Geotechnical Engineering: Unsaturated and Saturated Soils

Geotechnical Engineering:

Unsaturated and Saturated Soils

Jean-Louis Briaud

WILEY

Cover image: © Art Koenig, Photographer/Artist
Cover design: Wiley

This book is printed on acid-free paper.

Published by John Wiley & Sons, Inc., Hoboken, New Jersey
Published simultaneously in Canada

Library of Congress Cataloging-in-Publication Data:

Briaud, J.-L.
Introduction to geotechnical engineering : unsaturated and saturated soils / Jean-Louis Briaud.
pages cm
''Published simultaneously in Canada''—Title page verso.
Includes bibliographical references and index.
ISBN 978-0-470-94856-9 (cloth : acid-free paper); 978-1-118-41574-0 (ebk.); 978-1-118-41826-0 (ebk.)
1. Geotechnical engineering–Textbooks. 2. Soil mechanics–Textbooks. I. Title.
TA705.B75 2013
624–dc23

2013004684

Printed in the United States of America

10 9 8 7 6 5 4 3 2

CONTENTS

ACKNOWLEDGMENTS

One of the greatest joys in writing this book was working as a team with all my PhD students. From 2010 to 2013, they contributed tremendously to making this book possible. The leader of the team was Ghassan Akrouch. I thank them all very sincerely for their magnificent help. The beautiful memories of our work together on this huge project will be with me as a source of strength and friendship forever.

- Ghassan Akrouch (Lebanon)
- Alireza Mirdamadi (Iran)
- Deeyvid Saez (Panama)
- Mojdeh Asadollahipajouh (Iran)
- Congpu Yao (China)
- Stacey Tucker (USA)
- Negin Yousefpour (Iran)
- Oswaldo Bravo (Peru)
- DoHyun Kim (Korea)
- Axel Montalvo (Puerto Rico)
- Gang Bi (China)
- Mohsen Madhavi (Iran)
- Seung Jae Oh (Korea)
- Seok Gyu Kim (Korea)
- Mohammad Aghahadi (Iran)
- Yasser Koohi (Iran)
- Carlos Fuentes (Mexico)

My colleagues also provided advice on many topics:

- Marcelo Sanchez (Texas A&M University)
- Don Murff (Exxon)
- Jose Roesset (Texas A&M University)
- Giovanna Biscontin (Texas A&M University)
- Chuck Aubeny (Texas A&M University)
- Zenon Medina Cetina (Texas A&M University)
- Vincent Drnevich (Purdue)
- Chris Mathewson (Texas A&M University)

One person stands out as a major helper in this book project by her dedication to the task and her relentless denial of the impossible: my assistant Theresa Taeger, who took care of the hundreds of illustration permission requests in record time.

I also want to thank all those who share their knowledge and intellectual property online. Without the Internet as a background resource, this work would have taken much longer.

CHAPTER 1

Introduction

1.1 WHY THIS BOOK?

> "Things should be made as simple as possible but not a bit simpler than that."
>
> *Albert Einstein (Safir and Safire 1982)*

Finding the Einstein threshold of optimum simplicity was a constant goal for the author when writing this book (Figure 1.1).

The first driving force for writing it was the coming of age of unsaturated soil mechanics: There was a need to introduce geotechnical engineering as dealing with true three-phase soils while treating saturated soil as a special case, rather than the other way around. The second driving force was to cover as many geotechnical engineering topics as reasonably possible in an introductory book, to show the vast domain covered by geotechnical engineering and its important contributions to society. Dams, bridges, buildings, pavements, landfills, tunnels, and many other infrastructure elements involve geotechnical engineering. The intended audience is anyone who is starting in the field of geotechnical engineering, including university students.

1.2 GEOTECHNICAL ENGINEERING

Geotechnical engineering is a young (~100 years) professional field dealing with soils within a few hundred meters of a planet's surface for the purpose of civil engineering structures. For geotechnical engineers, soils can be defined as loosely bound to unbound, naturally occurring materials that cover the top few hundred meters of a planet. In contrast, rock is a strongly bound, naturally occurring material found within similar depths or deeper. At the boundary between soils and rocks are intermediate geo-materials. The classification tests and the range of properties described in this book help to distinguish between these three types of naturally occurring materials.

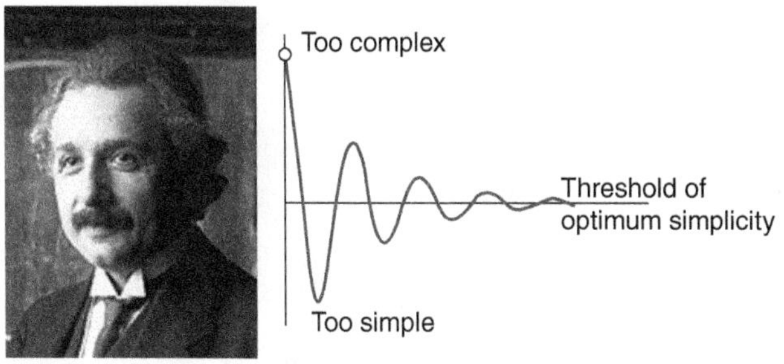

Figure 1.1 Einstein threshold of optimum simplicity. (Photo by Ferdinand Schmutzer)

Geotechnical engineers must make decisions in the best interest of the public with respect to safety and economy. Their decisions are related to topics such as:

- Foundations
- Slopes
- Retaining walls
- Dams
- Landfills
- Tunnels

These structures or projects are subjected to loads, which include:

- Loads from a structure
- Weight of a slope
- Push on a retaining wall
- Environmental loads such as waves, wind, rivers, earthquakes, floods, droughts, and chemical changes, among others

Note that current practice is based on testing an extremely small portion of the soil or rock present in the project area. A typical soil investigation might involve testing 0.001% of the soil that will provide the foundation support for the structure. Yet, on the basis of this extremely limited data, the geotechnical engineer must predict the behavior of the entire mass of soil. This is why geotechnical engineering is a very difficult discipline.

1.3 THE PAST AND THE FUTURE

While it is commonly agreed that geotechnical engineering started with the work of Karl Terzaghi at the beginning of the 20th century, history is rich in instances where soils and soils-related engineering played an important role in the evolution of humankind (Kerisel 1985; Peck 1985; Skempton 1985). In prehistoric times (before 3000 BC), soil was used as a building material. In ancient times (3000–300 BC), roads, canals, and bridges were very important to warriors. In Roman times (300 BC–300 AD), structures started to become larger and foundations could no longer be ignored. The Middle Ages (AD 300–1400) were mainly a period of war, in which structures became even heavier, including castles and cathedrals with very thick walls. Severe settlements and instabilities were experienced. The Tower of Pisa was started in 1174 and completed in 1370. The Renaissance (AD 1400–1650) was a period of enormous development in the arts, and several great artists proved to be great engineers as well. This was the case of Leonardo da Vinci and more particularly Michelangelo. Modern times (AD 1650–1900) saw significant engineering development, with a shift from military engineering to civil engineering. In 1776, Charles Coulomb developed his earth pressure theory, followed in 1855 by Henry Darcy and his seepage law. In 1857, William Rankine proposed his own earth pressure theory, closely followed by Carl Culman and his graphical earth pressure solution. In 1882, Otto Mohr presented his stress theory and the famous Mohr circle, and in 1885 Joseph Boussinesq provided the solution to an important elasticity problem for soils. From 1900 to 2000 was the true period of development of modern geotechnical engineering, with the publication of Karl Terzaghi's book *Erdbaumechanik* (in 1925), which was soon translated into English; new editions were co-authored with Ralph Peck beginning in 1948. The progress over the past 50 years has been stunning, with advances in the understanding of fundamental soil behavior and associated soil models (e.g., unsaturated soils), numerical simulations made possible by the computer revolution, the development of large machines (e.g., drill rigs for bored piles), and a number of ingenious ideas (e.g., reinforced earth walls).

Geotechnical engineering has transcended the ages because all structures built on or in a planet have to rest on a soil or rock surface; as a result, the geotechnical engineer is here to stay and will continue to be a very important part of humanity's evolution. The Tower of Pisa is one of the most famous examples of a project that did not go as planned, mostly because of the limited knowledge extant some 900 years ago. Today designing a proper foundation for the Tower of Pisa is a very simple exercise, because of our progress. One cannot help but project another 900 years ahead and wonder what progress will have been made. Will we have:

- complete nonintrusive site investigation of the entire soil volume?
- automated four-dimensional (4D) computer-generated design by voice recognition and based on a target risk?
- tiny and easily installed instruments to monitor geotechnical structures?
- unmanned robotic machines working at great depth?
- significant development of the underground?
- extension of projects into the sea?
- soil structure interaction extended to thermal and magnetic engineering?
- failures down to a minimum?
- expert systems to optimize repair of defective geotechnical engineering projects?
- geospace engineering of other planets?
- geotechnical engineers with advanced engineering judgment taught in universities?
- no more lawyers, because of the drastic increase in project reliability?

1.4 SOME RECENT AND NOTABLE PROJECTS

Among some notable geotechnical engineering projects and developments are the underpinning of the foundation of the Washington Monument in 1878 (Figure 1.2; Briaud et al.

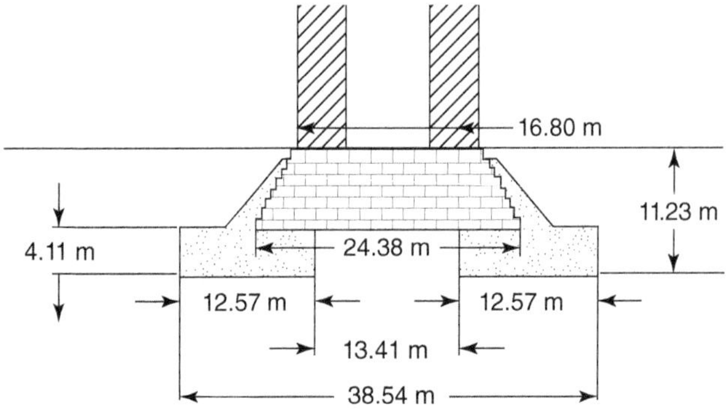

Figure 1.2 The Washington Monument.

Figure 1.3 Culebra cut of the Panama Canal, 1913. (a: Courtesy of Fernando Alvarado; b: Courtesy of United States Geological Survey)

2009); the Panama Canal (1913) and its slope stability problems (Figure 1.3; Marcuson 2001); the Tower of Pisa (1310) and its foundation repair in 1990 (Figure 1.4; Jamiolkowski 2001); the locks and dams on the Mississippi River and their gigantic deep foundations (Figure 1.5); and airports built offshore, as in the case of the Tokyo Haneda airport runway extension (Figure 1.6). Among the most significant milestones in the progress of geotechnical engineering are the discovery of the effective stress principle in saturated and then unsaturated soil mechanics; the development of laboratory testing and in situ testing to obtain fundamental soil properties; the combination of soil models with numerical methods to simulate three-dimensional behavior; the advent of geosynthetics and of reinforced soil, which is to geotechnical

Figure 1.4 The Tower of Pisa and its successful repair in 1995. (c: Courtesy of Dr. Gianluca De Felice (General Secretary), Opera Primaziale Pisana.)

Figure 1.5 Lock and Dam 26 on the Mississippi River in 1990. (a: Courtesy of United States Army Corps of Engineers, b: Courtesy of Thomas F. Wolff, St. Louis District Corps of Engineers, 1981. c: Courtesy of Missouri Department of Transportation.)

Figure 1.6 Extension of the Tokyo Haneda airport in 2010. (Courtesy of Kanto Regional Development Bureau, Ministry of Land, Infrastructure, Transport and Tourism, Japan.)

engineering what reinforced concrete is to structural engineering; and the development of instruments to monitor full-scale behavior of geotechnical engineering structures.

1.5 FAILURES MAY OCCUR

Failures do occur. The fact remains that it is not possible to design geotechnical engineering structures that will have zero probability of failure. This is because any calculation is associated with some uncertainty; because the geotechnical engineering profession's knowledge, despite having made great strides, is still incomplete in many respects; because human beings are not error free; and because the engineer designs the geotechnical engineering structure for conditions that do not include extremely unlikely events such as an asteroid hitting the structure at the same time as an earthquake, a hurricane, and a 100-year flood during rush hour.

Nevertheless, geotechnical engineers learn a lot from failures, because thorough analysis of what happened often points out weaknesses and needed improvement in our approaches. Some of the most notable geotechnical engineering failures have been the Transcona silo bearing capacity failure in 1913 (Figure 1.7), the Teton dam seepage failure in 1976 (Figure 1.8), and the failure of some of the New Orleans levees during Hurricane Katrina in 2005 (Figure 1.9).

1.6 OUR WORK IS BURIED

As Terzaghi is said to have noted, there is no glory in foundations. Indeed, most of our work is buried (Figure 1.10).

Figure 1.7 Transcona silo bearing capacity failure and repair (1913). (Courtesy of the Canadian Geotechnical Society.)

For example, everyone knows the Eiffel Tower in Paris, but very few know about its foundation (Figure 1.11; Lemoine 2006). The foundation was built by excavating down to the water level about 7 m deep—but the soil at that depth was not strong enough to support the 100 MN weight of the Tower, so digging continued. Because of the water coming from the River Seine, the deepening of the excavation had to be done using pressurized caissons (upside-down coffee cans, big ones!) so that the air pressure could balance the water pressure and keep it out of the excavation. Workers got into these 14 × 6 × 15 m caissons (Figure 1.12) and worked literally under pressure until they reached a depth where the soil was strong enough to support the Tower (about 13 m on the side closest to the river and about 8 m on the side away from the river).

1.7 GEOTECHNICAL ENGINEERING CAN BE FUN

Geotechnical engineering can be fun and entertaining, as the book by Elton (1999; Figure 1.13) on geo-magic demonstrates. Such phenomena as the magic sand (watch this movie: www.stevespanglerscience.com/product/1331?gclid=CNiW1uu-aICFc9J2godZwuiwg), water going uphill, the surprisingly strong sand pile (Figure 1.13), the swelling clay pie (Figure 1.13), and the suddenly very stiff glove full of sand will puzzle the uninitiated. Geotechnical engineering is seldom boring; indeed: the complexity of soil deposits and soil behavior can always surprise us with unanticipated results. The best geotechnical engineering work will always include considerations regarding geology, proper site characterization, sound fundamental soil mechanics principles, advanced knowledge of all the tools available, keen observation, and engineering judgment. The fact that geotechnical engineering is so complex makes this field an unending discovery process, which keeps the interest of its adepts over their lifetimes.

1.8 UNITS

In engineering, a number without units is usually worthless and often dangerous. On this planet, the unit system most commonly used in geotechnical engineering is the System International or SI system. In the SI system, the unit of mass is the *kilogram* (kg), which is defined as the mass of a platinum-iridium international prototype kept at the International Bureau of Weights and Measures in Paris, France. On Earth, the kilogram-mass weighs about the same as 10 small apples. The unit of length is the *meter*, defined as the length of the path travelled by light in vacuum during a time interval of 1/299,792,458 of a second. A meter is about the length of a big step for an average human. The *second* is the duration of 9,192,631,770 periods of the radiation corresponding to the transition between the two hyperfine levels of the ground state of the cesium 133 atom. Watches and clocks often have a hand ticking off the seconds. The unit of temperature is the

Figure 1.8 Teton Dam seepage failure (1976) (Photos by Mrs. Eunice Olson. Courtesy of Arthur G. Sylvester.)

Kelvin, defined as 1/273.16 of the difference in temperature between the absolute zero and the triple point of water. The *degree Celsius* (C) is also commonly used; it has the same magnitude as the degree Kelvin but starts at ~0°C (~273 K) for the freezing point of water and uses ~100°C (~373 K) for the boiling point of water. There are seven fundamental units in a unit system, but these four (kg, m, s, K) are the most commonly used in geotechnical engineering. The other fundamental units in the SI system are the mole (substance), the candela (light), and the ampere (electricity).

Other geotechnical engineering units are derived from these fundamental units. The unit of force is the *Newton*,

Figure 1.9 New Orleans levee failures during the Katrina hurricane in 2005. (Courtesy of United States Army Corps of Engineers.)

Figure 1.10 A rendition of the geotechnical engineering world. (Courtesy of Hayward Baker Inc., Geotechnical Contractor.)

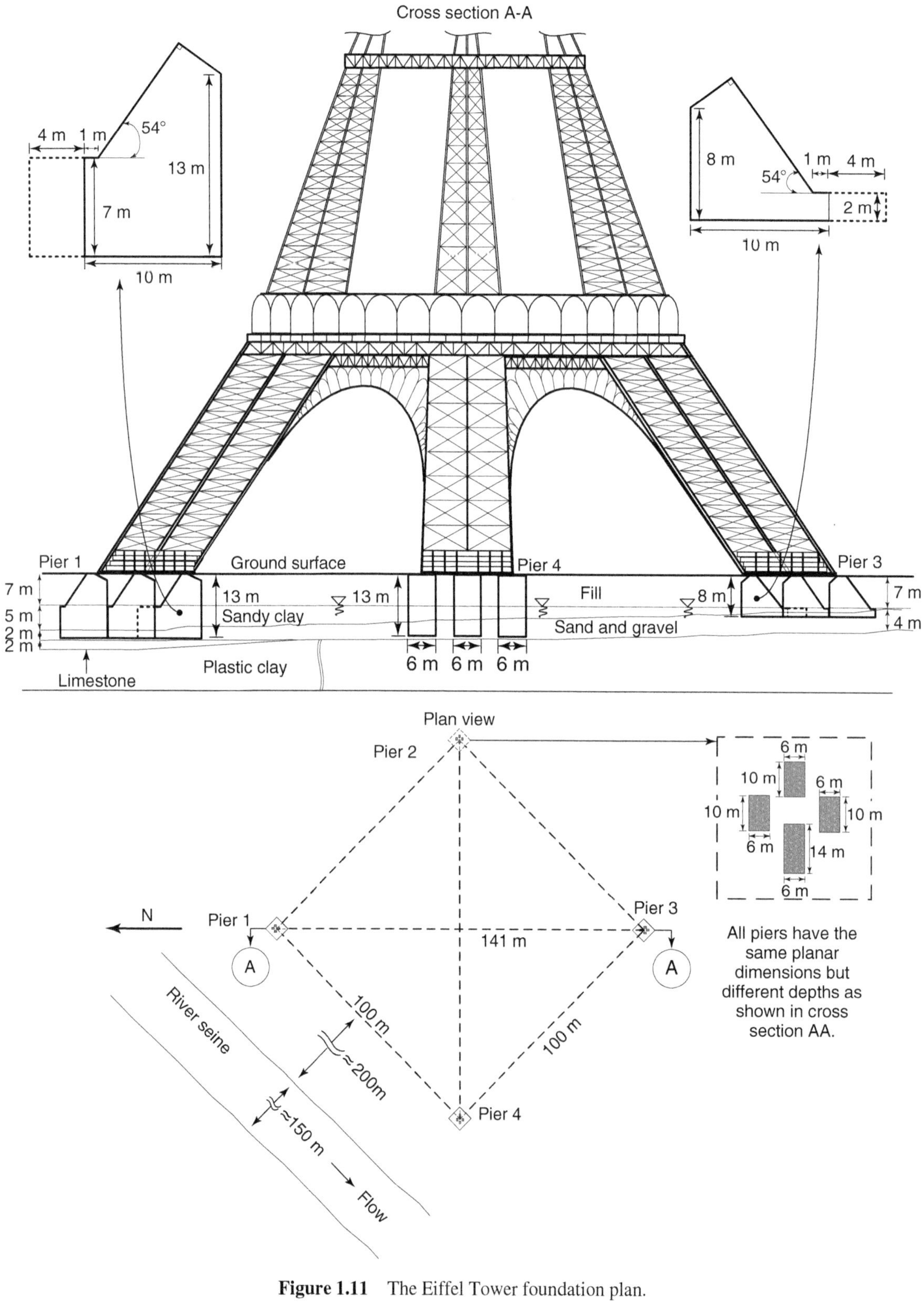

Figure 1.11 The Eiffel Tower foundation plan.

Figure 1.12 The Eiffel Tower foundation. (Photos b, c: Courtesy of the Musée d'Orsay, Paris.)

Figure 1.13 Soil magic. (Courtesy of David J. Elton.)

which is the force required to accelerate a mass of 1 kg to 1 m/s^2.

$$1\ \text{N} = 1\ \text{kg} \times 1\ \text{m/s}^2 \tag{1.1}$$

This force is about the weight of a small apple. Humans typically weigh between 600 and 1000 N. Most often the kilo-Newton (kN) is used rather than the Newton. The kilogram force is the weight of one kilogram mass. On Earth, the equation is:

$$1\ \text{kgf} = 1\ \text{kg} \times 9.81\ \text{m/s}^2 \tag{1.2}$$

The unit of stress is the kN/m^2, also called kilo-Pascal (kPa); there is about 20 kPa under your feet when you stand on both feet. Note that a kilogram force is the weight of a kilogram mass and depends on what planet you are on and even where you are on Earth. Other units are shown in a table at the beginning of this book.

Accepted multiples of units, also called SI prefixes, are:

terra	10^{12}
giga	10^{9}
mega	10^{6}
kilo	10^{3}
milli	10^{-3}
micro	10^{-6}
nano	10^{-9}
pico	10^{-12}
(An angstrom is 10^{-10} meter.)	

PROBLEMS

1.1 How would you decide if you have reached the threshold of optimum simplicity?

1.2 What was achieved by underpinning the 608 MN Washington Monument foundation from a 24.4 m square foundation to a 38.5 m square ring, as shown in Figure 1.2?

1.3 How would you go about deciding if the slopes of the Panama Canal are too steep?

1.4 What major geotechnical engineering problems come to mind for the extension of the Tokyo Airport?

1.5 Write a step-by-step procedure for the up-righting of the Transcona Silo.

1.6 For the 100 MN Eiffel Tower, calculate the average pressure under the foundation elements.

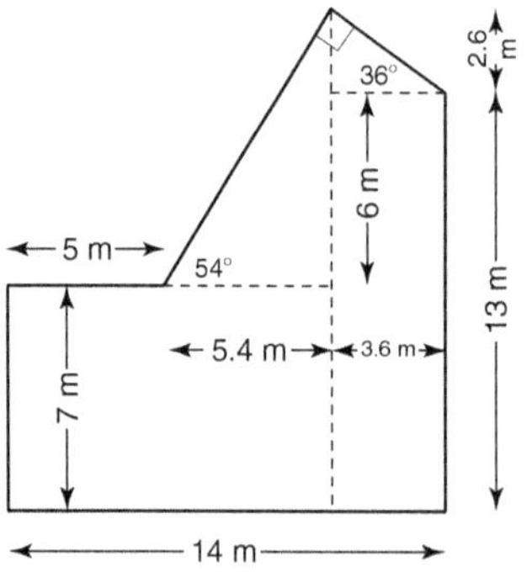

Figure 1.1s Foundation of the Eiffel Tower.

1.7 For the Tower of Pisa, calculate the pressure under the foundation, given that the foundation is a ring with a 19.6 m outside diameter and a 4.5 m inside diameter. Compare this pressure to the pressure obtained for the Eiffel Tower in problem 1.6.

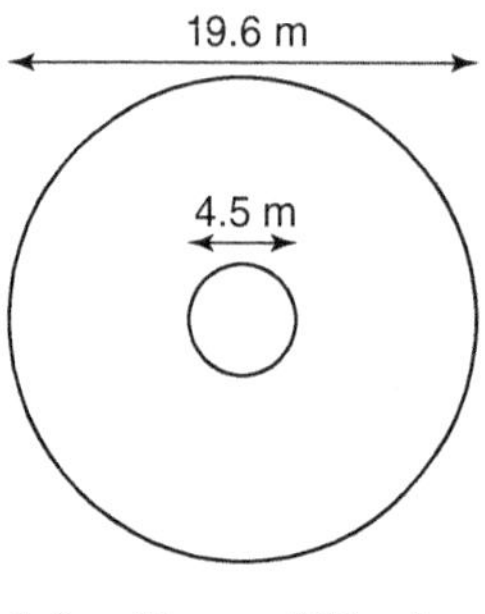

Figure 1.2s Tower of Pisa foundation.

1.8 Calculate the pressure under your feet.

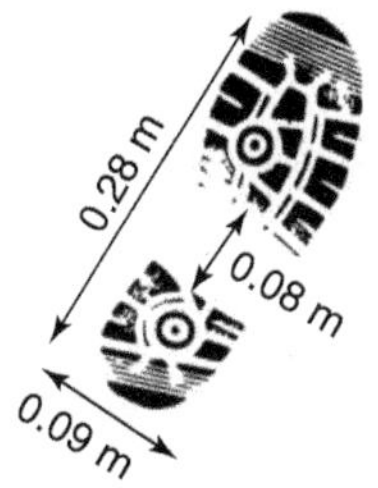

Figure 1.3s Feet geometry.

1.9 What do you think caused the failure of the Teton Dam? What do you think might have avoided this problem?
1.10 Explain the magic behind Figures 1.13d and 1.13e.
1.11 Are the following equations correct?

$$1 \text{ kgf} = 1 \text{ kg} \times 9.81 \text{ m/s}^2$$
$$1 \text{ N} = 1 \text{ kg} \times 1.0 \text{ m/s}^2$$
$$1 \text{ kgf} = 9.81 \text{ N}$$

1.12 What is the relationship between a kilopascal (kPa) and a pound per square foot (psf)? What is the net pressure in psf under the Eiffel Tower foundation?

Problems and Solutions

Problem 1.1

How would you decide if you have reached the threshold of optimum simplicity?

Solution 1.1

The threshold is not reached if:

- The solution seems too simple or too complicated.
- The solution is not used in practice.
- It costs too much time and money to obtain the solution.
- The solution leads to erroneous answers.
- The solution does not contain or address the essential elements of the problem.

The threshold is likely reached if:

- The solution seems reasonably simple and cannot be simplified further.
- The solution is used in practice.
- The cost of obtaining and implementing the solution is consistent with the budget of a large number of projects.
- The solution leads to reasonable answers.
- The solution is based on fundamental elements of the problem.

Problem 1.2

What was achieved by underpinning the 608 MN Washington Monument foundation from a 24.4 m square foundation to a 38.5 m square ring, as shown in Figure 1.2?

Solution 1.2

By increasing the area of the foundation, the pressure under the Washington Monument was decreased. This allowed the construction of the column to be completed with greatly reduced settlement and avoided the overturning or collapse of the structure that would likely have occurred if no underpinning had been done.

Problem 1.3

How would you go about deciding if the slopes of the Panama Canal are too steep?

Solution 1.3

I would draw a free-body diagram of the mass that would be likely to fail, I would show all the external forces, and I would check the equilibrium of the system.

I would also check the site and make observations of the slope as a function of time. If it had not already been built, I could observe neighboring slopes and make measurements.

Problem 1.4

What major geotechnical engineering problems come to mind for the extension of the Tokyo Airport?

Solution 1.4

Some of the problems associated with the extension of the Tokyo airport include:

- Soil failure in the form of rotational sliding at the edges of the embankment.
- Excessive settlement of the embankment, and in particular differential movements.
- Erosion problems during storms.
- Earthquake-induced problems, as the airport is in a high-seismicity area.

Problem 1.5

Write a step-by-step procedure for the up-righting of the Transcona Silo.

Solution 1.5

The following steps could be considered for the successful up-righting of the silo:

- Build footings on top of which hydraulic jacks can be installed to raise the structure. Make sure the footings can resist the force necessary to lift the structure.
- Lift the structure upward and start to backfill the failed soil. An alternative is to reinforce the existing failed soil.
- Complete the reinforcement of the key locations beneath the silo.
- Lower the jacks and allow the silo to rest on the reinforced earth.

Problem 1.6

For the 100 MN Eiffel Tower, calculate the average pressure under the foundation elements.

Solution 1.6

Pressure is force over area. The problem states that the Eiffel Tower exerts 100 MN of force on the foundation. From Figure 1.11, we know that the foundation of each leg of the Eiffel Tower is made of one rectangular foundation of 14 m by 6 m and three rectangular foundations of 10 m by 6 m. Therefore, the total area for the foundation of each leg is $14\text{ m} \times 6\text{ m} + 3(10\text{ m} \times 6\text{ m}) = 264\text{ m}^2$. Assuming that the load is evenly distributed among the four legs, the load per leg is 100 MN divided by 4, or 25 MN. The average pressure per foundation element is

$$\frac{25000}{264} = 94$$

Note that this pressure does not include the weight of the foundation.

Weight of the largest foundation element:

$$W = 25 \times \left(14 \times 7 + 3.6 \times 6 + \frac{2.6 \times 3.6}{2} + \frac{5.4 \times 8.6}{2}\right)$$
$$= 25 \times (98 + 21.6 + 4.68 + 23.22) \times 6 = 221$$

Average pressure due to the weight of this foundation is:

$$P_{\text{foundation}} = \frac{22125}{14 \times 6} = 263$$

which is much larger than the pressure due to the tower alone. Indeed, the weight of all the foundation elements is a lot more than the weight of the tower.

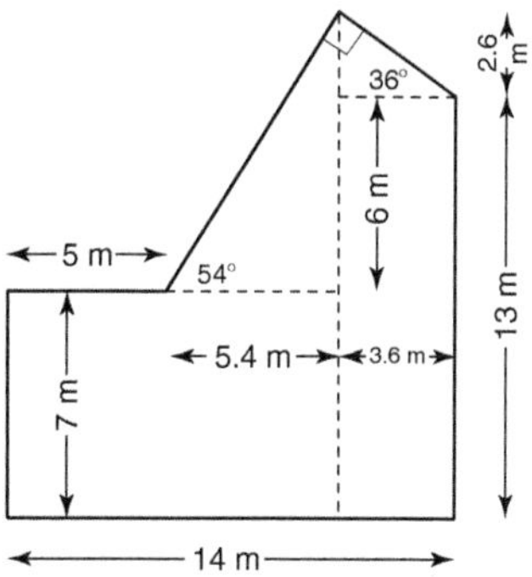

Figure 1.1s Foundation of the Eiffel Tower.

If we assume a total unit weight of soil of 20 kN/m^3, this pressure $P_{foundation}$ is equivalent to the pressure created by a height of soil equal to

$$h_{soil} = \frac{263}{20} = 1$$

Because 13 meters of soil were excavated, the weight of soil removed during the excavation was approximately equal to the weight of the foundation and the net pressure increase on the soil is $P_{net} = 94.6$ kPa. However, the actual pressure under the biggest foundation element is $P_{total} = 94.6 + 263 = 357$ kPa.

Problem 1.7

For the Tower of Pisa, calculate the pressure under the foundation, given that the foundation is a ring with a 19.6 m outside diameter and a 4.5 m inside diameter. Compare this pressure to the pressure obtained for the Eiffel Tower in problem 1.6.

Solution 1.7

$$\text{Pressure under the foundation} = \frac{142 \times 10^2}{285.21} = 469$$

If this pressure does not include the weight of the foundation, then $P_{net} = 496.8$ kPa is the net pressure. Net pressure under the Eiffel Tower foundation = 94.6 kPa. The net pressure under the Tower of Pisa is about five times higher than the net pressure under the Eiffel Tower.

Problem 1.8

Calculate the pressure under your feet.

Solution 1.8

Effective area for one foot $\approx (0.28 - 0.08) \times 0.09 = 0.018\ m^2$

$$\text{Average weight of a person} = 750\ \text{N}$$

$$\text{Pressure under two feet}: \frac{750 \times 10^{-3}}{2 \times 0.018} = 20$$

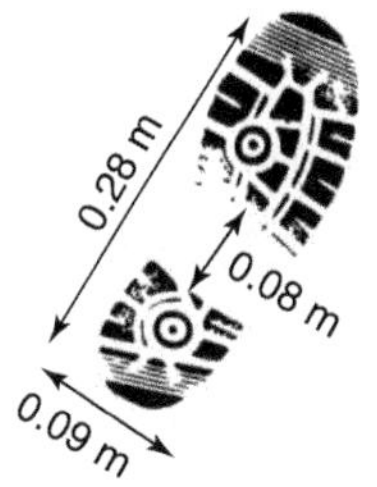

Figure 1.3s Feet geometry.

Problem 1.9

What do you think caused the failure of the Teton Dam? What do you think might have avoided this problem?

Solution 1.9

The failure of the Teton Dam was likely due to seepage at the boundary between the dam and the abutment. This seepage led to piping in the dam and ultimately to its breach. One way to avoid such a problem is to build a wall penetrating into the abutment, called a *key*, to minimize the seepage at that interface.

Problem 1.10

Explain the magic behind Figures 1.13d and 1.13e.

Solution 1.10

The swelling clay pie is made of smectite clay, which has a tremendous ability to attract water in the presence of a free water source. This is due to the chemical attraction between the water molecules and the smectite mineral ($Al_2Si_4O_{10}(OH)_2$ and x interlayers of H_2O). This clay type can swell an amount equal to its initial height or more. This is why the clay pie swelled to twice its height when subjected to a water source.

The sand pile at the top of the figure fails under the load applied (50 N) because the load exceeds the shear strength of the sand. The sand pile at the bottom of the figure is internally reinforced by sheets of toilet paper that are not visible from the outside. These paper sheets provide enough tension and increased shear strength in the sand for it to resist a much higher load (220 N) than the unreinforced sand pile.

Problem 1.11

Are the following equations correct?

Solution 1.11

$$1\ \text{kgf} = 1\ \text{kg} \times 9.81\ \text{m/s}^2: \text{Correct}$$

$$1\ \text{N} = 1\ \text{kg} \times 1.0\ \text{m/s}^2: \text{Correct}$$

$$1\ \text{kgf} = 9.81\ \text{N}: \text{Correct}$$

Problem 1.12

What is the relationship between a kilopascal (kPa) and a pound per square foot (psf)?

Solution 1.12

$$1\,\text{kPa} = 1000\,\text{N/m}^2 = \frac{\left(1000\,\text{N} \times \dfrac{0.22481\ \text{lb}}{1\ \text{N}}\right)}{\left(1\ \text{m}^2 \times \left(\dfrac{3.28\ \text{ft}}{1\ \text{m}}\right)^2\right)} = 20.9\,\text{psf}$$

What is the net pressure in psf under the Eiffel Tower foundation?

$$\text{Total weight} = 100\ \text{MN}$$

$$\text{Total area} = (14\ \text{m} \times 6\ \text{m} + 3(10\ \text{m} \times 6\ \text{m})) \times 4 = 10$$

$$\text{Pressure in kPa} = \frac{100\ \text{MN}}{1056\ \text{m}^2} = \frac{100 \times 10^6}{1056\ \text{N/m}^2}$$

$$= 94697\ \text{N/m}^2 = 94$$

$$\text{Pressure in psf} = 94.7\ \text{kPa} \times 20.9(\text{psf/kPa}) = 1975$$

CHAPTER 2

Engineering Geology

This chapter is intended to give readers a general overview of engineering geology. More detailed information should be sought in textbooks and other publications (Waltham 1994; Bell 2007).

2.1 DEFINITION

Geology is to geotechnical engineering what history is to humankind. It is the history of the Earth's crust. *Engineering geology* is the application of the science of geology to geotechnical engineering in particular and engineering in general. The same way we learn from history to avoid repeating mistakes in the future, we learn from engineering geology to improve geotechnical engineering for better design of future structures. Engineering geology gives the geotechnical engineer a large-scale, qualitative picture of the site conditions. This picture is essential to the geotechnical engineer and must always be obtained as a first step in any geotechnical engineering project.

2.2 THE EARTH

The age of the universe and of the Earth is a matter of debate. The most popular scientific views are that the universe started with a "big bang" some 15 billion years ago and that the Earth (Figure 2.1) began to be formed some 4.5 billion years ago (Dalrymple 1994), when a cloud of interstellar matter was disturbed, possibly by the explosion of a nearby star. Gravitational forces in this flat, spinning cloud caused its constituent material to coalesce at different distances from the Sun, depending on their mass density, and eventually to form planets. The Earth ended up with mostly iron at its center and silicates at the surface.

The Earth has a radius of approximately 6400 km (Jefferis 2008). The first layer, known as the crust (Figure 2.2), is about 100 km thick and is made of plates of hard silica rocks. The next layer, called the mantle, is some 2800 km thick and made of hot plastic iron silicates. The core is the third and last layer; it has a radius of 3500 km and is largely made of molten iron.

Early on, the planet was very hot and all earth materials were melted like they are on the Sun today. The cooling process started right away and has been progressing ever since. The present temperature gradient, shown in Figure 2.2, represents an average increase in temperature with depth of 15 degrees Celsius per kilometer in the crust, although the overall average is only 1 degree Celsius per kilometer. The gravity field is governed by the acceleration due to gravity ($9.81\ m/s^2$ on the average). This gravity field generates an increase in stress versus depth, which leads to an enormous pressure at the center of the Earth of about 340 GPa. The Earth's magnetic field is created by magma movement in the core and varies between 30 and 60 microteslas; it is strongest near the poles, which act as the two ends of the Earth dipole.

The Earth is a dynamic medium that changes and evolves through major events such as plate tectonics and earthquakes. The rock plates (about 100 km thick) that "float" on the semiliquid and liquid layers below accumulate strains at various locations where they run into each other. When the stress buildup is released abruptly, the result is an earthquake. Earthquakes and other movements allow the plates to move slowly (centimeters per year) yet significantly over millions of years. For example, on today's world map South America still looks like it could fit together with Africa—because in the distant past they were in fact joined (Figure 2.3).

2.3 GEOLOGIC TIME

Geologic time is a scale dividing the age of the earth (4600 million years) into 5 eras (Figure 2.4): Precambrian (4600 million years ago [MYA] to 570 MYA), Paleozoic (570 MYA to 245 MYA), Mesozoic (245 MYA to 65 MYA), Tertiary (65 MYA to 2 MYA), and Quaternary (2 MYA to the present) (Harland et al. 1989). Each era is subdivided into periods and then into epochs (Figure 2.5). The Quaternary era, for example, is divided into the Pleistocene period and the Holocene or Recent period.

Figure 2.1 The Earth. (Courtesy of NOAA-NASA GOES Project.)

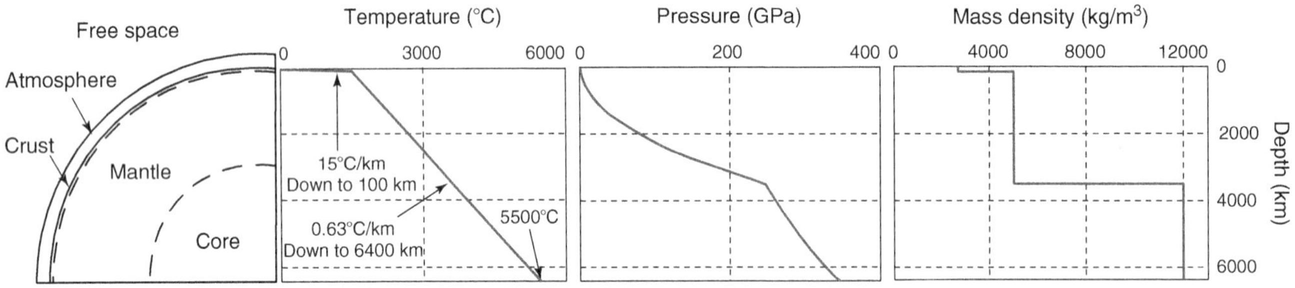

Figure 2.2 Earth temperature, pressure, and density.

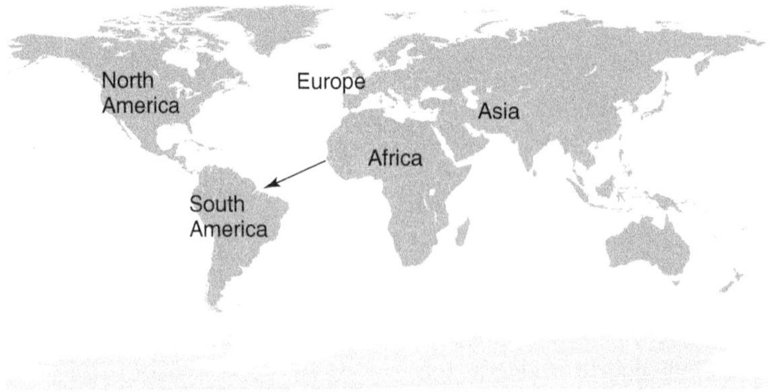

Figure 2.3 South America and Africa fit. (Courtesy of John Harvey.)

Typically, the older the earth material, the stronger it is. The last Ice Age occurred about 10,000 years ago at the beginning of the Holocene period. Glaciers, some of them 100 meters thick, covered the earth from the North Pole down to about the 40th parallel (St. Louis in the USA) and preloaded the soil. Because of this very heavy preloading, called overconsolidation or OC, those soil types (e.g., till) are very stiff and strong and do not settle much under load, but may erode quickly (as in the Schoharie Creek bridge failure disaster in 1987). When the glaciers melted, the soil surface rebounded; in some places this movement is still ongoing at a rate of about 10 mm per year.

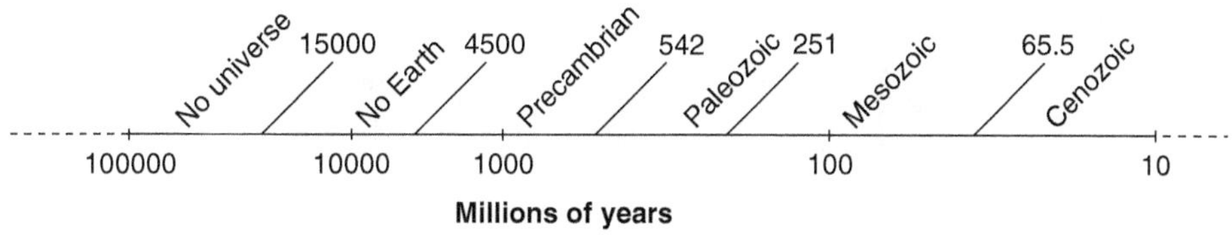

Figure 2.4 Geologic time (eras).

Cenozoic ERA			Mesozoic ERA	Paleozoic ERA
Periods		Epoch	Periods	Periods
Quaternary (Present – 2.6 My)		Holocene (present – 0.01 My)	Cretaceous (65.5 – 145.5 My)	Permian (251 – 299 My)
		Pleistocene (0.01 - 2.6 My)		Carboniferous (299 – 359 My)
Tertiary (2.6 – 65.5 My)	Neogene (2.6-23.0 My)	Pliocene (2.6-5.3 My)		Devonian (359 – 416 My)
		Miocene (5.3 – 23.0 My)		Silurian (416 – 444 My)
	Paleogene (23 – 65.5 My)	Oligocene (23 – 33.9 My)	Jurassic (145.5 – 201.6 My)	Ordovician (444 – 488 My)
		Eocene (33.9 – 55.8 My)	Triassic (201.6 – 251 My)	Cambrian (488 – 542 My)
		Paleocene (55.8 – 65.5 My)		

Figure 2.5 Geologic time (periods and epochs).

2.4 ROCKS

The Earth crust is 95% silica—and when silica cools, it hardens. This cooling creates the first kind of rocks: *igneous rocks*. Igneous rocks (e.g., granite, basalt, gneiss) are created by the crystallization of magma. *Sedimentary rocks* (e.g., sandstone, limestone, clay shales) are made of erosional debris on the Earth surface which was typically granular and recemented; they are created by wind erosion and water erosion, and are recemented by long-term high pressure or by chemical agents such as calcium. *Metamorphic rocks* (e.g., schist, slate) are rocks that have been altered by heat and/or pressure. The strength of rocks varies greatly, from 10 times stronger than concrete (granite) to 10 times weaker than concrete (sandstone). Older rocks are typically stronger than younger rocks. Figure 2.6 shows some of the main rock types.

2.5 SOILS

Soils are created by the exposure of rocks to the weather. This weathering can be physical (wetting/drying, thermal expansion, frost shatter) or chemical (solution, oxidation, hydrolysis). The elementary components of rocks and soils are minerals such as quartz and montmorillonite. Some minerals are easier to break down (montmorillonite) than others (quartz). As a result, the coarse-grained soils (sand, gravel) tend to be made of stable minerals such as quartz, whereas the fine-grained soils (silt and clay) tend to be made of less stable minerals such as montmorillonite. Organic soils may contain a significant amount of organic matter (wood, leaves, plants) mixed with the minerals, or may be made entirely of organic matter, such as the peat often found at the edges of swamps. Figure 2.7 shows some of those soils categories. Note that what the geotechnical engineer calls *soil* may be called *rock*

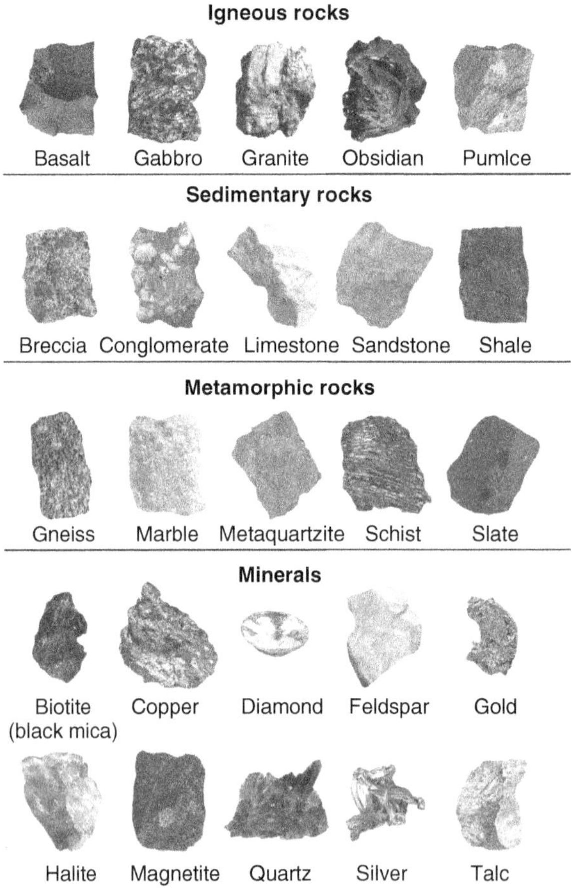

Figure 2.6 Main categories of rocks. (Courtesy of EDUCAT Publishers)

Figure 2.7 Main soil categories (crushed rock, gravel, sand, silt, clay).

by the engineering geologist; this can create confusion during discussion and interpretation.

2.6 GEOLOGIC FEATURES

The ability to recognize geologic features helps one to assess how the material at the site may be distributed. These features (Waltham 1994; Bell 2007) include geologic structures (faults, synclines, anticlines), floodplains and river deposits (alluviums, meander migration), glacial deposits (glacial tills and boulders left behind by a glacier), arid landforms (dunes, collapsible soils, shrink-swell soils), and coastal processes (shoreline erosion, sea-level changes).

The following list identifies some of the most common and important geological features that can affect geotechnical engineering projects.

Faults (Figure 2.8) are fractures in a rock mass that has experienced movement. They can lead to differences in elevation at the ground surface, differential erosion, contrasting visual appearance, and weaker bearing capacity of the fault material compared to the parent rock.

Outcrops show up at the ground surface when the rock layers are inclined. The area on the ground surface associated with an outcrop depends on the thickness of the layer and its *dip* or angle with the horizontal.

Escarpments are asymmetric hills formed when an outcrop is eroded unevenly or when the edge of rock layers is not flat. A *cliff* is an extreme case of an escarpment.

Folds (Figure 2.9) are created when rock layers are curved or bent by earth crust movement. *Synclines* are concave features (valleys), whereas *anticlines* are convex features (hills). Folds are best seen on escarpments.

Inliers and outliers are the result of erosion. Older rocks are typically below younger rocks. When an anticline erodes, the old rock appears at the surface between two zones of younger rocks (inlier). When a syncline erodes, it can lead to the reverse situation (outlier).

Figure 2.8 Example of rock fault. (Courtesy of USGS U.S. Geological Survey.)

Figure 2.9 Example of anticline–syncline combination. (Photo by R. W. Schlische.)

Figure 2.10 Examples of sinkholes. (*Left*: Courtesy of R.E. Wallace, United States Geological Survey, USA,; *Right*: Courtesy of International Association of Certified Home Inspectors, Inc.)

Karst is the underground landscape created when limestone is eroded or dissolved by groundwater. This process leads to holes in the limestone, called *sinkholes*, which can range from 1 meter to more than 100 meters in size and may become apparent while drilling during the site investigation (Figure 2.10).

Subsidence refers to settlement of the ground surface over large areas (in the order of square kilometers). Subsidence can be caused by pumping water out of the ground for irrigation or drinking purposes (Houston, Mexico City), pumping oil, digging large tunnels and mines, the presence of sinkholes, melting of the permafrost, and wetting of certain soils that collapse in the presence of water (called *collapsible soils*).

Meander migration occurs because rivers are dynamic features that change their contours by lateral erosion, particularly around bends or meanders. The soil forming the bank on the outside of the meander is eroded and is sent to the inside of the meander by the helical current of the river as it takes the meander turn. The inside of the meander then forms a sand bar (Figure 2.11).

Flood plain deposits occur when rivers experience flooding and the water spills over from the main channel into the floodplain. The main channel is a high-energy deposition environment, and only coarse-grained soils heavy enough

Figure 2.11 Example of meander migration.

not to be transported away are found there. In contrast, floodplains are a low-energy deposition environment where fine-grained soils are typically found. Floodplains and main channels can end up being buried or abandoned as the river migrates laterally and vertically. Abandoned floodplains are called *river terraces*.

Alluvium and *alluvial fans* are soil deposits transported to the bottom of a steep slope by the erosion of a river flowing down that steep slope (Figure 2.12).

Colluvial fans are deposits that form by gravity at the bottom of steep slopes when the slope fails.

Dunes are wind-blown sediments that accumulate over time to form a hill.

Permafrost is a zone of soil that remains frozen year round.

Figure 2.13 Example of geologic map. (Courtesy of National Park Service, NPS.)

2.7 GEOLOGIC MAPS

Geologic maps are very useful to the geotechnical engineer when evaluating the large-scale soil and rock environment to be dealt with in a project. These maps typically have a scale from 1:10,000 to 1:100,000 and show the base rock or geologic unit and major geologic features such as faults. Each rock area of a certain age is given a different color (Figure 2.13); soil is usually not shown on those maps. These maps can provide useful information regarding groundwater and hydrogeology, landslide hazards, sinkhole susceptibility, earthquakes, collapsible soils, flood hazards, and karst topography. Remember that what the geotechnical engineer calls *soil* may be called *rock* by the engineering geologist; to avoid confusion during discussion and interpretation, it is best to clarify the terminology.

Figure 2.12 Example of an alluvial fan. (Courtesy of Mike Norton.)

2.8 GROUNDWATER

Another important contribution of engineering geology to geotechnical engineering is a better understanding of how the groundwater is organized at a large scale. This field involves aquifer conditions, permeability of the rocks, and weather patterns (Winter et al. 1999). If you drill a hole in the ground, at some point you are likely to come to a depth where there is water. This water is called *groundwater* and it comes from infiltration from rain, rivers, springs, and the ocean. It may be stationary or flow slowly underground. If you go very deep (about 3 km or more), you will get to a point where there is no more water and the rocks are dry. The *groundwater table* (Figure 2.14) is the surface of the water within the soil or rock

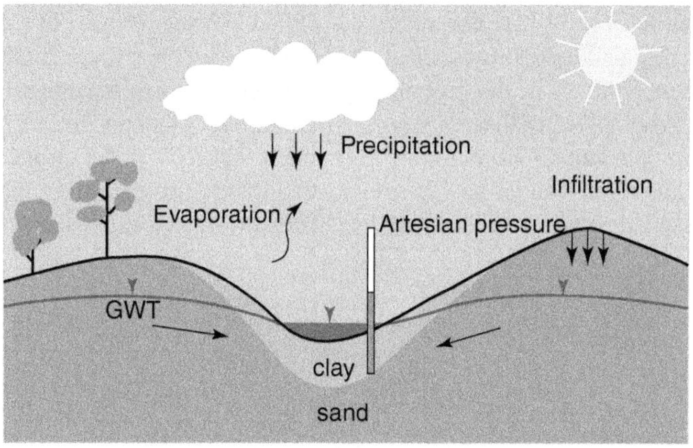

Figure 2.14 Groundwater.

where the water stress is equal to the atmospheric pressure (zero gauge pressure). Under natural conditions and in the common case, the groundwater table is close to being flat.

The *phreatic surface*, also called the *piezometric surface*, is the level to which the water would rise in a tube connected to the point considered in the soil mass. Most of the time, the groundwater table and the phreatic surface are the same. In some cases, though, they are different: *artesian pressure* refers to the case where the pressure in the water at some depth below the groundwater table is higher than the pressure created by a column of water equal in height to the distance between the point considered and the groundwater table. This can occur when a less permeable clay layer lies on top of a more permeable sand layer connected to a higher water source (Figure 2.14). Indeed, if you were to drill a hole through the soil down to a zone with artesian pressure, the water would rise above the level of the ground surface and could gush out into a spring (Figure 2.15).

Figure 2.15 Example of flow due to artesian pressure. (Courtesy of USGS U.S. Geological Survey.)

Perched water is a zone of water in the soil where the water appears at a certain depth in a boring and then disappears at a deeper depth; it acts as a pocket of water in the ground. *Aquifers* are typically deeper reservoirs of water that are supplied by surrounding water through a relatively porous rock. Aquifers are often pumped for human consumption. Their depletion can create kilometers-wide zones of settlement called *subsidence*, and in some instances the settlement can reach several meters in depth.

In geotechnical engineering, it is very important to know where the groundwater table is located, as it often affects many aspects of the project. Furthermore, it is important to identify irregularities in groundwater, such as artesian pressure or perched water.

PROBLEMS

2.1 Calculate the pressure at the center of the Earth.
2.2 Calculate the temperature at the center of the Earth
2.3 What is the depth of interest for most geotechnical engineering projects?
2.4 List the Tertiary and Quaternary epochs.
2.5 What happened about 10,000 years ago on the Earth? What are some of the consequences for soil and rock behavior today?
2.6 What are the three main categories of rocks, and what is the origin of each category?
2.7 What are the four main categories of soil sizes? How were each of these soils generated?
2.8 What engineering geology features can you look for when you visit a site for a geotechnical engineering project?
2.9 How can geologic maps be useful to the geotechnical engineer?
2.10 Define the following terms: groundwater level, perched water, phreatic surface, aquifer.

Problems and Solutions

Problem 2.1

Calculate the pressure at the center of the Earth.

Solution 2.1

To calculate the pressure at the center of the Earth, we will use Newton's law of universal gravitation. The force between two masses, m_1 and m_2, separated by a distance r, is:

$$F = G.\frac{m_1.m_2}{r^2}$$

where G is the gravitational constant $= 6.67 * 10^{-11}\ \mathrm{N\,m^2\,kg^{-2}}$

The density of soil layers varies with depth; the average density value for each layer is given in the following table:

Layer	Thickness (km)	Average Density (kg/m^3)
Crust	100	2700
Mantle	2800	5000
Core	3500	12000

Consider a small element of Earth dr thick and $rd\theta$ wide at a depth such that the distance from the center of the Earth is r (Figure 2.1s). This small element has a mass dm_1. The force acting on that element consists of three gravitational force components: the force due to mass Ma, which pulls the element away from the center; the force due to mass Mb, which pulls the element toward the center, and the force due to mass Mc, which also pulls the element toward the center. Newton showed that the forces due to mass *Ma* and *Mb* are equal and opposite so that the only force acting on the element is the force due to mass *Mc*. Therefore:

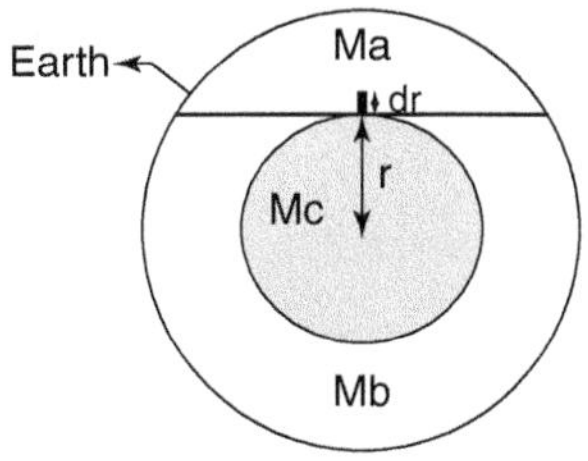

Figure 2.1s Parameters definition.

The pressure P is $P = \frac{F}{A}$ where A is the area of the element, so:

$$dP = \frac{dm_1}{A}.\frac{G.m_2}{r^2} = \frac{\rho.dV}{A}.\frac{G.m_2}{r^2} = \frac{\rho.dr.A}{A}.\frac{G.m_2}{r^2} = \rho.dr.\frac{G.m_2}{r^2}$$

$$P = \int \rho.G.\frac{m_2}{r^2}.dr, \text{ where } m_2 = \frac{4}{3}\pi r^3 \rho$$

$$P = \frac{4}{3}\pi.G.\int \rho^2.r.dr$$

Because the density of the Earth's layers is not constant (see Figure 2.2), the pressure at the center of the Earth is:

$$P = \frac{4}{3}\pi \times 6.67 \times 10^{-11}\left(\int_0^{3500\times1000} 12000^2\ \mathrm{rdr} + \int_{3500\times1000}^{6300\times1000} 5000^2\ \mathrm{rdr} + \int_{6300\times1000}^{6400\times1000} 2700^2\ \mathrm{rdr}\right.$$

$$P = 2.79 \times 10^{-4}\left(72r^2\Big|_0^{3.5\times10^6} + 12.5r^2\Big|_{3.5\times10^6}^{6.3\times10^6} + 3.645r^2\Big|_{3.3\times10^6}^{6.4\times10^6}\right) = 3.44\times10^{11}\frac{\mathrm{N}}{\mathrm{m}^2} = 344\ \mathrm{GPa}$$

Note that in geotechnical engineering we calculate the pressure, also called *vertical total stress*, at a given depth z as:

$$P = \sum \gamma_i \Delta Z_i$$

Where γ_i is the unit weight of the ΔZ_i thick i^{th} layer within the depth z. This is an approximation, as the unit weight $\Upsilon = \rho g$ is not constant and depends on the depth z (since g is a function of z). This approximation is very acceptable for the usual depth involved in a geotechnical project (a few hundred meters at most); indeed, this approximation only makes a difference of a small fraction of a percent.

Problem 2.2

Calculate the temperature at the center of the Earth.

Solution 2.2

The temperature gradient is 15° Celsius per kilometer in the crust and 0.63° Celsius per kilometer in the mantle and the core. Therefore, the temperature at the center of the Earth is:

$$T_{center} = 15 \times 100 + 0.63 \times 6300 = 5469^\circ$$

Problem 2.3

What is the depth of interest for most geotechnical engineering projects?

Solution 2.3

The depth of interest for most geotechnical engineers is a few hundred meters.

Problem 2.4

List the Tertiary and Quaternary epochs.

Solution 2.4

Holocene	0 to 10,000 years ago
Pleistocene	10,000 to 1.8 million years ago
Pliocene	1.8 to 5.3 million years ago
Miocene	5.3 to 23.8 million years ago
Oligocene	23.8 to 33.7 million years ago
Eocene	33.7 to 54.8 million years ago
Paleocene	54.8 to 65 million years ago

Problem 2.5

What happened about 10,000 years ago on the Earth? What are some of the consequences for soil and rock behavior today?

Solution 2.5

An ice age occurred about 10,000 years ago, at the beginning of the Holocene period. At that time, glaciers about 100 meters thick covered the earth from the North Pole down to about the 40th parallel and loaded the soil. This very heavy loading increased the density, stiffness, and strength of the soils below the glaciers. When the glaciers melted, they left behind these very dense, overconsolidated soils, called *glacial tills*. These soils do not settle much as long as the pressure does not exceed the pressure exerted by the Ice-Age glacier. (The glaciers also carried within them very large and heavy rocks, and deposited these boulders along their paths when they melted.) When the glaciers melted, the soil surface rebounded, and in some places this movement still goes on today at a rate of about 10 mm per year. An example of this is the landmass in England.

Problem 2.6

What are the three main categories of rocks, and what is the origin of each category?

Solution 2.6

The three main categories of rocks are:

- Igneous rocks, which come from the solidification and crystallization of magma. Common igneous rocks are granite, basalt, and gneiss.
- Sedimentary rocks, which are composed of rocks previously eroded through wind and hydraulic erosion and recemented by long-term high pressure or chemical agents (e.g., calcium). Common sedimentary rocks are sandstone, limestone, and clay shales.
- Metamorphic rocks, which have been altered by heat and/or pressure. Common types of metamorphic rocks are schist and slate.

Problem 2.7

What are the four main categories of soil sizes? How were each of these soils generated?

Solution 2.7

Soil class	Soil type	Size (by USCS)
Coarse-grained soil	Gravel	75 mm to 4.75 mm
	Sand	4.75 mm to 0.075 mm
Fine-grained soil	Silt	0.075 mm to 2 μm
	Clay	$<2\ \mu$m

Soils are generated by the exposure of rocks to the weather and other altering mechanisms. The weathering can be physical (wetting/drying, thermal expansion, frost shatter) or chemical (solution, oxidation, and hydrolysis). Erosion and deposition is another mechanism responsible for soil formation.

Problem 2.8

What engineering geology features can you look for when you visit a site for a geotechnical engineering project?

Solution 2.8

- Geologic structures (faults, synclines, anticlines)
- Floodplains and river deposits (alluviums, meander migration)
- Glacial deposits (glacial tills and boulders left behind after glacier melting)
- Arid landforms (dunes, collapsible soils, shrink-swell soils)
- Coastal processes (shoreline erosion, sea level changes)

Problem 2.9

How can geologic maps be useful to the geotechnical engineer?

Solution 2.9

Geologic maps help geotechnical engineers to evaluate the soil and rock in an area and to find specific geologic features such as faults.

Problem 2.10

Define the following terms: groundwater level, perched water, phreatic surface, aquifer.

Solution 2.10

Groundwater level: the level at which water is found in an open borehole.

Perched water: a zone of water in the soil where the water appears at a certain depth in a boring and then disappears at a deeper depth; it acts as a pocket of water in the ground.

Phreatic surface: the level where the water would rise in a tube connected to the point considered in the soil mass. Most of the time, the groundwater table and the phreatic surface are the same. Some exceptions include artesian pressure and water flow.

Aquifer: a deep reservoir of water created by infiltration of surrounding water through a porous soil or rock. Drinking water may come from an aquifer.

CHAPTER 3

Soil Components and Weight-Volume Parameters

3.1 PARTICLES, LIQUID, AND GAS

Soils are made of particles, gas (most often air), and fluid (most often water). Particles are also called *grains*. The space between the particles makes up the voids sometimes also called *pores*. If the voids are completely filled with air, the soil is called *dry*. If the voids are completely filled with water, the soil is called *saturated*. If the soil is filled partly with air and partly with water, the soil is called *unsaturated*. Figure 3.1 shows a soil sample and its graphical representation (the three-phase diagram discussed later in this chapter).

Note that in some cases, there is a subtle distinction between saying that a soil is dry and saying that a soil has no water. If a small sample of wet soft clay is left in the sun or in a low-humidity laboratory, it will become "dry" after a while and at the same time much stronger than when it was wet. This "dry" clay still has a tiny bit of water firmly bound between the particles. This water is in tension and sucks the particles together through a phenomenon called *suction* (explained in Chapter 10 on effective stress). This suction is responsible for the increase in strength of the clay. If the dried clay is ground into individual particles and placed in an oven at 100°C, then it will have no water and no strength. Thus, it becomes important to make a distinction between dried and no water; for example, a dried clay is a hard block of soil whereas a clay with no water may simply be a dry powder.

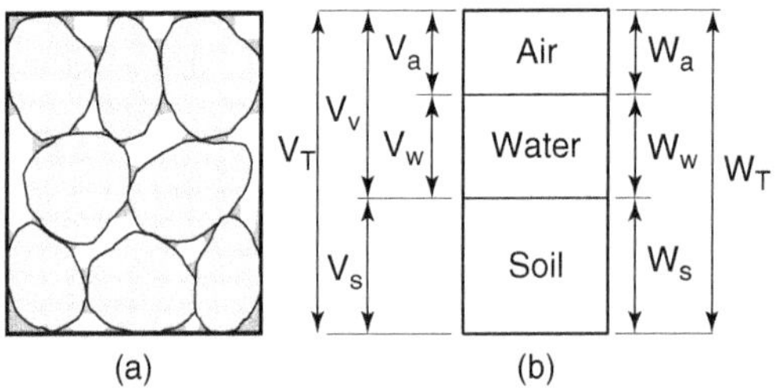

Figure 3.1 Three-phase diagram representation.

3.2 PARTICLE SIZE, SHAPE, AND COLOR

Depending on their size, soil particles are called gravel size particles, sand size particles, silt size particles, or clay size particles. Gravel, sand, and the coarser silt particles are typically made of quartz and are more rounded in shape. They can be seen with the naked eye or a simple microscope. Clay and the finer silt particles are too small to be seen with the naked eye; they are visible only with the use of electron microscopy or X-ray diffractometry. Figure 3.2 shows photos of soil particles.

Ranges of particle sizes are defined as:

Gravel-size particles: 20 mm to 4.75 mm
Sand-size particles: 4.75 mm to 0.075 mm
Silt-size particles: 0.075 mm to 0.002 mm
Clay-size particles: less than 0.002 mm

These ranges indicate a huge difference in size between a sand-size particle and a clay-size particle. For example, if the clay particle were a postage stamp, the sand particle would be a very large airplane. Soil particle sizes are so dramatically different that showing them on a natural scale is not very helpful (Figure 3.3); instead, a logarithmic scale is used which allows the very small particle to appear on the scale as well as the very large ones. Figure 3.4 shows such a scale and summarizes the main differences between soil particles.

There is also a big difference in shape between the gravel- and sand-size particles on the one hand and the silt- and clay-size particles on the other. Gravel, sand, and the larger silt particles tend to be rounded, whereas clays and the smaller silt size particles tend to be rodlike or platelike. This is because minerals such as quartz, which form the larger particles, are much more stable and resistant to weathering than the minerals, such as kaolinite (baby powder), that form the smaller particles. The surface of sand and gravel particles can present various degrees of roughness. At one end of the spectrum are the angular particles (freshly broken from the parent rock, for example) and at the other are the smooth,

Figure 3.2 Examples of cobbles, gravel-, sand-, silt-, and clay-size particles.

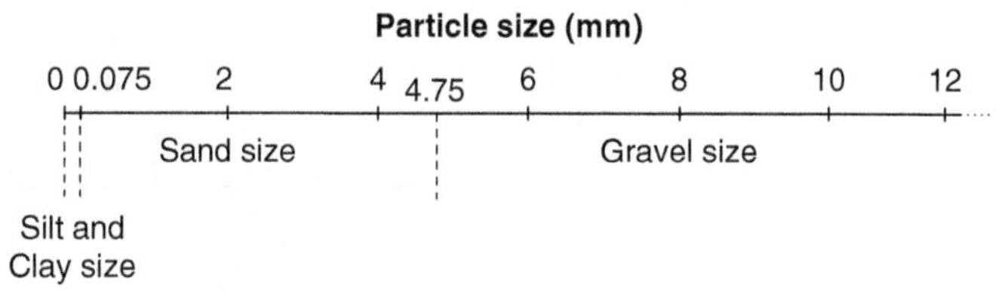

Figure 3.3 Particle sizes on a natural scale.

Particle size	Clay size		Silt size	Sand size	Gravel size
Particle size in mm on a log scale	10^{-4}	10^{-3}	10^{-2}	10^{-1} 10^{0}	10^{1} 10^{2}
Log of particle size on an arithmatic scale	−4	−3	−2	−1 0	1 2
	log (0.002) = −2.7		log (0.075) = −1.125	log (4.75) = 0.677	
Seeing the particle	Electron microscope		Microscope	Hand lens	Naked eye
Shape	Plate like-rod like			Sphere like	
Minerals	Smectite-montmorillonite-bentonite-illite-kaolinite			Mostly quartz-some felspar and mica	
Forces involved	Electostatic-electromagnetic-van der waals-intermolecular-gravity			Gravity	
Structure	Flocculated-dispersed			Loose-dense	
Other factors	Cation exchange capacity-suction			Surface roughness-suction	

Figure 3.4 Particle sizes on logarithmic scale and some characteristics of each size.

rounded particles (eroded by water over a long period of time, for example). Clays and silts are typically much smoother to the touch than sands and gravels.

Soil particles are grey, tan, brown, or reddish. The brown or reddish color may come from the presence of iron. The wetter the soil is, the darker the color will be; this may help in determining the location of the groundwater level when retrieving samples from a boring. A darker color may also indicate the presence of organic matter, although a foul smell is another and possibly better indicator.

3.3 COMPOSITION OF GRAVEL, SAND, AND SILT PARTICLES

Soil particles are made of mineral or organic matter. Mineral matter is inert matter such as silica, whereas organic matter is of biological origin (basically, anything that lives or has lived). Organic particles include leaves, plants, grasses, fibers, tree trunks, shells, and fossils. Most soil particles are made of minerals, which have a crystalline structure. The most common mineral is silica; indeed, silica makes up 95% of the Earth's crust. Minerals are to particles what bricks are to houses: they are the building blocks of the particle. The most stable minerals are framework minerals, which are resistant to erosion and weathering, and form the larger particles (gravel and sand). The least stable minerals are the sheet minerals which make up the clay particles. The most common constituent mineral in gravel, sand, and the coarser silt particles is quartz (SiO_2), but feldspar ($KAlSi_3O_8$), and mica (SiO_2, Al_2O_3) are also encountered. The behavior of gravel particles, sand particles, and the coarser silt particles is determined by the weight of the particle and associated friction. Other phenomena, such as electromagnetic and intermolecular forces, do exist, but in these coarser particles their effects are negligible compared to the weight. However, this is not the case for extremely small particles, such as clay particles or the finer silt particles.

3.4 COMPOSITION OF CLAY AND SILT PARTICLES

Note that silt particles are listed in the title of this section and the last section. The reason is that silt particles straddle the properties of coarse-grained particles and clay particles. Three major minerals make up clay particles. In decreasing order of size, they are kaolinite, illite, and smectite (Mitchell and Soga 2005). Montmorillonite and bentonite are subgroups of the smectite minerals. These minerals are composed of elementary sheets, which are the silica sheet (SiO_2), the gibbsite sheet ($Al_2(OH)_6$), and the brucite sheet ($Mg_3(OH)_6$).

The mineral kaolinite ($Al_2Si_2O_5(OH)_4$) is made of a stack of a silica sheet and a gibbsite sheet. Kaolinite makes up the larger clay particles with length on the order of 1000 nanometers (Figure 3.5), a thickness of about 100 nanometers, and a specific surface (particle surface per unit mass) of 10 m^2/g. Kaolinite is commonly used in baby powder. Smectite ($Al_2Si_4O_{10}(OH)_2$ and x interlayers of H_2O) is made of a gibbsite sheet sandwiched between two silica sheets. Smectite makes up the smaller clay particles with length on the order of 100 nanometers (Figure 3.5), a thickness on the order of 1 nanometer, and a specific surface (particle surface per unit mass) of 800 m^2/g. This remarkably high specific surface allows the smectite particle to absorb a significant amount of water between the elementary sheets. This leads to extreme swelling and shrinking potential for these clays

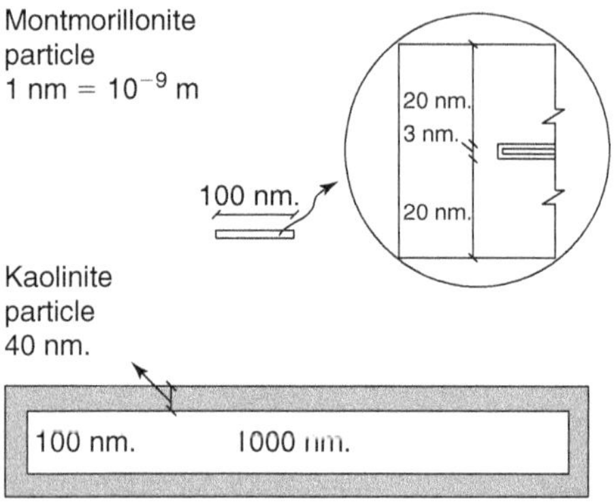

Figure 3.5 Approximate dimensions of montmorillonite and kaolinite particles.

(Figure 3.6). Montmorillonite and bentonite are subgroups of the smectite mineral group. Bentonite is sold commercially for drilling mud applications because it can form a nearly impervious cake on the wall of the borehole and keep groundwater from penetrating the borehole (see Chapter 6 on site investigation). The mineral illite has properties intermediate between those of kaolinite and smectite.

Cations are positive ions that are attracted to the surface of clay particles. Silicium (Si^{4+}) is a very common cation in soils. Because Si^{4+} has a high valence, a negative charge will be generated if it is replaced by another cation such as Al^{3+} or Mg^{2+} or Na^+. This cation exchange is called *isomorphous substitution* because the exchange cation has the same shape (*isomorphous* means "same shape" in Greek), allowing it to fit in the crystalline lattice, but a lower valence. This substitution will occur if an exchange cation is available when a Si^{4+} cation is not. The cation exchange capacity or CEC is a measure of how many cations a clay particle can catch; it is measured in milliequivalents per unit mass (meq/100 g). The milliequivalent is a unit of amount of substance and is related to the mole, the SI unit used to quantify the amount of substance. Kaolinite has a smaller CEC (~5 meq/100 g) than montmorillonite (~80 meq/100 g). As a result of isomorphous substitution, the surface of clay particles is negatively charged except at the ends of the particles, where positive charges may appear due to broken bonds. In this case, clay particles can be thought of as little magnets that attract or repel each other. The negative and sometimes positive electrical charges on the surface of clay particles influence the way the structure of the clay mass develops (flocculated or dispersed).

The water next to the clay particle surface is made of molecules that can be thought of as electrical dipoles (H^+ and OH^-). The H^+ end of the dipole is attracted to the negative charges on the clay particle surface and the water molecule adheres to the surface. Cations such as Na^+ may also be present in the water and will be attracted to the surface in an effort to neutralize the negative charge. The sodium adsorption ratio or SAR gives an indication of how

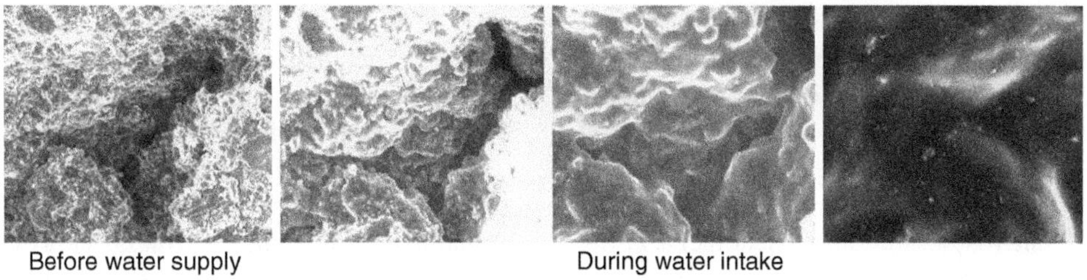

Figure 3.6 Absorption of water in bentonite. (Courtesy of Komine and Ogata, 2004.)

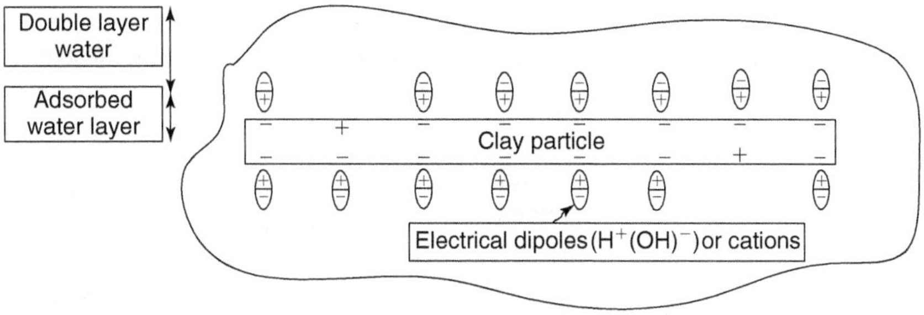

Figure 3.7 The electrical double layer of clay particles.

much sodium is available around the particles. It is defined as:

$$SAR = \frac{[Na^+]}{\left(\frac{[Ca^{2+}] + [Mg^{2+}]}{2}\right)^{0.5}} \tag{3.1}$$

where the value within brackets [] is the concentration of cations in meq/liter. This layer of bound water is called the *electrical double layer* (Figure 3.7) and its thickness is on the order of 1 to 50 nm, with the higher values found in very active clay particles such as montmorillonite and bentonite. The layer of water most closely bound to the particle surface within the electrical double layer is called the *adsorbed water layer* (Figure 3.7).

The attraction between clay particles is attributed to the Van der Waals forces that overcome the repulsion between two negatively charged particles. Van der Waals forces are intermolecular forces that give water its tensile strength, for example. The other important source of cohesion in a clay is the attraction between water and silica, which sucks the particles together. This phenomenon, called *suction*, and is discussed in Chapter 10 on effective stress.

3.5 PARTICLE BEHAVIOR

Gravels and sands are called coarse-grained soils, while silts and clays are called fine-grained soils. The weight of soil particles varies tremendously; for example, a gravel-size particle is about 10 billion times heavier than a clay-size particle. Coarse-grained soil particles tend to behave according to their weight. In contrast, the behavior of fine-grained, clay-size particles is significantly influenced by the electrostatic and electromagnetic forces that exist at the particle surface. These forces create attraction and repulsion much like small magnets would do. They give clays their consistency, which you might wish to think of as stickiness. The behavior of silt-size particles is intermediate between that of gravel and sand on the one hand and that of clay on the other.

In addition to the weight of the particle and the electrostatic/electromagnetic forces affecting the particles, water can strongly influence the behavior of an assembly of particles (Figure 3.8). First, the water can create buoyancy if the particle is below the groundwater level. This buoyancy reduces the effective weight of the particle (like when you go into

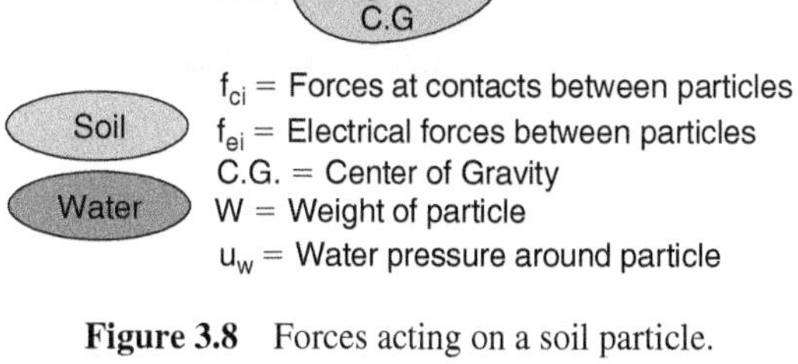

Figure 3.8 Forces acting on a soil particle.

a swimming pool) and therefore reduces the friction that it can generate when rubbing against other particles. Second, even above the groundwater level water is still present in the voids because of two fundamental phenomena: the attraction between water and the clay minerals (e.g., water is attracted to silica, which leads to capillary suction) and the attraction between water and salt (osmosis). Both phenomena allow the water to stay in the voids, go into tension, and suck the particles together. This "glue" between particles influences the behavior of the particles, contributes to soil plasticity (stickiness), and is responsible for the strength of a dry clay. This topic is developed in Chapter 10 on effective stress.

3.6 SOIL STRUCTURE

The *structure* of a soil refers to the arrangement of the soil grains. Loose or dense structures are found in coarse-grained soils, whereas flocculated and dispersed structures exist in fine-grained soils.

A loose soil structure is similar to the arrangement of the spheres shown in Figure 3.9. In this case, the contacts between particles are mostly at 90 and 180 degrees on the rosette of contacts. Shearing the mass would lead to a loss of volume of the mass, as the particles will tend to move toward a more stable arrangement. This soil would be called *contractive*. Such loose structures are found, for example, when the soil settles under water in a very low-energy environment and without vibration. This can be the case with hydraulic fills. A dense soil structure is similar to the arrangement of the spheres shown in Figure 3.10. In this case, the contacts between particles are mostly at 45 and 135 degrees on the rosette of contacts. Shearing the mass would lead to an increase in volume of the mass, as each particle will tend to ride on top of the next one. This soil would be called *dilatant*. Such dense structures are found in compacted soils for dams or pavements that are densified during placement by a combination of pressure and vibration.

In a *dispersed* structure, the particle arrangement is like a deck of cards (Figure 3.11). Such structures tend to be very stable and exhibit high stiffness. However, they have little strength against shearing that takes place in the direction of the "cards." The stacks of particles can, however, be organized in different ways within a single soil, and that will influence the overall behavior. In a *flocculated* structure, the particle arrangement looks like a card castle (Figure 3.11). Such structures tend to be unstable and can easily collapse. When a flocculated clay derives the strength of its particle contact from salt bonding, a quick clay may be formed. These quick clays (such as found in Norway and Canada) may liquefy if the salt is leached from the contacts by exposure to fresh water and/or if an event triggers the breaking of the bonds. The Risa event in Norway was a remarkable quick-clay landslide in which the clay literally turned into liquid—to the extent that houses floated down the hill. (This landslide was videotaped by an amateur, and the movie can be obtained by contacting the Norwegian Geotechnical Institute (NGI) in Oslo, Norway.) Most natural clays exhibit a mixture of dispersed and flocculated structures. Examples of clay and sand structures are shown in Figure 3.12 (Terzaghi et al. 1996).

Composite structures are associated with mixtures of coarse particles and fine particles. In a matrix structure, the fine particles are predominant and the coarse particles do not touch each other. In a void bound structure, the coarse particles touch each other and are bound together by the fine particles, which effectively act as a glue.

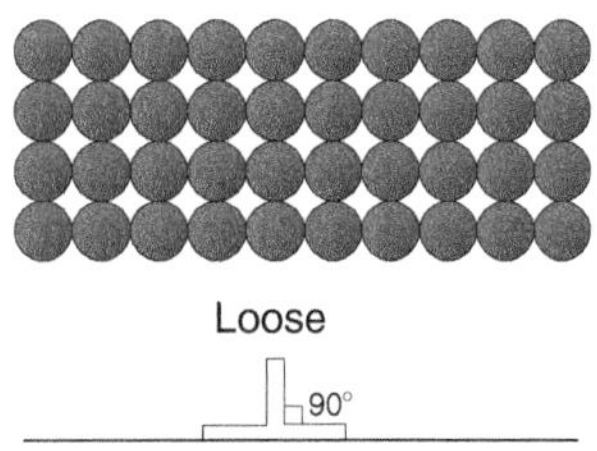

Figure 3.9 Sphere organized in a loose structure and associated rosette of contacts.

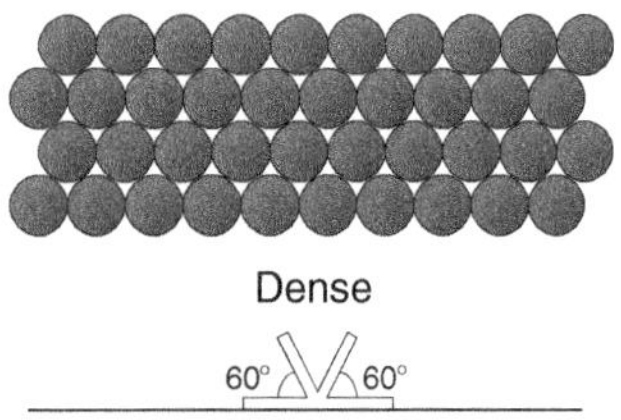

Figure 3.10 Sphere organized in a dense structure and associated rosette of contacts.

3.7 THREE-PHASE DIAGRAM

The three-phase diagram is a graphical representation of the soil components. Figure 3.1(a) shows a soil sample in its natural state, with particles, gas (most often air), and liquid (most often water) all mixed together. All the air can be regrouped into one volume V_a, and all the water can be regrouped into one volume V_w. The sum of V_a and V_w is the volume of voids V_v. Once all the air is in V_a and all the water is in V_w, then what is left are the particles regrouped into one volume V_s. This particle volume has no voids, because they have been sucked out into V_a and V_w; therefore, the volume V_s is a solid piece of rock with no voids. The unit weight of this solid piece of rock made with the particulate material has a unit weight called the *unit weight of solids*, γ_s, a ratio of the weight of solids W_s over the volume of solids V_s. The unit weight of solids varies depending on the mineral or organic matter of the particles, but for mineral matter it is in the range of 25.5 to 27 kN/m^3 and for organic matter it is 9 to 13 kN/m^3. The specific gravity of solids, *Gs*, is the ratio

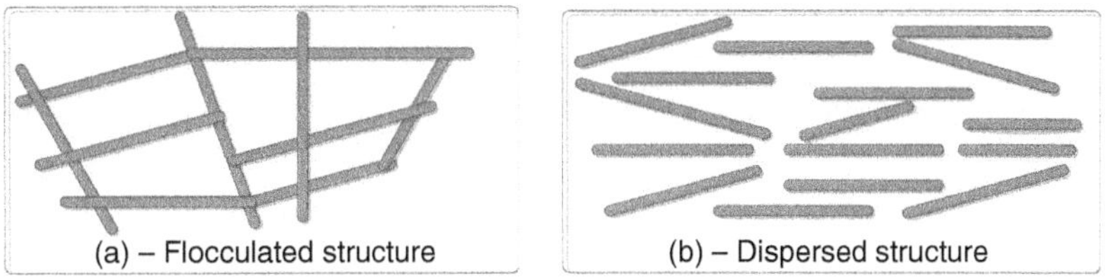

Figure 3.11 Flocculated and dispersed clay structures.

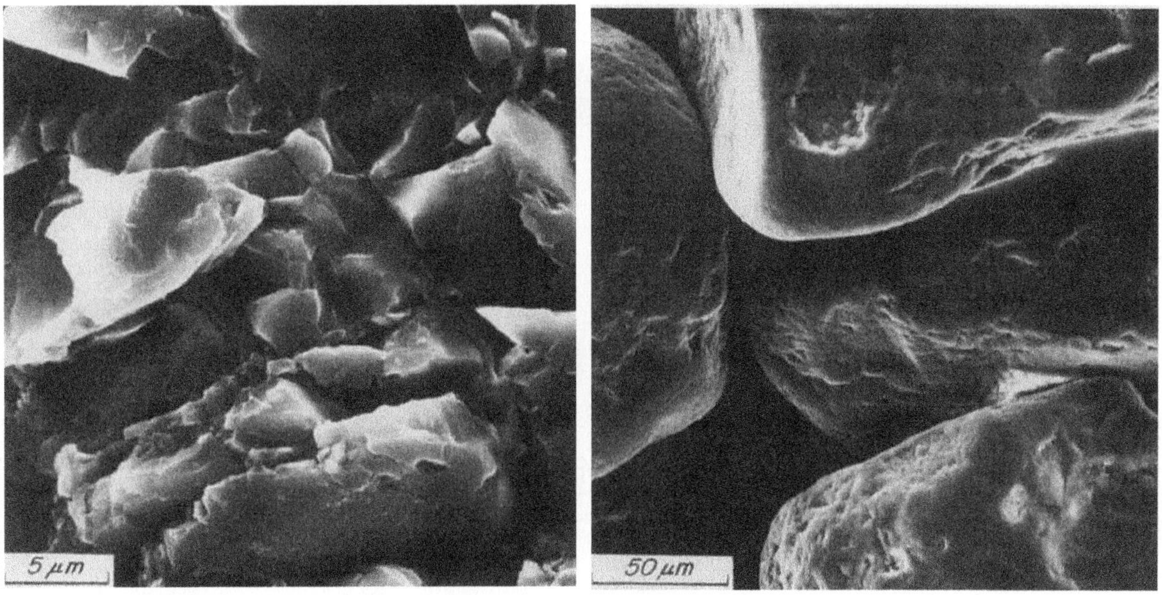

Figure 3.12 Example of clay and sand structure. (From Terzaghi et al. 1996. This material is reproduced with permission of John Wiley & Sons.)

between the unit weight of solids γ_s and the unit weight of water γ_w.

3.8 WEIGHT-VOLUME PARAMETERS

Some of the most important parameters describing the volume and weight of soils are the unit weight, the water content, the void ratio, the porosity, the degree of saturation, the specific gravity of solids, and the density index (Table 3.1).

The *natural unit weight* (total weight of soil W_t over total volume of soil V_t) is the unit weight of the soil as it is found in its natural environment. The natural unit weight is also called *total unit weight* or simply *unit weight*. Numbers between 17 and 22 kN/m^3 are common. The *dry unit weight* is the unit weight of the dry soil (weight of solids W_s over total volume of soil V_t). Numbers between 14 and 18 kN/m^3 are common. The *saturated unit weight* is the ratio of the weight of the soil when the voids are full of water or liquid over the total volume of the soil. The weight of the saturated soil is the weight of solids plus the weight of water necessary to fill the voids ($W_s + V_v\gamma_w$). Numbers between 18 and 22 kN/m^3 are common. The *submerged unit weight* is the difference between the saturated unit weight and the unit weight of water. The *effective unit weight* is equal to the total unit weight for a point in the soil mass above the groundwater level and equal to the submerged unit weight for a point below the groundwater level. The unit weight of solids γ_s is the unit weight of the particle itself. It is the ratio of the weight of solids W_s over the volume of solids V_s. The unit weight of solids varies depending on the composition of the particles (mineral or organic matter), but for mineral matter it is in the range of 25.5 to 27 kN/m^3 and for organic matter it is in the range of 9 to 13 kN/m^3. The specific gravity of solids G_s is the ratio between the unit weight of solids γ_s and the unit weight of water γ_w.

The water content w, also called *gravimetric water content*, is the ratio of the weight of water W_w over the weight of solids W_s or weight of dry soil. Although the water content is a ratio and should be used as such in most formulas, it is most often quoted as a percentage. Numbers around 10 to 40% are common, but the water content can be 0 for a dry soil and can reach 400%, as in the Mexico City silt, or even 2000% for some peaty soils (soil near swamps, made up mostly of grass and plants). Indeed, there is no theoretical upper limit to the water content. The gravimetric water content is the water content measure most widely used in geotechnical

Table 3.1 Weight-Volume Parameters and Typical Values

Parameter	Symbol	Definition	Typical range	Observation
Total unit weight	γ_t	W_t/V_t	17–22 kN/m^3	Unit weight of soil in natural state
Dry unit weight	γ_d	W_s/V_t	14–18 kN/m^3	Unit weight of dry soil
Maximum dry unit weight	γ_{dmax}	W_{smax}/V_t	15–19 kN/m^3	Densest state
Minimum dry unit weight	γ_{dmin}	W_{smin}/V_t	13–17 kN/m^3	Loosest state
Unit weight of solids	γ_s	W_s/V_s	25.5–27 kN/m^3 for mineral 9–13 kN/m^3 for organic	Unit weight of particles
Specific gravity of solids	G_s	γ_s/γ_w	2.6–2.7 for mineral 0.9–1.3 for organic	Dimensionless
Saturated unit weight	γ_{sat}	$(W_s + V_v\gamma_w)/V_t$	18–22 kN/m^3	Voids are full of water
Submerged unit weight	γ_{sub}	$\gamma_{sat}-\gamma_w$	8–12 kN/m^3	Buoyancy force accounted for
Effective unit weight	γ_{eff}	γ_t if above GWL $\gamma_{sat}-\gamma_w$ if below GWL	See γ_t and γ_{sub}	
Unit weight of water	γ_w	W_w/V_w	9.81 kN/m^3	
Water content (gravimetric)	w	W_w/W_s	10–40%	0–∞ theoretical range
Volumetric water content	θ_w	V_w/V_t	5–30%	0–1 theoretical range
Degree of saturation	S	V_w/V_v	50–100%	0–100% theoretical range
Porosity	n	V_v/V_t	25–50%	0–100% theoretical range
Void ratio	e	V_v/V_s	0.4–1	0–∞ theoretical range
Maximum void ratio	e_{max}	V_{vmax}/V_s	0.6–1.2	Loosest state
Minimum void ratio	e_{min}	V_{vmin}/V_s	0.3–0.9	Densest state
Density Index	I_d or D_r	$(e_{max}-e)/(e_{max}-e_{min})$	20–90%	0–100% theoretical range

engineering. Sometimes for unsaturated soils, the volumetric water content θ_w is used; θ_w is defined as the ratio of the volume of water V_w over the total volume V_t. Numbers between 5 and 30% are common; θ_w is zero for a dry soil and approaches 100% for extremely wet soils such as peat. The degree of saturation S is the volume of water V_w over the volume of voids V_v. Although the degree of saturation is a ratio and should be used as such in most formulas, it is most often quoted as a percentage. Numbers from 0 to 100% are found, although most soils below the groundwater level and some distance above it are saturated or nearly saturated. In many cases soils near the surface are unsaturated.

The porosity n is the ratio of the volume of voids V_v over the total volume V_t. Although the porosity is a ratio and should be used as such in most formulas, it is most often quoted as a percentage. Numbers in the range of 25 to 50% are common, and the porosity is always between 0 and 100%. The void ratio e is the ratio of the volume of voids V_v over the volume of solids V_s. It is most often quoted as a number. Numbers from 0.4 to 1 are common. Although the theoretical limits of the void ratio are 0 and infinity, the practical limits for a given soil are the minimum void ratio e_{min} and the maximum void ratio e_{max}. The minimum void ratio corresponds to the densest state of a given soil, and the maximum void ratio corresponds to the loosest state for a given soil. Both e_{min} and e_{max} are particularly useful in the case of coarse-grained soils and lead to the definition of the density index I_d (also designated as D_r), which is quoted as a percentage and expresses the density of a coarse-grained soil as a percentage between the two extreme states of density ($I_d = (e_{max} - e)/(e_{max} - e_{min})$). Also associated with the densest and loosest states are the maximum and minimum dry densities γ_{dmax} and γ_{dmin}. Note that γ_{dmax} corresponds to e_{min} and that γ_{dmin} corresponds to e_{max}. The density index can be expressed in terms of γ_d, γ_{dmax}, and γ_{dmin} as

$$I_d = \frac{\gamma_{d\,max}}{\gamma_d}\left(\frac{\gamma_d - \gamma_{d\,min}}{\gamma_{d\,max} - \gamma_{d\,min}}\right) \tag{3.2}$$

3.9 MEASUREMENT OF THE WEIGHT-VOLUME PARAMETERS

To obtain the natural or total unit weight of a soil, the sample is trimmed into a simple geometrical shape, the dimensions are measured to obtain the volume, the weight is measured, and the weight over volume is calculated (ASTM 2005b [ASTM D2937]). This test is possible only if the sample can keep a geometric shape long enough for the measurements to be made. If this is not possible, as in the case of a dry sand or gravel for example, then the unit weight is typically obtained by correlation with other measurements such as the blow count during a standard penetration test (SPT) (ASTM 2005a [ASTM D1586]). The water content is obtained by taking a small piece of the sample and measuring its wet weight (W_t), drying it in an oven at 100°C for 24 hours, and obtaining its dry weight W_s, and then calculating the water content $w = (W_t - W_s)/W_s$ (ASTM 2005c [ASTM D4959]). These two measurements, natural unit weight and water content, are the two most common measurements on a soil sample.

Unless the sample is dry or saturated, a third input parameter is necessary to obtain all the weight-volume parameters for a soil. This parameter is often the specific gravity of solids G_s. If it is known that the soil particles are mineral and not organic, then a reasonable assumption can be made for G_s, such as $G_s = 2.65$. If the composition is not known, or if a more precise value for G_s is needed, then G_s is determined by the specific gravity test (ASTM 2005d [ASTM D854]). This test consists of drying the soil in an oven, pulverizing it by grinding, placing the ground-up material in a container, and filling the container with water up to a chosen level. The container with water plus soil is weighed. Then the container is emptied, cleaned, and filled up to the same chosen level with water only and weighed. The weight of the container with water plus soil minus the weight of the container with water only gives the weight of the buoyant soil. The buoyancy force is the difference between the weight of the buoyant soil and the dry soil. The ratio of the dry weight over the buoyancy force is the specific gravity of the solids.

If the unit weight of the soil, its water content, and the specific gravity of solids are known, all other weight-volume parameters can be obtained by calculations (see section 3.10), including the dry unit weight, the saturated unit weight, the submerged unit weight, the effective unit weight, the degree of saturation, the porosity, and the void ratio.

Finding the density index of a coarse-grained soil requires two special tests in addition to the determination of the natural dry unit weight γ_d: one test to obtain the maximum dry unit weight γ_{dmax} (ASTM D4253) and one test to obtain the minimum dry unit weight γ_{dmin} (ASTM D4254). The maximum dry unit weight is obtained by pouring the dry sand or dry gravel into a container of known volume, placing a standard weight on top of the sample surface, and vibrating the soil and the container for a standard time. During the vibrations, the soil volume decreases and reaches equilibrium at the maximum dry unit weight. Measurements of weight and volume at that time allow one to calculate the maximum dry unit weight. The minimum dry unit weight is obtained by very gently pouring a dry sand or gravel sample into a container of known volume, measuring the weight, and calculating the dry unit weight. Once γ_d, γ_{dmax}, and γ_{dmin} are known, the density index I_d (or D_r) is calculated according to equation 3.2.

3.10 SOLVING A WEIGHT-VOLUME PROBLEM

Geotechnical engineers often encounter problems where some information related to the weight or volume of a soil is known but different weight-volume properties are required. The best way to solve such problems is to follow these steps:

1. Draw a three-phase diagram and indicate the known quantities. If the soil is dry or saturated, then only a two-phase diagram is necessary.
2. If no quantity is given (for example, you are given a unit weight but not a weight or a volume), assume a volume of solid of 1 m^3.
3. Using the information in the specific problem case, complete the weight and volume values for the different phases. If some information is missing, make reasonable assumptions (e.g., the unit weight of solids.) Also realize that the unit weight of water is known (9.81 kN/m^3).
4. Complete the calculations to derive the weight-volume parameters required.

The assumptions made in step 2 have no impact on the answers as long as the answers are in the form of ratios (unit weight, void ratio, porosity, degree of saturation); if a different volume of solids were assumed, the final answer would be the same. Although the step-by-step procedure described here is foolproof, it might be faster in some cases to use the relationships existing between weight-volume parameters. Table 3.2 shows some of these.

Table 3.2 Useful Relationships between Weight-Volume Parameters

$n = e/(1 + e)$
$e = n/(1 - n)$
$e = (\gamma_s - \gamma_d)/\gamma_d$
$Se = G_s w$
$W_s = W_t/(1 + w)$
$\gamma_t = \gamma_d\,(1 + w)$
$\gamma_t = \gamma_w\,(G_s(1 - n) + Sn)$
$\gamma_t = \gamma_w\,(G_s + Se)/(1 + e)$
$\gamma_d = \gamma_w\,G_s(1 - n)$

PROBLEMS

3.1 A sample of clay is brought back from the field, extruded from the Shelby tube, and trimmed to the following dimensions: height = 150 mm, diameter = 75 mm. It weighs 13.2 N. The water content has been determined to be 25% and the soil does not exhibit any signs of the presence of organic matter (e.g., the soil is not very dark and does not smell foul). Find the following parameters for the clay:

a. Natural unit weight
b. Degree of saturation
c. Porosity
d. Void ratio
e. Dry unit weight
f. Saturated unit weight

3.2 a. The sample from problem 3.1 shrinks by 10% when it dries. What is the difference between the dry unit weight and the unit weight of the dry soil?

b. The sample from problem 3.1 is placed under water and has swollen by 15% when it reaches its swell limit. What is the difference between the saturated unit weight and the unit weight of the soil at the swell limit?

3.3 A farmer wants to buy a 10 kg bag of fertilizer (organic soil). He has the choice between two merchants. Merchant A sells the 10 kg bag for $10 and the bag indicates that the fertilizer is completely dry. Merchant B sells the 10 kg bag for $8 and the bag indicates that the fertilizer has a water content equal to 20%. If the farmer wishes to buy the least expensive solid constituents, which merchant should he buy from? Show your calculations.

3.4 An airport runway is being extended into a bay and requires a 10 m high embankment above the bottom of the bay. Calculations indicate that, once constructed, the long-term settlement of the soil beneath the embankment will be about 1 m. The sand used to build the embankment is taken from a pit where the sand has a relative density of 40%. The maximum void ratio is 0.7; the minimum void ratio is 0.4. Once compacted in the embankment, the sand will have a relative density of 90%. What height of sand must be obtained from the borrow pit so that, a long time after completion, the embankment will be 10 m above the initial position of the bottom of the bay before construction started?

3.5 A shrink test is performed on a sample of clay. At time zero, the sample is 25 mm high, 75 mm in diameter, weighs 2.2 N, and is saturated. The sample is left on a laboratory table; this laboratory is at 20°C and 50% relative humidity. The sample dries and shrinks. It is weighed and the dimensions are measured with digital calipers as a function of time. At the end of the test, the sample is placed in the oven to obtain its dry weight, which comes out to be 1.8 N. The results of the test are shown in the following table.

Time (hr)	0	1	2	3	5	8	12	24
Height (mm)	25	24.932	24.662	24.490	24.315	24.138	23.958	23.958
Diameter (mm)	75	74.497	73.987	73.470	72.946	72.414	71.874	71.874
Weight (N)	2.200	2.160	2.115	2.079	2.034	1.989	1.944	1.872

Plot the curve of water content versus relative decrease in volume. Comment on the shape of that curve.

3.6 A 2.2 N sample of clay is 25 mm high and 75 mm in diameter and has a water content of 22.2% (same sample as in problem 3.5). It is placed in a stainless steel ring has the same dimensions as the sample, so the sample cannot expand laterally. The sample is inundated and allowed to swell vertically for several weeks until it reaches equilibrium. The height and time measurements are shown in the following table. At the end of this free swell test, the sample is taken out of the steel ring and weighed; it weighs 2.40 N. Plot the relative increase in volume of the sample versus time and calculate the swell limit for this clay. The *swell limit* is the water content at which the soil can no longer absorb any additional water.

Time (hr)	0	100	200	300	400	500
Height (mm)	25	28	29.5	30.25	30.75	31

Problem 3.2

a. The sample from problem 3.1 shrinks by 10% when it dries. What is the difference between the dry unit weight and the unit weight of the dry soil?
b. The sample from problem 3.1 is placed under water and has swollen by 15% when it reaches its swell limit. What is the difference between the saturated unit weight and the unit weight of the soil at the swell limit?

Solution 3.2

a. Shrinking case

The volume of the sample is $6.627 \times 10^{-4}\ \text{m}^3$.

The volume of the sample after the 10% reduction due to shrinkage is $V_{T\ (\text{Shrink})} = 6.627 \times 10^{-4}\ \text{m}^3 \times 0.90 = 5.96 \times 10^{-4}\ \text{m}^3$.

$$\gamma_{dried\ soil} = \frac{W_S}{V} = \frac{10.6}{5.96 \times 10^{-4}\ \text{m}^3} = 17.78 \times 10^3 \frac{N}{\text{m}^3} = 17.78 \frac{\text{kN}}{\text{m}^3}$$

Based on these results, the unit weight of the dry soil is higher than the dry unit weight and the difference is $17.78 - 15.94 = 1.84 \frac{\text{kN}}{\text{m}^3}$.

b. Swelling case

The volume of the sample after the 15% volume increase due to swelling is $6.627 \times 10^{-4}\ \text{m}^3 \times 1.15 = 7.62 \times 10^{-4}\ \text{m}^3$. It is assumed that during the swelling process the soil becomes completely saturated. Therefore, the volume of air in the original sample is replaced by a volume of water.

The increase in weight of the sample is equal to the weight of water corresponding to an increase in volume of water equal to $(7.62 - 6.627) \times 10^{-4}\ \text{m}^3$ plus the *volume* of water necessary to fill the air voids in the original sample.

$$\text{Additional weight of water: } (7.62 - 6.627) \times 10^{-4} \times 9.81 + 2.48 \times 10^{-6} \times 9.81 = 9.98 \times 10^{-4}\ \text{kN} = 0.998\ \text{N}$$

$$\gamma_{swollen\ soil} = \frac{W_T}{V} = \frac{13.2 + 0.998}{7.62 \times 10^{-4}\ \text{m}^3} = 18.63 \times 10^3 \frac{N}{\text{m}^3} = 18.6\,3 \frac{\text{kN}}{\text{m}^3}$$

Based on these results, the unit weight of the swollen soil at the swell limit is lower than the unit weight of the saturated soil in the previous problem and the difference is: $19.95 - 18.63 = 1.32 \frac{\text{kN}}{\text{m}^3}$.

Problem 3.3

A farmer wants to buy a 10 kg bag of fertilizer (organic soil). He has the choice between two merchants. Merchant A sells the 10 kg bag for \$10 and the bag indicates that the fertilizer is completely dry. Merchant B sells the 10 kg bag for \$8 and the bag indicates that the fertilizer has a water content equal to 20%. If the farmer wishes to buy the least expensive solid constituents, which merchant should he buy from? Show your calculations.

Solution 3.3

Case 1

Merchant A (fertilizer in completely dry condition). The three-phase diagram for the fertilizer from merchant A is shown in Figure 3.2s.

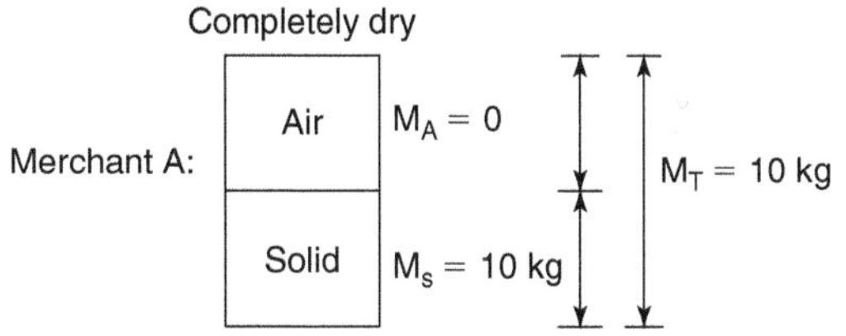

Figure 3.2s Three-phase diagram for the fertilizer from merchant A.

The unit price for the solid constituents of merchant A is $\frac{10\$}{10\ \text{kg}} = 1\$/\text{kg}$

Case 2

Merchant B (fertilizer with water content = 20%). The three-phase diagram for the fertilizer from merchant B is shown in Figure 3.3s.

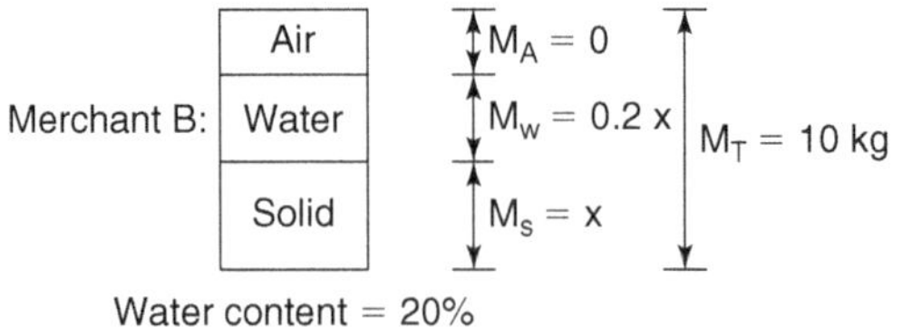

Figure 3.3s Three-phase diagram for the fertilizer from merchant B.

Assume that the mass of the solids is x; then the mass of the water is $0.2x$. The total mass of the fertilizer bag is $1.2x$, which is equal to 10 kg. So the mass of the solids can be obtained from the following equation: $1.2x = 10$.

$$\text{So}\, x = 8.33\ \text{kg}$$

The unit price for the solid constituents in merchant B's bag is: $\frac{8\$}{8.33\ \text{kg}} = 0.96\$/\text{kg}$

So, the farmer should buy the fertilizer from merchant B.

Problem 3.4

An airport runway is being extended into a bay and requires a 10 m high embankment above the bottom of the bay. Calculations indicate that, once constructed, the long-term settlement of the soil beneath the embankment will be about 1 m. The sand used to build the embankment is taken from a pit where the sand has a relative density of 40%. The maximum void ratio is 0.7; the minimum void ratio is 0.4. Once compacted in the embankment, the sand will have a relative density of 90%. What height of sand must be obtained from the borrow pit so that, a long time after completion, the embankment will be 10 m above the initial position of the bottom of the bay before construction started?

Solution 3.4

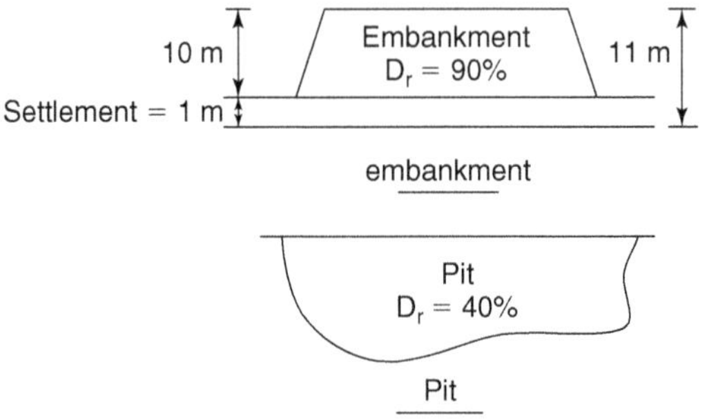

Figure 3.4s Illustration of embankment and pit.

The void ratio of the soil in the pit is obtained with the equation:

$$D_r = \frac{e_{max} - e_{40}}{e_{max} - e_{min}} = \frac{0.7 - e_{40}}{0.7 - 0.4} = 0.4$$

Therefore, $e_{40} = 0.58$. For the soil after compaction:

$$D_r = \frac{e_{max} - e_{90}}{e_{max} - e_{min}} = \frac{0.7 - e_{90}}{0.7 - 0.4} = 0.9$$

Therefore, $e_{90} = 0.43$

The three-phase diagram for the soil in both conditions is shown in Figure 3.5s.

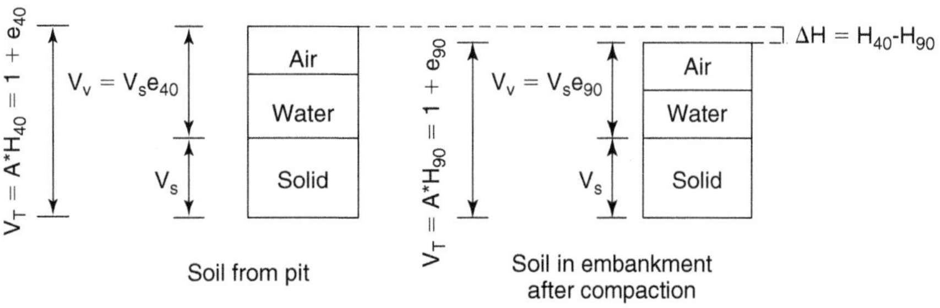

Figure 3.5s Three-phase diagram for the soil in the two conditions.

Based on the three-phase diagram in Figure 3.5s, we can write:

$$\frac{H_{40}}{H_{90}} = \frac{V_s(1 + e_{40})}{V_s(1 + e_{90})}$$

Knowing that the long-term settlement of the soil in the bay beneath the embankment will be 1 m, the total height of soil necessary is 11 m. We have to calculate the height of soil H_{40} that should be taken from the pit with a 40% relative density such that when compacted to 90% relative density, the height H_{90} will equal 11 m.

$$\frac{H_{40}}{H_{90}} = \frac{H_{40}}{11} = \frac{V_s(1 + e_{40})}{V_s(1 + e_{90})} = \frac{1 + 0.58}{1 + 0.43} \quad \text{or} \quad H_{40} = 12.15 \text{ m}$$

The height of the soil that must be taken from the pit is 12.15 m.

Problem 3.5

A shrink test is performed on a sample of clay. At time zero, the sample is 25 mm high, 75 mm in diameter, weighs 2.2 N, and is saturated. The sample is left on a laboratory table; this laboratory is at 20°C and 50% relative humidity. The sample dries and shrinks. It is weighed and the dimensions are measured with digital calipers as a function of time. At the end of the test, the sample is placed in the oven to obtain its dry weight, which comes out to be 1.8 N. The results of the test are shown in the following table.

Time (hr)	0	1	2	3	5	8	12	24
Height (mm)	25	24.932	24.662	24.490	24.315	24.138	23.958	23.958
Diameter (mm)	75	74.497	73.987	73.470	72.946	72.414	71.874	71.874
Weight (N)	2.200	2.160	2.115	2.079	2.034	1.989	1.944	1.872

Plot the curve of water content versus decrease in volume. Comment on the shape of that curve.

Solution 3.5

The water content and decrease in volume at each reading is given by the following equations:

$$w\% = \frac{(W_{Total} - W_{dry})}{W_{dry}} \times 100$$

$$\Delta V = \frac{\pi D_o^2}{4} H_o - \frac{\pi D^2}{4} H$$

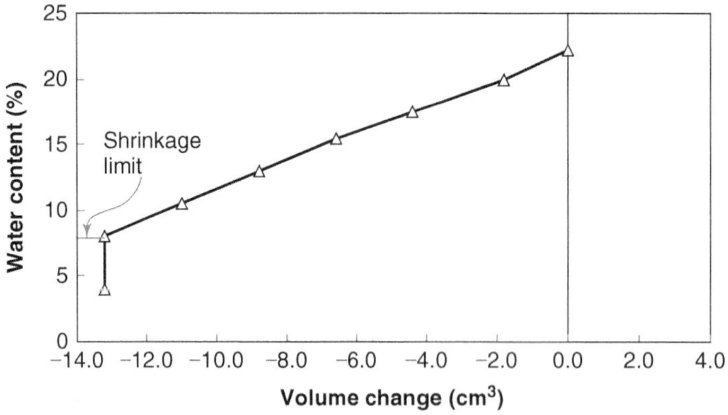

Figure 3.6s Water content versus decrease in volume.

We can make two observations:

1. The curve is almost linear for most of the test. This indicates that within that range of water content, the relative change in volume of the soil is linearly proportional to the change in water content.
2. At a water content of 8%, further drying does not lead to further reduction in volume. The soil has reached its shrinkage limit, which is 8% in this case. Note that this shrinkage limit is the shrinkage limit of the undisturbed soil, not the shrinkage limit of the Atterberg limit that would be obtained from a remolded sample.

Problem 3.6

A 2.2 N sample of clay is 25 mm high and 75 mm in diameter and has a water content of 22.2% (same sample as in problem 3.5). It is placed in a stainless steel ring has the same dimensions as the sample, so the sample cannot expand laterally. The sample is inundated and allowed to swell vertically for several weeks until it reaches equilibrium. The height and time measurements are shown in the following table. At the end of this free swell test, the sample is taken out of the steel ring and weighed; it weighs 2.40 N. Plot the relative increase in volume of the sample versus time and calculate the swell limit for this clay. The *swell limit* is the water content at which the soil can no longer absorb any additional water.

Time (hr)	0	100	200	300	400	500
Height (mm)	25	28	29.5	30.25	30.75	31

Solution 3.6

The volume, relative volume, and relative increase in volume are calculated in the following table:

Time (hr)	0	100	200	300	400	500
Height (mm)	25	28	29.5	30.25	30.75	31
Volume (cm^3)	110.45	123.70	130.33	133.64	135.85	136.95
Relative volume V/V_0	1	1.12	1.18	1.21	1.23	1.24
Relative increase in volume $\Delta V/V_0$	0	0.12	0.18	0.21	0.23	0.24
$t/(\Delta V/V_0)$		833	1111	1428	1739	2083

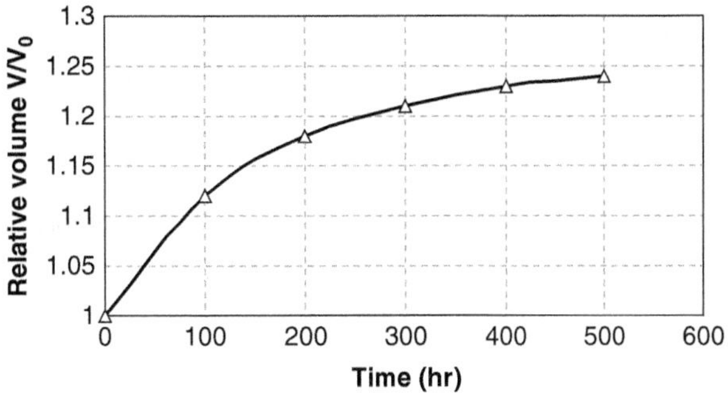

Figure 3.7s Relative volume variation with time.

As can be seen in Figure 3.7s, the sample did not reach the swell limit because at the end of the test its volume is still increasing slightly. One way to solve this issue is to use the hyperbolic extension method. To do so, we assume that the curve in the figure is a hyperbola with an equation:

$$\frac{\Delta V}{V} = \frac{t}{a + bt}$$

To determine the constants a and b, we write:

$$\frac{t}{(\Delta V/V)} = a + bt$$

Then we plot $\frac{t}{(\Delta V/V_0)}$ versus t and fit a straight line through the data as shown in Figure 3.8s.

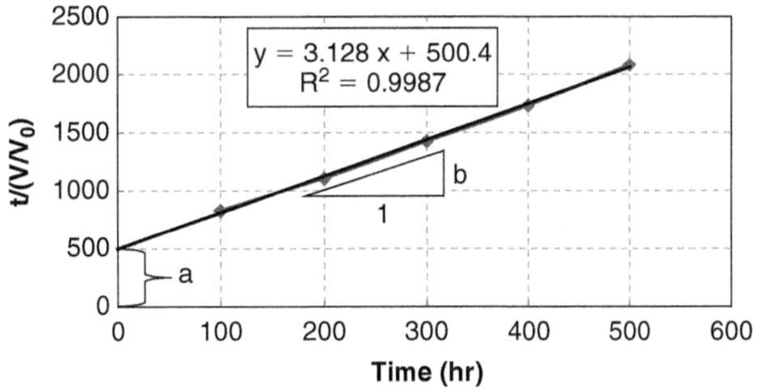

Figure 3.8s Graph showing the parameters a and b.

According to this graph, a = 500.4 hr and b = 3.13. The extended swell test curve is shown in Figure 3.9s.

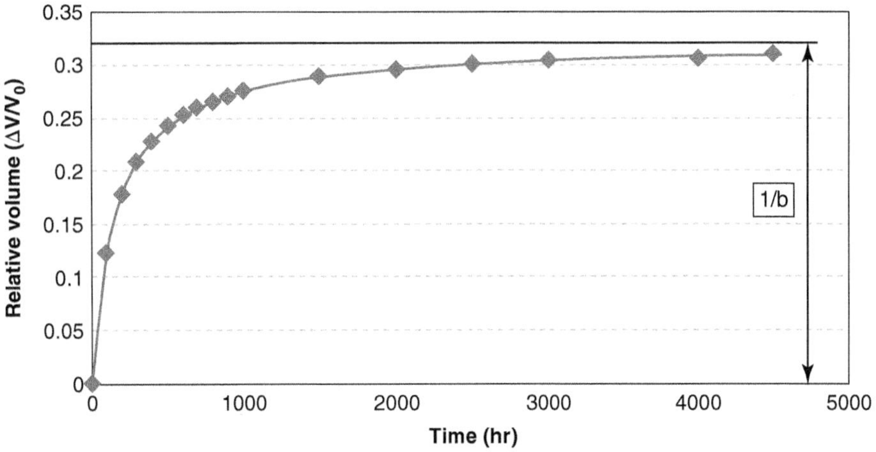

Figure 3.9s Extended swell test curve.

When t goes to infinity, $\Delta V/V$ goes to $1/b$.

$$\lim_{t\to\infty}\left(\frac{\Delta V}{V}\right) = \lim_{t\to\infty}\left(\frac{t}{a+bt}\right) = \frac{1}{b}$$

So, the limit value of $\Delta V/V = 1/b = 0.319$. The asymptotic volume of the sample at the swell limit is:

$$\frac{V_{\text{swell limit}}}{V_{initial}} = 1.319 \quad or$$

$$V_{\text{swell limit}} = 1.319 \times 2.5 \times \frac{7.5^2}{4} \times \pi = 145.67 \times 10^{-6}\ \text{m}^3$$

The volume of solids in the sample is:

$$W_{dry} = 1.8\ \text{N}$$

$$W_w = 2.2 - 1.8 = 0.4\ \text{N}$$

$$V_w = \frac{0.4}{9810} \times 10^6 = 40.77 \times 10^{-6}\ \text{m}^3$$

$$V_s = \overbrace{25 \times \frac{75^2}{4} \times \pi \times \frac{1}{1000}}^{\text{Total volume}} - 40.77 = 69.67 \times 10^{-6}\ \text{m}^3$$

At the swell limit, the water content of the sample is:

$$V_w = V_t - V_s = 145.67 - 69.67 = 76 \times 10^{-6}\ \text{m}^3$$

$$W_w = 76 \times 10^{-6} \times 9810 = 0.745\ \text{N}$$

$$w\% = \frac{W_w}{W_s} \times 100 = \frac{0.745}{1.8} \times 100 = 41.3\%$$

Problem 3.7

A silty sand is compacted in a mold. The volume of the mold is 9.46×10^{-4} m^3. The weight of compacted soil in the mold is 18.9 N and the water content is 8%. Assume that G_s is 2.65 and calculate the dry unit weight and the degree of saturation.

Solution 3.7

The volume of the sample is:

$$V_t = 9.46 \times 10^{-4}\ \text{m}^3$$

The weight of the sample is:

$$W_t = 18.9\ N$$

The weight of the solids is:

$$W_s = \frac{W_t}{1+\omega} = \frac{18.9}{1.08} = 17.5\ N$$

The weight of the water is:

$$W_w = W_t - W_s = 18.9 - 17.5 = 1.40\ \text{N}$$

Assuming that the density of solids is $G_s = 2.65$, the volume of solids is:

$$V_s = \frac{W_s}{G_s \times \gamma_w} = \frac{17.5}{2.65 \times 9810} = 6.74 \times 10^{-4}\ \text{m}^3$$

The volume of water is:

$$V_w = \frac{W_w}{\gamma_w} = \frac{1.40}{9810} = 1.43 \times 10^{-4}\ m^3$$

The volume of air is: $V_a = V_t - V_w - V_s = (9.46 - 6.74 - 1.42) \times 10^{-4} = 1.29 \times 10^{-4} m^3$

<table>
<tr><td rowspan="3">$V = 9.46 \times 10^{-4}$ m^3</td><td rowspan="2">$V_v = 2.69 \times 10^{-4}$ m^3</td><td>$V_A = (9.46–1.43–6.74) \times 10^{-4} = 1.29 \times 10^{-4}$ m^3</td><td>Air</td><td>$W_A = 0$</td><td rowspan="3">W = 18.91 N</td></tr>
<tr><td>$V_w = 1.4/9810 = 1.43 \times 10^{-4}\ m^3$</td><td>Water</td><td>$W_w = 0.08 \times 17.51 = 1.40$</td></tr>
<tr><td colspan="2">$V_s = 17.51/9810/2.65 = 6.74 \times 10^{-4}\ m^3$</td><td>Soil</td><td>$W_s = 18.91/1.08 = 17.51$ N</td></tr>
</table>

Figure 3.10s Three-phase diagram.

The degree of saturation can be calculated from the following formula:

$$S = \frac{V_w}{V_v} = \frac{1.43 \times 10^{-4}}{(1.29 + 1.43) \times 10^{-4}} \times 100 = 53\%$$

or

$$w.G_s = S.e$$

$$0.08 \times 2.65 = S \times \frac{1.29 + 1.43}{6.74}$$

$$S = 53\%$$

The dry unit weight of the sample is:

$$\gamma_d = \frac{W_s}{V_t} = \frac{17.5}{9.46 \times 10^{-4}} = 18.5\ \frac{kN}{m^3}$$

Problem 3.8

A consolidation test is performed on a sample of soft clay that is 25 mm high and 50 mm in diameter. The test consists of placing a disk of soil in a steel ring and applying load on the sample in a series of steps. The steps last 24 hours and measurements of vertical compression are obtained at the end of each step. The following table shows the time, load, and compression results of the test. Calculate the pressure and vertical strain for the sample at the end of each load step and plot the curve that links the pressure to the vertical strain (stress-strain curve). Why does this curve indicate an apparently surprising result, in that the more load applied to the sample, the stiffer the sample becomes? Can the sample fail?

Time (days)	0	1	2	3	4	5	6
Load (N)	0	294	589	1177	2453	4906	9812
Stress (kN/m^2)	0	149.73	299.98	599.45	1249.31	2498.61	4997.21
Height (mm)	25	24.62	24.25	23.62	22.87	21.47	19.7
Displacement (mm)	0	0.38	0.75	1.38	2.13	3.53	5.3
Strain	0	0.0152	0.03	0.0552	0.0852	0.1412	0.212

Solution 3.8

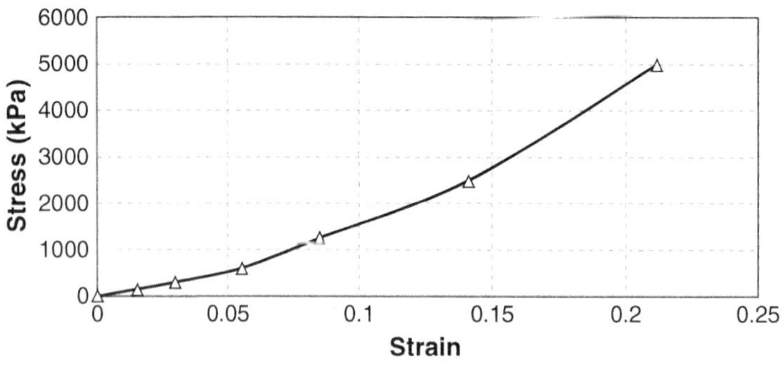

Figure 3.11s Stress-strain curve.

The stress-strain curve indicates that the soil become stiffer as the stress increases. Indeed, the ratio of stress increment to strain increment becomes increasingly larger. The reason is that as the stress increases, the influence of the steel ring becomes more important in providing confinement to the sample. With increased confinement, the sample becomes stiffer. The sample cannot fail unless the steel ring fails.

Problem 3.9

A silt has a unit weight of 20 kN/m^3 and a water content of 26%. What is the specific gravity of the particles?

Solution 3.9

The information given in this problem is not sufficient to solve for specific gravity; thus, the exact answer is that it is not possible to solve this problem. However, if we make one assumption, then it becomes possible. For example, assume that the sample is saturated.

$$w = 0.26, Y = 20\ kN/m^3, Y_w = 10\ kN/m^3, S = 1$$

$$\gamma = \frac{G_s\ \gamma_w(1+w)}{1+e}$$

$$S.e = G_s.w \quad \text{and} \quad S = 1 \quad \text{therefore} \quad e = G_s.w$$

$$\gamma = \frac{G_s\ \gamma_w(1+w)}{1+G_s.w} \rightarrow 20 = \frac{G_s \times 10(1+0.26)}{1+G_s\ .0.26} \rightarrow G_s = 2.7$$

Problem 3.10

A 5 m high embankment is made of a sand that has a void ratio of 0.55 and the following boundary void ratios: $e_{max} = 0.6$, $e_{min} = 0.4$. The embankment is subjected to an earthquake that creates a settlement of 0.32 m due to vibration without a change in lateral dimensions. Calculate the void ratio and the relative density of the sand after the earthquake.

Solution 3.10

Let's assume a reference volume of solids equal to 1 m^3:

$$e = \frac{V_v}{V_s} = \frac{V_v}{1} = V_v.$$

The total height of the embankment, H, is proportional to $1 + e$, so that

$$\frac{H(\text{before earthquake})}{H(\text{after earthquake})} = \frac{5}{5-0.32} = \frac{1+0.55}{1+e(\text{after earthquake})}$$

$$\text{Therefore } e(\text{after earthquake}) = \frac{(5-0.32)}{5} \times (1+0.55) - 1 = 0.45$$

The relative density is

$$D_r = \frac{e_{max} - e}{e_{max} - e_{min}} \times 100(\%) = \frac{0.6-0.45}{0.6-0.4} \times 100 = 75\%$$

Problem 3.11

Find the relationship between the dry unit weight, the void ratio, and the specific gravity of solids for a soil.

Solution 3.11

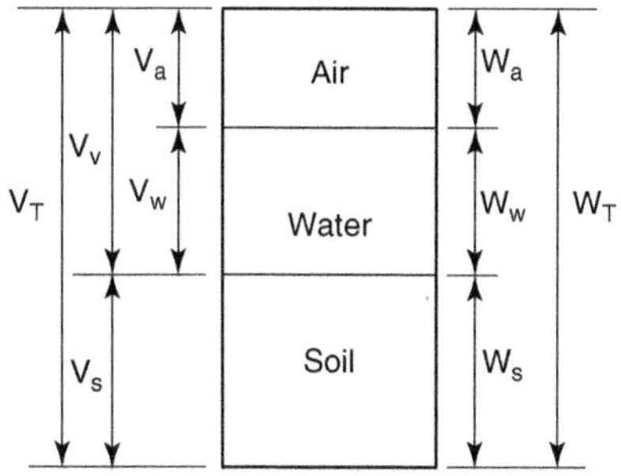

Figure 3.12s Three-phase diagram.

Definition:

$$S = V_w/V_v,\ e = V_v/V_s,\ G_s = W_s/(V_s \cdot \gamma_w),$$
$$w = W_w/W_s,\ \gamma_t = W_T/V_T, \text{ and } \gamma_d = W_s/V_T$$

Using the definition of S and e, the volume of water and air can be rewritten as:

$$V_a = V_v - V_w = e \cdot V_s - S \cdot e \cdot V_s = (1 - S) \cdot e \cdot V_s$$
$$V_w = S \cdot e \cdot V_s$$

Using the relationships and definitions, the dry unit weight is:

$$\gamma_d = \frac{W_s}{V_T} = \frac{W_s}{V_s + V_w + V_a} = \frac{G_s \cdot \gamma_w . V_s}{V_s + V_w + V_a}$$
$$= \frac{G_s . \gamma_w . V_s}{V_s + S \cdot e \cdot V_s + (1 - S) \cdot e \cdot V_s} = \frac{G_s \cdot \gamma_w . V_s}{V_s \cdot (1 + e)} = \frac{G_s \cdot \gamma_w}{1 + e}$$

Problem 3.12

Demonstrate that $S \cdot e = G_s \cdot w$.

Solution 3.12

Definition:

$$S = V_w/V_v,\ e = V_v/V_s,\ G_s = W_s/(V_s \cdot \gamma_w), \text{ and } w = W_w/W_s$$

$$G_s \cdot w = \frac{W_s}{V_s \cdot \gamma_w} \cdot \frac{W_w}{W_s} = \frac{W_w}{V_s \cdot \gamma_w} = \frac{V_w \cdot \gamma_w}{V_s \cdot \gamma_w} = \frac{V_w}{V_s}$$
$$S \cdot e = \frac{V_w}{V_v} \cdot \frac{V_v}{V_s} = \frac{V_w}{V_s}$$
$$S \cdot e = G_s \cdot w$$

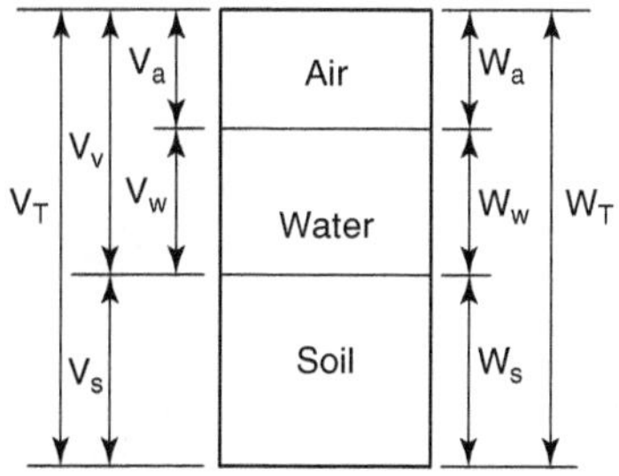

Figure 3.13s Three-phase diagram.

CHAPTER 4

Soil Classification

To classify a soil, tests are performed according to the American Society for Testing and Materials (ASTM) standards, and the results of these tests are used in a classification system recommended by the International Society for Soil Mechanics and Geotechnical Engineering (ISSMGE). The tests are the sieve analysis, the hydrometer analysis, and the Atterberg limits. The classification system is called the Unified Soil Classification System (USCS).

4.1 SIEVE ANALYSIS

Sieve analysis is used for the classification of gravels and sands, which are coarse-grained soils. It consists of taking a given weight of dry soil, breaking the clumps of soil down to individual particles (using a mortar and rubber-tipped pestle), washing the soil through the smallest sieve (sieve #200), drying what remains on the sieve #200, and then sieving that remainder by shaking it through a stack of sieves of decreasing openings (Figure 4.1), the last one being a retaining pan. Recording all the weights involved during this process leads to the percent of soil finer than a given particle size by weight versus the particle size; this is the particle size distribution curve (Figure 4.2).

A typical set of sieve numbers and sieve openings is given in Table 4.1. The sieve number corresponds to the number of openings per 25 mm. For example, the no. 200 sieve—the smallest sieve commonly used—has 200 openings per 25 mm; however, each opening is not equal to 25 mm divided by 200 because of the thickness of the wire between openings. In fact, the opening of the no. 200 sieve is 0.075 mm. This opening corresponds to the boundary between sand- and silt-size particles; this is why sieve analysis is limited to the classification of gravels and sands.

The sieve analysis proceeds as follows. First, each sieve is weighed empty. Then the dry soil sample is weighed, soil clumps are broken down, and the soil sample is placed on a sieve #200. The sample is washed under a gentle stream of water and the soil left on the sieve #200 is dried in the oven. The purpose is to wash out the fine particles that may adhere to the larger particles or form clumps. Sieves are stacked in order of increasing opening, with the largest-opening sieve at the top. The dry soil is placed on the top sieve, which is then covered so that no soil is ejected during shaking. The stack is shaken in a vibrator for a given period of time. At the end of shaking, each sieve is weighed with the soil retained on it.

Because the total weight of the dry sample is known, the proportion of the soil sample on each sieve is calculated as the weight of that sieve plus soil minus the weight of the empty sieve divided by the total weight of the sample. With this data, the particle size distribution curve can be obtained. This curve is a plot of the percent finer by weight (sum of the weight of soil passing a certain sieve divided by the total weight of the sample, expressed as a percentage) on the vertical axis and the sieve opening taken as the particle size on the horizontal axis (log scale). Figure 4.3 shows the sieves and the dry weight retained on each sieve. The sieve analysis calculations are shown in Table 4.2. Figure 4.4 gives examples of particle size distribution curves. Note that the particle size determined by sieving through a given sieve is the second largest dimension of the particle that can pass through the sieve opening.

Figure 4.1 Stack of sieves and shaker.

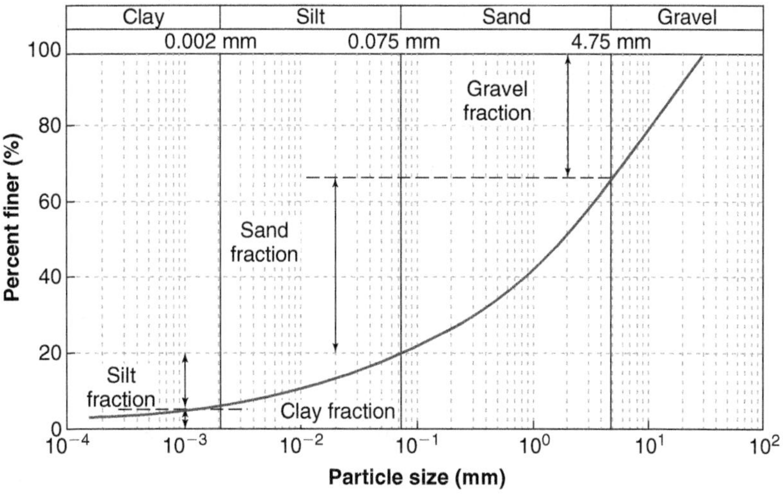

Figure 4.2 Particle size distribution curve.

Table 4.1 Sieve Numbers and Sieve Openings

Sieve Number	Sieve Opening
#1	25.4 mm
#4	4.75 mm
#10	2 mm
#20	0.85 mm
#40	0.425 mm
#80	0.18 mm
#200	0.075 mm

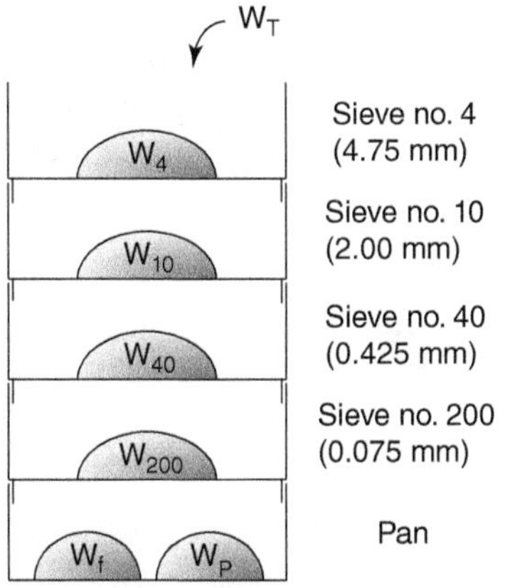

Figure 4.3 Dry weight retained on a stack of sieves.

4.2 HYDROMETER ANALYSIS

A *hydrometer* is an instrument made of glass (Figure 4.5) with a graduated stem on top of a bulb ballasted with lead beads so that it can float upright. It is used to measure the ratio of the density of the liquid in which it is immersed over the density of water. This ratio is read at the liquid surface (level of flotation of the hydrometer) on the graduated scale placed on the stem of the hydrometer. If the liquid being tested is very dense, the hydrometer does not sink very deep into the liquid, and vice versa. Therefore, the higher ratios are at the bottom of the stem.

Hydrometer analysis is used to obtain the particle size distribution curve of fine-grained soils: silts and clays. The

Table 4.2 Sieve Analysis Calculations

Initial weight of dry soil	W_t
Weight of dry soil retained on #200 after washing through #200	W_t (washed)
Weight of dry soil washed through #200	$W_{\text{fines}} = W_t - W_t$ (washed)
Dry weight retained on #4	W_4
Dry weight retained on #10	W_{10}
Dry weight retained on #40	W_{40}
Dry weight retained on #200	W_{200}
Dry weight retained on bottom pan	W_{p}
Percent finer than #4 (4.75 mm)	$((W_{10} + W_{40} + W_{200} + W_{\text{p}} + W_{\text{fines}})/W_t) \times 100$
Percent finer than #10 (2 mm)	$((W_{40} + W_{200} + W_p + W_{\text{fines}})/W_t) \times 100$
Percent finer than #40 (0.425 mm)	$((W_{200} + W_p + W_{\text{fines}})/W_t) \times 100$
Percent finer than #200 (0.075 mm)	$((W_p + W_{\text{fines}})/W_t) \times 100$

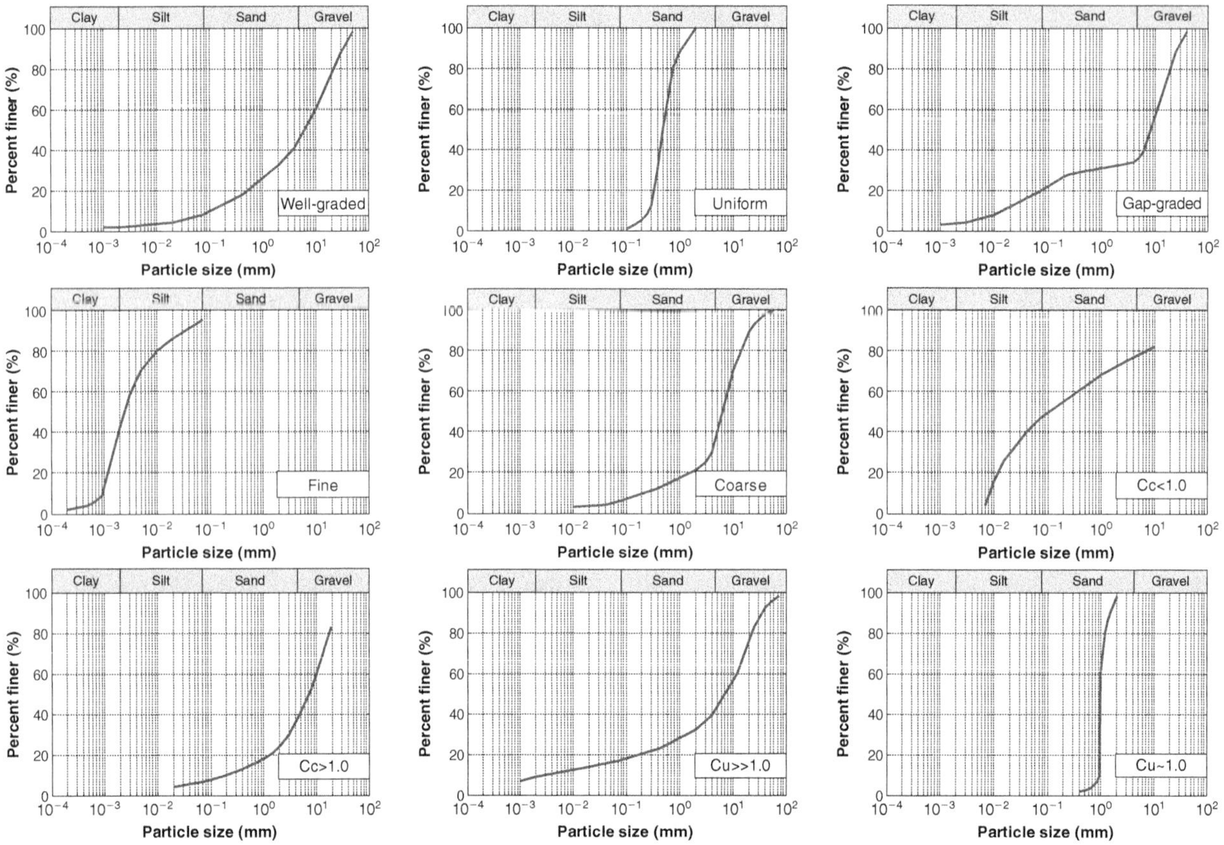

Figure 4.4 Examples of particle size distribution curves.

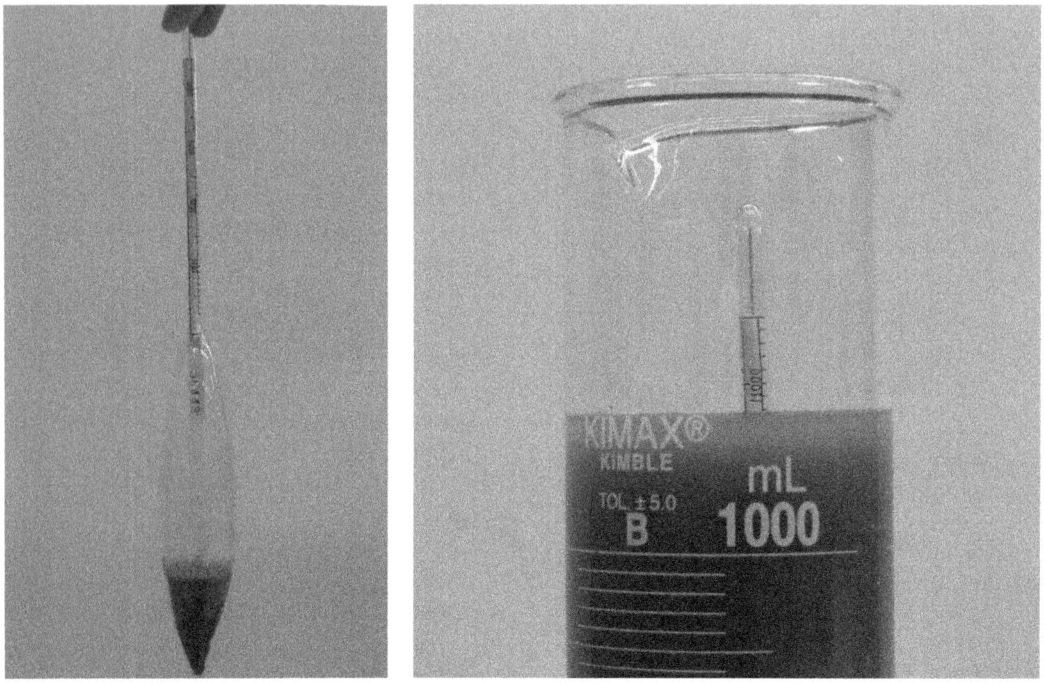

Figure 4.5 Hydrometer and hydrometer reading.

test consists of taking a given weight of dry soil, breaking it down into individual particles if clumps exist, mixing it with a dispersing agent (liquid), placing the wet mixture in a graduated cylinder, filling the container with water up to a known volume, shaking the cylinder to reach a uniform mixture, letting the soil particles settle, and recording the fall velocity at which the particles settle. The dispersing agent is used to ensure that the fine particles remain individually separated and do not form clusters. The fall velocity is obtained by measuring the unit weight of the soil-water mixture at a given depth z and at a given time t with a hydrometer. This unit weight decreases with time as the particles settle to the bottom of the container (Figure 4.6).

George Stokes was a British mathematician and physicist who made important contributions in fluid dynamics in the mid-1800s. Stokes's law relates the diameter of a sphere to its fall velocity in a liquid:

$$v = \left(\frac{\gamma_s - \gamma_f}{18\mu}\right) D^2 \quad (4.1)$$

where v is the fall velocity of the sphere, γ_s is the unit weight of the sphere, γ_f is the unit weight of the fluid (soil plus water), μ is the viscosity of the liquid, and D is the sphere diameter.

The depth z below the surface corresponding to the hydrometer reading ($r = \gamma_f/\gamma_w$) is the depth to the center of gravity of the hydrometer. At a time t after the beginning of the test, the smallest particles (equivalent spheres) which just passed the depth z have fallen at the velocity $v = z/t$.

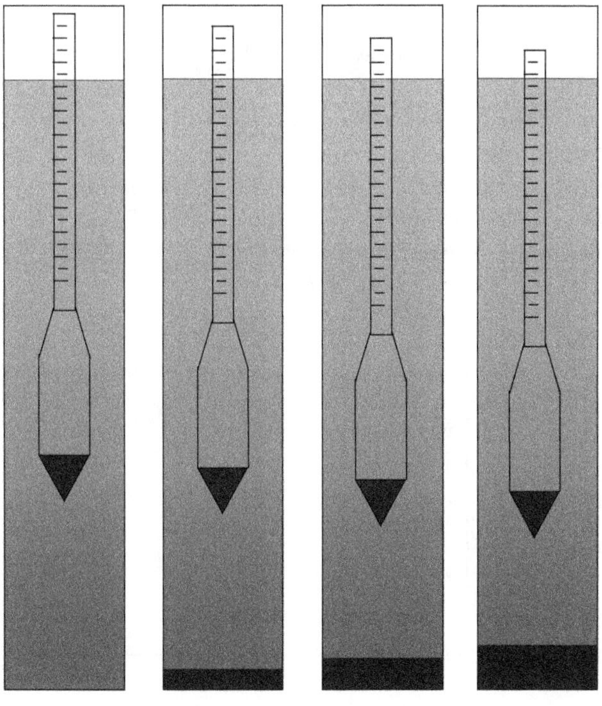

Figure 4.6 Different stages of the hydrometer analysis.

Knowing this velocity, plus the viscosity of water at the right temperature (e.g., 10^{-3} N.s/m^2 at 20°C), the unit weight of the sphere, and the unit weight of the liquid at time t as measured by the hydrometer, one can obtain the diameter D of this smallest particle (equivalent sphere) from Equation 4.1. The unit weight of the sphere is the unit weight of the soil particle (~26 kN/m^3 for mineral particles). The particle size determined by the hydrometer analysis is therefore the diameter D of a sphere made of the same material as the particle and falling at the same velocity as the particle.

Because the purpose of hydrometer analysis is to obtain the particle distribution curve, it is now necessary to obtain the percent finer P associated with the sphere diameter D. The unit weight measured by the hydrometer γ_f can be expressed as follows:

$$\gamma_f = \frac{W_{s(<D)} + \gamma_w(V - V_{s(<D)})}{V} \quad (4.2)$$

where $W_{s(<D)}$ is the weight of particles finer than the particle size D, γ_w is the unit weight of water, V is the total volume involved in the hydrometer measurement, and $V_{s(<D)}$ is the volume of particles finer than the particle size D within the volume V.

But

$$V_{s(<D)} = \frac{W_{s(<D)}}{\gamma_s} = \frac{PW_s}{G_s\gamma_w} \quad (4.3)$$

where γ_s is the unit weight of solids (~26 kN/m^3 for mineral particles), P the percentage by weight of particles finer than the particle size D, and G_s is the specific gravity of the particles.

Therefore

$$\gamma_f = \gamma_w + \frac{(G_s - 1)W_s P}{G_s V} \quad (4.4)$$

and

$$P = \frac{G_s V}{(G_s - 1)W_s}\gamma_w(r - 1) \quad (4.5)$$

where V is taken as the volume of water in the graduated cylinder (usually 1000 cubic centimeters), W_s is the total weight of dry soil placed into the cylinder, and r is the hydrometer reading ($r = \gamma_f/\gamma_w$). Hydrometer readings are taken at various times as the particles fall through the water column; the particle size D and associated percent finer P are calculated from these readings. This gives several points on the particle distribution curve.

Using the results of the hydrometer analysis, the particle size distribution curve can be obtained for particles ranging from 0.075 mm down to at least 0.001 mm (Figure 4.7). The hydrometer curve can be added to the sieve analysis curve so as to generate a curve from 0.001 mm up to 10 mm particle size. Note that there may be a discontinuity at the 0.075 mm size, as the sieve analysis and the hydrometer analysis do not strictly measure the same particle size (as explained earlier). The second largest dimension is measured in the sieve

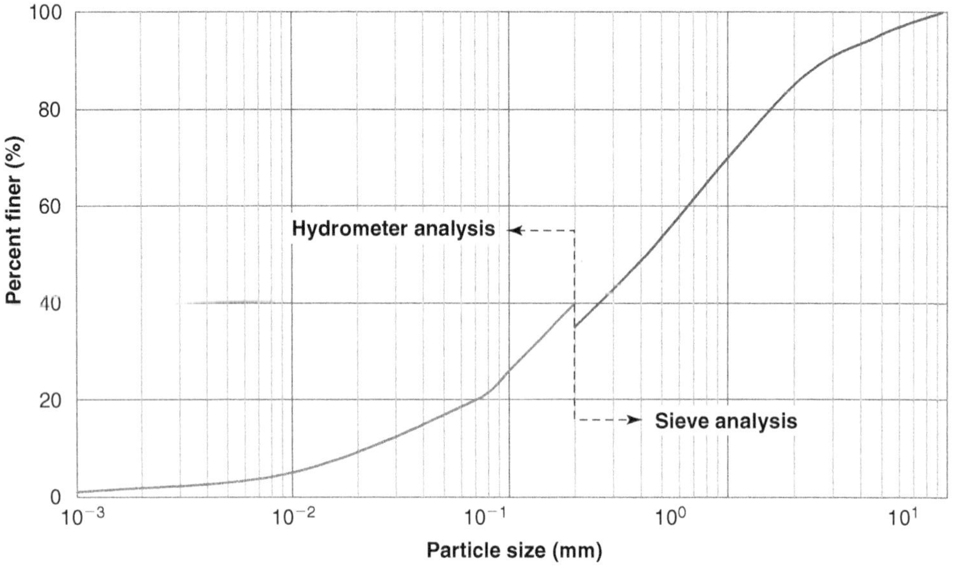

Figure 4.7 Combined sieve analysis and hydrometer analysis results.

analysis while the equivalent sphere diameter is measured in the hydrometer analysis. A discontinuity could also be due to other factors, such as the dispersing agent not working and fine particles clustering, thereby yielding a higher percentage of larger particles.

4.3 ATTERBERG LIMITS AND OTHER LIMITS

As mentioned earlier, particle size is not the main factor controlling the behavior of silts and clays. Instead, the behavior and therefore the classification are based on the ability of the soil to be deformed and stay together (*consistency*). This is measured by the *Atterberg limits*. Albert Atterberg was a Swedish chemist who worked in the field of agricultural science; he came up with what is now known as the Atterberg limits around 1910 as a means of classifying fine-grained soils. The Atterberg limits are water contents of remolded fine-grained soil. The limit tests are performed on remolded samples of silts or clays or more generally on the portion of a sample finer than sieve #40 (0.425 mm opening). These limits indicate the points at which the consistency of a fine-grained soil (Figure 4.8) changes from a liquid state to a plastic state (liquid limit), from a plastic state to a semisolid state (plastic limit), and from a semisolid state to a solid state (shrinkage limit).

The liquid limit w_L has a precise ASTM definition (ASTM D4318; ASTM 2004a). In short, it is the water content at which the remolded soil behaves like a soft paste (toothpaste consistency). This particular water content varies significantly depending on how fine the particles are. For example, very fine clay particles can have liquid limits approaching 100%, whereas silt particles may have liquid limits of around 30%. More precisely (Figure 4.9), the *liquid limit* is defined as the water content at which the two sides of a small amount of soil placed in a standard cup and grooved by a standard tool will flow together over a distance of 12.5 mm when hit by 25 blows in a standard liquid limit apparatus.

The plastic limit w_P also has a precise ASTM definition (ASTM D4318; ASTM 2004a). In short, it is the water content at which the remolded soil behaves like a hard paste (soft caramel). More precisely (Figure 4.10), the *plastic limit* is the water content at which a soil will begin to crumble when rolled into a thread 3.2 mm in diameter. The difference between the liquid limit and the plastic limit is the *plasticity index* or I_p. The plasticity index has been found to be related to a number of useful soil properties.

The shrinkage limit w_S is defined in ASTM D4943 (ASTM 2004b). The *shrinkage limit* is the water content corresponding to the amount of water necessary to fill all the voids of the dry soil after shrinkage. It is close to the lowest water content at which the remolded soil is still saturated during a drying process; any further drying leads to a degree of saturation less than 100%. The test consists of remolding the soil to a water

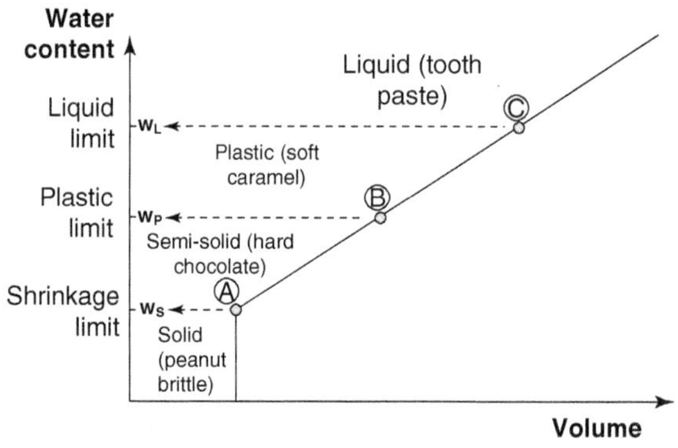

Figure 4.8 States of consistency and Atterberg limits.

Figure 4.9 Liquid limit apparatus and test.

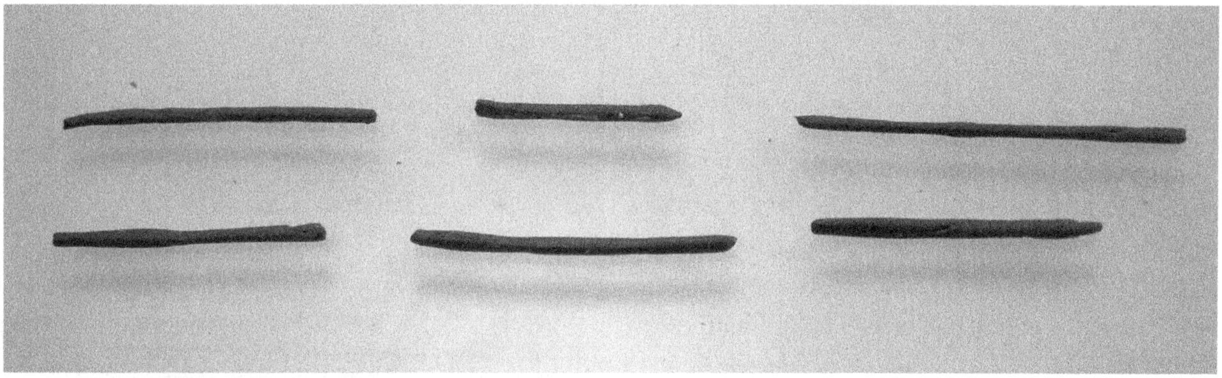

Figure 4.10 Plastic limit test and soil threads.

Figure 4.11 Shrinkage limit equipment and test.

content, w_o, above the shrinkage limit and filling a small cup of known volume, V_o, with the soil paste (Figure 4.11). The cup is weighed empty and then with the wet soil in it. The soil in the cup is then left to dry until it no longer shrinks. At this point, the cup plus dry soil is weighed and the weight of the dry soil is obtained (W_d). The dry sample is attached to a thread, dipped in hot wax, and pulled out. Once the film of wax now covering the sample has hardened, the sample is plunged into a graduated cylinder with water in it. The volume of the sample plus wax is measured by water displacement (V_{d+w}). The wax is removed and weighed; knowing the unit weight of the wax, the volume of wax V_w is calculated. The shrinkage limit is then:

$$w_S = w_o - \frac{(V_o - V_{d+w} + V_w)\gamma_w}{W_d}$$

where γ_w is the unit weight of water.

The liquid limit can also be determined by the fall-cone method (Figure 4.12). This method, developed in the early 1900 in Sweden, is now used in other countries such as France and the UK. In this test, a standard cone is brought to barely touch the surface of the soil and is released suddenly. The cone with a mass M and an apex angle θ penetrates into the soil a distance d. The liquid limit is reached when a

Figure 4.12 Fall-cone test for Atterberg limits.

Table 4.3 Fall-Cone Parameters for Liquid Limit Determination

Country	Cone mass	Cone apex angle	Cone penetration
Sweden	60 g	60°	10 mm
United Kingdom	80 g	30°	20 mm
France	80 g	30°	17 mm

Figure 4.13 Free swell test for swell limit of undisturbed sample.

chosen value of d is obtained. Table 4.3 shows the values of M, θ, and d in different countries. There is as yet no ASTM standard for the fall-cone test.

The plastic limit can also be determined by the fall-cone method, but in this case the distance d is much smaller and a value of 2.2 mm seems to be appropriate.

Although not associated with the Atterberg limits, the swell limit w_{SW} is important as well. It is defined as the water content at which a soil submerged in water can no longer absorb water. The test consists of placing a soil sample in a snug-fitting cylindrical container (Figure 4.13), inundating the soil, and measuring the vertical swell movement as a function of time. When the swelling stops, the water content is measured; this gives the soil's swell limit. This test is called a *free swell test* because no pressure is applied on top of the sample. Note that in this case the sample is undisturbed, whereas the Atterberg limits are performed on remolded samples.

Associated with the undisturbed-sample swell limit w_{SW} is the undisturbed-sample shrinkage limit w_{SH}. This shrinkage limit is obtained by performing a free shrink test (Figure 4.14). A sample of soil is trimmed in a cylinder, its dimensions are measured, and it is weighed. The initial volume V_o and the initial weight W_o are recorded. The sample is then left to dry while the dimensions and the weight are measured as a function of time. This gives the volume V(t) and weight W(t). When the sample is air-dried, it is placed in the oven to obtain the oven dry weight W_s. The average water content

Figure 4.14 Free shrink test for shrinkage limit of undisturbed sample.

of the sample at any time during the test is (W(t)−Ws)/Ws. A graph of the water content versus relative change in volume is plotted. The undisturbed-sample shrinkage limit w_{SH} is the water content corresponding to the point where the sample first stops decreasing in volume (point A on Figure 4.8).

Table 4.4 Summary of Water Content Limits

Atterberg liquid limit	w_L	Remolded soil, toothpaste consistency
Atterberg plastic limit	w_P	Remolded soil, soft toffee consistency
Atterberg shrinkage limit	w_S	Remolded soil, hard chocolate consistency
Swell limit	w_{SW}	Undisturbed soil, maximum natural water content
Shrinkage limit	w_{SH}	Undisturbed soil, highest water content at which further drying yields no more shrinkage

Sometimes point A is not clearly definable, particularly for high-plasticity soils (Figure 4.15), but there is always a distinct change in slope around the shrinkage limit. Table 4.4 summarizes the water content limits for soils.

4.4 CLASSIFICATION PARAMETERS

A number of reference particle sizes are determined from the particle size distribution curve. The parameter D_{50} is the particle size corresponding to a percent finer equal to

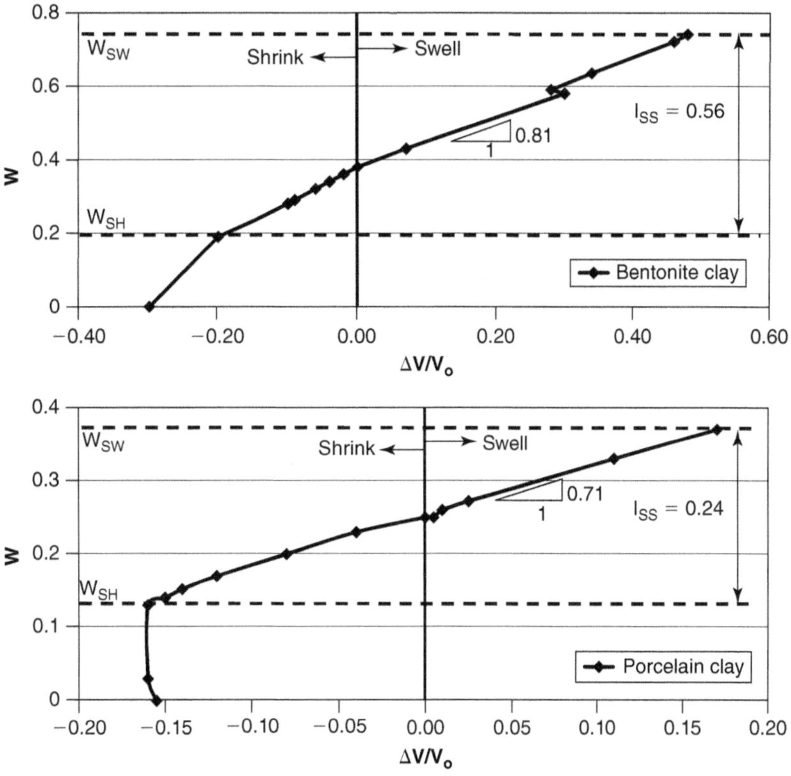

Figure 4.15 Shrink-swell test for porcelain clay and bentonite clay.

50 percent, D_{10} is the particle size corresponding to a percent finer equal to 10 percent, and so on. D_{10}, D_{30}, and D_{60} are used to calculate the coefficient of uniformity C_u and the coefficient of curvature C_c, as shown in Table 4.5.

Figure 4.4 illustrates some of these parameters. A coefficient of uniformity C_u close to 1 indicates that most of the particles in the soil have the same size. If C_u is large—say, larger than 6—then the soil contains particles that cover a wide range of sizes. If C_u is small, then the soil may be quite uniform with many particles of similar sizes. If the coefficient of curvature C_c is less than 1, then the particle size distribution has a downward curvature; if it is more than 3, then the particle size distribution has an upward curvature. If C_c is between 1 and 3, the particle size distribution curve will be reasonably straight and the soil is likely to contain particles with a wide range of sizes. Both C_u and C_c are used to classify coarse-grained soils. The size D_{50} is used extensively in erosion studies as a parameter that correlates well with the velocity at which a coarse-grained soil starts to erode. D_{15} and D_{85} are used in filter design for earth dams and other water-retaining structures. The particle size distribution curve is simple to obtain but very useful in geotechnical engineering as shown by these various applications.

A number of indices are determined from the Atterberg limits. The *plasticity index* I_P is the difference between the liquid limit w_L (quoted as a percent) and the plastic limit w_P (also quoted as a percent). The *shrinkage index* I_S is the difference between the plastic limit and the shrinkage limit, both in percent. The *liquidity index*, I_L, is defined in Table 4.5 and is quoted as a ratio or a percent; it indicates the relative position of the natural water content between the plastic limit and the liquid limit. The *shrink-swell index*, I_{SS}, is the difference between the undisturbed swell limit w_{SW} and the undisturbed shrinkage limit w_{SH}. It is very useful as an indicator of the shrink-swell potential of a soil. Other but less used indices are defined in Table 4.5.

Activity, A_c, is another parameter that helps describe a soil. This parameter is used for fine-grained soils, and is defined as the ratio between the plasticity index and the percent finer than 0.002 mm:

$$A_c = \frac{I_P}{\%\,finer\ than\ 0.002\ mm} \tag{4.6}$$

The values of A_c vary from less than 0.75 for relatively inactive soils (kaolinite) to more than 1.25 for very active soils (montmorillonite).

Table 4.5 Classification Parameters Definitions

Parameter Symbol	Name	Definition	Applications
D_X		Particle size corresponding to X% finer	Filter design, erosion of coarse-grained soils
C_u	Coefficient of uniformity	$C_u = \frac{D_{60}}{D_{10}}$	Classification of soils
C_c	Coefficient of curvature	$C_c = \frac{(D_{30})^2}{D_{60}D_{10}}$	Classification of soils
I_P	Plasticity index	$I_P = w_L - w_P$	Shrink-swell soil, fill specifications, correlations
I_{SS}	Shrink-swell index	$I_{SS} = w_{SW} - w_{SH}$	Shrink-swell potential
I_S	Shrinkage index	$I_S = w_P - w_S$	
I_L	Liquidity index	$I_L = \frac{(w - w_P)}{(w_L - w_P)}$	Correlations
I_C	Consistency index	$I_C = \frac{(w_L - w)}{(w_L - w_P)}$	
I_F	Flow index, slope of the water content vs. lg of number of blows in the liquid limit test	$I_F = \frac{(w_1 - w_2)}{(\lg N_1 - \lg N_2)}$	
I_T	Toughness index	$I_T = \frac{I_P}{I_F}$	
Ac	Activity	$A_c = \frac{I_P}{\%\,finer\ than\ 0.002\ mm}$	

4.5 ENGINEERING SIGNIFICANCE OF CLASSIFICATION PARAMETERS AND PLASTICITY CHART

The plasticity index I_P is definitely the index most used in practice, with the liquidity index a distant second. The others are rarely used. The I_P essentially relates to how small the clay particles are in the soil: the higher I_p is, the smaller the clay particles are. Table 4.6 shows the range of values that can be expected for common soils. Very high I_P values (60 or more) are associated with a predominance of very small clay particles, such as in montmorillonite; low I_P values (20 or less) are associated with a predominance of larger clay particles, such as in kaolinite. Thus, the I_P value gives an indication of some important properties of a soil. For example, a high I_P value indicates a soil that will be very difficult to compact, has a high shrink-swell potential, and has low permeability. A low I_P value is often required for fill material when good drainage is important, such as for pavement layers and retaining walls backfill.

From the point of view of soil strength, the friction between particles decreases with increasing I_P. Also, a comparison between the natural water content and the limits can give an indication of possible soil behavior. For example, if the natural water content is higher than the liquid limit, the soil is likely to be sensitive (it may lose significant strength when remolded). If the soil has a water content close to the shrinkage limit and a high shrink-swell index, beware of swelling problems if the soil can get wet.

The plasticity chart was developed by Arthur Cassagrande, an Austrian-born American civil engineer, around 1932. The plasticity chart is a plot of the plasticity index versus the liquid limit of a soil (Figure 4.16), and is used for the purpose of classifying fine-grained soils according to their plasticity. The A line is an empirically chosen line that splits the chart between clays above the A line and silts below the A line. The vertical line, corresponding to a liquid limit equal to 50%, separates high-plasticity fine-grained soils ($w_L > 50$) from low-plasticity fine-grained soils ($w_L < 50$). To classify a soil, the plasticity index and liquid limit of that soil are plotted on the chart; the region in which the point falls indicates what type of fine-grained soil it is or what kind of fines are encountered in a coarse-grained soil. The plasticity chart is the basis for the classification of fine-grained soils and of the fines fraction of coarse-grained soils.

Table 4.6 Range of Values for Atterberg Limits and Some Indices

Parameter	Low	Medium	High
Liquid limit	10–40	40–80	>80
Plastic limit	10–20	20–30	>30
Shrinkage limit	5–15	10–20	>20
Plasticity index	0–20	20–50	>50
Shrink-swell index	0–25	25–60	>60

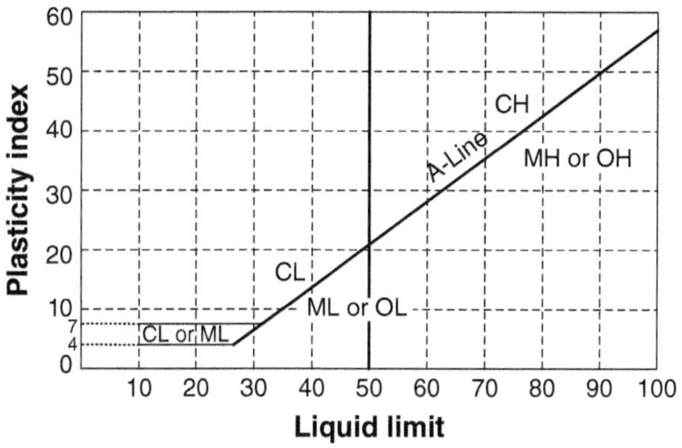

Figure 4.16 Plasticity chart.

4.6 UNIFIED SOIL CLASSIFICATION SYSTEM

The Unified Soil Classification System, or USCS, is the system used internationally to classify soils. Most commonly, it employs a two-letter symbol. The first letter indicates whether the soil is gravel (G), sand (S), silt (M), or clay (C). The letter for silt could not be S, as that letter was already used for sand, so the letter M was chosen; in Swedish *mjäla* means silt. The second letter gives additional information on the soil. For coarse-grained soils, the second letter can be M or C, indicating that the gravel or sand has a significant amount of silt or clay particles in it. For coarse-grained soils, the second letter can be W or P. W indicates that the gravel or sand is clean and well graded, meaning that all particle sizes are more or less represented. P indicates that the gravel or sand is clean and poorly graded, meaning that not all particle sizes are represented. For fine-grained soils, the second letter can be H, meaning high plasticity (high liquid limit and high I_p), or L for low plasticity (low liquid limit and low I_p).

An SC would be a soil with the majority of its particles in the sand-size range and Atterberg limits of the portion smaller than 0.425 mm, consistent with the Atterberg limits of clay. A GP would be a soil with the majority of its particles in the gravel-size range and poorly graded. An ML would be a low-plasticity silt, based on its Atterberg limits, and a CH would be a high-plasticity clay, again based on its Atterberg limits. The USCS two-letter symbols are understood throughout the world and help geotechnical engineers communicate with each other regardless of their native languages.

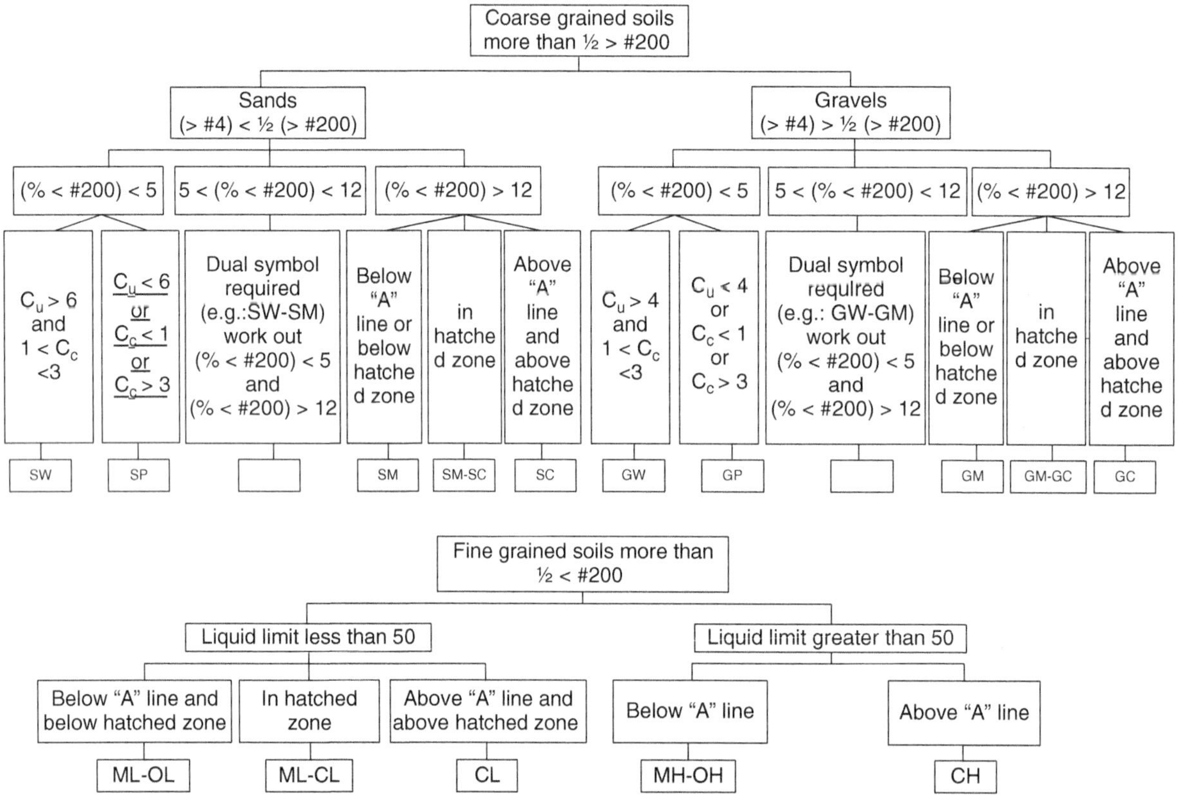

Figure 4.17 Flowchart to classify a soil by the USCS.

The exact process for classifying a soil consists of a series of steps organized in a decision tree as shown in Figure 4.17. The first decision is based on the percent passing the no. 200 sieve (#200), which has an opening of 0.075 mm. If the soil has more than 50% particles by weight larger than 0.075 mm (#200), the soil is a coarse-grained soil. If the soil has more than 50% by weight smaller than 0.075 mm (#200), the soil is a fine-grained soil. For coarse-grained soils, if the percent by weight of the gravel-size particles is larger than the percent by weight of the sand-size particles, the soil is a gravel and the first letter is G. If not, the soil is a sand and the first letter is S.

The second letter for a coarse-grained soil is W, P, M, or C. If the soil has less than 5% passing #200, it is clean and the second letter will be W or P, depending on the coefficient of uniformity C_u and the coefficient of curvature C_c obtained from the particle size distribution curve. If the coarse-grained soil has more than 12% passing #200, the soil is dirty and the second letter will be M or C, depending on the Atterberg limits of the portion smaller than 0.425 mm; M will be selected if the soil plots below the A line on the plasticity chart and C if it plots above. If the percent passing #200 is between 5% and 12%, then a dual symbol will be required, as the soil is intermediate between clean and dirty. In this instance, the classification for the <5% case and the >12% case are obtained and the soil ends up with a dual symbol (e.g., GP-GC or SW-SM). For fine-grained soils, the plasticity index and the liquid limit are plotted on the plasticity chart and the dual symbol is read from the quadrant of the chart where the point is situated.

PROBLEMS

4.1 Calculate the thickness of the wire in the no. 200 sieve.

4.2 A dry sample of soil weighs 5 N. It is shaken on a set of sieves: No. 4 (4.75 mm), No. 40 (0.425 mm), No. 200 (0.075 mm), and a pan. The weight retained on No. 4 is 2 N, on No. 40 is 1.5 N, and on No. 200 is 1 N. Calculate:

- The percent of coarse grain size particles by weight
- The percent of gravel-size particles by weight
- The percent of sand-size particles by weight
- The percent of fine grain size particles by weight
- The coefficient of uniformity and the coefficient of curvature

Based on these results, what would you call the soil?

4.3 Why is the particle size of the particle size curve plotted on a log scale? Plot the particle size curve of problem 2 as percent finer vs. particle size on a log scale and then as percent finer vs. log of particle size. Determine by calculations the position of a particle size equal to 0.075 mm and 4.75 mm on the particle size (log scale) axis and on the log of particle size axis.

4.4 Calculate how fast a particle of soil will settle in water if its equivalent diameter is 0.075 mm and then if its equivalent diameter is 0.002 mm.

4.5 A cylindrical hydrometer has a radius of 20 mm and weighs 2 N. It is lowered into water mixed with fine soil particles. If the hydrometer sinks and comes to floating equilibrium when it is 100 mm in the liquid, calculate the ratio of soil solids by volume that exists in the liquid. Assume that Gs = 2.65 if needed.

4.6 Explain the hydrometer analysis in your own words. Develop the equations necessary.

4.7 A soil has a natural water content of 22% and the following limits.

- Shrinkage limit = 13%
- Plastic limit = 25%
- Swell limit = 36%
- Liquid limit = 55%

Calculate the

- Plasticity index
- Liquidity index
- Shrink-swell index

4.8 Classify the following soils:

	S1 (% finer)	S2 (% finer)	S3 (% finer)	S4 (% finer)	S5 (% finer)
#4	52	52	63	98	100
#10	38	38	56	90	97
#40	18	18	42	47	82
#200	8	2	4	20	70
W_L	17	NP	NP	32	48
W_P	11	NP	NP	26	34

Problems and Solutions

Problem 4.1

Calculate the thickness of the wire in the no. 200 sieve.

Solution 4.1

The sieve number corresponds to the number of openings per 25 mm. For the sieve #200, the width of any opening is 0.075 mm; therefore, the total width of the openings in 25 mm of the #200 mesh is 200 × 0.075 = 15 mm. The total thickness of the wires in 25 mm of the #200 mesh is (25 − 15) = 10 mm, so the thickness of the wires in a sieve #200 is 10/200 = 0.05 mm (about the diameter of a human hair).

Problem 4.2

A dry sample of soil weighs 5 N. It is shaken on a set of sieves: No. 4 (4.75 mm), No. 40 (0.425 mm), No. 200 (0.075 mm), and a pan. The weight retained on No. 4 is 2 N, on No. 40 is 1.5 N, and on No. 200 is 1 N. Calculate:

- The percent of coarse grain size particles by weight
- The percent of gravel-size particles by weight

- The percent of sand-size particles by weight
- The percent of fine grain size particles by weight
- The coefficient of uniformity and the coefficient of curvature

Based on these results, what would you call the soil?

Solution 4.2

The percent of coarse grain size particles is $= \frac{2 + 1.5 + 1}{5} \times 100 = 90\%$

The percent of gravel-size particles is $= \frac{2}{5} \times 100 = 40\%$

The percent of sand-size particles is $= \frac{1.5 + 1}{5} \times 100 = 50\%$

The percent of fine grain size particles is $= \frac{0.5}{5} \times 100 = 10\%$

	Retained soil on sieve			Passing through sieve	
	Weight (N)	Accumulated weight (N)	Accumulated weight (%)	Weight (N)	Accumulated weight (%)
No. 4 (4.75 mm)	2	2	40	3	60
No. 40 (0.425 mm)	1.5	3.5	70	1.5	30
No. 200 (0.075 mm)	1	4.5	90	0.5	10
Pan	0.5	5	100	0	0

From these results, $D_{60} = 4.75$ mm, $D_{30} = 0.425$ mm, and $D_{10} = 0.075$ mm.

$$C_u = \frac{D_{60}}{D_{10}} = \frac{4.75}{0.075} = 63$$

$$C_c = \frac{D^2{}_{30}}{D_{10} \times D_{60}} = \frac{0.425^2}{0.075 \times 4.75} = 0.5$$

Based on these results, the soil has 90% coarse fraction, therefore the soil is a coarse-grained soil; furthermore, 50% of the soil is retained between sieves #40 and #200, so the soil is sand.

Problem 4.3

Why is the particle size of the particle size curve plotted on a log scale? Plot the particle size curve of problem 2 as percent finer vs. particle size on a log scale and then as percent finer vs. log of particle size. Determine by calculations the position of a particle size equal to 0.075 mm and 4.75 mm on the particle size (log scale) axis and on the log of particle size axis.

Solution 4.3

The range of particle sizes in soils is very large, so we use the logarithmic scale because this scale stretches out the particle size distribution in the very small range. This allows us to distinguish the small sizes as well as the large sizes. Figure 4.1s shows the particle size curve as percent finer vs. particle size on a log scale. Figure 4.2s shows the particle size curve as percent finer vs. log of particle size. For the 0.075 mm particle, $\log 0.075 = -1.125$; this point can easily be found on the linear scale of Figure 4.2s. The position of this point is the same on the scale of Figure 4.1s. The same approach applies to the 4.75 mm particle: $\log 4.75 = 0.677$.

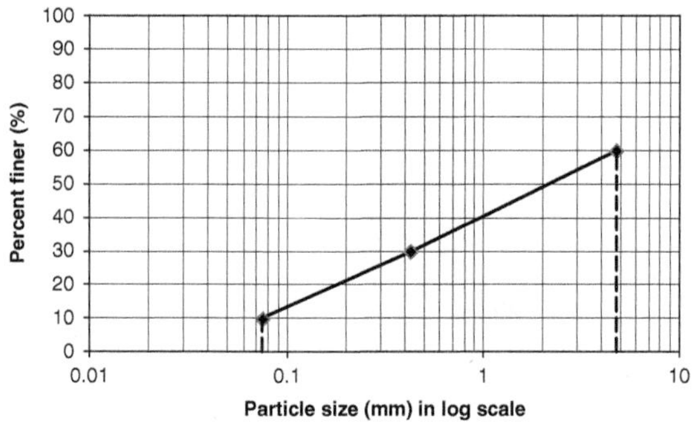

Figure 4.1s Percent finer vs. particle size on a log scale.

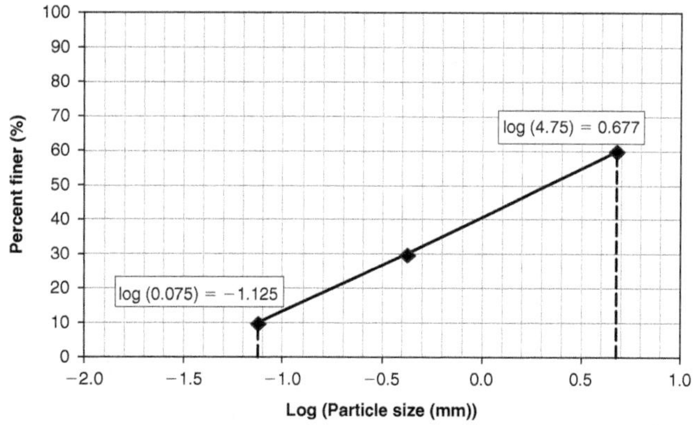

Figure 4.2s Percent finer vs. log of particle size on a normal scale.

Problem 4.4

Calculate how fast a particle of soil will settle in water if its equivalent diameter is 0.075 mm and then if its equivalent diameter is 0.002 mm.

Solution 4.4

Assume that:

- Water temperature = 20°C
- Specific gravity of particles is 2.65
- Viscosity of water is $\mu = 10^{-3}\ \mathrm{N \cdot s/m^2}$
- Unit weight of water is $\gamma_w = 9.79\ \mathrm{kN/m^3}$
- Unit weight of soil particles $\gamma_s = 2.65 \times 9.79\ \mathrm{kN/m^3} = 25.95\ \mathrm{kN/m^3}$

The fall velocity of a soil particle in water can be calculated using Stokes's law:

$$v = \left(\frac{\gamma_s - \gamma_f}{18\mu}\right) D^2$$

where $\gamma_s = 25.95\ \mathrm{kN/m^3}$ and $\gamma_f = \gamma_w = 9.79\ \mathrm{kN/m^3}$. For particles with D = 0.075 mm

$$v = \left(\frac{25.95 - 9.79}{18 \times 10^{-6}}\right) \times \left(\frac{0.075}{1000}\right)^2 = 0.0051(\mathrm{m/sec}) = 5.1(\mathrm{mm/sec})$$

For particles with D = 0.002 mm:

$$v = \left(\frac{25.95 - 9.79}{18 \times 10^{-6}}\right) \times \left(\frac{0.002}{1000}\right)^2 = 3.59 \times 10^{-6}(\mathrm{m/sec}) = 0.00359\,(\mathrm{mm/sec})$$

Problem 4.5

A cylindrical hydrometer has a radius of 20 mm and weighs 2 N. It is lowered into water mixed with fine soil particles. If the hydrometer sinks and comes to floating equilibrium when it is 100 mm in the liquid, calculate the ratio of soil solids by volume that exists in the liquid. Assume that Gs = 2.65 if needed.

Solution 4.5

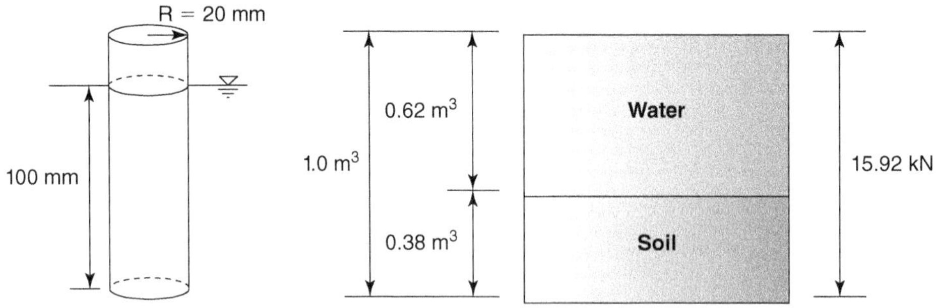

Figure 4.3s Hydrometer and three-phase diagram.

$$F_{\text{Buoyancy}} = W$$
$$V \times \gamma_{\text{mixture}} = W$$
$$\frac{\pi \times 0.04^2}{4} \times 0.1 \times \gamma_{\text{mixture}} = 2$$
$$\gamma_{\text{mixture}} = 15.92 \times 10^3\ \text{N/m}^3 = 15.92\ \text{kN/m}^3$$

Assuming 1 m^3 of the mixture and $G_s = 2.65$:

$$W_W + W_S = 15.92\ \text{kN}$$
$$V_W + V_S = 1\ \text{m}^3$$
$$\gamma_W V_W + \gamma_S V_S = 15.92\ \text{kN}$$
$$9.81 \times (1 - V_S) + (2.65 \times 9.81) \times V_S = 15.92\ \text{kN}$$
$$\therefore \quad V_S = 0.38\ \text{m}^3, \quad V_W = 0.62\ \text{m}^3$$

The volumetric percent of solids in the mixture is 38%.

Problem 4.6

Explain the hydrometer analysis in your own words. Develop the equations necessary.

Solution 4.6

See Section 4.2 in this chapter.

Problem 4.7

A soil has a natural water content of 22% and the following limits.

- Shrinkage limit = 13%
- Plastic limit = 25%
- Swell limit = 36%
- Liquid limit = 55%

Calculate the

- Plasticity index
- Liquidity index
- Shrink-swell index

Solution 4.7

- Plasticity index: PI = LL − PL = 55 − 25 = 30
- Liquidity index: LI = (w − PL)/PI = (22 − 25)/30 = −0.1
- Shrink-swell index: Iss = swell limit−shrinkage limit = 36 − 13 = 23

Problem 4.8

Classify the following soils:

	S1 (% finer)	S2 (% finer)	S3 (% finer)	S4 (% finer)	S5 (% finer)
#4	52	52	63	98	100
#10	38	38	56	90	97
#40	18	18	42	47	82
#200	8	2	4	20	70
w_L	17	NP	NP	32	48
w_P	11	NP	NP	26	34

Solution 4.8

The soils are classified based on the following criteria:

- Coarse grain size particles: retained on the no. 200 sieve (0.075 mm)
- Gravel-size particles: retained on the no. 4 sieve (4.75 mm)
- Sand-size particles: passing no. 4 sieve, retained on the no. 200
- Fine grain size particles: passing no. 200
- Plastic and liquid limit:

$$\text{Coefficient of uniformity } C_u = \frac{D_{60}}{D_{10}}$$

$$\text{Coefficient of curvature } C_c = \frac{D^2{}_{30}}{D_{10} \times D_{60}}$$

The particle size distribution curves are drawn on Figures 4.4s to 4.8s and theclassification of the 5 soils is presented in the Table below.

	S1 (% finer)	S2 (% finer)	S3 (% finer)	S4 (% finer)	S5 (% finer)
Sieve Opening (mm)			**Percent Finer**		
10	80	80	80	—	—
4.75	52	52	63	98	100
2	38	38	56	90	97
0.425	18	18	42	47	82
0.075	8	2	4	20	70
0.03 from hydrometer	—	—	—	—	9
Other properties					
w_L	17	NP	NP	32	48
w_P	11	NP	NP	26	34
I_p	6	NP	NP	6	14
Coarse fraction (%)	92	98	96	80	30
Fine fraction (%)	8	2	4	20	70
Gravel fraction (%)	48	48	37	2	0
Sand fraction (%)	44	50	59	78	30
D_{10} (mm)	0.11	0.19	0.098	0.036	
D_{30} (mm)	1.05	1.05	0.23	0.16	
D_{60} (mm)	7	7	3.2	0.7	
C_u	63.6	36.8	32.7	19.4	
C_c	1.4	0.8	0.2	1.0	
Classification	GW-(GC-GM)	SP	SP	SM	ML

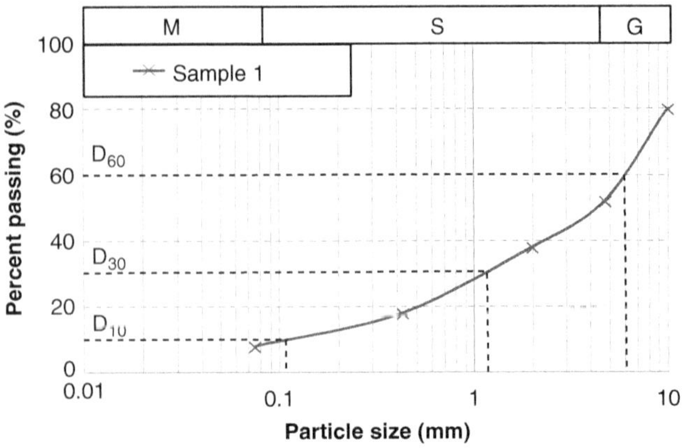

Figure 4.4s Percent finer vs. log of particle size of sample S1.

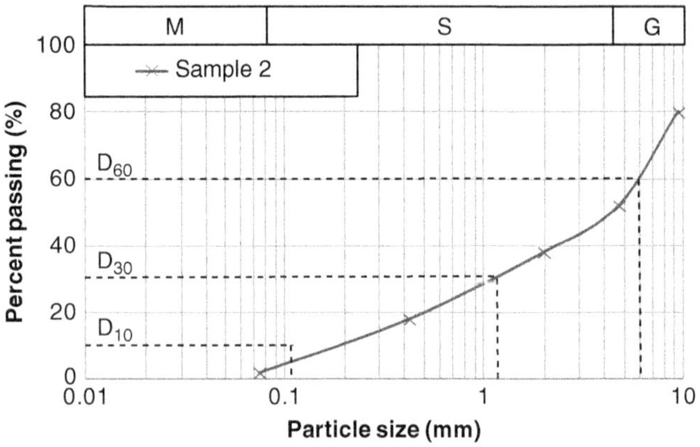

Figure 4.5s Percent finer vs. log of particle size of sample S2.

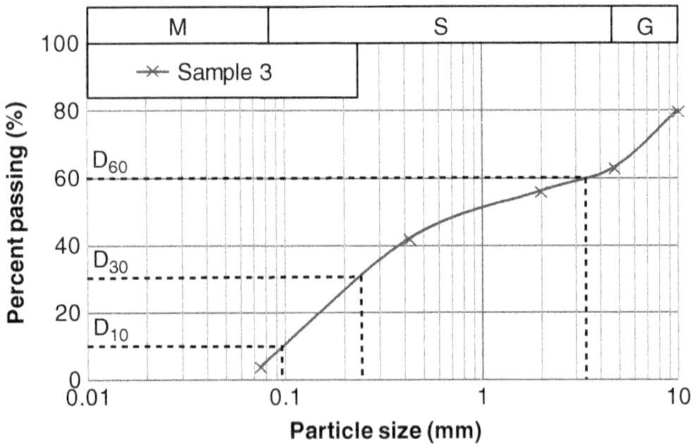

Figure 4.6s Percent finer vs. log of particle size of sample S3.

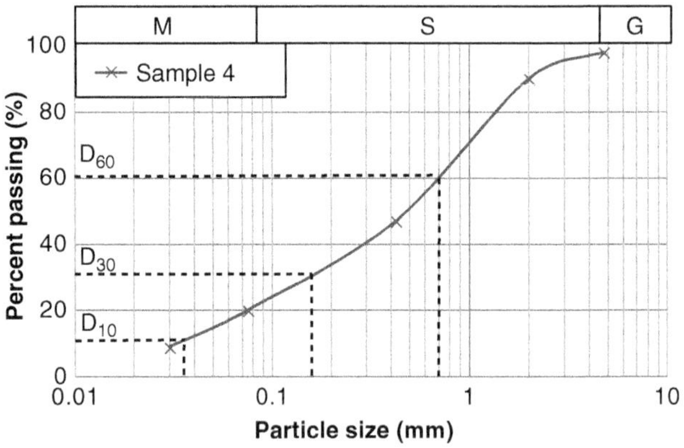

Figure 4.7s Percent finer vs. log of particle size of sample S4.

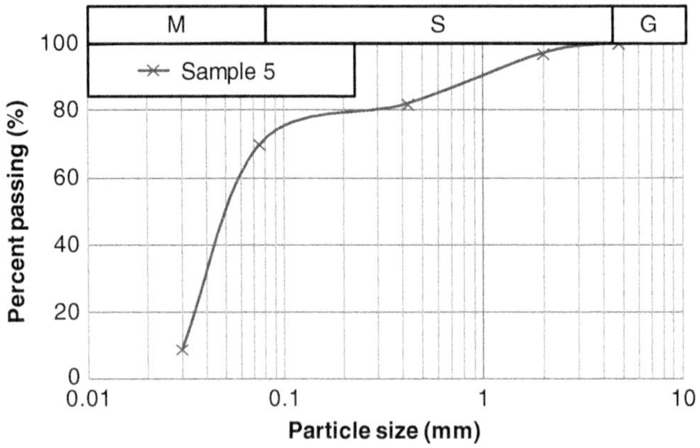

Figure 4.8s Percent finer vs. log of particle size of sample S5.

CHAPTER 5

Rocks

In many instances geotechnical engineers work on rock problems. For example, locating the depth of bedrock is often an important part of any soil investigation. Rock slopes, rock tunneling, rock excavations, rock fill in dams, and foundations on rock are other examples of projects requiring the expertise of the geotechnical engineer. This chapter is intended to give the reader an overview of rocks, rock properties, and rock engineering. Further information and more detailed coverage of the topic should be sought in textbooks and other publications such as Goodman (1989).

5.1 ROCK GROUPS AND IDENTIFICATION

A *rock* is a mixture of minerals (Sorrell and Sandström 2001). You may wish to think of minerals as being the building blocks of the various rocks. The primary mineral groups forming rocks are silicates (e.g., feldspar and mica), oxides (e.g., quartz), carbonates (e.g., dolomite and calcite), and sulfates (e.g., gypsum). Some of the rare minerals are topaz, jade, and emerald (silicate); ruby and sapphire (oxides); and turquoise (phosphate). Diamond is pure carbon, so it is a basic element rather than a mineral. From the point of view of their origin, rocks are classified as igneous, sedimentary, or metamorphic.

Igneous rocks (Figure 5.1) are formed by the cooling process of magma (i.e., granite and basalt). Granite is formed when viscous lava cools slowly. It is light in color and contains large elements such as quartz and feldspar. Basalt is formed by the rapid cooling of fluid lava. It is dark-colored and contains fine-grained elements undetectable by the naked eye.

Sedimentary rocks (Figure 5.2) are formed by the weathering of a parent rock, when the weathered materials are transported and redeposited into a different setting and lithified back into rock by some form of cementation, or pressure, or a heat process. They are divided into *clastic rocks* (rocks made from particles of other rocks) and *nonclastic rocks* (rocks formed by chemical precipitates, often calcite). Sandstone, siltstone, mudstone, marl, and shale are clastic rocks, whereas limestone, dolomite, gypsum, lignite, and coal are nonclastic rocks.

Metamorphic rocks (Figure 5.3) are formed when the constituents of sedimentary and igneous rocks are changed by tremendous heat and pressure, with the possible influence of water and gases. The two main types of metamorphism processes involve temperature and pressure or temperature alone. Pressure alone is uncommon. In order of decreasing strength, marble, gneiss, slate, and schist are all metamorphic rocks.

For identification purposes, the charts in Figure 5.4 and 5.5a are very useful. Figure 5.4 helps in identifying the minerals that form a rock. It proceeds through a series of testing steps, including use of a hand lens to observe the rock-forming mineral; use of a knife and one's fingernail to test the strength; and observation of the cleavage, the color, and the luster. Figure 5.5a helps in identifying the rock itself. It distinguishes between rocks with a crystalline texture, rocks that have no grains visible and are uniformly smooth, and rocks with a clastic texture.

5.2 ROCK MASS VS. ROCK SUBSTANCE

Rock mechanics makes a major distinction between rock substance and rock mass. *Rock substance* refers to a piece of intact rock with no fissures; *rock mass* refers to the entire mass of rock, including fissures and joints. There is usually a big difference between the tensile strength of an intact piece of rock (rock substance), and a weathered mass of rock (rock mass). In most cases the rock mass is much weaker than the rock substance. Therefore, a description of the joint pattern is very important, and should include joint spacing (less than 50 mm for very fractured rock to more than 3 m for solid rock), joint width, joint roughness, joint direction (using a rose diagram), and joint strength.

Although it is easiest to measure the properties of the rock substance through laboratory testing, it is often more important to determine the behavior of the rock mass. This is the case for rock slopes, foundations on or in rock, and seepage through rock. An exception is the behavior of rock fill and rip rap, where the properties of the rock substance are critical.

Figure 5.1 Igneous rocks: (*a*) granite, (*b*) feldspar, (*c*) basalt. (Courtesy of Mineral Information Institute, an affiliate of the SME Foundation.)

Figure 5.2 Sedimentary rocks: (*a*) sandstone, (*b*) slate, (*c*) limestone, (*d*) siltstone. (Courtesy of Mineral Information Institute, an affiliate of the SME Foundation.)

Figure 5.3 Metamorphic rocks: gneiss, marble. (Courtesy of Mineral Information Institute, an affiliate of the SME Foundation.)

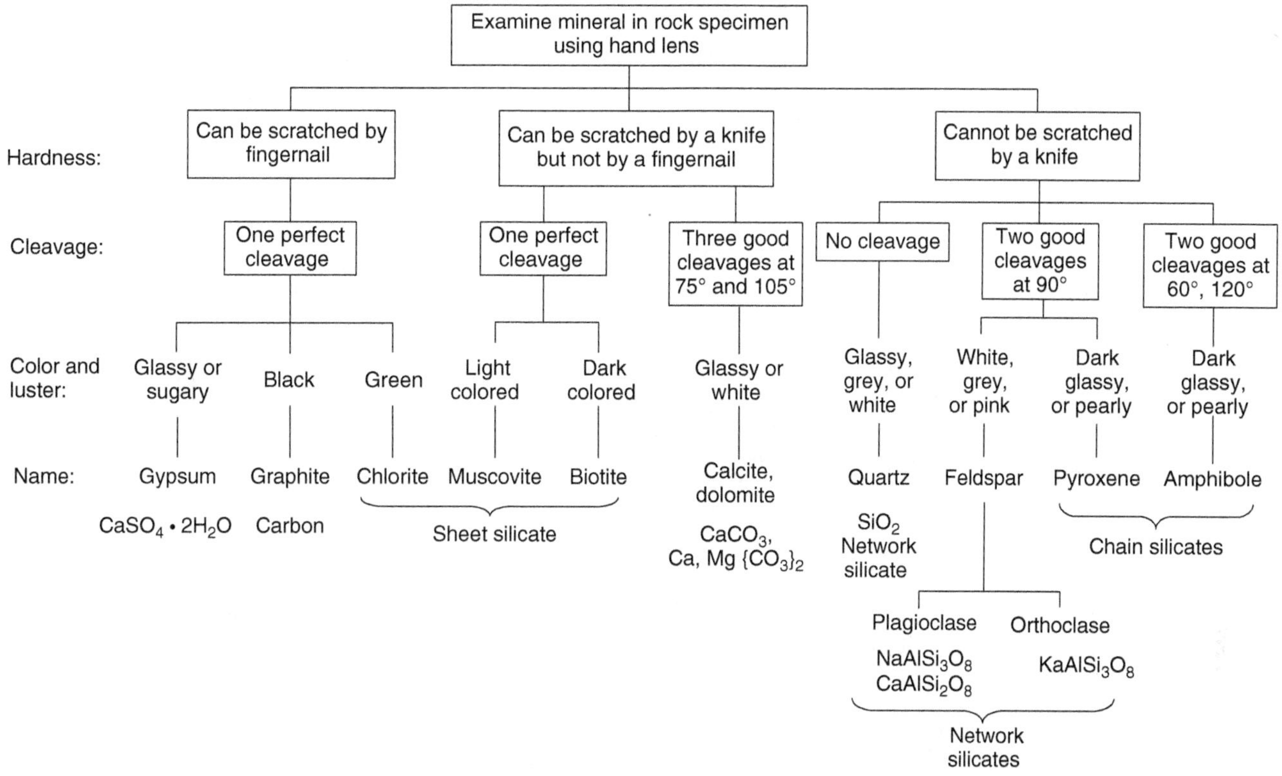

Figure 5.4 Identifying rock minerals. (From Goodman, 1989. Reprinted with permission of John Wiley & Sons, Inc.)

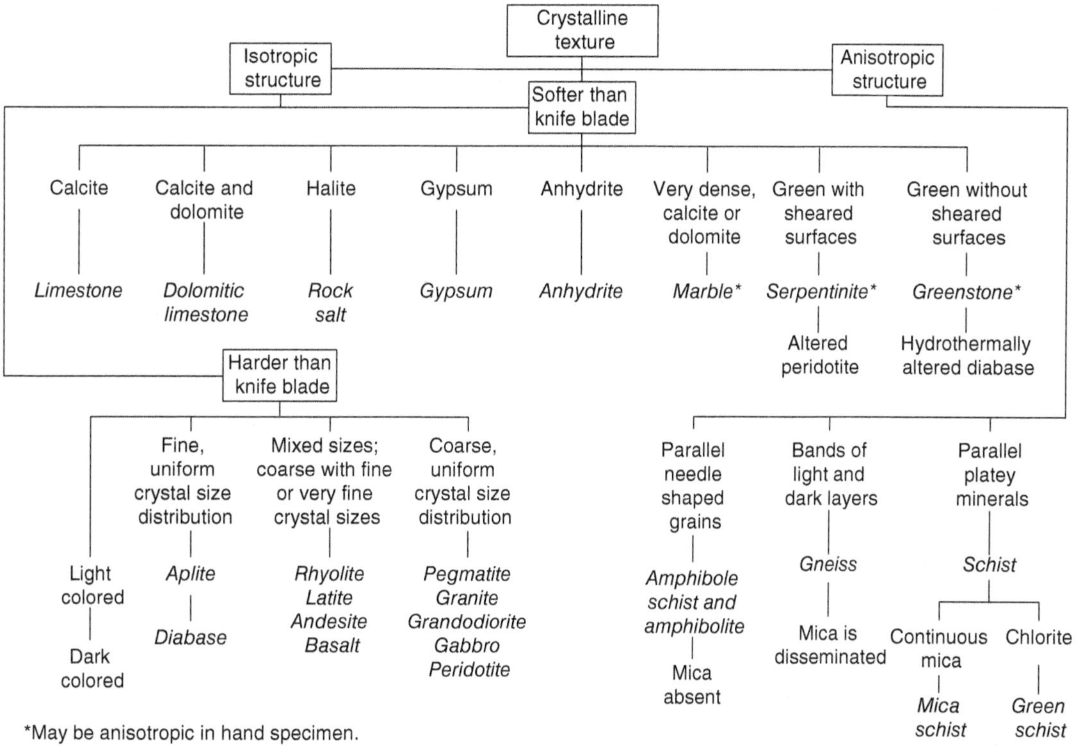

Figure 5.5a Identification of rocks with crystalline texture. (From Goodman, 1989. Reprinted with permission of John Wiley & Sons.)

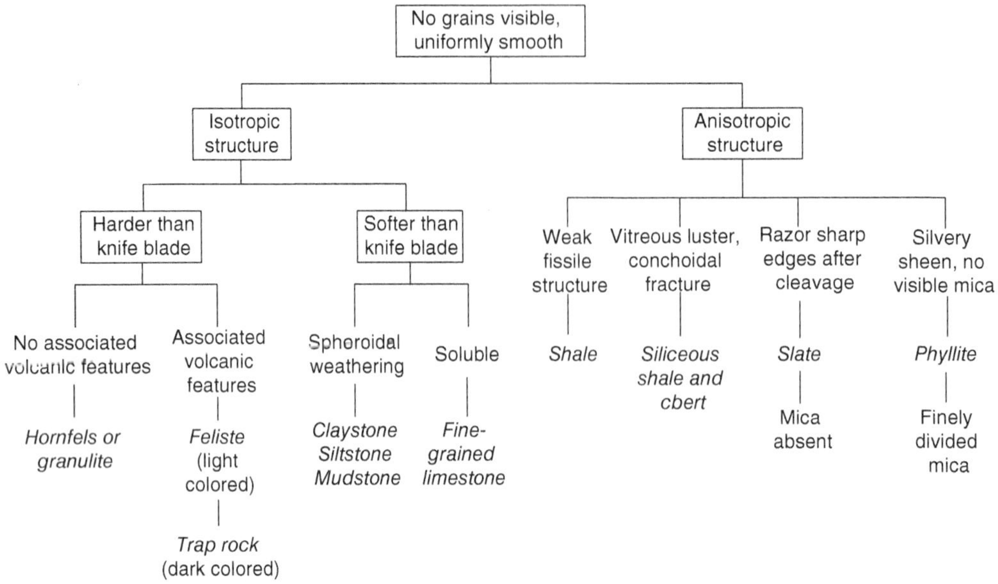

Figure 5.5b Identification of rocks with no grains visible. (From Goodman, 1989. Reprinted with permission of John Wiley & Sons.)

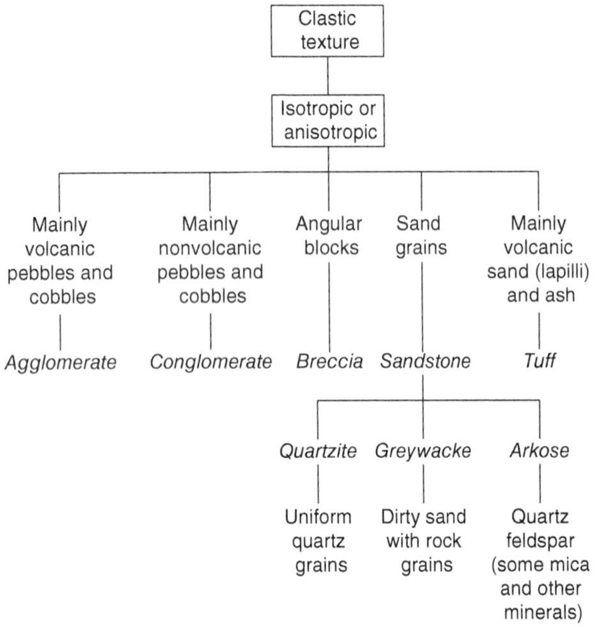

Figure 5.5c Identification of rocks with clastic texture. (From Goodman, 1989. Reprinted with permission of John Wiley & Sons.)

5.3 ROCK DISCONTINUITIES

Rocks usually exhibit a network of discontinuities that significantly affect the mass behavior. Many words exist to refer to these discontinuities: fissures, cracks, fractures, joints, and faults (Priest 1993). *Fissures* are the smallest and *faults* are the largest (Figure 5.6 and Figure 5.7). Nonetheless, the two main types are joints and faults. *Joints* are created over geologic time by bending of the rock mass, by vertical expansion, by horizontal stress relief (e.g., cliffs), by temperature differences, and sometimes by chemical action. Joints tend to exhibit a pattern. Faults are due to the movement of rock plates on a large scale and tend to be singular elements. These discontinuities introduce nonlinearities in behavior, stress dependency and anisotropy in properties, and weaknesses with regard to deformation and strength. Cementation in clastic rocks also significantly influences a rock's properties; often the properties of the binder control the behavior of the rock, much as cement controls the behavior of concrete. If the network of joints is random (rare), it weakens the rock evenly, but if the joints are directional (common), the weakness is accentuated in the direction of the joints in shear and reduces the shear strength to the strength of the joint surfaces. The tensile strength of the rock mass perpendicular to the joint direction is reduced to a small fraction of the intact rock strength. Compression perpendicular to joints increases deformation compared to the intact rock but has little influence on strength.

Another type of discontinuity is cavities and voids in the rock mass. These cavities most commonly form in limestone, dolomite, gypsum, and salt. Sinkholes in limestone occur in karst regions and can reach impressive dimensions.

5.4 ROCK INDEX PROPERTIES

Rock index properties include the dry unit weight of the rock substance and the porosity of the rock substance. The dry unit weight of the rock substance varies from a possible 21 kN/m^3 for a shale or a limestone to a possible 27 kN/m^3 for a marble or a granite. The most common values are between 25 and

Figure 5.6 Fissures and joints. (a: Courtesy of Lupin. c: Courtesy of Charles DeMets, University of Wisconsin-Madison. d: Courtesy of Alex Brollo.)

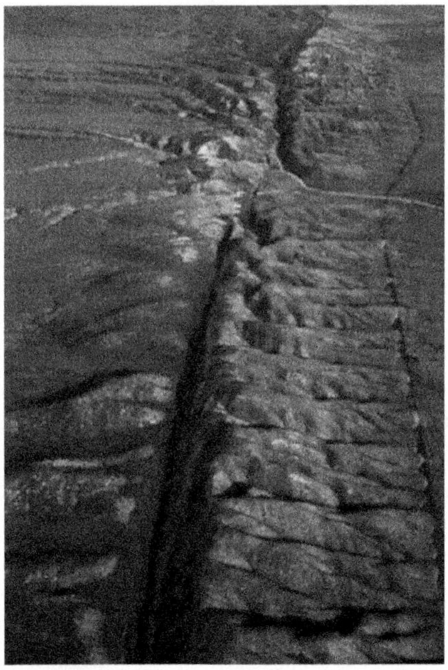

Figure 5.7 A fault. (Courtesy of The United State Geological Survey USGS, USA)

26 kN/m^3. The porosity of rock substance is at most a few percent; exceptions include shale, sandstone, and schist, for which the porosity can reach that of soils at several tens of percent. The degree of weathering significantly affects the rock mass unit weight and porosity, with the lowest unit weights and highest porosities for the highest degree of weathering.

5.5 ROCK ENGINEERING PROPERTIES

Engineering properties of the rock substance include durability, hardness, permeability, modulus, and strength (Waltham 1994). Although it is generally more important to know the properties of the rock mass, the first step is to find out the properties of the rock substance. An exception to this "rule" is when rip rap or rock fill has to be used for protection, as in scour or stability in rock-fill dams.

The *durability* of a rock is measured by a test called the *slaking durability test*. Ten pieces of rock are weighed and placed in a rotating drum lined with a 2 mm opening mesh. The drum is slowly rotated through a water bath for

10 minutes and the rock pieces remaining after the test are weighed again. The ratio in percent of the weight after and before the test is the slaking durability index I_{sd}. Rocks typically have I_{sd} values in excess of 90%. Values below 70 are undesirable for rip-rap applications.

Hardness is a measure of how hard a surface is. For rocks, it may refer to the hardness of the parent mineral or the rock surface. Talc is one of the softest minerals, whereas diamond is the hardest known mineral. On Mohs scale of hardness, talc has a rating of 1, gypsum 2, quartz 7, and diamond 10. The hardness of a rock surface can be measured by using a Schmidt hammer. The Schmidt hammer generates an impact on the rock surface and the mass that impacts the surface rebounds to a measured height. The rebound height divided by the maximum height is called the rebound value R. The rebound value has been correlated to the unconfined compression strength and the modulus of rocks.

The hydraulic conductivity, k, of a rock can be measured in the laboratory on an intact sample or in the field on the rock mass. The results are usually extremely different, with the field values being 10 to 100,000 times (or even more) larger than the laboratory values depending on the extent of the network of discontinuities in the rock mass. The densest intact rocks will have k values in the 10^{-10} to 10^{-15} m/s range, but volcanic intact rocks can have hydraulic conductivities in the range of 10^{-3} m/s. In the field, the hydraulic conductivity is drastically increased compared to the intact rock, as water could be gushing out of the joints of the rock mass. The k value can exhibit significant anisotropy depending on the direction of the joints.

The modulus of deformation, E, of the rock substance is measured on samples in the laboratory, most commonly using the unconfined compression test. In the field, the plate test, the half cylinder test, or the pressuremeter test can be used. Values of E for intact rock or rock substance are in the range of 2000 MPa to 100,000 MPa (concrete is around 20,000 MPa). The softer rocks include chalk and shale; the stiffer ones include granite and marble. The Poisson's ratio of rocks is relatively small, with values ranging from 0.15 to 0.3.

The strength of the intact rock, as measured by unconfined compression tests, can vary from more than 200 MPa for very hard rock to less than 10 MPa for very soft rock. Concrete has an unconfined compression strength of 20 MPa. Therefore, concrete is a soft to medium rock.

The ratio between the rock modulus of deformation E and the unconfined compression strength q_u is in the range of 150 to 600, with an average of 350. The lower values are found for the softer rocks (sandstone, shale), while the higher values are found for the harder rocks (marble, granite).

The tensile strength of a rock can be measured indirectly by using a special splitting test called the Brazilian test. The values range from less than 1 MPa for a shale up to about 15 MPa for granite. The shear strength of intact rocks leads to cohesion intercepts in the range of 5 to 40 MPa and friction angles in the range of 30 to 50 degrees.

5.6 ROCK MASS RATING

Rock masses are rated by using indices that help in evaluating the relationship between the rock substance properties and the rock mass properties.

Samples of rock are obtained by coring the rock, a process which consists of rotating an open steel tube or barrel with a coring bit (diamond) on the end of the steel tube wall. The tube is rotated into the rock at high speed while water is simultaneously injected for lubrication and cooling. Cores are retrieved and placed in core boxes. The recovery ratio (RR) is the ratio expressed in percent of the length of the core recovered divided by the length cored. The rock quality designation (RQD) is the ratio of the length obtained by adding all the pieces of core longer than 100 mm over the length cored. The *velocity index* I_v is also a useful index to evaluate the difference between the rock substance properties and the rock mass properties. It is defined as the ratio of the square of the compression-wave velocity of the rock mass in the field to the square of the compression-wave velocity of the intact rock in the laboratory. Rock mass quality is excellent for an RQD higher than 90% and a velocity index higher than 0.8. Rock mass quality is very poor for an RQD less than 25% and a velocity index less than 0.2.

The Unified Rock Classification System or URCS (Williamson 1984) was developed to parallel the Unified Soil Classification System (USCS). It provides a systematic and reproducible method of describing rock weathering, strength, discontinuities, and density in a manner directly usable by engineers. The URCS is described in ASTM D5878.

In 1989, Bieniawski proposed the rock mass rating (RMR) by combining several indicators of rock mass features. They include the strength of the rock substance (q_u), the rock quality designation (RQD), the joint spacing, the joint condition, the joint orientation, and the groundwater conditions. Table 5.1 shows the RMR categories. The RMR value is obtained by adding the ratings defined in each category. Rock mass classes I through V correspond to RMR values between 80–100, 60–80, 40–60, 20–40, and 0–20, respectively. A class I rock mass would be labeled a very good rock, whereas a class V rock mass would be considered very poor rock. Such classes can be correlated to estimated values of rock mass strength and safe bearing pressures, for example. Another and similar rock mass rating system exists and is called the Norwegian Q system. This system, created in 1974, is credited to Barton, Lien, and Lunde of the Norwegian Geotechnical Institute (1974). It is based primarily on the analysis of tunneling case histories and uses six parameters to assess the rock mass quality. The parameters are the RQD, the joint set number J_n, the roughness of the joints J_r, the degree of alteration and filling of the joints J_a, the water inflow J_w, and the stress reduction factor SRF. Using these six parameters, the Q factor is derived with the following equation:

$$Q = \frac{RQD \times J_r \times J_w}{J_n \times J_a \times SRF}$$

Table 5.1 Rock Mass Rating (RMR) Geomechanics System (Waltham 1994)

Parameter	Assessment of values and rating				
Intact rock USCS, MPa	>250	100–250	50–100	25–50	1–25
rating	15	12	7	4	1
RQD %	>90	75–90	50–75	25–50	>25
rating	20	17	13	8	3
Mean fracture spacing	>2 m	0.6–2 m	200–600 mm	60–200 mm	<60 mm
rating	20	15	10	8	5
Fracture conditions	Rough tight	Open <1 mm	Weathered	Gouge <5 mm	Gouge >5 mm
rating	30	25	20	10	0
Groundwater state	Dry	Damp	Wet	Dripping	Flowing
rating	15	10	7	4	0
Fracture orientation	Very favorable	Favorable	Fair	Unfavorable	Very unfavorable
rating	0	−2	−7	−15	−25

Rock mass rating (RMR) is sum of the six ratings. Note that orientation ratings are negative.

In 1994, Hoek introduced the geologic strength index (GSI) to rate jointed rock masses. The GSI takes into consideration the interlocking of rock pieces or "blockiness" of the rock mass on the one hand and the condition of the rock surfaces or joints on the other (Figure 5.8). Then the GSI is used to extrapolate from the intact rock strength and modulus to the strength and modulus of the rock mass.

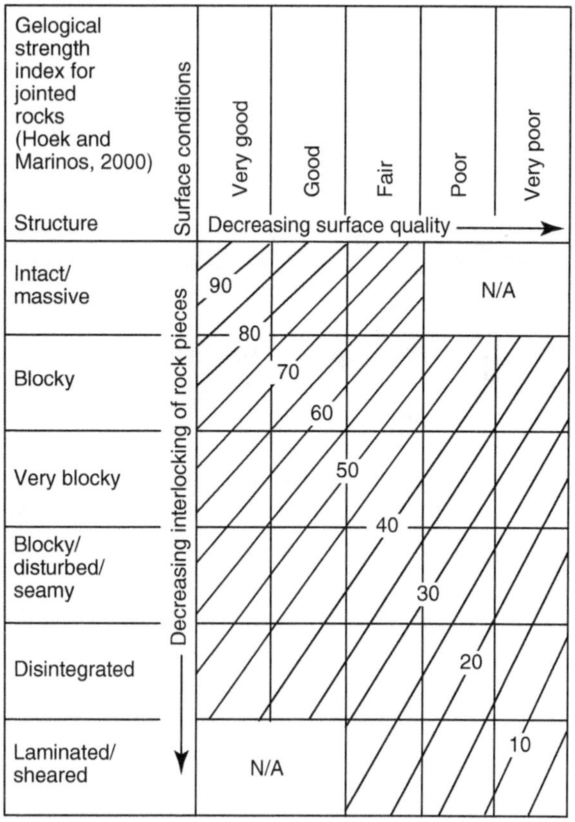

Figure 5.8 Geologic strength index (GSI) for jointed rock masses. (After Marinos and Hoek, 2000.)

5.7 ROCK ENGINEERING PROBLEMS

Some common rock engineering problems include allowable pressures for foundations, ultimate side shear and ultimate point pressure for bored piles, slope stability, tunneling, excavations, blasting, rippability, and scour.

The allowable pressure for shallow foundations in rock is sometimes used prescriptively. These values vary significantly and depend on the quality of the rock mass, which can be described by the RMR classes of the geomechanics system. Estimates of these allowable values are from 6000 to 10,000 kPa for a very good rock (Class I), 4000 to 6000 kPa for a good rock (Class II), 1000 to 4000 kPa for a fair rock (Class III), 200 to 1000 kPa for a poor rock (Class IV), and less than 200 kPa for a very poor rock (Class V). The settlement associated with these allowable pressures is usually calculated using elasticity theory; the main issue is obtaining the right modulus of deformation for the rock mass.

The columns of the World Trade Center towers were on shallow foundations on rock (Figures 5.9, 5.10, and 5.11). The towers weighed approximately 4500 MN each, were 417 m high, and had a footprint of 62 m × 62 m. The mica schist bedrock was found at a depth of about 21 m and exhibited inclined joints. Excavation took place so the shallow foundations could rest directly on the rock. The rock substance modulus was determined through laboratory tests that gave an average of 80,000 MPa. The rock mass was tested by a full-scale footing test which gave a rock mass modulus equal to 1400 MPa or 1/57 of the rock substance value, due to the presence of joints. The design pressure for the footings was approximately 3000 kPa and the maximum pressure applied during the full-scale footing test was well over 3000 kPa.

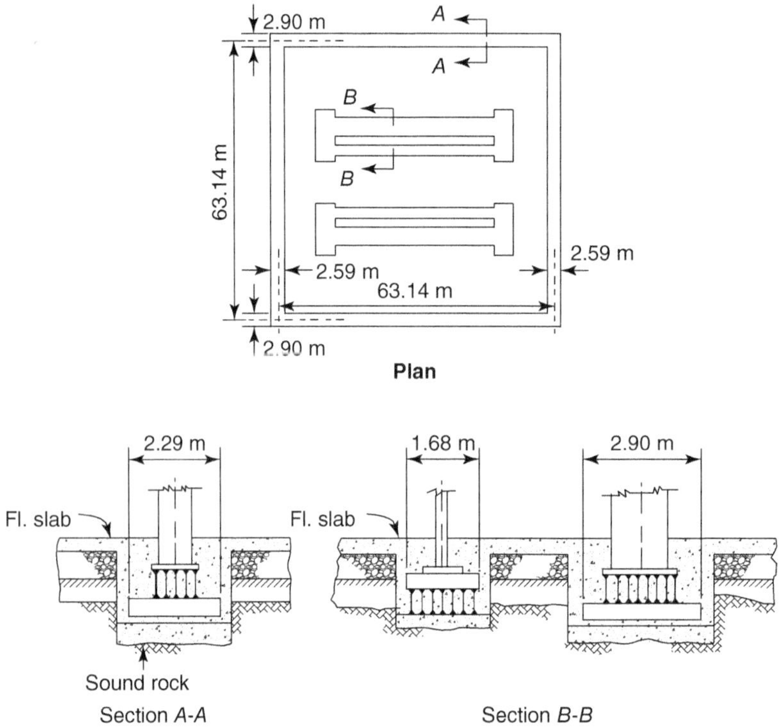

Figure 5.9 Foundation plan for the World Trade Center. (Courtesy of the Port Authority of New York and New Jersey.)

Figure 5.10 Photo of the foundation for the World Trade Center. (Courtesy of the Port Authority of New York and New Jersey.)

The shallow foundations are shown on Figure 5.8 and 5.9; the total area was 1426 m^2. The calculated settlement for the foundation elements using elasticity theory and the measured modulus was very small, varying from 6 to 12 mm. Most of this settlement is likely to have happened during construction.

The cliffs at the Pointe du Hoc site in Normandy, France, are made of interbedded layers of limestone and sandstone. These cliffs are eroded at their base by wave action from the sea; caverns develop at the base as a result of this wave action (Figure 5.12). When the caverns become deep enough, the overhanging rock mass fails. These failures allow back-calculation of the tensile strength of the rock mass. The tensile strength of the rock substance tested in the laboratory by the Brazilian test (Figure 5.13) gave an average tensile strength of 3400 kPa in the limestone and 4500 kPa in the sandstone. The average tensile strength of the rock mass back-calculated from the overhang failures (Figure 5.14) indicated 40 kPa tensile strength, or about 1/100 of the rock substance value.

Recommendations for the ultimate side shear values for a bored pile socketed in rock range from 300 kPa for a weak, fractured, decomposed rock (say, RQD = 20%) to 3000 kPa for a massive competent rock. Common formulas

Figure 5.11 Photo of the World Trade Center twin towers. (Courtesy of the Port Authority of New York and New Jersey.)

Figure 5.12 Jointed rock mass and caverns at Pointe du Hoc.

Figure 5.13 Brazilian tension test on Pointe du Hoc limestone.

equate the ultimate side shear with the square root of the unconfined compression strength diminished by additional factors that take the rock mass quality into consideration. The ultimate bearing pressure at the bottom of a bored pile or a driven pile to rock can range from 4000 kPa for a poor rock mass quality to 400,000 kPa for a massive competent rock with a high strength. Of course, the ultimate bearing capacity may be limited by the strength of the pile itself. The ultimate bearing pressure is usually given as proportional to the unconfined compression strength of the intact rock diminished by a coefficient that takes the rock mass quality into consideration.

In rock slope stability, the main influencing factors are the direction of the joints compared to the direction of the

Figure 5.14 Massive collapse of rock cliff at Pointe du Hoc.

potential failure surface, the shear strength of the joints, and the water pressures in the joints. Failure analyses usually use planar surfaces and wedges following the joints' contours. The failing mass is analyzed using fundamental laws and constitutive laws to give a factor of safety. The *factor of safety* is defined as the ratio of the resisting moment or resisting force in the direction of sliding over the driving moment or driving force in the direction of sliding.

Other rock engineering problems include tunneling, excavations, blasting, rippability, and scour.

5.8 PERMAFROST

In areas of the Earth where the mean annual temperature of the air does not get above 0° Celsius, the soil may be permanently frozen down to a certain depth. These areas include the North Pole, the South Pole, and any mountain above about 5000 m high (Figure 5.15). The permafrost can be shallow (a few meters) or deep (several hundred meters). Because permafrost is rich in ice, its properties are very much tied to the properties of the ice. This implies that, much like ice, the strength and modulus of permafrost increase when the temperature decreases and are rate dependent. Note that the influence of temperature is much more important than the influence of rate effect. Permafrost also exhibits creep under sustained loading. Because ice is the binder that strengthens permafrost, like cement for concrete, the higher the degree of saturation of permafrost, the stronger the permafrost is. For construction on permafrost, it is best to isolate the building or structure from the permafrost so as to minimize the temperature changes incurred in the permafrost. Indeed, when permafrost melts, it loses tremendous strength. This is why buildings and pipelines are elevated above permafrost ground through the use of piles (Figure 5.16).

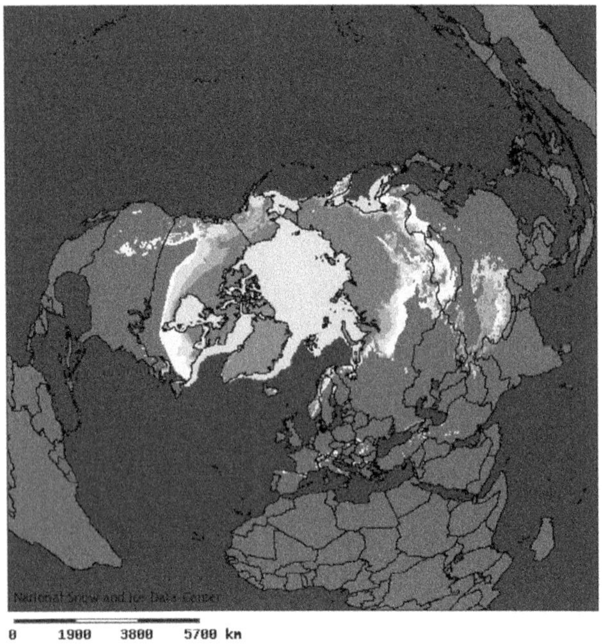

Figure 5.15 Zones of permafrost in the Northern hemisphere. (Courtesy of NSIDC.)

Figure 5.16 Elevated structure on permafrost. (Courtesy of Anadarko Petroleum Corporation.)

PROBLEMS

5.1 Answer the following questions:

a. What are the three main categories of rocks?

b. Are granite, feldspar, and basalt igneous, metamorphic, or sedimentary rocks?

c. Are sandstone, shale, limestone, and lignite igneous, metamorphic, or sedimentary rocks?

d. Are gneiss and marble igneous, metamorphic, or sedimentary rocks?

5.2 Is diamond a rock or a mineral? Is there any rock harder than diamond? Is there anything harder than diamond?

5.3 What is the difference between rock mass and rock substance, and how does this difference affect the engineering properties?

5.4 What is the typical range of values for the dry unit weight and porosity of a rock substance?

5.5 How is the durability of a rock substance measured? Describe the test.

5.6 What are the typical range of modulus and Poisson's ratio values for rock substance? How do this range and values compare with concrete?

5.7 What is the typical range of unconfined compression strength for rock substance and its ratio to the rock substance modulus? What is the typical range of tensile strength for rock substance? How does this compare with concrete?

5.8 Explain what the RR, the RQD, and the I_v are and use the words *excellent, good, fair, poor*, or *very poor* to qualify ranges of values of these various indices.

5.9 Explain what the RMR and the GSI are and how they are obtained.

5.10 Attempt a correlation between the safe pressure for a foundation on rock using the rock substance strength on the one hand and the RMR or GSI on the other.

5.11 Calculate the settlement of the foundation of the World Trade Center towers (Figure 5.1s).

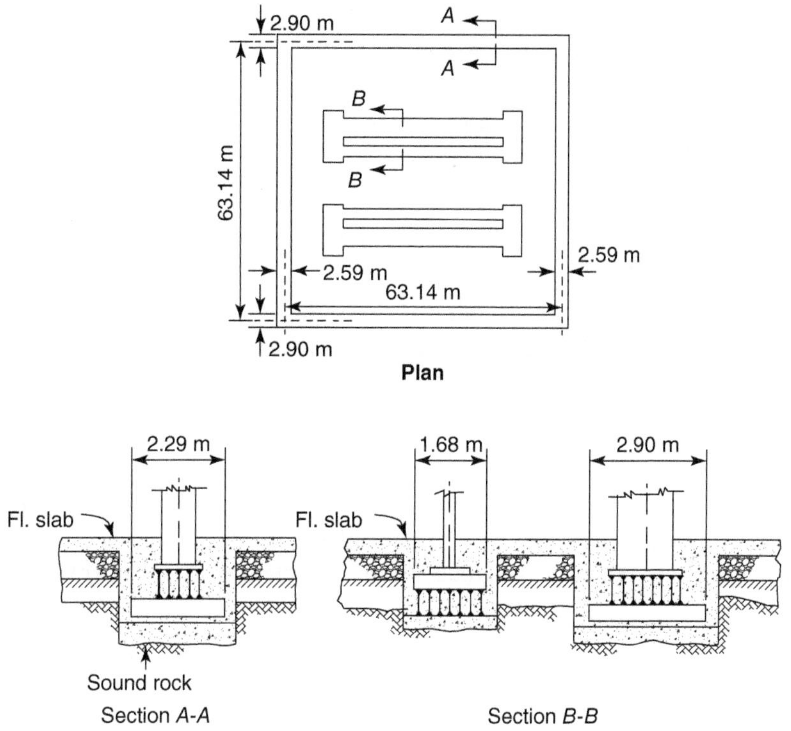

Figure 5.1s World Trade Center foundations section.

5.12 Calculate the stress distribution along the plane of failure for the cliff overhang in Figure 5.2s. Give the solution if the stresses are in the elastic range and assume that there is no failure due to tensile stress (no crack). Find the maximum stress when the cliff is 30 m high and the cave is 4 m deep and 3 m high.

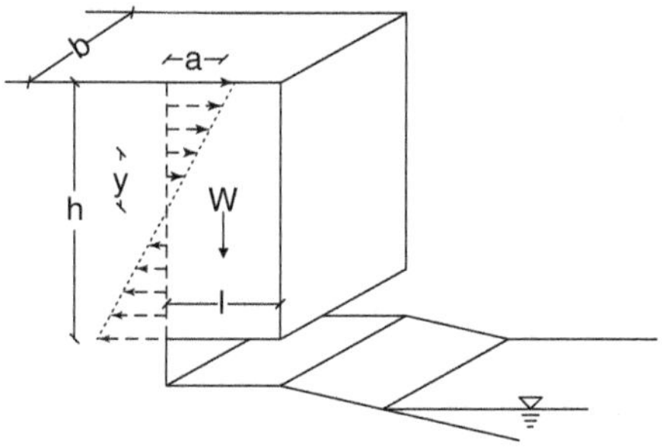

Figure 5.2s Geometry of the cliff.

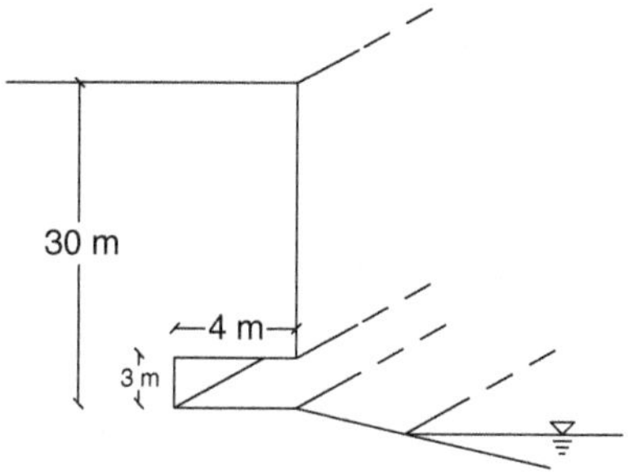

Figure 5.3s Dimensions of the cliff.

5.13 What is the best way to design a foundation on permafrost?

Problems and Solutions

Problem 5.1

Answer the following questions:

a. What are the three main categories of rocks?
b. Are granite, feldspar, and basalt igneous, metamorphic, or sedimentary rocks?
c. Are sandstone, shale, limestone, and lignite igneous, metamorphic, or sedimentary rocks?
d. Are gneiss and marble igneous, metamorphic, or sedimentary rocks?

Solution 5.1

a. From the point of view of their origin, rocks are classified as igneous, sedimentary, or metamorphic.
b. Granite, feldspar, and basalt are igneous rocks.
c. Sandstone, shale, limestone, and lignite are sedimentary rocks.
d. Gneiss and marble are metamorphic rocks.

Problem 5.2

Is diamond a rock or a mineral? Is there any rock harder than diamond? Is there anything harder than diamond?

Solution 5.2

Diamond is pure carbon, so it is a basic element and is classified as a mineral. Diamond is the hardest mineral on the Mohs scale of mineral hardness, based on its resistance to scratching. There is no rock or other natural material harder than diamond. A few manmade materials—all made of carbon—have been claimed to be harder than diamond.

Problem 5.3

What is the difference between rock mass and rock substance, and how does this difference affect the engineering properties?

Solution 5.3

Rock substance refers to a piece of intact rock with no fissures; *rock mass* refers to a large volume of rock, including the fissures and joints. In most cases, the rock mass is much weaker than the rock substance. The presence of fissures and joints weakens the rock mass and affects all its engineering properties.

Problem 5.4

What is the typical range of values for the dry unit weight and porosity of a rock substance?

Solution 5.4

The dry unit weight of rock substance varies from about 21 kN/m^3 (e.g., a shale or a limestone) to about 27 kN/m^3 (e.g., a marble or a granite). The most common values are between 25 and 26 kN/m^3. The porosity of rock substance is very low except for shale, sandstone, and schist, for which the porosity can reach that of soils (several tens of percent).

Problem 5.5

How is the durability of a rock substance measured? Describe the test.

Solution 5.5

The durability of a rock is measured by a test called the slaking durability test. Ten pieces of rock are weighed and placed in a rotating drum lined with a 2 mm opening mesh. The drum is slowly rotated through a water bath for 10 minutes and the dry weight of the rock pieces remaining after the test is measured again. The ratio in percent of the weight after and before the test is the slaking durability index I_{sd}. Rocks typically have I_{sd} values in excess of 90%. Values below 70% are undesirable for rip-rap applications.

Problem 5.6

What are the typical range of modulus and Poisson's ratio values for rock substance? How do this range and values compare with concrete?

Solution 5.6

Values of the modulus of deformation for rock substance range from 2000 MPa to 100,000 MPa and the Poisson's ratio values of rock substance range from 0.15 to 0.3. By comparison, concrete has a modulus of 20,000 MPa, equivalent to that of a soft to medium rock.

Problem 5.7

What is the typical range of unconfined compression strength for rock substance and its ratio to the rock substance modulus? What is the typical range of tensile strength for rock substance? How does this compare with concrete?

Solution 5.7

The typical range of unconfined compression strength for rock substance is from 10 MPa for very soft rock to 200 MPa for very hard rock. By comparison, concrete has an unconfined compression strength of 20 MPa, equivalent to the strength of a soft rock.

The ratio between the rock modulus of deformation E and the unconfined compression strength q_u is in the range of 150 to 600, with an average of 350.

The typical range of tensile strength for rock substance is 1 MPa to 15 MPa. The average tensile strength for concrete is about 2.5 MPa, so concrete is equivalent to a soft rock.

Problem 5.8

Explain what the RR, the RQD, and the I_v are and use the words *excellent, good, fair, poor*, or *very poor* to qualify ranges of values of these various indices.

Solution 5.8

The recovery ratio (RR) is the ratio of the length of the core recovered divided by the length cored, expressed in percent.

The rock quality designation (RQD) is the ratio of the length obtained by adding all the pieces with length longer than 100 mm over the length cored, expressed in percent.

The velocity index I_v is used to evaluate the difference between the rock substance properties and the rock mass properties. It is defined as the ratio of the square of the compression-wave velocity of the rock mass in the field to the square of the compression-wave velocity of the rock substance.

Rock Quality	RR	RQD (%)	I_v
Excellent	97–100	90–100	>0.8
Good	90–97	75–90	0.6–0.8
Fair	67–90	50–75	0.4–0.59
Poor	35–67	25–50	0.2–0.39
Very poor	<35	<25	<0.2

Problem 5.9

Explain what the RMR and the GSI are and how they are obtained.

Solution 5.9

The rock mass rating (RMR) is a value defined as the sum of ratings for several indicators of rock mass features. These indicators are the strength of the rock substance (q_u), rock quality designation (RQD), joint spacing, joint condition, joint orientation, and the groundwater conditions.

Rock Quality	RMR
Excellent	81–100
Good	61–80
Fair	41–60
Poor	21–40
Very poor	0–20

The geologic strength index or GSI is used to rate jointed rock masses, taking into consideration the interlocking of rock pieces or blockiness of the rock mass on the one hand and the condition of the rock surfaces or joints on the other.

The RMR and GSI can be used to extrapolate from the strength and modulus of the rock substance to the strength and modulus of the rock mass.

Problem 5.10

Attempt a correlation between the safe pressure for a foundation on rock using the rock substance strength on the one hand and the RMR or GSI on the other.

Solution 5.10

It has been suggested (Sjoberg 1997) that GSI can be related to RMR by GSI = RMR − 5 for rock masses with RMR larger than 25. Therefore, the following correlation between the safe pressure and the RMR or GSI may be developed:

Pressure (kPa)	RMR	GSI
6,000 ~ 10,000	Class I (81 ~ 100)	76 ~ 95
4,000 ~ 6,000	Class II (61 ~ 80)	56 ~ 75
1,000 ~ 4,000	Class III (41 ~ 60)	36 ~ 55
200 ~ 1,000	Class IV (21 ~ 40)	21 ~ 35
<200	Class V (0 ~ 20)	0 ~ 20

Sjoberg, J. 1997.

Problem 5.11

Calculate the settlement of the foundation of the World Trade Center towers.

Solution 5.11

The sections of the World Trade Center foundations are shown in Figure 5.1s.

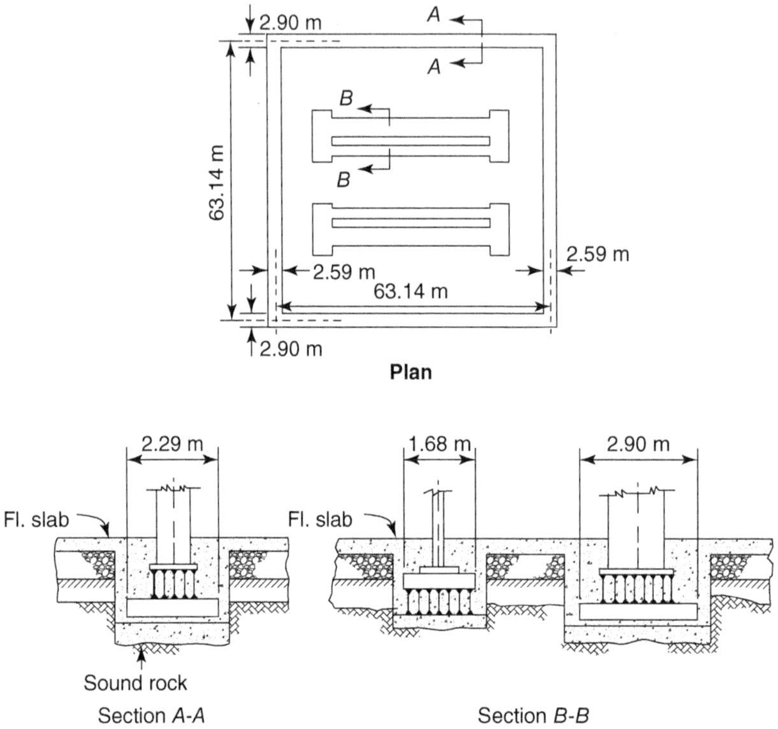

Figure 5.1s World Trade Center foundations section.

The pressure acting on the shallow foundation is $p = 3.0 \times 10^6 Pa$, and the rock mass modulus is $E = 1.4 \times 10^9 Pa$. The ratio between the effective length and the effective width of the foundation (*L'/B'*) is 10.0, and Poisson's ratio (ν) of the rock is 0.3. The settlement of the foundation, assuming that the thickness of the bedrock layer is infinite, is given by the following equation:

$$s = C_s pB \left(\frac{1-\nu^2}{E}\right)$$

where C_s is the shape factor (2 for a rigid foundation with an L/B = 10.0, Fang 1991). Therefore, the settlement is:

$$s(\text{mm}) = 2 \times 3 \times 10^6 \cdot B \cdot \left(\frac{1-0.3^2}{1.4 \times 10^9}\right) \times 1000 = 3.9 \cdot B$$

$$\text{For} \quad \text{B} = 2.29\ \text{m},\ \text{S}_1 = 8.93\ \text{mm}.$$

$$\text{For} \quad \text{B} = 1.68\ \text{m},\ \text{S}_2 = 6.55\ \text{mm}.$$

$$\text{For} \quad \text{B} = 2.9\ \text{m},\ \text{S}_3 = 11.31\ \text{mm}.$$

Problem 5.12

Calculate the stress distribution along the plane of failure for the cliff overhang in Figure 5.2s. Give the solution if the stresses are in the elastic range and assume that there is no failure due to tensile stress (no crack). Find the maximum stress when the cliff is 30 m high and the cave is 4 m deep and 3 m high.

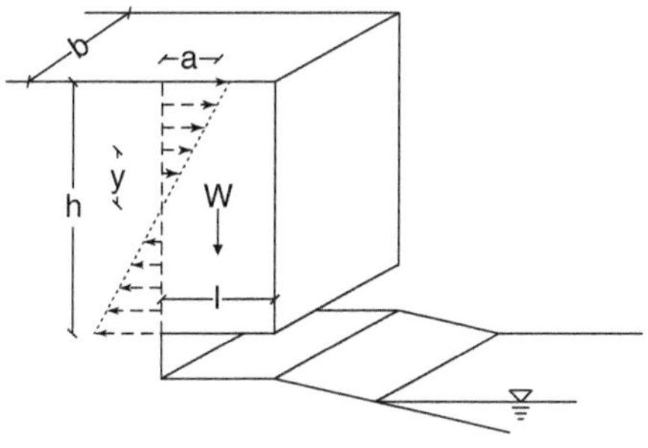

Figure 5.2s Geometry of the cliff.

Solution 5.12

The weight W of the rock overhang above the cave is:

$$W = \gamma \times b \times h \times l$$

where Y is the unit weight of the cliff rock, b is the unit width, h is height of the overhang, and l is the depth of the cave. The bending moment due to this mass of rock is:

$$M = W \times a$$

With $a = 1/2$

The stress distribution in the cliff due to the bending moment is:

$$\sigma = \frac{M \times y}{I}$$

where I is the moment of inertia of the cliff section and y is the vertical distance from the neutral axis (Figure 5.2s). The moment of inertia of the cliff section is:

$$I = \frac{b \times h^3}{12}$$

Combining the previous equations gives the equation for the normal stress distribution in the cliff section:

$$\sigma = \frac{12 \times \gamma \times l \times a}{h^2} \times y$$

Numerical application:

$$h = 27 \text{ m}$$
$$l = 4 \text{ m}$$
$$a = 2 \text{ m}$$
$$b = 1 \text{ m}$$
$$Y = 26 \times 10^3 \text{ N/m}^3$$

Weight per meter of cliff above the cave:

$$W = 26 \times 10^3 \times 27 \times 4 = 28.08 \times 10^5 \text{ N/m}$$

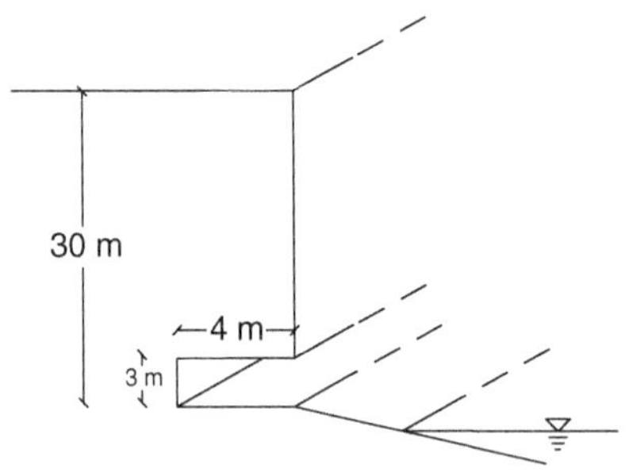

Figure 5.3s Dimensions of the cliff.

The moment due to the cliff mass above the cave:

$$M = 28.08 \times 10^5 \times 2 = 56.16 \times 10^5 \text{ N.m/m}$$

The moment of inertia of the cliff section:

$$I = 27^3/12 = 1640 \text{ m}^4/\text{m}$$

The stress at the top and bottom of the section is:

$$\sigma = \frac{12 \times 26 \times 10^3 \times 4 \times 2}{27^2} \times \frac{27}{2} = 46.22 \times 10^3 \text{ Pa} = 46.22 \text{ kPa}$$

The tension capacity of the intact rock varied between 2 and 20 MPa, yet the cliffs failed when the depth of the caverns reached about 4 m. Therefore, the rock mass tensile strength must have been about 46 kPa, or less than about 2% of the intact rock strength.

Problem 5.13

What is the best way to design a foundation on permafrost?

Solution 5.13

The best way to design a foundation on permafrost is to provide a space between the foundation slab and the soil layer (Figure 5.4s). That way air can circulate and prevent building-generated heat from thawing out the soil layer below and thereby dramatically decreasing the soil strength.

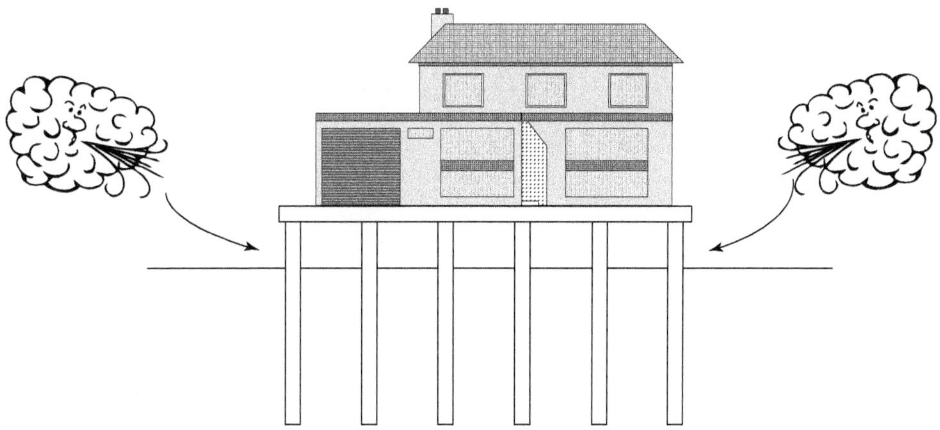

Figure 5.4s Solution for designing a foundation on permafrost.

CHAPTER 6

Site Investigation, Drilling, and Sampling

6.1 GENERAL

Site investigation is the first step in solving most geotechnical engineering problems (Figure 6.1). Indeed, when a geotechnical engineer is asked to solve a problem at a site, the first reflex is to go to the site, drill borings, take samples, and/or run in situ tests. Back in the laboratory, additional soil properties are determined and the problem is studied on the basis of the site-specific information already obtained. Note that laboratory tests and in situ tests are not mutually exclusive. The best site investigation features a combination of in situ tests and laboratory tests. Indeed, the advantages of laboratory tests and the advantages of in situ tests complement each other, as shown in Table 6.1. Boring logs add a very important component to the site investigation. The first part of this chapter deals with onshore site investigations, the second part with offshore site investigations.

A site investigation takes place in two steps: the preliminary investigation and the main investigation. Once the site investigation is completed, the geotechnical engineer makes appropriate calculations and recommendations to the project owner or representative. Sometimes additional site investigation allows the geotechnical engineer to optimize the design and propose less expensive options. For most projects, the cost of the soil investigation is a very small fraction of the cost of the project; it can be 0.1% for buildings up to 3% for dams. Yet it is extremely important that it be well carried out, as a poor site investigation can have disastrous consequences, generate great expenses, delay the project, and lead to litigation. For geotechnically complicated projects, it is very desirable for the geotechnical engineer to act as inspector of the work being done at the site.

Note that under current practice, only an extremely small portion of the soil involved in the project is tested. In a typical soil investigation, 0.001% of the soil involved in providing the foundation support for the structure might be tested. The proportion of soil tested is much smaller than the amount of testing done for the structure itself (concrete cylinder testing, for example).

6.2 PRELIMINARY SITE INVESTIGATION

The preliminary site investigation takes place in two steps: a paper study and a site visit. The paper study consists of obtaining documents related to the site information and history. In addition to maps, previous records of site uses are very helpful. Maps include geologic maps (e.g., http://ngmdb.usgs.gov/), aerial photographs (www.terraserver.com/, http://maps.google.com/), flood maps (www.fema.gov/hazard/map/flood.shtm), and seismicity maps (http://earthquake.usgs.gov/earthquakes/world/seismicity/). The site visit consists of going to the site, taking notes and photos of the site conditions, including the behavior of other projects in the vicinity. The site conditions include general topography, rig access, geologic features, stream banks exposing the stratigraphy, land use, water-flow conditions, and possibility of flood. A good site visit requires a keen eye and keeping a detailed record of what is found and observed at the site. In the case of environmentally related problems, special guidelines exist for what is called environmental site assessments (ESAs). The rest of this chapter describes the main site investigation.

6.3 NUMBER AND DEPTH OF BORINGS AND IN SITU TESTS

The word *sounding* is used in this section to refer to both borings and in situ tests. The number, location, and depth of soundings on the one hand and the type of samples and in situ tests on the other depend on several factors, including the type of geotechnical project, the stratigraphy of the site, the soil type, and the water table conditions.

About 2 to 6 soundings are performed for average size buildings and bridges. A common rule for a building is to perform 1 sounding per 250 m^2 of foundation surface area. For major bridges, a sounding is performed at each pier. For extended projects such as runways and highways, soundings are located anywhere from every 50 m for major runways to

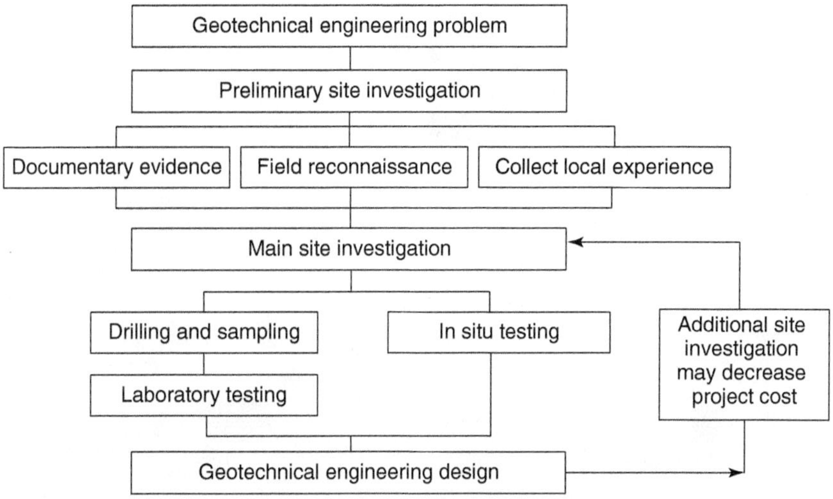

Figure 6.1 Flowchart for a geotechnical engineering project.

Table 6.1 Advantages and Drawbacks of Laboratory and In Situ Tests

Laboratory Testing		In Situ Testing	
Advantages	Drawbacks	Advantages	Drawbacks
Easier to analyze theoretically	Small-scale testing	Large-scale testing	Difficult to analyze theoretically
Drainage can be controlled	Time-consuming	Relatively fast to perform	Drainage difficult to control
Elementary parameters easier to obtain	Stresses must be simulated	Testing done under *in situ* stresses	Elementary paramaters harder to obtain
Soil identification possible	Some disturbance	Less disturbance for some tests	Soil identification rarely possible

every 500 m for secondary highways. For power lines and pipelines, soundings are performed for locations associated with difficult soil conditions and for special loading like corner towers. The depth of the soundings is typically equal to twice the foundation width below the foundation depth. Shallower borings may be accepted if a hard layer is found and confirmed to be thick enough for the project. Depths of soundings commonly vary from 5 m to 30 m.

It is critical to think about the zone of influence of the geotechnical project and ensure that the soil conditions are reasonably well known within that zone. For example, the zone of influence below the tip of a pile may be a few meters, but if 10,000 piles are driven with close spacing, the zone of influence of the foundation is related to the width of the pile group, not the width of a single pile. It is also critical to think about the cost-benefit ratio of the site investigation. The cost of an additional sounding is trivial compared to the cost of repair for most geotechnical projects.

6.4 DRILLING

The two most common methods of drilling for soil samples are the wet rotary method and the hollow stem auger.

6.4.1 Wet Rotary Drilling Method

The wet rotary method consists of drilling a borehole with a drill bit (Figure 6.2) while circulating drilling mud through the center of the rods. The drill bit is typically 75 to 150 mm in diameter and the rods 40 to 70 mm in diameter. The

Figure 6.2 Drill bits.

Figure 6.3 Wet rotary drilling method including mud pit. (*Right*: Courtesy of Wikimedia. See also this video: http://cee.engr.ucdavis.edu/faculty/boulanger/video.html)

drilling mud flows down the center of the rods while they rotate and back to the surface on the outside of the rods between the wall of the borehole and the exterior wall of the rods. This return flow carries the soil cuttings back to the surface by entrainment. The drilling mud arrives in the mud pit (Figure 6.3), where it is sucked back up to the top of the drilling rods by a pump. The connection between the hose carrying the drilling mud back to the top of the rods and the rods themselves is called the *water swivel*; this connection allows the hose to remain stationary while the rods keep rotating. The drill bit at the bottom end of the drill rods is typically either a drag bit or a roller bit (Figure 6.2). Drag bits tend to carve the soil with finger-like protrusions and are used for fine-grained soils. Roller bits are made of three rollers that roll against the soil and erode it or push it aside; they are used for drilling in gravel because the larger particles could get stuck between the fingers of a drag bit, damage it, and create excessive disturbance. In sand, either bit can be used; the bit itself is not that critical because the drilling proceeds by washing or eroding the sand with the mud flow in front of the bit. When the rods progress smoothly downward, the soil is likely a fine-grained soil; when the rods go down in a more jerky fashion, the soil is likely a coarse-grained soil. The grinding sound associated with drilling in gravel, cobbles, or rock can be easily identified. Once the borehole is advanced to the required depth, the rods and bit are withdrawn, the bit is uncoupled, and a sampling tube or an in situ test device is connected at the bottom of the rods.

6.4.2 Hollow Stem Auger Drilling Method

The hollow stem auger method (Figure 6.4 and 6.5) sometimes also called the continuous flight auger method consists of

Figure 6.4 Hollow stem augers. (Courtesy of C. Jeffries, Environmental Sampling Ltd.)

Figure 6.5 Hollow stem auger drilling method. (Courtesy of Geovil Ltd)

rotating hollow stem augers into the soil; these augers are 150 to 300 mm in diameter. The hollow center part of the augers gives access for sampling or any other testing device that can be lowered to the bottom of the hole. The hollow stem auger has the advantage of providing a casing against collapse of the side walls of the borehole, but is limited in penetration depth because it requires significant torque to advance the augers. The wet rotary method is much less limited by depth, but sometimes faces problems of borehole instability.

6.5 SAMPLING

6.5.1 Sample Disturbance

The objective in sampling a soil or rock deposit is to obtain samples that have the least amount of disturbance. This disturbance can come from:

1. Change in stress condition
2. Mechanical disturbance of the soil structure
3. Changes in water content and porosity
4. Chemical changes
5. Mixing and segregation of soil constituents

The goal is to minimize factors 1, 2, and 3, and to eliminate factors 4 and 5. Factor 1 recognizes that the sample follows a certain sequence of stress states as it goes from the intact field situation to testing in the laboratory. In the field, the sample exists under an at-rest effective vertical stress σ'_{ov} and an at-rest effective horizontal stress σ'_{oh}. During sampling, both stresses are likely to increase; σ'_{ov} because the friction between the sample wall and the inside of the sampler compresses the sample and σ'_{oh} because the thickness of the sampler creates horizontal displacement and associated compression. Upon extrusion of the sample from the sampling tube, both total stresses are decreased, but the effective stresses may or may not decrease as much as the total stresses. Indeed, as the sample tries to expand upon extrusion, the expansion may be limited by the inability of the air to get into the pores if the soil has a high degree of saturation and a low hydraulic conductivity. During transport, vibrations are likely and can affect the internal stress state. Figure 6.6 shows a possible scenario before and after testing the sample in the laboratory. It is important to remember this sequence to better understand sample behavior in the laboratory, especially at smaller strains. For one, it is clear that the sample does not start at a state of stress equal to the one it was at in the field. It is desirable to try to recreate this initial state of stress before starting the laboratory test. Similar observations can be made for in situ testing.

Factors 1 and 2 can be minimized by using samplers with a low area ratio. The area ratio, *AR*, is the ratio of the cross-sectional area of the tube wall over the cross-sectional area of the sample:

$$AR = \frac{\pi(D_0^2 - D_i^2)/4}{\pi D_i^2/4} = \frac{D_0^2}{D_i^2} - 1 \tag{6.1}$$

where D_o and D_i are the outside and inside diameter of the sampling tube, respectively.

Ratios less than 10% are desirable. Factor 3 in the preceding list can be minimized by sealing the samples as soon as they

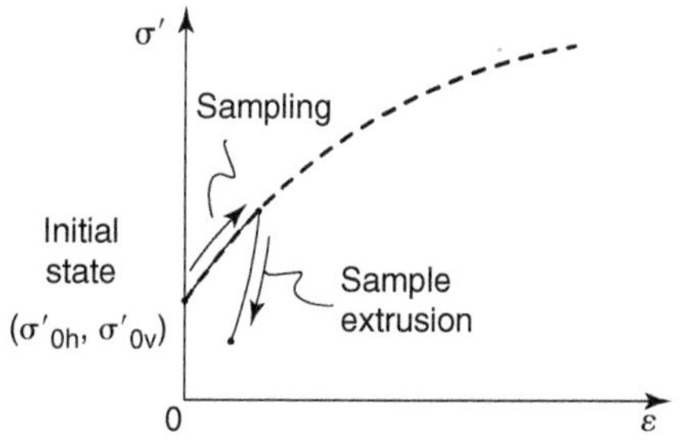

Figure 6.6 Sequence of stress strain behavior during sampling.

are extruded from the sampler. This can be done by pouring hot wax at each end or sealing them with a thin plastic film or foil wrap. Also, it is necessary to keep the vibrations during transport to a minimum, and store the samples in a humidity room as soon as possible. Note that humidity rooms are actually drying rooms, as the humidity level very rarely reaches 100%. Indeed, even at 95% humidity, significant suction exists that can draw water out of the soil into the air. Efforts should be made to seal the sample as well as possible for storage. Despite these best efforts, samples that have been in a humidity room for more than one month are likely to have been affected by drying.

The least disturbed samples of mineral soils are obtained when the sampler is pushed into the soil in one continuous motion. Driving creates much more disturbance. Repeated pushes are not acceptable either, as they create a series of compressions and extensions in the sample and disturbs it (Figure 6.7). There is one exception to this, for organic fibrous soils such as peats. In this case it is best to drive the sampler, because the driving action has a better chance of cutting the fibers rather than pushing them and simply compressing and disturbing the peat excessively. Further information on sample disturbance can be found in Hvorslev (1949).

Figure 6.8 Shelby tube sampling. (Courtesy of Leslie Kanat, 2010, http://kanat.jsc.vsc.edu/drh/.)

6.5.2 Common Sampling Methods

The two most common samplers are the *Shelby tube sampler* for clays and silts, also called *thin-wall steel tube*, and the *split spoon sampler* for sands and gravels.[1] The Shelby tube (Figure 6.8) is a seamless, thin-wall steel tube (e.g., 76.2 mm outside diameter, 73 mm inside diameter, 0.9 m long). The area ratio for the Shelby tube is 9%. This is a very low area ratio, so samples taken with the Shelby tube are considered undisturbed. The tube is pushed into the silt or clay at a steady pace under one continuous push. The tube is then pulled out of the soil and the sample stays in the tube by friction. At the surface, the sample is extruded, wrapped and sealed to prevent moisture loss, and then shipped to the laboratory for testing.

Note that the length of the sample recovered is rarely equal to the length pushed. One reason is that the friction that develops between the sample and the inner wall of the sampler increases as the sample enters the sampler. If the friction on the sample becomes larger than the ultimate bearing capacity of the silt or clay below the lower end of the sampler, the sampler becomes plugged and the soil ahead of the sampler experiences bearing-capacity failure, so no more soil enters the sampler. The length of sample required to plug

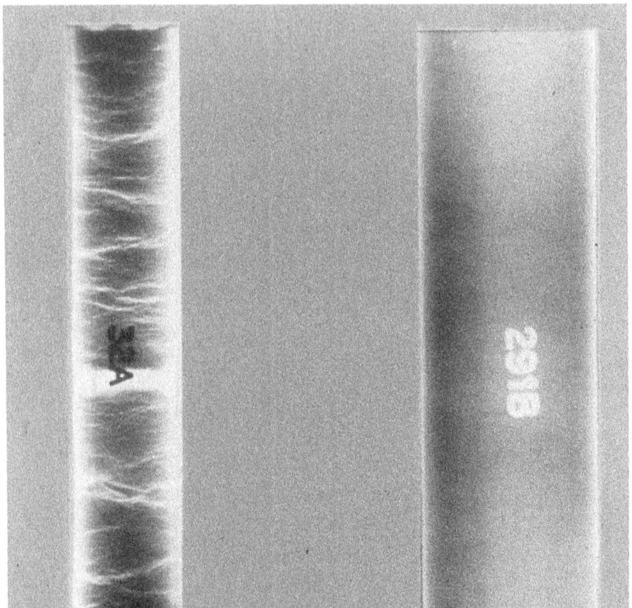

Figure 6.7 X-ray photographs of a driven sample (*left*) and pushed sample (*right*). (Courtesy of FUGRO Inc.)

[1]The name "Shelby" comes from the Shelby seamless steel tube company established in the late 1800s in Shelby, Ohio. The city of Shelby was named after General Isaac Shelby, a hero of the Revolutionary War and War of 1812 and first governor of Kentucky.

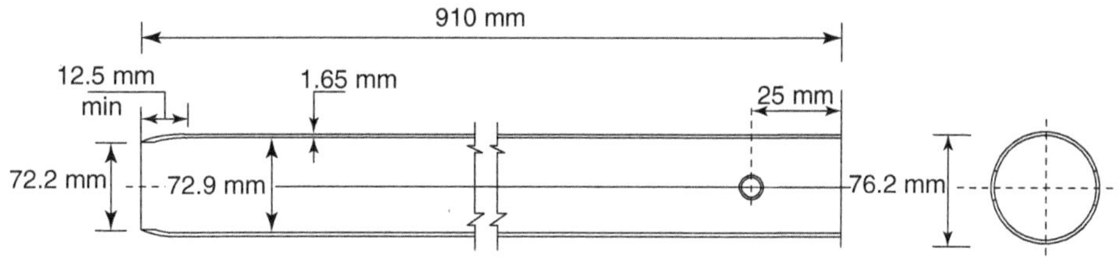

Figure 6.9 Shelby tube sampler cross section.

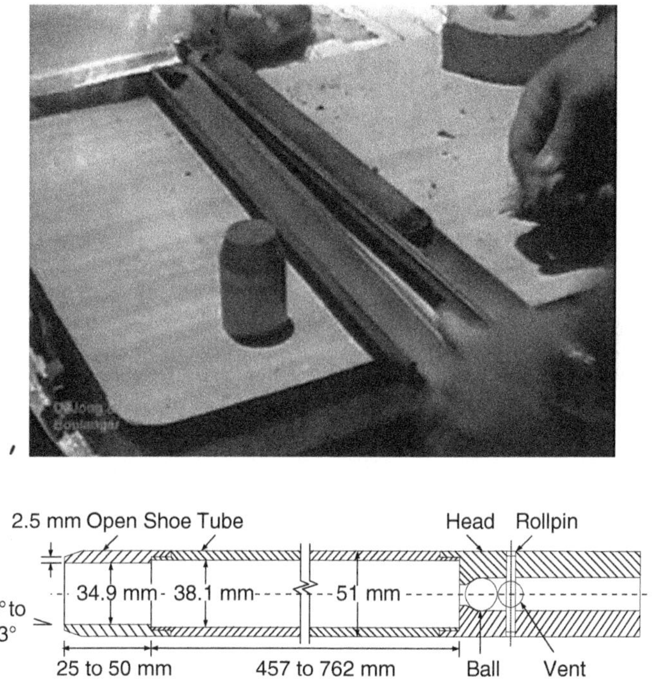

Figure 6.10 Split spoon sampler. (a: From DeJong and Boulanger 2000.)

the sampler depends on the soil and on the sampler, but a length equal to a few sampler diameters may be sufficient to plug the sampler. To minimize the friction between the sample and the inner wall of the sampler, Shelby tubes have an inward curl near the penetrating end (Figure 6.9).

The split spoon sampler (Figure 6.10) is a thick-wall steel tube (50.8 mm outside diameter, 34.9 mm inside diameter, about 0.6 m long) made of two half tubes kept together at the top and the bottom by rings. A core catcher in the bottom ring helps keep the sample in place upon retrieval. The area ratio of the split spoon sampler is 112%. This is a high area ratio, so the samples collected with a split spoon sampler are considered to be disturbed samples. The sampler is driven into the sand or gravel with a standard 623N hammer falling on an anvil at the top of the rods from a height of 0.76 m. This is called the standard penetration test (SPT). The driving process further contributes to the disturbance of the sample. The sampler is brought back to the surface, the tube is opened, and the sample is typically collected in glass jars.

The thin-wall steel tube sampler is used primarily with clays and silts and gives undisturbed samples well suited to many quality laboratory tests. By comparison, the split spoon sampler is used primarily with sands and gravels and gives disturbed samples well suited for soil identification and classification purposes. Other, more advanced samplers exist, such as the Osterberg piston sampler, Swedish foil sampler, Denison sampler, and Pitcher sampler (Hunt, 2005). Piston samplers have the advantage that they minimize part of the disturbance associated with open-tube samplers. Piston samplers have a piston blocking the opening at the bottom of the sampler (Figure 6.11). This piston is locked in place as the sampler is lowered to the desired depth. The piston is then held at that depth while the sampling tube is pushed past the piston into the soil. The vacuum that can develop at the top of the sample helps the soil enter the sampler and minimizes the plugging effect mentioned earlier.

6.6 GROUNDWATER LEVEL

The level of the groundwater at a site is a very important piece of information for any geotechnical investigation. This level can be found in a number of ways: existing information,

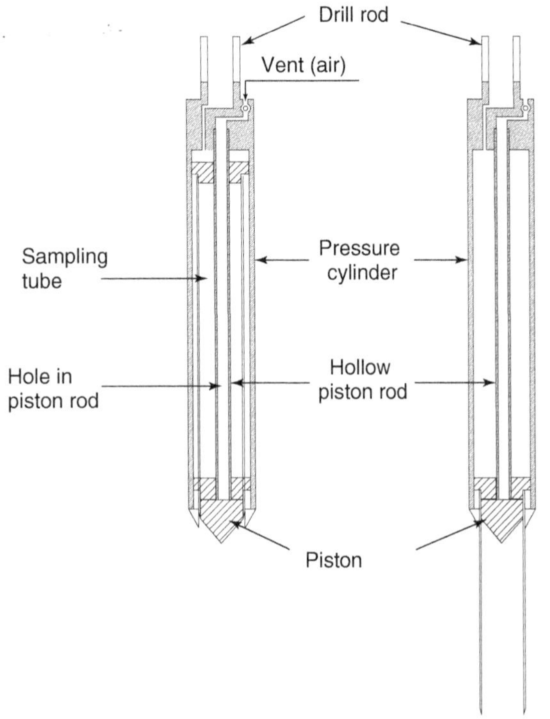

Figure 6.11 Piston sampler diagram.

water level in open borings, standpipe piezometer, and driven piezometers. Existing information can be found in records of water wells drilled in the area (Figure 6.12). These records are often kept by government agencies, such as those dealing with water resources, hydrology, natural resources, geology, and transportation. The well owners may have some very useful information on the seasonal fluctuation of the water level in the well.

Reading the water level in an open borehole is often done as part of a site investigation. At the end of the drilling process, the water or drilling mud is bailed out of the borehole and the water fills the borehole back up to the groundwater level. Water level readings are typically taken 24 hours after the boring is completed and recorded as such in the boring log. Note that 24 hours may not be long enough for the water level to come to equilibrium with the surrounding groundwater. Standpipe piezometers are made by preparing a special borehole (Figure 6.12). After the borehole is drilled, a plastic casing, slotted with holes at its bottom part and smaller in diameter than the borehole, is lowered to the bottom of the borehole. The annulus between the slotted casing and the borehole is filled with sand over the portion where the water pressure measurement is intended. The annulus above the sand-filled portion is filled with bentonite pellets to form an

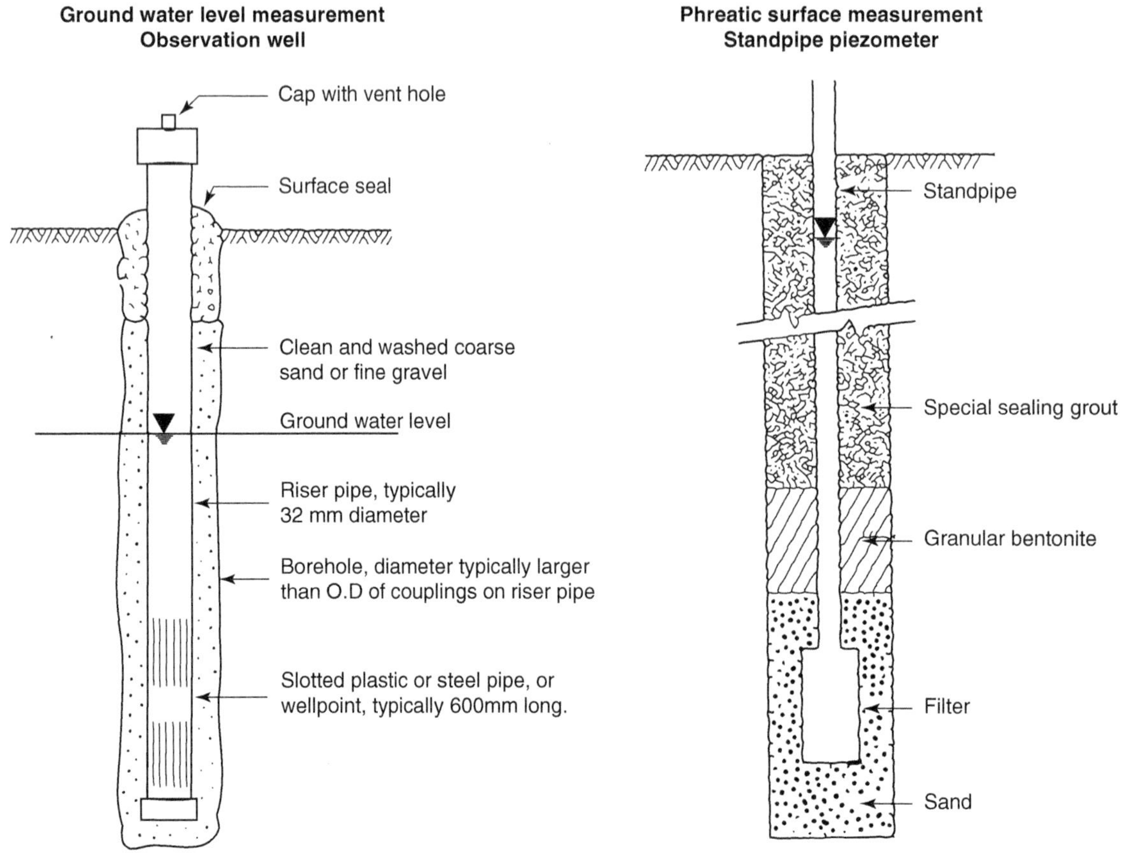

Figure 6.12 Measurement of groundwater level and phreatic surface: (a) groundwater level measurement; (b) phreatic surface measurement observation well. (Courtesy of FHWA.)

impervious plug and isolate the zone to be studied. Water is allowed to rise in the standpipe and the water level is measured when equilibrium is reached. This gives the pressure head at the depth of the slotted casing. Driven piezometers are pushed into the ground like a cone penetrometer and water-pressure measurements are made through pore-pressure transducers located at the bottom of the piezometer. After equilibrium is reached, the pressure measured is equal to the unit weight of water times the height to where the water level would rise in the pipe.

A distinction has to be made between the groundwater level and the phreatic surface. The *groundwater level* is the level to which the water rises in an open borehole. The *phreatic surface* is the line where the water would rise in a pipe (not an open borehole) connected to a point in the ground. These two definitions amount to the same thing unless there is a water pressure in the ground different from hydrostatic. This is the case with an artesian pressure, where the phreatic surface is higher than the groundwater level (Figure 2.14). Standpipe piezometers measure the phreatic surface, whereas open boreholes measure the groundwater level (Figure 6.12).

6.7 FIELD IDENTIFICATION AND BORING LOGS

The best way to identify the soil type is to classify the soil through proper laboratory tests, as described in Chapter 4 on soil classification. In the field, it is also possible to classify the soil through a series of simple tests (ASTM D2488).

Sands and gravels are easily identified, as the particle size is large enough to be seen with the naked eye. Sands will feel gritty when rubbed between your fingers. Dirty sands such as SM and SC tend to leave stains on your hands when wet, whereas SW and SP will have much less tendency to do so. If the sand is dry, taking a handful of sand and dropping it from a height of about 0.3 m will generate a cloud of fine particles for an SC or SM; very little dust will be observed for an SP or SW. Observations about obvious gradation gaps can help in deciding whether the sand is SP or SW.

The type of silt or clay is more difficult to identify. First, wet silts and clays will feel smooth when rubbed between your fingers. The tests described in this section help you decide whether the soil is an ML, MH, CL, or CH; this is the typical order from one extreme to another because in this sequence the soil particles become smaller and smaller and induce a progression in certain properties. The *wash hands test* simply refers to the fact that high-plasticity clays are very difficult to wash off your hands compared to low-plasticity clays and silts. High-plasticity clays tend to feel greasy and it requires quite a bit of rubbing to remove the soil from your skin. Also, when you wash your hands after handling a high-plasticity clay, the pores of your hands will tend to contract and your skin will feel tight after your hands dry. The *dry strength test* also helps you distinguish between high- and low-plasticity materials. Take a piece of soil and let it dry under the sun or in a field office. In a few hours, most soils will be dry enough that the dry strength test can be performed. High-plasticity clays will exhibit high dry strength: difficult to crush between your fingers, difficult to break a small piece by bending. Low-plasticity soils will crush easily or break easily by bending. Silts exhibit little cohesion. The *thread rolling test*, also called the *toughness test*, consists of taking the piece of soil used in the hand shaking test and trying to roll it into a thread as thin as 3 mm in diameter. If it is nearly impossible without the thread cracking, the soil is low plasticity; if it is possible, the soil is likely high plasticity.

The *hand shaking test*, also called the *dilatancy test*, can help in evaluating the hydraulic conductivity of a soil. Silts have a much higher hydraulic conductivity than clays because of the larger particles. The hand shaking test consists of taking a small piece of very wet soil, placing it in the bottom of your cupped hand such that it forms a mushroom-sized patty, and tapping your hand against your other hand to impart horizontal shaking blows to the soil. If the surface of the soil becomes glossy after a few blows, it means that water is coming to the surface and the soil has relatively high hydraulic conductivity (silt). If the soil surface stays matte after 10 blows, the soil has much lower hydraulic conductivity (clay). Organic soils have a distinct foul smell and dark color. Peat is fibrous when young and dark and smooth when decomposed.

Some simple tests can also be used to gauge the strength of the soil encountered. In clays and silts, the tests consist of taking a sample in your hand and trying to deform the sample with your thumb or fingernail. In sands and gravels, the tests consist of trying to push or drive a steel bar into the soil from the surface, as well as checking for footprints behind you. These tests and corresponding categories of strengths are presented in Table 6.2. Note that the tests for silts and clays can be performed on samples retrieved at depth, while the tests for sands and gravels are limited to the ground surface. The SPT blow count is used for evaluating the strength properties of sand and gravel at depth.

During drilling, the driller usually has a good idea what soil is being drilled through: the driller can hear the noise made by the drilling bit, the driller can observe the downward progress of the rods, and the driller can catch the cuttings coming back to the surface (in the mud pit, for example). Clays are carved by a drill bit without much noise and with smooth continuous penetration. Sands are washed by the mud flow and the downward movement of the rods is more erratic. Gravels make a grinding noise during drilling. The driller writes down the soil type that is encountered as the borehole advances; this is the field borehole log. (An example of a field log is shown in Figure 6.13.) While in the field the geotechnical engineer will take notes, collect samples, and perform simple tests. Back in the office, she or he will ask the lab technician to run classification tests and other engineering property tests best suited for the project. On the basis of the data collected, the engineer will prepare the final boring log corresponding to each borehole. Examples of boring logs are

Table 6.2 Simple Field Evaluation of Strength

Silts and Clays Strength			
Description	S_u (kPa)	N (bpf)	Simple field test[1]
Very soft	<12	<2	Squeezes between your fingers.
Soft	12–25	2–4	Easily penetrated by light thumb pressure.
Medium or firm	25–50	4–8	Penetrated by strong thumb pressure.
Stiff	50–100	8–15	Indented by strong thumb pressure.
Very stiff	100–200	15–30	Slightly indented by strong thumb pressure.
Hard	200–400	30–50	Slightly indented by thumbnail.
Very hard	>400	>50	Not indented by thumbnail.

Gravels and Sands Strength			
Description	Φ°	N (bpf)	Simple field test[2]
Very loose	$<28^\circ$	<4	12 mm diameter rebar pushed in 0.3 m by hand Shows definite marks of footsteps; hard to walk on
Loose	28°–30°	4–10	12 mm diameter rebar pushed in 0.1 m by hand Shows footsteps
Medium or compact	30°–36°	10–30	12 mm diameter rebar driven 0.3 m with carpenter hammer Footsteps barely noticeable
Dense	36°–41°	30–50	12 mm diameter rebar driven 0.1 m with carpenter hammer No marks or footsteps
Very dense	$>41^\circ$	>50	12 mm diameter rebar driven 0.03 m with carpenter hammer No marks of footsteps

[1]Note that these tests are performed on a sample of the soil.
[2]Note that these tests are performed at the ground surface of the gravel or sand deposit, not on a sample.

shown in Figure 6.14 and the key to soil type representation on boring logs is shown in Figure 6.15.

6.8 SOIL NAMES

In a natural soil, the four groups of particle sizes may exist side by side. A gravel will be a soil that has most of its particles in the gravel size range. A sand will be a soil that has most of its particles in the sand size range. Silts and clays are recognized according to their plasticity; that is, the ability of the soil to deform without breaking. Silts exhibit moderate plasticity, whereas clays can exhibit very high plasticity. Soils are classified as gravel, sand, silt, or clay according to a rigorous classification system described in Chapter 4.

Soils may also be given other names, such as those in the following list:

- *Adobe*: a local term describing silts and clays in semiarid regions
- *Aeolian soil*: soil deposited by wind such as loess
- *Alluvium*: soil carried by moving water and deposited when the water slows down
- *Bentonite*: a very fine clay with extreme swelling and shrinking properties; used with water as drilling mud
- *Calcareous sands*: sands formed by the shells of marine mollusks
- *Caliche*: soil cemented by calcium carbonate
- *Collapsible soils*: soils that exhibit sudden settlement (collapse of the structure) when placed under load and under water
- *Colluvium*: soil deposited by gravity at the bottom of a slope
- *Dispersive clays*: clays in which the particles separate from each other when exposed to water even when the water does not flow and the soil is not loaded
- *Expansive soils*: see shrink-swell soils
- *Lacustrine deposits*: soil deposited by settling in water under a low-energy environment such as a lake
- *Laterite*: soil rich in iron (red color) and found in hot and humid climates (tropics)

Project name:	Book										Borehole ID:	Borehole I
Project location	Texas A&M										Borehole length:	15 mm
											Date:	6 Feb. 2011
Depth (m)	Drilling method	Tube Dia. (mm)	SPT depth (m)	Blow count				Shelby tube depth (m)	Pocket penetrometer value (kPa)	Water observation	Description and notes	
				N1	N2	N3	N					
0 to 3	Auger	35	1	2	5	6	11			No	Brown silty sand	
1 to 2	Auger	35								No	Brown silty sand	
2 to 3	Auger	35	3	4	6	7	13			No	Brown silty sand	
3 to 4	Auger	35								Yes	Brown silty sand	
4 to 5	Auger	35	5	3	4	6	10			Yes	Brown silty sand	
5 to 6	Auger	35								Yes	Brown silty sand	
6 to 7	Auger	35	7	9	12	17	29			Yes	Brown silty sand	
7 to 8	Auger	35						8	70	Yes	Gray Plastic clay	
8 to 9	Auger	76								Yes	Gray Plastic clay	
9 to 10	Auger	76						10	120	Yes	Gray Plastic clay	
10 to 11	Auger	76								Yes	Gray Plastic clay	
11 to 12	Auger	76						12	130	Yes	Gray Plastic clay	
12 to 13	Auger	76								Yes	Gray Plastic clay	
13 to 14	Auger	76						14	135	Yes	Gray Plastic clay	
Generak notes:	Water found @ 3.4 m. After 24 hours still @ 3.4 m.											

Figure 6.13 Example of a driller's field log.

- *Loam*: a mixture of sand, clay, and decaying organic materials
- *Loess*: lightly cemented soil made mostly of silt, and deposited by wind
- *Marl*: very stiff clay of marine origin and with calcareous content
- *Montmorillonite*: a very fine clay with extreme swelling and shrinking properties
- *Organic clay or silt*: a clay or a silt with a significant amount of organic constituents
- *Peat*: organic soil made of live or decayed plant fragments
- *Quick clay*: clay that can liquefy when sheared excessively
- *Quick sand*: sand that turns into a liquid when subjected to a sufficiently strong upward flow of water
- *Residual soils*: soils created by intense weathering of crystalline rock (tropical regions)
- *Shale*: a very hard soil or soft rock made of silt and clay particles; can slake when subjected to wet-dry cycles
- *Shrink-swell soils*: soils above the groundwater level that shrink and swell when exposed to the seasonal cycles; also but less appropriately called *expansive soils*
- *Slickensided clay*: clay with fissures, the surfaces of which have been smoothened by repeated movement
- *Till*: soil created by glaciers and containing many particle sizes well distributed across the range from very small to very large; typically very strong
- *Tuff*: soil deposited by a volcanic explosion, usually silt size
- *Varved clay*: a clay made of thin alternating layers of silt and clay.

6.9 OFFSHORE SITE INVESTIGATIONS

Offshore structures (Figure 6.16) are built primarily to drill for oil, to collect any oil found, and to send it to shore through pipelines. The depth of offshore platforms can reach several thousand meters of water depth, and the foundation of these enormous structures requires proper site investigations. Other types of offshore structures requiring site investigation include windmills, pipelines, and bridges.

The site investigation is performed from boats, ships, or sometimes jack-up rigs. The size of the ship used depends on the water depth. Figure 6.17 shows some of the vessels available for various water depths. In shallow waters, ships are simply anchored. In deep waters, the most sophisticated ships have dynamic global positioning systems (GPSs) where motors on the hull of the ship are able to maintain

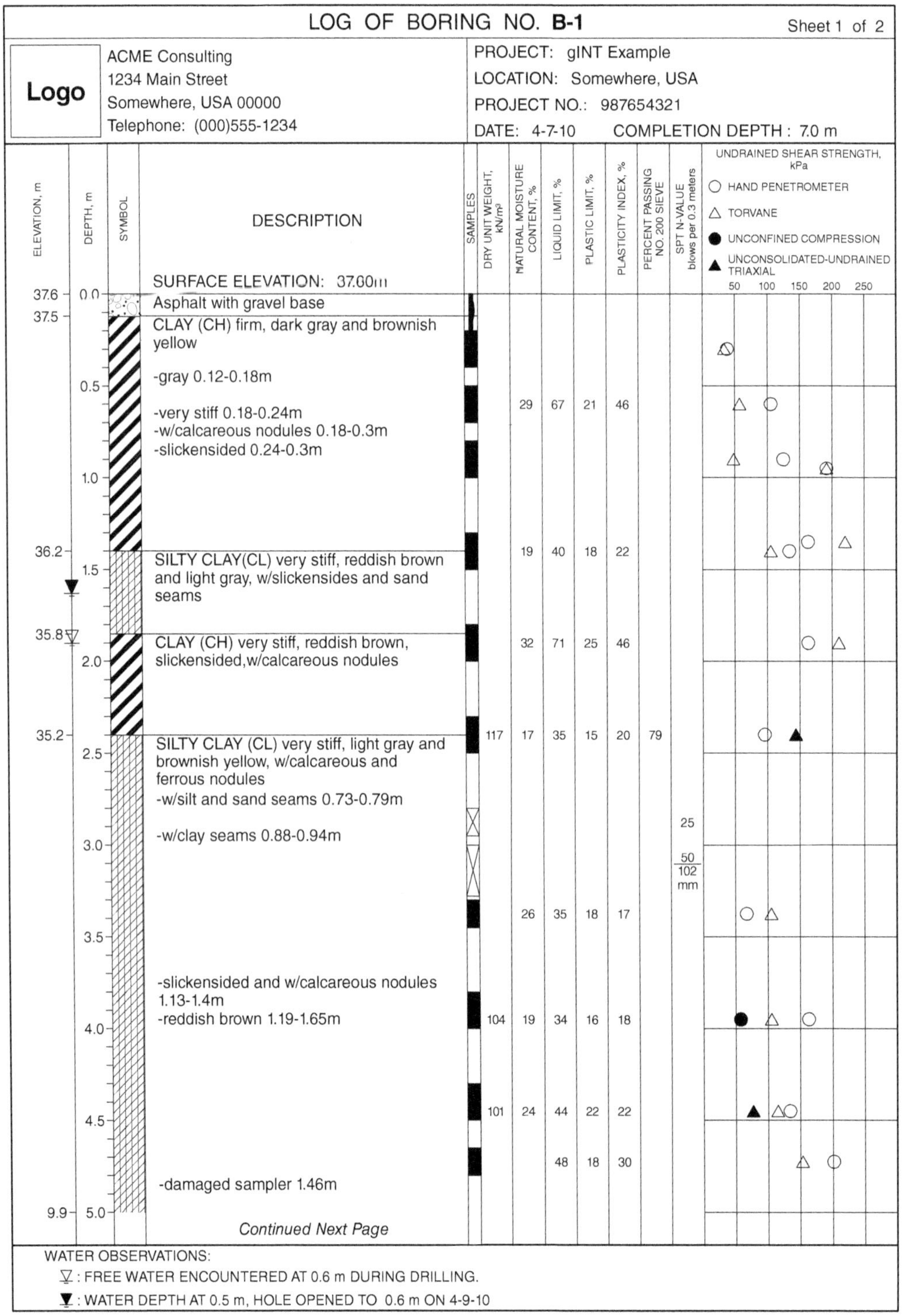

Figure 6.14 Examples of final borehole logs. (Copyright © 2011 Bentley Systems, Incorporated. All Rights Reserved.)

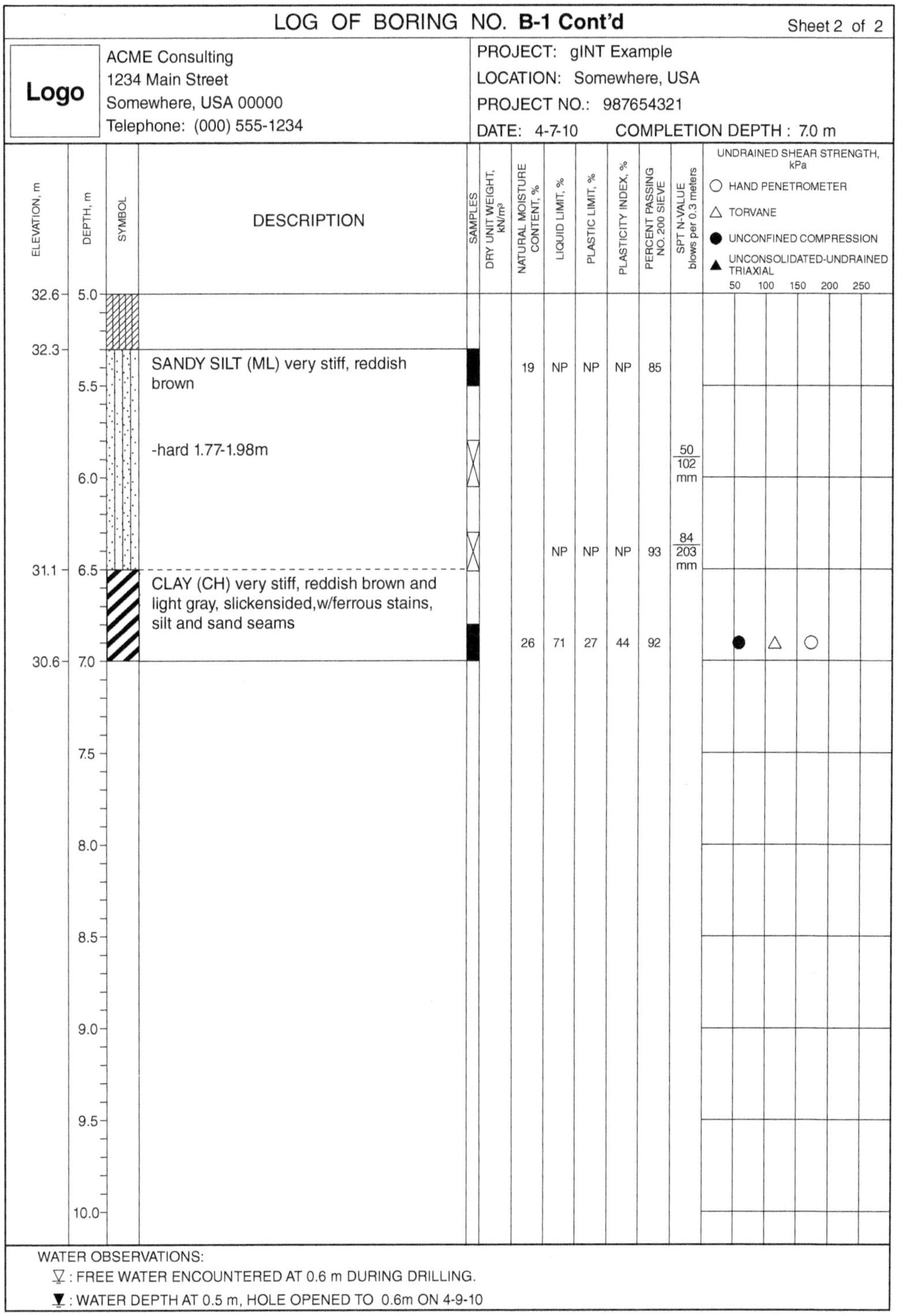

Figure 6.14 *(Continued)*

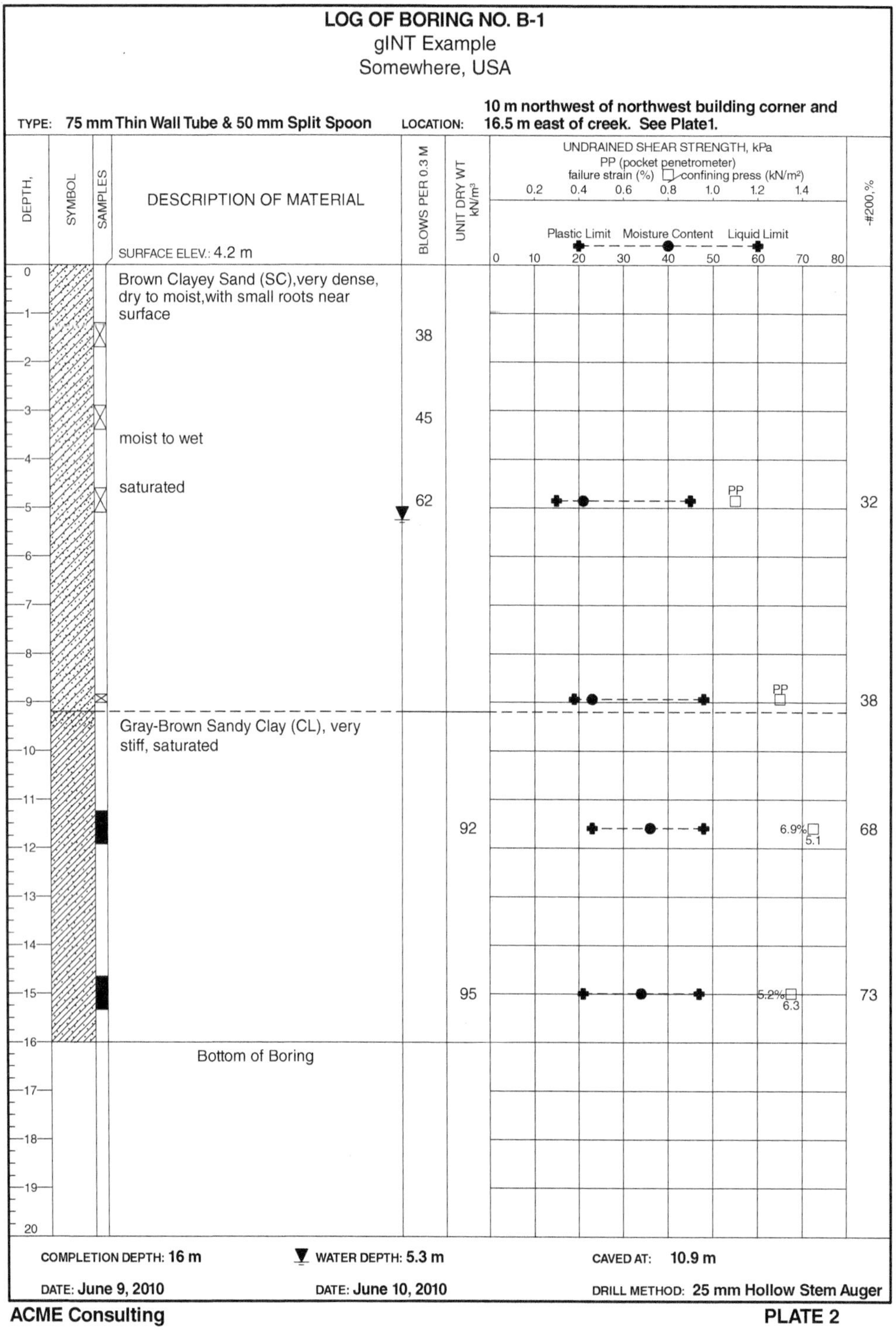

Figure 6.14 (*Continued*)

Majors divisions		Letter and symbol	Name
Coarse-grained soils	Gravel and gravelly soils	GW	Gravel or sandy gravel well graded
		GP	Gravel or sandy gravel poorly graded
		GM	Silty gravel or silty sandy gravel
		GC	Clayey gravel or clayey sandy gravel
	Sands and sandy soils	SW	Sand or gravelly sand well graded
		SP	Sand or gravelly sand poorly graded
		SM	Silty sand or silty gravelly sand
		SC	Clayey sand or clayey gravelly sand
Fine-grained soils	Silt and clay soils (low liquid limit)	ML	Silts, sandy silts, gravelly silts or diatomaceous soil
		CL	Lean clays, sandy clays, or gravelly clays
		OL	Organic silts or lean organic clays
	Silt and clay soils (high liquid limit)	MH	Micaceous silts, diatomaceous soils, or elastic silts
		CH	Fat clays
		OH	Fat organic clays
Fibrous organic soils		PT	Peat humus, and other organic swamp soils

Figure 6.15 Key to soil type representation.

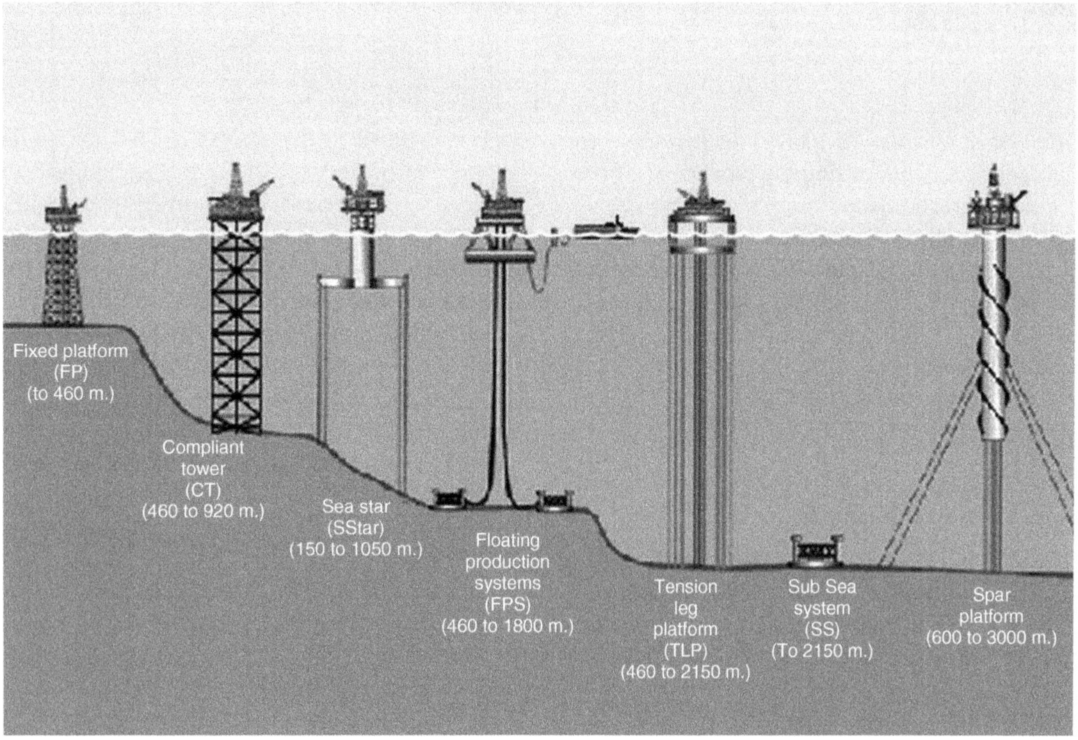

Figure 6.16 Types of offshore oil platforms. (Courtesy of Otis Armstrong and Greg Overton.)

Figure 6.17 Ships, jack-up rig, and remotely operated vehicle (ROV) for offshore investigations. (a, and c: Courtesy of FUGROSeacore., b and d: Courtesy of the ISSMGE Technical Committee on Offshore Geotechnics)

the ship in the same position (x and y) with respect to known satellite positions. Minimizing the movement of the drilling tool in the z direction due to ship movement is also important and is done through the use of heave compensators, described later in this chapter. Offshore geotechnical investigations include drilling, sampling, and in situ testing much like onshore investigations. The difference is a matter of scale, complexity, and cost. Another difference is the increased use of geophysical investigations for offshore work.

6.9.1 Offshore Geophysical Investigations

Offshore investigations rely on geophysics in addition to geotechnical investigations. The geophysics techniques most commonly used are seismic reflection and seismic refraction. Electrical resistivity methods are also used, but less frequently.

Seismic reflection systems use sound propagation energy generated by a device towed behind a ship. The device measures the travel time required for the acoustic energy or wave to travel to the seabed or an interface between two distinct soil layers below the seabed and be reflected to the same device or to a receiving array (Figure 6.18). The *seismic refraction* systems also use sound propagation energy, in this case generated by a device including an acoustic pulse generator and a line of hydrophones dragged on the sea bottom by a ship. The device measures the travel time required for the acoustic energy or wave to travel to an interface between two distinct soil layers below the seabed, refract critically along that interface, and send the wave back to the line of receivers or hydrophones. Seismic refraction is more often used for shallow penetration below the seafloor (pipelines, cables) and gives the thickness and the shear wave velocity of the material.

Precision of the measurement and penetration into the soil depend on the frequency and amplitude of the acoustic wave. A wave with a high frequency and low amplitude will give high resolution (good precision on the distance measured) but low penetration into the soil. A wave with a low frequency and high amplitude will give deeper penetration but lower resolution. Measurements of water depth or *bathymetry* (bathos

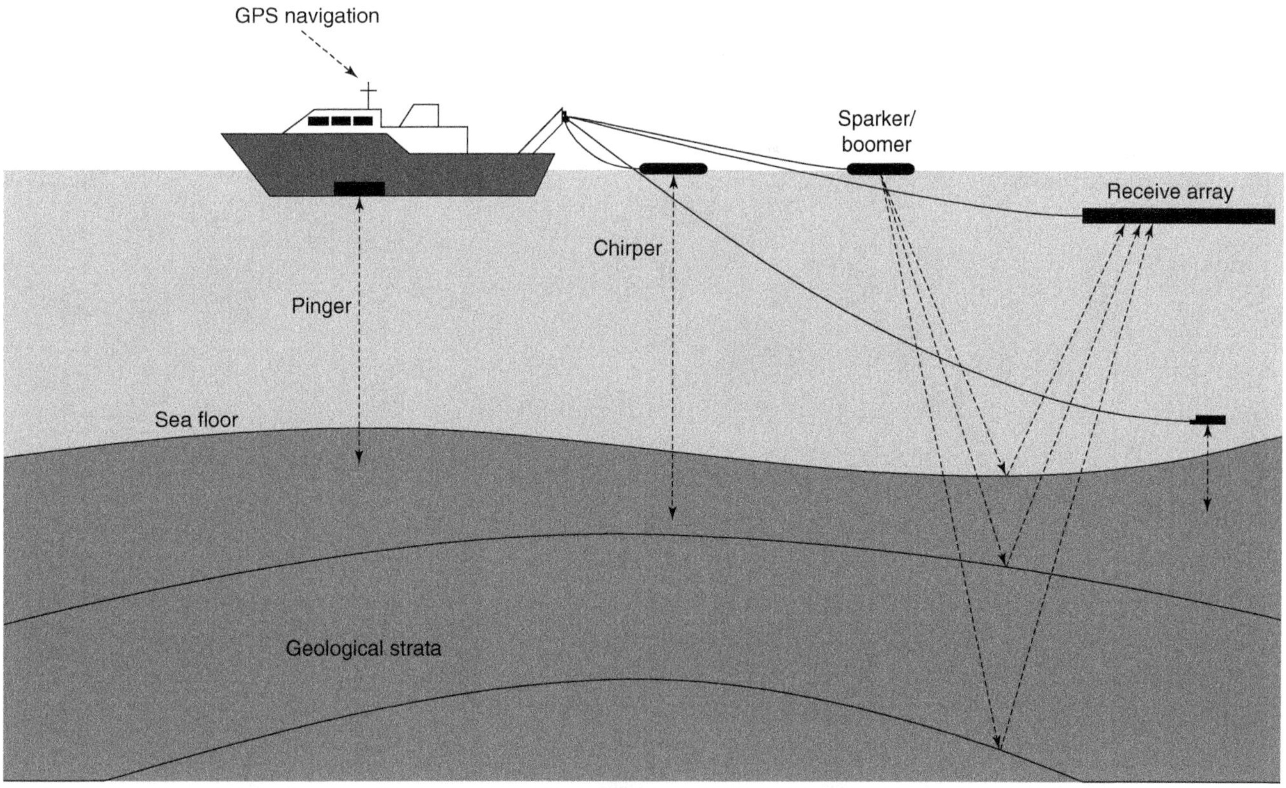

Figure 6.18 Offshore geophysics investigations: seismic reflection. (Courtesy of Ozcoast.)

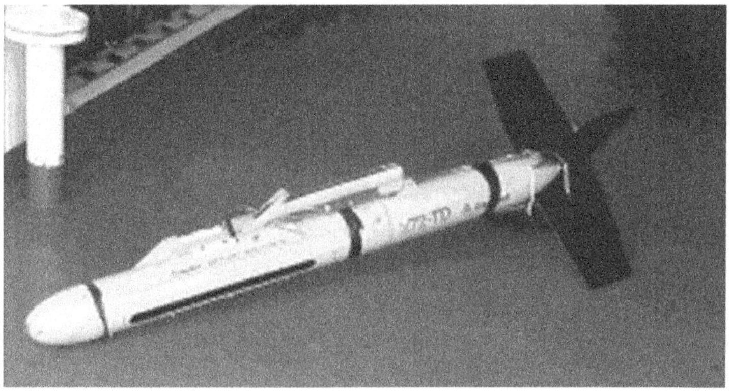

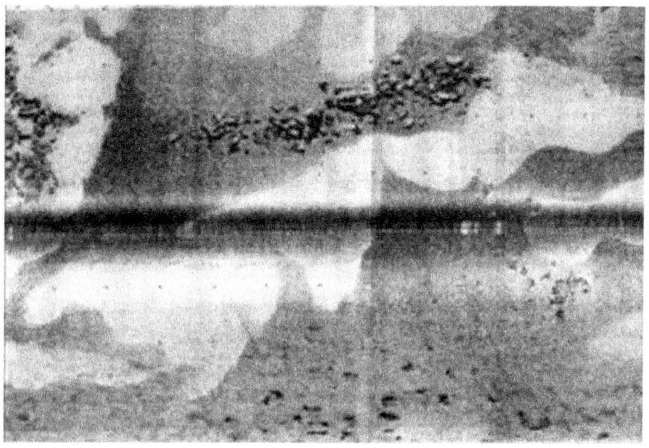

Figure 6.19 Sidescan sonar and sea-bottom image generated therefrom. (Courtesy of the ISSMGE Technical Committee on Offshore Geotechnics.)

means "depth" in Greek) are made using echosounders towed in the water column behind a boat. Pictures of the sea floor are obtained with *sidescan sonars* (Figure 6.19). These sonars aim sideways to capture a wide image of the sea floor. Images of the soil layers below the sea bottom are generated by using *sub-bottom profilers* such as pingers, chirpers, boomers, and sparkers (Figure 6.18). The frequencies generated by these devices vary from 0.5 kHz to 40 kHz and give penetration of the soil from 1 to 100 m with about a 1 to 10% resolution. Examples of the results obtained through geophysics tests are shown in Figure 6.20.

6.9.2 Offshore Geotechnical Drilling

Compared to onshore drilling, offshore drilling adds at least two complexities: larger depths, which can reach thousands of meters of combined water depth and penetration depth;

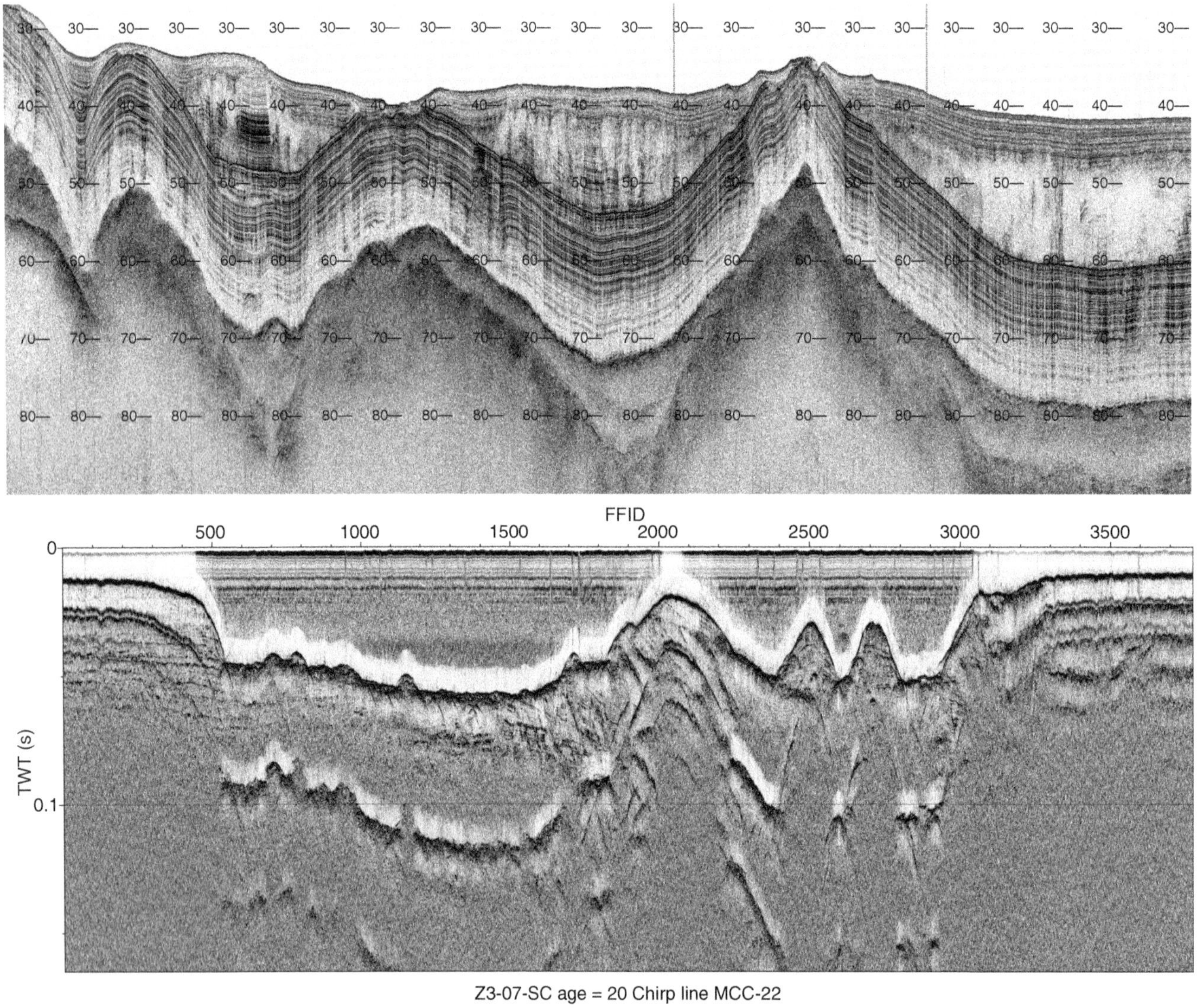

Figure 6.20 Examples of sub-bottom profiler results. (a: Courtesy of EdgeTech; image captured by EdgeTech 3100 Portable sub-bottom profiling system, b: Courtesy of United States Geological Survey, USA)

and vertical motion of the drill rig due to waves. The wet rotary drilling method is always used for offshore work, but it is very difficult to recirculate the drilling mud; this would require injecting the drilling mud from the ship down the drill pipes, and bringing it back to the ship. It would be necessary to have a double set of concentric drill pipes and would complicate the process dramatically while increasing the weight significantly. Instead, a single string of drill pipes is used and the biodegradable drilling mud is expended as waste on the sea floor. Many offshore sediments are very soft near the sea floor, so a casing is necessary to prevent collapse of the borehole. Furthermore, access to the borehole after drilling is necessary to take a sample or run an in situ test. For these reasons offshore drilling is done by rotating casing-size drill pipes, typically 127 mm outside diameter and 102 mm inside diameter. The drill bit at the end of the pipes is about 220 mm in outside diameter (Figure 6.21) and allows free access to the soil for various tools through the center of the pipes and the bit. Often the drill pipes will go through a support bottom platform (Figure 6.22) to guide the pipes and provide bottom support.

The ship movement must not be transmitted to the drill pipes or the drill bit would go up and down with the ship. This would lead to very poor borehole quality and could plug the bit. Heave compensators or motion compensators are instruments that minimize this problem (Figure 6.23). They can be passive (spring-and-dashpot system from which the top of the rods is hanging) or active (computer-controlled hydraulic jacks acting on the rods to compensate for measured motion).

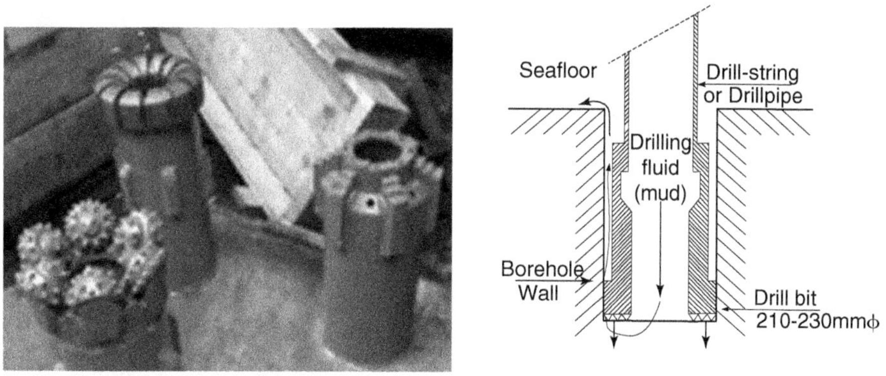

Figure 6.21 Bottom of drill pipes with drill bit. (*Left*: Courtesy of Rok Max Drilling Tools, Ltd.; *Right*: After Richards and Zuidberg 1985.)

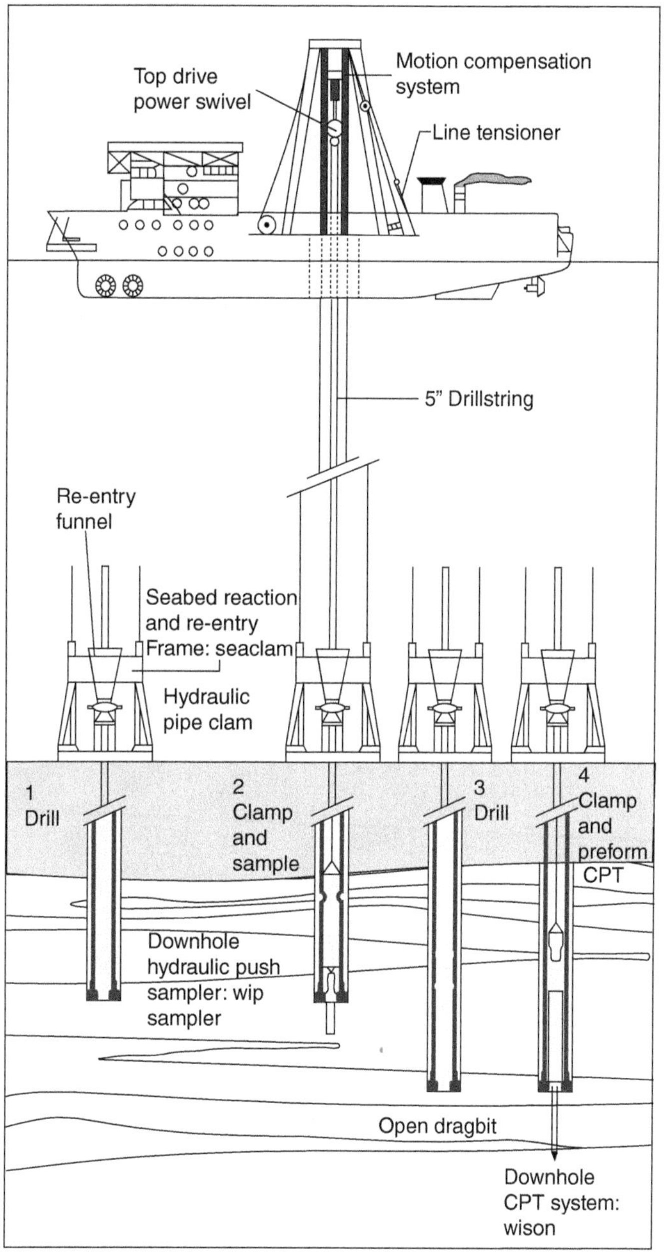

Figure 6.22 Drilling, sampling, and in situ testing through the drill string. (Courtesy of the ISSMGE Technical Committee on Offshore Geotechnics.)

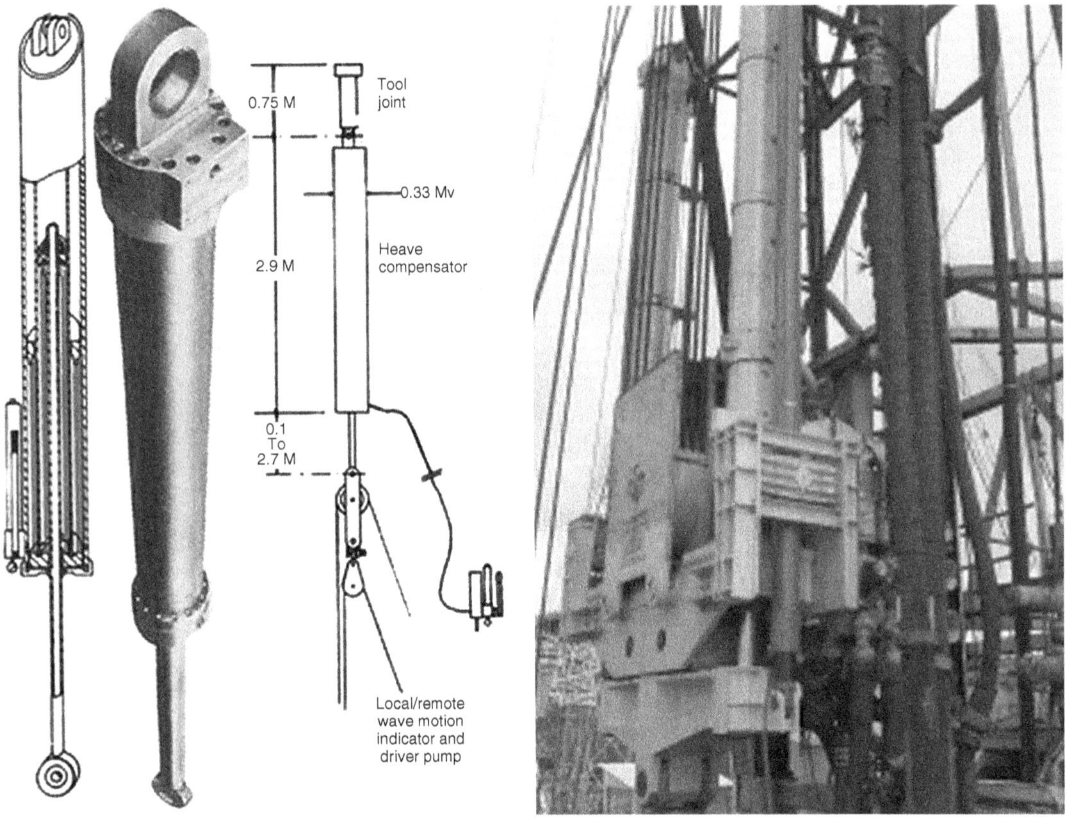

Figure 6.23 Drill derrick with heave compensator. (a: Courtesy of Pulse Guard; b: Courtesy of Integrated Ocean Drilling Program.)

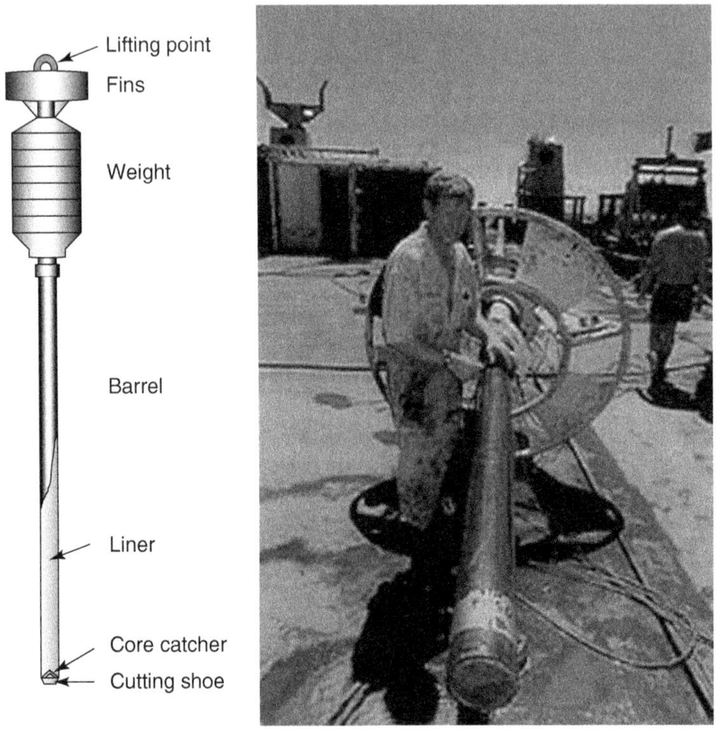

Figure 6.24 Drop core samplers. (a: After ISSMGE Technical Committee on Offshore Geotechnics, b: Courtesy of Offshore Magazine//PennWell Corp.)

6.9.3 Offshore Geotechnical Sampling

Sampling can be done remotely from a platform placed on the seabed (seabed mode) or through a drill pipe controlled from the ship deck (drilling mode). In the seabed mode, samples or in situ tests can be performed to a depth of 20 to 60 m below the sea floor, depending on the soil strength and the weight of the bottom platform. In the drilling mode, larger depths below the sea floor can be reached.

The simplest way to obtain soil samples offshore is by drop core sampling (Figure 6.24). These samples are taken by dropping a long, hollow tube (75 to 150 mm diameter) from a limited height above the sea floor. The length of these gravity samplers can reach tens of meters. Sometimes the process is aided by vibrating the sampler. In the seabed mode, the sampler is pushed hydraulically from a sea-bottom platform (Figure 6.25). In the drilling mode, the sampler is lowered through the drill pipes, locked in the bottom of the drill pipes by latches, and then pushed hydraulically out of the drill pipe into the soil by reaction against the weight of the drill pipes (Figure 6.22).

A piston sampler is preferred for softer, fine-grained soils. Otherwise, open tubes are used. In all cases, pushing is preferred to driving, although it may be necessary to drive the sampler into denser, coarse-grained soils to ensure penetration.

Further details on onshore site investigations can be found in Clayton, Simons, and Matthews (1982) and Hunt (2005). Further details on offshore investigations can be found in Poulos (1988) and a ISSMGE Technical Committee report (2010).

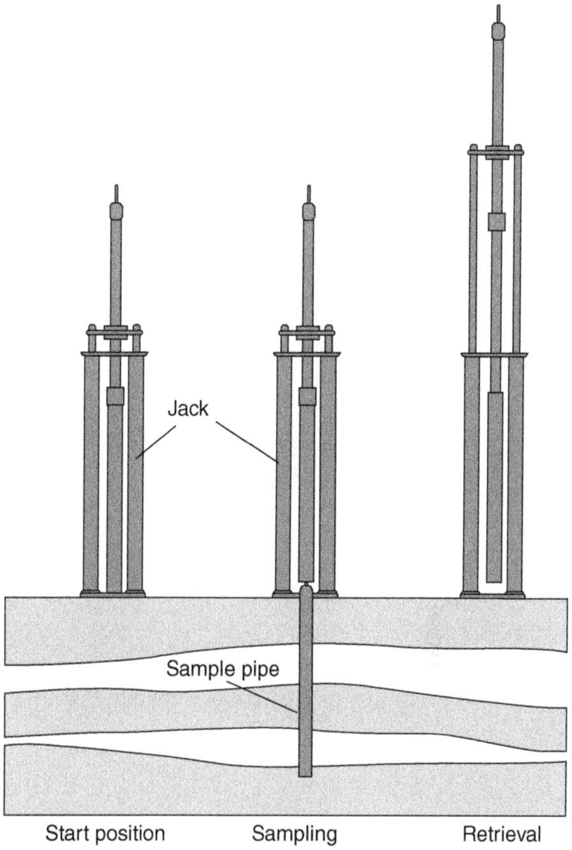

Figure 6.25 Push sampling from a seabed platform. (Courtesy of the ISSMGE Technical Committee on Offshore Geotechnics.)

PROBLEMS

6.1 A 70-story building has an imprint of 35 m by 25 m and will be supported on a mat foundation located at a depth of 10 m. How many borings would you propose and to what depth? Where would you place the borings on the building plan view?
6.2 For problem 6.1, estimate the ratio between the volume of soil that is tested over the volume of soil involved in supporting the building. Comment on the result.
6.3 What drill bit would you use for drilling in clay and which one would you use for drilling in gravel? Explain your choice.
6.4 Discuss and compare the wet rotary method and the hollow stem auger method. Make recommendations as to when to use one and when to use the other.
6.5 Give three sources of sample disturbance and calculate the area ratio for the Shelby tube sampler and the split spoon sampler.
6.6 Discuss when a sampler should be pushed and when it should be driven.
6.7 Calculate the length of clay sample necessary to plug a Shelby tube. (*Plugging* means that the friction between the sample and the inner wall of the sampler becomes equal to the ultimate bearing capacity of the soil below the sampler.) Give a parametric answer and do a few sample calculations to gauge the problem.

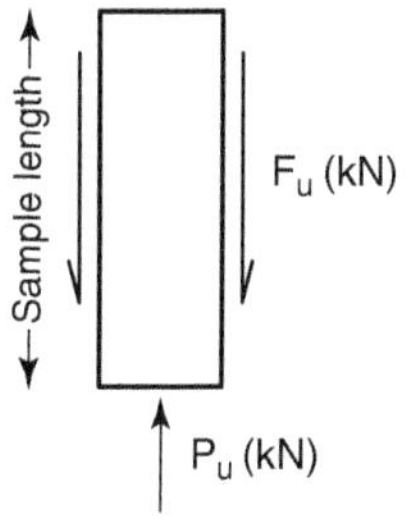

Figure 6.2s Free-body diagram of clay sample.

6.8 Describe the simple tests that would allow you to identify a soil in the field.
6.9 Explain the differences between drilling onshore and drilling offshore.
6.10 Explain the difference between the seismic reflection and the seismic refraction methods used for offshore investigations.
6.11 What is a piston sampler?

Problems and Solutions

Problem 6.1

A 70-story building has an imprint of 35 m by 25 m and will be supported on a mat foundation located at a depth of 10 m. How many borings would you propose and to what depth? Where would you place the borings on the building plan view?

Solution 6.1

A boring is required roughly every 250 m^2, so the minimum number of boreholes is

$$\frac{35 \times 25}{250} = 3.5$$

So, 4 or 5 borings are reasonable. The depth of the borings is usually one to two times the width of the foundation, with at least one boring extending to two times the width of the foundation below the foundation level. The depth of borings should be: $2B = 2 \times 25 = 50$ m.

For a rectangular mat, it is desirable to have a boring near each corner of the mat and one in the center. Therefore, a possible layout of the boring plan is shown in Figure 6.1s. Particular site specific soil conditions may affect this solution.

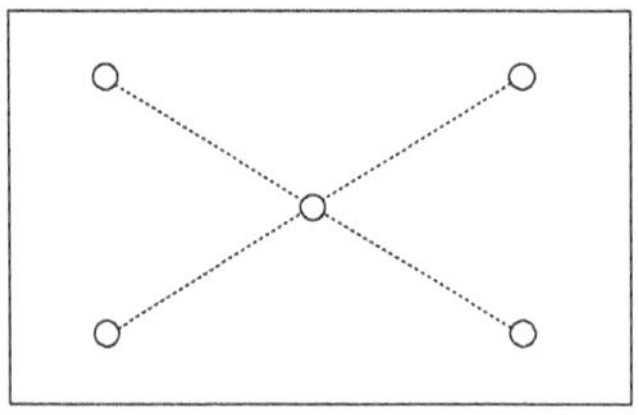

Figure 6.1s Borehole locations.

Problem 6.2

For problem 6.1, estimate the ratio between the volume of soil that is tested over the volume of soil involved in supporting the building. Comment on the result.

Solution 6.2

The depth of influence for the mat foundation can be considered to be $2B = 2 \times 25 = 50$ m. The volume of soil affected by the mat foundation of the building can be estimated as:

$$V_{Soil} = 35 \times 25 \times 50 = 43{,}750 \text{ m}^3$$

The volume of soil drilled (given a boring diameter of 100 mm) is:

$$V_{boreholes} = 5 \times \frac{\pi \times 0.1^2}{4} \times 50 = 1.96 \text{ m}^3$$

If it is assumed that the volume of soil tested is one-third of the volume of soil drilled, then:

$$V_{tested} = 0.33 \times 1.96 = 0.65 \text{ m}^3$$

The ratio of the volume of soil that is tested over the volume of soil involved in supporting the building is:

$$\frac{V_{tested}}{V_{soil}} = 1.48 \times 10^{-5} = 0.00148\,\%$$

This shows that the volume of soil tested in a typical soil investigation is extremely small. Add to this the fact that soils are known to be heterogeneous, and it is obvious that one must accept a significant degree of imprecision in geotechnical prediction.

Problem 6.3

What drill bit would you use for drilling in clay and which one would you use for drilling in gravel? Explain your choice.

Solution 6.3

A drag bit or finger bit is recommended for drilling in clay because it carves the soil with finger-like protrusions. This reduces the disturbance of the clay. Roller bits are used for drilling in gravel because it is made of three rollers that roll against the soil and erode it or push it aside. Finger bits are not used for drilling in gravel because the larger particles could get stuck between the fingers of the drag bit, damage it, and create excessive disturbance.

Problem 6.4

Discuss and compare the wet rotary method and the hollow stem auger method. Make recommendations as to when to use one and when to use the other.

Solution 6.4

The wet rotary method consists of drilling a borehole with a drill bit while circulating drilling mud through the center of the rods. The drill bit is typically 75 to 150 mm in diameter and the rods 40 to 70 mm in diameter. The drilling mud flows down the center of the rods while they rotate and back to the surface on the outside of the rods between the wall of the borehole and the exterior wall of the rods. This return flow carries the soil cuttings back to the surface by entrainment. The drilling mud arrives in the mud pit where it is sucked back up to the top of the drilling rods by a pump. The water swivel, which connects the hose carrying the drilling mud back to the top of the rods and the rods themselves, allows the hose to remain stationary while the rods keep rotating. The drill bit at the bottom end of the drill rods is typically either a drag bit or a roller bit.

The hollow stem auger method consists of rotating hollow stem augers into the soil; these augers are 150 to 300 mm in diameter. The hollow center part of the augers gives access for sampling or any other testing device that is to be lowered to the bottom of the hole. The hollow stem auger has the advantage of providing a casing against collapse of the side walls of the borehole, but is limited in penetration depth because it requires a significant torque to advance the augers. The wet rotary method has the advantage of being much less limited by depth, but sometimes faces problems of borehole instability.

Problem 6.5

(a) Give three sources of sample disturbance and (b) calculate the area ratio for the Shelby tube sampler and the split spoon sampler.

Solution 6.5

a. Three sources of sample disturbance:
   - Change in stress condition
   - Mechanical disturbance of the soil structure
   - Changes in water content and porosity
b. The equation to calculate the area ratio is:

$$AR = (\pi(D_o^2 - D_i^2)/4)/(\pi D_i^2/4)$$

   where D_o is the outside diameter of the sampling tube, and D_i is the inside diameter of the sampling tube.
   - For the Shelby tube sampler, $D_o = 76.2$ mm and $D_i = 72.9$ mm; therefore:

$$AR = (\pi(D_o^2 - D_i^2)/4)/(\pi D_i^2/4) = (D_o/D_i)^2 - 1 = (76.2/72.9)^2 - 1 = 0.092$$

     The area ratio for the Shelby tube sampler is 9.2%.
   - For the split spoon sampler, $D_o = 50.8$ mm and $D_i = 34.9$ mm; therefore:

$$AR = (\pi(D_o^2 - D_i^2)/4)/(\pi D_i^2/4) = (D_o/D_i)^2 - 1 = (50.8/34.9)^2 - 1 = 1.13$$

     The area ratio for the split spoon sampler is 113%.

Problem 6.6

Discuss when a sampler should be pushed and when it should be driven.

Solution 6.6

Samplers should usually be pushed in clays or silts to minimize soil disturbance and yield samples well suited for quality laboratory tests. Samplers are usually driven in sands or gravels because it is very difficult to push them into these soils without damaging the tube and therefore the sample. Driven samples are disturbed and only well suited for soil identification and classification purposes. For organic fibrous soils such as peats, it is best to drive the sampler, because the driving action has a better chance to cut the fibers rather than pushing them and simply compressing and disturbing the peat excessively.

Problem 6.7

Calculate the length of clay sample necessary to plug a Shelby tube. (*Plugging* means that the friction between the sample and the inner wall of the sampler becomes equal to the ultimate bearing capacity of the soil below the sampler.) Give a parametric answer and do a few sample calculations to gauge the problem.

Solution 6.7

The free-body diagram of the clay sample is shown in Figure 6.2s.

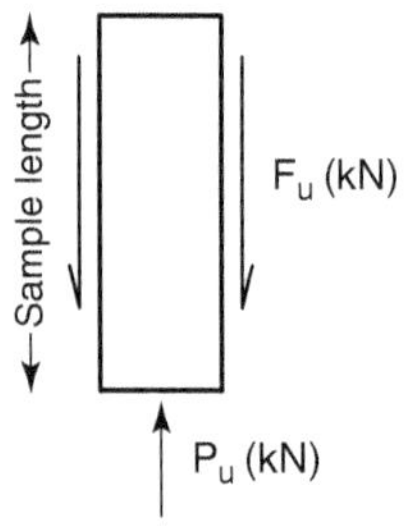

Figure 6.2s Free-body diagram of clay sample.

f_u: Unit side friction between Shelby tube and the soil
D: Diameter of the sample
L: Length of clay sample
S_u: Undrained shear strength
α: Friction coefficient
γ: Unit weight of the soil
Z: Depth of clay sample
N_c: Bearing capacity factor

$$F_u = f_u \pi DL = \alpha S_u \pi DL$$

$$P_u = (N_c S_u + \gamma z)\pi \frac{D^2}{4}$$

$$\text{Plugging occurs when: } \frac{P_u}{F_u} < 1 \rightarrow \frac{(N_c S_u + \gamma z)\pi \frac{D^2}{4}}{\alpha S_u \pi DL} < 1$$

$$\rightarrow \frac{(N_c S_u + \gamma z)D}{4\alpha S_u L} < 1$$

Sample calculations:

if $z = 0, \alpha = 1, N_c = 9$

$$\frac{9S_u D}{4S_u L} < 1 \rightarrow \frac{9D}{4L} < 1$$

$L > 2.25D$

if $z = 0, \alpha = 0.5, N_c = 9$

$$\frac{9S_u D}{4 \times 0.5 S_u L} < 1$$

$L > 4.5D$

if $\gamma z = N_c S_u, \alpha = 0.5, N_c = 9$

$$\frac{2 \times 9S_u D}{4 \times 0.5 S_u L} < 1$$

$L > 9D$

Problem 6.8

Describe the simple tests that would allow you to identify a soil in the field.

Solution 6.8

On cuttings:

1. Visual inspection
2. Feel the graininess or smoothness of the soil
3. Wash hands test
4. Dilatancy test (hand shaking test)
5. Dry strength test
6. Thread rolling test (toughness test)

On samples:

1. On clays and silts, the thumb or nail test for undrained shear strength
2. On the ground surface of a sand or gravel deposit, the 12 mm diameter steel bar test for strength

ASTM D2488, "Standard Practice for Description and Identification of Soils (Visual-Manual Procedure)," describes some of these tests.

Problem 6.9

Explain the differences between drilling onshore and drilling offshore.

Solution 6.9

Offshore geotechnical investigations include drilling, sampling, and in situ testing, much like onshore investigations. The difference is a matter of scale, complexity, and cost. Compared to onshore drilling, offshore drilling has at least two complexities: larger depths, which can reach thousands of meters of combined water depth and penetration depth; and vertical motion of the drill rig due to waves. Another difference is the increased use of geophysical investigations for offshore work.

Problem 6.10

Explain the difference between the seismic reflection and the seismic refraction methods used for offshore investigations.

Solution 6.10

Seismic reflection systems use sound propagation energy generated by a device towed behind a ship. The device measures the travel time required for the acoustic energy or wave to travel to the seabed or an interface between two distinct soil layers below the sea bed and be reflected to the same device or to a receiving array. Seismic refraction systems use sound propagation energy generated by a device including an acoustic pulse generator and a line of hydrophones dragged on the sea bottom by a ship. The device measures the travel time required for the acoustic energy or wave to travel to an interface between two distinct soil layers below the seabed, refract critically along that interface, and send the wave back to the line of receivers or hydrophones. Seismic refraction is more often used for shallow penetration below the seafloor (pipelines, cables) and gives the thickness and the shear wave velocity of the material.

Precision of the measurement and penetration into the soil depend on the frequency and amplitude of the acoustic wave. A wave with a high frequency and low amplitude will give high resolution (good precision on the distance measured) but low penetration into the soil while a wave with a low frequency and high amplitude will give deeper penetration but lower resolution.

Problem 6.11

What is a piston sampler?

Solution 6.11

A piston sampler has a piston blocking the opening at the bottom of the sampler (see Figure 6.10). This piston is locked in place as the sampler is lowered to the desired depth. The piston is then held at that depth while the sampling tube is pushed past the piston into the soil. The vacuum that can develop at the top of the sample helps the soil enter the sampler and minimizes the plugging effect.

CHAPTER 7

In Situ Tests

This chapter is devoted to the description of in situ tests and the test data that they generate. This chapter does not describe which soil properties can be inferred by correlation or other means from the test results; those correlations are discussed in Chapters 13, 14, and 15, dedicated to these soil properties. This chapter also does not describe the design methods that make use of in situ test results; this is covered in Chapters 17, 18, and 22, dedicated to design methods.

In situ tests are tests conducted on or in the soil at the site. They have been developed over the years as a complement to laboratory testing. Indeed, the drawbacks of laboratory tests are typically balanced by the advantages of in situ tests and vice versa (Table 6.1). Therefore, the best site investigation program uses a combination of in situ tests and laboratory tests. The most commonly used in situ tests are the standard penetration test, the field vane tests, the cone penetration test, the pressuremeter test, and the dilatometer test. Many other tests also exist, as shown in Figure 7.1 (Mayne et al. 2009).

7.1 STANDARD PENETRATION TEST

The standard penetration test (SPT) is the oldest of the in situ tests, and can be credited to Charles Gow in the United States who started developing it in 1902. After several decades of use, it was standardized in the mid-1930s (ASTM D1586). Today, the SPT (Figure 7.2) consists of driving a split spoon sampler into the soil using a standard 623N hammer falling from a height of 0.76 m onto an anvil at the top of the rods. The oldest hammer was the donut hammer, followed by the safety hammer and more recently the automatic hammer (Figure 7.3). For the donut hammer and the safety hammer, a person would raise the hammer with a rope. The rope would be wrapped around a cathead system (rotating drum) and the person would pull and release the rope to raise and drop the hammer in rhythm at about one blow per second. In the case of the automatic hammer, the hammer is raised automatically by a hydraulic jack. The rated energy of each blow is 623N × 0.76 m or 473 joules.

The rope-and-cathead system for the donut hammer and the safety hammer generate friction and other energy losses that decrease the amount of energy delivered to the split spoon sampler. Measurements have indicated that the mean energy actually delivered by these systems is around 285 J, or 60% of the maximum energy (ASTM D1586). Thus, the blow count N is often referred to as N_{60}. Because so much experience has been accumulated with these older systems, most correlations refer to N_{60}. However, automatic hammers may have much lower losses, so one should be careful about using the blow count N obtained with an automatic hammer without paying attention to this difference. The impact of the hammer on the anvil creates a compression wave in the steel rods which propagates at some 21,000 km/h (this, by the way, approaches the speed of the space shuttle in free space). The number of blows N_a necessary to drive the split spoon sampler 0.15 m into the soil is recorded. The SPT test continues and the number of blows N_b necessary to drive the sampler another 0.15 m is recorded. The SPT test continues and the number of blows N_c needed to drive the sampler yet another 0.15 m is recorded. The SPT blow count N (blows/0.3 m) is the sum of $N_b + N_c$, as N_a is considered to be a set of seating blows. A typical profile of SPT results is shown in Figure 7.4.

For design purposes, the N value is often corrected to account for influencing factors such as the energy level, the stress level, and the presence of silt (Table 7.1). Additional correction factors take into account the length of the rods, the diameter of the borehole, and whether or not the sampler has a liner. As explained earlier, the maximum energy that can be delivered by an SPT hammer system is 473 J (623 N × 0.76 m). If N_{measured} is the field value (N_b plus N_c, as explained earlier), N_{measured} corresponds to the energy (E) measured in the field, E_{measured}. To obtain the N_{60} value corresponding to 60% of the maximum energy that can be delivered by the system (0.6 × 435 J = 285 J), a linear interpolation is done as follows:

$$N_{60} = N_{\text{measured}} \left(\frac{E_{\text{measured}}\ (\text{J})}{285\ \text{J}} \right) \tag{7.1}$$

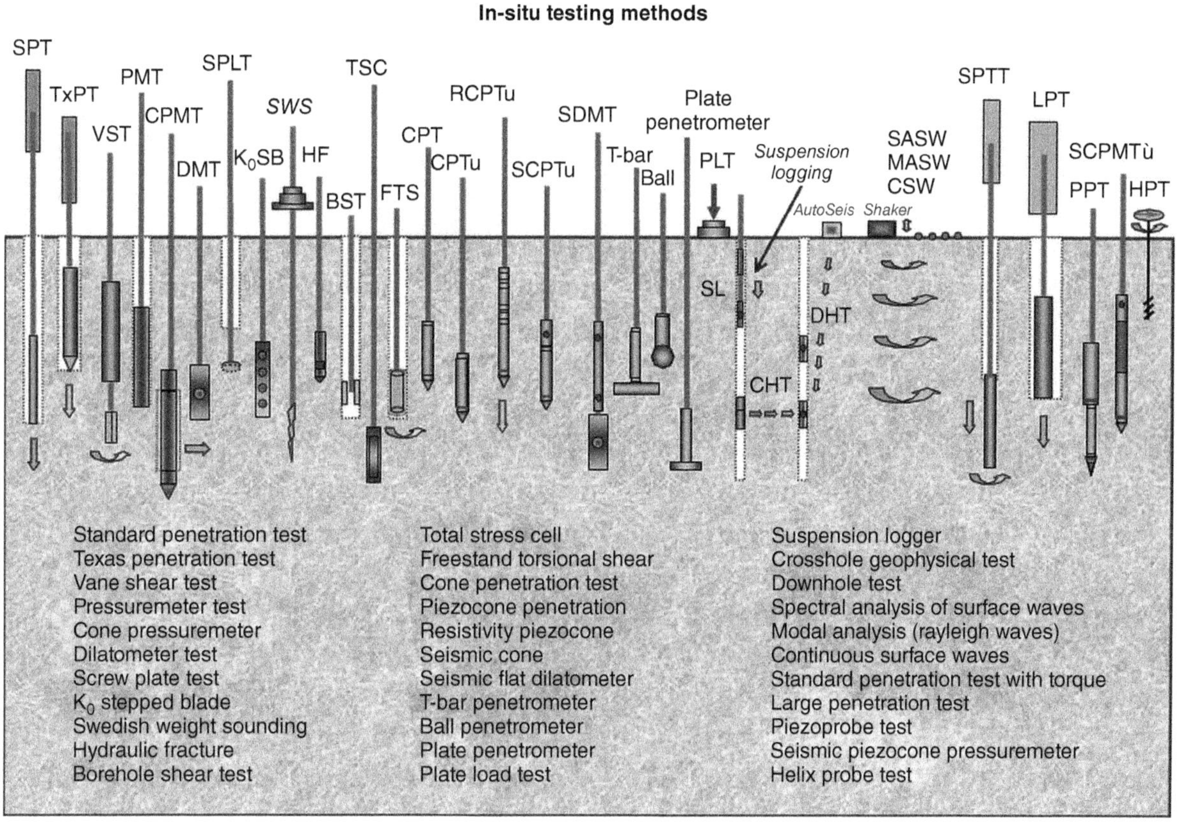

Figure 7.1 In situ tests. (Courtesy of Professor Paul Mayne, Georgia Institute of Technology, USA.)

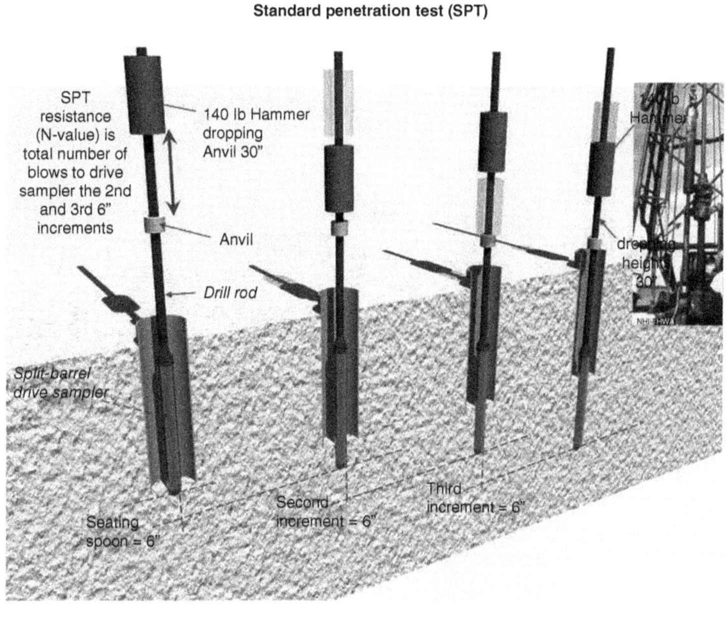

Figure 7.2 Standard penetration testing sequence. (Courtesy of Professor Kamal Tawfiq, Florida State University, USA.)

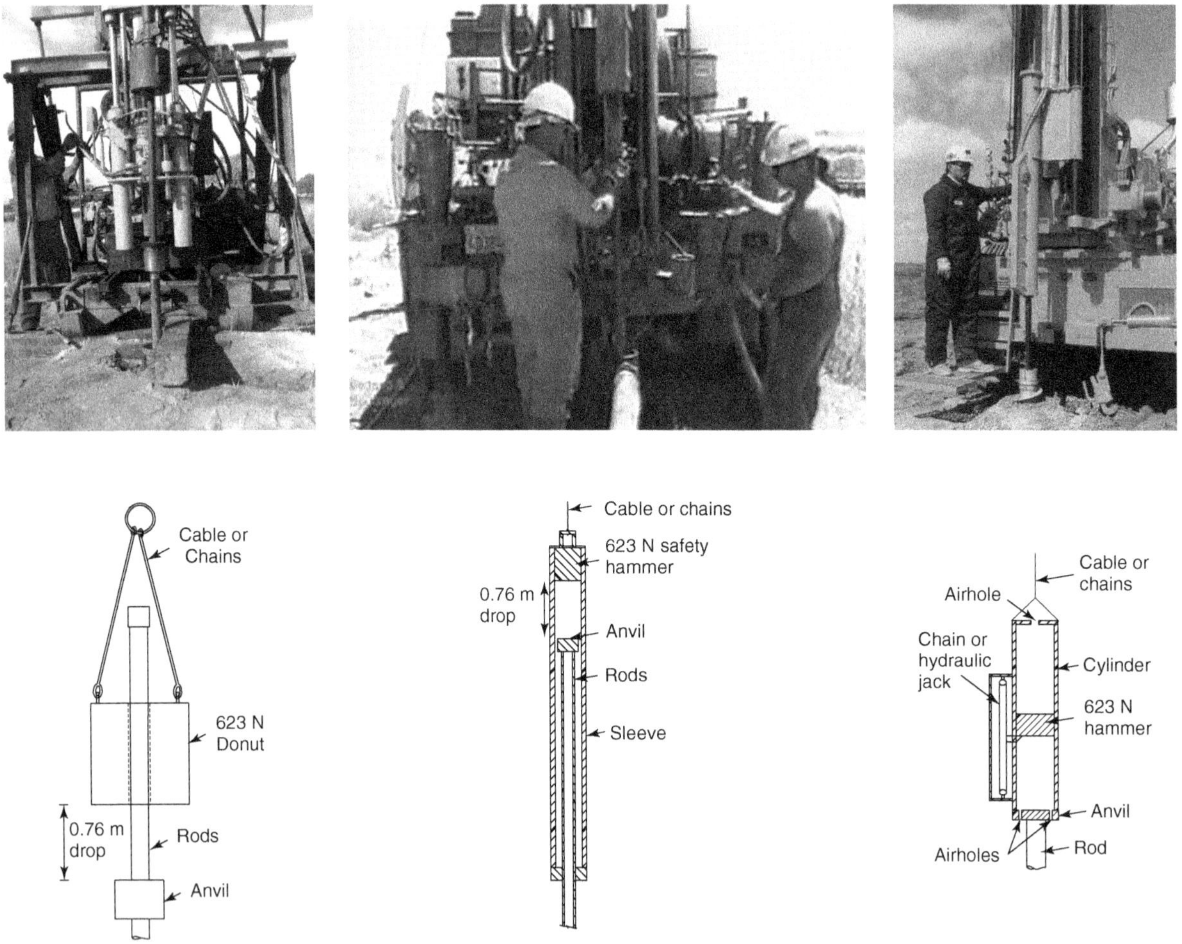

Figure 7.3 Standard penetration test hammers. (a: Courtesy of Fugro, b: Photo from Bray et al. 2001. Used by permission. c: Central Mine Equipment Co.)

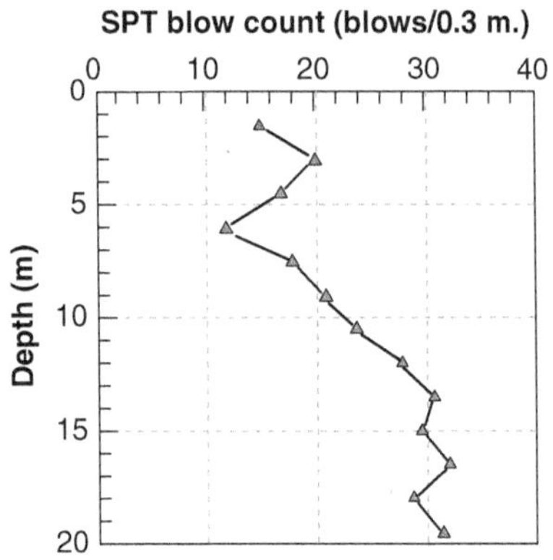

Figure 7.4 Example of SPT sounding result.

$N_{measured}$ also corresponds to the vertical effective stress at rest σ'_{vo} at the depth of the test. To obtain the N_1 value corresponding to a reference value of σ'_{vo} equal to 100 kPa, a power law interpolation is used:

$$N_1 = N_{measured}\left(\frac{100}{\sigma'_{vo}\,(\text{kPa})}\right)^{0.5} \tag{7.2}$$

Table 7.1 Correction of the SPT Blow Count Value N

Energy level	$N_{60} = N_{measured} \times \left(\frac{E_{measured}{}^{*}}{285\,\text{J}}\right)$
Stress level	$N_1 = N_{measured} \times \left(\frac{100\,\text{kPa}}{\sigma'_{vo}{}^{**}}\right)^{0.5}$ kPa
High silt content and effect of capillary	$N' = 15 + \left(\frac{N_{measured} - 15}{2}\right)$

$^{*}E_{measured}$ must be in joules
$^{**}\sigma'_{vo}$ must be in kPa

$N_{measured}$ is sometimes corrected for silt content as follows:

$$N' = 15 + \left(\frac{N_{measured} - 15}{2}\right) \tag{7.3}$$

Note that the decision to correct or not correct the N value requires engineering judgment. In general, N_{60} should always be used as a standardizing method, but this requires to measure the actual energy which is rarely done. If one needs to evaluate the friction angle φ of the soil, then N_1 should be used because N includes the effect of stress level, while φ does not. However, if one uses N in a direct design such as an ultimate bearing capacity equation of the form $p_u = kN + \gamma D$, then N should not be corrected for stress level, as the stress level is part of the soil resistance in both the SPT and the foundation capacity. Liquefaction charts do include most of the correction factors for N.

In the United States, the number N is used extensively in the design of structures over sand and gravel, but it is not used with silts and clays because it is felt that a better approach is possible, such as taking undisturbed samples. Some other countries, like Brazil, extend use of the SPT to silts and clays. Applications include settlement and ultimate bearing pressure for shallow and deep foundations, soil properties such as shear strength parameters and modulus values, and liquefaction potential. The advantages of the SPT include that it is a rugged test which can nearly always be performed and give results; that it is performed with the same drill rig used to collect samples; that it has been used for a long time and thus is well known and understood; and that it yields both an evaluation of strength and a sample for identification purposes at the same time. A primary drawback is that the amount of energy reaching the sample can vary quite a bit.

7.2 CONE PENETRATION TEST

The development of the cone penetration or penetrometer test (CPT) started in the early 1930s in the Netherlands, and can be credited to Pieter Barentsen, who performed the first CPT in 1932. At that time, a mechanical cone was used (Briaud and Miran 1992a; ASTM D3441), but in the mid-1950s electronic cones came into use (Mayne 2007a, b; ASTM D5778). Today, the CPT (Figure 7.5) consists of pushing a 35.6 mm diameter instrumented rod into the soil at 20 mm/s. A drill rig, or more commonly a truck, weighing as much as 200 kN provides the vertical reaction (Figure 7.6). At the bottom of the rods is the instrumented cone tip (Figure 7.7), which can be equipped with different sensors to make many measurements. The two primary measurements are the tip resistance q_c at the point of the cone and the sleeve friction f_s on a sleeve right behind the point. Note that the measured tip resistance i should be corrected for the influence of water pressure inside the cone to obtain the total cone tip resistance q_t (Mayne 2007a). Examples of continuous profiles obtained with the CPT are shown in Figure 7.8. Other possible measurements include water pressure measurements, shear wave velocity, electrical resistivity, inclination, sound level, lateral stress, camera, radio isotope, and temperature. The CPT can also be equipped with a soil and water sampler. The most common location for water pressure measurement is right behind the cone point (Figure 7.7); the measurement is made through a saturated porous element behind which a transducer senses the compression in the water as the cone advances. The shear wave velocity is typically measured between the surface and a geophone located in the rods and sensing the arrival of a shear wave generated at the ground surface (Figure 7.9). The electrical resistivity is measured between two electrodes mounted on the rods and separated by a nonconducting material to force the electrical current to go through the soil instead of through the rods.

The cone penetrometer point resistance q_t is influenced by the stress level surrounding the point where the test is performed and by the properties of the soil in the vicinity of that location. To use a cone parameter that is dependent only on an intrinsic soil property, it is desirable to

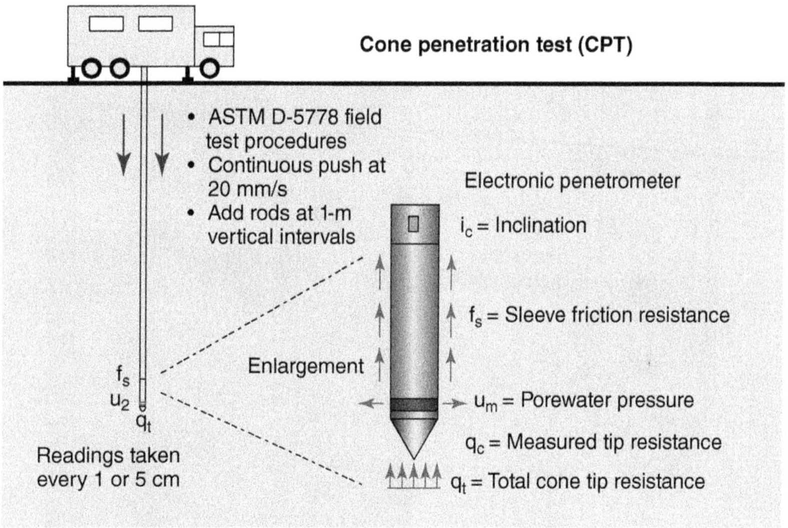

Figure 7.5 Cone penetrometer test. (From Mayne 2007a. Courtesy of Professor Paul Mayne, Georgia Institute of Technology)

Figure 7.6 Cone penetrometer truck. (From Mayne 2007a. Courtesy of Professor Paul Mayne, Georgia Institute of Technology.)

correct the q_t value for the stress level, as was done for the SPT (equation 7.2). The following corrections may be recommended.

For sands:

$$q_{t1} = \left(\frac{q_t}{\sigma_a}\right)\left(\frac{\sigma_a}{\sigma'_{vo}}\right)^{0.5} = \frac{q_t}{(\sigma'_{vo}\sigma_a)^{0.5}} \quad (7.4)$$

For clays:

$$q_{t1} = \frac{q_t - \sigma_{vo}}{\sigma'_{vo}} \quad (7.5)$$

where q_{t1} is the dimensionless corrected normalized CPT point resistance, q_t the total CPT point resistance, σ'_{vo} and σ_{vo} the vertical effective stress and vertical total stress at the depth of the cone respectively, and σ_a the atmospheric pressure used to nondimensionalize equation 7.4. The reason for not using Eq. 7.4 for clays is that the undrained shear strength of clays has been shown to be linearly proportional to the vertical effective stress. Alternatively, a progressive transition between the two soil types can be used through equation 7.6, which also includes a fine content influence factor K_c useful in liquefaction studies.

$$q_{t1} = \left(\frac{q_t}{\sigma_a}\right)\left(\frac{\sigma_a}{\sigma'_{vo}}\right)^n K_c \quad (7.6)$$

where n is 0.5 for sand, 0.7 for silty sand, 0.8 for silt, and 1 for clay, and K_c is a fine content factor gradually varying from 1 to 1.5 for clean sands, 1.5 to 3.5 for silty sands, and 3.5 to 6 for silts.

The most useful application of the CPT is stratigraphy, because the CPT penetration resistance profile gives the engineer a continuous display of the strength of the deposit. Note that the scale of the cone influences the thickness of the layer that can be detected, as well as the strength of that layer. If a layer is smaller than about 10 times the diameter of the cone, the tip resistance will not reach the value that would be obtained if the layer were infinitely deep. Associated with stratigraphy is the ability to classify the soil on the basis of the friction ratio, that is, the ratio of the sleeve friction over the

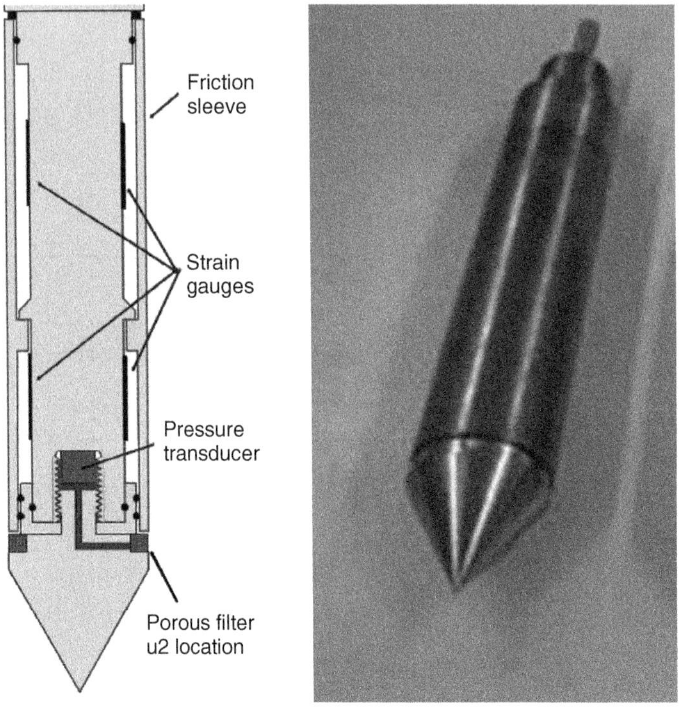

Figure 7.7 Cone penetrometers. (From Mayne 2007a. Courtesy of Professor Paul Mayne, Georgia Institute of Technology.)

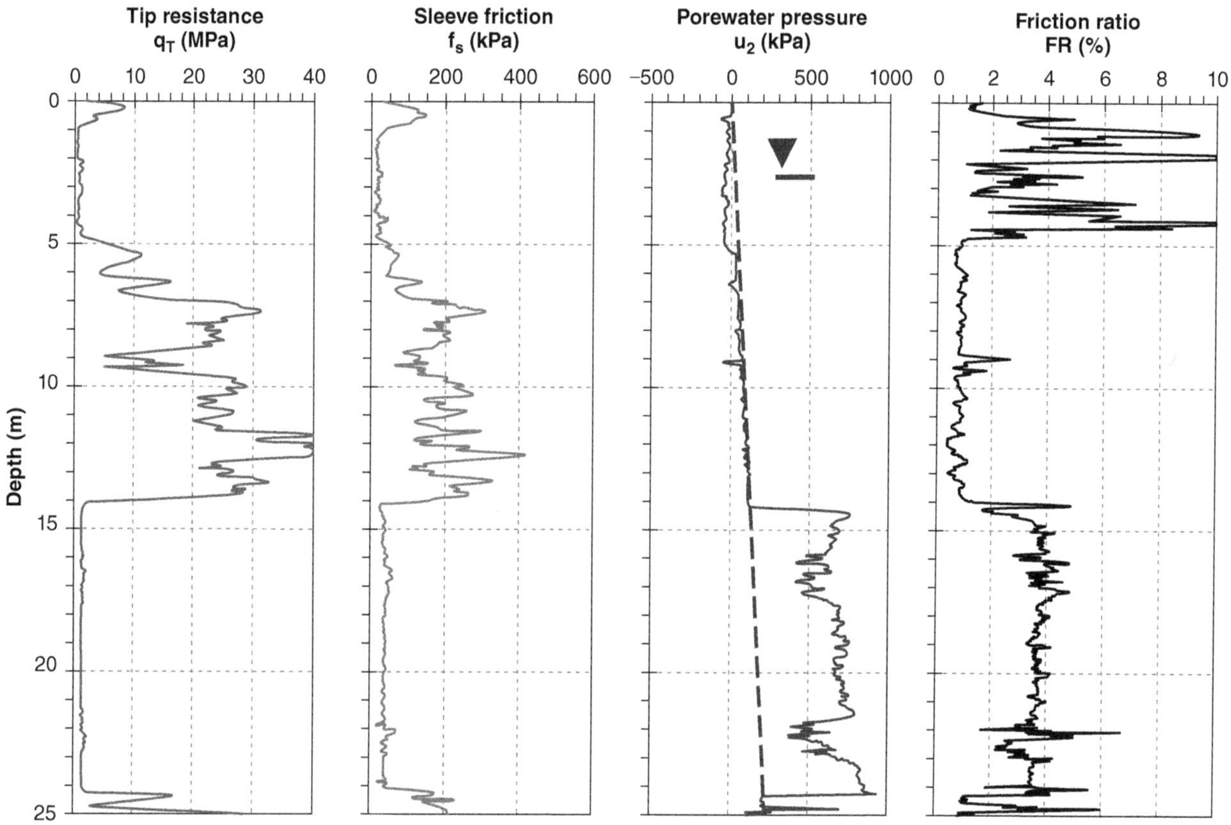

Figure 7.8 Examples of CPT profiles. (From Mayne 2007a. Courtesy of Professor Paul Mayne, Georgia Institute of Technology.)

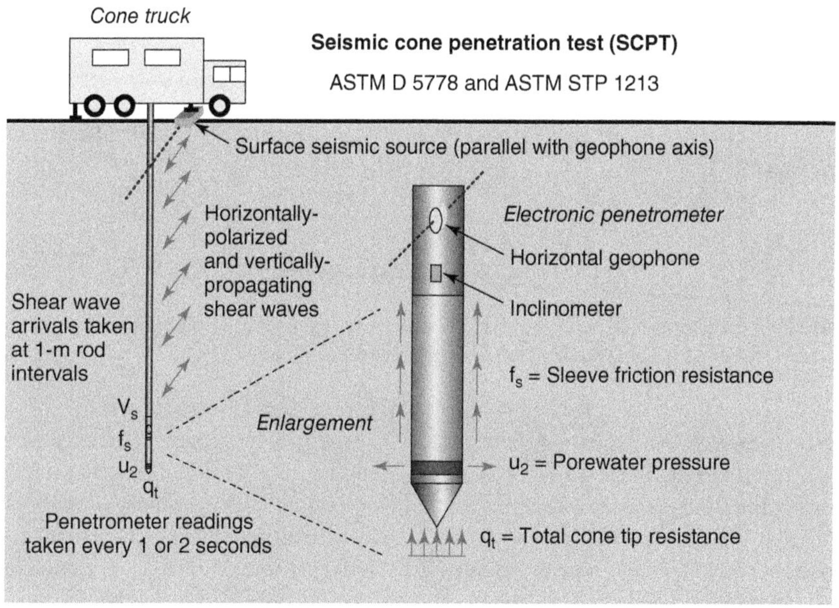

Figure 7.9 Seismic cone penetrometer test. (From Mayne 2007a. Courtesy of Professor Paul Mayne, Georgia Institute of Technology.)

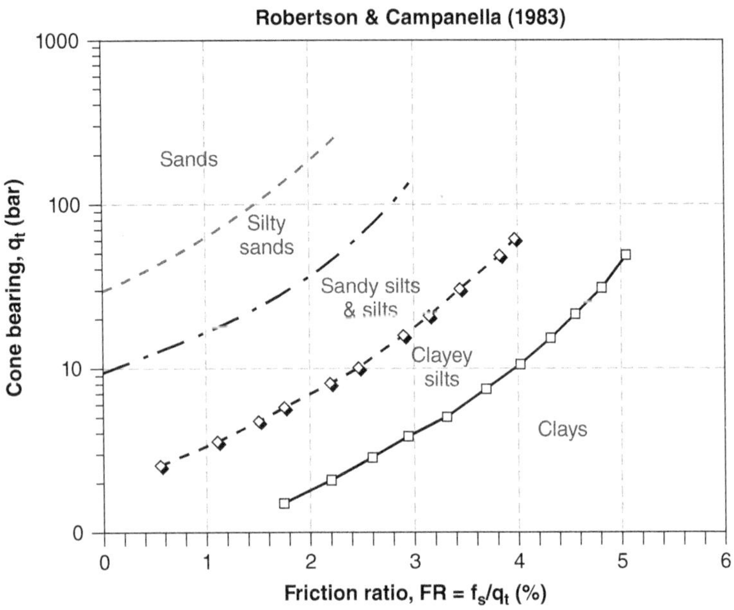

Figure 7.10 Robertson & Campanella (1983) soil classification using CPT results. (From Mayne 2007a. Courtesy of Professor Paul Mayne, Georgia Institute of Technology, USA.)

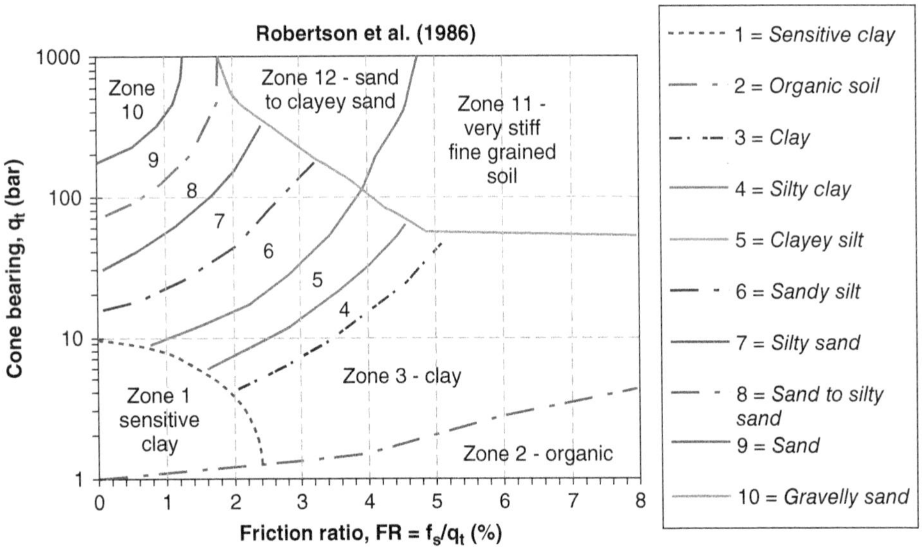

Figure 7.11 Robertson et al. (1986) soil classification using CPT results. (From Mayne 2007a. Courtesy of Professor Paul Mayne, Georgia Institute of Technology, USA.)

tip resistance (FR $= f_s/q_t$). Several classification schemes have been proposed; Figure 7.10 and 7.11 show two of them. The reason why it is possible to estimate the soil classification from the friction ratio is that the sleeve friction value does not change significantly between a sand and a clay, whereas the tip resistance changes dramatically. Maximum values of sleeve friction might be about 200 kPa for both sand and clay, whereas the maximum tip resistance may be 2000 kPa in a hard clay and 20,000 kPa in a dense sand. The friction ratio would be 10% for the clay and 1% for the sand.

The CPT parameters are used extensively in geotechnical engineering worldwide. Applications include obtaining soil properties such as shear strength parameters and modulus values, ultimate bearing pressure and settlement of shallow and deep foundations, and liquefaction potential. The advantages of the CPT include that it gives a rapid and continuous profile of soil strength; that it is much less operator dependent than other in situ tests; that it is relatively economical; that it does not create cuttings; and that it has a wide range of applications. For example, it is one of the best ways to obtain ultimate vertical pile capacity. One drawback is that the penetration depth is limited in stronger soils.

7.3 PRESSUREMETER TEST

There are three types of pressuremeters: the preboring pressuremeter, the self-boring pressuremeter, and the push-in or cone pressuremeter. In the preboring *pressuremeter test* (PMT), a borehole is drilled first, the drilling tool is removed, and the PMT probe is inserted in the open hole. For the self-boring PMT, the probe is equipped with its own drilling equipment and bores itself into the soil to avoid decompression of the soil due to preboring. For the push-in PMT, the probe is pushed into the soil and full displacement takes place during the insertion, as in the cone penetration test. This section discusses the preboring PMT, which is the most common of the three.

The pressuremeter test was developed in France in the late 1950s, and can be credited to Louis Menard, who conceived it as part of his university graduation project in 1957. The PMT (Figure 7.12; Briaud 1992; ASTM D4719) consists of boring a hole of a given diameter (e.g., 75 mm) down to the selected testing depth, withdrawing the drilling tool, lowering a cylindrical probe to the testing depth, and inflating the cylinder while recording the pressure necessary to do so and the corresponding increase in radius. The test result (Figure 7.13) is an in situ stress-strain curve that gives a number of useful soil parameters: the modulus E_0, called the first load modulus; the pressure p_{oh}, found at the beginning of the curve where the horizontal soil pressure is being reestablished; a yield pressure p_y; and a soil strength called the limit pressure p_L. Often an unload-reload loop is performed near p_y and a reload modulus E_r is determined. Typical profiles resulting from a PMT program are shown in Figure 7.14.

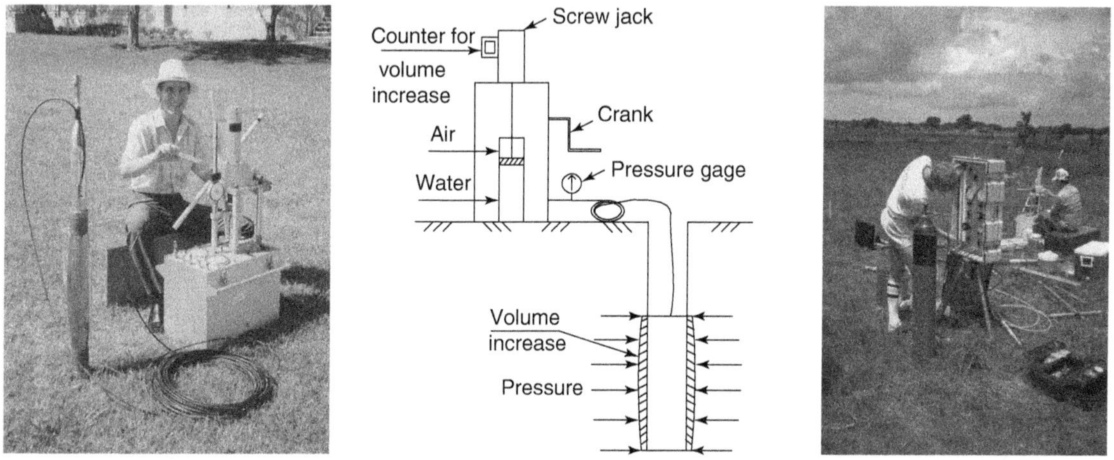

Figure 7.12 TEXAM and Menard pressuremeters.

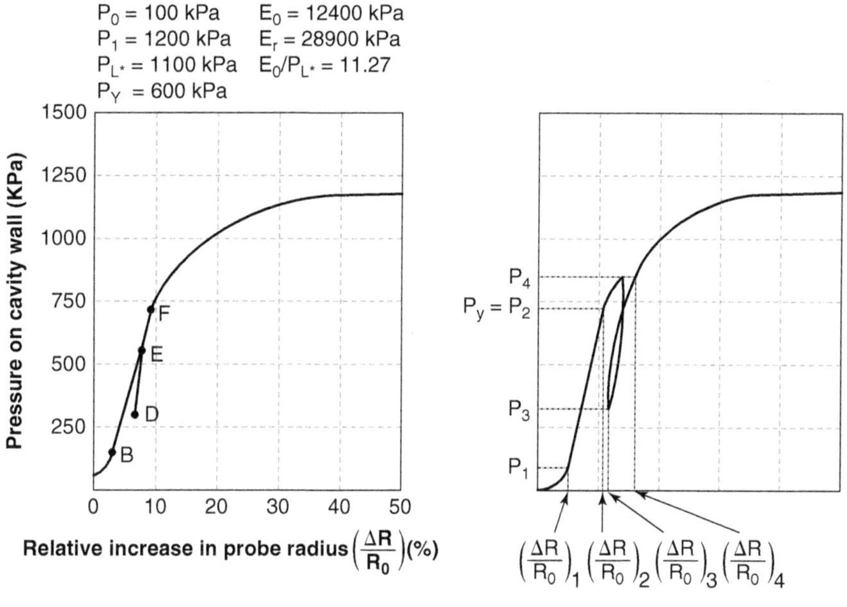

Figure 7.13 Pressuremeter test result.

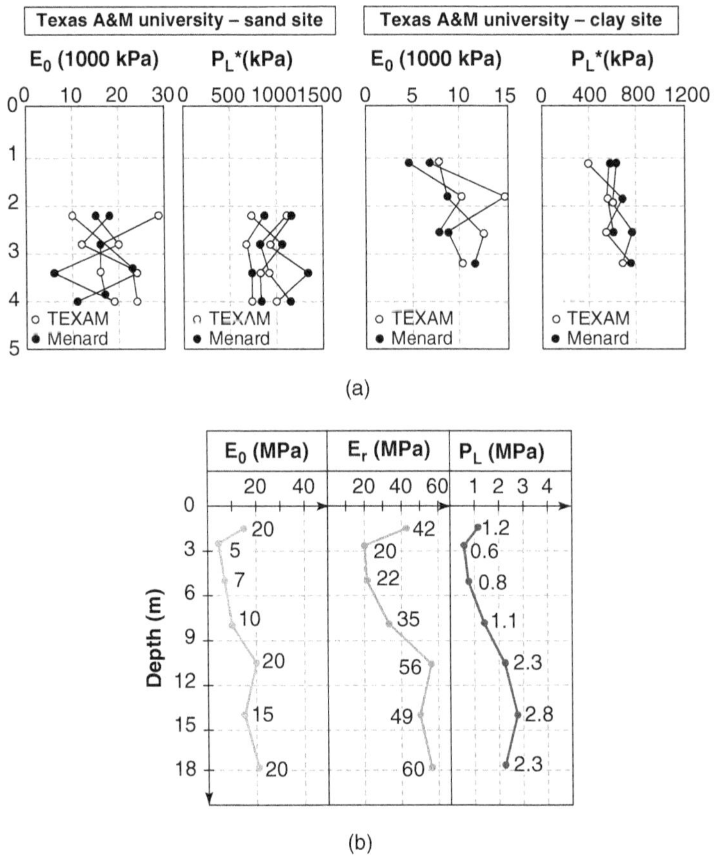

Figure 7.14 Typical PMT profile.

The most important part of a PMT is the preparation of the borehole in which to place the PMT probe. The disturbance of the walls of the borehole should be kept to a minimum and the diameter of the borehole should be only slightly larger than the PMT probe. If D_1, D_2, and D_3 refer to the diameter of the drilling tool, of the deflated probe, and of the borehole before inflation of the probe, respectively, then the following is recommended:

$$D_2 < D_1 < 1.03D_2 \tag{7.7}$$

$$1.03D_2 < D_3 < 1.20D_2 \tag{7.8}$$

The most commonly recommended method for preparing the borehole is the wet rotary method. In this case the rotation of the drill bit should be slow (about 60 rpm) and the circulation of the drilling mud should also be slow. The borehole should be advanced only as deep as necessary to perform one pressuremeter test at a time. The bottom of the borehole should be at least 1 m deeper than the PMT location, to allow any cuttings not transported up to the surface to settle at the bottom of the hole. The borehole should be prepared in one downward passage of the bit, followed by immediate retrieval of the bit; no multiple passages should be allowed, as they lead to an enlarged borehole. The borehole should be drilled to perform one PMT at a time. Other methods can also be used, as shown in Table 7.2.

The probe is calibrated to determine the amount of pressure p_c required to inflate the probe in the air. It is also calibrated to determine the amount of volume v_c necessary to inflate the probe in a tight-fitting thick steel tube. In the field and once the probe is in the borehole, the PMT is run in increments of either pressure or volume. Increases in volume have the advantage that they do not require a guess at the limit pressure. The test lasts about 10 minutes. Data reduction consists of converting the raw data into the actual pressure exerted against the wall of the borehole and the actual relative increase in borehole radius (Briaud 1992).

The modulus E_o and E_r are obtained from the portion of the curve between B and C, and D and E on Figure 7.13, respectively, by using linear elasticity. The equations to obtain E_o and E_r are:

$$E_0 = (1+\upsilon)(p_2 - p_1)\frac{\left(1+\left(\frac{\Delta R}{R_0}\right)_2\right)^2 + \left(1+\left(\frac{\Delta R}{R_0}\right)_1\right)^2}{\left(1+\left(\frac{\Delta R}{R_0}\right)_2\right)^2 - \left(1+\left(\frac{\Delta R}{R_0}\right)_1\right)^2} \tag{7.9}$$

$$E_r = (1+\upsilon)(p_4 - p_3)\frac{\left(1+\left(\frac{\Delta R}{R_0}\right)_4\right)^2 + \left(1+\left(\frac{\Delta R}{R_0}\right)_3\right)^2}{\left(1+\left(\frac{\Delta R}{R_0}\right)_4\right)^2 - \left(1+\left(\frac{\Delta R}{R_0}\right)_3\right)^2} \tag{7.10}$$

Table 7.2 Guidelines for PMT Borehole Preparation

Soil	Type	Rotary Drilling with Bottom Discharge of Prepared Mud	Pushed Thin Wall Sampler	Pilot Hole Drilling and Subsequent Sampler Pushing	Pilot Hole Drilling and Simul-taneous Shaving	Continuous Flight Auger	Hand Auger in the Dry	Hand Auger with Bottom Discharge of Prepared Mud	Driven or Vibro-Driven Sampler	Core Barrel Drilling	Rotary Percussion	Driven or Vibro-Driven or Pushed Slotted Tube
Clayey soils	Soft	2^B	2^B	2	2	NR	NR	1	NR	NR	NR	NR
	Firm to stiff	1^B	1	2	2	1^B	1	1	NR	NR	NR	NR
	Stiff to hard	1	2	1	1	1^B	NA	NA	NA	1^B	2^B	NR
Silty soils	Above GWL^C	1^B	2^B	2	2^B	1	1	2	2	NR	NR	NR
	Under GWL^C	1^B	NR	NR	2^B	NR	NR	1	NR	NR	NR	NR
Sandy soils	Loose and above GWL^C	1^B	NR	NR	2	2	2	1	2	NA	NR	NR
	Loose and below GWL^C	1^B	NR	NR	2	NR	NR	1	NR	NA	NR	NR
	Medium to dense	1^B	NR	NR	2	1	1	1	2	NR	2^B	NR
Sandy gravels or Gravelly sands below GWL	Loose	2	NA	NA	NA	NA	NA	NA	NR	NA	2	2
	Dense	NR	NA	NA	NA	NR	NA	NA	NR	NA	2	1^D
Weathered rock		1	NA	2^B	NA	1	NA	NA	1	2	2	NR

1 is first choice, 2 is second choice, NR is not recommended, and NA is nonapplicable.
B: Method applicable only under certain conditions.
C: GWL is groundwater level.
D: Pilot hole drilling required beforehand.
(After ASTM D4719.)

All parameter definitions are found in Figure 7.13. Note that the reload modulus E_r depends significantly on the amplitude of the unload-reload loop; therefore, unlike E_o, E_r is not unique. The parameter p_y is obtained by inspection as the point where the curve first departs from linearity. The limit pressure p_L is obtained by visual extrapolation of the data to a large value of $\Delta R/R_o$ equal to 0.40 or 40%. The pressure p_{oh} is found at the beginning of the curve at the point of maximum curvature during the reestablishment of the horizontal pressure that existed before placement of the PMT probe. The difference $p_L - p_{oh}$ is called the net limit pressure p_L*. Expected values of these PMT parameters are shown in Table 7.3; correlations to other soil properties are shown in Table 7.4 for sands and gravels and in Table 7.5 for silts and clays. The correlations in Tables 7.4 and 7.5 exhibit very large scatter and should be used for crude estimates only.

The applications of the PMT include the design of deep foundations under horizontal loads, the design of shallow foundations, the design of deep foundations under vertical loads, and the development of a modulus profile and the determination of other soil properties. The PMT is not very useful for slope stability and retaining structures. The advantages of the PMT are that it can be performed in most soils and rocks; that it stresses a larger soil mass than other tests; that it gives a complete stress-strain curve of the soil in situ, including cyclic loading and long-term loading; that it is relatively inexpensive; and that the quality of the test can be judged by the shape of the curve obtained. One drawback of the PMT is that the quality of the borehole influences the PMT parameters, in particular the first load modulus E_o.

7.4 DILATOMETER TEST

The dilatometer test (DMT) was developed in Italy in the mid-1970s and can be credited to Silvano Marchetti. The DMT (Marchetti 1975; Briaud and Miran 1992b; ASTM D6635) consists of pushing a flat blade located at the end of a series of rods (Figure 7.15) into a soil to a desired depth. The blade is 230 mm long, 95 mm wide, and 15 mm thick. Once the testing

Table 7.3 Expected Values of E_o and P_L in Soils

Clay					
Soil Strength	Soft	Medium	Stiff	Very Stiff	Hard
p^*_L(kPa)	0–200	200–400	400–800	800–1600	>1600
E_0(kPa)	0–2500	2500–5000	5000–12,000	12,000–25,000	>2500

Sand				
Soil Strength	Loose	Compact	Dense	Very Dense
p^*_L(kPa)	0–500	500–1500	1500–2500	>2500
E_0(kPa)	0–3500	3500–12,000	12,000–22,500	>22,500

Table 7.4 Correlations for Sand

Column A = number in table x row B

B A	E_0 (kPa)	E_R (kPa)	p^*_L (kPa)	q_c (kPa)	f_s (kPa)	N (bl/30 cm)
E_0(kPa)	1	0.125	8	1.15	57.5	383
E_R(kPa)	8	1	64	6.25	312.5	2174
p^*_L(kPa)	0.125	0.0156	1	0.11	5.5	47.9
q_c(kPa)	0.87	0.16	9	1	50	479
f_s(kPa)	0.0174	0.0032	0.182	0.02	1	9.58
N (bl/30 cm)	0.0026	0.00046	0.021	0.0021	0.104	1

Table 7.5 Correlations for Clay

Column A = number in table x row B

B A	E_0 (kPa)	E_R (kPa)	p^*_L (kPa)	q_c (kPa)	f_s (kPa)	s_u (kPa)	N (bl/30 cm)
E_0(kPa)	1	0.278	14	2.5	56	100	667
E_R(kPa)	3.6	1	50	13	260	300	2000
p^*_L(kPa)	0.071	0.02	1	0.2	4	7.5	50
q_c(kPa)	0.40	0.077	5	1	20	27	180
f_s(kPa)	0.079	0.0038	0.25	0.05	1	1.6	10.7
s_u(kPa)	0.010	0.0033	0.133	0.037	0.625	1	6.7
N (bl/30 cm)	0.0015	0.0005	0.02	0.0056	0.091	0.14	1

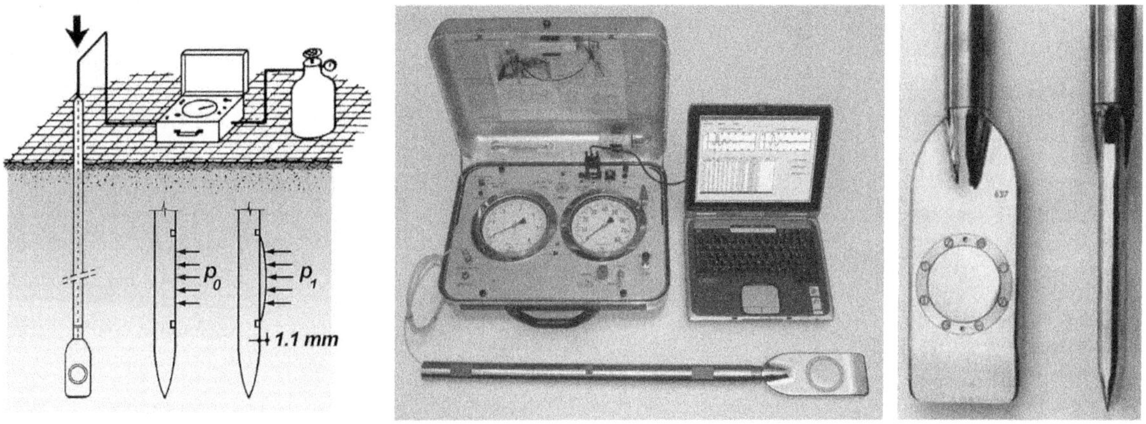

Figure 7.15 Dilatometer test and equipment. (Courtesy of Dr. Sylvano Marchetti, www.marchetti-dmt.it)

depth is reached, the operator uses gas pressure to expand horizontally into the soil a circular membrane located on one side of the blade. The membrane is 60 mm in diameter and expands 1.1 mm into the soil. Two pressures are recorded: p_o and p_1 : p_o is the pressure on the blade before expansion, and p_1 is the pressure required to produce the 1.1 mm expansion into the soil. A number of soil parameters are obtained from the DMT by using the formulas and correlations shown in Table 7.6.

The applications of the DMT include the design of foundations, the determination of soil properties, and soil classification (Figure 7.16). The advantages of the DMT include that it is fast, economical, easy to perform, and reproducible, giving a wealth of soil properties through correlations. A drawback is that it cannot be used in soils that are difficult to penetrate by pushing. Sample profiles are presented in Figure 7.17.

7.5 VANE SHEAR TEST

The vane shear test (VST) can be traced back to 1919 when it was first used in Sweden, but it is unclear if it can be credited to one person (Richards 1988). The VST (Figure 7.18) is used to determine the undrained shear strength of fine-grained soils (clays and silts). It can be performed either in the field with a field vane (ASTM D2573; Figure 7.19) or on the sample with a mini vane or a hand vane (ASTM D4648, Figure 7.20). The vane is made of two perpendicular blades, each having a 2-to-1 height-to-width ratio. The width of the field vanes varies from 38 to 92 mm; the larger vanes are used in softer soils. The width of the lab vanes varies from 10 to 20 mm. The VST consists of pushing a vane at the end of a rod into the soil until the desired depth is reached. Once the testing depth is reached, the vane is rotated at a slow rate (less than 1 degree per minute) while measuring the torque developed and the rotation angle (Figure 7.21). The peak value of the torque is recorded as T_{max}. Then the blade is rotated at least 10 times rapidly and a new maximum torque value, T_{res}, is measured.

The VST is used in saturated fine-grained soils to obtain the undrained shear strength s_u. The reason is that these soils have a low permeability and do not allow appreciable drainage during a test that typically lasts less than 10 minutes.

Table 7.6 Soil Parameters from Dilatometer Test

Symbol	Description	Basic Reduction Formulae	
p_0	Corrected first reading	$p_0 = 1.05\ (A - Zm + \Delta A) - 0.05\ (B - Zm - \Delta B)$	Zm = Gage reading when vented to atmosphere. However, if ΔA and ΔB are measured with the same gage used for current readings A & B, set Zm = 0 (Zm is compensated)
p_1	Corrected second reading	$p_1 = B - Zm - \Delta B$	
I_D	Material index	$I_D = (p_1 - p_0)/(p_0 - u_0)$	u_0 = pre-insertion pore pressure
K_D	Horizontal stress index	$K_D = (p_0 - u_0)/\sigma'_{V0}$	σ'_{V0} = pre-insertion overburden stress
E_D	Dilatometer modulus	$E_D = 34.7\ (p_1 - p_0)$	E_D is *not* a Young's modulus E. E_D should be used only *after* combining it with K_D (stress history). First obtain $M_{DMT} = R_M E_D$, then (e.g.) $E''0.8\ M_{DMT}$
K_0	Coefficient of Earth pressure in situ	$K_{0,DMT} = (K_D/1.5)^{0.47} - 0.6$	for $I_D < 1.2$
OCR	Overconsolidation ratio	$OCR_{DMT} = (0.5\ K_D)^{1.56}$	for $I_D < 1.2$
c_u	Undrained shear strength	$C_{u,DMT} = 0.22\,\sigma'_{V0}(0.5\ K_D)^{1.25}$	for $I_D < 1.2$
φ	Friction angle	$\varphi_{safe,DMT} = 28 + 14.6\ \log\ Kd - 2.1\ \log^2 Kd$	for $I_D > 1.8$
c_h	Coefficient of consolidation	$C_{h,DMTA} \approx 7cm^2/T_{flex}$	T_{flex} from A−log t DMTA−decay curve
k_h	Coefficient of permeability	$k_h = C_h \gamma_w / M_h (M_h \approx K_0 M_{DMT})$	
γ	Unit weight and description	(see chart)	
M	Vertical drained constrained modulus	$M_{DMT} = R_M E_D$ If $(I_D \leq 0.6)$ $R_M = 0.14 + 2.36\ \log\ Kd$ If $(I_D \geq 3)$ $R_M = 0.5 + 2\ \log\ Kd$ If $(0.6 < I_D < 3)$ $R_M = R_{M,0} + (2.5 - R_{M,0})\ \log\ Kd$ where $R_{M,0} = 0.14 + 0.15(I_D - 0.6)$ If Kd > 10 $R_{M}, = 0.32 + 2.18\ \log\ Kd$ If $R_M < 0.85$, set $R_M = 0.85$	
U_0	Equilibrium pore pressure	$U_0 = p_2 \approx C–Zm + \Delta A$	In freely draining soils

(Courtesy of Dr. Sylvano Marchetti, www.marchetti-dmt.it)

Therefore, in these saturated fine-grained soils, it is reasonable to assume that the shearing process is undrained and that the undrained shear strength s_u is the parameter being measured. For a rectangular vane, the following equation gives s_u from T_{max}:

$$T_{max} = \pi s_u D^2 \left(\frac{H}{2} + \frac{D}{6}\right) \tag{7.11}$$

where D is the diameter of the vane and H is the height of the vane. Proof of this equation is shown in the solution to problem 7.4. The residual undrained shear strength s_{ur} is obtained from the same formula using T_{res}:

$$T_{res} = \pi s_{ur} D^2 \left(\frac{H}{2} + \frac{D}{6}\right) \tag{7.12}$$

The VST can be used in coarse-grained soils, but no useful result can be obtained. These soils drain fast enough that one would not be measuring the undrained shear strength, but instead the drained or partially drained shear strength.

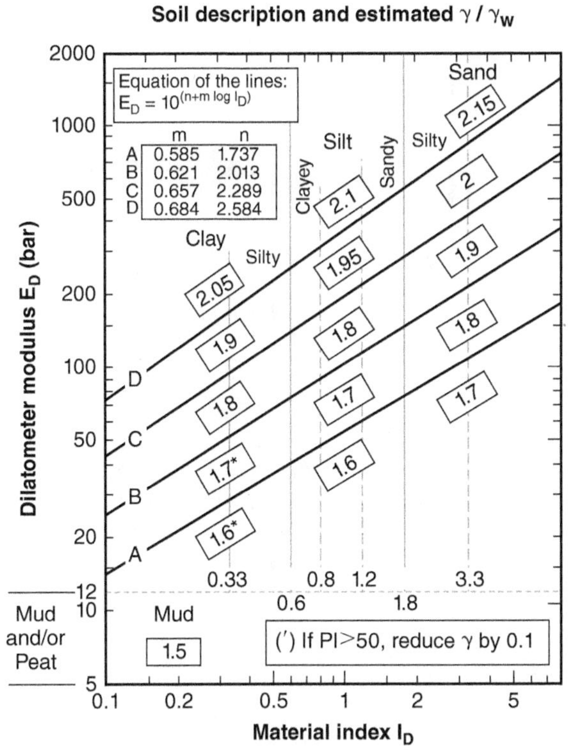

Figure 7.16 Soil classification using the DMT. (Courtesy of Dr. Sylvano Marchetti, www.marchetti-dmt.it)

Back-calculating the shear strength parameters from this test would require knowledge of the normal effective stress on the plane of failure in addition to T_{max}. This is not measured during the VST. The advantages of the VST include that it is fast, simple, economical, and useful for obtaining the undrained shear strength of fine-grained soils. A drawback is that it is limited to fine-grained soils where other methods are commonly used to obtain s_u. One exception is in offshore applications, where obtaining samples is very expensive and sample decompression can alter the true undrained strength of the soil in situ; in this case the VST is extremely useful.

7.6 BOREHOLE SHEAR TEST

The borehole shear test (BST) was developed in the USA in the 1960s and is credited to Richard Handy (Handy 1975, 1986). The BST (Figures 7.22 and 7.23) consists of drilling a borehole, removing the drilling tool, and inserting the borehole shear tester down to the testing depth. The device is made of two diametrically opposed grooved plates, which, once at the testing depth, are pushed horizontally against the wall of the borehole under a chosen total stress σ_h. After a proper time for dissipation of the pore pressures generated by the application of σ_h, the device is pulled upward to shear the soil along the side of the borehole. The force applied is measured as a function of time as it increases, and the peak force generated divided by the plates area gives the shear strength of the soil τ_f. If the shearing part of the test is performed slowly enough to ensure that no excess pore pressures arise, and if the soil has no effective stress cohesion intercept ($c = 0$), the ratio τ_f/σ_h is equal to $\tan\varphi'$ and φ' can be measured with the BST. If the shearing part of the test is performed slowly enough to ensure that no excess pore pressures arise, and if the soil has an effective stress cohesion intercept ($c' > 0$), a stage test can be performed where a second test at a higher value of σ_h follows the first one. The two tests give enough information to back-calculate c' and φ' for the soil (Figure 7.24). If, however, the test is performed rapidly, and does not allow any drainage to take place in the soil, an undrained shear strength s_u of the soil is obtained.

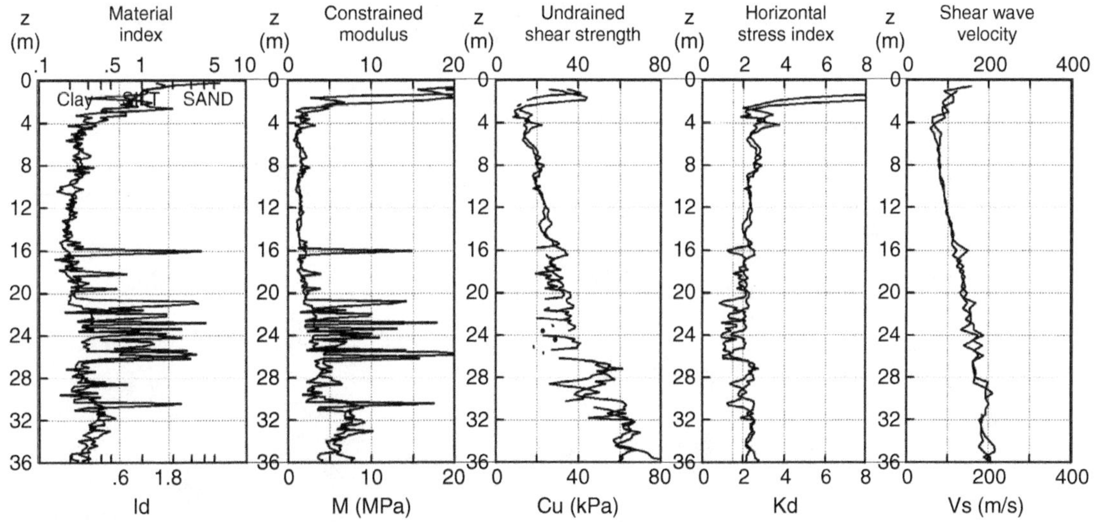

Figure 7.17 Example of dilatometer test results. (Courtesy of Dr. Sylvano Marchetti, www.marchetti-dmt.it)

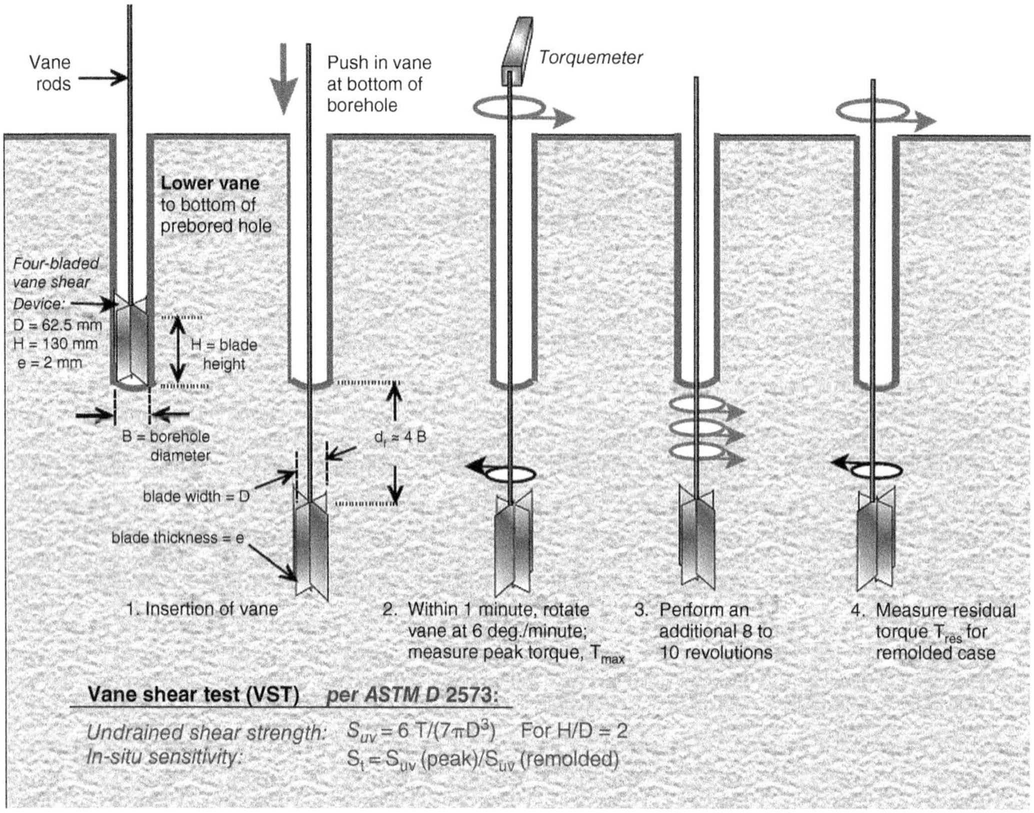

Figure 7.18 The vane shear test. (From Mayne et al. 2002. Courtesy of Professor Paul Mayne, Georgia Institute of Technology)

Figure 7.19 Field vane shear test. (Courtesy of Dr. Dimitrios P. Zekkos.)

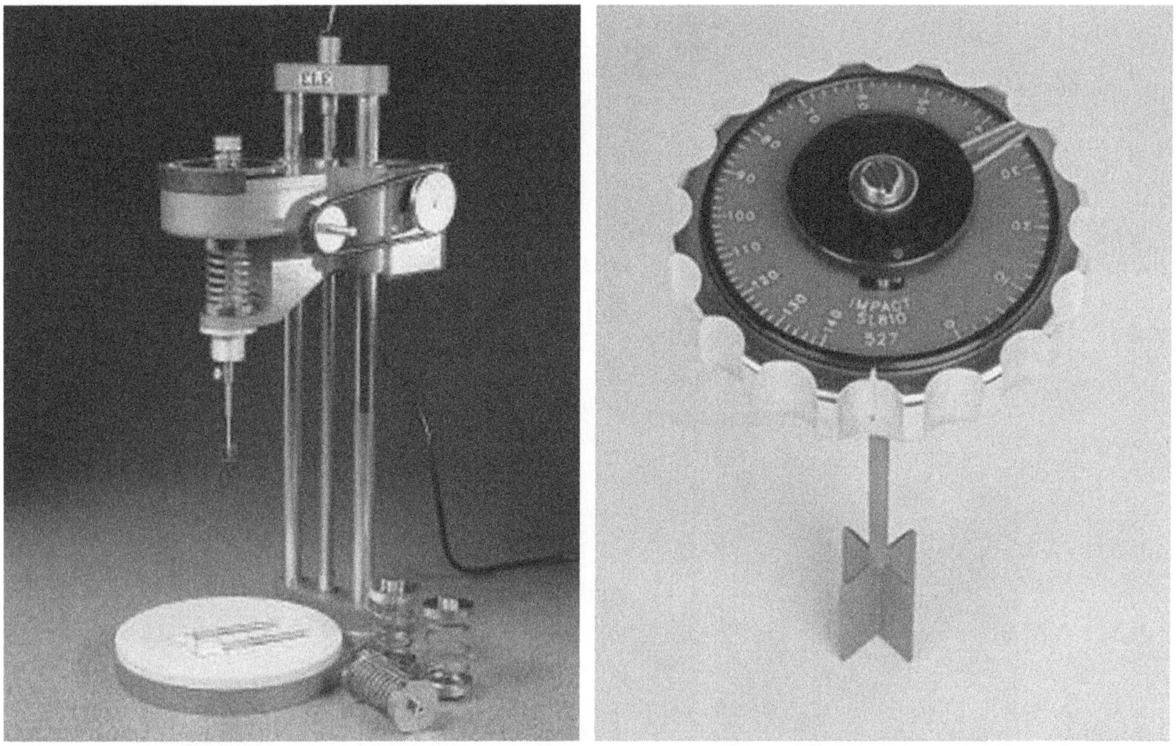

Figure 7.20 Laboratory vane shear test. (a: Courtesy of ELE International, b: Courtesy of Impact Test Equipment Ltd)

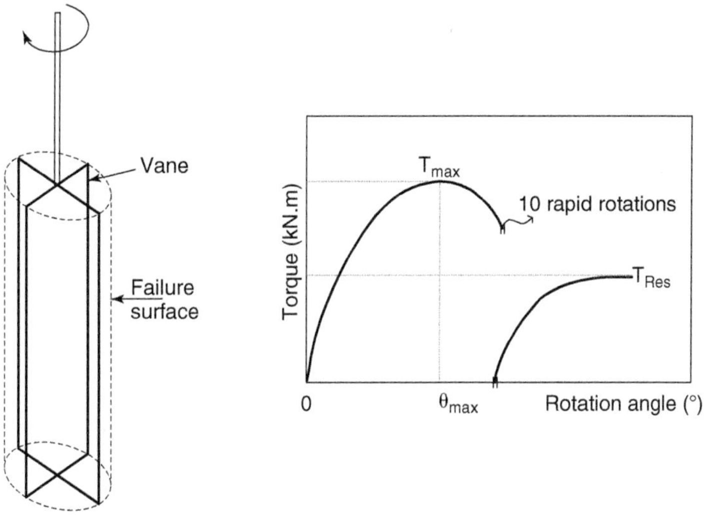

Figure 7.21 Vane shear test results.

The advantages of the BST are that it is simple, economical, and one of the best tools—if not the only tool—to obtain the friction angle of sands by direct measurements in the field. One drawback is that it is difficult to know exactly what pore pressures are generated. A pore pressure sensor on the plates helps in that respect. The phicometer developed by Philiponat (Philiponat, 1986, Philionat and Zerhouni, 1993) is a similar tool.

7.7 PLATE LOAD TEST

The plate load test or PLT (Figure 7.25; ASTM D1196 and D1195) is one of the simplest and oldest in situ tests. It consists of placing a circular plate with a diameter *D* on a prepared soil surface and loading the plate in steps until the desired pressure p is reached. The plate diameter is usually on the order of 0.3 m. Sometimes one or more unload-reload loops are performed during the test. All load

Figure 7.22 Borehole shear test device. (Courtesy of In-Situ Soil Testing, L.C.)

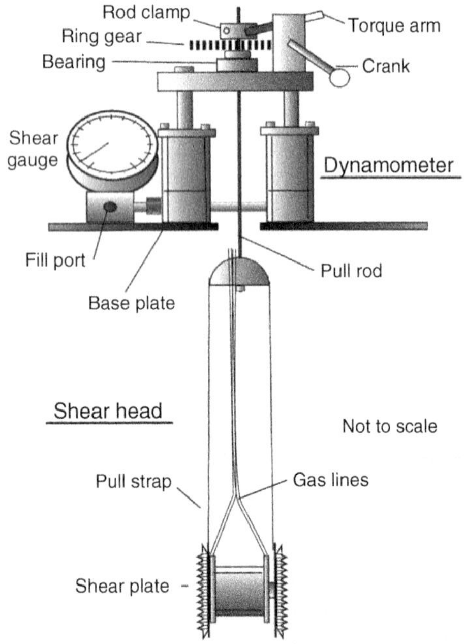

Figure 7.23 Borehole shear test device. (Courtesy of Professor Richard L. Handy, Handy Geotechnical Instruments, Inc.)

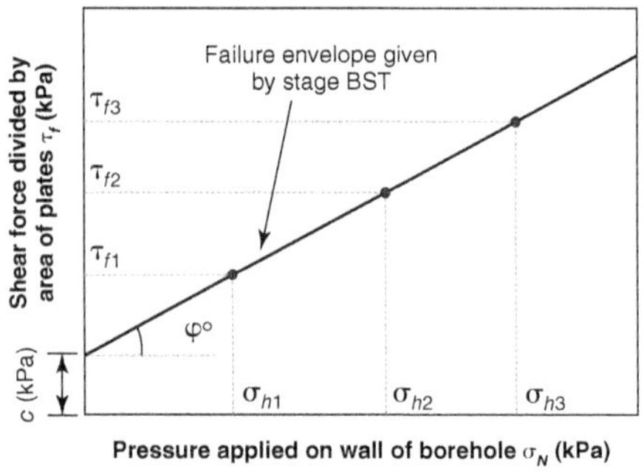

Figure 7.24 Results of a borehole shear test.

steps are held for the same period of time, during which readings of the plate settlement s are made as a function of time t. Loading the plate to soil failure is often desirable but not always possible. The load is measured with a load cell and the settlement is measured by using dial gages or electronic displacement devices (e.g., a linear variable differential transformer [LVDT]) attached to a settlement beam. It is critical that the supports of the settlement beam be far enough from the plate influence zone. Five plate diameters on each side seem appropriate.

The result of the test is a load Q versus displacement s curve (Figure 7.26), which can also be presented in normalized form as the ratio of the average pressure p under the plate over a measure of soil strength SS versus settlement of the plate s over the plate diameter D. The soil strength SS can be the ultimate bearing pressure under the plate p_u, the pressuremeter limit pressure p_L, the cone penetrometer point resistance q_c, the undrained shear strength s_u, the SPT blow

Figure 7.25 Plate load tests. (a: Photo by David Wilkins. Courtesy of Raeburn Drilling and Geotechnical (Northern) Limited; www.raeburndrillingnorthern.com. b: Courtesy of GEMTECH Limited, Fredericton, New Brunswick.)

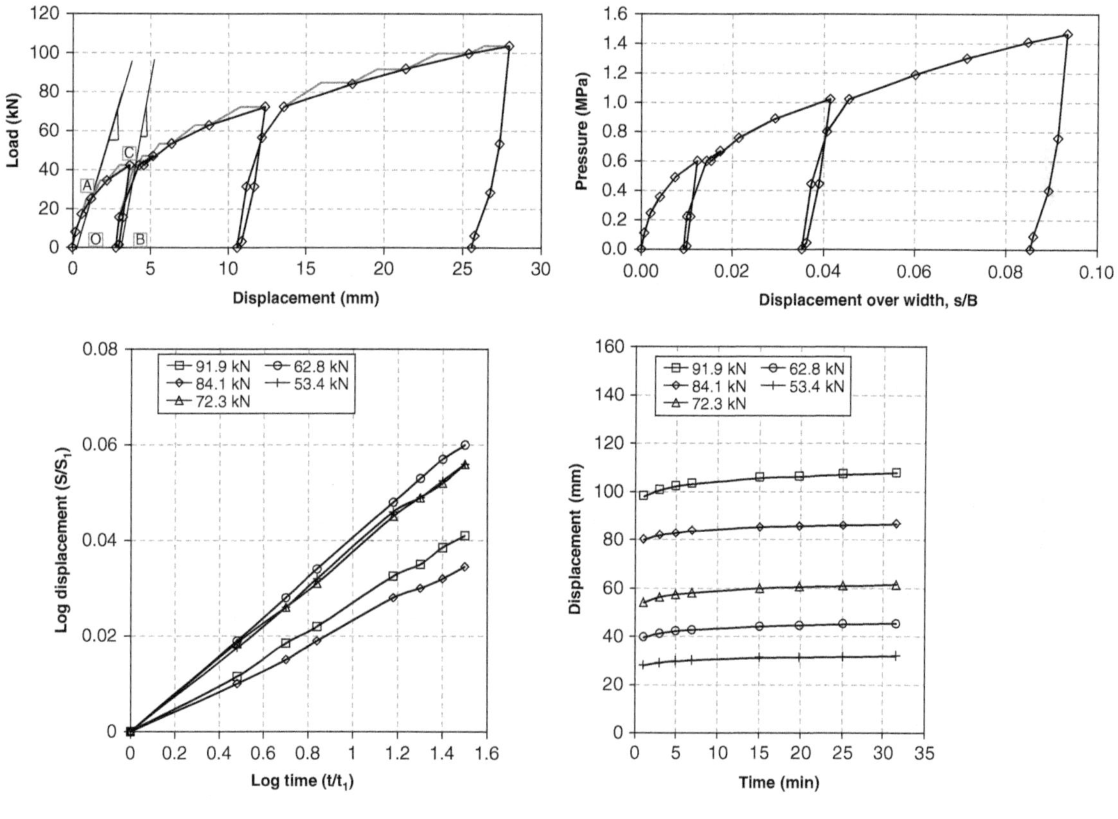

Figure 7.26 Results of load test for 0.3-m-diameter plate on medium dense silty sand.

count N, or another measure of soil strength. The ultimate bearing pressure p_u is often defined as the pressure reached when settlement of the plate is equal to 10% of the plate diameter. The advantage of plotting the results in this fashion (p/SS versus s/D) is that the results of the test become a property of the soil within the zone of influence of the plate and do not depend on the plate size (Briaud 2007). The soil modulus as measured during a plate test is obtained from the initial loading portion E_0 (O to A on Figure 7.26) or from the slope of the reloading part of the unload-reload loop E_r (B to C on Figure 7.26). The equations to be used for E_0 and E_r, if the plate can be assumed to be rigid, are:

$$E_0 = \frac{(1-\nu^2)\pi pD}{4s} \tag{7.13}$$

$$E_r = \frac{(1-\nu^2)\pi \Delta pD}{4\Delta s} \tag{7.14}$$

where E_0 is the initial modulus from a plate load test, ν is Poisson's ratio (to be taken as 0.5 if the plate test is fast enough that no drainage can take place during the test and 0.35 if the test is drained), p is the average pressure under the plate corresponding to the settlement s, D is the diameter of the plate in contact with the soil surface, E_r is the reload modulus from a plate load test, and Δp is the pressure increment during the reload loop corresponding to the settlement increment Δs.

In addition to obtaining the soil modulus, sometimes the modulus of subgrade reaction is calculated from the plate test, as follows:

$$K = \frac{p}{s} \text{ in kN/m}^3 \tag{7.15}$$

Note that K is not a soil parameter, since it depends on the size of the plate:

$$K = \frac{4E_0}{(1 - \upsilon^2)\pi D} \tag{7.16}$$

Therefore, the modulus of subgrade reaction K measured with a plate of a given diameter D cannot be used for plates or footings that have diameters significantly different from D.

It is also useful to plot the settlement of the plate s versus the time t for each load step on a log-log plot (Figure 7.26). The plot of log s versus log t is remarkably linear in most cases within the working load range. The slope of that line is called the *viscous exponent n* and allows one to predict by extrapolation the displacement at much longer times than the time taken to run the plate test, based on equation 7.17:

$$\frac{s_1}{s_2} = \left(\frac{t_1}{t_2}\right)^n \tag{7.17}$$

where s_1 is the settlement after a time t_1 and s_2 is the settlement after a time t_2 and n is the slope of the log s versus log t curve for the load step corresponding to s_1. Alternatively, the soil modulus E_0 or E_r can be written as:

$$\frac{E_1}{E_2} = \left(\frac{t_1}{t_2}\right)^{-n} \tag{7.18}$$

The advantage of the plate load test is that it is very simple and economical to perform. The drawback is that it only tests a zone of soil near the ground surface (one to two plate diameters deep), although larger depths can be reached by performing the test at the bottom of open pits.

7.8 CALIFORNIA BEARING RATIO TEST

The California bearing ratio test (CBR) is a form of plate test (Figure 7.27). It can be performed in the field or in the lab. In the field (ASTM D4429), it consists of placing a 254 mm diameter plate weighing 44.5 N on the ground surface and loading it until the settlement s is 2.5 mm. The load Q corresponding to a settlement s of 2.5 mm is divided by the plate area to get the pressure p. The California bearing ratio is the ratio between p and the pressure necessary to reach a

Figure 7.27 CBR test in the field. (Courtesy of A F Howland Company.)

settlement s of 2.5 mm on a reference soil (crushed California limestone). The pressure necessary to create 2.5 mm of settlement of the plate on the reference soil (crushed California limestone) has been measured to be 6900 kPa. So, the reference pressure is 6900 kPa and the CBR number is a percentage given by:

$$\text{CBR} = \frac{100 \times p(\text{kPa})}{6900} \tag{7.19}$$

This test is used primarily for pavement design, where the depth of influence of the plate is similar to the depth of influence of a truck tire. If the CBR value is less than 3%, the soil is too soft for road support without modification, values between 3% and 5% are average, and values from 5% to 15% are good. Crushed rock values are around 100%. Several correlations have been developed to link the CBR to soil properties, such as:

$$M_r \text{ (kPa)} = 10{,}000 \times \text{CBR} \tag{7.20}$$

$$s_u \text{ (kPa)} = 11 \times \text{CBR} \tag{7.21}$$

where M_r is the resilient modulus and s_u is the undrained shear strength.

7.9 POCKET PENETROMETER AND TORVANE TESTS

A number of simple tests can be performed on the sample in the field as soon as it is retrieved from the borehole. They are typically performed on the end of samples taken with a Shelby tube. These tests include the pocket penetrometer, the torvane, and the hand vane tests. The pocket penetrometer

Figure 7.28 Pocket penetrometer and torvane: (a) Pocket penetrometer (see also this video: www.encyclopedia.com/video/PBo0UDVWhSo-hand-penetrometer-test.aspx). (b) Torvane (see also this video: www.encyclopedia.com/video/9Su3ehhLfwc-torvane-test.aspx)

test (PPT) (Figure 7.28) consists of pushing by hand the end of a spring-loaded cylinder 6.35 mm in diameter until the ultimate bearing pressure is reached. The compression of the spring increases as the force increases and a floating ring on the body of the pocket penetrometer (PP) indicates how much force is exerted. The ultimate pressure is reached when the cylinder penetrates without further increase in the PP reading. The PP number ranges from 0 to 4.5 and has been correlated with the undrained shear strength of clays (s_u (kPa) $\sim$ 30 PP), but the scatter in this correlation is very large—not to mention the fact that the mass of soil tested is extremely small. The advantage of the PPT is that it is a very simple test that gives a quick indication of the soil strength. The drawback is that it tests only a very small zone of soil and thus must not be used in design. The torvane test (TVT) (Figure 7.28) consists of pushing a set of vanes about 6.5 mm into the face of the sample and then rotating the spring-loaded cap until the spring releases because the shear strength of the soil has been reached. A maximum value indicator stays at the maximum reading reached during the rotation and indicates the shear strength of the soil. The hand vane shear test (VST) (section 7.5, Figure 7.20) is also a simple and quick test that can be performed on the end of a Shelby tube sample. These three simple tests are mostly used on silts and clays. Of the three, the hand vane is the most reliable.

7.10 POCKET ERODOMETER TEST

The pocket erodometer test (PET) (Figure 7.29, Briaud, Bernhardt, and Leclair 2011) is to erosion resistance what the pocket penetrometer test is to shear resistance. The pocket erodometer (PE) is a regulated mini-jet-impulse-generating device. The water jet comes out of the nozzle at 8 m/s and is aimed horizontally at the vertical face of the sample. Verification that the velocity is 8 m/s when leaving the nozzle is achieved by aiming the jet from a height H (Figure 7.29), measuring the distance x where the water reaches the floor, and using the following equation:

$$v_{0x} = \frac{x}{\sqrt{\frac{2H}{g}}} \tag{7.22}$$

where v_{0x} is the velocity at the nozzle and g is the acceleration due to gravity. The depth of the hole in the surface of the sample created by 20 impulses of water is recorded. The depth of the hole is entered in the erosion chart (shown in Figure 7.30) to determine the erodibility category of the soil.

Figure 7.29 Pocket erodometer test.

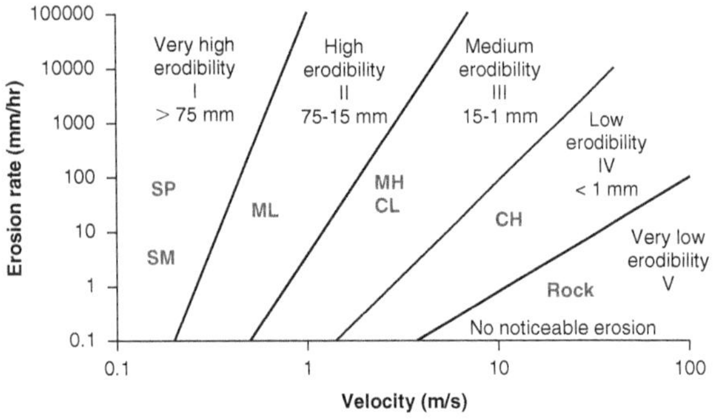

Figure 7.30 Erosion chart for various erosion depths from the PET.

This erosion category allows the engineer to make preliminary decisions in erosion-related work. The advantage of the PET is its simplicity; its drawback is that it tests a very small portion of the soil.

7.11 COMPACTION CONTROL TESTS

Soil compaction is one of many techniques of soil improvement and is discussed in Chapter 20. In short, the soil to be used at the site is tested in the laboratory where compaction tests are performed. The results of these tests are used to establish the target values (dry unit weight, modulus, water content) to be achieved during the compaction process in the field. In the field it becomes necessary to verify that the target value has been reached. These in situ tests include tests to measure the dry unit weight (e.g., sand cone method, rubber balloon method, nuclear density probe), water content (e.g., nuclear density probe, field oven test), and soil modulus (e.g., BCD, falling weight deflectometer).

7.11.1 Sand Cone Test

The sand cone test (SCT; Figure 7.31) consists of digging a hole in the ground, obtaining the weight and the volume of the soil excavated, drying the soil and obtaining the dry weight, and calculating the water content and the dry unit weight. More specifically, a standard steel plate with a 172 mm diameter hole through it is placed on the ground surface. A hole is dug into the ground through the hole in the steel plate to a depth of about 150 mm. The excavated soil is weighed, then dried, then weighed again. This gives the water content of the soil that was in the hole. As soon as the hole is excavated, an inverted funnel in the form of a cone is placed on top of the opening in the base plate and a bottle full of sand of known unit weight is connected to the top of the funnel. (The weight of the bottle full of sand is measured beforehand.) The valve between the bottle and the funnel is then opened and the sand of known unit weight flows out of the bottle until the hole in the ground and the funnel above it are full. The valve is closed, the bottle is disconnected, and the bottle is weighed again. The difference in weight of the bottle before and after filling the hole, divided by the known unit weight of the sand, gives the volume of the hole plus the funnel. Because the volume of the funnel is known, the volume of the hole can be deduced and the dry unit weight is obtained from the dry weight and the volume of the soil in the hole.

7.11.2 Rubber Balloon Test

The rubber balloon test (RBT; Figure 7.32) follows exactly the same procedure as the sand cone method except that the volume of the soil excavated is measured in a different way. The rubber balloon device is a cylinder filled with water up to a level indicated on a graduated scale. At the bottom of

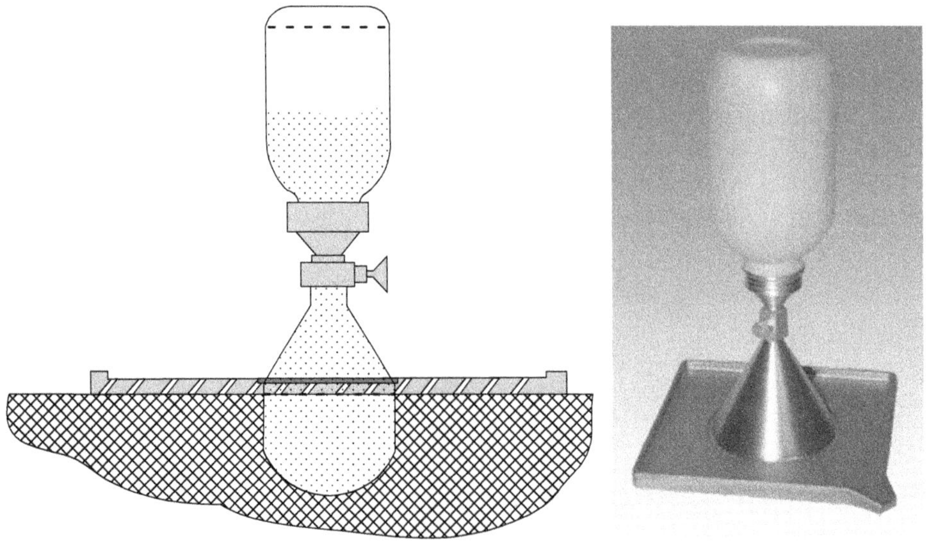

Figure 7.31 Field unit weight and water content by sand cone test. (b: Courtesy of Durham Geo Slope Indicator.)

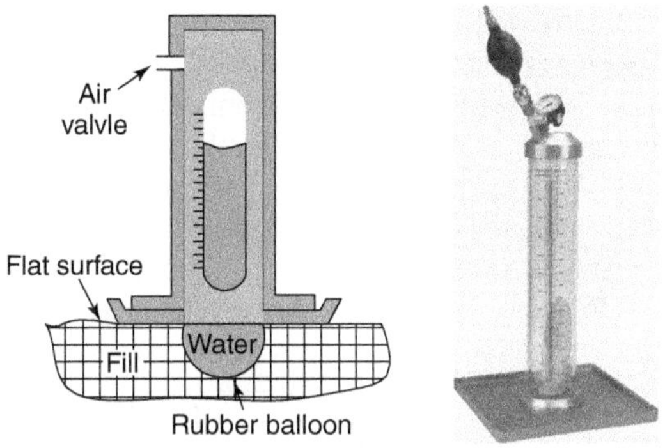

Figure 7.32 Field unit for testing weight and water content by rubber balloon. (b: Courtesy of Humboldt Mfg. Co.)

the cylinder is a rubber balloon that can be expanded into the hole below by pumping water into it. When the balloon fills the hole, the reading on the graduated scale on the cylinder gives the volume of the hole. The data reduction is the same as for the sand cone test.

7.11.3 Nuclear Density/Water Content Test

The nuclear density/water content test is a device to measure indirectly the density and water content of a soil at the soil surface. It consists of sending radiation from a source into the soil and counting the amount of radiation coming back to a detector. In the case of the nuclear density test, a source generating medium-energy gamma rays is used. These gamma rays send photons into the soil (photons are particles of light; see section 8.4.1). These photons go straight to the detector, or bump into the soil particles (Compton scattering) and deflect to arrive at the detector, or do not arrive at the detector. The gamma rays arriving at the detector are counted, and the gamma count is inversely proportional to density. In the case of the water content test, a source generating high-energy neutrons is used. The principle is that when a high-energy neutron hits a much heavier atomic nucleus, it is not slowed down significantly. However, if it hits an atomic nucleus that is about the same weight as the neutron, then the neutron is slowed down significantly. The hydrogen atom has a nucleus that is very comparable in weight to the neutron, and therefore is very good at slowing neutrons down. Because water has a lot of hydrogen, counting the number of slow neutrons coming back to a detector will indicate how much water is in the soil.

The test can be done in direct transmission or in back-scatter mode. In the direct transmission mode, the source rod penetrates into the soil anywhere from 75 mm to 220 mm (Figure 7.33); the detector is on the bottom side of the nuclear gage. This mode is preferred for density measurements. In the back-scatter mode, the nuclear gage sits on the soil surface and the source and detectors are on the bottom side of the gage (Figure 7.33). This is the mode used for water content determination. The nuclear gage is calibrated by the manufacturer initially and after any repair. The calibration consists of placing the gage on a sufficiently large block of material of known density and known water content.

7.11.4 Field Oven Test

The field oven (Figure 7.34) is a very simple instrument which is used to determine the water content of a soil in the field. A small piece of soil is carved from the soil surface; the sample is placed between the two plates of the field oven which looks like a waffle maker. A load cell located below the heating pad gives the weight of the sample. Then the two plates are closed and the oven dries the soil sample. After a

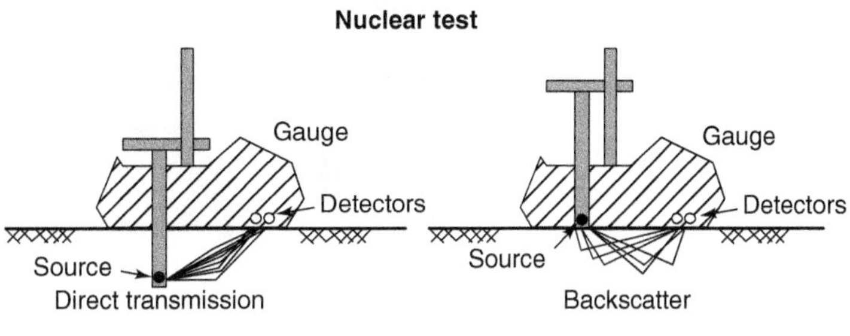

Figure 7.33 Nuclear density probe test for unit weight and water content.

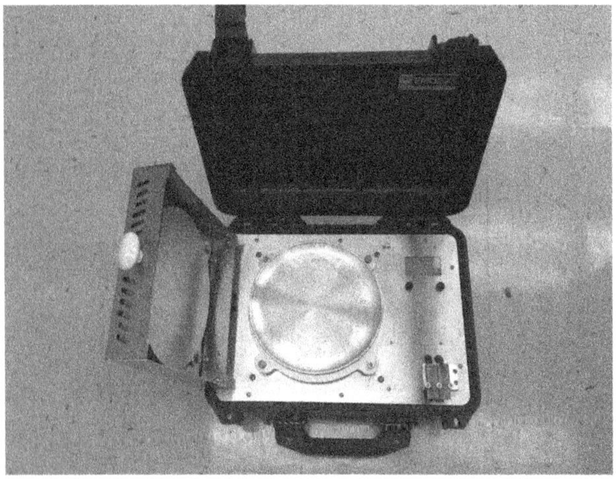

Figure 7.34 Field oven test (FOT) for water content.

few minutes, the soil is dry and the heating plates are opened. The load cell records the dry weight of the sample and the water content is displayed.

7.11.5 Lightweight Deflectometer Test

The lightweight deflectometer (LWD) test (Figure 7.35) (ASTM E2583) consists of dropping a weight guided along a rod from a chosen height onto a plate resting on the ground surface. The typical values for the LWD are a weight of 100 N, a drop height of 0.5 m, and a plate diameter of 0.2 m. A load cell located above the plate measures the force versus time signal and a geophone attached to the plate measures the deflection of the plate during the impact. The soil modulus is back-calculated from the knowledge of the peak force F and the peak deflection Δ. The soil modulus E is calculated using the theory of elasticity:

$$E = f(1 - \nu^2)\frac{4F}{\pi D\Delta} \tag{7.23}$$

where E is the soil modulus measured by the LWD, f is a plate rigidity factor (1 for flexible plates and 0.79 for rigid plates), ν is Poisson's ratio (range from 0.3–0.45, depending on soil type), F is the maximum force on the force versus time plot, D is the plate diameter, and Δ is the maximum displacement on the displacement versus time plot.

For example, referring to the flexible plate LWD test in Figure 7.36, the modulus would be calculated as:

$$E = 1(1 - 0.35^2)\frac{4 \times 7.5}{\pi \times 0.2 \times 0.55 \times 10^{-3}} = 76.3\,\text{MPa} \tag{7.24}$$

7.11.6 BCD Test

A modulus E can also be obtained with a device called the BCD (Figure 7.37). It consists of a 150 mm diameter, 2 mm thick flexible steel plate at the bottom of a rod with handles—a kind of scientific cane. Strain gages are mounted on the back of the plate to record the bending that takes place during the loading test. When the operator leans on the handle, the load on the plate increases and the plate bends. If the soil is soft (low modulus), the plate bends a lot. If the soil is hard (high modulus), the plate does not bend much. The amount of bending is recorded by the strain gages and is correlated to the modulus of the soil below.

The test is called the BCD test or BCDT (Briaud, Li, and Rhee 2006) and is performed as follows. First, the BCD plate is placed on top of the ground surface (Figure 7.37). Then the operator leans on the handles of the BCD and the vertical load increases. When the load goes through 223 N, a load sensor triggers the reading of the strain gages. The device averages the strain gage values, uses the internal calibration equation linking the strains to the modulus, and displays the modulus E. This evaluates the level of compaction achieved at that location.

The modulus obtained with the BCD corresponds to a reload modulus, to a mean stress level averaging about

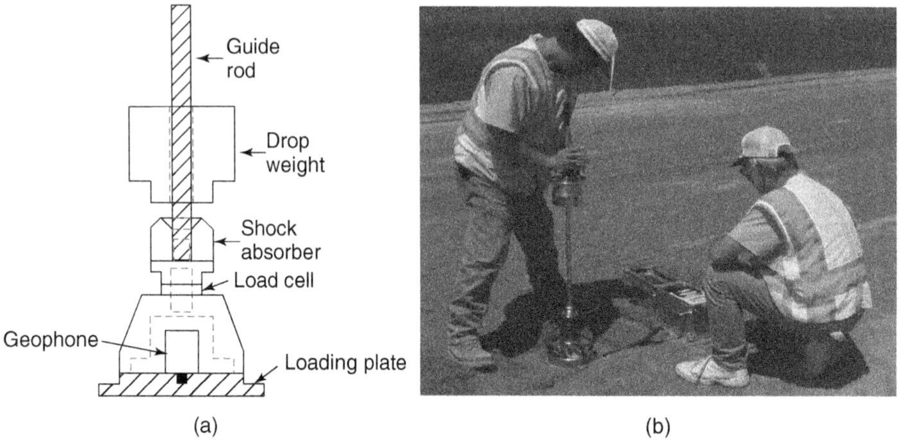

Figure 7.35 Falling weight deflectometer for soil modulus: (a) Principle. (b) Equipment. (b: Courtesy of Minnesota Department of Transportation.)

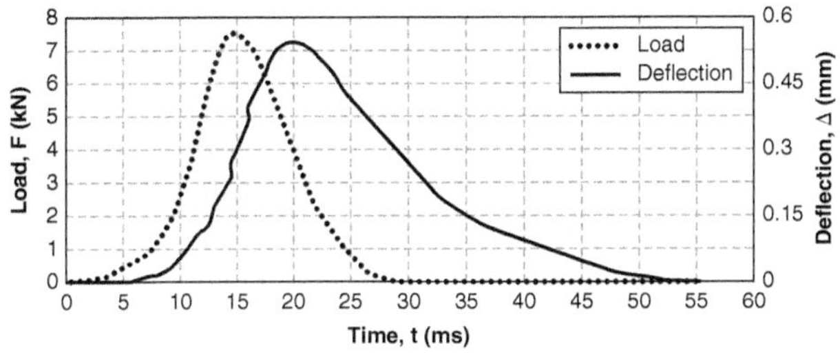

Figure 7.36 Falling weight deflectometer data.

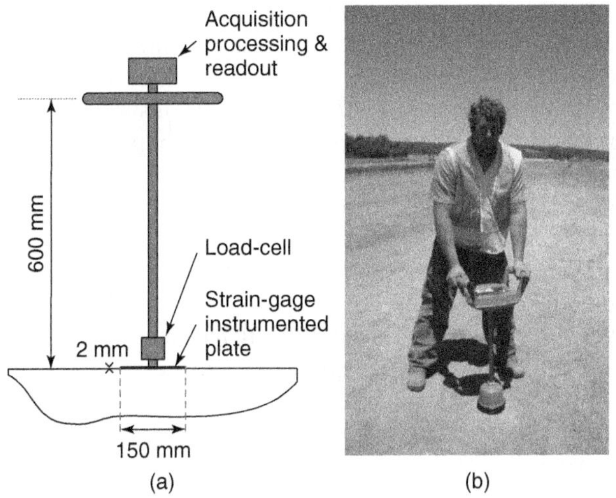

Figure 7.37 BCD test for soil modulus: (a) Principle. (b) Equipment.

50 kPa within the zone of influence, to a strain level averaging 10^{-3} within the zone of influence, and to a time of loading averaging about 2 s. The BCD test can also be performed in the laboratory on top of the compaction mold to obtain the modulus versus water content curve in parallel with the dry density versus water content curve (see chapter 20 section 20.2).

7.12 HYDRAULIC CONDUCTIVITY FIELD TESTS

The purpose of these hydraulic conductivity in situ tests is to measure the hydraulic conductivity k (m/s) of the soil. The soil can be either below the groundwater level (saturated), or above the groundwater level (saturated by capillary action or unsaturated). For saturated soils below the GWL, several tests exist, including the borehole tests (falling head test, rising head test, constant head tests), the pumping test, and the cone penetrometer dissipation test. For soils above the GWL, the tests include the sealed double-ring infiltrometer (SDRI) test and the two-stage borehole permeameter.

7.12.1 Borehole Tests

Borehole tests consist of drilling a borehole, changing the water level in the borehole, and recording the movement of the water level in the borehole as a function of time. Sometimes the borehole is cased to help in keeping the borehole stable. The data collected are used to back-calculate the hydraulic conductivity k. The equations to calculate k are based on developing the governing differential equation for the problem and then solving it while satisfying the boundary conditions. This is where the problem becomes quite complicated and requires charts or software. The following examples are cases in which the geometry is simple.

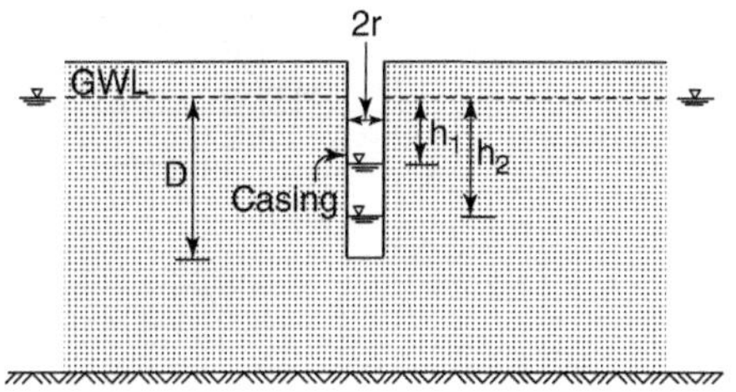

Figure 7.38 Inflow well test in deep uniform soil. (After Hunt 1984.)

When the soil layer is deep and uniform, when the casing goes down to the bottom of the borehole, and when the water is bailed out so that the water level starts far below the groundwater level outside of the casing (Figure 7.38), the hydraulic conductivity k is obtained from the equation:

$$k = \frac{2\pi r \ Ln\dfrac{h_1}{h_2}}{11(t_2 - t_1)} \tag{7.25}$$

where r is the radius of the casing, h_1 and h_2 are the distances from the groundwater level in the soil deposit outside of the casing to the level of the water in the casing, and t_1 and t_2 are the times at which h_1 and h_2 are measured. This equation applies when the depth D as shown in Figure 7.38 is between 0.15 m and 1.5 m.

In the case where the pervious soil layer to be tested is underlain by an impervious layer, where the uncased boring (or screened boring) penetrates through the entire pervious layer all the way to the impervious layer, and where the water level is maintained constant by pumping at a flow rate Q

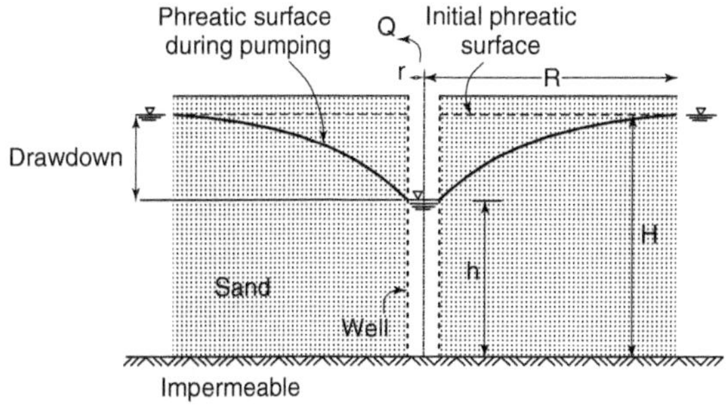

Figure 7.39 Pumping test in sand layer using one boring. (After Hunt 1984.)

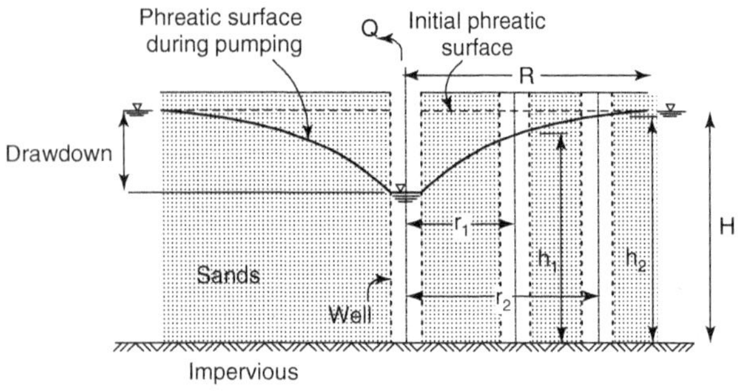

Figure 7.40 Pumping test in sand layer using three borings. (After Hunt 1984.)

(as shown in Figure 7.39), the hydraulic conductivity k is obtained from the equation:

$$k = \frac{Q \ Ln\dfrac{R}{r}}{\pi(H^2 - h^2)} \tag{7.26}$$

where Q is the flow rate pumped out of the well to maintain the water level constant in the well, r is the radius of the borehole, R is the radius of the zone of influence where the water table is depressed, H is the vertical distance between the bottom of the boring (impervious layer) and the groundwater level at or further than R, and h is the vertical distance between the bottom of the boring and the water level in the borehole. Note that for this equation to apply, a steady-state flow must be reached; this may take a time related to the hydraulic conductivity itself. Finding the value of R requires some borings down to the groundwater level away from the test boring.

To improve the precision of this test, observation borings can be drilled at radii r_1 and r_2 from the test boring and the vertical distances h_1 and h_2 between the bottom of the boring (impervious layer) and the water level in the observation borings recorded (Figure 7.40). Then equation 7.26 becomes:

$$k = \frac{Q \ Ln\dfrac{r_2}{r_1}}{\pi(h_2^2 - h_1^2)} \tag{7.27}$$

In the case where the pervious layer to be tested is sandwiched between two impervious layers, where the uncased boring (or screened boring) penetrates through the first two layers and stops at the top of the second impervious layer, and where the water level is maintained constant by pumping at a flow rate Q (as shown in Figure 7.41), the hydraulic conductivity k is obtained from the equation:

$$k = \frac{Q \ Ln\dfrac{R}{r}}{2\pi D(H - h)} \tag{7.28}$$

where Q is the flow rate pumped out of the well to maintain the water level constant in the well, r is the radius of the borehole, R is the radius of the zone of influence where the water table is depressed, H is the vertical distance between the bottom of the boring (top of the second impervious layer) and the groundwater level at or further than R, and h is the vertical distance between the bottom of the boring (top of the second impervious layer) and the water level in the borehole. Note that for this equation to apply, a steady-state flow must be reached; this may take a time related to the hydraulic conductivity itself. Finding the value of R requires some borings down to the ground-water level away from the test boring.

To improve the precision of this test, observation borings can be drilled at radii r_1 and r_2 from the test boring and the vertical distances h_1 and h_2 between the top of the second impervious layer and the water level in the borehole recorded (Figure 7.42). Then equation 7.28 becomes:

$$k = \frac{Q \ Ln\dfrac{r_2}{r_1}}{2\pi D(h_2 - h_1)} \tag{7.29}$$

Solutions for more complicated geometries are found in Mansur and Kaufman (1962) and in Cedergren (1967). The advantages of these tests are that they give a large-scale value of k in the field which includes the mass features of the soil deposit. Some of the drawbacks are the lack of control over problems such as filter cake development around the

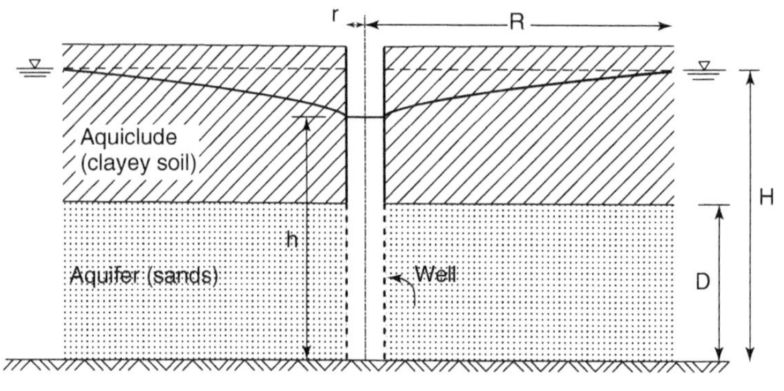

Figure 7.41 Pumping test in confined aquifer. (After Hunt 1984.)

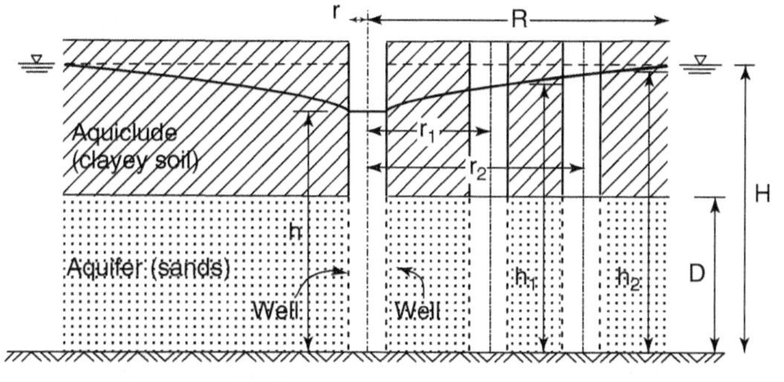

Figure 7.42 Pumping test in confined aquifer using three borings.

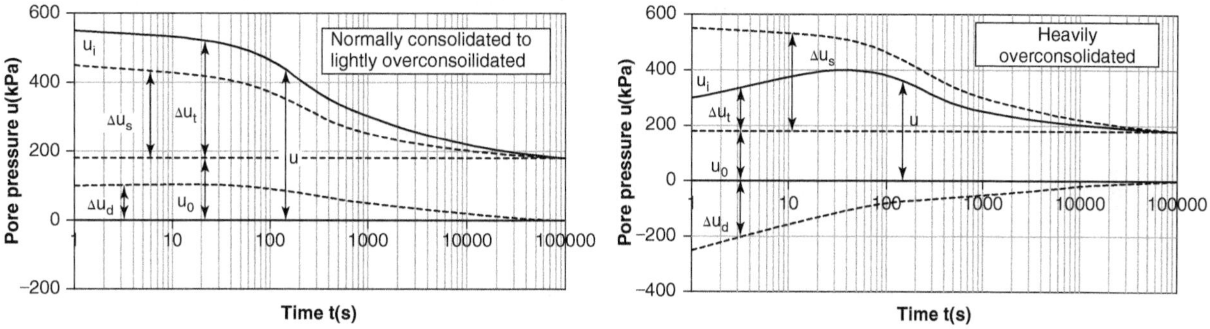

Figure 7.43 Decay of excess pore pressure in piezocone dissipation test.

wall of the borehole, and quick conditions development in high-gradient situations.

7.12.2 Cone Penetrometer Dissipation Test

The cone penetrometer dissipation test (CPDT) is performed during a CPT sounding and makes use of the cone point equipped with a pore pressure measuring sensor: a *piezocone*. The piezocone is pushed to a depth below the groundwater level where the measurement of k has to be made, the penetration stops, the initial excess pore pressure is read, and then the decay of excess pore pressure versus time is recorded. Two situations can arise: heavily overconsolidated soil or normally to lightly overconsolidated soil.

In the case of normally consolidated to lightly overconsolidated soil, the decay of excess pore pressure will be monotonic (Figure 7.43*a*). In the case of heavily overconsolidated soils, the response shows first an increase in excess pore pressure followed by a decrease (Figure 7.43*b*). The reason for this dual behavior is that the total excess pore pressure Δu_t has two components: one is due to the water stress response Δu_s to the mean all-around compression of the soil element (spherical stress tensor); the other is due to the water stress response Δu_d to the shearing of the soil element (deviatoric stress tensor). When the soil element is subjected to an all-around mean pressure, Δu_s is always positive, but when the soil element is subjected to a shear stress, Δu_d can be positive or negative depending on the change in volume of the element during shearing. If the soil element decreases in volume during shearing, it is called *contractive*, Δu_d is positive, and both Δu_s and Δu_d decrease as a function of time (Figure 7.43*a*). If, however, the soil element increases in volume during shearing, it is called *dilatant*, and Δu_d is negative. The combination of Δu_s decreasing with time and Δu_d increasing with time (becoming less negative) leads to a bump on the decay curve (Figure 7.43*b*).

The initial pore pressure when recording starts is u_i. Note that two u_i values exist depending on the location of the pore-pressure measuring device. In the case of a monotonic decay and for the pore-pressure measurement right behind the cone point (shoulder), Parez and Fauriel (1988) proposed a correlation between t_{50} and the hydraulic conductivity k (Figure 7.44), which is well represented by the equation:

$$k(\text{cm/s}) = \left(\frac{1}{251\ t_{50}\,(s)}\right)^{1.25} \tag{7.30}$$

Where k is the hydraulic conductivity in cm/s and t_{50} is the time in seconds to reach a decrease in water stress equal to 50% of the total decrease in water stress.

A typical example is shown in Figure 7.43a for a lightly overconsolidated clay. The time to 50% dissipation is found halfway between the initial value u_i ($t = 1$s in Figure 7.43*a*) and the equilibrium value corresponding to the hydrostatic pressure u_0. In the case of a decay curve exhibiting a rise followed by a decay (highly overconsolidated soil), obtaining the hydraulic conductivity k from the dissipation curve is more complicated (Burns and Mayne 1998).

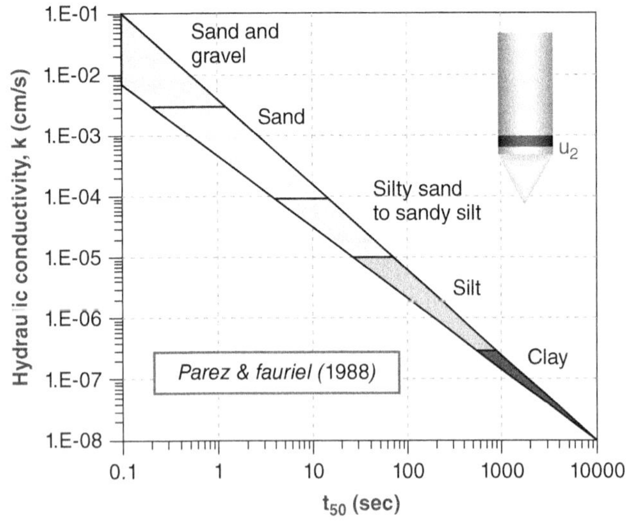

Figure 7.44 Relationship between t_{50} and hydraulic conductivity for piezocone dissipation test. (From Mayne, Christopher, Berg, and DeJong 2002. Courtesy of Professor Paul Mayne, Georgia Institute of Technology, USA.)

7.12.3 Sealed Double-Ring Infiltrometer Test

The sealed double-ring infiltrometer test (SDRIT) was developed in the late 1970s in the USA and is credited to Steve Trautwein and David Daniel (1994). The SDRIT aims at measuring the hydraulic conductivity at shallow depth in soils above the groundwater level. A typical situation is testing to obtain the hydraulic conductivity k of a 1 m thick clay liner above a free-draining layer of sand and gravel. The test setup starts by placing a square outer ring about 4 m in size in the soil surface and embedding and grouting the walls of the ring about 0.45 m below the surface (Figure 7.45). Then an inner ring is placed in the center of the outer ring and the walls are embedded and grouted about 0.15 m into the ground. The outer ring is open to the atmosphere while the inner ring is sealed. A tube goes from the inner ring to a deformable plastic bag, where it can be easily connected and disconnected. The bag is filled with water and weighed, and the entire system is saturated with water. The SDRIT is often used to test soils that are not saturated, in which case tensiometers are placed at different depths to measure the tension in the water within the layer being tested (see Chapter 10 on water stress for an explanation of how tension occurs in the soil water and Chapter 9 on laboratory tests for an explanation of how tensiometers work). As the water seeps through the unsaturated soil layer below the SDRI, the water fills the voids in the soil, thereby saturating the soil; a wetting front advances and the plastic bag loses water. The volume of water Q leaving the plastic bag and entering the soil is measured by weighing the bag as a function of time.

Reducing the data of an SDRIT requires knowledge of water flow through saturated and unsaturated soils (see chapter 13). Obtaining the hydraulic conductivity k from the SDRIT requires some assumptions: (1) steady-state seepage; (2) vertical, one-dimensional flow; and (3) saturated conditions. If the soil is unsaturated to start with, it will take time for the water to permeate through the soil layer thickness and saturate the soil. This time can be several weeks. To obtain the hydraulic conductivity k from the SDRIT data, the following equations are used:

$$v = k\,i \tag{7.31}$$

This is called *Darcy's law* and is explained in Chapter 13 on flow through soils; v is the discharge velocity; and i is the hydraulic gradient, defined as the loss of total head Δh_t of the flowing water per distance travelled Δz.

$$i = \frac{\Delta h_t}{\Delta z} \tag{7.32}$$

Conservation of mass leads to:

$$V_f = v_d A t C \tag{7.33}$$

where V_f is the volume of water that has infiltrated the soil in a time t, A is the plan view area of the inner ring, and v_d is the discharge velocity. This leads to an expression for k:

$$k = \frac{\dfrac{V_f}{At}}{\dfrac{\Delta h_t}{\Delta z}} \tag{7.34}$$

If the test is run long enough that the whole layer becomes saturated, then Δh_t is the vertical distance from the bottom of the layer to the level of the water in the outer ring and Δz is the thickness of the layer. The tensiometer readings help in deciding when this stage has been reached. If this assumption is made but the wetting front has not penetrated the whole layer, then i will be underestimated and the k value obtained will be lower than the true k value. If the test does not reach this stage and the water front has penetrated to a depth D_w below the top of the soil surface, the value of Δz is D_w and the value of Δh_t is:

$$\Delta h_t = H + D_w + h_p \tag{7.35}$$

where h_p is the tension in the water on the wetting front expressed in height of water. This value can be obtained from the tensiometer readings. Here two assumptions can be made: (1) h_p is given by the tensiometers, or (2) $h_p = 0$. In practice, the second assumption seems to give more acceptable results, especially as the test is often run to prove that the hydraulic conductivity of the soil layer is lower than 10^{-9} m/s (clay liner for waste disposals). Indeed, with assumption 2 ($h_p = 0$), Δh_t is underestimated and k is overestimated.

When the layer being tested swells, it is necessary to take the swelling into account. In this case some of the water leaving the plastic bag is stored in the swelling process while some of the water is seeping through the soil. Ignoring the

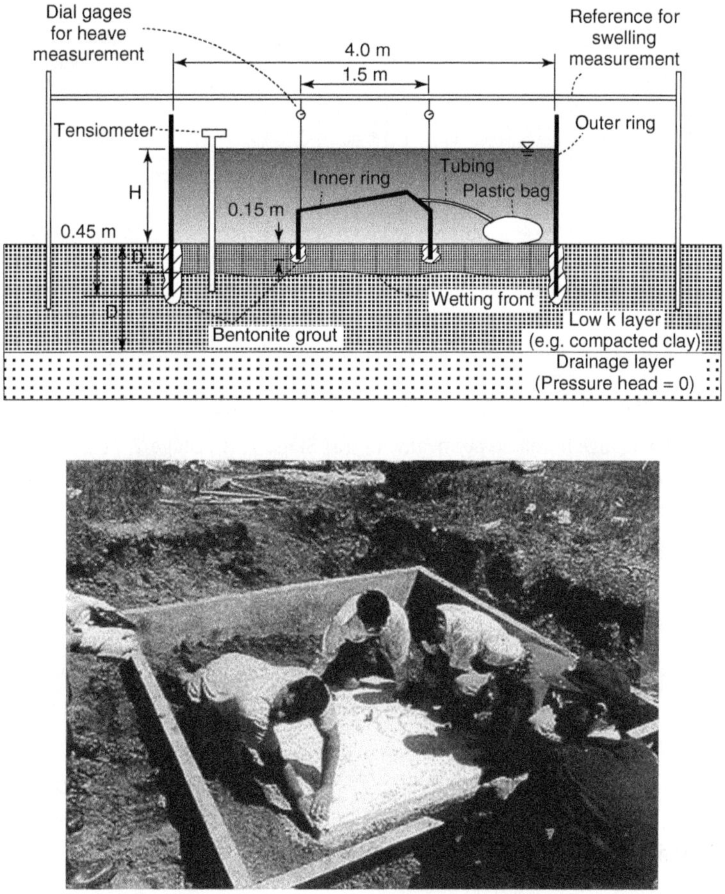

Figure 7.45 Sealed double-ring infiltrometer. (Courtesy of Professor Xiaodong Wang, University of Wisconsin, USA.)

swelling component would give an overestimated value of V_f and therefore an overestimated value of k. The volume of water V_s used to increase the volume of the soil through swelling is measured as follows: A reference beam is set up above the SDRI (Figure 7.44) and the vertical movement of the inner ring is recorded with respect to that beam (using dial gages, for example). The volume V_s corresponding to the vertical movement of the inner ring is subtracted from the volume of water V_t leaving the plastic bag to obtain the true volume V_f. flowing through the soil.

7.12.4 Two-Stage Borehole Permeameter Test

The two-stage borehole permeameter test (TSBPT) was developed in the USA in the 1980s and is credited to Gordon Boutwell (Boutwell and Derick 1986). The TSBPT aims at measuring the vertical and horizontal hydraulic conductivity at shallow depth in soils above the groundwater level. A typical situation is testing to obtain the hydraulic conductivity k of a 1 m thick clay liner above a free-draining layer of sand and gravel. The test takes place in two stages.

Stage 1 consists of drilling a hole (for example, 0.5 m deep and 0.1 m in diameter), inserting a permeameter (e.g., open PVC 75 mm diameter pipe with graduated cylinder above, Figure 7.46a) in the open hole, sealing the permeameter to the walls of the borehole by grouting, and keeping the bottom of the boring open and intact. Once the borehole is sealed, the test consists of filling the permeameter with water and letting the water seep into the soil through the bottom of the casing. The drop in water level in the graduated tube is recorded as a function of time. The hydraulic conductivity k_1 from stage 1 is calculated from the following equation (Hvorslev 1949):

$$k_1 = \frac{\pi d^2}{11D(t_2 - t_1)} Ln\frac{h_1}{h_2} \tag{7.36}$$

where d is the diameter of the graduated tube above the permeameter, D is the diameter of the permeameter, and h_1 and h_2 are the heights of water above the bottom of the casing recorded at times t_1 and t_2 respectively. The k_1 values are plotted as a function of time until steady state is reached. Note that this equation assumes that the material below the casing is uniform to a large depth. It is prudent to use it only if the depth to the next layer is at least 5 borehole diameters below the bottom of the boring.

Stage 2 consists of deepening the borehole (for example, 0.2 m deeper and 75 mm in diameter), and repeating the permeability test (falling head test). The hydraulic conductivity

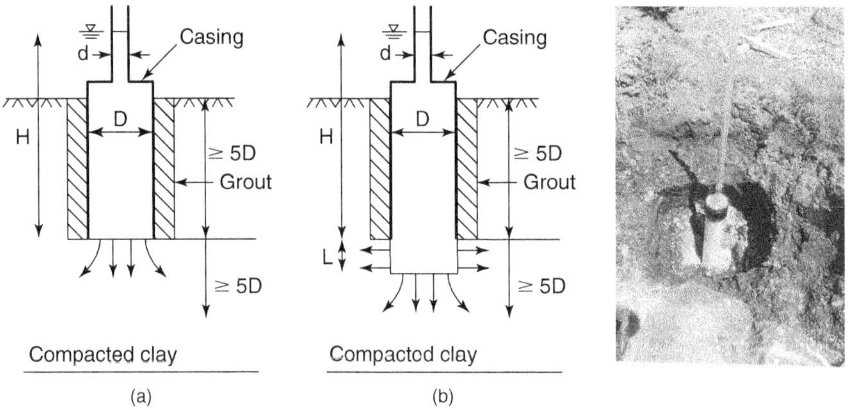

Figure 7.46 Two-stage borehole permeameter: (a) Stage 1; (b) Stage 2. (Third picture: Courtesy of Craig Benson, University of Wisconsin.)

k_2 from stage 2 is calculated from the following equations (Hvorslev 1949):

$$k_2 = \frac{A}{B} Ln \frac{h_1}{h_2} \tag{7.37}$$

with

$$A = d^2 \; Ln \left(\frac{L}{D} + \sqrt{1 + \left(\frac{L}{D} \right)^2} \right) \tag{7.38}$$

$$B = 8L(t_2 - t_1) \left(1 - 0.562 e^{-1.57 \frac{L}{D}} \right) \tag{7.39}$$

Note that A is in m^2 while B is in m.s. The k_2 values are plotted as a function of time until steady state is reached. Then the anisotropy can be taken into account by using the ratio k_2/k_1 and relating it to the ratio k_h/k_v. This is done by first defining m as:

$$m = \sqrt{\frac{k_h}{k_v}} \tag{7.40}$$

where k_h and k_v are the hydraulic conductivity in the horizontal and vertical directions respectively. Then k_2/k_1 is related to m through:

$$\frac{k_2}{k_1} = m \frac{Ln \left(\frac{L}{D} + \sqrt{1 + \left(\frac{L}{D} \right)^2} \right)}{Ln \left(\frac{mL}{D} + \sqrt{1 + \left(\frac{mL}{D} \right)^2} \right)} \tag{7.41}$$

In equation 7.41, all quantities are known except m, which can therefore be obtained. Alternatively, m can be found by using Figure 7.47, which presents k_2/k_1 versus k_h/k_v for L/D ratios of 1, 1.5, and 2. Once m is known, k_h and k_v can be found as follows (Daniel, 1989):

$$k_h = mk_1 \tag{7.42}$$

$$k_v = k_1/m \tag{7.43}$$

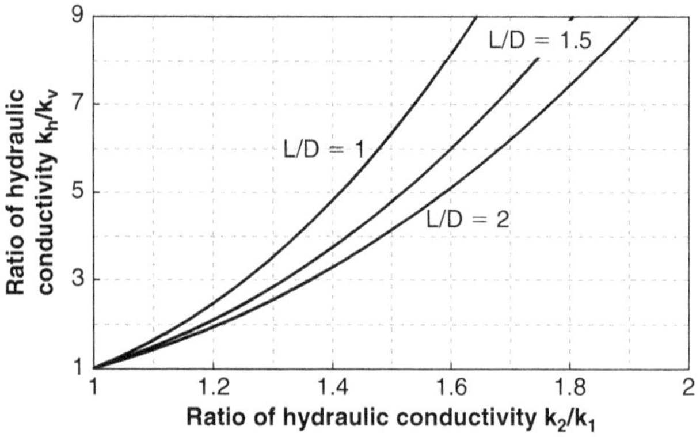

Figure 7.47 Relationship between k_1/k_2 and m for two-stage borehole permeameter. (After Daniel 1989.)

The analysis of both stage 1 and stage 2 presented here makes a number of limiting assumptions that may or may not be verified in the field (Daniel 1989).

7.13 OFFSHORE IN SITU TESTS

The in situ tests most commonly used offshore are the cone penetrometer test and the vane shear test. Other in situ tests used offshore include the pressuremeter test, the dilatometer test, and a number of geophysical tests (see Chapter 8).

The offshore CPT is used for stratigraphy, classification, undrained shear strength in fine-grained soils, and friction angle and relative density in coarse-grained soils. It is performed from the seabed or down a borehole. The seabed systems (Figure 7.48) are lowered to the seabed and provide the vertical reaction against which to push the CPT. A total push of 100 kN can be expected from these units. The rods are prestrung on the seabed unit. The downhole systems (Figure 7.49) consist of lowering the CPT system through the drill string that drilled the borehole, latching the CPT system to the bottom of the drill string, and pushing the CPT into the soil below by using the mud pressure in the drill string.

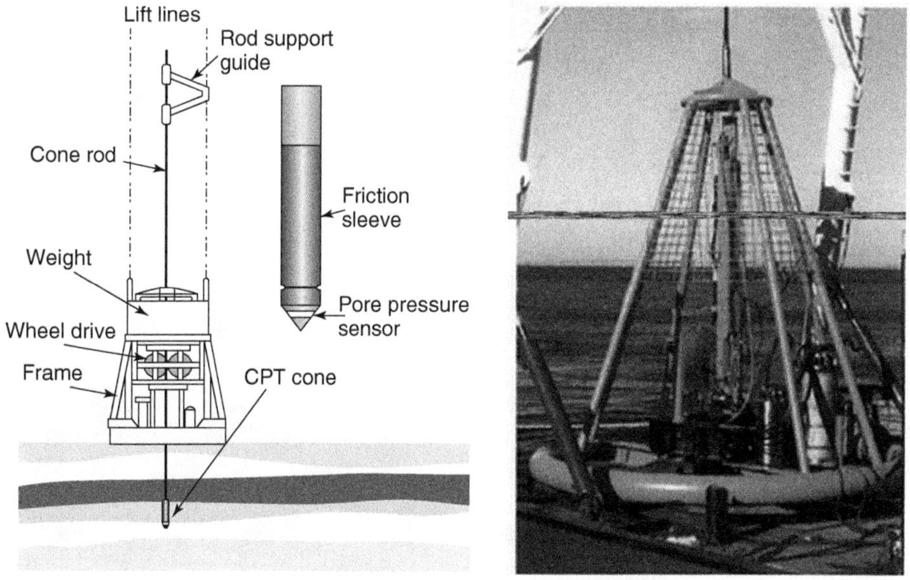

Figure 7.48 Seabed units to deploy the CPT offshore. (a and b: Image courtesy Swan Consultants Ltd., Copyright EFS Danson 2005.)

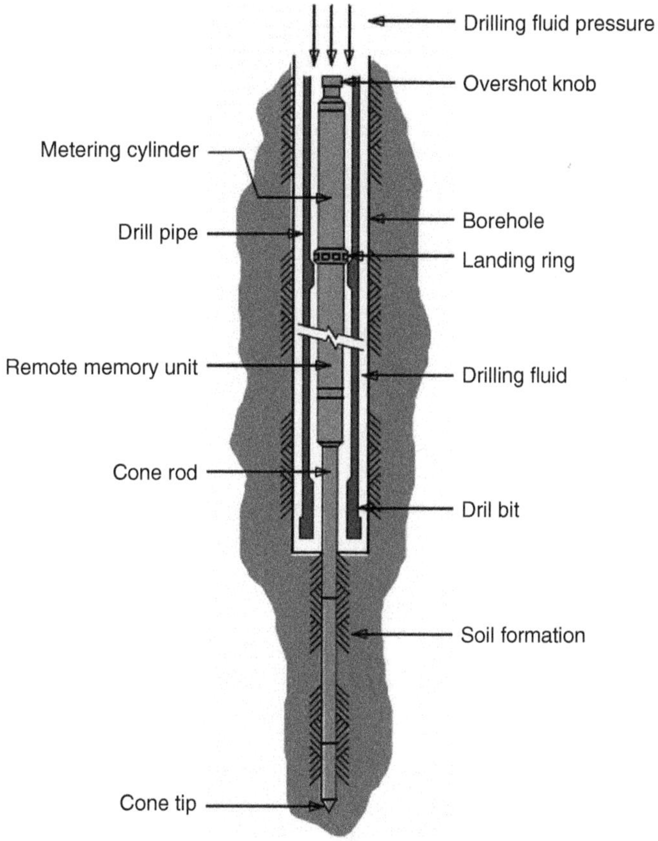

Figure 7.49 "Dolphin" downhole system to deploy the CPT offshore. (Courtesy of FUGRO Inc.)

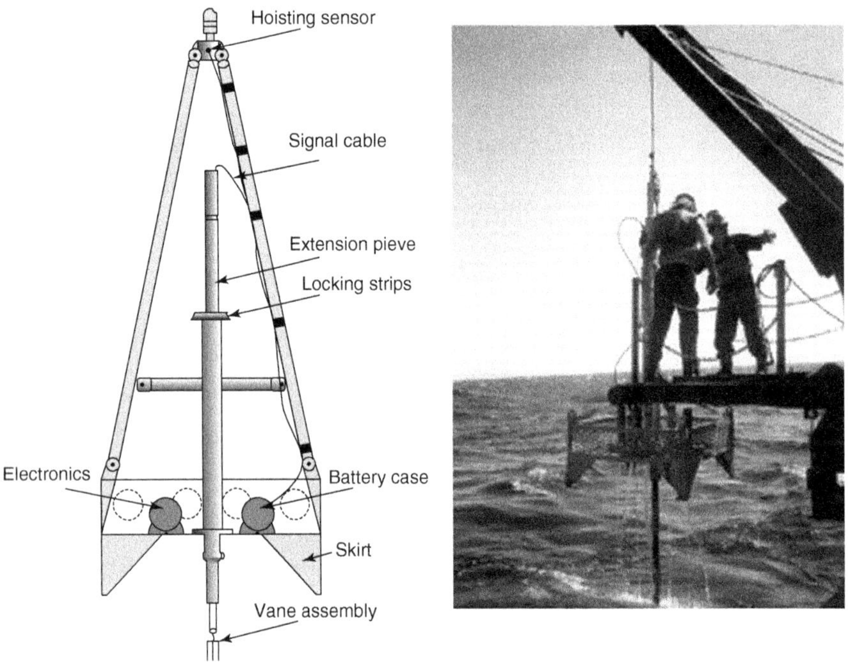

Figure 7.50 Seabed system for vane shear test. (Image courtesy Swan Consultants Ltd., Copyright EFS Danson 2005.)

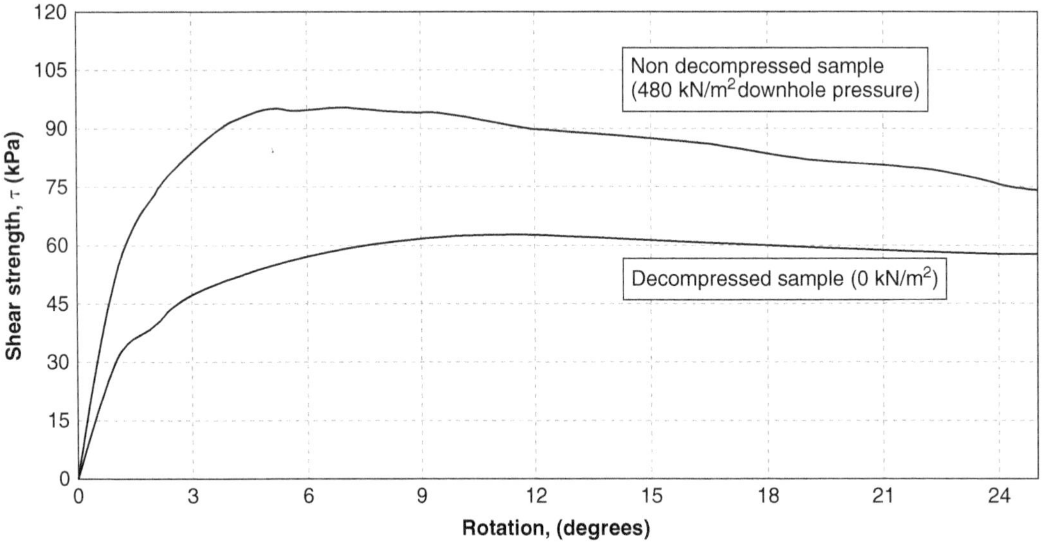

Figure 7.51 Influence of sample disturbance on vane shear results. (After Denk et al. 1981.)

The drill string is typically steadied by clamping the drill string to an external mass resting on the seabed.

The offshore vane shear test is used to measure the undrained shear strength of fine-grained soils. Like the CPT, the VST can be performed from a downhole tool (Figure 7.49) or from a seabed platform (Figure 7.50). Although samples can be taken, obtaining the undrained shear strength from such samples in the laboratory suffers from the decompression of the sample when it is brought back to the surface. In gassy soils, this decompression can be very significant and reduce the undrained shear strength by up to 40% (Figure 7.51; Denk et al. 1981). The VST measures the undrained shear strength in situ and therefore does not allow decompression. As a result, the value obtained is much more reliable.

PROBLEMS

7.1 Assume that the blow count profile shown in Figure 7.4 is an uncorrected blow count profile obtained for a silty sand. Assume further that the energy recorded during these SPT tests was 332 J, that the groundwater level was at the surface, and that the soil has a significant amount of silt. Create the corrected profile for energy level N_{60}, the corrected profile for stress level N_1, and the corrected profile for silt content N'. Then create the combined corrected profile for energy, stress level, and silt content, $N'_{1(60)}$.

7.2 A pressuremeter test gives the test curve shown in Figure 7.2s. Calculate the first load modulus E_0, the reload modulus of the first loop E_{r1}, the yield pressure p_y, the horizontal pressure p_{oh} corresponding to the reestablishment of the horizontal in situ stress, and the limit pressure p_L. What do you think each parameter can be used for?

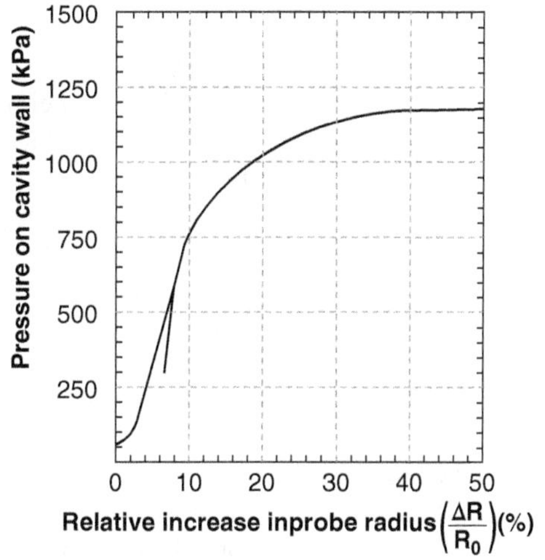

Figure 7.2s Pressuremeter test results.

7.3 Use the CPT profiles of Figure 7.8 to identify the main soil layers. Then classify the soil in each layer according to the CPT classification systems of Figure 7.10 and Figure 7.11.

7.4 Develop the equation for a rectangular vane that links the maximum torque T_{max} to the undrained shear strength s_u of a fine-grained soil.

7.5 Why is the vane test not used in coarse-grained soils? Develop a way, including placing instrumentation on the vane, that would allow the vane test to give the effective stress friction angle of a sand with no effective stress cohesion intercept.

7.6 A borehole shear test is performed in a saturated clay below the water level. The test is performed fast enough to ensure no drainage. When the horizontal pressure is applied, the plates penetrate 4 mm into the soil of the borehole wall. How long should the plates be for the end effect created by the resistance of the wedge at the leading edge of the plates to represent less than 10% of the shear force measured?

7.7 A plate test gives the load settlement curve shown in Figure 7.26. The plate is 0.3 m in diameter and the test is performed at the ground surface. Calculate the soil modulus from the early part of the plate test curve. Would you use this modulus to calculate the settlement of a 3 m by 3 m square footing? Explain.

7.8 Use the elastic settlement equation for a plate test to explain why the modulus of subgrade reaction K is not a soil property while the soil modulus E is. Which one would you rather use and why?

7.9 Calculate the settlement of a footing on sand after 50 years under a pressure of 100 kPa if the settlement after 1 hour under a pressure of 100 kPa during a load test is 10 mm. The soil has a viscous exponent $n = 0.04$.

7.10 Pocket erodometer tests (PETs) are performed on the end of Shelby tube samples retrieved from a levee. The average depth of the PET holes is 6 mm and the standard deviation is 2 mm. Estimate the rate of erosion if the mean velocity overflowing the levee will be 5 m/s. If the levee is subjected to overtopping for 2 hours (hurricane), how much erosion is likely to take place?

7.11 A sand cone apparatus is used to check the dry density of a compacted soil. The weight of dry sand used to fill the test hole and the funnel of the sand cone device is 8.7 N. The weight of dry sand used to fill the cone funnel is 3.2 N. The unit weight of the dry sand is calibrated to be 15.4 kN/m^3. The weight of the wet soil taken out of the test hole is 7.5 N and the water content of the soil from the test hole is 13.2%. Calculate the dry density of the compacted soil.

7.12 A lightweight deflectometer is used to obtain the modulus of the compacted soil. The plate is 200 mm in diameter and the results of the tests are shown in Figure 7.36. Calculate the modulus of deformation of the soil. What approximate stress level and strain level does it correspond to?

7.13 A borehole is drilled into a deep and uniform clay layer to a depth of 1.5 m. A 75 mm inside diameter casing is lowered to the bottom of the 100 mm diameter borehole and sealed to the borehole walls. The water is bailed out so that the water level starts 1 m below the groundwater level outside of the casing at time equal 0. Three days later the water level has risen 0.3 m in the casing. Calculate the hydraulic conductivity k of the clay layer.

7.14 A 10 m thick layer of silty sand is underlain by a deep layer of high-plasticity clay. The groundwater level is 2 m below the ground surface. A 100 mm diameter boring is drilled to a depth of 10 m and cased with a screen that allows the water to enter the borehole freely along the borehole walls. A pump is set up to pump the water out of the hole and reaches a steady state condition after 2 days; at that time it is able to maintain the water level in the hole at a depth of 6 m when the flow rate is 0.2 cubic meters per minute. Additional boreholes indicate that the radius of influence of the depressed water level is 9 m. Calculate the hydraulic conductivity of the silty sand layer.

7.15 A cone penetrometer dissipation test is performed at a depth of 15.2 m below the groundwater level in a silt deposit. The results of the tests are given in Figure 7.43a. Calculate the hydraulic conductivity of the silt layer.

7.16 A sealed double-ring infiltrometer is used to evaluate the field-scale hydraulic conductivity of a 1 m thick clay liner underlain by a free-draining layer of sandy gravel. The SDRI has a square outside ring that is 4 m by 4 m and an inside ring that is 1 m by 1 m. The wall of the outer ring is embedded and sealed 0.45 m below the ground surface and the wall of the inner ring is embedded and sealed 0.15 m below the ground surface. Water is poured into the infiltrometer to a height of 0.5 m above the ground surface and the inner ring is capped. After a period of one week, during which the liner below the infiltrometer becomes saturated and a steady-state flow develops, the daily volume of water flowing into the liner is 0.01 m^3 as measured by a plastic bag connected to the sealed inside ring. The soil swells, and vertical movement measurements of the inside ring indicate that this swelling amounts to 0.004 m^3 per day. Calculate the hydraulic conductivity of the liner.

7.17 A two-stage permeameter test is conducted to evaluate the vertical and horizontal hydraulic conductivity of a clay liner. In stage 1, a 0.1 m diameter borehole is drilled to a depth of 0.35 m. A 0.075 m inside diameter pipe is lowered to the bottom of the open borehole and sealed to the walls of the borehole. A 10 mm inside diameter graduated tube is placed on top of the 75 mm diameter pipe; then the pipe and the falling head permeameter fitted on top of it are saturated and the water seeps through the liner. After reaching a steady state, the following measurements are recorded. At time equal 0, the water is 0.6 m above the ground surface. After 30 minutes of infiltration, the water has dropped to a height of 0.5 m above the ground surface. In stage 2, a 75 mm borehole is advanced 0.2 m below the bottom of the stage 1 borehole (0.55 m below surface). The falling head permeameter test is repeated and the water level falls from 0.6 m above the ground surface at time equal 0 to 0.5 m above the ground surface in 5 minutes. Calculate the vertical and horizontal hydraulic conductivity of the clay liner.

7.18 Discuss the advantages and drawbacks of in situ tests versus laboratory tests.

Problems and Solutions

Problem 7.1

Assume that the blow count profile shown in Figure 7.4 is an uncorrected blow count profile obtained for a silty sand. Assume further that the energy recorded during these SPT tests was 332 J, that the groundwater level was at the surface, and that the soil has a significant amount of silt. Create the corrected profile for energy level N_{60}, the corrected profile for stress level N_1, and the corrected profile for silt content N'. Then create the combined corrected profile for energy, stress level, and silt content, $N'_{1(60)}$.

Solution 7.1

The corrections of the SPT values are shown in Table 7.1s and are based on the following formulas:

$$\text{Correction for energy level}: N_{60} = N_{\text{measured}} \times \left(\frac{E_{\text{measured}}\ (\text{J})}{285\ (\text{J})}\right)$$

$$\text{Correction for stress level}: N_1 = N_{\text{measured}} \times \left(\frac{100}{\sigma'_{v0}\ (\text{kPa})}\right)^{0.5}$$

$$\text{Correction for silt content}: N' = 15 + \left(\frac{N_{\text{measured}} - 15}{2}\right)$$

$$\text{Combined corrections}: N'_{1(60)} = 15 + \left(\frac{N_{60} \times \left(\frac{100}{\sigma'_{0v}}\right)^{0.5} - 15}{2}\right)$$

Table 7.1s Corrected SPT Values

Depth	Measured	Energy level		Stress level			Silt	Combination
	N_{measured}	E_{measured}	N_{60}	γ_{sat}	σ'_{ov}	N_1	N'	$N'_{1(60)}$
m	bpf	J	bpf	kN/m^3	kPa	bpf	bpf	bpf
1.5	15	332	17	19	14	40	15	31
3	20	332	23	19	28	38	18	30
4.5	17	332	20	19	41	26	16	23
6	12	332	14	19	55	16	14	17
7.5	18	332	21	19	69	22	17	20
9	21	332	24	19	83	23	18	21
10.5	24	332	28	19	96	24	20	22
12	28	332	33	19	110	27	22	23
13.5	31	332	36	19	124	28	23	24
15	30	332	35	19	138	26	23	22
16.5	32	332	37	19	152	26	24	23
18	29	332	34	19	165	23	22	21
19.5	31	332	36	19	179	23	23	21

The corrections of the SPT values are plotted on the graph shown in Figure 7.1s.

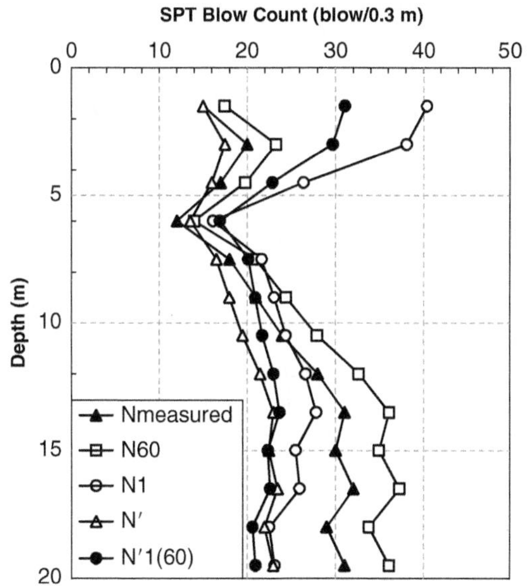

Figure 7.1s Corrected SPT values.

Problem 7.2

A pressuremeter test gives the test curve shown in Figure 7.2s. Calculate the first load modulus E_0, the reload modulus of the first loop E_{r1}, the yield pressure p_y, the horizontal pressure p_{oh} corresponding to the reestablishment of the horizontal in situ stress, and the limit pressure p_L. What do you think each parameter can be used for?

Solution 7.2

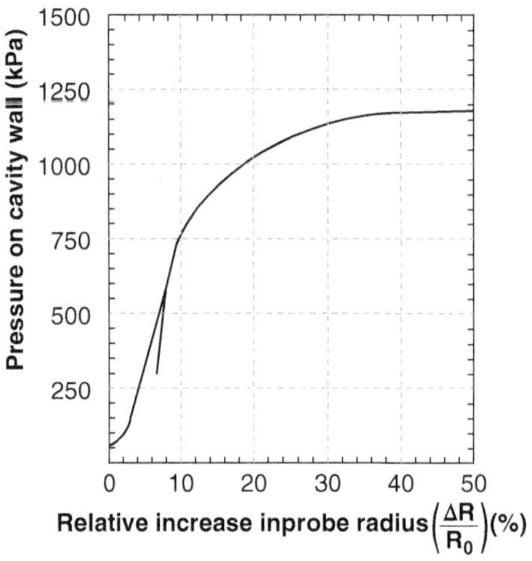

Figure 7.2s Pressuremeter test results.

According to the test results shown in Figure 7.2s, the following parameters are obtained:

- First load modulus $E_0 = (1 + 0.35)\dfrac{1500}{(0.18 - 0.017)} = 12423$ kPa
- The reload modulus of the first loop$E_{r1} = (1 + 0.35)\dfrac{1500}{(0.12 - 0.05)} = 28928$ kPa
- The yield pressure $p_y = 700$ kPa
- The horizontal pressure $p_{0h} = 120$ kPa
- The limit pressure $p_L = 1200$ kPa

The applications of the PMT include the design of deep foundations under horizontal loads, the design of shallow foundations, the design of deep foundations under vertical loads, and the determination of a modulus profile and other soil properties. The PMT is not very useful for slope stability and retaining structures.

The first load and reload modulus can be used in settlement analysis. The yield pressure can be used as an upper limit for the allowable foundation pressures. The limit pressure can be used to calculate the ultimate capacity of the foundation.

Problem 7.3

Use the CPT profiles of Figure 7.8 to identify the main soil layers. Then classify the soil in each layer according to the CPT classification systems of Figure 7.10 and Figure 7.11.

Solution 7.3

A total of 10 layers are identifiable from the CPT profiles of Figure 7.8 and are shown in Figure 7.3s and Table 7.2s. Furthermore, the porewater pressure profile can be extended back to zero pressure and indicates that the water level is at a depth of 2.5 m below the ground surface. The classifications of the soil layers based on Figures 7.10 and 7.11 are presented in Table 7.2s, Figure 7.4s, and Figure 7.5s. At a coarser level, the stratigraphy can be simplified as shown in Figure 7.6s.

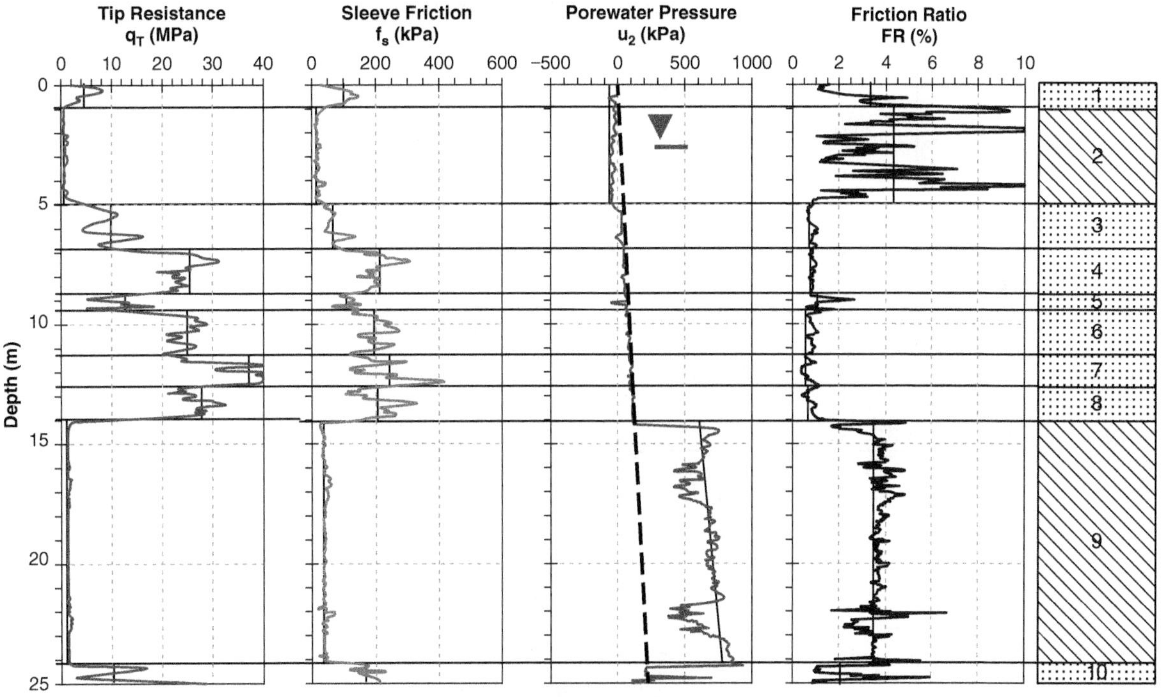

Figure 7.3s Soil layers. (Courtesy of Professor Paul Mayne, Georgia Institute of Technology)

Table 7.2s Classification of Soil Layers

Layer	Depth (m)	q_t (Mpa)	q_t (Bar)	f_s (kPa)	FR (%)	Figure 7.10	Figure 7.11
1	0.0–1.0	4.0	40	100	3.40	Sandy silts & silt	Silty sand
2	1.0–5.0	0.8	8	10	4.30	Clays	Clay
3	5.0–7.0	10.0	100	60	0.70	Sands	Sand to silty sand
4	7.0–8.8	26.0	260	210	0.70	Sands	Sand
5	8.8–9.5	13.0	130	100	1.10	Sands	Sand to silty sand
6	9.5–11.3	26.0	260	200	0.60	Sands	Gravelly sand
7	11.3–12.7	37.0	370	250	0.60	Sands	Gravelly sand
8	12.7–14	28.0	280	200	0.60	Sands	Gravelly sand
9	14–24.2	0.9	9	40	2.70	Clayey silts	Silty clay
10	24.2–25	10.0	100	160	2.00	Silty sands	Silty sand

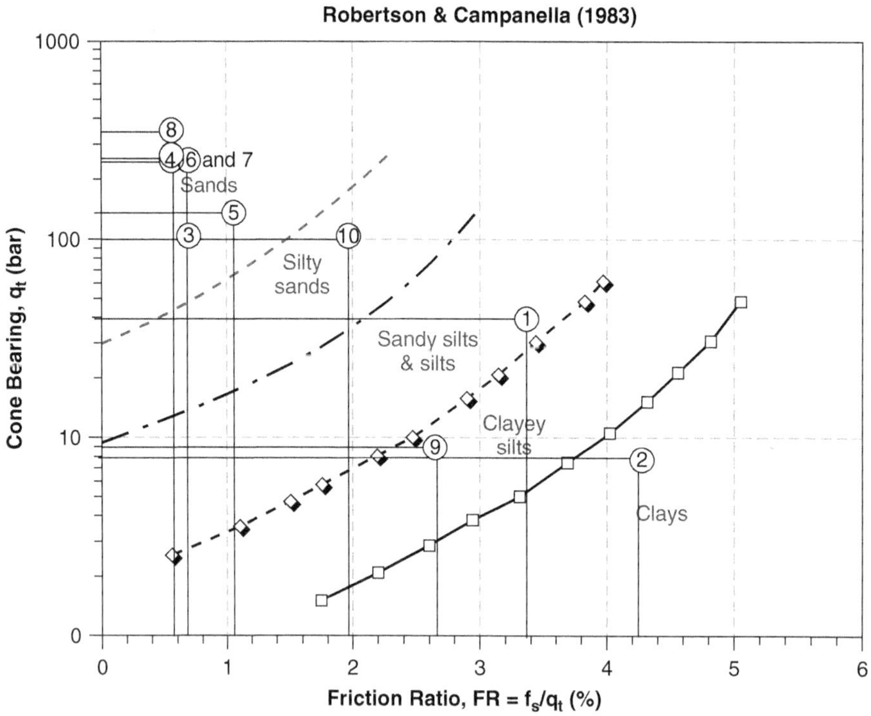

Figure 7.4s Soil classification based on CPT results. (Courtesy of Professor Paul Mayne, Georgia Institute of Technology)

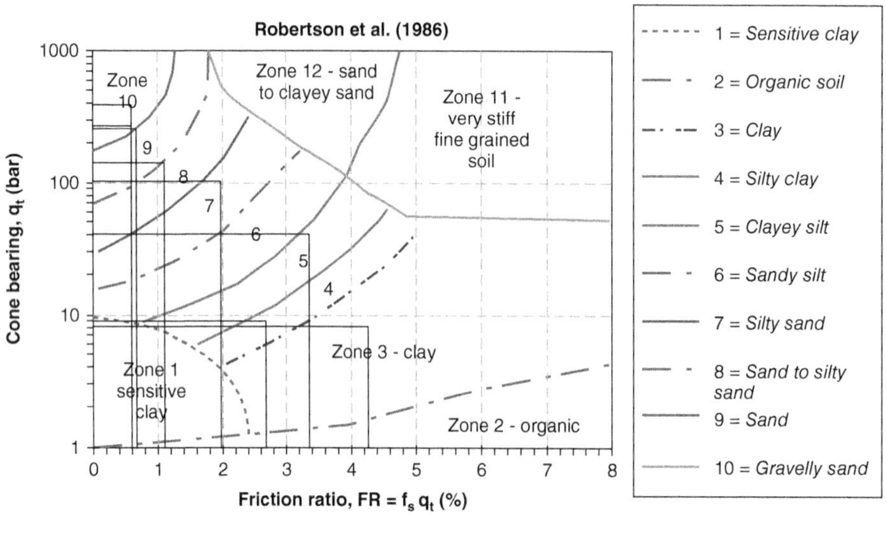

Figure 7.5s Soil classification based on CPT results. (Courtesy of Professor Paul Mayne, Georgia Institute of Technology)

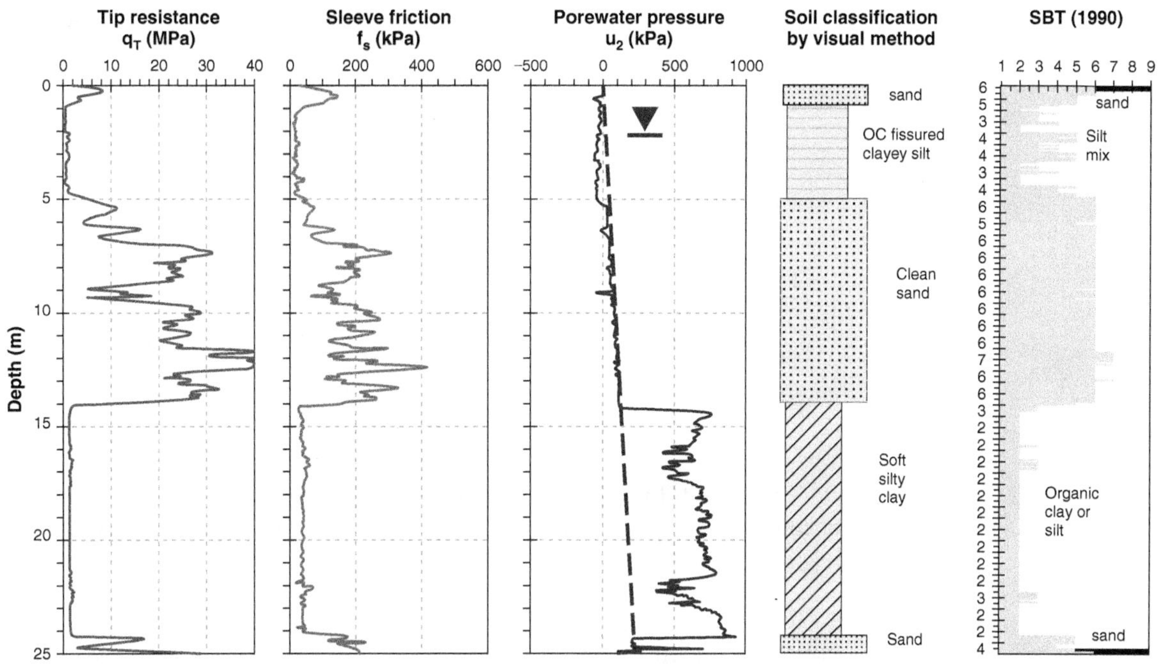

Figure 7.6s Simplified stratigraphy. (From Mayne 2007a, b, Courtesy of Professor Paul Mayne, Georgia Institute of Technology, USA.)

Problem 7.4

Develop the equation for a rectangular vane that links the maximum torque T_{max} to the undrained shear strength s_u of a fine-grained soil.

Solution 7.4

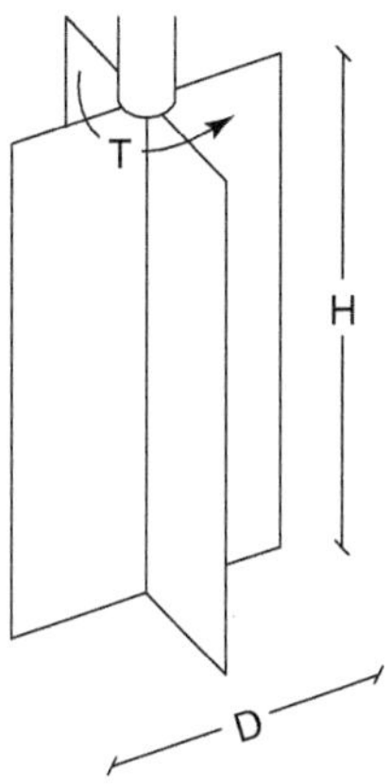

Figure 7.7s Vane subjected to torque.

The failure surface around the vane is a cylinder with a diameter D and a height H. The torque generated from the sides of the cylinder is:

$$T_1 = \pi D H s_u \frac{D}{2}$$

The torque generated by the top and bottom of the cylinder (ignoring the area occupied by the rod) is:

$$T_2 = \int_0^{\frac{D}{2}} 2.\pi.r.s_u.r.dr = 2.\pi.s_u\left(\frac{r^3}{3}\right)_0^{\frac{D}{2}} = \pi.s_u\frac{D^3}{12}$$

$$T = T_1 + 2T_2 = \pi DHs_u\frac{D}{2} + 2\pi.s_u.\frac{D^3}{12}$$

$$T = \pi s_u.D^2\left(\frac{H}{2} + \frac{D}{6}\right)$$

For vanes with H = 2D, the equation becomes:

$$T_1 = \frac{7}{6}\pi s_u D^3$$

Problem 7.5

Why is the vane test not used in coarse-grained soils? Develop a way, including placing instrumentation on the vane, that would allow the vane test to give the effective stress friction angle of a sand with no effective stress cohesion intercept.

Solution 7.5

The vane test gives one measurement: the torque at failure. It can easily be used to obtain the undrained shear strength of a fine-grained soil because in this case the strength is represented by one parameter, s_u. The vane test cannot be used easily to obtain the drained or effective stress parameters (c and ϕ) because we need three equations to solve for the three parameters involved: σ', c, and ϕ. The shear strength equation is:

$$\tau_f = c + \sigma' \tan\varphi$$

If c = 0, the shear strength equation becomes:

$$\tau_f = \sigma' \tan\varphi$$

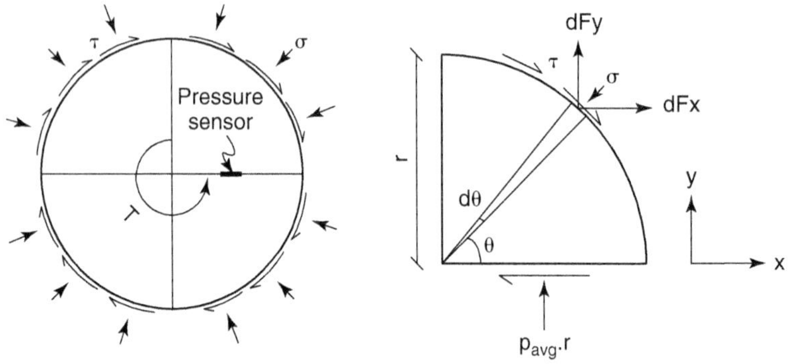

Figure 7.8s Applied stresses on vane.

To get ϕ from the vane test in this case, it is necessary to make two separate measurements. This can be accomplished by placing a pressure sensor on one of the blades, as shown in Figure 7.8s. A free-body diagram of a quadrant of the failing soil mass gives the following equations:

$$\begin{cases} dF_y = \sigma.r.d\theta.\sin\theta + \tau.r.d\theta.\cos\theta \\ dF_x = \sigma.r.d\theta.\cos\theta + \tau.r.d\theta.\sin\theta \end{cases}$$

Based on these equilibrium equations:

$$p.r = F_y = \int_0^{\frac{\pi}{2}} (\sigma.r.\sin\theta + \tau.r.\cos\theta)d\theta = -\sigma.r.\cos\theta + \tau.r.\sin\theta|_0^{\frac{\pi}{2}} = (\tau + \sigma)r$$

$$p = \tau + \sigma$$

At failure:

$$\tau_f = \sigma' \tan\varphi$$

$$\sigma' = p - \tau_f \rightarrow \tau_f = (p - \tau_f)\tan\varphi \rightarrow \tau_f = \frac{\tan\varphi}{1+\tan\varphi}p$$

From problem 7.4, we have:

$$T = \pi D H \tau_{side}\frac{D}{2} + \pi\tau_{top}\frac{D^3}{12} + \pi\tau_{bottom}\frac{D^3}{12}$$

$$T = \pi D H \frac{\tan\varphi}{1+\tan\varphi}p\frac{D}{2} + \pi\gamma' z\tan\varphi\frac{D^3}{12} + \pi\gamma'(z+H)\tan\varphi\frac{D^3}{12}$$

$$T = \pi\frac{D^2}{2}H\frac{\tan\varphi}{1+\tan\varphi}p + (2z+H)\gamma'\pi\tan\varphi\frac{D^3}{12}$$

$$(2z+H)\pi\gamma'\frac{D^3}{12}\tan^2\varphi + \left((2z+H)\pi\gamma'\frac{D^3}{12} + \frac{1}{2}\pi p D^2 H - T\right)\tan\varphi - T = 0$$

$$\tan\phi = \frac{-B+\sqrt{B^2+4AT}}{2A}$$

$$A = (2z+H)\pi\gamma'\frac{D^3}{12}$$

$$B = (2z+H)\pi\gamma'\frac{D^3}{12} + \frac{1}{2}\pi p D^2 H - T$$

T: torque applied to the vane
D: diameter of the vane
H: height of the vane
ϕ: internal friction angle of sand
p: pressure on the blade of the vane (which is measured by a sensor)
Y: unit weight of soil
z: depth of top of the vane

Problem 7.6

A borehole shear test is performed in a saturated clay below the water level. The test is performed fast enough to ensure no drainage. When the horizontal pressure is applied, the plates penetrate 4 mm into the soil of the borehole wall. How long should the plates be for the end effect created by the resistance of the wedge at the leading edge of the plates to represent less than 10% of the shear force measured?

Solution 7.6

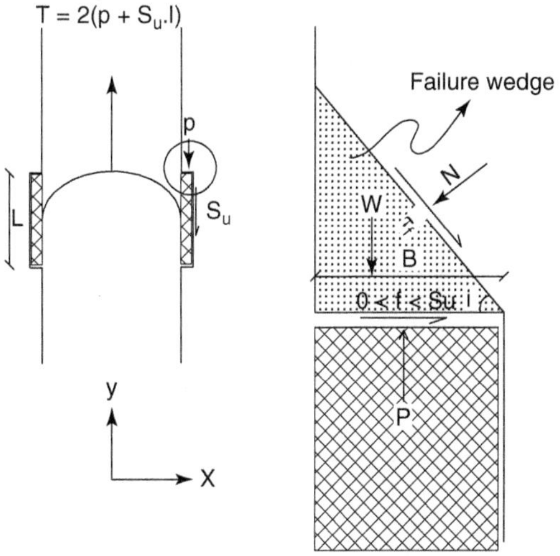

Figure 7.9s Borehole shear test.

$$\begin{cases} \sum F_x = 0 \to N \sin i = fB + \dfrac{\tau_f . B}{\cos i} \cos i \\ \sum F_y = 0 \to p = N \cos i + W + \dfrac{\tau_f . B}{\cos i} \sin i \end{cases}$$

$$N = \frac{f.B}{\sin i} + \frac{\tau_f B}{\sin i}$$

$$W = \frac{1}{2} \gamma B^2 \tan i$$

$$p = \left(\frac{f.B}{\sin i} + \frac{\tau_f B}{\sin i} \right) \cos i + \frac{1}{2} \gamma B^2 \tan i + \frac{\tau_f B}{\cos i} \sin i$$

$$p = \frac{f.B}{\tan i} + \frac{\tau_f B}{\tan i} + \frac{1}{2} \gamma B^2 \tan i + \tau_f B \tan i$$

Because B, the penetration of the blades into the soil, is typically very small (say, less than 10 mm), and because the weight of wedge W is a function of B^2, it is reasonable to neglect the influence of the weight of the wedge in calculating P:

$$p \sim \frac{f.B}{\tan i} + \frac{\tau_f B}{\tan i} + \tau_f B \tan i$$

By assuming $i = 45^\circ + \varphi/2$ and using Mohr-Coulomb theory, we have:

$$p \sim \frac{f.B}{\tan\left(45 + \frac{\phi}{2}\right)} + \frac{s_u B \cos \phi}{\tan\left(45 + \frac{\phi}{2}\right)} + s_u B \cos \phi \tan\left(45 + \frac{\phi}{2}\right)$$

$$p \sim \frac{f.B}{\tan\left(45 + \frac{\phi}{2}\right)} + 2 s_u B$$

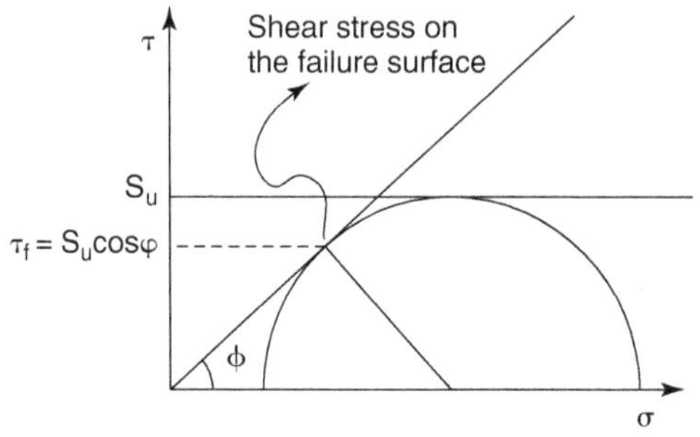

Figure 7.10s Stress envelope.

If $\Phi = 30^\circ$ for upper and lower limits of f, we will have:

$$\begin{cases} f = 0 \to p = 2 s_u B \\ f = s_u \to p = \left(2 + \dfrac{\sqrt{3}}{3}\right) s_u . B \end{cases}$$

$$2 s_u B < p < \left(2 + \frac{\sqrt{3}}{3}\right) s_u . B$$

P is the force needed to fail the wedge of soil above the borehole shear device. If this force must be less than 10% of the force measured by the borehole shear device, then:

$$T_{\text{measured}} = 2(p + s_u . l) \to \frac{2p}{T} < 10\% \to p < 0.1(p + s_u . l) \to l > \frac{0.9p}{0.1 s_u}$$

This assumes that the borehole shear device is associated with a plane strain failure, which is a simplifying assumption. In this case, the requirements on the length of the BSD to ensure that the end effect is less than 10% of the measured value are:

$$f = 0 \rightarrow l > 18B$$

$$f = s_u \rightarrow l > 23.2B$$

In the worst condition, which is ($f = s_u$), the length of plates must be longer than 23.2B. If $B = 4$ mm, for example, then $l > 92.8$ mm.

Problem 7.7

A plate test gives the load settlement curve shown in Figure 7.26. The plate is 0.3 m in diameter and the test is performed at the ground surface. Calculate the soil modulus from the early part of the plate test curve. Would you use this modulus to calculate the settlement of a 3 m by 3 m square footing? Explain.

Solution 7.7

The pressure versus displacement/width curve is shown in Figure 7.11s.

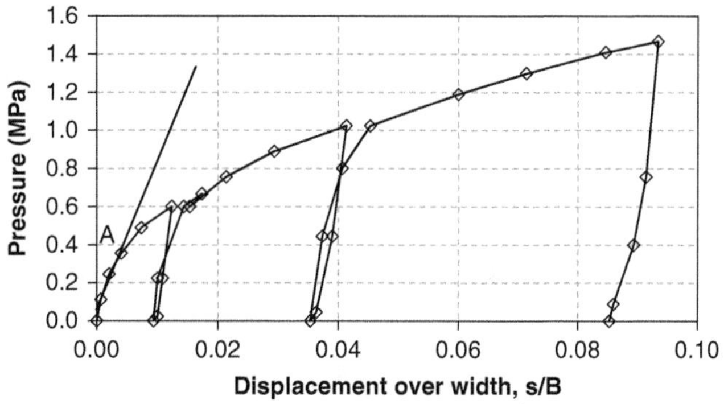

Figure 7.11s Pressure versus displacement/width curve.

The soil modulus is calculated based on point A in Figure 7.11s using the following equation:

$$E = \frac{\pi(1-\nu^2)pB}{4s} = \frac{\pi(1-\nu^2)p}{4 \times \frac{s}{B}} = \frac{\pi(1-0.35^2) \times 0.36}{4 \times 0.004} = 62\,\text{MPa}$$

The soil modulus obtained in this fashion from the plate test is 62 MPa.

I would not use this soil modulus to calculate the settlement of a 3 m by 3 m footing without checking the soil stratigraphy first. The plate bearing test can only give the response of the soil down to a depth of about twice the plate diameter, which is 0.6 m in this case. It cannot reflect the soil property beneath the 3 m by 3 m square footing unless they are the same.

Problem 7.8

Use the elastic settlement equation for a plate test to explain why the modulus of subgrade reaction K is not a soil property while the soil modulus E is. Which one would you rather use and why?

Solution 7.8

The elastic settlement equation for a plate load test is:

$$s = \frac{I(1-\nu^2)pB}{E}$$

Here, I is the shape factor, E is the soil modulus, p is the average pressure under the footing, B is the plate diameter, and ν is the Poisson's ratio. The modulus of subgrade reaction K is calculated as the ratio between the pressure and the settlement:

$$K = \frac{p}{s} = \frac{p}{\frac{I(1-\nu^2)pB}{E}} = \frac{E}{I(1-\nu^2)B}$$

Therefore, the modulus of subgrade reaction K is a function of the soil modulus E and the foundation size B. The larger the foundation is, the smaller the modulus of subgrade reaction is.

I would prefer to use the soil modulus E because it is a true soil property, whereas K is not. Indeed, as shown here, K depends on E and B. Any K value determined from a given size foundation test cannot be used directly for a different size without paying attention to the scale effect.

Problem 7.9

Calculate the settlement of a footing on sand after 50 years under a pressure of 100 kPa if the settlement after 1 hour under a pressure of 100 kPa during a load test is 10 mm. The soil has a viscous exponent $n = 0.04$.

Solution 7.9

Based on equation 7.14, the settlement s_2 of a footing after $t_2 = 50$ years under a pressure of 100 kPa based on the settlement s_1 of the same footing after $t_1 = 1$ hour is:

$$\frac{s_1}{s_2} = \left(\frac{t_1}{t_2}\right)^n$$

With $s_1 = 10$ mm, $t_1 = 1$ hr, $t_2 = 50$ years $= 50 \times 365 \times 24 = 438{,}000$ hr, and $n = 0.04$:

$$s_2 = \frac{s_1}{\left(\frac{t_1}{t_2}\right)^n} = \frac{10}{\left(\frac{1}{438000}\right)^{0.04}} = 16.8 \text{ mm}$$

So, the calculated settlement of the footing after 50 years under a pressure of 100 kPa is 16.8 mm.

Problem 7.10

Pocket erodometer tests (PETs) are performed on the end of Shelby tube samples retrieved from a levee. The average depth of the PET holes is 6 mm and the standard deviation is 2 mm. Estimate the rate of erosion if the mean velocity overflowing the levee will be 5 m/s. If the levee is subjected to overtopping for 2 hours (hurricane), how much erosion is likely to take place?

Solution 7.10

Using Figure 7.30 and a PET hole depth of 6 mm, the soil category is category III or medium erodibility. For this category, the PET hole varies between 1 mm and 15 mm, corresponding to erosion rates of 3 mm/hr and 2000 mm/hr respectively. For 6 mm, the erosion rate is estimated to be near the middle of the range on the logarithmic scale and an erosion rate of 80 mm/hr is selected (Figure 7.12s). With 2 hours of overtopping at this rate, 160 mm of erosion is estimated.

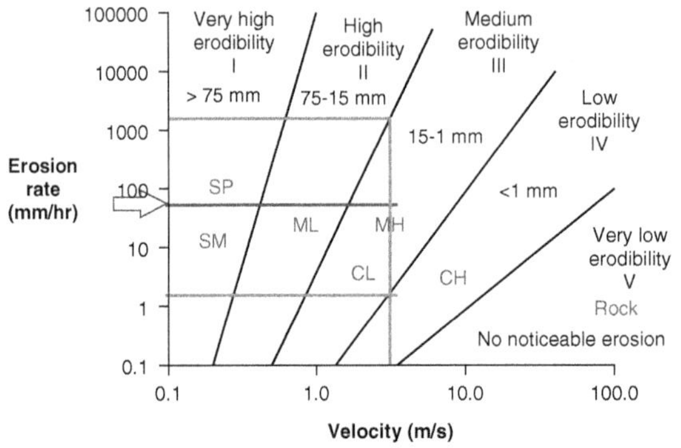

Figure 7.12s Erosion chart for various erosion depths from the PET.

Problem 7.11

A sand cone apparatus is used to check the dry density of a compacted soil. The weight of dry sand used to fill the test hole and the funnel of the sand cone device is 8.7 N. The weight of dry sand used to fill the cone funnel is 3.2 N. The unit weight of the dry sand is calibrated to be 15.4 kN/m^3. The weight of the wet soil taken out of the test hole is 7.5 N and the water content of the soil from the test hole is 13.2%. Calculate the dry density of the compacted soil.

Solution 7.11

The weight of dry sand used to fill the test hole is 8.7 N − 3.2 N = 5.5 N. The volume of the test hole is therefore 5.5×10^{-3} kN/15.4 kN/m^3 = 3.57×10^{-4} m^3. Therefore, the wet unit weight of the compacted soil is 7.5×10^{-3} kN/3.57×10^{-4} m^3 = 21 kN/m^3. Finally, the dry unit weight of the compacted soil is $21/(1 + 0.132) = 18.56$ kN/m^3.

Problem 7.12

A lightweight deflectometer is used to obtain the modulus of the compacted soil. The plate is 200 mm in diameter and the results of the tests are shown in Figure 7.36. Calculate the modulus of deformation of the soil. What approximate stress level and strain level does it correspond to?

Solution 7.12

The modulus of deformation of the soil is:

$$E = 1(1 - 0.35^2)\frac{4 \times 7.5}{\pi \times 0.2 \times 0.55 \times 10^{-3}} = 76.3\,\text{MPa}$$

This modulus of deformation corresponds to the stress level P:

$$P = \frac{4 \times 7.5}{\pi \times 0.2^2} = 238\,\text{kPa}$$

Problem 7.13

A borehole is drilled into a deep and uniform clay layer to a depth of 1.5 m. A 75 mm inside diameter casing is lowered to the bottom of the 100 mm diameter borehole and sealed to the borehole walls. The water is bailed out so that the water level starts 1 m below the groundwater level outside of the casing at time equal 0. Three days later the water level has risen 0.3 m in the casing. Calculate the hydraulic conductivity k of the clay layer.

Solution 7.13

In this case, equation 7.25 applies because the soil layer is deep and uniform, because the casing goes down to the bottom of the borehole, and because the water is bailed out to a depth far below the groundwater level outside of the casing (Figure 7.38). Therefore, the hydraulic conductivity k is obtained from:

$$k_{hyd} = \frac{2\pi r}{11(t_2 - t_1)} \ln \frac{h_1}{h_2}$$

where r is the radius of the casing (0.075 m), t_1 is 0, t_2 is 3 days, h_1 is the depth below the groundwater level at time t_1 (1 m), and h_2 is the depth below the groundwater level at time t_2 (0.7 m). Therefore, the solution is:

$$k_{hyd} = \frac{2\pi \times 0.075}{11(3 - 0)} \ln \frac{1}{0.7} = 5.08 \times 10^{-3}\ \text{m/day} = 5.87 \times 10^{-5}\ \text{mm/sec}$$

Problem 7.14

A 10 m thick layer of silty sand is underlain by a deep layer of high-plasticity clay. The groundwater level is 2 m below the ground surface. A 100 mm diameter boring is drilled to a depth of 10 m and cased with a screen that allows the water to enter the borehole freely along the borehole walls. A pump is set up to pump the water out of the hole and reaches a steady state condition after 2 days; at that time it is able to maintain the water level in the hole at a depth of 6 m when the flow rate is 0.2 cubic meters per minute. Additional boreholes indicate that the radius of influence of the depressed water level is 9 m. Calculate the hydraulic conductivity of the silty sand layer.

Solution 7.14

In this case, equation 7.26 applies because the pervious soil layer to be tested is underlain by an impervious layer, because the uncased boring (or screened boring) is penetrating through the entire pervious layer all the way to the top of the impervious layer, and because the water level is maintained constant by pumping at a flow rate Q, as shown in Figure 7.39. Therefore, the hydraulic conductivity k is obtained from:

$$k = \frac{Q\ Ln\frac{R}{r}}{\pi(H^2 - h^2)}$$

where Q is the flow rate pumped out of the well to maintain the water level constant in the well ($0.2\,m^3/\min = 288\,m^3/\text{day}$), r is the radius of the borehole (0.1 m), R is the radius of the zone of influence where the water table is depressed (9 m), H is the vertical distance between the bottom of the boring (impervious layer) and the groundwater level at or further than R (8 m), and h is the vertical distance between the bottom of the boring and the water level in the borehole (4 m). Therefore, the solution is:

$$k = \frac{288 \times \ln\frac{9}{0.1}}{\pi(8^2 - 4^2)} = 8.59\ ^{\text{m}}/_{\text{day}} = 9.94 \times 10^{-2}\ \text{mm/sec}$$

Problem 7.15

A cone penetrometer dissipation test is performed at a depth of 15.2 m below the groundwater level in a silt deposit. The results of the tests are given in Figure 7.43a. Calculate the hydraulic conductivity of the silt layer.

Solution 7.15

We can calculate the hydraulic conductivity of the silt layer using equation 7.30:

$$k(\text{cm/s}) = \left(\frac{1}{251 \times t_{50}\,(s)}\right)^{1.25}$$

with $t_{50} = 450$ sec, so $k = \left(\frac{1}{251\times450}\right)^{1.25} = 4.83 \times 10^{-7}\ \text{cm/sec} = 4.17 \times 10^{-4}\ ^{\text{m}}/_{\text{day}} = 4.83 \times 10^{-6}\ \text{mm/sec}$

Problem 7.16

A sealed double-ring infiltrometer is used to evaluate the field-scale hydraulic conductivity of a 1 m thick clay liner underlain by a free-draining layer of sandy gravel. The SDRI has a square outside ring that is 4 m by 4 m and an inside ring that is 1 m by 1 m. The wall of the outer ring is embedded and sealed 0.45 m below the ground surface and the wall of the inner ring is embedded and sealed 0.15 m below the ground surface. Water is poured into the infiltrometer to a height of 0.5 m above the ground surface and the inner ring is capped. After a period of one week, during which the liner below the infiltrometer becomes saturated and a steady-state flow develops, the daily volume of water flowing into the liner is 0.01 m^3 as measured by a plastic bag connected to the sealed inside ring. The soil swells, and vertical movement measurements of the inside ring indicate that this swelling amounts to 0.004 m^3 per day. Calculate the hydraulic conductivity of the liner.

Solution 7.16

The hydraulic conductivity of the clay layer for this test can be obtained by using equation 7.34:

$$k = \frac{v}{i} = \frac{\dfrac{V_f}{A \times t}}{\dfrac{\Delta h_t}{\Delta z}}$$

$$V_f = V_t - V_s = 0.01 - 0.004 = 6 \times 10^{-3}\ m^3$$

$$\Delta h_t = 1.5\ m$$

$$t = 1\ \text{day}$$

$$\Delta z = 1 \text{ m}$$

$$A = 1 \text{ m}^2$$

$$k = \frac{\dfrac{6 \times 10^{-3}}{1 \times 1}}{\dfrac{1.5}{1}} = 4 \times 10^{-3}\ \text{m/day} = 4.62 \times 10^{-5}\ \text{mm/sec}$$

Problem 7.17

A two-stage permeameter test is conducted to evaluate the vertical and horizontal hydraulic conductivity of a clay liner. In stage 1, a 0.1 m diameter borehole is drilled to a depth of 0.35 m. A 0.075 m inside diameter pipe is lowered to the bottom of the open borehole and sealed to the walls of the borehole. A 10 mm inside diameter graduated tube is placed on top of the 75 mm diameter pipe; then the pipe and the falling head permeameter fitted on top of it are saturated and the water seeps through the liner.

After reaching a steady state, the following measurements are recorded. At time equal 0, the water is 0.6 m above the ground surface. After 30 minutes of infiltration, the water has dropped to a height of 0.5 m above the ground surface. In stage 2, a 75 mm borehole is advanced 0.2 m below the bottom of the stage 1 borehole (0.55 m below surface). The falling head permeameter test is repeated and the water level falls from 0.6 m above the ground surface at time equal 0 to 0.5 m above the ground surface in 5 minutes. Calculate the vertical and horizontal hydraulic conductivity of the clay liner.

Solution 7.17

In the first stage, k_1 can be calculated using the following equation:

$$k_1 = \frac{\pi d^2}{11D(t_2 - t_1)} Ln\frac{h_1}{h_2} = \frac{\pi \times 0.01^2}{11 \times 0.075(30 - 0)} Ln\frac{0.6}{0.5} = 2.31 \times 10^{-6}\ \text{m/min} = 3.33 \times 10^{-3}\ \text{m/day}$$

$$k_1 = 3.85 \times 10^{-5}\ \text{mm/sec}$$

In the second stage, k_2 can be calculated using the following equation:

$$k_2 = \frac{A}{B} Ln\frac{h_1}{h_2}$$

with

$$A = d^2\ Ln\left(\frac{L}{D} + \sqrt{1 + \left(\frac{L}{D}\right)^2}\right) = 0.01^2\ Ln\left(\frac{0.2}{0.075} + \sqrt{1 + \left(\frac{0.2}{0.075}\right)^2}\right) = 1.70 \times 10^{-4}\ \text{m}^2$$

and

$$B = 8L(t_2 - t_1)\left(1 - 0.562e^{-1.57\frac{L}{D}}\right) = 8 \times 0.2 \times (5 - 0) \times \left(1 - 0.562e^{-1.57\frac{0.2}{0.075}}\right) = 7.93\ \text{m. min}$$

So

$$k_2 = \frac{A}{B} Ln\frac{h_1}{h_2} = \frac{1.70 \times 10^{-4}}{7.93} Ln\frac{0.6}{0.5} = 3.90 \times 10^{-6}\ \text{m/min} = 5.61 \times 10^{-3}\ \text{m/day} = 6.49 \times 10^{-5}\ \text{mm/sec}$$

$$\frac{k_2}{k_1} = 1.70$$

Based on Figure 7.47, $m = \sqrt{\frac{k_h}{k_v}} = \sqrt{4.84} = 2.2$

$$k_h = m \times k_1 = 2.2 \times 3.33 \times 10^{-3} = 7.32 \times 10^{-3}\ \text{m/day} = 8.47 \times 10^{-5}\ \text{mm/sec}$$

$$k_v = \frac{k_1}{m} = \frac{3.33 \times 10^{-3}}{2.2} = 1.51 \times 10^{-3}\ \text{m/day} = 1.74 \times 10^{-5}\ \text{mm/sec}$$

Problem 7.18

Discuss the advantages and drawbacks of in situ tests versus laboratory tests.

Solution 7.18

The advantages and drawbacks of in situ tests versus laboratory tests are summarized in Table 7.3s.

Table 7.3s Advantages and Drawbacks of In Situ and Laboratory Tests

Laboratory Testing Advantages	Drawbacks	In Situ Testing Advantages	Drawbacks
Easier to analyze theoretically			Difficult to analyze theoretically
	Small-scale testing	Larger-scale testing	
Drainage can be controlled			Drainage difficult to control
	Time consuming	Relatively fast to perform	
Elementary parameters easier to obtain			Elementary parameters harder to obtain
	In situ stresses must be simulated	Testing under in situ stresses	
Soil identification possible			Soil identification rarely possible
	Some disturbance	Less disturbance for some tests	

S waves propagate in solids more slowly than do P waves ($v_s \sim 0.6\, v_p$), so they arrive at the detector after the P waves:

$$v_s = \sqrt{\frac{G}{\rho}} \qquad (8.7)$$

where G is the shear modulus (see Chapter 12) and ρ is the mass density. For a homogeneous, isotropic, and elastic soil, the P wave velocity v_p and the S wave velocity v_s are related as follows:

$$v_p = v_s\sqrt{\frac{1-\nu}{0.5-\nu}} \qquad (8.8)$$

where ν is the Poisson's ratio.

In soils, P waves propagate both through the soil skeleton and through the water. S waves propagate through the soil skeleton only, as water cannot transmit shear waves. In solving geotechnical engineering problems, knowing the stiffness of the soil skeleton is often much more useful than knowing the stiffness of the water or the combined skeleton and water; therefore, shear waves are more useful than P waves in most cases except when trying to detect the depth of the ground water level.

Surface waves are also of two types: Rayleigh waves and Love waves. A large earthquake can create surface waves that travel around the Earth surface several times before dissipating. *Rayleigh waves*, sometimes called *ground rolls*, were discovered by Lord Rayleigh in the UK in 1885. Their propagation is analogous to the propagation you see when you drop a pebble into calm water. The wave displaces the particle along an ellipse in a plane perpendicular to the surface and in the direction of the wave as it passes through the soil. Rayleigh waves are slower than body waves ($v_R \sim 0.9\, v_s$). A good approximation of v_R is given by:

$$v_R \cong v_s\frac{0.87+1.12\nu}{1+\nu} \qquad (8.9)$$

where v_R and v_s are the Rayleigh wave and shear-wave velocities respectively and ν is the Poisson's ratio. Rayleigh waves have large amplitude, large wave length, and long duration, and propagate further than shear waves and P waves along the surface. Because their wave length is related to the depth being affected by the waves, different frequencies can be used to investigate the variation of soil properties with depth. *Love waves* are slightly faster than Rayleigh waves, and are named after Augustus Love in the UK who discovered them in 1911.

8.2.2 Seismic Reflection

Waves will reflect back to the surface (Figure 8.3) anytime they encounter a boundary separating two layers with a contrast in acoustic impedance I_a. *Acoustic impedance* is the product of the density ρ and the wave speed v:

$$I_a = \rho\, v \qquad (8.10)$$

The higher the impedance contrast is, the better the chance that it will be detected by seismic reflection. The acoustic impedance ratio R is defined as the ratio of the acoustic impedance of the lower layer over the acoustic impedance of the upper layer.

Seismic reflection consists of sending seismic waves down into the soil, receiving the reflected wave at a receiver, and identifying the time that it takes for the wave to travel down to the boundary and back to the surface (Figure 8.3).

The depth of the reflector or boundary is given by:

$$D = \frac{1}{2}\sqrt{(vt)^2 - L^2} \qquad (8.11)$$

where D is the depth of the boundary reflecting the wave, t is the measured travel time of the wave, v is the wave velocity, and L is the distance between the shock point and the geophone. The shock wave is usually created by hitting the ground and the receivers are usually geophones (instruments

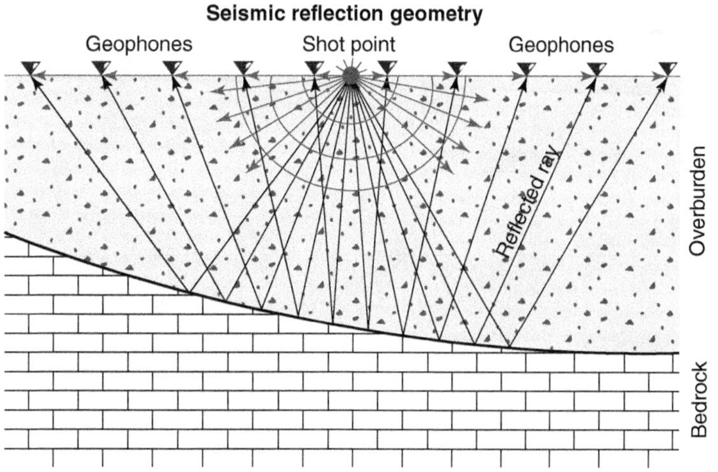

Figure 8.3 Seismic reflection test. (Courtesy of Timothy Bechtel, Enviroscan, Inc.)

that measure velocity of the point where they are located). The geophones are arranged in a line over a length related to the width and depth of the soil or rock boundaries to be tested. The distance between geophones is related to the required horizontal resolution: the closer they are, the higher the resolution is.

Seismic reflection typically makes use of P waves and has the following characteristics. The depth to be studied should be more than about 10 m; indeed, at shallower depths the surface waves arrive at about the same time as and with larger amplitude than the reflecting waves, making it difficult to distinguish them. At greater depths, the reflected waves arrive after the surface waves and thus can be detected more easily. Seismic reflection does not require a very long array of geophones because the waves simply reflect back to the surface. Nevertheless, seismic reflection tends to be 3 to 5 times more expensive than seismic refraction because the inversion and interpretation are more complex. The vertical resolution is between 5 and 10% of the depth, while the horizontal resolution is about 50% of the geophone spacing. The applications are the delineation of layer boundaries (such as finding the depth to bedrock), the discovery of fractures and faults, determination of water level, detection of cavities like tunnels or sinkholes, and determination of elastic modulus for soils and rocks.

8.2.3 Seismic Refraction

When a wave comes to a boundary with a distinct change of acoustic impedance (see section 8.2.2), part of the wave will be reflected (going back to the surface) and part of the wave will be refracted (going through to the next layer). The direction of the refracted wave will be at the angle of refraction, which follows Snell's law (Figure 8.4):

$$n_i \sin \alpha_i = n_r \sin \alpha_r \tag{8.12}$$

where n_i is the refractive index of the layer the wave is leaving, α_i is the incident angle between the wave direction and the normal to the boundary between the two layers, n_r is the refractive index of the layer the wave is entering, and α_r is the refractive angle between the wave direction and the normal to the two layers. Willebrord Snell was a Dutch physicist who made this contribution in 1621.

The *refractive index* is the ratio between the wave velocity in a reference medium and the wave velocity in the soil considered. Therefore, for seismic wave propagation at interfaces, Snell's law becomes:

$$\frac{\sin \alpha_1}{v_1} = \frac{\sin \alpha_2}{v_2} \tag{8.13}$$

where v_1 is the wave velocity in the upper layer and v_2 the wave velocity in the lower layer. Note that there is no change in wave frequency as the wave enters the next layer, only a change in wave direction, wave velocity, and wave length.

If v_2 is larger than v_1, there is an angle α_c such that:

$$\frac{\sin \alpha_c}{v_1} = \frac{\sin 90^\circ}{v_2} \tag{8.14}$$

The angle α_c is the critical angle at which the refracted wave propagates along the top of the second layer where the velocity is higher (Figure 8.4). At any time, the critically refracted wave that travels along the interface refracts back into the upper layer and strikes a geophone on the surface

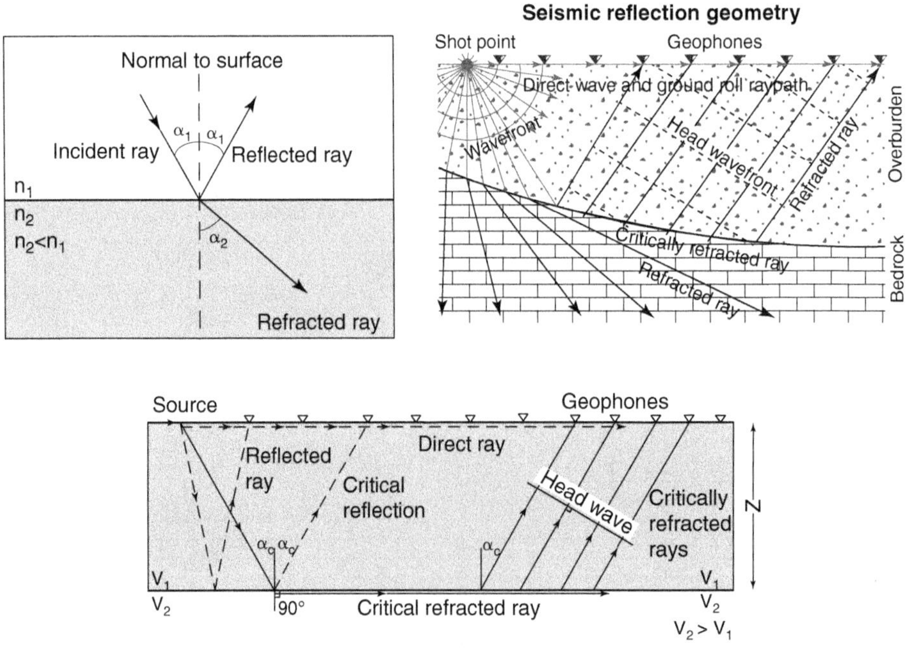

Figure 8.4 Seismic refraction test. (b: Courtesy of Timothy Bechtel, Enviroscan, Inc.)

which senses its arrival (Figure 8.4). The wave can travel directly along the surface from the source to the geophone or down to the lower layer and back to the surface. At the beginning of the recording, the waves travelling directly in the upper layer arrive first at a given geophone. After a while, the refracted waves arrive at the geophone before the reflected waves because the waves go faster in the lower layer if the lower layer has higher impedance (stiffer). The time at which this change occurs is called the crossover time t_c and corresponds to the cross-over distance x_c (Figure 8.5).

A plot of time of arrival versus distance between the detecting geophone and the source (Figure 8.5) shows two lines. The first is the arrival of the wave coming from direct propagation, which has a slope of $1/v_1$; the second line is the arrival of the refracted wave, which has a slope of $1/v_2$ (see problem solutions in this chapter for derivations). The intersection of these two lines gives the crossover time t_c and crossover distance X_c. The crossover time t_c can be used to obtain the depth of the lower layer:

$$Z = \frac{t_c v_1}{2} \frac{\sqrt{V_2 - V_1}}{\sqrt{V_2 + V_1}} \tag{8.15}$$

where Z is the depth to the lower layer and V_1 and V_2 are the velocities in the upper and lower layers, respectively. Therefore, seismic refraction can give the velocities V_1 and V_2 from the slope of the lines and the depth of the interface from t_c.

Seismic refraction typically makes use of P waves and has the following characteristics. The depth to be studied is typically up to 30 m; the length of the geophone array is on the order of 4 to 5 times the depth of the boundary to be detected. Although detection depths beyond 30 m are possible, they require very long geophone arrays and very large shock sources for the wave to be detected far away. Seismic refraction tends to be much less expensive than seismic reflection. The vertical resolution is about 15% of the depth studied and the horizontal resolution is about 50% of the geophone spacing.

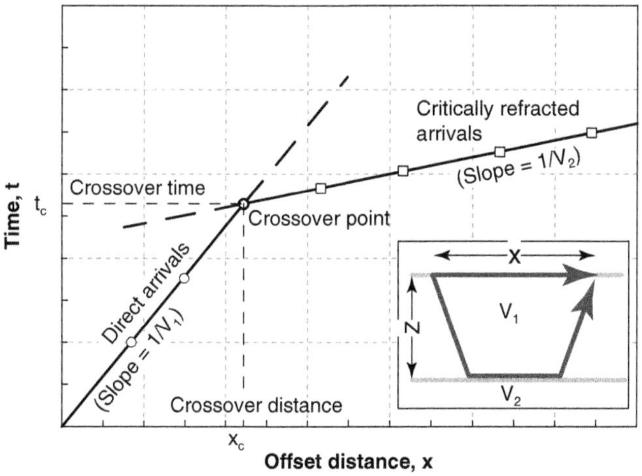

Figure 8.5 Interpreted signal from seismic refraction.

Seismic refraction is used primarily for determining the stratigraphy of soil layers, including depth to bedrock, as well as wave propagation characteristics of the layers penetrated. The critically refracted waves exist only if the soil or rock becomes stiffer or denser with depth, which is the most common case. However, a strong layer underlain by a weak layer will not produce critically refracted waves. The seismic refraction technique has been successfully applied to mapping depth to base of backfilled quarries, landfills, thickness of overburden, and the topography of groundwater.

8.2.4 Cross Hole Test, Seismic Cone Test, and Seismic Dilatometer Test

The cross hole test (CHT; ASTM D4428) (Figure 8.6) requires 2 borings separated by a distance L. This distance varies, but is typically between 3 to 6 m for geotechnical applications. Geophones are placed in boring 2 while the impact generator is placed in boring 1. Because shear waves isolate the behavior of the soil skeleton, they are more useful in geotechnical engineering than compression waves. Therefore, the source in boring 1 is usually one that generates a shear wave; this can be done by dropping an upper wedge on a lower wedge, for example. The time t required for the wave to travel from boring 1 to boring 2 is recorded and the shear-wave velocity is calculated as:

$$v_s = L/t \tag{8.16}$$

The distance L between borings may vary with depth, as the borings may not be perfectly parallel. For increased precision of L, it is desirable to run inclinometers in the borings to know the horizontal distance between borings at any depth with more accuracy. After the first CHT test, the depth of the source in boring 1 and the depth of the geophones in boring 2 are increased and the test is performed at each depth to obtain a shear-wave velocity profile (Figure 8.7). It is often desirable to use 3 borings, with the source in boring 1 and geophones in borings 2 and 3, because measurement of the wave travel time is easier to make in this case. The test can also be performed with the seismic cone penetrometer test by creating the shear-wave shock at the surface and recording the arrival of the shear wave at the depth where the cone penetrometer point (equipped with a geophone) is located. A similar test can be performed with the seismic dilatometer test.

The downhole technique consists of inserting a long probe with a source and a receiver on the same probe (Figure 8.8). The probe is inflated so that the source and receiver are in good contact with the wall of the borehole and the surrounding soil. With some probes, there is no direct contact; instead, transmission of the wave takes places through the liquid-filled borehole. The length of probe separating the source from the receiver is very flexible so that the wave propagating through the soil will arrive well before the one through the probe does. The wave travelling through the probe is purposely attenuated by the damping characteristic of that part

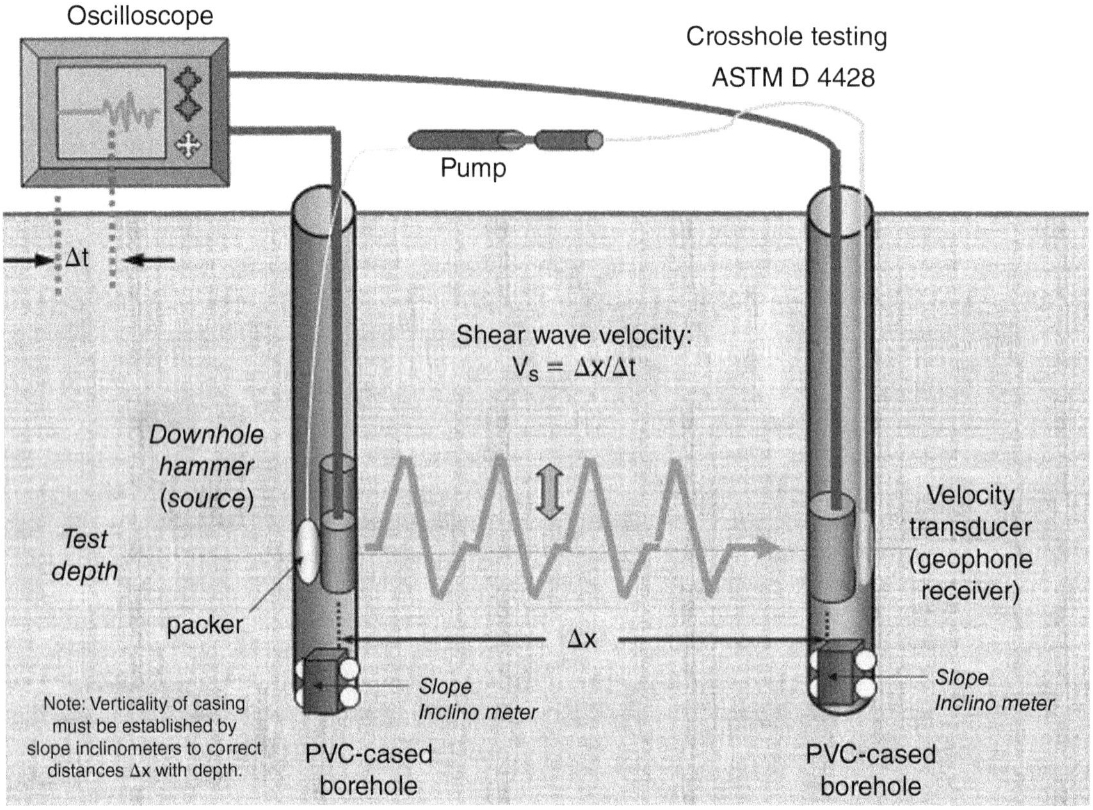

Figure 8.6 Cross hole test. (Courtesy of Professor Paul Mayne, Georgia Institute of Technology, USA.)

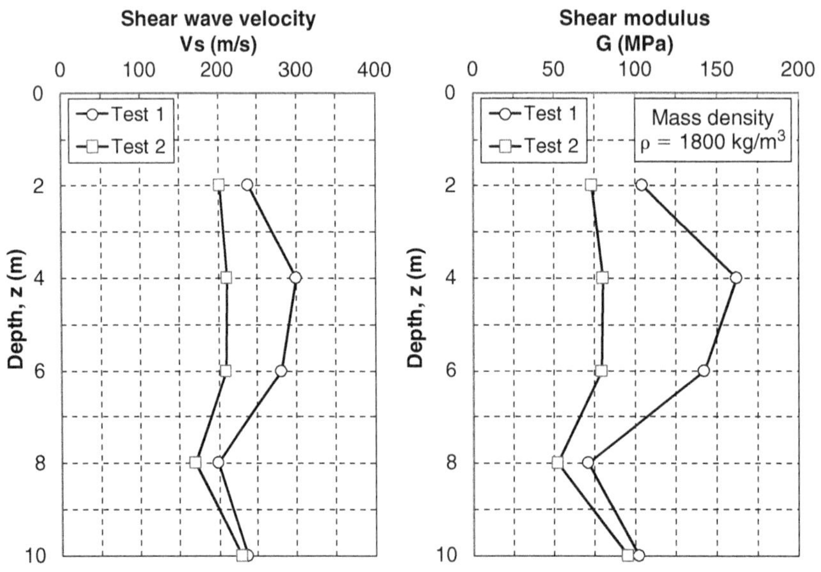

Figure 8.7 Profile of shear-wave velocities from a cross hole test.

of the probe. This type of equipment is used primarily in deep boreholes or for offshore investigations. This type of equipment is also used in the oil well industry for electrical resistivity logging, neutron logging, gamma logging, and caliper logging.

8.2.5 Spectral Analysis of Surface Waves

The technique of *spectral analysis of surface waves (SASW)* has evolved over the years, but it seems appropriate to give credit to Ken Stokoe in the USA for a major part of its early development during the 1970s (Stokoe, Joh, and Woods,

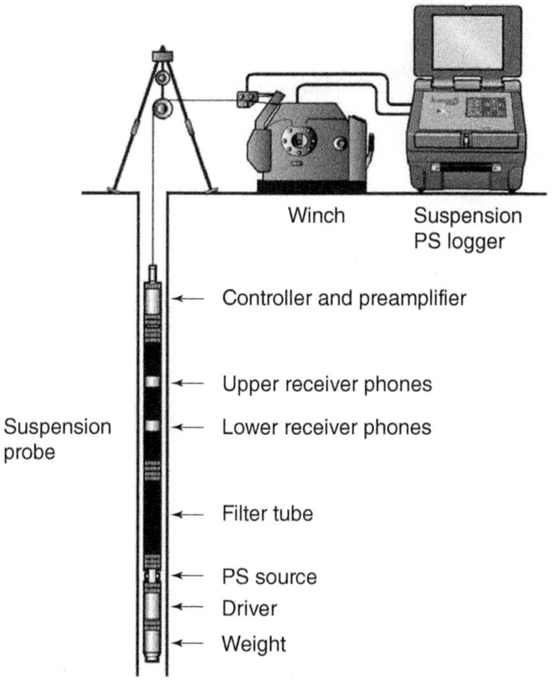

Figure 8.8 Downhole seismic test. (Courtesy of the OYO Corporation.)

2004). If the velocity of a wave travelling in a material depends only on the physical properties of the material, then the wave velocity is constant and independent of frequency. Such a material is called a *nondispersive material* and waves traveling through this medium will maintain a constant shape (light propagates in a nondispersive and nondissipating way; this is why it can propagate over astronomical distances). This would be the case of a wave travelling in a soil that has uniform properties independent of depth. However, almost all soils have properties that vary with depth, because of differences such as layering and variations in effective stress; therefore, soils are dispersive materials. As a result, when many waves with different frequencies travel through the material, the wave train contains a lot of waves with individual frequencies, and the shape of the wave train changes as the wave travels. Some waves within the wave train travel faster than the wave train (longer wave length) and die out as they approach the leading edge. Some waves within the wave train travel slower than the wave train (shorter wave length) and die out as they approach the trailing edge.

The *group velocity* v_g is the speed with which the wave train or wave envelope propagates; it is the travel speed of the energy carried by the wave. The *phase velocity* v_{ph}, in contrast, is the speed with which an individual wave of the wave train travels. The phase velocity depends on the frequency of the individual wave contributing to the overall wave train. In the case of nondispersive material, v_g and v_{ph} are the same and independent of frequency. In the case of dispersive material like soil, v_g and v_{ph} are different and v_{ph} depends on frequency. See http://physics.usask.ca/~hirose/ep225/animation/dispersion/anim-dispersion.html or http://paws.kettering.edu/~drussell/Demos/Dispersion/dispersion.html for an animation of the difference between dispersive and nondispersive waves and between group and phase velocities.

The SASW makes use of Rayleigh waves because they travel along the ground surface and because they attenuate a lot less than body waves: $1/\sqrt{r}$ instead of $1/r^2$. In fact, about two thirds of the seismic energy at shallow depth is made of Rayleigh waves. The SASW takes advantage of these concepts to link the frequency content of the wave train to the shear-wave velocity profile of the soil at a site. The high-frequency waves have short wave lengths and only penetrate the shallow layers of the soil deposit (Figure 8.9). Thus, they only give the shear-wave velocity of the shallow layers. The low-frequency waves have long wave lengths and penetrate much deeper in the soil deposit. Hence, they give the shear-wave velocity of the deeper layers. In the field, the test consists of placing receivers (geophones or accelerometers) on the ground surface at regular intervals away from where the shock is generated during the test. The receivers are placed along a single radial path from the impact location. These instruments measure the vertical movement of the soil as the waves pass by. A first set of data is collected at shallow depth by placing the receivers close to each other (Figure 8.10), generating the shock (by hammer blow, weight drop, explosive), and collecting the data at each receiver. A second set of data is collected after repositioning the receivers and doubling the distance between them to test the response of deeper layers. A third set of data is collected after again doubling that distance, and so on. This test is usually repeated about 6 to 8 times to obtain a shear-wave velocity versus depth profile. As the spacing between receivers increases, the impact source must generate larger amplitude and lower frequencies.

The data reduction involves the following sequence. First the amplitude versus time signal of the wave train, recorded at sequential receivers, is transformed from the time domain to

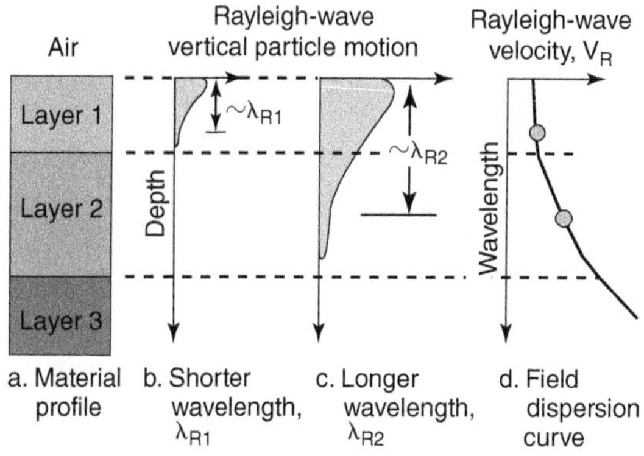

Figure 8.9 Principle of the SASW method. (Courtesy of Professor Kenneth Stokoe, University of Texas, USA)

Figure 8.10 Field SASW test. (Courtesy of Professor Kenneth Stokoe, University of Texas, USA)

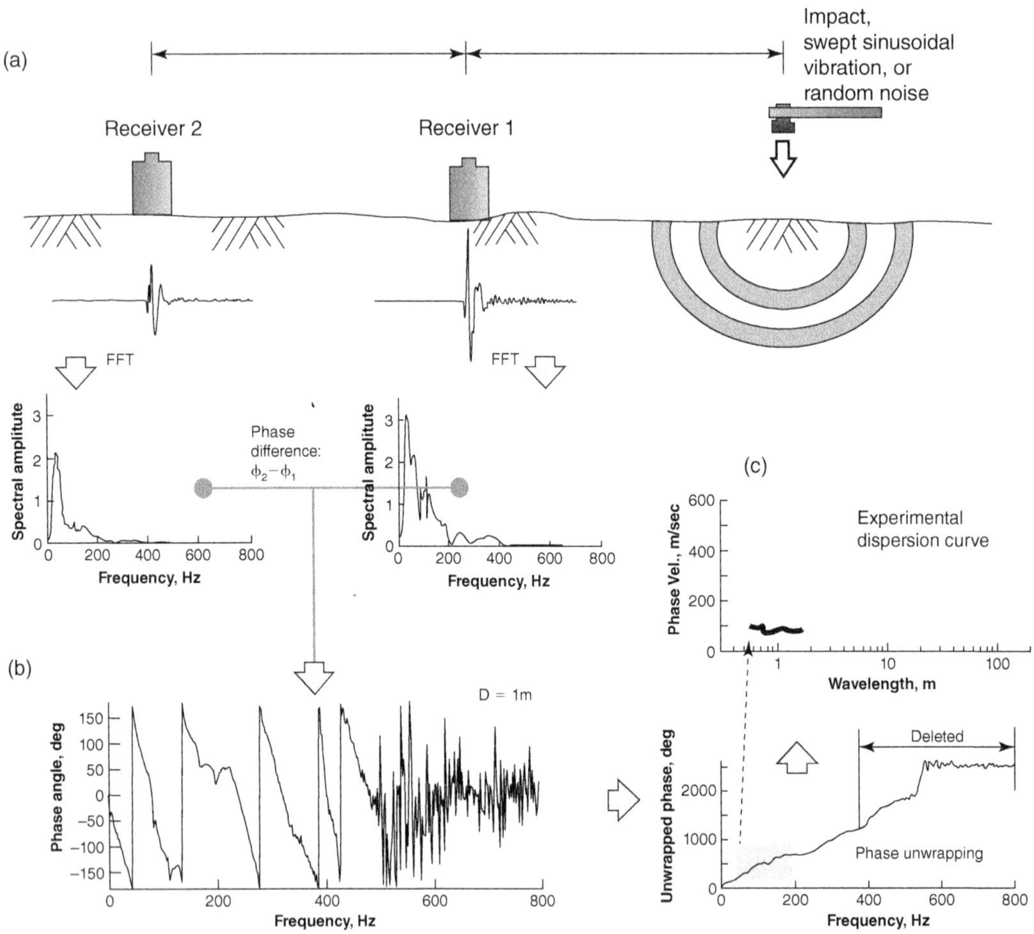

Figure 8.11 Calculation of phase velocity for the SASW. (Courtesy of Professor Kenneth Stokoe, University of Texas, USA.)

the frequency domain by performing Fourier transformation (Fourier 1822). The amplitude a versus frequency f plots and the phase versus frequency plots are obtained in this fashion (Figure 8.11). Then the phase angle versus frequency diagram is transformed into an unwrapped phase angle φ versus frequency f diagram. This means that rather than keeping the phase angle between 0 and 360 degrees, the phase angle keeps increasing instead of being re-zeroed at 360 degrees. Then the phase velocity v versus wave length λ diagram is obtained from each phase versus frequency plot, as the tester knows the frequency, the phase angle, and the distance between receivers. The phase velocity v is obtained from the distance s between receivers and the elapsed time t, while the wave length λ is obtained from the unwrapped phase angle φ. In simple terms, it consists of writing the following equations:

$$T = 1/f \qquad (8.17)$$

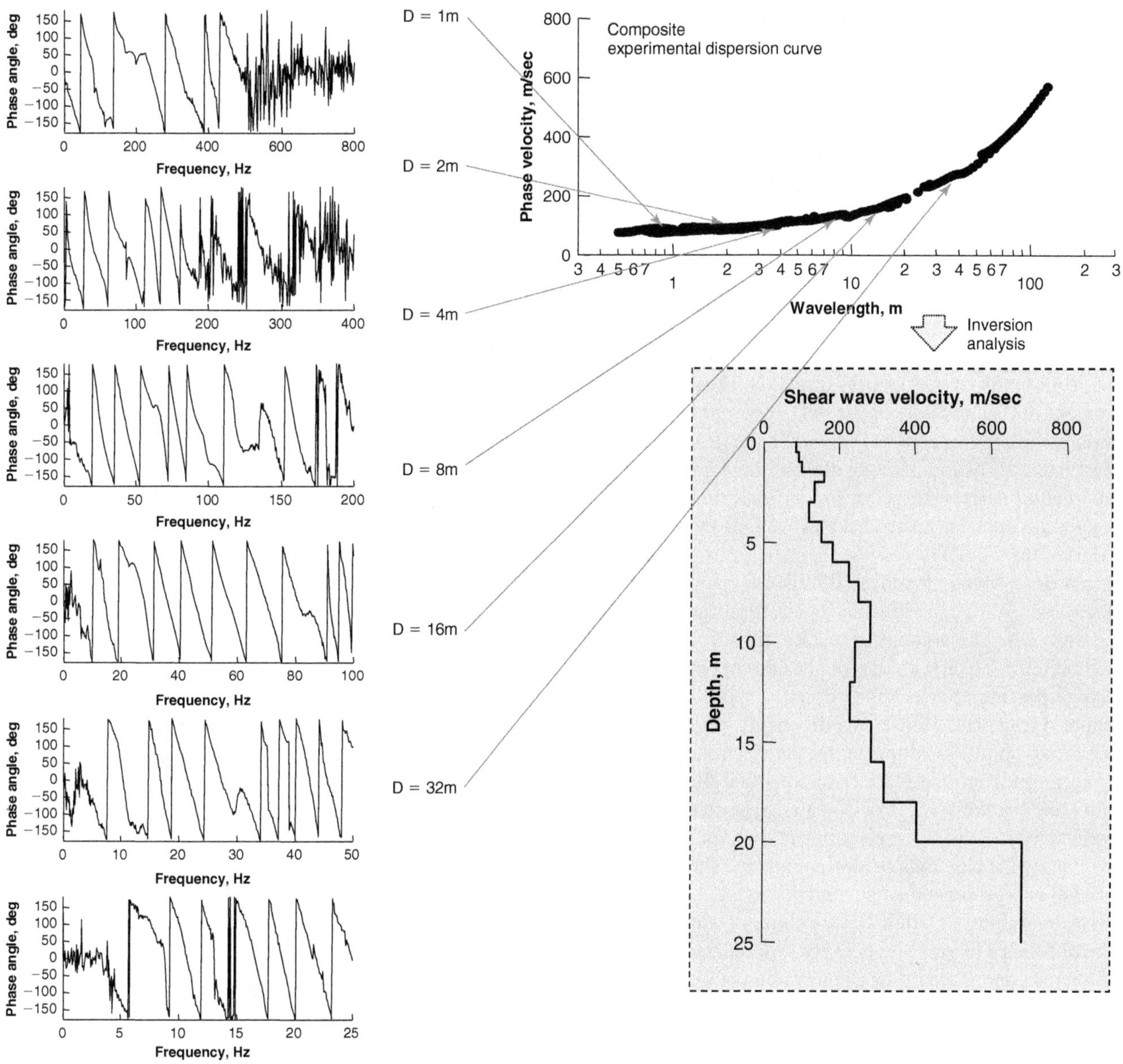

Figure 8.12 Determination of a dispersion curve and shear-wave velocity for the SASW. (Courtesy of Professor Kenneth Stokoe, University of Texas, USA.)

More generally:

$$t = \varphi/2\pi f \tag{8.18}$$

But

$$v = \lambda f = s/t \tag{8.19}$$

Therefore,

$$\lambda = 2\pi s/\varphi \tag{8.20}$$

where T is the period of the wave, f is the frequency, t is the time elapsed between the arrival of the wave at the first and second receivers, φ is the phase difference between the first and second receivers, v is the phase velocity, λ is the wave length, and s is the distance between the two receivers. This indicates that when the phase φ is known, the wave length λ can be calculated. The plot of phase velocity v versus wave length λ is the *dispersion curve* for a given receiver spacing (Figure 8.11). This procedure is repeated for all receiver spacings and the individual dispersion curves for each spacing are assembled into a single composite dispersion curve (Figure 8.12). Once the composite dispersion curve is generated for the site, an iterative forward modeling procedure or an inversion analysis algorithm is used to determine a shear-wave velocity profile by matching the field dispersion curve with the theoretically determined dispersion curve (Figure 8.9).

8.3 ELECTRICAL RESISTIVITY TECHNIQUES

8.3.1 Background on Electricity

Electricity is related to the organized movement of electrons (electronic conduction) or of ions (electrolytic conduction) in a medium. Electricity in metals, for example, is electronic conduction; in wet soils or human flesh, it is electrolytic conduction. Metals often have electrons that can be moved when subjected to a potential difference. Electrolytes (e.g., fluids) contain atoms that either have more protons than electrons or more electrons than protons. These atoms are charged and are called *ions*. When subjected to a potential difference, the positive ions move in one direction and the negative ions move in the opposite direction. The speed of this movement of electrons or ions is very low, but because the material is full of electrons or ions, when the first one moves the last one also begins moving almost immediately. Under an alternating current, the electrons or ions shake in place, but again the ones far away shake as well, as the shaking is transmitted very quickly because the material is packed with electrons or ions ready to move. In soils, the main conduction is electrolytic, although electronic conduction can also occur (e.g., iron ore). The amount of readily moving electrons or ions is called the electric charge Q. The current I is the amount of charge passing at a location per unit of time. Voltage or potential relates to the difference in energy per unit charge between two points. In simple terms, the electrons or ions are pushing to go from one place to another and the difference in "pressure" is the voltage or potential. The resistance R is the resistance to flow of the electrons or ions and depends on how strongly the electrons or ions are bound. The power P is the rate of energy consumed per unit of time. If an analogy is drawn to hydraulics, Q would be the volume of water, V would be the difference in pressure between two points, I would be the flow rate, and *R* would be a constriction in the pipe.

$$I = Q/t \tag{8.21}$$

$$V = E/Q \tag{8.22}$$

$$R = V/I \tag{8.23}$$

$$P = E/t = VI \tag{8.24}$$

where I is the current (amperes), Q the charge (coulombs), t the time (seconds), V the voltage (volts), E the energy (joules), R the resistance (ohms), and P the power (watts). When electricity goes through a wire, the resistance can be written as:

$$R = \rho L/A \tag{8.25}$$

where ρ is the resistivity (ohm.m), A is the cross-sectional area of the wire, and L the length of the wire. The resistivity ρ and its inverse, the conductivity ($\sigma = 1/\rho$), are properties of the material and independent of the dimensions. A low resistivity means very little resistance to an electrical current. Although soil and rock deposits are not wires, they also have resistivity values; Table 8.2 gives some of those values. The range is due in part to the significant influence of the water content in the soil or rock. Water has a very low resistivity, and a saturated soil will have a much lower resistivity than the same soil in the dry state. Porosity, degree of saturation, cation exchange capacity, temperature, and concentration of dissolved salts are other parameters that influence the resistivity of a soil.

Table 8.2 Example Values of Resistivity ρ for Soils and Rocks

Soil or Rock	Low Value (ohm.m)	High Value (ohm.m)
Groundwater	1	200
Seawater	0.2	1
Sea ice	20	1000
Permafrost	500	10,000
Intact igneous and metamorphic rocks	1000	100,000
Weathered igneous and metamorphic rocks	1	1000
Porous limestone	50	2000
Dense limestone	1000	100,000
Sandstone	50	5000
Shale	5	2000
Clay and silt	2	100
Sand	50	2000
Gravel	400	10,000

8.3.2 Resistivity Tomography

Figure 8.13 shows two electrodes placed at the surface of a homogeneous soil deposit. The current flows from electrode B to electrode A along the electrical flow lines. Perpendicular to the flow lines are the electrical equipotential lines.

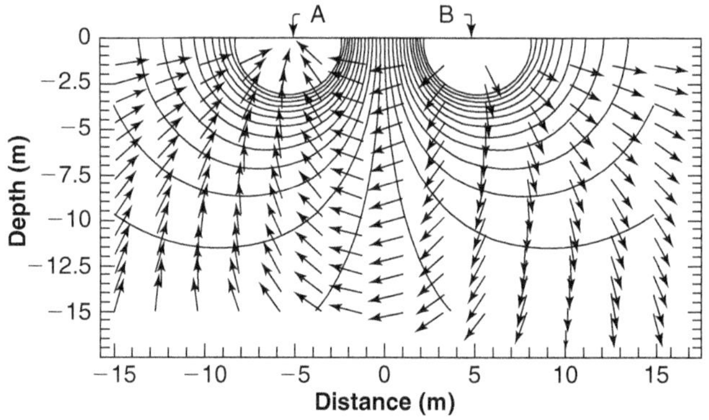

Figure 8.13 Current lines and equipotential lines for an electrical resistivity test in a homogeneous soil deposit. (From Herman, 2001, Courtesy of American Association of Physics Teachers.)

These two sets of lines are the graphical solution to the Poisson's differential equation that governs electrical flow in a homogeneous material.

$$\frac{d^2V}{dx^2} + \frac{d^2V}{dy^2} + \frac{d^2V}{dz^2} = 0 \tag{8.26}$$

where V is the voltage or potential and x, y, z are the Cartesian coordinates in three dimensions. In simple terms, from Eqs. 8.23 and 8.25 comes:

$$\rho_a = \frac{\Delta V}{I}\frac{A}{L} = RK \tag{8.27}$$

where $\Delta V/I$ is the resistance R, and A/L is the geometry coefficient K. The general solution to Eq. 8.26 depends on the placement of the electrodes and on the material in which they are placed, but the general form of Eq. 8.27 is maintained as:

$$\rho_a = \frac{\Delta V}{I}K \tag{8.28}$$

In the field, electrodes can be placed on the ground surface in a line, as shown in Figure 8.14. In this case, Eq. 8.28 becomes:

$$\rho_a = \frac{\Delta V}{I}\frac{2\pi}{\left(\frac{1}{r_1} - \frac{1}{r_2}\right) - \left(\frac{1}{r_3} - \frac{1}{r_4}\right)} \tag{8.29}$$

where ΔV is the difference of potential or voltage between the potential electrodes; I is the current existing between the current electrodes; and r_1, r_2, r_3, and r_4 are the distances between electrodes as shown in Figure 8.14. Different arrays have been proposed to optimize the arrangement of the electrodes. For the Wenner array shown in Figure 8.15, the geometric factor K becomes $2\pi a$ where a is the distance between electrodes. Other electrode spacings are being used in practice, such as the Schlumberger array and the dipole-dipole array. The best arrays for a field survey depend on the subsurface feature to be mapped, the sensitivity of the resistivity meter, and the background noise.

If the soil is made of two layers, with the lower layer having a lower electrical resistivity than the top layer, then the flow lines and equipotential lines are affected as shown in Figure 8.16. Furthermore, the resistivity obtained from the measurements is an equivalent or apparent electrical resistivity, as both layers are involved in the electrical response to the potential difference. The field test is generalized and many electrodes are placed on the ground surface at regular intervals. Alternatively, measurements may be made by using two electrodes as current electrodes and many others as potential electrodes. A mathematical inversion process is then used to back-calculate the electrical resistivity map that best fits the series of measurements. Such electrical resistivity maps are used for geotechnical engineering issues such as stratigraphy mapping, finding the depth of the water table, inferring the presence of leachates, and determining the depth of a landfill, the presence of cavities, and the depth of bedrock. The depth of investigation for resistivity tomography is about 20% of the length of the string of electrodes placed on the ground.

Figure 8.14 Current and potential electrodes placement. (*Bottom*: After Cardimona, 1993.)

8.4 ELECTROMAGNETIC METHODS

8.4.1 Electromagnetic Waves

There are basically two main types of waves: mechanical waves and electromagnetic waves. *Mechanical waves* can only propagate through a material; they cannot propagate in vacuum. Seismic waves are mechanical waves. *Electromagnetic waves* can propagate in both a material and a vacuum. Light is one example of an electromagnetic wave. You can create an electromagnetic wave by shaking an electron; the electron will create a wave that propagates as ripples across

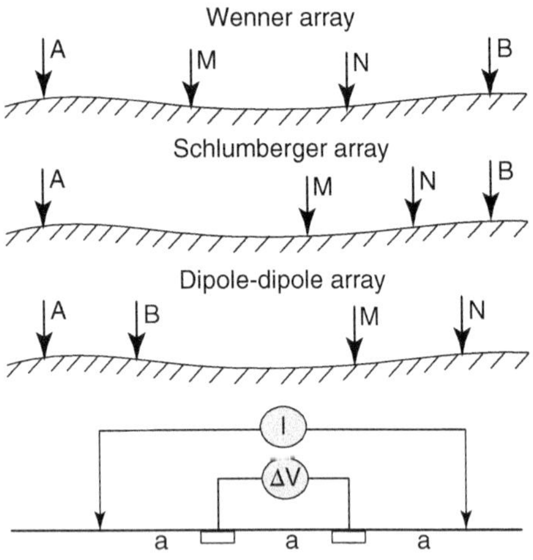

Figure 8.15 Common electrode arrays for field resistivity test. (After American Association of Physics Teachers.)

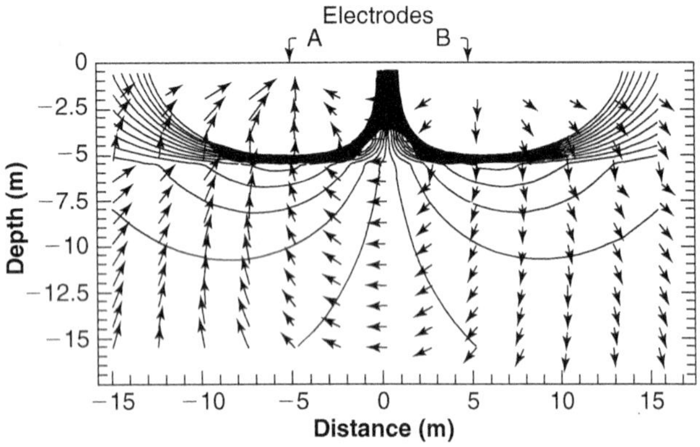

Figure 8.16 Current lines and equipotential lines for an electrical resistivity test in a two-layer soil deposit; the deeper layer has a lower resistivity than the top layer. (From Herman, 2001, Courtesy of American Association of Physics Teachers.)

the vacuum of space. When you shake the electron, the electromagnetic wave propagates as a *transverse wave*, a wave where the motion is perpendicular to the direction of propagation. What moves is a photon described as an electric field in the vertical direction and a magnetic field in the horizontal direction. Photons represent bundles of energy that can be equally considered as particles with zero mass or waves.

The elements of wave propagation described for seismic waves in section 8.2.1 (period, frequency, wave length, wave speed) also apply to electromagnetic waves. The speed of propagation of an electromagnetic wave is the speed of light. In a vacuum, that speed is approximately 300,000 km/s, which represents an accepted upper speed limit for our universe. The wave length λ of electromagnetic waves varies significantly from one end to the other of the spectrum (Figure 8.17). Radio waves ($\lambda = 3000$ to 0.3 m) are used in broadcasting, microwaves ($\lambda = 0.3$ to 3×10^{-4} m) are used to heat food and in communications, infrared light ($\lambda = 3 \times 10^{-4}$ to 4×10^{-7}m) is used for night vision and muscle therapy, visible light ($\lambda = 4 \times 10^{-7}$ to 7×10^{-7}m) is a very small range of the electromagnetic wave length spectrum, ultraviolet light (7×10^{-7} to 3×10^{-9}m) is used to detect forgery of paintings and in tanning salons, X-rays ($\lambda = 3 \times 10^{-9}$ to 3×10^{-11}m) are used to see through the human body and through sampling tubes, and gamma rays ($\lambda = 3 \times 10^{-11}$ to 3×10^{-13}m) are used to kill cancer cells in the human body.

8.4.2 Ground-Penetrating Radar

Ground-penetrating radar (GPR) uses electromagnetic waves in the radio-to-microwave range to penetrate the soil and give an image of the subsurface. The waves are generated by a source antenna that is in contact with the ground surface; the waves propagate in the soil, reflect from anomalies such as layer interfaces, cavities, and buried objects, and travel back to the surface where they are detected by a receiver antenna that is also in contact with the ground surface. *Antennas* are devices that transform electric current into electromagnetic waves and vice versa. Figure 8.18 shows a GPR test and a typical result. Note that the travel time from source to receiver is extremely short: Electromagnetic waves travel extremely fast, so this time is measured in nanoseconds. The electromagnetic waves are reflected any time they encounter a boundary between two materials with different dielectric constants. A dielectric material is a poor conductor of electricity and the *dielectric constant* is a measure of this property. Materials with relatively low dielectric constants, like air, glass, and ceramic, are good electric insulators. Materials with relatively high dielectric constants, like metal oxides, are good electric conductors. Soils having high electrical conductivity rapidly attenuate radar energy.

The depth of penetration of the GPR varies significantly depending on the soil type and on the frequency of the wave generated. The frequency used in GPR testing varies from as low as 20 MHz to as high as 2000 MHz. The user is often faced with a compromise between using a low frequency to penetrate deeply and a high frequency to obtain good definition. Indeed, as with mechanical waves, long wave lengths lead to deeper penetration, but short wave lengths lead to more precise definition of the objects encountered. Regarding the soil type, GPR works best in dry sand and gravel, where depths of tens of meters are possible with low-frequency antennas. However, in wet clays and saline soils the penetration is less than one meter. Figure 8.19 is a map of potential penetration with GPR in the USA. GPR is used for detecting pipes, tunnels, cavities, and unexploded ordinance, among other things.

8.4.3 Time Domain Reflectometry

Time domain reflectometry (TDR) makes use of the propagation of an electromagnetic wave in a cable. It was first used to

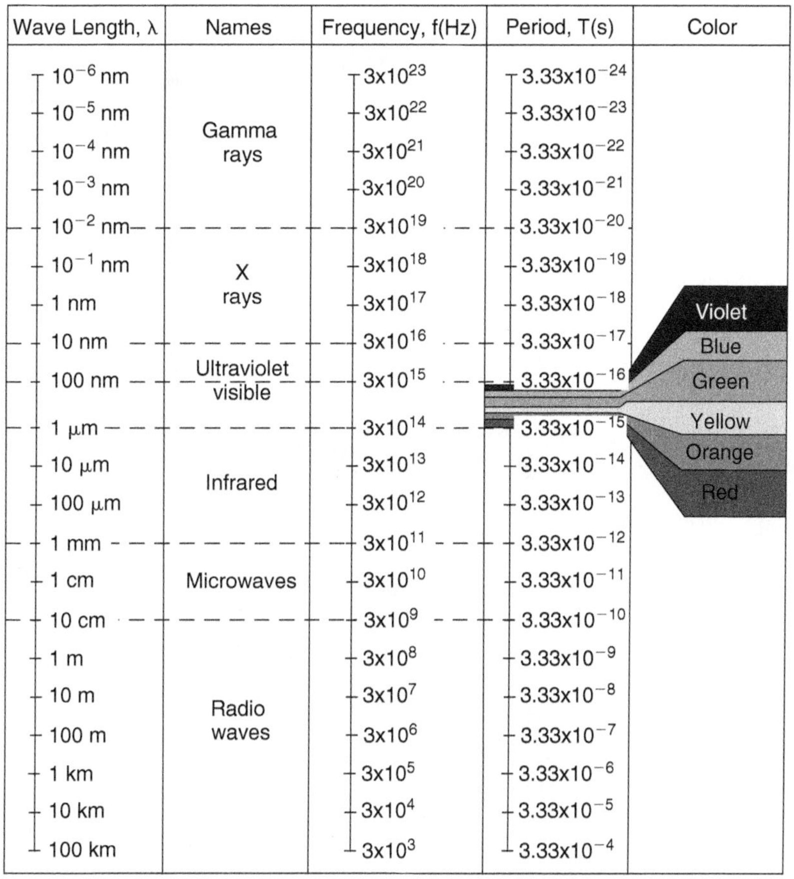

Figure 8.17 Electromagnetic wave length spectrum.

Figure 8.18 GPR testing and results. (Courtesy of Dig Smart of Maine)

find breaks in cables by measuring the travel time of the wave to the defect and back and using the travel speed to find out where the cable break was located. Although an electromagnetic wave travels at the speed of light in a vacuum, it travels at only at a fraction of that value in a cable. Nevertheless, the time measurements must still be in nanoseconds or even picoseconds. TDR was extended to soil water content and soil density measurements by using two rods pushed into the soil surface (Figure 8.20).

Materials are classified as conductors or insulators, also called dielectric, depending on their ability to conduct electricity. The dielectric permittivity ε of a soil is a measure

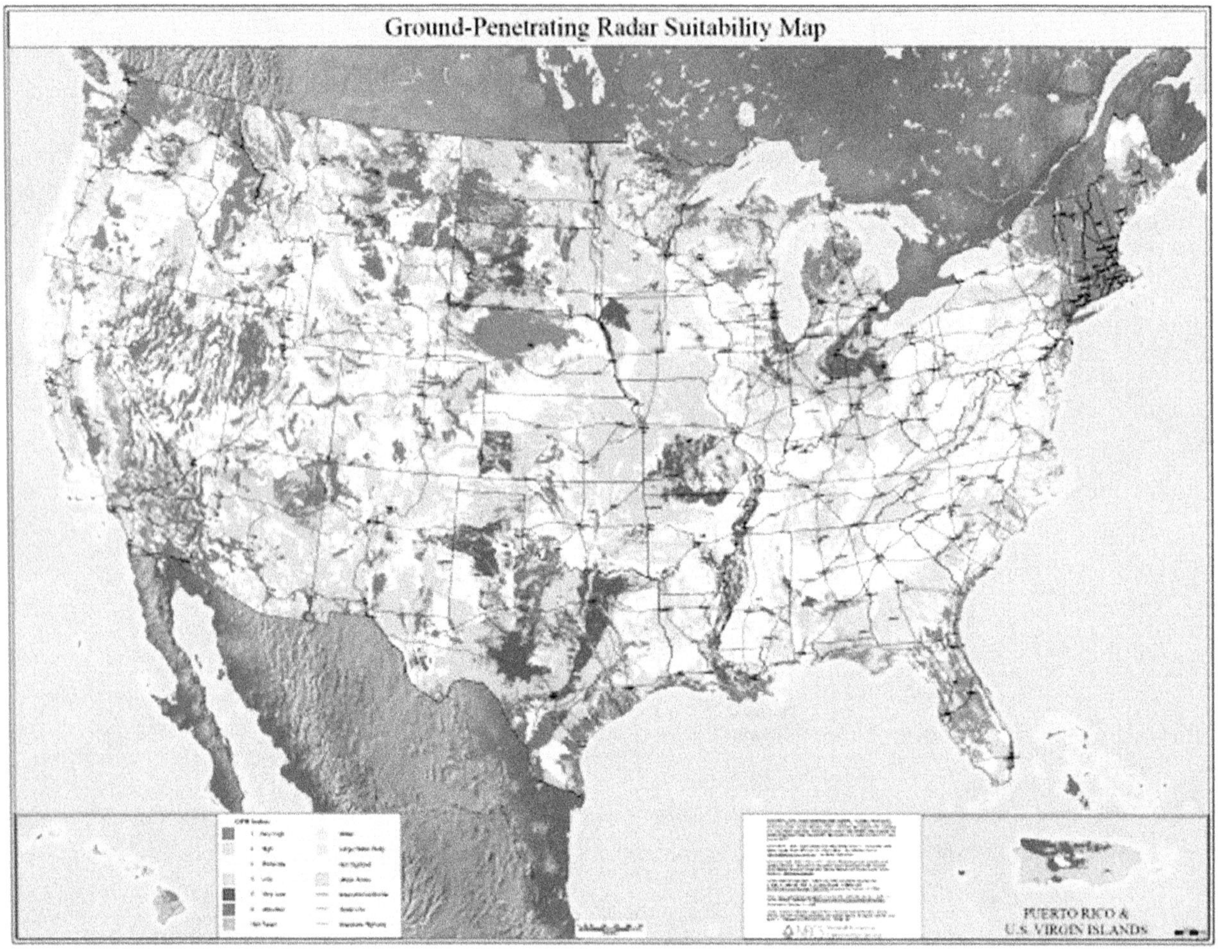

Figure 8.19 Suitability map for GPR testing. (Courtesy of NRCS [National Resources Conservation Services]).

Figure 8.20 Example of time domain reflectometry field probe. (Courtesy of Professor Vincent Drnevich, Purdue University)

of how fast an electromagnetic wave propagates through the soil:

$$\varepsilon = (c/v)^2 \tag{8.30}$$

where c and v are the velocity of the electromagnetic wave in the soil and in a vacuum (300,000 km/s) respectively. If a rod is embedded in a soil mass, the velocity of the electromagnetic wave propagating in the rod will be affected by the permittivity of the soil surrounding the rod. This velocity can be obtained by measuring the length of the rods embedded in the soil and the time required for the wave to travel down the rod and back; the soil dielectric permittivity can be obtained

Figure 8.21 Example of LIDAR instrument and result: (a) LIDAR instrument. (b) Image generated with LIDAR. (Courtesy of Professor Robert Warden, Texas A&M University)

from this measurement by use of equation 8.30. For example, a dry soil has a dielectric permittivity value of around 4, moist soils around 30, water about 80, and air close to 1. So water impacts the dielectric permittivity significantly, and testers take advantage of this fact to relate soil dielectric permittivity to the soil water content. Calibrations are necessary to obtain the best correlation equation. This technique is used to obtaining the soil density as well as the soil water content. Such measurements are particularly useful on compaction projects, such as the field performance of landfill covers.

8.5 REMOTE SENSING TECHNIQUES

8.5.1 LIDAR

LIDAR stands for light detection and ranging and is sometimes called *laser radar*. The LIDAR test consists of sending a laser (light amplification by stimulated emission of radiation) beam of electromagnetic waves (infrared, visible, ultraviolet) at an object and detecting the time required for that beam to reflect from the object and come back to the LIDAR receiver. The beam is sent in a series of wave pulses at a very high frequency. Because the beam travels at the speed of light in air (close to 300,000 km/s), the time involved is measured in nanoseconds or picoseconds. These very short times can be measured with instruments such as optoelectronic streak cameras. Knowing the time of flight and the speed of the wave, the tester can back-calculate the distance. LIDAR works like a camera, as it sweeps through the landscape it is aimed at and records the distance of all objects it is "seeing." The exact location of the instrument is obtained through the global positioning system (GPS), and the distances measured can be connected to elevations and coordinates. The result of a LIDAR test is a three-dimensional image of the landscape swept by the LIDAR equipment in which all points recorded are documented with coordinates.

LIDAR uses short wavelengths of the electromagnetic spectrum, typically in the ultraviolet, visible, or near-infrared range. This allows the LIDAR equipment to define objects within a few millimeters (or at least centimeters), as it is possible to image a feature or object only about the same size as the wavelength, or larger. This makes it difficult for LIDAR to see through aerosol, rain, snow, mist, fog, and smoke; LIDAR works best when the sky is clear without clouds, rain, or haze, and functions equally well day or night.

For applications from the ground, a LIDAR system (Figure 8.21) is composed of a laser scanning system and a global positioning system. GPS. A GPS is a space-based global navigation satellite system (GNSS) that provides reliable location and time information in all weather, at all times, and anywhere on or near the Earth when and where there is an unobstructed line of sight to four or more GPS satellites. It was established in 1973 by the United States government, which maintains it, and is freely accessible to anyone with a GPS receiver.

In addition, for LIDAR used from an airplane (airborne LIDAR), an inertial measuring unit (IMU) is required to take into account the speed of the airplane when calculating the coordinates of the points recorded. An IMU is an electronic device that measures the airplane's velocity and orientation using a combination of accelerometers and gyroscopes.

8.5.2 Satellite Imaging

Satellite imaging, also known as *radar satellite*, is based on the same principle as LIDAR but gives a picture of a much larger area than LIDAR. LIDAR is more applicable to smaller areas, whereas satellite imaging is more applicable to larger areas (Figure 8.22). For example, if the problem is to record the site contours of a levee breach after a flood, LIDAR is more applicable; if the problem is to record the subsidence over time of a large city due to water pumping, satellite imaging is faster and less time-consuming. The satellite imaging system

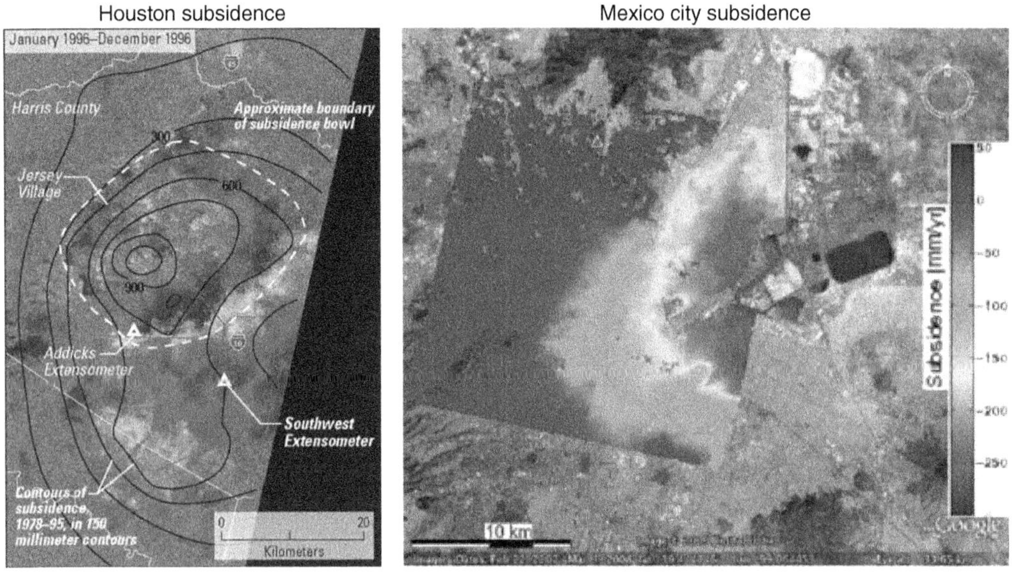

Figure 8.22 Example of satellite imaging results. (a) Houton subsidence. (b) Mexico City subsidence. (a: Courtesy of USGS. b: The ESA Envisat ASAR data is made available through the GEO Geohazards Supersite.)

requires the following parts: the optical system in the satellite, which views the area targeted; the internal processor, which collects and stores the data; the data transiting system; and the ground analysis and postprocessor. Satellite imaging is about as precise as LIDAR. Google Earth is a system based on satellite imaging.

PROBLEMS

8.1 Explain the difference between wave velocity and particle velocity.
8.2 If the shear-wave velocity in a soil is 250 m/s and the unit weight is 20 kN/m^3, what is the small-strain shear modulus of that soil?
8.3 If sound propagates in water at 5702 km/h, what is the constrained modulus of elasticity of water?
8.4 If sound propagates at 22000 km/h in steel, how long does it take for the wave to propagate down to the bottom of a 30 m long H pile, and what is the modulus of elasticity of steel? (Density of steel is 7850 kg/m^3 and the Poisson's ratio is 0.3.)
8.5 A wave has a wave length of 600 nm and a frequency of 5×10^{14} Hz. What kind of wave is it?
8.6 What is the difference in particle motion between a shear wave and a Rayleigh wave?
8.7 Explain the difference between seismic reflection and seismic refraction techniques.
8.8 Derive the crossover time equation for seismic refraction.
8.9 What is a dispersion curve?
8.10 Describe the basic concept of the SASW technique.
8.11 Describe the basic concept of the electrical tomography technique.
8.12 What is an electromagnetic wave, and what are its main properties?
8.13 Describe the basic concept of the GPR.
8.14 Describe the basic concept of the TDR.
8.15 Describe the basic concept of LIDAR.

Problems and Solutions

Problem 8.1

Explain the difference between wave velocity and particle velocity.

Solution 8.1

The wave velocity v is the speed at which the particle motion is propagated from one particle to the next. The particle velocity *u* is the speed at which the particle is moving around its own location.

Problem 8.2

If the shear-wave velocity in a soil is 250 m/s and the unit weight is 20 kN/m^3, what is the small-strain shear modulus of that soil?

Solution 8.2

$$\gamma = 20 \text{ kN/m}^3 = \rho g = \rho \times 9.81 \text{ m/s}^2$$

$$\rho = \frac{20000 \text{ N/m}^3}{9.81 \text{ m/s}^2} = 2039 \frac{\text{kg}}{\text{m}^3}$$

$$G = \rho \times v_s^2 = 2039 \times 250^2 = 127 \text{ MPa}$$

Problem 8.3

If sound propagates in water at 5702 km/h, what is the constrained modulus of elasticity of water?

Solution 8.3

$$v_p = 5702\frac{\text{km}}{\text{hr}} = 5702\frac{1000 \text{ m}}{3600 \text{ sec}} = 1584\frac{\text{m}}{\text{s}}$$

$$\left.\begin{aligned} v_p &= \sqrt{\frac{K + \frac{4}{3}G}{\rho}} \\ \text{water} &\rightarrow G = 0 \end{aligned}\right\} \Rightarrow K = v_p{}^2\rho$$

$$\left.\begin{aligned} v_p &= 1584 \frac{\text{m}}{\text{s}} \\ \rho &= 1000 \frac{\text{kg}}{\text{m}^3} \end{aligned}\right\} K = 2.51 \times 10^9 \frac{\text{N}}{\text{m}^2} = 2.51 \text{ GPa}$$

Problem 8.4

If sound propagates at 22000 km/h in steel, how long does it take for the wave to propagate down to the bottom of a 30 m long H pile, and what is the modulus of elasticity of steel? (Density of steel is 7850 kg/m^3 and the Poisson's ratio is 0.3.)

Solution 8.4

$$\left.\begin{aligned} &v_p = 22000 \frac{\text{km}}{\text{hr}} = 22000 \frac{1000 \text{ m}}{3600 \text{ sec}} = 6111 \frac{\text{m}}{\text{s}} \\ &\text{Pile Length L} = 30 \text{ m} \end{aligned}\right\} t = \frac{L}{v} = \frac{30}{6111} = 4.91 \times 10^{-3} \text{ sec}$$

$$v_p = \sqrt{\frac{E}{\rho}\frac{(1-\nu)}{(1+\nu)(1-2\nu)}} \Rightarrow E = v_p{}^2\rho\frac{(1+\nu)(1-2\nu)}{(1-\nu)}$$

$$\left.\begin{aligned} v_p &= 6111 \frac{\text{m}}{\text{s}} \\ \nu &= 0.3 \\ \rho &= 7850 \frac{\text{kg}}{\text{m}^3} \end{aligned}\right\} E = 6111^2 \times 7850\frac{(1+0.3)(1-2\times 0.3)}{(1-0.3)} = E = 2.17 \times 10^{11} \frac{\text{N}}{\text{m}^2} = 217 \text{ GPa}$$

Problem 8.5

A wave has a wave length of 600 nm and a frequency of 5×10^{14} Hz. What kind of wave is it?

Solution 8.5

Using the fundamental equation $\lambda = v/\mathrm{f}$, the velocity of the wave is:

$$v = \lambda \times f = (600 \times 10^{-9}) \times (5 \times 10^{14}) = 3 \times 10^{8}\ \mathrm{m/s}$$

3×10^{8} m/s is the speed of light. Therefore the wave is an electromagnetic wave.

Problem 8.6

What is the difference in particle motion between a shear wave and a Rayleigh wave?

Solution 8.6

In shear waves, the wave displaces the particle along a line perpendicular to the direction of the wave. In Rayleigh waves, which are surface waves, the wave displaces the particle along an ellipse in a plane that is in the direction of the wave and perpendicular to the surface.

Problem 8.7

Explain the difference between seismic reflection and seismic refraction techniques.

Solution 8.7

Seismic reflection consists of sending seismic waves down into the soil, receiving the reflected wave at a receiver, and identifying the time that it took for the wave to travel down to the boundary and back to the surface. It typically makes use of P waves.

Seismic refraction involves measuring the travel time of the component of seismic energy that travels down to the top of a layer boundary, is refracted along that boundary, and returns to the surface as a reflected wave. Seismic refraction typically makes use of P waves.

Problem 8.8

Derive the crossover time equation for seismic refraction.

Solution 8.8

$$t_1 = \frac{\overline{af}}{V_1} = \frac{X}{V_1}$$

$$t_2 = \frac{\overline{ac}}{V_1} + \frac{\overline{cd}}{V_2} + \frac{\overline{df}}{V_1}$$

$$\text{Here, } \overline{ac} = \overline{df} = \frac{Z}{\cos\alpha_c}$$

$$\text{and } \overline{cd} = X - 2Z\tan\alpha_c$$

Therefore,

$$t_2 = \frac{2Z}{V_1\cos\alpha_c} + \frac{X - 2Z\tan\alpha_c}{V_2} = \frac{2Z}{V_1\cos\alpha_c} - \frac{2Z\tan\alpha_c}{V_2} + \frac{X}{V_2}$$

$$= 2Z\left(\frac{1}{V_1\cos\alpha_c} - \frac{\tan\alpha_c}{V_2}\right) + \frac{X}{V_2} = 2Z\left(\frac{V_2 - V_1\sin\alpha_c}{V_1V_2\cos\alpha_c}\right) + \frac{Z}{V_2}$$

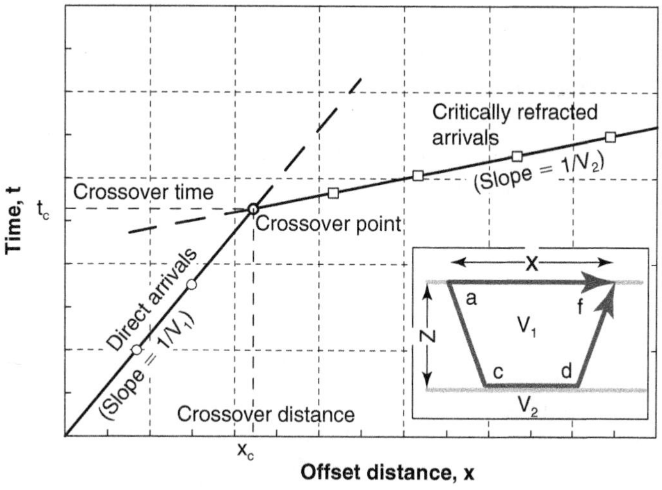

Figure 8.1s Illustration for seismic refraction test.

Based on Snell's law,

$$\sin\alpha_c = \frac{V_1}{V_2},$$

$$\cos\alpha_c = \sqrt{1-\sin^2\alpha_c} = \sqrt{1-\left(\frac{V_1}{V_2}\right)^2}$$

$$t_2 = 2Z\left(\frac{V_2 - V_1^2/V_2}{V_1V_2\sqrt{1-\left(\frac{V_1}{V_2}\right)^2}}\right) + \frac{Z}{V_2} = 2Z\left(\frac{V_2^2 - V_1^2}{V_1V_2^2\sqrt{1-\left(\frac{V_1}{V_2}\right)^2}}\right) + \frac{Z}{V_2} = 2Z\left(\frac{\sqrt{V_2^2-V_1^2}}{V_1V_2}\right) + \frac{X}{V_2}$$

Based on the definition of crossover time for seismic refraction,

$$t_1 = t_2 = t_c, \text{ meanwhile, } X = X_c$$

where t_c is the crossover time and x_c is the crossover distance.

Therefore,

$$\frac{X_c}{V_1} = 2Z\left(\frac{\sqrt{V_2^2-V_1^2}}{V_1V_2}\right) + \frac{X_c}{V_2},$$

$$X_c = 2Z\frac{\sqrt{V_2+V_1}}{\sqrt{V_2-V_1}},$$

$$t_c = \frac{X_c}{V_1} = \frac{2Z\sqrt{V_2+V_1}}{V_1\sqrt{V_2-V_1}}$$

$$\text{and } Z = \frac{t_cV_1}{2}\frac{\sqrt{V_2-V_1}}{\sqrt{V_2+V_1}}$$

Problem 8.9

What is a dispersion curve?

Solution 8.9

The plot of phase velocity v versus wave length λ is the dispersion curve for a given receiver spacing. This plot is obtained for all receivers at different spacing and the individual dispersion curves for each spacing are assembled into a single composite dispersion curve. Once the composite dispersion curve is generated for the site, an iterative forward modeling procedure or an inversion analysis algorithm is used to determine a shear-wave velocity versus depth profile by matching the field dispersion curve with the theoretically determined dispersion curve.

Problem 8.10

Describe the basic concept of the SASW technique.

Solution 8.10

The spectral analysis of surface waves technique is based on the concept that if the velocity of a wave travelling in a material depends only on the physical properties of the material, then the wave velocity is constant and independent of frequency. Waves traveling through such nondispersive material will maintain a constant shape. This would be the case of a wave travelling in a soil that had uniform properties independent of depth. However, soils have properties that vary with depth because of differences such as layering and variations in effective stress; therefore, soils are dispersive materials. As a result, when many waves with different frequencies travel through the material, the wave train contains a lot of waves with individual frequencies, and the shape of the wave train changes as the wave travels. Some waves within the wave train travel faster than the wave train (longer wave length) and die out as they approach the leading edge. Some waves within the wave train are slower than the wave train (shorter wave length) and die out as they approach the trailing edge. SASW makes use of Rayleigh waves because they travel along the ground surface and they attenuate a lot less than body waves. SASW links the frequency content of the wave train to the shear-wave velocity profile of the soil at a site. In the field, the test consists of placing receivers on the ground surface at regular intervals away from where the shock is generated during the test. The receivers are placed along a single radial path from the impact location. These instruments measure the vertical movement of the soil as the waves pass by. This procedure is repeated about 6 to 8 times to obtain a shear-wave velocity versus depth profile.

Problem 8.11

Describe the basic concept of the electrical tomography technique.

Solution 8.11

Electrical tomography is a geophysical technique in which a current is passed between metal electrodes inserted into the ground. For soils composed of different layers having different electrical resistivity, the resistivity obtained from the measurements is an equivalent or apparent electrical resistivity, depending on the layers involved in the electrical response to the potential difference. Because these measurements are made at different depths and in different directions, a three-dimensional image of the site can be obtained through an inversion process.

Problem 8.12

What is an electromagnetic wave, and what are its main properties?

Solution 8.12

Electromagnetic waves can propagate in both a material and a vacuum. Light is one example of an electromagnetic wave. You can create an electromagnetic wave by shaking an electron; the electron will create a wave that will propagate as ripples across the vacuum of space. When you shake the electron, the electromagnetic wave propagates as a transverse wave, a wave where the motion is perpendicular to the direction of propagation. What moves is an electric field in the vertical direction and a magnetic field in the horizontal direction called a photon. Photons represent bundles of energy that can be equally considered as particles with zero mass or waves.

Some of the properties are:

a. The direction of motion is perpendicular to the propagation of the wave.
b. The speed of propagation of an electromagnetic wave is the speed of light (300,000 km/s in vacuum)
c. The wave length λ of electromagnetic waves varies significantly. For example, the radio waves used in broadcasting have a wave length varying from 3000 to 0.3 m, the microwaves used to heat food have a wave length varying from 0.3 to 3×10^{-4}m, and the infrared light waves used for night vision and muscle therapy have a wave length varying from 3×10^{-4} to 4×10^{-7}m.

Problem 8.13

Describe the basic concept of the GPR.

Solution 8.13

Ground-penetrating radar is a nondestructive geophysical method that uses electromagnetic waves in the microwave range to penetrate the soil and give an image of the subsoil. The waves are generated by a source antenna in contact with the ground surface; the waves propagate in the soil, reflect from anomalies such as layer interfaces, cavities, and buried objects; and travel back to the surface, where they are detected by a receiver antenna also in contact with the ground surface. GPR can be used in a variety of media, including soil, rock, ice, pavement, and structures. It can also be used to detect changes in material properties and the presence of voids or cracks, among other things.

Problem 8.14

Describe the basic concept of the TDR.

Solution 8.14

Materials are classified as conductors or insulators, also called dielectric, depending on their ability to conduct electricity. The dielectric permittivity ε of a soil is a measure of how fast an electromagnetic wave propagates through the soil:

$$\varepsilon = (c/v)^2 \quad \text{(see equation 8.30)}$$

where c and v are the velocity of the electromagnetic wave in the soil and in a vacuum (300,000 km/s) respectively. If a rod is embedded in a soil mass, the velocity of the electromagnetic wave propagating in the rod will be affected by the permittivity of the soil surrounding the rod. This velocity can be obtained by measuring the length of the rods embedded in the soil and the time required for the wave to travel down the rod and back; then the soil dielectric permittivity can be obtained from this measurement. For example, a dry soil has a dielectric permittivity value of around 4, moist soils of around 30, water of about 80, and air close to 1. Obviously, water significantly affects the dielectric permittivity and advantage is taken of this fact to relate soil dielectric permittivity to the soil water content. Calibrations are necessary to obtain the best correlation equation. This technique is used to obtain the soil density as well as the soil water content. Such measurements are particularly useful on compaction projects, such as the field performance of landfill covers.

Problem 8.15

Describe the basic concept of LIDAR.

Solution 8.15

LIDAR stands for light detection and ranging and is sometimes called laser radar. The LIDAR test consist of sending a laser beam of electromagnetic waves (light) at an object and detecting the time required for that beam to reflect on the object and come back to the LIDAR receiver. The beam is sent in a series of wave pulses at a very high frequency. Because the beam travels at the speed of light in air (close to 300,000 km/s), the time involved is measured in nanoseconds or picoseconds. These very short times can be measured with instruments such as optoelectronic streak cameras. Knowing the time of flight and the speed of the wave, the distance can be back-calculated. LIDAR works like a camera as it sweeps through the landscape it is aimed at and records the distance of all objects it is seeing. The result of a LIDAR test is a three-dimensional image of the landscape swept by the LIDAR where all points recorded are documented with coordinates. LIDAR works equally well during the daytime and during the night.

CHAPTER 9

Laboratory Tests

9.1 GENERAL

Laboratory testing, in situ testing, and geophysical testing are the options a geotechnical engineer has to obtain the soil information necessary for a geotechnical engineering project. There are advantages and drawbacks to each one of these options (see Table 6.1). Among the advantages of laboratory tests are that they lend themselves to theoretical analysis, that the boundary drainage conditions can be controlled, and that the boundary loading conditions can also be controlled. Some of the drawbacks are the small scale of such testing and the influence of disturbance on the results. Many laboratory tests are available, as shown in Figure 9.1. They are typically classified in the following main categories:

1. Tests for index properties (e.g., water content, unit weight, particle size, Atterberg limits)
2. Tests for deformation properties (e.g., consolidation, triaxial, simple shear, resonant column)
3. Tests for strength properties (e.g., direct shear, unconfined compression, triaxial, lab vane)
4. Tests for flow properties (e.g., constant head permeameter, falling head permeameter, erosion tests)

In each category, one can also distinguish between static tests and dynamic tests. The measurements made during the tests include normal stress, shear stress, normal strain, shear strain, displacements, water compression stress, water tension stress, and air stress. The tests for index properties were presented in Chapter 4. This chapter presents some of the laboratory tests that are often used in practice. They include the consolidation test, the direct shear test, the simple shear test, the unconfined compression test, the triaxial test, the resonant column test, the lab vane test, the constant head permeameter test, the falling head permeameter test, and the erosion function apparatus test.

9.2 MEASUREMENTS

9.2.1 Normal Stress or Pressure

In laboratory testing, measurements of normal stress (Figure 9.2) are most often made by measuring the force and dividing by the area, although normal stress measurements can also be made by using a pressure cell. The measurement of force is done in a number of ways. The simplest way is to add weights on a hanger, as in the classic consolidation test. A proving ring is another device to measure force. It is a stiff steel ring inserted between a jack and the sample; the proving ring is deformed like a spring and the decrease in diameter of the ring is measured using a dial gage. A dial gage is made of a stem with indentations that make a wheel rotate as the stem moves up and down; this wheel rotates a needle on a graduated dial. Dial gages are precise down to a few micrometers. A load cell is the most common way to measure load; it consists of a deformable piece of steel (S shape or cylindrical) instrumented with strain gages. The most common and inexpensive are foil strain gages made of very thin metal strips glued to a surface and connected with an electrical circuit. A change in length of the strain gage created by the deformation of the piece to which the strain gage is glued induces a change in voltage, which is recorded. The change in voltage is correlated with the change in strain of the piece to which the strain gage is glued and therefore to the change in stress and then the change in force. Measurements of normal stress or pressure can also be made by using pressure cells. Such cells are circular and have a metallic membrane that deforms when it is in contact with the stressed soil. The deflection of the membrane is measured with strain gages glued to that membrane and related to the pressure on the membrane. Alternatively, the pressure gage is filled with a fluid and the pressure in the fluid is measured through a diaphragm further away.

9.2.2 Shear Stress

The simplest way to measure shear stress is to measure the shear force and divide by the corresponding area. This is done in the direct shear test. Alternatively, the shear stress can be measured by a shear stress transducer, an example of which is shown in Figure 9.3. In this example, two thin posts are equipped with strain gages to quantify the bending of the posts when a shear force is applied to the top platen.

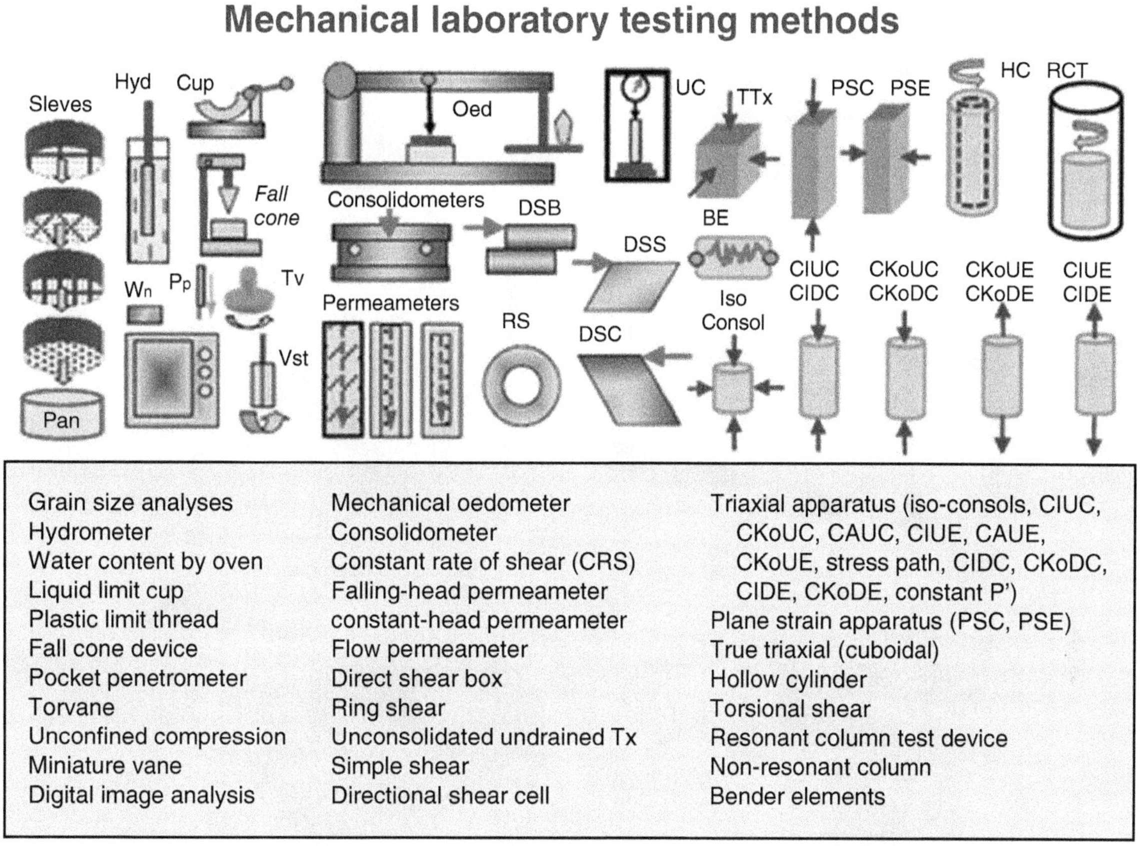

Figure 9.1 Laboratory tests. (From Mayne et al. 2009. Courtesy of Professor Paul Mayne, Georgia Institute of Technology, USA.)

Calibration of the transducer links the readings from the strain gages to the shear stress on the top platen.

9.2.3 Water Compression Stress

Water compression stress is also called *pore water pressure*. It can be measured through a manometer or through a pore-pressure transducer. A *manometer* or *standpipe* is simply a pipe connected to the point where the water compression stress is to be measured and open to the atmosphere at the other end. The pressure in the water makes the water rise in the manometer to the point of equilibrium. The water compression stress is then calculated as the vertical distance between the point of measurement and the water level in the manometer times the unit weight of water. A pore-pressure transducer measures the water pressure by letting that pressure deflect a membrane. A porous tip made of ceramic (Figure 9.4) is placed in contact with the soil where the water is in compression. This porous tip, which is saturated with de-aired water, allows water to come in but does not allow air to come in. This is called a *high air entry porous stone*. Behind the porous tip is a deformable body that responds to the pressure in the water. This could be a thin plate equipped with strain gages, although today it is most commonly a piezoelectric crystal. These crystals have the property of producing a voltage difference between the two sides of the crystal when they are subjected to deformations. So, by calibrating the crystal and measuring the voltage difference across the two sides, one can obtain the pressure.

9.2.4 Water Tension Stress

The tension stress in the water of a soil sample is generated by the suction potential. Suction has two components: matric suction and osmotic suction. Sometimes osmotic suction exists as a potential but is not realized as water tension (see Chapter 11 for more on this). Water tension and suction are usually measured in units of kPa, but sometimes water tension and suction are measured using the pF unit. A pF is defined as the decimal logarithm of the absolute value of the water tension stress or suction expressed in centimeters of water. For example, a water tension of −1000 kPa would correspond to −10,000 cm of water or 4 pF. Table 9.1 gives the equivalences for common values. The symbol *pF* reminds us of the chemical unit of pH which refers to the potential of hydrogen. The pF unit may be interpreted as the potential of flow because the tension in the water would create flow if water became available. Although the pF unit is not accepted

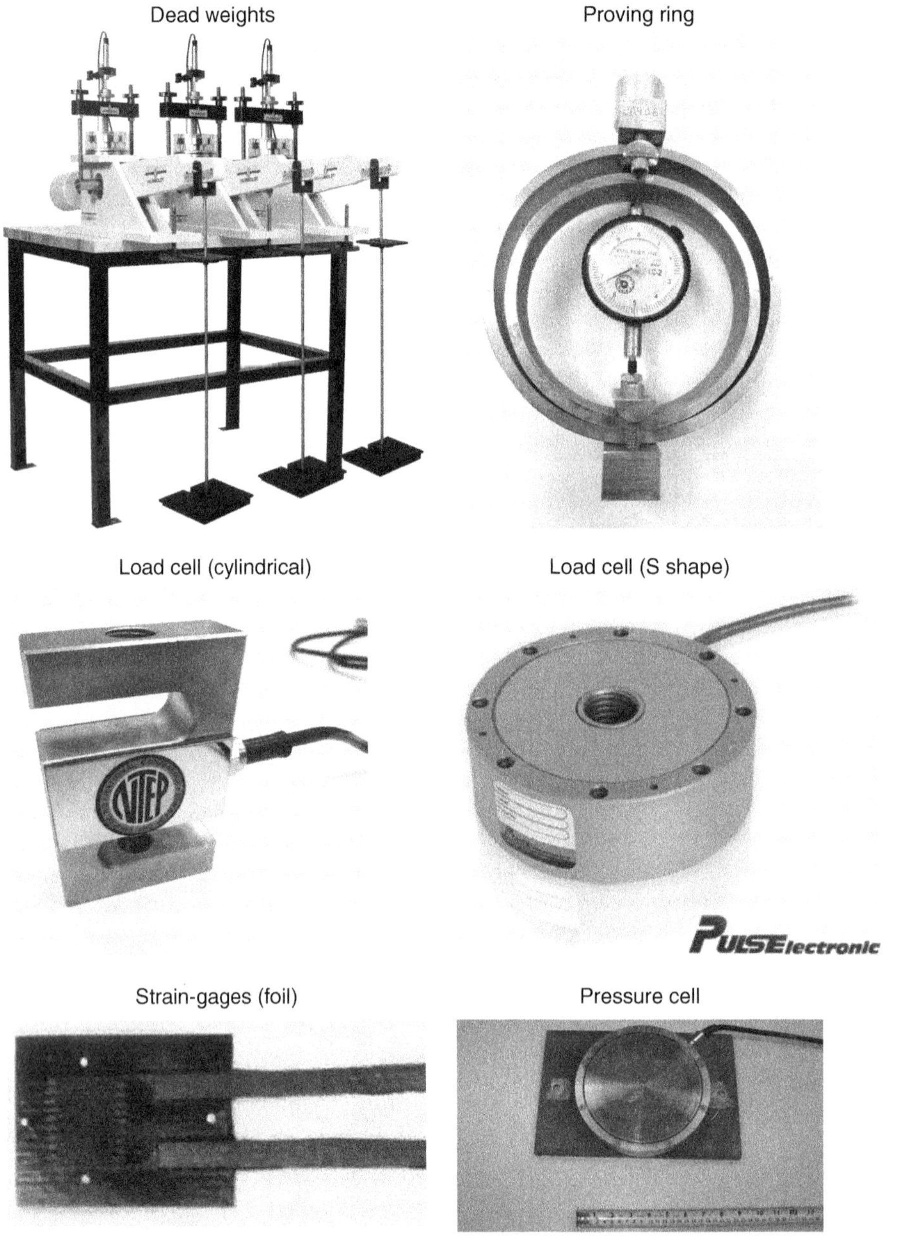

Figure 9.2 Devices to measure normal stress. (*a*) Dead weights. (*b*) Proving ring. (*c*) Load cell (S shape). (*d*) Load cell cylindrical. (*e*) Strain gages (foil). (*f*) Pressure cell. (*a*: Courtesy of Humboldt Mfg. Co.. *d*: Courtesy of Mediscale1.)

by the International Society for Soil Mechanics and Geotechnical Engineering, it conveys the message that using the log of the water tension is convenient.

Measuring water tension is not as easy as measuring water compression. As a result, there are many different measurement methods applicable to different ranges of water tension. These methods include filter paper, psychrometer, tensiometer, pressure plate apparatus, and salt solution equilibrium. Note that some devices or methods are geared toward measuring the natural water tension or suction in a sample, whereas others are geared toward forcing the sample to reach a chosen water tension. The second kind is often used to develop the soil water retention curve, also called the soil water characteristic curve, for a soil. Table 9.2 summarizes the range of application of these devices or methods.

Filter Paper Method

The filter paper method is the simplest of all. It consists of using a circular piece of filter paper (about 50 mm in diameter), weighing it dry, placing it either in contact with or above the soil sample, enclosing the filter paper and the sample in a sealed container until the filter paper comes

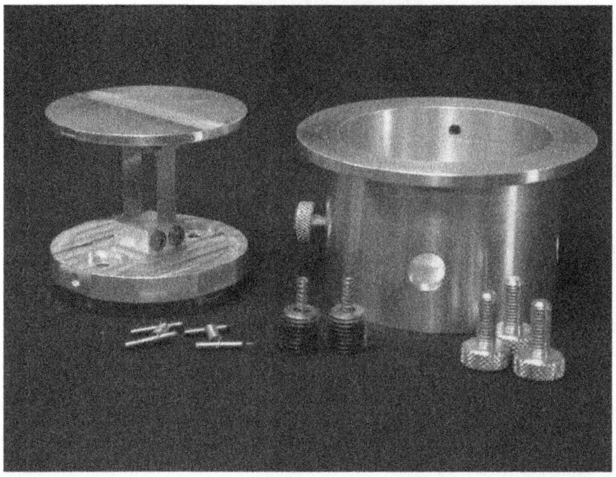

Figure 9.3 Shear stress sensor. (Courtesy of Department of Mechanical Engineering, University of Idaho, USA)

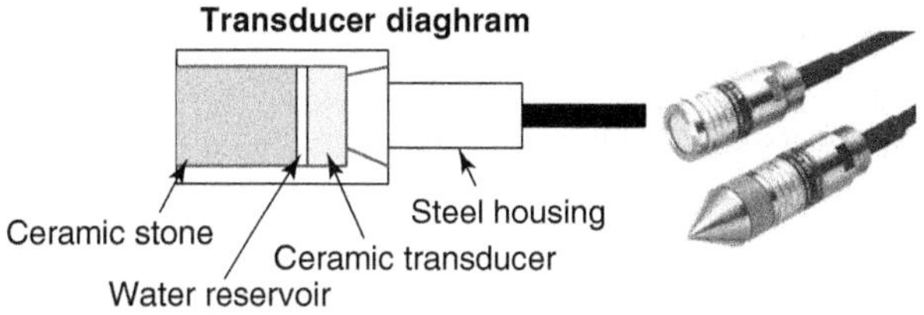

Figure 9.4 Water compression stress transducer or pore-pressure transducer. (*Right*: Courtesy of Bestech Australia provided, © 2010 Tokyo Sukki Kenkyujo Co., Ltd.)

into water tension equilibrium with the sample, retrieving the filter paper, and weighing it to obtain its water content (Bulut, Lytton, and Wray 2001). Because the soil sample is much larger than the filter paper, the water content of the sample remains unaffected by the amount of moisture drawn into the filter paper. The filter paper comes calibrated with a curve linking the filter paper moisture content and the water tension in the filter paper. Because the water tension is the same in the filter paper and the sample, the water tension of the sample is given in that fashion. Figure 9.5 shows the test in progress and a calibration curve. If the filter paper is in contact with the sample, the water drawn into the filter paper has the same chemistry as the water in the sample; therefore, the water tension due to osmotic suction is not distinguishable and only the water tension due to matric suction is measured. In contrast, if the filter paper is not in contact with the sample, then the water in the filter paper is pure water, while the water in the sample has its own chemistry. In this case the water tension due to osmotic suction is realized in addition to the matric suction and the water tension measured corresponds to the total suction. Note that the part of the procedure dealing with the weighing of the filter paper must be performed

Table 9.1 Equivalency between kPa and pF Units

Water Tension in kPa	Water Tension in pF (log(cm of H_2O))
−10	2
−100	3
−1000	4
−10000	5
−100000	6
−1000000	7

Table 9.2 Methods to Measure Water Tension Stress or Suction

Device or Method	Water Tension or Suction	Range (kPa)	Natural or SWRC*	Time Required	Comments	ASTM
Filter paper	Total	Entire range	Natural	1 to 2 weeks	May measure matric suction if in good contact	D5298
Thermocouple psychrometer	Total	100 to 8000	Natural	1 to 2 hours	Constant environment required	E337
Chilled mirror psychrometer	Total	1000 to 8000	Natural	10 minutes	Scatter at suction values less than 1000 kPa	D6836
Tensiometer	Matric	0 to 90	Natural	10 minutes	Difficulties with cavitation and diffusion through ceramic cup	D3404
Pressure plate	Matric	0 to 1500	Natural or SWRC*	1 to 5 days	Difficulties with pressures higher than 1500 kPa	D6836
Salt solution	Total	Entire range	SWRC*	1 to 2 weeks	Mainly used for calibrating other devices	None

*Soil water retention curve; also called soil water characteristic curve.

Figure 9.5 Filter paper method for water tension measurement. (a) Filter paper (matric suction only). (b) Filter paper (total suction). (c) High-precision scale with hood. (d) A calibration curve. (From Bulut et al., 2001. Courtesy of Dr. Bulut)

extremely carefully and quickly, as the weights involved are very small and the relative humidity of the air in the laboratory can influence the weight of the filter paper when it is transferred from the sample chamber to the scale chamber.

Thermocouple Psychrometers

Thermocouple psychrometers (*psykhros* means "cold" in Greek) can be used to give the total suction of a soil by measuring the relative humidity in the air phase of the soil pores or the region near the soil (Figure 9.6). They measure the total suction because the evaporation process creates pure water, while the water in the soil pores is not pure water. Hence, the osmotic suction is realized. Psychrometers give the relative humidity by measuring the difference in temperature between a nonevaporating surface and an evaporating surface. Imagine two thermometers, one with a dry bulb and the other with a wet bulb. The dry-bulb thermometer measures the ambient temperature, but the wet-bulb thermometer measures a temperature lower than ambient because the evaporation of the water on the bulb cools the bulb. The thermometers can be replaced by transistors in transistor psychrometers. If the air phase has a low relative humidity, the evaporation is faster, the cooling process is high, and the difference in temperature is larger. If the air phase has a high relative humidity, little evaporation takes place, the cooling process is limited, and the difference in temperature is smaller. The difference in temperature given by the two thermometers is related to the relative humidity, which in turn is related to the water tension or total suction. In the pores of a soil, there has to be a balance between the water tension in the air phase and in the water phase. See Chapter 11 for more details on these relationships. Because psychrometers work on the basis of precise temperature measurements, any exterior fluctuation in temperature will lead to poor precision. Therefore, psychrometers are not well suited for in situ measurements, because of the daily temperature cycle. It also takes a fair amount of time for equilibrium to be reached between the psychrometer and the air in the soil pores.

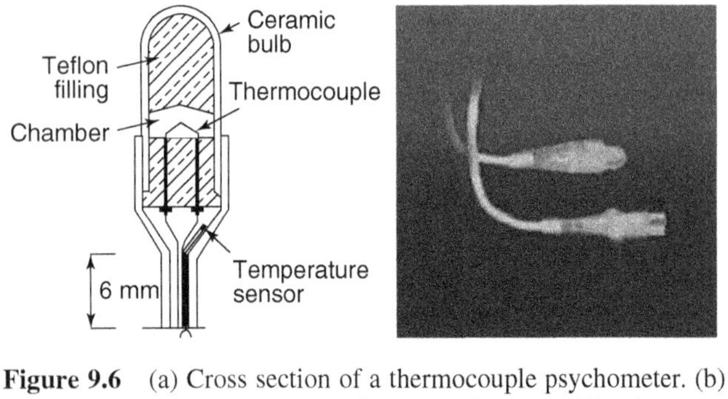

Figure 9.6 (a) Cross section of a thermocouple psychometer. (b) Thermocouple psychometer. (b: Courtesy of Wescor-Elitechgroup.)

Chilled Mirror Psychrometers

Chilled mirror psychrometers can be used to give the total suction of a soil (Figure 9.7). Much like the thermocouple psychrometers, they measure the relative humidity and then relate the relative humidity to the suction. The relative humidity in a chilled mirror psychrometer is obtained as follows: The soil is inserted into a small chamber that is sealed off from the outside air and has a mirror present. Facing the

Figure 9.7 Chilled mirror psychrometer. (Photo courtesy of Decagon Devices, Inc.)

mirror is a camera able to detect when dew forms on the mirror. The air in the chamber comes to relative humidity equilibrium with the air in the soil sample. Then the mirror is chilled down to the point where dew forms on the mirror and the temperature of the mirror at that point is recorded. The temperature of the soil is also recorded and the difference in temperature between the mirror at the dew point and the soil is related to the relative humidity in the soil. The suction is then obtained through its relationship with the relative humidity (see Chapter 11).

Tensiometers

Tensiometers can be used to measure the water tension or matric suction in a soil (Figure 9.8). A tensiometer consists of a high air entry porous ceramic tip (also called a ceramic cup) that is saturated with water and placed in good contact with the soil. In the tensiometer, the space behind the ceramic tip is filled with de-aired water and connected with a negative pressure measuring device. The stress slowly equalizes between the water tension in the tensiometer and the water tension in the soil pores. That tension is then measured either through a water-mercury manometer, a Bourdon-vacuum tube, or an electrical pressure transducer. The water tension that can be measured in a tensiometer is limited to approximately negative 90 kPa (2.95 pF) due to the possibility of water cavitation in the tensiometer above such a value.

Pressure Plate Apparatus (PPA)

The pressure plate apparatus (PPA) is a closed pressure chamber that can be used to increase the air pressure in the soil pores to the point where the air chases the water out of the pores (Figure 9.9). The sample is placed in the chamber on a high air entry ceramic disk. This disk, which is saturated with water, has the property of letting water go through but not air, up to a certain rated pressure, known as the air entry value of the disk. The air pressure is increased and the stress in the water is increased accordingly (decrease in tension). When the water tension becomes equal to zero, the water comes out and at that point, the air pressure is equal to the water tension. This technique is called the axis translation technique because it simply translates the origin of reference by applying an air pressure equal to the water tension (Figure 9.10).

The PPA can be used to determine the natural water tension or to generate a soil-water retention curve. If the soil sample is placed at its natural water content in the PPA, the air pressure that starts the water flow is the natural water tension. If the soil specimen starts as a saturated sample and the air pressure is increased in steps, each pressure step will drive

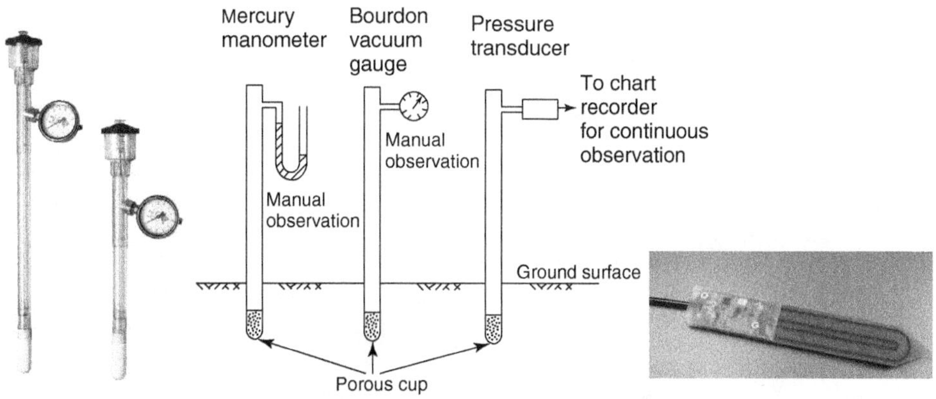

Figure 9.8 Tensiometers. (a) Tensiometer with pressure-vacuum gage. (b) Types of tensiometers. (c) Tensiometer with pressure transducer. (a: Courtesy of Envco Global. *c*: Courtesy of STEP Systems GmbH, www.stepsystems.de)

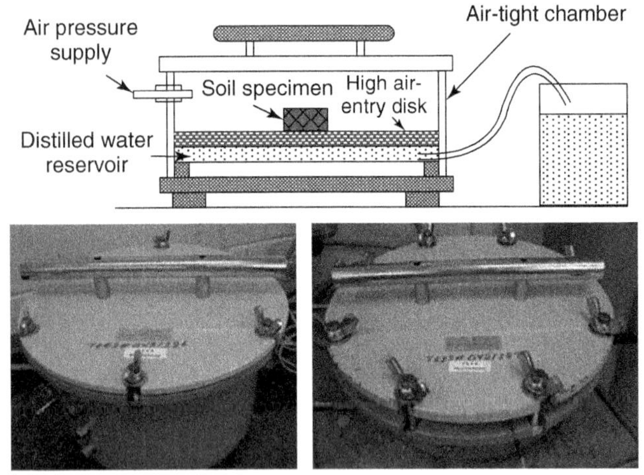

Figure 9.9 Pressure plate apparatus: (*a*) 500 kPa pressure plate. (*b*) 1500 kPa pressure plate.

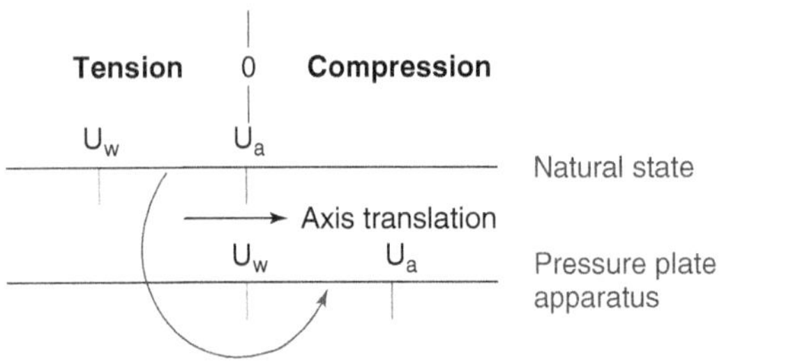

Figure 9.10 Axis translation for water tension determination.

water out of the sample until equilibrium is reached, and this will give the water tension corresponding to the water content of the sample. This water content can be measured separately by stopping the test or inferred from the water loss read on the burette connected to the PPA. The air pressure is increased in steps and each step gives the water tension and the corresponding water content. The soil-water retention curve (SWRC) is thus obtained. The range of application of the PPA is from 0 to about 1500 kPa (4.17 pF).

Salt Solution Equilibrium (SSE)

Salt solution equilibrium (SSE) is a water tension measurement technique which relies on the fact that salt solutions have significant osmotic suction. As explained in Chapter 11, osmotic suction comes from the fact that water molecules are attracted to salt molecules: more salt, more attraction. A closed chamber with a salt solution at its lower part (Figure 9.11) will generate a certain relative humidity in the air above it. The higher the salt concentration is, the lower the relative humidity above the solution in the chamber will be. If a soil sample is suspended in the air above the salt solution, it will dry and the water tension in the soil sample will come to equilibrium with the ambient relative humidity. At equilibrium, the water tension is given by the relative humidity in the air of the chamber. This relative humidity depends on the salt concentration in the solution and can be calculated from it (see Chapter 11). This relationship depends on the type of salt, the molality, and the temperature. Table 9.3 gives the osmotic suction for different salts and

Figure 9.11 Salt solution equilibrium containers for water tension determination.

Table 9.3 Osmotic Suction in kPa of Some Salt Solutions at 25°C

	Osmotic Suction in kPa at 25°C						
Molality (mol/kg)	N_aCl	KCl	NH_4Cl	Na_2SO_4	$CaCl_2$	$Na_2S_2O_3$	$MgCl_2$
0.001	5	5	5	7	7	7	7
0.002	10	10	10	14	14	14	14
0.005	24	24	24	34	34	34	35
0.010	48	48	48	67	67	67	68
0.020	95	95	95	129	132	130	133
0.050	234	233	233	306	320	310	324
0.100	463	460	460	585	633	597	643
0.200	916	905	905	1115	1274	1148	1303
0.300	1370	1348	1348	1620	1946	1682	2000
0.400	1824	1789	1789	2108	2652	2206	2739
0.500	2283	2231	2231	2582	3396	2722	3523
0.600	2746	2674	2671	3045	4181	3234	4357
0.700	3214	3116	3113	3498	5008	3744	5244
0.800	3685	3562	3558	3944	5880	4254	6186
0.900	4159	4007	4002	4384	6799	4767	7187
1.000	4641	4452	4447	4820	7767	5285	8249
1.200	5616	5354	5343	N/A	N/A	N/A	N/A
1.400	6615	6261	6247	N/A	N/A	N/A	N/A
1.500	N/A	N/A	N/A	6998	13391	7994	14554
1.600	7631	7179	7155	N/A	N/A	N/A	N/A
1.800	8683	8104	8076	N/A	N/A	N/A	N/A
2.000	9757	9043	9003	9306	20457	11021	22682
2.500	12556	11440	11366	11901	29115	14489	32776

(After Bulut et al. 2001.)

different molalities. *Molality*, in this case, is the number of moles of salt per kilogram of water. Note that in most cases, molarity is different from molality because *molarity* is the number of moles per liter of solvent.

The range of application for the SSE technique is very wide, from 0 to close to 100,000 kPa (6 pF). It is also a very inexpensive and very reliable technique. Hence, it is used as a reference to calibrate many other techniques. The drawback is that it is quite time consuming: The time necessary for equilibrium to be reached between the water tension in the soil sample and the relative humidity in the surrounding air can be a couple of weeks.

9.2.5 Normal Strain

A normal strain ε is defined in one direction as the change in length Δz divided by the initial length z between two points. A normal strain is measured either by measuring a displacement and a length ($\Delta z/z$) or by using a strain gage (ε). Measurements of length are done with a ruler or a set of calipers (Figure 9.12). Displacements are measured with mechanical devices such as dial gages (Figure 9.12) or electrical devices such as LVDTs, DCDTs, and potentiometers.

A linear variable differential transformer (LVDT) (Figure 9.13) has three solenoid coils arranged like three side-by-side donuts. A small metallic rod is attached to the point where the displacement is to be measured and the solenoids are attached to an immobile reference point. The small rod passes through the center of the three solenoids without touching them. An alternating current through the center solenoid creates a voltage in the side solenoids. The movement of the metallic rod creates a change in voltage that is linearly proportional to the movement of the rod. The change in voltage is transformed into a displacement measurement through calibration. A direct current differential transformer (DCDT) is an LVDT in which the current passing through the solenoids is a direct current instead of an alternating current. A potentiometer or pot is a resistor with three terminals. Two are fixed and one moves between the two fixed terminals. By sliding the moving terminal, the resistance offered by the potentiometer varies and so does the voltage. The rod connected to the point where the movement

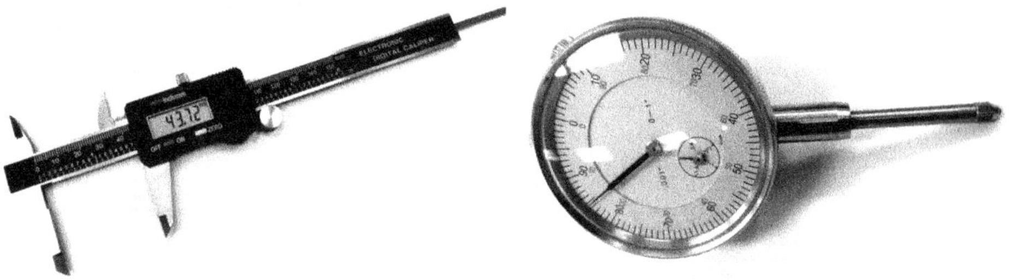

Figure 9.12 Mechanical devices to measure displacement: (a) Calipers. (b) Dial gage.

is to be measured is tied to the sliding terminal. The change in voltage induced by the movement of the rod is related to the movement through calibration.

Strain gages are of two main types: foil strain gages and vibrating wire strain gages. A foil gage is a thin sheet of metal (copper-nickel alloy is common) with a pattern (Figure 9.14) glued to the material that is deforming. Actually, a layer of flexible insulating material is first glued to the deforming material and then the foil gage is glued onto the insulator, so that the current passing through the gage only travels through the gage. When the material deforms, the foil length changes and so does its resistance. The voltage changes accordingly and the strain is related to the change in voltage through calibration. Vibrating wire strain gages consist of two small anchor blocks solidly connected to the material that is deforming. Between these two anchors is a high-tensile-strength wire brought taut to a chosen initial load. Around the wire is a cylinder that protects the wire and contains a permanent magnet and a plucking coil. When the wire is plucked, it vibrates at its natural frequency. If the material deforms, the end blocks move and the natural frequency of the vibrating wire changes. The change in natural frequency of the wire is related to the normal strain by theory and calibration.

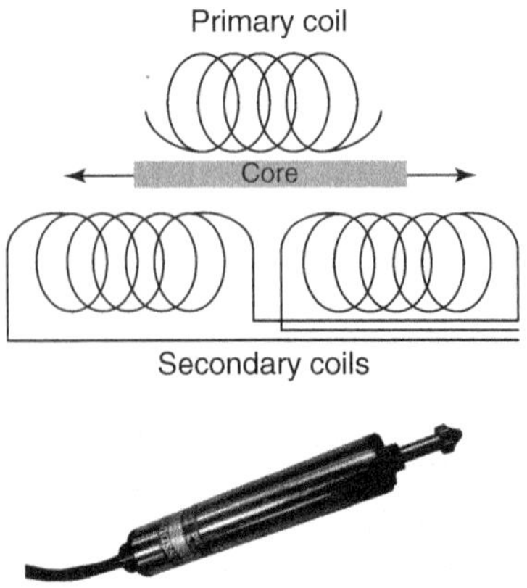

Figure 9.13 Linear variable differential transformers: (a) Principle. (b) Device.

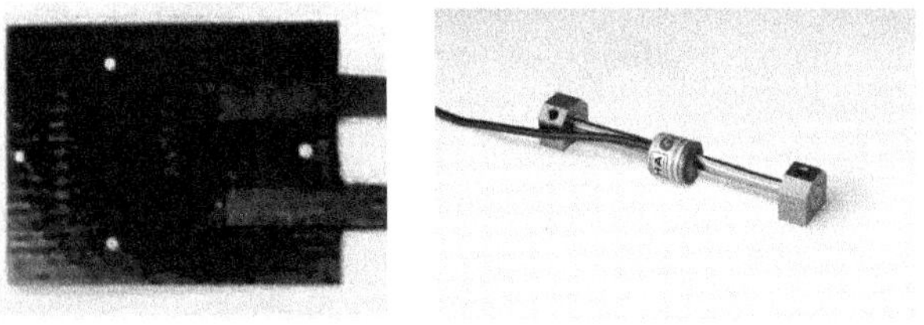

Figure 9.14 Foil and vibrating wire strain gages. (a) Foil strain gage. (b) Model 4000 Vibrating Wire strain gage. (*b*: Courtesy of Geokon, Inc.)

9.2.6 Shear Strain

A shear strain γ is defined for two perpendicular directions (x and y as shown in Figure 9.15). When the shear strain is small enough, the shear strain is equal to the change in angle γ expressed in radians between the two perpendicular directions due to the shearing process. Obtaining shear strain is most easily done by measuring the normal strain in two perpendicular directions (Figure 9.15). It can be shown (Chapter 10) that the shear strain in this case is given by:

$$\gamma_{xy} = \varepsilon_1 - \varepsilon_2 \tag{9.1}$$

9.2.7 Bender Elements

A *bender element* (Figure 9.16) is a small electromechanical device used to generate or sense bending waves. It is made of two thin piezoceramic plates glued together. Between the two plates and on the outside of the two plates are conducting surfaces. Because of the different polarizations of the two plates, when a voltage is driven through the plates, one shortens and the other lengthens; this forces the plates to shake in bending. If the small plates are buried in the soil,

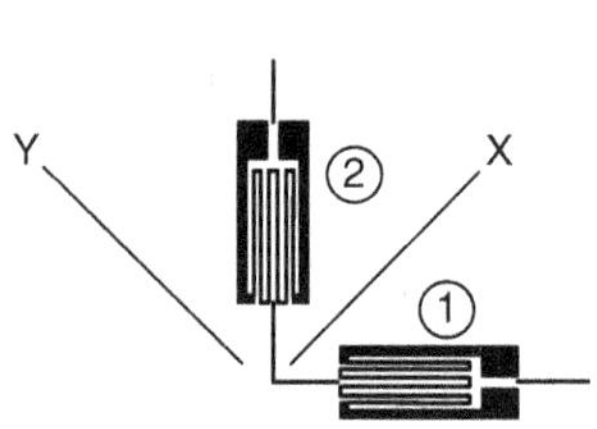

Figure 9.15 Getting shear strain from two normal strain gages.

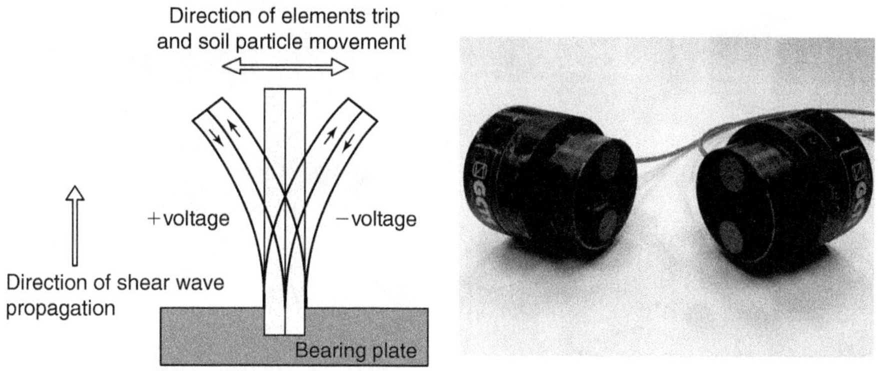

Figure 9.16 Bender elements: (a) Principle. (b) Device.

the repeated lateral motion of the plates generates a wave that propagates in shear through the soil. This is the wave generation function of a bender element. At the other end of the sample, a similar bender element is also buried in the soil and acts as a receiver. This receiver senses the arrival of the shear wave because that wave forces the two plates to move sideways. This bending movement shortens one and lengthens the other; this alternating tension and compression creates an electrical signal that can be measured. When the bender element generates a shear wave, the wave travels through the soil and reaches the bender element, which detects its arrival. Knowing the length of travel (sample length) and the time necessary for the wave to propagate from the generating bender element to the receiving bender element, one can calculate the shear-wave velocity v_s. Theory on shear-wave propagation in an elastic body tells us that the shear modulus G of the soil from measurement of shear-wave velocity v_S is given by:

$$G = \rho\,(v_S)^2 \tag{9.2}$$

where ρ is the mass density of the soil sample. Note that the shear modulus measured in this fashion is associated with very small shear strains.

9.3 COMPACTION TEST: DRY UNIT WEIGHT

9.3.1 Saturated Soils

Most of the time, the soil in a compaction test is unsaturated.

9.3.2 Unsaturated Soils

The compaction test dates back to the work of Ralph Proctor, an American civil engineer, in the early 1930s. Today, the test is actually two tests: the Standard Proctor Compaction Test (SPCT; ASTM D698) and the Modified Proctor Compaction Test, (MPCT; ASTM D1557). Proctor developed the SPCT, but in the late 1950s, as compaction machines became much bigger than in the 1930s, the MPCT was developed to better correspond to the higher energy generated by the larger roller compactors. In both cases, the result of the test is the dry unit weight γ_d vs. water content w curve (Figure 9.17).

The first step in the SPCT is to take a soil sample, dry it, break the clumps of soil down to individual particles (e.g., with a mortar and rubber-tip pestle), and measure its weight W_d. Then, calculate the weight of water W_w that must be added to the dry soil sample to reach a chosen water content w:

$$W_w = w\,W_s \tag{9.3}$$

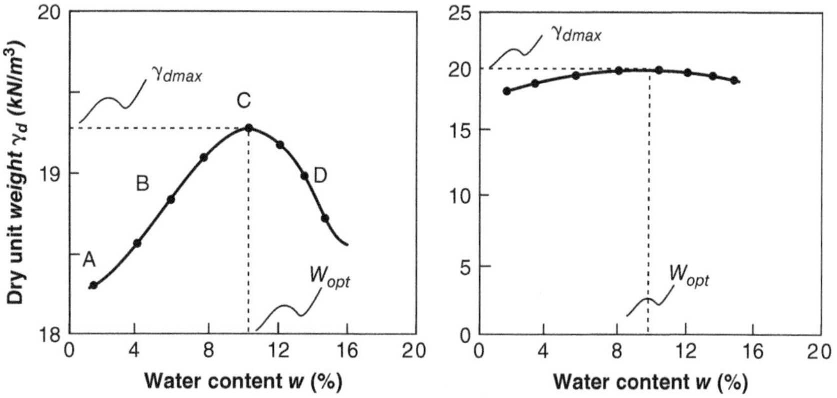

Figure 9.17 Compaction curve.

Figure 9.18 Compaction equipment and test: (a) Compaction mold. (b) Compaction test. (c) Compaction hammer. (a and c: Courtesy of Forney LP, Hermitage, PA.)

Add the water to the soil and mix thoroughly. Weigh the empty compaction mold to be used for the test. Using the prepared soil mixture, place a first layer in the compaction mold (Figure 9.18) and compact that layer of loose soil by dropping a standard compaction hammer a standard number of times. The blows should be distributed evenly across the soil layer to reach uniform compaction. Repeat this process for all layers and aim for the last layer to coincide with the top of the mold. Two mold sizes are used; Table 9.4 gives the detailed requirements. At the end, weigh the mold plus soil and calculate the soil weight W_t. The dry unit weight is obtained by:

$$\gamma_d = \frac{W_t}{V_t(1-w)} \tag{9.4}$$

where γ_d is the dry unit weight, W_t is the total weight of the soil sample in the mold, V_t is the total volume of the sample, and w is the water content of the sample. The combination of γ_d and w gives one point on the compaction curve. By repeating the SPCT for different water contents, the compaction curve is described point by point (Figure 9.17). Note that this curve has a well-defined bell shape because the vertical scale is concentrated around the range of values within which the dry unit weight varies. If the same curve is plotted at the full scale of the unit weight, the curve still has a bell shape but shows that the dry unit weight is not very sensitive to the water content.

Table 9.4 Compaction Requirements for Standard Proctor Compaction Test

102 mm diameter 116 mm high mold	152 mm diameter 116 mm high mold
3 soil layers	3 soil layers
25 blows per soil layer	56 blows per soil layer
Hammer weight 24.5 N	Hammer weight 24.5 N
Hammer drop height 305 mm	Hammer drop height 305 mm
Volume 9.43×10^{-4} m^3	Volume 21.2×10^{-4} m^3
Total energy 600 kN.m/m^3	Total energy 600 kN.m/m^3

The reason for this bell curve is that at point A on Figure 9.17 the soil is relatively dry and it is difficult for a given compaction energy to bring the particles closer together. At point B the water content is such that water tension exists between the particles and hinders the effectiveness of the compaction process. At point C, the water tension loses its effect and the primary role of the water is to lubricate the contacts between particles, thereby allowing the given compaction effort to reach a low void ratio and a high dry density. At point D the soil is nearing saturation and the added water simply increases the volume of the voids, which negates the benefit of the compaction.

The compaction curve is bounded on the right side by the saturation line for a degree of saturation equal to 1. Indeed, the relationship between the dry unit weight γ_d and the water content w is a function of the degree of saturation:

$$\gamma_d = \frac{SG_s\gamma_w}{S+G_s w} \tag{9.5}$$

This relationship can be demonstrated as follows:

$$\gamma_d = \frac{W_s}{V_t} = \frac{\gamma_s V_s}{V_v+V_s} = \frac{G_s\gamma_w}{1+\dfrac{V_v}{V_s}} = \frac{SG_s\gamma_w}{S+\dfrac{V_w}{V_v}\dfrac{V_v}{V_s}}$$

$$= \frac{SG_s\gamma_w}{S+\dfrac{V_w}{W_w}\dfrac{W_w}{W_s}\dfrac{W_s}{V_s}} = \frac{SG_s\gamma_w}{S+G_s w} \tag{9.6}$$

Equation 9.6 shows that the relationship between the dry unit weight and the water content for a given degree of saturation S is a hyperbola. This hyperbola is called the *saturation line* and corresponds to S (Figure 9.19). The

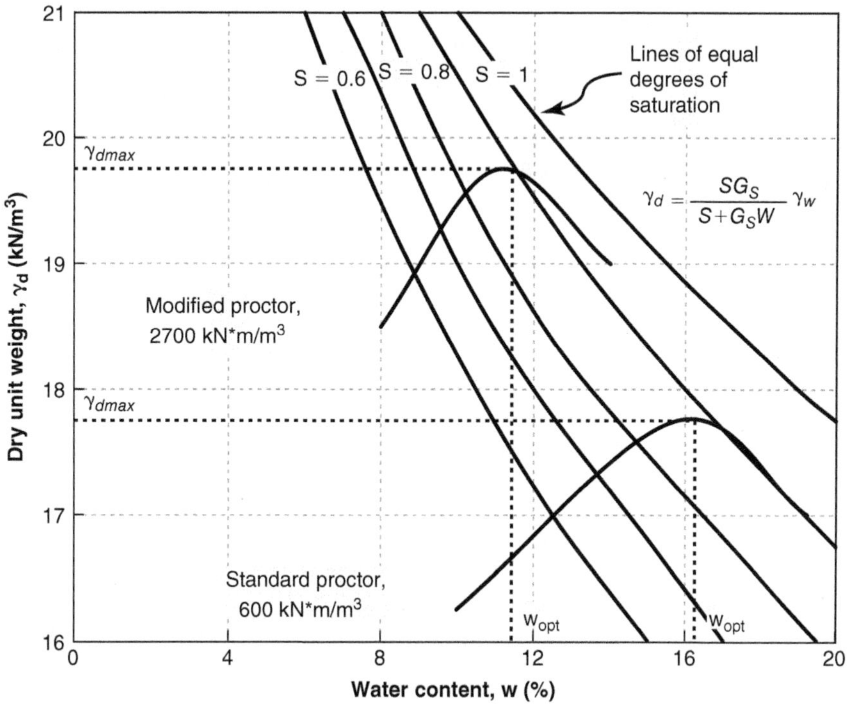

Figure 9.19 Compaction curve for Standard and Modified Proctor Compaction Tests.

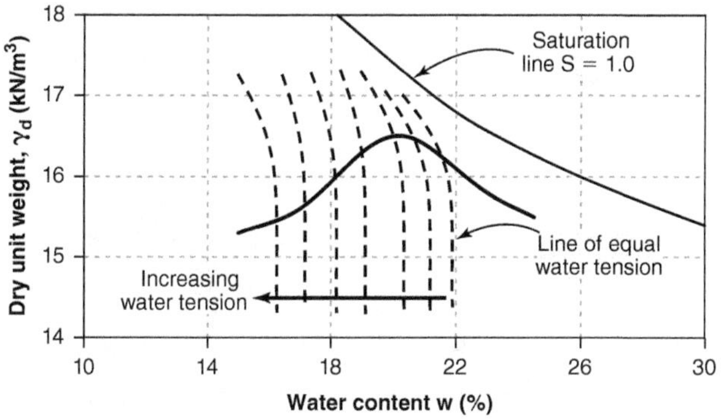

Figure 9.20 Compaction test and water tension lines.

Table 9.5 Compaction Requirements for Modified Proctor Compaction Test

102 mm diameter 116 mm high mold	152 mm diameter 116 mm high mold
5 soil layers	5 soil layers
25 blows per soil layer	56 blows per soil layer
Hammer weight 44.5 N	Hammer weight 44.5 N
Hammer drop height 457 mm	Hammer drop height 457 mm
Volume 9.43×10^{-4} m^3	Volume 21.2×10^{-4} m^3
Total energy 2700 kN.m/m^3	Total energy 2700 kN.m/m^3

saturation line for S = 1 is a bounding envelope for all compaction curves for that soil, called the *zero air void line*. It is also possible to draw the lines of equal water tension on the same graph as the compaction curve, as shown in Figure 9.20.

In 1958, a second compaction test, the Modified Proctor Compaction Test (MPCT), was developed as an ASTM standard. A higher compaction standard was necessary to better correspond to the larger and heavier compaction equipment, such as large vibratory compactors and heavier steam rollers. The MPCT is very similar to the SPCT except for the different requirements listed in Table 9.5. The data reduction is the same and the result is also the γ_d vs. w curve. The difference is that, due to the higher compaction effort (2700 kN.m/m^3 compared to 600 kN.m/m^3), the curve for the MPCT is located higher than the curve for the SPCT (Figure 9.19).

The peak of the curve has the coordinates γ_{dmax} and w_{opt}, called the *maximum dry density* and the *optimum water content* respectively. The specifications for field applications usually require that the water content be within ±x% of the optimum water content and that the dry density be at least y% of the maximum dry density. Then these requirements are checked by field testing at the compaction site (see Chapter 7 on in situ testing).

Note that the dry unit weight is used on the vertical axis of the compaction curve and not the total unit weight. The reason is best explained through the example of Figure 9.21. Both soil A and soil B have a total unit weight of 20 kN/m^3,

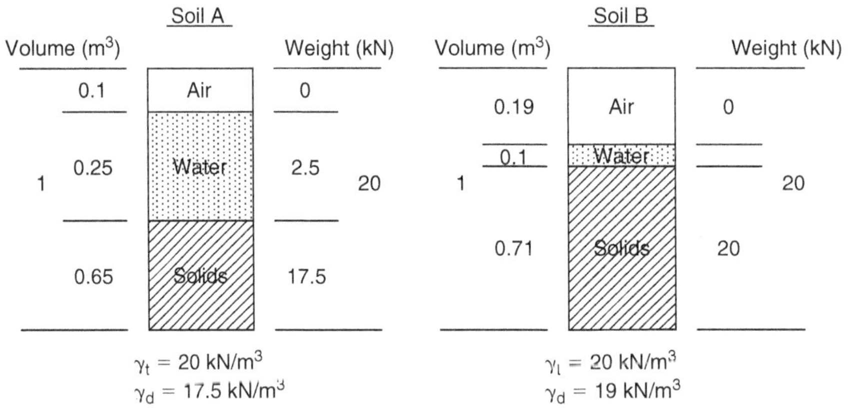

Figure 9.21 Three-phase diagram showing the usefulness of dry unit weight.

yet soil A has a dry unit weight of 17.5 kN/m^3 whereas soil B has a total and dry unit weight of 19 kN/m^3. Soil B has more solid constituents per unit volume and is therefore more compact. The selection of soil B over soil A can be made on the basis of the dry unit weight (19 vs. 17.5) but not on the basis of the total unit weight (20 vs. 20).

9.4 COMPACTION TEST: SOIL MODULUS

9.4.1 Saturated Soils

Most of the time, the soil in a compaction test is unsaturated.

9.4.2 Unsaturated Soils

The compaction test described in section 9.3 yields the dry unit weight γ_d vs. water content w curve. The soil modulus also plays a very important role in the field of compaction. Indeed, one of the major goals of compaction is to minimize deformation, so a sufficiently high modulus should be reached for compaction to be adequate. A modulus E vs. water content w curve can be generated in parallel with the γ_d vs. w curve by using a device called the BCD (Figure 9.22). It consists of a 150 mm diameter thin and flexible steel plate at the bottom of a rod with handles—a kind of scientific cane. Strain gages are mounted on the back of the plate to record the bending that takes place during the loading test. When the operator leans on the handle, the load on the plate increases and the plate bends. If the soil is soft (low modulus), the plate bends a lot. If the soil is hard (high modulus), the plate does not bend much. The amount of bending is recorded by the strain gages and is correlated to the modulus of the soil below.

This test, called the *BCD test* (Briaud et al. 2006), consists of the following steps. First, the BCD plate is placed on top of the sample in the 152 mm diameter compaction mold (Figure 9.22). The operator then leans on the handles of the BCD and the vertical load increases. When the load goes through 223 N, a load sensor triggers the reading of the strain gages. The device averages the strain gage values, uses the internal calibration equation linking the strains to the modulus, and displays the modulus E. This gives one point on the modulus vs. water content curve. By repeating this test for different water contents when the SPCT or MPCT is performed, a complete E vs. w curve can be obtained (Figure 9.23).

The modulus obtained with the BCD corresponds to a reload modulus, to a mean stress level averaging about 50 kPa

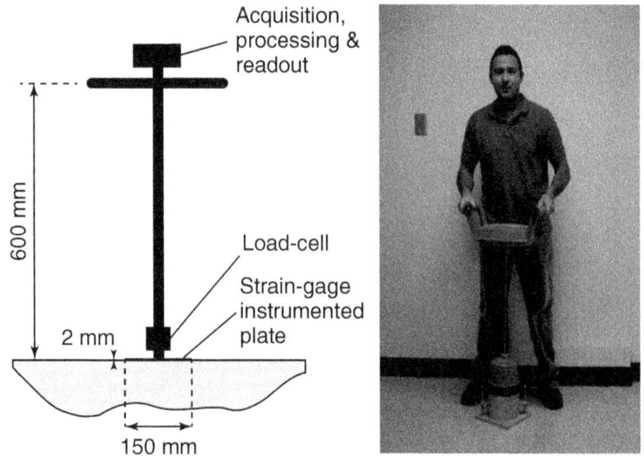

Figure 9.22 BCD apparatus to get soil modulus during a Proctor compaction test: (a) BCD principle. (b) BCD on Proctor mold.

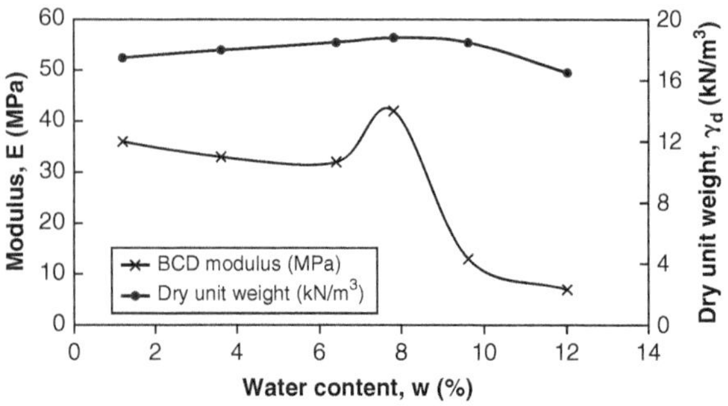

Figure 9.23 Compaction equipment and test.

within the zone of influence, to a strain level averaging 10^{-3} within the zone of influence, and to a time of loading averaging about 2 s.

9.5 CONSOLIDATION TEST

9.5.1 Saturated Soils

The consolidation test dates back to the early 1900s, and it may be appropriate to attribute its early development to Terzaghi, around 1925, with Cassagrande and Taylor making significant contributions as well. The consolidation test (ASTM D2435) is used mostly for determining the compressibility of saturated fine-grained soils. It consists of placing a disk of soil approximately 25 mm high and 75 mm in diameter in a steel ring of the same diameter and applying a vertical load on the sample while recording the decrease in thickness of the sample (Figure 9.24). Filter stones are placed at the top and bottom of the sample to allow the water squeezed out of the sample to drain at both ends. There are several loading procedures: incremental loading, constant rate of strain, and constant gradient.

The *incremental loading procedure* is the most popular and consists of placing a load on the sample for 24 hours while recording the decrease in sample thickness. The load creates a constant total normal stress σ on the surface of the sample. When σ is applied, the water stress u_w goes up because the water has difficulty escaping from the small soil pores quickly enough (Figure 9.25). It takes some time for the water stress u_w to decrease and come back to its original value. This decrease in u_w is associated with a corresponding

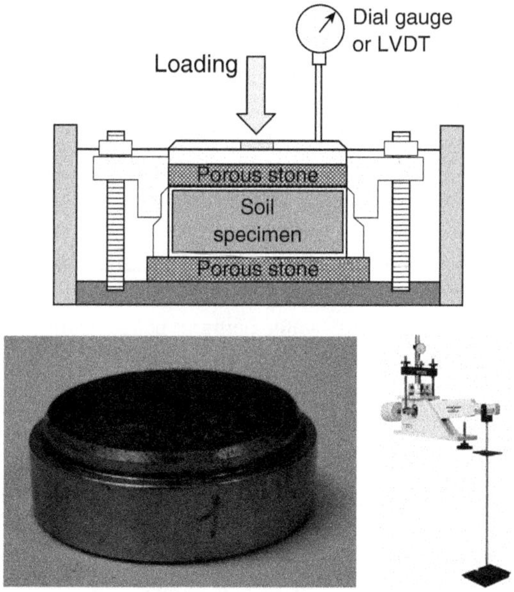

Figure 9.24 Consolidation test and equipment: (a) Principle. (b) Sample in ring. (c) Complete setup. (b: Courtesy of Lev Buchko, P.E. // Timely Engineering Soil Tests, LLC. c: Courtesy of Humboldt Mfg. Co.)

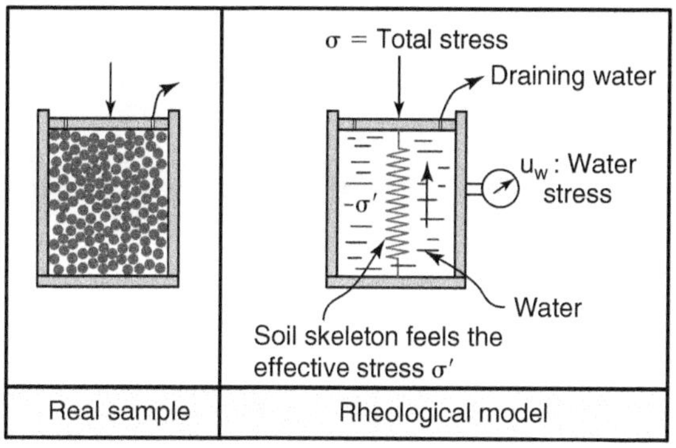

Figure 9.25 Consolidation model.

increase in effective stress ($\sigma' = \sigma - u_w$ in this case, because the soil is saturated) and a settlement of the soil; this is the process of consolidation (Figure 9.26).

The 24-hour loading step is considered to be sufficient in general for the water stress u_w to decrease back to zero. Therefore, it is assumed that at the end of each 24-hour loading step, the water stress is back to zero and the total normal stress σ is equal to the effective normal stress σ'. The loads and associated pressures are applied in a sequence where the load is doubled each time. A typical sequence is 12, 25, 50, 100, 200 kPa for σ.

The last point at the end of the 24-hour loading step curve (displacement vs. time, Fig. 9.26) gives one point (vertical effective stress σ' and vertical strain ε) on the consolidation test stress-strain curve (stress vs. strain, Figure 9.27*a*). The upward curvature of this stress-strain curve and the lack of maximum stress or failure stress or strength is due to the steel ring that confines the soil sample. The more load that is applied to the sample, the more the steel ring contributes to the resistance. Note that this curve is often presented as void ratio e versus the decimal logarithm of the effective stress $\log \sigma'$ (Figure 9.27*b*). The compression index C_c is defined as the slope of the linear portion of the $e - \log \sigma'$ curve past the initial rounded part of that curve (Figure 9.27). As such, C_c is:

$$C_c = \frac{\Delta e}{\Delta \log \sigma}, \tag{9.7}$$

During each 24-hour loading step, the decrease in sample height ΔH is recorded as a function of time t to be able to develop the ΔH vs. t curve. The vertical strain ε is obtained by dividing the change in height ΔH by the original height H_o of the sample. Figure 9.28 shows the ε vs. t curve for three loading steps.

The coefficient of consolidation c_v can be obtained from the ε vs. t curve of each load step through the formula:

$$c_v = T\frac{H^2}{t} \tag{9.8}$$

where T is the time factor, H the drainage length, and t the time elapsed. The drainage length H is equal to the height

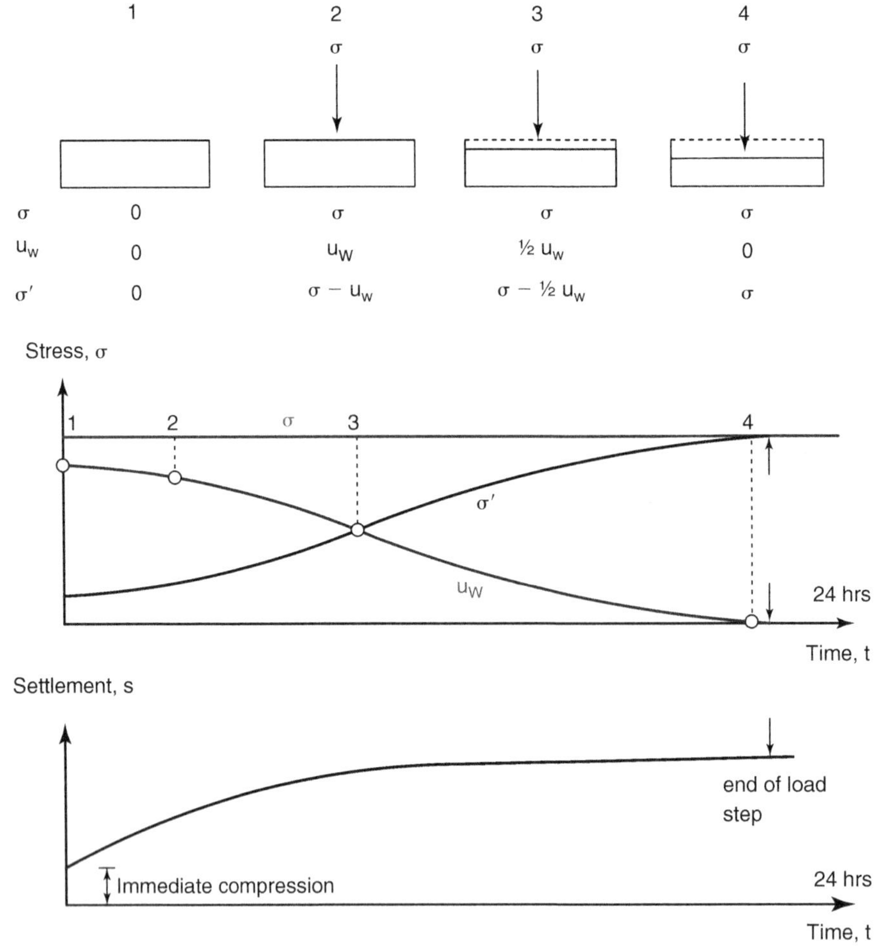

Figure 9.26 Consolidation process.

H_o of the sample if there is drainage only on one side of the sample (top only or bottom only) and equal to half the height of the sample, $H_o/2$, if there is drainage on the top and bottom of the sample. The time factor T comes from the solution of the governing differential equation for the one-dimensional consolidation theory (see Chapters 11 and 14 for more on consolidation theory). This time factor is linked to the average percent consolidation U, defined as:

$$U = \frac{s(t)}{s_{max}} = 1 - \frac{u_e}{u_{e\,max}} \tag{9.9}$$

where $s(t)$ is the settlement at time t, s_{max} is the settlement at a time equal to infinity, u_e is the excess water stress or pore pressure at time t, and $u_{e\,max}$ is the maximum excess water stress. The theoretical curve linking the average percent consolidation U to the time factor T is shown in Figure 9.29. This curve describes the normalized displacement vs. time curve for the sample according to the one-dimensional consolidation theory. It represents a normalized version of the settlement vs. time curve under a given load.

A value of c_v can be obtained for each load step by choosing a value of the percent consolidation U (50% or 90%, for example) and finding the corresponding time on the ε vs. t curve. Two methods are available to do this: the log time method developed by Cassagrande (1938) and the square root of time method developed by Taylor (1948). The log time (Cassagrande) method requires that ε_0 and ε_{100} be found on the ε vs. t curve (Figure 9.30). Note that ε_0 is not necessarily zero, as ε_0 refers to zero percent consolidation, not zero deformation. This is a subtle distinction, as the first part of the deformation may be elastic in nature and does not correspond to water being expelled from the pores (consolidation). Cassagrande proposed the following way to find ε_0 (Figure 9.30): Plot the ε vs. t curve as ε vs. log t; choose a point near the beginning of the ε vs. log t curve with coordinates ε_1 and log t_1; find the point with coordinates ε_2 and log t_2 = log 4 t_1; calculate the difference $\varepsilon_1 - \varepsilon_2$; and find ε_0 as:

$$\varepsilon_0 = \varepsilon_1 - (\varepsilon_2 - \varepsilon_1) = 2\,\varepsilon_1 - \varepsilon_2 \tag{9.10}$$

The basis for this technique is that, according to the theory, the beginning of the ε vs. t curve is a parabola, so that the

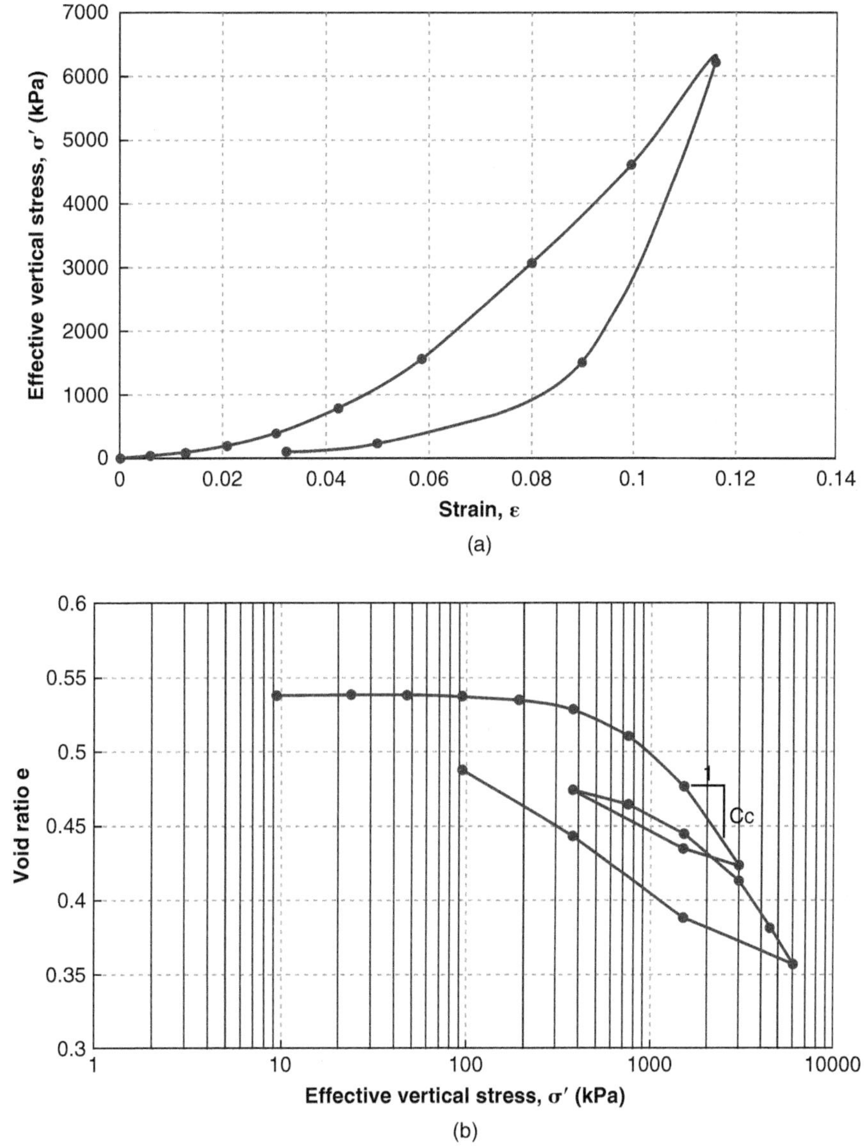

Figure 9.27 Consolidation test results: (*a*) Stress-strain curve. (*b*) e log σ' curve

beginning of the parabola satisfies equation 9.10. Once ε_0 is known, ε_{100} is found at the intersection of the two straight lines drawn on the ε vs. log t curve as shown in Figure 9.30. Then ε_{50} is read on the curve halfway between ε_0 and ε_{100}. The time t_{50} is read as the time corresponding to ε_{50} on the curve. Once t_{50} is obtained, Eq. 9.8 is used to calculate c_v. All other quantities are known, including $T_{50} = 0.197$, and the drainage length as described previously.

The square root of time (Taylor) method consists of plotting the ε vs. t curve as ε vs. $\sqrt{t}$ curve (Figure 9.31). Then a straight line is fitted to the early part of the curve (AB on Figure 9.31). A straight line with a slope equal to 1/1.15 times the slope of the first line is then drawn through point A (AC on Figure 9.31). The intersection of line AC with the ε vs. $\sqrt{t}$ curve corresponds to $\sqrt{t_{90}}$. Once t_{90} is known, Eq. 9.8 is used to calculate c_v. All other quantities are known, including $T_{90} = 0.848$, and the drainage length as described previously.

The preconsolidation pressure σ'_p is another important soil parameter that can be obtained from the consolidation test. It is the effective vertical stress before which the deformation of the soil is small and after which the deformation of the soil increases more rapidly. It can be thought of as a vertical yield stress, although failure does not necessarily happen at σ'_p. This effective stress corresponds to the highest long-term effective stress that the soil has been subjected to. The following procedure is recommended to obtain σ'_p from the consolidation test (Figure 9.32). Choose the point of highest curvature on the ε vs. log σ' curve (Point A on Figure 9.32); then draw a horizontal line through that point and a line tangent to the curve at that point. Then draw the bisectrice of the angle formed by these two lines. Draw the straight line

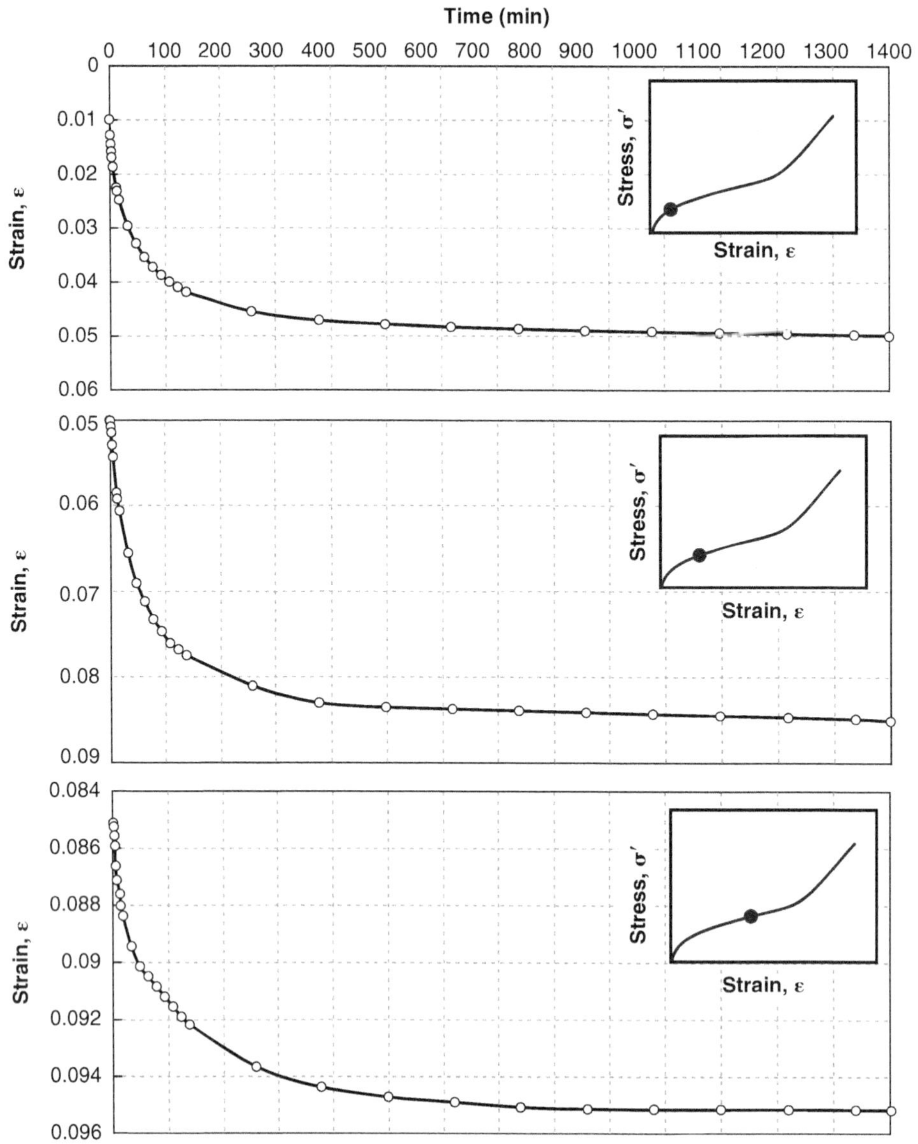

Figure 9.28 Vertical strain vs. time for three consolidation test loading steps.

that best fits the portion of the ε vs. log σ' curve past the σ'_p value. The intersection between this best-fit straight line and the bisectrice is a point that defines the preconsolidation pressure σ'_p (Figure 9.32).

The *constant rate of strain procedure* consists of the same procedure as the incremental loading procedure but with the following differences. The water is allowed to drain from the top of the sample but not from the bottom of the sample, where the water stress is measured. The sample is then deformed at a constant rate of displacement with time. This rate is chosen in such a way that the increase in water stress Δu_w at the bottom of the sample is kept at 5 to 10% of the vertical stress σ applied on the sample.

The *constant gradient procedure* consists of the same procedure as the constant rate of strain procedure but with the following differences. When the load is applied, a water stress (pore pressure) Δu_w develops throughout the sample. Soon the excess water stress at the top of the sample decreases to zero, because drainage is allowed but the bottom water stress remains close to Δu_w because the sample is not allowed to drain at the bottom. This creates a gradient between the top and bottom of the sample. This gradient is maintained constant as the load on the sample is slowly increased. However, at the end of each loading step, the water stress is allowed to dissipate to obtain an equilibrium compression of the soil.

Advantages of the consolidation test include its relative simplicity and its yield of the response of a soil sample to one-dimensional confined compression. A drawback is that the confinement provided by the steel ring around the sample

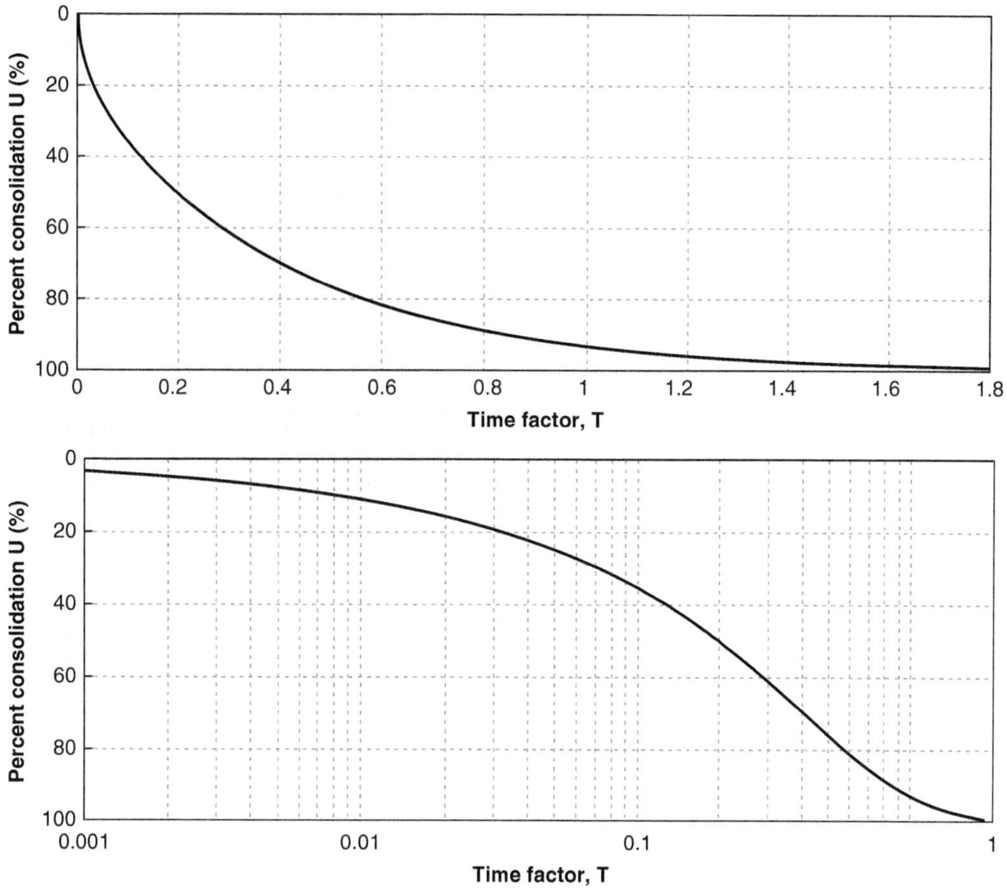

Figure 9.29 Percent consolidation vs. time factor.

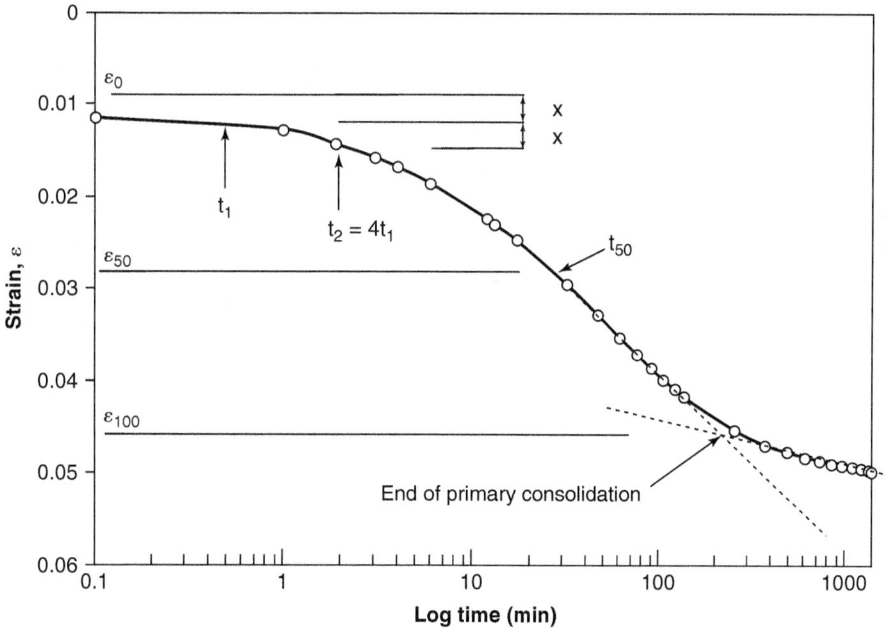

Figure 9.30 Log time method to obtain the coefficient of consolidation. (After Cassagrande 1938.)

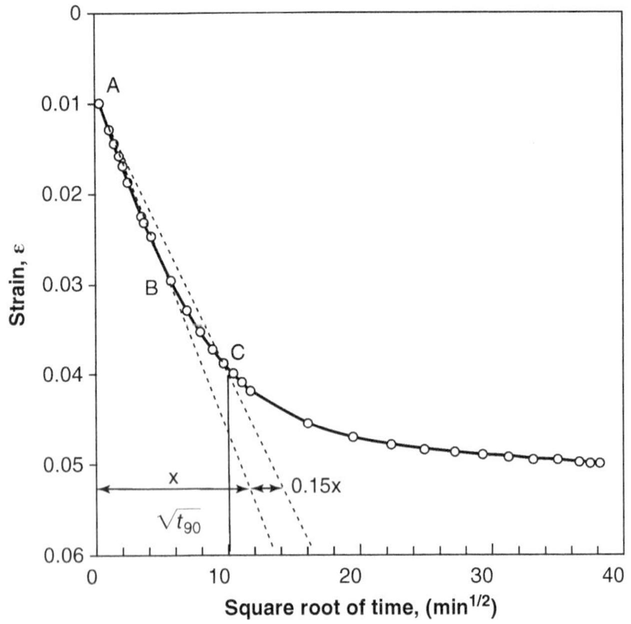

Figure 9.31 Square root of time method to obtain the coefficient of consolidation. (After Taylor 1948.)

prevents lateral deformations and may not represent the true deformation of the soil in the field.

9.5.2 Unsaturated Soils

If the soil is unsaturated, the test procedures are unchanged. However, the water is in tension initially, when the sample is placed in the consolidometer. The increase in vertical stress on the sample as the test proceeds may create enough of an increase in water stress that it goes from tension to compression. If the soil is saturated, it is implicitly assumed that at the end of each 24-hour loading step in the loading step procedure, the water stress is zero; that way the effective stress on the sample can be calculated for each step. In the case of unsaturated soils, it becomes more difficult to calculate the effective stress on the sample. The following expression can be used if the air stress is zero (see Chapter 10):

$$\sigma' = \sigma - \alpha \, u_w \tag{9.11}$$

where σ' is the effective stress, σ the total stress, α the water area ratio coefficient, and u_w the water tension stress. The coefficient α can be estimated as the degree of saturation S, but the error can be ±40% of the correct value. A better estimate consists of using the air entry value, as shown in Chapter 10. Either way, obtaining σ' requires that the water tension u_w be measured during the test. Most of the time, a soil in the saturated state with the water in compression is more compressible than the same soil in the unsaturated state with the water in tension. One exception is collapsible soils; with such materials, an unsaturated soil can experience significant and sudden compression when inundated (see section 9.8).

9.6 SWELL TEST

9.6.1 Saturated Soils

When soils absorb water, they may swell; some soils swell more than others. This is why it is important in many cases to measure how much swelling takes place when a soil has access to water. Consider a sample of dry, clean gravel in a container: When you add water to it, the water will fill the voids, but when the voids are full, no more water will be absorbed by the gravel. Clean gravel does not swell during wetting. Now consider a dry piece of montmorillonite clay

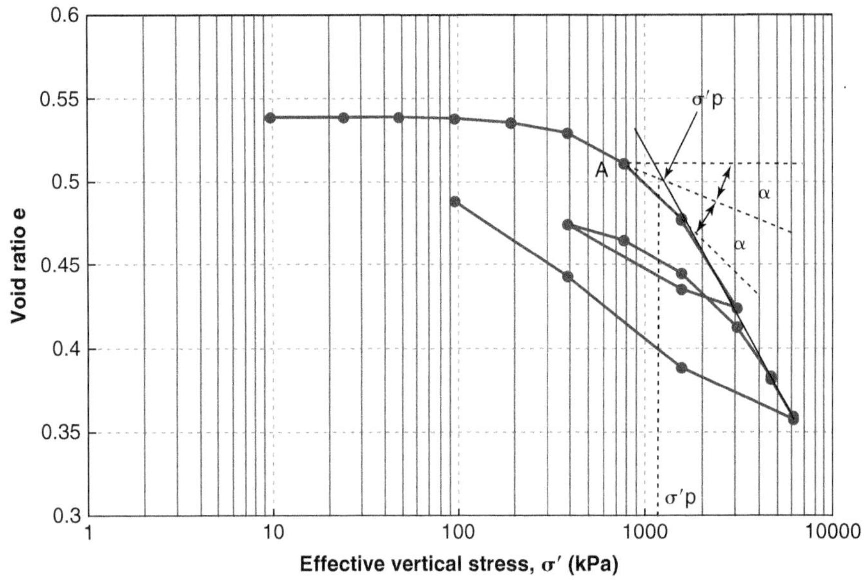

Figure 9.32 Method to determine the preconsolidation pressure from the consolidation test.

with a high dry density and place it at the bottom of a glass of water. The first thing that you will see through the wall is tiny explosions at the surface of the clay sample. The reason is that the water is drawn into the voids, but these voids are full of air that cannot escape because the water is coming in. This pressurizes the voids. The pressure increases until it overcomes the tensile strength of the dry clay, and a series of mini explosions is created. After a while the air finds a way to escape and the water enters the voids. The amount of swelling then depends on what the soil particles are made of. Montmorillonite minerals have a tremendous ability to attract water, so the swelling can be very significant for such clays and the sample may more than double in height. Swelling soils have very fine, highly plastic clay particles and are relatively dense. If they are located in regions where the water content of the soil varies significantly from one season to the next, they can create a lot of damage to structures, particularly light ones like houses, as they swell or shrink unevenly and distort those structures.

If the water in the voids is in compression (below the groundwater level), then no swelling will take place. If the water in the voids is in tension (above the groundwater level), then more water will be attracted into the voids. Thus, the swell test is more useful for soils above the groundwater level. These soils may be saturated or unsaturated. The procedure for the swell test is the same for both saturated soils and unsaturated soils and is described in section 9.6.2.

9.6.2 Unsaturated Soils

The *swell test* (Figure 9.33; ASTM 4546) consists of placing a soil sample in a snug-fitting cylindrical container (consolidometer ring), inundating the soil by placing it in a bath of water, and measuring the vertical swell movement (vertical strain) as a function of time (Figure 9.34). The vertical strain is the change in height of the sample divided by the initial height of the sample. Water access to the sample is provided by porous disks placed at the top and bottom of the sample. The swelling can take days or even weeks. If the top of the sample is not subjected to any vertical load, the test is called a *free swell test* (path AB on Figure 9.35, path CD on Figure 9.36). If a vertical load is applied, the

Figure 9.33 Swell test equipment.

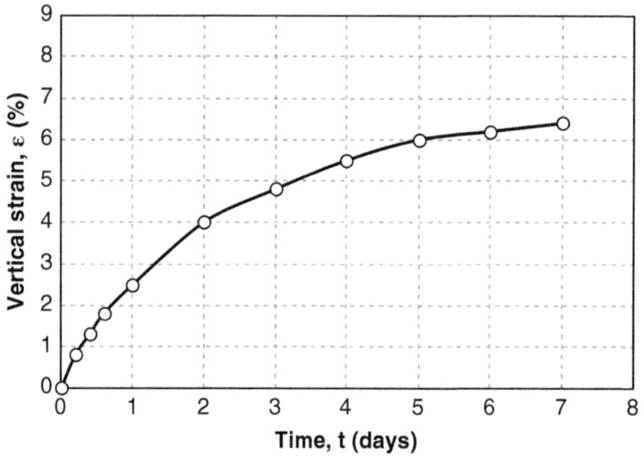

Figure 9.34 Swell test results: Vertical strain vs. time.

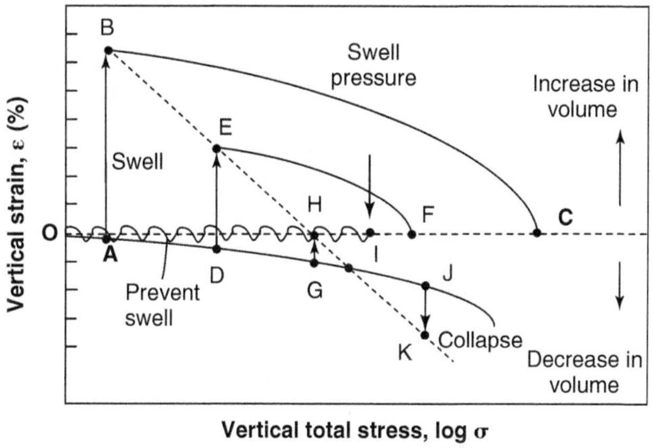

Figure 9.35 Shrink-swell test results: Vertical strain vs. vertical total stress.

test is simply called a swell test (path DE on Figure 9.35, path CE on Figure 9.36). Note that after swelling, a regular consolidation test can be performed on the sample (path BC and EF on Figure 9.35).

The free swell test gives the *swell limit*, which is the water content of the sample at the end of the free swell test (point B on Figure 9.35). The swell limit represents an upper limit of the water content that the soil can reach in the undisturbed state. When a vertical load is applied, it is usually applied before water is added on top of the sample and swelling starts. The magnitude of the load influences the swelling. It is often advantageous to apply a vertical stress on the sample equal to the stress that the soil will experience in the field (under the planned structure, for example). Sometimes the pressure is high enough that no swelling can take place, and settlement takes place instead.

Another way to run the swell test is to add the water first so that swelling can start and to increase the vertical stress on the sample gradually to prevent any swelling (path OI on Figure 9.35). During this test, the volume of the sample is maintained constant and equal to its initial volume. When

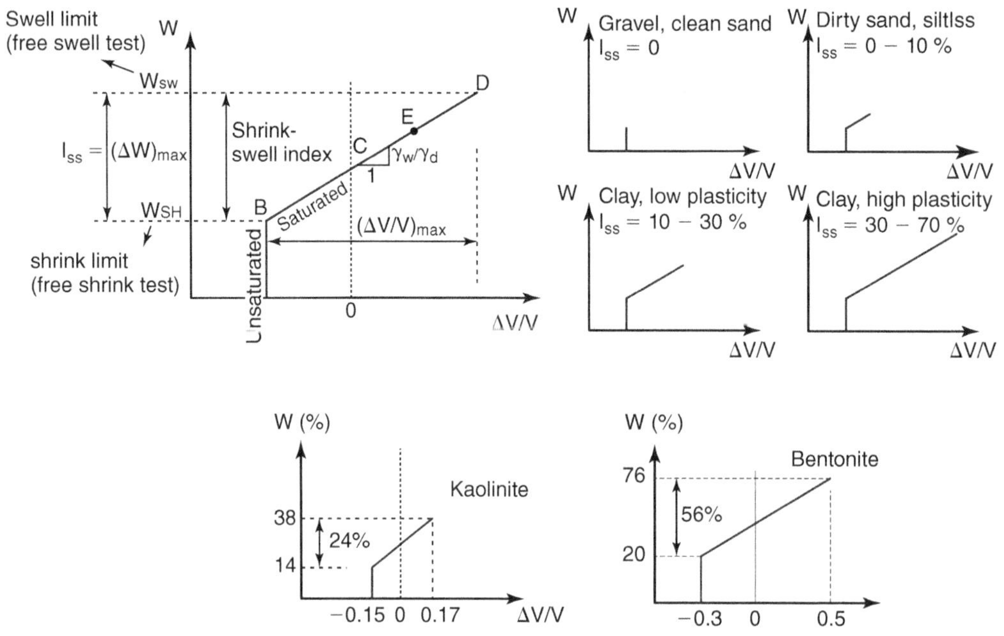

Figure 9.36 Shrink-swell test results: Water content vs. relative change in volume. (*a*) Idealized behavior. (*b*) Typical ranges. (*c*) Low-plasticity clay example.

the vertical stress reaches an equilibrium value, that stress is called the *swelling pressure*. Swelling pressures can reach 1000 kPa or higher for high-plasticity clays.

9.7 SHRINK TEST

9.7.1 Saturated Soils

The shrink test (Figure 9.37) consists of trimming a sample of soil into a cylindrical shape, measuring its dimensions, and recording its weight. The initial volume V_o and the initial weight W_o are recorded. Then the sample is left to dry while the dimensions and the weight are measured as a function of time. This gives the volume V(t) and weight W(t). When the sample is air-dried, it is placed in the oven to obtain the oven dry weight W_s. The average water content w of the sample at any time during the test is (W(t) – Ws)/Ws. The results of the test consist of a plot of the relative change in volume $(V(t) - V_0)/V_0 = \Delta V/V_0$ as a function of time t (Figure 9.38), and the water content w as a function of the relative change in volume $\Delta V/V_0$ (Figure 9.36). The undisturbed sample shrinkage limit w_{SH} is the water content corresponding to the point where the sample first stops decreasing in volume (point B on Figure 9.36(a)).

Figure 9.37 Free shrink test for shrinkage limit of undisturbed sample.

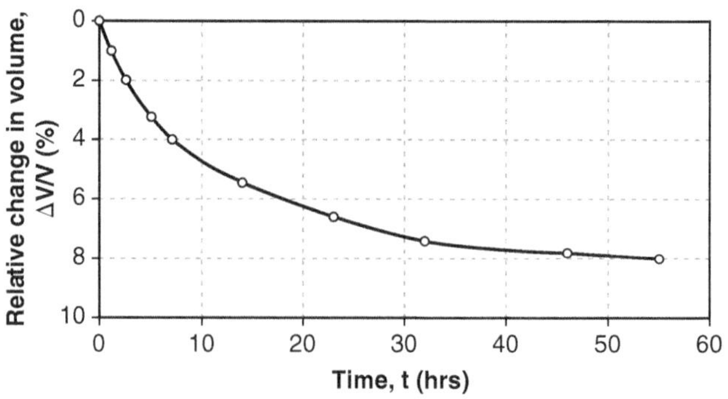

Figure 9.38 Free shrink test result: Relative change in volume vs. time.

As in the case of the swell test, the shrink test can be performed without any vertical load applied (free shrink test) or with vertical load applied (shrink test). The free shrink test is much more common.

9.7.2 Unsaturated Soils

The shrink test applies equally to saturated soils and unsaturated soils. In both cases the water is in tension throughout the test. The soil may start as a saturated soil, but, as it dries, it goes through the air entry value u_{we}, at which point it becomes unsaturated. The shape of the relative volume change vs. time curve for the free shrink test (Figure 9.38) is similar to the shape of the relative volume change vs. time curve for the free swell test (Figure 9.34). During the free shrink test, the weight of the sample is measured as a function of time, so it is possible to plot the water content as a function of relative volume change (BCD on Figure 9.36*a*). This curve indicates where the undisturbed shrink limit w_{sh} occurs. Note that the undisturbed shrink limit is more obvious for low-plasticity soils than for high-plasticity soils. The undisturbed shrink limit is different from the Atterberg shrink limit, which is obtained on a remolded sample.

9.8 COLLAPSE TEST

9.8.1 Saturated Soils

Consider a natural sample of dry silt with a low dry density and a reasonable strength. Place it in a steel ring and place some weight on top of the sample. In the dry state, the sample has no problem carrying the load without much deformation. Now add water on top of the sample: You will likely see a significant amount of compression take place due to collapse of the soil skeleton. What happens is that the small amount of water tension that exists at the contacts between the silt particles is lost when the water enters the voids and the loose structure of the silt collapses. It is important to check if a soil is collapsible; you can imagine the distress associated with any structure built on such soils if a significant amount of water permeates below the foundation.

Collapsible soils consist of loose, dry, low-density materials (say less than 16 kN/m^3) that decrease in volume (collapse and compact) with the addition of water. These soils are often found in arid regions, specifically in areas of wind-blown silty sediments (loess), young alluvial fans, and debris flow sediments. Soil collapse occurs within soils above the groundwater level. The process of saturation weakens or eliminates the clay bonds holding the soil grains together through water tension.

9.8.2 Unsaturated Soils

The collapse test (ASTM D5333) is the same for saturated and unsaturated soils. It is performed with the sample confined in a consolidometer ring. Typically, it consists of loading the soil sample to a vertical stress equal to the vertical total stress that the soil will experience at a chosen depth, recording the vertical strain vs. time curve (consolidation test), and then (once the compression is complete) inundating the sample while continuing to record the vertical strain vs. time curve. Once the collapse is completed, the consolidation test can be resumed by increasing the vertical stress. A sample vertical strain vs. vertical stress curve is shown in Figure 9.39.

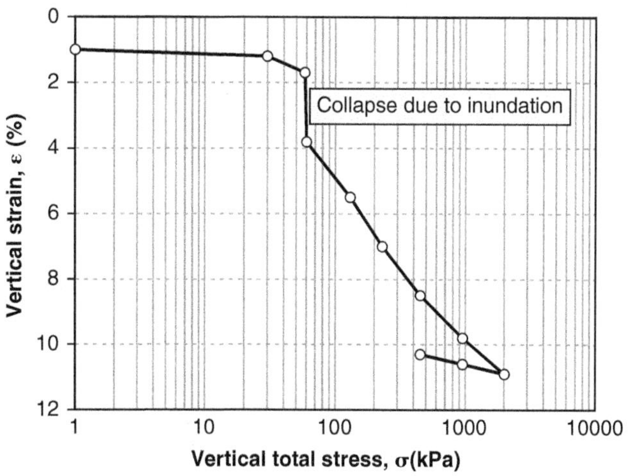

Figure 9.39 Collapse test: Vertical strain vs. vertical stress.

9.9 DIRECT SHEAR TEST

9.9.1 Saturated Soils

The direct shear test (ASTM D3080) is a simple test used to obtain the shear strength of a soil. A disk of soil is placed in a steel cylinder split horizontally at mid height (Figure 9.40). The cylinder is made of two rings stacked on top of each other. One filter stone is placed on top and one at the bottom of the sample so that the water can drain from the sample during the test. A vertical load is applied to the top of the sample and maintained constant during the test. This vertical load creates a total normal stress σ. Then the soil sample is sheared horizontally by pushing on the bottom ring while holding the top ring. This forces a shear plane to develop around the mid height of the sample. During the shearing process, the shear force is measured with a load cell, the horizontal displacement with an LVDT or dial gage, and the vertical displacement with an LVDT or dial gage. The result of a direct shear test is a shear stress vs. horizontal displacement curve and, if the vertical movement is also measured, a vertical movement vs. horizontal movement curve (Figure 9.41).

During the first part of the direct shear test, the soil sample is allowed to consolidate under the vertical stress applied, if such a stress is applied. The consolidation is monitored by recording the vertical movement of the sample as a function of time. When the settlement stops or becomes very small, it is assumed that the water stress has returned to zero and the shearing part of the test can start. During the second part of the test, the sample is sheared and shearing takes place along a thin horizontal band at mid height of the sample near the junction between the two steel rings. The shear stress versus horizontal movement curve is obtained point by point. The shear strength is the maximum shear stress on the shear stress versus horizontal movement curve. This shear strength is the undrained shear strength if the shearing part of the test is run quickly enough that water does not have time to drain; it is the drained shear strength if the test is run slowly enough that

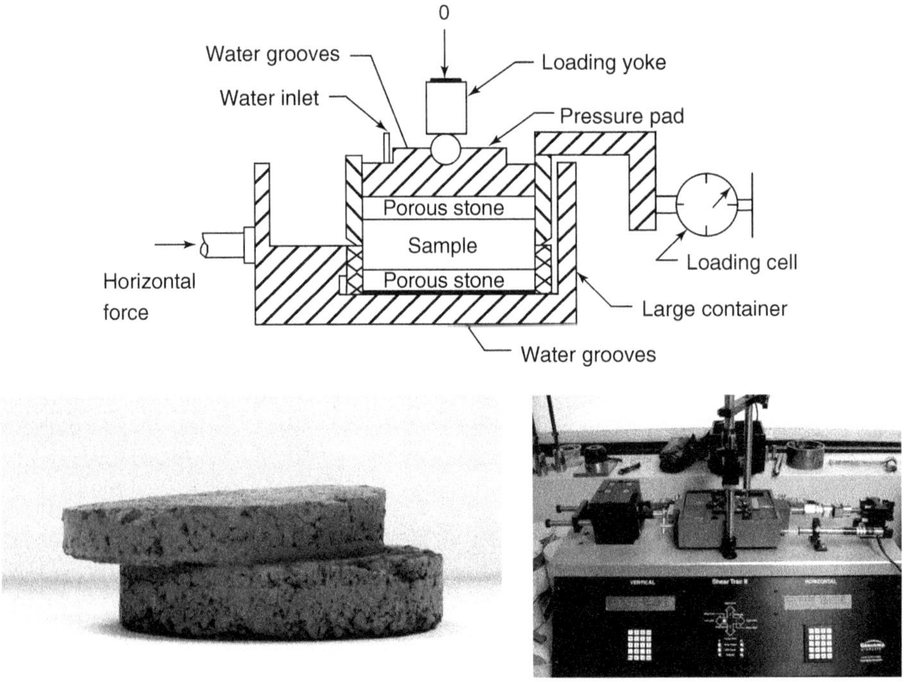

Figure 9.40 Direct shear test and equipment. (*a*) Principle. (*b*) Sample. (*c*) Complete setup. (*b*: Courtesy of Lev Buchko, P.E. // Timely Engineering Soil Tests, LLC.)

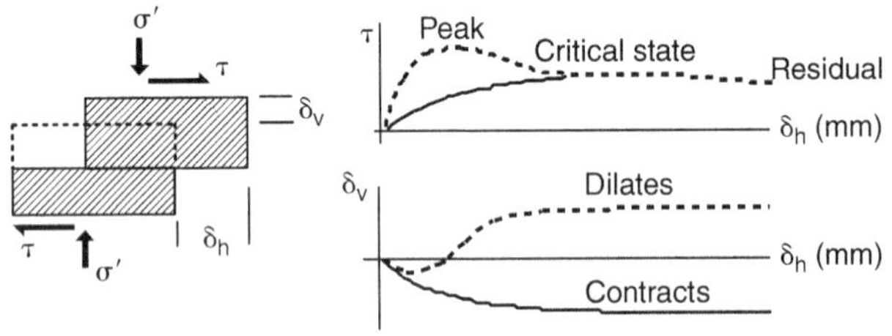

Figure 9.41 Direct shear test results stress-displacement curve.

the water stress remains zero. It is best also to measure the pore pressure or water stress, but that is not common with this simple test.

The shear strength measured in an undrained direct shear test is the undrained shear strength s_u. This undrained shear strength corresponds to the effective stress σ' generated at the end of the consolidation phase. This undrained shear strength also corresponds to the stress path followed in a direct shear test.

The shear strength measured in a drained direct shear test provides one point on the shear strength envelope. This envelope links the shear strength to the effective stress σ' normal to the plane of failure. As described in Chapter 15 on shear strength, the envelope is represented by the following equation:

$$s' = c + \sigma' \tan \phi' \tag{9.12}$$

This equation has two soil parameters: the effective stress cohesion c′ and the effective stress friction angle ϕ'. Because the drained direct shear test gives only one point on the envelope, it is necessary to run at least two direct shear tests to obtain c′ and ϕ' for a given soil (Figure 9.42).

When soils are subjected to shearing, they can increase in volume (dilate), decrease in volume (contract), or not change volume. If a soil dilates during shear, the shear strength

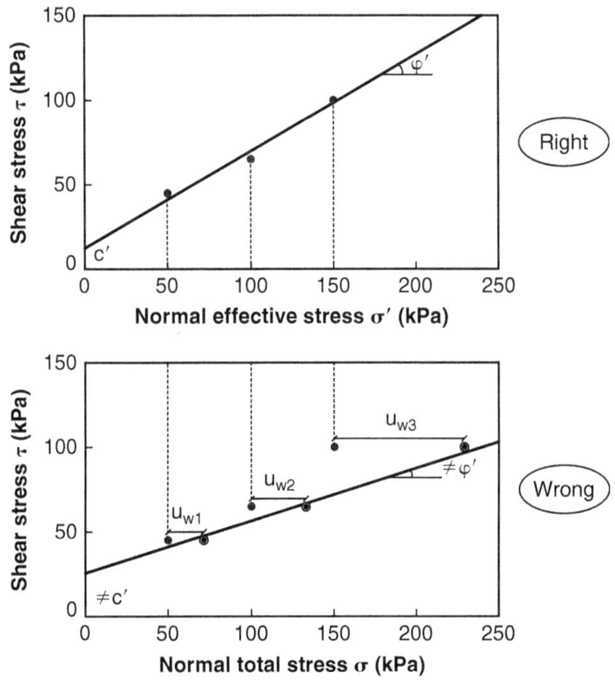

Figure 9.42 Example of direct shear test strength results for saturated soils.

increases compared to a soil that does not change in volume. The increase in shear strength is reflected by the dilation angle ψ (see Chapter 15 for more details). The dilation angle ψ can be estimated from a direct shear test as the slope of the curve linking the vertical movement z to the horizontal movement x. Because this curve is rarely a straight line, the equation is written in an incremental fashion.

$$\tan \psi = \Delta z / \Delta x \tag{9.13}$$

Advantages of the direct shear test include that it is easy to perform and gives a shear strength of the soil. A drawback of the direct shear test is that it cannot give the shear strain of the soil as it is sheared, because the thickness of the shearing zone is not known.

9.9.2 Unsaturated Soils

If the soil is unsaturated, or if the soil is saturated but the water in the voids is in tension (e.g., above the groundwater level), then the direct shear test requires measurement of the water tension stress (suction) to obtain the effective stress shear strength parameters c′ and φ'. Indeed, although the test procedure is the same for a soil with water in compression and for a soil with water in tension, the assumption that the water stress is zero when the test is performed slowly is not valid when the water is in tension. The reason is that if the water is in compression at the beginning of the direct shear test, the water compression stress is very small compared to the general stress level; in contrast, if the water is in tension, the water tension stress can be very large when the degree of saturation is low. The water tension stress u_w can be measured by any one of the methods described in section 9.2.4, but it is most often done with a tensiometer during the shear test. Once the water tension stress is known, the effective stress (assuming the air stress u_a is zero) is calculated as:

$$\sigma' = \sigma - \alpha\, u_w \tag{9.14}$$

where σ' is the effective stress, σ the total stress, α the water tension coefficient, and u_w the water tension stress. The coefficient α can be estimated as the degree of saturation S, but the error can be as large as ±40% of the correct value. A better estimate can be obtained by using the correlation to the air entry value u_{we} as shown in Chapter 10. It is assumed here that the air stress remains zero during the test.

The results are then plotted as shear strength vs. effective normal stress, as shown in Figure 9.43. If the results of direct shear tests on soils where the water is in tension are plotted as shear strength vs. total stress, the cohesion intercept will be much larger, as it includes the effect of the water tension on the soil strength (Figure 9.43). The apparent cohesion c_{app} is equal to:

$$c_{app} = -\alpha\, u_w \tag{9.15}$$

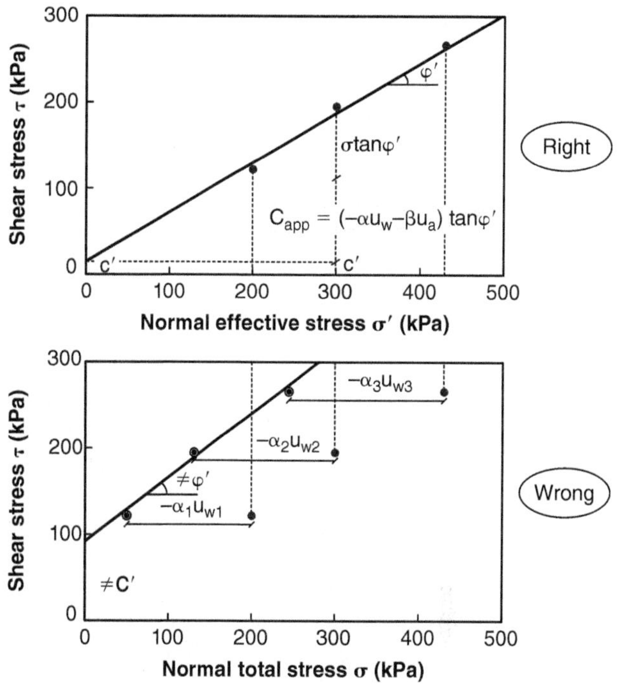

Figure 9.43 Example of direct shear test strength results for unsaturated soil.

However, c_{app} is not a constant for a given soil, because u_w depends on the water content of the sample. The apparent cohesion is called apparent rather than true cohesion because it is due to the effective stress created by the water tension and because it disappears if the soil is inundated (water tension goes to zero). In contrast, the parameter c′ is a characteristic of the soil that is constant and independent of the water content.

9.10 SIMPLE SHEAR TEST

9.10.1 Saturated Soils

The simple shear test (ASTM D6528) can be traced back to the mid 1960s with a publication by Bjerrum and Landva (1966). A disk of soil is placed in a flexible membrane with a porous stone on the top and on the bottom of the disk (Figure 9.44). A vertical load is applied to the top of the sample and maintained constant during the test. This vertical

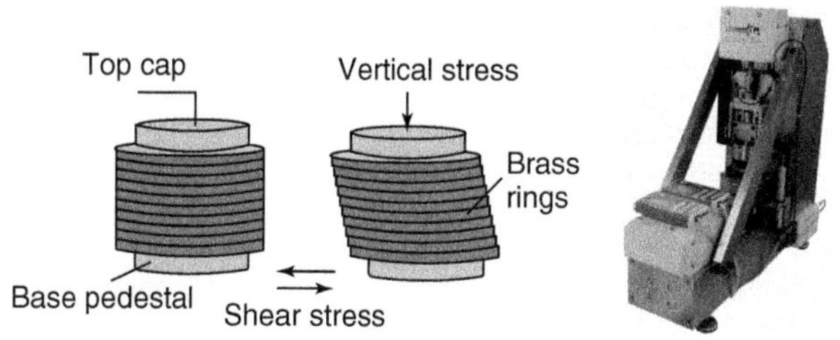

Figure 9.44 Simple shear test equipment: (a) Principle. (b) Complete setup. (b: Courtesy of GDS Instruments.)

load creates a total normal stress σ. Then the soil is sheared by holding one of the two platens and pushing the other one horizontally. The major difference between the direct shear test and the simple shear test is that in the direct shear test, the shearing takes place along a predetermined thin band of soil near the middle of the sample. In the simple shear test, the shearing takes places over the entire height of the sample. Therefore, the shearing strain γ can be measured in the simple shear test as:

$$\gamma = \Delta x / h_o \tag{9.16}$$

where Δx is the difference in horizontal movement between the top and the bottom of the sample and h_o is the initial height of the sample. The shear stress τ is measured as the shear force divided by the cross-sectional area of the sample. Thus, the simple shear test gives the shear stress-shear strain curve for the sample and therefore a shear modulus G.

During the first part of the simple shear test, the soil sample is allowed to consolidate (through drainage) under the vertical stress applied if such a stress is applied. The consolidation is monitored by recording the vertical movement of the sample as a function of time. When the settlement stops or becomes very small, it is assumed that the water stress has returned to zero and the shearing part of the test can start. During the second part of the test, the sample is sheared. The shear stress vs. shear strain curve is obtained point by point (Figure 9.45). The shear strength τ_f is the maximum shear stress on the shear stress vs. shear strain curve. This shear strength is the undrained shear strength if the shearing part of the test is run without allowing water to drain out of the sample; it is the drained shear strength if the test is run slowly enough that the water stress remains zero or if the water stress (pore pressure) is measured.

The shear strength τ_f is obtained in the same fashion as for the direct shear test, including the shear strength parameters c' and ϕ'. The shear modulus G is the slope of the τ vs. γ curve. Because the curve is typically nonlinear, G varies with γ and a G vs. γ curve can be generated. Therefore, an advantage of the simple shear test is that it can give the shear modulus G as a function of shear strain, in addition to the shear strength of the soil sample.

When soils are subjected to shearing, they can increase in volume (dilate), decrease in volume (contract), or not change volume. If a soil dilates during shear, the shear strength increases compared to a soil that does not change in volume. If a soil contracts during shear, the shear strength decreases compared to a soil with no change in volume. The increase or decrease in shear strength is reflected by the dilation angle ψ (see Chapter 15 on shear strength for more details). The dilatancy angle ψ can be estimated from a simple shear test as the slope of the curve linking the change in vertical movement Δz to the change in horizontal movement Δx (Eq. 9.13).

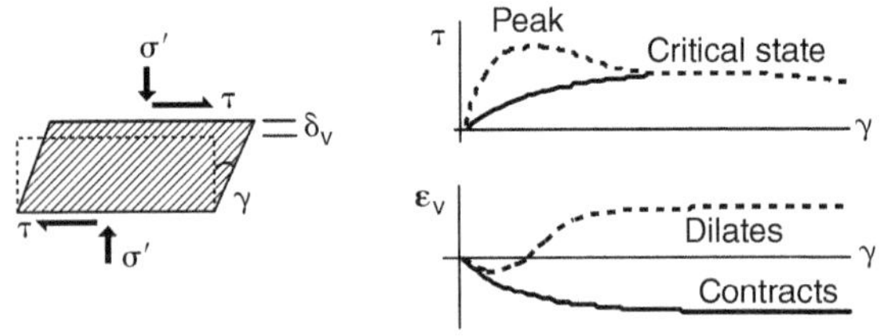

Figure 9.45 Simple shear test results.

9.10.2 Unsaturated Soils

If the soil is unsaturated, or if it is saturated but the water is in tension, the testing procedure is unchanged except for measurement of the water stress.The tensile stress in the water will typically require the use of a different measuring device, such as a tensiometer. The data reduction requires calculation of the effective stress, as discussed for the direct shear test.

9.11 UNCONFINED COMPRESSION TEST

9.11.1 Saturated Soils

The unconfined compression test (ASTM D2166) (Figure 9.46) is one of the simplest tests to perform if the soil can stand up under its own weight. In this test, the sample is a cylinder with a diameter d and a height h equal to about 2 times the diameter. The ratio h/d is about 2, to ensure that the oblique shear plane that typically develops during failure can propagate through the entire sample without intersecting the top or bottom platen. The sample remains unconfined during the test; therefore, the minor principal stress σ_3 is zero. A vertical load is applied to the sample by pushing up on the bottom platen at a constant rate of displacement while holding the top platen in a fixed position. The vertical total stress σ is calculated by dividing the vertical load by the cross-sectional area of the sample. Because it is assumed that there is no shear between the top of the sample and the bottom of the top platen, that stress is the major principal stress σ_1. The sample compresses and the vertical displacement Δh is measured with an LVDT or a dial gage. Knowing the initial height h of the sample, the vertical strain ε can be obtained as $\varepsilon = \Delta h/h$. The result of an unconfined compression test

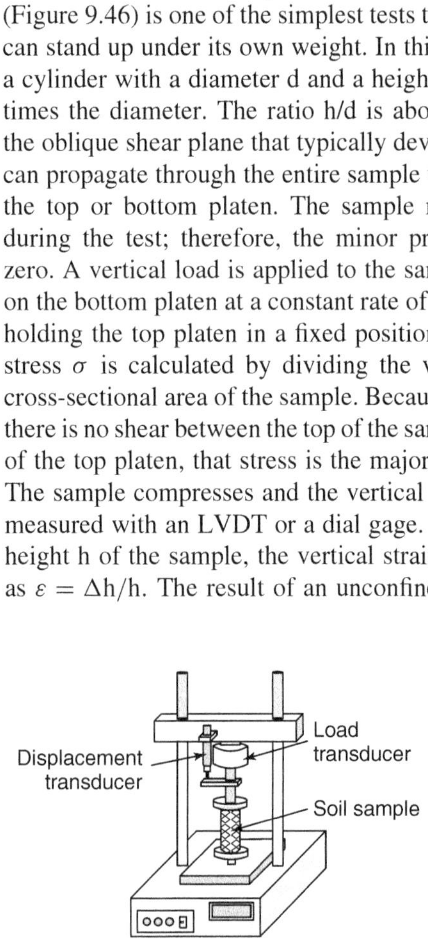

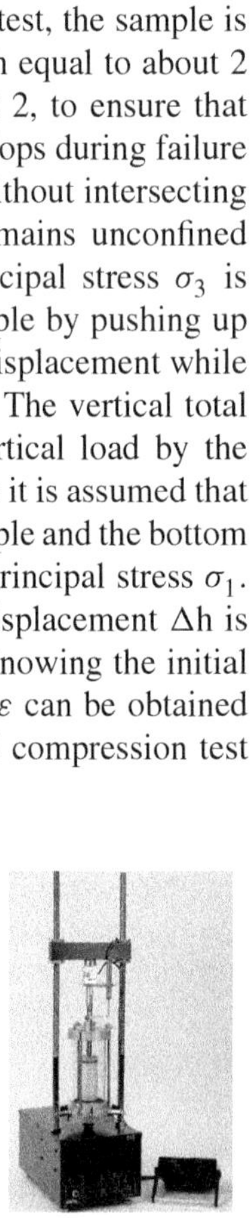

Figure 9.46 Unconfined compression test equipment: (a) Principle. (b) Complete setup. (*a*: After Ian Smith. *b*: Courtesy of ELE International.)

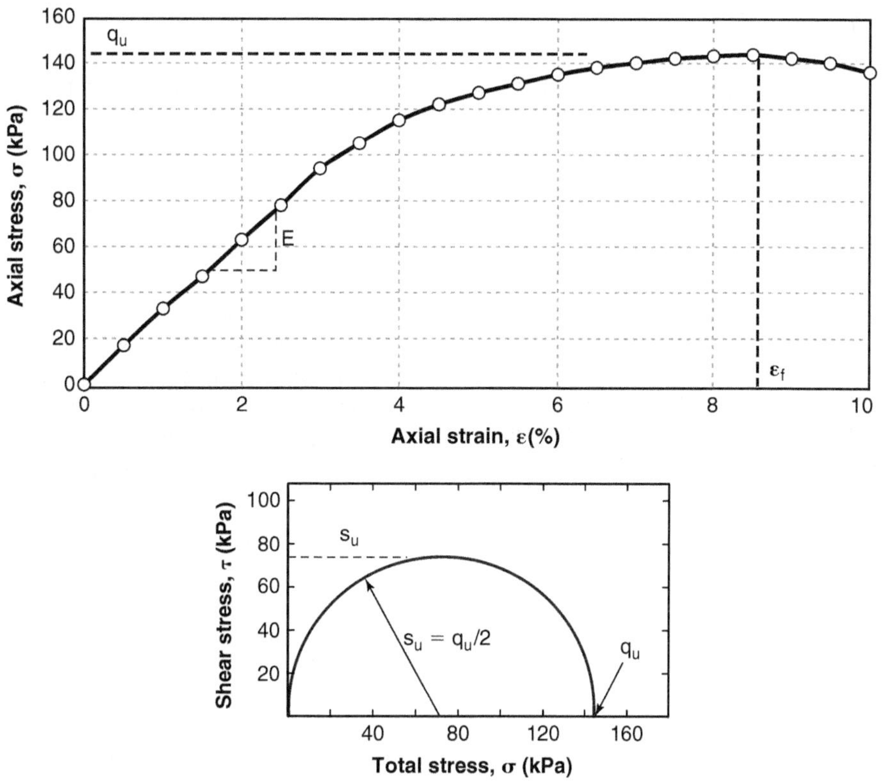

Figure 9.47 Unconfined compression test results.

is a complete total stress σ vs. strain ε curve for the soil sample under zero lateral confinement (Figure 9.47). The maximum stress on the curve is the unconfined compression strength q_u. Because the test is rather rapid, the shearing process is considered to be undrained for fine-grained soils. The undrained shear strength s_u is equal to $q_u/2$, as shown in Chapter 15.

$$s_u = q_u/2 \quad (9.17)$$

An unconfined compression modulus of deformation E can also be obtained from this test as:

$$E = \sigma_1/\varepsilon \quad (9.18)$$

Because the curve is often nonlinear, several moduli can be obtained depending on the chosen strain level. Advantages of the unconfined compression test are its simplicity and the fact that it gives both an undrained shear strength and a modulus of deformation for fine-grained soils.

9.11.2 Unsaturated Soils

If the soil is unsaturated, the test procedure is unchanged. Because the water stress is not measured in this test, there is also no difference in measurement and data reduction. One interesting observation is that the water tension can be estimated from the unconfined compression strength q_u. Indeed, the shear strength equation for unsaturated soils when the air stress u_a is assumed to be zero is:

$$s = c' + (\sigma - \alpha\, u_w) \tan \varphi' \quad (9.19)$$

In the unconfined compression test, the horizontal total stress is zero, therefore:

$$\sigma_h = 0 = \sigma'_h + \alpha\, u_w \text{ and therefore } \sigma'_h = -\alpha\, u_w \quad (9.20)$$

Meanwhile, the vertical total stress at failure is equal to q_u; therefore:

$$\sigma_v = q_u = \sigma'_v + \alpha\, u_w \text{ and therefore } \sigma'_v = q_u - \alpha\, u_w \quad (9.21)$$

The shear strength s is given by the point of tangency between the effective stress Mohr circle and the shear strength envelope (Figure 9.48). Triangle ACD on Figure 9.48 is such that:

$$\frac{CD}{AD} = \sin\varphi' = \frac{0.5((q_u - \alpha u_w) - (-\alpha u_w))}{0.5((q_u - \alpha u_w) + (-\alpha u_w)) + \dfrac{c'}{\tan\varphi'}} \quad (9.22)$$

Which leads to

$$u_w = \frac{0.5 q_u (\sin\varphi' - 1) + c' \cos\varphi'}{\alpha \sin\varphi'} \quad (9.23)$$

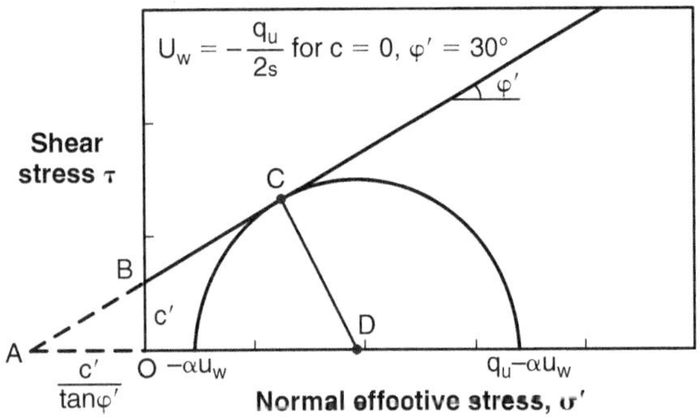

Figure 9.48 Water tension and unconfined compression strength relationship.

Eq. 9.23 gives the water tension at failure in the unconfined compression test. If it is further assumed that $c' = 0, \varphi' = 30^\circ$, and $\alpha = S$, then Eq. 9.23 becomes:

$$u_w = -\frac{q_u}{2S} \tag{9.24}$$

9.12 TRIAXIAL TEST

9.12.1 Saturated Soils

The triaxial test (ASTM D5311) (Figure 9.49) is similar to the unconfined compression test except that a chosen confining pressure is applied to the sample before compression takes place. The sample has a height equal to about two times the diameter to ensure that the oblique shear plane that typically develops during failure in compression can propagate through the entire sample without intersecting the top or bottom platen. Typical diameters range from 30 to 75 mm. First, porous disks (also called filter stones) are placed at the top and bottom of the sample. Then the sample is fit in an impervious rubber membrane and set on the pedestal of the triaxial cell. The top platen is placed, and the top of the triaxial cell is brought down to cover the sample. The shaft of the piston is lowered in contact with the top platen on one side and connected to the load cell or proving ring on the other. The cell is filled with liquid (water or oil) and the confining pressure is applied. Sometimes the cell is not filled with liquid and only air pressure is used. The triaxial cell is placed in a frame and the load is applied by moving the bottom of the frame upward and at a constant rate of displacement against the stationary top of the frame. The movement of the sample is typically obtained by measuring the movement of the shaft applying the load with respect to the triaxial cell. For more advanced testing, the movement measurements are taken between two rings directly tied to the sample. Pore-pressure measurements are an option and are typically made by placing a saturated porous stone at the base of the sample and measuring the pressure in the water through a pressure transducer tied to the base platen. Measuring the change in volume of the sample is also an option.

There are many different types of triaxial tests because of the possible combinations related to drainage and type and sequence of stress applications. However, nearly all triaxial tests start with a consolidation phase followed by a shearing phase. The consolidation phase is designed to bring the sample to a desired state of stress that is often intended to match the stress conditions that the sample would face in the field under the project conditions. During the consolidation phase, the cell pressure is increased to a chosen value of the confining pressure. This pressure confines the sample hydrostatically and represents the minor principal stress σ_3. During this phase of consolidation, drainage may or may not be allowed. If drainage is not allowed, the word “unconsolidated” is used in describing the triaxial test and the letter U is used in the acronym. If drainage is allowed and the water stress (pore pressure) generated by the application of σ_3 is allowed to dissipate back to zero, the word “consolidated” is used to describe the test and the letter C is used in the acronym.

During the shearing phase of the test, the vertical load Q on the sample is increased gradually and the stress in the

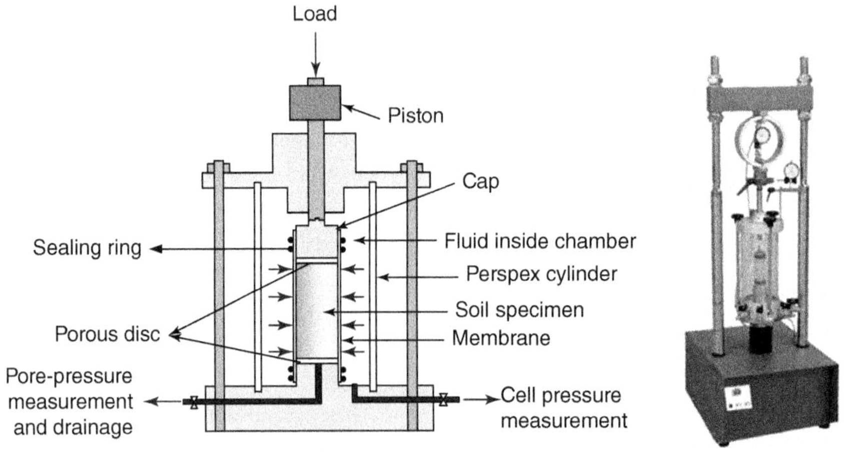

Figure 9.49 Triaxial test equipment: (a) Principle. (b) Equipment. (b: Courtesy of Geotechnical Testing Equipment Ltd., UK.)

vertical direction increases. This stress is the major principal stress σ_1:

$$\sigma_1 = \sigma_3 + Q/A \quad (9.25)$$

where σ_3 is the confining pressure, Q is the vertical load and A is the cross section of the sample. If drainage is not allowed during the shearing phase, the word "undrained" and the letter U are used. If drainage is allowed and the excess water stress (pore pressure) is kept equal to zero (very slow loading), then the word "drained" and the letter D are used. So, in the end, the following triaxial tests are possible:

1. UU test: unconsolidated undrained test
2. CU test: consolidated undrained test
3. CD test: consolidated drained test

A UD test is not possible, because allowing drainage during the shearing phase would also allow some consolidation under σ_3. UU tests are commonly performed to obtain the undrained shear strength, particularly in offshore studies where recompressing the sample to the high bottom pressures is important; UU tests are also simpler and faster than the other two. CD tests are quite time consuming, as loading must be slow enough not to generate water stresses (pore pressures), but they are simple to run. CU tests with water stress (pore pressure) measurements are faster to run, but require more sophisticated equipment because water stress (pore pressure) must be measured. Both CD tests and CU tests with water stress measurements are used to obtain the effective stress shear strength parameters c′ and ϕ'.

The result of a triaxial test is a stress-strain curve that typically links the deviator stress $(\sigma_1 - \sigma_3)$ to the vertical strain $(\varepsilon = \Delta h/h)$ where h is the initial height of the sample and Δh is the change in height of the sample. Figure 9.50 shows some results for two categories of soils: overconsolidated or dense soils on the one hand and normally consolidated or loose soils on the other. The first category exhibits a clear peak stress (maximum strength), followed by strain softening to reach a residual strength. The second category exhibits strain hardening, with the strength being reached at larger strain.

The peak stress value on this curve is the failure deviator stress $(\sigma_{1f} - \sigma_3)$. This failure stress, along with information on the water stress, is used to obtain the effective stress shear strength parameters c′ and ϕ'. This process requires use of the Mohr circle (see Figure 9.51 and Chapter 15). A *Mohr circle* is a circle in the shear stress vs. normal stress set of axes that describes the state of stress at a point when the principal stresses reduce from 3 stresses to 2 stresses. This is the case in the triaxial test where σ'_1 and σ'_3 are different and σ'_3 is equal to σ'_2. The points corresponding to the principal stresses σ'_1 and σ'_3 plot on the horizontal axis because they exist on planes with zero shear stress. The circle representing the state of stress in the triaxial sample at failure is drawn (Figure 9.51). Because the failure envelope is described by two parameters c′ and φ' (Eq. 9.12), a minimum of two triaxial tests at two different confining pressures (σ_3) must be performed to obtain the effective stress cohesion intercept c′ and the effective stress friction angle φ'. Figure 9.52 shows the difference between the Mohr circles in the effective stress set of axes and in the total stress set of axes.

A modulus of deformation E can also be obtained from the stress-strain curve as follows:

$$E = (\sigma_1 - 2\nu\sigma_3)/\varepsilon \quad (9.26)$$

where E is the total stress modulus of deformation of the soil, σ_1 and σ_3 are the major and minor principal total stresses respectively, ν is Poisson's ratio, and ε is the vertical strain. Note that because the stress-strain curve is rarely linear, many different moduli can be obtained depending on the strain level among other factors. The modulus defined in terms of effective stress is typically more useful and more fundamentally rooted:

$$E' = (\sigma'_1 - 2\nu\sigma'_3)/\varepsilon \quad (9.27)$$

where E′ is the effective stress modulus of deformation of the soil, and σ'_1 and σ'_3 are the major and minor principal effective stresses respectively.

The stress path describes the evolution of certain stresses during the test. Specifically, it tracks the path described by the points with p, q stress coordinates where p and q are defined as follows:

$$p = \frac{\sigma_1 + \sigma_3}{2} \quad \text{or} \quad p = \frac{\sigma_v + \sigma_h}{2} \quad (9.28)$$

$$q = \frac{\sigma_1 - \sigma_3}{2} \quad \text{or} \quad q = \frac{\sigma_v - \sigma_h}{2} \quad (9.29)$$

where σ_v and σ_v are the vertical and horizontal total stresses in a triaxial test, for example. The most useful stress paths are plotted in terms of effective stresses (p′ and q′):

$$p' = \frac{\sigma'_1 + \sigma'_3}{2} \quad \text{or} \quad p' = \frac{\sigma'_v + \sigma'_h}{2} \quad (9.30)$$

$$q' = \frac{\sigma'_1 - \sigma'_3}{2} = q \quad \text{or} \quad q' = \frac{\sigma'_v - \sigma'_h}{2} = q \quad (9.31)$$

where σ'_v and σ'_v are the vertical and horizontal total stresses in a triaxial test, for example. Examples of effective stress paths are shown in Figure 9.53 for different types of tests. In any lab test, it is most desirable to match the effective stress path followed by the soil in the field during the project construction and the project life.

9.12.2 Unsaturated Soils

If the soil is unsaturated, or if it is saturated and the water in the voids is in tension, the test procedure does not change, but the water and air stress measurements change. The water stress can be measured with a tensiometer and the air stress with a pressure transducer.

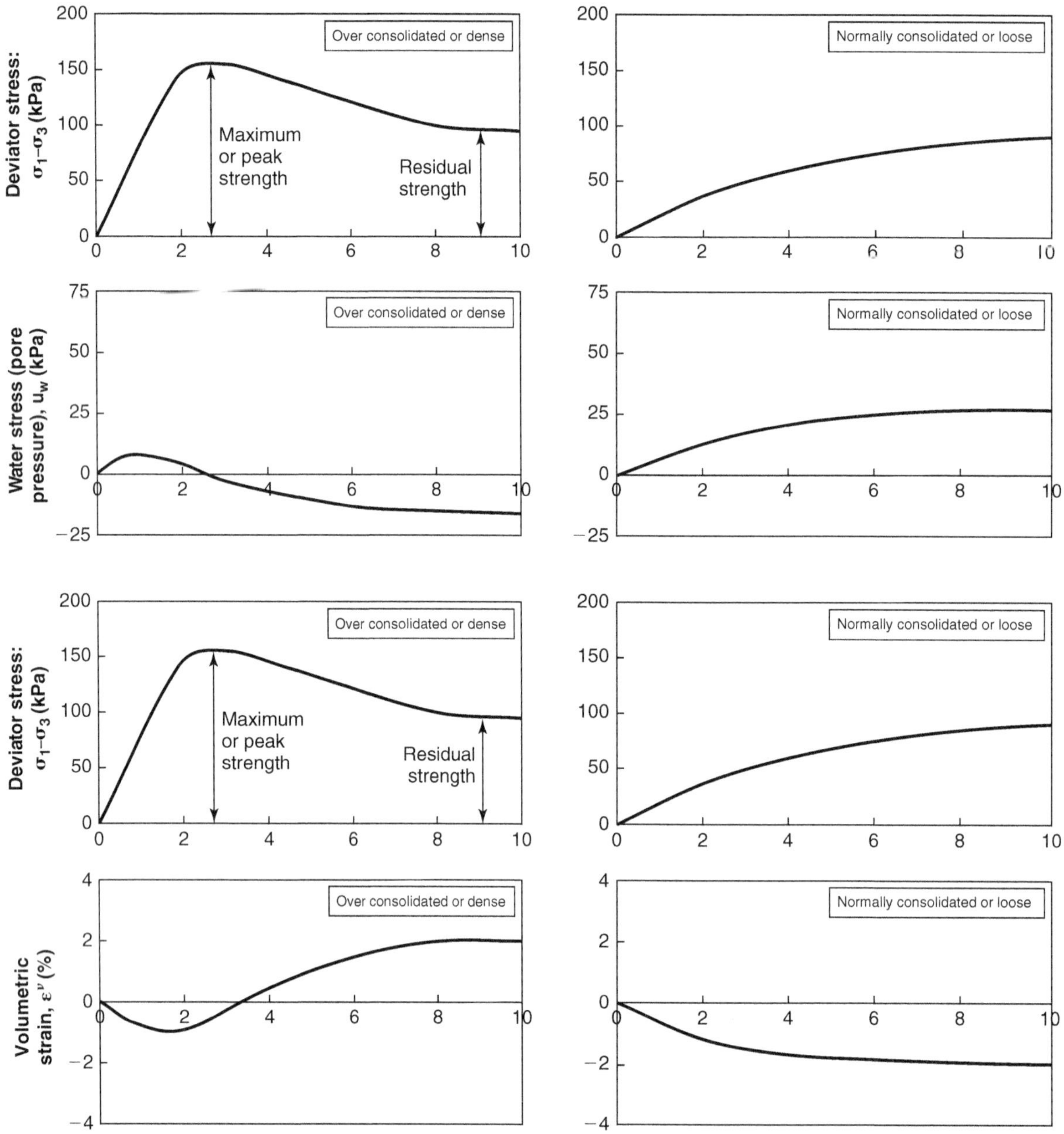

Figure 9.50 Triaxial test results (example stress-strain curves): (a) Consolidation, undrained test. (b) Consolidation, drained test.

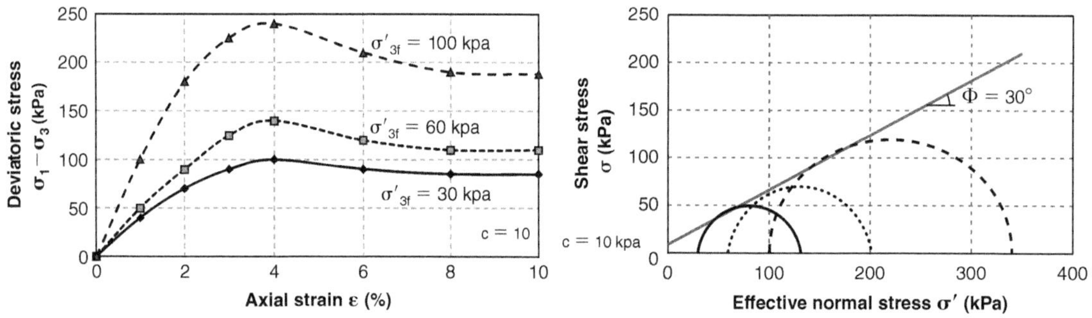

Figure 9.51 Triaxial test results: Examples of Mohr circles and strength envelope for saturated soils.

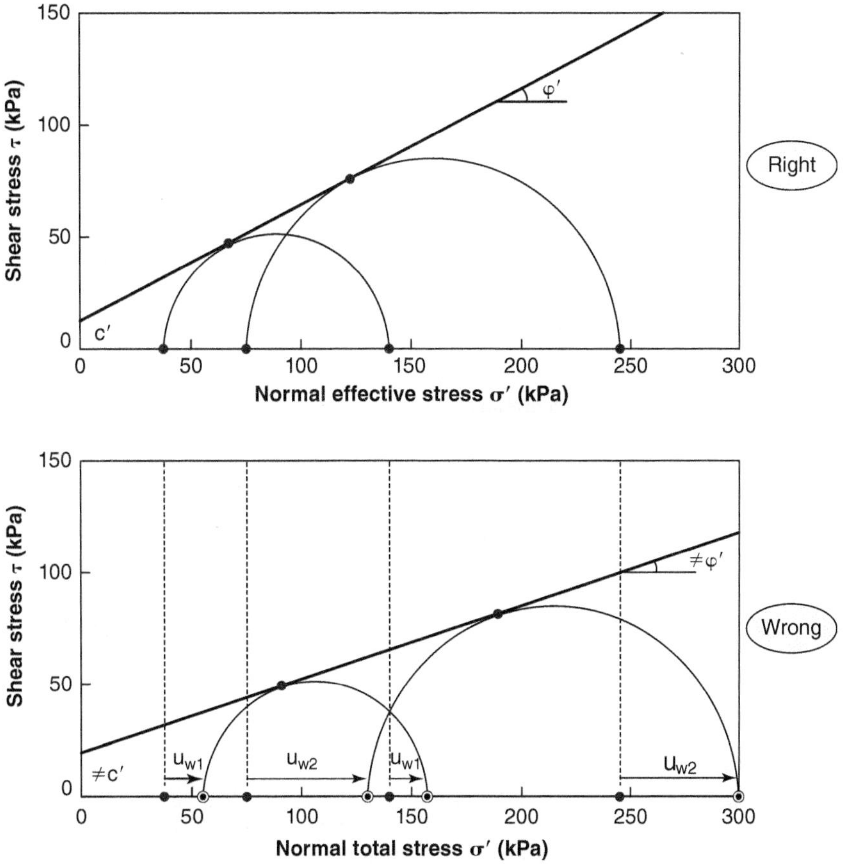

Figure 9.52 Triaxial test results: Mohr circles and strength envelope for saturated soils.

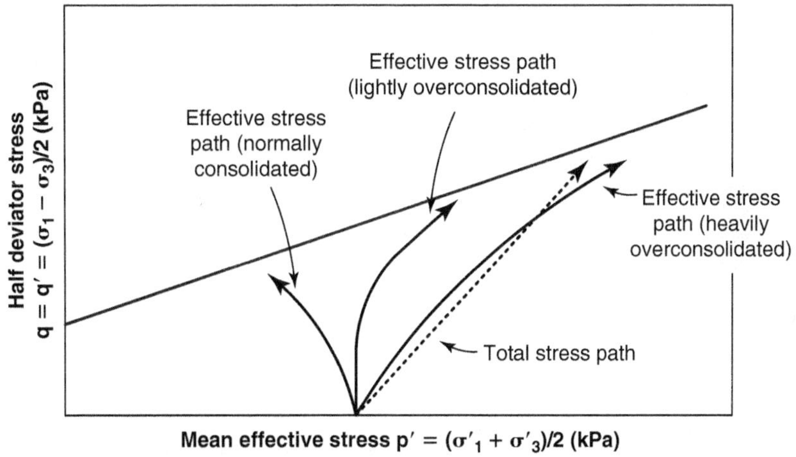

Figure 9.53 Triaxial test results: Stress paths.

The meaning of the tests that were described for saturated soils changes as well:

1. UU test: unconsolidated undrained test. For unsaturated soils, UU means that both the air and water are prevented from draining from the beginning to the end of the test. The air stress increases as the air compresses and the water stress increases (decrease in the absolute value of the water tension).
2. CU test: consolidated undrained test. For unsaturated soils, both air and water are allowed to drain during the consolidation phase. During the shearing phase, both are prevented from draining, so both pressures must be measured. Typically, the air stress and the water stress

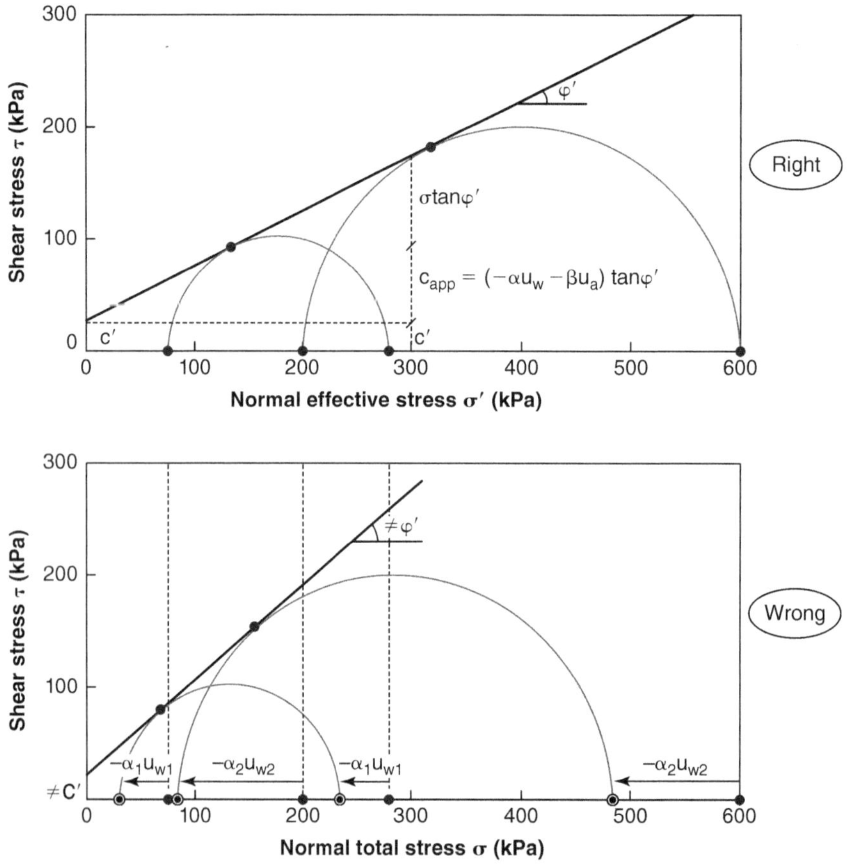

Figure 9.54 Triaxial test results: Mohr circles and strength envelope for unsaturated soils.

increase (decrease in water tension) during the shearing phase because the soil volume decreases (except for dilatant soils).

3. CD test: consolidated drained test. Both the air and the water are permitted to drain. The water tension can therefore be held constant throughout the test. The strain rate must be sufficiently slow to allow for flow of water from the soil through the high air entry disk.
4. CWC test: constant water content test. For unsaturated soils, it is also possible to conduct a test where the air can drain but not the water. Air drains much faster than water, so a judiciously chosen strain rate can achieve this condition.

The data reduction changes as well. The effective stress must be calculated according to the following formula (instead of $\sigma' = \sigma - u_w$):

$$\sigma' = \sigma - \alpha\, u_w - \beta\, u_a \qquad (9.32)$$

where σ' is the normal effective stress, σ the normal total stress, α the water area ratio parameter, u_w the water stress, β the air area ratio parameter, and u_a the air stress. This difference will affect the location of the Mohr circle on the shear stress τ vs. effective normal stress σ' graph. If instead the results are plotted in the shear stress τ vs. total normal stress σ graph, then the effective stress shear strength parameters c' and φ' cannot be obtained. The cohesion intercept c' in the shear stress τ vs. total normal stress σ graph is much larger than c', as it includes the effect of the water tension on the soil strength (Figure 9.54). The apparent cohesion c_{app} is equal to:

$$c_{app} = -\alpha\, u_w - \beta\, u_a \qquad (9.33)$$

This cohesion is called apparent cohesion rather than true cohesion because it is due to water tension and because it disappears if the soil is inundated (water and air stresses go to zero).

9.13 RESONANT COLUMN TEST

9.13.1 Saturated Soils

The resonant column test (ASTM D4015) is used to determine the dynamic small strain properties of a soil. Such results are applied in earthquake engineering and machine vibration, for example. A cylinder of soil with a height-to-diameter ratio of about 2 is placed in a cell where a confining pressure can be applied. The base of the sample is fixed to the bottom platen, which does not move. The top of the sample is mounted with a top platen having a mass m and able to generate

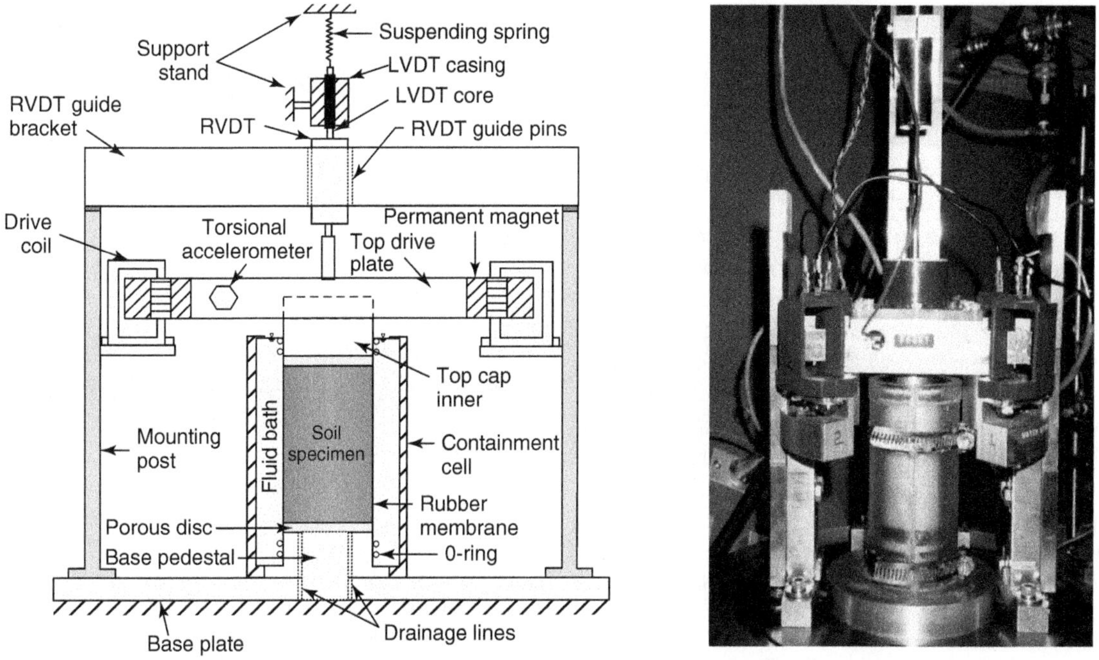

Figure 9.55 Resonant column test: (*a*) Principle. (*b*) Equipment. (*b*: Courtesy of Geotechnical Research Lab, Dept. of Civil Engineering, University of British Columbia.)

cyclic torsion (Figure 9.55). The test consists of applying a sinusoidal torque $T(\omega)$ to the top of the sample. This torque is generated through an electromagnetic drive system that controls the angular frequency ω of the sinusoidal torque application. The response of the sample is monitored by measuring (through LVDTs, for example) the rotation of the top of the sample. The water stress (pore pressure) is sometimes also measured during this test.

In a first step, a confining pressure is applied to the sample. Then the top of the sample is subjected to a chosen torque. The torque applied gives the shear stress τ imposed on the sample and the rotation θ is used to obtain the shear strain γ of the sample. The response is presented in term of loops linking τ to γ. The frequency of the sinusoidal torque is increased gradually while recording the strain in the sample. Resonance occurs when the frequency of the soil vibrations matches the frequency of the torque application (Figure 9.56). This

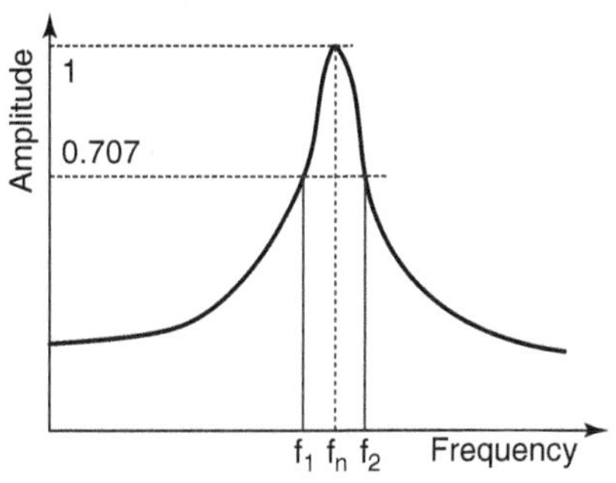

Figure 9.56 Rotation amplitude vs. frequency of induced vibration.

frequency is ω_n. At that point the sample rotation reaches its maximum value.

The data are used as follows to obtain the soil shear modulus G when the sample is fixed at the bottom and free at the top where the torque is applied. The mass polar moment of inertia of the sample J_s is:

$$J_s = M_s d_s^2/8 \tag{9.34}$$

where M_s is the sample mass and d_s is the sample diameter. The mass polar moment of inertia of the mass on top of the sample J_m is:

$$J_m = M_m d_m^2/8 \tag{9.35}$$

where M_m is the mass of the mass on top of the sample and d_m is the diameter of that mass. By using fundamental and constitutive equations, it can be shown that:

$$\frac{J_s}{J_m} = \frac{\omega_n L}{v_s} \tan\left(\frac{\omega_n L}{v_s}\right) \quad \text{or} \quad \frac{J_s}{J_m} = \frac{2\pi f_n L}{v_s} \tan\left(\frac{2\pi f_n L}{v_s}\right) \tag{9.36}$$

where J_s and J_m are the polar moments of inertia of the sample and of the mass on top of the sample respectively, ω_n is the resonant angular frequency, v_s is the shear-wave velocity in the sample, L is the length of the sample, and f_n is the natural frequency of the soil.

In Eq. 9.36, J_s, J_m, and L are known, f_n is measured in the test, and v_s can be back-calculated. Then the shear modulus is obtained from:

$$G = \rho v_s^2 \tag{9.37}$$

For the case where there is no mass at the top, $J_m = 0$, then $2\pi f_n L/v_s = \pi/2$, and then

$$G = \rho v_s^2 = 16 \rho f_n^2 L^2 \qquad (9.38)$$

Eq. 9.38 gives the shear modulus G for a given shear strain amplitude γ.

There are several ways to obtain the damping ratio, and each way has its own advantages and limitations. One way is to stop the excitation and let the sample vibration die out while recording the sample rotation as a function of time. This is called the *logarithmic decrement method.* The damping ratio *D* is defined as the ratio of the damping coefficient to the critical damping coefficient. The critical damping is the minimum amount of damping that results in the sample returning to its original position without oscillation. The damping ratio can be obtained from the decay curve (Figure 9.57) as follows. The amplitude of the first cycle is x_1 and the amplitude of the nth cycle is x_n, which is smaller than x_1. It can be shown that:

$$\frac{Ln x_1 - Ln x_n}{n-1} = \frac{2\pi D}{\sqrt{1-D^2}} \qquad (9.39)$$

In Eq. 9.39, all quantities are known except for D, the damping ratio. The damping obtained by this method includes the damping of the device, which must be accounted for separately. This method also requires stopping the test, and the strain level decreases during the vibration decay. Another way to obtain the damping ratio is to use the half-power bandwidth method. This method makes use of the amplitude vs. frequency plot (Figure 9.57) obtained during steady-state torsional vibration of the sample:

$$D = (f_2 - f_1)/2f_n \qquad (9.40)$$

where f_2, f_1, and f_n are defined in Figure 9.57. This method is best applied when the system is linear.

These curves can also be obtained from direct measurements of the shear stress and the shear strain. The maximum shear stress τ generated during the cycles is calculated as

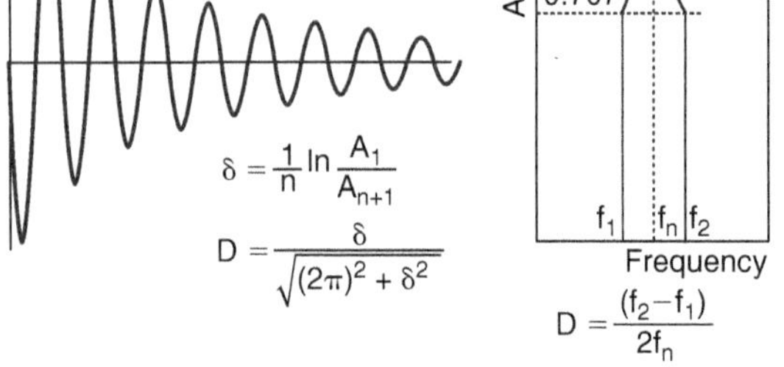

Figure 9.57 Method to obtain damping ratio from resonant column test: (a) Logarithmic decrement. (b) Half-power bandwidth.

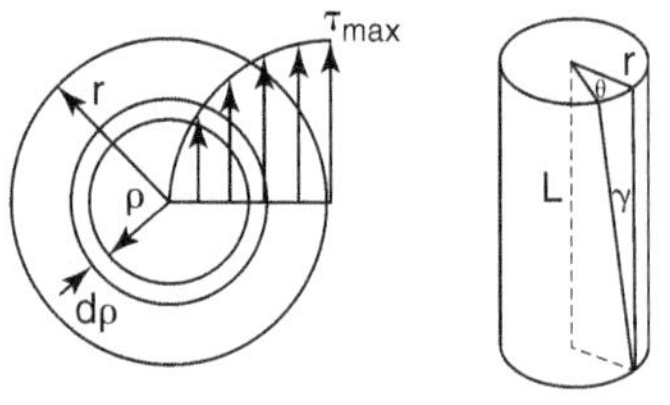

Figure 9.58 Shear stress and shear strain in a resonant column torsion test: (*a*) Shear stress. (*b*) Shear strain.

an average of the shear stress generated on the sample cross section. This shear stress is zero at the center of the sample ($\tau_{center} = 0$) and maximum at the edge (τ_{edge}) (Figure 9.58). The mean shear stress τ is related to the maximum torque T as follows:

$$\tau = 2T/\pi r_e^3 \qquad (9.41)$$

where r_e is the equivalent radius, which can be anywhere from 0.6r to 0.8r where r is the radius of the sample. The maximum shear strain during the cycle exists at the edge of the sample (γ_{edge}), while the shear strain is zero along the axis of the cylindrical sample ($\gamma_{center} = 0$) (Figure 9.58). The mean shear strain in the sample is usually taken as.

$$\gamma = r_e\theta/L \qquad (9.42)$$

where again r_e is the equivalent radius, often taken as 0.8r where r is the radius of the sample.

A typical τ vs. γ curve is shown in Figure 9.59. The shear modulus G is calculated as the slope of the line joining the two extremities of the loop. Alternatively, this curve can be generated by calculating the shear strain first, obtaining the shear modulus by the resonant frequency method, and then calculating the shear stress as $G\gamma$. The damping ratio D is defined from the curve as the ratio of the energy necessary to perform one cycle of torsion to the elastic energy expanded to load the sample to the peak of the cycle (Figure 9.59):

$$D = A_c/4\pi A_e \qquad (9.43)$$

where D is the damping ratio, A_c is the area inside the cycle, and A_e is the area inside the triangle shown in Figure 9.59.

The previous discussion identifies how G, D, and γ can be obtained for a given amount of torque applied at the top of the sample. This torque can then be increased to create a larger shear strain in the sample. The test is repeated and a new set of values of G, D, and γ are obtained. Point by point, the G vs. γ curve and the D vs. γ curve are described (Figure 9.60). The G vs. γ curve and the D vs. γ curve are the two results of a resonant column test. The strain that can be tested with this test typically ranges from 10^{-6} to 10^{-3}.

9.13.2 Unsaturated Soils

If the soil is unsaturated, or if the soil is saturated but the water is in tension, neither the test procedure nor the data

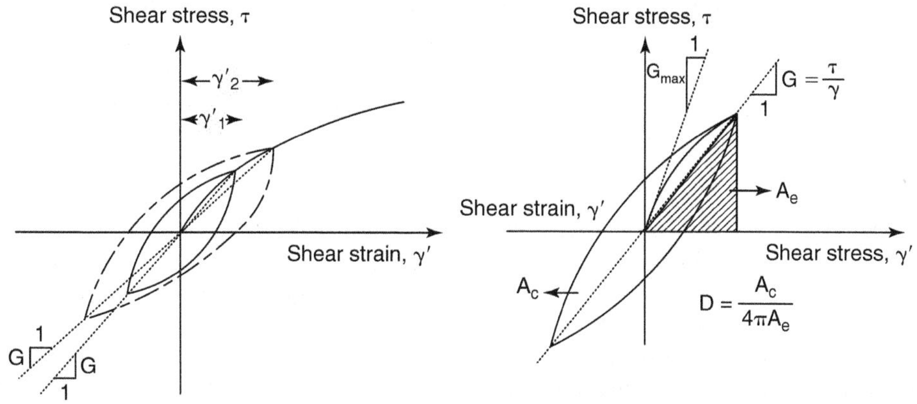

Figure 9.59 Shear stress-strain loops in resonant column test and damping ratio calculation: (*a*) Evolution of stress-strain loop. (*b*) Calculation of shear modulus and damping ratio.

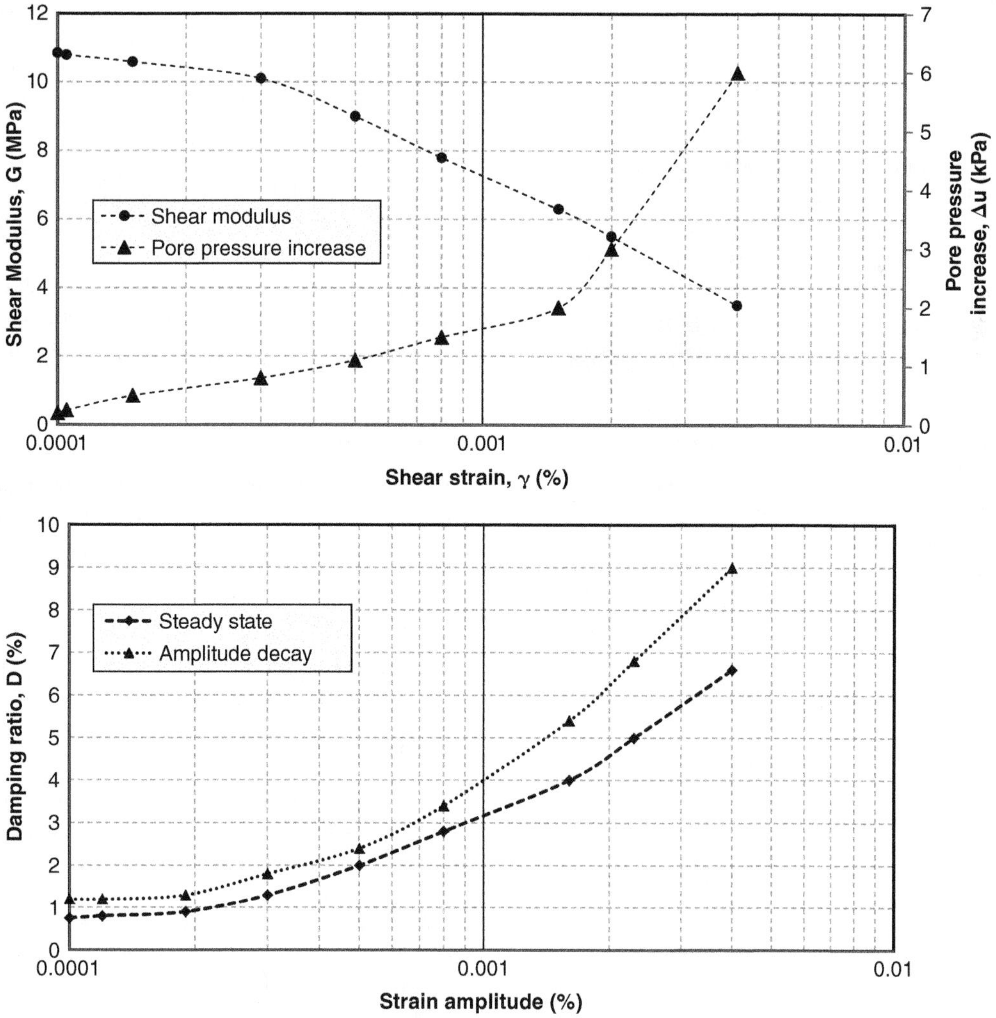

Figure 9.60 Shear modulus vs. shear strain and damping coefficient vs. shear strain: (*a*) Shearing modulus. (*b*) Damping ratio.

reduction changes. Indeed, the water stress is rarely measured during the resonant column test.

9.14 LAB VANE TEST

9.14.1 Saturated Soils

The lab vane test or VST (Figure 9.61) is used to determine the undrained shear strength of fine-grained soils (clays and silts). It can be performed either in the field with a field vane (ASTM D2573), or on the sample with a mini vane or a hand vane (ASTM D4648; Figure 9.61). The lab vane is made of two perpendicular blades, each having a 2-to-1 height-to-width ratio. The width of the blades varies from 12 to 25 mm; the larger vanes are used in softer soils. The vane is pushed perpendicularly into the end of a sample until the tops of the blades are one blade height below the surface of the sample. Then the vane is rotated at a slow rate (less than 1 degree per minute) while the testers measure the torque developed and the rotation angle (Figure 9.62). The peak value of the torque is recorded as T_{max}. The blade is then rotated at least 10 times rapidly and a new maximum torque value, T_{res}, is measured.

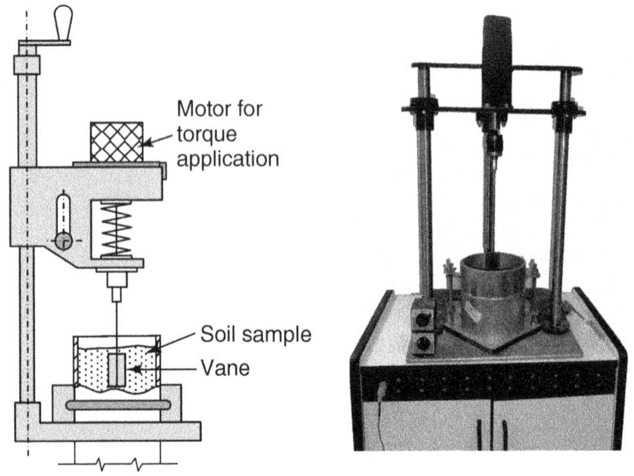

Figure 9.61 Lab vane test equipment: (*a*) Principle. (b) Equipment. (*a*: Adapted from BS 1377-7: 1990.)

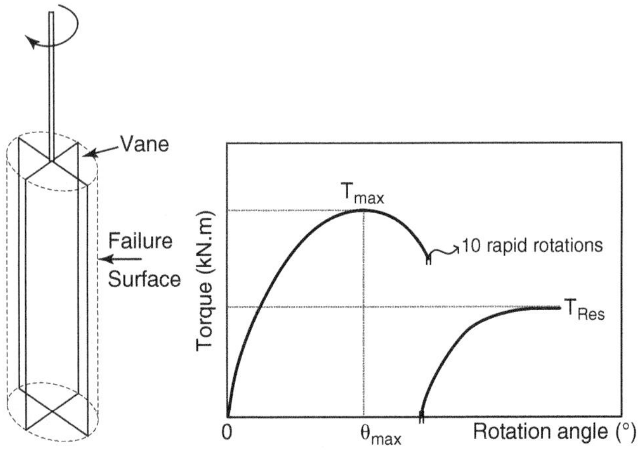

Figure 9.62 Lab vane test results.

The VST is used in saturated fine-grained soils to obtain the undrained shear strength s_u. The reason is that these soils have a low permeability and do not allow appreciable drainage during a test, which typically lasts less than 10 minutes. Therefore, for these saturated fine-grained soils, it is reasonable to assume that the undrained shear strength s_u is the parameter being measured. For a rectangular vane, the following equation gives s_u from T_{max}:

$$T_{max} = \pi s_u D^2 \left(\frac{H}{2} + \frac{D}{6} \right) \tag{9.44}$$

where D is the diameter of the vane and H is the height of the vane. Proof of this equation is shown in the solution to problem 7.4. The residual undrained shear strength s_{ur} is obtained from the same formula using T_{res}:

$$T_{res} = \pi s_{ur} D^2 \left(\frac{H}{2} + \frac{D}{6} \right) \tag{9.45}$$

The VST can be performed in coarse-grained soils, but no useful result can be obtained. These soils drain fast enough that one would not measure the undrained shear strength, but instead the drained or partially drained shear strength. Back-calculating the shear strength parameters from this test would require knowledge of the normal effective stress on the plane of failure in addition to T_{max}. This is not measured during the VST. The VST has the advantages of being fast, simple, economical, and useful for obtaining the undrained shear strength of fine-grained soils. Its drawbacks include that it is limited to fine-grained soils.

9.14.2 Unsaturated Soils

If the soil is unsaturated, or if the soil is saturated but the water is in tension, neither the test procedure nor the data reduction changes. Water stress is not measured during the vane test.

9.15 SOIL WATER RETENTION CURVE (SOIL WATER CHARACTERISTIC CURVE) TEST

9.15.1 Saturated Soils

The soil water retention curve (SWRC), also known as the soil water characteristic curve, is a property of the soil much like the shear strength parameters (Figure 9.63). It is a plot of the water content of the soil as a function of the water tension stress (suction) in the soil pores. It depends on many factors, including the particle size distribution, pore size distribution, soil structure, and soil texture.

During the drying process from a saturated state, the water tension in the soil will increase until it becomes large enough to force air into the soil pores. This water tension value is called the *air entry value* u_{wae}. Beyond the air entry value,

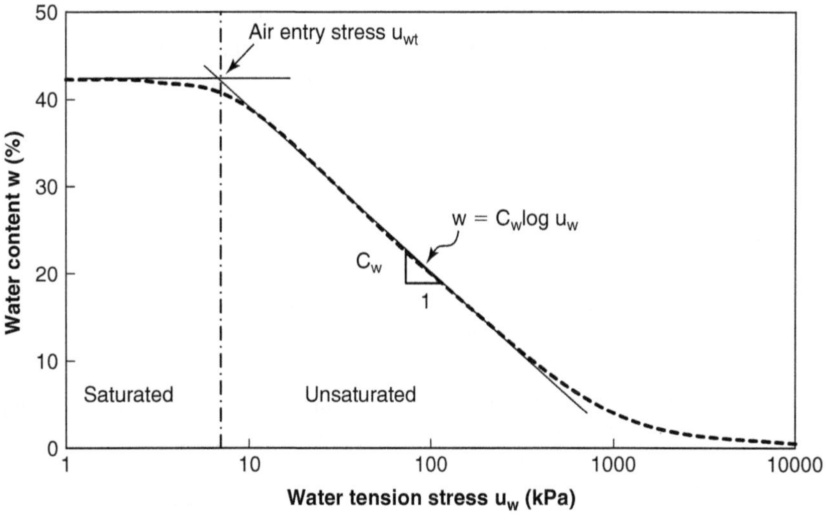

Figure 9.63 Soil water retention curve.

the decrease in water content is well approximated by a linear relationship between the water content and the log of the water tension, which can be written as:

$$dw = C_w d(\log_{10} u_w) \tag{9.46}$$

where w is in percent, u_w is in kPa (positive), and C_w is the slope of the SWRC.

The gravimetric water content is the most commonly used water content definition in geotechnical engineering, but for the SWRC the volumetric water content often is used. These are defined in the following equations:

$$\text{Gravimetric water content: } w = W_w/W_s \tag{9.47}$$

$$\text{Volumetric water content: } \theta_w = V_w/V \tag{9.48}$$

where W_w and V_w are the weight and volume of water respectively, W_s is the weight of solids, and V is the total volume. Example SWRCs are presented in Figure 9.64. It stands to reason that different soils will have different SWRCs: A sand will not retain water the same way a clay would. Imagine that you insert a straw into a sand; it would not take much sucking to get the water out of the sand. Now imagine that your straw is inserted into a clay; in this instance it would take a lot of sucking to get a little bit of water out. The suction or water tension that you would have to exert through the straw would be much higher for the clay than for the sand. This phenomenon is what the SWRC characterizes.

Soils under the groundwater level (GWL) are generally saturated and the water is in compression. Soils above the GWL can be saturated or unsaturated, but in both cases the water is in tension (suction). The SWRC is a property of a

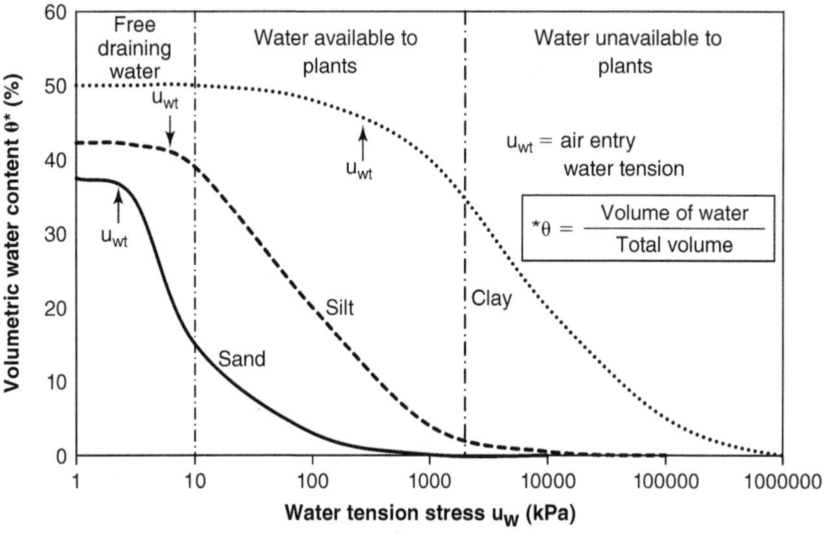

Figure 9.64 Example of soil water retention curve.

soil where the water is in tension. As such, the SWRC for a saturated soil refers to the case where the soil is saturated above the GWL by capillary action and other electrochemically based phenomena such as the affinity between water and clay minerals.

If a saturated soil sample is placed on a table top and is strong enough to stand by itself, it is likely held together by water tension unless it has some cementation (effective stress cohesion). As the soil dries, it initially shrinks while remaining saturated. The water tension increases and at a given water tension stress (suction), air enters the pores. The water tension at this point is called the *air entry value* (u_{we}). From this point on during the drying process, the soil is unsaturated. The procedure to determine the SWRC is the same below (saturated) and above (unsaturated) the air entry value. This procedure is detailed in section 9.15.2 related to unsaturated soils.

9.15.2 Unsaturated Soils

There are essentially two methods for obtaining the SWRC (ASTM D6836). The first consists of taking a saturated soil sample and measuring the water tension and the water content of the sample as a function of time as it dries up. The water content measurement was described in section 3.9; the water tension measurement was described in section 9.2.4 and summarized in Table 9.2. In this case, the two most common methods to measure water tension for the SWRC are the filter paper method and the chilled mirror psychrometer. For lower values of water tension, the hanging column method can also be used (ASTM D6836). As a guide, and for tests performed in an air-conditioned laboratory environment where the relative humidity is around 50%, a 25 mm high, 75 mm diameter sample is likely to become air-dry in about 24 hours. In these circumstances, a water content and water tension measurement every 1 to 2 hours is suitable to get a good description of the SWRC.

The second method of obtaining the SWRC is to use a saturated soil sample and force the sample to come to equilibrium at a selected series of water tension (suction), while measuring the water content for each one of those water tension values. The pressure plate apparatus can be used in this case (Figure 9.9); it makes use of the axis translation technique (Figure 9.10) and increases the air pressure to push the water out of the soil pores. The air pressure is equal to the water tension in the sample when the water starts moving out of the pores. The water content of the soil sample is measured when the water stops flowing. Such measurements are made at increasingly higher air pressures so as to describe the complete SWRC.

Yet another way is to use the salt solution equilibrium technique, in which "identical" samples are placed in different salt solution chambers (Figure 9.11) and left in the chamber until the water tension in the sample comes into equilibrium with the relative humidity created by the salt solution at the bottom of the chamber; reaching this equilibrium may take 1 or 2 weeks. The salt concentration in each chamber is different and is chosen to create a series of values of the relative humidity and therefore water tension, which gives a good description of the SWRC. After equilibrium is reached, the soil water content is measured in each chamber and the SWRC can be plotted.

The SWRC describes the fact that the water tension increases when the water content decreases but recognizes that this relationship is not the same when the soil is drying as when it is wetting; this is called the *hysteresis* in the SWRC. Figure 9.65 shows the difference between the drying curve

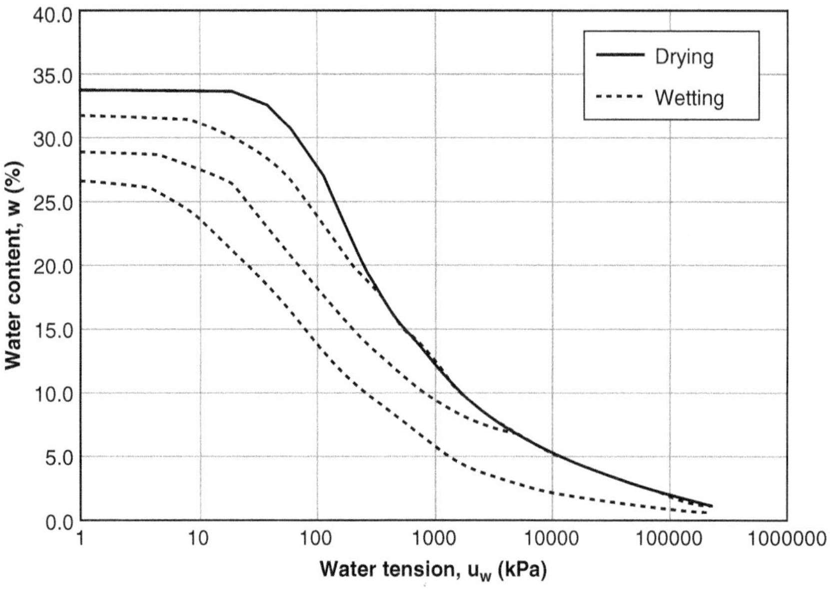

Figure 9.65 Drying and wetting hysteresis loop in the SWRC.

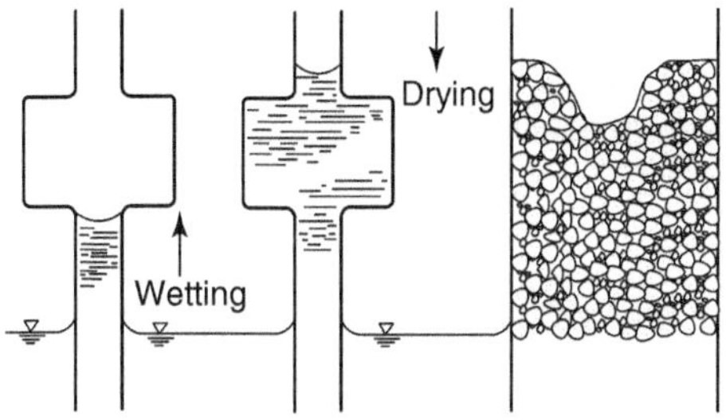

Figure 9.66 Geometrical explanation of drying and wetting hysteresis in the SWRC.

and the wetting curve. It is likely that this difference decreases as the number of drying and wetting cycles between the same values increases. The hysteresis effect may be attributed to several causes: the geometric nonuniformity of the individual pores, the pore fluid contact angle, entrapped air, and swelling, shrinking, or aging. The geometric nonuniformity of the pores can be explained as follows (Figure 9.66). When the soil is drying, the water level in the conduits formed by the voids between particles can drop down through a larger void cross section, as shown in Figure 9.66. However, if the soil is wetting, there is a limit to how large a cross section the water can move up, as the capillary force is limited. As a result, the loss of water is larger during drying than the gain of water during wetting and thus the wetting curve is below the drying curve (Figure 9.65). Several stages are identifiable in the drying or wetting process, as shown in Figure 9.67. During drying, at first the soil is saturated ($S = 1$) until the air entry value of the water tension u_{wae} is reached; then a linear semilog relationship exists between the water content and the water tension; and then the soil reaches a residual stage ($S = S_r$) where the water no longer forms continuous conduits in the pores, but rather exists only at the contacts between particles. The effective degree of saturation S_e is defined for a given degree of saturation S as:

$$S_e = \frac{S - S_r}{1 - S_r} \tag{9.49}$$

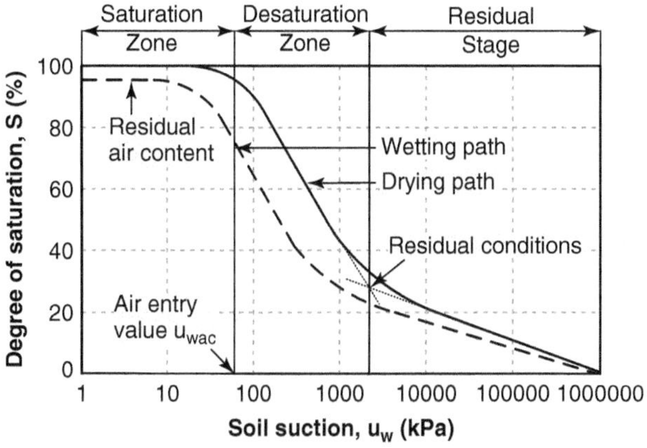

Figure 9.67 Various stages in the SWRC.

During the wetting process, a similar progression takes place in reverse and after the saturation phase, where again there is a linear semilog relationship between water content and water tension. The soil reaches a residual air content when the air is occluded and cannot be chased out of the voids through normal means.

Various empirical models have been proposed to describe the SWRC. Among the most common are:

Brooks and Corey (1964)
$$S_e = \begin{pmatrix} 1 \; if \; u_w \leq u_{wae} \\ \left(\frac{u_w}{u_{wae}}\right)^{-\lambda} \; if \; u_w \leq u_{wae} \end{pmatrix} \tag{9.50}$$

van Genuchten (1980)
$$S_e = \left(\frac{1}{1 + (\alpha u_w)^n}\right)^m \quad \text{with } m = 1 - 1/n \tag{9.51}$$

Fredlund and Xing (1994)
$$\theta = C(u_w)\theta_s \left(\frac{1}{Ln(e + (u_w/a)^n)}\right)^m \tag{9.52}$$

where S_e is the effective degree of saturation; u_w is the water tension (kPa); u_{wae} is the air entry value of the water tension (kPa); λ is a fitting parameter mostly influenced by the pore size distribution of the soil; α, n, and m are fitting parameters; θ is the volumetric water content (volume of water over total volume); θ_s is the volumetric water content at saturation; $C(u_w)$ is a correction factor that forces the model through a prescribed water tension value of 10^6 kPa at zero water content; a is a fitting parameter; and e is the logarithmic constant ($Ln\ e = 1$). More details on these models can be found in Lu and Likos (2004). ARA-ERES (2000) proposed a set of SWRCs (Figure 9.68) predicted on the basis of D_{60} in mm, the particle size for which 60% by weight is finer, and an index called the wPI. The wPI is defined as the product of the percent passing sieve number 200 as a decimal (ratio not percentage) and the plasticity index as a percent.

9.16 CONSTANT HEAD PERMEAMETER TEST

9.16.1 Saturated Soils

The constant head permeameter (CHP) (ASTM D2434; Figure 9.69) is used to obtain the coefficient of hydraulic conductivity k of saturated coarse-grained soils. The soil sample is placed in a cylinder about 75 mm in diameter and 150 mm high, with one filter stone at the top and another at the bottom. The top of the sample is connected by tubing to a container in which the water level is kept constant through an overflow regulator. The bottom of the sample is connected to another container in which the water level is also kept constant. The bottom container is kept lower than the top container and the flow Q (m^3/s) out of the bottom container is measured. The measurement simply consists of weighing the amount of water collected in the overflow container

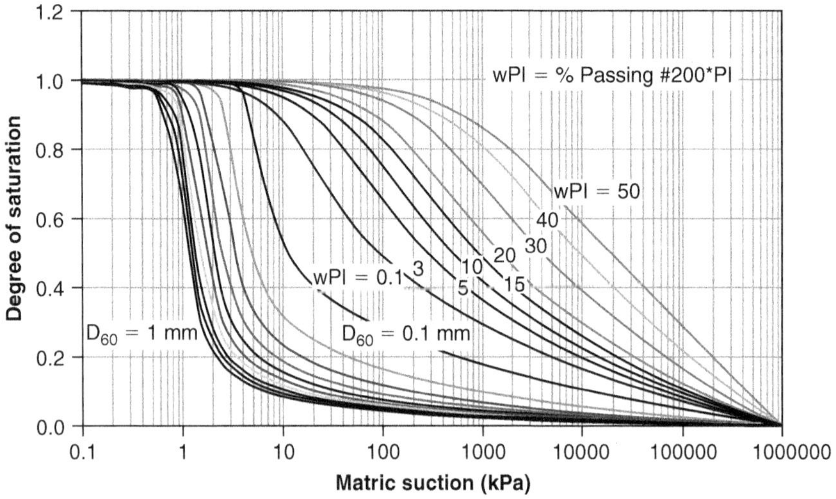

Figure 9.68 SWRC as a function of percent passing #200 and plasticity index. (Courtesy of NCHRP.)

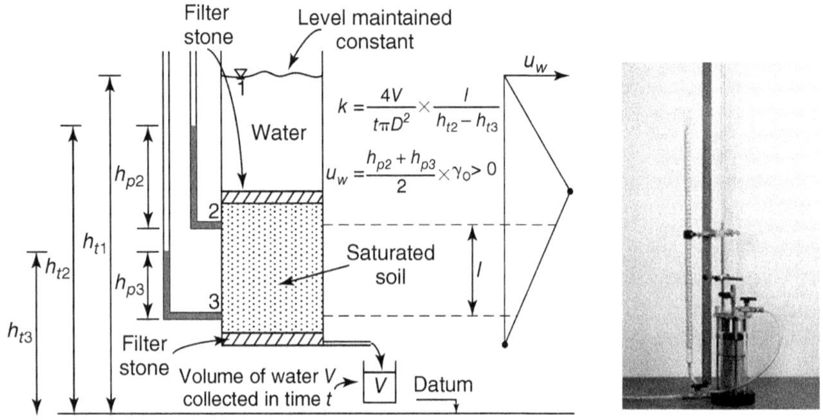

Figure 9.69 Constant head permeameter equipment: (*a*) Principle. (*b*) Equipment. (b: Courtesy of ELE International.)

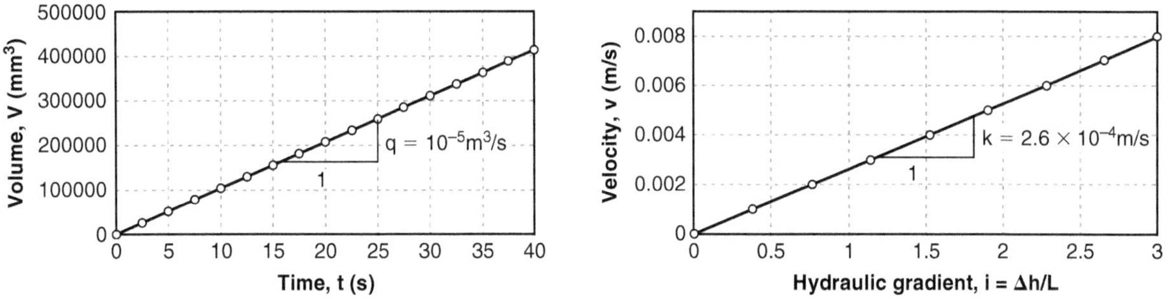

Figure 9.70 Constant head permeameter test results.

during a corresponding time. Typical results are shown in Figure 9.70. Often manometer tubes are connected to the side of the sample container at two points to give the water stress (pressure) at those two locations. (See Chapter 13 on flow through soils for an explanation of the following equations and parameters.) Darcy's law gives:

$$v = k\,i \tag{9.53}$$

where v is the discharge velocity through the sample, k is the hydraulic conductivity, and i is the hydraulic gradient. The hydraulic gradient in this case is given by:

$$i = h/l \tag{9.54}$$

where h is the loss of total head through the flow distance l (Figure 9.69). Also, conservation of mass gives:

$$Q = v\,A \tag{9.55}$$

where Q is the flow out of the sample, v is the discharge velocity, and A is the cross-sectional area of the sample. Note that A is the total cross-sectional area of the sample, not just the area of the pores through which the water is flowing. As a result, v is not the actual speed of the water molecules flowing through the pores (*seepage velocity*) but rather an equivalent speed called the *discharge velocity*. Combining equations 9.53 through 9.55 then gives the value of k:

$$k = Q\,l/h\,A \tag{9.56}$$

where Q is the discharge (m^3/s), l is the flow length between 2 points in the sample, h is the loss of total head between the same 2 points, and A is sample cross-sectional area. The discharge Q is the volume V collected in a time t divided by t. The cross-sectional area A is $\pi D^2/4$ where D is the sample diameter. The loss of total head h is $h_{t2} - h_{t3}$, as shown in Figure 9.69. Therefore, the hydraulic conductivity k is:

$$k = \frac{4V}{t\pi D^2}\frac{l}{h_{t2} - h_{t3}} \tag{9.57}$$

The advantage of the constant head permeameter test is that it is a very simple test to run; the drawback is that it is limited to measuring the hydraulic conductivity k of coarse-grained soils at a small scale. The k values typically measured with this test range from 10^{-1} to 10^{-6} m/s.

9.16.2 Unsaturated Soils

If the soil is unsaturated, things are quite different. The first thing to realize is that the hydraulic conductivity of an unsaturated soil is less than that of a saturated soil: Water goes through an unsaturated soil more slowly than through a saturated soil. The reason is that air is in the way of the flow, and the water is attracted to the walls of the tiny conduits formed by the particles. Of course, one must remember that in the equation giving the water velocity v (m/s) from the flow discharge Q (m^3/s) (Eq. 9.55), A is the total cross-sectional area of the sample, not the actual water flow area. Because the flow area is significantly reduced in the case of unsaturated flow, the actual water velocity is quite a bit higher than the velocity given by Eq. 9.55.

The steady-state permeameter test for unsaturated soils consists of the same equipment except for two differences: (1) The measurements of water compression are changed to measurements of water tension, and (2) a tube is connected to the center of the sample to control the air pressure in the sample. The measurement of the water tension is made at two locations, using tensiometers or other appropriate devices. Figure 9.71 shows the diagram for an unsaturated steady-state permeability test. The water level is maintained on the upstream side (point 1) and the water starts flowing. It arrives at the high air entry disk. This disk lets the water go through but not the air; that is a property of that disk. Then the water goes through the soil voids. One would think that it would flood the voids as it is attracted by the water tension (suction) in the water phase. But the air is in the way, and it has no way to escape because there is another high air entry disk at the other end of the sample. So the water is forced to flow through the continuous water phase around the air phase. The water tension is larger at point 3 than at point 2 (Figure 9.71) because the water loses energy as it drives through the soil. A friction force arises between the water molecules and the soil particles as the water drags through the voids. This force is called the *seepage force*. As a result of this force, there is an associated loss of pressure between points 2 and 3. Because the pressure at point 2 is negative (water tension), the pressure at point 3 is even more negative ($h_{p3} < h_{p2} < 0$).

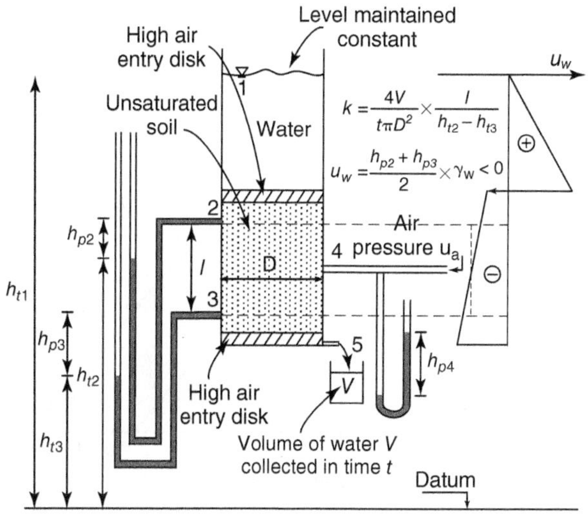

Figure 9.71 Constant head permeameter test for unsaturated soils.

The hydraulic conductivity depends on the water tension (Figure 9.72). As the water tension increases, the amount of water in the soil decreases, and it becomes harder and harder for the water to percolate through the soil: There is less room for the water to flow and a higher attraction between the water and the soil particles. The effect of the water tension on the hydraulic conductivity can be documented in this test by changing the air pressure through port 4. Applying an air pressure u_a different from zero changes the water tension u_w. This allows one to run the permeability test at different water tensions and establish the relationship between hydraulic conductivity and water tension.

From the calculations point of view, the hydraulic conductivity k is obtained as:

$$k = \frac{4V}{t\pi D^2}\frac{l}{(h_{t2} - h_{t3})} \tag{9.58}$$

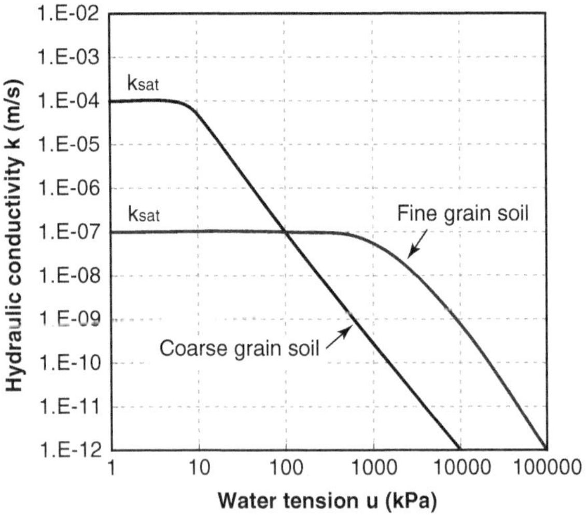

Figure 9.72 Constant head permeameter test results for unsaturated soils.

where V is the volume of water collected in a time t, D is the sample diameter, and l is the distance between the two points where the total heads h_{t2} and h_{t3} are measured. Note that Eq. 9.58 is the same as Eq. 9.57. The difference is that h_{t2} and h_{t3} are different, because the soil is unsaturated. The average water tension u_w associated with the hydraulic conductivity k of Eq. 9.58 is:

$$u_w = \frac{h_{p2} + h_{p3}}{2} < 0 \tag{9.59}$$

The combination (k, u_w) gives the coordinates of one point on the hydraulic conductivity vs. water tension curve. By testing the soil at different water content (water tension), one can get the complete curve (Figure 9.72).

Note that the chemistry of the water makes a difference when running a permeability test. If the water that seeps through the soil has a much different salt chemistry than the sample water, the osmotic suction could be activated and lead to a different water tension in the sample than if the water seeping through the sample had the same chemistry than the water in the sample.

9.17 FALLING HEAD PERMEAMETER TEST FOR SATURATED SOILS

The falling head permeameter (FHP) (Figure 9.73) is used to obtain the hydraulic conductivity k of saturated fine-grained soils. The soil sample is placed in a cylinder about 75 mm in diameter and 150 mm high with one filter stone at the top and another at the bottom. The top of the sample is connected to a tube with a much smaller diameter (say, 10 mm) filled with water. Unlike the constant head permeameter, the water level in this tube goes down with time. The bottom of the sample is connected to a container where the water level is kept constant through an overflow. The measurements consist of recording the time on one hand and the drop in height in the small tube on the other (Figure 9.74). Darcy's law gives:

$$v = k\,i \tag{9.60}$$

where v is the discharge velocity through the sample, k is the hydraulic conductivity, and i is the hydraulic gradient. The hydraulic gradient in this case is given by:

$$i = h/l \tag{9.61}$$

where h is the loss of total head through the flow distance l (Figure 9.73). Also, conservation of mass gives:

$$Q = v\,A = k\,i\,A = k\,A\,h/l \tag{9.62}$$

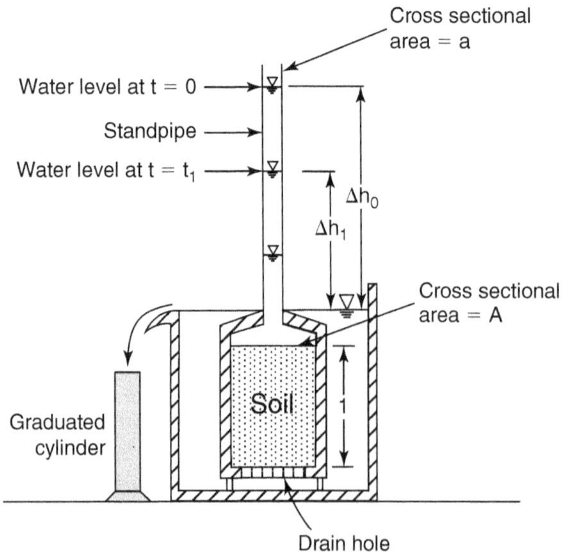

Figure 9.73 Falling head permeameter equipment: (a) Principle. (b) Equipment. (a: After FHWA. b: Courtesy of Gilson Company, Inc.)

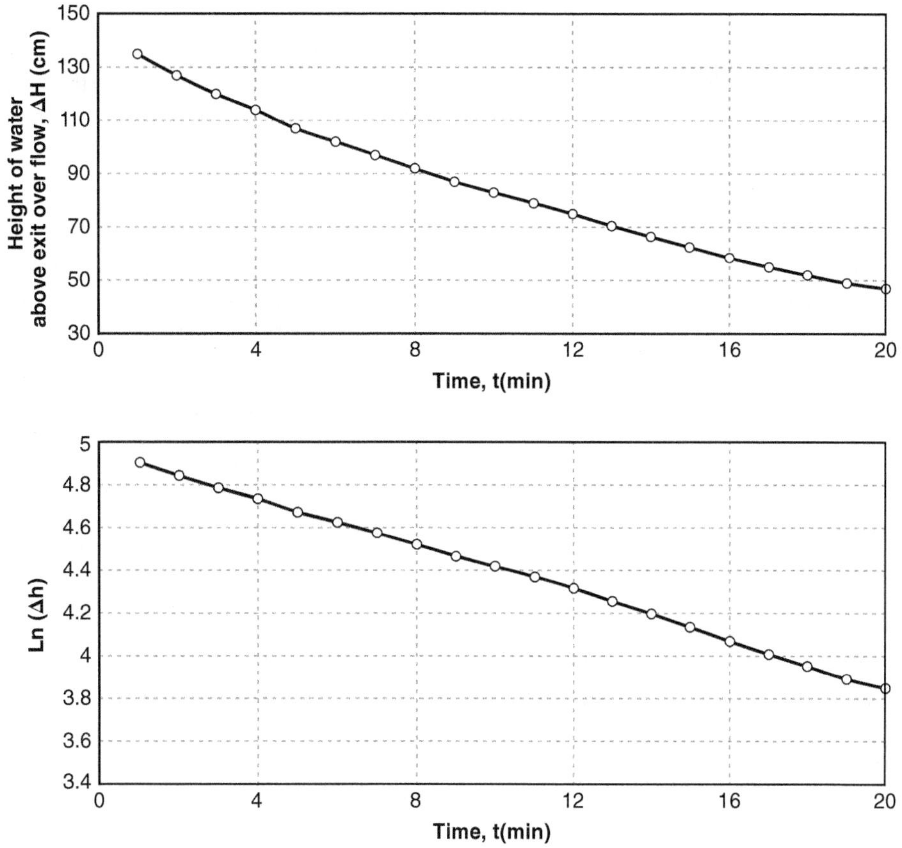

Figure 9.74 Falling head permeameter test results.

where Q is the flow out of the sample, v is the discharge velocity, and A is the cross-sectional area of the sample. The flow Q through the sample is also given by:

$$Q = -a\,dh/dt \tag{9.63}$$

where a is the cross-sectional area of the small tube, and dh is the drop in water level in the small tube during the time *dt*. The minus sign is necessary because *dh* is negative (the water level drops) while all other quantities are positive. Regrouping equations 9.58 and 9.59 gives:

$$dt = \frac{al}{Ak}\left(-\frac{dh}{h}\right) \tag{9.64}$$

After integration between the times 0 and t_1 corresponding to the losses of total head equal to Δh_0 and Δh_1 (Figure 9.73), the hydraulic conductivity is given by:

$$k = \frac{al}{At_1}Ln\frac{\Delta h_0}{\Delta h_1} = 2.3\frac{al}{At_1}\log\frac{\Delta h_0}{\Delta h_1} \tag{9.65}$$

The advantage of the falling head permeameter test is that it is very simple to run; however, it is limited to measuring the hydraulic conductivity *k* of fine-grained soils at a small scale. The *k* values typically measured with this test range from 10^{-7} to 10^{-11} m/s.

9.18 WETTING FRONT TEST FOR UNSATURATED SOILS

The wetting front test is used to measure the hydraulic conductivity of unsaturated soils as a function of the water tension or water content. In this test the water progresses through the soil and saturates (wets) the soil as it goes. Three methods exist: the instantaneous profile method, the capillary rise method, and the wetting front method. The instantaneous profile method is described here. The other methods are described in Li, Zhang, and Fredlund (2009).

The test setup is shown in Figure 9.75. The water enters the sample from the left and wets the sample progressively toward the right. The air is chased away in front of the wetting front and escapes through the filter stone at the right end of the sample. Tensiometers or psychrometers are placed at regular intervals along the sample to measure the water tension u_w. During the test, the water content of the sample increases progressively while the water tension decreases accordingly. Because the hydraulic conductivity *k* depends on the water tension u_w, during this single test the hydraulic conductivity varies significantly. The measurements of water tension give the water content through the SWRC and also the velocity of the water as a function of time. The result of this test consists of a curve linking the hydraulic conductivity to the water tension or water content.

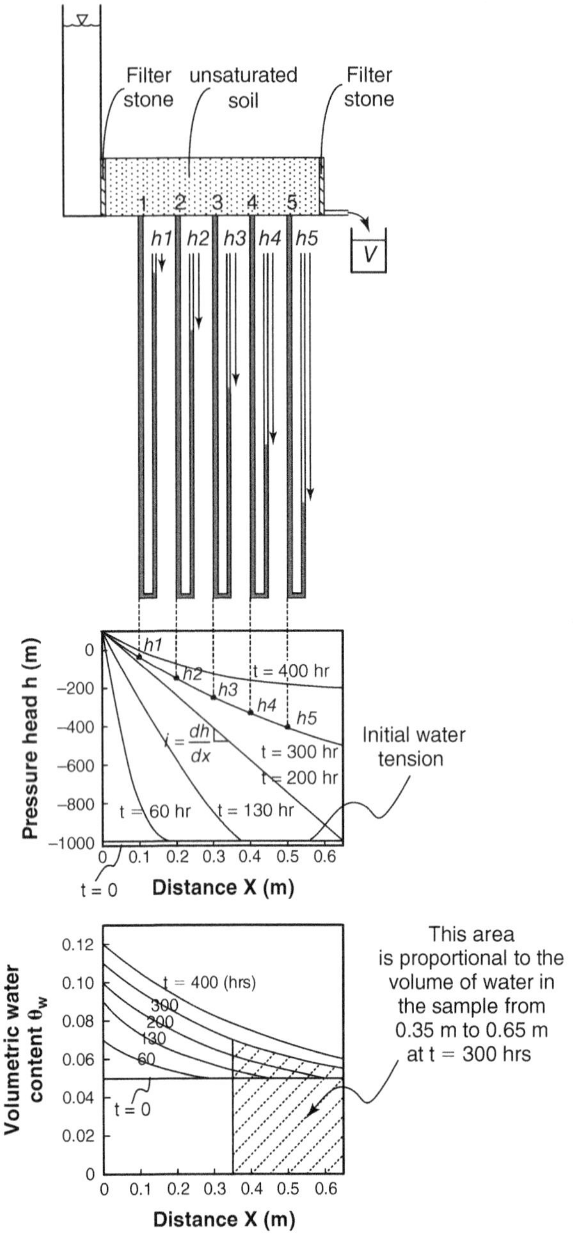

Figure 9.75 Wetting front test for unsaturated soils.

The data reduction proceeds as follows. The hydraulic gradient i varies as a function of time t and is given by the slope of the total head h_t versus distance x. Because the elevation head is constant, it is also the slope of the pressure head h_p versus distance x (Figure 9.75):

$$i_{(t)} = \frac{dh_t}{dx} = \frac{dh_p}{dx} \tag{9.66}$$

Note that the pressure head h_p is related to the water tension u_w and the unit weight of water γ_w as follows:

$$u_w = \gamma_w h_p \tag{9.67}$$

The volumetric water content θ_w may be obtained in the sample at any time and at any location by using the measured water tension at that time and at that location and the SWRC. In the center part of the curve, the relationship between θ_w and u_w can be approximated by:

$$\theta_w = C_w \log u_w + a \quad \text{or} \quad d\theta_w = C_w\, d(\log u_w) \tag{9.68}$$

where C_w is the slope of the θ_w vs. log u_w curve. Then the volume of water V_w in the sample between a given point j and the end of the sample is given by:

$$V_w = \int_{x_j}^{l} \theta_{w(x)} A dx \tag{9.69}$$

The velocity v_w of the water passing point j during an interval of time dt is given by:

$$v_w = \frac{dV_w}{Adt} \tag{9.70}$$

and the corresponding average hydraulic gradient is:

$$i_{ave} = \frac{1}{2}\left(i_{(t)} + i_{(t+dt)}\right) \tag{9.71}$$

Then the hydraulic conductivity $k(u_w)$ is found as the ratio of the velocity and the hydraulic gradient. The value of k depends on the water tension u_w; by using different corresponding pairs of pressure head vs. distance curves and volumetric water content vs. distance curves, the graph of $k(u_w)$ vs u_w can be generated with a single test.

9.19 AIR PERMEABILITY TEST FOR UNSATURATED SOILS

For unsaturated soils, it is sometimes necessary to measure the hydraulic conductivity of the soil to air flow k_a. One way to measure k_a is to use the apparatus shown in Figure 9.76. Air is supplied at a constant pressure p to the bottom of the unsaturated soil sample. The air flows through the soil and comes out at the top of the sample. The volume of air V_a collected during a time t is measured through a U-shaped, graduated burette. The air at the top of the sample is kept at

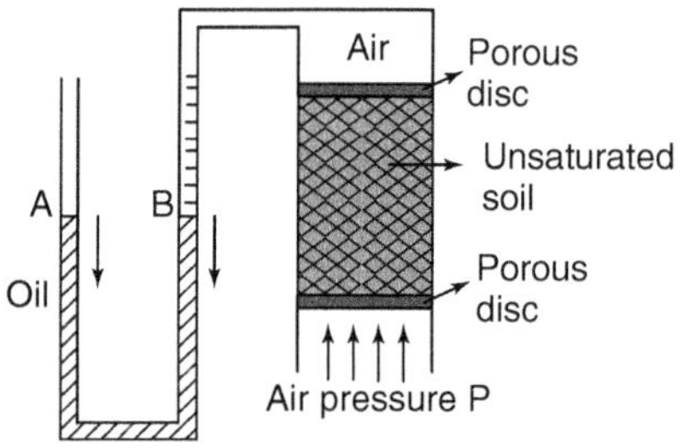

Figure 9.76 Measuring the hydraulic conductivity of air through an unsaturated soil.

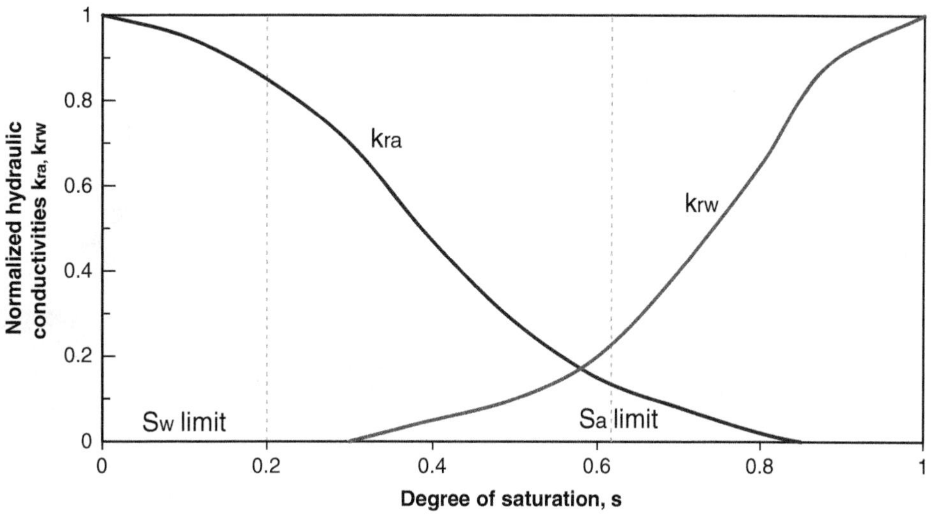

Figure 9.77 Relative hydraulic conductivity of water and air as a function of degree of saturation.

atmospheric pressure by adjusting the height of the burette so that the level of the oil remains the same on both sides (points A and B on Figure 9.76). Because point A is connected to the atmosphere, the pressure at B is also atmospheric.

The data reduction consists of the following. Darcy's law seems to describe the flow of air through soil reasonably well (Fredlund and Rahardjo 1993, p. 119). Therefore:

$$v_a = k_a i_a \tag{9.72}$$

where v_a is the air flow velocity, k_a is the hydraulic conductivity of air, and i_a is the hydraulic gradient for the air flow. Furthermore, the hydraulic gradient for the air is:

$$i_a = \frac{dh_a}{dx} = \frac{d\left(\frac{u_a}{\gamma_a}\right)}{dx} = \frac{1}{\gamma_a}\frac{du_a}{dx} \tag{9.73}$$

where h_a is the total head for air, x is the flow distance along the soil sample, u_a is the air pressure, and γ_a is the unit weight of air (0.0118 kN/m^3 at 20C). Note that the change in elevation head for air is typically negligible compared to the change in pressure head for air. This is why the change in total head is taken to be equal to the change in pressure head. The air pressure at the bottom of the sample is p and is maintained at 0 at the top of the sample, which has a length L. Therefore, the hydraulic conductivity of the air through the soil sample is given by:

$$k_a = \frac{\gamma_a V_a L}{Apt} \tag{9.74}$$

where k_a is the hydraulic conductivity of air, γ_a is the unit weight of air (0.0118 kN/m^3 at 20C), V_a is the volume of air flowing through the sample during a time t, L is the sample length, and p is the air pressure applied at the bottom of the sample.

The value of k_a depends on how dry the sample is as measured by the water content or water tension. The test described previously can be performed at different values of the water content or water tension by simply letting the sample dry and repeating the test at different water contents. The drier the sample is, the higher the value of k_a will be for a given soil. When the soil is dry, the value of k_a is maximum and equal to $k_{a(dry)}$. This trend is contrary to the trend for the hydraulic conductivity of water k_w. Indeed, k_w decreases when the soil gets drier; it is maximum when the soil is saturated and equal to $k_{w(sat)}$. Both hydraulic conductivities are often presented as normalized values as follows:

$$k_w = k_{rw} k_{w(sat)} \tag{9.75}$$

$$k_a = k_{ra} k_{a(dry)} \tag{9.76}$$

Figure 9.77 shows an example of the combined variation of both normalized hydraulic conductivity values k_{rw} and k_{ra} as a function of the degree of saturation S. Note that there is a limiting degree of saturation S_w (0.3 on Figure 9.77) where the water is no longer mobile and at the same time a limiting degree of saturation S_a (0.85 on Figure 9.77) where the air is no longer mobile.

9.20 EROSION TEST

9.20.1 Saturated Soils

The erosion function apparatus (EFA) test was developed in the early 1990s to measure the erodibility of soils and soft rocks (Figure 9.78; Briaud 2008). The principle is to go to the site where erosion is being investigated, collect samples within the depth of concern, bring them back to the laboratory and test them in the EFA. The 75 mm outside diameter sampling tube is placed through the bottom of the conduit where water flows at a constant velocity (Figure 9.78).

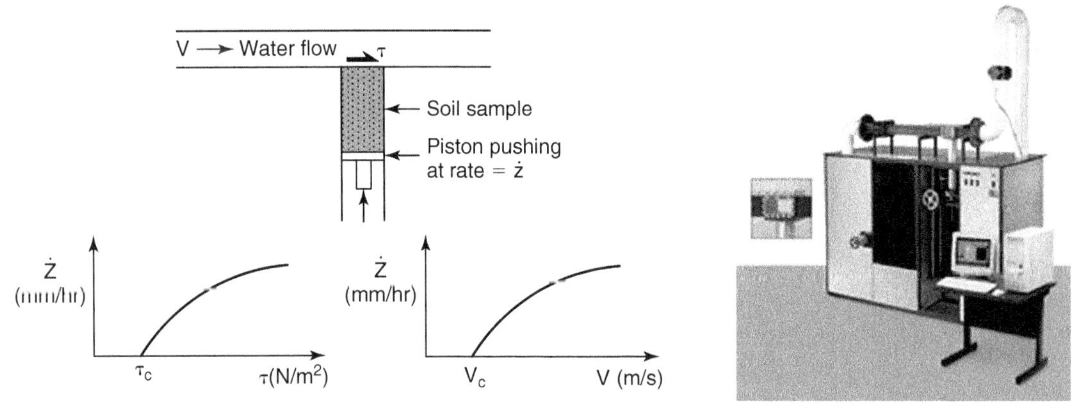

(a) (b)

Figure 9.78 Erosion function apparatus test equipment: (*a*) Principle. (*b*) Equipment.

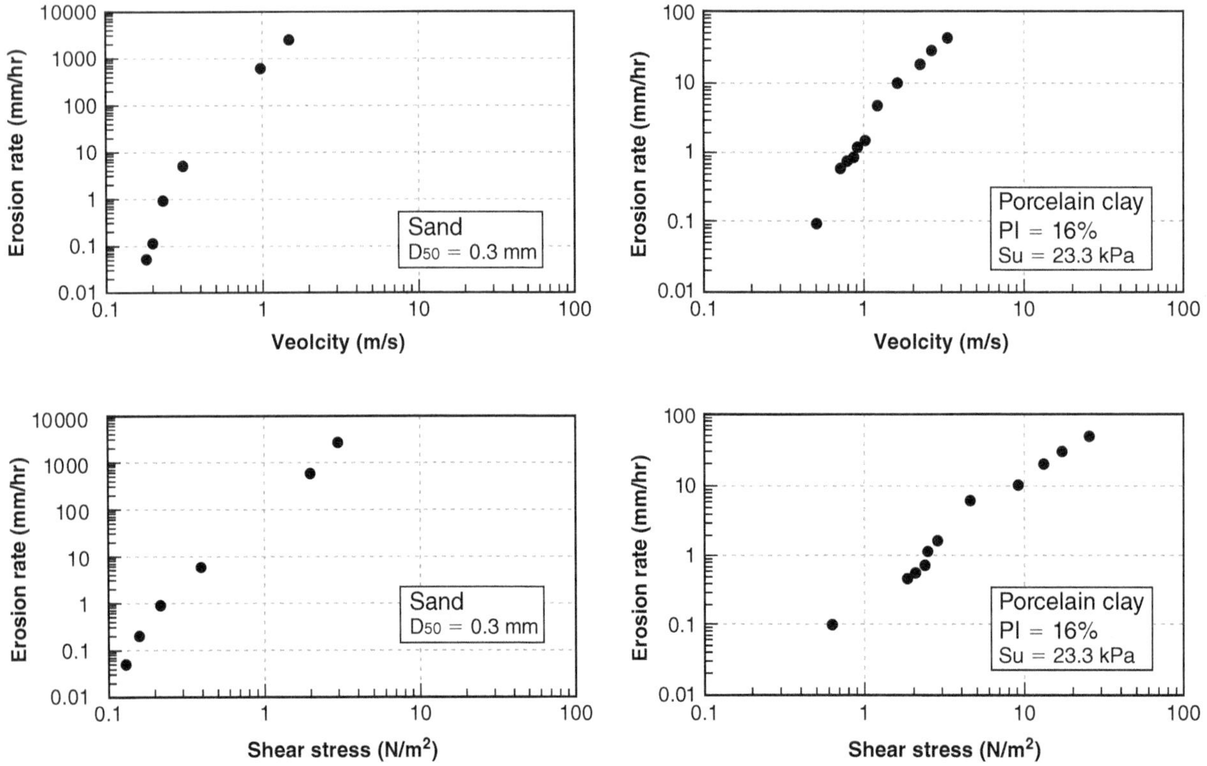

Figure 9.79 EFA test results: (*a*, *c*) A sand. (*b*, *d*) A clay.

The soil or rock is pushed by a piston out of the sampling tube only as fast as it is eroded by the water flowing over it.

The test result consists of the erosion rate $\dot{z}$ vs. shear stress τ curve and erosion rate $\dot{z}$ vs. mean flow velocity V curve (Figure 9.79). For each flow velocity V, the erosion rate $\dot{z}$ (mm/hr) is simply obtained by dividing the length h of sample eroded by the time t required to do so:

$$\dot{z} = \frac{h}{t} \tag{9.77}$$

The velocity V is obtained by measuring the flow Q and dividing by the flow area A. The shear stress τ is obtained by using the Moody Chart (Figure 9.80; Moody 1944) for pipe flows:

$$\tau = \frac{1}{8} f \rho V^2 \tag{9.78}$$

where τ is the shear stress on the wall of the pipe, f is the friction factor obtained from the Moody Chart (Figure 9.80), ρ is the mass density of water (1000 kg/m^3), and V is the mean flow velocity in the pipe. The friction factor f is a function of the pipe Reynolds Number R_e and the pipe roughness ε/D. The Reynolds Number is VD/v where D is the pipe diameter and v is the kinematic viscosity of water (10^{-6} m^2/s at 20°C).

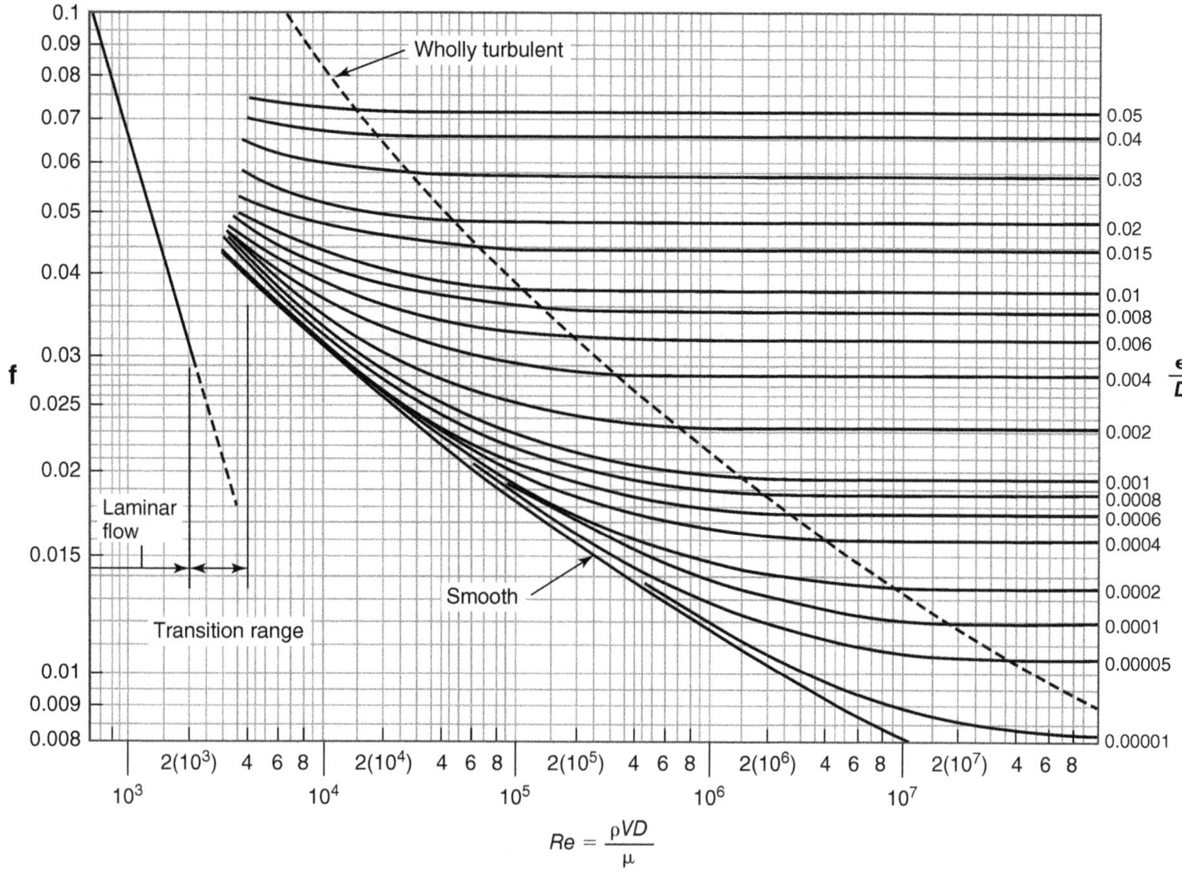

Figure 9.80 Moody Chart. (After Munson et al. 2012.)

Because the pipe in the EFA has a rectangular cross section, D is taken as the hydraulic diameter D = 4A/P where A is the cross-sectional flow area, P is the wetted perimeter, and the factor 4 is used to ensure that the hydraulic diameter is equal to the diameter for a circular pipe. For a rectangular cross-section pipe:

$$D = \frac{2ab}{a+b} \tag{9.79}$$

where a and b are the dimensions of the sides of the rectangle. The relative roughness ε/D is the ratio of the average height ε of the sample roughness over the pipe diameter D. The average height of the sample roughness ε is taken equal to $0.5D_{50}$ where D_{50} is the mean grain size for the soil. The factor 0.5 is used because it is assumed that the top half of the particle protrudes into the flow while the bottom half is buried in the soil mass. For fine-grained soils, the roughness is taken as one-half of the depth of the asperities on the sample surface.

For fine-grained and coarse-grained soils, ASTM standard thin-wall steel tube samples are favored. If such samples cannot be obtained (e.g., with coarse-grained soils), split spoon SPT samples are obtained and the coarse-grained soil is reconstituted in the thin-wall steel tube. Fortunately, in the case of erosion of uncemented coarse-grained soils, soil disturbance does not affect the results significantly. For erosion of rocks—if it is representative of the rock erosion process to test a 75 mm diameter rock sample—the rock core is placed in the thin-wall steel tube and tested in the EFA. Example erosion functions are shown in Figure 9.79 for a fine sand and for a low-plasticity clay. Note that for the same average velocity of 1 m/s in the EFA test conduit, the rate of erosion for the sand is about 1000 times faster than for the clay. This indicates that the rate of erosion can be very different for different soils.

Other devices have also been developed to evaluate how resistant earth materials are to water flow. These include the rotating cylinder to measure the erosion properties of stiff soils (e.g., Chapuis and Gatien 1986), the jet erosion test to evaluate the erodibility of soils (e.g., Hanson 1991), and the hole erosion test to measure the erosion properties of stiff soils (e.g., Wan and Fell 2004).

9.20.2 Unsaturated Soils

If the soil is unsaturated, the EFA test procedure and the data reduction are unchanged, except that a decision must be made on whether to let water stand on top of the sample for an extended period of time before starting the water flow. The decision should be based on which option best represents the field conditions. If the unsaturated state of the sample was created by drying, then the water should be left standing on top of the sample until saturation is recreated before testing. In ephemeral streams, for example, the soil dries in the summer, and a fine-grained soil may crack, thereby generating a thin crust of soil. This thin crust will be washed away as soon as significant water flow occurs. In this case it is necessary not to soak the sample before erosion testing. Note that in all cases, as the test progresses, the sample is likely to become nearly saturated.

PROBLEMS

9.1 What are the four main categories of laboratory tests? Give three examples for each category.

9.2 What device would you use to measure the following quantities? In each case explain the basic principle of the device.
   a. Force
   b. Pressure
   c. Shear stress
   d. Water compression stress

9.3 What devices and techniques would you use to measure the water tension stress (suction)? In each case explain the basic principle of the device.

9.4 What devices would you use to measure the following quantities? In each case explain the basic principle of the device.
   a. Displacement
   b. Normal strain
   c. Shear strain

9.5 Table 9.1s shows the results of a Standard Proctor Compaction Test for the same soil tested at different water contents. Calculate the dry unit weight and water content for each sample, plot a graph of dry unit weight versus water content, determine the maximum dry unit weight and optimum water content, and plot the saturation lines for S = 100%, S = 90%, and S = 80%.

Table 9.1s Results of a Standard Proctor Compaction Test

Volume of mold (m^3)	Unit weight (kN/m^3)	Weight wet soil + container (N)	Weight dry soil + container (N)	Weight of water (N)	Weight of container (N)	Weight of dry soil (N)
0.000943	18.48	0.194	0.182	0.012	0.048	0.134
0.000943	19.26	0.146	0.136	0.010	0.052	0.084
0.000943	19.80	0.204	0.184	0.020	0.047	0.137
0.000943	20.12	0.14	0.128	0.012	0.052	0.076
0.000943	20.21	0.273	0.238	0.035	0.054	0.184

9.6 A Modified Proctor Compaction Test is performed on a sample of silty sand in a 152 mm diameter mold. The maximum dry unit weight is 19.6 kN/m^3 and the optimum water content is 11%. If the specific gravity of solids is 2.65, draw the three-phase diagram of the sample in the mold and calculate all volumes and weight for that sample. What is the degree of saturation?

9.7 Referring to Figure 9.23, explain the following:
   a. Why is the dry unit weight vs. water content curve relatively flat compared to the modulus vs. water content curve?
   b. Why does the modulus vs. water content curve go downward as the water content increases from 1% to 6%?
   c. Why does the modulus vs. water content curve drop so significantly when the water content goes from 8% to 10%?

9.8 For the consolidation test, what is the difference between the incremental loading procedure and the constant rate of strain procedure?

9.9 For the consolidation test curve shown in Figure 9.27, calculate the compression index C_c and the recompression index C_r.

9.10 For the consolidation test curves shown in Figure 9.30 and Figure 9.31, calculate the coefficient of consolidation c_v according to the log time method and according to the square root of time method. Compare and comment.

9.11 For the consolidation curve shown in Figure 9.27, determine the preconsolidation effective stress.

9.12 Regarding deformation laboratory tests, discuss the differences between tests on saturated soils and tests on unsaturated soils (or, more precisely, tests on soils where the water is in compression and tests on soils where the water is in tension).

9.13 A direct shear test is performed on a sample of saturated clay. The sample is 25 mm high and 75 mm in diameter. The test cell is inundated such that the water stress is hydrostatic at the beginning of the test.

a. How would you run the test so as to measure the undrained shear strength of the clay?

b. How would you run the test so as to obtain the drained shear strength parameters for the clay?

9.14 A direct shear test is performed on a sample of dry sand. The sample is 50 mm in diameter and 25 mm high and is subjected to a vertical force of 100 N. At failure, the shear force applied is 60 N, the horizontal movement is 3 mm, and the vertical movement is 0.5 mm. Calculate the shear strength of the sand and the friction angle, and estimate the dilation angle.

9.15 A direct shear test is performed on a sample of saturated clay. The test is a quick test such that water does not have time to drain during the test. The vertical load on the sample induces a total normal stress of 50 kPa and at failure the shear force induces a shear stress of 100 kPa.

a. Calculate the undrained shear strength of the clay.

b. How is it possible for this clay to have such high shear strength, considering the low normal stress?

9.16 Two direct shear tests are performed on a sample of saturated clay. The tests are slow tests such that the water stress (pore pressure) remains equal to zero.

$$\text{Test 1: N} = 300\ \text{N, T} = 250\ \text{N, A} = 0.01\ \text{m}^2\text{, S} = 100\%\text{, u}_w = 0\ \text{kPa}$$

$$\text{Test 2: N} = 600\ \text{N, T} = 400\ \text{N, A} = 0.01\ \text{m}^2\text{, S} = 100\%\text{, u}_w = 0\ \text{kPa}$$

where N is the normal force, T is the shear force, A is the sample cross-sectional area, S is the degree of saturation, and u_w is the water stress. Calculate the effective stress cohesion and effective stress friction angle of the clay.

9.17 For strength laboratory tests, discuss the differences between tests on saturated soils and tests on unsaturated soils (or, more precisely, tests on soils where the water is in compression and tests on soils where the water is in tension).

9.18 Assume the conditions as in problem 9.16, but this time the soil is unsaturated and the readings are as follows:

$$\text{Test 1: N} = 600\ \text{N, T} = 1900\ \text{N, A} = 0.01\ \text{m}^2\text{, S} = 60\%\text{, u}_w = -400\ \text{kPa}$$

$$\text{Test 2: N} = 200\ \text{N, T} = 900\ \text{N, A} = 0.01\ \text{m}^2\text{, S} = 40\%\text{, u}_w = -300\ \text{kPa}$$

where N is the normal force, T is the shear force, A is the sample cross-sectional area, S is the degree of saturation, and u_w is the water tension stress. Calculate the effective stress cohesion and effective stress friction angle of the clay.

9.19 What are the differences between the direct shear test and the simple shear test? Explain your answers.

9.20 A simple shear test is performed on a sample of silt. The sample is 50 mm in diameter and 20 mm high. When the shear force applied is 200 N, the horizontal displacement of the top of the sample is 0.2 mm. Calculate the shear stress, the shear strain, and the shear modulus of the sample at that point on the stress-strain curve.

9.21 An unconfined compression (UC) test on a sample of clay gives the stress-strain curve shown in Figure 9.47. Calculate the undrained shear strength and the UC modulus for this sample. What geotechnical problem do you think this undrained shear strength and this modulus could be used for?

9.22 What are the two main phases in running a triaxial test? With respect to drainage during each one of these two phases, what are the different types of tests that can be run? For each type of test, what parameters can you obtain from the data?

9.23 A triaxial test is performed on a sample that is 50 mm in diameter and 100 mm high. The confining pressure is 30 kPa and at failure the vertical load on the sample is 118 N. Is the vertical total stress on the sample at failure expressed in N/m^2 equal to $\dfrac{118N}{\pi(25.10^{-3})^2}$? If yes, explain your answer. If not, what is it?

9.24 A triaxial test with water stress (pore-pressure) measurements is performed on a sample of saturated silty crushed rockfill and gives the results shown in Figure 9.50. The total confining stress is 35 kPa.

a. Calculate the total stress secant modulus E and the effective stress secant modulus E′ for a vertical strain equal to 0.2%, 0.5%, 1%, 2%, 3%, 4%, and 5%. Then plot the curve giving soil modulus both as a function of strain and as a function of log strain.

b. At failure, the vertical effective stress is 100 kPa. Calculate the effective stress friction angle of the sand if the effective stress cohesion c is zero.

9.25 Two CU triaxial tests are performed on a sample of saturated, overconsolidated, high-plasticity clay. At failure, the results are as follows:

$$\text{Test 1}: \sigma_3 = 30\ \text{kPa},\ \text{Q} = 0.45\ \text{kN},\ \text{A} = 0.01\ \text{m}^2,\ \text{u} = 10\ \text{kPa}$$

$$\text{Test 2}: \sigma_3 = 60\ \text{kPa},\ \text{Q} = 0.70\ \text{kN},\ \text{A} = 0.01\ \text{m}^2,\ \text{u} = 20\ \text{kPa}$$

where σ_3 is the total confinement stress, Q is the vertical load on the sample, A is the sample cross section, and u is the water stress (pore pressure).

a. Calculate σ_3, σ_1, σ'_3, and σ'_1 at failure.
b. Draw the Mohr circle at failure in the τ vs. σ' set of axes.
c. Draw the failure envelope and find the effective stress strength parameters c′ and φ'.

9.26 Two CU triaxial tests are performed on a sample of unsaturated clay. At failure, the results are as follows:

$$\text{Test 1}: \sigma_3 = 20\ \text{kPa},\ \sigma_1 = 190\ \text{kPa},\ \text{S} = 60\%,\ \text{u}_w = -100\ \text{kPa}$$

$$\text{Test 2}: \sigma_3 = 60\ \text{kPa},\ \sigma_1 = 450\ \text{kPa},\ \text{S} = 50\%,\ \text{u}_w = -300\ \text{kPa}$$

where σ_3 is the total confinement stress, σ_1 is the total vertical stress at failure, S is the degree of saturation, and u_w is the water stress.

a. Calculate σ_3, σ_1, σ'_3, and σ'_1 at failure.
b. Draw the Mohr circle at failure in the τ vs. σ' set of axes.
c. Draw the failure envelope and find the effective stress strength parameters c′ and φ'.

9.27 What is the stress path, and what shape does it typically have for the triaxial test?

9.28 A lab vane test is performed on a silty clay. At failure, the maximum torque is 5.7 N.m. The vane is 50 mm high and 25 mm in diameter. Calculate the undrained shear strength of the silty clay. The vane is rotated 10 times rapidly and the torque on the tenth revolution is measured to be 3.5 N.m. Calculate the residual undrained shear strength of the silty clay.

9.29 A silty sand is subjected to a constant head permeameter test. The flow collected at the downstream end is 221 mm^3/s; the sample is 75 mm in diameter and 100 mm high. The difference between the water level in the upstream overflow and the downstream overflow is 0.5 m. Calculate the hydraulic conductivity k of the silty sand.

9.30 A clay sample is tested in a falling head permeameter. The sample is 75 mm in diameter and 100 mm high. The small tube is 3 mm in diameter. The difference in height between the water level in the small tube above the sample and the downstream overflow is measured as a function of time. At time $t = 0$, the difference is 1.1 m and at time $t = 1$hr, the difference is 1.05 m. Calculate the hydraulic conductivity k of the clay.

9.31 This problem refers to Figure 9.69. A sample of unsaturated silt is tested in a constant head permeameter and the following parameters are measured: $D = 75$ mm, $l = 150$ mm, $V = 10\ cm^3$, $t = 1$ hour, $h_{p2} = -100$ mm, $h_{p3} = -200$ mm, $h_{p4} = 0$ mm. Calculate the unsaturated hydraulic conductivity k of the silt and the water tension u_w corresponding to that value.

9.32 A 0.65 m long, 75 mm diameter sample of unsaturated clay is tested in a wetting front permeameter. Initially the water tension in the sample is −1000 kPa. The results are shown in Figure 9.75. Use the results to develop the hydraulic conductivity k vs. water tension u_w curve for this clay.

9.33 This problem refers to Figure 9.76. A sample of unsaturated clayey sand has a degree of saturation of 40%. The sample length is 150 mm and the sample diameter is 75 mm. It is tested in a permeameter to determine the hydraulic conductivity of air through the sample. The air pressure at the base of the sample is 10 kPa and the volume of air collected at the top of the sample in one hour of testing is $10^{-3}\ m^3$. The top of the sample is kept at atmospheric pressure. Calculate the air hydraulic conductivity of the sample k_a and the air stress u_a associated with this hydraulic conductivity value.

9.34 A 1.8 m tall human being drinks one liter of water. Three hours later, this person goes to the bathroom and eliminates the liter of water. Is this case a constant head permeameter or a falling head permeameter? Calculate the hydraulic conductivity of the human body. Make reasonable assumptions when necessary.

9.35 A sample of fine sand is tested in the EFA. The mean diameter of the grains is $D_{50} = 1$ mm. When the velocity is set at 1 m/s, the piston below the sample of sand has to be raised at a rate of 16.7 mm/minute. The cross section of the conduit where the water is flowing is rectangular, with a width of 100 mm and a height of 50 mm. Calculate the shear stress at the interface between the water and the sand for the 1 m/s velocity.

9.36 A sample of low-plasticity clay is tested in the EFA. The surface of the clay sample is considered smooth. When the velocity is set at 3 m/s, the piston below the sample of sand has to be raised at a rate of 1 mm every 3 minutes. The cross section of the conduit where the water is flowing is rectangular, with a width of 100 mm and a height of 50 mm. Calculate the shear stress at the interface between the water and the sand for the 3 m/s velocity.

Problems and Solutions

Problem 9.1

What are the four main categories of laboratory tests? Give three examples for each category.

Solution 9.1

1. Tests for index properties
   a. Water content
   b. Unit weight
   c. Particle size
2. Tests for deformation properties
   a. Consolidation
   b. Triaxial
   c. Simple shear
3. Tests for strength properties
   a. Direct shear
   b. Unconfined compression
   c. Lab vane
4. Tests for flow properties
   a. Constant head permeameter
   b. Falling head permeameter
   c. Erosion tests

Problem 9.2

What device would you use to measure the following quantities? In each case explain the basic principle of the device.

a. Force
b. Pressure
c. Shear stress
d. Water compression stress

Solution 9.2

a. Force

A load cell is the most common way to measure load; it consists of a deformable piece of steel (S shape or cylindrical) instrumented with strain gages.

b. Pressure

Pressure cells are used to measure pressure. They are circular and have a metallic membrane that deforms when it is in contact with the stressed soil. The bending of the membrane is measured with strain gages glued to that membrane and the strains are related to the pressure on the membrane.

c. Shear stress

The simplest way to measure shear stress is to measure the shear force and divide by the corresponding area. This is the case with the direct shear test. Alternatively, shear stress can be measured by a shear stress transducer.

d. Water compression stress

A manometer can be used to measure water compression stress. A manometer or standpipe is simply a pipe connected to the point where the water compression stress must be measured and open to the atmosphere at the other end. The pressure in the water makes the water rise in the manometer to the point of equilibrium. The water compression stress is then calculated as the vertical distance between the point of measurement and the water level in the manometer times the unit weight of water.

A pore-pressure transducer can be also used to measure water pressure. The pore-pressure transducer measures the water pressure by letting that pressure deflect a membrane. A porous tip made of ceramic (Figure 9.4) is placed in contact with the soil where the water is in compression. This porous tip, which is saturated with de-aired water, allows water to come in but does not allow air to come in.

Problem 9.3

What devices and techniques would you use to measure the water tension stress (suction)? In each case explain the basic principle of the device.

Solution 9.3

Filter paper test

The filter paper test consists of using a circular piece of filter paper (about 50 mm in diameter), weighing it dry, placing it either in contact with or above the soil sample, enclosing the filter paper and the sample in a sealed container until the filter paper comes into water tension equilibrium with the sample, retrieving the filter paper, and weighing it to obtain its water content. Because the soil sample is much larger than the filter paper, the water content of the sample remains unaffected by the amount of moisture drawn into the filter paper. The filter paper comes calibrated with a curve linking the filter paper moisture content and the water tension in the filter paper. Because the water tension is the same in the filter paper and the sample, the water tension of the sample is determined in that fashion. The filter paper method, however, can measure matric suction only or matric suction plus osmotic suction, depending on whether or not the filter paper is in contact with the sample.

Thermocouple psychrometer

Psychrometers give the relative humidity by measuring the difference in temperature between a nonevaporating surface and an evaporating surface. Psychrometers measure the total suction because the evaporation process creates pure water, whereas the water in the soil pores is not pure.

Tensiometer

A tensiometer consists of a high air entry porous ceramic tip saturated with water and placed in good contact with the soil. In the tensiometer, the space behind the ceramic tip is filled with de-aired water and connected to a negative pressure measuring device. The stress slowly equalizes between the water tension in the tensiometer and the water tension in the soil pores. Then that tension is measured either through a water-mercury manometer, a Bourdon-vacuum tube, or an electrical pressure transducer. The water tension that can be measured in a tensiometer is limited to approximately negative 90 kPa (2.95 pF) due to the possibility of water cavitation in the tensiometer above such a value.

Pressure plate

A pressure plate is a closed pressure chamber that can be used to increase the air pressure in the soil pores to the point where the air drives the water out of the pores. The sample is placed in the chamber on a high air entry ceramic disk. This disk, which is saturated with water, has the property of letting water but not air go through up to a certain rated pressure (the air entry value of the disk). The air pressure is increased and the stress in the water is increased accordingly (decrease in tension). When the water tension becomes equal to zero, the water comes out; at that point, the air pressure is equal to the water tension. This technique is called the axis translation technique because it simply translates the origin of reference by applying an air pressure equal to the water tension.

Salt solution

Salt solution equilibrium is a water tension measurement that relies on the fact that salt solutions have significant osmotic suction. A closed chamber with a salt solution at its lower part will generate a certain relative humidity in the air above it. The higher the salt concentration is, the lower the relative humidity above the solution in the chamber will be. If a soil sample is suspended in the air above the salt solution, it will dry and the water tension in the soil sample will come to equilibrium with the ambient relative humidity. At equilibrium, the water tension is given by the relative humidity in the air of the chamber. This relative humidity depends on the salt concentration in the solution and can be calculated from it. This relationship depends on the type of salt, its molality, and the temperature.

Problem 9.4

What devices would you use to measure the following quantities? In each case explain the basic principle of the device.

a. Displacement
b. Normal strain
c. Shear strain

Solution 9.4

a. Displacement.

Displacement can be measured with a linear variable differential transformer. An LVDT has three solenoid coils, arranged like three side-by-side donuts. A metallic rod attached to the point where the displacement is to be measured passes through the center of the three solenoids without touching them. An alternating current through the center solenoid creates a voltage in the side solenoids. The movement of the metallic rod creates a change in voltage which is linearly proportional to the movement of the rod. The change in voltage is transformed into a displacement measurement through calibration.

b. Normal strain.

Normal strain can be measured using a foil strain gage. A foil gage is a thin sheet of metal (copper-nickel alloy is common) with a pattern glued to the material that will deform. Actually, a layer of insulating flexible material is first glued to the deforming material and then the foil gage is glued on the insulator so that the current passing through the gage only travels through the gage. When the material deforms, the foil length changes and so does its resistance. The voltage changes accordingly and the strain is related to the change in voltage through calibration.

c. Shear strain.

Shear strain can be measured using the same strain gage described for normal strain. Shear strain γ is defined for two perpendicular directions. When the shear strain is small enough, the shear strain is equal to the change in angle γ, expressed in radians between the two perpendicular directions due to the shearing process. Shear strain is most easily obtained by measuring the normal strain in two perpendicular directions. It can be shown that the shear strain measured in the x and y direction is given by: $\gamma_{xy} = \varepsilon_1 - \varepsilon_2$.

Problem 9.5

Table 9.1s shows the results of a Standard Proctor Compaction Test for the same soil tested at different water contents. Calculate the dry unit weight and water content for each sample, plot a graph of dry unit weight versus water content, determine the maximum dry unit weight and optimum water content, and plot the saturation lines for S = 100%, S = 90%, and S = 80%.

Table 9.1s Results of a Standard Proctor Compaction Test

Volume of mold (m^3)	Unit weight (kN/m^3)	Weight wet soil + container (N)	Weight dry soil + container (N)	Weight of water (N)	Weight of container (N)	Weight of dry soil (N)
0.000943	18.48	0.194	0.182	0.012	0.048	0.134
0.000943	19.26	0.146	0.136	0.010	0.052	0.084
0.000943	19.80	0.204	0.184	0.020	0.047	0.137
0.000943	20.12	0.14	0.128	0.012	0.052	0.076
0.000943	20.21	0.273	0.238	0.035	0.054	0.184

Solution 9.5

Assuming that Gs = 2.65, the dry unit weight and the water content for each sample are calculated in Table 9.2s.

Table 9.2s Results of a Standard Proctor Compaction Test

Volume of mold (m^3)	Unit weight (kN/m^3)	Weight wet soil + container (N)	Weight dry soil + container (N)	Weight of water (N)	Weight of container (N)	Weight of dry soil (N)	Water content (%)	Dry unit weight (kN/m^3)
0.000943	18.48	0.194	0.182	0.012	0.048	0.134	8.96	16.96
0.000943	19.26	0.146	0.136	0.01	0.052	0.084	11.90	17.21
0.000943	19.80	0.204	0.184	0.02	0.047	0.137	14.60	17.28
0.000943	20.12	0.14	0.128	0.012	0.052	0.076	15.79	17.37
0.000943	20.21	0.273	0.238	0.035	0.054	0.184	19.02	16.98

The plot of dry unit weight function of water content, and the saturation lines for S = 100%, S = 90%, and S = 80%, are shown in Figure 9.1s.

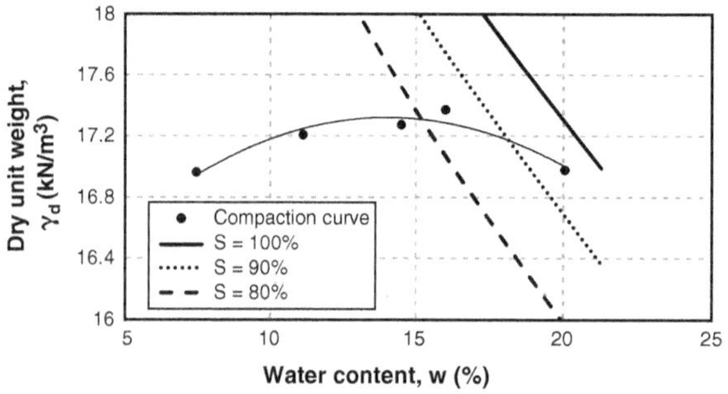

Figure 9.1s Plot of dry unit weight versus water content.

Problem 9.6

A Modified Proctor Compaction Test is performed on a sample of silty sand in a 152 mm diameter mold. The maximum dry unit weight is 19.6 kN/m^3 and the optimum water content is 11%. If the specific gravity of solids is 2.65, draw the three-phase diagram of the sample in the mold and calculate all volumes and weight for that sample. What is the degree of saturation?

Solution 9.6

Given that $\gamma_d = 19.6$ kN/m^3, $w_{opt} = 11\%$, $G_s = 2.65$, and D = 152 mm, we know the volume of the mold is 21.2×10^{-4} m^3 (Table 9.5).

The three-phase diagram of the sample is shown in Figure 9.2s.

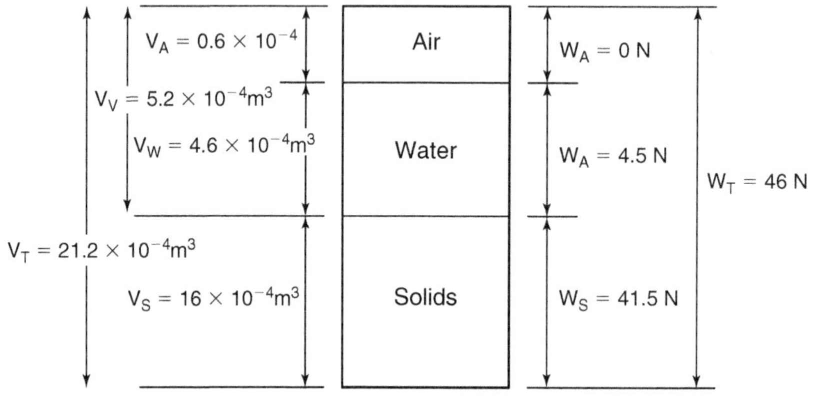

Figure 9.2s Three-phase diagram.

$$\gamma_d = \frac{w_s}{V_T} \rightarrow w_s = \gamma_d \times V_T = 19.6 \times 21.2 \times 10^{-4} = 4.15 \times 10^{-2}\ \text{kN} = 41.5\ \text{N}$$

$$V_s = \frac{w_s}{G_s \gamma_w} = \frac{41.5 \times 10^{-3}}{2.65 \times 9.81} = 1.6 \times 10^{-3}\ \text{m}^3$$

$$\gamma_d = \frac{\gamma_T}{1+w} \rightarrow \gamma_T = \gamma_d \times (1+w) = 19.6 \times (1+0.11) = 21.7\ \text{kN/m}^3$$

$$\gamma_T = \frac{W_T}{V_T} \rightarrow W_T = \gamma_T \times V_T = 21.7 \times 21.2 \times 10^{-4} = 46 \times 10^{-3}\ \text{kN} = 46\ \text{N}$$

$$\text{W}_\text{T} = \text{W}_s + \text{W}_w \rightarrow \text{W}_w = 46 - 41.5 = 4.5\ \text{N}$$

$$V_w = \frac{W_w}{\gamma_w} = \frac{4.5 \times 10^{-3}}{9.8} = 4.6 \times 10^{-4}\ \text{m}^3$$

$$\text{V}_\text{T} = \text{V}_\text{A} + \text{V}_\text{W} + \text{V}_\text{S} \rightarrow \text{V}_\text{A} = 0.6 \times 10^{-4}\ \text{m}^3$$

$$\text{V}_v = \text{V}_\text{A} + \text{V}_w = 6.2 \times 10^{-4}\ \text{m}^3$$

Degree of saturation:

$$S = \frac{V_w}{V_v} = \frac{4.6 \times 10^{-4}}{5.2 \times 10^{-4}} \times 100 = 88.5\%$$

Problem 9.7

Referring to Figure 9.23, explain the following:

a. Why is the dry unit weight vs. water content curve relatively flat compared to the modulus vs. water content curve?
b. Why does the modulus vs. water content curve go downward as the water content increases from 1% to 6%?
c. Why does the modulus vs. water content curve drop so significantly when the water content goes from 8% to 10%?

Solution 9.7

a. The dry unit weight vs. water content curve is relatively flat compared to the modulus vs. water content curve because the dry unit weight is much less sensitive to the water content than the modulus. The amount of dry particles in a given volume does not change much with increasing water content. However, the modulus will change much more dramatically because the effective stress and the skeleton structure of the soil change much more dramatically with the water content.
b. Increasing the water content from 1% to 6% decreases the tension stress in the water (suction), which weakens the soil; thus the BCD plate bends more, leading to lower E moduli. Beyond that point, the added water lubricates the particles and allows them to achieve a more compact arrangement for the given compaction effort. The soil becomes stiffer and the modulus becomes larger.
c. Near the optimum water content, the soil is approaching saturation. Increasing the water content beyond the optimum water content (8%) simply increases the size of the voids by filling them with water; thus, the soil becomes softer and the modulus drops significantly.

Problem 9.8

For the consolidation test, what is the difference between the incremental loading procedure and the constant rate of strain procedure?

Solution 9.8

The incremental loading procedure consists of placing a load on the sample for 24 hours while recording the decrease in sample thickness. Also, water is allowed to drain from both the top and the bottom of the sample. In the constant rate of strain procedure, the sample is deformed at a constant rate of displacement with time. Also, water is allowed to drain only from the top of the sample, while the tester ensures that the pore pressure does not rise above a set limit at the bottom of the sample.

Problem 9.9

For the consolidation test curve shown in Figure 9.27, calculate the compression index C_c and the recompression index C_r.

Solution 9.9

The compression index C_c and the recompression index C_r (Figure 9.3s) are calculated as follows:

$$C_c = \frac{\Delta e}{\log\left(\frac{\sigma_2'}{\sigma_1'}\right)} = \frac{0.48 - 0.36}{\log\left(\frac{6100}{1500}\right)} = 0.192$$

$$C_r = \frac{\Delta e}{\log\left(\frac{\sigma_2'}{\sigma_1'}\right)} = \frac{0.47 - 0.42}{\log\left(\frac{3000}{750}\right)} = 0.0823$$

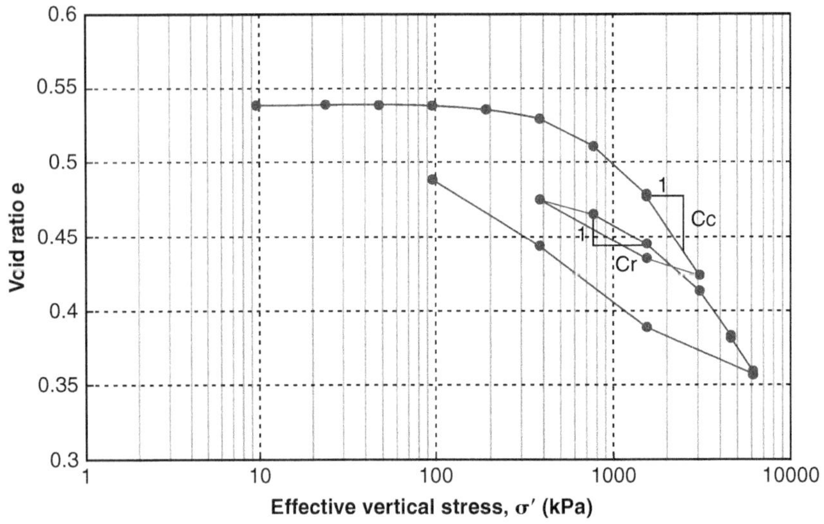

Figure 9.3s Consolidation test results.

Problem 9.10

For the consolidation test curves shown in Figure 9.30 and Figure 9.31, calculate the coefficient of consolidation c_v according to the log time method and according to the square root of time method. Compare and comment.

Solution 9.10

The coefficient of consolidation according to the log time method is calculated using Figure 9.4s. From the curve:

$$x = 0.0145 - 0.0115 = 0.003$$
$$\varepsilon_0 = 0.012 - 0.002 = 0.009$$
$$\varepsilon_{100} = 0.046$$
$$\varepsilon_{50} = (\varepsilon_0 + \varepsilon_{100})/2 = (0.009 + 0.046)/2 = 0.0275$$

The time corresponding to 50% of consolidation $t_{50} = 30$ min $= 1800$ sec.

Assuming that during the consolidation test the sample is double-drained, the coefficient of consolidation is calculated as follows:

$$c_v = T_{50}\left(\frac{H_{dr}^2}{t_{50}}\right) = 0.197\left(\frac{0.0125^2}{1800}\right) = 1.59 \times 10^{-8}\ \text{m}^2/\text{s}$$

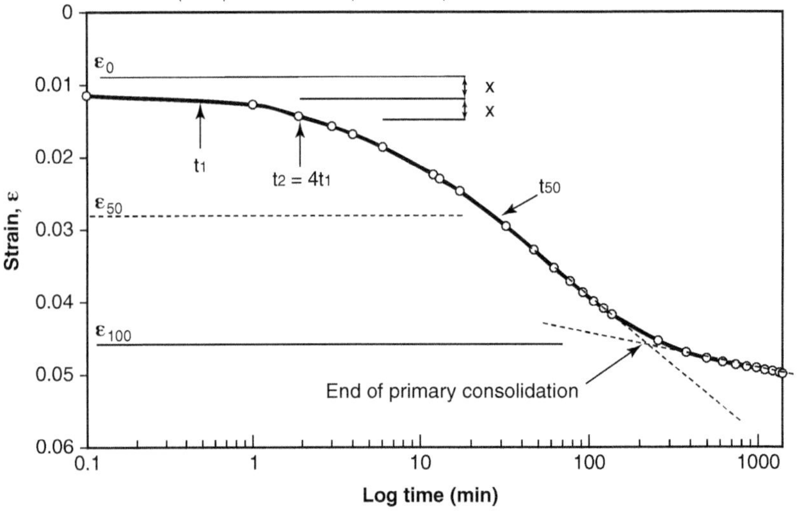

Figure 9.4s Log time method.

The coefficient of consolidation according to the square root of time method is calculated using Figure 9.5s. From the curve:

$$\sqrt{t_{90}} = 10 \Rightarrow t_{90} = 100\,\text{min} = 6000\,\text{sec}$$

Assuming that during the consolidation test the sample is double-drained, the coefficient of consolidation is calculated as follows:

$$c_v = T_{90}\left(\frac{H_{dr}^2}{t_{90}}\right) = 0.848\left(\frac{0.0125^2}{6000}\right) = 2.21 \times 10^{-8}\,\text{m}^2/\text{s}$$

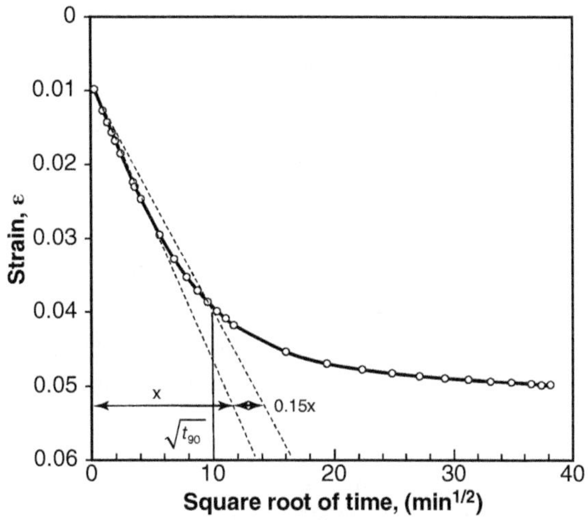

Figure 9.5s Square root of time method.

Problem 9.11

For the consolidation curve shown in Figure 9.27, determine the preconsolidation effective stress.

Solution 9.11

From the consolidation curve (Figure 9.6s) and using the Cassagrande construction, the preconsolidation effective stress is $\sigma'_p = 1020\,\text{kPa}$.

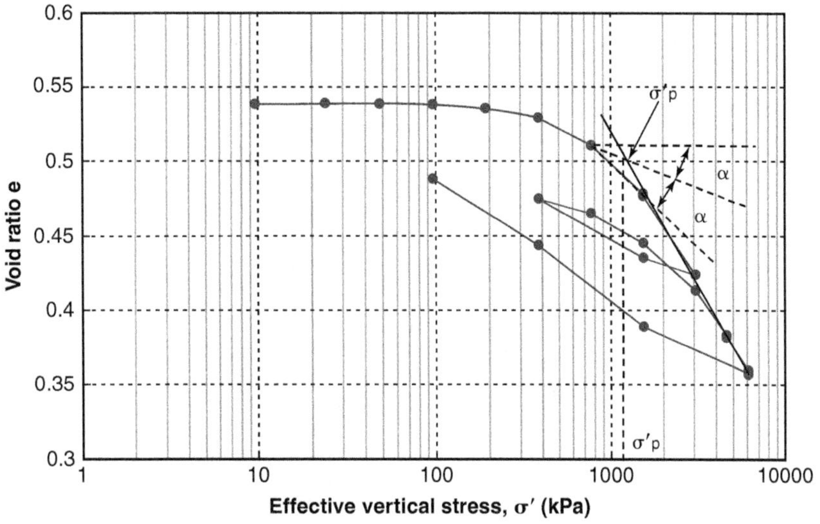

Figure 9.6s Consolidation test curve.

Problem 9.12

Regarding deformation laboratory tests, discuss the differences between tests on saturated soils and tests on unsaturated soils (or, more precisely, tests on soils where the water is in compression and tests on soils where the water is in tension).

Solution 9.12

For deformation laboratory tests on saturated and unsaturated soils, the test procedures are generally the same except for the measurement of water and air stress. For unsaturated soils, the effective stress on the sample should be calculated using the expressions:

$$\sigma' = \sigma - \alpha\, u_w \text{ or } \sigma' = \sigma - \alpha u_w - \beta u_a$$

if the air pressure is not zero. In these expressions, σ' is the effective stress, σ is the total stress, α is the water area coefficient, u_w is the water tension stress, β is the air area coefficient, and u_a is the air stress. The water tension stress u_w can be measured by several methods; most often this is done with a tensiometer during the test. The coefficient α can also be estimated as the degree of saturation S. Typically, saturated soils with the water in compression are more compressible than the same soils in the unsaturated state with the water in tension.

Problem 9.13

A direct shear test is performed on a sample of saturated clay. The sample is 25 mm high and 75 mm in diameter. The test cell is inundated such that the water stress is hydrostatic at the beginning of the test.

a. How would you run the test so as to measure the undrained shear strength of the clay?
b. How would you run the test so as to obtain the drained shear strength parameters for the clay?

Solution 9.13

a. The test has to be run quickly enough that the water does not have time to drain.
b. The test has to be run slowly enough that the water stress remains zero.

Problem 9.14

A direct shear test is performed on a sample of dry sand. The sample is 50 mm in diameter and 25 mm high and is subjected to a vertical force of 100 N. At failure, the shear force applied is 60 N, the horizontal movement is 3 mm, and the vertical movement is 0.5 mm. Calculate the shear strength of the sand and the friction angle, and estimate the dilation angle.

Solution 9.14

a. Shear strength of the sand:

$$A = \pi r^2 = 1.96 \times 10^3 \text{ m}^2$$

$$c = \frac{F_s}{A} = \frac{60\ N}{1.96 \times 10^{-3} \text{ m}^2} = 30600 \text{ Pa} = 30.6 \text{ kPa}$$

b. Friction angle of the sand:

$$\varphi = \tan^{-1}\left(\frac{F_s}{F_n}\right) = \tan^{-1}\left(\frac{60}{100}\right) = 31^\circ$$

c. Dilation angle of the sand (Figure 9.7s):

$$\psi = \tan^{-1}\left(\frac{\Delta y}{\Delta x}\right) = \tan^{-1}\left(\frac{0.5}{3}\right) = 9.5^\circ$$

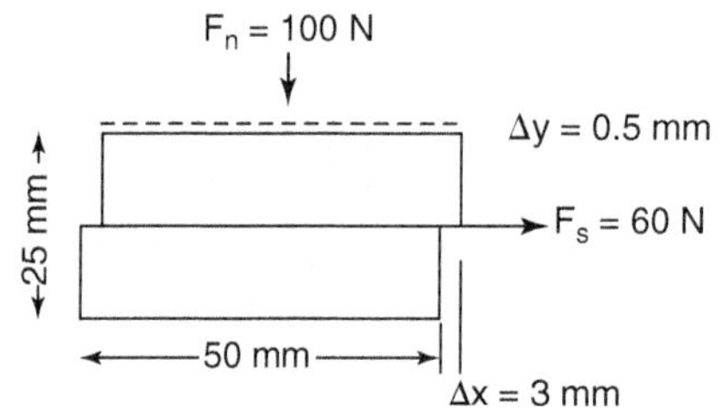

Figure 9.7s Direct shear test.

Problem 9.15

A direct shear test is performed on a sample of saturated clay. The test is a quick test such that water does not have time to drain during the test. The vertical load on the sample induces a total normal stress of 50 kPa and at failure the shear force induces a shear stress of 100 kPa.

a. Calculate the undrained shear strength of the clay.
b. How is it possible for this clay to have such high shear strength, considering the low normal stress?

Solution 9.15

a. The undrained shear strength of the saturated clay is associated with the shear force measured at failure. Therefore, the undrained shear strength is 100 kPa.
b. The shear strength equation is:

$$s = c + \sigma' \tan\varphi$$

We know that at failure the shear stress is 100 kPa, and we also know that the total normal stress is 50 kPa:

$$100 = c + (50 - \alpha u_w) \tan\varphi$$

The explanation is that either there is a lot of cohesion (cementation, for example), or there is a significant amount of tension in the water (highly negative u_w). The latter is more likely.

Problem 9.16

Two direct shear tests are performed on a sample of saturated clay. The tests are slow tests such that the water stress (pore pressure) remains equal to zero.

$$\text{Test 1: N} = 300\ \text{N, T} = 250\ \text{N, A} = 0.01\ \text{m}^2\text{, S} = 100\%\text{, u}_w = 0\ \text{kPa}$$

$$\text{Test 2: N} = 600\ \text{N, T} = 400\ \text{N, A} = 0.01\ \text{m}^2\text{, S} = 100\%\text{, u}_w = 0\ \text{kPa}$$

where N is the normal force, T is the shear force, A is the sample cross-sectional area, S is the degree of saturation, and u_w is the water stress. Calculate the effective stress cohesion and effective stress friction angle of the clay.

Solution 9.16

The effective stress cohesion and the effective stress friction angle can be calculated from the Mohr circle (Table 9.3s).

$$\sigma = \frac{N}{A} \quad \tau = \frac{T}{A} \quad \sigma' = \sigma - u_w = \sigma$$

Table 9.3s Calculation of Normal and Shear Stress for the Direct Shear Test

Test	Normal Force (N)	Shear Force (N)	Area (m^2)	S (%)	u$_w$ (kPa)	σ' (kPa)	τ (kPa)
1	300	250	0.01	100	0	30	25
2	600	400	0.01	100	0	60	40

Figure 9.8s shows the Mohr-Coulomb envelope for the two sample tests. Because the pore water pressure remains zero, the intersection of the envelope with the y_{axis} represents the effective cohesion of the soil (c′) and the slope of the envelope represents the effective friction angle of the soil (φ'). The plot shows a c′ and a φ' of 10 kPa and 26.6 degrees, respectively. The friction angle can also be calculated as:

$$\varphi' = \tan^{-1}\left(\frac{\Delta\tau}{\Delta\sigma'}\right) = \tan^{-1}\left(\frac{\tau_2 - \tau_1}{\sigma_2' - \sigma_1'}\right) = \tan^{-1}\left(\frac{40 - 25}{60 - 30}\right) = 26.6^\circ$$

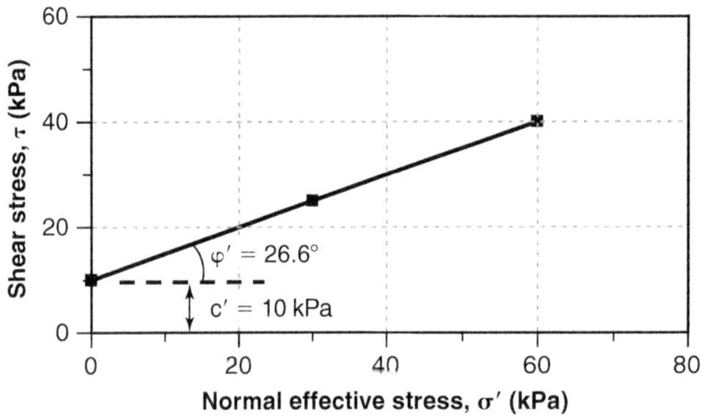

Figure 9.8s Shear stress and effective stress diagram.

Problem 9.17

For strength laboratory tests, discuss the differences between tests on saturated soils and tests on unsaturated soils (or, more precisely, tests on soils where the water is in compression and tests on soils where the water is in tension).

Solution 9.17

1. *Direct shear test*: When the soil is saturated (soil sample is inundated) there is no suction in the soil and the sample can be tested at low or high shearing rate. The high rate provides the undrained parameter of the soil and the low rate provides the drained parameters of the soil. If the soil is unsaturated, or if the soil is saturated but the water in the voids is in tension, then the direct shear test requires measurement of the water tension stress (suction) to obtain the effective stress shear strength parameters c′ and φ'.
2. *Simple shear test*: If the soil is unsaturated, or if it is saturated but the water is in tension, the testing procedure is the same as for the saturated case except for the measurement of the water stress.
3. *Unconfined compression test*: The test procedure and data reduction are the same for both saturated and unsaturated soils. The water stress is not measured in this test; however, it can be determined from the equations as shown in the text.
4. *Triaxial test*: If the soil is unsaturated, or if it is saturated and the water in the voids is in tension, the test procedure does not change from the saturated case; however, the water and air stress measurements change depending on whether the test is drained or undrained.
5. *Resonant column test*: The test procedure and data reduction are the same for both saturated and unsaturated soils. The water stress is rarely measured in this test.
6. *Lab vane test*: the test procedure and data reduction are the same for both saturated and unsaturated soils. The water stress is not measured in this test.

Problem 9.18

Assume the same conditions as in problem 9.16 but this time the soil is unsaturated and the readings are as follows:

$$\text{Test 1: N} = 600\ \text{N, T} = 1900\ \text{N, A} = 0.01\ \text{m}^2\text{, S} = 60\%\text{, u}_w = -400\ \text{kPa}$$

$$\text{Test 2: N} = 200\ \text{N, T} = 900\ \text{N, A} = 0.01\ \text{m}^2\text{, S} = 40\%\text{, u}_w = -300\ \text{kPa}$$

where N is the normal force, T is the shear force, A is the sample cross-sectional area, S is the degree of saturation, and u_w is the water tension stress. Calculate the effective stress cohesion and effective stress friction angle of the clay.

Solution 9.18

For Test 1 (at failure):

$$\sigma = \frac{N}{A} = \frac{600}{0.01} = 60000\ \text{Pa} = 60\ \text{kPa}$$

$$\sigma' = \sigma - \alpha u_w = \sigma - S \cdot u_w = 60 - 0.6 \times (-400) = 300\ \text{kPa}$$

$$\tau = \frac{T}{A} = \frac{1900}{0.01} = 190000\ \text{Pa} = 190\ \text{kPa}$$

For Test 2 (at failure):

$$\sigma = \frac{N}{A} = \frac{200}{0.01} = 20000 \text{ Pa} = 20 \text{ kPa}$$

$$\sigma' = \sigma - \alpha u_w = \sigma - S \cdot u_w = 20 - 0.4 \times (-300) = 140 \text{ kPa}$$

$$\tau = \frac{T}{A} = \frac{900}{0.01} = 90000 \text{ Pa} = 90 \text{ kPa}$$

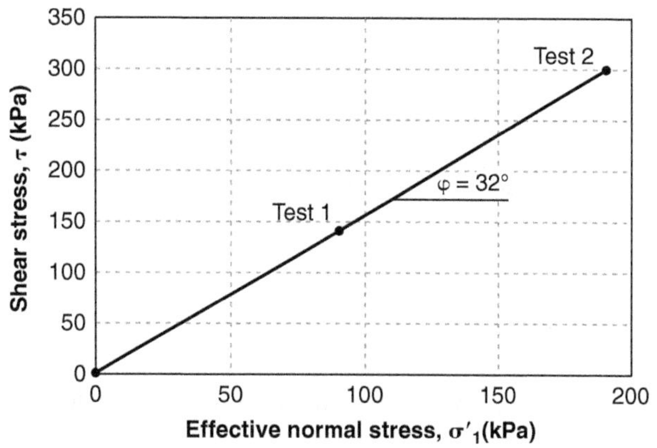

Figure 9.9s Failure envelope for the direct shear test on unsaturated soil.

The effective stress cohesion c' and effective stress friction angle φ' can be calculated by using the plotted curve of effective normal stress vs. shear stress:

$$\tan\phi' = \frac{\tau_1 - \tau_2}{\sigma_1' - \sigma_2'} = \frac{190 - 90}{300 - 140} = 0.625$$

Therefore, the effective friction angle φ' is 32°.
The effective stress cohesion can be calculated using the following equation:

$$\tau = \sigma' \cdot \tan\phi' + c'$$

$$90 = 140 \times \tan 32^\circ + c'$$

Therefore, the effective stress cohesion c' is 2.5 kPa.

Problem 9.19

What are the differences between the direct shear test and the simple shear test? Explain your answers.

Solution 9.19

In the direct shear test, the shearing takes place along a predetermined thin band of soil near the middle of the sample, whereas in the simple shear test the shearing takes places over the entire height of the sample. The shear strength can be obtained from both tests (including the strength parameters c′ and φ'). However, the simple shear test has the advantage of giving the shear modulus G as a function of shear strain in addition to the shear strength of the soil sample.

Problem 9.20

A simple shear test is performed on a sample of silt. The sample is 50 mm in diameter and 20 mm high. When the shear force applied is 200 N, the horizontal displacement of the top of the sample is 0.2 mm. Calculate the shear stress, the shear strain, and the shear modulus of the sample at that point on the stress-strain curve.

Solution 9.20

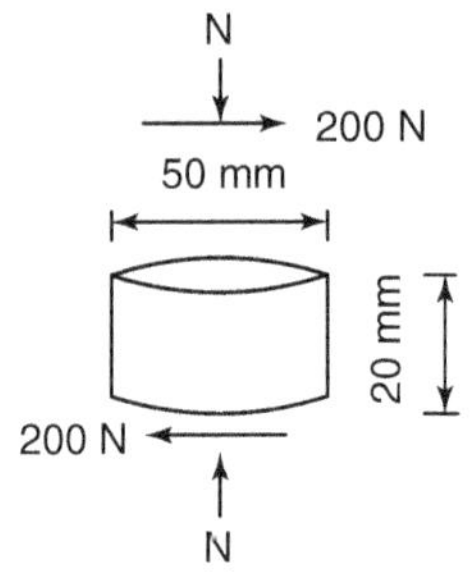

Figure 9.10s Silt sample.

Shear strain is:

$$\gamma = \frac{\Delta x}{h} = \frac{0.2 \text{ mm}}{20 \text{ mm}} = 0.01 = 1\%$$

Shear stress is:

$$\tau = \frac{F}{A} = \frac{200}{\frac{\pi}{4} \times (50 \times 10^{-3})^2} = 1.02 \times 10^5 \text{ Pa} = 102 \text{ kPa}$$

Assuming that τ and γ follow a linear relationship, then the shear modulus of the sample at that point is calculated:

$$G = \frac{\tau}{\gamma} = \frac{102 \text{ kPa}}{0.01} = 10.2 \text{ MPa}$$

Problem 9.21

An unconfined compression (UC) test on a sample of clay gives the stress-strain curve shown in Figure 9.11s. Calculate the undrained shear strength and the UC modulus for this sample. What geotechnical problem do you think this undrained shear strength and this modulus could be used for?

Solution 9.21

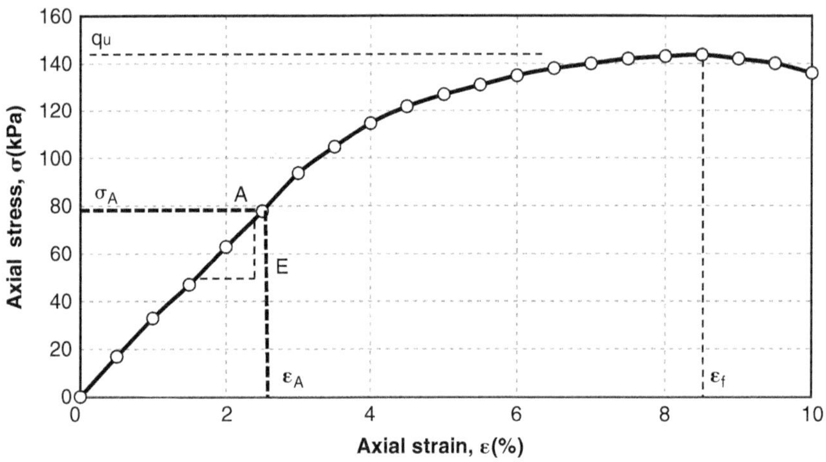

Figure 9.11s Stress-strain curve for unconfined compression test on clay.

From Figure 9.11s, we can determine that the maximum axial stress q_u is 144 kPa. The axial strain at failure is 8.5%. The undrained shear strength is half of the maximum axial stress: $s_u = \frac{q_u}{2} = \frac{144}{2}\text{kPa} = 72$ kPa. UC modulus can be obtained from the straight portion of the stress-strain curve for this sample. The following equation shows the result of UC modulus.

$$E = \frac{\sigma_A}{\varepsilon_A} = \frac{78 \text{ kPa}}{2.5\%} = 3120 \text{ kPa} = 3.12 \text{ MPa}$$

The results from the unconfined compression test can be used to estimate the short-term bearing capacity of fine-grained soils for foundation, estimate the short-term stability of slopes, and determine the stress-strain characteristics of a soil under fast (undrained) loading conditions.

Problem 9.22

What are the two main phases in running a triaxial test? With respect to drainage during each one of these two phases, what are the different types of tests that can be run? For each type of test, what parameters can you obtain from the data?

Solution 9.22

The two main phases in running a triaxial test are the consolidation phase and the shearing phase.

With respect to drainage during the two phases of the triaxial test, there are three different types of tests: UU test (unconsolidated undrained test), CU test (consolidated undrained test), and CD test (consolidated drained test).

From a UU test, we can obtain an undrained shear strength, the strain at failure, and undrained deformation moduli at different strains.

From a CU test with water stress measurement, we can obtain effective stress shear strength parameters: friction angle φ' and cohesion c′, and consolidated undrained deformation moduli at different strains.

From a CD test, we can obtain effective stress shear strength parameters: friction angle φ' and cohesion c′, and drained deformation moduli at different strains.

Problem 9.23

A triaxial test is performed on a sample that is 50 mm in diameter and 100 mm high. The confining pressure is 30 kPa and at failure the vertical load on the sample is 118 N. Is the vertical total stress on the sample at failure expressed in N/m^2 equal to $\frac{118\,\text{N}}{\pi(25.10^{-3})^2}$? If yes, explain your answer. If not, what is it?

Solution 9.23

The vertical total stress on the sample at failure expressed in N/m^2 is NOT equal to $\frac{118\text{ N}}{\pi(25\times10^{-3})^2\text{ m}^2}$. The confining pressure also contributes to the vertical total stress. Therefore, the vertical total stress on the sample at failure is equal to 30 kPa $+ \frac{0.118\text{ kN}}{\pi(25\times10^{-3})^2\text{ m}^2} = 90$ kPa. This answer does not consider the change in sample cross section during loading.

Problem 9.24

A triaxial test with water stress (pore-pressure) measurements is performed on a sample of saturated silty crushed rockfill and gives the results shown in Figure 9.50. The total confining stress is 35 kPa.

a. Calculate the total stress secant modulus E and the effective stress secant modulus E′ for a vertical strain equal to 0.2%, 0.5%, 1%, 2%, 3%, 4%, and 5%. Then plot the curve giving soil modulus both as a function of strain and as function of log strain.
b. At failure, the vertical effective stress is 100 kPa. Calculate the effective stress friction angle of the sand if the effective stress cohesion c is zero.

Solution 9.24

In the following calculation, Poisson's ratio is assumed to be 0.5 for the undrained modulus and 0.35 for the effective stress modulus. Figure 9.12s shows the soil sample in a triaxial test.

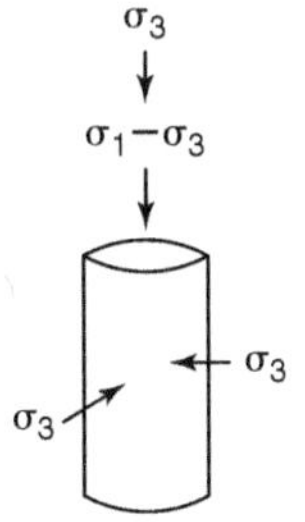

Figure 9.12s Illustration of soil sample in triaxial test.

From Figure 9.13s, we can determine that the deviator stresses $\sigma_1 - \sigma_3$ at the axial strains of 0.2%, 0.5%, 1%, 2%, 3%, 4% and 5% are 14 kPa, 40 kPa, 75 kPa, 145 kPa, 152 kPa, 142 kPa, and 131 kPa respectively. Given that $\sigma_3 = 35$ kPa, we can get both σ_1 and σ_3 for all the strain levels. The total stress secant modulus E can be obtained.

$$E = (\sigma_1 - 2\nu\sigma_3)/\varepsilon$$

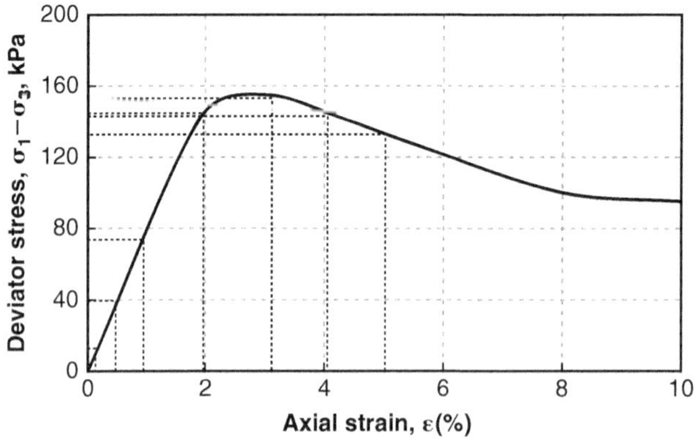

Figure 9.13s Deviator stress curve related to axial strain.

The results are shown in Table 9.4s. In Figure 9.14s, the water pressures can be read as 2 kPa, 5 kPa, 9 kPa, 11 kPa, 8 kPa, 4 kPa, and −2 kPa when the axial strain is 0.2%, 0.5%, 1%, 2%, 3%, 4%, and 5% respectively.

Table 9.4s Results of Calculation of Soil Modulus

ε	σ_3 (kPa)	$\sigma_1 - \sigma_3$ (kPa)	σ_1 (kPa)	E (MPa)	u (kPa)	σ_3' (kPa)	σ_1' (kPa)	E′(MPa)
0.2%	35	14	49	7.0	2	33	47	11.9
0.5%	35	40	75	8.0	5	30	70	9.8
1%	35	75	110	7.5	9	26	101	8.3
2%	35	145	180	7.2	11	24	169	7.6
3%	35	152	187	5.1	8	27	179	5.3
4%	35	142	177	3.5	4	31	173	3.8
5%	35	131	166	2.6	−2	37	168	2.8

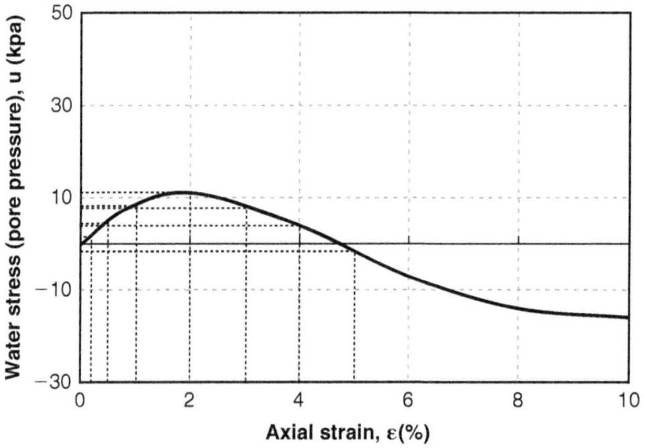

Figure 9.14s Water stress curve related to axial strain.

The effective stress can be calculated by using the total stress minus the water stress for this saturated soil. The effective stress secant modulus E' can then be obtained.

$$E' = (\sigma_1' - 2\nu'\sigma_3')/\varepsilon$$

The results are shown in Table 9.4s. Figure 9.15s shows the curves of total soil modulus and effective soil modulus as a function of strain, while Figure 9.16s shows the curve of total soil modulus and effective soil modulus as a function of logarithm of strain.

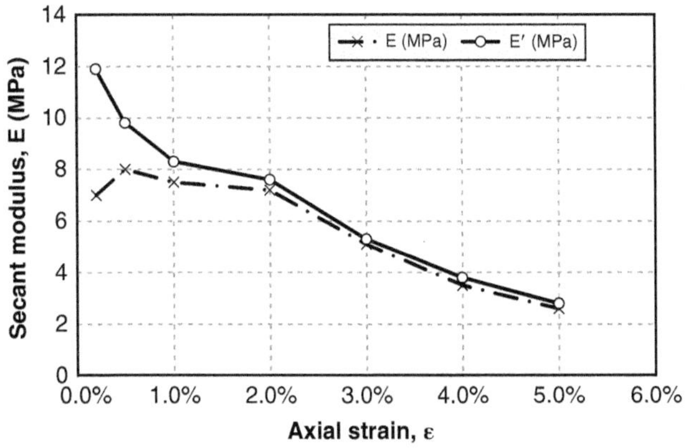

Figure 9.15s Curves of total soil modulus and effective soil modulus as a function of strain.

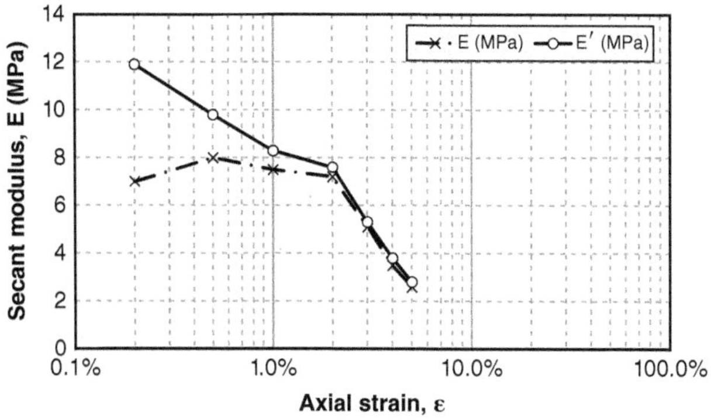

Figure 9.16s Curves of total soil modulus and effective soil modulus as a function of logarithm of strain.

Assuming that the effective cohesion is 0, the effective stress friction angle can be obtained from the Mohr circle using the effective principal stresses at failure (Figure 9.17s): $\varphi' = 47.5°$

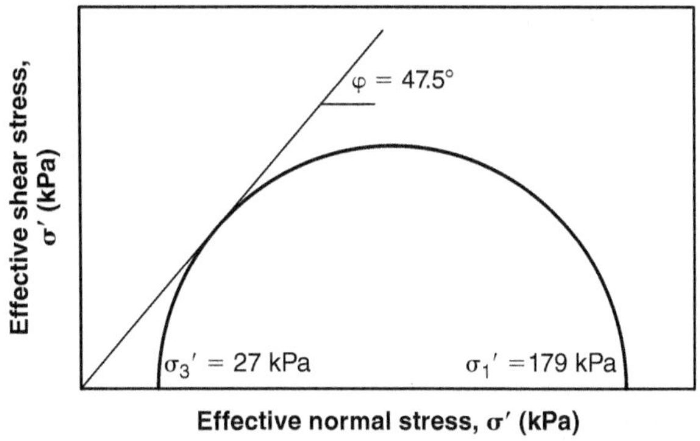

Figure 9.17s Mohr circle and friction angle.

Problem 9.25

Two CU triaxial tests are performed on a sample of saturated, overconsolidated, high-plasticity clay. At failure, the results are as follows:

$$\text{Test 1}: \sigma_3 = 30\text{ kPa}, Q = 0.45\text{ kN}, A = 0.01\text{ m}^2, u = 10\text{ kPa}$$

$$\text{Test 2}: \sigma_3 = 60\text{ kPa}, Q = 0.70\text{ kN}, A = 0.01\text{ m}^2, u = 20\text{ kPa}$$

where σ_3 is the total confinement stress, Q is the vertical load on the sample, A is the sample cross section, and u is the water stress (pore pressure).

a. Calculate $\sigma_3, \sigma_1, \sigma_3'$, and σ_1' at failure.
b. Draw the Mohr circle at failure in the τ vs. σ' set of axes.
c. Draw the failure envelope and find the effective stress strength parameters c′ and φ'.

Solution 9.25

a. At failure, the stresses $\sigma_3, \sigma_1, \sigma_3'$, and σ_1' are calculated as follows:

$$\sigma_1 = \sigma_3 + \frac{Q}{A} = 30 + \frac{0.45}{0.01} = 75 \text{ kPa}$$

$$\sigma_3 = 30 \text{ kPa}$$

$$\sigma_1' = \sigma_1 - \alpha u = 75 - 1 * 10 = 65 \text{ kPa}$$

$$\sigma_3' = \sigma_3 - \alpha u = 30 - 1 * 10 = 20 \text{ kPa}$$

Test 2:

$$\sigma_1 = \sigma_3 + \frac{Q}{A} = 60 + \frac{.70}{.01} = 130 \text{ kPa}$$

$$\sigma_3 = 60 \text{ kPa}$$

$$\sigma_1' = \sigma_1 - \alpha u = 130 - 1 * 20 = 110 \text{ kPa}$$

$$\sigma_3' = \sigma_3 - \alpha u = 60 - 1 * 20 = 40 \text{ kPa}$$

b. The Mohr circle at failure is shown in Figure 9.18s.

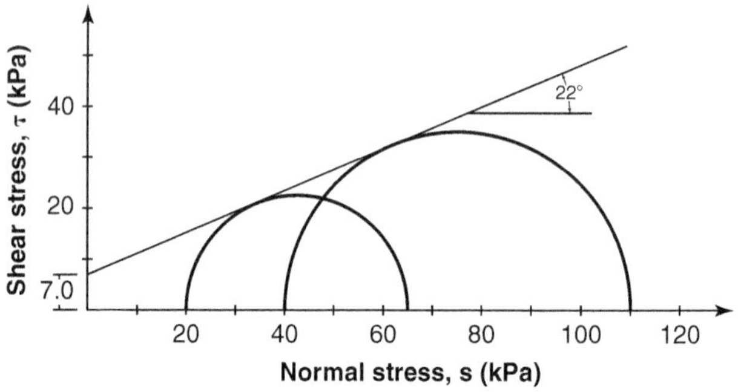

Figure 9.18s Mohr circle.

c. From the Mohr circle (Figure 9.18s), the cohesion c′ = 7 kPa and the friction angle $\varphi' = 22°$.

Problem 9.26

Two CU triaxial tests are performed on a sample of unsaturated clay. At failure, the results are as follows:

$$\text{Test 1: } \sigma_3 = 20 \text{ kPa}, \sigma_1 = 190 \text{ kPa}, \text{S} = 60\%, \text{u}_w = -100 \text{ kPa}$$

$$\text{Test 2: } \sigma_3 = 60 \text{ kPa}, \sigma_1 = 450 \text{ kPa}, \text{S} = 50\%, \text{u}_w = -300 \text{ kPa}$$

where σ_3 is the total confinement stress, σ_1 is the total vertical stress at failure, S is the degree of saturation, and u_w is the water stress.

a. Calculate $\sigma_3, \sigma_1, \sigma_3'$, and σ_1' at failure.
b. Draw the Mohr circle at failure in the τ vs. σ' set of axes.
c. Draw the failure envelope and find the effective stress strength parameters c′ and φ'.

Solution 9.26

a. Calculate σ_3, σ_1, σ_3', and σ_1' at failure.
For Test 1 (at failure):

$$\sigma_3 = 20\,\text{kPa}$$
$$\sigma_3' = \sigma_3 - \alpha u_w = \sigma_3 - S \cdot u_w = 20 - 0.6 \times (-100) = 80\,\text{kPa}$$
$$\sigma_1 = 190\,\text{kPa}$$
$$\sigma_1' = \sigma_1 - \alpha u_w = \sigma_1 - S \cdot u_w = 190 - 0.6 \times (-100) = 250\,\text{kPa}$$

For Test 2 (at failure):

$$\sigma_3 = 60\,\text{kPa}$$
$$\sigma_3' = \sigma_3 - \alpha u_w = \sigma_3 - S \cdot u_w = 60 - 0.5 \times (-300) = 210\,\text{kPa}$$
$$\sigma_1 = 490\,\text{kPa}$$
$$\sigma_1' = \sigma_1 - \alpha u_w = \sigma_1 - S \cdot u_w = 450 - 0.5 \times (-300) = 600\,\text{kPa}$$

b & c. Figure 9.19s shows the Mohr circle and failure envelope for the test. From the Mohr circle and failure envelope, we can get the value of c′ and φ' as 10 kPa and 27° respectively.

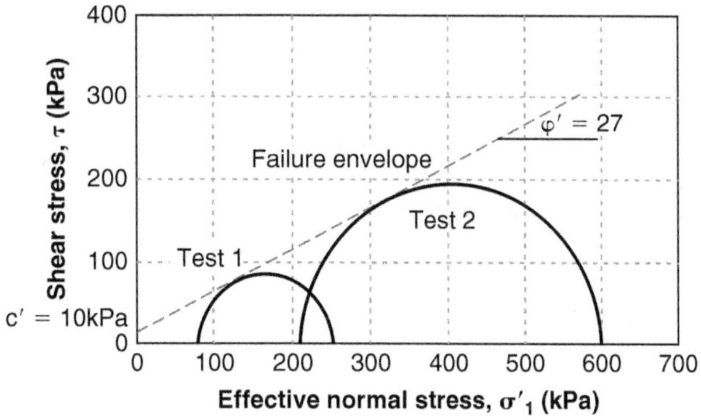

Figure 9.19s Mohr circle and failure envelope for the test.

Problem 9.27

What is the stress path and what shape does it typically have for the triaxial test?

Solution 9.27 (Figure 9.20s)

The stress path describes the evolution of certain stresses during the test. Specifically, it tracks the path described by the points with p, q stress coordinates where p and q are defined as follows:

Total stress

$$p = (\sigma_1 + \sigma_3)/2$$
$$q = (\sigma_1 - \sigma_3)/2$$

Effective stress

$$p' = (\sigma_1' + \sigma_3')/2$$
$$q' = (\sigma_1' - \sigma_3')/2$$

The most useful stress paths are plotted in terms of effective stresses (p′ and q′).

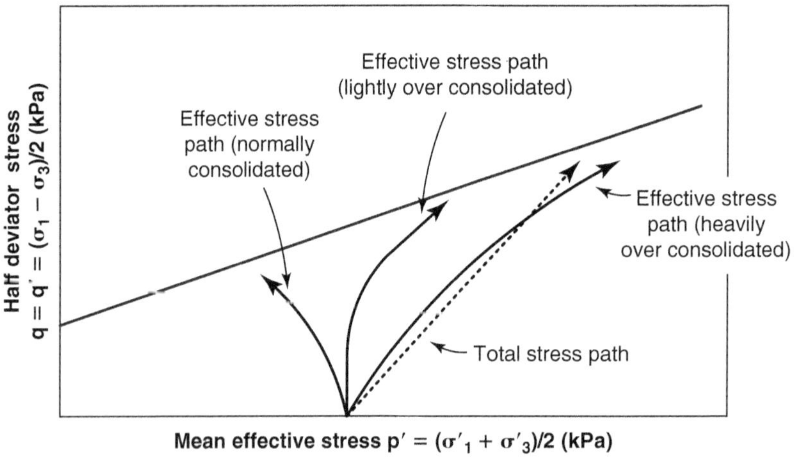

Figure 9.20s Stress paths.

Problem 9.28

A lab vane test is performed on a silty clay. At failure, the maximum torque is 5.7 N.m. The vane is 50 mm high and 25 mm in diameter. Calculate the undrained shear strength of the silty clay. The vane is rotated 10 times rapidly and the torque on the tenth revolution is measured to be 3.5 N.m. Calculate the residual undrained shear strength of the silty clay.

Solution 9.28

$$T = \pi s_u D^2 \left(\frac{H}{2} + \frac{D}{6}\right) \Rightarrow s_u = \frac{T}{\pi D^2 \left(\frac{H}{2} + \frac{D}{6}\right)}$$

$$T_{Max} = 5.7 \text{ N.m}$$

$$T_{res} = 3.5 \text{ N.m}$$

$$H = 0.05 \text{ m}$$

$$D = 0.025 \text{ m}$$

$$s_u = \frac{5.7}{\pi 0.025^2 \left(\frac{0.05}{2} + \frac{0.025}{6}\right)} = 99531 \text{ N/m}^2 \approx 100 \text{ kPa}$$

$$s_{ur} = \frac{3.5}{\pi 0.025^2 \left(\frac{0.05}{2} + \frac{0.025}{6}\right)} = 61115 \text{ N/m}^2 \approx 61 \text{ kPa}$$

Problem 9.29

A silty sand is subjected to a constant head permeameter test. The flow collected at the downstream end is 221 mm^3/s; the sample is 75 mm in diameter and 100 mm high. The difference between the water level in the upstream overflow and the downstream overflow is 0.5 m. Calculate the hydraulic conductivity k of the silty sand.

Solution 9.29

Given in the problem statement:

$$Q = 221 \text{ mm}^3/\text{s}$$

$$A = \pi d^2/4 = \pi (75 \text{ mm})^2/4 = 4417.8 \text{ mm}^2$$

$$i = h/l = 0.5 \text{ m}/0.1 \text{ m} = 5$$

Hydraulic conductivity:

$$k = (\mathrm{Q\,l})/(\mathrm{A\,h}) = 221 \times 0.1/(4417.8 \times 0.5) = 0.01\ \mathrm{mm/s}$$
$$k = 0.01\ \mathrm{mm/s} = 1.0 \times 10^{-5}\ \mathrm{m/s}$$

Problem 9.30

A clay sample is tested in a falling head permeameter. The sample is 75 mm in diameter and 100 mm high. The small tube is 3 mm in diameter. The difference in height between the water level in the small tube above the sample and the downstream overflow is measured as a function of time. At time $t = 0$, the difference is 1.1 m and at time $t = 1$ hr, the difference is 1.05 m. Calculate the hydraulic conductivity k of the clay.

Solution 9.30

Given in the problem statement:

Inner tube diameter, d = 3 mm
Sample diameter, D = 75 mm
Sample height = 100 mm

From test results:

$$\Delta t = 60\ \mathrm{min} = 3600\ \mathrm{s}$$
$$h_0 = 1100\ \mathrm{mm}$$
$$h_1 = 1050\ \mathrm{mm}$$

Calculations:

$$a = \pi \mathrm{d}^2/4 = \pi(3)^2/4 = 7.06\ \mathrm{mm}^2$$
$$A = \pi \mathrm{D}^2/4 = \pi(75)^2/4 = 4417.7\ \mathrm{mm}^2$$

Hydraulic conductivity:

$$k = 2.3\frac{al}{At}\log\frac{h_0}{h_1} = 2.3\frac{7.06 \times 100}{4417.7 \times 60 \times 60}\log\frac{1100}{1050} = 2.06 \times 10^{-6}\ \mathrm{mm/s} = 2.06 \times 10^{-9}\ \mathrm{m/s}$$

Problem 9.31

This problem refers to Figure 9.71. A sample of unsaturated silt is tested in a constant head permeameter and the following parameters are measured: $D = 75$ mm, l = 150 mm, V = 10 cm^3, $t = 1$ hour, $h_{p2} = -100$ mm, $h_{p3} = -200$ mm, $h_{p4} =$ 0 mm. Calculate the unsaturated hydraulic conductivity k of the silt and the water tension u_w corresponding to that value.

Solution 9.31

$$k = \frac{4V}{t\pi D^2}\frac{1}{(h_{t2} - h_{t3})}$$
$$h_{t2} - h_{t3} = |h_{p3}| + l - |h_{p2}| = 200 + 150 - 100 = 250\ \mathrm{mm}$$
$$k = \frac{4 \times 10 \times 10^3\ \mathrm{mm}^3}{60 \times 60 \times \pi \times 75^2\ \mathrm{mm}^2}\left(\frac{150}{250}\right)$$
$$= 3.77 \times 10^{-4}\frac{\mathrm{mm}}{s} = 3.77 \times 10^{-7}\frac{\mathrm{m}}{s}$$
$$u_w = \frac{h_{p2} + h_{p3}}{2}\gamma_w$$
$$u_w = \left(\frac{-100 + (-200)}{2}\right)10^{-3} \times 9.81 = -1.47\ \mathrm{kPa}$$

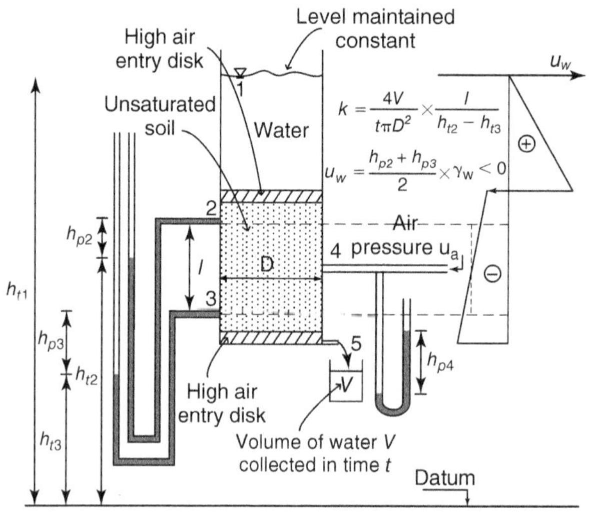

Figure 9.21s Constant head permeameter test for unsaturated soils.

Problem 9.32

This problem refers to Figure 9.75. A 0.65 m long 75 mm diameter sample of unsaturated clay is tested in a wetting front permeameter. Initially the water tension in the sample is −1000 kPa. The results are shown in Figure 9.22s. Use the results to develop the hydraulic conductivity k vs. water tension u_w curve for this clay.

Solution 9.32 (Figure 9.22s)

The hydraulic conductivity k is the ratio of water velocity v_w to the hydraulic gradient i, $k = v_w/\text{i}$. To develop the hydraulic conductivity k vs. water tension u_w curve, v_w and i are calculated from the plot on Figure 9.22s as follows.

The water velocity v_w is equal to the volume of water $d\text{V}_w$ passing through a point in a soil sample with a cross section A in a time interval dt:

$$v_w = \frac{d\text{V}_w}{dt} \times \frac{1}{A}$$

The volume of water that passes through the soil sample in a time interval $dt = t_2 - t_1$ is:

$$dV_w = V_{w,t_2} - V_{w,t_1}$$

where $V_{w,\text{ti}}$ is the volume of water present in the soil sample at time t_i between the distance x_J and the end of the sample. This volume can be calculated from the volumetric water content vs. distance curve and is equal to the area below the curve at a given time multiplied by the sample cross section A:

$$\text{V}_w \text{ (between point J and end of sample)} = \int_{xj}^{L} \theta_w(x)Adx$$

where $\theta_w = V_w/V$ is the volumetric water content and can be derived from the water tension as follows:

$$\theta_w = C_w \log|u_w| + a$$

where $u_w = \gamma_w\text{h}$ is the water tension in the soil and can be calculated from the pressure head vs. distance graph.

Let's call "a" the difference in area between two curves corresponding to two different times on the θ_w vs. distance diagrams. Then $d\text{V}_w$ is given as:

$$\text{dV}_w = a \times A$$

The water velocity is then:

$$v_w = \frac{d\text{V}_w}{dt} \times \frac{1}{A} = \frac{a \times A}{\Delta t} \times \frac{1}{A} = \frac{a}{\Delta t}$$

For each time interval, the hydraulic gradient is simply calculated from the pressure head vs. distance graph as follows:

$$\mathrm{i}_{\mathrm{avg}} = \frac{\Delta h}{\Delta x} = \frac{h_{x=0.6} - h_{x=0.0}}{0.60 - 0}$$

The hydraulic conductivity is then calculated as follows:

$$k = \frac{v_w}{i_{avg}} = \frac{a}{\Delta t} \times \frac{1}{i_{avg}}$$

Because the pressure head is variable through the section, the water tension u_w can be calculated as an average value in the soil sample for a given time from the pressure head vs. distance graph.

The following is an example of the calculations leading to one point on the k vs. u_w graph. This point is the one corresponding to the time interval $t = 300$ hr to 400 hr.

1. Calculation of average hydraulic gradient:

$$\mathrm{i}_{\mathrm{t=400}} = \frac{\Delta h}{\Delta x} = \frac{h_{x=0.60} - h_{x=0.0}}{0.60 - 0} = \frac{-190 - 0}{0.6 - 0} = -316$$

$$\mathrm{i}_{\mathrm{t=300}} = \frac{\Delta h}{\Delta x} = \frac{h_{x=0.6} - h_{x=0.0}}{0.6 - 0} = \frac{-420 - 0}{0.6 - 0} = -700$$

$$\mathrm{i}_{\mathrm{avg}} = \frac{\mathrm{i}_{\mathrm{t=400}} + \mathrm{i}_{\mathrm{t=300}}}{2} = -508$$

2. Calculation of "a":

The area S_1 below the $t = 400$ curve (assuming it is trapezoidal) is equal to:

$$\mathrm{S}_1 = \frac{(0.12 + 0.062) \times (0.6 - 0)}{2} = 0.0546$$

The area S_1 below the $t = 300$ curve (assuming it is trapezoidal) is equal to:

$$\mathrm{S}_2 = \frac{(0.11 + 0.056) \times (0.6 - 0)}{2} = 0.0498$$

$$\mathrm{a} = \mathrm{S}_1 - \mathrm{S}_2 = 4.8 \times 10^{-3}$$

3. Calculation of water velocity:

$$v_w = \frac{a \times L}{\Delta t} = \frac{4.8 \times 10^{-3} \times 0.6}{(400 - 300) \times 3600} = 8 \times 10^{-9}\ \mathrm{m/s}$$

4. The hydraulic conductivity k is equal to:

$$k = \frac{v_w}{i_{avg}} = \frac{8 \times 10^{-9}}{508} = 1.6 \times 10^{-11}\ \mathrm{m/s}$$

5. The average water tension corresponding to the hydraulic conductivity calculated in step 4 is:

$$u_{w,\mathrm{avg}} = \sum_{i=1}^{7} \frac{\gamma_w \mathrm{h_i}}{7} = \frac{9.81}{7} \times (0 - 20 - 60 - 100 - 150 - 170 - 190) = 967\ \mathrm{kPa}$$

More refined calculations can be done by using the actual pressure head vs. distance curve rather than a straight line and curved areas rather than trapezoidal shape assumptions for the volumetric water content vs. distance diagram.

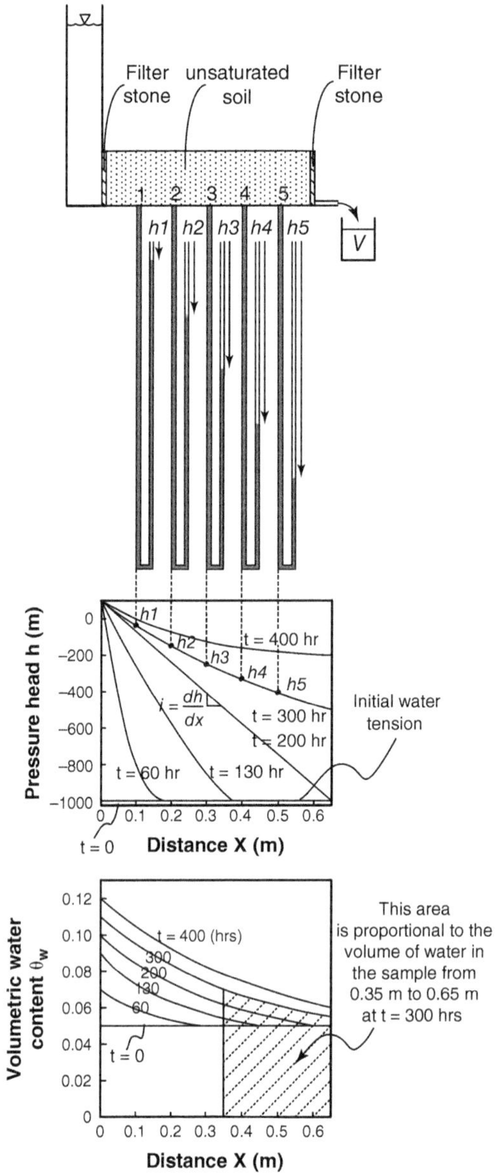

Figure 9.22s Wetting front test for unsaturated soils.

The hydraulic conductivity k vs. the water tension curve is shown in Figure 9.23s.

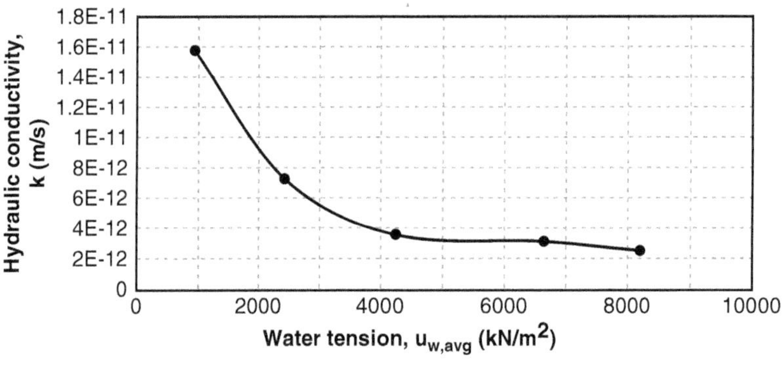

Figure 9.23s Hydraulic conductivity vs. water tension.

Problem 9.33

This problem refers to Figure 9.76. A sample of unsaturated clayey sand has a degree of saturation of 40%. The sample length is 150 mm and the sample diameter is 75 mm. It is tested in a permeameter to determine the hydraulic conductivity of air through the sample. The air pressure at the base of the sample is 10 kPa and the volume of air collected at the top of the sample in one hour of testing is 10^{-3} m^3. The top of the sample is kept at atmospheric pressure. Calculate the air hydraulic conductivity of the sample k_a and the air stress u_a associated with this hydraulic conductivity value.

Solution 9.33

Data:

$$\gamma_a = 0.0118 \text{ kN/m}^3 \text{ (assumed at 20°C)}$$
$$V_a = 10^{-3} \text{ m}^3$$
$$L = 0.15 \text{ m}$$
$$D = 0.075 \text{ m}$$
$$p_a = 10 \text{ kPa}$$
$$t = 1\text{hr} = 3600 \text{ sec}$$

a. The permeability of the soil to air can be calculated as:

$$k_a = \frac{\gamma_a V_a L}{Apt}$$
$$A = \frac{\pi D^2}{4} = \frac{\pi (0.075)^2}{4} = 0.004418 \text{ m}^2$$
$$k_a = \frac{\gamma_a V_a L}{Apt} = \frac{(0.0118 \text{ kN/m}^3) \times (1 \times 10^{-3} \text{ m}^3) \times (0.15 \text{ m})}{(0.0044818 \text{ m}^2) \times (10 \text{ kN/m}^2) \times (3600 \text{ sec.})} = 1.1 \times 10^{-8} \text{ m/sec.}$$

b. The air stress associated with the permeability k_a can be calculated as:
The air stress at the top of the sample is zero. The air stress at the bottom of the sample is 10 kPa. Thus, the average air stress in the sample associated with the measure of air hydraulic conductivity is $u_a = 0.5\ (0 + 10) = 5$ kPa.

Problem 9.34

A 1.8 m tall human being drinks one liter of water. Three hours later, this person goes to the bathroom and eliminates the liter of water. Is this case a constant head permeameter or a falling head permeameter? Calculate the hydraulic conductivity of the human body. Make reasonable assumptions when necessary.

Solution 9.34

This is a case of a falling head permeameter because the water level goes down with time. The equivalent hydraulic conductivity of the human body can be estimated using the following equation:

$$k = 2.3\frac{al}{At}\log\frac{h_0}{h_1}$$

Assumptions:

$$A = a;\ l = 0.21 \text{ m};\ t = 2 \text{ hrs};\ h_o = 1.3 \text{ m};\ h_1 = 1 \text{ m}$$
$$k = 2.3\frac{0.21}{2 \times 60 \times 60}\log\frac{1.3}{1} = 7.64 \times 10^{-6} \text{ m/s}$$

Problem 9.35

A sample of fine sand is tested in the EFA. The mean diameter of the grains is $D_{50} = 1$ mm. When the velocity is set at 1 m/s, the piston below the sample of sand has to be raised at a rate of 16.7 mm/minute. The cross section of the conduit where the water is flowing is rectangular, with a width of 100 mm and a height of 50 mm. Calculate the shear stress at the interface between the water and the sand for the 1 m/s velocity.

Solution 9.35

$$\tau = \frac{1}{8}\rho f v^2$$

Given in the problem statement are:

Flow velocity: $v - 1.0\ \mathrm{m/s}$
Mass density of water: $\rho = 1000\ \mathrm{kg/m^3}$
Mean grain size: $D_{50} = 1$ mm
Dimensions of test section = 50 mm by 100 mm
Hydraulic diameter, $D = 2ab/(a + b) = 2\ (50 \times 100)/(50 + 100) = 66.7$ mm = 0.0667 m
Viscosity of water: $\upsilon = 1.12 \times 10^{-6}\ \mathrm{m^2/s}$
Calculate friction factor, f, from Moody Chart
f is a function of roughness, ε, and Reynolds Number, Re
Roughness: $\varepsilon = D_{50}/2 = 1\ \mathrm{mm}/2 = 0.5$ mm
Reynolds Number: $Re = Dv/\upsilon = 0.0667 \times 1/1.12 \times 10^{-6} = 59553$
Friction factor read on Moody Chart: f = 0.032
Shear stress: $\tau = 1/8 \times 1000 \times 0.032 \times 1^2 = 4$ Pa

Problem 9.36

A sample of low-plasticity clay is tested in the EFA. The surface of the clay sample is considered smooth. When the velocity is set at 3 m/s, the piston below the sample of sand has to be raised at a rate of 1 mm every 3 minutes. The cross section of the conduit where the water is flowing is rectangular, with a width of 100 mm and a height of 50 mm. Calculate the shear stress at the interface between the water and the sand for the 3 m/s velocity.

Solution 9.36

$$\tau = \frac{1}{8}\rho f v^2$$

Given in problem statement:

Flow velocity: $v = 3.0\ \mathrm{m/s}$
Mass density: $\rho = 1000\ \mathrm{kg/m^3}$
Mean grain size: $D_{50} = 0$ mm—Smooth
Dimensions of test section = 50 mm by 100 mm
Hydraulic diameter, $D = 2ab/(a + b) = 2\ (50 \times 10)/(50 + 100) = 66.7$ mm = 0.0667 m
Viscosity of water: $\upsilon = 1.12 \times 10^{-6}\ \mathrm{m^2/s}$
Friction factor f is read on Moody Chart
f is a function of roughness, ε, and Reynolds Number, Re
Roughness is zero (smooth)
Reynolds Number: $Re = Dv/\upsilon = 0.0667 \times 3/1.12 \times 10^{-6} = 178660$
So, friction factor is read as f = 0.016
Shear stress: $\tau = 1/8 \times 1000 \times 0.016 \times 3^2 = 18$ Pa

CHAPTER 10

Stresses, Effective Stress, Water Stress, Air Stress, and Strains

10.1 GENERAL

A soil mass is subjected to internal and boundary forces due to loading by a building, bridge, dam, retaining wall, or even rain and evaporation near the ground surface. Under these forces, displacements take place. The objective of the design process is to ensure that the displacements are tolerable and safe for the structure. It is difficult to use forces and displacements as parameters in the design process because they are not normalized quantities and therefore cannot be compared between, for example, the full-scale behavior in the field and a small-scale test in the laboratory. The concept of stress and strain is used to normalize forces and displacements to the point where such comparisons can be made. Note that although using stresses and strains makes some problems easier to deal with, it may create some difficulties at the same time. For example, you cannot add stresses as you would add forces. Any time one wishes to compose stresses, it is much preferable to use the stresses to calculate the forces, then add the forces by conventional means to find the resultant, and then calculate the resultant stress.

10.2 STRESS VECTOR, NORMAL STRESS, SHEAR STRESS, AND STRESS TENSOR

A *stress* is a force divided by the area over which it applies. The force is not necessarily perpendicular or tangent to the area. Because the force is a vector, so is the stress. For a given point in a soil mass and for a given plane at that point, there is one stress vector:

$$\vec{t} = \lim_{A \to 0} \frac{\vec{F}}{A} \tag{10.1}$$

where t is the stress vector, F is the resultant force at the point considered, and A is the area of the plane on which F is acting. The stress vector, like the force, can always be decomposed into a normal stress σ and a shear stress τ. If the force is perpendicular to the area, the stress is a normal stress. If the force is tangential to the area, the stress is a shear stress:

$$\sigma = \frac{N}{A}, \quad \tau = \frac{T}{A}, \tag{10.2}$$

where σ is the normal stress, N is the force normal to the surface of area A, τ is the shear stress, and T is the force tangent to the surface of area A. For a given point in a soil mass, there is one resultant force but there is an infinity of stress vectors because, though there is only one force, one can choose an infinity of planes with different orientations through that point. By swiveling the plane around that point, one will find three planes where the shear stresses are zero. These planes are perpendicular to each other and are called the *principal planes*; the normal stresses on the principal planes are called *principal stresses* and are denoted σ_1, σ_2, and σ_3. The largest of the three is the *major principal stress* σ_1, the smallest is the *minor principal stress* σ_3, and σ_2 is called the *intermediate principal stress*.

The stress state at one point is usually represented by drawing a cube with axes in the x, y, and z directions. The stress vector on each face of the cube is decomposed into a normal stress (e.g., direction of x) and two shear stresses (e.g., directions of y and z). The definitions refer to the following labeling system:

- σ_{xx} is the stress on the plane perpendicular to x and in the direction of x; it is a normal stress.
- τ_{xy} is the stress on the plane perpendicular to x and in the direction of y; it is a shear stress.
- τ_{xy} is the stress on the plane perpendicular to x and in the direction of z; it is a shear stress.

Those three stresses are the decomposition of the stress vector t along the three orthogonal directions associated with the plane perpendicular to x (Figure 10.1).

For reasons of moment equilibrium, the shear stresses on two perpendicular planes must be equal ($\tau_{xy} = \tau_{yx}$). For reasons of symmetry and equilibrium, and because the cube is at the infinitesimal scale, stresses on opposite faces are equal and opposite. Therefore, while there are a total of 18 stresses

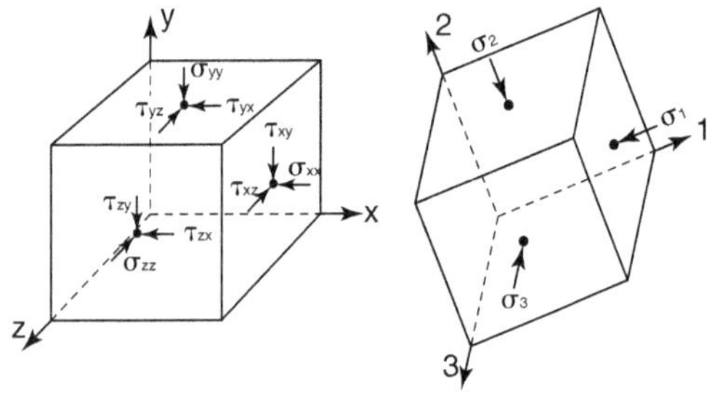

Figure 10.1 Stresses on an elementary cube.

(6 faces times 3 stresses), there are only 6 independent stresses (3 normal stresses and 3 shear stresses). These stresses are organized and presented in a stress tensor, which is a 3 × 3 matrix. That matrix has 9 elements, but is symmetric because the shear stresses on perpendicular planes are equal. Once the stress tensor is known at one point, all stresses are known at that point by simple geometric rules:

$$\Sigma = \begin{bmatrix} \sigma_{xx} & \tau_{xy} & \tau_{xz} \\ \tau_{yx} & \sigma_{yy} & \tau_{yz} \\ \tau_{zx} & \tau_{zy} & \sigma_{zz} \end{bmatrix} \tag{10.3}$$

The stress tensor Σ can be decomposed into the spherical tensor S and the deviatoric tensor *D*:

$$\Sigma = \begin{bmatrix} \sigma_{xx} & \tau_{xy} & \tau_{xz} \\ \tau_{yx} & \sigma_{yy} & \tau_{yz} \\ \tau_{zx} & \tau_{zy} & \sigma_{zz} \end{bmatrix} = S + D$$

$$= \begin{bmatrix} \sigma_M & 0 & 0 \\ 0 & \sigma_M & 0 \\ 0 & 0 & \sigma_M \end{bmatrix} + \begin{bmatrix} \sigma_{xx} - \sigma_M & \tau_{xy} & \tau_{xz} \\ \tau_{yx} & \sigma_{yy} - \sigma_M & \tau_{yz} \\ \tau_{zx} & \tau_{zy} & \sigma_{zz} - \sigma_M \end{bmatrix} \tag{10.4}$$

where

$$\sigma_M = \frac{1}{3}(\sigma_{xx} + \sigma_{yy} + \sigma_{zz}) \tag{10.5}$$

The spherical tensor represents a confinement effect at the point considered in the soil; it creates consolidation of the soil element with no shear. The deviatoric tensor represents the effect of various shear stresses on the soil element; it creates distortion with no mean normal stress.

10.3 SIGN CONVENTION FOR STRESSES AND STRAINS

Sign conventions are necessary in engineering because equations can differ for different conventions. Here, we will use compression stresses as positive because compression stresses are the most common case in soil mechanics. Note that in structures it is the contrary: there tension stresses are chosen to be positive normal stresses. Shear stresses are more complicated, so two cases must be considered.

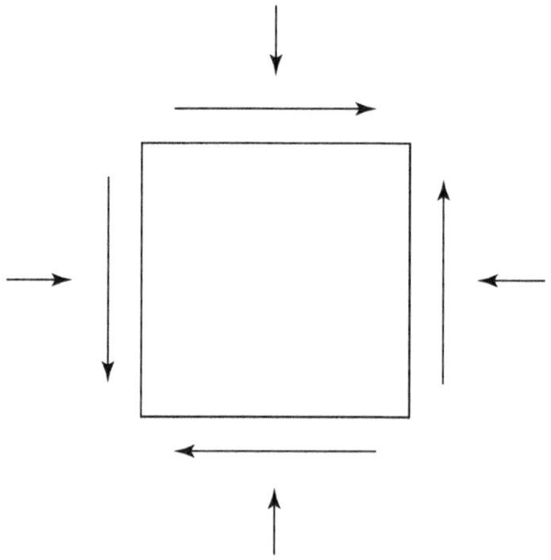

Figure 10.2 Positive sign convention for stress relationship equations.

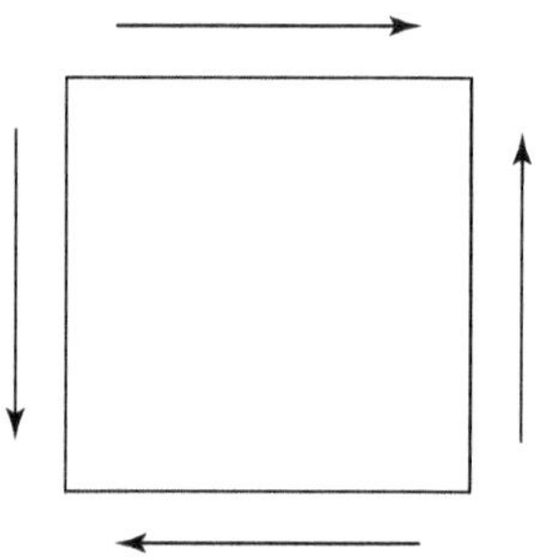

Figure 10.3 Positive sign convention for Mohr circle.

When dealing with the equations that relate stresses on two perpendicular planes to the stresses on an inclined plane, the positive convention for shear stress is as shown in Figure 10.2. However, when dealing with the Mohr circle representation of shear stresses, then the positive convention for shear stresses is as shown in Figure 10.3. Note that Figure 10.3 does not represent a feasible state of stress in a material, but simply the sign convention for the Mohr circle.

For normal strains, compressive strains will be considered positive. For shear strains, positive strains will be those that decrease an initially right angle.

10.4 CALCULATING STRESSES ON ANY PLANE: EQUILIBRIUM EQUATIONS FOR TWO-DIMENSIONAL ANALYSIS

At a specific point in the soil mass, and given the stresses on two perpendicular planes, the normal and shear stress on any other plane forming a wedge with the two perpendicular planes (Figure 10.4) can be related to the stresses on the two perpendicular planes as follows.

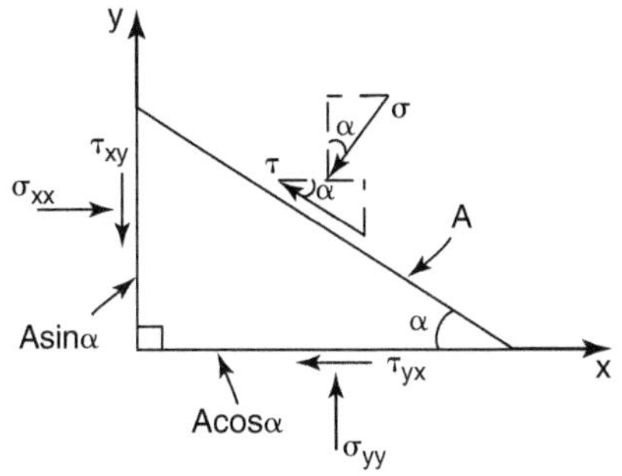

Figure 10.4 Wedge subjected to normal and shear stresses in equilibrium.

Referring to Figure 10.4, horizontal and vertical equilibrium of forces lead to equations 10.6 and 10.7:

$$\sigma_y \mathrm{A}\cos\alpha - \tau_{xy}\mathrm{A}\sin\alpha + \tau \mathrm{A}\sin\alpha - \sigma \mathrm{A}\cos\alpha = 0 \quad (10.6)$$

$$\sigma_x \mathrm{A}\sin\alpha - \tau_{yx}\mathrm{A}\cos\alpha - \tau \mathrm{A}\cos\alpha - \sigma \mathrm{A}\sin\alpha = 0 \quad (10.7)$$

where σ_y and σ_x are the normal stresses on the planes perpendicular to the y and x directions respectively, σ is the normal stress on the oblique surface, τ_{xy} and τ_{yx} are the shear stresses on the planes perpendicular to the x and y directions respectively, τ is the shear stress on the oblique surface, A is the area of the oblique surface, and α is the angle of the oblique surface as shown on Figure 10.4. From Eqs. 10.6 and 10.7 we get:

$$\sigma = \frac{\sigma_y + \sigma_x}{2} + \frac{\sigma_y - \sigma_x}{2}\cos 2\alpha - \tau_{xy}\sin 2\alpha \quad (10.8)$$

$$\tau = -\frac{\sigma_y - \sigma_x}{2}\sin 2\alpha - \tau_{xy}\cos 2\alpha \quad (10.9)$$

If the planes perpendicular to the *x* and *y* directions are principal planes (zero shear), then equations 10.8 and 10.9 become:

$$\sigma = \frac{\sigma_1 + \sigma_3}{2} + \frac{\sigma_1 - \sigma_3}{2}\cos 2\alpha \quad (10.10)$$

$$\tau = -\frac{\sigma_1 - \sigma_3}{2}\sin 2\alpha \quad (10.11)$$

where σ_1 and σ_3 are the major and minor principal stresses respectively, σ is the normal stress on the oblique surface, τ is the shear stress on the oblique surface, A is the area of the oblique surface, and α is the angle of the oblique surface as shown in Figure 10.4.

10.5 CALCULATING STRESSES ON ANY PLANE: MOHR CIRCLE FOR TWO-DIMENSIONAL ANALYSIS

In a set of coordinates where the shear stress on a plane is plotted on the vertical axis and the normal stress on the same plane is plotted on the horizontal axis, three circles bound the zone where the stress points are located (Figure 10.5). Indeed, all the stress points with τ, σ coordinates obtained for all the planes at that point fall in an area bounded by three circles centered on the horizontal axis. The reason why the center of the circles is on the horizontal axis goes back to the fact that shear stresses on perpendicular planes are equal. The circles intersect the normal stress axis at the principal stress values σ_1, σ_2, and σ_3; therefore, the circles have common points at the end of the diameter on the normal stress axis (Figure 10.5).

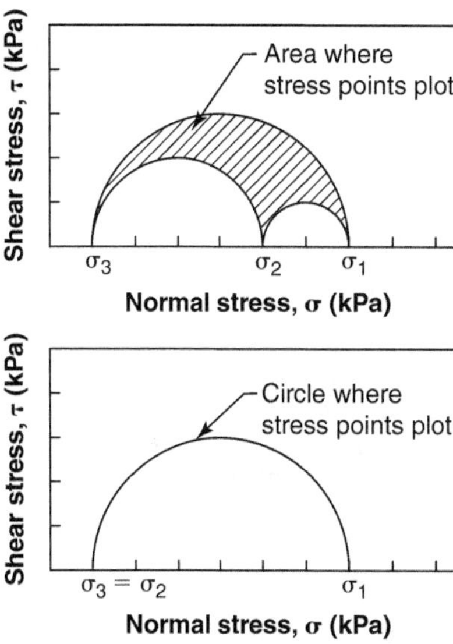

Figure 10.5 Shear stress vs. normal stress space and Mohr circle.

If the intermediate principal stress σ_2 is equal to the minor or the major principal stress, then there are only two principal stresses and the three circles collapse into one (Figure 10.5). This circle is called the *Mohr circle*. Otto Mohr was a German civil engineer who demonstrated in 1882 how this single circle could be used to find stresses on any plane at a point.

The case in which the intermediate principal stress is equal to the minor or the major principal stress occurs in a number of common situations (unconfined compression test, column loading, triaxial test, tension test). In this case, the zone representing all the stress points becomes the circle itself, and simple geometric constructions can be used to find the normal stress and the shear stress given a plane at that point in the soil mass (e.g., the Pole method). The Mohr circle can be defined as the graphical representation of the stresses at one point in a mass for the case where the intermediate principal stress is equal to the minor or the major principal stress. If one considers a different point, then the Mohr circle will be different. However, in the general case, there are three circles at one point and most stress points are not on the circles.

In the simpler case where the three principal stresses reduce to two, the following construction can be used to find the stresses on a randomly chosen plane (Figure 10.6). Although the problem can be posed in many different ways, the key, once the Mohr circle is known, is the relationship between:

1. a stress point on the Mohr circle for which we know the plane on which these stresses act
2. the direction of another plane in the two-dimensional (2D) space
3. the stresses on that other plane

If you known 1 and 2, you can find 3. If you know 1 and 3, you can find 2. The relationship is that if α is the angle between the two planes in space, the angle between the two stress points on the Mohr circle is 2α. This is due to equations 10.8 and 10.9, which have 2α in them. The angle 2α on the Mohr circle could be taken clockwise or counterclockwise from the known stress point, and that would lead to two different answers. The correct direction is such that if you go from the known plane to the plane where you seek the stresses by an angle α in space, you have to go from the known stress point to the unknown stress point through 2α in the same direction on the Mohr circle. Figure 10.6 illustrates the case for the triaxial test; Figure 10.7 illustrates the case for the direct shear test.

The Pole method is another popular method for solving the same problem. The *Pole* is a point on the Mohr circle such that a line on the Mohr circle passing through the stress point and parallel to the plane on which the stresses act will

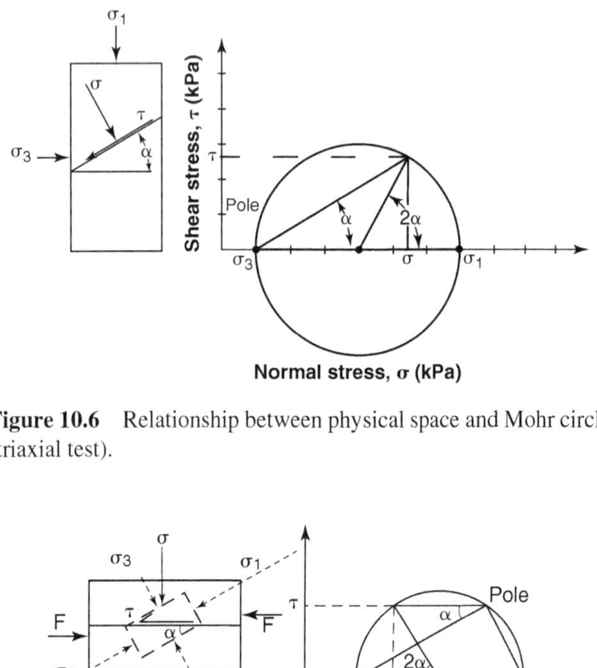

Figure 10.6 Relationship between physical space and Mohr circle (triaxial test).

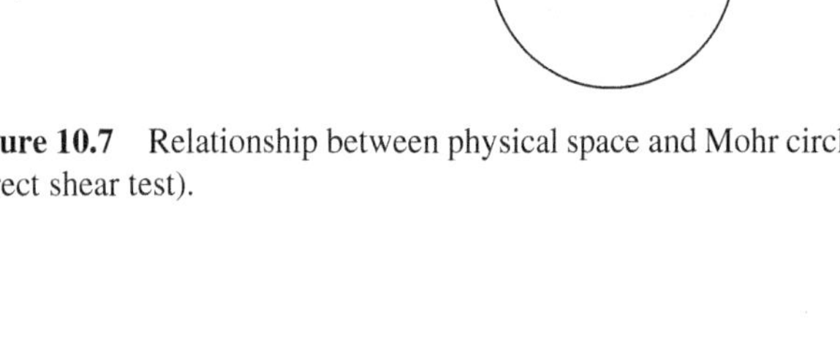

Figure 10.7 Relationship between physical space and Mohr circle (direct shear test).

intersect the Mohr circle at two points: the stress point and the Pole. The Pole method always has three components:

1. the Pole on the Mohr circle
2. the stress point on the Mohr circle
3. the plane on which the stresses act in space

You always have to know 2 of these 3 components to solve a problem. Typically, the first step is to find where the Pole is. For this you need to know the Mohr circle and a plane on which you know the stresses and therefore the stress point on the Mohr circle. The steps are as follows:

1. Draw a line from the known stress point on the Mohr circle parallel to the plane in the two-dimensional space on which the stresses act.
2. That line intersects the Mohr circle at 2 points: the stress point and the Pole. This gives the location of the Pole.
3. From the Pole on the Mohr circle, draw a line parallel to the plane on which the stresses are to be found.
4. That line intersects the Mohr circle at 2 points: the Pole and the stress point. The coordinates of this point are the stresses on the chosen plane, and the direction of these stresses are given by the sign convention discussed in section 10.3.

10.6 MOHR CIRCLE IN THREE DIMENSIONS

Section 10.5 dealt with the special case in which the intermediate principal stress σ_2 is equal to the minor principal stress σ_3 or to the major principal stress σ_1. In this case, there is only one Mohr circle and the stress points are on the circle. In the general case, the intermediate principal stress σ_2 is not equal to σ_1 or σ_3. As a result, there are three Mohr circles (Figure 10.5). In this general case, the stress point is located within the area bound by the three circles. The construction to find the stress point is more complicated than in the 2D case, as might be expected. It requires knowledge of the location of the point and plane considered in spherical coordinates, and the graphical solution defines the stress point at the intersection of three circles centered at the centers of the Mohr circles. Most advanced mechanics books describe this solution.

10.7 STRESS INVARIANTS

Stress invariants are combinations of stresses. There are three stress invariants:

$$I_1 = \sigma_1 + \sigma_2 + \sigma_3 \tag{10.12}$$

$$I_2 = \frac{1}{6}((\sigma_1 - \sigma_2)^2 + (\sigma_2 - \sigma_3)^2 + (\sigma_3 - \sigma_1)^2) \tag{10.13}$$

$$I_3 = \sigma_1\sigma_2\sigma_3 \tag{10.14}$$

where I_1, I_2, I_3 are the first, second, and third stress invariants, and σ_1, σ_2, σ_3 are the principal stresses. These stress invariants are quite useful in describing yield criteria for soils. For example, the Drucker-Prager yield criterion (see Chapter 12) is:

$$\sqrt{I_2} = A + BI_1 \tag{10.15}$$

where A and B are constants for a given material.

10.8 DISPLACEMENTS

Displacements take place as a result of many possible factors: loading change, temperature change, and water content changes are common. They can occur in the three directions x, y, and z. Any point A in a soil mass can experience displacements in the three directions x, y and z. These displacements can also be a function of time t. We will call the displacements u, v, and w, corresponding to the directions x, y, and z respectively. Figure 10.8 illustrates the displacements in a two-dimensional space. For a point B different from A but very close to A, the displacements will be slightly different, so the displacements are a function of the location of the point considered: $u(x,y)$, $v(x,y)$, $w(x,y)$ for the two-dimensional space of Figure 10.8.

Point A moves to A′ such that the displacements are $u(x,y)$ in the direction of x and $v(x,y)$ in the direction of y. Point B (Figure 10.8) is at a distance dx from A in the direction of x. Point B moves to B′ such that the displacements are $u(x + dx, y)$ and $v(x + dx, y)$. The displacement $u(x + dx, y)$ can be written as u(x,y) plus or minus a little bit. This little bit is expressed mathematically as $\frac{\partial u}{\partial x}dx$, which is the product of the partial derivative of u with respect to x times the distance dx. So:

$$u(x + dx, y) = u(x, y) + \frac{\partial u}{\partial x}dx \tag{10.16}$$

This equation can be understood by looking at the diagram of Figure 10.9. In the same way, v(x + dx, y) can be written as:

$$v(x + dx, y) = v(x, y) + \frac{\partial v}{\partial x}dx \tag{10.17}$$

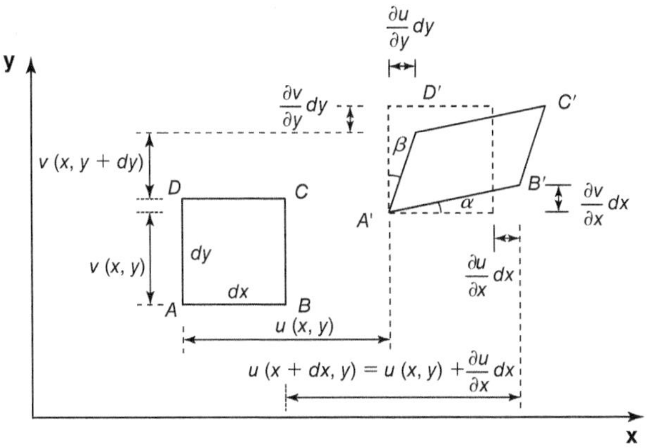

Figure 10.8 Illustration of displacements in a two-dimensional space.

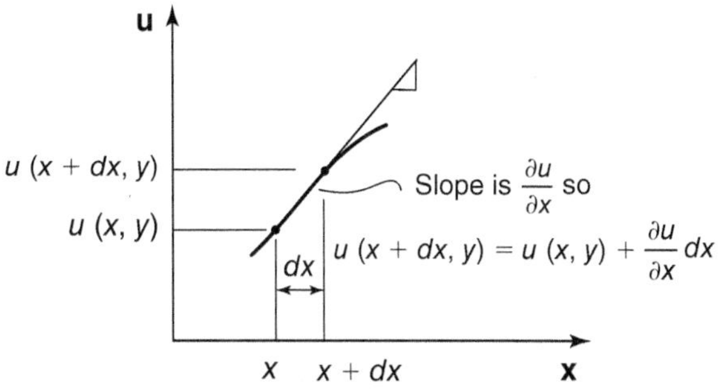

Figure 10.9 Visual illustration of equation 10.16.

Now consider point D on Figure 10.8. Point D is at a distance dy from A in the direction of y. Point D moves to D′ such that the displacements are $u(x, y + dy)$ and $v(x, y + dy)$. These displacements satisfy equations similar to 10.16 and 10.17, as follows:

$$u(x, y + dy) = u(x, y) + \frac{\partial u}{\partial y}dy \tag{10.18}$$

$$v(x, y + dy) = v(x, y) + \frac{\partial v}{\partial y}dy \tag{10.19}$$

10.9 NORMAL STRAIN, SHEAR STRAIN, AND STRAIN TENSOR

Strains are used to quantify the deformation of a material as a result of a loading process, a temperature change, a water content change, or some other change. Six strains are defined at one point: three normal strains and three shear strains. These six strains are defined from the knowledge of the three displacements (u, v, w) at a given point. Therefore, the six strains are not independent variables, and three strain relationships can be written linking the six strains to one another. *Normal strains* are used to quantify the change in length between two points. *Shear strains* are used to quantify the distortion of an angle.

Considering a point in a mass and an infinitesimal length in the x direction, the normal strain ε_{xx} at that point in the x direction is defined as the change in length of that infinitesimal length divided by the original length. The same definition applies for the normal strains in the y and z direction. More precisely, and referring to Figure 10.8, the normal strain ε_{xx} is defined as:

$$\varepsilon_{xx} = \frac{length\ A'B' - length\ AB}{length\ AB}$$
$$= \frac{dx + u(x + dx, y) - (dx + u(x, y))}{dx} = \frac{\partial u}{\partial x} \tag{10.20}$$

This equation assumes that the displacements are small and that the error in taking the length $A'B'$ equal to its projection on the x axis is very small. This is called the *small strain theory*. By the same reasoning in this theory, the two other normal strains are defined as:

$$\varepsilon_{yy} = \frac{\partial v}{\partial y} \tag{10.21}$$

$$\varepsilon_{zz} = \frac{\partial w}{\partial z} \tag{10.22}$$

Now consider the same point in the mass and two initially perpendicular directions x and y (DAB on Figure 10.8). In the deformed state, the right angle is deformed and becomes the angle formed by D′A′B′. The shear strain at point A is defined as one-half of the change in angle between DAB and D′A′B′ expressed in radians:

$$\varepsilon_{xy} = \frac{1}{2}(DAB - D'A'B') = \frac{1}{2}(\alpha + \beta)$$
$$= \frac{1}{2}(\tan\alpha + \tan\beta) = \frac{1}{2}\left(\frac{\partial v}{\partial x} + \frac{\partial u}{\partial y}\right) \tag{10.23}$$

This equation assumes that the displacements are small because the angles α and β in radians are taken to be equal to $\tan\alpha$ and $\tan\beta$ respectively, and that the projection of A′B′ on the x axis and the projection of A′D′ on the y axis are equal to dx and dy respectively. The other shear strains are then:

$$\varepsilon_{yz} = \frac{1}{2}\left(\frac{\partial w}{\partial y} + \frac{\partial v}{\partial z}\right) \tag{10.24}$$

$$\varepsilon_{zx} = \frac{1}{2}\left(\frac{\partial u}{\partial z} + \frac{\partial w}{\partial x}\right) \tag{10.25}$$

These six strains (Eqs. 10.20–10.25) form the strain tensor, which is a 3 × 3 matrix where the shear strains are repeated on either side of the diagonal. Mathematically, these six strains are defined from the knowledge of the three independent displacements; therefore, the six strains represent only three independent variables.

$$\varepsilon = \begin{bmatrix} \varepsilon_{xx} & \varepsilon_{xy} & \varepsilon_{xz} \\ \varepsilon_{yx} & \varepsilon_{yy} & \varepsilon_{yz} \\ \varepsilon_{zx} & \varepsilon_{zy} & \varepsilon_{zz} \end{bmatrix}$$
$$= \begin{bmatrix} \frac{\partial u}{\partial x} & \frac{1}{2}\left(\frac{\partial v}{\partial x} + \frac{\partial u}{\partial y}\right) & \frac{1}{2}\left(\frac{\partial w}{\partial x} + \frac{\partial u}{\partial z}\right) \\ \frac{1}{2}\left(\frac{\partial u}{\partial y} + \frac{\partial v}{\partial x}\right) & \frac{\partial v}{\partial y} & \frac{1}{2}\left(\frac{\partial w}{\partial y} + \frac{\partial v}{\partial z}\right) \\ \frac{1}{2}\left(\frac{\partial u}{\partial z} + \frac{\partial w}{\partial x}\right) & \frac{1}{2}\left(\frac{\partial v}{\partial z} + \frac{\partial w}{\partial y}\right) & \frac{\partial w}{\partial z} \end{bmatrix} \tag{10.26}$$

Note that there is a factor ½ in front of the shear strain. This is because that shear strain is an average of the shear strains in both directions. In engineering practice, the factor ½ is

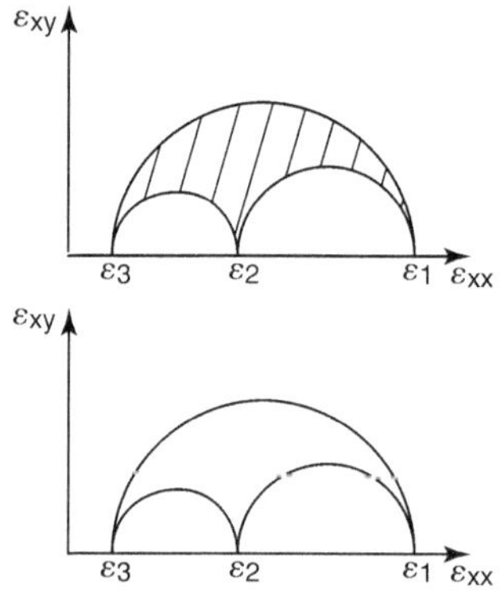

Figure 10.10 Mohr circle for strain.

not used and the engineering shear strains are defined as follows:

$$\gamma_{xy} = \frac{\partial v}{\partial x} + \frac{\partial u}{\partial y} \tag{10.27}$$

$$\gamma_{yz} = \frac{\partial w}{\partial y} + \frac{\partial v}{\partial z} \tag{10.28}$$

$$\gamma_{zx} = \frac{\partial u}{\partial z} + \frac{\partial w}{\partial x} \tag{10.29}$$

Note also that the same Mohr circle concepts apply to strains as apply to stresses. One can draw Mohr circles for strains on the shear strain vs. normal strains set of axes (Figure 10.10). Note too that the Mohr circle for strains applies to the ε values and not the γ values of shear strains.

10.10 CYLINDRICAL COORDINATES AND SPHERICAL COORDINATES

Sometimes the geometry of a problem makes it convenient to use cylindrical coordinates or even spherical coordinates to solve the problem. In cylindrical coordinates (Figure 10.11),

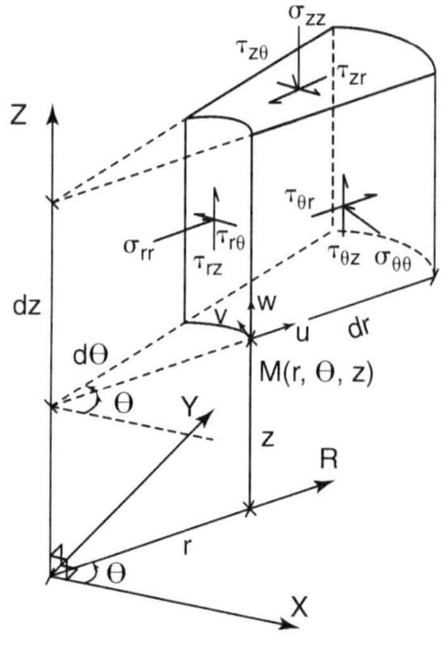

Figure 10.11 Stresses in cylindrical coordinates.

point M has coordinates r, θ, and z and the displacements of point M are u, v, and w in the directions of r, θ, and z respectively. The stresses are shown in Figure 10.11 and the stress tensor is:

$$\Sigma = \begin{bmatrix} \sigma_{rr} & \tau_{r\theta} & \tau_{rz} \\ \tau_{\theta r} & \sigma_{\theta\theta} & \tau_{\theta z} \\ \tau_{zr} & \tau_{z\theta} & \sigma_{zz} \end{bmatrix} \tag{10.30}$$

The strains definitions are:

$$\varepsilon_{rr} = \frac{\partial u}{\partial r} \tag{10.31}$$

$$\varepsilon_{\theta\theta} = \frac{u}{r} + \frac{1}{r}\frac{\partial v}{\partial \theta} \tag{10.32}$$

$$\varepsilon_{zz} = \frac{\partial w}{\partial z} \tag{10.33}$$

$$\gamma_{r\theta} = \frac{\partial v}{\partial r} + \frac{1}{r}\frac{\partial u}{\partial \theta} - \frac{v}{r} \tag{10.34}$$

$$\gamma_{\theta z} = \frac{1}{r}\frac{\partial w}{\partial \theta} + \frac{\partial v}{\partial z} \tag{10.35}$$

$$\gamma_{zr} = \frac{\partial u}{\partial z} + \frac{\partial w}{\partial r} \tag{10.36}$$

In spherical coordinates (Figure 10.12), point M has coordinates r, θ, and φ and the displacements are u, v, and w, in the directions of r, θ, and φ respectively. The stresses are shown in Figure 10.12 and the stress tensor is:

$$\Sigma = \begin{bmatrix} \sigma_{rr} & \tau_{r\theta} & \tau_{r\varphi} \\ \tau_{\theta r} & \sigma_{\theta\theta} & \tau_{\theta\varphi} \\ \tau_{\varphi r} & \tau_{\varphi\theta} & \sigma_{\varphi\varphi} \end{bmatrix} \tag{10.37}$$

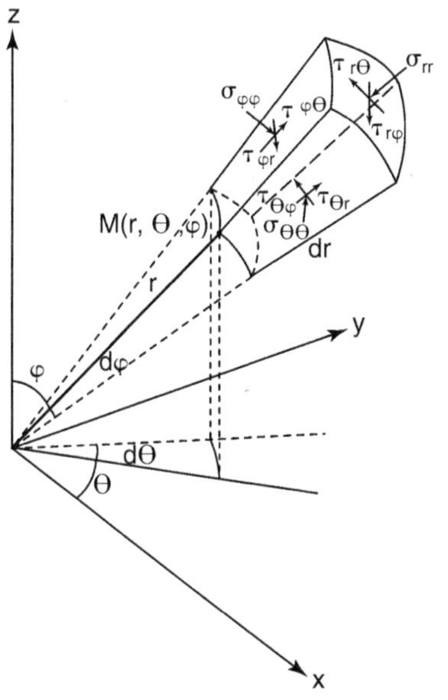

Figure 10.12 Stresses in spherical coordinates.

The strains definitions are:

$$\varepsilon_{rr} = \frac{\partial u}{\partial r} \tag{10.38}$$

$$\varepsilon_{\theta\theta} = \frac{u}{r} + \frac{1}{r}\frac{\partial v}{\partial \theta} \tag{10.39}$$

$$\varepsilon_{\varphi\varphi} = \frac{1}{r\sin\theta}\left(\frac{\partial w}{\partial \varphi} + u\sin\theta + v\cos\theta\right) \tag{10.40}$$

$$\gamma_{r\theta} = \frac{\partial v}{\partial r} + \frac{1}{r}\frac{\partial u}{\partial \theta} - \frac{v}{r} \tag{10.41}$$

$$\gamma_{\theta\varphi} = \frac{1}{r}\left(\frac{\partial w}{\partial \theta} + \frac{1}{\sin\theta}\frac{\partial v}{\partial \varphi} - w\cot\theta\right) \tag{10.42}$$

$$\gamma_{\varphi r} = \frac{1}{r\sin\theta}\frac{\partial u}{\partial \varphi} + \frac{\partial w}{\partial r} - \frac{w}{r} \tag{10.43}$$

10.11 STRESS-STRAIN CURVES

Stress-strain curves are often obtained when one tests a material in the laboratory or in the field. They usually relate one of the six stresses applied to an element of the material or to the mass to one of the six strains measured as a result of the stress applied. These stress-strain curves are very useful because they give fundamental soil properties that enter into the design process. As a result of this stress-strain curve relationship, one might be tempted to conclude that stresses and strains are intimately linked. However, there are exceptions to that intuitive statement. Take the example of the rails of high-speed trains. These rails have very few joints so that the very fast ride will be smooth. The rails change temperature during the daily temperature cycle; this temperature change would induce a change in length if such a length change were possible—but the anchors of the track do not permit such change and a stress develops because the strain is being suppressed. There is stress but no strain. Alternatively, consider a wire between two power-line poles. When the temperature increases, the wire gets longer but there is no change in stress. In this case there is strain but no stress. Nevertheless, in most cases stresses and strains are in fact intimately related.

10.12 STRESSES IN THE THREE SOIL PHASES

Concrete and steel are considered to be mono-phase materials (only one material). Soils, however, are three-phase materials, and stresses exist in each of the phases. The water can experience compression (also called *positive pore pressure*), or tension (also called *suction* or *negative pore pressure*). The air can also experience compression or tension. The shear stresses in the water and the air are neglected because they are very small compared to the shear stresses existing between the grains. The normal stress between the grains is very important because it has a significant influence on the shear strength and the compressibility of the soil. Note that failure in shear is the most common failure mechanism in soils.

10.13 EFFECTIVE STRESS (UNSATURATED SOILS)

Effective stress is a normal stress, and one of the most important parameters to know when dealing with soils. The effective stress equation gives the relationship between the various normal stresses that exist in the three phases. The derivation of this equation proceeds as follows. Consider a half space of soil in equilibrium and then within that half space consider an imaginary vertical cylinder. The top of the cylinder is the ground surface and the bottom of the cylinder is a generally horizontal plane that goes through the grain contacts and cuts through the voids (Figure 10.13). The external forces acting on that soil cylinder in the vertical direction are the total weight, including the grains, the water, and the air (acting downward); the vertical components of the contact forces between the grains on the bottom plane (acting upward); the vertical forces on the bottom plane corresponding to the water stress times the area of the water; and the vertical forces on the bottom plane corresponding to the air stress times the area of the air. The water area plus the air area plus the area of the contacts is equal to the total area. There are no vertical forces on the sides of the cylinder (shear forces) because there is no relative movement at that boundary.

Writing vertical equilibrium leads to the following equation:

$$F = \sum f_{ci} + \sum f_{wi} + \sum f_{ai} \tag{10.44}$$

where F is the weight of the soil mass plus any surcharge, and f_{ci}, f_{wi}, and f_{ai} are the vertical components of the forces between the grains, transmitted through the water, and transmitted through the air along the lower boundary of the free-body respectively. The forces f_{wi}, and f_{ai} are equal to:

$$f_{wi} = u_{wi} a_{wi} \tag{10.45}$$

$$f_{ai} = u_{ai} a_{ai} \tag{10.46}$$

where u_{wi} and u_{ai} are the water stress and air stress respectively, and a_{wi} and a_{ai} are the horizontal projections of the areas of water and air respectively on the bottom surface

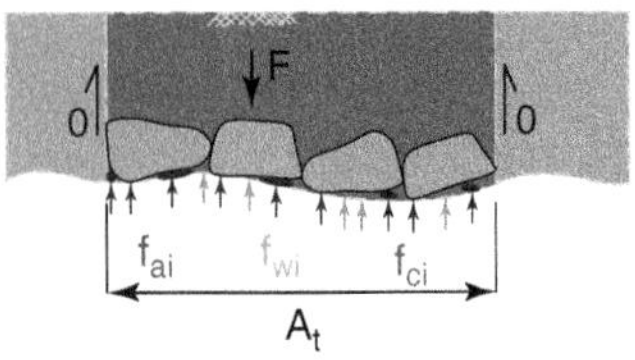

Figure 10.13 Free-body diagram for derivation of effective stress equation.

of the free body. It is further assumed that u_{wi} and u_{ai} are constant along the bottom surface and equal to u_w and u_a respectively. Therefore, equation 10.44 becomes:

$$F = \sum f_{ci} + u_w \sum a_{wi} + u_a \sum a_{ai} \tag{10.47}$$

Now divide both sides of the equation by the total horizontal projected area A_t of the bottom of the cylinder:

$$\frac{F}{A_t} = \frac{\sum f_{ci}}{A_t} + u_w \frac{\sum a_{wi}}{A_t} + u_a \frac{\sum a_{ai}}{A_t} \tag{10.48}$$

On the left-hand side, we get a quantity that is the total weight divided by the total area; this is called the *total (normal) stress* σ_t. On the right-hand side, the first term is the sum of the vertical components of the contact forces divided by the total area; this is the *effective (normal) stress* σ'. The second term is the water stress times the water area divided by the total area. This ratio of areas is lower than or equal to 1 and is called α. The third term is the air stress times the air area divided by the total area. This ratio of areas is lower than or equal to 1 and is called β. The total area can be written as:

$$A_t = \sum a_{ci} + \sum a_{wi} + \sum a_{ai} \tag{10.49}$$

Then

$$1 = \frac{\sum a_{ci}}{A_t} + \frac{\sum a_{wi}}{A_t} + \frac{\sum a_{ai}}{A_t} \tag{10.50}$$

And

$$1 = \frac{\sum a_{ci}}{A_t} + \alpha + \beta \tag{10.51}$$

where a_{ci} is the contact areas between particles. If it is assumed that Σa_{ci} is negligible compared to Σa_{wi} and Σa_{ai}, then:

$$\alpha + \beta = 1 \tag{10.52}$$

So, in summary:

$$\sigma = \sigma' + \alpha \mathrm{u_w} + \beta \mathrm{u_a} \tag{10.53}$$

Or

$$\sigma' = \sigma - \alpha \mathrm{u_w} - \beta \mathrm{u_a} \tag{10.54}$$

where $\sigma' = \frac{\sum f_{ci}}{A_t}$ is the effective stress, $\sigma = \frac{F}{A_t}$ is the total stress, α and β are the water and air area ratios ($\alpha + \beta = 1$), and u_w and u_a are the water stress and the air stress respectively.

Note that σ' is not the contact stress σ_c, which is the sum of the vertical components of the contact forces divided by the contact areas. This real stress σ_c is not used in geotechnical engineering because it is very difficult to know the area of the contacts. The contact stress σ_c is much higher than the effective stress σ'.

Note also that the effective shear stress is equal to the total shear stress, because the shear stress in the water τ_w and the shear stress in the air τ_a are neglected. The stresses τ_w and τ_a are not zero, however, and are responsible in part for the process of erosion (τ_w) and the drag force on airplanes (τ_a). Nevertheless, their order of magnitude is in N/m^2 rather than kN/m^2 as in the shear strength of soils.

$$\tau = \tau' \tag{10.55}$$

Again, the shear stress calculated is the shear force at the particle contacts divided by the total area rather than the contact area; therefore, it does not represent the shear stress at the contacts, but instead a much lower, well-defined value.

10.14 EFFECTIVE STRESS (SATURATED SOILS)

If the soil is saturated, Eq. 10.53 is simpler, as there is no air. The left-hand side is unchanged and equal to the total (normal) stress. The first term on the right-hand side is unchanged and equal to the effective (normal) stress. The second term reduces to u_w because the α value becomes equal to one, and the third term vanishes because there is no air in the soil:

$$\sigma' = \sigma - u_w \tag{10.56}$$

In unsaturated soils, the water stress can be significantly negative (high water tension); in this case the water stress can contribute significantly to increasing the effective stress between particles. For saturated soils with water in compression, that water stress detracts from the effective stress between particles. As in the case of unsaturated soils, however, the effective shear stress is the same as the total shear stress and Eq. 10.55 is equally valid for saturated soils and for unsaturated soils.

10.15 AREA RATIO FACTORS α AND β

In nature, the degree of saturation is either high enough that the air is occluded (air bubbles surrounded by water) or low enough that there is a continuous air path to the surface. The transition from occluded air to continuous path occurs at a degree of saturation approximately equal to 85%. If the air is occluded, the air stress u_a can be taken as being equal to the water stress u_w, as there is equilibrium at the bubble boundary between the air and the water. In this case, Eq. 10.54 reduces to Eq. 10.56 and the soil behaves as if it were saturated, except that the water phase is much more compressible due to the air bubbles. This increase in water compressibility can have a beneficial effect, as in reducing the potential for liquefaction. If the air phase is continuous, then the air path to the atmospheric pressure ensures that the air stress is zero and the term involving the air stress drops out. In this case, the effective stress equation expresses that the total stress is equal to the sum of the effective stress plus the product of the water stress (negative) by the ratio of the water area divided by the total area:

$$\sigma' = \sigma - \alpha u_w \tag{10.57}$$

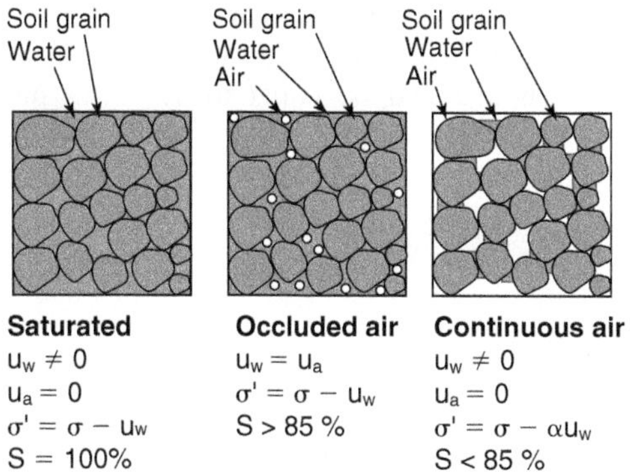

Figure 10.14 Effective stress equation for various common situations.

Therefore, for most common cases, the general effective (normal) stress equation is Eq. 10.57. Figure 10.14 summarizes these situations.

Note that in the case of occluded air, there can be a difference between u_w and u_a because of the contractile skin. Indeed, that membrane allows for a difference in pressure that can be obtained by writing equilibrium of the free-body diagram of half the bubble (Figure 10.15):

$$u_a \pi \frac{D^2}{4} = u_w \pi \frac{D^2}{4} + \pi D T \tag{10.58}$$

$$u_a - u_w = \frac{4T}{D} \tag{10.59}$$

Therefore, the expression of the effective stress for the case of the occluded air is an approximation. This approximation is reasonable, as the value of β is much smaller than the value of α in this case.

Because the effective stress has such a fundamental impact on the behavior of soils, it is very important to be able to evaluate the coefficient α in Eq. 10.57. This coefficient was first proposed by Bishop in the 1960s as the factor χ. This factor has been correlated with the degree of saturation S.

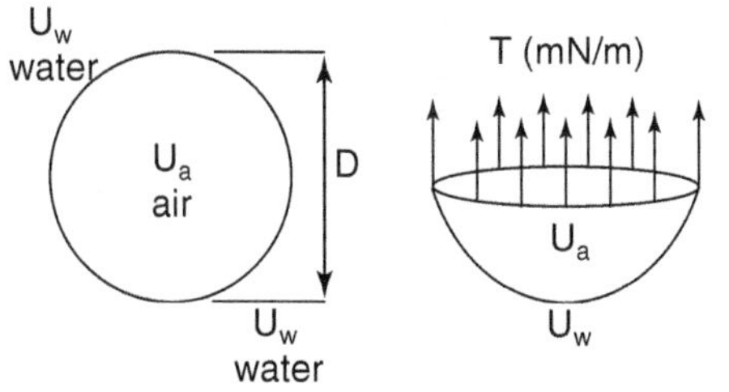

Figure 10.15 Pressure difference across an air-bubble boundary.

This makes some sense, because when the soil has no water (S = 0), α should also be zero, and when the soil is saturated (S = 1), α should also be equal to 1. Furthermore, if it is assumed that the area of the contacts A_c is negligible compared to the area of the voids A_v, then the definition of α becomes:

$$\alpha = \frac{A_w}{A_t} = \frac{A_w}{A_v} \tag{10.60}$$

Recall that the degree of saturation is defined as:

$$S = \frac{V_w}{V_v} \tag{10.61}$$

The analogy is tempting, but it must be said that the ratio of areas A_w/A_v is not likely equal to the ratio of volumes V_w/V_v, because the plane that cuts through the contacts in Figure 10.13 does not represent the general situation in the soil volume. As a result, there is quite a bit of scatter in the correlation between α and S (Figure 10.16).

Khalili and Khabbaz (1998) proposed a better relationship to predict α (Figure 10.17):

$$\alpha = \left(\frac{(u_a - u_w)}{(u_a - u_w)_{ae}}\right)^{-0.55} \tag{10.62}$$

which can be simplified without much loss of accuracy when u_a is zero as:

$$\alpha = \sqrt{\frac{u_{wae}}{u_w}} \tag{10.63}$$

where u_a is the air stress, u_w is the water stress, and $(u_a - u_w)_{ae}$ refers to the difference between u_a and u_w at the air entry value. At the beginning of the drying process of a saturated sample of soil, the water tension u_w increases (becomes more negative) as the water is evaporating out of the soil and into the surrounding air, but the soil remains saturated. As the drying continues, u_w continues to become more negative and gets to a point where air first enters the pores. This value of the water tension is called the *air entry value* u_{wae}. As the drying continues, the water tension continues to become more negative. The area ratio for air is β, and because $\alpha + \beta$ is equal to 1 ($A_c \sim 0$), once α is known so is β.

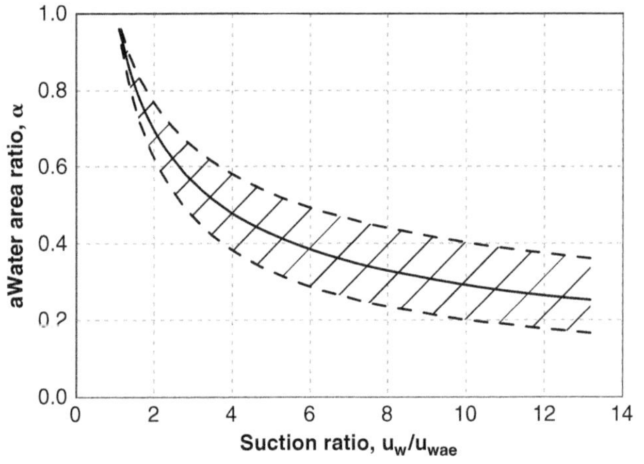

Figure 10.17 Water area ratio α vs. suction ratio. (After Khalili and Khabbaz 1998. Courtesy of Nasser Khalili)

10.16 WATER STRESS PROFILES

The water normal stress can be positive (pore pressure, compression) or negative (suction, tension). In the field, the groundwater level (GWL) is found at some depth below the ground surface (Figure 10.18).

In some cases that depth is very large (deserts); in others it is very shallow (regions close to oceans, lakes, or rivers). At the GWL, the water stress is zero. Below the GWL, the water is in compression (pore pressure) and, in the most common case, the water stress profile shows a linear increase with

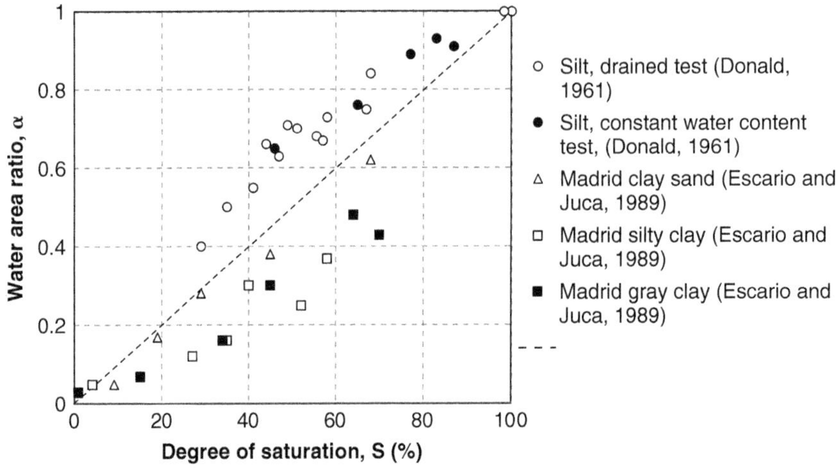

Figure 10.16 Water area ratio α vs. degree of saturation S. (After Lu and Likos 2004)

Figure 10.18 Groundwater level (GWL) and zones above the GWL. (Adapted from a photo of Art Koenig, reproduced with permission)

depth (hydrostatic pressure) and can be calculated as $\gamma_w z$ where γ_w is the unit weight of water and z is the depth below the GWL. Sometimes the water stress profile below the GWL is complicated by the presence of perched aquifers (water bodies sandwiched between dry soil) or artesian conditions (water body connected to a pressure higher than the local hydrostatic pressure). Figure 10.19 shows examples of such conditions.

Above the GWL, the water is in tension (suction). In the zone above the GWL and deep enough to be unaffected by the weather at the ground surface, the water stress is linear and given by $(-\gamma_w z)$ where z is the absolute value of the vertical distance above the water table. In the zone above the GWL and close enough to the ground surface that the weather can influence the water stress profile by evapotranspiration and rainfall (generally a few meters), the water stress profile becomes curved to reach an equilibrium between the weather and the soil (Figure 10.20). This part of the water stress profile is very difficult to calculate and varies daily with the weather.

10.17 WATER TENSION AND SUCTION

Water tension is the tension in the water expressed in kN/m^2. *Suction* is the potential that the water has to achieve a certain water tension; it is also expressed in kN/m^2. This suction potential is not always realized. If the suction potential is fully realized, the suction is equal to the water tension. If the suction is not fully realized, the suction is higher than the water tension. It is a bit like standing on top of a building but not jumping: you have potential energy, but you are not transforming it into velocity because you are not jumping. Later we will discuss cases in which the suction is not transformed into water tension. Although the suction is important, the water stress is the one that enters into most calculations. Note that suction is often defined as the difference between the air stress and the water stress $(u_a - u_w)$. Because the air stress is often zero in the field

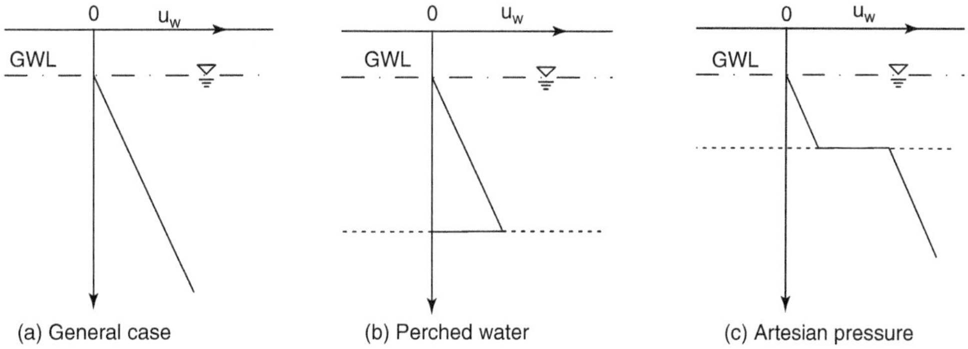

Figure 10.19 Examples of water stress profiles below the groundwater level.

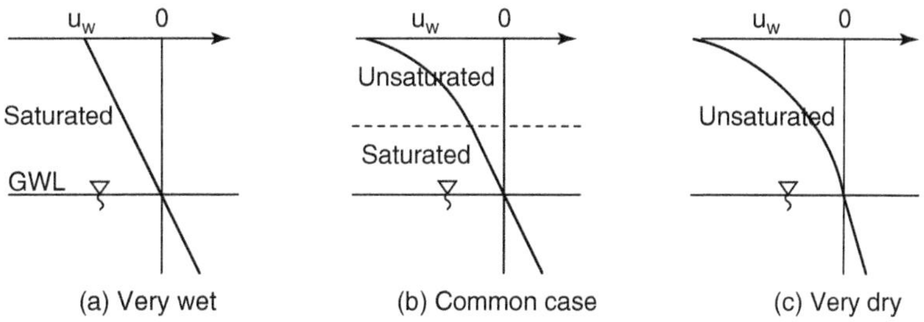

Figure 10.20 Examples of water stress profiles above the groundwater level.

(continuous air voids), suction is defined here as u_w. Note further that with $u_a - u_w$, suction is positive, whereas with u_w, suction is negative. Suction and water tension will always be negative in the rest of this book, as compression has been chosen as the positive sign convention for stresses.

The water tension and suction come from two different sources: attraction of water to the minerals in the soil particles and attraction of distilled water to salty water. The first one is called *matric suction*; the second one *osmotic suction*.

10.17.1 Matric Suction

Matric suction is due to the attraction between water molecules and the minerals in soil particles. If the mineral is silica, the phenomenon is called *capillary action*. The attraction between water and silica generates a force of 73 mN/m. Other minerals, such as smectite ($Al_2Si_4O_{10}(OH)_2$), can generate much higher attraction forces and therefore much higher water tension. Let's discuss capillary attraction first. The force of 73 mN/m is given per unit of length because it exists along the contact line of the meniscus interface between the water, the air, and the silica. Recall that one Newton is about the weight of a small apple, so 73 mN is a very small force, yet it is responsible for some major phenomena when dealing with very small scales. For example, when a very-small-diameter glass (silica) tube open at both ends is placed in water, that force lifts the water in the tube like one would pull up a sock. If the glass tube is small enough, the water can rise more than 10 m in the tube. Note that if the tube were made of a different mineral, the water would not rise to the same level. Also, if the tube were made of glass, but instead of water you had mercury in the container, the mercury would actually go down in the small tube rather than up, because there is a basic repulsion between mercury and silica.

The water rising in the silica tube does so up to a height where the volume of water lifted in the tube has a weight equal to the vertical component of the attraction force at the top of the column times the contact length of the meniscus. Equating the weight of the column of water to the vertical component of the attraction force leads to the height of the water column or capillary rise (Figure 10.21):

$$h_c \frac{\pi d^2}{4} \gamma_w = \pi d T \cos \alpha \tag{10.64}$$

Therefore,

$$h_c = \frac{4T \cos \alpha}{\gamma_w d} \tag{10.65}$$

It is clear that the capillary rise depends on the diameter of the tube; the capillary rise will be high in small-diameter tubes and small in larger-diameter tubes. If the tube has a diameter equal to the size of clay particles—say, 0.001 mm—Eq. 10.65 gives a height of capillary rise equal to 29.2 m (height of a 10-story building). The continuous voids in a soil play the role of the tiny glass tube because, like glass, many soil particles are made of silica. Continuous clay voids are similar to tiny tubes and the water can saturate the clay high above the groundwater level (15 m or more). In sands, the height to which the water can rise is more limited.

Let's study the water stress profile in the capillary tube (Figure 10.21). Below the water level, in the big container, the water is in compression and the water stress is positive. Above that level, in the tiny glass tube, the water is in tension because the water is pulled up into the tube by the force $\pi dT\cos\alpha$. The water tension increases (becomes more negative) linearly with the height in the tube, as shown in Figure 10.21. At the top of the column, the water tension is maximum and equal to $-h_c\gamma_w$. Yet in the air immediately above the water level in the small tube, the pressure is atmospheric or zero gage pressure. It is not possible for such a discontinuity to exist between two fluids unless there is a membrane separating the water from the air: this membrane is the *contractile skin*. It is similar to a car tire: the pressure in the tire is much higher than outside the tire, and this is made possible by the membrane represented by the tire. We will discuss the contractile skin a bit later.

Consider now two soil particles in the form of spheres (Figure 10.22). The soil is allowed to dry and the water

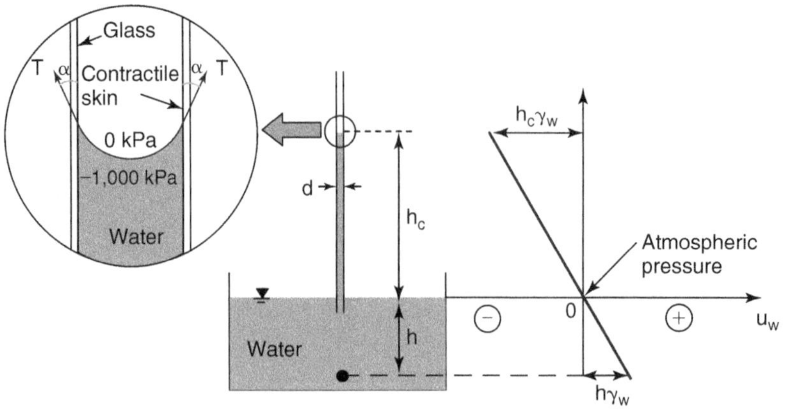

Figure 10.21 Capillary tube experiment.

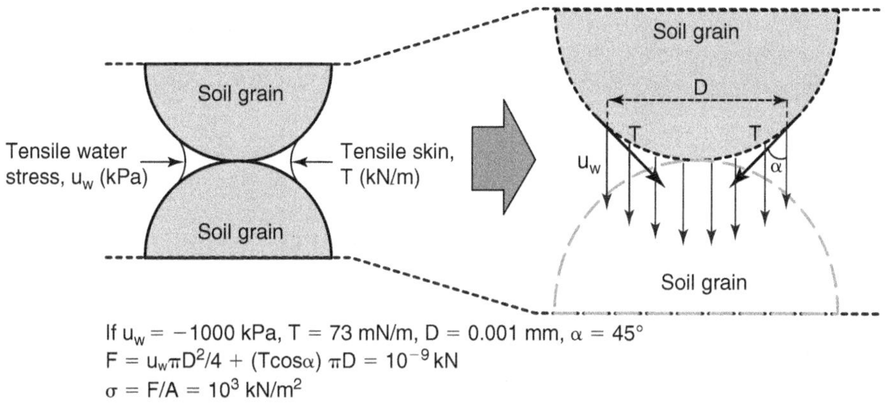

Figure 10.22 Water tension at the contact between two spherical particles.

between the particles evaporates. When the water is almost gone, the water is only found around the contact between the two particles (Figure 10.22). The water is in tension and the air is at atmospheric pressure. The contractile skin allows large stress difference to exist between the two fluids. Now let's calculate the force at the contact. We draw a free-body diagram of the upper particle and show the forces imparted by the water and the contractile skin on the particle. The water is under a tension stress u_w, so the water pulls on the particles above and below the contact area A with a force $u_w A$. The contractile skin is also in tension and pulls on the particle at an angle α. The calculations are shown in Figure 10.22. The force is a compression force equal to 10^{-6} N. Remember that 1 N is about the weight of a small apple, so the force is extremely small—yet the stress is very large (1000 kPa). These stresses develop when the soil dries and are the reasons why dry soils are a lot harder than saturated soils.

The preceding discussion focused on the case of water attraction to silica and the water tension that can be generated due to this phenomenon. Some clay minerals, such as smectite ($Al_2Si_4O_{10}(OH)_2$), can generate much higher attraction forces and therefore water tension which can reach 100,000 kPa or even 1,000,000 kPa (Figure 10.23). These water tension values correspond to soils that are very dry yet have a little bit of water between particles. It is not clear in these cases whether the water is still in liquid form, or in viscous form, or possibly approaching solid form.

10.17.2 Contractile Skin

The membrane called the *contractile skin* exists at the interface between the water and the air. The existence of this membrane is rooted in the Van der Waals forces, which are elementary attractive forces between molecules. In the water, these forces act in all three directions and give water its tensile strength. This tensile strength can be measured by placing water in a cylinder and pulling on the piston until the water breaks in tension, at about 20 MPa. This is remarkably large, approaching the strength of concrete in compression.

At the interface between the water and the air, the molecules of water attract those that are below the interface but are unable to attract water molecules above the surface, as there are none available. Instead, the water molecules enhance their attraction in the horizontal direction, thereby creating a membrane. Figure 10.24 shows a water strider resting on that contractile skin. (So it is possible to walk on water, at least for the water strider.)

This water membrane is able to generate 73 mN of force for every meter of linear contact with silica. This represents

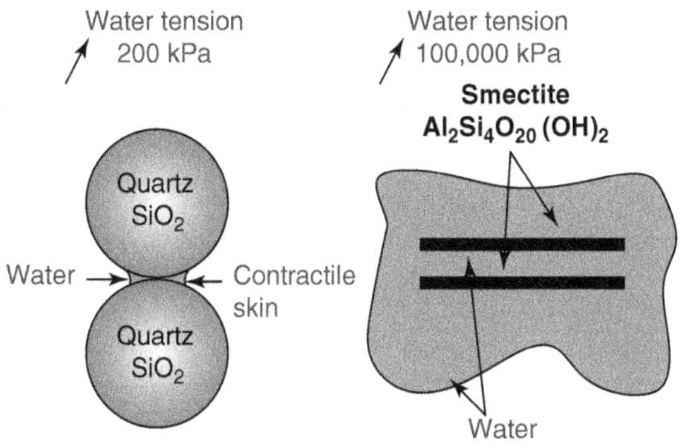

Figure 10.23 Water tension between soil particles.

Figure 10.24 Water strider resting on contractile skin.

a very small force, but the membrane is extremely thin. Its thickness is estimated at 20 to 30 nanometers; therefore, the stress in the contractile skin under 73 mN/m is larger than 20 MPa.

10.17.3 Osmotic Suction

There is a second reason why water can go into tension in a soil: osmotic suction. Osmotic suction is due to the basic attraction that exists between water and salt. The phenomenon can be explained as follows. Imagine a container with two sides (Figure 10.25). On one side is distilled water, and on the other side is water with salt in it. Imagine that there is an imaginary screen separating the two sides that allows water molecules to travel across it but not salt molecules. This imaginary screen therefore prevents the two water bodies from mixing. In this experiment, the distilled water will be attracted to the salt water and therefore a difference in elevation will be generated, as shown in Figure 10.25. This difference is a suction potential called the osmotic suction. To help you remember that the distilled water goes towards the salt water, just remember that when you eat salty food, you get thirsty!

Osmotic suction depends on the salt concentration in the water on the right side and on the type of salt in that water. Osmotic suction exists in a soil if the soil contains dissolved salts. This suction exists as a potential and is realized into a water tension if there is a change in salt concentration between two locations. This can happen when a sprinkler system is installed in the backyard of a home. In the majority of real situations, the osmotic suction is much smaller than the matric suction. The sum of the matric suction plus the osmotic suction is the total suction. Figure 10.26 shows values of total suction or water tension for a range of conditions.

If the salt concentration is high, as would be the case in a prepared solution, the osmotic suction can be very high. Table 10.1 shows the values of osmotic suction associated with various concentrations and various salt types. Note that osmotic suction exists in saturated soils as well as in unsaturated soils, as it is related only to the chemistry of the pore fluid.

10.17.4 Relationship between Total Suction and Relative Humidity

If you place water at the bottom of a container with air above it and then you close the container, the humidity of the air in the container will increase or decrease until it comes to an equilibrium. This equilibrium depends on the pressure and temperature in the container. At atmospheric pressure and at a temperature of 25°C, dry air consists of nitrogen (~78% by volume), oxygen (~21% by volume), and a few other

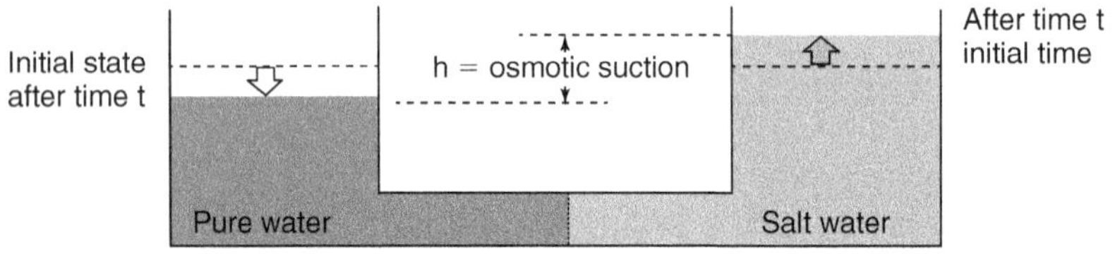

Figure 10.25 Osmotic suction experiment.

Water state	Examples	Water stress pF	cm	kPa	Degree of saturation	Water content	Swell	Shrink
Tension	Oven dry	7	-10^7	-10^6	0	0	Yes	No
Tension	Air dry	6	-10^6	-10^5			Yes	No
Tension	Shrinkage limit	4	-10^4	-10^3	Near 100 %	8 to 15 %	Yes	No ↕ Yes
Tension	Swell limit	2	-10^2	-10^1		25 to 50 %	Yes ↕ No	Yes
		0	0	0	100 %		No	Yes
Compression	Large river		10^3	10^2			No	Yes
Compression	Deepest offshore platforms		10^5	10^4			No	Yes
Compression	Bottom of deepest ocean		10^9	10^8			No	Yes

Figure 10.26 Range of water tension and water compression for various conditions.

Table 10.1 Osmotic Suction in kPa of Some Salt Solutions at 25°C

	Osmotic Suction in kPa at 25°C						
Molality (mol/kg)	N_aCl	KCl	NH_4Cl	Na_2SO_4	$CaCl_2$	$Na_2S_2O_3$	$MgCl_2$
0.001	5	5	5	7	7	7	7
0.002	10	10	10	14	14	14	14
0.005	24	24	24	34	34	34	35
0.010	48	48	48	67	67	67	68
0.020	95	95	95	129	132	130	133
0.050	234	233	233	306	320	310	324
0.100	463	460	460	585	633	597	643
0.200	916	905	905	1115	1274	1148	1303
0.300	1370	1348	1348	1620	1946	1682	2000
0.400	1824	1789	1789	2108	2652	2206	2739
0.500	2283	2231	2231	2582	3396	2722	3523
0.600	2746	2674	2671	3045	4181	3234	4357
0.700	3214	3116	3113	3498	5008	3744	5244
0.800	3685	3562	3558	3944	5880	4254	6186
0.900	4159	4007	4002	4384	6799	4767	7187
1.000	4641	4452	4447	4820	7767	5285	8249
1.200	5616	5354	5343	N/A	N/A	N/A	N/A
1.400	6615	6261	6247	N/A	N/A	N/A	N/A
1.500	N/A	N/A	N/A	6998	13391	7994	14554
1.600	7631	7179	7155	N/A	N/A	N/A	N/A
1.800	8683	8104	8076	N/A	N/A	N/A	N/A
2.000	9757	9043	9003	9306	20457	11021	22682
2.500	12556	11440	11366	11901	29115	14489	32776

*All suction values are in kPa.
(After Bulut et al. 2001)

gasses. If such a dry air is in the container, there is plenty of room for water molecules to become part of the air, thereby increasing the relative humidity of the air. Part of the liquid water at the bottom of the container will become vaporized, and join the air phase by fitting vaporized water molecules between the molecules of nitrogen and oxygen. This process will continue until an equilibrium is reached.

Each gas component in the air has a partial pressure, and the partial pressures add up to the total pressure, according to the ideal gas law:

$$p_{air} = p_{nitrogen} + p_{oxygen} + p_{water} + \cdots \quad (10.66)$$

At a certain relative humidity, the air has a corresponding partial water vapor pressure p_{water}. At 100% relative humidity, the partial water vapor pressure p_{water} equals the saturated water vapor pressure $p_{water,sat}$. This pressure is 3.17 kPa for conditions of atmospheric pressure (101.3 kPa) and a temperature of 25°C. The general equation for the saturated partial vapor pressure of water in air $p_{water,sat}$ at atmospheric pressure for different temperatures is (Tetens 1930):

$$p_{water,sat}(\text{kPa}) = 0.611\, e^{\left(17.27 \frac{T(^oC)}{T(^oC)+237.2}\right)} \quad (10.67)$$

where T is the temperature in degree Celsius. The relative humidity of the air is defined as the ratio:

$$R_H = \frac{p_{water}}{p_{water,sat}} \quad (10.68)$$

The relationship between the relative humidity R_H of the air in the void of an unsaturated soil and the suction potential ψ is given by Kelvin's equation (Fredlund and Rahardjo 1993; Lu and Likos 2004):

$$\Psi = \frac{\rho_w RT}{M} Ln\, R_H \quad (10.69)$$

where ψ is the suction potential in Pa, ρ_w is the mass density of the water (1000 kg/m^3), M is the molecular weight of water

(0.01802 kg/mol), T is the absolute temperature in Kelvin, R is the universal gas constant (8.314 N m/mol K), and R_H is the relative humidity expressed as a ratio rather than a percent. This suction potential in the void of the unsaturated soil can develop into a water tension, in which case:

$$u_w = \frac{\rho_w RT}{M} Ln\, R_H \tag{10.70}$$

At 20°C and given the same constants used earlier, the relationship is:

$$u_w(\text{kPa}) = 135000\, Ln R_H \tag{10.71}$$

where R_H is taken as a fraction. This equation is shown in Figure 10.27. It indicates, among other interesting observations, that a humidity room at 95% relative humidity has a water tension potential of almost 7000 kPa and therefore is a drying room.

10.17.5 Trees

Water is drawn up to the top of trees through suction. Osmotic suction in the tree is due to the difference in mineral concentration of the water in the tree and of the water in the soil. Capillary suction is due to the very small size of the tiny tubes (*xylem conduits*) that exist through the stem or tree trunk; water is attracted to the walls of the xylem conduits much like water is attracted to the glass (silica) wall of a capillary tube. In trees, suction or water tension can reach 2000 kPa. Evaporation takes place from the leaf surfaces and a continuous flow of water is generated in this fashion. This flow can reach 1 m^3 per day.

The tree absorbs carbon dioxide (CO_2) from the air and pumps water (H_2O) from the ground. It then uses the energy from the sun (photosynthesis) to combine the carbon dioxide with the water to make sugar ($C_6H_{12}O_6$) and release oxygen (O_2). Sugar is the essential basis for all plant growth. Trees and plants in general are extremely important to humankind because they absorb what we exhale (CO_2) and produce what we inhale (O_2).

10.18 PRECISION ON WATER CONTENT AND WATER TENSION

Water tension is more complicated to measure than water content. Water content also typically varies much less than water tension. A typical range of water content variation is 5 to 50%, whereas the typical range for water tension is −10 to −1,000,000 kPa. In an experiment conducted by Garner (2002, unpublished), three samples were sent to eight laboratories in Texas requesting that the water content and the suction be measured. Most laboratories used the filter paper method for the suction determination. The results were collected and an error band was created for each sample. The results are shown in Figure 10.28. They confirm that the arithmetic value of the suction varies a lot more than the water content. They also indicate that the error band for identically prepared samples is much larger for the determination of suction than for water content. If the log of the suction is used instead of the arithmetic value, then the error band of log(suction) approaches the error band of water content (Figure 10.28).

10.19 STRESS PROFILE AT REST IN UNSATURATED SOILS

The total vertical stress at rest σ_{ov} at any depth z in a uniform soil is equal to the total unit weight of the soil γ_t times the depth z:

$$\sigma_{ov} = \gamma_t\, z \tag{10.72}$$

If the soil above the depth z is made of n layers, the total vertical stress at rest σ_{ov} at depth z is:

$$\sigma_{ov} = \sum_{i=1}^{n} \gamma_{ti} h_i \tag{10.73}$$

where γ_{ti} is the total unit weight of layer i and h_i is the thickness of layer i. Note that if there is water above the

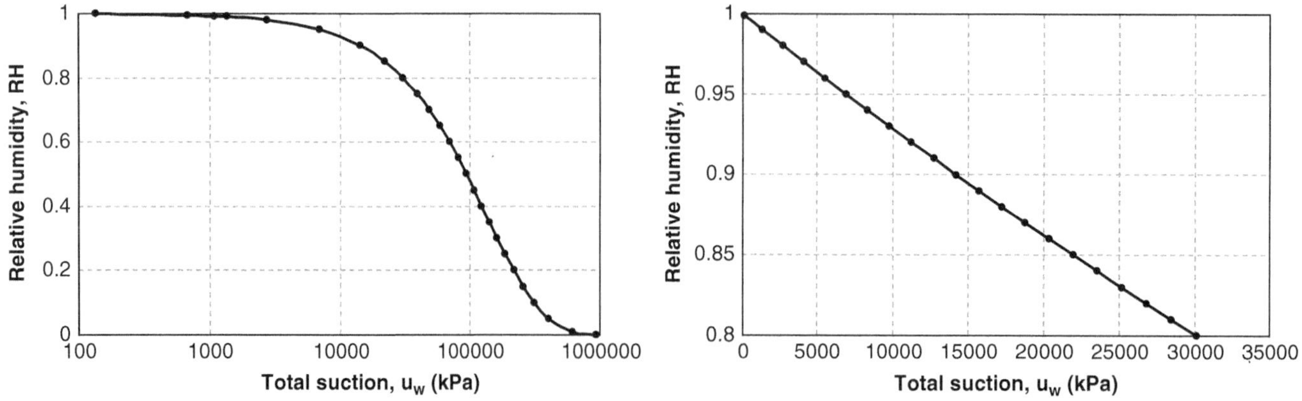

Figure 10.27 Water tension vs. relative humidity: *(a)* Relative humidity 0–100%. *(b)* Relative humidity 80-100%.

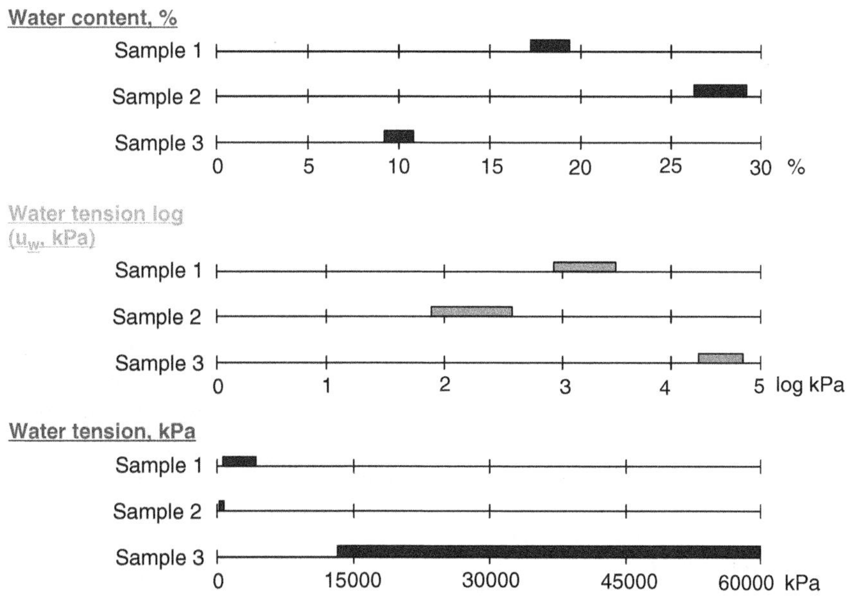

Figure 10.28 Error bands for suction and water content determination by eight different laboratories in Texas for three identically prepared samples. (After Garner 2002)

ground surface, the water must be included as a layer to calculate the total vertical stress. This is the case with a river, a lake, or an ocean. At the bottom of the deep oceans, the total vertical stress is very large and compresses any object tremendously. For example a Styrofoam coffee cup going to 3000 m of water depth comes back the size of a thimble.

Below the groundwater level, the water stress at rest u_{wo} is calculated under normal circumstances as:

$$u_{wo} = \gamma_w z_w \tag{10.74}$$

where γ_w is the unit weight of water and z_w is the depth below the GWL. Note that this water stress acts equally in all directions (hydrostatic), as it is assumed that water has no shear strength. This stress is a compressive stress. If an artesian condition exists, then information about the water stress in the artesian layer must be known or inferred from the global aquifer analysis. If a perched GWL condition exists, then information must be gathered where the groundwater layer ends.

Above the GWL, in the zone saturated by capillary action, the water stress u_w is calculated as:

$$u_{wo} = -\gamma_w z_w \tag{10.75}$$

where z_w is positive and represents the vertical distance above the GWL.

The water stress in this case is a tensile stress. Close to the surface, the water tension no longer exhibits a linear profile (Figure10.29). In that zone there is a power struggle between the soil particle minerals, which tend to attract the water, and the low relative humidity in the soil pores caused by the sun, which tends to draw the water away from the particles. The water is pulled hard in both directions, so high tensile stresses develop. Quantifying the variation of u_w with depth within that region requires advanced computations and depends on many factors, including rainfall, wind speed, solar radiation, temperature, soil hydraulic conductivity, extent of the cracks in the soil, and so on. Such computations are beyond the scope of this book.

Once the total vertical stress at rest σ_{ov} is known, and once the water stress at rest u_{wo} is known, the vertical effective stress at rest σ'_{ov} is calculated as:

$$\sigma'_{ov} = \sigma_{ov} - \alpha\, u_{wo} \tag{10.76}$$

where α is the water area ratio estimated as the degree of saturation or obtained from Eq. 10.63.

One of the important initial steps in solving a geotechnical problem is to prepare the profile of vertical stresses at rest for the site. This is done in the following steps:

1. Identify the layers and their thicknesses for the deposit considered.
2. Determine the total unit weight of each layer.
3. Determine the location of the GWL and any irregularity associated with the water regime (artesian pressure, perched water table).
4. Identify the points of discontinuity versus depth. These points include boundaries between two layers and depth to the GWL.
5. Calculate the total vertical stress at rest, σ_{ov}, at each discontinuity using Eq. 10.73.
6. Calculate the water stress at rest, u_w, at each discontinuity using Eq. 10.74 below the GWL and Eq. 10.75 above the GWL).

7. Calculate the water area ratio α at each discontinuity, estimated as the degree of saturation or by using Eq. 10.63 (α will be 1 under the GWL and in the zone saturated by capillary action).
8. Calculate the effective vertical stress at rest, σ'_{ov}, using Eq. 10.76.
9. Plot the values of σ'_{ov} on a graph at the depths corresponding to the discontinuities and join these points by straight lines.

Figure 10.29 is an illustration of this step-by-step procedure under the following conditions:

1. The soil is uniform.
2. The total unit weight of the soil is equal to 20 kN/m^3, and the unit weight of water is taken as 10 kN/m^3.
3. The GWL is at a depth of 5 m.
4. The points of discontinuity are the bottom of the profile (z = 7 m), the GWL (z = 5 m), the top of the capillary zone (z = 3 m), and the ground surface (z = 0 m).
5. The total vertical stress at rest at the bottom of the profile is equal to $\sigma_{ov} = 20 \times 7 = 140$ kN/m^2. At the top of the profile, it is $\sigma_{ov} = 0$. Because there is no discontinuity in total unit weight between these two discontinuities, the profile is a straight line between the two values.
6. The water stress at rest at the bottom of the profile is equal to $u_{wo} = 2 \times 10 = 20$ kN/m^2. At the GWL, the water stress u_{wo} is zero. At the top of the capillary zone, the water stress is $u_{wo} = -2 \times 10 = -20$ kN/m^2. The profile of water stress in the unsaturated zone above the top of the capillary zone is estimated as shown in Figure 10.29.
7. The water area ratio is equal to 1 in the zone where the soil is saturated. In the unsaturated zone, a linear decrease of α from 1 at the top of the capillary zone to 0 at the ground surface is assumed in this case.
8. The effective stress is calculated according to Eq. 10.76. For the point at the bottom of the profile, $\sigma'_{ov} = 140 - 20 = 120$ kN/m^2. For the point at the GWL, it is $\sigma'_{ov} = 100 - 0 = 100$ kN/m^2. For the point at the top of the saturated capillary zone, it is $\sigma'_{ov} = 60 - 1 \times (-20) = 80$ kN/m^2. In the unsaturated zone above the capillary zone, the profile is obtained by using Eq. 10.76.
9. The values of effective stress are plotted in Figure 10.29. Note that the effective stress decreases linearly as the depth decreases when the soil is saturated, but increases as the depth continues to decrease in the unsaturated zone.

We can then calculate the shear strength of the soil on horizontal planes by multiplying the vertical effective stress by the tangent of the friction angle, assuming that the soil has no effective stress cohesion: $s = \sigma'_{ov} \tan\varphi$. Therefore, the shear strength profile has the same shape as the effective stress profile. The increase in effective stress, and therefore strength close to the surface due to higher water tension, often leads to a crust that can be a few meters thick.

10.20 SOIL WATER RETENTION CURVE

The soil water retention curve (SWRC), also known as the soil water characteristic curve, is a property of the soil much like the shear strength parameters (Figure 10.30). It is a plot of the water content of the soil as a function of the water tension stress (suction) in the soil pores.

Figure 10.30 is a SWRC on a semilog plot; the water content is on a natural scale while the water tension is on a log scale. From point A to point B on Figure 10.30, the soil remains nearly saturated while the water tension increases. At the air entry value (point B), the water content decreases while the water tension increases. Up to point C on Figure 10.30, the water content is usually well represented

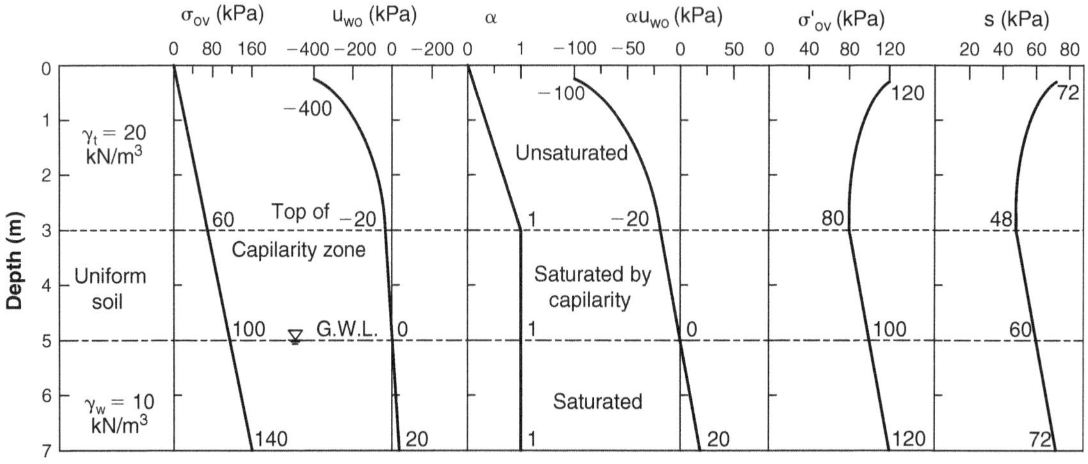

Figure 10.29 Stress profiles in a soil deposit.

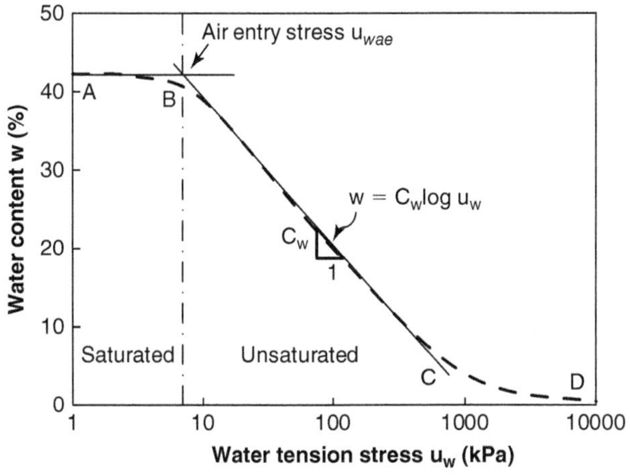

Figure 10.30 Soil water retention curve.

by a straight line and the slope of that line is the coefficient C_w:

$$\Delta w = C_w \log \frac{u_w}{u_{wae}} \tag{10.77}$$

where Δw is the change in water content, C_w is the slope of the SWRC, u_w is the water tension, and u_{wae} is the air entry value of the water tension. From C to D, the water content continues to decrease while the water tension continues to increase, but at a much higher rate.

If a saturated soil sample is placed on a table top and is strong enough to stand by itself, it is likely held together by water tension unless it has some cementation (effective stress cohesion). As the soil dries, it initially shrinks while remaining saturated. The water tension increases, and at a given water tension stress (suction), air enters the pores. This water tension is called the *air entry value* (u_{wae}). From this point on during the drying process, the soil is unsaturated. By definition, the water content at the air entry value is the undisturbed shrinkage limit because, during the shrinkage process, it is the last water content where the soil is saturated.

The gravimetric water content is the water content definition most commonly used in geotechnical engineering, but for the SWRC, the volumetric water content is often used. They are defined as follows:

$$\text{Gravimetric water content: } w = W_w/W_s \tag{10.78}$$

$$\text{Volumetric water content: } \theta_w = V_w/V \tag{10.79}$$

When the term *water content* is used in this book, it means gravimetric water content. Example SWRCs are presented in Figure 10.31. Different soils have different SWRCs; for instance, a sand will not retain water the same way a clay would. Imagine that you insert a straw into a sand. It would not take much sucking to get the water out of the sand. Now imagine that your straw is inserted into a clay. In this case it would take a lot of sucking to get a little bit of water out. The suction or water tension that you would exert through the straw would be much higher for the clay than for the sand. This phenomenon is what the SWRC characterizes.

Soils under the groundwater level are generally saturated and the water is in compression. Soils above the GWL can be saturated or unsaturated, but in both cases the water is in tension (suction). The SWRC is a property of a soil where the water is in tension. Thus, the SWRC for a saturated soil refers to the case where the soil is saturated above the GWL by capillary action and other electrochemically

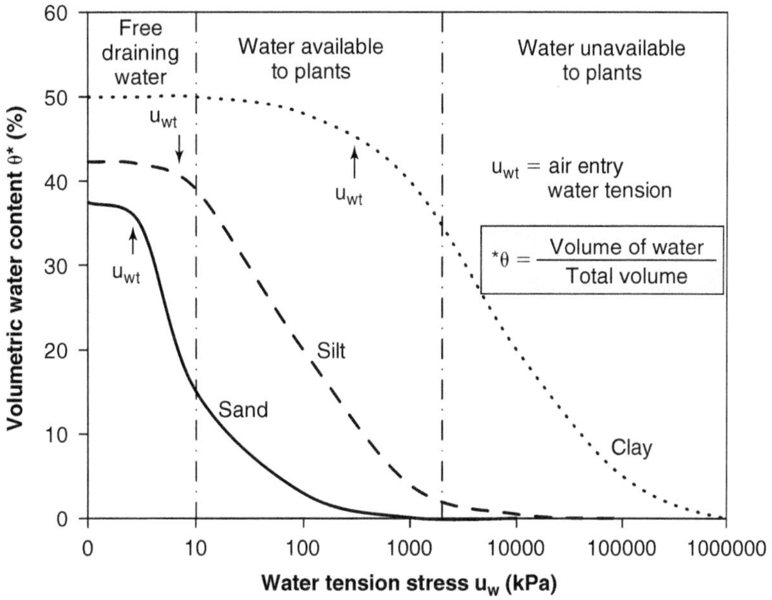

Figure 10.31 Example of soil water retention curve.

based phenomena such as the affinity between water and clay minerals (point A to B on Figure 10.30). Beyond point B the soil is unsaturated.

10.21 INDEPENDENT STRESS STATE VARIABLES

Effective stress σ', as defined in Eq. 10.54, is:

$$\sigma' = \sigma - \alpha \mathrm{u}_w - \beta \mathrm{u}_\mathrm{a} \qquad (10.80)$$

Effective stress is defined on the basis of three stresses (σ, u_w, $\mathrm{u_a}$) and two soil properties (α, β). Therefore, it depends not only on the state of stress in a soil, but also on the soil properties. Hence, it cannot be considered an independent stress variable, even if it is a very useful stress in solving many soil problems. Equation 10.54 can be rewritten as follows:

$$\begin{aligned}\sigma' &= \sigma - \alpha u_w - \beta u_a - u_a + u_a \\ &= (\sigma - u_a) - \alpha u_w + (1-\beta)u_a \\ &= (\sigma - u_a) + \alpha (u_a - u_w) \qquad (10.81)\end{aligned}$$

In this form, it becomes clear that two independent stress state variables are necessary to describe the effective stress: the net normal total stress in excess of air stress ($\sigma - u_a$) and the net water tension with respect to the air stress ($u_a - u_w$).

In terms of total stresses, the stress tensor at a point is defined in Eq. 10.3. This stress tensor does not include information on the water stress or the air stress. Keeping in mind that shear stresses are unaffected by water or air stress, the stress state in a soil can be fully described by the following two stress tensors, which include all the stress information necessary to solve an unsaturated soil problem.

$$\Sigma_1 = \begin{bmatrix} \sigma_{xx} - u_a & \tau_{xy} & \tau_{xz} \\ \tau_{yx} & \sigma_{yy} - u_a & \tau_{yz} \\ \tau_{zx} & \tau_{zy} & \sigma_{zz} - u_a \end{bmatrix} \qquad (10.82)$$

$$\Sigma_2 = \begin{bmatrix} u_a - u_w & 0 & 0 \\ 0 & u_a - u_w & 0 \\ 0 & 0 & u_a - u_w \end{bmatrix} \qquad (10.83)$$

In the case of a saturated soil, only one tensor is necessary:

$$\Sigma_1 = \begin{bmatrix} \sigma_{xx} - u_w & \tau_{xy} & \tau_{xz} \\ \tau_{yx} & \sigma_{yy} - u_w & \tau_{yz} \\ \tau_{zx} & \tau_{zy} & \sigma_{zz} - u_w \end{bmatrix} \qquad (10.84)$$

PROBLEMS

10.1 A wedge has applied stress vectors on two faces as shown in Figure 10.1s*a* and Figure 10.1s*b*. Calculate the stress on the third face in both cases. Hint: You can compose forces, but you cannot compose stresses unless they act on the same area.

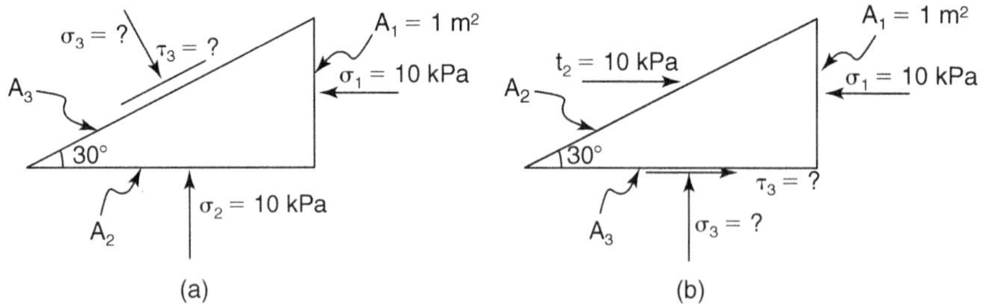

Figure 10.1s Stress vectors on wedge faces.

10.2 In a triaxial test, the confining stress (minor principal stress) σ_3 is 50 kPa, and the vertical stress (major principal stress) σ_1 is 150 kPa.

a. Form the total stress tensor shown in Eq. 10.3. Decompose the tensor into the deviatoric and spherical tensor forms shown in Eq. 10.4.

b. The soil is saturated, and under the given stresses, the water stress is 20 kPa. Form the stress tensor in terms of effective stress.

c. The soil is unsaturated, and under the given stresses, the air stress is 30 kPa and the water tension is −1000 kPa. Form the two tensors describing the state of stress in the sample in terms of independent stress state variables.

10.3 A simple shear test is performed in a plane strain condition. The vertical normal stress on the plane of failure is 80 kPa, the horizontal normal stress is 40 kPa, and the shear stress is 30 kPa on the horizontal plane. The Poisson's ratio for the soil is 0.35. Form the total stress tensor (Eq. 10.3). Decompose this tensor into the deviatoric and spherical tensor forms shown in Eq. 10.4.

10.4 A sample of cohesionless silt is tested in a direct shear test. At failure, the vertical normal stress is 100 kPa and the shear stress on the horizontal plane where failure occurs is equal to 40 kPa. The water stress is 20 kPa.

a. Calculate the effective principal stresses by using the equilibrium equations approach in two dimensions.

b. Calculate the effective principal stresses by using the Mohr circle approach in two dimensions.

10.5 For problem 10.4, use the Pole method to locate the planes where the principal stresses act.

10.6 For the sample in Figure 10.6s:

a. Find the stresses on the plane shown.

b. On what plane does the maximum shear stress exist?

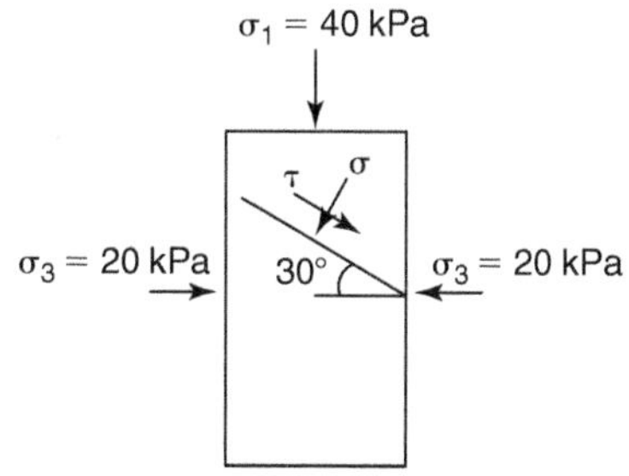

Figure 10.6s Stress state.

10.7 What happens to the Pole method when the diagram of a stress element in space is rotated by an angle θ? Does the Mohr circle change? Do the stresses on any plane change? Does the Pole location change?

10.8 In a simple shear test, the horizontal displacement at the top of the sample is 1 mm and the vertical displacement is a reduction in height of 0.5 mm. The original height of the sample is 25 mm.

a. Calculate the shear strain and the vertical normal strain.

b. Is the sample dilating or contracting?

10.9 In a triaxial test, the sample has an initial height of 150 mm and an initial diameter of 75 mm. During the loading in the vertical direction, the vertical displacement is 3 mm and the increase in diameter is 2 mm.

a. Calculate the normal strains ε_{zz} and ε_{rr}.

b. Form the strain tensor.

c. Calculate the shear strain on a 45-degree plane.

10.10 Consider the sphere-shaped soil particles shown in Figure 10.10s. The degree of saturation S is 1, the porosity n is 0.4, and the ratio between the sum of the contact areas and the total area (A_c/A_t) is 0.01. Calculate the following quantities and show the relationship between the total stress and the effective stress if the water stress is +40 kPa.

a. The average effective normal stress

b. The average normal stress at the contacts

c. The average normal total stress

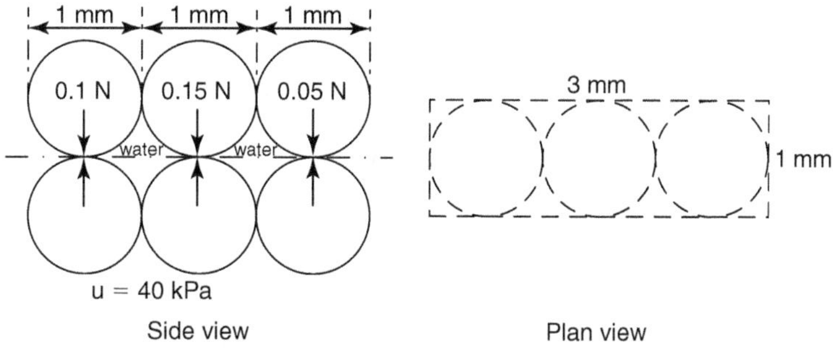

Figure 10.10s Sphere-shaped soil particles.

10.11 The surface tension of water is T = 73 mN/m, the diameter of a glass tube plunged into water is 0.002 mm, and the contact angle between the wall of the clean glass and the water is $\alpha = 10$ degrees. Find the height to which the water will rise in the small tube.

10.12 Consider the sphere-shaped soil particles shown in Figure 10.11s. The porosity n is 0.4, the ratio between the sum of the contact areas and the total area (A_c/A_t) is 0.01, and the ratio between the sum of the areas of water and the total area (A_w/A_t) is 0.1. Calculate the following quantities and show the relationship between the total stress and the effective stress if the water stress is −6000 kPa.

a. The average normal effective stress
b. The average normal stress at the contacts
c. The average normal total stress

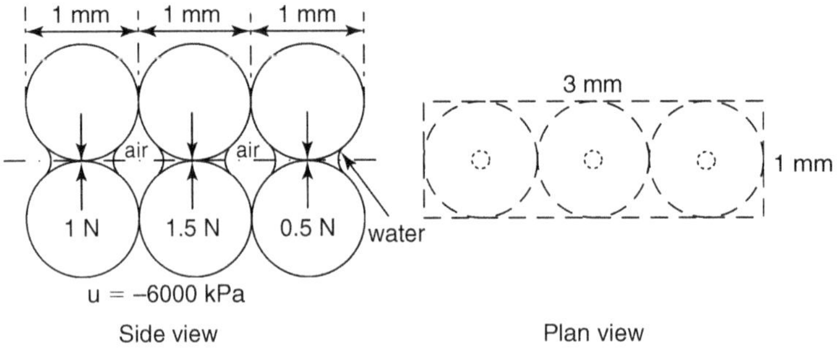

Figure 10.11s Sphere-shaped soil particles.

10.13 A soil has a degree of saturation of 92%. The air is occluded and the bubbles are 1 mm in diameter. Knowing that the water tension can reach 73 mN/m, what is the maximum difference in pressure that can exist between the water stress and the air stress?

10.14 A soil has a degree of saturation of 35%, an air entry value of −150 kPa, and a water tension stress of −1500 kPa at a depth of 2 m. Estimate the vertical effective stress at rest at a depth of 2 m below the ground surface, assuming that the unit weight of the soil is 19 kN/m^3.

10.15 Draw the three profiles (σ_{ov}, u_o, σ'_{ov}) for the layered system shown in Figure 10.12s.

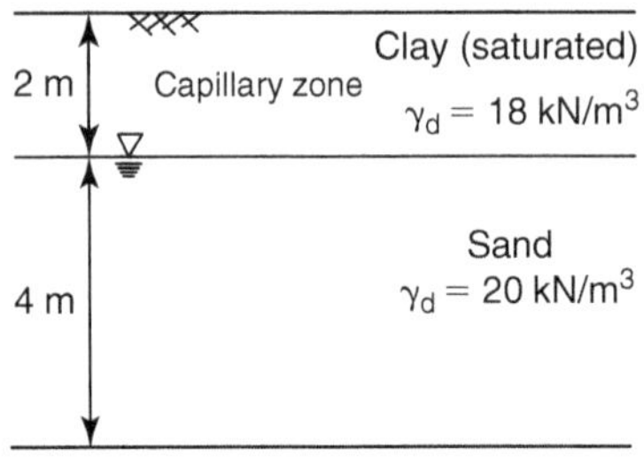

Figure 10.12s Soil profile

10.16 Draw the effective stress profiles (σ'_{ov}) for the layered system shown in Figure 10.14s.

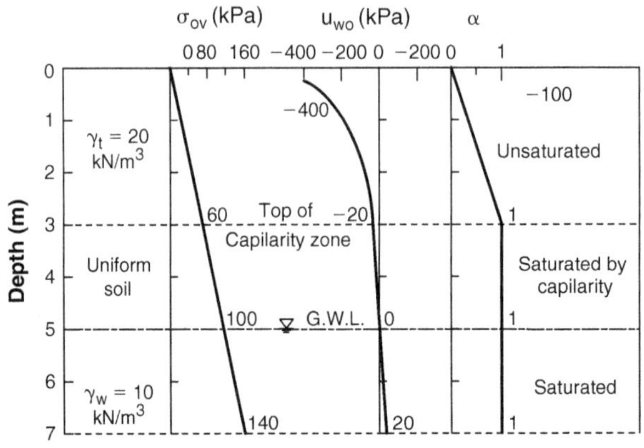

Figure 10.14s Soil and stress profile.

10.17 Draw the three profiles (σ_{ov}, u_w, σ'_{ov}) at the center of the river for the layered system shown in Figure 10.16s.

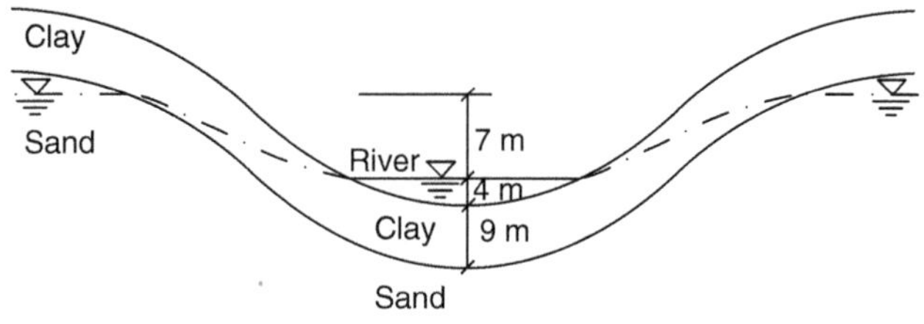

Figure 10.16s River profile.

10.18 An insect has 4 legs and is able to walk on water. The depression created under each foot is a sphere, as shown in Figure 10.18s. What is the maximum possible weight of the insect?

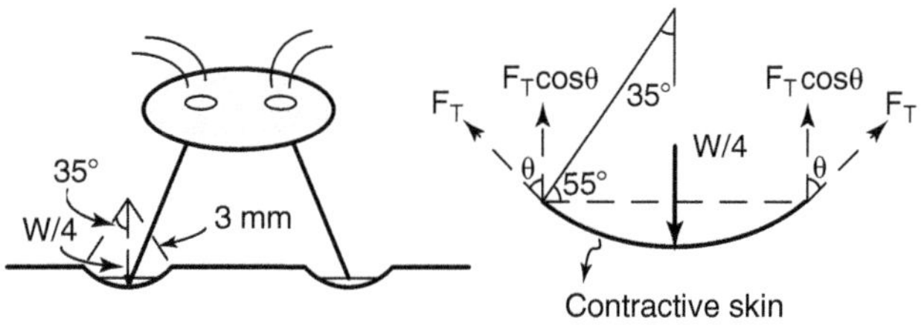

Figure 10.18s Free-body diagram of insect.

10.19 A soil has a water content of 42% and an air entry value of −8 kPa. If the slope of the soil water retention curve is 0.2 per log cycle of water tension in kPa, calculate the water tension for a water content of 10%.

10.20 A tree's root system occupies a volume equal to 1000 m^3. How much water is available to that tree if it is rooted in the three soils described by the retention curves of Figure 10.31?

Problems and Solutions

Problem 10.1

A wedge has applied stress vectors on two faces as shown in Figure 10.1s*a* and Figure 10.1s*b*. Calculate the stress on the third face in both cases. Hint: You can compose forces, but you cannot compose stresses unless they act on the same area.

Solution 10.1

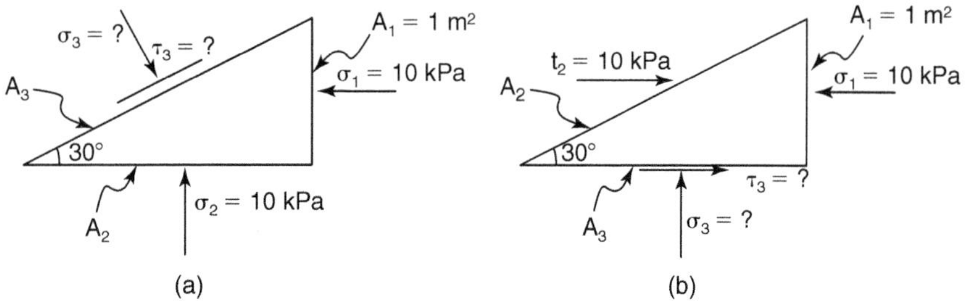

Figure 10.1s Stress vectors on wedge faces.

Part a:

$$F_1 = \sigma_1 . A_1 = 10 \text{ (kN)}$$

$$A_2 = \frac{A_1}{\tan\theta} = \frac{1}{\tan 30^\circ} = 1.732 \text{ (m}^2\text{)}$$

$$F_2 = \sigma_2 . A_2 = 10 * 1.732 = 17.32 \text{ (kN)}$$

$$A_3 = \frac{A_1}{\sin\theta} = \frac{1}{\sin 30^\circ} = 2\ (\text{m}^2)$$

$$\sum_{i=1}^{3} F_{xi} = 0 \rightarrow F_{x3(shear)} \cdot \cos 30 + F_{x3(normal)} \cdot \sin 30 - 10 = 0 \rightarrow (I)$$

$$\sum_{i=1}^{3} F_{yi} = 0 \rightarrow F_{y3(shear)} \cdot \sin 30 - F_{y3(normal)} \cdot \cos 30 + 17.32 = 0 \rightarrow (II)$$

$$(I)\&(II) \rightarrow \begin{cases} F_{S3} = 0\,(\text{kN}) \\ F_{N3} = 20\ (\text{kN}) \end{cases} \rightarrow \begin{cases} \tau_3 = 0\ (\text{kPa}) \\ \sigma_3 = 10\ (\text{kPa}) \end{cases}$$

Part b:

$$F_1 = \sigma_1.A_1 = 10\ (\text{kN})$$

$$A_2 = \frac{A_1}{\sin\theta} = \frac{1}{\sin 30^\circ} = 2\ \text{m}^2$$

$$A_3 = \frac{A_1}{\tan\theta} = \frac{1}{\tan 30^\circ} = 1.732\ (\text{m}^2)$$

$$\sum_{i=1}^{3} F_{xi} = 0 \rightarrow -F_1 + t_2 A_2 + \tau_3 A_3 = 0 \rightarrow \tau_3 = -5.77\ \text{kPa}$$

$$\sum_{i=1}^{3} F_y = 0 \rightarrow \sigma_3 = 0$$

Problem 10.2

In a triaxial test, the confining stress (minor principal stress) σ_3 is 50 kPa, and the vertical stress (major principal stress) σ_1 is 150 kPa.

a. Form the total stress tensor shown in Eq. 10.3. Decompose the tensor into the deviatoric and spherical tensor forms shown in Eq. 10.4.
b. The soil is saturated, and under the given stresses, the water stress is 20 kPa. Form the stress tensor in terms of effective stress.
c. The soil is unsaturated, and under the given stresses, the air stress is 30 kPa and the water tension is −1000 kPa. Form the two tensors describing the state of stress in the sample in terms of independent stress state variables.

Solution 10.2

a.

$$\Sigma = \begin{bmatrix} \sigma_{xx} & \tau_{xy} & \tau_{xz} \\ \tau_{yx} & \sigma_{yy} & \tau_{yz} \\ \tau_{zx} & \tau_{zy} & \sigma_{zz} \end{bmatrix} = \begin{bmatrix} 50 & 0 & 0 \\ 0 & 50 & 0 \\ 0 & 0 & 150 \end{bmatrix} (\text{kPa})$$

$$\sigma_M = \frac{(\sigma_{xx} + \sigma_{yy} + \sigma_{zz})}{3} = \frac{(50 + 50 + 150)}{3} = 83.33\ (\text{kPa})$$

$$\Sigma = \begin{bmatrix} \sigma_{xx} & \tau_{xy} & \tau_{xz} \\ \tau_{yx} & \sigma_{yy} & \tau_{yz} \\ \tau_{zx} & \tau_{zy} & \sigma_{zz} \end{bmatrix} = \begin{bmatrix} \sigma_M & 0 & 0 \\ 0 & \sigma_M & 0 \\ 0 & 0 & \sigma_M \end{bmatrix} + \begin{bmatrix} \sigma_{xx} - \sigma_M & \tau_{xy} & \tau_{xz} \\ \tau_{yx} & \sigma_{yy} - \sigma_M & \tau_{yz} \\ \tau_{zx} & \tau_{zy} & \sigma_{zz} - \sigma_M \end{bmatrix}$$

$$\Sigma = \begin{bmatrix} 83.33 & 0 & 0 \\ 0 & 83.33 & 0 \\ 0 & 0 & 83.33 \end{bmatrix} + \begin{bmatrix} 50 - 83.33 & 0 & 0 \\ 0 & 50 - 83.33 & 0 \\ 0 & 0 & 150 - 83.33 \end{bmatrix} (\text{kPa})$$

$$\Sigma = \begin{bmatrix} 83.33 & 0 & 0 \\ 0 & 83.33 & 0 \\ 0 & 0 & 83.33 \end{bmatrix} + \begin{bmatrix} -33.33 & 0 & 0 \\ 0 & -33.33 & 0 \\ 0 & 0 & 66.67 \end{bmatrix} (\text{kPa})$$

b.

$$\sigma' = \sigma - u$$

$$\Sigma = \begin{bmatrix} \sigma'_{xx} & \tau'_{xy} & \tau'_{xz} \\ \tau'_{yx} & \sigma'_{yy} & \tau'_{yz} \\ \tau'_{zx} & \tau'_{zy} & \sigma'_{zz} \end{bmatrix} = \begin{bmatrix} 50-20 & 0 & 0 \\ 0 & 50-20 & 0 \\ 0 & 0 & 150-20 \end{bmatrix} = \begin{bmatrix} 30 & 0 & 0 \\ 0 & 30 & 0 \\ 0 & 0 & 130 \end{bmatrix} \text{(kPa)}$$

c.

$$\sigma' = (\sigma - u_a) + \alpha(u_a - u_w)$$

$$\Sigma_1 = \begin{bmatrix} \sigma_{xx} - u_a & \tau_{xy} & \tau_{xz} \\ \tau_{yx} & \sigma_{yy} - u_a & \tau_{yz} \\ \tau_{zx} & \tau_{zy} & \sigma_{zz} - u_a \end{bmatrix} = \begin{bmatrix} 50-30 & 0 & 0 \\ 0 & 50-30 & 0 \\ 0 & 0 & 150-30 \end{bmatrix} \text{(kPa)}$$

$$\Sigma_1 = \begin{bmatrix} 20 & 0 & 0 \\ 0 & 20 & 0 \\ 0 & 0 & 120 \end{bmatrix} \text{(kPa)}$$

$$\Sigma_2 = \begin{bmatrix} u_a - u_w & 0 & 0 \\ 0 & u_a - u_w & 0 \\ 0 & 0 & u_a - u_w \end{bmatrix} = \begin{bmatrix} 30-(-1000) & 0 & 0 \\ 0 & 30-(-1000) & 0 \\ 0 & 0 & 30-(-1000) \end{bmatrix} \text{(kPa)}$$

$$\Sigma_2 = \begin{bmatrix} 1030 & 0 & 0 \\ 0 & 1030 & 0 \\ 0 & 0 & 1030 \end{bmatrix} \text{(kPa)}$$

Problem 10.3

A simple shear test is performed in a plane strain condition. The vertical normal stress on the plane of failure is 80 kPa, the horizontal normal stress is 40 kPa, and the shear stress is 30 kPa on the horizontal plane. The Poisson's ratio for the soil is 0.35. Form the total stress tensor (Eq. 10.3). Decompose this tensor into the deviatoric and spherical tensor forms shown in Eq. 10.4.

Solution 10.3

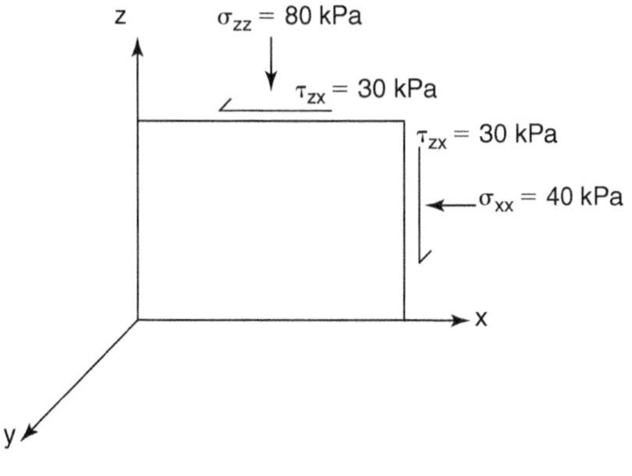

Figure 10.2s Stresses during the simple shear test.

Eq. 10.4:

$$\sum = \begin{pmatrix} \sigma_{xx} & \tau_{xy} & \tau_{xz} \\ \tau_{yx} & \sigma_{yy} & \tau_{yz} \\ \tau_{zx} & \tau_{zy} & \sigma_{zz} \end{pmatrix} = S + D = \begin{pmatrix} \sigma_m & 0 & 0 \\ 0 & \sigma_m & 0 \\ 0 & 0 & \sigma_m \end{pmatrix} + \begin{pmatrix} \sigma_{xx} - \sigma_m & \tau_{xy} & \tau_{xz} \\ \tau_{yx} & \sigma_{yy} - \sigma_m & \tau_{yz} \\ \tau_{zx} & \tau_{zy} & \sigma_{zz} - \sigma_m \end{pmatrix} \text{kPa}$$

where $\sigma_m = \frac{1}{3}(\sigma_{xx} + \sigma_{yy} + \sigma_{zz})$.

From the problem statement: $\sigma_{xx} = 40$ kPa, $\sigma_{zz} = 80$ kPa, $\tau_{xz} = \tau_{zx} = 30$ kPa, and $\tau_{xy} = \tau_{yz} = 0$ due to the plane strain condition. The value of σ_{yy} is found using the plain strain condition:

$$\varepsilon_{yy} = 0 = \frac{1}{E}(\sigma_{yy} - \nu(\sigma_{xx} + \sigma_{zz}))$$

so that

$$\sigma_{yy} = \nu(\sigma_{xx} + \sigma_{zz}) = 0.35(40 + 80) = 42 \text{ kPa}$$

Therefore,

$$\sigma_m = \frac{1}{3}(40 + 12 + 80) - 54 \text{ kPa.}$$

The deviatoric and spherical tensor forms are:

$$\sum = \begin{pmatrix} \sigma_{xx} & \tau_{xy} & \tau_{xz} \\ \tau_{yx} & \sigma_{yy} & \tau_{yz} \\ \tau_{zx} & \tau_{zy} & \sigma_{zz} \end{pmatrix} = S + D = \begin{pmatrix} 54 & 0 & 0 \\ 0 & 54 & 0 \\ 0 & 0 & 54 \end{pmatrix} + \begin{pmatrix} -14 & 0 & 30 \\ 0 & -12 & 0 \\ 30 & 0 & 26 \end{pmatrix} \text{ kPa}$$

Problem 10.4

A sample of cohesionless silt is tested in a direct shear test. At failure, the vertical normal stress is 100 kPa and the shear stress on the horizontal plane where failure occurs is equal to 40 kPa. The water stress is 20 kPa.

a. Calculate the effective principal stresses by using the equilibrium equations approach in two dimensions.
b. Calculate the effective principal stresses by using the Mohr circle approach in two dimensions.

Solution 10.4

a. The effective principal stresses are related to the shear and normal stress on the failure plane through the equilibrium equations (Eq. 10.10 and Eq. 10.11):

$$\sigma' = \frac{\sigma_1' + \sigma_3'}{2} + \frac{\sigma_1' - \sigma_3'}{2}\cos 2\alpha$$

$$\tau = -\frac{\sigma_1' - \sigma_3'}{2}\sin 2\alpha$$

Because the sample is at failure and the silt is cohesionless, the shear strength equation can be written as:

$$\tau = \sigma' \tan \varphi'$$

This also means that:

$$\sin \varphi' = \frac{\sigma_1' - \sigma_3'}{\sigma_1' + \sigma_3'}$$

The last four equations, together with the given values of $\sigma' = 80$ kPa and $\tau = 40$ $\tau = 40$ kPa, give the values of the four unknowns: φ', σ_1', σ_3', and α. The solution is $\varphi' = 26.6^{\circ}$, $\sigma_1' = 144.4$ kPa, $\sigma_3' = 55.4$ kPa, and $\alpha = 59^{\circ}$.

b. The effective principal stresses can be found using the Mohr circle as shown in Figure 10.3s.

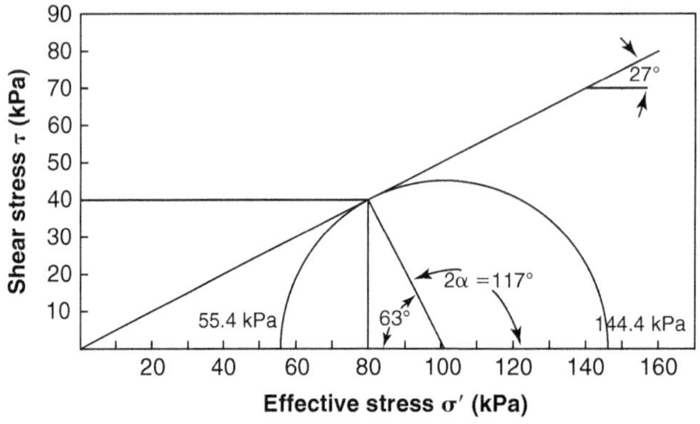

Figure 10.3s Mohr circle for direct shear test.

Problem 10.5

For problem 10.4, use the Pole method to locate the planes where the principal stresses act.

Solution 10.5

First we draw the failure stress point on the shear stress vs. effective normal stress set of axes ($\tau = 40$ kPa, $\sigma = 80$ kPa). This point is on the failure envelope, and because the soil has no cohesion intercept, the failure envelope can be drawn through the origin and the failure point. The Mohr circle is found tangent to the failure envelope at the failure stress point. According to the Pole method, the line parallel to the plane on which the stresses act (horizontal plane) intersects the Mohr circle at two points: the stress point and the Pole. This allows us to find the Pole (Figure 10.4s). Knowing the Pole, we draw the lines that join the Pole to the two principal stress points σ_1' and σ_3'. These lines define the directions of the planes on which the principal stresses are acting (Figure 10.5s).

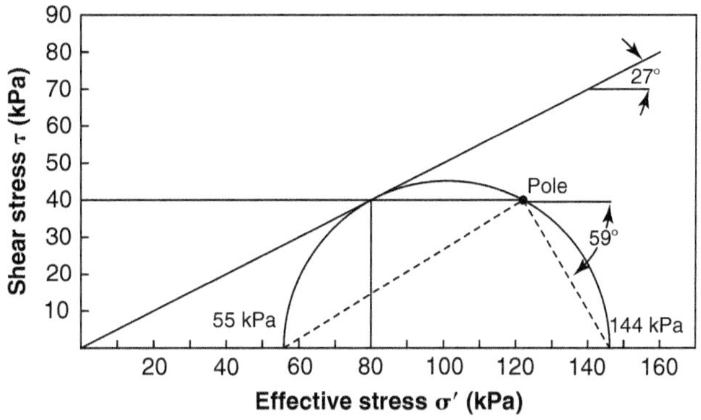

Figure 10.4s Pole method.

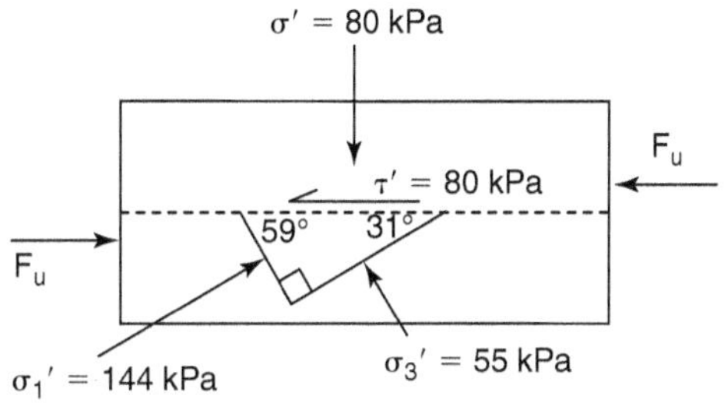

Figure 10.5s Principal planes in direct shear test.

Problem 10.6

For the sample in Figure 10.6s,

a. Find the stresses on the plane shown.
b. On what plane does the maximum shear stress exist?

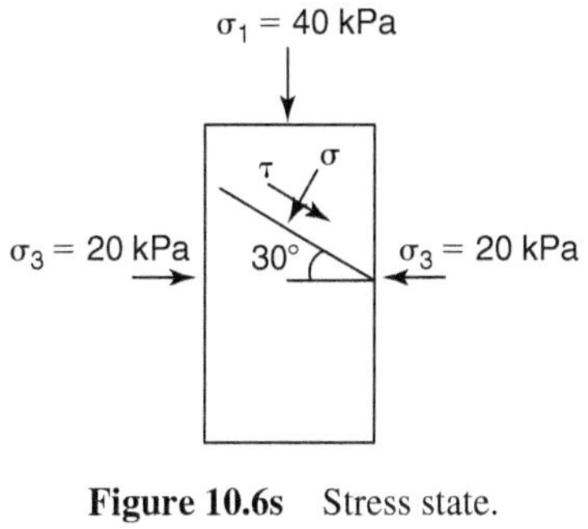

Figure 10.6s Stress state.

Solution 10.6

a. We can solve this problem with the equilibrium equations or with the Mohr circle. Recall Eqs 10.10 and 10.11 from the text:

$$\sigma = \frac{\sigma_1 + \sigma_3}{2} + \frac{\sigma_1 - \sigma_3}{2}\cos 2\alpha$$

$$\tau = -\frac{\sigma_1 - \sigma_3}{2}\sin 2\alpha$$

By using these equations, we obtain:

$$\sigma = \frac{40 + 20}{2} + \frac{40 - 20}{2}\cos(2 \times 30^\circ) = 35 \text{ kPa}$$

$$\tau = -\frac{40 - 20}{2}\sin(2 \times 30^\circ) = -8.67 \text{ kPa}$$

We then confirm the solutions by use of the Mohr circle (Figure 10.7s).

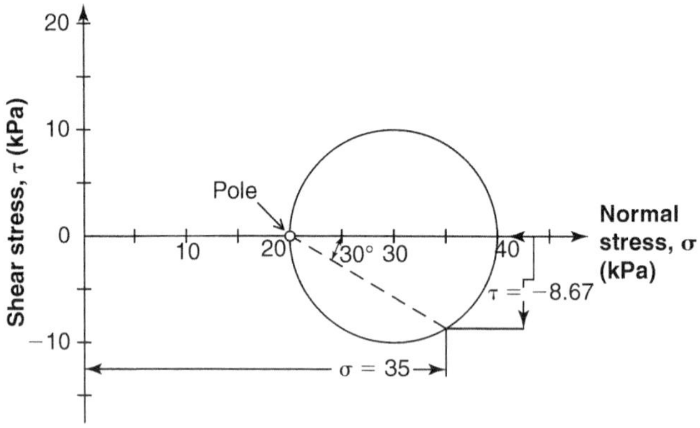

Figure 10.7s Mohr circle.

b. To find the plane where the maximum shear stress acts, we use the Pole method. We first find the Pole by drawing a line parallel to the plane where σ_1 acts (horizontal). That line intersects the Mohr circle at two points: the σ_1 stress point and the Pole. Then we join the Pole to the largest shear stress point of the Mohr circle. That line is a 45-degree line and gives the plane on which the highest shear stress acts (Figure 10.8s).

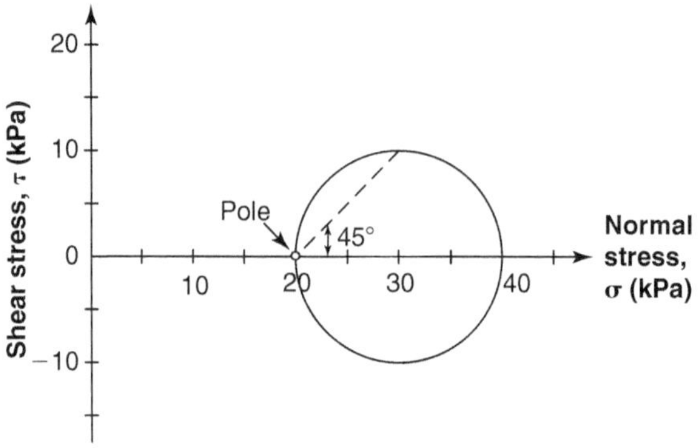

Figure 10.8s Pole method.

Problem 10.7

What happens to the Pole method when the diagram of a stress element in space is rotated by an angle θ? Does the Mohr circle change? Do the stresses on any plane change? Does the Pole location change?

Solution 10.7

When the diagram of the stress element in space is rotated by an angle θ, the Mohr circle does not change because the principal stresses do not change; accordingly, the stresses on any plane do not change either. However, the location of the Pole on the Mohr circle rotates with the diagram to maintain the rule of parallelism.

Problem 10.8

In a simple shear test, the horizontal displacement at the top of the sample is 1 mm and the vertical displacement is a reduction in height of 0.5 mm. The original height of the sample is 25 mm.

a. Calculate the shear strain and the vertical normal strain.
b. Is the sample dilating or contracting?

Solution 10.8

a. The shear strain and the vertical normal strain are (Figure 10.9s):

$$\varepsilon_{shear} = \tan^{-1}\frac{1}{25} = 0.04 \quad \text{or 4\% shear strain}$$

$$\varepsilon_{normal} = \frac{0.5}{25} = 0.02 \quad \text{or 2\% compression normal strain}$$

b. The sample is contracting.

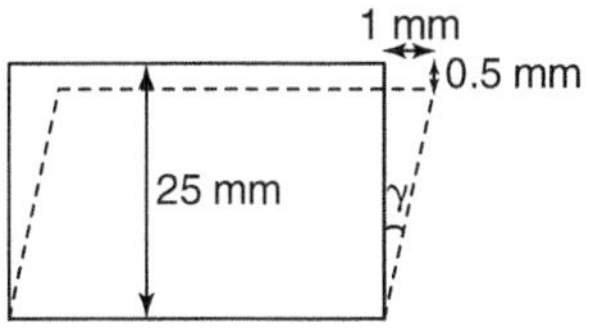

Figure 10.9s Normal and shear strain.

Problem 10.9

In a triaxial test, the sample has an initial height of 150 mm and an initial diameter of 75 mm. During the loading in the vertical direction, the vertical displacement is 3 mm and the increase in diameter is 2 mm.

a. Calculate the normal strains ε_{zz} and ε_{rr}.
b. Form the strain tensor.
c. Calculate the shear strain on a 45-degree plane.

Solution 10.9

a. The normal strains ε_{zz} and ε_{rr} are:

$$\varepsilon_{zz} = \frac{3}{150} = 0.02, \varepsilon_{rr} = \frac{2}{75} = -0.027$$

b. The strain tensor is:

$$\begin{bmatrix} \varepsilon_{rr} & \frac{1}{2}\gamma_{rz} \\ \frac{1}{2}\gamma_{zr} & \varepsilon_{zz} \end{bmatrix} = \begin{bmatrix} -0.027 & 0 \\ 0 & 0.02 \end{bmatrix}$$

c. The shear strain on a 45-degree plane is:

$$\gamma_{45} = (\varepsilon_{zz} - \varepsilon_{rr}) \sin 2\alpha = (0.02 - (-0.027)) * \sin(90) = 0.047$$

Problem 10.10

Consider the sphere-shaped soil particles shown in Figure 10.10s. The degree of saturation S is 1, the porosity n is 0.4, and the ratio between the sum of the contact areas and the total area (A_c/A_t) is 0.01. Calculate the following quantities and show the relationship between the total stress and the effective stress if the water stress is +40 kPa.

a. The average effective normal stress
b. The average normal stress at the contacts
c. The average normal total stress

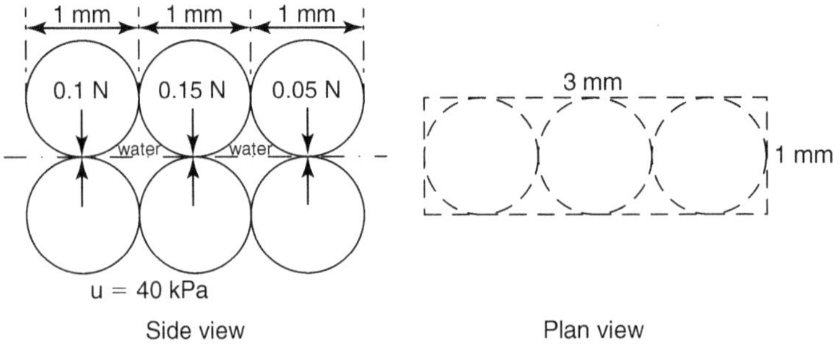

Figure 10.10s Sphere-shaped soil particles.

Solution 10.10

$$A = 3 \times 1 \times 10^{-6} = 3 \times 10^{-6}\ \text{m}^2$$

The average effective normal stress is:

$$\sigma'_{aver} = \frac{(0.1 + 0.15 + 0.05) \times 10^{-3}}{3 \times 10^{-6}} = 100\ \text{kPa}$$

The average normal stress at the contacts is:

$$\sigma_{c-aver} = \frac{(0.1 + 0.15 + 0.05) \times 10^{-3}}{3 \times 10^{-6} \times 0.01} = 10000\ \text{kPa}$$

The average normal total stress is:

$$\sigma_{aver} = \frac{(0.1 + 0.15 + 0.05) \times 10^{-3} + 40 \times (1 - 0.01) \times 3 \times 10^{-6}}{3 \times 10^{-6}} = 139.6\ \text{kPa}$$

By definition, the relation between the total stress and the effective stress is:

$$\sigma = \sigma' + u = 100 + 40 = 140 \approx 139.6\ \text{kPa}$$

Problem 10.11

The surface tension of water is T = 73 mN/m, the diameter of a glass tube plunged into water is 0.002 mm, and the contact angle between the wall of the clean glass and the water is $\alpha = 10$ degrees. Find the height to which the water will rise in the small tube.

Solution 10.11

$$d = 0.002\ \text{mm}$$

$$T = 73\ \text{mN/m}$$

$$\alpha = 10\ \text{deg}$$

$$h_c = \frac{4T\cos\alpha}{d\gamma_w} \rightarrow h_c = \frac{4 \times 73 \times 10^{-6} \times \cos(10)}{0.002 \times 10^{-3} \times 10} = 14.37\ \text{m}$$

Problem 10.12

Consider the sphere-shaped soil particles shown in Figure 10.11s. The porosity n is 0.4, the ratio between the sum of the contact areas and the total area (A_c/A_t) is 0.01, and the ratio between the sum of the areas of water and the total area (A_w/A_t) is 0.1. Calculate the following quantities and show the relationship between the total stress and the effective stress if the water stress is −6000 kPa.

a. The average normal effective stress
b. The average normal stress at the contacts
c. The average normal total stress

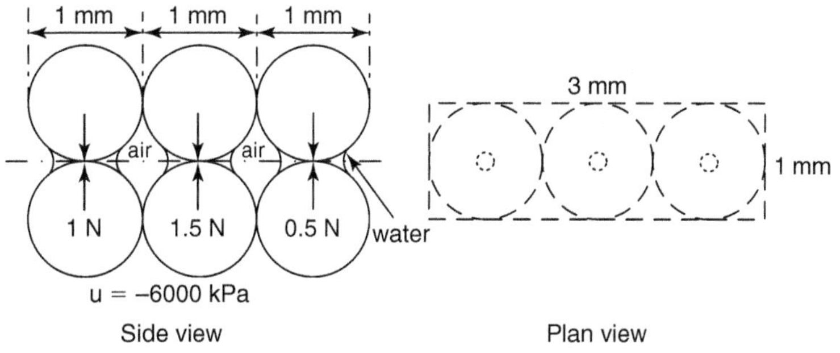

Figure 10.11s Sphere-shaped soil particles.

Solution 10.12

$$A = 3 \times 1 \times 10^{-6} = 3 \times 10^{-6}\text{m}^2$$

a. The average normal effective stress is:

$$\sigma'_{aver} = \frac{(1 + 1.5 + 0.5) \times 10^{-3}}{3 \times 10^{-6}} = 1000\text{ kPa}$$

b. The average normal stress at the contacts is:

$$\sigma_{c-aver} = \frac{(1 + 1.5 + 0.5) \times 10^{-3}}{3 \times 10^{-6} \times 0.01} = 100000\text{ kPa}$$

c. The average normal total stress is:

$$\sigma_{aver} = \frac{(1 + 1.5 + 0.5) \times 10^{-3} - 6000 \times 0.1 \times 3 \times 10^{-6}}{3 \times 10^{-6}} = 400\text{ kPa}$$

The relation between the total stress and the effective stress is:

$$\sigma' = \sigma - \alpha u = 400 - 0.1 \times (-6000) = 1000\text{ kPa}$$

Problem 10.13

A soil has a degree of saturation of 92%. The air is occluded and the bubbles are 1 mm in diameter. Knowing that the water tension can reach 73 mN/m, what is the maximum difference in pressure that can exist between the water stress and the air stress?

Solution 10.13

It seems reasonable that the air is occluded, as the degree of saturation of the soil is 92%, which is larger than 85%. Based on the equilibrium of the free-body diagram of half the bubble, and knowing that the water tension is 73 mN/m, we have (Eq. 10.59):

$$u_a - u_w = \frac{4T}{D} = \frac{4 \times 73 \times 10^{-6}\text{ kN/m}}{1 \times 10^{-3}\text{ m}} = 0.29\text{ kPa}$$

Problem 10.14

A soil has a degree of saturation of 35%, an air entry value of −150 kPa, and a water tension stress of −1500 kPa at a depth of 2 m. Estimate the vertical effective stress at rest at a depth of 2 m below the ground surface, assuming that the unit weight of the soil is 19 kN/m^3.

Solution 10.14

The soil has a degree of saturation of 35%, which means the soil is unsaturated and the relationship between the effective stress and total stress is:

$$\sigma' = \sigma - \alpha\, u_w$$

Here, α can be obtained based on Khalili and Khabbaz (1998):

$$\alpha = \sqrt{\frac{u_{wae}}{u_w}} = \sqrt{\frac{-150}{-1500}} = 0.316$$

Therefore, at the given depth of 2 m below the ground surface, the vertical effective stress at rest is:

$$\sigma' = \sigma - \alpha\ u_w = \gamma\ Z - \alpha\ u_w = 19 \times 2 - 0.316 \times (-1500) = 512 \text{ kPa}$$

Problem 10.15

Draw the three profiles (σ_{ov}, u_o, σ'_{ov}) for the layered system shown in Figure 10.12s.

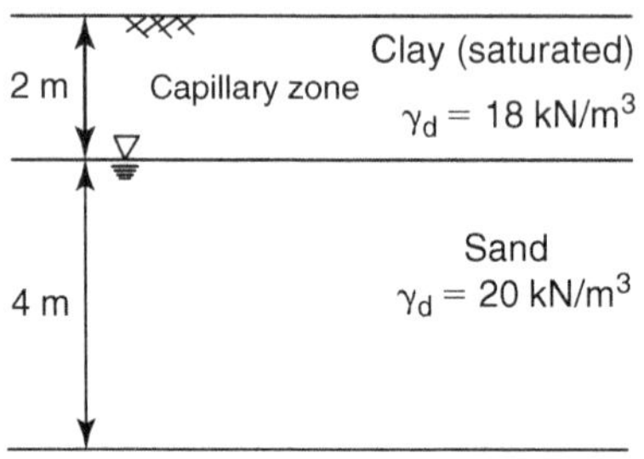

Figure 10.12s Soil profile.

Solution 10.15

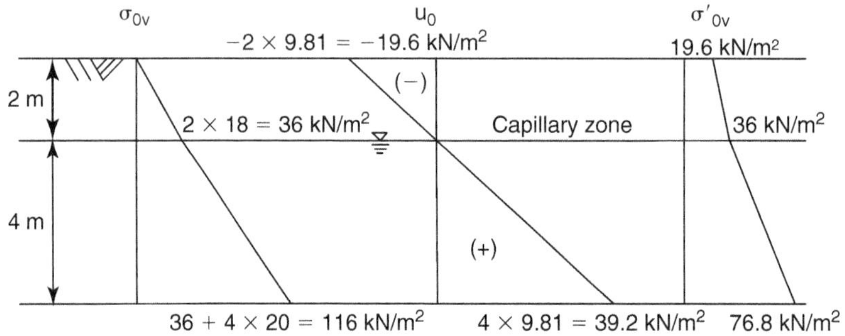

Figure 10.13s Stress profiles.

Problem 10.16

Draw the effective stress profiles (σ'_{ov}) for the layered system shown in Figure 10.14s.

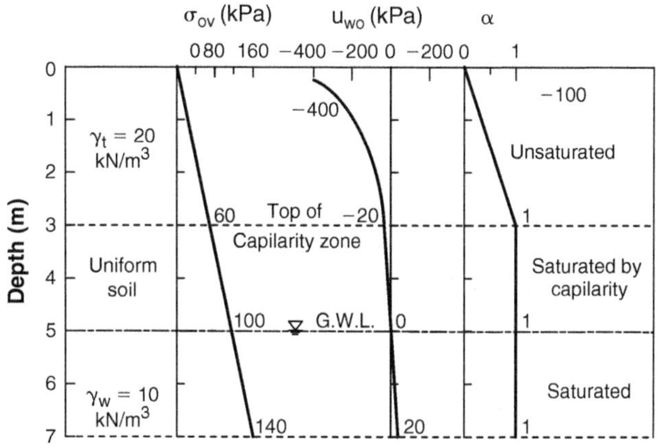

Figure 10.14s Soil and stress profile.

Solution 10.16

The effective vertical stress σ'_{ov} profile is shown in Figure 10.15s.

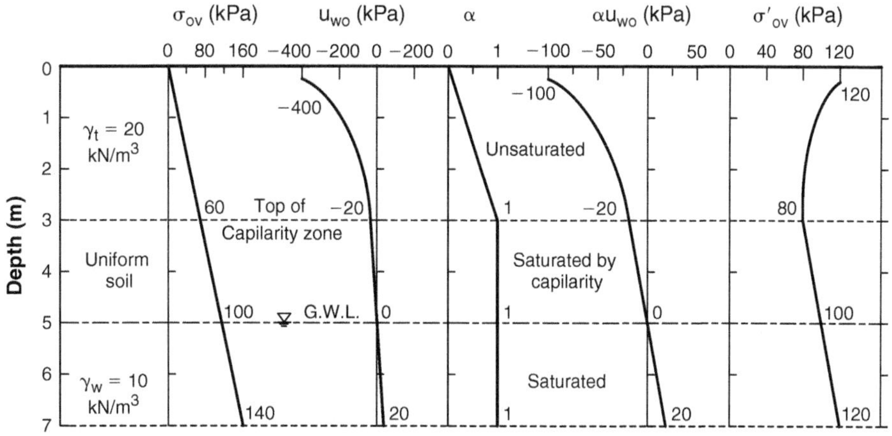

Figure 10.15s Stress profiles.

Problem 10.17

Draw the three profiles (σ_{ov}, u_w, σ'_{ov}) at the center of the river for the layered system shown in Figure 10.16s.

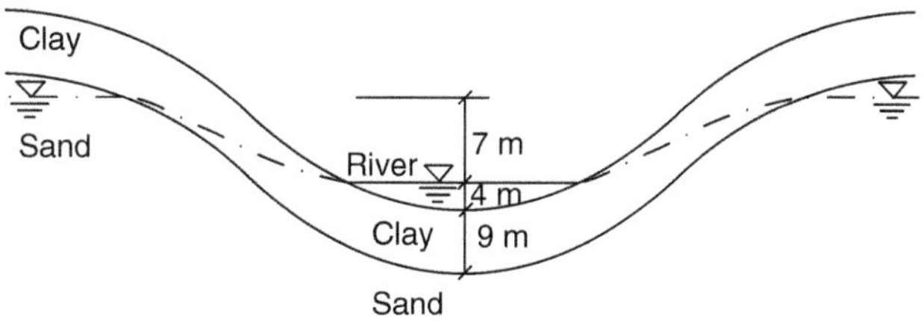

Figure 10.16s River profile.

Solution 10.17

The stress profile for σ_{ov}, u_w, σ'_{ov} is shown in Figure 10.17s:

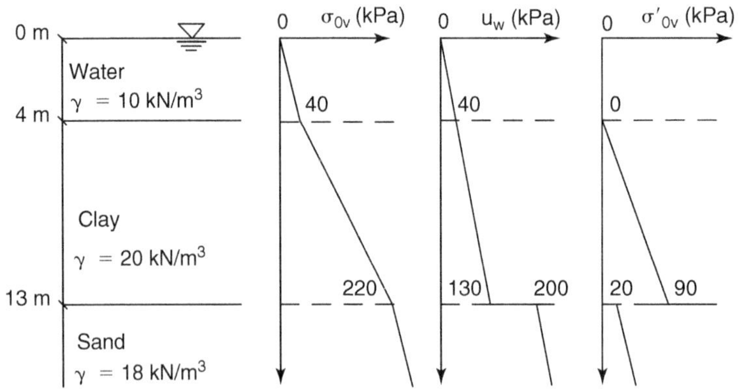

Figure 10.17s Stress profile in the river.

Problem 10.18

An insect has 4 legs and is able to walk on water. The depression created under each foot is a sphere, as shown in Figure 10.18s. What is the maximum possible weight of the insect?

Solution 10.18

The contact radius of the insect foot with the contractile skin is:

$$r = 3\text{ mm} \times \sin 35^\circ = 1.72\text{ mm}$$

For a water temperature of 20°C, the surface tension (σ_T) is 73 mN/m.

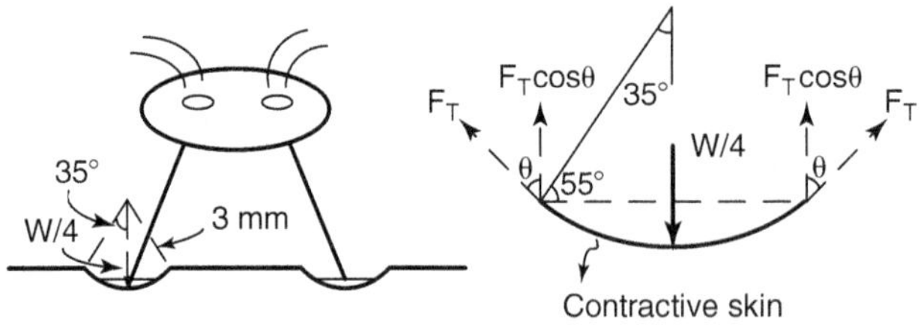

Figure 10.18s Free-body diagram of insect.

$$\sum F_v = 0$$

$$F_T \times 2\pi r \cos\theta - W/4 = 0$$

$$W = 8\pi r F_T \cos\theta$$

$$W = 8\pi(0.00172\text{m})(73\text{ mN/m})\cos 55^\circ = 1.79\text{ mN}$$

Problem 10.19

A soil has a water content of 42% and an air entry value of −8 kPa. If the slope of the soil water retention curve is 0.2 per log cycle of water tension in kPa, calculate the water tension for a water content of 10%.

Solution 10.19

Given values:

$$C_w = -0.2$$
$$u_{wae} = -8 \text{ kPa}$$
$$w_1 = 42\%$$
$$w_2 = 10\%$$
$$\Delta w = 0.42 - 0.10 = -0.2 \log\left(\frac{-8}{u_w}\right)$$
$$\log\left(\frac{-8}{u_w}\right) = \frac{0.32}{-0.2} = -1.6$$
$$10^{-1.6} = \frac{-8}{u_w}$$
$$u_w = \frac{-8}{10^{-1.6}}$$
$$u_w = -318.5 \text{ kPa}$$

Problem 10.20

A tree's root system occupies a volume equal to 1000 m^3. How much water is available to that tree if it is rooted in the three soils described by the retention curves of Figure 10.31?

Solution 10.20

From the portion in the graph referring to "Water available to plants":
Change in volumetric water content (Θ):

1. Clay: $\Delta\Theta = 0.50 - 0.34 = 0.16$
2. Silt: $\Delta\Theta = 0.40 - 0.02 = 0.38$
3. Sand: $\Delta\Theta = 0.15 - 0 = 0.15$

$\Delta\Theta \times$ Volume of soil in root system = Water available to the tree

1. Clay: $0.16 \times 1000 \text{ m}^3 = 160 \text{ m}^3$
2. Silt: $0.38 \times 1000 \text{ m}^3 = 380 \text{ m}^3$
3. Sand: $0.15 \times 1000 \text{ m}^3 = 150 \text{ m}^3$

CHAPTER 11

Problem-Solving Methods

11.1 GENERAL

There are three main types of problems in geotechnical engineering: failure load problems, deformation problems, and flow problems. Each problem can be solved by performing experimental modeling, by doing theoretical modeling, or by using experience. The best solutions are those that have a theoretical framework, are calibrated against and correlated with experimental measurements, and are verified by experience at full scale. Experience is obtained by years of practice. As the saying goes, good judgment comes from experience, but experience comes from bad judgment. An attempt can be made at teaching experience by letting engineers, who have been practicing successfully for a long time, discuss case histories—including failures—in a classroom environment. Theoretical modeling includes continuum mechanics closed-form solutions, numerical simulations, dimensional analysis, probabilistic analysis, and risk analysis. Experimental modeling includes the use of scaled models, centrifuge models, and/or full-scale models. In all cases, fundamental laws and constitutive laws help in solving the problem.

11.2 DRAWING TO SCALE AS A FIRST STEP

One very important first step in solving a geotechnical engineering problem (or any engineering problem in general) is to always start by making a drawing to scale of the problem. If this step is not taken, the engineer may not get a proper sense of the issues at hand. For example, if one is designing a pile foundation under a building with the pile tips bearing into a sand layer, making a drawing to scale helps the engineer evaluate whether the sand layer is thick enough to prevent serious compression of the layers below. Failing to make that drawing properly, and instead drawing only a sample single pile bearing into the sand layer, may give the false impression of a thick sand layer (Figure 11.1). Also, if you draw a driven pile as a thick, short vertical member instead of the actual slender member, the issue of buckling will not come to your attention. Embankments typically have side slopes of 2 to 1 or 3 to 1, yet when they are sketched on a piece of paper, these

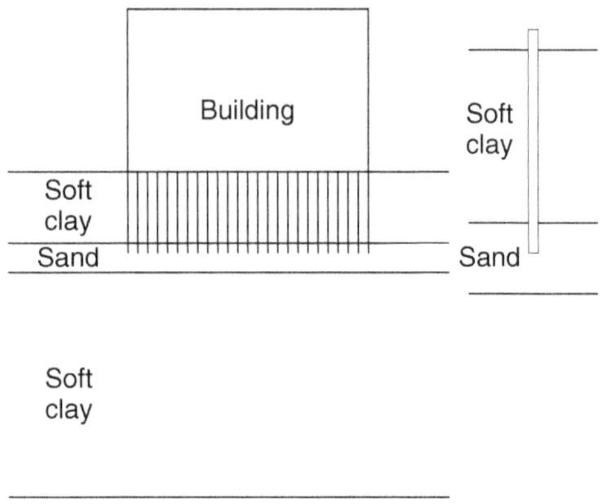

Figure 11.1 Make a drawing to scale.

slopes are often drawn too steep. By making a drawing to scale, you give yourself a better chance of recognizing some of the problems associated with the physical dimensions of the project. Always make a drawing to scale as a first step in solving an engineering problem.

11.3 PRIMARY LAWS

Two main types of laws are used to solve problems: fundamental laws and constitutive laws. *Fundamental laws* are valid no matter what material is being considered. They apply equally to soil, concrete, steel, or marshmallow. Fundamental laws include, for example, force and moment equilibrium, conservation of energy, and conservation of mass. The *constitutive laws* describe the behavior of the material. They are different for each material, whether it is soil, concrete, steel, or peanut butter. Constitutive laws include, for example, elasticity, plasticity, and viscoelasticity. Shear strength laws such as the Mohr-Coulomb criterion belong to the class of constitutive laws. Most theoretical problems are solved by making use of a combination of fundamental laws and constitutive laws. Other laws exist, such as the similitude laws used in dimensional analysis and the probabilistic laws used in risk analysis.

11.4 CONTINUUM MECHANICS METHODS

The basic and general steps in developing a theoretical solution to a soil problem are to describe the problem precisely, identify the variables, write the applicable equations, and solve for the unknowns. If there are more equations than unknowns, then one must choose which equations are most important to satisfy. If there are more unknowns than equations, it is time to make reasonable assumptions to generate new equations. The reasonableness of the assumptions should then be verified by experimentation at model scale or (even better) at full scale. In soil mechanics, there are three main types of problems: failure problems, deformation problems, and flow problems.

11.4.1 Solving a Failure Problem: Limit Equilibrium, Method of Characteristics, Lower and Upper Bound Theorems

A typical solution to the problem of finding a failure load (e.g., ultimate bearing capacity of a footing) or a failure moment (e.g., slope stability) is to use the limit equilibrium analysis. In such a failure analysis, the step-by-step process advances as follows:

1. Assume a reasonable failure mechanism. If such a failure mechanism is not obvious, an experiment can be performed to observe the failure mechanism.
2. Draw a free body diagram of the failing body (soil mass) and identify the external forces and external moments applied to the failing body.
3. Write the applicable fundamental equations. These are equations that are valid for all problems and independent of the type of material involved. They include equilibrium equations (three forces, three moments), conservation of mass, and conservation of energy, among others.
4. Write the applicable constitutive equations. These are the equations describing the behavior of the material under load. Constitutive laws include, for example, elasticity, plasticity, and viscoelasticity. The shear strength equation for soils, which states that the shear strength is a function of the effective stress on the failure plane, is another example of a constitutive law.
5. Count the number of equations and the number of unknowns. If there are as many equations as there are unknowns, proceed to the next step. If there are more equations than unknowns (rare), choose which equations are most important to satisfy. If there are more unknowns than equation, formulate assumptions that lead to additional equations. These assumptions should be based on engineering judgment, or experience, or experimental observations. The reasonableness of these assumptions should be verified by comparing the solution to observed full-scale behavior.
6. Combine all equations and solve for the unknown (failure load or failure moment).

There can be as many solutions as there are assumed failure mechanisms, so obviously it becomes important to choose the failure mechanism that most closely duplicates the real one. This is where observation of full-scale failures becomes very useful.

Another solution is to use the method of characteristics. *Characteristics* are lines in the physical soil mass where the partial differential equation collapses into an ordinary differential equation. The equilibrium equations at the element level typically lead to partial differential equations. The method of characteristics simplifies these equations to the point where the problem is easier to solve. The method of characteristics can help to calculate failure loads for simple geometries.

Yet another solution is to use the bound theorems and apply them to soil masses. There are two such theorems: the lower bound theorem and the upper bound theorem. The lower bound theorem states that if any stress distribution throughout the soil mass can be found which is everywhere in equilibrium internally, does not violate the yield condition, and balances the external loads, the soil mass will safely carry the external loads. The upper bound theorem states that if an estimate of the failure load of a soil mass is made by equating internal rate of energy to the rate at which external forces do work in any postulated but kinematically admissible mechanism of deformation of the soil mass, the estimate will be either high or correct. In short, the lower bound theorem involves guessing a stress field that leads to a lower bound of the failure load; the upper bound theorem involves guessing a velocity or displacement field that leads to an upper bound estimate of the failure load.

11.4.2 Examples of Solving a Failure Problem

The first exemple problem is to find the ultimate pressure p_u that a strip footing of width B (Figure 11.2) can exert on the surface of a saturated clay that has a shear strength s equal to the undrained shear strength s_u because the loading is rapid. The steps described in section 11.4.1 for the limit equilibrium method are followed.

1. A cylindrical failure surface, as shown in Figure 11.3*a*, seems reasonable. This failure mechanism has been observed in many old silo failures.
2. The failing soil mass is the half cylinder shown in Figure 11.3*b* together with its free body diagram. All external forces and stresses are shown on the diagram, including the weight of the mass. Note that the weight is always an external force.

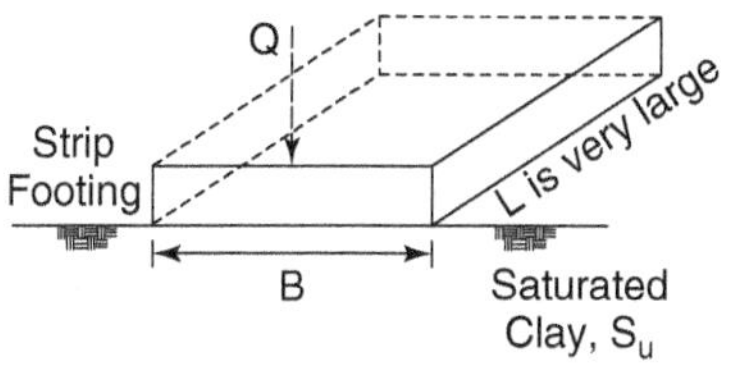

Figure 11.2 Strip footing example.

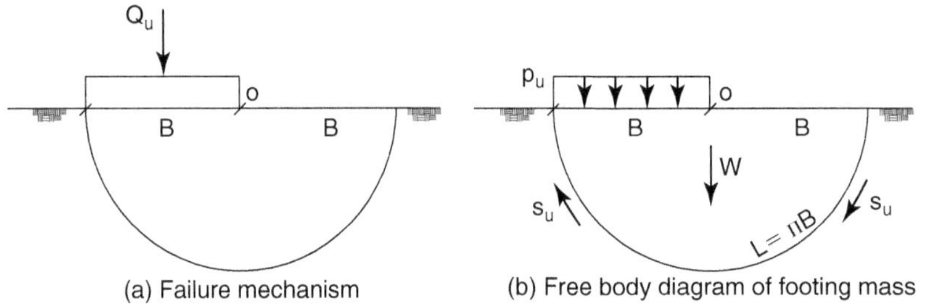

Figure 11.3 Failure load for a strip footing.

3. The most useful fundamental equation in this case is moment equilibrium around point O on Figure 11.3*b*.

$$M@O = 0 = p_u B \frac{B}{2} - s\,\pi B\,B \tag{11.1}$$

4. The constitutive equation in this case is the shear strength equation, which states that the shear strength s is equal to the undrained shear strength of the clay.

$$s = s_u \tag{11.2}$$

5. There are two unknowns (p_u and s) and two equations, so the problem can be solved.
6. Now we combine the equations and obtain:

$$p_u = 2\pi s_u \tag{11.3}$$

Other failure mechanisms are plausible and would lead to slightly different estimates of p_u.

The second problem is the one of a vertical wall with a height H supporting a clean, dry sand backfill with a friction angle φ (Figure 11.4). It is assumed that there is no friction between the wall and the backfill. The wall exerts a horizontal load P against the sand. As the wall moves very slightly away from the sand, the load P decreases and there is a point where the sand behind the wall starts to fail. At that point, the load is P_a and the question is to find the load P_a corresponding to impending failure of the sand. Note that the problem is a plane strain problem; therefore, all the loads will be line loads expressed in kN/m.

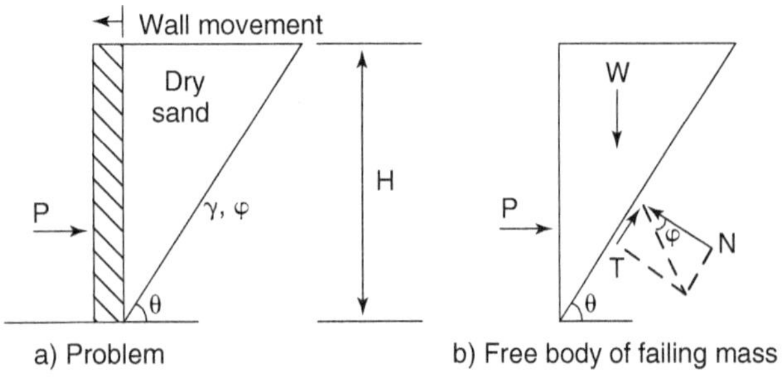

Figure 11.4 Example of a wall moving away from the backfill.

The steps described in section 11.4.1 for the limit equilibrium method are followed.

1. The soil is assumed to fail as a wedge making an angle θ with the horizontal, as shown in Figure 11.4. This failure mechanism has been observed in model scale and centrifuge experiments.
2. The failing soil mass is the wedge; its free-body diagram is shown in Figure 11.4. All external forces are shown on the diagram, including the action of the wall P, the weight of the soil mass W, the normal force N, and the shear force T on the failure plane. Note that the shear force T is acting uphill because the wedge is falling down along that plane and the soil outside of the wedge is resisting that tendency.
3. The most useful fundamental equations in this case are vertical and horizontal equilibrium of forces:

$$\sum F_v = 0 = W - N\cos\theta - T\sin\theta \tag{11.4}$$

$$\sum F_h = 0 = P - N\sin\theta + T\cos\theta \tag{11.5}$$

4. The constitutive equations in this case are the shear strength equation of the sand and the expression of the weight of the wedge. The shear strength equation states that the ultimate shear force T comes from the friction generated by the normal force N on the failure plane. The weight of the wedge is equal to the area of the wedge times the unit weight of the sand γ:

$$T = N\tan\varphi \tag{11.6}$$

$$W = \frac{\gamma H^2}{2\tan\theta} \tag{11.7}$$

5. There are four unknowns (W, N, T, P) and four equations, so the problem can be solved.
6. Now we combine the equations and obtain:

$$P = \frac{\gamma H^2}{2}\left(\frac{\sin\theta\cos\theta - \tan\varphi\cos^2\theta}{\sin\theta\cos\theta + \tan\varphi\sin^2\theta}\right) \tag{11.8}$$

There is one more issue to resolve. The load P depends on θ, yet there is a unique value of θ associated with the failure load P_a. This is the load at which the wedge fails behind the

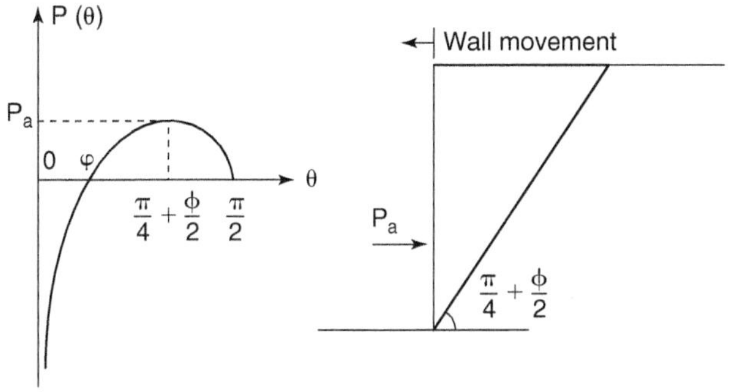

Figure 11.5 Load P as a function of the wedge angle θ; wall moves away from backfill.

wall, and this load corresponds to the θ value that maximizes P (Figure 11.5). In other words, the wedge that needs the maximum support will fail first. The maximum value of P is obtained by setting $\frac{dP}{d\theta} = 0$. This derivative is:

$$\frac{dP}{d\theta} = \frac{\gamma H^2}{2}\left(\frac{\sin\theta\cos\theta - \sin(\theta-\varphi)\cos(\theta-\varphi)}{\sin^2\theta\cos^2(\theta-\varphi)}\right) = 0 \tag{11.9}$$

There are two solutions to Eq. 11.9: one is $\varphi = 0$, which is not realistic, and the other one is:

$$\theta = \frac{\pi}{4} + \frac{\varphi}{2} \tag{11.10}$$

The load P_a can then be obtained from Eq. 11.8.

$$P_a = \frac{\gamma H^2}{2}\left(\frac{1-\sin\varphi}{1+\sin\varphi}\right) \tag{11.11}$$

This problem is repeated but now with the wall being pushed into the sand (Figure 11.6) instead of pulled away from the sand as in the previous case. The question is to find the load P_p that corresponds to the failure of the soil mass. Only one thing changes in the derivation: the direction of the shear force T on the failure plane. Because the wedge will now move up along the failure plane, the soil outside the wedge will exert a shear force acting toward the bottom of the wedge. Therefore, in the equations T is replaced by $-T$ and the problem leads to the situation shown in Figure 11.7. The failure load P_p is the load corresponding to the value of θ that minimizes P; that is, the wedge offering the minimum resistance is the wedge that will fail first.

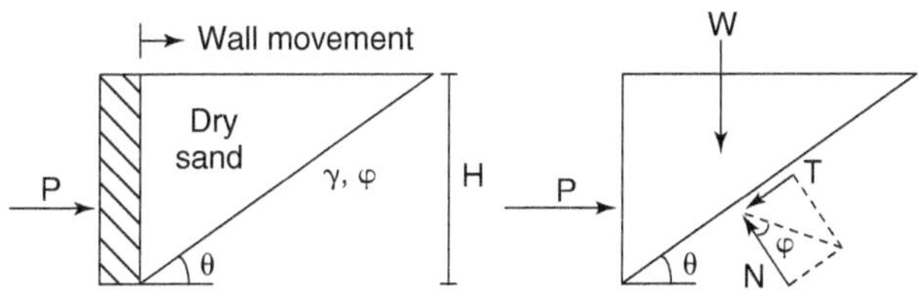

Figure 11.6 Example of a wall moving toward the backfill.

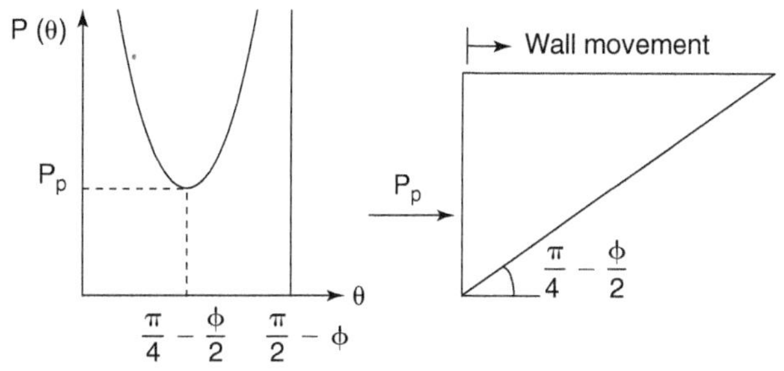

Figure 11.7 Load P as a function of the wedge angle θ; wall moves into backfill.

$$\theta = \frac{\pi}{4} - \frac{\varphi}{2} \tag{11.12}$$

$$P_p = \frac{\gamma H^2}{2}\left(\frac{1+\sin\varphi}{1-\sin\varphi}\right) \tag{11.13}$$

11.4.3 Solving a Deformation Problem

A typical solution to a deformation problem proceeds through the following steps:

1. Zoom in at the infinitesimal element level. This element has dimensions expressed in differential lengths.
2. The knowns and unknowns (e.g., loads, displacements, stresses, and strains) are identified on the element, including their variation from one side of the element to the other. This variation involves derivatives expressing the rate of change of the variable in one direction over a small distance.
3. The fundamental equations are written using the knowns and unknowns identified in step 2. These are equations that are true for all materials. They include equilibrium equations (three forces and three moments), conservation of mass, and conservation of energy, among others.
4. The constitutive equations are written using the knowns and unknowns identified in step 2. These equations describe the behavior of the material involved in the deformation. They include elasticity equations, plasticity equations, and viscosity equations, among others.
5. All equations are regrouped into the governing differential equations (GDEs).
6. The boundary conditions are expressed mathematically. If the problem is a dynamic problem, the boundary conditions involve both space and time.
7. The GDEs are solved in closed-form solutions if they are simple enough and through numerical solutions such as the finite difference method if they are too complicated. The boundary conditions are used to define the constants involved in the solution.

11.4.4 Example of Solving a Deformation Problem

The exemple problem is to find the horizontal displacement y(z) of a pile as a function of z if the pile is loaded in

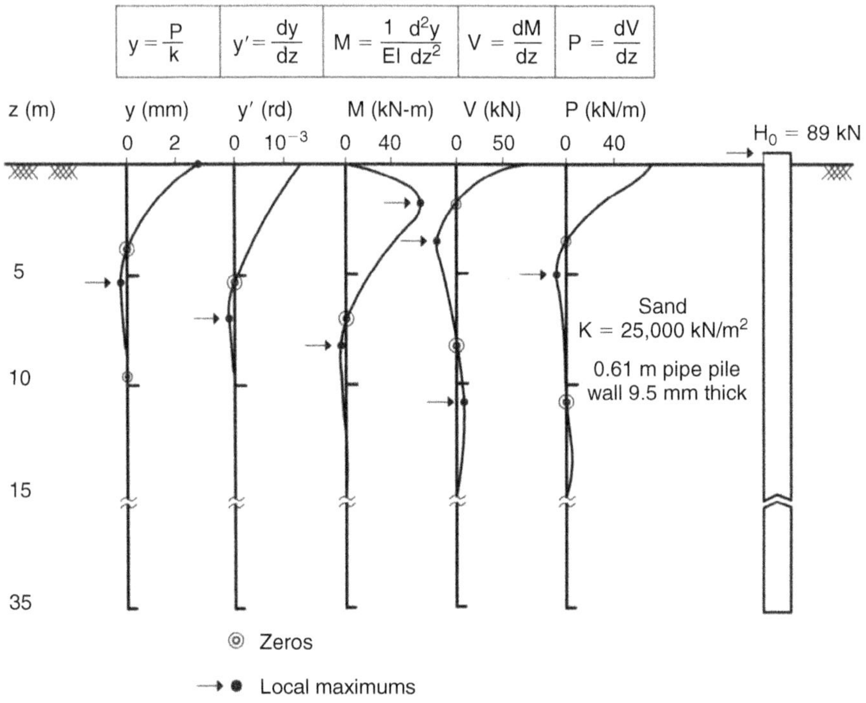

Figure 11.8 Horizontally loaded pile example.

overturning by a horizontal load H_o and an overturning moment M_o applied at the ground surface (Figure 11.8). For this simple example, the influence of the axial load will be ignored.

The solution proceeds by following the steps described in section 11.4.3.

1. Zoom in at the element level. In this case, we will select an element of the pile that is dz long (Figure 11.9).
2. The forces and moments acting on the element are shown on the element (Figure 11.9). These actions are the shear V (kN) and moment M (kN.m) at both ends of the element, and the soil resistance P (kN/m) as a line load. Some of these quantities change by a little bit from one end of the element to the other. This little bit is expressed mathematically as $\frac{\partial V}{\partial z}dz$ for the shear force V, for example, expressing that the change is equal to the rate of change of V times the distance dz. Because V is dependent only on z, $\frac{\partial V}{\partial z}dz$ can be simply written as dV.
3. The fundamental equations that are most useful in this case are horizontal equilibrium and moment equilibrium. Let's write horizontal equilibrium first (Figure 11.9):

$$\sum F_H = 0 = Pdz + V - (V + dV) \tag{11.14}$$

or

$$P = \frac{dV}{dz} \tag{11.15}$$

So, horizontal equilibrium states that the line load P on a pile is equal to the first derivative of the shear V. Now

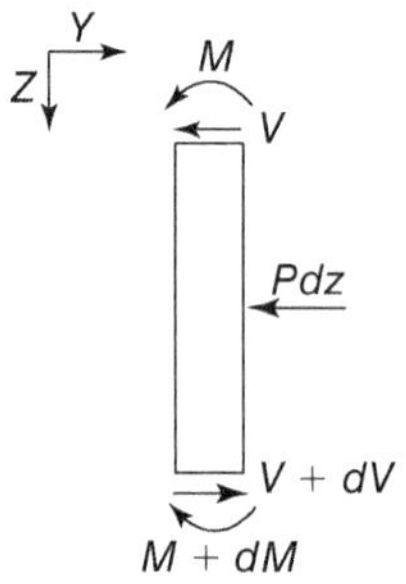

Figure 11.9 Element of horizontally loaded pile.

let's write moment equilibrium around point O (Figure 11.9). Again, Because M is only a function of z, $\frac{\partial M}{\partial z}dz$ can be simply written as dM.

$$\sum M@o = 0 = M + dM - M - V\frac{dz}{2} - (V + dV)\frac{dz}{2} \tag{11.16}$$

Neglecting the higher-order term, we are left with:

$$V = \frac{dM}{dz} \tag{11.17}$$

So, moment equilibrium states that the shear in a pile is equal to the first derivative of the bending moment.

4. The constitutive equations describe the behavior of the pile and of the soil. The pile behavior is described by relating the bending moment M applied to the pile to the curvature generated in the pile. This curvature is expressed by the

second derivative of y. The proportionality constant between the moment and the curvature is the bending stiffness E_pI, where E_p is the modulus of the pile material and I is the moment of inertia of the pile cross section around the axis of the moment. Again, because y is only a function of z, $\frac{\partial^2 y}{\partial z^2}$ can be more simply written as $\frac{d^2 y}{dz^2}$.

$$M = E_p I \frac{d^2 y}{dz^2} \tag{11.18}$$

Note that the unit of $\frac{d^2 y}{dz^2}$ is 1/m because $d^2 y$ is a little piece of y and dz^2 is the square of a little piece of z. This extends to the nth derivative; the unit of $\frac{d^n y}{dz^n}$ is $1/m^{(n-1)}$ because $d^n y$ is still a little piece of y while dz^n is the nth power of a little piece of z. For the constitutive equation describing the soil behavior, a simple linear relationship is used between the line load P (kN/m) characterizing the soil resistance and the deflection y (m) of the soil-pile interface. The proportionality constant is a spring constant K (kN/m^2) that characterizes the stiffness of the soil:

$$P = -Ky \tag{11.19}$$

The minus sign is there because P and y are in opposite directions (Figure 11.9).

5. The governing differential equations can now be assembled by regrouping the fundamental and constitutive equations:

$$P = -Ky = \frac{dV}{dz} = \frac{d^2 M}{dz^2} = E_p I \frac{d^4 y}{dz^4} \tag{11.20}$$

or

$$y + \frac{E_p I}{K} \frac{d^4 y}{dz^4} = 0 \tag{11.21}$$

6. The boundary conditions are stated for both ends of the pile. To simplify the solution of the differential equation, it is assumed that the pile is infinitely long and that the deflection is zero at the infinite end. At the top of the pile, the horizontal force and the overturning moment are known. The boundary conditions are:

a. z = infinity, y = 0
b. z = 0, $M = M_o$,
c. z = 0, $V = H_o$

7. Now we need to solve the differential equation. The solution y(z) has to be a function that becomes the same function when differentiated four times. This means a combination of exponential and trigonometric functions. The solution is therefore of the general form:

$$y(z) = e^{-\frac{z}{l_o}} \left(a \sin \frac{z}{l_o} + b \cos \frac{z}{l_o} \right) + e^{\frac{z}{l_o}} \left(c \sin \frac{z}{l_o} + d \cos \frac{z}{l_o} \right) \tag{11.22}$$

The l_o parameter is required because of the need to match the factor E_pI/K in the differential equation 11.21. Applying boundary condition 6a gives c = d = 0. Applying boundary conditions 6b and 6c requires that the expressions of V and M be derived using Eqs. 11.17 and 11.18:

$$y(z) = e^{-\frac{z}{l_o}} \left(a \sin \frac{z}{l_o} + b \cos \frac{z}{l_o} \right) \tag{11.23}$$

$$\frac{dy}{dz} = \frac{1}{l_o} e^{-\frac{z}{l_o}} \left(-(a+b) \sin \frac{z}{l_o} + (a-b) \cos \frac{z}{l_o} \right) \tag{11.24}$$

$$\frac{d^2 y}{dz^2} = -\frac{2}{l_o^2} e^{-\frac{z}{l_o}} \left(-b \sin \frac{z}{l_o} + a \cos \frac{z}{l_o} \right) \tag{11.25}$$

$$\frac{d^3 y}{dz^3} = \frac{2}{l_o^3} e^{-\frac{z}{l_o}} \left((a-b) \sin \frac{z}{l_o} + (a+b) \cos \frac{z}{l_o} \right) \tag{11.26}$$

$$\frac{d^4 y}{dz^4} = -\frac{4}{l_o^4} e^{-\frac{z}{l_o}} \left(a \sin \frac{z}{l_o} + b \cos \frac{z}{l_o} \right) \tag{11.27}$$

It can be seen from Eq. 11.23 and Eq. 11.27 that:

$$\frac{d^4 y}{dz^4} = -\frac{4}{l_o^4} y \tag{11.28}$$

which compared to Eq. 11.21 leads to:

$$l_o = \left(\frac{4E_p I}{K} \right)^{\frac{1}{4}} \tag{11.29}$$

Now boundary condition 6b can be written as:

$$\frac{d^2 y}{dz^2}_{@z=0} = \frac{M_o}{E_p I} = -\frac{2}{l_o^2} e^{-\frac{0}{l_o}} \left(-b \sin \frac{0}{l_o} + a \cos \frac{0}{l_o} \right) \tag{11.30}$$

and

$$a = -\frac{M_o l_o^2}{2E_p I} \tag{11.31}$$

Then:

$$\frac{d^3 y}{dz^3}_{@z=0} = \frac{H_o}{E_p I} = \frac{2}{l_o^3} e^{-\frac{0}{l_o}} \left((a-b) \sin \frac{0}{l_o} + (a+b) \cos \frac{0}{l_o} \right) \tag{11.32}$$

and

$$a + b = \frac{H_o l_o^3}{2E_p I} \tag{11.33}$$

so

$$b = \frac{H_o l_o^3}{2E_p I} + \frac{M_o l_o^2}{2E_p I} \tag{11.34}$$

Now we can build the equations for the deflection y, the slope y′, the bending moment M, the shear force V, and the line load P:

$$y(z) = e^{-\frac{z}{l_o}}\left(-\frac{M_o l_o^2}{2E_p I}\sin\frac{z}{l_o} + \left(\frac{H_o l_o^3}{2E_p I} + \frac{M_o l_o^2}{2E_p I}\right)\cos\frac{z}{l_o}\right) \tag{11.35}$$

But

$$E_p I = \frac{K l_o^4}{4} \tag{11.36}$$

Therefore, finally:

$$y(z) = \frac{2H_o}{l_o K}e^{-\frac{z}{l_o}}\cos\frac{z}{l_o} + \frac{2M_o}{l_o^2 K}e^{-\frac{z}{l_o}}\left(\cos\frac{z}{l_o} - \sin\frac{z}{l_o}\right) \tag{11.37}$$

$$y'(z) = -\frac{2H_o}{l_o^2 K}e^{-\frac{z}{l_o}}\left(\cos\frac{z}{l_o} + \sin\frac{z}{l_o}\right) - \frac{4M_o}{l_o^3 K}e^{-\frac{z}{l_o}}\cos\frac{z}{l_o} \tag{11.38}$$

$$M(z) = H_o l_o e^{-\frac{z}{l_o}}\sin\frac{z}{l_o} + M_o e^{-\frac{z}{l_o}}\left(\cos\frac{z}{l_o} + \sin\frac{z}{l_o}\right) \tag{11.39}$$

$$V(z) = H_o e^{-\frac{z}{l_o}}\left(\cos\frac{z}{l_o} - \sin\frac{z}{l_o}\right) - \frac{2M_o}{l_o}e^{-\frac{z}{l_o}}\sin\frac{z}{l_o} \tag{11.40}$$

where y(z) is the pile displacement at a depth z, y′(z) is the pile slope at z, M(z) is the bending moment at z, and V(z) is the shear force at z; H_o and M_o are the shear and moment at the ground surface, K is the soil spring constant, and l_o is the transfer length given by Eq. 11.29. The profiles of y(z), y′(z), M(z), and V(z) corresponding to Eq. 11.37 to 11.40 are shown in Figure 11.8 as a function of depth for a real pile.

11.4.5 Solving a Flow Problem

A typical solution to a flow problem proceeds through the following steps:

1. Zoom in at the infinitesimal element level. This element has dimensions expressed in differential lengths.
2. The knowns and unknowns (flow velocities, volumes, total head, and water stress, for example) are identified on the element, including their variation from one side of the element to the other. This variation involves derivatives expressing the rate of change of the variable in one direction over a small distance.
3. The fundamental equations are written using the knowns and unknowns identified in step 2. These equations are true for all materials. The most useful in this case is the conservation of mass equation.
4. The constitutive equations are written using the knowns and unknowns identified in step 2. These equations describe the behavior of the material involved in the flow. The main equation in this case is Darcy's law in the three dimensions.
5. All equations are regrouped into the governing differential equations.
6. The boundary conditions are expressed mathematically. These boundary conditions are usually in the form of total head or flow conditions. If the problem is a transient flow problem, the initial conditions are also stipulated.
7. The governing differential equations are solved in closed-form solutions if they are simple enough, and through numerical solutions such as the finite difference method if they are too complicated. The boundary conditions are used to define the constants involved in the solution.

11.4.6 Example of Solving a Flow Problem

One example of a flow problem is the flow of water out of a saturated soil layer when it is loaded by a long embankment (Figure 11.10*a*). Before loading, the layer is under an at-rest state of stress with a vertical effective stress σ'_{ov} and an initial water stress u_{wo}. Both σ'_{ov} and u_{wo} vary with the depth z. When the vertical stress is increased by $\Delta\sigma$ due to the embankment loading, the water stress increases by an amount called the *excess pore pressure* u_{we}. The excess pore pressure u_{we} is high at first and decreases as a function of time while the water drains out. The settlement takes place as a result of this water drainage (Figure 11.10*b*). The problem is to predict the variation of the excess pore pressure u_{we} as a function of time t and the settlement ΔH of the embankment as a function of time t.

The following simplifying assumptions are made:

a. The soil is saturated with water
b. The water is incompressible
c. The soil skeleton is linear elastic (linear stress-strain relation)
d. The soil particles are incompressible
e. Darcy's law governs the flow of water through the soil
f. The water drains at the top and at the bottom of the layer
g. The flow is in the vertical direction only

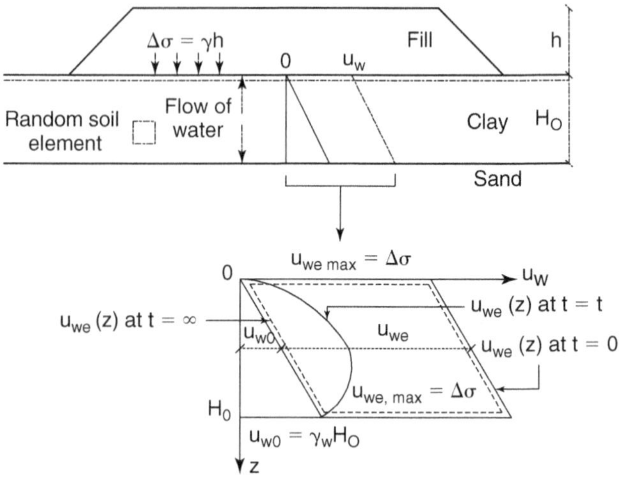

Figure 11.10 Embankment example.

h. The increase in stress $\Delta\sigma$ in the layer due to the embankment is constant within the layer
i. The excess water stress u_{we} increases by $\Delta\sigma$ when the embankment is placed
j. No lateral soil movement takes place

With these assumptions, the solution proceeds as follows:

1. Zoom in at the element level. In this case, we will select an element of soil with an elementary volume V equal to dx dy dz (Figure 11.11).

2. Considering the element of Figure 11.11, the water velocity in the z direction is v_z when it enters the element and $v_z + dv_z$ when it exits the element. It is assumed that the water does not flow in the y direction because of the plain strain condition induced by the infinitely long embankment. It is also assumed that there is no flow in the x direction because the total head gradient is much higher in the z direction than in the x direction. Because the water velocity is proportional to the total head gradient (Darcy's law), most of the water goes in the vertical direction. Also shown on the element is the change of volume dV of the element during a time dt. This change of volume corresponds to the water loss and also to the compression of the element, given that the soil is saturated.

3. The fundamental equation in this case is the conservation of mass equation expressing that, during a time dt, the volume of water entering the element plus the water squeezed out of the element due to the stress applied is equal to the volume of water exiting the element. Use is made of the flow equation (Qdt = vAdt):

$$v_z\,dx\,dy\,dt + dV = (v_z + dv_z)\,dx\,dy\,dt \tag{11.41}$$

$$\frac{dV}{Vdt} = \frac{dv_z}{dz} \tag{11.42}$$

Another fundamental equation is conservation of energy, which leads to the relationship between the total head h_t, the elevation head h_e, and the pressure head h_p. Note that the velocity head is neglected because water flows very slowly through soils:

$$h_t = h_e + h_p \tag{11.43}$$

and by differentiation

$$dh_t = dh_e + dh_p \tag{11.44}$$

Note that for the element, the elevation head h_e is constant and therefore $dh_e = 0$. Note also that, by definition:

$$h_p = \frac{u_{wo} + u_{we}}{\gamma_w} \tag{11.45}$$

Because u_{wo} is constant:

$$dh_p = \frac{du_{we}}{\gamma_w} \tag{11.46}$$

Combining the previous observation, we get:

$$dh_t = dh_p = \frac{du_{we}}{\gamma_w} \tag{11.47}$$

The effective stress in the element is:

$$\sigma' = \sigma - (u_{wo} + u_{we}) \tag{11.48}$$

By differentiation and noting that both σ and u_{wo} are constant during the loading and subsequent drainage:

$$d\sigma' = -du_{we} \tag{11.49}$$

4. The first constitutive equation describes how fast the water flows through the soil (Darcy's law):

$$v_z = ki = -k\frac{dh_t}{dz} \tag{11.50}$$

and by taking the first derivative of v_z with respect to z:

$$\frac{dv_z}{dz} = -k\frac{d^2h_t}{dz^2} \tag{11.51}$$

The second constitutive equation describes how much the soil compresses under stress (stress-strain relationship):

$$d\sigma' = M\frac{dV}{V} \tag{11.52}$$

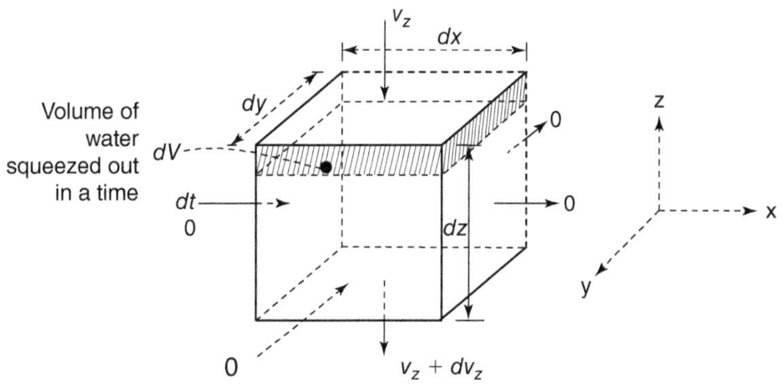

Figure 11.11 Element of soil under the embankment.

The strain in this case is the volumetric strain ($\varepsilon_v = dV/V$) and M is the constrained modulus because the soil is not allowed to expand laterally.

5. By regrouping Eqs. 11.42, 11.47, 11.49, 11.51, and 11.52, the governing differential equation is obtained:

$$\frac{1}{V}\frac{dV}{dt} = \frac{dv_z}{dz} = \frac{1}{M}\frac{d\sigma'}{dt} = -\frac{1}{M}\frac{du_{we}}{dt} = -k\frac{d^2h_t}{dz^2} = -\frac{k}{\gamma_w}\frac{d^2u_{we}}{dz^2} \quad (11.53)$$

$$\frac{du_{we}}{dt} = \frac{k\,M}{\gamma_w}\frac{d^2u_{we}}{dz^2} \quad (11.54)$$

The coefficient of consolidation c_v is expressed in m^2/s and is defined as:

$$c_v = \frac{k\,M}{\gamma_w} \quad (11.55)$$

and the governing differential equation for this problem is:

$$\frac{du_{we}}{dt} = c_v\frac{d^2u_{we}}{dz^2} \quad (11.56)$$

6. Now we need to organize the space and time boundary conditions. The space boundary conditions state that the excess water stress u_{we} at the ground surface is zero because the water can drain freely at that location. Also, the excess water stress u_{we} is zero at the bottom of the layer because the water can drain freely at that depth:

$$u_{we@z=0} = 0 \quad at\ any\ time\ t \quad (11.57)$$

$$u_{we@z=H_o} = 0 \quad at\ any\ time\ t \quad (11.58)$$

The time boundary conditions state that the excess water stress u_{we} is equal to the increase in total stress $\Delta\sigma$ at time $t = 0$ and then equal to 0 at time $t =$ infinity:

$$u_{we@t=0} = \Delta\sigma \quad at\ any\ depth\ z \quad (11.59)$$

$$u_{we@t=\infty} = 0 \quad at\ any\ depth\ z \quad (11.60)$$

7. This is the step where we solve the governing differential equation (11.56) and apply the boundary conditions. To simplify the mathematical process, it is convenient to use the following transformation into dimensionless variables:

$$Z = \frac{z}{H_d} \quad (11.61)$$

$$U = 1 - \frac{u_{we}}{u_{we(\max)}} \quad (11.62)$$

$$T = \frac{c_v t}{H_d^2} \quad (11.63)$$

where z is the depth below ground surface, H_d is the longest drainage path, U is the degree of consolidation at depth z and time t, u_{we} is the excess water stress at depth z and time t, $u_{we(\max)}$ is the maximum excess water stress at time $t = 0$ at any depth taken as equal to $\Delta\sigma$, T is the time factor, and t is the time. Note that the maximum drainage length is equal to the layer thickness H_o if the water can only drain on one side (top or bottom of the layer), but is equal to $0.5H_o$ if the water can drain at both ends. With these transformed variables, the GDE (Eq. 11.56) becomes:

$$\frac{dU}{dT} = \frac{d^2U}{dZ^2} \quad (11.64)$$

The solution to this partial differential equation, together with the space and time boundary conditions, is a Fourier series expansion of the form (Terzaghi 1943):

$$U = 1 - \sum_{m=0}^{m=\infty}\frac{2}{M}\sin(MZ)\exp(-M^2T) \quad with \quad M = \frac{\pi}{2}(2m+1) \quad (11.65)$$

The graphical representation of U as a function of Z and T is shown in Figure 11.12.

It is also useful to define the average degree of consolidation U_{av}:

$$U_{av} = 1 - \frac{\int_0^H u_{we}dz}{\int_0^H u_{we\max}dz} = 1 - \sum_{m=0}^{m=\infty}\frac{2}{M^2}\exp(-M^2T) \quad with \quad M = \frac{\pi}{2}(2m+1) \quad (11.66)$$

The average degree of consolidation represents the ratio of the area under the excess water stress profile at time t over the same area at time $t = 0$ (Figure 11.13).

The graphical representation of U_{av} as a function of T is shown in Figures 11.14 and 11.15.

Equation 11.52 indicates that the volumetric strain dV/V in the layer is linearly proportional to the increase in effective stress $d\sigma'$. Because the soil is assumed not to move laterally, the volumetric strain is also the vertical strain dH/H. Also, because the total stress is constant, the increase in effective stress is equal to the decrease in excess water stress (Eq. 11.49). Therefore, the average degree of consolidation U_{av} can be rewritten as:

$$U_{av} = 1 - \frac{u_{we}(average)}{u_{we(\max)}(average)} = \frac{u_{we(\max)}(av) - u_{we}(av)}{u_{we(\max)}(av)} = \frac{\Delta\sigma'(av)}{\Delta\sigma'_{\max}(av)} = \frac{M\frac{\Delta H}{H}}{M\frac{\Delta H_{\max}}{H}} = \frac{\Delta H}{\Delta H_{\max}} \quad (11.67)$$

This means that U_{av} represents the settlement of the structure divided by the maximum settlement. In contrast, because T is a function of the time t, the complete settlement vs.

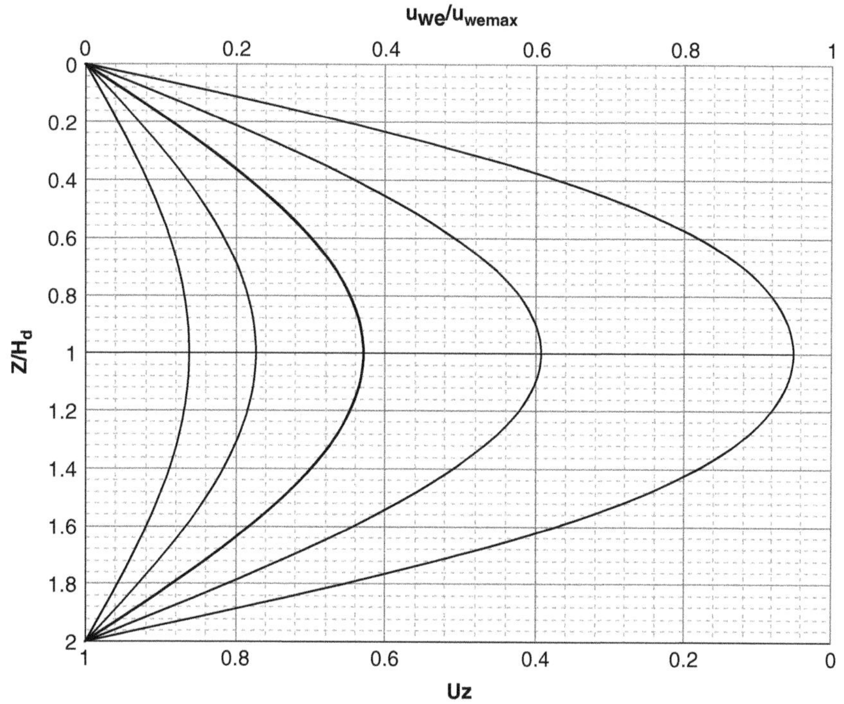

Figure 11.12 Degree of consolidation and excess water stress as a function of depth and time.

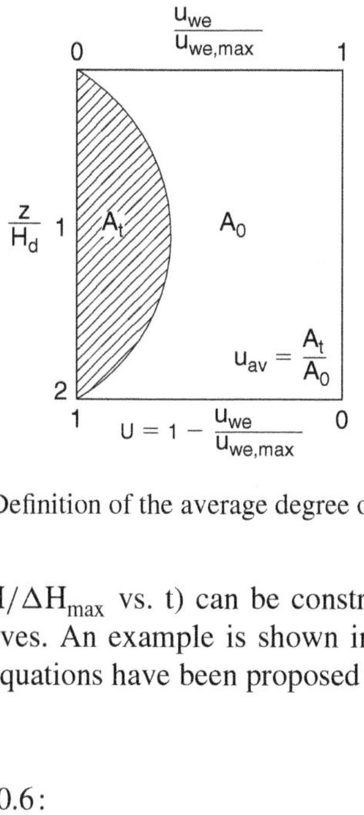

Figure 11.13 Definition of the average degree of consolidation.

time curve ($\Delta H/\Delta H_{max}$ vs. t) can be constructed by using the U vs. T curves. An example is shown in Figure 11.16. The following equations have been proposed to approximate Eq. 11.66:

For $U_{av} < 0.6$:

$$T = \frac{\pi}{4} U_{av}{}^{2} \quad or \quad \Delta H = \Delta H_{max} \frac{2}{H_d} \sqrt{\frac{c_v t}{\pi}} \tag{11.68}$$

For $U_{av} > 0.6$:

$$T = -0.933 \log(1 - U_{av}) - 0.085 \quad or$$

$$\Delta H = \Delta H_{max} \left(1 - 10^{-\left(\frac{\frac{c_v t}{H_d^2} + 0.085}{0.933} \right)} \right) \tag{11.69}$$

11.5 NUMERICAL SIMULATION METHODS

Numerical solutions typically require the use of a computer because of the complexity and amount of the mathematics involved. They tend to work as follows. The soil space or the foundation is discretized into many small elements (linear, surface, or volume). The points forming the geometry of these elements are the *nodes*. The unknowns (e.g., stresses, strains, displacements, forces, moments, flow velocity, head) have to be calculated at all the nodes. The governing differential equations are transformed into algebraic equations that must be written as many times as there are nodes in the discretized soil space. This usually yields a large number of equations organized in matrix form. From this matrix equation, the unknowns must be extracted and solved for; this often requires an inversion process of the main matrix and can only be done by computers. The output of these numerical solutions is in the form of large tables that give the calculated values of the unknowns at each node within the soil mass. Numerical methods (Jing and Hudson 2002; Bobet 2010) include the finite difference method (FDM), the finite element method (FEM), the boundary element method (BEM), and the discrete element method (DEM).

11.5.1 Finite Difference Method

The finite difference method is very powerful in solving differential equations. The main idea is to replace the differential equation by incremental algebraic equations. This is done by using algebraic expressions of the derivatives of the functions involved in the governing differential equation. In Figure 11.17, the function y(z) has values $y_{i-2}, \ldots, y_{i+2}$

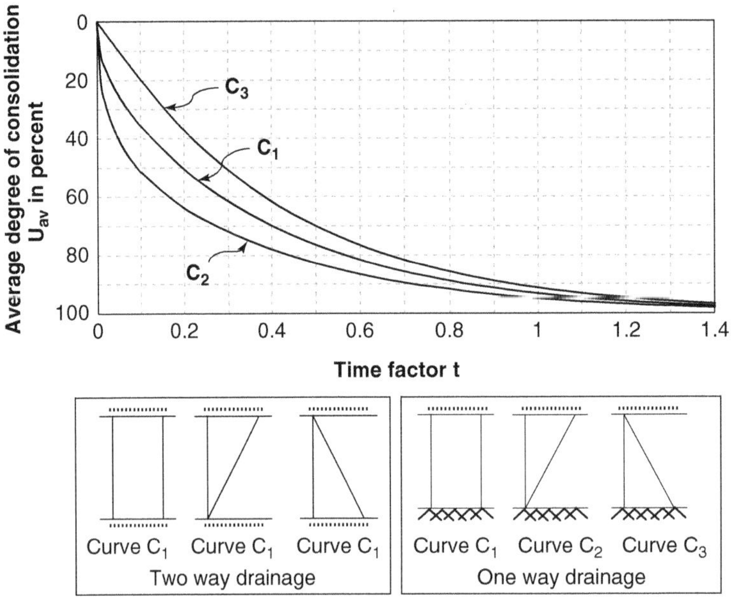

Figure 11.14 Average degree of consolidation U_{av} vs. time factor T on natural scale for different stress increase profiles.

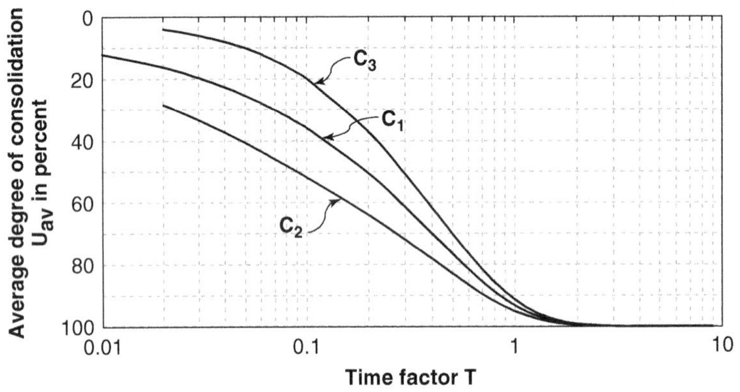

Figure 11.15 Average degree of consolidation U_{av} vs. time factor T on semilog scale.

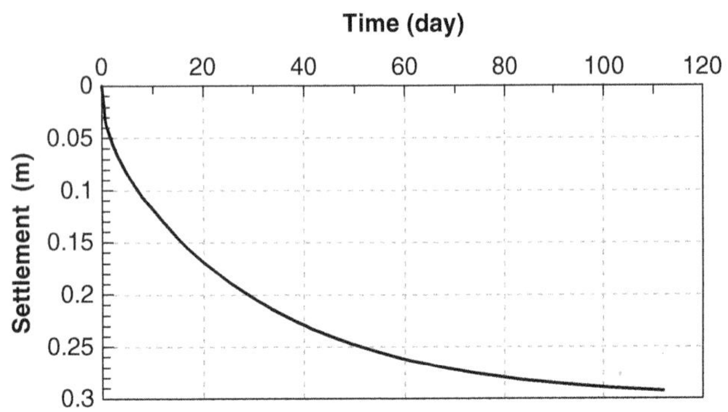

Figure 11.16 Example of settlement vs. time curve obtained from the U_{av} vs. T curve.

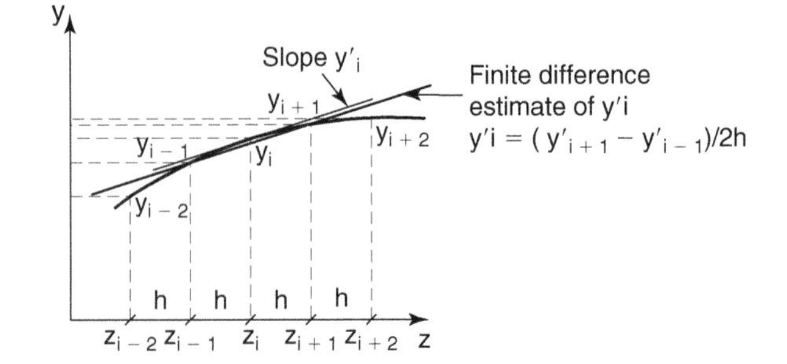

Figure 11.17 Derivative expressed by the central finite difference formulation.

corresponding to values of z equal to $z_{i-2}, \ldots, z_{i+2}$ respectively. The values of z are separated by a constant distance h. The first derivative of y evaluated at $z = z_i$ can be expressed as the slope of the tangent at z_i:

$$\frac{dy}{dz}_{@z_i} = y_i' = \frac{y_{i+1} - y_{i-1}}{2h} \tag{11.70}$$

This expression is called the *central difference expression* of the derivative, as it balances the influence of both sides of the function with respect to point i (Figure 11.17).

The forward difference would be:

$$\frac{dy}{dz}_{@z_i} = y_i' = \frac{y_{i+1} - y_i}{h} \tag{11.71}$$

and the backward difference would be:

$$\frac{dy}{dz}_{@z_i} = y_i' = \frac{y_i - y_{i-1}}{h} \tag{11.72}$$

The second derivative can be expressed using the same approach. Indeed, the second derivative is the first derivative of the first derivative. This gives the following expression, using a forward and a backward formulation for y′ to end up with a centered formulation of y″.

$$\frac{d^2y}{dz^2}_{@z_i} = \frac{dy'}{dz}_{@z_i} = y_i'' = \frac{y_{i+1}' - y_i'}{h} = \frac{\dfrac{y_{i+1} - y_i}{h} - \dfrac{y_i - y_{i-1}}{h}}{h} = \frac{y_{i+1} - 2y_i + y_{i-1}}{h^2} \tag{11.73}$$

Using the same process, the third derivative can be expressed as:

$$\frac{d^3y}{dz^3}_{@z_i} = \frac{dy''}{dz}_{@z_i} = y_i''' = \frac{y_{i+2} - 2y_{i+1} + 2y_{i-1} - y_{i-2}}{2h^3} \tag{11.74}$$

and the fourth derivative:

$$\frac{d^4y}{dz^4}_{@z_i} = \frac{dy'''}{dz}_{@z_i} = y_i'''' = \frac{y_{i+2} - 4y_{i+1} + 6y_i - 4y_{i-1} + y_{i-2}}{h^4} \tag{11.75}$$

A typical finite difference solution proceeds through the following steps:

1. The structure or soil mass involved is broken down into small elements of chosen finite dimensions. Each element has a number and each node at the boundaries of these elements has a number.
2. The knowns and unknowns (loads, displacements, stresses, strains, velocities, and heads, for example) are identified for each node and given a subscript corresponding to the node number.
3. The governing differential equation is written in algebraic finite difference form as many times as there are nodes in the structure or soil mass.
4. The space and time boundary conditions are also expressed in terms of the algebraic expressions of the variables.
5. All equations are regrouped into a matrix equation.
6. The matrix equation is solved to extract the unknown quantities. This usually requires that the matrix be inverted. Considering the size of these matrices, a computer is required for this step.
7. The solution is presented in the form of a table that gives the sought quantities at all the nodes.

11.5.2 Examples of Finite Difference Solutions

The example is to solve the governing differential equation by using the FDM for the problem of section 11.4.4: a pile subjected to a horizontal force H_o and an overturning moment M_o applied at the ground surface. The GDE is (Eq. 11.21):

$$y + \frac{E_p I}{K}\frac{d^4y}{dz^4} = 0 \tag{11.76}$$

The solution to this problem is the function y(z) describing the horizontal deflection of the pile as a function of the depth z. The process consists of the following steps:

1. The pile is discretized into elements as shown in Figure 11.18. The displacement at node i is y_i. There are a total of n + 1 unknown values of the horizontal displacement y (y_0 to y_n).

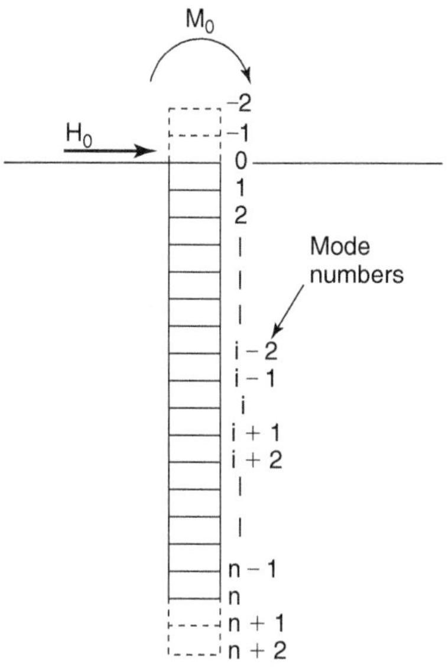

Figure 11.18 Pile discretized into numbered elements and nodes.

2. The GDE is written for each node using the expressions of the derivatives presented in section 11.5.1:

$$y_i + \frac{E_p I}{K}\frac{d^4y}{dz^4}_{@z_i} = y_i + \frac{E_p I}{K}\left(\frac{y_{i+2} - 4y_{i+1} + 6y_i - 4y_{i-1} + y_{i-2}}{h^4}\right) = 0 \quad (11.77)$$

or

$$y_{i+2} - 4y_{i+1} + \left(6 + \frac{Kh^4}{E_p I}\right) y_i - 4y_{i-1} + y_{i-2} = 0 \quad (11.78)$$

Because there are n + 1 nodes along the pile (0 to n), Eq. 11.78 theoretically could be written n + 1 times. That is not the case here, because Eq. 11.78 involves 5 nodal values of the displacement y, so in fact Eq. 11.78 can only be written n − 3 times. Because there are n + 1 values of the horizontal displacement y, we are missing four equations. Can the boundary conditions help us?

3. The boundary conditions are that the horizontal load is H_o at the ground surface and zero at the bottom of the pile and that the moment is M_o at the ground surface and zero at the bottom of the pile. To express these four boundary conditions, additional and fictitious nodes are created. These are nodes −1 and −2 at the top of the pile and nodes n + 1 and n + 2 at the bottom of the pile (Figure 11.18). The fact that the moment is M_o at the ground surface and zero at the bottom of the pile is written as:

$$M_{@z=0} = E_p I \frac{d^2y}{dz^2}_{@z=0} = E_p I\left(\frac{y_1 - 2y_0 + y_{-1}}{h^2}\right) = M_o \quad (11.79)$$

$$M_{@z=L} = E_p I \frac{d^2y}{dz^2}_{@z=L} = E_p I\left(\frac{y_{n+1} - 2y_n + y_{n-1}}{h^2}\right) = 0 \quad (11.80)$$

The fact that the shear force is H_o at the ground surface and zero at the bottom of the pile is written as:

$$V_{@z=0} = E_p I \frac{d^3y}{dz^3}_{@z=0} = E_p I\left(\frac{y_2 - 2y_1 + 2y_{-1} - y_{-2}}{h^3}\right) = H_o \quad (11.81)$$

$$V_{@z=L} = E_p I \frac{d^3y}{dz^3}_{@z=L} = E_p I\left(\frac{y_{n+2} - 2y_{n+1} + 2y_{n-1} - y_{n-2}}{h^3}\right) = 0 \quad (11.82)$$

The boundary conditions lead to four new equations, but we have also created four new unknowns (y_{-2}, y_{-1}, y_{n+1}, and y_{n+2}). Thus, the new count is n + 5 unknowns and n + 1 equations. The extra four equations are created because the additional nodes allow the GDE to be written four more times. Now we have n + 5 unknowns and n + 5 equations. These n + 5 equations are written in matrix form as:

$$[K][Y] = [C] \quad (11.83)$$

where $[K]$ is an n + 5 by n + 5 matrix of the coefficients of the y_i values in the algebraic equations corresponding to the GDE and the boundary conditions, $[Y]$ is a n + 5 long column matrix of the y values (y_{-2} to y_{n+2}), and $[C]$ is a n + 5 column matrix of the constants in the n + 5 GDE equations. Because the y values are the unknowns to be solved for, the $[K]$ matrix must be inverted and the solution is:

$$[Y] = [K]^{-1}[C] \quad (11.84)$$

This solution is illustrated by solving for the deflection and pressure distribution for a retaining wall as shown in Figure 11.19.

The units for this problem are not stated, because as long as the units are consistent the solution is independent of the units. The bending stiffness of the wall is 10,000 and the element height is 1. The soil reaction curves at each node must be prepared (Figure 11.20). The reaction curves represent the relationship between the line load P on the wall and the horizontal displacement y of the wall. A number of simplifying assumptions will be made to facilitate the solution.

At node 0, the reaction curve shows that the line load P_o is equal to zero for all y values:

$$P_0 = 0 \quad (11.85)$$

At node 1, the reaction curve is taken as a constant equal to 60. In fact, the reaction curve at node 1 should reflect the mobilization of the active pressure if the wall moves away from the soil and of the passive pressure if the wall moves into the soil. However, because the active pressure is the pressure that will be mobilized considering the problem, and because the active pressure requires very little movement to be mobilized, it is reasonable to assume that the movement will be large enough that the line load on the wall will

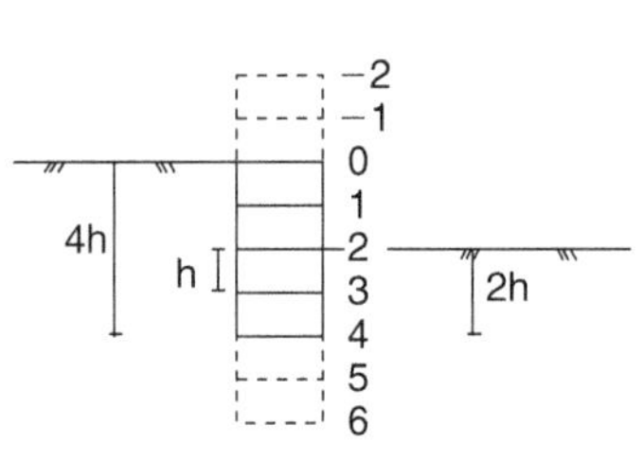

Figure 11.19 Wall discretized into numbered elements and nodes.

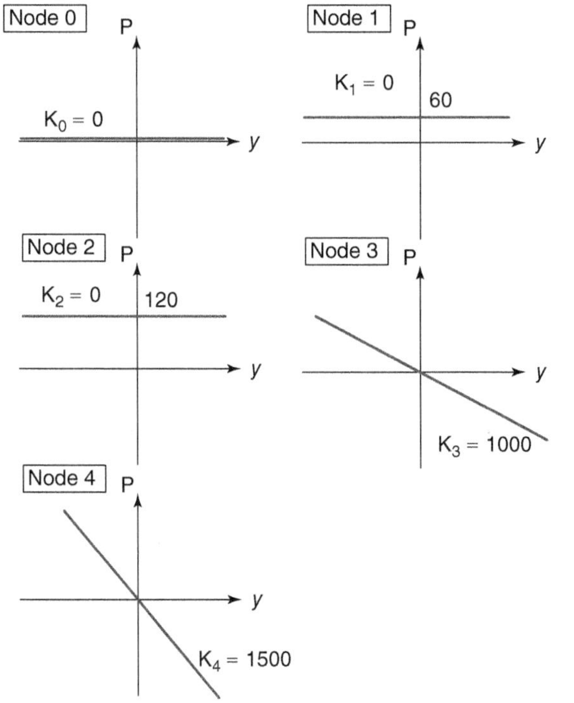

Figure 11.20 Reaction curves for the wall at each node.

correspond to the active pressure for a large range of lateral displacements:

$$P_1 = 60 \tag{11.86}$$

The same reasoning applies to the reaction curve at node 2, where the pressure is twice as high and the line load is equal to 120:

$$P_2 = 120 \tag{11.87}$$

At node 3, the reaction curve is as shown in Figure 11.20. It indicates that the line load is linearly proportional to the lateral displacement of the wall. Again, this reaction curve should reflect the influence of the active and passive pressures on both sides of the wall. The simplifying assumption in this case is that the passive resistance dominates the behavior of the wall below the excavation level. Knowing that it takes much larger displacements to mobilize the passive resistance than the active pressure, it is likely that below the excavation depth the wall will be in the range of displacement where a linear assumption is reasonable. Therefore, the reaction curve at node 3 is characterized by:

$$P_3 = -K_3 y = -1000y \tag{11.88}$$

The reason for the minus sign is that when the deflection increases to the right ($\Delta y > 0$), the line load decreases ($\Delta P < 0$). The same reasoning applies for the reaction curve at node 4, but with a higher stiffness K_4, as node 4 is deeper in the soil and therefore likely stiffer:

$$P_4 = -K_4 y = -1500y \tag{11.89}$$

As you can see, this problem has been greatly simplified compared to the real problem. The reason is that without such simplifications, the mathematics would become quite complicated.

Now the problem is clearly defined and we can proceed with the step-by-step procedure:

1. The wall has been discretized as shown in Figure 11.19.
2. The line loads and the horizontal displacements are numbered from 0 at the top of the wall to 4 at the bottom of the wall.
3. The GDE is the same as the one for the horizontally loaded pile (Eq. 11.21):

$$P - E_p I \frac{d^4 y}{dz^4} = 0 \tag{11.90}$$

Expressed in finite difference formulations, it becomes:

$$P_i - E_p I \left(\frac{y_{i+2} - 4y_{i+1} + 6y_i - 4y_{i-1} + y_{i-2}}{h^4} \right) = 0 \tag{11.91}$$

4. The boundary conditions are that the moment and the shear force are zero at both ends of the wall. This requires adding two fictitious nodes at both ends of the wall, as shown in Figure 11.19. The equations for the shear and moment are:

$$M = E_p I \left(\frac{y_{n+1} - 2y_n + y_{n-1}}{h^2} \right) = 0 \tag{11.92}$$

$$V = E_p I \left(\frac{y_{n+2} - 2y_{n+1} + 2y_{n-1} - y_{n-2}}{h^3} \right) = 0 \tag{11.93}$$

5. Now all the equations can be written and assembled in a matrix:

$$\begin{bmatrix} -1 & 2 & 0 & -2 & 1 & 0 & 0 & 0 & 0 \\ 0 & 1 & -2 & 1 & 0 & 0 & 0 & 0 & 0 \\ 1 & -4 & 6 & -4 & 1 & 0 & 0 & 0 & 0 \\ 0 & 1 & -4 & 6 & -4 & 1 & 0 & 0 & 0 \\ 0 & 0 & 1 & -4 & 6 & -4 & 1 & 0 & 0 \\ 0 & 0 & 0 & 1 & -4 & 6.1 & -4 & 1 & 0 \\ 0 & 0 & 0 & 0 & 1 & -4 & 6.15 & -4 & 1 \\ 0 & 0 & 0 & 0 & 0 & 1 & -2 & 1 & 0 \\ 0 & 0 & 0 & 0 & -1 & 2 & 0 & -2 & 1 \end{bmatrix}$$

$$\times \begin{bmatrix} y_{-2} \\ y_{-1} \\ y_0 \\ y_1 \\ y_2 \\ y_3 \\ y_4 \\ y_5 \\ y_6 \end{bmatrix} = \begin{bmatrix} 0 \\ 0 \\ 0 \\ 0.006 \\ 0.012 \\ 0 \\ 0 \\ 0 \\ 0 \end{bmatrix} \tag{11.94}$$

Table 11.1 Results of the Finite Difference Solution for the Simulated Wall

Node number (depth)	Horizontal deflection y of wall	Line load P on wall
1	2.72	0
2	1.95	60
3	1.18	120
4	0.42	−420
5	−0.32	480

The first two equations in the matrix equation are the two equations for the moment and shear boundary conditions at the top of the wall; then there are five GDEs written at five nodes; and the last two equations are the two equations for the moment and shear boundary conditions at the bottom of the wall. Now it is time to invert the matrix to obtain the $[Y]$ matrix as the solution to the problem.

6. The computer does that for us, and the deflections y at each node are calculated. The line loads on the wall are obtained by using the relationship between the load and the deflection given by the reaction curves of Eqs. 11.85 to 11.89. The results of this finite difference solution are presented in Table 11.1.

The deflection profile y(z) and the line load profile P(z) are shown in Figure 11.21. The profile P(z) shows that the wall is in horizontal equilibrium because the area under the left side of the profile is equal to the areas under the right side of the profile. This is the way it should be, as horizontal equilibrium was one of the fundamental equations we used in setting up the solution.

11.5.3 Finite Element Method

The finite element method (FEM) is another powerful numerical method to solve geotechnical problems (Clough 1960; Desai and Abel 1972; Zienkiewicz et al. 2005). The output, like most numerical methods, will be in the form of tables giving the unknown quantities at discrete locations in the soil mass. The general steps in developing a solution to a finite element simulation are as follows:

1. Discretize the soil mass into finite elements connected by nodes.
2. Choose the functions describing the variation of the unknowns across each element and between its nodes.
3. Write the strain-displacement equations.
4. Write the stress-strain equations for the soil.
5. Derive the equations governing the behavior of the soil element.
6. Assemble the element equations into the global matrix equation.
7. Introduce the boundary conditions into the global matrix equation.
8. Solve the global matrix equation for the unknowns.

Each step is discussed in more detail here.

1. *Discretize the soil mass into finite elements connected by nodes.* In this step the soil mass is subdivided into a number of small elements (Figure 11.22). The sides of the elements intersect at the nodes. Each element and each node is numbered in sequence. The size of the elements is influenced by a number of factors, including how fast the stress changes from one point to another of the soil mass (stress gradient). Various shapes of elements exist: lines, triangles, quadrilaterals, parallelepipeds, or brick elements. One of the big advantages of the FEM is that irregular boundaries do not present a big problem.

2. *Choose the functions describing the variation of the unknowns across each element and between its nodes.* These are called *interpolation functions* or *shape functions*. The solution of the FEM will give the answers at the nodes (Fig. 11.23), but we need to be able to calculate the unknowns everywhere in the mass to establish the general equations. The interpolation functions relate for example the displacement anywhere in the element to the displacements at the nodes. These interpolation functions are typically in the form of polynomials. It is more convenient, however, to write them in the following form:

$$u_x(x, y) = H_1 u_{x1} + H_2 u_{x2} + H_3 u_{x3} + H_4 u_{x4} = \sum_{i=1}^{\#nodes} H_i u_{xi} \tag{11.95}$$

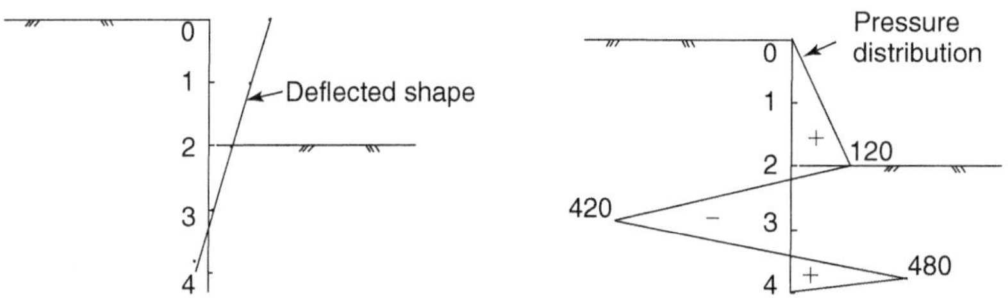

Figure 11.21 Wall deflection and line load.

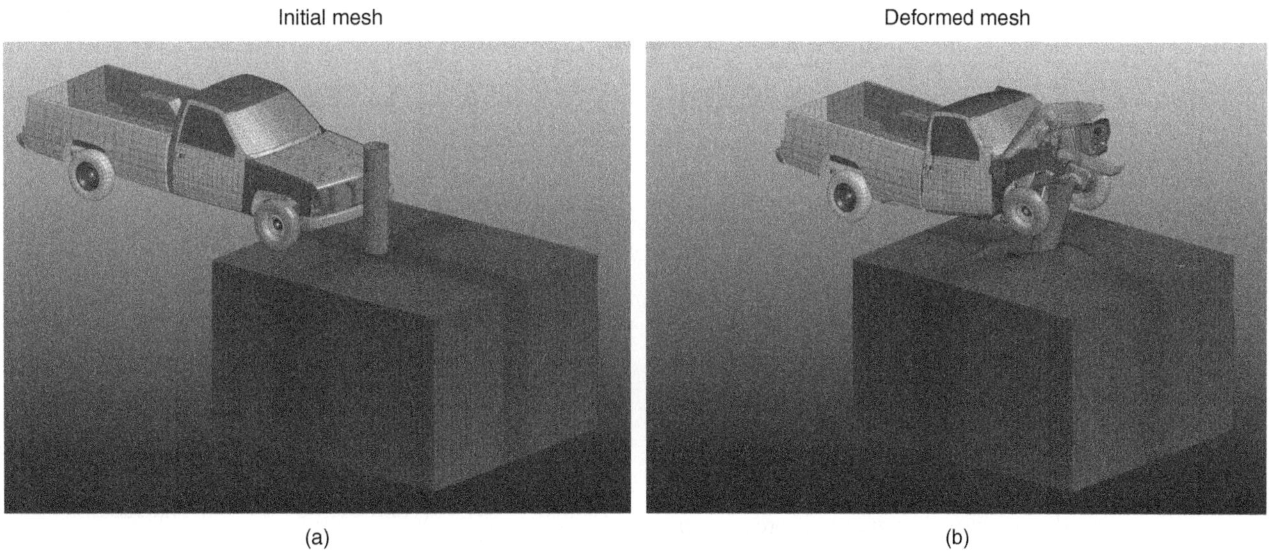

Figure 11.22 Example of finite element mesh: (*a*) Initial mesh. (*b*) Deformed mesh.

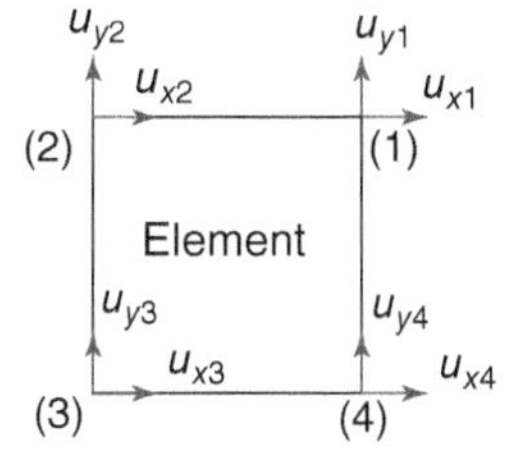

Figure 11.23 Element in plane strain.

$$u_y(x, y) = H_1 u_{y1} + H_2 u_{y2} + H_3 u_{y3} + H_4 u_{y4} = \sum_{i=1}^{\#nodes} H_i u_{yi} \tag{11.96}$$

where $u_x(x, y)$ is the displacement in the x direction of any point within the element with coordinates x and y, u_{xi} is the displacement in the x direction of node i, $u_y(x, y)$ is the displacement in the y direction of any point within the element with coordinates x and y, u_{yi} is the displacement in the y direction of node i, and the H_is are the interpolation functions. Equations 11.95 and 11.96 would be for an element with four nodes and plain strain condition in the z direction. They describe the shape of the displacement surface across the element.

In matrix form:

$$\begin{bmatrix} u_x \\ u_y \end{bmatrix} = \begin{bmatrix} H_1 & 0 & H_2 & 0 & H_3 & 0 & H_4 & 0 \\ 0 & H_1 & 0 & H_2 & 0 & H_3 & 0 & H_4 \end{bmatrix} \begin{bmatrix} u_{x1} \\ u_{y1} \\ u_{x2} \\ u_{y2} \\ u_{x3} \\ u_{y3} \\ u_{x4} \\ u_{y4} \end{bmatrix}$$

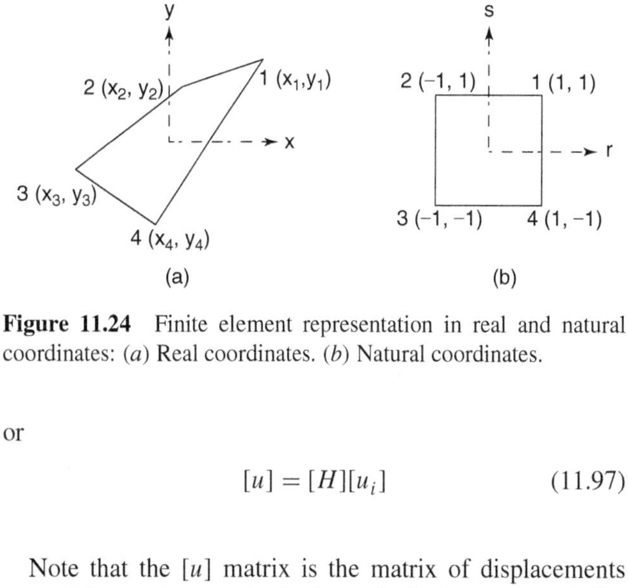

Figure 11.24 Finite element representation in real and natural coordinates: (*a*) Real coordinates. (*b*) Natural coordinates.

or

$$[u] = [H][u_i] \tag{11.97}$$

Note that the $[u]$ matrix is the matrix of displacements as variables, whereas the $[u_i]$ matrix is the matrix of displacements at the nodes. The $[H]$ matrix is called the *shape function matrix*. Note also that these matrices are written for the element and not for the entire soil mass.

Regarding the coordinates x and y, it is more convenient to use natural coordinates (Figure 11.24). As can be seen, regardless of the element's original shape, the transformation leads to a set of coordinates varying from −1 to +1 along each face. Also, the element is square. The interpolation functions for a four-node element in natural coordinates are:

$$H_1 = \frac{1}{4}(1 + r)(1 + s) \tag{11.98}$$

$$H_2 = \frac{1}{4}(1 - r)(1 + s) \tag{11.99}$$

$$H_3 = \frac{1}{4}(1-r)(1-s) \quad (11.100)$$

$$H_4 = \frac{1}{4}(1+r)(1-s) \quad (11.101)$$

where r and s are the natural coordinates (Figure 11.24).

In the general case, coordinates can be expressed in terms of interpolation functions as follows:

$$x = \sum_{i=1}^{\#nodes} H_i x_i \quad (11.102)$$

$$y = \sum_{i=1}^{\#nodes} H_i y_i \quad (11.103)$$

3. *Write the strain-displacement equations.* There are typically 9 equations: 3 force equilibrium equations and 6 constitutive equations linking the stresses to the strains. The other equations are the 3 moment equilibrium equations, but they simply lead to the fact that shear stresses on perpendicular planes are equal and in opposite directions so they have already been used up. However, there are 15 unknowns: 6 stresses, 6 strains, and 3 displacements. So we are short 6 equations. What saves the day is that the 6 strains are defined on the basis of the 3 displacements, so this adds 6 strain-displacement equations. In the end we have 15 unknowns and 15 equations.

Recalling Eq. 11.95, the normal strain in the x direction is ε_{xx}:

$$\varepsilon_{xx} = \frac{\partial u_x(x,y)}{\partial x} = \left[\frac{\partial H}{\partial x}\right][u_{xi}] \quad (11.104)$$

The same equation holds true for ε_{yy}:

$$\varepsilon_{yy} = \frac{\partial u_y(x,y)}{\partial y} = \left[\frac{\partial H}{\partial y}\right][u_{yi}] \quad (11.105)$$

For the shear strain γ_{xy}, the equation becomes:

$$\gamma_{xy} = \frac{\partial u_x}{\partial y} + \frac{\partial u_y}{\partial x} = \left[\frac{\partial H}{\partial y}\right][u_{xi}] + \left[\frac{\partial H}{\partial x}\right][u_{yi}] \quad (11.106)$$

or, in matrix form:

$$[\varepsilon] = [B][u_i] \quad (11.107)$$

where $[\varepsilon]$ is the strain matrix (3 × 1 vector for a two-dimensional problem), $[B]$ is the matrix containing the derivatives of the interpolation functions H_i (3 × 8 for a two-dimensional problem), and $[u_i]$ is the matrix of nodal displacements (8 × 1 for a two-dimensional problem).

$$\begin{bmatrix}\varepsilon_{xx}\\ \varepsilon_{yy}\\ \gamma_{xy}\end{bmatrix} = \begin{bmatrix}\frac{\partial H_1}{\partial x} & 0 & \frac{\partial H_2}{\partial x} & 0 & \frac{\partial H_3}{\partial x} & 0 & \frac{\partial H_4}{\partial x} & 0\\ 0 & \frac{\partial H_1}{\partial y} & 0 & \frac{\partial H_2}{\partial y} & 0 & \frac{\partial H_3}{\partial y} & 0 & \frac{\partial H_4}{\partial y}\\ \frac{\partial H_1}{\partial y} & \frac{\partial H_1}{\partial x} & \frac{\partial H_2}{\partial y} & \frac{\partial H_2}{\partial x} & \frac{\partial H_3}{\partial y} & \frac{\partial H_3}{\partial x} & \frac{\partial H_4}{\partial y} & \frac{\partial H_4}{\partial x}\end{bmatrix} \times \begin{bmatrix}u_{x1}\\ u_{y1}\\ u_{x2}\\ u_{y2}\\ u_{x3}\\ u_{y3}\\ u_{x4}\\ u_{y4}\end{bmatrix} \quad (11.108)$$

Because the interpolation functions H_i are defined in natural coordinates, the derivatives $\partial H_i/\partial x$ and $\partial H_i/\partial y$ can be related to $\partial H_i/\partial r$ and $\partial H_i/\partial s$ through the Jacobian matrix [J] as follows:

$$\begin{bmatrix}\frac{\partial H_i}{\partial x}\\ \frac{\partial H_i}{\partial y}\end{bmatrix} = J^{-1}\begin{bmatrix}\frac{\partial H_i}{\partial r}\\ \frac{\partial H_i}{\partial s}\end{bmatrix} \quad (11.109)$$

where the Jacobian matrix $[J]$ is described as follows:

$$J = \begin{bmatrix}\frac{\partial x}{\partial r} & \frac{\partial y}{\partial r}\\ \frac{\partial x}{\partial s} & \frac{\partial y}{\partial s}\end{bmatrix} \quad (11.110)$$

Recalling Eq. 11.102 and Eq. 11.103, the components of this Jacobian matrix are written as follows:

$$\frac{\partial x}{\partial r} = \sum \frac{\partial H_i}{\partial r}.x_i \quad (11.111)$$

$$\frac{\partial x}{\partial s} = \sum \frac{\partial H_i}{\partial s}.x_i \quad (11.112)$$

$$\frac{\partial y}{\partial r} = \sum \frac{\partial H_i}{\partial r}.y_i \quad (11.113)$$

$$\frac{\partial y}{\partial s} = \sum \frac{\partial H_i}{\partial s}.y_i \quad (11.114)$$

4. *Write the stress-strain equations for the soil.* These are the constitutive equations, the ones that are specific to the soil involved. One of the simplest constitutive laws is the case where the stresses are linearly related to the strains (elasticity):

$$[\sigma] = [C][\varepsilon] \quad (11.115)$$

where $[\sigma]$ is the stress matrix, which is a 3 × 1 matrix for a two-dimensional problem and a 6 × 1 matrix for a three-dimensional problem; $[\varepsilon]$ is the strain matrix, which is a 3 × 1 matrix for a two-dimensional problem and a 6 × 1 matrix for a three-dimensional problem; and $[C]$ is the soil stiffness matrix, which is a 3 × 3 matrix for a two-dimensional problem and a 6 × 6 matrix for a three-dimensional problem. In elasticity and for three dimensions, Eq. 11.115 is written as:

$$\begin{bmatrix} \sigma_{xx} \\ \sigma_{yy} \\ \sigma_{zz} \\ \tau_{xy} \\ \tau_{yz} \\ \tau_{zx} \end{bmatrix} = \frac{E}{(1-2\nu)(1+\nu)} \begin{bmatrix} 1-\nu & \nu & \nu & 0 & 0 & 0 \\ \nu & 1-\nu & \nu & 0 & 0 & 0 \\ \nu & \nu & 1-\nu & 0 & 0 & 0 \\ 0 & 0 & 0 & \frac{1}{2}-\nu & 0 & 0 \\ 0 & 0 & 0 & 0 & \frac{1}{2}-\nu & 0 \\ 0 & 0 & 0 & 0 & 0 & \frac{1}{2}-\nu \end{bmatrix} \times \begin{bmatrix} \varepsilon_{xx} \\ \varepsilon_{yy} \\ \varepsilon_{zz} \\ \varepsilon_{xy} \\ \varepsilon_{yz} \\ \varepsilon_{zx} \end{bmatrix} \quad (11.116)$$

In the case of two dimensions, the C matrix becomes:

$$C = \frac{E}{(1-2\nu)(1+\nu)} \begin{bmatrix} 1-\nu & \nu & 0 \\ \nu & 1-\nu & 0 \\ 0 & 0 & \frac{1}{2}-\nu \end{bmatrix} \quad (11.117)$$

5. *Derive the equations governing the behavior of the soil element.* You may have noticed that we have not yet written any fundamental equations such as the equilibrium equations. We do that in this step, but it is done by using another technique called a *variational principle* that includes the equilibrium equations. More specifically, we use the minimum total potential energy principle (MTPE) along with the virtual work technique. The MTPE principle states that the actual displacement solution of a deformable body is the solution that renders the TPE functional Π minimum, meaning that the derivative of Π is equal to zero. The two types of energies involved in the TPE are the work done by the external forces W and the internal strain energy U of the deformable soil mass. The TPE is minimum (system in equilibrium) when the change in work done by the external forces δW is equal to the change in internal strain energy δU:

$$\delta \Pi = 0 \quad \rightarrow \quad \delta U = \delta W \quad (11.118)$$

The increment of virtual internal strain energy δU for a bar is:

$$\delta U = \sigma A \, \delta\varepsilon \, dx = \delta\varepsilon \, \sigma \, dV \quad (11.119)$$

where σ is the axial stress, A is the cross-sectional area, $\delta\varepsilon$ is a virtual axial strain, and dx and dV are an infinitesimal length and volume of the bar. This is generalized for the three-dimensional soil element as:

$$\delta U = \int_V \delta[\varepsilon]^T [\sigma] dV \quad (11.120)$$

The change in work is calculated by assuming that the soil mass is subjected to virtual small displacements (virtual work). The increment in virtual external work δW for a bar is:

$$\delta W = F_{body} \delta u + F_{boundary} \delta u = bdV\delta u + tdA\delta u \quad (11.121)$$

where F_{body} is the body force, δu is a virtual displacement, F is the boundary force, b is the body force density (e.g., unit weight), t is the boundary tractions (e.g., pressure), and dV and dA are an infinitesimal volume and area of the bar. This is generalized for the three-dimensional soil element as:

$$\delta W = \int_V \delta[u]^T [b] dV + \int_V \delta[u]^T [t] dA \quad (11.122)$$

Then the principle of virtual work states that the expressions in Eq. 11.120 and 11.122 are equal:

$$\int_V \delta[\varepsilon]^T [\sigma] dV = \int_V \delta[u]^T [b] dV + \int_V \delta[u]^T [t] dA \quad (11.123)$$

Using Eqs. 11.107 and 11.115, we get:

$$\int_V \delta[u]^T [B]^T [C][B] \delta[u] dV = \delta[u]^T \left([F_{body}] + [F_{boundary}] \right) = \delta[u]^T [F] \quad (11.124)$$

The element stiffness matrix K^e is defined as:

$$[K^e] = \int_V [B]^T [C][B] dV \quad (11.125)$$

To calculate the integral on the right side of Eq. 11.125, we select integration points where all the components of the B and C matrices are evaluated. In the special case of a plane strain problem, the components of the stiffness matrix can be reduced to the following expression:

$$t \iint_{Area} f_{mn}(x, y) dx dy = \sum_{i=1}^{2} \sum_{j=1}^{2} t \times f_{mn}(r_i, s_j) . \det J . w_i . w_j \quad (11.126)$$

where t is the thickness of the element (1 in plane strain cases), $f_{mn}(x, y)$ is the function found at the intersection of the m row and n column of the B^tCB matrix of Eq. 11.125 expressed in real coordinates, $f_{mn}(r, s)$ is the same function but expressed in natural coordinates, i and j are the running indices identifying the location of the integration point, r_i and s_j are the natural coordinates of the chosen integration points on the element, w_i and w_j are the weighting factors that depend on the number and location of the integration points, and det J is the determinant of the Jacobian matrix. In the general case, the thickness is not a constant and must be calculated at each integration point by using the interpolation functions (see problem 11.7). Figure 11.25 shows an example of four integration points.

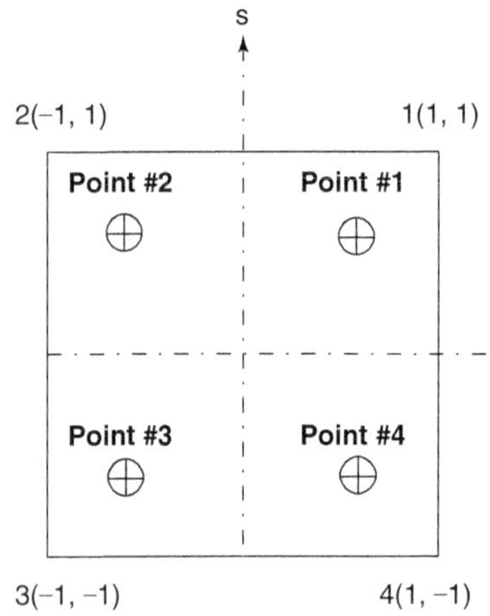

Figure 11.25 Four integration points.

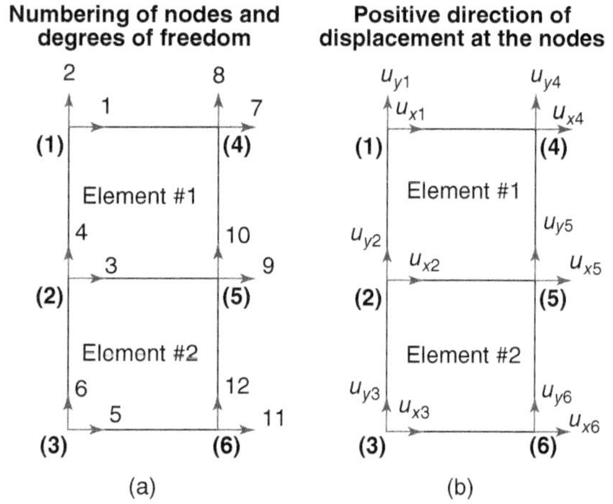

Figure 11.26 Two 2D FEM elements and numbering the degrees of freedom: (*a*) Number of nodes and degrees of freedom. (*b*) Positive direction of displacements at nodes.

Because Eq. 11.124 must be satisfied for any kinematically admissible virtual displacement field $[u]$, we must have:

$$[K^e][u] = [F] \tag{11.127}$$

In Eq. 11.127, most of the displacements u are unknown and most of the forces are either zero or known. This is the equation governing the behavior of the element. If the element were a spring, K would be the spring constant, but in the case of the three-dimensional element, K is a square matrix.

6. *Assemble the element equations into the global matrix equation.* Equation 11.126 is the equation for one element. There are as many such matrix equations as there are elements in the mesh. They must be assembled to form the stiffness matrix for the entire soil mass. To do so, we specify that the body must remain continuous during the deformation. This means that each node can have only one displacement vector common to all elements containing this node. At each node, we also have only one body force and one external force value. The following example illustrates how the global matrix is assembled.

Consider the two elements of Figure 11.26. The stiffness matrices for the 2 elements and their assembly into the global matrix of the soil mass of the 2 elements are shown in Figure 11.27. As can be seen, the coefficients of the individual element matrices are labeled K^i_{jk}. The index i designates the element number, j refers to the node number corresponding to the force F_j, and k refers to the number of the node where a displacement u_k contributes an additional displacement at node j. With these definitions for the indices, the stiffness coefficients for adjacent elements are simply added when they refer to the same j and k values while coming from different elements i. This simple example is extended to all nodes in the mesh to form the global stiffness matrix $[K]$. Then the global governing equation for the entire soil mass is:

$$[K][u] = [F] \tag{11.128}$$

Figs. 11.26 and 11.27 show how to assemble the global matrix for two four-node elements.

7. *Introduce the boundary conditions into the global matrix equation.* Equation 11.128 describes how the soil mass will behave in general terms. The boundary conditions make the problem specific. These boundary conditions (also called *constraints*) are given in the way of specified values of displacements, forces, temperatures, or any other parameters that affect the problem. In dynamics, these conditions involve the same types of parameters, but all of them are associated with a specific time. Examples of boundary conditions include requiring no movement at a node ($u^i_x = u^i_y = u^i_z = 0$), no external force at a node ($F^i_x = F^i_y = F^i_z = 0$), or movement at a node allowed only in one direction, or a single force applied at a node. The specified values of displacement and forces go directly into the $[u]$ and $[F]$ matrices. Of course, for problems other than deformation problems, the boundary conditions are different and can be in terms of specified flow velocities, heat flux, and so on.

8. *Solve the global matrix equation for the unknowns.* The matrix equation to be solved is:

$$[K][u] = [F] \tag{11.129}$$

In a three-dimensional problem, the $[K]$ matrix is a 3n × 3n matrix where n is the number of nodes; the $[u]$ matrix is a 3n × 1 matrix; and the $[F]$ matrix is also a 3n × 1 matrix. The reason it is 3n is that there are 3 directions at each node with 3 associated displacements and 3 associated forces. The displacement vectors and the force vectors will be:

$$\begin{bmatrix} u^1_x \\ u^1_y \\ u^1_z \\ \cdot \\ \cdot \\ \cdot \\ u^n_x \\ u^n_y \\ u^n_z \end{bmatrix} \quad \begin{bmatrix} F^1_x \\ F^1_y \\ F^1_z \\ \cdot \\ \cdot \\ \cdot \\ F^n_x \\ F^n_y \\ F^n_z \end{bmatrix}$$

Stiffness Matrix of Element #1

$$\begin{bmatrix} K^1_{77} & K^1_{78} & K^1_{71} & K^1_{72} & K^1_{73} & K^1_{74} & K^1_{79} & K^1_{7,10} \\ & K^1_{88} & K^1_{81} & K^1_{82} & K^1_{83} & K^1_{84} & K^1_{89} & K^1_{8,10} \\ & & K^1_{11} & K^1_{12} & K^1_{13} & K^1_{14} & K^1_{19} & K^1_{1,10} \\ & & & K^1_{22} & K^1_{23} & K^1_{24} & K^1_{29} & K^1_{2,10} \\ & & & & K^1_{33} & K^1_{34} & K^1_{39} & K^1_{3,10} \\ & SYM & & & & K^1_{44} & K^1_{49} & K^1_{4,10} \\ & & & & & & K^1_{99} & K^1_{9,10} \\ & & & & & & & K^1_{10,10} \end{bmatrix}$$

(a)

Stiffness Matrix of Element #2

$$\begin{bmatrix} K^2_{99} & K^2_{9,10} & K^2_{93} & K^2_{94} & K^2_{95} & K^2_{96} & K^2_{9,11} & K^2_{9,12} \\ & K^2_{10,10} & K^2_{10,3} & K^2_{10,4} & K^2_{10,5} & K^2_{10,6} & K^2_{10,11} & K^2_{10,12} \\ & & K^2_{33} & K^2_{34} & K^2_{35} & K^2_{36} & K^2_{3,11} & K^2_{3,12} \\ & & & K^2_{44} & K^2_{45} & K^2_{46} & K^2_{4,11} & K^2_{4,12} \\ & & & & K^2_{55} & K^2_{56} & K^2_{5,11} & K^2_{5,12} \\ & SYM & & & & K^2_{66} & K^2_{6,11} & K^2_{6,12} \\ & & & & & & K^2_{11,11} & K^2_{11,12} \\ & & & & & & & K^2_{12,12} \end{bmatrix}$$

(b)

Assembled Global Stiffness Matrix

$$\begin{array}{c} 1 \\ 2 \\ 3 \\ 4 \\ 5 \\ 6 \\ 7 \\ 8 \\ 9 \\ 10 \\ 11 \\ 12 \end{array}
\begin{bmatrix} K^1_{11} & K^1_{12} & K^1_{13} & K^1_{14} & 0 & 0 & K^1_{17} & K^1_{18} & K^1_{19} & K^1_{1,10} & 0 & 0 \\ & K^1_{22} & K^1_{23} & K^1_{24} & 0 & 0 & K^1_{27} & K^1_{28} & K^1_{29} & K^1_{2,10} & 0 & 0 \\ & & K^1_{33}+K^2_{33} & K^1_{34}+K^2_{34} & K^2_{35} & K^2_{36} & K^1_{37} & K^{11}_{38} & K^1_{39}+K^2_{39} & K^1_{3,10}+K^2_{3,10} & K^2_{3,11} & K^2_{3,12} \\ & & & K^1_{44}+K^2_{44} & K^2_{45} & K^2_{46} & K^1_{47} & K^1_{48} & K^1_{49}+K^2_{49} & K^1_{4,10}+K^2_{4,10} & K^2_{4,11} & K^2_{4,12} \\ & & & & K^2_{55} & K^2_{56} & 0 & 0 & K^2_{59} & K^2_{5,10} & K^2_{5,11} & K^2_{5,12} \\ & & & & & K^2_{66} & 0 & 0 & K^2_{69} & K^2_{6,10} & K^2_{6,11} & K^2_{6,12} \\ & & & & & & K^1_{77} & K^1_{78} & K^1_{79} & K^1_{7,10} & 0 & 0 \\ & & & & & & & K^1_{88} & K^1_{89} & K^1_{8,10} & 0 & 0 \\ & & SYM & & & & & & K^1_{99}+K^2_{99} & K^1_{9,10}+K^2_{9,10} & K^2_{9,11} & K^2_{9,12} \\ & & & & & & & & & K^1_{10,10}+K^2_{10,10} & K^2_{10,11} & K^2_{10,12} \\ & & & & & & & & & & K^2_{11,11} & K^2_{11,12} \\ & & & & & & & & & & & K^2_{12,12} \end{bmatrix}$$

(c)

Figure 11.27 Assembling the global stiffness matrix: (*a*) Stiffness of matrix of element #1. (*b*) Stiffness matrix of element #2. (*c*) Assembled global stiffness matrix.

In these vectors most of the unknowns are the displacements at the nodes, except for the displacement boundary conditions. However, most of the forces at the nodes are known and are zero. Remember that we are talking about the external forces, not the internal forces. The soil experiences stresses all over its mass, but the external forces at the nodes are zero except at supports or at boundary conditions. This distinction between internal forces and external forces is critically important and can be illustrated as follows.

Consider a simply supported beam resting on rigid supports at both ends. Place a heavy load in the center of the beam. If the beam is in equilibrium, the external moment is zero everywhere along the beam, but the internal moment (bending moment) is significant along most of the beam. You know the displacement at both ends (zero), but you do not know the force (support reaction). Along the rest of the beam, you do not know the displacement, but you know the force, which is zero except in the center where the force is equal to the applied external load.

The same principle applies to the finite element method and Eq. 11.128. The displacement matrix $[u]$ is largely unknown and the external force matrix $[F]$ is largely known. Therefore, because we want to know $[u]$, it will be necessary to invert the stiffness matrix $[K]$ to get the displacements:

$$[u] = [K]^{-1}[F] \tag{11.130}$$

Because the global stiffness matrix is very large, this operation can require a lot of time when the mesh has many elements. Techniques for optimizing this operation have been developed in mathematics, including matrix banding. This banding is affected by the numbering of the nodes and it is always desirable to ensure that neighboring nodes do not have very different numbers.

One issue arises with a boundary condition that specifies a displacement: say, $u_i = \delta$. An example may be a support where no movement is allowed. In this case, the displacement is zero but the force is unknown. To solve the matrix problem (Eq. 11.130), all unknowns must be in the displacement matrix and all values in the force matrix must be known. To satisfy this mathematical need, the following trick is applied. The known displacement is entered in the displacement matrix as an unknown u_i. The corresponding force is entered as the value of the known displacement δ to form the modified force matrix F' and the corresponding row (row i) in the K matrix is set to be all zeroes except for the diagonal value, which is 1. The same applies to column i, because the matrix is symmetrical. That way the ith equation simply says that $u_i = \delta$. This is repeated for all such cases and gives rise to a new matrix K′. The matrix K' is inverted and all displacements at all nodes are found by:

$$[u] = [K']^{-1}[F'] \tag{11.131}$$

Then the complete force matrix F is found as the matrix product Ku:

$$[F] = [K][u] \tag{11.132}$$

Once the displacement matrix is obtained, the strains and stresses can be obtained by using the strain-displacements relationships (Eq. 11.107) and the stress-strain relationships (Eq. 11.115).

11.5.4 Example of Finite Element Solution

Use the FEM to solve the deformation field for a test performed on an elastic soil. The height of the sample is 0.1 m, the width is 0.05 m, and the length is infinite. The major principal stress is 300 kPa and the minor principal stress is 100 kPa. The modulus is 40 MPa and the Poisson's ratio is 0.35. Consider a plane strain geometry and use two four-noded elements. Use numerical integration with four points to construct the stiffness matrix.

Step 1: Discretize the soil mass into finite elements connected by nodes

The elements are shown in Figure 11.28. The element dimensions are $a = 0.05$ m and $b = 0.05$ m; the soil properties are $E = 40{,}000$ kPa and $\mu = 0.35$.

Step 2: Choose the interpolation functions in natural coordinates

Recalling Eqs. 11.98 to 11.101, these functions are considered:

$$H_1 = \frac{1}{4}(1+r)(1+s) \tag{11.133}$$

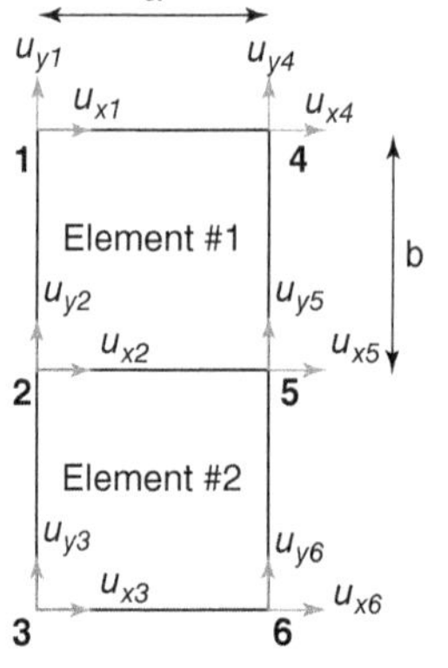

Figure 11.28 Triaxial test in plane strain.

$$H_2 = \frac{1}{4}(1-r)(1+s) \tag{11.134}$$

$$H_3 = \frac{1}{4}(1-r)(1-s) \tag{11.135}$$

$$H_4 = \frac{1}{4}(1+r)(1-s) \tag{11.136}$$

Step 3: Write the strain-displacement equations

$$[\varepsilon] = [B][u_i] \tag{11.137}$$

$$\begin{bmatrix} \varepsilon_{xx} \\ \varepsilon_{yy} \\ \gamma_{xy} \end{bmatrix} = \begin{bmatrix} \frac{\partial H_1}{\partial x} & 0 & \frac{\partial H_2}{\partial x} & 0 & \frac{\partial H_3}{\partial x} & 0 & \frac{\partial H_4}{\partial x} & 0 \\ 0 & \frac{\partial H_1}{\partial y} & 0 & \frac{\partial H_2}{\partial y} & 0 & \frac{\partial H_3}{\partial y} & 0 & \frac{\partial H_4}{\partial y} \\ \frac{\partial H_1}{\partial y} & \frac{\partial H_1}{\partial x} & \frac{\partial H_2}{\partial y} & \frac{\partial H_2}{\partial x} & \frac{\partial H_3}{\partial y} & \frac{\partial H_3}{\partial x} & \frac{\partial H_4}{\partial y} & \frac{\partial H_4}{\partial x} \end{bmatrix} \times \begin{bmatrix} u_{x1} \\ u_{y1} \\ u_{x2} \\ u_{y2} \\ u_{x3} \\ u_{y3} \\ u_{x4} \\ u_{y4} \end{bmatrix} \tag{11.138}$$

Constructing the [B] Matrix.

a. Calculate the inverse of the Jacobian matrix used in the transformation from natural coordinates to real coordinates.

$$J = \begin{bmatrix} \frac{\partial x}{\partial r} & \frac{\partial y}{\partial r} \\ \frac{\partial x}{\partial s} & \frac{\partial y}{\partial s} \end{bmatrix} = \begin{bmatrix} \frac{a}{2} & 0 \\ 0 & \frac{b}{2} \end{bmatrix} = \begin{bmatrix} 0.025 & 0 \\ 0 & 0.025 \end{bmatrix} \tag{11.139}$$

Therefore:

$$\det J = 6.25 * 10^{-4}$$

and

$$J^{-1} = \left(\frac{1}{\det J}\right) \cdot \begin{bmatrix} \frac{b}{2} & 0 \\ 0 & \frac{a}{2} \end{bmatrix} = \begin{bmatrix} 40 & 0 \\ 0 & 40 \end{bmatrix} \tag{11.140}$$

b. Obtain the relation between the derivatives of the interpolation functions in real coordinates and in natural coordinates:

$$\begin{bmatrix} \frac{\partial H_i}{\partial x} \\ \frac{\partial H_i}{\partial y} \end{bmatrix} = J^{-1} \begin{bmatrix} \frac{\partial H_i}{\partial r} \\ \frac{\partial H_i}{\partial s} \end{bmatrix} \tag{11.141}$$

$$\begin{bmatrix} \dfrac{\partial H}{\partial x} \\ \dfrac{\partial H}{\partial y} \end{bmatrix} = \begin{bmatrix} \dfrac{b}{2} & 0 \\ 0 & \dfrac{a}{2} \end{bmatrix} \cdot \begin{bmatrix} \dfrac{\partial H}{\partial r} \\ \dfrac{\partial H}{\partial s} \end{bmatrix} = \begin{bmatrix} \dfrac{b}{2} \cdot \dfrac{\partial H}{\partial r} \\ \dfrac{a}{2} \cdot \dfrac{\partial H}{\partial s} \end{bmatrix} \tag{11.142}$$

c. Select the natural coordinates of integration points r and s for a four-node element. This information is found in most FEM books (e.g., Zienkiewicz 2005).

$$r = \begin{bmatrix} \dfrac{1}{\sqrt{3}} & -\dfrac{1}{\sqrt{3}} \\ -\dfrac{1}{\sqrt{3}} & \dfrac{1}{\sqrt{3}} \end{bmatrix} \tag{11.143}$$

$$s = \begin{bmatrix} \dfrac{1}{\sqrt{3}} & \dfrac{1}{\sqrt{3}} \\ -\dfrac{1}{\sqrt{3}} & -\dfrac{1}{\sqrt{3}} \end{bmatrix} \tag{11.144}$$

d. Compute the components of the matrix [B] at the four integration points (Figure 11.29):

Point #1. Recalling Eqs. 11.133 to 11.136, the derivatives of the interpolation function are:

$$\frac{\partial H}{\partial r} = \left[\frac{1}{4}(1+s) \quad -\frac{1}{4}(1+s) \quad -\frac{1}{4}(1+s) \quad \frac{1}{4}(1+s)\right] \tag{11.145}$$

$$\frac{\partial H}{\partial s} = \left[\frac{1}{4}(1+r) \quad \frac{1}{4}(1+r) \quad -\frac{1}{4}(1+r) \quad -\frac{1}{4}(1+r)\right] \tag{11.146}$$

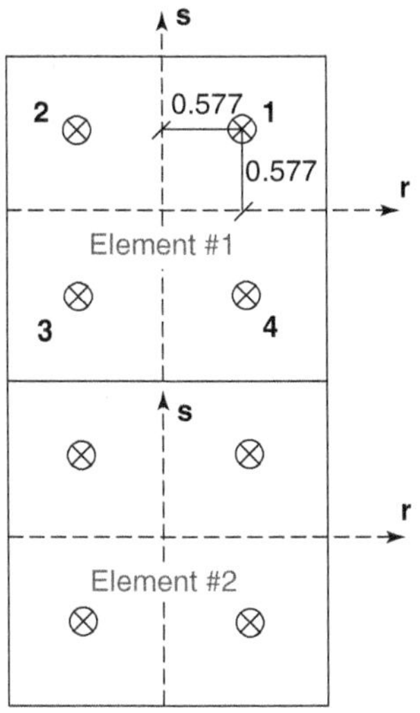

Figure 11.29 The integration points.

For integration point #1. the natural coordinates are:

$$r = \frac{1}{\sqrt{3}}$$

$$s = \frac{1}{\sqrt{3}}$$

$$\frac{\partial H}{\partial r} = \left[\frac{1}{4}\left(1+\frac{1}{\sqrt{3}}\right) \quad -\frac{1}{4}\left(1+\frac{1}{\sqrt{3}}\right) \quad -\frac{1}{4}\left(1+\frac{1}{\sqrt{3}}\right) \right.$$
$$\left. \times \frac{1}{4}\left(1+\frac{1}{\sqrt{3}}\right)\right]$$
$$= [0.394 \quad -0.394 \quad -0.105 \quad 0.105] \tag{11.147}$$

$$\frac{\partial H}{\partial s} = \left[\frac{1}{4}\left(1+\frac{1}{\sqrt{3}}\right) \quad \frac{1}{4}\left(1+\frac{1}{\sqrt{3}}\right) \quad -\frac{1}{4}\left(1+\frac{1}{\sqrt{3}}\right) \right.$$
$$\left. -\frac{1}{4}\left(1+\frac{1}{\sqrt{3}}\right)\right]$$
$$= [0.394 \quad 0.105 \quad -0.105 \quad -0.394] \tag{11.148}$$

Point #2.

$$r = -\frac{1}{\sqrt{3}}$$

$$s = \frac{1}{\sqrt{3}}$$

$$\frac{\partial H}{\partial r} = \left[\frac{1}{4}\left(1+\frac{1}{\sqrt{3}}\right) \quad -\frac{1}{4}\left(1+\frac{1}{\sqrt{3}}\right) \quad -\frac{1}{4}\left(1-\frac{1}{\sqrt{3}}\right) \right.$$
$$\left. \times \frac{1}{4}\left(1-\frac{1}{\sqrt{3}}\right)\right]$$
$$= [0.394 \quad -0.394 \quad -0.105 \quad 0.105] \tag{11.149}$$

$$\frac{\partial H}{\partial s} = \left[\frac{1}{4}\left(1-\frac{1}{\sqrt{3}}\right) \quad \frac{1}{4}\left(1+\frac{1}{\sqrt{3}}\right) \quad -\frac{1}{4}\left(1+\frac{1}{\sqrt{3}}\right) \right.$$
$$\left. -\frac{1}{4}\left(1-\frac{1}{\sqrt{3}}\right)\right]$$
$$= [0.105 \quad 0.394 \quad -0.394 \quad -0.105] \tag{11.150}$$

Point #3.

$$r = -\frac{1}{\sqrt{3}}$$

$$s = -\frac{1}{\sqrt{3}}$$

$$\frac{\partial H}{\partial r} = \left[\frac{1}{4}\left(1-\frac{1}{\sqrt{3}}\right) \quad -\frac{1}{4}\left(1-\frac{1}{\sqrt{3}}\right) \quad -\frac{1}{4}\left(1+\frac{1}{\sqrt{3}}\right) \right.$$
$$\left. \times \frac{1}{4}\left(1+\frac{1}{\sqrt{3}}\right)\right]$$
$$= [0.105 \quad -0.105 \quad -0.394 \quad 0.394] \tag{11.151}$$

$$\frac{\partial H}{\partial s} = \left[\frac{1}{4}\left(1-\frac{1}{\sqrt{3}}\right) \ \frac{1}{4}\left(1+\frac{1}{\sqrt{3}}\right) \ -\frac{1}{4}\left(1+\frac{1}{\sqrt{3}}\right) \times -\frac{1}{4}\left(1-\frac{1}{\sqrt{3}}\right)\right]$$

$$= \begin{bmatrix} 0.105 & 0.394 & -0.394 & -0.105 \end{bmatrix} \quad (11.152)$$

Point #4.

$$r = \frac{1}{\sqrt{3}}$$

$$s = -\frac{1}{\sqrt{3}}$$

$$\frac{\partial H}{\partial r} = \left[\frac{1}{4}\left(1-\frac{1}{\sqrt{3}}\right) \ -\frac{1}{4}\left(1-\frac{1}{\sqrt{3}}\right) \ -\frac{1}{4}\left(1+\frac{1}{\sqrt{3}}\right) \times \frac{1}{4}\left(1+\frac{1}{\sqrt{3}}\right)\right]$$

$$= \begin{bmatrix} 0.105 & -0.105 & -0.394 & 0.394 \end{bmatrix} \quad (11.153)$$

$$\frac{\partial H}{\partial s} = \left[\frac{1}{4}\left(1+\frac{1}{\sqrt{3}}\right) \ \frac{1}{4}\left(1-\frac{1}{\sqrt{3}}\right) \ -\frac{1}{4}\left(1-\frac{1}{\sqrt{3}}\right) \ -\frac{1}{4}\left(1+\frac{1}{\sqrt{3}}\right)\right]$$

$$= \begin{bmatrix} 0.394 & 0.105 & -0.105 & -0.394 \end{bmatrix} \quad (11.154)$$

Now Eqs. 11.138, 11.140, and Eqs. 11.145 to 154 are combined to create the B matrix. For example, the top right element of matrix B is 40 × 0.394 = 15.773.

$$B_{\substack{i=1\\j=1}} = \begin{bmatrix} 15.773 & 0 & -15.773 & 0 & -4.226 & 0 & 4.226 & 0 \\ 0 & 15.773 & 0 & 4.226 & 0 & -4.226 & 0 & -15.773 \\ 15.773 & 15.773 & 4.226 & -15.773 & -4.226 & -4.226 & -15.773 & 4.226 \end{bmatrix}$$

$$B_{\substack{i=1\\j=2}} = \begin{bmatrix} 15.773 & 0 & -15.773 & 0 & -4.226 & 0 & 4.226 & 0 \\ 0 & 4.226 & 0 & 15.773 & 0 & -15.773 & 0 & -4.226 \\ 4.226 & 15.773 & 15.773 & -15.773 & -15.773 & -4.226 & -4.226 & 4.226 \end{bmatrix}$$

$$B_{\substack{i=2\\j=1}} = \begin{bmatrix} 4.226 & 0 & -4.226 & 0 & -15.773 & 0 & 15.773 & 0 \\ 0 & 4.226 & 0 & 15.773 & 0 & -15.773 & 0 & -4.226 \\ 4.226 & 4.226 & 15.773 & -4.226 & -15.773 & -15.773 & -4.226 & 15.773 \end{bmatrix} \quad (11.155)$$

$$B_{\substack{i=2\\j=2}} = \begin{bmatrix} 4.226 & 0 & -4.226 & 0 & -15.773 & 0 & 15.773 & 0 \\ 0 & 15.773 & 0 & 4.226 & 0 & -4.226 & 0 & -15.773 \\ 15.773 & 4.226 & 4.226 & -4.226 & -4.226 & 15.773 & -15.773 & 15.773 \end{bmatrix}$$

Step 4. Write the stress-strain equations for the soil and obtain the constitutive matrix

Recalling Eq. 11.116:

$$C = \frac{E(1-\mu)}{(1+\mu)(1-2\mu)} \begin{bmatrix} 1 & \frac{\mu}{(1-\mu)} & 0 \\ \frac{\mu}{(1-\mu)} & 1 & 0 \\ 0 & 0 & \frac{(1-2\mu)}{2(1-\mu)} \end{bmatrix}$$

$$= 10^4 * \begin{bmatrix} 6.419 & 3.457 & 0 \\ 3.457 & 6.419 & 0 \\ 0 & 0 & 1.481 \end{bmatrix} \quad (11.156)$$

Step 5. Derive the equations governing the behavior of the soil element

Recalling Eqs. 11.124 and 11.126:

$$[K^e] = \int_V [B]^T[C][B]dV \quad (11.157)$$

$$[K^e][u] = [F] \quad (11.158)$$

and recalling the numerical integration from Eq. 11.125:

$$K_e = \int_v B^T CBdv = \sum_{i=1}^{2}\sum_{j=1}^{2} B_{ij}^T C_{ij} B_{ij} \det J.w_i.w_j.t \quad (11.159)$$

For two-point Gauss integration, w_i, and w_j are equal to 1. In the case of plane strain, the thickness t of the elements is taken as 1. Therefore, the stiffness matrix for each element is as follows:

$$K^e = 10^4 \times \begin{bmatrix} 2.63 & 1.23 & -1.89 & 0.49 & -1.32 & -1.23 & 0.58 & -0.49 \\ & 2.63 & -0.49 & 0.57 & -1.23 & -1.31 & 0.49 & -1.89 \\ & & 2.63 & -1.23 & 0.58 & 0.49 & -1.31 & 1.23 \\ & & & 2.63 & 0.49 & -1.89 & 1.23 & -1.31 \\ & & & & 2.63 & 1.23 & -1.89 & 0.49 \\ & \mathit{SYM} & & & & 2.63 & -0.49 & 0.58 \\ & & & & & & 2.63 & -1.23 \\ & & & & & & & 2.63 \end{bmatrix} \quad (11.160)$$

Step 6. Assemble the element equations into the global matrix equation

The global stiffness matrix equation K_g is based on the connected degrees of freedom shown in Figure 11.28, and is assembled as:

$$K_g = 10^4 \times \begin{bmatrix} 2.63 & -1.23 & 0.58 & 0.49 & 0 & 0 & -1.89 & 0.49 & -1.32 & 1.23 & 0 & 0 \\ & 2.63 & -0.49 & -1.89 & 0 & 0 & 0.49 & 0.57 & 1.23 & -1.32 & 0 & 0 \\ & & 5.27 & 0 & 0.58 & 0.49 & -1.32 & -1.23 & 0.39 & 0 & -1.32 & 1.23 \\ & & & 5.27 & -0.49 & -1.89 & -1.23 & -1.32 & 0 & 1.15 & 1.23 & -1.32 \\ & & & & 2.63 & 1.23 & 0 & 0 & -1.32 & -1.23 & -1.89 & 0.49 \\ & & & & & 2.63 & 0 & 0 & -1.23 & -1.32 & -0.49 & 0.57 \\ & & & & & & 2.63 & 1.23 & 0.57 & -0.49 & 0 & 0 \\ & & & & & & & 2.63 & 0.49 & -1.89 & 0 & 0 \\ & & \mathit{SYM} & & & & & & 5.27 & 0 & 0.57 & -0.49 \\ & & & & & & & & & 5.27 & 0.49 & -1.89 \\ & & & & & & & & & & 2.63 & -1.23 \\ & & & & & & & & & & & 2.63 \end{bmatrix} \quad (11.161)$$

Step 7. Introduce the boundary conditions into the global matrix equation

Referring to Figure 11.28, the degrees of freedom of the triaxial sample at nodes (3) and (6) should be constrained in both directions. Moreover, nodes (1) and (4) can only deform vertically. Thus, the rows and columns associated with those degrees of freedom should be zero.

Step 8. Solve the global matrix equation for the unknowns

The triaxial sample is subjected to a confining pressure σ_3 and a vertical pressure σ_1. For this problem, $\sigma_3 = 100$ kPa, and $\sigma_1 = 300$ kPa. The force components applied at the nodes due to the confining pressure and the vertical stress are:

At nodes 2 and 5

$$P_{horizontal} = \sigma_3 \times \frac{b}{2} \times 2 = 100 \times \frac{0.05}{2} \times 2 = 5\,\text{kN/m} \quad (11.162)$$

At nodes 1 and 4

$$P_{vertical} = \sigma_1 \times \frac{a}{2} = 300 \times \frac{0.05}{2} = 7.5\,\text{kN/m} \quad (11.163)$$

Now the force matrix is assembled as:

$$F' = \begin{bmatrix} F'_{x1} \\ F_{y1} \\ F_{x2} \\ F_{y2} \\ F'_{x3} \\ F'_{y3} \\ F'_{x4} \\ F_{y4} \\ F_{x5} \\ F_{y5} \\ F'_{x6} \\ F'_{y6} \end{bmatrix} = \begin{bmatrix} 0 \\ -7.5 \\ 5 \\ 0 \\ 0 \\ 0 \\ 0 \\ -7.5 \\ -5 \\ 0 \\ 0 \\ 0 \end{bmatrix} (\text{kN}) \quad (11.164)$$

Note that in fact the forces F_{x1}, F_{x3}, F_{y3}, F_{x4}, F_{x6}, and F_{y6} are actually unknown, but they are set equal to zero because of the mathematical trick mentioned at the end of section 11.5.3 and because the corresponding displacements are zero. Note also that the matrix K' will have zeroes on rows corresponding to the displacement boundary conditions, except the diagonal, which will have a 1. The same applies to the corresponding columns. The 12 × 12 matrix K' is inverted by the computer and the displacement vector u is found as $K'^{1} \times F'$:

$$u = \begin{bmatrix} u_{x1} \\ u_{y1} \\ u_{x2} \\ u_{y2} \\ u_{x3} \\ u_{y3} \\ u_{x4} \\ u_{y4} \\ u_{x5} \\ u_{y5} \\ u_{x6} \\ u_{y6} \end{bmatrix} = \begin{bmatrix} 0 \\ -0.5134 \\ -0.0428 \\ -0.2567 \\ 0 \\ 0 \\ 0 \\ -0.5134 \\ 0.0428 \\ -0.2567 \\ 0 \\ 0 \end{bmatrix} \text{(mm)} \tag{11.165}$$

Then we can obtain the force vector by $K \times u$:

$$F = \begin{bmatrix} F_{x1} \\ F_{y1} \\ F_{x2} \\ F_{y2} \\ F_{x3} \\ F_{y3} \\ F_{x4} \\ F_{y4} \\ F_{x5} \\ F_{y5} \\ F_{x6} \\ F_{y6} \end{bmatrix} = \begin{bmatrix} 3.62 \\ -7.5 \\ 5 \\ 0 \\ 3.62 \\ 7.5 \\ -3.62 \\ -7.5 \\ -5 \\ 0 \\ -3.62 \\ 7.5 \end{bmatrix} \text{(kN)} \tag{11.166}$$

11.5.5 Boundary Element Method

The FDM and the FEM model the continuum by discretizing the entire body of the soil mass. The boundary element method (BEM) (Crouch and Starfield 1983; Brebbia et al. 1984) is different in that it models the continuum by discretizing only the boundaries of the continuum (Figure 11.30). The mathematical technique for the BEM consists of replacing the governing differential equations valid over the entire soil mass by integral equations that consider only the boundary values. If the soil mass extends to infinity, the FEM requires a boundary at some distance from the imposed loading or deformation. No artificial boundaries are needed in the BEM; this is an advantage of the BEM over the FEM and the FDM. Another advantage is that for a 3D problem, only the boundary surface need be discretized; this reduces the problem from a 3D problem (volume) to a 2D (surface) problem. This is attractive if the boundary surface is small compared to the volume of soil to be simulated. The BEM is particularly well suited to addressing static continuum problems with small boundary-to-volume ratios, with elastic behavior, and with stresses or displacements applied to the boundaries (Bobet 2010).

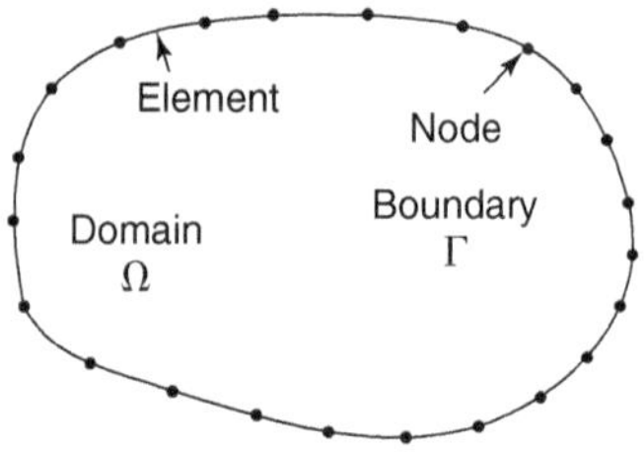

Figure 11.30 Discretization with the boundary element method.

11.5.6 Discrete Element Method

The discrete element method (DEM), also called the distinct element method (Cundall and Strack 1979; Ghaboussi and Barbosa 1990) differs from the finite element method in that it does not assume that the soil mass is a continuum; rather, it treats the soil mass as an assembly of particles of various sizes (Figure 11.31). Obviously, this is an improvement that gets us closer to reality for soils. The DEM addresses three issues during the calculations: the representation of the contacts, the representation of the solid material, and the detection and revisions of the contacts during deformation. Each soil particle is subjected to the forces transmitted at the contacts by adjacent particles and to its own body forces (gravity). The representation of the contact is usually handled through the use of spring and dashpot models (Figure 11.32). The springs have a stiffness k_n for the normal force and k_s for the shear force. The dashpots have damping factor c_n for the normal force and c_s for the shear force.

The solution proceeds in small time steps and the finite difference method (FDM) is used in the solution (see sections 11.5.1 and 11.5.2). The steps are:

1. The state of all the particles in the soil mass is known at time t. This includes contact forces, displacements, velocities, and accelerations.

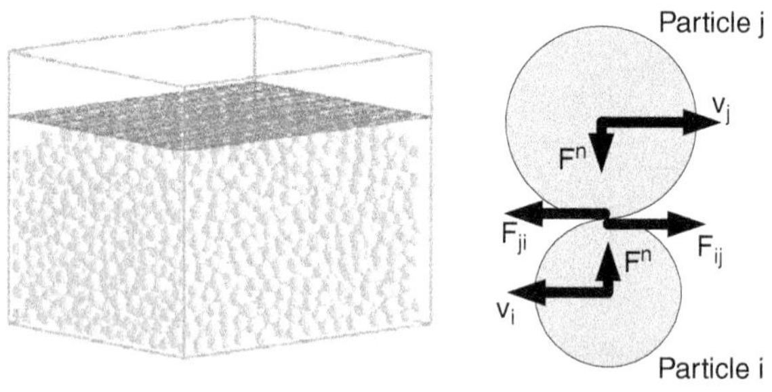

Figure 11.31 Distinct element method: (a) DEM domain. (b) Particle interaction. (*a*: Courtesy of C. Couroyer PhD Thesis, 2000, University of Surrey, Guildford, Surrey, UK.)

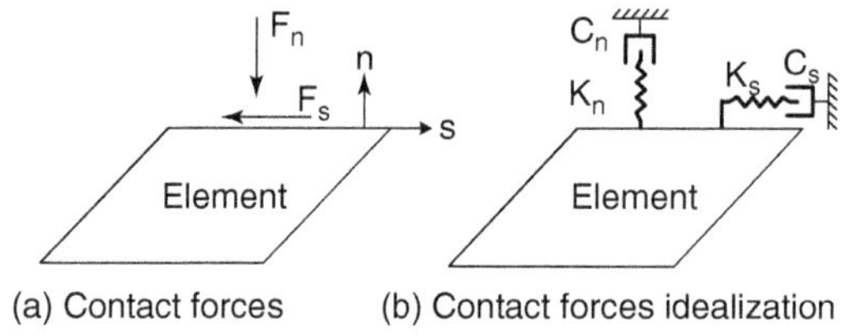

Figure 11.32 DEM element and idealized contact models.

2. A time increment Δt is considered. This time increment has to be small enough for the solution to be numerically stable. The following condition can be used (Hart et al. 1998):

$$\Delta t < 0.1\sqrt{\frac{m_{\min}}{2k_{\max}}} \tag{11.167}$$

where m_{min} is the smallest particle mass and k_{max} is the largest stiffness of all contacts. In the DEM, time comes into play for both dynamic and static problems. Even in a static problem, it takes time for the deformations to take place.
3. The differential equations of motion are then used to obtain the displacement and rotation of the particles at time $t + \Delta t$. The accelerations of the particles are calculated assuming that the forces and moments are constant over Δt:

$$\ddot{u}_i^t = \frac{\sum F_i^t}{m_i} \tag{11.168}$$

$$\ddot{\theta}_i^t = \frac{\sum M_i^t}{I_i} \tag{11.169}$$

where $\ddot{u}_i^t$ and $\ddot{\theta}_i^t$ are the linear and angular acceleration of particle i at time t respectively, F_i^t and M_i^t are the resultant force and resultant moment on particle i at time t respectively, and m_i and I_i are the mass and the moment of inertia of particle i respectively. Then the velocities of the particles are calculated assuming that the accelerations are constants over $-\Delta t/2$ and $+\Delta t/2$:

$$\dot{u}_i^{t+\frac{\Delta t}{2}} = \dot{u}_i^{t-\frac{\Delta t}{2}} + \ddot{u}_i^t \Delta t \tag{11.170}$$

$$\dot{\theta}_i^{t+\frac{\Delta t}{2}} = \dot{\theta}_i^{t-\frac{\Delta t}{2}} + \ddot{\theta}_i^t \Delta t \tag{11.171}$$

where $\dot{u}_i$ and $\dot{\theta}_i$ are the linear and angular velocities respectively. Then the displacements and rotations of the particles are calculated assuming that the velocities are constant over Δt:

$$u_i^{t+\Delta t} = u_i^t + \dot{u}_i^{t+\frac{\Delta t}{2}} \Delta t \tag{11.172}$$

$$\theta_i^{t+\Delta t} = \theta_i^t + \dot{\theta}_i^{t+\frac{\Delta t}{2}} \Delta t \tag{11.173}$$

where u_i and θ_i are the displacement and the rotation respectively.
4. The equations representing the behavior of the contacts are then used to update the forces and moments. Figure 11.32 gives a common model for the contact normal forces F_n and the contact shear forces F_s:

$$F_n^{t+\Delta t} = k_n \Delta u_n^{\Delta t} + c_n \Delta \dot{u}_n^{\Delta t} \tag{11.174}$$

$$F_s^{t+\Delta t} = k_s \Delta u_s^{\Delta t} + c_s \Delta \dot{u}_s^{\Delta t} \tag{11.175}$$

where k_s and k_n are the stiffnesses in the normal and shear directions respectively, c_n and c_s are the damping factors in the normal and shear directions respectively, Δu_n and Δu_s are the incremental displacements in the normal and shear directions respectively, and $\Delta \dot{u}_n$ and $\Delta \dot{u}_s$ are the incremental velocities in the normal and shear directions. The shear force F_s cannot exceed the shear strength of the soil, so the following condition is checked at each increment:

$$F_s^{t+\Delta t} \leq c' A_c + F_n^{t+\Delta t} \tan \varphi' \tag{11.176}$$

where c' is the effective stress cohesion intercept, A_c is the contact area, and φ' is the effective stress friction angle.
5. The cycle of calculations in 1 through 4 is repeated many times. The final solution is obtained when a chosen tolerance in the difference between two consecutive sets of calculations is achieved.

The DEM is quite efficient with these calculations. The calculations are done through a straightforward process solving one equation at a time, and no large matrix has to be inverted. Where the computing power and storage capacity are required is in recognizing and keeping track of all the contacts between elements from one step to the next. The DEM is very useful for soils and fissured rock masses.

11.6 PROBABILITY AND RISK ANALYSIS

All the methods discussed so far are deterministic in nature, which means that they give one precise answer for one problem. Considering the fact that uncertainty exists in every step taken in arriving at a solution, it makes sense to calculate the uncertainty associated with the solution or predicted value. This is called the *probabilistic approach.*

11.6.1 Background

This subsection reviews some basic concepts of statistics because they are useful in the steps described for the general procedure. When many values of a certain variable are collected—such as the undrained shear strength s_u of a clay at a site and at a given depth, for example—they will vary and can be organized in a table from the lowest to the highest value (Table 11.2). These values s_{ui} can then be regrouped into sets of increments or ranges, as shown in Table 11.2. A *histogram* is a plot of the number of times the variable is found in each increment as a function of the value of the variable (Figure 11.33a and b). Note that a different histogram is generated if a different increment magnitude is selected.

A distinction is made between the variable X and the values of that variable x_i. The mean μ of a set of values

Table 11.2 Values of Undrained Shear Strength and Histogram Input

Undrained strength value (kPa)	Number of values (10 kPa increments)	Number of values (20 kPa increments)
49	1 value between 40 and 50	1 value between 40 and 60
62	2 values between 60 and 70	6 values between 60 and 80
67		
73	4 values between 70 and 80	
75		
76		
79		
81	3 values between 80 and 90	4 values between 80 and 100
85		
86		
93	1 value between 90 and 100	
105	1 value between 100 and 110	1 value between 100 and 120

($x_1, x_2, x_3, \ldots, x_n$) is defined as follows and is called the *expected value* E(X) of X:

$$\mu = \frac{x_1 + x_2 + \cdots + x_n}{n} = \frac{\sum_{i=1}^{n} x_i}{n} = E(X) \quad (11.177)$$

The standard deviation σ is a measure of the deviation of the values with respect to the mean. It is given by:

$$\sigma = \sqrt{\frac{(x_1 - \mu)^2 + (x_2 - \mu)^2 + \cdots + (x_n - \mu)^2}{n - 1}} = \sqrt{\frac{\sum_{i=1}^{n} (x_i - \mu)^2}{n - 1}} \quad (11.178)$$

The reason for using the squares is that the difference ($x_i - \mu$) can be positive or negative and might cancel out during summation, thereby not giving a true rendition of the scatter around the mean. We could have used the absolute values of the difference, but that is not what is chosen in practice. Also, the reason for using (n − 1) rather than n is the fact that only (n − 1) values of ($x_i - \mu$) are independent, as the sum of the n values of ($x_i - \mu$) is equal to zero. This is called the *Bessel correction*. The square of the standard deviation σ^2 is called the *variance* v and the ratio of the standard deviation to the mean is the *coefficient of variation* CoV. The CoV is a measure of the scatter in the data. The CoV of structural dead loads may be around 0.05, whereas the CoV of soil data may be around 0.3:

$$CoV = \frac{\sigma}{\mu} \quad (11.179)$$

For normal distributions, the inverse of the CoV is the *reliability index* β. The reliability index tells us how many standard deviations the mean is from the zero origin. It is very useful in reliability analysis and engineering code calibration. In this case, the variable is the difference between the resistance R and the load L and the reliability index β tells us how many standard deviations $\sigma_{(R-L)}$ the mean $\mu_{(R-L)}$ is from failure (R − L = 0). It serves as an indication of the safety level (reliability index).

$$\beta = \frac{\mu}{\sigma} \quad (11.180)$$

For distributions different from normal distributions, the generalized reliability index is still used, but is defined differently.

If the number of values of x_i increases, the histogram becomes smoother; if the number becomes infinity, a smooth function is obtained. This function is $f(x)$ and is called the *probability density function* (PDF) (Figure 11.33c). It is defined as the function $f(x)$ that satisfies:

$$P(a < X < b) = \int_a^b f(x)dx \quad (11.181)$$

where $P(a < X < b)$ is the probability that X will be between a and b. The curves on Figure 11.34 are examples of the function $f(x)$. The area under the curve between two values a and b is the probability that X will fall between those two values. The function also satisfies:

$$P(-\infty < X < +\infty) = \int_{-\infty}^{+\infty} f(x)dx = 1 \quad (11.182)$$

Recall that for the histogram, the distribution depended on the increment selected for the variable. The same happens for f(x): Different functions will be obtained depending on the units used for the variable axis. However, the integral in Eq. 11.181 will be the same because it is a relative measure. The *cumulative distribution function* (CDF) gives the value:

$$P(X < x) = \int_{-\infty}^{x} f(x)dx \quad (11.183)$$

One of the most commonly used PDFs is the normal distribution. The normal distribution function is:

$$f(x) = \frac{1}{\sigma\sqrt{2\pi}} e^{-\frac{1}{2}\left(\frac{x-\mu}{\sigma}\right)^2} \quad (11.184)$$

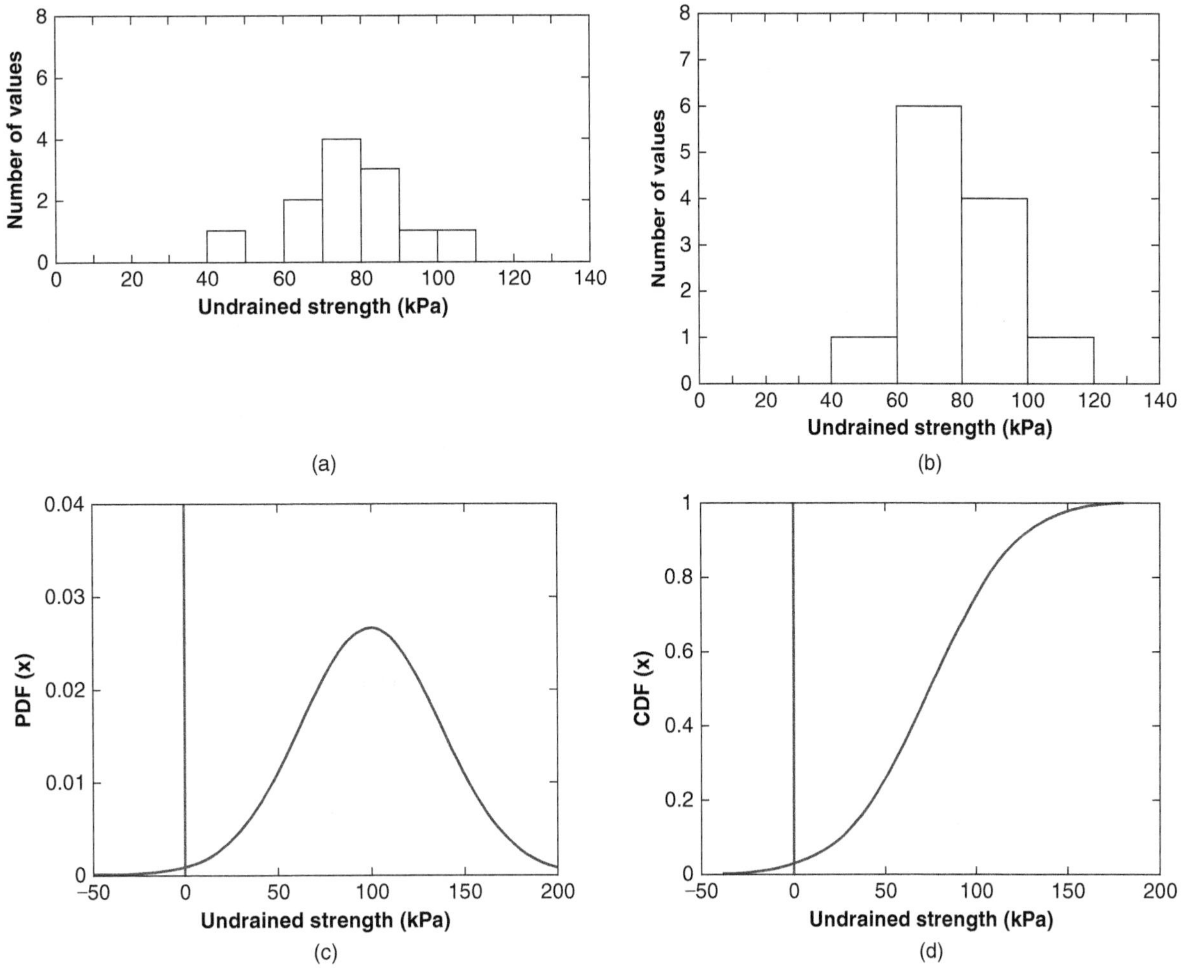

Figure 11.33 Histogram for two values of the variable increment.

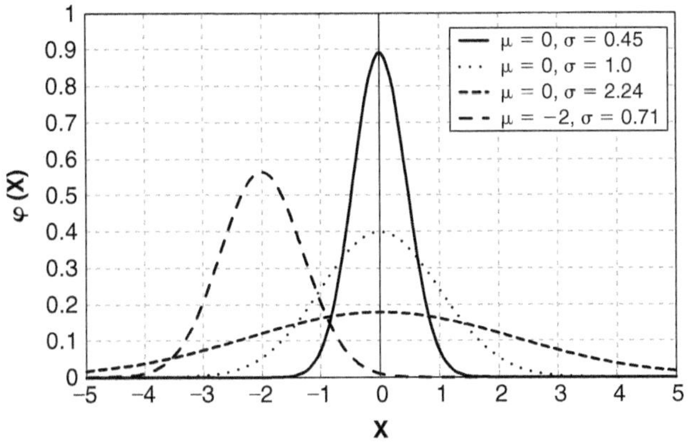

Figure 11.34 Examples of probability density function for normal distributions.

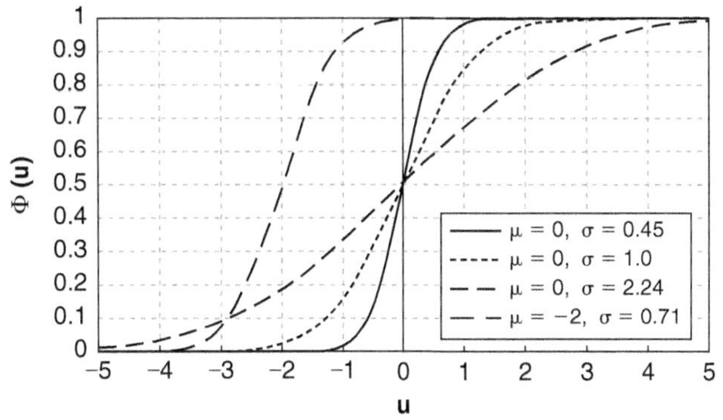

Figure 11.35 Examples of cumulative distribution function for normal distributions.

The corresponding CDF is:

$$F(x) = \frac{1}{2}\left(1 + erf\left(\frac{x-\mu}{\sigma\sqrt{2}}\right)\right) \tag{11.185}$$

The "erf" function is called the *error function*; it does not have a closed-form expression, but can be tabulated. Figure 11.34 shows normal distributions and Figure 11.35 shows cumulative distributions.

It is often advantageous to normalize the variable. The standard normal variable (SNV) is denoted u:

$$u = \frac{x - \mu}{\sigma} \tag{11.186}$$

Therefore, the mean and the standard deviation of the SNV are 0 and 1 respectively. The PDF and CDF for the SNV are:

$$\text{PDF} \quad \varphi(u) = \frac{1}{\sqrt{2\pi}} e^{-\frac{u^2}{2}} \tag{11.187}$$

$$\text{CDF} \quad \Phi(u) = \frac{1}{2}\left(1 + erf\left(\frac{u}{\sqrt{2}}\right)\right) \tag{11.188}$$

Values of the CDF function $\Phi(u)$ for the Standard Normal Variable u are presented in Table 11.3. So, if you wish to find out the probability $P(X < x)$ that a normally distributed variable X is less than a chosen value x, the steps are:

1. Obtain the mean μ and standard deviation σ of X
2. Calculate the value of the standard normal variable $u = (x - \mu)/\sigma$
3. Look in Table 11.3 to find $\Phi(u)$
4. Then

$$\begin{aligned}\Phi(u) = P(U < u) &= P\left(\frac{X - \mu}{\sigma} < \frac{x - \mu}{\sigma}\right) \\ &= P(X < x)\end{aligned} \tag{11.189}$$

5. Remember that $\Phi(u)$ has the following properties:

$$P(U < u) = 1 - P(U < -u) \quad so$$

$$\Phi(u) = 1 - \Phi(-u) \tag{11.190}$$

$$P(U < u) = P(U > -u) \tag{11.191}$$

Figure 11.36 shows some useful areas under the normal distribution.

Another distribution that is very commonly used is the *lognormal distribution* (Figures 11.37 and 11.38). This distribution of a variable X is defined as a distribution such that the LnX (natural logarithm) is normally distributed. The probability density function of the lognormal distribution is therefore:

$$f(x) = \frac{1}{x\sigma_{Lnx}\sqrt{2\pi}} e^{-\frac{1}{2}\left(\frac{Lnx - \mu_{Lnx}}{\sigma_{Lnx}}\right)^2} \tag{11.192}$$

Note that the distribution differs slightly from the normal distribution equation. This is because the function is $f(x)$ rather than $f(Lnx)$. The function $f(Lnx)$ would have the same expression as Eq. 11.184, but $f(x)$ is equal to $f(Lnx)$ times the derivative of Lnx with respect to x, which brings about the additional $1/x$. In this case, the mean and standard deviation of the lognormal distribution are:

$$\mu_{Lnx} = Ln\left(\frac{\mu_x{}^2}{\sqrt{\mu_x{}^2 + \sigma_x{}^2}}\right) \tag{11.193}$$

$$\sigma_{Lnx} = \sqrt{Ln\left(1 + \frac{\sigma_x{}^2}{\mu_x{}^2}\right)} \tag{11.194}$$

Table 11.3 can be used to obtain the probability $P(X < x)$ that a lognormal distributed variable X is smaller than a chosen value x. The process takes place as follows:

1. Obtain the mean μ_X and the standard deviation σ_X of X
2. Obtain the mean μ_{Lnx} and standard deviation σ_{Lnx} of LnX. This can be done by using Eqs. 11.193 and 11.194 once μ_X and σ_X are known.
3. Calculate the value of the standard normal variable $u = \left(\frac{Lnx - \mu_{Lnx}}{\sigma_{Lnx}}\right)$
4. Look in Table 11.3 to find $\Phi(u)$Then

$$\begin{aligned}\Phi(u) = P(U < u) &= P\left(\frac{LnX - \mu_{Lnx}}{\sigma_{Lnx}} < \frac{Lnx - \mu_{Lnx}}{\sigma_{Lnx}}\right) \\ &= P(LnX < Lnx) = P(X < x)\end{aligned} \tag{11.195}$$

5. Remember that $\Phi(u)$ has the following properties:

$$P(U < u) = 1 - P(U < -u) \quad so$$

$$\Phi(u) = 1 - \Phi(-u) \tag{11.196}$$

and

$$P(U < u) = P(U > -u) \tag{11.197}$$

11.6.2 Procedure for Probability Approach

A method of calculating the uncertainty associated with a predicted value usually proceeds as follows:

1. First, the uncertainty associated with each variable involved in the solution is quantified. This quantification process often requires that the mean μ and standard deviation σ of each variable be determined, or that the mean μ and the coefficient of variation $CoV = \sigma/\mu$ be determined. Soil properties tend to have coefficients of variation on the order of 0.3 to 0.4.
2. Deterministic approaches may use mean values of the variables to obtain the mean value of the predicted value. In probabilistic approaches, a second set of equations is organized dealing with the relationship between standard deviations. There are special mathematical rules of operation to combine the standard deviations of the contributing variables and obtain the standard deviation of the variable to be predicted. If the expression of the variable to be predicted as a function of the contributing variables is too complicated, one may have to use numerical probabilistic simulations such as the Monte Carlo simulation.

Table 11.3 Values of the Areas under the Distribution of the Standard Normal Variable

The table gives the cumulative probability up to the standarized normal value of x

$$P[X < x] = \int_{-\infty}^{x} \frac{1}{\sqrt{2\pi}} \exp\left(\frac{1}{2} X^2\right) dX$$

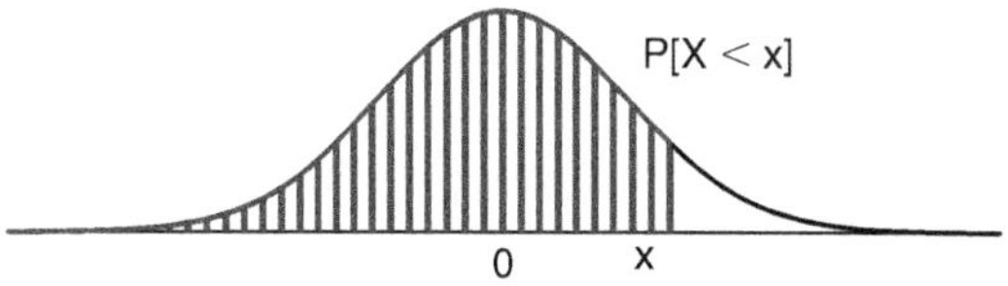

x	0.00	0.01	0.02	0.03	0.04	0.05	0.06	0.07	0.08	0.09
0.0	0.5000	0.5040	0.5080	0.5120	0.5159	0.5199	0.5239	0.5279	0.5319	0.5359
0.1	0.5398	0.5438	0.5478	0.5517	0.5557	0.5596	0.5636	0.5675	0.5714	0.5753
0.2	0.5793	0.5832	0.5871	0.5910	0.5948	0.5987	0.6026	0.6064	0.6103	0.6141
0.3	0.6179	0.6217	0.6255	0.6293	0.6331	0.6368	0.6406	0.6443	0.6480	0.6517
0.4	0.6554	0.6591	0.6628	0.6664	0.6700	0.6736	0.6772	0.6808	0.6844	0.6879
0.5	0.6915	0.6950	0.6985	0.7019	0.7054	0.7088	0.7123	0.7157	0.7190	0.7224
0.6	0.7257	0.7291	0.7324	0.7357	0.7389	0.7422	0.7454	0.7486	0.7517	0.7549
0.7	0.7580	0.7611	0.7642	0.7673	0.7704	0.7734	0.7764	0.7794	0.7823	0.7854
0.8	0.7881	0.7910	0.7939	0.7967	0.7995	0.8023	0.8051	0.8078	0.8106	0.8133
0.9	0.8159	0.8186	0.8212	0.8238	0.8264	0.8289	0.8315	0.8340	0.8365	0.8389
1.0	0.8413	0.8438	0.8461	0.8485	0.8508	0.8531	0.8554	0.8577	0.8599	0.8621
1.1	0.8643	0.8665	0.8686	0.8708	0.8729	0.8749	0.8770	0.8790	0.8804	0.8830
1.2	0.8849	0.8869	0.8888	0.8907	0.8925	0.8944	0.8962	0.8980	0.8997	0.9015
1.3	0.9032	0.9049	0.9066	0.9082	0.9099	0.9115	0.9131	0.9147	0.9162	0.9177
1.4	0.9192	0.9207	0.9222	0.9236	0.9251	0.9265	0.9279	0.9292	0.9306	0.9319
1.5	0.9332	0.9345	0.9357	0.9370	0.9382	0.9394	0.9406	0.9418	0.9429	0.9441
1.6	0.9452	0.9463	0.9474	0.9484	0.9495	0.9505	0.9515	0.9525	0.9535	0.9545
1.7	0.9554	0.9564	0.9573	0.9582	0.9591	0.9599	0.9608	0.9616	0.9625	0.9633
1.8	0.9641	0.9649	0.9656	0.9664	0.9671	0.9678	0.9686	0.9693	0.9699	0.9706
1.9	0.9713	0.9719	0.9726	0.9732	0.9738	0.9744	0.9750	0.9756	0.9761	0.9767
2.0	0.9773	0.9778	0.9783	0.9788	0.9793	0.9798	0.9803	0.9808	0.9812	0.9817
2.1	0.9821	0.9826	0.9830	0.9834	0.9838	0.9842	0.9846	0.9850	0.9854	0.9857
2.2	0.9861	0.9865	0.9868	0.9871	0.9874	0.9878	0.9881	0.9884	0.9887	0.9890
2.3	0.9893	0.9896	0.9898	0.9901	0.9904	0.9906	0.9909	0.9911	0.9913	0.9916
2.4	0.9918	0.9920	0.9922	0.9924	0.9927	0.9929	0.9931	0.9932	0.9934	0.9936
2.5	0.9938	0.9940	0.9941	0.9943	0.9945	0.9946	0.9948	0.9949	0.9951	0.9952
2.6	0.9953	0.9955	0.9956	0.9957	0.9959	0.9960	0.9961	0.9962	0.9963	0.9964
2.7	0.9965	0.9966	0.9967	0.9968	0.9969	0.9970	0.9971	0.9972	0.9973	0.9974
2.8	0.9974	0.9975	0.9976	0.9977	0.9977	0.9978	0.9979	0.9980	0.9980	0.9981
2.9	0.9981	0.9982	0.9982	0.9983	0.9984	0.9984	0.9985	0.9985	0.9986	0.9986
x	3.00	3.10	3.20	3.30	3.40	3.50	3.60	3.70	3.80	3.90
P	0.9986	0.9990	0.9993	0.9995	0.9997	0.9998	0.9998	0.9999	0.9999	1.0000

3. The Monte Carlo simulation consists of drawing values of the contributing variables at random from the range of possible values (using a random number generator), respecting the distribution of these variables, and calculating the value of the function to be predicted. This drawing process is repeated thousands of times and the values obtained are organized into a distribution for the predicted function from which a mean and a standard deviation are calculated.

4. Once the standard deviation of the predicted function is known, one can find out the probability that the predicted function value will be higher or lower than a chosen target.

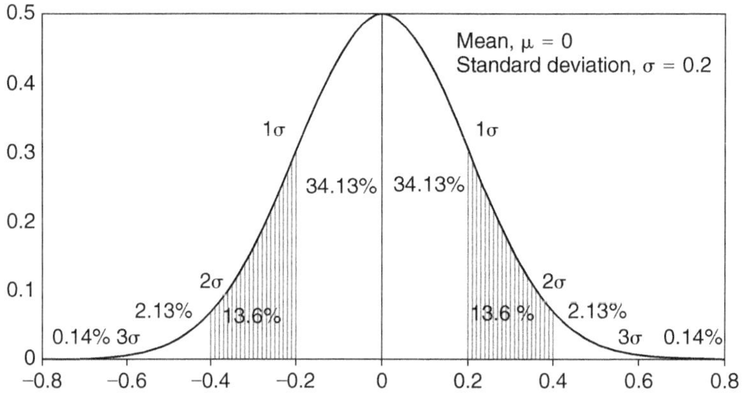

Figure 11.36 Useful areas under the normal distribution.

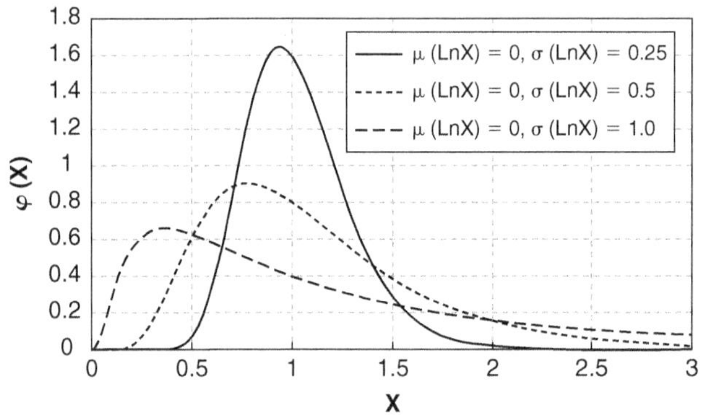

Figure 11.37 Examples of probability density function for lognormal distributions.

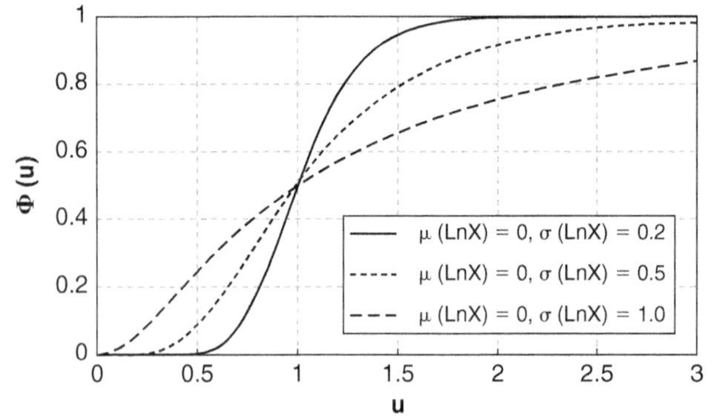

Figure 11.38 Examples of cumulative distribution function for lognormal distributions.

11.6.3 Risk and Acceptable Risk

There is a very important distinction to be made between probability of failure and risk. The *probability of failure* is simply the probability that something is going to collapse (e.g., a bridge, a slope, a building, a dam). *Risk* is defined as the probability of failure multiplied by the value of the consequence. It uses units of the value of the consequence, typically fatalities or dollars lost:

$$R = P_{(F<1)}C \tag{11.198}$$

where R is the risk, P is the probability of failure, and C the value of the consequence. For example, if a slope exists at a very steep angle, it likely would have a high probability of failure. If it were located in the middle of a deserted area and it failed, no one would die and the economic loss would be minimal; therefore the risk would be small. If the same slope were in the middle of a busy city with many buildings and people around it, the number of people dying and the economic loss from destroyed building, utilities, and transportation facilities would be significant; therefore the risk would be very high even though the probability of failure was the same. That illustrates the distinction between probability of failure and risk.

This brings up the point of what is an acceptable risk. First of all, it is not possible to design a structure (for example, a tunnel or an earth dam) that has zero risk associated with its engineering life. This is due to the facts that any calculation is associated with some uncertainty; that the engineering profession's knowledge, though having made great strides, is still incomplete in many respects; that human beings are not error free; and that the engineer designs the structure for conditions that do not include extremely unlikely events such as a falling satellite hitting the structure at the same time as an earthquake, a hurricane, and a 500-year flood during rush hour.

Most modern codes have been written with an accepted probability of failure of about 1 chance in 1000 (structural engineering); it may be estimated that geotechnical engineering operates at a somewhat higher risk than that. In any case, the choice of an acceptable risk is difficult because so many factors enter into the decision. One of those factors is the evaluation of how many fatalities are acceptable. Though few people are prepared to say that any fatality is acceptable,

Table 11.4 Approximate Probability of Human Death

Activity	Probability of death*
Heart disease	0.25
Cancer	0.23
Stroke	0.036
Car	0.012
Suicide	0.009
Fire	0.0009
Airplane	0.0002
Bicycle	0.0002
Lightning	0.00001
Earthquake	0.000009
Flood	0.000007

*These numbers represent the number of deaths due to that activity in one year in the USA divided by the total number of death in the USA during that same year. *Sources*: New York Times, Center for Disease Control, National Safety Council.

it is a matter of public record that some fatalities do occur because of civil engineering decisions. These fatalities can be due to malpractice or to unforeseen events. The choice of an acceptable risk involves other disciplines beyond geotechnical engineering, including philosophy, politics, and social sciences. One of the very difficult steps required in estimating an acceptable risk is what price to put on human life. It is not uncommon to use a number like \$1 million, because that is an average life insurance value for many people. The probabilities of death in the USA for various human activities (Table 11.4) help frame the acceptable numbers in this domain.

Note that the statistics do not always match the human perception. For example, the probability of dying in a car accident is much higher than the probability of dying in an airplane accident, yet people tend to be much more afraid of flying than of driving their cars. Figure 11.39 shows the annual risk associated with various activities in geotechnical engineering and in everyday life. The annual probability of failure (PoF) is on the vertical axis, and there are two scales on the horizontal axis: lives lost or fatalities per year (F) and dollars lost per year (D). Because the two do not necessarily correspond, the activities are shown as bubbles rather than precise points on the graphs. Since the risk is the product of the probability times the value of the consequence, two risk values can be defined:

$$R(fatalities) = PoF \times F \tag{11.199}$$

$$R(dollars\ lost) = PoF \times D \tag{11.200}$$

Therefore, the annual risk is constant on diagonals in Figure 11.39. The red, blue, and green lines correspond to a high, medium, and low annual risk. The numbers are shown in Table 11.5. These data indicate that 0.001 fatalities per year and \$1000 US per year may be acceptable target risk values.

A more advanced way to formulate the risk is:

$$R = T \times V \times C \tag{11.201}$$

where T is the threat, V the vulnerability, and C the value of the consequence.

As can be seen in this case, the probability of failure is split into two components. The *threat* is the probability that a certain event will occur (big flood or big earthquake), whereas the *vulnerability* is the probability that failure will occur if the event occurs. Vulnerability is the part of the system where

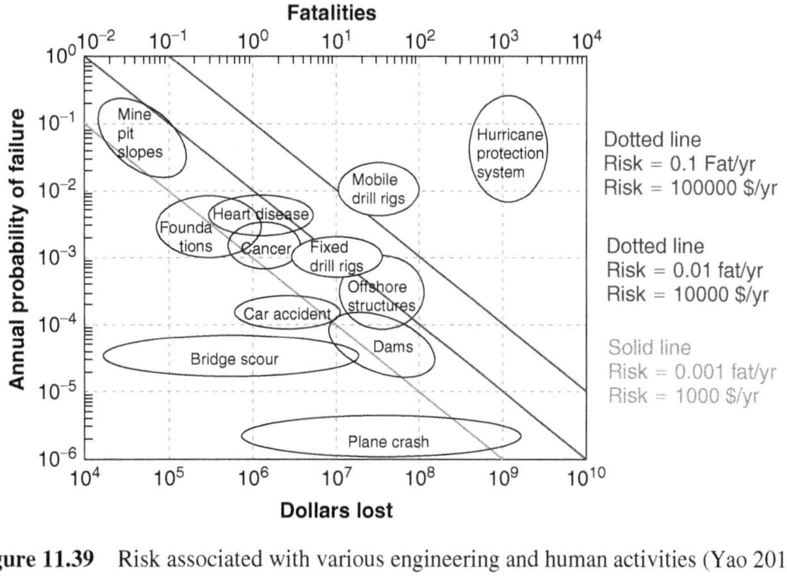

Figure 11.39 Risk associated with various engineering and human activities (Yao 2013).

Table 11.5 Annual Risks for the USA (risk = PoF x value of the consequence)

Annual risk level	Fatalities/year in USA	Dollars lost/year in USA
Low	0.001	1000
Medium	0.01	10000
High	0.1	100000

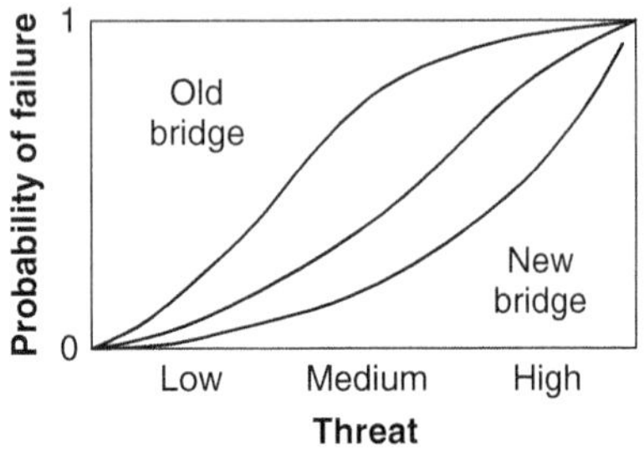

Figure 11.40 Fragility curves.

one has the most control. Fragility curves (Figure 11.40) link the probability of failure to the severity of the threat; they quantify the vulnerability function V.

11.6.4 Example of Probability Approach

A slope stability analysis is used as an example of the probability approach. In a *deterministic analysis*, a single factor of safety is calculated. In a *probabilistic analysis*, a mean factor of safety is calculated from the mean values of the soil parameters, the slope geometry, and the water stress conditions. Then a standard deviation of the factor of safety is obtained from the standard deviations of the parameters involved in the calculations. This is done either by mathematical calculations from the individual standard deviations of the parameters involved in the factor of safety of the slope, if the problem is simple enough; or by numerical simulations, such as the Monte Carlo simulation, if the problem involves several layers or complicated geometry and water stress conditions. Knowing the standard deviation of the factor of safety, one can calculate the probability that the calculated factor of safety will be below 1; this is the probability of failure of the slope.

The deterministic approach gives only one factor of safety, whereas the probabilistic approach gives a mean factor of safety and a probability of failure. This added information can be very valuable for the engineer who must accept or reject the calculated value of the factor of safety. Indeed, one could have the same mean factor of safety but drastically different probabilities of failure depending on whether the soil parameters are known with good precision (low coefficient of variation) or with poor precision (high coefficient of variation). For example, a mean factor of safety of 1.5 with a coefficient of variation of 0.5 would likely be unacceptable, whereas a mean factor of safety of 1.5 with a coefficient of variation of 0.05 would likely be acceptable. Yet no distinction could be made on the basis of the mean value alone.

Let us say that a slope has a mean factor of safety equal to 1.5 ($\mu_F = 1.5$) and a standard deviation equal to 0.45 ($\sigma_F = 0.45$). The coefficient of variation is 0.3 ($CoV_F = 0.3$). Let's further assume that F follows a lognormal distribution. The question is what is the probability of failure P(F < 1)? We follow the steps of section 11.6.2:

1. The mean and standard deviation of F are 1.5 and 0.45 respectively.
2. The mean and standard deviation of LnF are calculated as follows:

$$\mu_{Lnx} = Ln\left(\frac{1.5^2}{\sqrt{1.5^2 + 0.45^2}}\right) = 0.362 \qquad (11.202)$$

$$\sigma_{Lnx} = \sqrt{Ln\left(1 + \frac{0.45^2}{1.5^2}\right)} = 0.294 \qquad (11.203)$$

3. Calculate the value of the standard normal variable U for F = 1:

$$u = \frac{LnF - \mu_{LnF}}{\sigma_{LnF}} = \frac{Ln1 - 0.362}{0.294} = -1.231 \qquad (11.204)$$

4. Table 11.3 does not gives the value of $\Phi(-1.231)$, but it gives the value of $\Phi(1.231) = 0.8907$. Because $\Phi(u) = 1 - \Phi(-u)$, then $\Phi(-1.231) = 1 - 0.891 = 0.109$. Therefore, the probability of failure is 0.109.

This process can be repeated a number of times for different values of the factor of safety, and a plot of the factor of safety versus the probability of failure can be generated (Figure 11.41). Using that plot, if we wish to operate at a

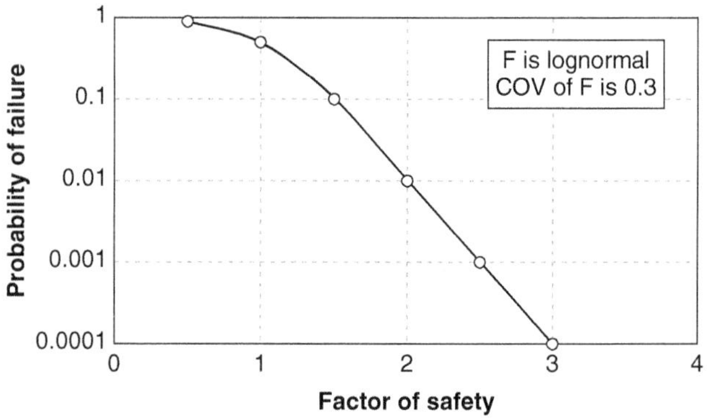

Figure 11.41 Probability of failure vs. mean factor of safety for a slope.

Figure 11.42 Slope and consequence of failure.

probability of failure of 0.001, then we would have to use a factor of safety of 2.5. Now let's say that the slope failure would have some serious consequences (Figure 11.42), such as 10 fatalities and \$5 million. For a factor of safety of 1.5, we would calculate a risk of 0.109 × 10 = 1.09 fatalities and 0.109 × 5M\$ = \$545,000. If we wish to operate at a risk level of 0.01 fatalities/year (Figure 11.39), then we need to have: Risk = 0.01 fatalities = P(F < 1) × 10 fatalities or P(F < 1) = 0.001; therefore F = 2.5 (Figure 11.41). If we wish to operate at a risk level of \$10,000/year, then we need to have: Risk = \$10000 = P(F < 1) × 5M\$ or P(F < 1) = 0.002 and F = 2.35 (Figure 11.41). In this case, the fatality-risk criterion controls.

11.7 REGRESSION ANALYSIS

Let's say that we have data presented on an x—y scatter plot (Figure 11.43) representing n x_i values and n corresponding y_i values. It is often desirable to find the best fit line for the data presented so that y can best be predicted for any value of x. This is *regression analysis*. The basic concepts are presented here for a linear regression where the best fit line to be found is a straight line y = ax + b. The first step is to define what is meant by *best fit*. The most common definition is that the sum of the squares of the differences d_i (Figure 11.43) between the predicted values of y and the measured value of y is minimum. The sum of the squares is:

$$f(a,b) = \sum_{i=1}^{n} d_i^2 = \sum_{i=1}^{n} (y_i - ax_i - b)^2 \qquad (11.205)$$

This function f(a,b) is minimum when the partial derivatives with respect to a and to b are zero:

$$\frac{\partial f(a,b)}{\partial a} = 0 = a\sum x_i^2 + b\sum x_i - \sum x_i y_i \qquad (11.206)$$

$$\frac{\partial f(a,b)}{\partial b} = 0 = a\sum x_i + bn - \sum y_i \qquad (11.207)$$

These two equations give a and b as:

$$a = \frac{\sum x_i \sum y_i - n\sum x_i y_i}{\left(\sum x_i\right)^2 - n\sum x_i^2} \qquad (11.208)$$

$$b = \frac{\sum x_i^2 \sum y_i - \sum x_i \sum x_i y_i}{n\sum x_i^2 - \left(\sum x_i\right)^2} \qquad (11.209)$$

If you look at a scatter plot and "eyeball" the regression line, you tend to minimize the normal distance (NM on Figure 11.44) between the data points and the best fit line rather than the vertical distance. This is called an *orthogonal regression*.

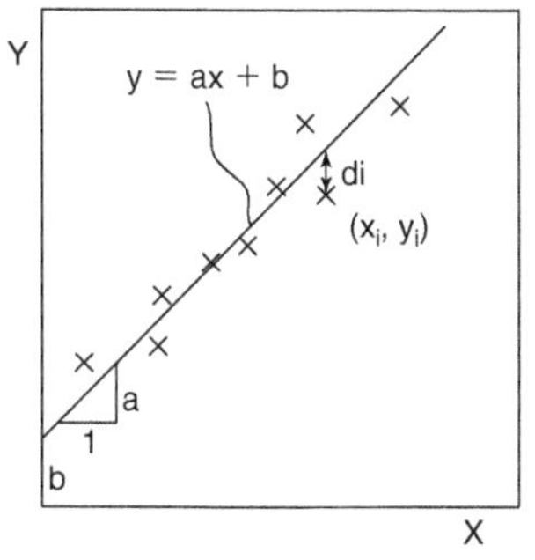

Figure 11.43 Regression minimizing the vertical distance.

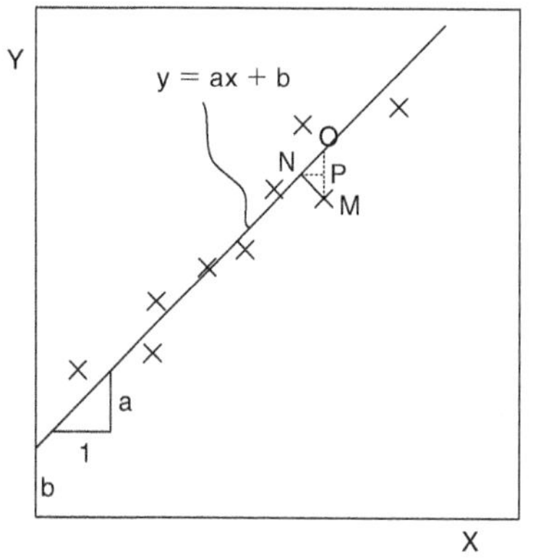

Figure 11.44 Regression minimizing the orthogonal distance.

In this case the expression for the sum of the squares of the distances becomes:

$$f(a,b) = \sum_{i=1}^{n} d_i{}^2 = \sum_{i=1}^{n}((x_{iM} - x_{iN})^2 + y_{iM} - y_{iN})^2)$$

$$= \sum_{i=1}^{n} \frac{1}{1+a^2}(y_{iM} - ax_{iM} - b)^2 \qquad (11.210)$$

Then the derivative with respect to b gives:

$$\frac{\partial f(a,b)}{\partial b} = 0 \quad or \quad \sum (y_i - ax_i - b) = 0 \qquad (11.211)$$

or

$$b = \frac{\sum y_i}{N} - a\frac{\sum x_i}{N} = \overline{y} - a\overline{x} \qquad (11.212)$$

Eliminating b from Eq. 11.210 gives:

$$f(a) = \frac{\sum (y_i - \overline{y})^2 - 2a\sum (x_i - \overline{x})(y_i - \overline{y}) + a^2 \sum (x_i - \overline{x})^2}{1+a^2} \qquad (11.213)$$

Then we set:

$$\frac{\partial f(a)}{\partial a} = 0 \qquad (11.214)$$

which gives the following equation:

$$a^2 \sum (x_i - \overline{x})(y_i - \overline{y}) - a\left(\sum (y_i - \overline{y})^2 - \sum (x_i - \overline{x})^2\right)$$
$$- \sum (x_i - \overline{x})(y_i - \overline{y}) = 0 \qquad (11.215)$$

which leads to the solution for a:

$$a = \frac{\sum (y_i - \overline{y})^2 - \sum (x_i - \overline{x})^2 + \sqrt{\left(\sum (y_i - \overline{y})^2 - \sum (x_i - \overline{x})^2\right)^2 + 4\left(\sum (x_i - \overline{x})(y_i - \overline{y})\right)^2}}{2\sum (x_i - \overline{x})(y_i - \overline{y})} \qquad (11.216)$$

Note that if the regression line is forced to go through the origin, then:

$$b = 0 \quad and \quad a = \frac{\sum y_i}{\sum x_i} \qquad (11.217)$$

The coefficient of regression r^2 is a measure of how well the regression equation predicts the data. It is given by:

$$r^2 = 1 - \frac{\sum (y_i - ax_i - b)^2}{\sum (y_i - \mu_y)^2} \qquad (11.218)$$

It tells us how well the regression line predicts the data compared to a simple average. Values close to 1 indicate that y = ax + b is a very good predictor of the data, whereas values close to zero indicate that y = ax + b is a very poor predictor of the data and that you might as well use the mean regardless of the value of x.

11.8 ARTIFICIAL NEURAL NETWORK METHOD

The artificial neural network (ANN) (De Wilde 1996; Schalkoff 1997) gets its name from the human brain, where neurons interact with each other to process information and make decisions. ANN can be thought of as a very sophisticated regression analysis where a data set is input and, after calculations through a number of neuron layers involving mathematical functions, is converted into a desired output (Figure 11.45). For example, there are many bridges in the USA for which the foundation type and depth are unknown. Let's say that you wish to predict the type and depth of the foundation on the basis of related information, such as the dead load of the bridge, the number of lanes, the length of the span, the foundation depth of neighboring bridges, the dates when the bridge was designed and built, the soil type and the soil strength if borings are available, and so on. This input becomes a set of numbers fitting in a layer of initial neurons and related in some fashion to the type and depth of the unknown foundation. A set of mathematical functions to be chosen by the user are placed in the next layer of neurons waiting for the arrival of the input data; these mathematical functions will transform the input values into a new set of values that is in turn sent to the next layer of neurons. Each time the data set goes through a new layer of neurons, it is mathematically transformed and then sent to the next set of neurons, where it undergoes new mathematical transformation. The output layer of neurons for this example would contain the type and depth of the foundation.

The neurons in a layer are connected only to the previous layer and to the next layer of neurons. Any given neuron is connected to some of the neurons in the previous layer and some of the neurons in the following layer but not necessarily

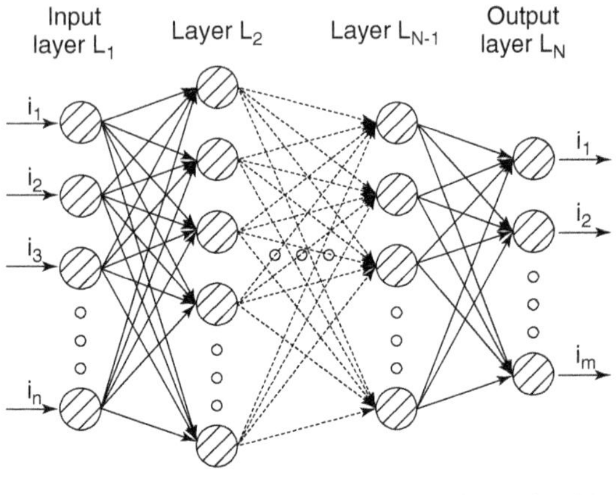

Figure 11.45 Artificial neural network. (After Bobet 2010.)

to all of them. The mathematical functions f that operate the transformation in a neuron are for example (Bobet 2010):

$$i_{jk} = f\left(\sum_{h\in L_{k-1}} \left(w_{hj} i_{h(k-1)} + \theta_{jk}\right)\right) = f(o_j) \quad with \quad j \in L_k \tag{11.219}$$

where i_{jk} is the information to be calculated and be stored in neuron j of layer k, often called the *state* of neuron jk; $i_{h(k-1)}$ is the known information stored in neuron h of layer (k − 1); w_{hj} is the weight factor associated with the connection between neuron h in layer k − 1 and neuron j in layer k (note that w_{jh} does not exist, as there is no connection back from neuron h to neuron j); θ_{jk} is the bias or threshold value associated with neuron jk; and o_j is the argument associated with neuron jk in the function f. Although these functions are chosen by the user at will, certain functions are more popular than others. This is the case of the sigmoidal function:

$$f(o_j) = \frac{1}{1 + e^{-o_j}} \tag{11.220}$$

Once the functions are in place, the ANN must be "trained," which means that the constants in the functions must be determined. This is done by minimizing the error E between the input data and the output predictions through a process similar to the regression analysis discussed in section 11.7:

$$E = \sum_{m\in L_N} (d_m - f(o_m))^2 \tag{11.221}$$

where d_m is the data for neuron m of layer n and $f(o_m)$ is the predicted value for neuron m of layer n. Once the ANN is trained, it can be used to make predictions concerning the type of data that was used to train it. However, the accuracy is tied to the quality of the ANN and the experience of the developer. Using ANN outside of the range of values used to train it can lead to serious errors.

11.9 DIMENSIONAL ANALYSIS

Units are essential to quantify engineering parameters. Unfortunately, there are several unit systems, and this often makes it difficult to communicate across countries using different systems. The most common system in the world is the SI unit system (Système International), but the U.S. customary unit system is still used in the USA. These systems were developed in the late 1700s (SI system in France) and the early 1800s (Imperial system in the UK). Although there are seven units in a system of units (Chapter 1), four are used commonly in engineering: length, mass, time, and temperature. In the SI system, the practice is to use the meter, the kilogram, the second, and the degree Celsius. In the U.S. customary system, the practice is to use the foot, the pound, the second, and the degree Fahrenheit. These four units are called *primary units*, from which derived units can be obtained. The Newton is a unit of force; it is not a primary unit but rather a derived unit, as it is a combination of mass, length, and time ($F = ma = \text{mass} \times \text{length}/\text{time}^2$). Stress is also not a primary unit, but rather a derived unit, as it uses mass, length, and time. One way to avoid worrying about units is to nondimensionalize the parameters used in a problem. Strains are an example of such nondimensional quantities. Strains are the same regardless of the system of units used. In geotechnical engineering, we tend to use dimensional parameters, whereas in hydraulic engineering the trend is toward using nondimensional parameters. As a result, the difference in unit systems does not affect hydraulic engineering as much as geotechnical engineering.

11.9.1 Buckingham Π Theorem

Dimensional analysis is a very useful tool when dealing with mechanics problems. It goes back to the work of Newton and Fourier, but culminated with Buckingham (an American physicist born in 1867) and his famous Π theorem in 1915. This theorem states that a function describing a relationship among n quantities, x_i, such as $f_1(x_1, x_2, x_3, \ldots, x_n) = 0$. where m primary units are required to express the x_i quantities, can be reduced to the form $f_2(\Pi_1, \Pi_2, \Pi_3, \ldots, \Pi_{n-m}) = 0$, where Π_i are nondimensional products of powers of the x_i of the form $\pi_i = x_1{}^a x_2{}^b \ldots x_n{}^c$. This means that the number n of variables necessary to describe a function can be reduced by the number m of primary units necessary to describe these variables. For example, if we have 5 variables (n = 5) with units of mass, length, and time (m = 3), then only 2 variables enter the function and are necessary to describe the solution.

The advantages of dimensional analysis include:

- Forcing us to think through a problem at the front end
- Providing insight about a solution
- Reducing the number of required experiments or simulations
- Providing a basis for direct scaling from model tests to prototype predictions
- Helping in memorizing formulas
- Helping in transforming empirical formulas from one system of units to another
- Detecting errors in equations revealed by lack of dimensional homogeneity
- Providing a mechanism for converting a formula from one unit system to another
- Interpreting the behavior of scale models
- Guiding the selection of experiments
- Obtaining partial solutions to complex problems

The procedure for applying the Buckingham Π theorem is as follows:

1. Identify all the n independent variables influencing the solution to the problem.
2. Identify the m primary units involved in these n independent variables and form m primary unit groups. In

each primary unit group, list all the variables containing that primary unit.
3. Select one variable from each group as a repeating variable. Do not select the variable that is to be predicted and do not select the same variable from each group.
4. Form the (n − m) Π terms as products of the repeating variables and each one of the nonrepeating variable in turn. Each variable is raised to a power exponent.
5. Determine the exponents of the power such that the products are dimensionless.
6. Write the function that links the Π terms and formulate the expression of the solution.

A certain degree of art and experience is associated with judicious use of the Π theorem, and it does require some trial and error, but as the following example will show it is worth the effort.

11.9.2 Examples of Dimensional Analysis

The problem is to use dimensional analysis to find the general expression of the function giving the lateral displacement at the top of an infinitely long pile subjected to a horizontal load applied at the top of the pile and placed in an elastic soil. Figure 11.46 shows the problem and the variables.

1. The independent variables are shown in Table 11.6 with their dimensions. There are 5 independent variables.

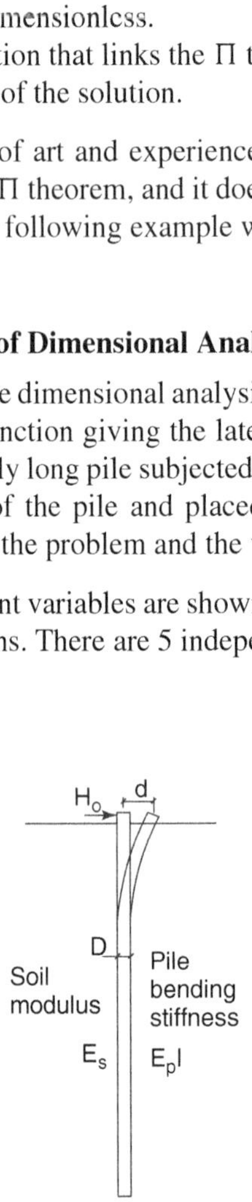

Figure 11.46 Laterally loaded pile problem.

Table 11.6 Variables and Their Dimensions

Quantity	Symbol	Dimension
Displacement	d	L
Force	H_o	F
Soil modulus	E_s	F/L^2
Pile bending stiffness	E_pI	FL^2
Pile diameter	D	L

2. There are 2 primary units, as listed in Table 11.6. We therefore form 2 primary unit groups of variable. For example,
   a. L group: d, E_s, E_pI, D
   b. F group: H_o, E_s, E_pI
3. We select one variable in each group, for example D in the L group and E_s in the F group. These are the repeating variables.
4. Because there are 5 variables and 2 primary units, we have $5 - 2 = 3$ Π terms. To obtain the 3 Π terms, we form the power product of the 2 repeating variables plus 1 of the remaining variables. The Π terms are:
   a. $\Pi_1 = D^a E_s{}^b d^c$
   b. $\Pi_2 = D^d E_s{}^e H_o{}^f$
   c. $\Pi_3 = D^g E_s{}^h E_p I^i$
5. Now we need to find the exponent of the powers in the Π terms such that they are dimensionless.
   a. For $\Pi_1 = D^a E_s{}^b d^c$ in terms of dimensions $L^a(F/L^2)^b L^c$
      For this term to be dimensionless, we must have $b = 0$ and $a - 2b + c = 0$
      This gives $b = 0$ and $a = -c$. We then set one exponent to a convenient value: say, $a = 1$ and Π_1 becomes $\Pi_1 = d/D$.
   b. For $\Pi_2 = D^d E_s{}^e H_o{}^f$ in terms of dimensions $L^d(F/L^2)^e F^f$
      For this term to be dimensionless, we must have $e + f = 0$ and $d - 2e = 0$
      This gives $e = -f$ and $d = 2e = -2f$. We chose $f = 1$ for this example, so Π_2 becomes $\Pi_2 = H_o/(E_s D^2)$.
   c. For $\Pi_3 = D^g E_s{}^h E_p I^i$ in terms of dimensions $L^g(F/L^2)^h(FL^2)^i$
      For this term to be dimensionless, we must have $h + i = 0$ and $g - 2h + 2i = 0$
      This gives $h = -i$ and $g = -4i$. We chose $i = 1$ for this example, so Π_3 becomes $\Pi_3 = E_p I_p/(E_s D^4)$
6. Then we can say that $g(\Pi_1, \Pi_2, \Pi_3) = 0$ or $f_1(d/D, H_o/(E_s D^2), E_p I_p/(E_s D^4)) = 0$. This can be rewritten as $d/D = f_2(H_o/(E_s D^2), E_p I_p/(E_s D^4))$. Because the problem is linear (linear soil and linear pile), we can write:

$$\frac{d}{D} = \frac{H_o}{E_s D^2} f_3\left(\frac{E_p I}{E_s D^4}\right) \tag{11.222}$$

Although the pile displacement cannot be calculated with this function, the result is still very helpful. For example, if we wish to find the function f_3, all we need to do is vary $E_pI/(E_sD^4)$. Without the dimensional analysis, we would have to vary many combinations of the four variables. Discovering the general expression of the solution saves a lot of research time in this case.

11.10 SIMILITUDE LAWS FOR EXPERIMENTAL SIMULATIONS

11.10.1 Similitude Laws

Experiments play a very important role in geotechnical engineering. From laboratory testing to in situ testing, from scaled models to centrifuge testing, all contribute to a better understanding of the problem. This section deals with similitude laws or scaling laws as they are used in scaled models and centrifuge tests. When facing a geotechnical problem where few established design procedures exist, the engineer may elect to perform scaled model tests to predict the behavior of the full-scale prototype. These scaled tests may be done by using models tested under one gravity (1g tests) or centrifuge tests.

A *geotechnical centrifuge* is a large rotating arm at the end of which is a swinging bucket (Figure 11.47). In that bucket is a model of the real problem (slope, retaining wall, foundation). The rotating arm spins at high speeds (e.g., 60 rpm); the swinging bucket first swings upward and then flies nearly horizontally. The centrifugal acceleration artificially increases the stresses in the sample. These high stresses make the sample behave as if it were much larger than it truly is, so the full-scale structure can be simulated and studied. Similitude laws must be evaluated and satisfied to ensure that a true similitude exists between the model scale (simulation) and the full scale, also called *prototype scale* (reality). If such a similitude is satisfied, the results from the scaled model can easily be extrapolated to the full-scale behavior.

To achieve similitude, each dimensionless term (Π terms from section 11.9) must be equal in the prototype and in the model:

$$\Pi_i \text{ model} = \Pi_i \text{ prototype} \quad (11.223)$$

11.10.2 Example of Similitude Laws Application for a Scaled Model

We wish to run a model test to predict the ultimate load P_u of a square footing of size B embedded at a depth d in clay with an undrained shear strength s_u and a unit weight γ. There are 5 parameters and 2 primary units. Therefore, there are 3 Π terms and the dimensional analysis gives:

$$f(\Pi_1, \Pi_2, \Pi_3) = f\left(\frac{P_u}{B^2 s_u}, \frac{d}{B}, \frac{s_u}{B\gamma}\right) = 0 \quad (11.224)$$

The subscript m will be used for the model parameters, and the subscript p will be used for the prototype. To simplify the experiment, we would like to use the same clay as the one found at the site for the prototype:

$$s_{um} = s_{up} \quad (11.225)$$

The scaled model will be n times smaller than the prototype:

$$n\, B_m = B_p \quad (11.226)$$

To satisfy the similitude, we now have to ensure that all Π terms are equal for the model and for the prototype. First we check the Π_1 term:

$$\Pi_{1m} = \Pi_{1p} \quad or \quad \frac{P_{um}}{{B_m}^2 s_{um}} = \frac{P_{up}}{{B_p}^2 s_{up}} \quad (11.227)$$

$$P_{um} = \frac{P_{up}}{n^2} \quad (11.228)$$

Therefore, we can expect the prototype ultimate load to be n^2 times larger than the load measured in the scaled model. Now let's look at the Π_2 term:

$$\Pi_{2m} = \Pi_{2p} \quad or \quad \frac{d_m}{B_m} = \frac{d_p}{B_p} \quad (11.229)$$

This is satisfied by geometric scaling. Finally let's look at the Π_3 term:

$$\Pi_{3m} = \Pi_{3p} \quad or \quad \frac{s_{um}}{B_m \gamma_m} = \frac{s_{up}}{B_p \gamma_p} \quad (11.230)$$

$$\gamma_m = n\gamma_p \quad (11.231)$$

Therefore, to satisfy similitude we will have to find a clay with the same undrained shear strength but with a unit weight n times larger than the unit weight of the prototype soil. This is very difficult to achieve, but there is an artificial way to do this using the centrifuge.

11.10.3 Example of Similitude Laws Application for a Centrifuge Model

Let's continue the example of section 11.10.2 and recognize that:

$$\gamma = \rho g \quad (11.232)$$

where ρ is the mass density of the clay and g is the acceleration due to gravity (9.81 m/s^2). Satisfying the Π_3 term leads to:

$$g_m = n g_p \quad (11.233)$$

Therefore, we can satisfy all similitude requirements by using a gravitational field n times larger for the model. This can be achieved in a geotechnical centrifuge (section 11.10.1) by spinning the bucket of the centrifuge fast enough to generate a centrifugal acceleration equal to ng. This is very useful, because in geotechnical engineering body forces play an important role, unlike in structures. These body forces affect the stability of slopes, tunnels, and mines, for example. Also, the strength of soils is greatly affected by the stress level; indeed, the shear strength depends on the effective stress on the plane of failure. Thus, the centrifuge plays an important role in solving geotechnical engineering problems. It is not without difficulties, however, as shown in the next example.

Consider the problem of flow through soil. The constitutive law for the soil is Darcy's law.

$$v = ki = k\frac{dh_t}{dx} \tag{11.234}$$

where v is the discharge velocity, k is the soil hydraulic conductivity, i is the hydraulic gradient, and dh_t is the change in total head over the distance dx. The hydraulic conductivity is dependent not only on the soil but also on the fluid. Indeed, as you can imagine for the same hydraulic gradient, water would flow faster through sand than thick oil. In fact, the hydraulic conductivity k is expressed as:

$$k = \frac{\gamma_w K}{\mu} \tag{11.235}$$

where γ_w is the unit weight of the flowing fluid, K is a property of the soil, and μ is the dynamic viscosity of the fluid. We can write:

$$\frac{k_m}{k_p} = \frac{\dfrac{\gamma_{wm} K_m}{\mu_m}}{\dfrac{\gamma_{wp} K_p}{\mu_p}} \tag{11.236}$$

Because we wish to use the same soil and the same fluid in the model and the prototype, $K_m = K_p$ and $\mu_m = \mu_p$. However, $\gamma_{wm} = n\,\gamma_{wp}$ because of the gravitational field, so we get:

$$k_m = n\,k_p \tag{11.237}$$

This means that the soil should be n times more pervious in the model than in the prototype—a conflict, as we wish to use the same soil.

Now consider the coefficient of consolidation c_v. The expression comes from Eq. 11.55:

$$c_v = \frac{k\,M}{\gamma_w} \tag{11.238}$$

where M is the constrained modulus. Because we are using the same soil in the model and in the prototype, $M_m = M_p$ and we can write:

$$\frac{c_{vm}}{c_{vp}} = \frac{\dfrac{k_m M_m}{\gamma_{wm}}}{\dfrac{k_p\,M_p}{\gamma_{wp}}} = \frac{\dfrac{nk_p M_m}{n\gamma_{wp}}}{\dfrac{k_p\,M_p}{\gamma_{wp}}} = 1 \tag{11.239}$$

So the coefficient of consolidation is the same. Now let's look at the time factor T (Eq. 11.63), which is one of the dimensionless terms to be satisfied:

$$T_m = T_p = \frac{c_{vm} t_m}{H_{dm}^2} = \frac{c_{vp} t_p}{H_{dp}^2} \tag{11.240}$$

We know that c_v is unchanged and that the drainage length H_d will scale geometrically, so:

$$t_m = \frac{t_p}{n^2} \tag{11.241}$$

Figure 11.47 Geotechnical centrifuge. (Courtesy of the Center for Geotechnical Modeling, University of California-Davis, USA.)

Therefore, the time scale in the model is n^2 times faster than in the prototype. If the model is 100 times smaller, the consolidation time will be 10,000 times faster. Note that the scaling of time is not always n^2 and depends on the problem. For example, the scaling of time for a dynamic event is n, not n^2.

11.11 TYPES OF ANALYSES (DRAINED–UNDRAINED, EFFECTIVE STRESS–TOTAL STRESS, SHORT-TERM–LONG-TERM)

With respect to water and air drainage, geotechnical engineering analyses can be

- Effective stress or total stress analyses
- Drained or undrained analyses
- Long-term or short-term analyses

An effective stress analysis is the best approach in all geotechnical engineering problems, but it is not always the simplest, and sometimes the added complexity may not be necessary. In the effective stress analysis, the soil is considered to be made of three distinct phases (water, air, solids) and the stresses in the three phases are handled separately. The effective stress analysis is always appropriate and applicable, but is often difficult because it requires knowledge of the water stress and even the air stress. For example, it is perfectly appropriate to do an effective stress analysis to solve a problem involving the undrained behavior of a saturated clay, but the water stress must be known in the soil mass.

A total stress analysis consists of considering that the soil is monophase. This is the approach taken when dealing with concrete or steel. For soils, such an analysis is appropriate when dealing with the undrained behavior of saturated clays, for example, because in this case the two phases involved (solids and water) remain bound together, as there is no water movement. An undrained analysis is simply the analysis of a soil that does not drain; it is a common analysis for clays that are loaded rapidly. In extreme cases, the liquefaction of sands during earthquakes can also be considered an undrained behavior.

In a drained analysis, the water stress remains equal to hydrostatic; it is a common assumption for the slow loading of sands or the very slow loading of clays. A long-term analysis is similar to a drained analysis because in the long term all soils become drained. A short-term analysis may be an undrained analysis for some soils (clays) and a drained analysis for others (static loading of clean sands).

PROBLEMS

11.1 A vertical wall is supporting a clean, dry sand backfill with a unit weight γ and effective angle of internal friction φ' (Figure 11.1s). It is assumed that there is no friction between the wall and the backfill. The wall exerts a horizontal load P against the sand. As the wall is pushed into the sand, the load P increases and there is a point where the sand behind the wall fails. At that point, the load is P_p corresponding to the passive earth pressure and the question is to find the load P_p corresponding to impending failure of the sand. Note that the problem is a plane strain problem.

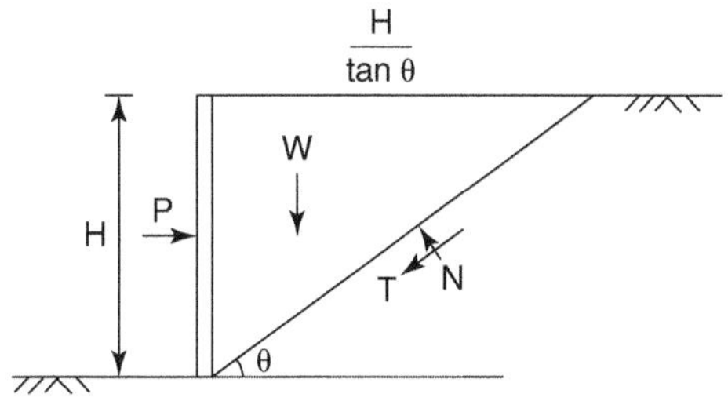

Figure 11.1s Free-body diagram of the failing soil mass.

11.2 A slope is made of a saturated clay with a total unit weight γ, and an undrained shear strength s_u. The slope makes an angle β with the horizontal. Choose 2 circles along which the slope could fail and calculate the factor of safety of that slope against rotation failure along the 2 circles. Why are the 2 factors of safety not the same? Describe how you would find the minimum factor of safety for this slope.

11.3 A pile has a diameter D, a length L, and a modulus E_p. It is subjected to a vertical load Q. The soil generates a constant pile soil friction f. At the pile point the soil generates a point pressure $p = k_p w$, where w is the vertical displacement of the point and k_p is a constant.

a. Develop the governing differential equation.
b. Find the expression for the top displacement by the finite difference method.

11.4 Develop the closed-form solution for the expansion of an infinitely long cylindrical cavity in an elastic soil space. The soil is weightless and has a Poisson's ratio ν and a modulus E. The cavity has an initial radius r_o. The goal is to generate the curve that gives the radial stress σ_r as a function of the relative increase in cavity radius $\Delta r/r_o$.

11.5 Develop the closed-form solution for the expansion of a spherical cavity in an elastic soil space. The soil is weightless and has a Poisson's ratio ν and a modulus E. The cavity has an initial radius r_o. The goal is to generate the curve that gives the radial stress σ_r as a function of the relative increase in cavity radius $\Delta r/r_o$.

11.6 Develop the solution for the flow of water through a saturated soil sample in a constant head permeameter. The goal is to find the excess water stress anywhere and at any time in the sample.

11.7 Use the finite element method to construct the global stiffness matrix for triaxial test performed on an elastic soil. The major principal stress is 300 kPa and the minor principal stress is 100 kPa. The modulus is 40 MPa and the Poisson's ratio is 0.35. The height and diameter of the sample are 0.1 m and 0.05 m respectively. Consider an axisymmetric geometry and use two four-noded elements.

11.8 Two weightless particles of fine sand have a diameter of 1 mm and are placed in the corner of a container as shown in Figure 11.2s. The vertical load applied on the top particle is 0.4 kN. Find all forces between the particles, the wall, and the ground surface. Calculate the contact stress between the two particles if the contact area is 0.005 mm^2. The angles θ_1 and θ_2 are equal to 45°.

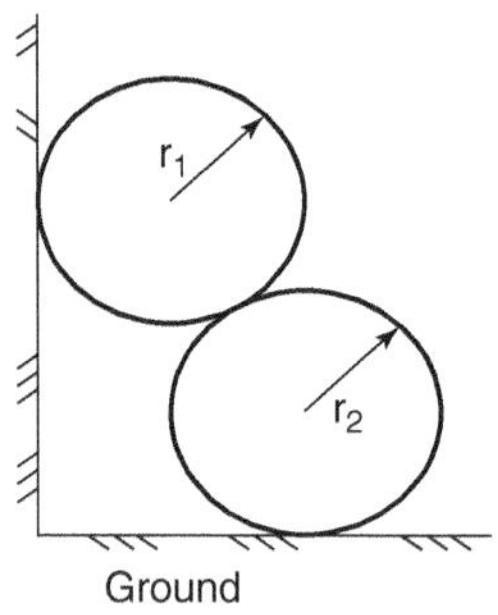

Figure 11.2s Discrete element problem.

11.9 A slope is to be designed for a target probability of failure of 0.001. Plot the mean factor of safety μ versus the coefficient of variation CoV_F in the following cases:
   a. F follows a normal distribution.
   b. F follows a lognormal distribution.

11.10 A levee system is to be designed to meet a risk of 0.001 fatalities/yr and \$1000/yr. It protects a city where 500,000 people could die and where the potential economic loss is \$200 billion if the system fails. What would you recommend for the design annual probability of failure of the levee system?

11.11 A levee system is to be designed to meet a risk of 0.001 fatalities/yr and \$1000/yr. It protects farmland where 100 people and a few cows could die and where the total potential economic loss is \$200 million. What would you recommend for the design probability of failure of the levee system?

11.12 The set of data (y, x) shown in Table 11.1s is plotted and a linear regression (y = ax + b) is performed. Calculate the values of a and b by:
   a. Minimizing the vertical distance between the measured and predicted y values.
   b. Minimizing the normal distance between the measured data and the regression line.
   c. Compare the results.

Table 11.1s Data Set

Data point number	x value	y value
1	2.1	7.4
2	4.5	10.1
3	4.8	11.7
4	5.3	12.4
5	5.7	13.1
6	6.2	16.7
7	7.8	23.4

11.13 Use consistent units to find the relationship between the shear wave velocity v_s, the mass density ρ, and the shear modulus of elasticity G.

11.14 The following empirical equations are used in sands to obtain the ultimate pressure p_u under a driven pile point and the ultimate friction f_u on a driven pile side. Use normalization to give these formulas with p_u and f_u in the U.S. customary system.

$$p_u(\text{kPa}) = 1000\ (N(\text{bl/ft}))^{0.5}$$
$$f_u(\text{kPa}) = 5\ (N(\text{bl/ft})^{0.7}$$

11.15 Perform a dimensional analysis for a square footing embedded at a depth d in a clay with an undrained shear strength s_u. The footing size is B and the failure load is Q_u.

Problems and Solutions

Problem 11.1

A vertical wall is supporting a clean, dry sand backfill with a unit weight γ and effective angle of internal friction φ' (Figure 11.1s). It is assumed that there is no friction between the wall and the backfill. The wall exerts a horizontal load P against the sand. As the wall is pushed into the sand, the load P increases and there is a point where the sand behind the wall fails. At that point, the load is P_p corresponding to the passive earth pressure and the question is to find the load P_p corresponding to impending failure of the sand. Note that the problem is a plane strain problem.

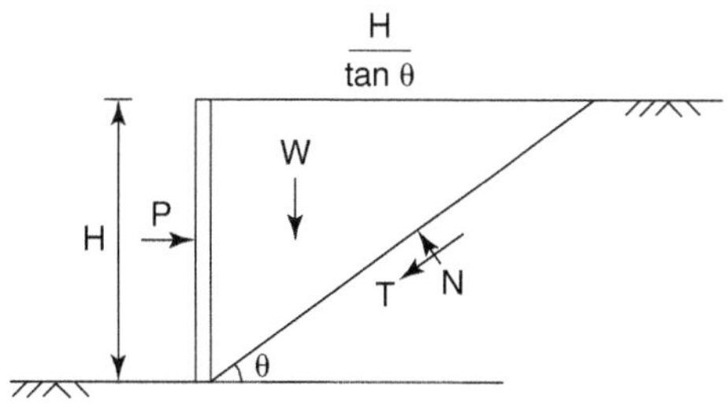

Figure 11.1s Free-body diagram of the failing soil mass.

Solution 11.1

The free-body diagram of the failing soil mass is shown in Figure 11.1s. All the external forces are shown on the diagram, including the weight of the soil mass W, the normal force N, and the shear force T on the failure plane. Also, the force generated by the wall is shown in the diagram as P. Here the failure plane is assumed to be at an angle θ from the horizontal plane. Note that the direction of the shear force T is acting toward the bottom of the wedge, because the soil has the tendency to move upward along the failure surface. The equilibrium equations are set up as follows:

$$W - N\cos\theta + T\sin\theta = 0$$

$$P - N\sin\theta - T\cos\theta = 0$$

The constitutive equations in this case are the shear strength equation of the sand and the expression of the weight of the wedge:

$$W = \frac{\gamma H^2}{2\tan\theta}$$

$$T = N\tan\varphi$$

We can then obtain N and P as:

$$N = \frac{W}{\cos\theta - \tan\varphi\sin\theta} = \frac{\gamma H^2}{2\tan\theta\,(\cos\theta - \tan\varphi\sin\theta)}$$

and

$$P = \frac{\gamma H^2(\sin\theta\cos\theta + \tan\varphi\cos^2\theta)}{2(\sin\theta\cos\theta - \tan\varphi\sin^2\theta)}$$

The maximum value of P, which is P_p, is obtained by setting $\frac{dP}{d\theta} = 0$:

$$\frac{dP}{d\theta} = \frac{\gamma H^2}{2} \times \frac{(-\cos 2\theta + \tan\varphi\sin 2\theta)\tan\varphi}{(\sin\theta\cos\theta - \tan\varphi\sin^2\theta)^2} = \frac{\gamma H^2(\sin 2\theta\sin^2\varphi - \cos 2\theta\sin\varphi\cos\varphi)}{2\sin^2\theta\ \cos^2(\theta+\varphi)} = 0$$

There are two solutions to this equation: one is $\varphi = 0$, which is not realistic, and the other one is:

$$\theta = \frac{\pi}{4} - \frac{\varphi}{2}.$$

The load P_p can then be expressed as:

$$P_p = \frac{\gamma H^2}{2}\left(\frac{1+\sin\varphi}{1-\sin\varphi}\right).$$

Problem 11.2

A slope is made of a saturated clay with a total unit weight γ, and an undrained shear strength s_u. The slope makes an angle β with the horizontal. Choose 2 circles along which the slope could fail and calculate the factor of safety of that slope against rotation failure along the 2 circles. Why are the 2 factors of safety not the same? Describe how you would find the minimum factor of safety for this slope.

Solution 11.2

Case 1: The circle is chosen as shown in Figure 11.3s. The center of the circle is 20 m horizontally away from the edge of the slope, and 10 m vertically above the top of the slope. The radius of the circle is 20 m.

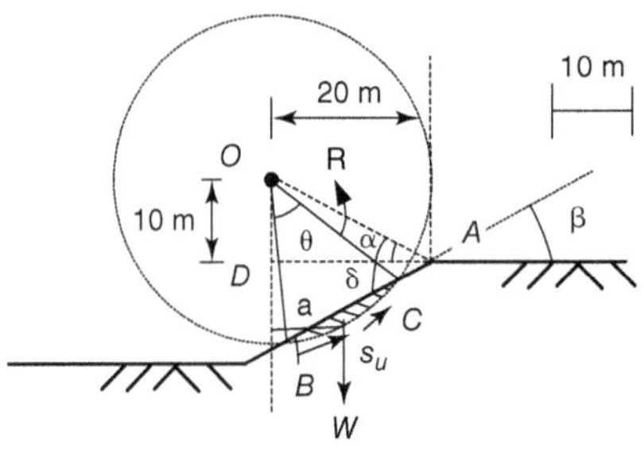

Figure 11.3s Illustration of the slope potential failure surface (case 1).

$$F.S. = \frac{M_r}{M_d} = \frac{s_u \cdot l \cdot R}{W \cdot a} = \frac{s_u \cdot \theta \cdot R^2}{W \cdot a} \quad \text{(a unit width of the soil slice is analyzed)}$$

Here, θ is in radians, a is the arm of the weight of the failure area (hatched), W is the weight of the failing soil, s_u is the undrained shear strength of the soil, and R is the radius of the circle.

In triangle ODA,

$$|OA| = \sqrt{|OD|^2 + |DA|^2} = \sqrt{10^2 + 20^2} = 22.4 \text{ m}$$

$$\alpha = \arctan\left(\frac{10}{20}\right) = 26.6^\circ$$

Therefore,

$$\delta = \alpha + \beta = 26.6 + 30 = 56.6^\circ$$

In triangle OCA (OBA),

$$|OC|^2 = |OA|^2 + |AC|^2 - 2 \times |OA| \times |AC| \times \cos\delta$$

and

$$20^2 = 22.4^2 + |AC|^2 - 2 \times 22.4 \times |AC| \times \cos 56.6^\circ$$

Therefore,

$$|AC| = 5.2 \text{ m}$$

and with the same method:

$$|AB| = 19.4 \text{ m}$$

$$|BC| = 14.2 \text{ m}$$

In triangle OBC,

$$\cos\theta = \frac{|OC|^2 + |OB|^2 - |BC|^2}{2 \times |OC| \times |OB|} = \frac{20^2 + 20^2 - 14.2^2}{2 \times 20 \times 20} = 0.748$$

$$\theta = 42^\circ = 0.733 \text{ rad}$$

The weight of the circular segment can be calculated as:

$$W = \gamma \cdot A \cdot 1 = \gamma \cdot \frac{R^2}{2}(\theta - \sin\theta) \cdot 1 = 19 \times \frac{20^2}{2}(0.733 - \sin 0.733) \times 1 = 242.8 \text{ kN}$$

$$a = \frac{4R\sin^3\left(\frac{\theta}{2}\right)}{3(\theta - \sin\theta)} \times \sin\beta = \frac{4 \times 20 \times \sin^3\left(\frac{0.733}{2}\right)}{3(0.733 - \sin 0.733)} \times \sin 30^\circ = 9.6 \text{ m}$$

The safety of factor can be obtained as follows:

$$F.S. = \frac{M_r}{M_d} = \frac{s_u \cdot l \cdot R \cdot 1}{W \cdot a} = \frac{s_u \cdot \theta \cdot R^2 \cdot 1}{W \cdot a} = \frac{50 \times 0.733 \times 20^2 \times 1}{242.8 \times 9.6} = 6.3$$

Case 2: The circle is chosen as shown in Figure 11.4s. The center of the circle is at 24.2 m distance horizontally away from the edge of the slope, and 16.2 m distance vertically above the top surface of the slope. The radius of the circle is defined to be 36.5 m.

$$F.S. = \frac{M_r}{M_d} = \frac{s_u \cdot l \cdot R \cdot 1}{W_1 \cdot a + W_2 \cdot b} = \frac{s_u \cdot \theta \cdot R^2 \cdot 1}{W_1 \cdot a + W_2 \cdot b} \quad \text{(unit width of the soil slice is analyzed)}$$

Here, θ is in radians, a is the moment arm of the weight of the failure area (circular segment), W_1 is the weight of the circular segment, b is the moment arm of the weight of triangle ABC, W_2 is the weight of triangle ABC, s_u is the undrained shear strength of the soil, and R is the radius of the circle.

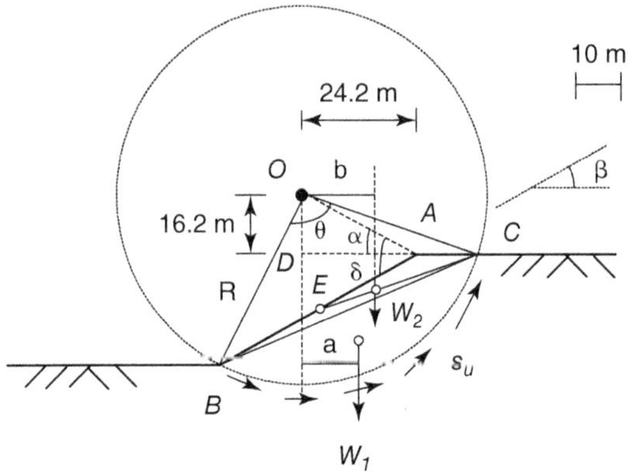

Figure 11.4s Illustration of the slope potential failure surface (case 2).

In triangle ODA,

$$|OA| = \sqrt{|OD|^2 + |DA|^2} = \sqrt{16.2^2 + 24.2^2} = 29.1 \text{ m}$$

$$\alpha = \arctan\frac{16.2}{24.2} = 33.8^\circ$$

Therefore,

$$\delta = \alpha + \beta = 33.8 + 30 = 63.8^\circ$$

In triangle OAC,

$$|OC|^2 = |OA|^2 + |AC|^2 - 2 \times |OA| \times |AC| \times \cos(180^\circ - \alpha)$$

$$36.5^2 = 29.1^2 + |AC|^2 - 2 \times 29.1 \times |AC| \times \cos(180^\circ - 33.8^\circ)$$

Therefore,

$$|AC| = 8.5 \text{ m}$$

In triangle OBA,

$$|OA|^2 + |AB|^2 - 2 \times |OA| \times |AB| \times \cos\delta = |OB|^2$$

$$29.1^2 + |AB|^2 - 2 \times 29.1 \times |AB| \times \cos 63.8^\circ = 36.5^2$$

$$|AB| = 38.4 \text{ m}$$

In triangle ABC,

$$|BC| = \sqrt{|AB|^2 + |AC|^2 - 2 \times |AB| \times |AC| \times \cos 150^\circ} = \sqrt{38.4^2 + 8.5^2 - 2 \times 38.4 \times 8.5 \times \cos 150^\circ} = 46 \text{ m}$$

$$\cos\angle ACB = \frac{8.5^2 + 46^2 - 38.4^2}{2 \times 8.5 \times 46} = 0.913$$

$$\angle ACB = 24.1^\circ$$

In triangle OBC,

$$\cos\theta = \frac{|OB|^2 + |OC|^2 - |BC|^2}{2 \times |OC| \times |OB|} = \frac{36.5^2 + 36.5^2 - 46^2}{2 \times 36.5 \times 36.5} = 0.206$$

$$\theta = 78^\circ = 1.36 \text{ rad}$$

The weight of the circular segment can be calculated as:

$$W_1 = \gamma \cdot A \cdot 1 = \gamma \cdot \frac{R^2}{2}(\theta - \sin\theta) = 19 \times \frac{36.5^2}{2}(1.36 - \sin 1.36) = 4836 \text{ kN/m}$$

$$a = \frac{4R\sin^3\left(\frac{\theta}{2}\right)}{3(\theta - \sin\theta)} \times \sin\angle ACB = \frac{4 \times 36.5 \times \sin^3\left(\frac{1.36}{2}\right)}{3(1.36 - \sin 1.36)} \times \sin 24.1^\circ = 12.9 \text{ m}$$

E is the center point of segment AB:

$$|AE| = \frac{1}{2}|AB| = \frac{1}{2} \times 38.4 = 19.2 \text{ m}$$

$$|CE| = \sqrt{|AE|^2 + |AC|^2 - 2 \times |AE| \times |AC| \times \cos 150^\circ} = \sqrt{19.2^2 + 8.5^2 - 2 \times 19.2 \times 8.5 \times \cos 150^\circ} = 26.9 \text{ m}$$

$$\cos\angle ACE = \frac{8.5^2 + 26.9^2 - 19.2^2}{2 \times 8.5 \times 26.9} = 0.934,$$

$$\angle ACE = 20.9^\circ$$

$$b = |DA| + |AC| - \frac{2}{3}|CE|\cos\angle ACE = 24.2 + 8.5 - \frac{2}{3} \times 26.9 \times \cos 20.9^\circ = 15.9 \text{ m}$$

$$W_2 = \gamma \cdot A \cdot 1 = \gamma \cdot \frac{1}{2}|AC| \cdot |AB| \cdot \sin 150^\circ \cdot 1 = 19 \times \frac{1}{2} \times 8.5 \times 38.4 \times \sin 150^\circ \times 1 = 1550.4 \text{ kN}$$

The safety of factor can be obtained as follows:

$$F.S. = \frac{M_r}{M_d} = \frac{s_u \cdot l \cdot R \cdot 1}{W_1 \cdot a + W_2 \cdot b} = \frac{s_u \cdot \theta \cdot R^2 \cdot 1}{W_1 \cdot a + W_2 \cdot b} = \frac{50 \times 1.36 \times 36.5^2 \times 1}{4836 \times 12.9 + 1550.4 \times 15.9} = 1.04$$

The two results are different because different potential failure circles are chosen to perform the calculation. The minimum factor of safety should be obtained by repeating this trial-and-error process until the minimum factor of safety is found. Choose different locations for the center of the circle and different radii, and perform the calculations following the procedure used previously until the minimum factor of safety is found. It is recommended that you use a software program to minimize the time spent on the calculations!

Problem 11.3

A pile has a diameter D, a length L, and a modulus E_p. It is subjected to a vertical load Q. The soil generates a constant pile soil friction f. At the pile point the soil generates a point pressure $p = k_p w$, where w is the vertical displacement of the point and k_p is a constant.

a. Develop the governing differential equation
b. Find the expression for the top displacement by the finite difference method.

Solution 11.3 (Figure 11.5s)

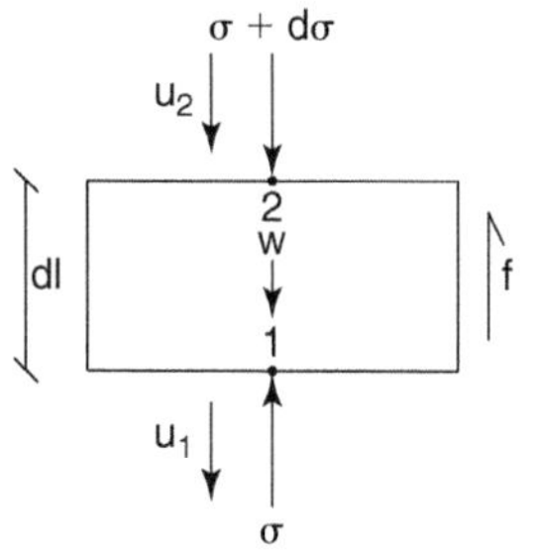

Figure 11.5s Pile element.

a.

$$\left.\begin{array}{l}\text{Equilibrium}: A_s.d\sigma + \gamma.dl.A_s = f.dl.p_p \Rightarrow \dfrac{d\sigma}{dl} = \dfrac{f}{A_s}.p_p - \gamma \\ \text{Constitutive}: \varepsilon = \dfrac{\overbrace{u_2 - u_1}^{du}}{dl} = \dfrac{\sigma}{E} \Rightarrow \dfrac{d\sigma}{dl} = \dfrac{d^2u}{dl^2}E\end{array}\right\} \dfrac{d^2u}{dl^2}E = \dfrac{f}{A_s}p_p - \gamma$$

A_s: Area of pile section σ: Compressive stress
f: friction stress p_p: Perimeter of pile
E: Young's modulus of pile Y: Unit weight of pile

$$\text{Governing differential equation}: \frac{d^2u}{dl^2} = \frac{f}{EA_s}p_p - \frac{\gamma}{E}$$

b. Figure 11.6s Finite difference application:

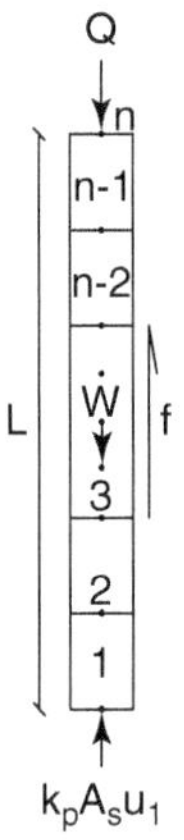

Figure 11.6s Pile discretization.

$$\frac{u_{i+1} - 2u_i + u_{i-1}}{\Delta l^2} = \frac{f}{EA_s}p_p - \frac{\gamma}{E} \rightarrow u_{i+1} = \left(\frac{f}{EA_s}p_p - \frac{\gamma}{E}\right).\Delta l^2 + 2u_i - u_{i-1}$$

Boundary condition:

$$\sigma_1 = k_p u_1 \rightarrow \frac{u_2 - u_1}{\Delta l} = \frac{k_p u_1}{E} \rightarrow u_2 = \left(1 + \frac{k_p \Delta l}{E}\right)u_1$$

Equilibrium:

$$k_p.A_s.u_1 = Q + W - fL.p_p \rightarrow u_1 = \frac{Q + W - f.L.p_p}{k_p.A_s}$$

$$u_3 = \left(\frac{f}{EA_s}p_p - \frac{\gamma}{E}\right).\Delta l^2 + 2u_2 - u_1 \rightarrow u_3 = \left(\frac{f}{EA_s}p_p - \frac{\gamma}{E}\right).\Delta l^2 + 2\left(1 + \frac{k_p\Delta l}{E}\right)u_1 - u_1$$

$$= \left(\frac{f}{EA_s}p_p - \frac{\gamma}{E}\right).\Delta l^2 + \left(1 + 2\frac{k_p\Delta l}{E}\right)u_1$$

$$u_4 = \left(\frac{f}{EA_s}p_p - \gamma\right).\Delta l^2 + 2u_3 - u_2 \rightarrow u_4$$

$$= \left(\frac{f}{EA_s}p_p - \frac{\gamma}{E}\right).\Delta l^2 + 2\left[\left(\frac{f}{EA_s}p_p - \frac{\gamma}{E}\right).\Delta l^2 + \left(1 + 2\frac{k_p\Delta l}{E}\right)u_1\right] - \left(1 + \frac{k_p\Delta l}{E}\right)u_1$$

$$= 3\left(\frac{f}{EA_s}p_p - \frac{\gamma}{E}\right).\Delta l^2 + \left(1 + 3\frac{k_p \Delta l}{E}\right)u_1$$

$$u_5 = \left(\frac{f}{EA_s}p_p - \frac{\gamma}{E}\right).\Delta l^2 + 2u_4 - u_3 \rightarrow u_5$$

$$= \left(\frac{f}{EA_s}p_p - \frac{\gamma}{E}\right).\Delta l^2 + 2\left[3\left(\frac{f}{EA_s}p_p - \frac{\gamma}{E}\right).\Delta l^2 + \left(1 + 3\frac{k_p \Delta l}{E}\right)u_1\right]$$

$$- \left[\left(\frac{f}{EA_s}p_p - \frac{\gamma}{E}\right).\Delta l^2 + \left(1 + 2\frac{k_p \Delta l}{E}\right)u_1\right]$$

$$= 6\left(\frac{f}{EA_s}p_p - \frac{\gamma}{E}\right).\Delta l^2 + \left(1 + 4\frac{k_p \Delta l}{E}\right)u_1$$

$$u_n = \frac{(n-1)(n-2)}{2}\left(\frac{f}{EA_s}p_p - \frac{\gamma}{E}\right).\Delta l^2 + \left[1 + (n-1)\frac{k_p \Delta l}{E}\right]u_1$$

$$\Delta l = \frac{L}{n-1}$$

$$u_n = \frac{(n-2)}{2(n-1)}\left(\frac{f}{EA_s}p_p - \frac{\gamma}{E}\right).L^2 + \left[1 + \frac{k_p L}{E}\right]\left(\frac{Q + W - f.L.p_p}{k_p.A_s}\right)$$

$$n \rightarrow \infty \quad u_{top} = \frac{Q.L}{EA_s} - \frac{f.p_p.L^2}{2EA_s} + \frac{W.L}{2EA_s} + \left(\frac{Q + W - f.L.p_p}{k_p A_s}\right)$$

Problem 11.4

Develop the closed-form solution for the expansion of an infinitely long cylindrical cavity in an elastic soil space. The soil is weightless and has a Poisson's ratio ν and a modulus E. The cavity has an initial radius r_o. The goal is to generate the curve that gives the radial stress σ_r as a function of the relative increase in cavity radius $\Delta r/r_o$.

Solution 11.4

The geometry of the problem indicates that this is an axisymmetric problem and a plane strain problem in the vertical direction. The initial state of stress is σ_{ov} in the vertical direction and σ_{oh} in the radial direction at any point in the soil space. After applying the pressure p at the cavity surface, the stresses in the mass become:

$$\sigma_{rr} = \sigma_{oh} + \Delta\sigma_{rr}$$

$$\sigma_{\theta\theta} = \sigma_{oh} + \Delta\sigma_{\theta\theta}$$

$$\sigma_{zz} = \sigma_{ov} + \Delta\sigma_{zz}$$

where σ_{rr} and $\sigma_{\theta\theta}$ are the radial stress and the hoop stress respectively at a distance r from the axis of the cylinder, and $\Delta\sigma_{rr}$ and $\Delta\sigma_{\theta\theta}$ are the increments of the radial and hoop stress above the at-rest stress value.

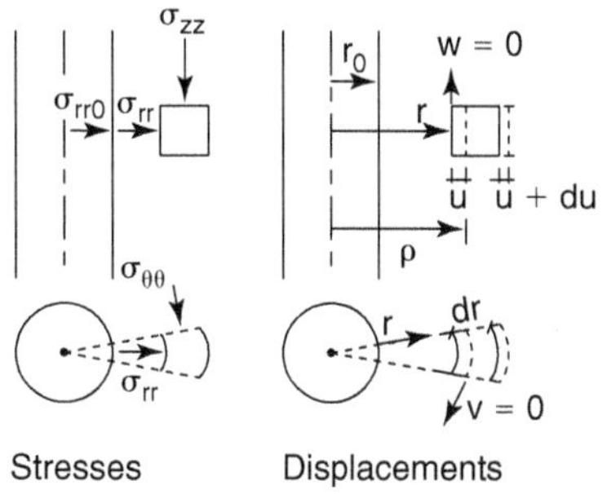

Figure 11.7s Element of soil around an expanding cylindrical cavity.

The radial displacement u is the only displacement type in this problem, as there are no displacements in the hoop direction or in the vertical direction. The radial strain is ε_{rr}, the hoop strain is $\varepsilon_{\theta\theta}$, and the vertical strain is ε_{zz}, which is zero because of the plane strain condition in the z direction. The relationships between the displacement and the strains for small strain theory are given in the following. Note that the minus sign is there to keep compression positive, because u decreases when r increases:

$$\varepsilon_{rr} = -\frac{du}{dr} \tag{11.1s}$$

$$\varepsilon_{\theta\theta} = -\frac{u}{r} \tag{11.2s}$$

$$\varepsilon_{zz} = 0 \tag{11.3s}$$

The equations of equilibrium reduce to:

$$\frac{d\sigma_{rr}}{dr} + \frac{\Delta\sigma_{rr} - \Delta\sigma_{\theta\theta}}{r} = 0 \tag{11.4s}$$

The constitutive equations are:

$$\varepsilon_{rr} = \frac{1}{E}(\Delta\sigma_{rr} - \nu(\Delta\sigma_{\theta\theta} + \Delta\sigma_{zz})) \tag{11.5s}$$

$$\varepsilon_{\theta\theta} = \frac{1}{E}(\Delta\sigma_{\theta\theta} - \nu(\Delta\sigma_{zz} + \Delta\sigma_{rr})) \tag{11.6s}$$

$$\varepsilon_{zz} = \frac{1}{E}(\Delta\sigma_{zz} - \nu(\Delta\sigma_{rr} + \Delta\sigma_{\theta\theta})) \tag{11.7s}$$

By combining Eqs. 11.1s to 11.7s, the governing differential equation is obtained as:

$$r^2\frac{d^2u}{dr^2} + r\frac{du}{dr} - u = 0 \tag{11.8s}$$

Assume that $u = r^n$. Then, by plugging it into Eq. 11.4s, we can get $n = 1, n = -1$.

$$\text{So, } u = \frac{A}{r} + Br$$

From boundary conditions

$$u = 0 \text{ when } r = \infty, \text{ we get } B = 0$$

$$u = u_0, \text{ when } r = r_0, \text{ we get } A = u_0 r_0$$

Therefore,

$$u = \frac{u_0 r_0}{r}$$

The strains are:

$$\varepsilon_r = -\frac{du}{dr} = \frac{u_0 r_0}{r^2}$$

$$\varepsilon_\theta = -\frac{u}{r} = -\frac{u_0 r_0}{r^2}$$

Note that from Eqs. 11.1s to 11.7s:

$$\Delta\sigma_r = \frac{E(1-\nu)}{(1+\nu)(1-2\nu)}\left[\frac{u_0 r_0}{r^2} - \frac{\nu}{1-\nu}\frac{u_0 r_0}{r^2}\right] = \frac{E}{1+\nu}\frac{u_0 r_0}{r^2}$$

Therefore,

$$\Delta\sigma_{r(r=r_0)} = \frac{E}{1+\nu}\frac{u_0 r_0}{{r_0}^2} = \frac{E}{1+\nu}\frac{u_0}{r_0}$$

and

$$\sigma_{rr} = \sigma_{oh} + 2G\frac{u_o r_o}{r^2} \qquad \sigma_{\theta\theta} = \sigma_{oh} - 2G\frac{u_o r_o}{r^2} \qquad \sigma_{zz} = \sigma_{ov}$$

In a pressuremeter test, the relative increase in radius ($u_o/r_o = \varepsilon_{\theta o}$) of the cavity is measured along with the pressure exerted on the cavity wall σ_{rro}. Therefore, the pressuremeter curve is a direct plot of a stress-strain curve of the soil.

Problem 11.5

Develop the closed-form solution for the expansion of a spherical cavity in an elastic soil space. The soil is weightless and has a Poisson's ratio ν and a modulus E. The cavity has an initial radius r_o. The goal is to generate the curve that gives the radial stress σ_r as a function of the relative increase in cavity radius $\Delta r/r_o$.

Solution 11.5

1. The equilibrium equation in spherical coordinates is:

$$\sigma_r(r d\theta)(r d\phi) - \left(\sigma_r + \frac{\partial \sigma_r}{\partial r}dr\right)(r + dr)^2 d\theta d\phi + \left(\sigma_\theta + \sigma_\theta + \frac{\partial \sigma_\theta}{\partial \theta}\partial\theta\right)\frac{d\theta}{2} r d\phi dr$$
$$+ \left(\sigma_\phi + \sigma_\phi + \frac{\partial \sigma_\phi}{\partial \phi}\partial\phi\right)\frac{d\phi}{2} r d\theta dr = 0$$

Ignoring the higher terms:

$$-\frac{\partial \sigma_r}{\partial r} - \frac{2\sigma_r}{r} + \frac{\sigma_\theta}{r} + \frac{\sigma_\phi}{r} = 0$$

For $\sigma_\theta = \sigma_\varphi$, it becomes:

$$\frac{\partial \sigma_r}{\partial r} + \frac{2}{r}(\sigma_r - \sigma_\theta) = 0$$

2. Obtain stress-strain relationships (constitutive equations) in the elastic range.

Due to symmetry, $\begin{matrix}\sigma_\theta = \sigma_\phi \\ \varepsilon_\theta = \varepsilon_\phi\end{matrix}$ and σ_θ, σ_ϕ, and σ_r are principal stresses.

Constitutive Equations

$$\varepsilon_r = \frac{1}{E}(\sigma_r - \nu(\sigma_\theta + \sigma_\phi)) = -\frac{du}{dr}$$
$$\varepsilon_\theta = \frac{1}{E}(\sigma_\theta - \nu(\sigma_r + \sigma_\phi)) = -\frac{u}{r}$$
$$\varepsilon_\phi = \frac{1}{E}(\sigma_\phi - \nu(\sigma_r + \sigma_\theta)) = -\frac{u}{r}$$

Again, the minus signs are there to keep compression positive. So:

$$\varepsilon_r = \frac{1}{E}(\sigma_r - 2\nu\sigma_\theta) = -\frac{du}{dr}$$
$$\varepsilon_\theta = \varepsilon_\phi = \frac{1}{E}((1-\nu)\sigma_\theta - \nu\sigma_r) = -\frac{u}{r}$$

Obtain

$$\sigma_r = f(u, r, du, dr)$$
$$\sigma_\theta = g(u, r, du, dr)$$

Solve for $u = F(r)$ with the appropriate boundary conditions.
Because

$$\varepsilon_r = -\frac{du}{dr} = \frac{1}{E}(\sigma_r - 2\nu\sigma_\theta)$$

we have:

$$-E\frac{\mathrm{d}u}{\mathrm{d}r} + 2\nu\sigma_\theta = \sigma_r$$

$$-E\frac{u}{r} = (1-\nu)\sigma_\theta - \nu\left[-E\frac{\mathrm{d}u}{\mathrm{d}r} + 2\nu\sigma_\theta\right]$$

$$-E\frac{u}{r} = (1-\nu)\sigma_\theta + \nu E\frac{\mathrm{d}u}{\mathrm{d}r} - 2\nu^2\sigma_\theta$$

$$-E\left(\frac{u}{r} + \frac{\nu \mathrm{d}u}{\mathrm{d}r}\right) = \sigma_\theta(1-\nu-2\nu^2)$$

$$\sigma_\theta = -\frac{E}{(1-\nu-2\nu^2)}\left[\frac{u}{r} + \nu\frac{\mathrm{d}u}{\mathrm{d}r}\right]$$

Solve for

$$\sigma_r = -E\frac{\mathrm{d}u}{\mathrm{d}r} - \frac{2\nu E}{(1-\nu-2\nu^2)}\left[\frac{u}{r} + \nu\frac{\mathrm{d}u}{\mathrm{d}r}\right]$$

$$\sigma_r = -E\frac{\mathrm{d}u}{\mathrm{d}r}\left[\frac{1-\nu-2\nu^2+2\nu^2}{(1-\nu-2\nu^2)}\right] - \frac{2\nu E}{(1-\nu-2\nu^2)}\frac{u}{r}$$

$$\sigma_r = -\frac{E(1-\nu)}{(1-\nu-2\nu^2)}\frac{\mathrm{d}u}{\mathrm{d}r} - \frac{2\nu E}{(1-\nu-2\nu^2)}\frac{u}{r}$$

$$\frac{\partial\sigma_r}{\partial r} = -\frac{E(1-\nu)}{(1-\nu-2\nu^2)}\frac{\mathrm{d}^2u}{\mathrm{d}r^2} + \frac{2\nu E}{(1-\nu-2\nu^2)}\frac{u}{r^2} - \frac{\mathrm{d}u}{\mathrm{d}r}\frac{2\nu E}{(1-\nu-2\nu^2)}\frac{1}{r}$$

$$(\sigma_r - \sigma_\theta)\frac{2}{r} = -\frac{2}{r}\left\{\frac{E}{(1-\nu-2\nu^2)} \times \left[(1-\nu)\frac{\mathrm{d}u}{\mathrm{d}r} + 2\nu\frac{u}{r} - \frac{u}{r} - \frac{\nu \mathrm{d}u}{\mathrm{d}r}\right]\right\}$$

$$= -\frac{2E}{r(1-\nu-2\nu^2)} \times \left[(1-2\nu)\frac{\mathrm{d}u}{\mathrm{d}r} + (2\nu-1)\frac{u}{r}\right]$$

So

$$\frac{\partial\sigma_r}{r} + \frac{2}{r}(\sigma_r - \sigma_\theta) = -\frac{E}{(1-\nu-2\nu^2)}\left\{(1-\nu)\frac{\mathrm{d}^2u}{\mathrm{d}r^2} - 2\nu\frac{u}{r^2} + \frac{2(1-2\nu)}{r}\frac{\mathrm{d}u}{\mathrm{d}r} + 2(2\nu-1)\frac{u}{r^2} + \frac{1}{r}2\nu\frac{\mathrm{d}u}{\mathrm{d}r}\right\}$$

$$= -\frac{E}{(1-\nu-2\nu^2)}\left\{(1-\nu)\frac{\mathrm{d}^2u}{\mathrm{d}r^2} - 2(1-\nu)\frac{u}{r^2} + \frac{2(1-\nu)}{r}\frac{\mathrm{d}u}{\mathrm{d}r}\right\} = 0$$

and

$$\frac{\mathrm{d}^2u}{\mathrm{d}r^2} - 2\frac{u}{r^2} + \frac{2}{r}\frac{\mathrm{d}u}{\mathrm{d}r} = 0$$

or

$$r^2\frac{\mathrm{d}^2u}{\mathrm{d}r^2} + 2r\frac{\mathrm{d}u}{\mathrm{d}r} - 2u = 0$$

Solve the differential equation:

$$u = r^n$$

$$r^2(n)(n-1)r^{n-2} + 2r(n)r^{n-1} - 2r^n = 0$$

$$\left.\begin{aligned} n(n-1) + 2n - 2 = 0 \\ n(n-1) + 2(n-1) = 0 \end{aligned}\right\} n = -1, n = 1$$

$$u = Ar + \frac{B}{r^2}$$

Apply the boundary conditions. When

$$r = \infty \rightarrow u = 0$$
$$A = 0$$
$$u = \frac{B}{r^2}$$

When

$$r = r_0, u = u_0$$
$$B = u_0 r_o^2$$
$$u = \frac{u_0 r_0^2}{r^2}$$

The strains can be expressed as:

$$\left.\begin{aligned} \varepsilon_\theta &= -\frac{u}{r} = -\frac{u_0 r_0^2}{r^3} = \varepsilon_\phi \\ \varepsilon_r &= -\frac{du}{dr} = 2\frac{u_0 r_0^2}{r^3} \end{aligned}\right\} \frac{\Delta V}{V} = \varepsilon_r + 2\varepsilon_\theta = 2\frac{u_0 r_0^2}{r^3} - 2\frac{u_0 r_0^2}{r^3} = 0$$

Recall the equilibrium equation:

$$\frac{\partial \sigma_r}{\partial r} + \frac{2}{r}(\sigma_r - \sigma_\theta) = 0$$

The preceding derivation assumes an unstressed initial state. If the soil is under a hydrostatic initial state of stress equal to $p_o(\sigma_\phi = \sigma_r = \sigma_\theta = p_0)$, then the preceding solutions are in terms of stress increments as follows:

$$\sigma_r = p_0 + \Delta\sigma_r$$
$$\sigma_\theta = p_0 + \Delta\sigma_\theta$$
$$\sigma_\phi = p_0 + \Delta\sigma_\phi$$

Thus, the equilibrium equation can be written as:

$$\frac{\partial \sigma_r}{\partial r} + \frac{2}{r}(\Delta\sigma_r - \Delta\sigma_\theta) = 0$$

As seen previously, we have:

$$\Delta\sigma_r = -\frac{E}{(1 - \nu - 2\nu^2)}\left[(1-\nu)\frac{du}{dr} + 2\nu\frac{u}{r}\right]$$
$$\frac{du}{dr} = -2\frac{u_0 r_0^2}{r^3}$$
$$\frac{u}{r} = \frac{u_0 r_0^2}{r^3}$$

Therefore,

$$\begin{aligned} \Delta\sigma_r &= -\frac{E}{(1-\nu-2\nu^2)}\frac{u_0 r_0^2}{r^3}(-2(1-\nu) + 2\nu) = -\frac{2(2\nu-1)E}{(1-\nu-2\nu^2)}\frac{u_0 r_0^2}{r^3} \\ &= -\frac{2(2\nu-1)}{(1+\nu)(1-2\nu)}E\left(\frac{u_0 r_0^2}{r^3}\right) \end{aligned}$$
$$\Delta\sigma_r = \frac{2E}{(1+\nu)}\frac{u_0 r_0^2}{r^3}$$

At

$$r = r_0 \rightarrow \Delta\sigma_r = \frac{2E}{1+\nu}\frac{u_0}{r_0}$$

Therefore,

$$\Delta\sigma_{r(r=r_0)} = \frac{2E}{1+\nu}\frac{u_0 r_0^2}{r_0{}^3} = \frac{2E}{1+\nu}\frac{u_0}{r_0} = -4G\varepsilon_{\theta 0}$$

and by the same process:

$$\Delta\sigma_{\theta(r=r_0)} = 2G\varepsilon_{\theta 0}$$

Problem 11.6

Develop the solution for the flow of water through a saturated soil sample in a constant head permeameter. The goal is to find the excess water stress anywhere and at any time in the sample.

Solution 11.6

Conservation of mass law: Qdt = vAdt

Darcy's law: vdl = kdh

$$dh = \frac{Q}{kA}dl$$

$$\int_{h_1}^{h_2} dh = \frac{Q}{kA}\int_{l_1}^{l_2} dl \rightarrow \Delta h = \frac{Q}{kA}\Delta l$$

$$\Delta u_w = \gamma_w \Delta h = \gamma_w \frac{Q}{kA}\Delta l$$

Note that the total head h decreases through the sample; therefore so will u_w. See Figure 9.69.

Problem 11.7

Use the finite element method to construct the global stiffness matrix for a triaxial test performed on an elastic soil. The major principal stress is 300 kPa and the minor principal stress is 100 kPa. The modulus is 40 MPa and the Poisson's ratio is 0.35. The height and diameter of the sample are 0.1 m and 0.05 m respectively. Consider an axisymmetric geometry and use two four-noded elements.

Solution 11.7

Refer to section 11.5.3 for the equations used in this problem

Step 1: The selected elements are shown in Figure 11.8s.
The element dimensions are $a = 0.025$ m and $b = 0.05$ m. The soil properties are $E = 40000$ kPa and $\mu = 35$.

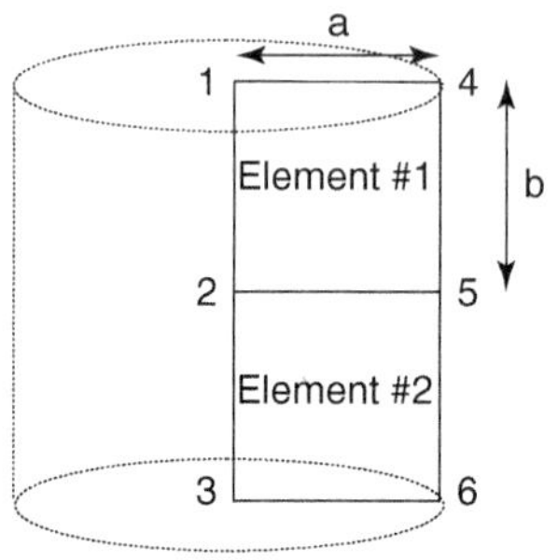

Figure 11.8s Triaxial test mesh.

Step 2: Choose the interpolation or shape functions. The equations for these functions are:

$$H_1 = \frac{1}{4}(1+r)(1+s)$$
$$H_2 = \frac{1}{4}(1-r)(1+s)$$
$$H_3 = \frac{1}{4}(1-r)(1-s)$$
$$H_4 = \frac{1}{4}(1+r)(1-s)$$

Step 3: Write the strain-displacement equations:

$$[\varepsilon] = [B][u_i]$$

$$\begin{bmatrix} \varepsilon_{xx} \\ \varepsilon_{yy} \\ \gamma_{xy} \end{bmatrix} = \begin{bmatrix} \frac{\partial H_1}{\partial x} & 0 & \frac{\partial H_2}{\partial x} & 0 & \frac{\partial H_3}{\partial x} & 0 & \frac{\partial H_4}{\partial x} & 0 \\ 0 & \frac{\partial H_1}{\partial y} & 0 & \frac{\partial H_2}{\partial y} & 0 & \frac{\partial H_3}{\partial y} & 0 & \frac{\partial H_4}{\partial y} \\ \frac{\partial H_1}{\partial y} & \frac{\partial H_1}{\partial x} & \frac{\partial H_2}{\partial y} & \frac{\partial H_2}{\partial x} & \frac{\partial H_3}{\partial y} & \frac{\partial H_3}{\partial x} & \frac{\partial H_4}{\partial y} & \frac{\partial H_4}{\partial x} \end{bmatrix} \begin{bmatrix} u_{x1} \\ u_{y1} \\ u_{x2} \\ u_{y2} \\ u_{x3} \\ u_{y3} \\ u_{x4} \\ u_{y4} \end{bmatrix}$$

Construct the [B] matrix:

a. Calculate the inverse of the Jacobian matrix used in the transformation from natural coordinates to real coordinates:

$$J = \begin{bmatrix} \frac{\partial x}{\partial r} & \frac{\partial y}{\partial r} \\ \frac{\partial x}{\partial s} & \frac{\partial y}{\partial s} \end{bmatrix} = \begin{bmatrix} \frac{a}{2} & 0 \\ 0 & \frac{b}{2} \end{bmatrix} = \begin{bmatrix} 0.0125 & 0 \\ 0 & 0.025 \end{bmatrix}$$

Therefore,

$$\det J = 3.125 * 10^{-4}$$

and

$$J^{-1} = \left(\frac{1}{\det J}\right) \cdot \begin{bmatrix} \frac{b}{2} & 0 \\ 0 & \frac{a}{2} \end{bmatrix} = \begin{bmatrix} 80 & 0 \\ 0 & 40 \end{bmatrix}$$

$$\frac{\partial H}{\partial r} = \begin{bmatrix} \frac{1}{4}(1+s) & -\frac{1}{4}(1+s) & -\frac{1}{4}(1+s) & \frac{1}{4}(1+s) \end{bmatrix}$$

$$\frac{\partial H}{\partial s} = \begin{bmatrix} \frac{1}{4}(1+r) & \frac{1}{4}(1+r) & -\frac{1}{4}(1+r) & -\frac{1}{4}(1+r) \end{bmatrix}$$

b. Obtain the relation between the derivatives of the interpolation functions in real coordinates and in natural coordinates:

$$\begin{bmatrix} \frac{\partial H_i}{\partial x} \\ \frac{\partial H_i}{\partial y} \end{bmatrix} = J^{-1} \begin{bmatrix} \frac{\partial H_i}{\partial r} \\ \frac{\partial H_i}{\partial s} \end{bmatrix}$$

$$\begin{bmatrix} \frac{\partial H}{\partial x} \\ \frac{\partial H}{\partial y} \end{bmatrix} = \begin{bmatrix} \frac{b}{2} & 0 \\ 0 & \frac{a}{2} \end{bmatrix} \cdot \begin{bmatrix} \frac{\partial H}{\partial r} \\ \frac{\partial H}{\partial s} \end{bmatrix} = \begin{bmatrix} \frac{b}{2} \cdot \frac{\partial H}{\partial r} \\ \frac{a}{2} \cdot \frac{\partial H}{\partial s} \end{bmatrix}$$

c. Select the natural coordinates r and s of the integration points for a four-node element:

$$r = \begin{bmatrix} \frac{1}{\sqrt{3}} & -\frac{1}{\sqrt{3}} \\ -\frac{1}{\sqrt{3}} & \frac{1}{\sqrt{3}} \end{bmatrix}$$

$$s = \begin{bmatrix} \frac{1}{\sqrt{3}} & \frac{1}{\sqrt{3}} \\ -\frac{1}{\sqrt{3}} & -\frac{1}{\sqrt{3}} \end{bmatrix}$$

d. Compute the components of the matrix [B] at the four integration points:

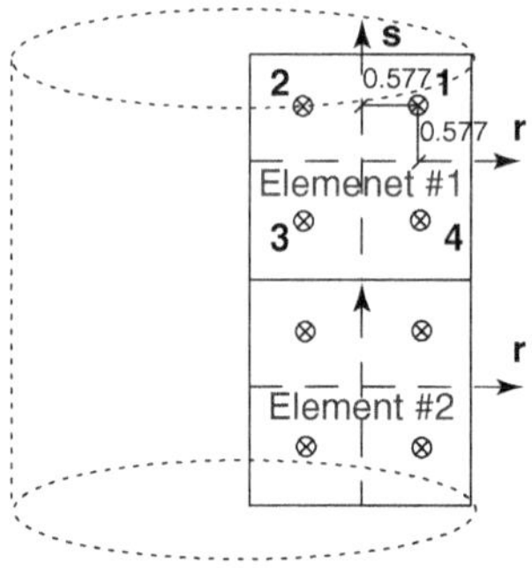

Figure 11.9s The integration points.

Intg. Point #1. The derivatives of the interpolation function are:

$$\frac{\partial H}{\partial r} = \begin{bmatrix} \frac{1}{4}(1+s) & -\frac{1}{4}(1+s) & -\frac{1}{4}(1+s) & \frac{1}{4}(1+s) \end{bmatrix}$$

$$\frac{\partial H}{\partial s} = \begin{bmatrix} \frac{1}{4}(1+r) & \frac{1}{4}(1+r) & -\frac{1}{4}(1+r) & -\frac{1}{4}(1+r) \end{bmatrix}$$

For integration point #1, the natural coordinates are $r = \frac{1}{\sqrt{3}}$ and $s = \frac{1}{\sqrt{3}}$

$$\frac{\partial H}{\partial r} = \begin{bmatrix} \frac{1}{4}\left(1+\frac{1}{\sqrt{3}}\right) & -\frac{1}{4}\left(1+\frac{1}{\sqrt{3}}\right) & -\frac{1}{4}\left(1+\frac{1}{\sqrt{3}}\right) & \frac{1}{4}\left(1+\frac{1}{\sqrt{3}}\right) \end{bmatrix}$$
$$= \begin{bmatrix} 0.394 & -0.394 & -0.105 & 0.105 \end{bmatrix}$$

$$\frac{\partial H}{\partial s} = \begin{bmatrix} \frac{1}{4}\left(1+\frac{1}{\sqrt{3}}\right) & \frac{1}{4}\left(1+\frac{1}{\sqrt{3}}\right) & -\frac{1}{4}\left(1+\frac{1}{\sqrt{3}}\right) & -\frac{1}{4}\left(1+\frac{1}{\sqrt{3}}\right) \end{bmatrix}$$
$$= \begin{bmatrix} 0.394 & 0.105 & -0.105 & -0.394 \end{bmatrix}$$

In the case of plane strain (Section 11.5.4), the thickness t of the elements was 1. However, in the case of axisymmetric geometry, the thickness varies across the element and must be evaluated at each integration point. If you look at the element in plan view, it looks like a piece of pizza with an angle θ. For convenience, we take a value of 1 radian for this angle. The thickness t of the element at a radius x_i is equal to x_i times θ. Because θ is 1 rd, the thickness is simply equal to x_i. Therefore, the equation for the thickness t is:

$$t = [H][x] = \begin{bmatrix} H_1 & H_2 & H_3 & H_4 \end{bmatrix} \begin{bmatrix} x_1 \\ x_2 \\ x_3 \\ x_4 \end{bmatrix} = H_1.x_1 + H_2.x_2 + H_3.x_3 + H_4.x_4$$

where x_i represents the real coordinates of the nodes. For elements 1 and 2, the x matrices are:

$$element\ \#1 \rightarrow [x] = \begin{bmatrix} 0.025 \\ 0 \\ 0 \\ 0.025 \end{bmatrix} \quad \text{and } element\ \#2 \rightarrow [x] = \begin{bmatrix} 0.025 \\ 0 \\ 0 \\ 0.025 \end{bmatrix}$$

$$H_1 = \frac{1}{4}(1+r)(1+s) = \frac{1}{4}(1+\frac{1}{\sqrt{3}})(1+\frac{1}{\sqrt{3}}) = 0.622$$

$$H_2 = \frac{1}{4}(1-r)(1+s) = \frac{1}{4}\left(1-\frac{1}{\sqrt{3}}\right)(1+\frac{1}{\sqrt{3}} = 0.1667$$

$$H_3 = \frac{1}{4}(1-r)(1-s) = \frac{1}{4}\left(1-\frac{1}{\sqrt{3}}\right)\left(1-\frac{1}{\sqrt{3}}\right) = 0.0446$$

$$H_4 = \frac{1}{4}(1+r)(1-s) = \frac{1}{4}\left(1-\frac{1}{\sqrt{3}}\right)\left(1-\frac{1}{\sqrt{3}}\right) = 0.1667$$

$$t = \begin{bmatrix} 0.622 & 0.1667 & 0.0446 & 0.1667 \end{bmatrix} \begin{bmatrix} 0.025 \\ 0 \\ 0 \\ 0.025 \end{bmatrix} = 0.0197\ (\text{m})$$

Intg Point #2: For integration point #2, the natural coordinates are $r = -\frac{1}{\sqrt{3}}$ and $s = \frac{1}{\sqrt{3}}$

$$\frac{\partial H}{\partial r} = \left[\frac{1}{4}\left(1+\frac{1}{\sqrt{3}}\right) \quad -\frac{1}{4}\left(1+\frac{1}{\sqrt{3}}\right) \quad -\frac{1}{4}\left(1-\frac{1}{\sqrt{3}}\right) \quad \frac{1}{4}\left(1-\frac{1}{\sqrt{3}}\right)\right]$$
$$= \begin{bmatrix} 0.394 & -0.394 & -0.105 & 0.105 \end{bmatrix}$$

$$\frac{\partial H}{\partial s} = \left[\frac{1}{4}\left(1-\frac{1}{\sqrt{3}}\right) \quad \frac{1}{4}\left(1+\frac{1}{\sqrt{3}}\right) \quad -\frac{1}{4}\left(1+\frac{1}{\sqrt{3}}\right) \quad -\frac{1}{4}\left(1-\frac{1}{\sqrt{3}}\right)\right]$$
$$= \begin{bmatrix} 0.105 & 0.394 & -0.394 & -0.105 \end{bmatrix}$$

$$H_1 = \frac{1}{4}(1+r)(1+s) = \frac{1}{4}\left(1-\frac{1}{\sqrt{3}}\right)\left(1+\frac{1}{\sqrt{3}}\right) = 0.1667$$

$$H_2 = \frac{1}{4}(1-r)(1+s) = \frac{1}{4}\left(1+\frac{1}{\sqrt{3}}\right)\left(1+\frac{1}{\sqrt{3}}\right) = 0.622$$

$$H_3 = \frac{1}{4}(1-r)(1-s) = \frac{1}{4}\left(1+\frac{1}{\sqrt{3}}\right)\left(1-\frac{1}{\sqrt{3}}\right) = 0.1667$$

$$H_4 = \frac{1}{4}(1+r)(1-s) = \frac{1}{4}\left(1-\frac{1}{\sqrt{3}}\right)\left(1-\frac{1}{\sqrt{3}}\right) = 0.0446$$

$$t = \begin{bmatrix} 0.1667 & 0.622 & 0.1667 & 0.0446 \end{bmatrix} \begin{bmatrix} 0.025 \\ 0 \\ 0 \\ 0.025 \end{bmatrix} = 0.0053\ (\text{m})$$

Intg. Point #3: For integration point #3, the natural coordinates are $r = -\frac{1}{\sqrt{3}}$ and $s = -\frac{1}{\sqrt{3}}$

$$\frac{\partial H}{\partial r} = \left[\frac{1}{4}\left(1-\frac{1}{\sqrt{3}}\right) \quad -\frac{1}{4}\left(1-\frac{1}{\sqrt{3}}\right) \quad -\frac{1}{4}\left(1+\frac{1}{\sqrt{3}}\right) \quad \frac{1}{4}\left(1+\frac{1}{\sqrt{3}}\right)\right]$$
$$= \begin{bmatrix} 0.105 & -0.105 & -0.394 & 0.394 \end{bmatrix}$$

$$\frac{\partial H}{\partial s} = \left[\frac{1}{4}\left(1-\frac{1}{\sqrt{3}}\right) \quad \frac{1}{4}\left(1+\frac{1}{\sqrt{3}}\right) \quad -\frac{1}{4}\left(1+\frac{1}{\sqrt{3}}\right) \quad -\frac{1}{4}\left(1-\frac{1}{\sqrt{3}}\right)\right]$$

$$= \left[0.105 \quad 0.394 \quad -0.394 \quad -0.105\right]$$

$$H_1 = \frac{1}{4}(1+r)(1+s) = \frac{1}{4}\left(1-\frac{1}{\sqrt{3}}\right)\left(1-\frac{1}{\sqrt{3}}\right) = 0.0446$$

$$H_2 = \frac{1}{4}(1-r)(1+s) = \frac{1}{4}\left(1+\frac{1}{\sqrt{3}}\right)\left(1-\frac{1}{\sqrt{3}}\right) = 0.1667$$

$$H_3 = \frac{1}{4}(1-r)(1-s) = \frac{1}{4}\left(1+\frac{1}{\sqrt{3}}\right)\left(1+\frac{1}{\sqrt{3}}\right) = 0.622$$

$$H_4 = \frac{1}{4}(1+r)(1-s) = \frac{1}{4}\left(1-\frac{1}{\sqrt{3}}\right)\left(1+\frac{1}{\sqrt{3}}\right) = 0.1667$$

$$t = \left[0.0446 \quad 0.1667 \quad 0.622 \quad 0.1667\right]\begin{bmatrix} 0.025 \\ 0 \\ 0 \\ 0.025 \end{bmatrix} = 0.0053 \text{ (m)}$$

Intg. Point #4: For integration point #4, the natural coordinates are $r = \frac{1}{\sqrt{3}}$ and $s = -\frac{1}{\sqrt{3}}$

$$\frac{\partial H}{\partial r} = \left[\frac{1}{4}\left(1-\frac{1}{\sqrt{3}}\right) \quad -\frac{1}{4}\left(1-\frac{1}{\sqrt{3}}\right) \quad -\frac{1}{4}\left(1+\frac{1}{\sqrt{3}}\right) \quad \frac{1}{4}\left(1+\frac{1}{\sqrt{3}}\right)\right]$$

$$= \left[0.105 \quad -0.105 \quad -0.394 \quad 0.394\right]$$

$$\frac{\partial H}{\partial s} = \left[\frac{1}{4}\left(1+\frac{1}{\sqrt{3}}\right) \quad \frac{1}{4}\left(1-\frac{1}{\sqrt{3}}\right) \quad -\frac{1}{4}\left(1-\frac{1}{\sqrt{3}}\right) \quad -\frac{1}{4}\left(1+\frac{1}{\sqrt{3}}\right)\right]$$

$$= \left[0.394 \quad 0.105 \quad -0.105 \quad -0.394\right]$$

$$H_1 = \frac{1}{4}(1+r)(1+s) = \frac{1}{4}\left(1+\frac{1}{\sqrt{3}}\right)\left(1-\frac{1}{\sqrt{3}}\right) = 0.1667$$

$$H_2 = \frac{1}{4}(1-r)(1+s) = \frac{1}{4}\left(1-\frac{1}{\sqrt{3}}\right)\left(1-\frac{1}{\sqrt{3}}\right) = 0.0446$$

$$H_3 = \frac{1}{4}(1-r)(1-s) = \frac{1}{4}\left(1-\frac{1}{\sqrt{3}}\right)\left(1+\frac{1}{\sqrt{3}}\right) = 0.1667$$

$$H_4 = \frac{1}{4}(1+r)(1-s) = \frac{1}{4}\left(1+\frac{1}{\sqrt{3}}\right)\left(1+\frac{1}{\sqrt{3}}\right) = 0.622$$

$$t = \left[0.1667 \quad 0.0446 \quad 0.1667 \quad 0.622\right]\begin{bmatrix} 0.025 \\ 0 \\ 0 \\ 0.025 \end{bmatrix} = 0.0197 \text{ (m)}$$

Then we assemble the B matrix:

$$B_{\substack{i=1 \\ j=1}} = \begin{bmatrix} 31.547 & 0 & -31.547 & 0 & -8.453 & 0 & 8.453 & 0 \\ 0 & 15.773 & 0 & 4.266 & 0 & -4.266 & 0 & -15.773 \\ 15.773 & 31.547 & 4.266 & -31.547 & -4.266 & -8.453 & -15.773 & 15.773 \\ 31.547 & 0 & 31.547 & 0 & 8.453 & 0 & 8.453 & 0 \end{bmatrix}$$

$$B_{\substack{i=1\\j=2}} = \begin{bmatrix} 31.547 & 0 & -31.547 & 0 & -8.453 & 0 & 8.453 & 0 \\ 0 & 4.226 & 0 & 15.733 & 0 & -15.733 & 0 & -4.226 \\ 4.226 & 31.547 & 15.733 & -31.547 & -15.733 & -8.453 & -4.226 & 8.453 \\ 8.453 & 0 & 117.735 & 0 & 31.547 & 0 & 2.2650 & 0 \end{bmatrix}$$

$$B_{\substack{i=2\\j=1}} = \begin{bmatrix} 8.453 & 0 & -8.453 & 0 & -31.547 & 0 & 31.547 & 0 \\ 0 & 4.226 & 0 & 15.773 & 0 & -15.773 & 0 & -4.226 \\ 4.226 & 8.453 & 15.773 & -8.453 & -15.773 & -31.547 & -4.226 & 31.547 \\ 2.2650 & 0 & 31.547 & 0 & 117.735 & 0 & 8.453 & 0 \end{bmatrix}$$

$$B_{\substack{i=2\\j=2}} = \begin{bmatrix} 8.453 & 0 & -8.453 & 0 & -31.547 & 0 & 31.547 & 0 \\ 0 & 15.773 & 0 & 4.266 & 0 & -4.266 & 0 & -15.773 \\ 15.773 & 8.453 & 4.266 & -8.453 & -4.266 & -31.547 & -15.773 & 31.547 \\ 8.453 & 0 & 8.453 & 0 & 31.547 & 0 & 31.547 & 0 \end{bmatrix}$$

Step 4: Write the stress-strain equations for the soil using the constitutive matrix.

$$C_{4X4} = \frac{E(1-\mu)}{(1+\mu)(1-2\mu)} \begin{bmatrix} 1 & \frac{\mu}{(1-\mu)} & \frac{\mu}{(1-\mu)} & 0 \\ \frac{\mu}{(1-\mu)} & 1 & \frac{\mu}{(1-\mu)} & 0 \\ \frac{\mu}{(1-\mu)} & \frac{\mu}{(1-\mu)} & 1 & 0 \\ 0 & 0 & 0 & \frac{1-2\mu}{2(1-\mu)} \end{bmatrix}$$

$$= 10^4 * \begin{bmatrix} 6.419 & 3.457 & 3.457 & 0 \\ 3.457 & 6.419 & 3.457 & 0 \\ 3.457 & 3.457 & 6.419 & 0 \\ 0 & 0 & 0 & 1.418 \end{bmatrix}$$

Step 5: Derive the equations governing the behavior of the soil element.

$$[K^e] = \int_V [B]^T[C][B]dV$$

$$[K^e][u] = [F]$$

The numerical integration equation is:

$$K_e = \int_v B^T CBdv = \sum_{i=1}^{2}\sum_{j=1}^{2} B_{ij}^T C_{ij} B_{ij} \det J.w_i.w_j.t$$

For a 2-point Gauss quadrature integration, the weighing factors w_i, w_j are equal to 1 and the element stiffness matrix is:

$$K^e = 10^3 \times \begin{bmatrix} 1.12 & 0.8 & -0.42 & -.44 & -0.48 & -0.52 & 0.11 & 0.16 \\ & 1.02 & -0.26 & -0.54 & -0.45 & -0.55 & -0.08 & 0.07 \\ & & 0.92 & 0.19 & 0.43 & 0.05 & -0.05 & 0.09 \\ & & & 0.46 & 0.16 & 0.2 & 0.16 & -0.11 \\ & & & & 1.2 & 0.53 & -0.27 & -0.24 \\ & SYM & & & & 0.75 & -0.05 & -0.39 \\ & & & & & & 0.57 & 0.135 \\ & & & & & & & 0.45 \end{bmatrix}$$

Now the global stiffness matrix K_g can be assembled:

$$K_g = 10^3 \times \begin{bmatrix} 0.92 & 0.12 & 0.43 & 0.046 & 0 & 0 & -0.42 & -0.26 & -0.05 & 0.09 & 0 & 0 \\ & 0.45 & 0.17 & 0.2 & 0 & 0 & -0.45 & -0.54 & 0.16 & -0.11 & 0 & 0 \\ & & 2.12 & 0.65 & 0.43 & 0.04 & -0.48 & -0.45 & -0.68 & -0.5 & -0.05 & 0.09 \\ & & & 1.2 & 0.16 & 0.2 & -0.5 & -0.5 & -0.5 & -0.93 & 0.15 & -0.11 \\ & & & & 1.2 & 0.52 & 0 & 0 & -0.48 & -0.45 & -0.27 & -0.27 \\ & & & & & 0.75 & 0 & 0 & -0.5 & -0.55 & -0.05 & -0.39 \\ & & & & & & 1.12 & 0.8 & 0.11 & 0.16 & 0 & 0 \\ & & & & & & & 1.02 & -0.08 & 0.06 & 0 & 0 \\ & & SYM & & & & & & 1.67 & 0.79 & 0.11 & 0.16 \\ & & & & & & & & & 1.47 & -0.08 & 0.06 \\ & & & & & & & & & & 0.54 & -0.015 \\ & & & & & & & & & & & 0.44 \end{bmatrix}$$

Problem 11.8

Two weightless particles of fine sand have a diameter of 1 mm and are placed in the corner of a container as shown in Figure 11.2s. The vertical load applied on the top particle is 0.4 kN. Find all forces between the particles, the wall, and the ground surface. Calculate the contact stress between the two particles if the contact area is 0.005 mm^2. The angles θ_1 and θ_2 are equal to 45°.

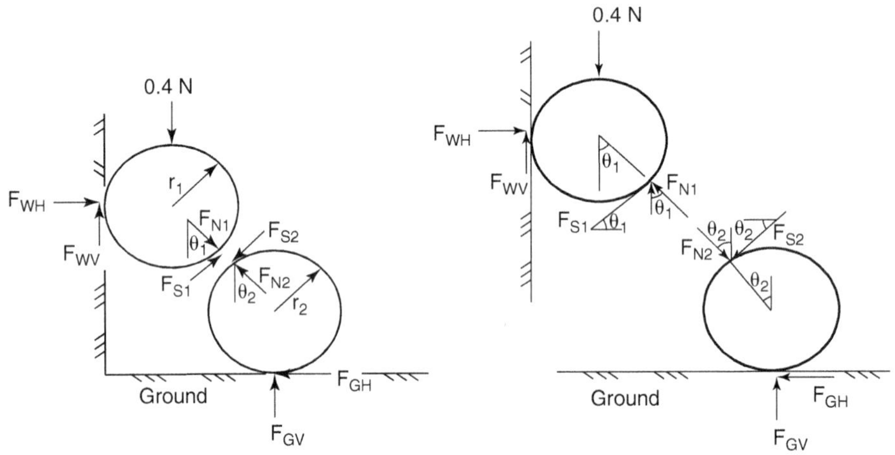

Figure 11.2s Discrete element problem.

Solution 11.8

Ball 1:

$$\sum F_V = F_{WV} + F_{S1}\sin\theta_1 + F_{N1}\cos\theta_1 - Q = 0$$

$$\sum F_H = F_{WH} + F_{S1}\cos\theta_1 - F_{N1}\sin\theta_1 = 0$$

$$\sum M_{center} = F_{S1}r_1 - F_{WV}r_1 = 0$$

Ball 2:

$$\sum F_v = F_{GV} - F_{N2}\cos\theta_2 - F_{S2}\sin\theta_2 = 0$$

$$\sum F_H = F_{N2}\sin\theta_2 - F_{S2}\cos\theta_2 - F_{GH} = 0$$

$$\sum M_{center} = F_{S2}r_2 - F_{GH}r_2 = 0$$

Also, at the contact between the two balls:

$$F_{S1} = F_{S2}$$
$$F_{N1} = F_{N2}$$

There are 8 unknown forces and 8 equations:

$$F_{WV} = F_{S1}$$
$$F_{WV} + F_{WV} \sin\theta_1 + F_{N1}\cos\theta_1 - Q = 0$$
$$F_{N1} = \frac{Q - F_{WV}(1 + \sin\theta_1)}{\cos\theta_1}$$
$$F_{S2} = F_{GH}$$
$$F_{N2}\sin\theta_2 - F_{S2}\cos\theta_2 - F_{S2} = 0$$
$$F_{N2}\sin\theta_2 = F_{S2}(\cos\theta_2 + 1)$$
$$F_{N2} = \frac{F_{S2}(\cos\theta_2 + 1)}{\sin\theta_2}$$

For Q = 0.4 kN, $r_1 = r_2 = 0.5$ mm, $\theta_1 = \theta_2 = 45°$

$$F_{WV} = 0.117\text{ kN}$$
$$F_{S1} = F_{S2} = 0.117\text{ kN}$$
$$F_{N1} = F_{N2} = 0.283\text{ kN}$$
$$F_{WH} = 0.117\text{ kN}$$
$$F_{GH} = 0.117\text{ kN}$$
$$F_{GV} = 0.283\text{ kN}$$

If the two balls touch over an area with a unit width of 0.05 mm, the stress distribution is:

$$P = \frac{F_N}{A} = \frac{0.283}{0.05 \times 10^{-6}} = 5660\text{ kPa}$$

Problem 11.9

A slope is to be designed for a target probability of failure of 0.001. Plot the mean factor of safety μ versus the coefficient of variation CoV_F in the following cases:

a. F follows a normal distribution.
b. F follows a lognormal distribution.

Solution 11.9

a. Normal distribution (Figure 11.10s)
   Probability of failure = 0.001. The mean of F is μ and the standard deviation is σ

$$P(F < 1) = 0.001 \Rightarrow P\left(\frac{F-\mu}{\sigma} < \frac{1-\mu}{\sigma}\right) = 0.001 \Rightarrow \frac{1-\mu}{\sigma} = -3.1$$

$$\sigma = \mu.COV$$

$$1 - \mu = -3.1\mu.COV \rightarrow \mu = \frac{1}{(-3.1COV + 1)}$$

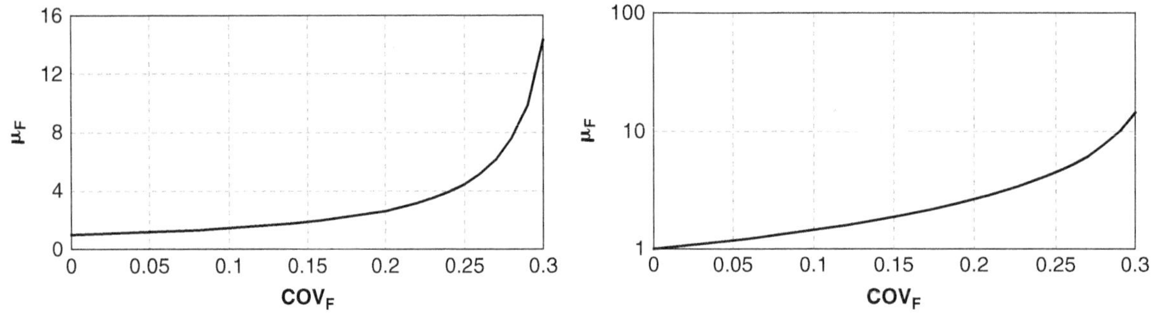

Figure 11.10s μ_F vs. CoV_F for a probability of failure of 0.001 when F follows a normal distribution.

b. Lognormal distribution (Figure 11.11s)

$$P(F < 1) = 0.001 \Rightarrow P\left(\frac{\ln F - \mu_{\ln F}}{\sigma_{\ln F}} < \frac{\ln 1 - \mu_{\ln F}}{\sigma_{\ln F}}\right) = 0.001 \Rightarrow \frac{\ln 1 - \ln\left(\dfrac{\mu^2}{\sqrt{\mu^2 + \mu^2.COV^2}}\right)}{\sqrt{\ln(1 + COV^2)}} = -3.1$$

$$\frac{\ln\left(\dfrac{\mu}{\sqrt{1 + COV^2}}\right)}{\sqrt{\ln(1 + COV^2)}} = 3.1 \rightarrow \ln(\mu) = 3.1\sqrt{\ln(1 + COV^2)} + \ln\sqrt{1 + COV^2}$$

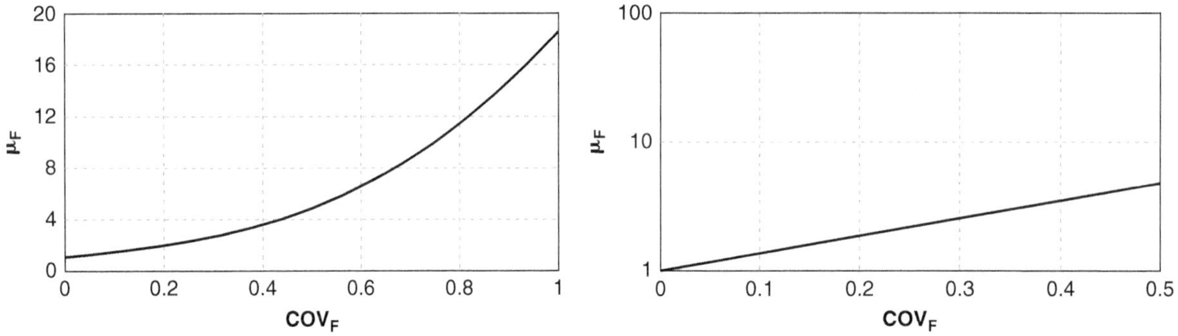

Figure 11.11s μ_F versus CoV_F for a probability of failure of 0.001 when F follows a lognormal distribution.

Problem 11.10

A levee system is to be designed to meet a risk of 0.001 fatalities/yr and \$1000/yr. It protects a city where 500,000 people could die and where the potential economic loss is \$200 billion if the system fails. What would you recommend for the design annual probability of failure of the levee system?

Solution 11.10

R(fatalities) = PoF × F

R(dollars lost) = PoF × D

R: risk

PoF: annual probability of failure

F: lives lost or fatalities if failure occurs = 500,000

D: dollars lost if failure occurs = 200×10^9 \$.

R(fatalities) = 0.001 fatalities/yr

R(dollars lost) = 1000 \$/yr

$$\frac{0.001 \text{ fatality/yr}}{500{,}000 \text{ people}} = 2 \times 10^{-9}$$

$$\frac{1000 \text{ dollars/yr}}{200 \times 10^{9}} = 5 \times 10^{-9}$$

The recommended annual probability of failure is 2×10^{-9}, as it is the most demanding of the two: fatalities control.

Problem 11.11

A levee system is to be designed to meet a risk of 0.001 fatalities/yr and \$1000/yr. It protects farmland where 100 people and a few cows could die and where the total potential economic loss is \$200 million. What would you recommend for the design probability of failure of the levee system?

Solution 11.11

R(fatalities) = PoF × F

R(dollars lost) = PoF × D

R: risk

PoF: annual probability of failure

F: lives lost or fatalities if failure occurs = 100

D: dollars lost if failure occurs = 100×10^{6} \$.

R(fatalities) = 0.001 fatalities/yr

R(dollars lost) = 1000 \$/yr

$$\frac{0.001 \text{ fatality/yr}}{100 \text{ people}} = 1 \times 10^{-5}$$

$$\frac{1000 \text{ dollars/yr}}{200 \times 10^{6}} = 5 \times 10^{-6}$$

The recommended annual probability of failure is 5×10^{-6}, as it is the most demanding of the two: economic loss controls.

Problem 11.12

The set of data (y, x) shown in Table 11.1s is plotted and a linear regression (y = ax + b) is performed. Calculate the values of a and b by:

a. Minimizing the vertical distance between the measured and predicted y values.
b. Minimizing the normal distance between the measured data and the regression line.
c. Compare the results.

Table 11.1s Data Set

Data point number	x value	y value
1	2.1	7.4
2	4.5	10.1
3	4.8	11.7
4	5.3	12.4
5	5.7	13.1
6	6.2	16.7
7	7.8	23.4

Solution 11.12

a. Vertical distance:

$$a = \frac{\sum xi \sum yi - n \sum xiyi}{\left(\sum xi\right)^2 - n \sum xi^2} = 2.77$$

$$b = \frac{\sum xi^2 \sum yi - \sum xi \sum xiyi}{n \sum xi^2 - \left(\sum xi\right)^2} = -0.86$$

b. Normal distance:

$$a = \frac{\sum (y_i - \overline{y})^2 - \sum (x_i - \overline{x})^2 + \sqrt{\left(\sum (y_i - \overline{y})^2 - \sum (x_i - \overline{x})^2\right)^2 + 4\left(\sum (x_i - \overline{x})(y_i - \overline{y})\right)^2}}{2\sum (x_i - \overline{x})(y_i - \overline{y})} = 2.06$$

$$b = \frac{\sum y_i}{N} - a\frac{\sum x_i}{N} = \overline{y} - a\overline{x} = 2.82$$

c. The plot of the linear regression and the orthogonal regression are shown in Figure 11.12s.

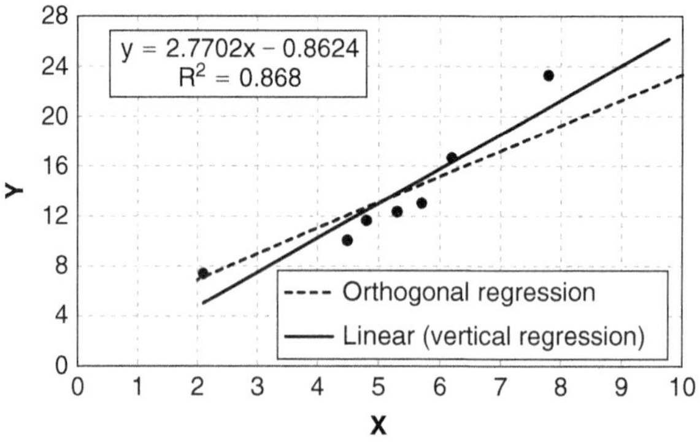

Figure 11.12s Regression plots.

Problem 11.13

Use consistent units to find the relationship between the shear wave velocity v_s, the mass density ρ, and the shear modulus of elasticity G.

Solution 11.13

If we use length, time, and force as primary units, we have

Variable	Dimension
V_s	L/T
G	F/L^2
ρ	FT^2/L^4

We hypothesize that $v_s = f(G, \rho)$. Thus, G and ρ must appear to cancel the force dimension, because F does not appear in v_s. Let's try the ratio $G/\rho = (F/L^2)/(FT^2/L^4) = L^2/T^2$. This ratio has the units of velocity squared, so a reasonable guess is $(v_s)^2 = G/\rho$ or $v_s = (G/\rho)^{0.5}$. This is, of course, correct.

Problem 11.14

The following empirical equations are used in sands to obtain the ultimate pressure p_u under a driven pile point and the ultimate friction f_u on a driven pile side. Use normalization to give these formulas with p_u and f_u in the U.S. customary system.

$$p_u(\text{kPa}) = 1000\ (N(\text{bl/ft}))^{0.5}$$

$$f_u(\text{kPa}) = 5\ (N(\text{bl/ft})^{0.7}$$

Solution 11.14

We first normalize the right-hand term.

$$p_u(\text{kPa}) = 1000\left(\frac{N\ (\text{bl/ft})}{1(\text{bl/ft})}\right)^{0.5}$$

Now the blow count N is normalized and the factor 1000 is in kPa. It can be changed to tsf, for example, by recalling that 100 kPa = 1.0443 tsf. In that case:

$$p_u(\text{tsf}) = 1000\ \text{kPa}\left(\frac{1.0443\ \text{tsf}}{100\ \text{kPa}}\right)\left(\frac{N\ (\text{bl/ft})}{1(\text{bl/ft})}\right)^{0.5} = 10.443\ (N(\text{bl/ft}))^{0.5}$$

We repeat the process for the friction:

$$f_u(\text{kPa}) = 5\left(\frac{N\ (\text{bl/ft})}{1(\text{bl/ft})}\right)^{0.7}$$

$$f_u(\text{tsf}) = 5\ \text{kPa}\left(\frac{1.0443\ \text{tsf}}{100\ \text{kPa}}\right)\left(\frac{N\ (\text{bl/ft})}{1(\text{bl/ft})}\right)^{0.7} = 0.0522\ (N(\text{bl/ft}))^{0.7}$$

Problem 11.15

Perform a dimensional analysis for a square footing embedded at a depth d in a clay with an undrained shear strength s_u. The footing size is B and the failure load is Q_u.

Solution 11.15 (Figure 11.13s)

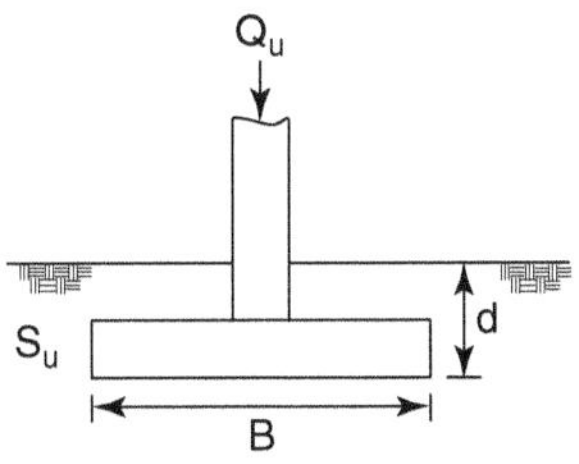

Figure 11.13s Square footing on clay.

The independent variables are shown in Table 11.2s with their dimensions. There are four independent variables.

Table 11.2s Independent Variables and Dimensions

Quantity	Variable	Dimension
Load	Q_u	F
Embedment	d	L
Footing size	B	L
Undrained shear strength	S_u	F/L^2

There are two primary units, as listed in Table 11.2s. We therefore form two primary unit groups of variables:

a. L group: d, B, S_u
b. F group: Q_u, S_u

We select one variable from each group; for example, B in the L group and S_u in the F group. These are the repeating variables.

Because there are 4 variables and 2 primary units, we have $4 - 2 = 2\ \pi$ terms. To obtain the $2\ \pi$ terms, we form the power:

a. $\pi_1 = B^a S_u^b d^c$
b. $\pi_2 = B^a S_u^b Q_u^c$

Now, we find the exponent and we determine that the $2\ \pi$ terms are:

$$\pi_1 = \frac{d}{B}$$

$$\pi_2 = \frac{Q_u}{B^2 S_u}$$

Then we can say that $g(\pi_1, \pi_2) = 0$, or:

$$f\left(\frac{d}{B}, \frac{Q_u}{B^2 S_u}\right) = 0$$

We can also write this expression as:

$$Q_u = \frac{d}{B} f_1(B^2 S_u)$$

Notice that if the embedment $d = 0$, then the expression becomes:

$$Q_u = C_1 B^2 S_u$$

where C_1 is a constant that is approximately 6.0.

CHAPTER 12

Soil Constitutive Models

A *soil model* is a mathematical representation of the behavior of the soil under load. The model typically relates the stresses applied to the strains experienced by the soil as a result. The simplest of these relationships is the theory of elasticity.

12.1 ELASTICITY

12.1.1 Elastic Model

The theory of *elasticity* states that stresses and strains are linearly related (Figure 12.1).

Because there are 6 stresses and 6 strains, the matrix relating the stresses to the strains is made of 36 constants. Satisfying isotropy and symmetry reduces those 36 constants to only 2: the *modulus of elasticity* E (also called *Young's modulus*), and the *Poisson's ratio* ν. The equations are:

$$\varepsilon_{xx} = \frac{1}{E}(\sigma_{xx} - \nu(\sigma_{yy} + \sigma_{zz})) \quad (12.1)$$

$$\varepsilon_{yy} = \frac{1}{E}(\sigma_{yy} - \nu(\sigma_{zz} + \sigma_{xx})) \quad (12.2)$$

$$\varepsilon_{zz} = \frac{1}{E}(\sigma_{zz} - \nu(\sigma_{xx} + \sigma_{yy})) \quad (12.3)$$

$$\varepsilon_{xy} = \frac{1+\nu}{E}\tau_{xy} = \frac{\gamma_{xy}}{2} \quad (12.4)$$

$$\varepsilon_{yz} = \frac{1+\nu}{E}\tau_{yz} = \frac{\gamma_{yz}}{2} \quad (12.5)$$

$$\varepsilon_{zx} = \frac{1+\nu}{E}\tau_{zx} = \frac{\gamma_{zx}}{2} \quad (12.6)$$

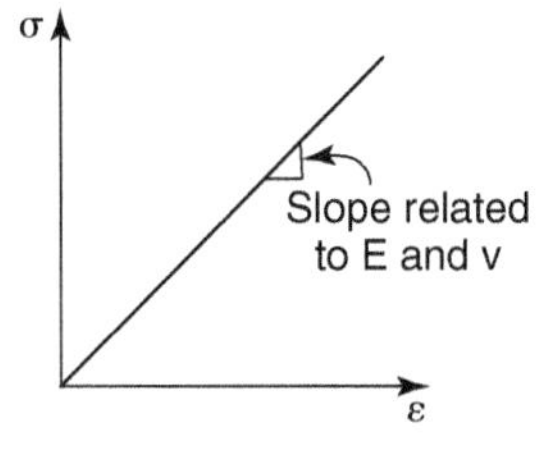

Figure 12.1 Linear elasticity stress-strain curve.

where σ_{ii} is the normal stress on the plane perpendicular to i in the direction of i, τ_{ij} is the shear stress on the plane perpendicular to i in the direction of j, ε_{ii} is the normal strain associated with the normal stress σ_{ii}, ε_{ii} is the shear strain associated with the shear stress τ_{ij}, γ_{ij} is the engineering shear strain associated with the shear stress τ_{ij}, E is Young's modulus or modulus of elasticity, and ν is Poisson's ratio. Young's modulus is named after Thomas Young, a British physician and physicist who made his contribution around the turn of the 1800s. Poisson's ratio is named after Simeon Poisson, a French mathematician and physicist who lived around the turn of the 1800s and had Lagrange and Laplace as his doctoral advisors at the École Polytechnique in Paris. In matrix form, the elasticity equations are:

$$\begin{bmatrix} \varepsilon_{xx} \\ \varepsilon_{yy} \\ \varepsilon_{zz} \\ \varepsilon_{xy} \\ \varepsilon_{yz} \\ \varepsilon_{zx} \end{bmatrix} = \frac{1}{E}\begin{bmatrix} 1 & -\nu & -\nu & 0 & 0 & 0 \\ -\nu & 1 & -\nu & 0 & 0 & 0 \\ -\nu & -\nu & 1 & 0 & 0 & 0 \\ 0 & 0 & 0 & 1+\nu & 0 & 0 \\ 0 & 0 & 0 & 0 & 1+\nu & 0 \\ 0 & 0 & 0 & 0 & 0 & 1+\nu \end{bmatrix}\begin{bmatrix} \sigma_{xx} \\ \sigma_{yy} \\ \sigma_{zz} \\ \tau_{xy} \\ \tau_{yz} \\ \tau_{zx} \end{bmatrix} \quad (12.7)$$

or

$$\begin{bmatrix} \sigma_{xx} \\ \sigma_{yy} \\ \sigma_{zz} \\ \sigma_{xy} \\ \sigma_{yz} \\ \sigma_{zx} \end{bmatrix} = \frac{E}{(1+\nu)(1-2\nu)} \times \begin{bmatrix} 1-\nu & \nu & \nu & 0 & 0 & 0 \\ \nu & 1-\nu & \nu & 0 & 0 & 0 \\ \nu & \nu & 1-\nu & 0 & 0 & 0 \\ 0 & 0 & 0 & 1-2\nu & 0 & 0 \\ 0 & 0 & 0 & 0 & 1-2\nu & 0 \\ 0 & 0 & 0 & 0 & 0 & 1-2\nu \end{bmatrix}\begin{bmatrix} \varepsilon_{xx} \\ \varepsilon_{yy} \\ \varepsilon_{zz} \\ \varepsilon_{xy} \\ \varepsilon_{yz} \\ \varepsilon_{zx} \end{bmatrix} \quad (12.8)$$

Note that the normal strain in one direction is affected by the normal stress in that direction and also by the normal stresses in the other two directions–yet this is not true for the shear strains. Indeed, the shear strain is affected by the shear stress in that direction, but not by the shear stresses in the other two directions. Note also that although ε_{xy} is the shear strain, the engineering shear strain γ_{xy} (=2 ε_{xy}) is often used in practice.

Other elasticity moduli have been defined from E and ν. They are the shear modulus G, the bulk modulus K, and the constrained modulus M. The *shear modulus G* can be obtained by performing a simple shear test; it is defined as the ratio of the shear stress τ over the corresponding engineering shear strain γ. The *bulk modulus K* is obtained when a soil sample is subjected to an all-around (hydrostatic) pressure σ; it is defined as the ratio of the pressure σ over the volumetric strain generated $\varepsilon_v = \Delta V/V$. The *constrained modulus M* is obtained when a soil sample is subjected to a vertical normal stress in a cylinder that prevents any lateral movement; it is defined as the ratio of the normal stress applied over the vertical strain obtained. The relationships are as follows:

$$\text{Shear modulus} \quad G = \frac{\tau_{xy}}{\gamma_{xy}} = \frac{\tau_{xy}}{2\varepsilon_{xy}} = \frac{E}{2(1+\nu)} \tag{12.9}$$

$$\text{Bulk modulus} \quad K = \frac{\sigma}{\Delta V/V} = \frac{\frac{1}{3}(\sigma_{xx}+\sigma_{yy}+\sigma_{zz})}{\varepsilon_{xx}+\varepsilon_{yy}+\varepsilon_{zz}} = \frac{E}{3(1-2\nu)} \tag{12.10}$$

$$\text{Constrained modulus} \quad M = \frac{\sigma_{xx}}{\varepsilon_{xx}} = \frac{E(1-\nu)}{(1+\nu)(1-2\nu)} \tag{12.11}$$

The term *plane strain* means that the normal strain in one direction is zero throughout the soil mass. The term *plane stress* means that the normal stress in one direction is zero throughout the soil mass. Such conditions lead to an additional equation, as setting the normal strain in one direction equal to zero (for example) gives a relationship between the normal stresses in the three directions.

One of the advantages of the elastic model is the associated superposition principle, which is possible because the equations are linear. Table 12.1 indicates some of the possible superposition operations. The superposition principle is not applicable to nonlinear theories, such as the theory of plasticity.

Table 12.1 Superposition Principle Operations

Force	Stress	Strain	Displacement
F_1	σ_1	ε	u_1
λF_1	$\lambda\sigma_1$	$\lambda\varepsilon_1$	λu_1
F_2	σ_2	ε_2	u_2
$F_1 + F_2$	$\sigma_1 + \sigma_2$	$\varepsilon_1 + \varepsilon_2$	$u_1 + u_2$

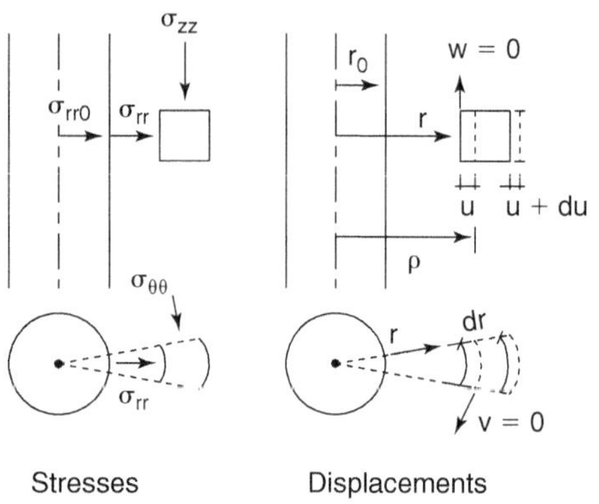

Figure 12.2 Element of soil around an expanding cylindrical cavity.

12.1.2 Example of Use of Elastic Model

The problem is to solve the expansion of an infinitely long cylinder subjected to a pressure p in an elastic soil space (Figure 12.2). The geometry of the problem indicates that this is an axisymmetric problem and a plane strain problem in the vertical direction. The initial state of stress is σ_{ov} in the vertical direction and σ_{oh} in the radial direction at any point in the soil space. After applying the pressure p at the cavity surface, the stresses in the mass become:

$$\sigma_{rr} = \sigma_{oh} + \Delta\sigma_{rr} \tag{12.12}$$

$$\sigma_{\theta\theta} = \sigma_{oh} + \Delta\sigma_{\theta\theta} \tag{12.13}$$

$$\sigma_{zz} = \sigma_{ov} + \Delta\sigma_{zz} \tag{12.14}$$

where σ_{rr} and $\sigma_{\theta\theta}$ are the radial stress and the hoop stress respectively at a distance r from the axis of the cylinder, and $\Delta\sigma_{rr}$ and $\Delta\sigma_{\theta\theta}$ are the increments of the radial and hoop stress above the at-rest stress value.

The radial displacement u is the only displacement type in this problem, as there are no displacements in the hoop direction or in the vertical direction. The radial strain is ε_{rr}, the hoop strain is $\varepsilon_{\theta\theta}$, and the vertical strain is ε_{zz}, which is zero because of the plane strain condition in the z direction. The relationships between the displacement and the strains for small-strain theory are:

$$\varepsilon_{rr} = -\frac{du}{dr} \tag{12.15}$$

$$\varepsilon_{\theta\theta} = -\frac{u}{r} \tag{12.16}$$

$$\varepsilon_{zz} = 0 \tag{12.17}$$

The minus sign is used in Eqs. 12.15 and 12.16 because compression has been chosen to be positive. In fact, u is positive but decreases with radial distance; hence, if the minus sign were not there, du/dr would be negative and associated with compression considering the loading for this problem. The equations of equilibrium reduce to:

$$\frac{d\sigma_{rr}}{dr} + \frac{\Delta\sigma_{rr} - \Delta\sigma_{\theta\theta}}{r} = 0 \tag{12.18}$$

The constitutive equations are:

$$\varepsilon_{rr} = \frac{1}{E}(\Delta\sigma_{rr} - \nu(\Delta\sigma_{\theta\theta} + \Delta\sigma_{zz})) \quad (12.19)$$

$$\varepsilon_{\theta\theta} = \frac{1}{E}(\Delta\sigma_{\theta\theta} - \nu(\Delta\sigma_{zz} + \Delta\sigma_{rr})) \quad (12.20)$$

$$\varepsilon_{zz} = \frac{1}{E}(\Delta\sigma_{zz} - \nu(\Delta\sigma_{rr} + \Delta\sigma_{\theta\theta})) \quad (12.21)$$

By combining equations 12.12 to 12.21, the governing differential equation is obtained as:

$$r^2\frac{d^2u}{dr^2} + r\frac{du}{dr} - u = 0 \quad (12.22)$$

The boundary conditions are:

$$u = 0 \text{ at } r = \text{infinity}$$

$$u = u_o \text{ at } r = r_o$$

The solution that satisfies Eq. 12.22 and the boundary conditions is:

$$u = \frac{u_o r_o}{r} \quad (12.23)$$

$$\varepsilon_{rr} = \frac{u_o r_o}{r^2} \quad \varepsilon_{\theta\theta} = -\frac{u_o r_o}{r^2} \quad \varepsilon_{zz} = 0 \quad (12.24)$$

$$\sigma_{rr} = \sigma_{oh} + 2G\frac{u_o r_o}{r^2} \quad \sigma_{\theta\theta} = \sigma_{oh} - 2G\frac{u_o r_o}{r^2} \quad \sigma_{zz} = \sigma_{ov} \quad (12.25)$$

At the wall of the cylindrical cavity, the equations become:

$$u_o = u_o \quad (12.26)$$

$$\varepsilon_{rro} = \frac{u_o}{r_o} \quad \varepsilon_{\theta\theta o} = -\frac{u_o}{r_o} \quad \varepsilon_{zzo} = 0 \quad (12.27)$$

$$\sigma_{rro} = \sigma_{oh} + 2G\frac{u_o}{r_o} \quad \sigma_{\theta\theta o} = \sigma_{oh} - 2G\frac{u_o}{r_o} \quad \sigma_{zzo} = \sigma_{ov} \quad (12.28)$$

In a pressuremeter test, the relative increase in radius ($u_o/r_o = \varepsilon_{\theta\theta o}$) of the cavity is measured along with the pressure exerted on the cavity wall σ_{rro}. Therefore, the pressuremeter curve is a direct plot of a stress-strain curve of the soil (Figure 12.3).

In the case of large-strain theory, the large-strain definitions require use of the current radius ρ, the initial radius r, and the displacement u:

$$\rho = r + u \quad (12.29)$$

Then the strains can be defined as:

$$\text{Radial strain} \quad \alpha_r = \frac{1}{2}\left(\frac{d\rho^2 - dr^2}{d\rho^2}\right) \quad (12.30)$$

$$\text{Hoop strain} \quad \alpha_\theta = \frac{1}{2}\left(\frac{\rho^2 - r^2}{\rho^2}\right) \quad (12.31)$$

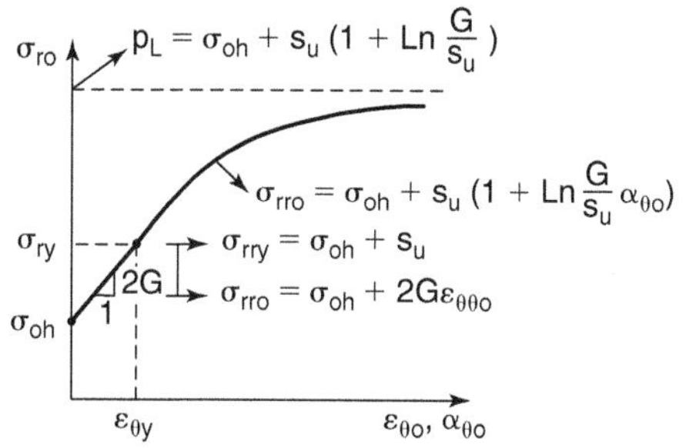

Figure 12.3 Expansion of a cylindrical cavity.

and the solution becomes:

$$\sigma_{rr} = \sigma_{oh} + 2G\alpha_\theta = \sigma_{oh} + G\left(\frac{\rho^2 - r^2}{\rho^2}\right) = \sigma_{oh} + G\left(\frac{\Delta V}{V}\right) \quad (12.32)$$

where ΔV is the increase in volume of the cylinder having an initial radius r and V is the current volume of the cylinder having a current radius ρ and an initial radius r.

12.2 LINEAR VISCOELASTICITY

When load is applied to a linear elastic material, the stresses, strains, and displacements occur instantaneously and remain constant with time. *Viscoelasticity* introduces the influence of time on the deformation process (Figure 12.4). *Linear viscoelasticity* further simplifies the phenomenon by allowing superposition of the elastic deformation and the time-dependent deformation. A good way to understand viscoelasticity is to start by studying simple models.

12.2.1 Simple Models: Maxwell and Kelvin-Voigt Models

Simple, one-dimensional models help to understand the potential use of linear viscoelasticity (Figure 12.5). These models make use of mechanical elements such as a spring and a *dashpot* (also called *damper*). The spring behavior is governed by $\sigma = k\,\varepsilon$ where σ is the axial stress applied, k is the spring stiffness, and ε is the axial strain. The dashpot behavior is governed by $\sigma = \eta\,(d\varepsilon/dt)$ where η is the viscosity and $d\varepsilon/dt$ the strain rate. The dashpot behavior is very similar to the behavior of a shock absorber in a car suspension. If you load it fast, it generates a stiff response; if you load it

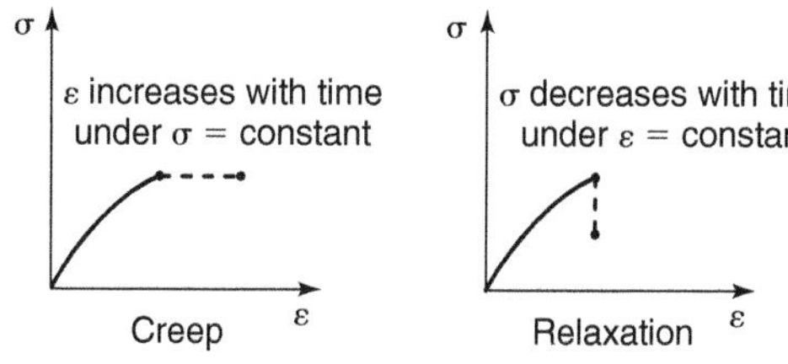

Figure 12.4 Creep and relaxation of viscous models.

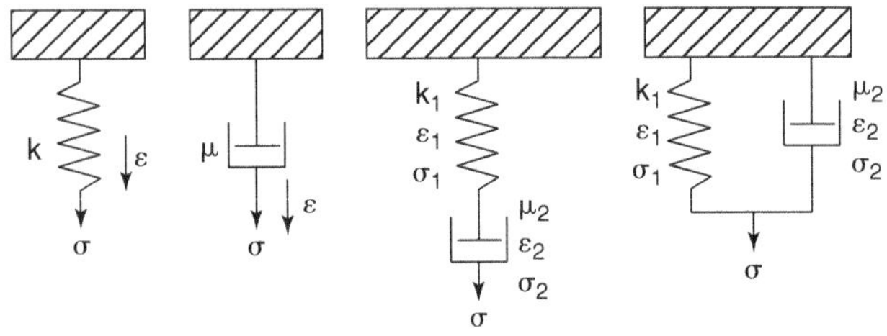

Figure 12.5 One-dimensional viscoelastic models: *(a)* Spring, $\sigma = k\,\varepsilon$. *(b)* Dashpot, $\sigma - \eta\,(d\varepsilon/dt)$. *(c)* Maxwell, $\sigma - \sigma_1 - \sigma_2$, $\varepsilon - \varepsilon_1 + \varepsilon_2$. *(d)* Kelvin-Voigt, $\sigma = \sigma_1 + \sigma_2$, $\varepsilon = \varepsilon_1 = \varepsilon_2$.

slowly, it offers little resistance. These mechanical elements can be combined to represent a more complex behavior. The Maxwell model is made of a spring and a dashpot in series, whereas the Kelvin-Voigt model is made of a spring and a dashpot in parallel. The Maxwell model is named after James Maxwell, a British physicist and mathematician of the mid-1800s. The Kelvin-Voigt model is named after William Thompson, First Baron Kelvin, a British physicist and engineer of the late 1800s; and Woldemar Voigt, a German physicist of the late 1800s.

Two basic phenomena can be investigated with these models: creep and relaxation (Figure 12.4). *Creep* refers to the increase in strain as a function of time when a constant stress is applied. For example, creep could occur in the soil under a high embankment. *Relaxation* refers to the decrease in stress as a function of time when a constant strain is applied. For example, relaxation of the horizontal total stress could occur against the side of a pile after driving. To find out how the Maxwell model creeps and relaxes, we write (Figure 12.5):

$$\sigma = \sigma_1 = \sigma_2 \tag{12.33}$$

$$\varepsilon = \varepsilon_1 + \varepsilon_2 \tag{12.34}$$

Therefore, the governing equation for the Maxwell model is:

$$\frac{d\varepsilon}{dt} = \frac{1}{k}\frac{d\sigma}{dt} + \frac{\sigma}{\eta} \tag{12.35}$$

Creep occurs under a constant stress σ_o. If that stress is applied instantaneously, only the spring deflects and the initial value of the strain is $\varepsilon_o = \sigma_o/k$. Therefore:

$$\int_{\sigma_o/k}^{\varepsilon} d\varepsilon = \frac{\sigma_o}{\eta}\int_0^t dt \quad and \quad \varepsilon = \frac{\sigma_o}{k} + \frac{\sigma_o}{\eta}t \tag{12.36}$$

which shows that the Maxwell model creeps linearly. This does not fit well with observed soil behavior. Relaxation occurs under constant strain ε_o. If that strain is applied instantaneously, only the spring deflects and the initial value of the stress is $\sigma_o = k\,\varepsilon_o$, Therefore:

$$\int_{k\varepsilon_o}^{\sigma} \frac{d\sigma}{\sigma} = -\frac{k}{\eta}\int_0^t dt \quad and \quad \sigma = k\varepsilon_o e^{-\frac{t}{\eta/k}} \tag{12.37}$$

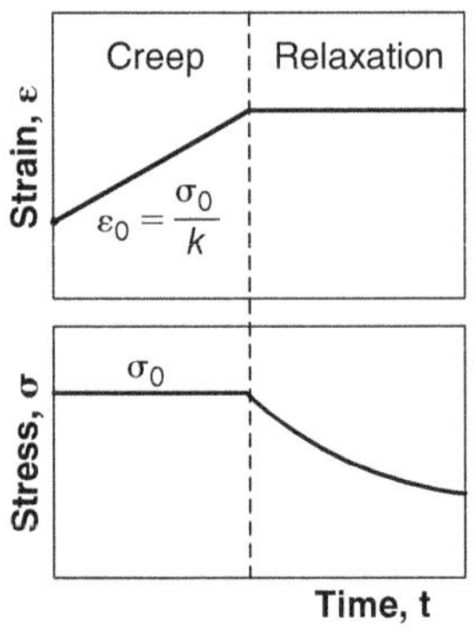

Figure 12.6 Creep and relaxation of the Maxwell model.

Equation 12.37 shows a relaxation process that is closer to what one would expect in actual soils. Figure 12.6 summarizes the behavior of the Maxwell model.

To find out how the Kelvin-Voigt model creeps and relaxes, we write (Figure 12.5):

$$\sigma = \sigma_1 + \sigma_2 \tag{12.38}$$

$$\varepsilon = \varepsilon_1 = \varepsilon_2 \tag{12.39}$$

Therefore, the governing equation for the Maxwell model is:

$$\sigma = k\varepsilon + \eta\frac{d\varepsilon}{dt} \tag{12.40}$$

Creep occurs under a constant stress σ_o. If that stress is applied instantaneously, the dashpot is infinitely stiff, all the stress is carried by the dashpot, no strain occurs initially under σ_o, and $\varepsilon_o = 0$. After an infinite time, however, the dashpot carries no load, the stress is entirely carried by the spring, and $\varepsilon_{t = \text{infinity}} = \sigma_o/k$. Therefore:

$$\int_0^{\varepsilon} \frac{d\varepsilon}{\varepsilon - \frac{\sigma_o}{k}} = -\frac{k}{\eta}\int_0^t dt \quad and \quad \varepsilon = \frac{\sigma_o}{k}\left(1 - e^{-\frac{t}{\eta/k}}\right) \tag{12.41}$$

which shows that the Kelvin-Voigt model creeps in a way consistent with what can be expected for actual soils. Relaxation occurs under constant strain ε_o, therefore there is no contribution from the dashpot and the stress is simply:

$$\sigma = k\varepsilon_o \tag{12.42}$$

The Kelvin-Voigt model does not relax. Figure 12.7 summarizes the behavior of the Kelvin-Voigt model.

12.2.2 General Linear Viscoelasticity

The simple models from section 12.2.1 indicate that stress behavior over time is related to the strain through a function called the *relaxation modulus function G(t)*. Similarly, the strain behavior over time of a viscoelastic material is related to the stress through a function called the *creep*

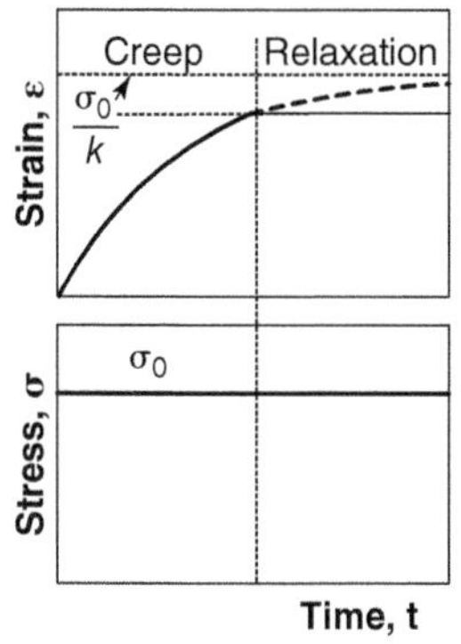

Figure 12.7 Creep and relaxation of the Kelvin-Voigt model.

compliance function $J(t)$. For example, Eq. 12.37 indicates that the relaxation modulus function $G(t)$ for the Maxwell model is:

$$G(t) = \frac{\sigma(t)}{\varepsilon_o} = ke^{-\frac{t}{\eta/k}} \tag{12.43}$$

and that the creep compliance function $J(t)$ for the Kelvin-Voigt model (Eq. 12.41) is:

$$J(t) = \frac{\varepsilon(t)}{\sigma_o} = \frac{1}{k}\left(1 - e^{-\frac{t}{\eta/k}}\right) \tag{12.44}$$

Ludwig Boltzmann, an Austrian physicist of the late 1800s, generalized these observations by proposing a superposition principle that can be explained as follows. At time $t_1' = 0$, a constant stress σ_1 is applied and the strain induced is (Figure 12.8):

$$\varepsilon_1(t) = J(t)\sigma_1 \tag{12.45}$$

Then at a time t_2' an increment of stress $(\sigma_2 - \sigma_1)$ is imposed and the strain increase is:

$$\varepsilon_2(t) = J(t - t_2')(\sigma_2 - \sigma_1) \tag{12.46}$$

Note here that the function J is the same as in equation 12.41 and independent of the stress level. This is the property of linear viscoelasticity. Again, at a time t_3', an increment of stress $(\sigma_3 - \sigma_2)$ is imposed and the strain increase is:

$$\varepsilon_3(t) = J(t - t_3')(\sigma_3 - \sigma_2) \tag{12.47}$$

And so on, so that in the end the total strain is:

$$\varepsilon(t) = \sum_{i=1}^{n} \varepsilon_i(t) = \sum_{i=1}^{n} J(t - t_i')(\sigma_i - \sigma_{i-1}) \tag{12.48}$$

For a continuous stress function $\sigma(t)$, Eq. 12.48 becomes:

$$\varepsilon(t) = \int_0^t J(t - t')\frac{d\sigma(t')}{dt'}dt' \tag{12.49}$$

This represents the viscous part of the strain, to which should be added the elastic part. So, in the end, the general form of the model is:

$$\varepsilon_{ij}(t) = \varepsilon_{ij(elastic)} + \int_0^t J(t - t')\frac{d\sigma_{ij}(t')}{dt'}dt' \tag{12.50}$$

Similarly, for the relaxation modulus the equation is:

$$\sigma_{ij}(t) = \sigma_{ij(elastic)} + \int_0^t G(t - t')\frac{d\varepsilon_{ij}(t')}{dt'}dt' \tag{12.51}$$

12.3 PLASTICITY

One way to model a soil is to consider that it behaves elastically at first, then reaches a yield point, and then continues to deform plastically until it reaches failure. Beyond the yield point, the soil can strain harden, strain soften, or be perfectly plastic (Figure 12.9).

If the material is perfectly plastic beyond the yield point, the yield criterion and the failure criterion are the same. If the material strain hardens, they are not the same; and if the material strain softens, the yield criterion and the failure criterion are the same but postyield behavior requires further calculations. It is accepted that strain can be decomposed into an elastic component and a plastic component. Furthermore, because plasticity is primarily a nonlinear theory, the calculations involve strain increments $d\varepsilon_{ij}$:

$$d\varepsilon_{ij} = d\varepsilon_{ij}^e + d\varepsilon_{ij}^p \tag{12.52}$$

where $d\varepsilon_{ij}^e$ is the elastic part of the strain increment, and $d\varepsilon_{ij}^p$ is the plastic part of the strain increment. There are four elements to any plasticity method (Potts and Zdravkovic 1999; Davies

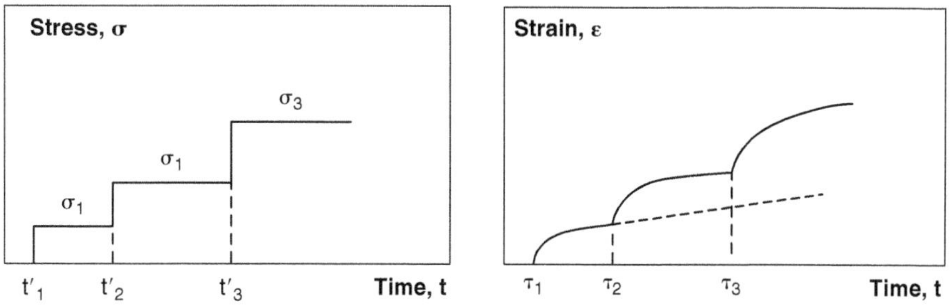

Figure 12.8 Boltzmann superposition principle.

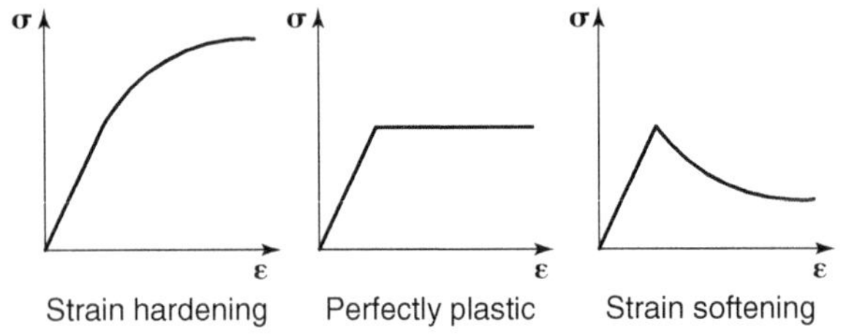

Figure 12.9 Plastic models.

and Selvadurai, 2002): coincidence of axes, yield function, plastic potential function, and hardening or softening rule. The coincidence of axes is a common assumption stating that the axes of the accumulated principal stress vectors (σ_1, σ_2, σ_3) and the axes of the plastic principal strain increment vectors ($d\varepsilon_1^p, d\varepsilon_2^p, d\varepsilon_3^p$) coincide. This is an extension of what is used in elasticity, but in plasticity it applies to the stress and the corresponding strain increment and not to the stress increment and the corresponding strain increment. The yield function and associated yield criterion give the combination of stresses that lead to yielding of the soil. The plastic potential function gives the direction of the plastic strain increments through a flow rule, and the hardening or softening rule gives the magnitude of the plastic strain increments.

12.3.1 Some Yield Functions and Yield Criteria

The combination of stresses that create yielding of the soil are given by the yield function, which is set equal to zero to give the yield criterion. The yield function involves a state parameter k:

$$Y(\sigma_{ij}, k) = 0 \tag{12.53}$$

The two most common yield criteria in soil mechanics are the Tresca yield criterion and the Mohr-Coulomb yield criterion. The *Tresca yield criterion* is named after Henri Tresca, a French mechanical engineer, who proposed it in 1864. When applied to soil mechanics and the undrained behavior of fine-grained soils, it states that yield will occur when the difference between the major principal stress and the minor principal stress reaches a value equal to two times the undrained shear strength s_u (Figure 12.10):

$$\sigma_1 - \sigma_3 - 2s_u = 0 \tag{12.54}$$

As can be seen in the Tresca criterion, s_u is the state parameter. It corresponds to the Mohr circle plotted in the shear stress vs. total stress set of axes reaching the undrained shear strength failure envelope.

The *Mohr-Coulomb yield criterion* is named after Otto Mohr, a German civil engineer of the late 1800s, and Charles de Coulomb, a French civil engineer of the late 1700s. It states that yield will occur when the Mohr circle reaches the line corresponding to the shear strength equation (Figure 12.11):

$$\tau_f - c' - \sigma' \tan \varphi' = 0 \tag{12.55}$$

As can be seen in the Mohr-Coulomb criterion, c' and φ' are the state parameters. This can be rewritten in terms of major and minor principal stresses by using the rectangular

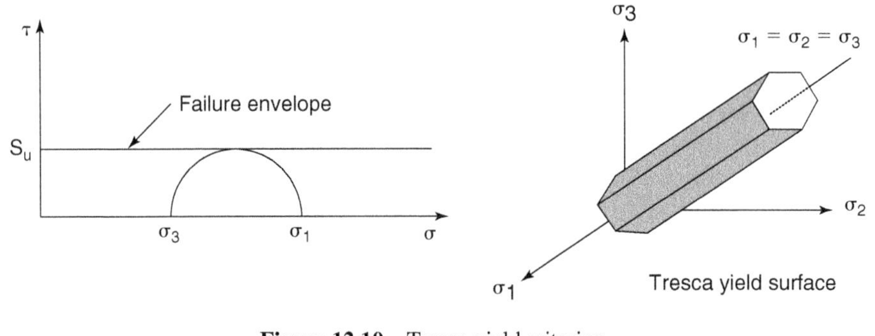

Figure 12.10 Tresca yield criterion.

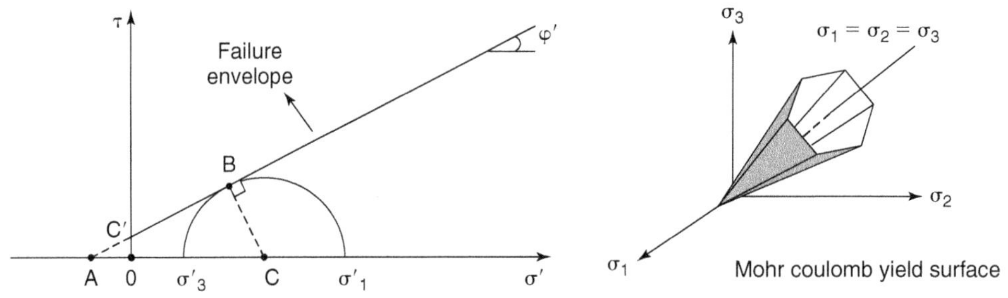

Figure 12.11 Mohr-Coulomb yield criterion.

triangle ABC in Figure 12.11.

$$\sin\varphi' = \frac{\dfrac{\sigma_1' - \sigma_3'}{2}}{\dfrac{c'}{\tan\varphi'} + \dfrac{\sigma_1' + \sigma_3'}{2}} \quad (12.56)$$

or

$$\sigma_1' - \sigma_3' - 2c'\cos\varphi' - (\sigma_1' + \sigma_3')\sin\varphi' = 0 \quad (12.57)$$

The Mohr circle starts at a stress state that corresponds to the soil equilibrium in situ. As the soil is loaded, it deforms elastically at first until the circle reaches the yield criterion envelope (shear strength equation). At that point, the circle cannot grow past the envelope, but it can grow along the envelope (strain hardening) or decrease in size along the envelope (strain softening). Note that in sand ($c' = 0$), the Mohr-Coulomb yield criterion simplifies to:

$$\frac{\sigma_1'}{\sigma_3'} - \frac{1+\sin\varphi}{1-\sin\varphi} = 0 \quad (12.58)$$

Yet another yield criterion is the *Von Mises criterion*, named after Richard Von Mises, an Austrian engineer in the early 1900s:

$$\sqrt{J_2} - k = 0 \quad (12.59)$$

where J_2 is the second stress invariant of the deviatoric tensor (Section 10.7) and k is a constant to be determined experimentally. The *Drucker-Prager criterion* is a generalization of the Von Mises criterion named after two American engineers of the mid-1900s. It introduces the influence of the mean stress on the strength of soils:

$$\sqrt{J_2} - A - BI_1 = 0 \quad (12.60)$$

where I_1 is the first invariant of the stress tensor.

These four yield criteria are compared (Figure 12.12) on the π plane, the plane perpendicular to the bisectrice of the three-dimensional space $\sigma_1 - \sigma_2 - \sigma_3$. This bisectrice has the equation $\sigma_1 = \sigma_2 = \sigma_3$.

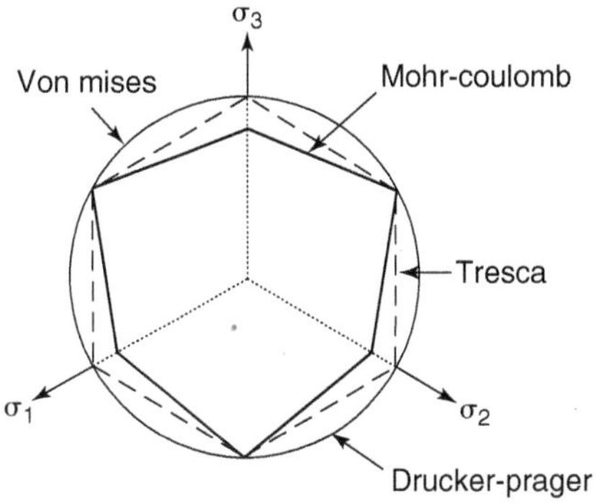

Figure 12.12 Yield criteria compared on the π plane.

12.3.2 Example of Use of Yield Criteria

Let's go back to the expansion of an infinite cylinder subjected to an internal pressure p and find out at what pressure the soil first yields. We will first use the Tresca criterion (undrained fine-grained soil behavior). The radial stress at the cavity wall σ_{rro} increases with p, because it is equal to p, and represents the major principal stress σ_1. The hoop stress at the cavity wall $\sigma_{\theta\theta o}$ decreases as much as the radial stress increases (Eq. 12.28), and represents the minor principal stress σ_3. The difference $\sigma_1 - \sigma_3$ increases as p increases, and when p reaches a value where the Tresca criterion is first satisfied, the soil yields.

$$\sigma_1 - \sigma_3 = 2s_u \quad (12.61)$$

where s_u is the undrained shear strength of the soil. A plastic zone is initiated around the cylindrical cavity and grows as the pressure continues to increase (Figure 12.13).

Using Eqs. 12.12 and 12.13 plus the observation that the increase in stress $\Delta\sigma$ in the radial direction is equal to the decrease in stress $\Delta\sigma$ in the hoop direction, we write:

$$\sigma_{oh} + \Delta\sigma - (\sigma_{oh} - \Delta\sigma) = 2s_u \quad or \quad \Delta\sigma = s_u \quad (12.62)$$

Therefore, the yield pressure p_y will be:

$$p_y = \sigma_{oh} + s_u \quad (12.63)$$

If we use the Mohr-Coulomb criterion for a soil with $c' = 0$ (drained behavior of a coarse-grained soil, for example), then we write that:

$$\frac{\sigma_1'}{\sigma_3'} = \frac{1+\sin\varphi}{1-\sin\varphi} = K_p \quad (12.64)$$

Using Eqs. 12.12 and 12.13 plus the observation that the increase in stress in the radial direction is equal to the decrease

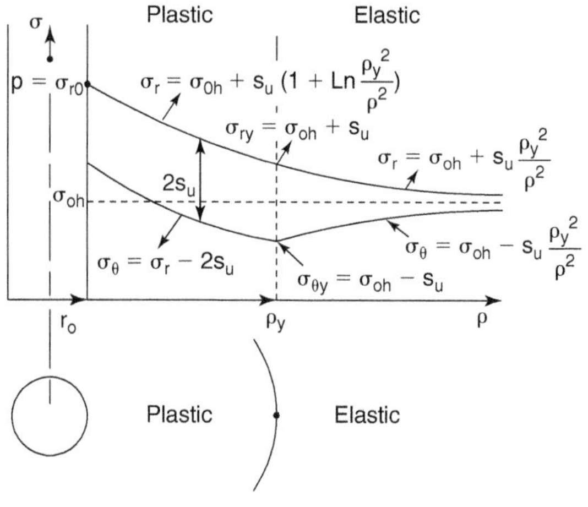

Figure 12.13 Elastic zone and plastic zone around an expanding cylindrical cavity.

in stress in the hoop direction, we write:

$$\frac{\sigma'_{oh} + \Delta\sigma'}{\sigma'_{oh} - \Delta\sigma'} = \frac{1 + \sin\varphi}{1 - \sin\varphi} \quad or \quad \Delta\sigma' = \sigma'_{oh}\sin\varphi \qquad (12.65)$$

Therefore, the yield pressure p_y will be:

$$p_y = \sigma'_{oh}(1 + \sin\varphi) \qquad (12.66)$$

Note that this solution is presented in terms of effective stresses; thus, the water stress would have to be taken into account to obtain the total stress.

12.3.3 Plastic Potential Function and Flow Rule

Because the behavior in the plastic domain is nonlinear, the relationship is written in terms of strain increments $d\varepsilon_{ii}$. It is accepted that the strain increment can be separated into an elastic portion and a plastic portion (Figure 12.14):

$$d\varepsilon_{ij} = d\varepsilon^e_{ij} + d\varepsilon^p_{ij} \qquad (12.67)$$

Now we need a way to predict the direction and magnitude of the plastic strain increments in the plastic region as we stress the soil beyond the yield point (if that is possible). As will be seen, the plastic potential gives the direction of the plastic strain increment, while the flow rule gives its magnitude. Von Mises proposed the existence of a plastic potential $P(\sigma_{ij}, m)$ function of the stress state at one point and material parameters m. This plastic potential is used to define a flow rule such that:

$$d\varepsilon^p_{ij} = \lambda\frac{dP(\sigma_{ij}, m)}{d\sigma_{ij}} \qquad (12.68)$$

where λ is a proportionality constant. If $P(\sigma_{ij}, m)$ is set equal to zero, the equation defines a surface in the stress space and $\frac{dP}{d\sigma_{ij}}$ is a vector perpendicular to that surface. This is called the *normality rule*, indicating that the increment of plastic strain $d\varepsilon^p_{ij}$ is perpendicular to the plastic potential surface. Figure 12.15 shows plastic potential contours in the $q - p'$ plane where q is the deviator stress ($q = \sigma_1 - \sigma_3$ for a triaxial test) and p′ is the mean normal effective stress ($p' = 0.33(\sigma'_1 + 2\sigma'_3)$ for a triaxial test).

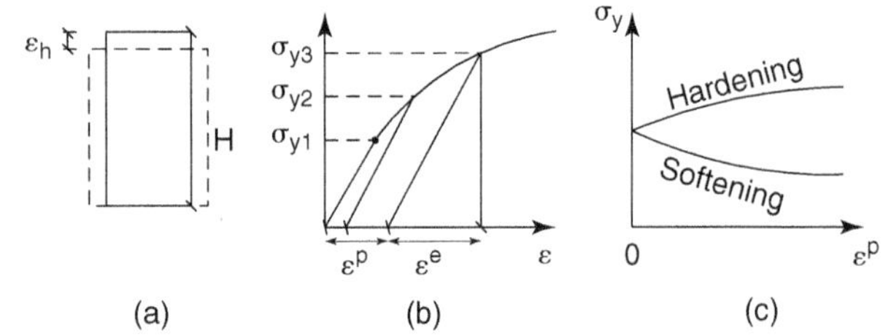

Figure 12.14 Elastic and plastic strains.

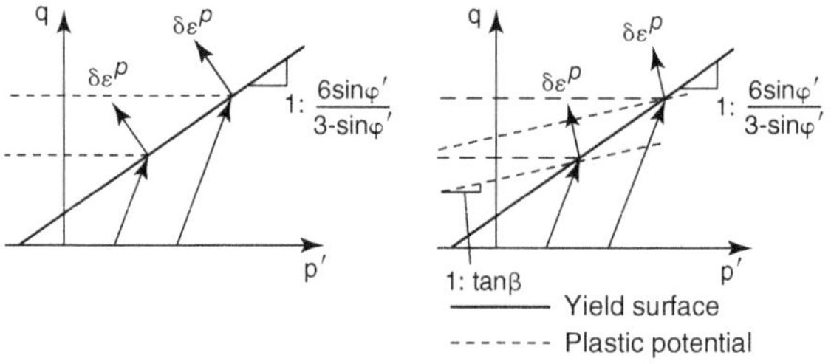

Figure 12.15 Plastic potential, yield surface, and normality rule.

A flow rule is said to be an *associated flow rule* if the plastic potential is equal to the yield function (Figure 12.15*a*):

$$P(\sigma_{ij}, m) = Y(\sigma_{ij}, k) \qquad (12.69)$$

So, in the case of an associated flow rule, the plastic strain increment is perpendicular to the yield surface, because:

$$d\varepsilon^p_{ij} = \lambda\frac{dY}{d\sigma_{ij}} \qquad (12.70)$$

For example, if we use the Tresca yield criterion, the plastic potential would be:

$$P(\sigma_{ij}, m) = Y(\sigma_{ij}, k) = \frac{\sigma_1 - \sigma_3}{2} - s_u \qquad (12.71)$$

the derivatives $\frac{dY}{d\sigma_{ij}}$ would be:

$$\frac{dY}{d\sigma_1} = \frac{1}{2}, \quad \frac{dY}{d\sigma_2} = 0, \quad \frac{dY}{d\sigma_3} = -\frac{1}{2} \qquad (12.72)$$

and the flow rule would be:

$$\begin{bmatrix} d\varepsilon^p_1 \\ d\varepsilon^p_2 \\ d\varepsilon^p_3 \end{bmatrix} = d\lambda \begin{bmatrix} +0.5 \\ 0 \\ -0.5 \end{bmatrix} \qquad (12.73)$$

If the plastic potential $P(\sigma_{ij}, m)$ is different from the yield function $F(\sigma_{ij}, k)$, then the flow rule is said to be nonassociated (Figure 12.15*b*). Associated flow rules work well for pressure-nonsensitive soils (undrained behavior of fine-grained soils) but nonassociated flow rules work better for pressure-sensitive soils (effective stress approach for soils). Nonassociated flow rules require more complicated calculations and increase the computing time.

12.3.4 Hardening or Softening Rule

Now we know the direction of the plastic strain increment vector, because it has to be normal to the plastic potential surface. We need to determine its magnitude, which is done by using a hardening or softening rule. As mentioned before, beyond the yield point the soil can strain harden, strain soften, or be rigid plastic (Figure 12.9). The hardening or softening rule defines what happens to the yield function beyond yield. If the hardening/softening is due to the plastic strains, it is called *strain hardening/softening*; if it is due to the plastic work done, it is called *work hardening/softening*. The hardening/softening rule describes how the state parameters k vary with plastic strain. This relationship can then be used in Eq. 12.68 or 12.70 as appropriate. Figure 12.14 illustrates how the hardening rule can be obtained for a simple axial compression test.

12.3.5 Example of Application of Plasticity Method

Let's go back to the expansion of the infinite cylinder subjected to an internal pressure p and find out the relationship between stress and strain in the plastic domain beyond the yield pressure p_y. We will first take the case of the undrained behavior of fine-grained soils and use the Tresca criterion. Because the cavity expands beyond small-strain theories, we need to use large-strain definitions (section 12.1.2) to still be able to form a valid strain tensor. In the plastic zone (Figure 12.13), the constitutive law has changed from elasticity to plasticity, but the equilibrium equation is still valid:

$$\frac{d\sigma_{rr}}{d\rho} + \frac{\Delta\sigma_{rr} - \Delta\sigma_{\theta\theta}}{\rho} = 0 \tag{12.74}$$

The Tresca criterion gives:

$$\begin{aligned}\sigma_1 - \sigma_3 = \sigma_{rr} - \sigma_{\theta\theta} &= \sigma_{oh} + \Delta\sigma_{rr} - (\sigma_{oh} + \Delta\sigma_{\theta\theta}) \\ &= \Delta\sigma_{rr} - \Delta\sigma_{\theta\theta} = 2s_u \end{aligned} \tag{12.75}$$

This leads to the solution:

$$d\sigma_{rr} = -2s_u\frac{d\rho}{\rho} \quad \textit{and} \quad \sigma_{rr} = -2s_u Ln\,\rho + A \tag{12.76}$$

The constant A is defined by the boundary condition, which states that at the boundary between the elastic region and the plastic region the radial stress is equal to p_y, as given by Eq. 12.63. Therefore, A is found as:

$$A = \sigma_{oh} + s_u(1 + Ln\,\rho_y^2) \tag{12.77}$$

and the radial stress σ_{rr} in the plastic zone at a radial distance ρ from the axis of the cylinder (Figure 12.13) is given by:

$$\sigma_{rr} = \sigma_{oh} + s_u\left(1 + Ln\frac{\rho_y^2}{\rho^2}\right) \tag{12.78}$$

This equation gives the value of the radial stress anywhere in the plastic zone. At the cavity wall, the pressure is therefore:

$$\sigma_{rro} = \sigma_{oh} + s_u\left(1 + Ln\frac{\rho_y^2}{\rho_o^2}\right) \tag{12.79}$$

Now we want to evaluate the maximum pressure that can be resisted by the soil at the cavity wall, called the *limit pressure* p_L. This limit pressure p_L is reached when the entire soil mass has reached the yield criterion—in other words, when ρ_y becomes infinite. Therefore, we are looking for the limit:

$$\lim_{\rho_y \to \infty} \frac{\rho_y^2}{\rho_o^2} \tag{12.80}$$

For this we need a flow rule. Because we are dealing with the undrained behavior of a fine-grained soil, it makes sense to assume that there will be no volume change in the soil mass (a simple flow rule). Thus, the volume increase at radius r_o has to be the same as the volume increase at radius r_y, so that the soil mass in between the two radii does not change volume:

$$\rho_o^2 - {r_o}^2 = \rho_y^2 - r_y^2 \tag{12.81}$$

or

$$\alpha_{\theta o}\rho_o^2 = \alpha_{\theta y}\rho_y^2 \quad \text{and} \quad \frac{\alpha_{\theta o}}{\alpha_{\theta y}} = \frac{\rho_y^2}{\rho_o^2} \tag{12.82}$$

At the boundary between the elastic and plastic regions, both the yield criterion and the elastic solution must be satisfied. Using Eqs. 12.32 and 12.63:

$$\text{Elastic side of the boundary} \quad \sigma_{rry} = \sigma_{oh} + G\left(\frac{\rho_y^2 - r_y^2}{\rho_y^2}\right) \tag{12.83}$$

$$\text{Plastic side of the boundary} \quad \sigma_{rry} = p_y = \sigma_{oh} + s_u \tag{12.84}$$

Then

$$\frac{\rho_y^2 - r_y^2}{\rho_y^2} = \frac{s_u}{G} = \alpha_{\theta y} \tag{12.85}$$

and Eq. 12.79 becomes:

$$\sigma_{rro} = \sigma_{oh} + s_u\left(1 + Ln\frac{G}{s_u}\alpha_{\theta o}\right) \tag{12.86}$$

This equation gives the curve linking the radial stress at the cavity wall vs. the hoop strain at the cavity wall (the pressuremeter curve) (Figure 12.3). The limit of $\alpha_{\theta o}$ when ρ_o goes to infinity (limit pressure) is 1 because r_o^2 becomes negligible compared to ρ_o^2. Then the limit pressure p_L can be given from Eq. 12.86 as:

$$p_L = \sigma_{oh} + s_u\left(1 + Ln\frac{G}{s_u}\right) \tag{12.87}$$

12.4 COMMON MODELS

12.4.1 Duncan-Chang Hyperbolic Model

The *Duncan-Chang model* or DC model (Duncan and Chang 1970) is a nonlinear stress-dependent model where the

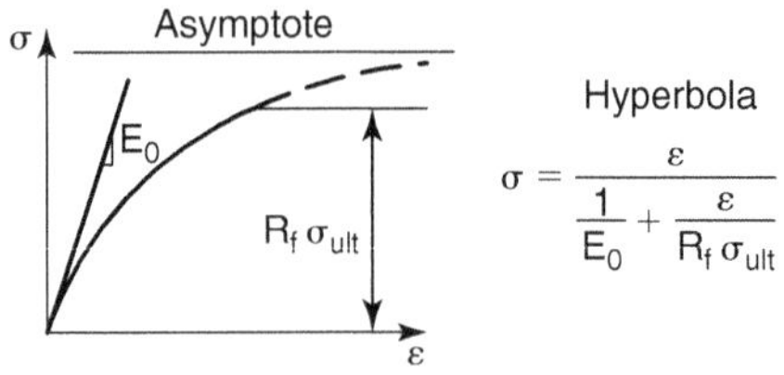

Figure 12.16 Duncan-Chang model.

stress-strain curve is described by a hyperbola (Figure 12.16):

$$\sigma = \frac{\varepsilon}{\frac{1}{E_o} + \frac{\varepsilon}{R_f \sigma_{ult}}} \tag{12.88}$$

where σ is typically taken as the deviator stress, ε is the axial strain, E_o is the initial tangent modulus, which depends on the stress level, σ_{ult} is the asymptotic value of the deviator stress, and R_f is a reduction factor such that R_F times σ_{ult} is the soil strength. The initial tangent modulus E_o increases when the mean confinement stress increases:

$$E_o = E_{o@p_a}\left(\frac{\sigma_m}{p_a}\right)^n \tag{12.89}$$

where $E_{o@p_a}$ is the initial tangent modulus for the reference pressure p_a (often taken as the atmospheric pressure), σ_m is the mean principal stress ($\sigma_m = 0.33(\sigma_1 + \sigma_2 + \sigma_3)$), and n is a soil-specific stress influence exponent. The nonlinearity of the model also recognizes the decrease in modulus with increase in strain. The volume change is characterized by a Poisson's ratio model dependent on the log of the confining stress. An unload-reload modulus E_{ur} is used to characterize the unload-reload path. The DC model uses the Mohr-Coulomb criterion as the failure criterion with a friction angle dependent on the confining stress level but does not directly include dilatancy. The soil parameters needed for the Duncan-Chang model are easily obtained from triaxial tests. Although this model does not have a plasticity framework, it is a very practical model.

12.4.2 Modified Cam Clay Model

Roscoe, Schofield, and Wroth (1958) at Cambridge University (UK) used the theory of plasticity to develop a complete stress-strain model for normally consolidated and lightly overconsolidated saturated clays, which they called the *Cam Clay model* (named after the River Cam, which passes through the campus of Cambridge University). This model was modified in 1965 (Roscoe and Burland 1968) and became known as the *Modified Cam Clay (MCC) model*. The MCC model is an elastic plastic strain hardening model based on critical-state soil mechanics (CSSM) theory, which makes the assumption that all stress paths end up at failure on the critical state line (CSL on Figure 12.17). On the critical state line (CSL), there is no more change in volume or stress. This line exists in the e–$Ln\,p'$ set of axes and in the $q - p'$ set of axes (Figure 12.17).

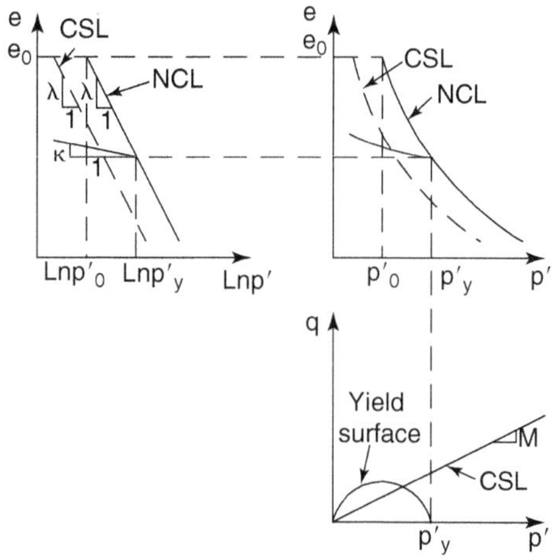

Figure 12.17 Modified Cam Clay model.

Recall that q is the deviator stress ($q = \sigma_1 - \sigma_3$ for the triaxial test) and p′ is the mean normal stress ($p' = 0.33(\sigma_1' + 2\sigma_3')$ for the triaxial test). The parameters defining these two lines are shown in Figure 12.17. In addition, a shear modulus G is necessary, as well as the initial state of the soil described by its initial void ratio e_o, its initial effective stress p_o', and its initial overconsolidation ratio (OCR).

Note that for the consolidation test, the axial strain ε and the change in void ratio Δe from e_o to e are linked by:

$$\varepsilon = \frac{\Delta e}{1 + e_o} \tag{12.90}$$

The normal compression line (NCL) describes the stress-strain curve as a straight line in the $e - Ln\ p'$ set of axes where e is the void ratio and p′ is the mean effective stress ($p' = 0.33(\sigma_1' + \sigma_2' + \sigma_3')$):

$$e = e_o - \lambda Ln\frac{p'}{p_o'} \tag{12.91}$$

where e_o is the initial void ratio corresponding to the initial mean effective stress p_o', e is the void ratio corresponding to the current mean effective stress p', and λ is the isotropic logarithmic compression index (slope of the line). The line in the e–Lnp' set of axes corresponding to the critical state (the critical state line or CSL) links the critical void ratio e_c to Lnp' and is assumed to have the same slope as the NCL. Recall that the critical void ratio is obtained at the end of loading when the soil reaches a state where no more volume change and no more stress increase or decrease occurs:

$$e_c = e_{co} - \lambda Ln\frac{p'}{p_o'} \tag{12.92}$$

The part of the strain recovered upon unload is the elastic component of the strain, e_e. This unload-reload line is considered to be a straight line in the e–Lnp' set of axes and is expressed as:

$$e^e = e_y - \kappa Ln \frac{p'}{p'_y} \quad (12.93)$$

where e_y is the void ratio corresponding to the yield stress p'_y, and κ is the swelling index (slope of the line). The critical state line in the $q - p'$ plot is:

$$q = M\, p' \quad (12.94)$$

where q is the deviator stress ($q = \sigma_1 - \sigma_3$ for a triaxial test), M is the critical state parameter, and p' is the mean confining stress ($p' = 0.33(\sigma'_1 + 2\sigma'_3)$ for a triaxial test). The plastic potential is the same as the yield function because the MCC model uses an associated flow rule. The yield function f is an ellipse (Figure 12.18) with the following equation:

$$f = q^2 - M^2(p'(p'_y - p')) \quad (12.95)$$

where f is the plastic potential and the yield function; it becomes the yield surface for $f = 0$. The direction of the plastic strain is perpendicular to the yield surface and the magnitude is given by the hardening rule (Figure 12.18). For the Cam Clay model, the hardening rule is an isotropic hardening rule. It is obtained from the increase in yield stress and from the recognition that the strain is made of the elastic part related to the swelling line and the plastic part associated with the difference between the normal compression line and the swelling line (Figure 12.19):

$$de^p = (\lambda - \kappa)\frac{dp'}{p'} \quad (12.96)$$

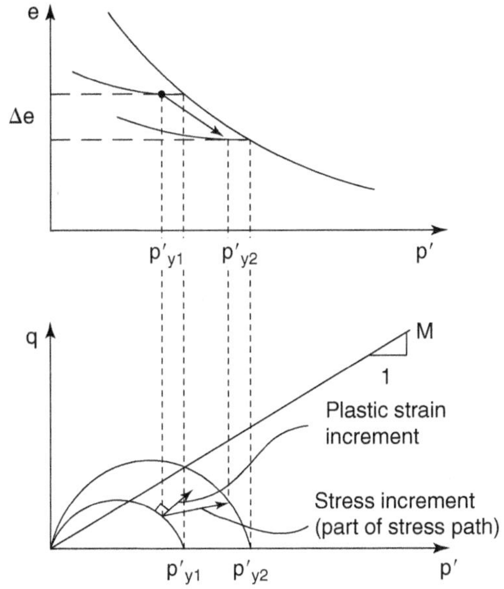

Figure 12.18 Strain hardening for Cam Clay model.

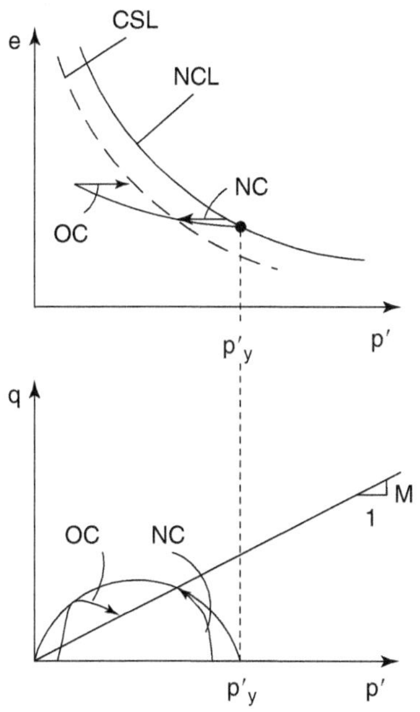

Figure 12.19 Evolution of stress path for NC and OC clays.

12.4.3 Barcelona Basic Model

Alonso, Gens, and Josa (1990) proposed a model to describe the behavior of unsaturated soils. It has become known as the Barcelona Basic Model (BBM), named after the city where the researchers' university is located. The BBM is an elastic plastic strain hardening model that makes use of two stress variables: the net normal stress ($p^* = \sigma - u_a$) and the net water tension or suction ($s = u_w - u_a$). The model is based on several observations of the behavior of unsaturated soils, including reversible swelling and shrinking at low confining pressures, collapse at high pressures, and increase in yield stress (preconsolidation pressure) with increase in net water tension. BBM becomes equal to the Modified Cam Clay model when the suction is equal to zero. Like the MCC model, the BBM uses a linear relationship between the void ratio e and the natural logarithm of the net normal stress p^*, called the normal compression loading (NCL) curve (Figure 12.20*a*). The BBM adds another NCL curve with a linear relationship between the void ratio e and the natural logarithm of the net water tension or suction s. The reference NCL curve in the $e - Lnp^*$ set of axes corresponds to a suction equal to zero (Figure 12.20) and has the same equation as in the MCC model except that the stress is now the net mean normal stress p* instead of the mean effective stress p':

$$e = e_o - \lambda_o Ln \frac{p^*}{p^*_o} \quad (12.97)$$

where e is the void ratio corresponding to p^*, e_o is the initial void ratio corresponding to p^*_o, and λ_o is the compression index for zero suction. Then the NCL curve for a suction s different from zero is:

$$e = e_{so} - \lambda Ln \frac{p^*}{p_o^*} \quad (12.98)$$

where e_{so} is the initial void ratio corresponding to p_o^*, and λ is the compression index for a suction s.

The unload-reload line is also considered to be a straight line and is expressed as:

$$e^e = e_y - \kappa Ln \frac{p^*}{p_y^*} \quad (12.99)$$

where e^e is the void ratio after elastic rebound swelling, e_y is the void ratio corresponding to the yield stress p_y^*, and κ is the swelling index (slope of the line). The NCL curve is also presented in the $e - Lns$ set of axes (Figure 12.20*b*) and the equations are similar to those of the NCL curve in the $e - Lnp^*$ set of axes. However. p^* is replaced by s and the slopes λ_s and κ_s are defined as the compression index with respect to suction s and the swelling index with respect to suction s, respectively.

The yield stress p_y^* depends on the suction s (Figure 12.20*a*), and the curve linking the two is called the *loading-collapse curve* or *LC curve* (Figure 12.20*c*). The LC curve is a yield curve and the equation of this curve is given by:

$$Ln \frac{p_y^*}{p_r} = \left(\frac{\lambda_o - \kappa}{\lambda_\infty - (\lambda_\infty - \lambda_o)\, e^{-\beta s} - \kappa} \right) Ln \frac{p_{yo}^*}{p_r} \quad (12.100)$$

where p_y^* is the yield pressure at a suction s, $p_{yo}{}^*$ is the yield pressure at a suction $s = 0$, p_r is a reference pressure (atmospheric pressure, for example), λ_o is the compression index for a suction $s = 0$, λ_∞ is the compression index at very high suction, β is a coefficient controlling the rate of compression index λ with suction, and κ is the swelling index.

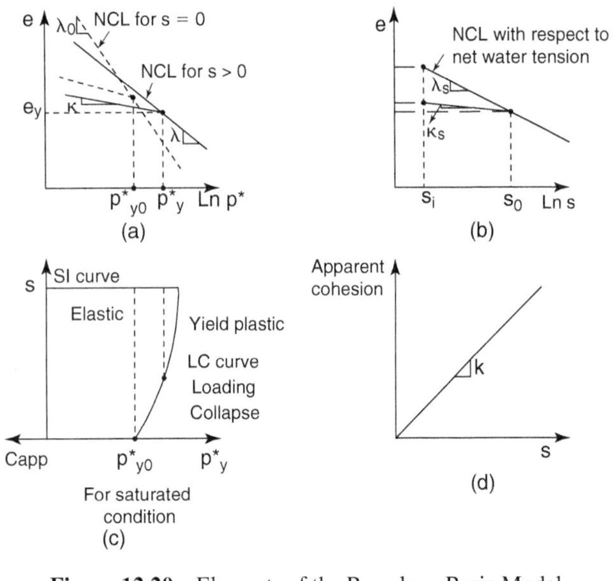

Figure 12.20 Elements of the Barcelona Basic Model.

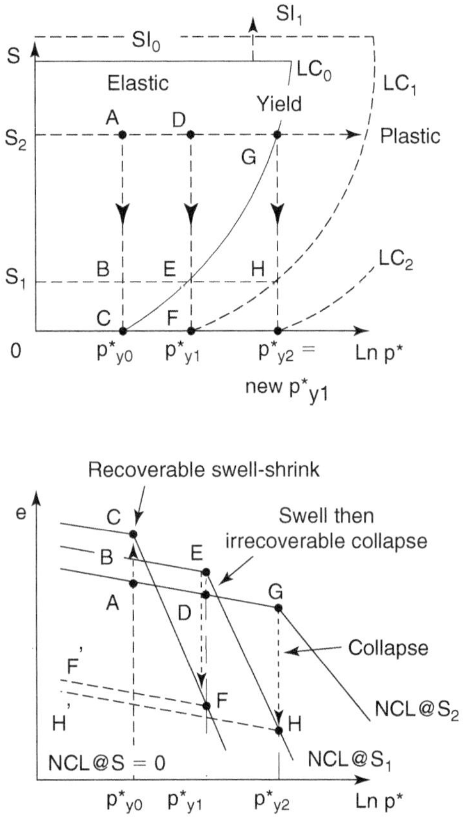

Figure 12.21 Loading collapse curve.

The LC curve dictates whether the sample will swell, shrink, or collapse. Figure 12.21 illustrates this point. To the left of the LC curve in Figure 12.21a, the soil behaves elastically; it yields on the LC curve and with strain hardening it deforms plastically to the right of the LC curve. If the soil is far inside the elastic domain, the soil will swell upon wetting (suction decreases) and the volume change will be reversible (path ABC on Figure 12.21b). If the soil is inside the elastic zone but not far enough from the LC curve, the soil can swell upon wetting (suction decreases) at first and then collapse (path DEF on Figure 12.21b). If the soil is outside the LC curve, the soil will collapse upon wetting (suction decreases) (path GH on Figure 12.21b). In all three cases, the suction decreases under constant total stress; this means that the water stress increases (e.g., from −1000 kPa to −100 kPa) and therefore the effective stress decreases. These three cases show that for unsaturated soils, there is not a single relationship between the volume change of the soil and the effective stress, unlike for saturated soils as postulated by Terzaghi. The inability of effective stress to explain this dual behavior has been the biggest obstacle in the development of a single effective stress model for unsaturated soils.

In the case of the swelling-collapse path DEF in Figure 12.21b, the soil moves from the NCL@s_2 curve on Figure 12.21a to the NCL@s_1 curve and finally comes to rest on the NCL@s = 0 curve. During that process, the value of the net normal stress does not change and remains equal to the yield stress p^*_{y1}; it starts associated with s_2 and ends up associated with $s = 0$. Therefore, the starting point of the LC curve for the soil after collapse is p^*_{y1}, which represents the new value of p^*_{y0}. This shows how the LC curve can evolve as the material experiences wetting or drying.

The LC curve is capped by the suction increase curve or SI curve (Figure 12.21), indicating that during drying the soil will reach a maximum suction value. This maximum suction is nearly independent of the net stress, and a horizontal line is chosen to represent this yield limit. Much like the LC curve, the SI curve can evolve with straining, wetting, or drying of the soil (Figure 12.21).

The critical state line (CSL) failure envelope in the $q - p'$ plot is the same as in the MCC model, but the suction increases the apparent cohesion c_{app} by a value linearly related to the suction s:

$$c_{app} = ks \tag{12.101}$$

where k is a constant of proportionality. The CSL equation (Eq. 12.90) is modified as follows:

$$q = Mp^* + ks \tag{12.102}$$

The shape of the yield surface is kept as an ellipse but is modified to include the contribution of the suction on the apparent cohesion (Figure 12.22):

$$q^2 - M^2(p^* + k\,s)(p^*_y - p^*) = 0 \tag{12.103}$$

The postyield behavior for the BBM is described through the strain hardening of the yield function. Much like in the MCC model, the incremental plastic strain is given by:

$$de^p = (\lambda - \kappa)\frac{dp^*}{p^*} \tag{12.104}$$

What is different with the BBM is that the flow rule that gives the direction of the incremental plastic strain is non-associated and given by a plastic potential:

$$G = \alpha q^2 - M^2(p^* + k\,s)(p^*_y - p^*) \tag{12.105}$$

where α is a parameter determined from the condition that the flow rule predicts zero lateral strains in a K_0 stress path (Alonso et al. 1990).

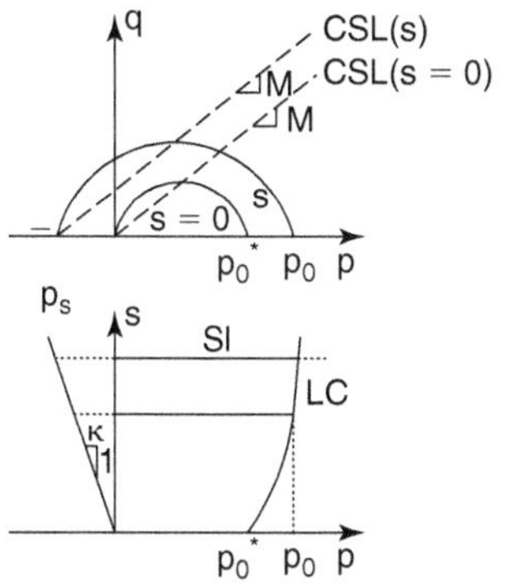

Figure 12.22 Increase in yield surface with increase in suction.

12.4.4 Water Stress Predictions

The prediction of water stresses for saturated and unsaturated cases in numerical methods can be classified in four categories:

1. **No water**. This is the case where the soil has no water. In this case the numerical simulations proceed on the basis of total stress or effective stress without distinction, as there is no difference between the two.
2. **Saturated and drained**. This is the case where the soil is saturated but the loading is slow enough that no water stress in excess of hydrostatic is generated. In this case the solution proceeds in terms of effective normal stress and the water stress is added at the end to obtain the total normal stress.
3. **Total stress approach**. In this case the numerical simulation proceeds in terms of total stresses regardless of the water regime. This is not a theoretically satisfying approach, as it does not recognize the basic and separate behavior of the soil skeleton and the water in the soil. This type of analysis can be accepted in the case of undrained behavior of saturated soils or in the case of high-water-tension soils. In both cases, it is approximately acceptable to consider the soil as a one-phase material, as it is likely that there is very little movement of the water in the soil mass. An exception is the liquefaction of loose sands.
4. **Saturated or unsaturated with water stress formulation**. This is the best and most appropriate way to simulate soil behavior, but it is also the most complicated. This approach requires one to formulate the flow of water through soil in the case of either a saturated soil or an unsaturated soil. This is the topic of Chapter 13.

PROBLEMS

12.1 Develop the expression for the bulk modulus K (hydrostatic compression) and the constrained modulus M (no lateral strain) by using the equations of elasticity linking the stresses and the strains.

12.2 A triaxial test is performed on an elastic soil and the result is plotted as major principal stress σ_1 versus axial strain ε_1. Is the slope of the line equal to the modulus E? If not, what is it? Give the expression of Poisson's ratio in terms of the stresses σ_1 and σ_3, and strains ε_1 and ε_3, for this test. What measurements would you have to make to back-calculate the modulus and Poisson's ratio from such a test?

12.3 Find the ultimate pressure that can be resisted by a soil subjected to a cylindrical expansion in the following case. The cylinder is infinitely long and the initial radius is r_o. The soil is a clay that behaves as a rigid plastic material with a yield criterion $\sigma_1 - \sigma_3 = 2s_u$. Beyond the yield criterion, the soil deforms without changing volume (undrained behavior of the clay).

12.4 Find the ultimate pressure that can be resisted by a soil subjected to a spherical expansion in the following case. The sphere has an initial radius equal to r_o. The soil is a clay that behaves as a rigid plastic material with a yield criterion $\sigma_1 - \sigma_3 = 2s_u$. Beyond the yield criterion, the soil deforms without changing volume (undrained behavior of the clay).

12.5 Find the ultimate pressure that can be resisted by a soil subjected to a cylindrical expansion in the following case. The cylinder is infinitely long and the initial radius is r_o. The soil is a sand that behaves as a rigid plastic material with a yield criterion $\sigma_1/\sigma_3 = K_p$. Beyond the yield criterion, the soil deforms without changing volume. (Although "no volume change" is not a common case in sand, it drastically simplifies the mathematics of this problem.)

12.6 Find the ultimate pressure that can be resisted by a soil subjected to a spherical expansion in the following case. The sphere has an initial radius r_o. The soil is a sand that behaves as a rigid plastic material with a yield criterion $\sigma_1/\sigma_3 = K_p$. Beyond the yield criterion, the soil deforms without changing volume. (Although "no volume change" is not a common case in sand, it drastically simplifies the mathematics of this problem.)

12.7 A Duncan-Chang (DC) model soil has an initial tangent modulus E_o equal to 100 MPa, a strength ratio R_f equal to 0.9, and a stress exponent n equal to 0.5. This DC soil is tested in a triaxial test with a confinement stress $\sigma_3 = 60$ kPa. The cohesion intercept if 5 kPa and the friction angle 34°.

a. Generate the complete σ_1 vs. ε_1 curve.
b. Derive the equation for the modulus as a function of stress level and strain level.

Problems and Solutions

Problem 12.1

Develop the expression for the bulk modulus *K* (hydrostatic compression) and the constrained modulus *M* (no lateral strain) by using the equations of elasticity linking the stresses and the strains.

Solution 12.1

The bulk modulus K:

$$K = \frac{\sigma}{\frac{\Delta V}{V}} = \frac{\frac{1}{3}\left(\sigma_{xx} + \sigma_{yy} + \sigma_{zz}\right)}{\varepsilon_{xx} + \varepsilon_{yy} + \varepsilon_{zz}} = \frac{E}{3(1 - 2\upsilon)}$$

$$\varepsilon_{xx} = \frac{1}{E}\left(\sigma_{xx} - \upsilon\left(\sigma_{yy} + \sigma_{zz}\right)\right)$$

$$\varepsilon_{yy} = \frac{1}{E}\left(\sigma_{yy} - \upsilon\left(\sigma_{xx} + \sigma_{zz}\right)\right)$$

$$\varepsilon_{zz} = \frac{1}{E}\left(\sigma_{zz} - \upsilon\left(\sigma_{xx} + \sigma_{yy}\right)\right)$$

$$\varepsilon_{xx} + \varepsilon_{yy} + \varepsilon_{zz} = \frac{1}{E}\left[\left(\sigma_{xx} + \sigma_{yy} + \sigma_{zz}\right) - 2\upsilon\left(\sigma_{xx} + \sigma_{yy} + \sigma_{zz}\right)\right]$$

$$\varepsilon_{xx} + \varepsilon_{yy} + \varepsilon_{zz} = \frac{1}{E}\left(\sigma_{xx} + \sigma_{yy} + \sigma_{zz}\right)(1 - 2\upsilon)$$

If substituted in *K* formula:

$$K = \frac{\frac{1}{3}\left(\sigma_{xx} + \sigma_{yy} + \sigma_{zz}\right)}{\varepsilon_{xx} + \varepsilon_{yy} + \varepsilon_{zz}} = \frac{\frac{1}{3}\left(\sigma_{xx} + \sigma_{yy} + \sigma_{zz}\right)}{\frac{1}{E}\left(\sigma_{xx} + \sigma_{yy} + \sigma_{zz}\right)(1 - 2\upsilon)} = \frac{E}{3(1 - 2\upsilon)}$$

The constrained modulus M:

$$M = \frac{\sigma_{xx}}{\varepsilon_{xx}} = \frac{E(1-\upsilon)}{(1+\upsilon)(1-2\upsilon)}$$

Because there is no lateral strain:

$$\varepsilon_{yy} = \frac{1}{E}\left(\sigma_{yy} - \upsilon\left(\sigma_{xx} + \sigma_{zz}\right)\right) = 0 \rightarrow \sigma_{yy} = \upsilon\left(\sigma_{xx} + \sigma_{zz}\right) \quad \text{(I)}$$

$$\varepsilon_{zz} = \frac{1}{E}\left(\sigma_{zz} - \upsilon\left(\sigma_{xx} + \sigma_{yy}\right)\right) = 0 \rightarrow \sigma_{zz} = \upsilon\left(\sigma_{xx} + \sigma_{yy}\right) \quad \text{(II)}$$

$$\text{(I)} + \text{(II)} \rightarrow \left(\sigma_{yy} + \sigma_{zz}\right) = 2\upsilon\sigma_{xx} + \upsilon\left(\sigma_{yy} + \sigma_{zz}\right)$$

$$\left(\sigma_{yy} + \sigma_{zz}\right) = \frac{2\upsilon\sigma_{xx}}{(1-\upsilon)}$$

By substituting in M formula:

$$M = \frac{\sigma_{xx}}{\varepsilon_{xx}} = \frac{\sigma_{xx}}{\frac{1}{E}\left(\sigma_{xx} - \upsilon\left(\sigma_{yy} + \sigma_{zz}\right)\right)} = \frac{\sigma_{xx}}{\frac{1}{E}\left(\sigma_{xx} - \frac{2\upsilon^2\sigma_{xx}}{(1-\upsilon)}\right)} = \frac{E\sigma_{xx}}{\sigma_{xx}\left(1 - \frac{2\upsilon^2}{(1-\upsilon)}\right)}$$

$$M = \frac{E\sigma_{xx}}{\sigma_{xx}\left(\frac{1-\upsilon-2\upsilon^2}{(1-\upsilon)}\right)} = \frac{E(1-\upsilon)}{(1+\upsilon)(1-2\upsilon)}$$

Problem 12.2

A triaxial test is performed on an elastic soil and the result is plotted as major principal stress σ_1 versus axial strain ε_1. Is the slope of the line equal to the modulus E? If not, what is it? Give the expression of Poisson's ratio in terms of the stresses σ_1 and σ_3, and strains ε_1 and ε_3, for this test. What measurements would you have to make to back-calculate the modulus and Poisson's ratio from such a test?

Solution 12.2

No. The slope of the line is not the modulus E and is given by the following expression.

In a triaxial test, $\sigma_2 = \sigma_3$

$$\varepsilon_1 = \frac{1}{E}(\sigma_1 - 2\nu\sigma_3)$$

$$\frac{\sigma_1}{\varepsilon_1} = \frac{E\sigma_1}{\sigma_1 - 2\nu\sigma_3}$$

If σ_3 is equal to zero (unconfined compression test) then the slope is E. The Poisson's ratio is calculated as follows.

$$\varepsilon_1 = \frac{1}{E}(\sigma_1 - 2\nu\sigma_3)$$

$$\varepsilon_3 = \frac{1}{E}[\sigma_3 - 2\nu(\sigma_1 + \sigma_3)]$$

$$\nu = \frac{\varepsilon_3\sigma_1 - \varepsilon_1\sigma_3}{2\varepsilon_3\sigma_3 - \varepsilon_1(\sigma_1 + \sigma_3)}$$

The following measurements should be made to back-calculate the modulus and Poisson's ratio: confining pressure (σ_3), deviatoric stress ($\sigma_1 - \sigma_3$), axial strain (ε_1) or (ε_a) and radial strain (ε_3) or (ε_r).

Problem 12.3

Find the ultimate pressure that can be resisted by a soil subjected to a cylindrical expansion in the following case. The cylinder is infinitely long and the initial radius is r_o. The soil is a clay that behaves as a rigid plastic material with a yield criterion $\sigma_1 - \sigma_3 = 2s_u$. Beyond the yield criterion, the soil deforms without changing volume (undrained behavior of the clay).

Solution 12.3

Step 1

The elastic solution is summarized as follows:

$$\frac{d\sigma_r}{dr} + \frac{\sigma_r - \sigma_\theta}{r} = 0$$

where $\sigma_r = \sigma_{oh} + \Delta\sigma_r, \sigma_\theta = \sigma_{oh} + \Delta\sigma_\theta$

$$\varepsilon_r = \frac{1}{E}(\Delta\sigma_r - \nu(\Delta\sigma_\theta + \Delta\sigma_z)) = -\frac{du}{dr}$$
$$\varepsilon_\theta = \frac{1}{E}(\Delta\sigma_\theta - \nu(\Delta\sigma_r + \Delta\sigma_z)) = -\frac{u}{r}$$
$$\varepsilon_z = \frac{1}{E}(\Delta\sigma_z - \nu(\Delta\sigma_r + \Delta\sigma_\theta)) = 0$$

The governing differential equation is

$$r^2\frac{d^2u}{dr^2} + r\frac{du}{dr} - u = 0$$

By applying the boundary conditions, we have:

$$u = \frac{u_0 r_0}{r}$$

Then the strains are:

$$\varepsilon_r = \frac{du}{dr} = \frac{u_0 r_0}{r^2}$$
$$\varepsilon_\theta = \frac{u}{r} = -\frac{u_0 r_0}{r^2}$$
$$\Delta\sigma_{r(r=r_0)} = \frac{E}{1+\nu}\frac{u_0 r_0}{r_0^2} = \frac{E}{1+\nu}\frac{u_0}{r_0} = 2G\varepsilon_{\theta 0}$$

We know that:

$$\varepsilon_{\theta 0} = -\frac{u_0}{r_0} = -\frac{2\pi r_0 u_0}{2\pi r_0 r_0} = -\frac{1}{2}\frac{\Delta V}{V}$$

Therefore,

$$\sigma_{r(r=r_0)} = \sigma_{oh} - 2G\varepsilon_{\theta 0} = \sigma_{oh} + G\frac{\Delta V}{V} \text{ and } \sigma_\theta = \sigma_{oh} - G\frac{\Delta V}{V}$$

Step 2

In plasticity and for this problem, the yield criterion is Tresca:

$$\sigma_1 - \sigma_3 = 2s_u$$
$$\sigma_r - \sigma_\theta = 2s_u$$

We know that (equilibrium using the current radius ρ):

$$\frac{d\sigma_r}{d\rho} + \frac{\sigma_r - \sigma_\theta}{\rho} = 0$$
$$\frac{d\sigma_r}{d\rho} + \frac{2s_u}{\rho} = 0$$
$$d\sigma_r = -2s_u\frac{d\rho}{\rho}$$
$$\int d\sigma_r = \int -2s_u\frac{d\rho}{\rho}$$
$$\sigma_r = -2s_u \ln\rho + A$$

Boundary conditions:

$$\sigma_r = p_F @ \rho = \rho_F$$
$$A = p_F + 2s_u \ln \rho_F$$
$$\sigma_r = p_F - s_u \ln \frac{\rho^2}{\rho_F^2}$$

Compatibility equations at the plastic-elastic boundary:

$$\sigma_r = p_F - s_u \ln \frac{\rho_F^2}{\rho_F^2}$$
$$\sigma_r = p_F$$

When assuming no volume change, $\Delta V = const$, $\sigma_\theta = \sigma_{oh} - G\frac{\Delta V}{V}$, and $\sigma_r = \sigma_{oh} + G\frac{\Delta V}{V}$, so:

$$\sigma_r - \sigma_\theta = 2G\frac{\Delta V}{V} = 2s_u$$

at the interface., so:

$$p_F = \sigma_{oh} + s_u \text{ (for Tresca criterion)}$$

Step 3

Find p_L in plasticity condition (Figure 12.1s).

To get the limit pressure:

$$p_L = p_F - s_u \ln \frac{\rho_0^2}{\rho_F^2}$$

Let us look at no volume change, $\frac{\partial \Delta V}{\partial r} = 0$

$$\Delta V_0 = \Delta V_F = const$$
$$\pi \rho_0^2 - \pi r_0^2 = \pi \rho_F^2 - \pi r_F^2$$
$$\rho_F^2 - r_F^2 = \rho_0^2 - r_0^2 = const$$
$$\frac{\rho_F^2 - r_F^2}{\rho_F^2} = \frac{\Delta V_F}{V_F}$$
$$\frac{\rho_0^2 - r_0^2}{\rho_0^2} = \frac{\Delta V_0}{V_0} = \frac{\rho_F^2 - r_F^2}{\rho_0^2}$$

When $\rho_F \to \infty$, we have $\frac{\Delta V_0}{V_0} \to 1$

$$\frac{\rho_0^2}{\rho_F^2} = \frac{\Delta V_F}{V_F}$$

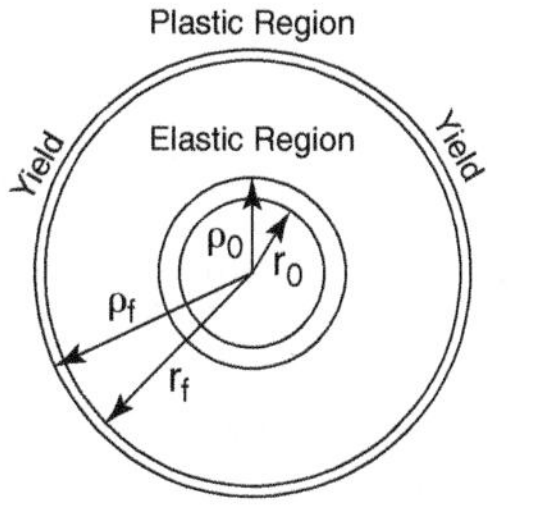

Figure 12.1s r and ρ definition.

We already know that:

$$G\frac{\Delta V}{V} = s_u$$

$$p_L = p_F - s_u \ln\frac{s_u}{G} = p_F + s_u \ln\frac{G}{s_u}$$

$$p_L = \sigma_{oh} + s_u\left(1 + s_u \ln\frac{G}{s_u}\right)$$

Problem 12.4

Find the ultimate pressure that can be resisted by a soil subjected to a spherical expansion in the following case. The sphere has an initial radius equal to r_o. The soil is a clay that behaves as a rigid plastic material with a yield criterion $\sigma_1 - \sigma_3 = 2s_u$. Beyond the yield criterion, the soil deforms without changing volume (undrained behavior of the clay).

Solution 12.4

Step 1

The elastic solution is summarized as follows (problem 11.5) Equilibrium in spherical space ($\sigma_\theta = \sigma_\phi$) gives:

$$\frac{d\sigma_r}{dr} + 2\frac{\sigma_r - \sigma_\theta}{r} = 0$$

where $\sigma_r = p_o + \Delta\sigma_r$, $\sigma_\theta = p_o + \Delta\sigma_\theta$, and p_o is the initial hydrostatic stress at rest in the soil.

$$\varepsilon_r = \frac{1}{E}(\Delta\sigma_r - 2\nu\Delta\sigma_\theta) = -\frac{du}{dr}$$

$$\varepsilon_\theta = \varepsilon_\phi = \frac{1}{E}[(1-\nu)\Delta\sigma_\theta - \nu\Delta\sigma_r] = -\frac{u}{r}$$

or

$$\Delta\sigma_\theta = -\frac{E}{(1+\nu)(1-2\nu)}\left(\frac{u}{r} + \nu\frac{du}{dr}\right)$$

$$\Delta\sigma_r = -\frac{E}{(1+\nu)(1-2\nu)}\left[2\nu\frac{u}{r} + (1-\nu)\frac{du}{dr}\right]$$

The governing differential equation is:

$$r^2\frac{d^2u}{dr^2} + 2r\frac{du}{dr} - 2u = 0$$

By applying the boundary conditions we get:

$$u = \frac{u_0 r_0^2}{r^2}$$

The strains are:

$$\varepsilon_r = -\frac{du}{dr} = 2\frac{u_0 r_0^2}{r^3}$$

$$\varepsilon_\theta = -\frac{u}{r} = -\frac{u_0 r_0^2}{r^3}$$

$$\Delta\sigma_r = \frac{2E}{1+\nu}\frac{u_0 r_0^2}{r^3}$$

Therefore,

$$\Delta\sigma_{r(r=r_0)} = \frac{2E}{1+\nu}\frac{u_0}{r_0} = -4G\varepsilon_{\theta 0}$$

$$\frac{\Delta V}{V} = \frac{4\pi r^2 u}{\frac{4}{3}\pi r^3} = 3\frac{u}{r} = -3\varepsilon_\theta \rightarrow \Delta\sigma_r = -4G\varepsilon_\theta = \frac{4}{3}G\frac{\Delta V}{V}$$

By the same process:

$$\Delta\sigma_\theta = 2G\varepsilon_\theta = -\frac{2}{3}G\frac{\Delta V}{V}$$

Step 2

In plasticity and for this problem, the yield criterion is Tresca:

$$\sigma_r - \sigma_\theta = 2s_u$$

We know that (using the current radius ρ):

$$\frac{d\sigma_r}{d\rho} + 2\frac{\sigma_r - \sigma_\theta}{\rho} = 0$$

$$\frac{d\sigma_r}{d\rho} + \frac{4s_u}{\rho} = 0$$

$$d\sigma_r = -4s_u\frac{d\rho}{\rho}$$

$$\int d\sigma_r = \int -4s_u\frac{d\rho}{\rho}$$

$$\sigma_r = -4s_u \ln\rho + A$$

The boundary conditions are:

$$\sigma_r = p_F @ \rho = \rho_F$$

$$A = p_F + 4s_u \ln\rho_F$$

$$\sigma_r = p_F + 4s_u \ln\frac{\rho_F}{\rho}$$

From elasticity theory, we have already proved that:

$$\sigma_r = p_0 + 4G\frac{u_0 r_0^2}{r^3}$$

$$\sigma_\theta = p_0 - 2G\frac{u_0 r_0^2}{r^3}$$

At yield:

$$\sigma_r = p_F = p_0 + 4G\frac{u_0 r_0^2}{r_f^3} = p_0 + \frac{4}{3}G\frac{\Delta V}{V}$$

$$\sigma_\theta = p_0 - 2G\frac{u_0 r_0^2}{r_f^3}$$

At the boundary between the elastic zone and the plastic zone and using the Tresca criterion gives

$$\sigma_r = p_F$$

$$\sigma_r - \sigma_\theta = 6G\frac{u_0 r_0^2}{r_f^3} = 2G\frac{\Delta V}{V}$$

$$G\frac{\Delta V}{V} = s_u$$

$$p_F = p_0 + \frac{4}{3}s_u$$

Step 3

Find p_L in a plasticity condition:

$$p_L = p_F + 4s_u \ln\frac{\rho_F}{\rho_0}$$

$$\Delta V_0 = \Delta V_F = const$$

Furthermore

$$\frac{\rho_F^3 - r_F^3}{\rho_F^3} = \frac{\Delta V_F}{V_F}$$

$$\frac{\rho_0^3 - r_0^3}{\rho_0^3} = \frac{\Delta V_0}{V_0} = \frac{\rho_F^3 - r_F^3}{\rho_0^3}$$

$$\frac{\rho_F^3}{\rho_0^3} = \frac{\Delta V_0}{V_0}\frac{V_F}{\Delta V_F}$$

when $\rho_F \to \infty$ we have $\dfrac{\Delta V_0}{V_0} \to 1$ so at infitnite expansion

$$\frac{\rho_F^3}{\rho_0^3} = \frac{V_F}{\Delta V_F}$$

$$\frac{\Delta V}{V} = \frac{s_u}{G}$$

$$\frac{\rho_F}{\rho_0} = \left[\frac{G}{s_u}\right]^{1/3}$$

$$p_L = p_F + \frac{4}{3}s_u \ln\frac{G}{s_u}$$

$$p_L = p_0 + \frac{4}{3}s_u\left(1 + \ln\frac{G}{s_u}\right)$$

Problem 12.5

Find the ultimate pressure that can be resisted by a soil subjected to a cylindrical expansion in the following case. The cylinder is infinitely long and the initial radius is r_o. The soil is a sand that behaves as a rigid plastic material with a yield criterion $\sigma_1/\sigma_3 = K_p$. Beyond the yield criterion, the soil deforms without changing volume. (Although "no volume change" is not a common case in sand, it drastically simplifies the mathematics of this problem.)

Solution 12.5

Step 1

The elasticity constitutive model (problem 12.3, step 1) gives:

$$\sigma_r = p_0 + G\frac{\Delta V}{V} \text{ and } \sigma_\theta = p_0 - G\frac{\Delta V}{V}$$

Step 2

In plasticity and for this problem, the yield criterion is Mohr-Coulomb:

$$\frac{\sigma_\theta + \dfrac{c}{\tan\phi}}{\sigma_r + \dfrac{c}{\tan\phi}} = \frac{1-\sin\phi}{1+\sin\phi} = k_a \tag{12.1s}$$

From elasticity theory, we have already proved that:

$$\sigma_r = p_0 + 2G\frac{u_0 r_0}{r^2}$$

$$\sigma_\theta = p_0 - 2G\frac{u_0 r_0}{r^2}$$

At the boundary between the elastic and the plastic region, we have:

$$\sigma_r = p_F = p_0 + 2G\frac{u_0 r_0}{{r_F}^2} \tag{12.2s}$$

$$\sigma_\theta = p_0 - 2G\frac{u_0 r_0}{{r_F}^2} \tag{12.3s}$$

From Eq. 12.2s, we can get

$$r_F^2 = \frac{2Gu_0 r_0}{p_F - p_0}$$

Therefore,

$$\sigma_\theta = p_0 - \frac{2Gu_0 r_0}{2Gu_0 r_0}(p_F - p_0)$$

$$\sigma_\theta = 2p_0 - p_F \tag{12.4s}$$

$$\sigma_r = p_F \tag{12.5s}$$

Plugging Eqs. 12.4s and 12.5s into Eq. 12.1s, we get:

$$\frac{2p_0 - p_F + \dfrac{c}{\tan\phi}}{p_F + \dfrac{c}{\tan\phi}} = \frac{1-\sin\phi}{1+\sin\phi}$$

so

$$p_F = p_0 + p_0\sin\phi + c\ \cos\phi$$

Step 3

Find p_L in a plasticity condition.

From Eq. 12.1s, we get:

$$\sigma_\theta = \left(\sigma_r + \frac{c}{\tan\phi}\right)k_a - \frac{c}{\tan\phi} \tag{12.6s}$$

Plugging Eq. 12.6s into the equilibrium equation, we get:

$$\frac{d\sigma_r}{dr} + \frac{\sigma_r - \left(\sigma_r + \dfrac{c}{\tan\phi}\right)k_a + \dfrac{c}{\tan\phi}}{r} = 0$$

$$\frac{d\sigma_r}{dr} + \frac{\sigma_r(1-k_a)}{r} + \frac{c}{\tan\phi}(1-k_a)\frac{1}{r} = 0$$

$$r\frac{d\sigma_r}{dr} + \sigma_r(1-k_a) = \frac{c}{\tan\phi}(k_a - 1) \tag{12.7s}$$

The solution for σ_r includes a general solution and a particular solution. The general solution of σ_r is:

$$r\frac{d\sigma_r}{dr} + \sigma_r(1 - k_a) = 0$$
$$\sigma_r = A\rho^{(k_a-1)}$$

The particular solution of σ_r is:

$$\sigma_r^* = -\frac{c}{\tan\phi}$$

The solution of σ_r is:

$$\sigma_r = A\rho^{(k_a-1)} - \frac{c}{\tan\phi}$$

Based on the boundary conditions:

$$\rho = \rho_F, \quad \sigma_r = p_F$$

therefore,

$$p_F = A\rho_F^{k_a-1} - \frac{c}{\tan\phi}$$

So,

$$A = \left(p_F + \frac{c}{\tan\phi}\right)\left(\frac{1}{\rho_F}\right)^{k_a-1}$$

Therefore,

$$\sigma_r = \left(p_F + \frac{c}{\tan\phi}\right)\left(\frac{\rho}{\rho_F}\right)^{k_a-1} - \frac{c}{\tan\phi}$$

The no volume change condition gives:

$$\Delta V_0 = \Delta V_F = const$$
$$\pi\rho_0^2 - \pi r_0^2 = \pi\rho_F^2 - \pi r_F^2$$
$$\rho_F^2 - r_F^2 = \rho_0^2 - r_0^2 = const$$
$$\frac{\rho_F^2 - r_F^2}{\rho_F^2} = \frac{\Delta V_F}{V_F}$$
$$\frac{\rho_0^2 - r_0^2}{\rho_0^2} = \frac{\Delta V_0}{V_0} = \frac{\rho_F^2 - r_F^2}{\rho_0^2}$$

when $\rho_F \to \infty$, we have $\dfrac{\Delta V_0}{V_0} \to 1$, Therefore at the limit pressure we have

$$\frac{\rho_0^2}{\rho_F^2} = \frac{\Delta V_F}{V_F}$$

In elasticity, $\sigma_r = p_0 + G\dfrac{\Delta V}{V}$; therefore at the elastic-plastic boundary,

$$\frac{\Delta V_F}{V_F} = \frac{p_F - p_0}{G} = \frac{p_0 + p_0\sin\phi + c\cos\phi - p_0}{G} = \frac{p_0\sin\phi + c\cos\phi}{G}$$

The limit pressure p_L corresponds to $\frac{\Delta V_o}{V_o} = 1$; therefore,

$$p_L = \left(p_F + \frac{c}{\tan\phi}\right)\left(\frac{\rho}{\rho_f}\right)^{k_a - 1} - \frac{c}{\tan\phi}$$

$$p_L = \left(p_0 + p_0 \sin\phi + c\,\cos\phi + \frac{c}{\tan\phi}\right)\left(\frac{G}{p_0 \sin\phi + c\,\cos\phi}\right)^{\frac{1-k_a}{2}} - \frac{c}{\tan\phi}$$

Problem 12.6

Find the ultimate pressure that can be resisted by a soil subjected to a spherical expansion in the following case. The sphere has an initial radius r_o. The soil is a sand that behaves as a rigid plastic material with a yield criterion $\sigma_1/\sigma_3 = K_p$. Beyond the yield criterion, the soil deforms without changing volume. (Although "no volume change" is not a common case in sand, it drastically simplifies the mathematics of this problem.)

Solution 12.6

Step 1

The elasticity solution is presented in problem 12.4, step 1.

Step 2

In plasticity and for this problem, the yield criterion is Mohr-Coulomb:

$$\frac{\sigma_r + \frac{c}{\tan\phi}}{\sigma_\theta + \frac{c}{\tan\phi}} = \frac{1 + \sin\phi}{1 - \sin\phi} = k_p = \frac{1}{k_a} \tag{12.8s}$$

From elasticity theory, we have already proved that:

$$\sigma_r = p_0 + 4G\frac{u_0 r_0^2}{r^3}$$

$$\sigma_\theta = p_0 - 2G\frac{u_0 r_0^2}{r^3}$$

At yield:

$$\sigma_r = p_F = p_0 + 4G\frac{u_0 r_0^2}{r_f^3} = p_0 + \frac{4}{3}G\frac{\Delta V}{V} \tag{12.9s}$$

$$\sigma_\theta = p_0 - 2G\frac{u_0 r_0^2}{r_f^3} = p_0 - \frac{2}{3}G\frac{\Delta V}{V}$$

Combining Eq. 12.8s and 12.9s, we get.

$$p_F = \frac{3p_0(1 + \sin\phi) + 4c\cos\phi}{3 - \sin\phi} \tag{12.10s}$$

Step 3

Find p_L in plasticity. From Eq. 12.8s, we get:

$$\sigma_\theta = \left(\sigma_r + \frac{c}{\tan\phi}\right)k_a - \frac{c}{\tan\phi} \tag{12.11s}$$

Plugging Eq. 12.11s into the equilibrium equation, we get:

$$\frac{d\sigma_r}{dr} + 2\frac{\sigma_r - \left(\sigma_r + \frac{c}{\tan\phi}\right)k_a + \frac{c}{\tan\phi}}{r} = 0$$

$$\frac{d\sigma_r}{dr} + 2\frac{\sigma_r(1-k_a)}{r} + \frac{c}{\tan\phi}(1-k_a)\frac{2}{r} = 0$$

$$\frac{r}{(1-k_a)}\frac{d\sigma_r}{dr} + 2\sigma_r = -2\frac{c}{\tan\phi} \tag{12.12s}$$

The solution for σ_r includes a general solution and a particular solution. The general solution of σ_r is:

$$r\frac{d\sigma_r}{dr} + 2\sigma_r(1-k_a) = 0$$

$$\sigma_r = A\rho^{2(k_a-1)}$$

The particular solution of σ_r is:

$$\sigma_r^* = -\frac{c}{\tan\phi}$$

The solution of σ_r is:

$$\sigma_r = A\rho^{2(k_a-1)} - \frac{c}{\tan\phi}$$

At the elastic-plastic boundary, we have:

$$\rho = \rho_f,\ \sigma_r = p_F$$

therefore,

$$p_F = A\rho_f^{2(k_a-1)} - \frac{c}{\tan\phi}$$

So,

$$A = \left(p_F + \frac{c}{\tan\phi}\right)\rho_f^{-2(k_a-1)}$$

Therefore,

$$\sigma_r = \left(p_F + \frac{c}{\tan\phi}\right)\left(\frac{\rho}{\rho_f}\right)^{2(k_a-1)} - \frac{c}{\tan\phi}$$

by using p_F from Eq. 12.10s:

$$P_L = \sigma_r = (c + p_0\tan\phi)\left(\frac{3(1+\sin\phi)\cos\phi}{(3-\sin\phi)\sin\phi}\right)\left(\frac{r}{r_f}\right)^{2(k_a-1)} - \frac{c}{\tan\phi} \tag{12.13s}$$

Using the no volume change condition leads to

$$\Delta V_0 = \Delta V_F = const$$

$$\rho_F^3 - r_F^3 = \rho_0^3 - r_0^3$$

$$\frac{\rho_F^3 - r_F^3}{\rho_F^3} = \frac{\Delta V_F}{V_F}$$

$$\frac{\rho_0^3 - r_0^3}{\rho_0^3} = \frac{\Delta V_0}{V_0} = \frac{\rho_F^3 - r_F^3}{\rho_0^3}$$

$$\frac{\rho_F^3}{\rho_0^3} = \frac{\Delta V_0}{V_0}\frac{V_F}{\Delta V_F}$$

when $\rho_F \to \infty$, we have $\dfrac{\Delta V_0}{V_0} \to 1$

$$\frac{\rho_F^3}{\rho_0^3} = \frac{V_F}{\Delta V_F} \qquad (12.14s)$$

From Eqs. 12.9s and 12.10s:

$$G\frac{\Delta V}{V} = \frac{3p_0 \sin\phi + 3C\cos\phi}{3 - \sin\phi} \qquad (12.15s)$$

By using Eqs. 12.13s, 12.14s, and 12.15s, we get:

$$p_L = (c + p_0 \tan\phi)\left(\frac{3(1+\sin\phi)\cos\phi}{(3-\sin\phi)\sin\phi}\right)\left[\frac{G}{c + p_0\tan\phi}\left(\frac{3-\sin\phi}{3\cos\phi}\right)\right]^{\frac{4}{3}\left(\frac{\sin\phi}{1+\sin\phi}\right)} - \frac{c}{\tan\phi}$$

Problem 12.7

A Duncan-Chang (DC) model soil has an initial tangent modulus E_o equal to 100 MPa, a strength ratio R_f equal to 0.9, and a stress exponent n equal to 0.5. This DC soil is tested in a triaxial test with a confinement stress $\sigma_3 = 60$ kPa. The cohesion intercept if 5 kPa and the friction angle 34°. Generate the complete $\sigma_1 - \sigma_3$ vs. ε_1 curve.

Solution 12.7

Given: $E_o = 100$ MPa, $R_f = 0.9$, n = 0.5, c = 5 kPa, $\varphi = 34°$, and $\sigma_3 = 60$ kPa, we can get the σ_1 vs. ε_1 curve using the DC formulation:

$$\sigma_1 - \sigma_3 = \frac{\varepsilon_1}{\dfrac{1}{E_o} + \dfrac{\varepsilon_1}{R_f \sigma_{ult}}}$$

Based on the information from the triaxial test and the Mohr Coulomb failure criterion, the soil strength in term of deviator stress is computed as follows:

$$(\sigma_1 - \sigma_3)_f = \frac{2c\cos\phi + 2\sigma_3\sin\phi}{1 - \sin\phi}$$

$$(\sigma_1 - \sigma_3)_f = \frac{2\times 5\times \cos 34 + 2\times 60\times \sin 34}{1 - \sin 34}$$

$$(\sigma_1 - \sigma_3)_f = 171 \text{ kPa}$$

The asymptotic value is given by

$$(\sigma_1 - \sigma_3)_{ult} = \frac{(\sigma_1 - \sigma_3)_f}{R_f} = \frac{171 \text{ kPa}}{0.9} = 190 \text{ kPa}$$

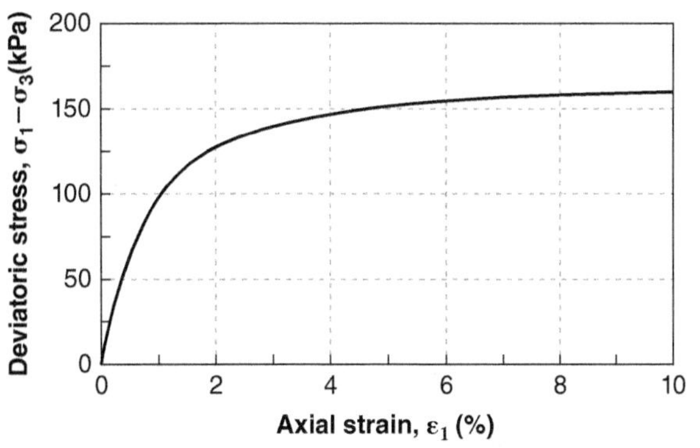

Figure 12.2s Stress-strain curve.

CHAPTER 13

Flow of Fluid and Gas Through Soils

13.1 GENERAL

The fluid and gas phase can either flow through the voids between the soil particles or stay static in the voids. The flow is affected by the resistance offered by friction between the soil particles, by the size of the voids, by the blockage posed by the gas phase to the fluid flow, by the blockage posed by the fluid phase to the gas flow, and by the energy gradient in the fluid or gas. Most of the time the fluid is water and the gas is air, so from this point on we will talk about water and air. If other fluids or gasses are involved, the viscosity of the fluid or gas will change from that of water and air; additionally, any chemical reaction that may occur between the fluid and the particles can increase or decrease the size of the voids.

Some of the output quantities of interest in a flow problem are the water stress, the air stress, the water velocity and its direction, the air velocity and its direction, and the quantity of water flowing per unit time. In geoenvironmental studies, the future location of a moving body of a contaminant is of interest in predicting the extent of contamination. The soil can be saturated or unsaturated. The flow of water in a saturated soil is the simplest case, so we will start with that.

13.2 FLOW OF WATER IN A SATURATED SOIL

13.2.1 Discharge Velocity, Seepage Velocity, and Conservation of Mass

One of the two main equations used to solve flow problems in soils is the conservation of mass equation, which in this case states that the flow Q in m^3/s is equal to the cross-sectional area A times the water velocity v:

$$Q = vA \tag{13.1}$$

One distinguishes between the discharge velocity and the seepage velocity (Figure 13.1). The *seepage velocity* v_s is the actual velocity of the water. In other words, if you were riding on the water molecule, what you would read on the speedometer would be the seepage velocity. Also, if you put a dye in the water and you could see through the soil, you would see the colored water propagate at the seepage velocity. The cross section associated with that velocity is the actual cross section of the voids A_v.

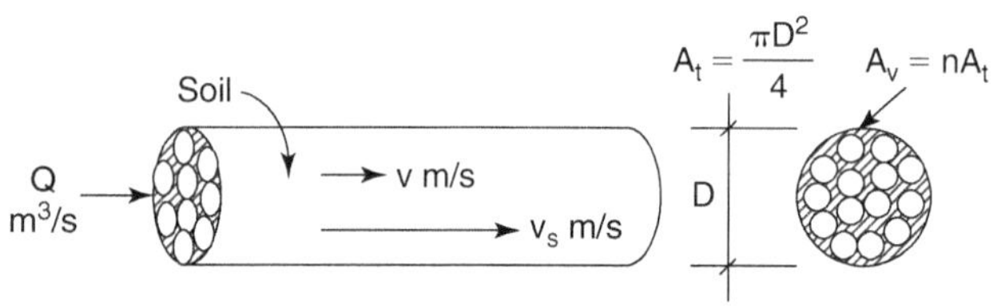

Figure 13.1 Steady flow of water through soil.

$$Q = v_s A_v \tag{13.2}$$

Because A_v is difficult to estimate, the discharge velocity v is used instead in almost all calculations. The *discharge velocity* is the ratio of the flow Q divided by the total cross-sectional area A_t of the soil being traversed by the water (voids plus grains). The discharge velocity is not the actual water velocity, but it is a convenient value to use for calculation purposes:

$$Q = vA_t \tag{13.3}$$

Using Eqs. 13.2 and 13.3 gives the relationship between the seepage and discharge velocity as

$$vA_t = v_s A_v \quad \text{or} \quad v = nv_s \tag{13.4}$$

where n is the porosity of the soil. This shows that the seepage velocity is higher than the discharge velocity (2 to 3 times higher). Although v_s is the actual water velocity, in most geotechnical problems we will use the discharge velocity v. One exception concerns the propagation of contaminated plumes, where it is important to know where the contaminated water is going as a function of time; in this case the seepage velocity must be used. In any case, switching from one to the other can be achieved simply by using Eq. 13.4.

13.2.2 Heads

The energy level in the water is measured in height of water or *head*. The total head h_t represents the total energy available to the water to drive through the soil voids. The pressure head h_p represents the energy stored as pressure in the water. The elevation head h_e represents the potential energy, and the velocity head h_v represents the kinetic energy. The total head is the sum of the pressure head plus the elevation head plus the velocity head:

$$h_t = h_p + h_e + h_v \tag{13.5}$$

Because water flows very slowly through soils (mm per second at most), the velocity head is assumed to be zero for all practical purposes. The pressure head times the unit weight of water γ_w gives the water stress u_w or pore pressure:

$$u_w = h_p \gamma_w \tag{13.6}$$

The elevation head is measured with reference to an arbitrarily chosen datum. This arbitrary choice does not affect the results because all calculations involve changes in quantities, not absolute quantities. The total head of a water molecule at the surface of a lake is the same as the total head of a water molecule at the bottom of that lake (Figure 13.2). Indeed, at the surface, the pressure head is zero but the elevation head is the water depth (if the lake bottom is chosen as the datum), whereas at the bottom of the lake the elevation head is zero but the pressure head is equal to the water depth (water pressure divided by unit weight).

As the water drives through the voids of the soil, it burns energy or total head. The loss of energy is due to the friction that exists between the water molecules and the soil particles. This friction force is called the *seepage force* S. At any point M in the soil, the elevation head can be obtained as the vertical distance between M and the arbitrarily chosen datum (Figure 13.3). At any point M in the soil, the pressure head can be measured by placing a standpipe connected to M at one end and to the atmosphere at the other and measuring the vertical distance between M and the water level in the pipe. The standpipe does not have to be vertical or even straight as long as there is a clear water path from M to the atmospheric pressure. Figure 13.4 shows an example of head diagrams from a constant head permeameter and a falling head permeameter.

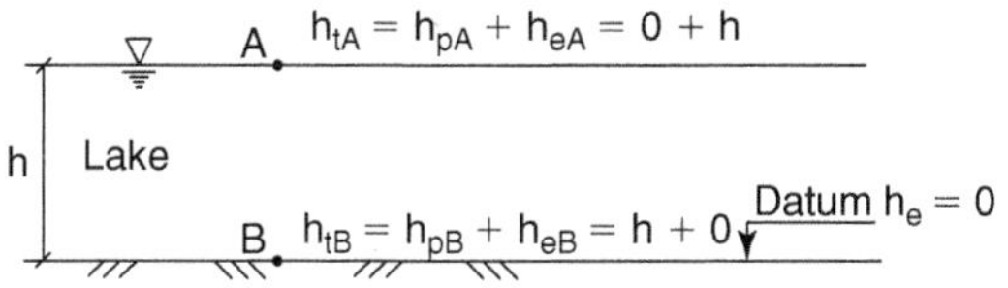

Figure 13.2 Heads in a lake with no flow.

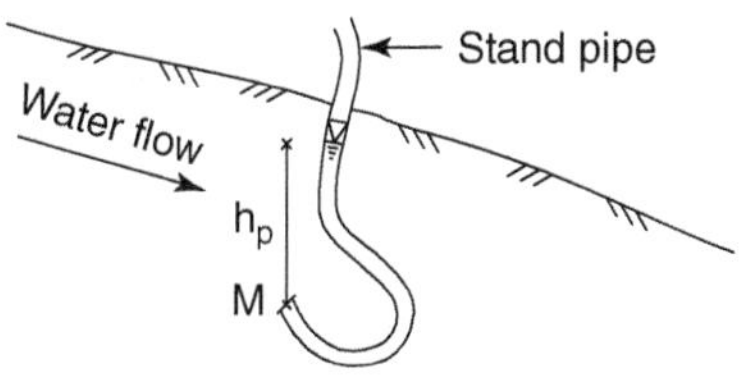

Figure 13.3 Pressure head at M.

13.2.3 Hydraulic Gradient

The *hydraulic gradient* is defined between two points A and B along the path of water travel, called the *flow path*. The hydraulic gradient is the ratio of the loss of total head between A and B over the actual distance traveled by the water to go from A to B (not always the straight line joining the two points):

$$i_{AB} = \frac{h_{tA} - h_{tB}}{l_{AB}} = \frac{\Delta h_{tAB}}{l_{AB}} \tag{13.7}$$

where i_{AB} is the hydraulic gradient between A and B, Δh_{tAB} is the loss of total head between A and B, and l_{AB} is the length of the flow path from A to B. The hydraulic gradient is unitless and varies from about 0.1 to 2 in the field. The hydraulic gradient represents a rate of energy consumption. It is similar in concept to gas consumption for a car. Gas in the tank of a car is energy that is burned when traveling from one point to another; the amount of gas burnt per actual distance traveled on the highway is the gas consumption. If you wish to figure your car consumption from one point to another, you use the actual distance travelled, not the straight-line distance between the two cities. The same applies for the hydraulic gradient. The hydraulic gradient between D and C in Figure 13.4*a* is:

$$i_{DC} = \frac{h_{tD} - h_{tC}}{l_{DC}} = \frac{1.12 - 0.96}{0.2} = 0.8 \tag{13.8}$$

The hydraulic gradient between C and D on Figure 13.4*b* is:

$$i_{CD} = \frac{h_{tC} - h_{tD}}{l_{CD}} = \frac{0.8 - 0.4}{0.2} = 2 \tag{13.9}$$

13.2.4 Darcy's Law: The Constitutive Law

This law is named after Henry Darcy, a French engineer, who discovered it in 1855 as he was working on a problem with the public fountains in Dijon, France (yes, that's the mustard city). The experiment that Darcy set up is essentially the same as the constant head permeameter shown in Figure 13.4*a*. Darcy varied the water level and the length of the sample while measuring the flow coming out of the sample. He found that there was a linear relationship between the water velocity and the hydraulic gradient (Figure 13.5). The slope of that line is called the *hydraulic conductivity* k and Darcy's law states that the discharge velocity v is equal to the hydraulic conductivity k times the hydraulic gradient i:

$$v = ki \tag{13.10}$$

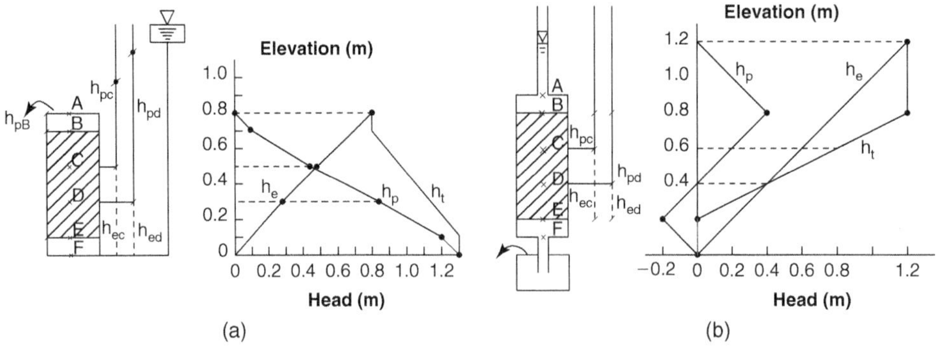

Figure 13.4 Head diagram examples: (a) Constant heat permeameter. (b) Falling head permeameter.

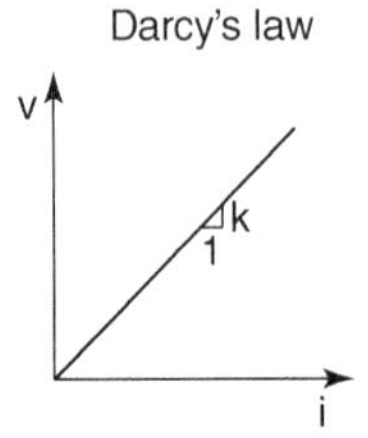

Figure 13.5 Darcy's law.

Going back to the analogy with the car and its gas consumption, it would mean that the gas consumption of the car is linearly proportional to the speed of the car. Darcy's law is the most important constitutive law related to the flow of fluids through soils.

In Figure 13.4*a*, if the hydraulic conductivity of the sand is 10^{-5} m/s, and considering the hydraulic gradient of Eq. 13.8, the discharge velocity is:

$$v = 10^{-5} \times 0.8 \text{ m/s} = 0.008 \text{ mm/s} = 691 \text{ mm/day} \tag{13.11}$$

In Figure 13.4*b*, if the hydraulic conductivity of the clay is 10^{-10} m/s, the discharge is:

$$v = 10^{-10} \times 2 \text{ m/s} = 0.0000002 \text{ mm/s} = 6.3 \text{ mm/year} \tag{13.12}$$

Darcy's law applies to fluid flow and is parallel to Fourier's law for heat flux, to Ohm's law for electrical flux, and to Fick's law for diffusive flux. All these laws express that the propagation of a phenomenon is linearly related to a gradient of potential through a conductivity constant specific to the material through which the propagation is taking place. The analogy can be taken further, as shown in Table 13.1.

13.2.5 Hydraulic Conductivity

The hydraulic conductivity of saturated soils varies widely, from about 10^{-2} m/s for some gravels to about 10^{-11} m/s for some clays. It is measured in the laboratory with a constant head permeameter for sands and gravels, or with a falling head permeameter for silts and clays (see Chapter 9). In the field, it is measured by using a borehole and performing either a pumping test or an infiltration test (see Chapter 7). It can also be obtained by using the piezocone penetrometer test through pore-pressure decay as a function of time (see Chapter 7). The hydraulic conductivity obtained by laboratory tests can be 10 times to 100 times lower than the field value, because the lab test may be testing the intact soil between fissures while the field test may include a network of fissures.

The hydraulic conductivity of saturated soils depends on many factors, including the void ratio, the shape and roughness of the particles, the structure of the soil skeleton, and

Table 13.1 Equivalency between Hydraulic, Heat, and Electricity

Parameter	Hydraulic	Heat	Electricity
Law	Darcy	Fourier	Ohm
Soil property	Hydraulic conductivity	Thermal conductivity	Electrical conductivity
Quantity	Volume V(m^3)	Heat Q(J)	Charge q (C)
Potential	Head (m)	Temperature (K)	Potential (V)
Gradient	Hydraulic gradient (m/m)	Temperature gradient (K/m)	Potential gradient (V/m)
Flux	Flow rate (m^3/s)	Heat transfer rate (J/s)	Current I(A)
Flux density	Velocity v(m/s)	Heat flux (W/m^2)	Current density j(A/m^2)

the fluid properties (viscosity and unit weight). To separate the influence of the fluid from that of the soil skeleton on the hydraulic conductivity k, the intrinsic hydraulic conductivity or simply *permeability*, K is used:

$$K = k\frac{\mu_f}{\gamma_f} \tag{13.13}$$

where μ_f is the dynamic viscosity of the fluid, and γ_f is the unit weight of the fluid. The value of μ_f for water is 10^{-6} kPa.s at 20°C, and γ_f for water is 9.79 kN/m^3 at 20°C. Note that K is in units of m^2.

Some empirical relationships have been proposed over the years for estimating the hydraulic conductivity of coarse-grained soils. Hazen (1892), working on sand filters, proposed a formula relating the hydraulic conductivity k to the D_{10} particle size corresponding to 10% finer on the particle size distribution curve. The Hazen formula seems to work best for sands with D_{10} values between 0.1 and 1 mm.

$$k(\text{m/s}) = C(D_{10}(\text{mm}))^2 \tag{13.14}$$

where C is a constant usually taken as 0.01 but with reported values from 0.1 to 0.001. For D_{10} values above 1 mm, the power of 2 for D_{10} decreases. Kozeny (1927) and Carman (1938) proposed a semi-empirical formula also for sands:

$$k(\text{m/s}) = \frac{\gamma e^3}{\mu C S_o^2(1+e)} \tag{13.15}$$

where γ is the unit weight of the permeating fluid (kN/m^3), μ is its dynamic viscosity (kN.s/m^2), e is the void ratio of the soil, C is a constant usually taken equal to 5×10^6, and S_o is the specific surface of the particles (1/m). The *specific surface* is the ratio of the particle surface area over the volume of the particle. For a sphere it would be:

$$S_o = \frac{\pi D^2}{\pi D^3/6} = \frac{6}{D} \tag{13.16}$$

The factor 6 in Eq. 13.16 goes up to 8.5 for very angular coarse-grain particles. Using a value of 5×10^6 for C, a value of 10 kN/m^3 for the unit weight of water, and a value of 10^{-6} kN.s/m for the dynamic viscosity of water at 20°C, the formula becomes:

$$k(\text{m/s}) = \frac{2e^3}{S_o^2(\text{mm}^{-2})(1+e)} \tag{13.17}$$

The specific surface can be measured or estimated from the particle size distribution curve (Carrier 2003). Some equations have been proposed to estimate the hydraulic conductivity of remolded clays from index properties (Carrier and Beckman 1984):

$$k(\text{m/s}) = \frac{0.0174}{1+e}\left(e - \frac{0.027\left(w_L - 0.242PI\right)}{PI}\right)^{4.29} \tag{13.18}$$

Table 13.2 Approximate Range of Hydraulic Conductivity of Soils

Soil type	Hydraulic conductivity (m/s) for water flow in saturated soils
Gravel	10^{-4} to 10^{-2}
Sand	10^{-6} to 10^{-4}
Silt	10^{-8} to 10^{-6}
Clay	10^{-11} to 10^{-8}

where e is the void ratio, w_L is the liquid limit, and PI is the plasticity index. Table 13.2 gives a range of possible hydraulic conductivity values for saturated water flow through soils.

13.2.6 Field vs. Lab Values of Hydraulic Conductivity

One of the difficult issues in soil hydraulic conductivities is the different values obtained in the laboratory at small scale, k_{lab}, and in the field at large scale, k_{field}, particularly for fine-grained soils. The difference can be several orders of magnitude, as shown conceptually in Figure 13.6. Among the reasons for this large difference is the lack of representativeness of the small samples. Indeed, often the small samples do not reflect the influence of the large-scale features of a soil deposit on the hydraulic conductivity k. These features include cracks and fissures formed through successive drying and wetting or simply bending of the soil mass over geologic time. Hence, the lower values given in Table 13.2 may represent the lab values, while the higher values may represent the large-scale field values.

13.2.7 Seepage Force

Seepage force is a drag force that develops at the interface between flowing water and soil particles. It is due to the viscous friction between the two elements. If the water and

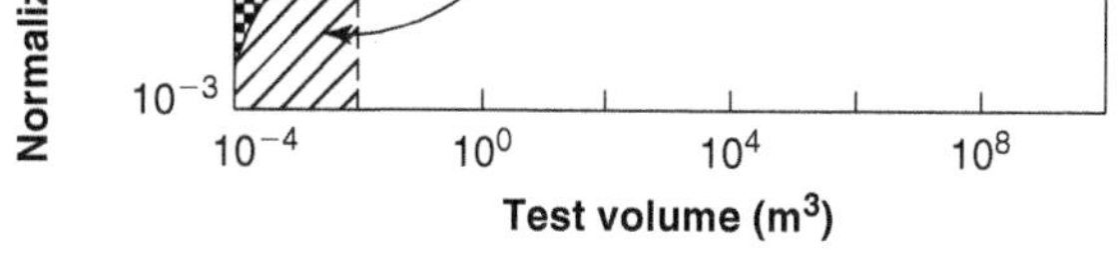

Figure 13.6 Conceptual difference between small-scale and large-scale hydraulic conductivity.

the soil particles are considered together in a free-body diagram, the seepage force is an internal force and does not enter into the equilibrium of the free body. Indeed, only the external forces influence the equilibrium of the free body. However, if the soil skeleton made only of the particles is considered as the free body and the water is external, then the seepage force must be considered in the equilibrium of the free body, as it is now an external force. Similarly, if the water only is considered as the free body, then again the seepage force is an external force.

Let's see how we can calculate the magnitude of this seepage force. For this, we consider the sketch of Figure 13.7 in which water flows from A to B in a cylinder of diameter D filled with soil. As explained earlier, we must consider either the free body of the soil skeleton alone or the free body of the water alone to make the seepage force appear in the equilibrium of the free body. It is easier in this case to consider the free body of the water in the cylinder between point A and point B. So, imagine the body of water with all the soil particles removed; it looks like Swiss cheese. The external forces are the seepage force, which is the summation of all the small friction or drag forces the soil particles exert on the water; the oblique upward force at point A due to the water pressure at that point; the oblique downward force at point B due to the water pressure; and the weight of the water. The weight of the particles is another external force, but it is carried by the container.

The equilibrium of the free body of water in the flow direction is written as:

$$h_{pA}\gamma_w A - h_{pB}\gamma_w A - S - W\sin\alpha = 0 \tag{13.19}$$

where h_{pA} and h_{pB} are the pressure head at A and B respectively, γ_w is the unit weight of water, A is the total cross-sectional area $(\pi D^2/4)$, S, S is the seepage force, W is the weight of the water body, and α is the angle between the flow direction and the horizontal. Note that A is not the correct area to use, as there are holes in this "Swiss cheese" water body. We consider this area, even though it is wrong, for the same reason that we consider the cross-sectional area, even though it is wrong, for the determination of

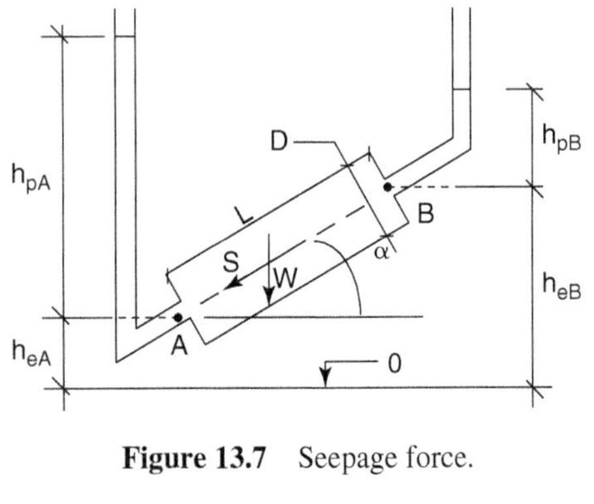

Figure 13.7 Seepage force.

discharge velocity (Section 13.2.1). The term $W\sin\alpha$ can be expressed as:

$$W\sin\alpha = AL\gamma_w \frac{h_{eA} - h_{eB}}{L} \tag{13.20}$$

where h_{eA} and h_{eB} are the elevation head at A and B respectively and A is the total cross-sectional area. Again, this is not the correct area to use, as there are holes in the "Swiss cheese" water body, but it is consistent with the area chosen for the equilibrium equation (Eq. 13.19). Then, using Eqs. 13.19 and 13.20, the seepage force per unit of soil volume (S/AL) can be written as:

$$\frac{S}{AL} = \frac{h_{pA}\gamma_w - h_{pB}\gamma_w}{L} + \gamma_w \frac{h_{eA} - h_{eB}}{L} = \frac{h_{tA} - h_{tB}}{L}\gamma_w = i_{AB}\gamma_w \tag{13.21}$$

where i_{AB} is the hydraulic gradient between A and B. Therefore, the seepage force s per unit volume of soil is given by $i\gamma_w$ and exists in the direction of the hydraulic gradient. This is where the choice of the "wrong" area—total cross-sectional area—becomes useful; if we had used the correct area, we would have obtained the seepage force per unit of water volume, which would be more difficult to calculate than the simpler volume of soil:

$$\vec{s} = \gamma_w \vec{i} \tag{13.22}$$

For example, if the hydraulic gradient is 1 (a rather high but not unusual hydraulic gradient for common flow problems), the seepage force for one cubic meter of volume is 10 kN, or about one ton. This is a significant force. Note that the seepage force is zero when there is no flow ($i = 0$) and therefore does not include the buoyancy force, which is always in the vertical upward direction.

13.2.8 Quick Sand Condition and Critical Hydraulic Gradient

If the flow is upward and in the vertical direction, the seepage force adds to the buoyancy force to lighten the soil particles, and can become high enough to make the soil particles weightless (Figure 13.8).

This is called a *quick sand condition* and the corresponding hydraulic gradient is called the *critical hydraulic gradient* i_c. Referring to Figure 13.9, the buoyancy force F_b on the soil volume of cross-sectional area A and length L is equal to:

$$F_b = AL\gamma_w \quad or \quad f_b = \frac{F_b}{AL} = \gamma_w \tag{13.23}$$

where f_b is the buoyancy force per unit volume. In the case where the flow is vertical upward, the equilibrium of forces per unit volume when the quick condition is reached is:

$$s + f_b = \gamma_w i_c + \gamma_w = \gamma_{sat} \quad or \quad i_c = \frac{\gamma_{sat}}{\gamma_w} - 1 \tag{13.24}$$

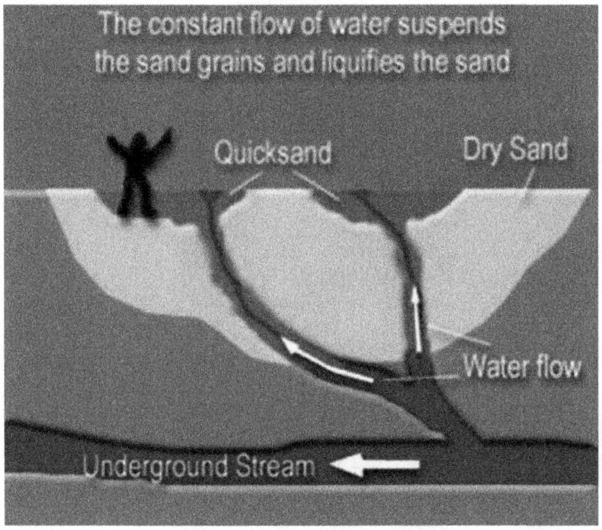

Figure 13.8 Quick sand condition. (Courtesy of Lee Krystek)

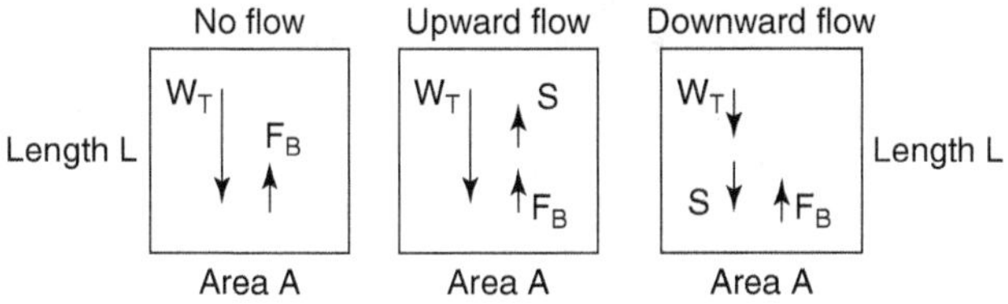

Figure 13.9 Seepage force for upward and downward flow.

Another way to arrive at this result for the critical gradient i_c is to consider the experiment of Figure 13.4 and ask when the total head difference between A and B will be sufficient to generate an effective stress equal to zero at the bottom of the sample:

$$\sigma' = \sigma - u_w = \gamma_{sat}L + \gamma_w h_{pB} - \gamma_w h_{pE} = 0 \quad (13.25)$$

$$\gamma_{sat}L + \gamma_w(h_{tB} - h_{eB}) - \gamma_w(h_{tE} - h_{eE})$$
$$= \gamma_{sat}L - \gamma_w(h_{tE} - h_{tB}) - \gamma_w L = 0 \quad (13.26)$$

$$i_c = \frac{h_{tE} - h_{tB}}{L} = \frac{\gamma_{sat}}{\gamma_w} - 1 \quad (13.27)$$

Note that because γ_{sat}/γ_w is about 2, the critical gradient is about 1. If you fall into a quick sand, it is like falling into a very thick liquid. This liquid has the unit weight of the soil ($\sim$20 kN/m^3), which is typically equal to about two times the unit weight of the human body ($\sim$10 kN/m^3). Therefore, theoretically you should sink halfway into the quick sand until the buoyancy force counterbalances your weight. One problem is that if you do not stay still, you will go down, as there is no bearing capacity under your feet, and it will be difficult to go back up as this heavy liquid can develop friction resisting your movement upward. So, if you fall in such a quick sand, stay still and hope that the theory is correct! In contrast, if the flow is downward, the seepage force increases the weight of the particles artificially and the bearing capacity is improved compared to a no-flow condition. Figure 13.9 illustrates these conditions.

13.2.9 Quick Clay

A quick clay is something completely different from a quick sand condition. A *quick clay* is a clay that is a solid in its natural state, but turns into a liquid when disturbed. This disturbance can come from shearing because of loading. Such clays typically were slowly deposited in a seawater, offshore environment and, through geologic aging, are now in an onshore environment. One mechanism is loading by glaciers, which were as thick as 300 m some 10,000 years ago but have now melted, thereby allowing the offshore clays to rebound and become above ground in the process. The slow offshore deposition can lead to a card-castle type of structure for the clay particles (edge-to-face contacts) with a high porosity and abundant salt content in the pore water and at the contacts. Then, in the onshore environment, the clay has been permeated by rainwater (distilled water) or groundwater (usually low-salinity water) and the salt has been washed away from the voids in the clay, leaving only some salt at the contacts between the clay particles and the low-salinity water in the voids. The salt strengthens the particle contacts and therefore the structure because it decreases the repulsion that typically exists between electrically charged clay platelets. The intact clay may have an undrained shear strength of 25 kPa and a water content of 30% in the undisturbed case, for example. The low-salinity water leaches the salt away and weakens the bond between particles. If this fragile structure is disrupted by shearing or vibrations, for example, the structure collapses and the mixture of water from the voids and clay particles becomes a thick liquid (Figure 13.10). If salt is then added to the thick liquid and mixed by stirring, the clay regains some strength (Figure 13.10).

13.2.10 Sand Liquefaction

The phenomenon of *sand liquefaction* should be distinguished from the quick sand condition. Quick sand conditions are due to sufficiently rapid upward flow, whereas sand liquefaction is typically related to earthquake shaking. During such violent, repeated shaking, the water in the saturated sand does not have time to escape the pores (undrained behavior), so the pressure in the water goes up. If the water stress u_w becomes so high as to equal the total stress σ, then the effective stress $\sigma'(\sigma' = \sigma - u_w)$ becomes equal to zero and the sand liquefies. This heavy liquid can flow to the surface and create sand boils, which are often found at the ground surface after a severe earthquake (Figure 13.11).

13.2.11 Two-Dimensional Flow Problem

Some of the structures involving problems associated with steady-state flow through saturated soils include earth dams, cofferdams, spillways, cutoff walls, retaining walls, and

(a) (b) (c) (d)

Figure 13.10 Rissa landslide clay, Norway 1978: (a) Intact quick clay. (b) Remolded clay. (c) Adding salt to remolded clay. (d) Remolded clay strengthened by salt. (Pictures/images are from the film *The Quick Clay Landslide in Rissa, Norway*, Made by Norwegian Geotechnical Institute [NGI])

Figure 13.11 Sand boil. (Courtesy of USGS.)

slopes. Some of the questions that must be answered in design are:

1. What is the water stress (pore pressure) at any point in the soil mass?
2. What is the amount of water flowing through the soil?
3. What is the uplift force exerted on a structure buried in the soil?
4. What is the factor of safety against a quick sand condition developing under or near a structure?
5. What happens when the hydraulic conductivity is different in two directions?
6. What happens if the soil is layered rather than being uniform?

To answer these questions, it is necessary to solve the flow problem. The following assumptions are made:

1. The soil is uniform with the same hydraulic conductivity in all directions.
2. The soil is saturated with water and the water is in compression.
3. The water is incompressible.
4. Darcy's law governs the water flow through the soil.
5. The flow is in two directions only (x and z, but no flow in the y direction).
6. The flow is independent of time: steady-state flow.

To solve the problem with these assumptions, we use the problem-solving method outlined in section 11.4.5:

1. Zoom in at the element level. We select an element of soil that has an elementary area dx dz with a dimension of 1 in the y direction (Figure 13.12).
2. Considering the element of Figure 13.12, the water velocity is v_x in the x direction and v_z in the z direction when it enters the element and $v_x + \frac{\partial v_x}{\partial x}dx$ and $v_z + \frac{\partial v_z}{\partial z}dz$ when it exits the element. It is assumed that the water does not flow in the y direction because of the plane strain assumption.

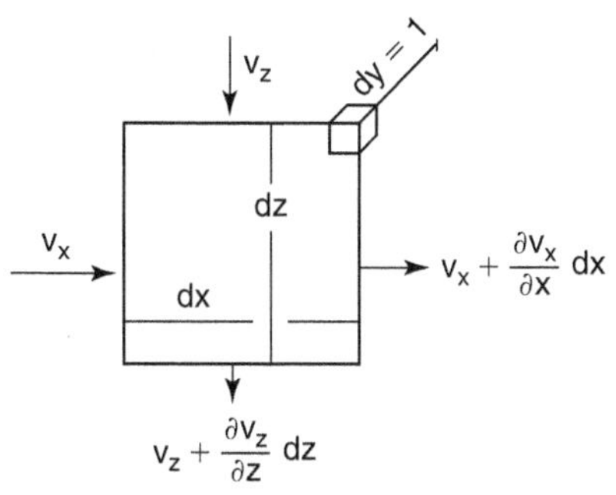

Figure 13.12 Element of soil in the flow mass.

3. The fundamental equation in this case is the conservation of mass equation, expressing that the flow of water in the element is equal to the flow of water out of the element. Use is made of the flow equation (Q = vA):

$$v_x dz \times 1 + v_z dx \times 1 = \left(v_x + \frac{\partial v_x}{\partial x}dx\right) dz \times 1 + \left(v_z + \frac{\partial v_z}{\partial z}dz\right) dx \times 1 \quad (13.28)$$

$$\frac{\partial v_x}{\partial x} + \frac{\partial v_z}{\partial z} = 0 \quad (13.29)$$

4. The constitutive equation describes how fast the water flows through the soil (Darcy's law):

$$v_x = ki_x = k\frac{dh_t}{dx} \quad and \quad v_z = ki_z = k\frac{dh_t}{dz} \quad (13.30)$$

5. The governing differential equation is obtained by combining Eq. 13.29 and the first derivative of the terms in Eq. 13.30:

$$\frac{d^2h_t}{dx^2} + \frac{d^2h_t}{dz^2} = 0 \quad (13.31)$$

6. This form of differential equation is called the *Laplace equation* and the solutions are called *harmonic functions*.
7. The complexity of the solution is brought about by the complexity of the boundary conditions. These boundary conditions describe the flow conditions at the geometric boundaries of the flow. Equation 13.31 can be solved mathematically or graphically. The most common solution to this problem is a graphical solution called the flow net.

The preceding solution is based on the assumption that the soil mass is uniform, meaning that the hydraulic conductivity k_h is equal to the vertical hydraulic conductivity k_v. If k_h is very different from k_v, then Eq. 13.31 becomes:

$$k_h\frac{d^2h_t}{dx^2} + k_v\frac{d^2h_t}{dz^2} = 0 \quad (13.32)$$

13.2.12 Drawing a Flow Net for Homogeneous Soil

A *flow net* is a graphical solution to the governing differential equation for a steady-state flow of water through a pervious soil. The flow net is made of two sets of lines: the flow lines and the equipotential lines. The flow lines describe the path of the water molecules. The equipotentials are lines of equal potential or total head h_t. A *flow channel* is the soil conduit between two consecutive flow lines (Figure 13.13*d*). A *flow field* is the geometric shape between two consecutive flow lines and two consecutive equipotentials (Figure 13.13*d*). A flow net is a map of the total head h_t giving the value of h_t for any point in the flow net with an x and z coordinate.

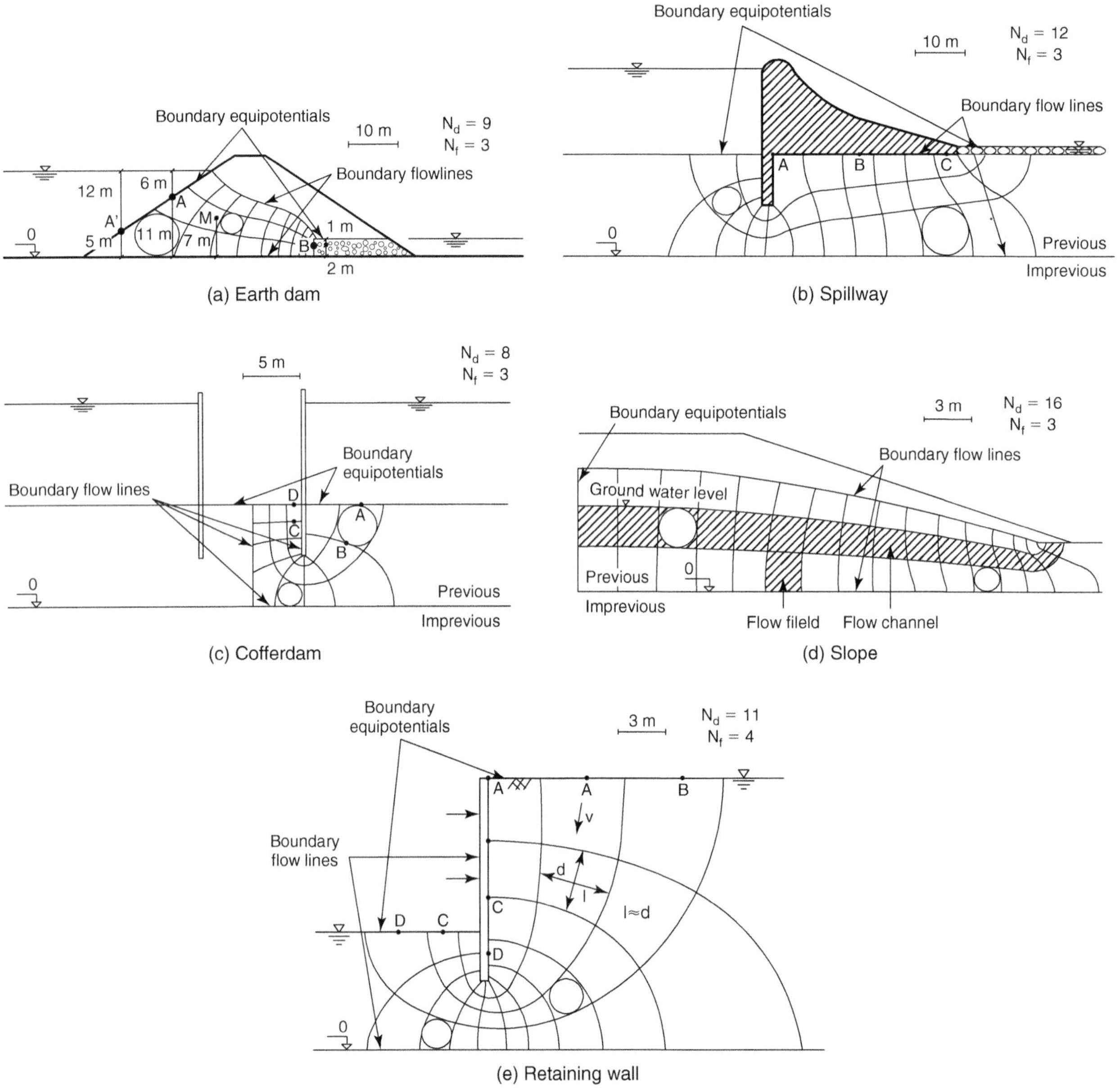

Figure 13.13 Sample flow nets.

Figure 13.13 gives examples of flow nets. To draw a flow net, proceed as follows:

1. Draw the cross section of the flow problem to scale.
2. Draw the boundary flow lines. These are flow lines such that the total flow through the flow net occurs between these flow lines.
3. Draw the boundary equipotential lines (also called boundary equipotentials). These lines define the total head at the beginning of the flow net $h_{t(beg)}$ and the total head at the end of the flow net $h_{t(end)}$.
4. Draw an additional two to three flow lines between the boundary flow lines.
5. Draw the equipotentials such that they cross the flow lines at a right angle; this is the condition expressed by the governing differential equation. Choose the equipotentials in such a way that the flow fields are squares. This is the case if the flow fields are very small, but at the scale at which most flow nets are drawn, this condition should be replaced by: choose the equipotentials in such a way that a circle can be inscribed in each flow field and is tangent to all four sides.
6. Adjust the flow lines and the equipotentials until the conditions of step 5 (perpendicularity and circle inscribed) are satisfied. This step usually takes the longest time and requires some experience.

13.2.13 Properties of a Flow Net for Homogeneous Soil

Two of the most important properties of the flow net are:

1. The potential drop or drop in total head Δh_t from one equipotential to the next is the same across any of the equipotentials.
2. The flow is the same through any of the flow channels.

One of the first things to do when working with a flow net is to choose the datum: the horizontal line where the elevation is equal to zero. The chosen datum is usually located at the location of the bottom impervious layer, so that all elevations will be positive, but theoretically it can be set at any level within the diagram. The total head at the beginning of the flow net $h_{t(beg)}$ and at the end of the flow net $h_{t(end)}$ can be calculated as follows: The elevation heads $h_{e(beg)}$ and $h_{e(end)}$ are simply measured to scale on the diagram (e.g., $h_{e(beg)(A)} = 11$ m and $h_{e(end)(B)} = 2$ m on Figure 13.13*a*). The pressure heads $h_{p(beg)}$ and $h_{p(end)}$ are readily available from the free water body connected to the beginning and the end of the flow net (e.g., $h_{p(beg)(A)} = 6$ m and $h_{p(end)(B)} = 1$ m on Figure 13.13*a*). The value of $h_{t(beg)}$ and $h_{t(end)}$ can then be found (e.g., $h_{t(beg)(A)} = 11 + 6 = 17$ m and $h_{t(end)(B)} = 2 + 1 = 3$ m on Figure 13.13*a*). Note that $h_{t(beg)}$ and $h_{t(end)}$ are constant on the equipotential; indeed, if you try a different point on that equipotential you will find the same value (e.g., $h_{t(beg)(A')} = 5 + 12 = 17$ m on Figure 13.13*a*).

The number of flow channels is N_f and the number of equipotential drops is N_d (Figure 13.13*d*). One of the properties of the flow net is that the drop of total head Δh_t across two consecutive equipotential lines is the same for all flow fields; it is given by:

$$\Delta h_t = \frac{(h_{t(beg)} - h_{t(end)})}{N_d} \tag{13.33}$$

Once $h_{t(beg)}$ and $h_{t(end)}$ are known, the total head $h_{t(M)}$ can be found for any point in the flow net by interpolation:

$$h_{t(M)} = h_{t(beg)} - n_d \frac{(h_{t(beg)} - h_{t(end)})}{N_d} \tag{13.34}$$

where n_d is the number of drops to go from the beginning of the flow net to the point considered. For example, consider point *M* in the flow net of Figure 13.13*a*. The elevation head at point M is 7 m and the total head is:

$$h_{t(M)} = 17 - 1.7\frac{17-3}{9} = 14.35 \text{ m} \tag{13.35}$$

The hydraulic gradient *i* in any flow field is:

$$i = \frac{\Delta h_t}{l} \tag{13.36}$$

where Δh_t is the loss of total head in the flow field (a constant for all flow fields) and l is the flow path across the flow field (varies from one flow field to another). Therefore, the hydraulic gradient varies throughout the flow net and is inversely proportional to the length of the flow field. Because the velocity is linearly related to the hydraulic gradient through the hydraulic conductivity of the soil, the water velocity increases when the size of the flow field decreases. To illustrate, imagine that you can ride a water molecule in Figure 13.13*e* and that you have the choice between molecule at point A and molecule at point B; which molecule should you choose if you wish to win the flow net race? Molecule A is your best bet because it is associated with smaller flow fields, higher gradients, and therefore higher velocities. Note that although A will burn the same amount of energy (total head) as B to travel through the flow net, it has a shorter trip to travel and can afford to step on the gas and have a higher energy consumption per meter travelled (hydraulic gradient). Thus, molecule A will get to C before molecule B gets to D. You can check this race by using a colored dye in the water at the upstream face of the flow net in a laboratory experiment.

13.2.14 Calculations Associated with Flow Nets

Quantity of Flow

How much water will go through a flow net per unit time? This is important for a dam, for example. The flow q through one flow channel is:

$$q = kiA = kid \times 1 \tag{13.37}$$

where *A* is the cross section through which the water flows. *A* is equal to d × 1 where *d* is the width of the flow field perpendicular to the flow and 1 represents the unit width of the flow net perpendicular to the flow net (Figure 13.13*e*). Because the flow field is a square, its width d is equal to its length l over which the total head drops by Δh_t. Therefore, Eq. 13.37 can be rewritten as:

$$q = k\frac{\Delta h_t}{l} l \times 1 = k\Delta h_t \tag{13.38}$$

Furthermore, the flow through one flow field is the same as the flow through all flow fields in one flow channel, because no water crosses over into other flow channels. If we go back to the car traffic analogy, in water flow no one changes lanes; everybody stays in their own lane, but the highway is totally congested (saturated soil), so all lanes carry the same traffic flow. The flow q is also the same in all flow channels. Because there are N_f flow channels, the flow per unit width of flow net is $N_f \times q$. If the length perpendicular to the plane of the flow net over which the flow takes place is L, the total flow Q through or under the structure is $N_f \times q \times L$. Using Eq. 13.38 leads to the formula for the total flow:

$$Q = k\frac{N_f}{N_d} L(h_{t(beg)} - h_{t(end)}) \tag{13.39}$$

Water Stress

What is the water stress u_w at any point in the flow net? This is important for calculating the effective stress at any point or even calculating the uplift force on a buried solid structure like a spillway. The procedure is as follows:

1. Calculate the total head at the point M considered by using Eq. 13.34.
2. Subtract the elevation head to obtain the pressure head $h_{p(M)}$.
3. Get the water stress by:

$$u_w = \frac{h_{p(M)}}{\gamma_w} \tag{13.40}$$

where γ_w is the unit weight of water. Note that u_w includes the hydrostatic stress, because the pressure head is the level at which the water would rise in a pipe connected to M. If there were no flow, that height would correspond to the hydrostatic height.

Uplift Force on a Buried Structure

What is the upward force generated by the water pressure under a solid structure buried in the flow net? This force F_{up} is the result of the water pressure acting on the bottom of the structure, as in the case of the spillway shown in Figure 13.13*b*. The procedure for determining upward force is as follows:

1. Select a few points in the flow net along the bottom of the structure. A minimum of four points is recommended (A, B, C, D).
2. Calculate the water stress u_A, u_B, u_C, and u_D at A, B, C, and D.
3. Calculate the average water stress u_{av} under the structure.
4. Calculate the uplift force as:

$$F_{up} = u_{av} BL \tag{13.41}$$

where B is the width of the structure (dam) and L is the length.

Exit Gradient

What is the highest hydraulic gradient on the exit face of the flow net? This is called the *exit gradient*. Because the drop in total head is the same for any two consecutive equipotentials, the highest hydraulic gradient on the exit face (exit gradient) is associated with the smallest flow field on the exit face (Eq. 13.36). Because the exit face is often a horizontal plane, the exit gradient i_{exit} is compared to the critical hydraulic gradient i_{crit} to avoid a critical condition (quick sand). A large factor of safety F is usually used:

$$i_{exit} = \frac{i_{crit}}{F} \quad or \quad \frac{\Delta h_t}{l} = \frac{1}{F}\left(\frac{\gamma_{sat}}{\gamma_w} - 1\right) \tag{13.42}$$

If the required factor of safety is not satisfied, the flow must be modified to satisfy the required factor of safety. This can be done by using cutoff walls, deepening barriers, or preventing the flow altogether.

Heave and Critical Block

A calculation similar to the comparison between the exit gradient and the critical gradient can be performed in the case of retaining structures, as shown in Figure 13.14. In this case a block of soil is identified where the flow is upward and could create a quick sand condition. The free body considered is the soil particles only, with the water as an external body. In this case, the weight of the soil particles is the buoyant weight and the seepage force is an external force acting vertically and upward on the soil particles. The factor of safety against heave is the ratio of the buoyant weight of the particles divided by the seepage force:

$$F = \frac{W'}{S} = \frac{(\gamma_{sat} - \gamma_w) D \times \frac{D}{2}}{\left(\frac{h_{t(A)} - h_{t(B)}}{D}\right)\gamma_w D \times \frac{D}{2}} = \frac{(\gamma_{sat} - \gamma_w) D}{(h_{t(A)} - h_{t(B)})\gamma_w} \tag{13.43}$$

13.2.15 Flow Net for Hydraulically Anisotropic Soil

The procedure described in section 13.2.12 is used for a soil that has the same hydraulic conductivity in the vertical and horizontal directions. If the horizontal hydraulic conductivity k_h is significantly different from the vertical hydraulic conductivity k_v, then the flow net is distorted because Eq. 13.32 applies and the flow lines and equipotential lines no longer intersect at right angles. A change of variable can bring Eq. 13.32 back to Eq. 13.31:

$$x = \alpha x' \quad and \quad z = z' \tag{13.44}$$

$$\frac{k_h}{\alpha^2 k_v}\frac{d^2 h_t}{dx'^2} + \frac{d^2 h_t}{dz'^2} = 0 \tag{13.45}$$

which shows that if:

$$\alpha = \sqrt{k_h / k_v} \tag{13.46}$$

then the flow net can be drawn for the anisotropic soil with a proper scale transformation in the x direction. The steps to

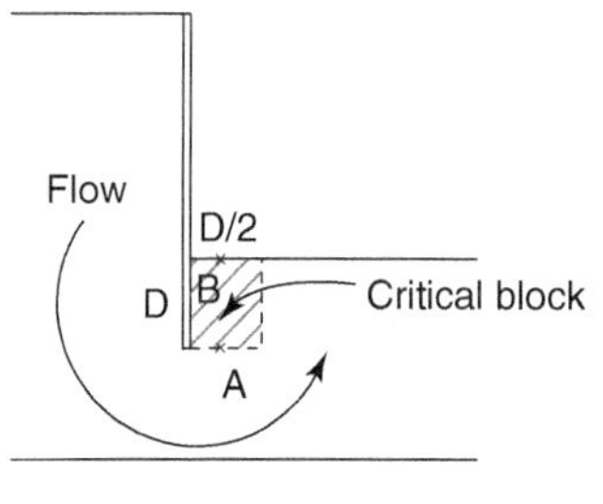

Figure 13.14 Heave of critical block.

draw the flow cross section to scale in the section 13.2.12 procedure are modified as follows:

1. Select a scale for the vertical z direction.
2. Select a scale for the horizontal x direction such that the horizontal scale is equal to $\sqrt{k_h/k_v}$ times the vertical scale.
3. Draw the flow net according to the procedure of section 13.2 12.
4. If needed, use that flow net to go back to the untransformed set of axes and draw the resulting flow net in that space; understand that in that space, the flow lines and equipotential lines will not intersect at right angles and the flow fields will not be squares.

For example, if the hydraulic conductivity k_h was 4×10^{-8} m/s in the horizontal direction and k_v was 10^{-8} m/s in the vertical direction, the transformed cross section of the scaled diagram would be shrunk by a factor of 2 in the horizontal direction while it was kept unchanged in the vertical direction (Figure 13.15). Then the flow net would be drawn as if the soil were uniform. Note that the quantity of flow equation would become:

$$Q = \sqrt{k_h k_v}\frac{N_f}{N_d}L(h_{t(beg)} - h_{t(end)}) \tag{13.47}$$

13.2.16 Flow and Flow Net for Layered Soils

If the flow goes from layer 1 with a hydraulic conductivity k_1 to a layer 2 with a hydraulic conductivity k_2, then the flow lines and the equipotential lines are deflected. If the approach angle of the flow line coming from layer 1 to the interface is θ_1 (Figure 13.16), the angle with which that flow line leaves the interface into layer 2 is θ_2 and is different from θ_1. The angles are linked by the following equations:

$$\frac{k_1}{k_2} = \frac{\tan\theta_1}{\tan\theta_2} \tag{13.48}$$

If the flow is either parallel to the interface or perpendicular to the interface, then the flow lines are not deflected and an equivalent hydraulic conductivity k_e can be found.

In the case where the flow is parallel to the interface of two layers (Figure 13.17), the hydraulic gradient across two equipotentials is the same in both layers:

$$i_1 = i_2 = \frac{h_{tB} - h_{tA}}{L} = i_e \tag{13.49}$$

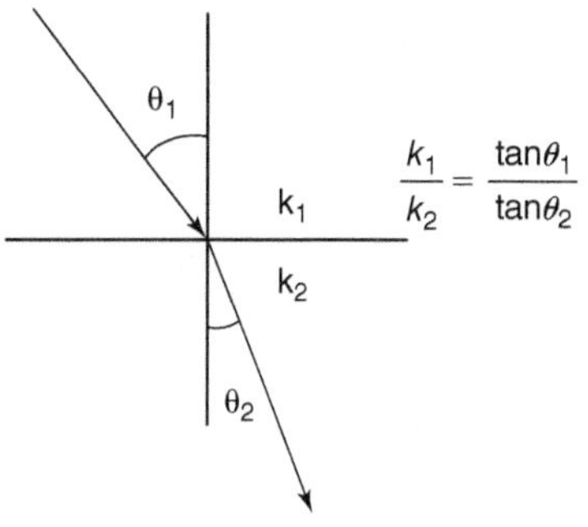

Figure 13.16 Flow line crossing layer interface.

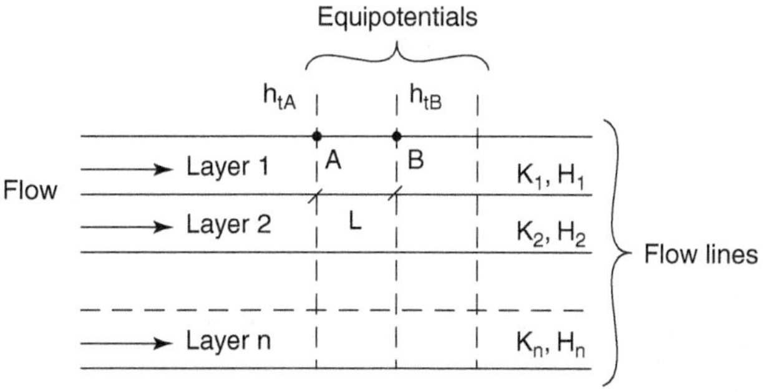

Figure 13.17 Flow parallel to parallel layers.

The flow is additive:

$$q = q_1 + q_2 = v_1 H_1 \times 1 + v_2 H_2 \times 1 \tag{13.50}$$

or

$$k_e i_e (H_1 + H_2) = k_1 i_1 H_1 + k_2 i_2 H_2 \tag{13.51}$$

Therefore:

$$k_e = \frac{k_1 H_1 + k_2 H_2}{H_1 + H_2} \tag{13.52}$$

This result can be generalized for n layers.

$$k_e = \frac{\sum_{i=1}^{n} k_i H_i}{\sum_{i=1}^{n} H_i} \tag{13.53}$$

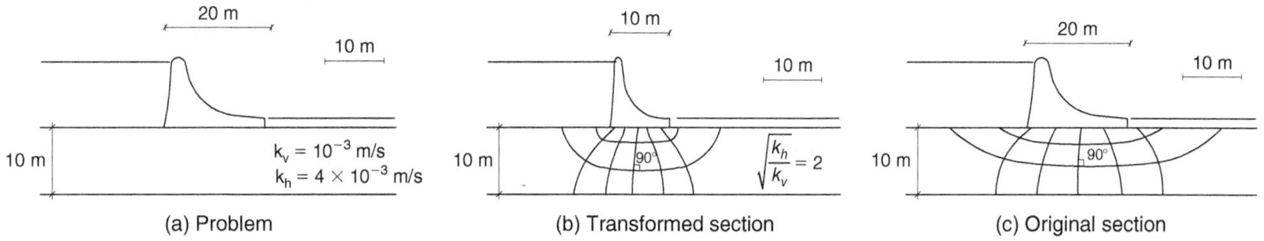

Figure 13.15 Flow net for anisotropic soil.

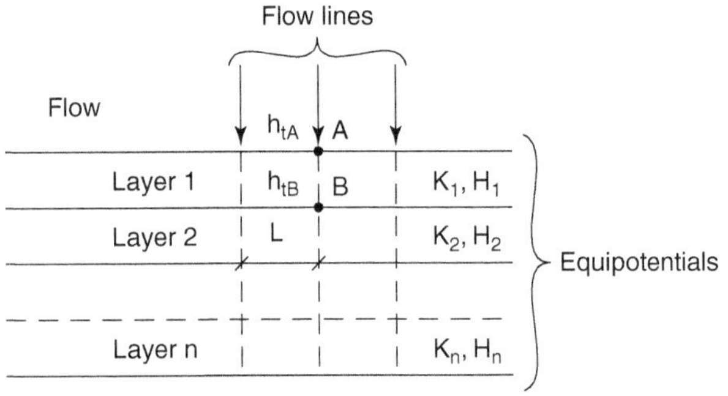

Figure 13.18 Flow perpendicular to parallel layers.

In the case where the flow is perpendicular to two layers (Figure 13.18), the flow across two equipotentials is the same in both layers:

$$q = q_1 = q_2 \quad and$$
$$q = k_e i_e L \times 1 = k_1 i_1 L \times 1 = k_2 i_2 L \times 1 \quad (13.54)$$

The loss of total head, however, is additive:

$$\Delta h_t = \Delta h_{t1} + \Delta h_{t2} \quad (13.55)$$

but

$$i_e = \frac{\Delta h_t}{H_1 + H_2} \quad and \quad i_1 = \frac{\Delta h_{t1}}{H_1} \quad and \quad i_2 = \frac{\Delta h_{t2}}{H_2} \quad (13.56)$$

Therefore,

$$\frac{H_1 + H_2}{k_e L} q = \frac{H_1}{k_1 L} q_1 + \frac{H_2}{k_2 L} q_2 \quad (13.57)$$

and

$$k_e = \frac{H_1 + H_2}{\frac{H_1}{k_1} + \frac{H_2}{k_2}} \quad (13.58)$$

This result can be generalized for n layers:

$$k_e = \frac{\sum_{i=1}^{n} H_i}{\sum_{i=1}^{n} \frac{H_i}{k_i}} \quad (13.59)$$

13.3 FLOW OF WATER AND AIR IN UNSATURATED SOIL

13.3.1 Hydraulic Conductivity for Water and for Air

There is a need to distinguish between the soil hydraulic conductivity for water k_w and the soil hydraulic conductivity for air k_a; k_w expresses how fast water travels through the water phase and k_a expresses how fast air travels through the air phase. One of the fundamental observations regarding water flow in unsaturated soils is that the hydraulic conductivity of water decreases compared to saturated soils. You might think that as the soil becomes drier, there is more room for the water to go through, but that is not the case, because air occupies the voids and cannot get out of the way unless you chase it out somehow. Instead, the water has to go through what is left of water in the soil. The air may be thought of as blocking the flow like particles do. In this sense, the area blocking the flow has increased from the area associated with the particles (solid phase) in the case of a saturated flow to the area associated with the particles plus the air phase in the case of unsaturated flow. This means that the cross-sectional area decreases and that the drag force increases because the water is bound more tightly to the particles. Thus, the hydraulic conductivity decreases because:

1. Cross-sectional area of water flow decreases
2. Tortuosity increases
3. Drag forces increase

Note also that the difference between the discharge velocity v and the seepage velocity v_s (actual water molecule velocity) is increased. Recall that for saturated flow we had:
Saturated flow:

$$vA_t = v_s A_v \quad or \quad v = nv_s \quad (13.60)$$

where A_t is the total cross-sectional area of the soil where the water flows, A_v is the area of voids where the water can flow in the saturated case, and n is the porosity.

In the case of unsaturated flow, the equation becomes:
Unsaturated flow:

$$vA_t = v_s A_w \quad or \quad v = Snv_s \quad (13.61)$$

This gives an indication that the degree of saturation will have a significant influence on the hydraulic conductivity of the water. When the degree of saturation decreases, so does the water content, and the water tension increases in the soil. Therefore, the higher the water tension, the lower the hydraulic conductivity of water is for an unsaturated soil. The reverse is observed for the hydraulic conductivity of air, which increases as the water content decreases and the water tension increases.

Figure 13.19 illustrates what happens to the water hydraulic conductivity k_w when a coarse-grained soil and a fine-grained soil desaturate, which means that they are subjected to higher and higher water tension. At low water tension (~ saturated), the coarse-grained soil has a much higher k_w (e.g., 10^{-4} m/s) than the fine-grained soil (e.g., 10^{-7} m/s). Indeed, the water travels a lot faster through a saturated coarse-grained soil than through a saturated fine-grained soil. The water tension corresponding to the air entry value for the coarse-grained soil (e.g., $u_{wae} = 10$ kPa) is much lower than for the fine-grained soil (e.g., $u_{wae} = 1000$ kPa) because it is a lot easier

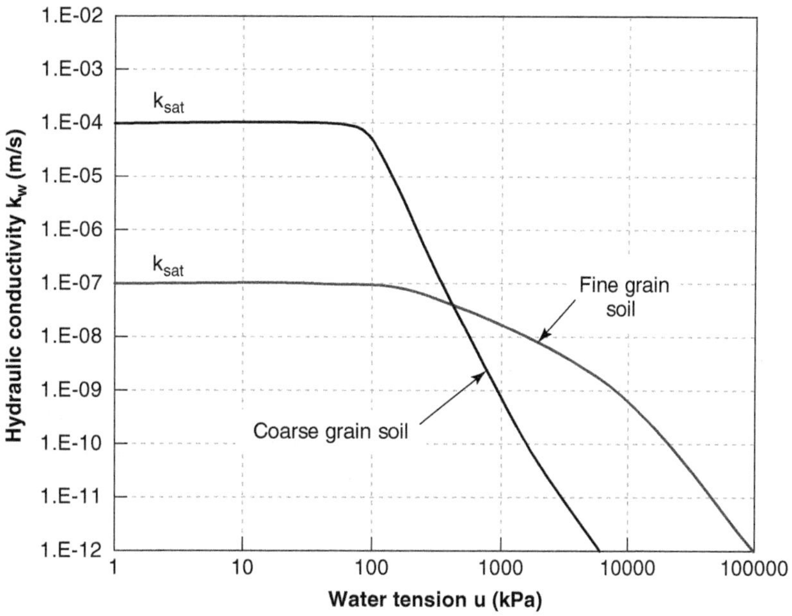

Figure 13.19 Constant head permeameter test results for unsaturated soils.

for the air to enter the large pores than the small pores. The crossover occurs because it does not take much of an increase in water tension to desaturate the pores of a coarse-grained soil compared to what is required to drive the water out of the pores of a fine-grained soil. The fine-grained soil retains more water longer while the water travels through its pores, compared to the coarse-grained soil. As a result, at high water tension (e.g., 5000 kPa), the k_w value of the coarse-grained soil (e.g., 10^{-11} m/s) can be much lower than the k_w value of the fine-grained soil (e.g., 10^{-8} m/s).

When the soil is dry, the value of air hydraulic conductivity k_a is maximum and equal to $k_{a(dry)}$. This trend is contrary to the trend for the hydraulic conductivity of water k_w. Indeed, k_w decreases when the soil gets drier; it is maximum when the soil is saturated and equal to $k_{w(sat)}$. Both hydraulic conductivities are often presented as normalized values as follows:

$$k_w = k_{rw} k_{w(sat)} \tag{13.62}$$

$$k_a = k_{ra} k_{a(dry)} \tag{13.63}$$

Figure 13.20 shows an example of the combined variation of both normalized hydraulic conductivity values k_{rw} and k_{ra} as a function of the degree of saturation S. Note that there is a limiting degree of saturation S_w (0.3 on Figure 13.20) where the water is no longer mobile (bound water) and at the same time a limiting degree of saturation S_a (0.85 on Figure 13.20) where the air is no longer mobile (occluded air). These two stages correspond to the residual stages.

A number of models have been proposed to describe the variation of the hydraulic conductivity as a function of water content, or water tension, or degree of saturation. Among the most popular are:

Averjanov (1950)

$$k_w = k_{ws} S_e^n \tag{13.64}$$

LaLiberte and Correy (1966)

$$k_w = k_{ws} \left(\frac{u_{wae}}{u_w} \right)^n \tag{13.65}$$

Gardner (1958)

$$k_w = \frac{k_{ws}}{1 + a u_w^n} \tag{13.66}$$

where k_w is the hydraulic conductivity to water, k_{ws} is the hydraulic conductivity to water when the soil is saturated, S_e is the effective degree of saturation, u_w is the water tension, u_{wae} is the water tension at the air entry, and a and n are fitting parameters.

The hydraulic conductivity of unsaturated soils depends on many factors, including the degree of saturation, the void ratio, the shape and roughness of the particles, the structure of the soil skeleton, and the fluid properties (viscosity and unit weight). To separate the influence of the fluid from that of the soil skeleton on the hydraulic conductivity k, the intrinsic hydraulic conductivity, or simply permeability K, is used:

$$K = k \frac{\mu_f}{\gamma_f} \tag{13.67}$$

where μ_f is the dynamic viscosity of the fluid, and γ_f is the unit weight of the fluid. At 20°C and one atmosphere,

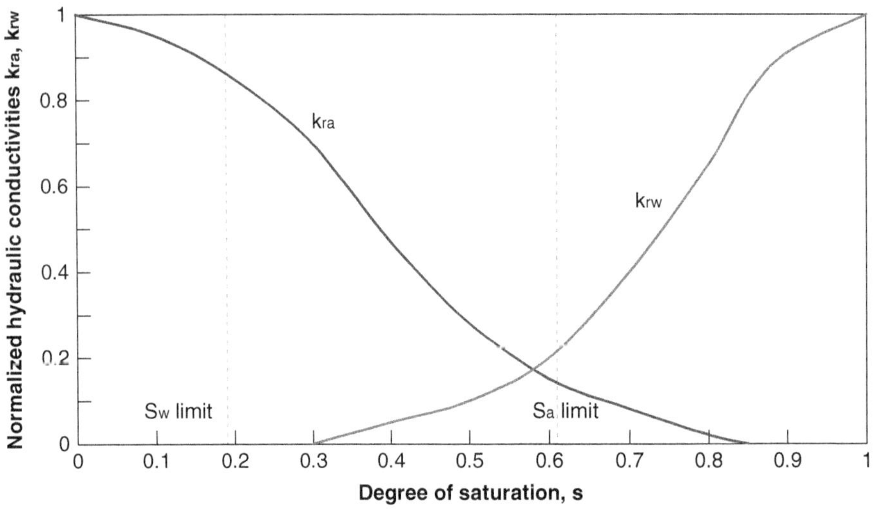

Figure 13.20 Relative hydraulic conductivity of water and air as a function of degree of saturation.

the value of μ_f for water is 10^{-6} kPa.s and γ_f for water is 9.79 kN/m^3 At 20°C and one atmosphere, the value of μ_f for air is 1.82 x 10^{-5} Pa.s and γ_f for air is 11.8 N/m^3. Note that K is in units of m^2.

13.3.2 One-Dimensional Flow

Let's now study the problem of a soil layer sitting in the sun and drying from the inside out or sitting in the rain and getting wet from the outside in. The question here is: What is the change in water stress as a function of time and depth in the soil layer? The assumptions are:

1. The soil is uniform with the same hydraulic conductivity in all directions.
2. The soil is unsaturated, with both water and air present
3. The water is incompressible, but the volume of water in a given soil element can change with time.
4. The soil is not changing volume.
5. Darcy's law governs the water flow through the soil.
6. The change in elevation head is negligible compared to the change in pressure head (water tension).
7. The flow is in one direction only (flow in the z direction, but no flow in the x or y directions).
8. The flow is transient (dependent on time).

The problem is solved (after Aubeny and Lytton 2004) by following a process similar to the case of the saturated soil and as described in the problem-solving method of section 11.4.5.

1. Zoom in at the element level. We select an element of soil that has an elementary area dx dz with a dimension of 1 in the y direction (Figure 13.21).
2. Considering the element of Figure 13.21, the water velocity is v_z in the z direction when it enters the element and $v_z + dv_z$ when it exits the element. It is assumed that the water does not flow in the x and y directions (one-dimensional flow).

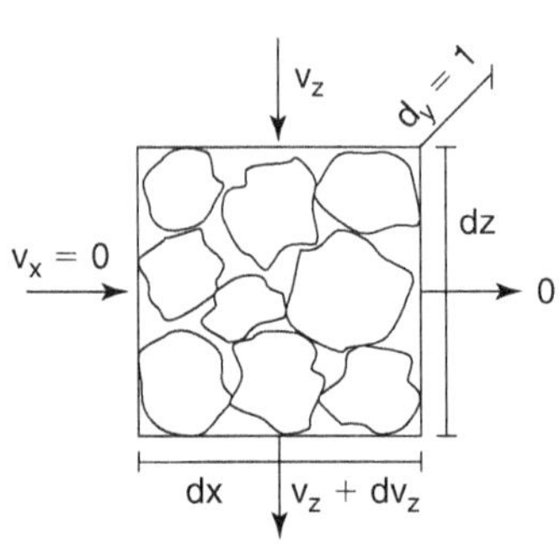

Figure 13.21 Element of unsaturated soil in the flow mass.

3. The fundamental equation in this case is the conservation of mass equation, expressing that the flow of water coming out of the element minus the flow of water entering the element is equal to the time rate of change of the volume of water in the element:

$$q(out) - q(in) = \frac{dV_w}{dt} \tag{13.68}$$

$$\frac{dv_z}{dz}\, dx\, dy\, dz = \frac{dV_w}{dt} \quad or \quad \frac{dv_z}{dz} = \frac{dV_w}{V\, dt} \tag{13.69}$$

4. The first constitutive equation links the water velocity v_z to the hydraulic conductivity k and the hydraulic gradient i_z (Darcy's law):

$$v_z = ki_z = k\frac{dh_t}{dz} \tag{13.70}$$

where h_t is the total head, which is equal to the pressure head h_p plus the elevation head h_e. It is assumed that the change in elevation head is negligible compared to the change in pressure head (water tension):

$$\boxed{dh_e \ll dh_p}$$

Then Eq. 13.70 can be written:

$$v_z = k\frac{dh_p}{dz} \quad (13.71)$$

5. The second constitutive equation describes how the hydraulic conductivity varies with the water stress (suction). Here the Laliberte and Corey model is used, with an exponent n equal to 1, which is not unreasonable but is particularly convenient mathematically:

$$k = \frac{k_0 h_{p0}}{h_p} \quad (13.72)$$

Therefore,

$$v_z = k_0 h_{p0}\frac{dh_p/h_p}{dz} = k_0 h_{p0}\frac{d(\log_e h_p)}{dz} \quad (13.73)$$

6. The third constitutive equation describes how the water content varies with the water stress (suction):

$$dw = c\, d(\log_{10} h_p) = cd(0.434 \log_e h_p) \quad (13.74)$$

where c is the slope of the soil water retention curve (SWRC).

7. Then Eqs. 13.69 to 13.74 are regrouped while making use of a change of variable and phase relationships to obtain the governing differential equation:

Change of variable

$$u = \log_{10} h_p = 0.434 \log_e h_p \quad (13.75)$$

Phase relationship

$$V_w = \frac{W_w}{\gamma_w} = \frac{W_w}{\gamma_w}\frac{W_s}{W_s} = w\frac{\gamma_d}{\gamma_w}V \quad (13.76)$$

$$\frac{dv_z}{dz} = 2.3k_0 h_{p0}\frac{d^2u}{dz^2} = \frac{dw}{dt}\frac{\gamma_d}{\gamma_w} = c\frac{du}{dt}\frac{\gamma_d}{\gamma_w} \quad (13.77)$$

$$\frac{du}{dt} = \frac{2.3k_0 h_{p0}\gamma_w}{c\gamma_d}\frac{d^2u}{dz^2} \quad (13.78)$$

With the diffusivity

$$\alpha = \frac{2.3k_0 h_{p0}\gamma_w}{c\gamma_d} \quad (13.79)$$

Then

$$\frac{du}{dt} = \alpha\frac{d^2u}{dz^2} \quad (13.80)$$

Note that this is the same equation as the one-dimensional consolidation equation for a saturated soil, except that u is the $\log_{10}$ of the water stress expressed as a height of water rather than being the excess water stress itself.

8. Now the space and time boundary conditions must be addressed. Let's assume that the entire semi-infinite layer is at an initial water tension stress u_{wi} at a time equal to zero and that the top of the layer is suddenly subjected to a wet condition that permanently imposes a much lower water tension $u_{w(z=0)}$ at that boundary (ground surface).
9. We define the degree of wetting U at any depth z as:

$$U = \frac{u_{w(z,t)} - u_{wi}}{u_{w(z=0,t)} - u_{wi}} \quad (13.81)$$

where $u_{w(z,t)}$ is the water tension at a depth z and a time t, u_{wi} is the initial water tension throughout the layer, and $u_{w(z=0,t)}$ is the wetting value of the water tension permanently applied at the ground surface. We also define the time factor T as:

$$T = \alpha\frac{t}{z^2} \quad (13.82)$$

The solution to the governing differential equation is given in this case by the complementary error function, as follows:

$$U = erfc\left(\frac{1}{2\sqrt{T}}\right) \quad (13.83)$$

Figure 13.22 shows that function.

The average degree of wetting represents the ratio of the area under the water tension (in excess of the wetting value) vs. depth profile at time t over the same area at time t = 0 (Figure 13.22). Another way to present the results is shown in Figure 13.23, where the evolution of the water tension toward the imposed wet value at the surface is shown as a function

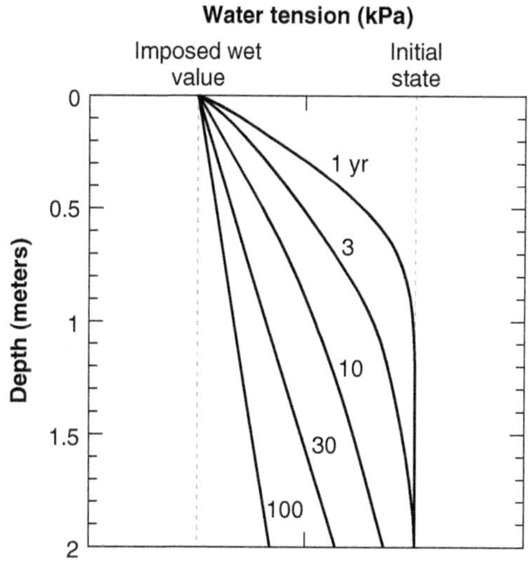

Figure 13.22 Decrease in water tension with depth in an initially high-water-tension soil layer subjected to wetting at the ground surface. (After Aubeny and Lytton 2004)

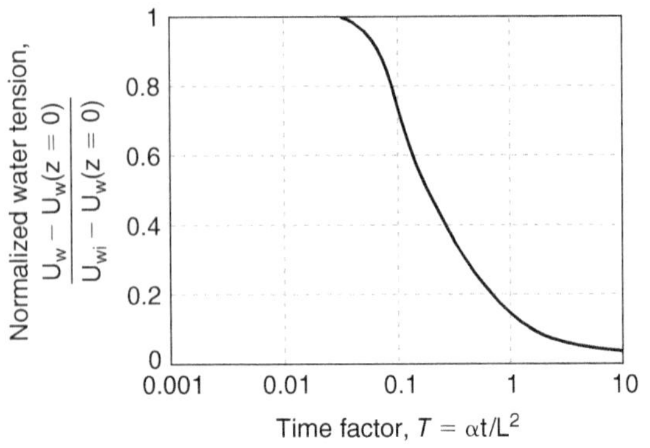

Figure 13.23 Decrease in water tension with time at a chosen depth. (After Aubeny and Lytton 2004)

of time for a given depth z. In that figure the water tension is normalized as:

$$U' = \frac{u_{w(z,t)} - u_{w(z=0,t)}}{u_{wi} - u_{w(z=0,t)}} \tag{13.84}$$

13.3.3 Three-Dimensional Water Flow

The previous example was of a one-dimensional flow of water perpendicular to the surface of an unsaturated soil. Let's look now at what happens in a three-dimensional case. The assumptions are:

1. The flow is in three directions (x, y, z).
2. The flow is transient (dependent on time).
3. The soil is uniform with different hydraulic conductivities in the x, y, and z directions.
4. The soil is unsaturated, with both water and air present.
5. The water is incompressible, but the volume of water in a given soil element can change with time.
6. The soil is not changing volume.
7. Darcy's law governs the water flow through the soil.
8. The hydraulic conductivity k is a function of the water tension u_w.

The problem is solved by following a process similar to the case of the saturated soil and as described in the problem-solving method of section 11.4.5:

1. Zoom in at the element level. We select an element of soil that has an elementary volume V = dx dy dz.
2. The water velocity is v_x in the x direction when it enters the element and $v_x + \frac{\partial v_x}{\partial x} dx$ when it exits the element. The same applies in the y and the z directions (three-dimensional flow).
3. The fundamental equation in this case is the conservation of mass equation, expressing that the flow of water coming out of the element minus the flow of water entering the element is equal to the time rate of change of the volume of water in the element:

$$q(out) - q(in) = \frac{dV_w}{dt} \tag{13.85}$$

Or, using Q = vA on all faces of the element,

$$\frac{\partial v_x}{\partial x} + \frac{\partial v_y}{\partial y} + \frac{\partial v_z}{\partial z} = \frac{dV_w}{V\,dt} \tag{13.86}$$

4. The first constitutive equation links the water velocity v_x to the hydraulic conductivity k_{wx} and the hydraulic gradient i_x (Darcy's law):

$$v_x = k_{wx} i_x = k_{wx} \frac{dh_t}{dx} \tag{13.87}$$

where h_t is the total head equal to the pressure head h_p plus the elevation head h_e, which is the coordinate z. Then Eq. 13.87 can be written:

$$v_x = k_{wx} \frac{d(h_p + z)}{dx} = k_{wx} \frac{dh_p}{dx} \quad and \quad v_y = k_{wy} \frac{dh_p}{dy}$$

but

$$v_z = k_{wz} \frac{d(h_p + z)}{dz} = k_{wz} \left(\frac{dh_p}{dz} + 1 \right) \tag{13.88}$$

5. The second constitutive equation describes how the hydraulic conductivity varies with the water stress (suction). Here several models could be selected, but in general suffice to say that:

$$k_w = k_w(h_p) \tag{13.89}$$

6. The third constitutive equation describes how the water content varies with the water stress (suction). If a linear semilog model is accepted for this part of the soil water retention curve, then:

$$dw = c\, d(\log_{10} u_w) \tag{13.90}$$

where c is the slope of the SWRC. Using the phase relationship of Eq. 13.76, the following expression is obtained for the term on the right-hand side of Eq. 13.86:

$$\frac{dV_w}{V\,dt} = \frac{\gamma_d}{\gamma_w} \frac{dw}{dt} \tag{13.91}$$

7. Then Eqs. 13.86 to 13.91 are regrouped to obtain the governing differential equation. In this process the pressure head is transformed into the water tension by using:

$$u_w = \gamma_w h_p \tag{13.92}$$

The governing differential equation is then

$$\frac{\partial}{\partial x}\left(k_{wx}\left(u_w\right)\frac{\partial u_w}{\partial x}\right)+\frac{\partial}{\partial y}\left(k_{wy}\left(u_w\right)\frac{\partial u_w}{\partial y}\right)+\frac{\partial}{\partial z}\left(k_{wz}\left(u_w\right)\left(\frac{\partial u_w}{\partial z}+1\right)\right)=c\gamma_d\frac{\partial(\log_{10}u_w)}{\partial t} \quad (13.93)$$

8. Now the space and time boundary conditions must be addressed and the differential equation can be solved. The solution is the function that describes the water tension u_w for any location (x, y, z) and any time t.

13.3.4 Three-Dimensional Air Flow

Section 13.3.3 gave the steps for generating the governing differential equation for the water flow in an unsaturated soil. Now we need to repeat the process for the flow of air in the unsaturated soil. As shown in Figure 13.17, the water hydraulic conductivity decreases when the water tension increases. At the same time, the air hydraulic conductivity increases when the water tension increases, because more of the void space is occupied by air. For the flow of water we made a distinction between the discharge velocity v and the seepage velocity v_s, which were related as follows in the case of unsaturated flow:

$$v_w A_t = v_{ws} A_w \quad or \quad v_w = Snv_{ws} \quad (13.94)$$

For the air flow the relationship becomes:

$$v_a A_t = v_{as} A_a \quad or \quad v_a = (1-S)nv_{as} \quad (13.95)$$

For a degree of saturation of about 85% or more, the air is usually occluded, so we cannot talk about air flow; rather, we address diffusion of the air mass through the water in the soil voids. For a degree of saturation of 20% or less, the air hydraulic conductivity approaches its maximum value.

Blight (1971) and Fredlund and Rahardjo (1993) showed that Darcy's law is applicable to the flow of air in soils, and related the air velocity to the gradient of the total head in the air by:

$$v_{ax} = k_a \frac{\partial h_{ta}}{\partial x} \quad (13.96)$$

The pressure head h_{pa} is related to the air pressure u_a by:

$$u_a = \gamma_a h_{pa} \quad (13.97)$$

Note that the unit weight of air γ_a varies with temperature and pressure (Table 13.3); at 20°C and at atmospheric pressure, γ_a is 11.8 N/m^3.

To develop the solution for the three-dimensional flow of air in soil, we follow the same procedure as in the case of water:

Table 13.3 Unit Weight of Air

Temperature (°C)	Pressure (atm)	Unit weight of air (N/m^3)	Mass density (kg/m^3)
−10	1	13.17	1.341
0	1	12.67	1.316
10	1	12.23	1.247
20	1	11.81	1.204
30	1	11.43	1.164
40	1	11.05	1.127

1. Zoom in at the element level. We select an element of soil that has an elementary volume $V = dx\,dy\,dz$.

2. The air velocity is v_{ax} in the x direction when it enters the element and $v_{ax} + \frac{\partial v_{ax}}{\partial x}dx$ when it exits the element. The same applies in the y and the z directions (three-dimensional air flow).

3. Writing the conservation of mass principle for air poses a problem a bit different from the conservation of mass for water. Because water is considered incompressible at usual pressures, conservation of mass is also conservation of volume, which is what we used for the fundamental equation for water. However, because air is very compressible, the mass of air in a given volume could be very different depending on temperature and pressure. We write that the air mass exiting the element minus the air mass entering the element is equal to the change in air mass corresponding to a decrease in volume of the soil pores of the element over time:

$$\frac{\partial(\rho_a v_{ax})}{\partial x}+\frac{\partial(\rho_a v_{ay})}{\partial y}+\frac{\partial(\rho_a v_{az})}{\partial z}=\frac{1}{V}\frac{\partial(\rho_a V_a)}{\partial t} \quad (13.98)$$

where ρ_a is the mass density of air, V_a is the volume of air in the element, and V is the volume of the soil element.

4. The first constitutive equation is Darcy's law for air flow, which is written for each direction:

$$v_{ax} = \frac{k_{ax}}{\gamma_a}\frac{\partial u_a}{\partial x} \quad \text{and} \quad v_{ay} = \frac{k_{ay}}{\gamma_a}\frac{\partial u_a}{\partial y}$$

$$\text{but} \quad v_{az} = \frac{k_{az}}{\gamma_a}\left(\frac{\partial u_a}{\partial z}+\rho_a g\right) \quad (13.99)$$

where u_a is the air pressure. The second term in the z direction indicates the influence of gravity on the air flow. Note that:

$$\frac{k_{ax}}{\gamma_a} = \frac{K_{ax}}{\mu_a} \quad (13.100)$$

where K is the intrinsic hydraulic conductivity and μ_a is the viscosity of air.

5. The term on the right-hand side of Eq. 13.98 can be rewritten as follows by using phase relationships:

$$\frac{1}{V}\frac{\partial(\rho_a V_a)}{\partial t} = \frac{\partial(\rho_a n(1-S))}{\partial t} \tag{13.101}$$

6. The ideal gas law is an additional constitutive equation describing how the air density ρ_a varies with the air pressure u_a and temperature T:

$$\rho_a = \frac{\omega_a}{RT}u_a \tag{13.102}$$

where ω_a is the molecular weight of air (kg/mol), R is the universal gas constant (J/mol.K), and T is the absolute temperature (K).

7. Then Eqs. 13.98 to 13.102 are regrouped to obtain the governing differential equation for the flow of air through a soil:

$$\frac{\partial\left(\frac{\omega_a}{RT}u_a\frac{K_{ax}}{\mu_a}\frac{\partial u_a}{\partial x}\right)}{\partial x} + \frac{\partial\left(\frac{\omega_a}{RT}u_a\frac{K_{ay}}{\mu_a}\frac{\partial u_a}{\partial y}\right)}{\partial y} + \frac{\partial\left(\frac{\omega_a}{RT}u_a\frac{K_{az}}{\mu_a}\left(\frac{\partial u_a}{\partial z}+\frac{\omega_a}{RT}u_a g\right)\right)}{\partial z} = \frac{\partial\left(\frac{\omega_a}{RT}u_a n(1-S)\right)}{\partial t} \tag{13.103}$$

8. Now the space and time boundary conditions must be addressed and the differential equation can be solved. The solution is the function that describes the air pressure u_a for any location (x, y, z) and any time t.

PROBLEMS

13.1 A soil has a porosity of 40%.
   a. The soil is saturated and water flows through the soil. Calculate the ratio between the discharge velocity v and the seepage velocity v_s.
   b. The soil is unsaturated, with a degree of saturation equal to 35%. Calculate the ratio between the discharge velocity v and the seepage velocity v_s.

13.2 Water is flowing through three soil layers as shown in Figure 13.1s. The cross section is a square with sides of 100 mm. The hydraulic conductivity of each soil layer is given in Table 13.1s.
   a. What is the equivalent hydraulic conductivity of the three layers?
   b. Determine the flow rate exiting the system.
   c. Determine the elevation head diagram, the total head diagram, and the pressure head diagram from point A to point D.

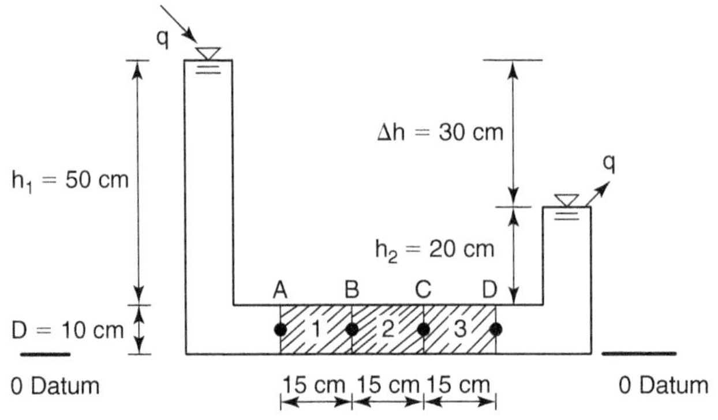

Figure 13.1s Three-layer permeameter.

Table 13.1s Hydraulic Conductivity of Three Soil Layers

Soil	Hydraulic conductivity (m/s)
1	1×10^{-4}
2	5×10^{-6}
3	3×10^{-5}

13.3 Water is flowing through three soil layers as shown in Figure 13.2s. The cross-section is a square with sides of 100 mm. The hydraulic conductivity of each soil layer is given in Table 13.1s.

a. What is the equivalent hydraulic conductivity of the three layers?
b. Determine the flow rate exiting the system.
c. Determine the elevation head diagram, the total head diagram, and the pressure head diagram from point A to point D.

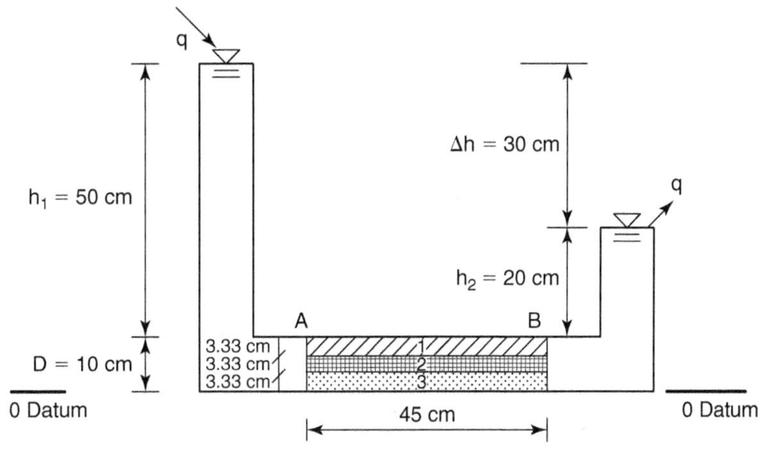

Figure 13.2s Three-layer permeameter.

13.4 Use the uplift force equation (Eq. 13.41) to calculate the uplift force on a ship and demonstrate Archimedes' principle.

13.5 Referring to Figure 13.3s, calculate the following quantities:

a. Elevation head, total head, and pressure head at point M on Figure 13.3s*a*
b. The quantity of water seeping through the dam of Figure 13.3s*a* per day
c. Elevation head, total head, and pressure head at points A, B, and C on Figure 13.3s*b*
d. The uplift force on the bottom of the concrete dam in Figure 13.3s*b*
e. The hydraulic gradient between points A and B and then between points C and D on Figure 1.3s*c*
f. The factor of safety against a quick condition on the exit face of the cofferdam (Figure 13.3s*c*) by the exit gradient method and the critical block method
g. The seepage force applied by the water on a soil grain on the exit face of the slope if the grain has a volume of 1 mm^3 (Figure 13.3s*d*)
h. The water pressure distribution behind the retaining wall of Figure 13.3s*e*

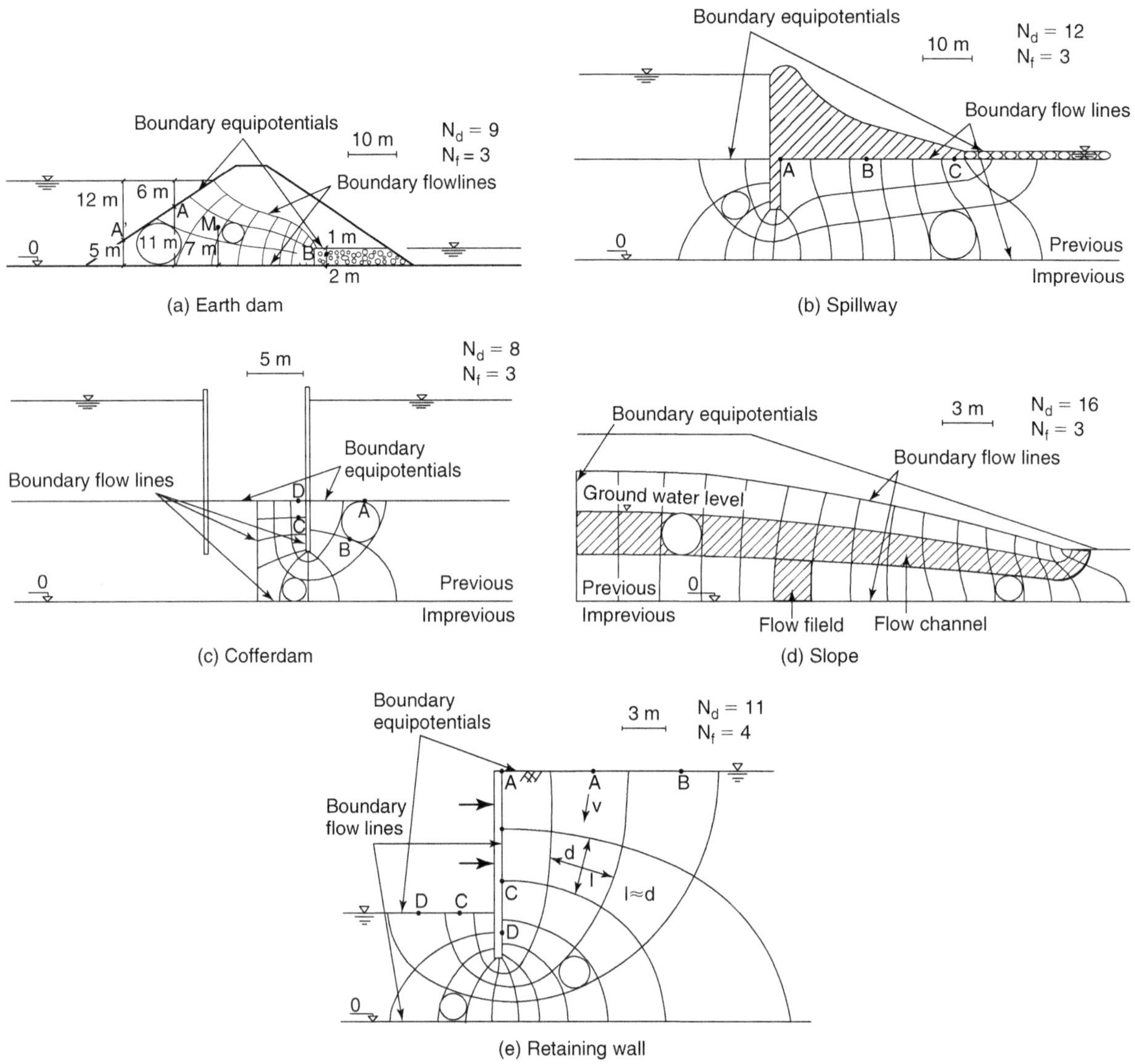

Figure 13.3s Flow nets.

13.6 A tube is filled with a relatively dry soil at a water tension corresponding to a pressure head h_0 and a volumetric water content θ_0. Water is made available at one end of the tube (Figure 13.4s). As a result, a wetting front is created and advances from left to right on the figure. The wetted soil has a water tension corresponding to a pressure head of h_1 and a volumetric water content of θ_1. How fast will the wetting front propagate across the sample?

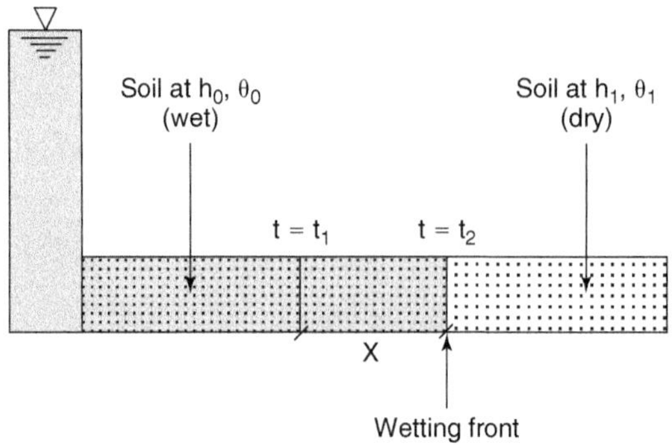

Figure 13.4s Horizontal wetting front propagation.

13.7 A soil sample has a saturated hydraulic conductivity k_{sat} equal to 10^{-8} m/s. Estimate the hydraulic conductivity of the sample if it dries to a degree of saturation equal to 0.9 and then 0.5. Use Figure 13.5s to estimate k_{unsat}.

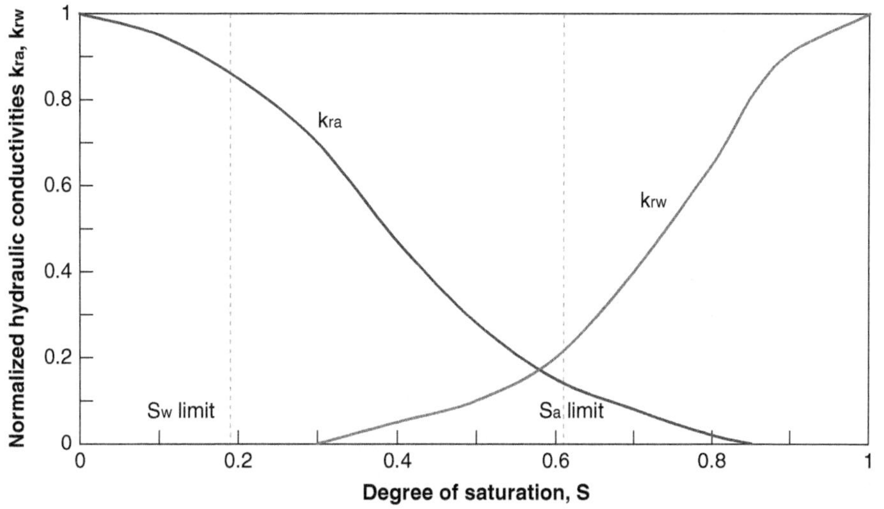

Figure 13.5s Relative hydraulic conductivity of water and air as a function of degree of saturation.

Problems and Solutions

Problem 13.1

A soil has a porosity of 40%.

a. The soil is saturated and water flows through the soil. Calculate the ratio between the discharge velocity v and the seepage velocity v_s.
b. The soil is unsaturated, with a degree of saturation equal to 35%. Calculate the ratio between the discharge velocity v and the seepage velocity v_s.

Solution 13.1

a. For a saturated soil, the relation between seepage and discharge velocity is:

$$v = nv_s$$

Given $n = 40\%$, we have:

$$v = 0.4v_s$$
$$\frac{v}{v_s} = 0.4$$

b. For an unsaturated soil, the relation between seepage and discharge velocity is:

$$v_w = Snv_{ws}$$

Given n = 40% and $S = 35\%$, then:

$$\frac{v_w}{v_{ws}} = Sn = 0.35 \times 0.4 = 0.14$$

Problem 13.2

Water is flowing through 3 soil layers as shown in Figure 13.1s. The cross section is a square with sides of 100 mm. The hydraulic conductivity of each soil layer is given in Table 13.1s.

a. What is the equivalent hydraulic conductivity of the three layers?
b. Determine the flow rate exiting the system.
c. Determine the elevation head diagram, the total head diagram, and the pressure head diagram from point A to point D.

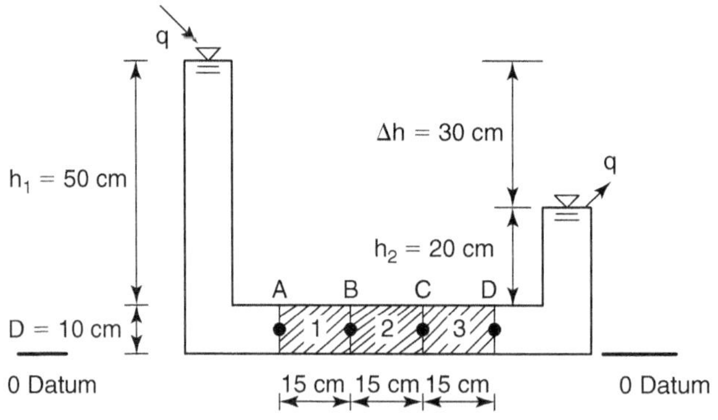

Figure 13.1s Three-layer permeameter.

Table 13.1s Hydraulic Conductivity of Three Soil Layers

Soil	Hydraulic conductivity (m/s)
1	1×10^{-4}
2	5×10^{-6}
3	3×10^{-5}

Solution 13.2

This is a problem about water flowing perpendicularly to the soil layers. The equivalent hydraulic conductivity is calculated as:

$$k_{eq} = \frac{\Sigma H_i}{\Sigma \frac{H_i}{k_i}} = \frac{0.45\text{ m}}{\frac{0.15\text{ m}}{k_1} + \frac{0.15\text{ m}}{k_2} + \frac{0.15\text{ m}}{k_3}} = \frac{0.45\text{ m}}{\frac{0.15\text{ m}}{1 \times 10^{-4}\text{ m/s}} + \frac{0.15\text{ m}}{5 \times 10^{-6}\text{ m/s}} + \frac{0.15\text{ m}}{3 \times 10^{-5}\text{ m/s}}} = 1.23 \times 10^{-5}\text{ m/s}$$

The flow rate is calculated as:

$$q = vA = k_{eq} iA = k_{eq} \frac{\Delta h}{L} D^2 = 1.23 \times 10^{-5}\text{ m/s} \times \frac{0.3\text{ m}}{0.45\text{ m}} \times 0.1^2\text{ m}^2 = 8.2 \times 10^{-8}\text{ m}^3/\text{s}$$

The zero datum is set at the bottom of the soil layer. The elevation heads at points A, B, C, and D are the same:

$$h_{eA} = h_{eB} = h_{eC} = h_{eD} = 5\text{ cm}$$

The total head at point A is calculated as:

$$h_{tA} = h_e + h_p = 5\text{ cm} + 55\text{ cm} = 60\text{ cm}$$

The total head at point B is calculated as:

$$q = vA = k_1 iA = k_1 \frac{\Delta h_{AB}}{L_{AB}} D^2 = 1 \times 10^{-4}\,\text{m/s} \times \frac{\Delta h_{AB}}{0.15\,\text{m}} \times 0.1^2\,\text{m}^2 = 8.2 \times 10^{-8}\,\text{m}^3/\text{s}$$

$$\Delta h_{AB} = 0.012\,\text{m} = 1.2\,\text{cm}$$

Because $h_{tA} = 60$ cm from the previous calculation, the total head at point B is:

$$h_{tB} = h_{tA} - \Delta h_{AB} = 60\text{ cm} - 1.2\text{ cm} = 58.8\text{ cm}$$

The total head at point C is calculated as:

$$q = vA = k_2 iA = k_2 \frac{\Delta h_{BC}}{L_{BC}} D^2 = 5 \times 10^{-6}\text{ m/s} \times \frac{\Delta h_{BC}}{0.15\text{ m}} \times 0.1^2\text{ m}^2 = 8.2 \times 10^{-8}\text{ m}^3/\text{s}$$

$$\Delta h_{BC} = 0.246\text{ m} = 24.6\text{ cm}$$

Because $h_{tB} = 58.8$ cm from the previous calculation, the total head at point C is:

$$h_{tC} = h_{tB} - \Delta h_{BC} = 58.8\text{ cm} - 24.6\text{ cm} = 34.2\text{ cm}$$

The total head at point D is calculated as:

$$q = vA = k_3 iA = k_3 \frac{\Delta h_{CD}}{L_{CD}} D^2 = 3 \times 10^{-5}\text{ m/s} \times \frac{\Delta h_{CD}}{0.15\text{ m}} \times 0.1^2\text{ m}^2 = 8.2 \times 10^{-8}\text{ m}^3/\text{s}$$

$$\Delta h_{CD} = 0.041\text{ m} = 4.1\text{ cm}$$

Because $h_{tc} = 34.2$ cm from the previous calculation, the total head at point D is:

$$h_{tD} = h_{tC} - \Delta h_{CD} = 34.2\text{ cm} - 4.1\text{ cm} = 30\text{ cm}$$

The pressure head can be obtained by subtracting the elevation head from the total head at each point:

$$h_{pA} = h_{tA} - h_{eA} = 60\text{ cm} - 5\text{ cm} = 55\text{ cm}$$
$$h_{pB} = h_{tB} - h_{eB} = 58.8\text{ cm} - 5\text{ cm} = 53.8\text{ cm}$$
$$h_{pC} = h_{tC} - h_{eC} = 34.2\text{ cm} - 5\text{ cm} = 29.2\text{ cm}$$
$$h_{pD} = h_{tD} - h_{eD} = 30\text{ cm} - 5\text{ cm} = 25\text{ cm}$$

The elevation head diagram, the total head diagram, and the pressure head diagram from point A to point D are plotted in Figure 13.6s.

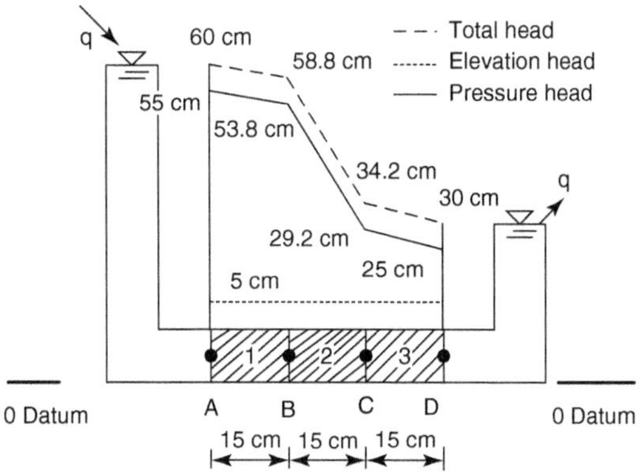

Figure 13.6s Elevation head, total head, and pressure head diagram from point A to point D.

Problem 13.3

Water is flowing through three soil layers as shown in Figure 13.2s. The cross section is a square with sides of 100 mm. The hydraulic conductivity of each soil layer is given in Table 13.1s.

a. Determine the flow rate exiting the system.
b. Determine the elevation head diagram, the total head diagram, and the pressure head diagram from point A to point D.
c. What is the equivalent hydraulic conductivity of the three layers?

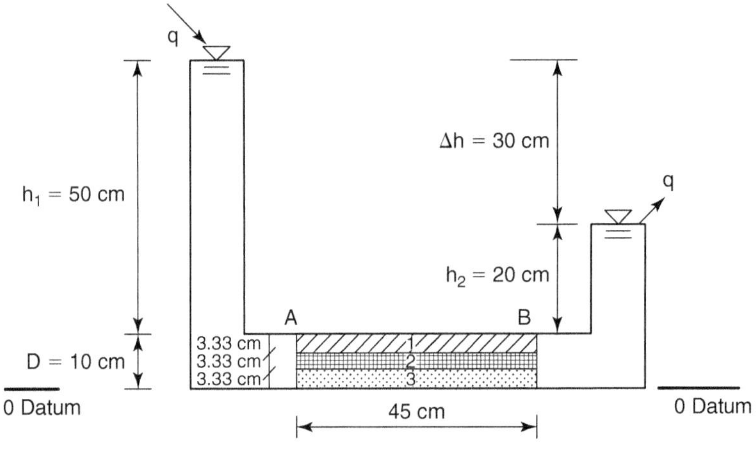

Figure 13.2s Three-layer permeameter.

Solution 13.3

a. This is a problem about water flowing in the direction of the soil layer boundaries. The equivalent hydraulic conductivity is calculated as:

$$k_{eq} = \frac{k_1 + k_2 + k_3}{3} = \frac{(100 + 5 + 30)}{3} \times 10^{-6} = 4.5 \times 10^{-5} \text{ m/s}$$

b. The flow rate is calculated as:

$$q = vA = k_{eq} iA = k_{eq} \frac{\Delta h}{L} D^2 = 4.5 \times 10^{-5} \text{ m/s} \times \frac{0.3 \text{ m}}{0.45 \text{ m}} \times 0.1^2 \text{ m}^2 = 3 \times 10^{-7} \text{ m}^3/\text{s}$$

c. The zero datum is set at the bottom of the soil layer. The total head at point A is calculated as:

$$h_{tA} = h_e + h_p = 5 \text{ cm} + 55 \text{ cm} = 60 \text{ cm}$$

The total head at point B is calculated as:

$$q = vA = k_{eq} iA = k_3 \frac{\Delta h_{AB}}{L_{AB}} D^2 = 4.5 \times 10^{-5} \text{ m/s} \times \frac{\Delta h_{AB}}{0.45 \text{ m}} \times 0.1^2 \text{ m}^2 = 3 \times 10^{-7} \text{ m}^3/\text{s}$$

$$\Delta h_{AB} = 0.3 \text{ m} = 30 \text{ cm}$$

$$h_{tB} = h_{tA} - \Delta h_{AB} = 60 \text{ cm} - 30 \text{ cm} = 30 \text{ cm}$$

The total head at point B can be found by simply using right side condition.

The elevation head diagram, the total head diagram, and the pressure head diagram from point A to point B are plotted in Figure 13.7s.

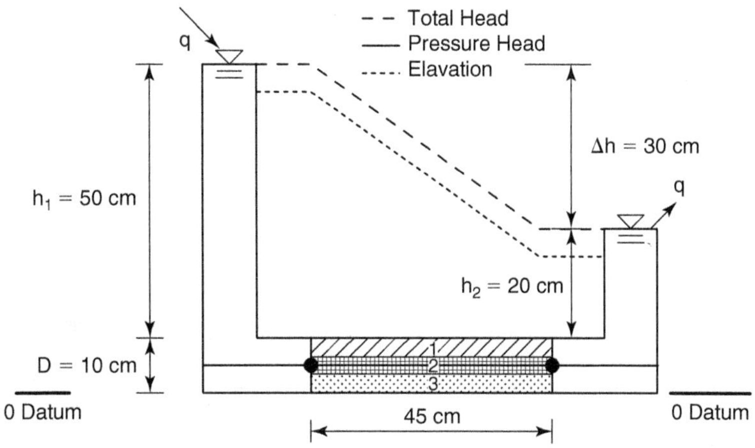

Figure 13.7s Total head, pressure head, and elevation diagram.

Problem 13.4

Use the uplift force equation (Eq. 13.41) to calculate the uplift force on a ship and demonstrate Archimedes' principle.

Solution 13.4

Consider that the height of the ship under water is Z and the total height of the ship is H. The width of the ship is B and the length is L.

Uplift water pressure on the bottom of the ship is:

$$u_{av} = \gamma_w Z$$

Uplift force would be:

$$F_{up} = u_{av} BL \Rightarrow F_{up} = \gamma_w ZBL$$

This is Archimedes' principle, which states that the upward buoyant force exerted on a body immersed in a fluid is equal to the weight of the fluid displaced by the body.

Problem 13.5

Referring to Figure 13.3s, calculate the following quantities:

a. Elevation head, total head, and pressure head at point M on Figure 13.3s*a*
b. The quantity of water seeping through the dam of Figure 13.3s*a* per day
c. Elevation head, total head, and pressure head at points A, B, and C on Figure 13.3s*b*
d. The uplift force on the bottom of the concrete dam in Figure 13.3s*b*
e. The hydraulic gradient between points A and B and then between points C and D on Figure 13.3s*c*
f. The factor of safety against a quick condition on the exit face of the cofferdam (Figure 13.3s*c*) by the exit gradient method and the critical block method
g. The seepage force applied by the water on a soil grain on the exit face of the slope if the grain has a volume of 1 mm^3 (Figure 13.3s*d*)
h. The water pressure distribution behind the retaining wall of Figure 13.3s*e*

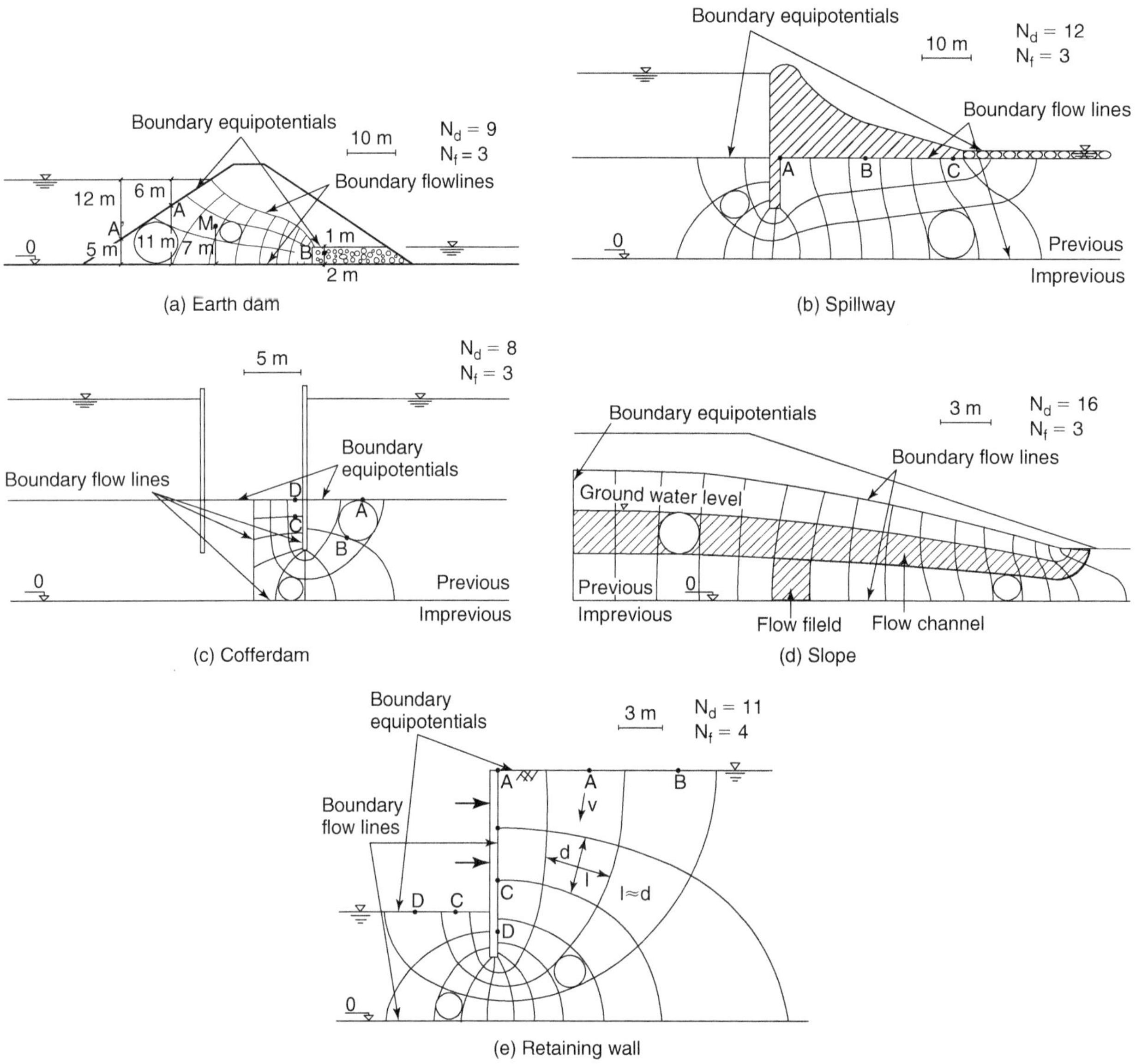

Figure 13.3s Flow nets.

Solution 13.5

a. Elevation head, total head, and pressure head at point M on Figure 13.3s*a*

The head loss between each equipotential line:

$$\Delta h = \frac{\Delta H}{N_d} = \frac{17 - 3}{9} = 1.56 \text{ m}$$

The total head:

$$h_t = \Delta H_t - \Delta h \times (N_d)_B = 17 - 1.56 \times 2 = 13.9 \text{ m}$$

The elevation head at M:

$$h_e = 7 \text{ m}$$

The pressure head at M:

$$(h_p)_B = h_t - h_e = 13.9 - 7 = 6.9 \text{ m}$$

The pore water pressure at M:

$$u_B = h_p \times \gamma_w = 6.9 \times 9.81 = 67.69 \text{ kN/m}^2 = 67.69 \text{ kPa}$$

b. The quantity of water seeping through the dam of Figure 13.3s*a* per day if the hydraulic conductivity of the soil is 10^{-8} m/s:

$$q = k\Delta H \frac{N_f}{N_d} = 10^{-8} \times (17 - 3) \times \frac{3}{9} = 4.67 \times 10^{-8} \text{ m}^3/\text{ sec /m}$$

$$Q = q \times L = \left(k\Delta H \frac{N_f}{N_d}\right) L = 4.67 \times 10^{-8} \times 1 \text{ m} = 4.67 \times 10^{-8} \text{ m}^3/\text{ sec}$$

$$Q_{1day} = 4.67 \times 10^{-8} \times (24 \times 60 \times 60) = 0.00403 \text{ m}^3$$

c. Elevation head, total head, and pressure head at points A, B, and C on Figure 13.3s*b*
The head loss between each equipotential line:

$$\Delta h = \frac{\Delta H}{N_d} = \frac{16 - 2}{12} = 1.17 \text{ m}$$

The total head at each point:

$$(h_t)_A = \Delta H_t - \Delta h \times (N_d)_B = (20 + 16) - 1.17 \times 5.5 = 29.57 \text{ m}$$

$$(h_t)_B = \Delta H_t - \Delta h \times (N_d)_B = (20 + 16) - 1.17 \times 7.5 = 27.23 \text{ m}$$

$$(h_t)_C = \Delta H_t - \Delta h \times (N_d)_B = (20 + 16) - 1.17 \times 10.5 = 23.72 \text{ m}$$

The elevation head at A, B, and C:

$$h_e = 20 \text{ m}$$

The pressure head at each point:

$$(h_p)_A = h_t - h_e = 29.57 - 20 = 9.57 \text{ m}$$

$$(h_p)_B = h_t - h_e = 27.23 - 20 = 7.23 \text{ m}$$

$$(h_p)_B = h_t - h_e = 23.72 - 20 = 3.72 \text{ m}$$

The pore water pressure at each point:

$$u_A = h_p \times \gamma_w = 9.57 \times 9.81 = 93.88 \text{ kN/m}^2 = 93.88 \text{ kPa}$$

$$u_B = h_p \times \gamma_w = 7.23 \times 9.81 = 70.93 \text{ kN/m}^2 = 70.93 \text{ kPa}$$

$$u_C = h_p \times \gamma_w = 3.72 \times 9.81 = 36.49 \text{ kN/m}^2 = 36.49 \text{ kPa}$$

d. The uplift force on the bottom of the concrete dam in Figure 13.3s*b*
The pressure head at each point:

$$h_p = h_t - h_e = \Delta H - N_d \times \Delta h - h_e$$

$$u = h_p \times 9.81$$

$$u_A = 9.57 \times 9.81 = 93.88 \text{ kPa}$$

$$u_B = 7.23 \times 9.81 = 70.93 \text{ kPa}$$

$$u_C = 3.72 \times 9.81 = 36.49 \text{ kPa}$$

$$h_e \text{ at end of wall} = (20 - 10) \text{ m}$$

$$(h_p)_{end\ of\ wall} = (16 + 20) - 4 \times 1.17 - (20 - 10) = 21.32 \text{ m}$$

$$\therefore u_{end\ of\ wall} = 21.32 \times 9.81 = 209.15 \text{ kPa}$$

The resultant uplift force is:

$$F_{up} = (A_{wall} \times u_{wall}) + A_{AB} \times u_{AB} + A_{BC} \times u_{BC}$$
$$= ((2 \times 1) \times 209.15) + (16 \times 1) \times \frac{93.88 + 70.93}{2} + (16 \times 1) \times \frac{70.93 + 36.49}{2}$$
$$= 2596.14 \text{ kN/m}$$

e. The hydraulic gradient between points A and B and then between points C and D on Figure 13.3s*c*
The head loss between each equipotential line:

$$\Delta h = \frac{\Delta H}{N_d} = \frac{8}{8} = 1 \text{ m}$$

The total head:

$$(h_t)_A = \Delta H_t - \Delta h \times (N_d)_A = (8 + 8) - 1 \times 0 = 16 \text{ m}$$
$$(h_t)_B = \Delta H_t - \Delta h \times (N_d)_B = (8 + 8) - 1 \times 1 = 15 \text{ m}$$
$$(h_t)_C = \Delta H_t - \Delta h \times (N_d)_C = (8 + 8) - 1 \times 7 = 9 \text{ m}$$
$$(h_t)_D = \Delta H_t - \Delta h \times (N_d)_D = (8 + 8) - 1 \times 8 = 8 \text{ m}$$

Hydraulic gradient between points A and B:

$$i = \frac{\Delta h}{L_{AB}} = \frac{16 - 15}{3.5} = 0.29$$

Hydraulic gradient between points C and D:

$$i = \frac{\Delta h}{L_{CD}} = \frac{9 - 8}{1.5} = 0.67$$

f. The factor of safety against a quick condition on the exit face of the cofferdam (Figure 13.3s*c*) by the exit gradient method and the critical block method

Exit Gradient

The critical hydraulic gradient (i_c):

$$i_c = \frac{\gamma_{sat} - \gamma_w}{\gamma_w} = \frac{20 - 9.81}{9.81} = 1.04$$

Determine the factor of safety against quicksand and explain it

$$FOS = \frac{i_c}{i} = \frac{1.04}{0.67} = 1.55 < 4$$

Heave and Critical Block

Taking flow field CD as the critical block

$$FOS = \frac{W'}{S} = \frac{(\gamma_{sat} - \gamma_w)d}{(h_{t(C)} - h_{t(D)})\gamma_w} = \frac{(20 - 9.81) \times 1.5}{(9 - 8) \times 9.81} = 1.6$$

g. The seepage force applied by the water on a soil grain on the exit face of the slope if the grain has a volume of 1 mm^3 (Figure 13.3s*d*) The drop in total head in the last flow field on the exit face is

$$\Delta h_t = \frac{4}{16} = 0.25 \text{ m}$$

The corresponding hydraulic gradient and the force on the soil grain are

$$i = \frac{\Delta h_t}{l} = \frac{0.25}{0.625} = 0.4$$

$$F_s = i\,\gamma_w V = 0.4 \times 9.81 \times 1 \times 10^{-9} = 3.92 \times 10^{-9}\ \text{kN}$$

h. The water pressure distribution behind the retaining wall of Figure 13.3se
   The head loss between each equipotential line:

$$\Delta h = \frac{\Delta H}{N_d} = \frac{9.25}{11} = 0.841\ \text{m}$$

The total head:

$$h_{t(A)} = H_{t(beg)} - \Delta h \times (N_d)_A = 16.5 - 0.841 \times 0 = 16.5\ \text{m}$$

$$h_{t(C)} = 16.5 - 0.841 \times 2 = 14.82\ \text{m}$$

$$h_{t(D)} = 16.5 - 0.841 \times 3.5 = 13.56\ \text{m}$$

$$h_{t(bottom\ of\ wall)} = 16.5 - 0.841 \times 5.8 = 11.62\ \text{m}$$

The water pressure:

$$u_{(A)} = (h_{t(A)} - h_{e(A)}) \times \gamma_w = (16.5 - 16.5) \times 9.81 = 0\ \text{kPa}$$

$$u_{(C)} = (14.82 - 9.37) \times 9.81 = 53.46\ \text{kPa}$$

$$u_{(D)} = (13.56 - 5.87) \times 9.81 = 75.44\ \text{kPa}$$

$$u_{(bottom\ of\ wall)} = (11.62 - 4.10) \times 9.81 = 73.77\ \text{kPa}$$

Problem 13.6

A tube is filled with a relatively dry soil at a water tension corresponding to a pressure head h_0 and a volumetric water content θ_0. Water is made available at one end of the tube (Figure 13.4s). As a result, a wetting front is created and advances from left to right on the figure. The wetted soil has a water tension corresponding to a pressure head of h_1 and a volumetric water content of θ_1. How fast will the wetting front propagate across the sample?

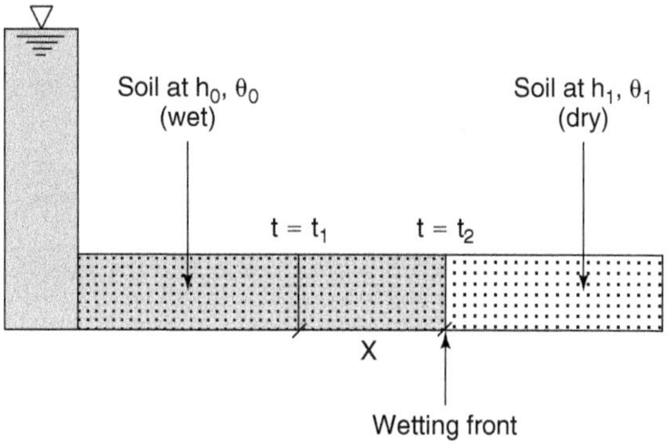

Figure 13.4s Horizontal wetting front propagation.

Solution 13.6

Consider the position of the wetting front at time t and then at time $t + dt$. The volume of water dV_w which has filled the voids during that interval of time is:

$$dV_w = (\theta_o - \theta_1)dV_t = (\theta_o - \theta_1)A\,dx$$

where A is the tube cross section and dx is the advance of the wetting front over the time dt. The corresponding flow rate is:

$$Q = \frac{dV_w}{dt} = \frac{(\theta_o - \theta_1)Adx}{dt}$$

The velocity can be obtained from the flow rate and also from Darcy's law:

$$v = \frac{Q}{A} = \frac{(\theta_o - \theta_1)dx}{dt} = k_o \frac{h_o - h_1}{x}$$

or

$$xdx = k_o \frac{h_o - h_1}{\theta_o - \theta_1} dt$$

Then the distance x is given as a function of time, as:

$$x = \sqrt{2k_o \frac{h_o - h_1}{\theta_o - \theta_1} t}$$

Problem 13.7

A soil sample has a saturated hydraulic conductivity k_{sat} equal to 10^{-8} m/s. Estimate the hydraulic conductivity of the sample if it dries to a degree of saturation equal to 0.9 and then 0.5. Use Figure 13.5s to estimate k_{unsat}.

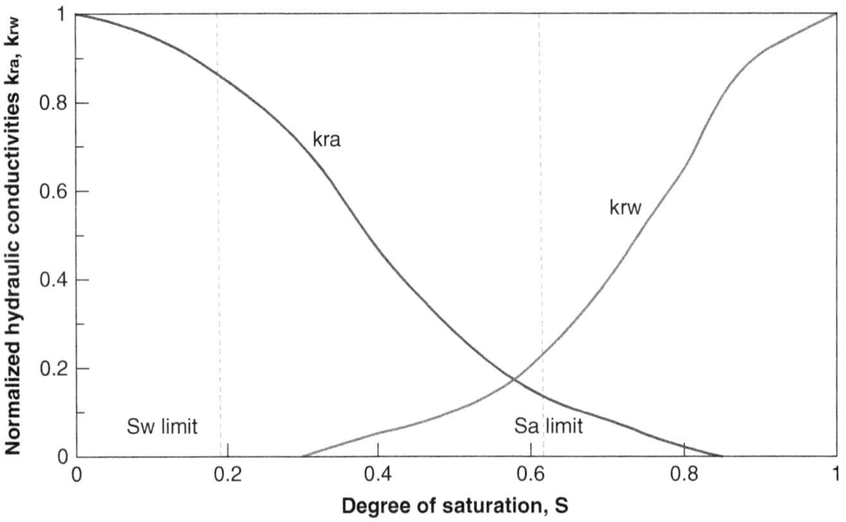

Figure 13.5s 5s Relative hydraulic conductivity of water and air as a function of degree of saturation.

Solution 13.7

From Figure 13.5s, we can find k_{rw} and use it in Eq. 13.62 to calculate the hydraulic conductivity of the soil sample in unsaturated conditions.

For S = 0.9, $k_{rw} = 0.93$, $k_{unsat} = k_{rw} x\ k_{sat}. = 0.93 \times 10^{-8}$ m/s $= 9.3 \times 10^{-9}$ m/s

For S = 0.5, $k_{rw} = 0.1$, $k_{unsat} = k_{rw} x\ k_{sat}. = 0.1 \times 10^{-8}$ m/s $= 1 \times 10^{-9}$ m/s

CHAPTER 14

Deformation Properties

Two of the major reasons for soil to deform are mechanical loading (settlement or rebound) and environmental changes (water content, temperature, chemistry). Deformations due to loading are governed by an appropriate stress-strain curve of the soil, and deformations due to environmental changes by corresponding constitutive relationships. For shrinking and swelling, for example, those relationships would be the water content-strain curve or the water tension-strain curve.

The choice of an appropriate laboratory test or in situ test and appropriate parameters to solve a particular deformation problem in the field is not easy. One of the important concepts in making that choice is to favor a laboratory test or an in situ test that duplicates the deformation condition of the soil in the field at the element level or the global level. For example, if the problem is a wide embankment over a comparatively thin layer of compressible soil, the consolidation test makes sense and is often used. Although the rigid vertical boundaries in the consolidation test do not exist in the field, the confinement created by the friction between the thin clay layer and the stronger top and bottom layers minimize the lateral movement, just as the consolidation ring does. The remaining difference leads to the need for a correction factor in the calculations. As another example, if the problem is a foundation over a deep deposit, the lateral squeezing of the soil is well represented by the horizontal deformation around the pressuremeter; thus, the pressuremeter is well suited to predicting the settlement in such a case. This approach also requires correction factors to compensate for the lack of complete correspondence.

There are several ways to quantify the deformation characteristics of a soil. One of the simplest is through a modulus of deformation.

14.1 MODULUS OF DEFORMATION: GENERAL

The shape of the stress-strain curve for a soil is typically nonlinear and depends on a number of factors. The early part of that curve can be approximated by a straight line, where the theory of linear elasticity becomes very useful. The slope of that line is related to the modulus of elasticity E (Young's modulus) and to Poisson's ratio ν. E is called *Young's modulus* after Thomas Young, a British physician and physicist who made his contribution around the turn of the 1800s. Poisson's ratio is named after Simeon Poisson, a French mathematician and physicist who lived around the turn of the 1800s and had Lagrange and Laplace as his doctoral advisors at the École Polytechnique in Paris.

Elasticity refers to the ability of a material to regain its original shape when deformed by load. That is not the case with soils, as they experience irrecoverable (plastic) deformations even at low stresses. *Linear elasticity* refers to the fact that the stress-strain curve is linear. That also is not the case with soils, as they exhibit nonlinear behavior very early in the stress-strain curve. Nevertheless, a modulus can be calculated from a soil stress-strain curve by using the secant line from the origin to the point considered on the stress-strain curve (*first load modulus*) or the slope of an unload-reload cycle loop (*cyclic modulus*). Note that the slope of the line calculated as the stress increment divided by the strain increment is not generally the modulus. This is true only if the loading is unconfined, as in an unconfined compression test. The theory of elasticity is presented in section 12.1.1. In the general case, the modulus E is given by applying the elasticity equations (Eqs. 12.1 to 12.3). In the case of a triaxial test (Figure 14.1) for example, the modulus is given by:

$$E = \frac{\sigma_1 - 2\nu\sigma_3}{\varepsilon_1} \tag{14.1}$$

where σ_1 and σ_3 are the major and minor principal stresses, ν is Poisson's ratio, and ε_1 is the major principal strain. As can be seen, the slope s is equal to E only if σ_3 is zero (unconfined compression test).

$$E = \frac{\sigma_1}{\varepsilon_1} \quad \textit{only for unconfined compression test} \tag{14.2}$$

The Poisson's ratio ν is also obtained by applying the elasticity equations (Eqs. 12.1 to 12.3). In the case of the

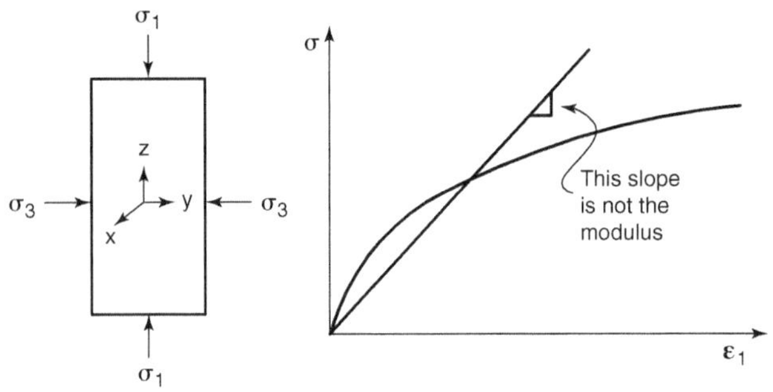

Figure 14.1 Case of a triaxial test.

triaxial test, and if the radial strain ε_3 measurements are made (rare), then:

$$\nu = \frac{-\varepsilon_3\sigma_1 + \varepsilon_1\sigma_3}{\varepsilon_1\sigma_1 + \varepsilon_1\sigma_3 - 2\varepsilon_3\sigma_3} \tag{14.3}$$

If σ_3 is zero (unconfined compression test), the Poisson's ratio is given by.

$$\nu = -\frac{\varepsilon_3}{\varepsilon_1} \quad \textit{only for unconfined compression test} \tag{14.4}$$

The minus sign indicates that when ε_1 is in compression, ε_3 is in tension and the Poisson's ratio is positive.

The stress-strain curve of a soil, and therefore the soil modulus, is influenced by state factors and by loading factors. The state factors include the soil density, the soil structure, the soil water content, the soil stress history, and any cementation between the particles. The loading factors include the stress level, the strain level, the strain rate, the number of cycles, and the drainage conditions. The modulus typically increases when the density increases, when the water content decreases, when the soil has been prestressed by overburden or desiccation, when cementation increases, when the mean stress level increases, when the strain level decreases, when the strain rate increases, when the number of cycles decreases, and when better drainage takes place.

14.2 MODULUS: WHICH ONE?

Because soils do not exhibit a linear stress-strain curve, many moduli can be defined from triaxial test results, for example. In section 14.1, it was pointed out that the slope of the stress-strain curve is not the modulus of the soil. However, the slope of that curve is related to the modulus, and it is convenient to associate the slope of the stress-strain curve with a modulus. This gives a simple image tied to the modulus value; note, however, that in the figures the slope is never labeled as modulus E, but rather as slope S. Referring to Figure 14.2, if the slope is drawn from the origin to a point on the curve (O to A in Figure 14.2), the secant slope S_s is obtained and the secant modulus E_s is calculated from it. One would use such a modulus for predicting the movement due to the first

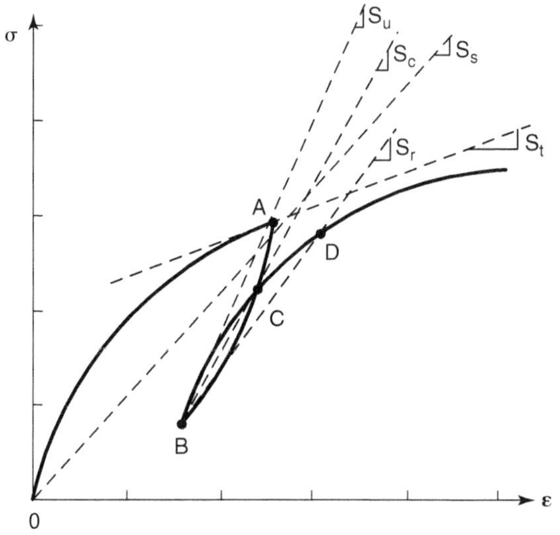

Figure 14.2 Definition of soil modulus.

application of a load, as in the case of a spread footing. If the slope is drawn as the tangent to the point considered on the stress-strain curve, then the tangent slope S_t is obtained and the tangent modulus E_t is calculated from it. One would use such a modulus to calculate the incremental movement due to an incremental load, as in the case of the movement due to one more story in a high-rise building. If the slope is drawn as the line joining points A and B in Figure 14.2, then the unloading slope S_u is obtained and the unloading modulus E_u is calculated from it. One would use such a modulus when calculating the heave at the bottom of an excavation or the rebound of a pavement after the loading by a truck tire (*resilient modulus*). If the slope is drawn from point B to point D in Figure 14.2, then the reloading slope S_r is obtained and the reload modulus E_r is calculated from it. One would use this modulus to calculate the movement at the bottom of an excavation if the excavated soil or a building of equal weight is placed back in the excavation, or to calculate the movement of the pavement under reloading by the same truck tire. If the slope is drawn from point B to point C in Figure 14.2, then the cyclic slope S_c is obtained and the cyclic modulus E_c is calculated from it. One would use such a modulus, and its evolution as a function of the number of cycles, to calculate the movement of a pile foundation subjected to repeated wave loading.

Regardless of which modulus is defined and considered, the state of the soil at any given time will affect that modulus. Section 14.3 describes some of the main state parameters influencing soil moduli.

14.3 MODULUS: INFLUENCE OF STATE FACTORS

The state factors include particle packing and organization, water content, past stress history, and cementation.

How closely packed are the particles? If the particles are closely packed, the modulus tends to be high. This is measured by the dry density (ratio of the weight of solids over the total volume of the wet sample) of the soil, for example; it can also be measured by the porosity (ratio of the volume of voids over the total volume of the wet sample).

How are the particles organized? This factor refers to the structure of the soil. For example, a coarse-grained soil can have a loose or dense structure; a fine-grained soil can have a dispersed or flocculated structure. Note that two soil samples can have the same dry density yet different structures and therefore different soil moduli. This is why taking a disturbed sample of a coarse-grained soil in the field and reconstituting it to the same dry density and water content in the laboratory can lead to differing laboratory and field moduli.

What is the water content? This parameter has a major impact because at low water contents the water binds the particles (especially for fine-grained soils) and increases the effective stress between the particles through the water tension (suction) phenomenon. Therefore, in this case low water contents lead to high soil moduli. This is why a clay shrinks and becomes very stiff when it dries. At the same time, at very low water contents the compaction of coarse-grained soils is not as efficient as it is at higher water contents, because the lubrication effect of water is not present. Therefore, in this case very low water contents lead to low moduli. As the water content increases, water lubrication increases the effect of compaction and the modulus increases as well. However, if the water content rises beyond an optimum value, the water occupies more and more room and gets to the point where it pushes the particles apart, thereby increasing compressibility and reducing the modulus.

What has the soil been subjected to in the past? This is referred to as the *stress history factor*. If the soil has been prestressed in the past, it is called *overconsolidated*. This prestressing could come, for example, from a glacier that was 100 meters thick 10,000 years ago and has now totally melted. Prestressing can also come from the drying and wetting cycles of the seasons in semiarid parts of the world. If the soil has not been prestressed in the past—in other words, if today's stress is the highest stress ever experienced by the soil—and if the soil is at equilibrium under this stress, the soil is normally consolidated. An overconsolidated (OC) soil will generally have higher moduli than the same normally consolidated (NC) soil, because the OC soil is on the reload part of the stress-strain curve whereas the NC soil is on the first loading part. Some soils are still in the process of consolidating under their own weight. These so-called *underconsolidated* soils are those such as the clays deposited offshore of the Mississippi Delta, where the deposition rate is faster than the rate that would allow the pore water pressures induced by deposition to dissipate. These clays have very low moduli.

What about cementation? *Cementation* refers to the "glue" that may exist at the contacts between particles. As discussed earlier, low water contents in fine-grained soils can generate water tension strong enough to simulate a significant "glue effect" between particles. This effect is temporary, as an increase in water content will destroy it. Another glue effect is due to the chemical cementation that can develop at the contacts. This cementation can be due to the deposition of calcium at the particle-to-particle contacts, for example. Such cementation leads to a significant increase in modulus.

These are some of the most important factors related to the state of the soil and influencing its modulus. Section 14.4 discusses the factors associated with the loading process.

14.4 MODULUS: INFLUENCE OF LOADING FACTOR

In this section it is assumed that the state factors for the soil considered are fixed (unchanging). In other words, the discussion of each of the following factors can be prefaced by saying "all other factors being equal." Also, in this section the secant modulus is used.

What is the mean stress level in the soil? The loading process induces stresses in the soil. These stresses can be shear stresses or normal stresses or a combination of both. At any given point and at any given time in a soil mass, there is a set of three principal normal stresses. The mean of these three stresses has a significant influence on the soil modulus, called the *confinement effect*. Figure 14.3*a* shows two sample stress-strain curves at two different confinement levels. As common sense would indicate, the higher the confinement is, the higher the soil modulus will be. A common model for quantifying the influence of confinement on the soil modulus is given in Figure 14.3*a* and is usually attributed to the work of Kondner. According to this model, the modulus is proportional to a power law of the confinement stress. The modulus E_0 is the modulus obtained when the confinement stress is equal to the atmospheric pressure p_a. A common value for the power exponent a in Figure 14.3*a* is 0.5.

What is the strain level in the soil? The loading process induces strains in the soil mass. Because soils are nonlinear materials, the secant modulus depends on the mean strain level in the zone of influence. In most cases the secant modulus will decrease as the strain level increases, because the stress-strain curve has a downward curvature. Note that an exception to this downward curvature occurs when the results of a consolidation test are plotted as a stress-strain curve on arithmetic scales for both axes. In this case the stress-strain curve exhibits an upward curvature, because the increase in confinement brought about by the steel ring is more influential than the decrease in modulus due to the increase in strain in the soil. In the triaxial test, the stress-strain curve can be fitted with a hyperbola up to the peak value; the associated model for this modulus is shown on Figure 14.3*b*. This hyperbolic model is usually attributed to the work of Duncan. In this model (Figure 14.3*b*), E_0 is the initial tangent modulus, also equal to the secant modulus for a strain of zero. The parameter

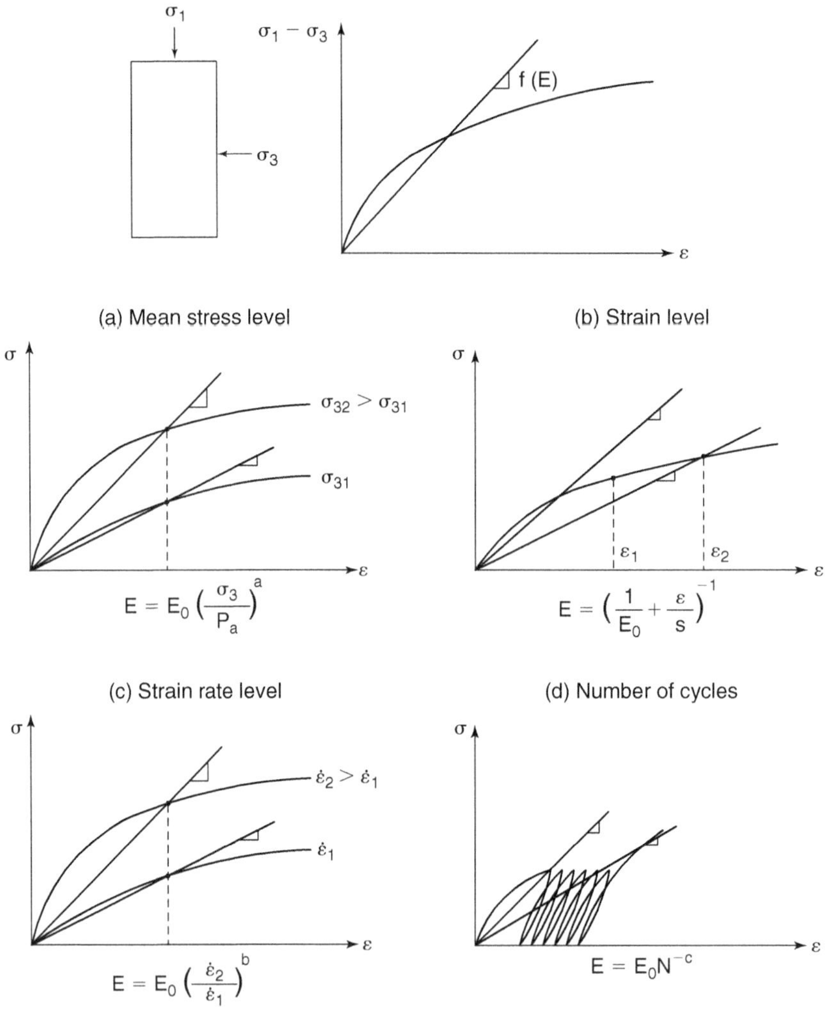

Figure 14.3 Loading factors for soil moduli.

s is the asymptotic value of the stress for a strain equal to infinity. In that sense it is related to the strength of the soil.

What is the strain rate in the soil? Soils, like many other materials, are viscous. This means that the faster a soil is loaded, the stiffer it is and therefore the higher the modulus is. In some instances, though, the reverse behavior is observed. Figure 14.3*c* shows an example of two stress-strain curves obtained by loading the soil at two drastically different strain rates. *Strain rate* is defined as the strain accumulated per unit of time. The modulus usually varies as a straight line on a log-log plot of modulus vs. strain rate. The slope of that line is the exponent b in Figure 14.3*c*. In clays, common values of this exponent vary from 0.02 for stiff clays to 0.1 for very soft clays. In sands, common values of b vary from 0.01 to 0.03. The modulus E_0 is the modulus obtained at a reference strain rate. Much of the work on this model has been done at Texas A&M University.

What is the number of cycles experienced by the soil? If the loading process is repeated a number of times, the number of cycles applied will influence the soil modulus. Again referring to the secant modulus, the larger the number of cycles, the smaller the modulus becomes. This is consistent with the accumulation of movement with an increasing number of cycles. The model used to describe this phenomenon is shown in Figure 14.3*d*. The exponent c in the model is negative and varies significantly. The most common values are on the order of −0.1 to −0.3. Much of the work on this model has also been done at Texas A&M University.

Is there time for the water to drain during the loading process? Two extreme cases can occur: drained or undrained loading. The undrained case may occur if the drainage valve is closed during a laboratory test or if the test is run sufficiently quickly in the field. The time required to maintain an undrained behavior or to ensure that complete drainage takes place depends mainly on the soil type. For example, a 10-minute test in a high-plasticity clay is probably undrained, whereas a 10-minute test in a clean sand is probably drained. The Poisson's ratio is sensitive to whether or not drainage takes place. For example, if no drainage takes place during loading in a clay, it is common to assume a Poisson's ratio equal to 0.5 (no volume change). In contrast, if complete drainage takes place (excess pore pressures are kept equal to zero), then a Poisson's ratio value of 0.35 may be reasonable. The difference between the two calculated moduli is the difference between the undrained modulus and the drained modulus. Note that the shear modulus remains theoretically constant when the drainage varies, because the effective shear stress is equal to the total shear stress. Note also that the Poisson's ratio can be larger than 0.5 if the soil dilates during shear associated with compression.

14.5 MODULUS: DIFFERENCES BETWEEN FIELDS OF APPLICATION

The modulus is useful in many fields of geotechnical engineering, but the modulus required for one field may be significantly different from the modulus for another field.

In the case of shallow foundations, the mean stress level applied under the foundation is often between 100 and 200 kPa. The normal strain level in the vertical direction is about 0.01 or less and is typically associated with a movement of about 25 mm. The rate of loading is extremely slow because that strain occurs first at the construction rate, and then the load is sustained over many years. The number of cycles is one unless cycles due to seasonal variations or other cyclic loading (such as compressor foundations or wind loads) are included. Example values of the modulus in this case are 10,000 to 20,000 kPa.

In the case of deep foundations, the mean stress level varies because the side friction on the piles occurs over a range of depth, whereas the point resistance occurs at a relatively large depth. The strain level at the pile point is usually smaller than in shallow foundations because a percentage of the load dissipates in friction before getting to the pile point. The strain rate is similar to the case of shallow foundations, with rates associated with months of construction and years of sustained loads. Some of the highest strain rates occur in the case of earthquake or wave loading. Cycles can be a major issue for earthquake loading of buildings and bridges or for wave loading of offshore structures. Because deep foundations are used in very different types of soils and for very different types of loading, the moduli vary over a much wider range of values than do the moduli for shallow foundations.

In the case of slope stability and retaining structures, movements are associated with the deformation of the soil mass essentially under its own weight. Therefore, the stress level corresponds to gravity-induced stresses. The strains are usually very small and the strain rate is again associated with the rate during initial construction and then the long-term deformation rate during the life of the slope or of the retaining structure. Cycles may occur due to earthquakes or other cyclic phenomena. For properly designed slopes and retaining structures, the moduli tend to be higher than in foundation engineering because the strain levels tend to be smaller.

In the case of pavements, the mean stress level in the subgrade is relatively low. The pressure applied to the pavement is on the order of 200 kPa for car tires, 500 kPa for truck tires, and 1700 kPa for airplane tires. However, the vertical stress at the top of the subgrade under a properly designed pavement may be only one-tenth of the tire pressure applied at the surface of the pavement. The strain level is very low because the purpose of the pavement is to limit long-term movements so that they do not exceed a few tenths of a millimeter. Typical strain levels are 0.001 or less at the top of the subgrade. The rate of loading is very high and associated with the passing of a traveling vehicle. The loading time is on the order of milliseconds for a car traveling at 100 km/h, but is measured in hours for an airplane parked at the gate. The number of cycles is tied to the number of vehicles traveling on the pavement during the life of the pavement. This number varies drastically from less than a million vehicle cycles for small roads to tens of millions for busy interstates. Typical modulus values for the subgrade range from 20,000 kPa to 150,000 kPa.

14.6 MODULUS, MODULUS OF SUBGRADE REACTION, AND STIFFNESS

The modulus of deformation E was defined in Figure 14.1. It is measured in units of force per unit area (kN/m^2). The stiffness K is defined here as the ratio of the force Q applied on a boundary through a loading area divided by the displacement s experienced by the loaded area. It is notated in units of force per unit length (kN/m). The loaded area is typically a plate, which can be square or circular. There is a relationship between the modulus E and the stiffness K. For the case of a circular plate having a diameter B, the elastic settlement s of the plate is given by:

$$s = I_1 \frac{Q}{EB} \tag{14.5}$$

Where I_1 is a constant. Therefore, the relationship between K and E is:

$$K = \frac{EB}{I_1} \tag{14.6}$$

This relationship shows that, if the modulus is a soil property, *the stiffness is not a soil property*, because it depends on the size of the loaded area. Therefore, for an elastic material, the stiffness measured with one test will be different from the stiffness measured with another test if the loading areas are different. Yet, for the same elastic material, the modulus obtained from both tests would be the same. In that sense, the stiffness is not as convenient as the modulus, so the use of the modulus is preferred.

Similar considerations apply to the modulus of subgrade reaction k. The *modulus of subgrade reaction* is defined here as the ratio of the pressure p applied to the boundary through a loading area divided by the displacement s experienced by the loaded area. It is noted in units of force per unit volume (kN/m^3). The loaded area can be a footing (coefficient of vertical subgrade reaction) or a horizontally loaded pile (coefficient of horizontal subgrade reaction). There is a relationship between the modulus E and the coefficient of subgrade reaction k. Eq. 14.5 can be rewritten as:

$$s = I_2 \frac{pB}{E} \tag{14.7}$$

Therefore, the relationship between k and E is:

$$k = \frac{E}{I_2 B} \tag{14.8}$$

where B is the footing width or the pile width or diameter. This relationship shows that, if the modulus is a soil property, *the coefficient of subgrade reaction is not a soil property*, because it depends on the size of the loaded area. Therefore, if a coefficient of subgrade reaction k is derived from load tests on a footing or a pile of a certain dimension, the value of k cannot be used directly for other footing or pile sizes. Indeed, in this case careful considerations of size and scale must be addressed. The modulus is not affected by this problem. In that sense, the coefficient of subgrade reaction is not as convenient as the modulus, so the use of the modulus is preferred.

14.7 COMMON VALUES OF YOUNG'S MODULUS AND POISSON'S RATIO

Considering all those factors, it is clear that the modulus of a soil is not a unique number. Therefore, when one says that

Table 14.1 Range of Quoted Modulus and Poisson's Ratio Values for Clays

Clay	Modulus E (MPa) Quoted ranges	Poisson's Ratio v Quoted ranges	Undrained Strength (kPa)
Very soft clay	2.5–15	0.4	<12
	2–15	0.35–0.45	
	3	0.4–0.5 undrained	
	1–3	0.1–0.3 unsaturated	
	2–4	0.2–0.3 sandy clay	
	0.5–5		
Soft clay	2–25	0.15–0.25	12–25
	1.8–3.5	0.4–0.5 undrained	
	5–20	0.1–0.3 unsaturated	
		0.2–0.3 sandy clay	
Medium or firm clay	15–50	0.3	25–50
	15–50	0.3–0.35	
	7	0.2–0.5	
	5–10	0.4–0.5 undrained	
	20–50	0.1–0.3 unsaturated	
		0.2–0.3 sandy clay	
Stiff clay	15–50	0.1–0.3	50–100
	2.5–5	0.4–0.5 undrained	
	8–19	0.1–0.3 unsaturated	
	4.2–8	0.2–0.3 sandy clay	
	50–100		
Very stiff clay	50–100	0.4–0.5 undrained	100–200
		0.1–0.3 unsaturated	
		0.2–0.3 sandy clay	
Hard clay	50–100	0.25	200–400
	14	0.4–0.5 undrained	
	8–19	0.1–0.3 unsaturated	
	6–14	0.2–0.3 sandy clay	
Very hard clay	100–200	0.4–0.5 undrained	>400
		0.1–0.3 unsaturated	
		0.2–0.3 sandy clay	

Table 14.2 Range of Quoted Modulus and Poisson's Ratio Values for Silts

Silt	Modulus E (MPa) Quoted ranges	Poisson's Ratio v Quoted ranges
Silt	2–20	0.3–0.35
	3–10	0.3–0.35
	2–19	0.3–0.35
	2–20	
	2–20	
Soft silt	2–5	
	0.5–3	
	4–8	
Firm silt	5–20	

the modulus of a soil is 10,000 kPa, for example, the very next question should be: What are the conditions associated with this number? It is also clear that the best way to obtain an appropriate modulus for a soil is to measure it directly with a test that reproduces the situation that the soil will undergo during the deformation process. Hence, tests like the pressuremeter test, the triaxial test, and the consolidation test are among the best for such a measurement. These tests are not always available or even within the budget of small projects. The next best way to obtain an appropriate modulus is to use correlations to other results or tests, such as the undrained shear strength s_u or the standard penetration test blow count N. The last resort for estimating a modulus is to use tables that give ranges of typically encountered values. Tables 14.1 through 14.7 are a collection of ranges quoted in various publications for the values of soil and rock moduli. It is not always clear from these publications

Table 14.3 Range of Quoted Modulus and Poisson's Ratio Values for Sands

Sand	Modulus E (MPa) Quoted ranges	Poisson's Ratio v Quoted ranges	SPT Blow Count N (bpf)
Very loose			<4
Loose	8–12 (fine)	As low as 0.1	4–10
	10–28	0.25 (fine)	
	10–25	0.2–0.36	
	10–30	0.2	
	15	0.2–0.35	
	10–21	0.35–0.4	
	20–80	0.2–0.4	
	10–29		
Medium or compact	12–20 (fine)	0.25 (fine)	10–30
	30–50	0.3–0.35	
	50–150	0.25–0.4	
	29–48		
Dense	20–30 (fine)	0.25 (fine)	30–50
	50–80	0.3–0.4	
	35–70	0.3–0.45	
	50–81	0.3	
	80	0.3–0.36	
	52–83	0.3–0.4	
	49–78	0.25–0.3	
	48–77	Up to 1	
Very dense			>50

Table 14.4 Range of Quoted Modulus and Poisson's Ratio Values for Gravels

Gravel	Modulus E (MPa) Quoted ranges	Poisson's Ratio v Quoted ranges	SPT Blow Count N (bpf)
Loose	50–150	0.2–0.35	4–10
	50–150	0.2	
	100		
	29–77		
	30–80		
Medium or compact	80–100		10–30
Dense	100–200	0.3–0.4	30–50
	100–200	0.3	
	150		
	102–204		
	96–192		

Table 14.5 Range of Quoted Modulus and Poisson's Ratio Values for Other Soils

Other Soils	Modulus E (MPa) Quoted ranges	Poisson's Ratio v Quoted ranges
Peat	0.1–0.3 (pure)	
	0.4–1	
	0.8–2 (some clay)	
Loess	14–60	–0.3
	15–60	0.1–0.3
	14–58	
Glacial till	10–150 (loose)	
	150–720 (dense)	
	500–1440 (Very dense)	
Clay shale	100–200	0.25–0.33
	150–5000	
	10000–40000 (intact)	

Table 14.6 Range of Quoted Modulus and Poisson's Ratio Values for Some Rocks

Rocks	Modulus E (MPa) Quoted ranges	Poisson's Ratio ν Quoted ranges
Dolomite	110000–121000	0.3
Gneiss	83000–118000	0.15–0.2
Granite	73000–86000	0.23–0.27
	31000–57000	0.15–0.24
	7000–14000 (partially decomposed)	
Limestone	87000–108000	0.27–0.3
	21000–48000	0.16–0.23
Marble	87000–108000	0.27–0.3
Mica schist	79000–101000	0.15–0.2
Quartzite	82000–97000	0.12–0.15
Rock salt	35000	0.25
Slate	79000–112000	0.15–0.2
Sandstone	38000–76000	0.25–0.33
Coal	10000–20000	

what conditions are associated with these modulus values (stress level, strain level, rate of loading, number of cycles, undrained or drained). Nevertheless, they offer some guidance for the overall range of possible values. Among the sources of these ranges of values are Lambe and Whitman (1979), Hunt (1986), USACE (1990), Bowles (1996), AASHTO (2007), and FHWA (2010). In summary, common values of soil moduli vary between 1 MPa and 150 MPa; Poisson's ratio is about 0.3 to 0.35 for drained behavior and unsaturated soils and close to 0.5 for undrained behavior of saturated soils. Poisson's ratio can be higher than 0.5 for dilatant soils because the volume can increase during compression.

Table 14.7 Range of Quoted Modulus and Poisson's Ratio Values for Other Materials

Other Material	Modulus E (MPa) Quoted ranges	Poisson's Ratio v Quoted ranges
Steel	1220000	
Steel	200000	0.28–0.29
		0.33
Aluminum	55000–76000	0.34–0.36
Concrete	20000–40000	0.15
Wood	11000–14500	
Glass	65000	
Plastic (polyethylene)	13000	
Ice	7000	0.36
Water	2200 (bulk modulus)	
Air	0.1 (bulk modulus)	

14.8 CORRELATIONS WITH OTHER TESTS

Correlations have been developed between soil modulus and the results of soil tests, mostly in situ tests. The correlations are presented in Tables 14.8 through 14.10. Among the sources for these correlations are Bowles (1996), FHWA (2010), Briaud and Miran (1992a, 1992b), Briaud (1992), and Mayne (2007a, 2007b). Note that most of these correlations lead to a modulus that would be associated with foundation settlements at working loads. There is one exception to this statement for the relationship between the soil modulus of clays and the undrained shear strength. In this case the modulus refers to the elastic immediate settlement, a higher modulus than would be used for the long-term settlement. Figure 14.4 shows a more detailed relationship between the undrained modulus E_u and the undrained shear strength s_u as a function of the overconsolidation ratio (OCR). Recall that the OCR is the ratio between the effective preconsolidation pressure σ'_p and the vertical effective stress at rest σ'_{ov}. The pressure σ'_p is found on the stress-strain curve of the consolidation test around the maximum curvature of the semilog plot (see Figure 9.32).

14.9 MODULUS: A COMPREHENSIVE MODEL

To acknowledge the influence of the various loading factors on the modulus, it is useful to regroup them into one single model. The model should include the influence of:

- Stress confinement level
- Strain level
- Rate of loading
- Number of cycles

First, the influence of the confinement level on the modulus is quantified. For this, the initial tangent modulus E_i is selected to isolate the influence of the strain level, which is to be included separately. This confinement is due to the mean normal stress σ_M:

$$\sigma_M = \frac{\sigma_1 + \sigma_2 + \sigma_3}{3} \tag{14.9}$$

where σ_1, σ_2, and σ_3 are the principal stresses. The influence of the confinement is quantified through the Kondner model:

$$E_{\sigma i} = E_{ai}\left(\frac{\sigma_M}{p_a}\right)^m \tag{14.10}$$

where $E_{\sigma i}$ is the initial tangent modulus, E_{ai} is the value of E_i for a confinement equal to the atmospheric pressure, σ_M is the mean confining stress, p_a is the atmospheric pressure used as a reference, and m is the stress level exponent. Then the

Table 14.8 Correlations between Soil Modulus and Soil Test Results for Sands and Gravels

Soil Types	Correlation
Silts, sandy silts, slightly cohesive mixtures	$E(kPa) = 400\,N(bpf)^*$
Clean fine to medium sands and slightly silty sands	$E(kPa) = 700\,N(bpf)^*$
Coarse sand and sand with little gravel	$E(kPa) = 1000\,N(bpf)^*$
Sandy gravels and gravels	$E(kPa) = 1200\,N(bpf)^*$
	$E(kPa) = 7000\,(N(bpf))^{0.5}$
Sand (normally consolidated)	$E(kPa) = (15000 \text{ to } 22000)\log_e(N(bpf))$
	$E(kPa) = 500\,(N(bpf)^* + 15)$
Sand (saturated)	$E(kPa) = 250\,(N(bpf)^* + 15)$
Sand (overconsolidated)	$E(kPa) = 40000 + 1050\,N(bpf)^*$
Gravelly sand	$E(kPa) = 1200\,(N(bpf)^* + 6)$
	$E = 2\,q_c^{**}$
Sandy soils (normally consolidated)	$E = (2.5 \text{ to } 3.5)\,q_c^{**}$ recent < 100 yrs
	$E = (3.5 \text{ to } 6)\,q_c^{**}$ old > 3000 yrs
Sand (normally consolidated)	$E = (1 + Dr^2)\,q_c$ Dr is relative density as ratio
Sand (overconsolidated)	$E = (6 \text{ to } 10)\,q_c^{**}$
Sand: $q_c < 5\,MPa$	$E = 2\,q_c^{**}$
Sand: $q_c > 10\,MPa$	$E = 1.5\,q_c^{**}$

*SPT blow count N in bpf, blows per 0.3 m
**CPT point resistance in units of pressure

Table 14.9 Correlations between Soil Modulus and Soil Test Results for Clays and Silts

Soil Types	Correlation
Normally consolidated sensitive clay	$E = (200 \text{ to } 500)\,s_u^{***}$ for immediate undrained settlement
Normally consolidated insensitive and lightly overconsolidated clay	$E = (750 \text{ to } 1200)\,s_u^{***}$ for immediate undrained settlement
Heavily overconsolidated clay	$E = (1500 \text{ to } 2000)\,s_u^{***}$ for immediate undrained settlement
Clays of low plasticity (CL)	$M^* = (1 \text{ to } 2.5)\,q_c^{**}$ for $q_c > 2\,MPa$
	$M^* = (2 \text{ to } 5)\,q_c^{**}$ for $0.7 < q_c < 2\,MPa$
	$M^* = (3 \text{ to } 8)\,q_c^{**}$ for $q_c < 0.7\,MPa$
Silts of low plasticity (ML)	$M^* = (3 \text{ to } 6)\,q_c^{**}$ for $q_c > 2\,MPa$
	$M^* = (1 \text{ to } 3)\,q_c^{**}$ for $q_c < 2\,MPa$
High-plasticity silts and clays (MH, CH)	$M^* = (2 \text{ to } 6)\,q_c^{**}$ for $q_c < 2\,MPa$
Organic silt (OL)	$M^* = (2 \text{ to } 8)\,q_c^{**}$ for $q_c < 1.2\,MPa$
Peat and organic clay (Pt, OH, $q_c < 0.7\,MPa$)	$M^* = (1.5 \text{ to } 4)\,q_c^{**}$ for $50 < w < 100$
	$M^* = (1 \text{ to } 1.5)\,q_c^{**}$ for $100 < w < 200$
	$M^* = (0.4 \text{ to } 1)\,q_c^{**}$ for $w > 200$
Chalk	$M^* = (2 \text{ to } 4)\,q_c^{**}$ for $2 < q_c < 3\,MPa$

*M is the constrained modulus: $M = E(1-v)/((1+v)(1-2v))$
**CPT point resistance in units of pressure
***Undrained shear strength

Table 14.10 Correlations between Pressuremeter Modulus and Other Data

(a) Correlations for Sand

Column A = number in table × row B

B / A	E_0 (kPa)	E_R (kPa)	p^*_L (kPa)	q_c (kPa)	f_s (kPa)	N (bl/30 cm)
E_0 (kPa)	1	0.125	8	1.15	57.5	383
E_R (kPa)	8	1	64	6.25	312.5	2174
p^*_L (kPa)	0.125	0.0156	1	0.11	5.5	47.9
q_c (kPa)	0.87	0.16	9	1	50	479
f_s (kPa)	0.0174	0.0032	0.182	0.02	1	9.58
N (bl/30 cm)	0.0026	0.00046	0.021	0.0021	0.104	1

(b) Correlations for Clay

Column A = number in table × row B

B / A	E_0 (kPa)	E_R (kPa)	p^*_L (kPa)	q_c (kPa)	f_s (kPa)	s_u (kPa)	N (bl/30 cm)
E_0 (kPa)	1	0.278	14	2.5	56	100	667
E_R (kPa)	3.6	1	50	13	260	300	2000
p^*_L (kPa)	0.071	0.02	1	0.2	4	7.5	50
q_c (kPa)	0.40	0.077	5	1	20	27	180
f_s (kPa)	0.079	0.0038	0.25	0.05	1	1.6	10.7
s_u (kPa)	0.010	0.0033	0.133	0.037	0.625	1	6.7
N (bl/30 cm)	0.0015	0.0005	0.02	0.0056	0.091	0.14	1

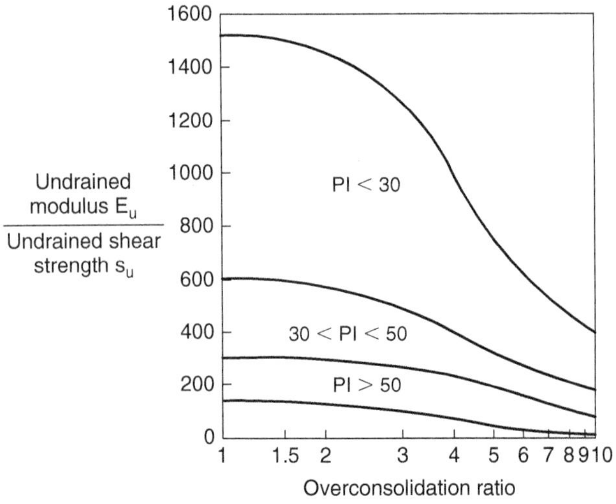

Figure 14.4 Modulus of clays correlated to undrained shear strength (After Duncan and Buchignani, 1976)

influence of the strain level is included by using the Duncan hyperbolic model. This model states that the stress-strain curve is well described by a hyperbola:

$$\sigma = \frac{\varepsilon}{a + b\varepsilon} \tag{14.11}$$

Note that when ε goes to zero, the ratio σ/ε goes to $1/a$, therefore $1/a$ is associated with the initial tangent modulus E_i. Also, when ε goes to infinity, σ goes to $1/b$, which represents the ultimate strength of the soil σ_{ult}. Equation 14.11 can be rewritten in terms of modulus variation as:

$$E_{\sigma\varepsilon} = \left(\frac{1}{E_{\sigma i}} + \frac{\varepsilon}{\sigma_{ult}}\right)^{-1} \tag{14.12}$$

Now we can include the rate effect on the modulus. This is done by using a rate effect exponent model (Briaud and Garland 1985) that quantifies the modulus increase when the time of loading t decreases or the strain rate $\dot{\varepsilon}$ increases:

$$E_t = E_{t0}\left(\frac{t}{t_o}\right)^{-n} = E_{t0}\left(\frac{\dot{\varepsilon}}{\dot{\varepsilon}_o}\right)^{n} \quad (14.13)$$

where t_o is a reference time of loading, $\dot{\varepsilon}_o$ is a reference strain rate, and n is the rate effect exponent for the soil. Figure 14.5 shows a correlation between the exponent n and the undrained shear strength s_u. The best-fit equation is:

$$n = 0.12(s_{u\,ref}(\text{kPa}))^{-0.22} \quad (14.14)$$

The number of loading cycles N is included by using a power law (Briaud 1992):

$$E_N = E_1 N^{-p} \quad (14.15)$$

where E_N and E_1 are the secant moduli to the top of the Nth cycle and the first cycle respectively and p is the cyclic degradation exponent. By combining all effects, the general model becomes:

$$E_{\sigma\varepsilon tN} = \left(\frac{1}{E_{ai}\left(\frac{\sigma_M}{p_a}\right)^m} + \frac{\varepsilon}{\sigma_{ult}}\right)^{-1}\left(\frac{t}{t_o}\right)^{-n} N^{-p} \quad (14.16)$$

Values around 0.5 are common for the stress level exponent m and for drained or unsaturated conditions ($S < 0.8$). The value of m becomes very low and even zero for the undrained behavior of fine-grained soils. In clays, common values of the rate effect exponent n vary from 0.02 for stiff clays to 0.1 for very soft clays. In sands, common values of n vary from 0.01 to 0.03. The cyclic degradation exponent p varies widely depending on how close to failure the soil is loaded. At working loads, this exponent is generally less than 0.1. The time t_o that serves as a reference for strain rate effect can correspond to the typical length of a soil test and may be taken as 10 minutes. The time t can vary from 75 years for the typical design life of a bridge to 10 milliseconds for a car impact on a guardrail post or the passage of a vehicle on a pavement. The number of cycles can vary from 1 for a building, to about 1000 for hurricane wave loading on an offshore platform, and to millions for a pavement. As can be seen in Eq. 14.16, the modulus E_{ai} is very important. It is the reference modulus for all other calculations and represents the initial tangent modulus (zero strain) at a stress level corresponding to atmospheric pressure (100 kPa), at a loading time of possibly 10 minutes, and for the first loading (one cycle).

14.10 INITIAL TANGENT MODULUS G_o OR G_{max}

This modulus is typically referred to as G_o or G_{max}. This is because the shear modulus G is more convenient than the Young's modulus E. Indeed, the shear modulus G does not require knowledge of the Poisson's ratio, whereas E does. The subscript "o" or "max" refers to the fact that it is the modulus at the origin and also the maximum shear modulus value one can expect for the soil. Several expressions for G_{max} have been formulated. Hardin and Drnevich (1972) and Hardin (1978) proposed, for all soil types:

$$\frac{G_{max}}{p_a} = \frac{625}{0.3 + 0.7e^2}(OCR)^k\left(\frac{\sigma'_M}{p_a}\right)^n \quad (14.17)$$

where p_a is the atmospheric pressure, e is the void ratio, OCR is the overconsolidation ratio, σ'_M is the mean effective normal stress, and k and n are exponents. Then Jamiolkowski et al. (1991) proposed:

$$\frac{G_{max}}{p_a} = \frac{625}{e^{1.3}}(OCR)^k\left(\frac{\sigma'_M}{p_a}\right)^n \quad (14.18)$$

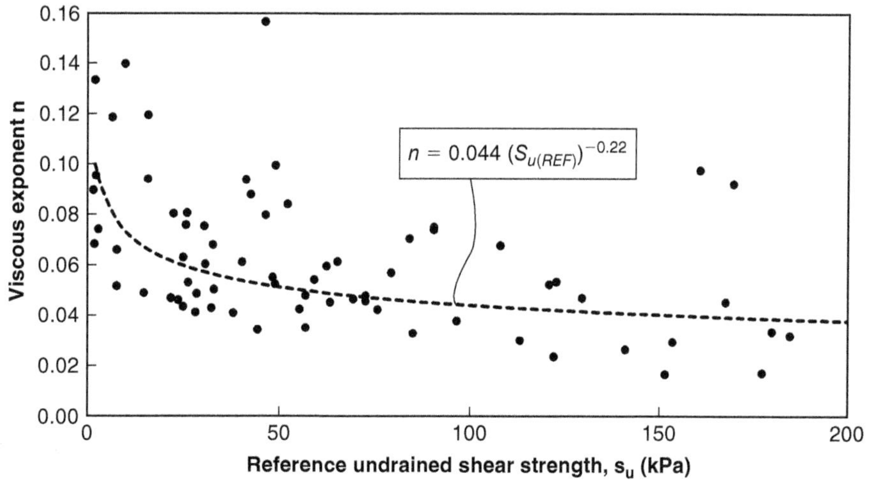

Figure 14.5 Rate effect exponent for clays.

The exponent n is usually taken equal to 0.5, and k is given in Table 14.11.

For sands, Seed and Idriss (1970) proposed:

$$\frac{G_{max}}{p_a} = 22.4K_{2,max}\left(\frac{\sigma'_M}{p_a}\right)^{0.5} \tag{14.19}$$

where p_a is the atmospheric pressure and $K_{2,max}$ is a modulus number given in Table 14.12 for sands. For gravel, $K_{2,max}$ is higher than for sands, ranging between 80 and 180.

For fine-grained soils, Kramer (1996) suggests relating G_{max} to the OCR and the undrained shear strength s_u measured in a CU triaxial test (Table 14.13).

The value of G_{max} has also been correlated to the results of in situ tests, in particular the SPT and the CPT. Ohta and Goto (1976) and Seed et al. (1986) proposed, for sands:

$$\frac{G_{max}}{p_a} = 447N^{0.33}\left(\frac{\sigma'_M}{p_a}\right)^{0.5} \tag{14.20}$$

where N is the SPT blow count corrected for 60% of maximum energy and corrected to 100 kPa of pressure (see Chapter 7). Rix and Stokoe (1991) proposed a correlation for quartz sands with the CPT point resistance q_c (Chapter 7):

$$\frac{G_{max}}{p_a} = 290\left(\frac{q_c}{p_a}\right)^{0.25}\left(\frac{\sigma'_M}{p_a}\right)^{0.375} \tag{14.21}$$

Table 14.11 Overconsolidation Exponent *k*

Plasticity Index	Value of k
0	0.00
20	0.18
40	0.30
60	0.41
80	0.48
100	0.50

(After Hardin and Drnevich 1972; Kramer 1996)

Table 14.12 Values of $K_{2,max}$

Void ratio	$K_{2,max}$	Relative density (%)	$K_{2,max}$
0.4	70	30	34
0.5	60	40	40
0.6	51	45	43
0.7	44	60	52
0.8	39	75	59
0.9	34	90	70

(After Seed and Idriss 1970; Kramer 1996.)

Table 14.13 Values of G_{max}/s_u

	Overconsolidation ratio, OCR		
Plasticity index	1	2	5
15–20	1100	900	600
20–25	700	600	500
035–45	450	380	300

(After Kramer 1996.)

Table 14.14 Common Values of G_{max} for Different Soils Based on Shear Wave Velocity

Type of Soil	Small-Strain Shear Wave Velocity, v_s (m/s)	Initial Shear Modulus, G_{max} (MPa)
Soft clay	40–90	3–14
Firm clay	65–140	7–36
Loose sand	125–270	29–144
Dense sand and gravel	270–400	72–360

For clay, Mayne and Rix (1993) proposed:

$$\frac{G_{max}}{p_a} = 100\left(\frac{q_c}{p_a}\right)^{0.695} e^{-1.13} \tag{14.22}$$

In the end, however, the best way to obtain G_{max} is through testing. In the field, the tests that can be used are the cross hole test and the SASW, as described in Chapter 7. Some common values are shown in Table 14.14. In the laboratory, the best test is the resonant column test (Chapter 9), yet sample disturbance may lead to lower values of G_{max} compared to the field values. The field values come from testing a large, undisturbed mass of soil through wave propagation, whereas disturbance has a much more pronounced effect on the small scale of the lab test. The contrary is likely true for weathered rocks, where the sample is likely to be much stiffer than the rock mass.

14.11 REDUCTION OF G_{max} WITH STRAIN: THE G/G_{max} CURVE

As discussed earlier (section 14.9, and section 9.13.1 in Chapter 9), the modulus decreases with an increase in strain. As a result, the G_{max} value, which corresponds to zero strain, is the highest shear modulus attainable. When the strain

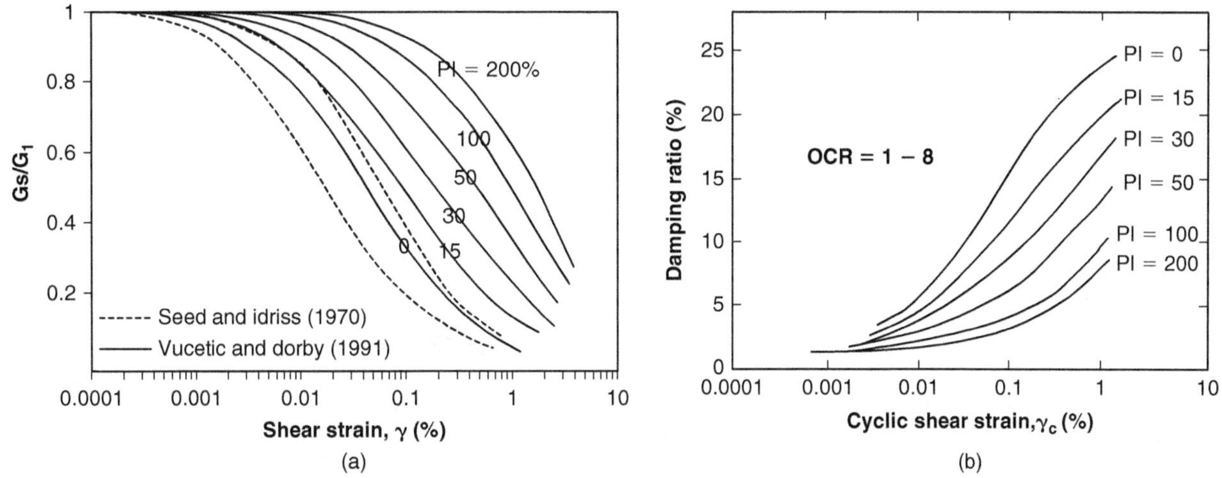

Figure 14.6 Degradation of the shear modulus and damping ratio with shear strain: (a) Modulus. (b) Damping ratio. (After Vucetic and Dobry 1996)

increases, G decreases and the ratio G/G_{max} becomes less than 1. The G/G_{max} ratio is plotted against the shear strain γ in a semilog plot: G/G_{max} on a natural scale and γ on a log scale. Figure 14.6 shows such a plot, proposed by Vucetic and Dobry (1991), indicating the influence of the plasticity index PI on the G/G_{max} curve. As can be seen from Figure 14.6, the high PI clays maintain the G_{max} value over a larger range of strain than the low PI clays. Kramer (1996) cautiously suggests that the $PI = 0$ curve could also be used for sands.

Parallel to this variation is the evolution of the damping ratio as defined in section 9.13.1. The damping ratio quantifies the loss of energy in the deformation process. During cyclic loading, the loop of the shear stress vs. shear strain curve widens, and it takes more and more energy to deform the soil. This leads to an increase in the damping ratio, as shown in Figure 14.6.

Mayne (2007a; 2007b) summarizes the data of several authors using the following equation, which is based on the mobilized strength rather than the strain as the variable influencing the reduction in shear modulus:

$$\frac{G}{G_{max}} = 1 - \left(\frac{\tau}{\tau_{max}}\right)^g \tag{14.23}$$

where τ is the applied shear stress, τ_{max} is the maximum shear stress or shear strength of the soil, and g is the reduction exponent (which has a range of 0.2 to 0.4 for common soils). Figure 14.7 shows plots of Eq. 14.23 with data for various soils.

Again, testing is the best way to obtain the G/G_{max} degradation curve, but no single test can be used to give the entire curve. The problem is that the strain range associated with the G/G_{max} degradation curve goes from shear strains of 10^{-6} to 10^{-1}or 10^{-4}percent to 10^{1}percent. To cover this wide range of strains, several tests are used, as shown in Figure 14.8.

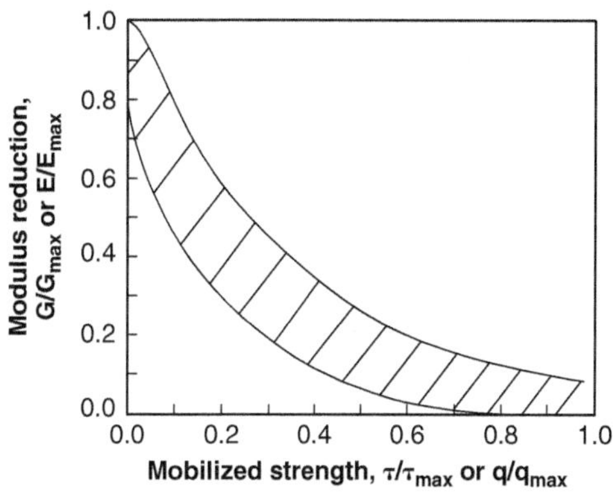

Figure 14.7 Degradation of the shear modulus with mobilized stress. (After Mayne 2007a; 2007b)

14.12 PRECONSOLIDATION PRESSURE AND OVERCONSOLIDATION RATIO FROM CONSOLIDATION TEST

One of the oldest tests in geotechnical engineering is the consolidation test (see Chapters 9 and 11). This test yields a stress-strain curve and a set of strain-time curves from which soil properties can be obtained regarding the magnitude of deformation and the time rate of deformation. The preconsolidation pressure σ'_p at a depth z is the effective normal stress found from the effective stress vs. strain consolidation curve on a sample from depth z by the technique described in section 9.5 and Figure 9.32. The vertical effective stress at depth z where the sample was taken is σ'_{ov}. Soils can be classified according to whether σ'_{ov} is larger than, smaller than, or equal to σ'_p. Three categories are identified:

- Normally consolidated soils: $\sigma'_{ov} = \sigma'_p$
- Overconsolidated soils: $\sigma'_{ov} < \sigma'_p$
- Underconsolidated soils: $\sigma'_{ov} > \sigma'_p$

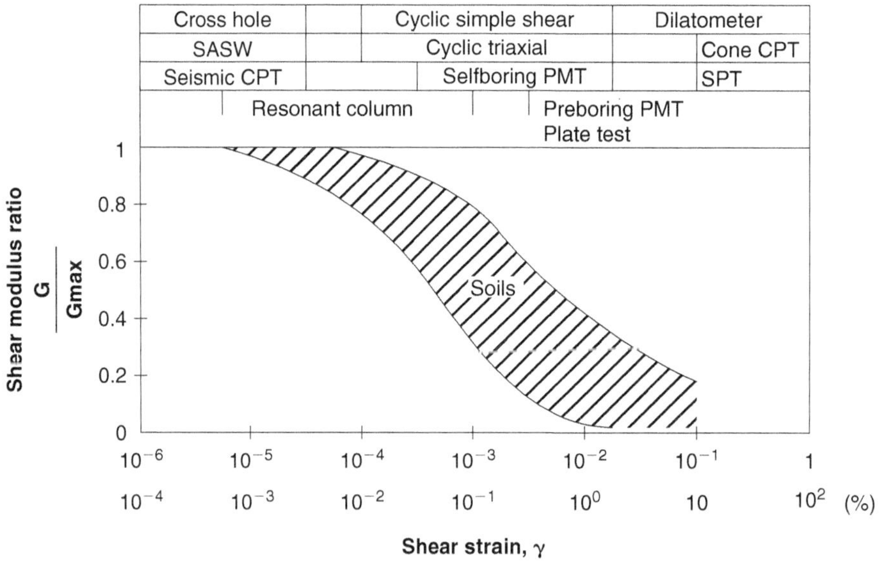

Figure 14.8 Range of strains covered by various soil tests to generate the G/G_{max} vs. γ curve.

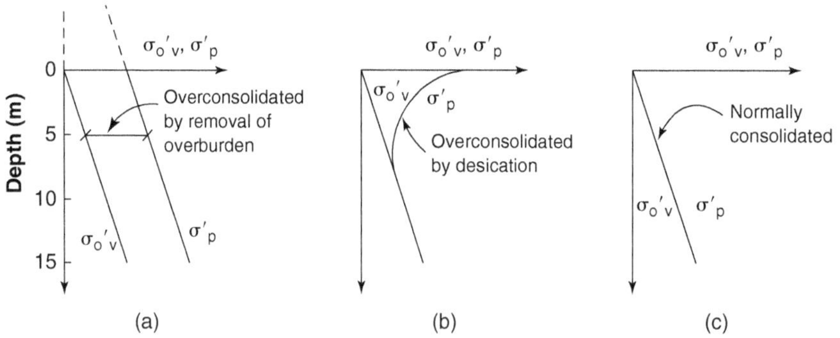

Figure 14.9 Overconsolidation profiles.

The pressure σ'_p is very important for many reasons; it essentially plays the role of a yield stress for soils when the consolidation test applies to the soil problem being studied. For example, the settlement of a structure or embankment will likely be small if the pressure under the foundation on an overconsolidated soil is kept below σ'_p. This pressure can be due to a number of phenomena, including overburden removal and desiccation. Indeed, if a soil exists for a long time with a high overburden pressure and then this overburden pressure is significantly reduced, the soil will keep the "memory" of the high overburden pressure and will not exhibit large deformation until the high overburden pressure is surpassed during the soil loading process. For example, if a 100 m thick glacier was covering an area 10,000 years ago but has completely melted today, the soil below the glacier will have a preconsolidation pressure equal to the pressure of the large glacier (100 m × 9 kN/m^3 = 900 kPa). One would expect small settlements for pressures less than 900 kPa but larger settlements for pressures above 900 kPa. This stress relief would be felt throughout the depth of the deposit, and the profile of preconsolidation pressure would look like the one in Figure 14.9*a*. If, in contrast, the soil had not been subjected to erosion, but had been subjected to a series of wetting and drying cycles, it would exhibit an overconsolidation profile, as shown in Figure 14.9*b*. Indeed, repeated wetting and drying cycles induce and release a significant effective stress that prestresses the soil. This effect is typically localized near the ground surface where the wetting and drying cycles are prevalent. If the soil had not been subjected to any significant erosion, loading, wetting, or desiccation, the current vertical effective stress would be the same as the preconsolidation pressure and the soil would be normally consolidated (Figure 14.9*c*). Underconsolidated soils are soils that are still consolidating. This is the case, for example, with the soils in the delta of the Mississippi River, where the sediments drained from the plains of the USA through erosion are loading the bottom of the Gulf of Mexico faster than the soft clays can consolidate.

The overconsolidation ratio (OCR) is the ratio of the preconsolidation pressure σ'_p over the vertical effective stress σ'_{ov}:

$$OCR = \frac{\sigma'_p}{\sigma'_{ov}} \tag{14.24}$$

The OCR varies from 1 for normally consolidated soils up to 5 for heavily overconsolidated soils, with values between 1.5 and 2.5 being relatively common.

14.13 COMPRESSION INDEX, RECOMPRESSION INDEX, AND SECONDARY COMPRESSION INDEX FROM CONSOLIDATION TEST

The *compression index* C_c is defined as the slope of the linear portion of the e–logσ' curve after the initial curved part (Figure 14.10):

$$C_c = \frac{\Delta e}{\Delta \log \sigma}, \tag{14.25}$$

The value of C_c for common saturated soils varies between 0.2 and 1. Terzaghi and Peck (1967) gave the following empirical equation:

$$C_c = 0.009(LL - 10) \tag{14.26}$$

where *LL* is the liquid limit expressed as a percentage. This equation has a reliability range of ±30% and should not be used for clays with a sensitivity greater than 4, or a liquid limit greater than 100, or a large percentage of organic matter (Holtz et al. 2011). The compression index increases with the initial void ratio and the water content for saturated soils. Among other correlations for saturated clays are:

$$C_c = 1.15(e_o - 0.35) \tag{14.27}$$

$$C_c = w \tag{14.28}$$

Table 14.15 Compression Index C_c Values for Saturated Soils

Soil	Compression Index C_c
Peat	10–15
Organic clays	2–8
Sensitive clays	1–4
High-plasticity clays	0.5–0.9
Low-plasticity clays	0.15–1.2

where e_o is the initial void ratio and *w* is the natural water content expressed as a ratio rather than a percentage. The compression index does increase with the initial void ratio and the water content for saturated soils. Table 14.15 gives some ranges of observed values of C_c for various soils.

The *recompression index* C_r is the slope of the unload-reload loop performed during a consolidation test. The problem is that the value of C_r depends on both the point at which the unloading is started and the extent of the unloading. This is why it is important, during the test, to reproduce the loading path experienced by the soil. Typically, the higher the stress at which the unloading starts, the higher the C_r; conversely, the larger the unloading stress, the smaller the C_r is.

The secondary compression index C_α (Figure 14.11) is associated with the void ratio vs. log time curve. This curve is the one obtained during each load step in the conventional consolidation test. Toward the end of the consolidation process, once the pore pressures have dissipated, the void ratio vs. log time curve tends toward a straight line. At that point, the soil deformation is called *secondary consolidation* or *creep*. The slope of the later part of the *e* versus. log *t* curve

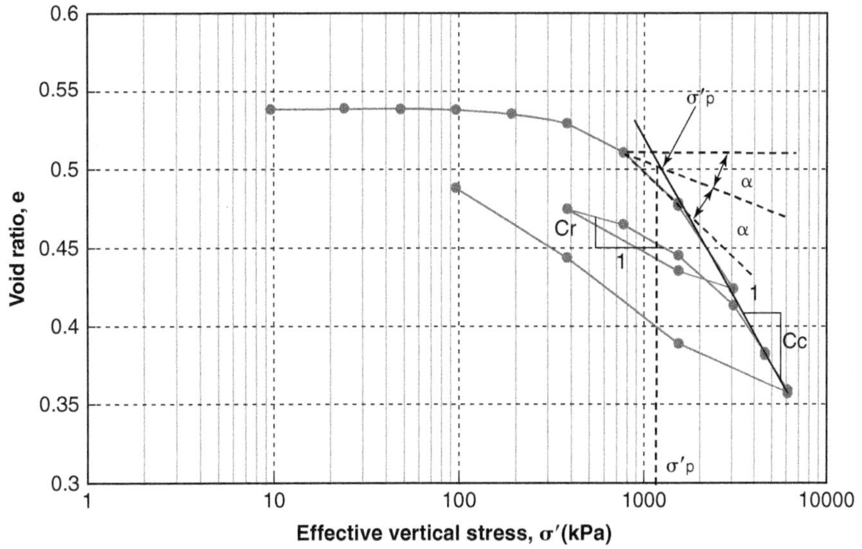

Figure 14.10 Definition of σ'_p, the compression index C_c, and the recompression index C_r.

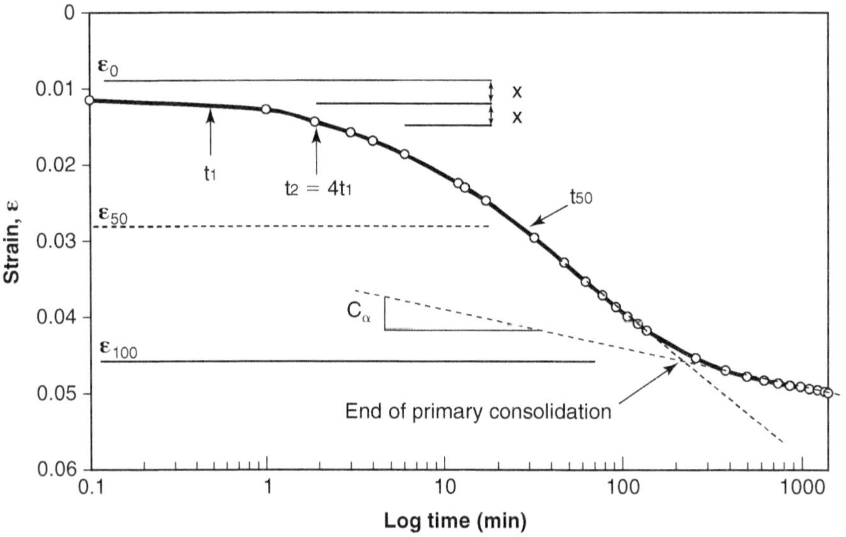

Figure 14.11 Definition of the secondary compression index C_α.

is the secondary compression index C_α:

$$C_\alpha = \frac{\Delta e}{\Delta \log t} \tag{14.29}$$

C_α tends to increase with an increase in organic content in the soil and decrease with an increase in sand content in fine-grained soils. Terzaghi et al. (1996) state that the secondary compression index C_α is linked to the compression index C_c and propose Table 14.16.

Note that a modulus E can be obtained from the consolidation test. By definition, the constrained modulus M is the ratio between the vertical stress σ_z and the vertical strain ε_z when any lateral expansion is restrained. The constrained modulus M is related to the Young's modulus E by:

$$M = \frac{\sigma_z}{\varepsilon_z} = E\frac{1-\nu}{(1+\nu)(1-2\nu)} \tag{14.30}$$

Figure 14.12 shows consolidation curves plotted on axes with natural scale and associated constrained modulus M. Note that as the strain increases, the value of M increases, because the steel ring constraining the soil is playing an increasing role.

Table 14.16 Ratio of Secondary Compression Index C_α over Compression Index C_c for Various Soils

Soil	C_α/C_c
Granular soils	0.02 ± 0.01
Inorganic clays and silts	0.04 ± 0.01
Organic clays and silts	0.05 ± 0.01
Peat	0.06 ± 0.01

(After Terzaghi et al. 1996)

14.14 TIME EFFECT FROM CONSOLIDATION TEST

The rate of deformation can be quantified by the coefficient of consolidation c_v. As demonstrated in Chapter 11 (section 11.4.6), the time rate of settlement can be predicted by the following equation, provided the assumptions associated with that derivation are satisfied:

$$t = \frac{TH^2}{c_v} \tag{14.31}$$

where t is the time required for $U\%$ of the settlement to take place, T is the unitless time factor corresponding to $U\%$ of settlement (Figure 14.13), H is the drainage length (m), and c_v is the coefficient of consolidation (m^2/s). The drainage length H is equal to the thickness H_o of the consolidating layer if drainage can take place on only one side (top or bottom) of the layer and equal to half the thickness of the layer $H_o/2$ if drainage can take place on both sides of the layer (top and bottom) (Figure 14.14). The assumptions required for this equation to be applicable are:

a. The soil is saturated with water.
b. The water is incompressible.
c. The soil skeleton is linear elastic (linear stress-strain relation).
d. The soil particles are incompressible.
e. Darcy's law governs the flow of water through the soil.
f. The water drains through one or both of the horizontal boundaries.
g. The flow is in the vertical direction only.

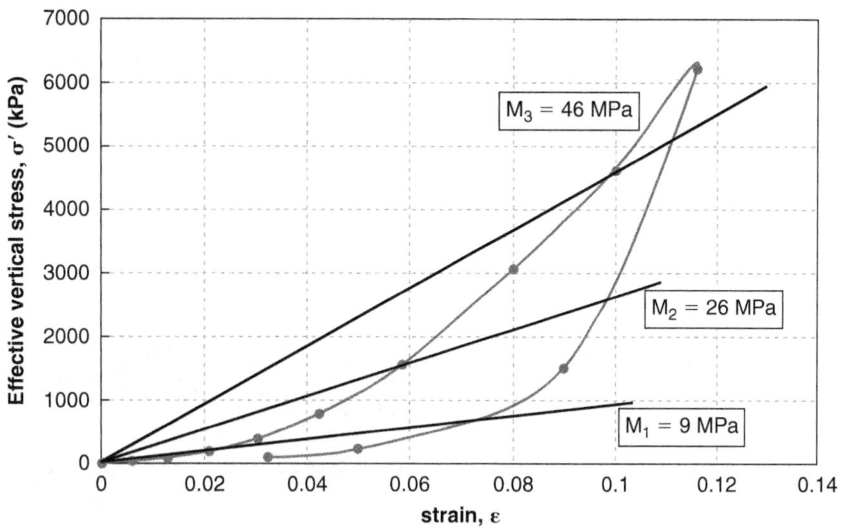

Figure 14.12 Stress-strain curve from consolidation test.

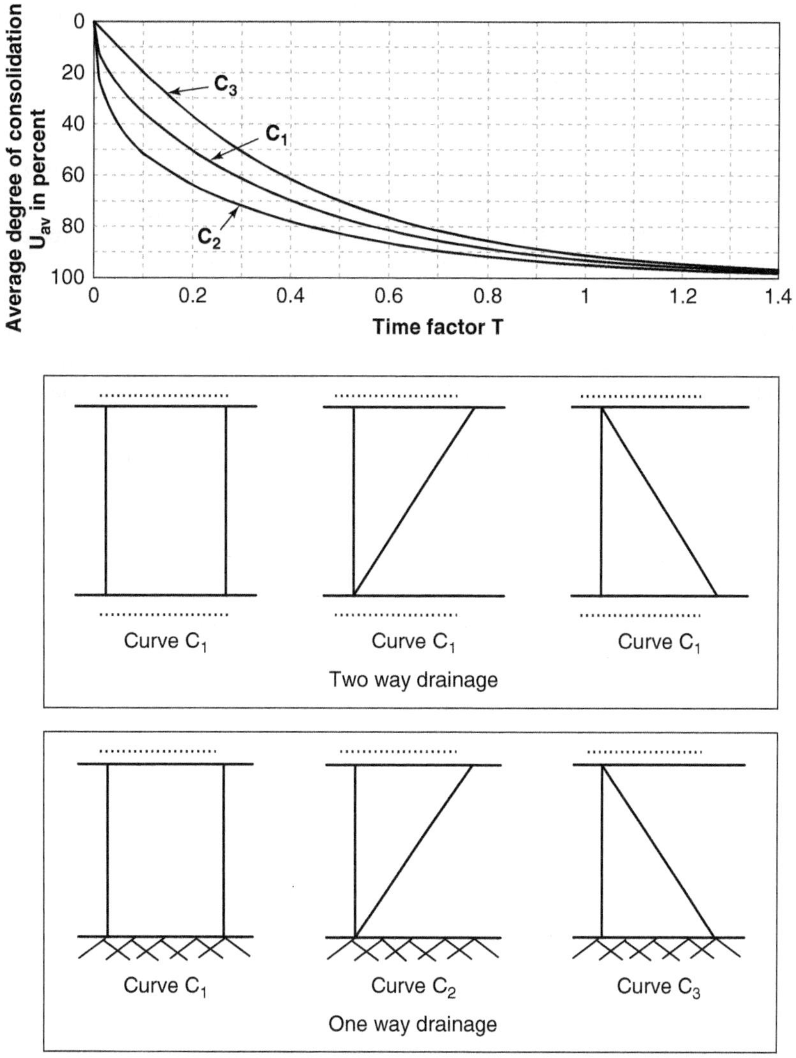

Figure 14.13 Average degree of consolidation U_{av} vs. time factor T on natural scale for different stress increase profiles.

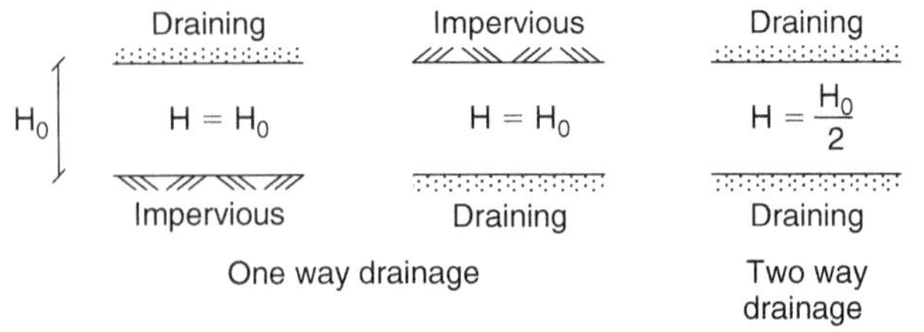

Figure 14.14 Drainage length.

h. The increase in stress $\Delta\sigma$ in the layer due to the embankment is constant within the layer.
i. The excess water stress u_{we} increases by $\Delta\sigma$ when the embankment is placed.
j. No lateral soil movement takes place.

The coefficient of consolidation c_v can be obtained from the consolidation test, as explained in Chapter 9 (section 9.5.1). Typical values of c_v for fine-grained soils range from 10^{-3} m^2/day to 10^{-1} m^2/day.

14.15 MODULUS, TIME EFFECT, AND CYCLIC EFFECT FROM PRESSUREMETER TEST

One of the tools with which one can measure a soil modulus in situ is the pressuremeter (PMT). There are three types of pressuremeters: the preboring pressuremeter, the self-boring pressuremeter, and the push-in or cone pressuremeter. With a preboring PMT, a borehole is drilled first; then the drilling tool is removed and the PMT probe is inserted in the open hole. With a self-boring PMT, the probe is equipped with its own drilling equipment, and bores itself into the soil to avoid decompression of the soil due to preboring. With a push-in PMT, the probe is pushed into the soil and full displacement takes place during the insertion, as in the cone penetrometer test. This section refers to tests done with the preboring pressuremeter, which is by far the most common of the three.

The stress tensor in the soil can be decomposed into the spherical component (confinement) and the deviatoric component (shearing). The consolidation test mostly generates an increase in confinement, and the spherical part of the stress tensor dominates the deformation process measured in that test. The pressuremeter test, in contrast, mostly generates an increase in shear around the probe, while the mean confining stress remains relatively unchanged. Thus, for any deformation process in which shearing dominates, the PMT is a good candidate test.

A PMT curve is shown in Figure 14.15. The pressure p_y is the yield pressure, and it represents an important threshold of pressure. If the soil is loaded at pressures below this value, the creep deformation will be small. Such creep deformation will increase gradually and significantly beyond p_y. The modulus E_o is the modulus obtained from the first loading part of the PMT curve below p_y (see Chapter 7, section 7.3). This modulus corresponds to about 1% strain and to pressure levels below p_y associated with ordinary foundation work. It is measured in minutes and with the first cycle of loading. Thus, it is a relatively low modulus that tends to give reasonable to conservative values of settlement when used in elastic equations. Typical values and correlations for the PMT modulus are presented in Chapter 7 (see Tables 7.3 to 7.5).

The influence of time on the modulus can be measured if a PMT is run by holding the pressure constant while measuring the relative increase in radius (strain) as a function of time (Figure 14.16). The model described in Eq. 14.13 is used and the exponent n is back-calculated directly from the measurements taken during the test. This is done by plotting the modulus E_o as a function of time on a log-log plot and

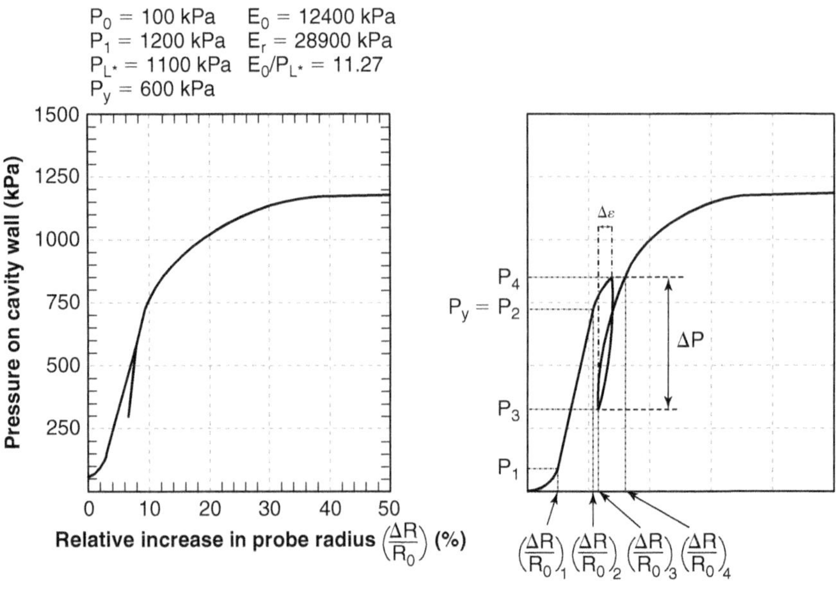

Figure 14.15 Pressuremeter curve and modulus.

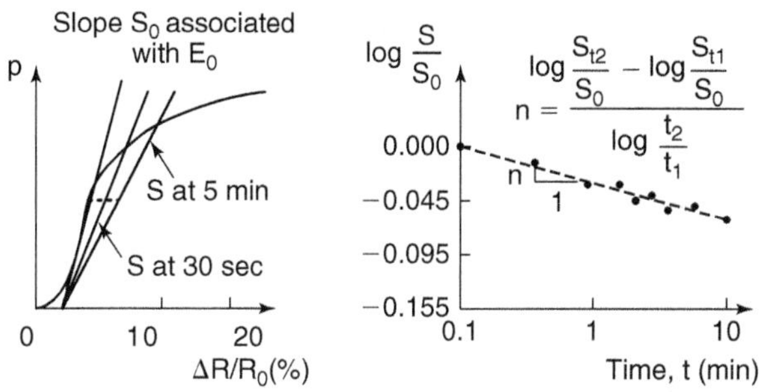

Figure 14.16 PMT modulus as a function of time.

obtaining the slope of that line (Figure 14.16). That way a modulus corresponding to the building settlement after 50 years can be calculated. For example, if E_o is 30 MPa as measured in the PMT in 1 minute; and if n is 0.03, also as measured in the PMT, then the modulus to use for the settlement at 50 years would be:

$$E_{o(50\text{ years})} = E_{o(1\text{ min})}\left(\frac{50 \times 365 \times 24 \times 60}{1}\right)^{-0.03} = 0.6E_{o(1\text{ min})} \tag{14.32}$$

and the settlement at 50 years would be 1.67 times larger than the settlement at 1 minute.

The unload-reload modulus E_r comes from an unload-reload loop performed around the value of p_y. The value of E_r depends on the amplitude of the unload-reload cycle ($\Delta p = p_4 - p_3$; Figure 14.15). The larger Δp is, the smaller E_r will be. The strain amplitude $\Delta\varepsilon$ corresponding to the pressure amplitude Δp can be controlled and the unload-reload modulus can be associated with that strain amplitude as a way to obtain a modulus as a function of strain:

$$E_{\sigma\varepsilon} = \left(\frac{1}{E_{\sigma i}} + \frac{\Delta\varepsilon}{p_L}\right)^{-1} \tag{14.33}$$

$$\Delta\varepsilon = \left(\frac{\Delta R}{R_0}\right)_4 - \left(\frac{\Delta R}{R_0}\right)_3 \tag{14.34}$$

where $E_{\sigma\varepsilon}$ is a modulus corresponding to the strain amplitude $\Delta\varepsilon$, $E_{\sigma i}$ would be obtained from a very small unload-reload loop at p_4, $\Delta\varepsilon$ is given by Eq. 14.34 and Figure 14.15, and p_L is the PMT limit pressure. In practice, the unload-reload loop is performed by unloading the pressure around the p_y value (p_4; Figure 14.15) down to a pressure equal to about one-half of that value ($\Delta p = p_4/2$). The reload modulus E_r obtained in this fashion and used with elastic equations seems to give reasonable to optimistic values of settlement.

A PMT modulus can also be obtained as a function of the number of cycles by repeating the unload-reload loop (Figure 14.17). As can be seen, the secant modulus E_{sN} to the top of the cycle will decrease as the number of cycles increases, but the cyclic modulus E_{cN} obtained from the slope of the unload-reload loop will increase—at least at low stress-to-strength ratios. In both cases, the cyclic exponent can be obtained from the evolution of the modulus as a function of cycles. It remains important to think whether or not the loading process around the pressuremeter (stress path) is analogous to the loading process in the geotechnical problem at hand.

$$E_{sN} = E_{s1}N^{-n_c} \tag{14.35}$$

$$E_{cN} = E_{c1}N^{n_c} \tag{14.36}$$

14.16 RESILIENT MODULUS FOR PAVEMENTS

The *resilient modulus* is used in pavement engineering to quantify the deformation characteristic of the various layers involved in the response of a pavement to the cyclic loading of traffic. It is measured in a triaxial test and is defined as the ratio of the applied cyclic stress to the recoverable strain after many cycles of repeated loading (Figure 14.18):

$$M_R = \frac{\Delta\sigma_c}{\Delta\varepsilon_c} \tag{14.37}$$

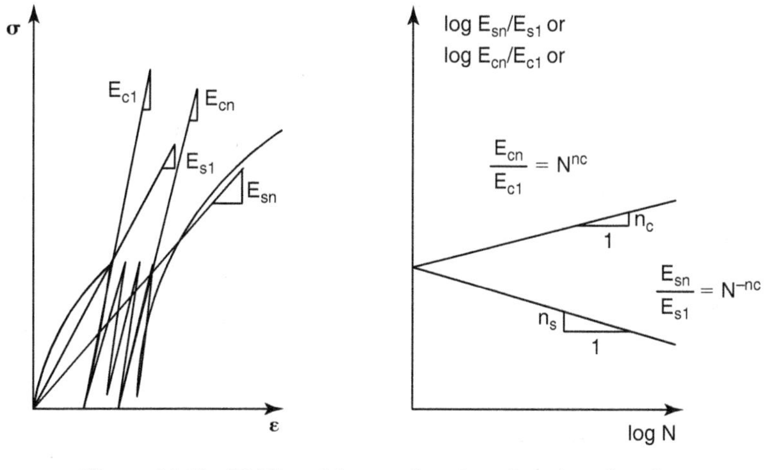

Figure 14.17 PMT modulus as a function of number of cycles.

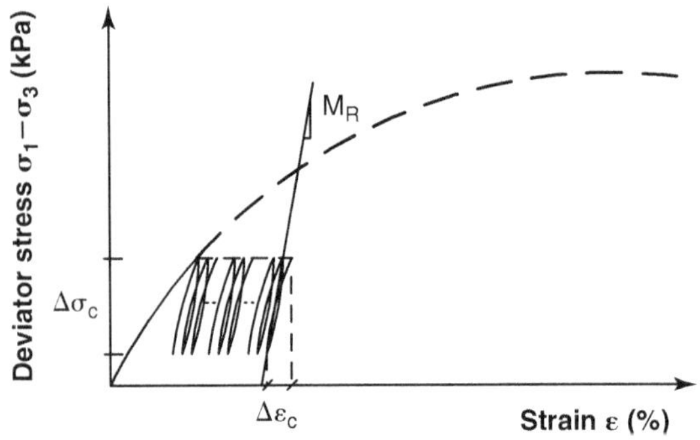

Figure 14.18 Resilient modulus test result.

where $\Delta\sigma_c$ is the amplitude of the deviatoric cyclic stress $(\sigma_1 - \sigma_3)$ over which the cycles are performed (Figure 14.18) and $\Delta\varepsilon_c$ is the corresponding strain amplitude after many cycles. The number of cycles should be sufficient for the value of M_R to reach a constant value for several cycles in a row. Generally, the value of M_R will increase (stiffening) as the number of cycles increases, at least if the peak stress is much lower than the strength of the soil. However, if the peak cyclic stress (top of the cycle) is too close to the strength of the soil sample, the value of M_R will decrease as the number of cycles increases, and the sample may fail at a cyclic strength value less than the static strength value.

Table 14.17 Default MR Values for Unbound Granular and Subgrade Materials at Unsoaked Optimum Moisture Content and Density Conditions

USCS Soil Class	Range of resilient modulus (MPa)	Typical value (MPa)
GW	2725–2898	2829
GP	2449–2760	2622
GM	2277–2898	2656
GC	1656–2587	2070
GW-GM	2449–2794	2656
GP-GM	2070–2760	2484
GW-GC	1932–2760	2380
GP-GC	1932–2691	2346
SW	1932–2587	2208
SP	1656–2277	1932
SM	1932–2587	2208
SC	1483–1932	1656
SW-SM	1656–2277	1932
SP-SM	1656–2277	1932
SW-SC	1483–2070	1759
SP-SC	1483–2070	1759
ML	1173–1759	1380
CL	931–1656	1173
MH	480–1207	793
CH	345–931	480

(After FHWA 2006.)

Table 14.18 Typical Poisson's Ratio Values for Geomaterials in Pavements

Material Description	Poisson's ratio ν Range	ν Typical
Clay (saturated)	0.4–0.5	0.45
Clay (unsaturated)	0.1–0.3	0.2
Sandy clay	0.2–0.3	0.25
Silt	0.3–0.35	0.325
Dense sand	0.2–0.4	0.3
Coarse-grained sand	0.15	0.15
Fine-grained sand	0.25	0.25
Bedrock	0.1–0.4	0.25

(After FHWA 2006.)

The confinement stress and the amplitude of the deviator stress should be chosen to match the expected values during the pavement loading. The confinement stress is fairly small for pavements, as the layers are not very thick and the depth is shallow. Typical confinement for pavement layers varies from 30 to 200 kPa. The amplitude of the deviator stress also depends on the depth below the rolling surface, and may be 50 to 200 kPa for cars, 100 to 500 kPa for trucks, and 350 to 1500 kPa for airplanes. Table 14.17 gives some range and typical values of the resilient modulus and Table 14.18 gives some ranges and typical values of the Poisson's ratio for soils in pavement layers.

14.17 UNSATURATED SOILS: EFFECT OF DRYING AND WETTING ON THE MODULUS

When a soil dries, it becomes stiffer, because the water tension that develops acts as glue by increasing the effective stress between particles; thus, the soil skeleton becomes stiffer. This has been studied particularly for pavement geotechnics, where the compaction curve helps to document the influence of the water content on the dry density and on the modulus of a soil. Figure 14.19 shows the variation of the soil modulus as a function of the water content. This modulus curve was obtained with the BCD (see Chapter 9, section 9.4). As can be seen, the modulus decreases drastically on the wet side of optimum, but the variation is not as severe on the dry side of optimum. Note that this variation of modulus with water content is associated with a remolded soil that was prepared by compaction at each water content. It does not represent the increase in modulus as the soil dries.

Some models have been proposed to document the increase in modulus with an increase in water tension and a decrease

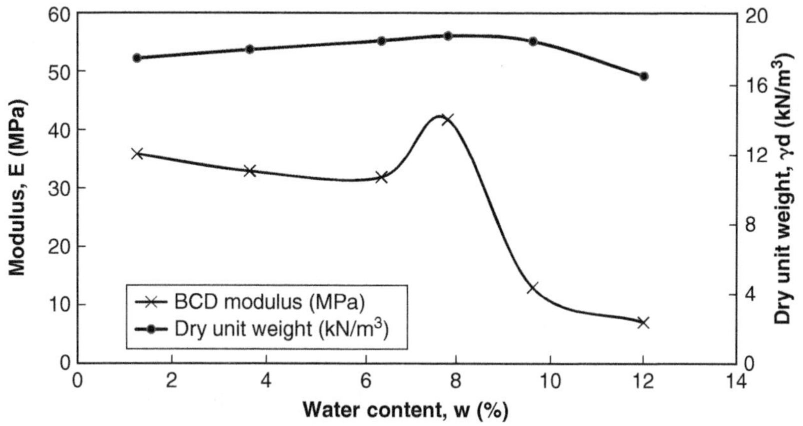

Figure 14.19 Variation of soil modulus measured in the Proctor Compaction Test.

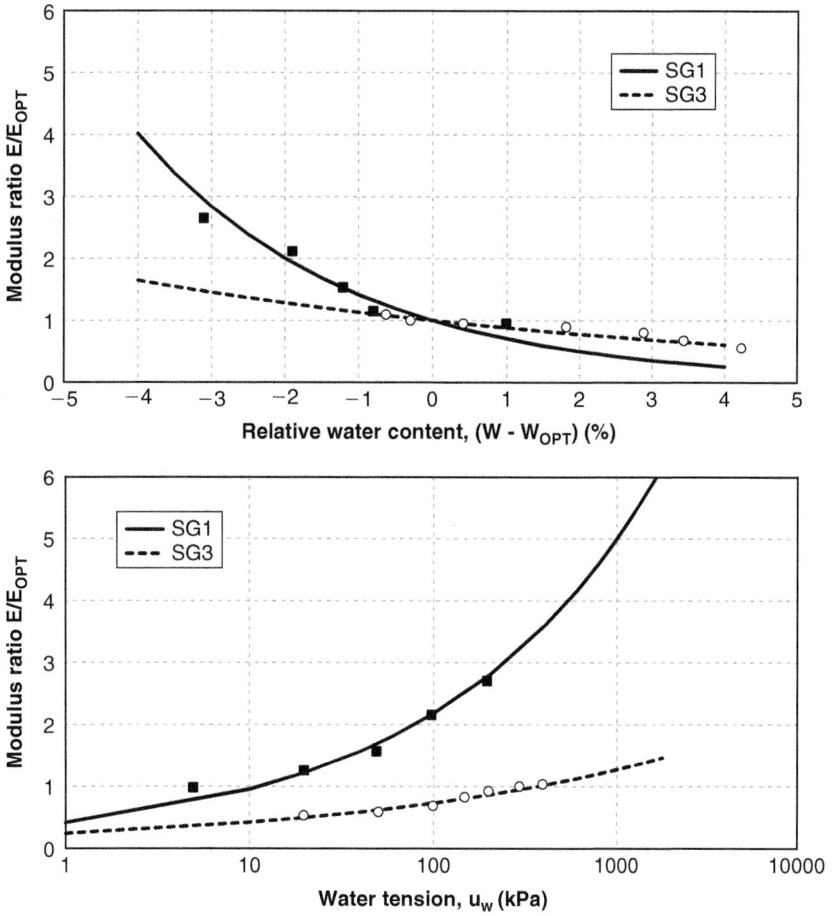

Figure 14.20 Soil modulus as a function of water content and water tension. (After Kim et al. 2006)

in water content. They use the optimum water content w_{opt} as the reference:

$$\frac{E}{E_{opt}} = 10^{k_1(w-w_{opt})} \tag{14.38}$$

where E and E_{opt} are the modulus at the water content w and w_{opt} (in percent) respectively and k_1 is a dimensionless constant. Kim et al. (2006) measured values of k_1 ranging from −0.05 to −0.15 and present the relation shown in Figure 14.20. In terms of water stress or suction:

$$\frac{E}{E_{opt}} = k_2\left(\frac{u_w}{u_{w\,opt}}\right)^{k_3} \tag{14.39}$$

where E and E_{opt} are the modulus at the water tension u_w and the water tension $u_{w\,opt}$ corresponding to the optimum water content respectively, and k_2 and k_3 are dimensionless constants. Kim et al. (2006) measured values of k_2 close to 1 and values of k_3 ranging from 0.25 to 0.35 (Figure 14.20).

14.18 SHRINK-SWELL DEFORMATION BEHAVIOR, SHRINK-SWELL MODULUS

Now let's talk about deformations due to changes in water content or water tension in soils that are not collapsible soils. When the unstressed soil has access to as much water as it can absorb, the soil swells and the water content ends up reaching an equilibrium value called the *free swell limit* w_{SW}. It is not possible for the water content w to be higher than this value under natural conditions. As the soil dries, it shrinks and decreases in volume. This happens because the loss of water brings the particles closer together, as the water no longer occupies that space. During this shrinking process, the soil remains saturated until it reaches the shrinkage limit w_{SH}. Air then enters the voids, the water content continues to decrease, and the soil stops changing volume, or at least the volume change is drastically decreased.

The soil remains saturated from the swell limit to the shrink limit, but the water tension gradually increases. At the shrink limit, the water tension reaches the air entry value u_{wae}, air first enters the soil pores, the soil starts to lose saturation, and the volume stops decreasing (see Chapter 4, Figure 4.15). You can think of this shrinking process as the soil particles coming closer and closer together. At the shrink limit, they touch each other and therefore the volume change stops—but the soil can still lose water if air comes into the voids to replace the water. For high-plasticity clays, this is not quite true: the soil continues to decrease in volume past the shrink limit, but at a much slower rate (Figure 4.15). This is because water is bound to the surface of the particle to various degrees and the transition is not as clearly defined as in the case of a sand or a gravel.

It turns out that the slope of the water content w vs. the relative decrease in volume $\Delta V/V$ (volumetric strain) between the swell limit and the shrinkage limit is well approximated by a straight line (Figure 14.21). During the seasons, the soil within a few meters of the ground surface shrinks and swells along this straight line. The double slope line linking the water content to the volumetric strain is to shrink-swell soils what the stress-strain curve is to compressible soils: a constitutive law. In this analogy, the water content plays the role of the stress. Note that the water content could be replaced by the log of the water tension to define the same type of curve. The slope of the straight line from the shrink limit to the swell limit is called the *shrink-swell modulus* E_{SS}, by analogy with the loading problem.

$$E_{SS} = \frac{\Delta w}{\Delta(\Delta V/V)} \tag{14.40}$$

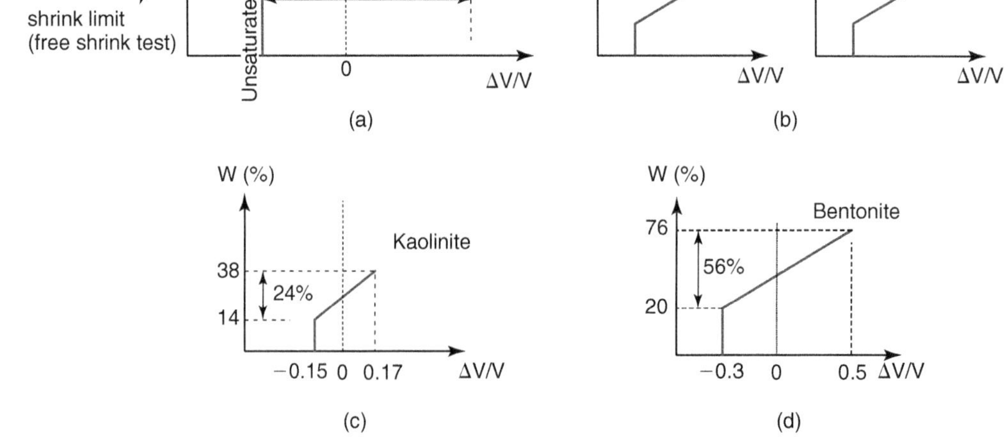

Figure 14.21 Water content versus relative change in volume: *(a)* Idealized behavior. *(b)* Typical ranges. *(c)* Low-plasticity clay, example. *(d)* High-plasticity clay, example.

Recall that the soil remains saturated along the straight line. Therefore, the change in soil volume ΔV along that line is also the change in volume of water ΔV_w and thus:

$$E_{SS} = \frac{\Delta w}{\Delta(\Delta V/V)} = \frac{\Delta W_w}{W_s} \times \frac{V_t}{\Delta V_w} = \frac{\gamma_w}{\gamma_d} \quad (14.41)$$

Equation 14.41 shows that the shrink-swell modulus E_{SS} is equal to the ratio of the unit weight of water over the dry unit weight of the soil. Unlike the Young's modulus E, the shrink-swell modulus is a constant for a given soil and does not vary much from soil to soil, with values in the range of 0.5 to 1 and an average of 0.7. The shrink-swell modulus E_{SS} can be obtained in the laboratory by performing a free shrink test (see Chapter 9, section 9.6) or a free swell test (section 9.7). The much simpler free shrink test gives the shrink limit, but the free swell test is necessary to obtain the swell limit. The shrink-swell modulus E_{SS} is independent of the stress level. What happens is that if a vertical stress is applied, it will decrease the amount of swelling $\Delta V/V$ and the associated change in water content Δw, but it will not change the ratio that is E_{SS}. The vertical stress can reach a value large enough to prevent swelling altogether; that vertical stress is called the *swelling pressure* p_{SW} of the soil. By the same token, the swell limit depends on the vertical stress and the swell limit w_{SW} refers to the free swell limit, the one obtained when there is no pressure on top of the soil.

If you dug a hole in the soil under your feet, you would likely encounter three zones (Figure 14.22). The first zone would be unsaturated (degree of saturation less than 1). This zone corresponds to water contents below the shrink limit in Figure 14.21; therefore, this zone can swell all the way from the shrink limit to the swell limit if enough water becomes available, but this zone has very little shrinkage potential. The weather or a leaking pipe, for example, could affect the water content and water tension in that zone and create significant swelling. The water tension in this zone is quite high and above the air entry value, as the soil is unsaturated. The thickness of the first zone can vary drastically, from nearly nonexistent in wet regions to hundreds of meters in arid and desert regions.

If you kept on digging, you would encounter a second zone, a zone of saturated soil called the *capillary zone*. There would be no standing water in the hole because you would be above the groundwater level and the water would be held in the soil by water tension. The soil in this zone is at a natural water content somewhere between the swell limit and the shrink limit. Therefore, it can shrink or swell according to any change in water regime. The weather, for example, could affect the water content and water tension in that zone and create significant movements. The thickness of this second zone can also vary drastically, from nearly nonexistent near the coast or near rivers where the groundwater level is high to tens of meters of thickness. The thickness of this zone increases when the soil particles become finer; indeed, the height above the groundwater level to which the soil can draw water up and saturate the zone depends on the size of the soil particles (see Chapter 10, section 10.17.1).

If you kept on digging, you would encounter a third zone, a zone of saturated soil where, given enough time, water would collect in the hole and stand at a constant level called the *groundwater level* (GWL). The water content in this zone is at the swell limit, because the soil has access to all the water it can absorb and has been in this condition for a long time. Note that this swell limit is less than the free swell limit and depends on the vertical stress at rest applied at that depth. In this zone, the soil cannot swell, because it is at the swell limit, but it can shrink should water migrate upward (during a drought, for example). This would be associated with a lowering of the groundwater level.

One good way to tell if a soil is sensitive to shrink-swell deformation is to measure the water content difference between the swell limit and the shrinkage limit. This difference in percent is called the *shrink-swell index* I_{SS}. Shrink-swell

Figure 14.22 The three soil zones. (Courtesy of Art Koenig)

Table 14.19 Shrink-Swell Potential and Various Soil Parameters

Shrink-Swell Potential	Potential Volume Change[1]	Shrink-Swell Index[2]	Plasticity Index[3]	Percent Passing #200[4]	Swell Pressure(kPa)[5]	Slope of SWRC[6]
Very low	<5	<15	<10	<10	<50	>−5
Low	5–10	15–30	10–20	10–30	50–150	−7.5 to −5
Medium	10–20	30–45	20–30	30–60	150–250	−10 to −7.5
High	20–30	45–60	30–40	60–95	250 1000	−17 to −10
Very high	>30	>60	>40	>95	>1000	<−17

[1]Potential volume change ΔV/V in percent from dry to swell limit in free swell test.
[2]Difference between swell limit and shrink limit in percent.
[3]Difference between the liquid limit and the plastic limit in percent.
[4]Percent passing sieve number 200 with opening equal to 0.075 mm.
[5]Lowest pressure necessary to prevent swelling of an inundated sample.
[6]Slope of the soil water retention curve: $\Delta w = C_w \Delta(\log_{10}|u_w|)$ where Δw is the change in water content expressed as a percent and u_w is the water tension in kPa or any other unit, and C_w is the slope of the SWRC; it is negative because when the water content increases, the absolute value of the water tension decreases.

indices lower than 0.2 indicate a soil without much shrink-swell deformation potential. Shrink-swell indices with values above 0.6 indicate a soil with very high shrink-swell deformation potential. Other indices to evaluate the shrink-swell potential of a clay include the plasticity index, the percent passing sieve #200, the swelling pressure, and the slope of the soil water retention curve. Table 14.19 gives some guidance on the shrink-swell potential of clays.

It is the movements in these three zones that accumulate to create the shrink-swell movement of the ground surface. Shrink-swell deformations take place when two conditions exist: (1) there is a water content change between the shrinkage limit and the swell limit, and (2) the soil is sensitive to such water content changes. The natural water content cannot be higher than the free swell limit, by definition of the free swell limit. If the water content changes but remains below the shrink limit, there is either no change or very little change in soil volume, again by definition of the shrink limit. Because the soil is saturated between the shrink limit and the swell limit, most of the volume change of a soil takes place when the soil is saturated. The volume change is given by:

$$\frac{\Delta V}{V} = \frac{\Delta w}{E_{SS}} \tag{14.42}$$

Often, one is not interested in the volume change but in the change in height. Free shrink tests indicate that the shrinkage is about the same in all directions and therefore:

$$\frac{\Delta H}{H} = \frac{\Delta V}{3V} = \frac{1}{3}\frac{\Delta w}{E_{SS}} \tag{14.43}$$

Water content and water tension are tied together through the soil water retention curve. The main part of the SWRC leads to a linear relationship between the water content w and the decimal log of the water tension u_w (see Chapter 9, Figure 9.63):

$$\Delta w = C_w \Delta(\log_{10} u_w) \tag{14.44}$$

Therefore,

$$\frac{\Delta H}{H} = \frac{\Delta V}{3V} = \frac{1}{3}\left(\frac{C_w \Delta\left(\log_{10} u_w\right)}{E_{SS}}\right) \tag{14.45}$$

14.19 COLLAPSE DEFORMATION BEHAVIOR

Collapsible soils consist of loose, dry, low-density materials (say, less than 16 kN/m^3) that decrease in volume (collapse and compact) under the addition of water. These soils are often found in arid regions, specifically in areas of wind-blown silty sediments (loess), young alluvial fans, and debris flow sediments. Soil collapse can occur in these soils when they are above the groundwater level. The process of saturation weakens or eliminates the clay bonds holding the soil grains together through water tension.

Here are some indicators to help recognize if a soil is collapsible (USACE 1990):

1. Liquid limit below 45
2. Plasticity index below 25
3. Dry unit weight between 10 and 17 kN/m^3
4. Porosity between 40 and 60%

Table 14.20 Guidelines to Determine Collapsible Soils

Criterion	Source
$\gamma_d(\text{kN}/m^3) < \dfrac{25.5}{1+0.026\text{LL}(\%)}$	Gibbs and Bara, 1962
or	
$e_o > \dfrac{2.6LL(\%)}{100}$	
$\dfrac{\dfrac{w_o}{S_o} - PL(\%)}{PI(\%)} > 0.85$	Feda, 1966

γ_d = dry unit weight, LL = liquid limit, e_o = natural void ratio, w_o = natural water content, S_o = natural degree of saturation, PL = plastic limit, PI = plasticity index

Table 14.20 gives additional criteria to determine if a soil can be collapsible or not.

Once it is recognized that a soil may be collapsible using these criteria, the extent of the collapse and its severity can be gauged by the scale proposed by Jennings and Knight (1975). Their scale is based on the collapse potential index CP:

$$CP = \frac{e_o - e_c}{1 + e_o} \times 100 \qquad (14.46)$$

where e_o is the void ratio of the soil at its natural water content under 200 kPa of vertical pressure in the consolidation test before wetting, and e_c is the void ratio after soaking under 200 kPa of vertical pressure. Table 14.21 gives a severity scale for the collapse.

The best way to determine the amount of collapse that may occur is to perform a consolidation test and simulate what would happen to the soil in the field. For this, the sample at its natural water content is placed in the consolidometer, the sample height h is recorded, and the vertical pressure p is increased in steps. For each step, the change in height Δh of the sample is recorded every 30 minutes and the curve of stress p vs. strain $\Delta h/h$ is plotted (Figure 14.23). Each pressure is kept on the sample until the rate of strain is less than 0.1%/hour. When the vertical pressure p reaches the pressure p_f anticipated in the field (under the foundation, for example) and at the end of that load step, the sample is inundated and the readings of strain continue during the collapse as a function of time. The end of the collapse step is when the strain has become less than 0.1%/hour. The next pressure step is applied, and so on, until the curve is completed (Figure 14.23). For collapse pressures less than p_f, a line is drawn from A to C on the curve and used to estimate the collapse strain for intermediate values of p_f.

Table 14.21 Severity of Collapsible Soils Scale

CP, Collapse Potential in Percent	Severity
0–1	Negligible
1–5	Moderate trouble
5–10	Trouble
10–20	Severe trouble
> 20	Very severe trouble

(After Jennings and Knight 1975.)

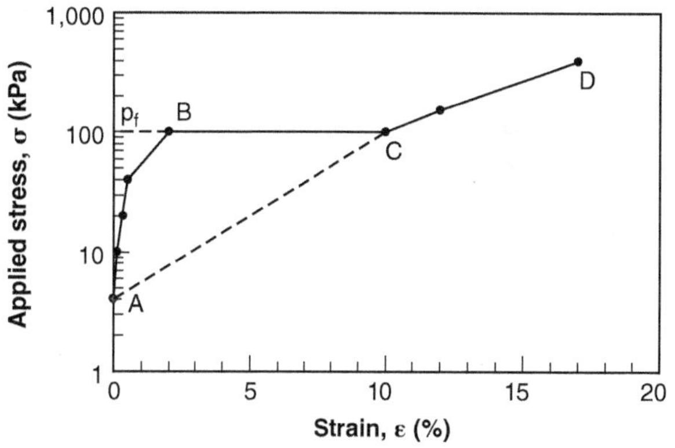

Figure 14.23 Stress-strain curve for a collapsible soil.

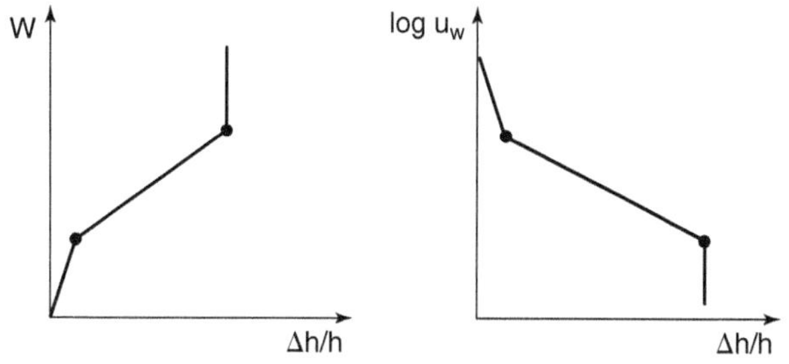

Figure 14.24 Water content and water tension vs. collapse strain.

In this consolidation test, the sample is inundated and the collapse occurs under the extreme situation in which the sample can absorb all the water it is able to absorb. In reality, there could be a limit to the amount of water available to the soil (the rainstorm stops, for example) and the collapse might be partial. A model to predict the collapse strain under partial wetting links the water content to the strain or the water tension to the strain (Figure 14.24) (Pereira and Fredlund 2000).

PROBLEMS

14.1 Consider the stress-strain curve from a triaxial test shown in Figure 14.1s:
   a. Why is $\varepsilon_r < 0$ when $\varepsilon_z > 0$?
   b. Calculate Poisson's ratio.
   c. Calculate the soil modulus between 0 and A.
   d. Calculate the ratio between this soil modulus and the modulus of concrete.

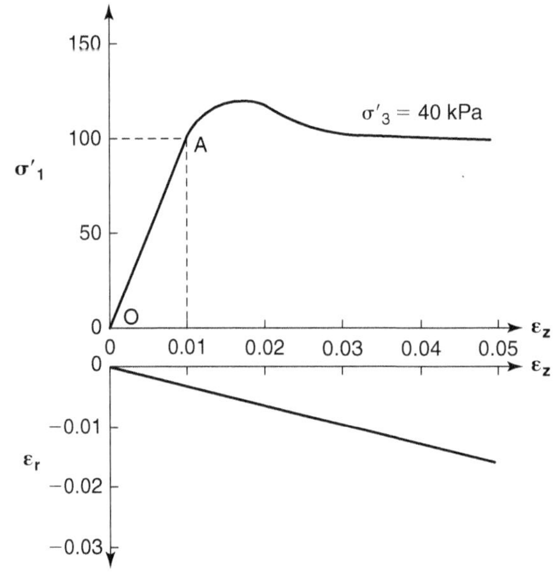

Figure 14.1s Triaxial test results.

14.2 Given that Figure 14.2s is the result of an unconfined compression test, calculate the secant modulus (OA), the unload modulus (AB), the unload-reload modulus (BC), the reload modulus (BD), and the tangent modulus at A. Which is the smallest modulus? Which is the largest modulus? Which is the right modulus?

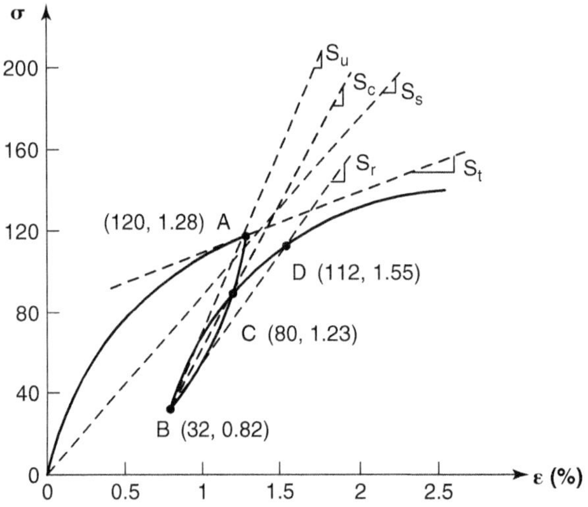

Figure 14.2s Modulus values.

14.3 Explain the difference between the modulus, the stiffness, and the modulus of subgrade reaction. Comment on which one is a true soil property and why.

14.4 Equation 14.12 gives the secant modulus for any confinement level, any strain level, any time of loading, and any number of cycles. *If* E_{ai} is equal to 1000 MPa, n is 0.5, σ_{ult} is 100 kPa, t_o is 1 minute, m is 0.03, and p is 0.1:

a. Plot the initial tangent modulus E_i as a function of the confinement level σ_M for the reference loading time t_o and for monotonic loading ($N = 1$).

b. Plot the secant modulus E_s as a function of the strain level ε for a confinement of 50 kPa, for the reference loading time t_o, and for monotonic loading ($N = 1$).

c. Plot the secant modulus E_s as a function of the time of loading t for a confinement stress of 50 kPa, a strain of 0.5%, and for monotonic loading ($N = 1$).

d. Plot the secant modulus E_s as a function of the number of cycles N for a confinement of 50 kPa, an initial strain of 0.5%, and the reference time t_o.

$$E_{\sigma \varepsilon t N} = \left(\frac{1}{E_{ai}\left(\frac{\sigma_M}{p_a}\right)^n} + \frac{\varepsilon}{\sigma_{ult}} \right)^{-1} \left(\frac{t}{t_o}\right)^{-m} N^{-p} \tag{14.12}$$

14.5 A soil sample has a void ratio e = 0.6, an OCR = 2, a PI = 20%, and a shear strength of 40 kPa at a confining pressure of 70 kPa. Use an equation similar to Eq. 14.12 for the shear modulus G and prepare two plots of G/G_{max} versus γ. The first one is G/G_{max} versus γ on natural scales and the second one is G/G_{max} on the vertical natural scale and γ on the horizontal decimal log scale. What other influencing factors are missing from this classical G-γ curve?

14.6 Given the log of vertical effective stress vs. vertical strain curve (Figure 14.3s), find the preconsolidation pressure σ'_p and calculate the compression index C_c.

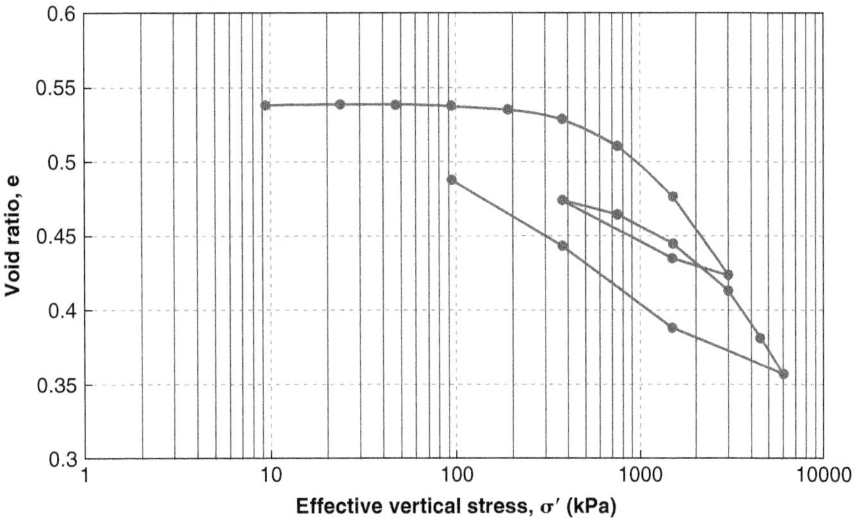

Figure 14.3s Strain vs. log of stress consolidation curve.

14.7 Given a straight-line relationship (Figure 14.4s) between the vertical effective stress and the vertical strain ($\sigma' = \sigma'_{ov} + 40{,}000\ \varepsilon$), a vertical effective stress at rest of 100 kPa, and an initial void ratio e_o of 1, draw the log of vertical stress vs. void ratio curve, and find the preconsolidation pressure σ'_p and the compression index C_c from that curve. Discuss.

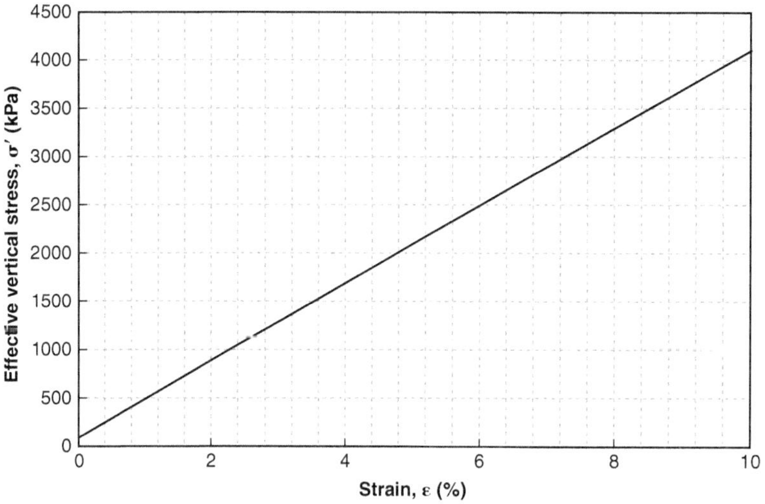

Figure 14.4s Effective vertical stress vs. strain.

14.8 Given the three vertical strain versus time curves of Figure 14.5s from a consolidation test with drainage top and bottom, and the original height of the sample of 14.2 mm, calculate the coefficient of consolidation c_v by the t_{50} method and by the log time method.

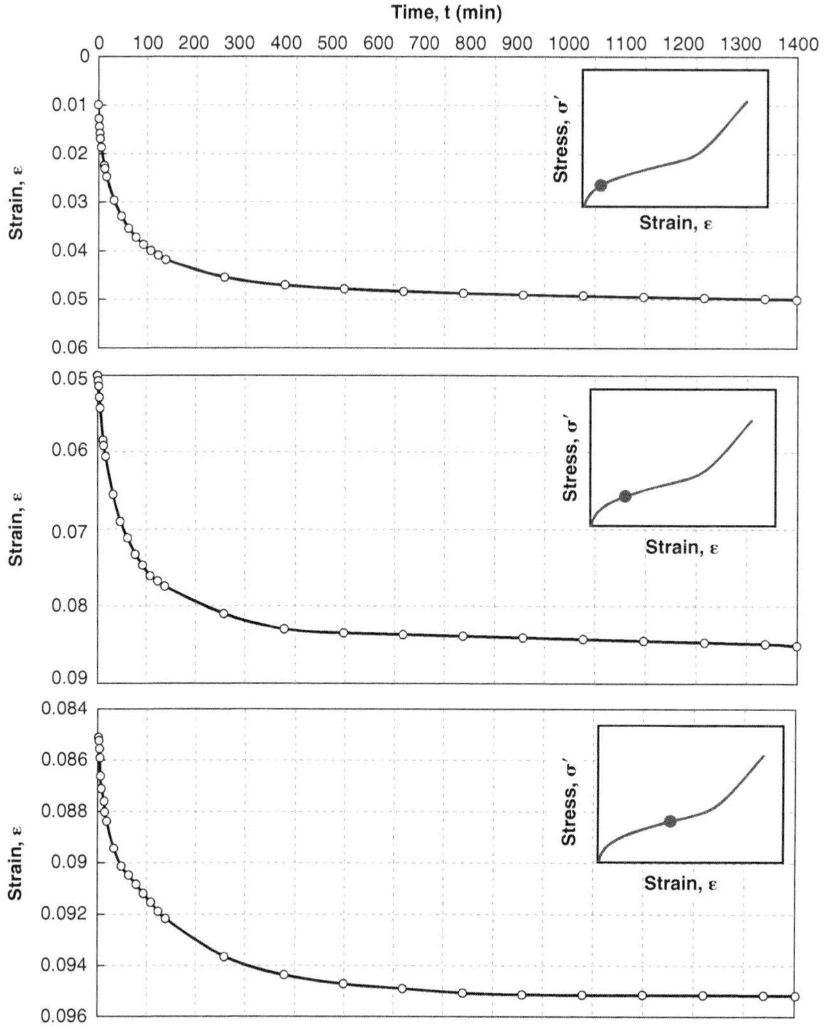

Figure 14.5s Strain vs. time consolidation curves.

14.9 Given the strain versus. time curve of Figure 14.6s, and knowing that the initial void ratio e_o is 0.7, calculate the secondary compression index C_α.

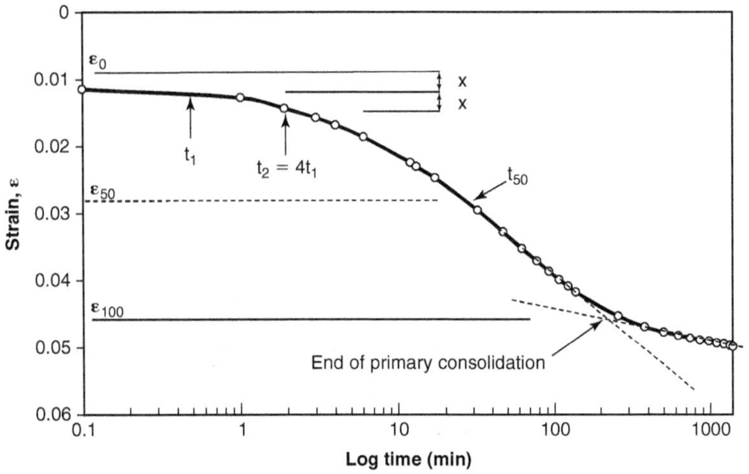

Figure 14.6s Strain vs. log time consolidation curve.

14.10 Devise a pressuremeter test procedure that allows you to measure as many parameters as possible for equation 14.12.

14.11 Give the range of shrink-swell modulus that can be expected for soils. Use that range and the range of shrink-swell indices in Table 14.19 to give the range of expected relative volume change in shrink-swell soils.

14.12 Which of the following two soils is the most likely to collapse upon wetting?

a. Silt with a dry unit weight of 14 kN/m^3, a liquid limit of 40%, a plastic limit of 20%, a porosity of 50%, and a natural water content of 10%

b. Clay with a dry unit weight of 16 kN/m^3, a liquid limit of 55, a plastic limit of 20, a porosity of 35%, and natural water content of 20%

Problems and Solutions

Problem 14.1

Consider the stress-strain curve from a triaxial test shown in Figure 14.1s:

a. Why is $\varepsilon_r < 0$ when $\varepsilon_z > 0$?
b. Calculate Poisson's ratio.
c. Calculate the soil modulus between 0 and A.
d. Calculate the ratio between this soil modulus and the modulus of concrete.

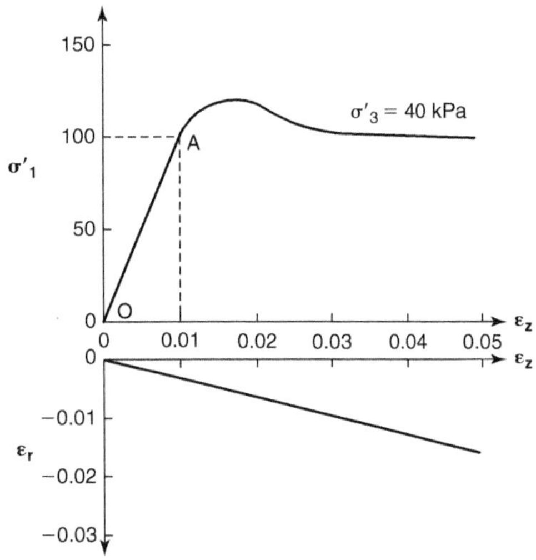

Figure 14.1s Triaxial test results.

Solution 14.1

The Poisson's ratio is obtained from the elasticity equations (Eq. 14.3). In the principal directions, the Poisson's ratio is given by:

$$\nu = \frac{-\varepsilon_3\sigma_1 + \varepsilon_1\sigma_3}{\varepsilon_1\sigma_1 + \varepsilon_1\sigma_3 - 2\varepsilon_3\sigma_3} \quad (14.1s)$$

Considering the geometry of the triaxial test:

$$\varepsilon_z = \varepsilon_1 \quad and \quad \varepsilon_r = \varepsilon_3 \quad (14.2s)$$

At point A, the values of the strains are:

$$\varepsilon_z = \varepsilon_1 = 0.01 \quad and \quad \varepsilon_r = \varepsilon_3 = -0.004 \quad (14.3s)$$

At point A, the values of the stresses are:

$$\sigma_z = \sigma_1 = 100\,\text{kPa} \quad and \quad \sigma_r = \sigma_3 = 40\,\text{kPa} \quad (14.4s)$$

Then the Poisson's ratio is calculated as:

$$\nu = \frac{-(-0.004) \times 100 + 0.01 \times 40}{0.01 \times 100 + 0.01 \times 40 - 2 \times (-0.004) \times 40} = 0.47 \quad (14.5s)$$

Note that ν is different from the ratio of $-\varepsilon_r/\varepsilon_z$, which would be 0.4. The soil modulus E (Eq. 14.1) is given by:

$$E = \frac{\sigma_1 - 2\nu\sigma_3}{\varepsilon_1} = \frac{100 - 2 \times 0.47 \times 40}{0.01} = 6240\,\text{kPa} \quad (14.6s)$$

Note that E is different from the ratio σ_1/σ_3, which would be 10,000 kPa. The ratio between this soil modulus and the modulus of concrete (20,000 MPa) is:

$$\frac{E_{Concrete}}{E_{Soil}} = \frac{20000 \times 10^3}{6240} = 3205 \quad (14.7s)$$

Problem 14.2

Given that Figure 14.2s is the result of an unconfined compression test, calculate the secant modulus (OA), the unload modulus (AB), the unload-reload modulus (BC), the reload modulus (BD), and the tangent modulus at A. Which is the smallest modulus? Which is the largest modulus? Which is the right modulus?

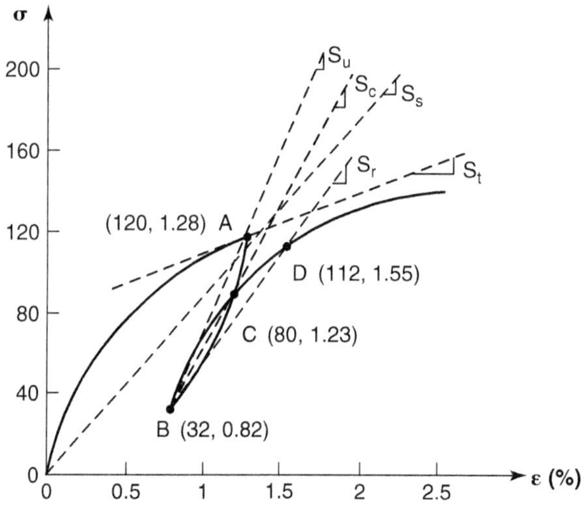

Figure 14.2s Modulus values.

Solution 14.2

Because the test is an unconfined compression test, each modulus can be calculated directly from the slopes in Figure 14.2s. From the secant slope OA (S_s), the secant modulus is:

$$E_s = \frac{\Delta\sigma}{\Delta\varepsilon} = \frac{120 - 0}{0.0128 - 0} = 9375\ \text{kPa} \tag{14.8s}$$

From the unload slope AB (S_u), the unload modulus is:

$$E_u = \frac{\Delta\sigma}{\Delta\varepsilon} = \frac{120 - 32}{0.0128 - 0.0082} = 19130\ \text{kPa} \tag{14.9s}$$

From the cyclic slope BC (S_c), the unload-reload modulus is:

$$E_c = \frac{\Delta\sigma}{\Delta\varepsilon} = \frac{80 - 32}{0.0123 - 0.0082} = 10707\ \text{kPa} \tag{14.10s}$$

From the reloading slope BD (S_r), the reload modulus is:

$$E_r = \frac{\Delta\sigma}{\Delta\varepsilon} = \frac{112 - 32}{0.0155 - 0.0082} = 10960\ \text{kPa} \tag{14.11s}$$

From the tangent slope (S_t), the tangent modulus is:

$$E_t = \frac{\sigma}{\varepsilon} = \frac{167 - 80}{0.03 - 0} = 2900\ \text{kPa} \tag{14.12s}$$

- The smallest modulus is the tangent modulus.
- The largest modulus is the unload modulus.
- There is no "right" modulus. Each slope or modulus is used for different situations or applications.

Problem 14.3

Explain the difference between the modulus, the stiffness, and the modulus of subgrade reaction. Comment on which one is a true soil property and why.

Solution 14.3

The modulus of deformation (kN/m^2) is defined by the equations of elasticity and as the slope of the line of a stress-strain curve of a material in the case of an unconfined compression test. Stiffness (kN/m) is the ratio of a force Q applied on a boundary through a loading area divided by the displacement s experienced by the loaded area (square or circular shape). The modulus of subgrade reaction (kN/m^3) is the ratio of pressure p applied to the boundary through a loading area divided by the displacement s experienced by the loaded area. Only the modulus of deformation is a true soil property, because stiffness and modulus of subgrade reaction depend on the size of the loaded area. The results of stiffness and modulus of subgrade reaction in one test will be different from the results of other tests with different areas. The modulus of deformation for the same material is not affected by the size of the loaded area.

Problem 14.4

Equation 14.12 gives the secant modulus for any confinement level, any strain level, any time of loading, and any number of cycles. If E_{ai} is equal to 1000 MPa, n is 0.5, σ_{ult} is 100 kPa, t_o is 1 minute, m is 0.03, and p is 0.1:

a. Plot the initial tangent modulus E_i as a function of the confinement level σ_M for the reference loading time t_o and for monotonic loading ($N = 1$).
b. Plot the secant modulus E_s as a function of the strain level ε for a confinement of 50 kPa, for the reference loading time t_o, and for monotonic loading ($N = 1$).
c. Plot the secant modulus E_s as a function of the time of loading t for a confinement stress of 50 kPa, a strain of 0.5%, and for monotonic loading ($N = 1$).

d. Plot the secant modulus E_s as a function of the number of cycles N for a confinement of 50 kPa, an initial strain of 0.5%, and the reference time t_o.

$$E_{\sigma\varepsilon tN} = \left(\frac{1}{E_{ai}\left(\frac{\sigma_M}{p_a}\right)^n} + \frac{\varepsilon}{\sigma_{ult}} \right)^{-1} \left(\frac{t}{t_o}\right)^{-m} N^{-p} \tag{14.12}$$

Solution 14.4

a. For the initial tangent modulus, for the reference time t_o, and for monotonic loading ($N = 1$), the general equation becomes as shown in Eq. 14.13s, and the results are plotted in Figure 14.7s.

$$E_{\sigma\varepsilon tN}(\text{MPa}) = E_{ai}\left(\frac{\sigma_M}{p_a}\right)^n = 1000\left(\frac{\sigma_M\ (\text{kPa})}{100}\right)^{0.5} \tag{14.13s}$$

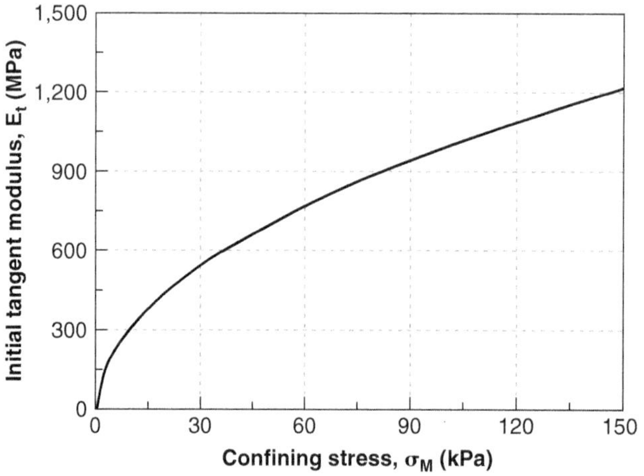

Figure 14.7s Initial tangent modulus vs confining stress.

b. For the secant modulus E_s as a function of the strain level ε for a confinement of 50 kPa, for the reference loading time t_o, and for monotonic loading ($N = 1$), the equation becomes as shown in Eq. 14.14s and the results are plotted in Figure 14.8s:

$$E_{\sigma\varepsilon tN}(\text{MPa}) = \left(\frac{1}{E_{ai}\left(\frac{\sigma_M}{p_a}\right)^n} + \frac{\varepsilon}{\sigma_{ult}} \right)^{-1} = (1.414 \times 10^{-3} + 10\varepsilon)^{-1} \tag{14.14s}$$

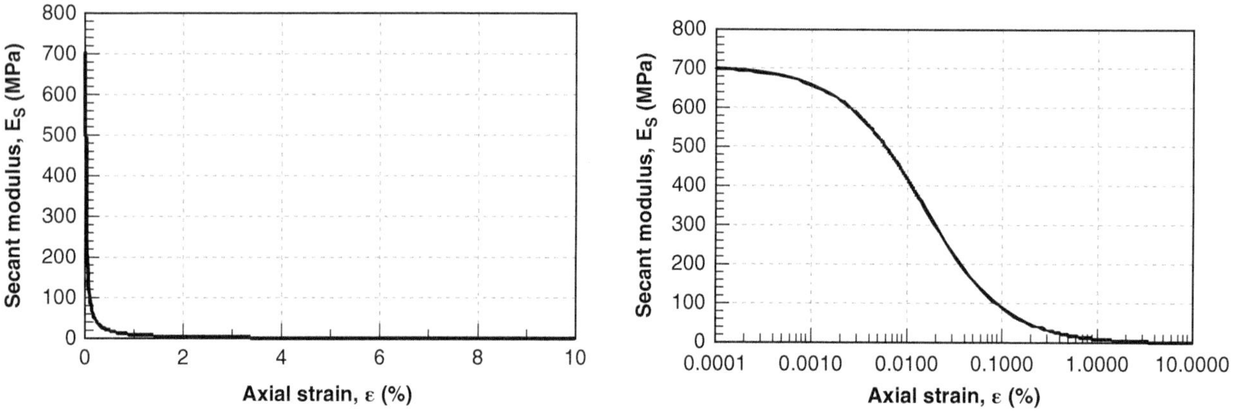

Figure 14.8s Secant modulus vs. axial strain.

c. For the secant modulus E_s as a function of the time of loading t for a confinement stress of 50 kPa, a strain of 0.5%, and for monotonic loading ($N = 1$), the equation becomes as shown in Eq. 14.15s and the results are plotted in Figure 14.9s:

$$E_{\sigma\varepsilon tN}(\text{MPa}) = \left(\frac{1}{E_{ai}\left(\frac{\sigma_M}{p_a}\right)^n} + \frac{\varepsilon}{\sigma_{ult}} \right)^{-1} \left(\frac{t}{t_o}\right)^{-m} = 19.45 \times (t(\text{min}))^{-0.03} \quad (14.15\text{s})$$

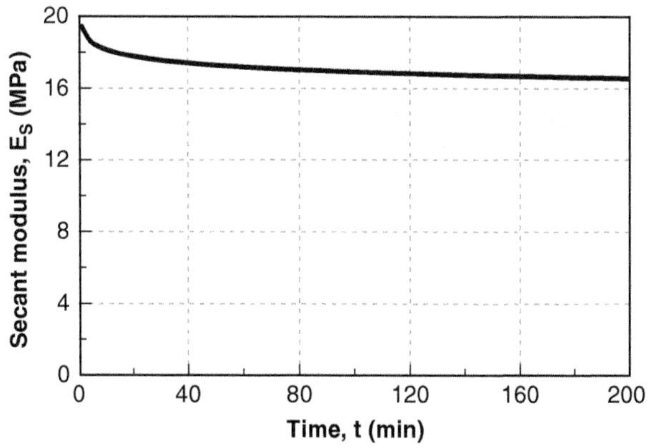

Figure 14.9s Secant modulus vs time

d. For the secant modulus E_s as a function of the number of cycles N for a confinement of 50 kPa, an initial strain of 0.5%, and the reference time t_o, the equation becomes as shown in Eq. 14.16s and the results are plotted in Figure 14.10s:

$$E_{\sigma\varepsilon tN} = \left(\frac{1}{E_{ai}\left(\frac{\sigma_M}{p_a}\right)^n} + \frac{\varepsilon}{\sigma_{ult}} \right)^{-1} N^{-p} = 19.45 \times N^{-0.1} \quad (14.16\text{s})$$

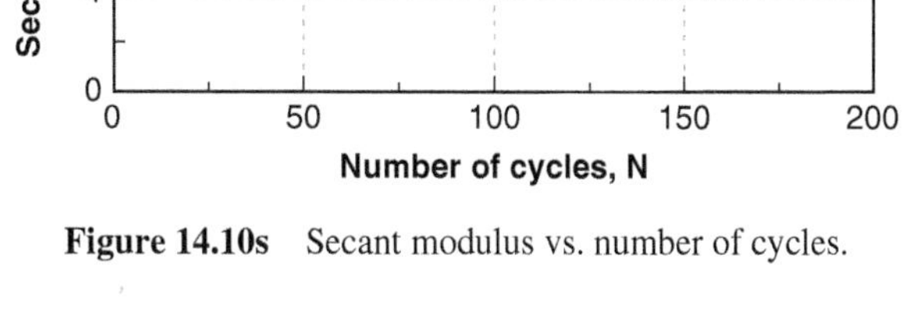

Figure 14.10s Secant modulus vs. number of cycles.

Problem 14.5

A soil sample has a void ratio $e = 0.6$, an OCR $= 2$, a $PI = 20\%$, and a shear strength of 40 kPa at a confining pressure of 70 kPa. Use an equation similar to Eq. 14.12 for the shear modulus G and prepare two plots of G/G_{max} versus γ. The first one is G/G_{max} versus γ on natural scales and the second one is G/Gmax on the vertical natural scale and γ on the horizontal decimal log scale. What other influencing factors are missing from this classical G-γ curve?

Solution 14.5

We select an equation of the form

$$G = \left(\frac{1}{G_{max}} + \frac{\gamma}{s}\right)^{-1} \tag{14.17s}$$

We are given $e = 0.6$, $OCR = 2$, $PI = 20\%$, $p_a = 101.325$ kPa, s = 40 kPa, and $\sigma_M = 70$ kPa. The equation proposed by Jamiolkowski (1991) can be used to estimate the maximum shear modulus (G_{max}). The overconsolidation exponent k is 0.18 for $PI = 20\%$ (Table 14.11).

$$\frac{G_{max}}{P_a} = \frac{625}{e^{1.3}}(OCR)^k\left(\frac{\sigma'_M}{P_a}\right)^{0.5} \tag{14.18s}$$

$$G_{max} = 101.325\left(\frac{625}{(0.6)^{1.3}}(2)^{0.18}\left(\frac{70}{101.325}\right)^{0.5}\right) = 115844 \text{ kPa} \tag{14.19s}$$

Then the equation for G/G_{max} is:

$$\frac{G}{G_{max}} = \frac{1}{G_{max}}\left(\frac{1}{G_{max}} + \frac{\gamma}{s}\right)^{-1} = \frac{1}{115844}\left(\frac{1}{115844} + \frac{\gamma}{40}\right)^{-1} \tag{14.20s}$$

The plots of G/G_{max} versus γ are shown in Figure 14.11s. This plot includes the effect of strain level and confinement level, but not rate effect or the influence of cycles.

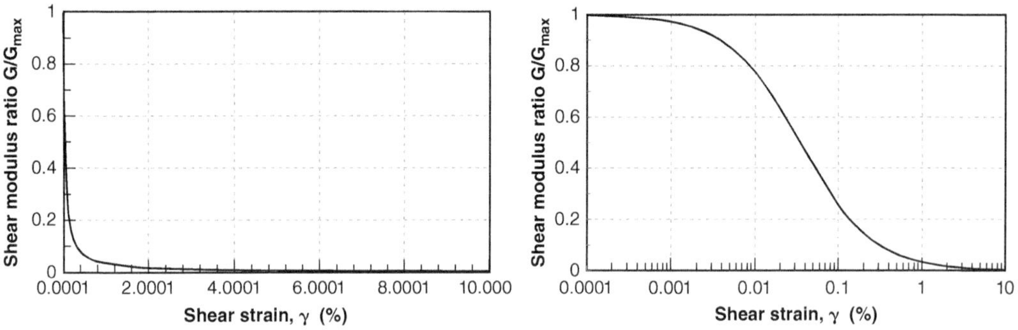

Figure 14.11s G/G_{max} vs. shear strain.

Problem 14.6

Given the log of vertical effective stress vs. vertical strain curve (Figure 14.3s), find the preconsolidation pressure σ'_p and calculate the compression index C_c.

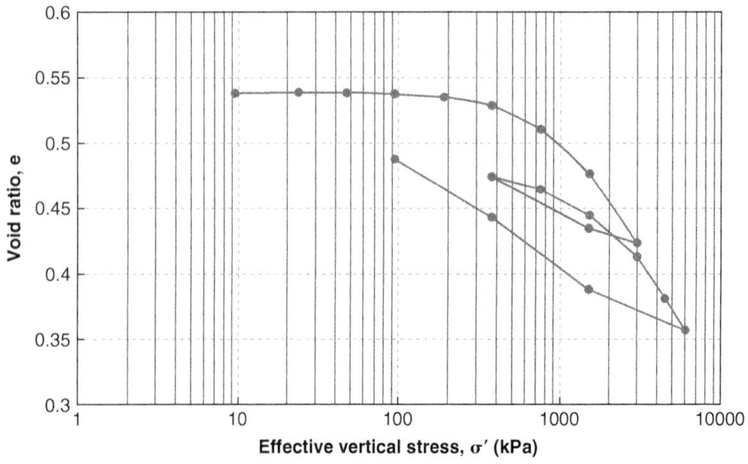

Figure 14.3s Strain vs. log of stress consolidation curve.

Solution 14.6

- Based on Cassagrande's method shown in Figure 14.10:

$$\sigma_p' = 1200(\text{kPa})$$

- The compression index C_c is:

$$C_c = \frac{\Delta e}{\log \sigma_1 - \log \sigma_2} = \frac{0.52 - 0.30}{\log\left(\frac{10000}{1000}\right)} = 0.22 \quad (14.21\text{s})$$

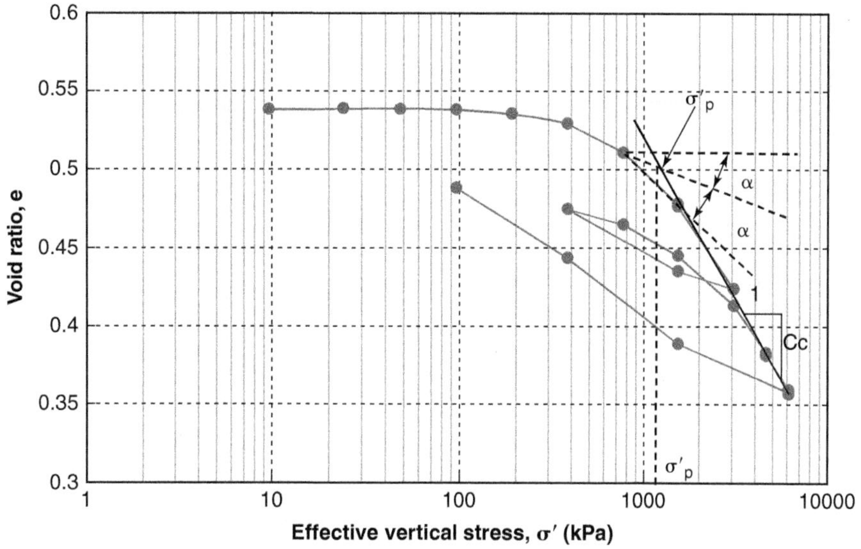

Figure 14.12s Void ratio vs. effective vertical stress.

Problem 14.7

Given a straight-line relationship (Figure 14.4s) between the vertical effective stress and the vertical strain ($\sigma' = \sigma_{ov}' + 40{,}000\,\varepsilon$), a vertical effective stress at rest of 100 kPa, and an initial void ratio e_0 of 1, draw the log of vertical stress versus void ratio curve, and find the preconsolidation pressure σ_p' and the compression index C_c from that curve. Discuss.

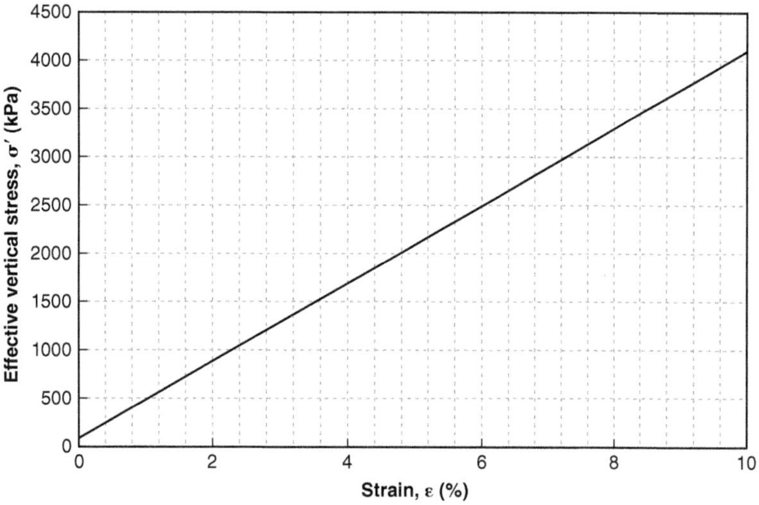

Figure 14.4s Effective vertical stress vs. strain.

Solution 14.7

The relationship between the strain and the void ratio is $\varepsilon = (e - e_o)/(1 + e_o)$. By substituting in the stress-strain equation, we get:

$$\sigma' = \sigma'_{ov} + 40{,}000\left(\frac{e - e_o}{1 + e_o}\right) = 100 + 20{,}000(e - 1) \tag{14.22s}$$

If we plot this equation as $\log \sigma'$ versus e, we get Figure 14.13s. Using this curve and Cassagrande construction, we find a preconsolidation pressure of the order of 2000 kPa (Figure 14.13s). Then we can calculate the compression index as:

$$C_c = \frac{\Delta e}{\log \sigma_1 - \log \sigma_2} = \frac{0.965 - 0.80}{\log\left(\frac{4000}{2000}\right)} = 0.548$$

As can be seen from Figure 14.13s, a preconsolidation pressure can be found for a soil that obviously does not have one (Figure 14.14s), as it is linear. The distortion created by the semilog plot creates the apparent preconsolidation pressure in this case.

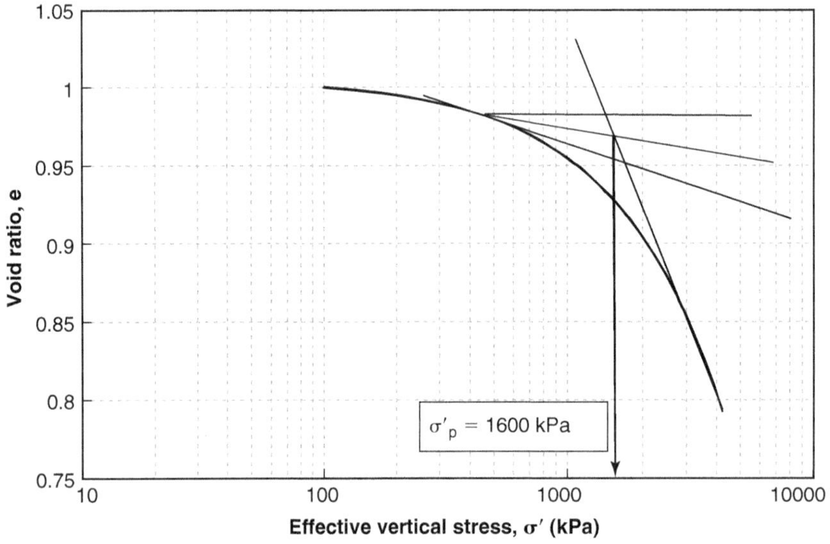

Figure 14.13s Preconsolidation pressure.

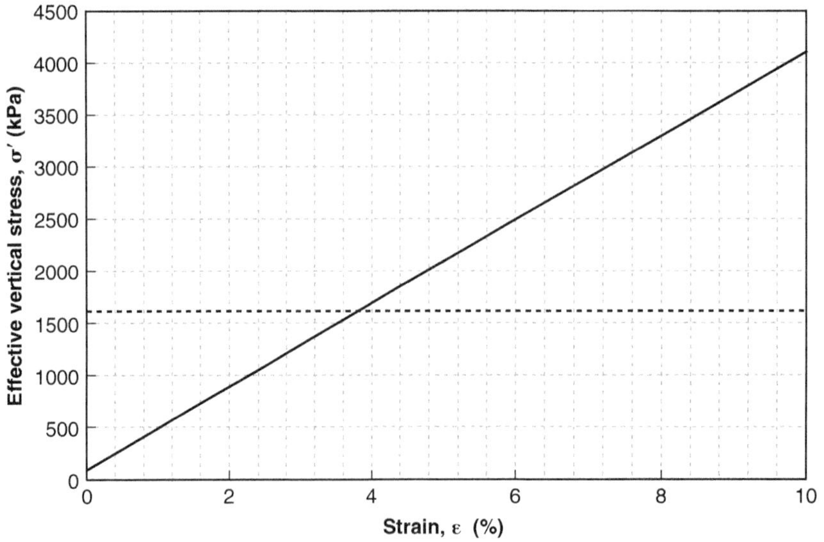

Figure 14.14s Effective vertical stress vs. strain.

Problem 14.8

Given the three vertical strain versus time curves of Figure 14.5s from a consolidation test with drainage top and bottom, and the original height of the sample of 14.2 mm, calculate the coefficient of consolidation c_v by the t_{50} method and by the log time method.

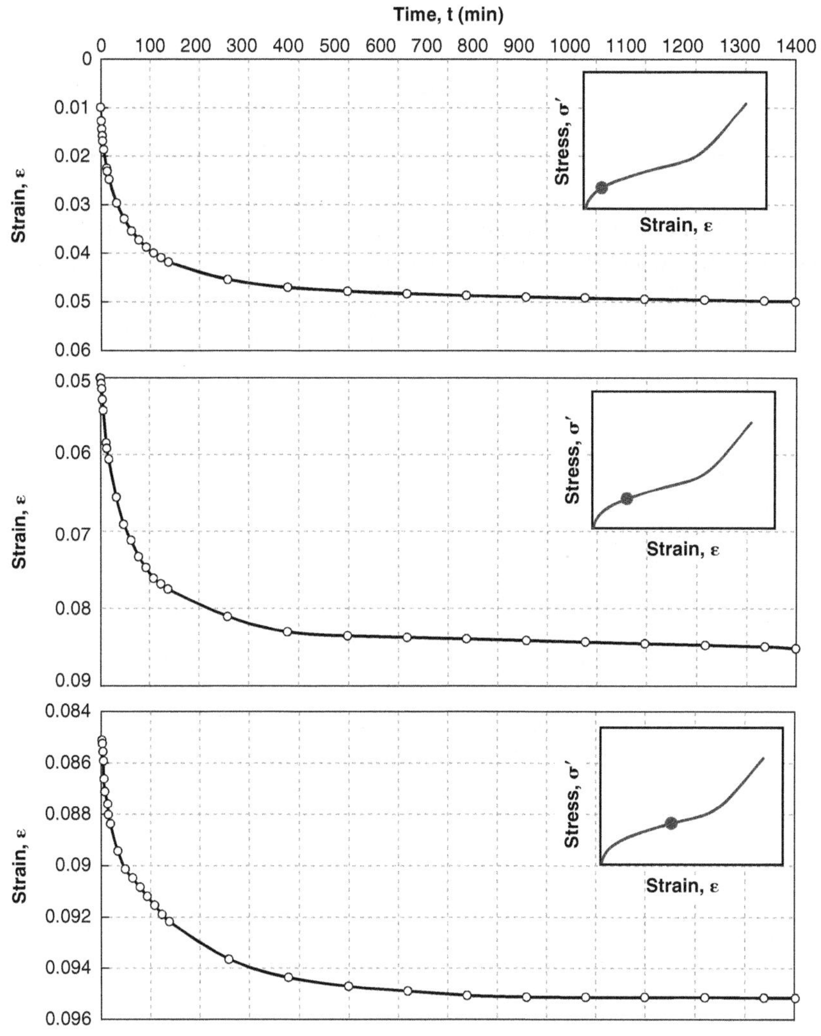

Figure 14.5s Strain vs. time consolidation curves.

Solution 14.8

a. $\sqrt{t}$ Method

1. Plot the sample height H versus $\sqrt{t}$
2. Draw the tangent to the initial part of the curve
3. Choose a point M at an arbitrary but convenient $\sqrt{t_1}$ value and a height H_1 on that tangent
4. Plot a point N with coordinates $\sqrt{t_2} = 1.15\sqrt{t_1}$ and H_1
5. Connect N to the start of the curve
6. The intersection with the curve gives $\sqrt{t_{90}}$
7. Calculate C_v from the equation:

$$C_v = \frac{0.848\, d^2}{t_{90}} \tag{14.23s}$$

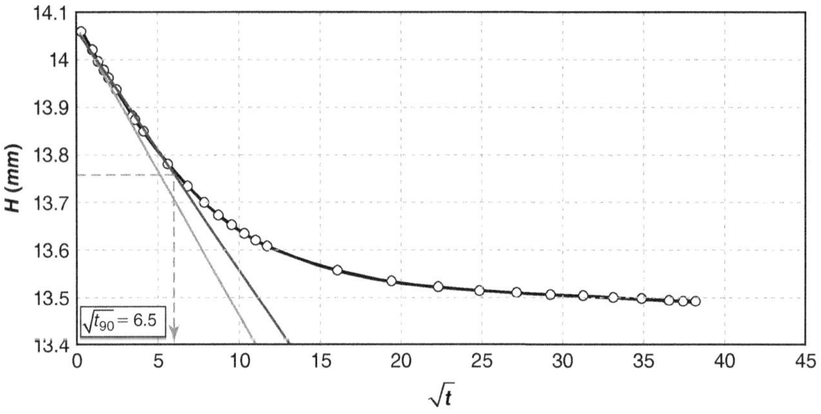

Figure 14.15s Square root of time method.

$$C_{v1} = \frac{0.848d^2}{t_{90}} = \frac{0.848^*(14.2/2)^2}{6.5^2} = 1.0118(\text{mm}^2/\text{ min})$$

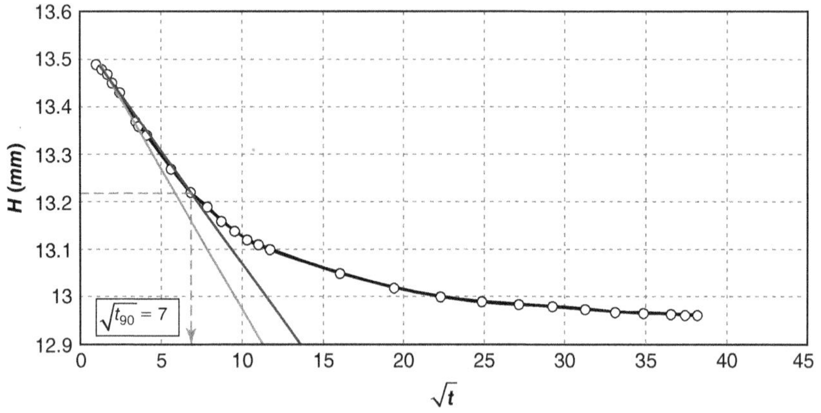

Figure 14.16s Square root of time method.

$$C_{v2} = \frac{0.848d^2}{t_{90}} = \frac{0.848^*(14.2/2)^2}{7^2} = 0.8724(\text{mm}^2/\text{ min})$$

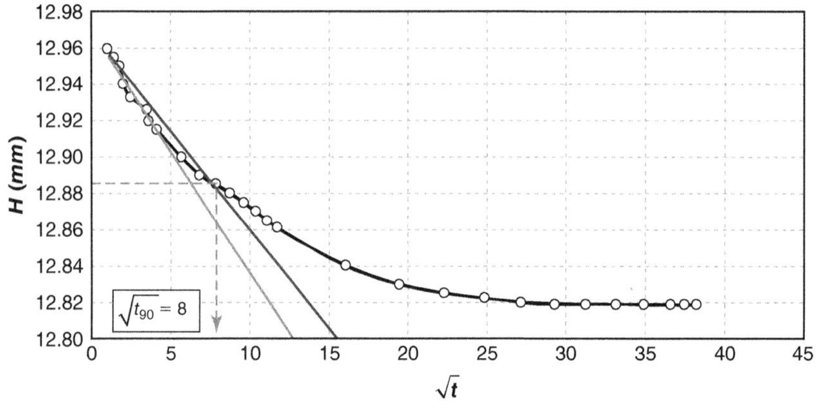

Figure 14.17s Square root of time method.

$$C_{v1} = \frac{0.848d^2}{t_{90}} = \frac{0.848 * (14.2/2)^2}{8^2} = 0.6679(\text{mm}^2/\text{ min})$$

b. Log time method (Figure 14.18s)
   1. Plot H vs. log time
   2. Find H_{100} and H_0 (as shown in Figure 14.18s), then calculate H_{50} from the equation:

$$H_{50} = \frac{H_0 + H_{100}}{2} \tag{14.24s}$$

   3. Find t_{50} from the plot
   4. Calculate C_v as follows:

$$C_v = \frac{0.197d^2}{t_{50}} \tag{14.25s}$$

$$H_0 = 14.08(\text{ mm})$$

$$H_{100} = 13.55(\text{mm})$$

$$H_{50} = 13.815(\text{mm}) \rightarrow t_{50} = 20\,\text{min}$$

$$C_v = \frac{0.197d^2}{t_{50}} = \frac{0.197(14.2/2)^2}{20} = 0.50\ (\text{mm}^2/\text{min})$$

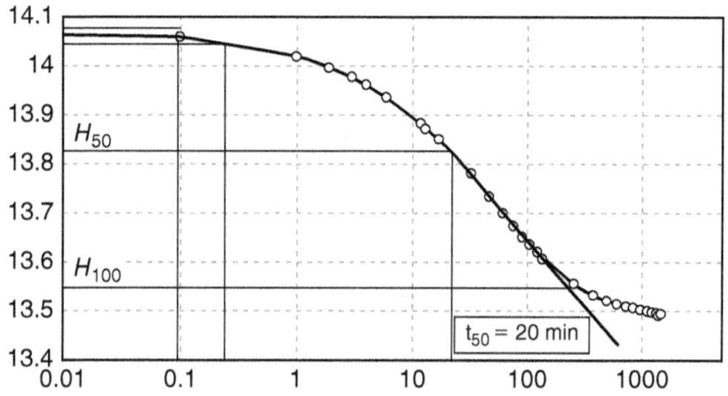

Figure 14.18s Log time method.

Problem 14.9

Given the strain versus time curve of Figure 14.6s, and knowing that the initial void ratio e_o is 0.7, calculate the secondary compression index C_α.

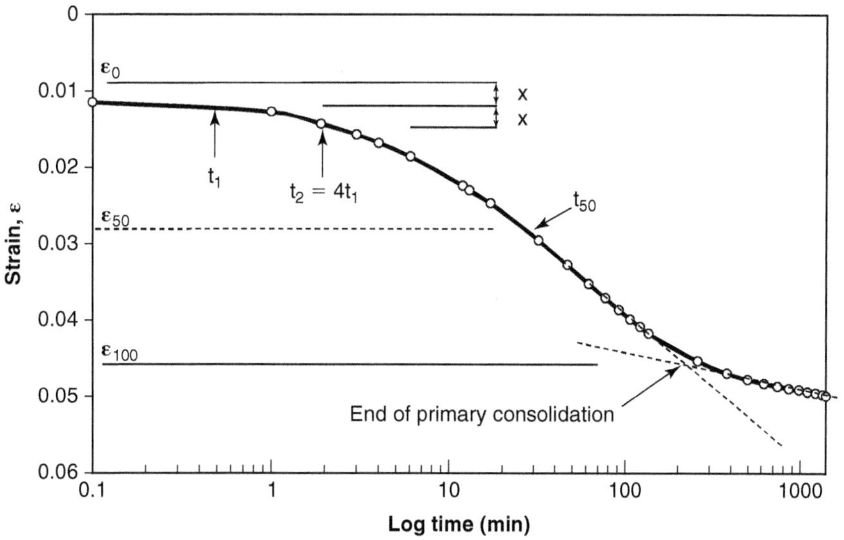

Figure 14.6s Strain vs. log time consolidation curve.

Solution 14.9

$$C_\alpha = \frac{\Delta e}{\Delta \log t} \tag{14.26s}$$

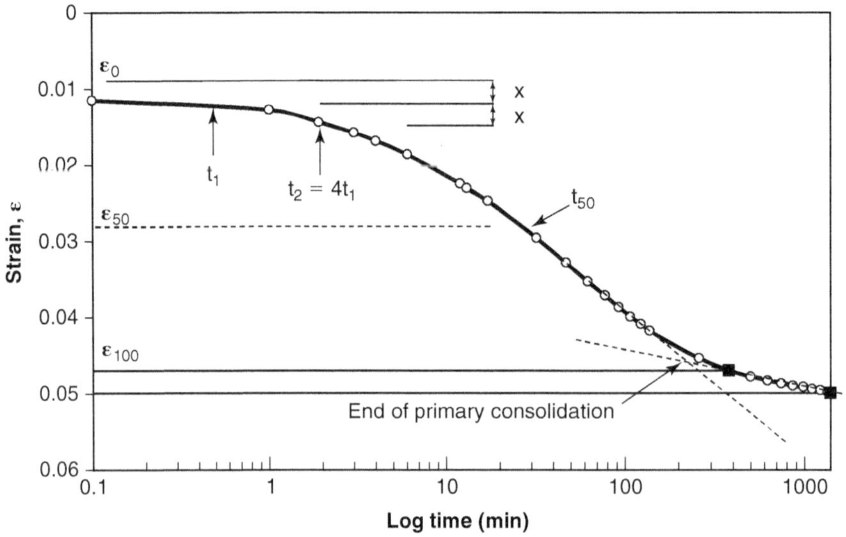

Figure 14.19s Strain vs. log time consolidation curve.

From Figure 14.19s:

$$\varepsilon_1 = 0.04696$$
$$t_1 = 377(\text{min})$$
$$\varepsilon_2 = 0.0499$$
$$t_2 = 1400(\text{min})$$

$$\frac{\Delta H}{H_0} = \frac{\Delta e}{1+e_0} \rightarrow \Delta H = \varepsilon_1 H_0 - \varepsilon_2 H_0 \rightarrow \frac{\Delta H}{H_0} = \Delta\varepsilon \tag{14.27s}$$

$$\Delta\varepsilon = \frac{\Delta e}{1+e_0} \tag{14.28s}$$

$$(0.0499 - 0.04696) = \frac{\Delta e}{1+0.7} \rightarrow \Delta e = 0.004998$$

Recalling Eq. 14.26s:

$$C_\alpha = \frac{0.004998}{\log(1400) - \log(377)} = 0.008772 \qquad (\text{in } 1/\log(\text{minutes}))$$

Problem 14.10

Devise a pressuremeter test procedure that allows you to measure as many parameters as possible for equation 14.12.

Solution 14.10

Strain level influence: A PMT can be run by performing unload-reload cycles around a chosen mean borehole pressure. The cycles would be of increasing amplitude to vary the associated strain, but always around the same mean pressure so that the mean stress level would not change and the influence of the strain amplitude would be isolated. This would quantify the influence of the strain level.

Stress level influence: A PMT can be run by performing unload-reload cycles with the same strain amplitude but at different stress levels. The loops of the cycles would have the same amplitude of $\Delta(\Delta R/R_o)$, but would take place at increasingly higher pressure over limit pressure ratio. This would isolate the influence of the stress level. One must be cautious here and realize that because the pressuremeter test is primarily a shear test, the influence of the stress level would not be the influence of the confinement level.

Time influence: A PMT can be run by holding a chosen pressure p and recording the relative increase in probe radius $\Delta R/R_0$ as a function of time t. This $\Delta R/R_0$ vs. t curve will give information on the time dependency of the soil deformation.

Cycle influence: A PMT can be run by performing cycles between two chosen pressure levels. The evolution of the relative increase in radius $\Delta R/R_o$ with the number of cycles N will give a quantification of the sensitivity of the soil to cyclic loading.

It is very important, when running these kinds of PMTs, to keep in mind the analogy or difference between the stress path and deformation process around the pressuremeter and in the geotechnical project. The closer the analogy, the more useful the information.

Problem 14.11

Give the range of shrink-swell modulus that can be expected for soils. Use that range and the range of shrink-swell indices in Table 14.19 to give the range of expected relative volume change in shrink-swell soils.

Solution 14.11

The shrink-swell modulus (E_{ss}) is a constant for a given soil and does not vary much from soil to soil, with values in the range of 0.5 to 1. Table 14.1s shows the expected range of relative volume change, $\Delta(\Delta V/V)$, for the corresponding range of shrink-swell indices (I_{ss}):

$$\Delta\left(\frac{\Delta V}{V}\right) = \frac{\Delta w}{E_{ss}} = \frac{I_{ss}}{E_{ss}} \tag{14.29s}$$

Table 14.1s Relative Change in Volume of a Soil in Percent for Various Values of Shrink-Swell Modulus and Shrink-Swell Index

		Shrink-swell index				
		0–15	15–30	30–45	45–60	>60
Shrink-swell modulus	**0.5**	0–30	30–60	60–90	90–120	>120
	0.6	0–25	25–50	50–75	75–100	>100
	0.7	0–21.4	21.4–42.9	42.9–64.3	64.3–85.7	>85.7
	0.8	0–18.8	18.8–37.5	37.5–56.3	56.3–75	>75
	0.9	0–16.7	16.7–33.3	33.3–50	50–66.7	>66.7
	1.0	0–15	15–30	30–45	45–60	>60

Problem 14.12

Which of the following two soils is the most likely to collapse upon wetting?

a. Silt with a dry unit weight of 14 kN/m^3, a liquid limit of 40%, a plastic limit of 20%, a porosity of 50%, and a natural water content of 10%
b. Clay with a dry unit weight of 16 kN/m^3, a liquid limit of 55, a plastic limit of 20, a porosity of 35%, and natural water content of 20%

Solution 14.12

Use the USACE (1990) indicators to determine if a soil is likely to collapse.

a. Silt:
   1. $LL = 40\%$ (less than 45%)
   2. $PI = 40\% - 20\% = 20\%$ (less than 25%)

3. $\gamma_d = 14\ \text{kN/m}^3$ (between 10 and 17)
4. $n = 50\%$ (between 40% and 60%)

b. Clay

1. $LL = 55\%$ (not less than 45%)
2. $PI = 55\%\text{–}20\% = 25\%$ (equals 25%)
3. $\gamma_d = 16\ \text{kN/m}^3$ (between 10 and 17)
4. $n = 35\%$ (not between 40% and 60%)

According to these guidelines, the silt is more likely to collapse than the clay.

CHAPTER 15

Shear Strength Properties

15.1 GENERAL

Three strengths are usually considered for a material: compressive strength, tensile strength, and shear strength. *Compressive strength* is tested by applying an all-around pressure (hydrostatic loading) on a sample and recording the pressure at which the sample fails. In general, soils are very strong in all-around compression. Exceptions include soils with a very loose structure and a slight cementation such as calcareous sands; under such loading, these soils can collapse on themselves and crush with a drastic reduction in volume. (For comparison purposes, other materials that are weak in compression are puffed rice and marshmallow.)

Tension strength is tested by pulling on a sample at both ends. In general, soils are very weak in tension. This mode of failure does not often control the behavior of soils, however, because tensile stresses between the grains are rare, due in part to gravity stresses that impose a natural prestressing in the deposit. If tensile stresses develop between the grains, they first correspond to a decrease in compression rather than true tension. Tensile cracks do develop at the top of failing slopes or in shrinking soils near the ground surface.

Shear strength can be tested by moving the top part of a sample with respect to the bottom part of a sample in the direction of the plane separating the top from the bottom. Most often, the shear strength is what controls the ultimate loads in geotechnical engineering projects. This is why it is so important to the geotechnical engineer. As an added example to convince you, think of the unconfined compression test. The soil sample is loaded vertically and has no lateral pressure applied. When the vertical stress becomes too high, the sample fails along a diagonal where the shear stress reaches the shear strength. So, even though the loading is compression, the failure is in shear. By comparison with concrete and steel, the strength of soil is very small (Table 15.1).

Table 15.1 Strength of Soils, Concrete, and Steel

Material	Shear Strength	Unconfined Compression Strength	Tensile Strength
Soil	5 kPa to 500 kPa	10 to 1000 kPa	0 to 100 kPa
Concrete	750* to 1100* kPa	20,000 to 40,000 kPa	2000 to 4000 kPa
Steel	230,000 kPa	250,000 kPa	400,000 kPa

*This low value is explained in section 15.3.

Where does the shear resistance come from in a soil mass? It cannot be from the shearing resistance of the air or the water, because these shear resistances are negligible. In fact, it comes from the shearing resistance at the particle-to-particle contacts. The particles are pressed against each other by normal forces and the shear resistance is due in large part to the friction at the contacts. The normal stress between particles is quantified by the effective stress, and therefore one component of the shear strength is the product of the effective stress on the plane of failure times the coefficient of friction of the interface. The second component of the shearing resistance at the contacts is the glue that may exist at the contacts. This glue may be real, as in the case of calcium cementation, or apparent, as in the case of water tension between the particles that pulls the particles together when the soil dries. The apparent cohesion is in fact a part of the friction resistance, as the effective stress is enhanced by the tension in the water.

As will be seen, many factors can affect the shear strength of a soil. The best way to obtain the shear strength of a soil is to measure it directly by laboratory test or in situ test and by reproducing in the test the same stress conditions as those anticipated in the field. Any shear strength parameter should be quoted by explaining how it was measured and over what stress range the soil was tested.

15.2 BASIC EXPERIMENTS

15.2.1 Experiment 1

If a block of concrete with a weight N is placed on a concrete floor (Figure 15.1), the force F necessary to initiate motion

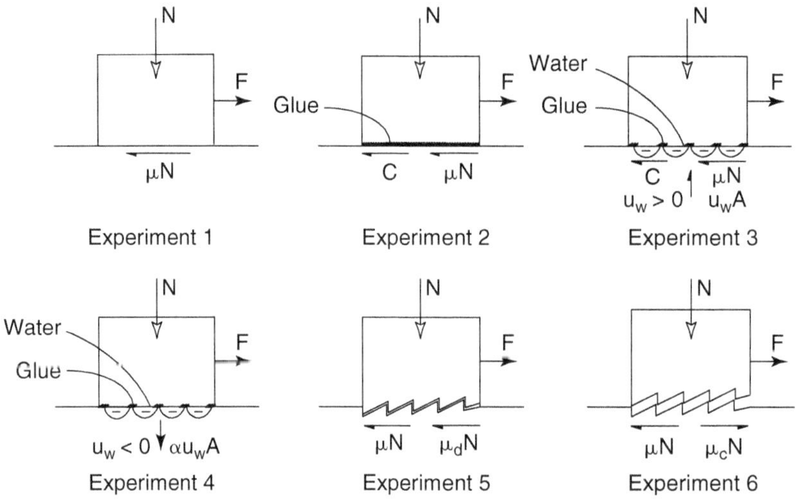

Figure 15.1 Basic experiments.

by dragging the concrete block on the concrete floor is given by:

$$F = \mu N \quad (15.1)$$

where μ is the coefficient of friction of the concrete-to-concrete interface, F is the shear force, and N is the normal force. By dividing both sides of Eq. 15.1 by the interface contact area A, and replacing μ by $\tan\varphi$, the equation becomes:

$$\frac{F}{A} = \frac{N}{A}\tan\varphi \quad \text{or} \quad \tau_f = \sigma \tan\varphi \quad (15.2)$$

15.2.2 Experiment 2

Imagine that before I place the concrete block on the concrete floor, I paint a layer of glue on the concrete floor (Figure 15.1) and then I place the concrete block on the glue and I let it set. Now I repeat the experiment and exert a force F, higher than in the first case because of the glue, to drag the block. Then Eq. 15.1 becomes:

$$F = C + \mu N \quad (15.3)$$

where C is the force required to break the glue. If I divide again by the total area A and use $\tan\varphi$ instead of μ, I get:

$$\frac{F}{A} = \frac{C}{A} + \frac{N}{A}\tan\varphi \quad \text{or} \quad \tau_f = c + \sigma\tan\varphi \quad (15.4)$$

15.2.3 Experiment 3

Imagine now that I make some small holes on the concrete floor, that I paint the glue only on the top of the bumps between holes, and that I flood the holes with water before I place the block (Figure 15.1). When I place the concrete block on top of the concrete floor, two things happen: the glue sets and the water is squeezed between the two surfaces. If the water saturates the holes and if the water cannot escape, there will be a water compression stress $u_w (u_w > 0)$ under the block and an associated uplift force $u_w \times A$, which decreases the normal force on the sliding plane. Equation 15.1 then becomes:

$$F = C + \mu(N - u_w A) \quad (15.5)$$

If I divide by the total area A, I get:

$$\frac{F}{A} = \frac{C}{A} + \frac{(N - u_w A)}{A}\tan\varphi \quad \text{or} \quad \tau_f = c + (\sigma - u_w)\tan\varphi \quad (15.6)$$

15.2.4 Experiment 4

Let's repeat that last experiment, but this time, before I place the concrete block, I dry up some of the water in the holes such that the little amount of water that is left is held in the holes by tension in the water ($u_w < 0$) (Figure 15.1). This creates a suction force between the block and the concrete floor that increases the force F necessary to move the block. This force is equal to the water tension times the area over which the water exists. This area is a fraction α of the total area A and is represented by αA where α is less than one. Equation 15.1 becomes:

$$F = C + \mu(N - \alpha u_w A) \quad (15.7)$$

If I divide by the total contact area A, I get

$$\frac{F}{A} = \frac{C}{A} + \frac{(N - \alpha u_w A)}{A}\tan\varphi \quad \text{or} \quad \tau_f = c + (\sigma - \alpha u_w)\tan\varphi \quad (15.8)$$

15.2.5 Experiment 5

Let's go back to experiment 1, but this time we design some special grooves in the concrete floor and matching grooves at the bottom of the concrete block. These grooves are inclined as shown in Figure 15.1, such that to move the concrete block,

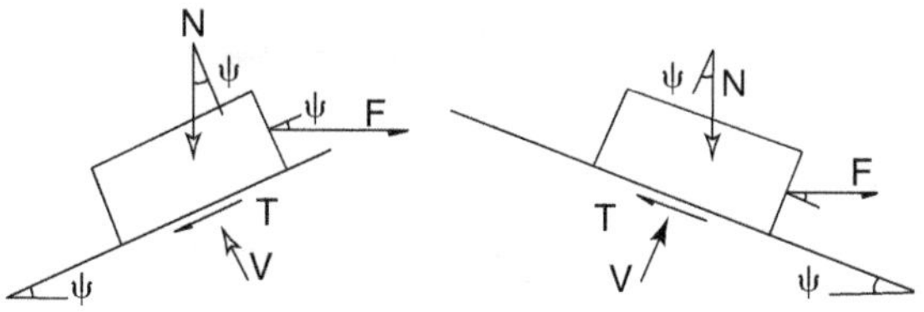

Figure 15.2 Concrete block on slopes.

the block has to be pushed sideways and upward. This type of interface increases the shear force F necessary to move the block as follows. The friction force μN is still necessary, but a force $\mu_d N$ has to be added to overcome the roughness of the upward grooves. The subscript d is for dilation. The coefficient μ_d is equal to $\tan\psi$ where ψ is the angle of the groove with the horizontal. Here is why. Referring to Figure 15.2, the friction force T is equal to the normal force $N\cos\psi$ times the coefficient of friction $\tan\varphi$; this is the constitutive law:

$$T = N\cos\psi\tan\varphi \tag{15.9}$$

Then equilibrium in the direction of the force T can be written; this is the fundamental law:

$$T + N\sin\psi = F\cos\psi \tag{15.10}$$

By combining Equations 15.9 and 15.10, we get:

$$F = N\tan\varphi + N\tan\psi \tag{15.11}$$

If we divide by the total contact area, we get:

$$\begin{aligned}\tau_f &= c + \sigma\tan\varphi + \sigma\tan\psi \\ &= c + \sigma\tan(\varphi+\psi)(1-\tan\varphi\tan\psi)\end{aligned} \tag{15.12}$$

Note that if ψ is relatively small, the term $(1 - \tan\varphi\tan\psi)$ is close to 1.

15.2.6 Experiment 6

Let's repeat experiment 5, but now with the grooves slanted in the other direction (Figure 15.1). This time the downward slope creates a force that decreases the value of F. Equation 15.11 becomes:

$$F = N\tan\varphi - N\tan\psi \tag{15.13}$$

If we divide by the total contact area, we get:

$$\begin{aligned}\tau_f &= c + \sigma\tan\varphi - \sigma\tan\psi \\ &= c + \sigma\tan(\varphi-\psi)(1+\tan\varphi\tan\psi)\end{aligned} \tag{15.14}$$

Again, if ψ is relatively small, the term $(1 + \tan\varphi\tan\psi)$ is close to 1.

15.3 STRESS-STRAIN CURVE, WATER STRESS RESPONSE, AND STRESS PATH

The stress-strain curve of a soil depends on a number of factors, including the soil stress history, the current stress level, the structure of the soil, and others. Two types of curves are usually encountered. The first exhibits a peak followed by a strain softening region; the second does not exhibit a peak but simply an increase toward a plateau at large strains (Figure 15.3).

Overconsolidated soils, hard soils, and dense soils have curves exhibiting peaks (brittle), whereas normally consolidated soils, soft soils, and loose soils have curves exhibiting no peak (ductile). For the same soil, under the same confinement, but for an overconsolidated and normally consolidated case, both curves tend to reach a common strength at large strain (Figure 15.3). This point is called the *critical state*. At that point the soil does not change volume while shearing.

The water stress exhibits two different types of behavior for these two distinct types of curves. In the case of the curve with no peak, the soil compresses throughout the shearing process and the water goes into compression, thereby reducing the effective stress. In the case of the curve with a peak, the water goes into compression initially (reduction in effective stress) and then the soil starts to dilate; the associated increase in volume creates a decrease in water stress that ends up as tension. As a result, the effective stress increases. Note that water stress is not always measured during such tests. Nevertheless, the water stress is necessary for proper reduction of the data in terms of effective stress.

The stress path in two dimensions is the path described by the top of the Mohr circle. It describes the evolution of certain stresses throughout the loading of the sample. Specifically, it tracks the path described by the points with p, q stress coordinates where p and q are defined as:

$$\mathrm{p} = \frac{\sigma_1+\sigma_3}{2} \quad \text{or} \quad \mathrm{p} = \frac{\sigma_v+\sigma_h}{2} \tag{15.15}$$

$$\mathrm{q} = \frac{\sigma_1-\sigma_3}{2} \quad \text{or} \quad \mathrm{q} = \frac{\sigma_v-\sigma_h}{2} \tag{15.16}$$

where σ_v and σ_h are the vertical and horizontal total stresses in a triaxial test, for example. The most useful stress paths are plotted in terms of effective stresses (p′ and q′)::

$$\mathrm{p}' = \frac{\sigma_1'+\sigma_3'}{2} \quad \text{or} \quad \mathrm{p}' = \frac{\sigma_v'+\sigma_h'}{2} \tag{15.17}$$

$$\mathrm{q}' = \frac{\sigma_1'-\sigma_3'}{2} \quad \text{or} \quad \mathrm{q}' = \frac{\sigma_v'-\sigma_h'}{2} \tag{15.18}$$

where σ_v' and σ_h' are the vertical and horizontal total stresses in a triaxial test, for example. Examples of effective stress paths for different types of tests are shown in Figure 15.4. In any lab test, it is most desirable to match the effective stress path followed by the soil in the field during the project construction and the project life. Examples of field stress

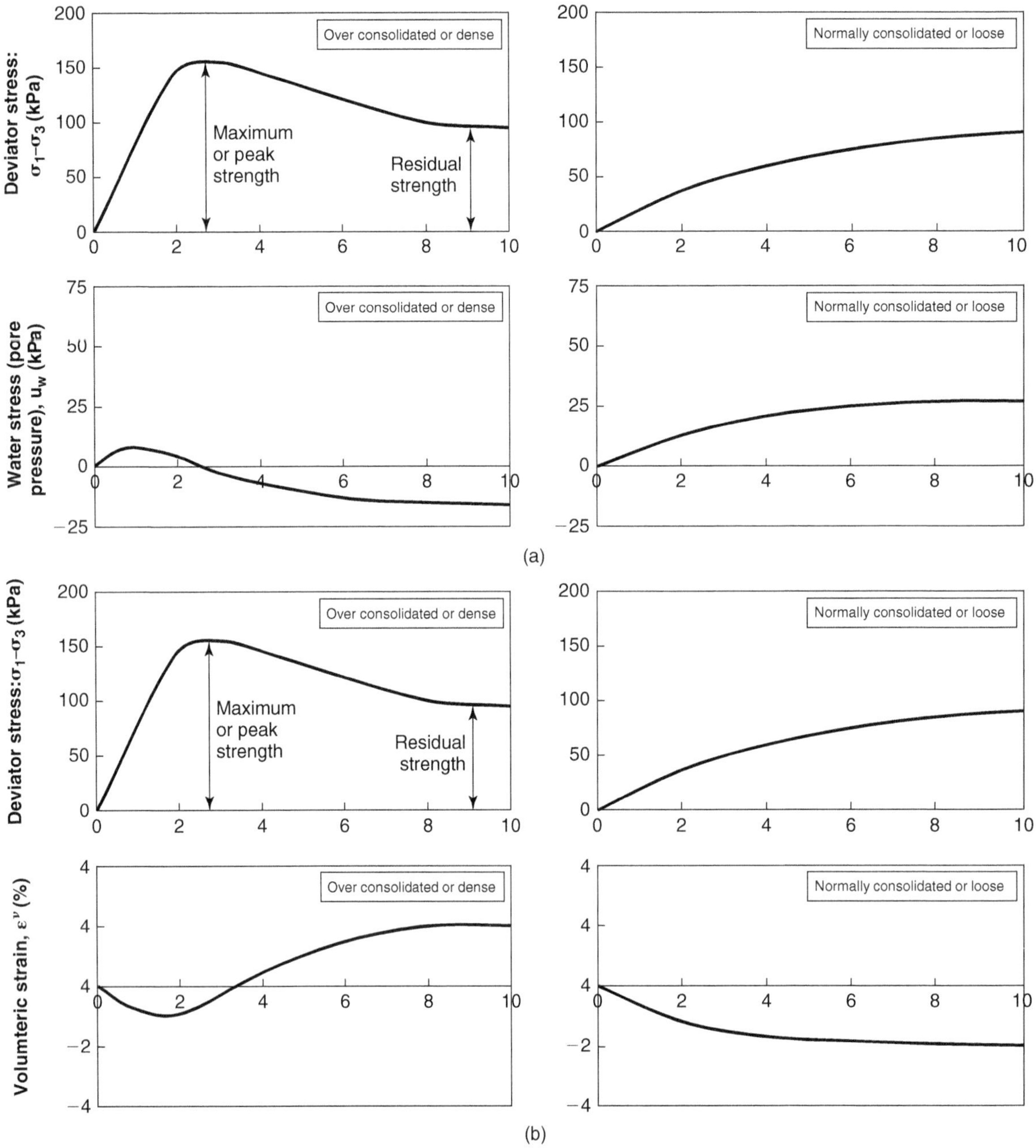

Figure 15.3 Stress-strain curves in soils: *(a)* Consolidated undrained test. *(b)* Consolidated drained test.

paths are shown in Figure 15.5. Stress path OA would be the case of the wetting of an unsaturated soil or the filling of an earth dam reservoir. Stress path OB might be associated with a slow excavation process. Stress path OC would correspond to a rapid embankment construction. Stress path OD would be the case of a slow embankment construction. As can be seen from Figure 15.5, stress paths OC and OB are those that will approach the failure envelope the fastest, because they go toward the strength envelope with the shortest distance.

The *shear strength* τ_f of a soil is defined as the highest shear stress the soil can resist. For the curve with a peak, it will be the shear stress corresponding to the peak of the curve, known as the *peak shear strength*. For the curve with no peak, it is the shear stress at large strain; a value of 10% strain is often used when no obvious plateau is reached. The residual shear strength of a soil is defined only when the curve has a peak. In this case, the value of the shear stress corresponding to the post-peak plateau is the *residual shear strength*. The *remolded shear strength* is the shear strength of the remolded soil. The remolded shear strength can be equal to the residual shear strength, but more often it is less than the residual shear strength.

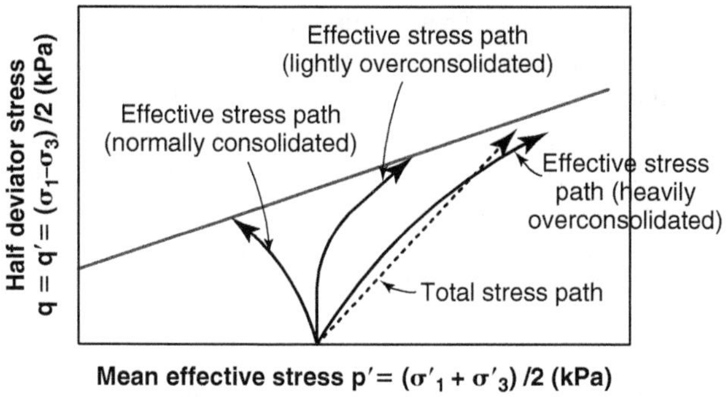

Figure 15.4 Stress paths for an overconsolidated and a normally consolidated soil.

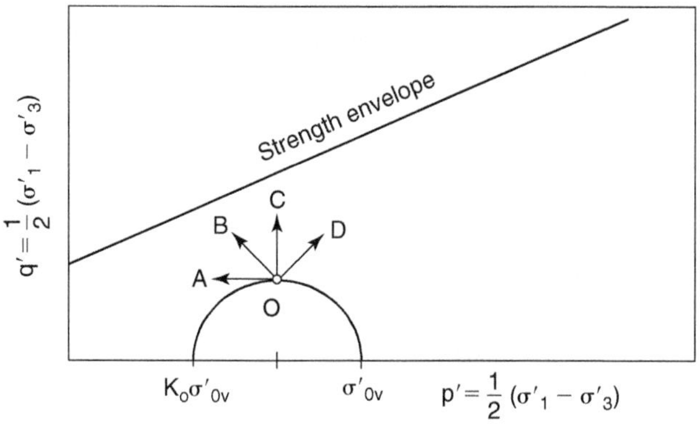

Figure 15.5 Example of field stress paths.

15.4 SHEAR STRENGTH ENVELOPE

15.4.1 General Case

Each one of the experiments described in section 15.2 has a parallel for soils. Imagine now that the interface, instead of concrete on concrete, is a plane in a soil with no water. The shear strength that can be generated by the soil will definitely have a component due to friction, as explained in experiment 1:

$$\tau_f = \sigma \tan \varphi \tag{15.19}$$

where τ_f is the soil shear strength, σ is the normal stress on the plane of failure, and φ is the friction angle. Recall that $\tan \varphi$ is a coefficient of friction and as such is often between 0 and 1, although we will see later that it could actually be higher than 1. The glue added in experiment 2 refers to any cohesion that may exist between the soil particles. This cohesion is relatively rare, and when it is not zero, it is quite small (5 to 20 kPa). The cohesion plus the friction give:

$$\tau_f = c + \sigma \tan \varphi \tag{15.20}$$

The water added in experiment 3 refers to the case where the voids between the soil particles are full of water or 100% saturation. In this case the water is under a certain amount of pressure u_w (compression below the groundwater level, GWL, or tension above the GWL) that changes the effect of the normal stress. The normal stress σ becomes the effective normal stress σ': the difference between the total normal stress σ and the water stress u_w. Also, the cohesion becomes the effective stress cohesion c′ and the friction angle becomes the effective stress friction angle φ':

$$\tau_f = c' + (\sigma - u_w) \tan \varphi' = c' + \sigma' \tan \varphi' \tag{15.21}$$

In experiment 4, the water no longer filled the voids and covered a fraction α of the total area. As a result, Eq. 15.21 is modified because the expression of the effective stress σ' has changed:

$$\tau_f = c' + (\sigma - \alpha u_w) \tan \varphi' = c' + \sigma' \tan \varphi' \tag{15.22}$$

Experiment 5 conveys an important message regarding soil shear strength: the concept of dilatancy. When a very dense soil is sheared, it tends to increase in volume or dilate. This is due to each particle having to climb over the one in front of it during shearing. This increase in volume is associated with a lifting effect similar to that of the concrete block and increases the shear strength compared to a no-volume-change situation. The shear strength equation becomes:

$$\tau_f = c' + \sigma' \tan \varphi' + \sigma' \tan \psi' \tag{15.23}$$

where ψ' is the effective stress dilatancy angle. If ψ' is small, Eq. 15.23 can be rewritten as:

$$\tau_f = c' + \sigma' \tan(\varphi' + \psi') \tag{15.24}$$

In experiment 6, the problem of dilatancy became a problem of compression and the term $\sigma' \tan \psi'$ had to be subtracted rather than added. In geotechnical engineering, it is common instead to use a negative value of ψ' and keep Eq. 15.24 the same. In the general case, the shear strength of soils is measured and the effects of dilatancy or compression are absorbed in the value of φ'. The general equation for the shear strength of soils is therefore:

$$\tau_f = c' + \sigma' \tan \varphi' \tag{15.25}$$

where τ_f is the shear strength, c′ is the effective stress cohesion intercept, σ' is the effective stress normal to the plane of failure ($\sigma - \alpha u_w$), and φ' is the effective stress friction angle. This equation works for all soils in all situations, including saturated or unsaturated, drained or undrained, dilative or compressive. If c′ and φ' are considered to be constants, then Eq. 15.25 is a straight line on the τ vs. σ' set of axes (Figure 15.6) and is referred to as the *strength envelope*. Any stress point below or on the envelope is possible, but it is not possible for any stress point to plot above that line. Thus, any failure Mohr circle will have to be tangent to the failure envelope.

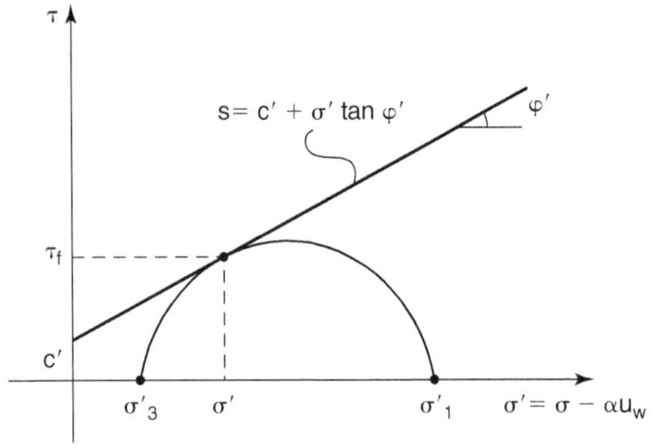

Figure 15.6 Shear strength envelope for soils.

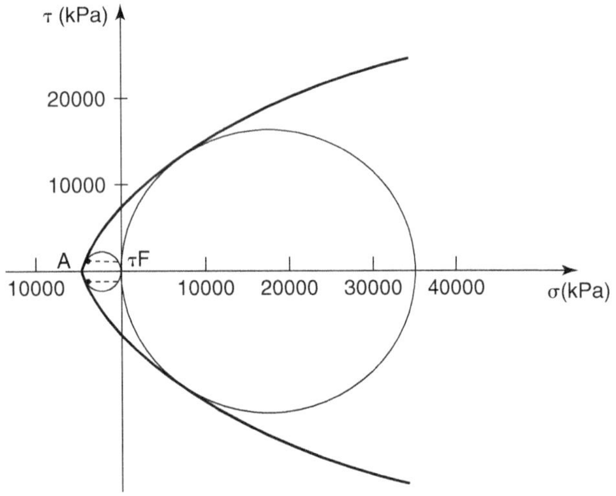

Figure 15.7 Strength envelope for concrete.

15.4.2 The Case of Concrete

Concrete has a very large cohesion intercept compared to that of soils. Figure 15.7 shows the Mohr circle for an unconfined compression test on concrete ($f_c = 35000$ kPa) on one side and for an unconfined tension test on concrete ($f_t = 3500$ kPa) on the other. The shear strength envelope for concrete is also shown conceptually. The value of the shear strength used in the code is given by the equation s (kPa) = $5.25(f_c(kPa)^{0.5}) = 982$ kPa. This value is shown on the strength envelope (Figure 15.7) and is associated with a significant tension. The reason is that in concrete beam design, shear typically occurs in sections near the supports where tension is large. It would be like using Eq. 15.25 with a large tension for the normal stress; this would decrease the shear strength significantly. This is why the shear strength of concrete in Table 15.1 is quite small—much less than one-half the unconfined compression strength.

15.4.3 Overconsolidated Fine-Grained Soils

A special case occurs with overconsolidated soils where the shear strength envelope does not quite follow the straight line of Eq. 15.25. These soils exhibit a preconsolidation pressure σ'_p as measured in the consolidation test. For stresses less than σ'_p, deformations are small; for stresses higher than σ'_p, deformations are much larger for the same increase in effective stress. The preconsolidation pressure can be thought of as a *yield stress* on the consolidation stress-strain curve. This yield stress also affects the shear strength envelope. Indeed, when the effective stress σ' on the plane of failure is less than σ'_p, the cohesion intercept found in many overconsolidated clays is apparent. However, when the effective stress σ' on the plane of failure is larger than σ'_p, the cohesion intercept is destroyed by the stress level that destructures the soil, and the envelope goes through the origin (Figure 15.8). Others have proposed that the envelope be curved as shown in Figure 15.8. Mesri and Abdelghafar (1993) proposed an empirical equation, for stresses less than σ'_p, that takes into account the overconsolidation ratio on the drained shear strength, as follows:

$$\tau_f = \sigma' \tan \varphi' \left(\frac{\sigma'_p}{\sigma'}\right)^{1-m} \tag{15.26}$$

where m is a shear strength coefficient given in Table 15.2.

15.4.4 Coarse-Grained Soils

A special case also arises for coarse-grained soils where the shear strength envelope does not quite follow the straight line of Eq. 15.25. These soils tend to dilate during shear at

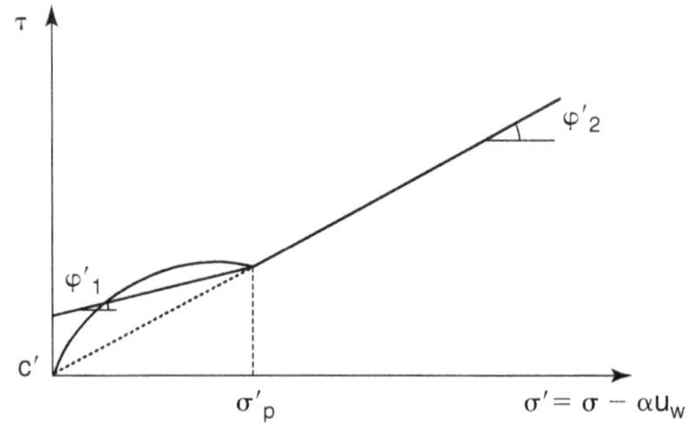

Figure 15.8 Strength envelope for overconsolidated fine-grained soils.

Table 15.2 Parameter m for Equation 15.22

Soil	m for intact soil	m for destructured soil
Cemented soft clays	0.4–0.5	0.5–0.7
Stiff clays and shales	0.5–0.6	0.6–0.8
Soft clays	0.6–0.7	0.7–0.9

(After Terzaghi et al. 1996.)

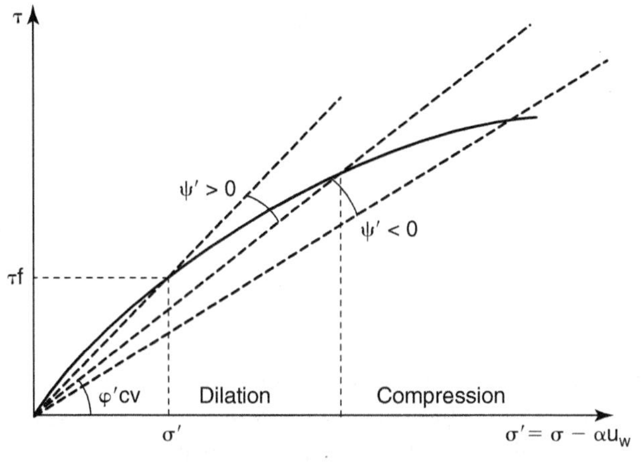

Figure 15.9 Strength envelope for coarse-grained soil.

a low confinement level and compress at higher confinement levels. The stress level at which the change between dilation and compression occurs depends on the relative density of the coarse-grained soil: the higher the density, the larger the stress range over which the soil dilates. All soils end up compressing during shear at some level of stress. Recall the simplified equation including the dilation angle ψ':

$$\tau_f = c' + \sigma' \tan(\varphi' + \psi') \tag{15.27}$$

Because ψ' is positive at smaller stresses (dilation) and becomes negative at higher stresses (compression), the sum $\varphi' + \psi'$ is larger at smaller effective stresses than it is at higher effective stresses, and the shear strength envelope is curved (Figure 15.9).

When the soil dilates, a distinction is made between the friction angle φ'_{peak} associated with the peak of the stress-strain curve and the post-peak large strain friction angle φ'_{cv} at which the soil reaches a point where shearing takes place at constant volume. The difference between the two is the dilation angle ψ':

$$\varphi'_{peak} = \varphi'_{cv} + \psi' \tag{15.28}$$

For dense soils, φ'_{peak} is larger than φ'_{cv} and ψ' is positive; for loose soils, φ'_{peak} is smaller than or equal to φ'_{cv} and ψ' can be negative. In most tests, the angle φ'_{peak} is the one measured. The angle ψ' can be inferred from the post-peak large strain shear strength that gives φ'_{cv} and then using Eq. 15.28.

15.5 UNSATURATED SOILS

For unsaturated soils, the effective stress can be calculated as explained in section 10.13. The most general expression for the effective stress σ' is:

$$\sigma' = \sigma - \alpha u_w - \beta u_a \tag{15.29}$$

Therefore, the general equation for the shear strength of a soil (unsaturated or saturated) is:

$$\tau_f = c' + \sigma' \tan\varphi' = c' + (\sigma - \alpha u_w - \beta u_a)\tan\varphi' \tag{15.30}$$

where σ' is the normal effective stress on the plane of failure, σ is the normal total stress on the plane of failure, α is the fraction of the total area of the failure plane covered by the water, β is the fraction of the plane covered by the air, u_w is the water stress, u_a is the air stress, and φ' is the effective stress friction angle. As explained in section 10.13, when the soil is saturated or when the air is occluded, Eq. 15.29 becomes:

$$\sigma' = \sigma - u_w \tag{15.31}$$

If the air is not occluded, there is a path for the air to be connected directly to the atmosphere and the air stress is atmospheric or zero gauge pressure. Then the most general expression of the effective stress in soils covering all real cases is:

$$\sigma' = \sigma - \alpha u_w \tag{15.32}$$

Therefore, in all real cases for unsaturated soils ($u_a = 0$ or $u_a = u_w$) and saturated soils, the equation for the shear strength τ_f is:

$$\tau_f = c' + (\sigma - \alpha u_w)\tan\varphi' \tag{15.33}$$

The parameter α can be estimated by taking it equal to the degree of saturation S (Figure 10.16) or by using a slightly modified version of the Khalili and Khabbaz (1998) equation (Figure 10.17):

$$\alpha = S \tag{15.34}$$

$$\alpha = \sqrt{\frac{u_{wae}}{u_w}} \tag{15.35}$$

where u_{wae} is the air entry value of the water tension and u_w is the water tension.

Shear strength equations other than Eq. 15.30 have been proposed, such as the one of Fredlund and Rahardjo (1993):

$$\tau_f = c' + (\sigma - u_a)\tan\varphi' + (u_a - u_w)\tan\varphi^b \tag{15.36}$$

where φ^b is an angle indicating the rate of increase in shear strength relative to the matric suction $u_a - u_w$. Equation 15.36 can be reorganized as follows:

$$\tau_f = c' + \left(\sigma - \frac{\tan\varphi^b}{\tan\varphi'}u_w - \left(1 - \frac{\tan\varphi^b}{\tan\varphi'}\right)u_a\right)\tan\varphi' \tag{15.37}$$

Comparison of Eq. 15.37 with Eq. 15.30 shows that the two equations are identical if:

$$\alpha = \frac{\tan\varphi^b}{\tan\varphi'} \quad \text{and} \quad \beta = 1 - \frac{\tan\varphi^b}{\tan\varphi'} \tag{15.38}$$

You may recall from section 10.13, Eq. 10.52, that $\alpha + \beta = 1$; therefore, both conditions are satisfied automatically and the ratio $\tan \varphi^b / \tan \varphi'$ can be estimated through Eqs. 15.34 and 15.35.

15.6 EXPERIMENTAL DETERMINATION OF SHEAR STRENGTH (LAB TESTS, IN SITU TESTS)

There are many ways to determine the effective stress shear strength parameters of soils. Because many factors influence the shear strength, it is best to aim at reproducing the initial stress conditions and the stress path during loading, while matching the drainage conditions to be encountered in the field. In the laboratory, the most common tests are the unconsolidated undrained triaxial test (UUT), the consolidated undrained triaxial test (CUT), the unconsolidated undrained direct shear test (UUDS), the consolidated undrained direct shear test (CUDS), the unconsolidated undrained simple shear test (UUSS), and the consolidated undrained simple shear test (CUSS). *Unconsolidated* means that no drainage is allowed when the confining pressure is applied; *consolidated* means that drainage is allowed during application of the confining pressure until the excess water stress has come back down to zero. The second letter in the acronym refers to the loading process; for example, a consolidated undrained test means that the loading process is done while allowing no drainage. Table 15.3 shows which test and test requirements are applicable to determining which shear strength parameters for saturated and unsaturated soils. Note that if the water in the soil voids is in tension (saturated or unsaturated), additional

Table 15.3 Laboratory Tests for Shear Strength Determination of Saturated and Unsaturated Soils

Test	Measurements	Shear Strength	Comments
Direct shear test, Unconsolidated Undrained	Normal stress, shear stress	s_u	Effective stress $\sigma' =$ existing σ' in sample
Direct shear test, Consolidated Undrained	Normal stress, shear stress	s_u	Effective stress $\sigma' =$ chosen σ' for confinement
Direct shear test, Consolidated Drained	Normal stress, shear stress	c', φ'	Estimate of dilatancy angle ψ' if horizontal and vertical displacements measured. If water is in tension, measurements of water tension, air entry water tension, and water content are also necessary.
Simple shear test, Unconsolidated Undrained	Normal stress, shear stress, displacement	s_u and complete stress-strain curve	Effective stress $\sigma' =$ existing σ' in sample
Simple shear test, Consolidated Undrained	Normal stress, shear stress, displacement	s_u and complete stress-strain curve	Effective stress $\sigma' =$ chosen σ' for confinement
Simple shear test, Consolidated Drained	Normal stress, shear stress, displacement	c', φ', and complete stress-strain curve	Estimate of dilatancy angle ψ' if horizontal and vertical displacements measured. If water is in tension, measurements of water tension, air entry water tension, and water content are also necessary.
Triaxial test, Unconsolidated Undrained	Vertical stress, confinement stress, displacement	s_u, complete stress-strain curve, and c', φ' if water stress measured	Effective stress $\sigma' =$ existing σ' in sample
Triaxial test, Consolidated Undrained	Vertical stress, confinement stress, displacement	s_u, complete stress-strain curve, and c', φ' if water stress measured	Effective stress $\sigma' =$ chosen σ' for confinement
Triaxial test, Consolidated Drained	Vertical stress, shear stress, displacement	c', φ' and complete stress-strain curve	Estimate of dilatancy angle ψ' if volume change measured. If water is in tension, measurements of water tension, air entry water tension, and water content are also necessary.

measurements are necessary to obtain the effective stress shear strength parameters. These additional measurements include the measurement of the water tension u_w, the air entry water tension u_{wae}, and the water content w or degree of saturation S. The reason is that the equation for the shear strength is:

$$\tau_f = c' + (\sigma - \alpha u_w) \tan \varphi' \qquad (15.39)$$

which requires estimating α as S or $\sqrt{u_{wae}/u_w}$

The undrained shear strength s_u is simply read as the peak shear stress reached during an undrained test. The effective stress shear strength parameters (c', φ') require plotting the results on the shear stress τ vs. effective normal stress σ', as shown in sections 9.9, 9.10, and 9.12.

In situ tests (see Chapter 7) can also be used to obtain the shear strength of soils. The most direct tests are the vane shear test (VST) and the borehole shear test (BHST). The VST is simple and can be used to obtain the undrained shear strength of fine-grained soils. The BHST is a bit more complicated, but can be used to obtain the effective stress friction angle of coarse-grained soils. The BHST can also be used for the undrained shear strength of saturated fine-grained soils by conducting a rapid test, and the effective stress shear strength parameters of saturated soils by conducting a test slow enough not to generate water stress. Water stress is not typically measured during the BHST or the VST. Other tests such as the standard penetration test (SPT) and the cone penetration test (CPT) can be used to obtain shear strength parameters through correlations. For example, the blow count N of the SPT and the point resistance q_c of the CPT have been used to estimate the friction angle of coarse-grained soils, as well as the undrained shear strength of fine-grained soils.

15.7 ESTIMATING EFFECTIVE STRESS SHEAR STRENGTH PARAMETERS

The parameters referred to in this section are the effective stress cohesion intercept c', the effective stress friction angle φ', and the effective stress dilation/compression angle ψ'.

15.7.1 Coarse-Grained Soils

For coarse-grained soils, the *effective stress cohesion intercept* c′ is considered to be equal to zero, which often leads to coarse-grained soils being called cohesionless soils. The parameter φ' controls the shear strength of these soils, along with the normal effective stress on the plane of failure. The friction angle φ' for coarse-grained soils varies between 25 and 50 degrees. Recall that $\tan \varphi'$ is the coefficient of friction μ, which varies correspondingly between 0.5 and 1.2. A coefficient of friction higher than 1 is possible in soils because of the dilatancy effect, which combines friction and lifting. Tables 15.4 and 15.5 as well as Figure 15.10 give suggested values of the friction angle for coarse-grained soils. These are values of φ' typically obtained in a triaxial test or a direct shear test. Note that the value of φ' obtained in a plane strain test is about 10% higher than the one obtained in a triaxial compression test. The reason is that in the plane strain deformation process, the particles are forced to move in a restricted two-dimensional domain and cannot find the path of least resistance. Thus, the resistance is slightly higher and so is the friction angle. An application of this observation is in the difference between the friction angle for a strip footing and for a circular or square footing:

$$\varphi'_{plane\ strain} \simeq 1.1 \times \varphi'_{triaxial\ compression} \qquad (15.40)$$

The dilation/compression angle ψ' is typically included in the measurement of the friction angle φ'. Therefore, it should not be added to the measured value of φ'. The following relationship between the two angles has been used:

$$\psi' = \varphi' - 30 \qquad (15.41)$$

Houlsby (1991) presents a plot (Figure 15.11) indicating that Eq. 15.41 should be modified to:

$$\psi' = \varphi' - 34 \qquad (15.42)$$

In any case, the angle ψ' varies between −5 for very loose soils to +15 degrees for very dense soils.

Table 15.4 Range of Values for φ'

Soil	Friction angle in degrees		Coefficient of friction	
	Loose	Dense	Loose	Dense
Gravel with sand	35	50	0.7	1.2
Sand, angular grains, well graded	33	45	0.65	1.0
Sand, round grain, uniform	27.5	34	0.52	0.67
Silty sand	27 to 33	30 to 34	0.51 to 0.65	0.58 to 0.67
Inorganic silt	27 to 30	30 to 35	0.51 to 0.58	0.58 to 0.7

(After Terzaghi and Peck 1967)

Table 15.5 Guide for Values for φ'

Gravels and Sands Strength			
Description	φ'°	N (bpf)	Simple field test*
Very loose	$<28^{\circ}$	<4	12 mm diameter rebar pushed in 0.3 m by hand. Shows definite marks of footsteps; hard to walk on.
Loose	28°–30°	4–10	12 mm diameter rebar pushed in 0.1 m by hand. Shows footsteps.
Medium or compact	30°–36°	10–30	12 mm diameter rebar driven 0.3 m with carpenter hammer. Footsteps barely noticeable.
Dense	36°–41°	30–50	12 mm diameter rebar driven 0.1 m with carpenter hammer. No marks of footsteps.
Very dense	$>41^{\circ}$	>50	12 mm diameter rebar driven 0.03 m with carpenter hammer. No marks of footsteps.

*Note that these tests are performed at the ground surface of the gravel-sand deposit, not on a sample.

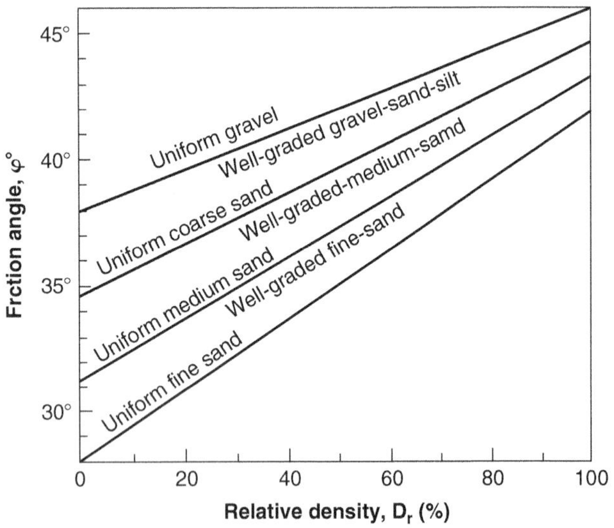

Figure 15.10 Friction angle vs. relative density. (From Schmertmann 1975)

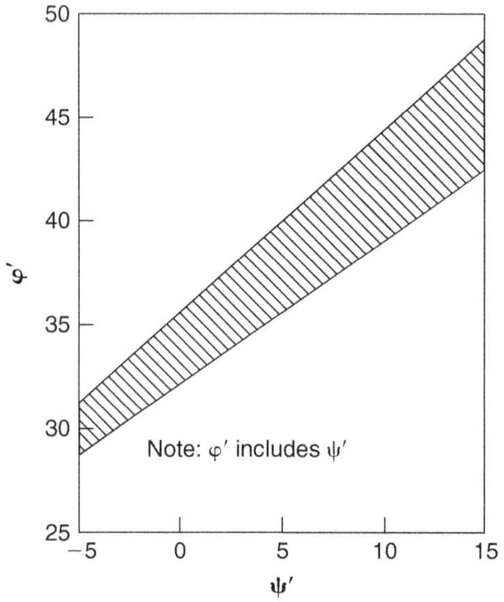

Figure 15.11 Peak friction angle vs. dilation/compression angle. (After Houlsby 1991)

The parameter φ' can be measured directly in situ by using the BHST. The BHST may be the only tool that can give a direct measure of φ' for coarse-grained soils (see section 7.6). The parameter φ' has also been correlated with in situ test results including the SPT blow count N and the CPT point resistance q_c. It is not recommended to use the PMT limit pressure p_L to obtain the friction angle. Using N or q_c to obtain φ' requires understanding the following. The shear strength of a coarse-grained soil is expressed as:

$$s = \sigma' \tan \varphi' \tag{15.43}$$

Therefore, there are two components involved in the soil response to the SPT or CPT: the effective stress level σ' at the depth of the test and the frictional characteristics of the soil $\tan \varphi'$. Hence, it is important to extract the influence of σ' from N and q_c before correlating them with φ'. The corrections for the influence of σ' on N were discussed in section 7.2:

$$N_1 = N_{measured} \times \left(\frac{\sigma'_{ov}}{p_a}\right)^{-0.5} \tag{15.44}$$

where N_1 and $N_{measured}$ are the corrected and uncorrected values of the SPT blow count respectively, σ'_{ov} is the vertical effective stress at the depth of the test, and p_a is the atmospheric pressure used for normalization.

There are other ways to include the influence of the stress level in the correlation. The following is a correlation between N and φ' that incorporates the stress level influence separately; it was proposed by Schmertmann (1975) and formulated into an equation by Kulhawy and Mayne (1990):

$$\tan \varphi' = \left(\frac{N}{12.2 + 20.3 \frac{\sigma'_{ov}}{p_a}} \right)^{0.34} \tag{15.45}$$

Terzaghi and Peck (1967) proposed the simple correlation shown in Figure 15.12.

The cone penetrometer point resistance q_c should also be corrected for the stress level before attempting correlation with the friction angle φ' (Figure 15.13). Much like the correction for N, the correction for q_c is:

$$q_{c1} = q_{c\,measured} \times \left(\frac{\sigma'_{ov}}{p_a} \right)^{-0.5} \tag{15.46}$$

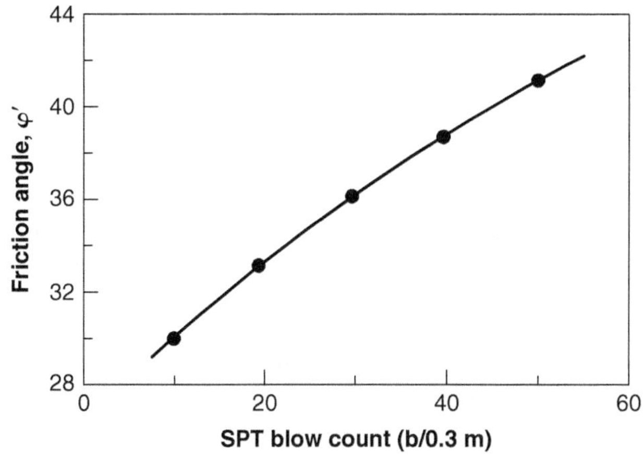

Figure 15.12 Correlation between the SPT blow count N and the friction angle φ' for coarse-grained soils. (After Terzaghi and Peck 1967)

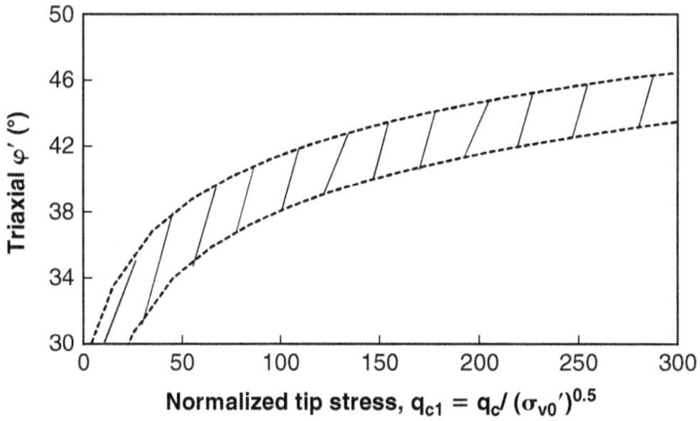

Figure 15.13 Correlation between the CPT point resistance q_c and the friction angle φ' for coarse-grained soils. (After Mayne 2007a, 2007b)

Then the following correlations exist between q_{c1} and φ' (Mayne 2007a, 2007b):

$$\varphi'_{deg} = 17.6 + 11 \times \log \left(\frac{q_{c1}}{p_a} \right)$$

$$= 17.6 + 11 \times \log \left(\frac{q_{c\,measured}}{\sqrt{\sigma'_{ov} p_a}} \right) \tag{15.47}$$

15.7.2 Fine-Grained Soils

Normally consolidated fine-grained soils have no cohesion, but some overconsolidated fine-grained soils do exhibit true cohesion c'. It is obtained by drawing a straight line (the shear strength envelope) through the failure points from shear strength tests and determining the intercept at $\sigma' = 0$. Sometimes fine-grained soils are called *cohesive soils*, but this is misleading, as the friction component of the shear strength still dominates. In fact, it is safe to ignore the cohesion c′ for most geotechnical problems. What creates the c′ value ? The phenomenological reason for any "glue" between particles can be attributed to electrical forces between fine particles and to cementation that may develop through chemical reaction. These bonds are sometimes called *diagenetic bonds*. This parameter c′ is called *true cohesion* and is not to be confused with the apparent cohesion c_{app} which comes from water tension in the voids. In fact, c_{app} is part of the friction term in Eq. 15.33:

$$c_{app} = -\alpha u_w \tan \varphi' \tag{15.48}$$

where α is the water area ratio, u_w is the water tension, and φ' is the effective stress friction angle. Because u_w has a negative value, c_{app} is positive and can be significant if the soil dries enough to generate significant water tension. This water tension can reach 10,000 kPa; therefore, c_{app} can reach hundreds of kPa. The value of c', in comparison, is rarely higher than 25 kPa. Table 15.6 gives some possible values for different soils.

The friction angle φ' corresponding to the peak shear strength for overconsolidated fine-grained soils and to the large strain strength for normally consolidated fine-grained

Table 15.6 Range of Possible Values for the Effective Stress Cohesion c′ of Fine-Grained Soils

Soil	Cohesion c′ in kPa
Coarse-grained soils	0
Silts, low plasticity	0
Silts, high plasticity, overconsolidated	5 to 10
Clays, normally consolidated	0
Clays, overconsolidated, low plasticity	10 to 15
Clays, overconsolidated, high plasticity	15 to 20

Table 15.7 Range of Possible Values for the Effective Stress Friction Angle φ' of Fine-Grained Soils

Soil	Friction Angle φ' in degrees
Silts, low plasticity	30 to 38
Silts, high plasticity	18 to 30
Clays, low plasticity	23 to 31
Clays, high plasticity	16 to 26

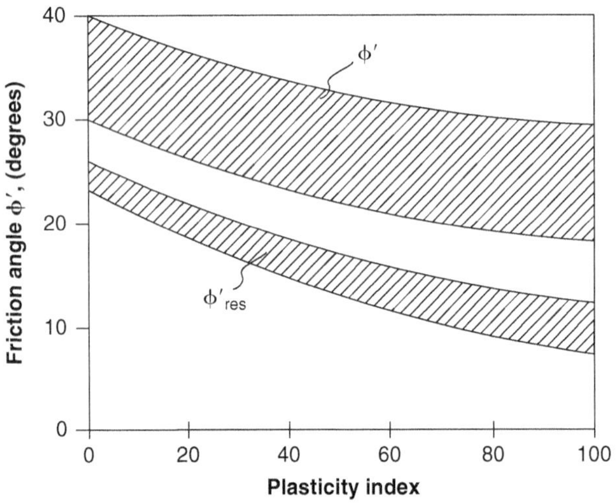

Figure 15.14 Effective stress friction angle φ' versus plasticity index for fine-grained soils.

soils is lower than the one for coarse-grained soils and varies from 20 to 35 degrees. Table 15.7 shows some possible values of φ' for various fine-grained soils. In general, the friction angle φ' decreases when the plasticity index increases. You will realize this if you wash your hands after handling a kaolinite clay (baby powder) and then after handling a bentonite clay. The bentonite will feel a lot more slippery than the kaolinite. Figure 15.14 shows general trends of φ' with the plasticity index I_p. The effective stress parameters for fine-grained soils are not obtained from in situ tests because it is difficult to ensure that the test is a drained test. One exception to this statement is the use of the borehole shear test, which is essentially a direct shear tests on the wall of the borehole; in this case the test must be performed slowly enough during the consolidation phase and the shearing phase that the assumption of no excess water stress can be made.

15.8 UNDRAINED SHEAR STRENGTH OF SATURATED FINE-GRAINED SOILS

A particular case arises when a soil is loaded fast enough that the water does not have time to drain during the loading time or if drainage is prevented in a laboratory test. In this case the shear strength is called the *undrained shear strength* and designated as s_u. This undrained case occurs rarely for most construction problems concerning coarse-grained soils, but it is often encountered with construction problems involving fine-grained soils. For example, if it takes a month to build an embankment, a clean sand would have time to fully drain, but a high-plasticity clay would not. As will be shown next, during the undrained loading of a fine-grained soil, the effective stress does not increase significantly and therefore the shear strength does not increase significantly either. Instead, the water stress increases significantly. So the controlling design case for loading on a fine-grained soil is often the *undrained case*, also called the *short-term case*. Indeed, at that time the load is maximum and the shear strength is minimum. As time goes by, the water stress decreases due to water drainage, the shear strength increases accordingly, and the factor of safety against failure increases. The critical time in the case of a fine-grained soil is immediately after loading. This is why the undrained shear strength of fine-grained soils is so important: It controls the stability design of many geotechnical structures.

As pointed out before, the general equation (Eq. 15.25) applies in all cases, including the undrained case, and s_u can be expressed as:

$$s_u = c' + \sigma' \tan \varphi' \tag{15.49}$$

The problem is that it is often difficult to obtain the effective stress on the plane of failure σ'. One of the important factors in this case is how compressible the soil skeleton is compared to water (Figure 15.15).

15.8.1 Weak Soil Skeleton: Soft, Normally Consolidated Soils

When a load is applied rapidly to a soft, normally consolidated soil, the water picks up the entire load because the soil skeleton is too weak to contribute. Therefore, the increase in normal stress $\Delta\sigma$ on the soil due to loading is equal to the increase in water stress Δu_w. The effective stress before loading σ'_b is equal to:

$$\sigma'_b = \sigma_b - u_{wb} \tag{15.50}$$

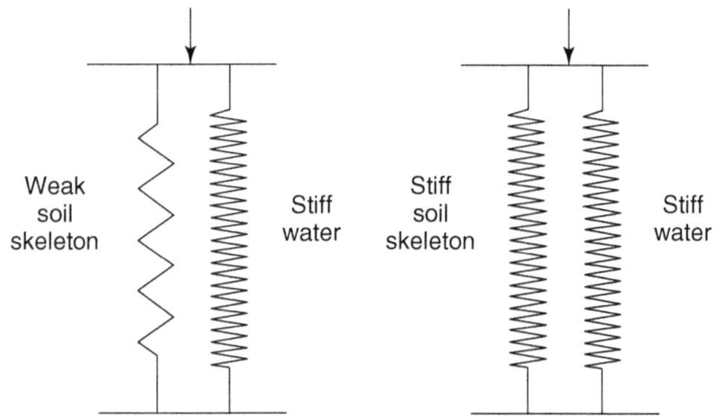

Figure 15.15 Model of saturated soil skeleton and water.

Where σ_b is the total stress before loading and u_{wb} is the water stress before loading. The effective stress immediately after loading σ'_a is equal to:

$$\begin{aligned}\sigma'_a &= \sigma_b + \Delta\sigma - (u_{wb} + \Delta u_w)\\ &= \sigma_b + \Delta\sigma - (u_{wb} + \Delta\sigma)\\ &= \sigma_b - u_{wb} = \sigma'_b\end{aligned} \tag{15.51}$$

As can be seen, the effective stress has not increased and therefore the shear strength has not increased. The undrained shear strength of a saturated, fine-grained soil with a weak skeleton is a constant s_u. This statement must be qualified by adding the following: provided that the stress level (confinement) is the same, the stress history is the same (OCR), and the stress path followed to go from the initial state to failure is the same. Indeed, all three factors can influence s_u and selecting the correct s_u is more complex than often thought (Ladd, 1991).

When a fine-grained soil with a weak skeleton is loaded in an undrained test, the Mohr circle is as shown in Figure 15.16*a* in the effective stress set of axes (Mohr circle 1) and as shown in Figure 15.16*b* in the total stress set of axes (Mohr circle 2). The difference between the total stress and the effective stress is the water stress u_w. If a second undrained test is performed on the same soil but after increasing the confining pressure by $\Delta\sigma$ (Mohr circle 3), then the water stress also increases by $\Delta\sigma$, the effective stresses do not change, and the effective stress Mohr circle does not change (still Mohr circle 1 on Figure 15.16*a*). The reason why the undrained shear strength is a constant independent of the total stress is because the effective stress Mohr circle remains the same regardless of the total stress.

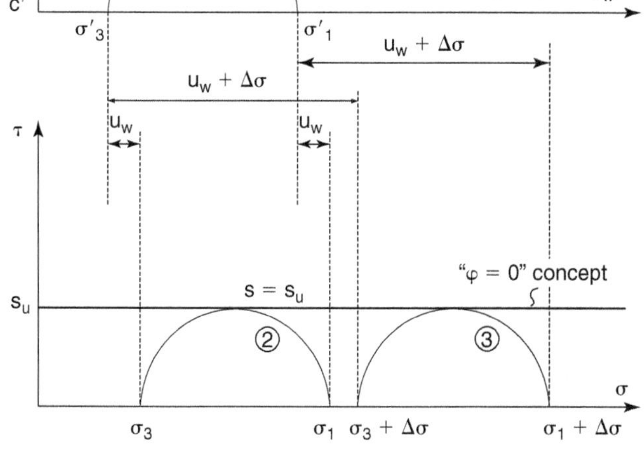

Figure 15.16 Undrained shear strength for weak soil skeleton.

This is often called the $\varphi = 0$ concept because the envelope on the shear strength vs. total stress set of axes is horizontal. This is not to say that such a soil is frictionless ($\mu = \tan\varphi = 0$). It simply means that the envelope is horizontal. In reality, the soil always has a nonzero friction angle, but it shows up in the shear strength vs. effective stress set of axes only ($\mu' = \tan\varphi' \neq 0$). Recall that the intergranular stress is represented by the effective stress, and in that set of axes, the soil friction is identified.

15.8.2 Strong Soil Skeleton: Overconsolidated Soils

In the case of an overconsolidated soil with a strong skeleton, when the load is applied rapidly, the soil skeleton is able to resist part of the load and the water picks up the rest of the load. The increase in water stress Δu_w is not as large as the increase in normal stress $\Delta\sigma$ on the soil, and is equal to f × $\Delta\sigma$ where f is smaller than 1. The effective stress before loading σ'_b is equal to:

$$\sigma'_b = \sigma_b - u_{wb} \tag{15.52}$$

Where σ_b is the total stress before loading and u_{wb} is the water stress before loading. The effective stress immediately after loading σ'_a is equal to:

$$\begin{aligned}\sigma'_a &= \sigma_b + \Delta\sigma - (u_{wb} + \Delta u_w)\\ &= \sigma_b + \Delta\sigma - (u_{wb} + f\Delta\sigma)\\ &= \sigma_b - u_{wb} + (1 - f)\Delta\sigma > \sigma'_b\end{aligned} \tag{15.53}$$

As can be seen, the effective stress has increased and therefore the shear strength has increased. The undrained shear strength s_u of a saturated, fine-grained soil with a strong skeleton increases somewhat with the total stress because the effective stress increases somewhat. Again, factors like stress level reached under drained conditions (confinement), the stress history (OCR), and the stress path followed to go from the initial state to failure influence the value of s_u (Ladd 1991).

When a fine-grained soil with a strong skeleton is loaded in an undrained test, the Mohr circle is as shown in Figure 15.17*a* in the effective stress set of axes (Mohr circle 1) and as shown in Figure 15.17*b* in the total stress set of axes (Mohr circle 2). The difference between the total stress and the effective stress is the water stress u_w. If a second undrained test is performed on the same soil, but after increasing the confining pressure by $\Delta\sigma$ (Mohr circle 4 in the total stress set of axes), then the water stress increases by a fraction f x $\Delta\sigma$ of $\Delta\sigma$, the effective stress increases somewhat, and the effective stress Mohr circle moves (Mohr circle 3 on Figure 15.17*a*). The reason why the undrained shear strength increases slightly with an increase in total stress is that the effective stress increases slightly.

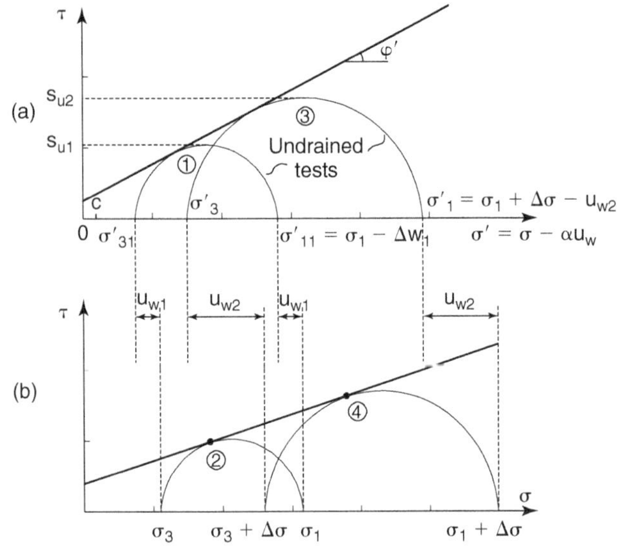

Figure 15.17 Undrained shear strength for strong soil skeleton.

15.8.3 Rate of Loading Effect on the Undrained Strength

Soils, like many other materials, are viscous: They increase in strength when the loading rate increases. The reason is attributed to the difference in water stress being developed at slower rates and at higher rates. At higher rates, the soil grains do not have time to move by finding the path of least resistance and more dilation is generated, thereby inducing higher effective stresses and shear strength. Also, the water in the voids has viscosity of its own. Indeed, water and air are viscous as well. They are Newtonian fluids and therefore are linearly viscous. They obey the following law:

$$\tau = \eta \dot{\gamma} \tag{15.54}$$

where τ is the shear stress, η is the dynamic viscosity of the material, and $\dot{\gamma}$ is the strain rate. The dynamic viscosity of water at 20°C is 10^{-6} kPa.s and the dynamic viscosity of air at 20°C is 1.8×10^{-8} kPa.s. The kinematic viscosity ν takes units of m^2/s and is defined as:

$$\nu = \frac{\eta}{\rho} \tag{15.55}$$

where ρ is the mass density of the material. For water, ρ is 1000 kg/m^3 and for air it is 1.2 kg/m^3 at the Earth's surface.

Soils are much less viscous than water and air. Equation 15.54 states that if the strain rate is doubled, the shear stress resistance will also double. In soils, if the strain rate is doubled, the shear stress resistance will be increased by a few percentage points. You might think: "Then why worry about it?" The issue is that sometimes the strain rate can be multiplied by factors of 1000 or more, and in such cases the increase or decrease can be significant. Briaud and Garland (1985) proposed the following model for the undrained shear strength of fine-grained soils:

$$\frac{s_{u1}}{s_{u2}} = \left(\frac{t_1}{t_2}\right)^{-n} \tag{15.56}$$

where s_{u1} and s_{u2} are the undrained shear strengths measured in time to failure t_1 and t_2 respectively and n is the viscous exponent for the fine-grained soil. This exponent was correlated to the reference undrained shear strength s_{uref} (Figure 15.18). The exponent n was also correlated with other soil parameters as follows:

$$n = 0.044\left(\frac{s_{u\,ref}}{p_a}\right)^{0.22} \tag{15.57}$$

$$n = 0.028 + 0.00060\, w \tag{15.58}$$

$$n = 0.035 + 0.00066\, PI \tag{15.59}$$

$$n = 0.036 + 0.046\, LI \tag{15.60}$$

where n is the soil viscous exponent in Eq. 15.56, s_{uref} is the reference undrained shear strength taken as the one obtained with a time to failure equal to one hour, p_a is the atmospheric pressure, w is the natural water content in percent, PI is the plasticity index in percent, and LI is the liquidity index (as a fraction, not a percent). The scatter in those correlations is significant, as shown in Figure 15.18. All in all, the most common values of the exponent n vary from 0.03 to 0.06, with 0.03 occurring for a high-strength, low-plasticity clay and 0.06 for a low-strength, high-PI clay.

15.9 THE RATIO S_U/σ'_{OV} AND THE SHANSEP METHOD

The undrained shear strength, like any soil shear strength, depends on the effective stress on the failure plane at the time of failure. A measure of this effective stress level is the vertical effective stress at rest σ'_{ov} in the field at the depth z considered. The ratio s_u/σ'_{ov} has been used to try to normalize the variation of s_u with depth and with effective stress level. For normally consolidated, saturated, fine-grained soils, the ratio has been found to vary between 0.2 and 0.35, increasing slightly with the plasticity index. Holtz et al. (2011) propose that for normally consolidated, saturated, fine-grained soils:

$$\left(\frac{s_u}{\sigma'_{ov}}\right)_{NC} = 0.23 \pm 0.04 \tag{15.61}$$

When the overconsolidation ratio (OCR) increases above 1, the ratio s_u/σ'_{ov} of the overconsolidated soil becomes higher than the ratio s_u/σ'_{ov} of the normally consolidated soil. This increase is not linear, and the following relationship has been proposed (Ladd et al., 1977)

For overconsolidated, saturated, fine-grained soils:

$$\left(\frac{s_u}{\sigma'_{ov}}\right)_{OC} = S(OCR)^m \tag{15.62}$$

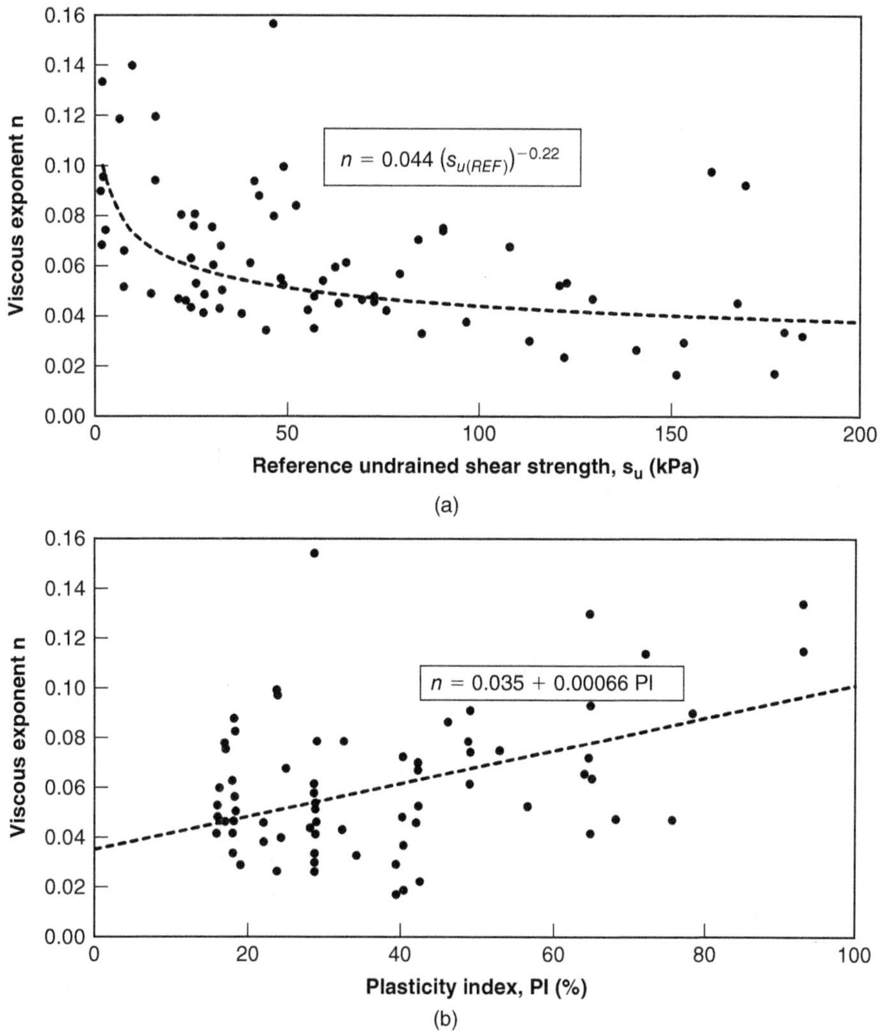

Figure 15.18 Viscous exponent n for fine-grained soils: *(a)* Influence of strength. *(b)* Influence of plasticity.

where S is the ratio for a normally consolidated soil (Eq. 15.61) and m is estimated to be 0.8.

For overconsolidated saturated fine grained soils:

$$\left(\frac{s_u}{\sigma'_{ov}}\right)_{OC} = 0.23 \times (OCR)^{0.8} \qquad (15.63)$$

Several factors influence the value of the undrained shear strength, one of which is the disturbance of the sample. Several methods have been proposed for "healing" a sample from its disturbance. One is to do a drained recompression of the sample to the in situ effective stress σ'_{ov}. This approach tends to give too high an s_u value, as the recompression decreases the sample volume and water content below its natural state. Another is the stress history and normalized soil engineering properties (SHANSEP) method developed by Ladd and Foott (1974). The method consists of four steps:

1. Determine the preconsolidation pressure σ'_p from consolidation tests.
2. Test samples of the soil in consolidated undrained tests (preferably under K_o consolidation) at confining pressures well beyond σ'_p to destructure the clay and obtain the normally consolidated behavior. These tests give the value of S in Eq. 15.62.
3. Obtain the influence of OCR by overconsolidating the sample, reducing the vertical stress, and measuring s_u at that point. These tests give the value of m in Eq. 15.62 and both S and m are therefore known.
4. Use Eq. 15.62 to develop the s_u profile for the consolidation pressure σ'_{vc} to be encountered under the structure (e.g., foundation or embankment):

$$\frac{s_u}{\sigma'_{vc}} = S\left(\frac{\sigma'_p}{\sigma'_{vc}}\right)^m \qquad (15.64)$$

As was shown, many factors affect the undrained shear strength of a soil. Therefore, any undrained shear strength value should be quoted by explaining how it was measured and over what stress range the soil was tested.

15.10 UNDRAINED SHEAR STRENGTH FOR UNSATURATED SOILS

The undrained shear strength for unsaturated soils is obtained by shearing the soil while preventing any drainage of air or water during the test. A distinction must be made among four categories of soils:

1. Soils where the water is in tension and the air has a continuous path to the ground surface or the boundary (typical degree of saturation $S < 0.85$)
2. Soils where the water is in tension and the air is occluded (typical degree of saturation $0.85 < S < 1$)
3. Soils where the water is in tension and the soil is saturated
4. Soils where the water is in compression and the soil is saturated

Sections 15.8 and 15.9 discussed results applicable to categories 2, 3, and 4 in the preceding list. This section discusses the undrained shear strength of soils in category 1: soils that are unsaturated and where the air has a continuous path to the boundary. In this case, the model in Figure 15.19 shows that part of the total stress applied to the soil will be transferred to the soil skeleton (effective stress) because the air spring is very compressible and must be compressed before stress is transferred to the water spring. The amount of total stress transferred to the water depends on the degree of saturation of the soil. For soils with very low degrees of saturation, most of the total stress will be transferred to the soil skeleton, whereas for soils with degrees of saturation close to about 0.85, most of the total stress will be transferred to the water.

This has a big impact on the undrained shear strength. Indeed, if most of the total stress imposed is carried by the soil skeleton (low degree of saturation S), then the effective stress increases nearly as much as the total stress imposed and the shear strength increases with the total stress (Figure 15.20*a*). If, in contrast, most of the total stress imposed is carried by the water (S approaching 0.85), then the effective stress does not

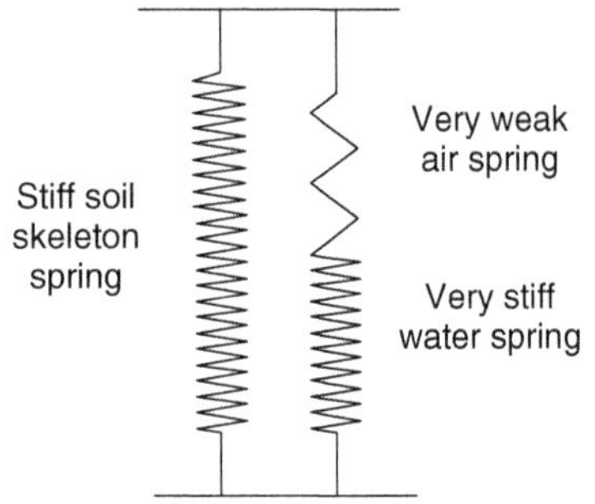

Figure 15.19 Model of unsaturated soil skeleton, air, and water.

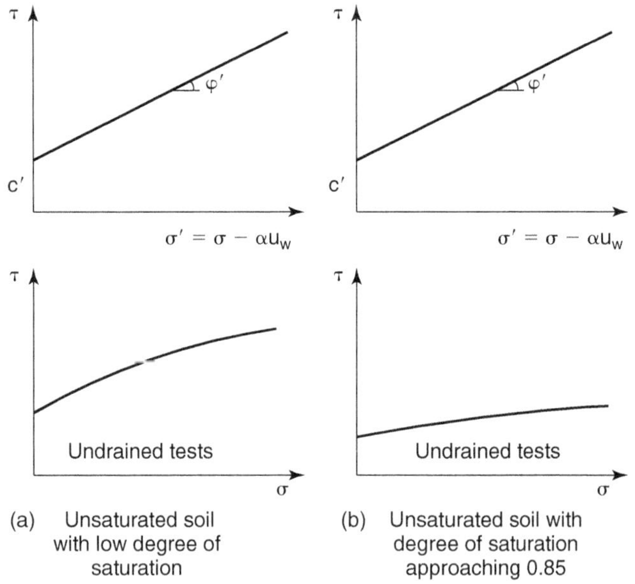

Figure 15.20 Strength envelopes for unsaturated soils.

increase much and the shear strength is nearly independent of the total stress (Figure 15.20*b*). If the degree of saturation is low but the confining stress is high enough to compress the air, including bringing it into solution, then the initially low-saturation soil will start behaving more like a saturated soil (Figure 15.20*a*). As a result, the undrained shear strength can be highly variable for unsaturated soils, depending on the degree of saturation and the total stress level. Note that as the fine-grained soil becomes drier, the water tension that is generated thereby increases the effective stress between particles and therefore the undrained shear strength.

In the field, cases in which an unsaturated soil would be loaded in an undrained fashion are rare, and are limited to high-rate dynamic loading. The concept of undrained shear strength should be handled with care for unsaturated soils, as the total stress level influences the value, especially for soils with a low degree of saturation. For these soils, the undrained shear strength case does not support the simplifying assumption that it offers for soft saturated soils, where the undrained shear strength can be considered independent of the total stress.

15.11 PORE-PRESSURE PARAMETERS A AND B

Pore-pressure parameters have been found convenient to quantify the variation of the water stress in response to undrained loading. Skempton (1954) and Bishop and Henkel (1962) proposed the following equation linking the change in water stress Δu_w due to a variation in the major principal stress $\Delta\sigma_1$ and a variation in the minor principal stress $\Delta\sigma_3$:

$$\Delta u_w = B[\Delta\sigma_3 + A(\Delta\sigma_1 - \Delta\sigma_3)] \quad (15.65)$$

where B is the pore-pressure parameter associated with an increase in confining stress $\Delta\sigma_3$ and A is the pore-pressure parameter associated with an increase in deviator stress $\Delta\sigma_1 - \Delta\sigma_3$. For saturated soils, B is close to one and A depends on the overconsolidation ratio. At failure, A_f is about 1 for normally consolidated soils, decreases with OCR, and can be negative for heavily overconsolidated soils. In practice, the coefficient $\overline{B}$ is sometimes used:

$$\Delta u_w = \overline{B}\,\Delta\sigma_v \tag{15.66}$$

where $\Delta\sigma_v$ is the increase in vertical stress. The coefficient $\overline{B}$ can be assumed in the design calculations, say 0.5, and then construction is monitored with piezometers to ensure that the water stress does not rise above $\overline{B}\,\Delta\sigma_v$. If it does, construction is halted until the water stress recedes sufficiently below that value.

For unsaturated soils, there is a need to distinguish between the response of the water and that of the air. Fredlund and Rahardjo (1993) propose:

$$du_w = B_w\,(d\sigma_3 + A_w d(\sigma_1 - \sigma_3)) \tag{15.67}$$

$$du_a = B_a\,(d\sigma_3 + A_a d(\sigma_1 - \sigma_3)) \tag{15.68}$$

Note that B_w, A_w, B_a, and A_a all depend on the degree of saturation of the soil. Also note that all pore-pressure parameters are like moduli, in that they depend on the strain level and strain rate at which they are defined.

15.12 ESTIMATING UNDRAINED SHEAR STRENGTH VALUES

There are many ways to estimate the undrained shear strength of fine-grained soils. The problem is that the value of s_u is not unique and depends on many factors. Nevertheless, fine-grained soils are often categorized by their undrained strength, as shown in Table 15.8.

The best way to determine s_u is to test the fine-grained soil in the laboratory using high-quality samples and to reproduce during the tests the initial stress conditions and the stress path during loading, while assuring no drainage. As discussed earlier, common laboratory tests available to obtain s_u include the:

1. unconsolidated undrained triaxial test (UUT)
2. consolidated undrained triaxial test (CUT)
3. unconsolidated undrained direct shear test (UUDS)
4. consolidated undrained direct shear test (CUDS)
5. unconsolidated undrained simple shear test (UUSS)
6. consolidated undrained simple shear test (CUSS)

In situ tests can also be used, including the vane shear test (VST), the borehole shear test (BHST), the cone penetrometer test (CPT), the pressuremeter test (PMT), and the standard penetration test (SPT). The VST (see Section 7.5) is the best in situ test to obtain s_u and is particularly useful offshore, where sample decompression upon retrieval from deep-water boreholes can decrease the undrained shear strength by up to 40% (Denk et al. 1981). Bjerrum (1972) used 14 case histories to back-calculate the full-scale undrained shear strength s_u (field) from embankment failures and compare it to s_u (VST) obtained from the VST performed at the sites. Because the values did not correspond, Bjerrum proposed a correction factor μ (Figure 15.21) as a function of the plasticity index I_p:

$$s_{u(Field)} = \mu\, s_{u(VST)} \tag{15.69}$$

Ladd et al. (1977) collected additional failure case histories and confirmed the trend. As can be seen, the correction factor indicates that s_u (VST) is larger than s_u (field); this is attributed to the facts that the rate of shearing is much higher in the VST than in the failure of the embankment and that this rate effect is more prominent in high-plasticity clays than in low-plasticity clays. Differences in the influence of anisotropy and plane strain conditions between the VST and the embankment are also contributing factors.

Table 15.8 Classification of Fine-Grained Soils by Undrained Shear Strength

Silts and Clays Strength			
Description	s_u (kPa)	N (bpf)	Simple field test*
Very soft	<12	<2	Squeezes between your fingers
Soft	12–25	2–4	Easily penetrated by light thumb pressure
Medium or firm	25–50	4–8	Penetrated by strong thumb pressure
Stiff	50–100	8–15	Indented by strong thumb pressure
Very stiff	100–200	15–30	Slightly indented by strong thumb pressure
Hard	200–400	30–50	Slightly indented by thumbnail
Very hard	>400	>50	Not indented by thumbnail

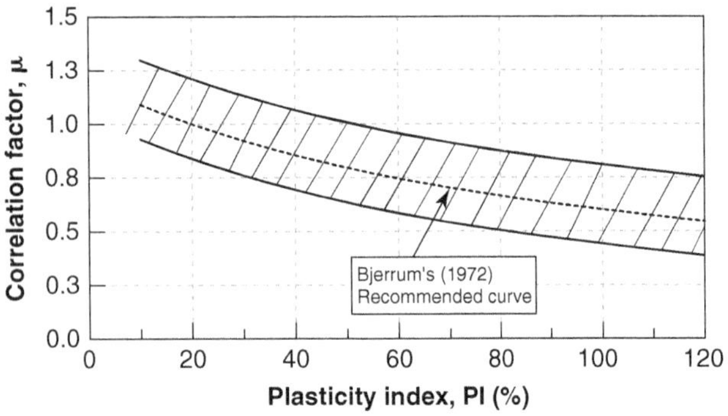

Figure 15.21 Bjerrum correction factor for undrained shear strength from vane test. (After Bjerrum 1972; Ladd et al. 1977)

The borehole shear test (see section 7.6) can be used to obtain a value of s_u in situ by direct measurement as long as the shearing is performed rapidly to ensure undrained behavior. Because the normal total stress (horizontal) on the plane of failure (vertical) can be varied in the BHST, the influence of the total normal stress on s_u (as discussed in sections 15.8 to 15.10) can be evaluated.

The cone penetrometer test (see section 7.2) has also been used to obtain s_u. The equation used is:

$$s_{u(CPT)} = \frac{q_c - \sigma_{vo}}{N_k} \tag{15.70}$$

where q_c is the point resistance of the CPT, σ_{vo} is the vertical total stress at the depth where q_c is measured, and N_k is the cone factor. This equation comes from the ultimate bearing pressure p_u under a pile point:

$$p_u = N_c s_u + \sigma_{vo} \tag{15.71}$$

where N_c is a bearing capacity factor usually taken equal to 9 for deep localized failure. One would therefore expect that N_k would be 9. However, many differences between a pile point and the CPT lead N_k to be quite different from 9 and quite variable. The differences include the rate of loading effect, the scale effect, and the installation procedure. The penetration of the CPT goes much faster than the pile penetration during a typical load test ($N_k > 9$). The cone is much smaller in size than the pile; as a result, the cone detects thinner layers than the pile, which averages the soil resistance over a larger zone ($N_k > 9$); also, the cone is pushed in, whereas the pile is either driven or drilled in place. All in all, the value of N_k seems to average 14 ± 5 for s_u being determined from Eq. 15.70 (Figure 15.22), but correlations have led to values varying from 5 to 70. The main problem is that, as discussed earlier, s_u is not unique, so no general correlation can be proposed. The best way to approach the problem is to run a few lab tests to obtain the right s_u value needed for the project, run parallel CPT soundings, generate a local correlation to obtain a site value of N_k from s_u and q_c, and then extend the results by running additional CPTs.

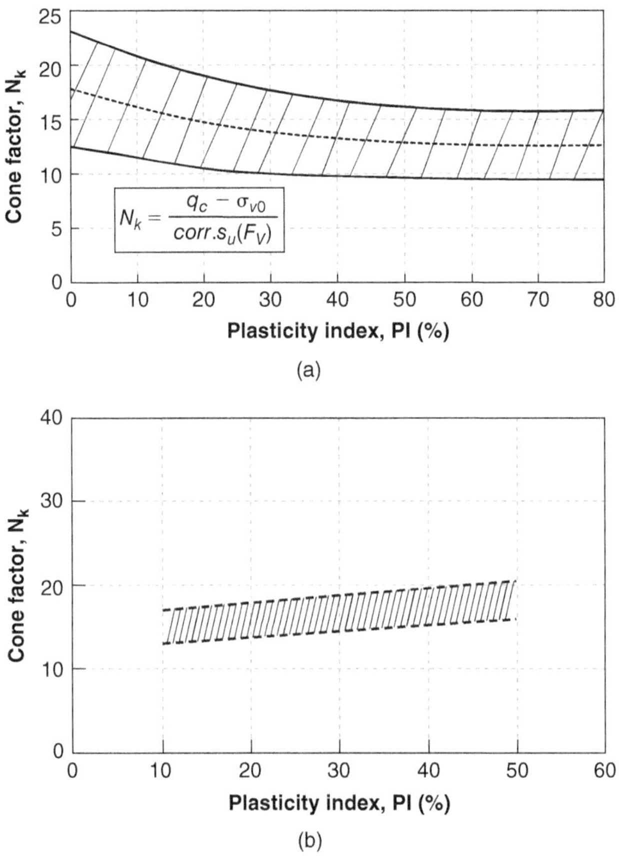

Figure 15.22 N_k factor for obtaining s_u from CPT q_c value: *(a)* based on data from Baligh et al. 1980; Lunne and Kleven 1981; s_u mostly from vane shear tests. *(b)* based on data from Aas et al. 1986; s_u mostly from vane shear tests.

The undrained shear strength s_u can also be obtained from a pressuremeter test (PMT; see section 7.3). In this case the limit pressure p_L is used as follows:

$$s_{u(PMT)} = \frac{p_L}{N_p} \tag{15.72}$$

where N_p is the pressuremeter factor. This factor can be taken as 7.5 in first approximation, but the relationship is nonlinear and Briaud (1992) proposed:

$$s_{u(PMT)}(\text{kPa}) = 0.67(p_L(\text{kPa}))^{0.75} \tag{15.73}$$

Figure 15.23 shows this relationship compared to two databases. The likely reason for this nonlinearity is that for lower values of s_u, the fine-grained soils tend to have stress-strain curves exhibiting no peak (strain hardening behavior), whereas at higher s_u values the fine-grained soils tend to exhibit peak strength and post-peak softening down to a residual strength. Because the limit pressure involves large strains near the cavity and smaller strain at some distance from the cavity, the strength mobilized is an average between the two. This average will tend to be higher for strain hardening

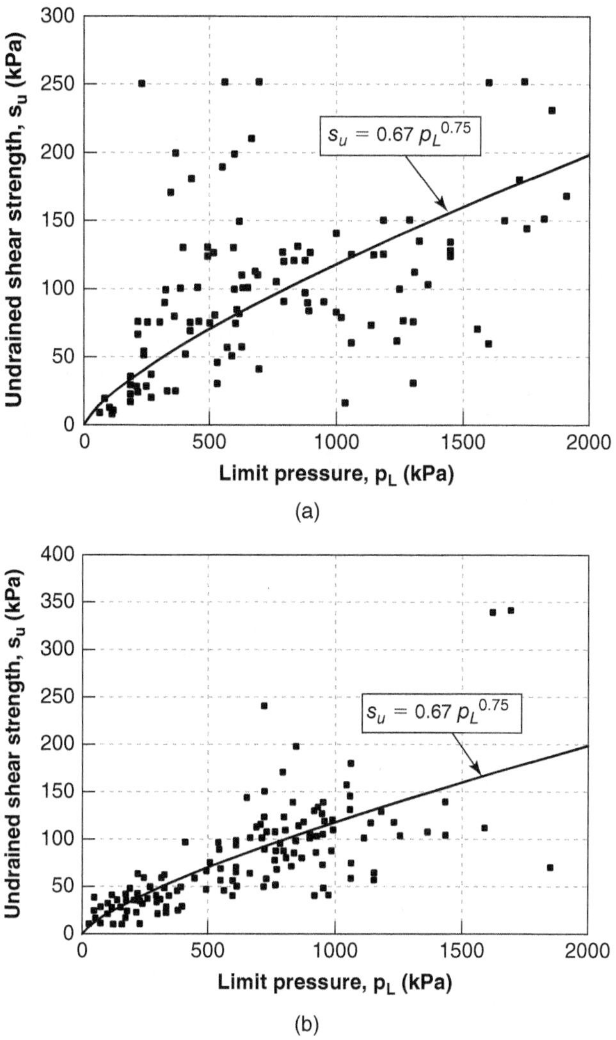

Figure 15.23 Correlation between s_u and the pressuremeter limit pressure: (a) based on data from Briaud 1992; s_u mostly from unconfined compression tests. (b) based on data from Baguelin et al. 1978; s_u from laboratory tests and vane tests.

soils (low s_u) than for strain softening soils (high s_u). The advantage of using p_L to get s_u is that the PMT involves a larger mass of soil than most other tests in the response to the expansion; as such, it can bridge over microfissures and other small-scale features and is more representative of the mass strength. The drawback is that the test is typically more expensive than the vane test, for example.

The standard penetration test (SPT; see section 7.1) and its blow count N have also been used to obtain the undrained shear strength s_u. Such correlations should be used as a last resort, however. Terzaghi et al. (1996) propose the following relationship to obtain a relatively conservative value of s_u:

$$s_{u(SPT)}(\text{kPa}) = 4.4N_{60} \tag{15.74}$$

where N_{60} is the blow count (blows per foot) corrected to 60% of maximum energy (see section 7.1). Terzaghi et al.

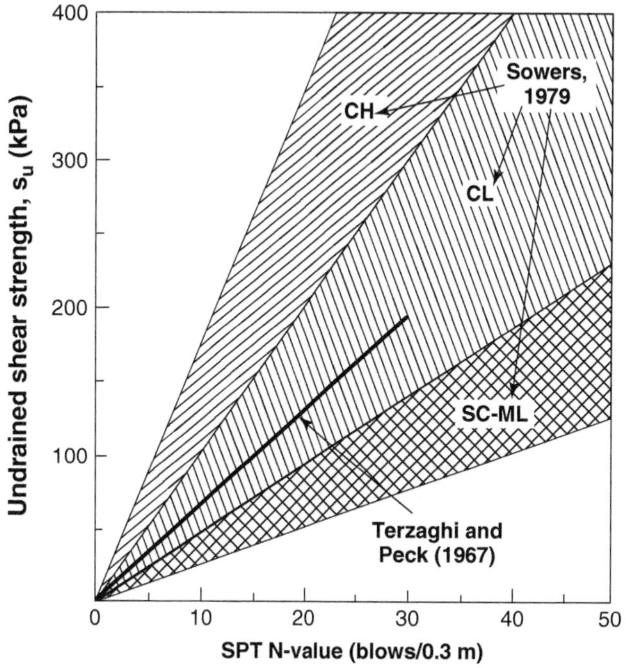

Figure 15.24 Correlation between s_u and the SPT blow count N. (After Sowers 1979; Terzaghi and Peck 1967)

point out that for low-plasticity, fine-grained soils, the factor 4.4 in Eq. 15.74 can go up to 7. Terzaghi and Peck (1967) proposed:

$$s_{u(SPT)}(\text{kPa}) = 6.7N \tag{15.75}$$

Sowers (1979) presents his experience in a figure relating N and s_u (Figure 15.24).

15.13 RESIDUAL STRENGTH PARAMETERS AND SENSITIVITY

The *residual strength* of a soil is the strength at very large strains long after the peak strength. It exists for the effective stress shear strength and for the undrained shear strength. The residual effective stress cohesion can be taken as zero and the residual effective stress friction angle is reduced:

$$\tau_{f\ res} = \sigma' \tan \varphi'_{res} \tag{15.76}$$

where $\tau_{f\ res}$ is the residual shear strength, σ' is the effective normal stress on the plane of failure, and φ'_{res} is the residual friction angle. The amount of reduction from φ' to φ'_{res} depends on the soil type. Loose coarse-grained soils and normally consolidated, saturated, low-plasticity, fine-grained soils do not exhibit much reduction between the friction angle and the residual friction angle. The reduction for soils with higher plasticity is more significant, as exemplified by Figure 15.25 based on data from Stark and Eid (1994) and Lupini et al. (1981).

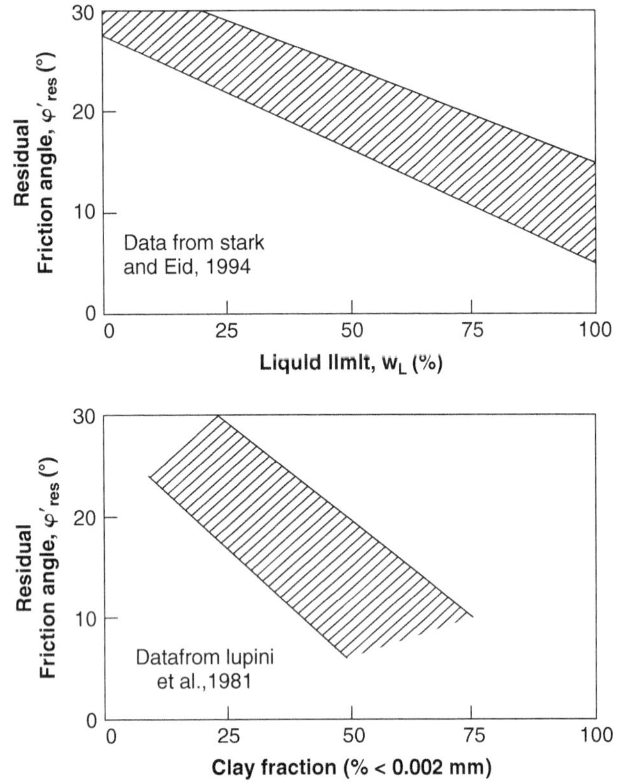

Figure 15.25 Correlation between effective stress residual friction angle and soil properties. (After Stark and Eid 1994; Lupini et al. 1981.)

The residual undrained shear strength $s_{u\,res}$ is best measured directly, either in the laboratory or in the field. In the laboratory, the best apparatus is the ring shear apparatus, which consists of a split-donut type of device. In this apparatus the top half of the donut is rotated one way while the bottom half is held in place. In this fashion, very large strains can be reached until the shear strength reaches the residual strength plateau. In the field, the vane shear test can be used. The vane is rotated until the peak shear strength $s_{u\,peak}$ is obtained and then rotation continues while recording the torque. When the torque stabilizes, the residual undrained shear strength $s_{u\,res}$ is reached. ASTM recommends that after reaching the residual shear strength, the remolded shear strength be obtained by rapidly rotating the vane 5 to 10 times. The remolded undrained shear strength $s_{u\,rem}$ is then obtained by repeating the vane test immediately after the rapid rotations.

The sensitivity S_t of a fine-grained soil is defined as:

$$S_t = \frac{s_{u\ peak}}{s_{u\ rem}} \tag{15.77}$$

where $s_{u\ peak}$ is the peak undrained shear strength and $s_{u\ rem}$ is the remolded undrained shear strength. Some clays are not sensitive and some are very sensitive. For example, a low-plasticity, soft kaolinite clay is unlikely to be very sensitive ($S_t < 2$), but a quick clay may have a sensitivity in excess of 20. These quick clays do have some strength when undisturbed, say $s_u = 25$ kPa, but become a thick liquid when disturbed (see section.13.2.9). A soil with a sensitivity of less than 4 would be qualified as a low-sensitivity soil; from 4 to 10 would be medium sensitivity; 10 to 20 would be highly sensitive; and above 20 would be quick.

15.14 STRENGTH PROFILES

The strength profile of a soil deposit can give a lot of information about the deposit. This strength can be measured by the CPT point resistance q_c, or by the SPT blow count N, or by the PMT limit pressure p_L, or by the undrained shear strength s_u for a fine-grained soil. If the profile shows a linear increase with depth with a zero value at the surface, the deposit could be a normally consolidated, soft, fine-grained soil, as would be expected in a city like New Orleans (Figure 15.26*a*). If the profile goes through zero at the surface but increases nonlinearly with depth with a downward curvature, then the deposit could be a dry sand deposit (Figure 15.26*b*). If the profile increased linearly with depth but had a definite nonzero value at the surface, the deposit could be a fine-grained soil overconsolidated by overburden removal through erosion or through the melting of a glacier (Figure 15.26*c*). If the profile indicated a constant strength with depth (Figure 15.26*d*), the deposit could be an unsaturated silty sand where the total stress increased with depth but the water tension decreased with depth, thereby maintaining the effective stress constant and the shear strength constant with depth. It could also be an underconsolidated soft clay. If the profile showed a curved decrease with depth near the surface followed by

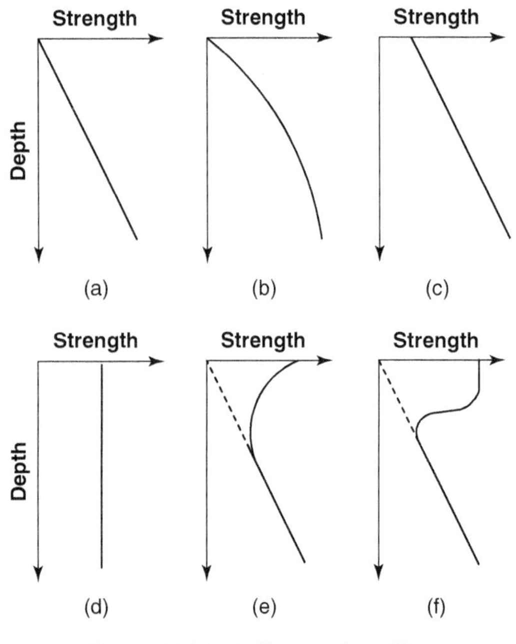

Figure 15.26 Soil strength profiles.

an increase at larger depths, the deposit could be a high-plasticity, fine-grained soil overconsolidated by desiccation near the surface but becoming normally consolidated at depth where the seasonal shrink-swell cycles no longer have an influence (Figure 15.26*e*). If the profile shows a strong layer near the surface and a softer layer at some depth, it could indicate the presence of a crust of the softer layer below (Figure 15.26*f*).

Note that these strength profiles, if rotated 90 degrees counterclockwise, represent shear strength envelopes in the case of uniform soil. Indeed, after rotation, the vertical axis represents a measure of the strength and the horizontal axis a measure of the total stress. The horizontal axis can be transformed in a measure of the effective stress if the water stresses are known (such as a hydrostatic condition, for example).

15.15 TYPES OF ANALYSES

In this chapter we have talked about effective stress, total stress, undrained strength, and drained strength. Each strength is associated with a type of analysis in design and it is important to understand which analysis is used for what strength. The types of strength analyses typically performed in geotechnical engineering include: effective stress analysis, total stress analysis, undrained analysis, drained analysis, short-term analysis, and long-term analysis.

Effective stress analysis: an analysis in which the soil is considered to be made of particles, water, and air. It is the most theoretically sound analysis, but it is also the most complicated analysis, because it requires knowledge of the total stress, the water stress, and the air stress (unless it can be assumed to be zero). It is applicable to all design cases.

Total stress analysis: an analysis in which the soil is considered to be made of one material, without distinguishing between particles, water, and air. It is the easiest of the analyses because the number of variables is decreased significantly. It is also the most likely to be erroneous, because the fundamental principles are not respected, except in a few cases like the undrained behavior of fine-grained soils where the undrained strength can be considered constant.

Undrained analysis: an analysis in which the water and air are not allowed to drain during loading. This analysis can be performed by using effective stress models and in a few specific cases total stress models. One of the difficulties in using this analysis together with an effective stress model is prediction of the water stress and possibly the air stress for unsaturated soils.

Drained analysis: an analysis in which the water and air are allowed to drain until any excess water stress and any excess air stress have gone back to zero. It is one of the simplest of all effective stress analyses, but its usefulness is limited because it only applies to long loading times.

Short-term analysis: an analysis of the behavior of the soil in the short term. A short-term analysis can be a drained analysis for a clean, coarse-grained soil and an undrained analysis for a fine-grained soil. It tends to control the design of structures that will load fine-grained soils.

Long-term analysis: an analysis of the behavior of the soil in the long term. A long-term analysis is similar to a drained analysis because in the long term—sometimes in the very long term—the soil will drain and excess water and air stresses will vanish. This analysis tends to control the design of excavations.

15.16 TRANSFORMATION FROM EFFECTIVE STRESS SOLUTION TO UNDRAINED STRENGTH SOLUTION

The results of an effective stress analysis can be transformed into the results of an undrained analysis when the undrained strength is constant and the $\varphi = 0$ concept applies. In this case the transformation consists of using the following correspondence principles:

1. Effective unit weight becomes total unit weight

$$\gamma_{eff} \rightarrow \gamma_t \tag{15.78}$$

2. Effective stress becomes total stress

$$\sigma' \rightarrow \sigma \tag{15.79}$$

3. Effective stress cohesion becomes undrained shear strength

$$c' \rightarrow s_u \tag{15.80}$$

4. Effective stress friction angle becomes zero

$$\varphi' \rightarrow 0 \tag{15.81}$$

In this fashion, for example, the shear strength changes as follows:

$$s = c' + \sigma' \tan \varphi' \rightarrow s = s_u \tag{15.82}$$

The ultimate bearing pressure changes as follows (see Chapter 17):

$$p_u = c'N_c + \frac{1}{2}\gamma_{eff} B N_\gamma + \gamma D N_q \rightarrow p_u = N_c s_u + \gamma D \tag{15.83}$$

because for $\varphi = 0$, $N_\gamma = 0$, and $N_q = 1$.

The passive earth pressure equation changes as follows (see Chapter 21):

$$\sigma'_{ph} = K_p \sigma'_{ov} + 2c'\sqrt{K_p} \rightarrow \sigma_{ph} = \sigma_{ov} + 2s_u \tag{15.84}$$

because for $\varphi = 0$, $K_p = 1$.

PROBLEMS

15.1 It is well known that a car with wider tires can take corners faster than the same car with narrower tires. That is to say, the shearing resistance of the car with wider tires is larger than the shearing resistance of the car with narrower tires. This seems counterintuitive when one considers that in both cases the weight of the car is the same, and therefore the friction should be the same regardless of the width of the tires. Explain why the car with wider tires develops more resistance to shear in the corners than the car with narrower tires.

15.2 A medium dense sand deposit has a dry unit weight of 17 kN/m^3, a saturated unit weight of 20 kN/m^3, and a friction angle of 32 degrees. Calculate the shear strength on a horizontal plane at a depth of 10 m if:

   a. The groundwater level is much deeper than 10 m and the sand has no water.
   b. The groundwater level is at the ground surface.
   c. The groundwater level is at 12 m and the sand is saturated by capillary action.

15.3 In a simple shear test on a dense sand with no water, the normal stress is 100 kPa and the shear stress at failure is 80 kPa. At failure also, the vertical displacement is 0.5 mm and the horizontal displacement is 5 mm.

   a. Calculate the friction angle φ' and the dilation angle ψ'.
   b. Calculate the shear strength of the sand if the normal stress increases to 200 kPa and the angles φ' and ψ' remain the same.

15.4 A soft clay has formed a crust near the ground surface due to drying under the sun. At the ground surface the relative humidity has been 40% for a long time. A sample of the surface clay gives a unit weight of 17.5 kN/m^3 and a water content of 10%. Estimate the shear strength of the clay at the ground surface if the effective stress friction angle is 27 degrees and G_s is 2.7. What is the apparent cohesion of that clay?

15.5 A medium-stiff clay is tested in an undrained triaxial test. At failure, the effective stress on the failure plane is 230 kPa and the shear stress on the failure plane is 122 kPa. Calculate the undrained shear strength of this clay.

15.6 A soft, saturated clay is tested in an unconsolidated undrained direct shear test with a normal stress of 50 kPa; the shear strength obtained is 20 kPa. An identical sample is tested, also in an unconsolidated undrained direct shear test, but this time the normal stress is 100 kPa. What would you expect the shear strength to be?

15.7 A sand layer has an SPT blow count of 27 bpf and a CPT point resistance of 13.5 MPa. Both measurements come from a depth of 12 m. The groundwater level is at a depth of 5 m. What is your best estimate of the friction angle for this sand at that depth?

15.8 The undrained shear strength of a medium-stiff clay is 46 kPa when sheared in a time to failure equal to 3 minutes in a vane shear test. The medium-stiff clay has a water content of 35% and a plasticity index of 30% Solve the following two problems:

   a. A guardrail post is placed in this clay on the side of the road to arrest cars upon impact. The rise time of the force during the impact is anticipated to be 20 milliseconds. What shear strength value should you use?
   b. An embankment is placed on that clay. In the design process it is assumed that if a failure occurs, the failure of the embankment would be very slow and take place in about 6 hours. What undrained shear strength should be used in calculating the factor of safety against embankment failure?

15.9 Use average and associated ranges of rate effect viscous exponent to generate a curve similar to the Bjerrum correction factor for the vane shear test, undrained strength. Assume that the vane reaches the peak undrained shear strength in 3 minutes and that the embankment reaches failure in half a day.

15.10 A clay has an overconsolidation ratio equal to 2.5. Use the SHANSEP method and reasonable values of the parameters to estimate the undrained shear strength of that clay at a depth of 20 m. The clay is offshore at the bottom of the North Sea in 300 m of water.

15.11 An unsaturated sample of clay is tested in a simple shear test. At failure the total normal stress on the failure plane is 70 kPa and the shear stress is 175 kPa.

   a. Is that possible?
   b. After testing, the water content on the plane of failure is measured and the soil water retention curve gives a water tension of 1450 kPa. The water content coupled with the measurement of the unit weight and the assumption that Gs is 2.7 leads to a degree of saturation of 20%. If the clay has no effective stress cohesion, calculate the effective stress friction angle.

15.12 A lightly overconsolidated clay has a CPT point resistance of 1100 kPa, an OCR of 1.7, a PMT limit pressure of 590 kPa, an SPT blow count N of 13 bpf, and a unit weight of 18 kN/m^3. Estimate the undrained shear strength of that clay if the data comes from a depth of 6 m with the groundwater level being at a depth of 2 m.

15.13 You are at the beach lying on dry uniform sand. You take a handful of sand and let it fall from your hand onto a 0.3 m by 0.3 m wide plate. The sand pile on the plate has the shape of a pyramid and the angle of the pyramid with the horizontal is β. Demonstrate that β is equal to the friction angle φ'. You then take that same pile of sand and add a bit of water. Now you are able to mold the sand pile into a cylinder standing vertically. Where does the sand strength come from? Is it cohesion or friction?

Problems and Solutions

Problem 15.1

It is well known that a car with wider tires can take corners faster than the same car with narrower tires. That is to say, the shearing resistance of the car with wider tires is larger than the shearing resistance of the car with narrower tires. This seems counterintuitive when one considers that in both cases the weight of the car is the same, and therefore the friction should be the same regardless of the width of the tires. Explain why the car with wider tires develops more resistance to shear in the corners than the car with narrower tires.

Solution 15.1

The weight of the car is the same in both cases, so the friction should be the same (in theory). However, we need to consider the force generated by the cohesion or "glue" between the tire and the asphalt. In equation form, $F = \mu N + C$, where μ is the friction coefficient, N is the normal force, and C is the cohesion force. The cohesion force C depends on the contact area, whereas the normal force N does not. The area of a wide tire is larger than the area of a narrow tire. A wide tire will thus provide more area to resist the force between the tire and the pavement to turn around a corner. Direct shear tests between a piece of pavement and a piece of rubber from a tire would be required to demonstrate this possible explanation.

Problem 15.2

A medium dense sand deposit has a dry unit weight of 17 kN/m^3, a saturated unit weight of 20 kN/m^3, and a friction angle of 32 degrees. Calculate the shear strength on a horizontal plane at a depth of 10 m if:

a. The groundwater level is much deeper than 10 m and the sand has no water.
b. The groundwater level is at the ground surface.
c. The groundwater level is at 12 m and the sand is saturated by capillary action.

Solution 15.2

a. The groundwater level is much deeper than 10 m and the sand has no water:

$$\tau_f = c' + (\sigma - \alpha u_w) \tan \varphi'$$
$$\tau_f = 0 + (10 \times 17 - 0 \times 0) \tan 32 = 106.2 \text{ kPa}$$

b. The ground-water level is at the ground surface:

$$\tau_f = c' + (\sigma - \alpha u_w) \tan \varphi'$$
$$\tau_f = 0 + (10 \times 20 - 1 \times 10 \times 9.81) \tan 32 = 63.7 \text{ kPa}$$

c. The ground-water level is at 12 m and the sand is saturated by capillary action.

In this case, there is suction in the soil. Equation 15.8 is used and the pore water pressure is negative:

$$\tau_f = c' + (\sigma - \alpha u_w) \tan \varphi'$$
$$\tau_f = 0 + (10 \times 20 - 1(-2 \times 9.81)) \tan 32 = 137.2 \text{ kPa}$$

Problem 15.3

In a simple shear test on a dense sand with no water, the normal stress is 100 kPa and the shear stress at failure is 80 kPa. At failure also, the vertical displacement is 0.5 mm upward and the horizontal displacement is 5 mm.

a. Calculate the friction angle φ' and the dilation angle ψ'.
b. Calculate the shear strength of the sand if the normal stress increases to 200 kPa and the angles φ' and ψ' remain the same.

Solution 15.3

a. Using the shear strength equation and knowing that the effective stress cohesion of the dense sand is zero:

$$\tau_f = c' + (\sigma - \alpha u_w) \tan \varphi'$$

$$80 = 0 + (100 - 0 \times 0) \tan \varphi' \quad \text{or} \quad \varphi' = 38.66 \text{ degrees}$$

The tangent of the dilation angle is given by the ratio of vertical to horizontal displacement:

$$\tan \psi' = \frac{0.5}{5} \quad \text{or} \quad \psi' = 5.71 \text{ degrees}$$

b. Again using the shear strength equation:

$$\tau_f = 0 + (200 - 0 \times 0) \tan 38.66 = 160 \text{ kPa}$$

The dilation angle is not used because it is included in the friction angle φ'.

Problem 15.4

A soft clay has formed a crust near the ground surface due to drying under the sun. At the ground surface the relative humidity has been 40% for a long time. A sample of the surface clay gives a unit weight of 17.5 kN/m^3 and a water content of 10%. Estimate the shear strength of the clay at the ground surface if the effective stress friction angle is 27 degrees and G_s is 2.7. What is the apparent cohesion of that clay?

Solution 15.4

Based on the Kelvin equation (Eq. 10.69; see Chapter 10), we can calculate the water tension at the ground surface:

$$u(\text{kPa}) = 135000 \times \ln(RH) = 135022 \times \ln(0.4) = -123719 \text{ kPa} \tag{15.1s}$$

Based on the three-phase soil relationships, the void ratio is linked to the unit weight of solids, the water content, and the soil unit weight by:

$$e = \frac{\gamma_s(1+\omega)}{\gamma} - 1$$

Given

$$G_s = 2.7$$

and

$$\omega = 10\%, \gamma = 17.5 \text{ kN/m}^3$$

we can obtain:

$$e = \frac{\gamma_s(1+\omega)}{\gamma} - 1 = \frac{2.7 \times 9.81 \times (1+0.1)}{17.5} - 1 = 0.66$$

Another useful equation links the degree of saturation to Gs, w, and e:

$$S = \frac{\omega G_s}{e}$$

We can obtain:

$$S = \frac{\omega \gamma_s}{e \gamma_w} = \frac{0.1 \times 2.7}{0.66} = 0.41$$

Therefore, a can be estimated as S (i.e., 0.41). At the ground surface, the shear strength of the clay can now be calculated as:

$$\tau = c' + (\sigma - \alpha u_w)\tan\varphi' = 0 + (0 - 0.41 \times (-123719)) \times \tan 27^\circ + 0 = 25846 \text{ kPa}$$

The apparent cohesion of the clay is also 25,846 kPa, because both c′ and σ are zero.

Problem 15.5

A medium-stiff clay is tested in an undrained triaxial test. At failure, the effective stress on the failure plane is 230 kPa and the shear stress on the failure plane is 122 kPa. Calculate the undrained shear strength of this clay.

Solution 15.5

$$s_u = 122 \text{ kPa}$$

Problem 15.6

A soft, saturated clay is tested in an unconsolidated undrained direct shear test with a normal stress of 50 kPa; the shear strength obtained is 20 kPa. An identical sample is tested, also in an unconsolidated undrained direct shear test, but this time the normal stress is 100 kPa. What would you expect the shear strength to be?

Solution 15.6

Because the soil sample is being tested in an undrained condition, and because the soil skeleton is weak (soft clay), the increase in normal stress is taken up by the water and there is no increase in effective stress. Therefore, the expected undrained shear strength is the same as in the first test: 20 kPa.

Problem 15.7

A sand layer has an SPT blow count of 27 bpf and a CPT point resistance of 13.5 MPa. Both measurements come from a depth of 12 m. The groundwater level is at a depth of 5 m. What is your best estimate of the friction angle for this sand at that depth?

Solution 15.7

The vertical effective stress at the point of measurement of the SPT and the CPT is computed as:

$$\sigma'_{ov} = 20 \times 12 - 9.81 \times 7 = 171.3 \text{ kPa}$$

The SPT blow count is corrected for stress level:

$$N_1 = N_{measured} \times \left(\frac{\sigma'_{ov}}{p_a}\right)^{-0.5} = 27 \times \left(\frac{171.3}{101.3}\right)^{-0.5} = 20.8 \text{ blows/0.3 m}$$

Then the CPT point resistance is corrected for stress level:

$$q_{c1} = q_{c\,measured} \times \left(\frac{\sigma'_{ov}}{p_a}\right)^{-0.5} = 13.5 \times \left(\frac{171.3}{101.3}\right)^{-0.5} = 10.4 \text{ MPa}$$

The friction angle φ' can be evaluated in a number of ways. Using Mayne's recommendation:

$$\varphi' = 17.6 + 11 \times \log\left(\frac{q_{c1}}{p_a}\right) = 17.6 + 11 \times \log\left(\frac{10400}{101.3}\right) = 39.7 \text{ degrees}$$

Using an equation from Schmertmann (1975) and Kulhawy and Mayne (1990):

$$\tan\varphi' = \left(\frac{N}{12.2 + 20.3\dfrac{\sigma'_{ov}}{p_a}}\right)^{0.34} = \left(\frac{27}{12.2 + 20.3 \times \dfrac{171.3}{101.3}}\right)^{0.34} \quad \text{and} \quad \varphi' = 39.7 \text{ degrees}$$

Using the Terzaghi and Peck (1967) figure relating the friction angle to the blow count, we get:

$$\Phi' = 35 \text{ degrees}$$

Considering all the values collected, a cautiously conservative estimate of the friction angle might be 36 degrees.

Problem 15.8

The undrained shear strength of a medium-stiff clay is 46 kPa when sheared in a time to failure equal to 3 minutes in a vane shear test. The medium-stiff clay has a water content of 35% and a plasticity index of 30%. Solve the following two problems:

a. A guardrail post is placed in this clay on the side of the road to arrest cars upon impact. The rise time of the force during the impact is anticipated to be 20 milliseconds. What shear strength value should you use?
b. An embankment is placed on that clay. In the design process it is assumed that if a failure occurs, the failure of the embankment would be very slow and take place in about 6 hours. What undrained shear strength should be used in calculating the factor of safety against embankment failure?

Solution 15.8

The rate effect equation for the undrained shear strength of a clay is:

$$\frac{s_{u1}}{s_{u2}} = \left(\frac{t_1}{t_2}\right)^{-n}$$

The viscous exponent n is related to the water content by:

$$n = 0.028 + 0.0006\, w\% \text{ so } n = 0.049$$

The viscous exponent n is related to the plasticity index by:

$$n = 0.035 + 0.00066 \text{ PI\%} \text{ so } n = 0.0548$$

Use an average n value of $n_{Avg} = 0.0519$

a. In this case $s_{u1} = 46$ kPa, $t_1 = 180$ sec, $t_2 = 0.02$ sec, $s_{u2} = ?$

$$s_{u2} = s_{u1}\left(\frac{t_1}{t_2}\right)^{n}$$

$$s_{u2} = 46\left(\frac{180}{0.02}\right)^{0.0519} = 73.8 \text{ kPa}$$

b. In that case $s_{u1} = 46$ kPa, $t_1 = 180$ sec, $t_2 = 21600$ sec, $s_{u2} = ?$

$$s_{u2} = 46\left(\frac{180}{21600}\right)^{0.0519} = 35.9 \text{ kPa}$$

Problem 15.9

Use average and associated ranges of rate effect viscous exponent to generate a curve similar to the Bjerrum correction factor for the vane shear test, undrained strength. Assume that the vane reaches the peak undrained shear strength in 3 minutes and that the embankment reaches failure in half a day.

Solution 15.9

Use is made of the rate effect equation and of the correlation between the rate exponent n and the plasticity index PI in percent:

$$\frac{s_{u1}}{s_{u2}} = \left(\frac{t_1}{t_2}\right)^{-n}$$

$$n = 0.035 + 0.00066 \times \text{PI\%}$$

The strength s_{u1} is the field value of s_u, the strength s_{u2} is the vane test value of s_u, the time t_1 is the time for the embankment failure or half a day (720 min), the time t_2 is the time for the vane test or 3 min. Therefore, the equation for the Bjerrum factor is:

$$\mu = \frac{s_{u,field}}{s_{u,VST}} = 240^{-(0.035+0.00066\ \text{PI\%})}$$

To take into account the scatter in the n vs. PI correlation, the calculated μ value is bracketed between the following two expressions.

$$\mu = \frac{s_{u,field}}{s_{u,VST}} = 240^{-(0.01+0.00066\ \text{PI\%})}$$

$$\mu = \frac{s_{u,field}}{s_{u,VST}} = 240^{-(0.065+0.00066\ \text{PI\%})}$$

Figure 15.1s shows the range of the function μ vs. PI based on the rate effect model. It appears that the rate effect model explains much of the correction factor except at low PI values:

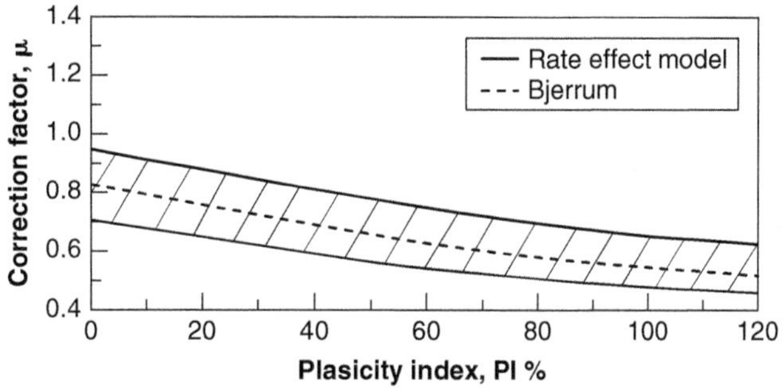

Figure 15.1s Correction factor vs. plasticity index.

Problem 15.10

A clay has an overconsolidation ratio equal to 2.5. Use the SHANSEP method and reasonable values of the parameters to estimate the undrained shear strength of that clay at a depth of 20 m. The clay is offshore at the bottom of the North Sea in 300 m of water.

Solution 15.10

$$\gamma_{sat} = 19\ \text{kN/m}^3$$

at depth 20 m below the sea floor in 300 m of water:

$$\sigma'_{ov} = 300 \times 9.81 + 20 \times 19 - 320 \times 9.81 = 183.8\ \text{kN/m}^2$$

For normally consolidated saturated fine grained soil:

$$\left(\frac{S_u}{\sigma'_{ov}}\right)_{NC} = 0.23$$

and for an overconsolidated fine grained soils with OCR = 2.5

$$\left(\frac{S_u}{\sigma'_{ov}}\right)_{OC} = \left(\frac{S_u}{\sigma'_{ov}}\right)_{NC} (OCR)^{0.8} = 0.23 \times 2.5^{0.8} = 0.48$$

Therefore

$$S_u = 0.48 \times 183.8 = 88.2\ \text{kN/m}^2$$

Problem 15.11

An unsaturated sample of clay is tested in a simple shear test. At failure the total normal stress on the failure plane is 70 kPa and the shear stress is 175 kPa.

a. Is that possible?
b. After testing, the water content on the plane of failure is measured and the soil water retention curve gives a water tension of 1450 kPa. The water content coupled with the measurement of the unit weight and the assumption that G_s is 2.7 leads to a degree of saturation of 20%. If the clay has no effective stress cohesion, calculate the effective stress friction angle.

Solution 15.11

a. Yes, it is possible. In unsaturated soils, the shear strength can be higher than the total normal stress because of the water tension increases the effective stress.
b.

$$\sigma = 70 \text{ kPa}$$

$$\tau_f = 175 \text{ kPa}$$

$$u_w = -1450 \text{ kPa}$$

$$S = 20\% \Rightarrow \alpha = 0.2$$

$$c' = 0$$

$$\tau_f = c + (\sigma - \alpha u_w)\tan\phi' = 0 + (70 - 0.2 \times (-1450))\tan\phi' = 175 \text{ kPa}$$

$$\tan\phi' = 0.486 \Rightarrow \phi' = 25.9^\circ$$

Problem 15.12

A lightly overconsolidated clay has a CPT point resistance of 1100 kPa, an OCR of 1.7, a PMT limit pressure of 590 kPa, an SPT blow count N of 13 bpf, and a unit weight of 18 kN/m^3. Estimate the undrained shear strength of that clay if the data comes from a depth of 6 m with the groundwater level being at a depth of 2 m.

Solution 15.12

$$\gamma_T = 18 \text{ kN/m}^3$$

$$\sigma_{ov} = 18 \times 6 = 108 \text{ kPa}$$

$$\sigma'_{ov} = \sigma_{ov} - \alpha u_w = 108 - 1 \times (6 - 2) \times 9.81 = 69 \text{ kPa}$$

The CPT data equation gives:

$$s_{u(CPT)} = \frac{q_c - \sigma_{ov}}{N_k}$$

Assuming an average value of N_k equal to 14, the equation becomes:

$$s_{u(CPT)} = \frac{1100 - 108}{14} = 70.9 \text{ kPa}$$

The PMT data equation gives:

$$s_{u(PMT)} = \frac{p_L}{N_P}$$

Assuming an average value of N_p equal to 7.5:

$$s_{u(PMT)} = \frac{590}{7.5} = 79 \text{ kPa}$$

The recommendations also give:

$$\frac{s_{u(PMT)}}{p_a} = 0.21 \times \left(\frac{p_L}{p_a}\right)^{0.75}$$

Using 101.3 kPa for the atmospheric pressure:

$$s_{u(PMT)} = 0.21 \times \left(\frac{p_L}{p_a}\right)^{0.75} \times p_a = 0.21 \times \left(\frac{590}{101.3}\right)^{0.75} \times 101.3 = 79.8 \text{ kPa}$$

The SPT data equation gives:

$$s_{u(SPT)} = 4.4 N_{60}$$

Assuming that N = N_{60}, then:

$$s_{u(SPT)} = 4.4 \times 13 = 57.2 \text{ kPa}$$

Given all the data, a cautiously conservative estimate of the undrained shear strength is 65 kPa.

Problem 15.13

You are at the beach lying on dry uniform sand. You take a handful of sand and let it fall from your hand onto a 0.3 m by 0.3 m wide plate. The sand pile on the plate has the shape of a pyramid and the angle of the pyramid with the horizontal is β. Demonstrate that β is equal to the friction angle φ'. You then take that same pile of sand and add a bit of water. Now you are able to mold the sand pile into a cylinder standing vertically. Where does the sand strength come from? Is it cohesion or friction?

Solution 15.13

The angle of the pyramid with the horizontal, β, is known as the *angle of repose*. When the sand falls, it comes to rest at the maximum angle possible; therefore, the slope of the sand pile is at impending failure. You can check that by tilting the plate slightly to one side: The side slope of the dry sand pyramid will fail to retain the same angle with the horizontal. If the slope is at impending failure, an element of soil as shown in Figure 15.2s is subjected to shear strength τ_f. The equilibrium of the element leads to the following equations, which show that the slope angle is the friction angle of the sand:

The shear force on the failure plane is:

$$T = W \sin \beta$$

The normal force on the failure plane is:

$$N = W \cos \beta$$

Therefore:

$$T = N \tan \beta$$

The maximum resisting force on the failure plane is:

$$S = N \tan \varphi'$$

At failure:

$$T = S \quad \text{therefore} \quad \beta = \varphi'$$

If we add a bit of water to the sand, water tension develops in the voids of the fine sand. This water tension pulls the particles against each other and creates an effective stress equal and opposite to the water tension. The shear strength of the sand is due to the friction related to the effective stress created by the water tension. This shear strength is often called *apparent cohesion* because the sand "sticks" together—yet the real mechanism is friction.

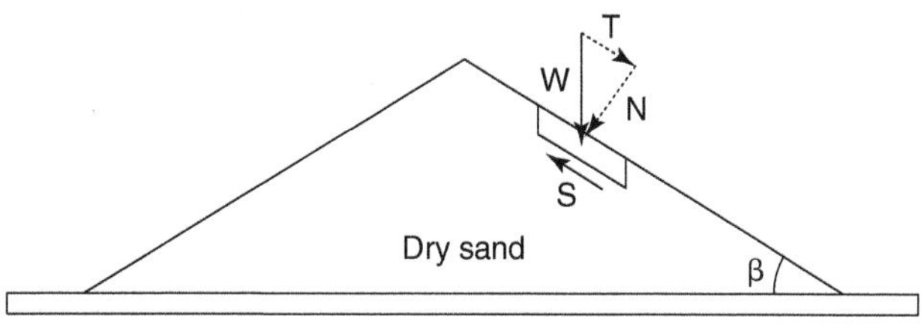

Figure 15.2s Dry sand pile and angle of repose.

CHAPTER 16

Thermodynamics for Soil Problems

16.1 GENERAL

Heat flow in soils involves several different phenomena: convection, radiation, and conduction. *Convection* takes place when a fluid flows over a solid that is at a different temperature than the fluid. When you set up a fan to blow air toward your body and cool yourself down in the summer, you use convection heat transfer. *Radiation* refers to the fact that all bodies continuously emit energy because of their temperature. This energy propagates to other nearby fluids or bodies through electromagnetic waves. Hot radiators that you may use in the winter to warm yourself up operate by radiation heat transfer. *Conduction* is a heat transfer mechanism whereby energy moves from a region of high temperature to a region of lower temperature. The phenomenon is due to the motion and impact of molecules, which increase as the temperature rises. Conduction of heat in soil is very similar to the flow of water through soil and is the most important mechanism of heat transfer through soils.

16.2 DEFINITIONS

Because of the analogy between temperature propagation and water flow, it is useful to draw a parallel between the parameters used in both fields of geotechnical engineering. Heat, Q, is a quantity of energy measured in joules (N × m). It is named after James Prescott Joule (1818–1889), an English physicist. The heat Q is equivalent to the volume of water V (m^3) in flow problems.

Temperature, T, is a measure of how hot a material is; it is sometimes measured in degrees Kelvin (K), but more commonly in degrees Celsius (C). The Kelvin is named after the British engineer and physicist William Thomson, First Baron Kelvin (1824–1907). The Celsius is named after the Swedish astronomer Anders Celsius (1701–1744). The Kelvin scale starts at absolute zero temperature, which is $-273°C$. There is in fact a lower bound to the temperature scale: It corresponds to the point where none of the molecules are moving. There is no known upper bound to the temperature scale. The temperature T is equivalent to the total head h_t(m) in flow problems.

The *temperature gradient* i_t is defined between two points in the soil mass; it is the ratio between the change in temperature dT over the distance dx separating the two points and is expressed in K/m. It corresponds to the hydraulic gradient i for the flow problem:

$$i_t = \frac{dT}{dx} \quad in \quad \text{K/m} \tag{16.1}$$

The *heat transfer rate* H is the amount of heat transferred per amount of time and is expressed in joules per second or watts, named after the Scottish engineer James Watt (1736–1819) (J/s or W). The heat transfer rate is equivalent to the flow rate Q (m^3/s) in flow problems:

$$H = \frac{dQ}{dt} \quad in \quad \text{J/s} \tag{16.2}$$

The *heat flow* q is the amount of heat dQ per unit time dt and per unit area A or the heat transfer rate H per unit area A. It is expressed in watts per meter square (W/m^2) or in joules per second and per meter square (J/s.m^2). It is equivalent to the velocity v (m/s) in the flow problem:

$$q = \frac{dQ}{dt} \times \frac{1}{A} = \frac{H}{A} \quad in \quad \text{J/s.m}^2 \tag{16.3}$$

The *thermal conductivity* k_t is a property of the soil. It takes units of J/s.K.m and is defined through Fourier's law (section 16.3) as the ratio between the heat flow and the thermal gradient:

$$k_t = \frac{q}{\frac{dT}{dx}} \quad in \quad \text{J/s.K.m} \tag{16.4}$$

The thermal conductivity is an indication of the speed with which the heat flows through the soil under a given temperature gradient. It is equivalent to the hydraulic conductivity for the flow problem.

The *specific heat* c is a property of the soil and takes units of J/kg.K. It is defined as:

$$c = \frac{1}{m}\frac{dQ}{dT} \quad in \quad \text{J/kg.K} \tag{16.5}$$

where m is the mass of the soil element considered, and dQ is the increase in heat stored in the element when the temperature is raised by dT. In the flow problem, the compressibility of the soil skeleton plays the role of the inverse of the specific heat. The inverse of the specific heat tells you how much heat you can squeeze out of the soil for a given change in temperature, much like the compressibility tells you how much water you can squeeze out the soil if you apply a change in effective stress.

The *diffusivity* α appears in the governing differential equation. It is in m^2/s and is defined as:

$$\alpha = \frac{k}{\rho c} \quad in \quad \text{m}^2/\text{s} \tag{16.6}$$

The diffusivity gives the speed with which the temperature will decay in a soil. It is closely linked to the thermal conductivity k_t, but is also influenced by the specific heat, which indicates how much heat can be squeezed out of the soil for a given change in temperature. In other words, you could have two soils with the same thermal conductivity but different specific heats. In this instance the heat would flow at the same speed in both soils for the same thermal gradient, but if the heat source stopped, the one with the highest specific heat would cool down the slowest because it would be harder to squeeze the heat out of the soil.

Table 16.1 Equivalency between Thermal Conductivity and Hydraulic Conductivity

Parameter	Flow of water	Flow of heat
Quantity	Volume V (m^3)	Heat Q (J)
Potential	Head h_t(m)	Temperature T (K)
Gradient	Hydraulic gradient i_h(unitless)	Temperature gradient i_t(K/m)
Flux	Flow rate Q (m^3/s)	Heat transfer rate H (J/s)
Flux density	Velocity v (m/s)	Heat flow q (J/s.m^2)
Conductivity	Hydraulic conductivity k_h(m/s)	Thermal conductivity k_t(J/s.K.m)
Law	Darcy	Fourier
Storage	Compressibility	Specific heat c (J/kg.K)
Decay coefficient	Coefficient of consolidation c_v(m^2/s)	Thermal diffusivity α(m^2/s)

Table 16.1 summarizes the equivalency between soil thermal flow problems and soil hydraulic flow problems.

16.3 CONSTITUTIVE AND FUNDAMENTAL LAWS

Fourier's law is the constitutive law for heat flow. It is named after Joseph Fourier (1768–1830), a French mathematician and physicist. Fourier's law (Fourier 1822) states that the heat flow q is linearly related to the temperature gradient through the thermal conductivity k_t:

$$q = -k_t i_t = -k_t \frac{dT}{dx} \tag{16.7}$$

where q is the heat flow, k_t is the thermal conductivity, i_t is the temperature gradient, T is the temperature, and x is the length in the direction of the heat flow. Therefore, the units of thermal conductivity are J/s.K.m. The minus sign indicates that heat flows in the direction of decreasing temperatures. Fourier's law is equivalent to Darcy's law in the flow problem. By the way, the R rating of house insulation comes from Eq. 16.7 and is based on very much the same concept as the resistance of an electrical conductor:

$$R = \frac{dT}{q} = \frac{dx}{k_t} \quad or \quad dT = R\,q \tag{16.8}$$

The fundamental law is the conservation of energy. For the purpose of this chapter, this law states that during an amount of time dt, the amount of heat dQ_{in} flowing into an element of soil is equal to the amount of heat dQ_{out} flowing out of the element plus the heat stored or extracted dQ_{stored} from the element.

$$\frac{dQ}{dt}_{in} = \frac{dQ}{dt}_{out} + \frac{dQ}{dt}_{stored} \tag{16.9}$$

16.4 HEAT CONDUCTION THEORY

Let's first address the problem of one-dimensional heat conduction. An example is the penetration of frost into a surface layer of soil due to low air temperature in the winter months. To solve this problem, we follow the normal steps (see section 11.4):

1. Consider an element of soil dx wide, dy long, and with a unit length perpendicular to the page (Figure 16.1).

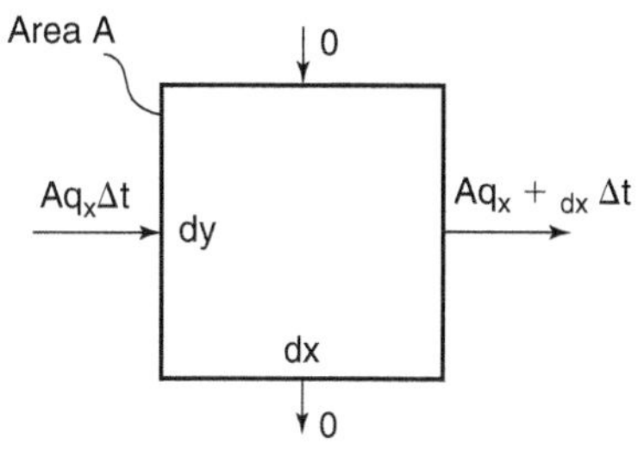

Figure 16.1 Element of soil.

2. The heat flows through the element volume, which has a cross-sectional area $A(dy \times 1)$ and a length dx. During a time dt, the quantity of heat entering the element is $Aq_x dt$, whereas the quantity of heat leaving the element is $Aq_{x+dx}dt$.
3. Conservation of energy allows us to state that the difference $(Aq_x dt - Aq_{x+dx}dt)$ is equal to the stored heat in the element.

$$dQ = A\, q_x dt - A\, q_{x+dx} dt \qquad (16.10)$$

4. The constitutive law is Fourier's law:

$$q(x,t) = -k\frac{dT(x,t)}{dx} \qquad (16.11)$$

where q is the heat flow (J/s.m^2), k is the thermal conductivity (J/s.K.m), T is the temperature (K), and x is the length (m) in the direction of the heat flow.
5. The second constitutive law is associated with the definition of specific heat. The amount of heat dQ will generate an increase in temperature dT in the element of mass m such that:

$$dQ = m\ c\ dT = A\ dx\ \rho\ c\ dT \qquad (16.12)$$

where ρ is the mass density of the material (kg/m^3) and c is the specific heat of the material (J/kg.K).
6. Regrouping Eqs. 16.10 and 16.12 gives:

$$Aq_x dt - Aq_{x+dx} dt = A\ dx\ \rho\ c\ dT \qquad (16.13)$$

Or, in partial derivative form:

$$-\frac{\partial q}{\partial x} = \rho\ c\ \frac{\partial T}{\partial t} \qquad (16.14)$$

Combining Eqs. 16.11 and 16.14, we get:

$$k\frac{\partial^2 T}{\partial x^2} = \rho\ c\frac{\partial T}{\partial t} \qquad (16.15)$$

If we define the thermal diffusivity α as:

$$\alpha = \frac{k}{\rho c} \qquad (16.16)$$

where α is the diffusivity in m^2/s, then the governing differential equation for one-dimensional conduction heat is:

$$\frac{\partial T}{\partial t} = \alpha\frac{\partial^2 T}{\partial x^2} \qquad (16.17)$$

In three dimensions, it becomes:

$$\frac{\partial^2 T}{\partial x^2} + \frac{\partial^2 T}{\partial y^2} + \frac{\partial^2 T}{\partial z^2} = \frac{1}{\alpha}\ \frac{\partial T}{\partial t} \qquad (16.18)$$

7. Now the boundary and initial conditions have to be expressed. This depends on the problem at hand. The complexity of the solution depends on the complexity of the boundary conditions, but numerical methods can always be used to solve such problems. Note that Eq. 16.17 is identical to Eq. 11.56 for the consolidation theory, where the temperature T is replaced by the excess water stress u_e. Therefore, the solutions are identical for identical boundary conditions. Jumikis (1977) presents the solution for a sinusoidal temperature fluctuation input at the ground surface to replicate seasonal variations.

16.5 AXISYMMETRIC HEAT PROPAGATION

In the case of an axisymmetric geometry, Eq. 16.18 becomes:

$$\frac{1}{r}\frac{\partial T}{\partial r} + \frac{\partial^2 T}{\partial r^2} = \frac{1}{\alpha}\frac{\partial T}{\partial t} \qquad (16.19)$$

where r is the radial distance from the axis, t is time, T is temperature, and α is the thermal diffusivity of the soil. Carslaw and Jaeger (1947) solved this problem in the case of an infinitely long cylindrical heat source of radius R_o maintained at a temperature T_o at the center of a full space, which was initially at a temperature equal to zero. In this case the time t required for a given temperature T_m to reach a certain distance R into the soil is given by:

$$t = T_F\frac{R_o^2}{\alpha} \qquad (16.20)$$

where T_F is the time factor (Figure 16.2), and is a function of the ratio R/R_o and T_m/T_o. This equation is very similar to the consolidation equation, which yields the time for excess water stress dissipation around a pile. At first glance, Eq. 16.20 seems to indicate that t increases with $R_o{}^2$. But in fact, t decreases as R_0 increases, because T_F decreases with R_o faster than R_o^2 increases.

The following reasoning illustrates this point. In Eq. 16.20, if R_o is multiplied by $\sqrt{10}$, the time t is not multiplied by 10 because the time factor T_F is not the same in both cases. If t was multiplied by 10, it would mean that it would take 10 times longer for the temperature to reach a value T_m at a distance $R-R_o$ from the boundary in the case of the large-radius heat source ($\sqrt{10}R_o$) than for the same temperature T_m to be reached at the same distance $R - R_o$ in the case of the smaller-radius heat source (R_o). This does not make sense: Because the heat source is larger, it should take less time—and indeed it does, because the time factor T_F decreases more than by a ratio of 10 in this case (Figure 16.2). Therefore, as R_o increases, t in fact decreases nonlinearly.

For example, consider a hot cone penetrometer with a radius R_o of 20 mm that is kept at a temperature T_o of 100°C in a soil with an initial temperature of 20°C and a diffusivity of 1 mm^2/s. Let's calculate the time it will take for the temperature to reach 40°C at a distance of R equal to R_o + 100 mm = 120 mm. Considering that the temperature of the soil is at 20°C initially, the ratio of net

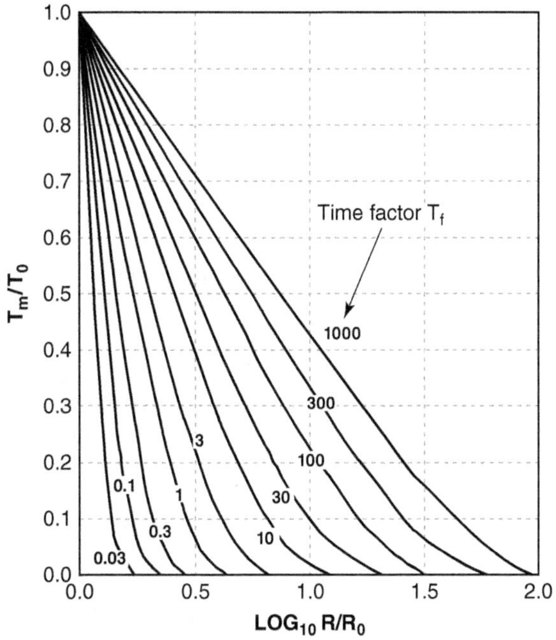

Figure 16.2 Time factor. (After Carslaw and Jaeger 1947)

temperature increase is $(40 - 20)/(100 - 20) = 0.25$. For this temperature ratio and a radius ratio of $120/20 = 6$, Figure 16.2 yields a time factor T_F equal to 32, and the time for the temperature to reach 40°C at $R = 120$ mm is:

$$t = 32\frac{20^2}{1} = 12800 \text{ s} = 3.55 \text{ hours} \tag{16.21}$$

Now consider a hot oil conductor in the bottom of the Gulf of Mexico with a radius R_o of 500 mm, that is kept at a temperature T_o of 100°C. Let's calculate the time it will take for the temperature to reach a temperature of 40°C at a distance of R equal to $R_o + 100$ mm = 600 mm. Considering that the temperature of the soil is at 20°C initially, the ratio of net temperature increase is $(40 - 20)/(100 - 20) = 0.25$. For this temperature ratio and a radius ratio of $600/500 = 1.2$, Figure 16.2 yields a time factor T equal to 0.02, and the time for the temperature to reach 40°C at $R = 600$ mm is:

$$t = 0.02\frac{500^2}{1} = 5000 \text{ s} = 1.39 \text{ hours} \tag{16.22}$$

16.6 THERMAL PROPERTIES OF SOILS

Any material can be found in solid, liquid, or gas form. For water, the transition from solid to liquid is at 0°C and the transition from liquid to gas is at 100°C. These temperatures correspond to 1 atmosphere of pressure, but would be different at different pressure levels. Figure 16.3 shows the pressure-temperature phase diagram for water and its triple point. By the way, the *latent heat* of a material is the heat necessary to change the phase of the material (solid to gas, for example).

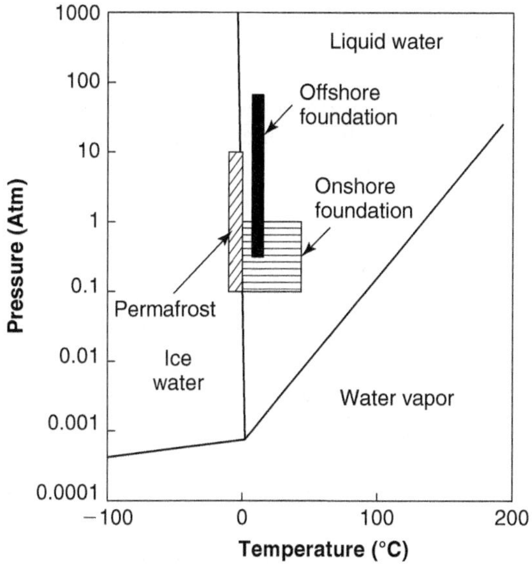

Figure 16.3 Temperature phases for water.

The temperature on the Earth varies from about −50°C to about +50°C. The temperature in the Earth varies from −50°C on the surface to 5500°C at the center of the Earth. Rocks and soil particles melt at a temperature varying between 600°C and 1200°C. The temperature gradient in the Earth varies and may be taken as 15°C per km over the first 100 km of depth. The deepest types of projects involving the geotechnical engineer may be offshore platforms and the associated retrieval of oil. The water depth in which the largest platforms are constructed reaches several kilometers. At the bottom of such oceans, the temperature is only a few degrees Celsius. The oil reservoir may be at a depth of 15 km; thus, the temperature of the oil can easily be 100°C when it comes back up to the surface (Figure 16.4). So, for

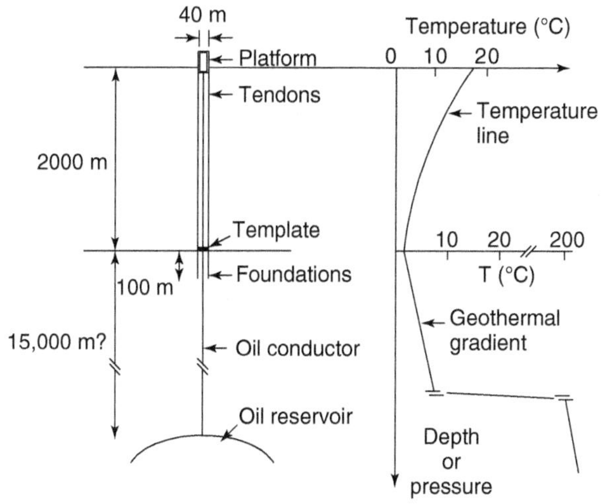

Figure 16.4 Temperature gradient for an offshore platform. (After Briaud and Chaouch 1997)

the geotechnical engineer, soil particles and rocks remain in solid form. However, within the range of Earth surface temperatures, water can be in liquid or in solid form (frozen), and the thermal properties of the soil may differ depending on whether the soil is frozen or not.

The thermal properties of interest are the thermal conductivity k(J/s.m.°C), the specific heat C(J/kg.°C), and the diffusivity α(m^2/s). A high value of thermal conductivity means that heat travels easily through the material; a high value of specific heat means that it takes a lot of heat to raise the temperature of the material; and a high value of diffusivity means that it will take little time for the temperature to rise in the material. These thermal properties depend on a number of factors, among which are the temperature level T, the pressure level p, the moisture content w, and the density ρ. Table 16.2 shows an estimate of the range of values one can expect for those thermal properties at ordinary temperature and pressure levels. The range of values in this table helps one to understand the factors affecting the thermal properties. For example, a dry soil will have a thermal conductivity lower than the same soil once saturated, because air has a lower thermal conductivity than water. Also, sand in a very dense state will have a higher thermal conductivity than the same sand in a very loose state, because soil particles have a higher thermal conductivity than air or water.

16.7 MULTILAYER SYSTEMS

Heat can flow through a layered system, such as an asphalt-concrete pavement over soil in the heat of the summer or a layer of snow covering the soil surface in the winter. Consider the case in which heat flows parallel to the interface of the two layers (Figure 16.5), where the starting temperature and the ending temperature are maintained at T_A and T_B.

These temperatures exist at two points separated by a horizontal distance L. Layer 1 is h_1 thick and layer 2 is h_2 thick. The thermal gradient is the same in both layers:

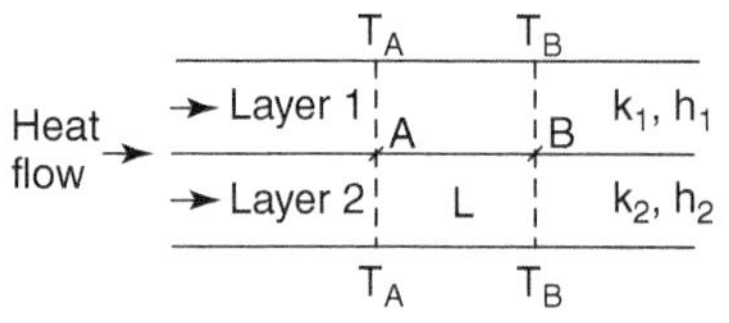

Figure 16.5 Horizontal heat flow through two layers.

$$i_1 = i_2 = \frac{T_B - T_A}{L} = i_e \tag{16.23}$$

However, the total heat transfer rate H is the sum of the heat transfer rate H_1 in layer 1 plus the heat transfer rate H_2 in layer 2:

$$H = q_e(h_1 + h_2) \times 1 = H_1 + H_2 = q_1 h_1 \times 1 + q_2 h_2 \times 1 \tag{16.24}$$

where q_e is the total heat flow, q_1 is the heat flow in layer 1, and q_2 is the heat flow in layer 2. Using Fourier's law gives:

$$k_e i_e (h_1 + h_2) \times 1 = k_1 i_1 h_1 \times 1 + k_2 i_2 h_2 \times 1 \tag{16.25}$$

where k_e is the equivalent thermal conductivity, k_1 is the thermal conductivity of layer 1, k_2 is the thermal conductivity of layer 2, i_e is the equivalent gradient, i_1 is the gradient in layer 1, and i_2 is the gradient in layer 2. Therefore:

$$k_e = \frac{k_1 h_1 + k_2 h_2}{h_1 + h_2} \tag{16.26}$$

This result can be generalized for n layers:

$$k_e = \frac{\sum_{i=1}^{n} k_i h_i}{\sum_{i=1}^{n} h_i} \tag{16.27}$$

Table 16.2 Thermal Properties for Various Earth Materials at Standard Conditions of Temperature and Pressure

Material	Density ρ(kg/m^3)	Specific Heat c(J/kg.°C)	Thermal Conductivity k(J/s.m.°C)	Thermal Diffusivity α(mm^2/s)
Air	1 to 1.4	1000 to 1050	0.02 to 0.03	13 to 30
Water	960 to 1000	4190 to 4220	0.5 to 0.8	0.13 to 0.17
Ice	917 to 920	1960 to 2110	2.0 to 2.6	1.24 to 1.52
Clay (unfrozen)	1400 to 1800	750 to 920	0.8 to 2.8	0.1 to 1.66
Clay (frozen)	1400 to 1800	650 to 800	1.0 to 3.6	0.15 to 2.3
Sand (unfrozen)	1500 to 2200	630 to 1460	2.3 to 3.8	0.87 to 3.0
Sand (frozen)	1500 to 2200	500 to 1200	2.9 to 4.7	1.2 to 4.2
Rock	2200 to 3000	710 to 920	2 to 6	1.1 to 3.0

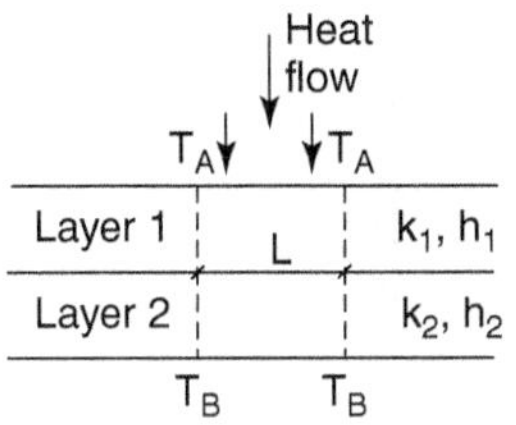

Figure 16.6 Vertical heat flow through two horizontal layers.

Now consider the case in which heat flows perpendicular to two layers (Figure 16.6), where the starting temperature and the ending temperature are maintained at T_A and T_B.

Layer 1 is h_1 thick and layer 2 is h_2 thick. The temperatures T_A and T_B exist at two points separated by a distance $(h_1 + h_2)$. The heat transfer rate is the same in both layers:

$$H = H_1 = H_2 \quad and$$

$$H = k_e i_e L \times 1 = k_1 i_1 L \times 1 = k_2 i_2 L \times 1 \tag{16.28}$$

The change in temperature, however, is additive:

$$\Delta T = \Delta T_1 + \Delta T_2 = T_A - T_B \tag{16.29}$$

But

$$i_e = \frac{\Delta T}{h_1 + h_2} \quad and \quad i_1 = \frac{\Delta T_1}{h_1} \quad and \quad i_2 = \frac{\Delta T_2}{h_2} \tag{16.30}$$

Therefore

$$\frac{h_1 + h_2}{k_e L} H = \frac{h_1}{k_1 L} H_1 + \frac{h_2}{k_2 L} H_2 \tag{16.31}$$

and

$$k_e = \frac{h_1 + h_2}{\dfrac{h_1}{k_1} + \dfrac{h_2}{k_2}} \tag{16.32}$$

This result can be generalized for n layers:

$$k_e = \frac{\sum_{i=1}^{n} h_i}{\sum_{i=1}^{n} \dfrac{h_i}{k_i}} \tag{16.33}$$

16.8 APPLICATIONS

Let's consider a soil deposit in a cold country (Figure 16.7). The question is: How deep will the frost penetrate during a very cold period? At depth, where the soil is not frozen, the temperature is T_d. The air is at a temperature T_a, much lower than 0° Celsius. It is assumed that the temperature of the soil surface T_s is the same as the air temperature T_a. The temperature at the bottom of the frozen soil is assumed to be 0° Celsius (C).

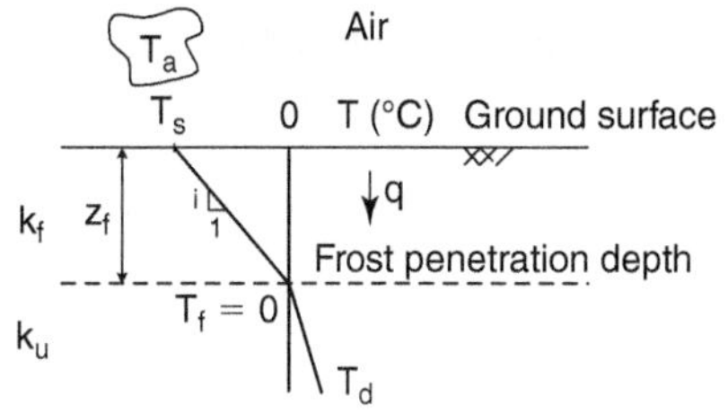

Figure 16.7 Frost penetration depth for a uniform soil.

The gradient of temperature in the frozen layer is i and is associated with a heat flow q and a thermal conductivity k_f. Therefore, the depth of the frozen soil is:

$$z_f = \frac{0 - T_s}{i} = -\frac{T k_f}{q} \tag{16.34}$$

Now let's consider that a layer of snow covers the ground surface (Figure 16.8). The question here is: Would the depth of the frozen soil z_f be the same? The thickness of the snow cover is h_s and the thermal conductivity of the snow is k_s. The air temperature is T_a, and the temperature of the snow surface is T_s and is assumed equal to T_a. The thermal conductivity of the frozen soil is k_f.

The difference in temperature between the bottom of the frozen soil layer at 0°C and the surface of the snow layer at T_s can be written as:

$$0 - T_s = 0 - T_1 + T_1 - T_s \tag{16.35}$$

where T_1 is the temperature at the interface between the bottom of the snow layer and the soil surface (Figure 16.8). By using the definition of the temperature gradient and then Fourier's law, Eq. 16.35 can be rewritten as:

$$0 - T_s = i_{sn} h_{sn} + i_f z_f = \frac{q_{sn}}{k_{sn}} h_{sn} + \frac{q_f}{k_f} z_f \tag{16.36}$$

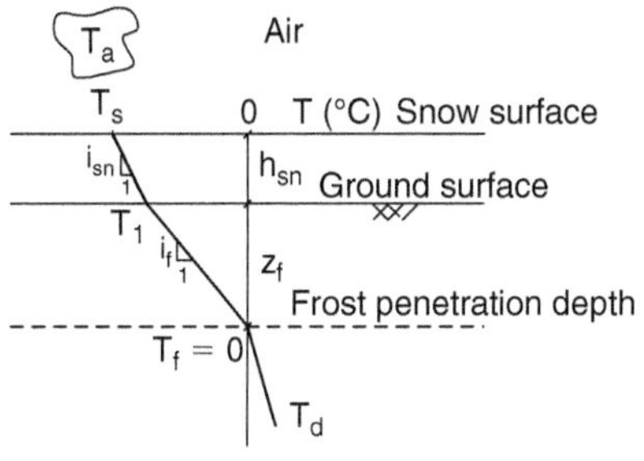

Figure 16.8 Frost penetration depth for a two-layer system.

For continuity purposes, though, the heat flow has to be the same in the snow and the frozen part of the soil layer. Then:

$$q_{sn} = q_f = q \tag{16.37}$$

and the frost penetration depth z_f is:

$$z_f = \frac{-T_s k_f}{q} - h_{sn}\frac{k_f}{k_{sn}} \tag{16.38}$$

As can be seen, the snow cover reduces the frost penetration depth by $h_{sn}\frac{k_f}{k_{sn}}$.

16.9 FROZEN SOILS

The general term *frozen soils* regroups problems of freezing soils, frozen soils, and thawing soils. Frozen soils are usually classified in three categories: soils with nonvisible ice (N), soils with visible ice and ice lenses less than 25 mm thick (V), and soils with visible ice with ice lenses larger than 25 mm thick (ICE). *Permafrost* is a term indicating that the ground, including soil and rock, is at or below 0° Celsius for more than two consecutive years. The temperature at which the water in the voids will freeze depends on many factors, including the salt content. The more salt there is, the lower the temperature has to be before the water will freeze. Generally, freezing starts at around −1°C, and at −20°C most soils are completely frozen. Figure 16.9 shows typical temperature profiles in frozen soils. It indicates that close to the surface there is usually a zone that freezes and thaws each year, called the *active zone*.

The water very close to the mineral surface of a particle can be tightly bound to the particle, especially for very small particles. This adsorbed water layer practically never freezes. Therefore, clays tend to resist freezing more than sands. The water film the furthest away from that boundary is the first one to freeze. Figure 16.10 shows conceptually the evolution of the water content of a soil as the temperature plunges below zero. As can be seen, the equilibrium frozen water content is higher for clayey soils than for sandy soils.

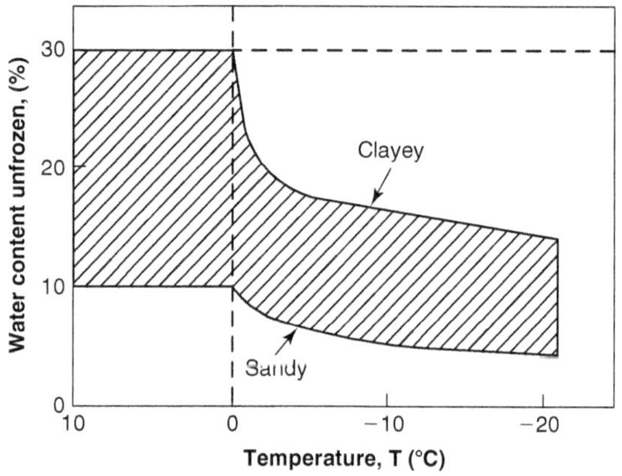

Figure 16.10 Evolution of water content with temperature.

Table 16.3 Frost Susceptibility and Soils

Soil Type	Frost Susceptibility
High-plasticity clays	Negligible
Low-plasticity clays, clays with sand and gravel	Moderate
Silty clays	Moderate to severe
Silts, silty sands, very fine sands	Severe
Gravels and sands with fines	Moderate
Clean sands and gravels	Negligible

Frost susceptibility is smallest for clean gravels and clean sands, on the one hand, and for high-plasticity clays on the other. The most frost-susceptible soils are silts, as shown in Table 16.3. The reason is that frost heave requires the soil to have the ability to lift water by capillary action and let the water flow through its voids. Clean gravels and clean sands have high hydraulic conductivity but little ability for capillary action; in other words, it is easy for the water to move, but the water has no energy to go anywhere. High-plasticity clays, in contrast, have a very high ability for capillary action but a very low hydraulic conductivity; in other words, the water has plenty of energy, but it is very hard to move through

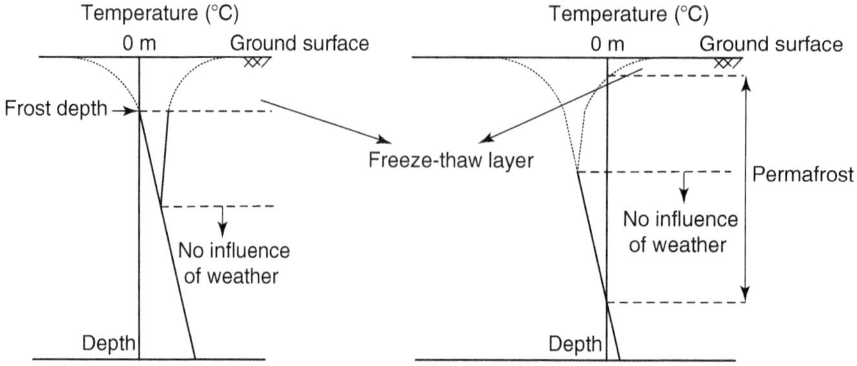

Figure 16.9 Typical temperature profiles in frozen soils.

the clay. Silts optimize the two requirements of capillary potential and water flow and are therefore some of the most frost-susceptible soils.

Note that the unit weight of ice is about 10% less than the unit weight of water. Therefore, if a certain weight of water becomes ice, it will occupy about 10% more volume. This is why icebergs float with only one-tenth of the iceberg mass showing up above the water level and 90% below it (hence the expression "this is only the tip of the iceberg"). If a soil becomes frozen, it will expand according to the increase in volume of the water becoming ice. These *ice lenses*, once started, continue to attract water and become thicker by something called the *cryosuction process*. Such ice lenses have significant lifting potential; the uplift pressures can be several hundreds of kPa and can reach 2000 kPa if the heave is confined. Heave magnitudes of 50 to 75 mm are common. The frozen soil can also develop an "adfreeze" bond with neighboring objects such as foundation piles. This bond can generate shear stresses from 50 to 150 kPa.

Frozen soils have four phases instead of three. Note that nearly all frozen soils contain liquid water. The phase diagram is shown in Figure 16.11. For the water content, a distinction must be made between the unfrozen water content and the frozen water or ice content. They are defined as:

Total water content $$w = \frac{W_w + W_i}{W_s} \quad (16.39)$$

Unfrozen water content $$w_u = \frac{W_w}{W_s} \quad (16.40)$$

Frozen water (ice) content $$w_i = \frac{W_i}{W_s} \quad (16.41)$$

$$w = w_u + w_i \quad (16.42)$$

where W_w is the weight of water, W_i is the weight of ice, and W_s is the weight of solids (Figure 16.11). In all other index parameters it is necessary to state what is included and what is not. For example, the degree of saturation, the void ratio, and the porosity can be defined by including or not including the ice.

The thermal properties of a frozen soil are the combination of the properties of the water, the ice, the air, and the soil skeleton. Table 16.2 shows these properties for each material

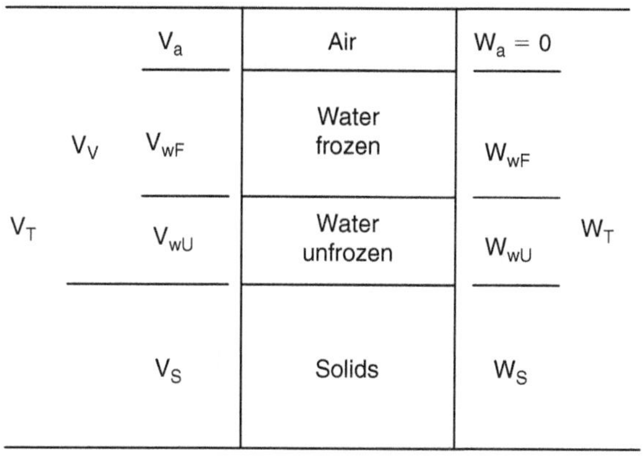

Figure 16.11 Phase diagram for a frozen soil.

individually and the impact they have on the soil. As can be seen, the frozen soil will have a higher thermal conductivity, a lower specific heat, and a higher thermal diffusivity. In other words, the heat will flow faster in the frozen soil, and it will be easier to squeeze the heat out of the frozen soil.

The mechanical properties will also be affected. The shear strength will increase significantly, as the ice will contribute to increasing the cohesion intercept. The stiffness will also increase, as the ice essentially increases the amount of solids in the soil. However, the creep component of the settlement will be increased as the ice content increases. Indeed, ice exhibits creep properties that depend on the ice temperature; the lower the ice temperature, the less it will creep. The viscous exponent mentioned in Eq. 15.56 varies in the range of 0.1 to 0.5 for ice. Recall that the same exponent for unfrozen clays was 0.02 to 0.05 (see Figure 15.18). As a result, a frozen soil will creep more than the unfrozen soil under constant load, but the initial movement will be less. The hydraulic conductivity will decrease, as there is less area for the water to flow through. In that sense, frozen soils act according to the same principles as unsaturated soils. The best way to obtain the mechanical properties of frozen soils is to perform a laboratory or in situ test that duplicates the conditions under which the soil will be stressed in the project at hand.

There is a close analogy between frozen soils and unsaturated soils, and more interaction between these two fields is likely to be very rewarding.

PROBLEMS

16.1 A house is built on a frozen soil layer. The house generates heat such that it maintains a temperature of 20°C in the house. If the thermal conductivity of the frozen soil is $k_{frozen} = 1.3$ W/m.K, if the thermal conductivity of thawed-out soil is $k_{unfrozen} = 1.1$ W/m.K, and the temperature gradient in the frozen soil $i_{frozen} = -15°C/m$, what thickness of soil will thaw out?

16.2 A building is to be built with a geothermal foundation in a soil with a thermal diffusivity α equal to 5×10^{-7} m^2/s. The energy piles are 0.4 m in diameter and water circulates up and down the piles to take advantage of the beneficial effect of the soil temperature (hotter in the winter and cooler in the summer). The energy piles operate 8 months out of the year, and for optimum operation performance, the increase in temperature in adjacent energy piles due to the operation of one energy pile must not exceed 10% of the initial temperature difference between pile and soil. Calculate the minimum spacing between energy piles.

16.3 A cylindrical soil sample ($D = 0.075$ m, $L = 0.150$ m) is put in an oven where the temperature is kept at $T_f = 45°C$. The initial temperature of the soil sample is $T_i = 25°C$. The soil sample thermal conductivity k is 1.2 W/m.K and the volumetric heat capacity C is 1.2×10^6 J/m^3.K. Using the literature, find the solution that gives the increase in temperature at the center of the cylindrical soil sample as a function of time and calculate how long it will take for the center of the sample to reach 30°C, 35°C, and 40°C.

16.4 A two-layer system is made of a concrete pavement overlaying a sandy subgrade. The thermal properties of the two layers are shown in Figure 16.1s.

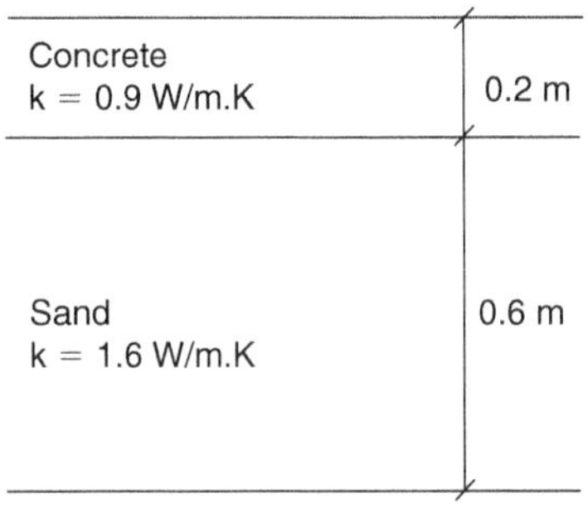

Figure 16.1s Two-layer system.

a. What is the equivalent thermal conductivity of the system if the heat flows horizontally?
b. What is the equivalent thermal conductivity of the system if the heat flows vertically?

16.5 Calculate the change in volume of a saturated soil with a water content of 30% and unit weight of 18 kN/m^3 if 90% of the water by weight becomes frozen.

16.6 Add a column to Table 16.1 dealing with electricity. Write Ohm's law and compare it to Darcy's and Fourier's laws.

16.7 A 0.3 m diameter, 10 m long probe is pushed into a clay. The clay has a thermal conductivity equal to 1.2 W/m.K and a thermal diffusivity equal to 2×10^{-6}m^2/s. The probe is at 2500°C and the intent is to bake the clay in place to create a baked-in-place pile with a wall thickness equal to 0.1 m. If the clay becomes permanently solidified at 1700°C and the initial clay temperature is 0°C, how long will it take before the pile is cooked and the probe can be removed to bake the next pile?

Problems and Solutions

Problem 16.1

A house is built on a frozen soil layer. The house generates heat such that it maintains a temperature of 20°C in the house. If the thermal conductivity of the frozen soil is $k_{frozen} = 1.3$ W/m.K, if the thermal conductivity of thawed-out soil is $k_{unfrozen} = 1.1$ W/m.K, and the temperature gradient in the frozen soil $i_{frozen} = -15$°C/m, what thickness of soil will thaw out?

Solution 16.1

Based on the principle of continuity of heat flow, and assuming that the surface temperature is T_s, the freezing temperature is T_f (equal to 0°C), and the thawing depth is x, we can write:

$$q_{unfrozen} = q_{frozen}$$

$$k_{unfrozen} \times A \times \frac{T_f - T_s}{x} = k_{frozen} \times A \times i_{frozen}$$

$$k_{unfrozen} \times A \times \frac{T_f - T_s}{x} = k_{frozen} \times A \times i_{frozen}$$

$$x = \frac{k_{unfrozen}}{k_{frozen}} \times \frac{T_f - T_s}{i_{frozen}}$$

The thawing depth is thus:

$$x = \frac{1.1}{1.3} \times \frac{0 - 20}{(-15)} = 1.13 \text{ m}$$

Problem 16.2

A building is to be built with a geothermal foundation in a soil with a thermal diffusivity α equal to 5×10^{-7} m^2/s. The energy piles are 0.4 m in diameter and water circulates up and down the piles to take advantage of the beneficial effect of the soil temperature (hotter in the winter and cooler in the summer). The energy piles operate 8 months out of the year, and for optimum operation performance, the increase in temperature in adjacent energy piles due to the operation of one energy pile must not exceed 10% of the initial temperature difference between pile and soil. Calculate the minimum spacing between energy piles.

Solution 16.2

From the problem data and using Figure 16.2 from the text, $T_m/T_o = 0.1$. The time factor T_F is calculated using Eq. 16.20:

$$T_F = \frac{t \times \alpha}{R_o^2} = \frac{8 \times 30 \times 24 \times 3600 \times 5 \times 10^{-7}}{0.2^2} = 259.2$$

From Figure 16.2, $\log_{10} R/R_0 = 1.3$; therefore, $R/R_0 = 19.95$ and the minimum distance between energy piles should be $R = R_o \times 19.95 = 4$ m.

Problem 16.3

A cylindrical soil sample (D = 0.075 m, L = 0.150 m) is put in an oven where the temperature is kept at $T_f = 45°C$. The initial temperature of the soil sample is $T_i = 25°C$. The soil sample thermal conductivity k is 1.2 W/m.K and the volumetric heat capacity C is 1.2×10^6 J/m^3.K. Using the literature, find the solution that gives the increase in temperature at the center of the cylindrical soil sample as a function of time and calculate how long it will take for the center of the sample to reach 30°C, 35°C, and 40°C.

Solution 16.3

Carslaw and Jaeger (1947) developed the solution for the temperature increase at the center of a cylindrical sample as a function of time. The percentage increase or decrease in soil sample temperature U can be plotted versus the normalized time factor T as shown in Figure 16.1s, for both a finite-length sample and an infinite-length sample.

$$U = \frac{T - T_{\min}}{T_{\max} - T_{\min}}$$

$$T = \frac{\alpha(m^2/s) \times t(s)}{D^2(m^2)}$$

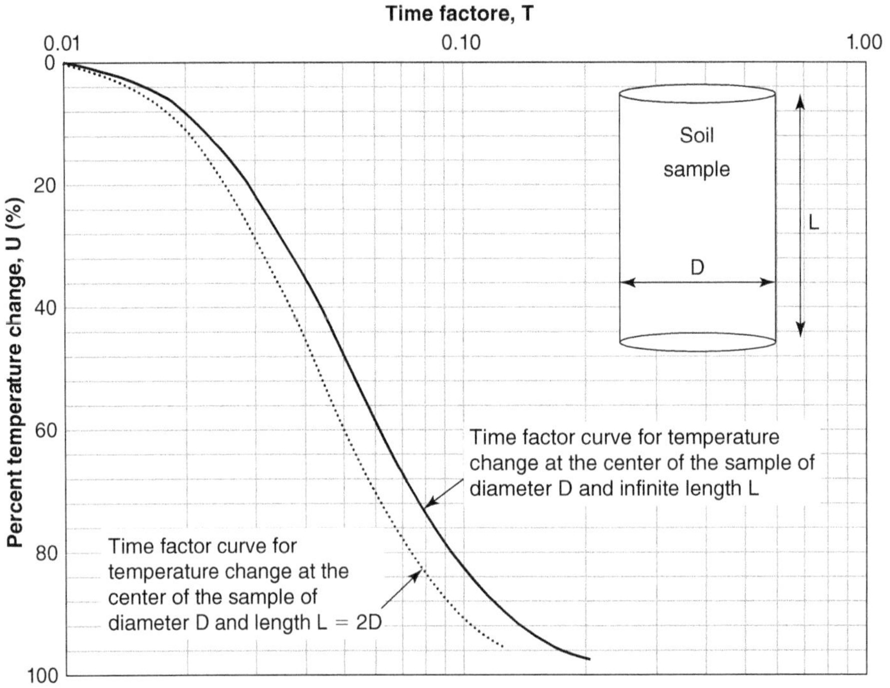

Figure 16.2s Percent temperature change U versus time factor T.

The thermal diffusivity α of the soil sample is:

$$\alpha(m^2/s) = \frac{k}{C} = \frac{1.2}{1.2 \times 10^6} = 10^{-6}$$

When the temperature reaches 30°C, $U = 25\%$; from Figure 16.1s, $T = 0.03$:

$$t(s) = \frac{T \times D^2}{\alpha} = \frac{0.03 \times 0.075^2}{10^{-6}} = 168 \text{ sec}$$

When the temperature reaches 35°C, $U = 50\%$; from Figure 16.1s, $T = 0.04$:

$$t(s) = \frac{T \times D^2}{\alpha} = \frac{0.04 \times 0.075^2}{10^{-6}} = 225 \text{ sec}$$

When the temperature reaches 40°C, $U = 75\%$; from Figure 16.1s, $T = 0.07$:

$$t(s) = \frac{T \times D^2}{\alpha} = \frac{0.07 \times 0.075^2}{10^{-6}} = 393 \text{ sec}$$

Problem 16.4

A two-layer system is made of a concrete pavement overlaying a sandy subgrade. The thermal properties of the two layers are shown in Figure 16.2s.

a. What is the equivalent thermal conductivity of the system if the heat flows horizontally?
b. What is the equivalent thermal conductivity of the system if the heat flows vertically?

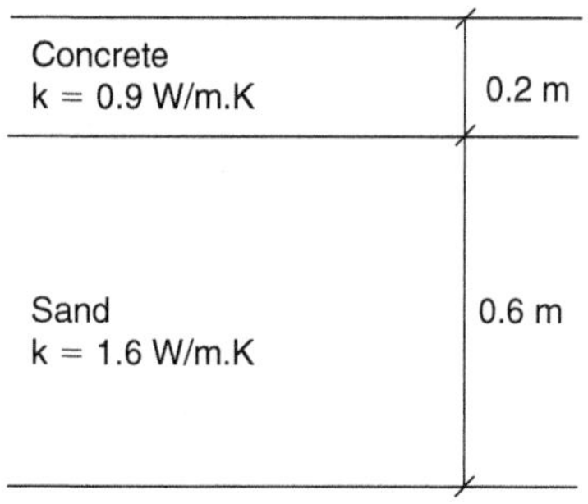

Figure 16.1s Two-layer system.

Solution 16.4

a. The equivalent thermal conductivity of the system if the heat flows horizontally can be calculated using Eq. 16.26:

$$k_e = \frac{k_1 h_1 + k_2 h_2}{h_1 + h_2} = \frac{0.9 \times 0.2 + 1.6 \times 0.6}{0.2 + 0.6} = 1.42 \text{ W/m.K}$$

b. The equivalent thermal conductivity of the system if the heat flows horizontally can be calculated using Eq. 16.32:

$$k_e = \frac{h_1 + h_2}{\dfrac{h_1}{k_1} + \dfrac{h_2}{k_2}} = \frac{0.2 + 0.6}{\dfrac{0.2}{0.9} + \dfrac{0.6}{1.6}} = 1.34 \text{ W/m.K}$$

Problem 16.5

Calculate the change in volume of a saturated soil with a water content of 30% and unit weight of 18 kN/m^3 if 90% of the water by weight becomes frozen.

Solution 16.5

First we have to calculate the weight of water in unfrozen conditions. The total unit weight γ is:

$$\gamma_t = \frac{W_w + W_s}{V_t} = 18 \text{ kN/m}^3$$

The water content is 30%; therefore, $W_w = 0.3W_s$. Assuming a soil unit volume of 1 m^3, the water weight W_w is 4.15 kN and the solid weight W_s is 13.85 kN.

The volume of water in the unfrozen condition is:

$$\gamma_w = \frac{W_w}{V_w} = 10 \text{ kN/m}^3 \quad \text{or} \quad V_w = 0.415 \text{ m}^3$$

If 90% of water weight becomes frozen, then the weight of ice W_i is 3.735 kN. The unit weight of ice γ_i is 10% less than the unit weight of water; therefore, the volume of ice is:

$$V_i = \frac{W_i}{\gamma_i} = \frac{3.735}{9} = 0.415 \text{ m}^3$$

The volume of remaining unfrozen water is:

$$V_w = \frac{W_w}{\gamma_w} = \frac{(4.15 - 3.735)}{10} = 0.0415 \text{ m}^3$$

The total volume of water and ice when 90% of water mass becomes frozen is 0.4565 m^3; therefore, the change in volume of the soil is 0.0415 m^3 or 4.15% of the original volume.

Problem 16.6

Add a column to Table 16.1 dealing with electricity. Write Ohm's law and compare it to Darcy's and Fourier's laws.

Solution 16.6

Parameter	Flow of water	Flow of heat	Flow of current
Quantity	Volume $V\,(m^3)$	Heat Q (J)	Electric charge (C)
Potential	Head $h_t\,(m)$	Temperature T (K)	Voltage (V)
Gradient	Hydraulic gradient i_h (unitless)	Temperature gradient $i_t\,(K/m)$	Electric field gradient E(V/m)
Flux	Flow rate Q (m^3/s)	Heat transfer rate H (J/s)	Electric current flow (C/s)
Flux density	Velocity v (m/s)	Heat flow q $(J/s.m^2)$	Electrical flux density (C/m^2)
Conductivity	Hydraulic conductivity $k_h\,(m/s)$	Thermal conductivity $k_t\,(J/s.K.m)$	Electric conductivity, $\sigma\,(S/m)$
Law	Darcy	Fourier	Ohm
Storage	Compressibility	Specific heat c (J/kg.K)	Capacitance
Decay coefficient	Coefficient of consolidation $c_v\ (m^2/s)$	Thermal diffusivity $\alpha\,(m^2/s)$	Electrical diffusivity $D(m^2/s)$

Problem 16.7

A 0.3 m diameter, 10 m long probe is pushed into a clay. The clay has a thermal conductivity equal to 1.2 W/m.K and a thermal diffusivity equal to $2 \times 10^{-6} m^2/s$. The probe is at 2500°C and the intent is to bake the clay in place to create a baked-in-place pile with a wall thickness equal to 0.1 m. If the clay becomes permanently solidified at 1700°C and the initial clay temperature is 0°C, how long will it take before the pile is cooked and the probe can be removed to bake the next pile?

Solution 16.7

The probe can be considered an infinite cylindrical heat source, because the length-to-diameter ratio is very large. The increase in temperature that must be achieved at a radial distance $R = R_o + 0.1 = 0.25$ m is 1700°C. Using Figure 16.2:

$$T_m/T_o = 1700/2500 = 0.68$$

$$R/R_o = 0.25/0.15 = 1.67$$

$$\log(R/R_o) = 0.222$$

From Figure 16.2, $T_f = 1$. Therefore:

$$T_F = 1 = \frac{t \times \alpha}{R_o^2} \Rightarrow t = \frac{R_o^2}{\alpha} = \frac{0.15^2}{2 \times 10^{-6}} = 11250\,\text{sec.} = 3.125\ \text{hours}$$

CHAPTER 17

Shallow Foundations

17.1 DEFINITIONS

Shallow foundations (Figure 17.1) are those placed close to the ground surface, typically at a depth less than one times the width of the foundation. A 1 m thick, 3 m by 3 m foundation under a column, placed at a depth of 1.5 m, would be a shallow foundation called a *spread footing*. Spread footings can be square, circular, or very long compared to their width, in which case they are called *strip footings*. A 3 m thick, 40 m by 40 m square foundation, placed at a depth of 10 m, would be considered a particular type of shallow foundation called a *mat foundation*. A 0.1 m thick, 15 m by 15 m foundation stiffened with 1 m deep beams 3 m apart in both directions would be a shallow foundation called a *stiffened slab on grade*.

17.2 CASE HISTORY

This case history illustrates the behavior of shallow foundations. Five tests of spread footings were performed at the National Geotechnical Experimentation Site at Texas A&M University. The soil at the site is a medium-dense, fairly uniform, silty fine silica sand with the following average properties near the footings and within the top 5 meters: mean grain size $D_{50} = 0.2$ mm, SPT (standard penetration test) blow count 18 blows per 0.3 m, CPT (cone penetrometer test) point resistance 6 MPa, PMT (pressuremeter test) limit pressure 800 kPa, PMT modulus 8.5 MPa, DMT (dilatometer test) modulus 30 MPa, borehole shear test friction angle 32°, estimated total unit weight 15.5 kN/m^3, and cross hole shear wave velocity 240 m/s. The water table is 4.9 m deep. Additional data can be found in Briaud and Gibbens (1999; 1994). Geologically, the top layer of sand is a flood plain deposit of Pleistocene age about 3 m thick with a high fine content. The next layer of sand is a river channel deposit of Pleistocene age about 3 m thick, clean and uniform. The third layer is a mixed unit with an increasing amount of clay seams and gravel layers; it is also of Pleistocene age and was deposited by a stream of fluctuating energy. Below these 200,000-year-old sand layers and about 10 m below the ground surface is the 45-million-year-old Eocene bedrock; this bedrock is a dark gray clay shale that was deposited in a series of marine transgressions and regressions. Erosion of the Eocene marine clay took place before the Pleistocene river sediments were deposited.

The test setup is shown in Figure 17.2. The 5 footings were square with a side dimension equal to 1 m, 1.5 m, 2.5 m, 3 m, and 3 m. They were embedded 0.75 m into the sand and were 1.2 m thick. They were loaded in load step increments, each one lasting 30 minutes, while settlement was recorded every minute during the load step. All footings were pushed downward until the settlement reached 0.15 m. Figure 17.3 shows an example of the load settlement curve obtained for the 3 m by 3 m north footing, as well as the log of the settlement vs. the log of time for several load steps. The pressure vs. settlement curves for all footings are shown in Figure 17.4. These curves were normalized by dividing the pressure by the limit pressure of the pressuremeter and the settlement by the width of the footing. Figure 17.4 indicates that this normalization makes the footing size disappear: The p/p_L vs. s/B curve becomes a property of the soil, much like a stress-strain curve. Tell tales and inclinometers were placed below and on the side of the footing, respectively. They indicated the depth to which the soil was compressed and the lateral movement of the soil during the load application. Figure 17.5 shows the soil movement as a function of depth for four of the footings and the lateral movement for the 3 m north footing. The data show that most of the settlement and lateral movement occurs within one footing width below the footing.

17.3 DEFINITIONS AND DESIGN STRATEGY

The most important considerations in foundation design are to ensure:

1. The safety of the foundation against soil failure (ultimate limit state)

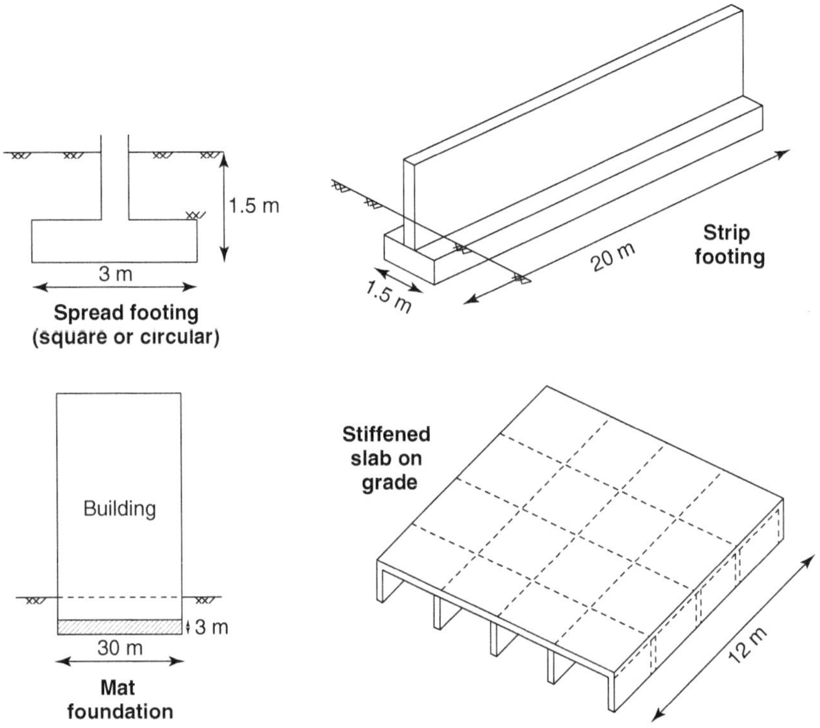

Figure 17.1 Types of shallow foundations.

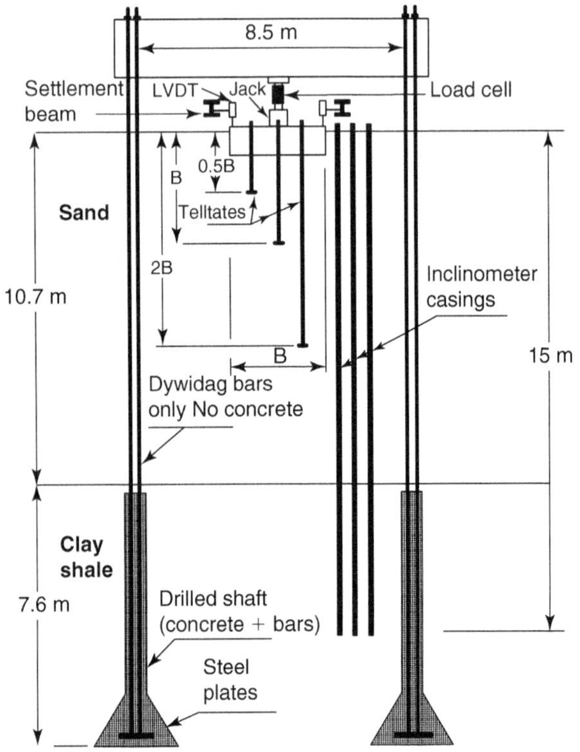

Figure 17.2 Load test setup: (a) Load settlement curve with 30-minute load steps. (b) Settlement time curves for each load step.

2. The functionality of the foundation and the structure above by minimizing the foundation movement and distortion (serviceability limit state)
3. The safety of the foundation against structural failure

Item 3 is handled primarily by the structural engineer and is not covered in this book. Items 1 and 2 in the preceding list are primarily geotechnical engineering considerations involving soil shear strength and the soil increase and decrease in volume when loaded. They are the topic of this chapter for shallow foundations and of Chapter 18 for deep foundations.

The geotechnical design of a shallow foundation consists of estimating the size and depth of the foundation. The depth is chosen on the basis of several factors, including profile of soil strength and compressibility, depth of the zone that shrinks and swells, depth of frost penetration, groundwater level, and ease of construction. The size is typically chosen once the depth is chosen.

No foundation can be designed to ensure zero probability of failure. This is because any calculation is associated with some uncertainty; because the engineering profession's knowledge, while having made great strides, is still incomplete in many respects; because human beings are not error free; because budgets are limited; and because the engineer designs the bridge or building for conditions that do not include extremely unlikely events, such as a big airplane hitting the bridge at the same time as an earthquake, a hurricane,

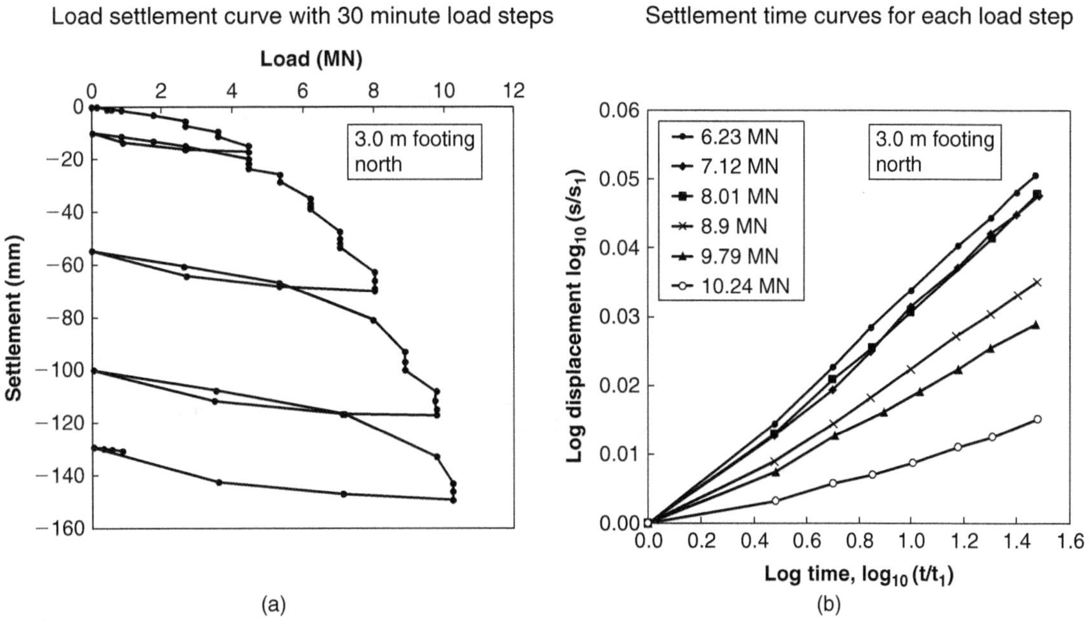

Figure 17.3 Result for the 3 m by 3 m north footing: (a) Pressure-settlement curve. (b) Normalized curves.

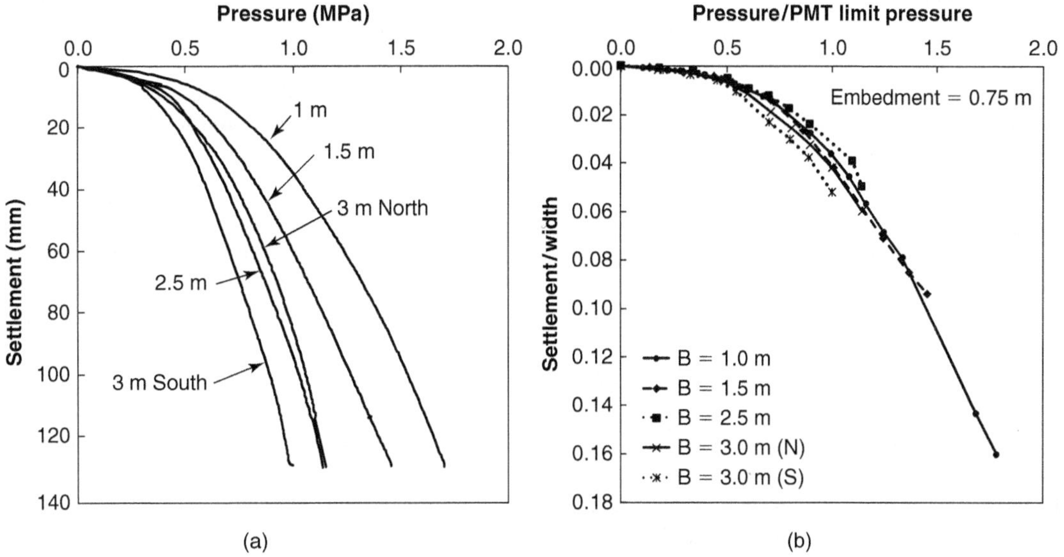

Figure 17.4 Pressure vs. settlement curve for all footings and normalized curves: (a) Pressure-settlement curve. (b) Normalized curves.

and a 500-year-flood during rush hour. The engineer and the public must accept a certain level of probability of failure. This acceptable level of probability of failure is tied to the number of deaths that the public accepts on a daily basis (fatalities) and to the amount of money that it can afford to spend (economy). In geotechnical engineering and in structural engineering, this acceptable probability of failure is typically less than 1 chance in 1000 (10^{-3}).

Design procedures have been developed to calculate a foundation size that meets these low probabilities of failure. These procedures involve:

1. Selecting the design issues (limit states)
2. Selecting load factors and resistance factors that are consistent with the low target probability of not meeting the design criterion
3. Determining the minimum size of the foundation that satisfies the low probability of not meeting the design criterion

For example, let's go back in time to the year 1100 and design the foundation of the Tower of Pisa, but with today's knowledge. The load is calculated to be 150,000 kN. The uncertainty about this load is small because the

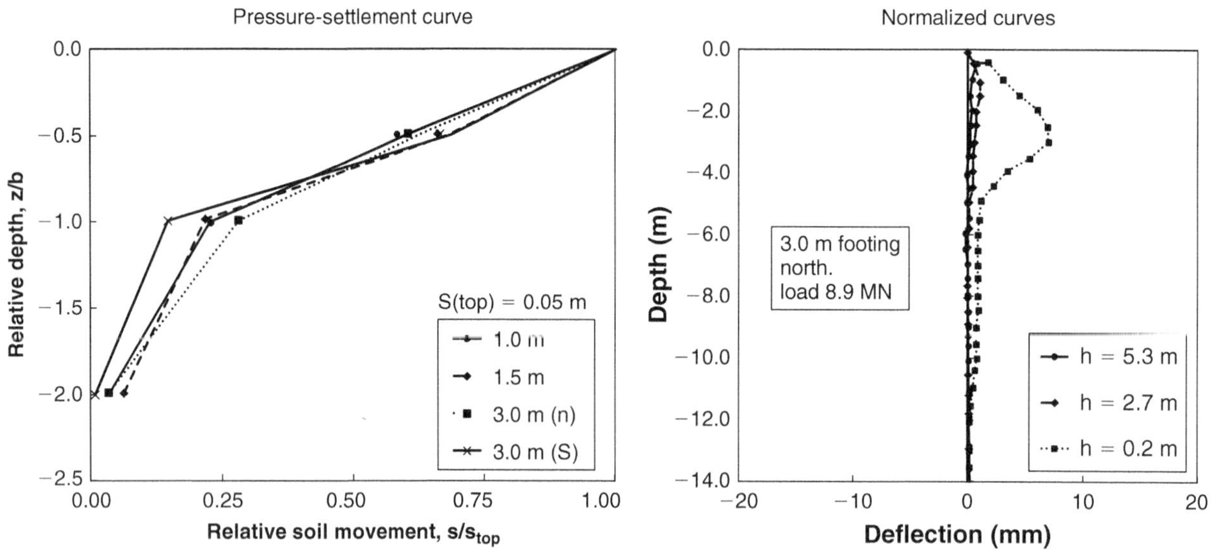

Figure 17.5 Vertical and horizontal movement vs. depth.

dimensions of the structure are on the plans. Nevertheless, a load factor of 1.2 is used to obtain the factored load of $1.2 \times 150{,}000 = 180{,}000$ kN, which lowers the probability of exceeding the load. The resistance is the ultimate bearing pressure of the soil below the tower. It is calculated as 6 times the undrained shear strength s_u of the soil within the depth of influence of the foundation (section 17.6.1). From the borings, in situ tests, and laboratory tests, a value of 80 kPa is selected for s_u. This leads to an ultimate bearing pressure of 480 kPa. The uncertainty associated with the undrained shear strength and the calculation model is not negligible, so a resistance factor of 0.6 is selected. The factored resistance is $0.6 \times 480 = 288$ kPa, which lowers the probability of not having the necessary resistance. The load factor 1.2 and the resistance factor 0.6 are based on the probability distribution of the load and of the resistance, and on ensuring that the probability that the difference between the factored load and the factored resistance is negative (failure) is less than approximately 10^{-3}. The difference between the load and the resistance is called the *limit state function*. We decide to place the 15 m diameter Tower of Pisa on a circular mat foundation 1 m thick with a diameter B. Now the ultimate limit state equation is written as:

$$1.2 \times 150000 < 0.6 \times 480 \times \pi\, B^2/4 \qquad (17.1)$$

which leads to B > 28 m. The actual, as-built foundation was less than 15 m in diameter and the soil below the foundation failed. The design should also include other considerations such as the serviceability limit state, but this simple example illustrates the design process and the concept of load and resistance factors.

More specifically, the design process proceeds as follows:

1. Decide on the foundation depth.
2. Make a reasonable estimate of the foundation size.
3. Calculate the ultimate bearing pressure of the foundation, p_u.
4. Check if the ultimate bearing pressure satisfies the safety criterion under the given load (ultimate limit state).
5. Repeat steps 1 through 3 until the safety criterion is satisfied and obtain the safe foundation pressure p_s, which is the unfactored load divided by the foundation area.
6. Under the safe foundation pressure p_s, check that the foundation satisfies the serviceability limit state by calculating the movement of the foundation and ensuring that it is less than the allowable movement.
7. If the calculated movement is larger than the acceptable movement s_a, increase the foundation size and/or the foundation stiffness and repeat step 6.
8. If the movement is acceptable, the design is complete, as the pressure applied is safe and allows only acceptable movement.

In addition to the preceding steps concerning soil strength and compressibility, the foundation must be well designed structurally. For example, one must ensure that the column will not punch through the spread footing, or that the mat foundation will not bend excessively. The structural aspect of foundation design is not covered in this book.

Shallow foundations are typically less expensive than deep foundations. Therefore, it is economically prudent, in most cases, to start with a shallow foundation solution. Only if it is shown to be insufficient or inappropriate should the design proceed with deep foundations.

17.4 LIMIT STATES, LOAD AND RESISTANCE FACTORS, AND FACTOR OF SAFETY

Limit states are the loading situations and the associated equations that are considered during the design of a foundation. They must be satisfied to yield a proper design. There are two major limit states: the ultimate limit state and the service limit state. In foundation engineering, *ultimate limit state* involves calculations of ultimate capacity using primarily the shear strength of the soil. Satisfying the ultimate limit state ensures that the foundation will meet a chosen level of safety against failure. The *service limit state* involves calculations of movements using deformation parameters. Satisfying the service limit state ensures that the foundation will meet a chosen degree of confidence against excessive movement or distortion of the structure.

The ultimate limit state refers to satisfaction of equations ensuring that the foundation will function far enough away from failure of the soil. This requires the choice of load factors γ and resistance factors φ that will achieve the chosen level of probability of success. These equations are of the form:

$$\gamma L < \varphi R \qquad (17.2)$$

where γ is the load factor, L is the load, φ is the resistance factor, and R is the resistance. The resistance here is meant to be the ultimate resistance of the soil. In the case of complex loading and multiple resistances, Eq. 17.2 becomes:

$$\sum \gamma_i L_i < \sum \varphi_i R_i \qquad (17.3)$$

where γ_i is the load factors, L_i is the loads, φ_i is the resistance factors, and R_i is the resistances. The load factors and the resistance factors make it possible to address separately the uncertainties associated with each load and each resistance. The term $\sum \gamma_i L_i$ also makes it possible to select the most appropriate combination(s) of loads that the soil has to resist. An example of an ultimate limit state equation is:

$$1.25\ DL + 1.75\ LL < 0.5\ R_u \qquad (17.4)$$

where DL is the dead load and permanent live load on the foundation, LL is the nonpermanent live load on the foundation, and R_u is the ultimate resistance of the foundation from the soil point of view. Typical load factors for ultimate limit state are shown in Table 17.1; typical resistance factors for ultimate limit state are shown in Table 17.2. Note that there are two choices for the resistance side. The first one consists of applying a factor φ to the resistance (resistance factor); the second one consists of applying factors to the individual material properties such as the components of the shear strength (material factors). The Eurocode gives designers the choice to use either of the approaches (not both), whereas the AASHTO specifications only use the resistance factors. The selection of the soil parameter is a very important step. The AASHTO specifications tend to use mean values of the parameters, whereas the Eurocode uses "cautious estimates" of the soil parameters. This affects the selection of the resistance and material factors.

These factors γ and φ are developed by using the following procedure:

1. The unbiased estimates or best estimates or true values or measured values of the ultimate resistance and the load are R_m and L_m. The nominal values or design

Table 17.1 Typical Load Factors for Ultimate Limit State

Type of Loading	Load Factor γ (AASHTO) For bridges	Load Factor γ (ASCE 7) For buildings	Load Factor γ_E (Eurocode 7) For buildings
Dead load and permanent live load	1.25	1.2	1.35
Other live load	1.75	1.6	1.5
Extreme events (earthquake, hurricane, etc.)	1	1	1

Table 17.2 Typical Resistance Factors for Ultimate Limit State and Shallow Foundations

Type of Soil Testing	Resistance Factor φ (AASHTO)	(Eurocode 7) Material Factor $\gamma_M = 1/\varphi$	(Eurocode 7) Resistance Factor $\gamma_R = 1/\varphi$
Many high-quality tests	0.5 to 0.6	1.25–1.4	1.1 to 1.7 (footings)
Ordinary quantity and quality of tests	0.4 to 0.5	(may be reduced for	1.1 to 1.6 (piles)
Extreme events (earthquake, hurricane, etc.)	1	extreme events)	(may be reduced for extreme events)

values or predicted values of the resistance and the load are R_p and L_p.

2. Obtain the probability distribution of the load L_m and of the ultimate resistance R_m. Note that L_m and R_m are probabilistic. Each follows a certain distribution (for example, lognormal) with specified means (μ_{Rm} and μ_{Lm}) and standard deviation (σ_{Rm} and σ_{Lm}).
3. Write the likelihood function as $g = R_m - L_m$. Because R_m and L_m are random, g is also random.
4. Compute, using reliability software such as FERUM (2001):
   a. the probability $P(g \leq 0)$
   b. the corresponding value of the generalized reliability index β
   c. the coordinates of the failure point (R_m^*, L_m^*)
5. Choose a target reliability index β_{target}, usually 2.33 for redundant systems and 3 for nonredundant systems.
6. Compare the β from step 4 with the β_{target} from step 5. If the β from step 4 is equal to the β_{target} from step 5, then the central resistance factor $\overline{\varphi}$ and the central load factor $\overline{\gamma}$ can be calculated as:

$$\overline{\varphi} = R_m^*/\mu_{Rm}$$
$$\overline{\gamma} = \mathrm{L}_m^*/\mu_{\mathrm{Lm}}$$

7. Otherwise, increase or decrease μ_{Rm} and repeat steps 1 through 5.
8. Calculate the nominal resistance factor φ and the nominal load factor γ as follows:

$$\varphi = \overline{\varphi}\frac{\mu_{Rm}}{\mu_{Rp}}$$
$$\gamma = \overline{\gamma}\frac{\mu_{Lm}}{\mu_{Lp}}$$

For normal distributions, the reliability index β is the inverse of the coefficient of variation and tells us how many standard deviations the mean of $R_m - L_m$ is from the zero origin. For more complex distributions, this definition does not hold true. Typical β values are 2.33 for redundant systems and 3 for nonredundant systems. These β values correspond to probabilities of failure equal to $10^{-2}(\beta = 2.33)$ and $10^{-3}(\beta = 3.0)$.

The service limit state involves calculations of movements using deformation parameters. Satisfying the service limit state ensures that the foundation will meet a chosen degree of confidence against excessive movement or distortion of the structure. The equations have the same format as the ultimate limit state equations. The load factors are applied to the loads to be considered for movement calculations and the resistance factors are applied to the predicted movement or the soil deformation parameters. Typically, however, the load factors and resistance factors are taken as equal to 1. The nonpermanent live loads are not included in the loads considered for calculating settlements that take a long time to develop, such as consolidation settlements in saturated clays.

For example, the service limit state in terms of loads for a spread footing can be written as follows:

$$\gamma_1 DL + \gamma_2 LL \leq \varphi \frac{s_{all}BE}{I(1-\nu^2)} \tag{17.5}$$

where $\mathrm{s_{all}}$ is the allowable settlement of the foundation, B is the width of the spread footing, E is the modulus of the soil below the footing, I is a shape factor, and ν is the Poisson's ratio of the soil. The term $\frac{s_{all}BE}{I(1-\nu^2)}$ on the right-hand side of Eq. 17.5 is the load that generates the allowable settlement of the footing on an elastic half space; it is the resistance of the system at the service limit state. As mentioned earlier, the load factors and the resistance factors are usually taken as equal to 1. Furthermore, if the settlement will take place over a long period of time, the live load is not included in the settlement calculations except for the permanent live load.

Before the development of the load and resistance factor design (LRFD) approach, also called limit state design (LSD), the working stress design (WSD), also called the allowable stress design (ASD), approach was used. WSD consists of applying a global factor of safety against the ultimate bearing capacity of the soil in order to obtain the safe load. The equation is:

$$L < R/F \tag{17.6}$$

where L is the applied load to be safely carried, R is the ultimate resistance, and F is the global factor of safety. The factor of safety varies depending on the type of design (shallow foundation, deep foundation, slope stability, retaining wall) and is typically between 1.5 and 3 (Table 17.3). For the ultimate bearing pressure under a shallow foundation obtained by calculations, it is 3. The settlement is calculated using the dead loads and permanent live load without applying any factors.

One is always tempted, when comparing the WSD and LRFD approaches, to compare the global factor of safety with the ratio of the load factor divided by the resistance factor. Indeed, from Eqs. 17.2 and 17.6 comes $F = \gamma/\varphi$.

Table 17.3 Typical Global Factors of Safety against Soil Failure

Type of Geotechnical Application	Global Factor of Safety F
Shallow foundations	2.5 to 3
Deep foundations	2 to 2.5
Retaining wall	1.5 to 2
Slope stability	1.3 to 1.5

Using this expression and the extreme values of the load factors (dead load) and resistance factors gives a global factor of safety ranging from 1.3 to 4.2. This is a larger range than the values in Table 17.3 and shows that not all geotechnical methods give the same degree of precision on the predicted resistance. The LRFD approach takes this factor clearly into account.

One very important issue is how the geotechnical design parameters are selected from the borings, tests results, and soundings resulting from the site investigation. For example, the issue is to know which value to select from an undrained shear strength profile or a blow count profile or a cone penetrometer point resistance profile. This value is called the *characteristic value*, and its selection obviously will have a major impact on the uncertainty associated with predictions of the resistance. The Eurocode 7 defines the characteristic value as "a cautious estimate of the value affecting the occurrence of the limit state." So, in this case the selection is tied to the limit state itself.

Design methods can be classified into three categories: design by theory, design by empiricism, and design by analogy. Design methods by theory rely on theoretical derivations for recommending the design equations. Design methods by empiricism rely on experimental data and correlations for recommending the design equations. Design methods by analogy rely on the close analogy between the mode of deformation in the soil test and under the foundation. Generally speaking, the best methods include—and accumulate the advantages of—all three, by using a close analogy, experimental data, and a solid theoretical background.

17.5 GENERAL BEHAVIOR

In a load test on a shallow foundation (say, a 3 m by 3 m spread footing), the load on the foundation is increased in steps (jacking against an anchored frame or accumulating dead weight) and the corresponding downward movement is recorded. The load settlement curve is plotted and usually shows a relatively linear part at lower loads (elastic behavior), followed by a curved part, followed by a part where the movement accelerates faster than the load (Figure 17.6).

Load tests on silts and clays often plunge; load tests on sands and gravel rarely do, with the load increasing steadily with more deflection (Figure 17.6). The reason for the difference is that the fine-grained soils tend to shear in an undrained mode during a load test that may last a few hours, whereas coarse-grained soils likely shear in a drained mode. The undrained shear strength of a clay does not vary much with the stress and confinement level (s_u = constant), so when the load on the footing increases, the shear strength does not increase and the failure is clearly defined. The drained shear strength of a sand depends on the stress and confinement level ($s = \sigma' \tan \varphi'$); thus, when the load increases, so does the stress level and therefore the shear strength. Hence, the ultimate resistance of the sand increases as more load is applied and the failure is ill

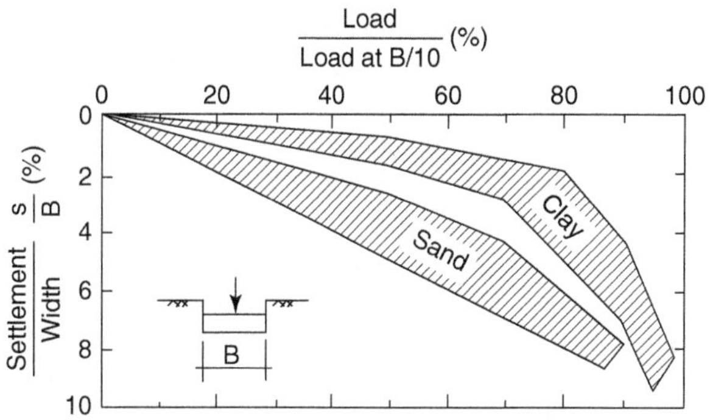

Figure 17.6 Typical shape of load test results on shallow foundations.

defined. In such a case, the ultimate load can be defined as the load corresponding to a movement equal to one-tenth of the foundation width. The true ultimate resistance of a footing on sand or gravel does exist, but at much larger displacements. These displacements are on the order of the width of the footing, as can be shown by the cone penetrometer test.

One important part of shallow foundation behavior is the movement of the foundation under sustained load, because most foundations are loaded with a static load for the life of the structure, which may be several decades or more. During the load test, the load can be maintained for a period of time and the movement can be observed as a function of time during that period.

17.6 ULTIMATE BEARING CAPACITY

The ultimate bearing capacity p_u is one of the critical values to be estimated when designing a shallow foundation. It is defined as the highest pressure the soil can resist. As explained in section 17.5, p_u corresponds to a plunging load in fine-grained soils, but to a load at large displacement (such as one-tenth of the footing width or B/10) in coarse-grained soils, because of the shape of the load settlement curve. Thus, the ultimate bearing capacity tends to control the design of shallow foundations on clay, whereas settlement tends to control the design of shallow foundations on sand. The value of p_u can come from an empirical formula (pressuremeter test, cone penetrometer test, or standard penetration test), from a formula based on theory (general bearing capacity equation), or from a load test. Load tests on shallow foundations are rare.

17.6.1 Direct Strength Equations

Direct strength equations rely on the average value of the strength of the soil within the depth of influence of the foundation below the foundation level. They are generally of the form:

$$p_u = ks + \gamma D \tag{17.7}$$

where k is the bearing capacity factor, γ is the effective unit weight of the soil, D is the embedment depth, and s is a measure of the soil strength averaged over the depth of influence. This depth of influence is typically taken as one foundation width below the foundation level for a uniform soil. The case of layered soils is addressed in section 17.6.3.

The first direct strength equation was proposed by Skempton (1951); it addresses the problem of the undrained ultimate bearing capacity of a shallow foundation on a fine-grained soil. The equation makes use of the average undrained shear strength s_u within the depth of influence below the footing. The theoretical background for this equation is rooted in the information presented in section 11.4.2. The equation is:

$$p_u = N_c s_u + \gamma D \quad (17.8)$$

where N_c is the bearing capacity factor (Figure 17.7) proposed by Skempton after calibration against field data, γ is the total unit weight of the soil above the foundation depth, and D is the depth of embedment. Note that N_c is higher for square footings than for strip footings. The reason is that the square footing can develop a relatively larger failure surface, because the failure surface can develop in four directions, whereas the failure surface for the strip footing is confined to only two directions. The N_c values for the square footing and the strip footing are related by:

$$N_{c(square)} = 1.2 N_{c(strip)} \quad (17.9)$$

Note also that N_c gradually increases with the relative depth of embedment, due to the gradual increase in the length of the failure surface with embedment. The values of N_c peak at:

$$N_{c(square)\max} = 9 \quad and \quad N_{c(strip)\max} = 7.5 \quad (17.10)$$

The second direct strength equation was proposed by Menard (1963a; 1963b); it addresses the problem of the

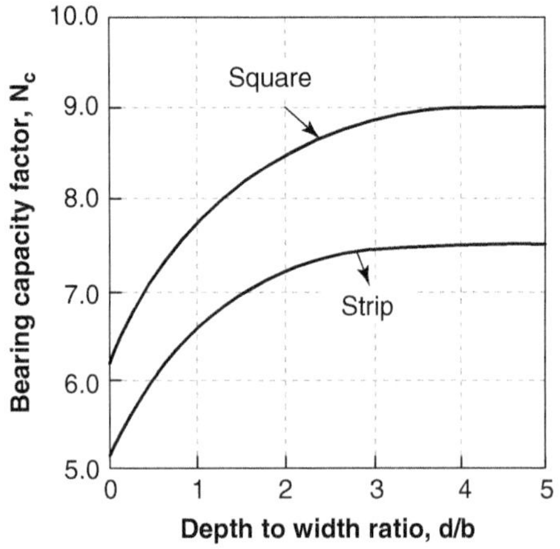

Figure 17.7 Skempton chart for N_c. (Skempton 1951)

ultimate bearing capacity of any soil in which the pressuremeter test can be performed. The theoretical background of this equation is rooted in the solution to the expansion of a cylindrical cavity. The equation is:

$$p_u = k_p p_L^* + \gamma D \quad (17.11)$$

where k_p is the pressuremeter bearing capacity factor, γ is the total unit weight of the soil above the footing depth, D is the depth of embedment, and p_L* is the net limit pressure equal to the PMT limit pressure p_L minus the horizontal total stress at rest σ_{oh}:

$$p_L^* = p_L - \sigma_{oh} \quad (17.12)$$

The PMT bearing capacity factor k_p is given in two steps (Frank, 1999, 2013, Norme Francaise AFNOR P94-261), first a soil classification step (Table 17.4) and then an equation for each soil category (Eqs. 17.13 to 17.18).

Clay and silt—strip footing:

$$k_p = 0.8 + \left(0.2 + 0.02\frac{D}{B}\right)\left(1 - e^{-1.3\frac{D}{B}}\right) \quad (17.13)$$

Clay and silt—square footing:

$$k_p = 0.8 + \left(0.3 + 0.02\frac{D}{B}\right)\left(1 - e^{-1.5\frac{D}{B}}\right) \quad (17.14)$$

Clay and silt—rectangular:

$$k_{p(B/L)} = k_{p(B/L=0)}\left(1 - \frac{B}{L}\right) + k_{p(B/L=1)}\frac{B}{L} \quad (17.15)$$

Sand and gravel—strip footing:

$$k_p = 1 + \left(0.3 + 0.05\frac{D}{B}\right)\left(1 - e^{-2\frac{D}{B}}\right) \quad (17.16)$$

Sand and gravel—square footing:

$$k_p = 1 + \left(0.22 + 0.18\frac{D}{B}\right)\left(1 - e^{-5\frac{D}{B}}\right) \quad (17.17)$$

Sand and gravel—rectangular:

$$k_{p(B/L)} = k_{p(B/L=0)}\left(1 - \frac{B}{L}\right) + k_{p(B/L=1)}\frac{B}{L} \quad (17.18)$$

where B and L are the width and length of the footing respectively, and D is the depth of embedment. These rules are primarily based on load tests with 1 m by 1 m square footings. As can be seen, the k_p factor varies within a typical

Table 17.4 Soil Classification for the PMT and CPT Foundation Rules (After Frank, 2013)

Soil Type	Strength	PMT p_L * (MPa)	CPT q_c(MPa)	SPT N(bpf)	Shear Strength s_u(kPa)
Clay, Silt	Very soft to soft	<0.4	<1		<75
	Firm	0.4 to 1.2	1 to 2.5		75 to 150
	Stiff	1.2 to 2	2.5 to 4		150 to 300
	Very stiff	>2	>4		>300
Sand, Gravel	Very loose	<0.2	<1.5	<3	
	Loose	0.2 to 0.5	1.5 to 4	3 to 8	
	Medium dense	0.5 to 1	4 to 10	8 to 25	
	Dense	1 to 2	10 to 20	25 to 42	
	Very dense	>2	>20	42 to 58	

After Norme Francaise AFNOR P94-261 as presented in Frank, 2013

range of 0.9 to 1 for clay and 1.2 to 1.4 for sands. The following simpler rule seems conservative in most cases:

$$p_u = k_p p_L \quad with \quad k_p = 0.9 \; for\; clay \quad and \quad k_p = 1.2 \; for\; sand \tag{17.19}$$

Interestingly, it can be shown that the horizontal resistance of a soil is the major component of the vertical resistance. Referring to Figure 17.8, consider a circular footing with a diameter D founded on the ground surface. The soil is a saturated clay layer with a thickness equal to $2D$. According to Eq. 17.8, the vertical ultimate bearing capacity p_u of that footing is $6.2 s_u$. To calculate how much of p_u comes from the horizontal soil resistance, let's remove that lateral support. In this case the footing sits on top of a large sample loaded in an unconfined compression test; thus, the ultimate bearing capacity that it generates is equal to $2s_u$. The difference between the two diagrams on Figure 17.8 gives the contribution of the horizontal strength of the soil to the vertical ultimate bearing pressure: $4.2s_u$. Therefore, 68% of the vertical ultimate bearing pressure is due to horizontal resistance. For sand, the percent contribution of the horizontal resistance is even larger, as the unconfined compression resistance of a sand is very small.

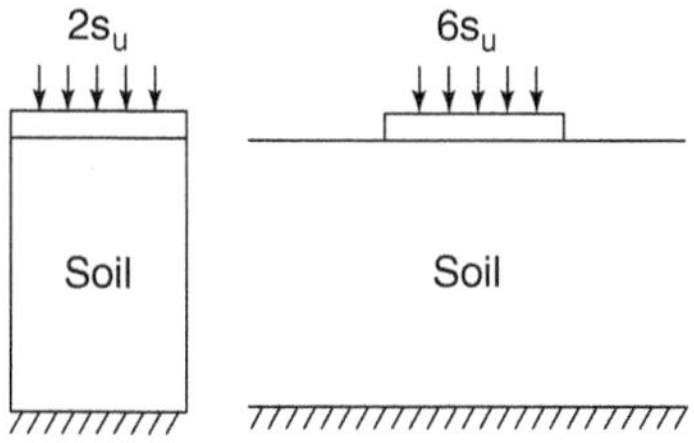

Figure 17.8 Lateral support as main contributor to vertical capacity. (Baguelin et al. 1978)

The third direct strength equation makes use of the cone penetrometer point resistance q_c; it addresses the problem of the ultimate bearing capacity of any soil into which the cone penetrometer can be pushed. The theoretical background of this equation is rooted in the solution to the expansion of a spherical cavity. There is one equation for clays and another one for sands. For clays, the equation is based on the work of Tand et al. (1986):

$$p_u = k_c q_c + \gamma D \tag{17.20}$$

where k_c is the cone penetrometer bearing capacity factor (Figure 17.9), q_c is the average point resistance within one footing width below the footing, γ is the total unit weight of the soil above the footing, and D is the depth of embedment. All in all, it appears that a k_c value of 0.35 is a reasonable estimate for shallow foundations on clay. In sand, a value of

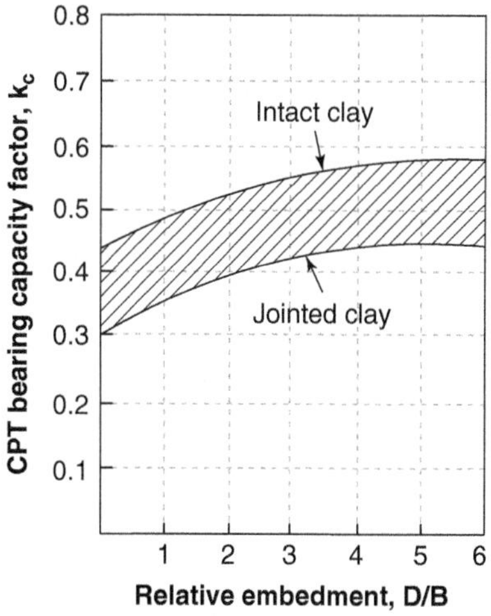

Figure 17.9 Chart for the CPT bearing capacity factor k_c. (After Tand et al. 1986)

k_c equal to 0.23 has been proposed by Briaud and Gibbens (1999). So, in summary:

$$\text{Clays} \qquad p_u = 0.40q_c + \gamma D \tag{17.21}$$

$$\text{Sands} \qquad p_u = 0.20q_c + \gamma D \tag{17.22}$$

The Norme Francaise AFNOR P94-261 as presented in Frank (2013) gives the following recommendations for k_c:

Clay and silt—strip footing:

$$k_c = 0.27 + \left(0.07 + 0.007\frac{D}{B}\right)\left(1 - e^{-1.3\frac{D}{B}}\right) \tag{17.23}$$

Clay and silt—square footing:

$$k_c = 0.27 + \left(0.1 + 0.007\frac{D}{B}\right)\left(1 - e^{-1.5\frac{D}{B}}\right) \tag{17.24}$$

Sand and gravle—strip footing:

$$k_c = 0.09 + \left(0.04 + 0.006\frac{D}{B}\right)\left(1 - e^{-2\frac{D}{B}}\right) \tag{17.25}$$

Sand and gravel—square footing:

$$k_c = 0.09 + \left(0.03 + 0.02\frac{D}{B}\right)\left(1 - e^{-5\frac{D}{B}}\right) \tag{17.26}$$

where B and L are the width and length of the footing respectively, and D_e is the depth of embedment. For the case of a rectangular footing, Eq. 17.15 is used. As can be seen, the k_c factor recommended by AFNOR varies within a typical range of 0.30 to 0.35 for clay and 0.10 to 0.14 for sands.

The fourth direct strength method makes use of the SPT blow count N; it addresses the problem of the ultimate bearing capacity of any soil in which the standard penetration test can be performed. There is one equation for sands and another one for clays. The form of the equation is:

$$p_u = k_N N\, p_a + \gamma D \tag{17.27}$$

where k_N is the SPT bearing capacity factor, N is the average blow count within one footing width below the footing, p_a is the atmospheric pressure used for normalization, γ is the total unit weight of the soil above the footing, and D is the depth of embedment. For sands, the k_N value is based on the work of Briaud and Gibbens (1999) and for clay the k_N value is back-calculated using Eq. 17.8 and the correlation between the blow count and the undrained shear strength. Note that calculating p_u based on the SPT blow count is probably the least accurate of all direct methods. So, in summary:

$$\text{Sands} \qquad p_u = 0.60N\, p_a + \gamma D \tag{17.28}$$

$$\text{Clays} \qquad p_u = 0.35N\, p_a + \gamma D \tag{17.29}$$

17.6.2 Terzaghi's Ultimate Bearing Capacity Equation

This equation is called the *general bearing capacity equation*. The assumptions made in deriving this equation are that the soil has no water, that it has a constant friction angle and cohesion c, and that it has a constant unit weight. As such, it corresponds to a soil strength profile that increases linearly with depth (Figure 17.10). If the soil strength profile does not meet this requirement, this equation should not be used, as it will give erroneous values of p_u.

The Terzaghi equation also assumes that a failure mechanism develops with a shear plane under the foundation (Figure 17.11) and that the soil mass is pushed sideways to allow for the foundation penetration. This was not observed in the large footing tests by Briaud and Gibbens (Figure 17.5).

The general bearing capacity equation for a strip footing is:

$$p_u = c'N_c + \frac{1}{2}\gamma BN_\gamma + \gamma DN_q \tag{17.30}$$

where p_u is the ultimate bearing capacity of the soil; c' is the effective stress cohesion intercept; N_c, N_γ, and N_q are bearing capacity factors function of the effective stress friction angle φ'; γ is the effective unit weight; B is the width of the foundation; and D is the depth of embedment of the foundation. The assumption of constant φ' and constant γ implies that the shear strength profile increases linearly with depth. If this matches the soil strength profile observed at the site, the equation is applicable. However, most field situations do not exhibit such simple linear profiles. In this case, the empirical equations give a more representative estimate of p_u. Note that the general bearing capacity equation is to be used with effective stress parameters (c', φ') and

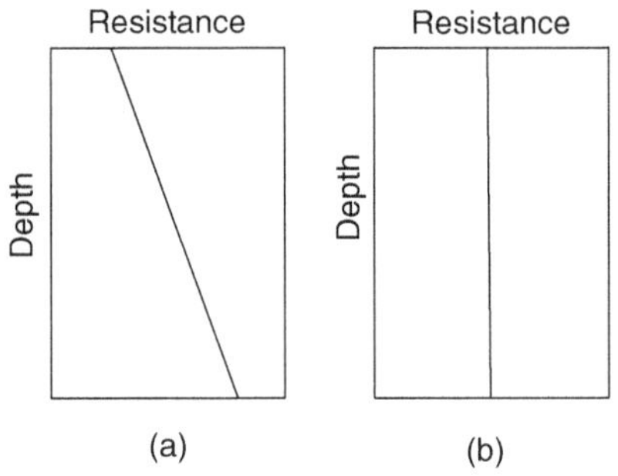

Figure 17.10 Soil strength profiles.

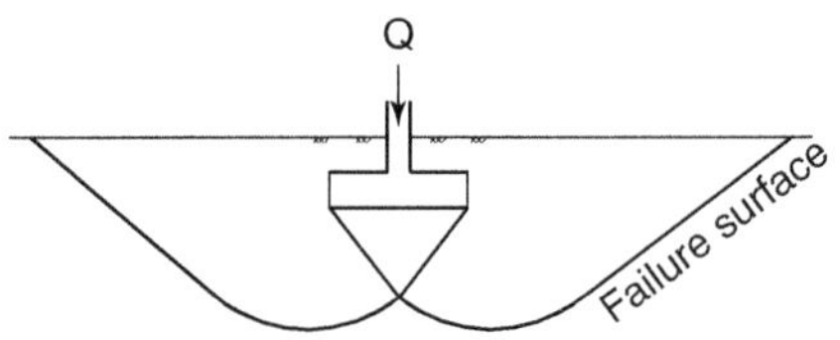

Figure 17.11 Bearing capacity failure mechanism.

drained conditions. It gives the long-term capacity of fine-grained soils and the short- and long-term capacity of coarse-grained soils. The undrained ultimate bearing capacity of fine-grained soils is given by Eq. 17.8.

The following derivation is an illustration of how the bearing capacity factors N_c, N_γ, and N_q can be obtained. The footing is a strip footing, which ensures a plane strain condition. The step-by-step procedure explained in section 11.4.1 is followed to obtain the failure load.

1. The failure mechanism of Figure 17.11 is assumed.

2. The free-body diagram of the wedge below the footing is drawn (Figure 17.12) and the reasoning is carried out on half of the wedge because of symmetry (OAB in Figure 17.12). The angle of the side of the wedge with the vertical is the angle of the failure plane. It is considered to be $45 + \varphi'/2$ because that is the angle of the failure plane in a triaxial test (see section 9.12.1) and with a passive pressure type of failure (see Chapter 21). All external forces are shown; they include the ultimate load Q_u at the soil-foundation interface ($Q_u(\text{kN/m}) = p_u \times B$), the weight W of the half wedge, the cohesion force C along the face AB, and the passive earth pressure force P_p (also on face AB).

3. Vertical equilibrium of forces is the fundamental equation used. Note that the forces are in force per unit length, as this is a plane strain problem:

$$\frac{Q_u}{2} = P_p \cos\left(45 - \frac{\varphi'}{2}\right) + C\cos\left(45 - \frac{\varphi'}{2}\right) - W \tag{17.31}$$

where φ' is the soil friction angle.
Referring to Figure 17.12, the weight W of the half wedge is:

$$W = \frac{1}{2}\gamma\frac{B}{2}H = \frac{1}{4}\gamma B\frac{B}{2}\tan\left(45 + \frac{\varphi'}{2}\right) = \frac{1}{8}\gamma B^2\tan\left(45 + \frac{\varphi'}{2}\right) \tag{17.32}$$

where γ is the unit weight of the soil.

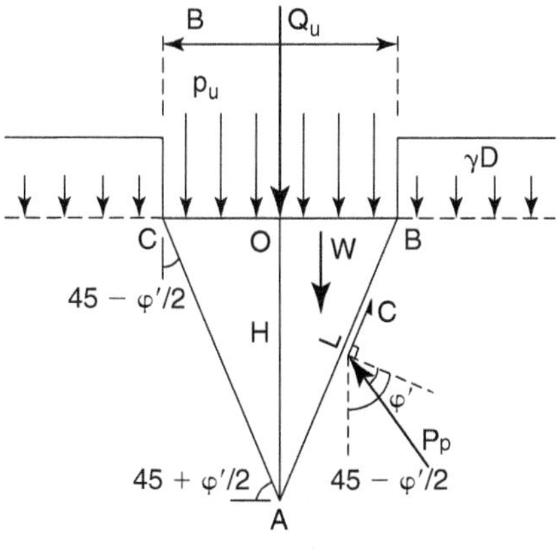

Figure 17.12 Free-body diagram of soil wedge in bearing capacity failure.

The cohesion force is:

$$C = c'L = c'\frac{B}{2\sin\left(45 - \frac{\varphi'}{2}\right)} \tag{17.33}$$

where c' is the soil cohesion intercept.
The passive resistance P_p is given by an equation presented in Chapter 21:

$$P_p = \frac{1}{2}K_p\gamma H^2 + 2c'H\sqrt{K_p} + \gamma DHK_p \tag{17.34}$$

where K_p is the passive earth pressure coefficient (see Chapter 21). This coefficient depends on φ'. Regrouping Eqs. 17.31 to 17.34 gives:

$$p_u = \frac{Q_u/2}{B/2} = c'\left(1 + 2\sqrt{K_p}\cos\left(45 - \frac{\varphi'}{2}\right)\right)\tan\left(45 + \frac{\varphi'}{2}\right) + \frac{1}{2}\gamma B\left(\frac{K_p\cos\left(45 - \frac{\varphi'}{2}\right)}{2\tan\left(45 - \frac{\varphi'}{2}\right)} - \frac{1}{2}\right)\tan\left(45 + \frac{\varphi'}{2}\right) + \gamma DK_p\tan\left(45 + \frac{\varphi'}{2}\right) \tag{17.35}$$

This can be rewritten as:

$$p_u = c'N_c + \frac{1}{2}\gamma BN_\gamma + \gamma DN_q \tag{17.36}$$

and the expressions of the bearing capacity factors N_c, N_γ, and N_q become clear. In Eq. 17.36, p_u is the ultimate bearing pressure the soil can resist, c' is the effective stress cohesion, γ is the soil effective unit weight, B is the foundation width, D is the depth of embedment, and N_c, N_γ, and N_q are the bearing capacity factors.

4. Note that the constitutive equation is buried in Eq. 17.34, which makes use of the shear strength equation of the soil. This is discussed in Chapter 21. The problem now is to obtain the expression of K_p as a function of φ'. Taking the expression that comes from Chapter 21 is not appropriate, because the assumptions for the retaining walls dealt with in Chapter 21 are not applicable to the extreme inclination of the "retaining wall" associated with plane AB in Figure 17.12 and Figure 17.13. In Chapter 21, a plane

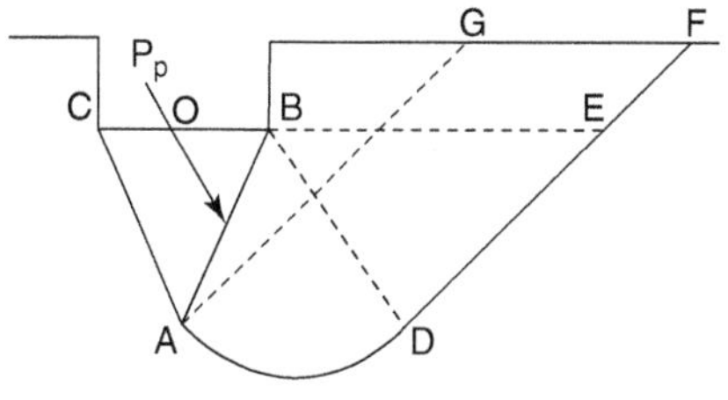

Figure 17.13 Evaluation of passive resistance.

is assumed as a failure surface (line AG in Figure 17.13), whereas a different shape failure surface is assumed for the bearing capacity failure (line ADE in Figure 17.13). Different assumptions have been made for line ADEF, and each one leads to a different set of bearing capacity factors N_c, N_γ, and N_q. Some assume a circle for line AD, some assume a log spiral, some assume that line DF stops at E, some go all the way to F, and some use a wedge ABC that is not a triangle.

5. The solution originally proposed by Terzaghi (1943) was decomposed into three superposed states:

a. State I, soil with cohesion and friction but no weight and no surcharge
b. State II, soil with friction and surcharge but no weight and no cohesion
c. State III, soil with weight and friction but no surcharge and no cohesion

Then each State is solved with separate failure envelopes and the solutions for each State are added in superposition of all States, to end up with Eq. 17.36. Although such a superposition principle is not theoretically correct in plasticity (or any other nonlinear) theory, the error appears to be small.

Many different bearing capacity factors have been proposed by various authors. All in all, the N_c, N_γ, and N_q factors most commonly used are those shown in Figure 17.14. They come from the work of Reissner (1924) for N_c and N_q and from the work of Meyerhof (1955) for N_γ.

The general bearing capacity equation requires that the soil be rigid enough to push the whole soil wedge from the footing to the ground surface. This may be the case when the soil is very dense or very stiff, but not when it is loose or soft. This also requires a very large amount of movement. To alleviate this limitation, Terzaghi and Peck (1963) recommended a correction that consists of reducing the value of the friction angle to $0.67\varphi'$ for loose and soft soils.

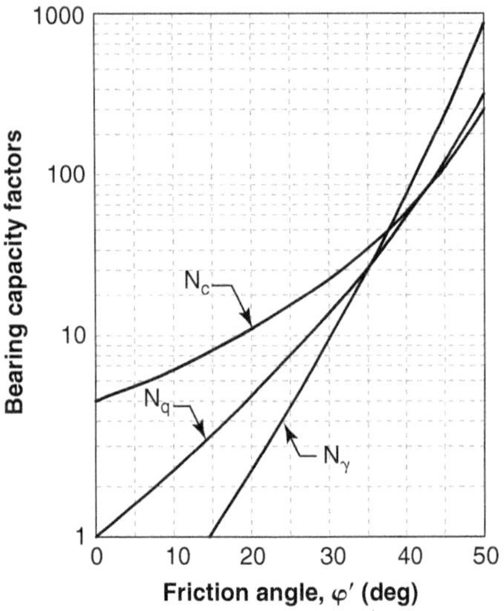

Figure 17.14 Bearing capacity factors.

Recall that one of the assumptions for the development of Eq. 17.36 is that the soil has no water. If the groundwater level (GWL) is within the depth of influence of the footing (1B below the footing), the unit weight in Eq. 17.36 should be the effective unit weight:

If the soil is below the GWL $\quad \gamma_{eff} = \gamma_t - \gamma_w \quad (17.37)$

If the soil is above the GWL $\quad \gamma_{eff} = \gamma_t \quad (17.38)$

For example, if the GWL is at the level of the foundation, the γ value for the third term in Eq. 17.36 should be γ_t, because that term refers to the soil above the foundation level, but the γ value for the second term in Eq. 17.36 should be $\gamma_t - \gamma_w$ because it refers to the soil below the foundation level.

17.6.3 Layered Soils

The previous two subsections dealt with relatively uniform soils. If the strength profile indicates that a layered system is involved in the responses to the foundation loading, modifications to the equations are necessary. The following simple examples show how this can be done for a strip footing.

Hard clay over soft clay. The first step is always to find a reasonable failure mechanism. Referring to Figure 17.15, it seems reasonable to assume that if the thickness H of the hard layer is large enough, the ultimate bearing pressure will be the one of the hard layer, $p_{u(hard)}$. If the thickness of the hard layer is negligible, then the ultimate bearing pressure will be $p_{u(soft)}$. If the thickness of the hard layer is intermediate, then the foundation will punch through the hard layer into the soft layer. This is very similar to punching through the ice layer when you walk across a frozen lake, if the ice is not thick enough.

Vertical equilibrium of forces for the failing mass (ABCD in Figure 17.15*a*) gives:

$$p_u B + \gamma_{(hard)} HB = 2F + p_{u(soft)} B = 2s_{u(hard)} H + (N_{c(soft)} s_{u(soft)} + \gamma_{(hard)} H)B \quad (17.39)$$

Or

$$p_u = N_{c(soft)} s_{u(soft)} + 2s_{u(hard)} \frac{H}{B} \quad (17.40)$$

where p_u is the ultimate bearing pressure of the foundation, N_c is the bearing capacity factor from Figure 17.7 for a depth of embedment of H/B, $s_{u(soft)}$ and $s_{u(hard)}$ are the undrained shear strength of the soft layer and hard layer respectively, $\gamma_{(hard)}$ is the unit weight of the hard layer, H is the thickness of the hard layer, and B is the width of the footing. Note that in Eq. 17.40 all forces are in kN/m, because they are calculated per unit length of footing perpendicular to the page. The p_u values for both layers taken independently are:

$$p_{u(soft)} = N_{c(soft)} s_{u(soft)} + \gamma_{(hard)} H \quad (17.41)$$

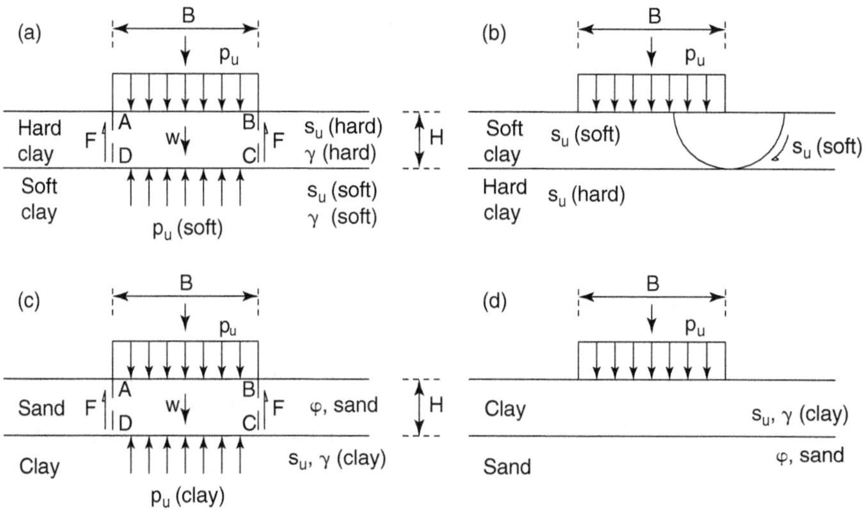

Figure 17.15 Layered systems.

$$p_{u(hard)} = N_{c(hard)} s_{u(hard)} \tag{17.42}$$

Then the critical height ratio, H_c/B, where the failure changes from a punching failure of the layered system to failure in the hard layer alone, can be found by writing that at that point the value of p_u for the layered system is equal to the p_u value for the hard layer:

$$N_{c(hard)} s_{u(hard)} = N_{c(soft)} s_{u(soft)} + 2 s_{u(hard)} \frac{H}{B} \tag{17.43}$$

Note that a distinction must be made between $N_{c(hard)}$ and $N_{c(soft)}$ because of the different depth of embedment for the foundation on top of the hard layer and the foundation on top of the soft layer. Because the top of the hard layer is at the ground surface, $N_{c(hard)}$ is equal to 5.14. Then the expression for H_c/B is:

$$\frac{H_c}{B} = \frac{5.14 s_{u(hard)} - N_{c(soft)} s_{u(soft)}}{2 s_{u(hard)}} = 2.57 - \frac{N_{c(soft)}}{2} \frac{s_{u(soft)}}{s_{u(hard)}} \tag{17.44}$$

Because $N_{c(soft)}$ depends on H/B, Eq. 17.44 has to be solved by iteration. Figure 17.16 illustrates the variation of p_u with an increase in H/B. As can be inferred from Eq. 17.44, the critical depth H_c varies from about 2 for significant strength contrast between the two layers to about 1 when the strength contrast is not very significant.

Soft clay over hard clay. In this case, the failure mechanism is different from the one for the hard clay over the soft clay. If the soft clay layer is thick enough, the failure will occur in the soft clay and p_u is equal to $p_{u(soft)}$. If the thickness of the soft layer is negligible, then it should be removed and p_u is equal to $p_{u(hard)}$. If the thickness of the soft layer is intermediate, then the failure mechanism is that the soft layer squeezes out

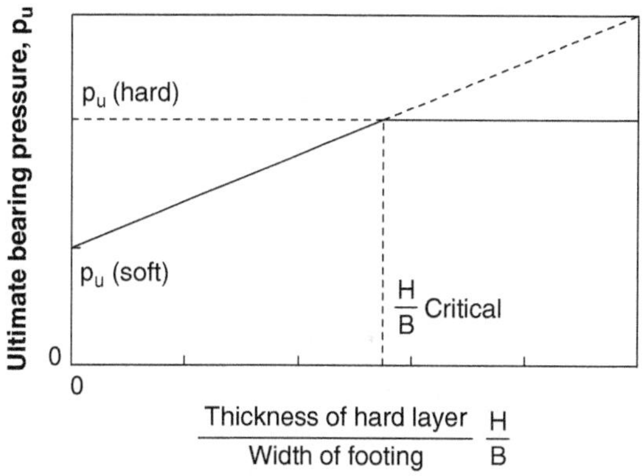

Figure 17.16 Ultimate bearing capacity for a layered system.

on the side of the footing. More scientifically put, a local failure occurs in the soft layer as shown in Figure 17.15*b*. Therefore, for a soft layer over a hard layer, the ultimate bearing pressure is always $p_{u(soft)}$.

Sand over clay. If the sand is very loose and the clay is very hard, a local failure in the sand layer can occur. Most of the time, in the case of a hard layer over a soft layer, the punching mechanism is likely to apply. If the thickness H of the sand layer is large enough, the ultimate bearing pressure will be the one of the sand layer, $p_{u(sand)}$. If the thickness of the sand layer is negligible, then the ultimate bearing pressure will be $p_{u(clay)}$. If the thickness of the sand layer is intermediate, then the foundation will punch through the hard layer into the soft layer. The force F in this case is equal to the horizontal force P_p times the coefficient of friction $\tan \varphi'$. The horizontal force P_p is the resultant force corresponding to the passive earth pressure distribution on the vertical plane BC (Figure 17.15c). Indeed, this plane is pushed sideways into the soil and generates the passive earth pressure at ultimate load. This force P_p is given by:

$$P_p = \frac{1}{2} K_{p(sand)} \gamma_{(sand)} H^2 \quad (17.45)$$

where $K_{p(sand)}$ is the coefficient of passive earth pressure for the sand. From Chapter 21 we get:

$$K_{p(sand)} = \frac{1 + \sin\varphi'}{1 - \sin\varphi'} \quad (17.46)$$

Vertical equilibrium of forces for the failing mass (ABCD in Figure 17.15c) gives:

$$\begin{aligned} p_u B + \gamma_{(sand)} HB &= 2F + p_{u(clay)} B \\ &= 2\frac{1}{2} K_{p(sand)} \gamma_{(sand)} H^2 \tan\varphi' \\ &\quad + (N_c s_{u(clay)} + \gamma_{(clay)} H) B \end{aligned} \quad (17.47)$$

Therefore, the ultimate bearing pressure is:

$$p_u = K_{p(sand)} \gamma_{(sand)} \frac{H^2}{B} \tan\varphi' + N_c s_{u(clay)} \quad (17.48)$$

where p_u is the ultimate bearing pressure of the foundation, $K_{p(sand)}$ is the coefficient of passive earth pressure for the sand, $\gamma_{(sand)}$ and $\gamma_{(clay)}$ are the unit weight of the sand and of the clay respectively, H is the thickness of the sand layer, B is the width of the foundation, φ' is the friction angle of the sand, N_c is the bearing capacity factor from Figure 17.7 for a depth of embedment of H/B, and $s_{u(clay)}$ is the undrained shear strength of the clay. Then the critical height ratio, H_c/B, where the failure changes from a punching failure of the layered system to failure in the sand layer, can be found by writing that at that point the value of p_u for the layered system is equal to the $p_{u(sand)}$ value for the sand layer, which is given by an equation of the form of Eq. 17.7.

Other combinations of layered systems should be addressed by considering the most likely failure mechanism and using the procedure outlined in section 11.4.1 to obtain p_u. If several failure mechanisms are possible, p_u should be calculated for each one and the minimum value should be retained, because the soil will fail at the lowest failure load encountered.

17.6.4 Special Loading

Most of the solutions for ultimate bearing pressure presented so far have been for simple cases. However, shallow foundations can be more complex (Figure 17.17) including:

1. Influence of the foundation shapes (rectangular, square, circular, strip), i_ε
2. Influence of the depth of embedment, i_d
3. Influence of the load eccentricity, i_e
4. Influence of the load inclination, i_i
5. Influence of a nearby slope, i_β

An increase in the depth of embedment tends to increase the ultimate bearing pressure p_u, while the eccentricity, the inclination, and the slope presence tend to decrease p_u. In each case, an influence factor must be added in front of the equation for the base case. Such factors have been proposed for the pressuremeter method, the cone penetrometer method, and the general bearing capacity method. The influence factors for the cone penetrometer method are the same as the ones for the pressuremeter method.

Pressuremeter method. These factors are recommended by Frank (1999) and Norme Francaise AFNOR P94-261 (2013) and are as follows. Note that the influence of the foundation shape and of the depth of embedment are already included in the formulas for the bearing capacity factor k_p and k_c (Eqs. 17.13 to 17.18 and Eqs. 17.23 to 17.26). If the load applied to a $B \times L$ footing has an eccentricity e_B along the width B and e_L along the length L, the influence of the eccentricity is taken into account by using a rule attributed to Meyerhof. This rule consists of reducing the footing size as follows:

$$B' = B - 2e_B \quad and \quad L' = L - 2e_L \quad (17.49)$$

Then the design rules are applied to the reduced-size $B' \times L'$ footing, but the final recommendation is a $B \times L$ footing.

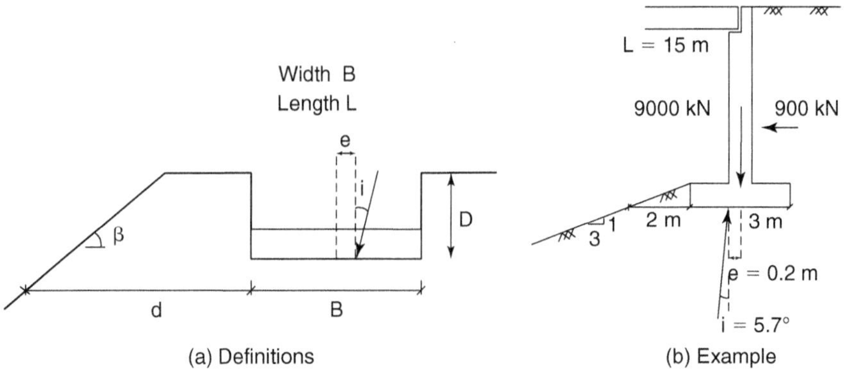

Figure 17.17 Complex loading cases for a shallow foundation.

If a footing is subjected to a centered inclined load making an angle α with the vertical, the influence factor i_i is given by Figure 17.18. Note that in Figure 17.18, the upper curve is for fine-grained soils, whereas the three lower curves are for coarse-grained soils and for three different relative depths of embedment D/B.

If a footing is located close to a slope and subjected to a centered vertical load, the presence of the slope reduces the ultimate bearing pressure. The influence factor i_β is given by Figure 17.19 as a function of d/B where d is the horizontal distance between the front edge of the bottom of the footing to the slope and B is the footing width. Each curve on Figure 17.19 corresponds to a slope angle β. Note that this figure corresponds to zero embedment depth. A simplified straight line relationship is also shown on Figure 17.19.

It is common practice to multiply the influence factors when several conditions are present at the same time.

General bearing capacity method. Several recommendations have been made for the influence factors to apply to the general bearing capacity equation. They take into account the foundation shape, the load eccentricity, the load inclination, and the presence of a nearby slope. They can be found in many manuals, including the Canadian foundation manual, the NAVFAC manual, the AASHTO bridge specifications, the API RP2A manual, the Norme Francaise AFNOR as presented by Frank (1999), and many others.

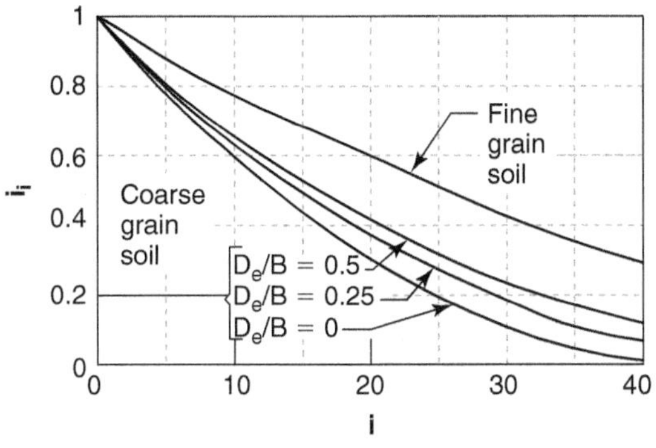

Figure 17.18 Influence of inclination. (After Frank, 1999)

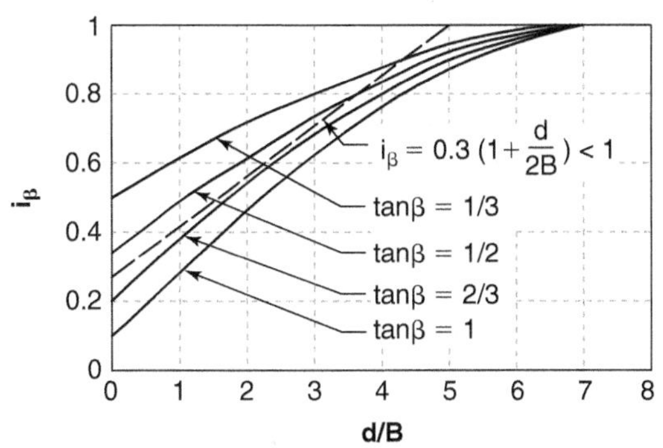

Figure 17.19 Influence of nearby slope. (After Frank, 1999)

The recommendations vary, but a review of these factors leads to the factors shown in Table 17.5, which represent reasonable averages. Note that there is a different factor for each of the three terms in Eq. 17.36. The subscript c is used for the cN_c term, the subscript γ is used for the term $0.5\gamma BN\gamma$, and the subscript q is used for the term γDN_q. Thus, the general formula is:

$$p_u = i_{cs}i_{ce}i_{ci}i_{c\beta}cN_c + i_{\gamma s}i_{\gamma e}i_{\gamma i}i_{\gamma\beta}\frac{1}{2}\gamma BN_\gamma + i_{qs}i_{qe}i_{qi}i_{q\beta}\gamma DN_q \quad (17.50)$$

17.6.5 Ultimate Bearing Capacity of Unsaturated Soils

Unsaturated soils and saturated soils with water in tension generally have higher ultimate bearing capacity p_u than the same soils with water in compression. Indeed, the water tension increases the effective stress and therefore the shear strength, which affects the value of p_u.

In the case of the direct equations, nothing changes because the change in strength is directly taken into account because the test itself takes the increase in strength into account. The PMT limit pressure, the CPT point resistance, the SPT blow count, and the undrained shear strength all reflect the impact of water tension on these soil parameters. Therefore, if one is using a direct method such as Eqs. 17.8, 17.11, 17.21, 17.22. 17.28, or 17.29, there is no need to change anything in the approach to be taken. Nevertheless, one must be aware of the fact that if the strength test is performed when the soil is very dry (high water tension), as is often the case in the summer, the predicted value of p_u will be high. If the soil loses that water tension in the winter, then the value of p_u will become much smaller. It is very possible for the water tension to vary significantly from one season to the next down to a depth of 3 m below the surface. Because shallow foundations are often placed within that depth, it is desirable to test the soil when it is in its wet state. If this is not possible, experience should be used from prior comparisons between summer and winter strength to reduce the strength accordingly before computing p_u.

In the case of the general bearing capacity equation, it is important to understand the role of each of the three terms. The first term, $c'N_c$, refers to the contribution made by the effective stress cohesion of the soil along the failure plane. The second term, $0.5\gamma BN\gamma$, refers to the contribution made by the friction along the failure plane due to the effective stress below the foundation but without a surcharge. The third term, γDN_q, refers to the contribution made by the friction along the failure plane due to the presence of the surcharge γD. It is relatively common practice to calculate the bearing capacity of soils with water tension (unsaturated or saturated) by increasing the cohesion c′ to include the apparent cohesion $c_{app} = \alpha\, u_w \tan\varphi$ in the value of c'. Then the equation is:

$$p_u = (c' - \alpha u_w \tan\varphi')N_c + \frac{1}{2}\gamma BN_\gamma + \gamma DN_q \quad (17.51)$$

Table 17.5 Influence Factors for the General Bearing Capacity Equation

	i_c for cN_c term	i_γ for $0.5\gamma BN_\gamma$ term	i_q for γDN_q term
Shape*	1 + 0.2 (B/L)	1 − 0.3(B/L)	1
Eccentricity	Meyerhof rule	Meyerhof rule	Meyerhof rule
Inclination**	$(1 - \alpha/90)^2$	$(1 - \tan\alpha)^{2.5}$	$(1 - \tan\alpha)^{1.5}$
Nearby slope***	0.3(1 + (d/2B))	0.3(1 + (d/2B))	0.3(1 + (d/2B))

*B is the footing width and L is the footing length
**α is the angle of inclination of the load
***For slope angles between 2 to 1 and 3 to 1, d is the horizontal distance from the footing edge to the slope, B is the footing width

This practice does not recognize the fact that the apparent cohesion is due to an increase in effective stress through the water tension and not to an increase in "glue" between the grains. It appears more appropriate to include this increase in effective stress in the second term. The expression $0.5\gamma B$ represents the effective stress σ'_{ov} for a "no water" condition at a depth of 0.5B below the foundation level in the case of no surcharge. This expression should be replaced by the effective stress at that same location but after consideration of the water tension. The bearing capacity for soils with water tension (unsaturated or saturated) would then be:

$$p_u = c'N_c + \frac{1}{2}(\gamma B - \alpha u_w)N_\gamma + \gamma DN_q \qquad (17.52)$$

Unfortunately, there are no known large-scale footing tests in which water tension was measured during a load test so as to provide verification for either approach. In any case, it is recommended that the direct method equations be used rather than the general bearing capacity equation, because the former methods are not restricted by the shape of the soil strength profile and have been extensively calibrated against footing load tests, particularly the PMT and CPT methods.

17.7 LOAD SETTLEMENT CURVE APPROACH

The design of a shallow foundation, much like the design of a deep foundation, is split into two steps. One addresses the ultimate bearing capacity, the other the movement at working loads. The load settlement curve (LSC) method (Jeanjean 1995; Briaud 2007) is used to predict the entire load settlement curve of the shallow foundation, rather than being limited to predicting only two points on that curve. It was developed in part after testing five large-scale footings (Figures 17.3 and 17.4). During these tests, inclinometer casings placed vertically at the edge of the footings gave the lateral deformation of the soil below the footings (Figure 17.5). These lateral deformation profiles never indicated that a plane of failure was developing as assumed in Figure 17.11. Instead, it showed that a "barreling" effect was progressively increasing in the same shape as the one created by the pressuremeter test. This is why the PMT curve was chosen as the curve to use and transform it into the footing load settlement curve. So, the LSC method is a way to transform the pressuremeter curve into the load settlement curve for a footing (Figure 17.20). During these large-scale tests, it was also observed that the normalized curve, plotted as pressure on the footing divided by the soil strength (PMT limit pressure) versus the settlement divided by the footing width, was independent of the footing size and essentially a unique property of the soil (Figure 17.4).

The transformation of the PMT curve into the footing curve is based on two equations as follows:

$$\frac{s}{B} = 0.24\frac{\Delta R}{R_o} \qquad (17.53)$$

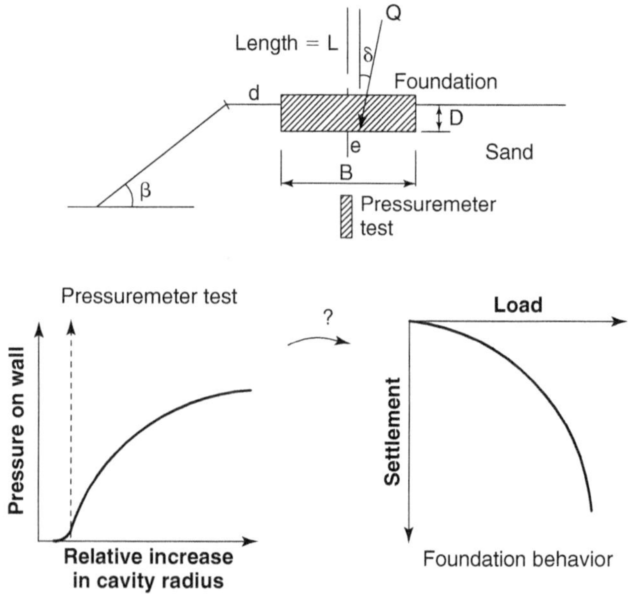

Figure 17.20 The load settlement curve (LSC) method. (Briaud 2007.)

$$p_f = f_{L/B} f_e f_\delta f_{\beta,d} \Gamma p_p \quad (17.54)$$

where s is the footing settlement, B is the footing width, R_o is the initial radius of the pressuremeter cavity, ΔR is the increase in cavity radius, p_f is the footing pressure corresponding to the settlement s, p_p is the pressuremeter pressure corresponding to $\Delta R/R_o$, and $f_{L/B}$, f_e, f_δ, and $f_{\beta,d}$ are the factors to include the influence of the footing shape, the load eccentricity, the load inclination, and the presence of a slope.

Equation 17.53 serves as a strain compatibility equation because it matches the strains at the ultimate values, which are $s/B = 0.1$ for the footing (a typical reference) and $\Delta R/R_o$ equal to 0.414 for the PMT (corresponding to the definition of the limit pressure). The value of 0.24 in Eq. 17.53 is the ratio of 0.1/0.414. In Eq. 17.54, Γ is a function of s/B (or 0.24 $\Delta R/R_o$), which represents the ratio between the footing pressure p_f and the PMT pressure p_p for the reference case of a centered vertical load on flat ground. Figure 17.21 shows the data from many sites used to generate the average Γ function and the design Γ function of Figure 17.22. The design Γ function is one standard deviation below the mean Γ function with respect to the data shown on Figure 17.21 and is recommended for design. The precision of the method can be gauged by the scatter on Figure 17.21.

The equations for the influence factors came mostly from numerical simulations (Hossain 1996; Briaud 2007):

$$\text{Shape} \quad f_{L/B} = 0.8 + 0.2\frac{B}{L} \quad (17.55)$$

$$\text{Load eccentricity} \quad f_e = 1 - 0.33\frac{e}{B} \quad \text{for the center} \quad (17.56)$$

$$\text{Load eccentricity} \quad f_e = 1 - \left(\frac{e}{B}\right)^{0.5} \quad \text{for the edge} \quad (17.57)$$

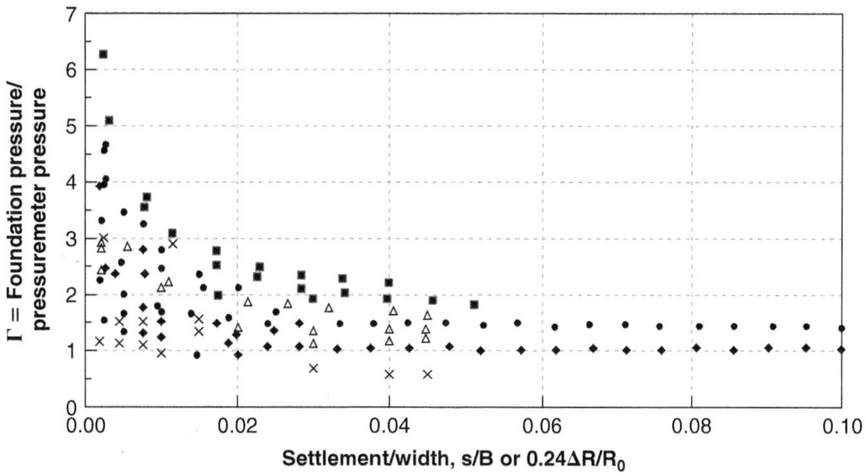

Figure 17.21 Data accumulated to generate the Γ function.

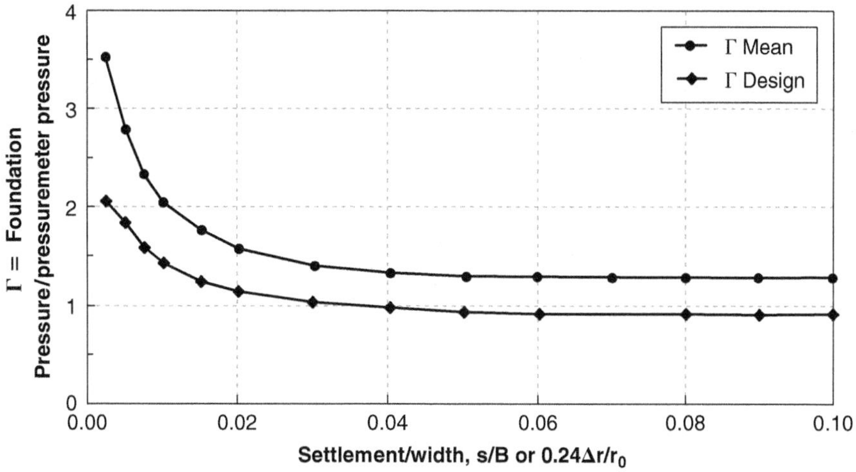

Figure 17.22 The Γ function for the load settlement curve method.

Load inclination $f_i = 1 - \left(\frac{i}{90}\right)^2$ for the center (17.58)

Load inclination $f_i = 1 - \left(\frac{i}{360}\right)^{0.5}$ for the edge (17.59)

Near a slope $f_{\beta,d} = 0.8\left(1 + \frac{d}{B}\right)^{0.1}$ for a 3 to 1 slope (17.60)

Near a slope $f_{\beta,d} = 0.7\left(1 + \frac{d}{B}\right)^{0.15}$ for a 2 to 1 slope (17.61)

where B and L are the footing width and length respectively, e is the load eccentricity, i is the inclination angle of the load, and d is the horizontal distance from the edge of the footing to the slope surface (Figure 17.17). The influence factor for the influence of a nearby slope is given for two common highway slopes: a 3 to 1 slope has a β angle with the horizontal of 18.4 degrees, and a 2 to 1 slope has a β angle with the horizontal of 26.6 degrees.

During the large footing tests discussed in section 17.2, the load was held for 30 minutes at each load level (Figure 17.3) and the settlement s was recorded as a function of time t. Figure 17.3 shows the relationship between the log of settlement and the log of time for each load step. The settlement $s(t)$ is normalized by the settlement value at the beginning of that load step $s(t_1)$ and the time t is normalized by a time t_1 equal to one minute. As can be seen from Figure 17.3, the relationship is linear in the log space; therefore, the model is a power model with an exponent n equal to the slope of the line in the log space:

$$\frac{s(t)}{s(t_1)} = \left(\frac{t}{t_1}\right)^n \quad (17.62)$$

The exponent n can be measured in a pressuremeter test where the pressure is held at an appropriate pressure level while the relative increase in radius $\Delta R/R_o$ is recorded as a function of time t. Equation 17.62 is then applied to the PMT data and n is back-calculated. The n value tends to be between 0.01 and 0.03 for sands and between 0.02 to 0.05 for stiff to hard clays.

The step-by-step procedure for the load settlement curve method is as follows:

1. Perform preboring pressuremeter tests within the zone of influence of the footing.
2. Plot the PMT curves as pressure p_p on the cavity wall versus relative increase in cavity radius $\Delta R/R_o$ for each test. Extend the straight-line part of the PMT curve to zero pressure and shift the vertical axis to the value of $\Delta R/R_o$ where that straight line intersects the horizontal axis; re-zero that axis. This is done to correct the origin for the initial expansion of the pressuremeter to allow it to come into contact with the borehole wall.
3. Develop the mean pressuremeter curve of all the PMT curves within the depth of influence of the footing. To do so, choose a value of $\Delta R/R_o$ and average the corresponding pressures of all the PMT curves; in doing so, give more weight to the shallower PMT curves, which will have more influence on the settlement than the deep PMT curves (Briaud 2007).
4. Transform the PMT curve point by point into the footing curve by using Eqs. 17.53 and 17.54.
5. Generate the short-term load settlement curve for the footing from the normalized curve.
6. Generate the long-term load settlement curve by multiplying all settlement values by the factor $(t/t_1)^n$ where t is the design life of the structure, t_1 is 1 hr, and n is the time exponent obtained from PMT tests or set equal to 0.03 as the default value.

Figure 17.23 is an example of the LSC method.

17.8 SETTLEMENT

17.8.1 General Behavior

Once the ultimate bearing capacity has been calculated and once the dimensions of the footing have been established such that the ultimate limit state (safety criterion) is satisfied, the settlement under the foundation pressure is calculated. This is the *service limit state*. Typically in this case, the load factors and resistance factors are taken as equal to 1. The nonpermanent live loads are not included in the loads considered for calculating settlements that take a long time to develop, such as consolidation settlements in saturated clays. The settlement of a structure is often decomposed into an elastic component (elastic settlement), then a time-delayed component associated with water stress dissipation (consolidation), then a time-delayed component associated with the slow movement of particles as a function of time (creep settlement). The settlement of a structure can also be decomposed into the settlement induced by the deviatoric stress tensor (shearing) and by the spherical stress tensor (compression). In cases where the settlement is concentrated in a thin (relative to the width of the foundation) layer, the settlement due to the spherical part of the tensor dominates. This would be the case of a wide embankment on top of a thin layer of soft clay. If, in contrast, the soil layer is deep (relative to the width of the foundation), the settlement due to the deviatoric tensor dominates. This would be the case of a tall building on top of a mat foundation underlain by a deep deposit of very stiff clay.

The pressure distribution under a shallow foundation depends on the flexibility of the foundation (Figure 17.24). For flexible foundations, the pressure is constant but the settlement is not. The settlement at the center $s_{flex\ (center)}$ is

Problem: A bridge abutment rests on a shallow foundation 15 m long and 3 m wide. The foundation is subjected to a vertical and centered load equal to 9000 kN. The lateral earth pressure generates a load of 900 kN on the back of the abutment. The resultant of the two forces has an eccentricity equal to 0.2 m. The soil is a sand characterized by the average pressuremeter curve shown.

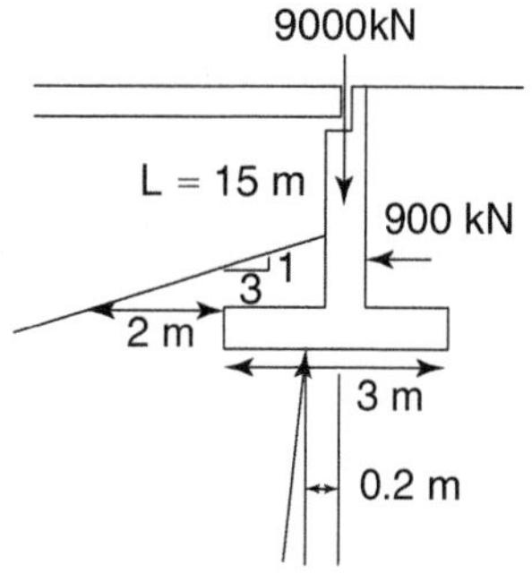

Solution Load Settlement Curve

$f_{L/B} = 0.8 + 0.2x3/15 = 0.84$

$f_e = 1-0.33x0.2/3 = 0.978$

$f_\delta = 1-$ (Arctan(900/9000) / 90)^2
$= 0.996$

$f_{\beta,d} = 0.8x(1+2/3)^{0.1} = 0.842$

$f = f_{L/B} f_e f_\delta f_{\beta,d} = 0.689$

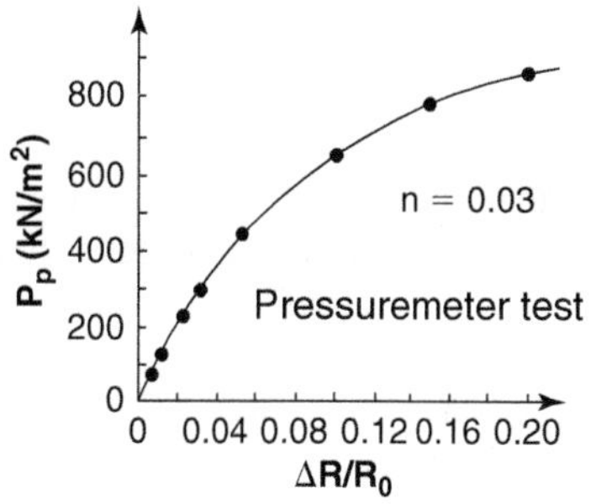

$\Delta R/R_0$	P_p (kN/m²)	s/B	s (mm)	Γ	f	P_f (kN/m²)	Q (MN)
0.006	75	0.00144	4.32	2.25	0.689	116.3	5.23
0.012	120	0.00288	8.64	2.00	0.689	165.4	7.44
0.024	220	0.00576	17.28	1.60	0.689	242.5	10.91
0.032	300	0.00768	23.04	1.50	0.689	310.0	13.95
0.055	450	0.0132	39.6	1.30	0.689	403.1	18.14
0.10	650	0.0240	72.0	1.10	0.689	492.6	22.17
0.15	775	0.036	108	1.00	0.689	534.0	24.03
0.20	850	0.048	144.0	0.95	0.689	556.4	25.04

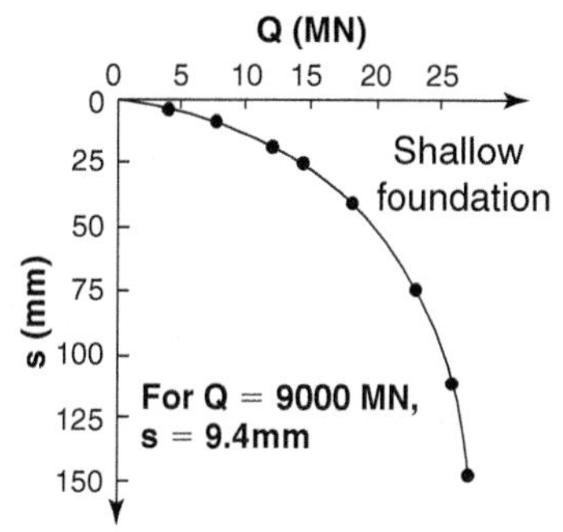

Figure 17.23 Example of the load settlement curve method.

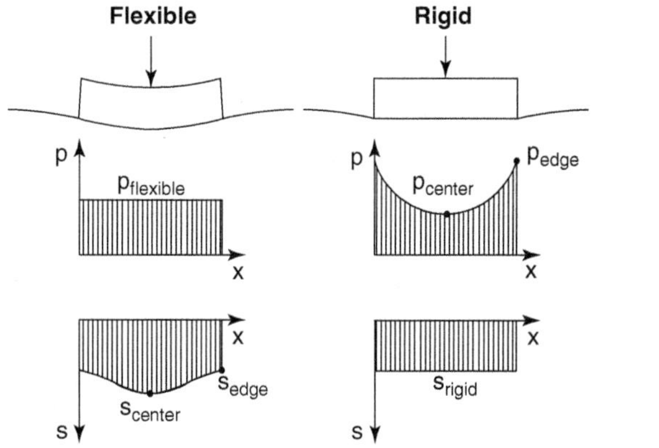

Figure 17.24 Settlement and pressure distribution below footings.

larger than the settlement at the edge $s_{flex(\text{edge})}$. For rigid foundations, the settlement s_{rigid} is constant but the pressure is not—at least initially. The following is an approximate relationship between the settlements:

$$s_{flex\ (center)} \simeq 2s_{flex\ (edge)} \simeq 1.33s_{rigid} \tag{17.63}$$

In other words, the settlement at the center of a flexible footing is about twice as large as the settlement at the edge of a flexible footing, and the settlement of a rigid footing is about the average of the center and the edge of a flexible footing. These observations are based on the theory of elasticity. Also in elasticity, the pressure near the edge of a rigid footing is very large and the pressure in the center of that footing is much smaller (Figure 17.24); in fact, it is about one-half the mean pressure. As will be discussed in section 17.8.7, the soil tends to develop a constant pressure under the foundation in the long term even if the foundation is very rigid.

There are a number of methods for performing settlement calculations:

1. Elasticity approach
2. Load settlement curve method (see section 17.7)

3. Chart approach
4. General layered soil approach
5. Consolidation settlement approach

17.8.2 Elasticity Approach for Homogeneous Soils

Soils are not elastic, as they do not recover all the deformation they experience when strained. Soils are not linear either, as their stress-strain curve are not a straight line. Nevertheless, if a foundation is loaded with a certain load Q and experiences a settlement s as a result, there is always a modulus E that, when combined with Q, can give the right s value. The elasticity equations also have a significant advantage in that they are simple to use. They have a drawback in that they require a very sensible and often very difficult choice of the soil modulus. The best way to obtain the modulus is to run a test that closely reproduces what the soil will be subjected to under the structure. At the preliminary design stage, one may wish to use estimated values as presented in Chapter 14.

The equation for the elastic settlement s of a shallow foundation is:

$$s = I(1 - \nu^2)\frac{pB}{E} \tag{17.64}$$

where I is an influence factor for any deviation from a footing on the ground surface subjected to a centered vertical load; ν is Poisson's ratio, usually taken as 0.35 for drained conditions and 0.5 for undrained conditions; p is the average pressure at the foundation level; B is the width of the foundation; and E is the soil modulus of deformation. The factor I can be written as:

$$I = I_s I_e I_h \tag{17.65}$$

where I_s is the factor for the influence of the shape of the footing, I_e is the factor for the influence of the embedment depth, and I_h is the factor for the presence of a hard layer at depth. Table 17.6 gives the values of I_s and shows that the strip footing settles a lot more than the square footing.

Table 17.6 Values of the Elastic Influence Factor I_s for Foundation Shape

		Influence Factor for Shape, I_s.		
	Length-to-Width		Flexible	
Shape	Ratio L/B	Rigid	Center	Corner
Circular	1	0.79	1	0.64
	1	0.88	1.12	0.56
	1.5	1.07	1.36	0.68
Rectangular	2	1.21	1.53	0.77
	3	1.42	1.78	0.89
	5	1.7	2.1	1.05
	10	2.1	2.54	1.27

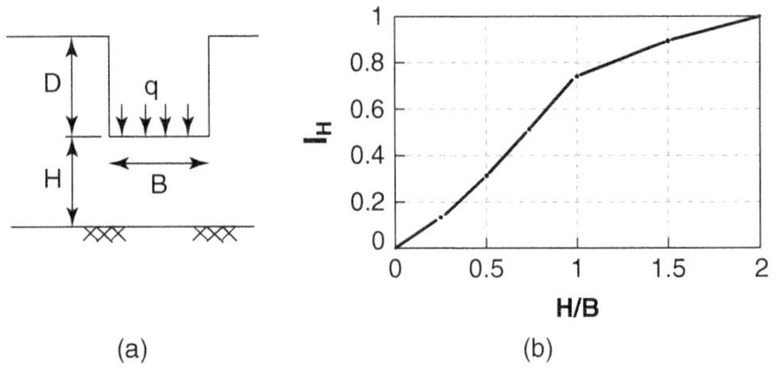

Figure 17.25 Influence factor for hard layer within depth of influence: (a) Hard layer at depth H. (b) Influence factor I_h. (After Christian and Carrier 1978)

This is due to a much larger depth of influence for the strip footing compared to the square footing. The factor I_e reduces settlement compared to a surface footing, because of the beneficial effect of having more mass to deform for a deeper footing than a shallower footing. The factor I_e can be estimated for footings with a relative depth of embedment (D/B) less than 1 (shallow foundations) by:

$$I_e = 1 - 0.1\frac{D}{B} \tag{17.66}$$

The maximum reduction for larger values of D/B is 15% $(I_e = 0.85)$. The factor I_h is a reduction factor when there is a hard layer within the depth of influence of the footing. Figure 17.25 gives the values of I_h when it is assumed that beyond 2B the hard layer has no reduction influence on the settlement and that the hard layer is incompressible.

The previous method assumes that the soil has a modulus which is constant with depth. If the soil has a modulus profile that increases linearly with depth (Figure 17.26), a correction factor I_G can be used. The equation for the soil modulus profile is:

$$E = E_o + E_1\left(\frac{z}{B}\right) \tag{17.67}$$

where E is the soil modulus at a depth z, E_o is the soil modulus at the ground surface, and E_1 is the rate of increase of the soil modulus as a function of the normalized depth (z/B). The influence factor I_G takes the modulus profile into account and is defined as:

$$I_G = \frac{s_1}{s_o} \tag{17.68}$$

where s_1 is the settlement calculated using E from Eq. 17.67 and s_o is the settlement calculated from Eq. 17.64 using a constant modulus E_o with depth $(E_1 = 0)$. Figure 17.26 shows the influence factor I_G as a function of the ratio E_1/E_o.

17.8.3 Elasticity Approach for Layered Soils

Another way to use elasticity to solve a settlement problem is to decompose the depth of influence z_i into several soil layers H_i thick and calculate the compression ΔH_i of each

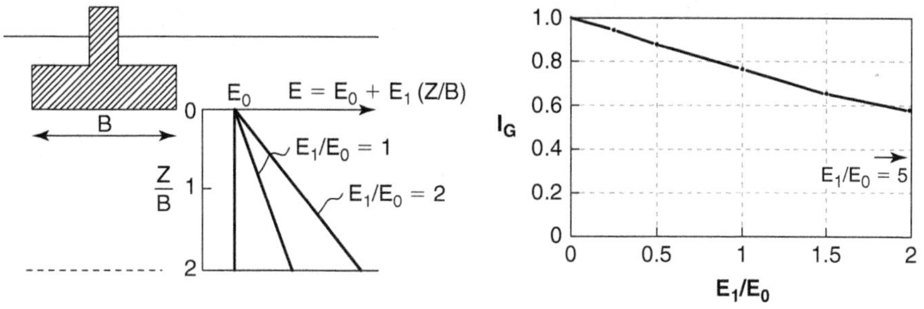

Figure 17.26 Influence of modulus increase vs. depth (Gibson soil).

layer. The vertical strain in each layer is ε_i and is related to the increase in stress $\Delta\sigma_i$ in the middle of that layer. The settlement s is expressed as:

$$s = \sum_{i=1}^{n} \Delta H_i = \sum_{i=1}^{n} \varepsilon_i H_i = \sum_{i=1}^{n} \frac{\Delta\sigma_i}{E_i} H_i \tag{17.69}$$

where s is settlement, n is the number of layers within the depth of influence, i refers to the ith layer, H_i is the thickness of the ith layer, ΔH_i is the compression of the ith layer, ε_i is the mean vertical strain of the ith layer, and $\Delta\sigma_i$ is the increase in stress in the center of the ith layer. Equation 17.69 assumes that the relationship between ε_i and $\Delta\sigma_i$ is given by:

$$\varepsilon_i = \frac{\Delta\sigma_i}{E_i} \tag{17.70}$$

This relationship ignores the influence of confinement on the strain and therefore is an approximation. This assumption is conservative, as taking the confinement into account would reduce the strain. How to obtain the magnitude of $\Delta\sigma_i$ in the middle of each layer is discussed in section 17.8.7.

Schmertmann (1970; 1978) used this approach and proposed a method to calculate the settlement s of footings on sand:

$$s = C_1 C_2 \Delta p \sum_{i=1}^{n} \frac{I_{zi}}{E_i} H_i \tag{17.71}$$

where C_1 takes into account the beneficial effect of the embedment, C_2 takes into account the increase in settlement with time, Δp is the net bearing pressure expressed as the difference between the footing pressure p (load over area) minus σ'_{ov} (the vertical effective stress in the soil at the level of the foundation near the footing), I_{zi} is called the strain influence factor, E_i is the soil modulus, and H_i is the thickness of the ith layer. The coefficient C_1 is:

$$C_1 = 1 - 0.5\frac{\sigma'_{ov}}{\Delta p} \geq 0.5 \tag{17.72}$$

where σ'_{ov} is the vertical effective stress in the soil at the level of the foundation near the footing, and Δp is the net increase in pressure expressed as the difference between the footing pressure p (load over area) minus σ'_{ov}. The coefficient C_2 is:

$$C_2 = 1 + 0.2\log\left(\frac{t\ (years)}{0.1}\right) \tag{17.73}$$

where t is the time in years.

The strain influence factor I_{zi} is such that $I_{zi} \times \Delta p$ represents $\Delta\sigma_i$ in Eq. 17.69. It is shown in Figure 17.27. In that figure, I_z increases first and then decreases. The peak value

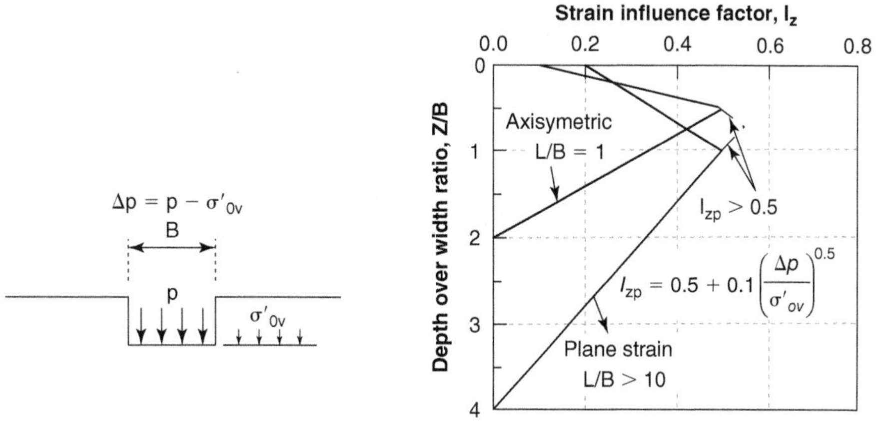

Figure 17.27 Strain influence factor. (After Schmertmann 1970)

Table 17.7 Conversion from CPT to SPT Values for Sands

Soil	q_c (kPa)/N (bpf)
Silts, sandy silts, slightly cohesive silt-sand	200
Clean, fine to medium sands and slightly silty sands	350
Coarse sands and sands with little gravel	500
Sandy gravel and gravel	600

of I_z is called I_{zp}. It is shown as 0.5 on Figure 17.27 but in fact it is given by:

$$I_{zp} = 0.5 + 0.1\left(\frac{\Delta p}{\sigma'_{Izp}}\right)^{0.5} \tag{17.74}$$

where σ'_{Izp} is the vertical effective stress at the location of I_{zp}. The soil modulus E_i is recommended by Schmertmann as follows:

For circular or square footings $E = 2.5q_c$ (17.75)

For strip footings (L/B > 10) $E = 3.5q_c$ (17.76)

where q_c is the CPT point resistance. Schmertmann adds the conversion values of Table 17.7 between q_c and N.

17.8.4 Chart Approach

The chart approach consists of simplifying the problem sufficiently so that the calculations are minimized and a chart can be read for the answer. Such a chart approach was developed by Terzaghi and Peck (1963) for footings on sand (Figure 17.28). This chart is only for footings on sands, and it gives the pressure that satisfies both the ultimate bearing pressure criterion and the settlement criterion of 25 mm. This chart was developed before LRFD was developed and as such

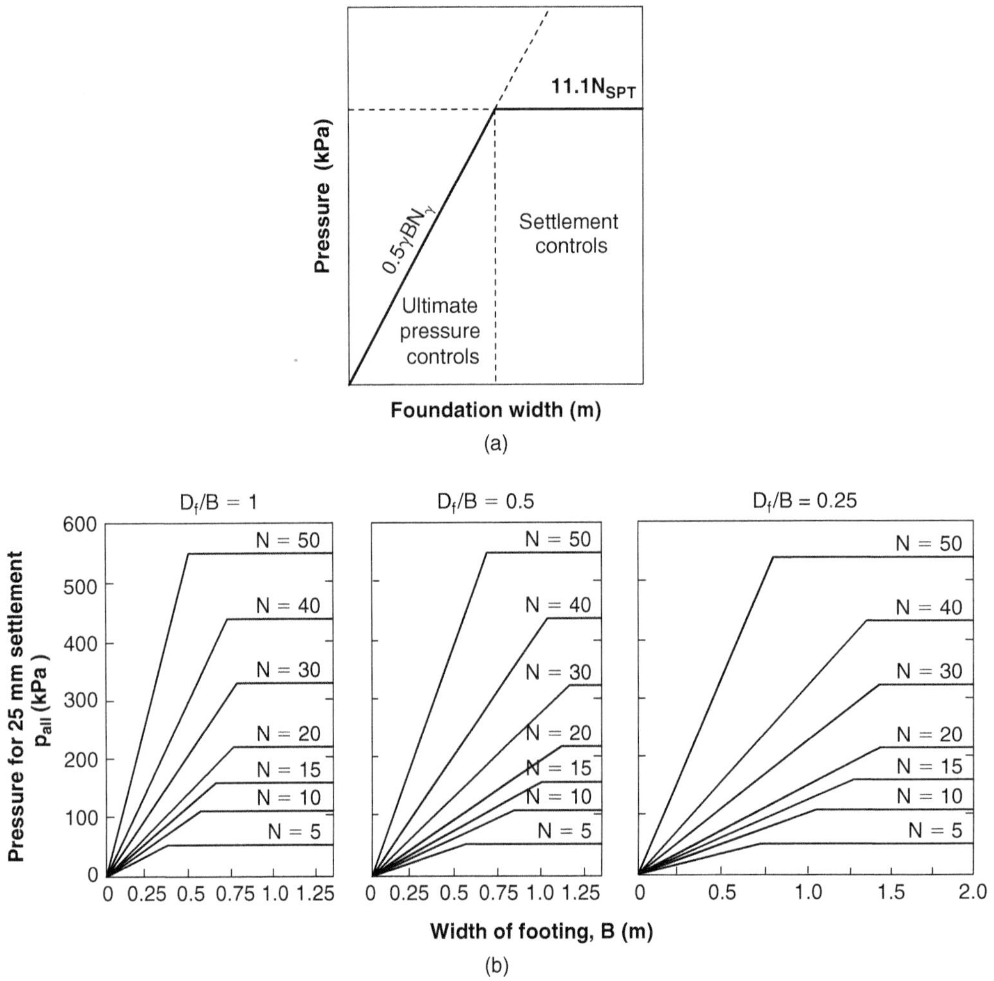

Figure 17.28 Chart for pressure leading to 25 mm settlement of footings on sand. (Terzaghi and Peck 1963)

is based on the following. The safe pressure criterion ensures that a reasonable factor of safety is applied to the ultimate bearing pressure:

$$p_{safe} = \frac{p_u}{F} \tag{17.77}$$

where p_{safe} is the safe bearing pressure, p_u is the ultimate bearing pressure, and F is the factor of safety. The allowable pressure criterion ensures that the settlement will be less than 25 mm in this case:

$$p_{allowable} = p \; for \; 25 \text{ mm } settlement \tag{17.78}$$

The chart of Figure 17.28 gives the minimum of p_{safe} and $p_{allowable}$. The first part of the design curves on the chart increases linearly with the width B of the footing for the following reason. For small values of B, it turns out that the ultimate bearing pressure criterion controls the design, and since there is no cohesion for sands, it is expressed as:

$$p_u = \frac{1}{2}\gamma B N_\gamma \tag{17.79}$$

As a result, p_{safe} increases linearly with B. The influence of the depth of embedment D is included by having several charts for different relative depths of embedment D/B. For the settlement s of the footing, Terzaghi and Peck found that s was proportional to the SPT blow count as follows:

$$p_{allowable}(\text{kPa}) \; for \; 25 \text{ mm } settlement = 11.1N(blows/0.30 \text{ m}) \tag{17.80}$$

This indicates that $p_{allowable}$ is not a function of B and therefore it shows up as a horizontal line on Figure 17.28. As a result, the ultimate pressure criterion controls for small footings and the settlement criterion controls for larger footings. If Eq. 17.80 is extended to other settlement values, and assuming linear behavior, the equation becomes:

$$s \text{ (mm)} = 2.3\frac{p(\text{kPa})}{N(\text{bpf})} \tag{17.81}$$

17.8.5 General Approach

The general approach to calculating the settlement of a structure is valid in all cases and proceeds as follows:

1. Determine the depth of influence z_i.
2. Divide that depth into an appropriate number n of layers (4 is a minimum), each layer being H_i thick.
3. Calculate the vertical effective stress σ'_{ovi} in the middle of each layer i before any load is applied.
4. Calculate the increase in stress $\Delta\sigma_{vi}$ in the middle of each layer i due to load.
5. Calculate the vertical effective stress $\sigma'_{ovi} + \Delta\sigma_{vi}$ in the middle of each layer i long after loading.
6. Obtain the vertical strain ε_{bi} before any load is applied, corresponding to the stress σ'_{ovi}.
7. Obtain the vertical strain ε_{ai} long after the load application corresponding to the stress $\sigma'_{ovi} + \Delta\sigma_{vi}$.
8. Calculate the compression ΔH_i of each layer i as:

$$\Delta H_i = (\varepsilon_{ai} - \varepsilon_{bi})H_i \tag{17.82}$$

9. Calculate the settlement ΔH as:

$$\Delta H = \sum_{i=1}^{n} \Delta H_i = \sum_{i=1}^{n} (\varepsilon_{ai} - \varepsilon_{bi})H_i \tag{17.83}$$

This general approach requires some other steps, which are addressed in the next sections. These steps are where one determines the zone of influence z_i (step 1), finds the increase in stress $\Delta\sigma_v$ (step 4), and obtains the strains ε_{bi} and ε_{ai} given the stresses σ'_{ov} and $\sigma'_{ov} + \Delta\sigma_v$ (steps 6 and 7).

17.8.6 Zone of Influence

The zone of influence z_i below a loaded area can be defined in one of two ways:

1. The depth at which the stress increase in the soil $\Delta\sigma_v$ has decreased to 10% of the stress increase p at the foundation level. This depth is called $z_{0.1\sigma}$.
2. The depth at which the downward movement of the soil becomes equal to 10% of the downward movement at the surface. This depth of influence is called $z_{0.1s}$.

Although the stress-based definition is the most commonly used in geotechnical engineering, the movement-based definition seems more reasonable because it ensures that 90% of the settlement is being calculated. Multiplying the answer by 1.11 will then give the full value of settlement.

The value of $z_{0.1\sigma}$ is typically taken as 2 times the width B of the footing for square and circular footings, and as 4 times the width B of the footing for long strip footings. These values are based on the elastic analysis of a uniform soil. Interpolation based on the ratio of width over length (B/L) is used for rectangular footings:

$$\frac{z_{0.1\sigma}}{B} = 4 - 2\left(\frac{B}{L}\right) \tag{17.84}$$

The value of $z_{0.1s}$ is the same as the value of $z_{0.1\sigma}$ if the soil modulus is constant with depth. If the soil modulus increases with depth, the value of $z_{0.1\sigma}$ does not change, but that of $z_{0.1s}$ does. The increase in modulus with depth is characterized by Eq. 17.67. Figure 17.29 shows the variation of $z_{0.1s}$ for a strip footing and for various values of the increase in modulus with depth characterized by E_1/E_o. For very small values of E_1/E_o (constant modulus with depth), the value of 4B is confirmed, but for high values of E_1/E_o the zone of influence based on settlement criterion $z_{0.1\sigma}$ is much smaller than $z_{0.1s}$. It decreases from 4B to 1B and reaches 1B for a modulus which is zero at the surface and increases linearly with depth. This phenomenon is explained as follows. When the soil is uniform and the soil modulus is constant with depth, the zone of influence is relatively deep (4B). At the other extreme,

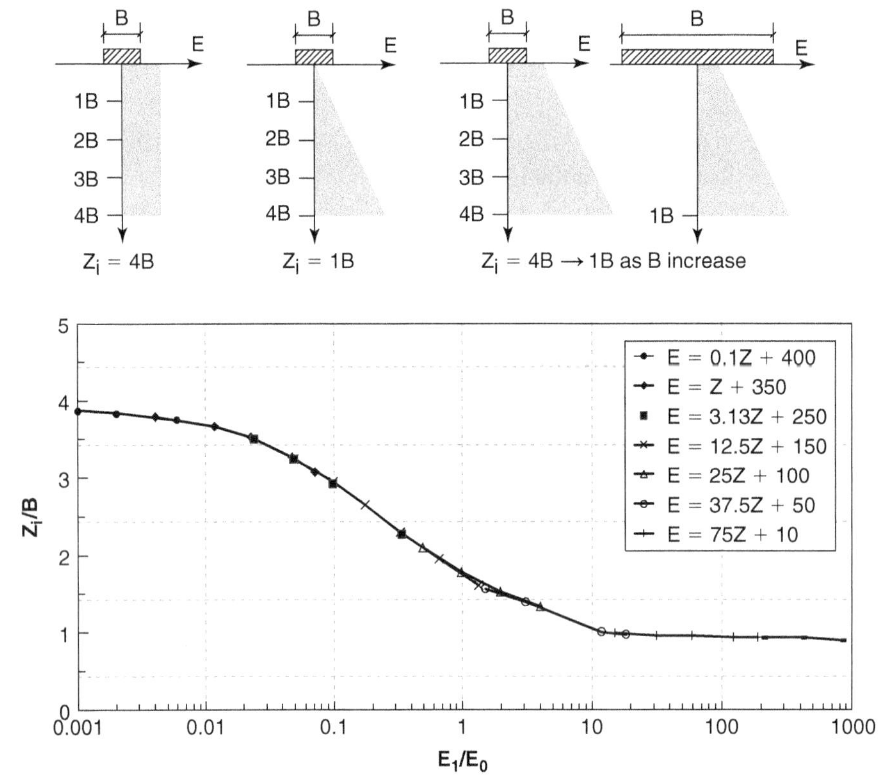

Figure 17.29 Zone of influence based on settlement criterion. (Briaud et al. 2007)

when the soil modulus increases with depth from a value of zero at the surface, the deeper layers are stiffer and do not compress as much as the shallower layers, which are softer. As a result, 90% of the settlement takes place within a much shallower depth and $z_{0.1s}$ is only 1B. For intermediate modulus profiles, the depth of influence depends on B and varies from 4B for small B values to 1B for large B values.

The procedure for finding the zone of influence below a foundation based on the settlement criterion using Figure 17.29 is as follows:

1. Fit the soil modulus profile with a straight line.
2. Determine the ratio E_1/E_o.
3. Obtain the depth of influence $z_{0.1s}$ from Figure 17.29 knowing the footing width B.
4. Calculate the settlement within that depth.
5. Multiply the answer by 1.11 to obtain the total settlement.

This approach has not been verified at full scale and is based solely on numerical simulations.

17.8.7 Stress Increase with Depth

2 to 1 method: One simple way to calculate the increase in stress below a foundation is the 2 to 1 method. This method consists of spreading the load with depth, as shown in Figure 17.30. The foundation is B wide and L long and is subjected to a load Q. At a depth z, the area over which the load is applied is increased by $z/2$ on both sides and becomes

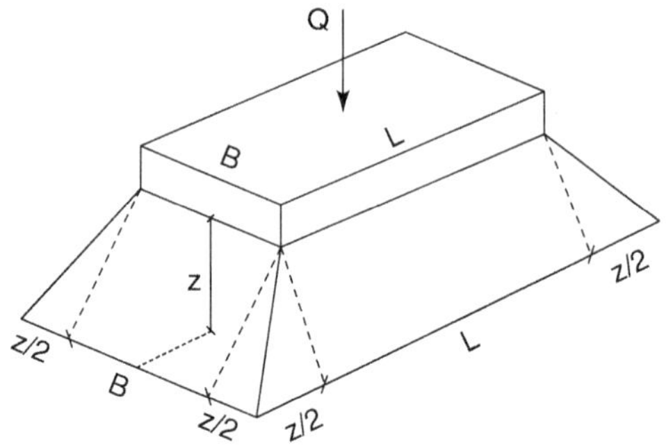

Figure 17.30 2 to 1 method for stress increase calculations.

$B + z$ and $L + z$. The average increase in stress at depth z is given by:

For a rectangular foundation $$\Delta\sigma_v = \frac{Q}{(B+z)(L+z)} \tag{17.85}$$

If the foundation is circular, then the diameter is increased by $z/2$ all around and the average increase in stress is given by:

For a circular foundation $$\Delta\sigma_v = \frac{4Q}{\pi(D+z)^2} \tag{17.86}$$

If the foundation is infinitely long (strip footing or embankment, for example), the load is defined as a line load (kN/m).

Furthermore, the load cannot spread in the direction of the length L, so the expression becomes:

$$\text{For a strip foundation} \quad \Delta\sigma_v = \frac{Q}{B+z} \tag{17.87}$$

Note that this method aims at estimating the average increase in vertical stress under the foundation. In elasticity, the increase in stress at the edge of a rigid foundation is different from the increase in stress at the center.

Bulbs of pressure: A more precise way to obtain the increase in stress at depth is the bulb of pressure chart (Figure 17.31). This chart gives the increase in stress below a square foundation and below a strip foundation for a uniform elastic soil. By using this chart, you can get the increase in stress at any location in the soil mass in the vicinity of the foundation. It is particularly useful for obtaining the increase in stress at the edge and at the center of the footing because this difference can affect the distortion of the foundation. Note that Figure 17.31 is for a flexible foundation where the pressure is uniform at the foundation level. Although foundations can be rigid, using the flexible solution in all cases is recommended for the following reason. Full-scale measurements (Focht, Khan, and Gemeinhardt 1978) indicate that the initially uneven pressure distribution under relatively rigid foundations redistributes itself and becomes close to the constant pressure under a flexible footing. This is attributed to the inability of the soil to sustain a large stress gradient for a long period of time. Therefore, the long-term settlement

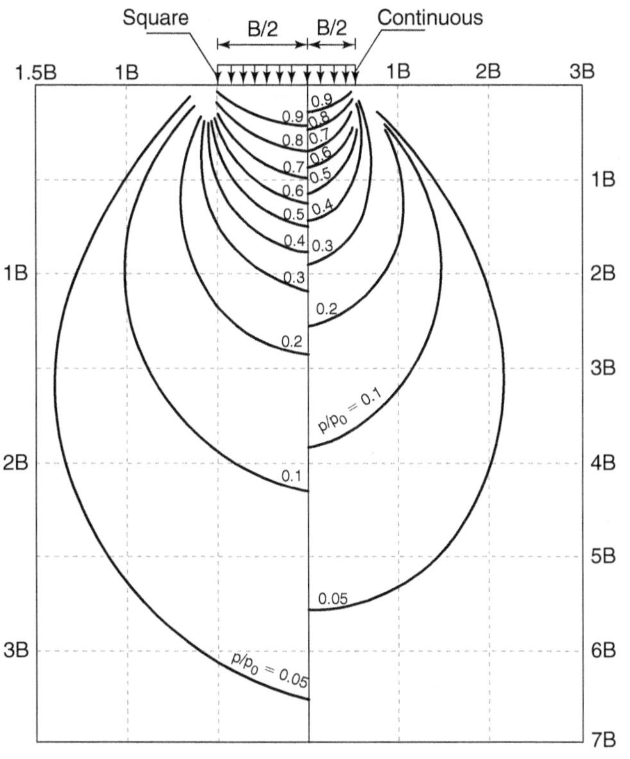

Figure 17.31 Bulbs of pressure based on Boussinesq elastic solution for a flexible foundation.

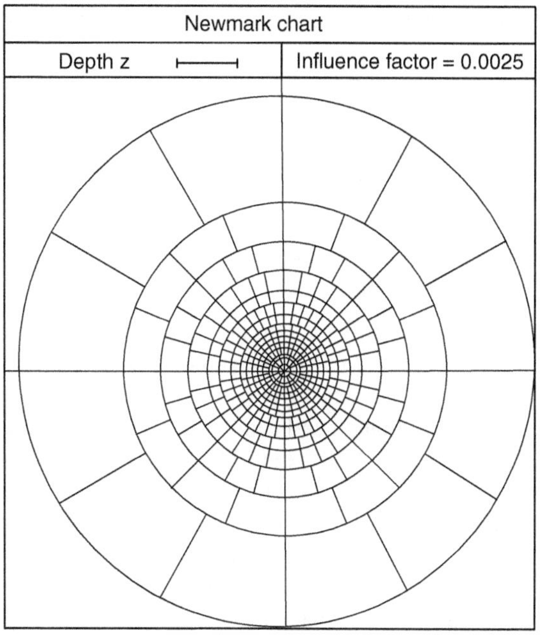

Figure 17.32 Newmark's chart.

of a foundation should be calculated on the basis of the stress distribution below a flexible distribution, regardless of whether the foundation is actually rigid or flexible.

Newmark's chart: The bulbs of pressure method gives the increase in stress under a square foundation or a strip foundation. If the foundation is more complicated, one possible solution is Newmark's chart (Figure 17.32). This chart is also for a uniform elastic soil and gives the increase in stress at any location below a foundation of any shape.

The best way to use the Newmark's chart is to make a transparency of the chart. If you do not have a transparency, then the drawing of the foundation has to be made on the original Newmark's chart, making it hard to reuse that chart. The transparency allows you many uses.

The procedure to use Newmark's chart is as follows:

1. Choose the depth z at which the stress increase $\Delta\sigma_v$ is required. Set the scale on the Newmark's chart (AB on Figure 17.32) equal to z. This gives the scale to be used for step 2.
2. Draw the foundation to the scale determined by z = AB.
3. Choose the point C in plan view on the foundation drawing under which $\Delta\sigma_v$ is required.
4. Overlay the transparent Newmark's chart on the foundation drawing such that point C of the foundation drawing is at the center of the Newmark's chart.
5. Count the number n of squares or fields of the Newmark's chart covered by the foundation drawing.
6. Calculate the increase in stress $\Delta\sigma_v$ as:

$$\Delta\sigma_v = n\, I\, p \tag{17.88}$$

where n is the number of squares, I is the influence factor of the Newmark's chart indicated on the chart, and p is the mean foundation pressure.

All of the solutions described in this subsection are limited to a uniform soil. If a layered soil such as a pavement is involved, or if the modulus is not constant with depth, the finite element method may be a good solution for finding the increase in stress with depth.

17.8.8 Choosing a Stress-Strain Curve and Setting Up the Calculations

Possibly the most difficult step in settlement calculations is to select the best and most applicable stress-strain curve to link the stress increment to the strain increment. This step requires a lot of thought and engineering judgment based on experience. Sections 17.8.2 and 17.8.3 described the elasticity approach, in which the stress-strain curve is a straight line and the modulus E is used to define the slope. Choosing such a modulus is a very difficult task; the content of Chapter 14 can help in that respect. The best and most applicable stress-strain curve is usually the one that most closely duplicates what the soil is being subjected to in the field, before and during the construction and then during the life of the structure. This includes the stress path, the strain path, the weather, the soil profile, the load level, and many more factors.

In general, if the structure is wide compared to the thickness of the compressing soil layer, then a test such as the consolidation tests will duplicate the soil deformation process quite closely. This would be true in the case of a wide embankment on a relatively thin, compressible layer, for example, because in this instance the friction between the embankment and the soil generates a natural resistance against lateral expansion of the soil much like the steel ring in the consolidation test. In contrast, if the soil deposit is deep compared to the width of the structure, then a test such as the pressuremeter test or the triaxial test duplicates the soil deformation process quite closely. Indeed, in this instance the soil below the foundation tends to barrel out in the same fashion as the soil around the pressuremeter or in the triaxial test.

Once the stress-strain curve is chosen, the strains corresponding to the stresses can be determined. The strains before and long after the loading (ε_{bi} and ε_{ai}) are obtained from the curve for the corresponding stresses σ'_{ov} and $\sigma'_{ov} + \Delta\sigma_v$. Note that although in the field the strain ε_{bi} corresponding to σ'_{ov} was zero, it is unlikely to be zero when reading the stress-strain curve. This is attributed to possible disturbance and stress relief upon extrusion. The calculations are then set up in the form of a spreadsheet, as shown in Table 17.8, where i is the layer number, H_i is the thickness of layer i, σ'_{ovi} is the vertical effective stress in the middle of layer I before loading, $\Delta\sigma_{vi}$ is the increase in stress in the middle of layer i due to loading, ε_{bi} is the vertical strain corresponding to σ'_{ovi}, ε_{ai} is the vertical strain corresponding to $\sigma'_{ovi} + \Delta\sigma_{vi}$, and ΔH_i is the compression of layer i. The sum of the last column in the table corresponds to Eq. 17.83 and represents the settlement.

Table 17.8 Calculation of Settlement by the General Approach

i	H_i (m)	σ'_{ovi} (kPa)	$\Delta\sigma_{vi}$ (kPa)	$\sigma'_{ovi} + \Delta\sigma_{vi}$ (kPa)	ε_{bi}	ε_{ai}	ΔH_i
1							
2							
3							
4							

17.8.9 Consolidation Settlement: Magnitude

As pointed out in section 17.8.8, the consolidation test is well suited to predicting the settlement of structures when most of the settlement is due to vertical compression and very little is due to lateral deformation. This limited horizontal movement can be created by the friction on the top and the bottom of a thin layer under a wide load. In this case the consolidation test curve can be used as the stress-strain curve in the general method, and the strains can be obtained by reading the curve for the corresponding stresses (Figure 17.33). It is recommended that you read the curve directly rather than reconstructing an undisturbed curve. This should be an incentive for obtaining quality samples, as disturbance is likely to increase the settlement prediction. Note that the consolidation curve is made of points corresponding to equilibrium points at the end of the 24-hour test period under each load step. The settlement is then calculated as:

$$\Delta H = H_o \frac{(e_{before} - e_{after})}{1 + e_o} = H_o(\varepsilon_{after} - \varepsilon_{before}) \quad (17.89)$$

Or, when the compressing zone is divided into several layers:

$$\Delta H = \sum_{i=1}^{n} H_{oi} \frac{(e_{i\ before} - e_{i\ after})}{1 + e_{oi}} = \sum_{i=1}^{n} H_{oi}(\varepsilon_{i\ after} - \varepsilon_{i\ before}) \quad (17.90)$$

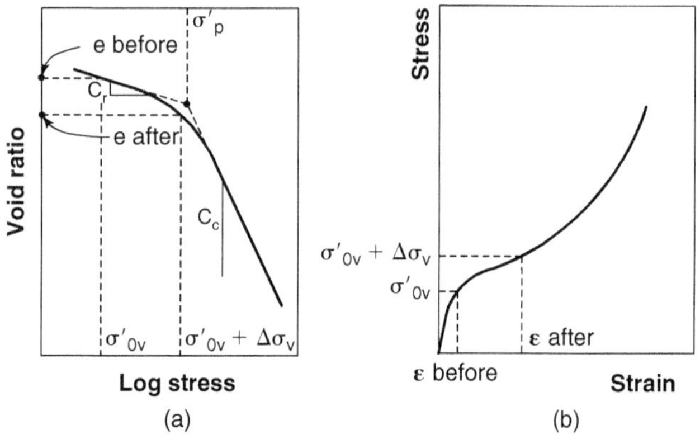

Figure 17.33 Obtaining the strains from the stresses for settlement calculations.

where ΔH is the settlement, H_{oi} is the thickness of each layer, $e_{ibefore}$ and e_{iafter} are the void ratios read on the consolidation curve (Figure 17.33) at σ'_{ov} and $\sigma'_{ov} + \Delta\sigma_v$ respectively, e_{oi} is the initial void ratio in each layer, and ε_{iafter} and $\varepsilon_{ibefore}$ are the strains read on the consolidation curve at $\sigma'_{ov} + \Delta\sigma_v$ and σ'_{ov} respectively.

If consolidation curves are not available, then the following equations can be used but the precision of the predictions will be affected. These equations correspond to the bilinear shape of the void ratio or vertical strain versus log of vertical effective stress (Figure 17.33a). Beyond the preconsolidation pressure σ'_p, the slope of the curve is the compression index C_c, defined as:

$$C_c = \frac{\Delta e}{\Delta \log \sigma'} = \frac{e_1 - e_2}{\log\left(\frac{\sigma'_1}{\sigma'_2}\right)} \quad (17.91)$$

Rough estimates of C_c can be obtained by correlation with index properties (see section 14.12). Before the preconsolidation pressure σ'_p, the slope of the curve is the recompression index C_r. It is defined as:

$$C_r = \frac{\Delta e}{\Delta \log \sigma'} = \frac{e_1 - e_2}{\log\left(\frac{\sigma'_1}{\sigma'_2}\right)} \quad (17.92)$$

The choice of the right equation for calculating the settlement is based on the relative magnitude of the effective stress before loading σ'_{ov} and the effective stress long after loading $\sigma'_{ov} + \Delta\sigma_v$, compared with the preconsolidation pressure σ'_p (see section 14.11). Normally consolidated (NC) soils have a vertical effective stress σ'_{ov} equal to the preconsolidation pressure σ'_p, and overconsolidated (OC) soils have a vertical effective stress σ'_{ov} smaller than the preconsolidation pressure σ'_p.

NC soils:

$$\Delta H = \frac{H_o}{1+e_o} C_c \log\left(\frac{\sigma'_{ov} + \Delta\sigma_v}{\sigma'_{ov}}\right) \quad (17.93)$$

OC soils and $\sigma'_{ov} + \Delta\sigma_v < \sigma'_p$

$$\Delta H = \frac{H_o}{1+e_o} C_r \log\left(\frac{\sigma'_{ov} + \Delta\sigma_v}{\sigma'_{ov}}\right) \quad (17.94)$$

OC soils and $\sigma'_{ov} + \Delta\sigma_v > \sigma'_p$

$$\Delta H = \frac{H_o}{1+e_o}\left(C_r \log\left(\frac{\sigma'_p}{\sigma'_{ov}}\right) + C_c \log\left(\frac{\sigma'_{ov} + \Delta\sigma_v}{\sigma'_p}\right)\right) \quad (17.95)$$

For Eq. 17.93, the curve is simply a single straight line in the e-log σ' set of axes, so there is only one term. For Eq. 17.94, the curve is bilinear, but the stresses σ'_{ov} and $\sigma'_{ov} + \Delta\sigma_v$ are both on the recompression part of the curve. In Eq. 17.95, the stresses σ'_{ov} and $\sigma'_{ov} + \Delta\sigma_v$ straddle the preconsolidation pressure; therefore, both C_c and C_r are involved and the equation has two terms. The first term represents the recompression from σ'_{ov} to σ'_p, the second term represents the virgin compression from σ'_p to $\sigma'_{ov} + \Delta\sigma_v$.

17.8.10 Consolidation Settlement: Time Rate

The time rate of settlement can be estimated by using the consolidation theory solution described in section 11.4.6. The time required for a given percentage of the settlement to take place is given by:

$$t_U = T_U \frac{H^2}{c_v} \quad (17.96)$$

where t_U is the time required for U% of the settlement to take place, T_U is the time factor (which comes from the theoretical solution and is obtained from Figure 17.34), H is the drainage length, and c_v is the coefficient of consolidation for the soil obtained from a consolidation test (see section 9.5.1). On Figure 17.34, curve C_1 represents the most common case. The parameter U is the average percent consolidation, which is a function of the time t and is defined as:

$$U(t) = \frac{\Delta H(t)}{\Delta H_{max}} \quad (17.97)$$

where $\Delta H(t)$ is the settlement after a time t and ΔH_{max} is the maximum settlement at time equal to infinity. ΔH_{max} is the settlement obtained from section 17.8.9. The drainage length (Figure 17.35) depends on the ability of the upper layer and the lower layer to drain the water away. If both the top and bottom layers are free draining (two-way drainage), then the drainage length H is equal to one-half the layer thickness H_o. This is because the furthest that a water molecule has to travel to get out of the compressing layer is one-half of the layer thickness. If only one of the two layers, top or bottom, is free draining (one-way drainage), then the drainage length H is equal to the layer thickness H_o. This is because the furthest that a water molecule has to travel to get out of the compressing layer is the layer thickness. Then, the complete settlement vs. time curve ($\Delta H(t)$ vs. t_U, Figure 17.36) can be created by using the combination of Eq. 17.96 and Eq. 17.97.

17.8.11 Creep Settlement

The consolidation settlement is associated with the dissipation of excess water stress by drainage of the soil mass. When the excess water stress has dissipated, the settlement may continue to occur due to creep in the soil. The creep settlement is attributed to the slow movement of particles with respect to each other with no change in water stress. This creep settlement can occur in saturated soils as well as in unsaturated soils. It is often slow and small, but can be significant in soft soils and soils with high organic content.

One can use consolidation test data to estimate the creep settlement. During each load step in a consolidation test,

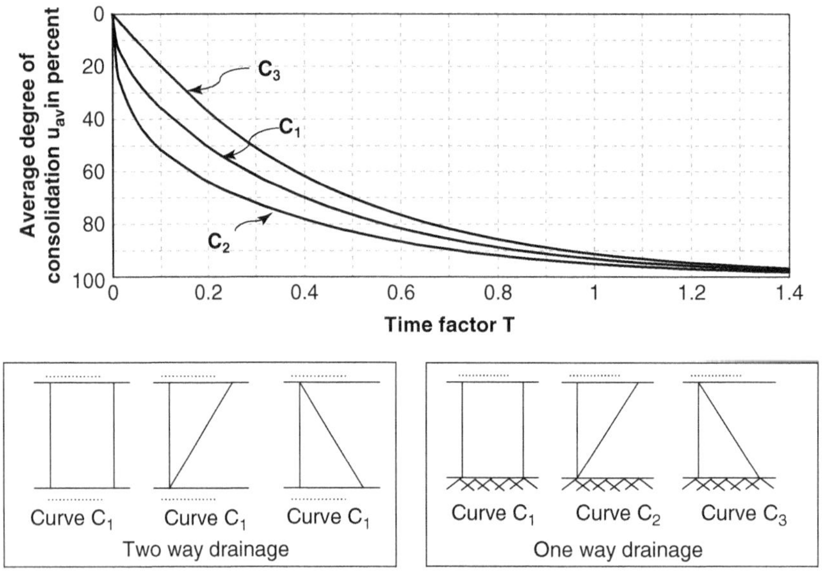

Figure 17.34 Time factor and percent consolidation curve.

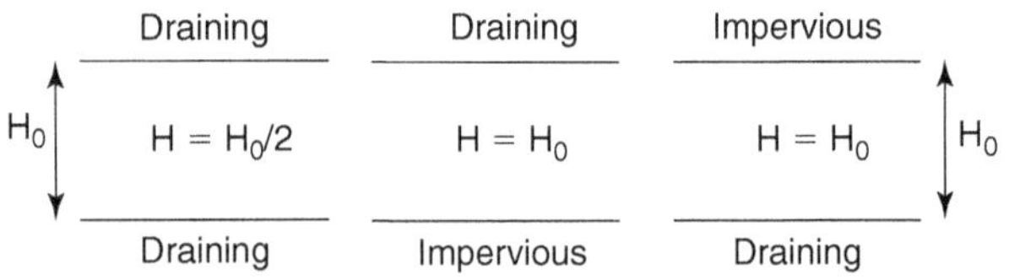

Figure 17.35 Drainage length.

the sample compression is recorded as a function of time (see section 14.12). The slope of the tail end of that curve corresponds to the creep settlement; because at that point on the curve, the excess water stress has dissipated. That slope C_α is called the *secondary compression index* and is defined as:

$$C_\alpha = \frac{\Delta e}{\Delta \log t} \tag{17.98}$$

where Δe is the change in void ratio between the start time t_{start} and the end time t_{end} and $\Delta \log t$ is the change in the log base 10 of the time. Then, the creep settlement is estimated as:

$$\Delta H = \frac{H_o}{1+e_o} C_\alpha \log\left(\frac{t_{end}}{t_{start}}\right) \tag{17.99}$$

where ΔH is the creep settlement, H_o is the layer thickness, e_o is the initial void ratio, C_α is the secondary compression index, t_{start} is the start time, and t_{end} is the end time.

To estimate the creep settlement, pressuremeter test data can be used, as discussed in section 17.7. The equation in this case is:

$$\frac{s(t_{end})}{s(t_{start})} = \left(\frac{t_{end}}{t_{start}}\right)^n \tag{17.100}$$

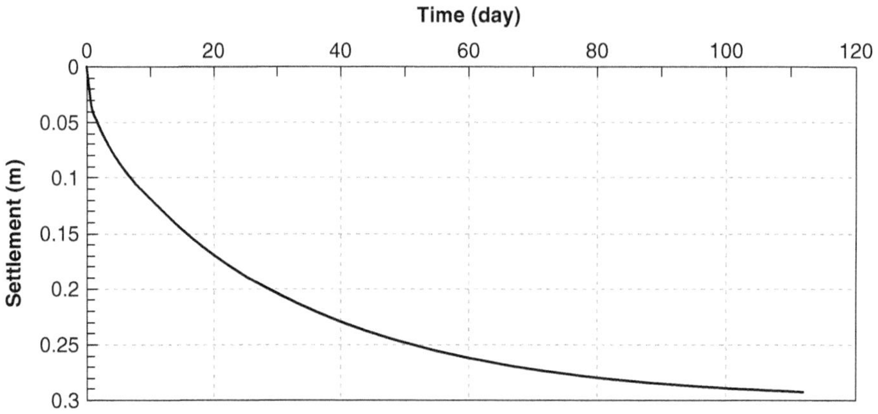

Figure 17.36 Settlement vs. time curve.

Table 17.9 Range of Possible Values for Footing Pressures

Category	Types of soils	Presumed allowable bearing value
Coarse-grained soils	Dense gravel or dense sand and gravel	$>600\ kN/m^2$
	Medium-dense gravel, or medium-dense sand and gravel	200 to 600 kN/m^2
	Loose gravel, or loose sand and gravel	$<200\ kN/m^2$
	Dense sand	$>300\ kN/m^2$
	Medium-dense sand	100 to 300 kN/m^2
	Loose sand	$<100\ kN/m^2$
Fine-grained soils	Very stiff and hard clays	300 to 600 kN/m^2
	Stiff clays	150 to 300 kN/m^2
	Firm clay	75 to 150 kN/m^2
	Soft clays and silts	$<75\ kN/m^2$
	Very soft clay	Not applicable

where $s(t_{end})$ and $s(t_{start})$ are the settlements at the times t_{end} and t_{start} respectively and n is the rate effect exponent. This exponent can be measured on a site-specific basis by performing the pressuremeter tests discussed in section 17.7. The typical range of values for n is 0.01 and 0.03 for sands and 0.02 to 0.05 for stiff to hard clays.

17.8.12 Bearing Pressure Values

The allowable pressure on a shallow foundation should always be calculated according to proper design procedures. It is useful to have an idea of what to expect as a range of possible pressure for various soils. Any pressure significantly outside of those ranges should be checked very carefully. Table 17.9 gives estimates of these ranges.

17.9 SHRINK-SWELL MOVEMENT

17.9.1 Water Content or Water Tension vs. Strain Curve

Most soils swell and shrink when they get wet and dry. Some soils are particularly prone to such movements, which must be taken into account in the foundation design. Such soils can be identified in a number of ways. One index is the *plasticity index*: the difference between the liquid limit and the plastic limit. The higher the plasticity index, the more prone to large shrink-swell movements the soil is. The reason is that a high plasticity index is indicative of a higher content of very small clay particles, and very small clay particles tend to absorb more water than larger clay particles. Another index is the *shrink-swell index*: the range of water content over which the soil freely shrinks and swells. The higher the shrink-swell index, the more prone to large shrink-swell movement the soil is. The reason is simply that the volume change is related to the water content and that a larger variation in water content leads to a larger change in volume.

The basic behavior of a soil with regard to water content changes can be shown by performing two simple tests: a free swell test and a free shrink test (see sections 9.6 and 9.7). For a shrink test, a disk of soil is placed on a table and allowed to dry; its weight and dimensions are recorded as a function of time. At the end of the test, the sample is oven dried and the dry weight is obtained. The test data give the water content versus volume change of the sample. For the swell test, a disk of soil is placed in a consolidometer ring, submerged, and allowed to swell. The thickness of the sample and the weight of the sample are kept constant so that the water content versus volume change curve can be plotted. The free shrink curve and the free swell curve are joined on the same graph to give the shrink-swell curve for the sample. The maximum water content that the soil can reach is the swell limit. As the soil dries, the soil shrinks along the water content vs. volume change curve, which is a straight line until the shrinkage limit is reached (Figure 17.37).

During shrinkage, the soil particles come closer and closer together until they can no longer get any closer; at that point, called the *shrinkage limit*, any further loss of water will no longer represent a loss of volume. In first approximation, it

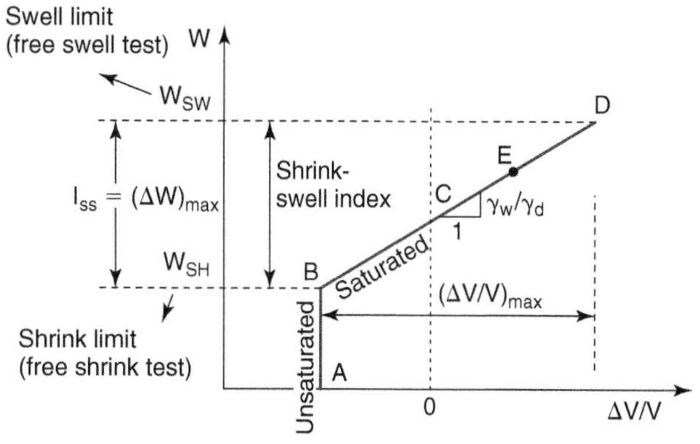

Figure 17.37 Water content vs. relative volume change.

can be said that above the shrinkage limit, the soil is saturated, and below the shrinkage limit, the soil is no longer saturated. The slope of the water content vs. volume change line is called the shrink-swell modulus E_w. Simple weight-volume relationships (see section 14.16) give E_w as:

$$E_w = \frac{\gamma_w}{\gamma_d} \tag{17.101}$$

where γ_w, and γ_d are the unit weight of water and the dry unit weight of the soil respectively.

17.9.2 Shrink-Swell Movement Calculation Methods

Several different types of methods are available to predict the shrink-swell movement of a soil: the potential vertical rise (PVR) method (McDowell 1956), the suction method, and the water content method. The PVR method consists of obtaining samples at the site, measuring the water content and the Atterberg limits, and using charts based on observations to calculate the maximum possible vertical movement. This movement corresponds to the case where the ground surface would be inundated for a very long time. This movement depends on the water content of the soil at the time of sampling. Therefore, a sample taken during the summer months will lead to a large predicted PVR and one taken during the winter months will lead to a small predicted PVR. The PVR only gives an indication of the swelling potential, not the shrinkage potential.

The suction method (Lytton 1994) relates the settlement to the log of the water tension. Lytton includes the settlement due to the change in mechanical stress in addition to the movement due to the change in water tension and proposes the following equation:

$$s = \sum_{i=1}^{n} f_i H_i \left(-\gamma_{hi} \log \frac{h_{fi}}{h_{ii}} - \gamma_{\sigma i} \log \frac{\sigma_{fi}}{\sigma_{ii}} - \gamma_{\pi i} \log \frac{\pi_{fi}}{\pi_{ii}} \right) \tag{17.102}$$

where i is the layer number, f_i is the crack fabric factor to convert the volumetric strain into vertical strain, H_i is the layer thickness, γ_{hi} is the matrix suction compression index for layer i, h_{fi} and h_{ii} are the final and initial values of the matric suction in layer i respectively, $\gamma_{\sigma i}$ is the mean principal stress compression index for layer i, σ_{fi} and σ_{ii} are the final and initial values of the mean principal stress in layer i respectively, $\gamma_{\pi i}$ is the osmotic suction compression index for layer i, and π_{fi} and π_{ii} are the final and initial values of the osmotic suction in layer i respectively. The three compression indices are obtained from correlations with index soil properties on samples from layer i, or from testing samples, and the boundary values of the matric and osmotic suction are based on experience. For the crack fabric factor f, Lytton recommends $f = 0.5$ when the soil is drying and $f = 0.8$ when the soil is wetting.

The water content method (Briaud et al. 2003) makes use of the water content vs. volume change curve recorded in a free shrink test to calculate the amplitude of the vertical movement. The equation used is:

$$s = \sum_{1}^{n} f_i \frac{\Delta w_i}{E_{wi}} H_i \tag{17.103}$$

where s is the vertical movement of the ground surface, n is the number of layers making up the depth of the active zone involved in the shrink-swell movement, f_i is the factor used to transform the volumetric strain into the vertical strain (0.33, according to Briaud et al. 2003), H_i is the thickness of the i th layer, Δw_i is the change of water content in the i th layer during the calculation period (expressed as a ratio, not a percentage), and E_{wi} is the shrink-swell modulus of the soil in the i th layer as given by Eq. 17.101 or measured in a shrink test or a swell test.

The number of layers involved in the calculations is given by the depth of the active zone, which is often obtained from local experience on water content profiles observed over several years. These profiles typically show large variations in water content near the surface, and a decrease in variation with depth down to a depth where the variation is negligible; that depth is the depth of the active zone. Typical values range between 3 and 5 m. The value of E_{wi} is obtained from measurements on samples (shrink test) from layer i or from using Eq. 17.101.

The value of Δw_i should not be taken as the difference between the swell limit and the shrink limit for the soil. This would assume that, during the life of the structure, the soil will shrink to the shrink limit and swell to the swell limit. This is extremely conservative and very unlikely. Instead, Δw is obtained from local experience as the amplitude of the water content variation at the chosen depth (middle of H_i) read on the water content profiles collected in an area over time. Briaud et al. (2003) collected more than 8000 water content measurements over a period of time and as a function of depth. They obtained values of the variation of water content over several seasons and found the amplitude of Δw for four cities in Texas (Figure 17.38). The samples came from right outside of the foundation as well as from under the foundation. The Δw values ranged from 0.05 to 0.08 for the samples outside of the foundation imprint. The Δw values for the samples directly under the foundation were lower:

$$\Delta w_{under} = 0.7 \Delta w_{outside} \tag{17.104}$$

17.9.3 Step-by-Step Procedure

Calculating the shrink-swell movement of a soil proceeds much like calculating the settlement of a building. The parallel is drawn in the following step-by-step procedure (Figure 17.39):

1. Determine the depth of the active zone H (the zone within which the movement takes place over time). This

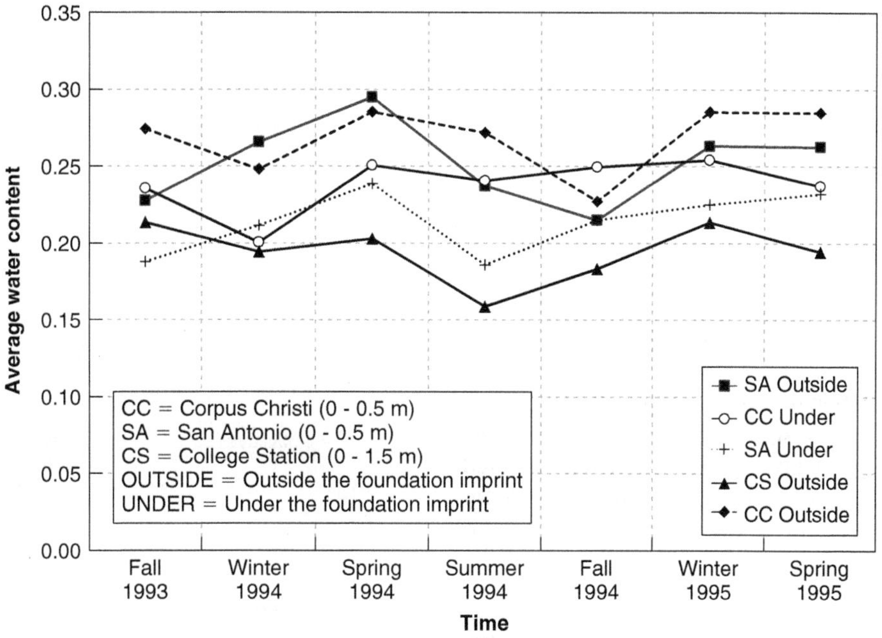

Figure 17.38 Water content variation for three cities in Texas. (Briaud et al. 2003)

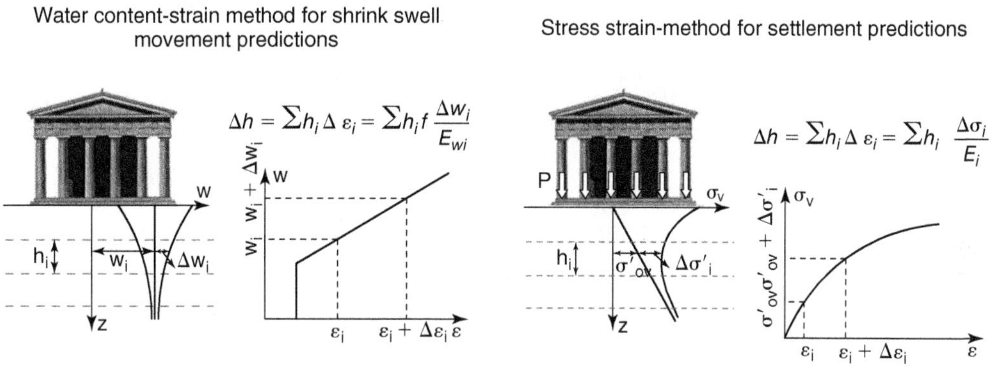

Figure 17.39 Parallel between shrink-swell and settlement methods.

is parallel to the zone of influence for the settlement case. The active zone is usually estimated from local experience or from water content or water tension profiles gathered over many years. Typical values vary between 3 and 5 m.

2. Decompose that zone into an appropriate number n of soil layers. This is the same step as in the settlement procedure.
3. Estimate the initial water content w_i in the center of each layer. The water content and associated water tension play the role of the effective stress in the settlement procedure.
4. Estimate the final water content w_f in the center of each layer. Again, the water content and associated water tension play the role of the effective stress in the settlement procedure. The final water content can be obtained from the soil boring files accumulated over time by a consulting company in a given geological area. This is what was done for Figure 17.38. This step is parallel to obtaining the increase in stress with depth by the elasticity method for the settlement case. Indeed, the water content in the shrink-swell calculations plays the role of the stress in the settlement calculations.
5. Obtain the relationship between the water content and the vertical strain by performing simple tests like the free shrink test or the free swell test. This relationship plays the role of the stress-strain curve in the settlement calculations.
6. Using the water content vs. vertical strain curve, obtain the strains ε_i and ε_f corresponding to the initial and final water content w_i and w_f. This is the same step as in the settlement calculations, but using water content instead of stress.

7. Calculate the shrink or swell movement using:

$$s = \sum_{1}^{n} f_i \frac{w_f - w_i}{E_{wi}} H_i \tag{17.105}$$

where s is the vertical movement of the ground surface, n is the number of layers making up the depth of the active zone involved in the shrink-swell movement, f_i is the factor used to transform the volumetric strain into the vertical strain (0.33, according to Briaud et al. 2003), H_i is the thickness of the ith layer, w_f and w_i are the final water and initial contents in the ith layer during the calculation period (expressed as a ratio, not a percentage), and E_{wi} is the shrink-swell modulus of the soil in the ith layer as given by Eq. 17.101 or measured in a shrink test or a swell test. Equation 17.105 for shrink-swell movement is the same as Eq. 17.83 for settlement calculations.

This procedure uses the water content as the main variable. The water content can be replaced by the water tension or suction in this procedure when using the suction-based shrink-swell movement method.

17.9.4 Case History

Four footings were placed at a site near Dallas, Texas, where the soil is a CL-CH (Figure 17.40). The soil below footings W1 and W2 was water injected, whereas the soil below footings RF1 and RF2 was left intact. The soil properties are shown in Figure 17.40; the groundwater level was about 4.5 m deep. The footings were constructed at the ground surface and were 2 m by 2 m by 0.6 m thick. The movement

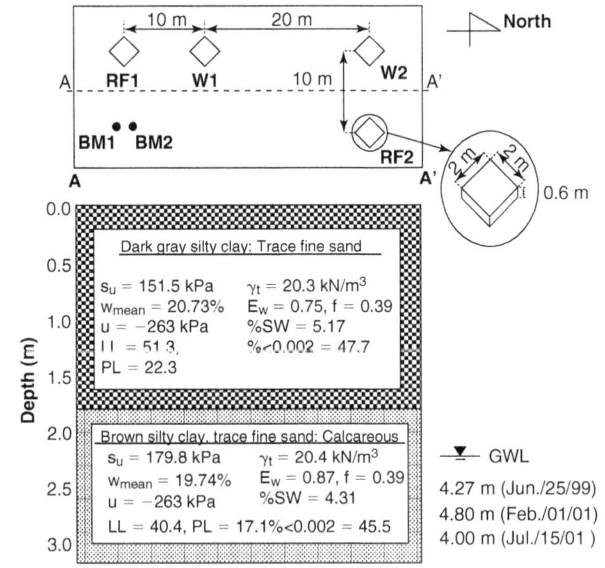

Figure 17.40 Footings and soil stratigraphy.

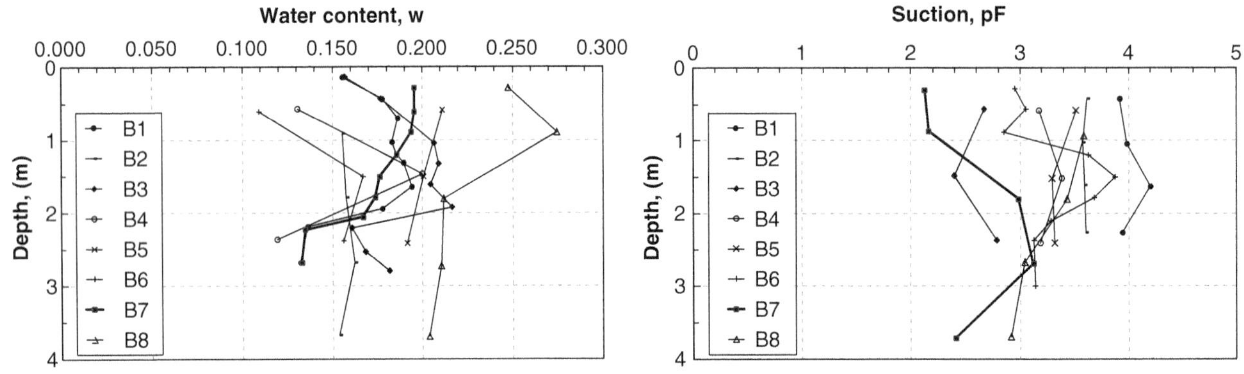

Figure 17.41 Water content and water tension variation over two years.

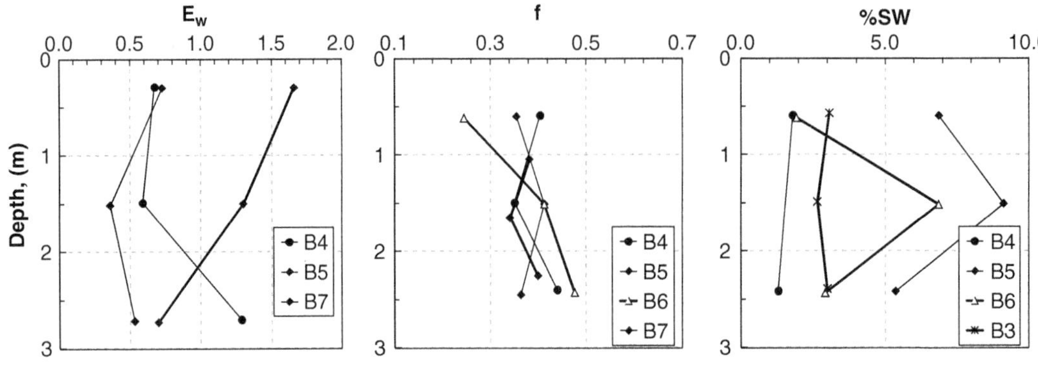

Figure 17.42 Soil properties.

of the footings was measured every month for 2 years and borings were drilled every 3 months. The samples were tested and gave the water content and water tension profiles shown in Figure 17.41. Additional properties, including the shrink-swell modulus E_w, the f factor to convert volumetric strain to vertical strain, and the maximum percent swell (%SW), are shown in Figure 17.42. The recorded movement of the footings is shown by Figure 17.43, and the temperature and rainfall variation during the same two years is shown in Figure 17.44. These data indicate the following:

1. The amplitude of movement of the footings on the water-injected soil is the same as the amplitude of the movement of the footings on the intact soil.
2. The footings on the water-injected soil swelled less and shrank more than the footings on the intact soil.
3. The movement was very small during the first year when the rainfall was very evenly distributed (Figure 17.44). During the second year, a three-month drought followed by three months of heavy rainfall created a lot of movement amplitude.

17.10 FOUNDATIONS ON SHRINK-SWELL SOILS

17.10.1 Types of Foundations on Shrink-Swell Soils

Predicting the vertical movement of the ground surface is useful but not a direct input to the design of foundations on shrink-swell soils. The problem with shrink-swell soils is that the soil shrinks and swells more at the edges of the building than under the center of the building. This tends to distort the building and damage it if the distortion exceeds the building's ability to deform. The best foundation systems are those that minimize building distortion even when the soil movement is very uneven. Foundations that have been used include (Figure 17.45):

1. Stiffened slab on grade for smaller structures (1 to 3 stories). These slabs consist of a thin (~0.1 m thick) slab on grade connected to deep beams (say, 1 to 1.2 m deep, 0.3 m wide, placed in both directions with a 3 to 5 m spacing center to center). This solution, sometimes called *waffle slab*, is typically economical (~\$100/m^2 in 2010) and very satisfactory if the slab is stiff enough.

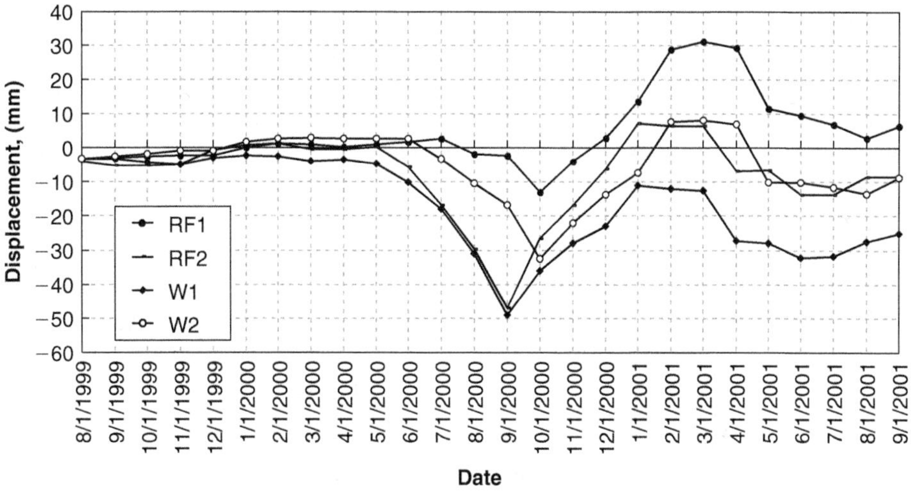

Figure 17.43 Observed movement over two years.

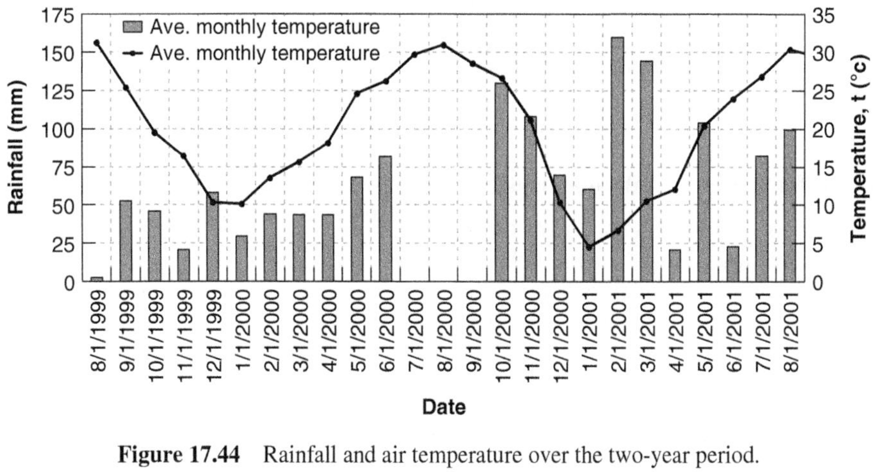

Figure 17.44 Rainfall and air temperature over the two-year period.

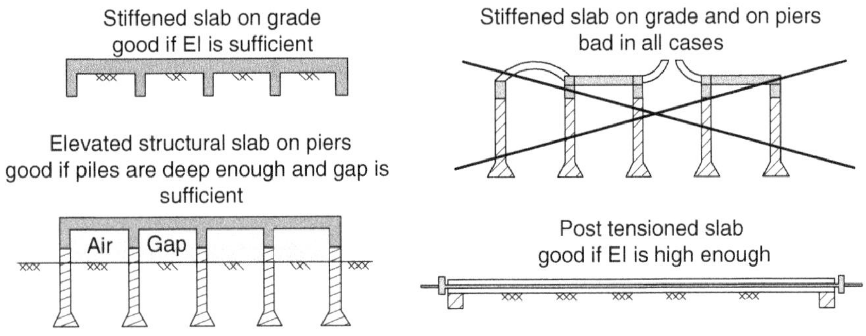

Figure 17.45 Types of foundations used for light buildings on shrink-swell soils.

2. Elevated structural slab for larger structures (3 to 15 stories). This system consists of a structural slab (a slab that can sustain the dead load and live load in free span) connected to piles in such a way that there is a sufficient gap (say, 0.3 m) between the ground surface and the bottom of the beams stiffening the slab. This solution is more expensive ($\sim\$200/\mathrm{m}^2$ in 2010) but very satisfactory provided the piles go deep enough below the zone of movement.
3. Anchored slab on grade. This system consists of a slab on grade and on piles. There is no gap between the ground surface and the slab that rests on it. This is a very undesirable system because when the soil swells, the bored piles prevent the slab from moving up, so the slab deforms and can break under the swell pressure. Alternatively, if the soil shrinks under the slab, the slab on grade becomes unsupported and breaks because it is not designed to carry the load in free span.
4. Posttensioned slabs are typically flat slabs that are posttensioned to keep the concrete in compression during bending. They are satisfactory systems provided they are stiff enough to minimize distortion. For equal stiffness, they are not as economical as a stiffened slab on grade. Thin posttensioned slabs on grade are undesirable for buildings because they are overly flexible. Although they minimize the cracking of the slab, they do not prevent distortion of the superstructure. Thin posttensioned slabs are a very good solution, however, for playing surfaces such as tennis or basketball courts on shrink-swell soils, because they minimize cracking.

17.10.2 Design Method for Stiffened Slabs on Grade

A stiffened slab on grade has deep beams (e.g., 1 m deep, 0.3 m thick), spaced relatively closely (e.g., 4 m) in both directions. These beams stiffen the slab, which is sometimes called a *waffle slab*. The stiffening limits the amount of distortion that the superstructure is subjected to in case of soil movement. In the summer, when the soil shrinks around the periphery of the structure and a gap develops between the ground surface and the edge of the foundation, the edges of the slab do not drop significantly, because of the rigidity of the slab and beams; this prevents excessive distortion of the superstructure. In the winter, when the soil swells around the periphery of the structure and lifts the foundation, the center of the foundation does not sag significantly, again because of the rigidity of the slab and beams; this prevents excessive deformations of the superstructure. The design of the foundation is therefore controlled by these two conditions, sometimes called *edge drop* and *edge lift*. The critical design parameters for these two conditions are the *cantilever edge distance* for the edge drop condition and the *free span distance* for the edge lift condition. These parameters depend on several factors, including weather, vegetation, soil shrink-swell sensitivity, soil stiffness, and slab stiffness. Several design procedures have been suggested over the years, including:

1. BRAB method (Building Research Advisory Board 1968)
2. PTI method (Post Tensioning Institute 2004)
3. WRI method (Wire Reinforcing Institute 1981)
4. Australian method (Australian Standard (AS) 2870, 1996)
5. TAMU-Slab method (Briaud et al. 2010)

The TAMU-Slab method is based on the use of charts. The details of the research work on which the method is based can be found in Abdelmalak (2007) and Briaud et al. (2010). The design parameters necessary to size the beams and their spacing are the maximum bending moment $M_{\max}$, the maximum shear force $V_{\max}$, and the maximum deflection $\Delta_{\max}$ of the slab (Figure 17.46).

In the TAMU-Slab method, these quantities are linked to the equations of $M_{\max}$, $V_{\max}$, and $\Delta_{\max}$ for an equivalent cantilever beam with a length L_{eqv}. These equations are applied to the design of the beams for the stiffened slab by using modification factors:

$$M_{\max} = \frac{1}{2} q L_{eqv}{}^2 \tag{17.106}$$

$$V_{\max} = F_v q L_{eqv} \tag{17.107}$$

$$\Delta_{\max} = \frac{q L_{eqv}^4}{F_{\Delta\max} EI} \tag{17.108}$$

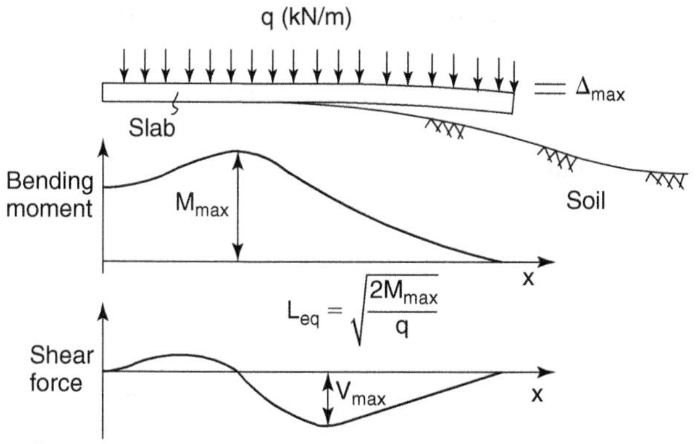

Figure 17.46 Half slab with deflection, bending moment, and shear.

where q (kN/m) is the distributed load on the cantilever beam, F_v is the modification factor for the shear force, $F_{\Delta\max}$ is the modification factor for the deflection, and $EI(\text{kN.m}^2)$ is the bending stiffness of the beam product of the modulus $E(\text{kN/m}^2)$ by the moment of inertia $I(\text{m}^4)$. For a true cantilever beam, the equivalent cantilever length L_{eqv} would simply be the length of the cantilever beam, the maximum shear modification factor F_V would be 1, and the maximum deflection factor $F_{\Delta\max}$ would be 8. For the stiffened slab on grade, these factors were obtained from numerical simulations. The equivalent length and the modification factors are given by charts for various slab thicknesses.

The three most important factors affecting the final choice of the beam depth are the weather, the soil, and the slab stiffness. The weather and the soil were found (Abdelmalak 2007) to be best characterized by the soil and weather index I_{SW}. This soil and weather index can be defined on the basis of the water tension or the water content:

$$I_{SW(Water\ Tension)} = I_{SS}\, H\, \Delta U_{edge} \tag{17.109}$$

$$I_{SW(Water\ Content)} = H\, \Delta w_{edge} = 0.5 I_{SW(Water\ Tension)} \tag{17.110}$$

where $I_{SW(Water\ Tension)}$ and $I_{SW(Water\ Content)}$ are the soil and weather indices on the basis of the water tension and the water content respectively; I_{SS} is the shrink-swell index, a soil parameter equal to the difference between the swell limit and the shrink limit; H is the depth of the active zone in meters; ΔU_{edge} is the change in the $\log_{10}$ of the water tension in kPa at the edge of the foundation over the period considered for the design; and Δw_{edge} is the change in water content expressed as a ratio at the edge of the foundation over the period considered for the design. The change in log10 of water tension is:

$$\Delta U_{edge} = \log_{10} \frac{u_w(final\ at\ edge)}{u_w(initial\ at\ edge)} \tag{17.111}$$

Based on testing of a number of clays, the relationship between $I_{SW(Water\ tension)}$ and $I_{SW(Water\ content)}$ was found to be:

$$I_{SW(Water\ Content)} = 0.5 I_{SW(Water\ Tension)}$$

because

$$\Delta w_{edge} = 0.5 I_{SS}\, \Delta U_{edge} \tag{17.112}$$

A few cities in the United States where the shrink-swell soil problem is acute were selected and the weather over the past 20 years was simulated to obtain estimates of the change in $\log_{10}$ of water tension (kPa), ΔU_{edge}. Simulations were performed for the free field and the edge of the foundation. It was found that the value at the edge was about one-half the value in the free field. The results are shown in Table 17.10.

The slab stiffness was represented by the slab equivalent depth d_{eq}, which represents the thickness of a flat slab having the same moment of inertia as the moment of inertia of a stiffened slab with a beam depth equal to D, a beam width equal to b, and a beam spacing equal to S. The slab equivalent depth can be calculated by:

$$S\, d_{eq}^{\ 3} = b\, D^3 \tag{17.113}$$

Step-by-step procedure for the water content method

1. Obtain the dimensions of the slab $B \times L$ and the loading pressure on the slab w (kPa).
2. Estimate the depth H of the active zone. This is best based on local practice and experience. In Texas, H is typically considered to be between 3 and 5 m.
3. Estimate the change in water content Δw_{edge} at the edge of the foundation. This is also best estimated from local practice and experience. Note that Δw_{edge} was found to be equal to one-half of the change in water content $\Delta w_{free\ field}$ in the free field. The borings

Table 17.10 Change in $\log_{10}$ of Water Tension in kPa for Six Cities in the USA

	College Station, TX	San Antonio, TX	Austin, TX	Dallas, TX	Houston, TX	Denver, CO
$\Delta U_{free\ field}$	0.788	1.392	0.866	1.295	1.283	1.374
ΔU_{edge}	0.394	0.696	0.433	0.648	0.642	0.687

(Abdelmalak 2007)

accumulated over time by a local company may come in very handy in estimating $\Delta w_{freefield}$ and therefore Δw_{edge}.

4. Choose a beam spacing s and a beam width b. Typical numbers for s are 3 to 5 m and for b are equal to 0.3 m.
5. Make a first assumption as to the beam depth D (say, 1 m).
6. Calculate the thickness of the equivalent slab d_{eq} by using Eq. 17.113 and the soil and weather index $I_{SW(water\ content)}$ by using Eq. 17.110.
7. Use the design charts of Figures 17.47 to 17.49 for the edge drop case (summer shrinkage) and Figure 17.50 to 17.52 for the edge lift case (winter swelling), and obtain the equivalent length L_{eq}, the shear factor F_v, and the deflection factor $F_{\Delta\max}$.
8. Use the distributed load on the beam q(kN/m) (q = $w \times$ s) to calculate the maximum bending moment M_{max}, maximum shear V_{max}, and maximum deflection Δ_{max} according to Eqs. 17.106 to 17.108.
9. Calculate the ratio $0.5L/\Delta_{max}$ and L_{eq}/Δ_{max}. Ratios larger than 500 typically lead to acceptable distortions. If this criterion is not reached, repeat steps 5 to 9 with a larger beam depth D.
10. If the deflection criterion of step 9 is met, use the maximum bending moment and maximum shear to design the beam reinforcement and the slab.

Step-by-step procedure for the water tension method

The steps for this method are the same as the steps for the water content method except for the following. In step 3, the change in $\log_{10}$ of the water tension in kPa is needed instead

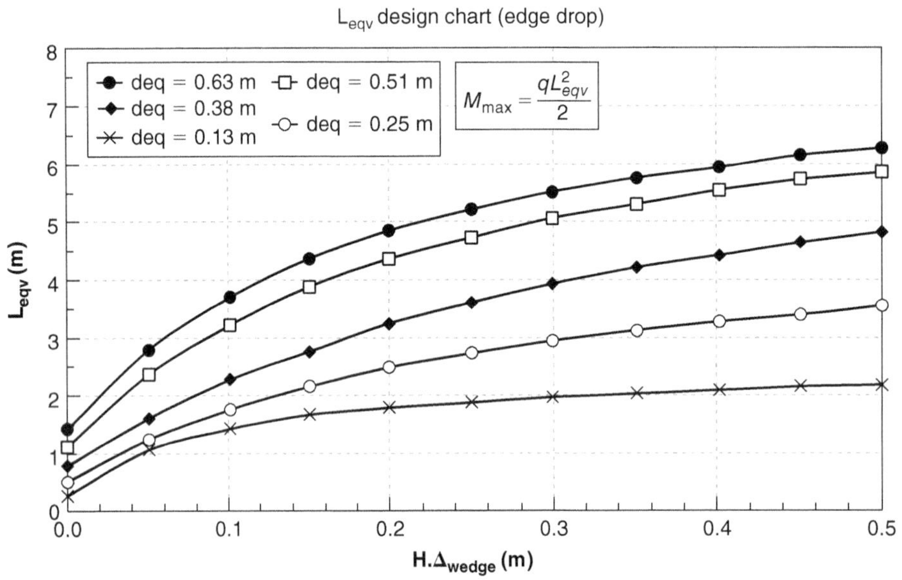

Figure 17.47 Equivalent cantilever length—water content method—edge drop case.

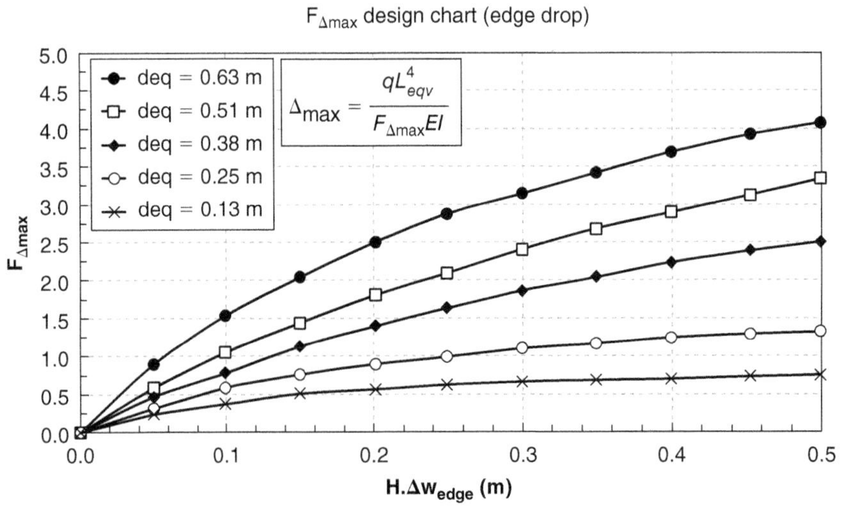

Figure 17.48 Maximum deflection factor—water content method—edge drop case.

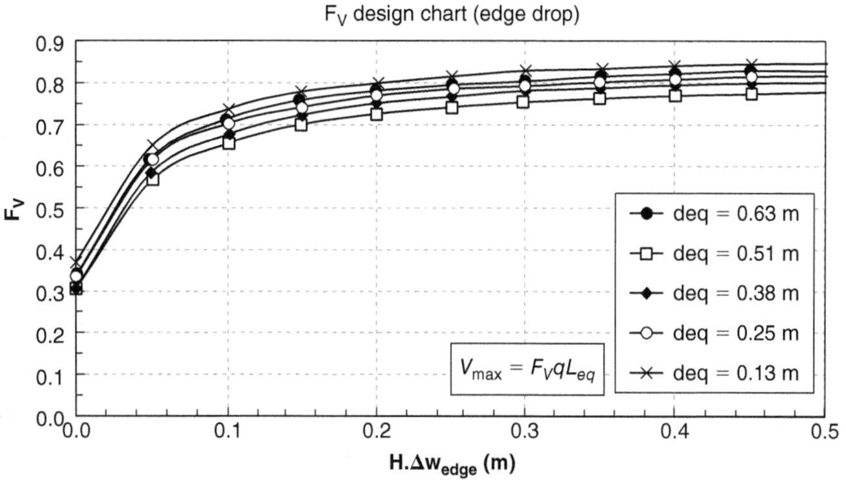

Figure 17.49 Maximum shear factor—water content method—edge drop case.

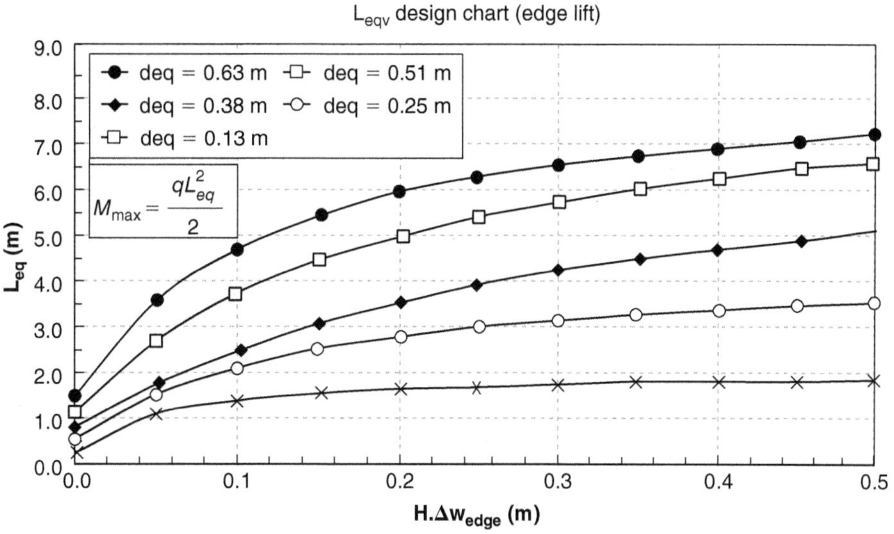

Figure 17.50 Equivalent cantilever length—water content method—edge lift case.

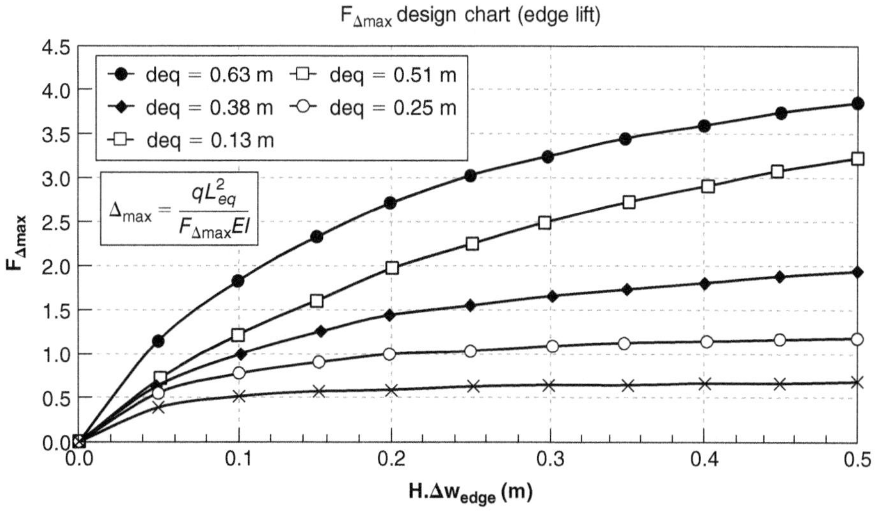

Figure 17.51 Maximum deflection factor—water content method—edge lift case.

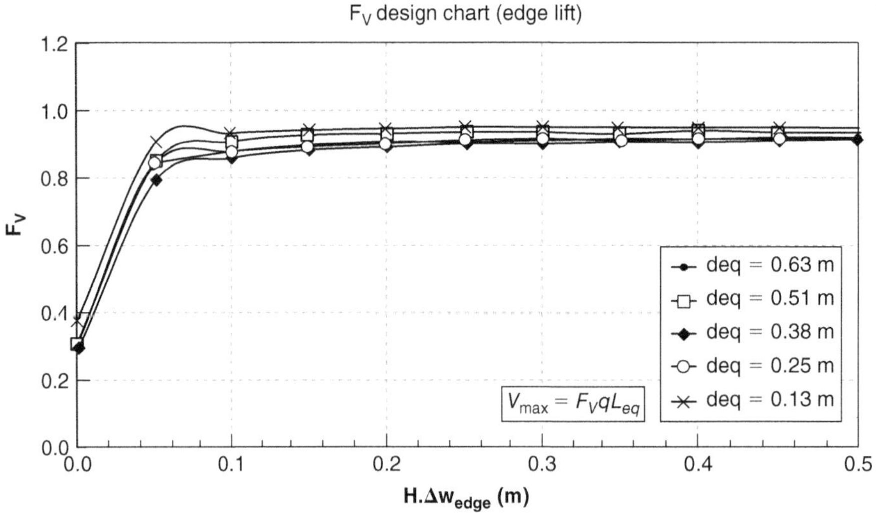

Figure 17.52 Maximum shear factor—water content method—edge lift case.

of the change in water content. In step 6, the soil and weather index $I_{SW(water\ tension)}$ is used instead of $I_{SW(water\ content)}$ based on Eq. 17.112. In step 7, the same charts are used after converting the $I_{SW(water\ tension)}$ into $I_{SW(water\ content)}$ based on Eq. 17.112.

17.11 TOLERABLE MOVEMENTS

Tolerable movements depend on the structure that is being built. Some embankments can tolerate 1 m of settlement as long as the pavement is built after the settlement takes place. Some sensitive facilities can tolerate only a few millimeters of settlement. Very tall buildings can settle anywhere from a few millimeters to 200 or 300 mm. One problem with a large total settlement is the connection to utilities outside the building, because typically the building settles or moves with respect to its surroundings. The total settlement s_t is one issue, but the differential settlement s_d is even more important in many cases. The differential settlement is rarely calculated, however, as there are rarely enough borings to make a settlement calculation at each building column or at each bridge pier. The practice is to calculate s_t and to assume that s_d is 3/4 of s_t:

$$s_d = 0.75 s_t \tag{17.114}$$

Bridges can tolerate a lot of differential movement, as documented in the study by Moulton et al. (1985). *Simply supported bridges* are bridges where each span is made of beams resting on top of the piers; the beams are not connected from one span to the next. For a *continuous bridge*, the beams continuously span several piers from one end of the bridge to the other. Simply supported bridges are easier and faster to construct, but continuous bridges make better use of the material and are thus lighter and therefore can be cheaper. Moulton surveyed more than 400 bridges and found that the supports (abutments and piers) moved vertically an average of 94 mm and horizontally an average of 68 mm. He found that bridges supported on shallow foundations had the same average movement as the ones founded on piles. From damage inspection, he concluded that total vertical and horizontal movements of up to 50 mm were tolerable. The *longitudinal distortion* is the ratio of the differential movement s_d between adjacent piers over the span length L. Moulton et al. (1985) and Barker et al. (1991) made recommendations for the limits of longitudinal angular distortion. In the end, it appears reasonable to accept 0.004 for continuous bridges and 0.008 for simply supported bridges. Simply supported bridges can sustain more differential movement than continuous bridges.

$$\text{For simply supported bridges} \quad \frac{L}{s_d} \geq 125 \tag{17.115}$$

$$\text{For continuous bridges} \quad \frac{L}{s_d} \geq 250 \tag{17.116}$$

where L is the span length and s_d is the differential vertical movement between adjacent piers. The amount of movement and distortion that buildings can tolerate has been studied by many researchers, including Skempton and MacDonald (1956), Polshin and Tokar (1957), Wahls (1994), and Zhang and Ng (2007). The tolerable amount of movement and distortion also depends on the level of damage that can be tolerated by the building, including the appearance and the function. It varies with many factors, such as the type, size, function, and properties of the structure; the soil type and properties; the method and time of construction; the type and stiffness of the foundation; and the rate and uniformity of the settlement. Zhang and Ng (2007) collected data for 380 buildings; the results are shown in Table 17.11. Many codes include tolerable values as well. All in all, it appears that, for buildings, vertical movements of 50 mm are generally tolerable and that

Table 17.11 Allowable Vertical Displacement and Angular Distortion for Buildings

	Allowable value (FS = 1.5)		Allowable value (95% percentile)	
Building category	Vertical displacement (mm)	Span length over differential movement	Vertical displacement (mm)	Span length over differential movement
Foundation type:				
Shallow foundation	145	245	49	833
Deep foundation	71	770	42	3333
Structural type:				
Frame buildings	92	323	29	1000
Load-bearing wall	60	417	22	1111
Soil type:				
Clay	113	333	44	1666
Sand	57		27	
Building use:				
Factory	141	263	53	526
Office	81	435	47	833

(Zhang and Ng 2007)

larger movements can be tolerated if they occur uniformly. It is common practice, however, to design buildings for 25 mm settlement. A span length over differential movement ratio of 500 also seems good guidance in most cases:

$$\text{For buildings} \quad \frac{L}{s_d} \geq 500 \qquad (17.117)$$

where L is the span length and s_d is the differential vertical movement between adjacent columns.

17.12 LARGE MAT FOUNDATIONS

17.12.1 General Principles

Mat foundations, also called *raft foundations*, are shallow foundations. A large mat may be used as the foundation for a tall building in an area where the soil strength does not increase significantly with depth. The design strategy is to place the foundation at a depth such that the weight of the excavated soil is nearly equal to the weight of the building. For this reason it is often called a *floating foundation*. The unit weight of soil (about 16 to 20 kN/m^3) is much larger than the unit weight of a building (about 2.5 to 5 kN/m^3). A story is about 3 m high; therefore, using a ratio of unit weight of soil over unit weight of building equal to 5, when a mat foundation is placed at a depth of 12 m, the weight of soil removed is equal to a 60 m high building with 20 stories (16 out of the ground). Placing such a building on such a foundation would lead to a postconstruction stress on the soil equal to the one in the soil prior to construction. Therefore, there would likely be very little problem with ultimate capacity and settlement. More precisely, the soil movement would be reduced to the unloading and reloading, which would take place during excavation of the soil and construction of the building. If a building taller than 20 stories were built, the ultimate capacity and settlement would have to be considered under the excess load beyond that of 20 stories. These are the basic geotechnical governing principles for the design of large mat foundations for tall buildings.

The design of the mat itself is controlled by the bending that it will undergo. The column loads represent point loads on the mat that must be transferred to the soil without punching or excessive bending. This requires an amount of concrete reinforcement dictated by the interaction between the stiff mat and the softer soil. The analysis can proceed in one of two ways: beam on elastic foundation approach or finite element approach. In the beam on elastic foundation approach, the stiffness of the soil comes from a stress-strain curve obtained, for example, from a consolidation test, and the stiffness of the mat is given by its bending stiffness value EI, where E is the modulus of concrete and I is the moment of inertia of the section. Because the stiffness of the soil is dependent on the strain experienced by the soil, and because the soil strain also depends on the mat stiffness, an iteration process develops where a run is made with the mat stiffness and a chosen soil stiffness; then the results are used to calculate the new soil strain and the next value of the soil stiffness. This process is repeated until the assumed soil stiffness and the calculated soil stiffness are within an acceptable tolerance. At that point, the bending moments in the mat are used to choose the amount of reinforcement necessary. With the more sophisticated FEM approach, the interaction is taken into account directly and the output gives the pressure distribution under the mat, the mat settlement profile, and the mat bending moment and shear. These mats often end up being of a uniform thickness equal to about 3 m. Pouring such mats is a large operation requiring many concrete trucks lined up one after the other; the concrete sets while developing very high temperatures (up to 80° C or more) due to the heat of hydration.

17.12.2 Example of Settlement Calculations

A large building weighing 400 MN is to be built on a deep deposit of very stiff clay (Figure 17.53). The building is to be placed at the bottom of a 15 m deep excavation, which will correspond to 4 levels of parking garages and 1 level for a mall. The height of the building is 180 m or 60 stories in addition to the 15 m of embedment. The footprint is a 30 m × 30 m square and the building will be founded on a thick mat foundation. The soil has a total unit weight of 20 kN/m^3 and the groundwater level is deeper than the zone of influence of the foundation. Pressuremeter tests gave a profile of first load modulus E_o and reload modulus E_r, as shown on Figure 17.53, with the equations:

$$\text{First load PMT modulus } E_o \text{ (MPa)} = 10 + 0.5z\text{(m)} \quad (17.118)$$

$$\text{Reload PMT modulus } E_r \text{ (MPa)} = 50 + 2z\text{(m)} \quad (17.119)$$

The strength has been deemed sufficient not to create problems of ultimate bearing capacity, but the settlement must be estimated.

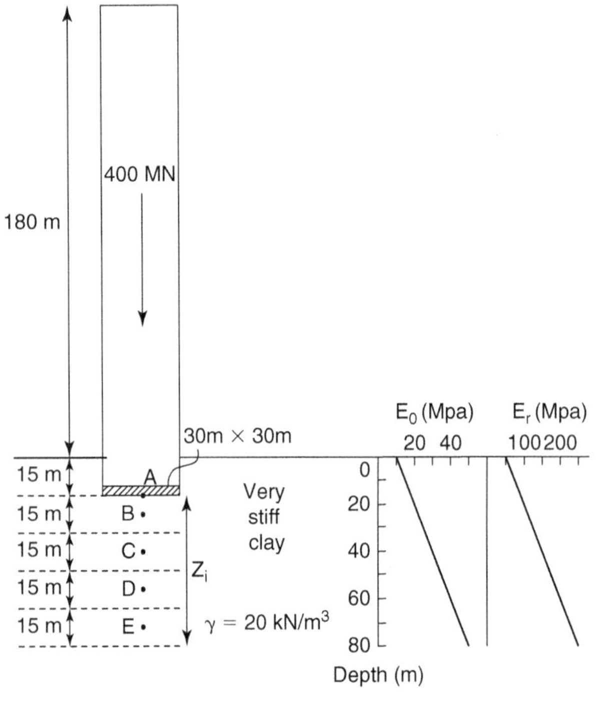

Figure 17.53 High-rise settlement: Problem definition.

It is first necessary to understand clearly what the soil will undergo at any depth below the building (Figure 17.54). Before any excavation, the vertical stress at a point in the soil is σ_{ov}. The 15 m deep excavation creates a decrease in stress equal to $\Delta\sigma_{exc}$ below the excavation level, such that after the excavation the stress has decreased to $\sigma_{ov} - \Delta\sigma_{exc}$. The construction of the building creates an increase in stress equal to $\Delta\sigma_{\text{bldg}}$ such that after the building is completed, the stress in the soil is $\sigma_{ov} - \Delta\sigma_{exc} + \Delta\sigma_{bldg}$. When the building is constructed starting at the bottom of the excavation, the soil first follows a reloading curve until the weight of the building becomes equal to the weight of soil removed. In other words, the recompression settlement S_{rel} should be calculated under the stress increment $\Delta\sigma_{exc}$ while using the PMT reload modulus E_r. Then, as construction continues, the soil is loaded in the virgin behavior and the settlement beyond the recompression settlement (sometimes called the *net settlement*, S_{net}) should be calculated under the stress increment $\Delta\sigma_{bldg} - \Delta\sigma_{exc}$ (sometimes called the *net increase in stress*) while using the PMT first load modulus E_o. This process is illustrated in Figure 17.54.

The steps outlined in section 17.8.5 are followed and are presented in Table 17.12.

1. The zone of influence z_i is taken as 2B because the building imprint is square: $z_i = 60$ m.
2. The zone of influence in this case is decomposed into 4 layers (Figure 17.53), each 15 m thick. This is column 1 in Table 17.12.
3. We calculate the initial stress σ_{ov} at the center of each layer. For example, the center of layer 1 is at a depth of 22.5 m and therefore the vertical stress at the center is $\sigma_{ov} = 22.5 \times 20 = 450$ kPa. This is column 2 in Table 17.12.
4. The first load modulus E_o is calculated in the center of each layer. For example, the modulus in the middle of layer 1 is $E_o = 10 + 0.5 \times 22.5 = 21.25$ MPa. This is column 3 in Table 17.12.
5. The reload modulus E_r is calculated in the center of each layer. For example, the modulus in the middle of layer 1 is $E_r = 50 + 2 \times 22.5 = 95$ MPa. This is column 4 in Table 17.12.
6. We calculate the decrease in stress $\Delta\sigma_{exc}$ in the middle of each layer due to the excavation. The total pressure decrease at the bottom of the excavation is $p_{\text{exc}} = 15 \times 20 = 300$ kPa. The bulb of pressure method is used to obtain the decrease in stress in the middle of each layer. For example, the decrease in stress $\Delta\sigma_{exc}$ in the middle of layer 1 is 0.85 $p_{exc} = 255$ kPa according to the bulb of pressure shown in Figure 17.31. This is column 5 in Table 17.12.
7. We calculate the increase in stress $\Delta\sigma_{bldg}$ in the middle of each layer due to the construction of the building. The total pressure increase at the foundation level is the weight of the building divided by the foundation area or $p_{bldg} = 400000/30 \times 30 = 444$ kPa. The bulb

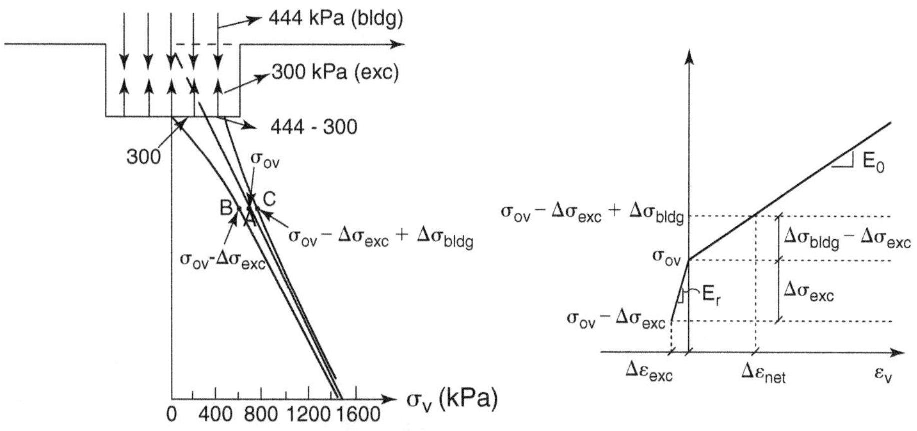

Figure 17.54 High-rise settlement: Stresses.

Table 17.12 High-Rise Settlement Calculations

1	2	3	4	5	6	7	8	9	10
H (m)	σ_{ov}(kPa)	E_o(kPa)	E_r(kPa)	$\Delta\sigma_{exc}$(kPa)	$\Delta\sigma_{bldg}$(kPa)	$\Delta\varepsilon_{exc}$	$\Delta\varepsilon_{net}$	ΔH_{rel}(mm)	ΔH_{net}(mm)
15	450	21250	95000	255	378	0.00268	0.00579	40	87
15	750	28750	125000	150	222	0.00120	0.00250	18	38
15	1050	36250	155000	75	111	0.00048	0.00099	7	15
15	1350	43750	185000	45	67	0.00024	0.00050	4	8
								69	148

Figure 17.55 Washington Monument and San Jacinto Monument on large mats.

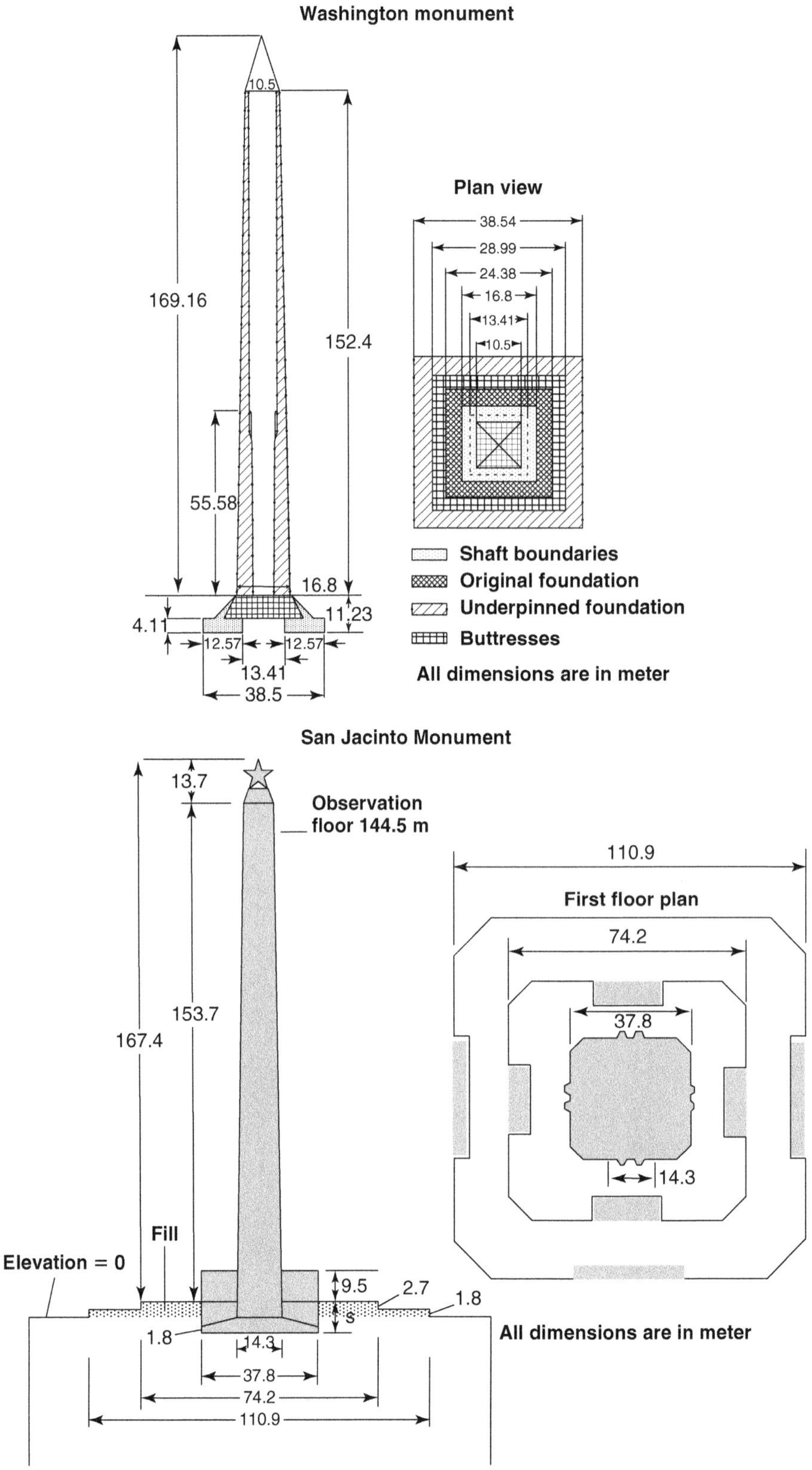

Figure 17.56 Washington Monument and San Jacinto Monument dimensions.

of pressure method is used to obtain the increase in stress in the middle of each layer. For example, the increase in stress $\Delta\sigma_{bldg}$ in the middle of layer 1 is $0.85 p_{exc} = 378$ kPa according to the bulb of pressure shown in Figure 17.31. This is column 6 in Table 17.12.

8. The weight of the soil excavated is equal to 20 × 30 × 30 × 15 = 270,000 kN. Because the building weighs 400,000 kN, the weight of soil excavated represents 67.5% of the total weight of the building. When the building has reached a weight of 270,000 kN, we assume that the soil will have been recompressed to its initial position. We calculate the strain increment $\Delta\varepsilon_{\rm rel}$ in the middle of each layer due to the reloading by the building construction from zero load to a load equal to the weight of the excavated soil as:

$$\Delta\varepsilon_{rel} = \frac{\Delta\sigma_{exc}}{E_r} \tag{17.120}$$

This is column 7 in Table 17.12. Equation 17.120 neglects the influence of the stress confinement on the strain (conservative).

9. The increase in stress $\Delta\sigma_{bldg}$ in the middle of each layer corresponds to the weight of the entire building. Because the strain increment corresponding to $\Delta\sigma_{exc}$ has already been calculated (Eq. 17.120), we now need to calculate the strain increment $\Delta\varepsilon_{net}$ due to $(\Delta\sigma_{bldg} - \Delta\sigma_{exc})$. This is done as follows:

$$\Delta\varepsilon_{net} = \frac{\Delta\sigma_{bldg} - \Delta\sigma_{exc}}{E_o} \tag{17.121}$$

10. Then the compression ΔH_{rel} of each layer due to $\Delta\varepsilon_{rel}$ is calculated as:

$$\Delta H_{rel} = H\Delta\varepsilon_{rel} \tag{17.122}$$

and the compression ΔH_{net} of each layer due to $\Delta\varepsilon_{net}$ is calculated as:

$$\Delta H_{net} = H\Delta\varepsilon_{net} \tag{17.123}$$

This corresponds to columns 9 and 10 in Table 17.12.

11. Finally, the settlement is calculated by adding the compression of the four layers. The settlement S_{rel} due to the reloading of the soil under the part of the weight of the building equal to the weight of soil excavated is:

$$S_{rel} = \sum_{i=1}^{4} H_i \Delta\varepsilon_{rel\ i} = \sum_{i=1}^{4} \Delta H_{rel\ i} = 69 \text{ mm} \tag{17.124}$$

Then the settlement S_{net} due to the weight of the building in excess of the weight of the excavated soil is:

$$S_{net} = \sum_{i=1}^{4} H_i \Delta\varepsilon_{net\ i} = \sum_{i=1}^{4} \Delta H_{net\ i} = 148 \text{ mm} \tag{17.125}$$

The total settlement S_{tot} of the building is:

$$S_{tot} = S_{rel} + S_{net} = 217 \text{ mm} \tag{17.126}$$

17.12.3 Two Case Histories

Two large mat case histories are presented here: the Washington Monument (Briaud et al. 2009) and the San Jacinto Monument (Briaud et al. 2007) (Figure 17.55). The Washington Monument (Washington, DC) was completed in 1884 and stands at 169.16 m tall above ground (Figure 17.56). In a first phase, it was built to a height of 55.6 m on a square mat 24.38 m by 24.38 m when construction stopped with a calculated settlement of 1.33 m. In a second phase, the mat was underpinned and extended to a square ring mat 38.54 m by 38.54 m on the outside and 13.41 m by 13.41 m on the inside. The monument experienced an additional measured settlement of 0.12 m while construction was completed. Additional data are shown in Table 17.13. The mat rests on an 8.3 m thick layer of sand and gravel with a blow count averaging 100 bpf underlain by a 11.7 m thick layer of very stiff clay with an average undrained shear strength of 100 kPa. Below the very stiff clay is the bedrock. The settlement during phase one was calculated, on the basis of available consolidation tests, to be 1.33 m (Briaud et al. 2009), whereas the settlement during phase two was only 0.17 m (Figure 17.57). The reason the settlement was so large during the first phase is that the bottom of the first mat foundation was shallow and rested

Table 17.13 Data for the Washington Monument and the San Jacinto Monument

Washington Monument	San Jacinto Monument
Total weight = 608 MN	Total weight = 313 MN
Weight of foundation = 184 MN	Weight of foundation = 133 MN
Pressure at foundation level = 465 kPa	Pressure at foundation level = 224 kPa
Net pressure = 252 kPa	Net pressure = 141 kPa
Calculated total settlement = 1.50 m	Calculated total settlement = 0.61 m
Measured settlement after underpinning = 0.17 m	Measured settlement after mat placed = 0.33 m

on a very compressible soft clay layer. The underpinning brought the foundation down to the strong sand and gravel layer. More details can be found in Briaud et al. (2009).

The San Jacinto Monument (Houston, Texas, USA) was completed in 1936 and stands at 171.9 m tall above ground; it is the tallest free standing column in the world. It rests on a square mat foundation 37.8 m by 37.8 m. Some of the data regarding weight and pressure are shown in Table 17.13. The mat rests on a deep deposit of very stiff clay with an average undrained shear strength equal to 100 kPa. The CPT point resistance is 1000 kPa at the ground surface and increases to 3000 kPa at 10 m depth. The pressuremeter limit pressure is 800 kPa at the ground surface, increasing to 3000 kPa at a depth of 40 m. The PMT first load modulus is 15 MPa at the ground surface, increasing to 60 MPa at a depth of 40 m. The PMT reload modulus is 2.1 times larger than the first load modulus on the average. The PMT viscous exponent n averages 0.045. The settlement was calculated, on the basis of available consolidation tests, to be 0.61 m (Briaud et al. 2007). The settlement measured after the mat was poured reached 0.33 m (Figure 17.58). More details can be found in Briaud et al. (2007).

These two tall, columnar structures on large mats settled significantly, yet both are as straight as possible, with no lean detectable to the naked eye. If heterogeneity had been an issue, these structures would likely have tilted. However, at the scale of a 38 m by 38 m mat, the soil is much more homogeneous than at the scale of a cone penetrometer, for example. This shows that heterogeneity is scale dependent and that tall structures can stand much larger settlement than might be thought. The weight of these simple structures is shown in Table 17.13. By comparison, the Eiffel Tower in Paris, France, weighs 94 MN; the Tower of Pisa in Pisa, Italy, weighs 142 MN; each tower of the World Trade Center in New York weighed 4500 MN; and the Burj Khalifa in Dubai weighs about 5000 MN.

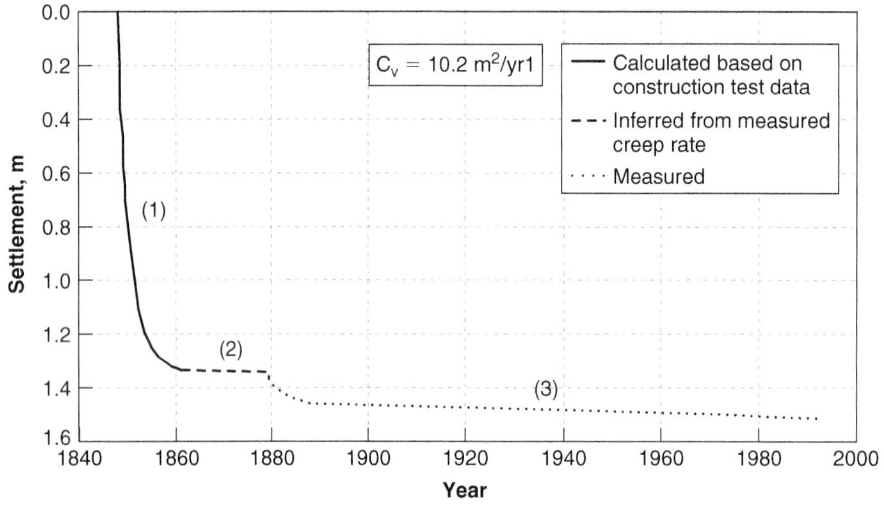

Figure 17.57 Settlement of the Washington Monument.

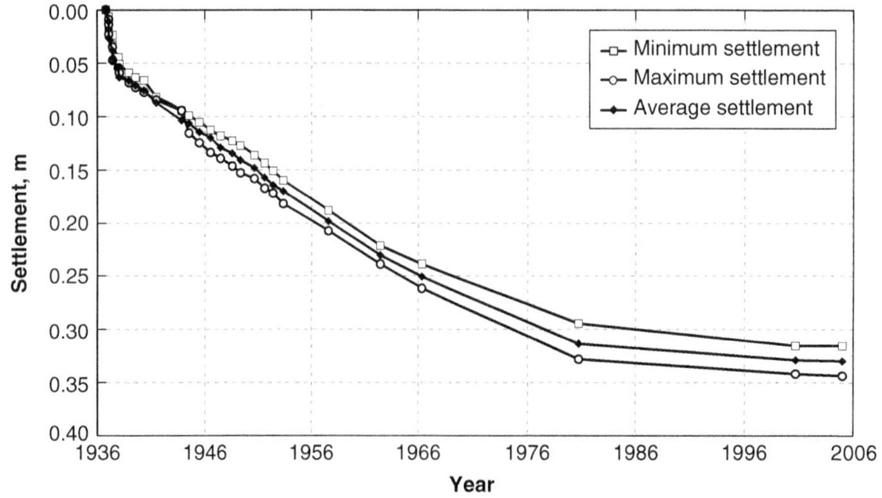

Figure 17.58 Settlement of the San Jacinto Monument.

PROBLEMS

17.1 If a shallow foundation test is performed on clay, there is a clear plunging load. If a shallow foundation test is performed on sand, the load continues to increase and a clear plunging load is not obvious. Explain why. How is the ultimate load defined for the load test on sand?

17.2 Calculate the ultimate bearing pressure for the footings and the soil described in section 17.2 by all applicable methods listed in section 17.6. Additional soil data can be obtained from Briaud and Gibbens (1999). If you had to give one answer what would you choose to do?

17.3 Calculate the ultimate bearing pressure (edge failure) for the mat of the San Jacinto Monument (section 17.12.2) by all applicable methods listed in section 17.6. If you had to give one answer, what would you choose to do?

17.4 Redo the example of Figure 17.23 but using the mean curve instead of the design curve for the Γ function.

17.5 Calculate the increase in stress under the center of the circular footing (Figure 17.1s) as a function of depth by all the methods presented in section 17.8.7. Show the profile of effective stress before construction and after construction. At what depth is $\Delta\sigma'(z)$ equal to 1/10 of $\Delta\sigma'(z = 0)$?

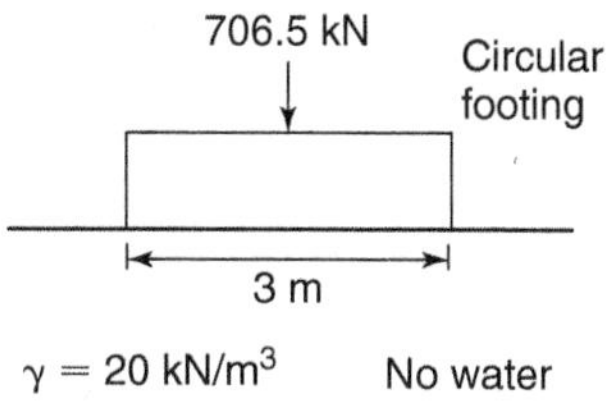

Figure 17.1s Circular footing.

17.6 Calculate the settlement of the footing shown in Figure 17.2s. If only 10 mm of settlement can be tolerated by the structure, what is the size of the footing required to carry the same load?

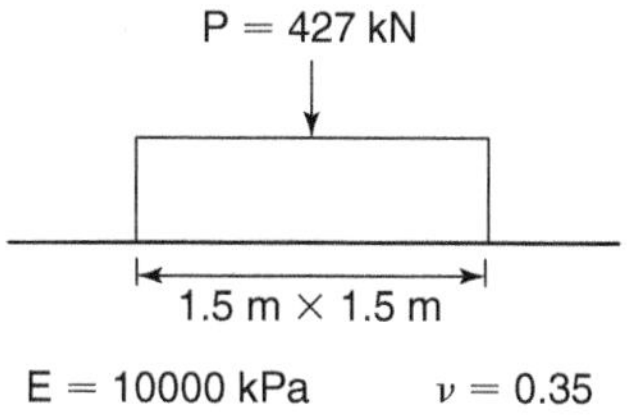

Figure 17.2s Square footing.

17.7 A square foundation is 3 m × 3 m and rests on a deep layer of sand at a depth of 1.5 m. The soil modulus at the ground surface is 10 MPa and increases linearly to 50 MPa at a depth of 10 m. What load can the footing carry if the allowable settlement is 25 mm?

17.8 Using the Schmertmann method, simplify the equation giving the settlement of a footing at the surface of a sand deposit when the soil is uniform with a constant value of E. Compare that equation to the elasticity equation.

17.9 A column load of 4000 kN is to be supported by a square spread footing on a medium-dense sand. Recommend the size and the embedment of the footing after addressing the issue of bearing capacity and settlement of the footing (25 mm is tolerable). Soil properties: N = 30 blows/ft, $q_c = 8$ MPa, $f_c = 70$ kPa, $p_L = 1500$ kPa, $E_o = 12$ MPa, $\gamma = 20$ kN/m^3. If you need additional properties, assume reasonable values.

17.10 A column load of 2000 kN is to be supported by a square spread footing on a very stiff clay. Recommend the size of the footing after addressing the issue of bearing capacity and settlement of the footing (25 mm is tolerable). Soil properties: $s_u = 100$ kPa, $q_c = 1.5$ MPa, $f_c = 70$ kPa, $p_L = 500$ kPa, $E_o = 7.5$ MPa, $C_c = 0.3$, $c_v = 10^{-4}$ cm^2/s, $\gamma = 18$ kN/m^3. If you need additional properties, assume reasonable values.

17.11 In 1955, an oil tank 10 m high and 38 m in diameter is built as shown in Figure 17.3s.

a. Calculate the settlement of the center of this tank (point C on Figure 17.3s) using the data from Figure 17.4s. Assume that the stress increase in the middle of the compressible layer is equal to the pressure under the tank because the layer is thin.

In 1975, this tank is removed; a year later, a new tank 15 m high and 76 m in diameter is built. The edge of the new tank goes through the center of the old tank.

b. Calculate the settlement of the edge of the new tank away from the old tank (point B on Figure 17.3s) using the data from Figure 17.4s. Assume that the stress increase at the edge of the new tank in the middle of the compressible layer is equal to one half of the pressure under the new tank.

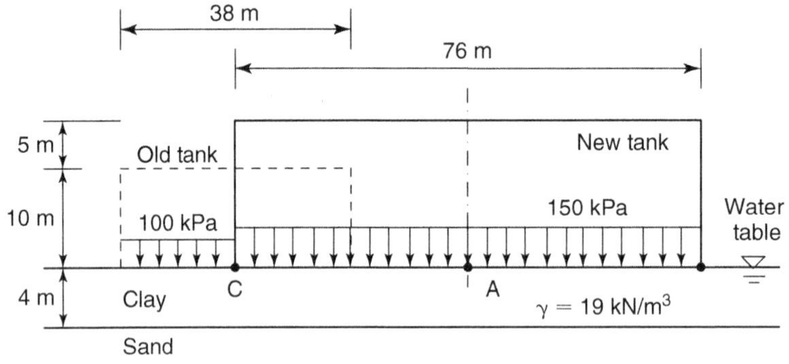

Figure 17.3s Old and new oil tanks.

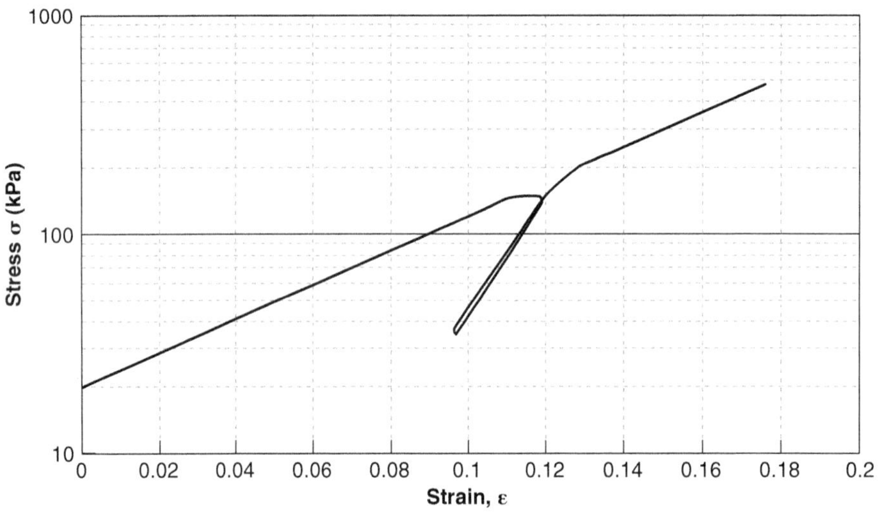

Figure 17.4s Stress-strain curve for oil tank problem.

c. Calculate the settlement of the edge of the new tank that passes over the center of the old tank (point C on Figure 17.3s) using the data from Figure 17.4s. Make the same assumption as in b.

d. Do you see any problem with the difference in settlement between C and B for the new tank?

17.12 Use the shrink-swell case history from section 17.9.4 to calculate the footing movements and compare your results with the measured movements.

17.13 The high-rise building shown in Figure 17.53 is subjected to a hurricane wind of 200 km/h. This wind creates a pressure of 3 kPa on the flat side of the building. Calculate the pressure diagram under the foundation.

17.14 The high-rise building shown in Figure 17.53 is placed on a stiff clay with the following properties: compression index C_c equal to 0.4, recompression index C_r equal to 0.1, initial void ratio e_o equal to 0.5, and total unit weight equal to 20 kN/m^3. The soil is lightly overconsolidated by overburden removal and has a preconsolidation pressure σ'_p150 kPa higher than the effective stress σ'_{ov}. The groundwater level is at the ground surface. Calculate the settlement of the building. How would you estimate the time required for the settlement to take place if c_v were known?

17.15 The annual drying and wetting condition of a site is shown in Figure 17.5s. Calculate the shrink and swell displacement at the center and at the edge of the building, and then calculate the differential movement between the two points. Hint: Use

$$\varepsilon = \frac{\Delta H_i}{H_i} = f\frac{\Delta w_i}{E_{wi}} \qquad \therefore \varepsilon = 0.33\frac{\Delta w_i}{(\gamma_w/\gamma_d)} \text{ and } w_i = 25\%, \qquad \gamma_d = 14\text{ kN/m}^3, \gamma_w = 10\text{ kN/m}^3$$

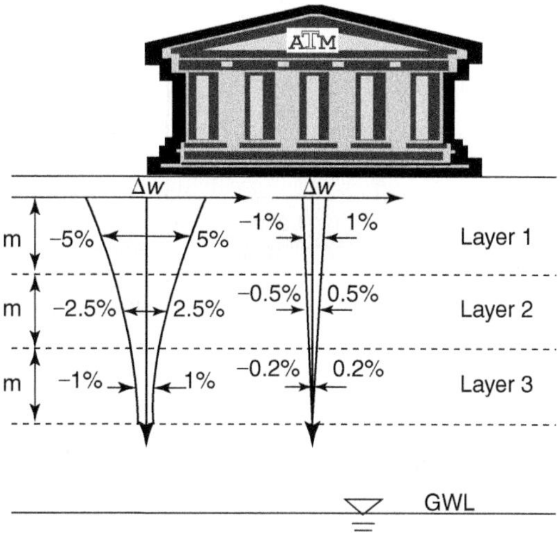

Figure 17.5s Annual drying and wetting condition.

17.16 A stiffened slab on grade for a two-story house is to be designed. The slab and site data are given as follows: slab dimensions 20 m by 20 m, beam spacing $s = 3.0$ m (for both directions), beam width $b = 0.3$ m, slab load $w = 10$ kPa, depth of movement zone $H = 3.0$ m, soil surface water content change $\Delta w_o = 20\%$. Recommend a beam depth that will minimize the distortion of the slab for the edge drop case to more than L/Δ = 500.

17.17 Calculate the settlement of the San Jacinto Monument using the pressuremeter data given in section 17.12.3.

Problems and Solutions

Problem 17.1

If a shallow foundation test is performed on clay, there is a clear plunging load. If a shallow foundation test is performed on sand, the load continues to increase and a clear plunging load is not obvious. Explain why. How is the ultimate load defined for the load test on sand?

Solution 17.1

The reason a shallow foundation test performed on sand shows no clear plunging load is that fine-grained soils tend to shear in an undrained mode during a load test, whereas coarse-grained soils likely shear in a drained mode. The undrained shear strength of a clay does not vary much with the stress and confinement level (s_u = constant), so when the load on the footing increases, the shear strength does not increase and the failure is clearly defined. The drained shear strength of a sand depends on the stress and the confinement level ($s = \sigma' \tan \varphi'$), so when the load increases the shear strength also increases. Therefore, the failure for the sand is ill defined, and no obvious plunging load is observed. The ultimate load on the sand can be defined as the load corresponding to a movement equal to one-tenth of the foundation width.

Problem 17.2

Calculate the ultimate bearing pressure for the footings and the soil described in section 17.2 by all applicable methods listed in section 17.6. Additional soil data can be obtained from Briaud and Gibbens (1999). If you had to give one answer what would you choose to do?

Solution 17.2

Figure 17.6s shows the illustration of the soil profile.

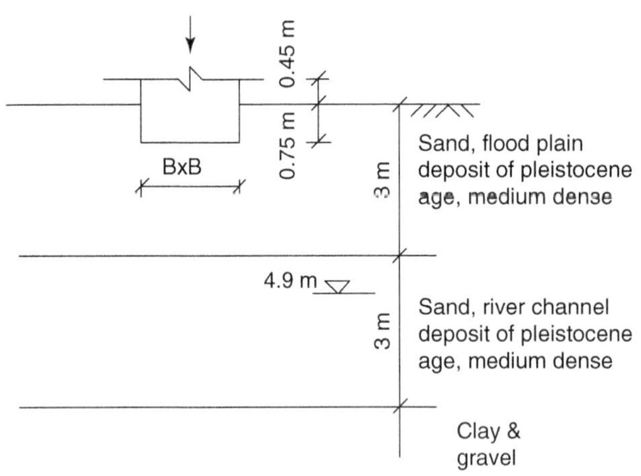

Figure 17.6s Footing and soil profile.

The methods include Skempton method, PMT method, CPT method, SPT method, and Terzaghi's general bearing capacity equation (GBE) method. The Skempton method is applicable only to fine-grained soil, so it is not discussed in this solution. The simple versions of the PMT method, CPT method, SPT method, and GBE method are used to solve the bearing capacity for the 1 m × 1 m footing, 1.5 m × 1.5 m footing, 2.5 m × 2.5 m footing, and 3 m × 3 m footing respectively. All footings are embedded 0.75 m. The soil unit weight is 15.5 kN/m^3. The following soil properties are selected within the zone of influence of the footings from the soil profiles presented in Figures 17.7s, 17.8s, 17.9s: the PMT limit pressure is 800 kPa, the CPT point resistance is 6000 kPa, and the SPT blow count is 18 bpf. Since the footing width does not appear in the ultimate bearing pressure p_u equations, and since the embedment depth is the same for all footings, then the value of p_u will be the same for all footings.

PMT method:

$$p_u = k_p p_L + \gamma D \tag{17.1s}$$

where k_p is the pressuremeter bearing capacity factor, p_L is the pressuremeter limit pressure, γ is the total unit weight of the soil above the footing depth, and D is the embedment of the footing.

Based on Equation 17.19, k_p is 1.2 and p_u is:

$$p_u = 1.2 \times 800 + 15.5 \times 0.75 = 972 \text{ kPa}$$

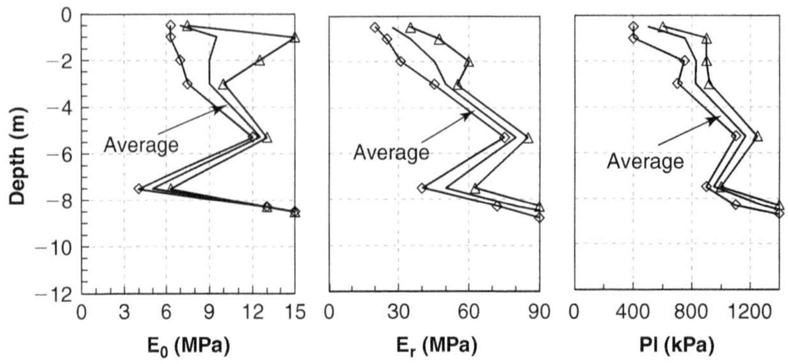

Figure 17.7s PMT profile.

CPT method:

$$p_u = k_c q_c + \gamma D \qquad (17.2s)$$

where k_c is the cone penetrometer bearing capacity factor, q_c is the average point resistance within one footing width below the footing, γ is the total unit weight of the soil above the footing depth, and D is the embedment of the footing. Based on Equation 17.22, k_c is 0.2 and p_u is:

$$p_u = 0.2 \times 6000 + 15.5 \times 0.75 = 1212 \text{ kPa}$$

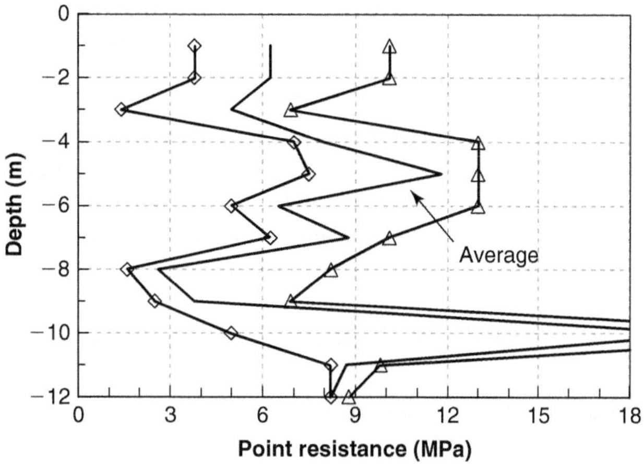

Figure 17.8s CPT profile.

SPT method:

$$p_u = k_N N p_a + \gamma D \qquad (17.3s)$$

where k_N is the SPT bearing capacity factor, N is the average blow counts within one footing width below the footing, p_a is the atmospheric pressure, γ is the total unit weight of the soil above the footing depth, and D is the embedment of the footing. Based on Equation 17.28, k_N is 0.6 and p_u is:

$$p_u = 0.6 \times 18 \times 101.3 + 15.5 \times 0.75 = 1106 \text{ kPa}$$

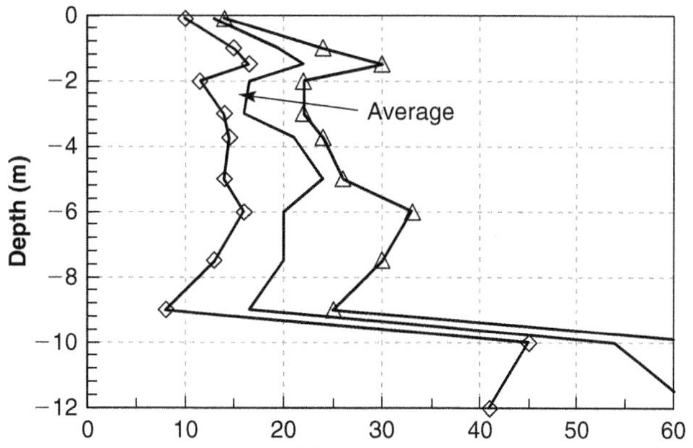

Figure 17.9s SPT profile.

Terzaghi's GBE:

This method is based on the following equation:

$$p_u = c' N_c + \frac{1}{2}\gamma_1 B N_r + \gamma_2 D N_q \qquad (17.4s)$$

where c' is the effective stress cohesion; N_c, N_r, and N_q are bearing capacity factors, γ_1 is the average effective unit weight of soil within the one footing width below the foundation, B is the width of the foundation, γ_2 is the effective unit weight of the soil above the foundation, and D is the depth of embedment of the foundation.

For this case, $N = 18$ bl/ft, $c' = 0$, $\phi' = 32.5°$ (Figure 15.12), $N_c = 39$, $N_r = 23$, and $N_q = 23$ (Figure 17.14):

$$p_u = 0 \times 39 + \frac{1}{2} \times 15.5 \times B \times 23 + 15.5 \times 0.75 \times 23 = 178.25B + 267.38$$

$$1 \times 1 \text{ m footing} : p_u = 446 \text{ kPa}$$
$$1.5 \times 1.5 \text{ m footing} : p_u = 535 \text{ kPa}$$
$$2.5 \times 2.5 \text{ m footing} : p_u = 713 \text{ kPa}$$
$$3 \times 3 \text{ m footing} : p_u = 802 \text{ kPa}$$

Based on this analysis, I would choose the average p_u value from the PMT method, CPT method, and SPT method as the ultimate bearing pressure of the footings. I would not use the general bearing capacity equation predictions because the soil profile does not correspond to the assumption made to derive that equation (linear strength increase with depth)

$$p_u = \frac{972 + 1212 + 1106}{3} = 1097 \text{ kPa}$$

Note that these footings were load tested (Briaud, Gibbens, 1999) individually and gave ultimate bearing pressures (pressure at one tenth of the footing width) equal to:

$$1 \times 1 \text{ m footing} : p_u = 1500 \text{ kPa}$$
$$1.5 \times 1.5 \text{ m footing} : p_u = 1500 \text{ kPa}$$
$$2.5 \times 2.5 \text{ m footing} : p_u = 1300 \text{ kPa}$$
$$3 \times 3 \text{ m footing (North)} : p_u = 1250 \text{ kPa}$$
$$3 \times 3 \text{ m footing (South)} : p_u = 1500 \text{ kPa}$$

Problem 17.3

Calculate the ultimate bearing pressure (edge failure) for the mat of the San Jacinto Monument (section 17.12.2) by all applicable methods listed in section 17.6. If you had to give one answer, what would you choose to do?

Solution 17.3

Skempton:

$$p_u = N_c s_u + \gamma D$$

$D = 9.1$ m (Figure 17.56), and $B = 37.8$ m, so the ratio $D/B = 0.24$.

From the Skempton chart in Figure 17.7, $N_c = 6.7$. The undrained shear strength below the monument is given as $S_u = 100$ kPa. Assuming a unit weight of 19 kN/m^3:

$$p_u = 100 \times 6.7 + 19 \times 9.1 = 843 \text{ kPa}$$

PMT Method (for clay):

$$p_u = k_p p_L^* + \gamma D$$

$D = 9.1$ m (Figure 17.56), $B = L = 37.8$ m, so the ratio $B/L = 1$, and $D/B = 0.24$:

$$p_u = 0.9 \times 800 + 19 \times 9.1 = 893 \text{ kPa}$$

CPT Method (for clay):

$$p_u = 0.40q_c + \gamma D$$

Because the profile of q_c increases from 1000 kPa at the surface to 3000 kPa at 10 m depth, and because there is 9.1 m embedment, it appears reasonable to select 3000 kPa as the cautious design value for q_c:

$$p_u = 0.4 \times 3000 + 19 \times 9.1 = 1373 \text{ kPa}$$

There are no SPT data, c', or φ' to use with the remaining methods.

If I had to choose one answer, I would choose 900 kPa as a conservative yet substantiated value (supported by two methods). The CPT method seems a bit optimistic in this case.

Problem 17.4

Redo the example of Figure 17.23 but using the mean curve instead of the design curve for the Γ function.

Solution 17.4

The f factor does not change when we use the mean curve instead of the design curve:

$\Delta R/R_o$	p_p (kN/m^2)	s/B	s (mm)	Γ_{Mean}	f	p_f (kN/m^2)	Q(kN)
0	0	0	0	0	0.689	0	0
0.006	75	0.00144	4.32	3.6	0.689	186.0	8370
0.012	120	0.00288	8.64	3.1	0.689	256.3	11533
0.024	220	0.00576	17.28	2.75	0.689	416.8	18756
0.032	300	0.00768	23.04	2.25	0.689	465.1	20929
0.055	450	0.0132	39.6	1.9	0.689	589.1	26509
0.1	650	0.024	72	1.5	0.689	671.8	30231
0.15	775	0.036	108	1.35	0.689	720.9	32440
0.20	850	0.048	144.0	1.3	0.689	761.3	34258

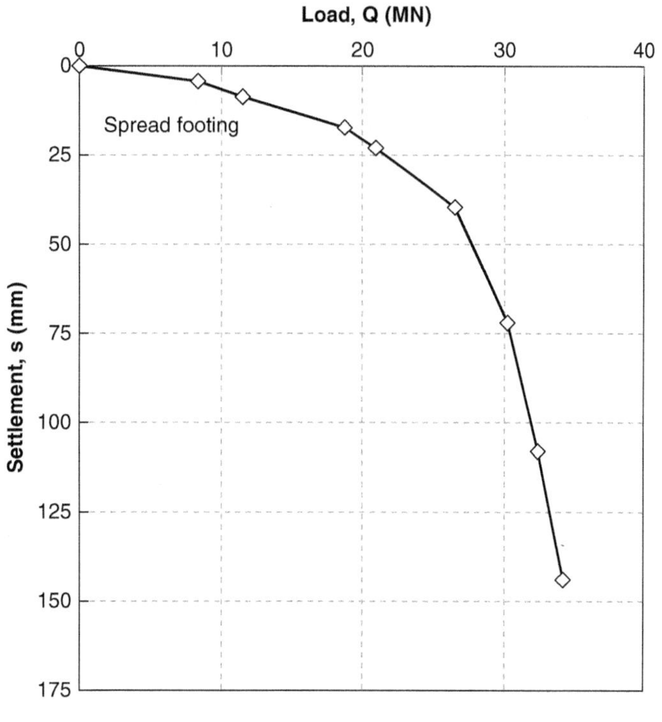

Figure 17.10s Load settlement curve.

Problem 17.5

Calculate the increase in stress under the center of the circular footing (Figure 17.1s) as a function of depth by all the methods presented in section 17.8.7. Show the profile of effective stress before construction and after construction. At what depth is $\Delta\sigma'(z)$ equal to 1/10 of $\Delta\sigma'(z = 0)$?

Solution 17.5

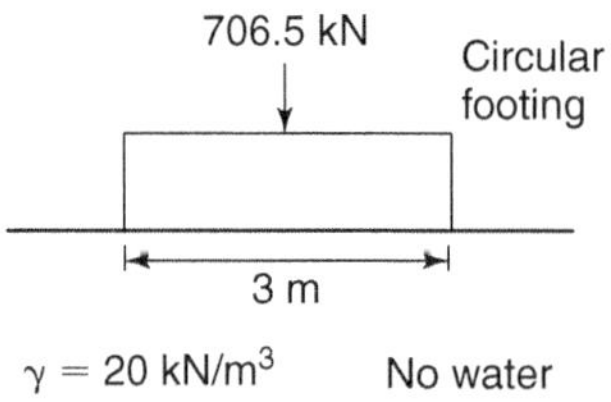

Figure 17.1s Circular footing.

2 to 1 method:

The average pressure under the footing is:

$$p_{ave} = \frac{Q}{\pi B^2/4} = \frac{706.5}{\pi(3)^2/4} = 100 \text{ kPa}$$

The increase in stress for a circular footing is:

$$\sigma_z = \frac{Q}{\pi(B+z)^2/4}$$

Depth (m)	$\Delta\sigma'_z$ (kN/m²)	Before construction (kN/m²)	After construction (kN/m²)
0	100	0.0	100
1	56.2	20.0	76.2
2	36.0	40.0	76.0
3	25.0	60.0	85.0
4	18.4	80.0	98.4
5	14.1	100.0	114.1
6	11.1	120.0	131.1
7	9.0	140.0	149.0
8	7.4	160.0	167.4
9	6.2	180.0	186.2
10	5.3	200.0	205.3

Bulbs of pressure (using Figure 17.31):

Depth (m)	Depth/diameter	Factor (from bulbs of pressure)	$\Delta\sigma'_z$ (kN/m²)	Before construction (kN/m²)	After construction (kN/m²)
0	0.00	1	100	0.0	100
1	0.33	0.8	80	20.0	100
2	0.67	0.57	57	40.0	97
3	1.00	0.35	35	60.0	95
4	1.33	0.23	23	80.0	103
5	1.67	0.17	17	100.0	117
6	2.00	0.12	12	120.0	132
7	2.33	0.09	9	140.0	149
8	2.67	0.07	7	160.0	167
9	3.00	0.06	6	180.0	186
10	3.33	0.04	5	200.0	204

Newmark's chart:

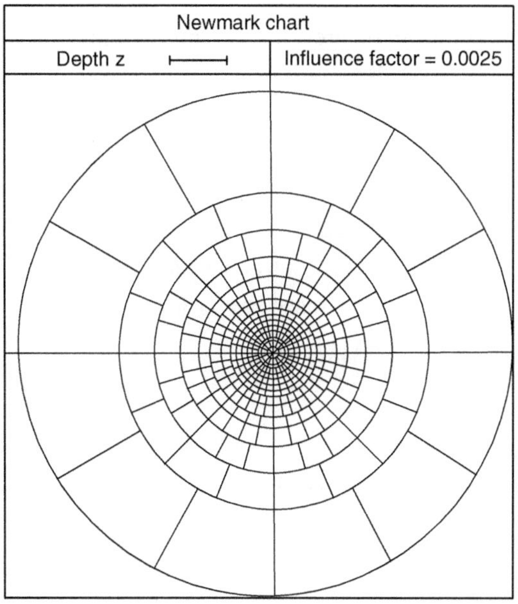

Figure 17.11s Newmark's chart.

Depth (m)	Number of squares	Factor	$\Delta\sigma'_z$ (kN/m^2)	Before construction (kN/m^2)	After construction (kN/m^2)
0		1.0	100	0.0	100
1	316	0.79	79	20.0	99
2	200	0.5	50	40.0	90
3	104	0.26	26	60.0	86
4	72	0.18	18	80.0	98
5	40	0.1	10	100.0	110
6	32	0.08	8	120.0	128
7	30	0.075	7.5	140.0	147.5
8	24	0.06	6	160.0	166.0
9	18	0.045	4.5	180.0	184.5
10	10	0.025	2.5	200.0	202.5

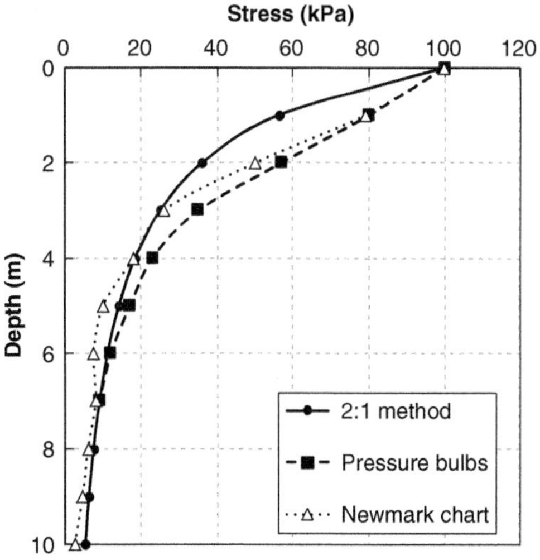

Figure 17.12s Difference between the three methods.

Problem 17.6

Calculate the settlement of the footing shown in Figure 17.2s. If only 10 mm of settlement can be tolerated by the structure, what is the size of the footing required to carry the same load?

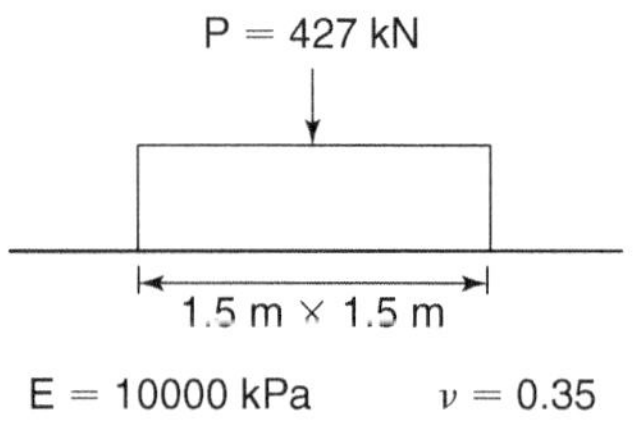

Figure 17.2s Square footing.

Solution 17.6

The settlement equation for an elastic soil gives (rigid square foundation: I = 0.88):

$$s = pB\frac{(1-\upsilon^2)}{E}I = \frac{427}{1.5 \times 1.5} \times 1.5 \times \frac{(1-0.35^2)}{10000} \times 0.88 = 0.022 \text{ m}$$

Footing size

$$s = pB\frac{(1-\upsilon^2)}{E}I = \frac{Q}{B^2}B\frac{(1-\upsilon^2)}{E}I$$

$$B = \frac{Q}{s}\frac{(1-\upsilon^2)}{E}I = \frac{427}{0.01}\frac{(1-0.35^2)}{10000}0.88 = 3.3 \text{ m}$$

The footing size should be 3.3 m by 3.3 m.

Problem 17.7

A square foundation is 3 m × 3 m and rests on a deep layer of sand at a depth of 1.5 m. The soil modulus at the ground surface is 10 MPa and increases linearly to 50 MPa at a depth of 10 m. What load can the footing carry if the allowable settlement is 25 mm?

Solution 17.7

$$s_o = pB\frac{(1-v^2)}{E_o} \times I$$

where:

s_o : reference settlement for uniform soil with E_o modulus
p : Pressure = ?
B : width of foundation = 3m
V : Poisson's ratio = 0.35
E_o : modulus of elasticity at the bottom of the foundation:

$$10 + \frac{1.5}{10}(50-10) = 16 \text{ MPa}$$

I : influence factor for square footing = 0.88

$$s_o = p \times 3 \times \frac{(1-0.35^2)}{16000} \times 0.88 = 1.448 \times 10^{-4} p$$

The settlement of the footing in the case of the increasing modulus with depth is s_1, such that:

$$I_G = \frac{s_1}{s_o}$$

where I_G is read on Figure 17.26 for the corresponding value of E_1/E_o, which is:

$$E = E_o + E_1\left(\frac{z}{B}\right) \quad or \quad \frac{E_1}{E_o} = \left(\frac{E}{E_o} - 1\right)\frac{B}{z}$$

We know that at a depth z equal to 8.5 m below the footing, the modulus E is 50 MPa, which gives an E_1/E_o ratio of:

$$\frac{E_1}{E_o} = \left(\frac{50}{16} - 1\right)\frac{3}{8.5} = 0.75$$

Figure 17.26 gives $I_G = 0.80$, so the settlement expression becomes:

$$s_1 = I_G \times pB\frac{(1-\nu^2)}{E_o} \times I = 0.80 \times p \times 3 \times \frac{(1-0.35^2)}{16000} \times 0.88 = 1.158 \times 10^{-4} p$$

Because s_1 must be 25 mm, then $p = 0.025/1.158 \times 10^{-4} = 216$ kPa.
The allowable footing load is then:

$$Q_{all} = 216 \times 3 \times 3 = 1944 \text{ kN}$$

Problem 17.8

Using the Schmertmann method, simplify the equation giving the settlement of a footing at the surface of a sand deposit when the soil is uniform with a constant value of E. Compare that equation to the elasticity equation.

Solution 17.8

The Schmertmann equation is:

$$s = C_1 C_2 p \Sigma \frac{I_{zi}}{E_i} H_i$$

The footing is placed on the ground surface, so:

$$\sigma'_{ov} = 0 \Rightarrow C_1 = 1 - 0.5\frac{\sigma'_{ov}}{\Delta p} = 1$$

Let's assume that the settlement occurs in 0.1 years, so:

$$C_2 = 1 + 0.2\log\left(\frac{0.1}{0.1}\right) = 1$$

E is a constant, so the final equation is:

$$s = \frac{p}{E}\Sigma H_i I_{zi}$$

The quantity $\Sigma H_i I_{zi}$ is the area under the strain influence factor curve on Figure 17.27. This area depends on the maximum value of I_{zp}, which is:

$$I_{zp} = 0.5 + 0.1\left(\frac{\Delta p}{\sigma'_{Izp}}\right)^{0.5}$$

A reasonable range for I_{zp} may be found when the ratio $\left(\frac{\Delta p}{\sigma'_{Izp}}\right)$ varies between 2 and 20 or a corresponding range for I_{zp} between 0.6 and 0.9 with an average of 0.75. For the value of 0.75, the area under the I_z diagram is:

$$0.5(0.1 + 0.75) \times 0.5B + 0.5 \times 0.75 \times 1.5B = 0.785B$$

and the final Schmertmann equation becomes:

$$s = 0.785\frac{pB}{E}$$

The elasticity equation is:

$$s = I(1 - \nu^2)\frac{pB}{E} = 0.88 \times (1 - 0.35^2)\frac{pB}{E} = 0.77\frac{pB}{E}$$

Problem 17.9

A column load of 4000 kN is to be supported by a square spread footing on a medium-dense sand. Recommend the size and the embedment of the footing after addressing the issue of bearing capacity and settlement of the footing (25 mm is tolerable). Soil properties: $N = 30$ blows/ft, $q_c = 8$ MPa, $f_c = 70$ kPa, $p_L = 1500$ kPa, $E_o = 12$ MPa, $\gamma = 20$ kN/m^3. If you need additional properties, assume reasonable values.

Solution 17.9

Let's assume that D = 0.5 m.

Ultimate bearing capacity: SPT

$$p_u\ (\text{kN/m}^2) = 60N + \gamma D = 60 \times 30 + 20 \times 0.5 = 1810\ \text{kPa}$$

$$p_{safe}\ (\text{kN/m}^2) = \frac{1810}{3} = 603\ \text{kPa} = \frac{Q}{B^2}$$

$$B = \sqrt{\frac{Q}{603}} = \sqrt{\frac{4000}{603}} = 2.58\ \text{m}$$

Ultimate bearing capacity: CPT

$$p_u\ (\text{kN/m}^2) = 0.2q_c + \gamma D = 0.2 \times 8000 + 20 \times 0.5 = 1610\ \text{kPa}$$

$$p_{safe}\ (\text{kN/m}^2) = \frac{1610}{3} = 537\ \text{kPa} = \frac{Q}{B^2}$$

$$B = \sqrt{\frac{Q}{537}} = \sqrt{\frac{4000}{537}} = 2.73\ \text{m}$$

Ultimate bearing capacity: PMT

$$p_u\ (\text{kN/m}^2) = 1.2p_L + \gamma D = 1.2 \times 1500 + 20 \times 0.5 = 1810\ \text{kPa}$$

$$p_{safe}\ (\text{kN/m}^2) = \frac{1810}{3} = 603\ \text{kPa} = \frac{Q}{B^2}$$

$$B = \sqrt{\frac{Q}{603}} = \sqrt{\frac{4000}{603}} = 2.58\ \text{m}$$

Settlement (25 mm tolerable)

$$s = pB(1 - \nu^2)\frac{I}{E}$$

I = shape factor (0.88 for square footing)

p = mean pressure under the foundation

B = foundation width

E = elasticity modulus of the soil (E = $2E_0$(sand) from Briaud (1992))

ν = Poisson's ratio

$$25 \times 10^{-3} = \frac{4000}{B^2} \times B(1 - 0.35^2)\frac{0.88}{2 \times 12000}$$

$$B = 5.15\ \text{m}$$

So the recommended foundation size is 5.2 m × 5.2 m and the settlement criterion controls the design.

Problem 17.10

A column load of 2000 kN is to be supported by a square spread footing on a very stiff clay. Recommend the size of the footing after addressing the issue of bearing capacity and settlement of the footing (25 mm is tolerable). Soil properties: $s_u = 100$ kPa, $q_c = 1.5$ MPa, $f_c = 70$ kPa, $p_L = 500$ kPa, $E_o = 7.5$ MPa, $C_c = 0.3$, $c_v = 10^{-4} cm^2/s$, $\gamma = 18\ kN/m^3$. If you need additional properties, assume reasonable values.

Solution 17.10

Let's assume that $D = 0.5$ m.

Ultimate bearing capacity: CPT

$$p_u\ (kN/m^2) = 0.4q_c + \gamma D = 0.4 \times 1500 + 18 \times 0.5 = 609\ kPa$$

$$p_{safe}\ (kN/m^2) = \frac{609}{3} = 203\ kPa = \frac{Q}{B^2}$$

$$B = \sqrt{\frac{Q}{203}} = \sqrt{\frac{2000}{203}} = 3.14\ m$$

Ultimate bearing capacity: Pressuremeter

$$p_u\ (kN/m^2) = 0.9p_L + \gamma D = 0.9 \times 500 + 18 \times 0.5 = 459\ kPa$$

$$p_{safe}\ (kN/m^2) = \frac{459}{3} = 153\ kPa = \frac{Q}{B^2}$$

$$B = \sqrt{\frac{Q}{153}} = \sqrt{\frac{2000}{153}} = 3.62\ m$$

Ultimate bearing capacity: Undrained shear strength

$$p_u\ (kN/m^2) = N_c S_u + \gamma D = 6.3 \times 100 + 18 \times 0.5 = 639\ kPa$$

$$p_{safe}\ (kN/m^2) = \frac{639}{3} = 213\ kPa = \frac{Q}{B^2}$$

$$B = \sqrt{\frac{Q}{213}} = \sqrt{\frac{2000}{213}} = 3.06\ m$$

Settlement (25 mm tolerable)

$$s = pB(1 - \nu^2)\frac{I}{E}$$

I = shape factor (0.88 for square footing)
p = mean pressure under the foundation
B = foundation width
E = elasticity modulus of the soil ($E = E_0$(clay) from Briaud (1992))
ν = Poisson's ratio

$$25 \times 10^{-3} = \frac{2000}{B^2} \times B(1 - 0.35^2)\frac{0.88}{7500}$$

$$B = 8.24\ m$$

So the recommended foundation size is 8.3 m × 8.3 m

Problem 17.11

In 1955, an oil tank 10 m high and 38m in diameter is built as shown in Figure 17.3s.

a. Calculate the settlement of the center of this tank (point C on Figure 17.3s) using the data from Figure 17.4s. Assume that the stress increase in the middle of the compressible layer is equal to the pressure under the tank because the layer is thin.

   In 1975, this tank is removed; a year later, a new tank 15 m high and 76 m in diameter is built. The edge of the new tank goes through the center of the old tank.

b. Calculate the settlement of the edge of the new tank away from the old tank (point B on Figure 17.3s) using the data from Figure 17.4s. Assume that the stress increase at the edge of the new tank in the middle of the compressible layer is equal to one half of the pressure under the new tank.
c. Calculate the settlement of the edge of the new tank that passes over the center of the old tank (point C on Figure 17.3s) using the data from Figure 17.4s. Make the same assumption as in b.
d. Do you see any problem with the difference in settlement between C and B for the new tank?

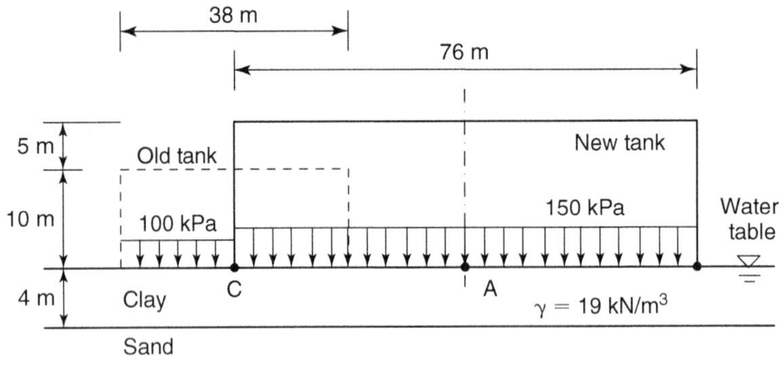

Figure 17.3s Old and new oil tanks.

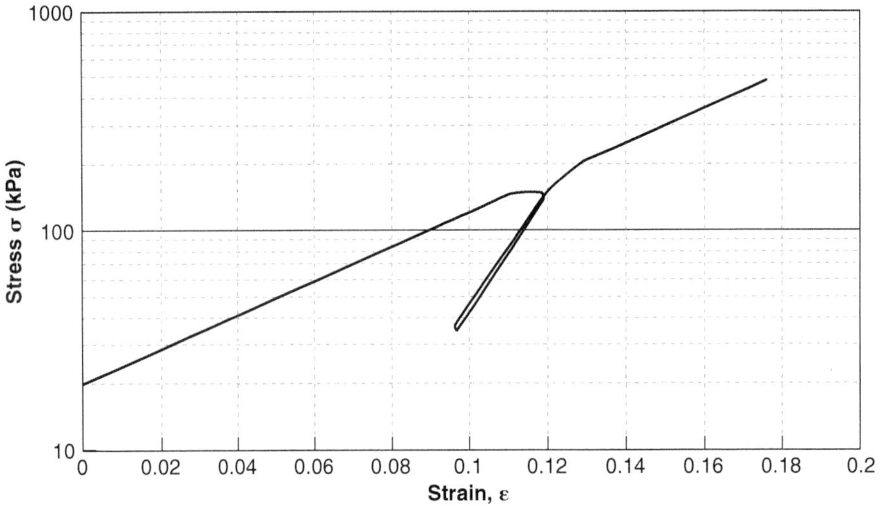

Figure 17.4s Stress-strain curve for oil tank problem.

Solution 17.11

$$\Delta H = \frac{\Delta e}{1+e_o} H = \frac{C_c H}{1+e_o} \log \frac{\sigma'_{ov} + \Delta\sigma'}{\sigma'_{ov}} = \Delta\varepsilon H$$

a. Settlement at point C for old tank

$$\sigma'_{ov} = 19 \times 2 - 9.81 \times 2 = 19.62 \text{ kPa}$$

$$\Delta\sigma = 100 \text{ kPa}$$

$$\sigma'_{ov} + \Delta\sigma = 119.62 \text{ kPa}$$

From Figure 17.4s, $\varepsilon_{119.6\text{kPa}} = 0.1$, $\varepsilon_{19.6\text{kPa}} = 0$, $\Delta\varepsilon = 0.1 - 0 = 0.1$

$$\Delta H = \Delta\varepsilon H = 0.1 \times 4 \text{ m} = 0.4 \text{ m}$$

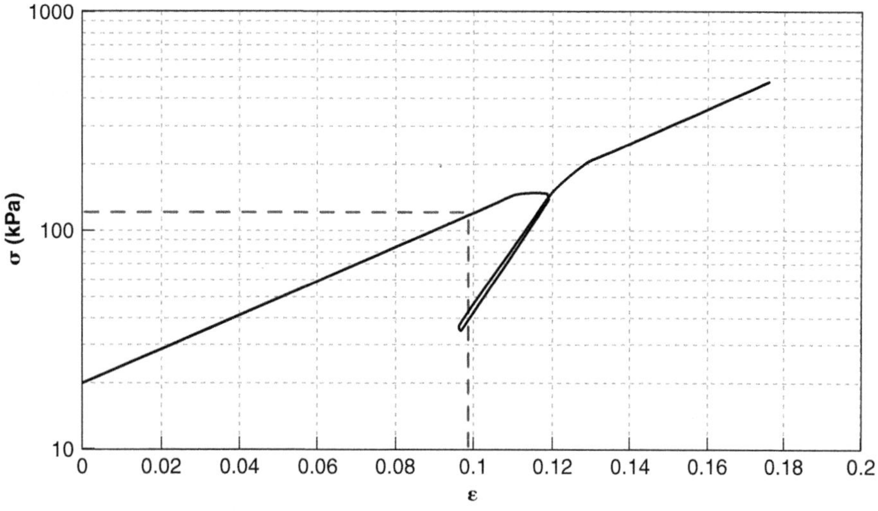

Figure 17.13s Settlement at point C for old tank.

b. Settlement at point B for new tank

$$\sigma'_{ov} = 19 \times 2 - 9.81 \times 2 = 19.62 \text{ kPa}$$
$$\Delta\sigma = 150 \times 1/2 = 75 \text{ kPa}$$
$$\sigma'_{ov} + \Delta\sigma = 94.62 \text{ kPa}$$

From Figure 17.4s, $\varepsilon_{94.6\text{kPa}} = 0.088$, $\varepsilon_{19.6\text{kPa}} = 0$, $\Delta\varepsilon = 0.088 - 0 = 0.088$

$$\Delta H = \Delta\varepsilon H = 0.088 \times 4 \text{ m} = 0.352 \text{ m}$$

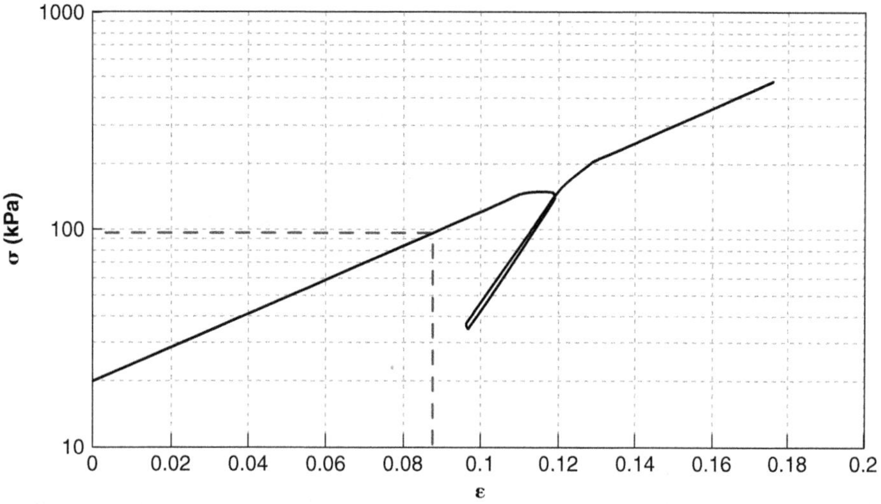

Figure 17.14s Settlement at point B for new tank.

c. Settlement at point C for new tank

$$\sigma'_{ov} = 19 \times 2 - 9.81 \times 2 = 19.62 \text{ kPa}$$
$$\Delta\sigma = 150 \times 1/2 = 75 \text{ kPa}$$
$$\sigma'_{ov} + \Delta\sigma = 94.62 \text{ kPa}$$

In 1955, the preconsolidation pressure was 119.6 kPa. Therefore:

$$\sigma'_{ov} + \Delta\sigma = 94.62 \text{ kPa} < \sigma'_c = 119.6 \text{ kPa} \quad \therefore \quad \textit{Overconsolidated clay}$$

From the rebound curve in Figure 17.4s:

$$\varepsilon_{94.6\text{kPa}} = 0.114, \varepsilon_{19.6\text{kPa}} = 0.086, \Delta\varepsilon = 0.114 - 0.086 = 0.028$$

$$\Delta H = \Delta\varepsilon H = 0.028 \times 4 \text{ m} = 0.112 \text{ m}$$

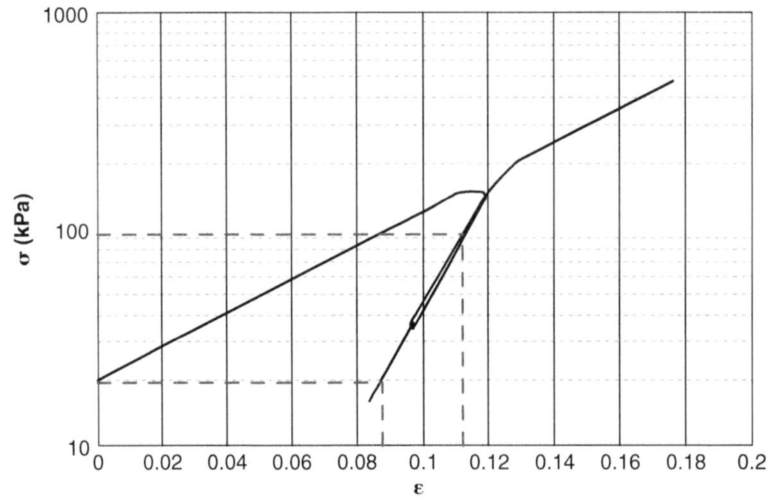

Figure 17.15s Settlement at point C for new tank.

d. The differential settlement between point B and point C:

$$\Delta H_{B-C} = \Delta H_B - \Delta H_C = 0.352 \text{ m} - 0.112 \text{ m} = 0.24 \text{ m}$$

This differential settlement between C and B is significant and will cause bending of the foundation and the new oil tank. This bending may create a problem with the sliding roof often used in such oil tanks.

Problem 17.12

Use the shrink-swell case history from section 17.9.4 to calculate the footing movements and compare your results with the measured movements.

Solution 17.12

The settlement time history was developed using the data presented in section 17.9.4. The change in water content was computed from the boring information and used to estimate the settlement. The analysis was conducted using a representative layer of 0.5 m for the dark gray silty clay and for the brown silty clay. The total settlement was estimated by the contribution of each layer. The results are presented in Table 17.2s and Table 17.3s. The average estimated settlement was compared with the average measured settlement from the four footings, as shown in the figure. The water content method seems to yield a reasonable prediction of the movement of the foundation:

$$s = \sum_{1}^{n} f_i \frac{w_f - w_i}{E_{wi}} H_i$$

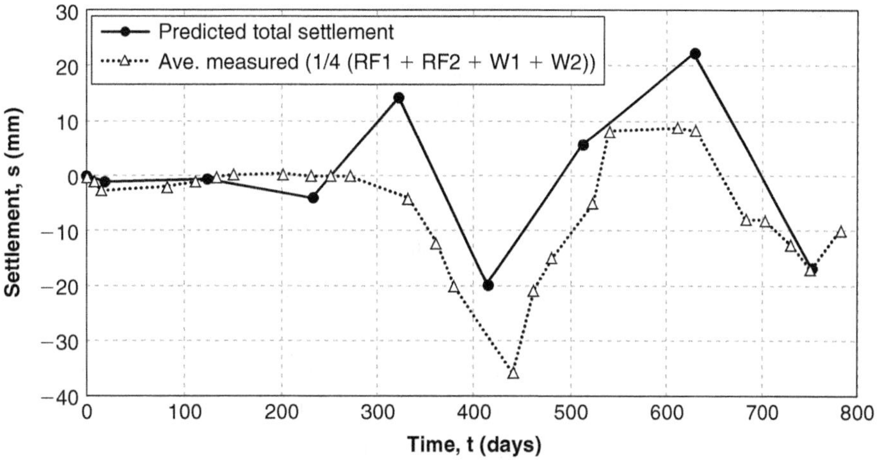

Figure 17.16s Measured vs. predicted settlement.

Table 17.2s Settlement of the Dark Gray Silty Clay

Boring	Boring Date	Elapsed time (days)	$w_i(\%)$	$\Delta w_i(\%)$	f_i	E_{wi}	$\Delta\varepsilon_i$	s_1(m)
B1	6/24/1999	0	0.19	—	0.33	0.752	0.0000	0.0
B2	7/13/1999	19	0.185	−0.005	0.33	0.752	−0.0026	−1.1
B3	10/25/1999	123	0.17	−0.02	0.33	0.752	−0.0104	−4.4
B4	2/11/2000	232	0.165	−0.005	0.33	0.752	−0.0026	−1.1
B5	5/11/2000	322	0.2	0.035	0.33	0.752	0.0182	7.7
B6	8/11/2000	414	0.14	−0.06	0.33	0.752	−0.0311	−13.2
B7	11/17/2000	512	0.17	0.03	0.33	0.752	0.0156	6.6
B8	3/13/2001	628	0.22	0.05	0.33	0.752	0.0259	11
B9	7/15/2001	752	0.17	−0.05	0.33	0.752	−0.0259	−13.0

Table 17.3s Settlement of Brown Silty Clay

Boring	Boring Date	Elapsed time (days)	$w_i(\%)$	$\Delta w_i(\%)$	f_i	E_{wi}	$\Delta\varepsilon_i$	s_2(m)	s_{total}(m)
B1	6/24/1999	0	0.15	—	0.33	0.869	0.0000	0.0	0.0
B2	7/13/1999	19	0.15	0	0.33	0.869	0.0000	0.0	−1.1
B3	10/25/1999	123	0.17	0.02	0.33	0.869	0.0090	3.8	−0.6
B4	2/11/2000	232	0.155	−0.015	0.33	0.869	−0.0067	−2.9	−4.0
B5	5/11/2000	322	0.19	0.035	0.33	0.869	0.0157	6.7	14.3
B6	8/11/2000	414	0.155	−0.035	0.33	0.869	−0.0157	−6.7	−19.8
B7	11/17/2000	512	0.15	−0.005	0.33	0.869	−0.0022	−0.9	5.7
B8	3/13/2001	628	0.21	0.06	0.33	0.869	0.0269	11.4	22.3
B9	7/15/2001	752	0.18	−0.03	0.33	0.869	−0.0135	−5.7	−16.7

Problem 17.13

The high-rise building shown in Figure 17.53 is subjected to a hurricane wind of 200 km/h. This wind creates a pressure of 3 kPa on the flat side of the building. Calculate the pressure diagram under the foundation.

Solution 17.13

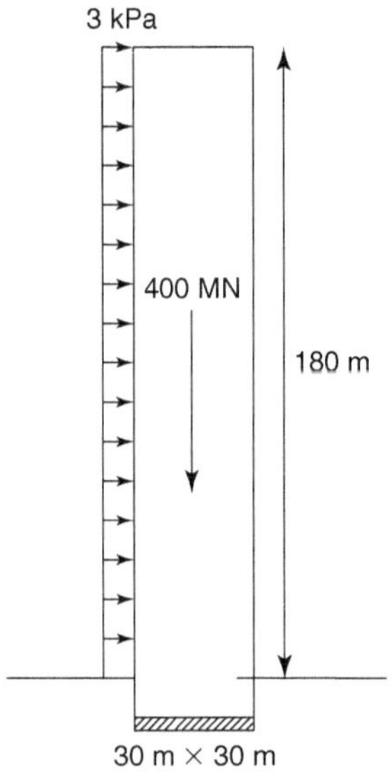

Figure 17.17s High-rise building.

The average pressure is:

$$p_{ave} = \frac{400000}{30 \times 30} = 444 \text{ kPa}$$

The wind creates a horizontal force equal to:

$$H = pA = 3 \times 30 \times 180 = 16200 \text{ kN}$$

The point of application of that force is at a height of 90 m above the ground surface. Therefore, the moment applied on the foundation is:

$$M = Hb = 16200 \times 90 = 1458000 \text{ kN} \cdot \text{m}$$

This moment will create a trapezoidal pressure distribution under the foundation such that the high pressure will be p_{max} and the low pressure p_{min}. The high pressure p_{max} is such that:

$$M = \frac{1}{2}(p_{max} - p_{ave})B \times \frac{B}{2} \times \frac{2}{3} \times \frac{B}{2} \times 2 = \frac{B^3}{6}(p_{max} - p_{ave})$$

But the average pressure is equal to the vertical load V divided by the foundation area A:

$$p_{ave} = \frac{V}{B^2}$$

and the eccentricity e is given by:

$$M = Ve$$

In the end,

$$p_{max} = p_{ave}\left(1 + \frac{6e}{B}\right)$$

and then:

$$p_{min} = p_{ave}\left(1 - \frac{6e}{B}\right)$$

Numerically:

$$p_{max} = 444\left(1 + \frac{6(1458000/400000)}{30}\right) = 768 \text{ kPa}$$

$$p_{min} = 444\left(1 - \frac{6(1458000/400000)}{30}\right) = 120 \text{ kPa}$$

The p_{max} value would have to be checked against the ultimate bearing capacity of the soil.

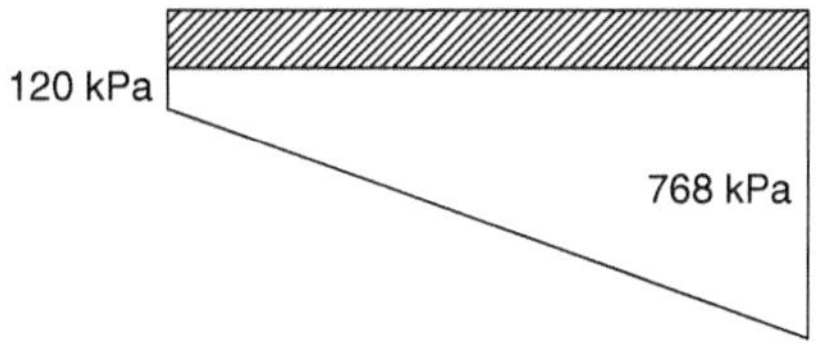

Figure 17.18s Stresses under foundation.

Problem 17.14

The high-rise building shown in Figure 17.53 is placed on a stiff clay with the following properties: compression index C_c equal to 0.4, recompression index C_r equal to 0.1, initial void ratio e_o equal to 0.5, and total unit weight equal to 20 kN/m^3. The soil is lightly overconsolidated by overburden removal and has a preconsolidation pressure σ'_p150 kPa higher than the effective stress σ'_{ov}. The groundwater level is at the ground surface. Calculate the settlement of the building. How would you estimate the time required for the settlement to take place if c_v was known?

Solution 17.14

The solution is the same as the one presented in Table 17.12 except that the consolidation test is used instead of the pressuremeter test. For the consolidation test, the equations change for the evaluation of the settlement due to the building pressure. The following equations cover all the possible cases at various depths:

$$\text{If } \sigma'_{ov} + \Delta\sigma_{bldg} < \sigma'_p \quad \text{use} \quad \Delta H = \frac{H_o}{1+e_o} C_r \log\left(\frac{\sigma'_{ov} + \Delta\sigma_{bldg}}{\sigma'_{ov}}\right)$$

$$\text{If } \sigma'_{ov} + \Delta\sigma_{bldg} > \sigma'_p \quad \text{use} \quad \Delta H = \frac{H_o}{1+e_o}\left(C_r \log\left(\frac{\sigma'_p}{\sigma'_{ov}}\right) + C_c \log\left(\frac{\sigma'_{ov} + \Delta\sigma_{bldg}}{\sigma'_p}\right)\right)$$

Note that from the statement of the problem:

$$\sigma'_p = \sigma'_{ov} + 150 \text{ kPa}$$

It is assumed that the excavation and subsequent construction are done under undrained conditions. Hence, the effective stress after excavation and therefore at the beginning of construction is the same as the effective stress before excavation begins. This assumption states that during the undrained behavior, the change in total stress due to excavation is the same as the change in water stress. The following table shows the calculations for each of the four layers of Figure 17.19s.

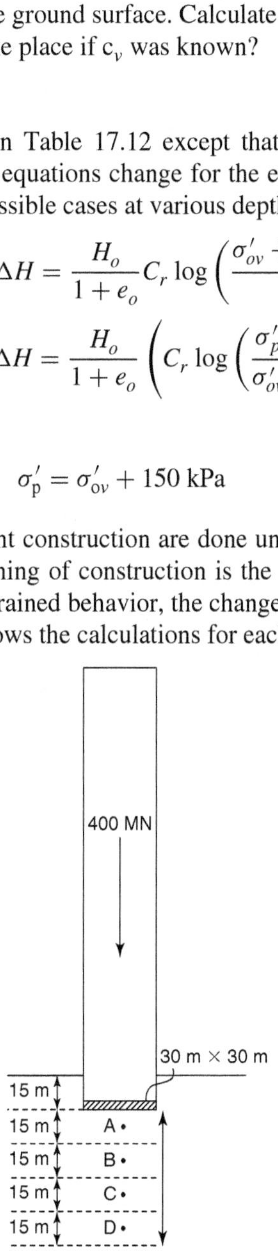

Figure 17.19s High-rise building.

Table 17.4s Calculations for the consolidation settlement of the highrise

1	2	3	4	5	6	7
Point	H (m)	Depth below ground surface (m)	Depth below foundation level (m)	σ_{ov}(kPa)	σ'_{ov}(kPa)	$\Delta\sigma_{bldg}$(kPa)
A	15	22.5	7.5	450	225	378
B	15	37.5	22.5	750	375	222
C	15	52.5	37.5	1050	525	111
D	15	67.5	52.5	1350	675	67

8	9	10	11	12	13	14
$\sigma'_{ov} + \Delta\sigma_{bldg}$(kPa)	σ'_p(kPa)	C_c	C_r	ΔH_R(mm)	ΔH_V(mm)	ΔH_T(mm)
603	375	0.4	0.1	221	825	1046
597	525	0.4	0.1	146	223	369
636	675	0.4	0.1	83	0	83
742	825	0.4	0.1	41	0	41

Summing column 14 gives a total settlement of 1539 mm. This is obviously not tolerable for such a building.

The time rate of settlement can be estimated by using the consolidation theory solution described in section 11.4.6. The time required for a given percentage of the settlement to take place is given by:

$$t_U = T_U \frac{H^2}{c_v}$$

where t_U is the time required for $U\%$ of the settlement to take place, T_U is the time factor (which comes from the theoretical solution and is obtained from Figure 17.34), H is the drainage length, and c_v is the coefficient of consolidation for the soil obtained from a consolidation test (see section 9.5.1). The parameter U is the average percent consolidation, which is a function of the time t and is defined as:

$$U(t) = \frac{\Delta H(t)}{\Delta H_{max}}$$

where $\Delta H(t)$ is the settlement after a time t and ΔH_{max} is the maximum settlement at time equal to infinity. ΔH_{max} is the settlement obtained from previous calculations.

The major question in this case is to find the drainage length H. This is done by carefully analyzing the stratigraphy to estimate the thickness of the compressing layer between two draining layers. Identifying the presence of sand seams in a clay deposit becomes very important in this case.

Problem 17.15

The annual drying and wetting condition of a site is shown in Figure 17.5s. Calculate the shrink and swell displacement at the center and at the edge of the building, and then calculate the differential movement between the two points. Hint: Use

$$\varepsilon_i = \frac{\Delta H_i}{H_i} = f\frac{\Delta w_i}{E_{wi}} \quad \therefore \varepsilon_i = 0.33\frac{\Delta w_i}{(\gamma_w/\gamma_d)}$$

and

$$w_i = 25\%, \quad \gamma_d = 14\,\text{kN/m}^3, \gamma_w = 10\,\text{kN/m}^3$$

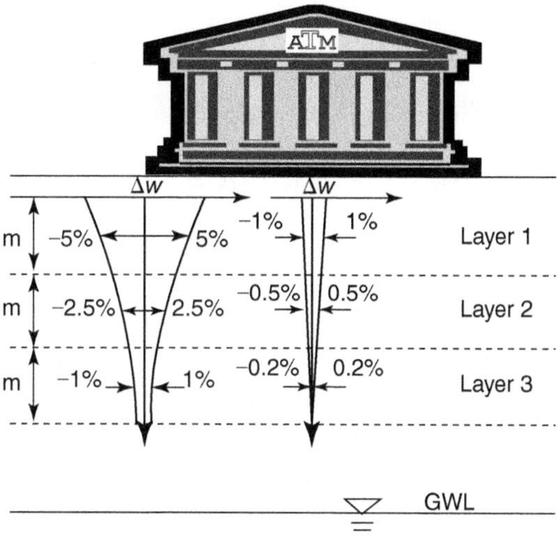

Figure 17.5s Annual drying and wetting condition.

Solution 17.15

$$\varepsilon = 0.33\frac{\Delta w_i}{(\gamma_w/\gamma_d)} = 0.33\frac{\Delta w_i}{10/14} = 0.462\Delta w_i$$

$$\varepsilon_i = \frac{\Delta H_i}{H_i} \qquad \therefore \Delta H_i = \varepsilon_i H_i$$

a. Center

Layer	w_i	Δw_i	ε_i	ΔH_i(mm)
1	0.25	±0.01	±0.00462	±4.62
2	0.25	±0.005	±0.00231	±2.31
3	0.25	±0.002	±0.00092	±0.92

The total displacement at the center of the foundation:

- Shrinking: −7.85 mm
- Swelling: +7.85 mm

b. Edge

Layer	w_i	Δw_i	ε_i	ΔH_i(mm)
1	0.25	±0.05	±0.0231	±23.1
2	0.25	±0.025	±0.01155	±11.55
3	0.25	±0.01	±0.00462	±4.62

The total displacement at the edge of the foundation:

- Shrinking: −39.27 mm
- Swelling: +39.27 mm

Problem 17.16

A stiffened slab on grade for a two-story house is to be designed. The slab and site data are given as follows: slab dimensions 20 m by 20 m, beam spacing s = 3.0 m (for both directions), beam width $b = 0.3$ m, slab load $w = 10$ kPa, depth of movement zone $H = 3.0$ m, soil surface water content change $\Delta w_o = 20\%$. Recommend a beam depth that will minimize the distortion of the slab for the edge drop case to more than $L/\Delta = 500$.

Solution 17.16

The design process advances by trial and error in the sense that the beam depth is assumed and then the resulting deflection is calculated and checked against the distortion criterion. If the deflection criterion is not met, a larger beam depth is assumed. Let's assume a beam depth of 1.2 m. The calculations then proceed with the soil-weather index I_{s-w} calculations:

$$\Delta w_{edge} = 0.5\ \Delta w_0 = 0.5 \times 0.2 = 0.1 \text{ or } 10\%$$
$$I_{s-w} = \Delta w_{edge} \times H = 0.1 \times 3 = 0.3 \text{ m}$$

and then the slab bending stiffness:

$$EI = E\ bh^3/12 = 2 \times 10^7 \times 0.3 \times 1.2^3/12 = 8.64 \times 10^5 \text{ kN.m}^2$$

which leads to the equivalent slab thickness:

$$b\,h^3/12 = s\ d_{eq}{}^3/12$$
$$d_{eq} = h(b/s)^{1/3} = 1.2\ (0.3/3)^{1/3} = 0.56 \text{ m}$$

The values of the design parameters are read on the water content charts for the edge drop case:

$$L_{eq} = 5.3 \text{ m for maximum moment}$$
$$L_{gap} = 3.6 \text{ m for information}$$
$$F_{\Delta\max} = 2.9 \text{ for maximum deflection}$$
$$F_v = 0.8 \text{ for maximum shear}$$

The maximum bending moment is calculated as:

$$q = 10 \times 3 = 30 \text{ kN/m line load on each beam}$$
$$M_{\max} = 0.5\ qL_{eq}{}^2 = 0.5 \times 30 \times 5.3^2 = 421.3 \text{ kN.m}$$

The maximum deflection is calculated as:

$$\Delta_{\max} = q\ L_{eq}{}^4/F_{\Delta\max}EI = 30 \times 5.34/2.9 \times 8.64 \times 10^5$$
$$\Delta_{\max} = 9.5 \times 10^{-3}\text{m}$$

The maximum shear force is calculated as:

$$V_{\max} = F_v q\ L_{eq} = 0.8 \times 30 \times 5.3 = 127.2 \text{ kN}$$

This results in a distortion of:

$$0.5L/\Delta_{\max} = 10/9.5 \times 10^{-3} = 1050$$
$$L_{eq}/\Delta_{\max} = 5.3/9.5 \times 10^{-3} = 558$$

Note that this example is an extreme case, as a Δw_0 of 20% corresponds to extreme weather conditions and a distributed pressure of 10 kPa is quite high for a house. This is why the beam depth is significant.

Problem 17.17

Calculate the settlement of the San Jacinto Monument using the pressuremeter data given in section 17.12.3.

Solution 17.17

The steps outlined in section 17.8.5 are used to solve this problem. The calculations are shown in Table 17.5s. The bulbs of pressure were used to obtain the change in stress at depth. The unit weight of the soil is 18 kN/m^3, the size of the foundation is 37.8 m × 37.8 m, the excavation depth is 4.5 m, the excavation pressure is −83 kPa, the contact pressure between the foundation and the soil is 224 kPa, and the following equations were assumed, using the values of the moduli given in section 17.12.3:

$$E_o(\text{MPa}) = 15 + 1.125\ \text{z(m)} = 15 + 1.125 \times 38\ (\text{z/B}) = E_o + E_1(z/B)$$

$$E_R(\text{MPa}) = 31.5 + 2.362\ \text{z(m)} = 31.5 + 2.362 \times 38(\text{z/B}) = E_{ro} + E_{r1}(z/B)$$

The release of pressure due to excavating 4.5 m of soil is:

$$\Delta p_{exc} = 4.5 \times 18 = 81\ \text{kPa}$$

The contact pressure under the building is:

$$\Delta p_{bldg} = 313000/(37.8 \times 37.8) = 219\ \text{kPa}$$

The total settlement of the monument is the sum of the last two columns or:

$$S_{total} = 41.5 + 114.1 = 155.6\ \text{mm}$$

Table 17.5s Calculations of the San Jacinto Monument settlement

H (m)	Depth to center of layer (m)	σ_{ov} (kPa)	E_o (kPa)	E_r (kPa)	Pressure factor	$\Delta\sigma_{exc}$ (kPa)	$\Delta\sigma_{bldg}$ (kPa)	$\Delta\varepsilon_{exc}$	$\Delta\varepsilon_{net}$	ΔH_{rel} (mm)	ΔH_{net} (mm)
19	14	252	30750	64575	0.88	71	193	0.001100	0.003970	20.9	75.4
19	33	594	52125	109462	0.50	40	110	0.000365	0.001343	16.9	25.5
19	52	936	73500	154350	0.25	20	55	0.000139	0.000476	2.6	9.0
19	71	1278	94875	199237	0.15	12	33	0.000060	0.000221	1.1	4.2
										$\Sigma = 41.5$	$\Sigma = 114.1$

Also:

- Elastic settlement (Equation 17.64)—Using an average modulus of 30 MPa and Poisson's ratio of 0.35:

$$s = I(1 - \nu^2)\frac{pB}{E} = 0.88(1 - 0.35^2)\frac{141 \times 38}{30} = 138\ \text{mm}$$

- Long-term settlement (Equation 17.100)—Where $s(t_o) = 138$ mm, $t_o = 5$ min, $t = 70$ years, and $n = 0.045$:

$$\frac{s(t)}{s(t_o)} = \left(\frac{t}{t_o}\right)^n$$

$$\frac{s(t)}{138} = \left(\frac{70 \times 365 \times 24 \times 60}{5}\right)^{0.045}$$

$$s(t) = 281.1\ \text{mm}$$

- Settlement using linear increase of modulus with depth (Equation 17.68)—Using the elastic settlement equation with the elastic modulus at the surface:

$$s_o = I(1 - \nu^2)\frac{pB}{E_o} = 0.88(1 - 0.35^2)\frac{141 \times 38}{15} = 276 \text{ mm}$$

From Figure 17.26, I_G is 0.5 (using $E_1/E_o = 2.85$):

$$\frac{E_1}{E_o} = \frac{1.125 \times 38}{15} = 2.85$$

From Equation 17.68, the settlement is:

$$I_G = \frac{s_1}{s_o}$$

$$s_1 = I_G s_o = 0.5(276) = 138 \text{ mm}$$

CHAPTER 18

Deep Foundations

18.1 DIFFERENT TYPES OF DEEP FOUNDATIONS

Shallow foundations are typically less expensive than deep foundations. Therefore, it is economically prudent in most cases to start by investigating whether a shallow foundation can be used. Only if it is shown to be insufficient should the design proceed with deep foundations. It is nearly always possible to use a shallow foundation to carry a vertical load, but the area required may be excessive or unavailable. For example, a building with a column spacing equal to s has a limited amount of room between columns to place the footing. Typically, if the area B^2 required for the footing is more than one-half of the area available (s^2), then it is better to use a mat foundation or a deep foundation (Figure 18.1). In other words:

$$B \leq 0.707s \tag{18.1}$$

where B is the footing width and s is the column spacing.

A typical deep foundation consists of a cluster of piles installed down to a certain depth in order to transfer the load to a more competent bearing layer or to distribute the load over a larger depth. Piles come in many different shapes and are made of many different materials. The cross section can be circular and full, tubular, square, or hexagonal. The diameter varies from 0.15 m for micropiles to 3 m for some of the bored piles and offshore pipe piles. The length may be as short as a few meters (bored piles for a house foundation) to more than 100 m for offshore piles (pipe piles to anchor offshore platforms). The material may be steel for pipe piles and H piles, concrete for bored piles or driven concrete piles, wood for timber piles, or even plastic (more recent installations). These piles may be prefabricated in a factory or cast in place. The installation process may consist of driving the piles into the soil with either impact hammers or vibratory hammers (driven piles), or the installers may proceed drilling a hole in the ground, lowering a reinforcing cage, and filling the hole with concrete (*bored piles*, also known as *drilled shafts* or *drilled piers*). There are many variations of these two basic installation techniques, but driven piles and bored piles remain the two major installation categories. The names *end-bearing piles* and *friction piles* are used to refer to the load distribution in the pile. End-bearing piles carry the load mostly at the pile point, whereas friction piles carry the load mostly in friction along the pile shaft. *Battered piles* are piles installed at an inclined angle in order to better resist horizontal loads.

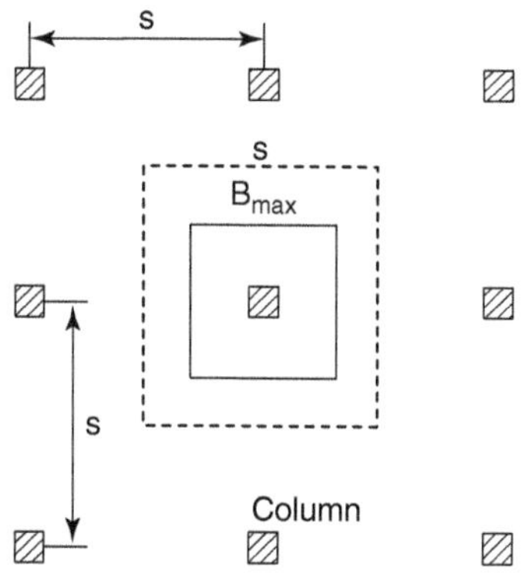

Figure 18.1 Maximum area for shallow foundation.

18.2 DESIGN STRATEGY

The design of a deep foundation consists of selecting the type of piles and calculating the length, size, and number of piles necessary to carry the load safely and within a tolerable settlement. The design also includes the planning of the installation process. Much like in the case of shallow foundations, deep foundations are now designed on the basis of the LRFD approach (see section 17.4). Here again two limit states are considered: the ultimate or strength limit state and the service limit state (section 17.4). In the LRFD approach, the limit state is written as:

$$\sum_{i=1}^{n} \gamma_i L_i \leq \sum_{j=1}^{m} \varphi_j R_j \tag{18.2}$$

where γ_i is the load factors for the loads L_i, and φ_j is the resistance factors for the resistances R_j. The load factors γ

are the same for shallow and deep foundations (Chapter 17, Table 17.1), but the resistance factors φ are different, as they are tied to specific design methods for calculating the pile capacity. These resistance factors for the ultimate limit state vary between 0.2 and 0.5 (AASHTO 2010) and will be discussed in sections 18.5.1 and 18.6.4. The ultimate limit state or strength limit state might look like this:

$$\gamma_1 DL + \gamma_2 LL \leq \varphi_1 R_{uf} + \varphi_2 R_{up} \tag{18.3}$$

where DL is the dead load, LL is the live load, R_{uf} is the pile ultimate friction resistance, and R_{up} is the pile ultimate point resistance. The service limit state can be presented as:

$$\gamma_3 DL + \gamma_4 LL \leq \varphi_3 R(s_{all}) \tag{18.4}$$

where s_{all} is the allowable settlement of the foundation, and $R(s_{all})$ is the pile load that generates the allowable settlement. For the service limit state, the load factors and the resistance factors are usually taken equal to 1. Furthermore, if the settlement will take place over a long period of time, the live load is not included in the settlement calculations except for the permanent live load.

Prior to the development of the load and resistance factor design approach (LRFD; also called limit state design or LSD), the working stress design (WSD; also called allowable stress design or ASD) approach was used. WSD consists of applying a global factor of safety against the ultimate bearing capacity of the soil in order to obtain the safe load. The equation is:

$$L < R_u / F \tag{18.5}$$

where L is the applied load to be safely carried, R_u is the ultimate resistance, and F is the global factor of safety. The factor of safety varied from 2.5 to 3 when R_u was based on calculations down to 2 when R_u was based on an appropriate number of load tests.

The type and size of piles selected for a project are often influenced by what is available locally; for example, in Hawaii steel pipe piles are rare, but concrete piles are common. Also, in given stratigraphies, some piles are easier to install than others; for example, in stiff clay with a water table at a large depth, bored piles drilled dry are very economical. In very soft soils, it is very difficult to drill and keep open a clean hole, so driven piles are preferred because they are easy to drive. Sometimes the size of the load to be carried dictates the pile type; for example, large, heavy loads can be carried more readily by a single large-diameter bored pile than by an equally large driven pile because it is easier to drill a large hole in the ground than to drive a large-diameter pile unless the soil is very soft.

The pile length can be determined by calculations. This is the case with offshore piles, where a required ultimate pile capacity is determined by using the load and resistance factors; then an ultimate pile capacity profile is generated as a function of depth, and the pile length is chosen to correspond to the depth where that ultimate capacity is first reached. In many instances, however, the pile length is chosen by inspecting the stratigraphy of the site. If a hard layer exists at some reasonable depth below the ground surface, the piles will be founded in that bearing layer and the pile length is fixed. If the stratigraphy is uniform and does not have a strong layer, or if the strong layer is too thin to support the pile group, the pile length may be dictated by the maximum length that can be transported without special permits, or the maximum length available; for example, timber piles are typically no longer than 20 m. In onshore practice and for driven piles, the length of the pile is often dictated by the pile blow count that is written into the specifications: this is called driving to a blow count. In simple terms, onshore you drive until you reach a set blow count; offshore you drive until you reach a set penetration.

If the pile size, type, and length are determined, then the design consists of finding the number of piles required to carry the load safely and within a tolerable settlement. Note that this load may be a vertical load, a horizontal load with or without overturning moment, or a combination of all of these. Combination loading is handled by considering the two load types separately and ignoring the interaction effect. The reason for this choice is that the resistance to vertical load tends to be mobilized at depth, whereas the resistance to horizontal loads tends to be mobilized close to the ground surface. The design strategy for either load case proceeds according to the following steps:

1. Choose the pile size, type, and length.
2. Calculate the ultimate bearing capacity of one pile (and maximum bending moment for horizontal loads).
3. Calculate the number of piles required to satisfy the ultimate limit state criterion under the given load.
4. Check the group effect.
5. Reiterate steps 1 through 4 until the ultimate limit state criterion is satisfied.
6. Under the foundation load, calculate the movement of the pile group and check that the service limit state is satisfied.
7. If the calculated movement is larger than the acceptable movement, the foundation must be modified (increase pile depth, pile size, use different pile type) and step 6 repeated.
8. If the movement is acceptable, the design is complete, as the ultimate limit state and the service limit state will have been satisfied.

Design methods for deep foundations can be classified into three categories: design by theory, design by empiricism, and design by analogy. Design methods by theory rely on theoretical derivations for recommending the design equations. Design methods by empiricism rely on experimental data and correlations for recommending the design equations. Design methods by analogy rely on the close analogy between the mode of deformation in the soil test and in the foundation case. Generally speaking, the best methods are those that combine the advantages of all three methods by including a close analogy, experimental data, and appropriate theoretical background.

18.3 PILE INSTALLATION

18.3.1 Installation of Bored Piles

Bored piles are also known as drilled shafts or drilled piers. Bored piles are installed by drilling a hole in the ground, removing the drilling tool, inserting the reinforcement cage, and filling the hole with concrete. In more detail, the sequence is as follows: First the hole is drilled with a drill rig. The diameter of the hole varies from 0.3 m all the way to 3 m. If the soil is free standing over the depth drilled, the hole is drilled dry. If not, slurry is placed into the hole to help prevent caving of the hole. The level of the slurry in the drilling hole must always be higher than the groundwater level, to ensure a positive flow from the drill hole to the soil through the borehole wall. Slurries can be mineral slurries or polymers. The most common type of mineral slurry is bentonite slurry, prepared by mixing bentonite particles with water. The consistency of this slurry is very liquid. When the slurry stands in the open hole, the slurry starts flowing horizontally into the soil; during this process, the bentonite particles accumulate on the wall and form a thin cake that seals the hole from incoming or outgoing water. This minimizes the sloughing of the soil into the hole that is often caused by entrainment of the incoming flow of water. Polymer slurries are viscous, but they do not form cakes on the wall; rather, they simply continue to flow into the soil, so new slurry must be added continuously. The unit weight of a bentonite slurry is between 3 and 10% higher than the unit weight of water, whereas the unit weight of a polymer slurry is less than 3% higher than that of water. The chemical composition of slurries should be checked before use (Brown et al. 2010).

If slurry is insufficient to keep the hole open, a steel casing can be lowered in the open hole and advanced as drilling progresses, to prevent collapse. After the hole is drilled, the casing may be left in place permanently or retrieved. Once the hole is opened, a steel reinforcement cage is lowered in the center of the hole. Then a tremie pipe is lowered to the bottom of the hole and concrete is poured into the hole from the bottom up. Because concrete is heavier than bentonite slurry, the concrete displaces the slurry upward, to overflow in a desanding pit, for example. It is very important to keep the bottom of the tremie pipe below the concrete level to prevent contamination of the concrete by the slurry. If the bottom of the tremie pipe is raised above the concrete-mud interface (called burping the tremie) during concreting, there is a possibility that some slurry or soil may become trapped in the concrete. This would create a weak inclusion or defect in the bored pile. Onshore, the slurry is recirculated, but for offshore drilled and grouted piles, the slurry is wasted on the ocean floor. Once the hole is full of concrete, it is allowed to cure; thereafter, the bored pile is complete.

In summary, there are three main procedures for placing a bored pile (Brown et al. 2010):

1. Dry method (Figure 18.2)
2. Casing method (Figure 18.3)
3. Wet method (Figure 18.4)

Note that often a bored pile is constructed by using a combination of two or three of the methods listed here. In addition to those methods, two other techniques are sometime used for bored piles: base grouting and underreams (also called bells). *Base grouting* consists of injecting grout under pressure at the base of the bored pile after the concrete is sufficiently hard (Figure 18.5). This increases the pressure at the base by reaction against the side friction of the bored pile. This increase in pressure stiffens and strengthens the soil under the pile point and actually prestresses the pile against the soil. This technique aims at decreasing the settlement and increasing the capacity of the pile under load. *Underreams* or *bells* are created by lowering a special drilling tool to

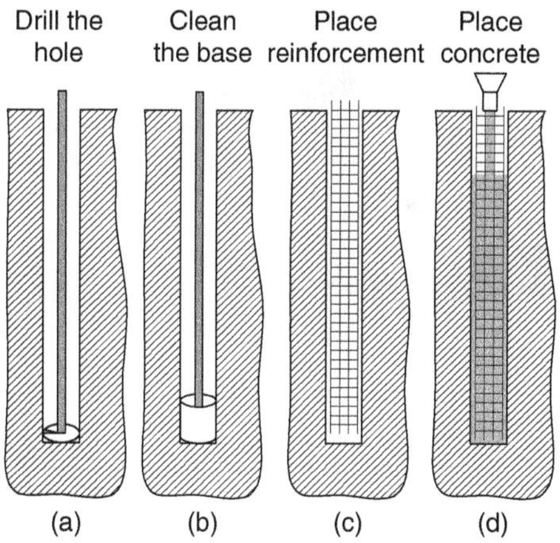

Figure 18.2 Installation of bored piles: Dry method (After Brown et al. 2010).

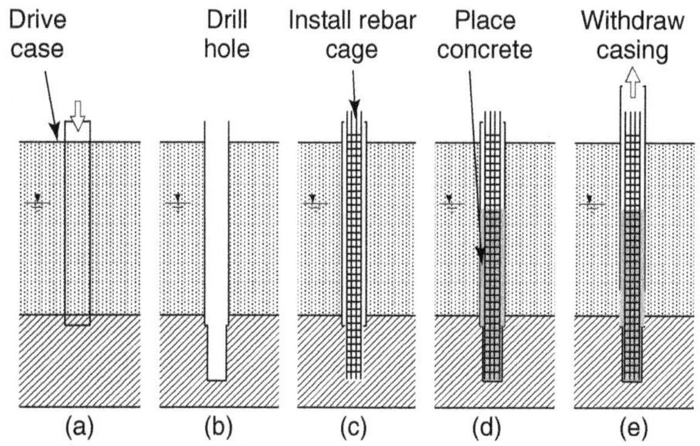

Figure 18.3 Installation of bored piles: Casing method (After Brown et al. 2010).

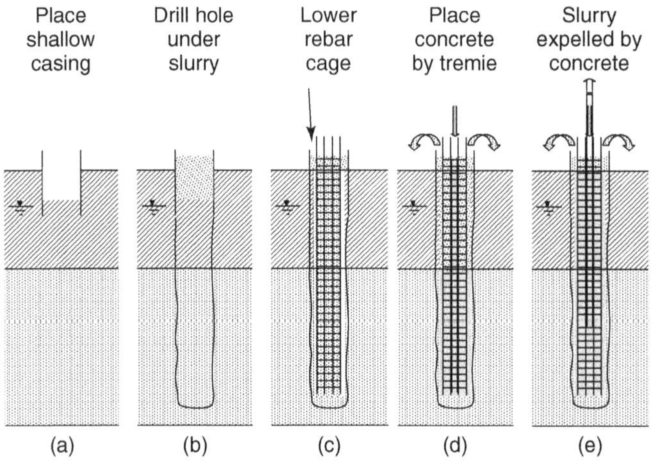

Figure 18.4 Installation of bored piles: Wet method (After Brown et al. 2010).

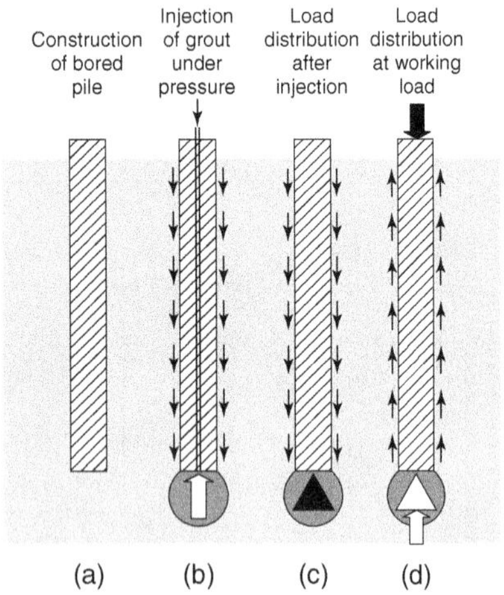

Figure 18.5 Installation of bored piles: Base grouting (After Brown et al. 2010).

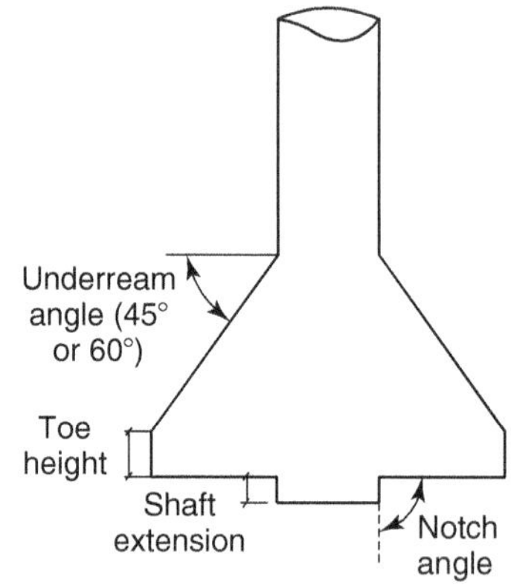

Figure 18.6 Installation of bored piles: Underream or bell (After Brown et al. 2010).

the bottom of the hole before concreting takes place. This tool expands sideways and creates a cone-shaped opening by rotation (Figure 18.6). The angle of the cone with the horizontal is commonly in the range of 45 to 60 degrees. The purpose of an underream or bell is to increase the point resistance of a bored pile or the uplift capacity without having to increase the diameter over the entire length of the pile.

Some of the important issues in bored piles installation are as follows (ADSC-DFI 2004):

1. For bored piles drilled dry, one must ensure that the concrete that falls in the hole is not segregated by hitting against the reinforcing bars.
2. One must also ensure that there is enough room between the outside rebar and the soil for the concrete aggregates to fit properly.
3. For bored piles constructed under slurry and in sand, it is important to de-sand the slurry as drilling progresses (settling pond). If not, sand will settle at the bottom of the hole and form a soft, compressible cushion that will hinder the settlement performance of the pile.
4. For bored piles constructed under slurry, it is also important not to keep the slurry in the hole too long. If it stays in the hole too long, the slick bentonite cake that forms on the wall of the hole will become very thick and will significantly decrease the friction capacity of the shaft.
5. In all cases, it is important to clean any loose soil from the bottom of the shafts just before concreting, to minimize settlement due to recompression of the excavated soil.

An experiment was conducted for two bored piles 1 m in diameter and 10 m long in sand (Briaud et al. 2000). In the first case, the contractor was asked to do the worst job possible (Pile 1) and in the second case the contractor was asked to do the best job possible (Pile 2). For Pile 1, the contractor did not de-sand the slurry, left the slurry in place in the finished hole for 72 hours, and did not clean the bottom of the pile. A 0.3 m thick cushion of soft sand was observed at the bottom of the pile and a 10 mm thick layer of bentonite

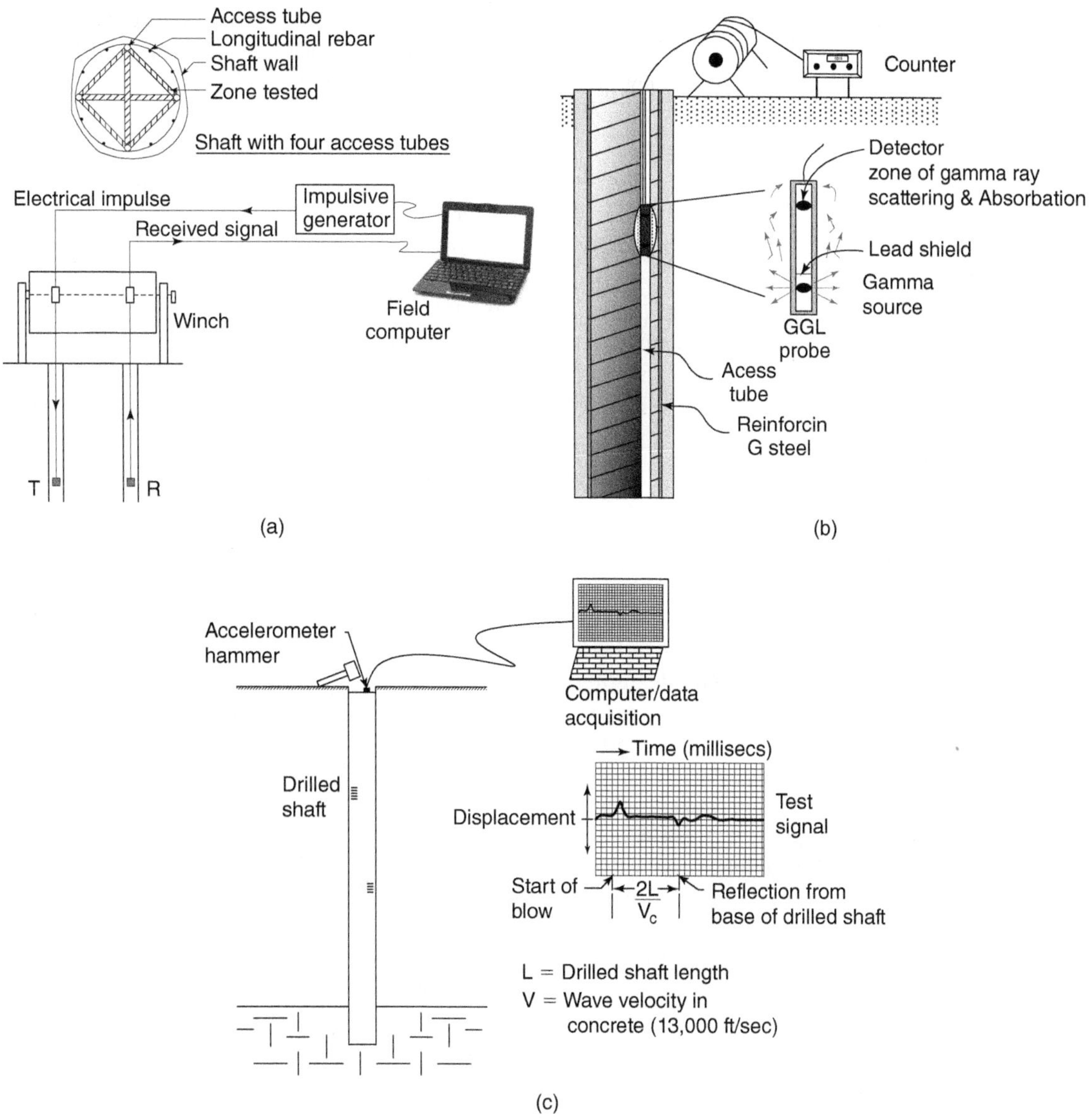

Figure 18.7 NDT techniques for bored piles: (a) Cross hole logging. (b) Gamma-gamma logging. (c) Sonic echo and impulse response (After Brown et al. 2010).

mud was measured on the walls of the bored pile hole by horizontal sampling. For Pile 2, the contractor was careful not to make any of those errors. Both piles were concreted, cured, and load tested to 150 mm of penetration. At 150 mm of penetration, Pile 1 carried only 1500 kN, whereas Pile 2 carried 4500 kN. The load distribution in the pile was measured with extensometers. Most of the difference in load came from the friction, which was reduced by a factor of 10; the point resistance was the same in both cases. This shows how important it is always to control the quality of foundation construction.

Drilled and grouted piles are also bored piles. These piles are installed by drilling a hole dry or under slurry; lowering a steel element such as an H beam, a steel pipe, or a rebar in the center; and filling the annulus between the reinforcement and the soil with grout. Offshore casings for oil wells are placed that way. *Micropiles* are small-diameter versions of bored piles or drilled and grouted piles. *Augercast piles* were initially installed in such a way that the drilling, lowering of the reinforcement, and grouting were done in a single down-and-up process. This was very efficient, but was limited to smaller diameters, and the reinforcement was also limited to a centralized bar or small casing placed through the center of the hollow stem auger used to drill the hole. Now augercast piles are drilled with larger diameter augers, the auger is retrieved, and a reinforcement cage with centralizers is lowered in the holefull of cement paste before the cement sets.

Deep soil mixing is yet another process leading to stronger elements placed in the ground to carry the load from a structure. This process consists of drilling the soil with a cement slurry and mixing the soil and cement in place while drilling. The cement and the soil harden into a soil-cement column made of a material with a strength intermediate between soil and concrete. The cement volume is around 20% of the soil volume and typical column diameters are around 1 m. The strength of this material varies significantly, but an unconfined compression strength equal to 2 MPa is not uncommon.

18.3.2 Nondestructive Testing of Bored Piles

Nondestructive testing (NDT) can be used on any deep foundation member. However, it is most often used in conjunction with the evaluation of onshore bored pile foundations. Several methods can be used (Figure 18.7):

1. Cross hole sonic logging
2. Gamma-gamma logging
3. Sonic echo
4. Impulse response

Cross Hole Sonic Logging

The cross hole sonic logging technique requires that at least two access holes and casings be left in the bored pile during construction. This is achieved by attaching the casings to the reinforcing cage. The casings are typically around 50 to 57 mm in diameter and should be very well connected to the bored pile to avoid loss of signal across the interface. A source transmitter is lowered in one of the access casings while a receiver is lowered to the same depth in another casing. The source emits a compression wave signal and the time t needed to receive the signal across the bored pile at the receiver is monitored. The compression wave speed is calculated as:

$$v = \frac{d}{t} \tag{18.6}$$

where v is the wave speed, d is the distance from the source to the receiver, and t is the travel time. The compression wave speed v in sound concrete is about 4000 m/s, in water is 1500 m/s, in air is 300 m/s, and in soils is anywhere from 400 to 2000 m/s. Therefore, any values much lower than 4000 m/s will be an indication of a problem with the bored pile. Table 18.1 gives an indication of how to rate concrete for various velocity readings. It is also possible to place the source and the receiver at different depths in the bored pile and across different horizontal paths. The data are then inverted to get a three-dimensional rendition of the bored pile. This is called cross hole tomography (Hollema and Olson 2002).

Table 18.1 Concrete Rating from Wave Speed

Compresion Wave Speed	Concrete Quality
3600 to 4000 m/s	Good
3200 to 3600 m/s	Questionable
< 3200 m/s	Poor/defective

(After Brown et al. 2010.)

Gamma-Gamma Logging

The gamma-gamma logging technique requires that an access tube be left in the bored pile during construction. A gamma ray source and a gamma ray detector are placed in the same cylindrical probe and lowered in the access tube. Gamma rays are beams of photons; some of the photons bounce back to the detector and are counted upon arrival. The gamma ray arrivals are recorded in counts per second (cps). There is a reasonably linear correlation between the concrete density and the log base 10 of the cps:

$$\gamma_{conc} = a \log(cps) + b \tag{18.7}$$

where γ_{conc} is the unit weight of concrete, cps is the gamma ray count recorded at the detector per second, and a and b are calibration constants. The radius of influence of the gamma ray test is about half the distance between the source and the detector on the probe. In most cases, the radius of influence is less than 0.2 m. The result of a gamma-gamma logging test is a profile of unit weight along the bored pile. This profile gives many values of the unit weight, thereby allowing one to calculate a mean and standard deviation. Any unit weight value that is less than 3 standard deviations below the mean reading for the pile is considered anomalous (Brown et al. 2010).

Sonic Echo Method

The sonic echo technique does not require any access tube in the bored pile. Thus, it can be used even if plans were not made ahead of time to NDT the bored piles. However, it is not as reliable as the more rigorous cross hole or gamma-gamma testing. The sonic echo test consists of hitting the top of the bored pile with a carpenter-size hammer and recording the return signal at a geophone glued to the top of the pile. The departure and arrival of the compression-tension wave are recorded and the distance travelled is calculated according to a known wave speed. If the wave encounters a necking defect (reduction in concrete cross section), then it returns as a tension wave. If the wave encounters a bulb defect (increase in cross section), then it returns as a compression wave. Hence, the sign of the return wave indicates whether the defect is a necking or a bulb.

The reason for the return wave being a compression or a tension wave is explained as follows: If you hit a set of billiard balls lined up in a row (Figure 18.8) with a hammer, the last ball will leave the row. The reason is that the compression wave you generate with the hammer propagates through the balls and gets to the end. Finding no resistance, it tries to go back as a tension wave, but because there is no tension capacity between the balls, the last ball leaves. Now, if the billiard balls are in line but the last one is against a wall, and

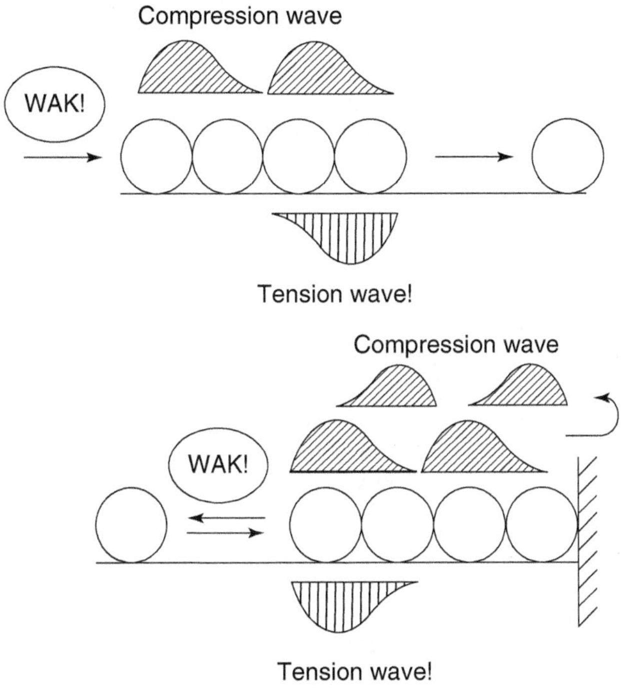

Figure 18.8 Impacts on billiard balls.

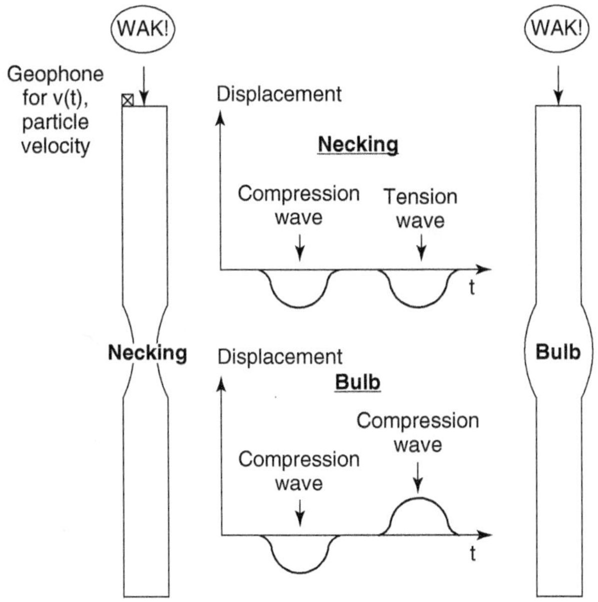

Figure 18.9 Sonic echo idealized signals.

if you hit the first one with a hammer, it is the first one that leaves the lineup. The reason is that the compression wave propagates through the balls, hits the wall, and returns as a compression wave back toward the first ball. There it finds no resistance and returns as a tension wave. Because there is no tension capacity between the balls, the first one leaves. The same thing happens in a bored pile. If the compression wave hits a necking defect (low resistance), it returns as a tension wave; if it hits a bulb defect (high resistance), it returns as a compression wave (Figure 18.9).

Some of the limitations of the sonic echo method are:

1. The soil strength affects the intensity of the return wave. Pile length-to-diameter ratios larger than 10 in rock are unlikely to give satisfactory returns. In soft soils, however, length-to-diameter ratios of up to 50 can give satisfactory returns.
2. The interface of soil layers with contrasting strengths can create some return waves that must be distinguished from defects in the bored pile.
3. The smallest defect that can be detected improved from about 50% in 1993 (Baker et al. 1993; Briaud et al. 2002) to 10% in 2001 (Iskander et al. 2001).
4. An important distinction must be made between an anomaly and a defect. What may be detected as an anomaly may not represent a defect that would make a bored pile unusable.

Impulse Response Method

The impulse response technique is similar to the sonic echo method, but in this case the head of the hammer is instrumented with a dynamic load cell. During the impact that generates the wave propagation, the force-time signal of the hammer is recorded through this load cell. In addition, the velocity is recorded at the pile top. The force-time signal and the velocity time signal are then transformed into the frequency domain to create the force spectrum F and the velocity spectrum V. The ratio V/F is called the *mobility* and is plotted against the frequency (Figure 18.10).

Interpretation of the mobility curve proceeds as follows (Finno and Gassman 1998). The slope of the initial part of the mobility curve gives the small strain stiffness of the bored pile–soil system. The distance between peaks on the mobility curve gives the pile length or the distance between anomalies:

$$\Delta f = \frac{v_{conc}}{2L} \tag{18.8}$$

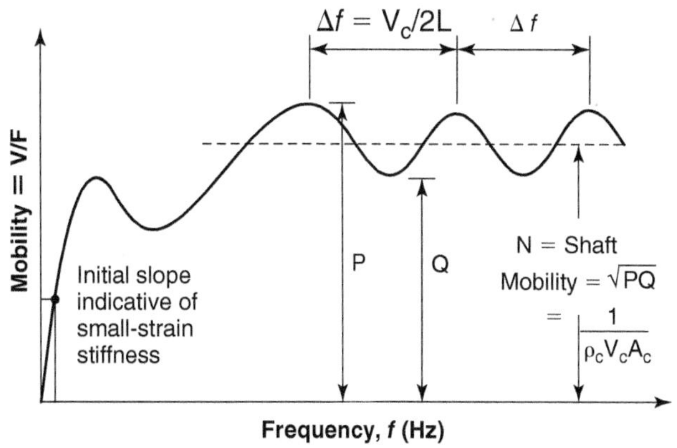

Figure 18.10 Mobility curve from an impulse response test.

where Δf is the distance between frequency peaks on Figure 18.10, v_{conc} is the compression wave velocity in concrete, and L is the length of the pile or the distance between anomalies. The mean value of the ratio V/F (Figure 18.10) is the inverse of the impedance:

$$\left(\frac{V}{F}\right)_{mean} = \frac{1}{I} = \frac{1}{\rho_c v_c A_c} \quad (18.9)$$

where $(V/F)_{mean}$ is the mean mobility from the mobility curve (Figure 18.10), I is the impedance of the system, ρ_c is the mass density of the concrete, v_c is the compression wave velocity in the concrete, and A_c is the cross-sectional area of the concrete pile.

Impedance Log Method

The impedance log technique is a derivative of the impulse response method. As mentioned regarding the impulse response method, the variation in impedance as a function of frequency can be generated. By comparing the mobility curve with the mobility curve for an infinitely long and constant-diameter bored pile, the variation in impedance as a function of depth can be generated (Hertlein 2009). This *impedance log* represents a two-dimensional rendition of the bored pile cross section as a function of depth (Figure 18.11). Note that in this rendition, because the impedance I is the product of $\rho_c v_c A_c$, all changes in mass density, wave velocity, and area are interpreted as changes in area. The impedance log has the advantage of giving a picture of the bored pile.

18.3.3 Installation of Driven Piles

A pile can be driven into the ground either by impact hammers or vibratory hammers. *Impact hammers* are big, heavy masses,

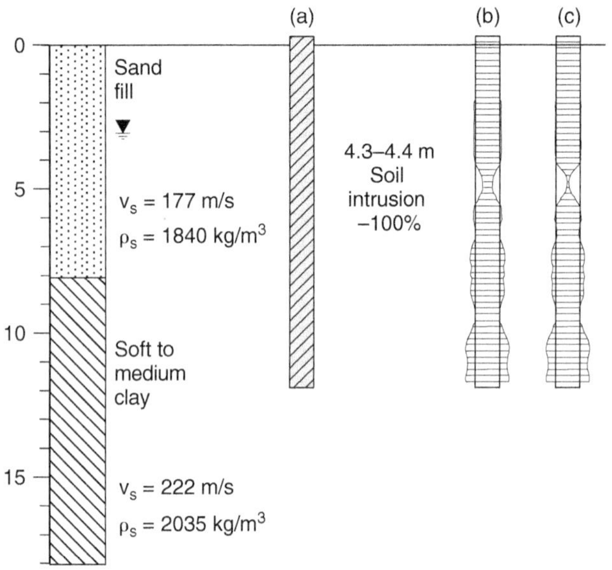

Figure 18.11 Example of impedance log of a bored pile (After Brown et al. 2010).

called *rams*, that are lifted and dropped repeatedly on the top of a pile to drive it into the ground. These hammers are of different types: steam hammers, diesel hammers, and hydraulic hammers. Steam hammers, the oldest types, use compressed steam to lift the ram; they may be single acting or double acting. In a single-acting hammer, the steam pressure lifts the ram which then falls under its own weight. In a double-acting hammer, the steam pressure also lifts the ram, but when the ram is ready to fall, the steam pressure acts on top of the ram to accelerate it downward, thereby increasing the force at impact. Diesel hammers, which use an explosion of ignited diesel fuel to lift the ram, can also be single acting or doubling acting. Hydraulic hammers use the hydraulic action of a piston. Hammers are rated in terms of maximum energy that can be delivered (drop height times weight of ram); these energies range from 20 kN.m or kJ to 800 kJ onshore and can reach up to 3000 kJ for offshore underwater hydraulic hammers. A cushion is placed between the hammer and the pile top to limit the stress generated in the pile material (soften the blow) by the hammer impact. Cushions made of wood are common and thicknesses can range from 25 to 100 mm. Sometimes a pile cap is also placed between the hammer and the pile. Pile driving formulas and the wave equation analysis (Lowery et al. 1967) are used to make calculations regarding drivability, hammer size, pile stresses, and pile capacity. Piles onshore are typically driven until a chosen blow count for a given hammer is reached. This blow count is usually around 75 blows per 0.3 meters of penetration. Offshore piles are usually driven to a penetration depth regardless of the blow count required.

Vibratory hammers grab the top of the pile and shake it vertically into the soil. The vibration is created by eccentrically rotating masses and the peak force is generated by the static weight of the hammer plus the centrifugal vibrating force. Although the frequency can vary from 10 to 100 Hz, the most common vibratory hammers operate at around 25 to 30 Hz. Resonance of the hammer-pile system is rarely reached, as it is typically higher than 30 Hz unless the pile is very long. Low-frequency, high-weight vibratory hammers (e.g., 1500 kN at 10 Hz) are used to drive large piles and caissons. Medium-frequency vibratory hammers (e.g., 250 kN at 25 Hz) are most common and are used for driving sheet pile and small piles. They work particularly well in sands where vibrations easily displace the soil particles. High-frequency vibratory hammers (e.g., 90 to 120 Hz) are rare and aim at reaching hammer-pile system resonance. Methods based on empirical formulas rooted in energy consideration as well as variances of the wave equation analysis are used to analyze these systems (Warrington 1992; Chua et al. 1987; Rausche 2002). The advantage of vibratory driving is that it is usually faster than impact driving, with a penetration rate that can be 10 times faster. The drawback is that it has limited penetration capability and does not develop residual stresses in the pile at the end of driving like impact-driven piles. As a result, a vibrodriven pile tends to exhibit more settlement at working loads than an impact-driven pile, although both may have the same ultimate load (Briaud et al. 1990). Sometimes vibratory-driven piles are impact driven at the end of penetration to benefit from the advantages of both methods.

18.3.4 Pile Driving Formulas

Pile driving analysis started by assuming that the pile motion into the soil under each blow was a rigid body motion. Under this assumption, the energy conservation equation gives:

$$R_{ud} = \frac{Wh}{s} \tag{18.10}$$

where W is the weight of the hammer, h is the drop height, R_{ud} is the ultimate pile capacity at the time of driving, and s is the penetration of the pile. This simple equation turns out to be riddled with problems:

1. The fall of the ram is not unimpeded (e.g., friction) and there are other energy losses (e.g., compression of the cushion), so that the energy delivered to the pile is not Wh but a fraction of Wh'. This can be written as eWh where e is the efficiency of the driving system; e values are very difficult to quantify unless special measurements are made during driving, and can vary from 0.3 to 0.9.
2. The pile is not a rigid body; it compresses and rebounds during each hammer blow. This compression and rebound uses up energy that is not used to advance the pile penetration. This elastic energy is often represented by a term equal to $R_{ud}c/2$ (Figure 18.12) and is added to the resistance side of the energy equation (Eq. 18.10).
3. The movement of the pile during the driving process is best represented by a wave propagation in the pile. This process is not consistent with a single energy equation such as Eq. 18.10. This process is better represented by what is called the *wave equation analysis*.
4. If the static ultimate capacity R_{us} is the quantity sought from Eq. 18.10, then the dynamic component of R_{ud} due to rate and inertia effects must be subtracted from R_{ud} to get R_{us}.

Nevertheless, various forms of Eq. 18.10 have been proposed and used. The incentive was clearly the simplicity and great usefulness of this equation. Referring to Figure 18.12, the energy used in driving the pile is equated to the effective energy delivered by the hammer, and an improved version of Eq. 18.10 is:

$$R_{ud} = \frac{eWh}{s + \dfrac{c}{2}} \tag{18.11}$$

where R_{ud} is the ultimate capacity of the pile at the time of driving; e is the efficiency of the driving system; W is the weight of the ram; h is the height of drop (Wh is the rated energy of the hammer); s is the net downward movement of the pile after the blow, often called the *permanent set*; and c is the elastic rebound of the pile (c is usually taken as 5 mm, based on experience). If the number of blows per 300 mm penetration is N, then the permanent set s is equal to $s = 300/N$. In order to include the effect of water stress dissipation and soil relaxation on the ultimate capacity, it is very desirable to use the blow count $N_{redrive}$ from redriving the pile a good while after the end of driving. Then Eq. 18.11 becomes:

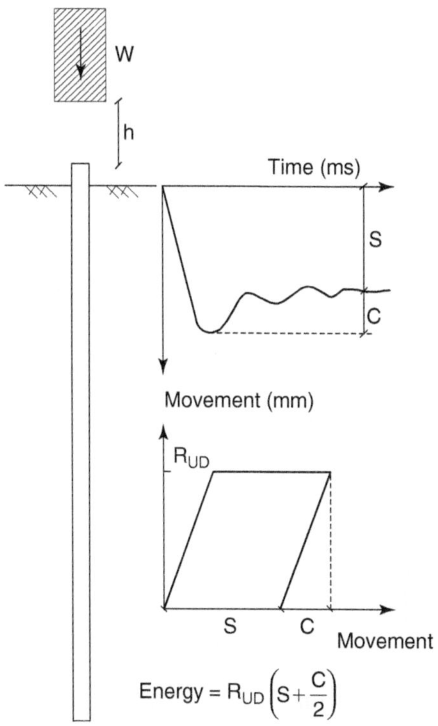

Figure 18.12 Pile driving event.

$$R_{ud} = \frac{eWh(\text{mm})}{\dfrac{300}{N_{redrive}} + 2.5} \tag{18.12}$$

This equation indicates how the pile resistance at the time of redriving is linked to the blow count. The R-N curve (Figure 18.13) gives the following information:

1. It gives the pile resistance at the time of driving R_{ud} for an observed value of the blow count N.
2. Alternatively, if R_{ud} is known, if a reasonable blow count N for the end of driving is selected, and if the efficiency e of the system can be estimated, then the rated hammer energy Wh required to drive the pile can be obtained.
3. Equation 18.12 is a hyperbola with an asymptotic value of the pile resistance at the time of driving $R_{ud(\max)}$ of:

$$R_{ud(\max)} = \frac{eWh(\text{mm})}{2.5} \tag{18.13}$$

This is the maximum resistance that can be overcome by the hammer.

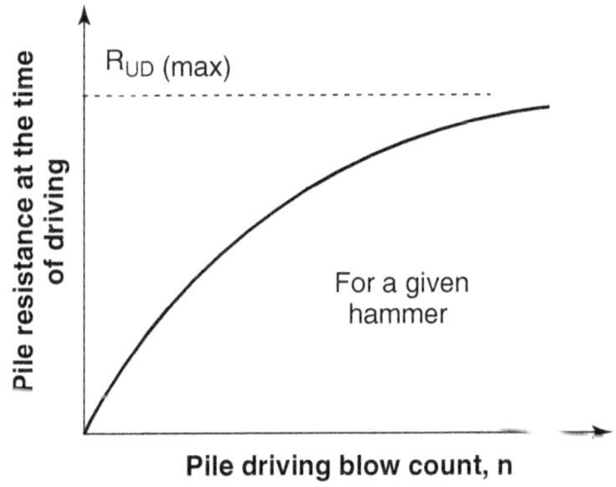

Figure 18.13 Pile driving R-N curve.

4. If a load test is performed to obtain the static ultimate capacity after driving R_{us}, the ratio K between R_{us} and R_{ud} can be calculated and used to evaluate R_{us} from the R_{ud} values on other piles.

One of the shortcomings of the pile driving equation is that it considers rigid body motion of the pile. In fact, a compression wave imparted by the hammer propagates down the pile and back up the pile in a time ranging from 5 to 20 milliseconds. This phenomenon must be accounted for to arrive at a more satisfactory analysis of the driving event.

18.3.5 Wave Propagation in a Pile

Before we talk about the wave equation for pile driving analysis, let's talk about wave propagation in a pile. We will apply the general principle to develop the solution for the displacement problem described in section 11.4.3.

1. We zoom in on an element of pile dz long (Figure 18.14).
2. We identify the knowns and unknowns, including the stresses in the pile element, the inertia force associated with the element mass, and the soil friction on the side of the element.
3. The fundamental equation in this case is the equation of motion:

$$\sum F = ma \quad or \quad (\sigma + d\sigma)A - \sigma A + fPdz = \rho A dz \frac{d^2u}{dt^2} \tag{18.14}$$

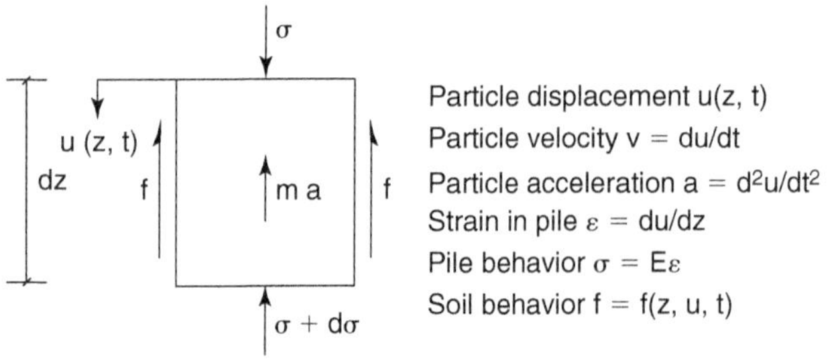

Figure 18.14 Element of pile.

where σ is the normal stress in the pile, A is the cross-sectional area of the pile, f is the shear stress at the interface, P is the pile perimeter, dz is the element length, ρ is the mass density of the pile, u is the pile particle displacement, and t is the time.

4. The constitutive equation for the pile is:

$$\sigma = E\varepsilon = E\frac{du}{dz} \tag{18.15}$$

where E is the pile modulus of elasticity, ε is the strain in the pile, and z is the depth. This equation ignores the influence of confinement on the pile. This confinement plays a minor role because the stress in the pile is typically much larger than the confining stress from the surrounding soil.

5. The constitutive equation for the soil links the shear stress f at the pile soil interface to the displacement u of the pile. The shear stress f is then a function of u, and of the depth z if the soil is not uniform, and also the time t:

$$f = f(u, z, t) \tag{18.16}$$

6. Equations 18.14 through 18.16 are regrouped to give the wave equation:

$$\frac{d^2u}{dz^2} - \frac{\rho}{E}\frac{d^2u}{dt^2} + \frac{P}{AE}f = 0 \tag{18.17}$$

7. The boundary and initial conditions are the hammer impact velocity at $z = 0$ and $t = 0$ and the point resistance from the soil at $z = \mathrm{L}$ and $t = 0$. The wave equation program solves this equation by stepping into time, as will be shown later.

The wave speed $c = dz/dt$ is different from the pile particle velocity $v = du/dt$. The wave speed c is typically thousands of times larger than the pile particle velocity. The wave travels down and up the pile during the few milliseconds of impact, whereas the particle only moves around its point of equilibrium. When a particle in the pile moves, it pushes its neighbor, which moves in turn. The very slight delay between the movements of these two neighbors is what creates the propagation of the wave. Of course, the stiffer the pile is, the faster the neighbor feels the push; thus, c is higher for stiffer materials. In contrast, the denser the pile is, the harder it is for the particle to move its neighbor, so c is lower for higher-density materials. The equation for the wave speed (compression) is derived as follows. We will use the case of a compression wave propagating in a pile without surrounding soil.

$$\frac{d^2u}{dz^2} = \frac{\rho}{E}\frac{d^2u}{dt^2} \tag{18.18}$$

In this instance it is convenient to change variables:

$$x = z + \sqrt{\frac{E}{\rho}}t \quad and \quad y = z - \sqrt{\frac{E}{\rho}}t \tag{18.19}$$

Then Eq. 18.18 becomes:

$$\frac{d^2u}{dxdy} = 0 \quad (18.20)$$

for which the solution is of the form:

$$u = f(x) + g(y) = f\left(z + \sqrt{\frac{E}{\rho}}t\right) + g\left(z - \sqrt{\frac{E}{\rho}}t\right) \quad (18.21)$$

Let's now consider the position of the wave at time t and $t + \Delta t$ (Figure 18.15). At time t, the wave is at a depth z, at time $t + \Delta t$ the wave is at a depth $z + \Delta z$, and the wave speed is $c = \Delta z/\Delta t$. The two functions f and g represent two waves, one coming down and one coming up in the pile. Let's consider one of the two waves represented by function f. The displacement $u(t)$ at time t and depth z is $u = f\left(z + \sqrt{\frac{E}{\rho}}t\right)$ and the displacement $u(t + \Delta t)$ at time $t + \Delta t$ and depth $z + \Delta z$ is $u = f\left(z + \Delta z + \sqrt{\frac{E}{\rho}}(t + \Delta t)\right)$. Because we have unimpeded propagation of the wave, the two values of u must be equal for all values of t and z. This requires that:

$$\Delta z = \sqrt{\frac{E}{\rho}}\Delta t \quad and \quad c = \sqrt{\frac{E}{\rho}} \quad (18.22)$$

where c is the wave speed.

The impedance I is defined as the ratio between the force and the velocity. The relationship is established as follows.

$$F = \sigma A = E\varepsilon A = E\frac{du}{dz}A = E\frac{du}{dt}\frac{dt}{dz}A = \frac{EA}{c}v = Iv \quad (18.23)$$

where F is the force generated by the impact, σ is the normal stress, A is the cross-sectional area, E is the modulus of elasticity, ε is the normal strain, u is the particle displacement, z is the depth, t is the time, c is the wave speed, v is the particle velocity, and I is the impedance equal to EA/c.

Equation 18.21 indicates that the particle velocity is made of the influence of a wave going down plus a wave going up in the pile. Similarly, the force F is made of a force going down plus a force going up:

$$F = AE\frac{du}{dz} = AE\left(\frac{df_{\downarrow}}{dz} + \frac{dg_{\uparrow}}{dz}\right) = F_{\downarrow} + F_{\uparrow} \quad (18.24)$$

The equation for the particle velocities is also similar:

$$v = \frac{du}{dt} = \frac{df_{\downarrow}}{dt} + \frac{dg_{\uparrow}}{dt} = v_{\downarrow} + v_{\uparrow} \quad (18.25)$$

Note that:

$$\frac{df_{\downarrow}}{dz} = c\frac{df_{\downarrow}}{dt} \quad and \quad \frac{dg_{\uparrow}}{dz} = -c\frac{dg_{\uparrow}}{dt} \quad (18.26)$$

Therefore, using the impedance I, defined as:

$$I = \frac{AE}{c} \quad (18.27)$$

we get the following relationships:

$$F_{\uparrow} = 0.5(F + Iv) = Iv_{\uparrow} \quad (18.28)$$

$$F_{\downarrow} = 0.5(F - Iv) = -Iv_{\downarrow} \quad (18.29)$$

18.3.6 Wave Equation Analysis

In 1960, Smith proposed a calculation scheme to include the wave propagation in the analysis. This was the beginning of the modern pile driving analysis. The equations proposed by Smith are based on a discretization of the hammer, the cushion, the pile, and the soil. This discretization breaks the pile into elements that have a mass M and a spring constant K (Figure 18.16). The hammer, the cap-block, the helmet,

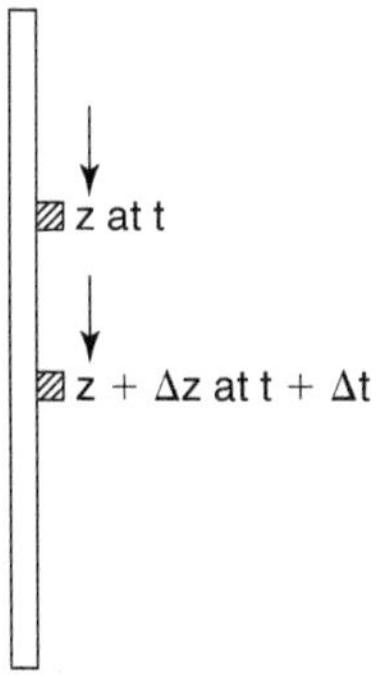

Figure 18.15 Wave location at time t and t + Δt.

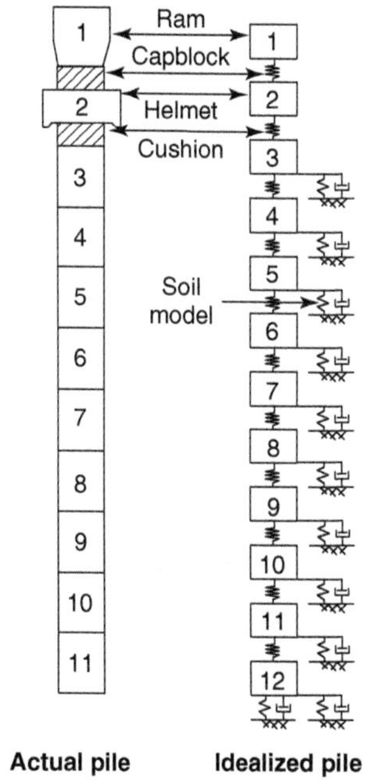

Figure 18.16 Discretization of the pile for wave equation analysis. (After Lowery et al. 1967).

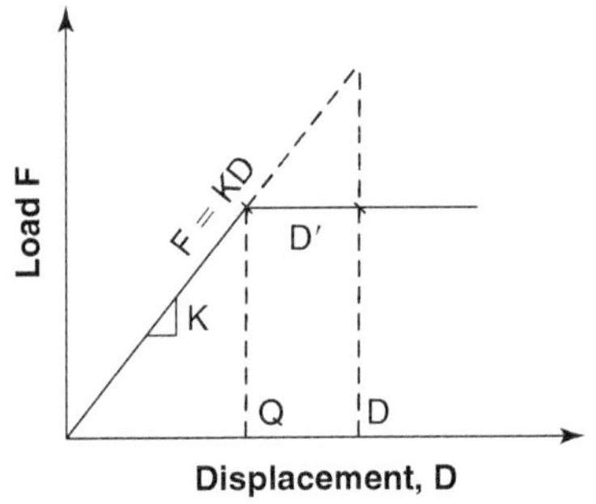

Figure 18.17 Soil model. (After Lowery et al. 1967).

and the cushion can also be represented by an element with a mass and a spring constant to represent their compressibility under load. The soil is represented as a series of springs tied to the pile elements with a dashpot for dynamic effects and a sliding block for maximum resistance. The soil model for the static resistance is an elastic, perfectly plastic model, as shown in Figure 18.17. The movement required to reach the plastic plateau is called the quake Q. The dynamic resistance of the soil is obtained from the static resistance by:

$$R_{DYN} = R_{STA}(1 + Jv) \tag{18.30}$$

where R_{DYN} and R_{STA} are the dynamic and static resistance of the soil respectively, v is the particle velocity of the pile element, and J is a damping coefficient. Table 18.2 gives the values for Q and J originally recommended by Coyle et al. (1973). Tables 18.3 and 18.4 give the values currently recommended in the GRLWEAP manual (2012). GRLWEAP makes additional comments regarding the quake values:

1. Nondisplacement piles are sheet piles, H piles, or open-ended pipe piles that are not plugging during driving.
2. Displacement piles are solid piles (concrete piles) or piles that plug during driving.
3. Typically, pipe piles with diameters larger than 900 mm will not plug, whereas H piles and pipe piles with diameters smaller than 500 mm will plug during driving.
4. For vibratory-driven piles in fine-grained soils, the quake value should be doubled.
5. For vibratory-driven piles in all soils, the damping values should be doubled.

The equations used to solve the wave equation are as follows:

$$D(m,t) = D(m,t-1) + V(m,t-1)\Delta t \tag{18.31}$$

where $D(m,t)$ is the displacement of mass number m at the time step number t, $V(m, \mathrm{t}-1)$ is the velocity of mass number m at time step number $t-1$, and Δt is the time

Table 18.4 Recommended Damping Coefficient J

Soil Type	Side Damping	Point Damping
Coarse-grained soils	0.16 s/m	
Fine-grained soils	0.65 s/m	
All soils		0.50 s/m

(GRLWEAP 2012.)

Table 18.2 Original Values of Quake Q and Damping Coefficient J

Soil Type	Side Damping	Point Damping	Side Quake	Point Quake
Clay	0.65 s/m	0.03 s/m	2.5 mm	2.5 mm
Silt	0.33 s/m	0.50 s/m	2.5 mm	2.5 mm
Sand	0.16 s/m	0.50 s/m	2.5 mm	2.5 mm

(Lowery et al. 1967)

Table 18.3 Recommended Quake Q

Soil Type	Pile Type or Size	Side Quake	Point Quake
All soil types	All pile types	2.5 mm	
All soil types, soft rock	Nondisplacement piles (unplugged)		2.5 mm
Very dense and hard soils	Displacement piles with diameter or width D (solid or plugged)		D(mm)/120
Loose or soft soils	Displacement piles with diameter or width D (solid or plugged)		D(mm)/60
Hard rock	All types		1 mm

(GRLWEAP 2012.)

increment. This time increment must be very small, as the entire event may take only 20 milliseconds. Samson et al. (1963) recommend that Δt be less than:

$$\Delta t \leq \frac{\Delta L}{\sqrt{\frac{E}{\rho}}} \tag{18.32}$$

where ΔL is the pile element length, E is the pile modulus of elasticity, and ρ is the mass density of the pile material. This ensures that the wave does not propagate past one element during one time step. Once the displacements of all the pile elements are calculated, the compression of the springs between the pile elements can be calculated:

$$C(m, t) = D(m, t) - D(m + 1, t) \tag{18.33}$$

where $C(m, t)$ is the compression of spring number m at time step number t, and $D(m, t)$ is the displacement of the mass number m at time step number t.

Once the spring compressions are known, the force in the spring can be calculated:

$$F(m, t) = C(m, t)K(\text{m}) \tag{18.34}$$

where $F(m, t)$ is the force in spring number m at time step number t and $K(\text{m})$ is the spring constant for spring number m.

The soil resistance is calculated as follows:

$$R(m, t) = (D(m, t) - D'(m, t))K'(\text{m})(1 + JV(m, t - 1)) \tag{18.35}$$

where $R(m, t)$ is the dynamic soil resistance on the side of mass m or under the last mass at time step number t; $D(m, t)$ is the displacement of mass number m at time step number t; $D'(m, t)$ is the displacement beyond the quake if $D(m, t)$ is larger than the quake (if not, $D'(m, t)$ is zero); $K'(\text{m})$ is the soil spring for mass number m; J is the soil damping factor; and $V(m, t - 1)$ is the velocity of mass number m at time step number $t - 1$. Then the velocity of the element is calculated as follows:

$$V(m, t) = V(m, t - 1) + (F(m - 1, t) - F(m, t) - R(m, t))\frac{g\Delta t}{W(\text{m})} \tag{18.36}$$

where $V(m, t)$ is the velocity of mass number m at time step number t, $F(m - 1, t)$ is the force in spring $m - 1$ above mass number m at time step number t, $F(m, t)$ is the force in spring m below mass number m at time step number t, $R(m, t)$ is the dynamic soil resistance on the side of mass number m at time step number t, g is the acceleration due to gravity, Δt is the time step, and $W(\text{m})$ is the weight of mass number m. In this string of equations, Eq. 18.34 is the constitutive equation for the pile, Eq. 18.35 is the constitutive equation for the soil, and Eq. 18.36 is the fundamental equation ($F = Ma$).

Several computer programs have been written to automate the calculations, which consist of stepping into time and making a set of calculations within each time step. These programs include MICROWAVE (Lowery 1993), GRLWEAP (2012), and TNOWAVE (2012). The first two use the equations described in this subsection; the last one uses the method of characteristics. The best way to understand the calculations is to go through a simple example.

Example of Wave Equation Calculations

A square concrete pile (Figure 18.18) has a cross section 0.3 m by 0.3 m and a length of 8 m. The concrete modulus E_{conc} is 2×10^7 kN/m^2 and the concrete unit weight is 25 kN/m^3. The hammer weighs 20 kN and strikes the pile at 3 m/s. Between the hammer and the pile is an oak cushion that has the same cross-sectional area as the pile, a thickness of 0.20 m, a modulus E_{cush} equal to 2×10^6 kN/m^2, and a unit weight of 7 kN/m^3. The pile is driven into a sand with a point resistance of 500 kN, a quake of 2.5 mm, and a damping coefficient of $J = 0.2$ s/m.

The following idealizations are made to simplify the calculations. Such extreme simplifications are not necessary when using one of the computer programs mentioned earlier. The hammer is idealized as a rigid mass with a weight W_1 equal to 20 kN. The cushion is idealized as a spring with no mass; the spring constant K_1 comes from the equation giving the compression C of a column:

$$C = \frac{FL}{AE} \quad or \quad K = \frac{F}{C} = \frac{AE}{L} \tag{18.37}$$

where F is the force in the column, L is the length of the column, A is the column cross section, and E is the modulus of the column material. Therefore:

$$K_1 = \frac{0.3 \times 0.3 \times 2 \times 10^6}{0.2} = 900 \text{ kN/mm} \tag{18.38}$$

The pile itself is decomposed into two elements. (In a computer program, at least 10 elements are recommended.) Both elements are 4 m long and have weights W_2 and W_3 equal to:

$$W_2 = W_3 = 0.3 \times 0.3 \times 4 \times 25 = 9 \text{ kN} \tag{18.39}$$

The elasticity or springiness of each pile element is characterized by a spring K_2, given by:

$$K_2 = K_3 = \frac{0.3 \times 0.3 \times 2 \times 10^7}{4} = 450 \text{ kN/mm} \tag{18.40}$$

The soil model is the point resistance model for the pile represented by an elastic, perfectly plastic model. The spring (Figure 18.12) is given by:

$$K' = \frac{500}{2.5} = 200 \text{ kN/mm} \tag{18.41}$$

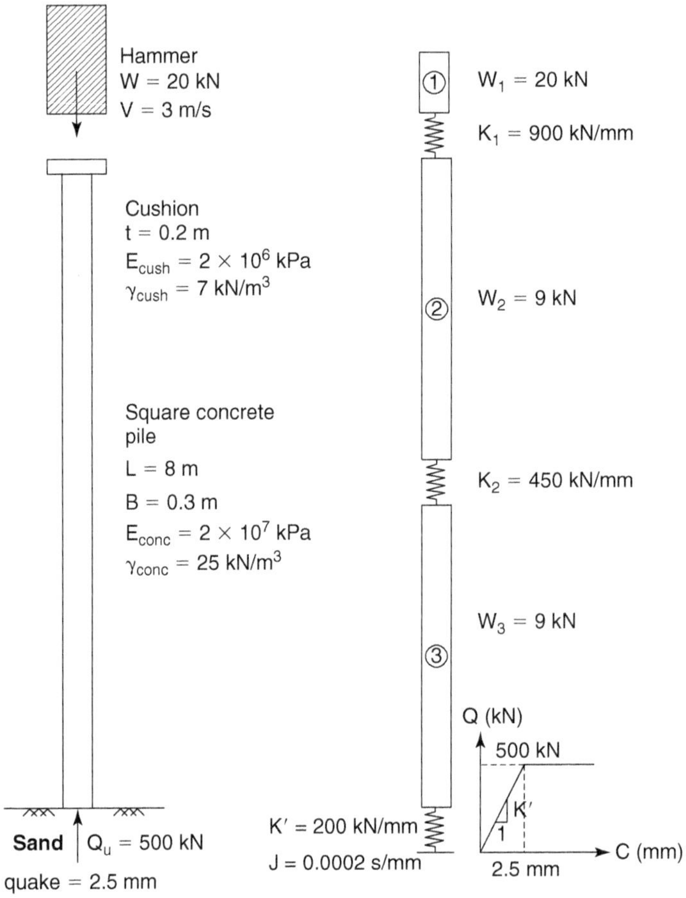

Figure 18.18 Example of wave equation analysis for a simple pile.

Figure 18.18 shows the real pile and the idealized pile. Note that I have placed the spring for pile element 2 under the pile segment, ignored the spring K_3, and kept only the spring K' under the pile point. If we wished to keep K_3, we would have to put it in series with K'. The time step is chosen by using Eq. 18.32:

$$\Delta t \leq \frac{4}{\sqrt{\frac{2 \times 10^7}{2500}}} = 0.045s \qquad (18.42)$$

This value of Δt is very large because the length of the pile element is very large, to simplify the calculations. We would normally have a pile element that is much smaller than 4 m. Nevertheless, we will use a time step of 0.0005 seconds. The results of all calculations for the first four time steps are shown in Table 18.5.

Residual stresses are generated in an impact-driven pile in the following way. During the last hammer blow, the pile goes down, say, 7 mm, and mobilizes the upward resistance of the soil in both friction and point resistance. Then the pile rebounds, and in doing so it reverses the direction of the friction stresses. Indeed, it takes very little movement for the friction to be mobilized (say, 2 mm). However, because it takes a lot more movement to totally decompress the pile point—say, 10 mm—the soil still pushes upward on the pile at the pile point. Equilibrium establishes itself at the end of the blow between the downward friction load and the upward point load (Figure 18.19). This creates a residual compression load in the pile. Thus, impact-driven piles end up being prestressed in the soil and their settlement is minimized because of this phenomenon. Briaud and Tucker (1984) showed how the wave equation can be used to simulate residual stresses. Under the first blow simulation, the pile is driven from a stress-free initial state, but the wave equation calculations end up with the residual stresses prediction. Using the residual loads from the output at the end of the first blow simulation as input to the second blow simulation allows one to simulate the influence of residual stresses. This influence makes pile driving easier, particularly for hard driving (e.g., Figure 18.20). Post grouting of bored piles is a way to establish beneficial residual stresses in bored piles and reduce potential settlement.

18.3.7 Information from Pile Driving Measurements (PDA, Case, CAPWAP)

Several methods are available to obtain the static capacity of the pile from dynamic measurements made at the top

Table 18.5 Wave Equation Calculations

	1	2	3	4	5	6	7
A	Time	s	0.000	0.0005	0.001	0.0015	0.002
B	D(1,t)	mm	0.000	1.500	2.834	3.897	4.672
C	D(2,t)	mm	0.000	0.000	0.368	1.296	2.708
D	D(3,t)	mm	0.000	0.000	0.000	0.045	0.241
E	C(1,t)	mm	0.000	1.500	2.467	2.601	1.964
F	C(2,t)	mm	0.000	0.000	0.368	1.250	2.467
G	F(1,t)	kN	0.000	1350.000	2219.923	2341.013	1767.614
H	F(2,t)	kN	0.000	0.000	165.544	562.706	1110.058
I	R(3,t)s	kN	0.000	0.000	0.000	9.022	48.211
J	R(3,t)d	kN	0.000	0.000	0.000	9.185	51.990
K	V(1,t)	mm/s	3000.000	2668.913	2124.476	1550.343	1116.836
L	V(2,t)	mm/s	0.000	735.750	1855.387	2824.564	3182.932
M	V(3,t)	mm/s	0.000	0.000	90.221	391.890	968.537

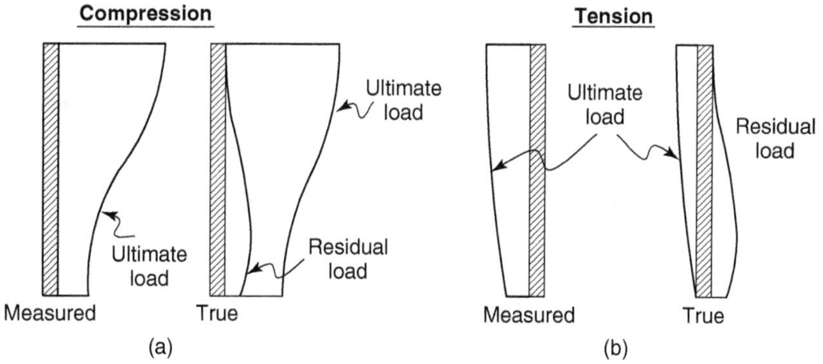

Figure 18.19 Residual load in an impact-driven pile.

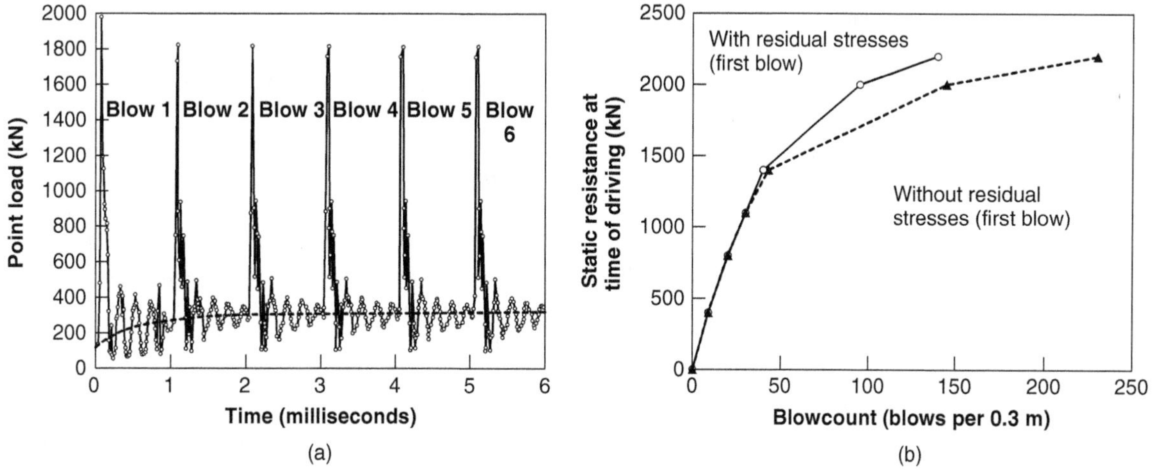

Figure 18.20 Pile driving R-N curve for a pile with and without residual stresses. (Briaud and Tucker 1984).

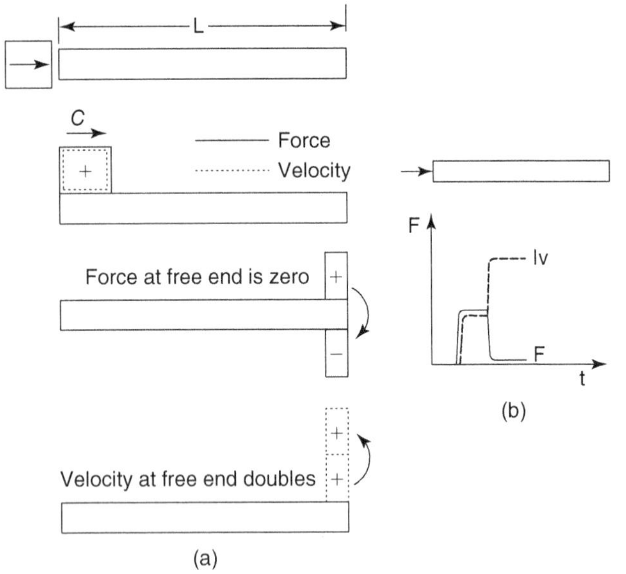

Figure 18.21 Free-end pile signal: (a) Pile. (b) Signal. (After Hannigan et al. 1998)

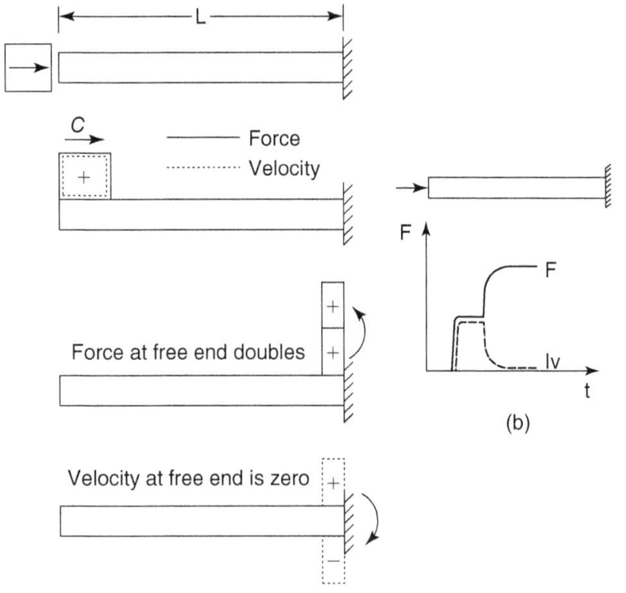

Figure 18.22 Fixed-end pile signal: (a) Pile. (b) Signal. (After Hannigan et al. 1998)

of the pile. The notion of making such measurements goes back a long way, but the credit for commercializing the idea goes to George Goble. Dynamic measurements made at the top of the pile usually include two strain gages and two accelerometers. ASTM has a standard for this test: D4945–89, entitled "Standard Method for High Strain Testing of Piles." The purpose of the strain gages is to obtain the force in the pile during the impact of the hammer and the purpose of the accelerometers is to obtain the acceleration as a function of time and then the velocity by integration of the acceleration signal. The pile driving analyzer or PDA (Likins and Hussein 1988) is a device used to record, digitize, and process the strain and acceleration signals measured at the pile head.

Understanding the signals is important. The following example helps in this process. Imagine a pile suspended horizontally from the ceiling and hit at one end (Figure 18.21). There is no soil surrounding it and the end of the pile is free. Then the compression force in the pile will be proportional to the particle velocity (F = Iv, from Eq. 18.23). Now the wave is racing along the pile at the wave speed c. When it gets to the end of the pile, the compression force F finds no resistance and reflects as a tension force, but the magnitude of the particle velocity doubles while the wave speed is unchanged (see Eqs. 18.24 to 18.26). The F and Iv signals are as shown in Figure 18.21. This would be close to a case of easy driving with very little point resistance. Now let's say that the pile is still suspended from the ceiling and it is hit at one end, but the other end is against a strong wall (Figure 18.22). When the compression wave gets to the wall, it cannot displace it. As a result, the compression force doubles and the velocity vanishes; the F and Iv signals are shown in Figure 18.22. Again, see Eqs. 18.24 to 18.26 for the mathematical reason. This would be close to hard driving into a strong bearing layer. Actual force and impedance times velocity signals for different driving conditions are shown in Figure 18.23 (Hannigan et al. 1998).

The Case Method

The Case Method (Likins and Hussein 1988) is a simple method for obtaining the dynamic and static pile capacity from the force and velocity signals. It is rooted in Eq. 18.43, which states that:

$$R_D = F - Ma \tag{18.43}$$

where R_D is the pile resistance, F is the force at the top of the pile, M is the mass of the pile, and a is the acceleration of the pile. This equation is based on rigid motion. To recognize the influence of the wave, Eq. 18.43 is modified empirically by taking average values of F and a during the time corresponding to the travel of the wave down and back up to the top of the pile:

$$R_D = 0.5(F_{(t_1)} + F_{(t_1+2L/c)}) - M\frac{(v_{(t_1+2L/c)} - v_{(t_1)})}{(t_1 + 2L/c - t_1)} \tag{18.44}$$

where R_D is the dynamic resistance of the pile, $F_{(t_1)}$ is the force at the top of the pile read at a time equal to t_1, t_1 is the time corresponding to the first peak of the force signal, $F_{(t_1+2L/c)}$ is the force at the top of the pile read at a time equal to $(t_1 + 2L/c)$, L is the length of the pile, c is the wave speed in the pile material, M is the mass of the pile, $v_{(t_1)}$ is the velocity at the top of the pile read at the time t_1, and $v_{(t_1+2L/c)}$ is the velocity at the top of the pile read at a time $(t_1 + 2L/c)$ (Figure 18.24). The term $2L/c$ corresponds

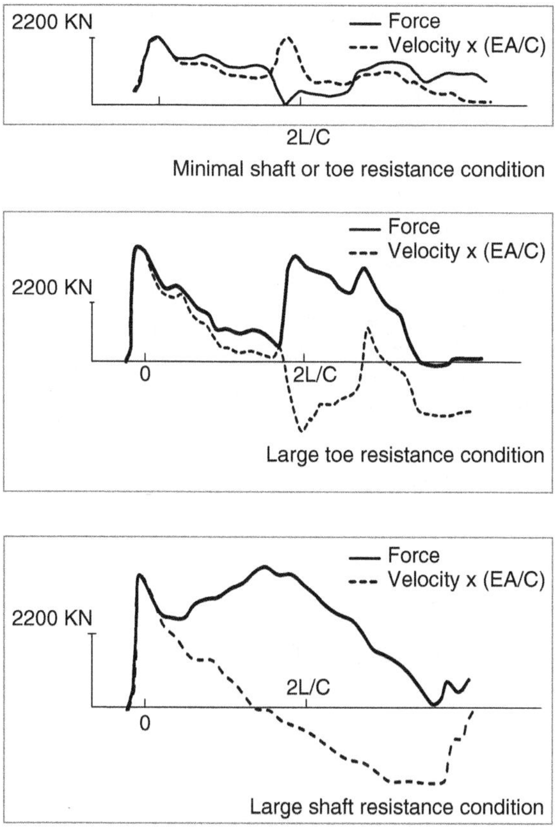

Figure 18.23 Actual force and impedance times velocity signals for different conditions. (After Hannigan 1990)

to the time necessary for the wave to travel to the bottom of the pile and back to the top, so $(t_1 + 2L/c)$ corresponds to the first return of the wave. Figure 18.24 shows a record of force-time signal at the top of a pile.

Eq. 18.44 can be rewritten as:

$$R_D = 0.5(F_{(t_1)} + F_{(t_1+2L/c)} + I(v_{(t_1)} - v_{(t_1+2L/c)})) \tag{18.45}$$

w because I, the impedance of the pile, is given by:

$$I = \frac{AE}{c} = \frac{Mc}{L} \tag{18.46}$$

In this equation A is the cross-sectional area of the pile, E is the modulus of the pile material, c is the wave speed, M is the mass of the pile, and L is the length of the pile. Figure 18.24 shows as a dashed line the signal obtained by multiplying the velocity measured at the top of the pile (actually integrated from accelerometer measurements) by the pile impedance I. The resistance R_D is the dynamic resistance of the pile. To get the static resistance of the pile R_S, a case damping factor J_c is used:

$$R_D = R_S + J_c I v \tag{18.47}$$

The velocity v is chosen to be the pile point velocity v_{point}, which can be obtained from wave propagation theory as:

$$v_{point(t_1)} = \frac{F_{(t_1)} + I v_{(t_1)} - R_D}{I} \tag{18.48}$$

where $v_{point(t1)}$ and $v_{(t1)}$ are the pile point and pile top velocities respectively, evaluated at the time t_1 corresponding to the first peak of the force signal; $F_{(t1)}$ is the force at the pile top at time t_1; I is the pile impedance; and R_D is the dynamic resistance at the time of driving. Regrouping Eqs. 18.45, 18.47, and 18.48 gives the static capacity of the pile as:

$$R_S = 0.5((1 - J_c)(F_{(t_1)} + I v_{(t_1)}) + (1 + J_c)(F_{(t_1+2L/c)} - I v_{(t_1+2L/c)})) \tag{18.49}$$

The recommended Case damping coefficients are shown in Table 18.6.

Table 18.6 Case Damping Coefficient J_c.

Soil Type	Case Damping Coefficient J_c
Clean sands	0.10 to 0.15
Silty sands	0.15 to 0.25
Silts	0.25 to 0.40
Silty clays	0.40 to 0.70
Clays	0.70 to 1.00

(Likins and Hussein 1988.)

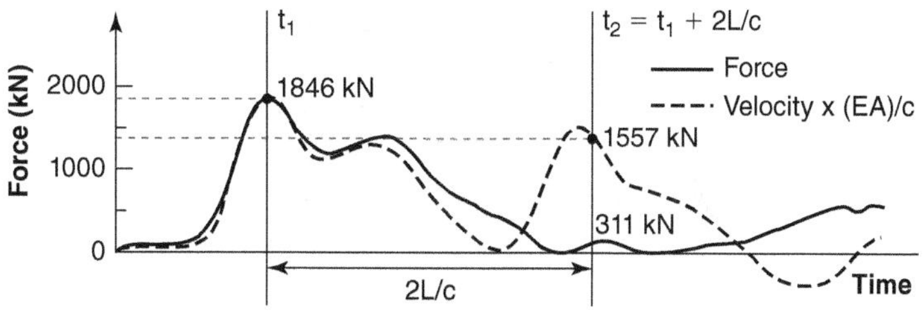

Figure 18.24 Force and impedance times velocity signals at the top of a pile. (After Likins and Hussein 1988.)

The CAPWAP Method

The CAPWAP method (Goble et al. 1993) makes use of the wave equation analysis described in section 18.3.5 and solves the inverse problem. It progresses by iterations to curve-fit the pile response determined in a wave equation model to the measured response of the actual pile during one hammer blow. The measured acceleration is used as input to the pile model and reasonable estimates are made for the soil resistance, quake, and damping parameters. The force-time signal at the pile head is calculated using a wave equation program and compared to the measured force-time signal. The input parameters, including the soil-resistance distribution, quake, and damping, are modified until the match between the measured and calculated signals is deemed satisfactory. Figure 18.25 shows an example of a comparison between measured and calculated force signals for a pile. Once an acceptable match is achieved, the solution yields an estimate of the ultimate static capacity, the distribution of soil resistance along the pile, and the quake and damping parameters. CAPWAP (PDI, CAse Pile Wave Analysis Program, 2012) and DLTWAVE (Dynamic Load Testing WAVE, TNO, 2012) are two programs that can be used.

Note that the soil resistance predicted from data collected during pile driving is tied to the blow count. Indeed, if the blow count is high, the pile penetration per blow is low and the associated pile capacity is low (Figure 18.26). However, if a much bigger hammer is brought in and the pile penetration per blow increases significantly, the soil resistance predicted from such measurement on the same pile can be much higher. For a given pile, the bigger the hammer, the larger the predicted soil resistance is. Figure 18.26 indicates the reason for this observation.

18.3.8 Suction Caissons

Suction caissons have become very popular in the last decade for the foundation of offshore platforms and offshore wind turbines. They consist of upside-down coffee cans (large ones!) pushed into the seafloor by sucking water out of the inside (Figure 18.27).

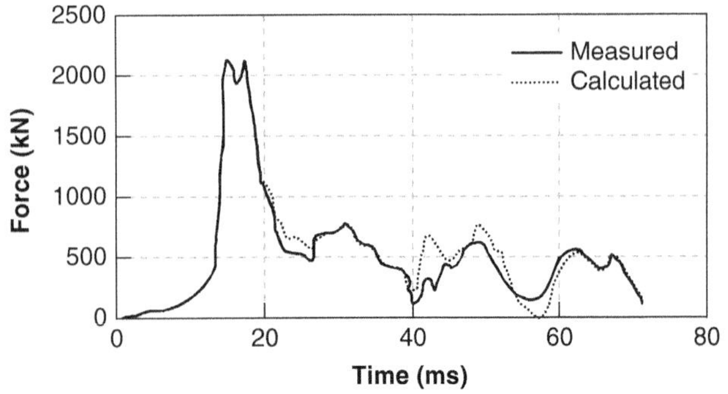

Figure 18.25 Signal matching using CAPWAP. (After Goble et al. 1993.)

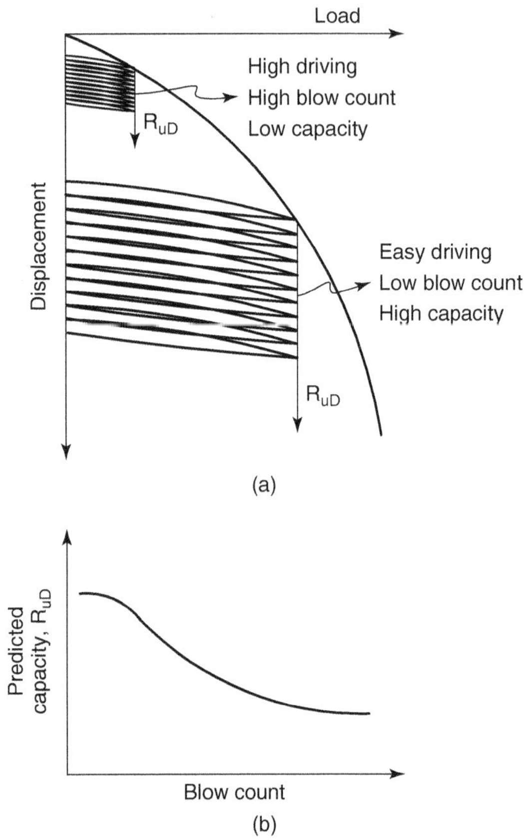

Figure 18.26 Predicted capacity depends on blow count and hammer size.

They are used primarily in clays, but occasionally have to penetrate through sand layers. In clays they have aspect ratios around 5 to 1, with diameters in the range of 3 to 6 m and lengths in the range of 15 to 30 m. Capacities in clays vary from 5 to 20 MN. Sand layers can offer high resistance to penetration; as a result, in sand, the aspect ratio is usually reduced, with values of around 2 to 1. The differential pressure between the inside and the outside of the caisson must be large enough to create a downward force that can overcome the soil resistance to penetration. In clays, this penetration resistance Q_{tot} is calculated as follows (API-RP 2SK, 2012):

$$Q_{tot} = Q_{side} + Q_{tip} = \alpha s_u A_{wall} + (N_c s_u + \gamma' z) A_{tip} \quad (18.50)$$

where Q_{side} is the soil friction resistance on the outside and the inside of the wall of the suction caisson, Q_{tip} is the area at the tip of the caisson corresponding to the thickness of the wall, α is the adhesion factor during installation (taken, for example, as the ratio of the residual shear strength over the peak shear strength), s_u is the average peak undrained shear strength obtained from direct simple shear tests, A_{wall} is the area of the inside wall plus the outside wall in contact with the soil, N_c is the bearing capacity factor (taken as 7.5), γ' is the effective unit weight of the soil, z is the final penetration depth, and A_{tip} is the area of the tip of the caisson corresponding to the cross-sectional area of the wall.

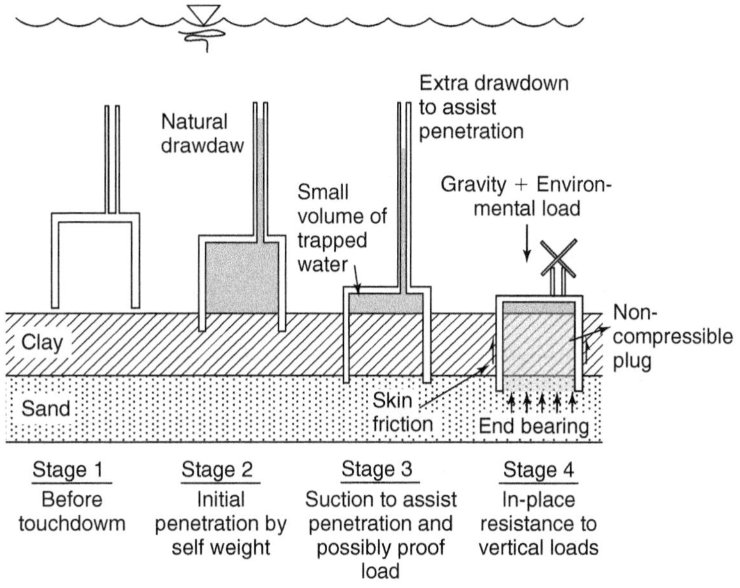

Figure 18.27 Installation of a suction caisson.

The underpressure Δu_{req} needed to enable installation of the suction caisson can then be calculated as:

$$\Delta u_{req} = \frac{Q_{tot} - W'}{A_{in}} \tag{18.51}$$

where Δu_{req} is the difference in pressure between the top and bottom of the roof of the suction caisson, Q_{tot} is the total resistance at full penetration, W' is the submerged weight of the suction caisson, and A_{in} is the area under the roof of the caisson where the underpressure is acting.

The amount of suction that can be generated has a limit which is set by any associated failure mechanism. For example, an excessive underpressure could create an inverse bearing capacity failure ($N_c s_u$) and excessive plug movement inside the wall of the caisson (αs_u). The failure underpressure can then be evaluated as follows:

$$\Delta u_{crit} = N_c s_u + \frac{\alpha s_u A_{wall}}{A_{in}} \tag{18.52}$$

where Δu_{crit} is the underpressure that would create inward failure of the soil, N_c is the bearing capacity factor (taken as 6.2 to 9 depending on the relative embedment; see Figure 17.7), s_u is the undrained shear strength measured in a direct simple shear test, α is the adhesion factor during installation (taken, for example, as the ratio of the residual shear strength over the peak shear strength), s_u is the average peak undrained shear strength obtained from direct simple shear tests, A_{wall} is the area of the inside wall in contact with the soil, and A_{in} is the area under the roof of the caisson where the underpressure is acting. Similar rules have been developed for suction caissons in sand (Andersen et al. 2008). The actual underpressure used is limited to Δu_{crit} divided by a factor of safety equal to 1.5, for operation at a safe level.

18.3.9 Load Testing (Static, Statnamic, Osterberg)

Static Load Tests

Static load tests are still the best way to obtain the load settlement curve for a pile (ASTM D1143). Most typically, these tests consist of installing two or more reaction piles on each side of the test pile, placing a beam across the two reaction piles, and pushing or pulling on the test pile with a jack (Figure 18.28). The load is measured by a load cell between the beam and the jack, while the settlement of the pile is measured by dial gages or LVDTs connected to a settlement beam with supports far away from the test pile. The result of the load test is a load settlement curve (Figure 18.29) up to the capacity of the test pile, the capacity of the reaction system, or the target load for proof tests. For more advanced load tests, the test pile is instrumented with strain gages or extensometers to measure the load in the pile at different depths. This is very useful when separate measurements are needed for the load carried in friction and the load carried in point resistance. For driven piles, the instrumentation should be read right after driving to obtain the distribution of the residual loads.

The ultimate load is obtained from the load settlement curve. For piles in fine-grained soils, the ultimate load is usually clearly identifiable as a plunging load. For coarse-grained soils, the ultimate load is much harder to identify because the curve tends to gradually curve without a plunging load. In this case, an ultimate load criterion is used. There are many such criteria; the most appropriate ultimate load definition seems to be the load corresponding to a settlement s (Figure 18.29) equal to:

$$s = \frac{B}{10} + \frac{PL}{AE} \tag{18.53}$$

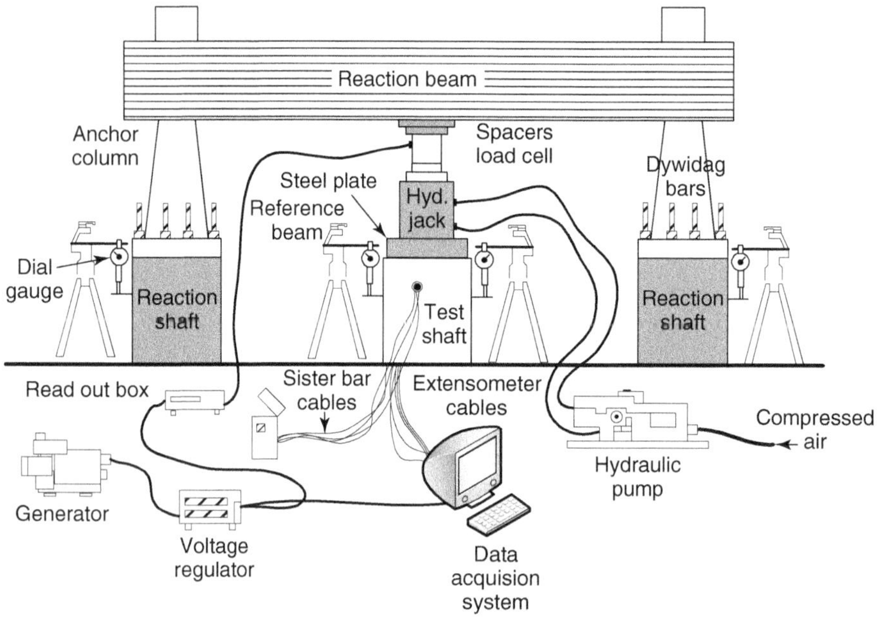

Figure 18.28 Load test setup.

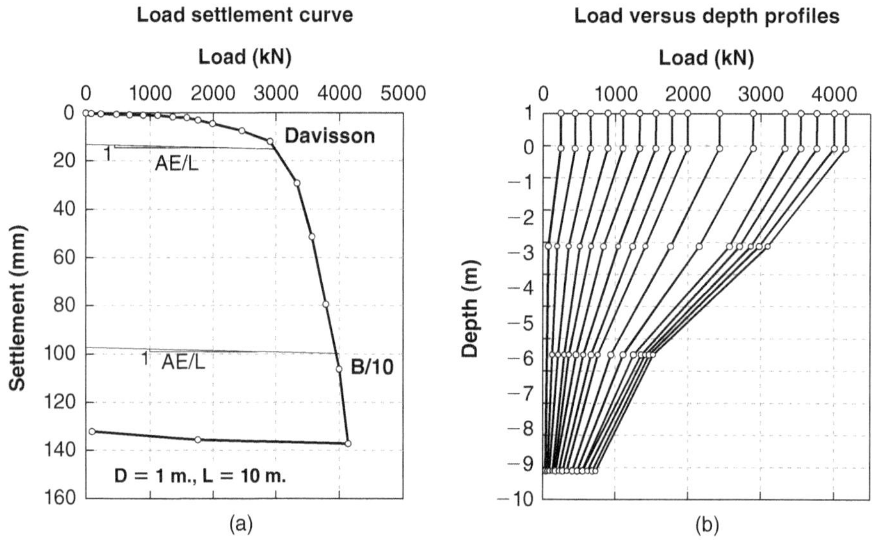

Figure 18.29 Results of an instrumented load test on a bored pile: (a) Load settlement curve. (b) Load versus depth profiles. (Briaud et al. 2000)

where B is the pile diameter, P is the load at the pile top, L is the pile length, A is the pile cross-sectional area, and E is the modulus of the pile material. The Davisson criterion (Figure 18.29) gives a load corresponding to a much smaller settlement:

$$s = 4\text{ mm} + \frac{B}{120} + \frac{PL}{AE} \tag{18.54}$$

For a 1 m diameter, 10 m long concrete pile loaded to 5000 kN, the two criteria give a settlement s of 103 mm (B/10) and 15 mm (Davisson). It is clear that the Davisson criterion corresponds to a much smaller settlement than the B/10 criterion.

Osterberg Load Cell Test

The Osterberg load cell test (Figure 18.30) is another form of static load test. The Osterberg cell or O-cell was developed by Jorge Osterberg (1984). The idea is to place a hydraulic flat jack at the bottom of the pile, so that after pile installation the jack can be inflated, thereby pushing the pile upward against the point resistance. During the test, the soil friction acts downward on the pile and the point resistance acts upward. The test ends when the ultimate friction load or the point resistance load is reached, whichever comes first. If the friction load is the smaller of the two, the ultimate friction load is determined, but only a lower bound of the

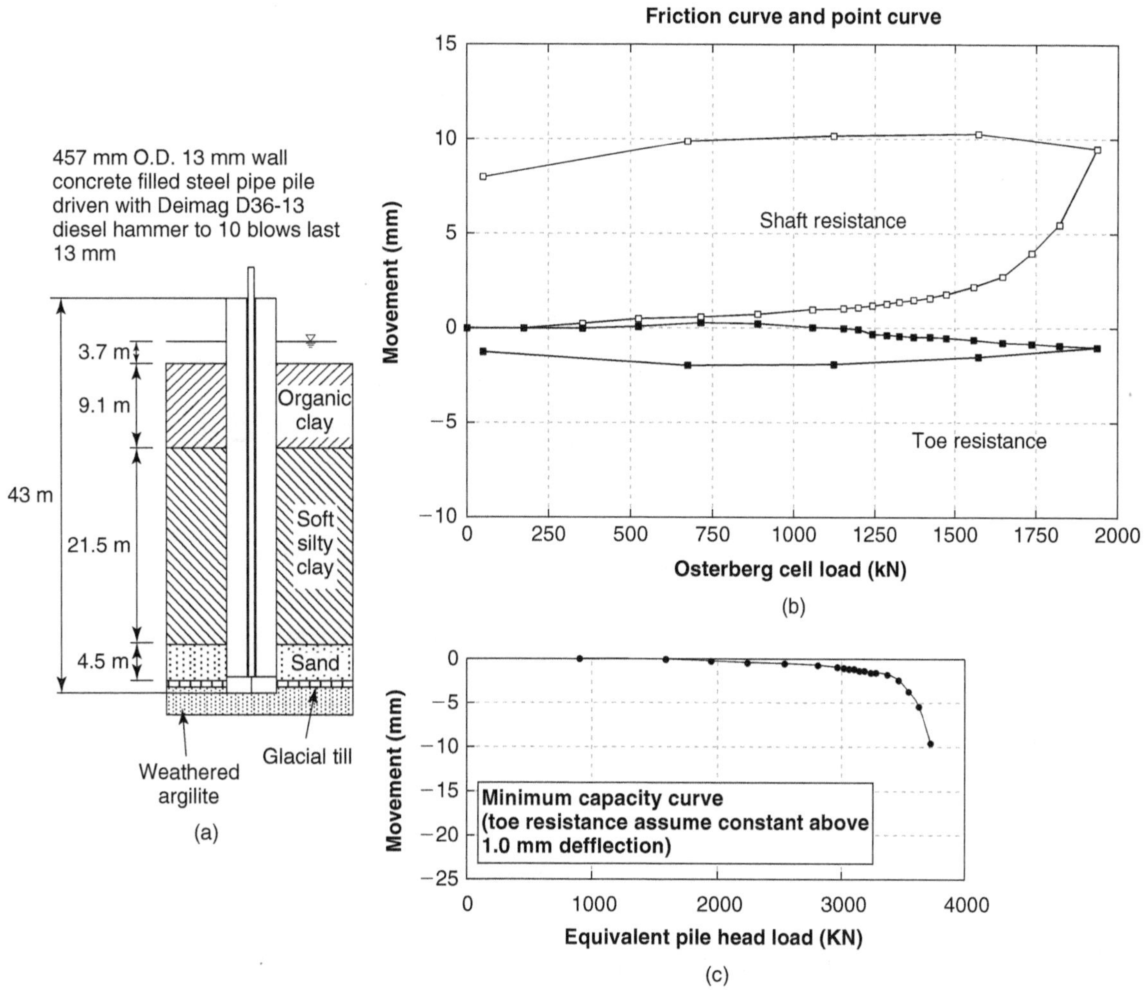

Figure 18.30 Osterberg load test. (a) Osterberg test. (b) Load test results, friction curve, and point curve. (c) Reconstructed top load top movement. (After Hannigan et al. 1998.)

ultimate point resistance is obtained. If the point resistance is the smaller of the two, the ultimate point load is determined, but only a lower bound of the ultimate friction resistance is obtained. In this case, the drawback can be mitigated by placing the O-cell along the pile at a location that balances the load above and the load below the O-cell position. Generally, ASTM Standard D1143 is followed, and loads as high as 300 MN have been generated on large piles. The cost of an O-cell test seems to be about one-third to two-thirds of the cost of a conventional load test, with more savings being realized as the maximum load necessary increases. Figure 18.30 shows an example of results from an O-cell test. There is one load settlement curve for the friction and one load settlement curve for the point resistance. These two curves are added to reconstruct the top load top settlement curve for the pile; however, this reconstructed curve assumes that either the friction or the point resistance remains constant at the end of testing (conservative).

The Statnamic Load Test

The Statnamic load test (Figure 18.31) was developed by Berminghammer (Bermingham and Janes 1989). It consists of placing a dynamic pressure chamber fueled by solid propellant on top of the pile and a large mass on top of the jack. When the fuel is ignited, the large mass is accelerated upward to about 20 g, and by reaction the same force acts downward on the pile. The loading event takes about 100 milliseconds. The load is measured through a dynamic load cell, the acceleration with an accelerometer, and the displacement through a laser beam on a target. Integrating the accelerometer data once for velocity and twice for displacement complements the data.

ASTM Standard D7383 is followed, and loads as high as 50 MN have been generated. Figure 18.32 shows an example of the data collected. The load F_{stn} measured during the Statnamic test is composed of the static resistance of the soil F_s, the rate effect component of the soil F_d, and the inertia

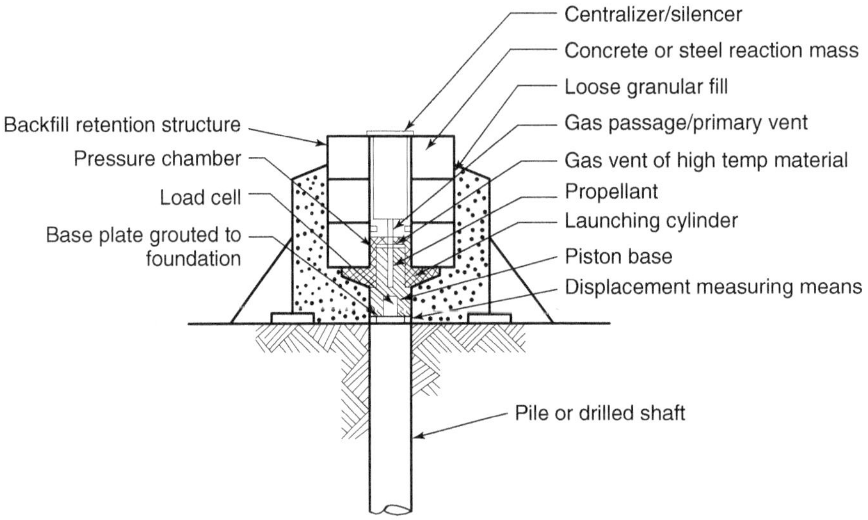

Figure 18.31 Original Statnamic set up. (Bermingham and Janes 1989)

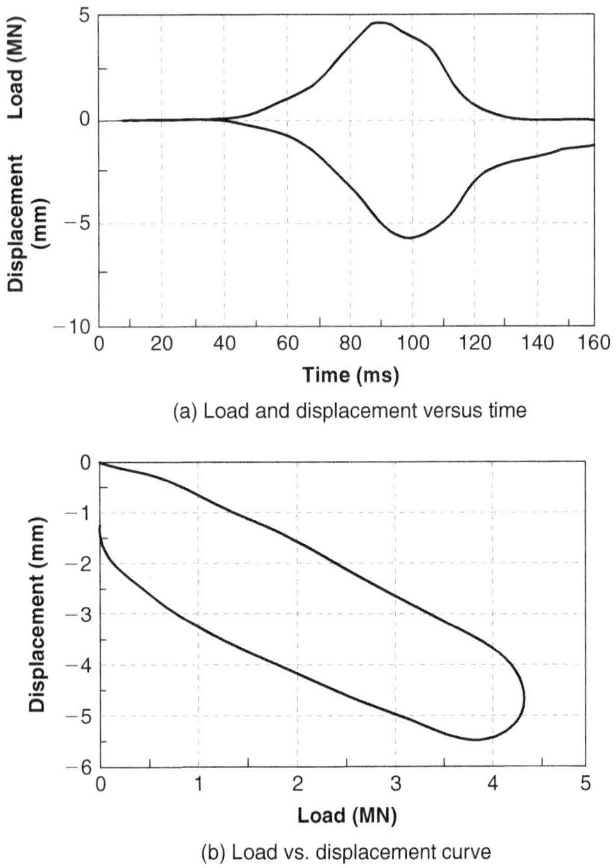

Figure 18.32 Statnamic load test data. (After Hannigan et al. 1998)

force F_i due to accelerating the pile and surrounding soil:

$$F_{stn}(t) = F_s(t) + F_d(t) + F_i(t) \quad (18.55)$$

The purpose of the test is to obtain the static resistance F_s. The rate effect force F_d is considered to be linearly proportional to the velocity:

$$F_d(t) = Cv(t) \quad (18.56)$$

where C is a damping factor similar to the J values for the wave equation analysis. Furthermore, the force F_i can be expressed as:

$$F_i(t) = Ma(t) \quad (18.57)$$

where M is the mass of the pile and $a(t)$ is the measured acceleration at the top of the pile. Here it is assumed that the entrained soil mass is negligible and that rigid body motion prevails. At the point of maximum displacement on the unloading part of the load settlement curve (point A in Figure 18.33), the velocity is zero and therefore the force F_d is zero. Then, if it is assumed that at A the displacement is large enough to mobilize the ultimate pile capacity, the load at point A is the sum of the static ultimate capacity of the pile F_{su} and the inertia force F_i:

$$F_{su}(t_A) = F_{stn}(t_A) - Ma(t_A) \quad (18.58)$$

where $F_{stn}(t_A)$ is the load measured in the Statnamic test at time t_A, t_A is the time corresponding to the largest displacement during the Statnamic test (point A in Figure 18.33), M is the mass of the pile, and $a(t_A)$ is the acceleration measured at the pile top at time t_A.

The damping coefficient C can be obtained by using the maximum load measured at point B on the load settlement curve as follows:

$$C = \frac{F_{stn}(t_B) - F_{su}(t_A) - Ma(t_B)}{v(t_B)} \quad (18.59)$$

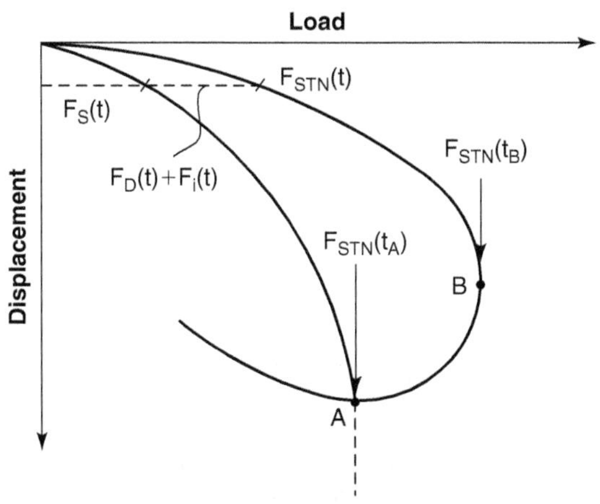

Figure 18.33 Obtaining the static curve from a Statnamic load test. (After Hannigan et al. 1998)

where $F_{stn}(t_B)$ is the load measured in the Statnamic test at time t_B, t_B is the time corresponding to the largest load during the Statnamic test (point B in Figure 18.33), M is taken as the mass of the pile, $a(t_B)$ is the acceleration measured at the pile top at time t_B, $F_{su}(t_A)$ is the static load obtained from Eq. 18.58, and $v(t_B)$ is the velocity measured at the top of the pile a time t_B. If it is further assumed that C is a constant during the load test, then point by point and for any given time t, the static load F_s versus displacement curve can be obtained from:

$$F_s(t) = F_{stn}(t) - Cv(t) - Ma(t) \tag{18.60}$$

Figure 18.33 shows the result of such a procedure.

18.4 VERTICAL LOAD: SINGLE PILE

If a pile is installed in the ground, is left to rest until it gains its full capacity, and is then load tested, the results of the load test could look like the load settlement curve of Figure 18.29. At the beginning of the load test, the load increases proportional to the settlement. This is the linear part of the behavior, and it is usually within this range of loads that the settlement of a single pile is calculated. When the load increases past that point, permanent deformations occur, nonlinear behavior becomes apparent, and, at high loads, a small increase in load leads to a large increase in settlement of the pile.

In clay, that part of the curve usually exhibits a plunging failure mode where the pile simply cannot sustain any increase in load. In sand, however, that part of the curve usually exhibits a continuous increase in load as a function of settlement. The reason for this difference is that during a typical load test, the behavior under the point of a pile in clay is undrained (for saturated soils), whereas the behavior in sand is drained. Under undrained conditions, the shear strength is nearly constant regardless of the total stress increase (plunging failure of the pile), whereas under drained conditions the shear strength increases with the total stress increase and so

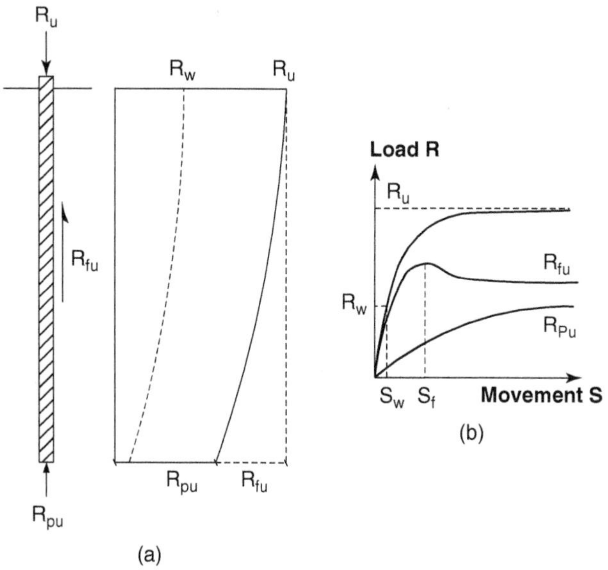

Figure 18.34 Load distribution in a single pile.

does the pile resistance. The distribution of the load in the pile is shown in Figure 18.34.

The issues to be discussed next—always in the simple case of a single pile—include how to estimate the ultimate load of the pile, estimate the settlement of the pile, negotiate downdrag problems, and handle shrink-swell situations.

18.4.1 Ultimate Vertical Capacity for a Single Pile

The ultimate load R_u of a single pile (Figure 18.35) is given by:

$$R_u = R_{uf} + R_{up} = \sum_{i=1}^{n} f_{ui}A_{si} + p_u A_p \tag{18.61}$$

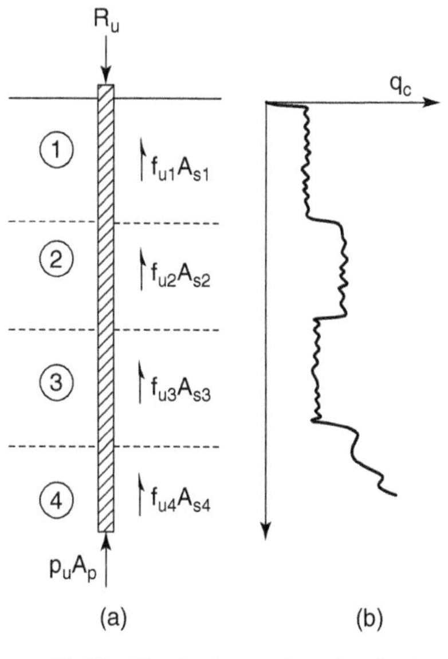

Figure 18.35 Vertical capacity of a single pile.

Table 18.7 Methods for Ultimate Capacity of Single Piles

	Bored Piles	Driven Piles
Fine-Grained Soils	LPC-PMT method LPC-CPT method FHWA, 2010	LPC-PMT method LPC-CPT method API RP2A Effective stress method
Coarse-Grained Soils	LPC PMT method LPC-CPT method FHWA, 2010	LPC-PMT method LPC-CPT method API RP2A Briaud-Tucker SPT method

where R_{uf} is the ultimate resistance in friction, R_{up} is the ultimate resistance in point bearing, A_{si} is the side area of the ith pile element, f_{ui} is the ultimate friction between the soil and the pile acting on the ith pile element, A_p is the area of the pile point, and p_u is the ultimate bearing pressure of the soil at the pile point.

Many methods exist for estimating the values of f_u and p_u. Most of them are empirical with some theoretical content. The methods listed in Table 18.7 have been selected for presentation in this book.

LPC-PMT Method

The LPC-PMT method (Frank, 1999, 2013, Norme Francaise AFNOR P94-262) was developed by the Laboratories des Ponts et Chaussees after 30 years of instrumented load testing of piles by Bustamante and his colleagues. It is not restricted to a pile type or a soil type. It makes recommendations of f_u and p_u for bored piles and driven piles and for fine-grained soils and coarse-grained soils. It is based on the limit pressure from the pressuremeter test (PMT). Although the PMT is not a very common test in many countries, it has the significant advantage over all other tests of being possible in almost all soils and rock. The steps to follow for the LPC-PMT method are:

1. Classify the soil according to Table 18.8.
2. Identify the pile type on Table 18.9 and Table 18.10.

Table 18.8 Soil Classification for LPC-CPT and LPC-PMT Methods

SOIL TYPE	STRENGTH	PMT $p_L{}^*$ (MPa)	CPT q_c (MPa)	SPT N (bpf)	Shear Strength s_u (kPa)
Clay, Silt	Very soft to soft	< 0.4	< 1		< 75
	Firm	0.4 to 1.2	1 to 2.5		75 to 150
	Stiff	1.2 to 2	2.5 to 4		150 to 300
	Very stiff	> 2	> 4		> 300
Sand, Gravel	Very loose	< 0.2	< 1.5	< 3	
	Loose	0.2 to 0.5	1.5 to 4	3 to 8	
	Medium dense	0.5 to 1	4 to 10	8 to 25	
	Dense	1 to 2	10 to 20	25 to 42	
	Very dense	> 2	> 20	42 to 58	
Chalk	Soft	< 0.7	< 5		
	Weathered	0.7 to 3	5 to 15		
	Intact	> 3	> 15		
Marl and Marly Limestone	Soft	< 1	< 5		
	Hard	1 to 4	5 to 15		
	Very hard	> 4	> 5		
Rock	Weathered	2.5 to 4			
	Fissured	> 4			

(After Frank 2013 and Norme Francaise AFNOR P94-262)

Table 18.9 Choosing the Friction Parameters for the LPC-PMT Method.

	Clay, Silt		Sand, Gravel		Chalk		Marl and Marly Limestone		Weathered rock	
Friction curve	Q1		Q2		Q3		Q4		Q5	
	α	f_{lim} (kPa)	α	f_{lim} (kPa)	α	f_{lim} (kPa)	α	f_{lim} (kPa)	α	f_{lim} (kPa)
Bored, dry	1.1	90	1.0	90	1.8	200	1.5	170	1.6	200
Bored, mud	1.25	90	1.4	90	1.8	200	1.5	170	1.6	200
Bored w. casing (left in place)	0.7	50	0.6	50	0.5	50	0.9	90	-	-
Bored w. casing (retrieved)	1.25	90	1.4	90	1.7	170	1.4	170	-	-
Driven concrete	1.1	130	1.4	130	1.0	90	0.9	90	-	-
Driven metal (closed end)	0.8	90	1.2	90	0.4	50	0.9	90	-	-
Driven metal (open end)	1.2	90	0.7	50	0.5	50	1.0	90	1.0	90
Driven H pile	1.1	90	1.0	130	0.4	50	1.0	90	0.9	90
Driven sheet pile	0.9	90	0.8	90	0.4	50	1.2	50	1.2	90
Micropiles (single injection)	2.7	200	2.9	380	2.4	320	2.4	320	2.4	320
Micropiles (repeated injections)	3.4	200	3.8	440	3.1	440	3.1	440	3.1	500

(After Frank 2013 and Norme Francaise AFNOR P94-262)

Table 18.10 Bearing Capacity Factor k_p for LPC-PMT Method

	Clay, Silt	Sand, Gravel	Chalk	Marl and Marly Limestone	Weathered rock
Bored, dry	1.15	1.1	1.45	1.45	1.45
Bored, mud	1.15	1.1	1.45	1.45	1.45
Bored w. casing (left in place)	1.15	1.1	1.45	1.45	1.45
Bored w. casing (retrieved)	1.15	1.1	1.45	1.45	1.45
Driven concrete	1.35	3.1	2.3	2.3	2.3
Driven metal (closed end)	1.35	3.1	2.3	2.3	2.3
Driven metal (open end)*	1.0	1.9	1.4	1.4	1.2
Driven H pile*	1.3	3.1	1.7	2.2	1.5
Driven sheet pile*	1.0	1.0	1.0	1.0	1.2
Micropiles (single injection)	1.15	1.1	1.45	1.45	1.45
Micropiles (repeated injections)	1.15	1.1	1.45	1.45	1.45

*For vibrodriven piles, use one half of these k_p values.
(After Frank 2013 and Norme Francaise AFNOR P94-262)

3. From Table 18.9, read the designation of the f_{soil} curve to be used (Q1 to Q5) as well as the α value and the maximum allowable value for f_{soil} called f_{lim}.
4. Enter the proper curve on Figure 18.36a and read the value of f_{soil} corresponding to the value of the net limit pressure p_L* defined as $p_L - \sigma_{oh}$ where p_L is the PMT limit pressure and σ_{oh} is the total horizontal stress at rest at the depth of the pressuremeter test that gave the value of p_L. The value of f_u to be used in Eq. 18.61 is given by:

$$f_u = \alpha f_{soil} \leq f_{\lim} \quad (18.62)$$

5. Repeat steps 1 through 4 to obtain the values of f_{ui} for all values of $p_{Li}*$ in the soil profile next to the pile.
6. The value of p_u is given by:

$$p_u = k_p p_L^* \quad (18.63)$$

where p_L* is the average net limit pressure within 1.5B below the pile point, and B is the pile point diameter or width.

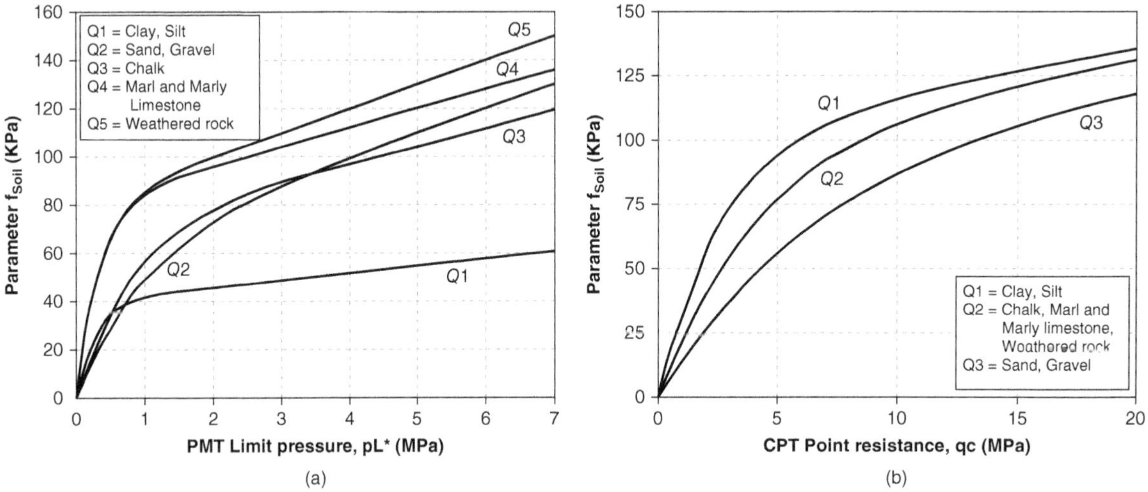

Figure 18.36 (a) Soil friction vs. PMT limit pressure curves and (b) vs. CPT point resistance curves. (After Frank 2013 and Norme Francaise AFNOR P94-262)

7. Obtain k_p for Eq. 18.63 from Table 18.10 and calculate p_u.
8. Calculate the ultimate vertical capacity R_u of the single pile according to Eq. 18.61.

LPC-CPT Method

The LPC-CPT method (Frank 1999, 2013; Norme Francaise AFNOR P94-262) was developed by the Laboratories des Ponts et Chaussees after 30 years of instrumented load testing of piles by Bustamante and his colleagues. It was developed in parallel with the LPC-PMT method as CPT soundings were performed at each load test site. It is not restricted to a pile type or a soil type. It makes recommendations of f_u and p_u for bored piles and driven piles and for fine-grained soils and coarse-grained soils. It is based on the point resistance of the cone penetrometer test (CPT). Although the CPT is a very popular test, which is becoming even more popular as time goes by, it is limited to soils that can be penetrated with a 200 kN truck to sufficient depth for pile design. The steps to follow for the LPC-CPT method are:

1. Classify the soil according to Table 18.8.
2. Identify the pile type on Table 18.11 and Table 18.12
3. From Table 18.11, read the designation of the f_{soil} curve to be used (Q1 to Q5) as well as the α value and the maximum allowable value for f_{soil} called f_{lim}.
4. Enter the proper curve on Figure 18.36b and read the value of f_{soil} corresponding to the value of the CPT point resistance q_c at the depth where f_u is required. The value of f_u to be used in Eq. 18.61 is given by:

$$f_u = \alpha f_{soil} \leq f_{lim} \quad (18.64)$$

where f_{lim} is the maximum permissible value of f_u.

5. Repeat steps 1 through 4 to obtain the values of f_{ui} for all values of q_c in the CPT sounding next to the pile.
6. The value of p_u is given by:

$$p_u = k_c q_c \quad (18.65)$$

where q_c is the average CPT point resistance within 1.5B below the pile point, and B is the pile point diameter or width and k_c a bearing capacity factor.

7. Obtain k_c for Eq. 18.65 from Table 18.12 and calculate p_u.
8. Calculate the ultimate vertical capacity R_u of the single pile according to Eq. 18.61.

FHWA Method for Bored Piles in Fine-Grained Soils

The FHWA method evolved from the initial recommendations of O'Neill and Reese (1999) and was modified by Brown et al. (2010). It gives the following recommendations for f_u and p_u:

From ground surface to depth of 1.5 m:

$$f_u = 0 \quad (18.66)$$

For the rest of the pile and $s_u < 150$ kPa:

$$f_u = 0.55 s_u \quad (18.67)$$

For the rest of the pile and $150 < s_u < 250$ kPa:

$$f_u = \left(0.55 - 0.1\left(\frac{s_u}{p_a} - 1.5\right)\right) s_u \quad (18.68)$$

where s_u is the undrained shear strength of the soil at the depth where f_u is calculated.

$$p_u = N_c s_u \quad (18.69)$$

where N_c is 9 unless the bored pile has length-to-diameter ratio less than 3, in which case N_c is given by Figure 17.7 (see section 17.6.1), s_u is the undrained shear strength of the soil averaged over 2B below the point of the bored pile, and B is the pile point diameter.

FHWA Method for Bored Piles in Coarse-Grained Soil

This FHWA method also evolved from the initial recommendations of O'Neill and Reese (1999) and was modified by

Table 18.11 Choosing the Friction Parameters for LPC-CPT Method

	Clay, Silt		Sand, Gravel		Chalk		Marl and Marly Limestone		Weathered Rock	
	Q1		Q3		Q2		Q2		Q2	
Friction curve	α	f_{lim} (kPa)	α	f_{lim} (kPa)	α	f_{lim} (kPa)	α	f_{lim} (kPa)	α	f_{lim} (kPa)
Bored, dry	0.55	90	0.7	90	0.8	200	1.4	170	1.5	200
Bored, mud	0.65	90	1.0	90	0.8	200	1.4	170	1.5	200
Bored w. casing (left in place)	0.35	50	0.4	50	0.25	50	0.85	90	-	-
Bored w. casing (retrieved)	0.65	90	1.0	90	0.75	170	0.13	170	-	-
Driven concrete	0.55	130	1.0	130	0.45	90	0.85	90	-	-
Driven metal (closed end)	0.4	90	0.85	90	0.2	50	0.85	90	-	-
Driven metal (open end)	0.6	90	0.5	50	0.25	50	0.95	90	0.95	90
Driven H pile	0.55	90	0.7	130	0.2	50	0.95	90	0.85	90
Driven sheet pile	0.45	90	0.55	90	0.2	50	1.25	50	1.15	90
Micropiles (single injection)	1.35	200	2.0	380	1.1	320	2.25	320	2.25	320
Micropiles (repeated injections)	1.7	200	2.65	440	1.4	440	2.9	440	2.9	500

(After Frank 2013, Norme Francaise AFNOR P94-262)

Table 18.12 Bearing Capacity Factor k_c for LPC-CPT Method

	Clay, Silt	Sand, Gravel	Chalk	Marl and Marly Limestone	Weathered Rock
Bored, dry	0.4	0.2	0.3	0.3	0.3
Bored, mud	0.4	0.2	0.3	0.3	0.3
Bored w. casing (left in place)	0.4	0.2	0.3	0.3	0.3
Bored w. casing (retrieved)	0.4	0.2	0.3	0.3	0.3
Driven concrete	0.45	0.4	0.4	0.4	0.4
Driven metal (closed end)	0.45	0.4	0.4	0.4	0.4
Driven metal (open end)*	0.35	0.25	0.15	0.15	0.15
Driven H pile*	0.4	0.4	0.35	0.2	0.20
Driven sheet pile*	0.35	0.15	0.15	0.15	0.15
Micropiles (single injection)	0.45	0.2	0.3	0.3	0.25
Micropiles (repeated injections)	0.45	0.2	0.3	0.3	0.25

*For vibrodriven piles, use one half of these k_c values.
(After Frank 2013, Norme Francaise AFNOR P94-262)

Brown et al. (2010). It gives the following recommendations for f_u and p_u:

$$f_u = K\tan\delta\,\sigma'_{ov} = \beta\,\sigma'_{ov} = (1-\sin\varphi')\left(\frac{\sigma'_p}{\sigma'_{ov}}\right)^{\sin\varphi'}\tan\varphi'\sigma'_{ov} \quad (18.70)$$

where K is taken as the at-rest horizontal earth pressure coefficient, δ is taken as the effective stress friction angle φ', σ'_{ov} is the vertical effective stress at rest in the soil at the depth where f_u is calculated, β is $K\tan\delta$, and σ'_p is the effective stress preconsolidation pressure of the soil. The value of σ'_p can be estimated by:

$$\sigma'_p(\text{kPa}) = 47(N_{60})^m \quad \text{(Mayne 2007a; 2007b)} \quad (18.71)$$

$$\sigma'_p(\text{kPa}) = 15N_{60} \quad \text{(Kulhawy and Chen 2007)} \quad (18.72)$$

where N_{60} is the SPT blow per 0.3 m corrected for 60% of maximum energy, and m is 0.6 for clean quartz sands and 0.8 for silty sands and sandy silts. O'Neill and Reese (1999) gave a value of β related to the depth z in meters at which f_u is calculated:

$$\beta = 1.5 - 0.244[z(\mathrm{m})]^{0.5} \quad with \quad 0.25 \leq \beta \leq 1.2 \quad (18.73)$$

The FHWA 2010 method retained the previous recommendation for p_u as:

$$p_u(\mathrm{kPa}) = 60N_{60} \quad (18.74)$$

where N_{60} is the SPT blow per 0.3 m corrected for 60% of maximum energy averaged over 2B below the point of the bored pile, with B being the pile point diameter.

API-RP2A Method for Driven Piles in Fine-Grained Soils

The API-RP2A method also evolved over the years and is now based on the work of Randolph and Murphy (1985). It gives the following recommendations for f_u and p_u:

$$f_u = \alpha s_u = 0.5\left(\frac{s_u}{\sigma'_{ov}}\right)^{-0.5} s_u \quad for \quad \frac{s_u}{\sigma'_{ov}} \leq 1 \quad (18.75)$$

$$f_u = \alpha s_u = 0.5\left(\frac{s_u}{\sigma'_{ov}}\right)^{-0.25} s_u \quad for \quad \frac{s_u}{\sigma'_{ov}} > 1 \quad (18.76)$$

where s_u is the undrained shear strength, and σ'_{ov} is the vertical effective stress at rest in the soil at the depth where f_u is calculated. Then p_u is obtained by:

$$p_u = 9s_u \quad (18.77)$$

Effective Stress Method for Driven Piles in Fine-Grained Soil

The effective stress method theoretically gives the long-term capacity of a pile in fine-grained soil because it uses the effective stress approach and the drained strength parameters of the soil. It is fundamentally correct, but the parameters that enter into the calculations are more difficult to obtain accurately than those for the total stress method (such as API-RP2A) for the same case. It appears that much precision is lost through adding steps and complexity in the calculations. At this time, it seems that simplicity wins over theoretical correctness. As research progresses, it is likely that theoretical correctness will prevail. Nevertheless, this method gives the following recommendations for f_u and p_u:

$$f_u = K \tan \delta \sigma'_{ov} = \beta \sigma'_{ov} \quad (18.78)$$

The problem is to obtain reliable and accurate values of β. The following values have been proposed (Jeanjean 2012):

$$\beta = (1 - \sin \varphi')\, OCR^{0.5} \tan \varphi' \quad (18.79)$$

where φ' is the effective stress friction angle of the soil, and OCR is the overconsolidation ratio (defined as the ratio of the preconsolidation pressure σ'_p over the vertical effective stress σ'_{ov}). Briaud and Tucker (1997) proposed the values shown in Table 18.13.

For the ultimate point pressure p_u, there is very little data available to make a recommendation, so one could use the ultimate bearing capacity of a shallow foundation. The following equation seems to fit the FHWA recommendations quite well (Hannigan et al. 1998).

Table 18.13 Proposed Values of β for Clays

Soil Type	β value
Soft clays and soft silts	0.2 to 0.25
Medium clays and medium silts	0.25 to 0.30
Stiff clays and stiff silts	0.3 to 0.35

(Briaud and Tucker 1997)

For φ' between 25 and 40 degrees:

$$p_u = N_q \sigma'_{ov} = 400(\tan \varphi')^6 \sigma'_{ov} \quad (18.80)$$

API-RP2A Method for Driven Piles in Coarse-Grained Soils

The API-RP2A method gives the following recommendations for f_u and p_u:

$$f_u = K \tan \delta \sigma'_{ov} \quad (18.81)$$

where K is the horizontal earth pressure coefficient (taken as 0.8), δ is the friction angle of the pile soil interface, and σ'_{ov} is the vertical effective stress at rest in the soil at the depth where f_u is calculated. Then p_u is obtained by:

$$p_u = N_q \sigma'_{ov} \quad (18.82)$$

where N_q is the bearing capacity factor. Recommendations for δ and N_q are presented in Table 18.14.

Briaud-Tucker SPT Method for Driven Piles in Coarse-Grained Soils

This method (Briaud and Tucker 1984) was developed for sands and gravels, and for driven piles only. It makes use of the SPT blow count N and is based on a database of pile load tests that included measured residual loads in the piles at the end of driving and before load testing. The values of f_u and p_u are given by:

$$f_u(\mathrm{kPa}) = 5\,(N)^{0.7} \quad (18.83)$$

$$p_u(\mathrm{kPa}) = 1000\,(N)^{0.5} \quad (18.84)$$

where N is the SPT blow count (blows per 0.3 m). The database used to develop these equations was populated with uncorrected N values, but it seems logical to use N_{60} if the energy is measured or can be estimated during the SPT. Indeed, most of the data came from U.S. pile load tests where drill rigs generate about 60% of maximum energy on the average.

18.4.2 Miscellaneous Questions about the Ultimate Capacity of a Single Pile

Minimum Thickness of the Bearing Layer

A question about the minimum thickness of the bearing layer arises when the layer in which the point of the pile ends has a finite thickness and is underlain by a weaker layer. The

Table 18.14 Recommended Values of δ and N_q

Soil Type	Density	Angle δ	Limiting Friction (kPa)	Bearing Factor N_q	Limiting Point Pressure (kPa)
Sand	Very loose	15	48	8	1900
Sand-silt	Loose				
Silt	Medium				
Sand	Loose	20	67	12	2900
Sand-silt	Medium				
Silt	Dense				
Sand	Medium	25	81	20	4800
Sand-silt	Dense				
Sand	Dense	30	96	40	9600
Sand-silt	Very dense				
Gravel	Dense	35	115	50	12000
Sand	Very dense				

(API-RP 2A 2000.)

question is: How thick should the bearing layer be to generate the full p_u capacity of that layer? One of the important factors is the difference in strength between the bearing layer and the underlying layer. One of the best ways to answer this question is to perform a failure load analysis (see section 17.6.3). Short of that, and if the difference in strength between the bearing layer and the underlying layer is not extreme (say, less than 4 to 1), then a thickness to pile width ratio of more than 4 may be appropriate.

$$\frac{H}{B} > 4 \tag{18.85}$$

where H is the distance between the pile point location and the bottom of the bearing layer and B is the pile diameter or width. One very important observation is that at 4 times the pile width, the thickness of the bearing layer may sustain the capacity of a single pile—but the question is very different for a pile group.

Which Area to Consider for A_s and A_p in Equation 18.61

For solid piles such as bored piles and concrete or timber driven piles, and also for closed-end pipe piles, the issue is clear: The side area should be the pile perimeter times the pile segment length. The question arises in the case of open-end pipe piles and H piles. These piles may plug or not plug during driving and later on during loading. Whether the pile plugs or not depends on the soil type, the pile diameter, and the loading. Some piles may not plug during driving, but most common-size piles plug during subsequent slow loading. It is common practice to calculate the ultimate load R_u (plug) corresponding to a plugged condition on the one hand, and then R_u (unplug) corresponding to an unplugged condition on the other, and to take the minimum of the two as the failure load R_u. For an open-end pipe pile, the plugged condition would be the addition of the outside friction plus the point capacity using the total area, whereas the unplugged condition would be the friction on the outside and the inside of the pile plus the point capacity on the thickness of the pipe wall.

Plugged case $$R_{u(plug)} = \sum f_{ui}\pi D\Delta L_i + p_u\frac{\pi D^2}{4} \tag{18.86}$$

Unplugged case $$R_{u(unplug)} = \sum 2f_{ui}\pi D\Delta L_i + p_u\pi Dt \tag{18.87}$$

Ultimate capacity $$R_u = Min(R_{u(plug)}, R_{u(unplug)}) \tag{18.88}$$

where D is the pile diameter, ΔL is the pile segment length, and t is the wall thickness. The recommendations in Table 18.15 are made by Frank 2013, Norme Francaise AFNOR P94-262.

Compression vs. Tension

An important thing to know is whether the friction in compression is the same as the friction in tension. One issue is

Table 18.15 Pile Areas to Be Used in Ultimate Capacity Calculations according to Frank (2013), Norme Francaise AFNOR P94-262

	Point	Side
Open-end pipe D = diameter	$A_p = \frac{\pi D^2}{4}$	$A_s = \pi DL$
H pile B = width of flange W = height of web	$A_p = BW$	$A_s = 2(B + W)L$

the Poisson's effect, which makes the pile expand laterally when in compression and contract laterally when in tension. This creates a lower lateral stress in tension than in compression, thereby leading to less friction capacity in tension than in compression. Counter to this is the fact that soil masses tend to relax and creep around piles and maintain horizontal stresses in the long term. This was measured by placing total stress pressure cells on driven piles and monitoring the horizontal pressure as a function of time (Briaud and Tucker 1989). Another problem is that the failure surface of the pile in compression may be different from the failure surface of the pile in tension. This was observed on 23 m long H piles in sand that had an ultimate load in friction-tension equal to one-half of the ultimate load in friction-compression (Briaud et al. 1984). The American Petroleum Institute recommends the same friction coefficient in compression and in tension for long pipe piles in clay and in sand. All in all, it appears that significantly different friction capacities in tension and in compression are the exception rather than the rule; however, they can exist and probably more so for short piles.

Should I Add γd or Not Add γd to the Expression Giving p_u?

The question about adding γd is related to the following two expressions:

$$p_u = ks \tag{18.89}$$

$$p_u = ks + \gamma d \tag{18.90}$$

where p_u is the ultimate point pressure, k is a bearing capacity factor, s is a soil strength measurement (CPT q_c, PMT p_L, SPT N, undrained shear strength s_u), γ is the soil unit weight, and d is the depth of the foundation below the nearby ground surface. The answer is that at the level of the foundation, the actual p_u value is given by Eq. 18.90. Thus, if Eq. 18.90 is used, then the weight of the foundation must be included on the load side of the ultimate limit state equation. If it is assumed that the pressure exerted by the weight of the foundation plus backfilling is equal to the pressure of the overburden on each side of the foundation, then the two cancel out and the term γd can be ignored. Generally, it is best to add γd and include the weight of the foundation on the load side, especially in more complex cases where the cancellation may not apply.

What about Buoyancy when the Foundation Is Under Water?

The problem of buoyancy is again associated with the term γd or the vertical total stress at the depth of the foundation σ_{ov}. If the total weight of the foundation is W and the water pressure under the foundation is u_w applied over the bottom surface A_p of the foundation, the buoyant weight is $W' = W - u_w A_p$. The point resistance $p_u A_p$ contains the term $\gamma d A_p$ (or better, $\sigma_{ov} A_p$) and therefore includes the buoyancy force of the foundation. Indeed, $\sigma_{ov} A_p$ is the sum of $\sigma'_{ov} A_p$ and $u_w A_p$, which is the buoyancy force. So there are two alternatives:

1. Use the buoyant weight W' of the pile and then use $\sigma'_{ov} A_p$.
2. Use the total weight W of the pile and then use $\sigma_{ov} A_p$.

It is usually best to use the total weight and the total vertical stress at the foundation level.

Rate of Loading Effect

Soils are somewhat viscous, so if the rate of loading or straining is not changed much, the difference in the pile ultimate capacity can be neglected. If, however, there is a drastic change in loading rate or strain rate, then the difference must be included in the calculations. For example, most of the calculation methods presented in section 18.4.1 are based on load test databases where the pile was pushed to large displacements in several hours. However, under a building or a bridge, the pile will experience the load for the design life of the structure, which may be 75 years, for example. In contrast, the rise time of a hurricane wave against an offshore structure may be only 3 seconds. In these two cases, the ultimate capacity would be significantly affected. The model used for the undrained shear strength of saturated clays (see section 15.8) is extended to pile capacities (Briaud and Garland 1985):

$$\frac{R_{u1}}{R_{u2}} = \left(\frac{t_1}{t_2}\right)^{-n} \tag{18.91}$$

where R_{u1} is the ultimate pile resistance when loaded to failure in a time t_1 and R_{u2} is the ultimate pile resistance when loaded to failure in a time t_2. Values of the viscous exponent n vary from 0.01 to 0.03 for sand and from 0.02 to 0.08 for clays (see Figure 15.18). So, if a pile has a capacity of 1000 kN according to usual methods associated with load tests averaging 3 hours, and if that pile is loaded in 3 s by a hurricane wave, then the load will be such that:

$$\frac{R_{u1}}{1000} = \left(\frac{3}{3 \times 3600}\right)^{-0.05} = 1.5 \tag{18.92}$$

The capacity R_{u1} of the pile during the hurricane is 50% larger than the capacity calculated by conventional methods. However, if Eq. 18.91 is applied to long-term loading under a 75-year-old building, the capacity becomes:

$$\frac{R_{u1}}{1000} = \left(\frac{75 \times 365 \times 24}{3}\right)^{-0.05} = 0.54 \tag{18.93}$$

The capacity R_{u1} of the pile after 75 years would be 50% smaller than the capacity calculated by conventional methods. The data upon which Eq. 18.91 is based are mostly populated with tests done in less than 3 hours; therefore, using this model for the long-term capacity of piles is not based on data.

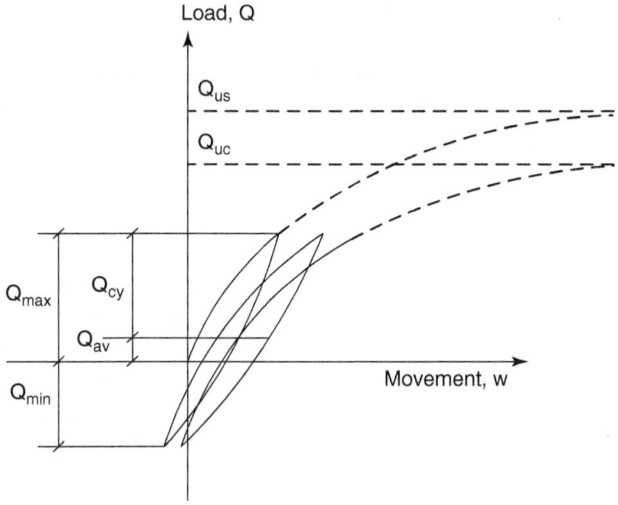

Figure 18.37 Cyclic loading of piles.

Cyclic Loading Effect

Soils are sensitive to cyclic loading and tend to weaken with the accumulation of cycles. Figure 18.37 shows the load settlement curve for a pile subjected to cyclic loading. Two main parameters are defined: the cyclic amplitude ratio R_{cy} and the average load ratio R_{av}:

$$R_{cy} = \frac{Q_{\max} - Q_{\min}}{2Q_{us}} \quad and \quad R_{av} = \frac{Q_{\max} + Q_{\min}}{2Q_{us}} \qquad (18.94)$$

where Q_{us} is the vertical ultimate capacity of the pile under static monotonic loading, and $Q_{\max}$ and $Q_{\min}$ are the maximum and minimum load applied respectively during cyclic loading. The ultimate cyclic capacity Q_{uc} is the maximum load that can be reached when performing a monotonic test to failure at the end of the cycles. Note also that Q_{cy} and Q_{av} are the cyclic load amplitude and the average load respectively. If a test reaches failure after N cycles, then Q_{uc} is equal to $Q_{\max}$ for N cycles. If the load does not change direction during loading (always compression, for example), it is called *one-way cyclic loading*. If it does change direction (compression to tension, for example), it is called *two-way cyclic loading*. The most severe loading seems to be symmetrical two-way cyclic loading where R_{av} is equal to zero.

Studies have been performed and recommendations made to quantify the influence of R_{cy} and R_{av} on the ratio Q_{uc}/Q_{us}. One of them is the work of Karlsrud et al. (1986) at NGI. The results, shown in Figure 18.38, indicate that for these tests and for full reversal of load ($Q_{av} = 0$), the cyclic capacity Q_{uc} is 35 to 50% of the static capacity Q_{us} depending on the number of cycles to failure.

Briaud and Felio (1986) studied the impact of cyclic vertical loading on the response of piles by quantifying the increase in vertical movement as a function of the number of cycles. After collecting a database of 16 studies on cyclic full-scale pile load tests and 10 studies on model pile load tests, they used the following power law model:

$$\frac{s(N)}{s(1)} = N^a \qquad (18.95)$$

where $s(N)$ and $s(1)$ are the movement of the pile top for the Nth and first cycle respectively, N is the cycle number, and a is the cyclic exponent. Figure 18.38 shows the range of a values obtained as a function of the cyclic load ratio $Q_{\max}/Q_{us}$.

Prediction Method vs. Design Method

A distinction should be made between a prediction method and a design method (Figure 18.39). The goal of a prediction

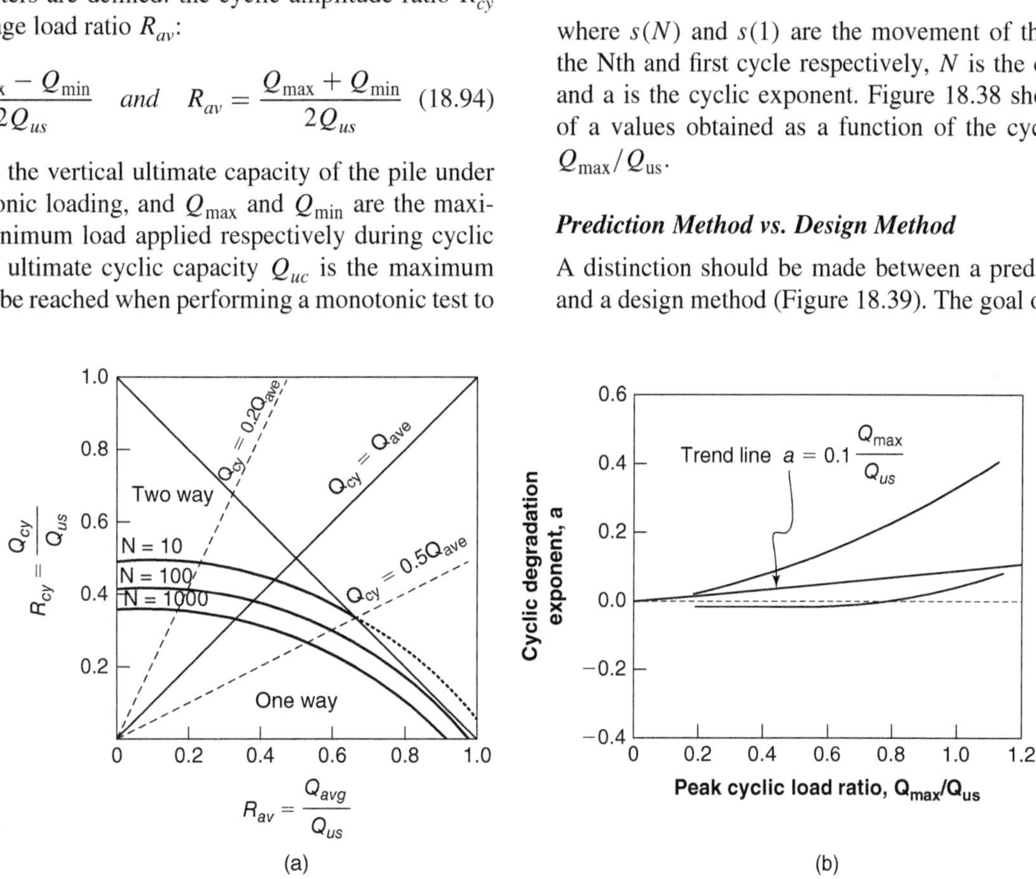

Figure 18.38 Results of two studies on cyclic loading of piles in clays: (a) Karlsrud et al. (1986). (b) Briaud and Felio (1986).

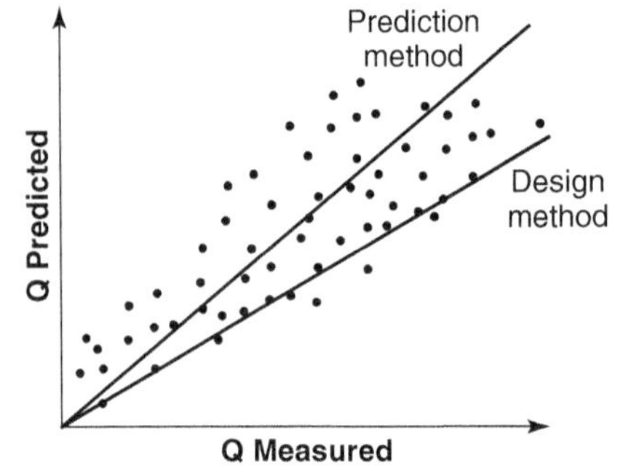

Figure 18.39 Difference between a prediction method and a design method.

method is for the calculated value to be as close as possible to the measured value on the average. The goal of a design method is to minimize the number of times the calculated value is unsafe. This is a big difference, and for ultimate pile capacity a design method will tend to give lower values than a prediction method. This also brings into play the issue of precision versus accuracy. An *accurate method* is a method that gives the right answer on the average. A *precise method* is a method that exhibits very little scatter around the mean. A precise and accurate method is a method that gives the right answer on the average with very little scatter around the mean; that is, very little uncertainty. An inaccurate but precise method is more desirable than an imprecise but accurate method. Indeed, it is easier to apply a calibration factor to the mean value of an inaccurate but precise method than to reduce the scatter of an imprecise but accurate method.

Resistance Factor

To use the LRFD approach, one must have the load factors γ and resistance factors φ in Eq. 18.3. These factors are given in the codes or guidelines specific to each method. For the load factors γ, see Table 17.1. For the resistance factors, some methods make a distinction between the side friction resistance factor φ_f and the point resistance factor φ_p. Some methods use one global resistance factor φ. Table 18.16 shows suggested ranges of global resistance factors as collected from various sources.

Length Effect on Ultimate Capacity

The length of the pile may have an effect on the pile capacity. This is particularly clear when the soil is overconsolidated and the pile is long. Here is why. Overconsolidated soils exhibit a peak shear strength followed by a lower to much lower residual shear strength. Long piles, when loaded, exhibit much more movement at the pile top than at the pile point because of the elastic compression of the pile (same in tension). At ultimate load, the displacement at the pile top will be large enough to be on the residual part of the strength curve while, at the pile point, the displacement will just reach the peak strength. As a result, the ultimate load will be lower than the one obtained by using the peak shear strength all along the pile for obtaining f_u. The best way to quantify the influence of pile length on ultimate load is to use a settlement analysis at large displacement. This topic is covered in section 18.4.3.

18.4.3 Settlement of a Single Pile

The settlement of the top of a pile s_{top} is equal to the settlement of the point of the pile s_{point} plus the compression of the pile:

$$s_{top} = s_{point} + \frac{PL}{A_{cs}E_p} \tag{18.96}$$

where P is the average load in the loaded portion of the pile, L is the length of pile under load, A_{cs} is the pile cross section, and E_p is the modulus of elasticity of the pile material. The load in the pile is Q_{top} at the top and Q_{point} at the point

Table 18.16 Estimated Resistance Factor φ for Ultimate Limit State in Pile Design

	Bored Piles		Driven Piles	
	Method	Resistance factor φ	Method	Resistance factor φ
Fine-grained soils	LPC-PMT method	0.5 to 0.6	LPC-PMT method	0.5 to 0.6
	LPC-CPT method	0.5 to 0.6	LPC-CPT method	0.5 to 0.6
	FHWA, 2010	0.35 to 0.45	API RP2A	0.6 to 0.7
			Effective stress method	0.3 to 0.4
Coarse-grained soils	LPC-PMT method	0.5 to 0.6	LPC-PMT method	0.5 to 0.6
	LPC-CPT method	0.5 to 0.6	LPC-CPT method	0.5 to 0.6
	FHWA, 2010	0.45 to 0.55	API RP2A	0.6 to 0.7
			Briaud-Tucker SPT method	0.35 to 0.45

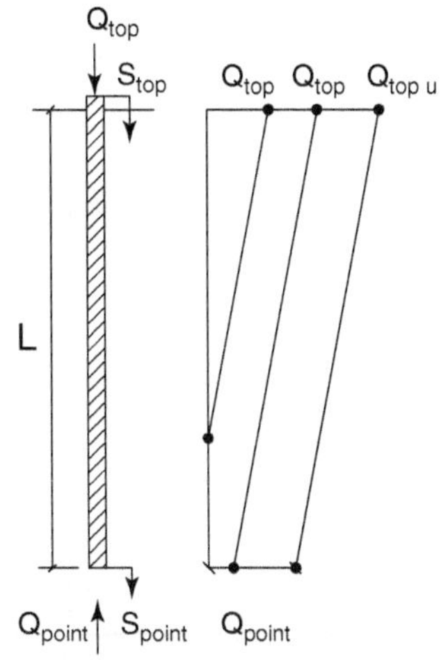

Figure 18.40 Load distribution in the pile.

(Figure 18.40). If the friction is constant between the top and the point of the pile, Eq. 18.96 becomes:

$$s_{top} = s_{point} + \frac{(Q_{top} + Q_{point})L}{2A_{cs}E_p} \quad (18.97)$$

As a preliminary estimate, $(Q_{top} + Q_{point})/2$ is often taken as $0.6Q_{top}$. Of course this is an estimated average. For friction piles where most of the pile capacity comes from the side friction, $0.6Q_{top}$ would likely be larger than the true average load, and in fact L is likely not the total length of the pile, as the load becomes zero along the pile length. For an end-bearing pile, though, $0.6Q_{top}$ may be too small, as much of the load is carried in point resistance. The settlement s_{point} is related to the load at the point through a load transfer curve that can be idealized as an elastic, perfectly plastic curve (Figure 18.41). The slope of the elastic part is a spring constant k_p such that:

$$Q_{point} = k_p s_{point} \quad (18.98)$$

Then Eq. 18.97 becomes:

$$s_{top} = \frac{Q_{point}}{k_p} + \frac{(Q_{top} + Q_{point})L}{2A_{cs}E_p} \quad (18.99)$$

The spring constant k_p has been related to the modulus of the soil under the point through theory. The relationship for a closed-end circular pile (Randolph and Wroth 1978) is:

$$Q_{point} = k_p s_{point} = \left(\frac{DE_s}{1 - \nu^2}\right) s_{point} \quad (18.100)$$

where D is the diameter of the pile point, and E_s and ν are the modulus and the Poisson's ratio of the soil below the pile point. In terms of pressure p under the pile point, the equation becomes:

$$p = \frac{4E_s}{\pi D(1 - \nu^2)} s_{point} \quad (18.101)$$

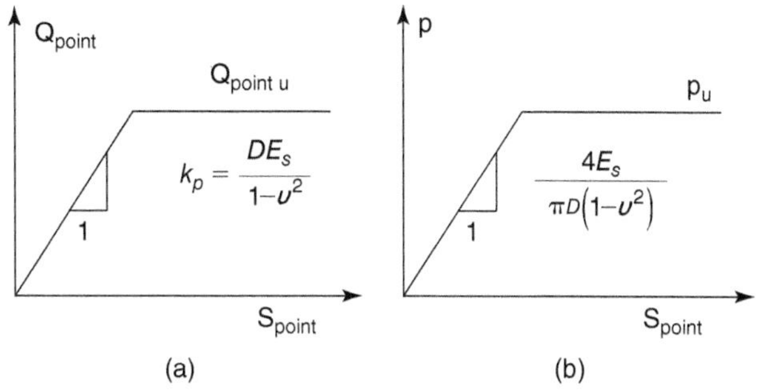

Figure 18.41 Point load transfer curve model.

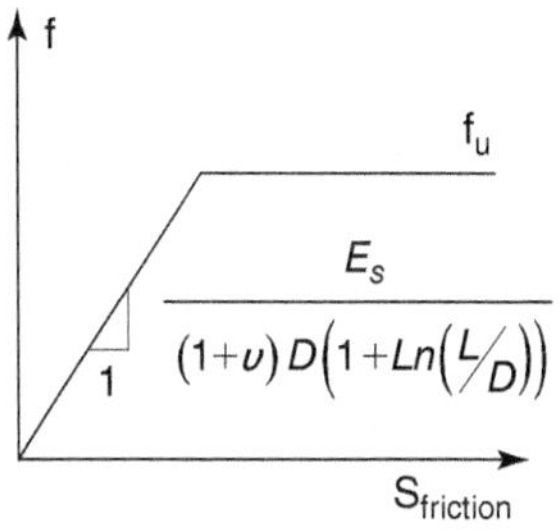

Figure 18.42 Friction load transfer curve model.

The friction load transfer curve can also be represented by an elastic, perfectly plastic curve (Figure 18.42). The slope of the elastic part of the friction transfer curve (Frank, Zhao 1982) is given by:

$$f = \frac{E_s}{(1 + \nu)(1 + Ln(L/D))D} s_{friction} \quad (18.102)$$

where f is the friction stress at the interface between the pile and the soil, E_s and ν are the soil modulus and the Poisson's ratio at depth z where the friction is generated, L is the embedded pile length, D is the pile diameter, and $s_{friction}$ is the downward movement of the pile at depth z.

We still do not know what the load distribution is in the pile for a given load at the top. Solving this problem requires a load transfer curve analysis, which is best explained through an example (Figure 18.43).

A 0.3 m diameter closed-end pipe pile with a 5 mm thick wall is driven 10 m below the ground surface. The steel has a modulus of 2×10^8 kPa. The soil is made of 9 m of a soft clay underlain by a thick layer of dense sand. The modulus of the soft clay is 5×10^3 kPa, and the modulus of the sand layer is 10^5 kPa. The drained Poisson's ratio is 0.35 for both soils. The ultimate friction f_u in the soft clay at the pile-soil interface is 20 kPa, and in the dense sand f_u is 80 kPa. The ultimate point pressure under the pile point in the dense sand is 10,000 kPa.

1. Divide the pile into a number of elements. We would typically use a minimum of 10 elements for the computer solution. For this hand calculation, we will use only 3 elements, as shown in Figure 18.43.

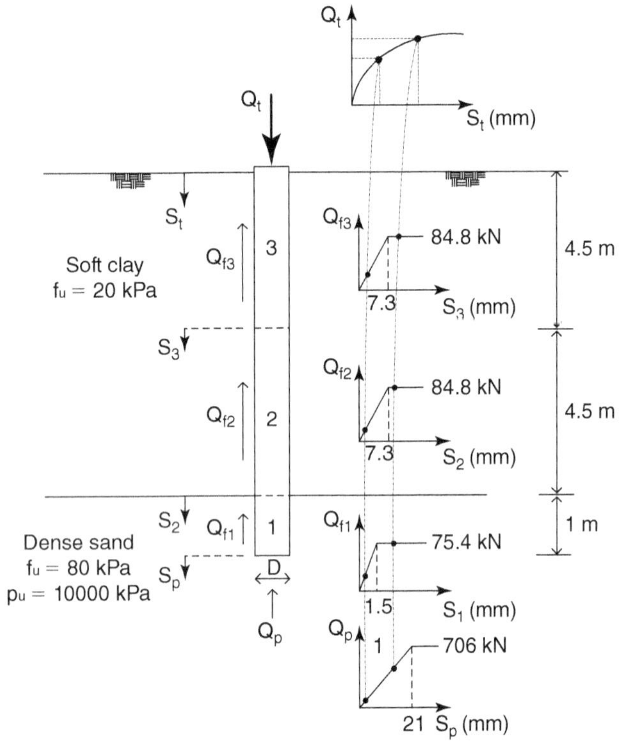

Figure 18.43 Pile problem and load transfer curves.

2. Prepare the load transfer curves for each element: three friction curves for elements 1, 2, and 3 and one point curve under element 1. To prepare these curves, the elastic slopes of Eqs. 18.101 and 18.102 are used together with the ultimate values of f_u and p_u. These curves are shown in Figure 18.43.
3. Assume a point movement of 1 mm and calculate the point load corresponding to that movement using Eq. 18.100:

$$Q_{point} = \left(\frac{0.3 \times 100000}{1 - 0.35^2}\right) \times 0.001 = 34.2 \text{ kN} \quad (18.103)$$

This corresponds to point 1 on the point load transfer curve in Figure 18.43.

4. Evaluate the friction f_1 mobilized in element 1 by reading the friction transfer curve for a movement of 1 mm (Eq. 18.102):

$$f_1 = \frac{100000}{(1 + 0.35)(1 + Ln(10/0.3))0.3} \times 0.001$$
$$= 54.8 \text{ kPa} < 80 \text{ kPa} \quad (18.104)$$

More correctly speaking, we should first calculate the movement at the midpoint in element 1 and then use that movement to obtain the friction value from the transfer curve. This would require an iteration. For simplicity in these hand calculations, we will use the movement at the bottom of the element to obtain the friction value in that element.

5. Calculate the load carried in friction in element 1:

$$Q_{friction1} = f_1 \pi D \Delta L_1 = 54.8 \times 3.14 \times 0.3 \times 1$$
$$= 51.6 \text{ kN} \quad (18.105)$$

6. Calculate the movement at the bottom of element 2:

$$s_2 = s_{point} + \frac{\left(Q_p + \frac{Q_{f1}}{2}\right) \Delta L_1}{A_{cs} E_p}$$
$$= 1 + \frac{\left(34.2 + \frac{51.6}{2}\right) 1000}{3.14 \times 0.3 \times 0.005 \times 2 \times 10^8}$$
$$= 1.063 \text{ mm} \quad (18.106)$$

7. Evaluate the friction f_2 mobilized in element 2 by reading the friction transfer curve for the movement s_2 (Eq. 18.102):

$$f_2 = \frac{5000}{(1 + 0.35)(1 + Ln(10/0.3))0.3} \times 0.001063$$
$$= 2.91 \text{ kPa} < 20 \text{ kPa} \quad (18.107)$$

8. Calculate the load carried in friction in element 2:

$$Q_{friction2} = f_2 \pi D \Delta L_2 = 2.91 \times 3.14 \times 0.3 \times 4.5$$
$$= 12.3 \text{ kN} \quad (18.108)$$

9. Calculate the movement at the bottom of element 3:

$$s_3 = s_2 + \frac{\left(Q_p + Q_{f1} + \frac{Q_{f2}}{2}\right) \Delta L_2}{A_{cs} E_p}$$
$$= 1.063 + \frac{\left(34.2 + 51.6 + \frac{12.3}{2}\right) 4500}{3.14 \times 0.3 \times 0.005 \times 2 \times 10^8}$$
$$= 1.502 \text{ mm} \quad (18.109)$$

10. Evaluate the friction f_3 mobilized in element 3 by reading the friction transfer curve for the movement s_3 (Eq. 18.102):

$$f_3 = \frac{5000}{(1 + 0.35)(1 + Ln(10/0.3))0.3} \times 0.001439$$
$$= 3.94 \text{ kPa} < 20 \text{ kPa} \quad (18.110)$$

11. Calculate the load carried in friction in element 3:

$$Q_{friction3} = f_3 \pi D \Delta L_3 = 3.94 \times 3.14 \times 0.3 \times 4.5$$
$$= 16.7 \text{ kN} \quad (18.111)$$

12. Calculate the movement at the top of element 3:

$$s_{top} = s_3 + \frac{\left(Q_p + Q_{f1} + Q_{f2} + \frac{Q_{f3}}{2}\right)\Delta L_3}{A_{cs}E_p}$$

$$= 1.502 + \frac{\left(34.2 + 51.6 + 12.3 + \frac{16.7}{2}\right)4500}{3.14 \times 0.3 \times 0.005 \times 2 \times 10^8}$$

$$= 2.01\text{ mm} \quad (18.112)$$

13. Calculate the load at the top of the pile:

$$Q_{top} = Q_p + Q_{f1} + Q_{f2} + Q_{f3}$$

$$= 34.2 + 51.6 + 12.3 + 16.7 = 114.8\text{ kN} \quad (18.113)$$

14. Now we have a point on the top load versus top movement curve for the vertically loaded pile. A second point can be generated by going back to step 3 and assuming a point movement of, say, 2 mm and repeating steps 3 to 13 to get the corresponding values of Q_{top} and s_{top}. Point by point, the load settlement curve for the pile is generated in this fashion.

Note that typically it takes very little displacement to mobilize the ultimate friction, whereas it takes more displacement to mobilize the ultimate point resistance. The displacement associated with full friction mobilization is often estimated to be 2.5 mm, while the displacement associated with full point resistance mobilization can be 10 mm or more.

18.5 VERTICAL LOAD: PILE GROUP

Piles are often installed in groups (Figure 18.45) to carry higher loads under columns of buildings, bridges, dams, and other structures. Again the questions of ultimate resistance and settlement arise.

18.5.1 Ultimate Vertical Capacity of a Pile Group

What we want to know is if a group of n piles, each having an isolated ultimate capacity R_{us}, will have an ultimate capacity of n times R_{us}. The first estimate of R_{ug} for the group can be written as:

$$R_{ug} = enR_{us} \quad (18.114)$$

where R_{ug} is the ultimate capacity of the pile group, e is the efficiency of the pile group, n is the number of piles in the group, and R_{us} is the ultimate capacity of one pile. The efficiency e of the group may be smaller than 1, but sometimes is larger than 1. Two cases are identified: sand and clay.

Sand

For sand, a further distinction is made between bored piles and driven piles. For bored piles, the current AASHTO recommendation (2010) is to use efficiencies as follows:

Bored pile groups in sand $\quad e = 0.67 \quad$ for $s/B = 2.5$ (18.115)

Bored pile groups in sand $\quad e = 1.0 \quad$ for $s/B > 4$ (18.116)

where e is the group efficiency, s is the center-to-center pile spacing, and B is the pile diameter. For values of s/B between 2.5 and 4, extrapolation is used. For driven pile groups in sand, it is reasonable to think that the efficiency depends on the relative density of the sand. In loose sands, driving piles in a group would densify the sand more than driving a single pile would. Thus, the efficiency of a driven pile group in loose sand should be higher than 1. In very dense sand, however, the pile driving could not make the sand any denser, so the efficiency of the driven pile group would not be enhanced.

A large-scale experiment was performed to check if the efficiency of driven pile groups in loose sand was larger than 1 (Briaud et al. 1989). A five-pile group was driven in place, as was a separate reference single pile (Figure 18.44). The piles were closed-end steel pipe piles with a diameter of 0.273 m, an embedded length of 9.15 m, and a wall thickness of 9.3 mm. The soil was a clean, fine sand hydraulic fill with the following properties: dry unit weight 15.7 kN/m^3, water content 23%, friction angle 35.4°, SPT blow count 15 blows per 0.3 m, CPT point resistance averaging 6200 kPa, and a PMT limit pressure averaging 500 kPa. The piles were instrumented with strain gages to obtain the residual load due to driving and (separately) the load carried in friction and in point resistance during load testing. The single pile and the pile group were load tested (pushed) to a penetration of 40 mm and exhibited plunging failure. At that penetration the single pile carried an ultimate load of 505 kN and the pile group an ultimate load of 2499 kN for an efficiency of 0.99. However, it is interesting to note that although the top load was the same for the single pile and each pile in the group, the distribution of point load and friction load was quite different. The ultimate point load was 360 kN for the single pile and 240 kN per pile in the group, leading to a point efficiency of 0.67. The ultimate load carried in friction was 150 kN for the single pile and 270 kN per pile in the group, leading to a friction efficiency of 1.8. The significant difference in point load between the single pile and the piles in the group was attributed to the difference in residual stresses. When a single pile is driven, it locks in a residual point load. Driving additional piles in the group next to the first pile releases the residual point load and decreases its beneficial effect. The significant difference in friction load between the single pile and the piles in the group was attributed to the difference in horizontal stresses. When a single pile is driven, it generates a certain horizontal stress. Driving additional piles in the group next to the first pile increases the horizontal stresses and therefore the pile friction. Based on these results, it may be best to write the efficiency equation for a pile group as:

$$R_{ug} = n(e_f R_{ufs} + e_p R_{ups}) \quad (18.117)$$

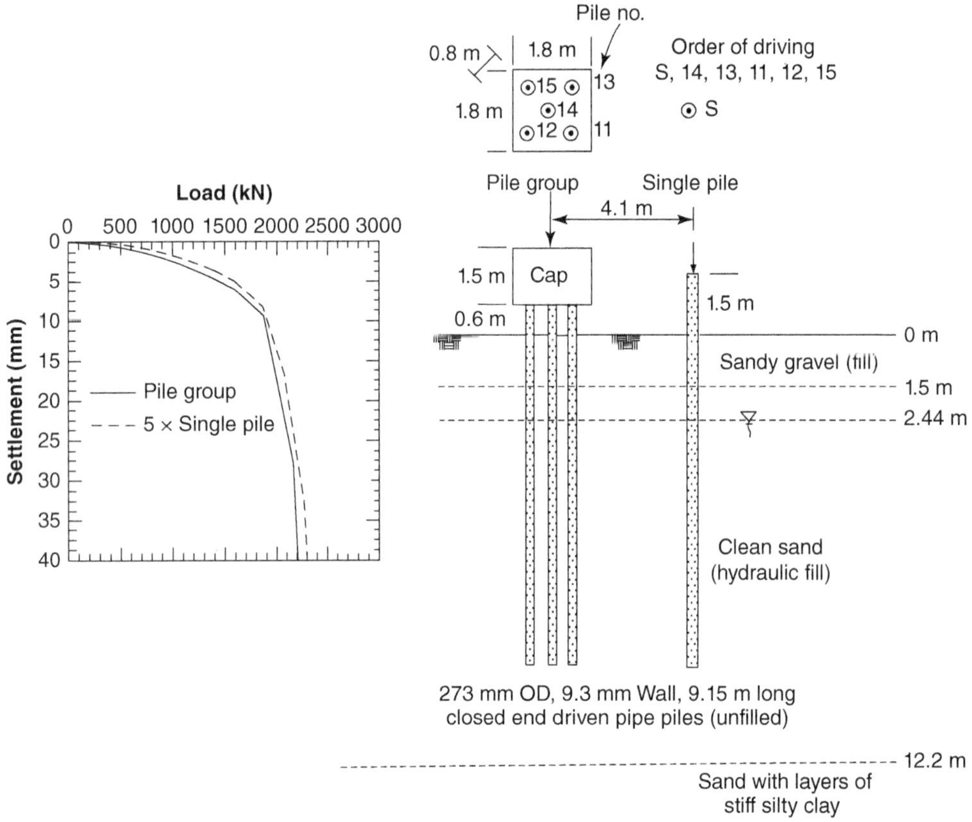

Figure 18.44 Pile group load test and results.

where R_{ug} is the ultimate capacity of the pile group, n is the number of piles in the group, e_f and e_p are the efficiencies of the pile group in friction and in point resistance respectively, and R_{ufs} and R_{ups} are the ultimate capacity of one pile in friction and in point resistance respectively. The pile group load test just described suggests e_f and e_p values of 0.67 and 1.8 respectively. In the case history just described, the increase in friction balanced the decrease in point resistance perfectly. Shorter piles that rely more on point resistance would have global efficiencies e lower than 1, and longer piles that rely more on friction resistance would have global efficiencies e higher than 1. In the absence of further evidence, it is suggested that the efficiency of driven piles in loose sand be taken as 1.

Clay

For clay, the group efficiency can be taken as 1, but it is very important also to check a second failure mechanism called *block failure*. This mechanism corresponds to the case where the pile group fails as a block, as shown in Figures 18.45 and 18.46. The ultimate capacity of the block is given by:

$$R_{ublock} = R_{ublock}(friction) + R_{ublock}(point) = 2(B_g + L_g)Df_u + B_g L_g p_u \tag{18.118}$$

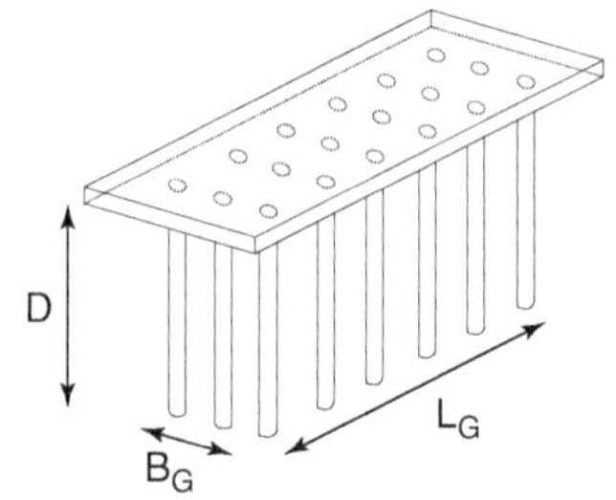

Figure 18.45 Pile group: (a) Closely spaced slender piles. (b) Strong layer not thick enough.

where the ultimate friction f_u is taken as the undrained shear strength s_u because the shearing will take place mostly in the clay, and p_u is $N_c s_u$ with N_c being obtained from the Skempton chart (see Figure 17.7). Note that for the single pile, N_c is likely equal to 9, because the relative embedment of the single pile, D/B, is often larger than 4; for the group, the relative embedment is much smaller, as it is equal to D/B_g. The ultimate capacity of the pile group is then the smaller of:

$$R_{ug} = Min(nR_{us}, R_{ublock}) \tag{18.119}$$

There are two cases in which R_{ublock} will be the smaller of the two (Figure 18.46). The first case is when the center-to-center pile spacing is small (say, 3 or less) and the piles are

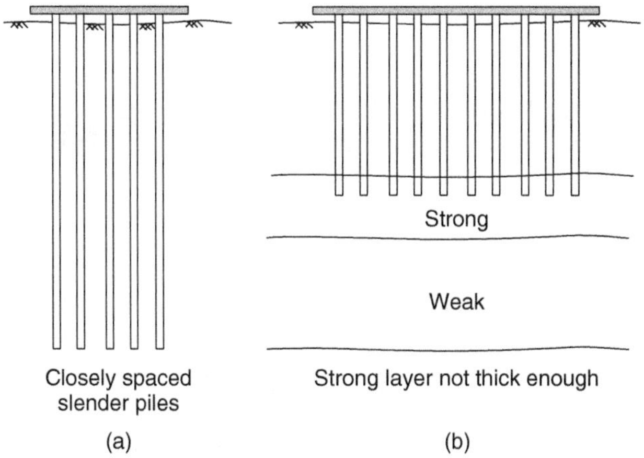

Figure 18.46 Cases where block failure is likely to control.

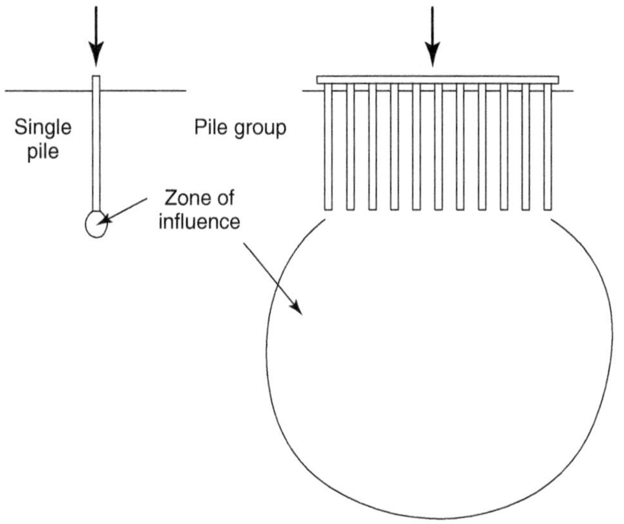

Figure 18.47 Difference in zone of influence between single pile and pile group.

long and slender (say, $D/B > 30$). The second case is when the pile point is in a strong layer that gives a high single-pile ultimate capacity but is not thick enough to prevent the pile group from punching through into a weaker layer below. In this case the R_{us} (point) would involve the strength of the strong layer, whereas the R_{ublock} (point) would involve the strength of the weak layer below.

18.5.2 Settlement of Pile Groups

The settlement of a pile group can be much larger than the settlement of a single pile. The major difference is the increase in the depth of influence under the pile group compared to the single pile (Figure 18.47) and the accumulation of load effect from all the piles. Many empirical equations have been proposed to relate the settlement of the group s_g to the settlement of the single pile s_s. The following one is recommended by O'Neill (1983):

$$\frac{s_g}{s_s} = \sqrt{\frac{B_g}{B_s}} \tag{18.120}$$

where s_s is the settlement of the single pile under the working load Q, s_g is the settlement of the group under nQ, n is the number of piles in the group, B_g is the width of the group, and B_s is the width of the single pile. This equation indicates that the settlement of the group does not increase linearly with B_g but rather with the square root of B_g. This is corroborated by limited data (O'Neill 1983).

For large groups of piles in a uniform soil deposit, Terzaghi proposed to calculate the settlement of the group by considering that the group was equivalent to a spread footing having the dimensions B_g by L_g and located at a depth equal to 2/3 of the pile embedded depth (Figure 18.48). If the piles penetrated through a weak layer into a strong layer (end-bearing piles), the equivalent footing is placed on top of the strong layer. If the penetration into the strong layer is significant, then the 2/3 rule would apply to the strong layer (Figure 18.48). Note that such an approach gives a linear increase of s_g with B_g which is more severe than Eq. 18.120. The Terzaghi approach has been found to be conservative in some cases.

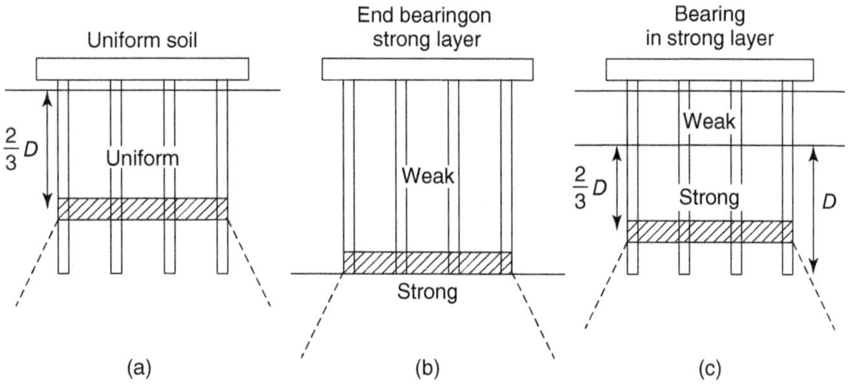

Figure 18.48 Terzaghi equivalent footing approach.

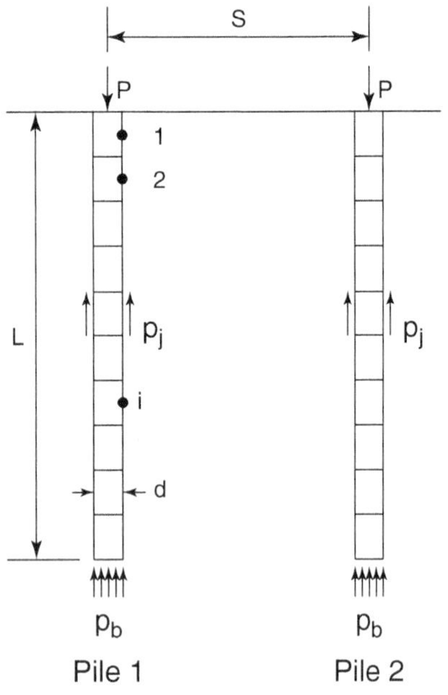

Figure 18.49 Influence of one pile on the other. (After Poulos and Davis, 1980)

A third approach to calculating the settlement of a pile group is to use a computer program that solves the problem of superposition of displacements induced by all piles (Poulos and Davis 1980). This is called the *interaction factor method.* Each pile stresses the soil surrounding it; therefore, each pile contributes to the global settlement at any other point. Each pile is first discretized as n pile elements versus depth (Figure 18.49). For a group of two identical and equally loaded piles, the movement of any node on pile 1 is given by:

$$[\rho_s] = \frac{d}{E_s}[I_1 + I_2][f] \qquad (18.121)$$

where $[\rho_s]$ is the vector of the displacements at all nodes i (1 to n) of pile 1 (same as pile 2), d is the pile diameter, E_s is the soil modulus, $[f]$ is the vector of shear stresses at all nodes j (1 to n) of pile 1 (same as pile 2), $[I_1 + I_2]$ is the $n + 1$ by $n + 1$ matrix ($n + 1$ because there are n friction elements and 1 point) of displacement influence factors containing elements I_{1ij} and I_{2ij}, and I_{1ij} and I_{2ij} are the displacement influence factors at element i on pile 1 caused by shear stress (friction) on element j of piles 1 and 2 respectively. These factors are obtained by integration of the Mindlin equation (1936). Note that the solution for pile 2 is the same, because the two piles are identical and equally loaded.

Of particular interest is the displacement at the top of one of the piles. By setting this displacement equal to 1, the distribution of shear stresses along the piles can be generated by solving Eq. 18.121. The interaction factor α is defined as:

$$\alpha = \frac{\text{additional settlement due to adjacent pile}}{\text{settlement of pile under its own load}} \qquad (18.122)$$

Figure 18.50 shows the range of expected values for the interaction factor depending on the spacing between piles, the slenderness of the pile, and the relative stiffness between the pile and the soil.

In the general case of n identical piles, the settlement of pile k can be expressed as:

$$\rho_k = \rho_1 \sum_{\substack{j=1 \\ j \neq k}}^{n} (P_j \alpha_{kj}) + \rho_1 P_k \qquad (18.123)$$

where ρ_k is the settlement of pile k, ρ_1 is the settlement of a single pile under a unit load, P_j is the load at the top of pile j, and α_{kj} is the interaction factor for spacing between piles k and j. Equation 18.123 can be written for all piles in the group, giving n equations. In addition, the load on the group can be written as:

$$P_G = \sum_{j=1}^{n} P_j \qquad (18.124)$$

The $n + 1$ equations thus assembled can be solved for the boundary conditions imposed by the pile cap. Two simple conditions exist:

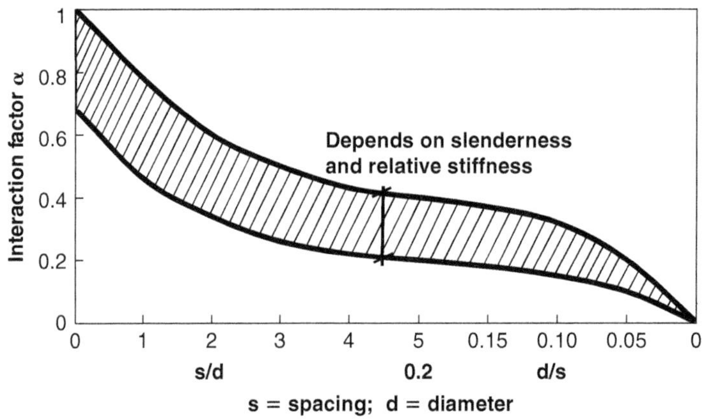

Figure 18.50 Interaction factor as a function of pile spacing. (After Poulos and Davis 1980)

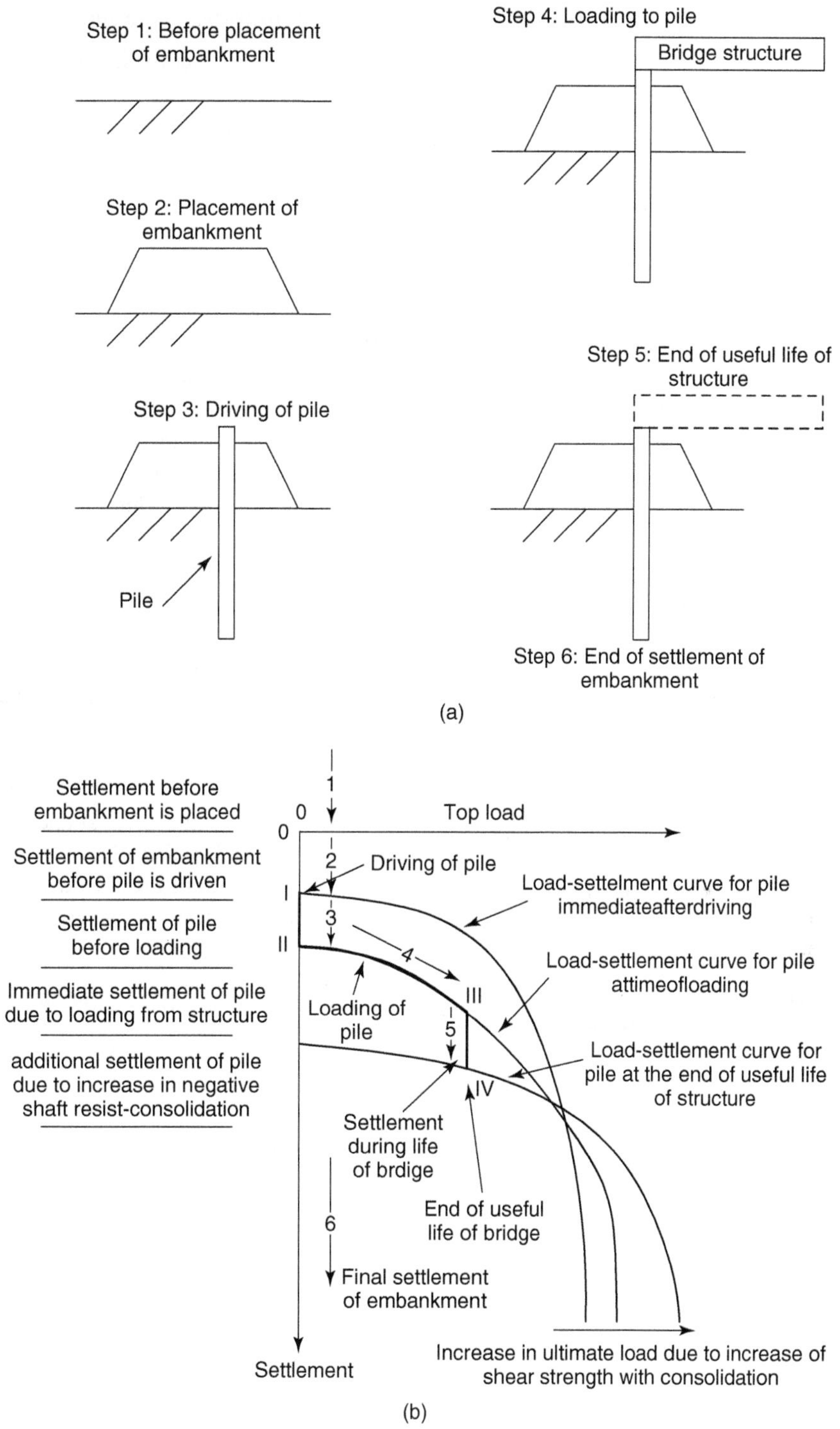

Figure 18.51 Construction sequence and settlement path for a case of downdrag: (a) Construction sequence. (b) Settlement path.

1. Equal load on all piles (perfectly flexible pile cap)
2. Equal settlement of all piles (perfectly rigid pile cap)

The program DEFPIG (Poulos 2012) automates these calculations. Other programs based on somewhat different approaches include PIGLET (Randolph 1980) and FLPIER (Hoit et al. 1997). Of course it is now possible to use the general finite element method to model a group of piles and the surrounding soil in three dimensions and with nonlinear soil behavior.

18.6 DOWNDRAG

18.6.1 Definition and Behavior

In the normal case, the loaded pile moves down more than the soil surrounding it. *Downdrag* refers to the special case where the soil around the upper part of the pile moves downward more than the pile. This occurs, for example, when a pile is driven through a compressible layer into a stronger layer, and then an embankment is placed on the compressible layer, creating significant settlement of the ground surface. A construction sequence leading to downdrag and the associated settlement path are shown in Figure 18.51. Table 18.17 gives some clues indicating when downdrag might occur. A guideline manual for downdrag on uncoated and bitumen coated piles should be consulted for further information on downdrag (Briaud and Tucker 1997; http://ceprofs.tamu.edu/briaud/).

Table 18.17 Some Situations in which Downdrag Can Occur

1	The total settlement of the ground surface will be larger than 100 mm
2	The settlement of the ground surface after the piles are installed will be larger than 10 mm
3	The height of the embankment to be placed on the ground surface exceeds 2 m
4	The thickness of the soft compressible layer is larger than 10 m
5	The water table will be drawn down by more than 4 m
6	The piles are longer than 25 m

18.6.2 Downdrag on a Single Pile

A shallow, soft layer loaded by an embankment, for example, settles more than the pile, which may rest in a strong deeper layer. In this instance the soil drags the pile down during the soil settlement. Because the pile does not move downward as much as the soil, at the ground surface it looks like the pile is coming out of the ground. So, when you see piles coming slowly out of the ground, it could be downdrag. This downdrag load extends to the point where the settlement of the soil becomes equal to the settlement of the pile (Figure 18.52); that point is called the *neutral point* (NP). In the case of no downdrag, the load in the pile decreases from the top down to the pile point (Figure 18.52). In the case of downdrag, the load increases to the NP and then decreases to the pile point. In extreme cases the NP is at the pile point. The evolution of the loads during and after construction is shown in Figure 18.53 for the case of downdrag and no downdrag.

As with other foundation design problems, the two limit states must be satisfied: serviceability limit state (allowable settlement) and ultimate limit state (ultimate capacity). What is special about downdrag cases is that the serviceability limit state (allowable settlement criterion) controls the design much more often than does the case of no downdrag. Indeed, downdrag induces larger settlement of the pile and settlement calculations must be performed with this in mind. Also, one must check that the load at the NP, which is most often the highest load in the pile, will not crush the pile material. Two main equations are the basis for finding the location of the neutral point. Vertical equilibrium gives the following expression:

$$Q_t + F_n = F_p + Q_p \tag{18.125}$$

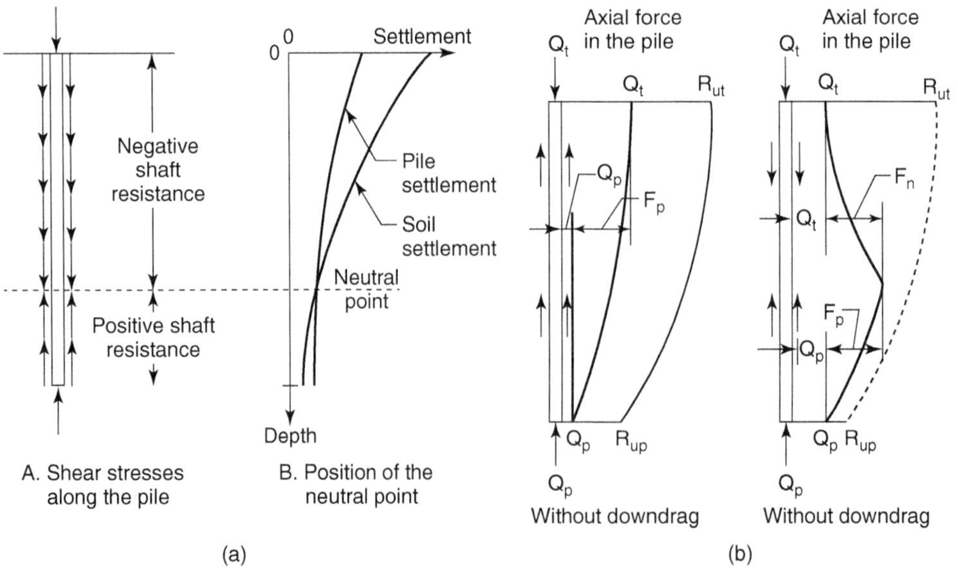

Figure 18.52 Settlement profile and load distribution: *(a)* Settlement profile. *(b)* Load distribution.

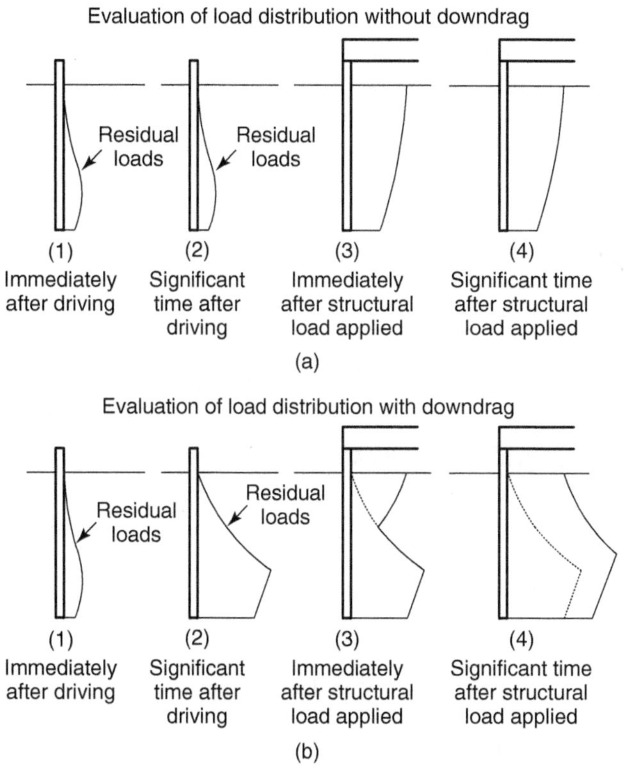

Figure 18.53 Evolution of the loads during and after construction.

where Q_t is the top load on the pile, F_n is the downdrag or negative friction load, F_p is the positive friction load, and Q_p is the point load (Figure 18.52). Compatibility of movement at the *NP* gives:

$$w_{NP(soil)} = w_{NP(pile)} \tag{18.126}$$

where $w_{NP(pile)}$ is the pile movement at the NP and $w_{NP(soil)}$ is the soil movement at the NP. These two equations are used together with an iteration procedure to find the depth of the *NP*. Although the necessary calculations have been automated in computer programs (e.g., PILNEG, available at http://ceprofs.tamu.edu/briaud/), these calculations are best illustrated through hand calculations.

18.6.3 Sample Downdrag Calculations

Consider the square concrete pile of Figure 18.54. It is 30 m long and 0.3 m × 0.3 m in cross section. The concrete has a modulus of elasticity of 2×10^7 kN/m^2. The soil develops an ultimate friction that is constant with depth (simplification) and equal to 25 kN/m^2. To simplify the problem further, it is assumed that the friction load transfer curve is rigid-perfectly plastic, so that any pile displacement generates the ultimate friction. The point resistance load transfer curve (Figure 18.54) is elastic-perfectly plastic, with an ultimate point resistance of 1000 kN and a 5 mm movement required to reach that ultimate value. The soil settlement profile shown in Figure 18.54 indicates that, at the pile point, the soil movement is zero (hard layer) and the ground surface settles 200 mm. The allowable settlement for the structure is 15 mm.

1. Find the ultimate capacity of the pile.

$$Q_u = 25 \times 1.2 \times 30 + 1000 = 1900 \text{ kN} \tag{18.127}$$

If the pile was not subjected to downdrag, we would apply the load and resistance factor design and end up

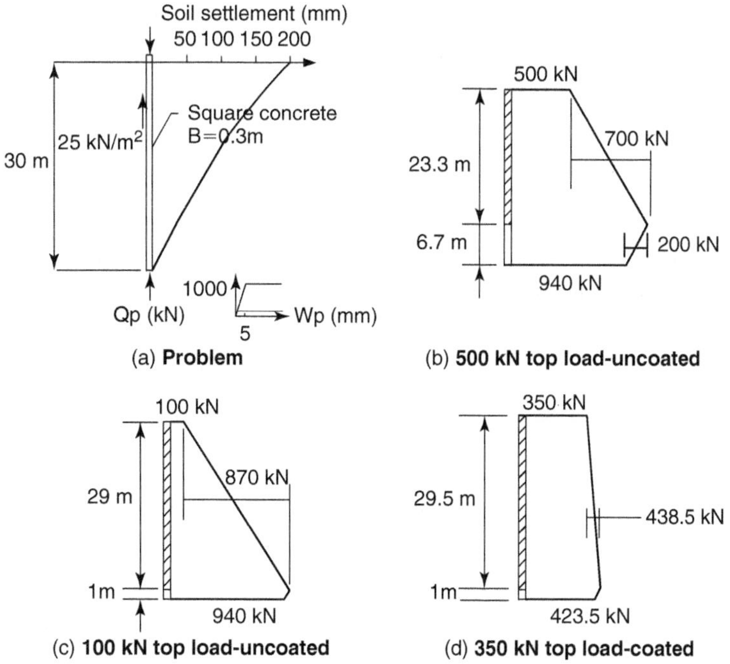

Figure 18.54 Example downdrag problem and results.

with a top load of around 800 kN for such a pile. Because we have downdrag, we need to reduce this load.

2. Try a top load of $Q_t = 500$ kN.
   a. Assume that the NP is at a depth of 20 m.
   From the soil settlement profile, we read a soil movement at the assumed depth of the NP (20 m) of:

$$w_{NP(soil)} = 50 \text{ mm} \tag{18.128}$$

Vertical equilibrium of the pile gives us the load in the pile at the NP, Q_{NP}, and the point load Q_p (Eq. 18.125):

$$Q_{NP} = 500 + 25 \times 1.2 \times 20 = 1100 \text{ kN} \tag{18.129}$$

$$Q_p = 500 + 25 \times 1.2 \times 20 - 25 \times 1.2 \times 10 = 800 \text{ kN} \tag{18.130}$$

We can now use the point load transfer curve together with the calculated point load to obtain the corresponding point movement w_p:

$$w_p = 4 \text{ mm} \tag{18.131}$$

Then we can calculate the pile movement at the location of the neutral point by adding the elastic compression of the pile between the pile point and the NP. First, the average load in the pile between the pile point and the NP is:

$$Q_{ave} = \frac{800 + 1100}{2} = 950 \text{ kN} \tag{18.132}$$

The pile movement at the NP is:

$$w_{NP(pile)} = 4 + \frac{950 \times 10}{0.3 \times 0.3 \times 2 \times 10^7} \times 1000 = 9.3 \text{ mm} \tag{18.133}$$

Comparing Eq. 18.128 with Eq. 2a shows that the movement of the soil at the NP is quite difference from the movement of the pile at the NP. Therefore, this cannot be the NP; our first guess was incorrect. Let's try another guess.

   b. *Assume that the NP is at a depth of 29 m.*
   From the soil settlement profile, we read a soil movement at the NP of:

$$w_{NP(soil)} = 5 \text{ mm} \tag{18.134}$$

Vertical equilibrium of the pile gives us the point load, Q_p (Eq. 18.125):

$$Q_p = 500 + 25 \times 1.2 \times 29 - 25 \times 1.2 \times 1 = 1340 \text{ kN} \tag{18.135}$$

This is not possible because the ultimate point load is only 1000 kN. What would happen in this case is that the pile point would settle into the soil until the settlement of the pile point became sufficient for equilibrium to be reached. This would happen when (Eq. 18.125):

$$500 + F_n = F_p + 1000 \tag{18.136}$$

Knowing that:

$$F_n + F_p = 25 \times 1.2 \times 30 = 900 \text{ kN} \tag{18.137}$$

Then (Figure 18.54):

$$F_n = 700 \text{ kN} \quad and \quad F_p = 200 \text{ kN} \tag{18.138}$$

This means that the neutral point would be at a depth of:

$$z_{NP} = \frac{700}{25 \times 1.2} = 23.3 \text{ m} \tag{18.139}$$

From the soil settlement profile, we get a settlement at the NP of:

$$w_{NP(soil)} = 35 \text{ mm} \tag{18.140}$$

The mean load in the pile between the top and the NP is:

$$Q_{ave} = \frac{500 + 1200}{2} = 850 \text{ kN} \tag{18.141}$$

and the settlement at the top of the pile is:

$$w_{top} = 35 + \frac{850 \times 23.3}{0.3 \times 0.3 \times 2 \times 10^7} \times 1000 = 46 \text{ mm} \tag{18.142}$$

This is more than the allowable settlement, not to mention that the pile point would be at failure. We need to reduce the top load on the pile.

3. Try a top load of $Q_t = 100$ kN.
   a. Assume that the NP is at a depth of 25 m.
   From the soil settlement profile, we read a soil movement at the NP of:

$$w_{NP(soil)} = 25 \text{ mm} \tag{18.143}$$

This is already larger than the allowable settlement, so we need to move the NP deeper.

   b. Assume that the NP is at a depth of 29 m.
   From the soil settlement profile, we read a soil movement at the NP of:

$$w_{NP(soil)} = 5 \text{ mm} \tag{18.144}$$

Vertical equilibrium of the pile gives us the load in the pile at the NP, Q_{NP}, and the point load Q_p (Eq. 18.125):

$$Q_{NP} = 100 + 25 \times 1.2 \times 29 = 970 \text{ kN} \tag{18.145}$$

$$Q_p = 100 + 25 \times 1.2 \times 29 - 25 \times 1.2 \times 1 = 940 \text{ kN} \tag{18.146}$$

We can now use the point load transfer curve together with the calculated point load to obtain the corresponding point movement w_p:

$$w_p = 4.7 \text{ mm} \tag{18.147}$$

Then we can calculate the pile movement at the location of the neutral point by adding the elastic compression of the pile between the pile point and the NP. First, the average load in the pile between the pile point and the NP is:

$$Q_{ave} = \frac{940 + 970}{2} = 955 \text{ kN} \tag{18.148}$$

The pile movement at the NP is:

$$w_{NP(pile)} = 4 + \frac{955 \times 1}{0.3 \times 0.3 \times 2 \times 10^7} \times 1000 = 5.2 \text{ mm} \tag{18.149}$$

The movement of the soil at the NP is very close to the movement of the pile at the NP. Let's calculate the movement at the top of the pile. The mean load in the pile between the NP and the top of the pile is:

$$Q_{ave} = \frac{100 + 970}{2} = 535 \text{ kN} \tag{18.150}$$

and the movement of the top of the pile is:

$$w_{top} = 5 + \frac{535 \times 29}{0.3 \times 0.3 \times 2 \times 10^7} \times 1000 = 13.6 \text{ mm} \tag{18.151}$$

Therefore, the settlement at the top of the pile is allowable. However, the pile point is close to failure, and this pile, which would typically carry about 800 kN of load with no downdrag, is reduced to carrying 100 kN under precarious conditions. What if we coated that pile with a friction reducer such as bitumen? Let's say that this friction reducer reduces the friction from 25 kN/m^2 to 2.5 kN/m^2 (common for properly selected bitumen).

4. Try a top load of 350 kN on the coated pile.
   a. *Assume that the NP is at a depth of 29.5 m.*
   From the soil settlement profile, we read a soil movement at the NP of:

$$w_{NP(soil)} = 2.4 \text{ mm} \tag{18.152}$$

Vertical equilibrium of the pile gives us the load in the pile at the NP, Q_{NP}, and the point load Q_p (Eq. 18.125):

$$Q_{NP} = 350 + 2.5 \times 1.2 \times 29.5 = 438.5 \text{ kN} \tag{18.153}$$

$$Q_p = 350 + 2.5 \times 1.2 \times 29.5 - 25 \times 1.2 \times 0.5 = 423.5 \text{ kN} \tag{18.154}$$

We can now use the point load transfer curve together with the calculated point load to obtain the corresponding point movement w_p:

$$w_p = 2.2 \text{ mm} \tag{18.155}$$

Then we can calculate the pile movement at the location of the neutral point by adding the elastic compression of the pile between the pile point and the NP. First, the average load in the pile between the pile point and the NP is:

$$Q_{ave} = \frac{423.5 + 438.5}{2} = 431 \text{ kN} \tag{18.156}$$

The pile movement at the NP is:

$$w_{NP(pile)} = 2.2 + \frac{431 \times 0.5}{0.3 \times 0.3 \times 2 \times 10^7} \times 1000 = 2.3 \text{ mm} \tag{18.157}$$

The movement of the soil at the NP is very close to the movement of the pile at the NP. Let's calculate the movement at the top of the pile. The mean load in the pile between the NP and the top of the pile is:

$$Q_{ave} = \frac{350 + 438.5}{2} = 394.2 \text{ kN} \tag{18.158}$$

and the movement at the top of the pile is:

$$w_{top} = 5 + \frac{394.2 \times 29.5}{0.3 \times 0.3 \times 2 \times 10^7} \times 1000 = 11.5 \text{ mm} \tag{18.159}$$

Therefore, the settlement is acceptable. Note that the ultimate capacity of the pile is reduced because of the coating:

$$Q_u = 2.5 \times 1.2 \times 29.5 + 25 \times 1.2 \times 0.5 + 1000 = 1103.5 \text{ kN} \tag{18.160}$$

18.6.4 LRFD Provisions

The preceding example showed calculations of settlement associated with the serviceability limit state. In the case of settlement, the unfactored dead load, the unfactored permanent live load, and the unfactored downdrag are included in the settlement calculation, but the transient live load is not. The transient live load is included when checking the ultimate limit state at the top of the pile, but not when the ultimate limit state is checked at the NP. The reason is that the transient live load does not last long enough to reverse the downdrag load.

Briaud and Tucker (1997) proposed to check the ultimate limit state at two locations along the pile: the pile top and the NP.

$$\text{At the pile top} \quad 1.25DL + 1.75PLL + 1.75TLL < 0.5R_u \tag{18.161}$$

$$\text{At the NP} \quad 1.25DL + 1.75PLL + 1.75F_n < 0.75(Q_{pu} + F_{pu}) \tag{18.162}$$

where DL is the dead load, PLL is the permanent live load, TLL is the transient live load, R_u is the ultimate capacity of the pile, F_n is the downdrag load, Q_{pu} is the ultimate point resistance, and F_{pu} is the friction resistance below the NP.

The resistance factors (0.5 and 0.75 on the right side of the equations) are for a high-quality static method of computing the resistances; they should be adjusted for other cases. Note that at the pile top the ultimate limit state is the same as the case of no downdrag, because the ultimate capacity of the pile is the same whether or not there is downdrag. The use of a higher resistance factor at the neutral point, 0.75, compared to 0.5 at the top, means that the consequence of exceeding the ultimate load below the neutral point is not as drastic as that of exceeding the ultimate load at the pile top. Indeed, if the ultimate capacity of the pile below the neutral point is exceeded, some settlement will take place, the neutral point will move up and the pile will find a new equilibrium; as long as the top load is less than the ultimate capacity, downdrag by itself cannot create plunging failure. In contrast, if the top load reaches the ultimate capacity of the pile, the pile will plunge, as there is no reserve in this case. AASHTO (2010) recommends checking the ultimate limit state at the neutral point as follows:

$$\text{If } F_n > F_{pu} \quad 1.25DL + 1.75PLL + 1.75(F_n - F_{pu}) < 0.5Q_{pu} \tag{18.163}$$

$$\text{If } F_n < F_{pu} \quad 1.25DL + 1.75PLL < 0.5(Q_{pu} + F_{pu} - F_n) \tag{18.164}$$

Going back to the example of section 18.6.3, case 4a, and assuming that the dead load is 300 kN and the permanent live load is 50 kN, then Eq. 18.162 gives:

$$1.25 \times 300 + 1.75 \times 50 + 1.75 \times 88.5 < 0.75(1000 + 15)$$

$$\text{or} \quad 617.4 < 761.2 \tag{18.165}$$

This Briaud-Tucker ultimate limit state at the NP is satisfied. The AASHTO guidelines give (Eq. 18.158):

$$1.25 \times 300 + 1.75 \times 50 + 1.75(88.5 - 15) < 0.5 \times 1000$$

$$\text{or} \quad 591.1 < 500 \tag{18.166}$$

This AASHTO limit state would not be satisfied.

18.6.5 Downdrag on a Group of Piles

The downdrag force on a group of n closely spaced piles is less than n times the downdrag force on an isolated single pile. The reason is that the soil tends to settle on the outside of the pile group but does not settle as much between piles inside the group, as illustrated in Figure 18.55. The full-scale case history by Okabe (1977) demonstrates the impact of this observation very clearly. Figure 18.56 shows the pile group configuration together with the load distribution for different piles in the group compared to a single pile. The center-to-center spacing for the piles in the group is approximately 2.1 diameters. It can be seen that the single pile experiences a large downdrag load (7000 kN), that the outer piles in the group carry about one-half of the single pile downdrag load (3500 kN), and that the interior piles in the group are subjected to only about 500 to 1000 kN of downdrag, or 7% to 14% of the single-pile downdrag load. The reduction of downdrag on the interior piles in the group is dramatic. After calibrating their three-dimensional nonlinear finite element simulation against Okabe's result, Jeong and Briaud (1994) performed a large parametric study to investigate the downdrag reduction in pile groups. Based on those results and other measurements,

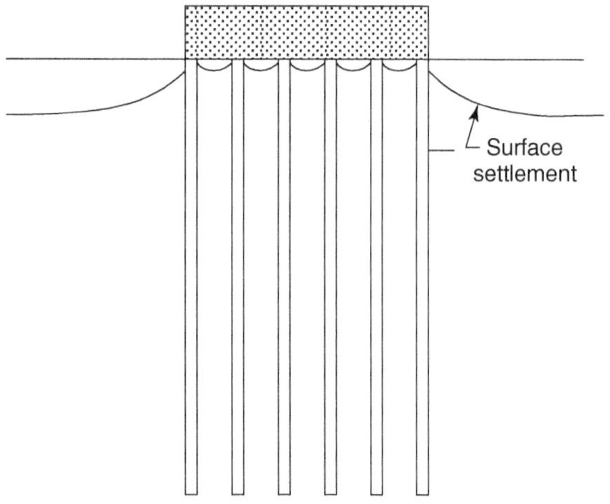

Figure 18.55 Soil settlement pattern around a pile group.

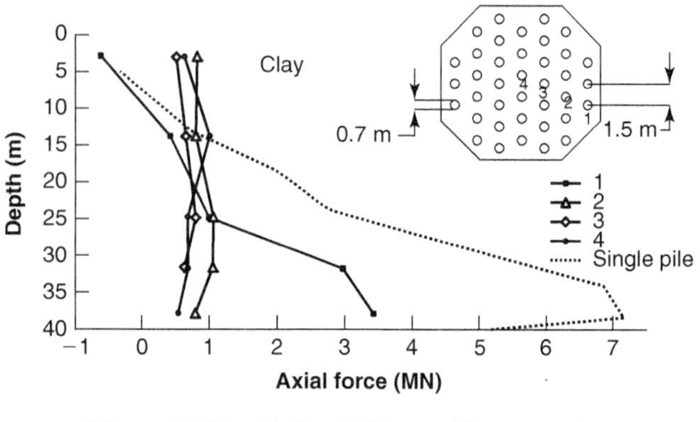

Figure 18.56 Okabe (1977) case history results.

Table 18.18 Downdrag Reduction Factors for Groups of Piles

S/D = 5		S/D = 2.5
$F_{n(corner)} = 0.9\ F_{n(single)}$ $F_{n(side)} = 0.8\ F_{n(single)}$ $F_{n(interior)} = 0.5\ F_{n(single)}$	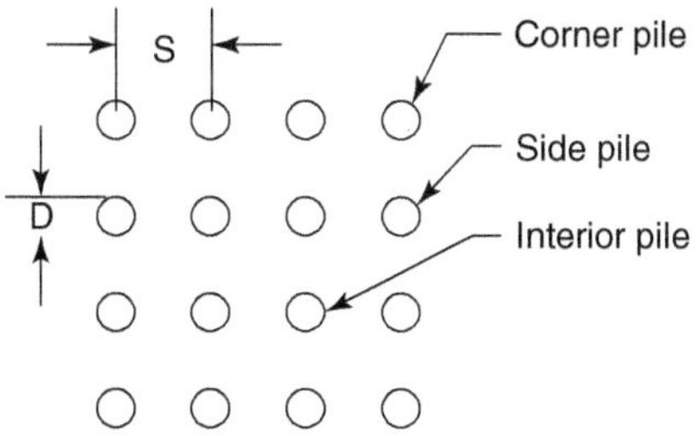	$F_{n(corner)} = 0.5\ F_{n(single)}$ $F_{n(side)} = 0.4\ F_{n(single)}$ $F_{n(interior)} = 0.15\ F_{n(single)}$

Definitions:
S = center-to-center spacing
D = pile diameter
$F_{n(single)}$ = downdrag force on the single pile
$F_{n(corner)}$ = downdrag force on a corner pile in the group
$F_{n(side)}$ = downdrag force on a side pile in the group
$F_{n(interior)}$ = downdrag force on an interior pile in the group
(Briaud and Tucker 1997)

Briaud and Tucker (1997) recommended the reduction factors listed in Table 18.18. Most often the piles in the groups are embedded into a rigid pile cap. The fact that the outside piles undergo more downdrag than the inside piles means that the outside piles pull down on the pile cap while the inside piles push up on it. This is the case of pile 1 in Okabe's experiment (Figure 18.56). Therefore, the connection between the pile cap and the outside piles should be designed for tension. The best way to simulate downdrag of a pile group is with the finite element method.

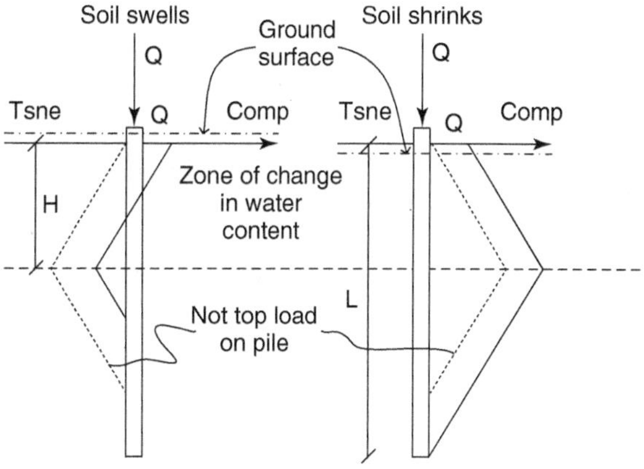

Figure 18.57 Load distribution in piles in shrink-swell soils.

18.7 PILES IN SHRINK-SWELL SOILS

Piles in shrink-swell soils are subjected to soil movements that may increase or decrease the compression load in the pile (Figure 18.57). These vertical soil movements take place during the year as the top part of the soil deposit changes water content from one season to another. The depth of the zone influenced by these movements, called the *active zone*, seems to be on the order of 3 to 5 m. If the soil shrinks, the soil moves down with respect to the pile; this is a case similar to downdrag, where excessive settlement and excessive load in the pile are the design issues. If the soil swells, the soil moves up with respect to the pile and could create excessive upward movement of the foundation if the pile is not deep enough.

18.7.1 The Soil Shrinks

The case in which the soil shrinks is very similar to the case of downdrag, except that the neutral point is found at the bottom of the soil shrinkage zone. If the soil is uniform and the depth of the active zone is H, a pile length equal to 2.5 H must be ignored in the calculation of load carrying capacity. This minimum pile length is used to ensure that the shrinkage of the soil pulling down on the pile does not create downward movement of the structure. The downward load is applied over the top H of the pile, the next H resists that movement, and the next 0.5 H is there as a safety factor. An additional length of pile beyond 2.5 H and/or the point resistance is necessary to safely carry the compressive structural load in friction and/or point resistance. Note that there is some uncertainty as to whether the shrinking soil can truly load the pile downward; indeed, the shrinkage is usually in all directions, including away from the pile in the radial direction. At the same time, the shrinking soil can be much stronger than the soil below the active zone, which does not shrink.

18.7.2 The Soil Swells

In the case of swelling soil, the soil swells against the pile and pulls it upward over the depth H of the active zone. If the soil is uniform and the depth of the active zone is H, again the minimum length of pile must be 2.5 H, excluding other loads. This minimum pile length is used to ensure that the swelling of the soil pulling up on the pile does not create upward movement of the structure. The uplift load is applied by the swelling soil over the top H, the next H resists that movement, and the next 0.5 H is there as a safety factor. If the structure applies a compressive load, this will counteract to some extent the uplift created by the soil. In this instance there is no need to lengthen the pile, unless the structural load is so large that the movement of the soil is overcome and the pile moves downward with respect to the soil over its entire length. In this extreme case, the swelling of the soil can be ignored and the pile is designed as an ordinary pile.

The friction load created by swelling $F_{u(swell)}$ or shrinking $F_{u(shrink)}$ of the soil over the depth of the active zone is calculated as:

$$F_{u(swell)} = F_{u(shrink)} = f_u PH \qquad (18.167)$$

where f_u is the ultimate friction, P is the pile perimeter, and H is the depth of the active zone. Reese et al. (1976) give some limiting values for the parameter f_u.

18.8 HORIZONTAL LOAD AND MOMENT: SINGLE PILE

18.8.1 Definitions and Behavior

This section deals with piles subjected to horizontal loads and overturning moments. Examples of such loading on pile foundations include wire tension at corner towers of power lines,

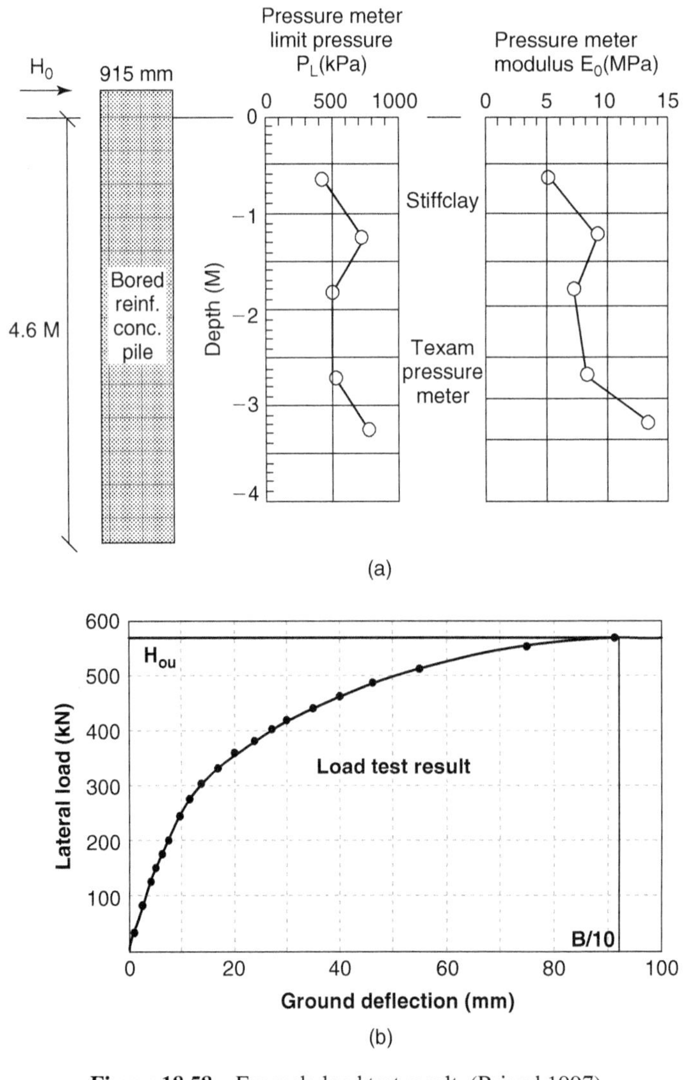

Figure 18.58 Example load test result. (Briaud 1997)

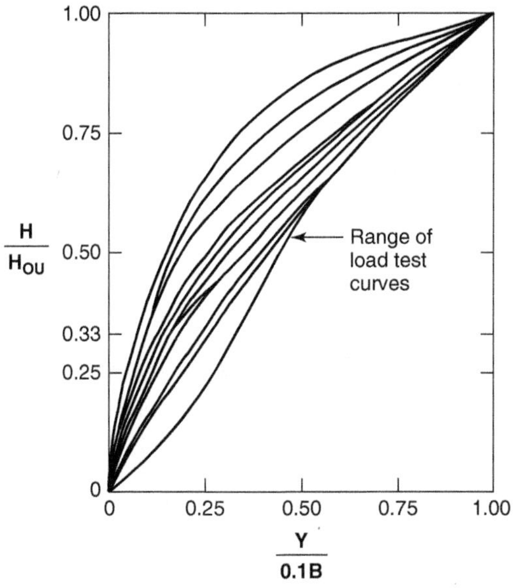

Figure 18.59 Normalized horizontal load test curves. (Briaud 1997)

hurricane waves on offshore platforms, ship impact on bridge piers, and earthquake shaking of a building and walls. A horizontal load test on a pile often consists of placing two piles at some distance from each other and either pulling them toward each other or pushing them apart. The resulting load displacement curve for one pile gives the horizontal load H_o versus the horizontal displacement y_o (Figure 18.58). From this curve, an ultimate load H_{ou} can be defined as the horizontal load H_o corresponding to a displacement equal to one-tenth of the pile diameter (B/10). With such a definition, the load test curve can be normalized as H_o/H_{ou} as a function of y_o/0. B. Using a database of 20 piles, Briaud (1997) generated such normalized horizontal load test curves (Figure 18.59). It was found that the curves with the least amount of curvature came from steel piles, whereas the curves with the largest amount of curvature came from concrete piles. The reason is that concrete piles gradually crack as they are bent; steel piles do not. The bending stiffness EI (E modulus of elasticity of the pile material, I moment of inertia around the bending axis) of concrete piles decreases due to cracking as the pile bends, inducing more and more curvature; by comparison, the EI of the steel piles does not change measurably within the elastic range.

18.8.2 Ultimate Capacity

The ultimate capacity H_{ou} can be determined by using the free-body diagram of the upper part of the pile from the ground surface down to the point of zero shear force or maximum bending moment (Figure 18.60). By writing horizontal equilibrium, we get:

$$H_{ou} = pBz_{\max} \tag{18.168}$$

where p is the mean pressure against the pile within that depth, B is the pile width, and $z_{\max}$ is the depth to zero shear

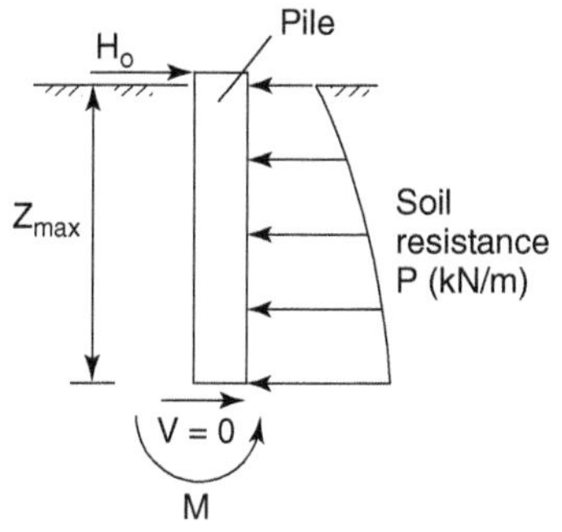

Figure 18.60 Free-body diagram of upper part of horizontally loaded pile.

force (maximum bending moment). By using the database of pile load tests and the associated pressuremeter data, Briaud found that p was equal to 0.75 p_L (pressuremeter limit pressure). Therefore, the ultimate load and the depth to zero shear are given by:

$$H_{ou} = \frac{3}{4} p_l B z_{\max} \quad \begin{cases} z_{\max} = \left(\frac{\pi}{4}\right) l_o & for \quad L > 3l_o \\ z_{\max} = \frac{L}{3} & for \quad L < l_o \\ l_o = \left(\frac{4E_p I}{K}\right)^{1/4} \\ K = 2.3E_o \end{cases} \tag{18.169}$$

where H_{ou} is the ultimate horizontal load (the load that "breaks" the soil around the pile, not the load that "breaks" the pile), p_L is the average limit pressure from pressuremeter tests within the depth $z_{\max}$, $z_{\max}$ is the depth to zero shear (maximum bending moment), B is the projected pile width, E_p is the modulus of the pile material, I is the pile moment of inertia, K is the soil stiffness, L is the length of the pile, l_o is the transfer length, and E_o is the pressuremeter first load modulus.

A comparison between predicted H_{ou} and measured H_{ou} is shown in Figure 18.61. The success of this methodology is attributed to the close analogy between the pressuremeter test and the lateral loading of the soil around the pile. It reminds us that when there is a close analogy between the test and the loading of the prototype, there is a very good chance for close predictions. Note that all piles in the database were pushed horizontally with no moment or very small moments applied at the ground surface. If a sizeable moment is applied, the value of $z_{\max}$ can change significantly (sections 18.8.3 and 18.8.4).

18.8.3 Displacement and Maximum Moment: Long Flexible Pile

The problem of predicting the behavior of a laterally loaded pile is solved in section 11.4.4 for a long flexible pile. A long

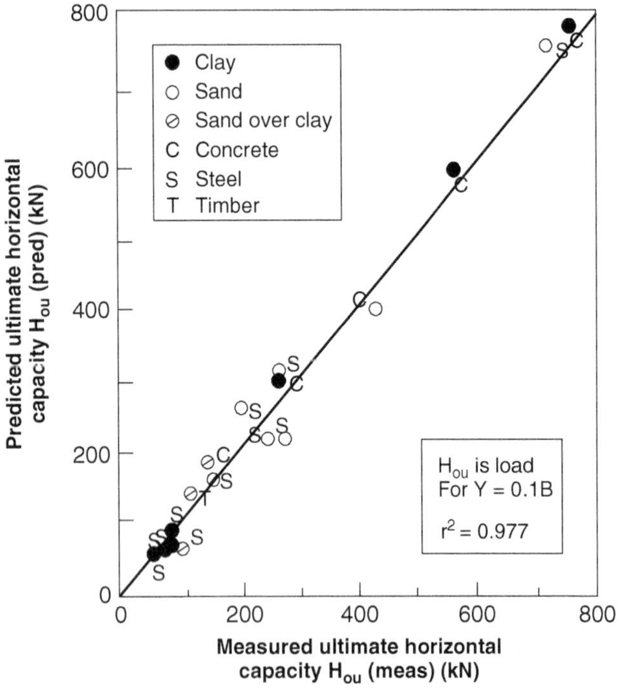

Figure 18.61 Predicted vs. measured ultimate horizontal capacity.

flexible pile corresponds to the case where:

$$L > 3l_o \quad with \quad l_o = \left(\frac{4E_pI}{K}\right)^{1/4} \tag{18.170}$$

where L is the length of the pile, l_o is the transfer length, E_p is the modulus of the pile material, I is the pile moment of inertia, and K is the soil stiffness. The soil stiffness K is taken as equal to 2.3 E_o where E_o is the pressuremeter first load modulus. The equations for the displacement $y(z)$, slope $y'(z)$, bending moment $M(z)$, shear $V(z)$, and line load $P(z)$ as a function of depth z are repeated here for convenience (Figure 18.62):

$$y(z) = \frac{2H_o}{l_oK}e^{-\frac{z}{l_o}}\cos\frac{z}{l_o} + \frac{2M_o}{l_o^2K}e^{-\frac{z}{l_o}}\left(\cos\frac{z}{l_o} - \sin\frac{z}{l_o}\right) \tag{18.171}$$

$$y'(z) = -\frac{2H_o}{l_o^2K}e^{-\frac{z}{l_o}}\left(\cos\frac{z}{l_o} + \sin\frac{z}{l_o}\right) - \frac{4M_o}{l_o^3K}e^{-\frac{z}{l_o}}\cos\frac{z}{l_o} \tag{18.172}$$

$$M(z) = H_ol_oe^{-\frac{z}{l_o}}\sin\frac{z}{l_o} + M_oe^{-\frac{z}{l_o}}\left(\cos\frac{z}{l_o} + \sin\frac{z}{l_o}\right) \tag{18.173}$$

$$V(z) = H_oe^{-\frac{z}{l_o}}\left(\cos\frac{z}{l_o} - \sin\frac{z}{l_o}\right) - \frac{2M_o}{l_o}e^{-\frac{z}{l_o}}\sin\frac{z}{l_o} \tag{18.174}$$

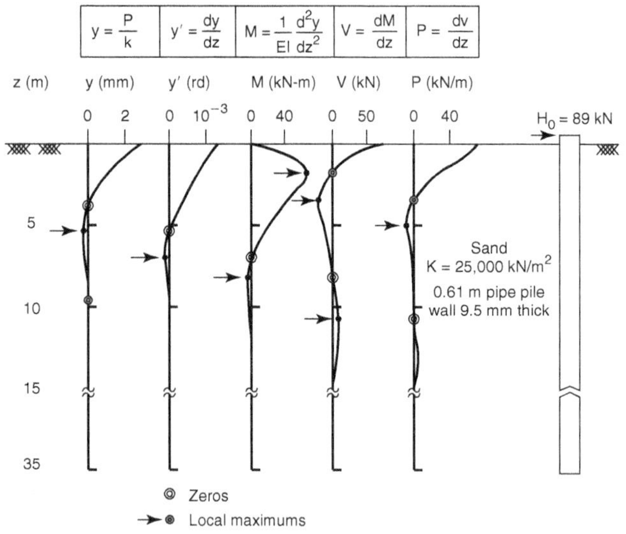

Figure 18.62 Displacement, slope, bending moment, shear, and line load profiles for a laterally loaded pile.

$$P(z) = -K\ y(z) \tag{18.175}$$

where H_o and M_o are the horizontal load and moment respectively applied at the ground surface, l_o is the transfer length given in Eq. 18.170, and K is the soil stiffness.

Using the same database of pile load tests and pressuremeter data as for the ultimate load equation, Briaud (1997) recommended that the soil stiffness K be taken as:

$$K = 2.3E_o \tag{18.176}$$

where E_o is the pressuremeter first load modulus. The important design quantities for the pile obtained from Eqs. 18.171 to 18.175 are the displacement at ground surface y_o, the pressure close to ground surface p_o, the slope at the ground surface y'_o, the depth to the maximum bending moment z_{max}, and the maximum bending moment M_{max}. The displacement at the ground surface comes from Eq. 18.171 for z = 0:

$$y_o = \frac{2H_o}{l_oK} + \frac{2M_o}{l_o^2K} \tag{18.177}$$

The pressure close to the ground surface is:

$$p_o = -\frac{Ky_o}{B} = -\frac{K}{B}\left(\frac{2H_o}{l_oK} + \frac{2M_o}{l_o^2K}\right) \tag{18.178}$$

where B is the diameter or width of the pile. This pressure should be compared to the yield pressure of the soil close to the ground surface. For the pressuremeter, this yield pressure p_y is on the order of 50% of the limit pressure p_L in clays and 33% of the limit pressure p_L in sands. Alternatively, a factor of safety can be applied to the limit pressure to ensure that the pressure p_o is acceptable:

$$p_o < \frac{p_L}{F} \quad or \quad p_o < p_y \tag{18.179}$$

The slope at the ground surface comes from Eq. 18.172 for z = 0:

$$y_o' = -\frac{2H_o}{l_o^2 K} - \frac{4M_o}{l_o^3 K} \tag{18.180}$$

The depth z_{max} to the location of the maximum bending moment M_{max} is found by setting the expression for the shear force (derivative of M) equal to zero and solving for z_{max}. This gives:

$$z_{max} = l_o \tan^{-1}\left(\frac{l_o H_o}{l_o H_o + 2M_o}\right) \tag{18.181}$$

Note that z_{max} must be calculated in radians; any other unit for angles, such as degrees or grades, will not give the right answer. Then M_{max} is calculated by using Eq. 18.173 and the calculated value of z_{max}.

Equation 18.177 was evaluated against the 20-pile database by comparing predicted versus measured values of y_o (Figure 18.63). As can be seen, the scatter is much larger than in the case of the ultimate load. This is in part due to the fact that the precision on the modulus is usually lower than the precision on the limit pressure or strength in general. Most of the piles in the database were flexible, but some were rigid. The case of a rigid pile is addressed in section 18.8.4.

18.8.4 Displacement and Maximum Moment: Short Rigid Pile

The case of a short rigid pile corresponds to:

$$L < l_o \quad with \quad l_o = \left(\frac{4E_p I}{K}\right)^{1/4} \tag{18.182}$$

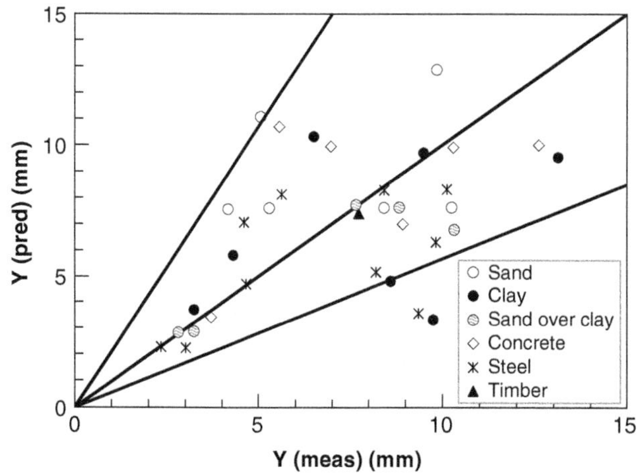

Figure 18.63 Predicted vs. measured horizontal displacements.

where L is the length of the pile, l_o is the transfer length, E_p is the modulus of the pile material, I is the pile moment of inertia, and K is the soil stiffness. In this case the constitutive law for the pile is no longer the relationship between the bending moment and the curvature of the pile, as the pile does not bend (rigid). Instead, the constitutive law for the pile expresses that the deflected shape is a straight line:

$$y = az + b \tag{18.183}$$

where y is the horizontal displacement of the pile and z is depth. The parameter a represents the first derivative of y with respect to z, which is the slope of the pile (y_o'), while b represents the horizontal displacement at the ground surface (y_o at $z = 0$). The solution is much like the solution for the case of the flexible pile.

The constitutive law for the soil is assumed to be a linear relationship between the line load on the pile and the horizontal displacement:

$$P = -Ky \tag{18.184}$$

where P is the line load on the pile, K is the soil stiffness (Eq. 18.176), and y is the horizontal displacement of the pile.

The shear force V at a depth z on the pile can be calculated by integration of the line load P as follows:

$$V = H_o - \int_0^z P d\zeta = H_o + Ka\frac{z^2}{2} + Kbz \tag{18.185}$$

where H_o is the horizontal load applied to the pile at the ground surface, z is the depth where V is calculated, and ζ is the running variable varying between 0 and z.

The bending moment at a depth z in the pile can be obtained by integration of the shear force as follows:

$$\begin{aligned} M &= M_o + H_o z - \int_0^z P(z - \zeta)d\zeta \\ &= M_o + H_o z + Ka\frac{z^3}{6} + Kb\frac{z^2}{2} \end{aligned} \tag{18.186}$$

We use the boundary conditions to find the values of a and b that represent the slope y_o' and the horizontal displacement y_o at the ground surface respectively:

$$for \quad z = 0, \quad V = H_o, \quad M = M_o \tag{18.187}$$

$$for \quad z = L, \quad V = 0, \quad M = 0 \tag{18.188}$$

The condition at $z = 0$ is already satisfied and the condition at z = L leads to:

$$y_o' = +\frac{6(H_o L + 2M_o)}{KL^3} = a \tag{18.189}$$

$$y_o = -\frac{2(2H_o L + 3M_o)}{KL^2} = b \tag{18.190}$$

The pressure close to the ground surface is:

$$p_o = -\frac{K y_o}{B} = 2\frac{K}{B}\left(\frac{2H_o L + 3M_o}{KL^2}\right) \quad (18.191)$$

where B is the diameter or width of the pile. This pressure should be compared to the yield pressure of the soil close to the ground surface. For the pressuremeter, this yield pressure p_y is about 50% of the limit pressure p_L in clays and 33% of the limit pressure p_L in sands. Alternatively, a factor of safety can be applied to the limit pressure to ensure that the pressure p_o is acceptable:

$$p_o < \frac{p_L}{F} \quad or \quad p_o < p_y \quad (18.192)$$

The depth z_{max} to the location of the maximum bending moment M_{max} is found by setting the expression for the shear force (derivative of M) equal to zero and solving for z_{max}. This gives:

$$z_{max} = \frac{H_o L^2}{3(H_o L + 2M_o)} \quad (18.193)$$

Then M_{max} is calculated by using Eq. 18 and the calculated value of z_{max}.

18.8.5 Modulus of Subgrade Reaction

There are three types of soil stiffness, as shown in the following equations 18.194 to 18.196:

Spring constant:

$$K_1(\text{kN}/m) = \frac{H(\text{kN})}{y(\text{m})} \quad (18.194)$$

Soil stiffness:

$$K_2(\text{kN/m}^2) = \frac{P(\text{kN}/m)}{y(\text{m})} \quad (18.195)$$

Modulus of horizontal subgrade reaction:

$$K_3(\text{kN/m}^3) = \frac{p(\text{kN/m}^2)}{y(\text{m})} \quad (18.196)$$

where H is the resultant force on the side of a given length of pile, y is the horizontal displacement of the pile, P is the line load on the pile, and p is the average pressure on the side of a given length of pile.

K_1 and K_3 contain foundation and soil properties, but K_2 is a soil property only. This can be illustrated by using the equation for the settlement of a square plate on an elastic soil:

$$y = I\frac{pB}{E_s} = I\frac{H}{BE_s} \quad (18.197)$$

where y is the settlement of the plate, I is an influence factor, p is the mean pressure under the plate, B is the width of the plate, E_s is the soil modulus, and H is the load on the plate. Using Eq. 18.197, the values of K_1, K_2, and K_3 can be obtained as:

Spring constant:

$$K_1 = \frac{BE_s}{I} \quad (18.198)$$

Soil stiffness:

$$K_2 = \frac{E_s}{I} \quad (18.199)$$

Modulus of horizontal subgrade reaction:

$$K_3 = \frac{E_s}{IB} \quad (18.200)$$

As can be seen, K_1 and K_3 have the width of the plate and the soil modulus in their expression, whereas K_2 only has the soil modulus. Therefore, one should use K_2 and not K_1 and K_3, especially if K_1 and K_3 are derived from tests performed at a scale very different from the field application.

18.8.6 Free-Head and Fixed-Head Conditions

A free-head condition exists when the loading at the top of the pile consists only of a horizontal load (no moment) (Figure 18.64):

$$M_o = 0 \quad (18.201)$$

A fixed-head condition exists when the loading at the top of the pile is such that the top of the pile remains vertical during the horizontal displacement:

$$y_o' = 0 \quad (18.202)$$

For the same horizontal load H_o, the displacement of the free-head pile will be larger than the displacement of the fixed-head pile. However, the fixed-head pile will develop a significant moment at the ground surface in the process. This moment is given in Eqs. 18.203 and 18.204 for a flexible pile and a rigid pile:

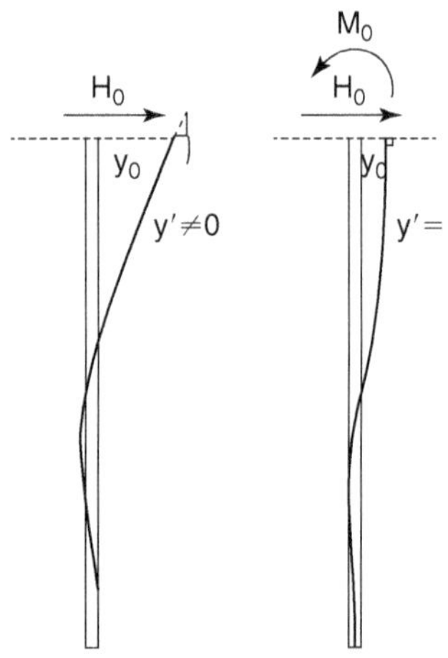

Figure 18.64 Free-head and fixed-head piles.

Table 18.19 Ground Surface Displacement for Horizontally Loaded Piles

	Free head	Fixed head
Long flexible pile	$y_o = \frac{2H_o}{l_o K}$ *for* $L > 3l_o$	$y_o = \frac{H_o}{l_o K}$ *for* $L > 3l_o$
Short rigid pile	$y_o = -\frac{4H_o}{LK}$ *for* $L < l_o$	$y_o = -\frac{H_o}{KL}$ *for* $L < l_o$

Fixed-head flexible pile:

$$M_o = -\frac{H_o l_o}{2} \tag{18.203}$$

Fixed-head rigid pile:

$$M_o = -\frac{H_o L}{2} \tag{18.204}$$

These moments also happen to be the maximum bending moments in the pile. Table 18.19 summarizes the equation giving the ground surface displacement y_o for flexible and rigid piles in free-head and fixed-head conditions.

18.8.7 Rate of Loading Effect

The rate of loading has an effect on the ultimate horizontal load and on the horizontal displacement at the ground surface. The model proposed by Briaud and Garland (1985) leads to the following relationships:

$$\frac{H_{ou}(t)}{H_{ou}(t_o)} = \left(\frac{t}{t_o}\right)^{-n} \tag{18.205}$$

$$\frac{y_o(t)}{y_o(t_o)} = \left(\frac{t}{t_o}\right)^{n} \tag{18.206}$$

where $H_{ou}(t)$ and $H_{ou}(t_o)$ are the ultimate horizontal load reached in a time t and t_o respectively, $y_o(t)$ and $y_o(t_o)$ are the horizontal displacements reached in a time t and t_o respectively, and n is the viscous exponent for the soil.

The value of n varies between 0.01 to 0.03 for sand and from 0.02 to 0.05 for clays, with values up to 0.08 or even 0.1 being reached for very soft, high-plasticity clays (see Figure 15.18). For example, if a retaining wall is founded on bored piles in stiff clay and is designed for 50 years of life, then the load $H_{ou}(t_o)$ obtained from Eq. 18.161 must be altered, because of the long-term sustained load. The reference time t_o is associated with the load tests used to calibrate the method. These tests are typically done in a few hours. If we say that t_o is equal to 2 hours, then Eq. 18.205 gives:

$$n = 0.02 \quad H_{ou}(50yrs) = H_{ou}(2hrs)\left(\frac{50 \times 365 \times 24}{2}\right)^{-0.02}$$
$$= 0.78 H_{ou}(2hrs) \tag{18.207}$$

$$n = 0.06 \quad H_{ou}(50yrs) = H_{ou}(2hrs)\left(\frac{50 \times 365 \times 24}{2}\right)^{-0.06}$$
$$= 0.48 H_{ou}(2hrs) \tag{18.208}$$

In contrast, if the pile is hit by a truck and the impact lasts 50 milliseconds, the results are:

$$n = 0.02 \quad H_{ou}(50ms) = H_{ou}(2hrs)\left(\frac{0.050}{2 \times 3600}\right)^{-0.02}$$
$$= 1.27 H_{ou}(2hrs) \tag{18.209}$$

$$n = 0.06 \quad H_{ou}(50ms) = H_{ou}(2hrs)\left(\frac{0.050}{2 \times 3600}\right)^{-0.06}$$
$$= 2.04 H_{ou}(2hrs) \tag{18.210}$$

The same model can be applied to the horizontal displacement $y_o(t)$. Note that the viscous exponent n can be measured directly and on a site-specific basis with a pressuremeter test. The PMT consists of holding the pressure in the probe for a chosen amount of time and recording the increase in radius as a function of time (Figure 18.65). Then n is given by the following equations:

$$\frac{E(t)}{E(t_o)} = \left(\frac{t}{t_o}\right)^{-n} \quad \text{or} \quad n = \frac{-\log\left(\frac{E(t)}{E(t_o)}\right)}{\log\left(\frac{t}{t_o}\right)} \tag{18.211}$$

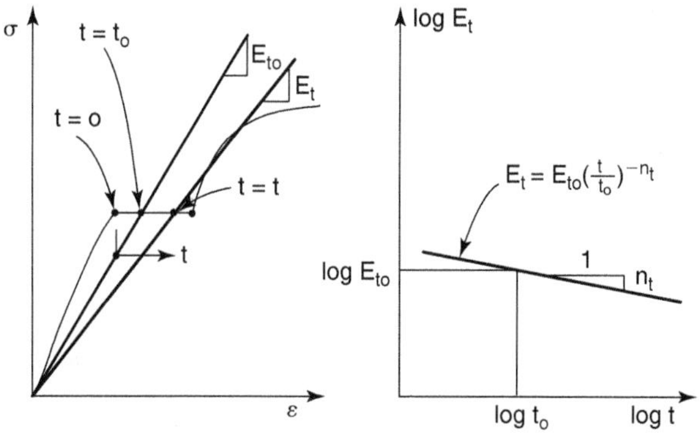

Figure 18.65 Obtaining the viscous exponent from a pressuremeter test.

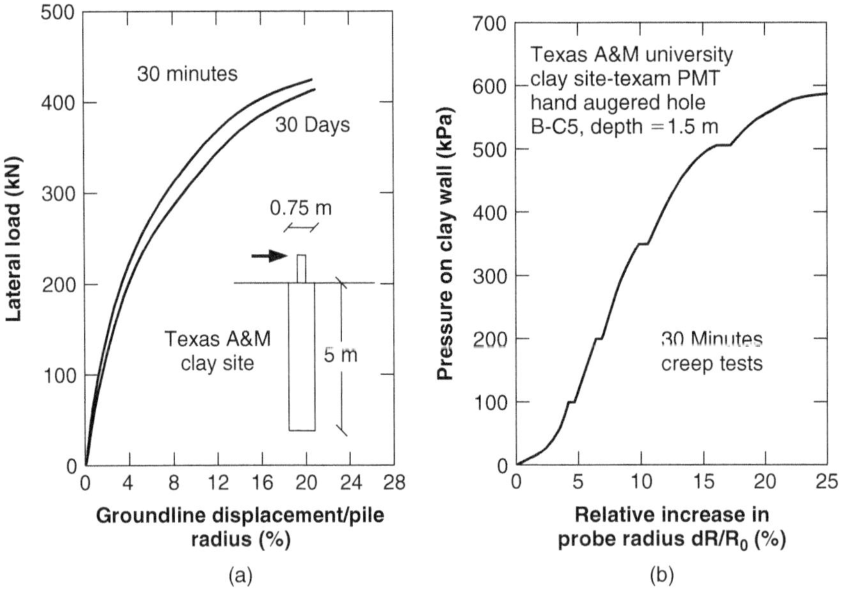

Figure 18.66 Horizontal pile load test and pressuremeter test with long loading steps: (a) Horizontal pile load test, 30-min and 30-day load steps. (b) Pressuremeter test, 30-min pressure steps.

where $E(t)$ and $E(t_o)$ are the first load PMT moduli at time t and t_o respectively. They are obtained from the slopes as shown in Figure 18.65.

Usually t_o is chosen as the reading at 1 minute after the start of the pressure holding step that lasts 10 minutes. The viscous exponent n should be obtained from a pressure holding step in the pressuremeter performed at a ratio p/p_L equal to the ratio H_o/H_{ou}.

Figure 18.66 shows the results of a horizontal pile load test in a stiff clay which was performed with 30-minute and 30-day-long load steps and a pressuremeter test that was performed next to the pile with 30-minute-long pressure steps. The parallel is striking.

18.8.8 Cyclic Loading Effect

The effect of cycles on the behavior of laterally loaded piles can be modeled as follows:

$$y_N = y_1 N^a \tag{18.212}$$

where y_1 and y_N are the ground surface horizontal displacement at the top of the first and the nth cycle respectively, N is the number of cycles, and a is the cyclic exponent.

A major distinction should be made between one-way cyclic loading and two-way cyclic loading (Figure 18.67). In one-way cyclic loading, the direction of the load is not reversed, whereas in the case of two-way cyclic loading the direction of the load is reversed. This distinction makes a difference in the response of the pile depending on the type of soil loaded. If the soil behaves in such a way that pushing the pile in one direction does not affect the behavior of the soil in the opposite direction, then there is little difference between one-way and two-way cyclic loading. This is typically the case in clay. There are cases, such as in dry sands, for example, where pushing the pile in one direction opens a gap behind the pile which fills up when the sand falls into it. Then, when the pile is pushed in the opposite direction, the pile is stiffer than it would have been had the sand not fallen into the open gap. This phenomenon can stiffen the pile during two-way cyclic loading and make two-way loading less detrimental to accumulation of displacement than one-way cyclic loading. The cyclic exponent a was collected from cyclic lateral load tests (Briaud 1992) and found to vary in the ranges shown in Table 18.20.

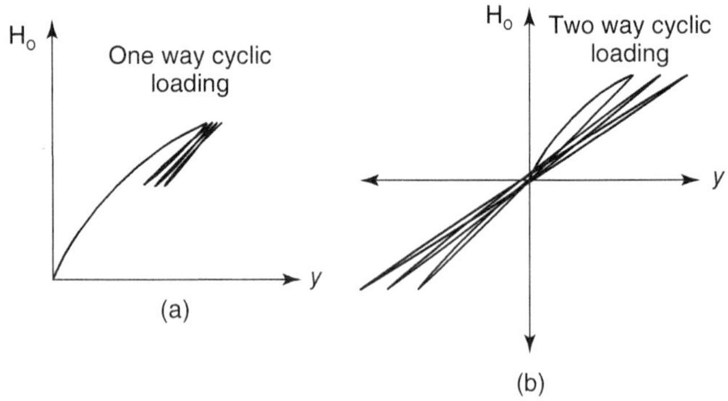

Figure 18.67 Difference between one-way and two-way cyclic loading: (a) One-way cyclic loading. (b) Two-way cyclic loading.

The pressuremeter test can be performed by including cycles of loading (Figure 18.68). The cyclic exponent can be

Table 18.20 Range of Measured Cyclic Exponent for Piles Subjected to Cyclic Horizontal Loads

Cyclic Loading Type	Soil Types	Range of Values of Cyclic Exponent a	Average
One-way and two-way	Clay	0.01 to 0.35	0.094
One-way	Sand	0.005 to 0.26	0.076
Two-way	Sand	−0.14 to 0.06	0.002

(Briaud 1992)

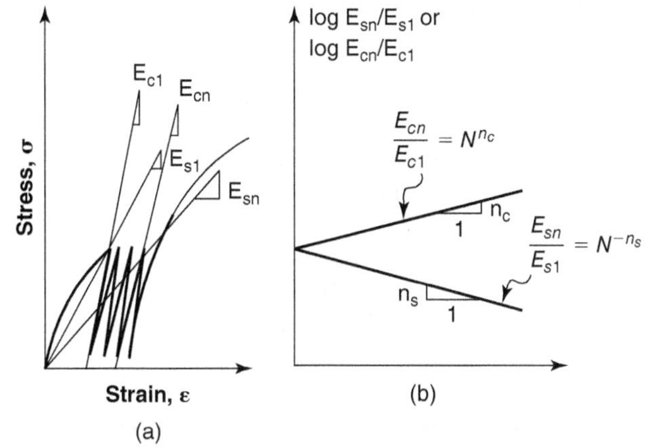

Figure 18.68 Obtaining the cyclic exponent from a cyclic PMT test.

obtained from the cyclic PMT. The cyclic exponent is given by the following equations:

$$\frac{E(N)}{E(1)} = \left(\frac{N}{1}\right)^a \quad \text{or} \quad a = \frac{\log\left(\frac{E(N)}{E(1)}\right)}{\log\left(\frac{N}{1}\right)} \tag{18.213}$$

where E(N) and E(1) are the first load PMT moduli corresponding to the nth cycle and the first cycle respectively. They are obtained from the slopes as shown in Figure 18.68.

The cycles should be performed by matching the anticipated cycles for the pile as closely as possible. In that respect, it is important to realize that the cyclic loading in a PMT can only be one-way cyclic loading. Indeed, in the pressuremeter test the soil is always in radial compression. Therefore, a cyclic pressuremeter test can be used for one-way and two-way cyclic loading for piles in clay where there does not seem to be any difference. However, it can only be used for one-way cyclic loading of sands. If the PMT is used for predicting the accumulation of movement as a function of the number of cycles for two-way cyclic loading in sand, the amount of movement will likely be overestimated. Figure 18.69 shows the results of a one-way horizontal cyclic loading test on a pile in sand. Figure 18.70 shows the results of a cyclic pressuremeter test in the same sand.

18.8.9 P-y Curve Approach

The previous approaches assume that the soil is uniform and that the pile has a constant cross section. In many cases, the actual soil is stratified with layers of different strength and stiffness. Also, sometimes the pile cross section varies as a

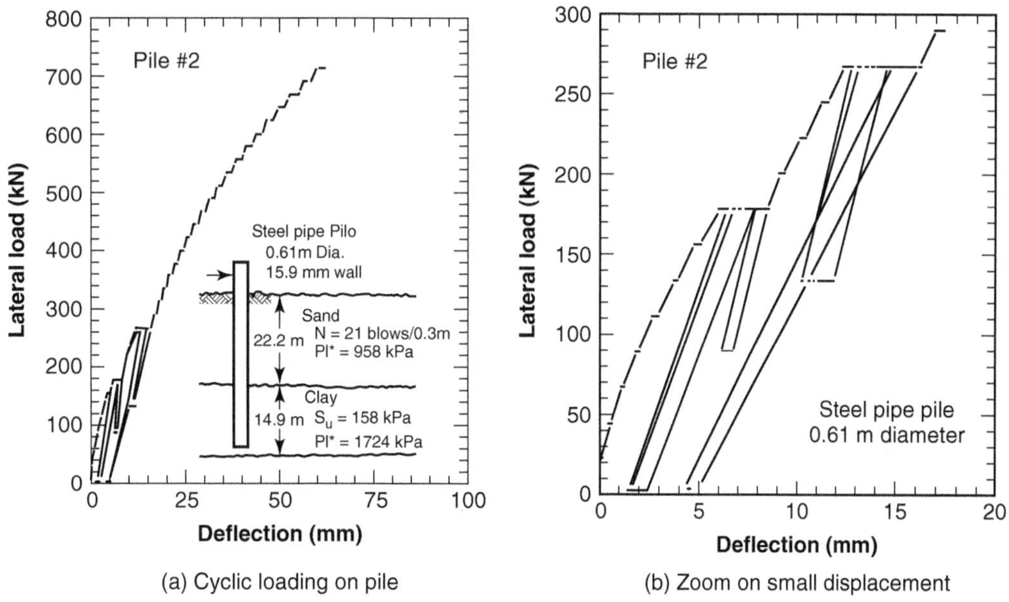

Figure 18.69 Cyclic lateral load test.

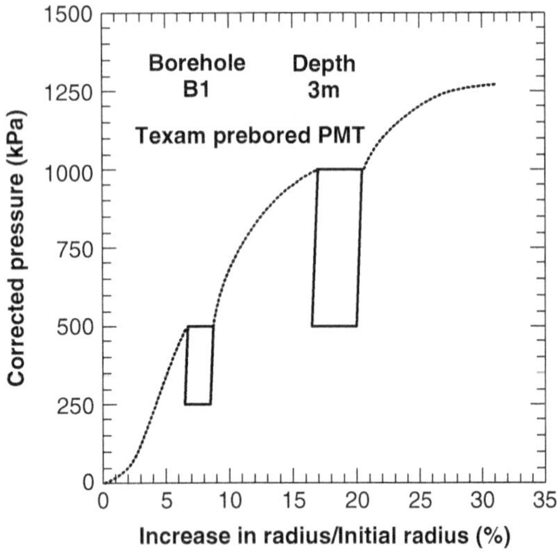

Figure 18.70 Cyclic pressuremeter test.

function of depth. In this case, one of the solutions is the *P*-*y* curve approach. In this approach, the soil resistance is described at any given depth by a nonlinear curve linking the line load *P* on the pile to the pile deflection into the soil *y*. Much of the early work on *P*-*y* curves was done by Matlock and Reese, who recommended a set of curves based on large-scale load tests, analytical developments, and software calibration. These curves are well documented in the offshore recommended practice API-RP 2A (2000), which includes the effect of cyclic loading. Briaud (1992) recommended P-y curves based on the pressuremeter curve. Frank (2013) and Norme Francaise AFNOR P94-262 (2012) also have recommendations as to how to construct *p*-*y* curves on the basis of pressuremeter data. The general solution for the *P*-*y* curve approach is the finite difference solution which is described in detail in Section 11.5.1 with a complete example in Section 11.5.2.

18.8.10 Horizontal Loading Next to a Trench

Sometimes there is a need to dig a trench in front of a laterally loaded pile. In this situation one often needs to know how far and how deep the trench can be dug and, if the trench is constructed within the zone of influence of the loading, how much the ultimate capacity will be reduced. If $H_{ou(no\ trench)}$ is the ultimate capacity when there is no nearby trench, and if $H_{ou(trench)}$ is the ultimate capacity when there is a trench, then the reduction factor λ is:

$$H_{ou(trench)} = \lambda H_{ou(no\ trench)} \quad (18.214)$$

Pressuremeter tests were conducted closer and closer to a trench in sand and then to a trench in clay (Briaud and Tucker 1987). The results, shown in Figure 18.71, indicate the weakening of the PMT curve as the distance from the PMT to the edge of the trench is decreased. These data were used to generate the reduction factor λ chart shown in Figure 18.72. Of course, before this chart can be used, one should first check that the trench is stable.

18.9 HORIZONTAL LOAD AND MOMENT: PILE GROUP

The resistance of pile groups to horizontal loading (Figure 18.73) includes several topics: resistance to overturning moments, ultimate loads, and movements at working loads.

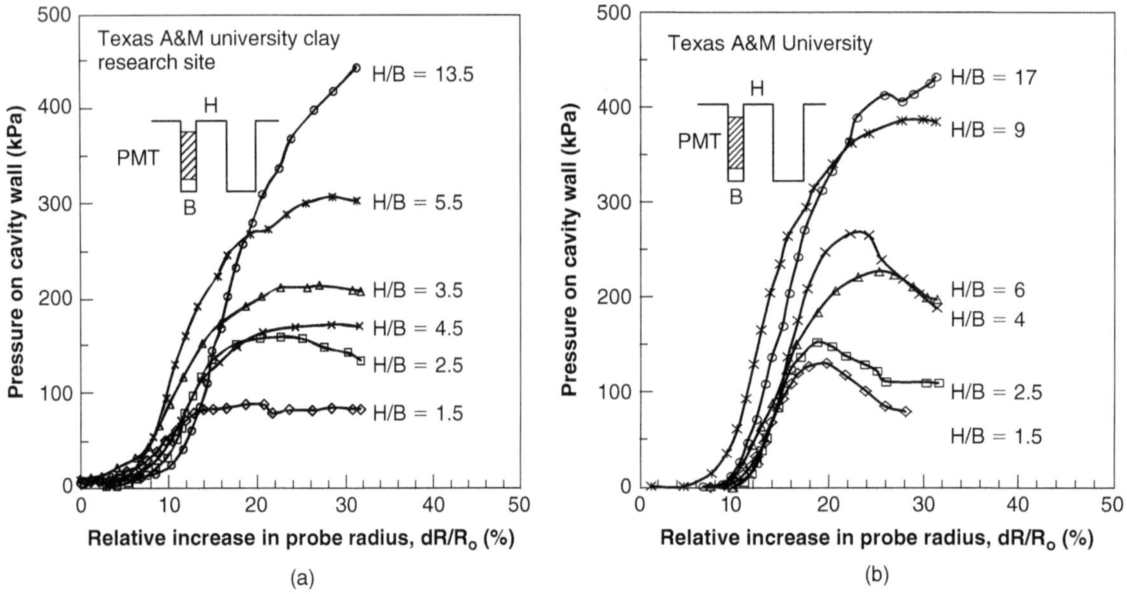

Figure 18.71 Pressuremeter tests near a trench: (a) PMT tests near a trench in clay. (b) PMT tests near a trench in sand.

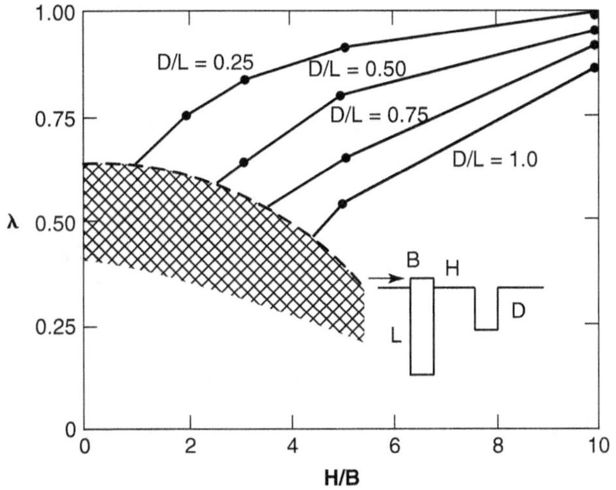

Figure 18.72 Horizontal capacity reduction factor for the presence of a trench.

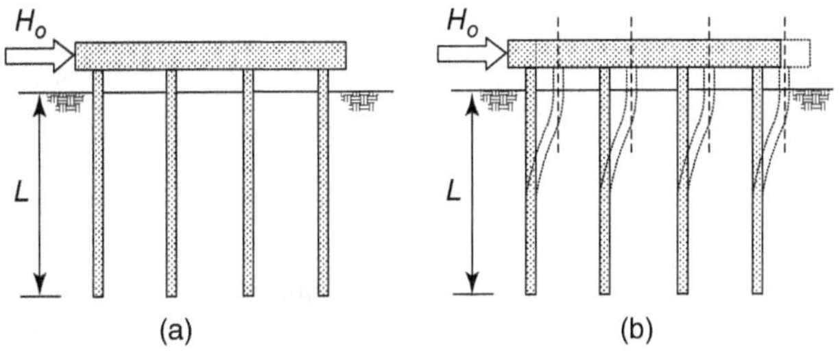

Figure 18.73 Horizontal loading of a pile group (cross section).

18.9.1 Overturning Moment

The resistance to overturning moment is usually taken through an increase in axial compression for the piles on one side of the group and a corresponding decrease in axial compression or possibly tension for the piles on the other side of the group (Figure 18.74).

Consider a rectangular group of piles with n piles in one direction and m piles in the other. The horizontal distances between the center of the group and individual piles in the group are a_i (Figure 18.74). The width of the group is B and the length is L. In this case, the change in load ΔQ_i in each pile due to the moment M is given by:

$$\Delta Q_i = \frac{a_i}{B/2} \Delta Q_{\max} \qquad (18.215)$$

where $\Delta Q_{\max}$ is the change in axial load in the pile located at the largest distance away from the center of the group ($B/2$). Then the resisting moment provided by the pile group is given by:

$$\Delta Q_{\max} = \frac{MB}{2m \sum_{i=1}^{n} a_i^2} \qquad (18.216)$$

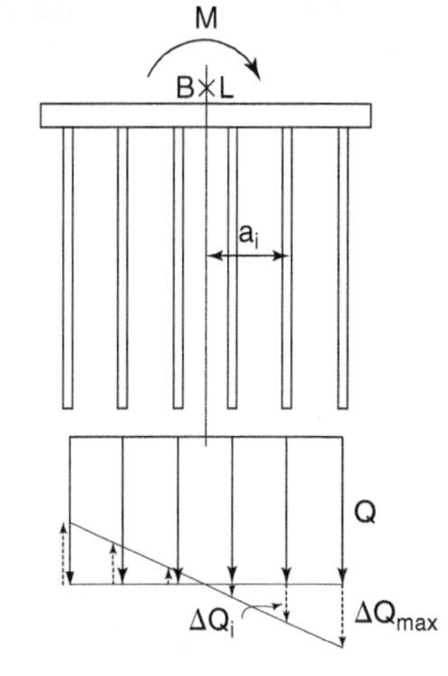

Figure 18.74 Overturning of a pile group.

where $\Delta Q_{\max}$ is the change in load in the piles at the edge of the group, M is the global moment applied, B is the width of the group, m is the number of piles in the length direction, n is the number of piles in the width direction, and a_i is the distance between pile i and the center axis around which the moment is applied. Once the value of ΔQ_i is known for each pile, the problem reverts to being a vertical load problem.

18.9.2 Ultimate Capacity

The ultimate horizontal load that can be applied to a pile group can be estimated as:

$$H_{ou(group)} = enH_{ou(single)} \qquad (18.217)$$

where $H_{ou(group)}$ and $H_{ou(single)}$ are the ultimate horizontal load for the group and for the single pile respectively, n is the number of piles in the group, and e is the efficiency of the group.

The load resisted by each pile in the group is not the same for all piles. The piles in the front of the group (leading piles) will develop more resistance than the piles behind them (trailing piles) (Figure 18.75).

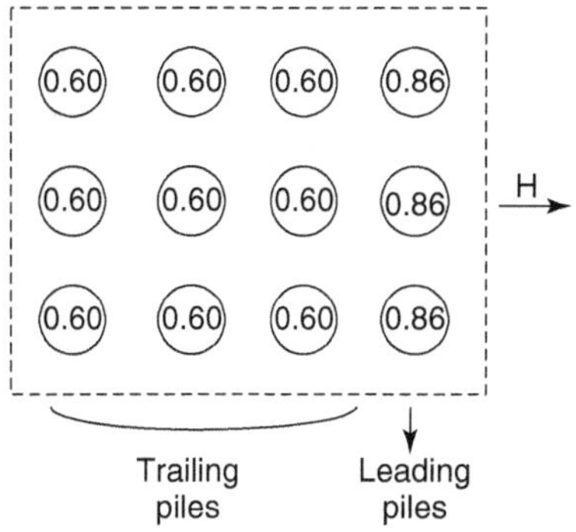

Figure 18.75 Horizontal load on a pile group (plan view).

One reasonable assumption is that all leading piles carry the same load and that all trailing piles carry the same load. Then the following influence factors can be defined for a leading pile and a trailing pile:

$$H_{ou(leading\ pile)} = e_{lp} H_{ou(single)} \tag{18.218}$$

$$H_{ou(trailing\ pile)} = e_{tp} H_{ou(single)} \tag{18.219}$$

where $H_{ou(leading\ pile)}$ and $H_{ou(trailing\ pile)}$ are the ultimate horizontal capacity of the leading pile and trailing pile respectively, and e_{lp} and e_{tp} are the efficiency factors for the leading pile and trailing pile respectively. The group ultimate capacity can then be assembled as:

$$H_{ou(group)} = (n_{lp} e_{lp} + n_{tp} e_{tp}) H_{ou(single)} \tag{18.220}$$

where n_{lp} and n_{tp} are the number of leading piles and trailing piles respectively.

Cox et al. (1983) measured the behavior of groups of in-line piles. They loaded these lines of piles in the direction of the line (in-line loading) and perpendicularly to that line (side-by-side loading). Their measurements were used to develop the global efficiency factors for line groups shown in Figure 18.76. The global efficiency factor is the ratio of the group ultimate capacity $H_{ou(group)}$ divided by n times the ultimate capacity of a single pile $H_{ou(single)}$ where n is the total number of piles in the group.

The measurements by Cox et al. (1983) also showed that all the trailing piles carry approximately the same load and that the leading pile carries more than the trailing piles. The ratio of the ultimate load of the leading pile over the ultimate load of the trailing pile depends on the spacing:

$$H_{ou(leading\ pile)} = \lambda H_{ou(trailing\ pile)} \tag{18.221}$$

The values of λ are shown in Figure 18.77. Therefore, another way to express the group capacity is:

$$H_{ou(group)} = \left(n_{lp} e_{lp} + n_{tp} \frac{e_{lp}}{\lambda} \right) H_{ou(single)} \tag{18.222}$$

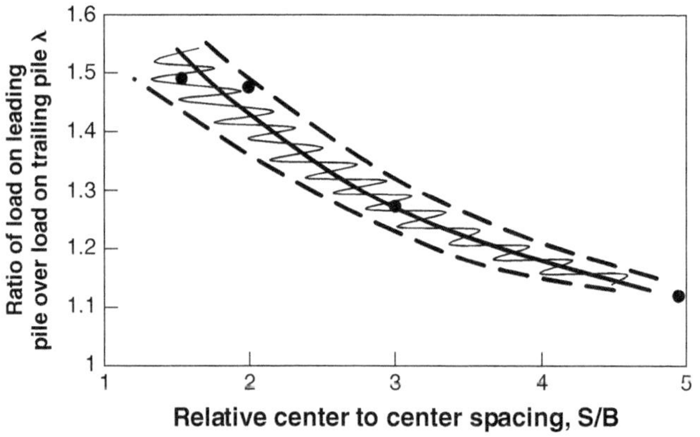

Figure 18.77 Ratio of load on leading pile over trailing pile.

As an example, consider the 3 by 4 pile group of Figure 18.75. The center-to-center spacing is equal to two times the pile diameter. The single-pile ultimate horizontal load $H_{ou(single)}$ has been calculated to be 100 kN. We now wish to obtain the group ultimate capacity if the horizontal load is applied in the direction perpendicular to the three-pile side as shown in Figure 18.75.

1. The leading-pile efficiency e_{lp} is obtained from Figure 18.76 for a pile spacing of 2. The value is 0.86. Therefore, because there are 3 leading piles, the contribution to the group capacity is $3 \times 0.86 = 2.58$.
2. The ratio λ between the capacity of the leading pile and the trailing pile is given by Figure 18.77. The value is 1.43 for a spacing of 2; therefore, the efficiency of the trailing piles is $0.86/1.43 = 0.60$. Because there are 9 trailing piles, the contribution to the group capacity is $9 \times 0.60 = 5.40$.
3. The contribution of the leading piles plus the trailing piles is then $2.58 + 5.40 = 7.98$. If the group was 100% efficient, it would carry 12 times the single-pile capacity, but in fact it carries 7.98 times the single-pile capacity. Therefore, the global efficiency of the group

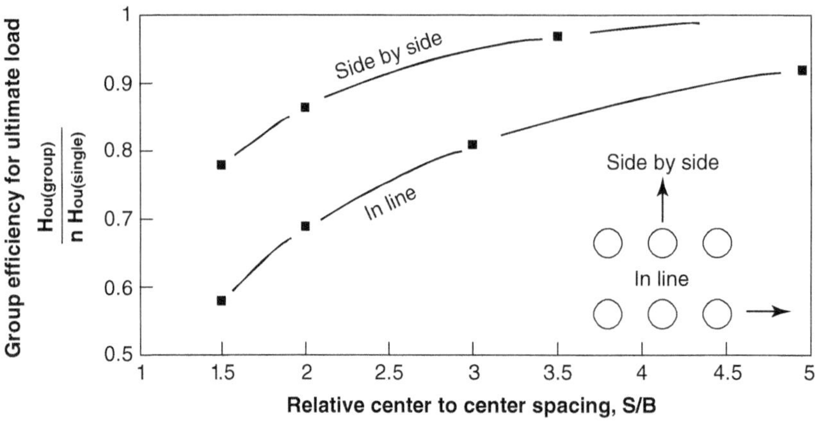

Figure 18.76 Efficiency for side-by-side and in-line groups (After Cox et al. 1983).

is 7.98/12 = 0.665. The ultimate horizontal capacity of the group is 0.665 × 100 × 12 = 798 kN.

18.9.3 Movement

The movement of a pile group is difficult to estimate by simple calculations. One way is to consider that the pile cap prevents any rotation of the individual piles, so that the displacement is the one associated with the case of fixed-head piles (Figure 18.73). The results in this case are given here and detailed in section 18.8.6. A moment M_o must develop between the pile and the pile cap to prevent rotation of the pile at the ground line. For flexible and rigid piles, this moment is given by:

$$\text{Fixed-head flexible pile} \quad M_o = -\frac{H_o l_o}{2} \qquad (18.223)$$

$$\text{Fixed-head rigid pile} \quad M_o = -\frac{H_o L}{2} \qquad (18.224)$$

where H_o is the horizontal load, l_o is the transfer length (Eq. 18.170), and L is the length of the pile.

These moments also happen to be the maximum bending moments in the pile. The displacement y_o at the ground surface in this case will be:

$$\text{Fixed-head flexible pile} \quad y_o = \frac{H_o}{l_o K} \qquad (18.225)$$

$$\text{Fixed-head rigid pile} \quad y_o = \frac{H_o}{LK} \qquad (18.226)$$

where K is the soil stiffness. This stiffness is recommended to be taken as 2.3 E_o for single piles, where E_o is the pressuremeter first load modulus.

In the case of pile groups, and because of the overlapping of soil stresses around the piles in the group, this number must be decreased by a factor indicative of the interaction between piles in the group.

A second approach for predicting the response of pile groups is to use the P-y curve approach (section 18.8.9) and soften the P-y curves to take into account the effect of overlapping stresses among piles. Given the P_s-y curve for a single pile, the P_g-y curve for the group is obtained simply by writing that for a given value of y (Figure 18.78):

$$P_g = mP_s \qquad (18.227)$$

where m is called the multiplier. Brown et al. (2010) gave recommendations for the values of m as shown in Figure 18.79.

The softened P-y curves are then used to simulate each pile in the group as a single pile while using a finite difference program to predict the deflection and maximum bending moment. Alternatively, programs such as FLPIER, which simulate the entire group on the basis of P-y curves, can be used. Ultimately, the finite element method in three dimensions is the best tool to predict the behavior of horizontally loaded pile groups, using programs such as ABAQUS or PLAXIS, but the computing time required is much larger.

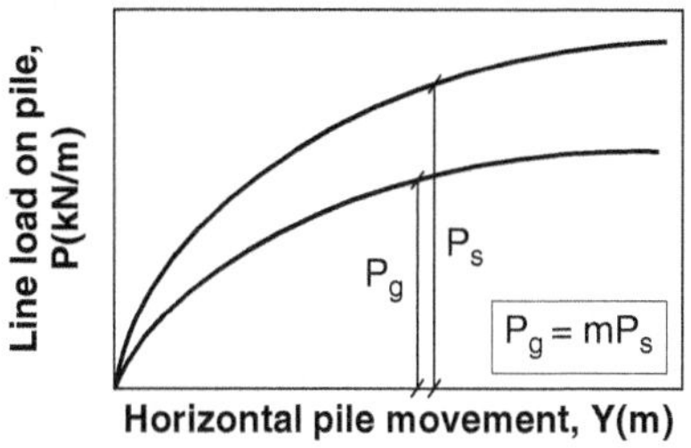

Figure 18.78 P-y curve for piles in groups.

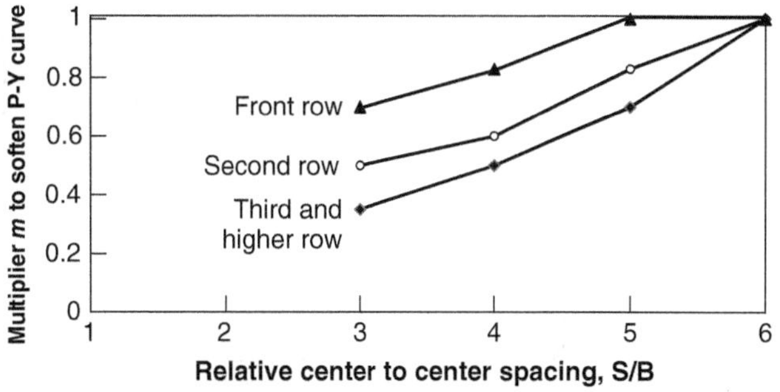

Figure 18.79 Multiplier to soften P-y curves. (After Brown et al. 2010)

18.10 COMBINED PILED RAFT FOUNDATION

A *combined piled raft foundation* (CPRF) is composed of a mat foundation with a number of piles underneath the mat. It is an intermediate between a mat foundation and a pile-group foundation (Figure 18.80). Guidelines for CPRF have been proposed by the ISSMGE technical Committee on Deep Foundations (Katzenbach, 2012). The difference between a CPRF and a pile-group foundation is twofold:

1. In the calculations of the pile-group foundation, the contribution of the pile cap or mat is ignored, whereas it is an integral part of the carrying capacity of the CPRF.
2. The CPRF has fewer piles and the piles are typically longer under the center of the mat than at the edges. The reason is that the mat foundation settles in the shape of a dish; the settlement tends to be larger under the center

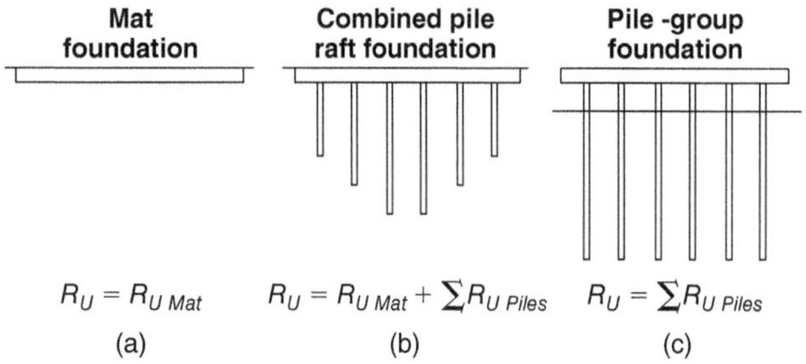

Figure 18.80 Difference between mat foundation, pile-group foundation, and combined piled raft foundation.

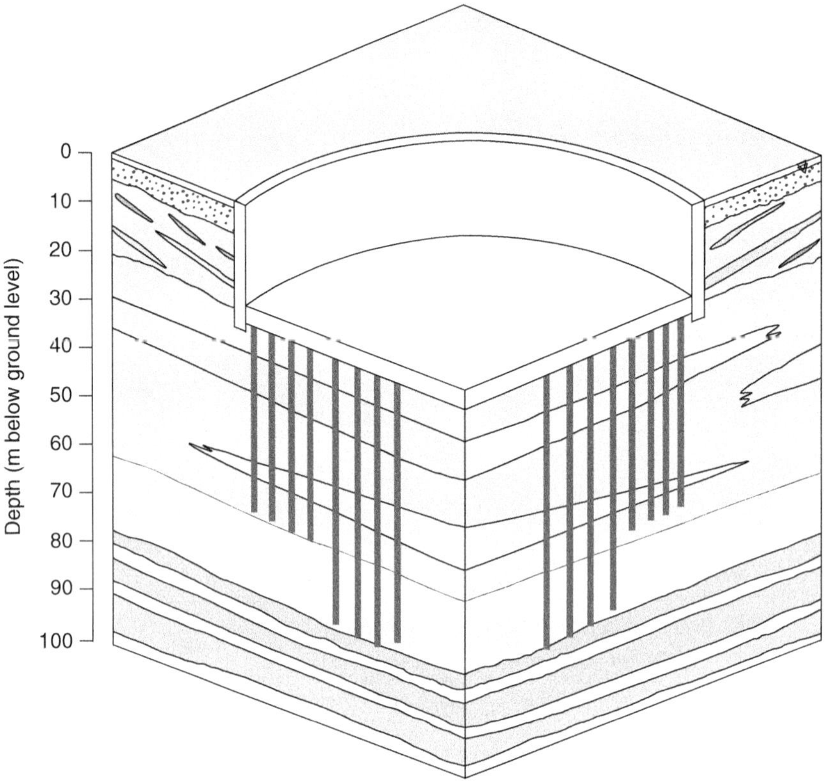

Figure 18.81 CPRF for a very high building in Dubai. (Courtesy of Chris Haberfield, Golder and Associates)

than under the edges. Thus, placing more piles under the center and fewer around the edges counterbalances the dishing tendency and reduces the bending moment in the mat.

An example of a CPRF for a very high building in Dubai is shown in Figure 18.81.

The CPRF offers a combined resistance to the building loads that comes from the pile point resistance (R_{pi}), pile friction resistance (R_{fi}), and raft or mat resistance (R_{ri}). The subscript i refers to pile i and to the raft resistance tributary to pile i (Figure 18.82). The CPRF ratio α is defined as the ratio of the load carried by the n piles divided by the total load R_t carried by the CPRF (Katzenbach 2012):

$$\alpha = \frac{\sum_{i=1}^{n}(R_{pi}(s) + R_{fi}(s))}{R_t(s)} \quad (18.228)$$

where s is the settlement of the CPRF. Values of the CPRF ratio around 0.5 are common.

At the ultimate limit state, the ultimate pile capacity can be estimated according to the guidelines presented in section 18.4 for single piles and section 18.5 for pile groups. The resistance R_{ri} contributed by the raft around pile i is given by:

$$R_{ri} = \iint_A \sigma(s, x, y)dxdy \quad (18.229)$$

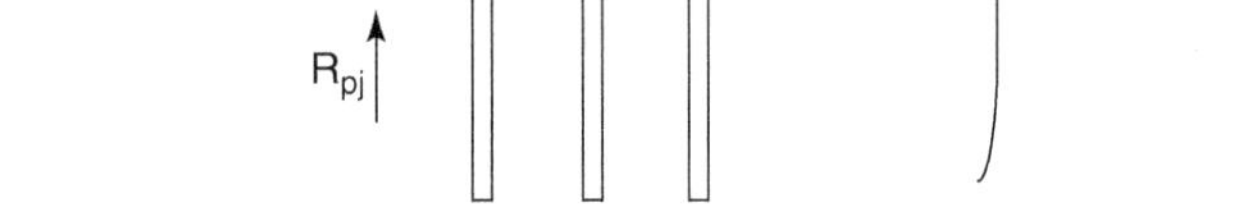

Figure 18.82 Free-body diagram of a CPRF. (After Katzenbach 2012)

where $\sigma(s, x, y)$ is the vertical and upward normal stress on the bottom of the raft around pile i, s is settlement, x and y are the coordinates in the horizontal plane, and A is the area domain of integration corresponding to the portion of the raft tributary to pile i.

It is very difficult to estimate the values of R_{pj}, R_{fj}, and R_{rj} by simple means because they all depend on the movement that takes place around the piles and under the mat. For this reason, the best way to predict the behavior of a CPRF is

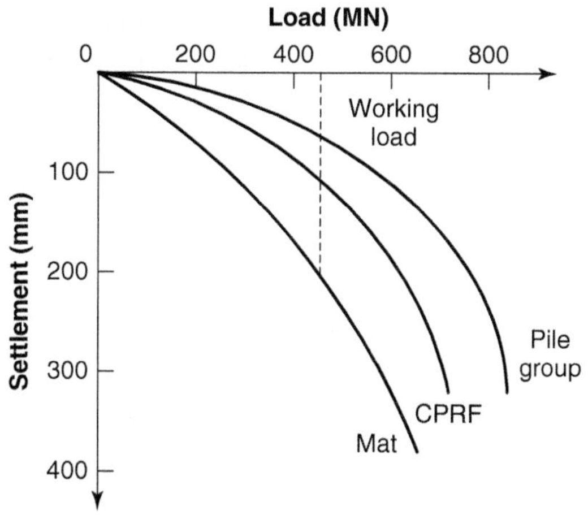

Figure 18.83 Load settlement curves of the structure for three types of foundations. (After Katzenbach, 2012)

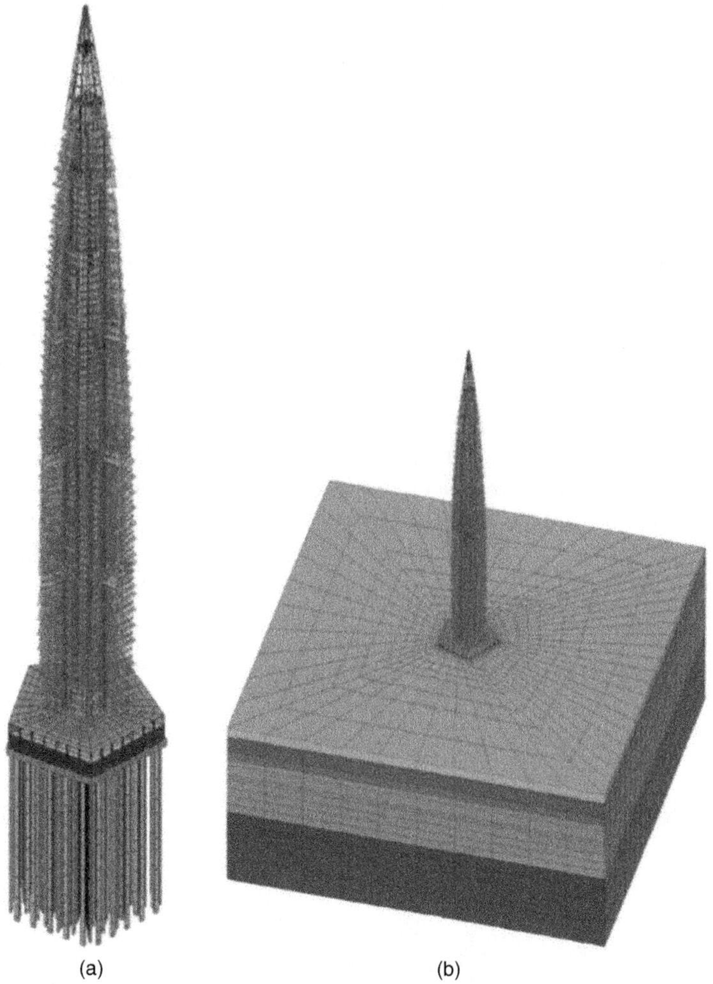

Figure 18.84 Complete soil-foundation-structure simulation: (a) Tower and foundation. (b) Tower, foundation, and mass. (Courtesy of Lisyuk and Ulitsky 2012)

through the finite element method. A typical approach for very large structures consists of the following steps:

1. Carry out load tests with group effect if possible.
2. Calibrate the FEM model to match the load test results.
3. Use the calibrated model to predict the behavior of the complete CPRF.
4. Monitor the construction of the structure to verify predictions and if needed make adjustments.

The result of such an approach appears in Figure 18.83, which shows the load settlement curve of the structure for three different foundation alternatives: mat, pile group, and CPRF. A parallel economic study can be used along with tolerable movements to decide which solution is both economical and safe.

The future of foundation engineering is in the use of the FEM with the goal of modeling the soil, the foundation, and the structure all in one model. An example of such approach is shown in Figure 18.84. Note that the unit weight of a building typically ranges between 2.5 and 5 kN/m^3 and is therefore a small fraction of the soil unit weight. As such, buildings are much lighter than soil for a given volume. This is why it is very advantageous to place basements in a building, as the weight of soil excavated for one story is equal to 5 to 10 stories of building. For example, a 20-story building with 3 levels of underground parking garages could well be as heavy as the soil removed to create the underground parking garages. In this case the settlement is limited to the recompression of the soil that expanded upon excavation.

PROBLEMS

18.1 In cross hole logging of a bored pile, the speed v of the compression wave is measured.
 a. If $v = 4000$ m/s how good is the concrete?
 b. If $v = 3500$ m/s how good is the concrete?
 c. If $v = 3000$ m/s how good is the concrete?

18.2 Draw a typical and clean record of velocity signal in a sonic echo test on a bored pile with a necking defect. Repeat the question for a bored pile with a bulb defect.

18.3 Use the pile driving equation to obtain the rated energy of a diesel hammer necessary to drive a pile with an ultimate resistance at the time of driving of 1000 kN to a penetration rate of 4 mm/blow. Assume that the diesel hammer has an efficiency of 0.5.

18.4 Calculate in km/h how fast the compression wave generated by the hammer blow from problem 18.3 propagates in a steel pile, in a concrete pile, and in a timber pile. If the pile is 20 m long, how much time does it take for the wave to go down to the bottom of the pile and back up to the top?

18.5 A hammer impacts a concrete pile with a 0.25 m^2 cross section and generates a particle velocity at the top of the pile equal to 3 m/s. Calculate the force and then the compressive stress in the concrete at the pile head.

18.6 The ultimate static soil resistance of a short, relatively rigid pile in silt is 800 kN with 500 kN of friction and 300 kN of point resistance. Calculate the ultimate dynamic soil resistance if the pile velocity is 4 m/s.

18.7 Show all calculations leading to the wave equation numbers populating Table 18.5.

18.8 Develop the theoretical expression of the residual load in a driven pile for the following conditions. The initial condition is the stress and load distribution in the pile at failure. The ultimate skin friction is f_u and the ultimate point pressure is p_u. The ultimate load at the top of the pile is R_u and the ultimate load at the point is R_{pu}. The unloading of the friction and point transfer curve is assumed to obey the linear elastic model shown in Figure 18.1s.

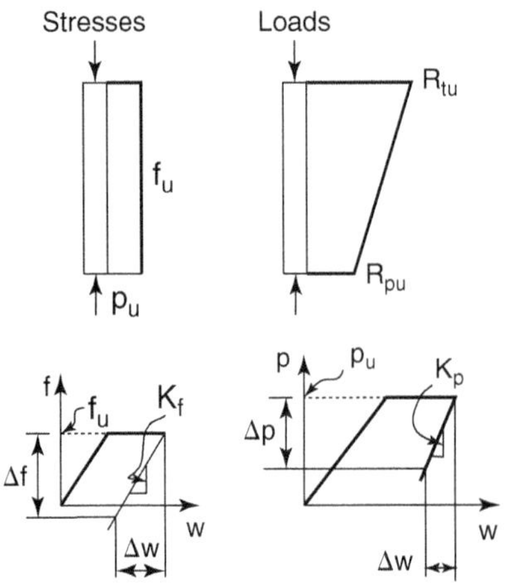

Figure 18.1s Initial conditions and models for residual loads.

18.9 Explain Figures 18.21 and 18.22 in your own words.

18.10 A pile is driven and the force (F) and particle velocity time impedance ($v \times EA/c$) at the top of the pile are measured as shown in Figure 18.2s. Calculate the dynamic resistance of the pile using the observations at times t_1 and t_2 separated by the down and up travel time of the wave ($2L/c$) and the Case method.

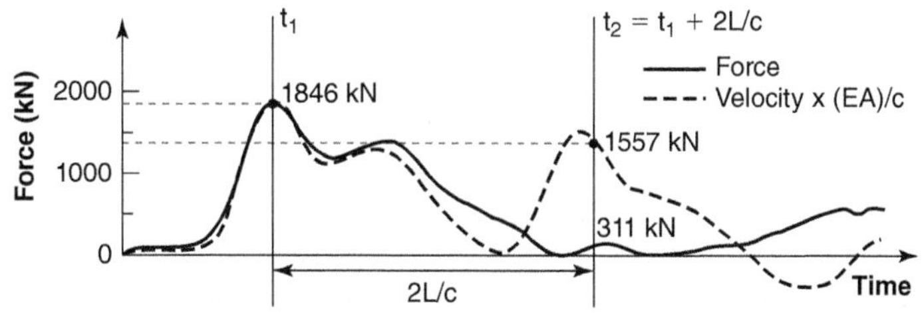

Figure 18.2s Force and velocity signal.

18.11 Calculate the static resistance of the pile in problem 18.10 by the Case method.

18.12 Is it possible to break a concrete pile in tension by driving it in the ground? If yes, explain how.

18.13 A suction caisson is 20 m long and 2 m by 2 m in cross section with a wall thickness of 20 mm. It is made of steel and is to be installed in a soft clay with an undrained shear strength of 20 kPa. Calculate the ultimate capacity of the caisson and the underpressure required to install it to full penetration. Check that this underpressure does not create inverse bearing capacity failure.

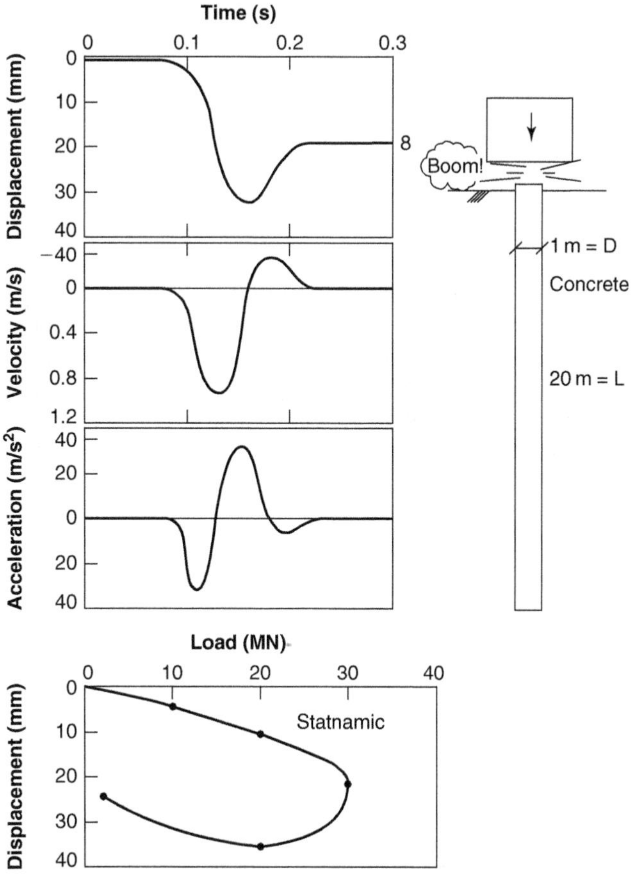

Figure 18.3s Statnamic test results.

18.14 Calculate the strain, the stress, and the friction on each segment of the pile shown in Figure 18.29 for the maximum load applied. How would you measure such values? The pile is a bored pile 1 m in diameter and made of concrete.

18.15 Calculate the slope of the load transfer curves for the axially loaded pile discussed in section 18.4.3 and verify the value of the movement necessary to reach the maximum friction and point resistance shown in Figure 18.43. Generate a spreadsheet to develop the complete load settlement curve of this axially loaded pile.

18.16 Calculate the ultimate static capacity of the pile subjected to the Statnamic test. The pile is a 1 m diameter, 20 m long bored concrete pile. The test results are summarized in Figure 18.3s.

18.17 Calculate the ultimate capacity of the two piles shown in Figure 18.4s by all possible methods. The pile in clay is a circular bored pile and the pile in sand is a square driven pile. At what depth along the pile would you place the Osterberg load cell to balance the load on both sides of the pile at ultimate load?

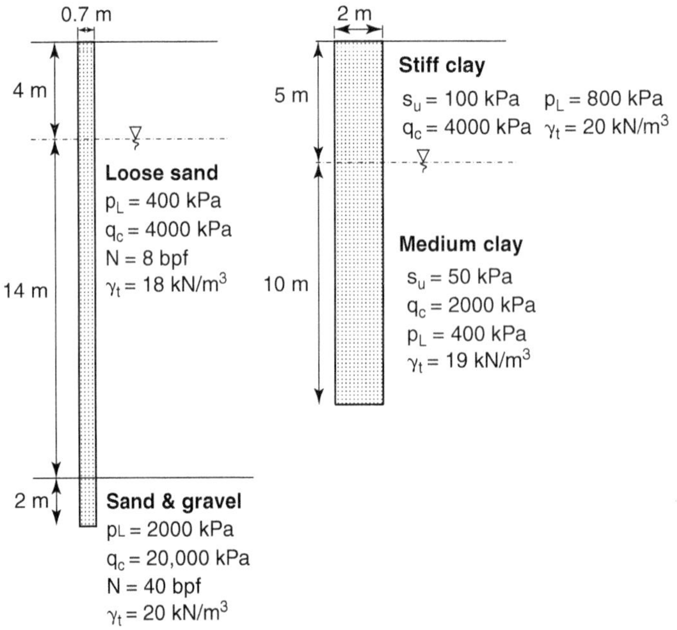

Figure 18.4s Ultimate capacity of two piles.

18.18 For the pile of Figure 18.43, find the top movement and the load distribution in the pile for a point movement of 5 mm.

18.19 A 16-story hospital weighs 1500 MN, and its imprint is 75 m by 75 m. The building rests on 10,000 timber piles, each 15 m long, 0.3 m in average diameter, and driven with a spacing of 0.75 m center to center. The soil is made of a clay layer down to 14.5 m ($s_u = 20$ kN/m^2, $e_o = 0.8$, and $C_c = 0.1$), then a sand layer down to 16.5 m (N = 30 bpf), and then clay again down to a depth of 100 m ($s_u = 30$ kN/m^2, $e_o = 0.7$, and $C_c = 0.06$). The water table is at the ground surface and the total unit weight of all soils is 20 kN/m^3. Calculate:

a. The capacity of one timber pile
b. The capacity of the pile group
c. The settlement of the hospital
d. Comment on this design.

18.20 Calculate the group efficiency for settlement using Poulos interaction factors for the case of a flexible pile cap (all piles carry the same load). The group is 4 by 4 with a 3-pile diameter center-to-center spacing.

18.21 If the uncoated pile subjected to downdrag in Figure 18.54 was pushed into the ground 100 mm at the pile top, what would be:

a. The new position of the neutral point?
b. The load at the top of the pile?
c. The load distribution in the pile?

18.22 A bored pile foundation is used for a house on a shrink-swell soil. The piles are 0.5 m in diameter, the load per pile is 50 kN, and the zone of active movement from one season to the next extends from the ground surface to a depth of 3 m. The soil is a very stiff clay with an undrained shear strength of 120 kPa and a total unit weight of 20 kN/m^3. The groundwater level is at a depth of 10 m. How deep should each bored pile be to minimize the potential uneven movement of the house?

18.23 For the long flexible pile shown in Figure 18.5s, calculate:

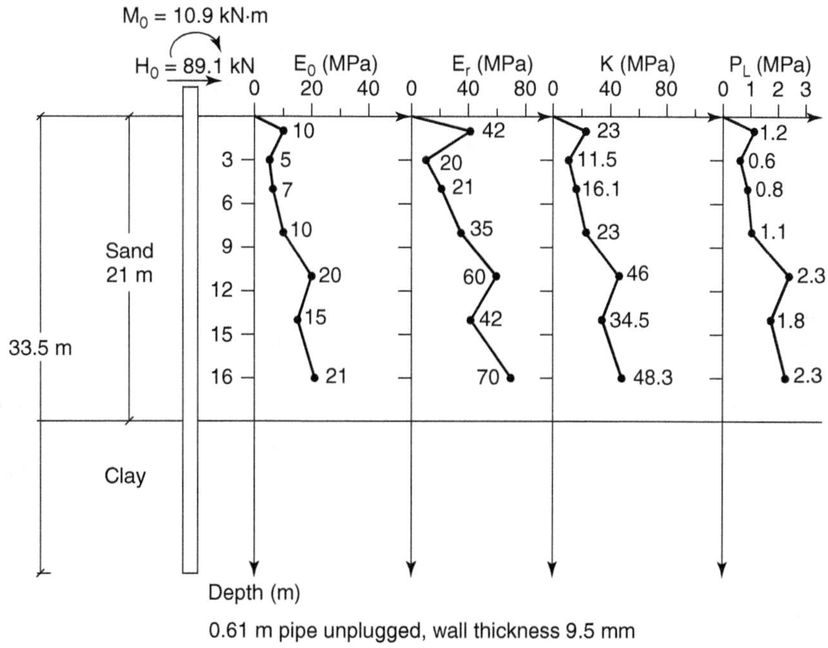

Figure 18.5s Long flexible pile loaded horizontally.

a. The ultimate load H_{ou}
b. The deflection and slope at the ground surface under the working load
c. The maximum bending moment under the working load
d. The factor of safety against yielding of the soil near the ground surface under the working load

18.24 For the short rigid pile shown in Figure 18.6s, calculate:

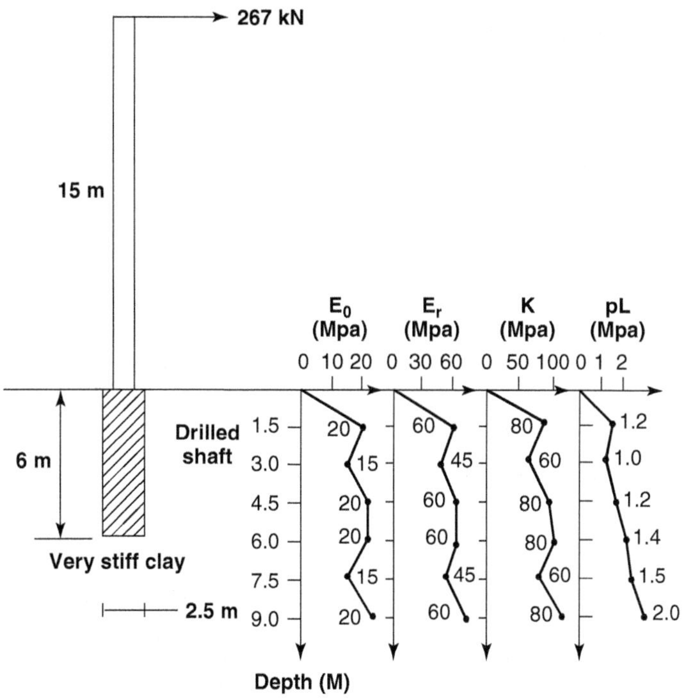

Figure 18.6s Short rigid pile loaded horizontally.

a. The ultimate load H_{ou}
b. The deflection and slope at the ground surface under the working load
c. The maximum bending moment under the working load
d. The factor of safety against yielding of the soil near the ground surface under the working load

18.25 Calculate z_{max} for a flexible pile and a rigid pile if the pile is subjected to a horizontal load only (H_o different from 0 but M_o equal to 0).

18.26 Calculate the ratio between the ground surface displacement for a free-head condition and for a fixed-head condition. Do the calculation first for a flexible pile and then for a rigid pile.

18.27 For the pile group shown in Figure 18.75, calculate the efficiency of the group if it is loaded horizontally in a direction perpendicular to the four-pile line.

18.28 The pile group of Figure 18.75 is subjected to an overturning moment of 10 MN.m in the direction of largest resistance to overturning of the group. The piles are 0.4 by 0.4 square concrete driven piles embedded 25 m in a loose sand with a blow count of 6 bpf. What will be the ratio between the applied tension load and the ultimate tension capacity of the most loaded pile in the group?

18.29 A steel pipe pile has a diameter D equal to 0.61 m and a wall thickness t equal to 9.5 mm. The pile is 33.5 m long and the steel has a modulus $\underline{E}$ equal to 200 GPa. The pile is loaded horizontally with a load H_o of 89 kN in fixed-head condition. The soil is characterized by stiffness coefficient K from pressuremeter tests equal to 25,000 kPa. Plot the profiles versus depth of the deflection, slope, shear, bending moment, and line load in the pile.

Problems and Solutions

Problem 18.1

In cross hole logging of a bored pile, the speed v of the compression wave is measured.

a. If $v = 4000$ m/s, how good is the concrete?
b. If $v = 3500$ m/s, how good is the concrete?
c. If $v = 3000$ m/s, how good is the concrete?

Solution 18.1

a. If $v = 4000$ m/s, the concrete is good.
b. If $v = 3500$ m/s, the concrete is questionable.
c. If $v = 3000$ m/s, the concrete is poor or there is a defect in the pile.

Problem 18.2

Draw a typical and clean record of velocity signal in a sonic echo test on a bored pile with a necking defect. Repeat the question for a bored pile with a bulb defect.

Solution 18.2

a. Necking defect

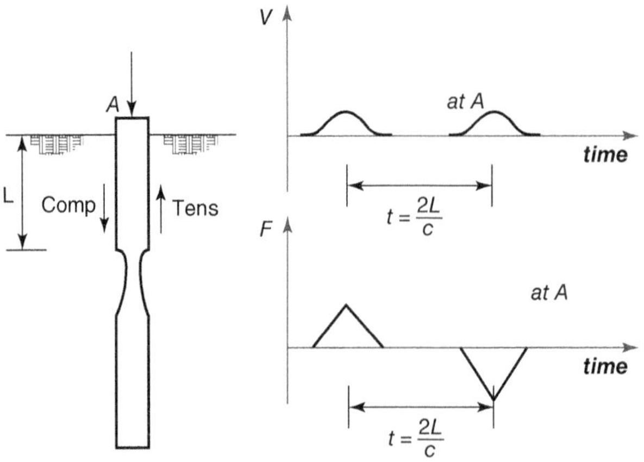

Figure 18.7s Necking defect.

b. Bulb defect

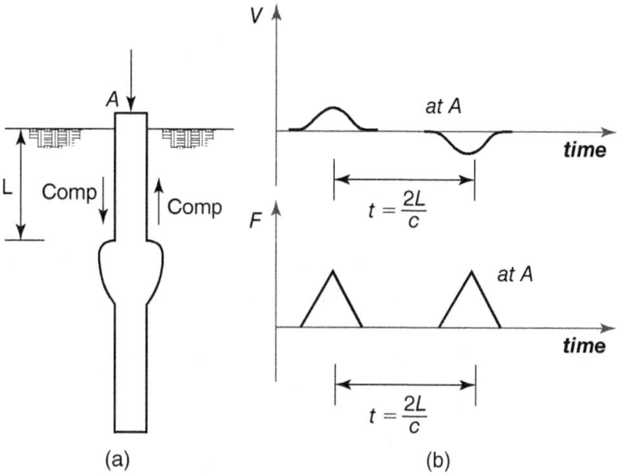

Figure 18.8s Bulb defect.

Problem 18.3

Use the pile driving equation to obtain the rated energy of a diesel hammer necessary to drive a pile with an ultimate resistance at the time of driving of 1000 kN to a penetration rate of 4 mm/blow. Assume that the diesel hammer has an efficiency of 0.5.

Solution 18.3

$$R_{ud} = \frac{eWh}{s + \frac{c}{2}} = \frac{eWh}{s + 2.5}$$

$$Wh = \frac{R_{ud} \times (s + 2.5)}{e}$$

$$Wh = \frac{1000 \times (4 + 2.5)}{0.5}$$

$$Wh = \frac{1000 \times (4 + 2.5)}{0.5}$$

$$Wh = 13000 \text{ kN.m} = 13000 \text{ kJ}$$

Problem 18.4

Calculate in km/h how fast the compression wave generated by a hammer blow propagates in a steel pile, in a concrete pile, and in a timber pile. If the pile is 20 m long, how much time does it take for the wave to go down to the bottom of the pile and back up to the top?

Solution 18.4

The wave speed for a given material can be calculated using the following formula:

$$c = \sqrt{\frac{E}{\rho}}$$

$$\text{For steel}: E = 2 \times 10^8 \text{ kPa} \quad \text{and } \rho = 7850 \text{ kg/m}^3$$
$$\text{For concrete}: E = 2 \text{ x } 10^7 \text{ kPa} \quad \text{and } \rho = 2400 \text{ kg/m}^3$$
$$\text{For wood}: E = 1 \text{ x } 10^7 \text{ kPa} \quad \text{and } \rho = 800 \text{ kg/m}^3$$

$$c_{steel} = \sqrt{\frac{2 \times 10^{11}}{7850}} = 5047.5 \text{ m/sec.} \quad \therefore \quad 18171.2 \text{ km/hr.}$$
$$c_{concrete} = \sqrt{\frac{2 \times 10^{10}}{2400}} = 2886.8 \text{ m/sec.} \quad \therefore \quad 10392.3 \text{ km/hr.}$$
$$c_{timber} = \sqrt{\frac{1 \times 10^{10}}{800}} = 3535.5 \text{ m/sec.} \quad \therefore \quad 12727.9 \text{ km/hr.}$$

The time it takes for the wave to go down to the bottom of the pile and back up to the top can be calculated as:

$$t = \frac{2L}{c}$$
$$t_{steel} = \frac{2 \times 20}{5047.5} = 0.00792 \text{ sec.}$$
$$t_{concrete} = \frac{2 \times 20}{2886.8} = 0.01386 \text{ sec.}$$
$$t_{timber} = \frac{2 \times 20}{3535.5} = 0.01131 \text{ sec.}$$

Problem 18.5

A hammer impacts a concrete pile with a 0.25 m^2 cross section and generates a particle velocity at the top of the pile equal to 3 m/s. Calculate the force and then the compressive stress in the concrete at the pile head.

Solution 18.5

Using the results of c = 2886 m/ sec . computed in problem 18.4 for a concrete pile, the force can be estimated using the following formula:

$$F = \frac{EA}{c} v$$
$$F = \frac{2 \times 10^7 \text{ kPa} \times 0.25 \text{ m}^2}{2886 \text{ m/sec.}} \times 3 \text{ m/sec.} = 5198 \text{ kN}$$

The compressive stress in the pile head is:

$$\sigma = \frac{F}{A} = \frac{5198 \text{ kN}}{0.25 \text{ m}^2} = 20792 \text{ kPa}$$

Problem 18.6

The ultimate static soil resistance of a short, relatively rigid pile in silt is 800 kN with 500 kN of friction and 300 kN of point resistance. Calculate the ultimate dynamic soil resistance if the pile velocity is 4 m/s.

Solution 18.6

The dynamic soil resistance of the soil is obtained using Eq. 18.30 and assuming rigid body motion of the pile:

$$R_{DYN} = R_{STA\,p}(1 + J_p v) + R_{STA\,f}(1 + J_f v)$$

where R_{STA} is 800 kN, v is 4 m/s, J is 0.65 m/s for fine-grained soils and side damping, and J is 0.50 m/s for fine-grained soils and point damping (from Table 18.2). The ultimate dynamic soil resistance is:

$$R_{DYN} = 500(1 + 0.65 \times 4) + 300(1 + 0.50 \times 4) = 2700 \text{ kN}$$

Problem 18.7

Show all calculations leading to the wave equation numbers populating Table 18.5.

Solution 18.7

Figure 18.18 gives the pile details. The calculations for all the numbers in Table 18.1s (Table 18.5 earlier in the chapter) are shown in this solution. Some conditions that must be satisfied for all calculations are:

- $F(1, t)$ is always ≥ 0, because no tension can be developed between the hammer and the pile head
- $F(3, t)_s$ is always ≤ 500, because of the maximum point resistance

Table 18.1s Wave Equation Calculations

	1	2	3	4	5	6	7
A	Time	s	0.0000	0.0005	0.0010	0.0015	0.0020
B	D(1, *t*)	mm	0.000	1.500	2.834	3.897	4.672
C	D(2, *t*)	mm	0.000	0.000	0.368	1.296	2.708
D	D(3, *t*)	mm	0.000	0.000	0.000	0.045	0.241
E	C(1, *t*)	mm	0.000	1.500	2.467	2.601	1.964
F	C(2, *t*)	mm	0.000	0.000	0.368	1.250	2.467
G	F(1, *t*)	kN	0.000	1350.000	2219.923	2341.013	1767.614
H	F(2, *t*)	kN	0.000	0.000	165.544	562.706	1110.058
I	R(3, *t*)s	kN	0.000	0.000	0.000	9.022	48.211
J	R(3, *t*)d	kN	0.000	0.000	0.000	9.185	51.990
K	V(1, *t*)	mm/s	3000.000	2668.913	2124.476	1550.343	1116.836
L	V(2, *t*)	mm/s	0.000	735.750	1855.387	2824.564	3182.932
M	V(3, *t*)	mm/s	0.000	0.000	90.221	391.890	968.537

At $t = 0$ s:

$$V(1, 0) = 3000 \text{ mm/s}$$

At $t = 0.0005$ s:

$$\mathrm{D}(1, 0.0005) = \mathrm{D}(1, 0) + \mathrm{V}(1, 0)\Delta\mathrm{t} = 0 + 3000 \times 0.0005 = 1.50 \text{ mm}$$

$$\mathrm{D}(2, 0.0005) = \mathrm{D}(2, 0) + \mathrm{V}(2, 0)\Delta\mathrm{t} = 0 \text{ mm}$$

$$\mathrm{D}(3, 0.0005) = \mathrm{D}(3, 0) + \mathrm{V}(3, 0)\Delta\mathrm{t} = 0 \text{ mm}$$

$$\mathrm{C}(1, 0.0005) = \mathrm{D}(1, 0.0005) - \mathrm{D}(2, 0.0005) = 1.50 \text{ mm}$$

$$\mathrm{C}(2, 0.0005) = \mathrm{D}(2, 0.0005) - \mathrm{D}(3, 0.0005) = 0.00 \text{ mm}$$

$$\mathrm{F}(1, 0.0005) = \mathrm{K}_1 \times \mathrm{C}(1, 0.0005) = 900\ (1.50) = 1350 \text{ kN}$$

$$\mathrm{F}(2, 0.0005) = \mathrm{K}_2 \times \mathrm{C}(2, 0.0005) = 450\ (0) = 0 \text{ kN}$$

$$\mathrm{F}(3, 0.0005)_s = \mathrm{K}' \times \mathrm{D}(3, 0.0005) = 200\ (0) = 0 \text{ kN}$$

$$\mathrm{F}(3, 0.0005)_d = \mathrm{F}(3, 0.0005)_s \times (1 + \mathrm{J}_s \times \mathrm{V}(3, 0)) = 0 \times (1 + 0.0002\ (0)) = 0 \text{ kN}$$

$$V(1, 0.0005) = V(1, 0) + (0 - F(1, 0.0005))\frac{g\Delta t}{W_H}$$

$$= 3000 + \left((0 - 1350) \times \frac{9.81 \times 0.0005}{20} \times 1000\right) = 2668.913 \text{ mm/s}$$

$$V(2, 0.0005) = V(2, 0) + (F(1, 0.0005) - F(2, 0.0005))\frac{g\Delta t}{W_1}$$

$$= 0 + \left((1350 - 0) \times \frac{9.81 \times 0.0005}{9} \times 1000\right) = 735.750 \text{ mm/s}$$

$$= 0 + \left((0 - 0) \times \frac{9.81 \times 0.0005}{9} \times 1000\right) = 0 \text{ mm/s}$$

At t = 0.0010 s:

$$D(1, 0.0010) = D(1, 0.0005) + V(1, 0.0005)\Delta t = 1.50 + 2668.9 \times 0.0005 = 2.834 \text{ mm}$$

$$D(2, 0.0010) = D(2, 0.0005) + V(2, 0.0005)\Delta t = 0 + 735.75 \times 0.0005 = 0.368 \text{ mm}$$

$$D(3, 0.0010) = D(3, 0.0005) + V(3, 0.0005)\Delta t = 0 \text{ mm}$$

$$C(1, 0.0010) = D(1, 0.0010) - D(2, 0.0010) = 2.83 - 0.37 = 2.467 \text{ mm}$$

$$C(2, 0.0010) = D(2, 0.0010) - D(3, 0.0010) = 0.368 \text{ mm}$$

$$F(1, 0.0010) = K_1 \times C(1, 0.0010) = 900\ (2.467) = 2219.923 \text{ kN}$$

$$F(2, 0.0010) = K_2 \times C(2, 0.0010) = 450\ (0.368) = 165.544 \text{ kN}$$

$$F(3, 0.0010)_s = K' \times D(3, 0.0010) = 200\ (0) = 0 \text{ kN}$$

$$F(3, 0.0010)_d = F(3, 0.0010)_s \times (1 + J_s \times V(3, 0.0005)) = 0 \times (1 + 0.0002\ (0)) = 0 \text{ kN}$$

$$V(1, 0.0010) = V(1, 0.0005) + (0 - F(1, 0.0010))\frac{g\Delta t}{W_H}$$

$$= 2668.913 + \left((0 - 2219.923) \times \frac{9.81 \times 0.0005}{20} \times 1000\right) = 2124.476 \text{ mm/s}$$

$$V(2, 0.0010) = V(2, 0.0005) + (F(1, 0.0010) - F(2, 0.0010))\frac{g\Delta t}{W_1}$$

$$= 735.750 + \left((2219.923 - 165.544) \times \frac{9.81 \times 0.0005}{9} \times 1000\right) = 1855.387 \text{ mm/s}$$

$$V(3, 0.0010) = V(3, 0.0005) + (F(2, 0.0010) - F(3, 0.0010))\frac{g\Delta t}{W_2}$$

$$= 0 + \left((165.544 - 0) \times \frac{9.81 \times 0.0005}{9} \times 1000\right) = 90.221 \text{ mm/s}$$

At $t = 0.0015$ s:

$$D(1, 0.0015) = D(1, 0.0010) + V(1, 0.0010)\Delta t = 2.834 + 2124.476 \times 0.0005 = 3.897 \text{ mm}$$

$$D(2, 0.0015) = D(2, 0.0010) + V(2, 0.0010)\Delta t = 0.368 + 1855.387 \times 0.0005 = 1.296 \text{ mm}$$

$$D(3, 0.0015) = D(3, 0.0010) + V(3, 0.0010)\Delta t = 0 + 90.221 \times 0.0005 = 0.045 \text{ mm}$$

$$C(1, 0.0015) = D(1, 0.0015) - D(2, 0.0015) = 3.896 - 1.296 = 2.601 \text{ mm}$$

$$C(2, 0.0015) = D(2, 0.0015) - D(3, 0.0015) = 1.296 - 0.045 = 1.250 \text{ mm}$$

$$F(1, 0.0015) = K_1 \times C(1, 0.0015) = 900\ (2.601) = 2341.013 \text{ kN}$$

$$F(2, 0.0015) = K_2 \times C(2, 0.0015) = 450\ (1.251) = 562.706\ \text{kN}$$

$$F(3, 0.0015)_s = K' \times D(3, 0.0015) = 200\ (0.045) = 9.022\ \text{kN}$$

$$F(3, 0.0015)_d = F(3, 0.0015)_s \times (1 + J_s \times V(3, 0.0010))$$
$$= 9.022 \times (1 + 0.0002\ (90.221)) = 9.185\ \text{kN}$$

$$V(1, 0.0015) = V(1, 0.0010) + (0 - F(1, 0.0015))\frac{g\Delta t}{W_H}$$
$$= 2124.476 + \left((0 - 2341.013) \times \frac{9.81 \times 0.0005}{20} \times 1000\right) = 1550.343\ \text{mm/s}$$

$$V(2, 0.0015) = V(2, 0.0010) + (F(1, 0.0015) - F(2, 0.0015))\frac{g\Delta t}{W_1}$$
$$= 1855.387 + \left((2341.013 - 562.706) \times \frac{9.81 \times 0.0005}{9} \times 1000\right) = 2824.564\ \text{mm/s}$$

$$V(3, 0.0015) = V(3, 0.0010) + (F(2, 0.0015) - F(3, 0.0015))\frac{g\Delta t}{W_2}$$
$$= 90.221 + \left((562.706 - 9.185) \times \frac{9.81 \times 0.0005}{9} \times 1000\right) = 391.890\ \text{mm/s}$$

At $t = 0.0020$ s:

$$D(1, 0.0020) = D(1, 0.0015) + V(1, 0.0015)\Delta t = 3.897 + 1550.343 \times 0.0005 = 4.672\ \text{mm}$$

$$D(2, 0.0020) = D(2, 0.0015) + V(2, 0.0015)\Delta t = 1.296 + 2824.564 \times 0.0005 = 2.708\ \text{mm}$$

$$D(3, 0.0020) = D(3, 0.0015) + V(3, 0.0015)\Delta t = 0.045 + 391.890 \times 0.0005 = 0.241\ \text{mm}$$

$$C(1, 0.0020) = D(1, 0.0020) - D(2, 0.0020) = 4.672 - 2.708 = 1.964\ \text{mm}$$

$$C(2, 0.0020) = D(2, 0.0020) - D(3, 0.0020) = 2.708 - 0.241 = 2.467\ \text{mm}$$

$$F(1, 0.0020) = K_1 \times C(1, 0.0020) = 900\ (1.964) = 1767.614\ \text{kN}$$

$$F(2, 0.0020) = K_2 \times C(2, 0.0020) = 450\ (2.467) = 1110.058\ \text{kN}$$

$$F(3, 0.0020)_s = K' \times D(3, 0.0020) = 200\ (0.241) = 48.211\ \text{kN}$$

$$F(3, 0.0020)_d = F(3, 0.0020)_s \times (1 + J_s \times V(3, 0.0015))$$
$$= 48.211 \times (1 + 0.0002\ (391.890)) = 51.990\ \text{kN}$$

$$V(1, 0.0020) = V(1, 0.0015) + (0 - F(1, 0.0020))\frac{g\Delta t}{W_H}$$
$$= 1550.343 + \left((0 - 1767.614) \times \frac{9.81 \times 0.0005}{20} \times 1000\right) = 1116.836\ \text{mm/s}$$

$$V(2, 0.0020) = V(2, 0.0015) + (F(1, 0.0020) - F(2, 0.0020))\frac{g\Delta t}{W_1}$$
$$= 2824.564 + \left((1767.614 - 1110.058) \times \frac{9.81 \times 0.0005}{9} \times 1000\right) = 3182.932\ \text{mm/s}$$

$$V(3, 0.0020) = V(3, 0.0015) + (F(2, 0.0020) - F(3, 0.0020))\frac{g\Delta t}{W_2}$$
$$= 391.890 + \left((1110.058 - 51.990) \times \frac{9.81 \times 0.0005}{9} \times 1000\right) = 968.537\ \text{mm/s}$$

Problem 18.8

Develop the theoretical expression of the residual load in a driven pile for the following conditions. The initial condition is the stress and load distribution in the pile at failure. The ultimate skin friction is f_u and the ultimate point pressure is p_u. The ultimate load at the top of the pile is R_u and the ultimate load at the point is R_{pu}. The unloading of the friction and point transfer curve is assumed to obey the linear elastic model shown in Figure 18.1s.

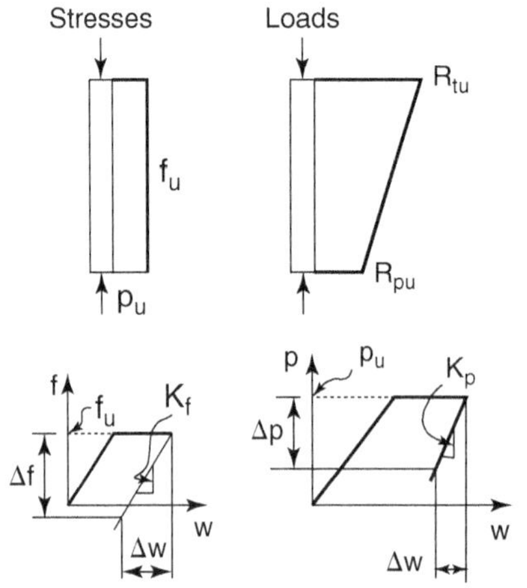

Figure 18.1s Initial conditions and models for residual loads.

Solution 18.8

Residual loads are loads that are locked in upon unloading after the pile has been brought to the ultimate soil resistance during driving or load testing. Therefore, the theoretical analysis takes, as an initial condition, the stress and load distribution in the pile at failure. The ultimate skin friction is f_u and the ultimate point resistance is p_u. The ultimate load at the pile head is R_{tu} and the ultimate load at the pile point is R_{pu}. The load anywhere in the pile is R_u. The unloading of the friction is assumed to obey the linear elastic model, which gives the following equation:

$$\Delta f = K_f \Delta w$$

in which Δf = decrease in pile-soil friction stress at depth z; K_f' = unloading stiffness in friction; and Δw = upward movement of the pile upon unloading at depth z. Similarly, the unloading of the point follows the equation:

$$\Delta p = K_p' \Delta w_p$$

in which Δp = decrease in point resistance; $K_p{}'$ = unloading stiffness for the point; and Δw_p = upward movement of the pile at the point upon unloading. The equilibrium equation of the elementary pile element can be written as follows:

$$-\frac{\partial \Delta\sigma}{\partial z} - \frac{P}{A}\Delta\tau = 0$$

in which $\Delta\sigma$ = normal stress decrease in the pile at depth z; A = cross-sectional area of the pile; and P = perimeter of the pile. The constitutive equation for the pile is:

$$\Delta\sigma = E_p \Delta\varepsilon = -E_p \frac{\partial \Delta w}{\partial z}$$

in which E_p = pile modulus of elasticity and $\Delta\varepsilon$ = change in normal strain at depth z due to stress change. The solution to the previous equation gives the residual load, R_r, in the pile at a depth z:

$$R_r = R_u - R_{tu}\left[\frac{\left(E_p\Omega + K_p'\right)e^{\Omega(L-z)} - (E_p\Omega - K_p')e^{-\Omega(L-z)}}{(E_p\Omega + K_p')e^{\Omega L} - (E_p\Omega - K_p')e^{-\Omega L}}\right]$$

in which L = length of the pile, z = depth at which R_r exists, and $\Omega = \sqrt{K_f' P / E_p A}$.

The residual point load, R_{pr}, is:

$$R_{pr} = R_{pu} - \frac{2R_{tu}}{\left(1 + \frac{E_p\Omega}{K'_p}\right)e^{\Omega L} + \left(1 - \frac{E_p\Omega}{K'_p}\right)e^{-\Omega L}}$$

Problem 18.9

Explain Figures 18.21 and 18.22 in your own words.

Solution 18.9

Imagine a pile suspended horizontally from the ceiling and hit at one end (Figure 18.21). There is no soil surrounding it and the end of the pile is free. In this case the compression force in the pile will be proportional to the particle velocity ($F = Iv$, from Eq. 18.23). Now the wave is racing along the pile at the wave speed c. When it gets to the end of the pile, the compression force F finds no resistance and reflects as a tension force, but the magnitude of the particle velocity doubles while the wave speed is unchanged (see Eqs. 18.24 to 18.26). This is similar to a case of easy driving with very little point resistance.

Now let's say that the pile is still suspended from the ceiling, but the other end is against a strong wall (Figure 18.22) and the pile is hit at one end. When the compression wave gets to the wall, it cannot displace it. As a result, the compression force doubles, the velocity vanishes, and the F and Iv signals are as shown in Figure 18.22. Equations 18.24 to 18.26 give the mathematical reason. This approximates hard driving into a strong bearing layer.

Problem 18.10

A pile is driven and the force (F) and particle velocity time impedance ($v \times EA/c$) at the top of the pile are measured as shown in Figure 18.2s. Calculate the dynamic resistance of the pile using the observations at times t_1 and t_2 separated by the down and up travel time of the wave ($2L/c$) and the Case method.

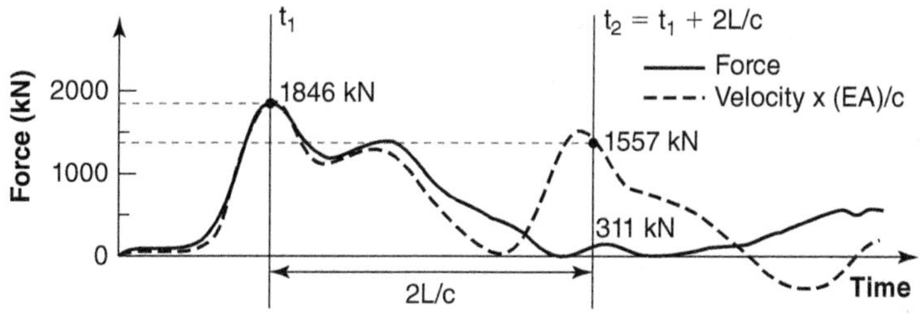

Figure 18.2s Force and velocity signal.

Solution 18.10

From the plot shown in Figure 18.2s:

$$t_1 = t_1$$
$$t_2 = t_1 + 2L/c$$
$$F_{(t_1)} = 1846 \text{ kN}$$
$$F_{(t_1+2L/c)} = 311 \text{ kN}$$
$$V_{(t_1)} \times \frac{EA}{c} = 1846 \text{ kN}$$
$$V_{(t_1+2L/c)} \times \frac{EA}{c} = 1557 \text{ kN}$$

The dynamic resistance of the pile can be computed as:

$$R_D = \frac{1}{2}(F_{(t_1)} + F_{(t_1+2L/c)}) + I(v_{(t_1)} - v_{(t_1+2L/c)})$$
$$R_D = \frac{1}{2}(F_{(t_1)} + F_{(t_1+2L/c)}) + \frac{EA}{c}(v_{(t_1)} - v_{(t_1+2L/c)})$$

$$R_D = \frac{1}{2}\left(F_{(t_1)} + F_{(t_1+2L/c)} + \frac{EA}{c}v_{(t_1)} - \frac{EA}{c}v_{(t_1+2L/c)}\right)$$

$$R_D = \frac{1}{2}(1846 + 311 + 1846 - 1557) = 1223 \text{ kN}$$

Problem 18.11

Calculate the static resistance of the pile in problem 18.10 by the Case method.

Solution 18.11

The static capacity of the pile can be computed as:

$$R_S = R_D - J_c I v$$

$$R_S = R_D - J_c I\left(\frac{F_{(t_1)} + Iv_{(t_1)} - R_D}{I}\right)$$

$$R_S = R_D - J_c(F_{(t_1)} + Iv_{(t_1)} - R_D)$$

$$R_S = R_D - J_c\left(F_{(t_1)} + \frac{EA}{c}v_{(t_1)} - R_D\right)$$

$$R_S = 1223 - 0.3(1846 + 1846 - 1223) = 482 \text{ kN}$$

Problem 18.12

Is it possible to break a concrete pile in tension by driving it in the ground? If yes, explain how.

Solution 18.12

Yes, it is possible to break a concrete pile in tension if it is not reinforced properly. Driving through a hard layer into a soft layer will generate tension in the pile when the pile point has very little resistance. The compression wave coming down the pile will turn back into a tension wave. The tension stresses created by the tension wave can be large enough to break the pile in tension. This condition should be considered during the design of the pile.

Problem 18.13

A suction caisson is 20 m long and 2 m by 2 m in cross section with a wall thickness of 20 mm. It is made of steel and is to be installed in a soft clay with an undrained shear strength of 20 kPa and an effective unit weight of 10 kN/m^3. Calculate the ultimate capacity of the caisson and the underpressure required to install it to full penetration. Check that this underpressure does not create inverse bearing capacity failure.

Solution 18.13

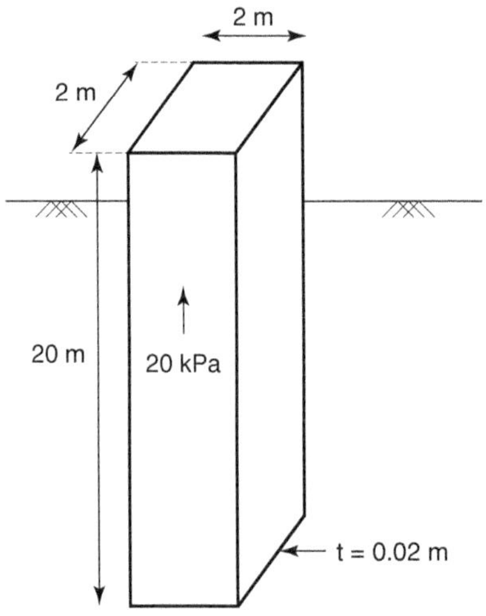

Figure 18.9s Suction caisson.

$$R_u = 4 \times 2 \times 20 \times 20 \times 2 + 4 \times 2 \times 0.02 \times (9 \times 20 + 10 \times 20) = 6460.8\ \text{kN}$$

$$W = 4 \times 2 \times 0.02 \times 20 \times 75 = 240\ \text{kN}$$

$$W' = 240 - 4 \times 2 \times 0.02 \times 20 \times 10 = 208\ \text{kN}$$

$$A_{in} = (2 - 0.04)^2 = 3.84\ \text{m}^2$$

$$\Delta u_{rq} = \frac{6460.8 - 208}{3.84} = 1628.33\ \text{kPa}$$

$$\Delta u_{crit} = N_c s_u + \frac{\alpha s_u A_{wall}}{A_{in}} = 9 \times 20 + \frac{1 \times 20 \times 8 \times 20}{3.84} = 1013.3\ \text{kPa}$$

Therefore it is unlikely that the suction caisson can be installed to the required depth of 20 m without some risk of inward bearing capacity failure

Problem 18.14

Calculate the strain, the stress, and the friction on each segment of the pile shown in Figure 18.29 for the maximum load applied. How would you measure such values? The pile is a bored pile 1 m in diameter and made of concrete.Solution 18.14

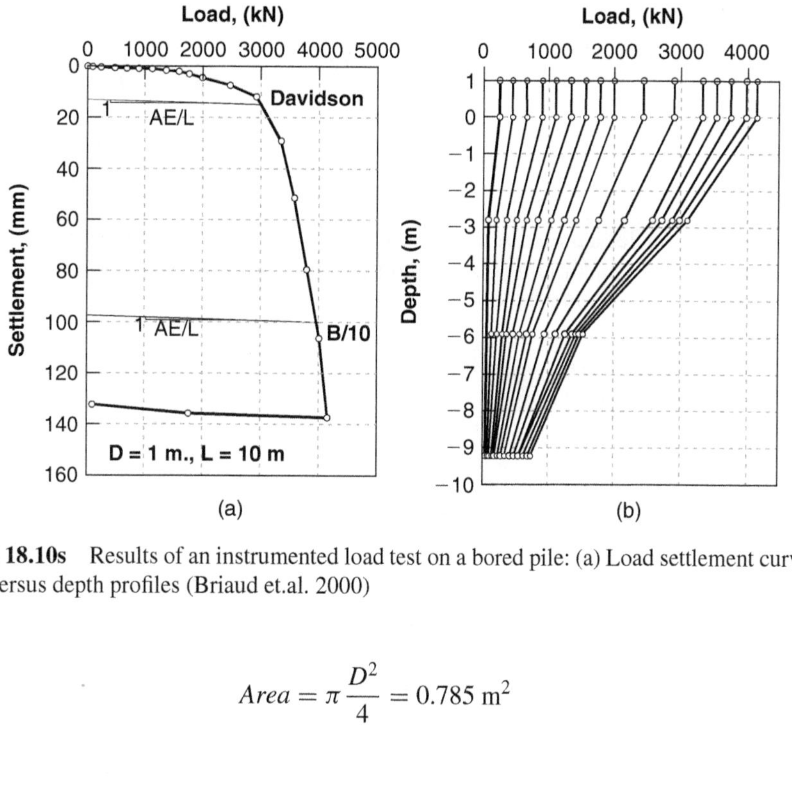

Figure 18.10s Results of an instrumented load test on a bored pile: (a) Load settlement curve. (b) Load versus depth profiles (Briaud et.al. 2000)

$$Area = \pi \frac{D^2}{4} = 0.785\ \text{m}^2$$

- Top load = 4200 kN
- Depth = 0 to 2.8 m, Average load = 3700 kN

$$\sigma_1 = \frac{F}{A} = \frac{3700}{0.785} = 4713 \left(\frac{\text{kN}}{\text{m}^2}\right)$$

$$\varepsilon_1 = \frac{\sigma}{E_{conc}} = \frac{4713}{2.5 \times 10^7} = 0.189 \times 10^{-3}$$

$$f_{u1} = \frac{F_{top} - F_{bot}}{P \times \Delta L} = \frac{4200 - 3200}{3.14 \times 1 \times 2.8} = 114 \left(\frac{\text{kN}}{\text{m}^2}\right)$$

- Depth = 2.8 to 5.8 m, Average load = 2400 kN

$$\sigma_2 = \frac{F}{A} = \frac{2400}{0.785} = 3057\left(\frac{\text{kN}}{\text{m}^2}\right)$$

$$\varepsilon_2 = \frac{\sigma}{E_{conc}} = \frac{3057}{2.5 \times 10^7} = 0.122 \times 10^{-3}$$

$$f_{u1} = \frac{F_{top} - F_{top}}{P \times \Delta L} = \frac{3200 - 1600}{3.14 \times 1 \times 3} = 167\left(\frac{\text{kN}}{\text{m}^2}\right)$$

- Depth = 5.8 to 9.2 m, Average load = 1175 kN

$$\sigma_3 = \frac{F}{A} = \frac{1175}{0.785} = 1497\left(\frac{\text{kN}}{\text{m}^2}\right)$$

$$\varepsilon_3 = \frac{\sigma}{E_{conc}} = \frac{1497}{2.5 \times 10^7} = 0.060 \times 10^{-3}$$

$$f_{u1} = \frac{F_{top} - F_{top}}{P \times \Delta L} = \frac{1600 - 750}{3.14 \times 1 \times 3} = 79.6\left(\frac{\text{kN}}{\text{m}^2}\right)$$

- Point load = 750 kN. Point pressure:

$$p_u = \frac{F_{point}}{\pi D^2/4} = \frac{750}{3.14 \times 1^2/4} = 955\left(\frac{\text{kN}}{\text{m}^2}\right)$$

Problem 18.15

Calculate the slope of the load transfer curves for the axially loaded pile discussed in section 18.4.3 and verify the value of the movement necessary to reach the maximum friction and point resistance shown in Figure 18.43. Generate a spreadsheet to develop the complete load settlement curve of this axially loaded pile.

Solution 18.15

The slope of the point load transfer curve is:

$$p_{point} = \frac{4E_s}{\pi D(1 - \nu^2)} s_{point}$$

For the pile of Figure 18.43, $D = 0.3$ m, $E_s = 100{,}000$ kPa, and $\nu = 0.35$, so the equation becomes:

$$p_{point} = \frac{4 \times 100{,}000}{\pi \times 0.3(1 - 0.35^2)} s_{point} = 483910 \times s_{point}$$

Because the ultimate point load is 706 kN, the ultimate pressure p_u is:

$$p_u = \frac{706}{\pi \times 0.3^2/4} = 9993 \text{ kPa}$$

Then the displacement necessary to mobilize p_u will be:

$$s_{point} = \frac{p_u}{483910} = 21 \text{ mm}$$

The slope of the friction load transfer curve is:

$$f = \frac{E_s}{(1 + \nu)(1 + Ln(L/D))D} s_{friction}$$

For the pile of Figure 18.43, $D = 0.3$ m, L = 10 m, $E_s = 100{,}000$ kPa for the dense sand layer, and $\nu = 0.35$, so the equation becomes:

$$f = \frac{100,000}{(1+0.35)(1+Ln(10/0.3))0.3} s_{friction} = 54790 \times s_{friction}$$

Because the ultimate friction load over the 1 m of pile in the dense sand layer is 75.4 kN, the ultimate friction f_u is:

$$f_u = \frac{75.4}{\pi \times 0.3 \times 1} = 80 \text{ kPa}$$

Then the displacement necessary to mobilize f_u will be:

$$s_{friction} = \frac{80}{54790} = 1.5 \text{ mm}$$

The same calculations apply to the two locations in the soft clay:

$$f = \frac{5000}{(1+0.35)(1+Ln(10/0.3))0.3} s_{friction} = 2739 \times s_{friction}$$

$$f_u = \frac{84.8}{\pi \times 0.3 \times 4.5} = 20 \text{ kPa}$$

$$s_{friction} = \frac{20}{2739} = 7.3 \text{ mm}$$

The results from an Excel spreadsheet for the load settlement curve are shown in Table 18.2s.

Table 18.2s Load Settlement Curve Results

Bottom Settlement (mm)	Top Settlement (mm)	Q_{top} (kN)
0.000	0.000	0.000
1.000	2.013	115.647
2.000	3.696	196.891
3.000	5.179	257.389
4.000	6.662	317.888
5.000	8.145	378.387
8.000	12.594	559.883
10.000	15.560	680.880
12.000	18.526	801.877
15.000	22.976	983.373
18.000	27.425	1164.869
20.000	30.391	1285.867
21.000	31.744	1333.455
25.000	36.191	1427.693
28.000	39.461	1471.174
30.000	41.628	1494.412
35.000	46.642	1496.362
38.000	49.642	1496.362
40.000	51.642	1496.362

The load settlement curve is shown in Figure 18.11s. You can note a slight bend in the load settlement curve when the friction is completely mobilized, at around 270 kN. This is often observed on the load settlement curve in load tests.

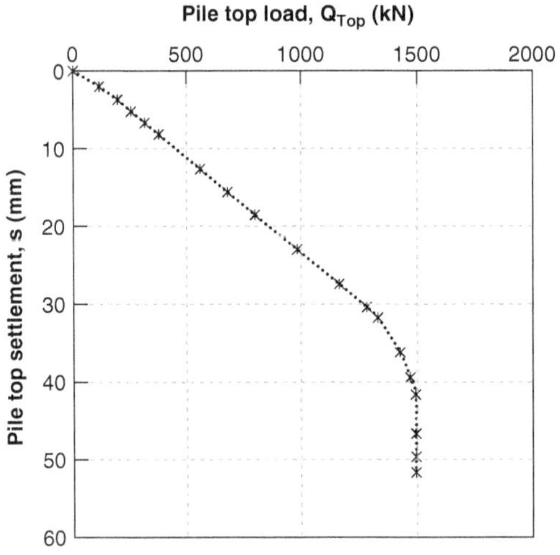

Figure 18.11s Load settlement curve.

Problem 18.16

Calculate the ultimate static capacity of the pile subjected to the Statnamic test. The pile is a 1 m diameter, 20 m long bored concrete pile. The test results are summarized in Figure 18.3s.

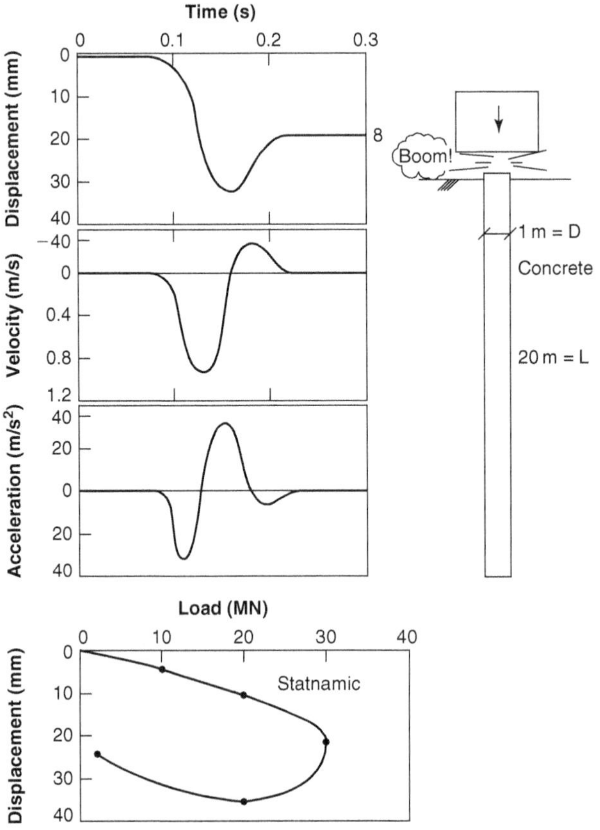

Figure 18.3s Statnamic test results.

Solution 18.16

$$F_{su}(t) = F_{stn}(t) - Ma(t)$$

From Figure 18.3s, the largest displacement during the Statnamic test is 35 mm occurring at time t equal to 0.16 s. At that point the velocity is zero, the acceleration is 33 m/s^2, and the force is:

$$F_{stn}(t) = 20 \text{ MN}$$
$$a(t) = 33 \text{ m/s}^2$$
$$M = \rho_{conc} V = 2450 \times 20 \times \pi \times (0.5)^2 = 38465 \text{ kg}$$
$$F_{su}(t) = 20000000 - 38465 \times 33 = 18.73 \text{ MN}$$

Problem 18.17

Calculate the ultimate capacity of the two piles shown in Figure 18.4s by all possible methods. The pile in clay is a circular bored pile and the pile in sand is a square driven pile. At what depth along the pile would you place the Osterberg load cell to balance the load on both sides of the pile at ultimate load?

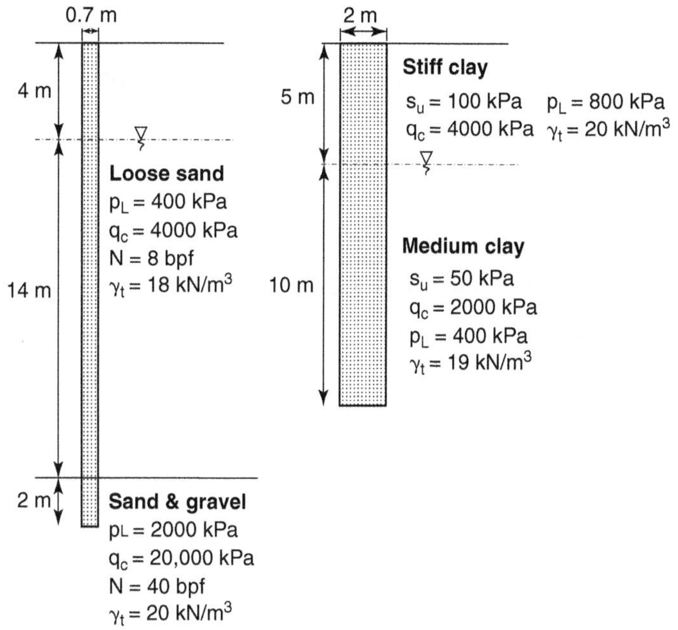

Figure 18.4s Ultimate capacity of two piles.

Solution 18.17

Case #1. Driven square concrete pile in sandy soil

Pile tip area: 0.7 m × 0.7 m
Layer 1: Loose sand / Layer 2: Sand & gravel

1. Briaud-Tucker SPT method:
   a. Calculate f_u:

$$f_{u1} = f_{u2} = 5(N)^{0.7} = 5(8)^{0.7} = 21.44 \text{ kPa}$$
$$f_{u3} = 5(N)^{0.7} = 5(40)^{0.7} = 66.1 \text{ kPa}$$

b. Calculate p_u :

$$p_u = 1000(N)^{0.5} = 1000(40)^{0.5} = 6324.6 \text{ kPa}$$

c. Ultimate bearing capacity calculation:

$$\begin{aligned} Q_u &= f_{u1}A_{s1} + f_{u2}A_{s2} + f_{u3}A_{s3} + p_{u2}A_p \\ &= (21.44)(4 \times 0.7 \times 4) + (21.44)(4 \times 0.7 \times 14) + (66.1)(4 \times 0.7 \times 2) + (6324.6)(0.7^2) \\ &= 4548 \text{ (kN)} \end{aligned}$$

2. LPC-PMT method:
   a. Classify the soil (Table 18.8):
      Layer 1, 2: Sand: loose
      Layer 3: Sand and gravel: dense to very dense
   b. Pile type being used (Table 18.9): Driven concrete pile, Curve Q2, $\alpha = 1.4$ and $f_{\text{lim}} = 130$ kPa for layers 1, 2, and 3.
   c. Calculations of f_u :
      Layer 1, 2 (Q2, $p_L = 400$ kPa): $f_u = \alpha f_{soil} = 1.4 \times 25 = 35$ kPa
      Layer 3 (Q2, $p_L = 2000$ kPa): $f_u = \alpha f_{soil} = 1.4 \times 74 = 103.6$ kPa
   d. The value of p_u:
      k_p value (Table 18.10): 3.1

$$p_u = k_p p_L = 3.1 \times 2000 = 6200 \text{ kPa}$$

   e. Ultimate bearing capacity calculation:

$$\begin{aligned} Q_u &= f_{u1}A_{s1} + f_{u2}A_{s2} + f_{u3}A_{s3} + p_{u2}A_p \\ &= (35)(4 \times 0.7 \times 4) + (35)(4 \times 0.7 \times 14) + (103.6)(4 \times 0.7 \times 2) + (6200)(0.7^2) \\ &= 5384 \text{ (kN)} \end{aligned}$$

3. LPC-CPT method:
   a. Classify the soil (Table 18.8):
      Layer 1, 2: Sand loose
      Layer 3: Sand and gravel dense to very dense
   b. Pile type being used (Table 18.11): Driven concrete, Curve Q2, $\alpha = 1.0$, and $f_{\text{lim}} = 130$ kPa for layers 1, 2, and 3.
   c. Calculations of f_u:
      Layer 1, 2 (Q3, $q_c = 4000$ kPa): $f_u = \alpha f_{soil} = 1.0 \times 47 = 47$ kPa Layer 3 (Q3, $q_c = 20000$ kPa): $f_u = \alpha f_{soil} = 1.0 \times 118 = 118$ kPa
   d. Calculate P_u:
      k_c value: 0.4 (Table 18.12)

$$p_u = k_c q_c = 0.4 \times 20000 = 8000 \text{ kPa}$$

   e. Ultimate bearing capacity calculation:

$$\begin{aligned} Q_u &= f_{u1}A_{s1} + f_{u2}A_{s2} + f_{u3}A_{s3} + p_{u2}A_p \\ &= (47)(4 \times 0.7 \times 4) + (47)(4 \times 0.7 \times 14) + (118)(4 \times 0.7 \times 2) + (8000)(0.7^2) \\ &= 6950 \text{ (kN)} \end{aligned}$$

4. API-RP2A method for driven piles in coarse-grained soils:

	σ'_{0v} (kPa)	k	δ	N_q	f_{max} (kPa)	f_u (kPa)	P_u (kPa)
Layer 1	36.0	0.8	15	8	48	7.7	
Layer 2	128.0	0.8	15	8	48	27.4	
Layer 3	194.0	0.8	25	20	81	72.4	
	204.0						4080.0

Ultimate bearing capacity calculation:

$$\begin{aligned} Q_u &= f_{u1}A_{s1} + f_{u2}A_{s2} + f_{u3}A_{s3} + p_{u2}A_p \\ &= (7.7)(4 \times 0.7 \times 4) + (27.4)(4 \times 0.7 \times 14) + (72.4)(4 \times 0.7 \times 2) + (4080)(0.7^2) \\ &= 3565.0 \text{ (kN)} \end{aligned}$$

Comparison of bearing capacity:

Estimation method	f_{u1} (kPa)	f_{u2} (kPa)	Q_{fu12} (kN)	f_{u3} (kPa)	Q_{fu3} (kN)	p_u (kPa)	Q_{pu} (kN)	Q_u (kN)
LPC-PMT	35	35	1764	104	582	6200	3038	5384
LPC-CPT	47	47	2369	118	661	8000	3920	6950
Briaud-Tucker	21.4	21.4	1079	66.1	370	6325	3099	4548
API-RP2A	7.7	27.4	1160	72.4	405	4080.0	1999	3564
Average			1593		505		3014	5112

The average ultimate capacity is 5112kN. Because the average friction capacity is 1593 + 505 = 2098 kN and the point capacity is 3014 kN, we have to place the O-cell at the bottom of the pile and will be limited to pushing the point capacity to the extent of the friction capacity. Nevertheless, we will be able to push the pile to 82% of its total capacity (2 × 2098/5112 = 0.82).

Case #2. Bored concrete pile in clayey soil

Pile diameter, $d = 2$ m
Bored by dry method
Layer 1: Stiff clay / Layer 2: Medium clay

1. LPC-PMT method:
   a. Classify the soil (Table 18.8):
      Layer 1: Clay firm
      Layer 2: Clay soft to firm
   b. Pile type being used (Table 18.9): Bored pile in the dry, Curve Q1, $\alpha = 1.1, f_{lim} = 90$ kPa
   c. Calculation of f_u:
      Layer 1 (Q1, $p_L = 800$ kPa): $f_u = \alpha f_{soil} = 1.1 \times 40 = 44$ kPa
      Layer 2 (Q1, $p_L = 400$ kPa): $f_u = \alpha f_{soil} = 1.1 \times 32 = 35.2$ kPa
   d. The value of P_u:
      k_p value (Table 18.10): 1.1

$$p_u = k_p p_L = 1.15 \times 400 = 460 \text{ kPa}$$

e. Ultimate bearing capacity calculation:

$$Q_u = f_{u1}A_{s1} + f_{u2}A_{s2} + p_{u2}A_p = (44)(\pi \times 2 \times 5) + (35.2)(\pi \times 2 \times 10) + 460\left(\frac{\pi}{4} \times 2^2\right)$$
$$= 5036 \text{ (kN)}$$

2. LPC-CPT method:
   a. Classify the soil (Table 18.8):
      Layer 1: Clay firm
      Layer 2: Clay soft to firm
   b. Pile type being used (Table 18.11): Bored pile in the dry, Curve Q1, $\alpha = 0.55, f_{lim} = 90$ kPa.
   c. Calculation of f_u:
      Layer 1 (Q1, $q_c = 4000$ kPa): $f_u = \alpha f_{soil} = 0.55 \times 87 = 47.8$ kPa
      Layer 2 (Q1, $q_c = 2000$ kPa): : $f_u = \alpha f_{soil} = 0.55 \times 55 = 30.2$ kPa
   d. Calculate P_u:
      k_c value: 0.4 (Table 18.12)

$$p_u = k_c q_c = 0.4 \times 2000 = 800 \text{ kPa}$$

   e. Ultimate bearing capacity calculation:

$$Q_u = f_{u1}A_{s1} + f_{u2}A_{s2} + p_{u2}A_p = (47.8)(\pi \times 2 \times 5) + (30.2)(\pi \times 2 \times 10) + 800\left(\frac{\pi}{4} \times 2^2\right)$$
$$= 5910 \text{ (kN)}$$

3. FHWA method for bored piles in fine-grained soils:
   a. Calculate f_u and p_u:
      Layer 1: $f_u = 0.55s_u = 0.55 \times 100 = 55$ kPa
      Layer 2: $f_u = 0.55s_u = 0.55 \times 50 = 27.5$ kPa

$$p_u = N_c s_u = 9 \times 50 = 450 \text{ kPa}$$

   (According to bearing capacity factors for clays, after Skempton)
   b. Ultimate bearing capacity calculation:

$$Q_u = f_{u1}A_{s1} + f_{u2}A_{s2} + p_{u2}A_p = (55)(\pi \times 2 \times 5) + (27.5)(\pi \times 2 \times 10) + 450\left(\frac{\pi}{4} \times 2^2\right)$$
$$= 4869.5 \text{ (kN)}$$

   c. Comparison of bearing capacity:

Estimation method	f_{u1} (kPa)	Q_{fu1} (kN)	f_{u2} (kPa)	Q_{fu2} (kN)	p_u (kPa)	Q_{pu} (kN)	Q_u (kN)
LPC-PMT	44	1381	35.2	2211	460	1444	5036
LPC-CPT	47.8	1501	30.2	1897	800	2512	5910
FHWA	55	1727	27.5	1727	450	1413	4867
Average		1536	30.9	1945		1789	5271

The average ultimate capacity is 5271 kN. The average friction capacity is 1536 + 1945 = 3481 kN and the average point capacity is 1789 kN. Therefore, because we have more friction capacity than point capacity, we can place the O-cell along the pile shaft to balance the loads. The position should be such that the point capacity plus some friction is equal to one-half of the total capacity:

$$1769 + 30.9 \times \pi \times 2 \times L_o = \frac{5271}{2} \quad or \quad L_o = 4.52\text{m}$$

Thus, we should place the O-cell 4.52 meters above the pile point to optimize our chances of reaching failure above and below the O-cell at the same time, thereby testing the pile to its full capacity.

Problem 18.18

For the pile of Figure 18.43, find the top movement and the load distribution in the pile for a point movement of 5 mm.

Solution 18.18

The point load for a point movement of 5 mm is is (Eq. 18.100):

$$Q_{point} = \left(\frac{0.3 \times 10^5}{1 - 0.35^2}\right) \times 0.005 = 170.9 \text{ KN}$$

The friction mobilized in element 1 (from Eq. 18.102) is:

$$f_1 = \frac{10^5}{(1 + 0.35)(1 + \ln(16.3/0.3)) \times 0.3} \times 0.005 = 247.2 \text{ KPa} > 80 \text{ KPa}$$

$$\rightarrow f_1 = 80 \text{ KPa}$$

The load carried in element 1 is:

$$Q_{f1} = f_1 \pi D \Delta L_1 = 80 \times \pi \times 0.3 \times 4 = 301.4 \text{ KN}$$

The movement at the bottom of element 2 is:

$$s_2 = s_p + \frac{\left(Q_p + \frac{Q_{f1}}{2}\right)\Delta L_1}{A_{cs}E_p} = 0.005 + \frac{\left(170.9 + \frac{301.4}{2}\right)4}{\pi \times 0.3 \times 0.005 \times 2 \times 10^8} = 0.00637 \text{ m} = 6.37 \text{ mm}$$

The friction mobilized in element 2 is:

$$f_2 = \frac{5000}{(1 + 0.35)(1 + \ln(16.3/0.3)) \times 0.3} \times 0.00637 = 4.72 \text{ KPa} < 20 \text{ KPa}$$

The load carried in element 2 is:

$$Q_{f2} = f_2 \pi D \Delta L_2 = 4.72 \times \pi \times 0.3 \times 7.3 = 32.46 \text{ KN}$$

The movement at the bottom of element 3 is:

$$s_3 = s_2 + \frac{\left(Q_p + Q_{f1} + \frac{Q_{f2}}{2}\right)\Delta L_2}{A_{cs}E_p} = 0.00637 + \frac{\left(170.9 + 301.4 + \frac{32.46}{2}\right)7.3}{\pi \times 0.3 \times 0.005 \times 2 \times 10^8} = 0.01016 \text{ m} = 10.16 \text{ mm}$$

The friction mobilized in element 3 is:

$$f_3 = \frac{5000}{(1 + 0.35)(1 + \ln(16.3/0.3)) \times 0.3} \times 0.01016 = 7.53 \text{ KPa} < 20 \text{ KPa}$$

The load carried in element 3 is:

$$Q_{f3} = f_3 \pi D \Delta L_3 = 7.53 \times \pi \times 0.3 \times 7 = 49.65 \text{ KN}$$

The top movement is:

$$s_t = s_3 + \frac{\left(Q_p + Q_{f1} + Q_{f2} + \frac{Q_{f3}}{2}\right)\Delta L_3}{A_{cs}E_p}$$

$$= 0.01016 + \frac{\left(170.9 + 301.4 + 32.46 + \frac{49.65}{2}\right)7}{\pi \times 0.3 \times 0.005 \times 2 \times 10^8} = 0.0141 \text{ m} = 14.1 \text{ mm}$$

The load at the top of the pile is:

$$Q_{top} = Q_p + Q_{f1} + Q_{f2} + Q_{f3} = 170.9 + 301.4 + 32.46 + 49.65 = 554.41 \text{ KN}$$

Problem 18.19

A16-story hospital weighs 1500 MN, and its imprint is 75 m by 75 m. The building rests on 10,000 timber piles, each 15 m long, 0.3 m in average diameter, and driven with a spacing of 0.75 m center to center. The soil is made of a clay layer down to 14.5 m ($s_u = 20$ kN/m^2, $e_o = 0.8$, and $C_c = 0.1$), then a sand layer down to 16.5 m ($N = 30$ bpf), and then clay again down to a depth of 100 m ($s_u = 30$ kN/m^2, $e_o = 0.7$, and $C_c = 0.06$). The water table is at the ground surface and the total unit weight of all soils is 20 kN/m^3. Calculate:

a. The capacity of one timber pile
b. The capacity of the pile group
c. The settlement of the hospital
d. Comment on this design.

Solution 18.19

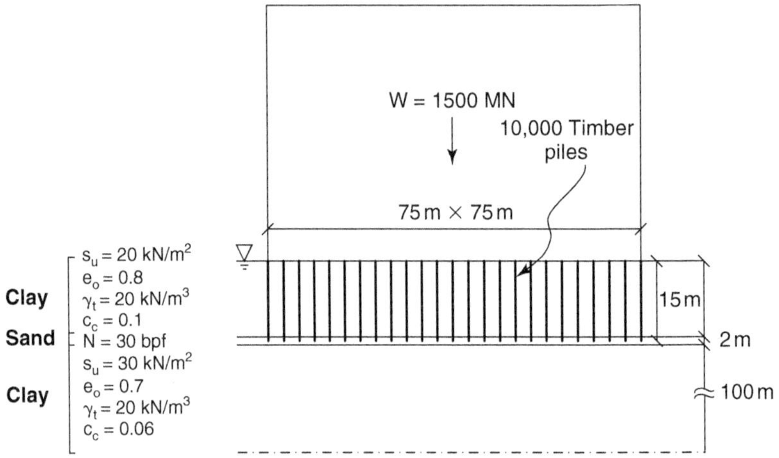

Figure 18.12s Building geometry and soil profile.

a. Bearing capacity of single pile

Side resistance with API-RP2A method

$$if \sigma'_{ov} = s_u \rightarrow z = \frac{20}{(18 - 9.8)} \approx 2.5 \text{ m}$$

$$f_u = \begin{cases} 0.5\left(\dfrac{s_u}{\sigma'_{ov}}\right)^{-0.5} s_u \; for \text{ z} \leq 2.5 \text{ m} \\ 0.5\left(\dfrac{s_u}{\sigma'_{ov}}\right)^{-0.25} s_u \; for \text{ 2.5m} < \text{z} < 14.5 \text{ m} \end{cases}$$

$$\rightarrow f_u = \begin{cases} 0.5\left(\dfrac{20}{(18 - 9.8) \times 1.25}\right)^{-0.5} 20 = 7.2 \text{ kPa} & \text{z} \leq 2.5 \text{ m} \\ 0.5\left(\dfrac{20}{(18 - 9.8) \times 8.5}\right)^{-0.25} 20 = 13.7 \text{ kPa} & 2.5\text{m} < \text{z} < 14.5 \text{ m} \end{cases}$$

$$R_{uf} = (7.2 \times 2.5 + 13.7 \times 12) \times 0.3\pi = 171.8 \text{ kN}$$

Point bearing capacity with Briaud-Tucker method:

$$p_u = 1000 \times (30)^{0.5} = 5477 \text{ kPa}$$

$$R_{up} = 5477 \times \frac{0.3^2\pi}{4} = 387 \text{ kN}$$

The capacity of one timber pile:

$$R_u = 171.8 + 387 = 558.8 \text{ kN}$$

b. Bearing capacity considering group effect:
Bearing capacity of 10,000 single piles

$$R_{u-group} = 558.8 \times 10000 = 5588 \times 10^3 \text{ kN} = 5588 \text{ MN}$$

Bearing capacity considering block failure:

$$R_{u-block} = 2(75 + 75) \times 14.5 \times 20 + 75 \times 75 \times (\text{N}_c \times 30)$$

The relative depth of embedment is $D/B = 15/75 = 0.2$

Skempton chart for D/B = 0.2 gives $N_c = 6.6$

$$R_{u-block} = 2(75 + 75) \times 14.5 \times 20 + 75 \times 75 \times (\text{N}_c \times 30)$$

$$R_{ug} = \text{Min } (5588, 1201) = 1201 \text{ MN}$$

c. Settlement calculations (Figure 18.13s)

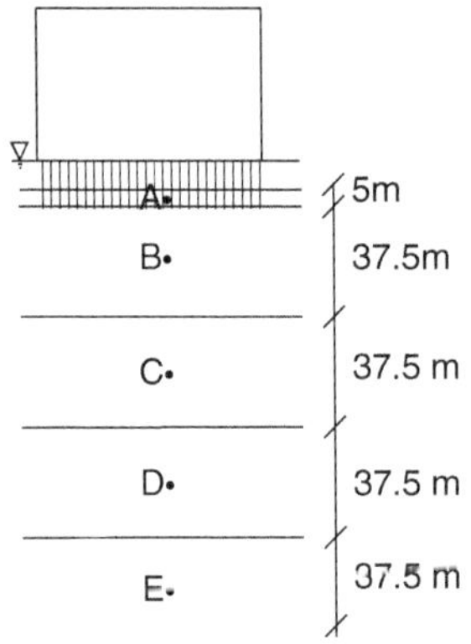

Figure 18.13s Settlement calculation points.

The average pressure under the building is:

$$p = \frac{F}{A} = \frac{1500000}{75 \times 75} = 267 \text{ kPa}$$

For settlement calculations, the large pile group is considered to be equivalent to a large footing located at 2/3 of the pile depth or 10 m below the ground surface. The depth of influence is considered to be 2 times the width of the foundation or 150 m. The thin sand layer is neglected; the layers involved are a 5 m layer of the upper clay and four layers of the lower clay each 37.5 m thick. The pressure factors are obtained from the bulb of pressure for a square foundation (see Figure 17.31). The settlement equation is:

$$\Delta H = \sum \frac{H_{oi}}{1+e_o} C_c \log\left(\frac{\sigma'_{ovi} + \Delta\sigma_{vi}}{\sigma'_{ovi}}\right) = \frac{C_c}{1+e_o} \sum H_{oi} \log\left(\frac{\sigma'_{ovi} + \Delta\sigma_{vi}}{\sigma'_{ovi}}\right)$$

Point	Depth (m)	H_o (m)	e_0	C_c	σ'_{ov} (kPa)	Pressure factor	$\Delta\sigma_v$ (kPa)	σ'_v (kPa)	ΔH (m)
A	12.5	5	0.8	0.1	125	1	267	392	0.138
B	33.75	37.5	0.7	0.06	337.5	0.8	213.6	551.1	0.282
C	71.25	37.5	0.7	0.06	712.5	0.45	120.1	832.6	0.090
D	108.75	37.5	0.7	0.06	1087.5	0.23	61.4	1148.9	0.032
E	146.25	37.5	0.7	0.06	1462.5	0.14	37.4	1499.9	0.015
									0.557

d. Comment

As can be seen, the ultimate load of the pile group is smaller than the weight of the building; therefore, the foundation is insufficient to carry the building weight safely. The settlement is also a great concern, because more than half a meter of settlement could lead to serious problems. This problem is actually close to the case of the New Orleans Charity Hospital built in 1939.

Problem 18.20

Calculate the group efficiency for settlement using Poulos interaction factors for the case of a flexible pile cap (all piles carry the same load). The group is 4 by 4 with a 3-pile diameter center-to-center spacing.

Solution 18.20

Because we have symmetry in piles location, we calculate the interaction factor only for three piles (a, b, and c in Figure 18.14s).

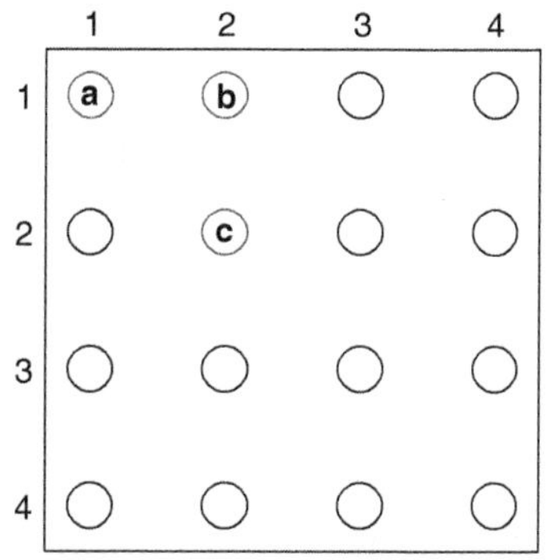

Figure 18.14s Piles location.

$$\rho_k = \rho_1 \sum_{\substack{j=1 \\ j \neq k}}^{n} (P_j \alpha_{kj}) + \rho_1 P_k$$

Flexible pile cap $\rightarrow P_i = P$

$$\rho_i = P \rho_1 \left(\sum_{\substack{j=1 \\ j \neq k}}^{n} (\alpha_{kj}) + 1 \right)$$

$$\alpha_{kj} = f\left(\frac{s}{d}\right) \rightarrow \text{Fig. 18.50}$$

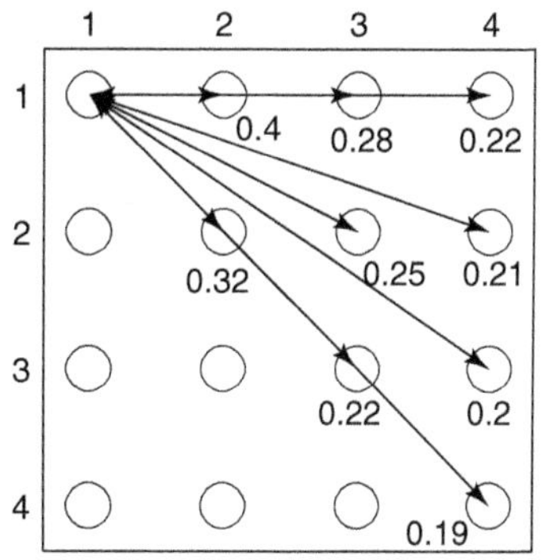

Figure 18.15s α values for pile a.

$$\rho_a = P\rho_1[2(0.4 + 0.28 + 0.22 + 0.25 + 0.21 + 0.2) + 0.32 + 0.22 + 0.19] = 3.85P\rho_1$$

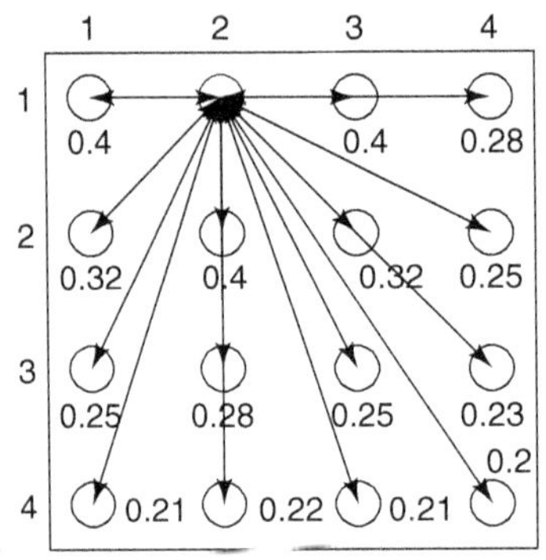

Figure 18.16s α values for pile b.

$$\rho_b = P\rho_1[3 \times 0.4 + 2 \times 0.32 + 2 \times 0.28 + 3 \times 0.25 + 0.23 + 2 \times 0.21 + 0.22 + 0.2] = 4.2P\rho_1$$

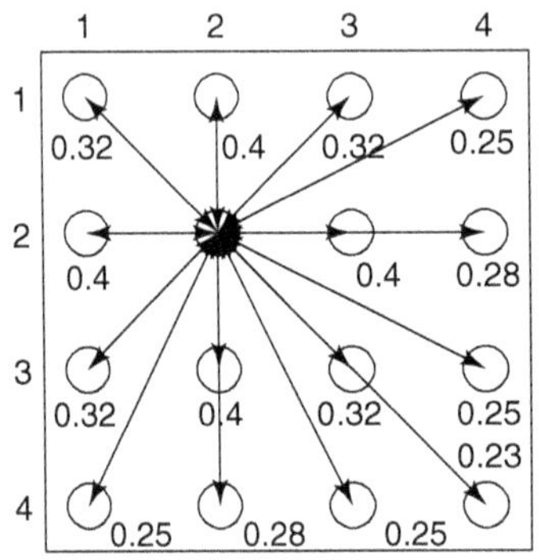

Figure 18.17s α values for pile c.

$$\rho_c = P\rho_1[4 \times 0.4 + 4 \times 0.32 + 2 \times 0.28 + 4 \times 0.25 + 0.23] = 4.67P\rho_1$$

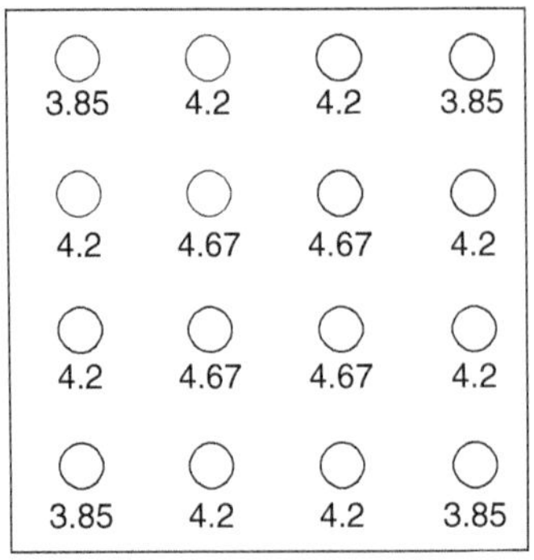

Figure 18.18s α values settlement factor for each pile in the group.

This gives an average of a group settlement equal to 4.23 times the settlement of one pile. The rule of thumb $\sqrt{B_G/B}$ would give $\sqrt{10/1} = 3.16$.

Problem 18.21

If the uncoated pile subjected to downdrag in Figure 18.54 was pushed into the ground 100 mm at the pile top, what would be:

a. The new position of the neutral point?
b. The load at the top of the pile?
c. The load distribution in the pile?

Solution 18.21

Initial assumption: $w_p > 5\text{ mm} \Rightarrow Q_p = 1000\text{ kN}$

$$w_{NP(soil)} = w_{NP(pile)} = w_{NP}$$

$$w_t = w_{NP} + \frac{(Q_{NP} + Q_t)}{2}\frac{z_{NP}}{EA} = 100\text{ mm} = 0.1\text{ m}$$

$$w_{NP(soil)} = \frac{0.2}{30}(30 - z_{NP})$$

$$Q_{NP} = Q_p + (L - z_{NP}) \times f \times A_p = 1000 + (30 - z_{NP}) \times 25 \times 1.2 = 1900 - 30z_{NP}$$

$$Q_t = Q_{NP} - z_{NP} \times f \times A_p = 1900 - 30z_{NP} - 25 \times 1.2 \times z_{NP} = 1900 - 60z_{NP}$$

$$w_t = \frac{0.2}{30}(30 - z_{NP}) + \frac{(1900 - 30z_{NP} + 1900 - 60z_{NP})}{2}\frac{z_{NP}}{0.3 \times 0.3 \times 2 \times 10^7} = 0.1\text{ m}$$

$$z_{NP} = 16.6\text{ m}$$

Check the initial assumption:

$$w_p = w_{NP} - \frac{(Q_{NP} + Q_p)}{2}\frac{30 - z_{NP}}{EA}$$

$$= \frac{0.2}{30}(30 - 16.6) - \frac{(1900 - 30 \times 16.6 + 1000)}{2}\frac{30 - 16.6}{0.3 \times 0.3 \times 2 \times 10^7} = 0.08 > 0.005$$

$$Q_t = 1900 - 60z_{NP} = 904\text{ kN}$$

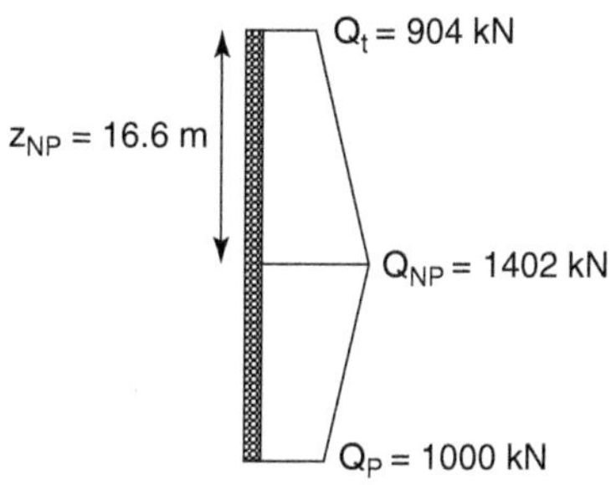

Figure 18.19s Load profile on pile.

Problem 18.22

A bored pile foundation is used for a house on a shrink-swell soil. The piles are 0.5 m in diameter, the load per pile is 50 kN, and the zone of active movement from one season to the next extends from the ground surface to a depth of 3 m. The soil is a very stiff clay with an undrained shear strength of 120 kPa and a total unit weight of 20 kN/m^3. The groundwater level is at a depth of 10 m. How deep should each bored pile be to minimize the uneven movement of the house?

Solution 18.22

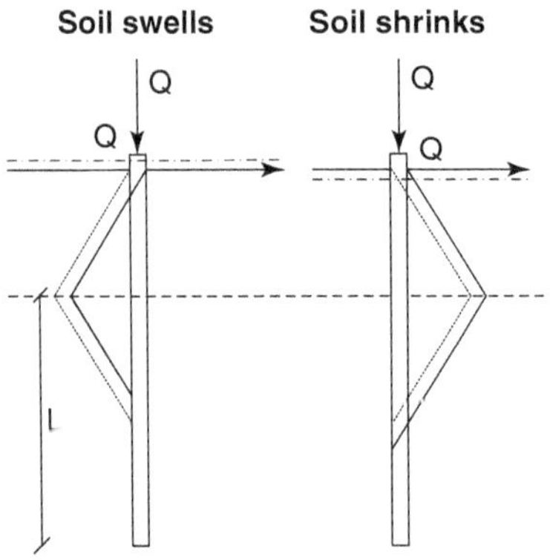

Figure 18.20s Load profile due to soil swelling and shrinking.

Swelling

Swelling and uplift load:

$$F_{sw} = \alpha s_u \pi D L_{sw} = 0.5 \times 120 \times \pi \times 0.5 \times 3 = 282.6 \text{ kN}$$

Maximum resisting load:

$$F_r = \alpha s_u \pi D L_r + Q = 0.5 \times 120 \times \pi \times 0.5 \times L_r + 50 = 94.2 L_r + 50 \text{ kN}$$

$$\frac{F_r}{F_{sw}} = SF = 2 \Rightarrow L = 5.47 \text{ m}$$

Shrinking

Shrinkage and downward load:

$$: F_{sh} = \alpha s_u \pi D L_{sh} + Q = 0.5 \times 120 \times \pi \times 0.5 \times 3 + 50 = 332.6 \text{ kN}$$

Maximum resisting load:

$$F_r = \alpha s_u \pi D L_r + 9 s_u A_p = 0.5 \times 120 \times \pi \times 0.5 \times L_r + 9 \times 120 \times \frac{0.5^2 \pi}{4} = 94.2 \text{L}_r + 212 \text{ kN}$$

$$\frac{F_r}{F_{sh}} = SF = 2 \Rightarrow L = 4.81 \text{ m}$$

$\text{L}_r = \text{Max}\,(5.47, 4.81) = 5.47$ m
Total length of the pile = 8.47 m

Problem 18.23

For the long flexible pile shown in Figure 18.5s, calculate:

a. The ultimate load H_{ou}
b. The deflection and slope at the ground surface under the working load
c. The maximum bending moment under the working load

d. The factor of safety against yielding of the soil near the ground surface under the working load

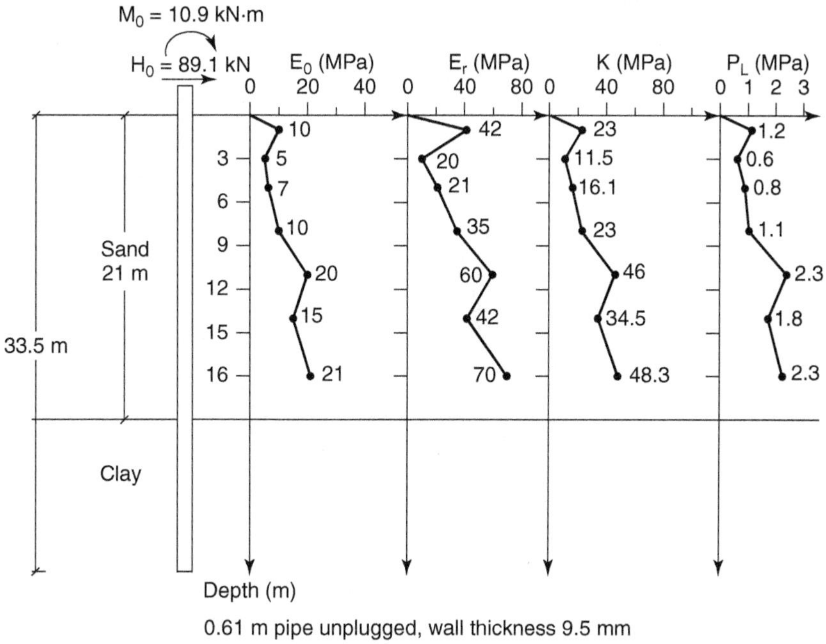

Figure 18.5s Long flexible pile loaded horizontally.

Solution 18.23

a. Select horizontal spring constant K:
   a. A value of K = 15,000 kPa is selected from the soil profile.
b. Calculate the transfer length l_o:

$$l_0 = \sqrt[4]{\frac{4EI}{K}}$$

$$I = \frac{\pi}{64}(D_0^4 - D_i^4) = \frac{\pi}{64}(0.61^4 - 0.591^4) = 8.08 \times 10^{-4}\ \text{m}^4$$

$$E = 2 \times 10^8\ \text{kPa}$$

$$l_0 = \sqrt[4]{\frac{4 \times 2 \times 10^8 \times 8.08 \times 10^{-4}}{15000}} = 2.56\ \text{m}$$

c. Check if pile is long and flexible or short and rigid:

$$\frac{L}{l_0} = \frac{33.5}{2.56} > 3,\ \text{flexible pile}$$

d. Deflection at the ground surface under the working load:

$$y_0 = \frac{2H_0}{l_0K} + \frac{2M_0}{l_0^2K} = \frac{2 \times 89.1}{2.56 \times 15000} + \frac{2 \times 10.9}{2.56^2 \times 15000} = 0.00464 + 0.00022 = 4.86\ \text{mm}$$

e. Slope at the ground surface under the working load:

$$y'_0 = \frac{2H_0}{l_0{}^2K} + \frac{4M_0}{l_0^3K}$$
$$= -\frac{2 \times 89.1}{2.56^2 \times 15000} - \frac{4 \times 10.9}{2.56^3 \times 15000}$$
$$= -1.81 \times 10^{-3} - 0.17 \times 10^{-3} = -1.98 \text{ radians}$$

f. Depth z_{max} to maximum bending moment M_{max}:

$$\tan \frac{z_{max}}{l_0} = \frac{1}{1 + \dfrac{2M_0}{l_0 H_0}}$$
$$= \frac{1}{1 + \dfrac{2 \times 10.9}{2.56 \times 89.1}} = 0.913 \quad and \quad \frac{z_{max}}{l_0} = 0.739$$
$$z_{max} = 1.89 \text{ m}$$

g. Maximum bending moment under the working load:

$$M_{max} = H_0 l_0 e^{-\frac{z_{max}}{l_0}} \sin\left(\frac{z_{max}}{l_0}\right) + M_0 e^{-\frac{z_{max}}{l_0}} \left(\cos\left(\frac{z_{max}}{l_0}\right) + \sin\left(\frac{z_{max}}{l_0}\right)\right)$$
$$M_{max} = 89.1 \times 2.56\, e^{-\frac{1.89}{2.56}} \sin\left(\frac{1.89}{2.56}\right) + 10.9\, e^{-\frac{1.89}{2.56}} \left(\cos\left(\frac{1.89}{2.56}\right) + \sin\left(\frac{1.89}{2.56}\right)\right)$$
$$= 73.4 + 7.4 = 80.8 \text{ kN.m}$$

h. Factor of safety against yielding of the soil near the ground surface under the working load:

$$P = Ky = 15000 \times 0.00486 = 72.9 \text{ kN/m}$$
$$p_a = \frac{P}{B} = \frac{72.9}{0.61} = 119.5 \text{ kPa}$$

p_L within the loaded depth is at least 0.6 MPa = 600 kPa
Safety factor:

$$F = \frac{p_l}{p_a} = \frac{600}{119.5} = 5.02$$

i. Ultimate horizontal load:

$$H_{ou} = \frac{3}{4} p_L B z_{max} = 0.75 \times 600 \times 0.61 \times 1.89 = 518.8 \text{ kN}$$

So the applied load of 89.1 kN is a safe load.

Problem 18.24

For the short rigid pile shown in Figure 18.6s, calculate:

a. The ultimate load H_{ou}
b. The deflection and slope at the ground surface under the working load
c. The maximum bending moment under the working load
d. The factor of safety against yielding of the soil near the ground surface under the working load

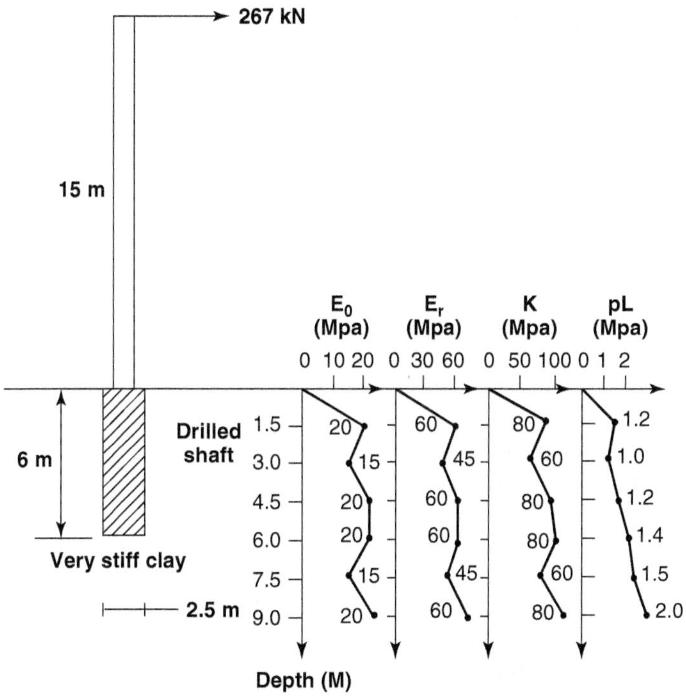

Figure 18.6s Short rigid pile loaded horizontally.

Solution 18.24

a. Select horizontal spring constant K:
   a. A value of $K = 60{,}000$ kPa is selected from the soil profile.
b. Calculate the transfer length l_o:

$$l_0 = \sqrt[4]{\frac{4EI}{K}}$$

$$I = \frac{\pi D^4}{64} = \frac{\pi 2.5^4}{64} = 1.92\ \text{m}^4$$

$$E = 2 \times 10^7\ \text{kPa}$$

$$l_0 = \sqrt[4]{\frac{4 \times 2 \times 10^7 \times 1.92}{60000}} = 7.1\ \text{m}$$

c. Check if the pile is long and flexible or short and rigid:
   a. $L = 6\text{m} < l_0 = 7.1$ m; therefore, the pile is rigid and short.
d. Deflection at the ground surface under the working load:

$$H_o = 267\ \text{kN}$$

$$M_o = 267 \times 15 = 4005\ \text{kN}$$

$$y_o = -\frac{2(2H_0L + 3M_0)}{KL^2}$$

$$= -\frac{2(2 \times 267 \times 6 + 3 \times 4005)}{60000 \times 6^2}$$

$$= -0.014\ \text{m}$$

e. Slope at the ground surface under the working load:

$$y'_o = \frac{6(H_0L + 2M_0)}{KL^3} = \frac{6(267 \times 6 + 2 \times 267 \times 15)}{60000 \times 6^3} = +0.0045 \text{ radians} = +0.26 \text{ degrees}$$

f. Depth zmax to maximum bending moment Mmax:

$$Z_{max} = -\frac{2y_0}{y'_0} - L = 0.22 \text{ m}$$

g. Maximum bending moment under the working load:

$$M_{max} = M_0 + H_0 Z_{max} - Ky'_o \frac{Z_{max}}{6} - Ky_o \frac{Z_{max}{}^2}{6} = 4071.6 \text{ kN.m}$$

h. Factor of safety against yielding of the soil
The load in the soil near the ground surface is:

$$P = Ky = 60000 \times 0.14 = 840 \text{ kN/m}$$

$$p_a = \frac{P}{B} = \frac{840}{2.5} = 336 \text{ kPa}$$

p_L at the top soil is ~1 MPa = 1000 kPa
Safety factor is $F = \frac{p_L}{p_a} = \frac{1000}{336} = 2.98 \approx 3$

i. Ultimate horizontal load:

$$H_{ou} = \frac{3}{4} p_L B z_{max} = 0.75 \times 1000 \times 2.5 \times 0.22 = 412.5 \text{ kN}$$

So, the applied load of 267 kN is safe, and the factor of safety is 412.5/267 = 1.54. Note that the large overturning moment affects the horizontal capacity by inducing horizontal movement of its own. If no moment was applied at the top, the z_{max} value would be $L/3$ and the ultimate horizontal load would be equal to 0.75 × 1000 × 2.5 × 2 = 3750 kN.

Problem 18.25

Calculate z_{max} for a flexible pile and a rigid pile if the pile is subjected to a horizontal load only (H_o different from 0 but M_o equal to 0).

Solution 18.25

Case 1: A flexible pile

The depth z_{max} to maximum bending moment is the value of z that gives zero shear. Therefore:

$$V(z) = H_o e^{-\frac{z}{l_o}} \left(\cos \frac{z}{l_o} - \sin \frac{z}{l_o} \right) - \frac{2M_o}{l_o} e^{-\frac{z}{l_o}} \sin \frac{z}{l_o} = 0$$

For $M_o = 0$, the equation simplifies to:

$$H_o e^{-\frac{z}{l_o}} \left(\cos \frac{z}{l_o} - \sin \frac{z}{l_o} \right) = 0 \quad or \quad \cos \frac{z}{l_o} = \sin \frac{z}{l_o}$$

This gives:

$$\frac{z_{max}}{l_o} = \frac{\pi}{4} \quad or \quad z_{max} = \frac{\pi}{4} \sqrt[4]{\frac{4EI}{K}}$$

Case 2: A rigid pile

The depth $z_{\max}$ to maximum bending moment is the value of z that gives zero shear. Therefore:

$$V = H_o + K\frac{6(H_oL + 2M_o)}{KL^3}\frac{z^2}{2} - K\frac{2(2H_oL + 3M_o)}{KL^2}z = 0$$

For $M_o = 0$, this equation simplifies to:

$$3z^2 - 4Lz + L^2 = 0$$

for which the positive root is L/3. Therefore:

$$z_{\max} = \frac{L}{3}$$

Problem 18.26

Calculate the ratio between the ground surface displacement for a free-head condition and for a fixed-head condition. Do the calculation first for a flexible pile and then for a rigid pile.

Solution 18.26

Case 1: A long flexible pile

Free-head condition

The displacement at the ground surface is calculated as:

$$y_o(\text{free head}) = \frac{2H_o}{l_oK}$$

Fixed-head condition

The displacement at the ground surface is calculated as:

$$\begin{cases} y_o\left(\text{fixed head}\right) = \dfrac{2H_o}{l_oK} + \dfrac{2M_o}{l_o{}^2K} \\ y'_o(\text{fixed head}) = -\dfrac{2H_o}{l_o{}^2K} - \dfrac{4M_o}{l_o{}^3K} = 0 \quad \Rightarrow M_o = -\dfrac{H_ol_o}{2} \end{cases}$$

Therefore,

$$y_o(\text{fixed head}) = \frac{2H_o}{l_oK} + \frac{2M_o}{l_o{}^2K} = \frac{2H_o}{l_oK} - \frac{H_o}{l_oK} = \frac{H_o}{l_oK}$$

Hence, for a long flexible pile, the ratio between the ground surface displacement for a free-head condition and for a fixed-head condition is:

$$\frac{y_o(\text{free head})}{y_o(\text{fixed head})} = \frac{\frac{2H_o}{l_oK}}{\frac{H_o}{l_oK}} = 2$$

Case 2: A short rigid pile

Free-head condition

The displacement at the ground surface is calculated as:

$$y_o(\text{free head}) = \frac{-2(2H_oL)}{KL^2} = \frac{-4H_o}{KL}$$

Fixed-head condition

The displacement at the ground surface is calculated as:

$$\begin{cases} y_o\left(\text{fixed head}\right) = \dfrac{-2(2H_oL + 3M_o)}{KL^2} \\ y'_o(\text{fixed head}) = \dfrac{6(H_oL + 2M_o)}{KL^3} = 0 \quad \Rightarrow M_o = -\dfrac{H_oL}{2} \end{cases}$$

Therefore,

$$y_o(\text{fixed head}) = \frac{-2(2H_oL + 3M_o)}{KL^2} = \frac{-2\left(2H_oL + 3\frac{-H_oL}{2}\right)}{KL^2} = \frac{-H_o}{KL}$$

Hence, for a short rigid pile, the ratio between the ground surface displacement for a free-head condition and for a fixed-head condition is:

$$\frac{y_o(\text{free head})}{y_o(\text{fixed head})} = \frac{\frac{-4H_o}{KL}}{\frac{-H_o}{KL}} = 4$$

Problem 18.27

For the pile group shown in Figure 18.75, calculate the efficiency of the group if it is loaded horizontally in a direction perpendicular to the four-pile line.

Solution 18.27

Figure 18.21s shows an illustration of the horizontal loading on the pile group. The center-to-center spacing S of the piles is two times the pile diameter B.

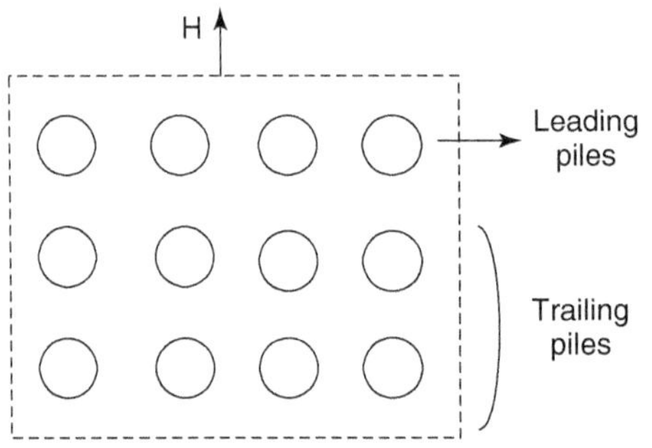

Figure 18.21s Horizontal loading of a pile group.

The leading pile efficiency e_{lp} is obtained from Figure 18.22s for a pile relative spacing of 2. The value is 0.86. Because there are 4 leading piles, the contribution to the group capacity is 4 × 0.86 = 3.44.

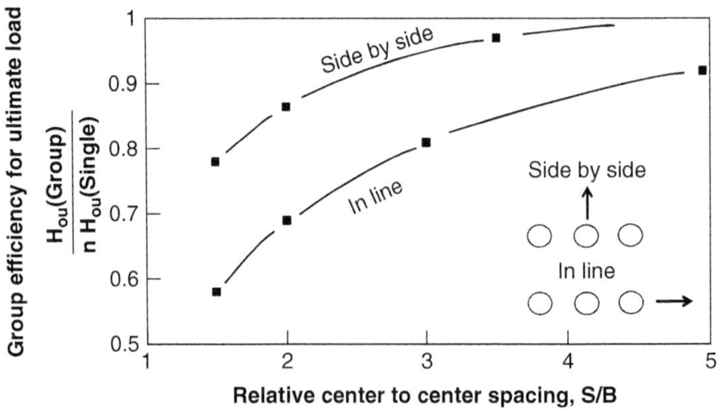

Figure 18.22s Efficiency for side-by-side and in-line groups.

The ratio λ between the capacity of the leading pile and the trailing pile is given by Figure 18.23s. The value is 1.43 for a relative spacing of 2. Therefore, the efficiency of each trailing pile is e_{tp} = 0.86/1.43 = 0.60. Because there are 8 trailing piles, the contribution to the group capacity is 8 × 0.6 = 4.8.

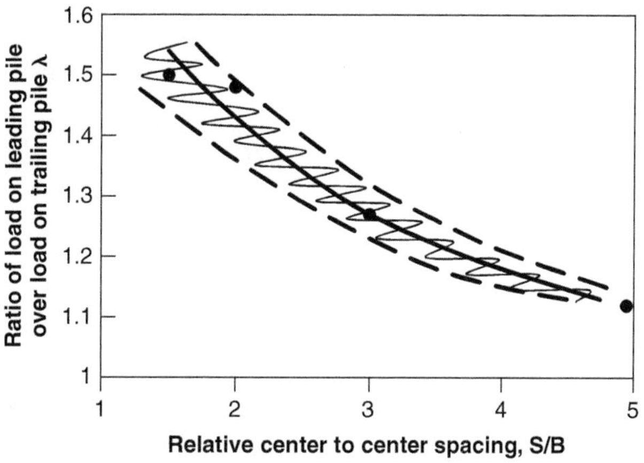

Figure 18.23s Ratio of load on leading pile over trailing pile.

The efficiency of the pile group is calculated as.

$$e = \frac{H_{ou(\text{group})}}{nH_{ou(\text{single})}} = \frac{n_{lp}e_{lp} + n_{tp}e_{tp}}{n} = \frac{4 \times 0.86 + 8 \times 0.6}{12} = 0.69$$

Problem 18.28

The pile group of Figure 18.75 is subjected to an overturning moment of 10 MN.m in the direction of largest resistance to overturning of the group. The piles are 0.4 by 0.4 square concrete driven piles embedded 25 m in a loose sand with a blow count of 6 bpf. What will be the ratio between the applied tension load and the ultimate tension capacity of the most loaded pile in the group?

Solution 18.28

The ultimate capacity of the piles in tension can be estimated by the Briaud-Tucker method:

$$Q_u = f_u PL = 5 \times 6^{0.7} \times 4 \times 0.4 \times 25 = 701 \text{ kN}$$

The maximum tension load on the outside of the group, in the absence of any compression load, is:

$$\Delta Q_{\max} = \frac{MB}{2\,\text{m}\sum_{i=1}^{n} a_i^2} = \frac{10000 \times 2.8}{2 \times 3((3 \times 0.4)^2 + (1 \times 0.4)^2 + (1 \times 0.4)^2 + (3 \times 0.4)^2)} = 1458 \text{ kN}$$

So, unless there is a significant compression load (coming from the structure deadweight, for example), the outside pile will fail.

Problem 18.29

A steel pipe pile has a diameter D equal to 0.61 m and a wall thickness t equal to 9.5 mm. The pile is 33.5 m long and the steel has a modulus E equal to 200 GPa. The pile is loaded horizontally with a load H_o of 89 kN in fixed-head condition. The soil is characterized by stiffness coefficient K from pressuremeter tests equal to 25 000 kPa. Plot the profiles versus depth of the deflection, slope, shear, bending moment, and line load in the pile.

Solution 18.29

Step 1: Define the type of pile

$$I = \frac{\pi(D^4 - d^4)}{64} = \frac{\pi(0.61^4 - (0.61 - 2 \times 9.5 \times 10^{-3})^4)}{64} = 8.08 \times 10^{-4} \text{ m}^4$$

Transfer length l_o is calculated as:

$$l_o = \sqrt[4]{\frac{4EI}{K}} = \sqrt[4]{\frac{4 \times 2 \times 10^{11} \times 8.08 \times 10^{-4}}{25 \times 10^6}} = 2.25 \text{ m}$$

Because $L > 3l_o$, the pile is defined as a long, flexible pile.

Step 2: Calculate the deflection, slope, shear, bending moment, and line load in the pile

The fixed-head flexible pile has a slope of zero at the top of the pile. Therefore,

$$y_o'(\text{fixed head}) = -\frac{2H_o}{l_o^2 K} - \frac{4M_o}{l_o^3 K} = 0$$

Hence, the bending moment at the ground surface can be calculated as:

$$M_o = -\frac{H_o l_o}{2} = -\frac{89 \times 2.25}{2} = -100 \text{ kN} \cdot \text{m}$$

With z as the depth along the pile, the deflection, slope, bending moment, shear, and line load in the pile can be calculated as:

Deflection $\quad y(z) = \frac{2H_o}{l_o K} e^{-\frac{z}{l_o}} \cos\frac{z}{l_o} + \frac{2M_o}{l_o^2 K} e^{-\frac{z}{l_o}} \left(\cos\frac{z}{l_o} - \sin\frac{z}{l_o}\right)$

Slope $\quad y'(z) = -\frac{2H_o}{l_o^2 K} e^{-\frac{z}{l_o}} \left(\cos\frac{z}{l_o} + \sin\frac{z}{l_o}\right) - \frac{4M_o}{l_o^3 K} e^{-\frac{z}{l_o}} \cos\frac{z}{l_o}$

Bending moment $\quad M(z) = H_o l_o e^{-\frac{z}{l_o}} \sin\frac{z}{l_o} + M_o e^{-\frac{z}{l_o}} \left(\cos\frac{z}{l_o} + \sin\frac{z}{l_o}\right)$

Shear $\quad V(z) = H_o e^{-\frac{z}{l_o}} \left(\cos\frac{z}{l_o} - \sin\frac{z}{l_o}\right) - \frac{2M_o}{l_o} e^{-\frac{z}{l_o}} \sin\frac{z}{l_o}$

Line load $\quad p(z) = -Ky(z)$

A Matlab program was written to plot these functions, as shown in Figure 18.24s.

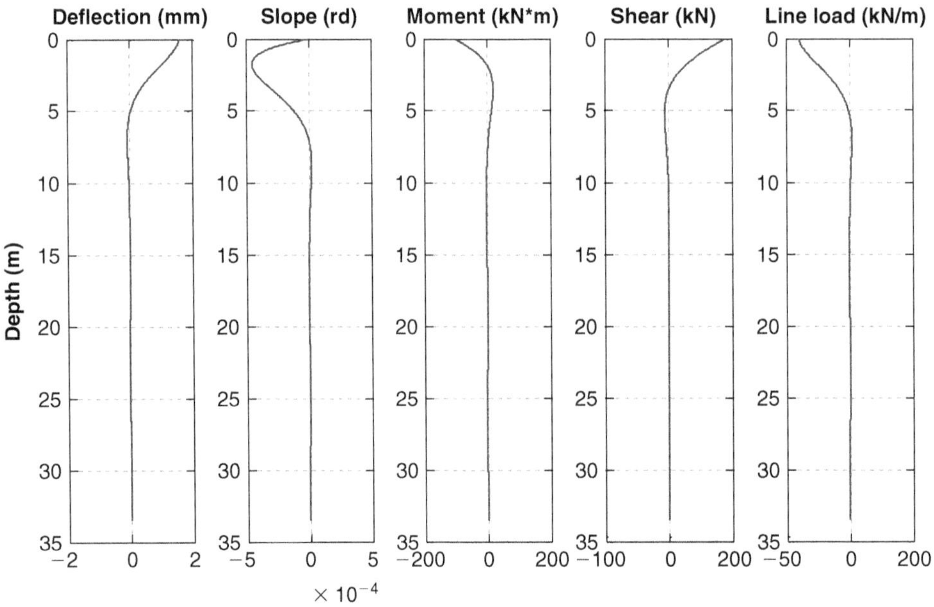

Figure 18.24s Deflection, slope, bending moment, shear, and line load in the fixed-head pile under a horizontal load H_o at the ground surface.

CHAPTER 19

Slope Stability

19.1 GENERAL

Slopes can be natural or manmade slopes (Figure 19.1). Natural slopes are found on the sides of mountains or at the edge of rivers, for example. Manmade slopes may be cut slopes, as in the case of an underpass for a road, or filled slopes, as in the case of an earth dam or a highway embankment. In all cases, the main parameter sought by the geotechnical engineer is the factor of safety against sliding failure of the slope.

Slopes fail along a failure surface. Most of the time, this surface can be approximated by a circle. However, the failure surface can take many other shapes, including a single plane, a series of planes, a log spiral, a sliding block, and others (Figure 19.2). Most analyses assume that the problem can be solved as if it were a plane strain problem in two dimensions. In three dimensions, the surface looks more like a spoon shape. Circular failure surfaces are the most common.

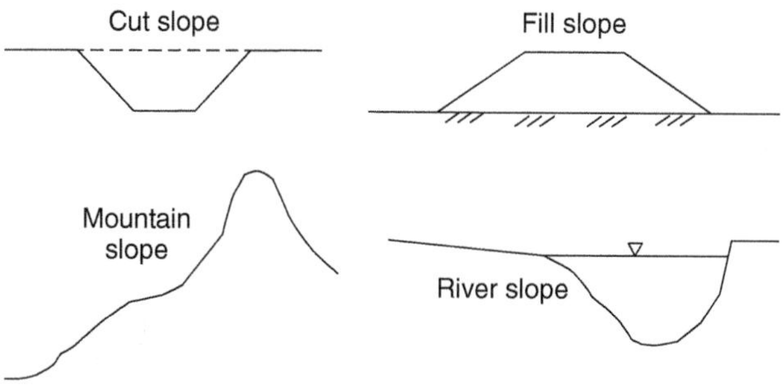

Figure 19.1 Natural slopes and manmade slopes.

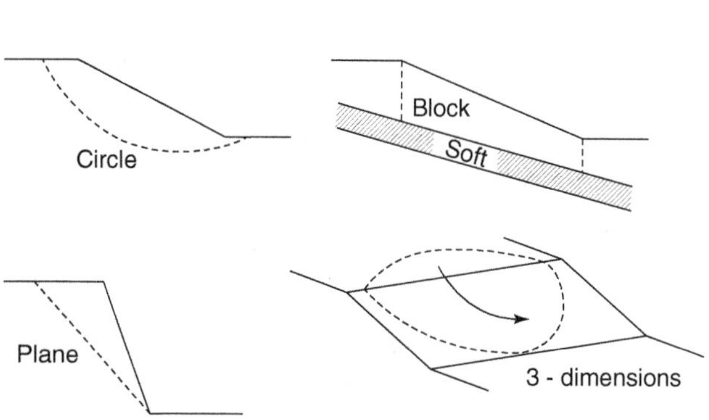

Figure 19.2 Failure surfaces.

19.2 DESIGN APPROACH

There are two aspects to slope stability: the safety against failure (ultimate limit state) and the movement under normal conditions (serviceability limit state). The movement under normal conditions is not often an issue and thus is rarely calculated; the best method for such estimates is the finite element method. The main issue is safety against failure; therefore, slope stability analysis consists of calculating the factor of safety *F*. Other issues include slope monitoring and slope stabilization methods. In the general case (circular failure surface), the factor of safety *F* is defined as (Figure 19.3):

$$F = \frac{\tau_{af}}{\tau_{am}} \tag{19.1}$$

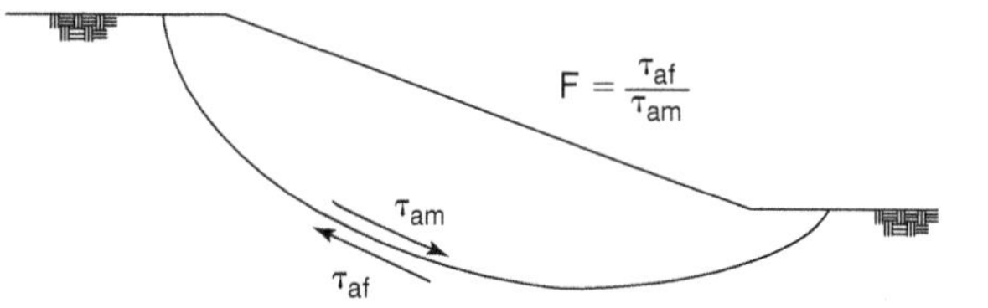

Figure 19.3 Factor of safety.

where τ_{af} is the average shear strength of the soil on the plane of failure and τ_{am} is the average shear stress mobilized on the plane of failure to keep the slope in equilibrium.

A simple example of how the factor of safety is obtained is shown in Figure 19.4. In this example the soil has a constant shear strength s along the length L of the failure plane, which is assumed to be an arc of a circle with a radius *R* and a center O. The weight of the failing soil mass is *W*, with a center of gravity generating a moment arm a around the center O. The factor of safety defined in Eq. 19.1 is also given in this

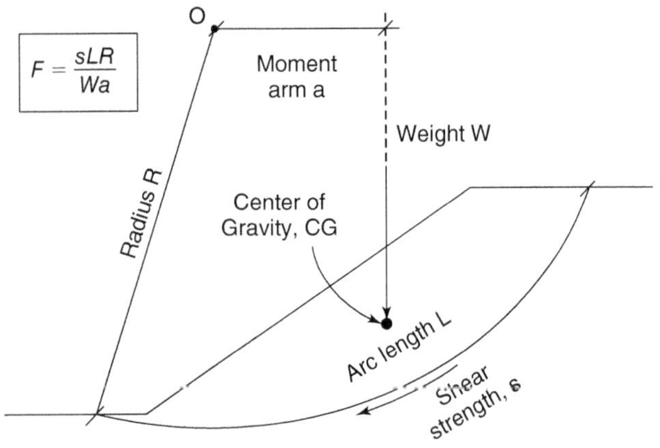

Figure 19.4 Simple slope stability problem.

case by the ratio of the maximum resisting moment over the driving moment around the center of the circle:

$$F = \frac{\tau_{af}}{\tau_{am}} = \frac{sLR}{Wa} \quad (19.2)$$

Note that because most of the time the slope stability problem is treated as a plane strain problem, the forces will be in kN/m and the moment in kN.m/m or kN. Typically the engineer will aim for a factor of safety between 1.25 and 1.5, depending on the application. As will be shown later, these values typically lead to a probability of failure that is higher than the probability of failure accepted in foundation engineering. Equation 19.2 is very simple; much more complexity is associated with slope stability analysis. The complexity arises from several issues:

1. The strength of the soil is not constant along the failure surface
2. The shape of the failure surface may vary (circle, plane, multilinear shape, log spiral)
3. One must find the failure surface corresponding to the lowest possible factor of safety
4. The boundary and external forces may be complex
5. The soil may be reinforced by inclusions

Equation 19.1 indicates how important it is to have a good estimate of the shear strength of the soil to obtain a good estimate of the factor of safety. The shear strength of soils was discussed in Chapter 15. The main equation is:

$$\tau_f = c' + (\sigma - \alpha u_w) \tan \varphi' \quad (19.3)$$

where τ_f is the shear strength, c′ is the effective stress cohesion intercept, σ is the total normal stress on the failure plane, α is the area ratio coefficient for the water phase, u_w is the water stress, and φ' is the effective stress friction angle. The parameters c′ and φ' can be obtained from a drained shear test (triaxial, direct shear, simple shear) or an undrained shear test with water stress measurements. The total stress σ can be calculated from the soil unit weight and any additional stress created by loading. The area ratio coefficient α can be estimated as the degree of saturation (α = S) or through a correlation to the air entry value u_{wae}($\alpha = (u_{wae}/u_w)^{0.5}$). If hydrostatic conditions exist, the water stress u_w can be estimated as follows. If the point considered is located under the groundwater level (GWL), u_w is the unit weight of water times the vertical distance from the point considered to the GWL. In this case the water stress is positive (compression). If the point considered is above the GWL in the zone that is saturated by capillary action, the water stress is also given as the unit weight of water times the distance between the point considered and the GWL, but this time the water stress is negative (tension). If the point considered is in the unsaturated zone above the zone saturated by capillary action, the water tension u_w can be estimated by its relationship with the water content through the soil water retention curve (SWRC). If the point considered is below the GWL but there is an excess water stress Δu_w, then the total water stress is the sum of the hydrostatic water stress and the excess water stress.

Because the factor of safety *F* involves the strength of the soil, it can be considered as an ultimate limit state where Eq. 19.1 is rewritten as:

$$\gamma \tau_{am} = \varphi \tau_{af} \quad (19.4)$$

where γ is the load factor and φ is the resistance factor. The values of γ and φ depend on how well the loading parameters and the shear strength parameters are known. The load factor in AASHTO (2007) for overall stability of slopes is taken as 1.0. The resistance factor φ proposed by AASHTO (2007) is:

- 0.75 if the geotechnical parameters are well defined and the slope does not support or contain a structural element
- 0.65 if the geotechnical parameters are not well defined or the slope supports or contains a structural element.

These resistance factors correspond to a probability of failure varying between 0.01 and 0.001.

19.3 INFINITE SLOPES

The simplest case of slope stability is the case where the slope is infinitely long (Figure 19.5). In this instance the

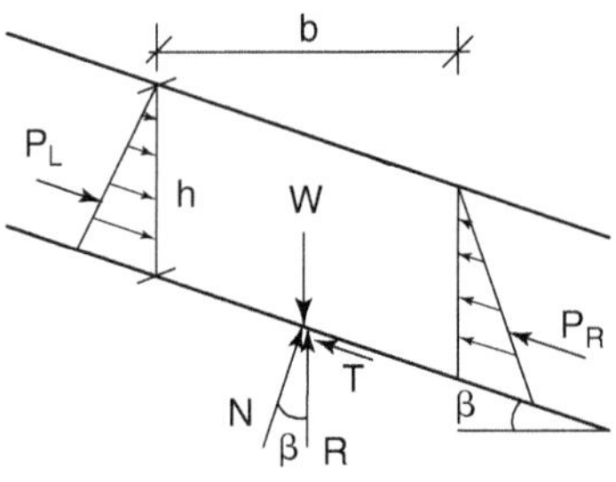

Figure 19.5 Infinite slope.

failure is assumed to be parallel to the ground surface. Several soil conditions, in order of increasing complexity, can be considered: dry sand, dry $c' - \varphi'$ soil, $c' - \varphi'$ soil with seepage, and $c' - \varphi'$ soil with unsaturated conditions.

19.3.1 Dry Sand

Consider a slice of the sheet of soil failing downslope. It is h high and b wide, and rests on the failure surface, which is at an angle β with the horizontal. The external forces acting on a free-body diagram of that slice include the weight of the slice *W*; the resistance R at the bottom of the slice, which can be decomposed into a normal force *N* and a shear force *T*; and the earth pressure force on the left P_L and the earth pressure force on the right P_R. The two forces P_L and P_R are equal, opposite, and in line with each other, so they simply cancel out of the equilibrium equation. The relationships between *W*, *N*, and *T* are:

$$N = W\cos\beta = \gamma bh\cos\beta \quad (19.5)$$

$$T = W\sin\beta = \gamma bh\sin\beta \quad (19.6)$$

where γ is the total unit weight of the dry sand. *N* and *T* represent the forces existing in the slope and mobilized to maintain the slope in equilibrium. The shear force *S* corresponding to the strength of the failure surface is not mobilized in the slope unless the slope is at failure. That force S represents the maximum value that *T* can have and is expressed as:

$$S = N\tan\varphi' = \gamma bh\cos\beta\tan\varphi' \quad (19.7)$$

The normal stress σ on the failure plane is:

$$\sigma = \frac{N}{A} = \frac{\gamma bh\cos\beta}{b/\cos\beta} = \gamma h\cos^2\beta \quad (19.8)$$

The shear stress τ on the plane of failure is:

$$\tau = \frac{T}{A} = \frac{\gamma bh\sin\beta}{b/\cos\beta} = \gamma h\sin\beta\cos\beta \quad (19.9)$$

The shear strength on the plane of failure is:

$$\tau_f = \frac{S}{A} = \frac{\gamma bh\cos\beta\tan\varphi'}{b/\cos\beta} = \gamma h\cos^2\beta\tan\varphi' \quad (19.10)$$

The factor of safety F is expressed as:

$$F = \frac{\tau_f}{\tau} = \frac{\gamma h\cos^2\beta\tan\varphi'}{\gamma h\sin\beta\cos\beta} = \frac{\tan\varphi'}{\tan\beta} \quad (19.11)$$

This is a very useful result, which says that a slope of dry sand cannot stand at an angle higher than the friction angle of the soil. This angle is usually around 30° for loose, dry sand. Next time you are at the beach, take a handful of dry sand, drop it gently on a flat surface, and measure the angle of the slope; this procedure will give you the friction angle of that loose sand, also called the *angle of repose*. Note that the factor of safety is independent of h, which means that all planes parallel to the ground surface are equally likely to be failure planes.

19.3.2 Dry $c' - \varphi'$ Soil

In the case of dry $c' - \varphi'$ soil, the only thing that changes is that the soil has a nonzero effective stress cohesion intercept c' in the expression of the shear strength (Eq. 19.10):

$$\tau_f = c' + \gamma h\cos^2\beta\tan\varphi' \quad (19.12)$$

Then the factor of safety becomes:

$$F = \frac{\tau_f}{\tau} = \frac{c' + \gamma h\cos^2\beta\tan\varphi'}{\gamma h\sin\beta\cos\beta} = \frac{c'}{\gamma h\sin\beta\cos\beta} + \frac{\tan\varphi'}{\tan\beta} \quad (19.13)$$

The factor of safety has increased compared to the dry sand case and depends on the depth h of the plane considered. Failure will occur on the plane defined by $F = 1$, called the *critical plane*, at a depth h_{crit}:

$$h_{cr} = \frac{c'}{\gamma\cos^2\beta(\tan\beta - \tan\varphi')} \quad (19.14)$$

Recall from section 15.16 that you can go from an effective stress solution to a total stress undrained solution by changing c' into s_u and taking φ' as equal to zero. Then the critical depth for the undrained case is.

$$h_{cr} = \frac{s_u}{\gamma\sin\beta\cos\beta} \quad (19.15)$$

19.3.3 $c' - \varphi'$ Soil with Seepage

In the case of $c' - \varphi'$ soil with seepage, the GWL is at the ground surface and the added complexity comes from having to take into account the influence of the water stress u_w on the shear strength. To obtain u_w, a flow net is drawn (Figure 19.6).

Recalling Eq. 19.8, the total normal stress on the failure plane is:

$$\sigma = \gamma_{sat} h\cos^2\beta \quad (19.16)$$

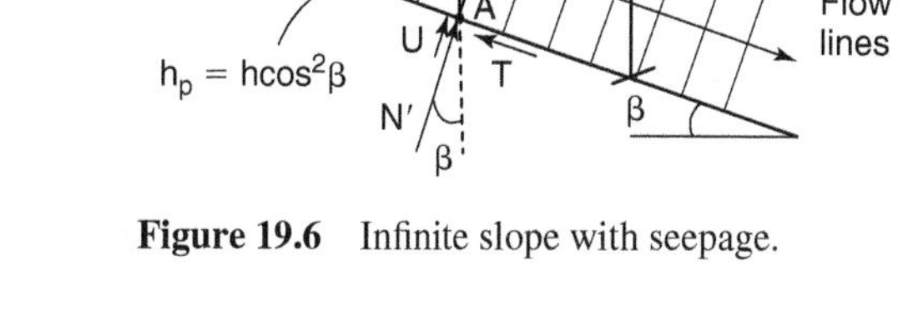

Figure 19.6 Infinite slope with seepage.

where γ_{sat} is the saturated unit weight of the soil. The water stress u_w on the failure is equal to:

$$u_w = h_p \gamma_w \tag{19.17}$$

where h_p is the pressure head on the failure plane.

We know from the flow net properties that the total head at A (Figure 19.6) is equal to the total head at B. We also know that the pressure head at B is zero; therefore, the pressure head at A is the difference in elevation between A and B:

$$h_{tA} = h_{eA} + h_{pA} = h_{tB} = h_{eB} + h_{pB} \tag{19.18}$$

Because h_{pB} is zero, then:

$$h_{pA} = h_{eB} - h_{eA} \tag{19.19}$$

Therefore, the vertical distance AC in Figure 19.6 is the pressure head at A, and from geometry we can calculate:

$$h_{pA} = h\cos^2\beta \quad \text{and} \quad u_w = \gamma_w h \cos^2\beta \tag{19.20}$$

Then the effective stress σ' is:

$$\sigma' = (\gamma_{sat} - \gamma_w) h \cos^2\beta \tag{19.21}$$

and the shear strength is:

$$\tau_f = c' + (\gamma_{sat} - \gamma_w) h \cos^2\beta \tan\varphi' \tag{19.22}$$

so the factor of safety becomes:

$$\begin{aligned} F &= \frac{\tau_f}{\tau} = \frac{c' + (\gamma_{sat} - \gamma_w) h \cos^2\beta \tan\varphi'}{\gamma_{sat} h \sin\beta\cos\beta} \\ &= \frac{c'}{\gamma_{sat} h \sin\beta\cos\beta} + \frac{(\gamma_{sat} - \gamma_w)}{\gamma_{sat}} \frac{\tan\varphi'}{\tan\beta} \end{aligned} \tag{19.23}$$

19.3.4 $c' - \varphi'$ Soil with Unsaturated Conditions

In the case of $c' - \varphi'$ soil with unsaturated conditions, the effective stress becomes:

$$\sigma' = \gamma_t h \cos^2\beta - \alpha u_w \tag{19.24}$$

The shear strength is now:

$$\tau_f = c' + (\gamma_t h \cos^2\beta - \alpha u_w) \tan\varphi' \tag{19.25}$$

Because the mobilized shear stress remains the same, the factor of safety is:

$$\begin{aligned} F &= \frac{\tau_f}{\tau} = \frac{c' + (\gamma_t h \cos^2\beta - \alpha u_w) \tan\varphi'}{\gamma_t h \sin\beta\cos\beta} \\ &= \frac{c'}{\gamma_t h \sin\beta\cos\beta} - \frac{\alpha u_w \tan\varphi'}{\gamma_t h \sin\beta\cos\beta} + \frac{\tan\varphi'}{\tan\beta} \end{aligned} \tag{19.26}$$

Comparison

The factors of safety corresponding to the various soil conditions can be compared. Assume that the soil is an overconsolidated silty clay with $c' = 5$ kPa and $\varphi' = 30°$, the slope has an angle of 20° with the horizontal, and the unit weight is 20 kN/m^3. The question is: What is the factor of safety against failure for a plane at a depth 2 m below the ground surface?

For the case of the dry soil, the factor of safety is:

$$\begin{aligned} F_{dry} &= \frac{5}{20 \times 2 \sin 20 \cos 20} + \frac{\tan 30}{\tan 20} \\ &= 0.389 + 1.586 = 1.975 \end{aligned} \tag{19.27}$$

For the case of the slope with seepage, the factor of safety is:

$$\begin{aligned} F_{seep} &= \frac{5}{20 \times 2 \sin 20 \cos 20} + \frac{(20 - 10)}{20} \frac{\tan 30}{\tan 20} \\ &= 0.389 + 0.793 = 1.182 \end{aligned} \tag{19.28}$$

For the case of the slope with an unsaturated condition, with a degree of saturation equal to 60% and a water tension equal to −1000 kPa, the factor of safety is:

$$\begin{aligned} F_{unsat} &= \frac{5}{20 \times 2 \sin 20 \cos 20} - \frac{0.6(-100) \tan 30}{20 \times 2 \sin 20 \cos 20} + \frac{\tan 30}{\tan 20} \\ &= 0.389 + 2.695 + 1.586 = 4.67 \end{aligned} \tag{19.29}$$

As can be seen, the factors of safety are organized as follows:

$$F_{unsat} > F_{dry} > F_{seep} \tag{19.30}$$

Note that F_{unsat} is much higher than the other factors of safety even though the water tension is quite modest. Therefore, water tension plays a very important role in slope stability. Note also that if there is no cohesion, F_{seep} will be equal to half of F_{dry}. Again the role of water proves to be very important in slope stability. These calculations explain why slope failures are more likely to happen after heavy prolonged rains, as is often reported in the news media.

19.4 SEEPAGE FORCE IN STABILITY ANALYSIS

The *seepage force* is the force exerted in friction by water flowing around soil particles and trying to drag them away. The forces shown on a free-body diagram are the external forces. The internal forces are resolved internally. The seepage force is an external force when the soil skeleton is considered the free body, but it is an internal force when the soil skeleton plus the water are considered the free body. Most slope stability analyses consider the soil skeleton plus the water as the free body. In those cases the seepage force must not be included in any slope stability calculations.

Figure 19.7 shows the two free-body diagram options. In the case where the free body is the soil particles plus the

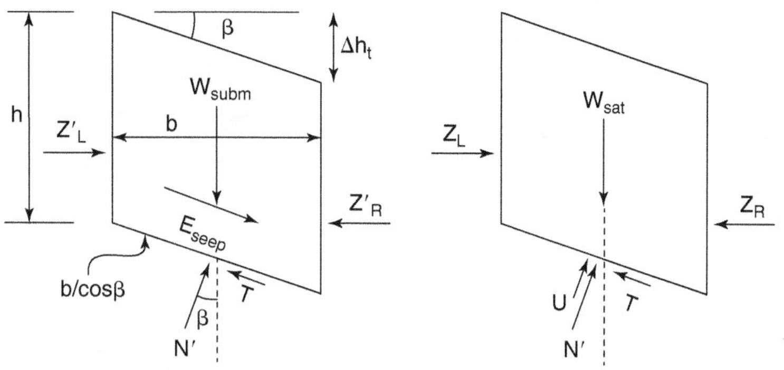

Figure 19.7 Soil skeleton and seepage force approach.

water, the weight W is the total weight including the solids and the water; the side forces are the total forces Z_R and Z_L; and the bottom forces are the effective normal force N′, the uplift force U, and the shear force T. In the case where the free body is the soil particles alone, the weight W_{subm} is the total weight minus the buoyancy force, the side forces are the effective components Z'_R and Z'_L, the bottom forces are the effective normal force N′ and the shear force T, and the seepage force E_u must be included.

Let's go back to the example of the infinite slope in the case of seepage through the slope. In the solution presented in the subsection concerning $c' - \varphi'$ soil with seepage, the free body considered was the soil skeleton and the water all together, as is usually done. As you recall, we did not consider the seepage force in that case. Indeed, it was an internal force because it was a force acting between the particles and the water, which are both part of the free body. Let's see what happens if we consider instead the soil skeleton alone to be the free body.

The forces are calculated as follows:

Submerged weight:

$$W_{subm} = (\gamma_{sat} - \gamma_w)bh \tag{19.31}$$

Normal force on bottom:

$$N' = W_{subm}\cos\beta = (\gamma_{sat} - \gamma_w)bh\,\cos\beta \tag{19.32}$$

Shear force on bottom:

$$T = W_{subm}\sin\beta + E_{seep} \tag{19.33}$$

Normal stress on bottom:

$$\sigma' = \frac{N'}{b/\cos\beta} = (\gamma_{sat} - \gamma_w)h\cos^2\beta \tag{19.34}$$

Uplift force on bottom (Eq. 9.20):

$$U = \gamma_w h\cos^2\beta \times \frac{b}{\cos\beta} \tag{19.35}$$

Seepage force:

$$E_{seep} = i\gamma_w bh \tag{19.36}$$

Hydraulic gradient:

$$i = \frac{\Delta h_t}{b/\cos\beta} = \sin\beta \tag{19.37}$$

Shear stress on bottom:

$$\tau = \frac{T}{b/\cos\beta} = (\gamma_{sat} - \gamma_w)h\sin\beta\cos\beta + \gamma_w h\sin\beta\cos\beta$$
$$= \gamma_{sat}h\sin\beta\cos\beta \tag{19.38}$$

Shear strength on bottom:

$$\tau_f = c' + \sigma'\tan\varphi' = c' + (\gamma_{sat} - \gamma_w)h\cos^2\beta\tan\varphi' \tag{19.39}$$

Then the factor of safety becomes:

$$F = \frac{\tau_f}{\tau} = \frac{c' + (\gamma_{sat} - \gamma_w)h\cos^2\beta\tan\varphi'}{\gamma_{sat}h\sin\beta\cos\beta}$$
$$= \frac{c'}{\gamma_{sat}h\sin\beta\cos\beta} + \frac{(\gamma_{sat} - \gamma_w)}{\gamma_{sat}}\frac{\tan\varphi'}{\tan\beta} \tag{19.40}$$

We get the same result as with Eq. 19.23, but after having started from a different free-body diagram (the free body of the soil skeleton with the water as an outside influence). The simplest approach in slope stability analysis is to consider the soil and the water together. When you do so, the seepage force is an internal force and does not enter into the calculations.

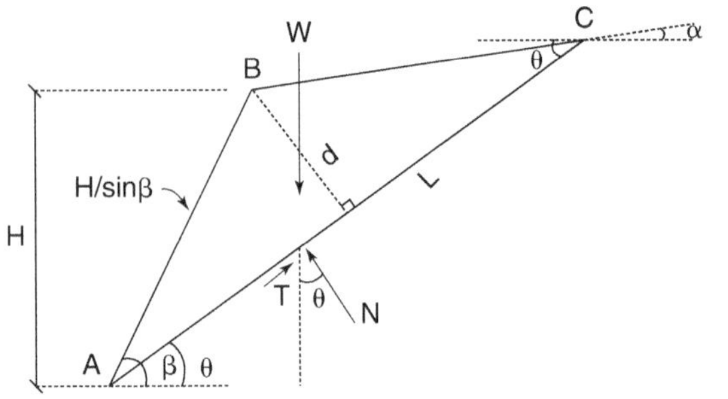

Figure 19.8 Plane failure along the bottom of a wedge.

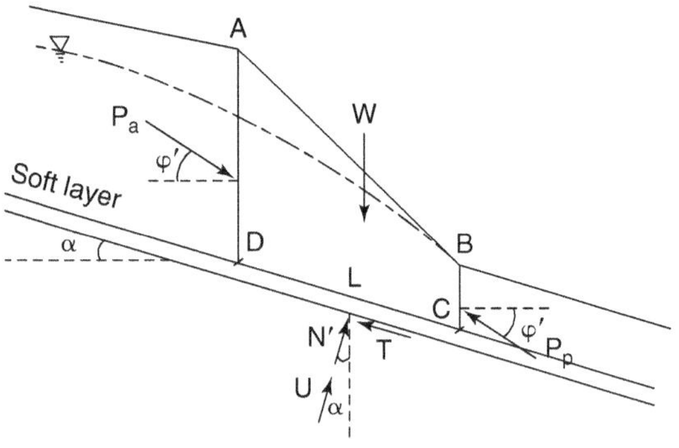

Figure 19.9 Block failure along a plane.

19.5 PLANE SURFACES

The infinite slope is a case of a plane surface. A plane surface can also be considered as the base of a wedge (Figure 19.8). In this case, the factor of safety can be calculated as follows. The length of the side AB of the triangle is H/sin β. Referring to Figure 19.8, the rule of sines in the triangle ABC gives:

$$\frac{H/\sin\beta}{\sin(\theta-\alpha)} = \frac{L}{\sin(\pi-\beta+\alpha)} \tag{19.41}$$

or:

$$L = \frac{H}{\sin\beta}\frac{\sin(\beta-\alpha)}{\sin(\theta-\alpha)} \tag{19.42}$$

Then the height d of the triangle is given by:

$$d = \frac{H}{\sin\beta}\sin(\beta-\theta) \tag{19.43}$$

The weight of the wedge is:

$$W = \frac{1}{2}\gamma L d = \frac{1}{2}\gamma H^2 \frac{\sin(\beta-\theta)\sin(\beta-\alpha)}{\sin^2\beta\sin(\theta-\alpha)} \tag{19.44}$$

The shear force T and normal force N necessary to keep the wedge in equilibrium are:

$$T = W\sin\theta \tag{19.45}$$

$$N = W\cos\theta \tag{19.46}$$

Then the factor of safety is:

$$F = \frac{S}{T} = \frac{c'L + W\cos\theta\tan\varphi'}{W\sin\theta} \tag{19.47}$$

where W is given by Eq. 19.44.

19.6 BLOCK ANALYSIS

Sometimes the most likely failure mechanism is a block of soil moving along a predetermined interface because of the presence of a weak layer along the bottom of the block (Figure 19.9). The stability analysis of block ABCD in Figure 19.9 is called a *block analysis*. In this case the driving shear force T along the potential failure plane DC is:

$$T = W\sin\alpha \tag{19.48}$$

The component of the active force P_a in the direction of sliding along DC is calculated according to the methodology described in Chapter 21. Along AD it is:

$$P_a\cos(\varphi'-\alpha) \tag{19.49}$$

At the same time, the component of the passive resistance P_p in the direction of sliding along BC is calculated according to the methodology described in Chapter 21. Along BC it is:

$$P_p\cos(\varphi'-\alpha) \tag{19.50}$$

The normal force N on the plane of failure is:

$$N = W\cos\alpha \tag{19.51}$$

and the uplift force U due to the average water stress u_w on the potential failure plane is:

$$U = u_w L \tag{19.52}$$

So, the maximum shear resistance on the potential plane of failure is:

$$S = c'L + (W\cos\alpha - u_w L)\tan\varphi' \tag{19.53}$$

The factor of safety against sliding of the block on plane DC can then be calculated as:

$$F = \frac{c'L + (W\cos\alpha - u_w L)\tan\varphi' + P_p\cos(\varphi'-\alpha)}{W\sin\alpha + P_a\cos(\varphi'-\alpha)} \tag{19.54}$$

19.7 SLOPES WITH WATER IN TENSILE CRACKS

Tensile cracks can develop at the top of a slope due either to impending failure or to desiccation. The depth of those cracks is highly variable. The depth of cracks due to desiccation is approximately equal to the horizontal distance between cracks on the ground surface. The depth of cracks due to active

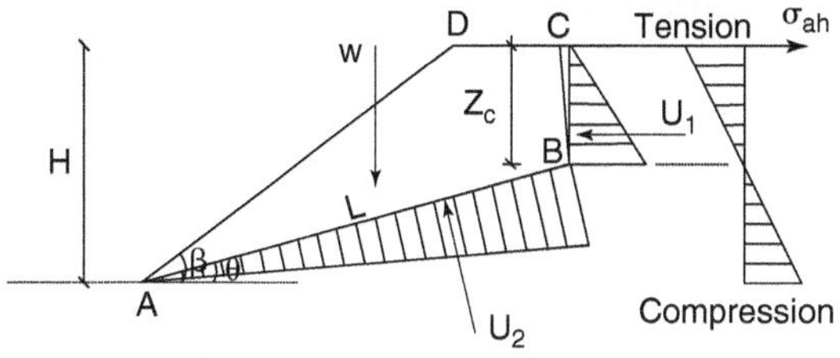

Figure 19.10 Influence of a tension crack at the top of a slope.

pressure failure can be calculated by using the active pressure expression and finding the depth where tension ends and the effective horizontal stress becomes zero (Figure 19.10). The expression for the active effective stress σ'_{ah} is (see Chapter 21):

$$\sigma'_{ah} = \sigma'_{ov} \tan^2\left(45 - \frac{\varphi'}{2}\right) - 2c' \tan\left(45 - \frac{\varphi'}{2}\right) \quad (19.55)$$

Setting Eq. 19.55 equal to zero gives the depth of the tension crack. This requires expressing σ'_{oh} as a function of the depth z. If the water table is at the ground surface, the depth of the crack is:

$$z_c = \frac{2c'}{(\gamma_{sat} - \gamma_w)\tan\left(45 - \frac{\varphi'}{2}\right)} \quad (19.56)$$

If the soil is unsaturated:

$$z_c = \frac{2c'}{\gamma_t \tan\left(45 - \frac{\varphi'}{2}\right)} + \frac{\alpha u_w}{\gamma_t} \quad (19.57)$$

Because u_w is negative, Eq. 19.57 gives a lower estimate of z_c than does Eq. 19.56. As it is rare to have the groundwater table at the ground surface near a slope, z_c is often estimated as:

$$z_c = \frac{2c'}{\gamma_t \tan\left(45 - \frac{\varphi'}{2}\right)} \quad (19.58)$$

Of course, engineering judgment always plays an important role in such decisions. Once an estimate of z_c is known, the slope stability analysis can proceed with the worst-case assumption that the crack is filled with water. Indeed, the water pressure pushes the slope horizontally. Figure 19.10 shows a planar surface analysis. In this case, the water forces U_1 and U_2 are:

$$U_1 = \frac{1}{2}\gamma_w {z_c}^2 \quad (19.59)$$

$$U_2 = \frac{1}{2}\gamma_w z_c L \quad (19.60)$$

The length L of segment AB is given by:

$$L = \frac{H - z_c}{\sin\theta} \quad (19.61)$$

The driving shear force on plane AB is:

$$T = W\sin\theta + U_1\cos\theta \quad (19.62)$$

The maximum resisting shear force on plane AB is:

$$S = c'L + (W\cos\theta - U_2)\tan\varphi' \quad (19.63)$$

The final expression of the factor of safety is then:

$$F = \frac{c'\dfrac{(H - z_c)}{\sin\theta} + \left(W\cos\theta - \dfrac{1}{2}\gamma_w z_c \dfrac{(H - z_c)}{\sin\theta}\right)\tan\varphi'}{W\sin\alpha + \dfrac{1}{2}\gamma_w z_c^2 \cos\theta} \quad (19.64)$$

19.8 CHART METHODS

When the soil is uniform and a circular failure surface is assumed, the problem is simple enough that the factor of safety can be determined from charts. These charts have been developed by various engineers, including Taylor (1948), Spencer (1967), Janbu (1968), and Morgenstern (1963), among others.

19.8.1 Taylor Chart

Taylor (1948) developed charts for two cases:

- $\varphi' = 0$, undrained shear strength s_u, and total stress analysis
- $\varphi' > 0$, $c' > 0$, no water

$\varphi = 0$, Undrained Shear Strength s_u, and Total Stress Analysis

The slope and its parameters are shown in Figure 19.11. This chart applies where the soil is uniform, can be represented by a constant undrained shear strength s_u, and has a total unit weight γ. Note that the concept $\varphi' = 0$ is not a true concept; however, it is mathematically convenient, and simply expresses the fact that the undrained shear strength is assumed to be independent of the normal total stress. In fact,

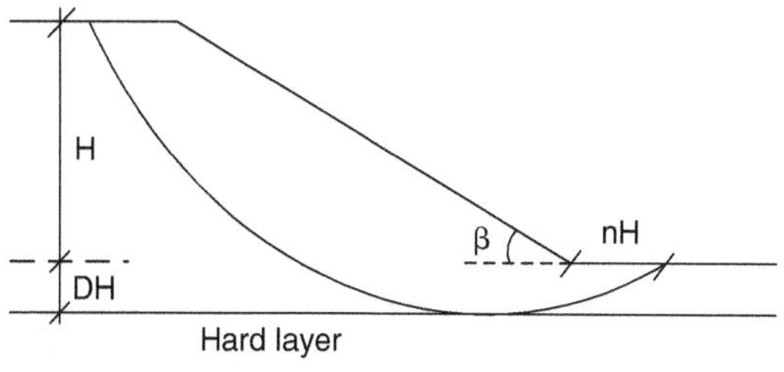

Figure 19.11 Slope parameters for Taylor chart.

φ' is always nonzero, as friction always exists between two materials. However, because the total stress changes and the effective stress does not, it looks like the friction angle is zero; this is why it is more appropriate to say that the total stress friction angle φ is zero whereas φ' is not.

The procedure is as follows:

1. Find the depth factor D, the height of the slope H, the total unit weight γ of the soil, the undrained shear strength s_u of the soil, and the slope angle β. The depth factor D (Figure 19.11) is the ratio between the vertical distance from the toe of the slope to the underlying hard layer and the height of the slope.
2. Knowing D and β, find the stability number N on the chart in Figure 19.12 by using the solid lines. The short dashed lines across the solid lines give the value of n, which is the ratio between the horizontal distance from the toe of the slope to the exit of the circle and the height of the slope. Once n is known, the circle can be identified, because it must be tangent to the hard layer.
3. The stability number N is defined as:

$$N = \frac{c_d}{\gamma H} \tag{19.65}$$

where c_d is the shear stress necessary to keep the slope in equilibrium. Using Eq. 19.65, calculate the value of c_d.

4. The factor of safety is given by:

$$F = \frac{s_u}{c_d} \tag{19.66}$$

5. If the geometry of the case at hand is such that the failure circle is most likely to be a toe circle, use the long dashed lines to find the stability number N.

$\varphi' > 0$, $c' > 0$, No Water, Effective Stress Analysis

This chart (Figure 19.13) applies to the case in which the soil is uniform, has a unit weight γ, has no water, and can be represented by an effective stress cohesion c' and an effective stress friction angle φ'. Note that the statement was that the soil has no water rather than that the soil was dry. Indeed, a dry soil can have enough water to develop very high water tension, which changes the shear strength significantly; this chart refers to the case of no water. For this chart, two factors of safety are defined:

$$F_{c'} = \frac{c'}{c'_d} \quad and \quad F_{\varphi'} = \frac{\tan \varphi'}{\tan \varphi'_d} \tag{19.67}$$

where c'_d and φ'_d are the fraction of c′ and φ' required to maintain the slope in equilibrium.

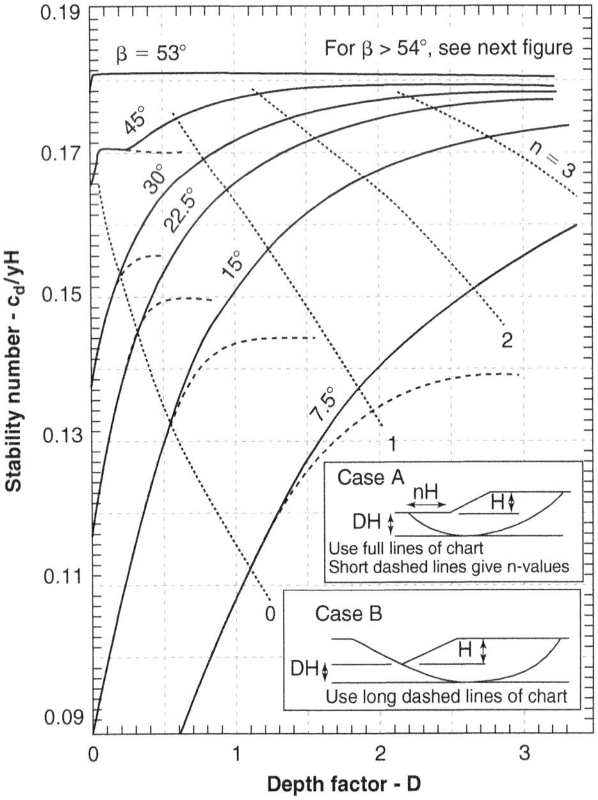

Figure 19.12 Taylor chart for $\varphi = 0$, undrained shear strength s_u soils (Taylor 1948). (This material is reproduced with permission of John Wiley & Sons, Inc.)

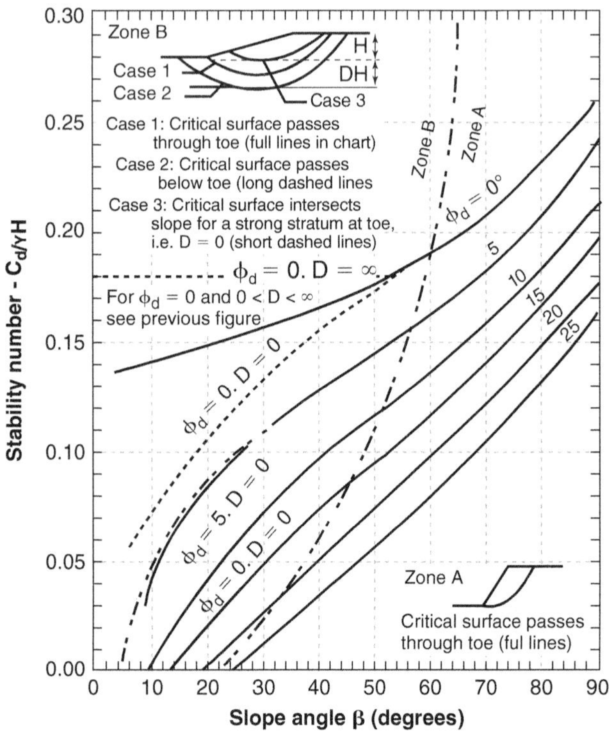

Figure 19.13 Taylor chart for $\varphi' > 0, c' > 0$. no water soils (Taylor 1948). (This material is reproduced with permission of John Wiley & Sons, Inc.)

The step-by-step procedure is as follows:

1. Choose an initial value of F'_c. A value of 1.5 is common.
2. Using Eq. 19.67, calculate the value of c'_d.
3. Calculate the depth factor D as defined in Figure 19.11 and the stability factor N as:

$$N = \frac{c'_d}{\gamma H} \tag{19.68}$$

4. Knowing the stability number N, the slope angle β, and the depth factor D, find φ'_d from the chart. Use the solid lines for the general case and the other lines as appropriate; check the chart for details.
5. Calculate F'_φ and compare to F'_c.
6. If F'_φ and F'_c are not equal or within a target tolerance, go back to step 1 and try a new value of F'_c until they are within that tolerance. It would be reasonable to use the mean of F'_c and F'_φ as the next guess.

19.8.2 Spencer Chart

Spencer (1967) developed charts for the case where the groundwater surface is within the slope circle (Figure 19.14). The soil strength is described by the effective stress parameters c' and φ'. The failure surface is considered to be circular and to go through the toe of the slope. The presence of the water in the slope is quantified by using the water stress ratio r_u:

$$r_u = \frac{u_w}{\sigma_{ov}} \tag{19.69}$$

where u_w is the water stress at the chosen point and σ_{ov} is the vertical total stress in the soil at the same point. Although r_u varies from one point to the next in the slope, a single value is used for the chart method. Referring to Figure 19.14 the average ratio r_u is estimated as:

$$r_u = \frac{\gamma_w}{\gamma_t} \times \frac{Area\ ABGEF}{Area\ ABCDEF} \tag{19.70}$$

where γ_w and γ_t are the unit weight of water and the total unit weight of the soil respectively. Note that the maximum value of r_u is about 0.5, because even if the slope is filled with water the ratio γ_w/γ_t is about 0.5. As a result, Spencer prepared charts for values of $r_u = 0$ (slope with no water), $r_u = 0.25$ (slope with water halfway up), and $r_u = 0.5$ (slope full of water). Note also that there is no water outside the slope.

The procedure for using Spencer's chart is as follows:

1. Choose an initial value of F'_c (Eq. 19.67). A value of 1.5 is common.
2. Using Eq. 19.67, calculate the value of c'_d.
3. Calculate the stability factor N as:

$$N = \frac{c'_d}{\gamma H} \tag{19.71}$$

4. Calculate the water stress ratio r_u.
5. Knowing the stability number N, the water stress ratio r_u, and the slope angle β, find φ'_d from the chart (Figure 19.15). If the ratio r_u is not exactly equal to 0, 0.25, or 0.5 as in the charts, the two closest cases of ratio r_u are calculated and interpolation on F'_φ is used.
6. Calculate F'_φ and compare to F'_c.
7. If F'_φ and F'_c are not equal or within a target tolerance, go back to step 1 and try a new value of F'_c until F'_φ and F'_c are within that tolerance. Using the mean of F'_c and F'_φ as the next guess for F'_c is reasonable.

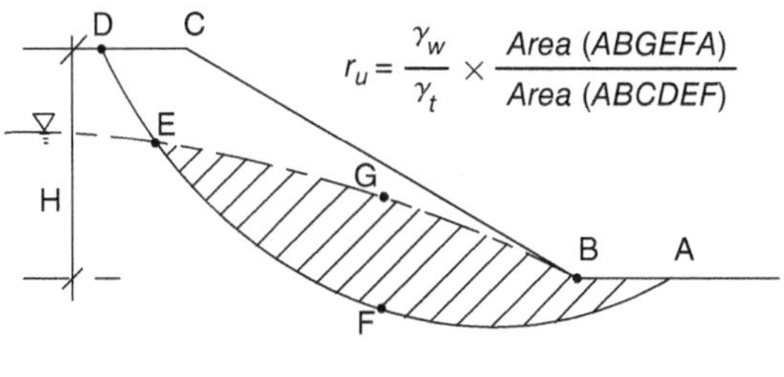

Figure 19.14 Slope parameters for Spencer chart.

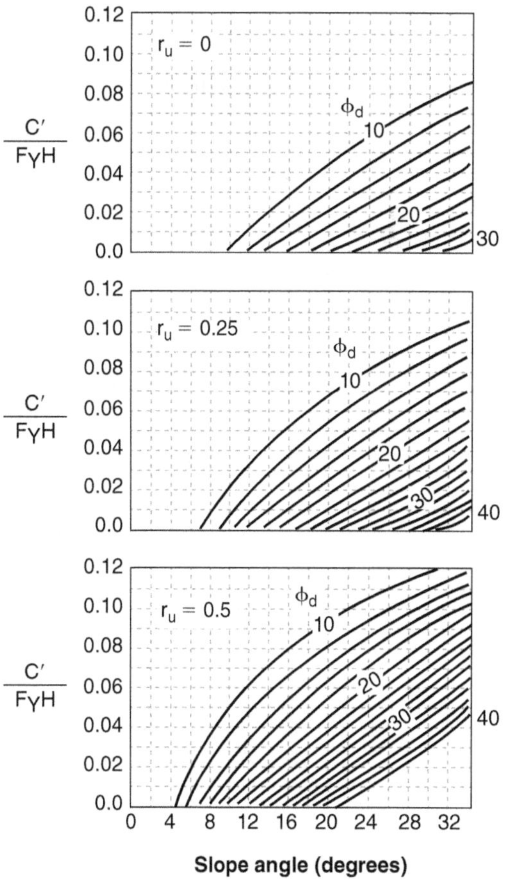

Figure 19.15 Spencer chart for $\varphi' > 0$, $c' > 0$. water in slope soils (Spencer 1967). (This material is reproduced with permission of John Wiley & Sons, Inc.)

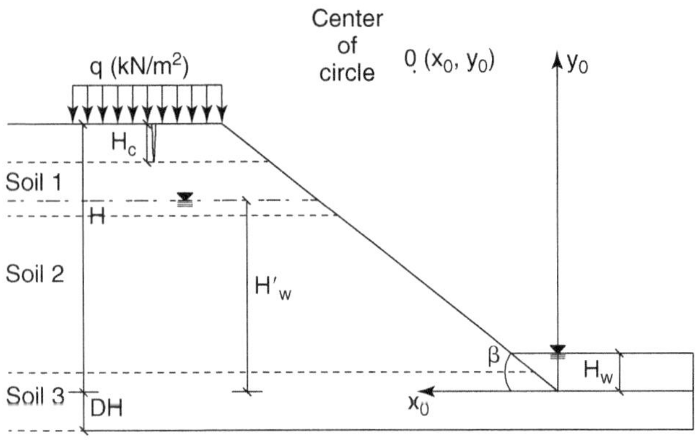

Figure 19.16 Slope parameters for Janbu chart.

19.8.3 Janbu Chart

Janbu (1968) developed an extensive set of charts covering many different cases. They include charts for the case where the soil has several layers, charts for the case where a surcharge exists on top of the slope, charts where a crack exists at the top of the slope, charts dealing with undrained short-term behavior, charts dealing with effective stress drained behavior, and charts dealing with different water levels outside the slope and inside the slope (Figure 19.16). These charts are detailed in Abramson et al. (2002) and in Duncan and Wright (2005). The chart dealing with a uniform soil, effective stress parameters c′ and φ', different water levels inside and outside the slope, and toe circles as failure surfaces is discussed here.

The procedure is as follows:

1. Calculate P_d as:

$$P_d = \frac{\gamma H + q - \gamma_w H_w}{\mu_q \mu_w \mu_t} \quad (19.72)$$

where P_d is a stress parameter characterizing the demand side of the slope stability; γ is the total unit weight of the soil; H is the height of the slope; q is the uniform surcharge at the top of the slope; γ_w is the unit weight of water; H_w is the height of water outside of the slope above the toe of the slope (Figure 19.16); and μ_q, μ_w, and μ_t are reduction factors for the surcharge, the submergence, and the tension crack respectively. In the case of no surcharge, μ_q is 1; in the case of no tension cracks, μ_t is 1 as well. The value of μ_w is found in the chart shown in Figure 19.17.

2. Calculate the effective stress parameter P_e as:

$$P_e = \frac{\gamma H + q - \gamma_w H'_w}{\mu_q \mu'_w} \quad (19.73)$$

where P_e is an effective stress parameter characterizing the average effective stress on the failure plane, γ is the total unit weight of the soil, H is the height of the slope, q is the uniform surcharge at the top of the slope, γ_w is the unit weight of water, H'_w is the height of water within the slope above the toe of the slope (Figure 19.16), and μ_q and μ'_w are the surcharge reduction factor and the seepage factor respectively. In the case of no surcharge, μ_q is 1 and the value of μ'_w is found in the chart shown in Figure 19.17.

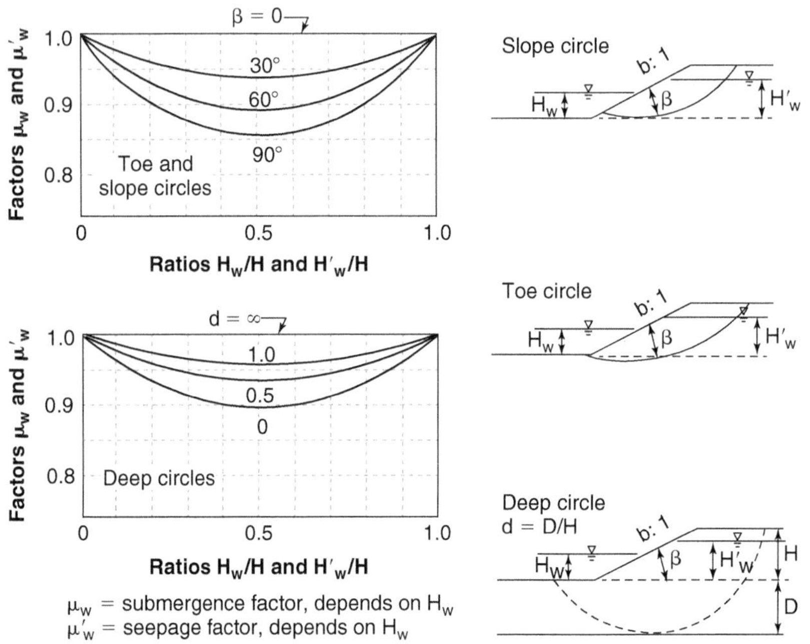

Figure 19.17 Janbu chart for the submergence reduction factor μ_w and μ'_w (Janbu 1968). (This material is reproduced with permission of John Wiley & Sons, Inc.)

3. Calculate $\lambda_{c\varphi}$ as:

$$\lambda_{c\varphi} = \frac{P_e \tan \varphi'}{c'} \tag{19.74}$$

where $\lambda_{c\varphi}$ is a parameter characterizing the ratio between the strength due to friction over the strength due to cohesion, and c' and φ' are the effective stress cohesion and friction angle respectively.

4. Using the chart in Figure 19.18, together with the slope angle β and the strength ratio $\lambda_{c\varphi}$, determine the stability number N_{cf}.

5. Calculate the factor of safety as:

$$F = N_{cf} \frac{c'}{P_d} \tag{19.75}$$

6. The location of the center of the failure circle is given by the chart in Figure 19.19. The chart gives the normalized values x_o and y_o of the coordinates:

$$X_o = x_o H \quad and \quad Y_o = y_o H \tag{19.76}$$

where X_o and Y_o are the actual coordinates in meters, and H is the height of the slope.

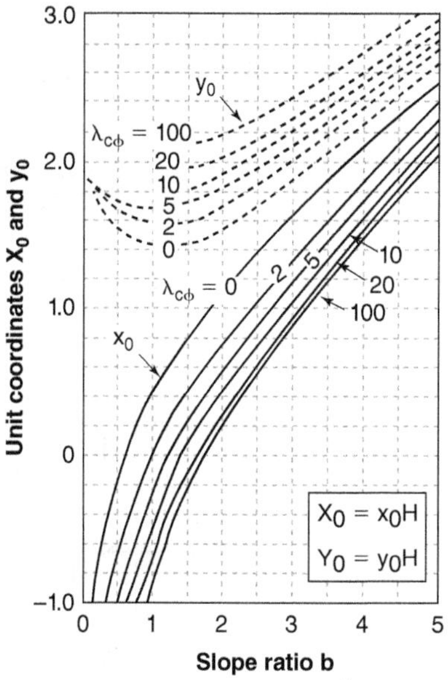

Figure 19.19 Janbu chart for locating the center of the critical circle. (After Janbu 1968. From J. Michael Duncan and Stephen G. Wright, *Soil Strength and Slope Stability*, Hoboken, NJ, John Wiley & Sons, 2005. This material is reproduced with permission of John Wiley & Sons, Inc.)

19.8.4 Morgenstern Chart

Morgenstern (1963) developed charts for the case of a rapid drawdown in a dam (Figures 19.20 and 19.21). The charts are for a uniform soil slope, effective stress parameters c′ and φ', soil total unit weight γ, a slope with a height H, and the

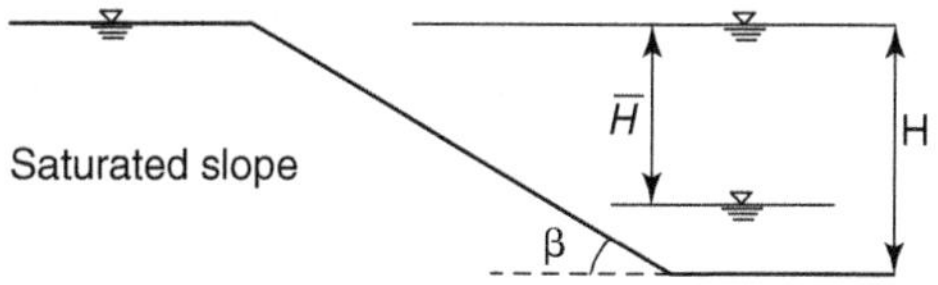

Figure 19.20 Slope parameters for Morgenstern chart: *(a)* $c'/\gamma H = 0.0125$. *(b)* $c'/\gamma H = 0.025$.

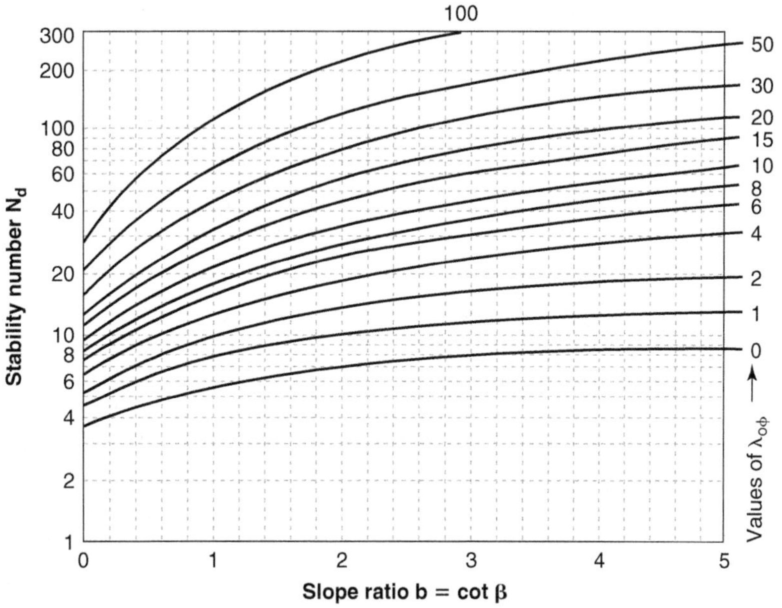

Figure 19.18 Janbu chart for obtaining the stability factor. (After Janbu 1968. From J. Michael Duncan and Stephen G. Wright, *Soil Strength and Slope Stability*, Hoboken, NJ, John Wiley & Sons, 2005. This material is reproduced with permission of John Wiley & Sons, Inc.)

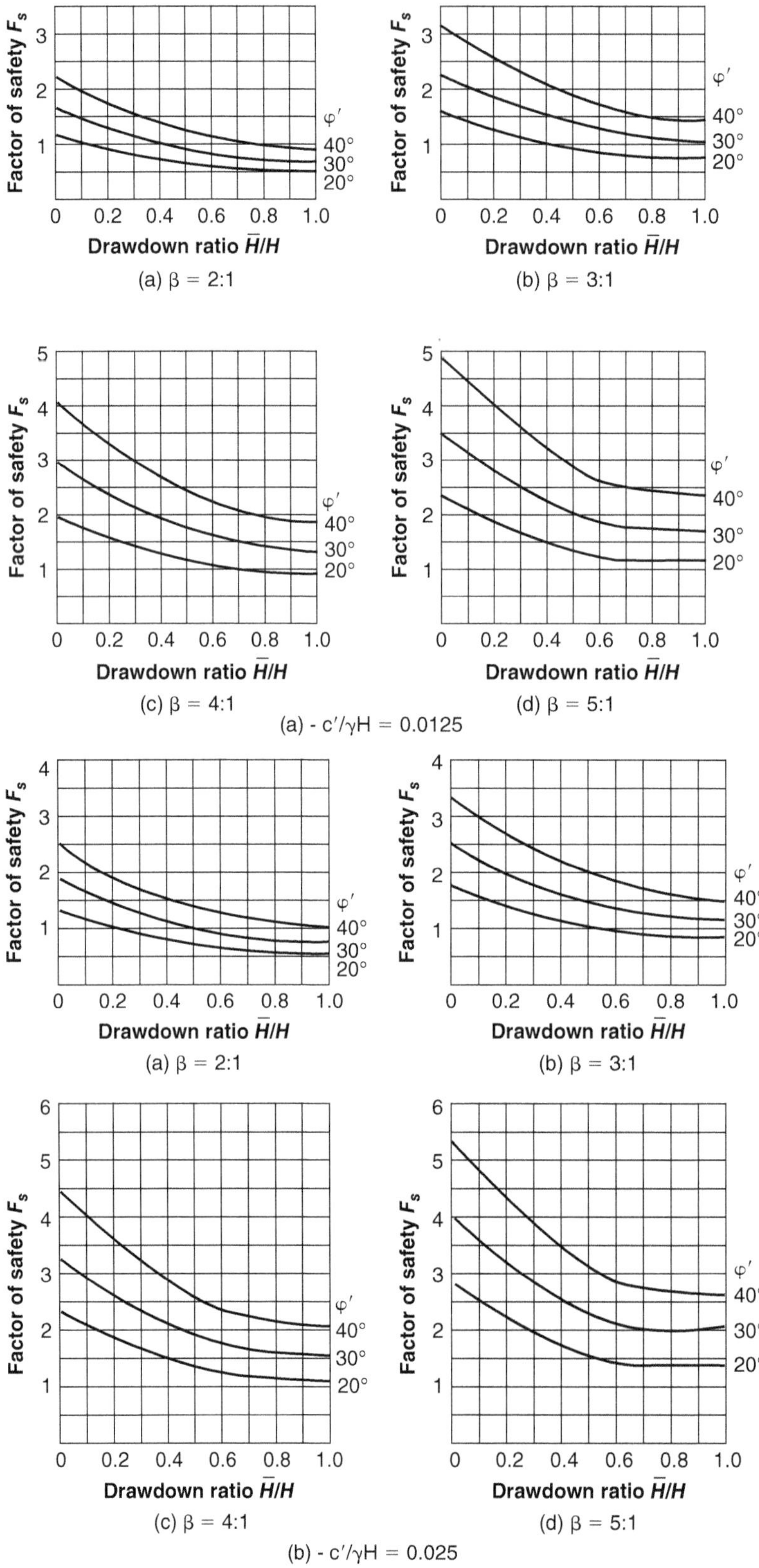

Figure 19.21 Morgenstern chart for rapid drawdown. (After Morgenstern 1963. *a* and *b*: From L. W. Abramson et al., *Slope Stability and Stabilization Methods* (1st ed.), Hoboken, NJ, John Wiley & Sons, 1996. This material is reproduced with permission of John Wiley & Sons, Inc.)

water level being drawn down an amount $\overline{H}$ from the top of the slope to a lower level. It is further assumed that the water stress in the soil does not have time to dissipate during the drawdown period.

The procedure is as follows:

1. Calculate the quantities $c'/\gamma H$ and $\tan\beta$.
2. Select the chart that corresponds to the correct $c'/\gamma H$, and the correct $\tan\beta$.
3. Using the values of $\overline{H}/H$ and φ', find the value of the factor of safety on the chart.

19.9 METHOD OF SLICES

The method of slices avoids some of the limitations associated with the chart methods. The method of slices is applicable to layered soils and to any water stress distribution. It is still associated with circular failure surfaces, although the concept can be applied to other shapes. The origin of the method goes back to the work of Fellenius (1927), a Swedish engineer. This problem-solving approach proceeds by breaking down the mass of soil into elements, drawing a free-body diagram of each element, writing the constitutive and fundamental equations at the element level, solving for the unknowns, and reassembling the pieces once the forces are known at the element level. Recall that what we really want to evaluate is the factor of safety of the slope F as defined in Eq. 19.1. In the method of slices, the soil mass is sliced as shown in Figure 19.22. Typically a minimum of 10 slices is necessary for reasonable accuracy. Figure 19.23 shows a slice with all parameters indicated. These parameters are defined in Table 19.1.

The number of unknowns and the number of equations available to find the values of the unknowns must be evaluated. The soil properties and the geometry of the slope are known quantities. The known forces are Q, U_β, W, k_h, k_v, and U_θ, whereas the known distances are b, h, and h_C. Furthermore, it is commonly assumed that the reactions N′ and U_θ are acting at the midpoint of the bottom of the slice while

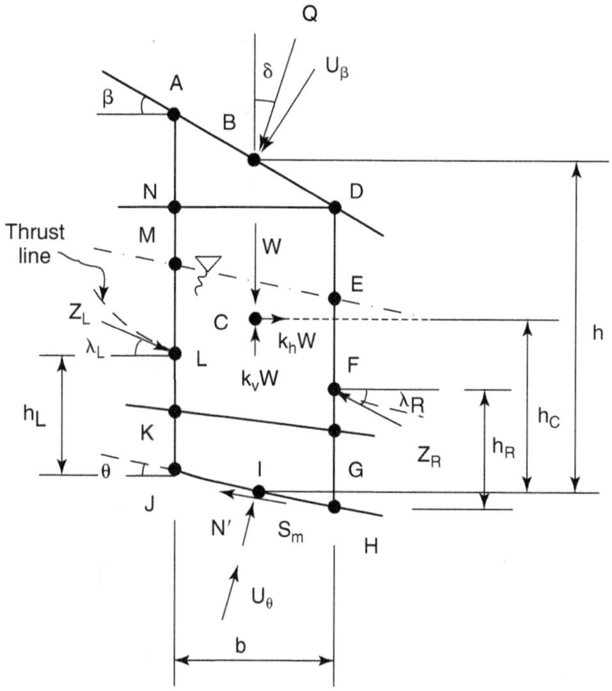

Figure 19.23 Slope slice i.

Q and U_β are applied at the middle of the top of the slice. The number of unknowns and the number of equations are shown in Table 19.2.

The total number of unknowns is 5n−2 and the total number of equations is 4n; therefore, there are n−2 unknowns in excess and the problem is statically indeterminate. It is necessary to make assumptions. Many assumptions have been made over time and each set of assumptions has been associated with one of the methods of slices. The assumptions and the associated names are presented in Table 19.3. That table shows the progress that took place over a period of 50 years in reducing the coarseness of the assumptions and increasingly satisfying the fundamental equations. The ordinary method of slices, the Bishop simplified method, and the generalized

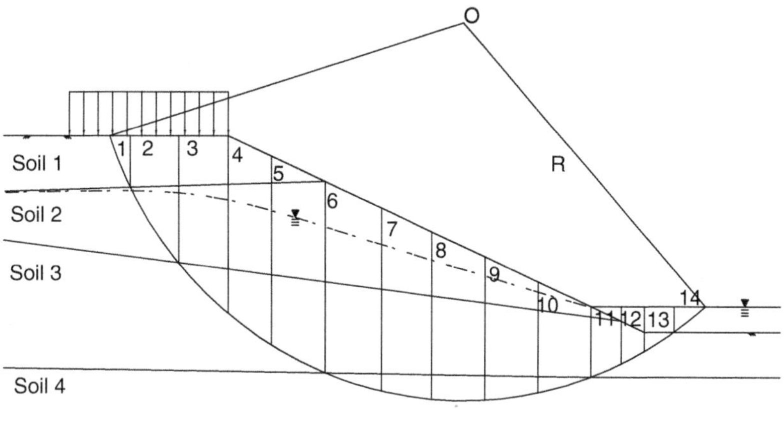

Figure 19.22 Sliced slope.

Table 19.1 Definition of Parameters in Figure 19.23

Q = force applied at the top of the slice	δ = angle of force Q with the vertical
U_β = water force applied at top of slice	θ = angle of bottom of slice with horizontal
W = total weight of slice	β = angle of top of slice with horizontal
$k_h W$ = horizontal static force due to earthquake	λ_L = angle of force Z_L with horizontal
$k_v W$ = vertical static force due to earthquake	λ_R = angle of force Z_R with horizontal
Z_L = earth pressure force on left side of slice	h_L = height of point of application of Z_L above bottom of left side of slice
Z_R = earth pressure force on right side of slice	h_R = height of point of application of Z_R above bottom of right side of slice
S_m = mobilized shear force at bottom of slice	h_C = height of center of gravity of slice above the middle of the bottom of slice
N′ = normal force on base of slice transmitted through the grains	h = height of slice from center of bottom to center of top
U_θ = water force applied at bottom of slice	b = width of slice

limit equilibrium method are detailed next. The other methods are presented in Abramson et al. (2002) and in Duncan and Wright (2005).

19.9.1 Ordinary Method of Slices

The assumption made by Fellenius (1927) is that the resultant of Z_L and Z_R (Figure 19.23) is equal to zero. This assumption decreases the number of unknowns by $3n - 3$ (Table 19.2), leaving $2n + 1$ unknowns and 4n equations. Therefore, the system is overdeterminate, meaning that not all equations can be satisfied. Fellenius chose to satisfy equilibrium of each slice in a direction perpendicular to the bottom of the slice. Referring to Figure 19.23, this leads to the following expression for slice i:

$$N_i' + U_{\theta i} + k_h W_i \sin\theta_i - W_i(1 - k_v)\cos\theta_i - U_{\beta i}\cos(\beta_i - \theta_i) - Q_i\cos(\delta_i - \theta_i) = 0 \quad (19.77)$$

Table 19.2 Unknowns and Equations for the Method of Slices

Unknowns	Equations
n values of S_m forces	n force equilibrium equations in x direction
n values of N′	n force equilibrium equations in y direction
n − 1 values of Z forces	n moment equilibrium equations
n − 1 values of the angles λ	n shear strength equation
n − 1 values of the location of the Z forces	
1 factor of safety	
TOTAL = $5n - 2$	TOTAL = $4n$

This equation gives the expression of N_i'.

The value of $U_{\theta i}$ is given by:

$$U_{\theta i} = \alpha_i u_{wi} \frac{b_i}{\cos\theta_i} \quad (19.78)$$

where α_i is the area ratio for the soil along the bottom of the slice (see section 15.5) and u_{wi} is the average water stress at the bottom of the slice. Obtaining the value u_{wi} is discussed in section 19.10. The weight of the slice W_i is calculated using the total unit weight of the soil. If the slice includes several soil layers, the weight is given by:

$$W_i = \sum_{j=1}^{m} \gamma_j A_j \quad (19.79)$$

where γ_j is the total unit weight of soil j within slice i and A_j is the area of soil j included within slice i.

The expression for the mobilized shear force at the bottom of slice i necessary to keep the slope in equilibrium S_{mi} is given by the shear strength equation and the factor of safety:

$$S_{mi} = \frac{c_i' \dfrac{b_i}{\cos\theta_i} + N_i' \tan\varphi_i'}{F} \quad (19.80)$$

where F is the global factor of safety for the slope. Then the global factor of safety F for the n slices in the slope is given by the ratio between the global maximum resisting moment $M_{R\max}$ around O, the center of the circle, and the global driving moment M_D around O:

$$F = \frac{M_{R\max}}{M_D} \quad (19.81)$$

The expressions for $M_{R\max}$ and M_D are:

$$M_{R\max} = \sum_{i=1}^{n} \left(c_i' \frac{b_i}{\cos\theta_i} + N_i' \tan\varphi_i' \right) R \quad (19.82)$$

Table 19.3 Methods of Slices, Authors, and Assumptions

Name of Method	Reference	Assumptions	Comment
Ordinary method of slices	Fellenius 1927	Resultant of Z forces on each slice is equal to zero.	Based on writing equilibrium perpendicular to base. Does not satisfy all equilibrium equations. Overdeterminate.
Janbu simplified method	Janbu 1954	Z forces are horizontal.	Does not satisfy all equilibrium equations. Overdeterminate.
Bishop simplified method	Bishop 1955	Shear forces on the side of all slices are zero (i.e., Z forces are horizontal).	Based on writing vertical force equilibrium. Does not satisfy all equilibrium equations. Overdeterminate.
Bishop rigorous method	Bishop 1955	Shear forces on the side of all slices are assumed.	Satisfies all equilibrium equations.
Lowe and Karafiath method	Lowe and Karafiath 1960	Z_R forces inclined at an angle equal to the average between the angle of the top and bottom of the slice.	Does not satisfy all equilibrium equations.
Morgenstern-Price method	Morgenstern and Price 1965	Inclination of Z forces given by a function of the horizontal distance multiplied by a scalar.	Satisfies all equilibrium equations.
Spencer method	Spencer 1967	Z forces have a constant but unknown inclination.	Satisfies all equilibrium equations.
Corps of Engineers method	U.S. Army Corps of Engineers 1970	Z forces inclined parallel to the ground surface or parallel to the line joining the beginning and the end of the failure circle.	Does not satisfy all equilibrium equations.
Janbu generalized method	Janbu 1973	Location of point of application of the Z forces on an assumed thrust line.	Does not satisfy all equilibrium equations.
Sarma method	Sarma 1973	Inclination of Z forces given by a function of the horizontal distance multiplied by a scalar.	Makes use of horizontal seismic coefficient. Satisfies all equilibrium equations.

$$M_D = \sum_{i=1}^{n} \left[\begin{array}{l} (W_i(1-k_v) + U_{\beta i}\cos\beta_i + Q_i\cos\delta_i) \\ \times R\sin\theta_i - (U_{\beta i}\sin\beta_i + Q_i\sin\delta_i) \\ \times(R\cos\theta_i - h_i) + k_h W_i (R\cos\theta_i - h_{ci}) \end{array} \right] \quad (19.83)$$

and the general expression of the factor of safety for the ordinary method of slices is:

$$F = \frac{\displaystyle\sum_{i=1}^{n} \left(\begin{array}{l} c_i' \dfrac{b_i}{\cos\theta_i} + (W_i(1-k_v)\cos\theta_i \\ +U_{\beta i}\cos(\beta_i - \theta_i) + Q_i\cos(\delta_i - \theta_i) - U_{\theta i} \\ -k_h W_i \sin\theta_i)\tan\varphi_i' \end{array} \right) R}{\displaystyle\sum_{i=1}^{n} \left[\begin{array}{l} (W_i(1-k_v) + U_{\beta i}\cos\beta_i + Q_i\cos\delta_i) \\ \times R\sin\theta_i - (U_{\beta i}\sin\beta_i + Q_i\sin\delta_i) \\ \times(R\cos\theta_i - h_i) + k_h W_i (R\cos\theta_i - h_{ci}) \end{array} \right]} \quad (19.84)$$

In the simple case where $k_h = k_v = U_{\beta i} = Q_i = 0$ (no earthquake, no water on top of ground surface, no structures on top of ground surface), the expression of the factor of safety becomes:

$$F = \frac{\displaystyle\sum_{i=1}^{n} \left(c_i' \frac{b_i}{\cos\theta_i} + \left(W_i\cos\theta_i - \alpha_i u_{wi} \frac{b_i}{\cos\theta_i} \right) \tan\varphi_i' \right)}{\displaystyle\sum_{i=1}^{n} W_i \sin\theta_i} \quad (19.85)$$

The sequential steps to be followed to obtain F correspond to the columns in Table 19.4.

Then the factor of safety is given by:

$$F = \frac{\textit{Sum of column } 14}{\textit{Sum of column } 6} \quad (19.86)$$

Table 19.4 Hand Calculations for the Ordinary Method of Slices (Simple Case of No Earthquake, No Water above Ground Surface, and No Structural Load on Ground Surface)

1	2	3	4	5	6	7	8	9	10	11	12	13	14
Slice no.	Area	Unit weight	W	θ	W $\sin\theta$	W $\cos\theta$	b/ $\cos\theta$	α	u_w	$\tan\phi'$	c′	7 − 8× 9 × 10	8 × 12+ 13 × 11
	m^2/m	kN/m^3	kN/m	°	kN/m	kN/m	m		kN/m^2		kN/m^2	kN/m	kN/m
1													
—													
i	$A_1, \ldots,$ $A_j, \ldots,$ A_m	$\gamma_1, \ldots,$ $\gamma_j, \ldots,$ γ_m											
—													
n													

The following notes are very important:

1. Make a drawing to scale of the slope, including the groundwater level and the external loads. This is necessary because the areas in column 2 are measured on the drawing.
2. Choose the circle to be analyzed.
3. Use a minimum of 10 slices and make the slices correspond to natural intersections with the chosen failure circle.
4. The unit weights in column 3 are total unit weights. This means that the seepage force is considered an internal force and must not be included in the calculations.
5. One way to handle a free water body (river, lake) on top of the ground surface is to let the circle cut through the water body, which then becomes part of the free-body diagram. Then, if there is water on top of a slice, the weight of that volume of water must be included in the total weight W in column 4. If water at the end of the circle is considered (e.g., Figure 19.22), then the last slice is a water slice with weight but zero values for c′ and φ'.
6. An alternative way to consider a free water body on top of the ground surface is to consider that the free body stops at the ground surface and to treat the water on top of this body as an external load with weight and direction. This is the way it is presented in Figure 19.23. External loads (due, for example, to structures on the slope surface) are handled in this fashion.
7. The angle θ in column 5 must carry a sign, which will affect the sign of the columns with θ in them. Column 6 will often be affected by the sign of θ; the negative sign of θ indicates that the slice decreases the driving moment of the soil mass.
8. Tan φ' and c′ must be the soil properties at the bottom of the slice, not the average properties of the soil within the slice. The reason is that the shear strength is being evaluated at the bottom of the slice.
9. The quantity in Column 13 for a given slice cannot be negative. If it is and if the calculations are correct, set it equal to zero.

19.9.2 Bishop Simplified Method

The assumption made by Bishop (1955) is that the Z forces (Z_L and Z_R in Figure 19.23) are horizontal. This assumption decreases the number of unknowns by $n-1$, because the angles λ are known (Table 19.2), leaving $4n-1$ unknowns and $4n$ equations. Therefore, the system is overdeterminate by one, meaning that not all equations can be satisfied. Bishop chose to satisfy equilibrium of each slice in the vertical direction. Referring to Figure 19.23, this leads to the following expression for slice i:

$$(N_i' + U_{\theta i})\cos\theta_i + S_{mi}\sin\theta_i - W_i(1-k_v) - U_{\beta i}\cos\beta_i - Q_i\cos\delta_i = 0 \tag{19.87}$$

The expression of S_{mi} remains the same as in the OMS (Eq. 19.80). By combining Eq. 19.87 with the expression of S_{mi} (Eq. 19.80), the following expression of N_i'. is obtained:

$$N_i' = \frac{1}{m_{\theta i}}\left(\begin{array}{c} W_i\,(1-k_v) - \dfrac{c_i'\dfrac{b_i}{\cos\theta_i}\sin\theta_i}{F} - U_{\theta i}\cos\theta_i \\ +U_{\beta i}\cos\beta_i + Q_i\cos\delta_i \end{array} \right) \tag{19.88}$$

with (Figure 19.24):

$$m_{\theta i} = \cos\theta_i\left(1 + \frac{\tan\theta_i\tan\varphi'}{F}\right) \tag{19.89}$$

The expressions of $U_{\theta i}$, W_i, F, M_{Rmax}, and M_D are given by the same equations as in the OMS (Eqs. 19.78, 19.79,

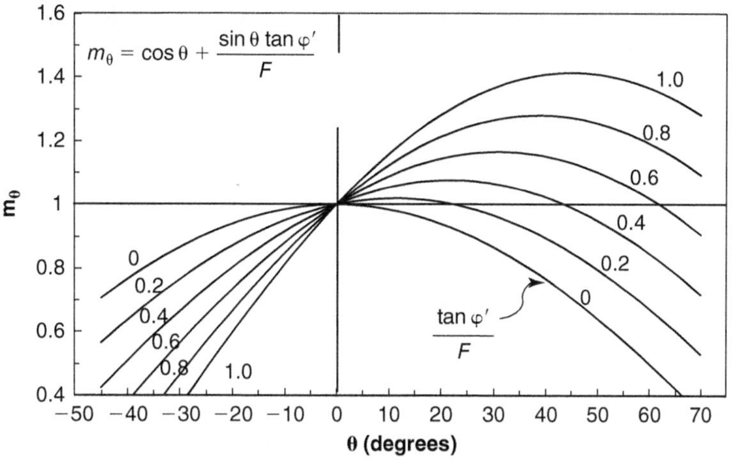

Figure 19.24 Graphical values of the parameters m_θ.

19.81, 19.82, and 19.83 respectively). The only thing that changes is the expression of N′. The final expression of the factor of safety is:

$$F = \frac{\sum_{i=1}^{n} \left(\begin{array}{c} c_i' \dfrac{b_i}{\cos\theta_i} + \left(\dfrac{1}{m_{\theta i}} \left(W_i\,(1-k_v) - \dfrac{c_i' b_i \tan\theta_i}{F} \right) \right) \\ -U_{\theta i}\cos\theta_i + U_{\beta i}\cos\beta_i + Q_i\cos\delta_i)) \tan\varphi_i' \end{array} \right) R}{\sum_{i=1}^{n} \left[\begin{array}{c} (W_i\,(1-k_v) + U_{\beta i}\cos\beta_i + Q_i\cos\delta_i) \\ R\sin\theta_i - (U_{\beta i}\sin\beta_i + Q_i\sin\delta_i) \\ \times (R\cos\theta_i - h_i) + k_h W_i (R\cos\theta_i - h_{ci}) \end{array} \right]} \tag{19.90}$$

In the simple case where $k_h = k_v = U_{\beta i} = Q_i = 0$ (no earthquake, no water on top of ground surface, no structures on top of ground surface), the expression of the factor of safety becomes:

$$F = \frac{\sum_{i=1}^{n} \frac{1}{m_{\theta i}} (c_i' b_i + (W_i - \alpha_i u_{wi} b_i) \tan\varphi_i')}{\sum_{i=1}^{n} W_i \sin\theta_i} \tag{19.91}$$

The sequential steps to be followed to obtain F correspond to the columns in Table 19.5.

Iterations are continued until two consecutive factors of safety fall within the target tolerance.

19.9.3 Generalized Equilibrium Method

Many other methods exist that make various assumptions about the side forces Z_L and Z_R, their inclination, and their location. The generalized equilibrium method (Abramson et al. 2002) exemplifies the general approach. In this method, the inclination angle λ of the side forces is assumed to be described by a function expressed as:

$$\lambda_i = \eta f(x_i) \tag{19.92}$$

where η is a scalar constant for the slope and $f(x_i)$ is the function with values between 0 and 1 describing the variation of the side forces angle λ_i as a function of the horizontal distance x_i along the slope. Examples of the function f(x) are shown in Figure 19.25.

Equation 19.92 decreases the number of unknowns by $n - 1$, as it gives the value of the interslice forces inclinations λ_i, but it does introduce one more unknown in η for a total reduction of unknowns of $n - 2$. This brings down the total number of unknowns to exactly 4n, which now corresponds exactly to the 4n number of equations available. Hence, the system is statically determinate. The equations are similar to those of the Bishop simplified method except that the side forces are now included. Force equilibrium parallel to the base gives n equations:

$$\begin{aligned} &S_{mi} + Z_{Li}\cos(\theta_i - \lambda_{Li}) - Z_{Ri}\cos(\theta_i - \lambda_{Ri}) \\ &\quad - W_i(1-k_v)\sin\theta_i - W_i k_h \cos\theta_i - U_{\beta i}\sin(\theta_i - \beta_i) \\ &\quad - Q_i \sin(\theta_i - \delta_i) = 0 \end{aligned} \tag{19.93}$$

Force equilibrium perpendicular to the base gives n equations:

$$\begin{aligned} &N_i' + Z_{Ri}\sin(\theta_i - \lambda_{Ri}) - Z_{Li}\sin(\theta_i - \lambda_{Li}) \\ &\quad - W_i(1-k_v)\cos\theta_i + W_i k_h \sin\theta_i + U_{\theta i} \\ &\quad - U_{\beta i}\cos(\theta_i - \beta_i) - Q_i\cos(\theta_i - \delta_i) = 0 \end{aligned} \tag{19.94}$$

Moment equilibrium around the point at the middle of the base leads to n equations:

$$\begin{aligned} &Z_{Li}\cos\lambda_{Li}\left(h_{Li} - \frac{b_i}{2}\tan\theta_i\right) - Z_{Ri}\cos\lambda_{Ri}\left(h_{Ri} + \frac{b_i}{2}\tan\theta_i\right) \\ &+ Z_{Li}\frac{b_i}{2}\sin\lambda_{Li} + Z_{Ri}\frac{b_i}{2}\sin\lambda_{Ri} - W_i k_h h_{ci} + U_{\beta i} h_i \sin\beta_i \\ &+ Q_i h_i \sin\delta_i = 0 \end{aligned} \tag{19.95}$$

Table 19.5 Hand Calculations for Bishop Simplified Method (Simple Case of No Earthquake, No Water above Ground Surface, and No Structural Load on Ground Surface)

1	2	3	4	5	6	7	8	9	10	11	12
Slice no.	Area	Unit weight	W	θ	W sin θ	b	α	u_w	tan φ'	c′	$11 \times 7 + (4 - 8 \times 9 \times 7) \times 10$
	m^2/m	kN/m^3	kN/m	°	kN/m	m		kN/m^2		kN/m^2	kN/m
1											
...											
i	$A_1, \ldots,$ $A_j, \ldots,$ A_m	$\gamma_1, \ldots,$ $\gamma_j, \ldots,$ γ_m									
...											
n											

13	14	15	16	17	18	19
Choose $F_1 = 1.5$?	$m_{\theta 1}$ = Eq. 19.85 or Figure 19.24	12/14	$F_2 = \Sigma 15/\Sigma 6$	$m_{\theta 2}$ = Eq. 19.85 or Figure 19.22 using F_2	12/17	$F_3 = \Sigma 18/\Sigma 6$

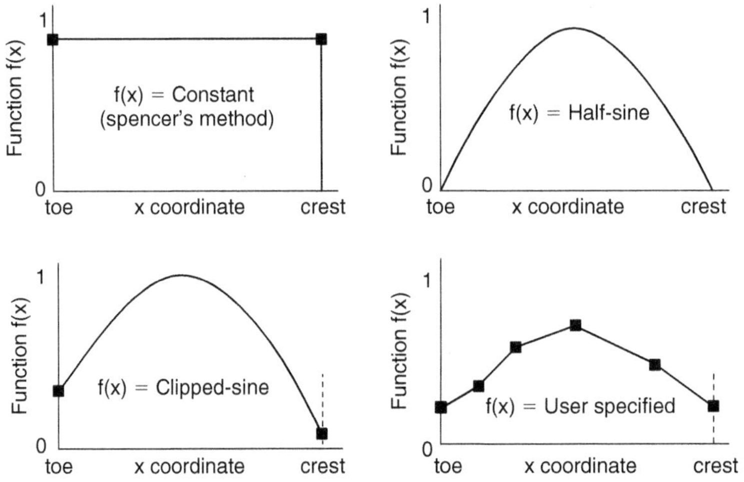

Figure 19.25 Example of functions f(x) (after Abramson et al., 2002).

Then one can write n shear strength equations:

$$S_{mi} = \frac{c_i' \dfrac{b_i}{\cos\theta_i} + N_i' \tan\varphi_i'}{F} \quad (19.96)$$

These 4n equations contain $4n$ unknowns, which are the n values of the S_m forces, the n values of the N' forces, the $n-1$ values of the Z forces, the n − 1 values of the location of the Z forces, the scalar η, and the factor of safety F. The system is solved for those variables and the factor of safety is found in that fashion.

19.9.4 Critical Failure Circle

The method of slices gives the factor of safety for a chosen failure circle. The trick is to find which circle will give the lowest possible factor of safety; this is called the *critical circle.* Because the center of the circle and the radius of the circle can both vary, there is a double infinity of possible circles. The search for the critical circle typically proceeds by choosing a center location and then varying the radius of the circle until the lowest factor of safety is found for that center. That center is then assigned the corresponding value of the factor of safety. Many different centers are tried and each time the radius is varied until the minimum factor of safety is found for that center. A map is prepared of the center locations and the associated factors of safety (Figure 19.26). This map describes the surface of the factor of safety F in two dimensions ($F = F(x, y)$).

Two options are available for a computer program to search for the minimum factor of safety: the automatic search and the grid approach. Some software programs have an automatic search mode, in which the slope of the surface F(x,y) is used to move the location of the center toward lower F values until a minimum is found. The problem with this approach is that the minimum could be a local minimum and not the absolute minimum. This is a bit like finding a low valley in a mountain range but not finding the deepest valley over the peak next to it. One way to alleviate this problem is to repeat the automatic search by starting the search at a different center location.

With the grid approach, the user inputs a grid of center locations and the program outputs the factor of safety surface, leaving the decision of where the minimum factor of safety might be up to the user. A broad grid is used at first and can be refined once the likely location is more precisely identified.

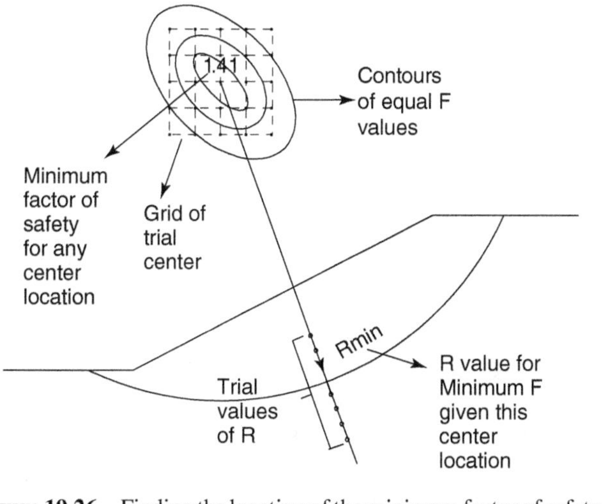

Figure 19.26 Finding the location of the minimum factor of safety.

19.10 WATER STRESS FOR SLOPE STABILITY

The water stress along the bottom of the failure surface has a significant influence on the factor of safety. The water stress at the bottom of a slice can be positive (below the groundwater level) or negative (above the groundwater level). A high positive water stress (compression) leads to a low factor of safety, whereas a high negative water stress (tension) leads to a high factor of safety. There are several ways to estimate the water stress in a slope: piezometric surface, water stress ratio value r_u, and grid of water stress values (Figure 19.27).

19.10.1 Piezometric and Phreatic Surface

A distinction must be made between the groundwater level, also called the *phreatic surface*, and the piezometric surface. If you drill a borehole in the ground, water will come to equilibrium at a certain level in the hole: this level corresponds to the phreatic surface or groundwater level. If you consider a point M in the ground and calculate the water stress at M as the product of the unit weight of water times the distance from M to a surface, then that surface is the piezometric surface. The groundwater level does not depend on the location of M, but the piezometric surface does. In most cases the piezometric surface is slightly below the phreatic surface, and using the phreatic surface as the piezometric surface will lead to a factor of safety slightly lower than the true factor of safety. The expression of the water stress u_w at point M is then:

$$u_w = \gamma_w h_p \quad (19.97)$$

where γ_w is the unit weight of water and h_p is the pressure head (positive or negative). The *pressure head* is the vertical

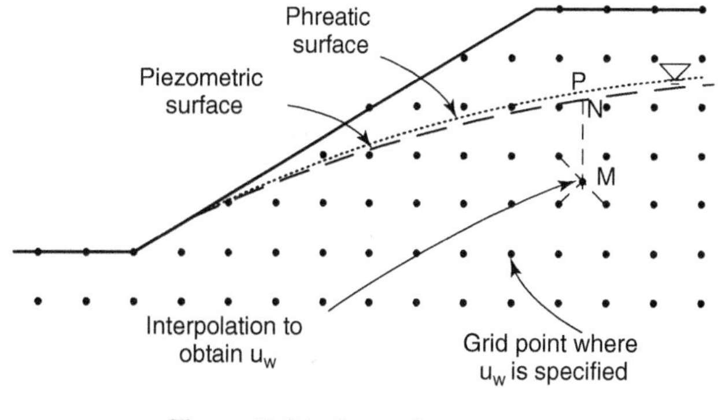

Figure 19.27 Input of water stress.

distance between point M and the piezometric surface. In the absence of a piezometric surface, the phreatic surface or groundwater level can be used as a first approximation. Note that if the point is above the groundwater level, the water stress will be negative, indicating water tension.

19.10.2 Water Stress Ratio Value

The water stress ratio r_u is defined as:

$$r_u = \frac{u_w}{\sigma_{ov}} \tag{19.98}$$

where u_w is the water stress at a point M and σ_{ov} is the total vertical stress at the same point M. Specifying a single value of r_u *for* a slope is very convenient, as it becomes simple to calculate the water stress at the bottom of the slice from the total vertical stress at the bottom of the slice. The problem is that the true r_u value may vary from one location along the failure surface to another; thus, this is a simplification, albeit a convenient one. Using a single r_u value is cruder than using a piezometric or phreatic surface, particularly for points of the failure surface that are above the groundwater level.

19.10.3 Grid of Water Stress Values

The approach that uses a grid of water stress values consists of inputting the water stress in the slope mass at grid points canvassing the slope. This input solution is more time-consuming than the two previously mentioned solutions, but it is also the most precise and versatile way to input the water stress. The grid size can vary, but should be fine enough that the interpolation between grid points leads to reasonable accuracy in the value of the water stress (Figure 19.27). Note that negative values (water tension) can be input with this solution at appropriate places in the grid.

19.10.4 Water Stress Due to Loading

If the slope is subjected to loading that induces water stress, the approach consists of calculating the hydrostatic water stress and the excess water stress separately:

$$u_w = u_{wo} + \Delta u_{we} \tag{19.99}$$

where u_{wo} is the hydrostatic component and Δu_{we} is the excess water stress. If the excess water stress is due to loading on the ground surface, it is generally calculated by first calculating the vertical normal total stress increase due to the load $\Delta\sigma_v$ in the soil mass (see section 17.8.7). Then the value of Δu_{we} is related to $\Delta\sigma_v$ by:

$$\overline{B} = \frac{\Delta u_{we}}{\Delta\sigma_v} \tag{19.100}$$

The value of the water stress parameter $\overline{B}$ is 1 in soft, saturated soil under the water table, but can be much smaller in stiff, overconsolidated soils. If $\overline{B}$ is not known, one solution is to assume a value (0.5, for example), calculate the factor of safety with that assumption, monitor the water stress in the slope with piezometers during construction, and stop construction if the water stress goes over the assumed value if that value is critical. One particular case in which such an approach is warranted is when an embankment is built over a soft clay.

19.10.5 Seepage Analysis

If water seeps through a slope, the water stress will be different from hydrostatic conditions. To calculate the water stress in this case, a flow net solution can be used (see sections 13.2.12 to 13.2.16). Figure 19.28 shows an example. In this case, the water stress at point M is given by equation 19.97, where h_p is the pressure head expressed as:

$$h_p = h_t - h_e \tag{19.101}$$

where h_t and h_e are the total head and elevation head at M respectively. Referring to Figure 19.28, the total head at M is the same as the total head at A, because they are on the same equipotential line. Because the pressure head at A is zero, the pressure head at M is expressed as follows:

$$h_{tM} = h_{tA} \quad and \quad h_{pA} = 0 \quad then \quad h_{pM} = h_{eA} - h_{eM} \tag{19.102}$$

and the pressure head at M is the difference in elevation between M and A. Thus, the piezometric line is slightly below the phreatic line. If the slope is relatively flat, as most soil slopes are, the difference is small, but if the slope is steep, the difference can be larger and using the phreatic line as the piezometric line can be excessively conservative.

19.11 TYPES OF ANALYSES

Several types of analyses can be performed, including:

1. Drained or undrained analysis
2. Effective stress or total stress analysis
3. Long-term or short-term analysis

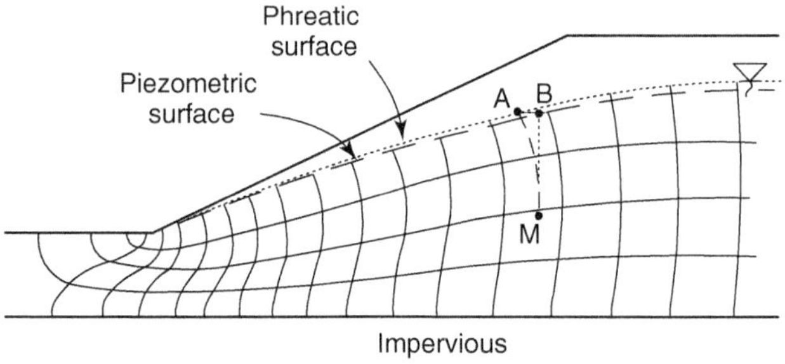

Figure 19.28 Flow net for slope stability.

In a *drained analysis*, the water stress is considered to be hydrostatic throughout the mass. The soil strength parameters associated with this analysis are the drained strength parameters or effective stress parameters.

An *undrained analysis* is used when the water does not have time to drain away. The soil strength parameter used in this case is the undrained shear strength. One must be careful to use the undrained shear strength corresponding to the stress path of the soil in the slope.

An *effective stress analysis* can be used in all cases. It makes no particular assumption regarding drainage and is based on sound fundamental principles. It makes use of the effective stress equation to obtain the shear strength of the soil based on the effective stress cohesion c′ and the effective stress friction angle φ'. It can be used for an undrained analysis, a drained analysis, a short-term analysis, or a long-term analysis. The difficulty with using an effective stress analysis is that the water stress in the soil mass must be known. This is particularly challenging in the case of an undrained analysis.

A *total stress analysis* considers that the soil is made of one material; it does not recognize the existence of the three components (particles, water, and air). Hence, one must be very careful when using such an analysis. A total stress analysis can be used in the case of a soil with no water and in the case of a soil where the shear strength is independent of rapid variations in total stress. One such case is the undrained behavior of soft, compressible, saturated soils under the water table right after loading by an embankment.

A *long-term analysis* considers that all water stresses induced by loading have had time to dissipate and are back to hydrostatic. In this regard, a long-term analysis is similar to a drained analysis.

A *short-term analysis* is used for a soil condition taking place shortly after loading. As such, it is often a drained analysis for fast-draining soils like free-draining sands and gravels, and an undrained analysis for slow-draining soils like silts and clays.

The effective stress analysis is the preferred analysis, but it is also often the most difficult to perform, because of the complexities associated with predicting water stresses in the soil mass due to loading and due to desaturation close to the ground surface. In all cases it is wise to perform both a short-term and a long-term analysis for any soil problem to ensure proper behavior in the short and long terms.

19.12 PROGRESSIVE FAILURE IN STRAIN-SOFTENING SOILS

An added complexity in selecting the shear strength to use in the failure analysis occurs when the soil exhibits strain-softening behavior. In this case there is a peak strength τ_{fmax} and a residual strength τ_{fres} after the peak (Figure 19.29). The complexity comes from the fact that the failing body is not rigid and moves differently along the failure surface. The largest displacements along the failure surface typically start at the bottom of the slope and move back until they reach the top of the slope. Therefore, the displacement could be large enough to be at the residual shear strength toward the bottom of the slope but still be at the peak shear strength toward the top of the slope. This is called *progressive failure* (Figure 19.29). In the case of progressive failure, the back-calculated shear strength from a slope failure could be between the peak shear strength and the residual shear strength. Progressive failure is most likely to occur for slopes excavated in overconsolidated, stiff, fine-grained soils; such slopes may exist for several years before failing.

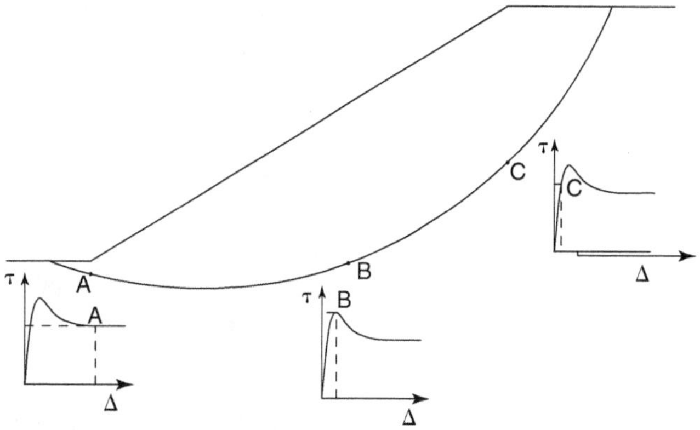

Figure 19.29 Progressive failure.

19.13 SHALLOW SLIDE FAILURES IN COMPACTED UNSATURATED EMBANKMENTS

Shallow slides may occur many years after an embankment is compacted at a water tension level that decreases with time, thereby weakening the soil strength. Aubeny and Lytton (2004) studied this problem and explained the phenomenon mathematically (see section 13.3.2) and experimentally. When embankments are built for freeway overpasses, for example, the approach embankment must be compacted and usually reaches a height of about 8 m. The side slopes are typically between 2 horizontal to 1 vertical and 3 horizontal to 1 vertical. The compaction takes place around the optimum water content (see Chapter 20), which corresponds to an unsaturated soil condition. Long after construction (e.g., 10 to 20 years), these embankments can experience shallow slide failures (Figure 19.30). The depth of these slides is about 1.5 to 2 meters. These failures take place because the water tension decreases as a function of time, as the as-built water tension is slowly reduced by repeated rainfalls. The drying and wetting process creates cracks that are typically as deep as they are horizontally spaced. This source of water at depth weakens the soil by decreasing the effective stress tied to the water tension—and the shallow slope fails.

Figure 19.30 Shallow slide in a compacted embankment. (Courtesy of Professor Charles Aubeny, Texas A&M University.)

The parts of the world where the rainfall and the temperature vary a lot during the seasons (tropics) are most likely to experience this problem. The solution is either to perform the slope stability analysis by using a wetted shear strength of the compacted soil or to prevent the water tension from being lost as a function of time. Geosynthetic covers may achieve this result.

19.14 REINFORCED SLOPES

19.14.1 Reinforcement Type

Many types of reinforcements can be used in a slope (Figure 19.31). They include rigid steel inclusions, geosynthetics, soil nails, stone columns, tieback anchors, and piles. Among these types of reinforcements, only tieback anchors are posttensioned to a preset tension force; all others are not. Of course, in most cases, the reinforcement ends up being in tension under working load conditions. Some reinforcement is considered to be rigid (e.g., soil nails), whereas other reinforcement is considered to be flexible (e.g., geosynthetics). This rigidity has an impact on the moment arm associated with the reinforcement.

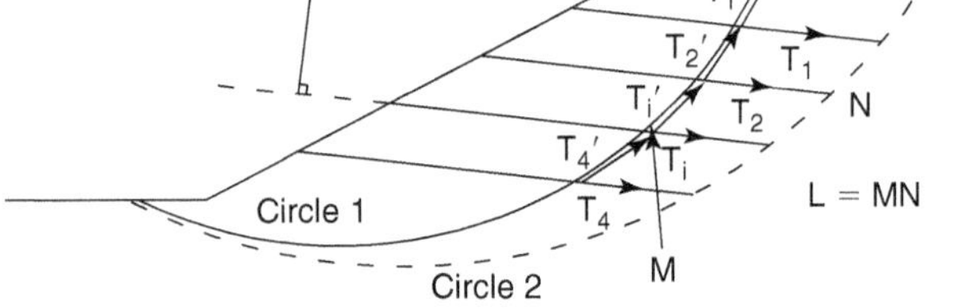

Figure 19.31 Reinforced slope.

19.14.2 Factor of Safety

The factor of safety for a slope and for a circular failure surface is defined in Eq. 19.81 as the ratio of the maximum resisting moment divided by the driving moment. Each layer of reinforcement increases the maximum resisting moment, so the factor of safety for a reinforced slope F_R is:

$$F_R = \frac{M_{R\,max(soil)} + M_{R\,max(reinforcement)}}{M_D} \tag{19.103}$$

where $M_{R\,max(soil)}$ is the maximum resisting moment provided by the soil along the failure circle considered, $M_{R\,max(reinforcement)}$ is the maximum resisting moment provided by the reinforcement, and M_D is the driving moment due to the soil weight and any other external loads. For an unreinforced slope, the factor of safety $F_{unreinforced}$ is:

$$F_U = \frac{M_{R\,max(soil)}}{M_D} \tag{19.104}$$

The expression of the maximum resisting moment provided by the reinforcement is:

$$M_{R\,max(reinforcement)} = \sum_{i=1}^{n} T_i d_i \tag{19.105}$$

where T_i is the maximum resistance of the ith reinforcement outside of the circle considered and d_i is the moment arm of the force T_i. The value of d_i depends on the flexibility of the reinforcement. If the reinforcement is rigid, such as soil nails, then the reinforcement will not bend along the potential plane of failure and the moment arm is the one associated with the direction of the reinforcement (d_i in Figure 19.31). If the reinforcement is flexible, then it will bend at the failure plane and follow the direction of the circle; in this case the moment arm of the reinforcement is the radius of the circle. The resistance force T_i is given by:

$$T_i = A_f f_u \tag{19.106}$$

where A_f is the contact area between the soil and the reinforcement (πDL for a cylindrical shape, 2(B + W)L for a rectangular shape), and f_u is the maximum shear stress that can be developed at the soil-reinforcement interface. The length L involved in calculating T_i is the length of reinforcement outside of the circle considered (MN in Figure 19.31). Note that for posttensioned reinforcement such as tieback anchors, the force T_i increases the compressive stresses on the failure plane, thereby increasing the shear strength along the plane of failure (see Chapter 21).

Another mode of failure in the case of a reinforced slope is for the failure circle to pass behind the reinforced zone (circle 2 in Figure 19.31). This circle is associated with a factor of safety F_2 (Eq. 19.104) which should be compared to F_1 of the reinforced slope (Eq. 19.103).

19.15 PROBABILISTIC APPROACH

In foundation engineering, uncertainty is included in the design through the use of load and resistance factor design (LRFD). In slope stability, LRFD is rarely used, but uncertainty does exist. The uncertainty associated with the calculated factor of safety in slope stability is quantified through a direct probability of failure calculation. This probabilistic approach has a great advantage over the deterministic approach, which gives only one value of the factor of safety. Imagine the following situations. You are given an average factor of safety equal to 1.5 and you feel comfortable about the safety of that slope. Then a probabilistic analysis is conducted to assess the probability of failure associated with the 1.5 factor of safety and yields a 0.2 or 1 chance in 5 of a failure occurring. Now you are not so comfortable about the 1.5 value. In contrast, if the assessed probability of failure turns out to be 0.001, you are very comfortable, as such a low probability of failure is well within the acceptable range for common civil engineering projects.

A background on probability is presented in section 11.6.1. The procedure for obtaining the probability of failure is outlined in section 11.6.2. A sample calculation of the probability of failure for a slope is given in section 11.6.4. The following simple examples illustrate the calculations to obtain the probability of failure.

19.15.1 Example 1

A slope exists as shown in Figure 19.32. It is made of clay with a normally distributed undrained shear strength s_u having a mean of 70 kPa and a standard deviation of 20 kPa. The failure circle has a radius of 16 m and the length of the arc is 24 m. The weight of the soil mass within the circle is 3200 kN per meter of length perpendicular to the page and the horizontal distance between the center of the circle and the center of gravity of the soil mass is 5.5 m.

The deterministic value of the factor of safety is:

$$F = \frac{RL}{Wa} s_u = \frac{16 \times 24}{3200 \times 5.5} 70 = 1.527 \tag{19.107}$$

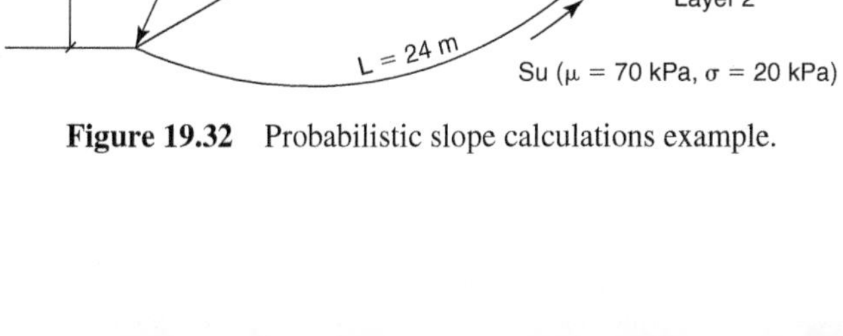

Figure 19.32 Probabilistic slope calculations example.

Note that if:

$$Y = aX \quad \text{then} \quad \mu_Y = a\mu_X \tag{19.108}$$

Therefore, the value of the factor of safety in Eq. 19.107 is also the mean of F, μ_F. Let's calculate the standard deviation of F. Again we note that if:

$$Y = aX \quad \text{then} \quad \sigma_Y = a\sigma_X \tag{19.109}$$

Therefore, the standard deviation of F is:

$$\sigma_F = \frac{RL}{Wa}\sigma_{S_u} = \frac{16 \times 24}{3200 \times 5.5} \times 20 = 0.436 \tag{19.110}$$

Failure occurs when $F < 1$ and the probability that $F < 1$ is $P(F < 1)$, which can be evaluated as follows. The first step is to transform F into the standard normal variable U:

$$U = \frac{F - \mu_F}{\sigma_F} = \frac{F - 1.527}{0.436} \tag{19.111}$$

The standard normal variable U has the following properties (see section 11.6.1):

$$P(U < u) = 1 - P(U < -u) \text{ and } P(U < u) = P(U > -u) \tag{19.112}$$

We are looking for the probability:

$$P(F < 1) = P\left(\frac{F - \mu_F}{\sigma_F} < \frac{1 - 1.527}{0.436}\right) = P(U < -1.21) = 1 - P(U < 1.21) \tag{19.113}$$

Table 11.3 gives:

$$P(U < 1.21) = 0.8869 \tag{19.114}$$

and the probability of failure for this case is:

$$P(F < 1) = 1 - 0.8869 = 0.1131 \tag{19.115}$$

For most civil engineering works, this would not be an acceptable probability of failure but the deterministic value of the factor of safety (1.527) would be.

19.15.2 Example 2

Consider the same slope geometry and the same circle, but now with the soil made of two layers. The top layer (crust) has a mean s_u value of 150 kPa, a standard deviation of 30 kPa, and a failure circle arc length of 6.5 m. The bottom layer has a mean s_u value of 70 kPa, a standard deviation of 20 kPa, and a failure circle arc length of 17.5 m. The weight and center of gravity of the soil mass are unchanged, with a weight of 3200 kN/m and a moment arm of 5.5 m.

The deterministic value of the factor of safety is:

$$F = \frac{R(L_1 s_{u1} + L_2 s_{u2})}{Wa} = \frac{16(6.5 \times 150 + 17.5 \times 70)}{3200 \times 5.5}$$
$$= 2.0 \quad (19.116)$$

Note that if:

$$Y = aX + bZ \quad \text{then} \quad \mu_Y = a\mu_X + b\mu_Z \quad (19.117)$$

Therefore, the value of the factor of safety in Eq. 19.116 is also the mean of F, μ_F. It makes sense that the factor of safety would be higher than in Example 1, because we have replaced part of the soil with a stronger layer. Let's calculate the standard deviation of F. Again we note that if:

$$Y = aX + bZ \quad \text{then} \quad \sigma_Y = \sqrt{a^2\sigma_X^2 + b^2\sigma_Z^2} \quad (19.118)$$

Therefore, the standard deviation of F is:

$$\sigma_F = \sqrt{\left(\frac{16 \times 6.5}{3200 \times 5.5}\right)^2 30^2 + \left(\frac{16 \times 17.5}{3200 \times 5.5}\right)^2 20^2} = 0.364 \quad (19.119)$$

We follow the same process as in Example 1:

$$U = \frac{F - \mu_F}{\sigma_F} = \frac{F - 2.0}{0.364} \quad (19.120)$$

Then, we are looking for the probability:

$$P(F < 1) = P\left(\frac{F - \mu_F}{\sigma_F} < \frac{1 - 2.0}{0.364}\right)$$
$$= P(U < -2.747) = 1 - P(U < 2.747) \quad (19.121)$$

Table 11.3 gives:

$$P(U < 2.747) = 0.997 \quad (19.122)$$

and the probability of failure for this case is:

$$P(F < 1) = 1 - 0.997 = 0.003 \quad (19.123)$$

This is an acceptable probability of failure in civil engineering. The main reason why the probability of failure has dramatically decreased (from 0.113 to 0.003) is that the mean factor of safety (2 instead of 1.527) is now further away from the failure value ($F = 1$).

19.15.3 Example 3

Let's repeat Example 1, but with two layers as in Example 2, except that these two layers are now identical and made of the Example 1 soil: mean $s_u = 70$ kPa and standard deviation of $s_u = 20$ kPa. The new calculations are as follows.

The deterministic value of the factor of safety is:

$$F = \frac{R(L_1 s_{u1} + L_2 s_{u1})}{Wa} = \frac{16(6.5 \times 70 + 17.5 \times 70)}{3200 \times 5.5}$$
$$= 1.527 \quad (19.124)$$

It makes sense that we find the same factor of safety as in Example 1. The standard deviation of F has changed, however; it is now:

$$\sigma_F = \sqrt{\left(\frac{16 \times 6.5}{3200 \times 5.5}\right)^2 20^2 + \left(\frac{16 \times 17.5}{3200 \times 5.5}\right)^2 20^2} = 0.339 \quad (19.125)$$

We follow the same process as in Example 1:

$$U = \frac{F - \mu_F}{\sigma_F} = \frac{F - 1.527}{0.339} \quad (19.126)$$

Then, we are looking for the probability:

$$P(F < 1) = P\left(\frac{F - \mu_F}{\sigma_F} < \frac{1 - 1.527}{0.339}\right)$$
$$= P(U < -1.555) = 1 - P(U < 1.555) \quad (19.127)$$

Table 11.3 gives:

$$P(U < 1.555) = 0.9400 \quad (19.128)$$

and the probability of failure for this case is:

$$P(F < 1) = 1 - 0.940 = 0.060 \quad (19.129)$$

This is about half the probability of failure calculated in Example 1, yet the soil conditions are the same except that we divided the soil into two identical layers. The reason for the decrease in the probability of failure is that if you randomly select a shear strength value from two identical distributions, you are very likely to make errors that tend to balance each other or reduce the error. The reason for this balancing error is that if you randomly pick a value that is too high for the first layer, you are more likely to pick a value that is too low for the second layer, as there are more values lower than your first guess. If there is only one layer, you have only your first guess for the calculations.

19.16 THREE-DIMENSIONAL CIRCULAR FAILURE ANALYSIS

The analysis of the circular failure in the preceding sections has assumed a plane strain condition. This means that the failing soil body has the shape of a cylindrical sector. Although in most cases this is a reasonable approximation, slope failures are always three dimensional (Figure 19.33). Three-dimensional or 3D slope failure analyses can be performed,

Figure 19.33 Three-dimensional slope failure. (Courtesy of Gordon W. Hunter, British Columbia Ministry of Transportation and Infrastructure.)

but are not done as commonly as plane strain analyses. One reason is that, most of the time, the assumption of a plane strain condition leads to a conservative factor of safety.

One way to perform a 3D slope stability analysis is to decompose the soil volume into a series of slices, each of which is considered to be a plane strain case (Figure 19.34). Many different assumptions can be made for such a mechanism, as was done for the 2D case. Some of them include a common axis of rotation for all circles and no forces between the circle slices. It can be seen in Figure 19.34 that if the deepest circle in the center of the volume is the critical circle for the 2D case, then all other circles will have a factor of safety higher than the 2D critical circle. From this observation, it follows that the global factor of safety for the 3D volume will be higher than for the 2D case. Though there are some exceptions to this statement, it is the general trend and has been documented

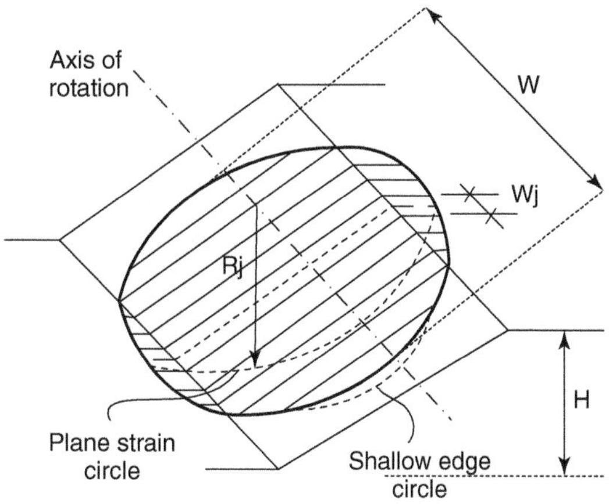

Figure 19.34 Decomposition of a 3D soil body into 2D soil slices.

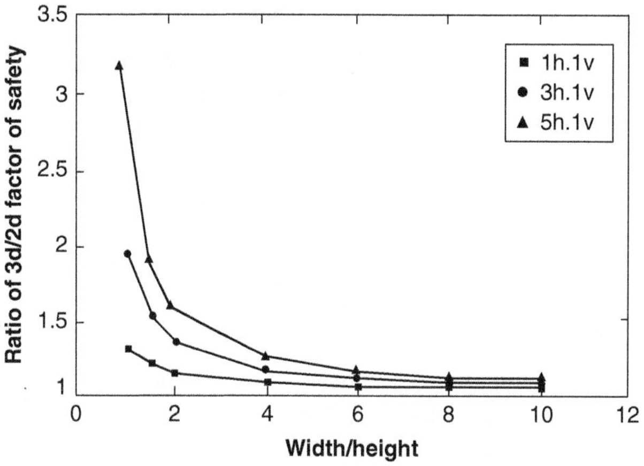

Figure 19.35 Comparison of 3D to 2D factor of safety. (After Stark 2003.)

by Stark (2003), for example (Figure 19.35). In Figure 19.35, H is the height of the slope and W is the width of slope considered in the analysis (sum of w_j on Figure 19.34). Three curves are shown for three different slope angles (1h to 1v, 3h to 1v, and 5h to 1v).

For the 2D case, the factor of safety of the jth circle (Bishop simplified) is modified after Eq. 19.91:

$$F_{2D,j} = \frac{\sum_{i=1}^{n} \frac{1}{m_{\theta ij}} (c'_{ij} b_{ij} + (W_{ij} - \alpha_{ij} u_{wij} b_{ij}) \tan \varphi'_{ij})}{\sum_{i=1}^{n} W_{ij} \sin \theta_{ij}} \quad (19.130)$$

where all parameters are defined in Table 19.1 and j is the number of the circle slice as shown in Figure 19.34. Then, if the axis of rotation is the same for all slices and if the forces between the circle slices are neglected, the factor of safety for the 3D volume becomes:

$$F_{3D} = \frac{\sum_{j=1}^{m} R_j w_j \sum_{i=1}^{n} \frac{1}{m_{\theta ij}} (c'_{ij} b_{ij} + (W_{ij} - \alpha_{ij} u_{wij} b_{ij}) \tan \varphi'_{ij})}{\sum_{j=1}^{m} R_j w_j \sum_{i=1}^{n} W_{ij} \sin \theta_{ij}} \quad (19.131)$$

where R_j and w_j are the radius and width of the circle slice j respectively.

The drastic assumptions associated with this equation limit its applicability to a first estimate. A number of computer programs are available to perform more sophisticated 3D analyses. The goal of the assumptions, as in the 2D case, is to make the problem a statically determinate problem and to satisfy equilibrium equations in all directions. In the end, the finite element method is again the best way to solve the 3D problem, because with this method all equilibrium equations will automatically be satisfied for all elements of soil.

19.17 FINITE ELEMENT ANALYSIS

The finite element method (FEM) can be used to analyze the stability of slopes. The mesh should be large enough that the boundaries have only a small and tolerable influence on the stability calculations. If the height of the slope is H, the mesh should be at least 3 H high. If the horizontal distance between the toe and the crest of the slope is L, the mesh should be at least 5 L long. The advantages of the FEM over the limit equilibrium method (LEM) are that (Griffiths and Lane 1999):

1. No assumptions need be made about the failure surface; the weakest surface will automatically be found through the stress field calculated as part of the solution.
2. All equilibrium equations are satisfied.
3. In addition to a factor of safety, information is obtained on the displacements of the slope. This information, of course, is only as good as the soil model and soil properties used to obtain it.
4. The FEM includes complex issues such as progressive failure up to complete failure.

The factor of safety is determined through the use of a strength reduction factor (SRF) which is applied to the strength parameters of the soil:

$$c'_r = \frac{c'}{SRF} \tag{19.132}$$

$$\tan \varphi'_r = \frac{\tan \varphi'}{SRF} \tag{19.133}$$

where c′ and φ' are the effective stress cohesion and friction angle of the soil respectively, and c'_r and φ'_r are the reduced effective stress cohesion and friction angle of the soil respectively. The FEM is performed repeatedly as the values of c'_r and φ'_r are gradually decreased by using an increasing SRF. The failure criterion can be defined in various ways (Abramson et al. 2002):

1. Bulging of the slope surface
2. Limiting shear stress reached on the failure surface
3. Nonconvergence of the solution

When an agreed-upon failure criterion is reached, the SRF is equal to the safety factor. Figure 19.36 shows an FEM output of a failed slope.

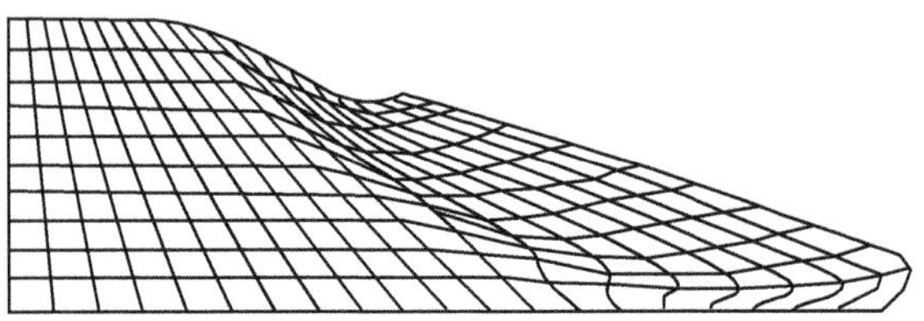

Figure 19.36 Failed slope in finite element method. (Courtesy of Griffiths and Lane 1999.)

19.18 SEISMIC SLOPE ANALYSIS

An earthquake can induce failure of a slope that is statically safe. The reason is twofold: The earthquake increases the driving moment, mostly through horizontal shaking; and in some soils it can decrease the shear strength of the soil by increasing the water compression stress through cyclic loading, possibly leading to liquefaction. There are several ways to include earthquake loading in slope stability analysis:

1. Pseudostatic method
2. Newmark's displacement method
3. Postearthquake stability method
4. Dynamic finite element method

19.18.1 Pseudostatic Method

The *pseudostatic method* is the most common and the simplest. It consists of adding a horizontal and vertical static force in the limit equilibrium analysis. These two forces are chosen to be equivalent to the effects of the inertia dynamic forces generated during shaking of the soil mass. They are assumed to be proportional to the weight W of the failing soil mass. The coefficients of proportionality are k_h and k_v for the horizontal and vertical direction respectively (Figure 19.37). Most commonly, the vertical seismic coefficient k_v is assumed to be zero and the horizontal seismic coefficient k_h depends on the severity of the shaking. Table 19.6 (Abramson

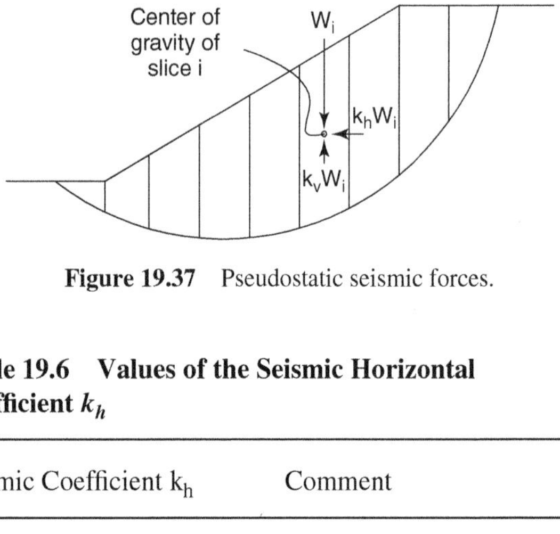

Figure 19.37 Pseudostatic seismic forces.

Table 19.6 Values of the Seismic Horizontal Coefficient k_h

Seismic Coefficient k_h	Comment
0.10	Major earthquake, U.S. Army Corps of Engineers, 1982
0.15	Great earthquake, U.S. Army Corps of Engineers, 1982
0.05 to 0.15	State of California
0.15 to 0.25	Japan
1/3 to 1/2 of peak ground acceleration (PGA)	Marcuson and Franklin 1983

(After Abramson et al. 2002.)

et al. 2002) is a summary of some common values for the seismic coefficient k_h.

The seismic force is usually placed at the center of gravity of the slice. Seismic analyses indicate that most of the time, the peak acceleration increases as the wave propagates from the bottom to the top of the slope. This would mean that the point of application of the seismic force should be above the center of gravity of the slice (CG); this would generate smaller overturning moments than if that force were placed at the CG. Therefore, placing the seismic force at the CG of the slices, as is usual practice, is conservative in most cases.

Another way to approach the problem is to find the horizontal seismic coefficient k_h that would lead to failure of the slope. This value of k_h, called the *yield horizontal seismic coefficient* k_y, corresponds to a factor of safety of 1 (Figure 19.38). Then the value of k_y can be compared to the peak ground acceleration (PGA) of the earthquake at the slope location. Abramson et al. (2002) suggest the observations in Table 19.7.

Note that a very important part of the pseudostatic analysis, as for any slope stability analysis, is to select the correct shear strength. The issue here is that the shear strength during shaking is likely to be reduced compared to the static case.

19.18.2 Newmark's Displacement Method

Newmark's displacement method is credited to Newmark (1965). Whereas most other slope stability methods aim at

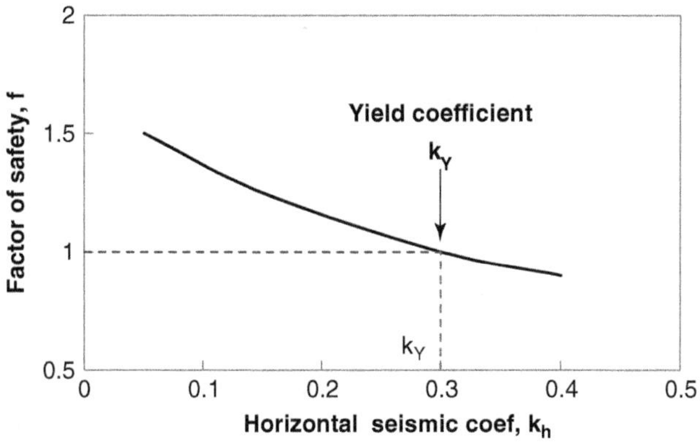

Figure 19.38 Yield horizontal seismic coefficient.

Table 19.7 Likelihood of Failure for Different Values of k_y

Relative Position of k_y and PGA	Observation
k_y > PGA	No failure likely
0.5 PGA < k_y < PGA	Minor damage possible
k_y < 0.5 PGA	Failure likely

(After Abramson et al. 2002.)

predicting the factor of safety, this method aims at predicting the accumulation of displacement of the slope during a series of acceleration cycles, as in an earthquake, for example.

The first step is to develop an acceleration history a(t) for the earthquake at the location of the slope (Figure 19.39). Then the yield acceleration a_y is found by using the pseudostatic method. The relationship between a_y and k_y is:

$$a_y = k_y g \tag{19.134}$$

where g is the acceleration due to gravity. Any acceleration above a_y will lead to movement as the slope fails during a short time increment (points B to D in Figure 19.39). By integration of the net acceleration $(a(t)-a_y)$ from B to D, the velocity of the soil mass is found (points B_1 to D_1 in Figure 19.39). Then the velocity decreases as the acceleration recedes below a_y and the shear strength slows the soil mass down (points D_1 to M_1 in Figure 19.39). By integrating the velocity from B_1 to M_1, the displacement of the slope mass is obtained (points B_2 to M_2 in Figure 19.39). At M_2, a permanent displacement has been accumulated; the displacement increases again when the acceleration exceeds the yield acceleration (point H in Figure 19.39). The process repeats itself until the earthquake is over. One of the key parts of this method is developing the acceleration history for the slope. This is discussed in Chapter 22.

Makdisi and Seed (1978) performed a parametric analysis using actual and hypothetical dams and embankments. Using the results, they simplified Newmark's method and presented it in the form of a chart (Figure 19.40). The acceleration ratio k_y/k_{max} is on the horizontal axis, where k_y is the yield horizontal seismic coefficient corresponding to failure of the slope and k_{max} is the maximum acceleration horizontal seismic coefficient corresponding to the maximum acceleration

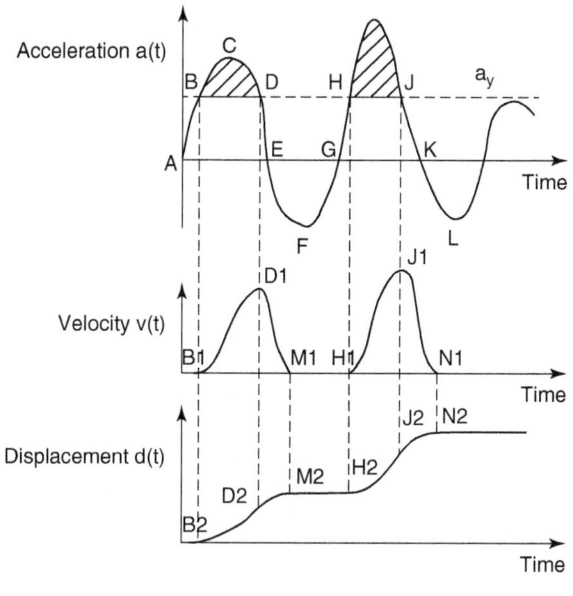

Figure 19.39 Newmark's displacement method (1965).

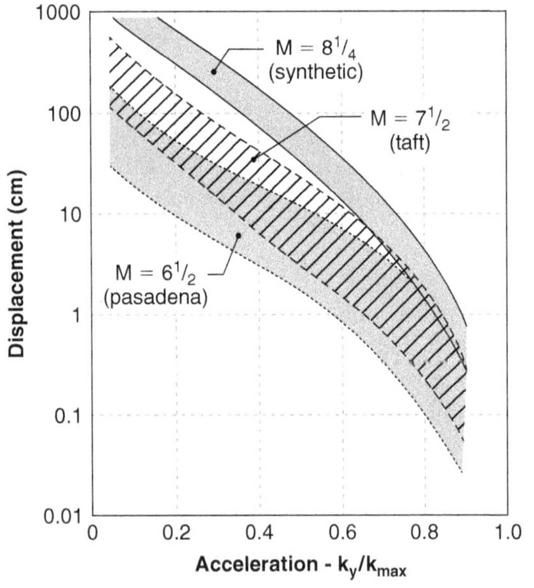

Figure 19.40 Makdisi and Seed (1978) chart. (This material is reproduced with permission of John Wiley & Sons, Inc.)

in the slope and defined as:

$$a_{max} = k_{max} g \qquad (19.135)$$

The earthquake magnitude M (see Chapter 22) is selected for the site and the range of displacement is read on the vertical axis using the acceleration ratio and the magnitude M.

19.18.3 Postearthquake Stability Analysis

Postearthquake stability analysis is a static analysis that considers the situation right after an earthquake. The main issue in this case is the proper selection of the shear strength existing after the earthquake. The steps for such an analysis (Duncan and Wright 2005) include:

1. Study whether the soil has liquefied or not (see Chapter 22)
2. Determine the reduced shear strength due to cyclic loading (see Chapter 22)
3. Use that shear strength for a conventional stability analysis

19.18.4 Dynamic Finite Element Analysis

Dynamic finite element analysis is a 2D or 3D dynamic analysis of the slope and its surroundings. The finite element mesh should be large enough that the effect of the boundaries does not significantly affect the stability calculations for the slope. The recommendations of section 19.17 for static analysis may not be sufficient, as earthquake shaking creates waves that propagate against the boundaries and are reflected toward the slope. The soil model should incorporate the evolution of the strength and deformation properties as a function of cycles. The earthquake motion is usually induced at the bottom of the mesh and propagates upward through the slope mesh. The fact that the soil model can more closely follow the soil behavior and the fact that the dynamic equilibrium of the elements is satisfied at all times are two major advantages of this approach. The complexity of the approach is its drawback.

19.19 MONITORING

Monitoring consists of making observations or measurements on a slope in order to evaluate its stability. Among the most useful parameters to observe or measure are:

1. Cracks, particularly on top of the slope
2. Movements of the slope surface or at depth
3. Groundwater and water stress conditions

Crack openings are indications that a slope is stressed (Figure 19.41). Cracks associated with instability on top of slopes are parallel to the slope crest and are several meters long. If the cracks are less than 25 mm wide, and if there is no difference in elevation between the two sides, the probability of failure is low. If the cracks are between 25 and 75 mm wide, with some difference in elevation between the two sides, the probability of failure is much higher. If the cracks are much larger than 100 mm with similar difference in elevation between the two sides, it is likely time to run. Instruments to measure cracks can be as simple as a tape measure or as sophisticated as an extensometer that monitors the distance between the two sides and sends readings to a remote monitoring station. It is very useful to monitor crack width b as a function of time t and plot the curve b vs. t. If the growth rate of the crack opening decreases with time, it indicates a trend toward stability, but if the growth rate increases steadily with time, failure may be imminent. There are cases in which the growth rate decreases but then reverses to an increase with time (Figure 19.42).

Movements of the slope surface can be tracked with useful and simple measurements. Tools can be as simple as surveying stakes driven in the ground and as advanced as GPS monitoring. Movement of the crest is a good indication, but swelling or heaving at the base of the slope is also an early sign. As in the case of cracks, the shape of the curve of the movement as a function of time is the best indication of potential failure.

Movement at depth in the slope is a very useful but more complicated measure, and is expensive to obtain. The most common method is to place a vertical inclinometer casing through the surface of the slope to a depth well below the potential failure surface (Figure 19.43). The bottom of the slope indicator casing should be in a soil zone that is not influenced by the slope movement. The reason is that the bottom readings will represent the zero position for the casing. The slope inclinometer casing is grooved and the probe has wheels (Figure 19.44) that fit in the grooves to keep

Figure 19.41 Example of cracks on top of slopes. (*a*: From Bray et al. 2001. Used by permission. *b*: Photo by and courtesy of Jonathan Wilkins. c: Courtesy of Dr. Ian West.)

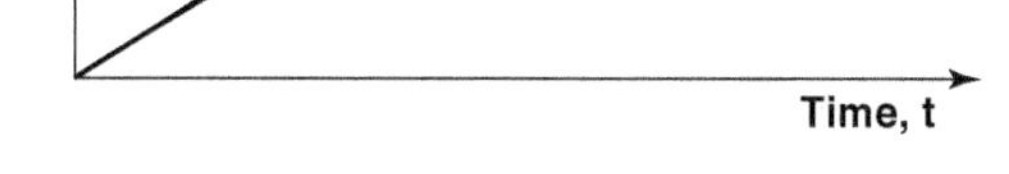

Figure 19.42 Monitoring crack width over time.

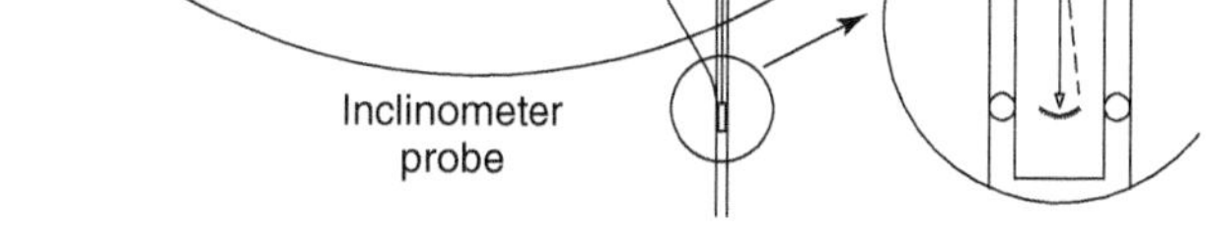

Figure 19.43 Inclinometer in a slope.

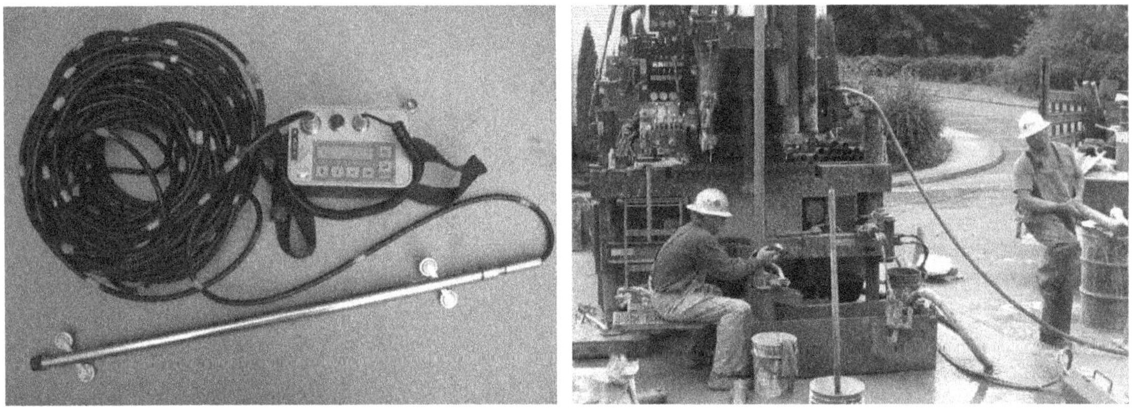

(a) (b)

Figure 19.44 Inclinometer instrument and casing installation. *(a)* Inclinometer. *(b)* Casing installation. (Courtesy of Landslide Technology.)

the direction of the probe constant during the readings. The probe is lowered to the bottom of the casing and then pulled up the casing while readings are taken at regular intervals. This interval is usually the length L of the probe between wheels (e.g., $L = 0.5$ m). The first reading R_1 is taken at the bottom of the casing at a depth z_1. The probe is pulled up an amount equal to L and the second reading R_2 at depth z_2 is obtained, and so on all the way to the top of the casing. The first set of readings in the casing is taken right after installation and provides a set of zero readings R_{oi}. It is assumed that the bottom of the casing is low enough below the slope potential failure surface that no movement takes place at the bottom of the casing; therefore, the readings at the bottom provide a reference for all the others. If there is doubt about whether the bottom is moving, then the top of the casing should be surveyed each time a set of inclinometer readings is taken.

Each reading represents the angle α between the probe direction and the vertical at that location. The instruments to measure this angle are servo-accelerometers located in the probe. A servo-accelerometer is essentially a mass placed at the end of a pendulum between two magnetic coils. When the pendulum begins to swing to a new position, a magnetic force is applied to keep the pendulum in the zero position. The current necessary for the magnetic force to keep the pendulum in the zero position is proportional to the angle that the pendulum would have taken had it not been restrained.

The reading R_i at depth z_i gives the angle of the inclinometer as:

$$\sin\theta_i = \frac{R_i}{C} \tag{19.136}$$

where C is a constant specific to each inclinometer and θ_i is the angle of the casing with the vertical at depth z_i.

Often readings will be taken in two opposite directions of the probe (0 and 180° in a horizontal plane) and the average of the two readings will be used. The horizontal distance d_i between the two points separating two consecutive readings (often the length between probe wheels) is given by (Figure 19.45):

$$d_i = L\sin\theta_i \tag{19.137}$$

where L is the increment of depth between readings. If the set of zero readings gave a value of d_i equal to d_{oi}, then the net horizontal distance is $(d_i - d_{oi})$. Because we want the overall position of the casing after deformation, it is necessary to add all net horizontal distances between all consecutive points from the bottom of the casing to the depth where the horizontal movement is required. If it is assumed that the

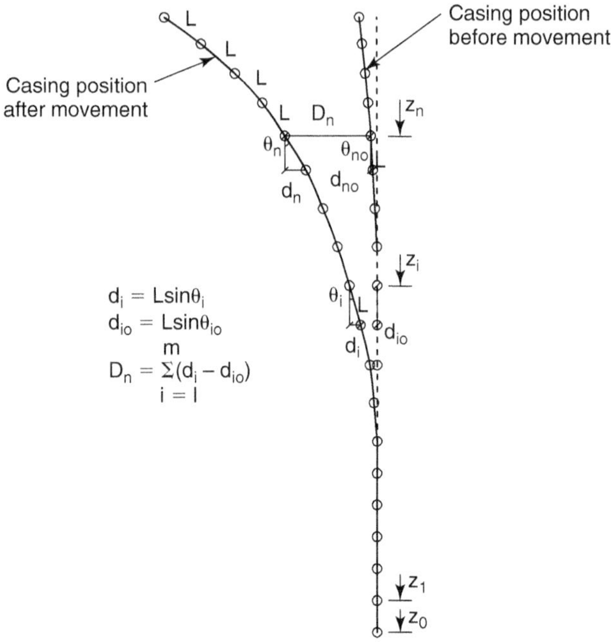

Figure 19.45 Inclinometer data reduction.

bottom of the casing is not moving, the horizontal movement of the casing at any depth z_n is (Figure 19.45):

$$D_n = \sum_{i=1}^{n}(d_i - d_{io}) = \frac{L}{C}\sum_{i=1}^{n}(R_i - R_{io}) \qquad (19.138)$$

Groundwater and water stress conditions are very important aspects of slope stability. The groundwater level can be measured by simply measuring the equilibrium water level in an open standpipe or by using a piezometer. If the water is in compression, the water stress can be measured with a piezometer. If the water is in tension, the water stress can be measured with a field tensiometer up to a water tension of −90 kPa. Above that value, a soil sample can be taken, the water content determined, and the soil water retention curve used to go from the water content to the water tension (see section 9.15).

19.20 REPAIR METHODS

There are essentially two ways to repair a slope that is getting close to failure (Figure 19.46):

1. Increase the resisting moment (e.g., soil improvement, inclusions)
2. Decrease the driving moment (e.g., shallower slope)

19.20.1 Increase the Resisting Moment

The shear strength s of the soil is:

$$s = c' + (\sigma - \alpha u_w)\tan\varphi' \qquad (19.139)$$

where c′ is the effective stress cohesion intercept, σ is the normal total stress on the plane of failure, α is the area ratio for the water phase, u_w is the water stress, and φ' is the effective stress friction angle. Therefore, increasing s may consist of increasing c', or σ, or $\tan\varphi'$, or decreasing αu_w. Increasing c′ can be done by chemical injection of cementing agents such as lime or cement. Increasing σ is usually not a good idea, as it also increases the driving moment. Increasing $\tan\varphi'$ is difficult, but can be done through densification by compaction or vibration. If the soil is saturated and if the water is in compression, the term αu_w becomes u_w and decreasing u_w consists of decreasing the water stress (through drainage, for example) if there are excess water stresses or by lowering the water level (by pumping, for example) if the water stress is hydrostatic. If the soil is saturated with water in tension or if the soil is unsaturated, decreasing u_w consists of drying or evaporation, for example. In this case u_w becomes more negative, but α also decreases, so the net result is not as efficient as a decrease in u_w alone. In some instances, the water tension is naturally decreased (less negative) during the life of the slope because of the weather. This may lead to failure, and one way to prevent such failures is to keep the water tension from changing by isolating the soil from the weather. Geosynthetic covers can achieve this goal.

The other way to increase the resisting moment is to insert inclusions in the slope and across the failure plane (Section 19.14). For existing slopes, soil nails or piles can be used. *Soil nails* are small-diameter inclusions that are placed most often by drilling and sometimes by driving. The drilling process consists of drilling a hole, removing the drilling tool, inserting a steel bar or cable with centralizers in the hole, and grouting the annulus between the bar and the soil. Soil nails have the advantage that they are relatively easy to place at any inclination, although they are most often placed nearly horizontally. Piles are placed vertically

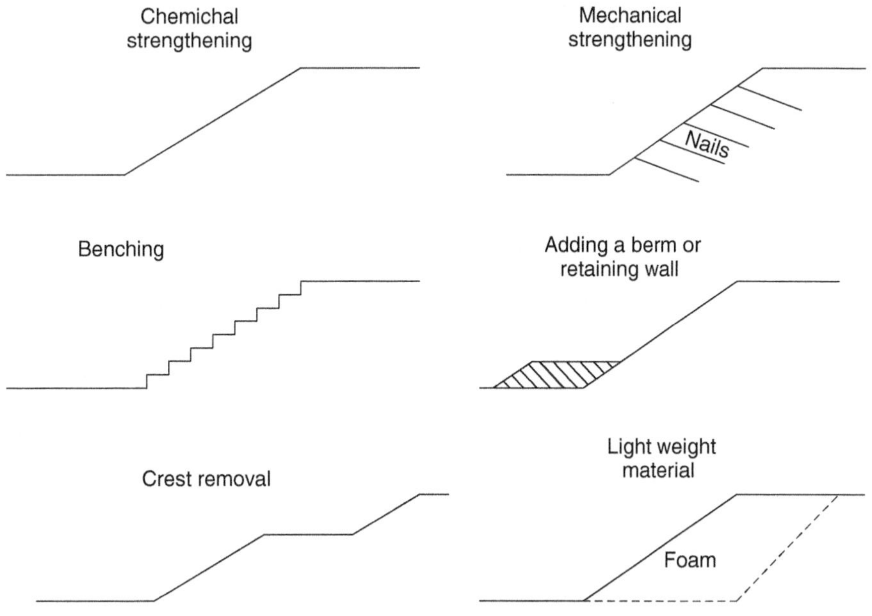

Figure 19.46 Slope repair methods.

or near vertically and have a larger diameter than nails. For new slopes like embankments, geosynthetic layers or reinforcing steel strips can be placed as reinforcement.

19.20.2 Decrease the Driving Moment

The driving moment M_d *for* the slope is:

$$M_d = Wa \quad (19.140)$$

where W is the weight of the failing soil mass and a is the horizontal distance from the center of the failure circle for a circular failure surface and the center of gravity of the failing soil mass. Therefore, decreasing M_d consists of decreasing W or a or both. To decrease W, lightweight material such as foam can be used for embankments. Also, the slope angle can be reduced by removing part of the crest, adding a berm at the bottom of the slope, or simply grading the slope to a flatter angle (Figure 19.46).

In the end, the choice of one method or another is based on effectiveness of the method, feasibility, and cost. For landfill slopes, see section 24.7.6. For slopes involving geosynthetics, see section 25.6.3.

PROBLEMS

19.1 Calculate the factor of safety for the slope shown in Figure 19.1s in the following cases:
   a. Slope alone
   b. Slope plus building
   c. Slope plus building and earthquake
   d. What is the yield coefficient k_y for that slope?

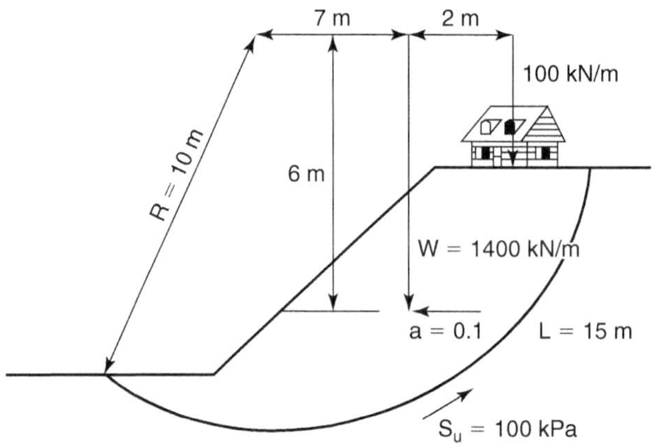

Figure 19.1s Slope with building and earthquake.

19.2 An infinite slope is made of sand with a friction angle of 32° and a unit weight of 20 kN/m^3. The slope angle is 2.5 horizontal to 1 vertical. Calculate the factor of safety in the summer when the slope has no water, then calculate the factor of safety in the spring when the slope is filled with water.

19.3 Derive the expression for the factor of safety of an infinite slope with a failure plane parallel to the ground surface at a depth h and with a groundwater level at a height mh above the failure plane ($m < 1$).

19.4 Design a safe slope angle for an excavated slope in a stiff clay to reach a 20 m deep deposit of lignite. The stiff clay has effective shear strength parameters of $c' = 10$ kPa and friction angle $\varphi' = 25°$. Consider the case where the water level is not within the slope failure zone and then the one where the water level follows the slope contour. Use the chart method. In practice, it is not uncommon to see such excavations with much steeper slopes than the answer you will get in this problem; although failures do occur, they do not occur too often. Why do you think that is?

19.5 Calculate the factor of safety by using Janbu's charts for the slope shown in Figure 19.2s in the following cases:
   a. Case 1 : $Hw = Hw' = 10$ m
   b. Case 2 : $Hw = 0$, $Hw' = 10$ m
   c. Case 3 : $Hw = Hw' = 0$

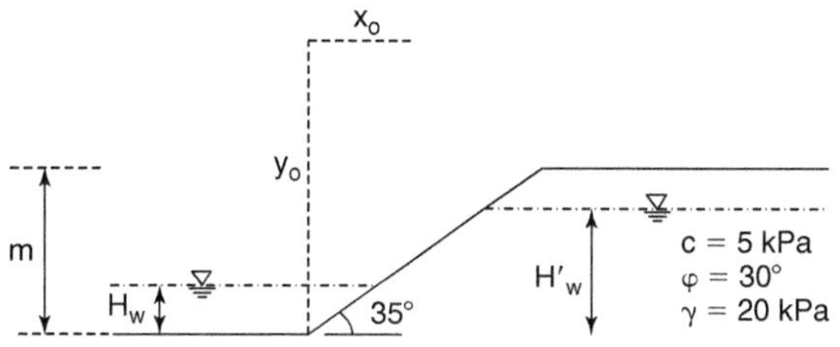

Figure 19.2s Slope with different water levels.

19.6 A 3 horizontal to 1 vertical slope is cut in a clay that has an effective stress cohesion of 5 kPa and an effective stress friction angle of 30°. The slope is 6 m high. Calculate:
   a. The factor of safety of the slope against long-term failure if the water table is below the critical circle.
   b. The factor of safety of the slope against long-term failure if the water table coincides with the slope contour.

19.7 Calculate the probability of failure of a slope that has a factor of safety with a mean of 1.5 and a coefficient of variation of 0.2 (assume that the factor of safety is normally distributed). If the acceptable probability of failure is 0.001, what must be the mean value of the factor of safety if the coefficient of variation remains equal to 0.2?

19.8 Which situation is more desirable? Explain and demonstrate.
   a. Mean factor of safety $F = 1.5$ and coefficient of variation of $F = 0.2$
   b. Mean factor of safety $F = 1.3$ and coefficient of variation of $F = 0.1$

19.9 Calculate the factor of safety for the slope shown in Figure 19.3s. Select your best estimate of the critical circle and calculate the factor of safety by the Bishop modified method of slices.

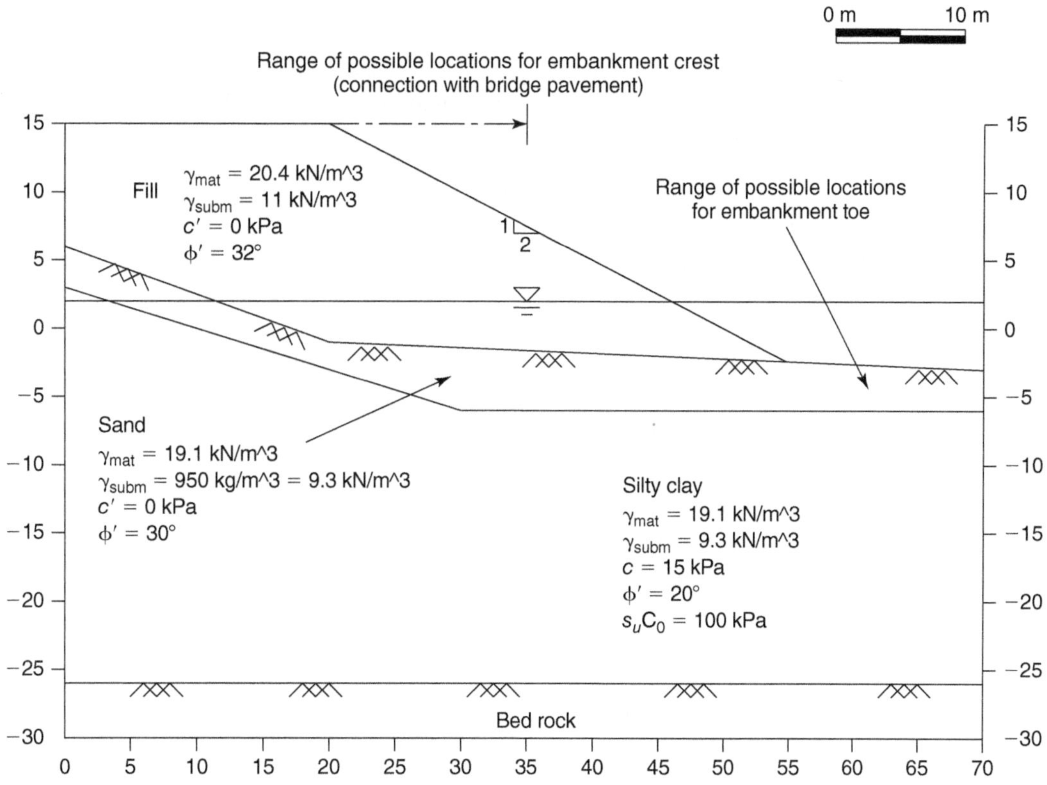

Figure 19.3s Slope of Fredericton embankment.

19.10 A slope is subjected to an acceleration history as shown in Figure 19.4s. The yield acceleration for that slope is 1.5 m/s^2. Calculate the displacement history of the slope according to Newmark's method.

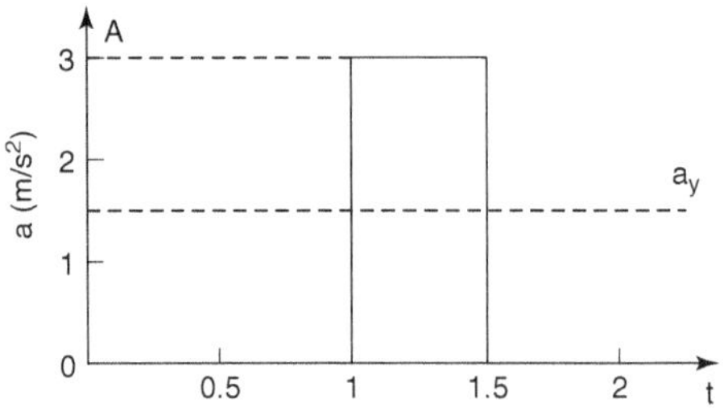

Figure 19.4s Acceleration history.

19.11 Consider the case of a $c' = 0$, $\varphi' > 0$ soil and demonstrate that for one slice, the factor of safety is the same for the ordinary method of slices (OMS) and for the Bishop simplified method of slices (BSMS). If it is true for one slice, why is it not true for n slices ($n > 1$)?

19.12 A 3D slope failure has a failure surface in the form of a sphere and a factor of safety F_{3D}. This sphere is sliced in a direction perpendicular to the crest. The slices have the same width b. The deepest slice in the center of the sphere has a factor of safety F_{min}. Each slice has a 2D factor of safety equal to F_i. What other assumptions must be made for the following equation to be true? $F_{3D} = \frac{1}{n}\sum_{i=1}^{n} F_i$

19.13 A dry fine sand slope has a factor of safety of 1.5 on the Earth.
a. Calculate and discuss the factor of safety for the same slope on the moon.
b. Assume that there could be water on the moon. Would the result of the dry case still hold?

19.14 Define the seepage force and discuss when the seepage force should be considered in a slope stability analysis. Why is it not usually considered?

19.15 Explain the difference between the following analyses, including what shear strength you would use: total stress analysis, effective stress analysis, undrained analysis, drained analysis, short-term analysis, long-term analysis.

19.16 An inclinometer casing is attached to a 10 m high retaining wall. Zero readings taken before the wall is backfilled indicate that the wall is perfectly vertical. The backfill is placed and compacted. At the end of construction, the inclinometer readings are taken again. Find what the readings are if:
a. The displacement y (m) of the wall obeys the equation $y = 0.01\ (z_{max} - z)$ where z_{max} is the maximum depth the inclinometer probe can reach in the casing (10 m) and z is the depth at which the reading is taken.
b. The displacement y (m) of the wall obeys the equation $y = 0.001\ (z_{max} - z)^2$ where z_{max} is the maximum depth the inclinometer probe can reach in the casing (10 m) and z is the depth at which the reading is taken.

Problems and Solutions

Problem 19.1

Calculate the factor of safety for the slope shown in Figure 19.1s in the following cases:

a. Slope alone
b. Slope plus building
c. Slope plus building and earthquake
d. What is the yield coefficient k_y for that slope?

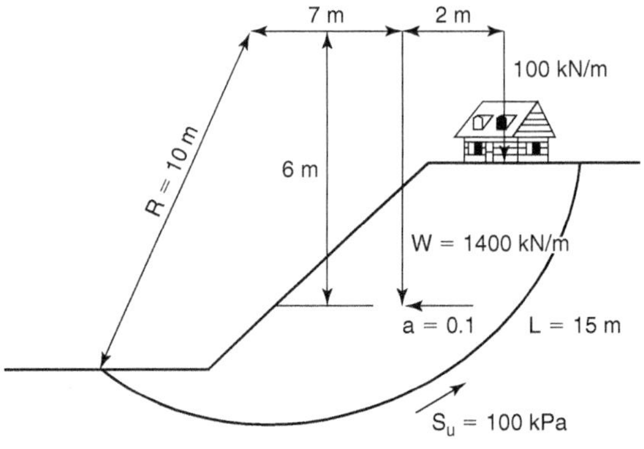

Figure 19.1s Slope with building and earthquake.

Solution 19.1

The factor of safety is the maximum resisting moment divided by the driving moment:

$$FS = \frac{\textit{Maximum resisting moment}}{\textit{Driving Moment}}$$

a. Slope alone:

$$FS = \frac{100 \times 15 \times 10}{1400 \times 7} = 1.53$$

b. Slope and building:

$$FS = \frac{100 \times 15 \times 10}{1400 \times 7 + 100 \times (7 + 2)} = 1.40$$

c. Slope plus building and earthquake:

$$FS = \frac{100 \times 15 \times 10}{1400 \times 7 + 100 \times (7 + 2) + 0.1 \times 1400 \times 6} = 1.30$$

d. The earthquake yield coefficient k_y for that slope:

$$FS = \frac{100 \times 10 \times 15}{1400 \times 7 + 100 \times (7 + 2) + k_Y \times 1400 \times 6} = 1.00 \Rightarrow k_Y = 0.51$$

Problem 19.2

An infinite slope is made of sand with a friction angle of 32° and a unit weight of 20 kN/m^3. The slope angle is 2.5 horizontal to 1 vertical. Calculate the factor of safety in the summer when the slope has no water, then calculate the factor of safety in the spring when the slope is filled with water.

Solution 19.2

For the case of sand with no water during the summer:

$$FS = \frac{\tan \varphi'}{\tan \beta} = \frac{\tan(32)}{1/2.5} = 1.56$$

For the case of the sand filled with water during the spring with no cohesion, and assuming a saturated unit weight of 22 kN/m^3 :

$$FS = \frac{(\gamma_{sat} - \gamma_w)}{\gamma_{sat}} \frac{\tan \varphi'}{\tan \beta} = \frac{(22 - 9.81)}{22} \frac{\tan(32)}{1/2.5} = 0.86$$

The presence of water significantly reduces the factor of safety of the slope.

Problem 19.3

Derive the expression for the factor of safety of an infinite slope with a failure plane parallel to the ground surface at a depth h and with a groundwater level at a height mh above the failure plane (m < 1).

Solution 19.3

Let's call γ_m the soil unit weight above the groundwater level and γ_{sat} the soil unit weight below the groundwater level. Referring to Figure 19.6 and the case of the infinite slope with seepage, the shear strength on the failure plane is:

$$\tau_f = c' + (\gamma_m(1 - m)h \cos^2\beta + \gamma_{sat} mh \cos^2\beta - \gamma_w mh \cos^2\beta) \tan \varphi'$$

$$\tau_f = c' + ((1 - m)\gamma_m + m(\gamma_{sat} - \gamma_w))h \cos^2\beta \tan \varphi'$$

The shear stress τ on the plane of failure is:

$$\tau = (\gamma_m(1-m)h\cos\beta + \gamma_{sat}mh\cos\beta)\sin\beta$$
$$\tau = ((1-m)\gamma_m + m\gamma_{sat})h\cos\beta\sin\beta$$

The factor of safety is:

$$FS = \frac{c' + ((1-m)\gamma_m + m(\gamma_{sat}-\gamma_w))h\cos^2\beta\tan\varphi'}{((1-m)\gamma_m + m\gamma_{sat})h\cos\beta\sin\beta}$$
$$FS = \frac{c'}{((1-m)\gamma_m + m\gamma_{sat})h\cos\beta\sin\beta} + \frac{((1-m)\gamma_m + m(\gamma_{sat}-\gamma_w))}{((1-m)\gamma_m + m\gamma_{sat})} \times \frac{\tan\varphi'}{\tan\beta}$$

Problem 19.4

Design a safe slope angle for an excavated slope in a stiff clay to reach a 20 m deep deposit of lignite. The stiff clay has effective shear strength parameters of $c' = 10$ kPa and friction angle $\varphi' = 25°$. Consider the case where the water level is not within the slope failure zone and then the one where the water level follows the slope contour. Use the chart method. In practice, it is not uncommon to see such excavations with much steeper slopes than the answer you will get in this problem; although failures do occur, they do not occur too often. Why do you think that is?

Solution 19.4

If the soil is uniform and a circular failure surface is assumed, chart methods can be used.

a. Case of No Water

a-1. Taylor (1948) (Figure 19.5s)

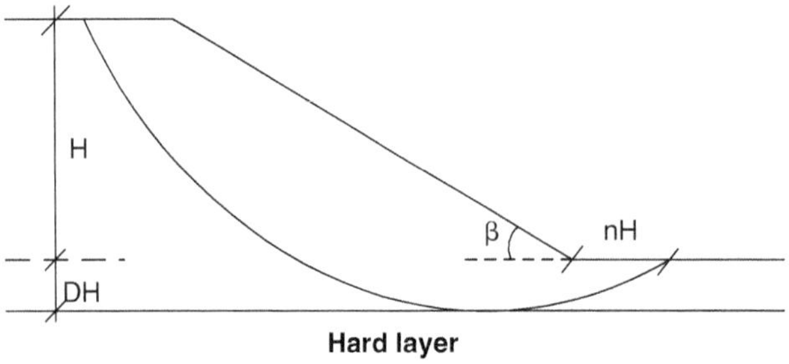

Figure 19.5s Sketch of problem 19.4.

Let's assume toe circles on Figure 19.13 with $c' = 10$ kPa, $\varphi' = 25°$, and $\gamma = 20$ kN/m^3. We start with an assumed factor of safety where c' equals 1.5.

The developed friction angle is calculated using:

$$\varphi_d = \tan^{-1}\left(\frac{\tan\varphi}{FS}\right) = \tan^{-1}\left(\frac{\tan 25}{1.5}\right) = 17.26°$$
$$F'_c = \frac{c'}{c'_d} \; and \; F_{\varphi'} = \frac{\tan\varphi'}{\tan\varphi'_d}$$
$$F'_c = \frac{c'}{c'_d} = 1.5 \rightarrow c'_d = \frac{10}{1.5} = 6.67 \text{ kPa}$$

The stability number $N = \dfrac{c'_d}{\gamma H} = \dfrac{6.67}{20 \times 20} = 0.0167 \rightarrow$ Figure 19.13 $\rightarrow \beta = 22°$

a-2. Spencer (1973)

$$r_u = \frac{u_w}{\sigma_{0v}} = 0 \quad \text{because there is no water}$$

$$F_c' = \frac{c'}{c_d'} \;\; and \;\; F_{\varphi'} = \frac{\tan \varphi'}{\tan \varphi_d'}$$

$$F_c' = \frac{c'}{c_d'} = 1.5 \rightarrow c_d' = \frac{10}{1.5} = 6.67 \text{ kPa} \;\; \text{and} \;\; N = \frac{c_d'}{\gamma H} = \frac{6.67}{20 \times 20} = 0.0167$$

Figure 19.15 $\rightarrow r_u = 0 \rightarrow \beta = 23^\circ$

b. Water Case

b-1. Taylor (1948), undrained: s_u = 70 kPa ($\varphi' = 0$)

Assuming a toe circle, a factor of safety of 1.5, and using $s_u = 70$ kPa:

$$FS = \frac{S_u}{c_d} \rightarrow c_d = \frac{S_u}{FS} = \frac{70}{1.5} = 46.67 \text{ kPa}$$

$$N = \frac{c_d}{\gamma H} = \frac{46.67}{20 \times 20} = 0.12 \rightarrow \text{Fig. 19.12 for n} = 0 \rightarrow \beta = 11^\circ$$

b-2. Spencer (1973), drained behavior: c', φ'

$$r_u = \frac{u_w}{\sigma_{0v}} = 0.5$$

Figure 19.15 $\rightarrow r_u = 0.5 \rightarrow \beta = 11^\circ$

Table 19.1s Slope Angle from Taylor and Spencer Methods

Slope Angle β	Taylor (1948)	Spencer (1973)
Dry case	22°	23°
Water case	11°	11°

Several factors come into play. First, the water in the soil is likely in tension, which increases the safe angle. Second, the excavation remains open for a limited amount of time and the soil behavior may be time dependent. The slopes may be well drained. The open-pit mine slope industry seems to accept a higher probability of slope failure as part of the economical optimization of the lignite mining process.

Problem 19.5

Calculate the factor of safety by using Janbu's charts for the slope shown in Figure 19.2s in the following cases:

a. Case 1 : $H_w = H_w' = 10$ m
b. Case 2 : $H_w = 0$, $H_w' = 10$ m
c. Case 3 : $H_w = H_w' = 0$

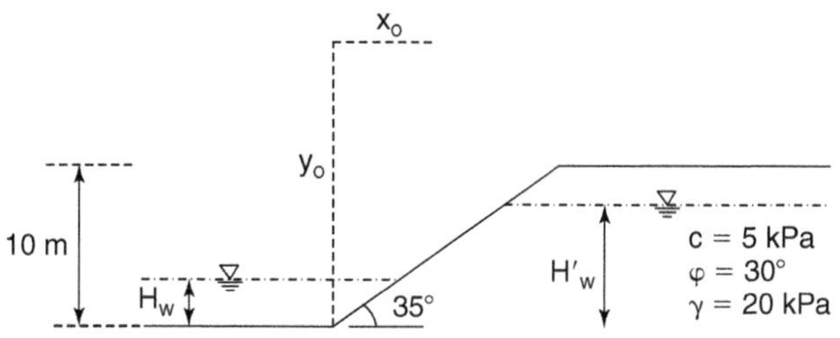

Figure 19.2s Slope with different water levels.

Solution 19.5

Janbu Chart

a. Case 1: $H_w = H'_w = 10$ m

$$\mu_w = \mu'_w = 1, \text{all other } \mu = 1$$

$$P_d = \frac{20 \times 10 - 10 \times 10}{1 \times 1 \times 1} = 100$$

$$P_e = \frac{20 \times 10 - 10 \times 10}{1 \times 1} = 100$$

$$\lambda_{c\phi} = \frac{100 \tan(30)}{5} = 11.55$$

For $\lambda_{c\phi} = 11.55$ and $\beta = 35^\circ \rightarrow \cot\beta = 1.43 \rightarrow N_{cf} = 27$

$$F = \frac{27 \times 5}{100} = 1.35$$

For $\lambda_{c\phi} = 11.55$ and $\beta = 35^\circ \rightarrow \cot\beta = 1.43 \rightarrow x_o = 0$ and $y_o = 1.82$
Therefore, $X_o = 0$ and $Y_o = 18.2$ m.

b. Case 2: $H_w = 0$, $H'_w = 10$ m

$$\mu_w = \mu'_w = 1, \text{all other } \mu = 1$$

$$P_d = \frac{20 \times 10}{1 \times 1 \times 1} = 200$$

$$P_e = \frac{20 \times 10 - 10 \times 10}{1 \times 1} = 100$$

$$\lambda_{c\phi} = \frac{100 \tan(30)}{5} = 11.55$$

For $\lambda_{c\phi} = 11.55$ and $\beta = 35^\circ \rightarrow \cot\beta = 1.43 \rightarrow N_{cf} = 27$

$$F = \frac{27 \times 5}{200} = 0.675$$

For $\lambda_{c\phi} = 11.55$ and $\beta = 35^\circ \rightarrow \cot\beta = 1.43 \rightarrow x_o = 0$ and $y_o = 1.82$
Therefore, $X_o = 0$ and $Y_o = 18.2$ m.

c. Case 3: $H_w = 0$, $H'_w = 0$

$$\mu_w = \mu'_w = 1, \text{all other } \mu = 1$$

$$P_d = \frac{20 \times 10}{1 \times 1 \times 1} = 200$$

$$P_e = \frac{20 \times 10}{1 \times 1} = 200$$

$$\lambda_{c\phi} = \frac{200 \tan(30)}{5} = 23.1$$

For $\lambda_{c\phi} = 23.1$ and $\beta = 35^\circ \rightarrow \cot\beta = 1.43 \rightarrow N_{cf} = 47$

$$F = \frac{47 \times 5}{200} = 1.175$$

For $\lambda_{c\phi} = 23.1$ and $\beta = 35^\circ \rightarrow \cot\beta = 1.43 \rightarrow x_o = -0.2$ *and* $y_o = 2$
Therefore, $X_o = -2$ m and $Y_o = 20$ m.

Problem 19.6

A 3 horizontal to 1 vertical slope is cut in a clay that has an effective stress cohesion of 5 kPa and an effective stress friction angle of 30°. The slope is 6 m high. Calculate:

a. The factor of safety of the slope against long-term failure if the water table is below the critical circle.
b. The factor of safety of the slope against long-term failure if the water table coincides with the slope contour.

Solution 19.6

The slope angle is: $\tan\beta = \frac{1}{3}$ *then* $\beta = 18.42^\circ$

a. Dry Case, No Water

a-1. Taylor (1948)

Assume depth factor as D = 0.5, $\beta = 18.5^\circ$, $c' = 5$ kPa, $\varphi' = 30^\circ$, and $\gamma = 20$ kN/m^3.

$$F'_c = \frac{c'}{c'_d} \quad and \quad F'_\varphi = \frac{\tan\varphi'}{\tan\varphi'_d}$$

Iteration #1, FS = 1.5:

$$F'_c = \frac{c'}{c'_d} \rightarrow 1.5 = \frac{5}{c'_d} \rightarrow c'_d = 3.33 \text{ kPa}$$

The stability number $N = \frac{c'_d}{\gamma H} = \frac{3.33}{20\times 6} = 0.0277 \rightarrow$ Figure 19.13 $\rightarrow \varphi'_d = 11^\circ$

$$F'_\varphi = \frac{\tan\varphi'}{\tan\varphi'_d} = \frac{\tan 30}{\tan 11} = 2.97$$

Iteration #2, FS = 2.2:

$$F'_c = \frac{c'}{c'_d} \rightarrow 2.2 = \frac{5}{c'_d} \rightarrow c'_d = 2.27 \text{ kPa}$$

The stability number $N = \dfrac{c'_d}{\gamma H} = \dfrac{2.27}{20 \times 6} = 0.019 \rightarrow$ Figure 19.13 $\rightarrow \varphi'_d = 14.5^\circ$

$$F'_\varphi = \frac{\tan\varphi'}{\tan\varphi'_d} = \frac{\tan 30}{\tan 14.5} = 2.23$$

The safety factor would be 2.21.

a-2. Spencer (1967)

$$r_u = \frac{u_w}{\sigma_{0v}} \text{ assumed equal to 0}$$

Iteration #1, FS = 1.5:

$$F'_c = \frac{c'}{c'_d} \quad and \quad F_{\varphi'} = \frac{\tan\varphi'}{\tan\varphi'_d}$$

$$F'_c = \frac{c'}{c'_d} \rightarrow 1.5 = \frac{5}{c'_d} \rightarrow c'_d = 3.33 \text{ kPa} \quad \text{and} \quad N = \frac{c'_d}{\gamma H} = \frac{3.33}{20 \times 6} = 0.0277$$

Figure 19.15 $\rightarrow r_u = 0 \rightarrow \varphi'_d = 12^\circ$

$$F'_\varphi = \frac{\tan\varphi'}{\tan\varphi'_d} = \frac{\tan 30}{\tan 12} = 2.7$$

Iteration #2, FS = 2.5:

$$F'_c = \frac{c'}{c'_d} \rightarrow 2.5 = \frac{5}{c'_d} \rightarrow c'_d = 2 \text{ kPa}$$

The stability number $N = \frac{c'_d}{\gamma H} = \frac{2}{20 \times 6} = 0.017 \rightarrow$ Figure 19.15 $\rightarrow \varphi'_d = 14^\circ$ $F'_\varphi = \frac{\tan \varphi'}{\tan \varphi'_d} = \frac{\tan 30}{\tan 14} = 2.3$
The average safety factor would be 2.4.

b. Water Case, water level at ground surface

b-1. Spencer (1967), Drained behavior: c′, φ′

$$r_u = \frac{u_w}{\sigma_{0v}} = 0.5$$

Iteration #1, FS = 1.5:

Figure 19.15 $\rightarrow r_u = 0.5, N = \frac{c'_d}{\gamma H} = \frac{3.33}{20 \times 6} = 0.0277 \rightarrow \varphi'_d = 23^\circ$

$$F'_\varphi = \frac{\tan \varphi'}{\tan \varphi'_d} = \frac{\tan 30}{\tan 23} = 1.36$$

Factor of safety is about 1.43.

Problem 19.7

Calculate the probability of failure of a slope that has a factor of safety with a mean of 1.5 and a coefficient of variation of 0.2 (assume that the factor of safety is normally distributed). If the acceptable probability of failure is 0.001, what must be the mean value of the factor of safety if the coefficient of variation remains equal to 0.2?

Solution 19.7

Probability of Failure for a Given Factor of Safety

Mean, $\mu = 1.5$
Coefficient of variation, $CoV = 0.2$
Acceptable probability of failure, $\text{PoF}_{\text{ac}} = 0.001$
Standard deviation σ:

$CoV = \frac{\sigma}{\mu}$

$\sigma = \mu \cdot CoV = 1.5 \times 0.2 = 0.3$
Standard normal variable U of F = 1:

$$u = \frac{F - \mu_F}{\sigma_F} = \frac{1 - 1.5}{0.3} = -1.67$$

$$P(F < 1) = P\left(\frac{F - \mu_F}{\sigma_F} < \frac{1 - \mu_F}{\sigma_F}\right) = P(U < -1.67)$$

Using Table 11.3,

$$\text{P}(\text{U} < 1.67) = 0.9525$$

$$\text{P}(\text{U} < \text{u}) = 1 - \text{P}(\text{U} < -\text{u})$$

$$\text{P}(\text{U} < -1.67) = 1 - 0.9525 = 0.0475 \text{ or } 4.75\% \text{ probability of failure}$$

Factor of Safety for a Given Probability of Failure

$P(F < 1) = 0.001$

$$P(F < 1) = P\left(\frac{F - \mu_F}{\sigma_F} < \frac{1 - \mu_F}{\sigma_F}\right) = P\left(\frac{F - \mu_F}{0.2\mu_F} < \frac{1 - \mu_F}{0.2\mu_F}\right) = P\left(U < \frac{1 - \mu_F}{0.2\mu_F}\right) = 0.001$$

Using Table 11.3, we get:

$$P(U < 3.1) = 0.999 \quad or \quad P(U < -3.1) = 0.001$$

Therefore, we must have:

$$\frac{1-\mu_F}{0.2\mu_F} = -3.1 \quad or \quad \mu_F = 2.63$$

Problem 19.8

Which situation is more desirable? Explain and demonstrate.

a. Mean factor of safety F = 1.5 and coefficient of variation of F = 0.2
b. Mean factor of safety F = 1.3 and coefficient of variation of F = 0.1

Solution 19.8

$$CoV = \frac{\sigma}{\mu}$$

$$\sigma_a = CoV.\mu = 0.2 \times 1.5 = 0.3$$

$$\sigma_b = CoV.\mu = 0.1 \times 1.3 = 0.13$$

$$P(F < 1) = P\left(\frac{F-\mu_F}{\sigma_F} < \frac{1-\mu_F}{\sigma_F}\right) = P\left(\frac{F-1.5}{0.3} < \frac{1-1.5}{0.3}\right)$$

$$= P(U < -1.67) = 1 - P(U < 1.67) = 1 - 0.9525 = 0.0475$$

$$P(F < 1) = P\left(\frac{F-\mu_F}{\sigma_F} < \frac{1-\mu_F}{\sigma_F}\right) = P\left(\frac{F-1.3}{0.13} < \frac{1-1.3}{0.13}\right)$$

$$= P(U < -2.31) = 1 - P(U < 2.31) = 1 - 0.9896 = 0.0104$$

Therefore, a mean factor of safety of 1.3 with a coefficient of variation of 0.1 (case b) is more desirable than a mean factor of safety of 1.5 and a coefficient of variation of 0.2 (case a). The reason is that the probability of failure is 1.04% in case b and 4.75% in case a.

Problem 19.9

Calculate the factor of safety for the slope shown in Figure 19.3s. Select your best estimate of the critical circle and calculate the factor of safety by the Bishop modified method of slices.

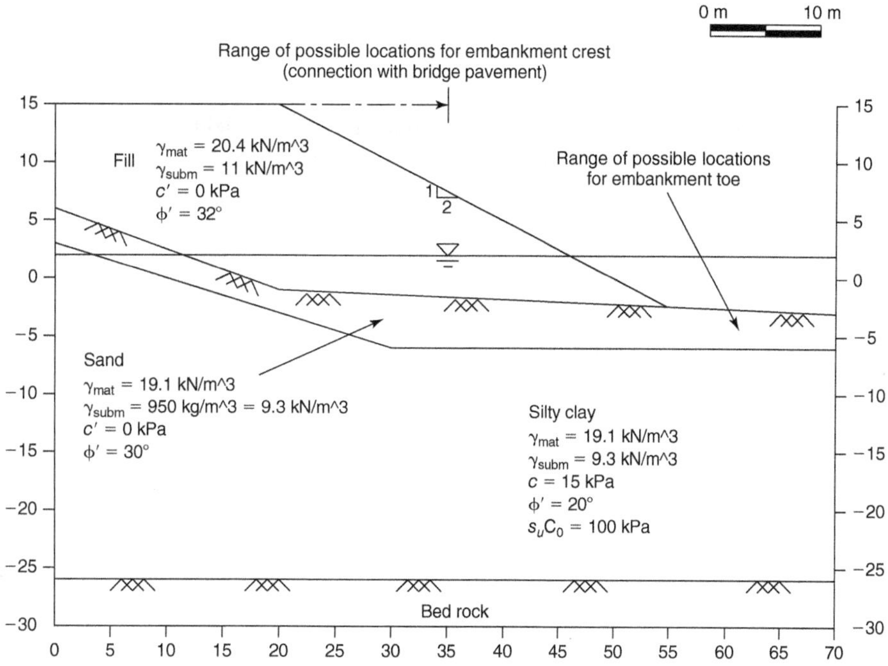

Figure 19.3s Slope of Fredericton embankment.

Solution 19.9

The sketch for the simplified method of slices is shown in Figure 19.6s. The results are shown in Table 19.2s.

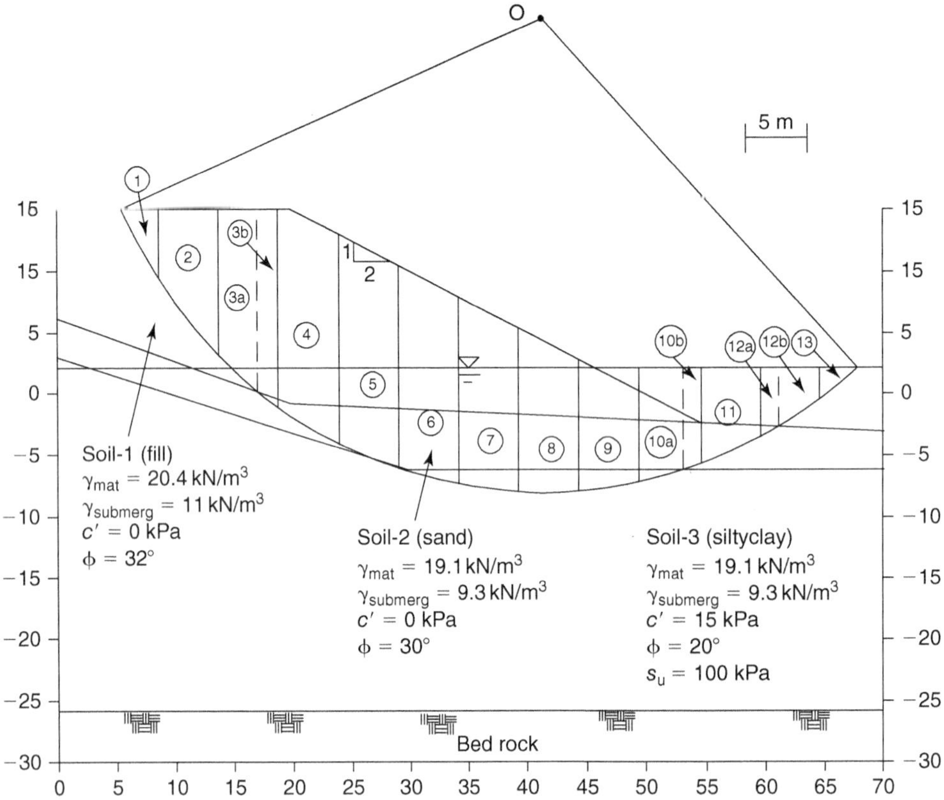

Figure 19.6s Bishop Simplified Method of Slices.

Problem 19.10

A slope is subjected to an acceleration history as shown in Figure 19.4s. The yield acceleration for that slope is 1.5 m/s^2. Calculate the displacement history of the slope according to Newmark's method.

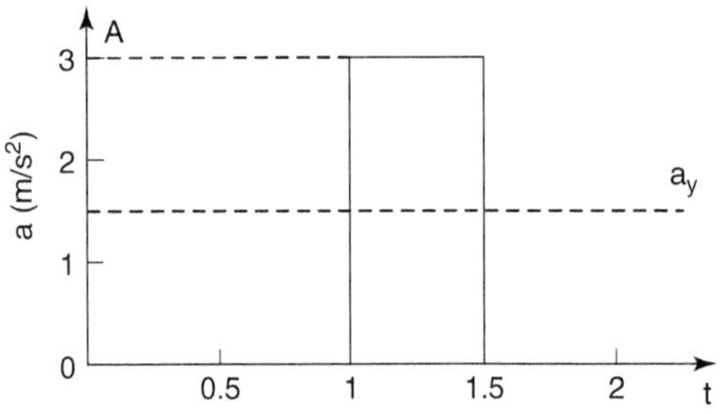

Figure 19.4s Acceleration history.

Table 19.2s Summary of the Results of the Bishop Simplified Method of Slices

Slice #	φ′	b (m)	Area fill (m2²)		Area total (m²)	Unit Weight (kN/m³)	Weight (kN/m)	Total Weight (kN/m)	θ	$\cos\theta_i$	Wsinθ (kN)	b (m)	hp (m)	u_w (kPa)	$u_w b_i$ (kN/m)	$(W-u_w b)\tan\varphi'$ (kN)	c′ (kN/m)	c′b/cosθ (kN/m)	m_θ	$[cb+(W-u_w b)\tan\varphi']/m_\theta$
1	32	3.2	Soil-1 (100%)	9.0	9.0	20.40	183.6	183.6	60.0	0.50	159.00	3.20	0.00	0.00	0.00	114.73	0.00	0.00	0.87	131.38
2	32	5.0	Soil-1 (100%)	43.5	43.5	20.40	887.4	887.4	51.1	0.63	690.61	5.00	0.00	0.00	0.00	554.51	0.00	0.00	0.96	575.61
3a	32	3.3	Soil-1 (95%)	66.3	66.3	20.40	1352.5	1352.5	42.7	0.73	917.22	3.25	0.90	8.83	28.69	827.22	0.00	0.00	1.03	805.34
3b	30	1.8	Soil-2 (5%)	3.5	3.5	19.10	66.9	66.9	36.6	0.80	39.86	1.75	2.60	25.51	44.64	12.83	0.00	0.00	1.04	12.33
4	30	5.0	Soil-1 (90%)	8.0	82.8	20.40	163.2	1591.9	30.1	0.87	798.34	5.00	5.00	49.05	245.25	777.48	0.00	0.00	1.06	730.14
			Soil-2 (10%)	74.8		19.10	1428.7													
5	30	5.0	Soil-1 (60%)	26.6	44.4	20.40	543.5	882.7	21.8	0.93	327.80	5.00	7.22	70.83	354.14	305.15	0.00	0.00	1.08	283.50
			Soil-2 (40%)	17.8		19.10	339.2													
6	20	5.0	Soil-1 (68%)	51.7	76.0	20.40	1054.7	1518.8	11.9	0.98	313.18	5.00	8.90	87.31	436.55	393.91	15.00	75.00	1.03	455.14
			Soil-2 (28%)	21.3		19.10	406.8													
			Soil-3 (4%)	3.0		19.10	57.3													
7	20	5.0	Soil-1 (57%)	40.6	71.2	20.40	828.2	1411.7	7.3	0.99	179.38	5.00	9.80	96.14	480.69	338.88	15.00	75.00	1.02	404.26
			Soil-2 (33%)	23.3		19.10	445.0													
			Soil-3 (10%)	7.3		19.10	138.5													
8	20	5.0	Soil-1 (49%)	29.1	59.8	20.40	593.6	1180.0	0.0	1.00	0.00	5.00	10.14	99.47	497.37	248.46	15.00	75.00	1.00	323.46
			Soil-2 (36%)	22.0		19.10	420.2													
			Soil-3 (15%)	8.7		19.10	166.2													
9	20	5.0	Soil-1 (36%)	17.5	48.7	20.40	357.0	920.4	-9.1	0.99	-145.57	5.00	9.70	95.16	475.79	161.83	15.00	75.00	0.95	249.89
			Soil-2 (43%)	20.9		19.10	399.2													
			Soil-3 (14%)	6.8		19.10	129.9													
			Water (7%)	3.5		9.81	34.3													
10a	20	3.5	Soil-1 (18%)	5.4	29.8	20.40	110.8	486.8	-15.1	0.97	-126.81	3.50	8.90	87.31	305.58	65.96	15.00	52.50	0.90	131.61
			Soil-2 (44%)	13.1		19.10	250.2													
			Soil-3 (5%)	1.6		19.10	30.6													
			Water (33%)	9.7		9.81	95.3													
10b	30	1.5	Soil-1 (5%)	0.5	11.8	20.40	10.8	167.7	-20.1	0.94	-57.63	1.50	8.00	78.48	117.72	28.86	0.00	0.00	0.80	35.97
			Soil-2 (43%)	5.0		19.10	96.1													
			Water (53%)	6.2		9.81	60.8													
11	30	5.0	Soil-2 (32%)	10.5	33.5	19.10	200.6	426.2	-23.7	0.92	-171.30	5.00	6.70	65.73	328.64	56.32	0.00	0.00	0.76	74.53
			Water (68%)	23.0		9.81	225.6													
12a	30	5.0	Soil-2 (17%)	0.7	4.2	19.10	13.4	48.0	-31.2	0.86	-24.86	5.00	5.23	51.31	256.53	0.00	0.00	0.00	0.65	0.00
			Water (83%)	3.5		9.81	34.6													
12b	0	3.5	Water (100%)	12.7	12.7	9.81	124.6	124.6	-60.0	0.50	-107.90	3.50	3.86	37.87	132.53	0.00	0.00	0.00	0.50	0.00
13	0	3.2	Water (100%)	4.1	4.1	9.81	40.2	40.2	-68.0	0.37	-37.29	3.20	1.50	14.72	47.09	0.00	0.00	0.00	0.37	0.00
										Σ	**2754.0**						Σ	**352.50**		**4213.17**
Note: the percentage number in the section of the area stand for the corresponding area for each of the slide in consideration.																			**F.S.=**	**1.53**

Solution 19.10

From Figure 19.4s: single rectangular acceleration, $A = 3$ m/s^2, yield acceleration, $a_y = 1.5$ m/s^2, relative acceleration $a_{rel}(t)$, relative velocity $v_{rel}(t)$ and relative displacement $d_{rel}(t)$.

At $t_0 \leq t \leq t_0 + \Delta t$ or 1 < t < 1.5 seconds, relative acceleration, relative velocity, and relative displacement are:

$$a_{rel}(t) = A - a_y = 3 - 1.5 = 1.5 \text{ m/s}^2$$

$$v_{rel}(t) = \int_{t_0}^{t} a_{rel}(t)dt = [A - a_y](t - t_0) = [3 - 1.5](t - 1) = 1.5t - 1.5 \text{ m/s}$$

$$d_{rel}(t) = \int_{t_0}^{t} v_{rel}(t)dt = \frac{1}{2}[A - a_y](t - t_0)^2 = \frac{1}{2}[3 - 1.5](t - 1)^2 - 0.75(t \quad 1)^2 \text{ m}$$

At $t_0 = t_0 + \Delta t$ or 1.5 seconds:

$$v_{rel}(t + \Delta t) = [A - a_y]\Delta t = [3 - 1.5] \times 0.5 = 0.75 \text{ m/s}$$

$$d_{rel}(t + \Delta t) = \frac{1}{2}[A - a_y]\Delta t^2 = \frac{1}{2}[3 - 1.5]0.5^2 = 0.1875 \text{ m}$$

At $t_0 + \Delta t \leq t \leq t_1$ or 1.5 < t < 2 seconds:

$$a_{rel}(t) = 0 - a_y = -1.5 \text{ m/s}^2$$

$$v_{rel}(t) = 0.75 + \int_{1.5}^{t} a_{rel}(t)dt = 0.75 + \int_{1.5}^{t} (-1.5)dt = -1.5t + 3 \text{ m/s}$$

$$d_{rel}(t) = 0.1875 + \int_{1.5}^{t} v_{rel}(t)dt = 0.1875 + \int_{1.5}^{t} (-1.5t + 3)dt = -0.75t^2 + 3t - 2.625$$

All results are plotted in Figure 19.7s.

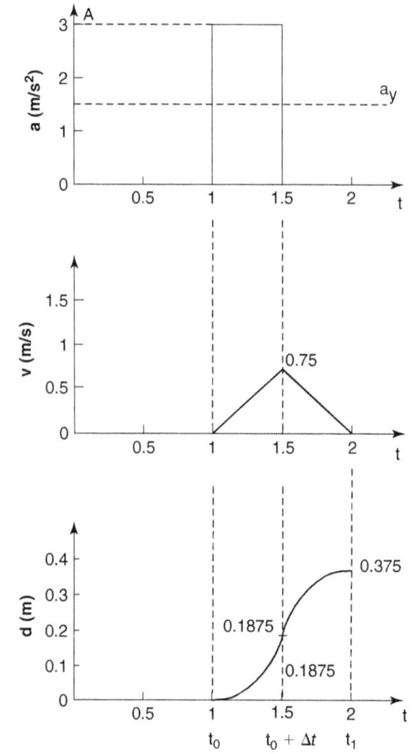

Figure 19.7s Acceleration, velocity, and displacement history of a slope.

Problem 19.11

Consider the case of a $c' = 0$, $\varphi' > 0$ soil and demonstrate that for one slice, the factor of safety is the same for the ordinary method of slices (OMS) and for the Bishop simplified method of slices (BSMS). If it is true for one slice, why is it not true for n slices ($n > 1$)?

Solution 19.11

OMS

$$F = \frac{W \cos\theta \tan\varphi}{W \sin\theta} = \frac{\tan\varphi}{\tan\theta}$$

BSMS

$$m = \cos\theta\left(1 + \frac{\tan\theta \tan\varphi}{F}\right)$$

$$F = \frac{\frac{1}{m}(W \tan\varphi)}{W \sin\theta} = \frac{\tan\varphi}{\cos\theta \sin\theta\left(1 + \frac{\tan\theta \tan\varphi}{F}\right)}$$

$$\frac{\tan\varphi}{F} = \cos\theta \sin\theta\left(1 + \frac{\tan\theta \tan\varphi}{F}\right)$$

$$\frac{\tan\varphi}{F}(1 - \cos\theta \sin\theta \tan\theta) = \cos\theta \sin\theta$$

$$F = \frac{\tan\varphi(1 - \sin^2\theta)}{\cos\theta \sin\theta} = \frac{\tan\varphi \cos^2\theta}{\cos\theta \sin\theta} = \frac{\tan\varphi}{\tan\theta}$$

The assumption in both methods is with respect to side forces. Because we have only one slice, there are no side forces between slices. Most importantly, there are as many unknowns as there are equations, so equilibrium equations can be written in any direction and will lead to the same answer. The three unknowns are S_m, N', and F; the three equations are vertical equilibrium, horizontal equilibrium, and the shear strength equation. Moment equilibrium is automatically satisfied because all forces go through the middle of the base of the slice. Thus, both methods should give identical safety factor. This is no longer true when we have more than one slice, because the assumptions are different for the side forces.

Problem 19.12

A 3D slope failure has a failure surface in the form of a sphere and a factor of safety F_{3D}. This sphere is sliced in a direction perpendicular to the crest. The slices have the same width b. The deepest slice in the center of the sphere has a factor of safety F_{min}. Each slice has a 2D factor of safety equal to F_i. What other assumptions must be made for the following equation to be true?

$$F_{3D} = \frac{1}{n}\sum_{i=1}^{n} F_i$$

Solution 19.12

As explained in section 19.16, in order to use this equation for the safety factor F_{3D}:

1. The axis of rotation must be the same for all slices (Figure 19.34)
2. The forces between circle slices must be negligible

Problem 19.13

A dry, fine sand slope has a factor of safety of 1.5 on the Earth.

a. Calculate and discuss the factor of safety for the same slope on the moon.
b. Assume that there could be water on the moon. Would the result of the dry case still hold?

Solution 19.13

a. The factor of safety for dry sand is $F = \frac{\tan\varphi'}{\tan\beta}$. It is independent of gravity acceleration (g), so it would be the same on the moon.

b. It would be if the water was at the surface of the slope, but it would not be if the slope was above the groundwater level and water tension developed in the slope. Because water tension is a chemically-based phenomenon and not a gravity-based phenomenon, it would be the same on the Earth and on the moon, but its ratio to gravity forces would be very different. Therefore, it would lead to different factors of safety. The same slope would be safer on the moon if water tension existed in both cases.

Problem 19.14

Define the seepage force and discuss when the seepage force should be considered in a slope stability analysis. Why is it not usually considered?

Solution 19.14

The seepage force is the force exerted in friction by water flowing around soil particles and trying to drag them away. The forces shown on a free-body diagram are the external forces. The internal forces are resolved internally. The seepage force is an external force when the soil skeleton is considered as the free body, but it is an internal force when the soil skeleton plus the water is considered as the free body. Most slope stability analyses consider the soil skeleton plus the water as the free body. In those instances, the seepage force must not be included in any slope stability calculations.

Problem 19.15

Explain the difference between the following analyses, including what shear strength you would use: total stress analysis, effective stress analysis, undrained analysis, drained analysis, short-term analysis, long-term analysis.

Solution 19.15

A total stress analysis considers that the soil is made of one material. During the analysis, the three components (particles, water, and air) are not recognized. This analysis can be used in the case of a soil with no water and in the case of a soil where the shear strength is independent of rapid variations in total stress.

An effective analysis can be used in all cases. It makes no particular assumption regarding drainage and is based on sound fundamental principles. It makes use of the effective stress equation to obtain the shear strength of the soil based on effective stress cohesion c′ and the effective stress friction angle φ' ($\tau = c' + \sigma' \tan\varphi'$). It can be used for an undrained analysis, a drained analysis, a short-term analysis, or a long-term analysis. The difficulty with this method is that the water stress in the mass must be known.

An undrained analysis is used in the case where the water is not allowed or does not have time to drain away. The soil strength parameter used in this case is the undrained shear strength (s_u).

In a drained analysis, the water stress is considered to be hydrostatic throughout the mass. The soil strength parameters used are the drained strength parameters or effective stress parameters ($\tau = c' + \sigma' \tan\varphi'$).

A short-term analysis considers a time shortly after loading. It is often a drained analysis for fast-draining soils like free-draining sands and gravels, and an undrained analysis for slow-draining soils like silts and clays. In the case of free-draining sands and gravels, the drained strength parameters are used. In the case of silts and clays, the undrained shear strength (s_u) is used.

A long-term analysis considers that all water stresses induced by loading have had time to dissipate and are back to hydrostatic condition. In this regard a long-term analysis is similar to a drained analysis. The soil strength parameters used are the drained strength parameters or effective stress parameters ($\tau = c' + \sigma' \tan\varphi'$).

Problem 19.16

An inclinometer casing is attached to a 10 m high retaining wall. Zero readings taken before the wall is backfilled indicate that the wall is perfectly vertical. The backfill is placed and compacted. At the end of construction, the inclinometer readings are taken again. Find what the readings are if:

a. The displacement y (m) of the wall obeys the equation $y = 0.01\ (z_{max} - z)$ where z_{max} is the maximum depth the inclinometer probe can reach in the casing (10 m) and z is the depth at which the reading is taken.

b. The displacement y (m) of the wall obeys the equation $y = 0.001\ (z_{max} - z)^2$ where z_{max} is the maximum depth the inclinometer probe can reach in the casing (10 m) and z is the depth at which the reading is taken.

The inclinometer has a calibration constant C = 20,000 and a wheel spacing of 0.5 m.

Solution 19.16

Assume that the constant parameter for the inclinometer is C = 20,000, and the length of probe between wheels is L = 0.5 m. Figure 19.8s shows the wall and inclinometer.

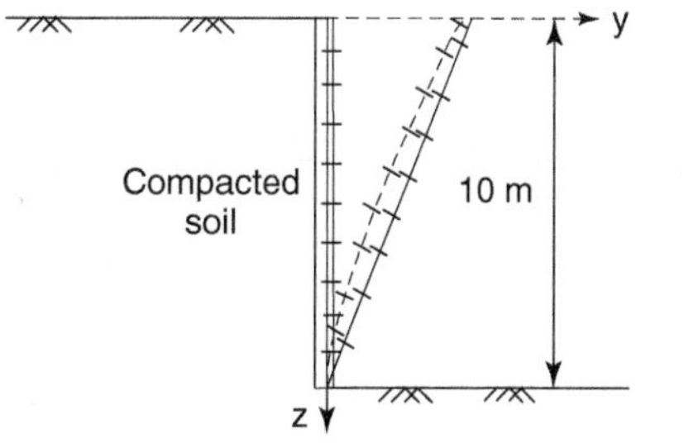

Figure 19.8s Illustration of the inclinometer.

The equation for the displacement D_n of the inclinometer casing is:

$$D_n = \sum_{i=1}^{n}(d_i - d_{io}) = \frac{L}{C}\sum_{i=1}^{n}(R_i - R_{io})$$

where D_n is the displacement at a depth z_n from the surface, z_n is the depth to the first (deepest) reading in the casing minus n times the distance L between readings, d_i is the difference in horizontal displacement between the i and $i-1$ reading points, d_{io} is the initial value of d_i, L is the length between readings, C is the inclinometer calibration constant, R_i is the reading at depth z_i, and R_{io} is the initial value of R_i.

Case a. In this case the wall deforms by simple rotation and the displacement y(m) of the wall is linear. The equation for D_n becomes:

$$D_n = \sum_{i=1}^{n} d_i = \frac{L}{C}\sum_{i=1}^{n} R_i$$

But D_n is also given in the problem as:

$$D_n = 0.01(10 - z)$$

Therefore, the difference between two consecutive readings is:

$$D_n - D_{n-1} = d_n = 0.01(10 - z) - 0.01(10 - z - 0.5) = 0.005 \text{ m}$$

The increment of displacement is constant and the angle of the wall is also constant:

$$\theta_n = \sin^{-1}\frac{d_n}{L} = \sin^{-1}\frac{0.005}{0.5} \simeq 0.01 \text{ rd}$$

Furthermore, D_o is equal to zero, because the bottom of the wall does not move. Now the reading is equal to:

$$R_n = \frac{C}{L}d_n = \frac{20000}{0.5} \times 0.005 = 200$$

So the reading of the inclinometer is constant equal to 200; Table 19.3s summarizes the results.

Case b. In this case, the displacement y(m) of the wall is nonlinear. The equation for D_n is still:

$$D_n = \sum_{i=1}^{n} d_i = \frac{L}{C}\sum_{i=1}^{n} R_i$$

But D_n is also given in the problem as:

$$D_n = 0.001(10 - z)^2$$

Therefore, the difference between two consecutive readings is:

$$D_n - D_{n-1} = d_n = 0.001(10 - z)^2 - 0.001(10 - z - 0.5)^2 = 0.001(9.75 - z)$$

The increment of displacement increases linearly with z and so does the angle of the wall:

$$\theta_n = \sin^{-1}\frac{0.001(9.75 - z)}{L} = \sin^{-1}\frac{d_n}{L}$$

Furthermore, D_o is equal to zero, because the bottom of the wall does not move. Now the reading is equal to:

$$R_n = \frac{C}{L}d_n = \frac{20000}{0.5} \times 0.001(9.75 - z) = 390 - 40z$$

Table 19.4s summarizes the results.

Table 19.3s Inclinometer Readings for Linear Displacement of Wall

Depth z (m)	Displacement y (m)	Inclined angle (radians)	Inclined angle (degree)	Inclinometer reading (R)
10	0	0.009999667	0.572938698	0
9.5	0.005	0.009999667	0.572938698	200
9	0.01	0.009999667	0.572938698	200
8.5	0.015	0.009999667	0.572938698	200
8	0.02	0.009999667	0.572938698	200
7.5	0.025	0.009999667	0.572938698	200
7	0.03	0.009999667	0.572938698	200
6.5	0.035	0.009999667	0.572938698	200
6	0.04	0.009999667	0.572938698	200
5.5	0.045	0.009999667	0.572938698	200
5	0.05	0.009999667	0.572938698	200
4.5	0.055	0.009999667	0.572938698	200
4	0.06	0.009999667	0.572938698	200
3.5	0.065	0.009999667	0.572938698	200
3	0.07	0.009999667	0.572938698	200
2.5	0.075	0.009999667	0.572938698	200
2	0.08	0.009999667	0.572938698	200
1.5	0.085	0.009999667	0.572938698	200
1	0.09	0.009999667	0.572938698	200
0.5	0.095	0.009999667	0.572938698	200
0	0.1	0.009999667	0.572938698	200

Table 19.4s Inclinometer Readings for Linear Displacement of Wall

Depth z (m)	Displacement y (m)	Inclined angle (radians)	Inclined angle (degree)	Inclinometer reading (R)
10	0	0	0	0
9.5	0.00025	0.0005	0.028648	10
9	0.001	0.0014997	0.085943	30
8.5	0.00225	0.0024995	0.143239	50
8	0.004	0.0034993	0.200535	70
7.5	0.00625	0.0044992	0.257832	90
7	0.009	0.0054990	0.315128	110
6.5	0.01225	0.0064988	0.372425	130
6	0.016	0.0074986	0.429722	150
5.5	0.02025	0.0084985	0.487020	170
5	0.025	0.0094983	0.544318	190
4.5	0.03025	0.0104982	0.601617	210
4	0.036	0.0114981	0.658916	230
3.5	0.04225	0.0124980	0.716216	250
3	0.049	0.0134978	0.773516	270
2.5	0.05625	0.0144978	0.830818	290
2	0.064	0.0154977	0.888120	310
1.5	0.07225	0.0164976	0.945423	330
1	0.081	0.0174976	1.002727	350
0.5	0.09025	0.0184976	1.060032	370
0	0.1	0.0194975	1.117338	390

CHAPTER 20

Compaction

20.1 GENERAL

Compaction refers to the densification of shallow soil layers by rollers. These rollers may be static cylindrical rollers, smooth or with protrusions; vibratory cylindrical rollers; or impact noncylindrical rollers. *Conventional compaction* refers to use of noninstrumented rollers, whereas *intelligent compaction* refers to use of instrumented rollers with feedback loops. *Dynamic compaction*, also discussed in this chapter, refers to dropping large weights from a given height onto the ground surface; this process creates a crater and compacts that material under the crater.

Compaction is required in many instances; examples include for the base layer of pavements, for embankment fills, for retaining wall backfills, for fill around pipes, and for landfills. Depending on the soil type and the size of the project, different compactors are used (Figure 20.1). For example, hand tampers (also called *jumping jacks*) are used in small areas around pipes, rollers are used for roadway compaction, and drop-weight compactors are used for dynamic compaction of large areas at larger depth.

The rollers are typically 50 to 150 kN in weight; the drums are 1 to 2 m in diameter and 2 to 3 m wide. The frequency of vibration for vibrating rollers is from 30 to 70 Hz. For dynamic compaction, the drop weight commonly varies from 50 to 250 kN and the drop height from 5 to 25 m. The depth over which the soil is compacted is up to 1 m for rollers and up to 10 m for dynamic compaction.

The compaction process typically takes the following steps:

1. Perform laboratory tests on the material to be compacted (Proctor test, for example) and establish the value of the soil property to be reached in the field work. These properties are most commonly the dry density and the water content. The modulus of deformation can also be used.
2. Write the field specifications, including the target dry density or target modulus within a chosen range of water content.
3. In the field, use compacting equipment to compact the soil in 0.15 m lifts after it is brought to the chosen water content.
4. In the field also carry out field tests to verify that the target values listed in the specifications have been reached.

20.2 COMPACTION LABORATORY TESTS

Laboratory tests are used to establish the characteristics of the soil to be compacted, to establish the target values to be achieved in the field, and to write the specifications for field work. The compaction process and the compaction curve associated with laboratory tests are described in detail in section 9.3; this section gives a brief summary.

The compaction curve links the dry density or a soil modulus to the water content. The dry density vs. water content curve is relatively flat, as the dry density is not very sensitive to the water content. Within the range of dry density variation, the curve has a bell shape (Figure 20.2). The reason for this is that at point A in Figure 20.2, the soil is relatively dry and it is difficult for a given compaction energy to bring the particles closer together. At point B the water content is such that water tension exists between the particles and hinders the effectiveness of the compaction process. At point C, the water tension loses its effect and the primary role of the water becomes to lubricate the contacts between particles, thereby allowing the given compaction effort to reach a low void ratio and a high dry density. At point D, the soil is nearing saturation and the added water simply increases the volume of the voids, which negates the benefit of the compaction. The maximum dry density γ_{dmax} and the optimum water content w_{opt} are two important parameters obtained from the curve (Figure 20.2). This curve is obtained in the laboratory with the Standard Proctor or Modified Proctor Compaction Test (Figure 20.3). These tests are described in detail in section 9.3.

The lines of equal degree of saturation can be presented on the same diagram as the dry density vs. water content curve (Figure 20.4). The equation for the saturation lines is:

$$\gamma_d = \frac{SG_s}{S + G_s w}\gamma_w \tag{20.1}$$

Figure 20.1 Compaction equipment: (a) Hand tamper. (b) Sheep-foot roller. (c) Smooth cylindrical roller. (d) Impact noncylindrical roller. (e) Drop-weight compactor. (a: Courtesy of Multiquip. b, c: Images courtesy of Caterpillar. d: Courtesy of LANDPAC. e: Courtesy of Serge Varaksin).

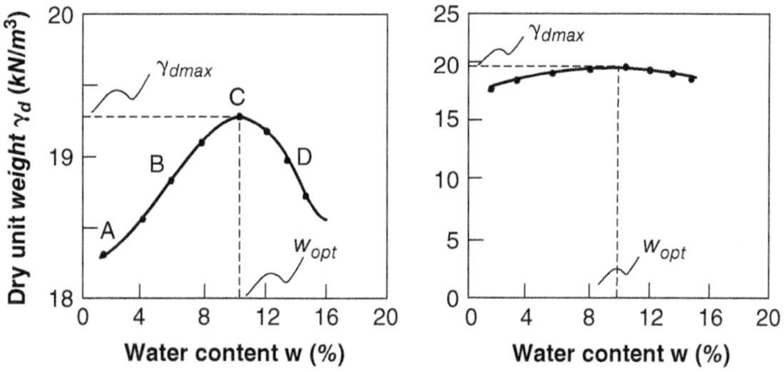

Figure 20.2 Compaction curve: dry density.

Figure 20.3 Proctor compaction laboratory test.

where γ_d is the dry density, S is the degree of saturation, G_s is the specific gravity of the solids, w is the water content, and γ_w is the unit weight of water. The derivation of this equation is shown in section 9.3.

In compaction control, the dry density can be replaced by the soil modulus as a governing parameter for specifications and quality control. The advantage of using the modulus is that the modulus is directly involved in the design calculations, whereas the dry density is not. The drawback is that the modulus depends on many factors (see section 14.2) and is not a single parameter for a given soil, whereas the dry density is. In the case of the modulus, the curve has the same bell shape as the dry density vs. water content curve, but is much more sensitive to the water content, especially on the wet side of the optimum water content (Figure 20.5). At low water contents, the modulus is influenced by the water tension that develops in the soil, whereas the dry density is not. As a result, the modulus curve can go back up at low water contents. The laboratory test to obtain the modulus vs. water content curve is the BCD test (Figure 20.6 and section 9.4), which is performed on the Proctor test sample. This is convenient because a dry density curve and a modulus curve can be obtained at the same time.

20.3 COMPACTION FIELD TESTS

The specifications indicate that the compacted soil must reach a dry density equal to a percentage of the maximum dry density measured in the laboratory (typically 95 to 100%) within a range of water content around the optimum water content. The specifications may also indicate that the compacted soil must reach a soil modulus equal to a percentage of the maximum soil modulus measured in the laboratory (typically 75% or so) within a range of water content around the optimum water content. Table 20.1 shows some possible target modulus values for pavement applications. Field tests are used to verify that the compaction work has been done according to specifications.

The field tests are divided into classic tests and new tests. The classic tests have been used for a long time and are relatively slow (15 to 30 minutes per test). They include the sand cone test for dry density, the rubber balloon test for dry density, and the nuclear density gage for dry density and water content. The new tests take only a few minutes to perform. They include the lightweight deflectometer, the BCD, and the field oven (Figure 20.7). All these tests are described in detail in section 7.11.

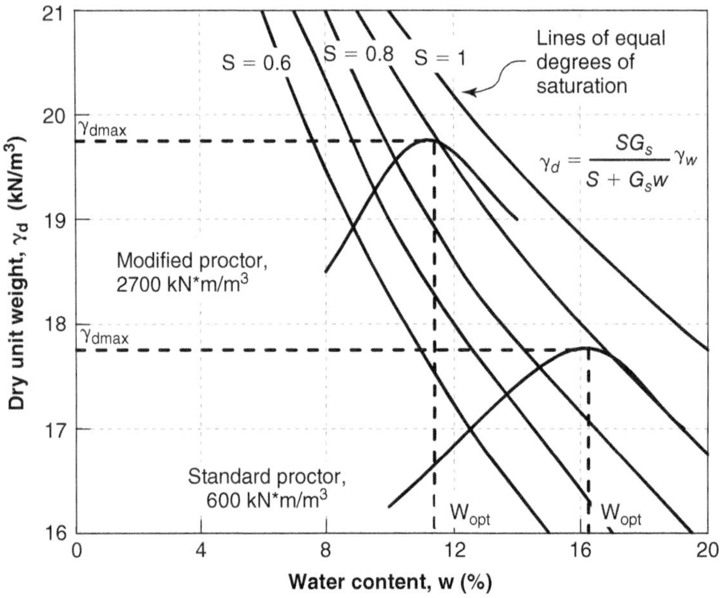

Figure 20.4 Compaction curves and saturation lines.

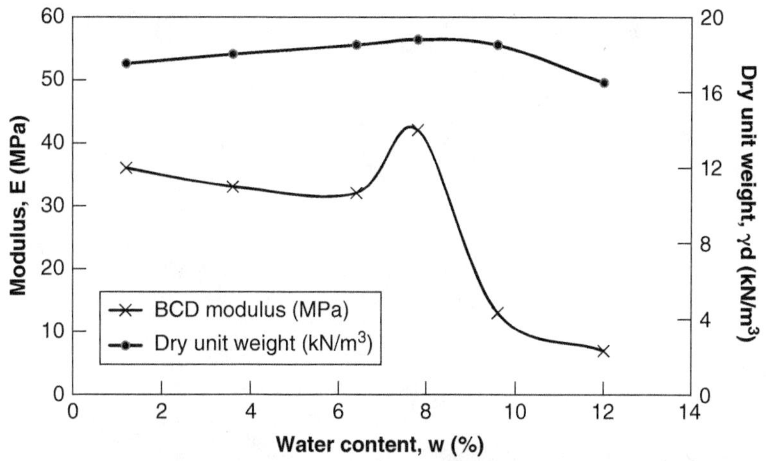

Figure 20.5 Compaction curve: modulus.

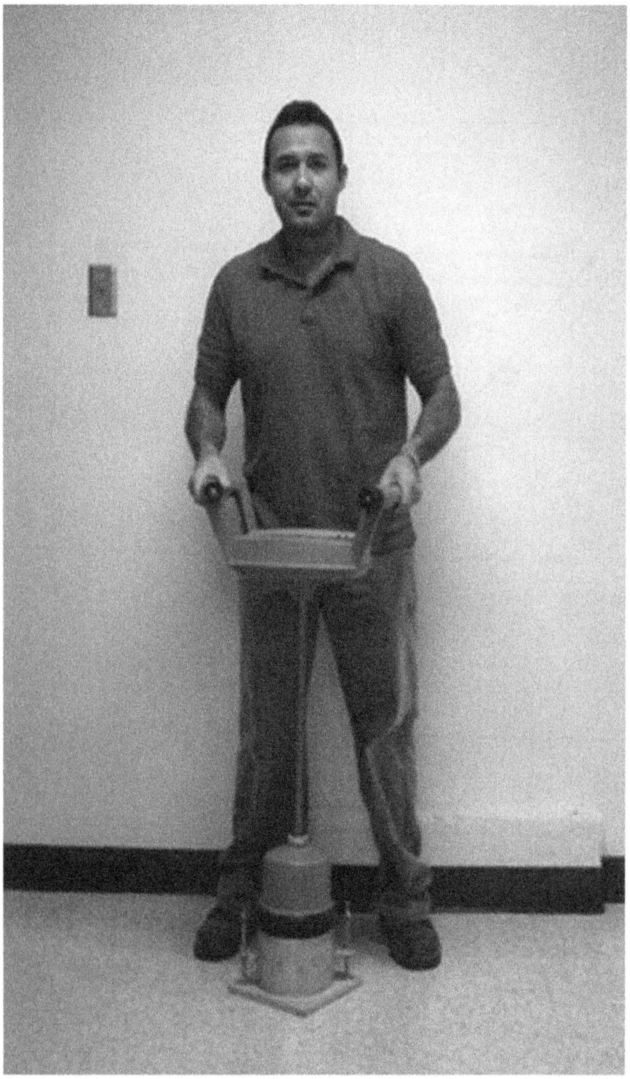

Figure 20.6 BCD-Proctor laboratory compaction test for modulus determination.

Table 20.1 Modulus Target Values for Pavements

Soil Layer	Plate Test Reload Modulus (MPa)	Lightweight Deflectometer Modulus (MPa)	BCD Reload Modulus (MPa)
Base course	100–150	100–150	55–85
Subgrade soil	45–80	45–80	25–45

20.4 COMPACTION AND SOIL TYPE

Different soils react differently to different compaction equipment. Coarse-grained soils are most effectively compacted through vibration combined with pressure. Pressure alone increases the effective stress and therefore the friction between particles, thereby preventing their sliding into a more compact position. Vibration breaks the friction bonds and lets the particles settle into a tighter arrangement. Fine-grained soils are most effectively compacted through kneading and pressure. Vibration may simply increase the water stress if the soil is saturated. Also, coarse-grained soils tend to reach optimum compaction at water contents lower than fine-grained soils. However, coarse-grained soils tend to reach maximum dry densities that are higher than those of fine-grained soils (Figure 20.8). Table 20.2 shows a rating of applications for various pieces of compaction equipment.

20.5 INTELLIGENT ROLLER COMPACTION

Continuous control compaction (CCC) refers to compaction with rollers that are instrumented, make measurements on the fly, and give an image of the complete compacted area

(a) (b) (c)

Figure 20.7 Compaction control tests in the field: (a) Nuclear gage. (b) Lightweight deflectometer. (c) BCD. (a: Photo by Lindsey D. Fields, Envirotech Engineering & Consulting, Inc. b: Courtesy of Minnesota Department of Transportation.)

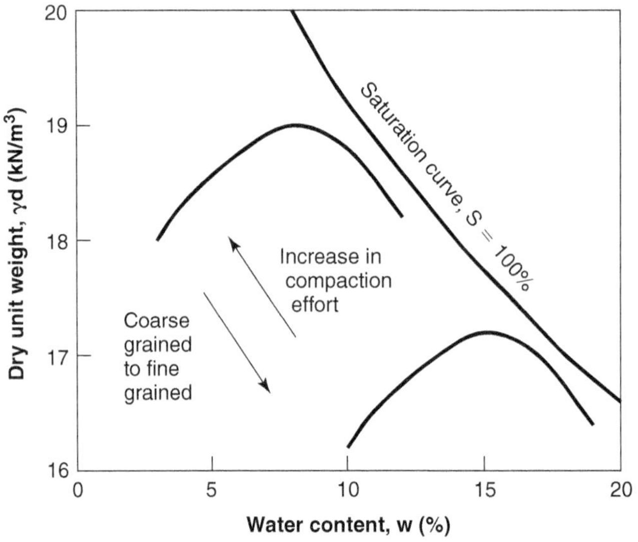

Figure 20.8 Influence of soil type and compaction effort on dry density curve.

Table 20.2 Applicability of Compaction Equipment for Various Soils

Soil Type	Static Sheep Foot Roller	Vibrating Cylindrical Roller	Impact Noncylindrical Roller	Dynamic Compaction
Gravel	Poor	Good	Good	Good
Sand	Poor	Very good	Good	Good
Silt	Good	Poor	Poor	Medium
Clay	Very good	Poor	Medium	Poor if saturated, good if unsaturated.
Domestic waste	Good	Poor	Good	Very good

with the values of the soil parameter measured (Figure 20.9). *Intelligent compaction* (IC) refers to CCC with the added feature that the roller is able to change its settings nearly instantaneously when it comes to a soft spot and to optimize the compaction process (Figure 20.10) while keeping track of the global position through GPS (Figure 20.11). In CCC and IC, the soil parameter most often measured is a soil modulus E.

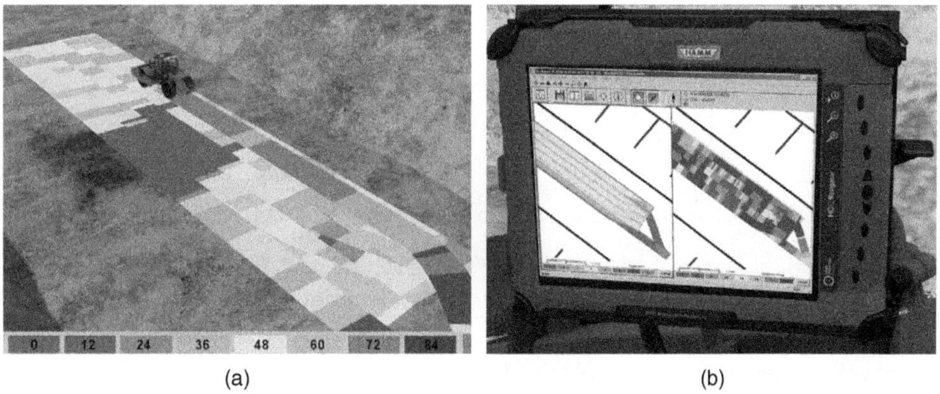

Figure 20.9 Continuous coverage in CCC and IC: (a) Continuous mapping of soil stiffness. (b) Screen display. (Courtesy of HAMM AG.)

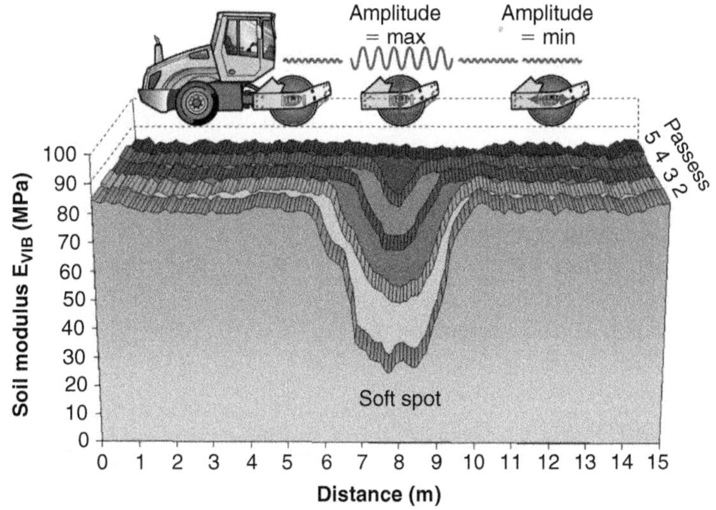

Figure 20.10 Intelligent compaction roller adjustment settings. (Courtesy of BOMAG.)

Figure 20.11 Intelligent rollers and readout equipment: *(a)* Roller. *(b)* Readout and control. (a: Courtesy of BOMAG; b: Courtesy of Ammann.)

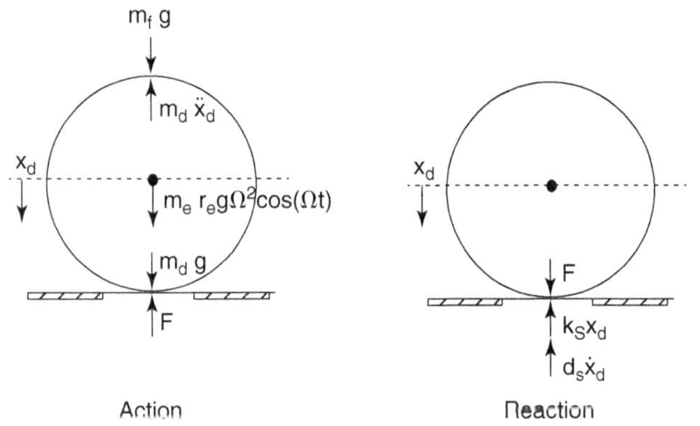

Figure 20.12 Forces acting on the drum of a vibrating roller (Anderegg 1997).

20.5.1 Soil Modulus from Vibratory Rollers

In the case of vibratory rollers, the modulus E is obtained from measurements of the acceleration of the roller (Anderegg 1997). A single degree of freedom model is used to represent the roller-soil interaction (Figure 20.12). Vertical equilibrium of the drum gives:

$$F = -m_d\ddot{x}_d + m_e r_e \omega^2 \cos(\omega t) + (m_f + m_d)g \qquad (20.2)$$

where F is the vertical force at the bottom of the drum, m_d is the mass of the drum, $\ddot{x}_d$ is the linear vertical acceleration of the drum, m_e is the eccentric mass creating the vibration, r_e is the radial distance at which m_e is attached, ω is the circular frequency of the rotating shaft, t is the elapsed time, m_f is the mass of the frame, and g is the acceleration due to gravity. All quantities on the right side of Eq. 20.2 come from the roller specifications and are known except for $\ddot{x}_d$, which is measured with an accelerometer on the drum axis. The vibration frequency of most rollers ranges from 30 to 70 Hz, their weight from 50 to 150 kN, their diameters from 1 to 2 meters, and their width from 2 to 3 meters.

The soil resistance F can also be obtained as follows if the soil is represented by a spring-and-dashpot model:

$$F = k_S x_d + c_S \dot{x}_d \qquad (20.3)$$

where k_s is the soil spring stiffness, x_d is the vertical displacement of the drum, c_s is the soil damping coefficient, and $\dot{x}_d$ is the vertical velocity of the soil boundary. In Eq. 20.3, k_s is the parameter to be solved for, x_d and $\dot{x}_d$ are obtained by integration of the acceleration signal, and c_s is typically assumed to be about 20% of critical damping. Numbers in the range of 50 to 100 MN/m for k_s and 150 to 250 kN s/m for c_s have been measured (Van Susante and Mooney 2008). An example of contact force vs. time and contact force vs. displacement is shown in Figure 20.13. Equations 20.2 and 20.3 are combined and the soil stiffness k_s can be obtained from the combined equation:

$$k_S = \frac{m_e r_e \omega^2 \cos(\omega t) + (m_f + m_d)g - m_d\ddot{x}_d - c_S\dot{x}_d}{x_d} \qquad (20.4)$$

As explained in section 14.6, the soil stiffness k_s is not an independent soil parameter because it depends on the size of the loaded area. The soil modulus E, in contrast, is an independent soil parameter; therefore, it is desirable to know how to obtain E from k_s. This problem was solved by Hertz in 1895 and further developed by Lundberg (1939):

$$k_s = \frac{F}{x_d} = \frac{\pi L E}{2(1-\nu^2)\left(2.14 + \frac{1}{2}Ln\left[\frac{\pi L^3 E}{16\left(1-\nu^2\right)(m_f + m_d)Rg}\right]\right)} \qquad (20.5)$$

where k_s is the soil stiffness, F is the force applied, x_d is the settlement of the drum, L is the drum width, υ is Poisson's ratio, Ln is the natural logarithm, m_f and m_d are the masses contributed by the frame and the drum of the

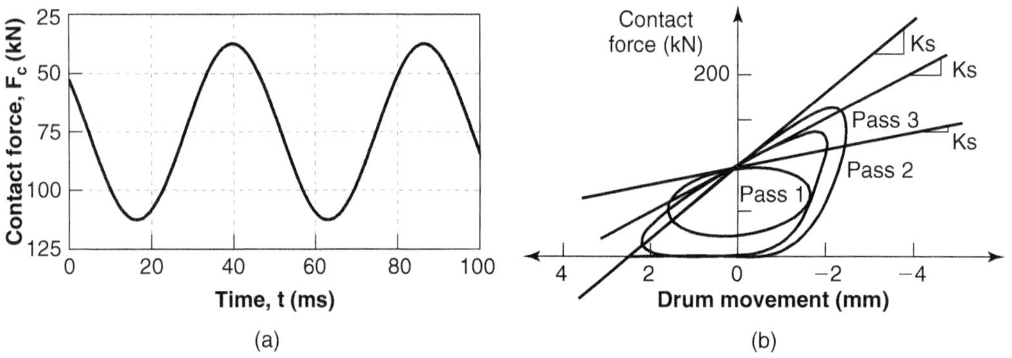

Figure 20.13 Contact force under the vibrating roller. *(a)* Force-time (Van Susante and Mooney 2008). *(b)* Force-displacement (Floss and Kloubert 2000)

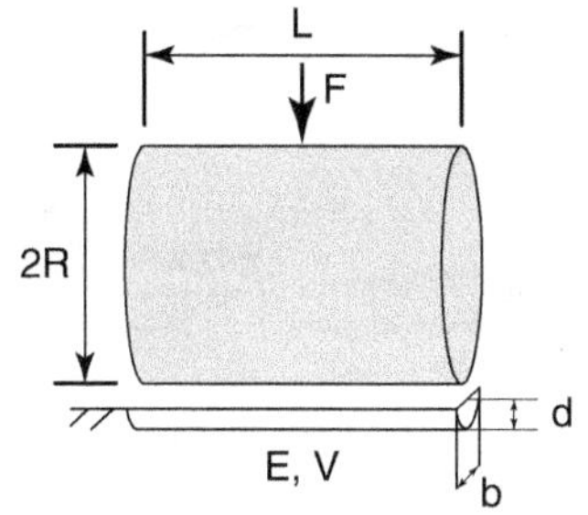

Figure 20.14 Drum on elastic soil problem. (After Lundberg 1939.)

roller respectively, R is the radius of the drum, and g is the acceleration due to gravity. Lundberg also gave the width b of the contact area between the drum and the soil:

$$b = \sqrt{\frac{16}{\pi}\frac{R(1-\upsilon^2)}{E}\frac{F}{L}} \tag{20.6}$$

As can be seen and as could be anticipated, the width b is inversely proportional to the soil modulus E. The roller soil contact area has a length L and a width b. Because the ratio L/b is very large, the loading is similar to a strip footing. Under static conditions, the depth of influence under a strip footing is $4b$. Thus, under the first pass, the width b is large because the soil is not very stiff and, as a result, the depth of influence is larger. Under subsequent passes, the soil stiffens and b decreases, and so does the depth of influence (Figure 20.14). The width b varies commonly between 200 mm on soft soils to 20 mm on very stiff soils.

20.5.2 Roller Measurements as Compaction Indices

The *machine drive power* (MDP) (White et al. 2005) is a roller index that can be used to evaluate the degree of compaction generated by any roller. The principle is that if the soil is soft, it will take more power for the roller to roll forward and compact the soil; if the soil is stiff, it will take less power for the roller to roll forward. The difference, as illustrated in Figure 20.15, is that the roller on soft soil has to overcome a lot more soil deformation energy than the roller on stiff soil. It is similar to the difference you feel when you run on loose sand compared to running on pavement; it is a lot more difficult (and takes more energy) to run on loose sand than on pavement.

The MDP is calculated as follows:

$$\text{MDP} = P_g - Wv\left[\sin\alpha + \frac{a}{g}\right] - (mv + b) \tag{20.7}$$

where MDP is the machine drive power (kJ/s), P_g is the gross power needed to move the machine (kJ/s), W is the roller weight (kN), v is the roller velocity (m/s), α is the slope angle (roller pitch from a sensor), a is the machine acceleration (m/s^2), g is the acceleration due to gravity (m/s^2), and m and b are the machine internal loss coefficient specific to a particular machine (kJ/m and kJ/s respectively). The second and third terms in Eq. 20.7 represent the machine power associated with a sloping grade and the internal machine loss respectively. The MDP represents only the machine power associated with the soil properties (White and Thompson 2008) and decreases as the soil becomes more compact.

For vibrating rollers, the compaction meter value (CMV) can be used. Some of the early work on continuous compaction control demonstrated that various indices incorporating drum acceleration amplitude and the amplitude of its harmonics could be linked to the stiffness of the underlying soil. Based on this early research, the CMV was proposed (Thurner and Sandström 1980). The CMV is a dimensionless compaction parameter that depends on roller dimensions (drum diameter and weight), roller operation parameters (frequency, amplitude, speed), soil mechanical properties (strength and stiffness), and soil stratigraphy. It is determined using the roller acceleration signal and calculated as:

$$CMV = C\frac{a_{2\Omega}}{a_\Omega} \tag{20.8}$$

where C is a constant (300), $a_{2\Omega}$ is the acceleration of the first harmonic component of the vibration, a_Ω is the acceleration of the fundamental component of the vibration, and Ω is the vibrating frequency of the roller.

If the soil is soft, the roller stays in contact with the soil and the roller and the soil move together; therefore the signal is sinusoidal and there is no other frequency content in the signal except for Ω, and CMV is thus zero. As the soil

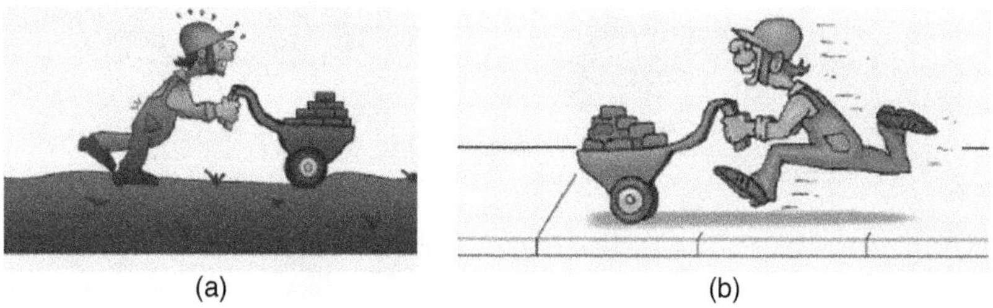

Figure 20.15 Principle of machine drive power: *(a)* Soft soil = hard to push. *(b)* Hard flat soil = easy to push.

becomes stiffer, the roller starts to jump and knock; this increases the frequency content of the signal, which becomes more complicated than just a sinusoidal signal, and the value of $a_{2\Omega}$ increases. A Fourier transform analysis of the time domain signal gives the frequency content and therefore the value of $a_{2\Omega}$. The CMV at a given point indicates an average value over an area with a width equal to the width of the drum and a length equal to the distance the roller travels in a set time period (0.5 seconds, for example).

The soil modulus E, or the MDP, or the CMV can be used to evaluate the degree of compaction achieved by the roller and a degree-of-compaction map of the area covered by the roller can be generated and located according to the GPS (Figure 20.9).

20.6 IMPACT ROLLER COMPACTION

Traditionally, compaction rollers have been cylindrical and have used their static weight, kneading action, or vibratory force to achieve the specific soil stiffness and soil strength. However, traditional rollers may have an energy capacity that is too low compared to the need. This might be the case for breaking the interparticle bonds of collapsible sands, for example. Impact compaction rollers were developed to alleviate this type of problem. They have noncircular drums (Figure 20.16) that rotate and fall to impact the ground surface. Such rollers tend to provide deeper compaction because the impact generates a wave that propagates at depth. Figure 20.17 demonstrates this point. It shows a freeze-frame picture of a numerical simulation movie describing the stress field in the soil as the roller passes over that spot (Kim 2010). The simulation compares the case of a cylindrical roller with one of a triangular impact roller. These types of simulations were used to generate the depth chart of Figure 20.18, which indicates that the depth of influence decreases as the soil becomes stiffer and as the roller becomes closer to a cylindrical roller. Here the depth of influence is defined as the depth at which the stress becomes equal to one-tenth of the stress under the roller at the ground surface. In that sense, impact rollers are more efficient; however, the biggest drawback is that they do not provide evenly compacted surfaces (Figure 20.16).

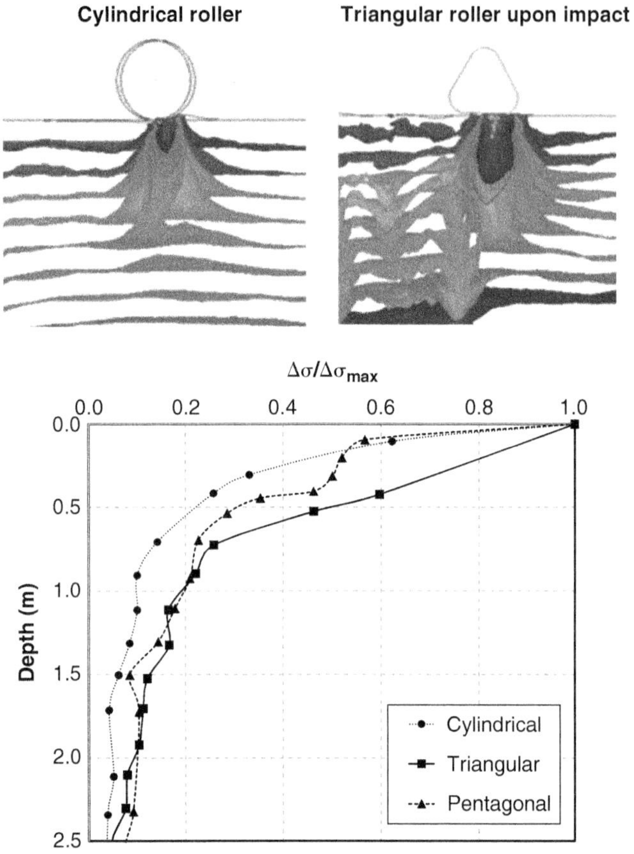

Figure 20.17 Stress field under rollers (Kim 2010).

Figure 20.16 Impact rollers. (a: Courtesy of LAND-PAC; b: Courtesy of Brooms)

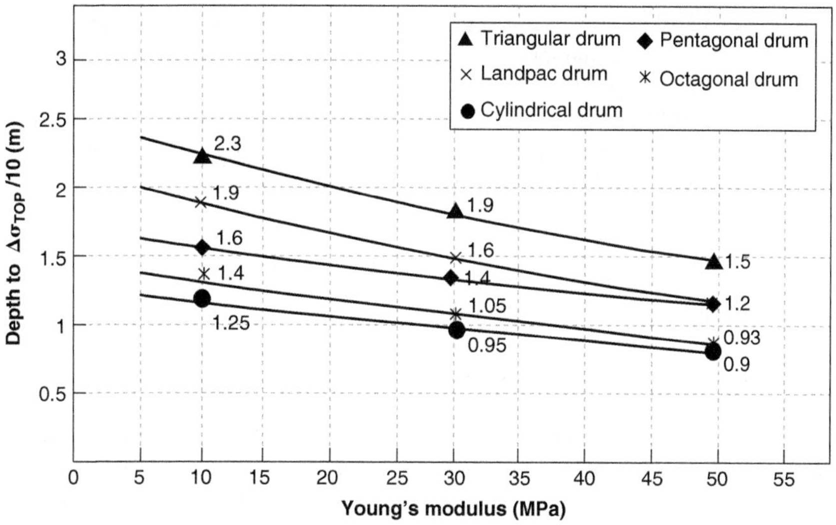

Figure 20.18 Depth of influence of rollers (Kim 2010).

The following comments summarize the situation with cylindrical and impact rollers:

1. The width of the contact area between the drum and the soil controls the depth of compaction. The softer the soil is, the deeper the roller sinks in the soil, the wider the contact area is, and the deeper the compaction is. Therefore, the depth of compaction depends on the stiffness of the soil. Hence, the depth of compaction decreases with the number of passes.
2. The surface pressure controls the degree of compaction. This pressure is higher for impact rollers than for cylindrical rollers due to the dynamic effect. However, the distribution of the pressure is much more uneven for impact rollers than for cylindrical rollers.
3. The depth of compaction is larger for impact rollers because they impart higher stresses that increase the penetration of the roller drum into the soil, thereby increasing both the width and depth of influence. The increased depth of influence is also due to wave propagation during the impact. These waves can propagate much deeper than the typical depth of influence for static loading.
4. If time and equipment allow it, it makes sense to compact first with an impact roller and use several passes to minimize the extent of the areas between impacts. Then, finish by using a cylindrical roller to provide a more evenly compacted surface and optimize the compaction of the shallow layers.
5. The process described in item 4 combines the benefits of both types of rollers: compaction of the deep layers (0.5 to 1.5 m) with the impact roller followed by evening out of the compaction of the shallow layers (0 to 0.5 m) with the cylindrical roller without disturbing the deep layers.

20.7 DYNAMIC OR DROP-WEIGHT COMPACTION

Dynamic compaction is often credited to Louis Menard (1975). It consists of lifting a heavy weight of mass M and dropping it from a preset height H so as to pound the soil and compact it in the process (Figure 20.19). The pounding is repeated at the same spot for a number of drops (say, 6 times) and thus creates a crater; then the crane moves to another

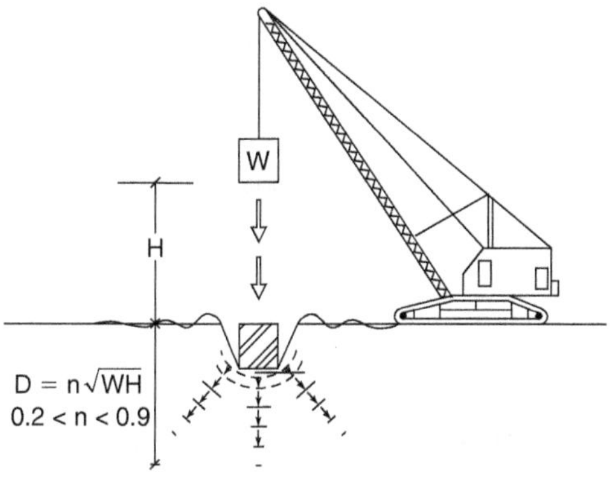

Figure 20.19 Dynamic compaction. (b: Courtesy of Menard Bachy Pty Ltd.)

location and repeats the process in a grid pattern. The spacing between impact points is about 2 times the diameter of the tamper. The crater should not be any deeper than 1.5 to 2 times the height of the tamper, to avoid collapse of the walls of the crater and associated difficulties in pulling the tamper out of the crater. The craters are typically backfilled with coarse-grained soil. After completing the first grid, the crane does a second pass by dropping the weight on the intermediate spots to complete the surface treatment. The drop weights commonly weigh 50 to 300 kN and drop from heights of up to 30 m, reaching velocities of 10 to 20 m/s at impact. During the final pass, called *ironing*, a flatter weight is dropped to smooth out the bumps.

Upon each drop, the energy generated by the impact propagates to the deeper layers by compression and shear wave propagation. Thus, the effectiveness of this compaction process depends on the dynamic response characteristics of the soil being compacted. Trials are usually run ahead of time to evaluate the potential results, but dynamic compaction works best for unsaturated coarse-grained soils and is not applicable to saturated fine-grained soils. The maximum depth D that can be compacted by dynamic compaction is influenced by many factors, including the soil properties, the groundwater level, the number of drops at each location, and the amount of time elapsed between the grids. The following equation is recommended by Lukas (1995):

$$D = n\sqrt{MH} \tag{20.9}$$

where n is a site factor less than 1 (Table 20.3), M is the mass of the tamper in tonnes (1000 kg), and H is the average drop height in meters.

Table 20.3 Recommended Values of n for Different Soils (Lukas 1995)

Soil Type	Degree of Saturation	Recommended n Value
Pervious soil deposits, granular soils	High	0.5
	Low	0.5–0.6
Semipervious soil deposits, primarily silts with plasticity index <8	High	0.35–0.4
	Low	0.4–0.5
Impervious deposits, primarily clayey soils with plasticity index of >8	High	Not recommended
	Low	0.35–0.40 Soil should be at water content less than the plastic limit

Table 20.4 Values of the Equipment Factor C (Chu et al. 2009)

Drop Method	Free Drop	Rig Drop	Mechanical Winch	Hydraulic Winch	Double Hydraulic Winch
Equipment factor C	1.0	0.89	0.75	0.64	0.5

For an applied energy of 1 to 3 MJ/m^2 and for a temper drop using a single cable with a free spool drum, Eq. 20.9 was modified by Varaksin as follows (Chu et al. 2009):

$$D = C\delta\sqrt{MH} \tag{20.10}$$

where C is an equipment factor given in Table 20.4, and δ is a soil factor equal to 0.9 for metastable soils, young fills, or very recent hydraulic fills, and equal to 0.4 to 0.6 for sands. Compaction depths of 10 m can be achieved with the heavier tampers (e.g., 20 tonnes dropping 20 m). The improvement ratio f, defined as the ratio of the strength after dynamic compaction over the strength before dynamic compaction, varies with depth and is typically measured by in situ testing (PMT, CPT, SPT). Varaksin proposes the following variation of f with depth below the tamper:

$$f = f_1 + (f_2 - f_1)\left(\frac{z}{D}\right)^2 \tag{20.11}$$

where f_1 and f_2 are the improvement ratios at the ground surface and at the depth D, respectively, and z is the depth at which f is evaluated.

The energy E input in the soil for each drop by dynamic compaction can be presented per unit of surface area compacted (E_2 in kJ/m^2) or per unit of soil volume compacted (E_3 in kJ/m^3). The energy per unit surface area compacted E_2 is:

$$E_2 = \frac{W \times H \times N \times P}{s^2} \tag{20.12}$$

where W is the weight of the tamper in kN, H is the height of drop in meters, N is the number of drops, P is the number of passes, and s is the grid spacing in meters for the pounding pattern.

The energy per unit volume of soil compacted E_3 is:

$$E_3 = \frac{E_2}{D} \tag{20.13}$$

where D is the depth of soil compacted. Lukas (1995) gives a list of typical energies used for different soil types (Table 20.5).

Table 20.5 Applied Energy Guidelines for E_3 (Lukas 1995)

Type of Deposit	Unit Applied Energy (kJ/m^3)	Percent Standard Proctor Energy
Pervious coarse-grained soil, Zone 1	200–250	33–41
Semipervious fine-grained soils, Zone 2; and clay fills above the water table, Zone 3	250–350	41–60
Landfills	600–1100	100–180

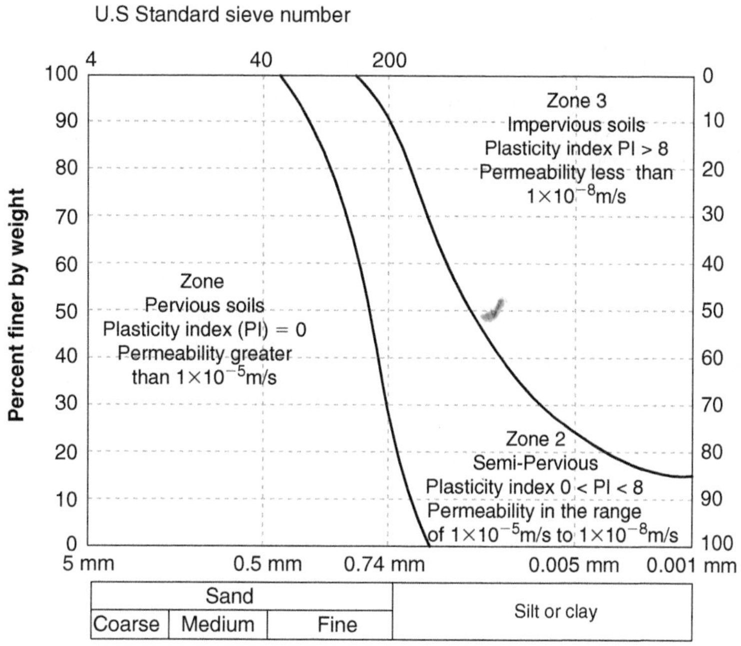

Note: Standard Proctor energy equals 600 kJ/m^3.

The degree of efficiency of dynamic compaction is measured by comparing the results of soil tests performed before and after the compaction process. The preferred test is the pressuremeter test (Figure 20.20), but the cone penetrometer test and the standard penetration test are also used. The *depth of compaction* is defined here as the depth to which the soil strength has increased compared to the initial state. This increase is not constant with depth, as seen in Figure 20.20. A typical use of dynamic compaction is to dynamically compact the soil deposit so that shallow foundations can be used instead of more expensive pile foundations. The decision is based on comparing the cost of a shallow foundation plus dynamic compaction to the cost of a deep foundation.

Dynamic compaction induces soil vibrations. These vibrations are typically measured in terms of the peak velocity of the soil particles or PPV. The PPV depends on a number of factors, primarily the energy of the impact and the distance from the impact. Mayne (1985) assembled a database

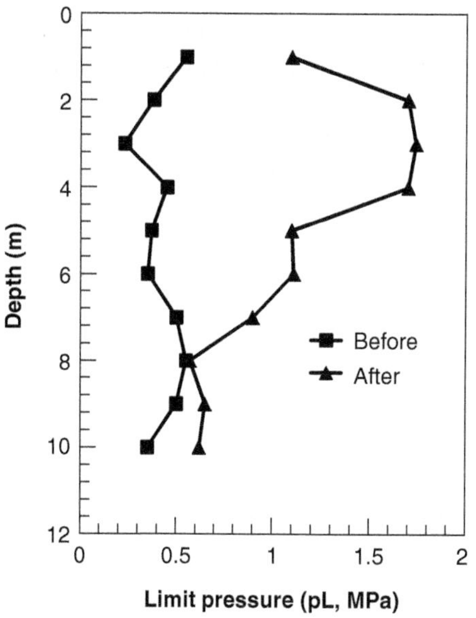

Figure 20.20 Improvement of soil strength due to dynamic compaction.

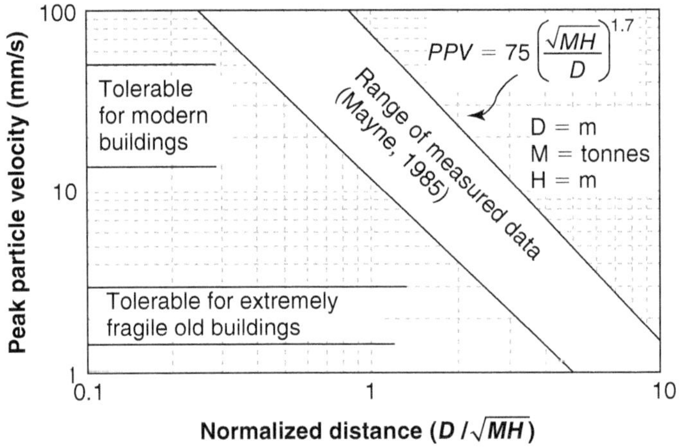

Figure 20.21 Peak particle velocity due to dynamic compaction. (After Mayne 1985.)

giving the range of values shown in Figure 20.21. The following equation gives an upper bound of the PPV generated by a dynamic compaction impact according to the data in Figure 20.21:

$$PPV = 75\left(\frac{\sqrt{MH}}{d}\right)^{1.7} \tag{20.14}$$

where PPV is the peak particle velocity in mm/s, M is the mass of the tamper in tonnes (1000 kg), H is the drop height in m, and d is the distance from the impact location in m. This value of the PPV can be compared with what is tolerable. Typical values of PPV for damage threshold vary from 1 to 3 mm/s for very old and fragile buildings to 20 to 50 mm/s for modern buildings (Figure 20.21).

PROBLEMS

20.1 Referring to Figure 20.2, give the maximum dry density and the optimum water content. What is the degree of saturation of the soil at that point if the specific gravity of solids is 2.7?

20.2 Use the data points from Figure 20.5 to draw a correlation between dry density and modulus. Find the R square value for that correlation. Discuss whether there should be or should not be a correlation between dry density and modulus.

20.3 Correlate the depth of the imprint that you can make with your thumb as a function of the soil modulus being compacted.

20.4 A vibratory intelligent roller weighs 140 kN; it has a drum diameter of 1.4 m and a drum length of 2.1 m. The eccentric weight generates a moment ($m_e r_e$ in Eq. 20.2) equal to 1.5 kg.m at an angular frequency of 200 rd/s. The drum weighs 30 kN and the added weight from the frame above the drum is 20 kN. The measured peak acceleration of the drum is + or −3g. Assume that the inertia force generated by the vibration of the frame is negligible compared to the one generated by the drum. Draw the acceleration signal, the velocity signal, and the displacement signal at the drum-soil contact point.

20.5 The vibratory roller from problem 4 rests on a soil that has a stiffness k_s to be determined. The damping coefficient of the soil is 200 kN s/m. Calculate the stiffness of the soil k_s, the modulus of the soil E, and the width b of the contact area.

20.6 A landfill must be compacted by dynamic compaction to improve its bearing capacity. The required depth of compaction is 10 m. Determine the weight of the tamper to be used and the drop height required to achieve the 10 m depth of compaction.

20.7 Regarding the landfill in problem 7, the closest building is located at 100 m from the edge of the compaction zone. Calculate the peak particle velocity that can be expected. Would this be normally tolerable for a recently constructed building?

Problems and Solutions

Problem 20.1

Referring to Figure 20.2, give the maximum dry density and the optimum water content. What is the degree of saturation of the soil at that point if the specific gravity of solids is 2.7?

Solution 20.1

The maximum dry unit weight and the optimum water content are obtained from the compaction curve. The values from Figure 20.2 are:

- Maximum dry unit weight, $\gamma_{dmax} = 19.3\ kN/m^3$
- Optimum water content, $w_{opt} = 10\%$
- Degree of saturation, S

Using the equation that links these quantities:

$$\gamma_d = \frac{G_s \gamma_w}{1 + \dfrac{G_s w}{S}}$$

$$S = \frac{G_s w}{\dfrac{G_s \gamma_w}{\gamma_d} - 1} = \frac{2.7 \times 0.10}{\dfrac{2.7 \times 9.81}{19.3} - 1} = 0.725 = 72.5\%$$

Problem 20.2

Use the data points from Figure 20.5 to draw a correlation between dry density and modulus. Find the *R* square value for that correlation. Discuss whether there should be or should not be a correlation between dry density and modulus.

Solution 20.2

The values of the modulus and dry density are as plotted in Figure 20.1s, and the *R* square value shows that there is not a good correlation between dry unit weight and soil modulus. The modulus depends on many other factors besides the amount of solids per unit volume. Factors such as structure, cementation, and stress history also affect the modulus. It is not surprising that there is no good correlation between dry density and modulus.

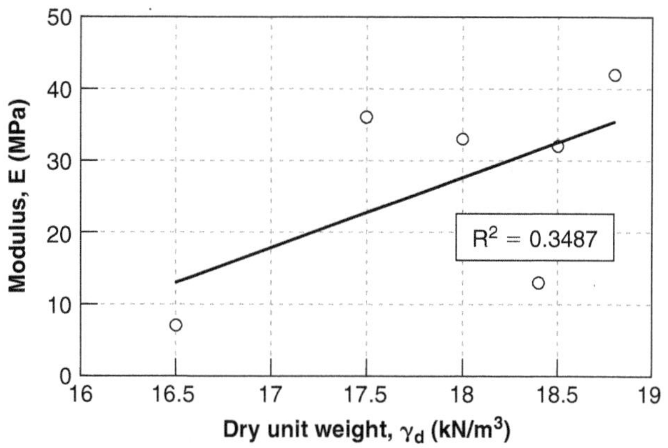

Figure 20.1s The correlation between dry unit weight and soil modulus.

Problem 20.3

Correlate the depth of the imprint that you can make with your thumb as a function of the soil modulus being compacted.

Solution 20.3 (Figure 20.2s)

υ = Poisson's ratio of the soil (assumed to equal 0.35)
S = settlement (m)
E = soil modulus

$$s = \frac{\pi}{4}(1 - \upsilon^2).p.\frac{B}{E}$$

Assuming that you can generate 0.1 kN with your thumb and that the area of contact between your thumb and the surface is 30 by 20 mm, the pressure under your thumb is:

$$p = \frac{F}{A} = \frac{100 \times 10^{-3}}{20 \times 30 \times 10^{-6}} = 166.67(\text{kN/m}^2)$$

The settlement can be calculated as:

$$s(m) = \frac{\pi}{4}(1 - \upsilon^2).p.\frac{B}{E} = \frac{\pi}{4}(1 - 0.35^2) \times 166.67 \times \frac{0.02}{E} = \frac{2.297}{E}$$

or, with E in kPa and s in mm:

$$\text{E (kPa)} = 2300/\text{s (mm)}$$

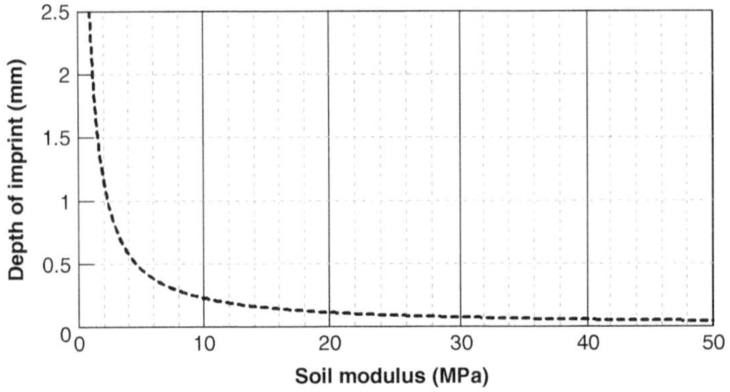

Figure 20.2s Depth of finger imprint vs. soil modulus.

Problem 20.4

A vibratory intelligent roller weighs 140 kN; it has a drum diameter of 1.4 m and a drum length of 2.1 m. The eccentric weight generates a moment ($m_e r_e$ in Eq. 20.2) equal to 1.5 kg.m at an angular frequency of 200 rd/s. The drum weighs 30 kN and the added weight from the frame above the drum is 20 kN. The measured peak acceleration of the drum is + or −3g. Assume that the inertia force generated by the vibration of the frame is negligible compared to the one generated by the drum. Draw the acceleration signal, the velocity signal, and the displacement signal at the drum-soil contact point.

Solution 20.4

$$\ddot{x} = a_{\max} \sin(\omega t) = 30 \sin(200t)$$

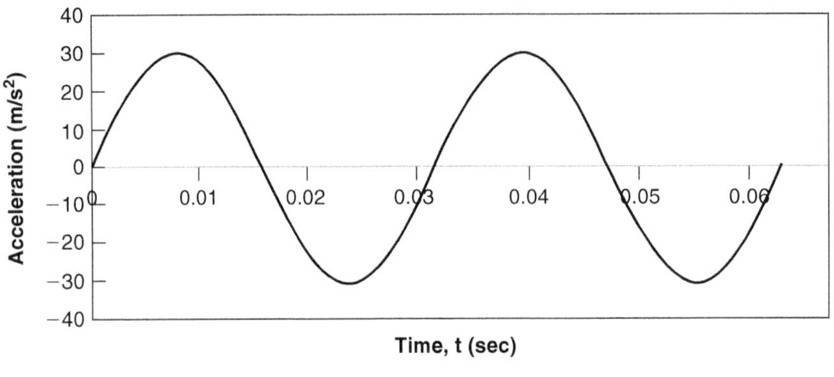

Figure 20.3s Acceleration signal.

$$\dot{x} = -\frac{a_{\max}}{\omega} \cos(\omega t) = -0.15 \cos(200t)$$

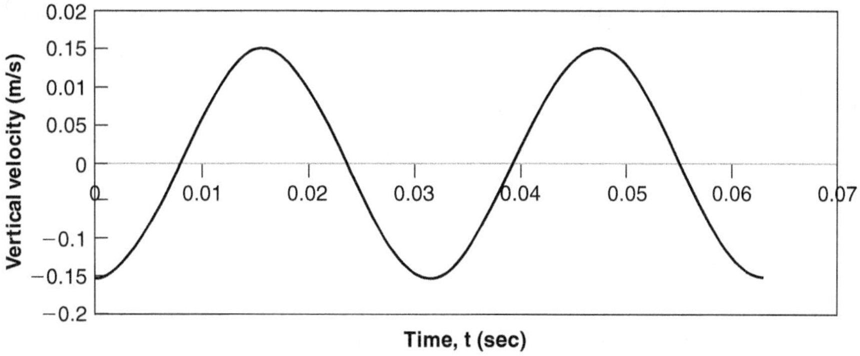

Figure 20.4s Velocity signal.

$$x = -\frac{a_{max}}{\omega^2} \sin(\omega t) = -0.75 \times 10^{-3} \sin(200t)$$

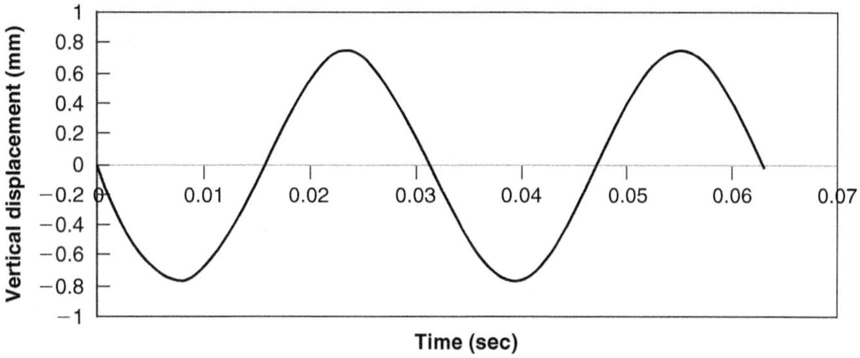

Figure 20.5s Vertical displacement signal.

Problem 20.5

The vibratory roller from problem 4 rests on a soil that has a stiffness k_s to be determined. The damping coefficient of the soil is 200 kN s/m. Calculate the stiffness of the soil k_s, the modulus of the soil E, and the width b of the contact area.

Solution 20.5 (Figures 20.6s, 20.7s)

$W_{roller} = 140$ kN
Drum radius, R = 0.7 m
Drum length, L = 2.1 m
$m_e r_e = 1.5$ kg.m
$\omega = 200$ rad/sec
$W_{drum} = 30$ kN
$c_s = 200$ kN s/m

Poisson's ratio, $\nu = 0.35$ (assumed)
The equations are:

$$\ddot{x}_d = \pm 3g \sin(\omega t)$$

$$\dot{x}_d = \mp \frac{3g}{\omega} \cos(\omega t)$$

$$x_d = \pm \frac{3g}{\omega^2} \sin(\omega t)$$

$$F = -m_d \ddot{x}_d + m_e r_e \omega^2 \cos(\omega t) + (m_f + m_d) g$$

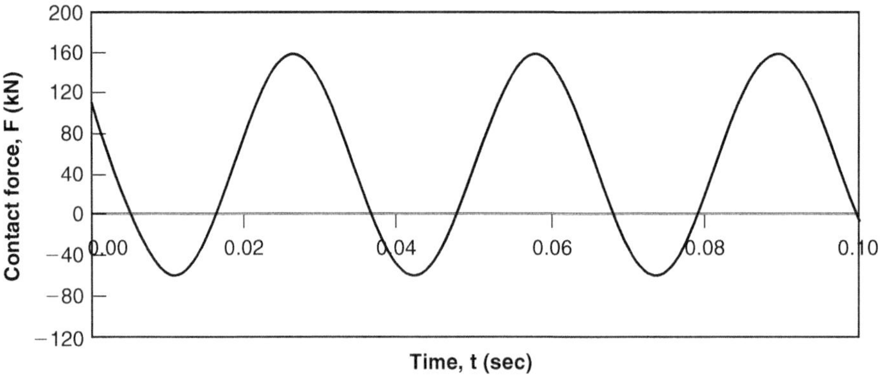

Figure 20.6s Contact force versus time.

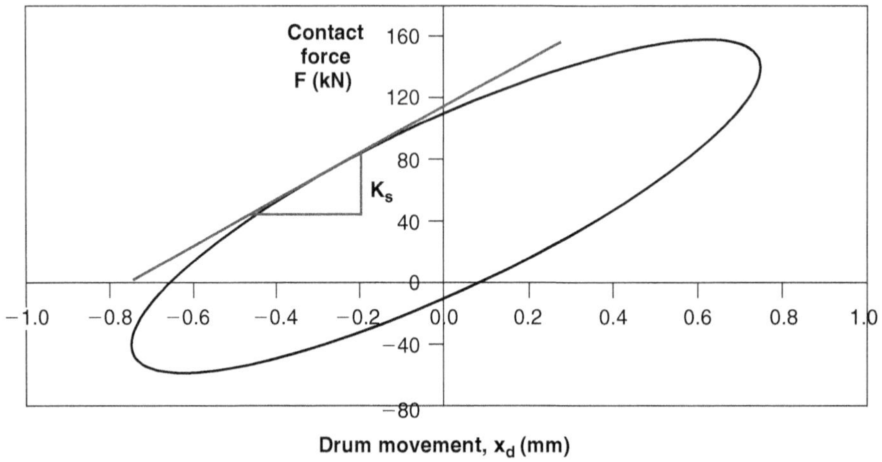

Figure 20.7s Contact force versus movement and soil stiffness.

The stiffness can be taken as the slope as shown in Figure 20.7s, or as the slope of the loop:

$$k_s = \frac{160\text{ kN}}{1.06\text{ mm}} = 150.9\text{ kN/mm}$$
$$k_s = 150900\text{ kN/m}$$

Then, the soil modulus is:

$$k_s = \frac{F_s}{x_d} = \frac{\pi L E}{2(1-\upsilon^2)\left(2.14 + \frac{1}{2}Ln\left[\frac{\pi L^3 E}{16\left(1-\upsilon^2\right)(m_f + m_d)Rg}\right]\right)}$$

$$k_s = \frac{\pi \times 2.1 \times E}{2(1-0.35^2)\left(2.14 + \frac{1}{2}Ln\left[\frac{\pi \times 2.1^3 \times E}{16\left(1-0.35^2\right) \times \left(\frac{20}{g} + \frac{30}{g}\right) 0.7 \times g}\right]\right)}$$

$$k_s = \frac{6.6 \times E}{1.755 \times \left(2.14 + \frac{1}{2}Ln\left[\frac{29.1 \times E}{491.4}\right]\right)}$$

$$40125.7 = \frac{E}{\left(2.14 + \frac{1}{2}Ln\left[\frac{29.1 \times E}{491.4}\right]\right)}$$

and solving for E gives $E = 281$ MPa.

To compute the contact width b between the roller and the soil at the time of the highest force, $F = 160$ kN, we use the following equation:

$$b = \sqrt{\frac{16}{\pi}\frac{R(1-\upsilon)}{E}\frac{F}{L}}$$

$$b = \sqrt{\frac{16}{\pi} \times \frac{0.7(1-0.35)}{280900} \times \frac{160}{2.1}} = 0.025 \text{ m}$$

$$b = 25 \text{ mm}$$

Problem 20.6

A landfill must be compacted by dynamic compaction to improve its bearing capacity. The required depth of compaction is 10 m. Determine the weight of the tamper to be used and the drop height required to achieve the 10 m depth of compaction.

Solution 20.6

The following equation is used to evaluate the depth of compaction, D:

$$D = \alpha\sqrt{MH}$$

From the problem statement, $D = 10$ m. Alpha is typically between 0.3 and 0.8; let's assume 0.5:

$$D/\alpha = 10/0.5 = 20$$

$$20 = \sqrt{MH}$$

$$M = 20 \text{ tonnes}$$

$$H = 20 \text{ m}$$

Problem 20.7

Regarding the landfill in problem 6, the closest building is located 100 m from the edge of the compaction zone. Calculate the peak particle velocity that can be expected. Would this be normally tolerable for a recently constructed building?

Solution 20.7

The peak velocity of the soil particles (PPV) (in mm/s) caused by the dynamic vibration is calculated as:

$$PPV = 75\left(\frac{\sqrt{MH}}{d}\right)^{1.7}$$

Where M is the mass of the tamper in tonnes, H is the drop height in meters, and d is the distance from the impact zone in meters. From problem 20.6, $M = 20$ tonnes and $H = 20$ m. Then, with $d = 100$ m, the PPV is:

$$PPV = 75\left(\frac{\sqrt{MH}}{d}\right)^{1.7} = 75 \times \left(\frac{20}{100}\right)^{1.7} = 4.86 \text{ mm/s}$$

The damage threshold for modern buildings is set to be $20 \sim 50$ mm/s; therefore, this PPV is tolerable for a recent building located 100 m away from the closest edge of the compaction zone.

CHAPTER 21

Retaining Walls

21.1 DIFFERENT TYPES (TOP-DOWN, BOTTOM-UP)

There are many different types of retaining walls, but they are generally classified into two main categories: bottom-up walls and top-down walls. *Bottom-up walls* are walls that are built before the soil is placed behind the wall. In this case the backfill is compacted in lifts from the bottom of the wall to the top of the wall, often with inclusions (e.g., metal strips, geosynthetics) being installed on the way up. *Top-down walls* are walls that are built in the ground; then the excavation in front of the wall takes place in stages, most often with inclusions (e.g., anchors, tiebacks, nails) being installed through the wall as excavation proceeds. Examples of bottom-up walls are gravity walls and mechanically stabilized earth (MSE) walls (Figure 21.1). Examples of top-down walls are cantilever walls, soil-nailed walls, and anchored walls (also known as tieback walls).

The design of retaining walls requires calculations regarding:

1. Earth pressure distribution behind the wall
2. Deflection of the wall
3. Drainage issues

The body of knowledge regarding the issue of earth pressure is much more developed than that on the issue of deflection.

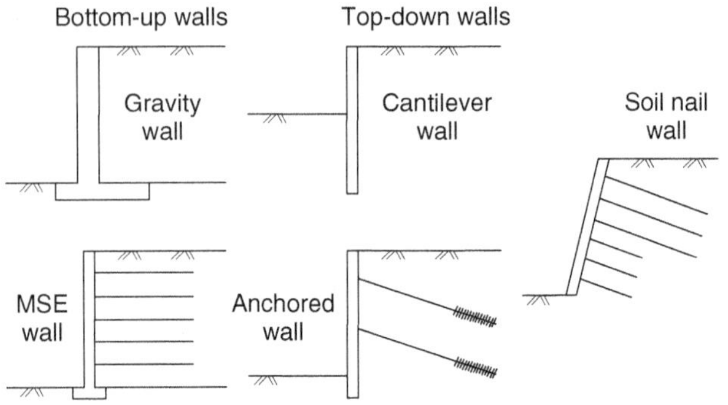

Figure 21.1 Types of retaining walls.

One of the reasons is that historically, earth pressure theories came first.

21.2 ACTIVE, AT REST, PASSIVE EARTH PRESSURE, AND ASSOCIATED DISPLACEMENT

Consider an imaginary wall in a lake. The water pressure u_w on both sides of the wall would be hydrostatic and equal to $\gamma_w z$ where u_w is the water pressure against the wall at depth z below the water surface, and γ_w is the unit weight of water. As a result, the pressure diagram is triangular and the resultant is located at two-thirds of the wall height from the top of the wall. Note that the water pressure is the same in all directions, including horizontal and vertical, because water has a negligible resistance to shear (the Mohr circle for water is a point). Now consider an imaginary wall in the ground (Figure 21.2). The at-rest earth pressure σ_{oh} exists on both sides of the wall. If you push the wall horizontally, the pressure will increase on the side that penetrates into the soil up to soil failure (passive pressure σ_{ph}) and decrease on the other side where the wall is moving away from the soil down to soil failure (active pressure σ_{ah}). Note that if you push the wall far enough and if the soil is strong enough because of true or apparent cohesion, the pressure may become zero on the side where the wall is moving away from the soil and a gap opens up.

On the passive side, the soil is pushed away and upward as a wedge of failing soil forms in front of the wall (Figure 21.3); as a result, the soil imposes an upward friction force on the

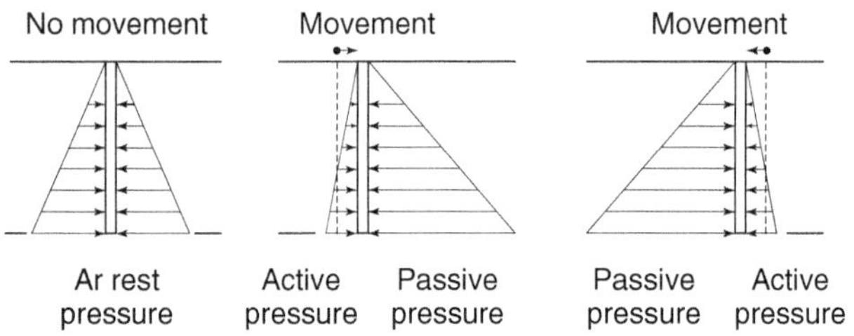

Figure 21.2 Imaginary wall and earth pressures.

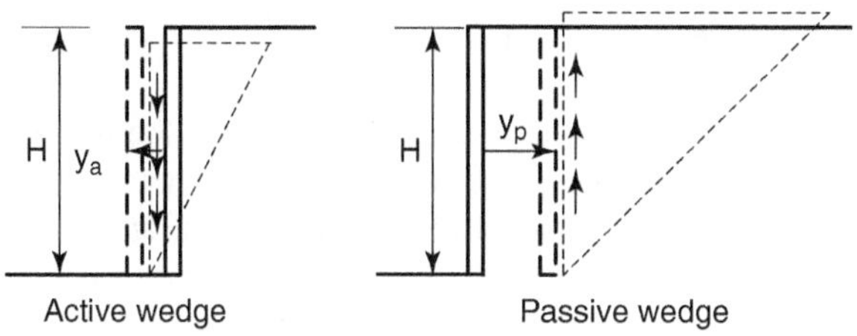

Figure 21.3 Earth pressures wedges.

wall. On the active side, the soil falls against the wall and downward as a wedge of failing soil forms behind the wall; as a result, the soil imposes a downward friction force on the wall. The passive wedge is much larger than the active wedge and requires more displacement to be mobilized. This is why the displacement required to mobilize the passive earth pressure is larger than the displacement required to mobilize the active earth pressure. The relationship between the soil pressure against the wall and the horizontal displacement of the wall is shown in Figure 21.4.

Now let's zoom in at the interface between the soil and the wall as shown in Figure 21.5. The soil particles contact the wall at several points where forces are transmitted between the soil and the wall. Between the particle contacts are the voids in the soil. These voids can be either completely filled with water (saturated soil) or filled with air and water (unsaturated soil). In the case of the saturated soil, the water will exert a pressure u_w against the wall. This water stress can be compression below the groundwater level (GWL) or tension within the capillary zone above the GWL. The water stress times the area of wall over which the water acts is the force transmitted by the water on the wall. The horizontal force on the wall is the sum of the forces at the particle contacts and the force contributed by the water stress u_w. Then we divide by the total area and, as in the case of vertical stress (see section 10.13), the total horizontal stress σ_h is

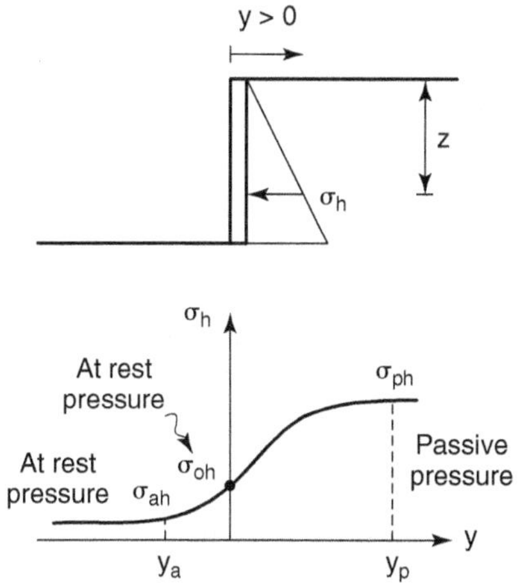

Figure 21.4 Earth pressure versus wall displacement.

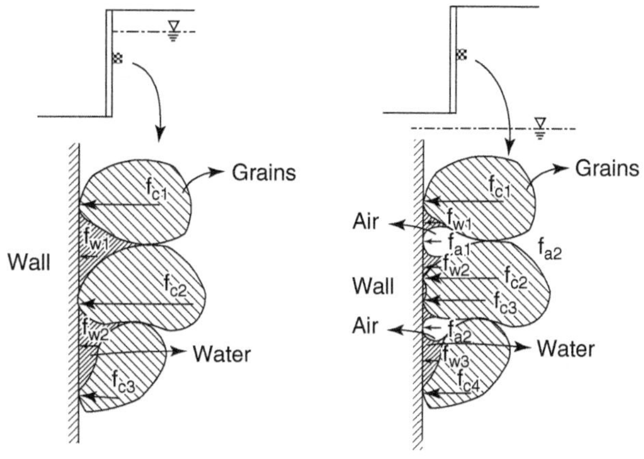

Figure 21.5 View of the soil-wall contact.

$\sigma'_h + u_w$ where σ'_h is the effective horizontal stress and u_w is the water stress (compression or tension).

If the soil is unsaturated, the horizontal force on the wall is the sum of the forces at the particle contacts, the forces transmitted through the water, and the forces transmitted through the air. If the air is occluded in the water phase and does not contact the wall, then σ_h is still equal to $\sigma'_h + u_w$, but the water phase is more compressible. Air tends to be occluded when the degree of saturation S is above 85%. If the air is not occluded ($S < 85\%$), there is a continuous air path to the ground surface and the air stress is atmospheric or zero gage pressure. In this case (see section 10.13), the horizontal stress σ_h is $\sigma'_h + \alpha u_w$ where σ'_h is the effective horizontal stress, α is the ratio of the water area in contact with the wall over the total area, and u_w is the water stress (which is in tension in this case). As pointed out in section 10.13, α can be estimated as the degree of saturation with a ±30% precision or by Khalili rule. The effect of the water tension in unsaturated soil will be to decrease the active horizontal pressure and increase the passive horizontal pressure compared to the case of the saturated soil with water in compression. Note that the active earth pressure and the passive earth pressure correspond to soil failure. Therefore, they should be thought of as strength rather than stress.

21.3 EARTH PRESSURE THEORIES

21.3.1 Coulomb Earth Pressure Theory

The earth pressure theories make the general assumption that the soil is at failure. In that sense, the earth pressures obtained by using these theories are similar to the concept of ultimate bearing capacity in foundation engineering; they represent strengths at failure rather than stresses at working loads. Coulomb, in 1776, was the first person to work on earth pressures. Charles Augustin de Coulomb was a French physicist who worked on this topic just before the French Revolution in the late 1700s, although he is better known for his work

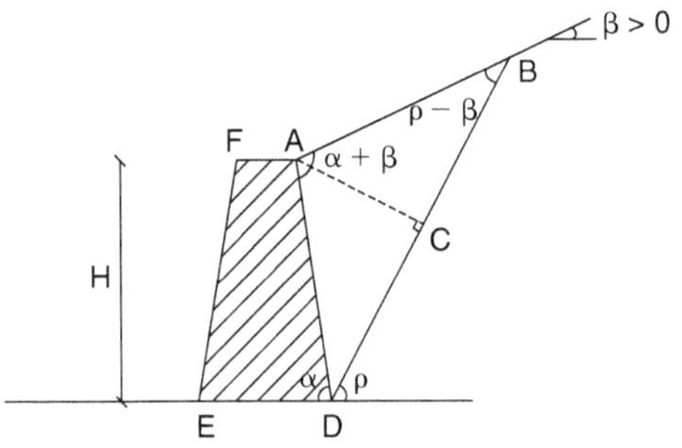

Figure 21.6 General geometry of the Coulomb soil wedge.

on electromagnetism. To develop his earth pressure theory, Coulomb made the following assumptions (Figure 21.6):

1. The problem is a plane strain problem
2. The soil has friction (φ') and cohesion (c')
3. The soil has no water
4. The failure wedge is a rigid body
5. The failure surface and the ground surface are planes
6. The friction coefficient between the wall and the soil wedge is $\tan\delta$

Let's first calculate the weight of the wedge W per unit length of wall. The area A of the triangle ABD is:

$$A = \frac{1}{2}BD \times AC = \frac{1}{2}AD\left(\frac{\sin(\alpha+\beta)}{\sin(\rho-\beta)}\right) \times AD\sin(180-\alpha-\rho) \quad (21.1)$$

Because

$$AD = \frac{H}{\sin\alpha} \quad (21.2)$$

then

$$A = \frac{H^2}{2\sin^2\alpha}\sin(\alpha+\rho)\left(\frac{\sin(\alpha+\beta)}{\sin(\rho-\beta)}\right) \quad (21.3)$$

and

$$W = \frac{\gamma H^2}{2\sin^2\alpha}\sin(\alpha+\rho)\left(\frac{\sin(\alpha+\beta)}{\sin(\rho-\beta)}\right) \quad (21.4)$$

In the case of the active earth pressure (Figure 21.7), the external forces acting on the wedge are the weight W, the active force P_a on the wall side AD, and the resultant force R on the soil side BD.

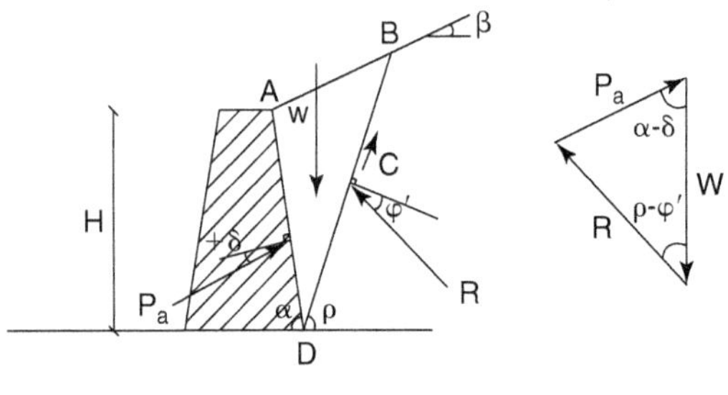

Figure 21.7 Free body of the active soil wedge.

The force P_a is inclined at an angle δ with the normal to the wall-soil interface. If the wall does not settle excessively, the wedge goes down with respect to the wall and the wall friction acts upward on the wedge (positive δ value for the active case). The resultant R is inclined at an angle φ' with the normal to the soil-soil failure plane at the back of the wedge. Because the wedge goes down with respect to the soil mass beyond the wedge, the friction force acts upward on the wedge. We will neglect the cohesion force at this time. Then the polygon of forces can be drawn (Figure 21.7) and the law of sines gives:

$$\frac{P_a}{\sin(\rho-\varphi')} = \frac{W}{\sin(180-\rho+\varphi'-\alpha+\delta)} \quad (21.5)$$

and

$$P_a = \frac{\gamma H^2}{2\sin^2\alpha}\sin(\alpha+\rho)\left(\frac{\sin(\alpha+\beta)}{\sin(\rho-\beta)}\right) \times \frac{\sin(\rho-\varphi')}{\sin(180-\rho+\varphi'-\alpha+\delta)} \quad (21.6)$$

Equation 21.6 shows that P_a is a function of a number of factors, including the angle ρ which is an unknown variable. The active earth pressure force will correspond to the value of ρ that leads to the lowest value of P_a, because that will be the first value reached as the wall is pulled away from the soil. Therefore, the ρ value corresponding to the active force is the one that minimizes P_a. For this we set:

$$\frac{\partial P_a}{\partial \rho} = 0 \quad (21.7)$$

and solve for ρ as was done in section 11.4.2. The final result for P_a is:

$$P_a = \frac{\gamma H^2}{2}\frac{\sin^2(\alpha+\varphi')}{\sin^2\alpha\sin(\alpha-\delta)\left[1+\sqrt{\dfrac{\sin(\varphi'+\delta)\sin(\varphi'-\beta)}{\sin(\alpha-\delta)\sin(\alpha+\beta)}}\right]^2} = \frac{1}{2}K_a\gamma H^2 \quad (21.8)$$

and the coefficient of active earth pressure K_a giving the magnitude of the vector P_a is:

$$K_a = \frac{\sin^2(\alpha+\varphi')}{\sin^2\alpha\sin(\alpha-\delta)\left[1+\sqrt{\dfrac{\sin(\varphi'+\delta)\sin(\varphi'-\beta)}{\sin(\alpha-\delta)\sin(\alpha+\beta)}}\right]^2} \quad (21.9)$$

Note that the direction of the force P_a is not horizontal, but rather acts at an angle 90—$\alpha + \delta$ with the horizontal. The horizontal component P_{ah} is:

$$P_{ah} = \frac{1}{2}K_a\gamma H^2\cos(90-\alpha+\delta) = \frac{1}{2}K_a\gamma H^2\sin(\alpha-\delta)$$
$$= \frac{1}{2}K_{ah}\gamma H^2 \tag{21.10}$$

Therefore, the coefficient of active earth pressure K_{ah} giving the horizontal component P_{ah} of the active push P_a is:

$$K_{ah} = \frac{\sin^2(\alpha+\varphi')}{\sin^2\alpha\left[1+\sqrt{\dfrac{\sin(\varphi'+\delta)\sin(\varphi'-\beta)}{\sin(\alpha-\delta)\sin(\alpha+\beta)}}\right]^2} \tag{21.11}$$

In the simpler case where the backfill is horizontal, the wall is vertical, and there is no soil-wall friction (conservative), then $\beta = \delta = 0$, $\alpha = 90°$, and K_a becomes:

$$K_a = \frac{1-\sin\varphi'}{1+\sin\varphi'} \tag{21.12}$$

In the case of the passive earth pressure (Figure 21.8), the external forces acting on the wedge are the weight W, the passive force P_p on the wall side AD, and the resultant force R on the soil side BD.

The force P_p is inclined at an angle δ with the normal to the wall-soil interface. As the wall pushes against the wedge, the wedge goes up with respect to the wall and the wall friction acts downward on the wedge (positive δ value for the passive case). The resultant R is inclined at an angle φ' with the normal to the soil-soil failure plane at the back of the wedge. Because the wedge goes up with respect to the soil mass beyond the wedge, the friction force acts downward on the wedge. We will neglect the cohesion force at this time. Then the polygon of forces can be drawn (Figure 21.8) and

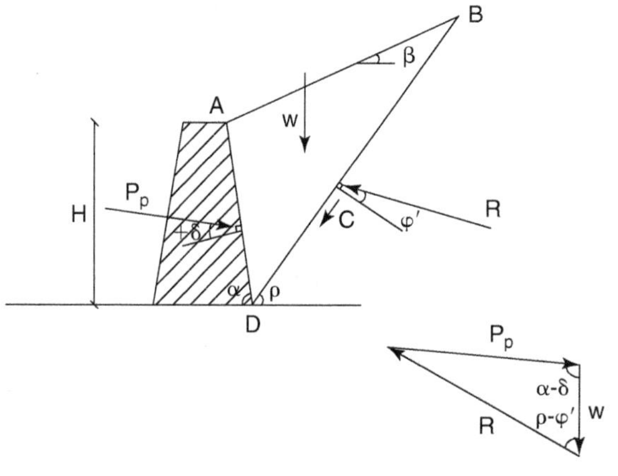

Figure 21.8 Free body of the passive soil wedge.

the derivation proceeds as for the active case. In the end, the equation for P_p is:

$$P_p = \frac{1}{2}K_p\gamma H^2 \tag{21.13}$$

The passive earth pressure coefficient giving the magnitude of the vector P_p is:

$$K_p = \frac{\sin^2(\alpha-\varphi')}{\sin^2\alpha\sin(\alpha+\delta)\left[1-\sqrt{\dfrac{\sin(\varphi'+\delta)\sin(\varphi'+\beta)}{\sin(\alpha+\delta)\sin(\alpha+\beta)}}\right]^2} \tag{21.14}$$

Note that the direction of the force P_p is not horizontal, but rather acts at an angle $\alpha + \delta - 90$ with the horizontal. The horizontal component P_{ph} is:

$$P_{ph} = \frac{1}{2}K_p\gamma H^2\cos(\alpha+\delta-90) = \frac{1}{2}K_p\gamma H^2\sin(\alpha+\delta)$$
$$= \frac{1}{2}K_{ph}\gamma H^2 \tag{21.15}$$

Therefore, the coefficient of passive earth pressure K_{ph} giving the horizontal component P_{ph} of the passive push P_p is:

$$K_{ph} = \frac{\sin^2(\alpha-\varphi')}{\sin^2\alpha\left[1-\sqrt{\dfrac{\sin(\varphi'+\delta)\sin(\varphi'+\beta)}{\sin(\alpha+\delta)\sin(\alpha+\beta)}}\right]^2} \tag{21.16}$$

In the simpler case where the backfill is horizontal, the wall is vertical, and there is no soil-wall friction (conservative), then $\beta = \delta = 0$, $\alpha = 90°$, and K_p becomes:

$$K_p = \frac{1+\sin\varphi'}{1-\sin\varphi'} \tag{21.17}$$

and the product $K_a \times K_p$ is equal to 1.

21.3.2 Rankine Earth Pressure Theory

In 1857, Rankine took a different approach to the same problem. William J. Rankine was a Scottish civil engineer, physicist, and mathematician. He made the following assumptions:

1. The problem is a plane strain problem
2. The soil has friction (φ') but no cohesion ($c' = 0$)
3. The soil has no water
4. The soil mass is in a state of plastic failure
5. The failure surface and the ground surface are planes
6. There is no friction between the soil and the wall

Coulomb considered the equilibrium of a rigid body wedge and reasoned in terms of equilibrium of forces, whereas Rankine considered the equilibrium of stresses at the element level in a failing mass. Rankine theory predates the work of Otto Mohr and the Mohr circle around 1882, but it is easiest to explain Rankine theory through the use of the Mohr circle, which will be done in section 21.3.3. The active and passive earth pressures are as follows (Figures 21.9 and 21.10):

$$\sigma_a = K_a \sigma_v = K_a \gamma z \tag{21.18}$$

$$\sigma_p = K_p \sigma_v = K_p \gamma z \tag{21.19}$$

where σ_a and σ_p are the active and passive earth stresses on the wall, K_a and K_p are the active and passive coefficients, γ is the soil unit weight, and z is the depth below the ground surface. Note that the stress vectors σ_a and σ_p are parallel to the ground surface and therefore inclined at an angle β with the horizontal (Figures 21.9 and 21.10). Rankine obtained the following expressions for K_a and K_p:

$$K_a = \cos\beta \frac{\cos\beta - \sqrt{\cos^2\beta - \cos^2\varphi'}}{\cos\beta + \sqrt{\cos^2\beta - \cos^2\varphi'}} \tag{21.20}$$

$$K = \cos\beta \frac{\cos\beta + \sqrt{\cos^2\beta - \cos^2\varphi'}}{\cos\beta - \sqrt{\cos^2\beta - \cos^2\varphi'}} \tag{21.21}$$

As can be seen from Eqs. 21.18 and 21.19, the stresses on the wall increase linearly with z. By integration of these two equations between 0 and H, the height of the wall, the active force P_a and the passive force P_p can be obtained and are given by Eqs. 21.8 and 21.13, but with different expressions

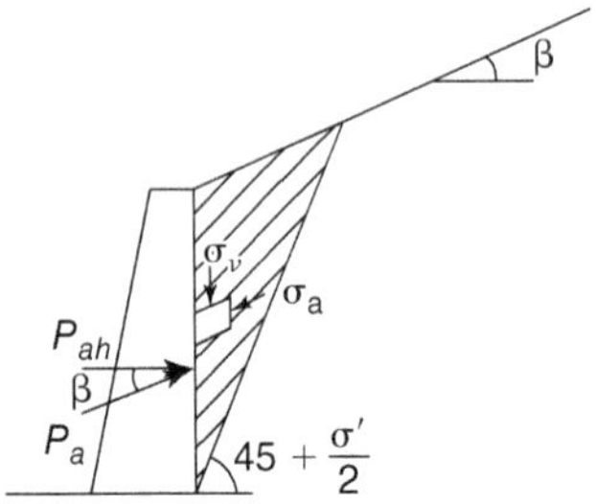

Figure 21.9 Active pressure mass (Rankine).

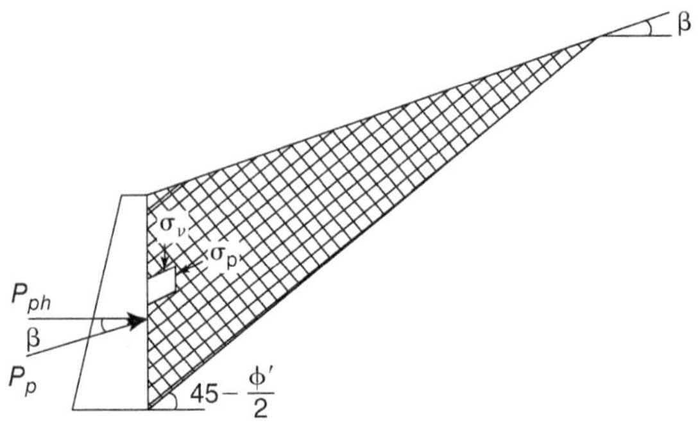

Figure 21.10 Passive pressure mass (Rankine).

for K_a and K_p given in Eqs. 21.20 and 21.21. Note also that the forces P_a and P_p are not horizontal, but rather parallel to the ground surface, which is at an angle β with the horizontal (Figures 21.9 and 21.10). The horizontal components P_{ah} and P_{ph} are:

$$P_{ah} = \frac{1}{2} K_a \gamma H^2 \cos\beta = \frac{1}{2} K_{ah} \gamma H^2 \tag{21.22}$$

$$P_{ph} = \frac{1}{2} K_p \gamma H^2 \cos\beta = \frac{1}{2} K_{ph} \gamma H^2 \tag{21.23}$$

Therefore, the coefficient of active earth pressure K_{ah} giving the horizontal component P_{ah} of the active force P_a and the coefficient of passive earth pressure K_{ph} giving the horizontal component P_{ph} of the passive force P_p are:

$$K_{ah} = \cos^2\beta \frac{\cos\beta - \sqrt{\cos^2\beta - \cos^2\varphi'}}{\cos\beta + \sqrt{\cos^2\beta - \cos^2\varphi'}} \tag{21.24}$$

$$K_{ph} = \cos^2\beta \frac{\cos\beta + \sqrt{\cos^2\beta - \cos^2\varphi'}}{\cos\beta - \sqrt{\cos^2\beta - \cos^2\varphi'}} \tag{21.25}$$

In the simple case where the backfill is horizontal, then $\beta = 0$, and K_a and K_p become:

$$K_a = \frac{1 - \sin\varphi'}{1 + \sin\varphi'} \tag{21.26}$$

$$K_p = \frac{1 + \sin\varphi'}{1 - \sin\varphi'} \tag{21.27}$$

So, should we use Coulomb or Rankine earth pressure coefficients? The Coulomb solution is a limit equilibrium solution giving upper-bound values because the chosen failure surface and mechanism is not necessarily the weakest one. In this context, Coulomb passive earth pressure coefficients tend to be very optimistic (too large). In contrast, the Rankine solution is an equilibrium of stresses solution that gives lower-bound values. Therefore, if a lower bound is conservative, one could choose Rankine; if an upper bound is conservative, one could choose Coulomb. Note that for extreme values of the geometry parameters, it is advisable to use engineering judgment, as the K_a and K_p values can become unreasonable. Note also that for the simple case of a vertical wall, no wall friction, and horizontal backfill, both theories give the same answers (Eqs. 21.12, 21.17, 21.26, and 21.27). The most common values vary from 0.25 to 0.40 for K_a and from 2.5 to 4 for K_p.

21.3.3 Earth Pressure Theory by Mohr Circle

Consider an element of soil behind a retaining wall (Figure 21.11). This element is in an at-rest state of stress to start with. The vertical effective stress is σ'_{ov}, the horizontal effective stress is σ'_{ov}, and the corresponding Mohr circle is shown in Figure 21.11. If the wall is pulled very slightly away from the soil, the horizontal effective stress will decrease until the

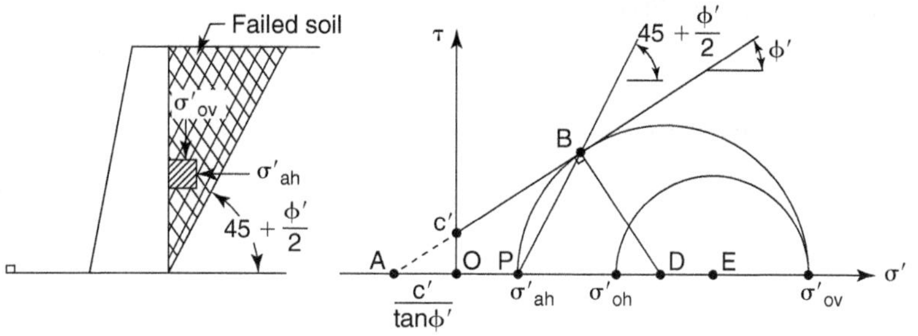

Figure 21.11 Element of soil and Mohr circle (active case).

Mohr circle touches the failure envelope. At that point the soil element will be in a state of failure: It will have mobilized all the shear strength it can offer to support itself, but will still need σ'_{ah} from the wall to avoid collapse. This value σ'_{ah} is the active earth pressure.

From triangle ABD in Figure 21.11, we can write that:

$$\sin\varphi' = \frac{BD}{AO + OD} = \frac{0.5(\sigma'_{ov} - \sigma'_{ah})}{\frac{c'}{\tan\varphi'} + 0.5(\sigma'_{ov} + \sigma'_{ah})} \tag{21.28}$$

which reduces to:

$$\sigma'_{ah} = \sigma'_{ov}\left(\frac{1 - \sin\varphi'}{1 + \sin\varphi'}\right) - 2c'\sqrt{\frac{1 - \sin\varphi'}{1 + \sin\varphi'}} \tag{21.29}$$

or:

$$\sigma'_{ah} = \sigma'_{ov}K_a - 2c'\sqrt{K_a} \quad \text{with} \quad K_a = \frac{1 - \sin\varphi'}{1 + \sin\varphi'} \tag{21.30}$$

The direction of the failure lines can be found by using the Pole method (see section 10.5). The stress point on the Mohr circle at σ'_{ah} corresponds to a stress acting on a vertical plane, so a vertical line will intersect the circle at two points: the stress point and the Pole. Because the vertical line is tangent to the circle, the two points are the same and the Pole is at point P on Figure 21.11. A line from the Pole to the failure point B gives the direction of the failure plane on the diagram. From geometry considerations, the angle of this plane with the horizontal is equal to $45 + \varphi'/2$. Because the entire mass is at failure, a set of parallel failure lines exists.

Now if the wall is pushed into the soil instead of pulled away (Figure 21.12), the horizontal effective stress will increase, pass the value of the vertical effective stress σ'_{ov}, and continue to increase until the Mohr circle touches the failure envelope. At that point the soil element will be in a state of failure: It will have mobilized all the shear strength it can offer to resist the wall push and σ'_{ph} will be generated. This value σ'_{ph} is the passive earth pressure. From triangle ABD in Figure 21.12, we can write that:

$$\sin\varphi' = \frac{BD}{AO + OD} = \frac{0.5(\sigma'_{ph} - \sigma'_{ov})}{\frac{c'}{\tan\varphi'} + 0.5(\sigma'_{ph} + \sigma'_{ov})} \tag{21.31}$$

which reduces to:

$$\sigma'_{ph} = \sigma'_{ov}\left(\frac{1 + \sin\varphi'}{1 - \sin\varphi'}\right) + 2c'\sqrt{\frac{1 + \sin\varphi'}{1 - \sin\varphi'}} \tag{21.32}$$

or

$$\sigma'_{ph} = \sigma'_{ov}K_p + 2c'\sqrt{K_p} \quad \text{with} \quad K_p = \frac{1 + \sin\varphi'}{1 - \sin\varphi'} \tag{21.33}$$

The direction of the failure lines can be found by using the Pole method (see section 10.5). The stress point on the Mohr circle at σ'_{ph} corresponds to a stress acting on a vertical plane, so a vertical line will intersect the circle at two points: the stress point and the Pole. Because the vertical line is tangent to the circle, the two points are the same and the Pole is at point P on Figure 21.12. A line from the Pole to the failure point B gives the direction of the failure plane on the diagram. From geometry considerations, the angle of this plane with the horizontal is equal to $45 - \varphi'/2$. Because the entire mass is at failure, a set of parallel failure lines exists. The conjugate failure lines on Figure 21.12 come from the failure point on the bottom part of the Mohr circle at failure that is not shown on the figure.

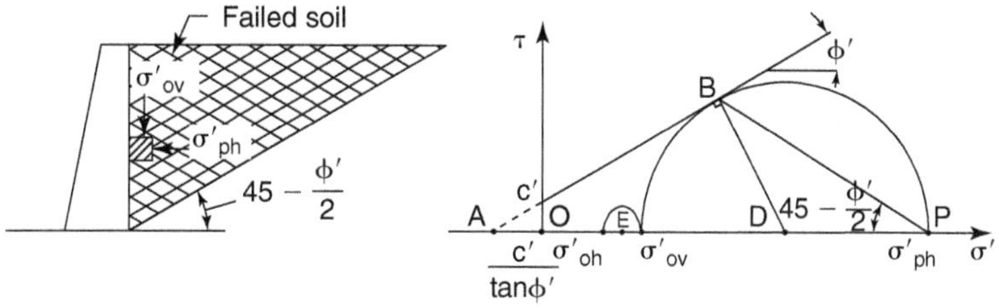

Figure 21.12 Element of soil and Mohr circle (passive case).

21.3.4 Water in the Case of Compression Stress (Saturated)

Up to this point we have calculated the effective horizontal stress for the active case and the passive case. The wall is subjected to the total horizontal stress. When the soil next to the wall is saturated and the water is in compression, the total active and passive earth pressures become:

$$\sigma_{ah} = \sigma'_{ov}K_a - 2c'\sqrt{K_a} + u_w \qquad (21.34)$$

$$\sigma_{ph} = \sigma'_{ov}K_p + 2c'\sqrt{K_p} + u_w \qquad (21.35)$$

The water stress u_w is obtained as follows:

$$u_w = \gamma_w h_p \qquad (21.36)$$

where γ_w is the unit weight of water, h_p is the distance from the groundwater level to the point considered if there is no flow, and h_p is the pressure head obtained from a flow net if there is flow.

Note that there is a big difference between the pressure against a wall that has to retain a soil without water and the pressure against a wall that has to retain a soil with a water level at the ground surface. For example, if a wall is 3 m high and retains a dry sand with a unit weight of 18 kN/m^3 and a friction angle of 30°, the active earth pressure behind the bottom of the wall will be:

$$\sigma_{ah} = 3 \times 18 \times 0.33 = 18\ \text{kN/m}^2 \qquad (21.37)$$

However, if the water rises to the top of the wall, increasing the unit weight of the soil to 20 kN/m^3, and if the water stress is hydrostatic, the active pressure behind the bottom of the wall becomes:

$$\sigma_{ah} = (3 \times 20 - 3 \times 10) \times 0.33 + 3 \times 10 = 40\ \text{kN/m}^2 \qquad (21.38)$$

As can be seen, the pressure doubles due to the presence of the water. If we had assumed that no water could be present and designed the wall for a factor of safety of 2, the wall would have been close to failure when the water accumulated behind it. It is extremely important to pay great attention to water when designing retaining walls.

21.3.5 Water in the Case of Tension Stress (Unsaturated or Saturated)

If the soil behind the wall is above the groundwater level, the water is in tension and the soil is either saturated or unsaturated. In both cases, the water stress u_w is negative. This increases the shear strength of the soil because it increases the effective stress. Thus, one would expect the active earth pressure to decrease and the passive earth pressure to increase. Equations 21.34 and 21.35 become:

$$\sigma_{ah} = \sigma'_{ov}K_a - 2c'\sqrt{K_a} + \alpha u_w \qquad (21.39)$$

$$\sigma_{ph} = \sigma'_{ov}K_p + 2c'\sqrt{K_p} + \alpha u_w \qquad (21.40)$$

where α is the water area ratio, which can be estimated as the degree of saturation S or by using the Khalili rule (see section 10.13). Note that the term αu_w is also embedded in σ'_{ov}. Regrouping gives:

$$\sigma_{ah} = \sigma_{ov}K_a - 2c'\sqrt{K_a} + (1 - K_a)\alpha u_w \qquad (21.41)$$

$$\sigma_{ph} = \sigma_{ov}K_p + 2c'\sqrt{K_p} + (1 - K_p)\alpha u_w \qquad (21.42)$$

Equations 21.41 and 21.42 show that water tension decreases the active earth pressure and increases the passive earth pressure. However, it is very important to consider if the water tension used in these equations will always be present or if it is a seasonal occurrence. Furthermore, it would be uncommon for the water tension to pull on the wall. In the case of unsaturated soils and for earth pressure calculations, it is therefore prudent to consider that the water stress is equal to zero.

21.3.6 Influence of Surface Loading (Line Load, Pressure)

Load is often applied at the top of a retaining wall (Figure 21.13) either during construction (e.g., compaction rollers) or after construction (e.g., bridge abutment, additional fill). In the case of a pressure p that covers the entire surface area at the top of the retaining wall, the active and passive earth pressures have an added term $K_a p$ and $K_p p$ respectively. The reason is that the pressure p simply adds to the total stress σ_{ov}.

In the case of a line load Q (kN/m) parallel to the wall crest and located at a perpendicular distance x from the wall, the increase in horizontal stress against the wall at a depth z below the top of the wall can be calculated by:

$$\Delta\sigma_h = \frac{4Q}{\pi}\frac{x^2 z}{(z^2 + x^2)^2} \qquad (21.43)$$

If the load is a point load P (kN) applied at a perpendicular distance x from the wall, the maximum increase in horizontal pressure against the wall at a depth z below the top of the wall can be calculated by:

$$\Delta\sigma_h = \frac{P}{\pi(z^2 + x^2)}\left(\frac{3x^2 z}{(z^2 + x^2)^{3/2}} - \frac{(z^2 + x^2)^{1/2}(1 - 2\nu)}{(z^2 + x^2)^{1/2} + z}\right) \qquad (21.44)$$

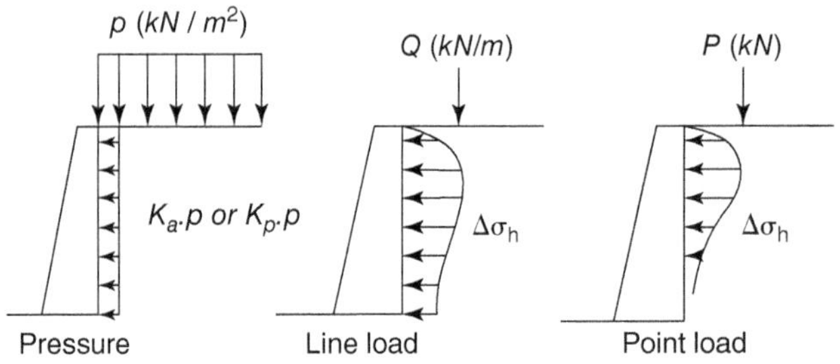

Figure 21.13 Horizontal pressures due to surface loading.

The values obtained from Eqs. 21.43 and 21.44 are added to both the active earth pressure and the passive earth pressure. Solutions for other surface loading can be found in the Canadian Foundation Engineering Manual (2007) and the AASHTO Bridge Specifications (2007).

21.3.7 General Case and Earth Pressure Profiles

In the general case, the total active and passive earth pressures are given by:

$$\sigma_{ah} = \sigma'_{ov} K_a - 2c' \sqrt{K_a} + \Delta\sigma_h + \alpha u_w \quad (21.45)$$

$$\sigma_{ph} = \sigma'_{ov} K_p + 2c' \sqrt{K_p} + \Delta\sigma_h + \alpha u_w \quad (21.46)$$

where σ_{ah} is the total active earth pressure on the wall at a depth z below the top of the wall, σ'_{ov} is the vertical effective stress at depth z, K_a is the coefficient of active earth pressure, $\Delta\sigma_h$ is the earth pressure due to surface loading, c' is the effective stress cohesion of the retained soil, α is the water area ratio, u_w is the water stress (tension or compression), σ_{ph} is the total passive earth pressure on the wall at a depth z below the top of the wall, and K_p is the coefficient of passive earth pressure.

These are the equations to use when calculating the active or passive earth pressure against the wall at a chosen depth z where σ'_{ov} and $\Delta\sigma_h$ exist. Keep in mind that these pressures or stresses may not be horizontal if the ground surface is not horizontal, the back of the wall is not vertical, or the wall friction is not assumed to be zero. A distinction is made in this respect between K_a and K_{ah} on the one hand and K_p and K_{ph} on the other (sections 21.3.1 and 21.3.2).

The next problem is to generate the complete profile of pressure against the wall versus depth. This is done by preparing a series of profiles using the following steps:

1. Profile of total vertical stress σ_{ov} versus depth
2. Profile of water stress u_w versus depth
3. Profile of water area ratio α versus depth
4. Profile of effective vertical stress ($\sigma'_{ov} = \sigma_{ov} - \alpha\ u_w$) versus depth
5. Profile of effective horizontal stress ($\sigma'_{ah} = \sigma'_{ov} K_a - 2c' K_a{}^{0.5}$) versus depth
6. Profile of horizontal stress due to surface loads ($\Delta\sigma_{ah}$) versus depth
7. Profile of total horizontal stress ($\sigma_{ah} = \sigma'_{ov} K_a - 2c' K_a{}^{0.5} + \Delta\sigma_{ah} + \alpha\ u_w$) versus depth

Figure 21.14 shows an example of the series of profile steps. The same sequence is followed for the passive earth pressure profiles.

If the soil is layered, the earth pressure has to be calculated twice at the depth of the layer boundary: once with the upper-layer soil parameters and once with the lower-layer soil parameters. As a result, there is typically a discontinuity in the earth pressure profile at the boundary between two soil layers (Figure 21.15).

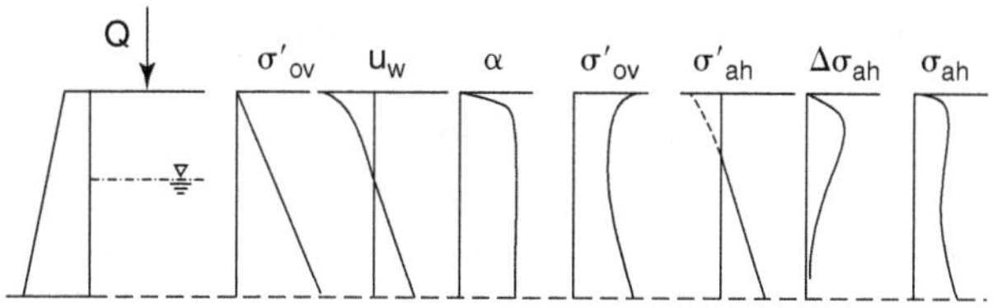

Figure 21.14 Series of profiles to generate earth pressure profile versus depth.

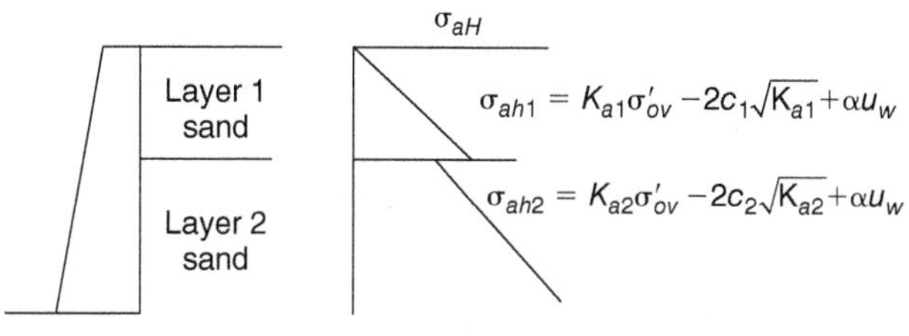

Figure 21.15 Active pressures at a soil layer boundary.

21.4 SPECIAL CASE: UNDRAINED BEHAVIOR OF FINE-GRAINED SOILS

As discussed in section 15.16, the equations for the undrained behavior of a fine-grained soil can be obtained from the effective stress equations by a simple transformation or correspondence principle:

1. Effective unit weight becomes total unit weight

$$\gamma_{eff} \rightarrow \gamma_t \quad (21.47)$$

2. Effective stress becomes total stress

$$\sigma\prime \rightarrow \sigma \quad (21.48)$$

3. Effective stress cohesion becomes undrained shear strength

$$c' \rightarrow s_u \quad (21.49)$$

4. Effective stress friction angle becomes zero

$$\varphi' \rightarrow 0 \quad (21.50)$$

Using this transformation on Eqs. 21.45 and 21.46, the following equations are obtained for the undrained behavior active and passive earth pressures:

$$\sigma_{ah} = \sigma_{ov} - 2s_u + \Delta\sigma_h \quad (21.51)$$

$$\sigma_{ph} = \sigma_{ov} + 2s_u + \Delta\sigma_h \quad (21.52)$$

where s_u is the undrained shear strength of the soil. These equations tend to give active earth pressures that are too low and passive pressures that are too high. One reason is that they assume that the soil is uniform with no fissures.

These equations should be used with great caution and proper judgment. For example, imagine that you have to design a wall for a clay that has an undrained shear strength of 100 kPa and a unit weight of 20 kN/m^3. Equation 21.51 says that no wall is needed until a depth of 10 m, as the active earth pressure is negative down to that depth. Now imagine that this clay has many fissures that are about 0.3 meters apart. The sample you tested was taken from one of the blocks between fissures and gave 100 kPa for s_u, but the soil mass is actually much weaker because of the fissures; the sample strength is not representative of the mass strength. If you dug a trench in such a material, it would be very surprising if you could dig down to 10 meters without a major collapse before that point. In contrast, if the material is truly uniform with no fissures (very rare), the theory says that you could dig to 10 m without support.

21.5 AT-REST EARTH PRESSURE

The *at-rest earth pressure* is the horizontal stress that exists in the soil under geostatic stresses and without displacement. The coefficient of at-rest earth pressure K_o is defined as:

$$K_o = \frac{\sigma'_{oh}}{\sigma'_{ov}} \tag{21.53}$$

where σ'_{ov} and σ'_{ov} are the horizontal and vertical effective stresses respectively. Note that K_o is the ratio of the effective stresses, not the total stresses; also, K_o does not involve the cohesion c', whereas the ratios K_a and K_p incorporate c' in their definition:

$$K_a = \frac{\sigma'_{ah}}{\sigma'_{ov}} + \frac{2c'\sqrt{K_a}}{\sigma'_{ov}} \tag{21.54}$$

$$K_p = \frac{\sigma'_{ph}}{\sigma'_{ov}} - \frac{2c'\sqrt{K_p}}{\sigma'_{ov}} \tag{21.55}$$

Thus, it is theoretically possible for K_o to have values higher than K_p and lower than K_a. For example, if $\sigma'_{ph} = 300$ kPa, $\sigma'_{ov} = 100$ kPa, and $c' = 20$ kPa, and if a high horizontal stress at rest is locked up tectonically at the value of σ'_{ph}, then K_p is 2.4 and K_o is 3.

In elasticity, the ratio of the horizontal stress to the vertical stress for a condition with no lateral movement (at-rest condition) is obtained in cylindrical coordinates from:

$$\varepsilon_h = \frac{1}{E}(\sigma'_{oh} - \nu(\sigma'_{ov} + \sigma'_{oh})) = 0 \tag{21.56}$$

where ε_h is the horizontal strain, E is a modulus of deformation of the soil, and ν is Poisson's ratio. Therefore

$$K_o = \frac{\nu}{1-\nu} \tag{21.57}$$

A commonly used value of Poisson's ratio for a drained case is 0.33; then K_o is equal to 0.5. However, measured K_o values have been reported in the range of 0.4 to more than 2. A value of 2 would require a Poisson's ratio equal to 0.67, which is possible for soils that dilate during compression, a well-known phenomenon. Such high K_o values are found in cases where high horizontal stresses have developed during geological events that densify or overconsolidate the soil. They may also be generated during compaction of shallow layers.

The coefficient of at-rest earth pressure is very difficult to measure, essentially because any instrument placed in the ground to measure K_o will create disturbance and change the at-rest state of stress. The best measurements are thought to be possible with a self-boring pressuremeter. However, even the self-boring pressuremeter creates significant disturbance due to shearing and side friction upon descent of the probe. Furthermore, the choice of zero volume of the probe can significantly affect the value of K_o obtained.

The early part of the preboring pressuremeter test offers another way to obtain an estimate of the horizontal stress. As the horizontal pressure applied by the pressuremeter probe on the borehole wall is increased, it goes through the threshold of pressure corresponding to the at-rest horizontal pressure. The curved line that describes the horizontal pressure versus increase in radius until the elastic portion of the curve is reached could be used. A construction much like the Casagrande construction for the preconsolidation pressure in the consolidation test would be needed, but calibration of such an idea has not been performed.

The step blade test consists of pushing a series of flat blades of increasing thickness into the soil while recording the horizontal stress on each blade. The idea was to extrapolate the horizontal stresses obtained on each blade back to a blade with zero thickness so as to find the at-rest horizontal stress. Although this idea was very clever, unfortunately the superposition of a penetration event and a lateral expansion event made the extrapolation unreliable.

One method consists of measuring the water tension developing in fine-grained soils upon extrusion of saturated samples. When the saturated sample comes out of the sampling tube, it decompresses and the total stress suddenly becomes zero—but the sample cannot readily expand because of the low hydraulic conductivity, and the water goes into tension to prevent any increase in volume. This results in a transfer from the mean effective stress to the water tension.

Sample at depth z:

$$\sigma_{mean} = \frac{1}{3}(\sigma'_{ov} + 2\sigma'_{oh}) + u_w \tag{21.58}$$

Sample extruded:

$$0 = \frac{1}{3}(\sigma'_{ov} + 2\sigma'_{oh}) + u_w \quad \text{or} \quad -u_w = \frac{1}{3}\sigma'_{ov}(1 + 2K_o) \tag{21.59}$$

Equation 21.59 shows that the water tension in the sample is a function of the horizontal effective stress. K_o can then be calculated knowing the vertical effective stress.

A K_o triaxial test can be used to obtain a value of K_o. This test consists of loading the sample vertically while increasing the horizontal stress (cell pressure) independently and in such a way that no lateral deformation will take place. During the test, the water stress is measured and the ratio between the horizontal effective stress (cell pressure minus water stress) and the vertical effective stress gives the K_o value. Alternatively, consolidometer tests with an instrumented ring can be used to obtain a value of K_o. The metal ring in which the sample is placed is instrumented with strain gages to measure the hoop strain in the metal, thereby giving the hoop stress that prevents lateral expansion. The radial stress is then obtained as:

$$\sigma_{oh} = \frac{\sigma_\theta t}{r} \tag{21.60}$$

where σ_{oh} is the radial or horizontal stress exerted by the soil on the metal ring that prevents expansion, σ_θ is the hoop stress in the metal obtained from the hoop strain measurements, t is the thickness of the metal ring, and r is the radius of the consolidometer. Knowing the vertical stress σ_{ov} imposed on the sample, and assuming that zero water stress is in the sample at the end of consolidation, gives data to calculate K_o. One of the difficulties with this approach is to ensure that the strain gages are sensitive enough to detect the strain in the metal ring under the relatively small radial stresses imposed by the soil.

Many correlations have also been proposed. The first one may be attributed to Jacky (1944), expressed as:

$$K_o = 1 - \sin\varphi' \tag{21.61}$$

This equation was later revised to include the effect of the overconsolidation ratio (OCR) for uncemented sands and clays of low to medium sensitivity:

$$K_o = (1 - \sin\varphi')OCR^{\sin\varphi'} \tag{21.62}$$

where φ' is the effective stress friction angle of the soil, and OCR is the overconsolidation ratio, defined as the ratio of the effective preconsolidation stress σ'_p over the current effective vertical stress. For clean quartz sand in chamber tests, Mayne (2007a, b) proposed:

$$K_o = 0.192\left(\frac{q_c}{\sigma_a}\right)^{0.22}\left(\frac{\sigma_a}{\sigma'_{ov}}\right)^{0.31}(OCR)^{0.27} \tag{21.63}$$

where q_c is the CPT point resistance, σ_a is the atmospheric pressure, σ'_{ov} is the vertical effective stress, and OCR is the overconsolidation ratio.

21.6 EARTH PRESSURE DUE TO COMPACTION

When soil is compacted behind bottom-up walls, the compaction process induces horizontal stresses that are higher than active earth pressures. This has been clearly documented

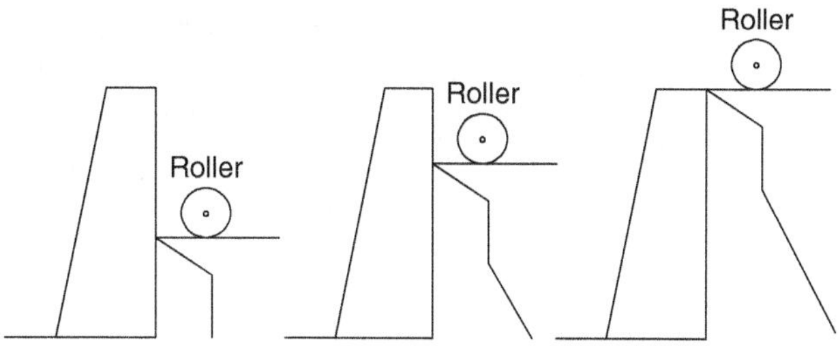

Figure 21.16 Compaction earth pressure during backfilling.

(Duncan and Seed 1986; Chen and Fang 2008). The compaction roller creates high vertical stresses, which in turn create high horizontal stresses during compaction. Because the soil does not return to an undeformed state after unloading (not elastic), and because the soil locks in plastic strains after unloading, high horizontal stresses remain after the roller moves on. This horizontal prestressing is actually very beneficial for improving the behavior of pavement base courses. For retaining walls, this means that designing for the active earth pressure case may not be prudent. At the same time, the depth of influence of the roller is limited and after several lifts of compaction have been completed the high stresses at depth (Figure 21.16) become smaller than the at-rest stresses at that depth.

The U.S. Navy (1982) made some recommendations for earth pressures due to compaction, which, considering more recent data, lead to the profile shown in Figure 21.17. The pressure diagram starts at a slope equal to the passive earth pressure coefficient. From the surface to a depth where the horizontal pressure reaches the value σ_h, the passive earth pressure profile, $K_p\gamma z$, is used. Then the pressure remains constant at a value of σ_h equal to:

$$\sigma_h = \frac{L}{a+L}\sqrt{\frac{2P\gamma}{\pi}} \tag{21.64}$$

where L is the length of the roller, a is the distance between the edge of the wall and the closest roller position, P is the

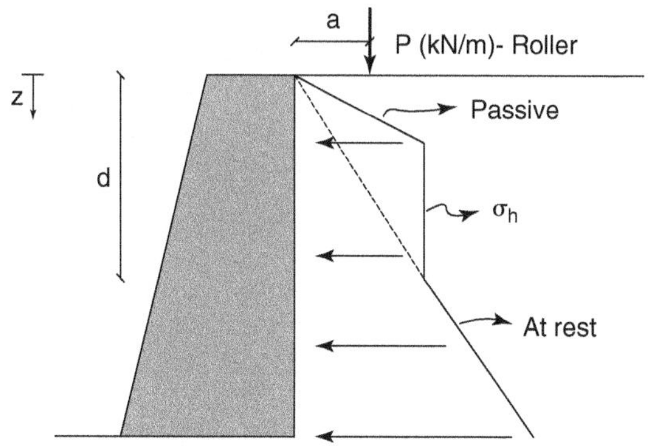

Figure 21.17 Wall pressure diagram including compaction stresses. (After U.S. Navy 1982.)

line load imposed by the roller (weight of the roller plus the centrifugal force for vibratory rollers divided by the length of the roller), and γ is the unit weight of the soil being compacted. At a depth d, the pressure diagram joins the at-rest earth pressure profile, $K_o\gamma z$, which is used beyond that point. That depth d is therefore equal to (Figure 21.17):

$$d = \frac{L}{K_o(a+L)}\sqrt{\frac{2P}{\pi\gamma}} \tag{21.65}$$

where K_o is the at-rest earth pressure coefficient.

21.7 EARTH PRESSURES IN SHRINK-SWELL SOILS

When the backfill of a bottom-up wall or the soil behind a top-down wall has a high plasticity index (Ip) or swell index (Is), it is necessary to consider the soil shrink-swell behavior in calculating the pressure diagram. Hong et al. (2010) studied this issue and made the following recommendation (Figure 21.18).

Three diagrams come into play in the resultant pressure diagram (Figure 21.18): the passive earth pressure diagram, the swell pressure diagram, and the at-rest earth pressure diagram. Both the passive and at-rest diagrams increase with depth according to K_p and K_o respectively. The swell pressure diagram, however, typically decreases with depth because the overburden pressure increases with depth and limits the swell pressure.

The pressure diagram starts at a slope equal to the passive earth pressure coefficient. Although the swell pressure is higher than the passive pressure within that zone, the soil fails in shear before it can reach the swell pressure. When the passive pressure profile reaches the swell pressure profile, the swell pressure limits the earth pressure against the wall and the pressure diagram follows the swell pressure profile. When the swell pressure profile reaches the at-rest pressure profile, the at-rest pressure is maintained against the wall because the swell pressure is smaller than that. As a result, the pressure diagram switches to the at-rest pressure profile. The coefficients K_p and K_o have been discussed in previous sections. The swell pressure profile can be obtained by performing swell tests on samples from the retained soil.

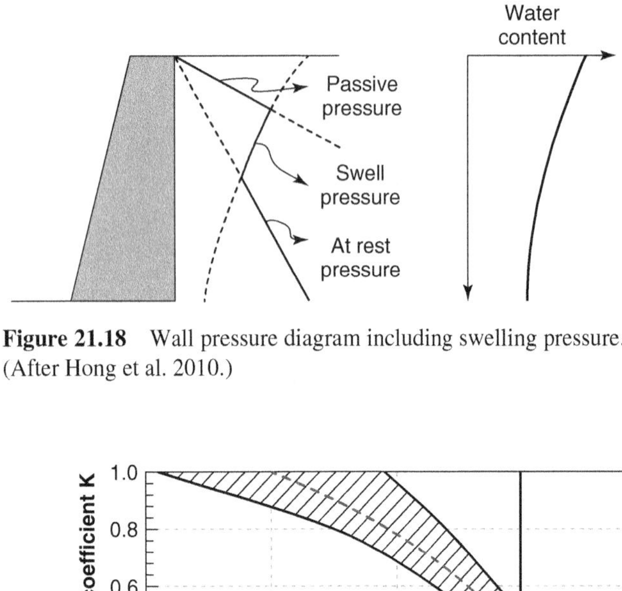

Figure 21.18 Wall pressure diagram including swelling pressure. (After Hong et al. 2010.)

21.8 DISPLACEMENTS

Figure 21.4 showed the general form of the earth pressure σ_h or p vs. displacement y curve. This curve, sometimes called a P-y curve, represents the plane strain behavior of the wall at a depth z. Figure 21.19 shows some values coming from measurement and numerical simulations (Briaud and Kim 1998). The vertical axis is a generalized earth pressure coefficient K, which is discussed further in section 21.12, and the horizontal axis is the horizontal displacement normalized by the height of the wall.

The amount of movement necessary to generate the active earth pressure σ_{ah} is y_a and the amount of movement necessary to generate the passive earth pressure σ_{ph} is y_p. Table 21.1 shows some possible values of y_a/H and y_p/H (H is the height of the wall) for different soil types. This means that if the wall is high, it will take more movement to mobilize the earth pressure than if the wall is low. The argument in favor

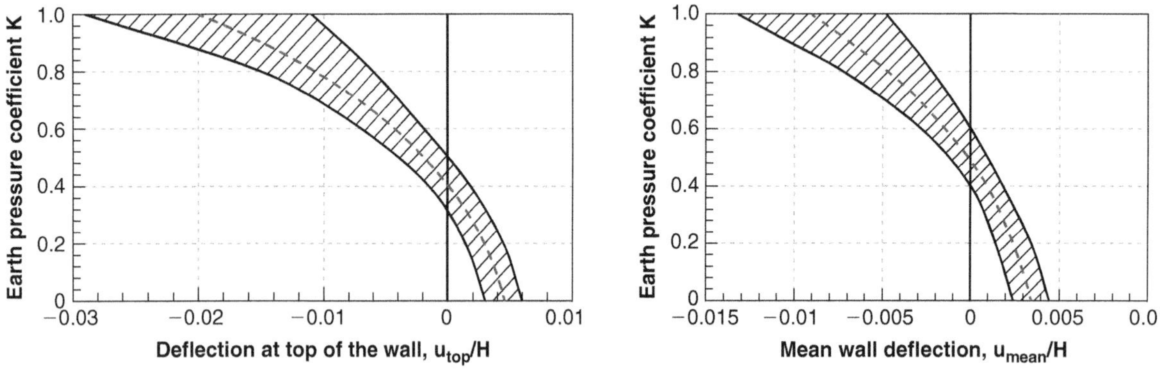

Figure 21.19 Measured earth pressure coefficient versus normalized displacement of a wall (Briaud and Kim 1998).

Table 21.1 Possible Range of Displacement to Generate Active and Passive Earth Pressures

Soil Type	Active, y_a/H	Passive, y_p/H
Loose sand	0.003 to 0.005	0.03 to 0.05
Dense sand	0.001 to 0.002	0.01 to 0.03
Soft clay	0.01 to 0.02	0.03 to 0.05
Stiff clay	0.005 to 0.01	0.01 to 0.03

of this concept is that if the wall is high, the earth pressure wedge will be large and it will take more movement to completely fail the wedge of soil behind a high wall compared to a low wall.

While it is clear that earth pressures depend on movement, and while it is also clear that predicting movements is important, our ability to make such predictions is not as good as our ability to calculate foundation settlement. Often the design of walls takes place solely on the basis of earth pressure distributions (ultimate limit state) rather than a combination of earth pressures and movements. Nevertheless, the trend in practice is toward increased inclusion of movement calculations in retaining wall design. Because the intact mass is the one deforming during such earth pressure problems, and because the overall strain level is quite small for well-designed systems, small strain moduli are most useful and can be obtained from cross hole sonic tests.

21.9 GRAVITY WALLS

Gravity walls are bottom-up walls usually made of reinforced concrete (Figure 21.20). In the early days they were heavy, massive blocks (concrete gravity wall), but such systems were soon replaced by walls that use less concrete weight and more backfill weight as dead weight to resist the soil push (cantilever gravity walls). In cantilever gravity walls, the slab under the retained portion of the backfill is subjected to the backfill weight, which increases the sliding resistance and the resistance to overturning. Cantilever gravity walls have to be heavily reinforced, as a high bending moment develops at the connection between the slab and the stem. The word *cantilever* is also used for a type of top-down wall; this is why the word *gravity* is added to *cantilever* to designate the wall shown in Figure 21.20.

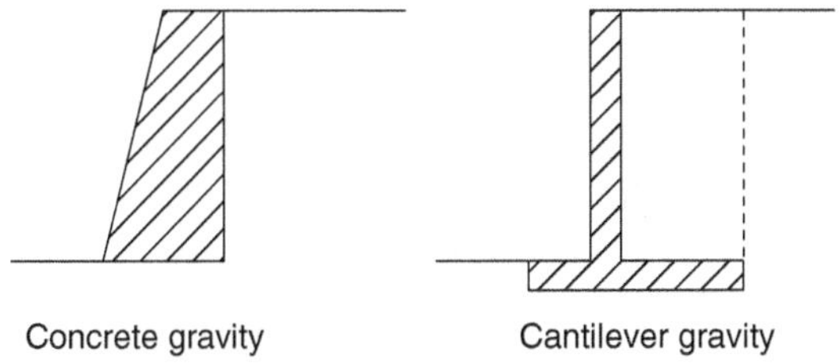

Figure 21.20 Types of gravity walls.

The geotechnical design of gravity walls consists of a number of steps aimed at ensuring the safety (low probability of failure) and functionality (low probability of intolerable movements) of the wall. The purpose of the design is to satisfy the ultimate limit state and the serviceability limit state of the wall as it is subjected to the earth pressures behind the wall and in front of the wall. For gravity walls, however, the serviceability limit state is rarely addressed, as movements are difficult to estimate and often small. The design steps include estimating the pressure distribution behind the wall (active pressure), the pressure distribution in front of the wall (passive pressure), the resultant active force and its location, the resultant passive force and its location, the sliding ultimate limit state, the overturning ultimate limit state, the bearing capacity ultimate limit state, the slope stability ultimate limit state, and the settlement serviceability limit state (rare). Figure 21.21 identifies the possible failure modes for a gravity wall.

1. Active pressure behind the wall σ_{ah}. For this, the steps in section 21.3.7 are followed and the profile of total active earth pressure is prepared. Special earth pressure conditions, such as compaction stresses, stresses due to shrink-swell soils, and stresses due to surface loading, are considered in arriving at the design active pressure diagram.
2. Passive pressure in front of the wall σ_{ph}. This refers to any embedded portion of the wall that could generate a passive resistance. Here again, the steps of section 21.3.7 are followed and the profile of passive earth pressure is prepared.
3. The resultant active push P_a (kN/m) is calculated as the area under the active earth pressure diagram (Figure 21.22):

$$P_a = \int_{z=0}^{z=H+D} \sigma_{ah}\, dz = \sum_{i=1}^{n} A_{ai} \tag{21.66}$$

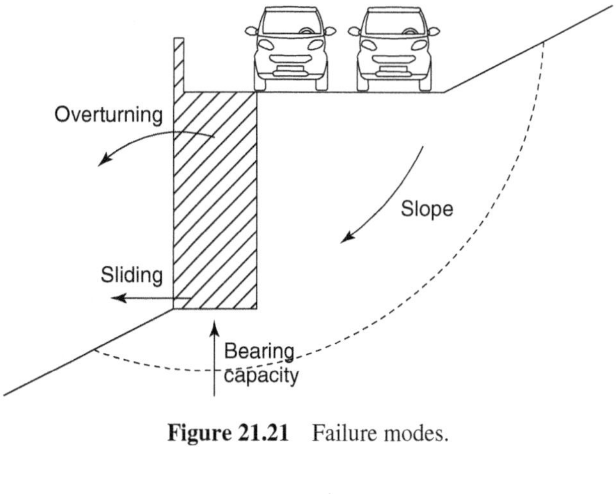

Figure 21.21 Failure modes.

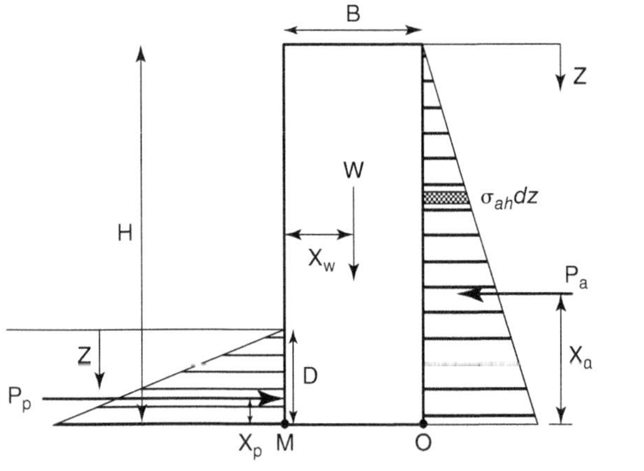

Figure 21.22 Forces acting on a gravity wall.

where σ_{ah} is the horizontal active earth pressure at depth z below the top of the wall, H is the height of the wall, D is the embedded depth, and A_a is the area under the active earth pressure diagram. If there is more than one soil layer, P_a is given by the sum of the areas A_{ai} (corresponding to layer i) under the active pressure diagram.

4. The resultant passive push P_p (kN/m) is calculated as the area under the passive earth pressure diagram (Figure 21.22):

$$P_p = \int_{z=0}^{z=D} \sigma_{ph}\, dz = \sum_{i=1}^{m} A_{pi} \tag{21.67}$$

where σ_{ph} is the horizontal passive earth pressure at depth z below the bottom of the wall, H is the height of the wall, D is the embedded depth, and A_p is the area under the passive earth pressure diagram. If there is more than one soil layer, P_p is given by the sum of the areas A_{pi} (corresponding to layer i) under the passive earth pressure diagram.

5. The point of application of P_a is found by writing that the moment around a chosen point (often the bottom of the wall, O in Figure 21.22) created by the active earth pressure diagram is the same as the moment created by the resultant P_a:

$$P_a x_a = \int_{z=0}^{z=H+D} \sigma_{ah}(H + D - z)dz = \sum_{i=1}^{n} A_{ai} a_{ai} \tag{21.68}$$

where x_a is the moment arm of P_a, and a_{ai} is the moment arm of the individual areas under the pressure diagram corresponding to A_{ai}. Of course, if the active earth pressure diagram is a simple triangle, then x_a is $0.33(H + D)$.

6. The point of application of P_p is found by writing that the moment around a chosen point (often the bottom of the wall, O in Figure 21.22) created by the passive earth pressure diagram is the same as the moment created by the resultant P_p:

$$P_p x_p = \int_{z=0}^{z=D} \sigma_{ph}(D - z)dz = \sum_{i=1}^{n} A_{pi} a_{pi} \tag{21.69}$$

where x_p is the moment arm of P_p, and a_{pi} is the moment arm of the individual areas under the pressure diagram corresponding to A_{pi}. Of course, if the passive earth pressure diagram is a simple triangle, then x_p is 0.33D.

7. Sliding ultimate limit state is checked by evaluating the following equation:

$$\gamma_1 P_{a1} + \gamma_2 P_{a2} \leq \varphi_1 W \tan\delta + \varphi_2 P_p \tag{21.70}$$

where γ_1 is the load factor associated with the active push P_{a1} due to soil weight, γ_2 is the load factor associated with the active push P_{a2} due to surcharge, φ_1 is the resistance factor for the resistance to sliding due to soil weight, φ_2 is the resistance factor for the resistance to sliding due to the passive earth pressure in front of the wall, and δ is the friction angle for the interface between the bottom of the wall and the soil on which it rests. The angle δ is usually taken as equal to the friction angle φ' of the soil for rough interfaces. The load factor γ_1 is typically taken as 1.5, and γ_2 as 1.75. The resistance factor for the sliding resistance due to the weight of the wall is in the range of 0.8 to 0.9, whereas the resistance factor for the sliding resistance due to the passive earth pressure is usually around 0.5.

8. Overturning ultimate limit state is checked by evaluating the following equation related to the moment around the front of the wall (point M in Figure 21.22):

$$\gamma_1 P_{a1} x_{a1} + \gamma_2 P_{a2} x_{a2} \leq \varphi_1 W x_w + \varphi_2 P_p x_p \tag{21.71}$$

where γ_{a1}, γ_{a2}, P_{a1}, and P_{a2} are as defined in step 7; x_{a1}, x_{a2}, and x_p are the moment arms of the forces P_{a1}, P_{a2} and P_p respectively; φ_1 and φ_2 are the same resistance factors as in step 7; W is the weight of the wall, and x_w is the corresponding moment arm. The values of the load and resistance factors for this ultimate limit state are the same as the values for step 7.

9. Bearing capacity ultimate limit state is checked as a shallow foundation subjected to the combination of W, P_a, and P_p (see section 17.4). This combination leads to the case of an inclined, eccentric load.
10. Slope ultimate limit state is checked in the same way as a slope with a wall loading the soil surface (see Chapter 19). The load and resistance factors were presented in section 19.2.
11. Serviceability limit state is usually not addressed in current practice.

The following comments may be made on the movement of gravity walls. For gravity walls founded on competent soil, the movement usually takes place by rotation around the bottom of the wall (point O in Figure 21.22). Most of the horizontal movement tends to occur during construction and corresponds to the order of magnitude given in Table 21.1 for the active case. Note that the main source of horizontal movement comes from rotation of the base under the overturning moment. Indeed, if the sliding ultimate limit state is satisfied, sliding movement should be very small. Vertical settlement of the wall will occur if the downdrag from the backfill and the high stresses under the front edge of the wall due to the applied moment compress the soil under the wall. Because this compression is uneven, with more settlement under the front edge, the wall will rotate with more horizontal movement at the top. To this extent, the settlement factors giving the settlement at the center and at the edge of the foundation (see section 17.7 on the load settlement curve approach) can be used to infer the rotation and movement of the wall. In that respect it is useful to study the case of a foundation subjected to a line load Q (kN/m) and an overturning moment M (kN.m/m) (Figure 21.23). The eccentricity e of the load Q is given by:

$$e = \frac{M}{Q} \tag{21.72}$$

The pressure diagram under the foundation is shown in Figure 21.23. The maximum pressure p_{max} and minimum pressure p_{min} under the foundation with a width B are given by:

$$p_{max} = \frac{Q}{B}\left(1 + \frac{6e}{B}\right) \tag{21.73}$$

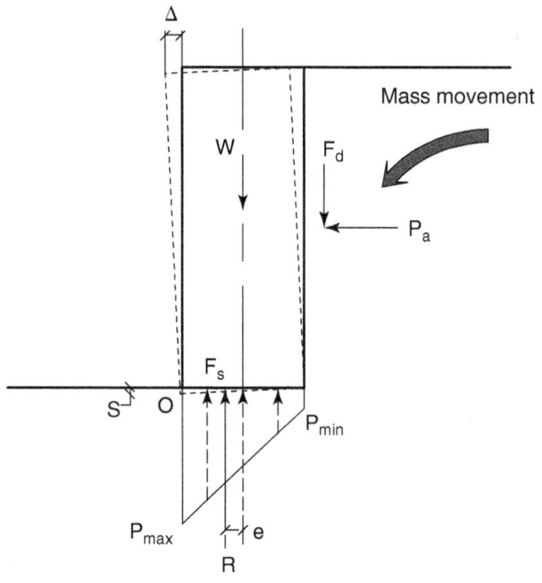

Figure 21.23 Pressure under a gravity wall.

$$p_{min} = \frac{Q}{B}\left(1 - \frac{6e}{B}\right) \tag{21.74}$$

Equation 21.74 indicates that p_{min} becomes zero when the eccentricity becomes equal to B/6. If p_{min} becomes zero, the instability of the wall is more likely, as the foundation cannot develop tensile resistance to overcome further increase in eccentricity. As long as the point of application of the resultant stays within a distance of $B/6$ from the axis of symmetry, the wall is more likely to be stable and experience limited movement. This is called the *rule of the middle third* as e can be $\pm$ $B/6$. Note that the wall can also move in the other direction (more horizontal movement at the bottom of the wall) if a slope stability problem exists.

21.10 MECHANICALLY STABILIZED EARTH WALLS

Mechanically stabilized earth (MSE) walls are bottom-up walls made mostly of soil with some reinforcement. Henri Vidal, a French engineer, is credited with inventing reinforced earth in 1957. This technology is to geotechnical engineering what reinforced concrete is to structural engineering. It consists of placing inclusions in the soil to give it significant tensile strength. These walls were called reinforced earth walls in the beginning and are now called mechanically stabilized earth walls (MSE walls). An MSE wall is built by placing a layer of soil (say, 0.7 m thick), then a layer of reinforcement, then a layer of soil, then a layer of reinforcement, and so on until the desired wall height is reached (Figure 21.24). Panels are placed in front of and attached to the reinforcement for esthetic purposes and to retain any soil that might fall between reinforcement layers close to the front. The pressure on the panels is very small, as most of the earth pressure is taken up by the reinforcement. The reinforcement can be galvanized steel strips, steel grids, or geosynthetics. The success of MSE walls is due to their lower cost compared to cantilever gravity walls, particularly for very high walls (Figure 21.25). Indeed, MSE walls built to 50 meters in height have performed very well.

The design of MSE walls includes an external stability design and an internal stability design.

21.10.1 External Stability

For this case, the MSE wall is considered to be a gravity wall consisting of the front panels, the reinforcement, and the soil between the reinforcement. This reinforced soil mass (ABCD in Figure 21.24) is the gravity wall and has to satisfy the design criterion of a gravity wall outlined in section 21.9. These include the sliding ultimate limit state, the overturning ultimate limit state, the bearing capacity ultimate limit state, the slope stability ultimate limit state, and the settlement serviceability limit state (rare). The design steps are identical to the steps detailed in section 21.9.

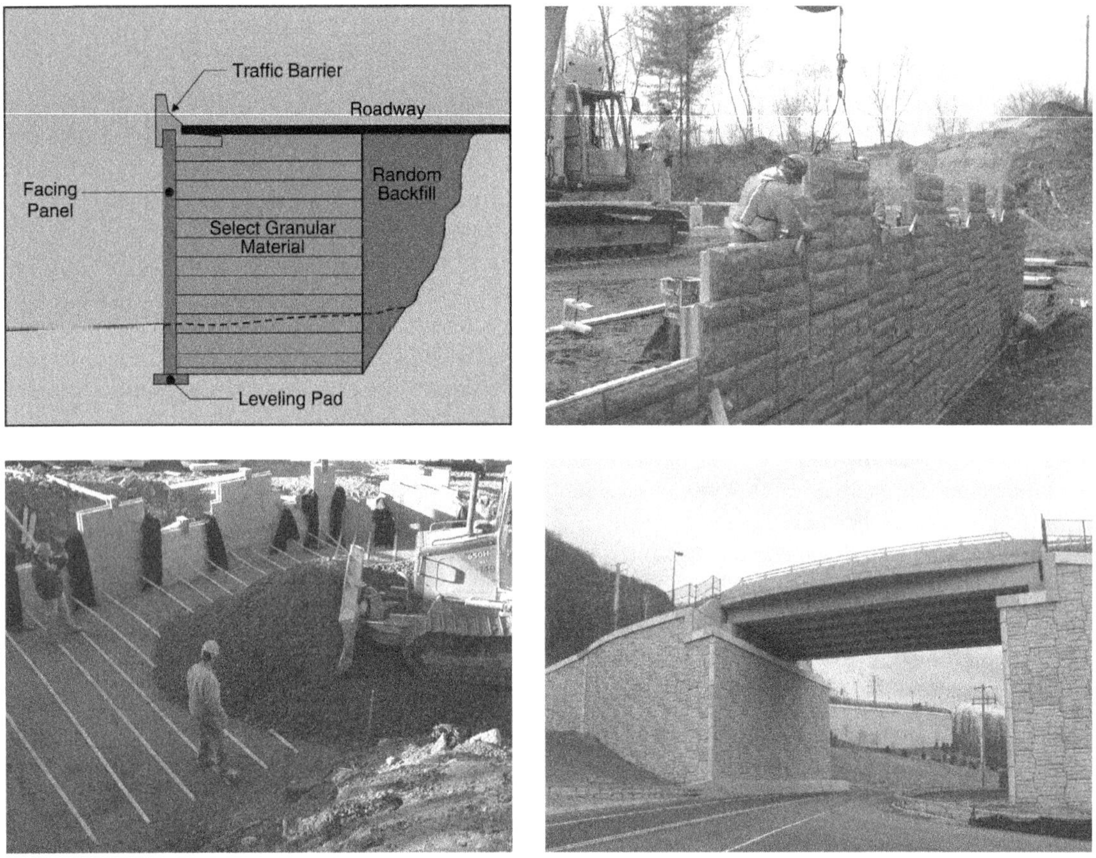

Figure 21.24 MSE wall. (Courtesy of The Reinforced Earth Company.)

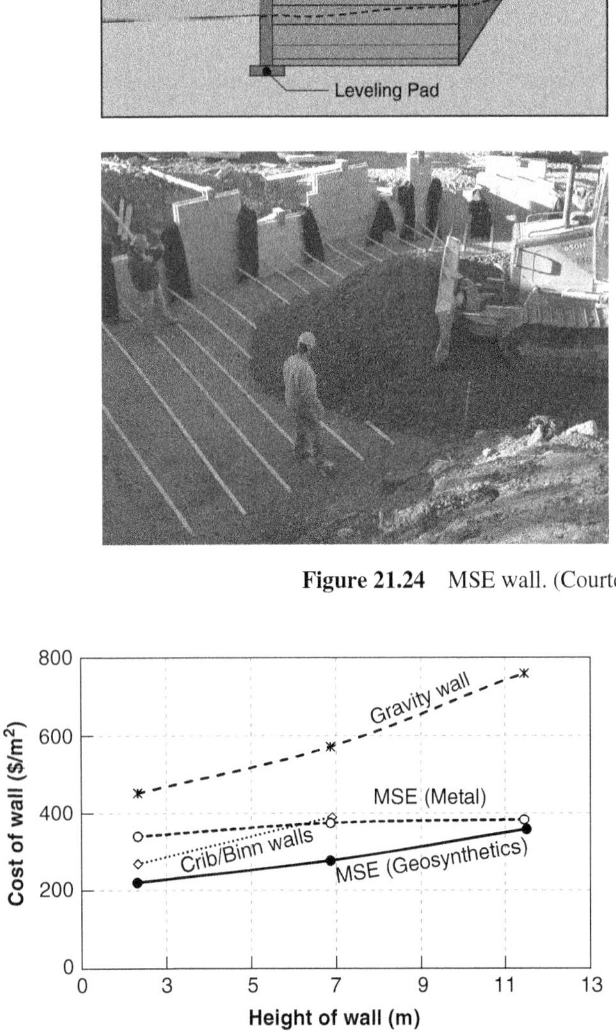

Figure 21.25 Cost of bottom-up walls (After Koerner and Soong 2001.)

Figure 21.26 Load in the reinforcement.

21.10.2 Internal Stability

Pull-out capacity and yield of the reinforcement are the two aspects of internal stability of an MSE wall. Let's address pull-out capacity first.

Pull-Out Design

This design consideration ensures that the load in the reinforcement will not be high enough to pull the reinforcement out of the soil. An understanding of the load distribution in the reinforcement is necessary. Figure 21.26 shows the variation of the tension load T (kN) in the reinforcement as a function of the distance from the front of the wall.

At the wall facing, the load T in the reinforcement is very small, and then it increases as the instability of the wedge of soil near the wall is transferred to the tension T (kN) in the reinforcement. At a distance L_{max} from the front, the tension T reaches a maximum T_{max}. Beyond T_{max}, the tension decreases as the load is transferred to the stable soil mass and reaches zero at a certain distance from the front. The true embedment or anchoring length L_a available to resist the active pressure force against the wall is $L - L_{max}$ where L is the total length of the reinforcement. The design requires knowledge of L_{max}, which is to be ignored in the length required to resist T_{max}. L_{max} is given in Figure 21.26; as can be seen for rigid inclusions, it is equal to 0.3 H in the top half of the wall and decreases to zero at the bottom of the wall. For flexible inclusions (geosynthetics), it is taken as the width of the active wedge. These recommendations are partially based on measurement and simulation data.

The force T_{max} is calculated as follows:

$$T_{max} = s_v s_h \sigma_h \qquad (21.75)$$

where T_{max} is the maximum line load (kN) to be resisted by the layer of reinforcement at depth z, s_v is the vertical spacing between reinforcement layers at depth z, s_h is the horizontal spacing between reinforcement inclusions at depth z, and σ_h is the total horizontal stress at depth z. The stress σ_h is calculated as:

$$\sigma_h = k_r \sigma_{ov} + \Delta\sigma_h \qquad (21.76)$$

where k_r is a coefficient of earth pressure defined in Figure 21.27 as a function of K_a. The reason that k_r is higher than K_a for rigid inclusions is that during compaction of the backfill, the rigid inclusions (e.g., steel strips) can lock in higher horizontal stresses. Flexible inclusions (geosynthetics) do not lock in additional compaction stresses. As a result, k_r is equal to K_a for flexible inclusions.

Now that we have calculated the load T_{max}, we need to find the length of reinforcement that will safely carry this load without pulling out of the soil. The pull-out capacity $T_{pullout}$

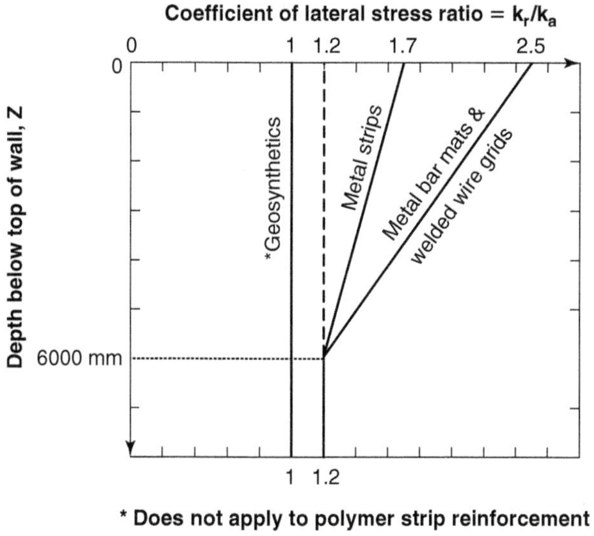

Figure 21.27 Earth pressure coefficient for load in the reinforcement.

(kN) of the reinforcement inclusion is given by:

$$T_{pull\ out} = 2 f_{max} b L_a \qquad (21.77)$$

where f_{max} is the maximum shear stress that can be developed on both sides of the interface between the reinforcement and the soil, b is the width of the inclusion, and L_a is the anchoring length beyond L_{max}, the width of the active failure zone. The shear stress f_{max} is evaluated as follows:

$$f_{max} = F^* \sigma'_v \alpha \qquad (21.78)$$

where F^* is the friction factor given in Figure 21.28; σ'_v is the vertical effective stress on the reinforcement; and α is a scale

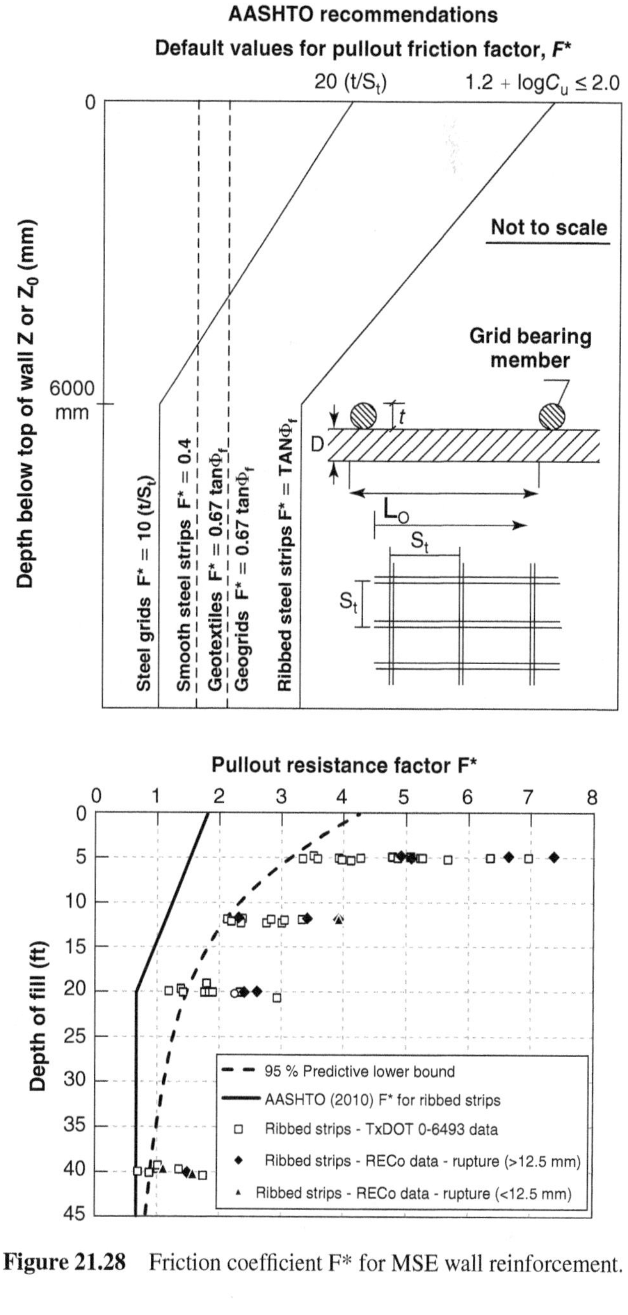

Figure 21.28 Friction coefficient F* for MSE wall reinforcement.

factor taken as 1 for steel reinforcement, 0.8 for geogrids, and 0.6 for geotextiles. Note that although the recommended F^* values can be as high as 2 at the ground surface, values of F^* much higher than 2 have been measured. The reason the coefficient of friction may be higher than 1 is that a combination of friction and bearing capacity is involved in the sliding-out of the reinforcement. The bearing capacity component comes from the protruding ribs for strips and from the transverse bars for grids.

Then the ultimate limit state for pull-out must be satisfied:

$$\gamma_1 T_{\max 1} + \gamma_2 T_{\max 2} = \phi T_{pull\ out} \tag{21.79}$$

where γ_1 is the load factor for the active earth pressure due to soil weight ($\gamma_1 = 1.35$), γ_2 is the load factor for the active earth pressure due to any surcharge on top of the wall ($\gamma_2 = 1.50$), φ is the resistance factor ($\varphi = 0.9$), $T_{\max 1}$ is the part of the load in the reinforcement due to the soil weight, $T_{\max 2}$ is the part of the load in the reinforcement due to any surcharge on top of the wall, and $T_{pullout}$ is the pull-out resistance calculated in Eq. 21.77. The required safe length L_a of the reinforcement is given by:

$$L_a = \frac{(\gamma_1 k_r \sigma'_{ov} + \gamma_2 \Delta\sigma_h) s_v s_h}{2\varphi F * \sigma'_{ov} \alpha b} \tag{21.80}$$

The total length L of the reinforcement is largest at the top of the wall, but because it is common practice to keep the reinforcement length L constant, L is given by:

$$L = L_a + L_{\max} = \frac{(\gamma_1 k_r \sigma'_{ov} + \gamma_2 \Delta\sigma_h) s_v s_h}{2\varphi F * \sigma'_{ov} \alpha b} + 0.3H \tag{21.81}$$

In the simple case, where $k_r = K_a = 0.33$, there is no surcharge, s_v and s_h are 0.75 m, F^* is 1, α is 1, and b is 0.05 m, then the length L_a is 2.8 m and independent of the wall height H. The reason is that the vertical stress contributes equally to the load and the resistance. The total length of reinforcement is $L = 2.8 + 0.3H$, where H is the height of the wall. For a 7 m high wall (common case of an overpass), L is approximately $0.7H$, which is a common recommendation. For higher walls, the 0.7 H rule is conservative, and for smaller walls a minimum of about 3 m reinforcement length is imposed.

Yield of the Reinforcement Design

We need to make sure that the reinforcement can safely carry the load $T_{\max}$ without yielding or rupturing. For this, we write the ultimate limit state as:

$$\gamma_1 T_{\max 1} + \gamma_2 T_{\max 2} = \phi T_{yield} \tag{21.82}$$

where γ_1 is the load factor for the active earth pressure due to soil weight ($\gamma_1 = 1.35$), γ_2 is the load factor for the active earth pressure due to any surcharge on top of the wall ($\gamma_2 = 1.50$), φ is the resistance factor ($\varphi = 0.75$ for strips, 0.65 for grids, and 0.9 for geosynthetics), $T_{\max 1}$ is the part of the load in the reinforcement due to the soil weight, $T_{\max 2}$ is the part of the load in the reinforcement due to any surcharge on top of the wall, and T_{yield} is the load corresponding to the yield strength of the reinforcement. T_{yield} for steel reinforcement is given by:

$$T_{yield} = \sigma_{yield} A \tag{21.83}$$

where σ_{yield} is the yield strength of the reinforcement and A is the cross-sectional area.

For geosynthetics, see section 27.6.2. For steel reinforcement, one issue is corrosion. This is addressed by using a thickness larger than required by the ultimate limit state for yield. Corrosion rates for nonaggressive soils are in the range of 0.005 to 0.015 mm/yr (AASHTO 2007). This means that a 1 mm excess thickness corresponds to a typical 75-year design life. Nonaggressive soils are recommended for backfill and are defined in Table 21.2.

Table 21.2 Characteristics of Nonaggressive Soils for Corrosion

pH	5 to 10
Resistivity	>3000 Ohm.cm
Chlorides	<100 ppm
Sulfates	<200 ppm
Organic content	<1%

(After AASHTO 2007.)

Movement

The movement of MSE walls is not typically calculated. If necessary, the settlement should be checked according to the procedures outlined in sections 17.7 and 17.8 and discussed in section 21.9, design step 11. The maximum horizontal movement $\Delta_{\max}$ of MSE walls during construction can be estimated for normal conditions and little or no surcharge as follows:

$$\text{For rigid inclusions} \quad \Delta_{\max} = 0.004 H \delta_r \tag{21.84}$$

$$\text{For flexible inclusions} \quad \Delta_{\max} = 0.013 H \delta_r \tag{21.85}$$

where H is the height of the wall and δ_r is given in Figure 21.29.

21.11 CANTILEVER TOP-DOWN WALLS

Cantilever walls are top-down walls, though they are sometimes confused with cantilever gravity walls. They are made, for example, of bored piles drilled side by side or sheet pile Z sections driven side by side (Figures 21.30 and 21.31). They can be used to retain soil up to a height of about 7 m; beyond that height anchored walls are more economical. The design

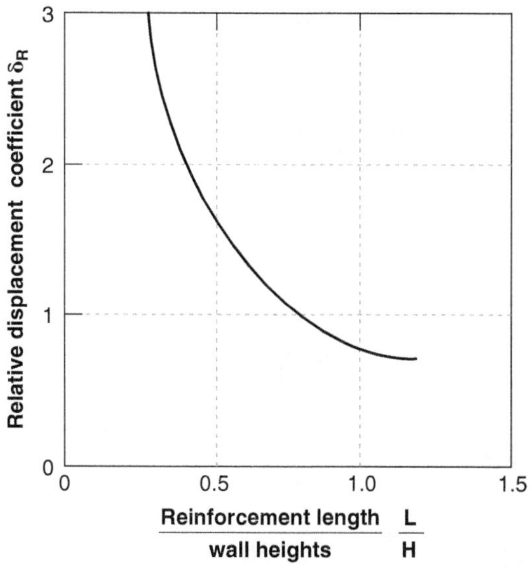

Figure 21.29 Movement parameter for estimating MSE wall horizontal displacement during construction (AASHTO 2010).

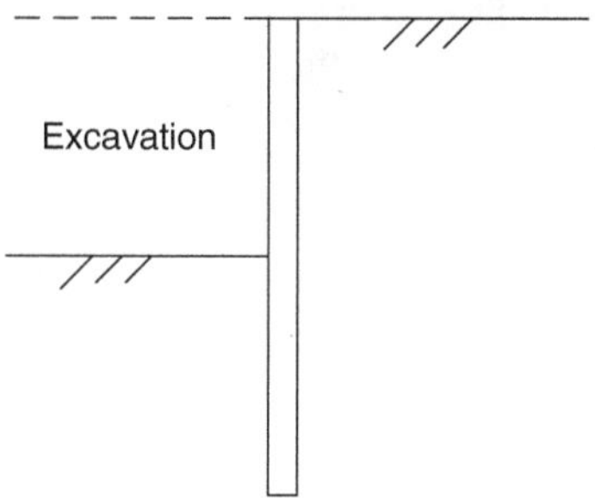

Figure 21.30 Cantilever top-down wall.

of such walls consists of satisfying the ultimate limit state (safety) and the serviceability limit state (limited movement). The parameters to be selected in the design are the depth of embedment D and the section of the wall to resist the maximum bending moment.

21.11.1 Depth of Embedment and Pressure Diagram

The pressure diagram is the first step. It is assumed that the wall will move into the excavation by an amount sufficient to generate the active earth pressure behind the wall, and that part of the passive earth pressure will be generated in front of the wall to resist the push (Figure 21.32). Of course, both the active push and the passive resistance depend on the depth of embedment D.

In the simplest case (no water, no surcharge, one uniform soil with no cohesion), the active push P_a (kN/m) is:

$$P_a = \frac{1}{2} K_a \gamma (H + D)^2 \tag{21.86}$$

where K_a is the active earth pressure coefficient, γ is the unit weight of the retained soil, H is the excavation height, and D is the depth of embedment. On the passive side, the passive pressure diagram is truncated at a depth where the passive pressure reaches half of the passive pressure at the embedment depth. This is done to acknowledge that the movement decreases with depth and may not be sufficient to generate the complete passive pressure at depth. This assumption brings into play the concept of both safety and serviceability, although it does not address that concept directly. As a result, the mobilized passive resistance P_{pm} is given by:

$$P_{pm} = \frac{3}{8} K_p \gamma D^2 \tag{21.87}$$

The point of application of P_a and P_{pm} are at a distance X_a and X_{pm} from the bottom of the wall respectively:

$$X_a = \frac{1}{3}(H + D) \tag{21.88}$$

$$X_{pm} = \frac{7}{18} D \tag{21.89}$$

Now we might be tempted to write horizontal equilibrium and we would find a depth D. The problem is that, even if we satisfied $P_a = P_{pm}$, the wall still could not be in moment equilibrium. For moment equilibrium to be satisfied, a force R is necessary at the bottom of the wall, and comes from the deflection pattern (Figure 21.32). By writing moment equilibrium around the bottom of the wall, we get the equation that leads to the value of D:

$$P_a X_a - P_{pm} X_{pm} = 0 \tag{21.90}$$

or

$$D = \frac{H}{\left(\frac{7}{8}\frac{K_p}{K_a}\right)^{0.33} - 1} \tag{21.91}$$

For a common ratio of K_p/K_a equal to 10, then:

$$D \approx H \tag{21.92}$$

This result shows that cantilever walls need an embedment at least equal to the excavation height. More detailed analysis shows that $D = 1.2H$ is more appropriate as a minimum for a uniform soil. Of course, more complex soil layering, surcharge, and water conditions will lead to a different result.

21.11.2 Displacement of the Wall, Bending Moment, and P-y Curves

The calculations shown in section 21.11.1 can give an estimate of the embedment depth D. Then the horizontal displacement of the wall and bending moment profile in the wall can be calculated by using a P-y curve analysis (see sections 18.6.8 and 11.4.4). The parameter P represents the load on the wall at depth z and the parameter y represents the horizontal deflection of the wall from the unloaded position. In the P-y curve analysis, a repeatable width of wall, usually one meter width,

Figure 21.31 Bored piles and sheet pile cantilever walls. (d: Courtesy of Associated Pacific Constructors, Inc.)

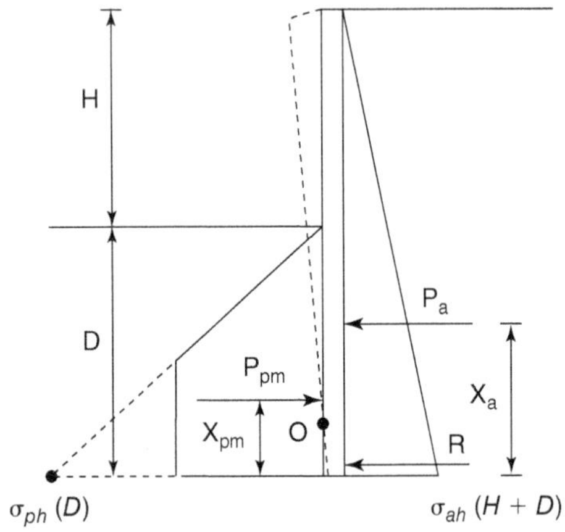

Figure 21.32 Simple approach for cantilever walls.

is simulated as a structural member and the soil mass is simulated by a series of nonlinear springs (*P-y* curves) tied to the wall and describing the response of the soil to the wall deflection. The first step in this analysis problem is to discretize the wall into elements (Figure 21.33); a minimum of 10 elements is recommended. The input to the problem includes.

1. Length of the wall (excavation height H plus depth of embedment D).
2. Length of the elements ($\Delta H < (H + D)/10$).
3. Bending stiffness EI of the wall (E modulus of elasticity, I moment of inertia) for the cross section corresponding to the repeatable wall section. This is usually a one meter width for continuous walls or the section tributary to one pile if a line of pile is involved.
4. P-y curves as a function of depth (one curve at each node).

The governing differential equation (GDE) and its finite difference method (FDM) solution are described in section

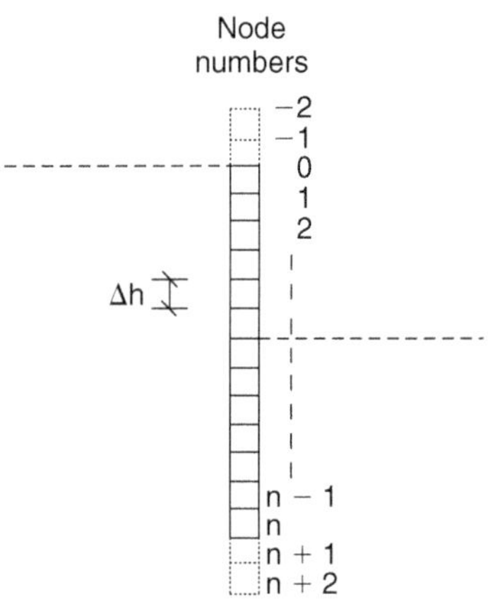

Figure 21.33 Cantilever wall discretized into elements.

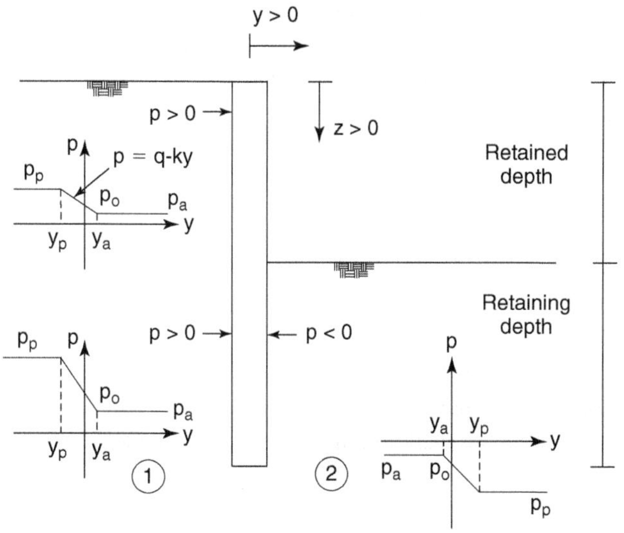

Figure 21.34 P-y curves for a cantilever top-down wall.

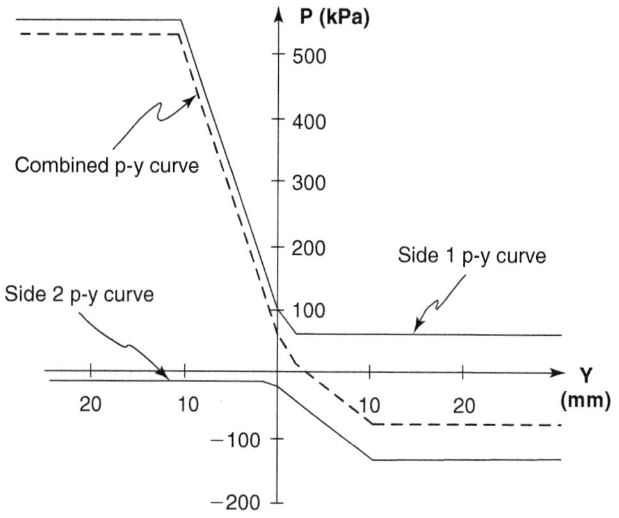

Figure 21.35 Combining P-y curves below the excavation level.

11.5.2 with a solved example. The *P-y* curves are constructed as follows (Figure 21.34). Above the excavation level in the retained soil zone, the soil is on only one side of the wall and the *P-y* curve is as shown in Figure 21.34. In that zone, if the wall moves toward the soil ($y < 0$), the *P* value increases from the P_o value corresponding to the at-rest earth pressure to the P_p value corresponding to the passive earth pressure when a displacement y_p is reached. In all cases the *P* values are given by:

$$P = \sigma_h \times b \times \Delta h \tag{21.93}$$

where σ_h is the horizontal stress (obtained as discussed in the previous sections in this chapter), Δh is the wall element length (vertical), and b is the width of the repeatable section (horizontal). The displacement y_p can be estimated by using Table 21.1. In the retained soil zone, the soil pushes in the chosen positive direction; therefore *P* is positive. In that zone also, if the wall moves away from the soil, the *P* value decreases from the P_o value corresponding to the at-rest earth pressure to the P_a value corresponding to the active earth pressure. The soil still pushes in the positive direction.

Below the excavation level, the soil is on both sides of the wall and there are two P-y curves: one for the retained soil side (Side 1 in Figure 21.34) and one for the retaining soil side (Side 2 in Figure 21.34). The *P-y* curve on Side 1 is similar to the one above the excavation except that the values of *P* are higher, because the depth is larger. The *P-y* curve on Side 2 is prepared as follows. If the wall moves toward Side 1 ($y < 0$), the magnitude of the *P* value decreases from the P_o value corresponding to the at-rest earth pressure to the P_a value corresponding to the active earth pressure. Because the soil pushes in a direction opposite to the chosen positive direction, *P* is negative. If the wall moves toward Side 2 ($y > 0$), the magnitude of the *P* value increases from the P_o value corresponding to the at-rest earth pressure to the P_p value corresponding to the passive earth pressure. Because the soil pushes in a direction opposite to the chosen positive direction, *P* is negative. The net *P-y* curve for the zone below the excavation level is constructed by combining the two curves for Side 1 and Side 2 (Figure 21.35).

Note that this *P-y* curve preparation is done for each node along the discretized wall. Then a finite difference program or spreadsheet is used and the solution gives the following parameters as a function of depth: wall deflection *y*(*z*), slope of the wall *y*′(*z*), bending moment in the wall *M*(*z*), shear force in the wall *V*(*z*), and pressure on the wall *p*(*z*). Sample outputs are shown in Figures 21.36 and 21.37. The deflection profile predicted by this method tends to underpredict the deflections observed in practice. The reason is that the mass movement of the retained soil is not included in the *P-y* curve in a theoretically sound manner. However, the bending moment profile predicted by this method and the maximum bending moment for design are much more consistently reliable than any hand calculation based on an assumed pressure distribution, such as shown in Figure 21.32 (Briaud and Kim 1998). For improved prediction of deflections including mass movement, the finite element method should be used; nevertheless, a problem remains concerning the quality of the input parameters and the selection of the soil model.

21.12 ANCHORED WALLS AND STRUTTED WALLS

Anchored walls (or tieback walls) and strutted walls are top-down walls (Figure 21.38). The wall portion may be a solid concrete wall built by the slurry wall method, a sheet pile wall, a bored pile wall, a deep soil mixing wall, or a soldier pile and lagging wall, to name a few. Concrete slurry walls are built by excavating the soil one rectangular panel at a time with a clamshell rig and under slurry if necessary, lowering the reinforcing cage into the slurry-filled hole, and placing the concrete in liquid paste through a tremie (tube that goes to the bottom of the hole) from the bottom of the panel to the top while displacing the slurry out of the rectangular hole

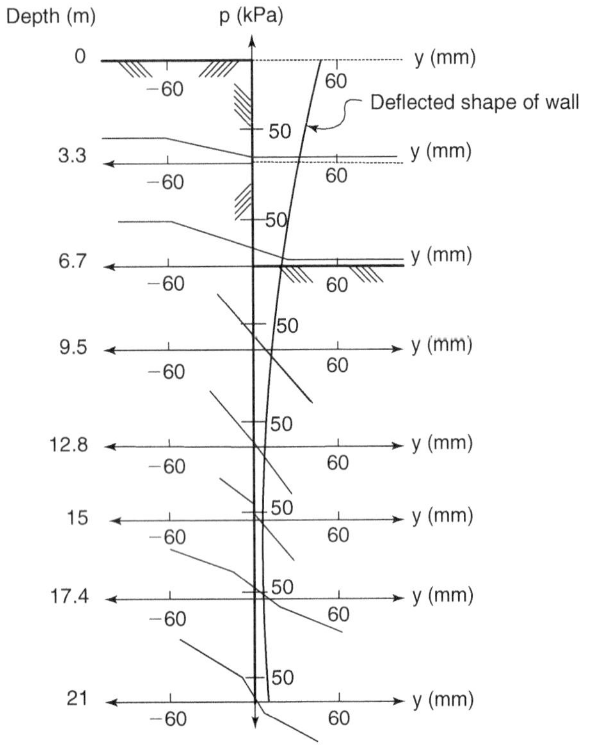

Figure 21.36 P-y curves and deflected shape of cantilever wall (Briaud et al. 1983).

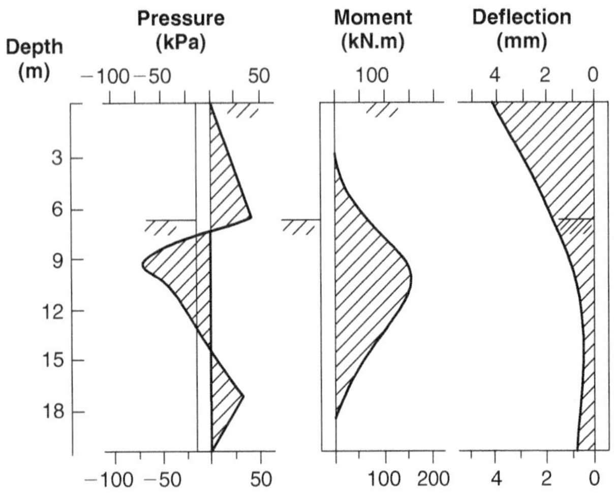

Figure 21.37 Pressure, bending moment, and deflection (Briaud et al. 1983).

(Figure 21.38). Bored pile walls and sheet pile walls were discussed in section 21.11. In this section they are anchored to be able to retain larger depth of soil. Deep soil mixing walls are like bored pile walls except that the bored piles are drilled by mixing the soil with about 20% cement; the resulting piles are not as strong, but they are less expensive. Soldier pile and lagging walls are constructed by driving or drilling piles in line on a 2 to 3 meter spacing and excavating in front of this line of piles while placing wood lagging to retain the soil between piles. Anchored and strutted walls are very convenient in tight settings like urban areas because they do not require much space for construction. On the one hand, struts clutter the excavation; on the other hand, anchors may hit underground utility lines.

The design of anchored walls and strutted walls includes many parts, with the main ones being estimating the pressure distribution behind the wall, calculating the anchor or strut loads, calculating the maximum bending moment in the wall, estimating the horizontal and vertical movements, and calculating the necessary length of anchors.

21.12.1 Pressure Distribution

Consider the pressure distribution behind a cantilever top-down retaining wall with the active pressure on the retained side (Figure 21.32). If you install an anchor within the excavated depth to hold the wall back, and if you stress that anchor in tension, the anchor head (plate) is going to press against the wall while you pull on the tendon, thereby increasing the local pressure (Figure 21.39). As a result, the pressure behind the anchor will be higher than the active pressure and will correspond to the prestressing load of the anchor or the strut.

It is very common to stress all the anchors to the same load, so that the pressure behind the wall in the retained soil depth (above excavation level) is nearly constant and equal to the sum of the anchor loads divided by the retained soil area. This is what led Terzaghi et al. (1996) to recommend a constant pressure diagram for strutted walls. Based on full-scale measurements, they recommended pressure diagrams for sand, for soft to medium clays, and for stiff fissured clays (Figure 21.40). The maximum total pressure σ_h is as follows:

For sands $\quad \sigma_h = 0.65 K_0 \sigma'_{ov} + u_w \quad (21.94)$

For soft to medium clays $\quad \sigma_h = \gamma H - 4\,\text{m} s_u \quad (21.95)$

For stiff fissured clay $\quad \sigma_h = 0.2\gamma H \text{ to } 0.4\gamma H + u_w(?) \quad (21.96)$

where K_a is the coefficient of earth pressure at rest, σ'_{ov} is the effective vertical stress on the retained soil (sand) side at the bottom of the excavation, γ is the total unit weight of the clay, H is the height of the excavation, s_u is the clay undrained shear strength, and m is a parameter that depends on the depth of the soft to medium clay layer below the excavation. It is taken as equal to 1 if the soft to medium clay layer stops at the bottom of the excavation, and as equal to 0.4 if the clay layer goes much deeper than the bottom of the excavation. Note that for sand, the analysis is an effective stress analysis and the water pressure must be added if water is present. For soft to medium clay, the analysis is an undrained analysis and the water pressure is included in γH. For stiff fissured clays, the coefficient 0.2 would correspond to less fissured clays and 0.4 to more fissured clays. Also, if the fissures are large enough that water will exert pressure on the wall, the water pressure must be added.

(a) Slurry wall digging the wall
(b) Slurry wall lowering cage
(c) Slurry wall final with anchors
(d) Deep soil mixing wall
(e) Soldier pile and lagging wall
(f) Strutted wall
(g)

Figure 21.38 Various anchored and strutted wall techniques. (a, b, c, d: Courtesy of Nicholson Construction; e, f, g: Courtesy of Schnabel Foundation Company.)

21.12.2 Pressure vs. Movement

Briaud and Kim (1998) collected a number of full-scale case histories on anchored walls and performed numerical simulations. For the case histories, the anchor loads were known, as were the horizontal deflections of the wall. The mean pressure σ_h behind the wall was calculated as the ratio of the sum of the individual anchor loads F_i divided by the total wall area A of soil retained by the anchors:

$$\sigma_h = \frac{\sum_{i=1}^{n} F_i}{A} \tag{21.97}$$

The mean pressure σ_h behind the wall was associated with the horizontal movement at the top of the wall u_{top} and the mean horizontal deflection u_{mean}. Note that one case history led to more than one combination of pressure and displacement, as the construction sequence included several excavation levels and several anchor installations. The earth pressure coefficient K was calculated as the ratio of the mean pressure σ_h over the vertical effective stress behind the wall at the bottom of the excavation:

$$K = \frac{\sigma_h}{\sigma'_{ov}(at\ z = \mathrm{H})} \tag{21.98}$$

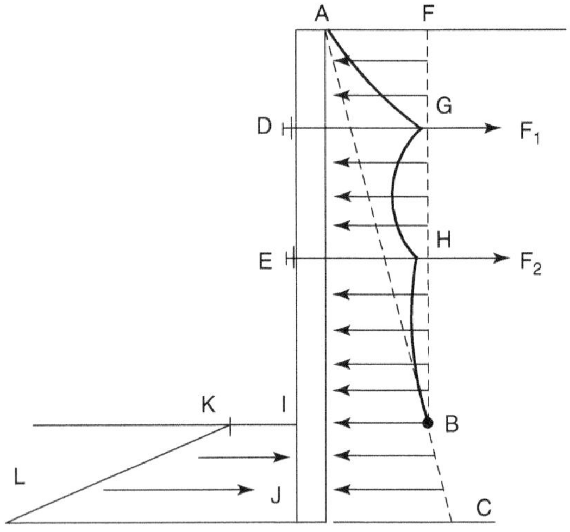

Figure 21.39 Influence of anchor stressing on pressure diagram.

Figure 21.41 shows the range of values of K versus u_{top}/H and K versus u_{mean}/H.

Terzaghi and Peck's earth pressure value of $0.65K_a\gamma\text{H}$ for strutted excavations in sand leads to a K value of 0.21 if the friction angle is 30° ($K_a = 0.33$). For a K value of 0.21, Figure 21.41 gives a range of u_{top}/H from 0.002 to 0.0045 and u_{mean}/H from 0.0015 to 0.0035. However, in the case of anchored walls, the engineer can choose the wall deflection by choosing the anchor loads. Indeed, if the anchor loads are very high, the wall could actually move back and go into passive resistance. In contrast, if the anchor loads are very low, there will be a lot of wall deflection toward the excavation. Figure 21.41 helps the engineer to select a K factor that will generate a targeted amount of wall movement. It appears that a K value of about 0.4 will lead to minimal displacements. For a given wall height H and for a chosen horizontal displacement u_{top} or u_{mean}, the total earth pressure σ_h at depth z is calculated according to:

$$\sigma_h = K\sigma'_{ov}(\text{at z} = \text{H}) + u_w \tag{21.99}$$

where K is read on Figure 21.41 at the corresponding relative displacement, σ'_{ov} is the vertical effective stress at the bottom of the wall, and u_w is the water pressure at depth z. Note that the term $K\sigma'_{ov}$ is a constant independent of depth, whereas u_w increases with depth (Figure 21.42).

21.12.3 Base Instability

In the case of clays, one concern is an inverted bearing capacity failure. In the case of sands, the concern is a loss of effective stress and the development of a quick condition at the bottom of the excavation. In clays, the bottom of the excavation may be unstable if the soil is not strong enough to sustain the lack of overburden on the excavated side. The factor of safety F against base instability is (Figure 21.43):

$$F = \frac{N_c s_u}{\sigma_{ov}(\text{at z} = \text{H})} \tag{21.100}$$

where N_c is a bearing capacity factor for a strip footing (Figure 17.7), s_u is the undrained shear strength, and $\sigma_{ov(\text{at } z = H)}$ is the vertical total stress behind the bottom of the wall.

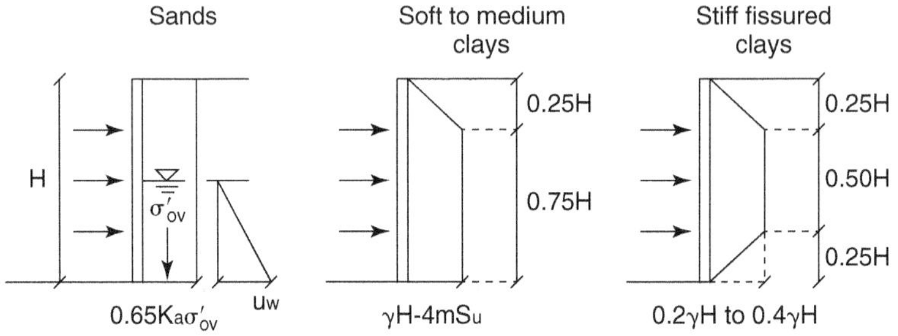

Figure 21.40 Pressure distribution for strutted walls. (After Terzaghi et al. 1996.)

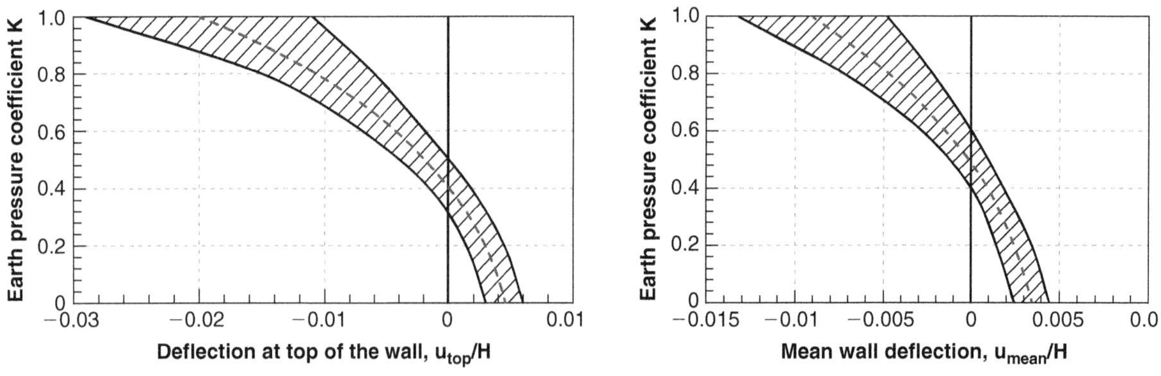

Figure 21.41 Measured earth pressure coefficient versus normalized displacement of the wall (Briaud and Kim 1998).

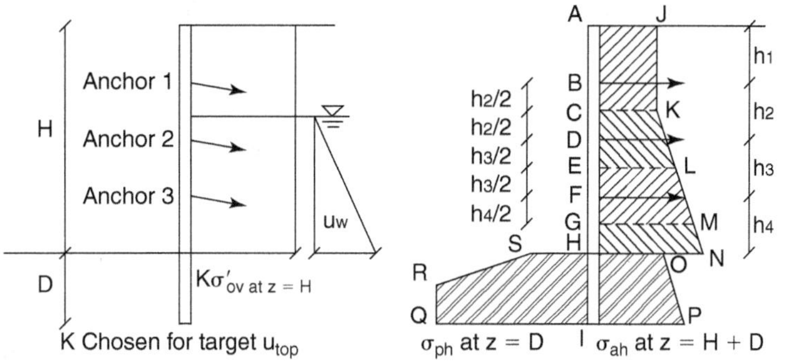

Figure 21.42 Pressure diagram for anchored walls.

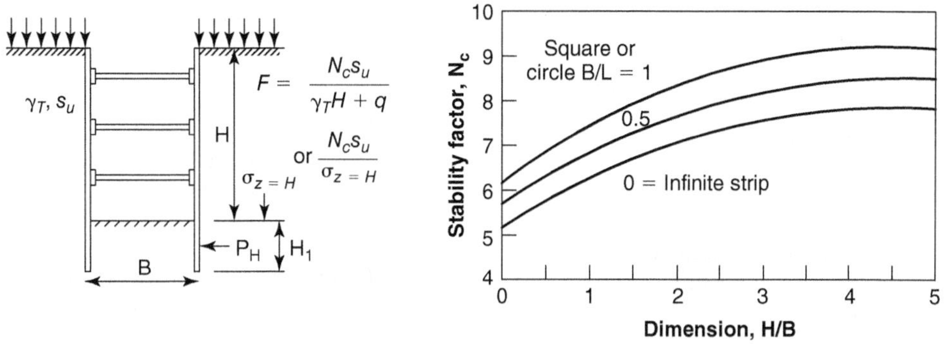

Figure 21.43 Base instability.

21.12.4 Movement of Wall and Ground Surface

The general shape of the deformed wall and adjacent ground surface has two components (Figure 21.44): a cantilever movement and a movement associated with deep deformations. The first one is associated with the lack of lateral support leading to the soil mass leaning into the excavation, and the second with the slope stability/bearing capacity type of deformation deeper in the soil mass. Predicting these displacements is not simple. Peck (1969) collected data on the movement of the ground surface near excavations and presented it in a very useful fashion (Figure 21.45). Peck divided the behavior according to soil type and to the value of the factor of safety *F* against base instability. He showed that the maximum settlement of the top of the wall can reach 0.01 H for excavations in sand or in soft to hard clay, it can reach 0.02 H for soft clays when *F* is larger than 1.3, and it can be larger than 0.02 H for soft clays when *F* is less than 1.3. Regarding the lateral extent over which the ground surface would be depressed, Peck found that it could be up to 2 H for excavations in sand or in soft to hard clay, it could be up to 4 H for soft clays when *F* is larger than 1.3, and larger than 4 H for soft clays when *F* is less than 1.3.

Clough and O'Rourke (1990) collected additional data and revised Peck's plots accordingly (Figure 21.46). In their work, Clough and O'Rourke also proposed a method to predict the maximum lateral movement of the wall depending on the relative stiffness *L* of the wall and the factor of safety *F*

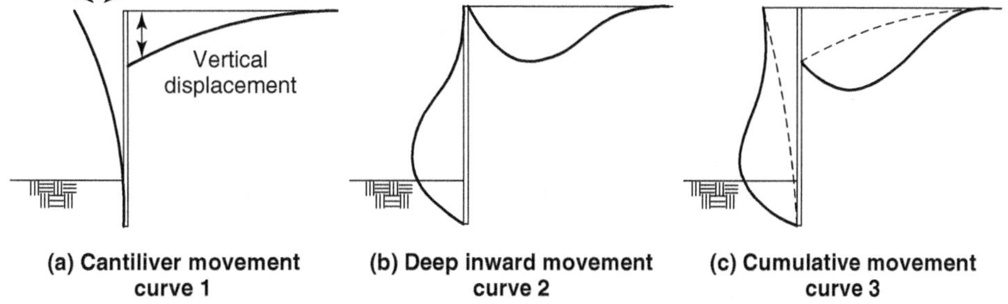

Figure 21.44 Components of excavation movements.

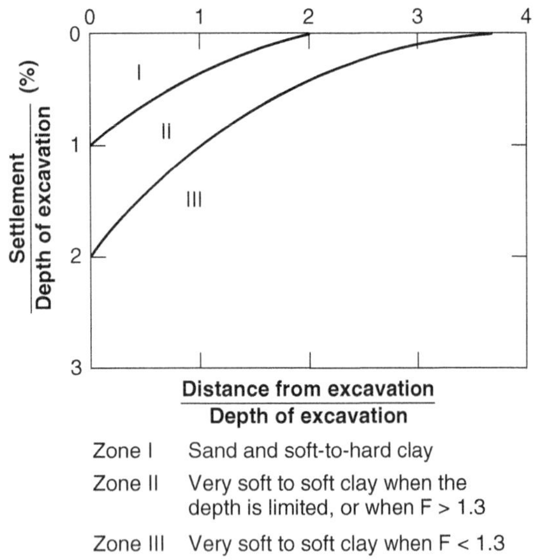

Figure 21.45 Peck diagram for ground surface settlement (Peck 1969).

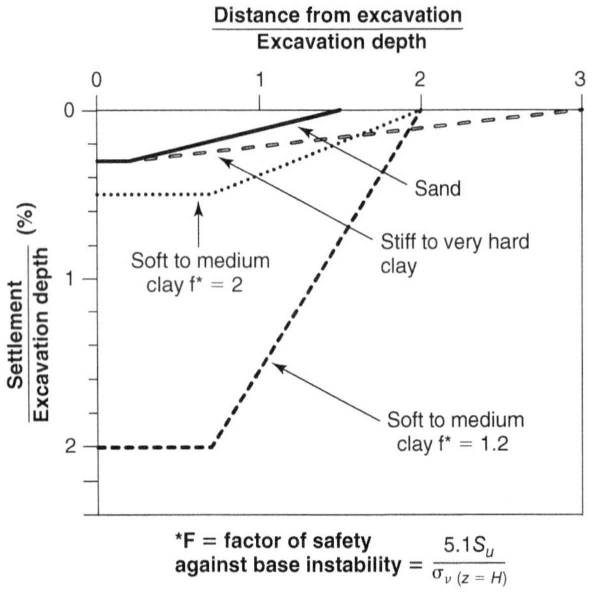

Figure 21.46 Clough and O'Rourke diagram for ground surface settlement. (After Clough and O'Rourke 1990.)

against base instability (Figure 21.47). The relative stiffness L(m) is defined as:

$$L = \frac{EI}{\gamma h^4} \tag{21.101}$$

where E is the modulus of elasticity of the wall material, I is its moment of inertia around the bending moment axis, γ is the unit weight of the soil, and h is the average vertical distance between anchors. For a given case, the relative stiffness L and the factor of safety for base instability are defined, the correct curve on Figure 21.47 is selected, and

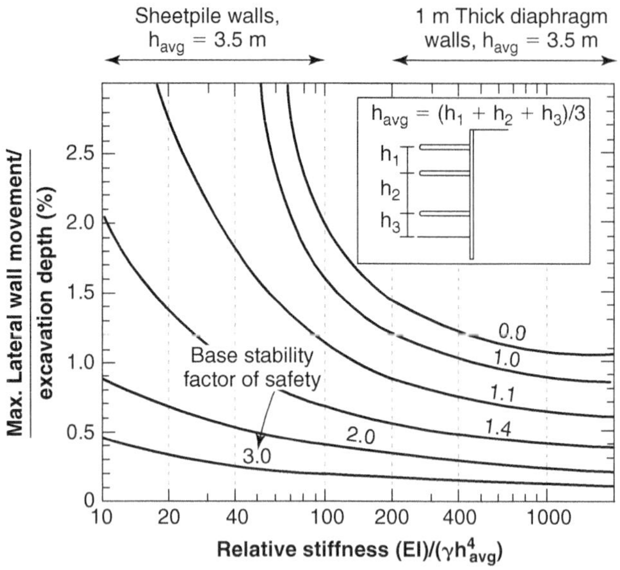

Figure 21.47 Clough and O'Rourke chart for maximum lateral movement. (After Clough and O'Rourke 1990)

the corresponding ratio between the maximum horizontal deflection and the excavation height is read on the vertical axis. In general, the vertical and horizontal displacements of excavations are on the same order of magnitude unless the soil is very dilatant or collapsible. Some of the ways to decrease movements are to place the first anchor as shallow and as early as reasonably possible and to use high anchor loads ($K = 0.4$ in Figure 21.41).

21.12.5 Anchors

Anchors can be constructed in different ways, but the most common way (Figure 21.48) is to drill a hole through the wall when the anchor depth is reached, insert a rod or multiple-strand cable in the open hole with centralizers, fill the hole with grout, wait for the grout to set, then tension the anchor, subject it to a proof test, and then lock the anchor at the design load. Sometimes a second injection of grout is performed through tubes left in place during the first injection to increase the anchor capacity. The rod or strand is in a bond-breaking sheath from the anchor head to a certain distance called the *tendon unbonded length* L_u. The sheath stops at L_u; the rest of the rod or strand is barren and is called the *tendon bond anchor length* L_b. The length of the anchor in the active wedge is called the *discounted anchor length* L_d. The rest of the anchor is called the *anchor bond length* L_a. The total length of the anchor is L_t:

$$L_t = L_u + L_b = L_d + L_a \tag{21.102}$$

The length L_d is taken as the length of the anchor within the active wedge behind the wall (Figure 21.48). An example

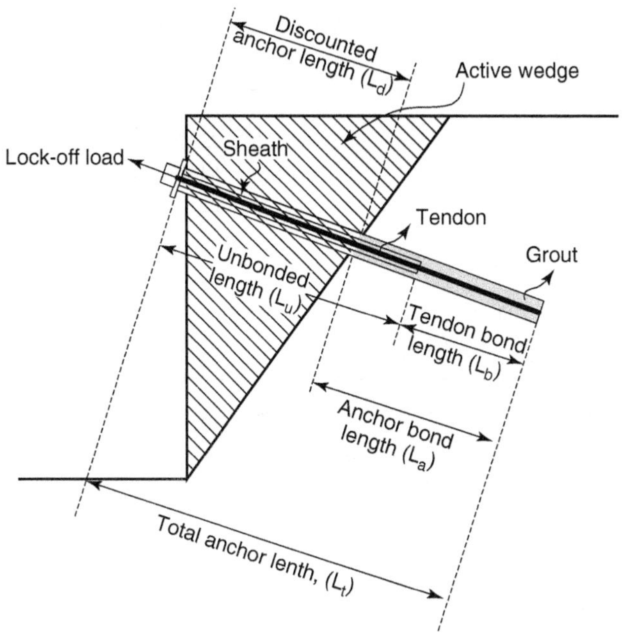

Figure 21.48 Anchor or tieback.

of load distribution in the anchor under tension is shown in Figure 21.49. A long unbonded length is best for anchors in tension because it maximizes the length of grout in compression (Briaud et al. 1998).

The design of anchors or tiebacks has two parts: calculating the anchor loads and calculating the required anchor capacity and associated length. The anchor load is determined by using the tributary method. Once the pressure diagram is obtained (Figure 21.42), the horizontal component A_{hi} of the load in anchor i is obtained by using the part of the pressure diagram tributary to anchor i. For example, the tributary area of the pressure diagram in Figure 21.42 for anchor 2 is CKLE. The expression is:

$$A_{hi} = p_i \left(\frac{h_i}{2} + \frac{h_{i+1}}{2} \right) s_h \tag{21.103}$$

where A_{hi} is the horizontal component of the anchor load A_i, p_i is the mean pressure behind the wall within the tributary depth, h_i is the anchor spacing above anchor i, h_{i+1} is the anchor spacing below anchor i, and s_h is the anchor spacing

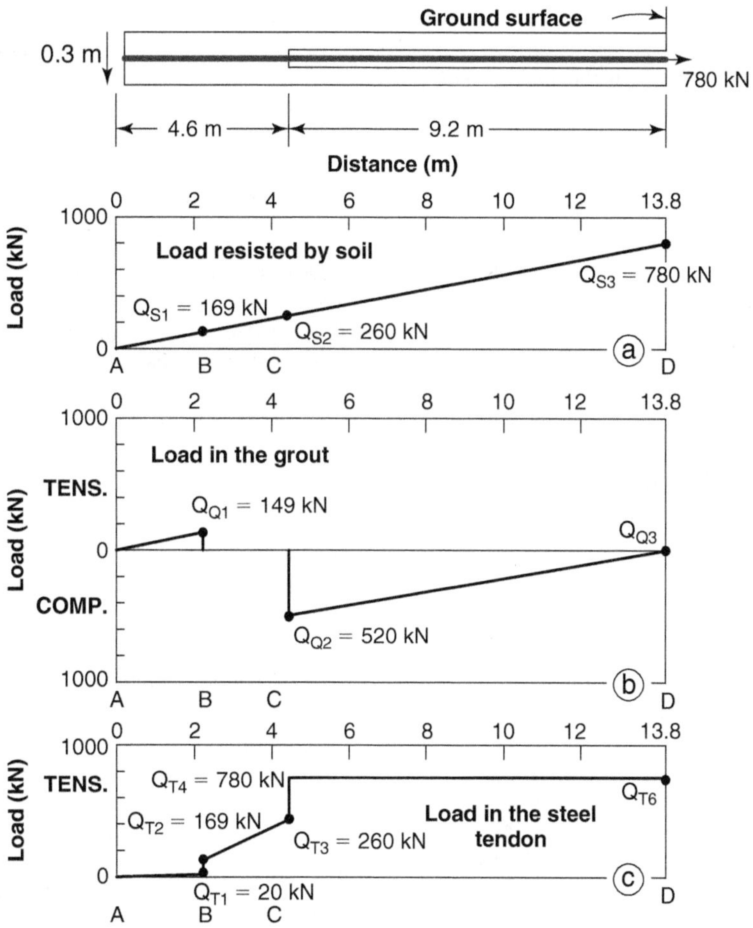

Figure 21.49 Example of load distribution in an anchor (Briaud et al. 1998).

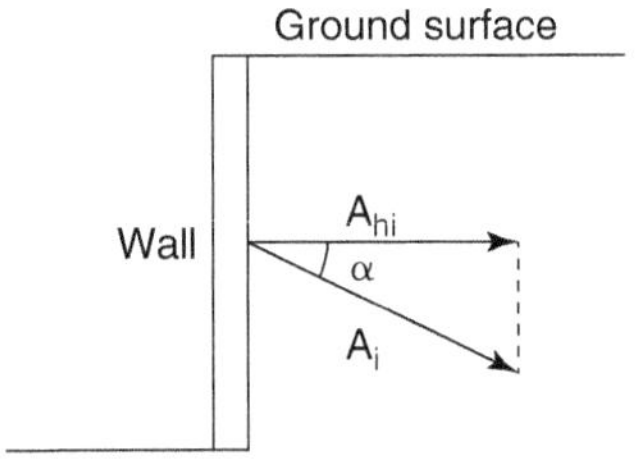

Figure 21.50 Anchor load components.

in the horizontal direction. Equation 21.103 applies to all anchor loads except the top anchor, where it becomes:

$$A_{h1} = p_1\left(h_1 + \frac{h_2}{2}\right)s_h \tag{21.104}$$

Often the anchor is not horizontal, but rather inclined at an angle α to the horizontal (15° to 30°). Thus, the anchor load A_i is (Figure 21.50):

$$A_i = \frac{A_{hi}}{\cos\alpha} \tag{21.105}$$

Once the anchor load is determined, the anchor resistance and length can be calculated. The LRFD equation gives:

$$\gamma A_1 = \varphi R_1 \tag{21.106}$$

where γ is the load factor ($\gamma = 1.35$), A_i is the anchor load, φ is the resistance factor, and R_i is the ultimate resistance of the anchor. If anchors are not proof tested, then φ is between 0.35 and 0.45. However, all anchors are usually proof tested and therefore there is little uncertainty as to the anchor capacity; in that case a resistance factor close to 1 can be used.

Once R_i is obtained, the length of anchor necessary to obtain R_i is calculated. The design is very similar to the case of a pile in tension, and R_i is given by:

$$R_i = \pi D L_a f_{\max} = F_{\max} L_a \tag{21.107}$$

where D is the diameter of the anchor, L_a is the anchor bond length, $f_{\max}$ is the shear strength of the soil-grout interface, and $F_{\max}$ is the maximum load that can be resisted per unit length of anchor. The parameter $f_{\max}$ is estimated as follows for various soils:

For sand and gravel $f_{\max} = \alpha_s \sigma'_{ov}$ with α_s from Table 21.3 (21.108)

For silts and clays $f_{\max} = \alpha_c s_u$ with α_c from Figure 21.51 (21.109)

Tables 21.4, 21.5, and 21.6 present some presumptive values of $f_{\max}$ as recommended by AASHTO (2007). Furthermore, the values of $F_{\max}$ in Table 21.7 can be used for anchors satisfying the following criteria:

- Diameter between 150 to 200 mm
- Grout pressure of about 1000 kPa

Table 21.3 Values of α_s Anchorage Factor for Sand and Gravel

	Relative Density		
Soil Type	Loose	Medium	Dense
Silt	0.1	0.4	1.0
Fine sand	0.2	0.6	1.5
Medium sand	0.5	1.2	2.0
Coarse sand, gravel	1.0	2.0	3.0

(Canadian Foundation Manual 2007.)

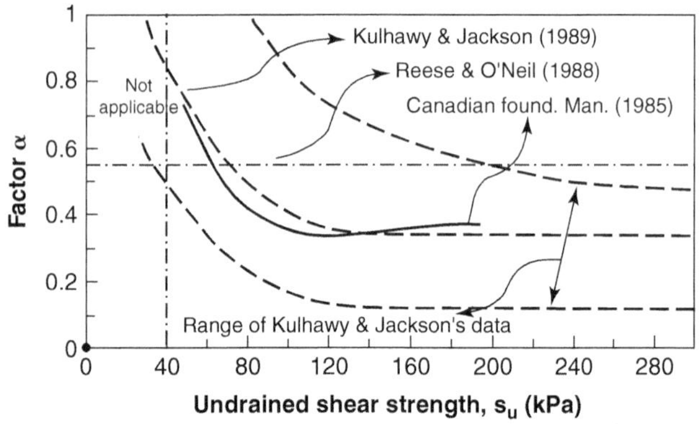

Figure 21.51 α factor for grouted anchors in clay (Briaud et al. 1998).

- Center-to-center spacing vertically and horizontally larger than 4 anchor diameter D

Design rules based on pressuremeter data for calculating the ultimate resistance of anchors also exist (Briaud 1992). These rules, established by the LCPC in France, make a distinction between several construction techniques for the anchors.

Typically, all anchors are tested after installation and curing time. These tests include proof tests, performance tests, creep tests, and 70-day load-hold tests (Briaud et al. 1998). The proof test is the most common and consists of increasing the load in steps up to 1.33 times the design load. In the United States, anchors are accepted if the creep movement at that load is less than 2 mm per log cycle of time. This creep movement is due to the creep in the steel tendon, the progressive cracking of the grout in tension, and the creep of the soil in shear. The loading history for the proof test and the result of a test are shown in Figure 21.52.

21.12.6 Embedment Depth and Downdrag

Another issue to be addressed is the embedment depth below the excavation level. You might think that the anchored wall would not need much embedment, since the anchors hold the soil back. The following reasoning shows the danger of having very little embedment (Briaud and Lim 1999).

Table 21.4 Presumptive f_{max} Values for Fine-Grained Soils

Anchor Type (Grout Pressure)	Soil Type	Shear Strength of Soil s_u (kPa)	Shear Strength of Soil-Grout Interface f_{max} (kPa)
Gravity grouted anchors (< 350 kPa)	Silt-clay mixtures	Stiff to very stiff 50 to 200 kPa	30 to 70 kPa
Pressure grouted anchors (350 to 2800 kPa)	High-plasticity clay	Stiff (50 to 120 kPa)	30 to 100 kPa
		Very stiff (120 to 200 kPa)	70 to 170 kPa
	Medium-plasticity clay	Stiff (50 to 120 kPa)	100 to 250 kPa
		Very stiff (120 to 200 kPa)	140 to 350 kPa
	Medium-plasticity sandy silt	Very stiff (120 to 200 kPa	280 to 380 kPa .

(After AASHTO 2007)

Table 21.5 Presumptive f_{max} Values for Coarse-Grained Soils

Anchor Type (Grout Pressure)	Soil Type	Relative Density and SPT N Value N (blows/0.3 m)	Shear Strength of Soil-Grout Interface f_{max} (kPa)
Gravity grouted anchors (< 350 kPa)	Sand or sand/gravel mixtures	Medium dense to dense (N = 10 to 50)	70 to 140 kPa
Pressure grouted anchors (350 to 2800 kPa)	Fine to medium sand	Medium dense to dense (N = 10 to 50)	80 to 380 kPa
	Medium to coarse sand with gravel	Medium to dense (N = 10 to 30)	110 to 670 kPa
		Dense to very dense (N = 30—50+)	250 to 950 kPa
	Silty sand		170 to 400 kPa
	Sandy gravel	Medium dense to dense (N = 10 to 40)	210 to 1400 kPa
		Dense to very dense (N = 40 to 50+)	280 to 1400 kPa
	Glacial till	Dense (N = 30 to 50)	300 to 520 kPa

(After AASHTO 2007)

Table 21.6 Presumptive f_{max} Values for Rock

Rock Type	Shear Strength of Soil-Grout Interface
Soft shale	200 to 800 kPa
Weathered sandstone	700 to 800 kPa
Sandstone	800 to 1700 kPa
Slate and hard shale	800 to 1400 kPa
Soft limestone	1000 to 1400 kPa
Dolomite limestone	1400 to 2100 kPa
Granite or basalt	1700 to 3100 kPa

(After AASHTO 2007)

When the excavation takes place, the soil mass behind the wall tends to move toward the excavation and downward (Figure 21.53). The downward movement drags the wall down, and if the embedment is insufficient, the downward movement can be significant. Even if the anchors are performing well, the wall can rotate; indeed, the anchors keep the soil from moving horizontally but not vertically. This rotation will generate horizontal movement as well. Therefore, to minimize horizontal movement, it is necessary to have well-designed anchors and a well-designed embedment depth to resist downdrag and the vertical component of the anchor loads.

The embedment depth must also resist the unbalanced lateral load from the bottom of the pressure diagram. This is

Table 21.7 Values of Anchor Load Transfer Capacity

Soil or Rock Type	Strength (SPT N Values)	Estimated Load Transfer (kN/m)
Sand and gravel	Loose (N = 4 to 10)	145
	Medium (N = 10 to 30)	210
	Dense (N = 30 to 50)	290
Sand	Loose (N = 4 to 10)	100
	Medium (N = 10 to 30)	145
	Dense (N = 30 to 50)	190
Sand and silt	Loose (N = 4 to 10)	75
	Medium (N = 10 to 30)	100
	Dense (N = 30 to 50)	130
Low-plasticity silt and clay	Stiff (N = 10 to 20)	30
	Hard (N = 20 to 40)	60
Soft shale		145
Slate and hard shale		360
Soft limestone		430
Sandstone		430
Dolomite limestone		580
Granite or basalt		730

(Canadian Foundation Engineering Manual 2007; FHWA 1984)

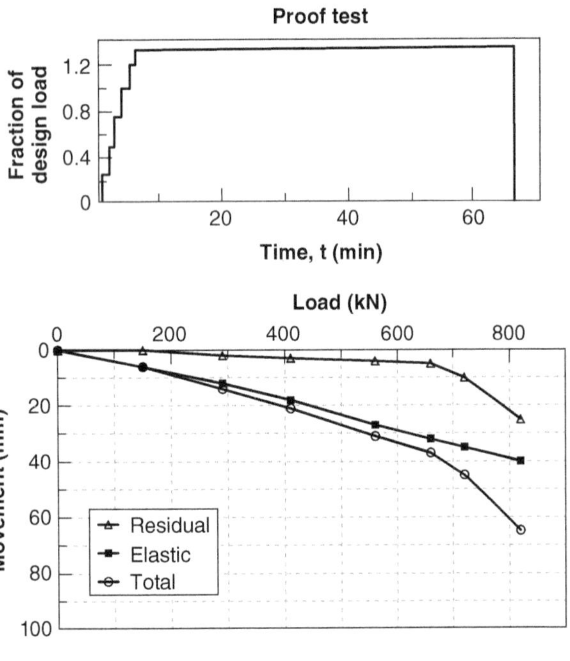

Figure 21.52 Loading history for an anchor proof test (Briaud et al. 1998).

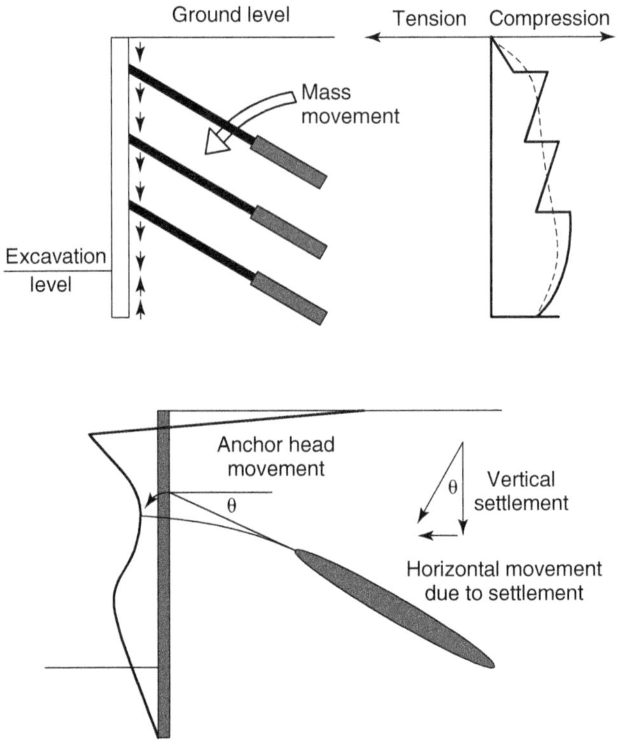

Figure 21.53 Downdrag creates horizontal movement (Briaud and Kim 1998).

area GMNH in Figure 21.42. The depth of embedment for lateral resistance is obtained by designing the bottom part of the wall (GHI in Figure 21.42) to resist the pressure from areas GMNH and HOPI with the factored passive resistance SHIQR. This design follows the approach described for the cantilever top-down wall. As a guide, the depth of embedment required for lateral resistance is on the order of 1.5 times the distance GH. The downdrag design requirement may be

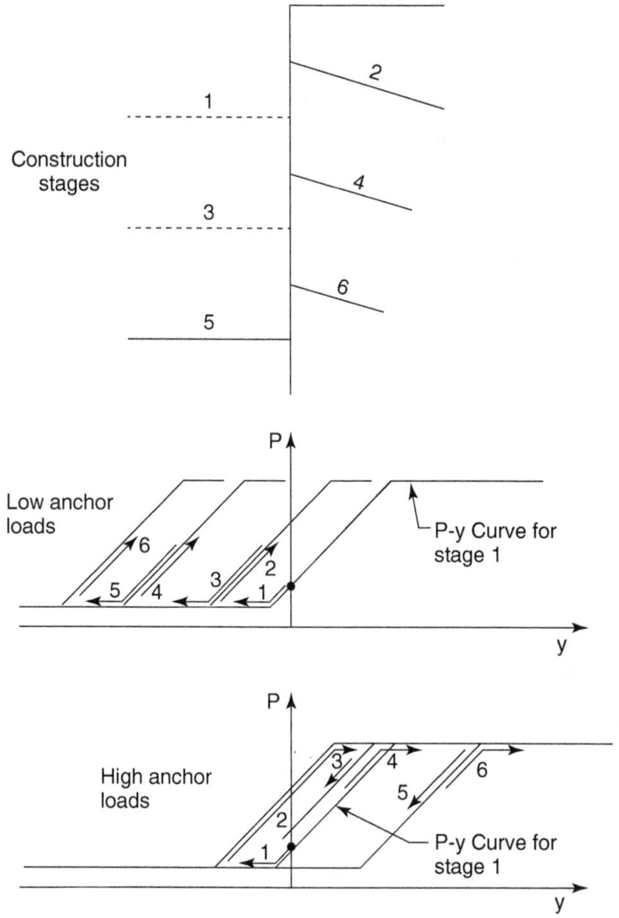

Figure 21.54 P-y curve path during construction sequence (Briaud and Kim 1998).

larger (see section 18.6). Note that two cases occur for the embedment depth: the case where the wall is continuous below the excavation level (e.g., slurry wall, sheet pile wall), and the case where only a row of piles exists below the excavation level (e.g., soldier pile and lagging wall). The depth of embedment of a system using a row of piles will have to be larger than that of a continuous wall system.

21.12.7 P-y Curve Approach and FEM Approach

The P–y curve approach described in section 21.11 is also applicable to anchored walls, and represents the best way to obtain the bending moment versus depth profile for the wall (Briaud and Kim 1998). In the process of preparing the P-y curves, it is possible to follow the construction sequence as shown in Figure 21.54. The anchors must have their separate P-y curves, as shown in Figure 21.55. A sample result for the P-y curve approach is shown in Figure 21.56. This is a comparison between the P-y curve predictions and the actual measurement for a full-scale instrumented wall at Texas A&M University (Figure 21.57). The P-y curve approach is not as reliable for predicting movements as it is for predicting bending moments. For better movement predictions, the FEM is preferred, provided quality soil parameters are obtained and a realistic soil model is selected (Briaud and Lim 1999).

Figure 21.58 shows a sample result for the FEM approach. This is a comparison of the FEM predictions with the same full-scale wall at Texas A&M University (Figure 21.57).

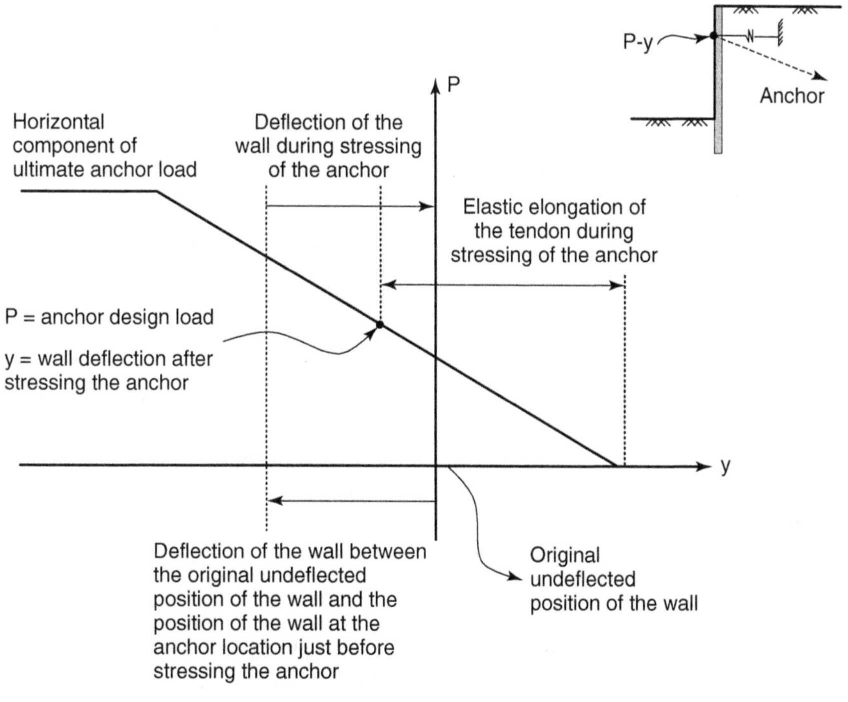

Figure 21.55 Anchor P-y curve (Briaud and Kim 1998).

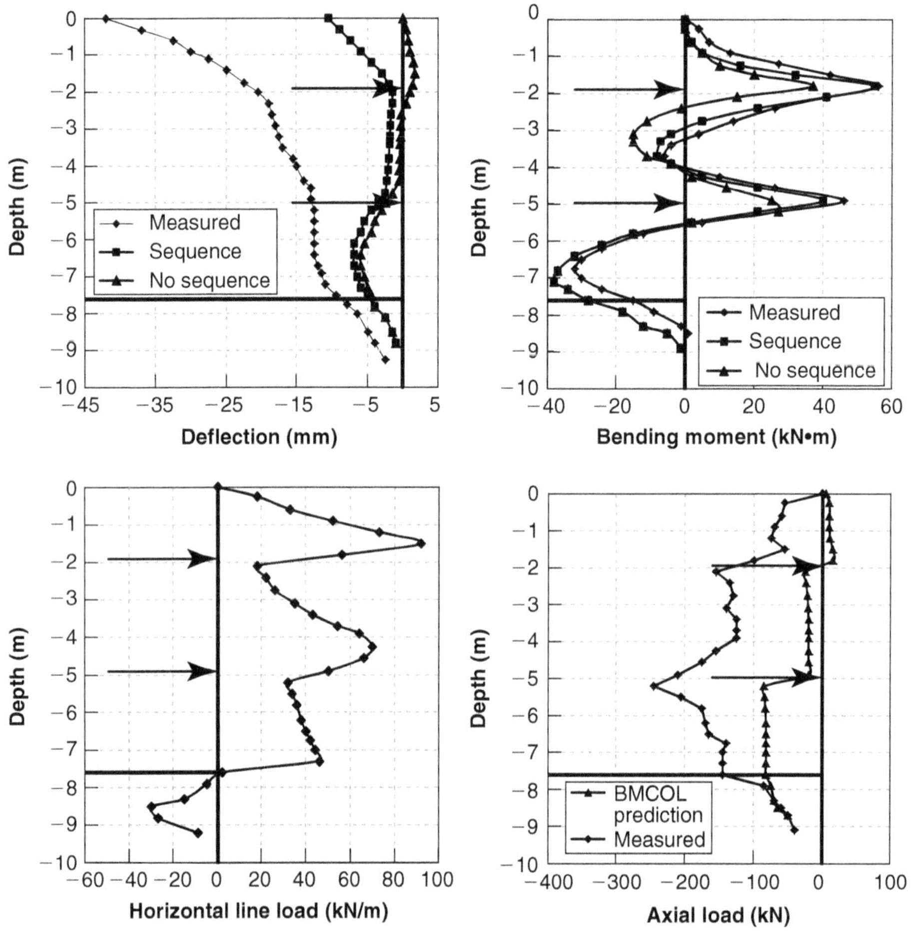

Figure 21.56 Predicted and measured result for the P-y curve method (Briaud and Kim 1998).

The predictions from the P-y curve approach and the FEM approach can be compared (Figure 21.56 and 21.58).

21.13 SOIL NAIL WALLS

Soil nail walls are top-down walls reinforced with rigid inclusions. They are to the top-down walls what MSE walls are to the bottom-up walls. *Soil nails* are rigid inclusions that are built much like anchors, by drilling a 100 to 300 mm diameter hole, inserting a steel bar with centralizers in the open hole, and backfilling the hole with grout. Unlike anchors, however, they are not posttensioned. The load in the nail develops as the soil mass deforms. The spacing between soil nails is typically quite a bit smaller than the spacing between anchors. Whereas all anchors and tiebacks are load tested, only a small percentage of soil nails are load tested. Soil nail walls are particularly suited for cases where the soil can stand unsupported for a height of 1 to 2 m long enough to place a row of nails (a few hours) and where the drill hole can stand open long enough for nail insertion and grouting. The front of the wall is typically covered with shotcrete to a thickness of 100 to 200 mm projected over a reinforcement mesh. Figures 21.59 and 21.60 show the construction sequence.

Much like the case of MSE walls, the design must consider external stability and internal stability as well as deformations.

21.13.1 External Stability

External stability includes global stability, sliding, and bearing capacity. Sliding and bearing capacity are handled in a fashion similar to the MSE wall design (section 21.12). The global stability, however, is different from the MSE wall approach. It is a slope stability type of analysis that considers failure along a surface through the nails. This surface can be a circle, two straight lines, or one line (Figure 21.61). The one-line solution is the simplest and is discussed here. Computer programs such as SNAIL (CALTRANS 1991) and GOLDNAIL (Golder, 1993) can be used to solve the problem for more complex failure surfaces.

Consider the soil nail wall of Figure 21.62. At equilibrium, the force resisted by the nails is T, the weight of the wedge is W, the surcharge force is Q, the shear force on the failure plane is S, and the normal force on the failure plane is N. The dimensions and angles involved are defined in Figure 21.62. The problem is to find the value of T to obtain a target factor

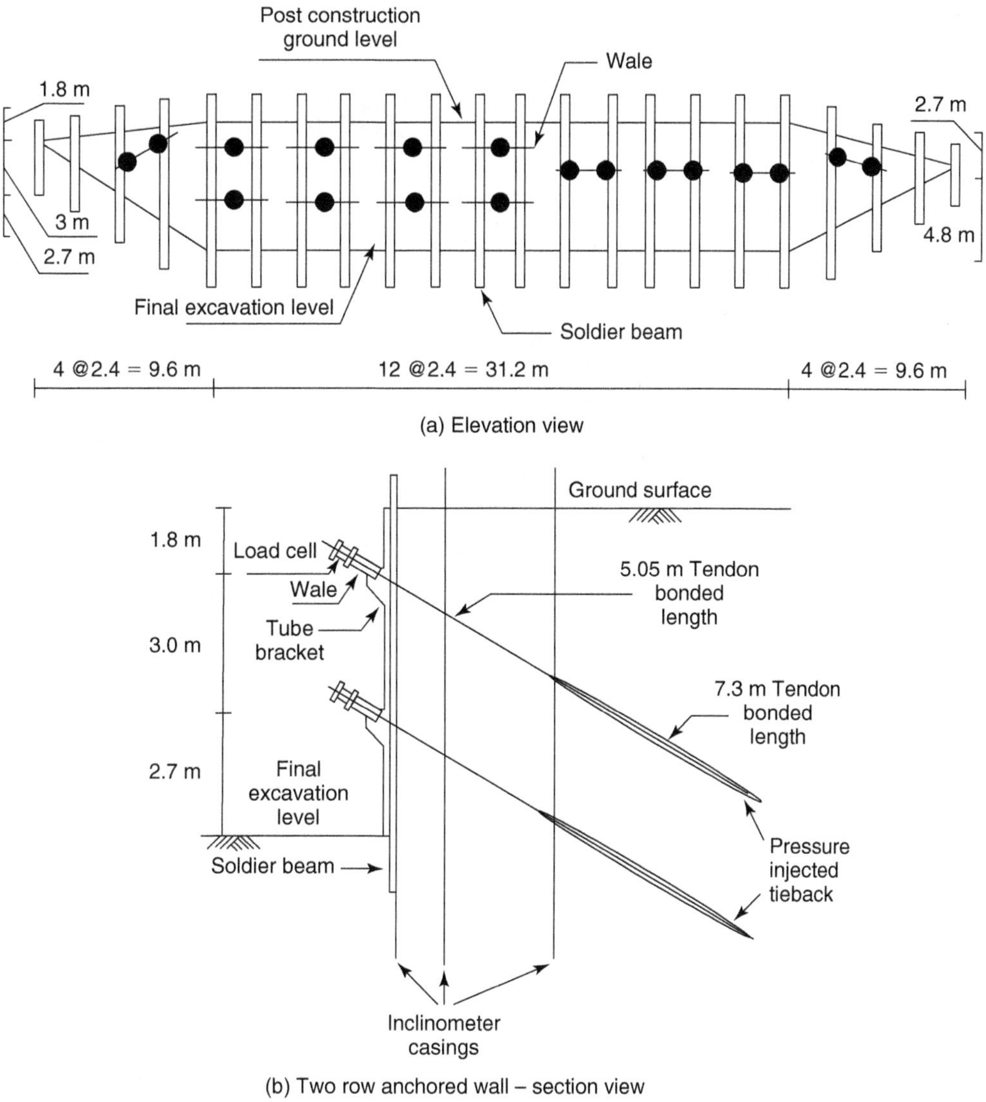

Figure 21.57 Full-scale instrumented wall at Texas A&M University (Briaud and Lim 1999).

of safety F (chosen value) on the ultimate shear resistance S. The factor of safety F is defined as:

$$F = \frac{\sum \text{maximum resisting shear forces on failure plane}}{\sum \text{driving shear forces on failure plane}} = \frac{\sum \text{R}}{\sum L} \tag{21.110}$$

Alternatively, the LRFD expression would be:

$$\gamma \sum L = \varphi \sum \text{R} \tag{21.111}$$

Writing equilibrium equations normal and along the plane of failure gives:

$$\sum \text{normal forces} = (W+Q)\cos\psi + T\sin(\psi+i) - N = 0 \tag{21.112}$$

$$\sum \text{tangent forces} = (W+Q)\sin\psi - T\cos(\psi+i) - S = 0 \tag{21.113}$$

The maximum value of the force S is S_{max} corresponding to the shear strength of the soil:

$$S = \frac{S_{max}}{F} = \frac{c'L + N\tan\varphi'}{F} \tag{21.114}$$

where c' and φ' are the effective stress cohesion and friction angle of the soil and F is the chosen factor of safety by design. The unknowns are N, S, and T and the three equations (21.112, 21.113, and 21.114) give the three quantities. Actually, the angle ψ corresponding to the lowest factor of safety is not known either, and must be found by trial and error. Once this is done, the load that must be safely carried by the nails is T.

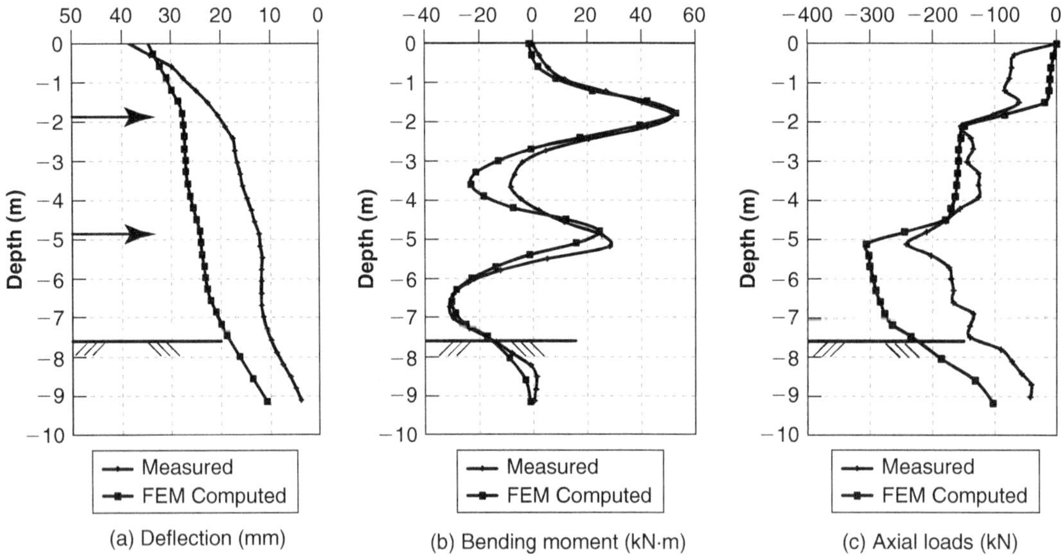

Figure 21.58 Predicted and measured result for the FEM method (Briaud and Lim 1999).

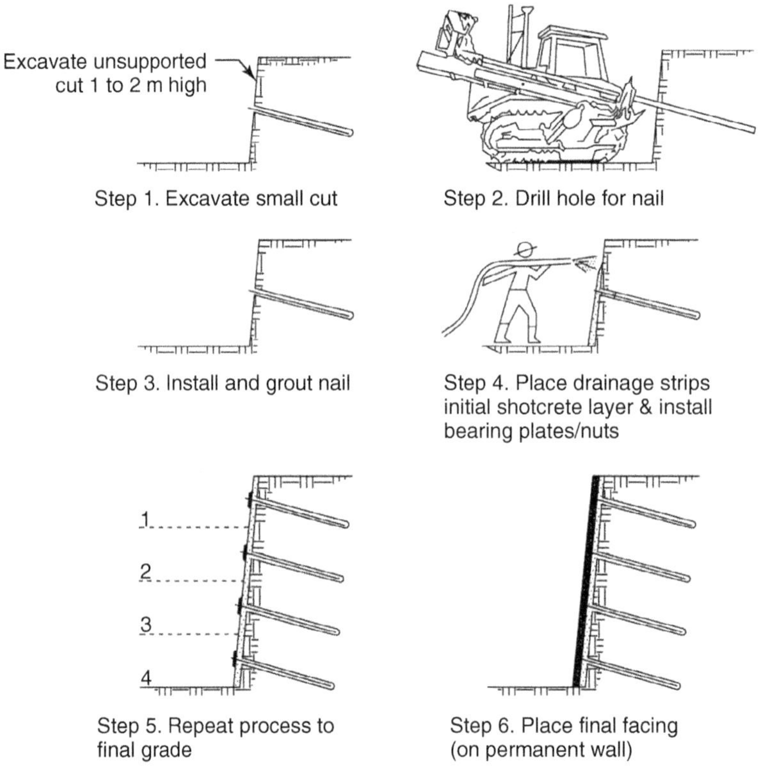

Figure 21.59 Soil nail wall construction sequence (FHWA 1998).

21.13.2 Internal Stability

Internal stability includes pull-out of the nails at the grout-soil interface, pull-out of the steel bar at the grout-bar interface, tensile yielding, and bending and shearing of the nails. The pull-out of the steel bar at the grout-bar interface is usually not controlling if threaded bars are used. Bending and shearing also do not appear to have a major influence on the behavior of the mass (Lazarte et al. 2003). Let's look first at pull-out at the grout-soil interface.

Pullout at Grout-Soil Interface

The equation for the ultimate axial resistance R of a nail in tension is:

$$R = \pi D L_p f_{\max} \tag{21.115}$$

Figure 21.60 Soil nail wall construction (Courtesy of FHWA www .fhwa.dot.gov/publications/publicroads/11septoct/alongroad.cfm).

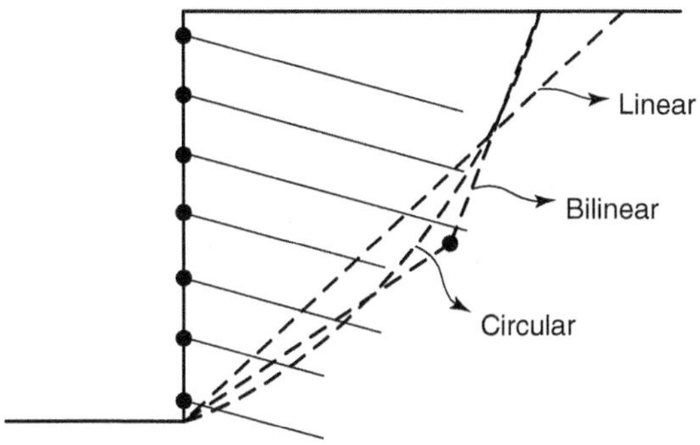

Figure 21.61 Soil nail wall failure surfaces.

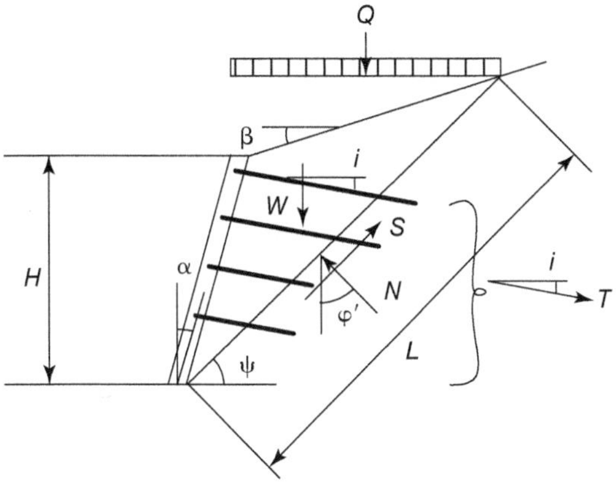

Figure 21.62 Soil nail wall global stability analysis.

where D is the diameter of the nail (drill hole), L_p is the useful length of the nail beyond the failure zone, and f_{max} is the shear strength at the grout-soil interface. Table 21.8 gives some estimated values of f_{max}.

Tensile Force Distribution in the Nail

Figure 21.63 shows a simplified distribution of the tension in a nail within the reinforced soil mass. As in the case of

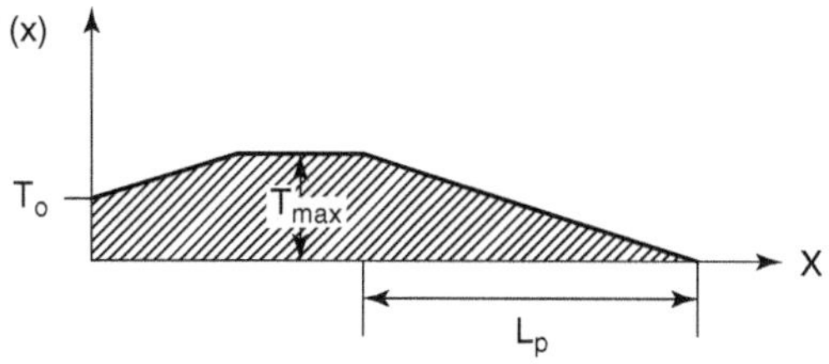

Figure 21.63 Tension load distribution in a soil nail.

the MSE wall, the tension load is lower at the wall face (T_o), then increases as the soil transfers the load to the nail until a maximum value is reached (T_{max}), and then decreases to zero as the load is transferred from the nail to the surrounding soil. At the wall face, the nail is usually connected to a plate pressed against the soil by a nut threaded on the nail steel bar. The load T_o varies from 0.6 to 1 times the maximum load T_{max}. The load T_{max} starts to decrease at a distance L_p from the end of the nail. The measured locus of T_{max}, the failure plane, and the distance L_p are shown in Figure 21.64. The load that must be globally carried by the nails is T, as calculated in the external stability analysis. The ultimate resistance that can be developed by individual nails over the length L_p is R_i, which must satisfy:

$$\gamma T = \varphi \sum_{i=1}^{n} R_i \tag{21.116}$$

The distribution of R_i among the nails is not precisely defined and experience plays a role in that determination. In general, shorter nails are placed at the bottom of the wall and longer ones at the top. A pattern such as the one shown in Figure 21.65 is not uncommon.

Length of Nails

The required length of each nail L_{pi} to resist R_i is calculated by using Eq. 21.115. The total length for nail i is L_{ti}; it is obtained by adding the length L_{pi} required to safely carry the

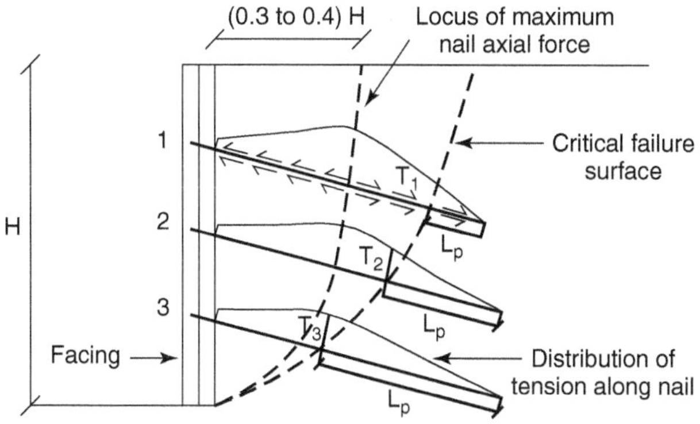

Figure 21.64 Load in the nails and available resisting length.

Table 21.8 Estimated Ultimate Grout-Soil Shear Strength, f_{max}

Material	Construction Method	Soil/Rock Type	Ultimate Grout-Soil Shear Strength, f_{max} (kPa)
Coarse-grained soils	Rotary drilling	Sand/gravel	100-180
		Silty sand	100-150
		Silt	60-75
		Piedmont residual	40-120
		Fine colluvium	75-150
	Driven casing	Sand/gravel	
		low overburden	190-240
		high overburden	280-430
		Dense moraine	380-480
		Colluvium	100-180
	Augered	Silty sand fill	20-40
		Silty fine sand	55-90
		Silty clayey sand	60-140
	Jet grouted	Sand	380
		Sand/gravel	700
Fine-grained soils	Rotary drilling	Silty clay	35-50
	Driven casing	Clayey silt	90-140
	Augered	Loess	25-75
		Soft clay	20-30
		Stiff clay	40-60
		Stiff clayey silt	40-100
		Calcareous sandy clay	90-140
Rock	Rotary drilling	Marl/limestone	300-400
		Phyllite	100-300
		Chalk	500-600
		Soft dolomite	400-600
		Fissured dolomite	600-1000
		Weathered sandstone	200-300
		Weathered shale	100-150
		Weathered schist	100-175
		Basalt	500-600
		Slate/Hard shale	300-400

(Elias and Juran 1991)

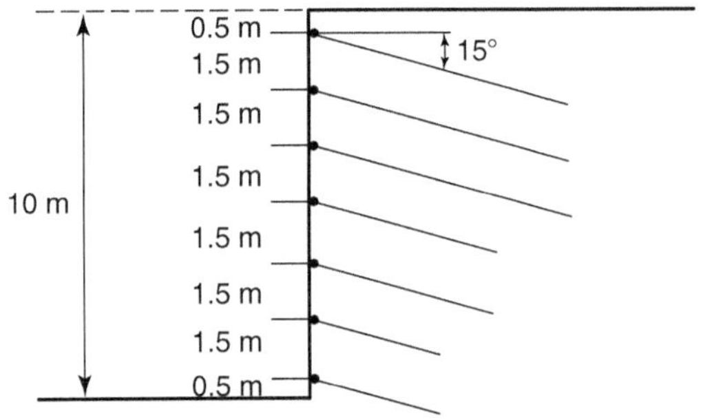

Figure 21.65 Typical pattern of nail length distribution.

required load R_i plus the discounted length L_{di} within the failure zone (Figure 21.64):

$$L_{ti} = L_{di} + L_{pi} \tag{21.117}$$

Tensile Yielding of Nails

The nails must be designed in such a way that the load applied does not break the nails. In calculating the tensile strength of the nail, the resistance of the grout is ignored and only the steel is considered. The area of the steel bar must satisfy:

$$\gamma T_{max} = \varphi A_t \sigma_y \tag{21.118}$$

where γ and φ are the load and resistance factors respectively, T_{max} is the highest load in the nail, A_t is the steel bar cross section, and σ_y is the yield strength of the steel. According to Briaud and Lim (1997) and Lazarte et al. (2003), the value of T_{max} is given by:

$$T_{max} = 0.65\ to\ 0.75\ K_a \gamma H s_v s_h \tag{21.119}$$

where K_a is the active earth pressure coefficient, γ is the total unit weight of the soil, H is the height of soil retained, and s_v and s_h are the vertical and horizontal nail spacing respectively. Note that Eq. 21.119 assumes that there is no water within the retained depth of soil or rock.

21.13.3 Wall Movement

The movement of soil nail walls is similar to the movement of anchored and strutted walls. According to Lazarte et al. (2003), for soil nail walls with ratios of length of nails to height of wall between 0.7 and 1.0, negligible surcharge loading, and typical load and resistance factors (safety factors), empirical data show that the maximum long-term horizontal and vertical wall displacements at the top of the wall, δ_h and δ_v, can be estimated by the values in Table 21.9 where H is the wall height. The parameter C helps to estimate the extent of the movement behind the wall (Figure 21.66):

$$D = CH(1 - \tan\alpha) \tag{21.120}$$

where D is the horizontal distance of influence of the excavation measured from the front of the wall where settlement of the ground surface takes place, C is the coefficient in Table 21.9, H is the height of the wall, and α is the batter of the wall (Figure 21.66).

21.13.4 Other Issues

Other issues include the details of the connection plates at the nail head, punching and bending of the wall cover at the front of the nail, corrosion resistance, and seismic loading. For more details on these matters, see Lazarte et al. (2003).

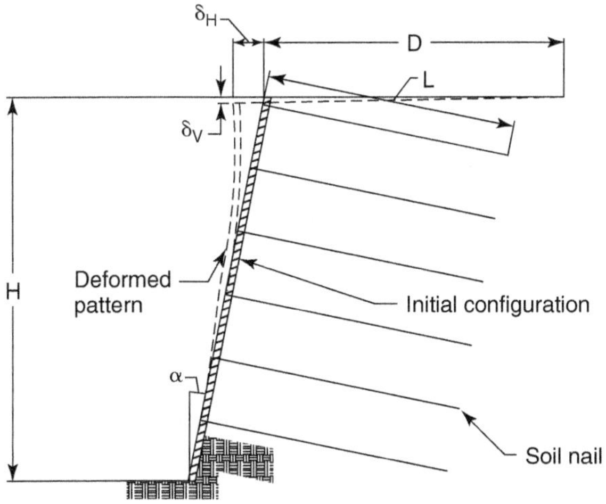

Figure 21.66 Deformation of soil nail walls.

Table 21.9 Estimates of Soil Nail Wall Movements

Variable	Weathered Rock or Stiff Soil	Sandy Soil	Fine-Grained Soil
δ_h/H and δ_v/H	0.001	0.002	0.003
C	1.25	0.8	0.7

(Lazarte et al. 2003.)

21.14 SPECIAL CASE: TRENCH

Trenches are narrow and fairly shallow excavations often used for placing utilities in congested areas. In the case of the undrained behavior of fine-grained soils, the relationship between the vertical total stress σ_{ov} and horizontal total stress σ_{oh} at failure of the trench is given by:

$$\sigma_{oh} = \sigma_{ov} - 2s_u \tag{21.121}$$

where s_u is the undrained shear strength of the soil.

The most stressed element in the trench is the soil element at the bottom of the trench, as shown in Figure 21.67. For that element, the vertical total stress σ_{ov} is:

$$\sigma_{ov} = \gamma h \tag{21.122}$$

where γ is the total unit weight of the soil and h is the depth of the trench. Initiation of failure of the trench corresponds to failure of the element shown in Figure 21.67. For this element, σ_{oh} is zero and the depth h_f at which initiation of failure starts is:

$$h_f = \frac{2s_u}{\gamma} \tag{21.123}$$

So, for example, if s_u is 100 kPa and γ is 20 kN/m^3, then h_f is 10 m and a safe depth might be 5 m. Would you go and work at the bottom of an open, unprotected trench 1 m wide and 5 m deep? You should not, and you should not allow anyone else to work in such a situation. The risk of collapse is too great, as evidenced by the number of related deaths every year. There is an average of 50 deaths per year due to trench accidents in the United States. Even going into a 1.2 m deep trench is not safe. You might think that as long as your head is above ground, you will be safe: Not true! If your head is above ground, you can open your mouth to

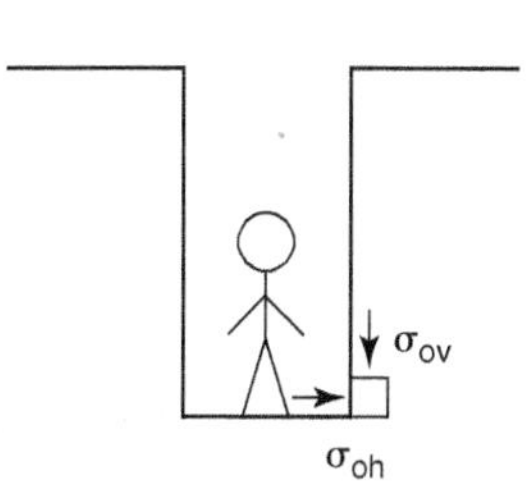

Figure 21.67 Initiation of failure in a trench.

take air in, but if your chest is below ground, you cannot expand your lungs, no air goes in, and you die. Do not go into an open, unsupported trench (Figure 21.68). You may go into a trench that is supported by what is called a trench box (Figure 21.69).

The theory leading to Eq. 21.101 is correct, but the assumptions may not match the reality. It is assumed that the soil is uniform and that every part of it has a minimum undrained shear strength s_u. This may not be the case in the field, for many reasons: the soil may be fissured and you tested the soil between fissures rather than at the fissures, which may control the mass strength; you may have tested the soil in the summer when it is harder (e.g., water tension higher) and the trench is opened in the winter; the undrained shear strength may not be the appropriate strength if the soil drains during and after the trench is open.

Figure 21.68 Do not go in there! (Courtesy of CDC's Public Health Image Library.)

Figure 21.69 Much safer when protected by a trench box. (Courtesy of www.cobletrenchsafety.com/jobprofile.php?id=95)

Additionally, if someone becomes partially buried in soil, do not try to pull that person out by rope and mechanical means. The tensile strength of the body is typically less than the pull-out capacity or force generated and you can imagine the result! Excavate around the body to free the person.

PROBLEMS

21.1 Show the pressure diagram, calculate the resultant push, and give its location for a 10 m high wall due to
  a. Water only
  b. Dry soil with unit weight of 20 kN/m^3 and a friction angle of 30° (active and passive)
  c. The same soil but with water to the top of the wall (active and passive)

21.2 Solve Coulomb's wedge analysis for the passive case of a soil with friction and cohesion. Write vertical and horizontal equilibrium and demonstrate equation 21.17.

21.3 Plot Coulomb and Rankine active and passive earth pressure coefficients for a vertical wall, no wall friction, as a function of the ground surface inclination β. Which one would you use?

21.4 Evaluate the influence of wall friction on the active and passive earth pressure coefficients by comparing Rankine value (no friction) and Coulomb values (varying friction angle from 0 to φ') for a vertical wall and horizontal backfill. Which one would you use?

21.5 Demonstrate that the direction of the plane of failure for the active pressure case (PB in Figure 21.11) is equal to 45 + $\varphi'/2$.

21.6 A 6 m high retaining wall has a backfill made of unsaturated sandy silt with a water tension equal to -1000 kPa and an area ratio (α) equal to 0.3. The total unit weight is 20 kN/m^3. The wall has no effective stress cohesion ($c' = 0$), and an effective stress friction angle equal to 30° ($\varphi' = 30$). The backfill is horizontal and the wall friction is neglected. Calculate the active and passive earth pressure diagram for this wall.

21.7 A wall is to be placed in a soil as described in Figure 21.1s. Prepare the active pressure diagram and the passive pressure diagram for that soil profile.

21.8 A 10 m high retaining wall has a horizontal backfill made of soil without water. The soil properties are $\gamma = 20$ kN/m^3, $c' = 0$, $\varphi' = 30°$. Draw the active pressure diagram against the wall due to the following surcharges at the top of the wall:
  a. Uniform surcharge equal to 20 kPa
  b. Line load of 20 kN/m at a distance of 1 m from the edge of the wall
  c. A point load of 20 kN at a distance of 1 m from the edge of the wall

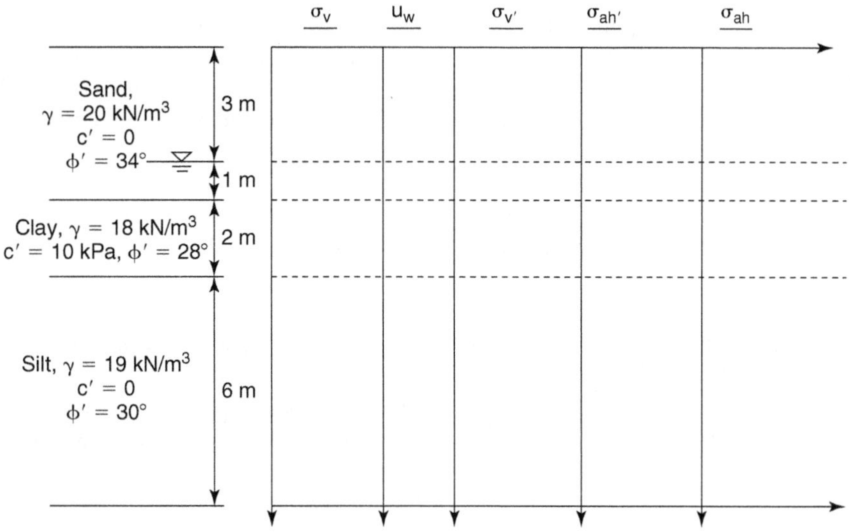

Figure 21.1s Soil profile.

21.9 How deep would you dig an unsupported trench in a stiff clay with an undrained shear strength of 75 kPa and a unit weight of 18 kN/m^3? The contract requires that you do the digging yourself while working at the bottom of the trench.

21.10 Plot the coefficient of earth pressure at rest K_o as a function of OCR for an overconsolidated clay with a friction angle φ' equal to 28°. On the same graph, plot K_a and K_p.

21.11 Draw the earth pressure diagram for a 7 m high gravity retaining wall with a backfill compacted with a vibratory roller. The roller weighs 150 kN, has a centrifugal force amplitude of 50 kN, is 2 m wide, and gets as close as 1 m to the top edge of the wall. The soil has a unit weight of 19 kN/m^3, a passive earth pressure coefficient equal to 3, and an at-rest earth pressure coefficient equal to 0.6.

21.12 An 8 m high top-down wall is retaining a shrink-swell soil with a swell pressure profile decreasing with depth from 500 kPa at the ground surface down to 50 kPa at the bottom of the wall. The soil has a friction angle φ' equal to 28° and no cohesion c'. K_o is 0.6. Draw the pressure diagram for the wall.

21.13 Draw the displacement y_a and y_p necessary to mobilize the active and passive earth pressure as a function of the wall height H for a dense sand.

21.14 Derive equations 21.73 and 21.74.

21.15 For the retaining wall shown in Figure 21.2s, calculate the pressure distribution against the wall, the resultant push, the factor of safety against sliding, and the factor of safety against overturning.

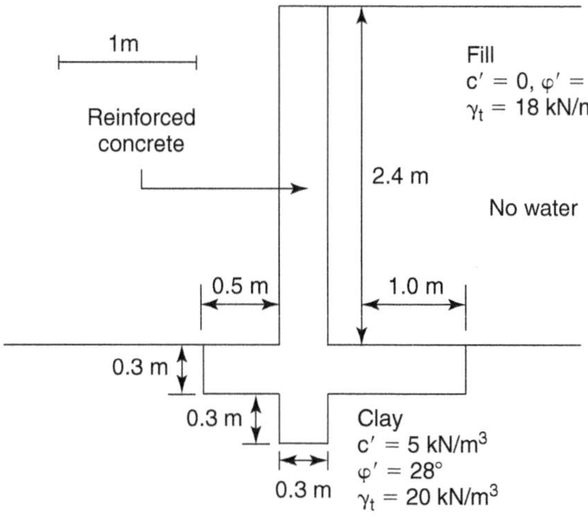

Figure 21.2s Retaining wall.

21.16 Design the soil reinforcing strips required for a 20 m high MSE wall. The precast concrete panels are 1.5 m by 1.5 m. The vertical and horizontal spacing between strips are 750 mm and 450 mm respectively. The unit weight of the backfill material is 19 kN/m^3 with an angle of internal friction of 34° and a coefficient of uniformity of 4.4. The location of the first layer of strips, measured from the finished grade, is 375 mm. Neglect the traffic surcharge.

21.17 A cantilever retaining wall is embedded 6 m below excavation level and retains 5 m of soil. An impervious layer exists 4 m below the bottom of the wall. The water level is at the ground surface on both sides of the wall and the soil deposit is uniform and deep. Draw the water pressure diagram against the wall on both sides of the wall, assuming that the water pressure is hydrostatic. Then draw a flow net and develop the water pressure diagram on both sides of the wall. Compare and comment.

21.18 Demonstrate Equation 21.91.

21.19 What is the depth of embedment d required for a cantilever wall retaining a height of sand H? Express the results as a function of H, K_p/K_a, and a factor of safety F applied to σ_p, the passive pressure. (Note: There is no water.)

21.20 For the anchored slurry wall shown in Figure 21.3s, calculate the pressure distribution on both sides of the wall for a deflection of 25 mm at the top of the wall. Calculate the anchor forces. How important is the vertical capacity of the wall? Explain your answer. What would happen if the water level rose on both sides of the excavation to the top of the wall? What would happen if the water level rose to the top of the wall on the retained-soil side of the excavation and to 2 m below that on the excavated side?

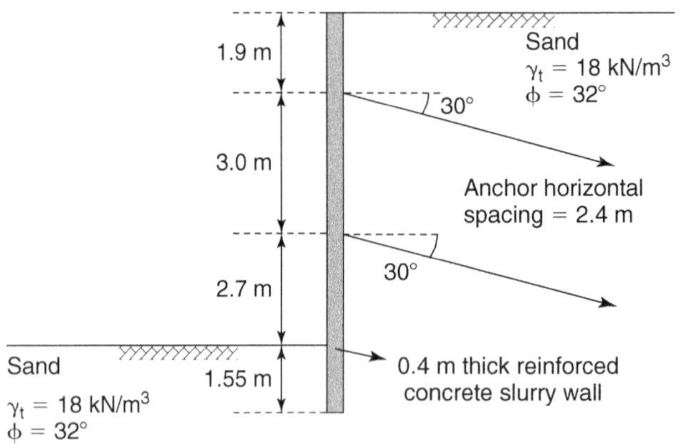

Figure 21.3s Anchored slurry wall.

21.21 Explain Figure 21.49.

21.22 Use Tables 21.4 and 21.5 and add a column giving the back-calculated alpha values.

21.23 For Figure 21.62, the height H is 9 m, α is 17°, β is 18°, ψ is 44°, and i is 10°. The stiff clay weighs 20 kN/m^3 with some cohesion c' (to be ignored), and a friction angle φ' of 32°. A uniformly applied surcharge of 10 kPa is to be considered on top of the wall. Calculate the required nail force T for a factor of safety against shear failure along the chosen plane to be 1.5. Distribute that force among the four nails and find the required length for each nail.

21.24 A 3 m wide strutted excavation is planned in a clay with an undrained shear strength equal to 40 kPa and a total unit weight of 19 kN/m^3. What depth of excavation corresponds to a factor of safety against base failure equal to 1.5?

Problems and Solutions

Problem 21.1

Show the pressure diagram, calculate the resultant push, and give its location for a 10 m high wall due to:

a. Water only
b. Dry soil with unit weight of 20 kN/m^3 and a friction angle of 30° (active and passive)
c. The same soil but with water to the top of the wall (active and passive)

Solution 21.1 (Figure 21.4s)

a. Water only

$$u_w = \gamma_w \times H$$
$$u_w = 9.81 \times 10 = 98.1 \text{ kPa}$$

The resultant push per unit length of wall is:

$$P_w = \frac{u_w \times H}{2} = \frac{\gamma_w \times H^2}{2}$$
$$P_w = \frac{9.81 \times (10)^2}{2} = 490.5 \text{ kN/m}$$

Location from the bottom of the wall:

$$z = \frac{h}{3} = 3.33 \text{ m}$$

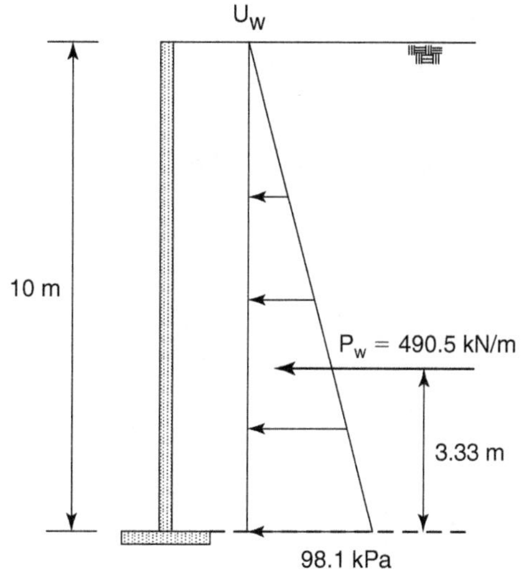

Figure 21.4s Pressure diagram for water only behind the wall.

b. Dry soil with unit weight of 20 kN/m^3 and a friction angle of 30° (active and passive)
The active force is:

$$\sigma'_{ov} = \gamma_d \times H = 20 \times 10 = 200 \text{ kPa}$$
$$P_a = \frac{K_a \times \gamma_d \times H^2}{2}$$
$$K_a = \frac{1 - \sin\varphi}{1 + \sin\varphi} = \frac{1 - \sin 30^\circ}{1 + \sin 30^\circ} = 0.33$$
$$P_a = \frac{0.33 \times 20 \times (10)^2}{2} = 333.3 \text{ kN/m}$$

The passive force is:

$$P_p = \frac{K_p \times \gamma_d \times H^2}{2}$$
$$K_p = \frac{1 + \sin\varphi}{1 - \sin\varphi} = \frac{1 + \sin 30^\circ}{1 - \sin 30^\circ} = 3.0$$
$$P_p = \frac{3 \times 20 \times (10)^2}{2} = 3000 \text{ kN/m}$$

The location of the active and passive force measured from the bottom of the wall (shown in Figure 21.5s) is:

$$z = \frac{h}{3} = 3.33 \text{ m}$$

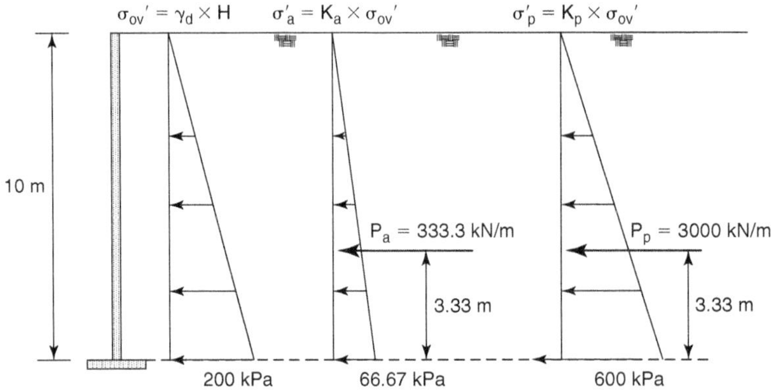

Figure 21.5s Pressure diagram for dry soil: Active and passive pressure profile.

c. Soil with water to the top of the wall (active and passive)
The vertical effective stress at the bottom of the wall is:

$$\sigma'_{ov} = \gamma_t \times H - \gamma_w \times H = 20 \times 10 - 9.81 \times 10 = 101.9 \text{ kPa}$$

The active force is

$$P_a = \frac{1}{2} H K_a \sigma'_{ov} + P_w$$

$$K_a = \frac{1 - \sin\varphi}{1 + \sin\varphi} = \frac{1 - \sin 30^\circ}{1 + \sin 30^\circ} = 0.33$$

$$P_a = \frac{10 \times 0.33 \times 101.9}{2} + 490.5 = 660.2 \text{ kN/m}$$

The passive force is:

$$P_a = \frac{1}{2} H K_p \sigma'_{ov} + P_w$$

$$K_p = \frac{1 + \sin\varphi}{1 - \sin\varphi} = \frac{1 + \sin 30^\circ}{1 - \sin 30^\circ} = 3.0$$

$$P_p = \frac{10 \times 3 \times 101.9}{2} + 490.5 = 2019 \text{ kN/m}$$

The location of the active and passive force measured from the bottom of the wall (shown in Figure 21.6s) is:

$$z = \frac{h}{3} = 3.33 \text{ m}$$

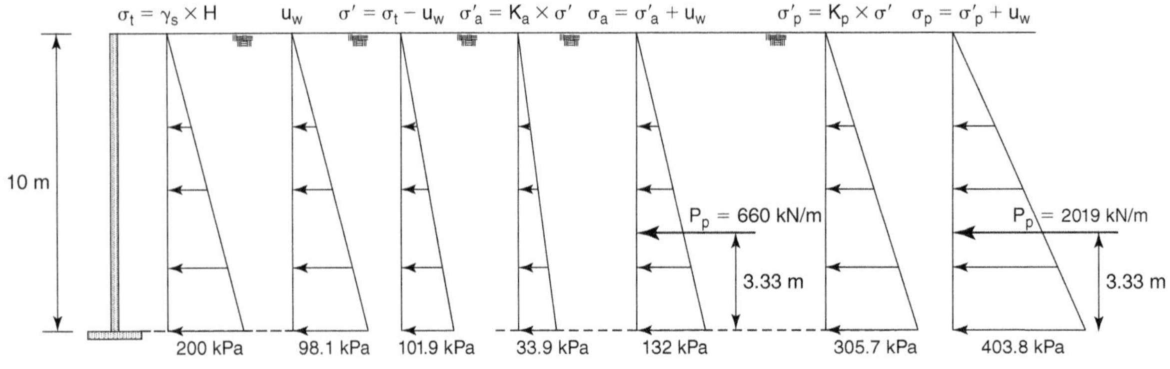

Figure 21.6s Pressure diagram for wall with water at ground surface.

Problem 21.2

Solve Coulomb's wedge analysis for the passive case of a soil with friction and cohesion. Write vertical and horizontal equilibrium and demonstrate equation 21.17.

Solution 21.2

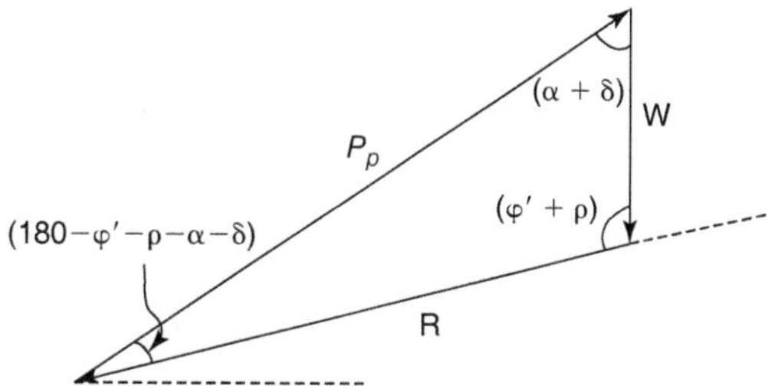

Figure 21.7s Coulomb wedge analysis for the passive case.

$$\frac{P_p}{\sin(\rho+\varphi')} = \frac{W}{\sin(180-\varphi'-\rho-\alpha-\delta)}$$

$$P_p = \frac{W\sin(\rho+\varphi')}{\sin(\varphi'+\rho+\alpha+\delta)}$$

$$W = \frac{\gamma H^2}{2\sin^2(\alpha)}\sin(\alpha+\rho)\left(\frac{\sin(\alpha+\beta)}{\sin(\rho-\beta)}\right)$$

$$P_p = \frac{\gamma H^2}{2\sin^2(\alpha)}\sin(\alpha+\rho)\left(\frac{\sin(\alpha+\beta)}{\sin(\rho-\beta)}\right)\frac{\sin(\rho+\varphi')}{\sin(\varphi'+\rho+\alpha+\delta)}$$

$$\frac{\partial P_p}{\partial \rho} = 0 \rightarrow \qquad P_p = \frac{\gamma H^2}{2}\frac{\sin^2(\alpha-\varphi')}{\sin^2(\alpha)\sin(\alpha+\delta)\left[1-\sqrt{\frac{\sin(\varphi'+\delta)\sin(\varphi'+\beta)}{\sin(\alpha+\delta)\sin(\alpha+\beta)}}\right]^2}$$

Problem 21.3

Plot Coulomb and Rankine active and passive earth pressure coefficients for a vertical wall, no wall friction, as a function of the ground surface inclination β. Which one would you use?

Solution 21.3

Coulomb theory

a. Figure 21.8s shows the wedge analysis in Coulomb theory for this case. As stated, the wall is vertical, and there is no wall friction. Therefore, the active earth force is acting horizontally. Note that β is the ground surface inclination, φ is the soil friction angle, and α is the failure plane inclination. H is the height of the wall, P_a is the maximum active force acting on the wall, W is the weight of the wedge, and R is the resultant force.

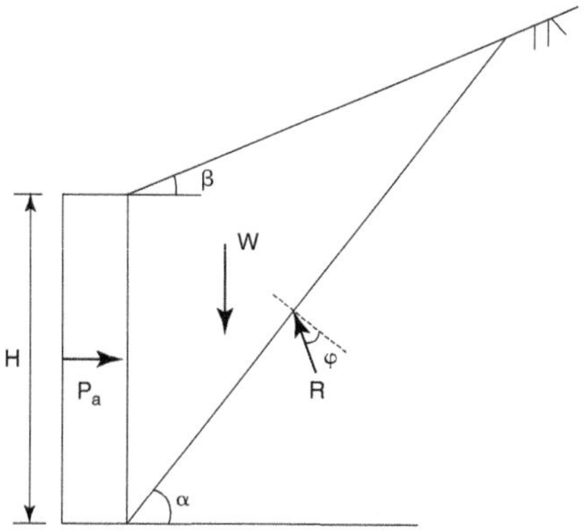

Figure 21.8s Illustration of active wedge analysis in Coulomb theory.

In this case, the expression of Coulomb active earth pressure coefficient simplifies to:

$$K_{ah} = \frac{\cos^2\varphi}{\left[1 + \sqrt{\dfrac{\sin\varphi \sin(\varphi - \beta)}{\cos\beta}}\right]^2}$$

To find the relationship between K_{ah} and β, a soil friction angle equal to 30° is assumed. The plot between K_a and β is shown in Figure 21.9s.

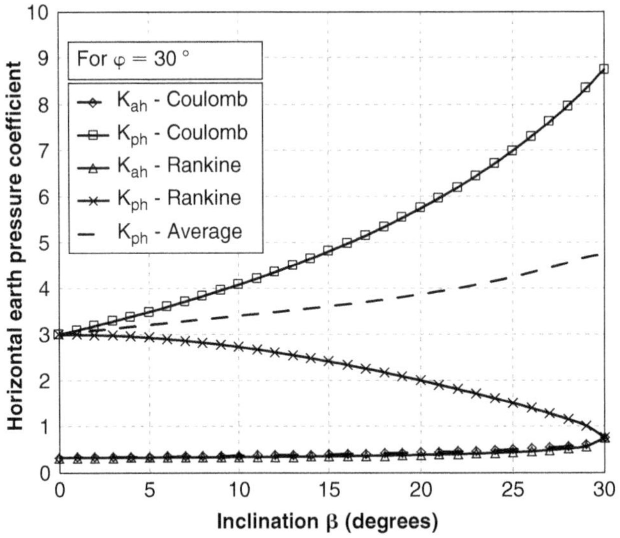

Figure 21.9s Plot of K_a and K_p vs. β using both Coulomb theory and Rankine theory.

b. Figure 21.10s shows the wedge analysis in Coulomb theory for this case. As stated, the wall is vertical, and there is no wall friction. Therefore, the passive earth force is acting horizontally. Note that β is the ground surface inclination, φ is the soil friction angle, and α is the failure plane inclination. H is the height of the wall, P_p is the maximum passive force acting on the wall, W is the weight of the wedge, and R is the resultant force.

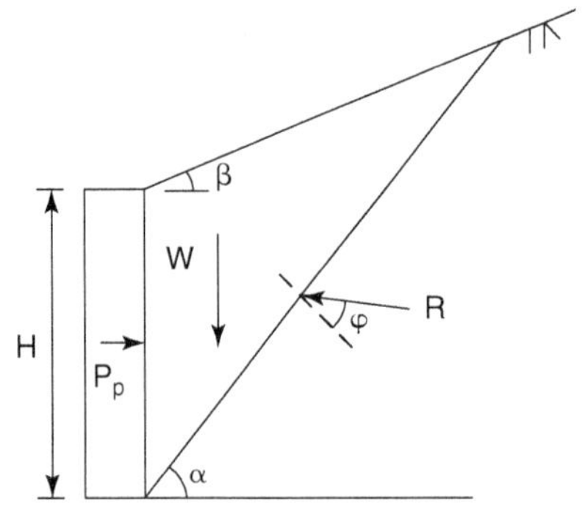

Figure 21.10s Illustration of passive wedge analysis in Coulomb theory.

In this case, the expression of Coulomb passive earth pressure coefficient simplifies to:

$$K_{ph} = \frac{\cos^2\varphi}{\left[1 - \sqrt{\frac{\sin\varphi \sin(\varphi+\beta)}{\cos\beta}}\right]^2}$$

To find the relationship between K_{ph} and β, a soil friction angle equal to 30° is assumed. The plot between K_{ph} and β is shown in Figure 21.9s.

Rankine theory

Fig 21.11s shows an illustration of the retaining wall analyzed using Rankine theory.

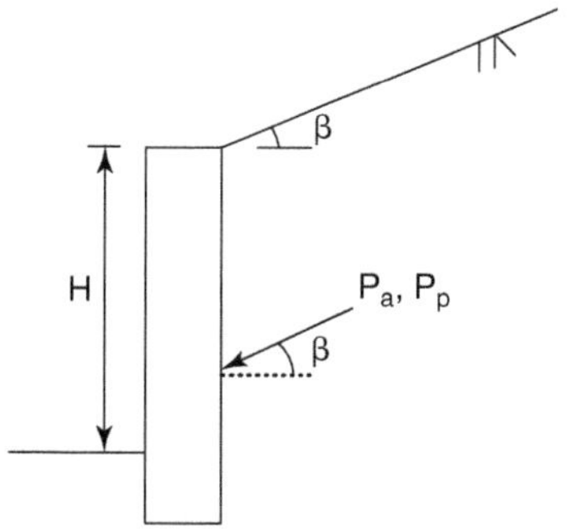

Figure 21.11s Retaining wall analyzed using Rankine theory.

Note that β is the ground surface inclination, φ is the soil friction angle, H is the height of the wall, P_p is the passive earth force, and P_a is the active earth force.

a. Based on Rankine theory, the active earth pressure coefficient that gives the horizontal component of P_a is:

$$K_{ah} = \cos^2\beta \frac{\cos\beta - \sqrt{\cos^2\beta - \cos^2\varphi}}{\cos\beta + \sqrt{\cos^2\beta - \cos^2\varphi}}$$

To find the relationship between K_{ah} and β, a soil friction angle equal to 30° is assumed. The plot between K_{ah} and β is shown in Figure 21.9s.

b. Based on Rankine soil theory, the passive earth pressure coefficient that gives the horizontal component of P_a is:

$$K_{ph} = \cos^2\beta \frac{\cos\beta + \sqrt{\cos^2\beta - \cos^2\varphi}}{\cos\beta - \sqrt{\cos^2\beta - \cos^2\varphi}}$$

To find the relationship between K_{ph} and β, a soil friction angle equal to 30° is assumed. The plot between K_{ph} and β is shown in Figure 21.9s.

Discussion

Figure 21.9s shows that for K_{ph} the Coulomb solution is a limit equilibrium solution giving upper-bound values, whereas the Rankine solution is an equilibrium-of-stresses solution that gives lower-bound values. Therefore, if a lower bound is conservative, one should choose Rankine theory; if an upper bound is conservative, one should choose Coulomb theory. To that end one would be tempted to use an average of the two values as a more reasonable estimate; however such an average is not based on any theoretical reasoning. Note that in the case of extreme values of the geometry parameters, it is advisable to use engineering judgment, as the K_a and K_p values can become unreasonable.

Problem 21.4

Evaluate the influence of wall friction on the active and passive earth pressure coefficients by comparing Rankine value (no friction) and Coulomb values (varying friction angle from 0 to φ') for a vertical wall and horizontal backfill. Which one would you use?

Solution 21.4

Active Pressure

For a vertical wall and horizontal backfill, the Coulomb value K_{ah} that gives the horizontal component P_{ah} of the active push P_a is:

$$K_{ah} = \frac{\sin^2(90 + \varphi')}{\left[1 + \sqrt{\frac{\sin(\varphi'+\delta)\sin\varphi'}{\sin(90-\delta)}}\right]^2}$$

and the Rankine value is:

$$K_{ah} = \frac{1 - \sin\varphi'}{1 + \sin\varphi}$$

Figure 21.12s shows the K_{ah} values versus friction angle for both the Coulomb and Rankine solutions. For Coulomb, different curves are presented for different values of the wall friction. As can be seen from the figure, the Coulomb value of K_{ah} decreases as the wall friction increases; the Rankine value does not change. The maximum value of Coulomb K_{ah} is reached for zero wall friction, which is equal to the Rankine value.

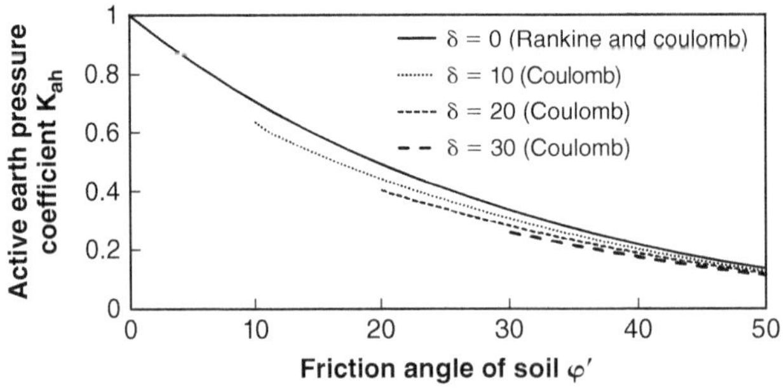

Figure 21.12s K_{ah} vs. soil friction angle for different wall friction angle.

Passive Pressure

Coulomb:

$$K_{ph} = \frac{\sin^2(90 - \varphi')}{\left[1 - \sqrt{\dfrac{\sin(\varphi' + \delta)\sin\varphi'}{\sin(90 + \delta)}}\right]^2}$$

Rankine:

$$K_{ph} = \frac{1 + \sin\varphi'}{1 - \sin\varphi}$$

Figure 21.13s shows the K_{ph} values versus friction angle for both the Coulomb and Rankine solutions. For Coulomb, different curves are presented for different values of the wall friction.

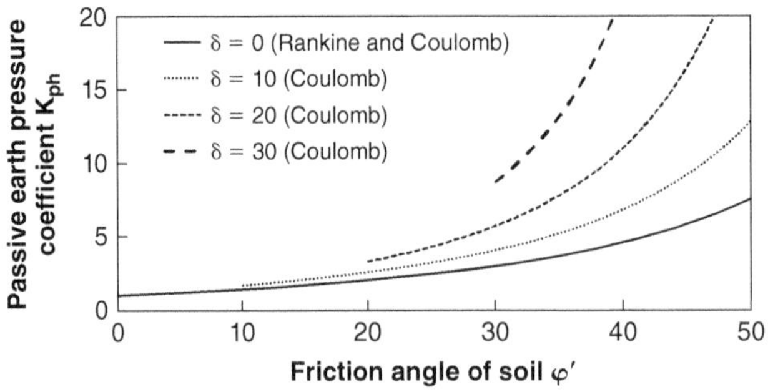

Figure 21.13s K_{ph} vs. soil friction angle for different wall friction angle.

Which one would you use? Rankine solution generally gives reasonable values. Coulomb theory is also reasonable, in that it takes into account wall friction in the case of the active earth pressure, but Coulomb's passive earth pressure values are quite optimistic and should not be used. The problem is that the failure surface is optimistically chosen as a straight line instead of a curved surface which would offer less resistance.

Problem 21.5

Demonstrate that the direction of the plane of failure for the active pressure case (PB in Figure 21.11) is equal to $45 + \varphi'/2$.

Solution 21.5

First we find the Pole on the failure circle in the active case (Figure 21.14s). The failure point is shown by point T on the failure circle. If we draw a line from the left of the failure circle M to T, the angle $\widehat{MTO}$ will be 90°. In the triangle TOM:

$$\widehat{TMO} = 180 - (90 + \varphi') = 90 - \varphi'$$

In the triangle TPM:

$$\widehat{MPT} = \widehat{MTP}$$

$$2 \times \widehat{MPT} + (90 - \varphi') = 180 \rightarrow \widehat{MPT} = 45 + \frac{\varphi'}{2}$$

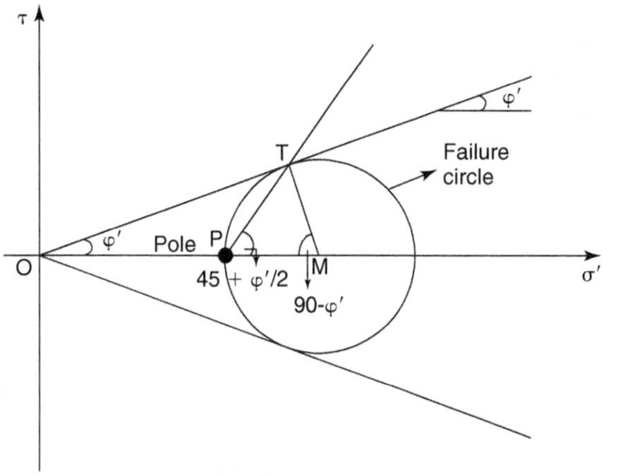

Figure 21.14s Pole method.

Problem 21.6

A 6 m high retaining wall has a backfill made of unsaturated sandy silt with a water tension equal to -1000 kPa and an area ratio (α) equal to 0.3. The total unit weight is 20 kN/m^3. The wall has no effective stress cohesion ($c' = 0$), and an effective stress friction angle equal to 30° ($\varphi' = 30$). The backfill is horizontal and the wall friction is neglected. Calculate the active and passive earth pressure diagram for this wall.

Solution 21.6

Use Rankine theory to solve this problem.

Active Earth Pressure

The active earth pressure coefficient for this problem is:

$$K_a = \frac{1 - \sin \varphi'}{1 + \sin \varphi'} = \frac{1}{3}$$

Because the soil behind the wall is uniform, we only need to choose two calculation points: point a and b, shown in Figure 21.15s.

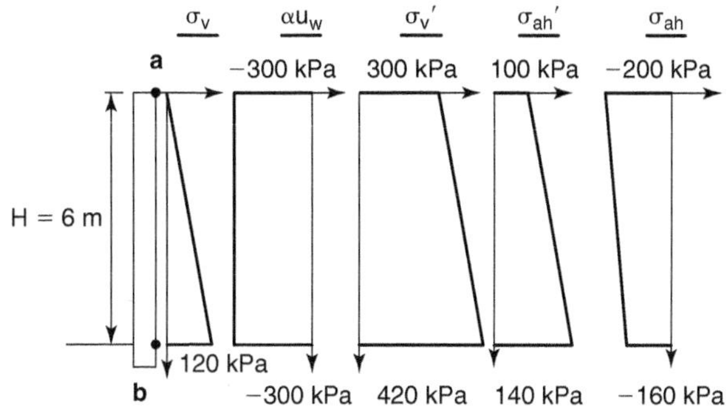

Figure 21.15s Active earth pressure diagram.

Point a: total vertical stress $\sigma_v = 0$

Therefore, effective vertical stress $\sigma'_v = \sigma_v - \alpha u_w = 0 + 0.3 \times (-1000) = 300$ kPa

Effective active horizontal stress is $\sigma'_{ah} = K_a \sigma'_v = \frac{1}{3} \times 300 = 100$ kPa

Total active horizontal stress is $\sigma_{ah} = \sigma'_{ah} + \alpha u_w = 100 + (-300) = -200$ kPa

Point b: total vertical stress $\sigma_v = \gamma_t H = 20 \times 6 = 120$ kPa

Therefore, effective vertical stress $\sigma'_v = \sigma_v - \alpha u_w = 120 - 0.3 \times (-1000) = 420$ kPa

Effective active horizontal stress is $\sigma'_{ah} = K_a \sigma'_v = \frac{1}{3} \times 420 = 140$ kPa

Total active horizontal stress is $\sigma_{ah} = \sigma'_{ah} + \alpha u_w = 140 + (-300) = -160$ kPa

The active earth pressure diagram is shown in Figure 21.15s. Practically, the suction should be ignored as it could disappear in the rainy season or cracks could develop in the backfill and the active earth pressure diagram would be the same as if the soil had no water.

Passive Earth Pressure:

The passive earth pressure coefficient for this problem is

$$K_p = \frac{1 + \sin \varphi'}{1 - \sin \varphi'} = 3$$

Point a: Effective passive horizontal stress is $\sigma'_{ph} = K_p \sigma'_v = 3 \times 300 = 900$ kPa

Total passive horizontal stress is $\sigma_{ph} = \sigma'_{ph} + \alpha u_w = 900 + (-300) = 600$ kPa

Point b: Effective passive horizontal stress is $\sigma'_{ph} = K_p \sigma'_v = 3 \times 420 = 1260$ kPa

Total passive horizontal stress is $\sigma_{ph} = \sigma'_{ph} + \alpha u_w = 1260 + (-300) = 960$ kPa

The passive earth pressure diagram is shown in Figure 21.16s. Practically, the suction would be ignored and the passive earth pressure diagram would be the same as if the soil had no water.

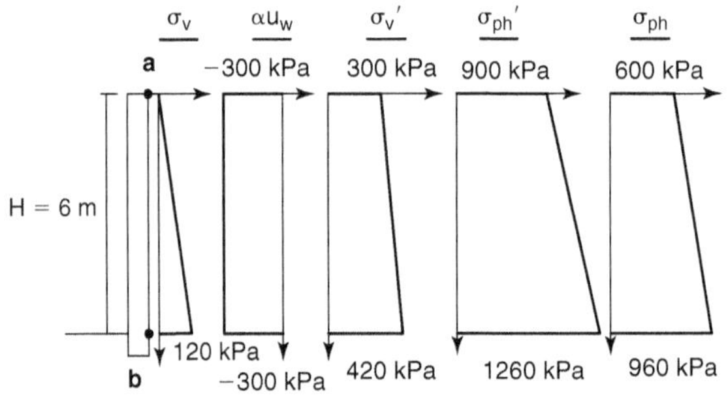

Figure 21.16s Passive earth pressure diagram.

Problem 21.7

A wall is to be placed in a soil as described in Figure 21.1s. Prepare the active pressure diagram and the passive pressure diagram for that soil profile.

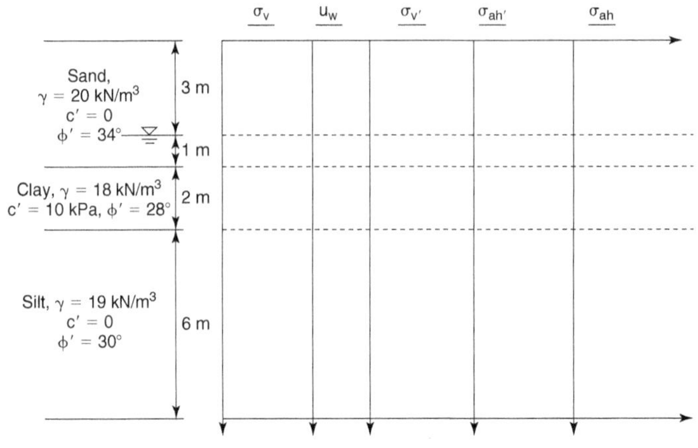

Figure 21.1s Soil profile.

Solution 21.7

Use Rankine theory to solve this problem.

For each layer, the active and passive earth pressure coefficients are calculated as follows:

$$\text{Sand}: K_a = \frac{1-\sin\varphi'}{1+\sin\varphi'} = 0.283,\ K_p = \frac{1+\sin\varphi'}{1-\sin\varphi'} = 3.54$$

$$\text{Clay}: K_a = \frac{1-\sin\varphi'}{1+\sin\varphi'} = 0.361,\ K_p = \frac{1+\sin\varphi'}{1-\sin\varphi'} = 2.77$$

$$\text{Silt}: K_a = \frac{1-\sin\varphi'}{1+\sin\varphi'} = 0.333,\ K_p = \frac{1+\sin\varphi'}{1-\sin\varphi'} = 3$$

Active Earth Pressure

Because several soil layers and a groundwater level are involved in this problem, 5 calculation points are chosen (Figure 21.17s).

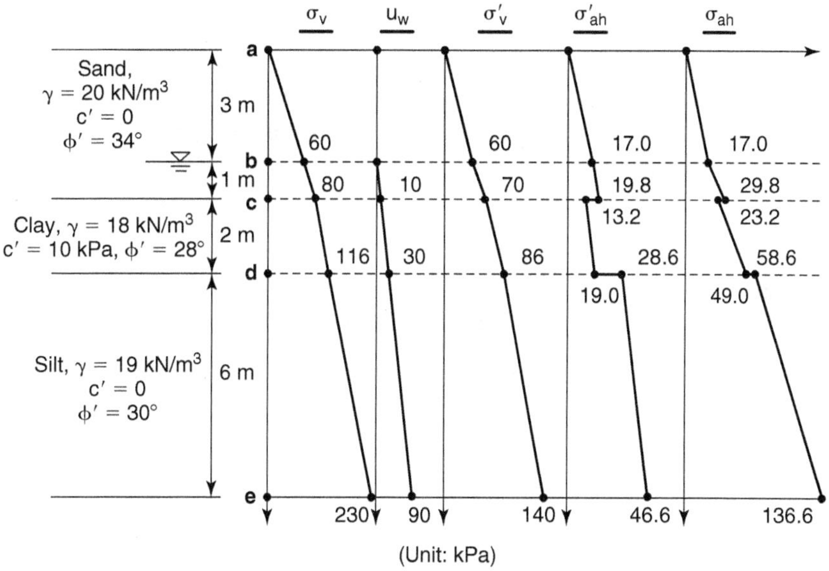

Figure 21.17s Active earth pressure diagram.

Point a: Total vertical stress: $\sigma_v = 0,\ u_w = 0$

Therefore, effective vertical stress:

$$\sigma'_v = \sigma_v - u_w = 0$$

Effective active horizontal stress is $\sigma'_{ah} = K_a\sigma'_v - 2c'\sqrt{K_a} = 0$

Total active horizontal stress is $\sigma_{ah} = \sigma'_{ah} + u_w = 0$

Point b: Total vertical stress:

$\sigma_v = \gamma h = 20 \times 3 = 60$ kPa, $u_w = 0$

Therefore, effective vertical stress:

$$\sigma'_v = \sigma_v - u_w = 60 \text{ kPa}$$

Effective active horizontal stress is $\sigma'_{ah} = K_a\sigma'_v - 2c'\sqrt{K_a} = 0.283 \times 60 - 2 \times 0 = 17.0$ kPa

Total active horizontal stress is $\sigma_{ah} = \sigma'_{ah} + u_w = 17.0$ kPa

Point c: Total vertical stress $\sigma_v = \gamma h = 60 + 20 = 80$ kPa, $u_w = 10 \times 1 = 10$ kPa

Therefore, effective vertical stress:

$$\sigma'_v = \sigma_v - u_w = 70 \text{ kPa}$$

Note that point c is on the interface between two different layers, so the effective active horizontal stress at that point should be calculated individually in each layer.

Point c, sand: Effective active horizontal stress is $\sigma'_{ah} = K_a\sigma'_v - 2c'\sqrt{K_a} = 0.283 \times 70 - 2 \times 0 = 19.8$ kPa
Total active horizontal stress is

$$\sigma_{ah} = \sigma'_{ah} + u_w = 19.81 + 10 = 29.8 \text{ kPa}$$

Point c, clay: Effective active horizontal stress is $\sigma'_{ah} = K_a\sigma'_v - 2c'\sqrt{K_a} = 0.361 \times 70 - 2 \times 10 \times \sqrt{0.361} = 13.2$ kPa
Total active horizontal stress is

$$\sigma_{ah} = \sigma'_{ah} + u_w = 13.2 + 10 = 23.2 \text{ kPa}$$

Point d: Total vertical stress: $\sigma_v = \sum \gamma h = 80 + 18 \times 2 = 116$ kPa, $u_w = 10 \times 3 = 30$ kPa
Therefore, effective vertical stress: $\sigma'_v = \sigma_v - u_w = 116 - 30 = 86$ kPa
Note that point d is on the interface between two different layers, so the effective active horizontal stress at that point should be calculated individually in each layer.
Point d, clay: Effective active horizontal stress is $\sigma'_{ah} = K_a\sigma'_v - 2c'\sqrt{K_a} = 0.361 \times 86 - 2 \times 10 \times \sqrt{0.361} = 19.0$ kPa
Total active horizontal stress is $\sigma_{ah} = \sigma'_{ah} + u_w = 19.0 + 30 = 49.0$ kPa
Point d, silt: Effective active horizontal stress is $\sigma'_{ah} = K_a\sigma'_v - 2c'\sqrt{K_a} = 0.333 \times 86 - 2 \times 0 = 28.6$ kPa
Total active horizontal stress is $\sigma_{ah} = \sigma'_{ah} + u_w = 28.6 + 30 = 58.6$ kPa
Point e: Total vertical stress: $\sigma_v = \sum \gamma h = 116 + 19 \times 6 = 230$ kPa, $u_w = 10 \times 9 = 90$ kPa
Therefore, effective vertical stress:

$$\sigma'_v = \sigma_v - u_w = 230 - 90 = 140 \text{ kPa}$$

Effective active horizontal stress is $\sigma'_{ah} = K_a\sigma'_v - 2c'\sqrt{K_a} = 0.333 \times 140 - 2 \times 0 = 46.6$ kPa
Total active horizontal stress is $\sigma_{ah} = \sigma'_{ah} + u_w = 46.6 + 90 = 136.6$ kPa
The active earth pressure diagram is shown in Figure 22.17s.

Passive Earth Pressure

Because several soil layers and a groundwater level are involved in this problem, 5 calculation points are chosen (Figure 21.18s.).

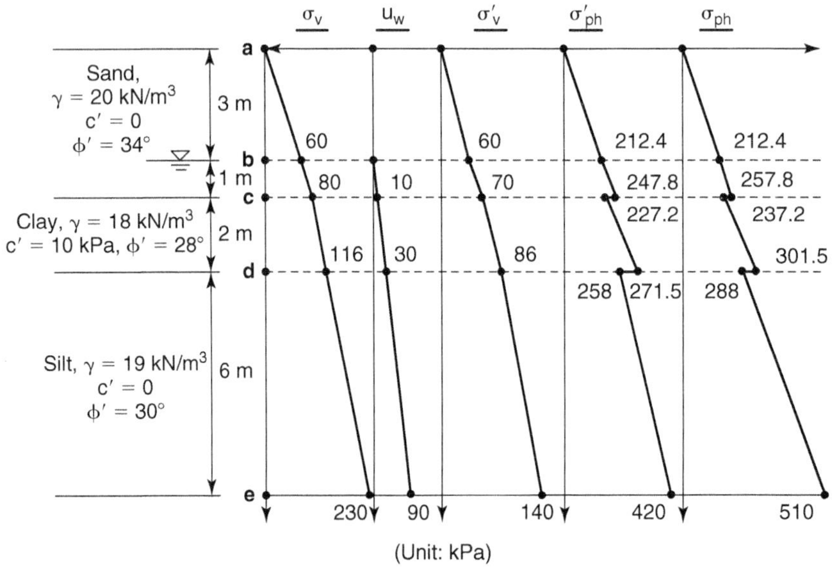

Figure 21.18s Passive earth pressure diagram.

For those calculation points, the vertical stresses, water stress, and effective vertical stresses are the same as for the active earth pressure. Here we only provide the calculation of effective passive horizontal stress and total passive horizontal stress at those five points.

Point a: Effective passive horizontal stress is

$$\sigma'_{ph} = K_p\sigma'_v + 2c'\sqrt{K_p} = 0$$

Total passive horizontal stress is

$$\sigma_{ph} = \sigma'_{ph} + u_w = 0$$

Point b: Effective passive horizontal stress is

$$\sigma'_{ph} = K_p\sigma'_v + 2c'\sqrt{K_p} = 3.54 \times 60 + 2 \times 0 = 212.4 \text{ kPa}$$

Total passive horizontal stress is

$$\sigma_{ph} = \sigma'_{ph} + u_w = 212.4 \text{ kPa}$$

Point c: Note that point c is on the interface between two different layers, so the effective passive horizontal stress at that point should be calculated individually in each layer.Point c, sand: Effective passive horizontal stress is

$$\sigma'_{ph} = K_p\sigma'_v + 2c'\sqrt{K_p} = 3.54 \times 70 + 2 \times 0 = 247.8 \text{ kPa}$$

Total passive horizontal stress is

$$\sigma_{ph} = \sigma'_{ph} + u_w = 247.8 + 10 = 257.8 \text{ kPa}$$

Point c, clay: Effective passive horizontal stress is

$$\sigma'_{ph} = K_p\sigma'_v + 2c'\sqrt{K_p} = 2.77 \times 70 + 2 \times 10 \times \sqrt{2.77} = 227.2 \text{ kPa}$$

Total passive horizontal stress is

$$\sigma_{ph} = \sigma'_{ph} + u_w = 227.2 + 10 = 237.2 \text{ kPa}$$

Point d: Note that point d is on the interface between two different layers, so the effective passive horizontal stress at that point should be calculated individually in each layer.

Point d, clay: Effective passive horizontal stress is

$$\sigma'_{ph} = K_p\sigma'_v + 2c'\sqrt{K_p} = 2.77 \times 86 + 2 \times 10 \times \sqrt{2.77} = 271.5 \text{ kPa}$$

Total passive horizontal stress is

$$\sigma_{ph} = \sigma'_{ph} + u_w = 271.5 + 30 = 301.5 \text{ kPa}$$

Point d, silt: Effective passive horizontal stress is

$$\sigma'_{ph} = K_p\sigma'_v + 2c'\sqrt{K_p} = 3 \times 86 + 2 \times 0 = 258 \text{ kPa}$$

Total passive horizontal stress is

$$\sigma_{ph} = \sigma'_{ph} + u_w = 258 + 30 = 288 \text{ kPa}$$

Point e: Effective passive horizontal stress is

$$\sigma'_{ph} = K_p \sigma'_v + 2c'\sqrt{K_p} = 3 \times 140 + 2 \times 0 = 420 \text{ kPa}$$

Total passive horizontal stress is

$$\sigma_{ph} = \sigma'_{ph} + u_w = 420 + 90 = 510 \text{ kPa}$$

The passive earth pressure diagram is shown in Figure 21.18s.

Problem 21.8

A 10 m high retaining wall has a horizontal backfill made of soil without water. The soil properties are $\gamma = 20$ kN/m^3, $c' = 0$, $\varphi' = 30°$. Draw the active pressure diagram against the wall due to the following surcharges at the top of the wall:

a. Uniform surcharge equal to 20 kPa
b. Line load of 20 kN/m at a distance of 1 m from the edge of the wall
c. A point load of 20 kN at a distance of 1 m from the edge of the wall

Solution 21.8 (Figure 21.19s)

$$K_a = \frac{1 - \sin 30}{1 + \sin 30} = \frac{1}{3}$$

$$\sigma_{ah} = \sigma'_{ov} K_a + \Delta\sigma_h$$

a.

$$\sigma_{ah} = \frac{1}{3} \times (20z + 20) = 6.67(z + 1)$$

b.

$$\sigma_{ah} = \frac{1}{3} \times 20z + \frac{4 \times 20}{\pi} \frac{1^2 z}{(z^2 + 1)^2} = 6.67z + \frac{25.46z}{(1 + z^2)^2}$$

c. Assume that $v = 0.35$

$$\sigma_{ah} = \frac{1}{3}z + \frac{20}{\pi(z^2 + 1)} \left(\frac{3z}{(z^2 + 1)^{\frac{3}{2}}} - \frac{(z^2 + 1)^{\frac{1}{2}}(1 - 2v)}{(z^2 + 1)^{\frac{1}{2}} + z} \right)$$

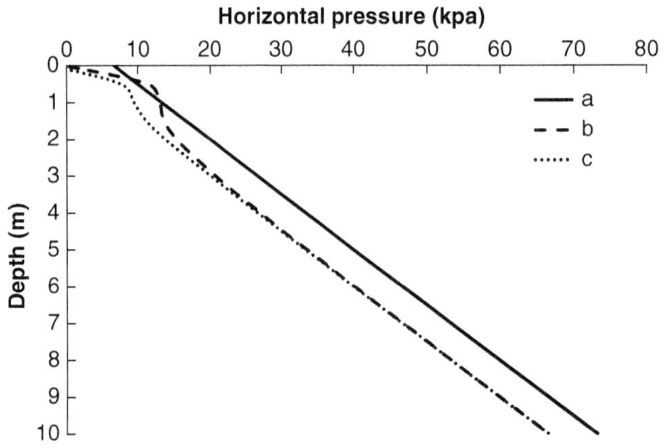

Figure 21.19s Horizontal pressure diagram.

Problem 21.9

How deep would you dig an unsupported trench in a stiff clay with an undrained shear strength of 75 kPa and a unit weight of 18 kN/m^3? The contract requires that you do the digging yourself while working at the bottom of the trench.

Solution 21.9

Assuming that the soil is truly uniform, with no fissures:

$$\sigma_{ah} = \sigma_{0v} - 2S_u$$
$$\sigma_{ah} = \gamma H - 2S_u = 18H - 2 \times 75 = 0 \rightarrow H = 4.16 \text{ m}$$

I would dig the trench to a depth of 1 m and stop there. Before going deeper, I would place a trench box to protect myself against trench collapse. Then I would dig further.

Problem 21.10

Plot the coefficient of earth pressure at rest K_o as a function of OCR for an overconsolidated clay with a friction angle φ' equal to 28°. On the same graph, plot K_a and K_p.

Solution 21.10 (Figure 21.20s)

$$K_0 = (1 - \sin\varphi')OCR^{\sin\varphi'}$$
$$K_a = \frac{1 - \sin\varphi}{1 + \sin\varphi} = \frac{1 - \sin 28^\circ}{1 + \sin 28^\circ} = 0.36$$
$$K_p = \frac{1 + \sin\varphi}{1 - \sin\varphi} = 2.77$$

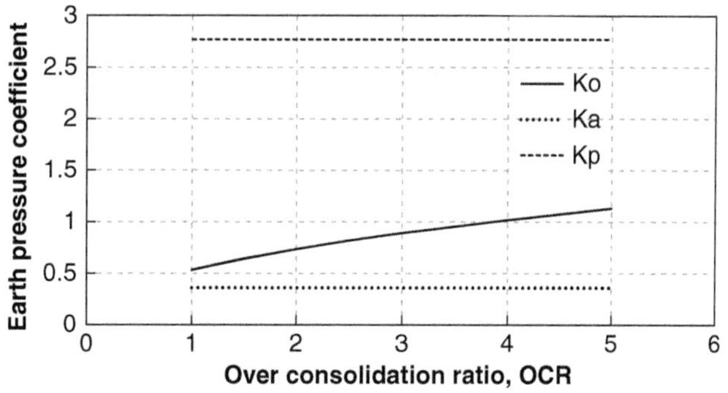

Figure 21.20s Earth pressure coefficients vs. OCR.

Problem 21.11

Draw the earth pressure diagram for a 7 m high gravity retaining wall with a backfill compacted with a vibratory roller. The roller weighs 150 kN, has a centrifugal force amplitude of 50 kN, is 2 m wide, and gets as close as 1 m to the top edge of the wall. The soil has a unit weight of 19 kN/m^3, a passive earth pressure coefficient equal to 3, and an at-rest earth pressure coefficient equal to 0.6.

Solution 21.11 (Figure 21.21s)

$$\sigma_h = \frac{L}{a + L}\sqrt{\frac{2P\gamma}{\pi}} = \frac{2}{1 + 2}\sqrt{\frac{2\left(\frac{150+50}{2}\right) \times 19}{\pi}} = 23.2 \text{ kN/m}^2$$
$$d = \frac{L}{K_o(a + L)}\sqrt{\frac{2P}{\pi\gamma}} = \frac{2}{0.6(1 + 2)}\sqrt{\frac{2\left(\frac{150+50}{2}\right)}{\pi \times 19}} = 2.0 \text{ m}$$
$$K_o\gamma z = 0.6 \times 19 \times 7 = 79.8 \text{ kN/m}^2$$

L: the length of the roller
a: the distance between the edge of the wall and the closest roller position
P: the line load imposed by the roller weight of the roller plus the centrifugal force for vibratory rollers
γ: the unit weight of the soil

K_0: the at-rest earth pressure coefficient
d: depth to which the pressure diagram is modified due to the roller

The depth z to reach the horizontal pressure equal to 23.2 kPa is such that $K_p\gamma z = 23.2$ kPa, therefore, z = 0.41 m.

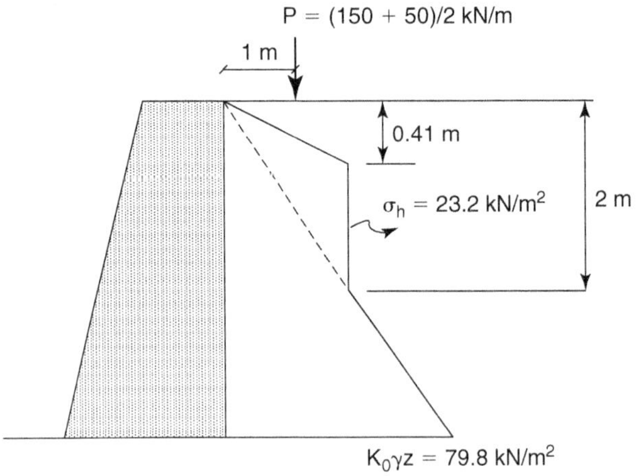

Figure 21.21s Earth pressure diagram.

Problem 21.12

An 8 m high top-down wall is retaining a shrink-swell soil with a swell pressure profile decreasing with depth from 500 kPa at the ground surface down to 50 kPa at the bottom of the wall. The soil has a friction angle φ' equal to 28° and no cohesion c'. K_o is 0.6. Draw the pressure diagram for the wall.

Solution 21.12 (Figure 21.22s)

$$K_0 = 0.6$$

$$K_p = \frac{1+\sin\varphi'}{1-\sin\varphi'} = \frac{1+\sin 28^\circ}{1-\sin 28^\circ} = 2.77$$

Assuming that the soil unit weight is γ = 18 kN/m^3:

$$K_o\gamma h = 0.6 \times 18 \times 8 = 86.4 \text{ kPa}$$

$$K_p\gamma h = 2.77 \times 18 \times 8 = 398.88 \text{ kPa}$$

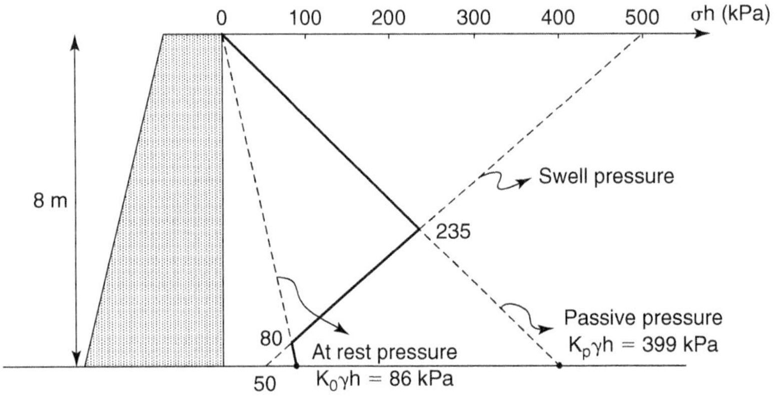

Figure 21.22s Earth pressure diagram.

Problem 21.13

Draw the displacement y_a and y_p necessary to mobilize the active and passive earth pressure as a function of the wall height H for a dense sand.

Solution 21.13 (Figures 21.23s, 21.24s)

From Table 21.1, the average displacements needed to generate active and passive earth pressures for different soil types are:

Loose sand	Soft clay
$\frac{y_a}{H} = 0.004, \frac{y_p}{H} = 0.04$	$\frac{y_a}{H} = 0.015, \frac{y_p}{H} = 0.04$
Dense sand	Stiff clay
$\frac{y_a}{H} = 0.0015, \frac{y_p}{H} = 0.02$	$\frac{y_a}{H} = 0.0075, \frac{y_p}{H} = 0.02$

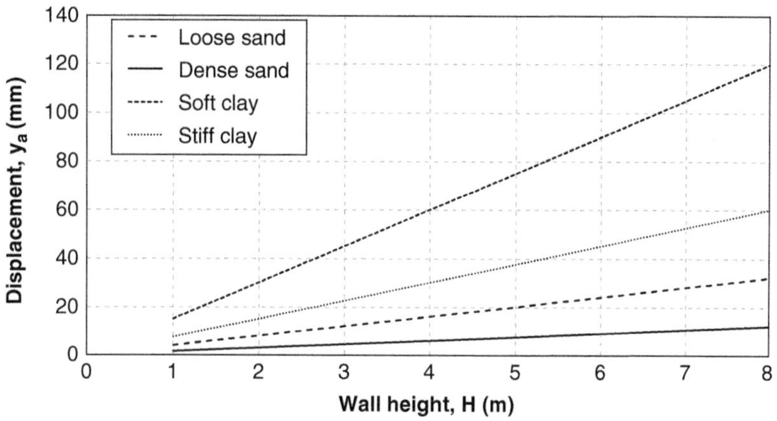

Figure 21.23s Active displacement vs. wall height.

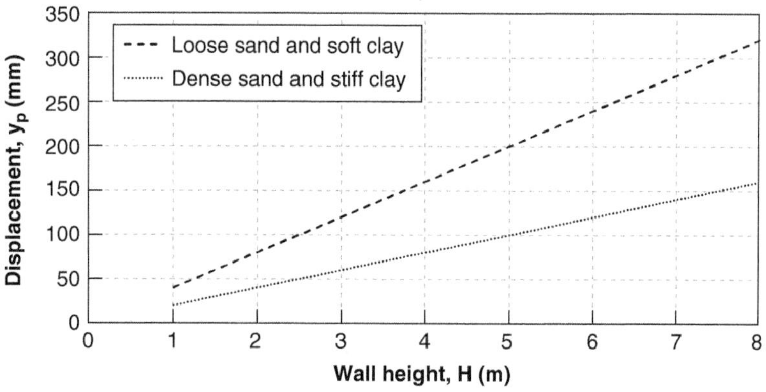

Figure 21.24s Passive displacement vs. wall height.

Problem 21.14

Derive equations 21.73 and 21.74.

Solution 21.14

The uniform soil pressure p_1 due to the line load is:

$$p_1 = \frac{Q}{B}$$

The soil pressure p_2 due to overturning moment is the maximum pressure at the edge of the triangular distribution under the foundation. The pressure distribution under the foundation must resist the moment. Writing the moment equilibrium gives:

$$\frac{1}{2}p_2\frac{B}{2}\times\frac{2}{3}\frac{B}{2}\times 2 = M \Rightarrow p_2 = \pm\frac{6M}{B^2}$$

$$p = \frac{Q}{B} \pm \frac{6M}{B^2} \& e = \frac{M}{Q} \Rightarrow \begin{cases} p_{\max} = \frac{Q}{B}\left(1+\frac{6e}{B}\right) \\ p_{\min} = \frac{Q}{B}\left(1-\frac{6e}{B}\right) \end{cases}$$

Problem 21.15

For the retaining wall shown in Figure 21.2s, calculate the pressure distribution against the wall, the resultant push, the factor of safety against sliding, and the factor of safety against overturning.

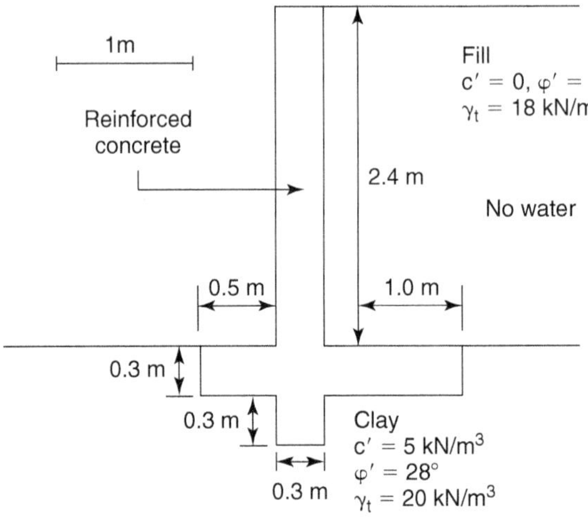

Figure 21.2s Retaining wall.

Solution 21.15 (Figures 21.25s, 21.26s)

Passive earth pressure, active earth pressure:

$$\sigma_{ph} = K_p\sigma'_{ov} + 2c'\sqrt{K_p} + \alpha u \qquad \sigma_{ah} = K_a\sigma'_{ov} - 2c'\sqrt{K_a} + \alpha u$$

$$K_p = \frac{1+\sin\varphi'}{1-\sin\varphi'} \qquad K_a = \frac{1-\sin\varphi'}{1+\sin\varphi'}$$

a. Calculate active earth pressure (active side):

$$K_{a1} = \frac{1-\sin 32}{1+\sin 32} = 0.307$$

$$K_{a2} = \frac{1-\sin 28}{1+\sin 28} = 0.361$$

$\sigma'_{ah} = 0.307 \text{ x } 43.2 = 13.26$ kPa at a depth of 2.4 m (in the fill)

$\sigma'_{ah} = 0.361 \text{ x } 43.2 - 2 \text{ x } 5\sqrt{0.361} = 9.59$ kPa at a depth of 2.4 m (in the clay)

$\sigma'_{ah} = 0.361 \text{ x } 55.2 - 2 \text{ x } 5\sqrt{0.361} = 13.92$ kPa at a depth of 3.0 m (in the clay)

Since there is no water $\sigma'_{ah} = \sigma_{ah}$

b. Calculate passive earth pressure (passive side):

$$K_p = \frac{1+\sin\varphi'}{1-\sin\varphi'} = \frac{1+\sin 28}{1-\sin 28} = 2.77$$

$$\sigma'_{ph} = K_P\sigma'_{ov} + 2c'\sqrt{K_p} = 16.64 \text{ kPa at the ground level}$$

$$\sigma'_{ph} = K_P\sigma'_{ov} + 2c'\sqrt{K_p} = 2.77 \times 12 + 10\sqrt{2.77} = 49.88 \text{ kPa at a depth of } 0.6\text{ m}$$

Since there is no water $\sigma'_{ph} = \sigma_{ph}$

c. Draw diagram (Figure 21.2s).

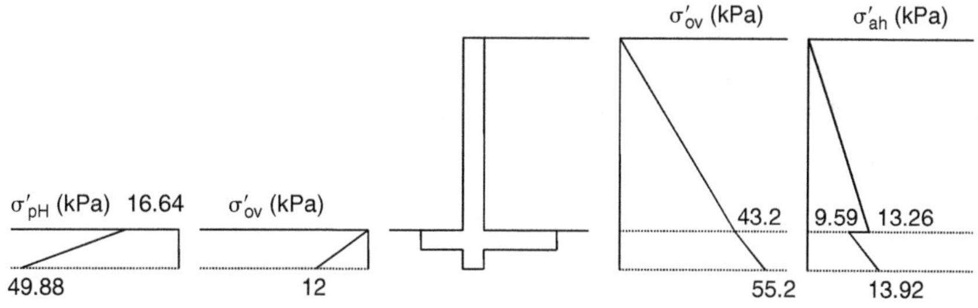

Figure 21.25s Earth pressure diagram.

$$P_{ah} = \frac{1}{2} \times 13.26 \times 2.4 + 9.59 \times 0.6 + \frac{1}{2} \times (13.92 - 9.59) \times 0.6 = 22.96\text{kN}$$

$$X_a = \frac{\frac{1}{2} \times 13.26 \times 2.4 \times 1.4 + 9.59 \times 0.6 \times 0.3 + \frac{1}{2} \times (13.92 - 9.59) \times 0.6 \times 0.2}{22.965} = \frac{24.26}{22.96} = 1.06\text{m}$$

$$P_{ph} = 16.64 \times 0.6 + \frac{1}{2} \times (49.88 - 16.64) \times 0.6 = 19.96\text{kN}$$

$$X_p = \frac{16.64 \times 0.6 \times 0.3 + \frac{1}{2} \times (49.88 - 16.64) \times 0.6 \times 0.2}{19.956} = \frac{4.99}{19.96} = 0.25\text{m}$$

$$W_{soil1} = 2.4\text{ m} \times 1\text{ m} \times 18\text{ kN/m}^2 = 43.2\text{ kN/m}$$

$$W_{soil2} = 0.5\text{ m} \times 0.3\text{ m} \times 20\text{ kN/m}^2 = 3\text{ kN/m}$$

$$W_{soil3} = 1\text{ m} \times 0.3\text{ m} \times 20\text{ kN/m}^2 = 6\text{ kN/m}$$

$$W_{stem} = (2.4\text{ m} + 0.3\text{ m}) \times 0.3\text{ m} \times 25\text{ kN/m}^2 = 20.25\text{ kN/m}$$

$$W_{base} = 1.8\text{ m} \times 0.3\text{ m} \times 25\text{ kN/m}^2 = 13.5\text{ kN/m}$$

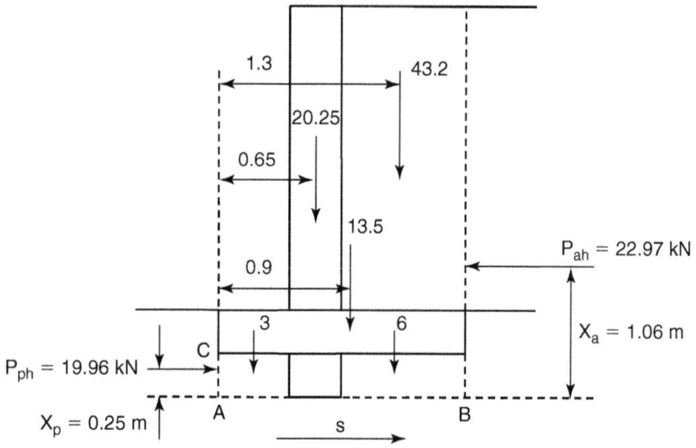

Figure 21.26s Forces diagram.

Considering a sliding failure along AB:

$$F_{\text{sliding}} = \frac{\sum W \tan\varphi' + P_{pH}}{P_{aH}} = \frac{(43.2 + 20.25 + 13.5 + 6 + 3) \times \tan 28^\circ + 19.96}{22.97}$$
$$= \frac{45.70 + 19.96}{22.97} = 2.84 > 2 \rightarrow OK$$

Considering a rotation failure around point A:

$$F_{\text{overturning}} = \frac{M_{\text{max,resist}}}{M_{\text{driving}}} = \frac{43.2 \times 1.3 + 20.25 \times 0.65 + 13.5 \times 0.9 + 19.96 \times 0.25}{22.97 \times 1.06}$$
$$= \frac{86.46}{24.35} = 3.55 > 2 \rightarrow OK$$

It is also reasonable to consider rotation failure around point C:

$$F_{\text{overturning}} = \frac{M_{\text{max,resist}}}{M_{\text{driving}}} = \frac{43.2 \times 1.3 + 20.25 \times 0.65 + 13.5 \times 0.9}{22.97 \times 0.76} = \frac{81.47}{17.46} = 4.67 > 2 \rightarrow OK$$

Problem 21.16

Design the soil reinforcing strips required for a 20 m high MSE wall. The precast concrete panels are 1.5 m by 1.5 m. The vertical and horizontal spacing between strips are 750 mm and 450 mm respectively. The unit weight of the backfill material is 19 kN/m^3 with an angle of internal friction of 34° and a coefficient of uniformity of 4.4. The location of the first layer of strips, measured from the finished grade, is 375 mm. Neglect the traffic surcharge.

Solution 21.16

$$\text{Panel section} = 1.5 \text{ m} \times 1.5 \text{ m}$$
$$s_v = 0.75 \text{ mm}$$
$$s_h = 450 \text{ mm}$$
$$\gamma_{\text{soil}} = 19 \text{ kN/m}^3$$
$$C_u = 4.4$$

a. Design for Pullout

The maximum line load ($T_{\max}$) to be resisted by the reinforcement inclusions at depth z can be computed as:

$$T_{\max} = s_v s_h \sigma_h$$

The horizontal stress σ_h can be calculated as:

$$\sigma_h = k_r \sigma_{ov} + \Delta\sigma_h$$
$$\sigma_h = k_r \sigma_{ov}$$

The coefficient of earth pressure k_r is computed using Figure 21.27s (AASHTO). The k_a value is computed as:

$$k_a = \frac{1 - \sin\varphi'}{1 + \sin\varphi'} = \frac{1 - \sin 34}{1 + \sin 34} = 0.283$$

Then the k_r value is computed as:
a-1. If z_i is less than 6 m, then:

$$\frac{k_r}{k_a} = 1.7 - \frac{z_i}{12}$$

a-2. If z_i is larger than 6 m, then:

$$\frac{k_r}{k_a} = 1.2$$

The calculation of T_{max} for the different strips is summarized in Table 21.1s.

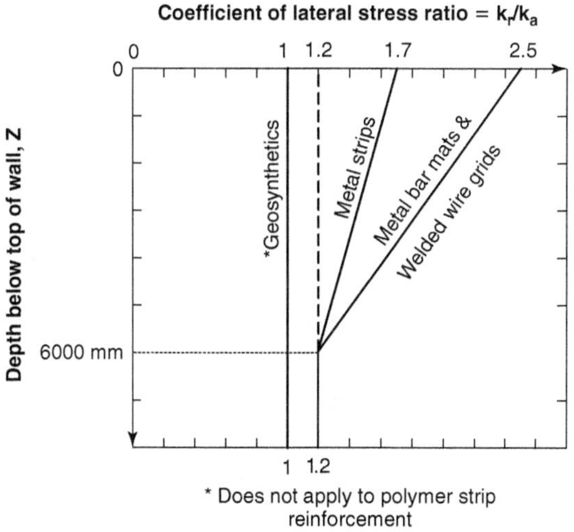

Figure 21.27s Coefficient of lateral stress ratio = k_r/k_a.

Table 21.1s Summary of Calculation of T_{max}

Strip No.	Depth (m)	k_a	k_r/k_a	k_r	σ_v (kPa)	σ_h (kPa)	T_{max} (kN)
1	0.375	0.283	1.67	0.472	7.1	3.365	1.14
2	1.125	0.283	1.61	0.455	21.4	9.716	3.28
3	1.875	0.283	1.54	0.437	35.6	15.564	5.25
4	2.625	0.283	1.48	0.419	49.9	20.907	7.06
5	3.375	0.283	1.42	0.402	64.1	25.747	8.69
6	4.125	0.283	1.36	0.384	78.4	30.082	10.15
7	4.875	0.283	1.29	0.366	92.6	33.913	11.45
8	5.625	0.283	1.23	0.348	106.9	37.240	12.57
9	6.375	0.283	1.20	0.340	121.1	41.134	13.88
10	7.125	0.283	1.20	0.340	135.4	45.973	15.52
11	7.875	0.283	1.20	0.340	149.6	50.813	17.15
12	8.625	0.283	1.20	0.340	163.9	55.652	18.78
13	9.375	0.283	1.20	0.340	178.1	60.491	20.42
14	10.125	0.283	1.20	0.340	192.4	65.331	22.05
15	10.875	0.283	1.20	0.340	206.6	70.170	23.68
16	11.625	0.283	1.20	0.340	220.9	75.009	25.32
17	12.375	0.283	1.20	0.340	235.1	79.848	26.95
18	13.125	0.283	1.20	0.340	249.4	84.688	28.58
19	13.875	0.283	1.20	0.340	263.6	89.527	30.22
20	14.625	0.283	1.20	0.340	277.9	94.366	31.85
21	15.375	0.283	1.20	0.340	292.1	99.206	33.48
22	16.125	0.283	1.20	0.340	306.4	104.045	35.12
23	16.875	0.283	1.20	0.340	320.6	108.884	36.75
24	17.625	0.283	1.20	0.340	334.9	113.724	38.38
25	18.375	0.283	1.20	0.340	349.1	118.563	40.01
26	19.125	0.283	1.20	0.340	363.4	123.402	41.65
27	19.875	0.283	1.20	0.340	377.6	128.241	43.28

Now that we have calculated the load T_{max}, we need to find the length of reinforcement that will safely carry that load without pulling out of the soil. The pull-out capacity $T_{pullout}$ (kN) of the reinforcement inclusion is given by:

$$T_{pullout} = 2 \times f_{max} \times b \times L_a$$
$$f_{max} = F^* \times \sigma'_{ov} \times \alpha$$

Using the ultimate limit state procedure, we have:

$$\gamma T_{max} = \phi T_{pullout}$$

The active length of the reinforcement strip required to resist the pullout load is:

$$T_{pullout} = \frac{\gamma T_{max}}{\varphi}$$
$$L_a = \frac{T_{pullout}}{2 \times f_{max} \times b}$$
$$L_a = \frac{(\gamma_1 k_r \sigma'_{ov}) \times s_v \times s_h}{2 \times \varphi \times F^* \times \sigma'_{ov} \times \alpha \times b}$$
$$L_a = \frac{(\gamma_1 k_r) \times s_v \times s_h}{2 \times \varphi \times F^* \times \alpha \times b}$$
$$L = L_a + L_{max} = \frac{(\gamma_1 k_r) \times s_v \times s_h}{2 \times \varphi \times F^* \times \alpha \times b} + 0.3H$$

The value of α is taken as 1.0 for strip reinforcements (Section 21.10.2). The resistance and load factors are taken as 0.9 and 1.35 respectively. The coefficient of friction (F^*) is computed according to AASHTO LRFD using Figure 21.28s.

If z_i is less than 6 m, then:

$$F^* = 1.2 + Log\ C_u = 1.8435\ at\ z = 0\ \text{m}$$
$$F^* = 0.6745\ at\ z = 6\ \text{m}$$
$$F^* = 1.8435 - 0.1948 \times z_i$$

If z_i is larger than 6 m, then:

$$F^* = \tan\phi = 0.6745$$

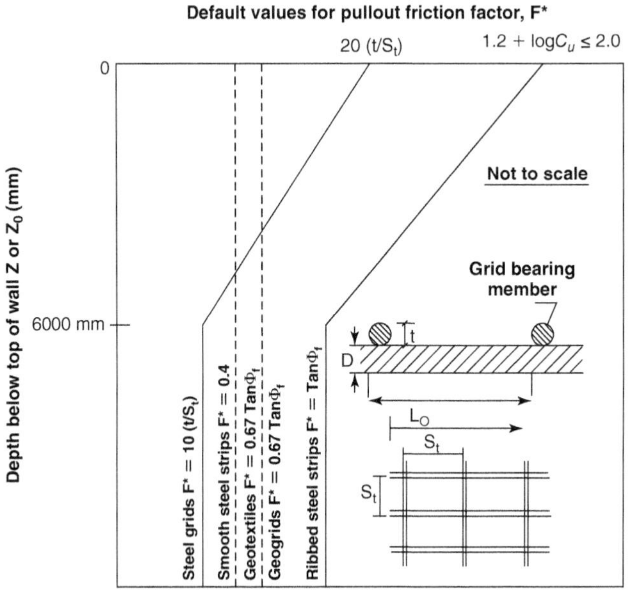

Figure 21.28s Friction coefficient F* for MSE wall reinforcement.

Table 21.2s Summary of Calculation of Total Strip Length (L)

Strip No.	Depth (m)	T_{max} (kN)	F*	f_{max} (kPa)	L_a (m)	L (m)
1	0.375	1.14	1.770	12.61	1.350	7.350
2	1.125	3.28	1.624	34.72	0.944	6.944
3	1.875	5.25	1.478	52.66	0.997	6.997
4	2.625	7.06	1.332	66.44	1.062	7.062
5	3.375	8.69	1.186	76.06	1.143	7.143
6	4.125	10.15	1.040	81.51	1.246	7.246
7	4.875	11.45	0.894	82.79	1.382	7.382
8	5.625	12.57	0.748	79.92	1.573	7.573
9	6.375	13.88	0.675	81.70	1.699	7.699
10	7.125	15.52	0.675	91.31	1.699	7.699
11	7.875	17.15	0.675	100.92	1.699	7.699
12	8.625	18.78	0.675	110.53	1.699	7.699
13	9.375	20.42	0.675	120.15	1.699	7.699
14	10.125	22.05	0.675	129.76	1.699	7.699
15	10.875	23.68	0.675	139.37	1.699	7.699
16	11.625	25.32	0.675	148.98	1.699	7.699
17	12.375	26.95	0.675	158.59	1.699	7.699
18	13.125	28.58	0.675	168.20	1.699	7.699
19	13.875	30.22	0.675	177.82	1.699	7.699
20	14.625	31.85	0.675	187.43	1.699	7.699
21	15.375	33.48	0.675	197.04	1.699	7.699
22	16.125	35.12	0.675	206.65	1.699	7.699
23	16.875	36.75	0.675	216.26	1.699	7.699
24	17.625	38.38	0.675	225.87	1.699	7.699
25	18.375	40.01	0.675	235.48	1.699	7.699
26	19.125	41.65	0.675	245.10	1.699	7.699
27	19.875	43.28	0.675	254.71	1.699	7.699

b. Design for Yielding

Using the ultimate limit state procedure, we have:

$$\gamma T_{max} = \phi T_{yield}$$

The resistance and load factors are taken as 0.75 and 1.35 respectively. The T_{yield} for steel reinforcement is given by:

$$T_{yield} = \sigma_{yield} \times A$$
$$T_{yield} = \sigma_{yield} \times b \times E_c$$

The value of A is the cross-sectional area of the strip after accounting for corrosion (AASHTO 2010). The structural thickness of the strip at the end of the service life is computed according to AASHTO LRFD as:

$$\textit{Service Life of Zinc Coating}\ (0.086\ \text{mm/year}) = 2\ \text{years} + \frac{0.086 - 2 \times 0.015}{0.004}\text{years}$$

$$\textit{Service Life of Zinc Coating}\ (0.086\ \text{mm/year}) = 16\ \text{years}$$

Use a strip thickness 50 mm wide and 5 mm thick:

$$E_c = 5\text{ mm} - 2E_s$$
$$E_c = 5\text{ mm} - 2 \times (75\text{ years} - 16\text{ years}) \times 0.12\text{ mm/year}$$
$$E_c = 3.58\text{ mm}$$

Then:

$$T_{yield} = 448159.2\text{ kPa} \times (0.00358\text{ mm} \times 0.05\text{ mm})$$
$$T_{yield} = 80.2\text{ kN}$$

Then, using the result of $T_{\max}$ at the bottom layer of strips where the maximum tension load is expected, we have:

$$0.75 \times 80.2\text{ kN} > 1.35 \times 43.3\text{ kN}$$
$$60.2\text{ kN} > 58.5\text{ kN} \quad \therefore\ OK$$

Detailed calculations for all the strips are shown in Table 21.3s.

Table 21.3s Summary of Calculations for Strip Resistance to Yielding

Strip No.	Depth (m)	T_{max} (kN)	R	φR	γT_{max}	Check
1	0.375	1.14	80.2	60.2	1.5	OK
2	1.125	3.28	80.2	60.2	4.4	OK
3	1.875	5.25	80.2	60.2	7.1	OK
4	2.625	7.06	80.2	60.2	9.5	OK
5	3.375	8.69	80.2	60.2	11.7	OK
6	4.125	10.15	80.2	60.2	13.7	OK
7	4.875	11.45	80.2	60.2	15.5	OK
8	5.625	12.57	80.2	60.2	17.0	OK
9	6.375	13.88	80.2	60.2	18.7	OK
10	7.125	15.52	80.2	60.2	20.9	OK
11	7.875	17.15	80.2	60.2	23.2	OK
12	8.625	18.78	80.2	60.2	25.4	OK
13	9.375	20.42	80.2	60.2	27.6	OK
14	10.125	22.05	80.2	60.2	29.8	OK
15	10.875	23.68	80.2	60.2	32.0	OK
16	11.625	25.32	80.2	60.2	34.2	OK
17	12.375	26.95	80.2	60.2	36.4	OK
18	13.125	28.58	80.2	60.2	38.6	OK
19	13.875	30.22	80.2	60.2	40.8	OK
20	14.625	31.85	80.2	60.2	43.0	OK
21	15.375	33.48	80.2	60.2	45.2	OK
22	16.125	35.12	80.2	60.2	47.4	OK
23	16.875	36.75	80.2	60.2	49.6	OK
24	17.625	38.38	80.2	60.2	51.8	OK
25	18.375	40.01	80.2	60.2	54.0	OK
26	19.125	41.65	80.2	60.2	56.2	OK
27	19.875	43.28	80.2	60.2	58.4	OK

Note: Bearing capacity and slope stability failure were not checked as part of this problem. However, they must be checked to ensure that the system is safe against these failure modes.

Problem 21.17

A cantilever retaining wall is embedded 6 m below excavation level and retains 5 m of soil. An impervious layer exists 4 m below the bottom of the wall. The water level is at the ground surface on both sides of the wall and the soil deposit is uniform and deep. Draw the water pressure diagram against the wall on both sides of the wall, assuming that the water pressure is hydrostatic. Then draw a flow net and develop the water pressure diagram on both sides of the wall. Compare and comment.

Solution 21.17

The hydrostatic water pressure diagram is shown in Figure 21.29s and the flow net in Figure 21.30s.

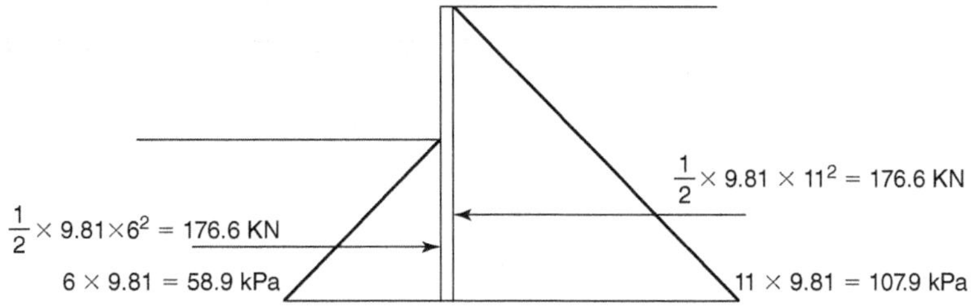

Figure 21.29s Water pressure in hydrostatic conditions.

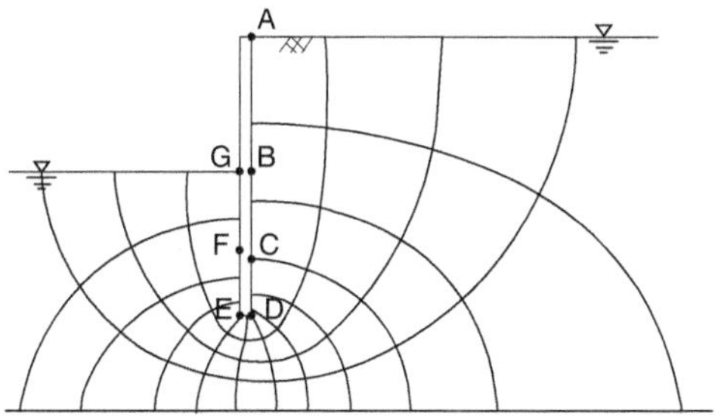

Figure 21.30s Flow net.

Water pressures are calculated at points A, B, C, D, E, F, and G on each side of the wall to generate the water stress profile. The loss of total head through the flow net is 5 m. The loss of total head through each flow field is 5/12 = 0.417 m. The total head $h_{t(M)}$ at any point M is calculated by:

$$h_{t(M)} = h_{t(beg)} - n_d \Delta h_t$$

where $h_{t(beg)}$ is the total head at the beginning of the flow net (11 m), n_d is the number of equipotential drops to go from the beginning of the flow net to point M, and Δh_t is the drop of total head across any flow field. Then the elevation head $h_{e(M)}$ is measured on the scaled drawing and the pressure head $h_{p(M)}$ is obtained as the difference between the total head and the elevation head (Table 21.4s.).

Table 21.4s

Point	Total Head (m)	Elevation Head (m)	Pressure Head (m)	WaterStress (kPa)
A	15	15	0	0
B	14.33	9.6	4.73	46.40
C	13.75	6.0	7.75	76.03
D	12.5	4	8.5	83.38
E	11.66	4	7.66	75.14
F	10.62	6.3	4.32	42.38
G	10	10	0	0

The water pressure diagram from the flow net is shown in Figure 21.31s together with the hydrostatic diagram. As can be seen, the hydrostatic diagram is more conservative.

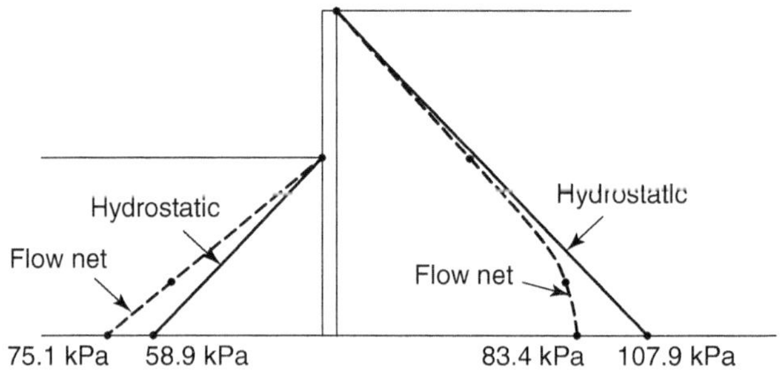

Figure 21.31s Water pressure under flow conditions.

Problem 21.18

Demonstrate Equation 21.91.

Solution 21.18

Equation 21.90 expresses moment equilibrium at the bottom of a wall:

$$P_a X_a - P_{pm} X_{pm} = 0$$

Using Equations 21.86–21.88, and 21.89 in Eq. 21.90, we get:

$$\left(\frac{1}{2}K_a\gamma(H+D)^2\right)\left(\frac{1}{3}(H+D)\right) - \left(\frac{3}{8}K_p\gamma D^2\right)\left(\frac{7}{18}D\right) = 0$$

$$\left(\frac{3}{8}K_p D^2\right)\left(\frac{7}{18}D\right) = \left(\frac{1}{2}K_a(H+D)^2\right)\left(\frac{1}{3}(H+D)\right)$$

$$\frac{21}{144}K_p D^3 = \frac{1}{6}K_a(H+D)^3$$

$$\frac{7}{8}K_p D^3 = K_a(H+D)^3$$

$$\sqrt[3]{\frac{7}{8}\frac{K_p}{K_a}D^3} = H + D$$

$$\left(\left(\frac{7}{8}\frac{K_p}{K_a}\right)^{0.33} D\right) - D = H$$

$$D\left(\left(\frac{7}{8}\frac{K_p}{K_a}\right)^{0.33} - 1\right) = H$$

$$D = \frac{H}{\left(\frac{7}{8}\frac{K_p}{K_a}\right)^{0.33} - 1}$$

Problem 21.19

What is the depth of embedment d required for a cantilever wall retaining a height of sand *H*? Express the results as a function of *H*, K_p/K_a, and a factor of safety *F* applied to σ_p, the passive pressure. (Note: There is no water.)

Solution 21.19

$$\frac{1}{2}K_a\gamma(H+D)^2 \times \frac{1}{3}(H+D) - \frac{\frac{1}{2}K_p\gamma D^2 \times \frac{1}{3}D}{F.S} = 0$$

$$K_a(H+D)^3 = \frac{1}{F.S}K_pD^3$$

$$\frac{H+D}{D} = \left(\frac{1}{F.S}\frac{K_p}{K_a}\right)^{0.33}$$

$$D = \frac{H}{\left(\frac{1}{F.S}\frac{K_p}{K_a}\right)^{0.33} - 1}$$

Problem 21.20

For the anchored slurry wall shown in Figure 21.3s, calculate the pressure distribution on both sides of the wall for a deflection of 25 mm at the top of the wall. Calculate the anchor forces. How important is the vertical capacity of the wall? Explain your answer. What would happen if the water level rose on both sides of the excavation to the top of the wall? What would happen if the water level rose to the top of the wall on the retained-soil side of the excavation and to 2 m below that on the excavated side?

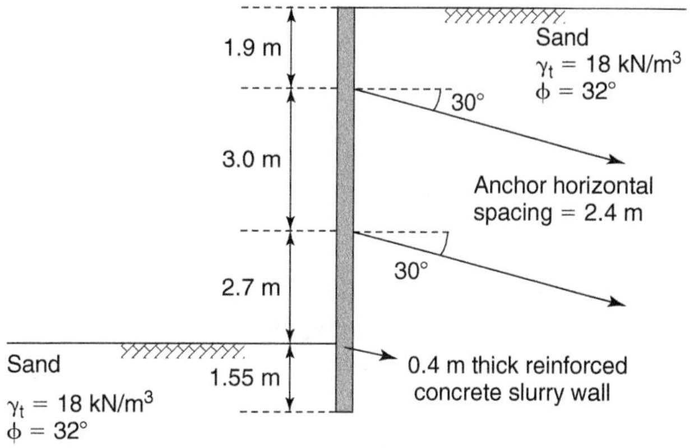

Figure 21.3s Anchored slurry wall.

Solution 21.20

From Figure 21.19 and $u_{top}/H = 0.025/7.6 = 0.003$, K behind the wall is 0.2 (average). Using Eq. 21.99, the constant pressure from $z = 0$ to $z = H$ is:

$$\begin{aligned}\sigma_h &= K\sigma'_{ov}(at\, z = H) + u_w \\ &= 0.2(18)(7.6) \\ &= 27.4\ \text{kN/m}^2\end{aligned}$$

For $z = H$, just below the constant pressure, the K_a active earth pressure is used:

$$\begin{aligned}\sigma_{ah} &= K_a\sigma_v \\ &= \frac{1-\sin(30)}{1+\sin(30)}(18)(7.6) \\ &= 45.5\ \text{kN/m}^2\end{aligned}$$

For $z = H + D$, the active pressure is:

$$\sigma_{ah} = K_a \sigma_v$$
$$= \frac{1}{3}(18)(9.15)$$
$$= 54.8 \text{ kN/m}^2$$

For $z = H$ on the excavation side, the passive pressure is 0 and at $z = H + D$, the passive pressure is:

$$\sigma_{ph} = K_p \sigma_v$$
$$= 3(18)(1.55)$$
$$= 83.7 \text{ kN/m}^2$$

Using the tributary area for the top anchor, the horizontal component for that anchor is:

$$F_{1h} = \sigma_h A_1 = 27.4 \times (1.9 + 1.5) \times 2.4 = 223.6 \text{ kN}$$

Using the tributary area for the bottom anchor, the horizontal component for that anchor is:

$$F_{2h} = \sigma_h A_2 = 27.4 \times (1.5 + 1.35) \times 2.4 = 187.4 \text{ kN}$$

Because the anchors are inclined at 30°, the actual loads in the anchors are:

$$F_1 = \frac{F_{1h}}{\cos\alpha} = \frac{223.6}{\cos 30} = 258.2 \text{ kN}$$

$$F_2 = \frac{F_{2h}}{\cos\alpha} = \frac{187.4}{\cos 30} = 216.4 \text{ kN}$$

The vertical capacity is important because the soil mass tends to move toward the excavation and downward. The downward movement imposes downdrag on the wall. If the vertical capacity is insufficient, the wall will move downward and rotate around the anchor. This will cause horizontal movement as well.

If water rises on both sides to the top of the wall, the water pressure on both sides will cancel out and the soil horizontal stress will decrease from the total stress ($K\gamma H$) to the effective stress ($K\gamma' H$). This would lead to a pressure on the wall of about one-half of the pressure with no water on either side. If there was a difference in level of 2 m, there would be a net water pressure equal to 2 m of water on the wall in addition to the $K\gamma' H$.

Problem 21.21

Explain Figure 21.49.

Solution 21.21

Figure 21.49 shows an example of load distribution in an anchor in tension. The load resisted by the soil increases steadily from the back of the anchor to the front of the anchor. The load in the tendon is constant and equal to the anchor load along the tendon unbonded length because the greased sheath that covers the anchor does not permit any load transfer. Then the load in the tendon drops off as the grout contributes to the load being resisted. Within the zone where the grout is in tension, the tendon is the only one carrying load, because the grout cracks and contributes no load to the resistance. Within the tensile strains where the grout can resist tension, some of the load is carried by the tendon and some by the grout. The grout has zero load at the ground surface and the load increases in compression over the unbonded tendon length because the grout moves with respect to the soil and is loaded in compression. Beyond the tendon unbonded length, the grout is in tension to such a level that it cracks and cannot contribute to the resistance. Then, in the back of the anchor, the tension load decreases to the point where the strains are low enough and the grout can resist some tension.

Problem 21.22

Use Tables 21.4 and 21.5 and add a column giving the back-calculated alpha values.

Solution 21.22 (Table 21.5s)

$$f_{\max} = \alpha_c s_u$$

$$\alpha_c = \frac{f_{\max}}{s_u}$$

Sample calculation: Stiff silt-clay mixture $s_u = 50$ kPa, $f_{\max} = 30$ kPa, $\alpha_c = \frac{30}{50} = 0.6$

Table 21.5s

Anchor Type (Grout Pressure)	Soil Type	Shear Strength of Soil s_u (kPa)	Shear Strength of Soil-Grout Interface f_{max} (kPa)	α_s
Gravity grouted anchors (< 350 kPa)	Silt-clay mixtures	Stiff to very stiff 50 to 200	30 to 70	0.35-0.6
Pressure grouted anchors (350 to 2800 kPa)	High-plasticity clay	Stiff (50 to 120)	30 to 100	0.6-0.83 0.58-0.85
		Very stiff (120 to 200)	70 to 170	
	Medium-plasticity clay	Stiff (50 to 120)	100 to 250	2-2.1 1.2-1.75
		Very stiff (120 to 200)	140 to 350	
	Medium-plasticity sandy silt	Very stiff (120 to 200)	280 to 380	2.3-1.9

Problem 21.23

For Figure 21.62, the height H is 9 m, α is 17°, β is 18°, ψ is 44°, and i is 10°. The stiff clay weighs 20 kN/m^3 with some cohesion c' (to be ignored), and a friction angle φ' of 32°. A uniformly applied surcharge of 10 kPa is to be considered on top of the wall. Calculate the required nail force T for a factor of safety against shear failure along the chosen plane to be 1.5. Distribute that force among the four nails and find the required length for each nail.

Solution 21.23 (Figure 21.32s)

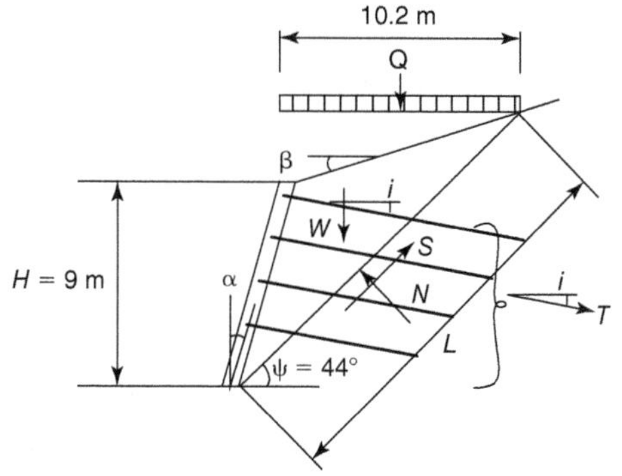

Figure 21.32s Nailed wall.

Equations 22.112, 22.113, and 22.114 are used to find the three unknowns N, S, and T:

$$(W + Q)\cos\psi + T\sin(\psi + i) - N = 0$$

$$(W + Q)\sin\psi - T\cos(\psi + i) - S = 0$$

$$S = \frac{S_{\max}}{F} = \frac{c'L + N\tan\varphi'}{F}$$

The weight of the soil is obtained by multiplying the unit weight of the soil by the area of the triangle. The area of the triangle is 41.4 m^2 per meter perpendicular to the page and W is 828 kN/m. Substituting these values in the equations:

$$(828 + 10.2 \times 10)\cos 44 + T\sin(44 + 15) - N = 0$$
$$(828 + 10.2 \times 10)\sin 44 - T\cos(44 + 15) - S = 0$$
$$\frac{N\tan 30}{1.5} - S = 0$$

Then:

$$669 + 0.857T - N = 0$$
$$646 - 0.515T - S = 0$$
$$0.385N - S = 0$$

After solving the system of equations, $N = 1065$ kN/m, $S = 410$ kN/m, and $T = 462$ kN/m. For simplification, if all the nails in the wall carry the same force, then the force T is divided by the four nails and the force at each nail is approximately 115.5 kN/m. The required length of the nails can be found using Eq. 22.115 plus a factor of safety F:

$$R_a = \frac{T}{n} = \frac{\pi D L_p f_{max}}{F}$$

where R_a is the allowable load on each nail, T is the total nail load, n is the number of nails, D is the diameter of the nail (drill hole), L_p is the useful length of the nail, and f_{max} is the shear strength at the grout-soil interface. The shear strength f_{max} depends on the soil and the construction method and is estimated using Table 22.8. For a stiff clay, an f_{max} of 50 kPa can be used and a diameter of 200 mm.

$$L_p = \frac{FT}{n\pi D f_{max}} = \frac{1.5 \times 462}{4\pi \times 0.2 \times 50} = 5.5 \text{ m}$$

The required length L_p of each nail is 4.4 m. The length of the nail inside the failure zone is the discounted length L_d. The total length of each nail is the sum of the required length and the discounted length. The total length of each nail is:

$$L_{ti} = L_{di} + L_{pi}$$
$$L_{t1} = L_{d1} + L_{p1} = 5.6 + 5.5 = 11.1 \text{ m}$$
$$L_{t2} = L_{d2} + L_{p2} = 4.3 + 5.5 = 9.8 \text{ m}$$
$$L_{t3} = L_{d3} + L_{p3} = 3 + 5.5 = 8.5 \text{ m}$$
$$L_{t4} = L_{d4} + L_{p4} = 1.8 + 5.5 = 7.3 \text{ m}$$

Problem 21.24

A 3 m wide strutted excavation is planned in a clay with an undrained shear strength equal to 40 kPa and a total unit weight of 19 kN/m^3. What depth of excavation corresponds to a factor of safety against base failure equal to 1.5?

Solution 21.24

The safety factor for the base failure can be calculated using the following equation:

$$F = \frac{N_c s_u}{\sigma_{ov(z=H)}}$$

Assume that $H = 3$m and $H/B = 1$. Then, using the Skempton chart, the $N_c = 6.4$

$$1.5 = \frac{6.4 \times 40}{19 \times H} \text{ and H} = 8.9 \text{ m}$$

Assume that $H = 9$ and $H/B = 3$. Then, using the Skempton chart, the $N_c = 7.3$

$$1.5 = \frac{7.3 \times 40}{19 \times H} \text{ and H} = 10.24 \text{ m}$$

Assume that $H = 10$ and $H/B = 3.33$. Then, using the Skempton chart, the $N_c = 7.3$

$$1.5 = \frac{7.3 \times 40}{19 \times H} \text{ and } H = 10.24 \text{ m}$$

$$H = 10 \text{ m}$$

CHAPTER 22

Earthquake Geoengineering

This chapter serves as an introduction to the large and complex field of geotechnical earthquake engineering. The book by Kramer (1996), the book by Towhata (2008), and the FHWA manual by Kavazanjian et al. (2011) are excellent references for further study.

22.1 BACKGROUND

Plate tectonics is the main reason for earthquakes on our planet. The Earth's crust is made of six continental plates (Figure 22.1) that have travelled large distances over geologic times. The plates move because of the thermal difference between the earth surface and the deeper layers. This thermal difference creates convection currents in the rock mass, which move the plates. The boundaries between plates are called *faults*. The problem is that the movement of the plates with respect to one another is not smooth. Indeed, the interface between plates or faults is rough and stresses accumulate along the fault over time. When the stress becomes equal to the strength of the fault surface, the fault shears in a dramatic motion known as an *earthquake* (Figure 22.2). A bit of lubricant would solve that problem!

An earthquake originates at some depth below the ground surface; this point is called the *hypocenter*. The point on the ground surface directly above the hypocenter is the *epicenter*. The distance between a site on the ground surface and the epicenter is the *epicentral distance*. An earthquake starts at one location, propagates along a fault, then propagates up into the rock mass and then into the soil mass (earthquake), and then sometimes into ocean (tsunami). This propagation is in the form of compression waves and shear waves (see section 8.2.1). Seismographs record the passage of the waves. A compression wave moves faster than the corresponding shear wave and will arrive first. The difference in time between the arrival of the compression wave and the shear wave can be used to determine the distance d between the location of the seismograph and the epicenter:

$$d = \frac{\Delta t_{p-s}}{\frac{1}{v_s} - \frac{1}{v_p}} \tag{22.1}$$

where $\Delta t_{p\text{-}s}$ is the difference in time between the arrival of the p wave and the s wave, v_s is the shear wave velocity, and v_p is the compression wave velocity. One seismograph can give the distance d but not the direction of the wave generating the signal; three seismographs are necessary to locate the epicenter (Foster 1988) (Figure 22.3).

22.2 EARTHQUAKE MAGNITUDE

The size of an earthquake can be quantified in several ways. The first and oldest way is the earthquake intensity, which is a qualitative description of the effect of the earthquake. The Mercalli scale (1883) is the best known and goes from I (not felt) to XII (total destruction). Giuseppe Mercalli was an Italian seismologist and volcanologist who proposed this scale in the late 1800s. It was revised a few times after that. The problem with the Mercalli scale is that it relies on human reactions and structural damage observations, both of which depend on more than just the size of the earthquake.

The Richter scale is the most well-known of the magnitude scales (Richter 1935). The Richter magnitude (M_L, with the subscript L used to designate local magnitude) is defined as the logarithm base 10 of the magnitude trace amplitude in micrometers recorded on a Wood-Anderson seismometer located 100 km from the epicenter of the earthquake. Seismic instruments were developed and installed around 1930 and are used extensively today to quantify earthquake magnitude. The Richter scale has been modified over the years and led to the use of the body wave magnitude and surface wave magnitude scales.

The body wave magnitude (m_b) is calculated from the amplitude of compression waves with periods of about 1 sec toward the beginning of the record. The surface wave magnitude (M_s) is calculated from the amplitude of Rayleigh waves with periods of about 20 sec. One limitation with these scales is that they are unable to recognize large earthquakes; this is called *saturation*. Saturation occurs at a magnitude of about 6.2 for m_b and 8 for M_s. Saturation is due to the fact that very large earthquakes release more of their energy at longer

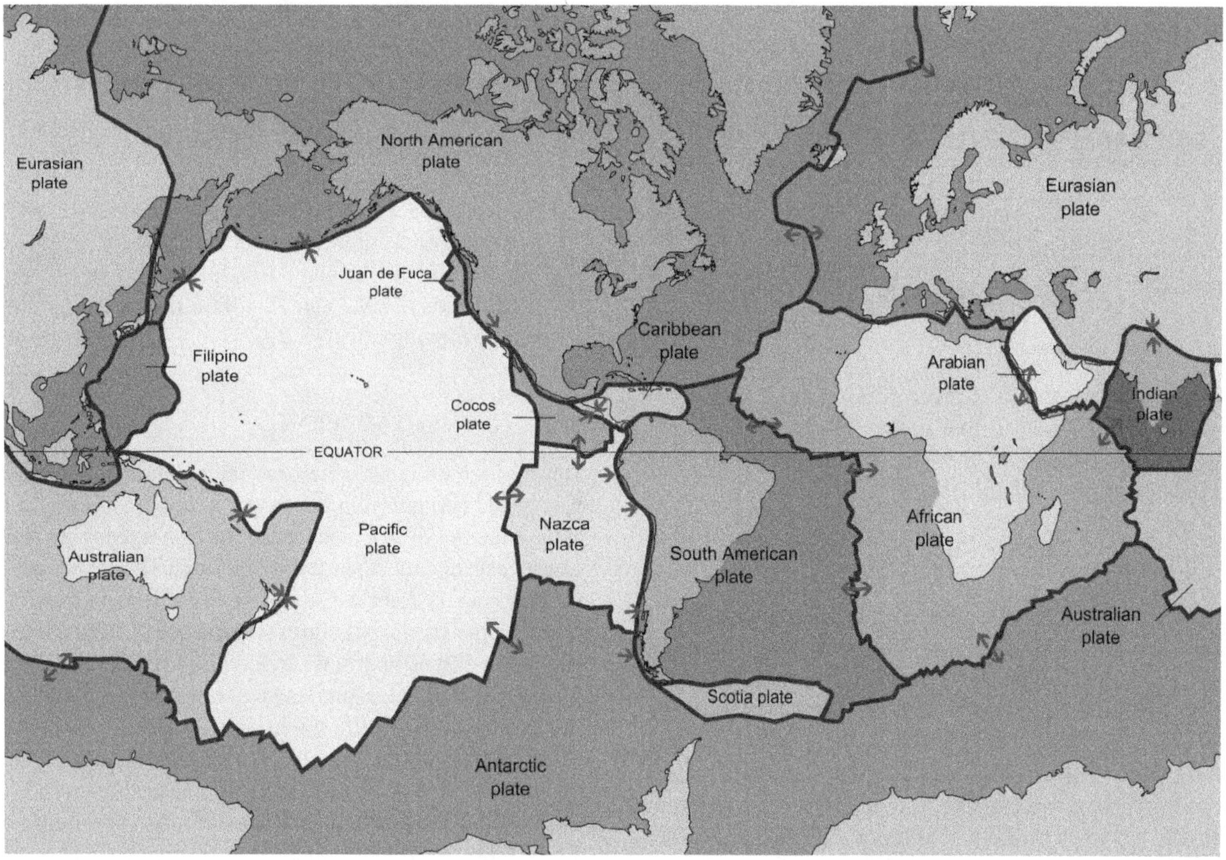

Figure 22.1 Tectonic plates on Earth. (Photo by United States Geologic Survey [USGS])

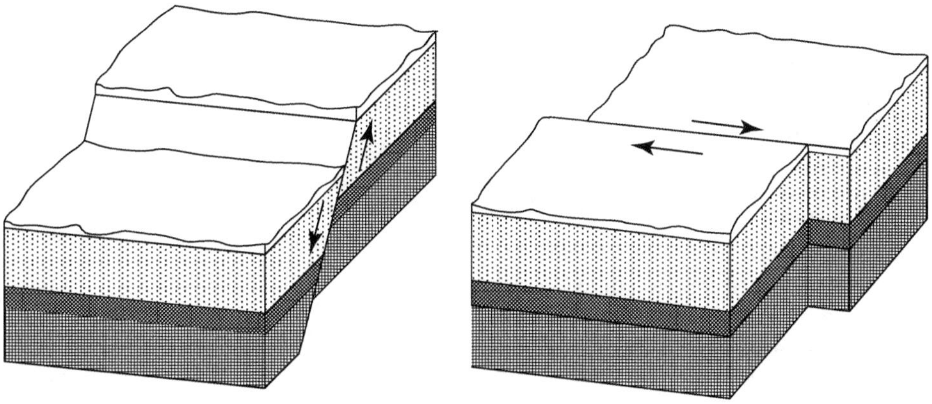

Figure 22.2 Movement of tectonic plates.

periods; because the periods associated with the m_b and M_s calculations are fixed, they cannot acknowledge higher periods and therefore larger earthquakes. Another limitation of these magnitude scales is that they do not address the amount of time associated with the shaking.

The moment magnitude (M_w) takes that aspect into account and is broadly used today. It is rooted in the seismic moment M_o associated with the work done by the earthquake along the fault:

$$M_o = GAD \tag{22.2}$$

where G is the shear modulus of the rock, A is the area over which the slip occurs, and D is the amount of slip movement. Because M_o is a very large number, and because

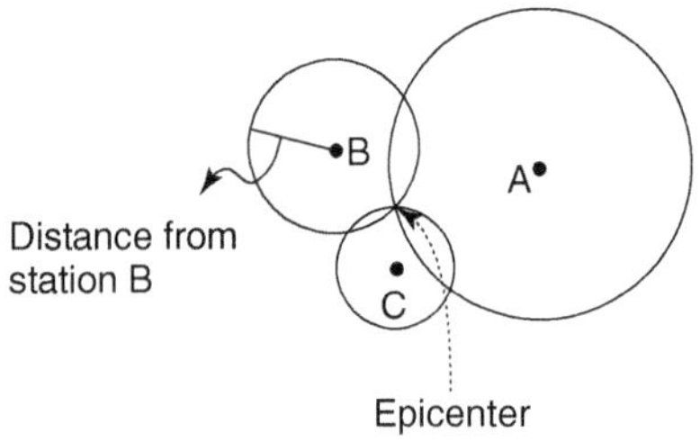

Figure 22.3 Locating the epicenter with three seismographs.

the public is used to the Richter scale, Kanamori (1977) proposed a transformation that makes M_w consistent with the other scales, including the Richter scale. The moment magnitude M_w is then obtained by:

$$M_w = 0.66 \log M_0(\text{N.m}) - 6.05 \quad (22.3)$$

As an example, let's calculate the moment magnitude of the December 26, 2004, Sumatra-Andaman earthquake. As reported by Lay et al. (2005), the fault surface was 1300 km long and 220 km deep, and the slip distance was 5 m. For a typical value for G of 3×10^{10} N/m^2, the seismic moment M_o was 4.7×10^{22} N.m. Then the moment magnitude is very close to 9 or a huge earthquake. Though this is the moment magnitude, the news media would report that the earthquake registered as 9 on the Richter scale. Note that all these scales are log scales, so that, for example, a magnitude 9 earthquake is 10 times larger than a magnitude 8 earthquake.

Yet another way to classify an earthquake is to calculate the energy released during the slip. Bath (1966) proposed to obtain the energy E from:

$$\log E = 5.24 + 1.44M \quad (22.4)$$

where E is the energy in N.m or joules and M is the magnitude. So if M is 9, then E is 1.6×10^{18} joules.

To give you an idea of how much energy this represents, it is enough to cover the electrical consumption of the entire United States for one month. So if we could harness that energy and turn it to good use, it would be very valuable, and unfortunately it seems to be renewable energy!

22.3 WAVE PROPAGATION

For specifics on wave propagation, see section 8.2.1.

22.4 DYNAMIC SOIL PROPERTIES

Dynamic soil properties have been discussed in previous chapters:

- See section 7.2 for the seismic CPT
- See section 7.11.5 for the lightweight deflectometer test
- See sections 8.2.2, 8.2.3, 8.2.4, and 8.2.5 for dynamic in situ tests based on wave propagation
- See section 9.13 for the resonant column test, shear modulus, and damping ratio
- See section 14.10 for the initial tangent shear modulus G_{max}
- See section 14.11 for the normalized shear modulus G/G_{max} and damping ratio vs. shear strain γ curves
- See sections 14.15 and 14.16 for the resilient modulus
- See sections 18.8.7 and 18.8.8 for the rate of loading and cyclic loading effects

22.5 GROUND MOTION

During an earthquake, the rock fault shears and sends shear waves and compression waves through to the ground surface. This shaking of the rock and soil mass can be recorded using instruments sensitive to motion. These are generally accelerometers that use the piezoelectric effect. They contain microscopic crystal structures (crystal quartz) that get stressed by inertia forces and react by creating a change in voltage. This voltage is measured and correlated by calibration to accelerations. While the soil motion created by an earthquake is in three directions, the horizontal motion is usually the one of most interest because it tends to cause the most damage. Figure 22.4 shows an acceleration record for an earthquake along with the velocity and the displacement. The velocity and the displacement are obtained by integrating once and then twice the acceleration versus time.

These time domain signals are quite complex, and there is a need to report simpler parameters to describe an earthquake. These parameters include information on the amplitude A, the frequency f, and the duration t of the acceleration a; velocity v; and displacement u. The amplitudes of a, v, and u can be characterized by the peak values, which are the highest values in the signal. The PGA is the peak ground acceleration, the PGV is the peak ground velocity, and the PGD is the peak ground displacement. The PGA, PGV, and PGD are indicated in Figure 22.4. A huge earthquake can generate 10 m/s^2 or 1 g acceleration, whereas acceleration of 0.1 m/s^2 or 0.01 g is associated with small earthquakes. Figure 22.5 shows a PGA map of the United States prepared by the United States Geologic Service (USGS) for two distinct return periods: 2275 years and 475 years. Also useful are the effective acceleration (acceleration closest to the structural response and damage of the structure), the sustained maximum acceleration (acceleration sustained for 3 or 5 cycles), and the effective design acceleration (peak acceleration after filtering accelerations above 8Hz).

A more detailed inspection of Figure 22.4 shows that the frequencies associated with the acceleration signal are higher than the frequencies associated with the velocity signal, which are themselves higher than the frequencies associated with the displacement signal. The frequency content is different and is best obtained by performing a Fourier transform

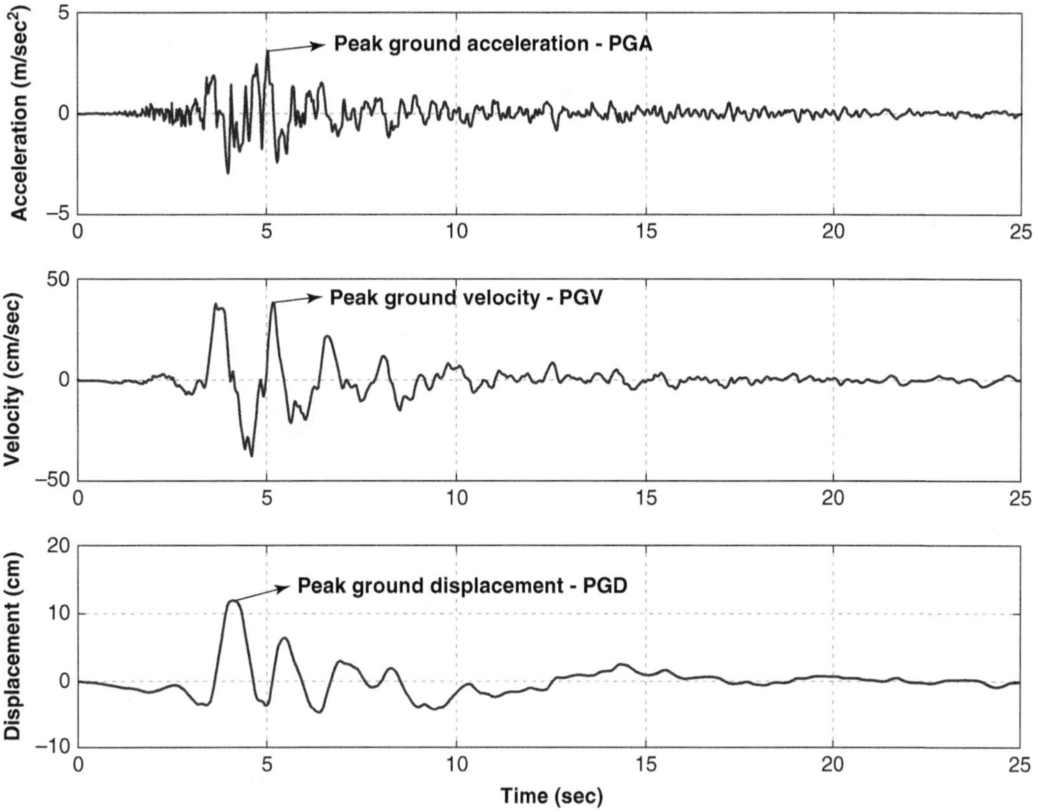

Figure 22.4 Ground motion for an earthquake (FHWA 1998).

analysis (Kramer 1996). This transformation is a mathematical transformation named after the work of Jean Baptiste Joseph Fourier, a French mathematician and physicist, who developed it around 1800. It transforms the signal from a plot of amplitude vs. time into a plot of amplitude vs. frequency (Figure 22.6) or from the time domain to the frequency domain. This amplitude vs. frequency plot is called a *Fourier spectrum*, so one will have a Fourier acceleration spectrum, a Fourier velocity spectrum, and a Fourier displacement spectrum. The Fourier spectra describe the ground motion.

Another spectrum is the response spectrum to a particular earthquake input motion. A *response spectrum* is a plot of the maximum response (a, v, or u) of a linear single degree of freedom (SDOF) system to an earthquake input motion versus the natural period T of the system for a given damping ratio β. The natural period T of an undamped SDOF system is given by:

$$T = 2\pi\sqrt{\frac{m}{k}} \tag{22.5}$$

where m is the mass of the system (kg) and k is the spring stiffness (N/m).

The damping ratio β in percent is given by:

$$\beta = \frac{c}{c_{crit}} \times 100 = \frac{c}{2\sqrt{mk}} \times 100 \tag{22.6}$$

where c is the damping coefficient of the dashpot (N.s/m) and c_{crit} is the critical damping.

The critical damping is what brings the system back to zero without oscillations. Nowadays many doors are equipped with critically damped pistons so they close back without oscillations. (Old saloon bar doors, for example, did not have critical dampers.) In earthquake engineering, a damping ratio equal to 5% is common. Three response spectra are typically created: one each for the acceleration (S_a), the velocity (S_v), and the displacement (S_d).

The process followed to obtain an acceleration response spectrum is illustrated in Figure 22.7 and an example is shown in Figure 22.8. The step-by-step procedure is as follows:

1. Choose the input motion signal for the earthquake.
2. Choose a value of the damping ratio β and the mass m for the SDOF.
3. Select a value of the stiffness k of the SDOF and excite the system with the selected earthquake motion.
4. Record the highest value (acceleration, velocity, displacement) of the output motion of the SDOF.
5. Repeat steps 3 and 4 for different values of k.
6. Plot the maximum values of step 4 (spectral value) versus the fundamental period of the SDOF. This is the response spectrum.

Note that structures have fundamental periods that increase with their height (Figure 22.9); these are generally in the

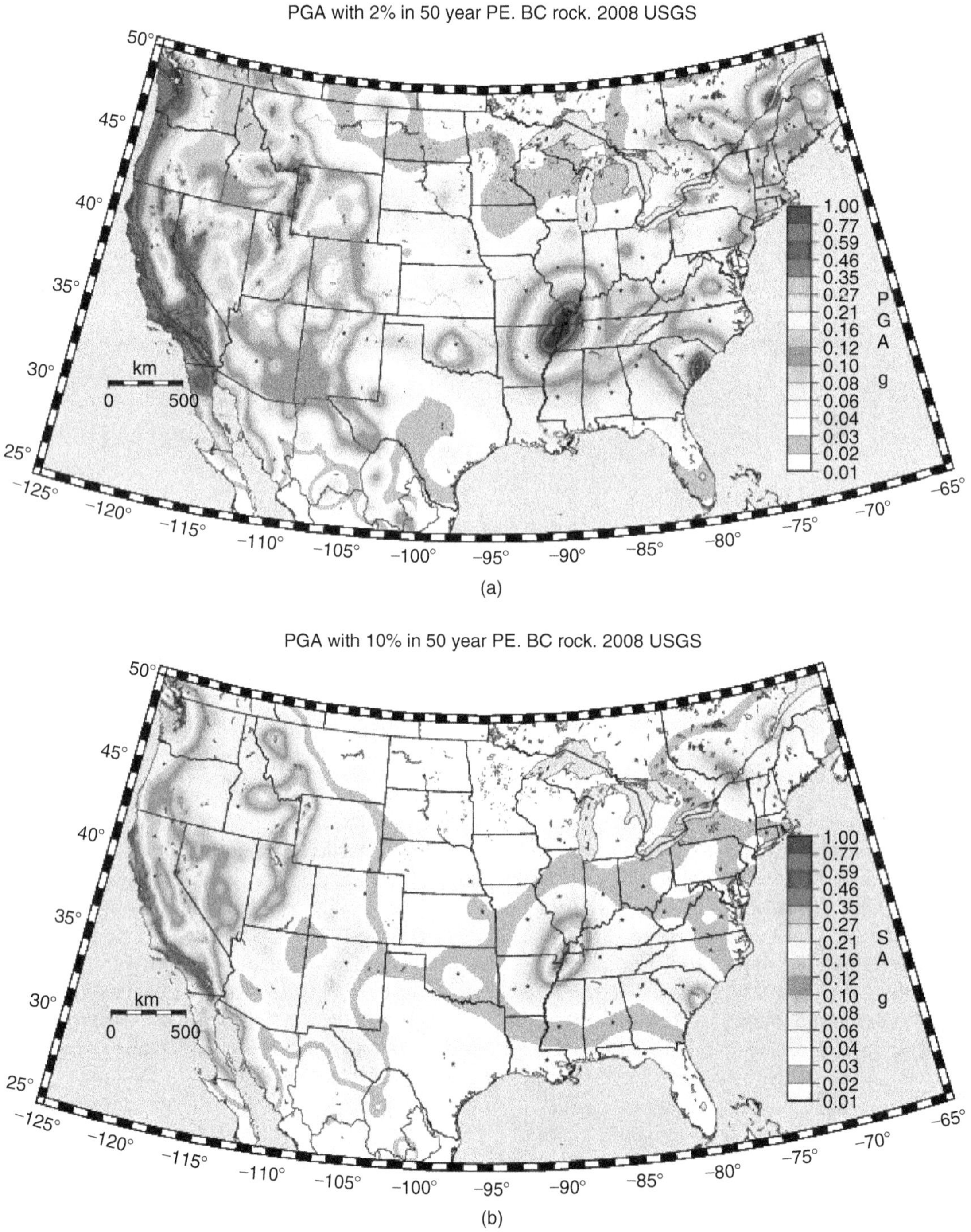

Figure 22.5 USGS maps of peak ground acceleration for 2% and 10% probability of exceedance over 50 years corresponding to 2275-year and 475-year return periods respectively. (Courtesy of USGS.)

range of 0.1 seconds for very small buildings to 10 seconds for extremely tall and flexible buildings. The spectral acceleration on the ordinate of the spectrum depends on the rock motion, the soil properties, the damping ratio, and the ratio of the SDOF stiffness over mass. It does not depend on k and m separately, because of the mathematics behind the problem. The spectrum itself—meaning the curve of a vs. T—is a function of the rock motion, the soil properties, and the damping ratio, but not of the stiffness and the mass of the SDOF. This unique property makes it possible to recommend a single design spectrum that can be used for any structure, given a damping ratio.

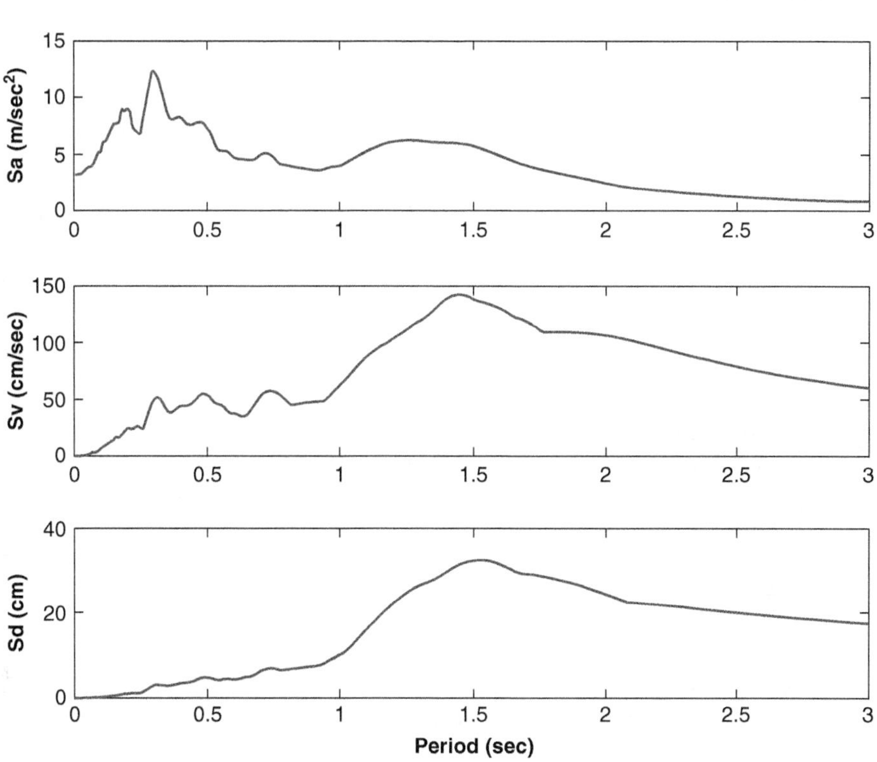

Figure 22.6 Fourier transforms of signals spectrum.

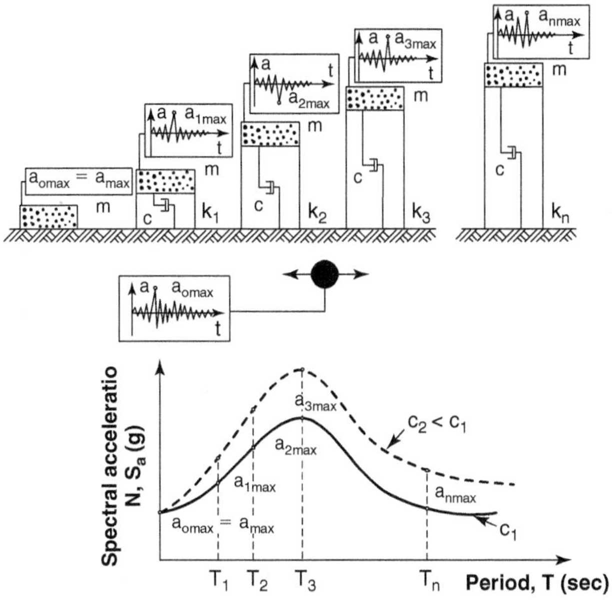

Figure 22.7 Development of an acceleration response spectrum. (After Matasovic 1993.)

The response spectrum is very useful because, for a building with a given fundamental period, this spectrum defines the highest spectral acceleration to which the structure is likely to be subjected. This acceleration times the mass of the building gives the inertia force to be resisted by the structure and the foundation. Table 22.1 shows orders of magnitudes of accelerations and velocities for the soil and the structures placed on it.

The duration of the earthquake's strong motion has a major influence on the amount of damage inflicted. The most common way to measure duration is to use the *bracketed duration*, which is defined as the time between the first and last exceedance of a chosen threshold of acceleration. This threshold is often taken as 0.05 g. Figure 22.10 shows an example in which the bracketed duration is 15 seconds.

22.6 SEISMIC HAZARD ANALYSIS

Now that we know how to characterize ground motion, we need to establish what parameters to consider for the site where the construction will take place, or where the stability must be evaluated, or where liquefaction is an issue. A distinction is made here between a deterministic analysis and a probabilistic analysis. In a seismic hazard deterministic analysis, the steps are as follows:

1. Identify all earthquake sources capable of creating significant ground motion at the site.
2. Determine the distance between the source and the site.
3. Select the controlling earthquake, that is, the earthquake most likely to produce the highest level of shaking at the site.

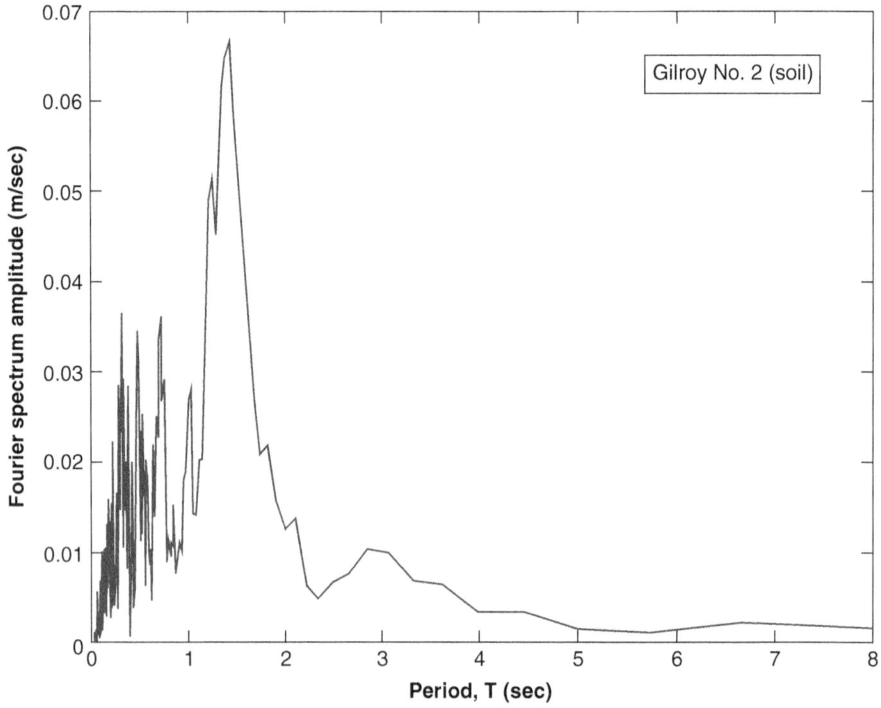

Figure 22.8 Examples of response spectrum.

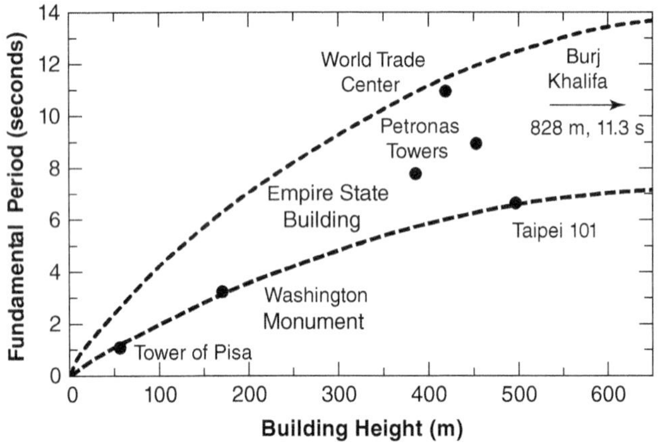

Figure 22.9 Relationship between fundamental period and height of a building.

Table 22.1 Order of Magnitude of Soil and Structure Horizontal Acceleration and Velocity for Different Earthquake Magnitudes

	Accelerations (Gravities)		Velocity (mm/second)	
Magnitude	Ground Motion	Structure	Ground Motion	Structure
8	0.60	0.33	740	410
7.5	0.45	0.22	560	280
7	0.30	0.15	360	180
5.5	0.12	0.1	150	130

(After Hall and Newmark 1977)

4. Determine the ground motion parameters at the site associated with the controlling earthquake: peak acceleration, peak velocity, response spectrum.

Step 3 requires the use of an *attenuation relationship*, a relationship that gives the decrease in acceleration, for example, as a function of the distance from the source. Such attenuation relationships have been developed based mostly on experimental data. These empirical equations (e.g., Cornell et al. 1979) are typically of the form:

$$\log(PGA) = A + BM - C\log(R + D) \qquad (22.7)$$

where PGA is the peak ground acceleration, M is the earthquake magnitude, R is the distance from the source to the site, and A, B, C, and D are calibration constants. Figure 22.11 shows a correlation by Boore et al. (1997), including the data points used.

In a seismic hazard probabilistic analysis, the steps are somewhat different and consist of the following:

1. Identify all earthquake sources capable of creating significant ground motion at the site. This is the same step as in a deterministic analysis, except that a probability distribution is associated with the location of the source to quantify that uncertainty.

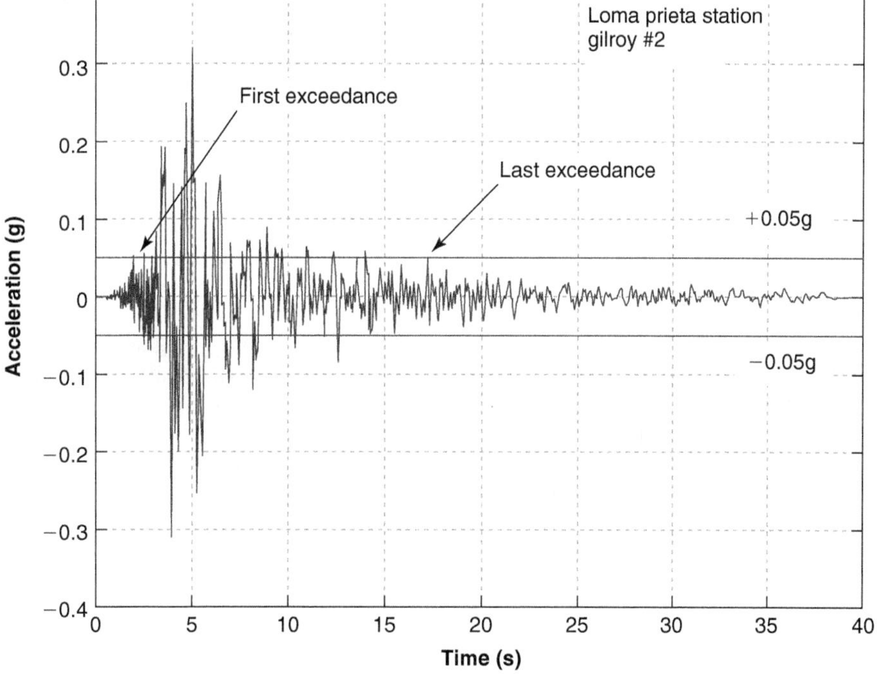

Figure 22.10 Example of bracketed duration.

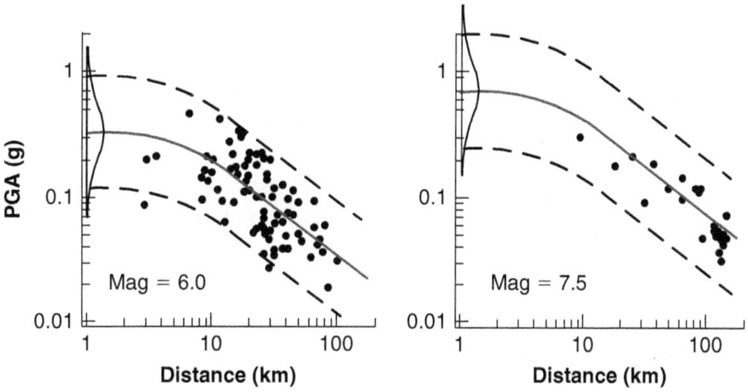

Figure 22.11 Attenuation of peak horizontal ground acceleration (Boore et al. 1997).

2. Determine the magnitude and recurrence of earthquakes from each source. Small earthquakes occur more often than large earthquakes. The magnitude-recurrence relationship gives the number N of earthquakes of a given magnitude or higher that may occur every year. Gutenberg and Richter (1944) proposed the following model:

$$\log N = a - bM \tag{22.8}$$

where N is the number of earthquakes per year of magnitude M or greater, and a and b are regional parameters.

This model has been revised (Figure 22.12) based on further measurements over the years and also by including geologic and geodetic data. Note that the reciprocal of N is called the *recurrence interval* or *return period.* In the building industry, the recurrence interval of the earthquake to use for insuring collapse prevention is the 2500-year earthquake; in the bridge industry, it is the 1000-year earthquake.

3. Determine the ground motion at the site by using an appropriate attenuation relationship. This is the same step as in a deterministic seismic hazard analysis, except that the uncertainty regarding the attenuation is now included as shown in Figure 22.11.
4. The uncertainties in steps 1 through 3 are combined to obtain the probability that the ground motion parameters will be exceeded over a chosen period of time and ensure that this probability meets a target value chosen as acceptable by design.

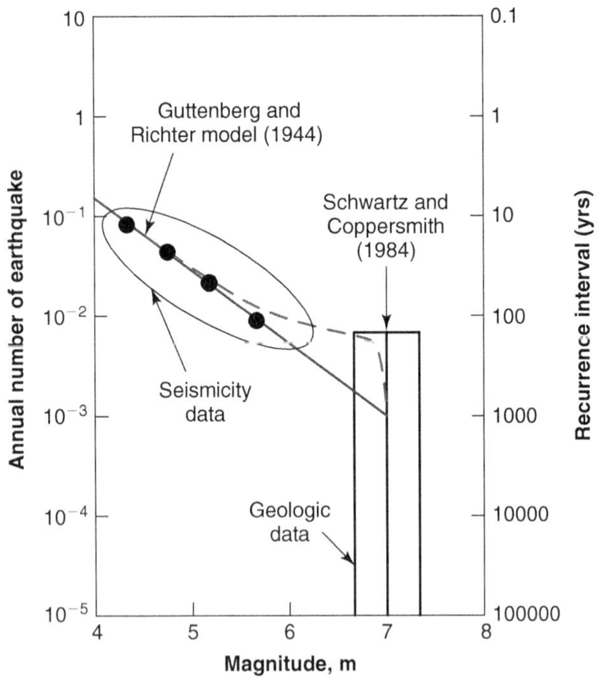

Figure 22.12 Magnitude-recurrence relationship. (Kavazanjian et al. 2011, based on Gutenberg and Richter 1944, and Schwartz and Coppersmith 1984.)

22.7 GROUND RESPONSE ANALYSIS

When a rock fault slides, it shakes the adjacent rock into motion. This motion is propagated in all directions, including upward toward the ground surface. Although most of the travel is through rock, the last few 100 meters may be through soil. The propagation through soil may have a significant impact on the motion of the ground surface, and this *ground response* is addressed in this section. Propagation analysis can be done as a one-dimensional (1D), two-dimensional (2D), or three-dimensional (3D) analysis. The theory for such analyses can get quite complex. Two simple cases of one-dimensional analysis are presented here. More advanced coverage is given in Kramer (1996).

22.7.1 One-Dimensional Solution for Undamped Linear Soil on Rigid Rock

In the case of undamped linear soil on rigid rock (Figure 22.13), the shaking of the rock generates, among other waves, a shear wave travelling in the soil at a shear wave velocity v_s and generating a horizontal movement equal to $u(t,z)$ where t is time and z is depth. The equation of motion for an element of soil (see section 18.3.4) is:

$$\frac{\partial^2 u}{\partial t^2} = {v_s}^2 \frac{\partial^2 u}{\partial z^2} \tag{22.9}$$

If the rock imposes a harmonic motion at the base of the soil layer which is H thick, two waves will be generated in

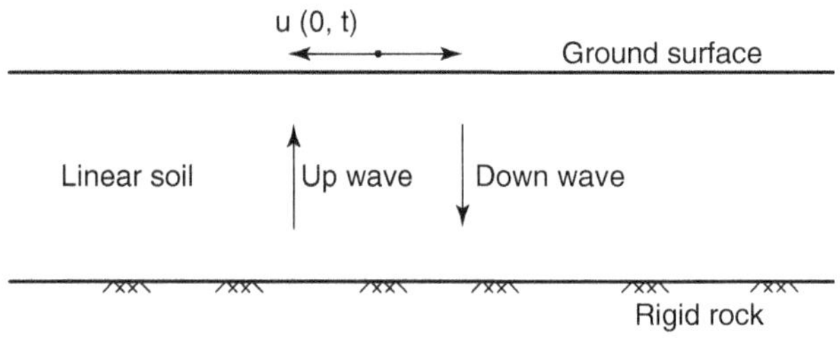

Figure 22.13 Linear soil on rigid rock.

the soil layer: one going up and one going down. The solution to this differential equation in complex notation reflects this decomposition and is of the form:

$$u(z,t) = ae^{i(\omega t + kz)} + be^{i(\omega t - kz)} \tag{22.10}$$

where a and b are the amplitude of the wave going up and the wave going down respectively, ω is the circular frequency of the harmonic motion, and k is a wave number given by:

$$k = \frac{\omega}{v_s} \tag{22.11}$$

Recall that:

$$T = \frac{1}{f} = \frac{2\pi}{\omega} \tag{22.12}$$

where T is the period (*s*), f is the frequency (Hz), and ω is the circular frequency (rd/s) of the harmonic motion. Because the ground surface is considered to be a free boundary, the shear stress $\tau(z,t)$ and shear strain $\gamma(z,t)$ must be zero on that boundary ($z = 0$). The shear strain $\gamma(0,t)$ must therefore satisfy:

$$\gamma(0,t) = \frac{\partial u(0,t)}{\partial z} = 0 \tag{22.13}$$

This leads to the condition that a = b and the final expression for u is:

$$u(z,t) = 2a\cos(kz)\ e^{i\omega t} \tag{22.14}$$

This represents a stationary wave (a wave that remains in a constant position) due to the superposition of the upward wave and the downward wave. Remember that we are interested in transforming the motion of the rock at the base of the soil layer into a motion at the ground surface. The transfer function $F(\omega)$ is therefore:

$$F(\omega) = \frac{u_{\max}(0,t)}{u_{\max}(H,t)} = \frac{1}{\cos(kH)} = \frac{1}{\cos\left(\dfrac{\omega H}{v_s}\right)} \tag{22.15}$$

While in the general case the transfer function will be a complex number, in this simple case it is a scalar. To obtain the horizontal displacement vs. time signal at the soil surface, the horizontal displacement vs. time at the rock boundary is simply multiplied by the transfer function for each time in the record. Note that if $\omega H/v_s$ is equal to $\pi/2$, Eq. 22.15 indicates that the transfer function becomes infinite and the soil is in resonance with the rock motion. Therefore, the natural period T of a soil layer with a height H and a shear wave velocity v_s, also called the *characteristic site period*, is:

$$\frac{\omega H}{v_s} = \frac{\pi}{2} \quad or \quad \frac{2\pi H}{T v_s} = \frac{\pi}{2} \quad or \quad T = \frac{4H}{v_s} \tag{22.16}$$

Equation 22.15 shows that the important factors in the response of a soil layer to an earthquake are the frequency of the rock motion, the thickness of the layer, and its shear wave velocity or small strain shear modulus, which are closely related (see section 8.2.1). In the United States, the natural period of soil deposits is on the order of 0.4 to 2 seconds.

22.7.2 One-Dimensional Solution for Damped Linear Soil on Rigid Rock

Let's assume that the damping in a soil layer can be represented by a Kelvin-Voigt model (see section 12.2.1). In the case of a damped linear soil on rigid rock, the governing differential equation becomes a bit more complicated (Kramer 1996):

$$\frac{\partial^2 u}{\partial t^2} = v_s{}^2 \frac{\partial^2 u}{\partial z^2} + \frac{\eta}{\rho}\frac{\partial^3 u}{\partial z^2 \partial t} \tag{22.17}$$

where η is the soil viscosity (N.s/m^2) and ρ is the mass density of the soil (kg/m^3). The solution is similar to the preceding undamped case except that the wave number k is now a complex number k^* with a real part k_1 and a complex part k_2:

$$u(z,t) = a e^{i(\omega t + k^* z)} + b e^{i(\omega t - k^* z)} \tag{22.18}$$

with:

$$k^* = \frac{\omega}{v_s}(1 - i\beta) \tag{22.19}$$

where β is the damping ratio, which is usually a small number between 0.05 and 0.1. The transfer function is:

$$F(\omega) = \frac{1}{\cos(k * H)} \tag{22.20}$$

and the modulus of that function or amplification function is the ratio of the movement at the ground surface over the movement at the rock level:

$$|F(\omega)| = \frac{u_{\max}(0,t)}{u_{\max}(H,t)} = \frac{1}{\sqrt{\cos^2\left(\frac{\omega H}{v_s}\right) + \sinh^2\beta\left(\frac{\omega H}{v_s}\right)}} \approx \frac{1}{\sqrt{\cos^2\left(\frac{\omega H}{v_s}\right) + \left(\beta\frac{\omega H}{v_s}\right)^2}} \tag{22.21}$$

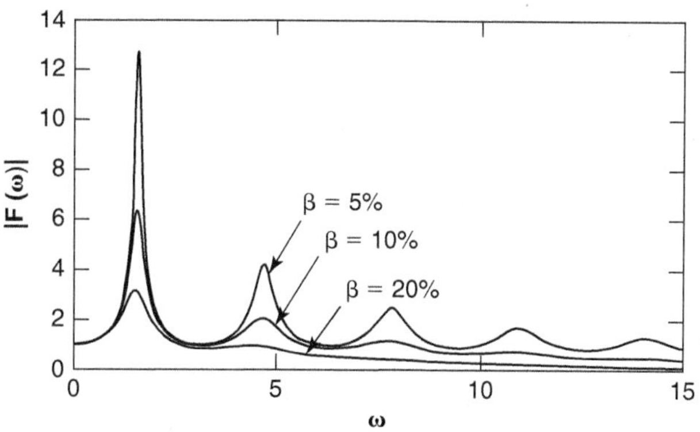

Figure 22.14 Amplification function.

This amplification function $|F(\omega)|$ is shown in Figure 22.14 for several values of the damping ratio β. As can be seen, the amplification is maximum for the lower frequencies.

Solutions can also be found for more complex situations, such as damped linear soil on elastic rock, layered damped soil on elastic rock, and nonlinear soil behavior (Kramer 1996).

22.7.3 Layered Soils

One very useful solution is that for a layered system. This solution was coded by Schnabel et al. (1972) with the program SHAKE and modified by Idriss and Sun (1992) with the program SHAKE91. It solves the problem of a soil deposit made of n layers ($i = 1$ to n), Δz thick, with a shear modulus G_i for each layer i. A column of soil is considered and each layer is represented by an element being deformed in simple shear (Figure 22.15).

The horizontal cross section of the column is 1 m × 1 m. Horizontal equilibrium of an element leads to:

$$\left(\tau + \frac{\partial \tau}{\partial z} dz\right) 1 \times 1 - \tau \times 1 \times 1 = ma = 1 \times 1 \times dz \times \rho \times \frac{\partial^2 u}{\partial t^2} \tag{22.22}$$

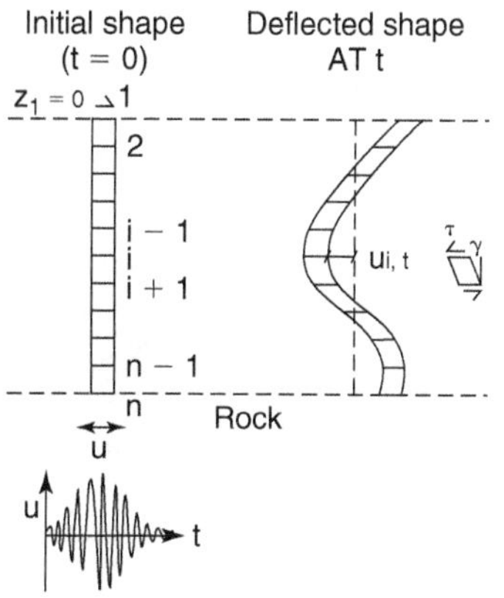

Figure 22.15 Column of soil elements deformed in simple shear during shear wave propagation.

or:

$$\frac{\partial \tau}{\partial z} = \rho \frac{\partial^2 u}{\partial t^2} \tag{22.23}$$

where τ is the shear stress on the horizontal plane, z is depth, ρ is the soil mass density, u is the horizontal displacement, and t is time. This equation is solved by the finite difference method (see section 11.5.1) and ends up for a centered expression as:

$$\frac{\tau_{i+1,t} - \tau_{i-1,t}}{\Delta z} = \rho \frac{u_{i,t+\Delta t} - 2u_{i,t} + u_{i,t-\Delta t}}{\Delta t^2} \tag{22.24}$$

where $\tau_{i,t}$ is the shear stresses at time t on element i, Δz is the increment of depth, ρ is the mass density of the soil, $u_{i,t}$ is the displacement of element i at time t, and Δt is the increment of time.

Boundary conditions exist at the bottom and at the top of the soil column. At the bottom, the displacement is equal to the input rock displacement at any time t. At the top, the shear stress is zero. The shear strain $\gamma_{i,t}$ is linked to the horizontal displacements of the nodes of the soil column as follows:

$$\gamma_{i,t} = \frac{u_{i+1,t} - u_{i,t}}{\Delta z} \tag{22.25}$$

The shear stress $\tau_{i,t}$ can then be calculated as:

$$\tau_{i,t} = G_i \gamma_{i,t} + \eta_i \frac{\partial \gamma_{i,t}}{\partial t} \tag{22.26}$$

where G is the shear modulus of the soil and η is the viscosity of the soil. It can be shown (Kramer 1996) that the viscosity η of the soil is linked to the damping ratio β by:

$$\eta = \frac{2G}{\omega}\beta \tag{22.27}$$

where G is the shear modulus and ω is the circular frequency of the motion.

The input of the problem consists of a shear modulus, a damping ratio, and a mass density for each element in the soil column. Then, the boundary conditions together with Eqs. 22.24, 22.25, 22.26, and 22.27 are written for all nodes in the soil column to solve for the unknown displacements in all elements. The solution consists of starting at $t = 0$ and stepping into time an amount Δt per step. The boundary conditions provide the first values of the displacements and shear stress, which are usually zero for most nodes except for the boundary condition nodes. At the ground surface, the shear stress is always zero, whereas the displacement at the rock level is set equal to the displacement of the bottom element. The displacement at the rock level at the beginning of the earthquake provides the first value.

More realistic analyses include the strain level dependency of the shear modulus and damping ratio and the influence of the confinement on the shear modulus. The process of *deconvolution* is the reverse process, where the ground surface motion is observed during an earthquake and the rock motion is back-calculated at the base of the soil column. The use of programs like SHAKE and other techniques has helped produce graphs like the one in Figure 22.16, which shows the acceleration at the ground surface for a given acceleration at the rock level.

22.8 DESIGN PARAMETERS

The design approach often considers two levels of earthquakes: a rare earthquake and an expected earthquake. A *rare earthquake* may be defined as an earthquake with a 2% probability of exceedance in 50 years, whereas an *expected earthquake* would correspond to a 10% probability of exceedance in 50 years (Figure 22.5). These definitions correspond approximately to a return period of 2500 years and 500 years respectively. The design parameters, including ground

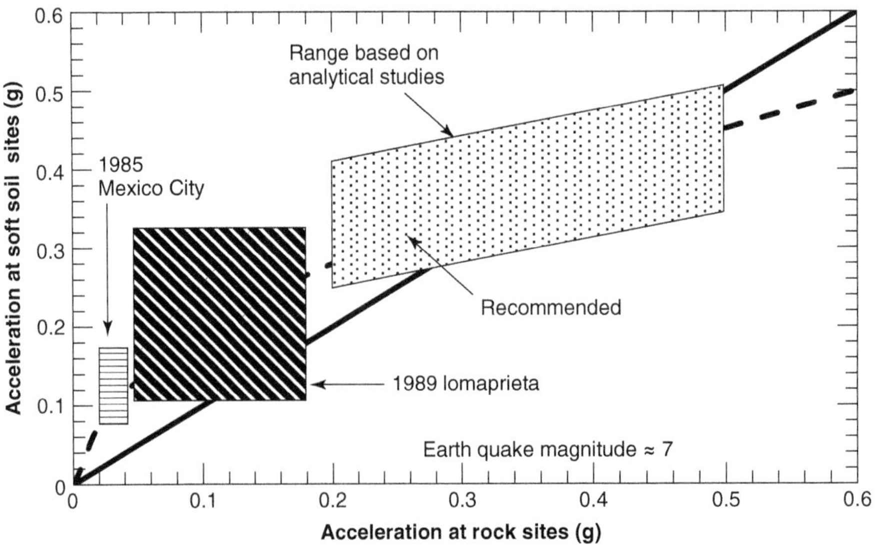

Figure 22.16 Amplification of rock motion at soil sites (After Idriss. 1990).

motion, are selected in one of two ways. The first way is to perform a site-specific analysis (as discussed in section 22.7) or by using recommendations outlined in building codes. This next section addresses the code approach.

22.8.1 Site Classes A–E for Different Soil Stiffness

The design spectrum depends on the soil at the site and in particular its stiffness. This is why *site classes* have been defined, ranging from site class A for hard rock to site class E for soft soil, as shown in Table 22.2. Site class F is a special category for which a site-specific dynamic site response analysis is recommended (section 22.7) instead of a code approach. The site classes use the average soil parameters within the top 30 m from the surface as a classification basis, because this depth is most influential in determining the dynamic response. The shear wave velocity is the parameter of choice, but the SPT blow count and the undrained shear strength are also helpful. Once the soil is classified according to the definitions listed in Table 22.2, the amplification factors can be selected to modify the acceleration spectrum.

22.8.2 Code-Based Spectrum

In the code approach, the acceleration response spectrum is constructed from the analysis of existing data, past experience, and engineering judgment. Such a spectrum is shown in Figure 22.17. First the reference spectrum is developed assuming that the soil at the site is rock (site class B) and then the values obtained for the reference spectrum are modified for the proper site class. The reference spectrum parameters are:

1. The spectral acceleration at a period equal to 0 seconds taken as the peak ground acceleration PGA. The PGA is used here because for a fundamental period of 0 seconds, the structure is infinitely stiff and the maximum acceleration of the structure is the same as the maximum acceleration from the ground.
2. The spectral acceleration at a short period equal to 0.2 seconds, called S_s.
3. The spectral acceleration at a long period equal to 1 second, called S_1.

These values come from the selection of the design earthquake (e.g., 1000-year or 2500-year return period) and the use

Table 22.2 Site Class Definitions

		Average Properties in Top 30 m		
Site Class	Soil Profile Name	Shear wave velocity, $\bar{v}_s$, (m/s)	Standard penetration resistance, N (blow/0.3m)	Undrained shear strength, $s_{u,}$ (kPa)
A	Hard rock	$\bar{v}_s > 1500$	N/A	N/A
B	Rock	$750 < \bar{v}_s \leq 1500$	N/A	N/A
C	Very dense soil and soft rock	$360 < \bar{v}_s \leq 750$	$N > 50$	$S_u \geq 100$
D	Stiff soil profile	$180 \leq \bar{v}_s \leq 360$	$15 \leq N \leq 50$	$50 \leq s_u \leq 100$
E	Soft soil profile	$\bar{v}_s < 180$	$N < 15$	$S_u < 50$
E		Any profile with more than 3 m of soil having the following characteristics: 1. Plasticity index PI > 20_s 2. Moisture content $\omega \geq 40\%$, and 3. Undrained shear strength $s_u < 25$ kPa		
F		Any profile containing soils having one or more of the following characteristics: 1. Soils vulnerable to potential failure or collapse under seismic loading, such as liquefiable soils, quick and highly sensitive clays, collapsible weakly cemented soils. 2. Peats and/or highly organic clays ($H > 3$ m of peat and/or highly organic clay where $H =$ thickness of soil) 3. Very high-plasticity clay ($H > 7.5$ m with plasticity index PI > 75) 4. Very thick soft/medium stiff clays ($H > 36$ m)		

(After Kavazanjian et al. 2011.)

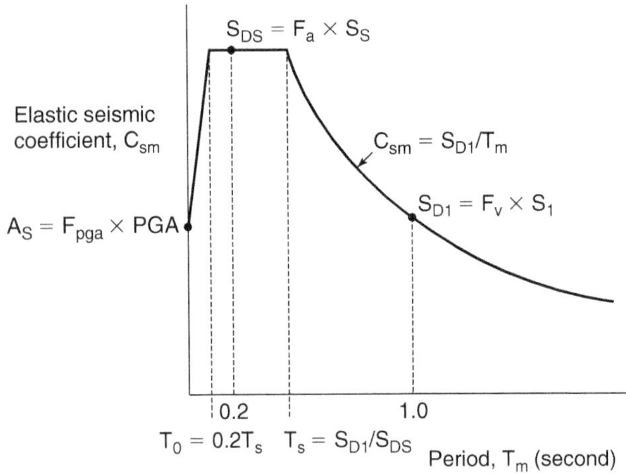

Figure 22.17 Design code acceleration response spectrum. (After Kavazanjian et al. 2011)

of the dedicated USGS web site, for example (Kavazanjian et al. 2011).

The site-specific spectrum is obtained from the values of the reference spectrum. The site-specific spectrum parameters are:

1. The spectral acceleration at a period equal to 0 seconds, called *As*:

$$A_S = F_{PGA} \times PGA \tag{22.28}$$

where F_{PGA} is the site factor for the *PGA* found in Table 22.3.

2. The spectral acceleration at a short period equal to 0.2 seconds, called S_{DS}:

$$S_{DS} = F_A \times S_S \tag{22.29}$$

where F_A is the site factor for S_S, found in Table 22.4.

3. The spectral acceleration at a long period equal to 1 second, called S_{D1}:

$$S_{D1} = F_V \times S_1 \tag{22.30}$$

where F_V is the site factor for S_1, found in Table 22.5.

4. The period T_S corresponding to the end of the spectrum plateau is given by:

$$T_S = \frac{S_{D1}}{S_{DS}} \tag{22.31}$$

5. The period T_o corresponding to the beginning of the spectrum plateau:

$$T_o = 0.2T_S \tag{22.32}$$

Table 22.3 Site Factor F_{PGA}

Site Class	Mapped Spectral Response Accelerations at Short Periods				
	PGA ≤ 0.1 g	PGA = 0.2 g	PGA = 0.3 g	PGA = 0.4 g	PGA ≥ 0.5 g
A	0.8	0.8	0.8	0.8	0.8
B	1.0	1.0	1.0	1.0	1.0
C	1.2	1.2	1.1	1.0	1.0
D	1.6	1.4	1.2	1.1	1.0
E	2.5	1.7	1.2	0.9	0.9
F	a	a	a	a	a

a: Site-specific geotechnical investigation and dynamic site response analysis are required in this case.
(After Kavazanjian et al. 2011)

Table 22.4 Site Factor F_A

Site Class	Mapped Spectral Response Accelerations at Short Periods				
	S_s ≤ 0.25 g	S_s = 0.50 g	S_s = 0.75 g	S_s = 1.00 g	S_s ≥ 1.25 g
A	0.8	0.8	0.8	0.8	0.8
B	1.0	1.0	1.0	1.0	1.0
C	1.2	1.2	1.1	1.0	1.0
D	1.6	1.4	1.2	1.1	1.0
E	2.5	1.7	1.2	0.9	0.9
F	a	a	a	a	a

a: Site-specific geotechnical investigation and dynamic site response analysis are required in this case.
(After Kavazanjian et al. 2011)

Table 22.5 Site Factor F_V

Site Class	Mapped Spectral Response Accelerations at 1-Second Periods				
	S_1 ≤ 0.1 g	S_1 = 0.2 g	S_1 = 0.3 g	S_1 = 0.4 g	S_1 ≥ 0.5 g
A	0.8	0.8	0.8	0.8	0.8
B	1.0	1.0	1.0	1.0	1.0
C	1.7	1.6	1.5	1.4	1.3
D	2.4	2.0	1.8	1.6	1.5
E	3.5	3.2	2.8	2.4	2.4
F	a	a	a	a	a

a: Site-specific geotechnical investigation and dynamic site response analysis are required in this case.
(After Kavazanjian et al. 2011)

In that fashion the design spectrum is completely defined (Figure 22.17) and can be used for structural or geotechnical analysis. In Figure 22.17, the elastic seismic coefficient C_{sm} is the ratio between the design horizontal shear force due to inertia and the effective weight of the structure.

22.8.3 Hazard Levels

The severity of an earthquake is described by the *hazard level*, which ranges from I to IV. Each level is tied to the S_{D1} or S_{DS} value. Recall that S_{D1} is the long-period (1 second) spectral acceleration, adjusted for the site factor, and S_{DS} is the short-period (0.2 second) spectral acceleration, also adjusted for the site factor. The hazard levels are defined in Table 22.6.

22.9 LIQUEFACTION

22.9.1 Phenomenon

When a loose coarse-grained soil under the groundwater level is shaken rapidly enough, it tends to decrease in volume. The decrease in volume causes the water to be pushed out of the pores. If the water cannot escape fast enough, the water stress u_w increases and can reach a value equal to the vertical total stress σ_{ov}. At that point the effective stress ($\sigma'_{ov} = \sigma_{ov} - u_w = 0$) becomes zero: The soil loses its strength and behaves like a thick liquid. This is the phenomenon of *liquefaction*. Loose sands under the groundwater level are particularly sensitive to this condition. Dense coarse-grained soils and fine-grained soils are much less sensitive. Liquefaction of the soil leads to flow slides, lateral spreading, loss of bearing capacity, increased earth pressures against retaining walls as the soil becomes a heavy liquid, and postearthquake settlement as the water stress dissipates.

22.9.2 When to Do a Liquefaction Study?

The need for liquefaction studies is tied first to the severity of the earthquake. This severity is described by hazard levels ranging from I to IV, as described in Table 22.6. For hazard levels I and II, a liquefaction study is not required. For hazard level IV, a liquefaction study is always required. For hazard level III, a liquefaction study is required unless:

1. The mean earthquake magnitude for the design event is less than 6, or
2. The mean magnitude is between 6 and 6.4, and $N_{1-60} >$ 20 (mean normalized SPT blow count; see section 7.1)
3. The mean magnitude is between 6 and 6.4, $N_{1-60} > 15$, and $S_{DS} < 0.35$ g

Table 22.6 Seismic Hazard Levels

Hazard Level	$S_{D1} = F_V S_1$	$S_{DS} = F_A S_S$
I	$S_{D1} < 0.15$	$S_{D1} < 0.15$
II	$0.15 < S_{D1} < 0.25$	$0.15 < S_{D1} < 0.35$
III	$0.25 < S_{D1} < 0.40$	$0.35 < S_{D1} < 0.60$
IV	$0.40 < S_{D1}$	$0.60 < S_{D1}$

Note: These hazard levels apply for site classifications A, B, C, and D. Further description and conditions apply for site classifications E and F (see Kavazanjian et al. 2011).

If the soil is resistant to liquefaction, a liquefaction study is not necessary even for hazard levels III and IV. Liquefaction-resistant soils include:

1. Bedrock
2. Fine-grained soils with more than 15% clay, liquid limit w_L higher than 35%, and water content lower than $0.9w_L$
3. Sands with $N_{1-60} > 30$ bpf or $q_{t1} > 160$ (mean corrected and normalized cone penetrometer resistance; see section 7.2)
4. Soils where the water table is deeper than 15 m below the ground surface

Note that quick clays should be considered as potentially liquefiable; however, the liquefaction is not due to the same process as the liquefaction of fine sands discussed here.

22.9.3 When Can a Soil Liquefy?

To predict whether a soil can liquefy, cyclic tests can be performed in the laboratory by cyclic triaxial testing, or (better) by cyclic simple shear testing, or (even better) by cyclic torsional shear testing. During such tests, the sample is subjected to an initial effective stress and then a chosen value of shear stress or deviator stress is applied cyclically in two-way symmetrical shearing. This means that the shear stress varies between $+\tau_c$ and $-\tau_c$. The frequency of the cycles is selected to be representative of earthquake frequencies (say, 1 to 10 Hz). Typical results for cyclic simple shear tests on saturated sand are shown in Figures 22.18 and 22.19. Liquefaction may or may not occur after a number of cycles or an amount of time consistent with typical earthquakes (less than 30 seconds for most cases and up to 2 minutes for a huge earthquake).

The cyclic stress ratio CSR is the ratio of the horizontal shear stress τ_c applied cyclically to a soil at a depth z over the vertical effective stress σ'_{vo} on the soil at the same depth. The lowest value of the CSR that triggers liquefaction is called the cyclic resistance ratio or CRR. Figure 22.20 shows the results of shaking table tests performed by De Alba et al. (1976); that figure indicates how the CRR decreases as the number of cycles increases. The goal of liquefaction studies is to calculate the CSR and the CRR within the depth of interest. Liquefaction is predicted if:

$$CRR < CSR \tag{22.33}$$

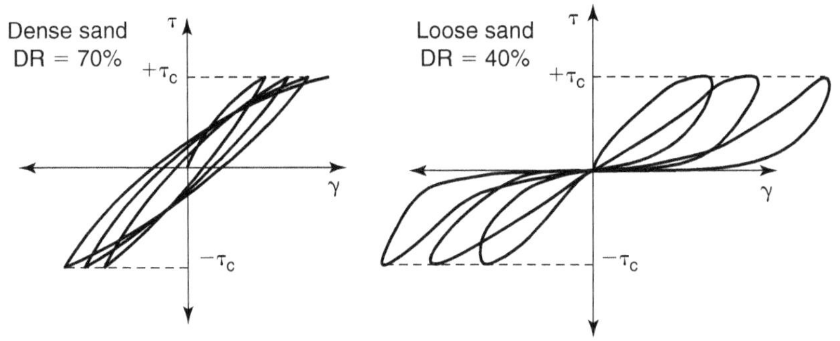

Figure 22.18 Undrained constant stress cyclic simple shear test results (stress-strain curves).

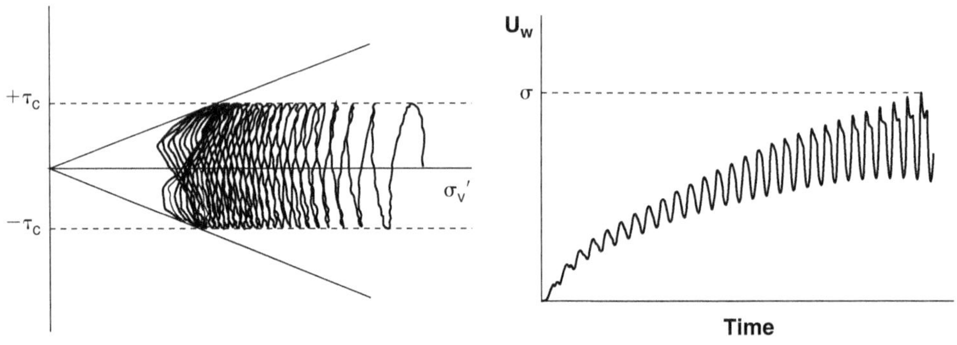

Figure 22.19 Undrained constant stress cyclic simple shear test results (stress path and water stress).

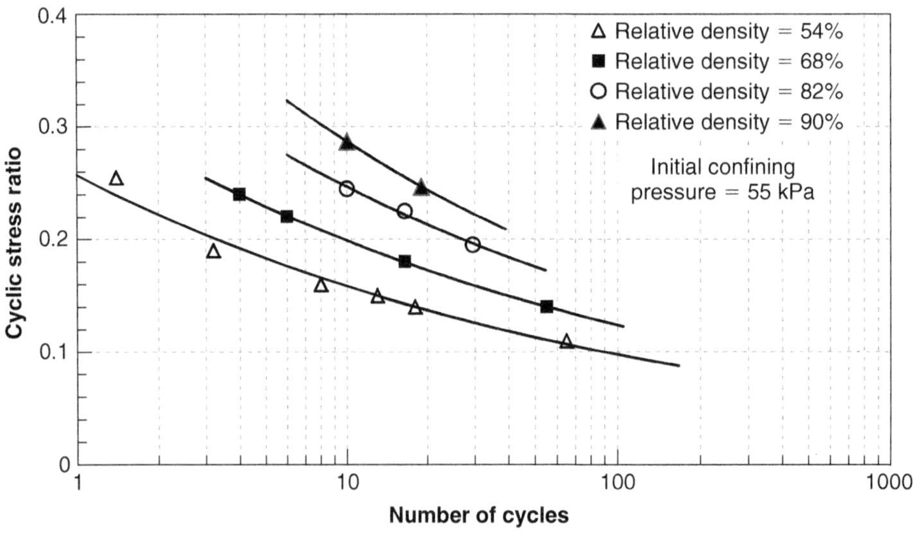

Figure 22.20 Cyclic stress ratio to reach liquefaction. (From De Alba et al. 1976. With permission from ASCE)

The drawbacks of using laboratory tests to predict liquefaction include sample disturbance, difficulty in reproducing in situ stresses, and difficulties in reproducing a true earthquake cyclic shear loading. As a result, the preferred approach in design has been to use earthquake case histories at sites where liquefaction did or did not occur.

The combination of the average shear stress τ_{av} due to the earthquake shaking, a measure of the soil strength, and the knowledge of whether the soil liquefied are used to produce design charts. The results of in situ tests are preferred in this approach to quantify the soil strength. The first charts, such as the one shown in Figure 22.21, were based on the SPT blow count. Additional charts were then proposed based on cone penetrometer data (Figure 22.22) and then shear wave velocity data (Figure 22.23). The chart based on the dilatometer is preliminary in nature (Figure 22.24).

In these charts, the vertical axis is the cyclic stress ratio CSR, defined as τ_{av}/σ'_{ov} where τ_{av} is the average shear stress generated during the design earthquake and σ'_{ov} is the vertical effective stress at the depth investigated and at the time of the in situ soil test. The shear stress τ_{av} is related to the maximum shear stress τ_{max}, which is obtained from a site response analysis (e.g., using the program SHAKE) for the design earthquake, or, more simply, by using the peak ground acceleration PGA obtained from maps such as the one show in Figure 22.5. If the PGA is used to obtain τ_{av} for a 7.5 magnitude, the expression is (Seed and Idriss 1971):

$$CSR = \frac{\tau_{av}}{\sigma'_{vo}} = 0.65\left(\frac{a_{max}}{g}\right)\left(\frac{\sigma_{vo}}{\sigma'_{vo}}\right) r_d \qquad (22.34)$$

where a_{max} is the PGA for the design earthquake, g is the acceleration due to gravity, σ_{vo} is the total vertical stress

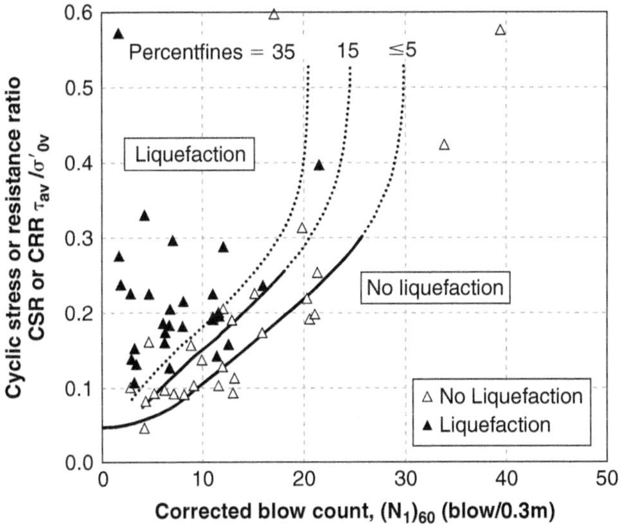

Figure 22.21 SPT-based liquefaction chart for magnitude 7.5. (After Youd and Idriss 1997)

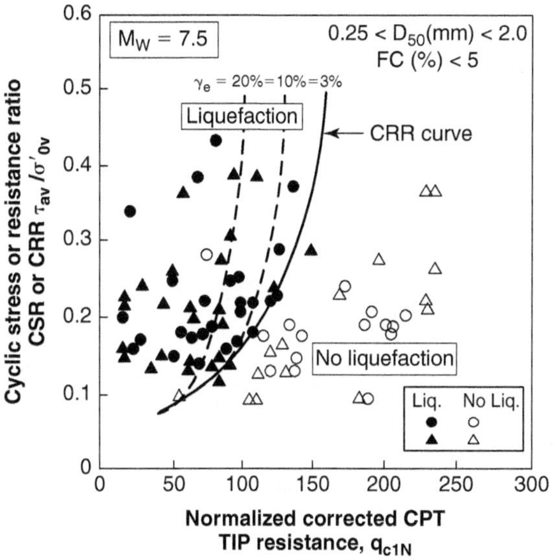

Figure 22.22 CPT-based liquefaction chart for magnitude 7.5. (After Robertson and Wride 1998)

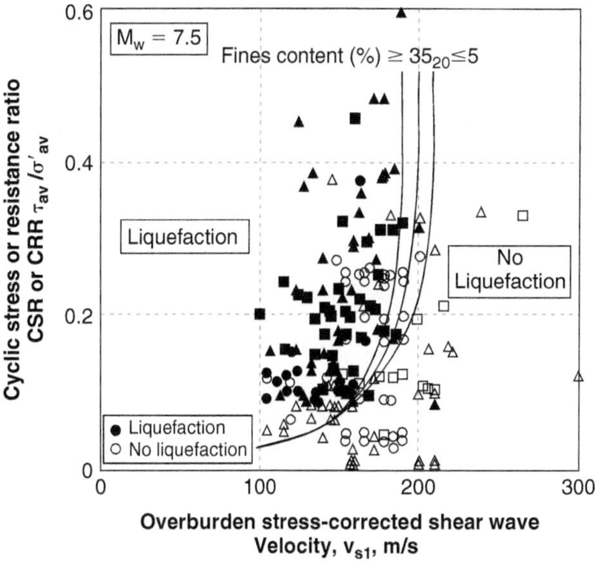

Figure 22.23 Shear wave velocity-based liquefaction chart for magnitude 7.5. (After Andrus and Stokoe 2000)

at the depth being investigated, σ'_{vo} is the effective vertical stress at the depth being investigated, and r_r is a flexibility factor. The flexibility factor depends on the depth at which the liquefaction is being evaluated. Such a factor is necessary because the PGA is acting at the ground surface while the possibility of liquefaction is evaluated at a depth z. Figure 22.25, after Seed and Idriss (1971), gives a range of values for r_d.

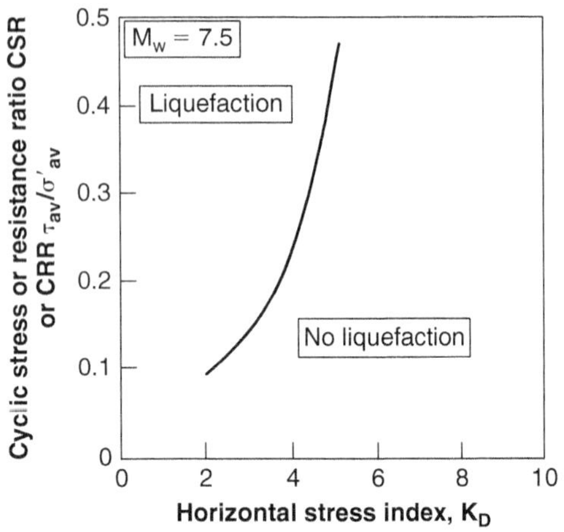

Figure 22.24 DMT-based liquefaction chart for magnitude 7.5. (After Monaco et al. 2005)

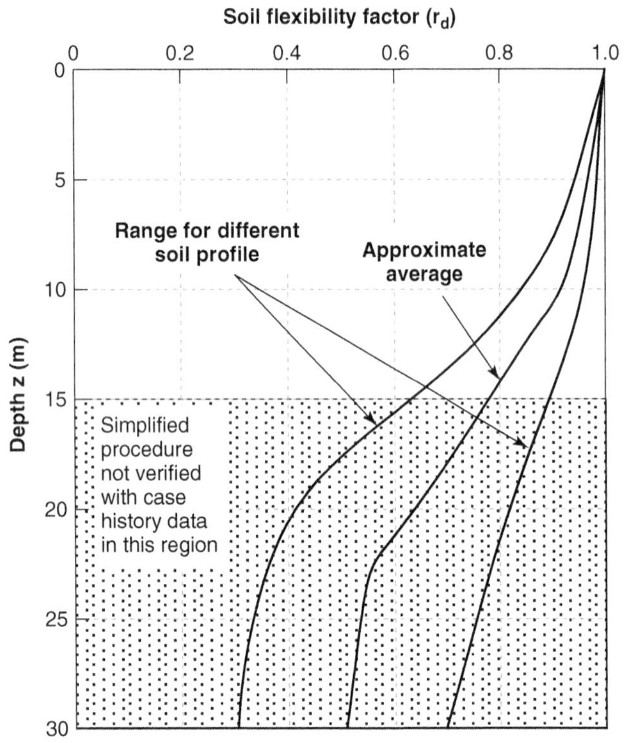

Figure 22.25 Soil flexibility factor to modify the PGA for a depth z (Seed and Idriss 1971)

On the horizontal axis of the charts in Figures 22.21 to 22.24 is the in situ test parameter normalized and corrected for the effective stress level in the soil at the time of the test and for fine content. The SPT blow count is N_{1-60} and the procedure to correct it for effective stress level is described in section 7.1. The correction for fine content is embedded in the chart of Figure 22.21. The CPT point resistance is q_{t1} and the procedure to correct for effective stress level and fine content is described in section 7.2. The shear wave velocity is v_{s1} and the procedure to correct for effective stress is:

$$v_{s1} = v_s \left(\frac{\sigma_a}{\sigma'_{vo}} \right)^{0.25} \tag{22.35}$$

where v_s is the shear wave velocity measured in the field, σ_a is the atmospheric pressure, and σ'_{vo} is the vertical effective stress at the depth investigated. Once the soil parameter is corrected, it is entered on the horizontal axis of the chart and the CRR is read on the liquefaction design curve of the corresponding chart. Note that the charts in Figures 22.21 to 22.24 give the cyclic resistance ratio (CRR) for an earthquake of magnitude 7.5. For earthquakes of different magnitude, a magnitude scaling factor (MSF) is applied as follows:

$$CRR_M = MSF \times CRR_{M=7.5} \tag{22.36}$$

Kavazanjian et al. (2011), building on the work of Youd and Idriss (1997), suggested that the hatched area in Figure 22.26 be used for MSF.

In summary, the way to use the charts is (Figure 22.27):

1. Obtain the soil parameter profile.
2. Correct the profile for stress level due to depth effects and fine content if necessary. Prepare a corrected soil parameter profile.
3. Enter the chart corresponding to the soil parameter and read the cyclic resistance ratio. Prepare a CRR profile.
4. Modify the CRR profile for a magnitude different from 7.5.

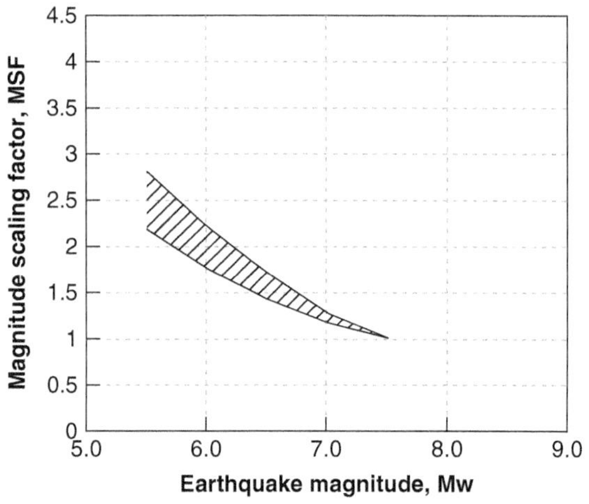

Figure 22.26 Magnitude scaling factor (MSF). (After Kavazanjian et al. 2011; Youd and Idriss 1997)

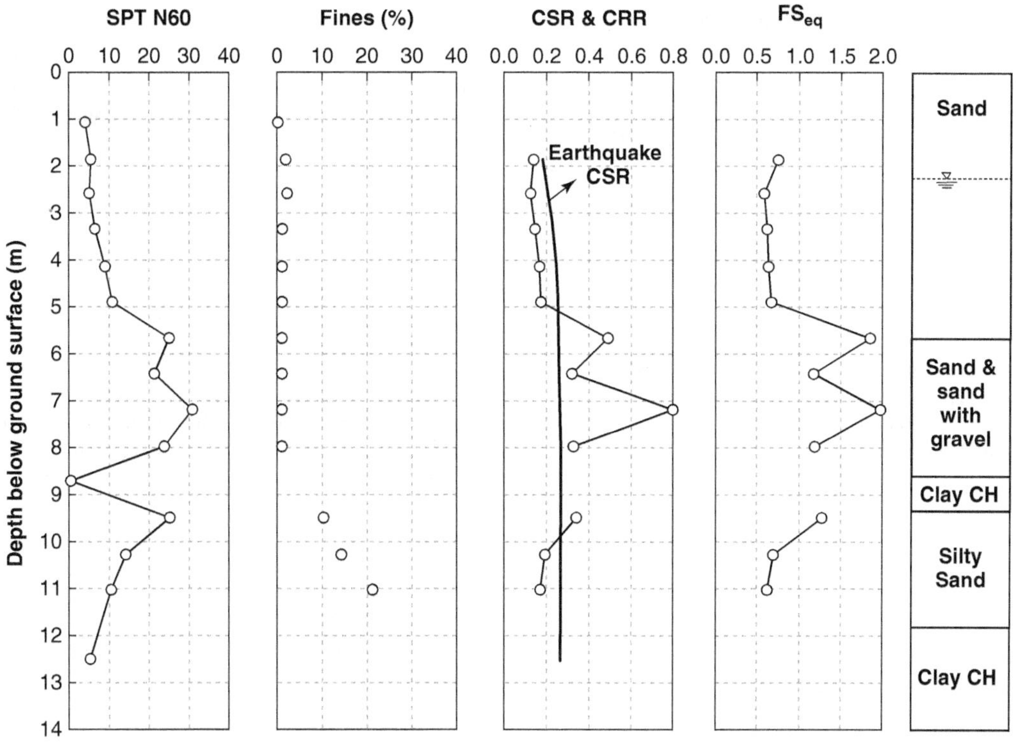

Figure 22.27 Profiles of liquefaction analysis (Idriss and Boulanger 2008).

5. Calculate the CSR generated by the design earthquake. Prepare the CSR profile.
6. Compare the CSR profile and the CRR profile.
7. The zone of potential liquefaction is the zone where CSR > CRR.

22.10 SEISMIC SLOPE STABILITY

Seismic slope stability was covered in section 19.18. The following summarizes the procedure to select the appropriate value of the horizontal seismic coefficient *k* for a pseudostatic analysis.

The step-by-step procedure that follows is as recommended by Kavazanjian et al. (2011):

1. Perform a static slope stability analysis without any earthquake loading to ensure that the slope is stable and that the factor of safety *F* is sufficient in the case of no earthquake (say, 1.5).
2. Using maps (USGS map, such as the ones shown in Figure 22.5, for example), obtain the peak ground acceleration PGA and the spectral acceleration at one second S_1 for site class B at the base of the slope for the design earthquake.
3. Select the site adjustment factor F_{PGA} from Table 22.3 and the factor F_V from Table 22.5 for the correct site class and the correct acceleration value.
4. Calculate the value of the maximum horizontal seismic inertia coefficient k_{max} as:

$$k_{max} = F_{PGA} \times PGA \tag{22.37}$$

5. Calculate the value of the average horizontal seismic inertia coefficient k_{av} as follows. The coefficient k_{av} is lower than k_{max} because the average horizontal acceleration over the slope mass is less than the PGA due to wave scattering:

$$k_{av} = \gamma\left(1 + 0.01H\left(0.5\frac{F_V S_1}{k_{max}} - 1\right)\right)k_{max} \tag{22.38}$$

where γ is equal to 1 for all site classes except for site classes A and B, where it is taken as 1.2; *H* is the height of the slope; and F_v is the site factor from Table 22.5.
6. If the slope can tolerate a movement of 25 to 50 mm, the value of k_{av} can be further reduced by a factor of 2. In the end, the factor k_h is given by:

$$k_h = 0.5\gamma\left(1 + 0.01H\left(0.5\frac{F_V S_1}{k_{max}} - 1\right)\right)k_{max} \tag{22.39}$$

7. Under the combined static and earthquake inertia loading, the target factor of safety should be at least 1.1.

22.11 SEISMIC DESIGN OF RETAINING WALLS

The design of retaining walls under static conditions is described in Chapter 21. This section addresses what happens under earthquake conditions. Gravity walls are considered first, followed by MSE walls, cantilever walls, and tieback walls.

22.11.1 Seismic Design of Gravity Walls

The horizontal force P_a per unit length of wall due to the active earth pressure behind a retaining wall when there is no earthquake and the water table is below the bottom of the wall (see sections 21.3.1 and 21.9) is given by:

$$P_a = \frac{1}{2} K_a \gamma H^2 \quad (22.40)$$

where γ is the soil unit weight, H is the wall height, and K_a is the active earth pressure coefficient expressed as (Figure 22.28):

$$K_a = \frac{\sin^2(\alpha + \varphi')}{\sin^2\alpha \sin(\alpha - \delta)\left[1 + \sqrt{\frac{\sin(\varphi' + \delta)\sin(\varphi' - \beta)}{\sin(\alpha - \delta)\sin(\alpha + \beta)}}\right]^2} \quad (22.41)$$

where α is the angle of the back of the wall with the horizontal, φ' is the effective stress friction angle of the soil behind the wall, δ is the angle of friction between the soil and the back of the wall, and β is the angle of the ground surface behind the wall with the horizontal.

In the case of earthquake loading on a gravity wall, the earth pressure is increased by the horizontal shaking of the soil and the associated horizontal inertia force. The vertical acceleration can also modify the weight of soil acting on the wall, but this vertical inertia force is usually ignored, mainly because it does not occur at the same time as the horizontal force; indeed, the horizontal and vertical accelerations are rarely in phase, so the peak horizontal and peak vertical accelerations do not occur simultaneously. The horizontal inertia force generated by the earthquake is written as $k_h W$ where k_h is the seismic coefficient and W is the weight of the soil wedge. The coefficient k_h is similar to the coefficient used for slope stability; it is taken as k_{av} (Eq. 22.38) if the wall cannot tolerate any movement and as k_h (Eq. 22.39) if a movement of 25 to 50 mm is tolerable.

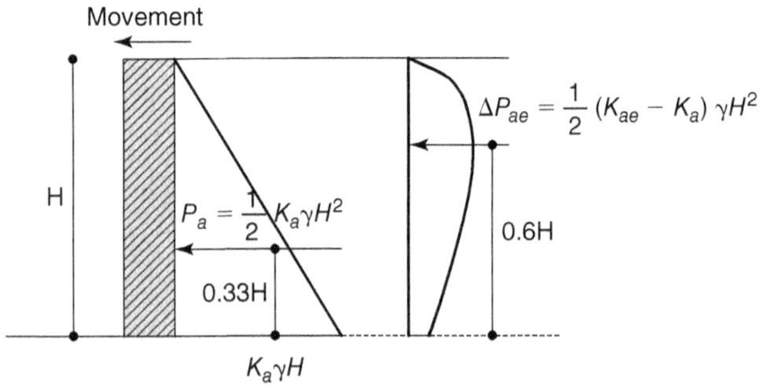

Figure 22.28 Gravity retaining wall with earthquake loading: active case.

In the case of an earthquake, the force P_a becomes P_{ae}, which is written as:

$$P_{ae} = \frac{1}{2} K_{ae} \gamma H^2 \quad (22.42)$$

where P_{ae} is the active force per unit length of wall due to the active earth pressure during an earthquake, K_{ae} is the active earth pressure coefficient in the earthquake case, γ is the soil unit weight, and H is the wall height. The coefficient K_{ae} is obtained in the same fashion as K_a (see section 21.3) except that the earthquake force $k_h W$ is added to the equilibrium equations.

The final expression of K_{ae} after finding the most critical wedge angle is credited to Mononobe and Okabe (Okabe 1926; Mononobe and Matsuo 1929):

$$K_{ae} = \frac{\sin^2(\alpha + \varphi' - \psi)}{\cos\psi \sin^2\alpha \sin(\alpha - \delta - \psi)\left[1 + \sqrt{\frac{\sin(\varphi' + \delta)\sin(\varphi' - \beta - \psi)}{\sin(\alpha - \delta - \psi)\sin(\alpha + \beta)}}\right]^2} \quad (22.43)$$

where α is the angle of the back of the wall with the horizontal, β is the angle of the ground surface behind the wall with the horizontal, δ is the angle of friction between the back of the wall and the soil, φ' is the friction angle of the soil, and ψ is the angle representing the earthquake inertia force through:

$$\psi = \tan^{-1}\left(\frac{k_h}{1 - k_v}\right) \quad (22.44)$$

where k_h and k_v are the horizontal and vertical seismic coefficients respectively. Note that k_v is often ignored (taken as equal to zero). The angle of the critical surface with the horizontal is flatter in the active earthquake case than in the static case (Kramer 1996). Note also that P_{ae} includes the static component P_a and a dynamic component ΔP_{ae} of the active push (Figure 22.28) and can be rewritten as:

$$P_{ae} = P_a + \Delta P_{ae} \quad (22.45)$$

For the passive earth pressure, the equations become:

$$P_{pe} = \frac{1}{2} K_{pe} \gamma H^2 \quad (22.46)$$

where P_{pe} is the passive force per unit length of wall due to the passive earth pressure during an earthquake, K_{pe} is the passive earth pressure coefficient in the earthquake case, γ is

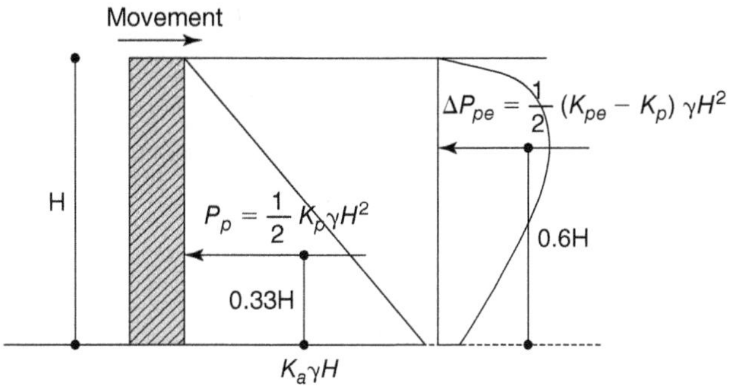

Figure 22.29 Gravity retaining wall with earthquake loading: passive case.

the soil unit weight, and H is the wall height. The expression of K_{pe} (Figure 22.29) is:

$$K_{pe} = \frac{\sin^2(\alpha - \varphi' + \psi)}{\cos\psi \sin^2\alpha \sin(\alpha + \delta + \psi)\left[1 - \sqrt{\frac{\sin(\varphi'+\delta)\sin(\varphi'+\beta+\psi)}{\sin(\alpha+\delta+\psi)\sin(\alpha+\beta)}}\right]^2} \tag{22.47}$$

Note that P_{pe} includes the static component P_p and dynamic component ΔP_{pe} of the passive push (Figure 22.29) and can be rewritten as:

$$P_{pe} = P_p + \Delta P_{pe} \tag{22.48}$$

The point of application of the static component of the active and passive forces, P_a and P_p, is located at 0.33 H (H is wall height) from the bottom of the wall in the simplest case of a uniform soil. Note that the static pressure distribution is triangular, but the pressure distribution associated with the earthquake inertia force is not triangular. Recall that K_{ae} was obtained from a Coulomb wedge analysis, which gives a global force solution, and not a Rankine stress analysis, which gives a pressure distribution. In fact, the point of application of the seismic component is different from the point of application of the static component. The point of application of the dynamic components of the active and passive forces, ΔP_{ae} and ΔP_{pe}, is higher than the one for the static components, because the amplitude of the soil movement due to the earthquake generally increases as the shear wave propagates upward. As a result, the point of application of ΔP_{ae} and ΔP_{pe} is located at 0.6 H from the bottom of the wall.

The Monobe-Okabe expressions of K_{ae} and K_{pe} in Eqs. 22.43 and 22.47 have the advantage of being simple to use. They also have shortcomings. One of them is that the failure surface is assumed to be the same for the static case and the dynamic case, which is not true. In the active case, the slope of the failure surface is flatter for the dynamic case than for the static case. Second, the effect of cohesion is not included, although it can reduce the effect of the dynamic part of the active pressure. Figure 22.30 shows the influence of the cohesion c' on K_{ae} for a friction angle of 35° and for different values of the horizontal seismic coefficient k_h. The cohesion c' is normalized in the figure by γH where γ is the soil unit weight and H is the wall height. Another shortcoming is that the wedge approach assumes a straight-line failure surface, which is not necessarily the weakest surface. This difference is particularly severe for the passive resistance P_{pe}, which can be seriously overestimated and should be used with caution if at all. A log spiral failure surface gives more conservative values for K_{pe} and should

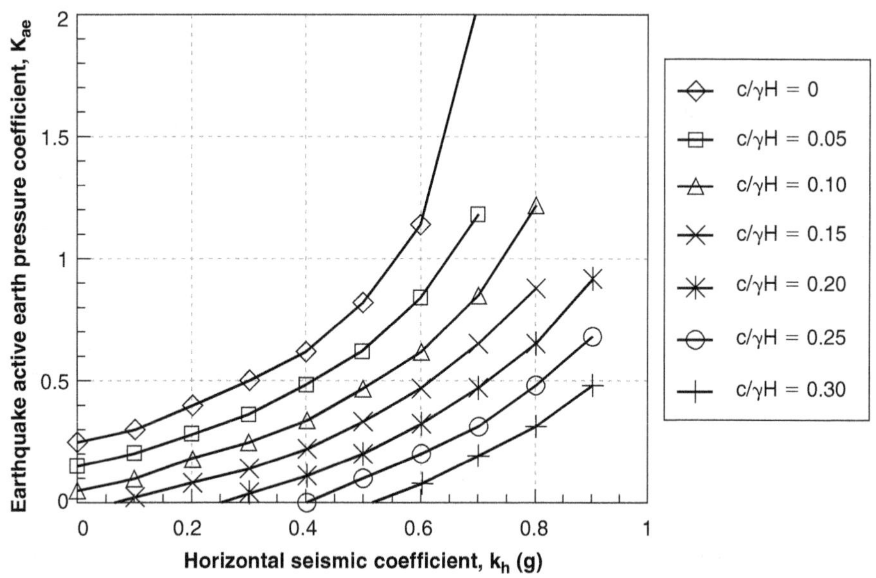

Figure 22.30 Influence of cohesion c' for $\varphi' = 35°$ on the earthquake active earth pressure coefficient K_{ae} for different values of the horizontal seismic coefficient k_h (NCHRP 2008).

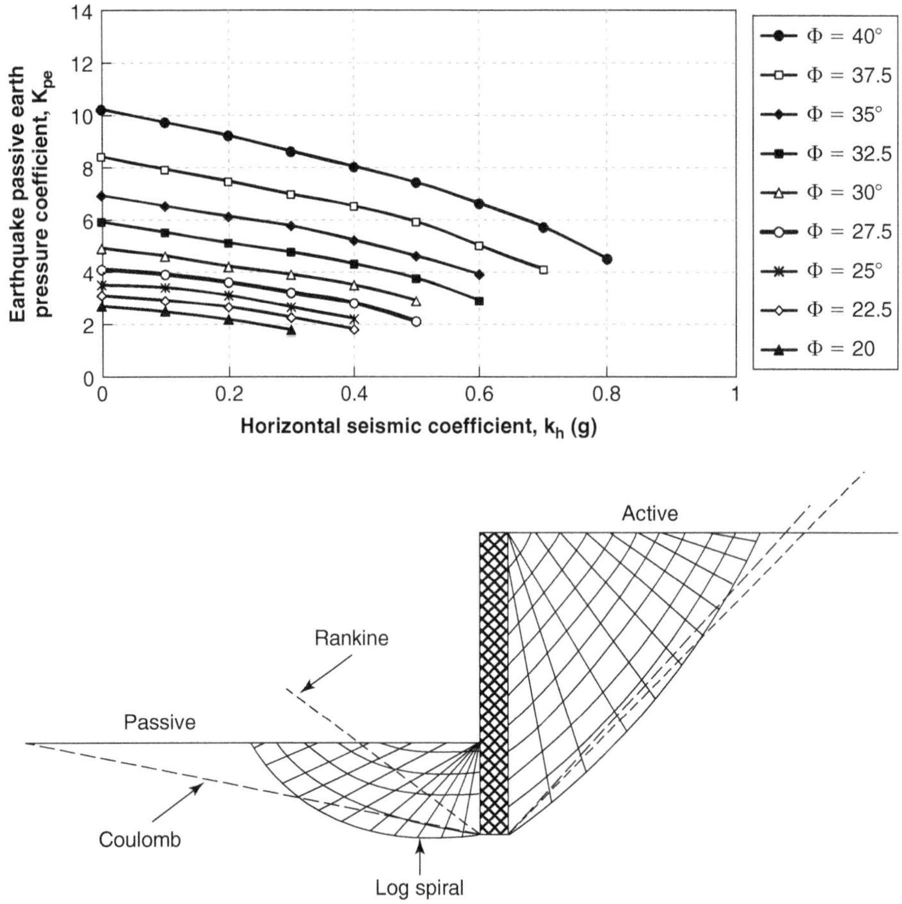

Figure 22.31 Coefficient of passive earth pressure in the case of earthquake loading for a log spiral failure surface and a wall friction angle equal to 2/3 of the soil friction angle (NCHRP 2008, Kavazanjian et al. 2011).

be preferred. Such values are shown in Figure 22.31 (after NCHRP 2008).

22.11.2 Water Pressures on Walls during Earthquake

It is generally desirable to ensure that the groundwater table is below the bottom of the retaining wall, as the water pressure significantly increases the active force. This is also true for walls in earthquake-prone areas. However, this may not be possible; a high water level is often encountered for walls in harbors or near shore. In such instances it is necessary to account for the water behavior during an earthquake in addition to the hydrostatic pressure associated with the static case.

Water on the Side That Has No Soil

If there is water on the side of the wall that has no soil (e.g., berthing wall in a harbor, earth dam), the pressure in the static case p_{wh} is hydrostatic and given by:

$$p_{wh} = \gamma_w z \tag{22.49}$$

where γ_w is the unit weight of water and z is the depth below the water level. The dynamic pressure during an earthquake Δp_{we} is given by Westergaard (1931):

$$\Delta p_{we} = \frac{7}{8} k_h \gamma_w \sqrt{z H_w} \tag{22.50}$$

where k_h is the horizontal seismic coefficient, z is the depth below the water level, and H_w is the total height of water against the wall (Figure 22.32). The assumptions made by Westergaard to develop this solution limit the application of this formula to the case where the earthquake frequency is below the fundamental frequency f_w of the water body. This frequency is given by:

$$f_w = \frac{v_p}{4H_w} \tag{22.51}$$

where v_p is the compression wave velocity. Note that the dynamic pressure works alternatively in both directions. The most detrimental condition for the retaining wall is likely to be when the dynamic pressure decreases the hydrostatic

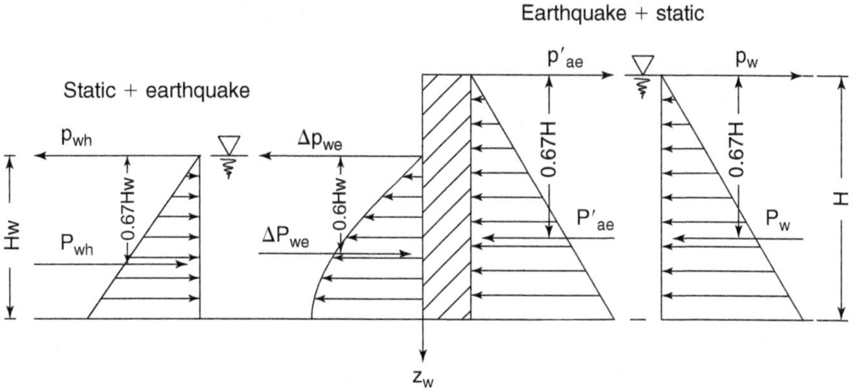

Figure 22.32 Water pressure on wall due to earthquake.

pressure, thereby decreasing the stabilization effect of the water. By integrating the expression in Eq. 22.50, we can obtain the resultant force ΔP_{we}:

$$\Delta P_{we} = \frac{7}{12} k_h \gamma_w H_w^{\;2} \tag{22.52}$$

The point of application of ΔP_{we} can be calculated by moment equilibrium and is found to be $0.6H_w$ below the water surface (Figure 22.32).

Water on the Retained-Soil Side

If there is water in the backfill, the problem becomes a bit more complicated, as the inertia force is proportional to the total unit weight γ and the shear resistance is proportional to the effective unit weight γ'. Therefore, Eq. 22.42 must be altered to reflect this dual effect. In the case where the water level in the backfill is at the ground surface and no excess water stress is generated, Towhata (2008) recommends the following approach:

1. Use γ' in the active earth pressure equation
2. Increase the horizontal seismic coefficient k_h to reflect the increase in inertia force
3. Add the hydrostatic pressure

The equation then becomes (Figure 22.32):

$$P'_{ae} = \frac{1}{2} K_{ae} \gamma' H^2 \tag{22.53}$$

However, the horizontal seismic coefficient k_h is increased to k'_h:

$$k'_h = \frac{\gamma}{\gamma'} k_h \tag{22.54}$$

This nearly doubles the value of k_h. Then the angle ψ used in the expression of K_{ae} is:

$$\psi = \tan^{-1}\left(\frac{k'_h}{1-k_v}\right) = \tan^{-1}\left(\frac{k_h}{1-k_v} \times \frac{\gamma}{\gamma'}\right) \tag{22.55}$$

The hydrostatic thrust must then be added:

$$P_w = \frac{1}{2} \gamma_w H^2 \tag{22.56}$$

$$P_{ae} = P_w + P'_{ae} = \frac{1}{2} \gamma_w H^2 + \frac{1}{2} K_{ae} \gamma' H^2 \tag{22.57}$$

Because of the triangular distribution of pressures, both P'_{ae} and P_w act at 0.67 H from the top of the wall in the simplest case of a uniform soil. Kramer (1996) gives recommendations for the more complex case where the water level behind the wall is not at the ground surface.

22.11.3 Seismic Design of MSE Walls

MSE walls retain the soil through a reinforced soil mass. The earthquake design of these types of walls follows the same approach as the static design (see section 21.10), except that the coefficient K_a is replaced by the coefficient K_{ae} in the calculation.

22.11.4 Seismic Design of Cantilever Walls

Cantilever walls retain the soil without anchors or strut simply by the resistance of their embedment into the foundation soil. The earthquake design of these types of walls follows the same approach as the static design (see section 21.11), except that the coefficients used for the earth pressure are K_{ae} and K_{pe} instead of K_a and K_p.

22.11.5 Seismic Design of Anchored Walls

Anchored walls retain the soil through the use of anchors or struts and through their depth of embedment. The earthquake design of these types of walls follows the same approach as the static design (see section 21.12), except that the coefficient K used for the earth pressure above the excavation level is increased by the ratio K_{ae}/K_a. Below the excavation level, the earth pressure coefficients are K_{ae} and K_{pe}.

22.12 SEISMIC DESIGN OF FOUNDATIONS

During an earthquake, a foundation and the soil around it will interact. Two kinds of interactions are identified: kinematic and inertial. *Kinematic interaction* refers to the interaction between the soil and the foundation as the foundation modifies the free field movement of the soil because of its presence. *Inertial interaction* refers to the interaction between the soil and the foundation as the foundation movement due to soil shaking generates accelerations throughout the building and associated inertia forces at the foundation level. In many instances, kinematic interaction can be neglected, and the foundation need only be designed to resist the inertia forces due to the inertial interaction.

The approaches used for the design of foundations to resist earthquake loading are the same for shallow foundations and deep foundations. There are two main categories: the design code approach and the dynamic analysis approach. Both approaches aim at obtaining the inertia forces and moments on the foundation and then designing the foundation to handle these forces on a pseudostatic basis.

In the design code approach, a response spectrum is specified, and then the fundamental period of the building is calculated. This fundamental period is entered on the horizontal axis of the spectrum and the corresponding spectral acceleration is obtained. The horizontal force H to be resisted by the foundation is the product of the spectral acceleration and the associated mass of the building. In this approach, the ductility of the structure and foundation are not considered. This ductility tends to reduce the inertia force and is included through the use of a reduction factor R_f. Table 22.7 shows such reduction factors for bridge substructures. The reduced force used for design purposes is H/R_f. This force is applied to the foundation and the ultimate limit state is checked to ensure safety against failure. Because earthquake is considered to be an extreme event, the load and resistance factors are close to 1 if not equal to 1.

Table 22.7 Force Reduction Factor R_f for Bridges (Kavazanjian et al. 2011)

Substructure	Importance Category		
	Critical	Essential	Other
Wall-type piers, larger dimension	1.5	1.5	2.0
Reinforced concrete pile bents			
• Vertical piles only	1.5	2.5	3.0
• With batter piles	1.5	1.5	2.0
Single columns	1.5	2.9	3.0
Steel or composite steel and concrete pile bents			
• Vertical pile only	1.5	3.5	5.0
• With batter piles	1.5	2.0	3.0
Multiple-column bents	1.5	3.5	5.0

In the dynamic analysis approach, the structure and foundation are simulated numerically. The foundation is usually simplified and represented by a system of translational and rotational springs and dashpots. The simulation gives the inertia forces and moments applied to the foundation. This approach has the advantage of including the ductility of the structure more directly. Again, this force is applied to the foundation and the ultimate limit state is checked to ensure safety against failure. Because earthquake is considered to be an extreme event, the load and resistance factors are close to 1 if not equal to 1.

There is a trend toward displacement base design (service limit state) rather than load-based design. In this approach the displacement due to the earthquake loads are calculated and allowance is made for what leads to no damage, medium damage, heavy damage but still standing, and total collapse.

In the case of deep foundations, it is possible for the liquefied soil to load the piles by flowing past them. The load generated by liquefied soil must be added to the inertia load. This brings in the importance of the shear strength of liquefied soils. Seed and Harder (1990) proposed a correlation of the liquefied soil shear strength to the corrected standard penetration test (SPT) blow count $(N_1)_{60}$ (see section 7.1). Further correction was added to the $(N_1)_{60}$ value for the presence of fines, which can be approximated as follows:

$$(N_1)_{60-cs} = (N_1)_{60} + \frac{P}{10} \tag{22.58}$$

where $(N_1)_{60}$ is the SPT blow count corrected for stress and energy level, P is the percent finer than 0.075 mm (expressed in percent), and $(N_1)_{60\text{-cs}}$ is the SPT blow count further corrected for the fine content. The correction increases the value of N to bring it back to the value that would have been obtained had the sand not contained fines (cs means clean sand). Olson and Stark (2002) further developed the original work of Seed and Harder, added data, and proposed the following equation on the basis of the corrected SPT blow count and the corrected CPT point resistance as follows (Figures 22.33 and 22.34):

$$\frac{s_{u-liq}}{\sigma'_{vo}} = 0.03 + 0.0075(N_1)_{60} \quad \text{for} \quad N_1 \leq 12 \text{ bpf} \tag{22.59}$$

$$\frac{s_{u-liq}}{\sigma'_{vo}} = 0.03 + 0.0143 \times q_{c1} \quad \text{for} \quad q_{c1} \leq 6.5 \text{ MPa} \tag{22.60}$$

where $s_{u\text{-}liq}$ is the shear strength of the liquefied soil, σ'_{vo} is the prefailure vertical effective stress in the soil, and $(N_1)_{60}$ and q_{c1} are the prefailure corrected SPT blow count and CPT point resistance respectively (see sections 7.1 7.2).

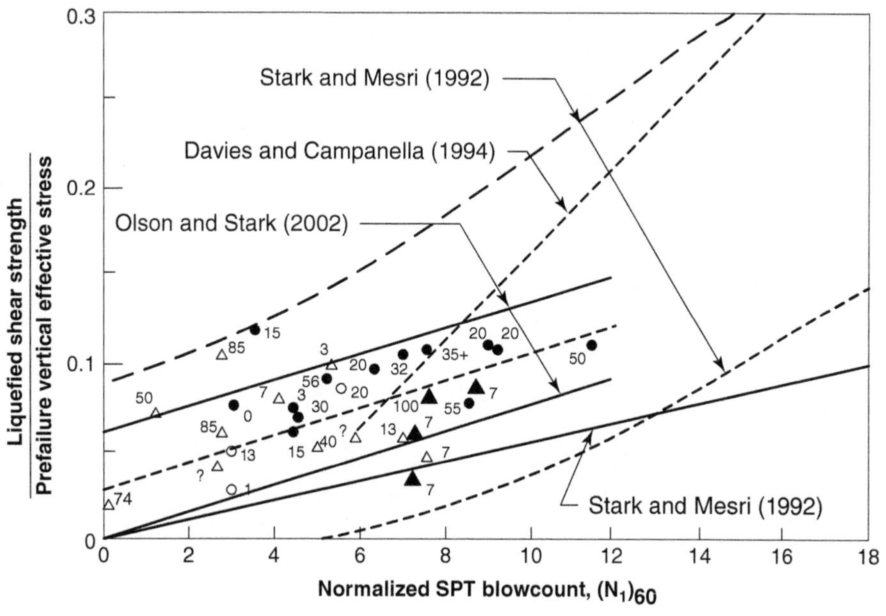

Figure 22.33 Shear strength of liquefied coarse-grained soils based on SPT blow count. (After Olson and Stark 2002)

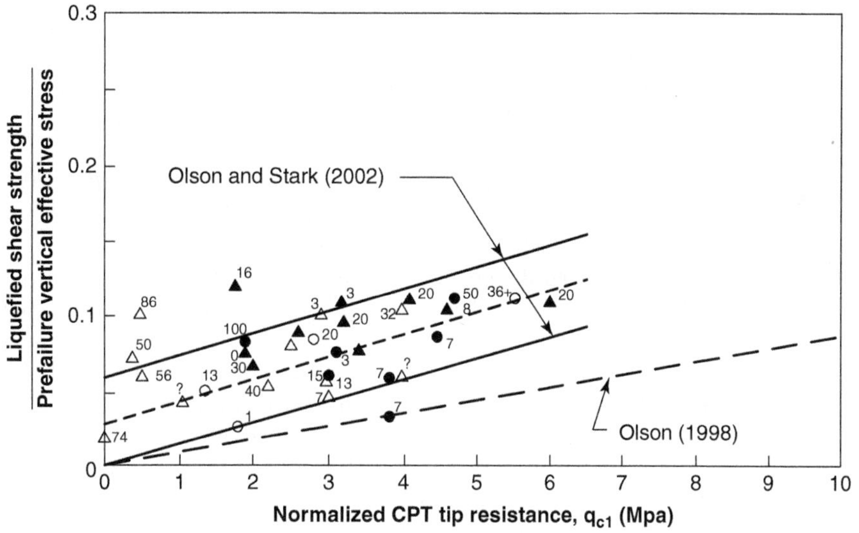

Figure 22.34 Shear strength of liquefied coarse-grained soils based on CPT point resistance. (After Olson and Stark 2002)

The pressure generated by the liquefied soil flowing past the pile can be estimated as 7 times the shear strength of the liquefied soil:

$$p_u = 7s_{u-liq} \tag{22.61}$$

where p_u is the pressure generated on the pile by the flowing soil and s_{u-liq} is the shear strength of the liquefied soil.

PROBLEMS

22.1 After an earthquake, a seismograph installed in the bedrock records the arrival of a compression wave and 10 seconds later the arrival of a shear wave. The rock has a compression wave velocity equal to 3000 m/s and a shear wave velocity equal to 1500 m/s. How far is the earthquake epicenter from the seismograph? How would you find the exact location of the epicenter?

22.2 An earthquake takes place along a fault and creates 2 m of relative displacement between two tectonic plates. The area over which the slip takes place is 500 km by 100 km and the shear modulus of the rock is 20 GPa. Calculate the seismic moment M_o, the moment magnitude M_w, and the energy E of the earthquake.

22.3 Search the Pacific Earthquake Engineering Research (PEER) Center web site (http://peer.berkeley.edu/nga/) and select an earthquake ground acceleration vs. time record. From this record, determine the peak ground acceleration. Then integrate the acceleration record to generate the velocity vs. time record and find the peak ground velocity. Then integrate the velocity record to generate the displacement vs. time record and find the peak ground displacement.

22.4 From the acceleration record of problem 22.3, find the bracketed duration for a threshold acceleration of 0.05 g and the sustained maximum acceleration for 3 cycles and then for 5 cycles.

22.5 An event with a return period *T* has a yearly probability of exceedance equal to 1/*T*. The equation linking the return period *T* of an event to the probability of exceedance *P* over a period of time *L* is:

$$P = 1 - (1 - 1/T)^L$$

Calculate (a) the return period for an earthquake that has a 2% probability of exceedance in 50 years and (b) the return period for an earthquake that has a 10% probability of exceedance in 50 years.

22.6 An 828 m tall tower weighs 6000 MN and has an equivalent stiffness of 200 MN/m.

a. Calculate the natural period of the tower. A one-story house weighs 1.4 MN and has a natural period of 0.15 seconds.
b. What is the equivalent stiffness of the house?

22.7 Search the PEER Center web site (http://peer.berkeley.edu/nga/) and select an earthquake ground acceleration vs. time record. For the acceleration record,

a. Develop the Fourier acceleration spectrum
b. Develop the response spectrum, for a damping ratio of 5%, by choosing *m* and varying *k*.
c. Choose a first set of values for *k* and m and find the spectral acceleration a_1, then find the spectral acceleration a_2 for a second set of values equal to 2k and 2m. Compare a_1 and a_2.

22.8 The PGA for a magnitude 6 earthquake is 0.5 g. What is the most likely PGA 50 km away?

22.9 What is the likely return period or recurrence interval for a magnitude 6 earthquake?

22.10 Calculate the natural period of a 20 m thick stiff soil layer if the soil shear wave velocity is 200 m/s. Then calculate the natural period of a 50 m thick soft soil layer if the shear wave velocity is 100 m/s.

22.11 What is the transfer function (amplification factor) for the displacement at the ground surface during an earthquake if the natural period of the deposit is 1 second and the depth of soil layer above rock level is 100 m? Assume an undamped linear soil on rigid rock. Redo the calculation for a damped linear soil on rigid rock if the damping ratio is 5%. The shear wave velocity of the soil is 250 m/s.

22.12 A soil has a shear wave velocity equal to 250 m/s and an SPT blow count equal to 30 bpf. The design earthquake corresponds to a PGA equal to 0.3 g. Develop the response spectrum according to Figure 22.17 if the reference spectrum has the following characteristics: spectral acceleration at 0.2 seconds = 0.5 g, spectral acceleration at 1 second = 0.2 g.

22.13 At a depth of 5 m below the ground surface, a saturated sand deposit has a corrected SPT blow count equal to 10 bpf, a CPT corrected and normalized point resistance of 90, and a corrected shear wave velocity of 170 m/s. The groundwater level is at the ground surface and the soil has a total unit weight of 18 kN/m^3. Will the soil liquefy in a magnitude 7.5 earthquake if the PGA is 0.6 g? What would be the highest magnitude for which the soil would not liquefy?

22.14 A slope is cut in a medium-stiff clay with an undrained shear strength s_u equal to 50 kPa. The site has a site class B, a PGA of 0.45 g, and a spectral acceleration at 1 second equal to 0.3 g. Calculate the horizontal seismic coefficient k_h to be used in the slope earthquake stability analysis.

22.15 Write the expression of the earthquake active earth pressure coefficient and the corresponding static active earth pressure coefficient. Plot the ratio versus k_h for $k_v = 0$, vertical back wall, horizontal backfill, frictionless wall, and a 30° friction angle for the backfill.

22.16 A 3 m high vertical gravity retaining wall has a dry horizontal backfill with a friction angle equal to 30° and a unit weight of 20 kN/m^3. It must be designed for a horizontal seismic coefficient equal to 0.2. Calculate:

a. Static coefficient of active and passive earth pressure, K_a and K_p
b. Seismic coefficient of active and passive earth pressure, K_{ae} and K_{pe}
c. The static component and dynamic component of the active push against the wall and their point of application, P_a, ΔP_{ae}, X_a, and X_{ae}
d. The static and dynamic components of the passive push against the wall if the wall was pushed into the soil backfill and their point of application, P_p, ΔP_{pe}, X_p, and X_{pe}

22.17 The wall of problem 22.16 has water on the no-soil side and water in the backfill up to the ground surface. The water depth on the no-soil side is 2 m. Calculate:

a. The hydrostatic pressure and the resultant water push on both sides of the wall, p_{w1}, p_{w2}, P_{w1}, and P_{w2}

b. The earthquake pressure and the resultant push on both sides of the wall if the horizontal seismic coefficient is 0.2

22.18 Demonstrate that the point of application of the dynamic water pressure in Eq. 22.50 is $0.6H_w$ from the top of the water level.

22.19 An anchored wall retains 10 m of sand with a blow count of 18 bpf and a unit weight of 20 kN/m^3. The water level is deeper than the excavation level. The design earthquake will generate a horizontal seismic coefficient equal to 0.25. Calculate:

a. The pressure p against the wall above the excavation in the case of no earthquake

b. The pressure p_e against the wall above the excavation in the case of an earthquake

c. The average load per anchor in both cases if the anchors are inclined at 15° and the vertical and horizontal spacing between anchors is 2.5 m.

22.20 A building is 60 m tall, weighs 500 MN, and has a horizontal stiffness of 400 MN/m. The design earthquake gives the response spectrum shown in Figure 22.1s. Calculate the horizontal force that must be resisted by the foundation.

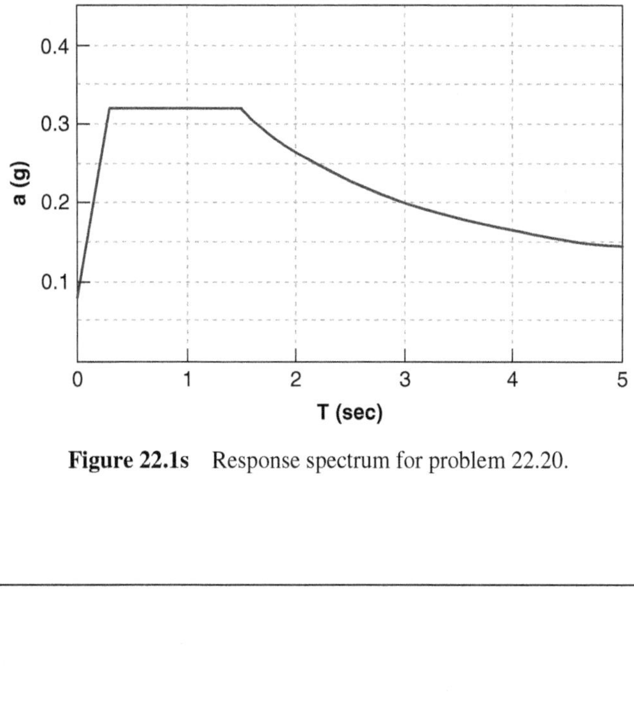

Figure 22.1s Response spectrum for problem 22.20.

Problems and Solutions

Problem 22.1

After an earthquake, a seismograph installed in the bedrock records the arrival of a compression wave and 10 seconds later the arrival of a shear wave. The rock has a compression wave velocity equal to 3000 m/s and a shear wave velocity equal to 1500 m/s. How far is the earthquake epicenter from the seismograph? How would you find the exact location of the epicenter?

Solution 22.1

The distance between the epicenter and seismograph is:

$$d = \frac{\Delta t_{\text{p-s}}}{\frac{1}{v_s} - \frac{1}{v_p}} = \frac{10}{\frac{1}{1500} - \frac{1}{3000}} = 30{,}000 \text{ m}$$

where $\Delta t_{p\text{-}s}$ is the arrival time difference of a shear wave and compression wave, v_s is the shear wave velocity, and v_p is the compression wave velocity. The earthquake epicenter is 30,000 m away from the seismograph. Three seismographs are needed to find the exact location of the epicenter: The intersection of the three circles gives the location.

Problem 22.2

An earthquake takes place along a fault and creates 2 m of relative displacement between two tectonic plates. The area over which the slip takes place is 500 km by 100 km and the shear modulus of the rock is 20 GPa. Calculate the seismic moment M_o, the moment magnitude M_w, and the energy E of the earthquake.

Solution 22.2

Seismic moment M_o:

$$M_o = GAD = (20 \times 10^9) \times (5 \times 10^{10}) \times (2) = 2 \times 10^{21}\ \text{N} \cdot \text{m}$$

where G is the shear modulus of the rock, A is the area over which the slip occurs, and D is the amount of slip movement.

Moment magnitude M_w:

$$M_w = 0.66 \log M_o(\text{N} \cdot \text{m}) - 6.05 = 0.66 \log(2 \times 10^{21}) - 6.05 = 8$$

Energy E:

$$\log E = 5.24 + 1.44M = 5.24 + 1.44 \times 8 = 16.76$$

Therefore, the energy E is $E = 10^{16.76} = 5.8 \times 10^{16}$ N.m $= 5.8 \times 10^{16}$ joules.

Problem 22.3

Search the Pacific Earthquake Engineering Research (PEER) Center web site (http://peer.berkeley.edu/nga/) and select an earthquake ground acceleration vs. time record. From this record, determine the peak ground acceleration. Then integrate the acceleration record to generate the velocity vs. time record and find the peak ground velocity. Then integrate the velocity record to generate the displacement vs. time record and find the peak ground displacement.

Solution 22.3

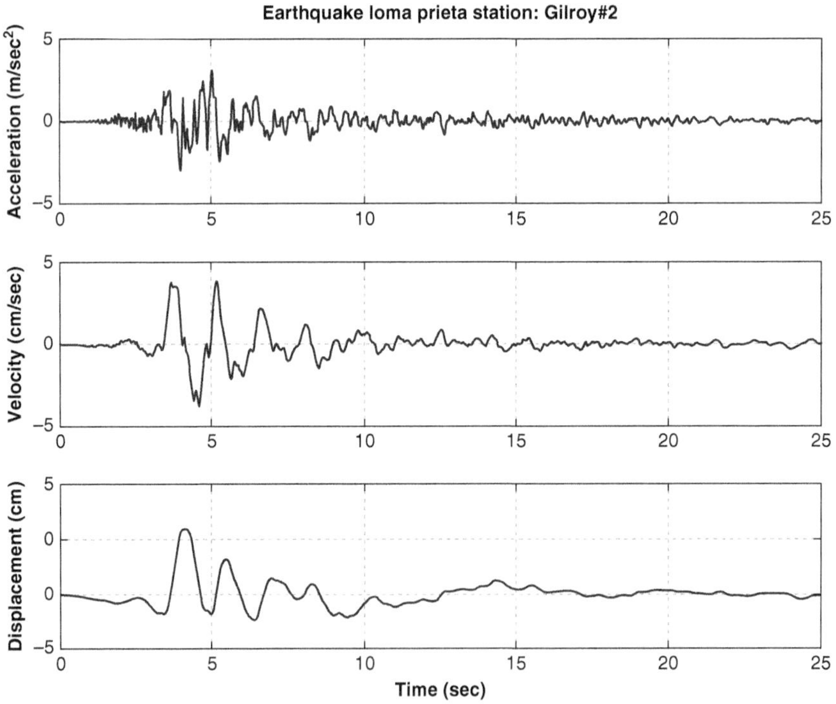

Figure 22.2s Acceleration, velocity, and displacement of an earthquake record.

A sample record chosen from the PEER web site is the Loma Prieta Station Gilroy #2 record. From Figure 22.2s:
Peak ground acceleration in gs (PGAg) = 0.322 g
Peak ground acceleration (PGA) = 3.159 (m/s^2)

Peak ground velocity (PGV) = 0.391 m/s
Peak ground displacement (PGD) = 0.121 (m)

Problem 22.4

From the acceleration record of problem 22.3, find the bracketed duration for a threshold acceleration of 0.05 g and the sustained maximum acceleration for 3 cycles and then for 5 cycles.

Solution 22.4

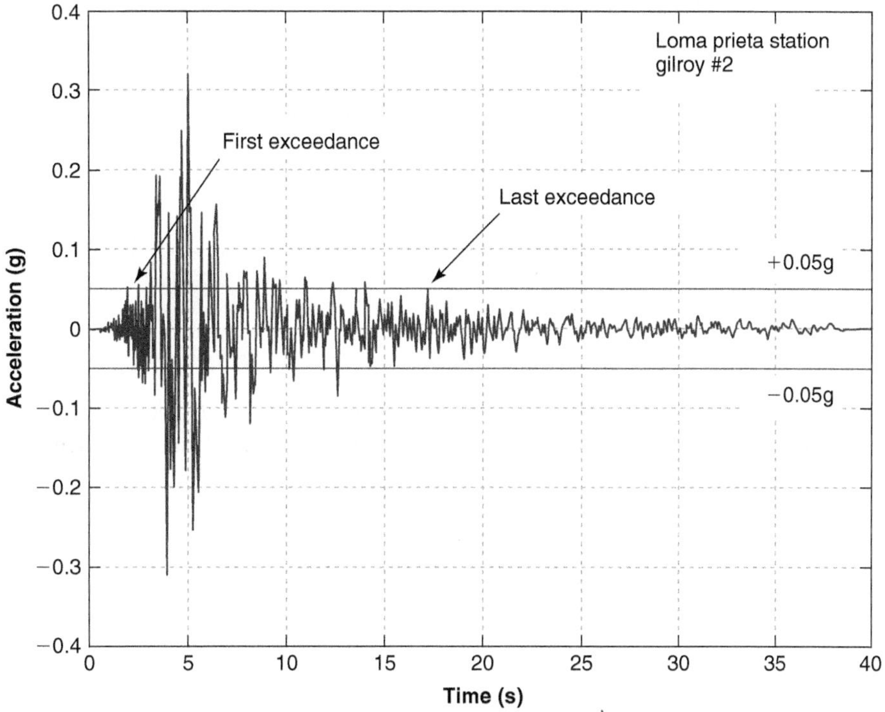

Figure 22.3s Bracketed duration of the ground acceleration.

The horizontal lines on Figure 22.3s show the threshold accelerations of ±0.05 g. The bracketed duration for this earthquake (the time between the first and last exceedance) is 15.26 seconds.

The maximum acceleration for three cycles is 0.145 g.

The maximum acceleration for five cycles is 0.13 g.

Problem 22.5

An event with a return period T has a yearly probability of exceedance equal to 1/T. The equation linking the return period T of an event to the probability of exceedance P over a period of time L is:

$$P = 1-(1-1/T)^L$$

Calculate (a) the return period for an earthquake that has a 2% probability of exceedance in 50 years and (b) the return period for an earthquake that has a 10% probability of exceedance in 50 years.

Solution 22.5

a. For an earthquake with a 2% probability of exceedance in 50 years, the return period is:

$$P = 1 - (1 - 1/T)^L$$
$$1 - P = (1 - 1/T)^L$$
$$(1 - P)^{1/L} = (1 - 1/T)$$
$$1/T = 1 - (1 - P)^{1/L}$$
$$T = \frac{1}{1 - (1 - P)^{1/L}}$$
$$T = \frac{1}{1 - (1 - 0.02)^{1/50}} = 2476 \text{ years}$$

b. For an earthquake with a 10% probability of exceedance in 50 years, the return period is:

$$T = \frac{1}{1 - (1 - P)^{1/L}}$$
$$T = \frac{1}{1 - (1 - 0.10)^{1/50}} = 475 \text{ years}$$

Problem 22.6

An 828 m tall tower weighs 6000 MN and has an equivalent stiffness of 200 MN/m. Calculate the natural period of the tower. A one-story house weighs 1.4 MN and has a natural period of 0.15 seconds. What is the equivalent stiffness of the house?

Solution 22.6

a. The natural period T of the tower is:

$$T = 2\pi\sqrt{\frac{m}{k}} = 2\pi\sqrt{\frac{W}{gk}}$$
$$T = 2\pi\sqrt{\frac{(6000/9.81)}{200}} = 10.98 \text{ sec}$$

b. Rearranging the natural period equation, the stiffness k of the house is:

$$k = 4\pi^2\frac{W}{gT^2} = 4 \times 3.14^2 \times \frac{1.4}{9.81 \times 0.15^2} = 250.1 \text{ MN/m}$$

Problem 22.7

Search the PEER Center web site (http://peer.berkeley.edu/nga/) and select an earthquake ground acceleration vs. time record. For the acceleration record:

a. Develop the Fourier acceleration spectrum
b. Develop the response spectrum, for a damping ratio of 5%, by choosing m and varying *k*.
c. Choose a first set of values for *k* and *m* and find the spectral acceleration a_1, then find the spectral acceleration a_2 for a second set of values equal to 2k and 2m. Compare a_1 and a_2.

Solution 22.7

The record selected for this example is the station Gilroy #2 on soil. The Fourier acceleration spectrum and the three response spectra (acceleration, velocity, and displacement) are given in Figures 22.4s and 22.5s.

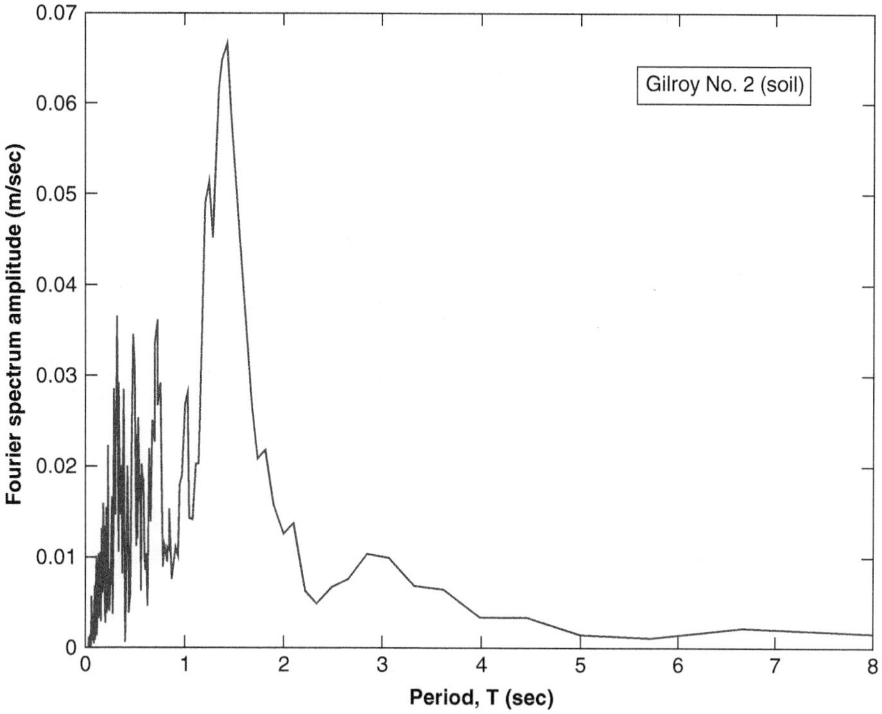

Figure 22.4s Fourier acceleration spectrum.

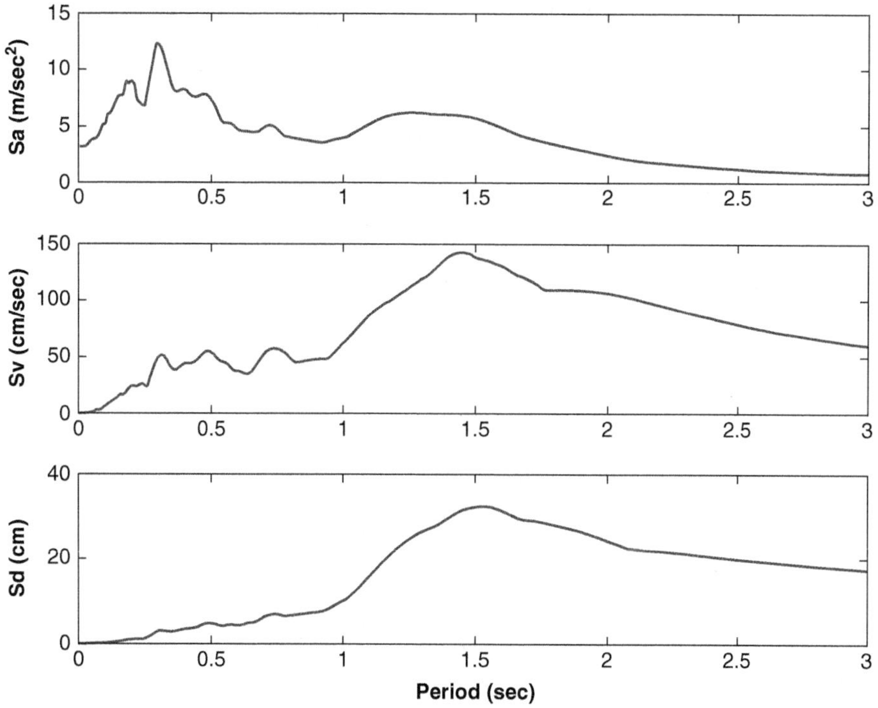

Figure 22.5s Fourier response spectrum.

Note: The response spectra are defined as the response of the SDOF with a natural period T. It is obtained by solving the equation of motion:

$$m\ddot{x} + c\dot{x} + kx = -ma(t)$$

By setting $\omega = \sqrt{k/m}$ and $c = 2\,m\omega\beta$, and then dividing by m, the equation becomes:

$$\ddot{x} + 2\beta\omega\dot{x} + \omega^2 x = -a(t)$$

Thus, by multiplying k and m by the same value, ω will not change and the response spectrum will not change, including the spectral acceleration. However, a change in β (damping) will change the response spectrum.

Problem 22.8

The PGA for a magnitude 6 earthquake is 0.5 g. What is the most likely PGA 50 km away?

Solution 22.8

The peak ground acceleration at a distance R(km) for a magnitude M earthquake can be estimated by using Figure 22.6s. A line is drawn parallel to the trend line starting at the PGA value of 0.5 g. Then the PGA value is read on that line at a distance of 50 km.

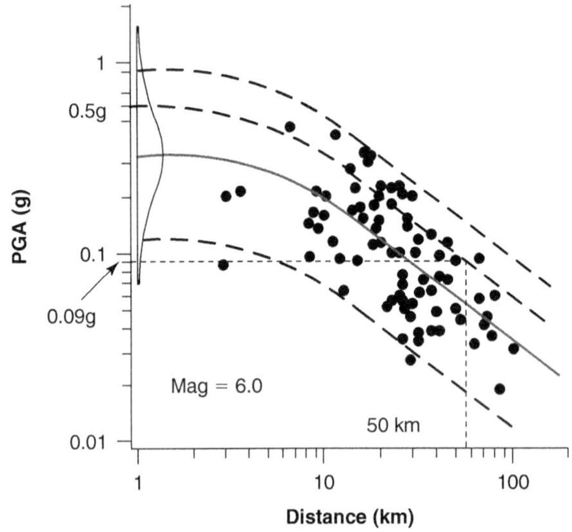

Figure 22.6s Attenuation of peak horizontal ground acceleration.

Problem 22.9

What is the likely return period or recurrence interval for a magnitude 6 earthquake?

Solution 22.9

According to Figure 22.12, the recurrence interval of a magnitude 6 earthquake is about 200 years.

Problem 22.10

Calculate the natural period of a 20 m thick stiff soil layer if the soil shear wave velocity is 200 m/s. Then calculate the natural period of a 50 m thick soft soil layer if the shear wave velocity is 100 m/s.

Solution 22.10

Using Eq. 22.6, the natural period is:

$$T = \frac{4H}{v_s}$$

Stiff soil:

$$T = \frac{4 \times 20}{200} = 0.4\,\text{sec}$$

Soft soil:

$$T = \frac{4 \times 50}{100} = 2\,\text{sec}$$

Problem 22.11

What is the transfer function (amplification factor) for the displacement at the ground surface during an earthquake if the natural period of the deposit is 1 second and the depth of soil layer above rock level is 100 m? Assume an undamped linear soil on rigid rock. Redo the calculation for a damped linear soil on rigid rock if the damping ratio is 5%. The shear wave velocity of the soil is 250 m/s.

Solution 22.11

The 1 second period is used to find ω:

$$T = \frac{2\pi}{\omega} = 1$$

Therefore, $\omega = 2\pi$. The first calculation is when the soil is undamped, $\beta = 0$. The transfer function is:

$$F(\omega) = \frac{1}{\sqrt{\cos^2\left(\frac{\omega H}{v_s}\right) + \left(\beta\frac{\omega H}{v_s}\right)^2}}$$

$$F(\omega) = \frac{1}{\sqrt{\cos^2\left(\frac{2\pi \times 100}{250}\right) + 0}} = 1.24$$

With damping, $\beta = 0.05$, the transfer function is:

$$F(\omega) = \frac{1}{\sqrt{\cos^2\left(\frac{2\pi \times 100}{250}\right) + \left(.05\frac{2\pi \times 100}{250}\right)^2}} = 1.22$$

Problem 22.12

A soil has a shear wave velocity equal to 250 m/s and an SPT blow count equal to 30 bpf. The design earthquake corresponds to a PGA equal to 0.3 g. Develop the response spectrum according to Figure 22.17 if the reference spectrum has the following characteristics: spectral acceleration at 0.2 seconds = 0.5 g, spectral acceleration at 1 second = 0.2 g.

Solution 22.12

The site-specific spectral parameters are found in Table 22.2. With the given soil parameters:

$$v_s = 250\text{ m/s} \quad \text{and} \quad N_{SPT} = 30 \Rightarrow \text{Soil clasification is “D”.}$$

From Table 22.3, with a PGA = 0.3 g and a site classification of D, $F_{PGA} = 1.2$.

$$A_S = F_{PGA} \times PGA$$
$$A_s = 1.2 \times 0.3g = 0.36\text{ g}$$

$S_{DS} = 0.5$ g and $S_{D1} = 0.2$ g (from the problem), therefore:

$$T_S = \frac{S_{D1}}{S_{DS}}$$
$$T = \frac{0.2\,g}{0.5\,g} = 0.4$$
$$T_o = 0.2T_S$$
$$T_o = 0.2 \times 0.4 = 0.08$$

The constants found using the site classification are used to develop the site-specific acceleration response spectrum, shown in Figure 22.7s.

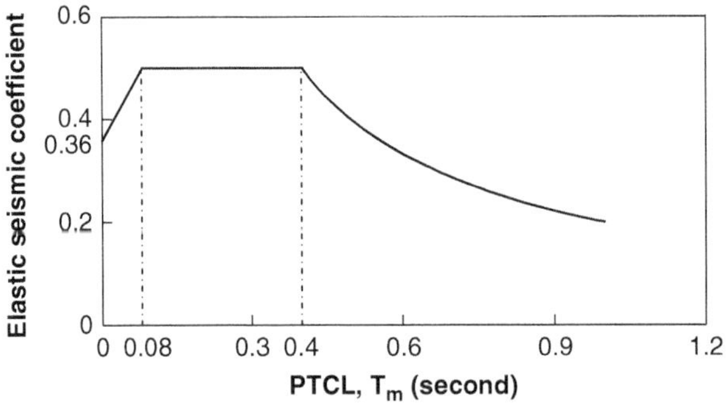

Figure 22.7s Design code acceleration response spectrum.

Problem 22.13

At a depth of 5 m below the ground surface, a saturated sand deposit has a corrected SPT blow count equal to 10 bpf, a CPT corrected and normalized point resistance of 90, and a corrected shear wave velocity of 170 m/s. The fine percentage is less than 5%. The groundwater level is at the ground surface and the soil has a total unit weight of 18 kN/m^3. Will the soil liquefy in a magnitude 7.5 earthquake if the PGA is 0.6g? What would be the highest magnitude for which the soil would not liquefy?

Solution 22.13

a. The cyclic stress ratio $CSR = \frac{\tau_{av}}{\sigma'_{vo}} = 0.65\left(\frac{a_{max}}{g}\right)\left(\frac{\sigma_{vo}}{\sigma'_{vo}}\right)r_d$

$$a_{max} = 0.6g, \sigma_{vo} = \gamma H = 18 \times 5 = 90(\text{kN/m}^2), \sigma'_{vo} = \gamma' H = (18 - 9.8) \times 5 = 41(\text{kN/m}^2)$$

Fig. 22.25 $\rightarrow r_d = 0.95$.

$$CSR = \frac{\tau_{av}}{\sigma'_{vo}} = 0.65(0.6)(2.195)0.95 = 0.81$$

Fig. 22.21 $\xrightarrow{N_{SPT}=10}$ *Liquefy*,

Fig. 22.22 $\xrightarrow{q=90}$ *Liquefy*,

Fig. 22.23 $\xrightarrow{v_s=170(\text{m/s})}$ *Liquefy*

b. $CSR = \frac{\tau_{av}}{\sigma'_{vo}} = 0.65\left(\frac{a_{max}}{g}\right)\left(\frac{\sigma_{vo}}{\sigma'_{vo}}\right)r_d$

Fig. 22.21 $\xrightarrow{CSR=0.1} a_{max} = 0.07$ g

Fig. 22.22 $\xrightarrow{CSR=0.13} a_{max} = 0.095$ g

Fig. 22.23 $\xrightarrow{CSR=0.13} a_{max} = 0.095$ g

Problem 22.14

A slope is cut in a medium-stiff clay with an undrained shear strength s_u equal to 50 kPa. The height of the slope is 10 m. The site has a site class B, a PGA of 0.45 g, and a spectral acceleration at 1 second equal to 0.3 g. Calculate the horizontal seismic coefficient k_h to be used in the slope earthquake stability analysis.

Solution 22.14

$$Class\ B,\ S_1 = 0.3\ \text{g} \xrightarrow{\text{Table 22.3}} F_{PGA} = 1.0$$

$$Class\ B,\ PGA = 0.45\ \text{g} \xrightarrow{\text{Table 22.5}} F_V = 1.0$$

$$k_{max} = F_{PGA} \times PGA = 0.45\ g$$

$$k_{av} = \gamma\left(1 + 0.01H\left(0.5\frac{F_V S_1}{k_{max}} - 1\right)\right)k_{max}$$

$$k_h = k_{av} = 1.2\left(1 + 0.01 \times 10\left(0.5\frac{1 \times 0.3\ \text{g}}{0.45\ \text{g}} - 1\right)\right)0.45\ \text{g} = 0.504\ \text{g}$$

Problem 22.15

Write the expression of the earthquake active earth pressure coefficient and the corresponding static active earth pressure coefficient. Plot the ratio versus k_h for $k_v = 0$, vertical back wall, horizontal backfill, frictionless wall, and a 30° friction angle for the backfill.

Solution 22.15

The expression for the active earth pressure coefficient in the earthquake case, K_{ae}, is found after finding the most critical wedge angle. K_{ae} is:

$$K_{ae} = \frac{\sin^2(\alpha + \varphi' - \psi)}{\cos\psi \sin^2\alpha \sin(\alpha - \delta - \psi)\left[1 + \sqrt{\dfrac{\sin(\varphi' + \delta)\sin(\varphi' - \beta - \psi)}{\sin(\alpha - \delta - \psi)\sin(\alpha + \beta)}}\right]^2}$$

where α is the angle of the back of the wall with the horizontal, β is the angle of the ground surface behind the wall with the horizontal, δ is the angle of friction between the back of the wall and the soil, φ' is the friction angle of the soil, and ψ is the angle representing the earthquake inertia force as:

$$\psi = \tan^{-1}\left(\frac{k_h}{1 - k_v}\right)$$

where k_h and k_v are the horizontal and vertical seismic coefficients respectively. The expression for the static active earth pressure coefficient, K_a, is:

$$K_a = \frac{\sin^2(\alpha + \varphi')}{\sin^2\alpha \sin(\alpha - \delta)\left[1 + \sqrt{\dfrac{\sin(\varphi' + \delta)\sin(\varphi' - \beta)}{\sin(\alpha - \delta)\sin(\alpha + \beta)}}\right]^2}$$

The ratio of K_{ae}/K_a for a vertical wall ($\alpha = 90$), no wall friction ($\delta = 0$), horizontal backfill ($\beta = 0$), and a 30° angle of friction for the backfill can be plotted as in Figure 22.8s.

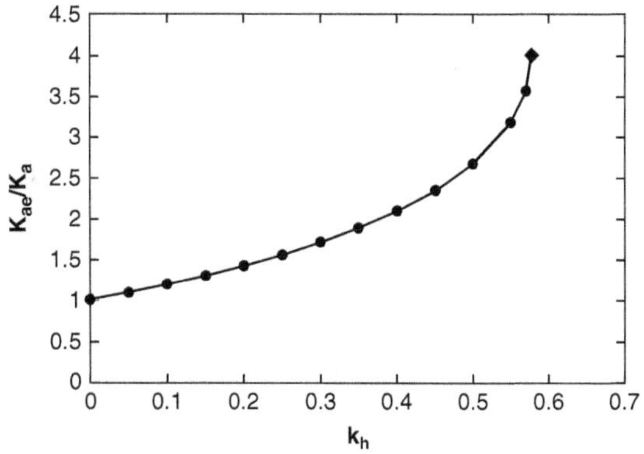

Figure 22.8s Ratio of the earthquake active earth pressure coefficient and the corresponding active earth pressure coefficient versus k_h.

Problem 22.16

A 3 m high vertical gravity retaining wall has a dry horizontal backfill with a friction angle equal to 30° and a unit weight of 20 kN/m^3. It must be designed for a horizontal seismic coefficient equal to 0.2. Calculate:

a. Static coefficient of active and passive earth pressure, K_a and K_p
b. Seismic coefficient of active and passive earth pressure, K_{ae} and K_{pe}
c. The static component and dynamic component of the active push against the wall and their point of application, P_a, ΔP_{ae}, X_a, and X_{ae}
d. The static and dynamic components of the passive push against the wall if the wall was pushed into the soil backfill and their point of application, P_p, ΔP_{pe}, X_p, and X_{pe}.

Solution 22.16

a. Static coefficient of active and passive earth pressure, K_a and K_p:

$$K_a = \frac{1-\sin\varphi'}{1+\sin\varphi'} = \frac{1-\sin(30)}{1+\sin(30)} = 0.333$$

$$K_p = \frac{1+\sin\varphi'}{1-\sin\varphi'} = \frac{1+\sin(30)}{1-\sin(30)} = 3$$

b. Seismic coefficient of active and passive earth pressure, K_{ae} and K_{pe}:

$$\psi = \tan^{-1}\left(\frac{k_h}{1-k_v}\right) = \tan^{-1}\left(\frac{0.2}{1-0}\right) = 11.3^\circ$$

The seismic coefficient of active earth pressure is:

$$K_{ae} = \frac{\sin^2(\alpha+\varphi'-\psi)}{\cos\psi\sin^2\alpha\sin(\alpha-\delta-\psi)\left[1+\sqrt{\dfrac{\sin(\varphi'+\delta)\sin(\varphi'-\beta-\psi)}{\sin(\alpha-\delta-\psi)\sin(\alpha+\beta)}}\right]^2}$$

$$K_{ae} = \frac{\sin^2(90+30-11.3)}{\cos(11.3)\sin^2(90)\sin(90-0-11.3)\left[1+\sqrt{\dfrac{\sin(30+0)\sin(30-0-11.3)}{\sin(90-0-11.3)\sin(90+0)}}\right]^2} = 0.473$$

The seismic coefficient of passive earth pressure is:

$$K_{pe} = \frac{\sin^2(\alpha-\varphi'+\psi)}{\cos\psi\sin^2\alpha\sin(\alpha+\delta+\psi)\left[1-\sqrt{\dfrac{\sin(\varphi'+\delta)\sin(\varphi'+\beta+\psi)}{\sin(\alpha+\delta+\psi)\sin(\alpha+\beta)}}\right]^2}$$

$$K_{pe} = \frac{\sin^2(90-30+11.3)}{\cos(11.3)\sin^2(90)\sin(90+0+11.3)\left[1-\sqrt{\dfrac{\sin(30+0)\sin(30+0+11.3)}{\sin(90+0+11.3)\sin(90+0)}}\right]^2} = 5.29$$

c. The static component of active push is:

$$P_a = \frac{1}{2}K_a\gamma H^2$$

$$P_a = \frac{1}{2}\times 0.33\times 20\times(3)^2 = 30\frac{\text{kN}}{\text{m}}$$

The dynamic component of active push is:

$$\Delta P_{ae} = \frac{1}{2}(K_{ae} - K_a)\gamma H^2$$

$$\Delta P_{ae} = \frac{1}{2}(0.473 - 0.333)(20)(3)^2 = 12.6\frac{\text{kN}}{\text{m}^3}$$

The point of application of P_a from the bottom of the wall is:

$$X_a = \frac{1}{3}H = \frac{1}{3}(3) = 1\text{ m}$$

The point of application of P_{ae} from the bottom of the wall is:

$$X_{ae} = 0.6H = 0.6(3) = 1.8\text{ m}$$

d. The static component of passive push is:

$$P_p = \frac{1}{2}K_p\gamma H^2$$

$$P_p = \frac{1}{2}(3)(20)(3)^2 = 270\frac{\text{kN}}{\text{m}^3}$$

The dynamic component of passive push is:

$$\Delta P_{pe} = \frac{1}{2}(K_{pe} - K_p)\gamma H^2$$

$$\Delta P_{pe} = \frac{1}{2}(5.29 - 3)(20)(3)^2 = 206.1\frac{\text{kN}}{\text{m}^3}$$

The point of application of P_a from the bottom of the wall is:

$$X_p = \frac{1}{3}H = \frac{1}{3}(3) = 1\text{ m}$$

The point of application of P_{ae} from the bottom of the wall is:

$$X_{pe} = 0.6\text{ H} = 0.6 \times 3 = 1.8\text{ m}$$

Problem 22.17

The wall of problem 22.16 has water on the no-soil side and water in the backfill up to the ground surface. The water depth on the no-soil side is 2 m. Calculate:

a. The hydrostatic pressure and the resultant water push on both sides of the wall, p_{w1}, p_{w2}, P_{w1}, and P_{w2}
b. The earthquake pressure and the resultant push on both sides of the wall if the horizontal seismic coefficient is 0.2

Solution 22.17

a. The hydrostatic pressure and the resultant water push on both sides of the wall, p_{w1}, p_{w2}, P_{w1}, and P_{w2}:
   The hydrostatic pressure on the no-soil side is:

$$p_{w1} = \gamma_w z_1$$

$$p_{w1} = 9.81(2) = 19.62\frac{\text{kN}}{\text{m}^2}$$

The resultant push for the hydrostatic pressure on the no-soil side is:

$$P_{w1} = \frac{1}{2}p_{w1}z_1 = \frac{1}{2}(19.62)(2) = 19.62\frac{\text{kN}}{\text{m}}$$

The hydrostatic pressure on the backfill side is:

$$p_{w2} = \gamma_w z_2$$

$$p_{w2} = 9.81(3) = 29.43\frac{\text{kN}}{\text{m}^2}$$

The resultant push for the hydrostatic pressure on the backfill side is:

$$P_{w2} = \frac{1}{2}p_{w2}z_2 = \frac{1}{2}(29.43)(3) = 44.15\frac{\text{kN}}{\text{m}}$$

b. The earthquake pressure and the resultant push on both sides of the wall if the horizontal seismic coefficient is 0.2:
The earthquake water pressure on the no-soil side is:

$$\Delta p_{we1} = \frac{7}{8}k_h\gamma_w\sqrt{zH_w}$$

$$\Delta p_{we1} = \frac{7}{8}(0.2)(9.81)\sqrt{(2)(2)} = 3.43\frac{\text{kN}}{\text{m}^2}$$

The resultant push for the hydrostatic pressure on the no-soil side is:

$$\Delta P_{we1} = \frac{7}{12}k_h\gamma_w {H_w}^2$$

$$\Delta P_{we1} = \frac{7}{12}(0.2)(9.81)(2)^2 = 4.58\frac{\text{kN}}{\text{m}}$$

The resultant water push on the no-soil side is:

$$P_{w1} + \Delta P_{we1} = 19.62 + 4.58 = 24.2\frac{\text{kN}}{\text{m}}$$

The earthquake push on the backfill side is obtained as follows:

$$\gamma' = \gamma - \gamma_h = 20 - 9.81 = 10.19\frac{\text{kN}}{\text{m}^3}$$

$$\psi = \tan^{-1}\left(\frac{k_h}{1-k_v} \times \frac{\gamma}{\gamma'}\right)$$

$$\psi = \tan^{-1}\left(\frac{0.2}{1-0} \times \frac{20}{10.19}\right) = 21.4^\circ$$

$$K_{ae} = \frac{\sin^2(\alpha + \varphi' - \psi)}{\cos\psi\sin^2\alpha\sin(\alpha - \delta - \psi)\left[1 + \sqrt{\dfrac{\sin(\varphi' + \delta)\sin(\varphi' - \beta - \psi)}{\sin(\alpha - \delta - \psi)\sin(\alpha + \beta)}}\right]^2}$$

$$K_{ae} = \frac{\sin^2(90 + 30 - 21.4)}{\cos(21.4)\sin^2(90)\sin(90 - 0 - 21.4)\left[1 + \sqrt{\dfrac{\sin(30 + 0)\sin(30 - 0 - 21.4)}{\sin(90 - 0 - 21.4)\sin(90 + 0)}}\right]^2} = 0.685$$

$$P'_{ae} = \frac{1}{2}K_{ae}\gamma' H^2$$

$$P'_{ae} = \frac{1}{2}(0.685)(10.19)(3)^2 = 31.41\frac{\text{kN}}{\text{m}^3}$$

The resultant push on the backfill side is:

$$P_{ae} = P_{w2} + P'_{ae}$$

$$P_{ae} = 44.15 + 31.41 = 75.6\frac{\text{kN}}{\text{m}}$$

Problem 22.18

Demonstrate that the point of application of the dynamic water pressure in Eq. 22.50 is $0.6H_w$ from the top of the water level.

Solution 22.18

By writing the moment equation:

$$\overline{Z}.\Delta P_{we} = \int_{z=0}^{z=H_W} (\Delta p_{we} \times z)dz$$

From Eqs. 22.50 and 22.52, we have:

$$\Delta p_{we} = \frac{7}{8} k_h \gamma_w \sqrt{zH_w}$$

$$\Delta P_{we} = \frac{7}{12} k_h \gamma_w H_w{}^2$$

By plugging Eqs. 22.50 and 22.52 into:

$$K_{ae} = \frac{\sin^2(90 + 30 - 11.3)}{\cos(11.3)\sin^2(90)\sin(90 - 0 - 11.3)\left[1 + \sqrt{\dfrac{\sin(30 + 0)\sin(30 - 0 - 11.3)}{\sin(90 - 0 - 11.3)\sin(90 + 0)}}\right]^2} = 0.473$$

we get:

$$\overline{Z} = \frac{\int_0^{H_w} \left(\frac{7}{8} k_h \gamma_w \sqrt{zH_w} z\right) dz}{\frac{7}{12} k_h \gamma_w H_w^2}$$

$$\overline{Z} = \frac{12 \int_0^{H_w} z^{\frac{3}{2}} dz}{8 H_w^{\frac{3}{2}}}$$

$$\overline{Z} = \frac{3}{2} \times \frac{1}{H_w^{\frac{3}{2}}} \times \frac{2}{5} H_w^{\frac{5}{2}} = \frac{6}{10} H_w$$

Problem 22.19

An anchored wall retains 10 m of sand with a blow count of 18 bpf and a unit weight of 20 kN/m^3. The wall is vertical, the backfill is horizontal, and the wall friction is zero. The water level is deeper than the excavation level. The allowable movement at the top of the wall is 30 mm. The design earthquake will generate a horizontal seismic coefficient equal to 0.25. Calculate:

a. The pressure p against the wall above the excavation in the case of no earthquake
b. The pressure p_e against the wall above the excavation in the case of an earthquake

Solution 22.19

1. The constant pressure p against the wall above the excavation in the case of no earthquake

$$\text{Fig. 21.19 } u_{top} = 30 \text{ mm}, H = 10000 \text{ mm} \rightarrow \frac{u_{top}}{H} = 0.003 \xrightarrow{\text{Fig. 21.19}} K = 0.2$$

$$p = K \times \gamma H = 0.2 \times 20 \times 10 = 40 \text{ kN/m}^2$$

2. The constant pressure p_e against the wall above the excavation in the case of the earthquake
   Calculate the coefficient of active earth pressure in the case of no earthquake
   Based on Fig. 15.12 → $\varphi' = 33^\circ$ $K_a = \dfrac{1 - \sin 33^\circ}{1 + \sin 33^\circ} = 0.3$
   Calculate the coefficient of active earth pressure in the case of earthquake

$$K_{ae} = \frac{\sin^2(\alpha + \varphi' - \psi)}{\cos\psi \sin^2\alpha \sin(\alpha - \delta - \psi)\left[1 + \sqrt{\dfrac{\sin(\varphi' + \delta)\sin(\varphi' - \beta - \psi)}{\sin(\alpha - \delta - \psi)\sin(\alpha + \beta)}}\right]^2}$$

Where $\psi = \tan^{-1}\left(\dfrac{k_h}{1 - k_v}\right) = \tan^{-1}\left(\dfrac{0.25}{1 - 0}\right) = 14^\circ$

$$K_{ae} = \frac{\sin^2(33 - 14)}{\cos(14)\sin^2(90)\sin(90 - 0 - 14)\left[1 + \sqrt{\dfrac{\sin(33 + 0)\sin(33 - 0 - 14)}{\sin(90 - 0 - 14)\sin(90 + 0)}}\right]^2} = 0.466$$

Calculate the pressure against the wall in the case of the earthquake.

$$p_e = \frac{K_{ae}}{K_a} \times K\gamma H = \frac{0.466}{0.3} \times 0.2 \times 20 \times 10 = 62.1\ \text{kN/m}^2$$

Problem 22.20

A building is 60 m tall, weighs 500 MN, and has a horizontal stiffness of 400 MN/m. The design earthquake gives the response spectrum shown in Figure 22.1s. Calculate the horizontal force that must be resisted by the foundation.

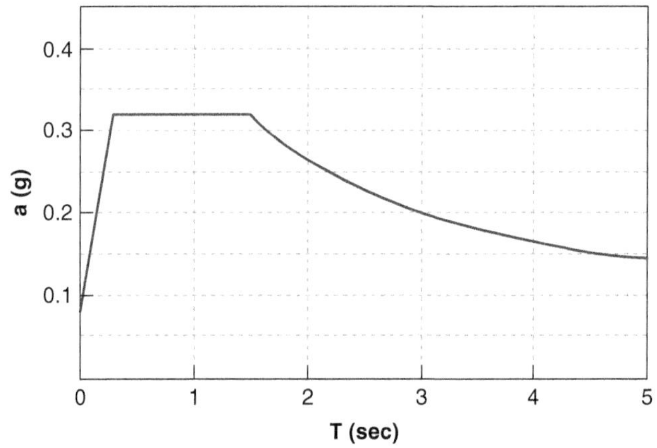

Figure 22.1s Response spectrum for problem 22.20.

Solution 22.20

Fundamental period: $T = 2\pi\sqrt{\dfrac{M}{K}} = 2\pi\sqrt{\dfrac{500/g}{400}} = 2.24\ \text{sec}$

$$T \xrightarrow{Spectrum} a = 0.245\ \text{g}$$

$$F = Ma = (500/9.81) \times 0.245 * 9.81 = 122.5\ \text{MN}$$

CHAPTER 23

Erosion of Soils and Scour Problems

23.1 THE EROSION PHENOMENON

An erosion problem always has three components: the soil or rock, the water, and the geometry of the obstacle that the water is interacting with. The resistance of the soil or rock is characterized by its erodibility, the water action is quantified by its velocity, and the geometry of the obstacle is quantified by its dimensions. Background on erosion from Briaud (2008) appears in this chapter, including the associated case histories.

Figure 23.1 shows a free-body diagram of a soil particle, a cluster of particles, or a rock block at the bottom of a lake. The water imposes a normal stress (hydrostatic pressure) around the soil particle or rock block. The normal stress is slightly higher at the bottom than at the top because the bottom is slightly deeper in the water column. This normal stress difference creates the buoyancy force, which reduces the weight of the soil particle or rock block.

Figure 23.2 shows the same particle, cluster of particles, or rock block at the bottom of a flowing river. Three things happen when water starts flowing. First, a drag force and associated shear stresses develop at the interface between the soil particle or rock block and the water flowing over it. This drag force is very similar to the seepage force. Second, the normal stress on top of the soil particle or rock block decreases because of the water flow. Indeed, as the velocity increases around the particle or the obstacle, the pressure drops to maintain conservation of energy according to Bernoulli's principle. This phenomenon is similar to the air flow on top of an airplane wing where the pressure is lower than below the wing, thereby developing the uplift force necessary for the plane to fly. Third, the normal stresses and shear stresses applied at the boundaries fluctuate with time because of the turbulence in the water. These fluctuations find their roots in the appearance and disappearance of eddies, vortices, ejections, and sweeps in the flowing water; they can contribute significantly to the erosion process, especially at higher velocities. In some cases they are the main reason for erosion. The contribution of turbulence fluctuations to the erosion process has been studied by several authors, including Croad (1981), Raudkivi (1998), Hoffmans and Verheij (1997), Bollaert (2002), and Hofland et al. (2005). The combination of the mean value and the fluctuations around the mean of the drag force and uplift force can become large enough to pluck and drag the soil particle, soil particle cluster, or rock block away and generate erosion.

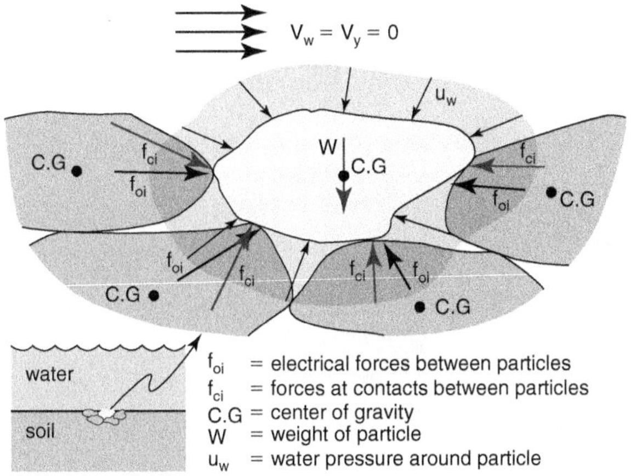

Figure 23.1 Free-body diagram of a soil particle or rock block for a no-flow condition (Briaud 2008).

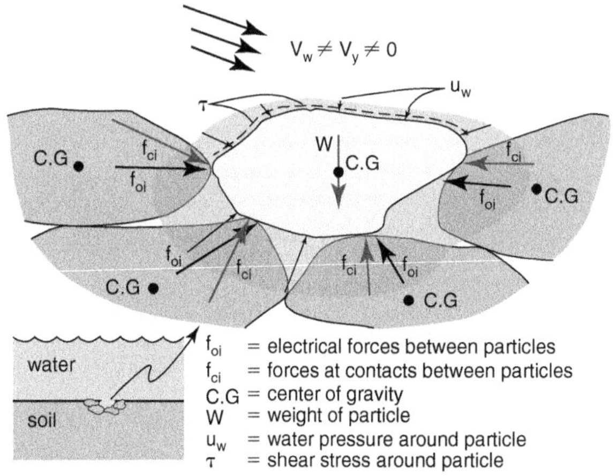

Figure 23.2 Free-body diagram of a soil particle or rock block when the water flows (Briaud 2008).

Note that in the case of unsaturated soils or saturated soils with water tension, the mechanical interparticle compressive forces (f_{ci} in Figures 23.1 and 23.2) can be significantly larger than in the case where the water is in compression. This apparent cohesion may increase the resistance to erosion, at least until the flow and presence of water destroy the water tension.

23.2 EROSION MODELS

The erodibility of a soil or rock can be defined as the relationship between the erosion rate $\dot{z}$ and the velocity of the water v near the soil-water interface. This definition is not very satisfactory because the velocity varies in direction and intensity in the flow field. In fact, strictly speaking, the water velocity is zero at the soil/rock-water interface. A more satisfactory definition is the relationship between the erosion rate $\dot{z}$ and the shear stress τ at the soil/rock-water interface:

$$\dot{z} = f(\tau) \qquad (23.1)$$

The erosion function described by Eq. 23.1 represents the constitutive law of the soil or rock for erosion problems, much like a stress-strain curve represents the constitutive law of the soil or rock for a settlement problem. Although a definition based on shear stress is an improvement over a velocity-based definition, it is still not completely satisfactory, as shear stress is not the only stress that contributes to the erosion rate. A more complete description of the erosion function is given by Eq. 23.2:

$$\frac{\dot{z}}{u} = \alpha\left(\frac{\tau - \tau_c}{\rho u^2}\right)^m + \beta\left(\frac{\Delta\tau}{\rho u^2}\right)^n + \gamma\left(\frac{\Delta\sigma}{\rho u^2}\right)^p \qquad (23.2)$$

where $\dot{z}$ is the erosion rate (m/s), u is the water velocity (m/s), τ is the hydraulic shear stress, τ_c is the threshold or critical shear stress below which no erosion occurs, ρ is the mass density of water (kg/m^3), $\Delta\tau$ is the turbulent fluctuation of the hydraulic shear stress, and $\Delta\sigma$ is the turbulent fluctuation of the net uplift normal stress.

All other quantities are parameters characterizing the soil being eroded. While this model is quite thorough, it is rather impractical at this time to determine the six parameters needed in Eq. 23.2 on a site-specific and routine basis. Today Eq. 23.3, which corresponds to the first term of Eq. 23.2, is widely accepted:

$$\frac{\dot{z}}{u} = \alpha\left(\frac{\tau - \tau_c}{\rho u^2}\right)^m \qquad (23.3)$$

As additional fundamental work is performed in erosion engineering, it is likely that Eq. 23.3 will evolve toward Eq. 23.2.

23.3 MEASURING THE EROSION FUNCTION

In the early 1990s, an apparatus was developed to measure the erosion function. This erosion function apparatus (EFA) was described in detail in section 9.20.1, including the data reduction (Briaud et al. 2001a). The principle is to go to the site where erosion is being investigated, collect samples within the depth of concern, bring them back to the laboratory, and test them in the EFA. A 75 mm outside diameter sampling tube containing the sample is placed through the bottom of the conduit where water flows at a constant velocity (Figure 23.3). The soil or rock is pushed out of the sampling tube only as fast as it is eroded by the water flowing over it. For each velocity, an erosion rate is measured and a shear stress is calculated using Moody's chart (Moody 1944). Thus the erosion function is obtained point by point.

Examples of erosion functions are shown in Figure 23.4 for a fine sand and Figure 23.5 for a low-plasticity clay. Note that for the same average velocity of 1 m/s in the EFA test conduit, the rate of erosion for the sand is about 1000 times faster than for the clay. This indicates that the rate of erosion can be very different for different soils.

Other devices have also been developed to evaluate how resistant earth materials are to water flow. These include the rotating cylinder to measure the erosion properties of stiff soils

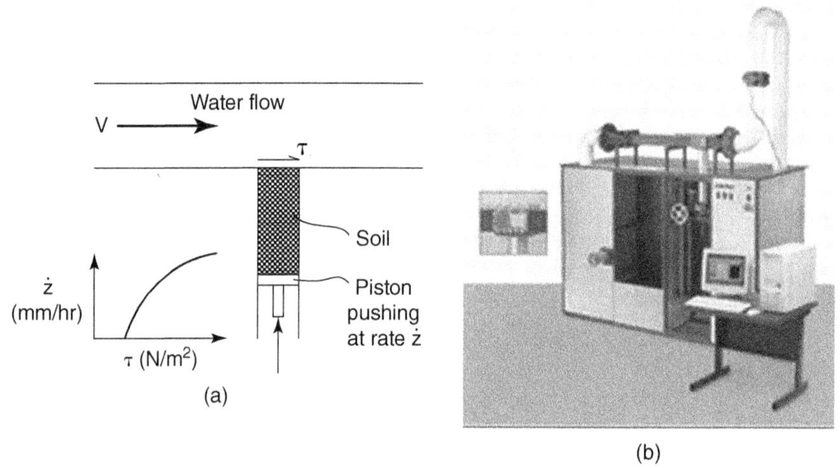

Figure 23.3 Erosion function apparatus to measure erodibility (Briaud et al. 1999).

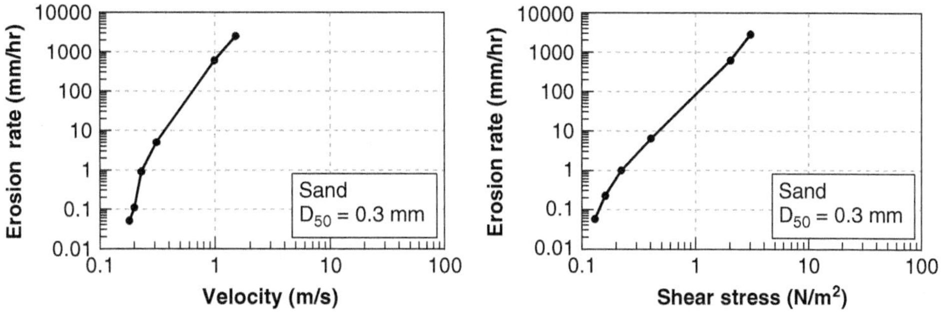

Figure 23.4 Erosion function for a fine sand as measured in the EFA.

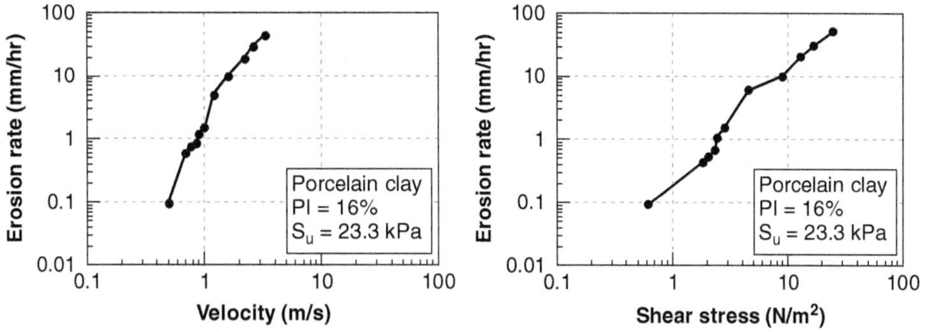

Figure 23.5 Erosion function for a low-plasticity clay as measured in the EFA.

(e.g., Chapuis and Gatien 1986), the jet test to evaluate the erodibility of soils (e.g., Hanson 1991), and the hole erosion test to measure the erosion properties of stiff soils (e.g., Wan and Fell 2004). More recently, a simple and inexpensive tool for field use has been developed called the pocket erodometer (Briaud et al. 2012). This tool is described in section 7.10. Tests with the pocket erodometer can be performed at the site on the end of a sample to get a first indication of the erodibility of the soil within minutes after sampling.

23.4 SOIL EROSION CATEGORIES

Categories are used in many fields of engineering: soil classification categories, hurricane strength categories, and earthquake magnitude categories, among others. Such categories have the advantage of quoting one number to represent a more complex condition. Erosion categories are proposed (Figure 23.6) to bring erodibility down in complexity from an erosion rate vs. shear stress function to a category number. Such a classification system can be presented in terms

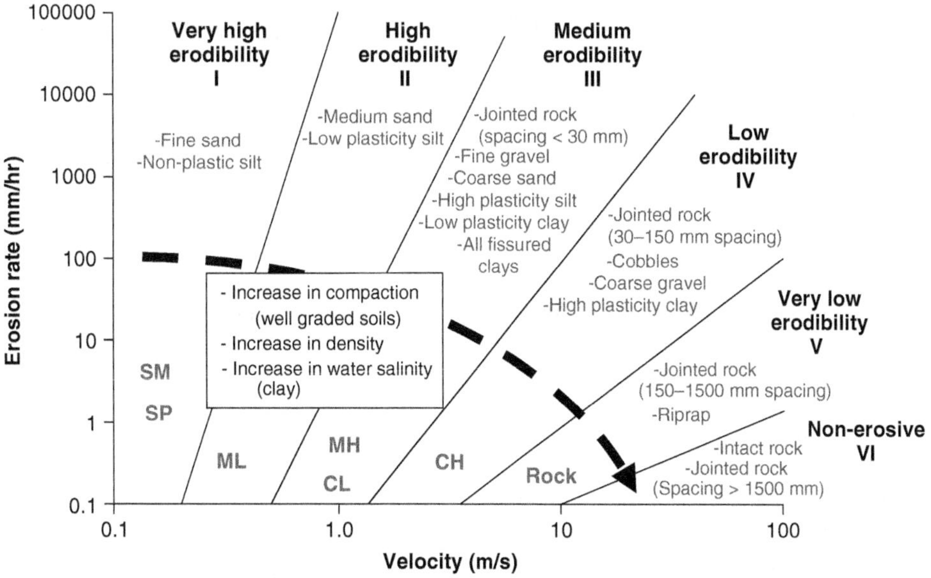

Figure 23.6 Proposed erosion categories for soils and rocks based on velocity.

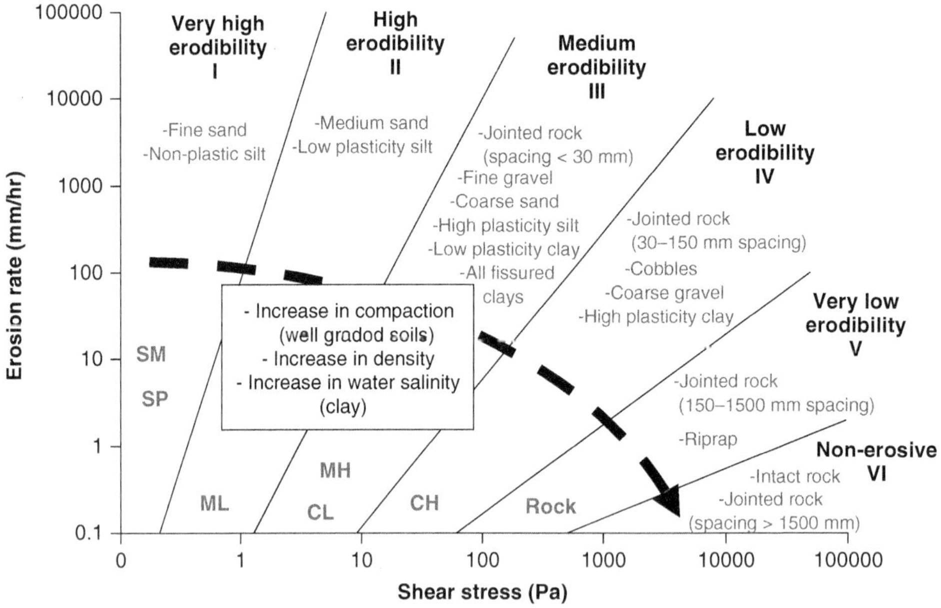

Figure 23.7 Proposed erosion categories for soils and rocks based on shear stress.

of velocity (Figure 23.6) or shear stress (Figure 23.7). The categories proposed are based on 20 years of erosion testing experience.

To classify a soil or rock, the erosion function is plotted on the category chart; the erodibility category number for the material tested is the number for the zone in which the erosion function fits. Note that, as discussed later, using the water velocity is less representative and leads to more uncertainty than using the shear stress; indeed, the velocity and the shear stress are not linked by a constant. The velocity chart has the advantage that it is easier to gauge a problem in terms of velocity. An erodibility classification chart developed on the basis of the pocket erodometer test is shown in Figure 7.30.

One of the most important soil parameters in erosion studies is the threshold of erosion. Below this threshold, erosion does not occur; above this threshold, erosion occurs. In terms of shear stress, this threshold is the critical shear stress τ_c; in terms of velocity, it is the critical velocity v_c. Figure 23.8 shows a plot of the critical velocity as a function of the mean grain size, and Figure 23.9 shows the same plot for the critical shear stress. The data come from measurements using the EFA as well as measurements published in the literature. As can be seen in Figures 23.8 and 23.9, the relationship between the critical value and the grain size has a V shape, indicating that the most erodible soils are fine sands with a mean grain size in the range of 0.1 to 0.5 mm. This V shape also points out that particle size controls the erosion threshold of coarse-grained soils, whereas particle size does not correlate with the erosion threshold of fine-grained soils. Note that Shields (1936) proposed a curve for coarse-grained soils in his doctoral work; his data are included in Figures 23.8 and 23.9. Shields's recommendations do not extend to fine-grained soils. Note also that Hjulstrom (1935) proposed such a curve for both coarse-grained soils and fine-grained soils, but his recommendations for fine-grained soils turned out to be too simple.

The erodibility of soils varies significantly from one soil to the next; therefore, erodibility depends on the soil properties. It depends also on the properties of the water flowing over the soil. For clays, the higher the salt concentration in the water, the more erosion-resistant the clay is (Cao et al. 2002; Croad 1981). The properties influencing erodibility are numerous; some of them are listed in Table 23.1. It appears reasonable to expect that a relationship would exist between common soil properties and erodibility—but erodibility is a function, not a number, so correlations can be made only with elements of that function, such as the critical shear stress or the initial slope of the erosion function. Such correlations have been attempted (Cao et al. 2002) and failed with very low coefficients of correlation. On the one hand, there should be a correlation; on the other hand, the correlation is complex and requires multiple parameters, all involved in the resistance of the soil to erosion. All in all, it is preferable to measure the erosion function directly with an apparatus such as the EFA.

23.5 ROCK EROSION

Soil erosion is not very well known, but rock erosion is even less known, so the engineer must exercise a great deal of engineering judgment when it comes to rock erosion. Nevertheless, many engineers and researchers have contributed to the advancement of knowledge in this relatively new field. They include Temple and Moore (1994), Annandale (1995), Kirsten et al. (1996), van Schalkwyk et al. (1995), Bollaert (2002), and Manso (2006).

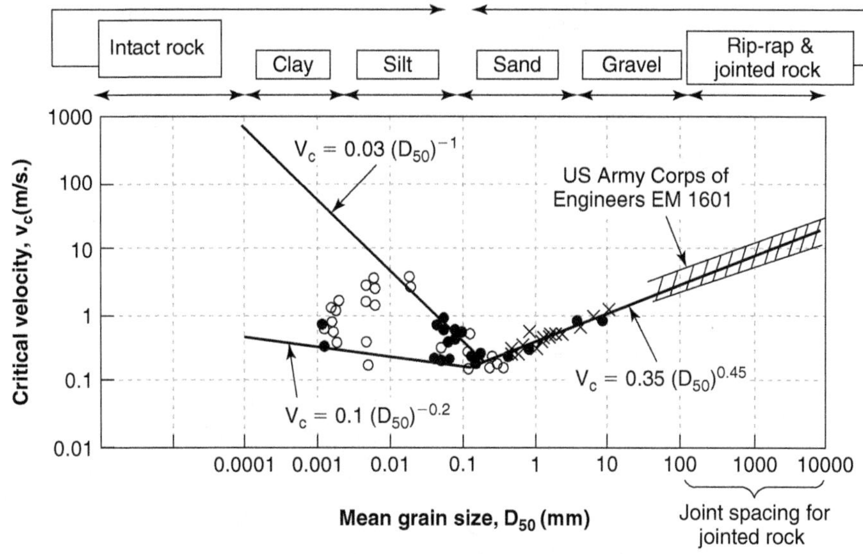

Legend:
- ● TAMU Data as reported by Briaud, J.-L. et. al. (2001). "Erosion Function Apparatus for Scour Rate Predictions." *J. Geotech. and Geoenvir. Engrg.,* ASCE, 127(2), 105–113.
- ○ TAMU Data as reported by Briaud, J.-L. (2006). "Erosion Tests on New Orleans Levee Samples." *Texas A&M University Internal Report.*
- x Data from Shields, Casey, US.WES, Gilbert, White as reported by Vanoni, V.A., ed. (1975). "Sedimentation Engineering." *ASCE manuals and reports on engineering practice,* ASCE, New York.

Figure 23.8 Critical velocity as a function of mean grain size.

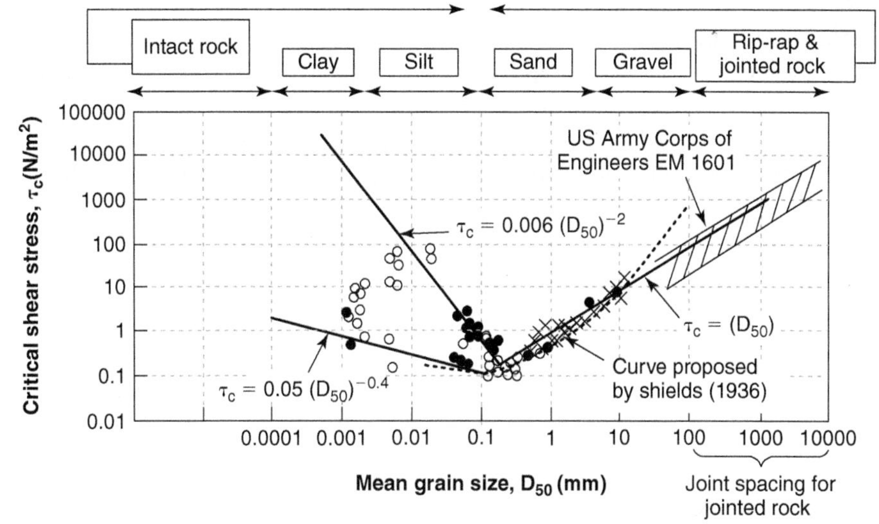

Legend:
- ● TAMU Data as reported by Briaud, J.-L. et. al. (2001). "Erosion Function Apparatus for Scour Rate Predictions." *J. Geotech. and Geoenvir. Engrg.,* ASCE, 127(2), 105–113.
- ○ TAMU Data as reported by Briaud, J.-L. (2006). "Erosion Tests on New Orleans Levee Samples." *Texas A&M University Internal Report.*
- x Data from Shields, Casey, US.WES, Gilbert, White as reported by Vanoni, V.A., ed. (1975). "Sedimentation Engineering." *ASCE manuals and reports on engineering practice,* ASCE, New York.

Figure 23.9 Critical shear stress as a function of mean grain size.

Table 23.1 Soil and Water Properties Influencing Erodibility

Soil water content	Soil clay minerals
Soil unit weight	Soil dispersion ratio
Soil plasticity index	Soil cation exchange cap
Soil undrained shear strength	Soil sodium absorption rate
Soil void ratio	Soil pH
Soil swell	Soil temperature
Soil mean grain size	Water temperature
Soil percent passing #200	Water salinity
	Water pH

Rock erodes through two main processes: rock substance erosion and rock mass erosion. *Rock substance erosion* refers to the erosion of the rock material itself, whereas *rock mass erosion* refers to the removal of rock blocks from the jointed rock mass. Rock substance erosion includes three submechanisms: erosion due to the hydraulic shear stress created by the water at the rock-water interface, erosion due to abrasion caused by sediments rubbing against the rock during the flow, and erosion from the impact of air bubbles that pit the rock surface due to cavitation at very high velocities. Rock mass erosion includes two submechanisms: erosion due to slaking, and erosion due to block removal between joints. Slaking can occur when a rock, such as a high-plasticity shale in an ephemeral stream, dries out and cracks during summer months; these small blocks are then removed by the next big flood. Block removal can occur if, during high-turbulence events, the difference in pressure between the top and the bottom of a rock block becomes large enough to overcome the weight and side friction on the block. Bollaert (2002) points out that brittle fracture and fatigue failure can contribute to breaking the rock into smaller pieces which are then carried away by the water. Note that most of the time, rock mass erosion will be the dominant process in rock erosion, with rock substance erosion occurring only rarely.

The critical velocity associated with rock erosion is much higher than the critical velocity associated with soil erosion in general. At the same time, the erosion rate for a given velocity is much lower for rock erosion than for soil erosion in general. Table 23.2 is an attempt at quantifying the critical velocity and the erosion rate of jointed rocks where rock mass erosion may control the process. This table is preliminary in nature and should be calibrated against field behavior. The critical velocities quoted in Table 23.2 refer to the velocity necessary to move a particle with a size equal to the spacing between joints; as such, they are likely lower bounds because they ignore any beneficial effect from the shear strength of the joints. Note that the orientation of the bedding of the rock mass is important, as shown in Figure 23.10. Engineering judgment must be used to increase or decrease the critical velocity when the bedding is favorable or unfavorable to the erosion resistance. In addition, it is highly recommended in all cases to measure the erosion function of the rock substance on core samples obtained from the site.

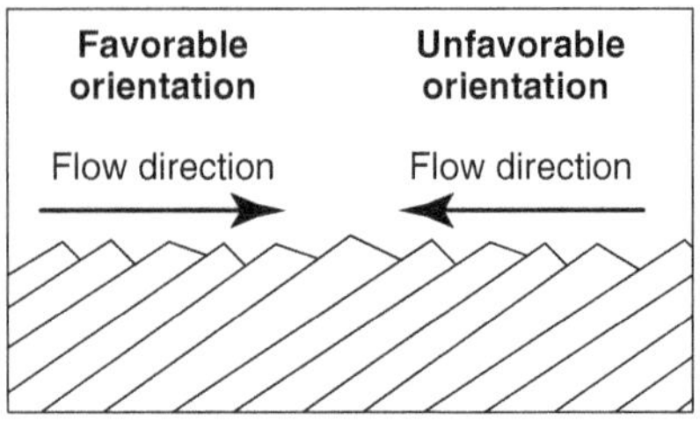

Figure 23.10 Effect of joint orientation on erosion resistance.

Examples of rock erosion rates can be collected from geology. For example, the Niagara Falls started about 12,000 years ago on the shores of Lake Erie, and have eroded back primarily through undercutting of the falls rock face to halfway between Lake Erie and Lake Ontario. This represents 11 km and an average rate of 0.1 mm/hr, through sandstones, shales, and limestone sedimentary rocks (http://en.wikipedia.org/wiki/Niagara_Falls). Another example is the Grand Canyon, where the Colorado River has generated 1600 m of vertical erosion through complex rock layers over an estimated 10 million years for an average rate of 0.00002 mm/hr (http://en.wikipedia.org/wiki/Geology_of_the_Grand_Canyon_area) as the Colorado Plateau was upheaving. These rates appear negligible at first glance, yet neglecting them would be to neglect the Grand Canyon or the

Table 23.2 Rock Mass Erosion

Joint Spacing (mm)	Critical Velocity (m/s)	Erosion Category	Orientation of Joints
<30	0.5–1.35	Category III Medium	Not applicable
30–150	1.35–3.5	Category IV Low	Evaluation needed
150–1500	3.5–10	Category V Very Low	Evaluation needed
>1500	>10	Category VI Nonerosive	Not applicable

Note: This table is preliminary in nature and should be calibrated against field behavior.

retreat of Niagara Falls. The lesson is clear: It is not only the rate of erosion that is important, but also the length of time over which that rate is applied.

23.6 WATER VELOCITY

Figure 23.11 shows the profile of water velocity as a function of flow depth. The water velocity is largest near the top of the water column and zero at the bottom. This has been measured repeatedly in hydraulic engineering. By comparison, the shear stress is highest at the bottom and near zero at the top of the water column. The relationship between the shear stress and the velocity can be established as follows. Because water is a Newtonian fluid, there is a linear relationship between the shear stress τ and the shear strain rate $d\gamma/dt$:

$$\tau = \mu\left(\frac{d\gamma}{dt}\right) \tag{23.4}$$

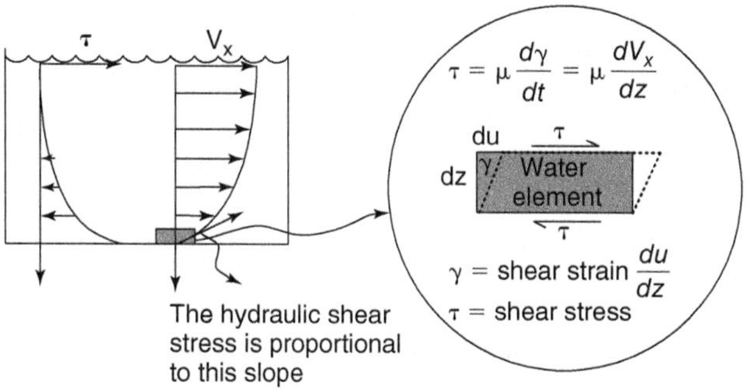

Figure 23.11 Velocity and shear stress profile versus flow depth.

where μ is the dynamic viscosity of the water (10^{-3} Pa.s at 20°C). This viscosity is different from the kinematic viscosity ν of water (10^{-6} m^2/s at 20°C) defined as $\nu = \mu/\rho$ where ρ is the mass density of water (1000 kg/m^3). Because, as shown in Figure 23.11, γ is du/dz, then $d\gamma/dt$ is dv/dz where v and u are the water shear velocity and horizontal displacement at a depth z respectively. Then the shear stress τ at depth z is given by:

$$\tau = \mu\left(\frac{dv}{dz}\right) \tag{23.5}$$

Therefore, the shear stress is proportional to the gradient of the shear velocity profile with flow depth, and the shear stress at the soil/rock-water interface is the slope of the profile at the interface. If the slope of the water velocity profile at the water-soil or water-rock interface (*interface shear stress*) is kept constant, and if the water depth is varied, it can be shown that the mean depth velocity will vary as well. This implies that there is no constant ratio between mean depth velocity and interface shear stress. This is one reason why velocity alone is not as good a predictor of erosion as shear stress. Thus, any erosion design tool presented in terms of velocity should be used with caution. Nevertheless, velocity is much easier for the engineer to gauge than shear stress, and this is why both velocity and shear stress are used in practice.

The magnitude of these shear stresses is very small and measured in N/m^2. They are much smaller than the shear stresses that the geotechnical engineer is used to calculating in foundation engineering, for example, which are in the range of kN/m^2. Figure 23.12 gives examples of the range of shear stresses associated with various fields of engineering.

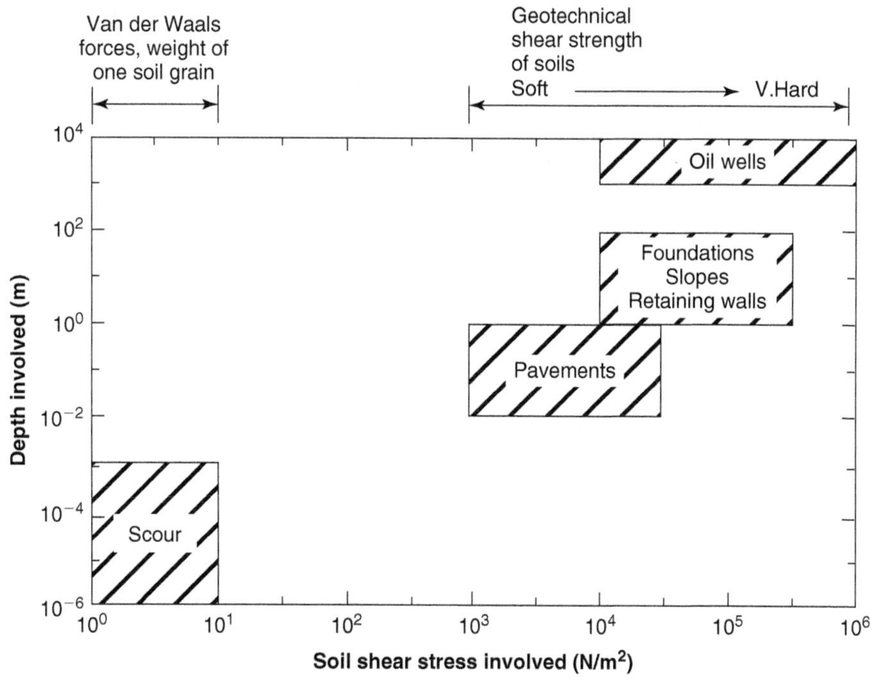

Figure 23.12 Range of shear stresses encountered in different engineering fields.

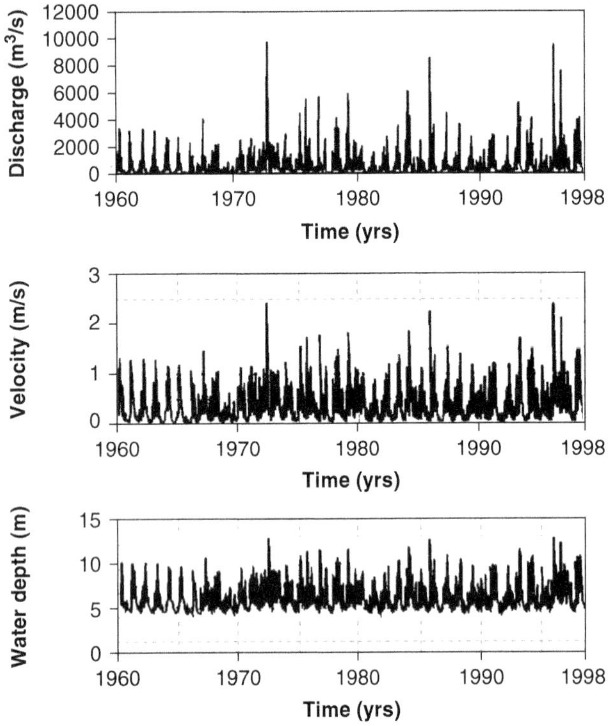

Figure 23.13 Discharge, velocity, and water depth hydrographs.

If the undrained shear strength is a reasonable measure of the strength of a clay for foundation engineering design, the critical shear stress is the "shear strength" of the same clay for erosion studies. The difference in magnitude of the stresses and the strengths between foundation engineering and erosion is that in erosion studies one looks at the resistance of one particle, or a small cluster of particles, whereas in foundation engineering one looks at the resistance of the soil mass associated with the scale of the foundation.

The water in a river does not flow at a constant velocity, so the velocity history over a period of time is a necessary input to many erosion problems. This velocity history or *hydrograph* is not usually readily available. Often, a discharge (m^3/s) hydrograph is available and must be transformed into a velocity (m/s) hydrograph and a water depth (m) hydrograph. This is commonly done by using software such as HEC-RAS (Brunner 2002). An example of the results of this transformation is shown in Figure 23.13. HEC-RAS solves the one-dimensional energy equation for gradually varied flow in natural or constructed channels and adds the one-dimensional momentum equation around hydraulic structures such as bridges, culverts, and weirs where the energy equation is no longer applicable.

The hydrograph can be used to determine the 100-year flood or the 500-year flood. One simple graphical method (e.g., Chow et al. 1988) consists of obtaining the yearly maximum flows from the hydrograph, ranking them in descending order of intensity, calculating for each flow the probability of exceedance as the rank divided by the total number of observations + 1, then plotting the flow versus the probability of exceedance on a semilog paper such as the one if Figure 23.14. Once the data are plotted, a linear regression is performed over, say, the first 20 to 30 years of data and extrapolated to the 0.01 probability of exceedance for the 100-year flood and to the 0.002 probability of exceedance for the 500-year flood. The return period is the inverse of the probability of exceedance. There are other and more refined ways of obtaining these design floods, but this simple graphical method helps one to understand the process and the meaning of a *100-year flood*: a flood that has a 1% chance of

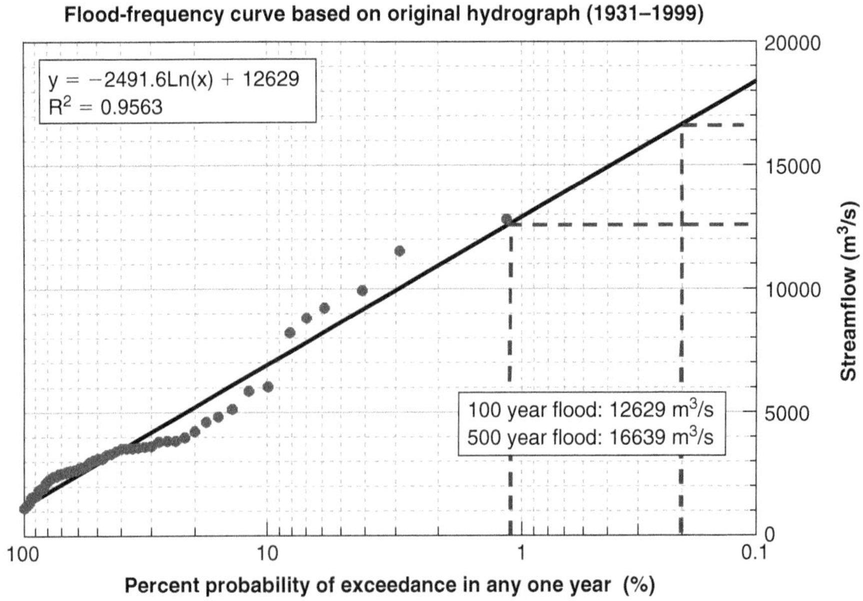

Figure 23.14 Flood frequency curve obtained from measured discharge hydrograph.

being exceeded in any one year. Figure 23.14 shows the result of an analysis for the hydrograph at the Woodrow Wilson bridge. As can be seen from that figure, the 100-year flood has a discharge of 12,600 m^3/s and the 500-year flood has a value of 16,600 m^3/s.

The probability of exceedance R of the design flood with a given return period T_r depends on the design life L_t of a structure:

$$R = 1 - (1 - 1/T_r)^{Lt} \tag{23.6}$$

If the design life of the bridge is 75 years, the probability that a flood with a return period of 100 years will be exceeded during the 75-year design life is 53%, according to Eq. 23.6; that probability is 14% for the 500-year flood. Only when one gets to the 10,000-year flood does the probability get lower than 1% (0.75%). Therefore, looking at those numbers alone, it seems desirable to use the 10,000-year flood for design purposes. This flood is used in design in the Netherlands for regions of the country deemed critical. The United States uses the 100- and 500-year floods for design purposes in hydraulic engineering; this leads to probabilities of exceedance in the tens of percent. By comparison, structural engineers use a probability of exceedance of about 0.1% for the design of bridge beams (LRFD target) and, judging from measured vs. predicted pile capacity databases (Briaud and Tucker 1988), geotechnical engineers use a probability of exceedance of the order of a few percent. While these numbers can be debated, it is relatively clear that these different fields of civil engineering operate at vastly different probability of exceedance levels. Note that risk is different from the probability of exceedance (see section 11.6.3), as it also involves the value of the consequence. Hence, the probability of exceedance target should vary with the consequence of the failure.

23.7 GEOMETRY OF THE OBSTACLE

The geometry of the obstacle encountered by the water influences the velocity of the water and the flow pattern, including turbulence intensity. When water approaches a pier in a river, it has to go around the pier. In doing so, it faces a restricted area and has to accelerate to maintain the flow rate. This acceleration results in a local mean depth velocity that can be 1.5 times higher than the approach mean depth velocity. If the approach velocity is lower than the critical velocity, but the local velocity around the pier reaches a value higher than the critical velocity, then *scour* occurs around the pier. This scour type is called *clear water scour*: that is, scour created by water that does not carry soil particles.

In contrast, if the approach velocity and the velocity around the pier are both higher than critical, then the scour type is *live bed scour*. This means that the water is carrying a significant amount of soil particles. The scour depth reached under live bed scour conditions is typically less than the scour depth reached under clear water scour conditions. The reason is that during live bed scour, some of the particles in suspension or rolling on the bottom fall down in the scour hole, thereby limiting the depth of the scour hole around the pier.

Figure 23.15*a* and *b* show results of numerical simulations of erosion created by water flow in a contracted channel. The CHEN 3D computer program (Chen et al. 1990; Chen 2002) was the program used.

23.8 BRIDGE SCOUR

Bridge scour refers to the erosion of the soil surrounding the foundation of bridge piers in rivers. The water flows at the approach velocity v, arrives at the bridge support, has to accelerate around that obstacle to maintain the flow rate, and thus has a higher potential to erode the river bottom around the foundation. Figure 23.16 shows the scour hole resulting from this erosion around a bridge pier. Bridge scour accounts for 60% of all bridge failures in the United States (Briaud 2006a). Figure 23.17 shows the progression of the scour depth as a function of time as a response to the flow history (hydrograph) at the bridge. It is important to know how deep the hole is going to be so that this scour depth can be ignored in the resistance of the foundation. The prediction of that scour depth requires knowledge of the soil erosion function, the water velocity, and the geometry of the obstacle. The obstacle can be a bridge pier, a bridge abutment, or the contraction of the river. As a result, we talk about pier scour, abutment scour, and contraction scour (Figure 23.18).

The simplest problem is that of a constant water velocity v flowing for an infinite time around a cylindrical pier of diameter B. Figure 23.19 shows a typical curve giving the scour depth as a function of time in this case. Experiments have shown that the scour depth z vs. time t curve is well described by a hyperbola:

$$z = \frac{t}{\dfrac{1}{\dot{z}_i} + \dfrac{t}{z_{max}}} \tag{23.7}$$

where z is the scour depth, $\dot{z}_i$ is the initial erosion rate at a time equal to zero under a velocity v, z_{max} is the scour depth at a time equal to infinity (asymptotic value) under a velocity v, and t is the time during which the water has been flowing at the velocity v. The scour depth z_{max} is called the maximum scour depth under v and would occur if a flood creating v lasted a long time. If instead the flood last a finite amount of time, say 24 hours, then the scour depth is called the final scour depth z_{final} at the end of the flood event, say 24 hours. If $\dot{z}_i$ is large, then z_{final} will quickly approach z_{max}, and one flood may be long enough to create z_{max}. This is the case with very erodible soils, such as sands, where a maximum scour depth analysis called z_{max} analysis is sufficient. If, however, $\dot{z}_i$ is small, then z_{final} is likely to be much lower than z_{max} and it is economical to perform the more complex z_{final} analysis. This is the case with fine-grained soils.

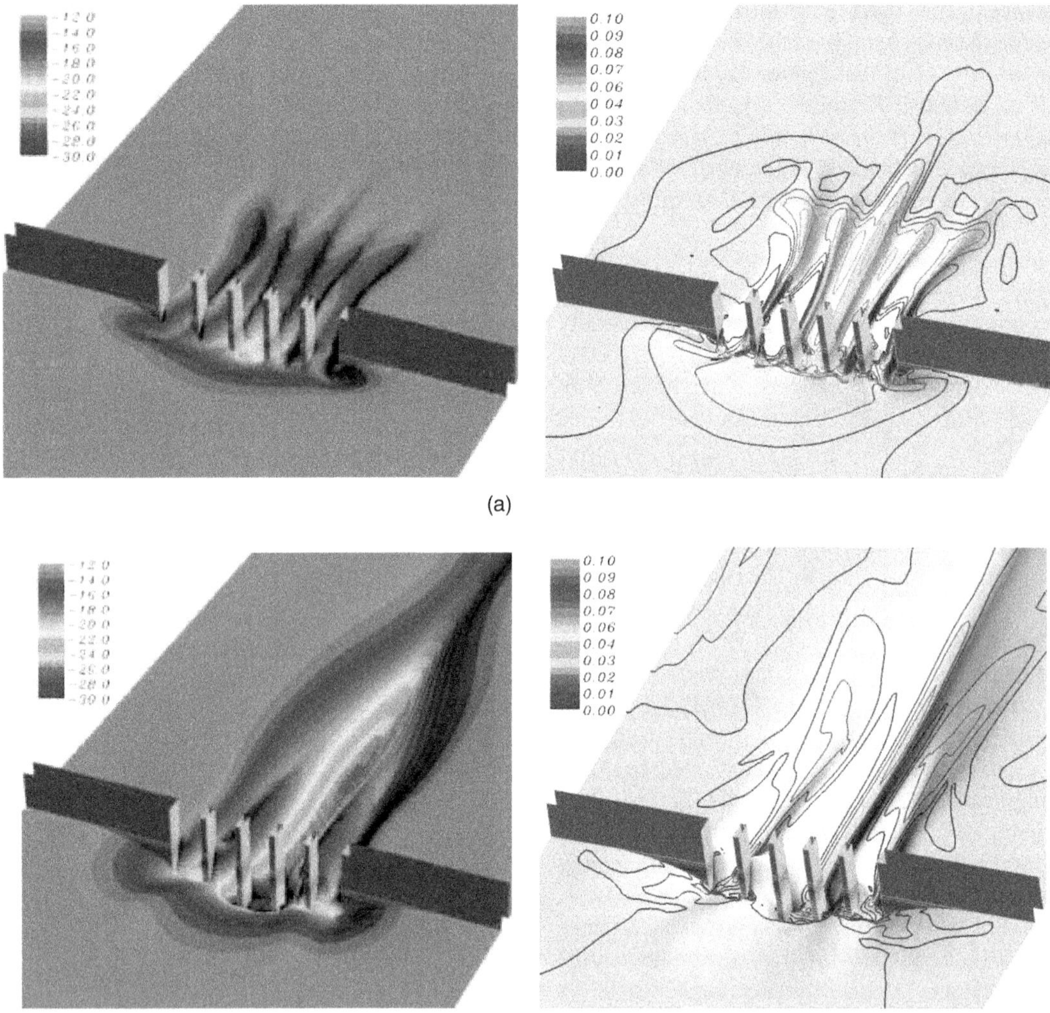

Figure 23.15 Predicted scour hole shape and streambed shear stresses around abutments and piers: *(a)* t = 2000 min. *(b)* t = 15,000 min. Scour depth and shear stress distributions at: *(a)* t = 2000 min and *(b)* t = 15,000 min. (From Chen 2002.)

Figure 23.16 Scour hole around bridge pier.

23.8.1 Maximum Scour Depth (z_{max}) Analysis

Pier Scour

The following equation gives the maximum scour depth for pier scour, that is to say the maximum depth of the hole that can form around the pier for a given set of parameters (Briaud 2012; Figure 23.18):

$$\frac{z_{\max(\text{Pier})}}{B'} = 2.2 \cdot K_{pw} \cdot K_{psh} \cdot K_{pa} \cdot K_{psp} \cdot (2.6 \cdot Fr_{(\text{pier})} - Fr_{c(\text{pier})})^{0.7} \qquad (23.8)$$

where $z_{\max(\text{pier})}$ is the maximum depth of pier scour, B' is the projected width of the pier perpendicular to the flow, K_{pw} is the water depth influence factor for pier scour depth, K_{psh} is

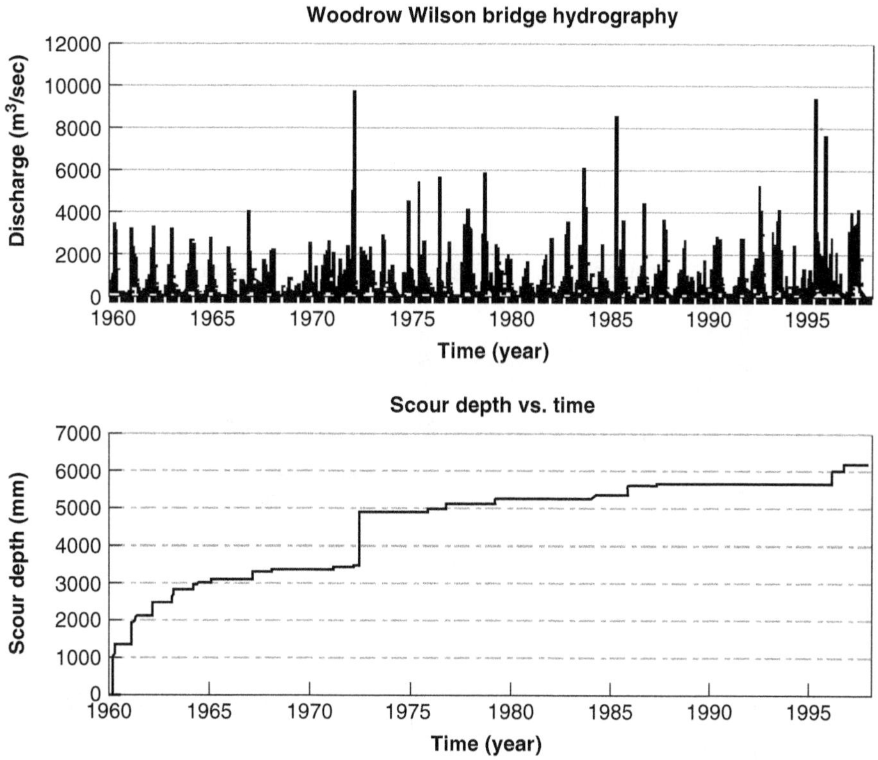

Figure 23.17 Increase in scour depth versus time as result of applied hydrograph.

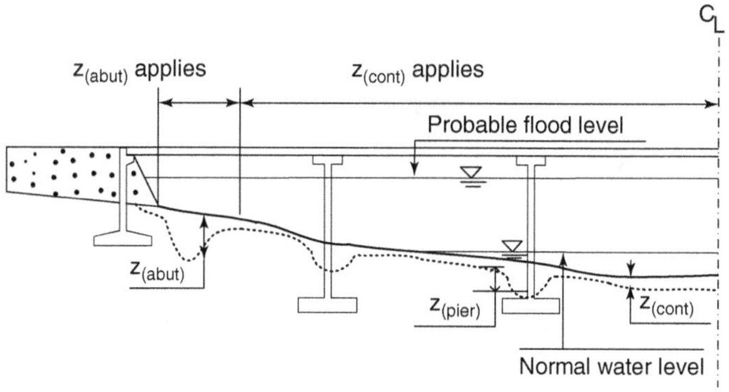

Figure 23.18 Pier scour, abutment scour, and contraction scour.

the pier shape influence factor for pier scour depth, K_{pa} is the aspect ratio influence factor for pier scour depth, the aspect ratio is L/B ratio of pier length L over pier width B, K_{psp} is the pier spacing influence factor for pier scour depth, $Fr_{(pier)}$ is the pier Froude Number (defined later) based on the approach velocity v_1 and pier width B', and $Fr_{c(pier)}$ is the critical pier Froude Number based on critical velocity v_c. The projected width B' (Figure 23.20) is given by:

$$B' = B\left(\cos\theta + \frac{L}{B}\cdot\sin\theta\right) \tag{23.9}$$

where B' is the projected width, B is the pier width, L is the pier length, and θ is the attack angle, which is the angle between the flow direction and the main direction of the pier.

The water depth influence factor K_{wa} corrects for the fact that the parenthetical expression on the right-hand side of Eq. 23.8 was developed for a pier in deep water. *Deep water* is defined as a water depth h_w larger than 1.43 B'. If the water depth is shallower than 1.43 B', the scour depth is reduced. The equation for K_{pw} is:

$$K_{pw} = \begin{cases} 0.89\left(\dfrac{h_w}{B'}\right)^{0.33}, & \text{for } \dfrac{h_w}{B'} < 1.43 \\ 1.0, & \text{else} \end{cases} \tag{23.10}$$

The pier shape influence factor K_{psh} is given in Table 23.3; it corrects for the fact that the parenthetical expression on the

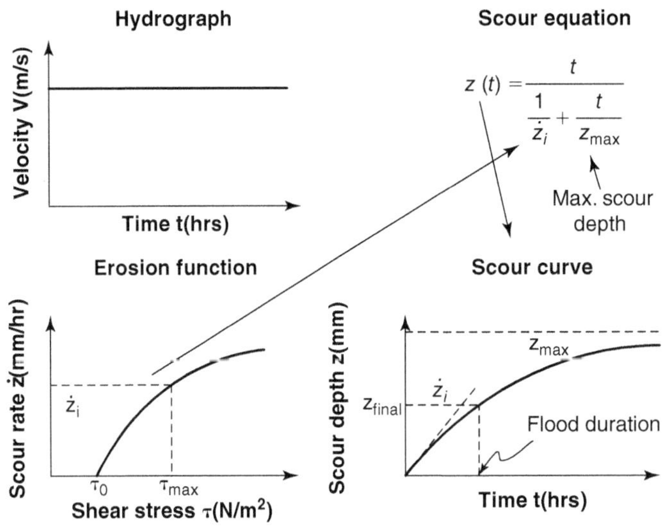

Figure 23.19 Scour depth vs. time curve for constant velocity.

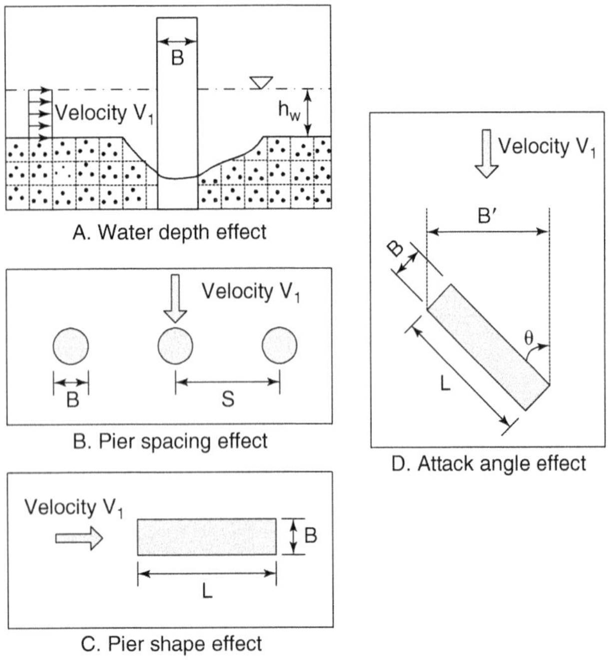

Figure 23.20 Definition of pier parameters.

right-hand side of Eq. 23.8 was developed for a cylindrical pier.

The aspect ratio influence factor K_{pa} corrects for the fact that the parenthetical expression on the right-hand side of Eq. 23.8 was developed for a cylindrical pier. This influence factor is taken care of by use of the projected width B' instead of B, so $K_{L/B}$ is always 1. The pier spacing influence factor K_{psp} corrects for the fact that the parenthetical expression on the right-hand side of Eq. 23.8 was developed for a single pier. If another pier is placed within the influence zone of the first one, the scour depth will be larger. The equation for K_{sp} is:

Table 23.3 Correction Factor for Pier Nose Shape (K_{psh})

Shape of Pier Nose	K_{psh}	Shape of Pier Nose	K_{psh}
Square nose	1.1	Circular cylinder	1.0
Round nose	1.0	Sharp nose	0.9

(Richardson Davis 2001)

$$K_{psp} = \begin{cases} 2.9\left(\frac{S}{B'}\right)^{-0.91}, & \text{for } \frac{S}{B'} < 3.42 \\ 1.0, & \text{else} \end{cases} \quad (23.11)$$

where S is the pier spacing and B' is the projected width. Equation 23.11 indicates that piers spaced more than 3.42 times the projected pier width from each other do not increase the scour depth at the pier. The pier Froude Number $Fr_{(pier)}$ is given by:

$$Fr_{(pier)}\left(= \frac{V_1}{\sqrt{g \cdot B'}}\right) \quad (23.12)$$

where V_1 is the water velocity at the location of the pier if the pier were not there, g is the acceleration due to gravity, and B' is the projected width of the pier. The critical pier Froude Number $Fr_{c(pier)}$ is given by:

$$Fr_{c(pier)} = \frac{V_c}{\sqrt{g \cdot B'}} \quad (23.13)$$

where V_c is the critical velocity for the soil.

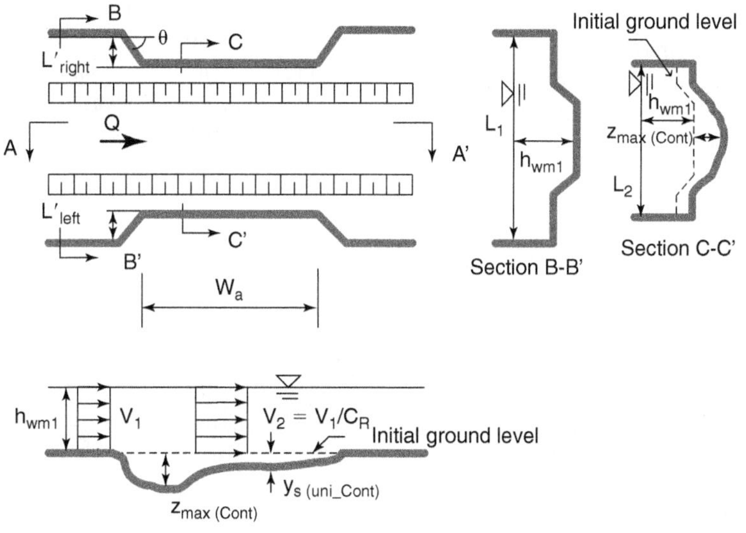

Figure 23.21 Definition of contraction scour parameters.

Contraction Scour

Contraction scour involves two regions of the river: the approach zone, called zone 1; and the contracted zone, called zone 2 (Figure 23.21).

The following equation gives the maximum scour depth for contraction scour, that is to say the maximum depth of scour that can develop in the contracted channel at the bridge location for a given set of parameters (Briaud 2012; Figure 23.21).

$$\frac{z_{max(Cont)}}{h_{wm1}} = 1.27(1.83Fr_{m2} - Fr_{mc}) \tag{23.14}$$

where $z_{max(cont)}$ is the maximum depth of contraction scour, h_{wm1} is the water depth in the main channel at the approach section, Fr_{m2} is the Froude Number for the main channel at the bridge in the contracted zone, and Fr_{mc} is the critical Froude Number for the main channel at the bridge. The Froude Number Fr_{m2} is given by:

$$Fr_{m2} = \frac{V_1/C_R}{\sqrt{gh_{wm1}}} \tag{23.15}$$

where V_1 is the velocity in the approach section, g is the acceleration due to gravity, h_{wm1} is the water depth in the main channel at the approach section, and C_R is the contraction ratio, defined as:

$$C_R = \frac{Q - Q_{block}}{Q} \tag{23.16}$$

where Q is the total discharge, and Q_{block} is the part of the discharge Q blocked by the approach embankments. The critical Froude Number Fr_{mc} is given by:

$$Fr_{mc} = \frac{V_{mc}}{\sqrt{gh_{wm1}}} = \frac{(\tau_c/\rho)^{0.5}}{gnh_{wm1}^{0.33}} \tag{23.17}$$

where V_{mc} is the critical velocity for the soil in the main channel, g is the acceleration due to gravity, h_{wm1} is the water depth in the main channel at the approach section, τ_c is the critical shear stress for the soil in the main channel, ρ is the mass density of the soil, g is the acceleration due to gravity, and n is the Manning's coefficient. Manning's coefficient characterizes the roughness of the river bottom. Estimated values are given in Table 23.4.

Table 23.4 Manning Coefficient n in $V = \frac{1}{n}R_h^{0.67}S_e^{0.5}$ *

Roughness	n (s.m$^{-0.33}$)	Roughness	n (s.m$^{-0.33}$)
Smooth clay surface	0.011	Gravel (D50 = 2 to 64 mm)	0.028 to 0.035
Sand (D50 = 0.2 mm)	0.012	Cobbles (D50 = 64 to 230 mm)	0.030 to 0.050
Sand (D50 = 0.4 mm)	0.020	Boulder (D50 > 230 mm)	0.040 to 0.070
Sand (D50 = 1 mm)	0.026		

*With V velocity in m/s, R_h hydraulic radius of channel in m, and S_e slope of the energy line (m/m).

Abutment Scour

The following equation gives the maximum scour depth for abutment scour, that is to say the maximum depth of scour that can develop around an abutment in the contracted channel at the bridge location for a given set of parameters (Briaud 2012; Figure 23.18):

$$\frac{z_{\max(Abut)}}{h_{wf1}} = 243 \times K_{ash} K_{ask} K_{al} K_{ag} \mathrm{Re}_{f2}^{-0.28} \times (1.65 Fr_{f2} - Fr_{fc}) \quad (23.18)$$

where $z_{\max(\text{abut})}$ is the maximum depth of abutment scour, h_{wf1} is the water depth in the flood plain in the approach flow next to the abutment, K_{ash} is the shape factor for abutment scour, K_{ask} is the skew angle influence factor for abutment scour, K_{al} is the influence factor taking into account the proximity of the abutment from the main channel, K_{ag} is the geometry of the channel influence factor for abutment scour, Re_{f2} is Reynolds Number around the toe of the abutment, Fr_{f2} is the Froude Number around the toe of the abutment, and Fr_{fc} is the critical Froude Number for the soil near the toe of the abutment.

The shape factor K_{ash} corrects for the fact that the parenthetical expression on the right-hand side of Eq. 23.18 was developed for a wing-wall abutment (Figure 23.22). The values of K_{ash} are:

$$K_{ash} = \begin{cases} 1.0 & \text{for wing-wall abutment} \\ 1.22 & \text{for vertical-wall abutment} \\ 0.73 & \text{for spill-through abutment with } 2:1 \text{ Slope} \\ 0.59 & \text{for spill-through abutment with } 3:1 \text{ Slope} \end{cases} \quad (23.19)$$

The skew angle factor K_{ask} corrects for the fact that the parenthetical expression on the right-hand side of Eq. 23.18 was developed for an approach embankment perpendicular to the river bank (Figure 23.23). If the embankment alignment is oblique to the river bank, the abutment scour depth is different. The equation for K_{ask} is:

$$K_{ask} = \begin{cases} 1.0 - 0.005\left(\left|\theta - 90^\circ\right|\right) & \text{for } 60^\circ \le \theta \le 120^\circ \\ 0.85 & \text{for other } \theta \text{ values} \end{cases} \quad (23.20)$$

where θ is the skew angle as shown in Figure 23.23.

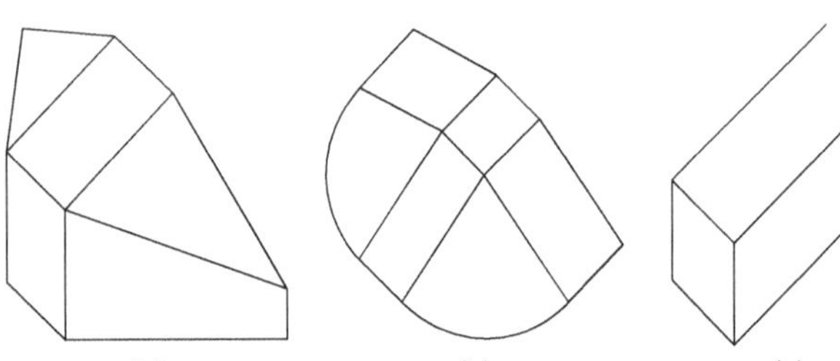

Figure 23.22 Abutment shapes: *(a)* Wing-wall abutment. *(b)* Spill-through abutment. *(c)* Vertical wall abutment.

The influence factor for the proximity of the abutment to the main channel K_{al} corrects for the fact that the parenthetical expression on the right-hand side of Eq. 23.18 was developed for an abutment far away from the bank of the main channel. When the abutment is close to the bank of the main channel, the abutment scour depth becomes larger. The equation for K_{al} is:

$$K_{al} = \begin{cases} -0.23\frac{(L_f - L_e)}{h_{wf1}} + 1.35 & \text{for } \frac{(L_f - L_e)}{h_{wf1}} < 1.5 \\ 1.0 & \text{otherwise} \end{cases} \quad (23.21)$$

where L_f is the length of the flood plain, L_e is the length of the embankment, and h_{wf1} is the water depth in the approach channel near the abutment.

The channel geometry influence factor K_{ag} corrects for the fact that the parenthetical expression on the right-hand side of Eq. 23.18 was developed for a compound channel geometry. For a rectangular channel geometry, the abutment scour depth is smaller. The values for K_{ag} are:

$$K_{ag} = \begin{cases} 1.0 & \text{for compound channel} \\ 0.42 & \text{for rectangular channel} \end{cases} \quad (23.22)$$

The Reynolds Number Re_{f2} is in the equation to respect the scaling laws and the influence of size. It is defined as:

$$\mathrm{Re}_{f2} = \frac{V_{f2} h_{wf1}}{\nu} \quad (23.23)$$

where h_{wf1} is the water depth in the approach channel near the abutment, ν is the kinematic viscosity of water (10^{-6} m^2/s at 20°C), and V_{f2} is the local velocity near the abutment in the flood plain, obtained as follows:

$$V_{f2} = \begin{cases} \frac{Q_{0.5}}{A_2}, & \text{for short setback } \left((L_f - L_e) \le 5h_{wm1}\right) \\ \frac{Q_{f1}}{A_{f2}}, & \text{for long setback } (L_e \le 0.25 L_f) \\ \text{otherwise use a linearly interpolated velocity between} & \\ \frac{Q_{0.5}}{A_2} & \text{for } (L_f - L_e) = 5h_{wm1} \text{ and} \\ \frac{Q_{f1}}{A_{f2}} & \text{for } L_e = 0.25 L_f \end{cases} \quad (23.24)$$

where $Q_{0.5}$ is the flow in half the channel defined as the sum of half the upstream flow in the main channel, 0.5 Q_{m1}, plus the flow in the flood plain immediately upstream of the abutment where the abutment is situated, Q_{f1}, h_{wm1} is the water depth in the main channel in the approach flow; A_2 is the cross sectional flow area in the contracted zone corresponding to the flow $Q_{0.5}$; A_{f2} is the cross sectional flow area on the floodplain at the contracted section; L_f is the width of the floodplain in the approach zone; and L_e is the length of embankment leading to the abutment.

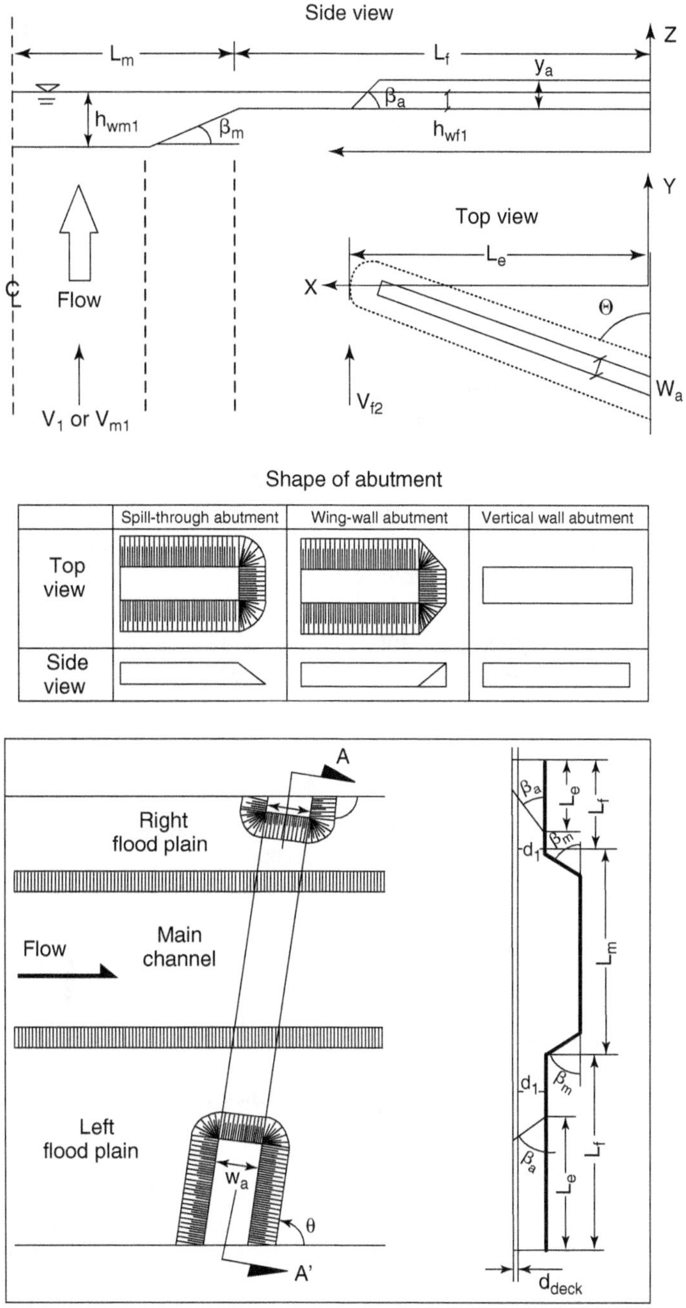

Figure 23.23 Abutment parameter definitions.

The Froude Number Fr_{f2} is calculated around the toe of the abutment and is given by:

$$Fr_{f2} = \frac{V_{f2}}{\sqrt{gh_{wf1}}} \tag{23.25}$$

where V_{f2} is defined in Eq. 23.24, g is the acceleration due to gravity, and h_{wf1} is the water depth in the approach flow near the abutment. The critical Froude Number Fr_{fc} is calculated around the toe of the abutment and is given by:

$$Fr_{fc} = \frac{V_c}{\sqrt{gh_{wf1}}} \tag{23.26}$$

where V_c is the critical velocity for the soil around the toe of the abutment, g is the acceleration due to gravity, and h_{wf1} is the water depth in the approach flow near the abutment.

23.8.2 Maximum Shear Stress at Soil–Water Boundary when Scour Begins

The maximum scour depth is the scour depth reached when the flood velocity v is applied long enough to reach z_{max}. If the flood velocity stops before z_{max} is reached, then only z_{final} is reached (Figure 23.19). To predict z_{final}, it is necessary to develop the relationship between scour depth z and time t. It was found that a hyperbolic equation would fit well with measured curves of z vs. t:

$$z = \frac{t}{\frac{1}{\dot{z}_i} + \frac{t}{z_{max}}} \tag{23.27}$$

where z is the scour depth, $\dot{z}_i$ is the initial erosion rate at a time equal to zero under a velocity v, z_{max} is the scour depth at a time equal to infinity (asymptotic value) under a velocity v, and t is the time during which the water flows at the velocity v.

The scour depth-time curve of Eq. 23.27 is defined once z_{max} and $\dot{z}_i$ are known. The maximum scour depth z_{max} is obtained as discussed in section 23.8.1. The initial erosion rate $\dot{z}_i$ is obtained from the erosion rate vs. shear stress curve measured in the EFA test or deduced from the soil classification and Figure 23.6. Therefore, it is necessary to know the maximum shear stress τ_{max} created by the water when it flows around the obstacle at the beginning of the scour process. The following equations were developed based on numerical simulations to calculate the shear stress τ_{max} for pier scour, contraction scour, and abutment scour.

Maximum Shear Stress for Pier Scour

For pier scour, the equation (Nurtjahyo 2003) is:

$$\tau_{max(Pier)} = k_{pw}k_{psh}k_{psk}k_{psp} \cdot 0.094\rho V_1^2 \left[\frac{1}{\log Re} - \frac{1}{10}\right] \tag{23.28}$$

where $\tau_{max(pier)}$ is the maximum shear stress for pier scour, k_{pw} is the water depth influence factor for pier scour shear stress, k_{psh} is the pier shape influence factor for pier scour shear stress, k_{psk} is the skew angle or angle of attack influence factor for pier scour shear stress, k_{psp} is the pier spacing influence factor for pier scour shear stress, ρ is the mass density of water, V_1 is the mean depth velocity of the water at the location of the pier if the pier were not there (also called upstream velocity in line with the pier), and Re is the pier Reynolds Number. The water depth influence factor corrects for the fact that the expression on the right-hand side of Eq. 23.28 excluding the influence factors was developed for a pier in deep water. At very shallow water depths, the shear stress τ_{max} increases significantly. The equation for k_{pw} is:

$$k_{pw} = 1 + 16e^{(-4h_w/B)} \tag{23.29}$$

where h_w is the water depth and B is the width of the pier.

The pier shape influence factor corrects for the fact that the expression on the right-hand side of Eq. 23.28 excluding the influence factors was developed for a circular pier. For square piers, the factor is 1.15; for rectangular piers, it depends on L/B where L is the pier length and B is the pier width. The equation for k_{psh} is:

$$k_{psh} = 1.15 + 7e^{(-4L/B)} \tag{23.30}$$

where L is the length of the pier and B is the width of the pier.

The skew angle or angle of attack influence factor k_{psk} corrects for the fact that the expression on the right-hand side of Eq. 23.28 excluding the influence factors was developed for a cylindrical pier. For square and rectangular piers with a length L and a width B, the factor k_{psk} is given by:

$$k_{psk} = 1 + 1.5\left(\frac{\theta}{90}\right)^{0.57} \tag{23.31}$$

where θ is the skew angle or attack angle, which is the angle between the flow direction and the main direction of the pier (Figure 23.20).

The pier spacing influence factor k_{psp} corrects for the fact that the expression on the right-hand side of Eq. 23.28 excluding the influence factors was developed for an isolated pier. For a line of piers, the pier spacing influence factor k_{psp} is given by:

$$k_{psp} = 1 + 5e^{(-1.1S/B)} \tag{23.32}$$

where S is the center-to-center spacing of the piers and B is the width of the pier (Figure 23.20).

Maximum Shear Stress for Contraction Scour

For contraction scour, the equation for $\tau_{max(Cont)}$ (Nurtjahyo 2003) is:

$$\tau_{max(Cont)} = k_{cr}k_{cl}k_{c\theta}k_{cw}\rho g n^2 V_1^{\,2} R_h^{\,-\frac{1}{3}} \tag{23.33}$$

where $\tau_{max(Cont)}$ is the maximum shear stress for contraction scour shear stress, k_{cr} is the contraction ratio influence factor for contraction scour shear stress, k_{cl} is the contraction length influence factor for contraction scour shear stress, $k_{c\theta}$ is the transition angle influence factor for contraction scour shear stress, k_{cw} is the water depth influence factor for contraction scour shear stress, ρ is the mass density of water, g is the acceleration due to gravity, n is Manning's coefficient, V_1 is the mean depth velocity of the water in the approach zone, and R_h is the hydraulic radius of the contracted channel.

The contraction ratio influence factor k_{cr} corrects for the fact that the velocity V_1 in the equation is the approach velocity, not the velocity in the contracted zone. It is given by:

$$k_{cr} = 0.62 + 0.38\left(\frac{A_1}{A_2}\right)^{1.75} \tag{23.34}$$

where A_1 is the cross-sectional flow area in the approach zone and A_2 is the cross-sectional flow area in the contracted zone.

Because A_2 is smaller than A_1, k_{cr} increases the shear stress in the contracted zone. The contraction length influence factor k_{cl} corrects for the fact that the main part of Eq. 23.33 (right hand side without correction factors) was developed for abutment widths that were larger than 0.7 times the length of the approach embankments. For abutments narrower than that, the k_{cl} factor is given by:

$$k_{cl} = \begin{cases} 0.77 + 1.36\left(\frac{W_a}{L_1-L_2}\right) - 1.98\left(\frac{W_a}{L_1-L_2}\right)^2, \\ \quad \text{for } \frac{W_a}{L_1-L_2} \le 0.35 \\ 1.0, \quad \text{otherwise} \end{cases} \tag{23.35}$$

where W_a is the width of the top of the abutment (Figure 23.23), L_1 is the width of the river in the approach zone, and L_2 is the width of the river in the contracted zone (Figure 23.21).

The transition angle influence factor $k_{c\theta}$ corrects for the fact that the main part of Eq. 23.33 (right hand side without correction factors) corresponds to no abutment ($\theta = 0$). If the abutment appears through a nonzero transition angle, then $k_{c\theta}$ must be used; it is given by:

$$k_{c\theta} = 1.0 + 0.9\left(\frac{\theta}{90}\right)^{1.5} \tag{23.36}$$

where θ is the transition angle (Figure 23.21).

The water depth influence factor for contraction scour shear stress k_{cw} was found to be equal to 1 in all conditions.

Maximum Shear Stress for Abutment Scour

For abutment scour, the equation for $\tau_{\max(\text{Abut})}$ is:

$$\tau_{\max(Abut)} = 12.5 k_{acr} k_{aar} k_{aw} k_{ash} k_{ask} k_{al} \rho V_1^2 \text{Re}^{-0.45} \tag{23.37}$$

where $\tau_{\max(\text{Abut})}$ is the maximum shear stress for abutment scour shear stress, k_{acr} is the contraction ratio influence factor for abutment scour shear stress, k_{ash} is the influence factor for the aspect ratio of the approach embankment for abutment scour shear stress, k_{aw} is the influence factor for Froude Number for abutment scour shear stress, k_{as} is the influence factor for abutment shape for abutment scour shear stress, k_{ask} is the influence factor for the skew angle of the abutment for abutment scour shear stress, k_{al} is the influence factor related to the location of the abutment in the flood plain for abutment scour shear stress, ρ is the mass density of water, V_1 is the mean depth velocity of the water in the approach zone, and Re is the abutment Reynolds Number.

The contraction ratio influence factor k_{acr} corrects for the fact that the velocity V_1 in the equation is the approach velocity and not the local velocity around the abutment. It is given by:

$$k_{acr} = 3.65\left(\frac{Q_{tot}}{Q_{tot} - Q_{block}}\right) - 2.91 \tag{23.38}$$

where Q_{tot} is the total discharge and Q_{block} is the part of the total discharge blocked by the approach embankments.

The influence factor k_{aar} takes into account the aspect ratio of the abutment. It is given by:

$$k_{aar} = 0.85\left(\frac{L_e}{W_a}\right)^{-0.24} \tag{23.39}$$

where L_e is the length of the approach embankment and W_a is the width of the top of the abutment (Figure 23.23).

The influence factor k_{aw} takes into account the water depth. It is given by:

$$k_{aw} = \begin{cases} 2.07Fr + 0.8 & for \quad Fr > 0.1 \\ 1 & for \quad Fr \le 0.1 \end{cases} \tag{23.40}$$

where Fr is the Froude Number, defined as:

$$Fr = \frac{V_{f2}}{\sqrt{g h_{wf1}}} \tag{23.41}$$

where V_{f2} is the water velocity in the approach zone in line with the abutment and h_{wf1} is the water depth in the approach zone in line with the abutment (Figure 23.23).

The influence factor k_{ash} takes into account the shape of the abutment. It is given by:

$$k_{ash} = \begin{cases} 1.0 & \textit{vertical-wall abutment} \\ 0.65 & \textit{wing-wall abutment} \\ 0.58 & \textit{spill-through abutment} \end{cases} \tag{23.42}$$

The influence factor k_{ask} takes into account the skew angle of the abutment. The reference case is the case when the embankment is perpendicular to the river bank with a skew angle equal to 90°. The skew angle can be smaller or larger than 90°, but was found to have little influence on the maximum bed shear stress and is conservatively taken as equal to 1:

$$k_{ask} = 1 \tag{23.43}$$

The influence factor k_{al} takes into account the location of the abutment in the flood plain. The factor k_{al} is different from 1 only when the abutment is near the edge of the main channel. It is given by:

$$k_{al} = \begin{cases} 1.0 & for \ (L_f - L_e)/h_{wf1} \le -2 \\ 1.2 + 0.1 \ (L_f - L_e)/h_{wf1} & for \ -2 \le (L_f - L_e)/h_{wf1} \le 0 \\ 1.2 - 0.2 \ (L_f - L_e)/h_{wf1} & for \ 0 \le (L_f - L_e)/h_{wf1} \le 1 \\ 1.0 & for \ (L_f - L_e)/h_{wf1} \ge 1 \end{cases} \tag{23.44}$$

23.8.3 Final Scour Depth (z_{final}) Analysis for Constant Velocity Flow and Uniform Soil

Once the maximum shear stress $\tau_{\max}$ is known, the erosion curve linking the erosion rate $\dot{z}$ to the shear stress τ is used to find the erosion rate $\dot{z}_i$ corresponding to $\tau_{\max}$. Equation 23.27 can then be used to find out what z_{final} is, as both $z_{\max}$ and $\dot{z}_i$ are known. The following example illustrates these calculations.

A round-nose pier, with a width of 2 m and a length of 6 m, is located in a river where the water depth is 7.89 m, the approach flow velocity is 1.4 m/s, and the attack angle is 0°

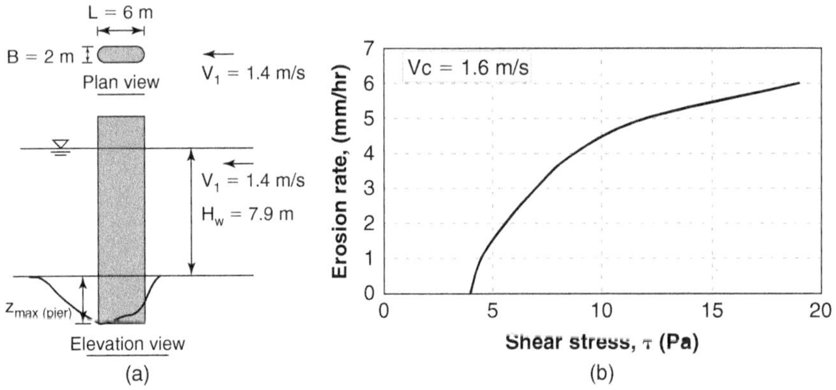

Figure 23.24 Data for example of bridge scour calculations: *(a)* Bridge pier geometry. *(b)* Erosion function of the soil.

(Figure 23.24). EFA tests were conducted on soil samples in the vicinity of the pier, and gave the average erosion function shown in Figure 23.24. The critical velocity of the soil V_c is 1.57 m/s and the duration of the flood is 48 hours. Find the pier scour depth after 48 hours of flood.

1. The maximum scour depth z_{max} is calculated first. The correction factors for water depth K_{pw}, pier shape K_{psh}, pier aspect ratio K_{pa}, and pier spacing K_{psp} are all equal to 1.0. The pier Froude Number needed in Eq. 23.8 is:

$$Fr_{(pier)} = \frac{V_1}{\sqrt{g \cdot B'}} = \frac{1.4}{\sqrt{9.81 \times 2}} = 0.316 \quad (23.45)$$

and the critical pier Froude Number is:

$$Fr_{c(pier)} = \frac{V_c}{\sqrt{g \cdot a}} = \frac{1.58}{\sqrt{9.81 \times 2}} = 0.356 \quad (23.46)$$

Therefore, the maximum pier scour depth $z_{max(pier)}$ is:

$$z_{max(Pier)} = 2.2 \times 1.0 \times 1.0 \times 1.0 \times 1.0 \times 2.0 \times (2.6 \times 0.316 - 0.356)^{0.7} = 2.58 \text{ m} \quad (23.47)$$

2. The maximum shear stress τ_{max} around the pier at the beginning of the scour process is calculated next. The correction factor for water depth k_{pw} is 1.0, for pier spacing k_{psp} is 1.0, for attack angle k_{pa} is 1.0, and for pier shape k_{psh} is 1.15 ($k_{psh} = 1.15 + 7e^{-4L/B} = 1.15 + 7e^{-12}$). The pier Reynolds Number Re is 2.8 × 10^6 $\left(\text{Re} = \frac{1.4 \times 2}{10^{-6}}\right)$. Therefore, the maximum shear stress around the pier is:

$$\tau_{max(pier)} = 1.0 \times 1.15 \times 1.0 \times 1.0 \times 0.094 \times 1000 \times 1.4^2 \left(\frac{1}{\log 2800000} - \frac{1}{10}\right) = 11.7 \text{ Pa} \quad (23.48)$$

3. The initial rate of scour $\dot{z}_i$ around the pier is read on the EFA curve (Figure 23.24) at $\tau = \tau_{max} = 11.7$ Pa, and gives 4.8 mm/hr.
4. The final depth of pier scour after 48 hours of flow can then be obtained from Eq. 23.27 as:

$$z_{final}(48h) = \frac{48}{\frac{1}{4.8} + \frac{48}{2580}} = 211 \text{ mm} \quad (23.49)$$

Therefore, the pier scour depth generated by the 48-hour flood is 8.2% of the maximum pier scour depth. Note that the erosion function used for this example corresponds to a soil with a medium resistance to erosion (Category 3) and that the flood is a relatively small flood (1.4 m/s). Major floods in rivers can reach 3 and 4 m/s. In very steep mountain torrents and at the bottom of levees during overtopping, the velocity can reach more than 10 m/s.

23.8.4 Final Scour Depth (z_{final}) Analysis for a Velocity Hydrograph and Layered Soil

Section 23.8.3 dealt with a uniform soil subjected to a constant velocity. However, in reality the flow velocity is not constant in a river, and the soil is likely to exhibit different layers versus depth. Let's look first at the velocity varying over time. The graph presenting the velocity as a function of time over many years is called a *velocity hydrograph* (Figure 23.17). This hydrograph represents an accumulation of events where the velocity v_i can be considered constant for a short period of time Δt_i. The solution (Briaud et al. 2001b) progresses by stepping into time with a time increment equal to δt_i for each iteration. For Δt_1, the velocity is V_1 and $z_{final\ 1}$ can be calculated. When the second velocity V_2 appears, the question is to know how to accumulate the second scour depth to the first one. The accumulation principle is as follows (Figure 23.25). The two scour depth z vs. time t curves for the velocities V_1 and V_2 are drawn separately. The scour depth $z_{final\ 1}$ is found on the V_2 curve and corresponds to the starting point for the scour depth increment for the second velocity.

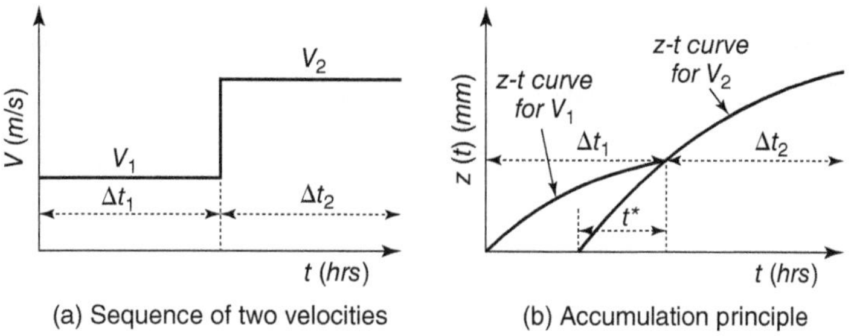

Figure 23.25 Accumulation of scour depth for two consecutive floods.

The time t^* is the time required for velocity V_2 to create $z_{\text{final 1}}$. Then the scour depth due to V_2 applied for δt_2 can be calculated by using the z vs. t curve for the velocity V_2 starting at t^*. More generally, the time t^* is the time required for velocity V_i to create the same scour depth as all the previous velocities. If that scour depth is larger than $z_{\max}$ for V_i, the velocity V_i does not increase the scour depth. This accumulation principle is applied for the entire hydrograph by stepping into time over the design life of the bridge.

For a layered soil system, the process is very similar (Figure 23.26). If the soil layer 1 is H_1 thick, the scour depth is predicted as a function of time by using the erosion function of soil 1 and the velocity accumulation principle. When the scour depth becomes equal to H_1, the erosion function is switched to that of soil layer 2 and the time t^* required for the first velocity impacting soil layer 2 to generate a scour depth equal to H_1 is found. After that, the calculations proceed using the erosion function of layer 2. These two algorithms have been automated in a program called SRICOS-EFA and are available at http://ceprofs.tamu.edu/briaud/.

The SRICOS-EFA method also allows the user to develop a probability of exceedance P vs. scour depth z curve so that the engineer can choose a scour depth corresponding to an acceptable probability of exceedance. The steps to develop the P-z curve are as follows (Brandimarte et al. 2006; Briaud et al. 2007a; Bolduc et al. 2008). First, the flow values in the hydrograph for the chosen period of time are organized in a log normal cumulative distribution function. Second, a random number generator is used to sample that distribution and create, say, 1000 equally likely future hydrographs. Third, for each of these 1000 future hydrographs, the final depth of scour, Z_{final}, is obtained according to the SRICOS-EFA method. Fourth, the 1000 values of Z_{final} are organized in a log normal distribution and presented as a cumulative density function referred to earlier as the P-z curve. This process is an integral part of the SRICOS-EFA computer program (Kwak et al. 2001; http://ceprofs.tamu.edu/briaud/).

The following case history gives an example of the calculation of scour depth, including probabilistic results.

23.8.5 The Woodrow Wilson Bridge Case History

The following case history (Kwak et al. 2002) describes the process followed to evaluate the scour depth around the main piers of the Old Woodrow Wilson Bridge, which carried I-95 across the Potomac River in Washington, D.C., from 1960 when it was built until 2005 when it was replaced.

Soil Erodibility

The soil stratigraphy is presented in Figure 23.27. It shows that at the location of the main pier in the main channel, the soil stratigraphy consists of a soft organic clay overlying a layer of hard plastic clay. Twelve ASTM Standard thin-wall steel tube samples were collected at the bottom of the Potomac River and sent to Texas A&M University for EFA testing. Examples of the erosion functions obtained for samples close to the main pier are shown in Figure 23.28. As can be seen, the soft layer has a much higher critical velocity than the hard

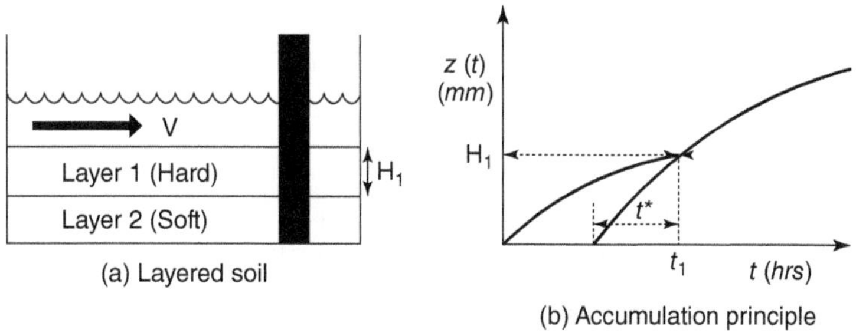

Figure 23.26 Scour depth for a layered soil system.

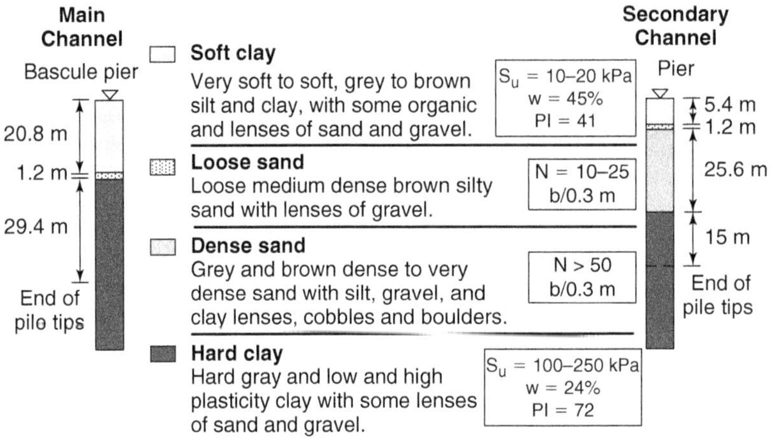

Figure 23.27 Soil stratigraphy at the location of the New Woodrow Wilson Bridge.

clay below, demonstrating yet again that critical velocity does not necessarily increase with shear strength.

Water Velocity

The nearest gage station (Gage Station 01646500; www.usgs.gov) on the Potomac River is located approximately 13 km upstream of the Woodrow Wilson Bridge and has a drainage area of 29,965 km^2. The discharge hydrograph from this gage station was multiplied by the drainage area ratio between the bridge location and the gage location (30742/29965) to obtain the discharge hydrograph at the bridge (Figure 23.29). The program HEC-RAS (Hydrologic Engineering Center's River Analysis System) (Brunner 2002) is a commonly used 1D flow analysis program. It was used to develop the relationship between the discharge and the velocity on the one hand and the relationship between the discharge and the water depth on the other (Figure 23.30). Note that the velocity in Figure 23.30 is the mean depth velocity of the water at the main pier location if the bridge were not there. That is the velocity, also called *approach velocity*, used in pier scour depth calculations. Using these relationships, the discharge vs. time curve was transformed into the water depth hydrograph and into the velocity hydrograph or velocity vs. time curve (Figure 23.31).

Geometry of the Obstacle

The old Woodrow Wilson Bridge was a bascule bridge and the obstacle to the flow considered for this case history was the main bascule pier for the bridge. The pier is square and 9.75 m by 9.75 m in plan view. The attack angle is zero, as the pier is in line with the flow.

Scour Depth Calculations

The time step for the 38-year period of observation (1960 to 1998) was chosen as one day, for a total of 13,870 time steps. The scour depth calculations progressed one day at

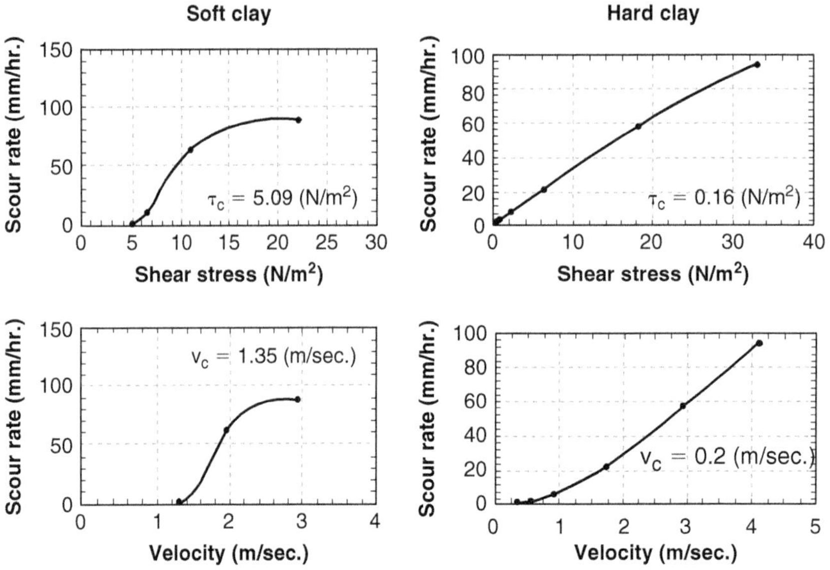

Figure 23.28 Erosion functions for the two main soil layers at the main pier location.

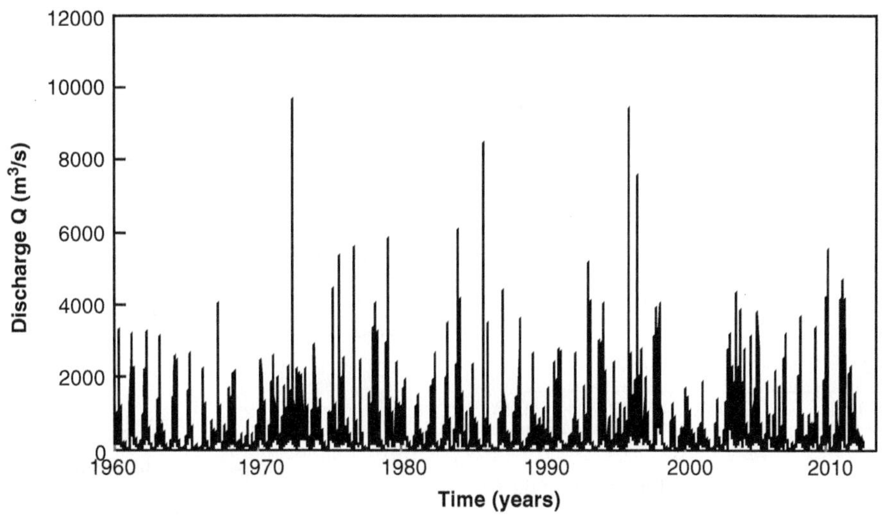

Figure 23.29 Discharge hydrograph (1960-2012) for the Potomac River at the Woodrow Wilson Bridge.

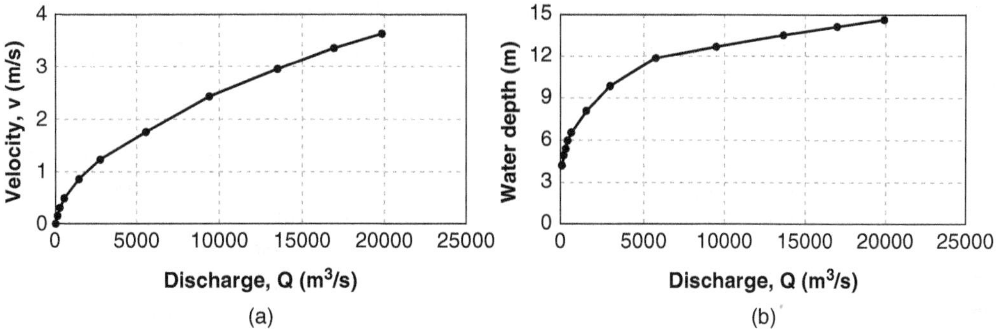

Figure 23.30 Calculated relationship between discharge, velocity, and water depth at the Woodrow Wilson Bridge if the bridge were not there.

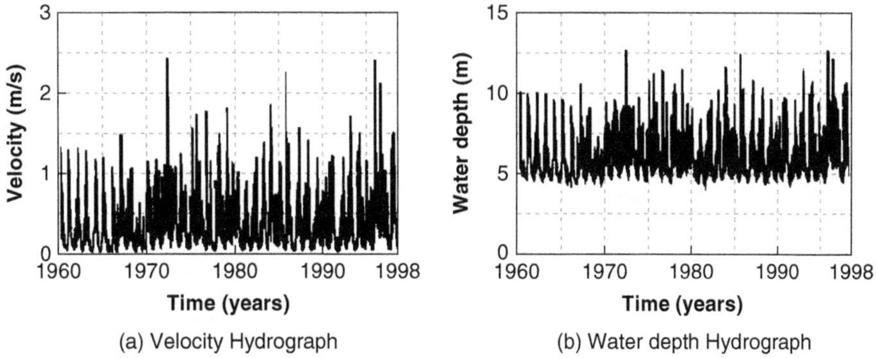

Figure 23.31 Velocity and water depth hydrograph.

a time by following the accumulation principle detailed in section 23.8.4. The program SRICOS-EFA (https://ceprofs.civil.tamu.edu/briaud/) was used with the soil erosion functions, the water velocity and water depth hydrographs, and the pier geometry as input. The resulting scour depth vs. time plot is shown in Figure 23.32.

The same procedure was repeated to predict the scour depths at the other piers of the old Woodrow Wilson Bridge

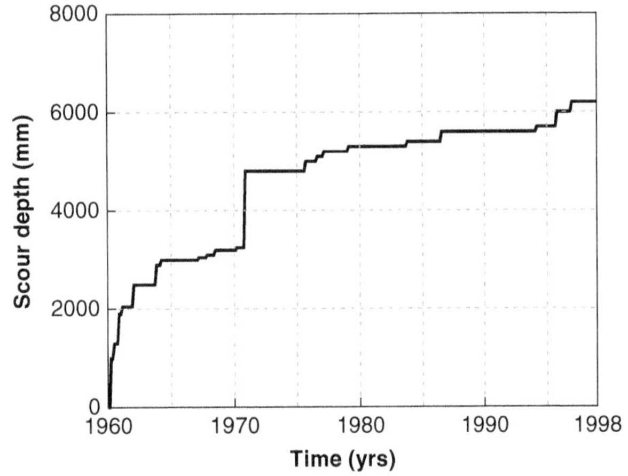

Figure 23.32 Predicted scour depth vs. time for pier 1E of the Old Woodrow Wilson Bridge.

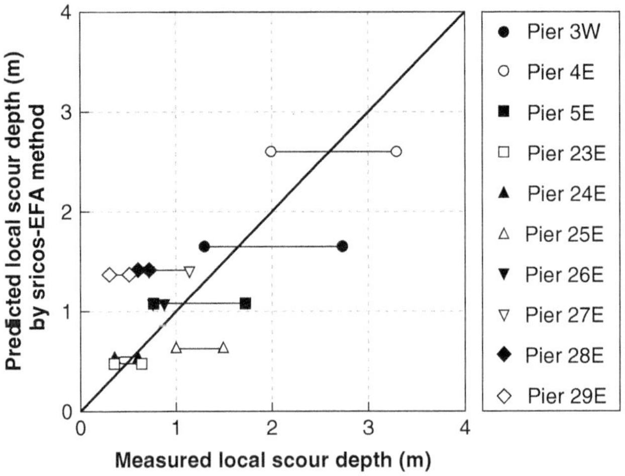

Figure 23.33 Predicted vs. measured scour depths at the old Woodrow Wilson Bridge.

where measured values were available (Hunt 2001). The comparison between predicted and measured values for all the piers that did not have rip-rap protection and where scour depth measurements were collected as a function of time is shown in Figure 23.33 (Kwak et al. 2002).

Probabilistic Scour Calculations

Figure 23.34 is an example of a probability vs. scour depth P-z curve for values of the design life L_t of the bridge. With this graph, the engineer can decide at what probability of exceedance to operate and choose the corresponding scour depth.

23.9 RIVER MEANDERING

23.9.1 Predicting River Meandering

Rivers are active system where meanders can move laterally several meters per year. This lateral migration of the main channel affects bridges, embankments, and other structures straddling the river. It is important to predict future meander movements to design remedial measures or move the structure. Many have contributed to the advancement of knowledge in this field, including Brice (1974), Hickin and Nanson (1984), Hooke (2001), Lagasse et al. (2001), and W. de Moor et al. (2007). Briaud et al. (2007a) developed the MEANDER method to predict the movement of a meander over time. It proceeds along the same steps followed to predict scour depth. First, the initial geometry of the river is described by fitting circles to the meander bends and placing straight-line tangents to the circles between circles. Second, the erosion function of the river banks is input. This can be done by using the results of EFA tests or by using the erosion classification charts of Figures 23.6 and 23.7 adjusted for the presence of vegetation, trees, or other erosion-retarding layers. Third, the velocity hydrograph is input from measurements at a nearby gage station. Fourth, the circles describing the meanders are moved according to erosion rules developed through a series of very large-scale laboratory meander experiments (in sand and then in clay) as well as numerical simulations (Briaud et al. 2007a; Wang 2006; Park 2007; Yeh

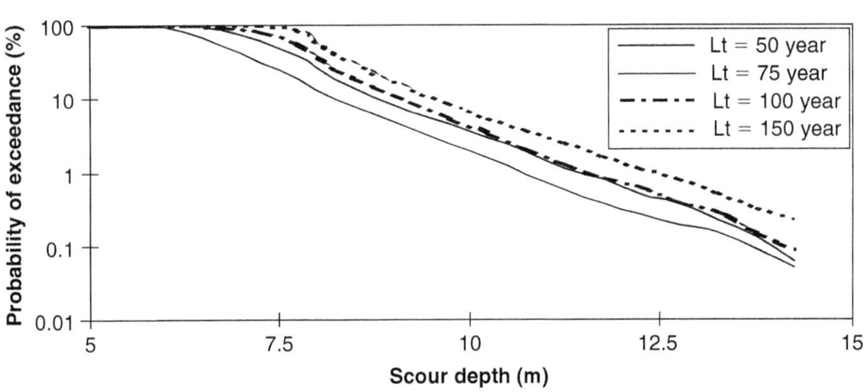

Figure 23.34 Probability of exceedance over the design life vs. scour depth curve for the bascule pier of the new Woodrow Wilson Bridge.

2008). This leads to a prediction of the location of the river after the period of time corresponding to the hydrograph.

The MEANDER method also allows the user to develop a map indicating the probability that the river will move a certain distance or more. The steps to develop that probabilistic river location are as follows (Briaud et al. 2007a). First, the flow values in the hydrograph for the chosen period of time T are organized in a log normal cumulative distribution function. Second, a random number generator is used to sample that distribution and create, say, 1000 equally likely future hydrographs. Third, for each of these 1000 future hydrographs lasting a time T, the final location of the river is obtained according to the MEANDER method. Fourth, the 1000 traces of the future river location are organized in a probabilistic map (Briaud et al. 2007b). This map gives the location of the river corresponding to the probability that the river will reach that location or go further after a time t. A conceptual example of this probabilistic map is shown in Figure 23.35 for a period of 20 years. This process is an integral part of the MEANDER computer program (http://ceprofs.tamu.edu/briaud/). The following case history illustrates the meander migration calculation process.

23.9.2 The Brazos River Meander Case History (Park 2007)

The river is the Brazos River in Texas, USA. The meander is located near Navasota, Texas (Figure 23.36) and the bridge carries highway SH105 over the Brazos River.

Observations

Records indicate that the meander has migrated significantly and rather steadily over a long period of time. Figure 23.36 shows the migration rate, which averages 4 m/yr. Observations at the site and large-scale laboratory experiments at Texas A&M University (Wang 2006; Park 2007; Yeh 2008) indicate that the process by which the meander progresses is

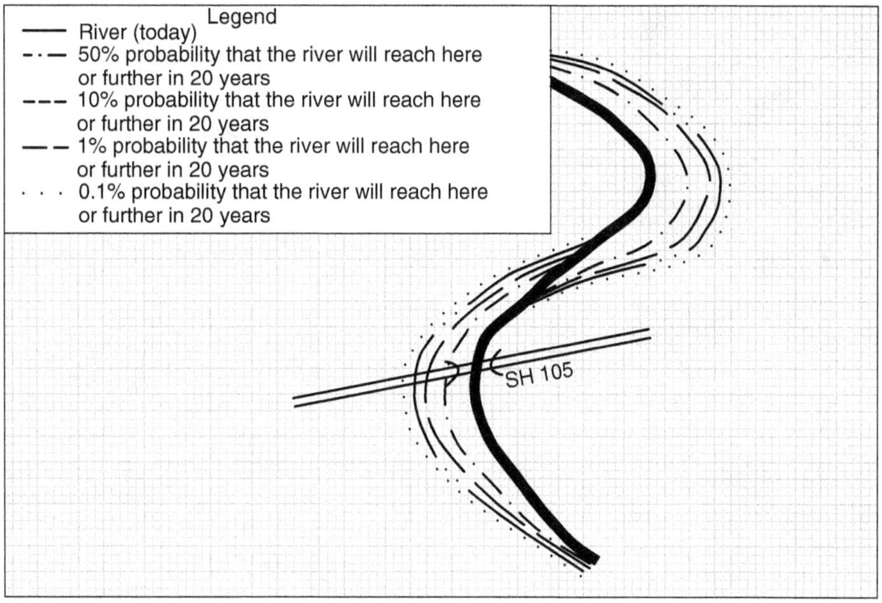

Figure 23.35 Conceptual presentation of the meandering risk for a river.

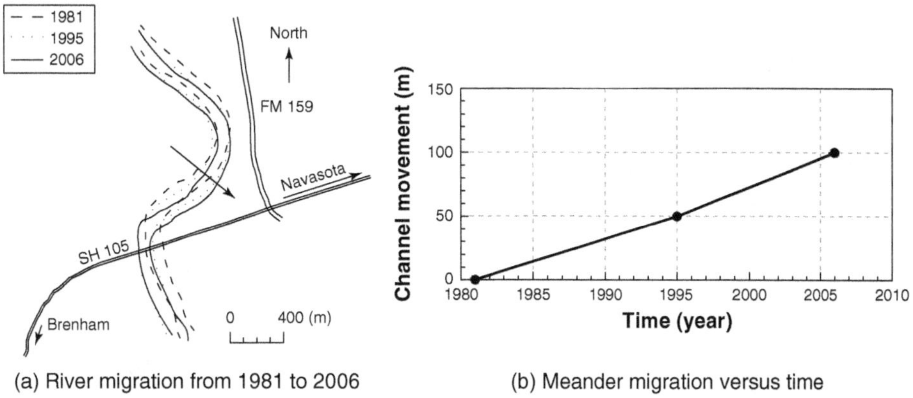

Figure 23.36 Measured migration of the meander over a 25-year period.

erosion of the base of the exterior river bank, which undercuts the steep slopes and leads to overhang failures of the banks. The material that falls into the flow is then moved to the other side of the main channel and slightly downstream. This cross-channel movement is due to the helical flow of the water in the meander. Such helical flow has been experimentally measured and numerically reproduced (Yeh 2008; Briaud et al. 2007a). This process leads to the formation of sand bars on the inside of the meander and to steep banks on the outside of the channel (Figure 23.37).

Soil Erodibility

Borings were done at the site of the meander from the top of the bank. The stratigraphy according to boring B-2 (Figure 23.38) shows 8 m of clay underlain by 7 m of sand. Thin-wall steel tube samples were collected and tested in

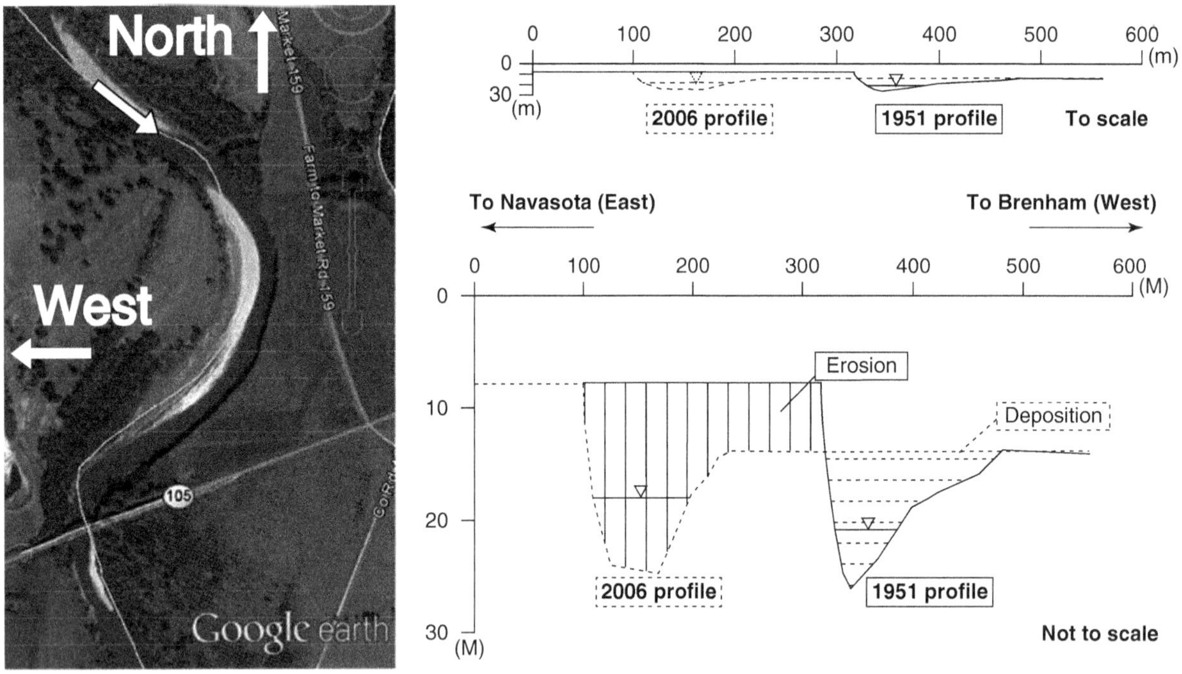

Figure 23.37 Lateral movement of the main channel between 1951 and 2006.

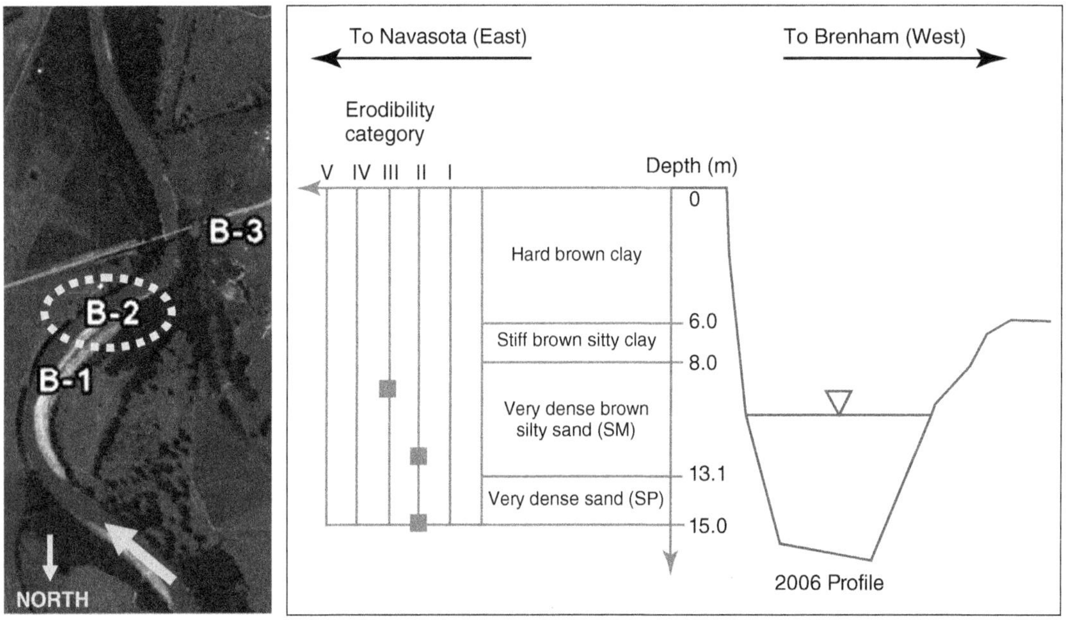

Figure 23.38 Soil stratigraphy at boring B-2.

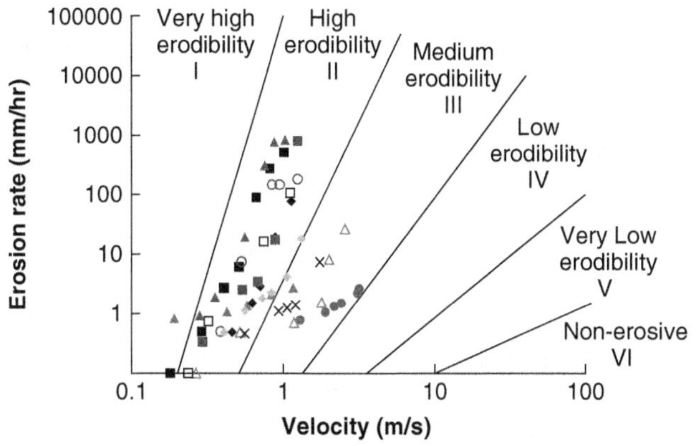

Figure 23.39 EFA test results on the soil from the meander bank.

the EFA. The results are shown in Figure 23.39. As could be predicted, the deeper layers were more erodible than the shallow ones. This means that the sand layer below will erode faster than the clay layer above. This will undercut the overhanging clay and lead to sloughing, as observed in the field. The prediction of meander migration was made using the erosion function of the deeper sand layer, as it was the controlling layer in this case.

Water Velocity

Gage Station ST #08110200 is located at the SH105 bridge over the Brazos River very close to the meander where the data were collected. This gage station worked from 1965 to 1987. To obtain the hydrograph over the prediction period 1958 to 2006, a process was developed (Park 2007) to make use of other nearby stations that had longer records (ST #08110200, ST #08108700, and ST #08109000). Then the relationship between discharge, velocity, and water depth was obtained from the actual measurements made during the period of 1965 to 1987 at gage ST #08110200. The velocity hydrograph of Figure 23.40 was finally obtained.

Geometry of the Obstacle

In this case, the obstacle is the shape of the meander, which is characterized primarily by its radius of curvature R and the width of the river channel W. To obtain R, a circle is fitted to the meander and the radius of the best-fit circle is retained as the value of R. The bend angle Φ is the angle to the center of that circle bounded by the beginning and the end of the meander on that circle. Any point M on the meander is then identified by the angle θ between the beginning and point M. Migration of the meander at point M is predicted as the movement over a period of time in the direction of the circle radius.

Meander Migration Calculations

The program MEANDER (http://ceprofs.tamu.edu/briaud/) was used to predict the migration of the meander over the period of time 1981 to 2006. The measured river centerline and the predicted river centerline are shown in Figure 23.41.

23.10 LEVEE OVERTOPPING

23.10.1 General Methodology

Levees or *dikes* are small dams built along a river or an ocean to prevent the water from inundating the land in case of flood. The top of the levee is set at a predetermined height corresponding to the water level for a chosen design flood. This flood corresponds to a certain return period, such as a 100-year flood. If the flood exceeds the design return period, water is likely to flow over the levee and generate potential erosion. One of the first observations is that if the water flows above a levee of height H, by the time the water reaches the bottom of the dry side of the levee it will have a velocity V, which can be very high. One simple way to evaluate that velocity is to write conservation of energy:

$$mgH = \frac{1}{2}mV^2 \quad or \quad V = \sqrt{2gH} \qquad (23.50)$$

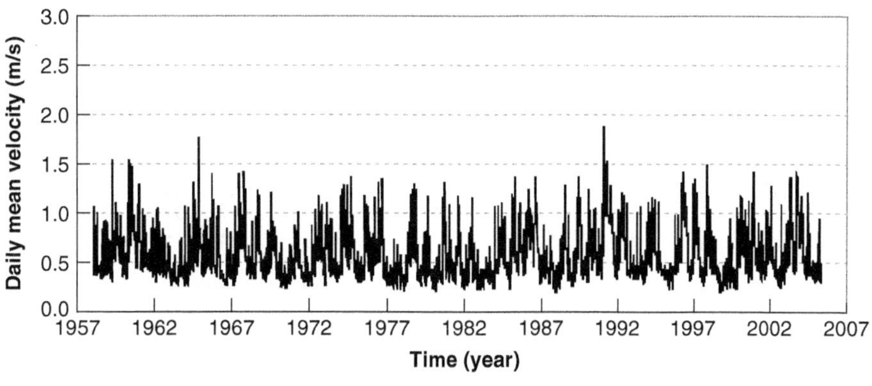

Figure 23.40 Velocity hydrograph for the Brazos River meander.

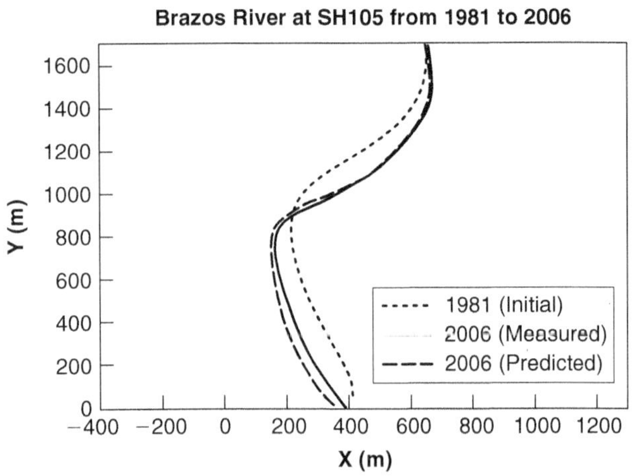

Figure 23.41 Predicted and measured migration of the Brazos River from 1981 to 2006.

For example, if the levee is 5 m high, the velocity V will be approximately 10 m/s. Of course, Eq. 23.50 does not take into account the energy lost in friction between the water and the levee surface, but it does indicate that the velocity range is much higher than typically encountered in rivers, where water rarely flows faster than 3 to 4 m/s. Furthermore, a distinction should be made between events such as hurricanes on one hand and river floods on the other. The major distinction is that hurricanes may overtop a levee for about 2 hours, whereas river floods may overtop a levee for 2 days. A levee-overtopping erosion chart developed for these two types of events is presented in Figure 23.42. It indicates which soil categories and associated erosion functions are likely to resist overtopping during a 2-hour and a 2-day overtopping. Recall that categories I to IV on the erosion chart are soils and categories V and VI are rocks. As can be seen, only the most erosion-resistant soils can resists 2 hours of overtopping without protection (Category IV), and no soil can sustain 2 days of overtopping without being totally eroded away. Armoring or vegetation satisfying strict criteria must be used to ensure that overtopping can be sustained for longer than 2 hours.

Vegetation can help significantly in retarding erosion. To be effective, though, this vegetation has to satisfy the following minimum requirements: It should have a mat-like appearance, have a sod-forming root system, be made of perennial grasses, have a dense consistent coverage, and have a minimum height of 0.3 m during flood season. Tree roots can be considered to help reinforce the levee slope, however, a the tree on a levee that is uprooted by a storm will create a major hole in the levee. Also, if the tree dies, the disappearance of the roots will leave channels for the water to seep through the levee. On the whole, trees on levees or near levees are not a good idea.

The following case history illustrates how the levee overtopping chart was generated and how it can be used.

23.10.2 Hurricane Katrina Levee Case History: New Orleans

On August 29, 2005, levee overtopping and associated erosion contributed significantly to the Katrina hurricane disaster in New Orleans, where some places are 6 m below the tops of the levees. This case history (Briaud 2006b) describes the process by which overtopped levees erode and discusses whether unprotected soils can resist overtopping erosion.

Soil Erodibility

Thin-wall steel tube samples and bag samples were obtained from the top of the levees at shallow depth (0 to 1 m). These samples were collected from locations S1 through S15 in Figure 23.43. The bag samples were reconstituted in a Shelby tube by recompacting the soil at a low and at a high compaction effort (Briaud 2006b). The soil type varied widely, from loose, uniform fine sand to high-plasticity stiff clay. EFA

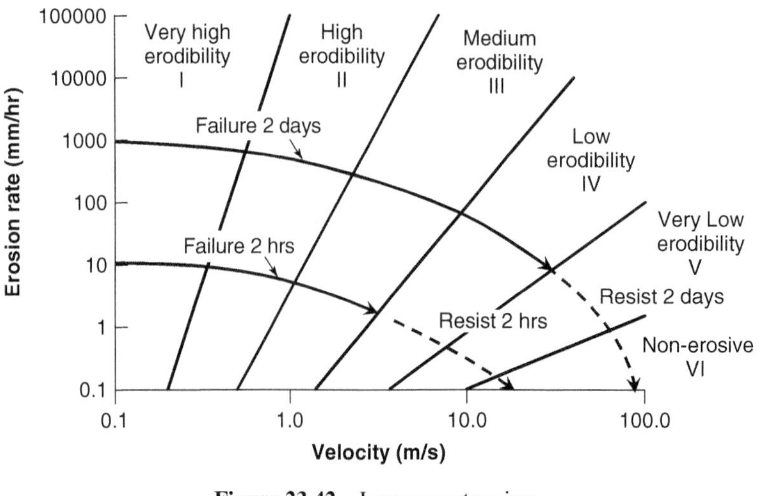

Figure 23.42 Levee overtopping.

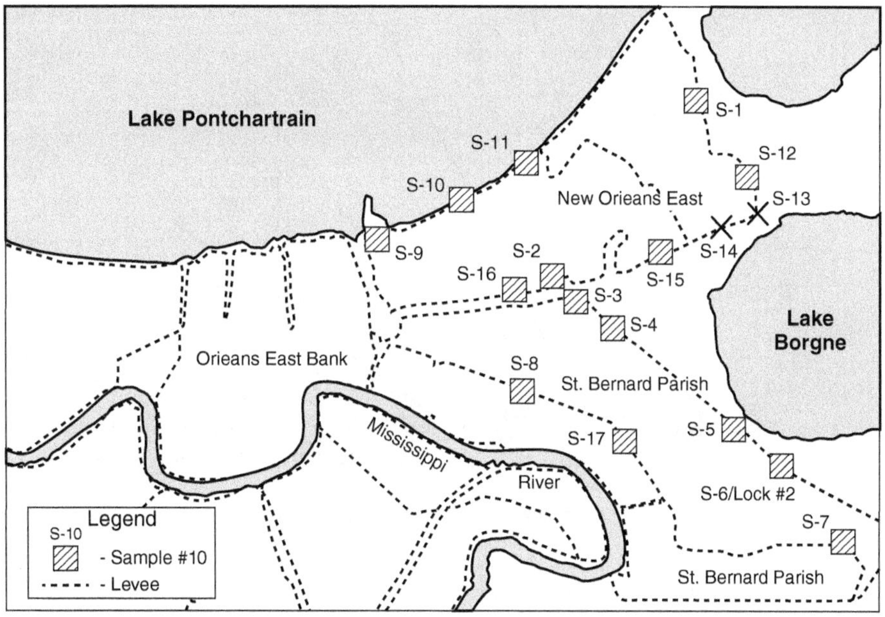

Figure 23.43 Location of shallow samples collected from the top of the levees.

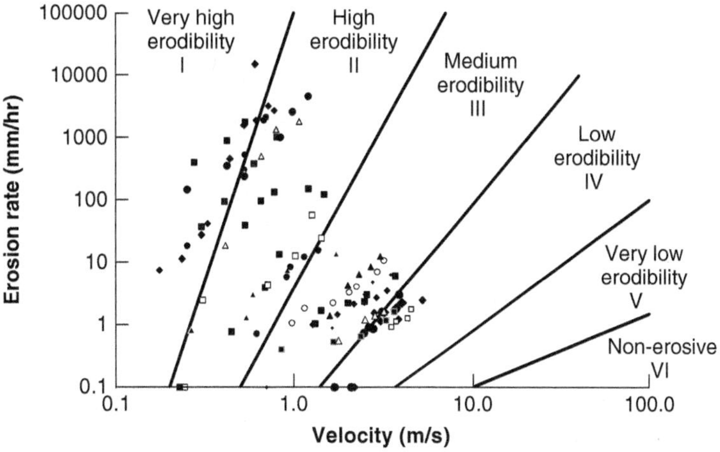

Figure 23.44 EFA test results in terms of velocity for some levee soils.

tests were performed on the samples. The results of all the tests are shown in Figures 23.44 and 23.45. One of the first observations from those figures is that the erodibility of the soils obtained from the New Orleans levees varies widely, all the way from very high erodibility (Category 1) to low erodibility (Category 4). This explains in part why some of the overtopped levees failed while other overtopped levees did not.

Water Velocity

Hurricanes are large rotating masses of moisture that can be 400 km in diameter. They travel relatively slowly at speeds of about 40 km/hr. Therefore, a hurricane takes about 10 hours to go over a levee or a bridge. The worst part of the storm, however, is only a fraction of that time. The friction generated by the wind at the air-water interface drags the water into a storm surge that can reach several meters above the mean sea level and kilometers in length. The surge associated with Katrina was about 8.5 m at Bay St. Louis, 4.6 m at Lake Borgne, and 3 m at Lake Pontchartrain. The storm surge was high enough to overtop some of the levees. As discussed earlier, the water velocity at the bottom of such levees can reach 10 m/s.

Geometry of the Obstacle

Most levees around New Orleans are between 3 and 6 m high. They have two main shapes. The first one consists of a flat top that is some 4 m wide with side slopes at about 5 horizontal to 1 vertical. Because the width of such a levee configuration takes a lot of space, the second shape consists of the same shape as the first, but at a reduced scale with a vertical wall

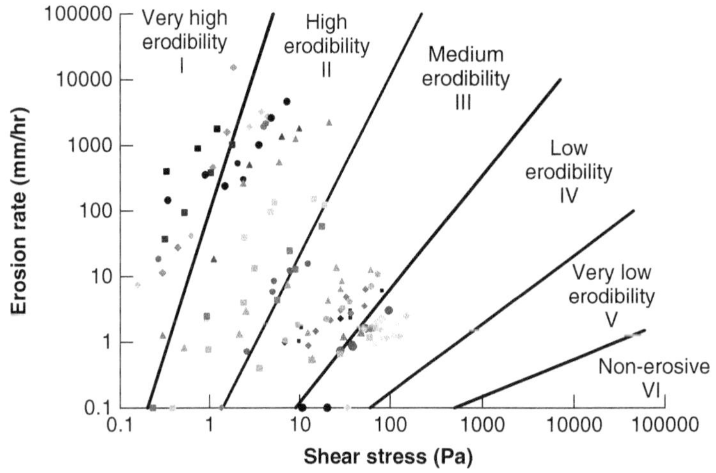

Figure 23.45 EFA test results in terms of shear stress for some levee soils.

extending from the top of the levee. The problem addressed here is limited to the first shape (no wall).

Predicting Levee-Overtopping Erosion

There was overwhelming evidence that the water overtopped the levees in many places; such evidence consisted mostly of ships being trapped on top of the levees when the water receded, but also included debris stuck in trees at levels higher than the top of the levees. Some levees resisted the overtopping well, whereas some levees were completely eroded. In Figure 23.46, the erodibility functions for the samples taken from levees that were overtopped and resisted well are plotted as open circles; the solid dots are for the samples of levees that were completely eroded. As can be seen, the eroded levees were made of soils in erodibility categories 1 and 2, whereas the levees that resisted well were made of soils in erodibility categories 3 and 4. This led to the levee overtopping chart shown in Figure 23.42.

23.11 COUNTERMEASURES FOR EROSION PROTECTION

Countermeasures for erosion protection include a number of solutions, the most prevalent of which is the use of rip rap (Figure 23.47). Rip rap can be sized by the following equation (USACE 1991):

$$d_{30} = H_w F\, C_{st} C_v C_t \left(\frac{V_{des}}{\sqrt{C_{sl}\,(G_s - 1)\, g H_w}} \right)^{2.5} \tag{23.51}$$

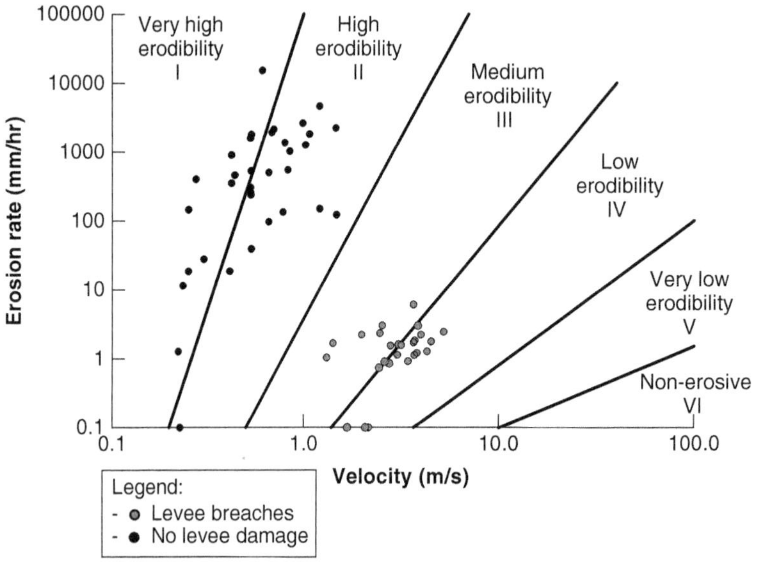

Figure 23.46 EFA test results for the soils of levees that failed and for the soils of levees that did not fail by overtopping erosion.

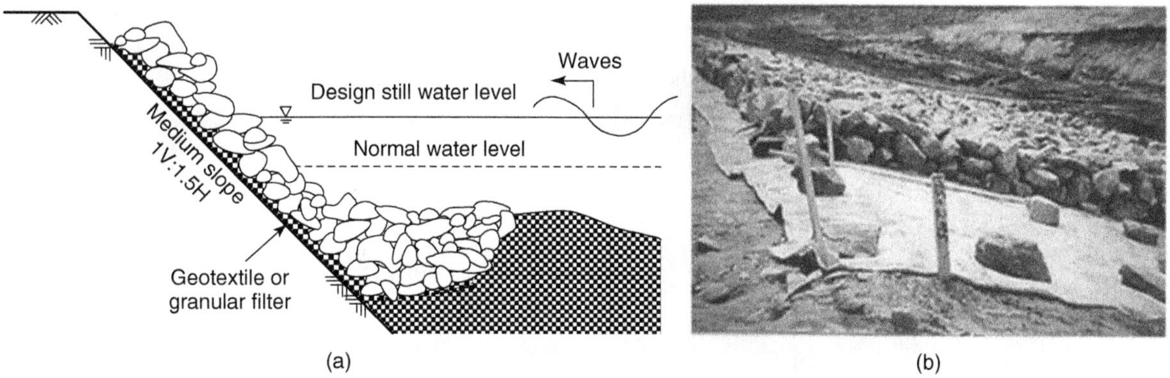

Figure 23.47 Rip rap with geosynthetic filter installation: *(a)* Design plan. *(b)* Field installation. (*Right picture*: Courtesy of FHWA.)

where d_{30} is the particle size of the rip-rap grain size distribution curve corresponding to 30% fines, H_w is the water depth, F is the factor of safety, C_{st} is the stability coefficient, C_v is the velocity distribution coefficient, C_t is the blanket thickness coefficient, V_{des} is the mean depth water velocity, C_{sl} is the side slope correction factor, G_s is the specific gravity of the rip rap, and g is the acceleration due to gravity (9.81 m/s^2).

The stability coefficient C_{st} takes into account the roughness of the rip-rap blocks; it is 0.3 for angular rock and 0.375 for round rocks. The velocity distribution coefficient C_v takes into account the fact that water tends to accelerate on the outside of river bends; it is 1 for straight channels and inside of bends, and 1.23 in most other cases. The blanket thickness coefficient C_t is a function of the rip-rap gradation, with a default value of 1 in the absence of additional data. The velocity V_{des} is the mean depth velocity for straight channels. For river bends it is given by:

$$V_{des} = V_{ave}\left(1.74 - 0.52\log\frac{R_c}{W}\right) \qquad (23.52)$$

where V_{ave} is the mean depth velocity upstream of the bend, R_c is the centerline radius of curvature of the river bend, and W is the river width at the water level.

The side slope coefficient C_{sl} is given by:

$$C_{sl} = \sqrt{1 - \left(\frac{\sin(\theta - 14^\circ)}{\sin 32^\circ}\right)^{1.6}} \qquad (23.53)$$

where θ is the bank angle in degrees. The specific gravity of solids G_s is usually taken as 2.65.

It is very important to place a filter between the soil to be protected and the rip-rap layer. Without a filter, the soil under the rip rap may continue to erode through the large voids in the rip rap. In the end, the rip rap may not move away, but may simply go down significantly as the underlying soil erodes away. The filter may be a sand filter or a geosynthetic filter (see Chapter 27). Design guidelines can be found in Heibaum (2004) for sand filters and in Koerner (2012) for geosynthetic filters.

Other countermeasures to prevent erosion include (Lagasse et al. 2009):

1. Flow deflectors such as spurs, jetties, dikes, and guide banks
2. Rigid armoring of the soil surface, such as soil-cement mixing and grouted mattresses
3. Flexible armoring, such as rip rap, gabions, and articulated blocks
4. Pier geometry modification, such as slender pier shape and debris deflectors
5. Vegetation such as woody mats and root wads
6. Fixed and portable instrumentation such as sonars and float-out devices
7. Periodic inspection

23.12 INTERNAL EROSION OF EARTH DAMS

23.12.1 The Phenomenon

It is estimated that 46% of earth dam failures occur due to internal erosion, and half of those failures occur during the first filling of the reservoir (Fell et al. 2005; Figure 23.48). Yet, handling of internal erosion of earth dams is still based primarily on engineering judgment and experience. Although guidelines and publications exist, much remains to be studied and researched in this field. For internal erosion of an earth dam to take place, the following are required:

1. A seepage flow path and a source of water
2. Erodible material that can be carried by the seepage flow within the flow path
3. An unprotected exit from which the eroded material may escape
4. For a pipe to form, the material must be able to form and support the roof of the pipe

Four different phenomena can lead to internal erosion of an earth dam (Figure 23.49):

1. Backward erosion
2. Concentrated leak

Figure 23.48 Internal erosion of an earth dam. (a) Blackman Creek Dam. (b) Teton Dam. (a: Photograph by Mark S. Harrison, Oklahoma Conservation Commission. Used by permission. b: Courtesy of Eunice Olson.)

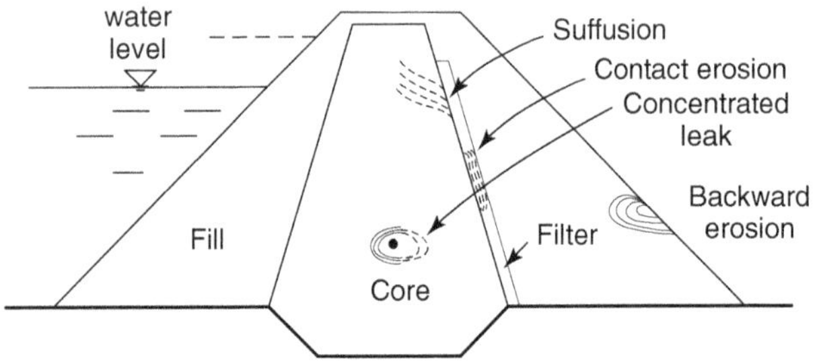

Figure 23.49 Mechanisms of internal erosion failures. (After Prezlmaier 2005)

3. Suffusion
4. Soil contact erosion

Backward erosion is initiated at the exit point of the seepage path when the hydraulic gradient is too high and the erosion gradually progresses backward, forming a pipe. A *concentrated leak* is internal to the soil mass; it initiates a crack or a soft zone emanating from the source of water and may or may not progress to an exit point. Erosion gradually continues and can create a pipe or a sinkhole. *Suffusion* develops when the fine particles of the soil wash out or erode through the voids formed by the coarser particles. This occurs when the amount of fine particles is smaller than the void space between the coarse particles. If, in contrast, the soil has a well-graded particle size distribution with sufficiently small voids, suffusion is unlikely. Soils are called internally unstable if suffusion takes place and internally stable if particles are not eroding under seepage flow. *Soil contact erosion* refers to sheet flow at interfaces between soil types. It may occur, for example, when water seeps down the back face of the core at the interface with the filter and then the stabilizing mass.

Earth dams deform during and after construction. This movement can be compression, extension, and/or shear distortion. Because typical dams are made of different zones playing different roles, they exhibit different deformation characteristics. This can lead to differential movement, resulting in cracks or soft zones where internal erosion can be initiated. Shrinkage can also create cracks that are prone to erosion if water comes to flow through them. Fell and Fry (2005) summarize the most likely locations where internal erosion can start in an earth dam (Figure 23.50).

23.12.2 Most Susceptible Soils

Coarse silt and fine sand are among the most erodible soils. Therefore, earth dams containing significant amounts of such materials will be more prone to internal erosion. Clays in general, and high-plasticity clays in particular, are more resistant to erosion as long as the electrical bonds between particles are not destroyed by chemicals. It seems that some core materials of glacial origin, such as glacial tills, can be particularly susceptible to internal erosion. Sherard (1979) gives a range of gradation of soils that can lead to internal erosion problems (Figure 23.51).

The soils that are most susceptible to suffusion are those where the volume of fines is less than the volume of the voids between coarse particles. In this case, the fines can move easily between the coarse particles and erode away to an exit face. After suffusion, such soils are devoid of fines and become very pervious clean gravel, for example. Fell and Fry (2005) indicate that gap-graded soils and coarsely graded soils with a flat tail of fines (Figure 23.52) are most susceptible to suffusion.

23.12.3 Criterion to Evaluate Internal Erosion Potential

One of the important criteria for evaluating erosion is to calculate the hydraulic gradient and compare it to the critical gradient. The critical gradient is given by:

$$i_{cr} = \frac{\gamma_{sat} - \gamma_w}{\gamma_w} \tag{23.54}$$

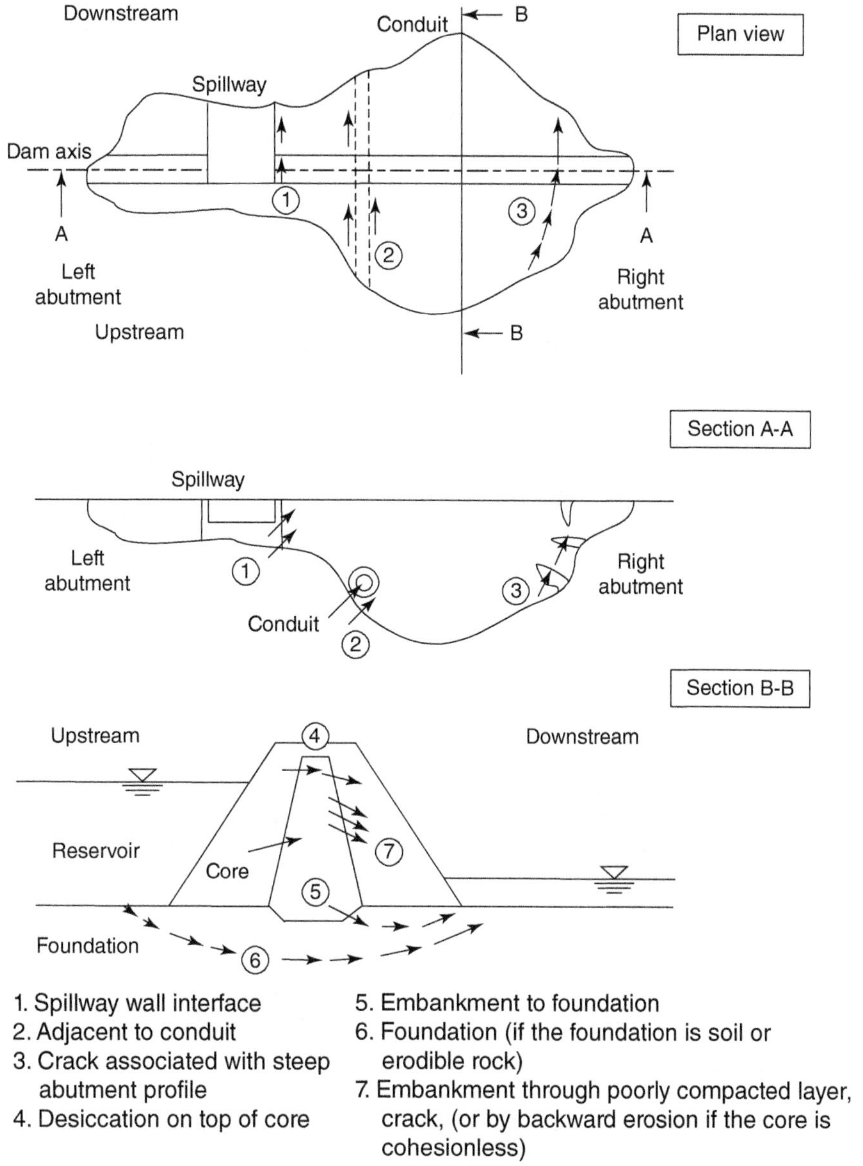

Figure 23.50 Possible locations of initiation of internal erosion. (After Fell and Fry 2005)

Values of i_{cr} typically vary in the range of 0.85 to 1.2. The hydraulic gradient in dams depends on many factors, including the difference in water level between the upstream and the downstream, the length of the drainage path, and the relative hydraulic conductivity of the various zones. The target maximum gradient in the flow must be kept much lower than the critical value, especially in areas where internal erosion is possible. Figure 23.53 shows ranges of hydraulic gradient values that are associated with initiation of internal erosion on the one hand and full development of piping on the other for unfiltered exit faces. Generally speaking, there is a trend toward higher-porosity soils beginning to erode at lower hydraulic gradients, even lower than 0.3. Yet, soils with plastic fines erode at higher gradients, and gap-graded soils begin to erode at lower gradients than nongap-graded soils with the same fine content. The U.S. Army Corps of Engineers uses a lower-bound value of the critical hydraulic gradient equal to 0.8 and allows a hydraulic gradient of up to 0.5 at the toe of levees, provided a number of conditions are met (USACE 2003). Another way to address the incipient motion of soil particles in internal erosion problems is to use the concept of critical velocity and charts such as Figure 23.8 and 23.9. However, these critical velocities were developed from sheet flow tests, and the critical velocity may be different from those initiating internal erosion.

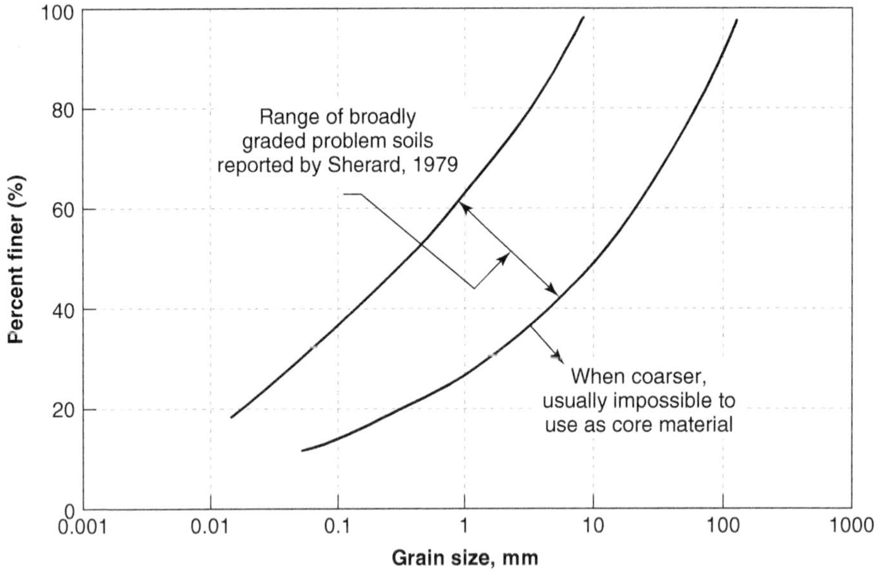

Figure 23.51 Range of problem soils for internal erosion. (After Sherard 1979)

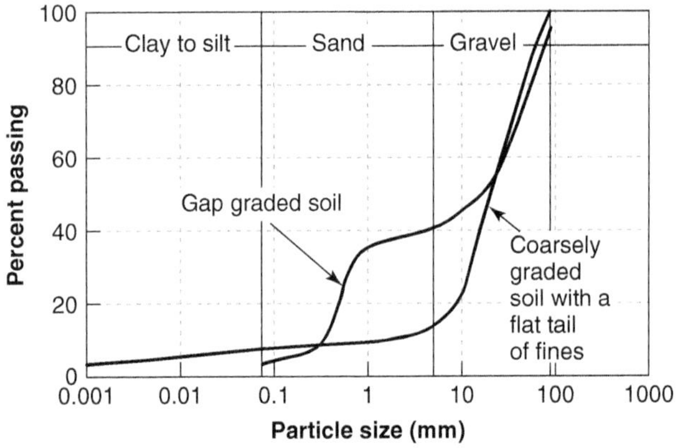

Figure 23.52 Range of problem soils for suffusion. (After Fell and Fry 2005)

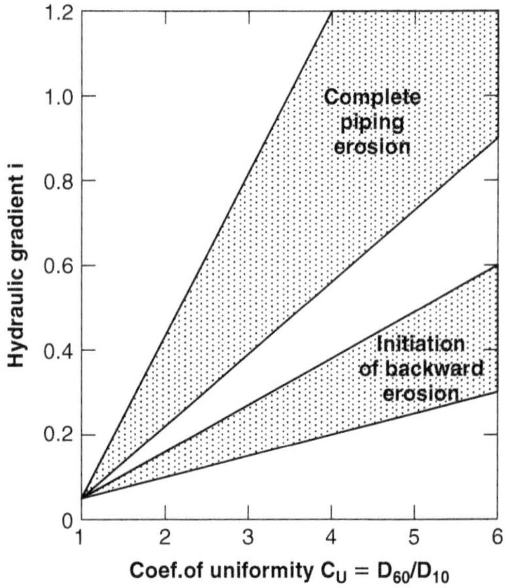

Figure 23.53 Range of hydraulic gradient values associated with internal erosion. (After Prezlmaier 2005)

Several methods, based in part on the analysis of the grain size curve, have been developed to evaluate the instability of soils in dams and their sensitivity to the suffusion phenomenon. They include Sherard (1979), Kenney and Lau (1986), Burenkova (1993), and Fell and Wan (2005).

23.12.4 Remedial Measures

Internal erosion of earth dams often occurs very quickly, leaving limited time for remedial action (Foster et al. 2000a, 2000b). Most of the time, complete breach occurs within 12 hours of first visual detection of internal erosion and sometimes in less than 6 hours. The majority of failures occur during the first filling or within 5 years after first filling. The process of suffusion tends to develop more slowly than the back erosion and piping processes.

One solution to many internal erosion problems is the use of quality filters. A *filter* is a layer of soil placed between a fine-grained soil and a coarse-grained soil to transition the flow without having the fines of the fine-grained soil erode through the voids of the coarse-grained soil. The grain size distribution curve of the soil filter layer is designed to provide this transition in a gradual fine-to-coarse fashion.

PROBLEMS

23.1 If a faucet drips on a pebble for 20 million years, will there be a hole in the pebble?

23.2 Water flows in a river at a mean depth shear velocity of 2 m/s. The gradient of the shear velocity at the bottom of the river is 7000 m/s per m of depth. Calculate the shear stress applied by the water to the bottom of the river. The soil particles at the bottom of the river are cubes 1 mm in size. They have a unit weight of 26.5 kN/m^3 and a friction angle equal to 35°. Calculate the shear stress necessary to move the soil grains. Compare this shear stress to the shear stress applied by the water; will there be erosion?

23.3 The particle of problem 2 is now a 1 mm diameter sphere that rests between two other spherical particles (Figure 23.1s). The particle is subjected to the same shear stress as in problem 2. Will the particle be able to roll over its neighbors and erode away?

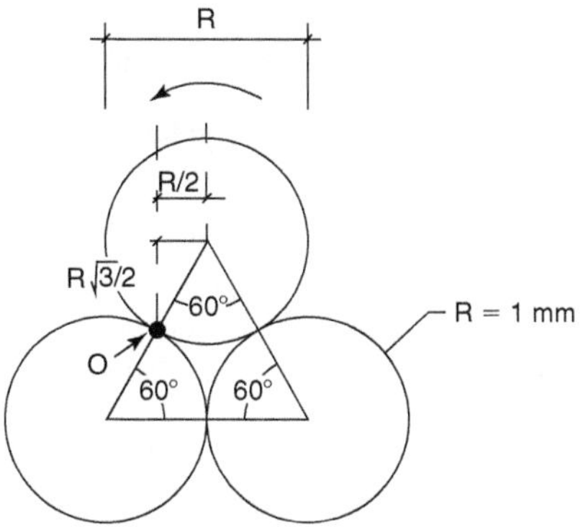

Figure 23.1s Soil particle.

23.4 The straight part of a river is at flood stage and experiences a 160-year flood. During the flood, the water depth is 6 m and the mean depth water velocity is 3 m/s. The bottom of the river is made of sand and the banks have a bank angle of 30°. Would you expect the sand to erode? If yes, what size rip rap would you recommend to place on top of the sand to prevent erosion? Would you place a geosynthetic filter between the sand and the rip rap? Explain.

23.5 A bridge is designed for a life of 50 years and you wish to design the bridge for a flood that has a probability of occurring or being exceeded of 0.001. What should the recurrence interval of the design flood be?

23.6 A round-nose pier is 3 m wide and 6 m long. The center-to-center spacing of the piers is 50 m. The water depth at the site is 10 m and the approach flow velocity of 3 m/s has an attack angle equal to 10° (Figure 23.2s). EFA tests were conducted; the average erosion function representing the soil is given in Figure 23.3s. The critical velocity of the soil is 1.6 m/s. The duration of the flood is 48 hours. Find the pier scour depth after 48 hours.

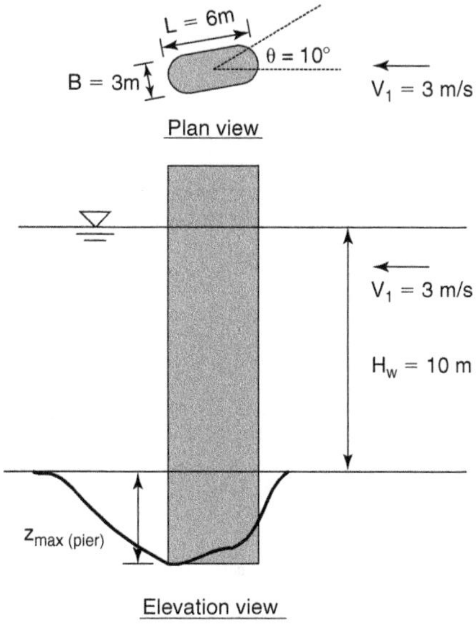

Figure 23.2s Pier scour problem.

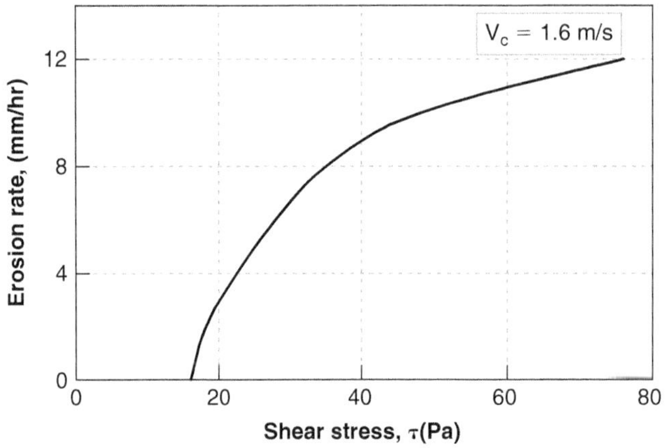

Figure 23.3s Erosion function.

23.7 Calculate the abutment and contraction scour depth after 48 hours of flood for the following case. The geometry of the channel and the bridge are given in Figure 23.4s. The compound channel is symmetrical, and the discharge during the flood is $Q = 2000\,m^3/s$. The critical velocity of the soil in the main channel and flood plain is 1.2 m/s. The erosion function of the soil from an EFA test is given in Figure 23.5s. The duration of flood is 48 hours, and the hydraulic data are as follows:

Mean velocity in the general approach cross section: $V_1 = 1.13\,m/s$
Mean velocity in the approach floodplain: $V_{f1} = 0.78\,m/s$
Mean velocity in the approach main channel: $V_{m1} = 1.4\,m/s$
Water depth in the approach flood plain: $H_{wf1} = 2.55\,m$
Water depth in the approach main channel: $H_{wm1} = 7.9\,m$
Mean velocity in the general contracted cross section: $V_2 = 1.75\,m/s$
Mean velocity in the contracted main channel: $V_{m2} = 1.83\,m/s$
Hydraulic radius in the approach main channel: $R_{h1} = 3.65\,m$
Find the abutment scour depth and the contraction scour depth after 48 hours of flood.

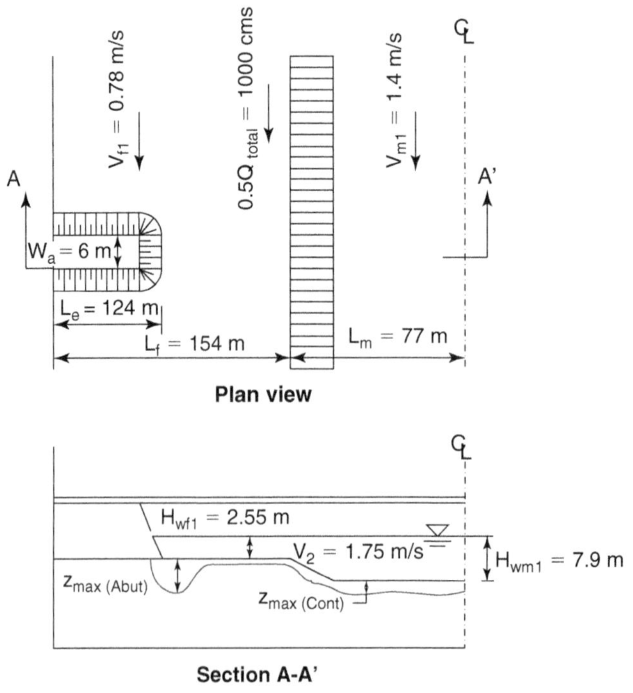

Figure 23.4s Channel geometry.

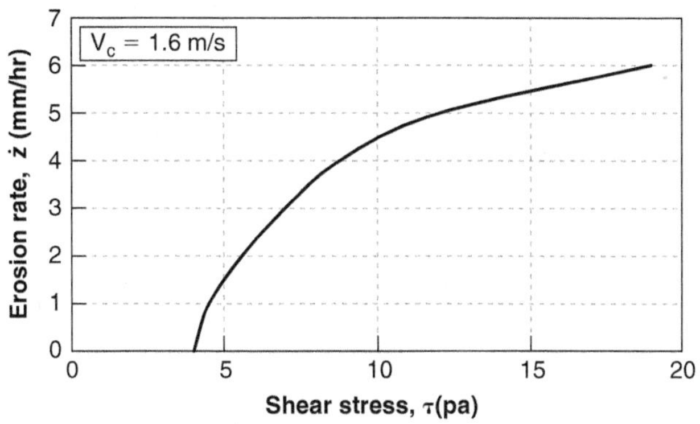

Figure 23.5s Erosion function.

23.8 Download the SRICOS-EFA program from the web site http://ceprofs.tamu.edu/briaud/ and run Example 1 from the list of examples.

23.9 Download the MEANDER program from the web site http://ceprofs.tamu.edu/briaud/ and run Example 1 from the list of examples.

23.10 A 5 m high levee is overtopped for 2 hours during a hurricane. The levee material and the soil below the levee are borderline between a high-plasticity clay CH and a low-plasticity clay CL. Draw a contour of the levee after 2 hours of overtopping.

23.11 Repeat problem 23.10 for a flood that lasts 72 hours.

Problems and Solutions

Problem 23.1

If a faucet drips on a pebble for 20 million years, will there be a hole in the pebble?

Solution 23.1

Common sense might lead you to say yes. Then the question might become: How is it possible for a stress level as small as the one created by a drop of water to destroy the bonds of the rock? The answer may be that any stress, no matter how small, can defeat any strength, no matter how large, provided the number of cycles is high enough. Experiments to check such a statement would be very valuable.

Problem 23.2

Water flows in a river at a mean depth shear velocity of 2 m/s. The gradient of the shear velocity at the bottom of the river is 7000 m/s per m of depth. Calculate the shear stress applied by the water to the bottom of the river. The soil particles at the bottom of the river are cubes 1 mm in size. They have a unit weight of 26.5 kN/m^3 and a friction angle equal to 35°. Calculate the shear stress necessary to move the soil grains. Compare this shear stress to the shear stress applied by the water; will there be erosion?

Solution 23.2

From the problem statement:

$$v = 2\ \text{m/s}$$

$$\frac{dv}{dz} = \frac{7000\ \text{m/s}}{\text{m}}$$

Particle size: 1 mm cube

$$\gamma_s = 26.5\ \text{kN/m}^3$$

$$\mu = 1 \times 10^{-3}\ \text{Pa} \cdot \text{s}$$

a. Shear stress applied by the water at the bottom of the river:

$$\tau_w = \mu\left(\frac{dv}{dz}\right) = (1 \times 10^{-3}\ \text{Pa}\cdot\text{s}) \times \left(7000\ \frac{\text{m/s}}{\text{m}}\right) = 7\ \text{Pa}$$

b. Shear stress necessary to move the grains:

$$\tau_s = \sigma_N \times \tan\varphi' = \left(\frac{\gamma_s \times V}{\text{A}}\right) \times \tan\varphi' = \left(\frac{26500 \times 10^{-9}}{10^{-6}}\right) \times \tan 35 = 18.56\ \text{Pa}$$

c. Comparison:

Stress necessary to move the grains ($\tau_s = 18.56$ Pa) is larger than the shear stress generated by the water ($\tau_w = 7$ Pa); therefore, there will be no erosion.

Problem 23.3

The particle of problem 2 is now a 1 mm diameter sphere that rests between two other spherical particles (Figure 23.1s). The particle is subjected to the same shear stress as in problem 2. Will the particle be able to roll over its neighbors and erode away?

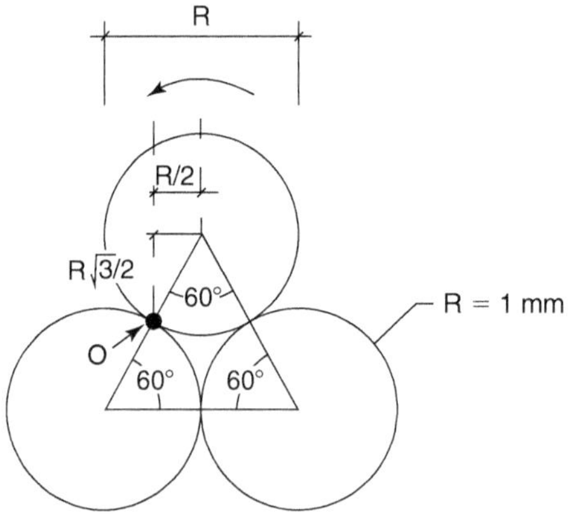

Figure 23.1s Soil particle.

Solution 23.3

The driving moment M_D and resisting moment M_R around point O in Figure 23.1s are calculated. For the driving moment, it is assumed that the drag force exerted by the shear stress acts on the projected surface of the spherical particle and that the normal stress applied by the water due to the flow is negligible:

$$M_D = \tau_w \frac{\pi D^2}{4}\left(R + \frac{\sqrt{3}R}{2}\right) = 7 * 10^{-6} * \frac{\pi}{4} * 1^2(1 + 0.87) = 10.25 * 10^{-6}\ \text{N.mm}$$

$$M_R = W\frac{R}{2} = \gamma_s V \frac{R}{2} = \frac{26500}{10^9} * \frac{4}{3}\pi D^3 * \frac{0.5}{2} = 27.74 * 10^{-6}\ \text{N.mm}$$

$M_R > M_D \rightarrow$ The particle won't be able to roll over its neighbors and erode away

Problem 23.4

The straight part of a river is at flood stage and experiences a 160-year flood. During the flood, the water depth is 6 m and the mean depth water velocity is 3 m/s. The bottom of the river is made of sand and the banks have a bank angle of 30°. Would you expect the sand to erode? If yes, what size rip rap would you recommend to place on top of the sand to prevent erosion? Would you place a geosynthetic filter between the sand and the rip rap? Explain.

Solution 23.4

Yes, one would expect the sand to erode. Indeed, the water velocity is 3 m/s and the critical velocity of the sand will be at most 1 m/s (Figure 23.8).

Rip rap can be sized as follows:

$$d_{30} = H_w F\, C_{st} C_v C_t \left(\frac{V_{des}}{\sqrt{C_{sl}\,(G_s - 1)\, g H_w}} \right)^{2.5}$$

The height of water is 6 m and we choose a factor of safety equal to 2. Also, we assume that the rip rap blocks are angular, so C_{st} is equal to 0.3.

The magnitude of C_{sl} is calculated with $\theta = 30^\circ$:

$$C_{sl} = \sqrt{1 - \left(\frac{\sin\left(\theta - 14^\circ\right)}{\sin 32^\circ} \right)^{1.6}} = \sqrt{1 - \left(\frac{\sin 16^\circ}{\sin 32^\circ} \right)^{1.6}} = 0.8$$

$$d_{30} = 6 \times 2 \times 0.3 \times 1.23 \times 1 \left(\frac{3}{\sqrt{0.8\,(2.65 - 1)\,9.81 \times 6}} \right)^{2.5} = 0.3 \text{ m}$$

It is very important to place a filter between the soil to be protected and the rip-rap layer. Without a filter, the soil under the rip rap may continue to erode through the large voids in the rip rap; in the end, the rip rap may not move away, but may simply go down significantly as the underlying soil erodes away.

Problem 23.5

A bridge is designed for a life of 50 years and you wish to design the bridge for a flood that has a probability of occurring or being exceeded of 0.001. What should the recurrence interval of the design flood be?

Solution 23.5

$$R = 1 - \left(1 - \frac{1}{T_R}\right)^{L_t}$$

where T_R is the return period and R is the probability of exceedance of the flood.

$$0.001 = 1 - \left(1 - \frac{1}{T_R}\right)^{50} \rightarrow \left(1 - \frac{1}{T_R}\right)^{50} = 0.999$$

$$50 \times \log\left(1 - \frac{1}{T_R}\right) = \log 0.999$$

$$\log\left(1 - \frac{1}{T_R}\right) = \frac{-4.34512 \times 10^{-4}}{50} = -8.690 \times 10^{-6}$$

$$1 - \frac{1}{T_R} = 0.99997999 \rightarrow T_R = 50000 \text{ yrs.}$$

The 50,000-year flood is the one to be considered.

Problem 23.6

A round-nose pier is 3 m wide and 6 m long. The center-to-center spacing of the piers is 50 m. The water depth at the site is 10 m and the approach flow velocity of 3 m/s has an attack angle equal to 10° (Figure 23.2s). EFA tests were conducted; the average erosion function representing the soil is given in Figure 23.3s. The critical velocity of the soil is 1.6 m/s. The duration of the flood is 48 hours. Find the pier scour depth after 48 hours.

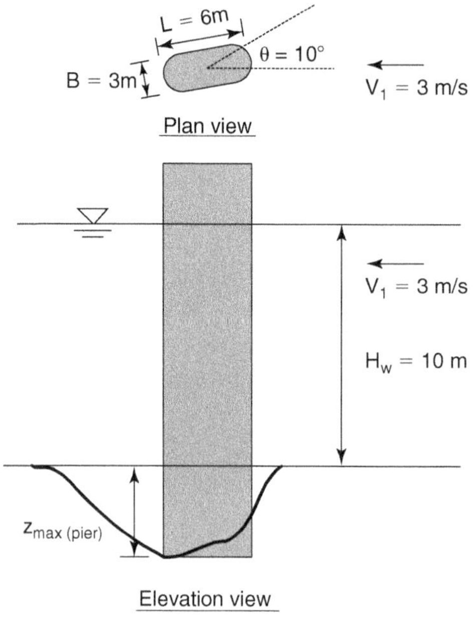

Figure 23.2s Pier scour problem.

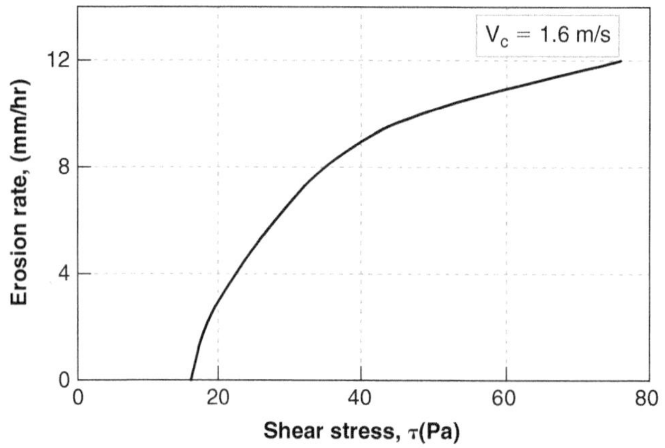

Figure 23.3s Erosion function.

Solution 23.6

The maximum scour depth and the maximum shear stress around the pier can be calculated as follows.

Maximum Scour Depth

The correction factors for water depth (K_{pw}), pier shape (K_{psh}), pier aspect ratio (K_{pa}), and pier spacing (K_{psp}) must be calculated first:

$$K_{pw} = \begin{cases} 0.89\left(\frac{h_w}{B'}\right)^{0.33}, & \text{for } \frac{h_w}{B'} < 1.43 \\ 1.0, & \text{else} \end{cases}$$

$$B' = B\left(\cos\theta + \frac{L}{B}\cdot\sin\theta\right) = 3\left(\cos 10 + \frac{6}{3}\sin 10\right) = 4 \text{ m}$$

$$\frac{h_w}{B'} = \frac{10}{4} = 2.5 > 1.43 \quad so \quad K_{pw} = 1$$

$K_{psh} = 1$ because the pier has a round nose

$K_{l/B} = 1$ because the projected width has been used.

$$K_{psp} = \begin{cases} 2.9\left(\frac{S}{B'}\right)^{-0.91} & , \text{ for } \frac{S}{B'} < 3.42 \\ 1.0 & , \text{ else} \end{cases}$$

$$\frac{S}{B'} = \frac{50}{4} = 12.5 > 3.42 \quad so \quad K_{psp} = 1$$

All correction factors are equal to 1.0. The Froude Number is calculated with the approach velocity and pier width:

$$Fr_{(\text{pier})} = \frac{V_1}{\sqrt{g \cdot B'}} = \frac{3}{\sqrt{9.81 \times 4}} = 0.48$$

The critical pier Froude Number is calculated:

$$Fr_{c(\text{pier})} = \frac{V_c}{\sqrt{g \cdot B'}} = \frac{1.6}{\sqrt{9.81 \times 4}} = 0.255$$

Therefore, the maximum pier scour depth in given condition is:

$$\begin{aligned} z_{\max(\text{Pier})} &= 2.2 \cdot K_{pw} \cdot K_{psh} \cdot K_{pa} \cdot K_{psp} \cdot a' \cdot (2.6Fr_{(\text{pier})} - Fr_{c(\text{pier})})^{0.7} \\ &= 2.2 \times 1 \times 1 \times 1 \times 1 \times 4 \times (2.6 \times 0.480 - 0.255)^{0.7} = 8.757 \text{ m} \\ &= 8757 \text{ mm} \end{aligned}$$

Maximum Shear Stress around Pier

The correction factors for water depth (k_{pw}) and for pier spacing (k_{psp}) are calculated and found equal to 1. For pier shape,

$$k_{psh} = 1.15 + 7e^{(-4L/a)} = 1.15 + 7e^{(-4\times6/2)} = 1.15$$

The angle of attack factor is:

$$k_{psk} = 1 + 1.5\left(\frac{\theta}{90}\right)^{0.57} = 1 + 1.5\left(\frac{10}{90}\right)^{0.57} = 1.429$$

The Reynolds Number based on pier width is:

$$R_e = \frac{VD}{\nu} = \frac{3 \times 4}{10^{-6}} = 12000000$$

Therefore, the maximum shear stress around the pier in the given condition is:

$$\begin{aligned} \tau_{\max(\text{Pier})} &= k_{pw} \times k_{psh} \times k_{psp} \times k_{psk} \times 0.094\rho V_1{}^2 \left(\frac{1}{\log R_e} - \frac{1}{10}\right) \\ &= 1 \times 1.15 \times 1 \times 1.429 \times 0.094 \times 1000 \times 3^2 \left(\frac{1}{\log 12000000} - \frac{1}{10}\right) \\ &= 57.36 \text{ Pa} \end{aligned}$$

Initial Rate of Scour

$\dot{z}_{i(\text{pier})}$ is read on the EFA curve at $\tau = \tau_{\max}$, and is 10.25 mm/hr (Figure 23.6s).

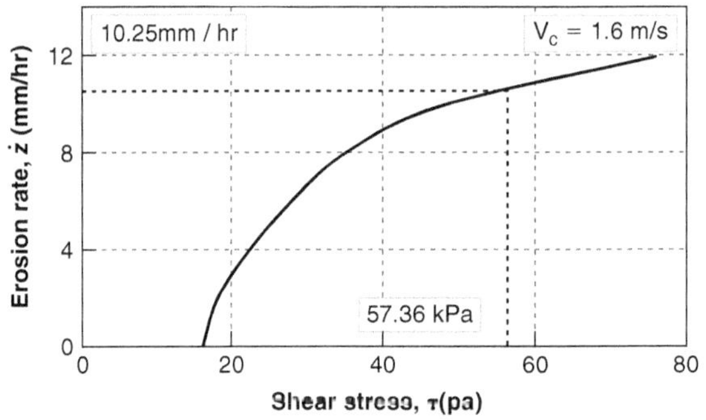

Figure 23.6s Erosion function and initial erosion rate for pier scour.

The depth of pier scour after the 48-hour duration of the 160-year flood can be calculated as:

$$z_{final}(t) = \frac{t\,(hrs)}{\dfrac{1}{\dot{z}_i} + \dfrac{t\,(hrs)}{y_{s(\max)}}} = \frac{48}{\dfrac{1}{10.2} + \dfrac{48}{8757}} = 463.7 \text{ mm}$$

Problem 23.7

Calculate the abutment and contraction scour depth after 48 hours of flood for the following case. The geometry of the channel and the bridge are given in Figure 23.4s. The compound channel is symmetrical, and the discharge during the flood is Q = 2000 m^3/s. The critical velocity of the soil in the main channel and flood plain is 1.2 m/s. The erosion function of the soil from an EFA test is given in Figure 23.5s. The duration of flood is 48 hours, and the hydraulic data are as follows:

Mean velocity in the general approach cross section: V_1 = 1.13 m/s
Mean velocity in the approach floodplain: V_{f1} = 0.78 m/s
Mean velocity in the approach main channel: V_{m1} = 1.4 m/s
Water depth in the approach flood plain: H_{wf1} = 2.55 m
Water depth in the approach main channel: H_{wm1} = 7.9 m
Mean velocity in the general contracted cross section: V_2 = 1.75 m/s
Mean velocity in the contracted main channel: V_{m2} = 1.83 m/s
Hydraulic radius in the approach main channel: R_{h1} = 3.65 m

Find the abutment scour depth and the contraction scour depth after 48 hours of flood.

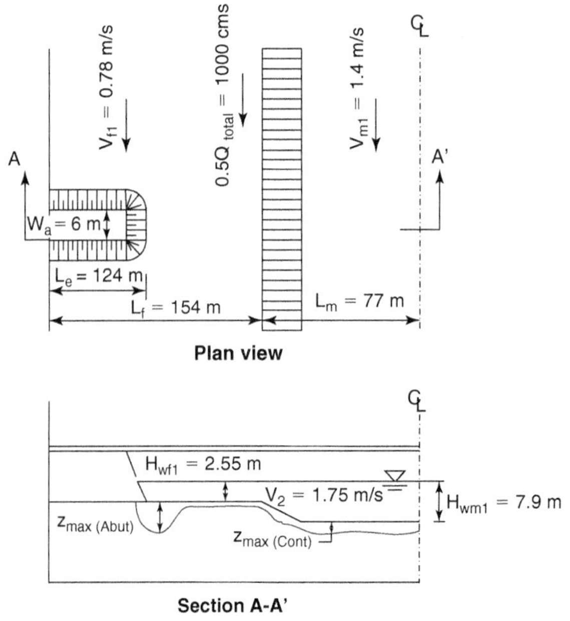

Figure 23.4s Channel geometry.

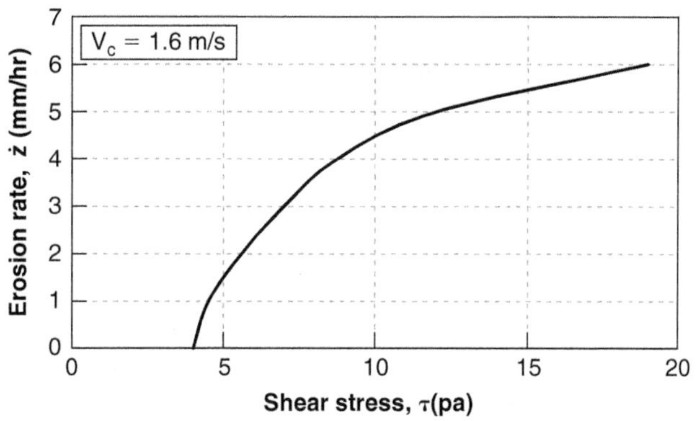

Figure 23.5s Erosion function.

Solution 23.7

Step 1. Calculate the maximum shear stress in the middle of the channel and around the abutment

Contraction scour

The maximum shear stress in the middle of the channel can be calculated as:

$$\tau_{\max(\text{Cont})} = k_{cr}k_{cl}k_{c\theta}k_{cw}\gamma n^2 {V_1}^2 {R_h}^{-1/3}$$

where k_{cr} is the correction factor for the contraction ratio, k_{cl} is the correction factor for the contraction length, $k_{c\theta}$ is the correction factor for the transition angle, and k_{cw} is the correction factor for the water depth.

For this case,

$$k_{cr} = 0.62 + 0.38\left(\frac{A_1}{A_2}\right)^{1.75} = 0.62 + 0.38 \times \left(\frac{V_2}{V_1}\right)^{1.75} = 0.62 + 0.38 \times \left(\frac{1.75}{1.13}\right)^{1.75} = 1.44$$

Width of the channel at approach section:

$$L_1 = (154 + 77) \times 2 = 462 \text{ m}$$

Width of the channel at contraction section:

$$L_2 = (30 + 77) \times 2 = 214 \text{ m}$$

$$k_{cl} = 0.77 + 1.36\left(\frac{6}{462 - 214}\right) - 1.98\left(\frac{6}{462 - 214}\right)^2 = 0.80$$

$$k_{c\theta} = 1 + 0.9\left(\frac{90}{90}\right)^{1.5} = 1.9$$

$$k_{cw} = 1$$

Therefore,

$$\tau_{\max(\text{Cont})} = k_{cr}k_{cl}k_{c\theta}k_{cw}\gamma n^2 {V_1}^2 {R_h}^{-1/3} = 1.44 \times 0.80 \times 1.9 \times 1 \times 9810 \times 0.018^2 \times 1.13^2 \times 3.65^{-1/3}$$
$$= 5.77 \text{ Pa}$$

Abutment scour

The maximum shear stress around the abutment can be calculated as:

$$\tau_{\max(\text{Abut})} = 12.45 \times k_{acr} \times k_{ash} \times k_{aw} \times k_{as} \times k_{ask} \times k_{al} \times \rho \times {V_1}^2 \times \text{Re}^{-0.45}$$

where k_{acr} is the contraction ratio influence factor for abutment scour shear stress, k_{ash} is the correction factor for aspect ratio of the approach embankment, k_{aw} is the correction factor for Froude Number, k_{as} is the correction factor for abutment shape, k_{ask} is the correction factor for the skew angle of the abutment, and k_{al} is the correction factor for abutment location in the flood plain.

For this case:

$$k_{acr} = 3.65\frac{q_2}{q_1} - 2.91 = 3.65 \times \frac{1.75}{1.13} - 2.91 = 2.74$$

$$k_{ash} = 0.85\left(\frac{L_e}{W_a}\right)^{-0.24} = 0.85 \times \left(\frac{124}{6}\right)^{-0.24} = 0.41$$

$$Fr = \frac{V_1}{\sqrt{gH_{wf1}}} = \frac{1.13}{\sqrt{9.81 \times 2.55}} = 0.23 > 0.1$$

Therefore,

$$k_{aw} = 2.07Fr + 0.8 = 2.07 \times 0.23 + 0.8 = 1.27$$

Because this is a spill-through abutment,

$$k_{as} = 0.58$$

$$k_{ask} = 1$$

Because

$$\frac{L_f - L_e}{H_{wf1}} = \frac{154 - 124}{2.55} = 11.76 > 1, k_{al} = 1$$

Therefore,

$$\begin{aligned}\tau_{\max(\text{Abut})} &= 12.45 \times k_{acr} \times k_{ash} \times k_{aw} \times k_{as} \times k_{ask} \times k_{al} \times \rho \times V_1{}^2 \times \text{Re}^{-0.45} \\ &= 12.45 \times 2.74 \times 0.41 \times 1.27 \times 0.58 \times 1 \times 1 \times 1000 \times 1.13^2 \times \left(\frac{1.13 \times 6}{10^{-6}}\right)^{-0.45} \\ &= 11.09 \text{ Pa}\end{aligned}$$

The initial rate of scour z_i for contraction scour and abutment scour are read on the EFA curve at $\tau = \tau_{\max}$: it is 2.01 mm/hr and 4.7 mm/hr respectively as shown in Figure 23.7s.

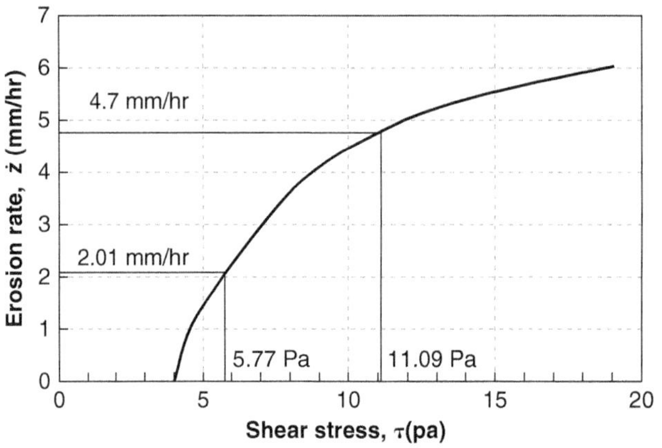

Figure 23.7s Erosion function and initial erosion rate for abutment scour and contraction scour.

Step 2. Calculate the maximum contraction scour depth and the maximum abutment scour depth

Contraction scour

The maximum contraction scour depth can be calculated as:

$$z_{\max(\text{Cont})} = 1.27(1.83Fr_{m2} - Fr_{mc}) \cdot H_{wm1}$$

where Fr_{m2} is the Froude Number for the main channel at the bridge in the contracted zone, Fr_{mc} is the critical Froude Number for the main channel at the bridge, and H_{wm1} is the water depth in the main channel at the approach section.

For this case,

$$Fr_{m2} = \frac{V_1/CR}{\sqrt{gH_{wm1}}} = \frac{V_2}{\sqrt{gH_{wm1}}} = \frac{1.75}{\sqrt{9.81 \times 7.9}} = 0.199$$

$$Fr_{mc} = \frac{V_{mc}}{\sqrt{gH_{wm1}}} = \frac{1.2}{\sqrt{9.81 \times 7.9}} = 0.136$$

Therefore,

$$y_{s(\text{Cont})} = 1.27(1.83Fr_{m2} - Fr_{mc}) \cdot H_{wm1} = 1.27 \times (1.83 \times 0.199 - 0.136) \times 7.9 = 2.29 \text{ m}$$

Abutment scour

The maximum abutment scour depth can be calculated as:

$$y_{s(\text{Abut})} = H_{wf1} \cdot K_{ash} \cdot K_{ask} \cdot K_{al} \cdot K_{ag} \cdot 243 \cdot \text{Re}_{f2}{}^{-0.28}(1.65Fr_{f2} - Fr_{fc})$$

where K_{ash} is the correction factor for the abutment shape, K_{ask} is the correction factor for the abutment skew, K_{al} is the influence factor that takes into account the proximity of the abutment to the main channel, K_{ag} is the geometry of the channel influence factor for abutment scour, Re_{f2} is the Reynolds Number around the toe of the abutment, Fr_{f2} is the Froude Number around the toe of the abutment, and Fr_{fc} is the critical Froude Number for soil near the toe of the abutment.

For this case, assume that it is a spill-through abutment with 2:1 slope, $K_{ash} = 0.73$

$$K_{ask} = 1 - 0.005|\theta - 90| = 1$$

For this compound channel, $K_{ag} = 1$

Because

$$\frac{L_f - L_e}{H_{wf1}} = \frac{154 - 124}{2.55} = 11.8 > 1.5$$

therefore, $K_{al} = 1$

Because

$$\frac{L_f - L_e}{H_{wm1}} = \frac{154 - 124}{7.9} = 3.8 < 5$$

it is a short setback condition. Therefore,

$$V_{f2} = \frac{0.5Q}{A_2} = 0.5 \times V_2 = 0.5 \times 1.75 = 0.875 \text{ m/s}$$

$$\text{Re}_{f2} = \frac{V_{f2} \cdot H_{wf1}}{\nu} = \frac{0.875 \times 2.55}{10^{-6}} = 2.23 \times 10^6$$

$$Fr_{f2} = \frac{V_{f2}}{\sqrt{gH_{wf1}}} = \frac{0.875}{\sqrt{9.81 \times 2.55}} = 0.175$$

$$Fr_{fc} = \frac{V_{fc}}{\sqrt{gH_{wf1}}} = \frac{1.2}{\sqrt{9.81 \times 2.55}} = 0.24$$

and thus

$$
\begin{aligned}
z_{\max(\mathrm{Abut})} &= H_{wf1} \cdot K_{ash} \cdot K_{ask} \cdot K_{al} \cdot K_{ag} \cdot 243 \cdot \mathrm{Re}_{f2}{}^{-0.28}(1.65 Fr_{f2} - Fr_{fc}) \\
&= 2.55 \times 0.73 \times 1 \times 1 \times 1 \times 243 \times (2.23 \times 10^6)^{-0.28}(1.65 \times 0.175 - 0.24) = 0.368 \text{ m}
\end{aligned}
$$

Step 3. Calculate the depth of contraction scour and abutment scour after 48 hours

$$
z_{(\mathrm{Cont})}(t) = \frac{t(hrs)}{\dfrac{1}{\dot{z}_i} + \dfrac{t(hrs)}{z_{\max(\mathrm{Cont})}}} = \frac{48}{\dfrac{1}{2.01} + \dfrac{48}{2290}} = 93 \text{ mm}
$$

$$
z_{\max(\mathrm{Abut})}(t) = \frac{t(hrs)}{\dfrac{1}{\dot{z}_i} + \dfrac{t(hrs)}{y_{s(\mathrm{Abut})}}} = \frac{48}{\dfrac{1}{4.7} + \dfrac{48}{368}} = 140 \text{ mm}
$$

Therefore, the contraction scour depth generated by the 48-hour flood is 4.1% of the maximum contraction scour depth, whereas the abutment scour depth generated by the same flood is 38% of the maximum abutment scour depth.

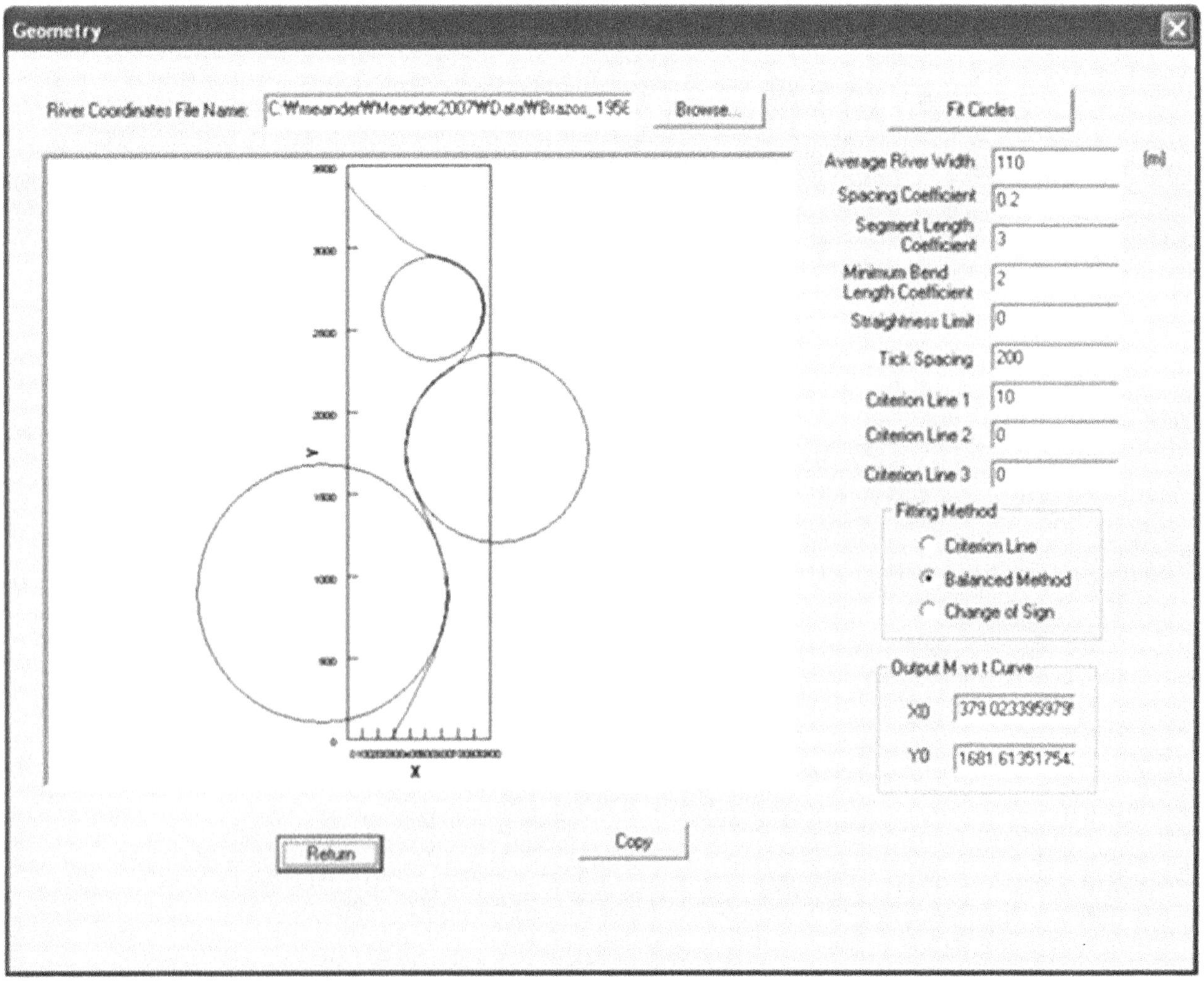

Figure 23.8s Input the geometry data.

Problem 23.8

Download the SRICOS-EFA program from the web site http://ceprofs.tamu.edu/briaud/ and run Example 1 from the list of examples.

Solution 23.8

See the web site.

Problem 23.9

Download the MEANDER program from the web site http://ceprofs.tamu.edu/briaud/ and run Example 1 from the list of examples.

Solution 23.9

1. Install MCRInstaller and run MEANDER.
2. Open the first example, Brazos1958C_const_NGP.meander, in the Data folder included with the program. This example is for migration with a constant discharge.
3. Choose between SI units or English units in Input > Units . . . . Choose the SI Units.
4. Open the *Geometry* window. This window lets you open the file with the initial coordinates of the river and fit circles that represent the meanders. Browse the geometry file Brazos_1958C_2006.dat included in the Data folder, which has the initial coordinates of the river.
5. The numbers in the *Geometry* window (Figure 23.8s) must be: Average River Width is 110 m, the Tick Spacing is 200, and the Criterion Lines 1, 2, and 3 are 10, 0, and 0 respectively. Click Fit Circles. Click Return after the circles are fitted.
6. The next window (Figure 23.9s) lets you input the data soil. Input the EFA curve on the *Soil Data* window and choose the sand option for the type of the soil.

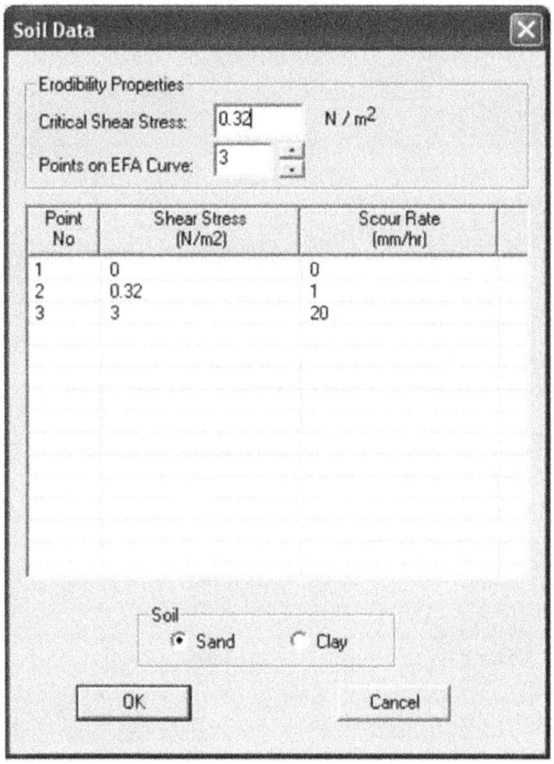

Figure 23.9s Soil data window.

7. Open the *Water Data* window (Figure 23.10s). The Critical Froude Number is 0.17 and the Time Step is 240. The speed of the program depends on this increment. The discharge, in the case of this example, is constant. The discharge units are in cubic meters per second. The time period is one year or 365 days. The discharge versus velocity and the discharge versus water depth have to be obtained from software such as HEC-RAS or TAMU-FLOW. These programs perform their analyses based on the cross section of the river. Click OK after you are done.
8. Before running the program, you may want to check the data again by clicking *Input Tables* and *Input Plots*. These two options let you review your data.
9. Once all the data are in, you can click the Run button. After the program finishes the calculations, click the *Output Plots* icon (Figure 23.11s). Click Center Line or One Bank to see the results of the meander migration (Figure 23.12s).

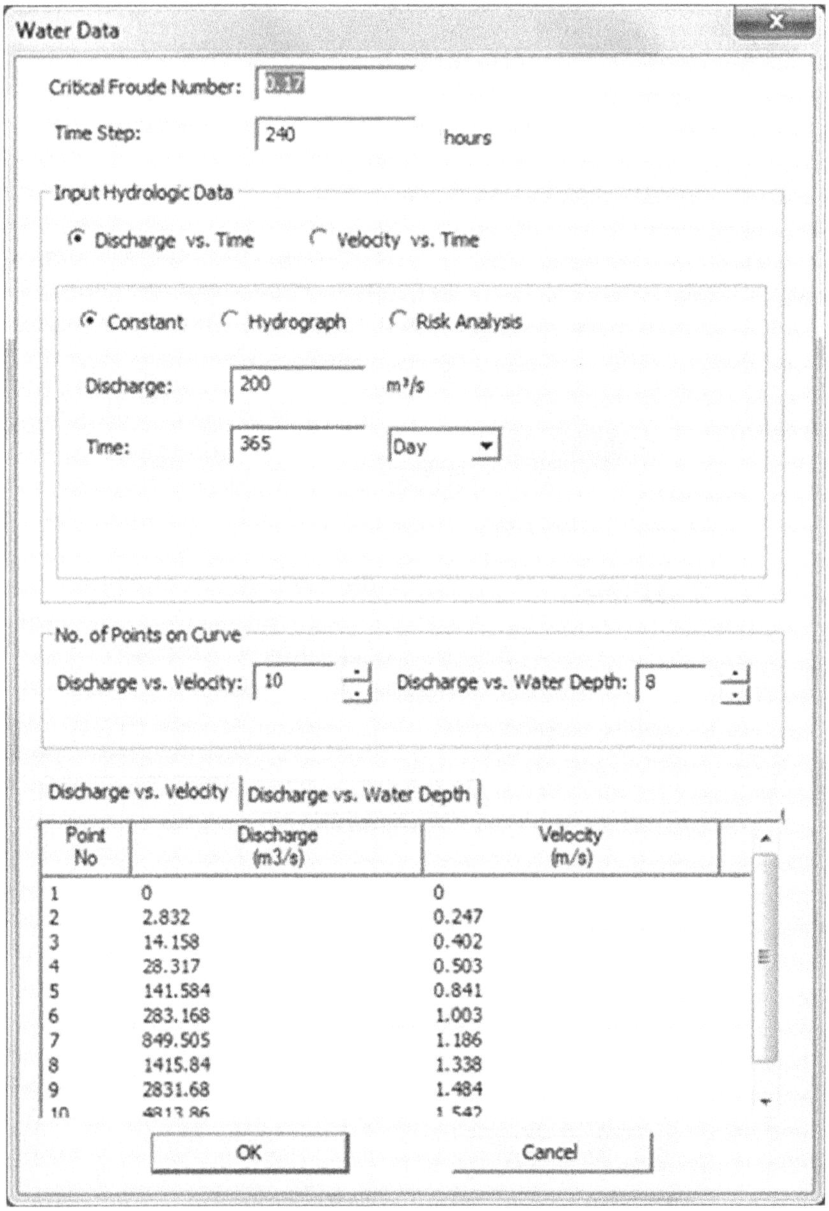

Figure 23.10s Water data input.

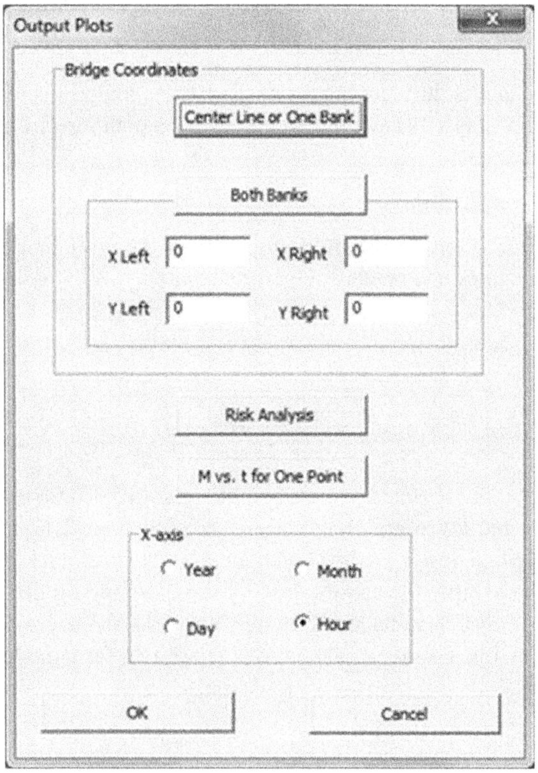

Figure 23.11s Output plots windows.

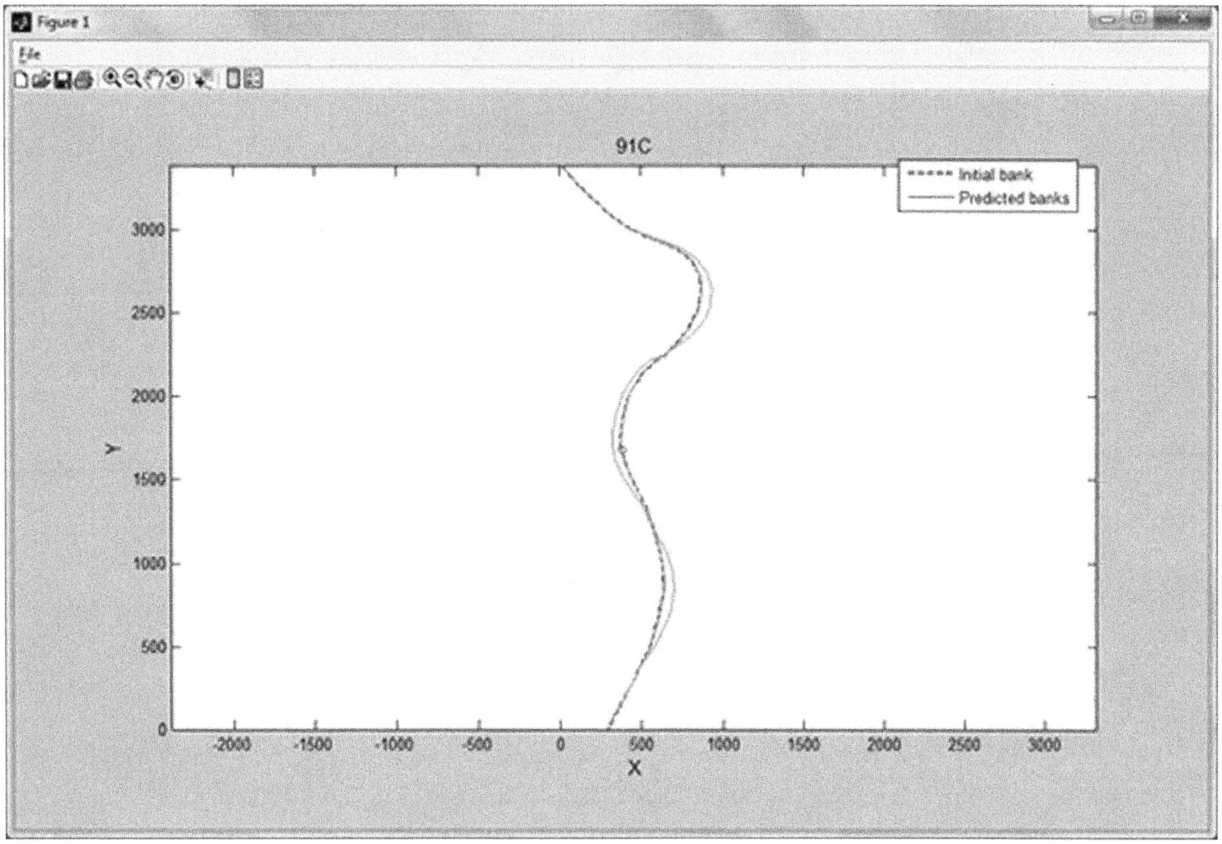

Figure 23.12s Results of the meander migration.

Problem 23.10

A 5 m high levee is overtopped for 2 hours during a hurricane. The levee material and the soil below the levee are borderline between a high-plasticity clay CH and a low-plasticity clay CL. Draw a contour of the levee after 2 hours of overtopping.

Solution 23.10

To draw the contour of the levee after 2 hours of overtopping erosion, we must first find where the erosion will start. When the water overtops the levee, it accelerates and reaches the critical velocity of the soil $Vc = 1.1$ m/s after some distance from the levee crest. This distance is such that:

$$V = \sqrt{2gH}$$
$$1.1 = \sqrt{2 \times 9.81 \times H} \rightarrow H = 0.06 \text{ m}$$

So, erosion will start once the water has reached a levee height equal to 5− 0.06 m = 4.94 m.

Then we select five points along the levee and compute the erosion after 2 hours (problem 23.10) and after 72 hours (problem 23.11). To calculate the erosion depth, we first calculate the water velocity, then find the corresponding erosion rate from the erosion function, and then multiply the erosion rate by the overtopping duration (2 or 72 hours). The levee soil is borderline between a high-plasticity clay CH and a low-plasticity clay CL, so the erosion function is selected as the boundary line on the erosion chart of Figure 23.13s. Example calculations are shown for a depth of 4 m below the crest of the levee.

$$H = 4 \text{ m}$$
$$V = \sqrt{2gH} = \sqrt{2 \times 9.81 \times 4} = 8.86 \text{ m/s} \rightarrow \text{erosion rate} = 58 \text{ mm/hr}$$
$$\text{Erosion depth} \quad z = 58 \times 2 = 116 \text{ mm}$$

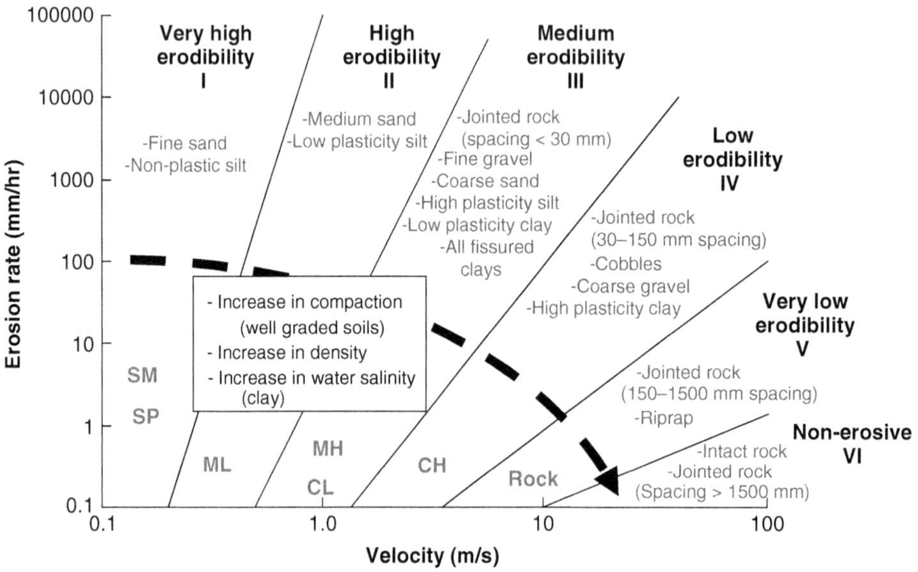

Figure 23.13s Proposed erosion categories for soils and rocks based on velocity.

Table 23.1s shows all the calculations and the contours of erosion after 2 hours and 72 hours of overtopping erosion are shown in Figure 23.14s.

Drop Height m	Velocity m/s	Erosion Rate mm/hr	Erosion Depth after 2 Hr m	Erosion Depth after 72 Hr m
0	0	0	0	0
0.06	1.1	0.1	0.0002	0.0072
1	4.43	6	0.0012	0.432
2	6.26	18	0.036	1.296
3	7.67	35	0.07	2.52
4	8.86	58	0.116	4.176
5	9.9	91	0.182	6.552

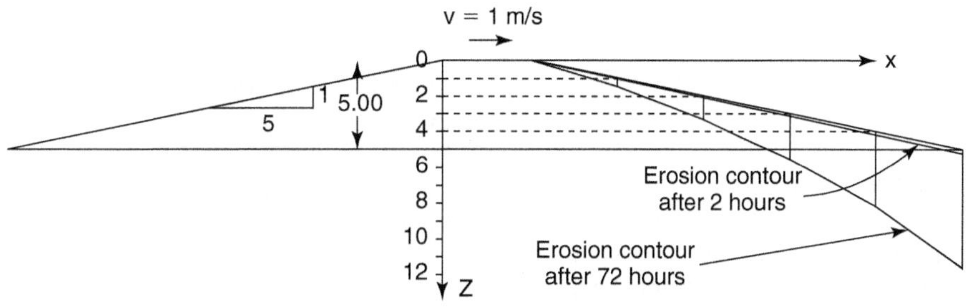

Figure 23.14s Contour of erosion after 2 and 72 hours.

Problem 23.11

Repeat problem 23.10 for a flood that lasts 72 hours.

Solution 23.11

Same approach but multiply by 72 hours instead of 2 hours.

CHAPTER 24

Geoenvironmental Engineering

This chapter introduces a relatively recent yet very important field: geoenvironmental engineering. The book by Sharma and Reddy (2004) is an excellent reference for further study.

24.1 INTRODUCTION

After World War II, there was a remarkable development of industries aimed at improving the quality of life. These chemical and manufacturing industries created a significant amount of waste in an unregulated environment. The advent of nuclear power plants in the late 1950s brought the problem of nuclear wastes, which can remain deadly for thousands of years. Disposal of such deadly wastes caught the attention of the public and emphasized the need for a more organized approach. The Love canal disaster occurred when an old canal that had been filled with toxic chemicals for many years started to leak and seriously affect the health of local residents in New York State in the 1970s. This broadly publicized disaster contributed to the development of laws and regulations aimed at ensuring the health and safety of the public by avoiding any environmental contamination.

24.2 TYPES OF WASTES AND CONTAMINANTS

Wastes are unwanted or useless materials. They come from many different human activities and take many different forms. Sources of wastes include dredging, mining, and farming: they can also be generated by the residential, commercial, institutional, industrial, nuclear power, and defense sectors. The waste generated by those human activities amounts to something like 100 billion kN or 10 billion m^3 per year worldwide. Wastes can be in gas form, liquid form, or solid form. In geotechnical engineering, the liquid and gas forms are of concern because they propagate through the soil that must be cleaned up; this falls under the topics of *contamination* and *remediation*. The solid wastes, hazardous or not, end up being stored in landfills or other facilities designed for that purpose. In the United States alone, the amount of waste generated each year is staggering, as shown in Figure 24.1.

Waste	Meganewtons per Year (MN/YR)
Municipal solid waste	2,500,000
Industrial solid waste	76,000,000
Hazardous waste	3,800,000
Radioactive waste	3,750
Medical waste	4,650

Wastes can be categorized in the following different types (Sharma and Reddy 2004):

1. *Solid wastes*. The term *solid waste* is misleading, as a solid waste can be a solid, a liquid, or a gas. These wastes include municipal solid waste (MSW), industrial solid wastes, and construction and demolition wastes. MSW includes household solid wastes and commercial solid wastes. Figure 24.2 shows the distribution of MSW in percent of the total quantity and its evolution over the past 50 years in the United States. Industrial solid wastes are nonhazardous wastes generated by manufacturing and industrial processes, including the production of furniture, apparel, machinery, busses, trucks, cars, airplanes, jewelry, shoes, and so on. Construction and demolition wastes include wood, concrete, bricks, and plumbing materials, among others.
2. *Hazardous wastes*. These are wastes that can cause death or serious illness or pose a substantial hazard to human health or the environment. In the United States, they are defined in subtitle C of the Resource Conservation and Recovery Act (RCRA). The Environmental Protection Agency (EPA) distinguishes three categories of hazardous wastes: nonspecific source wastes (e.g., solvents, dioxins), specific source wastes (e.g., sludge from petroleum refining and organic chemical manufacturing), and commercial chemical product source waste (e.g., creosote, acids, pesticides). Another important factor is the concentration level of the chemical in

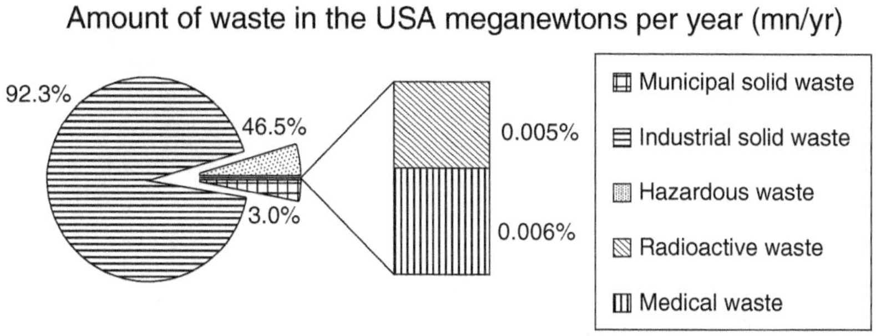

Figure 24.1 Amount of waste per year in the United States (After Sharma and Reddy 2004)

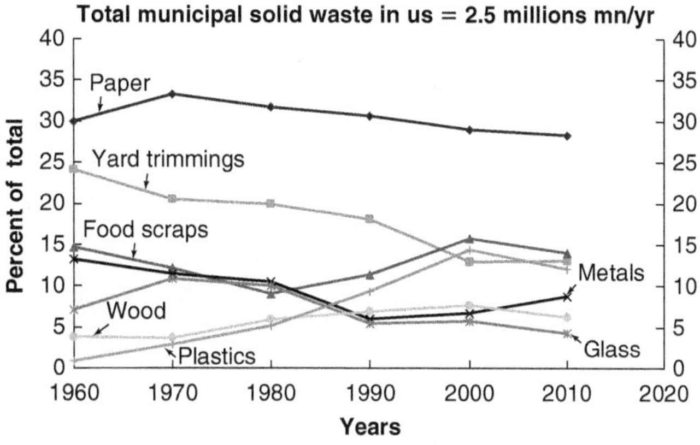

Figure 24.2 Distribution of MSW and its evolution.

the liquid (water); the acceptable level is defined in the regulations. For example, 0.05 milligram of mercury per liter would be acceptable but 0.3 mg/L would not; 1 mg of lead per liter would be acceptable but 7 mg/L would not.

3. *Radioactive waste*. These wastes are classified into four categories of their own:
   a. High-level wastes (HLW). These are liquid or solid wastes that are extremely dangerous and must not come into contact with humans. They come from defense or nuclear power plant activities. They require permanent isolation, as the radiation penetration remains lethal for 10,000 years.
   b. Transuranic wastes (TRU). These come from manufacture of nuclear weapons and processing of nuclear fuels. The radiation penetration remains lethal for 20 years. TRUs are relatively rare.
   c. Low-level wastes (LLW). These are the lowest-level radioactive wastes, which can be disposed of as regular waste after sufficient isotope decay.
   d. Mill tailings. These are the ore residues from mining uranium. There is a significant amount of it and it is typically stored, sometimes buried in large remote areas of a country.
4. *Medical waste*. These wastes come from hospitals and other health agencies. They include microbiological wastes (e.g., infectious cultures), human blood, pathological wastes (e.g., organs, body parts), contaminated animal wastes, isolation wastes (e.g., waste contaminated with highly communicable diseases), contaminated sharps (e.g., needles, scalpels), and uncontaminated sharps. Some MSW, such as disposable diapers and sanitary napkins, do contain pathogens, but usually not as much as medical wastes. One common solution for medical wastes is incineration to kill the disease-causing pathogens.

24.3 LAWS AND REGULATIONS

In the United States, once a bill is proposed, passes both Houses of Congress, and is signed by the president, it becomes a law (also known as an act or statute). To enforce the new act, the responsible agency—the Environmental Protection Agency in most cases for geoenvironmental engineering—develops regulations to implement the act. All federal government regulations are collected in enormous books of the *Code of Federal Regulations;* this series of books is divided into volumes called titles (Title 40 covers topics relating to geoengineering) which are further divided into parts and then sections. The EPA also develops Guidance Documents for technical issues and Policies for decision management to help industry comply with the regulations. Each state can use the federal regulations as is or enforce stricter, state-specific versions. The EPA office in each state is in charge of enforcing the regulations.

Laws have been passed on a wide variety of environmental topics. Some of the most important and with the widest range include:

- Solid Waste Disposal Act (SWDA; 1965, 1970)
- National Environmental Policy Act (NEPA; 1969)
- Occupational Safety and Health Act (OSHA; 1970)
- Clean Air Act (CAA; 1970, 1977, 1990)
- Clean Water Act (CWA; 1977, 1981, 1987)
- Safe Drinking Water Act (SDWA; 1974, 1977, 1986)
- Resource Conservation and Recovery Act (RCRA; 1976, 1980)
- Comprehensive Environmental Response, Compensation, and Liability Act (CERCLA; 1980)

These laws have positively impacted geoenvironmental engineering practice. In particular, CERCLA, also known as Superfund, generated a tremendous amount of work for geotechnical engineers regarding the cleaning of soil deposits. As society evolves and humankind creates new products, new laws are enacted and amendments are made to existing acts. This is the case of the Hazardous and Solid Waste Amendments to RCRA (HSWA; 1984) and the Superfund Amendments and Reauthorization Amendments to CERCLA (SARA; 1986). CERCLA/Superfund addressed the issue of cleaning up contaminated sites, whereas RCRA addressed the issue of landfill design.

Safety and health are the top priority regarding investigation and remediation at contaminated sites. The following levels of protection for humans working on contaminated sites have been formulated by the Occupational Safety and Health Administration (OSHA) in the United States. They address the level of protection required for eye, skin, and respiratory safety (Table 24.1).

1. *Level D*. Minimum protection including coveralls, gloves, chemical-resistant steel-toed boots, safety glasses, hard hat, escape mask, and face shield.
2. *Level C*. Moderate protection, including full-face or half-mask, air-purifying respirator, hooded chemical-resistant clothing, inner and outer chemical-resistant gloves, chemical-resistant boots and boot covers, hard hat, escape mask, and face shield.
3. *Level B*. Very high protection, including positive-pressure, full-face, self-contained breathing apparatus (SCBA), hooded chemical-resistant clothing, chemical-resistant inner and outer gloves, chemical-resistant steel-toed boots and boot covers, hard hat, and face shield.
4. *Level A*. Maximum protection, including positive-pressure, full-face, self-contained breathing apparatus (SCBA), totally encapsulating chemical-protective suit, chemical-resistant inner and outer gloves, chemical-resistant steel-toed boots and boot covers, and hard hat (under suit).

Table 24.1 Levels of Safety and Health Protection (OSHA)

Protection	Level A	Level B	Level C	Level D
Respiratory	Maximum	Maximum	Moderate	Minimum
Skin	Maximum	Very high	Moderate	Minimum
Eye	Maximum	Very high	Moderate	Minimum

24.4 GEOCHEMISTRY BACKGROUND

24.4.1 Chemistry Background

A discussion of geochemistry starts with a background on chemistry. Electrons, protons, and neutrons make up the universe. They are extremely small subatomic particles with masses of 1.66×10^{-27} kg for neutrons and protons and 9.1×10^{-31} kg for the much lighter electron. Electrons are negatively charged, protons are positively charged, and neutrons are electrically neutral. Atoms, which consist of electrons, protons, and neutrons in various combinations, are the basic building blocks of matter. Hydrogen, oxygen, and carbon, for example, are atoms. You can break them down into smaller pieces, but you will no longer have hydrogen, oxygen, or carbon. The nucleus of the atom is made of neutrons and protons, while electrons gravitate around the nucleus. An *element* is made of only one kind of atom. The periodic table (Figure 24.3) in chemistry gives the list of the different elements, 92 of which occur naturally (e.g., hydrogen, oxygen, carbon) and some two dozen others of which have been created by scientists. By changing the number of protons and electrons in the atom, you create different elements. For example, oxygen has 8 protons, but hydrogen has only 1. This is an element's *atomic number*. The periodic table gives the atomic number (number of protons in the nucleus) and the atomic mass, which is the sum of the weight of the protons and the neutrons times the Avogadro number. (Amadeo Avogadro was an Italian scientist who contributed significantly to molecular theory in the early 1800s.) Note that the weight of the electrons is negligible compared to the weight of the protons and neutrons. Stable atoms have the same number of electrons and protons.

The Avogadro number (6.022×10^{23}) is the number of atoms in 12 grams of carbon12. This number helps define the mole. In biology a *mole* is an underground rodent, but in chemistry it is the mass corresponding to 6.022×10^{23} (Avogadro number) molecules, atoms, or some other pure chemical substance unit. The periodic table gives the mass of one mole of each of the elements. For example, a mole of water (H_2O) has a mass of two moles of hydrogen (2×1.00794 g) plus one mole of oxygen (15.9994 g) or 18.01528 g. Note that, in chemistry, mass is calculated with a much larger number of significant figures than in geotechnical engineering. The

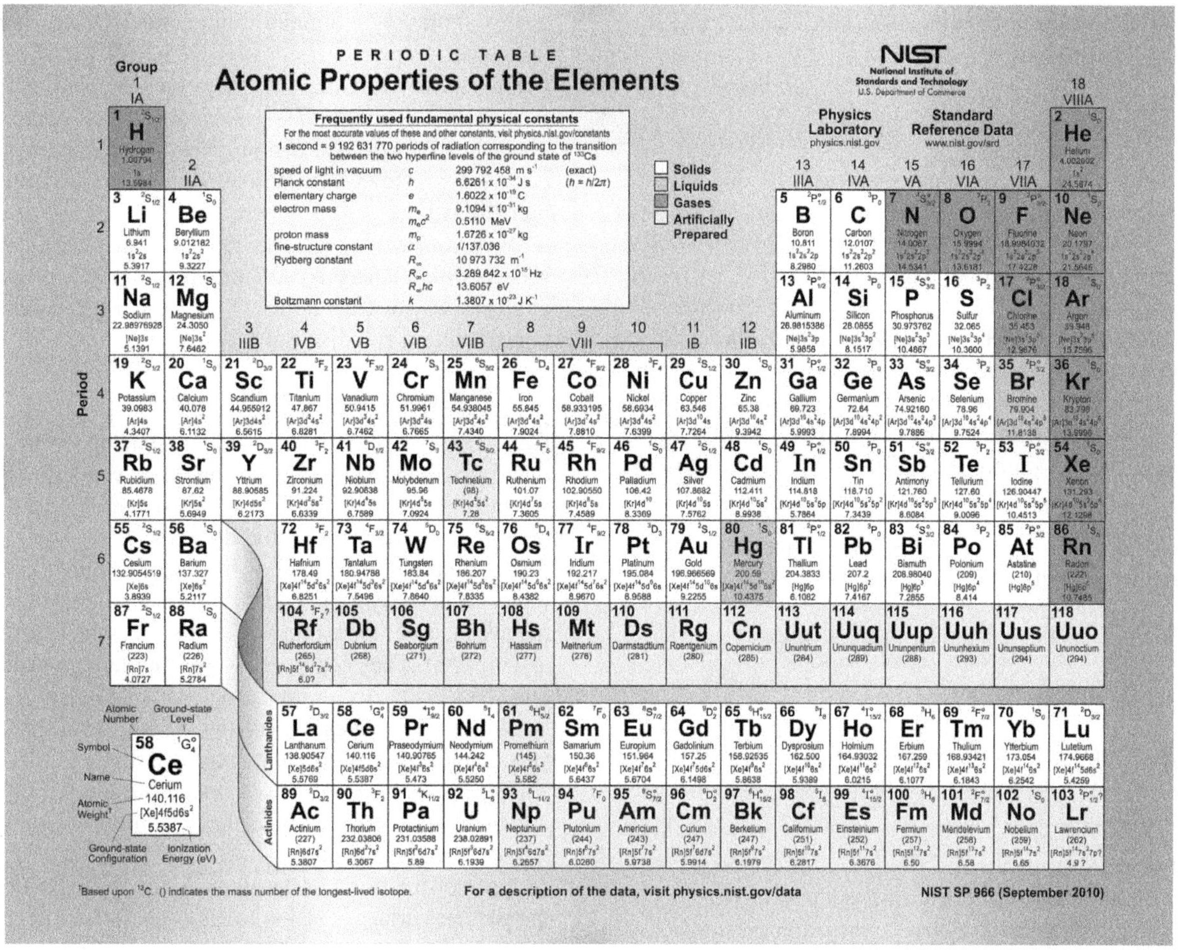

Figure 24.3 Periodic table of elements. (Courtesy of National Institute of Standards and Technology, U.S., Department of Commerce)

reason is the reproducibility of the results and the need to be that accurate.

Molecules are combinations of atoms. Two atoms of oxygen (O), for example, form the molecule O_2; two hydrogen atoms and an oxygen atom form a molecule of water (H_2O); a molecule of sugar is $C_{12}H_{22}O_{11}$. If an atom or a molecule loses or gains an electron on its outer orbital shell, it becomes an *ion*, either a cation or an anion. *Cations* have lost electron(s), are positively charged, and move toward the cathode because a cathode is negatively charged in an electrical circuit. Examples of cations are Na^+, Ca^{++}, and Al^{+++}. Anions have gained electron(s), are negatively charged, and move toward the anode because an anode is positively charged in an electrical circuit. Examples of anions are Cl^-, O^-, and N^-. *Radicals* are groups of atoms that are common to many molecules. Such radicals include hydroxyls (OH^-), carbonates ($CO_3{}^-$), sulfates ($SO_4{}^-$), and nitrates ($NO_3{}^-$). Molecules are bound by ionic bonds (attraction between ions having opposite charges) or covalent bonds (bound by shared electrons). The *valence* or oxidation number of an element in a molecule is the number of electrons transferred to or from an atom of the element with atoms of other elements in the molecule (e.g., O^{2-}, H^{1+}, C^{4+}). The *molecular mass* is the sum of the atomic masses of the elements in the molecule.

Concentration of a chemical in solution can be defined in one of three ways:

1. *Molar concentration*: number of moles per liter of solution.
2. *Mole fraction*: number of moles of a substance divided by the total number of moles in the solution.
3. *Mass concentration* (most commonly used in geoenvironmental engineering): mass of the element or substance in milligrams per liter (mg/L) of solution. If the solution is water (mass density = 1000 g/L), then the mass concentration is mg per 1,000,000 mg of solution or part per million (ppm). The unit of ppm is commonly used in geoenvironmental engineering. So, for the mass concentration of a chemical in water:

$$1 \text{ ppm} = 1 \text{ mg/liter} \tag{24.1}$$

The pH, a measure of acidity, gives the concentration of hydrogen ions by:

$$pH = -\log(H^+) \qquad (24.2)$$

where log is the logarithm base 10 and (H^+) is the concentration of hydrogen ions in moles per liter of solution—or, more accurately, the hydrogen ions' activity, which can be slightly different. The pH scale varies from 0 to 14, with a neutral solution (distilled water) having a pH of 7. A pH less than 7 is acidic and more than 7 is basic or alkaline. Example of pH values include battery acids pH = 1, vinegar and lemon juice 2.5, wine 3.2, beer 4, human blood 7, baking soda 8.5, soap 10, and ammonia 12. The p in pH is said to stand for *potential* and pH for *potential of hydrogen*.

Chemical reactions take place when reactants are transformed into products. For example:

$$NaOH + HCl \rightarrow NaCl + H_2O$$

$$\text{Sodium Hydroxide} + \text{Hydrochloric Acid} \rightarrow \text{Salt} + \text{Water} \qquad (24.3)$$

These chemical reactions must satisfy conservation of mass on both sides of the reaction equation. There are at least four types of chemical reactions:

1. *Acid-base reaction*. These reactions affect the pH of the soil and groundwater. Eq. 24.3 is an example.
2. *Precipitation-dissolution reactions*. Some chemicals are more soluble (sugar and salt) in water than others (oil). This is important in geoenvironmental engineering, as it can affect a remedial operation.
3. *Oxidation-reduction reactions* (redox reactions). *Oxidation* is the loss of electrons by a molecule, atom, or ion. *Reduction* is a gain of electrons. For example, oxidation of carbon can produce carbon dioxide, CO_2, but reduction of carbon can produce methane, CH_4. Both gases are produced by the long-term degradation of landfills.
4. *Complexation reactions*. These are chemical reactions that take place between a metal ion and a molecular or ionic entity known as a *ligand*. The properties of these complexes, including solubility, can be quite different from the properties of the metal itself, and such transformations can help in cleanup strategies.

Inorganic chemistry and organic chemistry are two very important branches of chemistry. The difference between organic and inorganic compounds is that most organic compounds contain carbon, whereas most inorganic compounds do not. Carbon is the fourth most abundant element on and in our planet, and organic compounds vastly outweigh (100 to 1?) inorganic compounds. Living organisms, petroleum and its derivatives, plastics, rubber, fat, sugar, proteins, and enzymes are examples of organic compounds. Metals and salt are examples of inorganic compounds. Nuclear chemistry is another branch of chemistry. Elements with high atomic numbers, like uranium, tend to be unstable and break down. During this process, these elements, called *radionuclides*, emit radiation (α rays, β rays, γ rays) with an intensity that decays very slowly. A radionuclide is characterized by its half-life, which is the time required for 50% of the atoms in a substance to decay into more stable substances. All these aspects of chemistry are important to the geoenvironmental engineer who wishes to select the best response possible to contamination and disposal problems.

24.4.2 Geochemistry Background

Geochemistry is the application of chemistry to the field of geoenvironmental engineering. It is concerned with the interaction between chemicals and soils at temperatures and pressures associated with soil deposits.

From the point of view of inorganic chemistry, contamination of soils by toxic metals is the main issue. These toxic metals include, for example, lead (Pb), mercury (Hg), and arsenic (As). They can be dissolved in the pore water (aqueous phase), attached to the particle surface (adsorbed phase), or stuck in the pores as separate solids (solid phase). The geochemical processes controlling the distribution of metals among the three phases include the four chemical reactions mentioned in section 24.4.2, plus adsorption and desorption. *Adsorption* is the accumulations of ions on the charged surface of soil particles; *desorption* is the decrease of ions on the particle surface. The impact of each of these geochemical processes on the contaminant and the soil should be carefully evaluated before any remediation decision is made. This can be done on samples in the laboratory or by computation and modeling.

The total concentration of metal in soil is obtained by washing the soil with an acid and using atomic absorption spectrophotometry (AAS), for example, to study the leachate. The toxicity characteristics leaching procedure (TCLP) simulates the leaching that a waste might undergo when disposed of in a landfill. The TCLP is used to determine if a waste is hazardous or not and to determine the necessary level of treatment if it is hazardous. The different types of metals present in the soil can be identified by sequential extraction, which consists of using solutions of increasing strength and analyzing the leachate. The metal concentration is the mass of metal divided by the mass of dry soil expressed in mg/kg or ppm (parts per million) or in μg/kg or ppb (parts per billion).

From the point of view of organic chemistry, contamination of soils by hydrocarbons is the main issue. These hydrocarbons include, for example, benzene, toluene, and xylene. Organic contaminants also include polychlorinated biphenyls (PCBs), and pesticides (aldrin, endrin). A commonly encountered group of hydrocarbon contaminants is the nonaqueous phase liquids (NAPLs), which do not mix with water (e.g., oil). NAPLs are further separated into those that float on water, called light NAPLs or LNAPLs; and those that sink through water, called dense NAPLs or DNAPLs. LNAPLs typically come from spills of fuels like gasoline, kerosene, or diesel, whereas DNAPLs come from degreasing, metal stripping, and pesticide manufacturing.

NAPLs are found in soils in the gas phase, in the liquid phase, and attached to the surface of the particle. Transformation from one phase to another involves volatilization (liquid to gas), dissolution (mixing in water), adsorption (attachment to particle surface), and biodegradation. *Biodegradation* is a redox reaction that is particularly suited to the action of microbes on NAPLs dissolved in water. The properties of NAPLs are studied in the laboratory and include density, viscosity, solubility, volatility, and surface tension. These properties all affect the optimization of the remedial measure. One way to quantify the amount of NAPLs in soils is to measure the NAPLs degree of saturation S_{NAPL}, defined as:

$$S_{NAPL} = \frac{V_{NAPL}}{V_v} \tag{24.4}$$

where V_{NAPL} is the volume of NAPLs in the voids and V_v is the total volume of voids. Another way to find out how many NAPLs are in the soil is to wash the soil with a solvent and then analyze the solution obtained by a process such as gas chromatography-mass spectrometry (GC-MS).

24.5 CONTAMINATION

24.5.1 Contamination Sources

Contamination can be due to sources on the ground surface, in the zone above the water table (vadose zone; *vadosus* means shallow in Latin), or in the zone below the water table. On the ground surface, sources include infiltration of contaminated surface water, land disposal of liquid or solid wastes, accidental spills, fertilizers, pesticides, disposal of sewage, wastewater treatment plant sludge, salt used on roads in icy conditions, animal feedlots, and fallout from automobile emissions. In the vadose zone, sources include landfills, surface impoundments, leakage from underground storage tanks (e.g., service station tanks and septic tanks), leakage from underground pipelines, and disposal at the bottom of shallow excavations. Below the groundwater level, sources include deep well injections, mines, abandoned oil wells, and disposal in deep excavations. Both the soil particles and the groundwater can potentially be contaminated, and the most serious contaminants are heavy metals, hydrocarbons, and radionuclides. Although the total contaminated land area may be a fraction of a percent of a country's total surface, it is important to remediate all sites. The cost of remediation can be very high, with an estimated average in the range of $1 million per site. As there are some hundreds of thousands of such sites, the cost could potentially reach hundreds of billions of dollars. Whatever the cost, and wherever the site, the cleanup or remediation process starts with detection of the contamination.

24.5.2 Contamination Detection and Site Characterization

The following are the steps in the remediation process:

1. Detection of the contamination
2. Establishment of the vertical and horizontal extent of the contamination
3. Identification of the contaminants
4. Assessment of the risk and impact
5. Choice and design of the remediation scheme
6. Execution of the remediation work
7. Verification of the solution

Environmental site assessments (ESAs) are part of the detection process. An ESA may be required when purchasing a piece of property in the United States. There are three levels:

ESA I. This phase consists of collecting information regarding previous ownership and prior use, using records of contaminated sites in the area, aerial photos, geologic and topographic maps, visit(s) to the site, and talking to neighbors. An ESA I indicates whether there are reasons to believe the site may be contaminated. If so, ESA II comes into play.

ESA II. This phase consists of testing the soil and the groundwater to find out if there is contamination and, if so, to what extent and to what level of severity (type of contaminants). If it is found that there is contamination requiring cleanup, ESA III comes into play.

ESA III. This phase consists of designing and implementing the remediation scheme, including verification that satisfactory level of cleanup has been achieved.

The plan for an environmental site characterization always includes a Safety and Health plan (S&H) and a Quality Assurance-Quality Control plan (QA-QC). The site characterization can make use of drilling and sampling methods, geophysical methods, or in situ testing methods. Drilling and sampling is described in Chapter 6 for noncontaminated sites. Drilling can be done by hollow stem auger drilling, wet rotary drilling, or air pressure rotary drilling, but the hollow stem auger is usually favored for contaminated sites. The reason is that it can be used dry, and minimizes the amount of contaminated fluid generated and the associated disposal cost. For most levels of contamination, the drill rig must be decontaminated after each boring. This is accomplished by pressure steam-washing the rig and washing the drilling and sampling tools with a strong detergent solution and rinsing with clean water. The purpose of this cleaning process is to avoid cross-contamination between borings. Another difference in the case of contaminated sites is that the drillers have to wear the appropriate level of protection (see section 24.3). The soil samples are the same as in the case of uncontaminated sites, but a much stricter chain of custody is followed for the samples. The *chain of custody* is a documentary trail that follows the sample through its entire life, including when

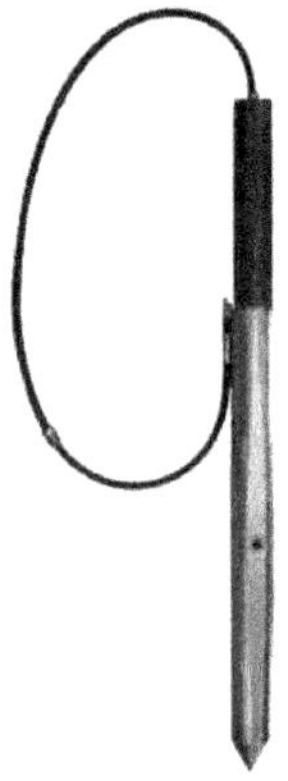

Figure 24.4 LED-induced fluorescence cone penetrometer probe. (Courtesy of Vertek, A Division of Applied Research Associates, Inc.)

and where it was taken, who was responsible for it, what happened to it, and everything else, all the way to final disposal. When two people are involved in the transfer of a sample, they both sign and date the chain of custody document.

Geophysical methods are described in Chapter 8 for noncontaminated sites. They are useful for determining the large-scale stratigraphy of the site and therefore the boundaries of the potential contamination. Surface geophysical methods are particularly convenient for contaminated sites because they are nonintrusive.

In situ testing methods are described in Chapter 7 for noncontaminated sites. In the case of contaminated sites, the cone penetrometer is particularly useful because it limits the amount of contaminated cuttings and contaminated water generated during testing. As a result, a number of techniques have been developed for using the CPT at such sites. One of them is the characterization of petroleum-contaminated sites with laser-induced fluorescence or LIF (ASTM D6187). In this test, the CPT probe is equipped with a side window (Figure 24.4) and a laser beam shines on the soil as the cone is penetrating at 20 mm/s. The laser beam causes the soil and the hydrocarbon to generate fluorescence, which is measured. Each type of hydrocarbon has a "fingerprint" or "signature" signal in terms of intensity and wave length, as shown in Figure 24.5 and Table 24.2. Side-by-side LIF soundings give the extent of the contaminated plume (Figure 24.6).

Another CPT technique adapted for contaminated sites is the BAT water sampler (Figure 24.7). BAT is the name of a company. In this case, the CPT probe is equipped with a porous filter that is obstructed until the CPT probe is pushed to the required depth. Then the filter is exposed and water is allowed to penetrate through the filter into a water sampling tube that can be removed through the CPT rods when full. Drawing the water into the sampling tube is accomplished by using the suction of a hypodermic needle.

Yet another CPT technique developed for contaminated sites is the MIP gas sampler (Figure 24.8). MIP stands for *membrane interface probe*. The cone penetrometer probe is

Table 24.2 Some Values of Fluorescence for Hydrocarbons

Aromatic Hydrocarbon	Molecular Weight (g/mole)	Fluorescence Range (nm)	Fluorescence Color
Toluene	92	270–310	Faint purple
Naphthalene	128	310–370	Blue
Anthracene	178	370–470	Blue-green
Benzo(a)	252	400–500	Green
Pyrene	252	440–530	Green
Perylene	228	470–580	Green-yellow

(Vertek, www.vertekcpt.com/hammerable-fuel-fluoresence-detection-cpt-hydrocarbon-probe)

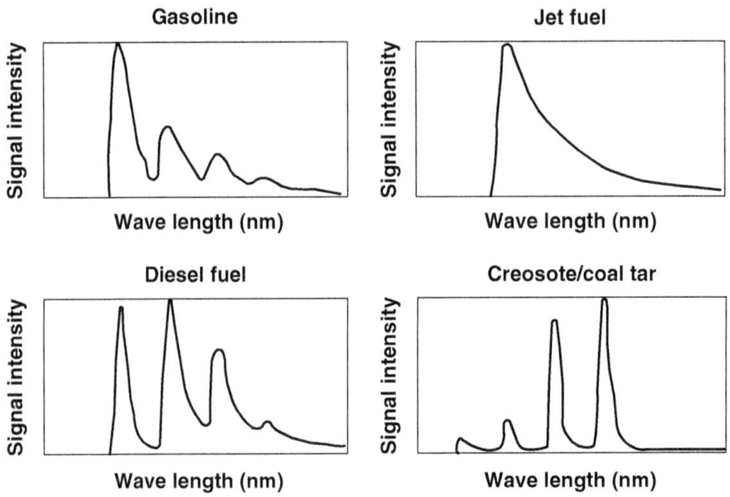

Figure 24.5 Fluorescence response of several hydrocarbons. (Courtesy of Fugro.)

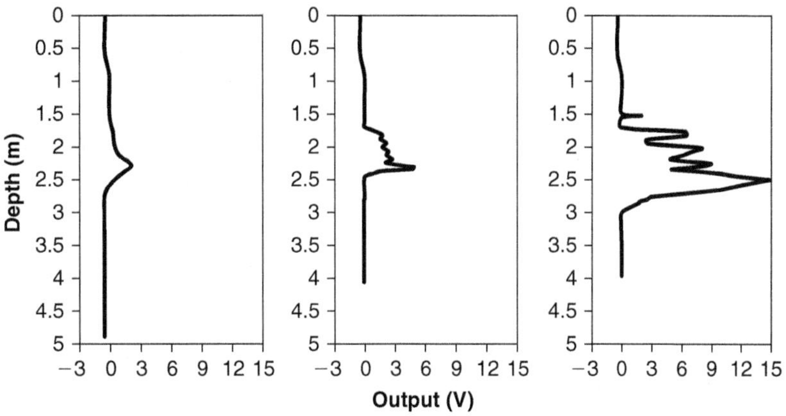

Figure 24.6 LIF CPT soundings and plume identification.

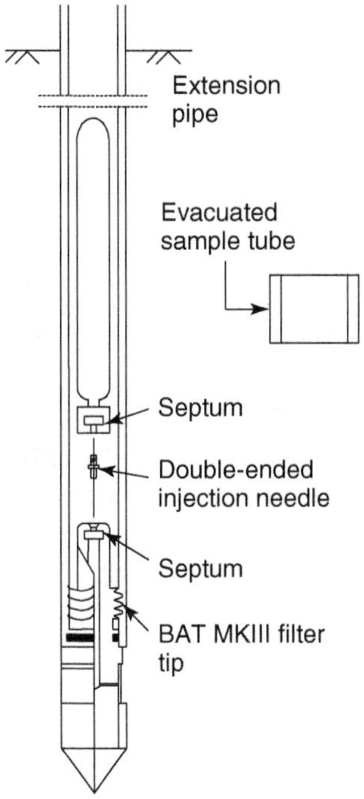

Figure 24.7 BAT CPT water sampler.

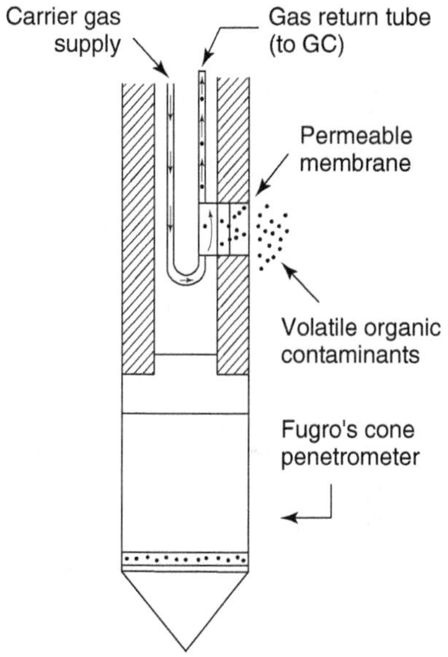

Figure 24.8 MIP CPT gas sampler.

fitted with a hydrophobic, semipermeable membrane and a heater. The heater is kept at a temperature higher than 100°C and vaporizes any volatile organic compound (VOC). The natural pressure gradient created by the heat source forces the gas to penetrate through the semipermeable membrane. Once through the membrane, the gas is swept by an inert carrier gas to the surface, where it is analyzed by a series of detectors.

The CPT soil sampler is also convenient for contaminated sites (Figure 24.9). The test consists of pushing a cone penetrometer to the depth where a sample is needed. At the required depth, the cone tip insert is disengaged, the CPT probe is advanced to collect the soil sample while the cone tip stays stationary as in a piston sampler, and then the CPT probe is pulled back to the surface. Such CPT samples are about 40 mm in diameter and up to 400 mm in length.

Hydrogeologic data are also very important, as contamination of the groundwater is of great concern to human life. There are two components to these studies. One deals with determining where the groundwater table is and which way the water is flowing (if at all). The other is a determination of the hydraulic conductivity of the soil. Determining the groundwater level is done through the use of monitoring wells and piezometers, as described in section 6.6; in situ hydraulic conductivity measurements are covered in section 7.12.

Finally, the chemical analysis of any soil samples and water samples collected has to be conducted under controlled conditions. This chemical analysis aims at identifying the type and concentration of the chemicals in the ground. The final report should include the geologic data, the soil data, the hydrogeologic data, and the chemical data. It should identify the type of contamination, the extent of the contamination, and the future movement of the contaminants.

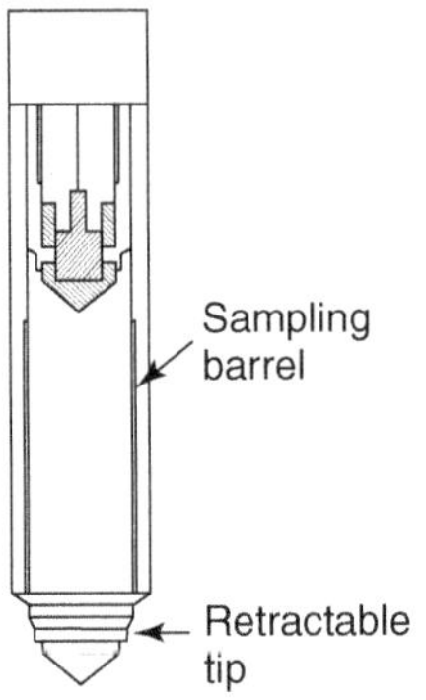

Figure 24.9 CPT soil sampler.

24.5.3 Contaminant Transport and Fate

The word *transport* refers to the flow of the contaminant and the word *fate* to the change in form and concentration of the contaminant through chemical reactions. For noncontaminated sites, the steady-state (no influence of time) flow of water through soil is governed by Darcy's law (constitutive law) and conservation of mass (fundamental law). These two equations, combined with boundary conditions and soil properties (hydraulic conductivity), give the solution to steady-state flow problems. This topic was covered in Chapter 13. For transient (influence of time) flow of water through soil, additional equations must be used. Consolidation of saturated layers is an example of transient flow; in that case the additional equation is the one linking the stress applied to the change in volume of the element. This change of volume (settlement) varies with time and adds to the volume of water flowing through the soil. This problem is presented in sections 11.4.6 and 14.14.

For contaminant flow through soil, an additional issue is the movement of the contaminant in terms of concentration. The contaminant may be found in the gas phase, in the liquid phase, or in the solid phase. This movement can take place through transport processes, chemical reaction processes, and biological processes.

Transport Processes

Transport processes include advection, diffusion, and dispersion.

Advection. *Advection* is the movement of the contaminant liquid under a hydraulic gradient. It is the same case as water flow, with Darcy's law and conservation of mass as the equations. One difference is that the seepage velocity v_s is used rather than the discharge velocity v (see section 13.2.1) because v_s represents the actual velocity of the contaminant movement:

$$Q = vA = v_s A_v \quad or \quad v = nv_s \tag{24.5}$$

where Q is the flow, A is the total cross-sectional area, A_v is the area of the voids, and n is the soil porosity.

The concentration C of the contaminant is defined as:

$$C = \frac{M}{V} \tag{24.6}$$

where M is the mass of contaminant (solute) within the volume V of the liquid carrying the contaminant (solvent). Associated with the concentration C is the contaminant mass flux F, defined as the mass of contaminant passing through a unit area of soil per unit time. The area is the area perpendicular to the flow direction:

$$F_{adv} = \frac{M}{At} = \frac{M}{V}\frac{x}{t} = Cv = nv_s C \tag{24.7}$$

where F_{adv} is the mass flux of contaminant due to advection (subscript *adv*), M is the mass of the contaminant, A is the total cross-sectional area perpendicular to the flow, t is the increment of time considered, and x is the distance travelled by the contaminant during t; other parameters are defined in Eqs. 24.5 and 24.6. Then the governing differential equation is:

$$\frac{dC}{dt} = -nv_s \frac{dC}{dx} \tag{24.8}$$

Diffusion. *Diffusion* refers to the change in concentration of a contaminant due to a chemical gradient. If you put a drop of colored food dye in a glass of water, you will observe diffusion of the dye in the water. This is due to the initial difference in concentration (gradient) of dye between locations in the glass. Another example is the intrusion of seawater into freshwater aquifers (Figure 24.10). This diffusion process is described by Fick's law, which states that the contaminant mass flux is linearly proportional to the change in mass concentration between two points:

$$F_{dif} = -nD^* \frac{dC}{dx} \tag{24.9}$$

where F_d is the mass flux of contaminant due to diffusion (subscript *dif*), n is the soil porosity, D^* is the diffusion

Figure 24.10 Example of diffusion: Seawater intrusion in aquifer. (Copyright © 1995-2013 GEO-SLOPE International Ltd. All Rights Reserved.)

coefficient, and dC/dx is the concentration gradient. Adolph Fick was a German-born physician who proposed this law in 1855. Remarkably, the value of D^* does not vary much, with a range of 1×10^{-9} to 2×10^{-9} m^2/s (Mitchell 1976). Using Fick's law and the equation of continuity gives the governing differential equation:

$$\frac{dC}{dt} = nD^* \frac{d^2C}{dx^2} \tag{24.10}$$

Note that this equation is the same as the governing differential equation for the consolidation theory (Chapter 11, Eq. 11.56), except that C is replaced by the excess water stress u_{we} in the consolidating layer and nD^* is replaced by the coefficient of consolidation c_v. The solution to this equation for simple boundary conditions was proposed by Crank (1956).

Dispersion. *Dispersion* refers to the fact that the velocity is not the same at all points in the flow field (Figure 24.11). This creates a problem similar to the diffusion component where the concentration C varies with the distance (longitudinal or transversal), and is written as:

$$F_{dsp} = -nD\frac{dC}{dx} = -n\alpha v_s \frac{dC}{dx} \tag{24.11}$$

where F_{dsp} is the mass flux of contaminant due to dispersion (subscript dsp), n is the soil porosity, D is the dispersion coefficient, α is the dispersivity, v_s is the seepage velocity, and dC/dx is the concentration gradient along the flow. Then the change in concentration with time is given by:

$$\frac{dC}{dt} = n\alpha v_s \frac{d^2C}{dx^2} \tag{24.12}$$

The total mass flux at one point is the sum of the three components. The soil parameters entering into the equations (n, D^*, α) can be measured through laboratory tests. One such test is the column test, in which the contaminant is injected at the top of the column and concentration measurements are made on the effluent at the bottom of the column. The soil parameters can also be obtained from in situ tests where a dye is injected at one location and the dye concentration is checked as a function of time in adjacent monitoring wells.

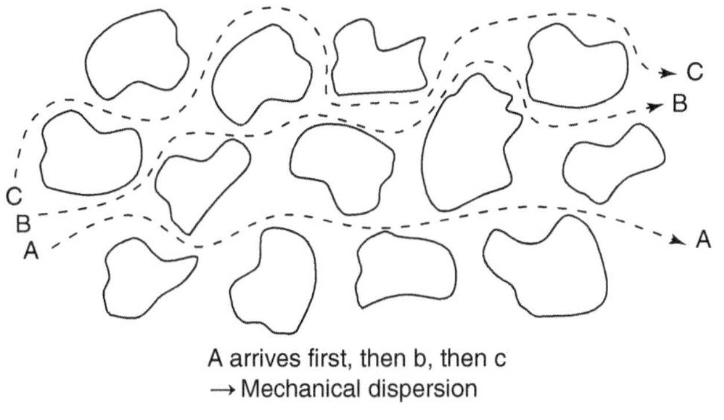

Figure 24.11 Mechanical dispersion.

Chemical Reaction Processes

Chemical reaction processes include many different types of chemical reactions.

Sorption and Desorption. *Sorption* is the process by which the contaminant becomes attached to the surface of the soil particle. *Desorption* is the process by which the contaminant gets detached from the surface of the particle. Empirical equations link the mass of contaminant sorbed per unit dry mass of soil S (e.g., mg/kg or ppm) to the concentration of contaminant in solution at equilibrium C (e.g., mg/L). Due to sorption, the contaminant velocity v_c is slower than the seepage velocity v_s; in other words, the contaminant movement is slowed down and the water goes faster than the contaminant. A retardation coefficient R (larger than 1) then links the seepage velocity v_s to the true contaminant velocity v_c:

$$v_c = \frac{v_s}{R} \tag{24.13}$$

The retardation coefficient is estimated by laboratory testing or correlation with the ratio S/C.

Precipitation and Dissolution. Precipitation and dissolution reactions involve the level of solubility of a contaminant in the carrier liquid (most often water). The degree of solubility varies from one contaminant to the next and is characterized by an equilibrium constant K, which is high for high solubility and vice versa. Sulfate salts and chlorides tend to be highly soluble, sulfide and hydroxides tend to be least soluble, and carbonates and silicates have intermediate solubility.

Oxidation and Reduction. These reactions are often called *redox reactions*. *Oxidation* is a loss of electrons. *Reduction* is a gain of electrons. Redox reactions are characterized by the redox potential Eh expressed in volts. The redox potential is a measure of the affinity of a substance for electrons (its *electronegativity*). The reference potential is that of hydrogen, which is set at 0 volt. Substances more strongly electronegative than hydrogen have positive redox potentials. These are substances that are capable of oxidizing; for example, oxygen has a redox potential of 1.23 V. In contrast, substances less strongly electronegative than hydrogen have negative redox potentials and are capable of reducing; for example, calcium has a redox potential of −2.87 V. Hydrogen peroxide is a powerful oxidant that is used in low concentrations to treat wounds because it releases oxygen, which kills bacteria.

Acid-Base Reactions. An *acid-base reaction* is a gain or loss of a proton (H^+) or the gain or loss of a hydroxyl group (OH^-). An acid is a proton donor, and a base is a proton taker. These reactions affect the pH of the soil and the groundwater and therefore the type of remediation strategy.

Other Reactions. Other reactions include complexation, ion exchange, hydrolysis, and volatilization. *Complexation* takes place when organic or inorganic ions or molecules combine in the dissolved phase. *Ion exchange* occurs when an ion is replaced by another one. This type of reaction, called *isomorphous substitution*, can take place at the surface of clay particles, with, for example, a Ca^{++} taking the place of an Na^{+}. *Hydrolysis* is the reaction between an organic molecule and water. *Volatilization* refers to the transformation of a liquid or solid into gas; it applies to volatile organics, for example, and is governed by Henry's law:

$$C_a = K_H C_w \tag{24.14}$$

where C_a is the concentration of the contaminant in the gas phase, K_H is Henry's constant for that contaminant, and C_w is the concentration of the contaminant in the liquid phase. William Henry was a British chemist who proposed this law in 1803. Henry's constant and many other constants associated with chemical reactions for typical contaminants in soils can be found in EPA publications such as *Subsurface Contamination Reference Guide* (U.S. EPA 1991) or in reference books such as Sharma and Reddy (2004).

Biodegradation

Biodegradation is a redox reaction mediated by microorganisms; it can be aerobic or anaerobic. Aerobic biodegradation takes place in the presence of oxygen, which acts as an electron acceptor from the substance; anaerobic biodegradation takes place in the absence of oxygen. Some microorganisms naturally occurring in soil have the remarkable ability to degrade and transform many compounds, including hydrocarbons, polychlorinated biphenyls (PCBs), polyaromatic hydrocarbons (PAHs), pharmaceutical substances, radionuclides, and metals. However, for degradation to take place, the microorganism must be matched to the contaminant to be degraded. The microorganisms facilitate the transfer of electrons from a donor to an acceptor and in the process transform the substance to which it is attached. Typical outputs of biodegradation are methane, carbon dioxide, and water.

Governing Differential Equation

The governing differential equation (GDE) can be obtained by adding the contributions from the previously discussed processes. In the simple case of a one-dimensional flow, the total mass of contaminant per unit volume of soil C_T is:

$$C_T = \rho_d C_s + \theta_w C_w + \theta_g C_g \tag{24.15}$$

where ρ_d is the dry density of the soil (mass of dry soil over volume of soil), C_s is the concentration of contaminant in the solid phase (mass of contaminant over mass of solids), θ_w is the volumetric water content (volume of water over volume of soil), C_w is the concentration of contaminant in the liquid phase (mass of contaminant over volume of water), θ_g is the volumetric gas content (volume of air over volume of soil), and C_g is the concentration of contaminant in the gas phase (mass of contaminant over volume of air). For a saturated flow, θ_g is zero, θ_w is equal to the porosity n, and Eq. 24.15 becomes:

$$C_T = \rho_d C_s + nC_w \tag{24.16}$$

The contaminant mass flux F_{adv} (mass of contaminant crossing a unit area of soil per unit time; see Eq. 24.7) due to water flow or advection is given by:

$$F_{adv} = C_w n v_s \tag{24.17}$$

where n is the soil porosity and v_s is the seepage velocity. The contaminant mass flux F_{dif} (mass of contaminant crossing a unit area of soil per unit time; see Eq. 24.9) due to diffusion is given by:

$$F_{dif} = -D^* n \frac{dC_w}{dx} \tag{24.18}$$

where D^* is the effective diffusion coefficient, n is the porosity, and x is the coordinate along the travel direction of the water.

The contaminant mass flux F_{dsp} (mass of contaminant crossing a unit area of soil per unit time; see Eq. 24.11) due to dispersion is given by:

$$F_{dsp} = -D_L n \frac{dC_w}{dx} = -\alpha_L v_s n \frac{dC_w}{dx} \tag{24.19}$$

where D_L is the dispersion coefficient in the longitudinal direction, n is the porosity, dx is the increment of x, and α_L is the longitudinal dispersivity.

If we combine the transport from advection, diffusion, and dispersion, we obtain, for saturated flow:

$$\begin{aligned} F_T &= F_{adv} + F_{dif} + F_{dsp} \\ &= nv_s C_w - D^* n \frac{dC_w}{dx} - \alpha_L v_s n \frac{dC_w}{dx} \end{aligned} \tag{24.20}$$

Then we write that for conservation of mass, the change in concentration of mass of contaminant with time has to be equal to the slope of the curve describing the flux vs. distance along the flow path:

$$\frac{dC_T}{dt} = -\frac{dF_T}{dx} \tag{24.21}$$

This leads to:

$$\begin{aligned} &\frac{\partial(\rho_d C_s)}{\partial t} + \frac{\partial(nC_w)}{\partial t} \\ &= -\frac{\partial}{\partial x}\left(nv_s C_w - \left(D^* n + \alpha_L v_s n\right)\frac{\partial C_w}{\partial x}\right) \pm S \end{aligned} \tag{24.22}$$

The term S is added to include any sources or sinks of contaminants, such as those due to chemical reactions. The partition coefficient K_d is defined as:

$$K_d = \frac{C_s}{C_w} \tag{24.23}$$

If we ignore sources and sinks, we get the equation:

$$\frac{\partial C_w}{\partial t} = -\frac{v_s}{\left(\frac{\rho_d K_d + n}{n}\right)} \frac{\partial C_w}{\partial x} + \frac{(D^* + \alpha_L v_s)}{\left(\frac{\rho_d K_d + n}{n}\right)} \frac{\partial^2 C_w}{\partial x^2} \quad (24.24)$$

After setting R_d, the retardation factor, and D_H, the hydrodynamic dispersion as:

$$R_d = \frac{\rho_d K_d + n}{n} = 1 + \frac{\rho_d K_d}{n} \quad (24.25)$$

$$D_H = D^* + \alpha_L v_s \quad (24.26)$$

Then the GDE for contaminant transport for this simplified case is:

$$\frac{\partial C_w}{\partial t} = -\frac{v_s}{R_d} \frac{\partial C_w}{\partial x} + \frac{D_H}{R_d} \frac{\partial^2 C_w}{\partial x^2} \quad (24.27)$$

Before we can solve this equation, we have to define the boundary conditions and the initial conditions. Let's take the simple case in which the water is flowing horizontally in a confined aquifer with a point source of contamination that remains constant versus time (Figure 24.12). In this case the boundary conditions and initial conditions are:

C_0 = constant contaminant concentration at point

$x = 0$ and $t = 0$ (24.28)

The solution was presented by Fetter (1992):

$$C_w(x, t) = \frac{C_0}{2} \, erfc \left(\frac{R_d x - v_s t}{\sqrt{4 R_d D_H t}} \right) \quad (24.29)$$

where *erfc* is the complementary error function. Propagation of the contamination is shown as normalized concentration (C/C_0) versus lateral extent $x = v_s t / R_d$ in Figure 24.13. Solutions for one-dimensional flow and more complicated boundary conditions can be found in Hemond and Fechner-Levy (2000). For two-dimensional and three-dimensional conditions, the GDE can be solved by the finite difference method or other numerical schemes.

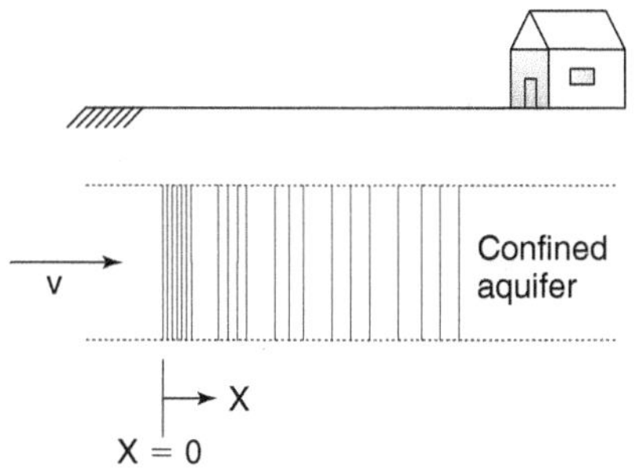

Figure 24.12 One-dimensional transport in a confined aquifer.

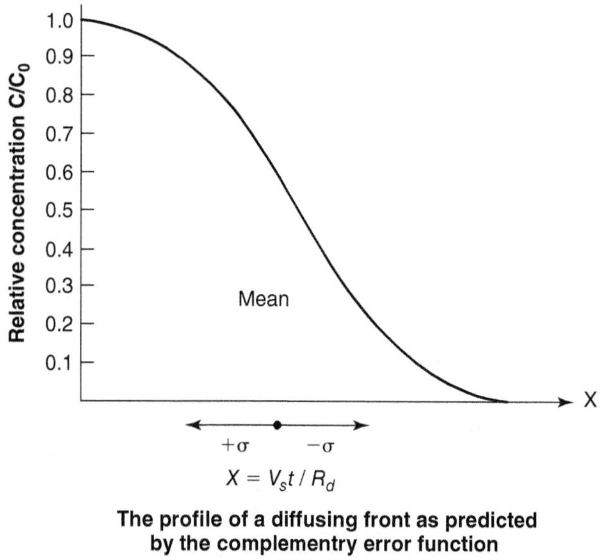

Figure 24.13 Diffusing front for one-dimensional contaminant transport. (After C. W. Fetter, *Contaminant Hydrogeology*, New York: Macmillan Publishing Company, 1992)

24.6 REMEDIATION

Remediation of contamination is the action of reducing the risk of detrimental effect on human life to an acceptable level. Once the type and extent of contamination have been identified, there are three common options: monitoring, containment, and remediation. Let's look first at risk assessment, as it influences the choice of remedial measure.

24.6.1 Risk Assessment and Strategy

Risk is the product of the probability of an event happening or being exceeded times the value of the consequence. The value of the consequence and therefore the risk can be expressed in cost units, in human or animal fatalities, or in number of people sick (among others). Acceptable risk is then established by considering the risk associated with normal life activities and accepting such levels as targets for contamination. In the United States, there are two risk assessment techniques: one general technique from the EPA and one technique for leaking petroleum tanks from ASTM.

EPA Procedure

The EPA procedure (U.S. EPA 2001) advances in four steps: data collection and evaluation, exposure assessment, toxicity assessment, and risk characterization. *Data collection* consists of identifying the contaminants, their concentration, the source of the contamination, and the soil in which the contaminant may propagate. *Exposure* is defined as the contact of a chemical or biological agent with the outer boundary of a human being. The amount of exposure is measured by the extent of contact with that outer boundary during a specified amount of time. Exposure assessment includes determining the exposure setting (e.g., weather, topography, geology, hydrogeology), the exposure pathways (e.g., likely transport routes, speed of propagation, proximity of human activities), and the exposure concentrations (e.g., current and future chemical concentration).

Toxicity is the adverse effect of a contaminant on human life (e.g., cancer, birth defect). Toxicity assessment includes determining if the chemical is a carcinogen or noncarcinogen. The toxicity of noncarcinogenic chemicals is quantified by the reference dose (RfD in mg/kg/day), which is the daily dose that would not create an appreciable risk of deleterious health effects during a lifetime. The toxicity of carcinogenic chemicals is quantified by the slope factor (SF in $(\text{mg/kg/day})^{-1}$) that is the upper bound (95% confidence) on the increased cancer risk from a lifetime exposure to the chemical. The EPA publishes tables of RfDs and SFs for many chemicals (Sharma and Reddy 2004). The *risk* quantification is done by using the hazard quotient *HQ* for noncarcinogens and the risk R for carcinogens, as follows:

$$HQ = \frac{E}{RfD} \tag{24.30}$$

$$R = CDI \times SF \tag{24.31}$$

where E is the chemical intake (mg/kg/day) and CDI is the chronic daily intake averaged over 70 years (mg/kg/day). The EPA considers a value of HQ higher than 1 and R larger than 10^{-6} to be unacceptable.

ASTM Procedure

The ASTM procedure (ASTM 1995) is aimed at the remediation of sites with leaking petroleum tanks, where the chemical contaminants may include benzene, toluene, and xylene, for example. It proceeds in three "tiers" or steps.

Tier 1, much like the EPA method, consists of collecting data, including the concentration of chemicals. These concentrations are then compared with risk-based screening levels (RBSLs) found in published tables. If the levels are below the RBSL for the most severe contamination propagation pathway, no action is required. If not, remediation can be planned to meet the tier 1 RBSL, or a more sophisticated Tier 2 analysis of the problem can be chosen.

In Tier 2, the points of compliance are selected and the site-specific target levels (SSTLs) are determined. Additional soil and water data are collected and simple calculations are made to predict the transport and fate of the chemical over time. If the STSL is met, no further action is necessary. If not, remediation to meet the STSL is undertaken, or a more refined Tier 3 evaluation takes place.

Tier 3 evaluation makes use of analyses more sophisticated than Tier 2 analyses and remediation takes place if the levels obtained after the Tier 3 analysis do not meet the target levels.

The general strategy, once the type and extent of contamination have been identified, is to choose among the following three common options: monitoring, containment, or remediation (Figure 24.14). Monitoring is selected if the

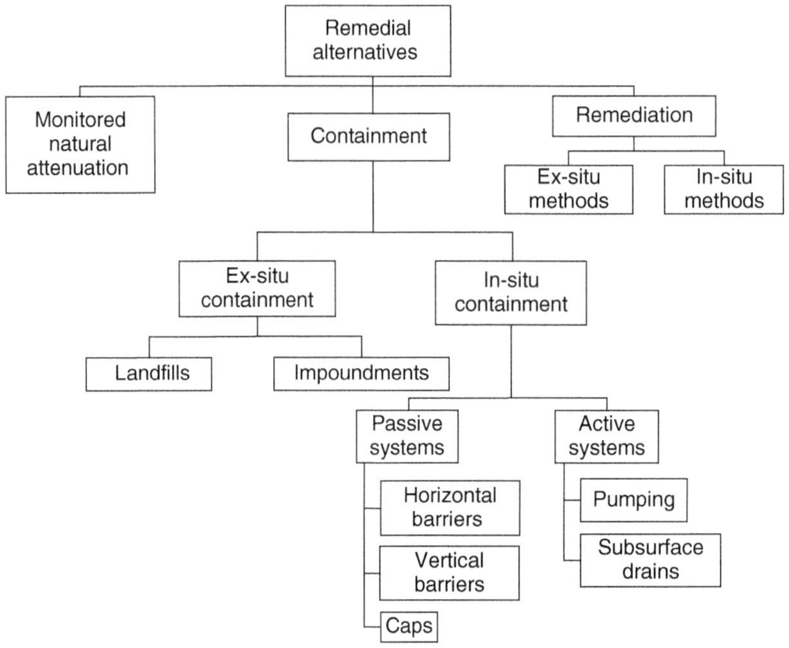

Figure 24.14 Remediation alternatives and decision tree. (After Sharma and Reddy, 2004. This material is reproduced with permission of John Wiley & Sons, Inc.).

level of contamination is minor and the risk is low; containment is most often a temporary measure; and remediation of soil and water is the long-term alternative.

24.6.2 In Situ Waste Containment

In situ waste containment is typically used as a temporary measure to prevent further propagation of the contaminant while a more permanent remediation solution is set up. Containment is sometimes used as a permanent measure if the cost of cleanup technologies is prohibitive or the technologies are impractical. Two types of systems exist: passive systems and active systems. Passive systems are barriers to enclose the waste and minimize spreading (vertical barriers, bottom barriers, or surface covers), whereas active systems are generally pumping wells and drains.

Vertical Barriers

Vertical barriers (Figure 24.15) are built to surround the waste. They are built from the surface down to a naturally impervious soil layer. They can be hanging barriers above an impervious layer if the contaminant will not propagate with depth, as in the case of LNAPLs that float on the water table. There are several types of vertical barriers: slurry trench barriers, grouted barriers, and steel sheet pile barriers.

Slurry Trench Barriers. Slurry trench barriers are the most common type of vertical barrier. They are constructed by excavating a narrow trench about 0.5 to 1 m wide to the depth required. During the excavation, the trench is filled with a liquid slurry made of water and bentonite clay (about 5% bentonite by weight). The role of the bentonite slurry is to seal the walls of the trench and provide a horizontal pressure that minimizes the chance of trench wall collapse. The slurry, which is at least 3% heavier than water, is kept level with the top of the trench, and is therefore higher than the water level in the soil. As a result, the slurry permeates out through the trench wall and deposits a film of very fine bentonite clay particles on the wall. This very low permeability film

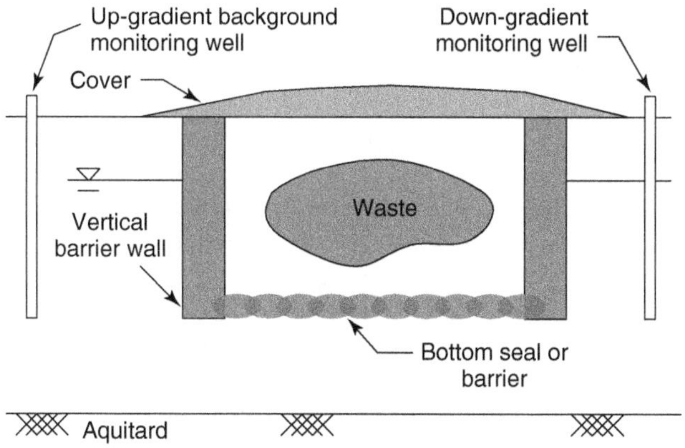

Figure 24.15 Vertical, bottom, and surface containment barriers. (After Sharma and Reddy, 2004. This material is reproduced with permission of John Wiley & Sons, Inc.).

can be 10 mm thick or more and seals the wall against water penetration in the trench.

When the trench is completed and full of slurry, the trench is backfilled with low-permeability backfill. This backfill may be a soil-bentonite mix (SB), or a cement-bentonite mix (CB), or it can involve a geomembrane. Soil-bentonite barrier backfill is typically a mixture of sand (excavated soil if possible), dry bentonite, and bentonite slurry. It has the consistency of wet concrete with 25 to 50% fines but no gravel. This low-permeability backfill is placed at the end of the slurry trench and displaces the slurry forward toward the ongoing excavation. The target hydraulic conductivity of the trench is 10^{-9} to 10^{-10} m/s. Cement-bentonite barrier backfill is similar to SB backfill except that the soil is replaced by cement. A typical mix would be 5% bentonite, 15% cement, and 80% water. CB barriers are stronger than SB barriers, but have lower permeability because the cement hinders the full expansion of the bentonite. Geomembrane barriers are installed by lowering a membrane into the open hole so that it seals the bottom and the side walls of the trench. Installation of a geomembrane can be a complex and difficult operation, but the permeability of the membrane is much lower than that of the other two systems.

One important aspect of slurry trench barriers is the design for stability of the trench. An earth pressure analysis must be conducted to calculate the global factor of safety against collapse of the trench wall when the trench is fully excavated. Such design is rooted in the content discussed in Chapter 22.

Grouted Barriers. Grouted barriers are prepared first by rotary drilling and then by grout injection. A column is constructed, then another one in line with the first one, and then another one in line with the first two, and so on in sequence so that in the end a wall of columns forms the barrier. Typically a first set of columns is built by skipping the intermediate columns, and then the intermediate columns are built when the first set of columns has hardened. Pressure grouting, jet grouting, and soil mixing are the techniques most commonly used. *Pressure grouting* consists of injecting grout under some pressure after a hole is drilled with diameters in the range of 1 to 1.5 m. The injection is done by the point injection technique or by the "tube a manchette" technique. *Jet grouting* consists of drilling a small hole and then rotating the drill rod upon withdrawal while jetting grout laterally under pressure to enlarge the hole. The pressure of the jet is in the range of 35 to 40 MPa. Grouted columns created with this technique can reach 3 m in diameter. *Soil mixing* consists of literally mixing the soil with grout (e.g., 20% by weight) as drilling takes place. In other words, the drilling mud is replaced by the soil-grout mixture. This minimizes the amount of cuttings generated. With grouted columns as barriers, it is very important to ensure overlapping of the columns to achieve a good seal against contaminant flow.

Steel Sheet Pile Barriers. To create a steel sheet pile barrier, steel sheet piles are driven one beside the other to

Table 24.3 Advantages and Drawbacks of Waste Containment Systems

System	Advantages	Drawbacks
Slurry barriers	Long-term, inexpensive, no maintenance required, well proven, available materials	Compatibility between slurry and contaminant; need for natural impervious layer; problems with boulders, caverns
Grout barriers	Injection of grout only requires small holes, can go very large depth, can fill caverns, can vary setting time	Difficult when soil is not pervious; holes and gaps more likely to jeopardize containment of liquids
Geomembrane barriers	Effective, compatible with many contaminants	Sealing between sheets is complex process; keying membrane in bottom layer is complex as well; expensive
Steel sheet pile barriers	No excavation needed, no maintenance required	Seal between sheet piles is not effective; corrosion problems; iron not compatible with many contaminants
Pumping	Less costly than barrier, design flexibility, control of pumping rates, common technology, depth not a problem	Requires frequent monitoring to limit propagation; maintenance required; capture zone limited
Subsurface drains	Economical to operate, drain location flexible, fairly reliable, simple and economical construction	Not for low-permeability soils; underdrains tough to place; monitoring required; not for deep contamination; potential clogging; excavation required

(After Sharma and Reddy 2004.)

form a barrier in the soil. The advantages of this technique are that no excavation is necessary and the barrier installation is rapid. One problem is the corrosion issue and the lack of a good seal across the joints between sheet piles.

Table 24.3 (after Sharma and Reddy 2004) summarizes the advantages and drawbacks of the various barrier techniques.

Bottom Barriers

Bottom barriers (Figure 24.15) are built to seal the bottom of a contaminated zone. This may be necessary when there is no natural low-permeability layer under the contaminated zone and the contaminant can propagate downward. A bottom barrier can be constructed by grouting or directional drilling. Grouting may be done by pressure grouting or jet grouting, but in both cases the injection pipe is driven or vibrodriven to the depth of the bottom barrier and a grout bulb is constructed. The operation is repeated until the overlapping bulbs form a bottom barrier. The drawback of this technique is that holes have to be punched through the waste or contaminated zone. Directional drilling can be used to reduce this problem: It consists of setting an inclined drill outside of the contaminated zone and reaching underneath that zone by drilling at an angle. Then the hole is grouted. Side-by-side holes are drilled and grouted to form the bottom barrier.

Surface covers (Figure 24.15) are built to cap the contaminated zone. They prevent the infiltration of running water and rain, thereby minimizing leaching; they prevent atmospheric contamination, reduce erosion, and improve aesthetics at the site. They are typically made of multiple layers, each with a specific purpose. The base layer or foundation layer creates a uniform surface on which to build other layers. Above that is the gas collection layer, made of coarse-grained soil and equipped with venting pipes. Above that is the barrier layer, made of compacted clay and a geomembrane or a geosynthetic clay liner (see Chapter 25) to prevent the surface water from entering the contaminated zone. This layer typically has a hydraulic conductivity of 10^{-9} m/s or less. Above that is the drainage layer, made of coarse-grained soil and /or geotextile to collect any water percolating down through the top layers. Above that is the surface and protection layer, made of topsoil and erosion control geosynthetic to prevent erosion and foster plant growth.

Pumping wells (Figure 24.16) are built to pump contaminated water out of the soil or introduce a hydraulic gradient that will force the plume to move in the desired direction. These wells are used when the contaminant is mixed into the water (soluble). Injection wells can be used at some distance from pumping wells to force the liquid to go toward the pumping wells. This strategy can be used when the contaminant is a liquid not miscible with water. Pumping and injection are most effective when the soil is coarse grained with high permeability. In the design of the well, the following issues must be addressed: depth, spacing, zone of influence, pumping rate, and number of wells. The treatment of the effluent must be addressed through an on-site or off-site treatment.

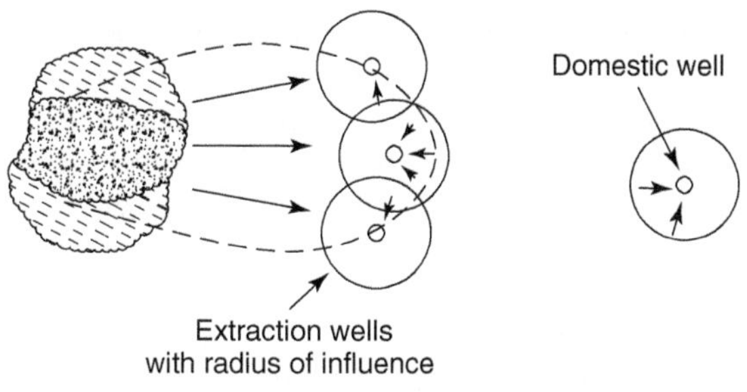

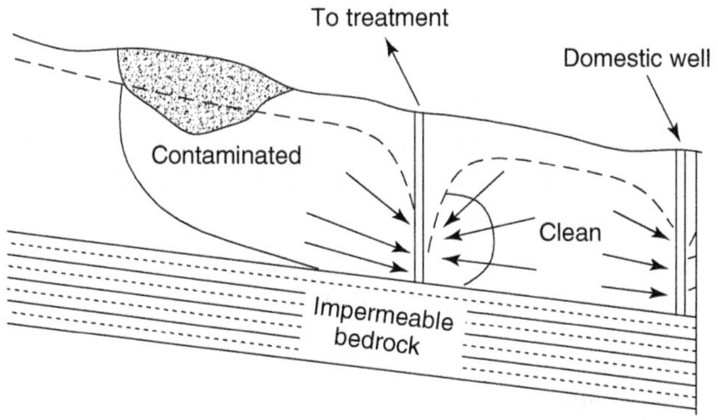

Figure 24.16 Pumping wells. (After Sharma and Reddy, 2004. This material is reproduced with permission of John Wiley & Sons, Inc.).

Subsurface drains (Figure 24.17) play essentially the same role as pumping wells, but they drain the contaminated zone by gravity instead of pumps. Drains have the advantages of being a more economical solution than pumping, and they can be used in low-permeability soils where pumping is not efficient. Drains can be placed horizontally as perforated pipes, or vertically as draining boreholes with a sump pump, or a combination of both. As in the case of pumping, the issues to be addressed are the location of the drains to take best advantage of gravity forces, depth, spacing, and zone of influence. Other issues specific to drains include pipe diameter, gradient of the slope, filters to prevent clogging, and size of the sump and pump.

24.6.3 Soil Remediation

The most commonly used methods of soil remediation are soil vapor extraction, soil washing, and solidification. However, many other techniques are available.

Soil Vapor Extraction

Soil vapor extraction (SVE), also called *soil venting*, *vacuum extraction*, and *aeration*, consists of sucking the contaminated air out of the voids in the unsaturated soil zone above the water table. It is applied mostly to the volatile organic compounds (VOCs) that form in petroleum-contaminated soils. Extraction wells are installed on a grid, the vapor is removed by the vacuum gradient, and the collected vapor is treated using carbon filters, for example. The best candidates for such treatment are soils that are highly permeable and gasses that have a vapor pressure larger than 70 Pa and a Henry's law constant higher than 0.01.

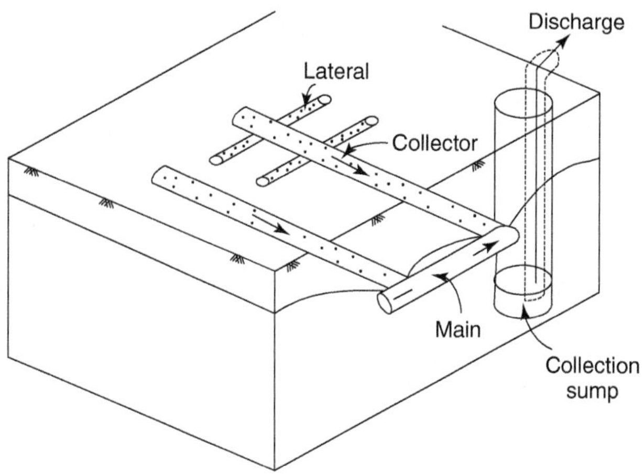

Figure 24.17 Subsurface drains. (After Sharma and Reddy, 2004. This material is reproduced with permission of John Wiley & Sons, Inc.).

Soil Washing

Soil washing (SW) consists of excavating the soil from the site and treating it on site with a soil scrubbing system. A chemically suitable washing fluid is selected on the basis of the contaminants to be removed. After excavation, the soil is washed by forcing it through an energetic scrubbing system where it is mixed with the washing fluid. The coarse fraction is usually easier to clean than the fine fraction, because of the size and chemical complexity of clay particles. The clean soil is returned to its initial location; the remaining contaminated soil and the contaminated effluent are sent elsewhere for further treatment or disposal. This technique becomes economically attractive when the amount of soil to be treated is large (say, more than 50 MN).

Stabilization and Solidification

Stabilization and solidification (S/S), also called *immobilization* or *fixation*, consists of treating the contaminant in such a way that it is bound to the soil particles and remains trapped at the site. The method can be done ex situ or in situ. For the ex situ method, the contaminated soil is excavated and mixed with a stabilizing agent that fixes the contaminant to the soil particles. The mix is cured, and then the stabilized soil is replaced at the site or disposed of in a landfill. For the in situ method, the stabilizing agent is injected or mixed with the in situ contaminated soil to prevent the contaminants from moving away from the site. Portland cement is an example of a stabilizing agent.

Electrokinetic Remediation

Electrokinetic (EK) remediation consists of applying a potential difference between two electrodes (e.g., steel bars) driven into the contaminated soil. The potential difference (say, 40 V/m) drives the contaminated fluid to the electrodes where it is collected and removed. The anode is positively charged and attracts negatively charged contaminants (anions); the cathode is negatively charged and attracts positively charged contaminants (cations). The flow rate (m^3/s) generated by EK remediation is proportional to the electrical potential between electrodes and the dielectric constant of the soil, but inversely proportional to the distance between electrodes and the viscosity of the fluid. EK remediation works well in fine-grained soils, which are otherwise difficult to clean.

Thermal Desorption

Thermal desorption (TD) consists of heating the contaminated soil to temperatures between 100 and 500°C. These high temperatures vaporize the contaminants, which are then removed by a vapor extraction system. This method works for volatile and semivolatile organic compounds, but not for metals. Note that the contaminants are not destroyed as they would be during incineration (much higher temperature). TD can be performed in situ or ex situ. For ex situ treatment, the soil is excavated, brought to the treatment plant, and subjected to the heating process. In situ, heating blankets are placed on the surface for shallow treatment and heating wells are installed for deeper zones. If the water content is too high (e.g., more than 15%), dewatering may be necessary as a first step.

Vitrification

Vitrification (VT) consists of melting the contaminated soil into glass. This requires a lot of heat, with temperatures of about 1800°C. At such a temperature, organics are either destroyed or vaporized and stable inorganic compounds are surrounded by the molten soil. Upon cooling, the mass turns to glass and the inorganic contaminants are fixed in place. The method can be applied in situ or ex situ. In situ electrodes are placed and very high voltage and very high current are applied (e.g., 4000 V, 4000 A) with a power requirement of 3 to 4 MW. Gas collection hoods are placed on top of the electrodes to evacuate the gasses created to a treatment system.

Bioremediation

Bioremediation (BR) involves microorganisms or microbes eating the contaminant, transforming it into nontoxic by-products through their digestive systems, and releasing those by-products to the atmosphere. The by-products are usually carbon dioxide and water or organic acids and methane. Appropriate microorganisms are found naturally in the soil and include yeast, fungi, and bacteria. The best contaminant food for them is petroleum hydrocarbons; other organic contaminants and inorganic contaminants are not as well suited to bioremediation. Microorganisms operate best in the presence of moisture, nutrients, and oxygen, so *biostimulation* consists of providing them with those three components to enhance the transformation process. *Bioaugmentation* consists of adding selected microorganisms to degrade a specific contaminant or supplement the work of the indigenous microorganisms.

Bioremediation may be aerobic (with oxygen) or anaerobic (without oxygen), but the aerobic process takes less time and is favored. Both in situ and ex situ treatment are possible. Either way, monitoring (e.g., for CO_2 and O_2) is necessary to adjust the stimulation process when appropriate. When stimulation involves providing more oxygen, it can be done in the form of bioventing (bringing air into the soil) or injection of hydrogen peroxide.

Phytoremediation

Phytoremediation (PR), from the word *phyto* in Greek which means "plant," is the natural soil remediation work done by plants through their root systems. The contaminant crosses the root membrane to enter the plant, which either degrades the contaminant or stores it in the plant tissue. Phytoremediation is best suited for sites with low levels of contamination at shallow depth (less than 3 m). It represents a final cleanup strategy rather than a main remediation method.

Table 24.4 summarizes the advantages and drawbacks of soil remediation methods.

24.6.4 Groundwater Remediation

Groundwater is a very important resource to humankind. It represents 40% of our drinking water and must be kept free of contaminants. Groundwater remediation includes several different methods: pump and treat, in situ flushing, permeable reactive barriers, in situ air sparging, monitored natural attenuation, and bioremediation.

Pump and Treat

The pump and treat (PT) technique consists of installing a well from which the water is pumped out of the contaminated soil, treating the contaminated water, and pumping the water back into the soil or to another appropriate location (Figure 24.18). A typical configuration is to have a row of pumping wells downstream of the contaminant flow and a series of recharging wells upstream of the contaminant flow. That way the pumping wells can also serve as monitoring wells for the efficiency of the treatment. The zone of influence and the depth of the wells, as well as the pumping rate, are part of the design.

In Situ Flushing

In situ flushing (ISF) consists of setting up the same kind of wells as in the pump and treat solution, but in this case the cleaning liquid is injected through the upstream wells, passes through the contaminated soil, cleans it, and is pumped out at the downstream wells. The cleaning liquid is carefully selected to remove the contaminant from the water without hurting the environment. The difference between PT and ISF is that with ISF, the cleaning is done in the soil rather than outside the soil as with PT.

Table 24.4 Advantages and Drawbacks of Soil Remediation Methods

Method	Advantages	Drawbacks
Soil vapor extraction	Easy installation, low disturbance, short time, economical	Not possible in low permeability areas; need air emission permits; only for unsaturated soils
Soil washing	Reduces volume of contaminated soil, excavation and efficient treatment on site, few permits required	Ineffective for soils with high fine content; relatively expensive; public exposure possible
Stabilization/solidification	Low cost, widely applicable, simple, high throughput rates	Contaminants remain; increased volume; volatiles created; limits future use
Electrokinetic remediation	Applicable to fine-grained soils, wide range of contaminants, less expensive	Changes pH; buried metal is a problem; stagnant zones between electrodes
Thermal desorption	Very rapid treatment, readily available equipment, very good for volatile organics	Dewatering may be necessary; not good for fine-grained soils; not usable for heavy metals; large space required
Vitrification	Long-term durability, wide applicability, reduction of volume, public acceptance, cost-effective for difficult sites	Difficult for very wet soils; limited depth (<7 m); not possible if >10% organics; high energy cost; dangerous in some cases
Bioremediation	Very good for organic contaminants, minimum equipment required, no excavation, low cost	Highly sensitive to local conditions; monitoring required; long treatment time
Phytoremediation	Less expensive, safe, in-place treatment	Shallow treatment (<7 m); slow; still experimental; potential contamination of food chain

(After Sharma and Reddy 2004.)

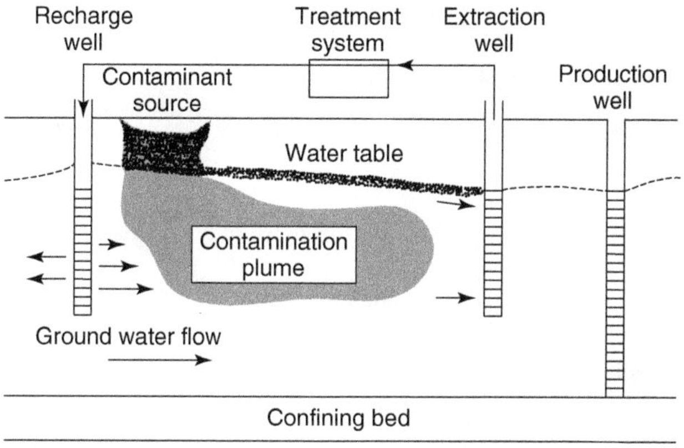

Figure 24.18 Pump and treat setup. (After Sharma and Reddy, 2004. This material is reproduced with permission of John Wiley & Sons, Inc.).

Permeable Reactive Barriers

Permeable reactive barriers (PRBs) are treatment walls placed in the soil; they let the water go through but not the contaminants, which are immobilized or degraded as the groundwater flows through the barrier. Typically a trench is built and filled with a carefully selected reactive agent. As the contaminated water flows through the PRB, the clean water comes out and the contaminant is transformed into nontoxic by-products.

In Situ Air Sparging

In situ air sparging (ISAS) consists of drilling injections wells through the contaminated soil to reach underneath the plume and injecting compressed air under the contaminated plume. Because it is lighter, the air flows upward through the contaminated water, entrains contaminants vapors which are then evacuated through soil vapor extraction. As a positive side effect and in the process the air brings oxygen which enhances the activity of microorganisms and bioremediation. ISAS is best suited to high permeability soils ($k > 10^{-5}$ m/s) and to the decontamination of volatile organic compounds, as in the case of leaking underground petroleum storage tanks.

Monitored Natural Attenuation

Monitored natural attenuation (MNA) is the "do nothing and monitor" or "watch and wait" solution. It consists of monitoring the process of natural remediation, including natural bioremediation, dilution, dispersion, and volatilization. There is no human intervention in this decontamination process.

Bioremediation

Bioremediation (BR) for contaminated groundwater works according to the same principles as bioremediation for contaminated soil. It is different from MNA in that there is human intervention to optimize the digestive process of the microorganisms that eat the contaminant. This intervention includes the injection of oxygen and nutrients in the water (biostimulation) or the injection of additional microorganisms (bioaugmentation). BR works best with low hydraulic gradients and permeable soils (k > 10^{-6} m/s).

24.7 LANDFILLS

Most landfills (Figures 24.19 and 24.20) are used as permanent repositories of municipal solid waste, which is the main topic of this section. In the United States, each person generates about 20 N (1 N is the weight of a small apple) of MSW per day (20 N/person/day). This number used to be 12 N in1960, reached 20 N in 1990, and has stabilized since then, but the population continues to grow, so landfills have to handle more and more MSW. The total amount of MSW per year in the United States is close to 2.5 million MN per year. The best ways to reduce waste, in order of preference, are:

1. Source reduction
2. Recycling and/or composting
3. Disposal in combustion facilities and landfills

Although the amount of waste being recycled has increased 10-fold over the past 40 years, today more than 50% of all MSW still ends up in a landfill. It is extremely important that these landfills be designed to keep the waste from contaminating soil and water and burdening future generations with unwanted problems.

24.7.1 Waste Properties

As mentioned in section 24.2 and Figure 24.2, municipal solid waste in landfills consists primarily of paper, plastic, and food scraps. However, you can also find the odd rusted refrigerator and car tires. It is difficult to come up with the friction angle or modulus of elasticity of an old fridge or a car tire, yet these are the type of properties we are accustomed to using. To complicate matters further, the waste can be in various stages of decomposition, which affect its engineering properties. The only way to answer this problem is by testing the site-specific waste at a large-enough scale. This has been the effort of many researchers and engineers, including Landva and his colleagues (Landva and Clark 1990). The following values are given to provide an order of magnitude of such properties, but the best approach consists of obtaining site-specific values of these parameters through testing at large scale—a scale large enough to be representative of the MSW behavior.

The *unit weight* γ of MSW has been measured in large pits and reported by many authors. The first observation is that γ is highly variable depending on the type of waste, the degree of compaction, the state of decomposition, the proportion of daily soil cover, and the depth of the landfill.

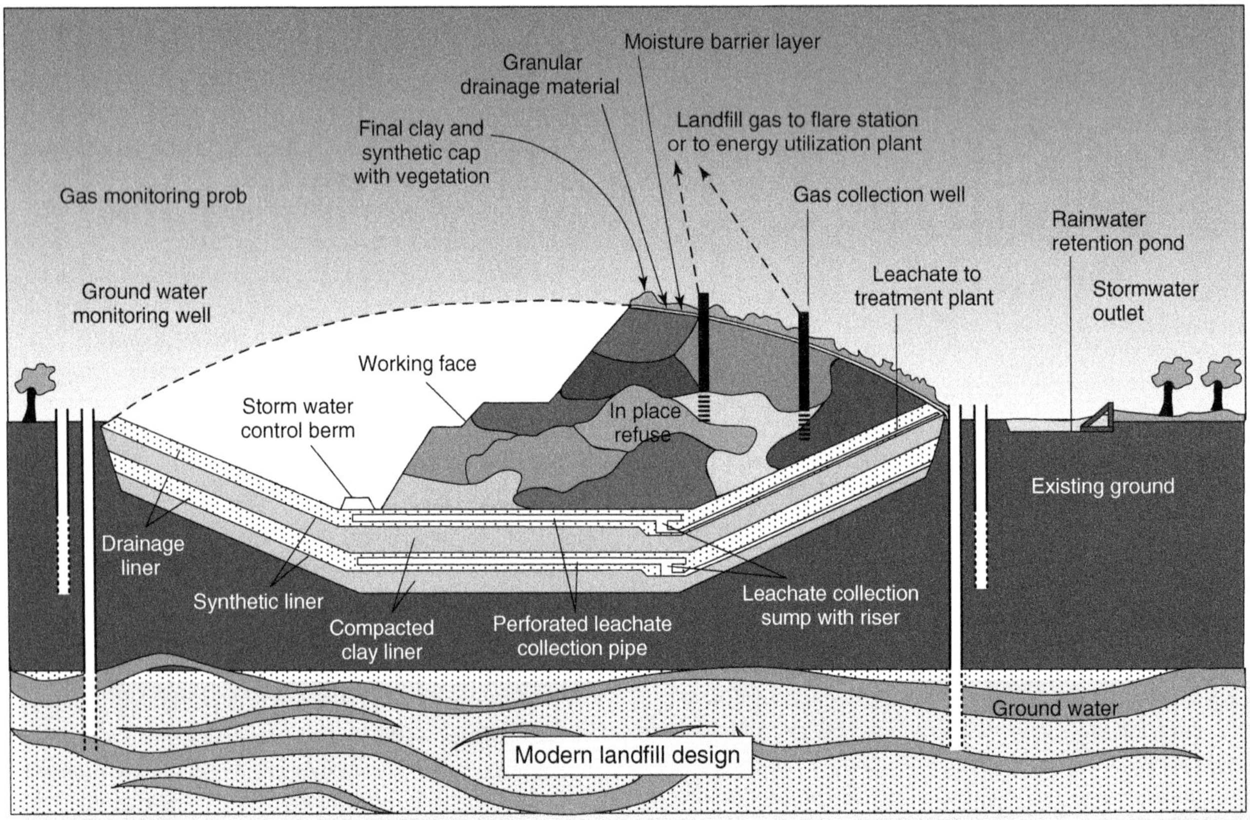

Figure 24.19 Cross section of a landfill

Figure 24.20 Landfill under operation. (Photo provided courtesy of the Texas Comptroller of Public Accounts from the Energy Report 2008, available at http://www.window.state.tx.us/specialrpt/energy/)

Numbers ranging from 3 to 14 kN/m^3 have been reported, with an average of 8 kN/m^3. The unit weight increases and the variability decreases as depth increases in the waste increases. Porosity is reported to vary between 0.4 and 0.6, void ratio between 0.67 and 1.5, and water content between 0.15 and 0.4 (Sharma and Reddy 2004). Field permeability measured in MSW pits gave a range of 10^{-5} to 4×10^{-3} m/s (Landva and Clark 1990). Shear strength data collected by many authors was reviewed by Kavazanjian (1999), who proposed a bilinear lower-bound envelope. The first part applies to normal stresses lower than 30 kPa and gives $c = 24$ kPa and $\varphi = 0$. The second part applies to normal stresses higher than 30 kPa and gives $c = 0$ and $\varphi = 33°$:

$$\text{For } \sigma < 30\,\text{kPa}, \qquad s = 24\,\text{kPa} \tag{24.32}$$

$$\text{For } \sigma > 30\,\text{kPa}, \qquad s = \sigma \tan 33 = 0.65\,\sigma \tag{24.33}$$

Kavazanjian (1999) suggested a shear wave velocity for MSW which varies from 150 m/s at the surface to 350 m/s at a depth of 60 m. For compressibility, most investigators favor the consolidation equation (see section 17.8.9):

$$\Delta H = H_o \frac{C_c}{1 + e_o} \log \left(\frac{\sigma'_{ov} + \Delta\sigma_v}{\sigma'_{ov}} \right) \tag{24.34}$$

where ΔH is the settlement, H_0 is the initial thickness of the waste layer, C_c is the compression index, e_o is the initial void ratio, σ'_{ov} is the effective vertical stress before loading, and $\sigma'_{ov} + \Delta\sigma_v$ is the effective stress long after loading. Values of $C_c/(1 + e_o)$ between 0.1 to 0.4 have been suggested (Navfac 1983), with the higher values corresponding to higher organic content. With MSW, a significant amount of delayed settlement (*creep*) can be expected over 10 to 15 years, with the magnitude of $\Delta H/H_0$ as much as 50% for new landfills and 15 to 20% for old landfills (Sharma and Reddy 2004). The creep settlement equation is written as:

$$\Delta H = H_o \frac{C_\alpha}{1 + e_o} \log \left(\frac{t_{end}}{t_{start}} \right) \tag{24.35}$$

where ΔH is the creep settlement, H_0 is the layer thickness, e_0 is the initial void ratio, C_α is the secondary compression index, t_{start} is the start time, and t_{end} is the end time. Values of $C_\alpha/(1 + e_0)$ have been reported (Sharma 2000) as varying from 0.1 to 0.4, with the higher values for higher organic content and higher degree of decomposition of the waste.

24.7.2 Regulations

The U.S. Resource Conservation and Recovery Act (RCRA) was passed in 1970 and amended in 1980 and 1984. Subtitle D of RCRA applies to MSW landfills, whereas subtitle C of RCRA applies to hazardous solid waste landfills. The issues covered are location, operation, design, monitoring, closure, and postclosure. Restrictions exist when landfill locations are proposed near airports, wetlands, floodplains, and fault areas. The surface area A required for a landfill in a city is calculated by:

$$A = \frac{WPt}{D\gamma} \tag{24.36}$$

where W is the weight of waste generated by a person per day, P is the total population of the city, t is the design period for the landfill, D is the depth of the landfill, and γ is the unit weight of the compacted landfill. The weight generated by one person per day is about 20 N. The unit weight of waste in a landfill varies widely, with an average of around 8 kN/m^3. The period t varies from 10 to 30 years, and the depth D is between 10 and 30 m.

There are many aspects to operating a landfill properly. First, a daily cover of about 0.3 m thick coarse-grained soil is required to cover the waste that was brought in that day. Other aspects include monitoring of the gas generated by the waste, control of public access, control of discharge and surface water, and recordkeeping regarding compliance.

One of the main components of the design of a landfill is the bottom composite liner, with a leachate collection system, a gas venting system, and a groundwater monitoring system; the top cover is another primary component. Closure takes place when the final cover is completed. Mandatory postclosure activities including maintenance of the top cover and of the leachate collection system, as well as monitoring of the gas generated and the groundwater, must continue for 30 years.

24.7.3 Liners

Liners are barriers constructed at the bottom and on the side of landfills. Their purpose is to keep the waste and any by-product(s) out of the surrounding soil and groundwater. For municipal solid waste landfills, the liner composition is specified by RCRA Subtitle D (40 C.F.R. 258), and consists of a series of layers performing different functions. Going from the top to the bottom of the bottom liner, the following layers (2 through 6 for the liner) are encountered (Figure 24.21):

1. Waste
2. Protective soil cover to minimize damage to the underlying geotextile
3. Geotextile layer that acts as a filter for any liquid coming down from the waste
4. Coarse-grained soil layer to serve as a leachate collection system
5. Geomembrane layer to prevent liquid penetration into the underlying layers; this geomembrane must be at least 0.75 mm thick for a flexible membrane liner and 1.5 mm thick for a high-density polyethylene (Figure 24.22)
6. Low-permeability soil layer ($k < 10^{-9}$ m/s) with a minimum thickness of 0.6 m
7. Natural soil

The liner should have a slope so that the leachate can drain naturally by gravity, be collected at a low point or sump, and be pumped and treated on a regular basis. The liner on

Figure 24.22 Installing a geomembrane in a bottom liner. (Courtesy of Layfield Environmental Systems, Layfield Group Limited, 11120 Silversmith Place, Richmond, British Columbia, Canada V7A 5E4.)

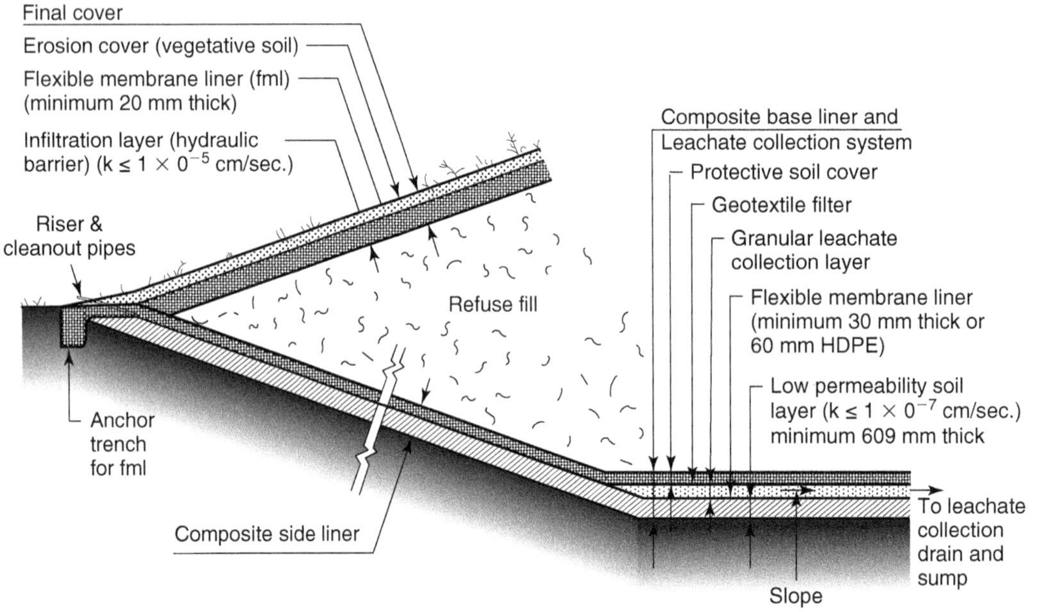

Figure 24.21 Landfill cover and bottom liner composition for municipal solid wastes. (After Sharma and Reddy, 2004. This material is reproduced with permission of John Wiley & Sons, Inc.).

the side slopes is the same as the bottom liner except that it does not typically have a leachate collection layer. The leachate naturally goes to the bottom of the landfill, where it is collected in the leachate collection layer.

For hazardous solid waste landfills, the liner composition is specified by RCRA Subtitle C (40 C.F.R. 244) with a series of layers as follows. Going from the top to the bottom of the bottom liner, the following layers (2 through 6 for the liner) are encountered:

1. Waste.
2. Protective soil cover (optional) to minimize damage to the underlying geomembrane.
3. Geomembrane to act as a barrier for any liquid coming down from the waste. This geomembrane must be at least 0.76 mm thick if there is a protective soil layer above it or at least 1.14 mm thick if there is no protective layer above. For HDPE liners, the minimum required thickness is larger, varying from 1.5 to 2.5 mm.
4. Coarse-grained soil layer to serve as a leachate collection system.
5. Geomembrane layer to serve as a barrier preventing liquid penetration into the underlying layers.
6. Low-permeability soil layer ($k < 10^{-9}$ m/s) with a minimum thickness of 0.9 m.
7. Natural soil.

As can be seen, a municipal solid waste liner is a single liner, whereas a hazardous solid waste liner is a double liner with the leachate collection system sandwiched between the two liners.

The geomembranes and geotextiles used in landfill liners are discussed in Chapter 25. The hydraulic conductivity k of the liner must be less than 10^{-9} m/s. The hydraulic conductivity k of soils is discussed in sections 13.2.5 and 13.2.6. The measurement of k in the laboratory is discussed in sections 9.16 to 9.19 and in the field in section 7.12. The k values of clays permeated by contaminated liquids may differ significantly from the values obtained with water because of the chemistry of the permeating fluid. Various experiments starting in the late 1980s (e.g., Bowders and Daniel 1987; Shackelford 1994) indicated that when a clay is permeated with different chemicals, the hydraulic conductivity changes—sometimes dramatically. For example, a high concentration of methanol or heptane or trichloroethylene in the fluid will increase k; this is because such chemicals decrease the thickness of the clay particle double layer. In contrast, diluted acid in the permeating fluid will tend to decrease the value of k because the acid can create precipitates that clog the clay pores and render flow more difficult; however, k will likely increase in the long term. One first step in gauging whether a chemical will alter the hydraulic conductivity of a soil is to investigate the change in Atterberg limits when the soil is mixed with the chemical; note, though, that the link between the effect on Atterberg limits and k is not always clear. Mitchell and Madsen (1987) concluded that permeation with hydrocarbons may affect k, but only if the concentration in the permeating fluid exceeds their solubility limit. Similar caution should be exercised for geosynthetic bentonite-clay liners (Shackelford 2000). In all cases, it is best to run site-specific tests with the clay from the site and the anticipated fluid, including the appropriate chemical concentration.

24.7.4 Covers

Covers (Figure 24.21) are placed on top of landfills that are full and must be closed. A cover has many purposes, including minimizing the infiltration of rainwater, decreasing the hydraulic head on the bottom liner, resisting surface erosion, keeping away rodents and insects, controlling gas emissions, and improving aesthetics. The typical cross section of a cover consists of a series of layers (1 through 5) as follows:

1. Vegetative layer for aesthetics and erosion protection.
2. Protective soil layer (optional).
3. Drainage layer to collect water, made of gravel and sand.
4. Barrier layer to stop water from penetrating into the waste. This layer may consist of a compacted clay layer, a geosynthetic clay liner (GCL), a geomembrane, or combinations thereof.
5. Drainage layer to collect gas generated by the waste, made of sand and gravel or geotextile.
6. Waste.

The specifications for covers of hazardous solid waste landfills (RCRA subtitle C) are more stringent than for covers of municipal solid waste landfills (RCRA subtitle D). For hazardous wastes, the required thickness of the layers is larger than for municipal wastes.

The final elevation of the top of a landfill is usually higher than the surrounding ground elevation (Figure 24.23). The side slopes of the final cover may be at 24° with the horizontal if the cover is made of soil layers, but it may be prudent to have the slopes at only 18° if a geomembrane is included in the cover, unless special measures to improve geomembrane roughness are taken. The top of the landfill is also sloped, but only at 2 to 5% on either side of the center to provide natural drainage.

24.7.5 Leachate Collection

The amount of leachate that would go through a single compacted clay liner is given by:

$$q = k \frac{\Delta h}{L} A \tag{24.37}$$

where q is the flow in m^3/s, k is the soil hydraulic conductivity in m/s, Δh is the change in total head when crossing the compacted clay layer, L is the length of the flow path through the liner (thickness), and A is the plan view area of the liner. The hydraulic conductivity k is required by design to be less than 10^{-9} m/s, so this is the number used in Eq. 24.37. The

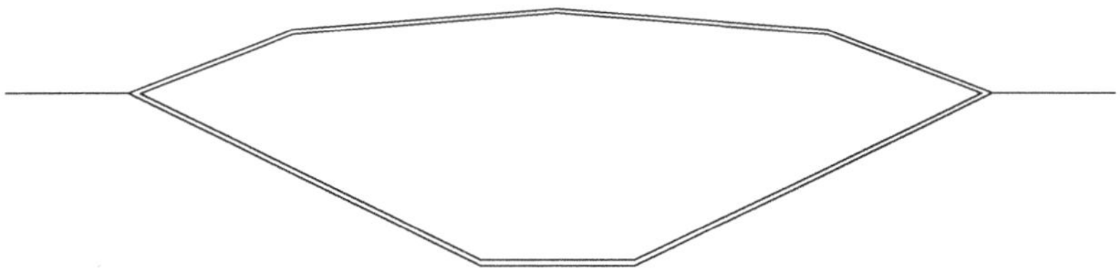

Figure 24.23 General cross section of a landfill.

change in total head Δh is usually taken as the sum of the height of liquid standing on top of the liner plus the thickness of the liner. This assumes that the total head under the liner is zero.

A *composite liner* is made of a geomembrane underlain by a compacted clay liner. The amount of leachate that would go through a composite liner was studied by Giroud and Bonaparte (1989), who recommended the following equation:

$$q = \frac{a^{0.1}k^{0.88}h_w A}{170000} \tag{24.38}$$

where q is the flow in m^3/s, a is the cumulative area of holes in the geomembrane in m^2 per acre (4047 m^2), k is the soil hydraulic conductivity in m/s, h_w is the height of liquid on top of the geomembrane in m, and A is the area of the bottom liner over which the flow of leachate q is calculated. Furthermore, Giroud and Bonaparte (1989) recommend assuming one hole of 3.2 mm^2 per 4047 m^2 of geomembrane under operating conditions, but a much larger hole for conservative sizing of the leachate collection system. For sizing purposes, they recommend a hole of 103 mm^2 per 4047 m^2 of geomembrane.

A cover is exposed to rain, runoff, and evaporation. The amount of leachate through a cover's top layer is calculated as follows:

$$I = P - R - E \pm \Delta S \tag{24.39}$$

where I is the infiltration, P is the precipitation, R is the runoff, E is the evapotranspiration, and ΔS is the change in water volume per unit time of the soil cover. All terms in Eq. 24.39 take the same units (m^3/yr, for example). If there is no cover on the waste, as is the case during operation, the amount of leachate reaching the bottom liner should be reflected by adding another term to Eq. 24.39, to represent the amount of liquid generated by the waste itself by compression or by chemical reaction.

The leachate collection system within covers and liners is built with a slope such that the leachate flows downward in the drainage layer toward a sump. At the sump, the leachate is collected and pumped to the surface, where it is analyzed and treated.

24.7.6 Landfill Slopes

The topic of slope stability is covered in Chapter 19. In the case of a landfill, slope stability comes into play in a number of instances (Figure 24.24), including the side slopes of the excavation, the stability of the side slope liner at the time of construction, the stability of the side slope liner when loaded unevenly by the waste pile, the stability of the waste when the landfilling operation advances through the landfill area, and the stability of the waste and cover upon closure of the landfill.

The stability of the side slope of the excavation can be addressed by using the methods described in Chapter 19. The stability of the waste, for the case of a failure in the waste itself, can also be addressed using conventional methods, except that the shear strength of the waste may or may not follow soil mechanics principles (section 24.7.1). The stability of the side slope liner is the case of a thin and long slope feature; it can be addressed by using the infinite slope method (see section 19.3). The stability of the side slope liner when loaded by the waste is usually a controlling factor in design because it is more severe than the case of the liner by itself. In this case, the most likely failure mechanism is a block failure along the side and bottom liner, because the liner may be the weak link in the resistance to shear. Thus, it is best not to have the front face of the waste at a steep slope. One important issue is the shear strength of the interface between the various materials making up the liner. Each interface should be checked and the associated factor of safety calculated. Factors of safety between 1.3 and 1.5 are common.

The interfaces involve the geomembrane, the geotextile, the geosynthetic clay liner (GCL), the drainage layer, the natural soil, and the waste. The geomembrane should be textured rather than smooth, to improve its interface shear strength,

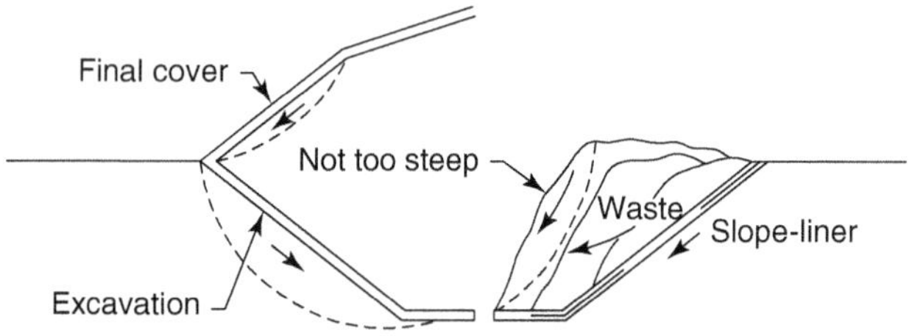

Figure 24.24 Slope stability design issues in a landfill

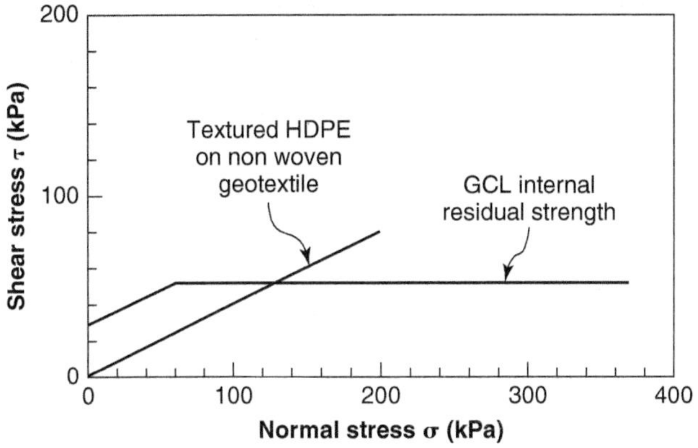

Figure 24.25 Strength envelopes of various interfaces (After Sharma et al. 1997)

and the GCL should be stitched to dramatically increase the shear strength of the bentonite layer. The best way to obtain design values for the interface shear strength is to perform direct shear strength tests (ASTM D5321 and D 6143) on site-specific materials under simulated field conditions. Some aspects of the behavior are important to document: the peak shear strength, the postpeak residual shear strength, the influence of the normal stress level, and the nonlinearity of the strength envelope. It is useful to place all shear strength envelopes on the same graph when the tests are completed (Figure 24.25) to find out which of the interfaces is the weak link for a given normal stress. A seismic slope analysis is also necessary (see section 19.18).

An additional problem may arise when during construction of the landfill, the excavation proceeds through a clay layer with an underlying sand layer under artesian pressure (Figure 24.26). Though this case is rare, it can be disastrous, because if the excavation is dug to a depth where the water pressure ($\gamma_w h_w$) at the top of the sand layer overcomes the downward pressure of the clay remaining on top of the sand (γh), the bottom of the excavation will blow up and a mixture of sand and water will run into the excavation. A factor of safety must be applied to the maximum depth of excavation to guard against such an event. The safe remaining thickness

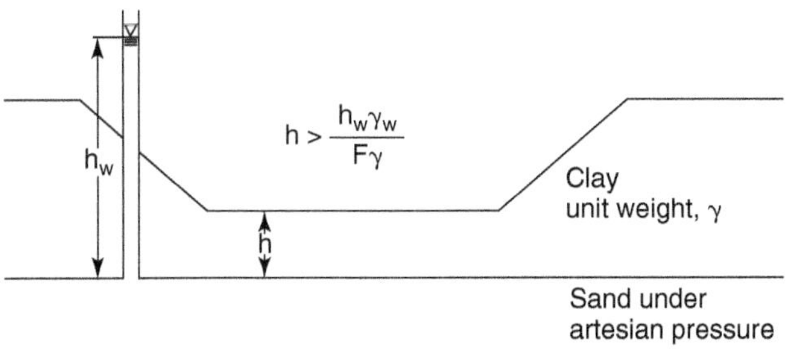

Figure 24.26 Blowout problem at bottom of excavation.

h of the clay layer is:

$$h = \frac{\gamma_w h_w}{F\gamma} \tag{24.40}$$

24.7.7 Gas Generation and Management

Landfills generate gas, mostly carbon dioxide and methane, through biodegradation. These gases are flammable, are toxic to humans, can create excessive deformation of the liners, and smell bad. Furthermore, methane is a greenhouse gas. The gas generation process is due to the work of bacteria that transform some of the waste through digestion. The product is approximately 50% carbon dioxide (CO_2) and 50% methane (CH_4). The carbon dioxide is usually generated first, followed by the methane. The intensity of this process depends on a number of factors, including the availability of nutrients for the bacteria, temperature, humidity, pH, and age of the waste. Landfill temperatures vary from 20 to 60°C. Higher temperature and higher water content of the waste are more favorable to gas generation, which can reach 10,000 m^3 per kN of waste over the life of the landfill. Gas generation in a landfill does have a finite life, which can vary from 20 years under favorable conditions where biodegradation is rapid (e.g., humid climates) to 100 years under unfavorable conditions where biodegradation is slow (e.g., arid climates).

The gas generated must be disposed of and the disposal process monitored. There are essentially three ways to dispose of gasses: vent to the atmosphere, vent and burn with no energy recovery, and vent and burn with energy recovery. The most common of the three is vent and burn without energy recovery, through the use of flares. Venting is achieved by placing gas wells into the waste, which facilitate gas migration to the surface where the gas is burned. Old landfills used open flame flares, which are the simplest kind, but modern landfills use enclosed flares because they allow for measurement of the gas coming out of the waste and yield better overall control. Wells typically consist of perforated pipes 50 to 300 mm in diameter that extend to 75% of the full depth of the landfill. The spacing varies from 15 to 100 m and averages 60 m. The energy recovery systems use the gas to power gas turbines or combustion engines to generate electricity, but the initial cost of such a system is worth the investment only for large landfills with more than 10 million kN of waste.

24.8 FUTURE CONSIDERATIONS

Lowering the generation of waste at the source is the first and best way to decrease the amount of waste generated by humankind. Recycling is the second best option. Recycling of household waste or municipal solid waste has become part of everyday life, and over the past 20 years has reduced the amount of waste going to landfills to about 50% of the MSW generated (Figure 24.27). The most successful programs have been recycling of aluminum and paper, because in both cases the cost-benefit ratio is favorable. Recycling does not stop at

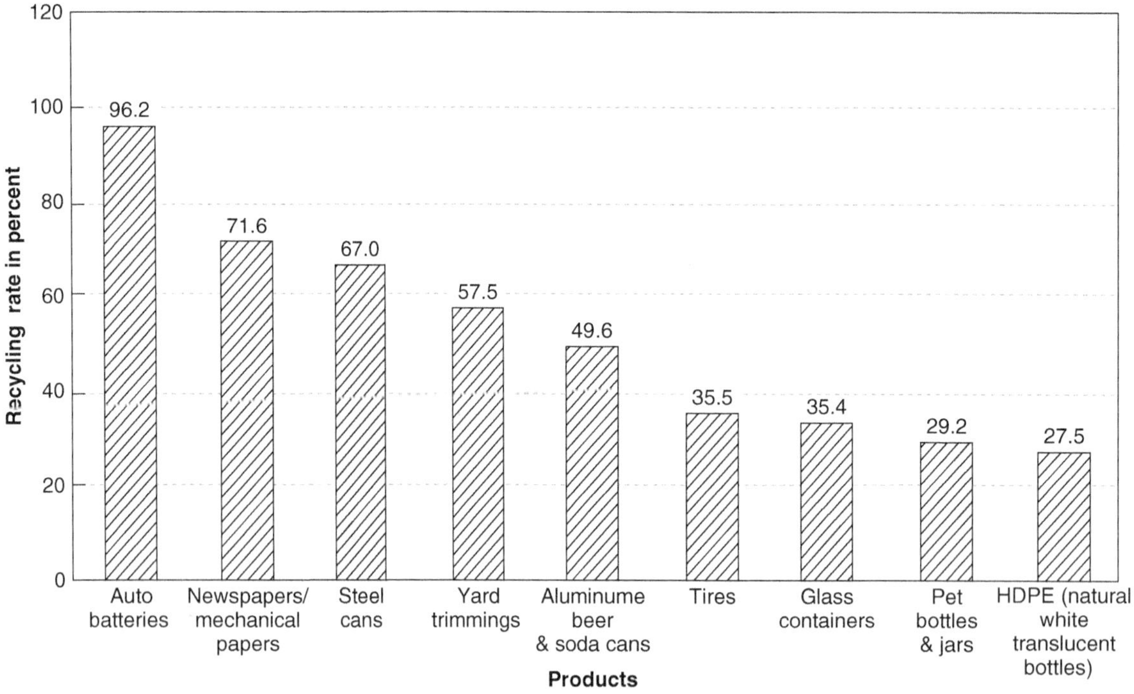

Figure 24.27 Recycling rate for various waste products in 2011 (Source: EPA).

Figure 24.28 Postclosure use of landfills.

household waste, but extends as well to the industrial sector, which is by far the largest generator of waste. Efforts for recycling fly ash, blast furnace slag, foundry sand, paper mill sludge, incinerator ash, glass, plastics, scrap tires, demolition and concrete debris, and wood waste are being made (Sharma and Reddy 2004). Note that once landfills are closed, the area can be used for various activities including parks, golf courses, airports, and sports stadiums (Figure 24.28).

PROBLEMS

24.1 How do you define waste?
24.2 What is the biggest generator of waste in the United States?
24.3 Are solid wastes solids?
24.4 What are the four main categories of waste?
24.5 How long can a high-level radioactive waste continue to be deadly?
24.6 What are RCRA and CERCLA and what do they regulate?
24.7 What does an OSHA level C mean and what does it require?
24.8 What is the name of the smallest piece of matter and what are its components?
24.9 What is the difference between an atom and a molecule?
24.10 What is the difference between an ion, an anion, and a cation?
24.11 In the case of concentration in water, how many parts per million (ppm) are in 1 mg/m^3?

24.12 What is the difference between organic and inorganic materials?

24.13 What is the difference between atomic absorption spectrophotometry (AAS) and gas chromatography-mass spectrometry (GC-MS)?

24.14 What are ESA I, II, and III, and when are they used?

24.15 What is the difference between the LIF CPT, the MIP CPT, and the BAT CPT?

24.16 In contaminant transport, what is the difference between concentration and flux?

24.17 Choose some reasonable values of the parameters in the solution to contamination propagation (Eq. 24.29) and draw the propagation plot. Then vary each parameter to understand the influence each one has on the propagation. For help with the solution, go tot www.lmnoeng.com/Groundwater/transportStep.htm

24.18 How can you form a bottom barrier at depth for a waste containment system?

24.19 A landfill has been closed for one day and the long-term settlement must be evaluated. The waste is 15 m deep, has a unit weight of 8 kN/m^3, and has a $C_\alpha/(1 + e_o)$ coefficient equal to 0.2. Calculate the creep settlement after 20 years.

24.20 Calculate the area of the landfill necessary to handle the municipal solid waste generated by a city of 1 million people over a period of 10 years. Each person in that city generates 20 N of MSW per day. The depth to the water table is 20 m and a high-plasticity clay layer exists at a depth of 15 m.

24.21 Calculate the flow of leachate through a 0.6 m thick clay liner covering a 200 m × 200 m area. The leachate level is 0.4 m above the top of the liner and the hydraulic conductivity of the clay meets the specification of 10^{-9} m/s.

24.22 Chloride dissolved in water is leaching through a liner and permeating into an aquifer-bearing 2 m thick layer of silty sand. The concentration of the dissolved chloride is 1500 mg/liter, the discharge velocity is 3.7×10^{-7} m/s, and the porosity of the silty sand is 0.25. Calculate the mass flux of chloride into the aquifer per unit area of landfill liner due to advection.

24.23 Calculate the flow of leachate through a composite liner with a 0.75 m thick compacted clay layer over an HDPE geomembrane. The clay has a hydraulic conductivity of 10^{-9} m/s and the HDPE membrane is 1.5 mm thick. The height of liquid above the geomembrane is 0.2 m. Give the answer for expected operating conditions first and then give a more conservative estimate for sizing the leachate pumping system.

24.24 An excavation is dug for a landfill in a 20 m thick stiff, high-plasticity clay layer underlain by a sand layer under artesian pressure. The unit weight of the clay is 19 kN/m^3 and that of the sand is 20 kN/m^3. The artesian pressure is such that a casing through the clay into the sand has water rising 10 m above the top of the clay layer (ground surface). How deep can the excavation be dug into the clay to maintain a minimum factor of safety of 1.5 against bottom blowout failure? Draw the effective vertical stress profile before and after excavation to that depth.

Problems and Solutions

Problem 24.1

How do you define waste?

Solution 24.1

Waste is unwanted or useless material.

Problem 24.2

What is the biggest generator of waste in the United States?

Solution 24.2

The industrial sector is the largest generator of waste.

Problem 24.3

Are solid wastes solids?

Solution 24.3

No. The term *solid waste* is misleading, as a solid waste can be a solid, a liquid, or a gas.

Problem 24.4

What are the four main categories of waste?

Solution 24.4

The four main categories of waste are: solid wastes, hazardous wastes, radioactive wastes, and medical wastes.

Problem 24.5

How long can a high-level radioactive waste continue to be deadly?

Solution 24.5

The radiation penetration from high-level wastes, which are generated by defense or nuclear power plant activities, remains lethal for 10,000 years.

Problem 24.6

What are RCRA and CERCLA and what do they regulate?

Solution 24.6

RCRA is the Resource Conservation and Recovery Act; it addresses the issue of landfill design. CERCLA is the Comprehensive Environmental Response, Compensation, and Liabilities Act; it addresses the issue of cleaning up contaminated sites.

Problem 24.7

What does an OSHA level C mean and what does it require?

Solution 24.7

OSHA stands for Occupational Safety and Health Administration. OSHA level C refers to moderate protection for humans working at contaminated sites. That is, it requires moderate protection including full-face or half-mask air-purifying respirator, hooded chemical-resistant clothing, inner and outer chemical-resistant gloves, chemical-resistant boots and boot covers, hard hat, escape mask, and face shield.

Problem 24.8

What is the name of the smallest piece of matter and what are its components?

Solution 24.8

An atom is the smallest piece of matter. Atoms are made of protons, electrons, and neutrons

Problem 24.9

What is the difference between an atom and a molecule?

Solution 24.9

Atoms consist of a nucleus containing protons and neutrons with electrons surrounding the nucleus; they are the basic building blocks of matter (e.g., hydrogen atom). Molecules are combinations of atoms bonded together. For example, two hydrogen atoms and an oxygen atom form a molecule of water (H_2O).

Problem 24.10

What is the difference between an ion, an anion, and a cation?

Solution 24.10

Ion is the general term for an atom that has lost or gained an electron on its outer orbital. More specifically, an ion can be a cation or an anion. *Cations* are neutral atoms that have lost one or more electrons, making them positively charged, such as Na^+, Ca^{++}, and Al^{+++}. *Anions* are the opposite: They have gained one or more electrons and thus have a net negative charge, such as Cl^-, O^-, and N^-.

Problem 24.11

In the case of concentration in water, how many parts per million (ppm) are in 1 mg/m^3?

Solution 24.11

For the mass concentration of a chemical in water, 1 ppm = 1 mg/liter and 1 liter = 0.001 m^3. Therefore, there are 0.001 ppm in 1 mg/m^3.

Problem 24.12

What is the difference between organic and inorganic materials?

Solution 24.12

The difference between organic and inorganic compounds is that most organic compounds contain carbon, whereas most inorganic compounds do not.

Problem 24.13

What is the difference between atomic absorption spectrophotometry (AAS) and gas chromatography-mass spectrometry (GC-MS)?

Solution 24.13

GC-MS is used to identify the components of a chemical mixture. AAS is used to measure the molar concentration of chemicals.

Problem 24.14

What are ESA I, II, and III, and when are they used?

Solution 24.14

Environmental site assessments or ESAs are part of the contamination detection process. They are often required when purchasing a piece of property in the United States. There are three levels:

ESA I: This phase consists of collecting information regarding previous ownership and prior use through records of contaminated sites in the area, aerial photos, geologic and topographic maps, visits to the site, and talking to neighbors. An ESA I indicates whether there are reasons to believe the site is contaminated. If so, ESA II comes into play.

ESA II: This phase consists of testing the soil and the groundwater to find out if there is contamination and, if there is, to what extent and to what level of severity (type of contaminants). If contamination that requires cleanup is found, ESA III comes into play.

ESA III: This phase consists of designing the remediation scheme and achieving it, including verification that a satisfactory level of cleanup has been realized.

Problem 24.15

What is the difference between the LIF CPT, the MIP CPT, and the BAT CPT?

Solution 24.15

LIF (laser-induced fluorescence) is a CPT technique used to determine the extent of plumes at petroleum-contaminated sites and the type of petroleum product contaminating the site. A laser beam is shone on the soil, which emits different fluorescence depending on the hydrocarbon present.

MIP (membrane interface probe) is a CPT technique used to identify the type of volatile organic compound by heating the soil and letting the gas permeate through a membrane located on the side of the CPT. Once in the CPT housing, the gas is swept by an inert carrier gas to the surface where it is analyzed.

BAT is a CPT technique used to collect groundwater. (BAT is the name of a company.) The CPT probe is equipped with a porous filter that is obstructed until the CPT probe is pushed to the required depth. Then the filter is exposed and water is allowed to penetrate through the filter into a water sampling tube, which can be removed through the CPT rods when full.

Problem 24.16

In contaminant transport, what is the difference between concentration and flux?

Solution 24.16

Concentration C is the mass of contaminant (solute) per volume of liquid carrying the contaminant (solvent); it is measured in kg/m^3.

Flux F is the mass of contaminant flowing through a unit area of soil per unit of time; it is measured in kg/m^{2}s.

They are related through F = Cv.

Problem 24.17

Choose some reasonable values of the parameters in the solution to contamination propagation (Eq. 24.29) and draw the propagation plot. Then vary each parameter to understand the influence each one has on the propagation. For help with the solution, go to www.lmnoeng.com/Groundwater/transportStep.htm

Solution 24.17

$$C_w\,(x,t) = \frac{C_0}{2}\,erfc\left(\frac{R_d x - v_s t}{\sqrt{4\,R_d D_H t}}\right)$$

C_0 = Constant contaminant concentration at point $X = 0$ and $t = 0$

X = Distance

V_s = Seepage velocity

R_d = Retardation factor $R_d = 1 + \frac{\rho_d K_d}{n}$ where ρ_d is dry density and n is total porosity

K_d = Partition coefficient

D_H = Hydrodynamic dispersion $D_H = D^* + \alpha_L v_s$ where α_L dispersivity varies from 0.1 to 100

$D*$ = Molecular diffusion coefficient; typical value 1×10^{-9} (m^2/s)

Propagation of the contamination is plotted as normalized concentration C_w/C_0 versus x for the following input parameters:

$$\alpha_L = 100,\ \rho_d = 1.6(\text{g/cm}^3),\ n = 35\ (\%),\ D^* = 1 \times 10^{-5}\ (\text{cm}^2/\text{s}),$$
$$v_s = 1.92e-5\ (\text{cm/s}),\ K_d = 0.1\ (\text{cm}^3\ \text{g}),\ t = 1000\ (\text{days})$$

- Propagation plot (Figure 24.1s)

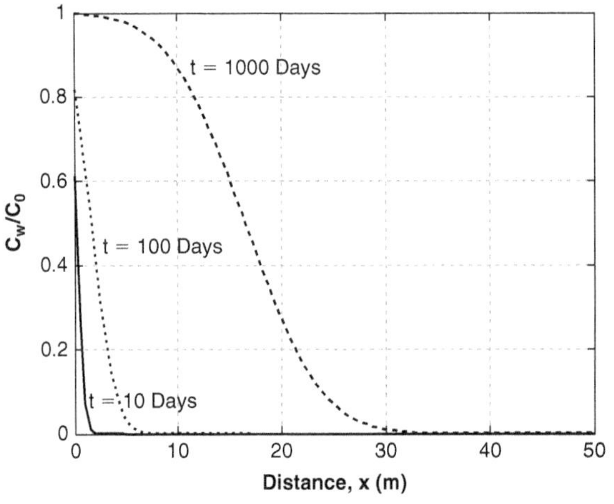

Figure 24.1s Propagation plot.

- Varying parameter α_L (Figure 24.2s)

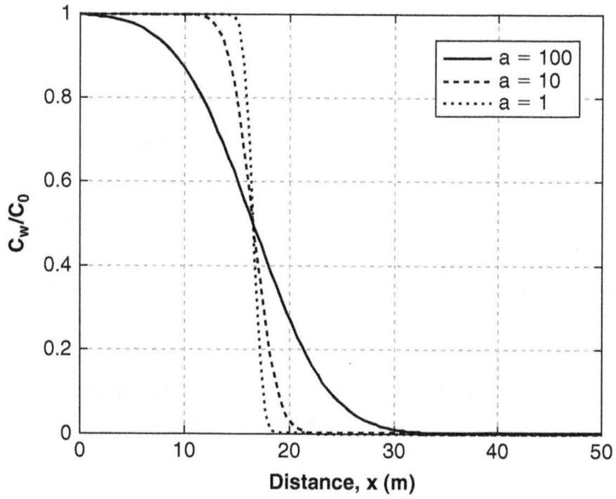

Figure 24.2s Propagation plot for different values of α_L.

- Varying parameter v_s (Figure 24.3s)

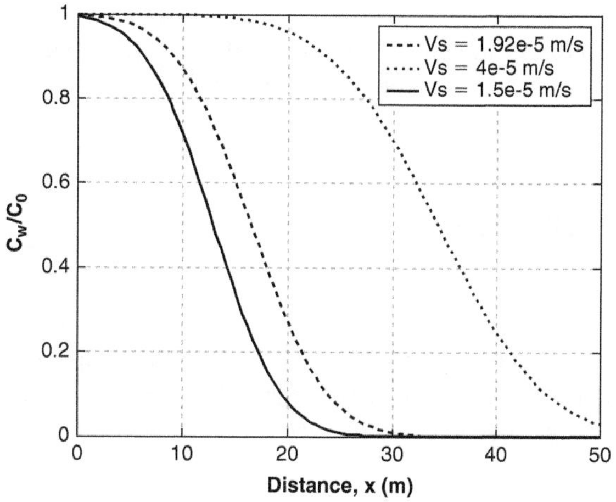

Figure 24.3s Propagation plot for different values of v_s.

- Varying parameter ρ_d (Figure 24.4s)

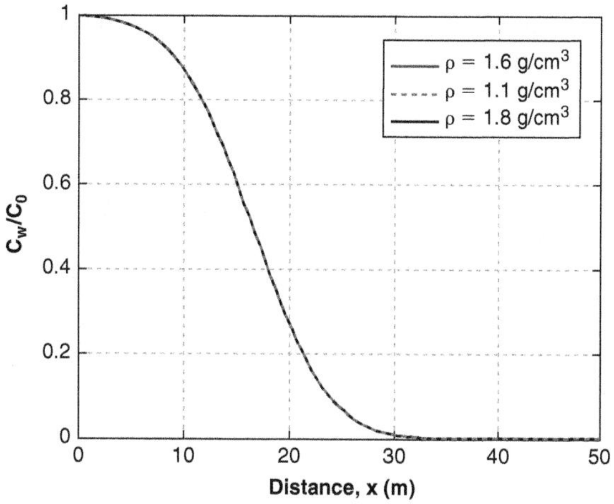

Figure 24.4s Propagation plot for different values of ρ_d.

Problem 24.18

How can you form a bottom barrier at depth for a waste containment system?

Solution 24.18

A bottom barrier can be constructed by grouting or directional drilling.

Grouting can be pressure grouting or jet grouting, but in both cases the injection pipe is driven or vibrodriven to the depth of the bottom barrier and a grout bulb is constructed. The operation is repeated until the overlapping bulbs form a bottom barrier. The drawback with this technique is that holes have to be punched through the waste or contaminated zone.

Directional drilling consists of setting an inclined drill outside of the contaminated zone and drilling at an angle to reach underneath that zone. Then the hole is grouted. Side-by-side holes are drilled and grouted to form the bottom barrier.

Problem 24.19

A landfill has been closed for one day and the long-term settlement must be evaluated. The waste is 15 m deep, has a unit weight of 8 kN/m^3, and has a $C_\alpha/(1 + e_o)$ coefficient equal to 0.2. Calculate the creep settlement after 20 years.

Solution 24.19

$$\Delta H = H_0 \frac{C_\alpha}{1 + e_0} \log\left(\frac{t_{end}}{t_{start}}\right) = 15 \text{ m} * 0.06 * \log\left(\frac{20^*365 \text{ day}}{1 \text{ day}}\right) = 3.48 \text{ m}$$

Problem 24.20

Calculate the area of the landfill necessary to handle the municipal solid waste generated by a city of 1 million people over a period of 10 years. Each person in that city generates 20 N of MSW per day. The depth to the water table is 20 m and a high-plasticity clay layer exists at a depth of 15 m.

Solution 24.20

The surface area A required for a landfill in a city is calculated by:

$$A = WPtD\gamma$$

where W is the weight of waste generated by a person per day, P is the total population of the city, t is the design period for the landfill, D is the depth of the landfill, and γ is the unit weight of the compacted landfill (estimated at 8 kN/m^3).

The depth of the water table is 20 m, and there is high-plasticity clay at depth 15 m, so the depth of landfill is selected as 15 m:

$$A = \frac{20 \text{ N/day} \times 10^6 \times 3650 \text{ day}}{15 \text{ m} \times 8 \text{ kN/m}^3} = \frac{7.3 \times 10^7 \text{ kN}}{1.2 \times 10^2 \text{ kN/m}^2} = 6.08 \times 10^5 \text{ m}^2$$

Problem 24.21

Calculate the flow of leachate through a 0.6 m thick clay liner covering a 200 m × 200 m area. The leachate level is 0.4 m above the top of the liner and the hydraulic conductivity of the clay meets the specification of 10^{-9} m/s.

Solution 24.21

The amount of leachate that would go through a single compacted clay liner is calculated by $q = K\frac{\Delta h}{L}A$, where q is the flow in m^3/s, k is the soil hydraulic conductivity in m/s, Δh is the change in total head when crossing the compacted clay layer, L is the length of the flow path through the liner (thickness), and A is the plan view area of the liner:

$$\Delta h = 0.6 + 0.4 = 1 \text{ m}$$

$$A = 200 \times 200 = 40000 \text{ m}^2$$

$$q = K\frac{\Delta h}{L} \text{A} = 10^{-9} \times \frac{1}{0.6} \times 40000 = 6.67 \times 10^{-5} \text{ m}^3/\text{s}$$

Problem 24.22

Chloride dissolved in water is leaching through a liner and permeating into an aquifer-bearing 2 m thick layer of silty sand. The concentration of the dissolved chloride is 1500 mg/liter, the discharge velocity is 3.7×10^{-7} m/s, and the porosity of the silty sand is 0.25. Calculate the mass flux of chloride into the aquifer per unit area of landfill liner due to advection.

Solution 24.22

$$F_{adv} = vC = nv_sC = 3.7 \times 10^{-7} \times 1500 \times 10^{-3} = 5.55 \times 10^{-7}\ \text{mg/m}^2\ \text{s}$$

Problem 24.23

Calculate the flow of leachate through a composite liner with a 0.75 m thick compacted clay layer over an HDPE geomembrane. The clay has a hydraulic conductivity of 10^{-9} m/s and the HDPE membrane is 1.5 mm thick. The height of liquid above the geomembrane is 0.2 m. Give the answer for expected operating conditions first and then give a more conservative estimate for sizing the leachate pumping system.

Solution 24.23

$$q = \frac{a^{0.1}k^{0.88}h_wA}{170000}$$

Case 1: Operating conditions

$$q = \frac{\left(\dfrac{3.2 \times 10^{-6}}{4047}\right)^{0.1} \times (10^{-9})^{0.88} \times 0.2 \times 1}{170000} = 1.73 \times 10^{-15}\ \text{m}^3/\text{s} = 0.55\ \text{cm}^3/\text{year}$$

Case 2: Pumping system design

$$q = \frac{\left(\dfrac{103 \times 10^{-6}}{4047}\right)^{0.1} \times (10^{-9})^{0.88} \times 0.2 \times 1}{170000} = 2.64 \times 10^{-15}\ \text{m}^3/\text{s} = 0.83\ \text{cm}^3/\text{year}$$

Problem 24.24

An excavation is dug for a landfill in a 20 m thick stiff, high-plasticity clay layer underlain by a sand layer under artesian pressure. The unit weight of the clay is 19 kN/m^3 and that of the sand is 20 kN/m^3. The artesian pressure is such that a casing through the clay into the sand has water rising 10 m above the top of the clay layer (ground surface). How deep can the excavation be dug into the clay to maintain a minimum factor of safety of 1.5 against bottom blowout failure?

Solution 24.24

The safe thickness h of the clay layer after excavation is:

$$h = \frac{\gamma_w\ \text{h}_w}{F\gamma} = \frac{9.81 \times 30}{1.5 \times 19} = 10.33\ \text{m}$$

Therefore, the safe excavation depth is 20—10.33 = 9.67 m.

CHAPTER 25

Geosynthetics

25.1 GENERAL

Geosynthetics have been to geotechnical engineering what computers have been to humankind in general: a revolution. The use of these planar synthetic materials in soils to reinforce, to drain, and to separate has grown remarkably over the past 50 years to the point where it is a huge industry today. According to ASTM D4439, a *geosynthetic* is a planar product manufactured from polymeric material (plastics) to be used with soil, rock, earth, or other geotechnical engineering-related materials as an integral part of a human-made project, structure, or system. There are many types of geosynthetics, including geotextiles, geomembranes, geogrids, geosynthetic clay liners, geofoam, geonets, geocells, geobags, and geocomposites. Geotextiles and geomembranes are the two largest groups of geosynthetics. In 2013, the cost of geosynthetics was between \$1 and \$7 m^2. The book by Koerner (2012) is an excellent reference on geosynthetics.

25.2 TYPES OF GEOSYNTHETICS

Geotextiles (Figure 25.1) are textiles made of synthetic fibers. The fibers are either woven together or tied together (nonwoven). Weaving consists of standard interlacing with textile machinery. In nonwoven fabrics, the fibers are tied together by heating, gluing, or needle-punching. In needle-punching, short needles with barbs are punched through the fabric to provide a mechanical interlocking. Geotextiles are flexible and porous to liquid flow. They are used mainly for separation, reinforcement, filtration, and drainage.

Geomembranes (Figure 25.1) are relatively thin, impervious sheets of plastic material. They are made by first preparing the polymer resin and its additives. The actual forming of the membrane takes place by extrusion through two parallel plates or rollers. The resulting sheet is between 1 and 3 mm thick and can be smooth or roughened. Geomembranes are used mostly as nearly impervious barriers to contain liquids or vapors.

Geogrids (Figure 25.1) are plastic grids that have a very open configuration; they have large holes between ribs. They are formed by bonding rods together, by weaving and then coating, or by stretching. Their main use is reinforcement.

Geosynthetic clay liners or GCLs (Figure 25.1) are made of a thin layer of bentonite clay sandwiched between two layers of geotextiles or geomembranes. GCLs are manufactured by feeding the bentonite on top of a conveyor-belt-style geosynthetic and covering it after the feed point by a top geosynthetic. The two geosynthetic layers are kept together by needle punching, stitching, or gluing. The bentonite will expand dramatically when wetted. GCLs are about 4 to 6 mm thick when the bentonite is hydrated at water contents of 10 to 35%. They are used mostly as nearly impervious barriers to contain liquids or vapors.

Geofoams (Figure 25.1) are extremely light blocks made of polymer bubbles. They are fabricated by thermal expansion and stabilization of polystyrene bubbles. The density of the blocks is about 2% of the density of soils, but 3 to 4 times more expensive per unit volume. They are stacked together to form lightweight fills, and are used as compressible layers behind retaining walls, as vibration dampers for seismic protection, and as thermal insulation in foundations.

Geonets (Figure 25.1), like geogrids, are open netting geosynthetics made of plastic. They are different from geogrids that they are thicker; they are sometimes called *spacers* as they provide space for fluid to flow within the structure. Also, the openings are more like diamonds than the squares of geogrid openings. They are used primarily for drainage purposes.

Geocells are a form of geogrid in the sense that they have a very open configuration, but their purpose is to reinforce by confining the soil within the cells. The cells may be 1 m × 1 m in plan view and 1 to 2 m high. The soil is placed within the cells, which provide lateral confinement and thereby significantly increase the bearing capacity of the soil layer.

Geobags are literally bags made of geosynthetic material; they are usually filled with sand and used for erosion protection in lieu of rip rap. Their size is in the range of rip rap, and can be as large as 5 m^3.

Figure 25.1 Examples of geosynthetics. (Photographs compliments of the Geosynthetic Institute.)

Geocomposites are combinations of the previous geosynthetics that are intended to maximize the usefulness of a geosynthetic layer. They are used as filter layers, for example.

Geosynthetics are useful in a number of geotechnical applications, as shown in Table 25.1

25.3 PROPERTIES OF GEOSYNTHETICS

The parameters used to characterize geosynthetics are much more numerous than and often different from those used for soils. The reason is that the material and the applications are quite different and more versatile than those associated with soils alone. Also, the field of geosynthetics is quite a bit younger than the field of geotechnical engineering. Although very significant progress has been made, some of the properties' definitions, the tests used to determine their value, and the design guidelines are still evolving.

25.3.1 Properties of Geotextiles

Physical Properties

The unit weight of typical plastics varies from 9 to 13 kN/m^3. The unit weight of dry, clean geotextiles is between 3 and 7 kN/m^3—but that is not the way it is typically given. Instead, it is quoted as mass per unit area (ASTM D5261) with values between 150 and 750 g/m^2 (Koerner 2012). The

Table 25.1 Applications for Some Geosynthetics

Geotextiles	Geogrids	Geomembranes	Geosynthetic Clay Liners	Geofoam
• Separation • Roadway reinforcement • Soil reinforcement • Filtration • Drainage	• Reinforcement • Roads • Slopes • Walls • Foundations	• Liners • Ponds • Canals • Landfills, dry • Landfills, wet • Landfill covers	• Liners	• Lightweight fill • Compressible inclusions • Thermal insulations • Drainage

thickness of commonly used geotextiles is between 0.5 to 4 mm (ASTM D5199).

Mechanical Properties

Stiffness is usually defined as the ratio between the force applied and the resulting displacement, as in stiffness of a spring. The stiffness of a geotextile is defined in a very different way; it is obtained from a laboratory test (ASTM D1388) in which a 25 mm wide strip of geotextiles is gradually pushed over the edge of the crest of a slope under controlled conditions (Figure 25.2). The slope is 41.5° with the horizontal and when the strip touches the slope, the length L of the overhanging strip is recorded. The stiffness of the geotextile is defined as:

$$G = M\left(\frac{L}{2}\right)^3 \tag{25.1}$$

where G is the flexural stiffness (g.m), M is the mass per unit area (g/m^2), and L is the overhang length (m). The G values for geotextiles are in the range of 0.01 to 1 g.m.

The average modulus of deformation of geotextile under tension stresses varies widely. It can be 60 MPa for some nonwoven, needle-punched geotextiles all the way to 400 MPa for some woven monofilament geotextiles (Koerner 2012). Because the evolution of the thickness during the test is not certain, this modulus is not commonly quoted for these products. Instead, it is more commonly presented as the ratio of the force per unit length of fabric over the normal strain generated. Numbers in the range of 30 kN/m to 150 kN/m are common.

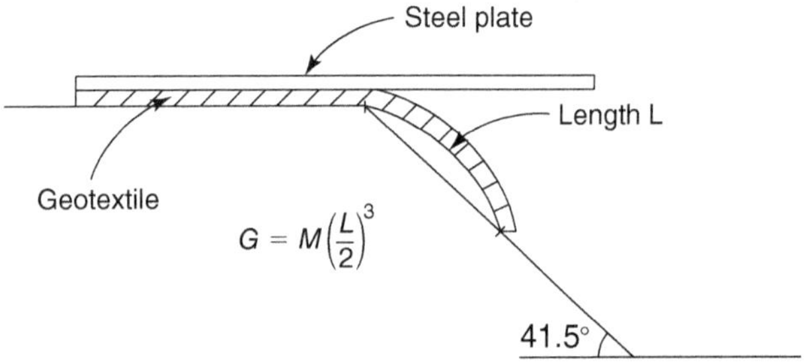

Figure 25.2 Flexure stiffness test for geotextile.

The average tensile strength of geotextiles (ASTM D4632) is in the range of 50 to 100 MPa; again, however, that is not the way it is typically cited. Instead, the average tensile strength is quoted as the force S_t per unit length of fabric that creates rupture; average numbers are in the range of 25 to 60 kN/m. One of the problems is that the thickness varies during elongation of the geotextiles. Another problem is that the strain to failure is much larger than in soils, with values around 25% for some woven fabrics and up to 70% for some nonwoven fabrics. As a result, the tensile strength is usually quoted together with a value of the strain at failure. The tensile strength of the seams (ASTM D4884) is typically 50 to 75% of the tensile strength of the intact fabric. The compressibility of geotextiles is generally not a concern except when they are used to convey water or other liquids in the in-plane direction. In this case it is important to make sure, by testing, that the small conveyance tubes within the geotextiles will not collapse under the in situ compression.

The puncture strength is important and may be quoted as an impact puncture strength or a static puncture strength. The impact puncture strength is tested by dynamically puncturing the fabric with a pendulum test (energy to puncture) (ASTM D256) or a drop cone (penetration distance). The impact puncture strength of a geotextile is quoted as the energy that leads to puncture. Common values of geotextile impact puncture strength vary from 25 Joules to 300 Joules. It is named after the English physicist James Prescott Joule who contributed in the middle to late nineteenth century. The static puncture strength (ASTM D6241) is determined by slowly pushing a 50 mm diameter beveled plunger into the fabric and recording the puncture failure load P. The following empirical relationship between the puncture load P (kN) and the tensile strength S_t (kN/m) has been proposed (Cazzuffi and Venezia 1986):

$$P = S_t \pi d \tag{25.2}$$

where d is the diameter of the punching plunger.

The interface shear strength between a geotextile and a soil can be very important in design and should be measured using

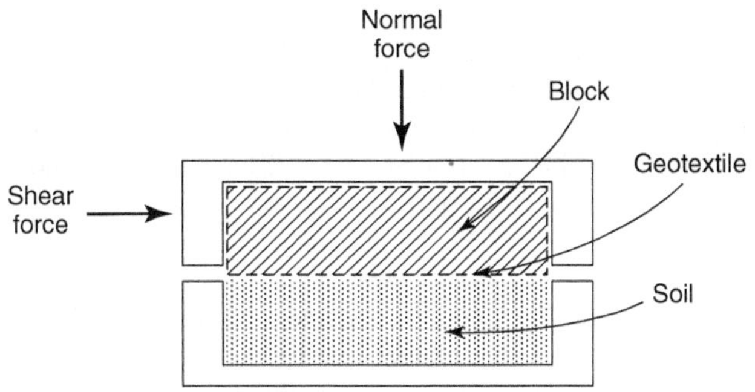

Figure 25.3 Direct shear test for soil-geotextiles interface shear strength.

site-specific materials. The accepted test (ASTM D5321) is a variant of the soil direct shear test in which the top part of the soil is replaced by a geotextile-covered block (Figure 25.3). The interface shear strength is typically in the range of 75% to 100% of the soil shear strength.

Creep or deformation under constant stress is also important for geotextiles, and creep tests are necessary for long-term applications under load (ASTM D5262). The general model used for soils can be extended to geotextiles:

$$\frac{\varepsilon_1}{\varepsilon_2} = \left(\frac{t_1}{t_2}\right)^n \tag{25.3}$$

where ε_1 and ε_2 are the strains reached in a time t_1 and t_2 respectively and n is the viscous exponent.

Table 25.2 gives estimates of the n values for several polymers based on the data from den Hoedt (1986). Note that the time-temperature equivalency, which has been used for a long time in the asphalt field, is also used for geotextiles for speeding up time in creep tests.

Hydraulic Properties

The percent open area of a geotextile can be measured by shining a light through the geotextile onto a poster board and measuring the illuminated area on the poster. Monofilament woven geotextiles have percent open areas in the range of 6 to 12%. The apparent opening size (AOS) of a geotextile is obtained through a test (ASTM D4751) in which glass beads of uniform diameter are placed on top of the geotextile and wet-sieved through the geotextile. The diameter for which 95% of the beads by weight are retained on the geotextile is the AOS, designated as O_{95}. Typical values range from 0.01 mm to 0.5 mm.

A distinction is made between the cross-plane hydraulic conductivity k_{cp} and the in-plane hydraulic conductivity k_{ip}. *Cross-plane* refers to the case where the liquid flows in a direction normal to the plane of the geotextile, this is called *filtration*. *In-plane* refers to the case where the liquid flows parallel to the plane or within the geotextile; this is called *drainage*. Typical values of the hydraulic conductivity of geotextiles (ASTM D4491) range from 10^{-3} m/s to 10^{-5} m/s for the stress-free product. This is in the range of gravel to coarse sand. Because the thickness of the geotextile can vary due to the in situ stress condition, the permittivity ψ (s^{-1}) is used for the cross-plane flow and the transmissivity Θ (m^2/s) is used for in-plane flow instead of the hydraulic conductivity. They are defined as follows:

$$\text{Permittivity} \quad \Psi = \frac{k_{cp}}{t} = \frac{q}{iAt} = \frac{q}{\frac{\Delta h}{t}At} = \frac{q}{\Delta h\ A} \tag{25.4}$$

$$\text{Transmissivity} \quad \Theta = k_{ip}t = \frac{q}{iA}t = \frac{q}{iWt}t = \frac{q}{iW} \tag{25.5}$$

where q is the flow, A is the flow area perpendicular to the flow, i is the hydraulic gradient, Δh is the loss of total head over the flow distance, t is the thickness, and W is the width of the geotextile involved in the flow.

The hydraulic conductivity, the permittivity, and the transmissivity should be tested under the compressive stress likely to be experienced in the field (ASTM D5493). While it is always important to do this testing, the values under load do not appear to be very different from the values under no load. As mentioned earlier, the cross-plane hydraulic conductivity k of geotextiles for water flow typically ranges from 10^{-3} to 10^{-5} m/s. If the fluid is not water, then the hydraulic conductivity and the permittivity should be corrected as follows:

$$\frac{k_f}{k_w} = \frac{\Psi_f}{\Psi_w} = \frac{\Theta_f}{\Theta_w} = \frac{\rho_f\,\mu_w}{\rho_w\,\mu_f} \tag{25.6}$$

Table 25.2 Viscous Exponent n for Several Polymers

Geotextile Polymer	Viscous Exponent n Value at 20% of Ultimate Strength	Viscous Exponent n Value at 60% of Ultimate Strength
Polyester (PET)	0 to 0.01	0 to 0.01
Polyamide (PA) (nylon)	0.02	—
Polypropylene (PE)	0.07	0.19 to 0.2
Polyethylene (PP)	0.08 to 0.14	0.12 to 0.19

(From data by den Hoedt 1986.)

where k_f and k_w are the hydraulic conductivity for the fluid and for water respectively, ψ_f and ψ_w are the permittivity for the fluid and for water respectively, Θ_f and Θ_w are the transmissivity for the fluid and for water respectively, ρ_f and ρ_w are the density of the fluid and of water respectively, and μ_f and μ_w are the viscosity of the fluid and of water respectively.

Other properties of geotextiles include resistance to abrasion from repeated action of gravel impacting the geotextile, soil retention and clogging (as in filters and silt fences), sunlight degradation from long-term exposure to ultraviolet rays, and degradation due to temperature, oxidation, chemical action, and biological action. Designers use reduction factors to take into account these factors, all of which affect the long-term strength and function of the geotextile.

25.3.2 Properties of Geomembranes

Physical Properties

Remember that geomembranes are solid and have no intended holes in them. The unit weight of dry, clean geomembranes is between 8.5 and 15 kN/m^3 depending on the polymer used to make them (ASTM D792). The high-density polyethylene (HDPE) membranes have a unit weight of about 9.2 kN/m^3. However, the density of geomembranes is usually given in terms of mass per unit area (ASTM D1910), in g/m^2. The thickness of commonly used smooth geomembranes is between 0.5 to 3 mm (ASTM D5199). The height of asperities for textured geomembranes can be 0.25 to 0.75 mm; these asperities do increase the interface shear strength.

Mechanical Properties

The stress-strain curve of geomembrane specimens tested in tension exhibits the same two types of shapes as soils: some have a peak strength followed by a residual strength, like overconsolidated soils; and some have a gradual increase in strength with no strain softening, like normally consolidated soils. One major difference is that the range of strains is drastically larger for geomembranes. Strains to failure for soils are in the 2 to 10% range, whereas strains to failure for geomembranes are in the 20 to 100% range. *Failure* refers to no more increase in resistance, but rupture may take as much as 1000% strain. The initial tangent modulus of deformation in tension can vary from 30 MPa for polyvinyl chloride (PVC) to 250 MPa for HDPE geomembranes. The peak tensile strength shows values in the range of 10 to 50 MPa depending on the type of polymer. The high peak strengths tend to lead to lower residual strengths, which can be 50 to 70% of the peak value. The tensile strength of the seams can be less than that of the parent material and should be tested. Seams are manufactured by overlapping two sheets and fusing them together, or by pinching two sheets and fusing them together. The tensile test for the overlapped seam is a shear test and the tensile test for the pinched seam is a peel test (e.g., ASTM D6392 and D882). Peeling tends to offer less resistance than shearing.

The interface shear strength between a geomembrane and the soil is important in many designs, especially for slope stability. It is measured with the same test as for geotextiles (Figure 25.3). A major difference exists between smooth membranes and textured membranes. Koerner (2012) quotes friction angles of 17° and 18° for fine sand and smooth HDPE, and 22° to 30° for fine sand and textured HDPE. Sometimes geomembranes are placed against geotextiles. Again Koerner (2012) quotes friction angles of 6° to 11° for geotextiles and smooth HDPE, and 19° to 32° for geotextiles and textured HDPE. The worst combination seems to be a woven monofilament on top of a smooth HDPE (6°); the best combination appears to be a nonwoven, needle-punched geotextile on top of a textured HDPE (32°).

The puncture strength of a geomembrane is important and can be quoted as impact puncture strength or static puncture strength. The impact puncture strength relates to the ability of the geomembrane to resist shocks from falling objects. It is tested by dropping a heavy object on the membrane (ASTM D3029) or through a pendulum test (energy to puncture) (ASTM D1822). The static puncture strength is related to the ability of the geomembrane to resists puncturing when the membrane is in contact with large aggregates under high pressures. Two alternatives exist to test static puncture strength: a small-scale test and a large-scale test. In the small-scale test (ASTM D4833), an 8 mm diameter beveled-edge piston is pushed through a geomembrane stretched over a 45 mm diameter empty mold. The puncture failure load P is expected to be in the range of 50 to 500 N for thin, nonreinforced geomembranes and 200 to 2000 N for reinforced geomembranes (Koerner 2012). The large-scale test consists of pressing the geomembrane against a bed of cones simulating aggregates (ASTM D5514).

Hydraulic Properties

Geomembranes are often used to prevent a fluid from passing from one side of the membrane to the other: this is called *separation*. Such geomembranes are essentially impervious. Nevertheless, nothing is truly and completely impervious. When the minute amount of fluid passing through the geomembrane must be known, the hydraulic properties of the geomembrane become important. Conventional hydraulic conductivity of geomembranes is in the range of 10^{-13} to 10^{-15} m/s, but, in the language of geosynthetics, terms such as *water-vapor transmission* and *permeance* (which actually designate other parameters) are often used. If solvents are to be retained instead of water, the hydraulic conductivity can increase drastically, by a factor of 100 or even 1000, depending on the nature of the chemical.

As in the case of geotextiles, other properties of geomembranes include sunlight degradation with long-term exposure to ultraviolet rays, radioactive degradation (limit of 10^6 to 10^7 rads), and degradation due to hot and cold temperatures, chemical action, and biological action. Designers use reduction factors to take into account these factors, all of which affect the long-term strength and function of the geomembrane.

25.3.3 Properties of Geogrids

Physical Properties

Geogrids are open-grid geosynthetics (Figure 25.1). The distance between ribs is in the range of 10 to 100 mm. They can be unidirectional (applied stress is in one direction) or bidirectional (direction of applied stress can be random). The percent open area (POA) is measured by shining a light through the geogrids onto a poster board where the illuminated area is measured. Most geogrids have POAs in the range of 40 to 95%. The mass per unit area varies quite a bit, from 200 to 1000 g/m^2.

Mechanical Properties

The flexural stiffness G was defined in Eq. 25.1. Stiff geogrids have G values above 10 g.m, whereas flexible geogrids have G values of less than 10 g.m. From the point of view of tensile strength, several strengths can be identified: the rib strength, the junction strength, and the wide width strength. The *rib strength* refers to the strength of the individual longitudinal elements. The *junction strength* refers to the strength of the connection between the longitudinal and transversal elements. The *wide width strength* F_{ug} is the strength of the geogrid at the field scale where all element strengths are integrated (ASTM D6637). The wide width tensile strength F_{ug} of many geogrids is in the range of 20 to 140 kN/m (force per unit length of fabric) reached at a wide range of strains from 5% to 30%. The tensile modulus of a geogrid is defined as the tensile load applied per unit length of geogrid (kN/m) divided by the corresponding strain of the geogrids. Numbers in the range 125 to 255 kN/m have been measured at small strains (1% to 5%) for stiff geogrids (Austin et al. 1993).

The interface shear strength between soil and geogrid can be measured in a direct shear test, as shown in Figure 25.3. In this test the geogrid is glued to a solid block that fits in the upper part of the direct shear box, which is 0.3 m by 0.3 m minimum (ASTM D5321). Results of such tests indicate that the ratio of the soil-geogrid shear strength over the soil-soil shear strength (called *efficiency*) is close to 1 in bidirectional loading and somewhat less (e.g., 75%) in unidirectional loading (Koerner 2012). Furthermore, Sarsby (1985) showed that if the geogrid aperture (distance between two ribs) is at least equal to 3.5 times the mean particle size d_{50} of the soil, it is likely that the efficiency of the geogrid will be 1, meaning that the shear strength of the soil-geogrid interface will be equal to the soil-soil shear strength.

The pull-out strength of geogrids embedded in soils is very important, as geogrids are most often used as reinforcement. There are two pull-out strengths: you can break the geogrid (F_{ug} given by the manufacturer), or you can break the soil (F_{us}). *Breaking the soil* means failure in shear at the interface between the soil and the geogrid. That ultimate pull-out load can be written as:

$$F_{us} = 2L_e K \sigma'_v \tan \varphi' \tag{25.7}$$

where F_{us} is the ultimate pull-out load per unit width of geogrid, L_e is the embedment length of the geogrid, K is a pull-out coefficient specific to the soil and geogrid involved, σ'_v is the vertical effective stress at the depth of the geogrids, and φ' is the soil effective stress friction angle.

The value of K should be obtained by testing. Pull-out tests can be carried out in the field on full-scale structures, or in the laboratory on large containers simulating the field conditions (Figure 25.4).

Note that two phenomena contribute to the ultimate load F_{us} and therefore the coefficient K: friction between the soil and the geogrid on one hand and penetration of the geogrid transversal elements into the soil. The first one can be calculated from friction laws while the second one calls for bearing capacity estimates. Consider a 2.5 mm thick geogrid made of 7 m long, 5 mm wide longitudinal ribs with a spacing of 100 mm and 3 mm wide transverse elements with a spacing of 200 mm. It is located 2 m below the ground surface of a soil weighing 20 kN/m^3. The interface shear strength between the geogrid and the soil is 18 kPa and the bearing capacity of the transverse element is 900 kPa, then the pull out load per unit width of geogrid is given by:

$$F_{us} = \text{friction longitudinal} + \text{friction transverse} + \text{bearing capacity transverse} \tag{25.8}$$

$$F_{us} = 2 \times 0.005 \times 7 \times 18 \times \frac{1}{0.1} + 2 \times 0.003 \times \left(1 - \frac{1}{0.1} \times 0.005\right) \times 18 \times \frac{7}{0.2} + 0.0025 \times \left(1 - \frac{1}{0.1} \times 0.005\right) \times 900 \times \frac{7}{0.2} \tag{25.9}$$

$$F_{us} = 12.6 + 3.59 + 74.81 = 91 \text{ kN/m} \tag{25.10}$$

As can be seen from this example, the bearing capacity on the ribs is the major contribution, but the friction on the longitudinal ribs is not negligible. Recall that it takes a lot less displacement to mobilize the friction than the bearing

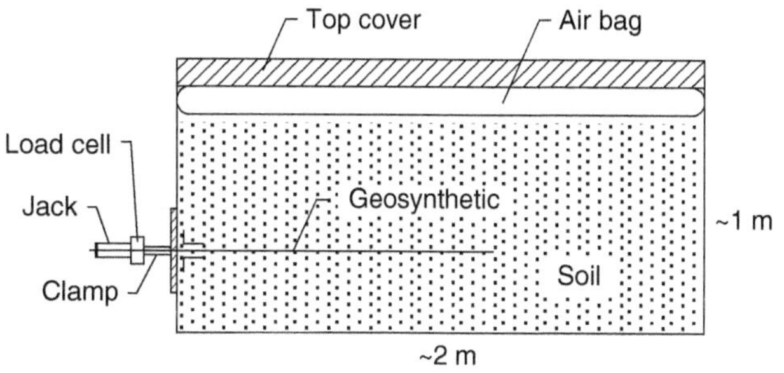

Figure 25.4 Laboratory container for pull-out tests on geogrid.

capacity, so the friction will be mobilized first and the bearing capacity last. Now we can calculate the global friction factor K in Eq. 25.7 knowing that the friction angle is 30° and the vertical effective stress at the depth of the geogrid is 40 kPa:

$$K = \frac{F_{us}}{2L_e\sigma'_v \tan\varphi'} = \frac{91}{2 \times 7 \times 40 \times \tan 30} = 0.281 \tag{25.11}$$

The geogrid covers an area A_g per meter of geogrid width, which is a very small fraction of the total area A_t:

$$A_g = 0.005 \times 7 \times \frac{1}{0.1} + 0.003 \times \left(1 - \frac{1}{0.1} \times 0.005\right) \times \frac{7}{0.2} = 0.35 + 0.10 = 0.45\ \text{m}^2 \tag{25.12}$$

The total area A_t is:

$$A_t = 7 \times 1 = 7\ \text{m}^2 \tag{25.13}$$

Therefore, the area covered by the geogrid is 6.4% (0.45/7) of the total area, yet it develops 28.1% (Eq. 25.11) of the total friction. Of course, in addition to the pull-out resistance of the geogrid itself, it is important to ensure that the connection can handle such a force.

Creep tension properties are also very important for geogrids, as they are usually subjected to constant tension during their design life. At low stress levels (low fraction of the tensile strength F_{ug}), the geogrid will exhibit strain that increases linearly with the log of time. The slope of that line is the constant strain rate (e.g., 0.5% strain per log cycle of time). This creep strain rate depends on the type of polymer, the temperature, and the stress level. At intermediate stress levels, the geogrid may exhibit a delayed failure, in which the creep strain rate is constant for a while but increases dramatically after a certain time, leading to failure (Figure 25.5). Delayed creep failure typically occurs within the range of 25% to 50% of the ultimate tension F_{ug}. Geogrids should be tested for creep response (ASTM D5292 and ASTM D6992). The time temperature superposition (TTS) principle can be used to shorten the testing time. In TTS, advantage is taken of the fact that a long time at low temperature is equivalent to a short time at high temperature.

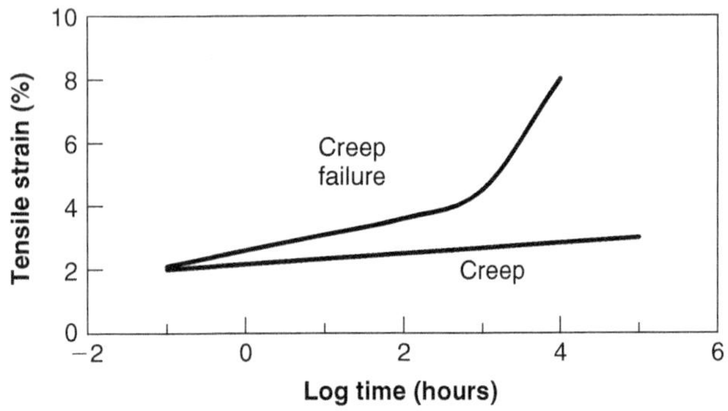

Figure 25.5 Creep behavior of geogrids.

As in the case of geotextiles and geomembranes, other properties of geogrids include temperature effects, chemical effects, biological effects, radioactive effects, and sunlight effects. Designers use reduction factors to take into account these factors, all of which affect the long-term strength and function of the geogrid.

25.3.4 Properties of Geosynthetics Clay Liners

Physical Properties

Geosynthetic clay liners (GCLs) are a recent innovation; they are made of a thin layer of bentonite clay sandwiched between two layers of geotextiles or geomembranes (Figure 25.6). They come in large flexible rolls and are used as containment barriers in the case of landfill liners and covers, for example. They are either nonreinforced or reinforced. The reinforcement solves the following problem. When hydrated, the bentonite clay is extremely slick and represents a weak shear plane that would initiate failure when placed in a slope. To remedy this situation, GCLs can be reinforced by fibers (needle-punched) or stitches (stitch bonds) that tie the two sides of the GCL together. Nonreinforced GCLs are used as barriers on flat ground, whereas the more common reinforced GCLs are used on sloping ground.

The clay type can be sodium bentonite or calcium bentonite. Sodium bentonite has the lowest hydraulic conductivity, but its availability worldwide is limited. The thickness of a GCL varies significantly because of the difference in hydrated and dry thicknesses. Furthermore, it is difficult to isolate the thickness of the clay layer from its boundaries. The hydrated thickness is more important, as it affects the hydraulic properties. The total hydrated thickness of GCLs typically varies between 10 and 30 mm. The mass per unit area of GCL is in the range of 5 to 6 kg/m^2, with 4 to 5 kg/m^2 of dry bentonite between the geosynthetic layers. Once hydrated, the GCL can easily become twice as heavy. The GCL is sold "dry," which means that it has a low initial water content of around 10%. When the bentonite hydrates, it can reach water contents well over 100%.

Hydraulic Properties

The hydraulic properties of GCLs are very important, as GCLs are mostly used as barriers. The chemistry of the

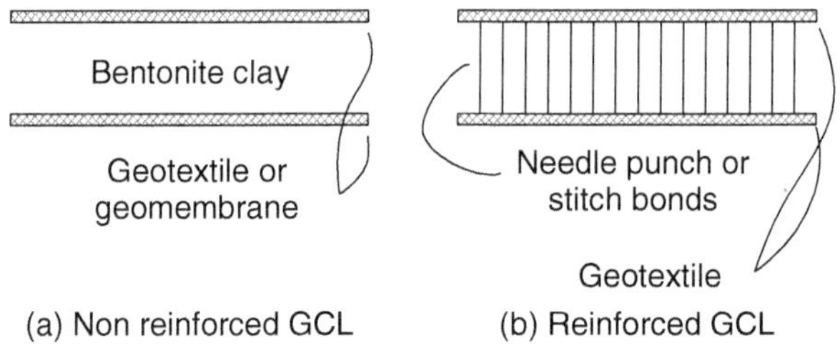

Figure 25.6 Geosynthetic clay liners cross section.

liquid hydrating the GCL can make a significant difference. Koerner (2012) reports on the difference between distilled water, tap water, mild landfill leachate, harsh landfill leachate, and diesel fuel. The results show that the swell movement is largest with distilled water and zero with diesel fuel; the other liquids lead to intermediate swell movement. The swell test consists of placing a "dry" sample of bentonite in a consolidometer, submerging it in water, and allowing it to swell under light vertical pressure. The bentonite will swell, reaching the maximum swell movement in a time that can vary from 2 weeks to 2 months.

The hydraulic conductivity k of a GCL can be measured in the laboratory using a flexible-wall triaxial permeameter (ASTM D5887). The in situ conditions should be reproduced as closely as possible, including the applied pressure and type of liquid. The value of k is obtained as follows:

$$q = kiA = k\left(\frac{\Delta h}{t}\right)A \tag{25.14}$$

where q is the flow rate, k is the hydraulic conductivity, I is the hydraulic gradient, A is the cross-sectional area through which the liquid flows, Δh is the loss of total head across the GCL, and t is the thickness to be permeated.

This thickness is very difficult to measure accurately and leads to inaccuracies in quoting the k value. Daniel et al. (1997) reported that the same GCL tested by many different laboratories yielded values between 2×10^{-11} and 2×10^{-12} m/s. This range can be expected for the k values of sodium bentonite GCLs and for water as permeate. Because of the difficulties associated with thickness measurements, the results of a GCL permeability test are usually given in terms of flow per unit area (q/A in $m^3/s.m^2$). At the junction between contiguous sheets of GCL, there is an overlap. A minimum overlap of 150 mm is recommended to maintain a hydraulic conductivity equal to that of the GCL itself. Sometime a layer of bentonite without GCL is added at the junction. From the long-term endurance point of view, freeze-thaw cycles and shrink-swell cycles do not seem to affect the hydraulic conductivity of GCLs significantly.

Mechanical Properties

The bentonite contributes very little to the wide width tension strength of a GCL. As a result, the tension strength of a GCL is estimated by using the values of the geotextile or geomembrane within which the bentonite is sandwiched.

The shear strength of the GCL depends on the interface considered: upper geosynthetic and soil or waste, bentonite clay layer with or without reinforcement, lower geosynthetic and soil or waste. The upper and lower interface shear strengths between the materials above or below the GCL are addressed by considering the type of geosynthetic involved. For geotextiles, see section 25.2.1; for geomembranes, see section 25.2.2. The shear strength of the bentonite clay layer is measured by a direct shear test (ASTM D6243). Koerner (2012) reports on tests where the shear strength parameters c and φ decrease dramatically upon hydration of the bentonite clay layer with water when tested in a relatively rapid direct shear test. This decrease in shear strength parameters is not as severe when the hydrating liquid is leachate, and no decrease was found when the hydrating liquid was diesel fuel. The shear strength of reinforced GCL is much higher than that of unreinforced GCL and larger displacements are required to reach failure.

One concern is the long-term shear strength of reinforced GCLs. This is related to the long-term strength of needle-punched fibers or stitch bonds. The long-term (100-year) internal shear strength of reinforced GCL is up to 50% of the short-term shear strength (Koerner 2012). The *peel strength* of reinforced GCLs refers to the maximum force per unit length that the upper and lower geosynthetic layers can resist when pulled away from each other at a 90° angle to the main direction of the GCL seam; it is measured in kN/m (ASTM D6496). Resistance to puncturing is also important and should be measured. The tests include ASTM D4883 and ASTM D6241. Squeezing of the bentonite layer away from a location by local pressure is avoided by placing a layer of sand, for example, above the GCL.

25.3.5 Properties of Geofoams

Physical Properties

Geofoams are blocks made of light yet hard polystyrene materials. They are used as light fills, as thermal insulations, and as compressible inclusions. The width and the height of the blocks vary from 0.3 to 1.2 m, and the length from 1.2 to 5 m. A distinction is made between expanded polystyrene (EPS) and extruded polystyrene (XPS). EPS is made from solid beads of polystyrene expanded by blowing gas through them. XPS consists of melted polystyrene crystals mixed with additives and a blowing agent and shaped by extrusion through a die; the white Styrofoam coffee cups are made of extruded polystyrene. EPS geofoam blocks are typically larger than XPS geofoam blocks. The unit weight of geofoams ranges from 0.1 to 0.5 kN/m^3, which is much smaller than the average unit weight of soil (~20 kN/m^3). Geofoams do not absorb much water, but are combustible and should not be exposed to temperatures in excess of 95°C.

Mechanical Properties

Because geofoams are often used as lightweight fill, the unconfined compressive strength is of interest. Figure 25.7 shows results from Negussey (1997), as presented by Koerner (2012).

This unconfined compression stress-strain data indicates that the geofoam exhibits a linear behavior with a modulus E until a yield strength σ_y, and then strain hardens at a modest rate. The yield strength σ_y is reached at around 2% compressive strain. The unit weight of the geofoam γ_{GF} has

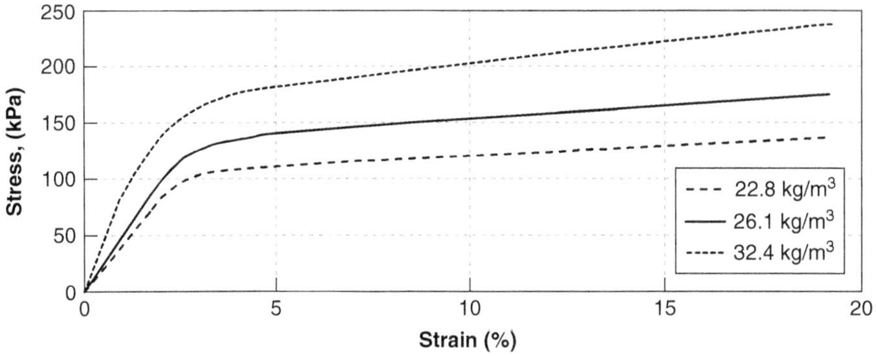

Figure 25.7 Stress-strain curves for geofoam. (After Negussey 1997.)

a direct impact on its mechanical properties and the following equations can be derived from Negussey's data:

$$\text{Modulus of EPS geofoam} \quad E\,(\text{MPa}) = 20\gamma_{GF}\,(\text{kN/m}^3) \tag{25.15}$$

$$\text{Modulus of XPS geofoam} \; E\,(\text{MPa}) = 60\gamma_{GF}\,(\text{kN/m}^3) \tag{25.16}$$

$$\text{Yield strength of EPS geofoam} \; \sigma_y\,(\text{kPa}) = 500\gamma_{GF}\,(\text{kN/m}^3) \tag{25.17}$$

$$\text{Yield strength of XPS geofoam} \; \sigma_y\,(\text{kPa}) = 800\gamma_{GF}\,(\text{kN/m}^3) \tag{25.18}$$

The internal shear strength of geofoams can be tested by following ASTM C253, and the shear strength between geofoam blocks can be tested by direct shear testing (ASTM D5321). The tensile strength of geofoams, σ_t, is much larger than that of soils. Using the data from Styropor (1993), the following equation can be proposed:

Tensile strength of EPS geofoam

$$\sigma_t\,(\text{kPa}) = 1250\gamma_{GF}\,(\text{kN/m}^3) \tag{25.19}$$

Creep properties are important because geofoams may be subjected to long-term loads (in lightweight embankments, for example). In creep testing of geofoams, the time temperature superposition can be used to shorten the time required to characterize the long-term behavior. Data from Negussey (1997) indicates that when the sustained compression stress is below 50% of the unconfined compression strength, the viscous exponent (Chapter 14, Eq. 14.9; section 15.8) of geofoam is within the range of values found in soils (n = 0.01 to 0.08).

Thermal Properties

Geofoams can be used as thermal insulation under buildings. Therefore, their thermal properties are important. The main property is the R value (see section 16.3). The R value is defined as:

$$R(^{\circ}\text{C.m}^2/\text{W}) = \frac{\Delta T(^{\circ}\text{C})}{q(\text{J/s.m}^2)} \tag{25.20}$$

where R is the R value or thermal resistance in °Celsius.m^2/Watt, ΔT is the difference in temperature on either side of the geofoam in °Celsius, and q is the heat flow in J/s.m^2. The higher the R value is, the more insulating the geofoam is. The R value per unit width of geofoam is R′ expressed in °Celsius.m/Watt. The R′ value of geofoams varies from 20 to 40 and increases with unit weight. It is typically higher for XPS than for EPS.

Other aspects to be addressed are the chemical resistance of geofoams, which are readily attacked by hydrocarbons such as gasoline; degradation due to long-term exposure to UV rays; and flammability. This is why it is best for geofoams to be covered by a soil backfill as soon as possible after installation.

25.3.6 Properties of Geonets

Geonets are open-grid geosynthetics very similar to geogrids; however, their purpose is to serve as spacers by providing flow conduits within their thickness. They are typically used in conjunction with a geotextile or geomembrane on top and bottom of the geonet. Whereas geogrids have a single layer of ribs typically perpendicular to each other, the ribs in geonets are stacked on top of each other (2 or 3 layers) and lined up in diagonals to facilitate flow. The mass per unit area varies from 0.8 to 1.6 kg/m^2 and the thickness from 4 to 8 mm. The mechanical properties of geonets are similar to those of geogrids, but the hydraulic properties are most important, as geonets are used primarily for drainage purposes.

The drainage capacity is quoted in flow per unit width of geonet. Values in the range of 10^{-3} to 10^{-4} m^3/s.m are common, but can decrease by 30% when the pressure increases to 1000 kPa. The drainage capacity of geonets may also be quoted in terms of transmissivity Θ (Eq. 25.5), which is related to the flow rate per unit width (*q/W*) by:

$$\Theta = \frac{1}{i} \times \frac{q}{W} \tag{25.21}$$

where i is the hydraulic gradient. The EPA has regulations indicating that a geonet must have a transmissivity of at least 3×10^{-5} m^2/s for landfills and 3×10^{-4} m^2/s for surface impoundments.

25.4 DESIGN FOR SEPARATION

Separation means that the two materials on each side of the geosynthetic cannot penetrate it. This is associated with failure mechanisms by impact, punching, or tear (Figure 25.8). *Impact* refers to the case where a stone falls on top of the geosynthetic. *Punching* refers to the case where the geosynthetic is pushed through an opening between large aggregates. *Tear* refers to the case where the geosynthetic is pulled apart by stones that are moving away from each other in the deformation process of the geotechnical structure.

Designing for impact first requires estimating the energy of the falling object. If the stone is represented by a sphere, the energy E_{stone} to be dissipated at impact is:

$$E_{stone} = Wh = \frac{\pi d^3}{6}\gamma h \tag{25.22}$$

where W is the weight of the stone, h is the height of drop, d is the stone diameter, and γ is the unit weight of the stone.

This energy is absorbed in part by the geosynthetic layer and in part by the soil immediately below the geosynthetic. If the soil is soft, the stone has a soft landing and the peak force in the geosynthetic is lower than if the soil is stiff but the deformation is large. If the soil is extremely weak, only the geosynthetic resists the impact. To take the soil support contribution into account, the value of E_{stone} is divided by a soil support factor F_s varying between 5 and 25 (Koerner 2012). This energy is then compared to the impact strength E_{geosyn} (Joules) of the geosynthetic (section 25.3.1). The impact strength of the geosynthetic is divided by a cumulative reduction factor F, which accounts for installation damage, creep, and chemical/biological degradation, for example. F_r varies from as low as 1.1 to as high as 9. The design ensures that:

$$\frac{E_{stone}}{F_{soil}} \le \frac{E_{geosyn}}{F_{reduc}} \tag{25.23}$$

Designing against puncture requires estimating the force F_{stone} generated by the stone protruding into the geosynthetic due to a pressure p applied. The pressure p may be applied by a rolling truck, for example. Koerner (2012) proposed:

$$F_{stone} = p\,d_a^{\,2} S_1 S_2 S_3 \tag{25.24}$$

where d_a is the diameter of the penetrating stone, and S_1, S_2, and S_3 are the protrusion factor, the scale factor, and the shape factor respectively. The product $S_1S_2S_3$ varies from 0.65 in the most severe condition (angular large stone) to 0.01 in the most favorable condition (rounded small particles). The value of F_{stone} is then compared to the strength of the geosynthetic (Eq. 25.2) divided by the cumulative reduction factor.

Designing against tear starts by calculating the tension force generated in the geosynthetic when squeezed between two layers of soil. When the upper and lower layers of soil are subjected to a rolling truck, for example, the layers deflect and bend locally under the wheel load. During this bending, the geosynthetic trapped between the two soil layers is subjected to a tension force $F_{tension}$, which is given by (Giroud 1981):

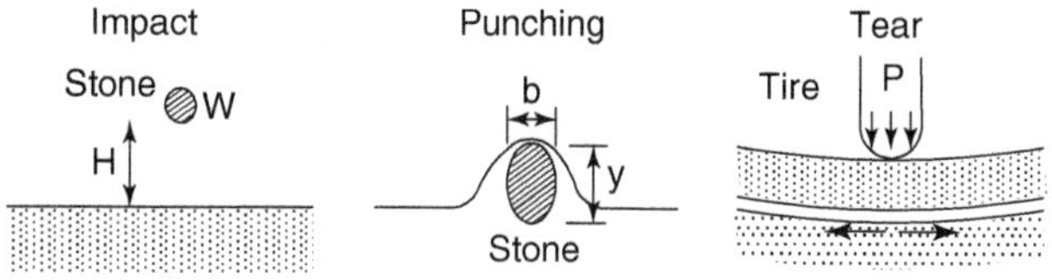

Figure 25.8 Modes of failure of geosynthetics in separation.

$$F_{tension} = 0.1pd_a^{\,2} f(\varepsilon) = 0.025pd_a^{\,2}\left(\frac{2y}{b} + \frac{b}{2y}\right) \tag{25.25}$$

where p is the pressure applied, d_a is the particle or stone diameter, $f(\varepsilon)$ is a function of the strain in the geosynthetic, y is the displacement into the stone void, and b is the width of the stone void (Figure 25.8). Then the value of $F_{tension}$ is compared to the strength of the geosynthetic (Eq. 25.2) divided by the cumulative reduction factor.

25.5 DESIGN OF LINERS AND COVERS

Liners are barriers placed at the bottom of landfills to prevent the waste and the liquid it generates from contaminating the soil and water below the waste. *Covers* are barriers placed on top of landfills to close them, prevent the waste from contaminating the surrounding environment, and prevent the gas it generates from escaping without control. Both liners and covers have evolved dramatically in the past 30 years, with most of the change taking place between 1980 and 1990.

Before 1980, only a compacted clay liner was required at the bottom of landfills (Figure 25.9). The leachate collection system was a layer of sand and gravel with perforated pipes; there was no leachate detection system and no secondary liner to decrease the probability of leaks through the liner. Nowadays, liners are double composite systems (Figure 25.10) with a leachate collection system, a primary liner, a leak detection system, and a secondary liner. The liner involves many layers: geosynthetic for the purpose of separation (geomembrane), barrier (geosynthetic clay liner), and drainage and leachate detection-collection system (geocomposite with geonet-geotextile-geomembrane) in addition

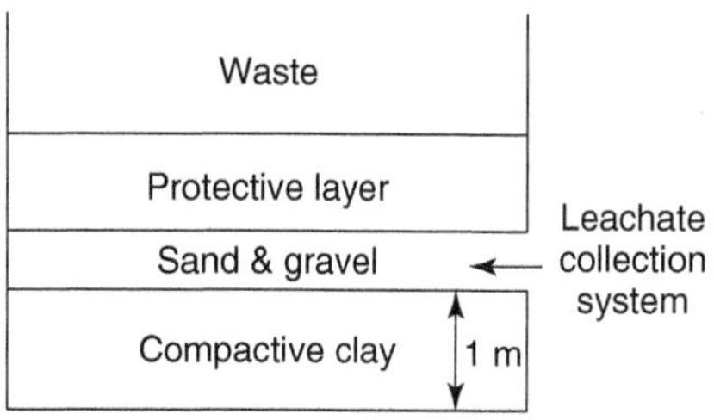

Figure 25.9 Early liner cross section.

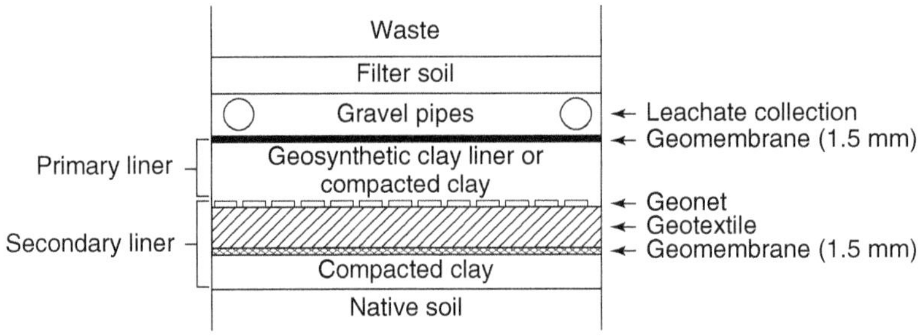

Figure 25.10 Example of a modern liner cross section.

to layers of compacted clay. The change from compacted clay liners to geosynthetic base liners was prompted by the fact that the compacted clay layer had to be quite thick (up to 1.5 m thick), thereby taking up space that could otherwise have been used for waste; and by the discovery that certain chemicals, such as organic solvent leachate, dramatically increase the hydraulic conductivity of clays. Given this, a compacted clay liner alone could not ensure that no leakage would occur.

The leakage through a liner is very rarely zero. It is measured in liters per hectares per day (lphd). A liter is 10^{-3} m^3 and an hectare is 10,000 m^2. The leakage varies with time as the landfill is being constructed and during its operation. Koerner (2012) defines stage 1 as the stage during construction, stage 2 as when considerable waste is placed, and stage 3 as when the final cover is placed. Furthermore, a distinction should be made between the different types of liners: geomembrane alone (GM), geomembrane over a compacted clay liner (GM/CCL), and geomembrane over a geosynthetic clay liners (GM/GCL). Based on the work of Othman et al. (1997) and Bonaparte et al. (2002), who gathered leakage rates for 289 landfills, Koerner gives the rates shown in Figure 25.11.

The geomembranes used in liners are typically made of high-density polyethylene (HDPE). Some global minimum recommendations for the survivability of geomembranes used in liners are presented in Table 25.3. Low severity refers to a careful manual placement with light loads on smooth ground, for example; very high severity refers to machine handling on rough, stiff ground under heavy loads.

Landfill covers are necessary so the waste does not contaminate the surrounding environment. These covers prevent rainwater from accessing the waste and keep the gas generated by the landfill from escaping into the atmosphere without control. An example of a cover cross section is shown in Figure 25.12. The layers involved include vegetation for a positive landscape impact, an erosion control geosynthetic, a top soil and cover soil layer for the plants to grow, a drainage layer (combination of geotextile/geonet/geomembrane), and then a second barrier (compacted clay or GCL), a gas collection layer, and the waste. The erosion control geosynthetic is often necessary because the top of the landfill is like a big hill, so the runoff water can erode the top soil in the cover. The drainage layer drains the rainwater away from the landfill. The barrier layer provides a second assurance that the water will not penetrate and also that the gas produced will not escape without control. The gas collection layer is necessary because most municipal landfills generate a lot of gas, mainly

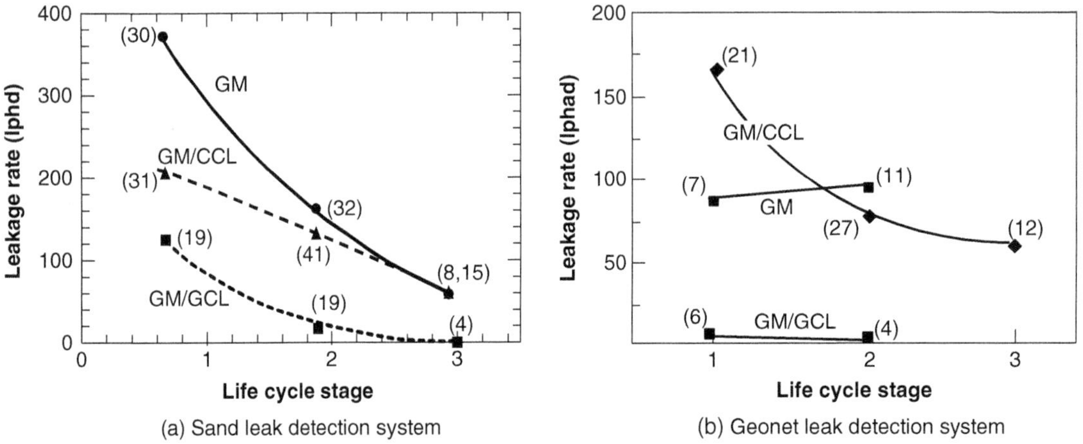

Figure 25.11 Leakage rates in landfill liners. (After Koerner 2012. Robert M. Koerner—Copyright Owner.)

Table 25.3 Minimum Requirements for Geomembrane Survivability

Property	Required Value Considering the Degree of Severity: Low	Medium	High	Very High
Thickness (mm)	0.63	0.75	0.88	1.00
Tensile strength (kN/m)	7	9	11	13
Tear resistance (N)	33	45	67	90
Puncture (N)	110	140	170	200
Impact resistance (J)	10	12	15	20

(After Koerner 2012)

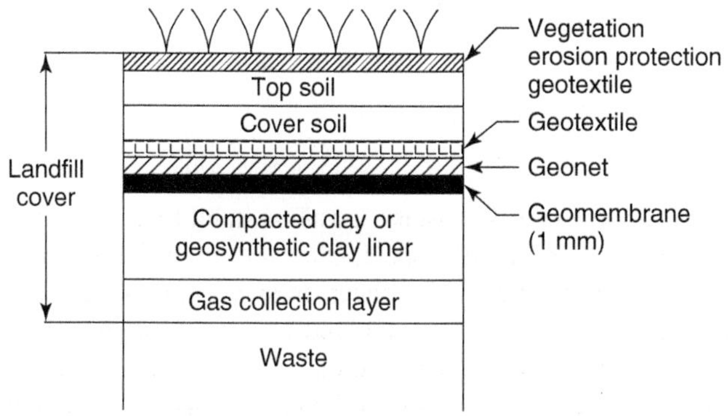

Figure 25.12 Example of a landfill cover. (After Koerner, 2012.)

methane (CH_4) and carbon dioxide (CO_2). This layer can be made of sand with perforated pipes that collect the gas; the perforated pipes are connected to risers (vertical unperforated pipes) that bring the gas to a collection point; alternatively, the gas may be burned and the combustion products released into the atmosphere (flare).

25.6 DESIGN FOR REINFORCEMENT

25.6.1 Road Reinforcement

The role of geosynthetics in road reinforcement is threefold: separation, reinforcement, and minimization of crack propagation. Geotextiles can be used for separation; geotextiles and geogrids can be used for reinforcement and mitigation of crack propagation. A distinction is made here between applications for unpaved roads, paved roads, and overlay of asphalt flexible pavements. Separation has already been addressed in section 25.4. It applies to the case of unpaved and paved roads, but rarely in the case of the overlay of asphalt flexible pavements.

Prevention of crack propagation applies to the case of asphalt flexible pavements only. Indeed, for overlay of asphalt pavements, the rolling surface may contain vertical cracks. Minimizing the chances that the crack will propagate from the lower cracked asphalt layer vertically through the overlay to the rolling surface can be achieved by using a thicker overlay asphalt layer or the combination of a geotextile and a thinner asphalt overlay. It is important to keep moisture from rising through the overlay. For this, the geotextile is first rendered impervious by impregnating it with bitumen; then it is placed on top of the old pavement; then the overlay is constructed. The concept is that the geotextile will provide horizontal reinforcement with significant tensile strength and contain the future increase of the crack growth.

The role of geosynthetics in road reinforcement is better suited to unpaved road than paved roads. The reason is that, on the one hand, geosynthetics tend to generate their resistance over a level of strain much larger than materials like asphalt and concrete and, on the other hand, unpaved roads deflect more under traffic load than paved roads. Thus, geosynthetics will contribute more to the capacity of unpaved roads than paved roads. The design concept is to calculate the pavement thickness with and without the geosynthetics layer and perform an economic analysis on the two options. The benefit of using the geosynthetic layer is derived by assuming that the pressure level on the rolling surface (tire pressure) can be increased due to the presence of the geosynthetics. Without the geosynthetics, the stress on the subgrade must be kept within the elastic limit, whereas with the geosynthetics, the stress on the subgrade can reach the bearing capacity of the subgrade, as failure will be prevented by the geosynthetics. Both geotextiles and geogrids can be used for this application. Details of these designs can be found in Koerner (2012).

25.6.2 Mechanically Stabilized Earth Geosynthetic Walls

Retaining walls (see Chapter 22) may be top-down walls, such as tieback walls; or bottom-up walls, such as gravity walls. Mechanically stabilized earth (MSE) walls are bottom-up walls, meaning that they are built starting at the bottom and going up until the top of the wall is completed. MSE walls are built by placing a layer of soil (say, 0.3 m thick), compacting it, then placing a layer of reinforcement, then a layer of soil and compacting it, placing a layer of reinforcement (geotextile or geogrids in this case), and so on to the top of the wall (Figure 25.13). These walls can be built in such

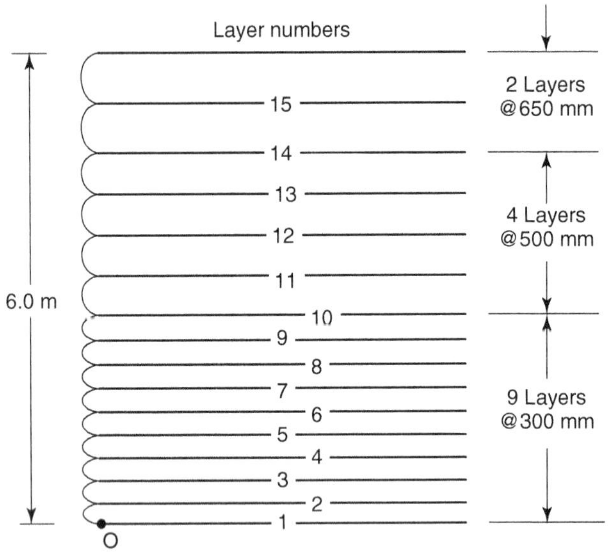

Figure 25.13 MSE wall with geosynthetics reinforcement.

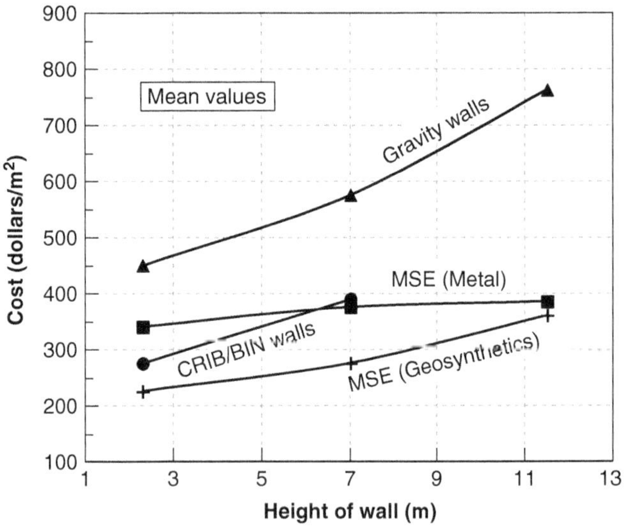

Figure 25.14 Cost of retaining walls. (From Koerner 2012.)

a fashion to heights reaching tens of meters and are less expensive than conventional gravity or cantilever retaining walls (Figure 25.14). The reinforcement can be made of rigid inclusions such as steel strips and steel wire mesh or flexible inclusions such as geosynthetics (geotextiles and geogrids). The front of the wall is covered with panels that are tied to the reinforcement. The design of geosynthetic MSE walls includes internal stability and external stability. The minimum length of reinforcement is set at 0.7 H where H is the height of the wall.

External Stability

Bearing capacity at the base of the wall, general slope stability of the wall and the slope within which it rests, sliding of the wall mass, and overturning of the wall mass are all external stability issues. Commonly used factors of safety for a global factor of safety approach and for each one of those failure modes are presented in Table 25.4. Average load and resistance factors for an LRFD approach are also shown in Table 25.4. Bearing capacity and slope stability are dealt with in Chapters 17 and 19 respectively. Sliding of the wall mass can be addressed through a factor of safety expressed as:

$$F_{sliding} = \frac{\sum \text{Resisting forces}}{\sum \text{Driving forces}} = \frac{W \tan \varphi'}{P_a} \tag{25.26}$$

where W is the weight of the wall mass per unit length of wall, φ' is the friction angle of the interface at the bottom of the wall, and P_a is the horizontal force per unit length of wall due to the active earth pressure against the back of the wall (see Chapter 22).

Alternatively, for the LRFD approach the equation is:

$$\gamma P_a = \varphi W \tan \varphi' \tag{25.27}$$

where γ is the load factor and φ is the resistance factor.

Overturning of the wall is addressed through a factor of safety expressed as a ratio of moments. The moments are

Table 25.4 Some Possible Load and Resistance Factors for External and Internal Stability of Geosynthetic MSE Wall

				Load Factor	
Stability	Design Issue	Factor of Safety	Resistance Factor	Dead	Live
External Stability	Bearing capacity	2	0.5	1.25	1.75
	Slope stability	1.5	0.7	1.25	1.75
	Sliding	1.5	0.85	1.25	1.75
	Overturning	2	0.7	1.25	1.75
Internal Stability	Pull-out	1.5	0.9	1.5	1.75
	Breakage	1.5	0.9	1.5	1.75

taken around the bottom of the front of the wall (point O on Figure 25.13):

$$F_{overturning} = \frac{\sum \text{Resisting moments}}{\sum \text{Driving moments}} = \frac{Wd}{P_a x_a} \quad (25.28)$$

where d is the moment arm of the weight W of the wall and x_a is the moment arm of the active earth pressure force P_a.

Alternatively, for the LRFD approach the equation is:

$$\gamma P_a x_a = \varphi W d \quad (25.29)$$

where γ is the load factor and φ is the resistance factor.

Internal Stability

Pull-out capacity and yield of the geosynthetic reinforcement are the two aspects of internal stability of an MSE wall with geosynthetic reinforcement. Commonly used factors of safety for a global factor of safety approach and for each one of those failure modes are presented in Table 25.4. Average load and resistance factors for an LRFD approach are also shown in Table 25.4. Pull-out capacity requires an understanding of the load distribution in the reinforcement. Figure 25.15 shows the variation of the tension load T (kN/m) in the reinforcement as a function of the distance from the front of the wall.

At the wall facing, the load T in the reinforcement is very small, and then it increases as the instability of the wedge of soil near the wall is transferred into the geosynthetic as a tension force T (kN/m). At a distance L_{max} from the front, the tension T reaches a maximum T_{max}. Beyond T_{max}, the tension decreases and reaches zero at a certain distance from the front. This distance must be less than the actual length L of the reinforcement, or significant deformations and possibly failure will occur. The true embedment or anchoring length L_a available to resist the active pressure force against the wall is $L - L_{max}$. The design requires a knowledge of L_{max}, which is to be ignored in the length required to resist T_{max}. The force T_{max} is calculated as:

$$T_{max} = s_v \sigma_{ah} \quad (25.30)$$

where T_{max} is the maximum line load (kN/m) to be resisted by the geosynthetic layer at a depth z, s_v is the vertical spacing between reinforcement layers at the depth z, and σ_{ah} is the total horizontal active stress at the depth z. The stress σ_{ah} is discussed in Chapter 22.

Now that we have calculated the load, we need to find the length of reinforcement that will safely carry the load without pulling out of the soil. The pull-out line capacity $T_{pullout}$ (kN/m) of the geosynthetic layer is given by:

$$T_{pull\ out} = 2 f_{max} L_a \quad (25.31)$$

where f_{max} is the maximum shear stress that can be developed on both sides of the interface between the geosynthetic and the soil, and L_a is the anchoring length. Recall that L_a is the length beyond the failure wedge. The shear stress f_{max} is evaluated as follows:

$$f_{max} = \sigma'_v \tan\delta \quad (25.32)$$

where σ'_v is the vertical effective stress on the geosynthetic layer at depth z (including any effect from surcharge or load at the ground surface), and $\tan\delta$ is the coefficient of friction between the soil and the geosynthetic. Then the ratio between the load T_{max} and the resistance $T_{pullout}$ must satisfy a factor of safety F (Table 25.4):

$$T_{pull\ out} = F \times T_{max} \quad (25.33)$$

and the required safe length L_a of the geosynthetic sheet is given by:

$$L_a = \frac{F s_v \sigma_{ah}}{2\sigma'_v \tan\delta} \quad (25.34)$$

In the simple case where $\sigma_{ah} = K_a \sigma'_v$, where $K_a = 0.33$, $F = 2$, and $\tan\delta = 0.5$, then L_a is equal to 0.66 s_v, which is quite small for normal vertical spacing of 0.3 to 0.5 meters.

A load and resistance factor approach would consist of replacing Eq. 25.34 by:

$$\gamma 2 L_a \sigma'_v \tan\delta = \varphi s_v \sigma_{ah} \quad (25.35)$$

where γ is the load factor and φ is the resistance factor.

Note that the anchoring length L_a is constant with depth. The reason is that as the load increases with depth, so does the resistance. In practice, the minimum embedment length is set at 1 m. The distance L_{max} required to develop the load

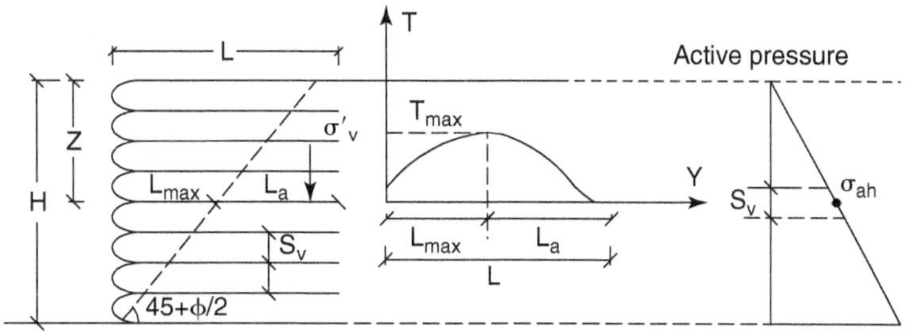

Figure 25.15 Load in the reinforcement. (After Theisen, 1992.)

in the reinforcement layer is taken as the width of the active wedge (see Chapter 22 and Figure 25.15):

$$L_{max} = (H - z)\tan\left(45 - \frac{\varphi'}{2}\right) \tag{25.36}$$

Therefore, the final length of the geosynthetic layer L is:

$$L = (H - z)\tan\left(45 - \frac{\varphi'}{2}\right) + \frac{\frac{\gamma}{\varphi} s_v \sigma_{ah}}{2\sigma_v' \tan\delta} \tag{25.37}$$

For construction simplicity, the length L is often kept constant for the entire wall. Because the length L is largest at the top of the wall, practically the length of reinforcement is taken as:

$$L = H\tan\left(45 - \frac{\varphi'}{2}\right) + \frac{\frac{\gamma}{\varphi} s_v \sigma_{ah}}{2\sigma_v' \tan\delta} \tag{25.38}$$

where σ_{ah} and σ_v' are calculated at the depth of the first reinforcement layer.

Yield of the geosynthetic layer is the next design issue. We must make sure that the geosynthetic can safely carry the load T_{max} without yielding or rupturing. For this, we need to find the allowable tensile resistance of the geosynthetic layer T_{allow}. This allowable tensile resistance T_{allow} is obtained from the measured ultimate tensile resistance T_{ult} and given by:

$$T_{allow} = \frac{T_{ult}}{RF_{ID} \times RF_{CR} \times RF_{CBD}} \tag{25.39}$$

where T_{allow} and T_{ult} are the allowable and ultimate resistance of the geosynthetic layer (kN/m); and RF_{ID}, RF_{CR}, and RF_{CBD} are strength reduction factors that take into account installation damage, creep, and chemical and biological factors.

These strength reduction factors vary between 1 and 2 depending on the application (Koerner 2012), and average 1.55, 2.15, and 1.32 respectively. Note that once combined, these reduction factors lead to using an allowable tension that is about 20% of the measured ultimate tensile strength T_{ult} of the geosynthetic. The required ultimate strength T_{ult} of the geosynthetic is such that:

$$T_{allow} = \frac{T_{ult}}{RF_{ID} \times RF_{CR} \times RF_{CBD}} = T_{max}\frac{\gamma}{\varphi} \quad \text{or}$$

$$T_{ult} = T_{max}\frac{\gamma}{\varphi} \times RF_{ID} \times RF_{CR} \times RF_{CBD} \tag{25.40}$$

We must also make sure that there is no slip at the location of the overlap between geosynthetic layers. Thus, the overlap distance must satisfy Eq. 25.35. Because the overlap is located near the wall where the tension load is less than T_{max}, a length equal to ½ L_a is typically used for the required overlap distance.

25.6.3 Reinforced Slopes

A distinction must be made here between manmade bottom-up slopes (e.g., embankments and dams) and natural slopes or top-down slopes (e.g., hillside slopes and cuts). In the first case, it is possible to install geosynthetic reinforcement layers (geotextiles or geogrids) as the slope is built. In the second case, it is not possible to use geosynthetics as reinforcement; however, natural slopes and cuts can be reinforced with geosynthetics by covering the slope with a geosynthetic layer and anchoring the cover deeply beyond the failure plane. For bottom-up slopes, the factor of safety of the slope with reinforcement is calculated as presented in section 19.14.

25.6.4 Reinforced Foundations and Embankments

Geosynthetics can be placed below a foundation or embankment to improve its carrying capacity and performance. Geotextiles and geogrids are most commonly used this way. The main design issues are bearing capacity, settlement, and anchoring length.

Bearing capacity. Bearing capacity is improved because the failure plane has to pull on the geosynthetic layer. The degree of improvement is analyzed by the same method used for slope stability analysis once a circular surface has been chosen. For example, consider a strip footing of width B at the surface of a soft clay with an undrained shear strength s_u (Figure 25.16). A geosynthetic layer is placed at a depth of $B/2$ with a tensile strength T kN/m. Let's find the value of T required to increase the bearing capacity by a factor of 2. At failure and for a circular failure surface as shown on Figure 25.16, moment equilibrium around O gives:

$$p_u B\frac{B}{2} = s_u \pi B B + TB \quad \text{or} \quad p_u = 2\pi s_u + 2\frac{T}{B} \tag{25.41}$$

Note that it is assumed here that the geosynthetic is flexible and that it will deform at the intersection with the failure surface in such a way that it will become tangential to the failure circle. For the geosynthetic layer to double the bearing capacity, we must have:

$$\frac{2\pi s_u + 2\frac{T}{B}}{2\pi s_u} = 2 \quad \text{or} \quad T = \pi s_u B \tag{25.42}$$

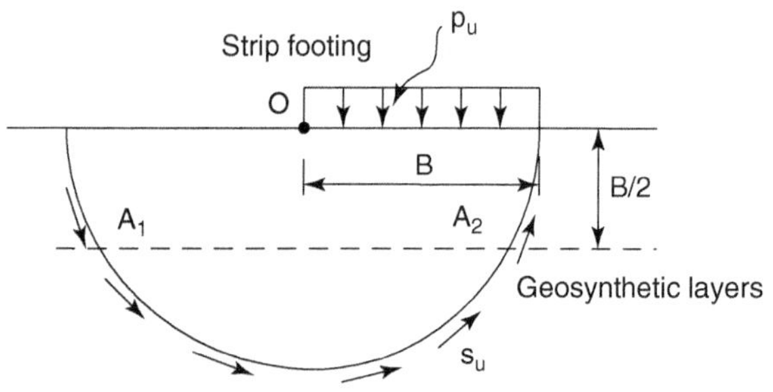

Figure 25.16 Foundation reinforcement.

For a 2 m wide strip footing and $s_u = 120\,\text{kPa}$, the value of T is 240 kN/m. This is a very high allowable tensile strength for a geosynthetic and several layers would have to be used to safely achieve this level of tensile strength. However, for $s_u = 40\,\text{kPa}$, the value of T is 80 kN/m, which can be achieved with one layer.

Settlement. Settlement is also affected by the presence of a geosynthetic layer. At small displacements, the contribution is limited, as the geosynthetic typically has to deform enough to make a difference. The magnitude of settlement necessary for this contribution to be significant is more consistent with embankments than foundations. Indeed, larger settlements are more readily accommodated by flexible embankments than by rigid foundations. Figure 25.17 shows an embankment with a width L and a geosynthetic layer at the bottom of it. If it is assumed that the embankment settles s at its center, that the deflected shape of the bottom of the embankment is an arc of a circle, and that the circle passes through the ends of the embankment, then the relationship between the settlement s and the radius R of the circle is:

$$(R - s)^2 + \frac{L^2}{4} = R^2 \text{ or (neglecting higher-order terms)}$$

$$R = \frac{L^2}{8s} \tag{25.43}$$

where L is the width of the embankment.

Then the geosynthetic stretches from an initial length L to a deformed length L':

$$L' = 2R\, Arc \sin \frac{L}{2R} \tag{25.44}$$

which leads to a strain ε in the geosynthetic of:

$$\varepsilon = \frac{Arc \sin(4s/L)}{4s/L} - 1 \tag{25.45}$$

and a tension T equal to:

$$T = E\varepsilon \tag{25.46}$$

where E is the geosynthetic modulus (section 25.3.1).

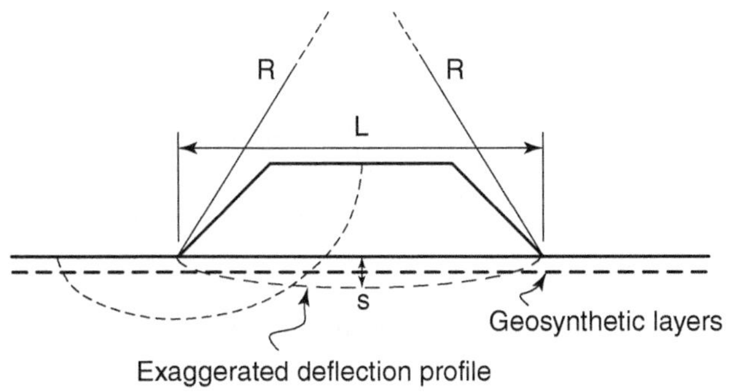

Figure 25.17 Embankment reinforcement.

The geotextile or geogrid required needs to have a tensile strength much higher than T because of the reduction factors (Eq. 25.39).

Anchoring length. The geosynthetic layer must extend far enough beyond the edges of the embankment or the foundation to ensure that it will not pull out when loaded. The anchor length is calculated as presented in Eq. 25.7. The anchor length can be shortened by wrapping the geosynthetic around large embedded stones or timber cribbing.

25.7 DESIGN FOR FILTRATION AND DRAINAGE

Filtration refers to the case in which water is flowing perpendicular to the plane of the geosynthetic; *drainage* is the case in which water flows in the direction of the geosynthetic. The design of a geosynthetic filter or drain (mostly nonwoven, needle-punched geotextile) has two aspects: water passage and soil retention. The problem is that for water conveyance, the geotextile should have large openings, whereas for soil retention it should have small openings. A compromise must be found.

For filtering, the required permittivity for water passage is calculated:

$$\Psi_{req} = \frac{k}{t} = \frac{q}{\Delta h A} \tag{25.47}$$

where ψ_{req} is the permittivity, k is the hydraulic conductivity, t is the thickness of the geotextile, q is the flow through the flow net to be handled by the geotextile, Δh is the drop of total head through the flow net, and A is the cross-sectional area perpendicular to the flow.

The steps include drawing a flow net, calculating the flow q through the flow net, determining the required permittivity from Eq. 25.47, and seeking the geotextile that satisfies this requirement. Note that the permittivity of the geotextile has to be corrected for reduction factors as follows:

$$\Psi_{allow} = \frac{\Psi_{ult}}{(RF_{SCB} \times RF_{CR} \times RF_{IN} \times RF_{CC} \times RF_{BC})} \tag{25.48}$$

where ψ_{allow} is the permittivity that can be used in design, ψ_{ult} is the permittivity quoted by the manufacturer, RF_{SCB} is the reduction factor for soil clogging and blinding, RF_{CR} is the reduction factor for creep reduction of void space, RF_{IN} is the reduction factor for adjacent materials intruding into the geotextile void space, RF_{CC} is the reduction factor for chemical clogging, and RF_{BC} is the reduction factor for biological clogging.

These reduction factors vary between 1 and 10 depending on the application, and average 4.41 (RF_{SCB}), 1.83 (RF_{CR}), 1.1 (RF_{IN}), 1.25 (RF_{CC}), and 2.2 (RF_{BC}). As can be seen, multiplying all these factors leads to using an allowable permittivity that is a very small fraction of the ultimate value. In addition, a regular factor of safety is applied as follows:

$$F = \frac{\Psi_{allow}}{\Psi_{req}} \tag{25.49}$$

For critical applications, this factor of safety should be as high as 5 to 10.

For filtering, the problem of soil retention occurs when the water flows from a soil with fines into a much coarser soil or an open space. In this case the fines may wash out through the coarser soil and follow the water flow. The filter ensures that the transition from fine-grained to coarse-grained is gradual and lets the water go through while retaining the fines of the soil. The design makes use of the geotextile AOS. (Recall from section 25.3.1 that the AOS is the apparent opening size, defined as the diameter of glass beads corresponding to 95% retained by weight.) Typical AOS values range from 0.01 mm to 0.5 mm. A simple criterion for the opening of the geotextile is (Carroll 1983):

$$O_{95} < 2.5D_{85} \tag{25.50}$$

where O_{95} is the AOS of the geotextile and D_{85} is the particle size corresponding to 85% passing by weight of the soil to be protected. More detailed criteria for geosynthetic filters have been developed (e.g., Luettich et al. 1992; Koerner 2012; Giroud 2010). In particular, Giroud (2010) proposed two new criteria based on porosity and thickness in addition to water conveyance and soil retention. The porosity criterion ensures that the geotextile has enough openings per unit area and makes a clear distinction between woven and nonwoven geotextiles. The thickness criterion recognizes that, unlike granular filters, the opening size of a geotextile filter depends on its thickness.

For drainage, the water flows in the direction of the geosynthetic. The design of a geosynthetic for drainage purposes follows an approach similar to that used in the design for filtering. Instead of permittivity, however, we use transmissivity in this case, defined as:

$$\Theta_{req} = kt = \frac{q}{iw} \tag{25.51}$$

where Θ_{req} is the transmissivity required, k is the in-plane hydraulic conductivity, t is the thickness of the geosynthetic, q is the water flow to be handled, i is the hydraulic gradient, and w is the width of the geosynthetic.

Although geosynthetic have very useful applications in filtering and drainage, they provide limited flow capacity for water conveyance (10^{-8} to 10^{-6} m^3/s per meter of geosynthetic).

25.8 DESIGN FOR EROSION CONTROL

Geosynthetics have been used for decades in the field of erosion control. There are several application domains, including geosynthetic filters under rip rap, geosynthetics to facilitate revegetation, and silt fences.

Filters

Rip rap is often placed to prevent erosion when high water velocities affect the ground surface. Sizing the rip rap consists of finding out the highest water velocity to be handled and choosing the rip-rap size accordingly. Figure 23.8 can be used for rip-rap size selection. In addition, one must check that the rock itself is not degradable over time when subjected to wet-dry cycles. Once the rip rap is chosen, it is very important to place a geosynthetic layer between the soil to be protected and the rip rap. If such a layer is not placed, the soil can erode from underneath the rip rap. The rip rap will not move downstream, but will sink into the soil below and not prevent erosion. The geosynthetic layer has two functions: soil retention to prevent the soil underneath from eroding away, and water conveyance to prevent compression water stresses from developing in the underlying soil. These water stresses would weaken the soil and lead to failure (e.g., slope instability). Therefore, the geosynthetic must be a filter, and geotextiles are best suited for this purpose. The design of the filter follows the same rules as those discussed in section 25.7.

Revegetation

Erosion on the slopes of embankments, dams, levees, and river banks can be minimized by strong and thick vegetation. The problem is that it takes time for appropriate vegetation to grow and become dense and deeply rooted. To help fix the vegetation, geosynthetics can be used. A distinction is made between temporary erosion and revegetation materials (TERMS) and permanent erosion and revegetation materials (PERMS). TERMS are completely biodegradable (hay straws, mulches) or partially biodegradable (hydraulic mulch geofibers, erosion control blankets). PERMS include turf reinforcement mats (TRMs) and vegetated geocellular containment systems (GCSs). The geosynthetics used in PERMS have openings to let the vegetation and roots grow through and some filtering capability. After the seeds are sown, the vegetation grows and gets entangled with the geosynthetic, which provides reinforcement to the root system. Figure 25.18 gives a range of velocities that can be resisted by various forms of

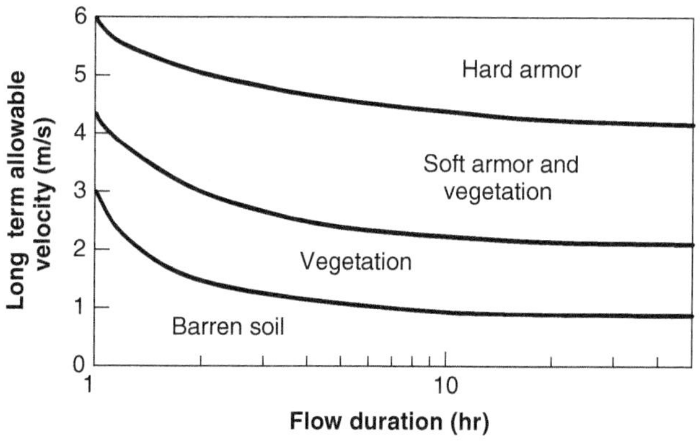

Figure 25.18 Allowable velocities for erosion control measures. (After Theisen 1992.)

armoring, including the soft armor with vegetation discussed here.

Silt Fences

Water flowing on barren soil along roadways or construction sites erodes the soil. To prevent this erosion, silt fences are often placed to let the water go through but stop and collect the silt-size particles that would otherwise flow downstream. Silt fences consist of a geosynthetic (most often woven geotextile) placed above ground by attaching it to vertical posts driven in the soil (Figure 25.19). The silt fence catches and retains the fine soil particles yet lets the water flow through. The water flow is typically quite shallow compared to a river flow, but the velocity can be high on steep slopes. The following design issues must be addressed: maximum length of slope between fences, runoff flow rate, sediment flow rate, height of fence, spacing and strength of fence posts, and geotextile selection. The maximum length of slope L_{max} that can be handled by one silt fence may be estimated by (Koerner 2012):

$$L_{max}(\text{m}) = 36.2\, e^{-11.1 \tan\alpha} \qquad (25.52)$$

where α is the slope angle.

If the length of slope to be protected is longer than that, a sequence of silt fences separated by a distance less than L_{max} is used. The runoff flow rate Q is tied to the recurrence interval of the rainfall selected (often taken as the 10-year flow) and is given by:

$$Q\,(\text{m}^3/\text{hr}) = C \times I\,(\text{m/hr}) \times A(\text{m}^2) \qquad (25.53)$$

where C is a dimensionless coefficient taken as 0.5 for barren soil, I is the rainfall intensity, and A is the drainage area. The weight of soil accumulated per unit area of soil drained and per unit time behind the silt fence can be estimated by using the Uniform Soil Loss Equation (USLE; Wishmeier and Smith 1960):

$$E\ (\text{kN/km}^2.\text{yr}) = 10 \times R \times K \times LS \times C \times P \qquad (25.54)$$

where R is the dimensionless rainfall coefficient, K is the dimensionless soil erodibility factor, LS is the dimensionless length of slope or gradient factor, C is the dimensionless vegetation cover factor, and P is the dimensionless conservation practice factor. These factors are given in Wishmeier and Smith (1960). The USLE equation has shortcomings and does not apply to channel and gully flow. Nevertheless, it provides a first estimate.

The height H of the silt fence can be calculated by finding out the volume V of water and soil that can be retained by the fence over a 1 m width of fence:

$$V(\text{m}^3) = Qt = H\left(\frac{H}{\tan\alpha}\right) \times 1\text{ m} \qquad (25.55)$$

where Q is obtained from Eq. 25.53, t is the duration of the rainstorm, H is the height of the fence, and α is the angle of the slope on which the water flows.

Silt fences are usually 0.3 to 0.9 m high. Then the spacing of the posts retaining the fence is chosen. This spacing is usually between 1 and 3 m. The load on the fence due to the water pressure can then be calculated to obtain the lateral load and maximum bending moment on the posts. This bending moment is in the range of 5 to 30 kN.m. The last step is to calculate the tensile load in the fence material. If it is assumed that the fence deflects an amount s in an arc of circle under the average water pressure of $0.5\gamma_w$H behind the fence, the tension in the geosynthetic is given by the following equation:

$$T = \frac{\gamma_w H L^2}{16s\left(\dfrac{Arc\sin\dfrac{4s}{L}}{\dfrac{4s}{L}}\right)} \quad \text{or } T = \frac{\gamma_w H L^2}{16s} \text{ if 4s/L is small} \qquad (25.56)$$

where T is the tension per meter of geotextile, γ_w is the unit weight of water, H is the height of the fence, L is the distance between posts, and s is the horizontal deflection of the fence at midspan (Figure 25.19). Typical values of T range from 5 to 30 kN/m.

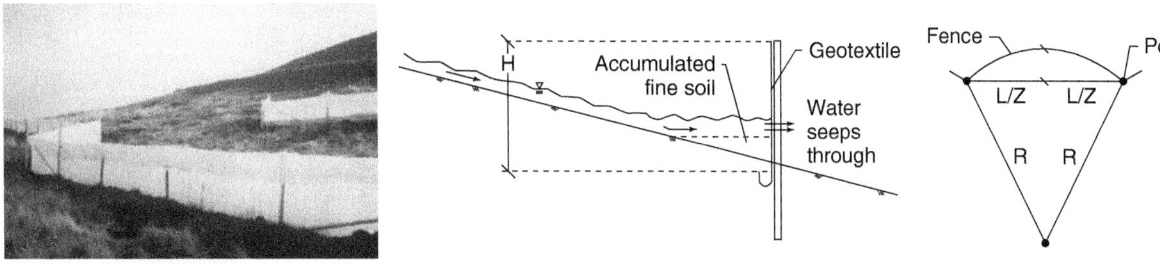

(a) Silt fences (b) Cross section (c) Plan view

Figure 25.19 Silt fences. (a: Courtesy of Robert Koerner, 2012)

25.9 OTHER DESIGN APPLICATIONS

25.9.1 Lightweight Fills

Lightweight fills are most commonly built of geofoam blocks (Horvath 1994; Saye et al. 2000). The unit weight of the geofoam blocks is at most 10% of the unit weight of soil. Therefore, the pressure on the native soil and the associated settlement can be reduced significantly. Note that a pavement layer still has to be constructed with heavier materials (granular base course for drainage and asphalt rolling layer) on top of a geofoam embankment. Of course, the compression of the geofoam must be added to the settlement of the soil below, but considering the typical application (embankment on soft soils), that compression is most of the time negligible compared to the settlement of the soil below. Overall, 80% reduction in settlement is not uncommon.

25.9.2 Compressible Inclusions

Another notable application of compressible inclusions is the case of geofoam blocks behind retaining walls to decrease the earth pressure. This type of solution is particularly useful for walls that cannot tolerate much lateral deflection without damaging the structure. This is the case of basement walls and bridge abutments in shrink-swell soil areas. The pressure-absorbing layer may be 50 to 600 mm thick and tends to decrease the pressure as shown in Figure 25.20. The thicker the geofoam layer is, the lower the pressure is likely to be. Note also that the pressure distribution is altered toward the bottom of the wall where the geofoam is most effective. Another application is to mitigate seismically induced pressures (Athanasopoulos 2007).

25.9.3 Thermal Insulation

Geofoams are among the best temperature insulators. Recall from Chapter 16, Eq. 16.8 that the R factor of an insulator is a measure of the resistance to temperature propagation and is given by:

$$R = \frac{dx}{k_t} \tag{25.57}$$

where dx is the thickness of the insulating layer in meters and k_t is the thermal conductivity of the insulating material in watts per degree Kelvin per meter (W/K.m). Therefore, the R rating is expressed in m^2.K/W. Because the degree Kelvin is equal to the degree Celsius, the R rating has the same value in m^2.K/W and in m^2.C/W.

Table 25.5 shows that geofoam blocks have some of the highest R ratings of any materials. Within the geofoam range of R values, extruded polystyrene (XPS) has higher R values than expanded polystyrene (EPS). Styrofoam coffee cups are made of XPS. Once the R factor is known, the heat flow can be calculated:

$$\frac{dQ}{dt} = k_t A \frac{dT}{dx} = \frac{A}{R} dT \tag{25.58}$$

where dQ is the amount of heat (J) flowing in a time dt (s), k_t is the thermal conductivity (J/s.K.m or W/K.m), A is the area perpendicular to the heat flow (m^2), dT is the change in temperature (K), dx is the length over which the change of temperature is occurring (m), and R is the thermal resistance or R factor (m^2.K/W).

The applications include insulation under a house on permafrost to avoid ground thawing, or under a refrigerated building to avoid ground freezing.

25.9.4 Geosynthetics and Landfill Slopes

Modern landfill liners are made of many different layers, including geosynthetics. These geosynthetic layers, particularly GCLs, can represent planes of lower shear strength where slope failure can develop. This issue should be addressed at the time the liner is designed, together with a plan and possible restrictions on where and how high the waste can be piled up at one location. This topic is addressed in Chapter 26.

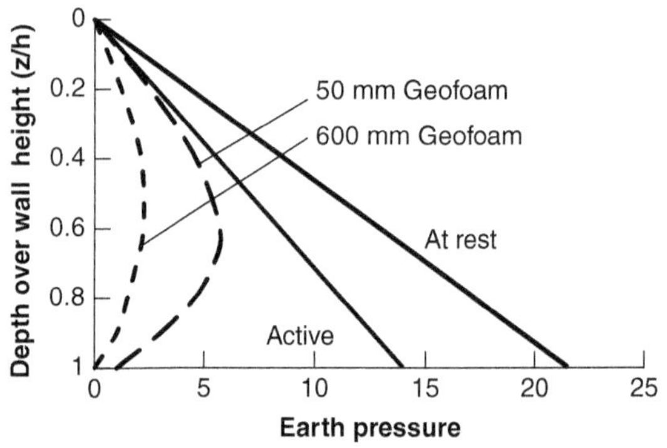

Figure 25.20 Decrease in earth pressure by compressible inclusions. (After Horvath 1997)

Table 25.5 R Factor for Various Materials

Material	R Factor (M^2.K/W or M^2.C/W)
Steel	0.022
Ice	0.45
Concrete	0.95
Glass	1.25
Water (25°C)	1.64
Glass wool	23.8
Air (25°C)	38.5
Geofoam blocks	25 to 40

(After Koerner 2012)

PROBLEMS

25.1 A geosynthetic is placed on the ground surface and stones are to be placed on top of it. The maximum diameter of the stones is 60 mm and the drop height from the truck is 1.5 m. The soil below the geosynthetic is medium stiff with a soil support reduction factor of $F_s = 15$; the cumulative reduction factor for the geosynthetic is $F_r = 5$. What is the impact strength required of the geosynthetic to safely handle the impact loading?

25.2 A 0.5 m thick layer of base course has been placed on top of a geotextile. Trucks with tire pressures equal to 600 kPa will travel on top of the base course during construction. The stones are 60 mm in diameter and fairly sharp, such that the product $S_1S_2S_3$ in Eq. 25.24 is equal to 0.3. If the geotextile strength reduction factor is 4.5 (Eq. 25.39), what is the required ultimate strength of the geotextile to safely avoid puncture?

25.3 A landfill owner is considering replacing a 1 m thick layer of compacted clay with a 15 mm thick GCL as part of the design of a new landfill liner. The landfill has an area of 7.5 hectares and the fee collected per cubic meter of waste is $90. How much additional income does the owner stand to collect from the saving in the thickness of the liner?

25.4 A geosynthetic clay liner and a compacted clay liner are being compared. The GCL is 15 mm thick and has a hydraulic conductivity of 10^{-11} m/s; the CCL is 500 mm thick and has a hydraulic conductivity of 10^{-9} m/s. The water level is 1 m above the top of the liner and the pressure head is assumed to be zero on the bottom side of the liner. Calculate the amount of water going through the GCL and the CCL.

25.5 A geosynthetic clay liner has a bentonite clay layer with the following shear strength characteristics: $c' = 0$ and $\varphi' = 10^o$. It is placed on the side slope of a landfill that has an 18° angle with the horizontal. The plan is to cover the GCL uniformly with 20 m of waste weighing 10 kN/m^3. What cohesion c' must be developed by needle-punching in the GCL to have a factor of safety of 1.5 against failure in the bentonite? Use the infinite slope equation from section 19.3.

25.6 Design a 6 m high geosynthetic-reinforced MSE wall. The vertical spacing between geosynthetic layers is 0.5 m, the backfill is sand with a unit weight of 20 kN/m^3 and a friction angle of 34°, a surcharge of 20 kN/m^2 is applied on the ground surface at the top of the wall, and the geosynthetic is a geogrid. The soil on which the wall is being built is a very stiff clay with an undrained shear strength s_u equal to 100 kN/m^2 and a friction angle of 25°. The soil behind the wall is a sandy clay with a unit weight of 20 kN/m^3 and a friction angle of 30°. Assume reasonable values for all other parameters needed for the design.

25.7 A layer of geotextile is placed 1 m below a 2 m wide strip footing. The footing rests on the surface of a loose sand with a friction angle equal to 30° and a pressuremeter limit pressure of 500 kPa within a depth equal to one footing width below the footing. The geogrid has an ultimate tensile strength of 100 kN/m.

a. Calculate the percent increase in ultimate bearing capacity between the case of no geogrid and the case with geogrid.

b. If the geogrid has a global friction factor K (Eq. 25.7) of 0.3, what length of geogrid is required to safely anchor the geogrid on each side of the footing?

25.8 A 30 m wide, 7 m high embankment is placed on soft clay with a geotextile between the surface of the soft clay and the embankment fill. The purpose of the geotextile is to increase the bearing capacity and reduce the settlement reduction, but it is also used for separation, drainage, and filtering. The bottom of the embankment settles along an arc of circle with 1 m of settlement at the center and a negligible amount at the edges. What will be the tension load in the geotextile if its modulus is 500 kN/m?

25.9 A geotextile has an ultimate tensile strength of 100 kN/m and a maximum flow rate capacity of 8×10^{-7} m^3/s per meter of geotextile. What are reasonable values of the allowable tensile strength and allowable flow rate for this geotextile?

25.10 A construction site has a 30 m long erodible slope with an angle of 6°. Silt fences are required.

a. How many silt fences are needed?

b. If the 10-year rainstorm generates 100 mm/hr, what is the flow rate to be handled per meter of width of the fence?

c. Calculate the height of the fence so that it can safely handle two 10-year storms each lasting 3 hours.

d. Posts are placed every 3 m and the fence is allowed to deflect 0.2 m at its center. Estimate the tension in the fence fabric.

25.11 Derive Eq. 25.56 for silt fences.

25.12 A 7 m high, 60 m wide embankment is to be built on a layer of soft clay with a water table at the ground surface. The soft clay is 5 m thick and the increase in stress in the clay layer can be taken as the pressure under the embankment because the clay layer is thin compared to the width of the embankment. The clay layer has the following consolidation characteristics: $e_o = 1.1$, $\gamma = 19$ kN/m^3, $C_c = 0.5$. Two options are considered for the embankment fill: soil fill and geofoam fill. The soil fill has a unit weight of 20 kN/m^3 and the geofoam fill 2 kN/m^3. What will be the settlement of the embankment in each case? Which fill type will have the shortest time to reach 90% consolidation?

25.13 A building refrigerated at -5°C is being designed on a soil with a high water table. The concern is the cost of the power (watts) to maintain the difference in temperature across the foundation. Two alternatives are considered. The first consists of a relatively inexpensive 100 mm thick concrete slab on grade on top of the soil. The second one consists of the same slab on grade on top of a 150 mm thick geofoam. The concrete has a thermal resistance R value equal to 0.9 m^2.C/W per meter of thickness and the geofoam 35 m^2.C/W per meter of thickness. Calculate the amount of power required in each case to maintain the difference in temperature at -5°C above the slab and 0°C on top of the soil.

25.14 A Styrofoam coffee cup holds coffee at 80°C. Your hand holding the coffee cup is at 30°C. Assuming a steady-state heat transfer in the cup wall, what is the R rating per meter of the Styrofoam if the amount of heat released from the coffee cup through the wall of the cup is 45 W? What is the thermal conductivity of the Styrofoam if the cup wall is 1.5 mm thick?

Problems and Solutions

Problem 25.1

A geosynthetic is placed on the ground surface and stones are to be placed on top of it. The maximum diameter of the stones is 60 mm and the drop height from the truck is 1.5 m. The soil below the geosynthetic is medium stiff with a soil support reduction factor of $F_s = 15$; the cumulative reduction factor for the geosynthetic is $F_r = 5$. What is the impact strength required of the geosynthetic to safely handle the impact loading?

Solution 25.1

We assume that the stone has a unit weight of 26 kN/m^3:

$$E_{stone} = Wh = \frac{\pi d^3}{6}\gamma h = \frac{\pi \times 0.06^3}{6} \times 26000 \times 1.5 = 4.41 \text{ J}$$

$$\frac{E_{stone}}{F_{soil}} = \frac{4.41}{15} = 0.294$$

We must satisfy:

$$\frac{E_{stone}}{F_{soil}} \leq \frac{E_{geosyn}}{F_{reduc}}$$

$$E_{geosyn} \geq 0.294 \times 5 = 1.47 \text{ J}$$

The geosynthetic must have an impact strength at least equal to 1.47 J.

Problem 25.2

A 0.5 m thick layer of base course has been placed on top of a geotextile. Trucks with tire pressures equal to 600 kPa will travel on top of the base course during construction. The stones are 60 mm in diameter and fairly sharp, such that the product $S_1 S_2 S_3$ in Eq. 25.24 is equal to 0.3. If the geotextile strength reduction factor is 4.5 (Eq. 25.39), what is the required ultimate strength of the geotextile to safely avoid puncture?

Solution 25.2

$$F_{stone} = p\, d_a{}^2 S_1 S_2 S_3 = (600 + 0.5 \times 20) \times 0.06^2 \times 0.3 = 0.659 \text{ kN}$$

$$S_t = \frac{P}{\pi d} = \frac{0.659}{0.06\pi} = 3.5 \frac{\text{kN}}{\text{m}}$$

$$T_{ultimate} = T_{allowable} \times RF = 3.5 \times 4.5 = 15.8 \frac{\text{kN}}{\text{m}}$$

In this situation, a geotextile rated at 15.8 kN/m is needed.

Problem 25.3

A landfill owner is considering replacing a 1 m thick layer of compacted clay with a 15 mm thick GCL as part of the design of a new landfill liner. The landfill has an area of 7.5 hectares and the fee collected per cubic meter of waste is \$90. How much additional income does the owner stand to collect from the saving in the thickness of the liner?

Solution 25.3

The change in height after replacing the compacted clay layer with the GCL:

$$\Delta H = 1 - 0.015$$
$$= 0.985 \text{ m}$$

For a landfill area of 7.5 hectares and a fee of \$90 per m^3:

$$\text{Additional income per hectare} = 90 \times 10{,}000 \times 0.985$$
$$= \$886{,}500/\text{ha}$$
$$\text{Total income} = \frac{\$886{,}500}{\text{ha}} \times 7.5 \text{ ha} = \$6{,}648{,}750$$

Problem 25.4

A geosynthetic clay liner and a compacted clay liner are being compared. The GCL is 15 mm thick and has a hydraulic conductivity of 10^{-11} m/s; the CCL is 500 mm thick and has a hydraulic conductivity of 10^{-9} m/s. The water level is 1 m above the top of the liner and the pressure head is assumed to be zero on the bottom side of the liner. Calculate the amount of water going through the GCL and the CCL.

Solution 25.4

Using Darcy's law, and assuming that the hydraulic gradient is the total head divided by the thickness of the GCL and a flow through a unit area, the amount of water through the GCL is:

$$q = kiA$$
$$= (1 \times 10^{-11})\left(\frac{1.015}{0.015}\right)(1 \times 1)$$
$$q = 7 \times 10^{-10} \text{ m}^3/\text{s}$$

Using the same procedure, the flow rate through the CCL is:

$$q = kiA$$
$$= (1 \times 10^{-9})\left(\frac{1.500}{0.500}\right)(1 \times 1)$$
$$q = 3 \times 10^{-9} \text{ m}^3/\text{s}$$

Problem 25.5

A geosynthetic clay liner has a bentonite clay layer with the following shear strength characteristics: $c' = 0$ and $\varphi' = 10°$. It is placed on the side slope of a landfill that has an 18° angle with the horizontal. The plan is to cover the GCL uniformly with 20 m of waste weighing 10 kN/m^3. What cohesion c' must be developed by needle-punching in the GCL to have a factor of safety of 1.5 against failure in the bentonite? Use the infinite slope equation from section 19.3.

Solution 25.5

Figure 25.1s shows the illustration of the infinite slope. Note that W is the weight of the wedge, T is the shear force, and N is the normal force. H is the height of the waste, L is the length of the wedge, and β is the inclination of the slope.

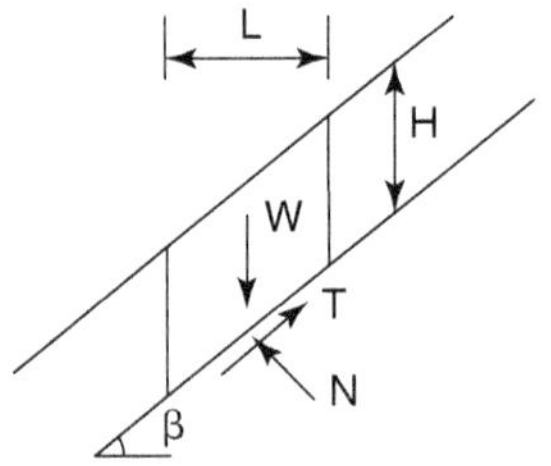

Figure 25.1s Illustration of infinite slope.

Based on the equilibrium condition and the definition of factor of safety, *FS* can be calculated:

$$FS = \frac{\tan \varphi'}{\tan \beta} + \frac{c'}{\gamma H \sin \beta \cos \beta}$$

Here, γ is the unit weight of the waste and φ' is the friction angle.
To achieve a factor of safety of 1.5, the cohesion developed by needle-punching has to satisfy the following equation:

$$1.5 = \frac{\tan 10^\circ}{\tan 18^\circ} + \frac{c'}{10 \times 20 \times \sin 18^\circ \cos 18^\circ}$$

Therefore, the cohesion developed by needle-punching must be $c' = 56\,\text{kPa}$.

Problem 25.6

Design a 6 m high geosynthetic-reinforced MSE wall. The vertical spacing between geosynthetic layers is 0.5 m, the backfill is sand with a unit weight of 20 kN/m^3 and a friction angle of 34°, a surcharge of 20 kN/m^2 is applied on the ground surface at the top of the wall, and the geosynthetic is a geogrid. The soil on which the wall is being built is a very stiff clay with an undrained shear strength s_u equal to 100 kN/m^2 and a friction angle of 25°. The soil behind the wall is a sandy clay with a unit weight of 20 kN/m^3 and a friction angle of 30°. Assume reasonable values for all other parameters needed for the design.

Solution 25.6 (Figure 25.2s)

$$H = 6\ \text{m}$$

$$S_v = 0.5\ \text{m}$$

$$\Delta\sigma_v = 20\ \text{kN/m}^2$$

Use a minimum reinforcement length $L = 4.2$ m as the length-to-height ratio of the reinforced wall (should be no less than 0.7).

Assume that the first layer of geosynthetics is placed at 0.25 m from the finished grade.

Consider the ultimate strength resistance of the geosynthetic (T_{ult}) as 170 kN/m.

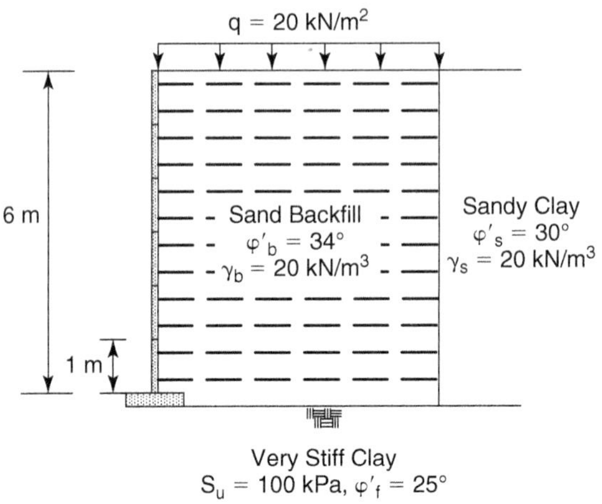

Figure 25.2s Retaining wall.

a. External stability, earth pressures.

$$k_a = \frac{1 - \sin \varphi_b}{1 + \sin \varphi_b} = \frac{1 - \sin 30}{1 + \sin 30} = 0.33$$

Then the active load generated by the horizontal soil pressure P_{a1} and the traffic surcharge P_{a2} can be computed as:

$$P_{a1} = \frac{K_a \times \gamma_s \times H^2}{2} = \frac{0.333 \times 20 \times (6)^2}{2} = 120 \text{ kN/m}$$

- Located 2 m above the bottom of the wall ($x_{a1} = 2$ m, as shown in Figure 25.3s).

$$P_{a2} = K_a \times q \times H = 0.333 \times 20 \times 6 = 40 \text{ kN/m}$$

- Located 3 m above the bottom of the wall ($x_{a2} = 3$ m as shown in Figure 25.3s).

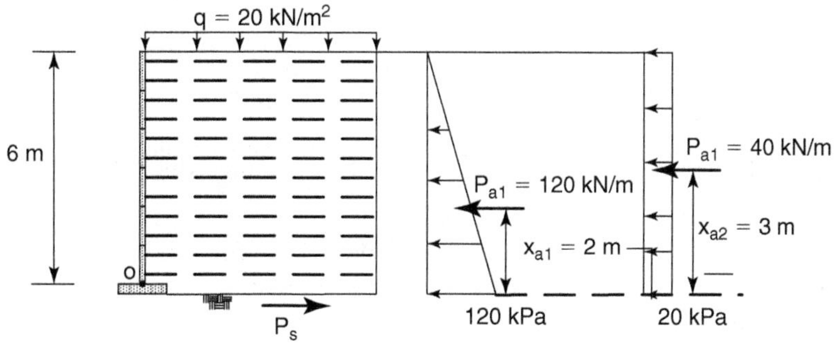

Figure 25.3s Pressure diagram on retaining walls.

We can now calculate the sliding and overturning stability (ignoring the traffic surcharge).

b. External stability, sliding analysis.
   Using the LRFD approach and no traffic surcharge:

$$\varphi W \tan \varphi'_f \geq \gamma P_{a1} \quad \text{or} \quad \varphi \gamma_b H L \tan \varphi'_f \geq \gamma P_{a1}$$

Using φ as 0.85, γ as 1.25, and L as 4.2 m, we have:

$$0.85 \times 20 \times 6 \times 4.2 \times \tan 25 \geq 1.25 \times 120$$

$$428.4 \text{ kN/m} \geq 150 \text{ kN/m} \quad \therefore \text{OK}$$

Using the LRFD approach and the traffic surcharge:

$$\varphi W \tan \varphi'_f \geq \gamma_1 P_{a1} + \gamma_2 P_{a2} \quad \text{or} \quad \varphi \gamma_b H L \tan \varphi'_f \geq \gamma P_{a1}$$

Using φ as 0.85, γ as 1.25 for the dead load and γ as 1.75 for the live load, and L as 4.2 m, we have:

$$0.85 \times 20 \times 6 \times 4.2 \times \tan 25 \geq 1.25 \times 120 + 1.75 \times 40$$

$$428.4 \text{ kN/m} \geq 220 \text{ kN/m} \quad \therefore \text{OK}$$

c. External stability, overturning analysis.

Overturning around the toe (point O) of the wall with no traffic surcharge:

$$\frac{\varphi WL}{2} \geq \frac{\gamma P_{a1} H}{3} \quad \text{or} \quad \frac{\varphi \gamma_b H L^2}{2} \geq \frac{\gamma P_{a1} H}{3} \quad \text{or} \quad \frac{0.85 \times 20 \times 6 \times 4.2^2}{2} \geq \frac{1.25 \times 120 \times 6}{3}$$

$$899.6 \text{ kN} \geq 300 \text{ kN} \quad \therefore \text{OK}$$

Overturning around the toe (point O) of the wall with traffic surcharge:

$$\frac{\varphi(W + qL)L}{2} \geq \frac{\gamma_1 P_{a1} H}{3} + \frac{\gamma_2 P_{a2} H}{2} \quad \text{or}$$

$$\frac{0.85(20 \times 6 \times 4.2 + 20 \times 4.2)4.2}{2} \geq \frac{1.25 \times 120 \times 6}{3} + \frac{1.75 \times 40 \times 6}{2}$$

$$1076 \text{ kN} \geq 510 \text{ kN} \quad \therefore \text{OK}$$

d. External stability, bearing capacity analysis.

The eccentricity of the wall applied forces can be calculated as:

$$W \times e + q \times L \times e = M_{ov} \quad \text{or} \quad (W + q \times L) \times e = P_{a1} \times x_{a1} + P_{a2} \times x_{a2}$$

$$e = \frac{P_{a1} \times x_{a1} + P_{a2} \times x_{a2}}{(W + q \times L)} = \frac{120 \times 2 + 40 \times 3}{(20 \times 6 \times 4.2 + 20 \times 4.2)} = 0.61 \text{ m}$$

This eccentricity cannot be outside of the central one-third of the footing, which is:

$$e \leq \frac{L}{6} = \frac{4.2}{6} = 0.7 \text{ m} \quad \therefore \text{OK}$$

This means that there is no tension underneath the footing.

The active length, according to Meyerhof's distribution, is:

$$L_{active} = L - 2 \times e = 4.2 - 2 \times 0.61 = 2.98 \text{ m}$$

The bearing pressure is:

$$p = (\gamma_b \times H + q) \times \frac{L}{L_{active}} = (20 \times 6 + 20) \times \frac{4.2}{2.98} = (169.1)_{weight} + (28.2)_{traffic} = 197.3 \text{ kPa}$$

The bearing capacity of the existing soil can be calculated according to the Skempton chart:

$$q_u = N_c S_u + \gamma_b D = 7.5 \times 100 = 750 \text{ kPa}$$

Checking for bearing capacity failure:

$$\varphi \times q_{bc} \geq \gamma_1 p_1 + \gamma_2 p_2 \quad \text{or} \quad 0.5 \times 750 \geq 1.25 \times 169.1 + 1.75 \times 28.2$$

$$375 \text{ kPa} \geq 260.7 \text{ kPa} \quad \therefore \text{OK}$$

e. Internal stability, pull-out failure.

No traffic surcharge:

$$T_{\max} = s_v \sigma_{ah} = s_v k_a \sigma'_v$$

$$k_a = \frac{1 - \sin \varphi_r}{1 + \sin \varphi_r} = \frac{1 - \sin 34}{1 + \sin 34} = 0.283$$

Note that for an MSE wall built with geosynthetics, the k_r and k_a ratio are the same according to AASHTO LRFD (Figure 25.4s).

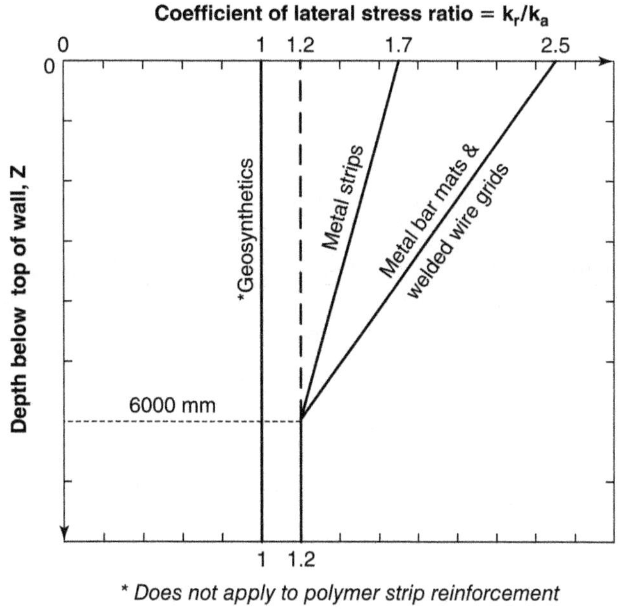

Figure 25.4s Coefficient of lateral stress ratio.

The results of T_{max} at different heights are shown in Table 25.1s. T_{max1} is due to the soil weight (s_v σ_{ah}, active earth pressure) and T_{max2} is due to the traffic surcharge (s_v $k_a \times 20\text{kN/m}^2$).

Table 25.1s Summary of Calculation of T_{max}

Layer No.	Depth (m)	k_a	k_r	σ_v (kPa)	σ_{ah} (kPa)	T_{max1} (kN/m)	T_{max2} (kN/m)	$T_{max\text{-}total}$ (kN/m)
1	0.25	0.283	0.283	5.0	1.414	0.71	2.83	3.53
2	0.75	0.283	0.283	15.0	4.241	2.12	2.83	4.95
3	1.25	0.283	0.283	25.0	7.068	3.53	2.83	6.36
4	1.75	0.283	0.283	35.0	9.895	4.95	2.83	7.77
5	2.25	0.283	0.283	45.0	12.722	6.36	2.83	9.19
6	2.75	0.283	0.283	55.0	15.549	7.77	2.83	10.60
7	3.25	0.283	0.283	65.0	18.376	9.19	2.83	12.02
8	3.75	0.283	0.283	75.0	21.204	10.60	2.83	13.43
9	4.25	0.283	0.283	85.0	24.031	12.02	2.83	14.84
10	4.75	0.283	0.283	95.0	26.858	13.43	2.83	16.26
11	5.25	0.283	0.283	105.0	29.685	14.84	2.83	17.67
12	5.75	0.283	0.283	115.0	32.512	16.26	2.83	19.08

Using the ultimate limit state procedure, we have:

$$\gamma_1 T_{\max 1} + \gamma_2 T_{\max 2} = \phi T_{pullout}$$

The active length of the reinforcement strip required to resist the pull-out load is:

$$T_{pullout} = \frac{\gamma_1 T_{\max 1} + \gamma_2 T_{\max 2}}{\varphi}$$

$$L_a = \frac{T_{pullout}}{2 \times f_{\max} \times b} = \frac{(\gamma_1 \sigma'_{ah} \times s_v) + (\gamma_2 k_a q \times s_v)}{2 \times \varphi \times \sigma'_v \times \tan\delta}$$

$$L_{\max} = (H - z) \times \tan\left(45 - \frac{\varphi}{2}\right)$$

$$L_{total} = L_{\max} + L_a = (H - z) \times \tan\left(45 - \frac{\varphi}{2}\right) + \frac{(\gamma_1 \sigma'_{ah} \times s_v) + (\gamma_2 k_a q \times s_v)}{2 \times \varphi \times \sigma'_v \times \tan\delta}$$

However, in construction practice, the length is often taken as constant throughout the height of the wall. The longest value of L_{total} is at the top of the wall ($z = 0$). Then:

$$L_{total} = H \times \tan\left(45 - \frac{\varphi}{2}\right) + \frac{(\gamma_1 \sigma'_{ah} \times s_v) + (\gamma_2 k_a q \times s_v)}{2 \times \varphi \times \sigma'_v \times \tan\delta}$$

The resistance (φ) and load factor (γ) are taken as 0.9 and 1.5, respectively. The coefficient of friction ($\tan\delta$) is computed according to AASHTO LRFD using Figure 25.5s. Based on this figure, the friction factor is equal to the tangent of the friction angle of the reinforced backfill (φ_r). Therefore:

$$F^* = \tan\delta = \tan\varphi_b = 0.6745$$

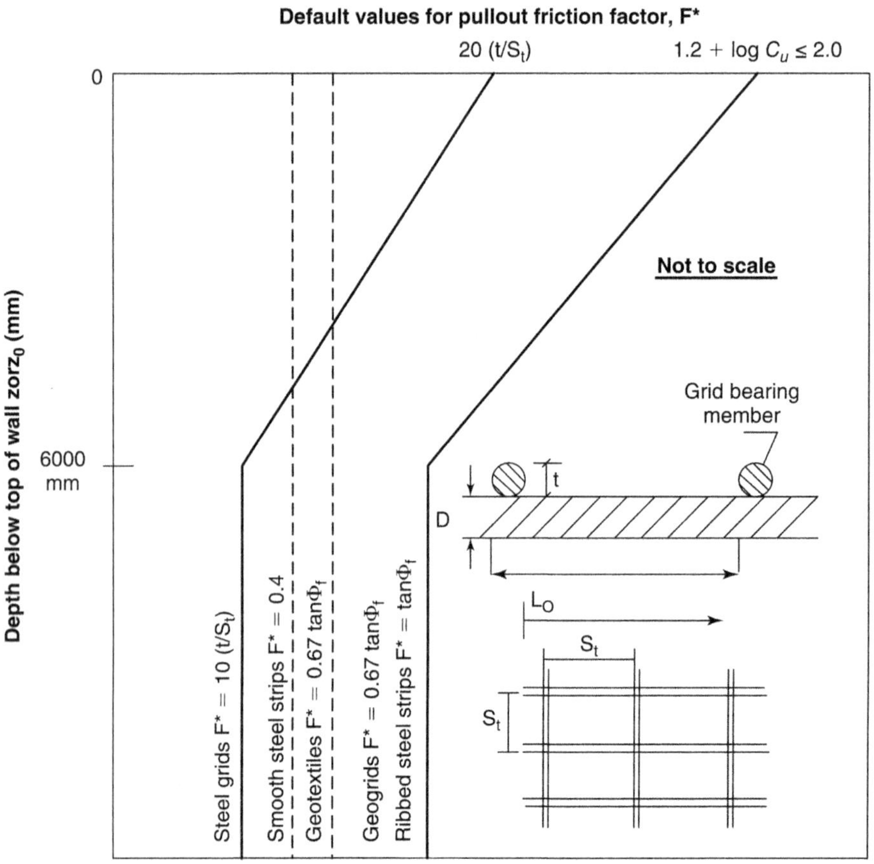

Figure 25.5s Default values for pull-out friction factor.

$$L_{total} = H \times \tan\left(45 - \frac{\varphi}{2}\right) + \frac{(\gamma_1 \sigma'_{ah} \times s_v) + (\gamma_2 k_a q \times s_v)}{2 \times \varphi \times \sigma'_v \times \tan\delta}$$

$$L_{total} = 6 \times \tan\left(45 - \frac{34}{2}\right) + \frac{(1.5 \times 0.283 \times 20 \times 0.25 \times 0.5) + (1.75 \times 0.283 \times 20 \times 0.5)}{2 \times 0.9 \times 5 \times 0.6745}$$

$$L_{total} = 3.19\text{ m} + 0.17\text{ m} + 0.81\text{ m} = 4.18\text{ m}$$

However, the required length of reinforced soil mass is 0.7 H or 4.2 m.

f. Internal stability, yield of reinforcement.

No traffic surcharge:

$$T_{allow} = \frac{T_{ult}}{RF_{ID} \times RF_{CR} \times RF_{CBD}}$$

Consider RF_{ID}, RF_{CR}, and RF_{CBD} as 1.55, 2.15, and 1.32, respectively:

$$T_{allow} = \frac{T_{ult}}{1.55 \times 2.15 \times 1.32} = \frac{T_{ult}}{4.4} = \frac{170\text{ kN/m}}{4.4} = 38.6\text{ kN/m}$$

$$\varphi\, T_{allow} = 34.7\text{ kN/m}$$

Using the ultimate limit state analysis, we have:

$$\gamma_1 T_{\max 1} + \gamma_2 T_{\max 2} = \phi T_{allow}$$

$$T_{allow} = \frac{\gamma_1 T_{\max 1} + \gamma_2 T_{\max 2}}{\varphi}$$

$$T_{allow} \geq \frac{1.5 \times s_v \times \sigma_{ah}}{0.9} + \frac{1.75 \times s_v \times k_a q}{0.9}$$

The maximum horizontal strength required is at the bottom of the wall, so we will check that layer of soil reinforcement:

$$T_{allow} \geq \frac{1.5 \times 0.5 \times 32.5}{0.9} + \frac{1.75 \times 0.5 \times 0.283 \times 20}{0.9} \quad \text{or} \quad T_{allow} \geq 27 + 5.5 = 32.5\text{ kN/m}$$

$$T_{allow} = 38.6\text{ kN/m} > 32.5\text{ kN/m}$$

$$\varphi T_{allow} = 34.7\text{ kN/m} > \gamma T_{\max\text{-}total} = 29.3\text{ kN/m}$$

Here we compare either the factored or the unfactored resistance to the factored or unfactored loads. Detail calculations are shown in Table 25.2s.

Table 25.2s Summary of Calculation for Strength

Layer No.	Depth (m)	$T_{max\text{-}total}$ (kN/m)	L_{total} (m)	T_{ult} (kN/m)	T_{allow} (kN/m)	φT_{allow} (kN/m)	$\gamma T_{max\text{-}total}$	Check
1	0.25	3.53	4.18	170.0	38.6	34.8	6.0	OK
2	0.75	4.95	3.64	170.0	38.6	34.8	8.1	OK
3	1.25	6.36	3.53	170.0	38.6	34.8	10.2	OK
4	1.75	7.77	3.48	170.0	38.6	34.8	12.4	OK
5	2.25	9.19	3.46	170.0	38.6	34.8	14.5	OK
6	2.75	10.60	3.44	170.0	38.6	34.8	16.6	OK
7	3.25	12.02	3.43	170.0	38.6	34.8	18.7	OK
8	3.75	13.43	3.42	170.0	38.6	34.8	20.9	OK
9	4.25	14.84	3.41	170.0	38.6	34.8	23.0	OK
10	4.75	16.26	3.41	170.0	38.6	34.8	25.1	OK
11	5.25	17.67	3.40	170.0	38.6	34.8	27.2	OK
12	5.75	19.08	3.40	170.0	38.6	34.8	29.3	OK

Problem 25.7

A layer of geotextile is placed 1 m below a 2 m wide strip footing. The footing rests on the surface of a loose sand with a friction angle equal to 30° and a pressuremeter limit pressure of 500 kPa within a depth equal to one footing width below the footing. The geogrid has an ultimate tensile strength of 100 kN/m.

a. Calculate the percent increase in ultimate bearing capacity between the case of no geogrid and the case with geogrid.
b. If the geogrid has a global friction factor K (Eq. 25.7) of 0.3, what length of geogrid is required to safely anchor the geogrid on each side of the footing?

Solution 25.7

Figure 25.6s shows the foundation without geogrid. The ultimate bearing capacity of the foundation without geogrid is equal to p_L:

$$p_{u1} = p_L$$

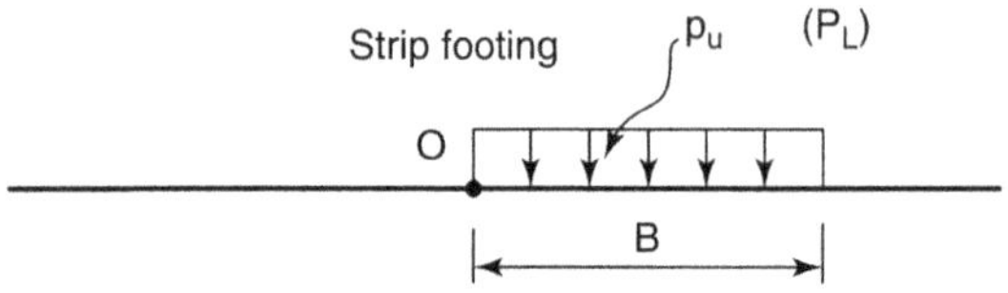

Figure 25.6s Illustration of foundation failure without geogrid.

Figure 25.7s shows the foundation with geogrid. The bearing capacity is calculated as:

$$p_{u2} = p_L + \frac{2T}{B}$$

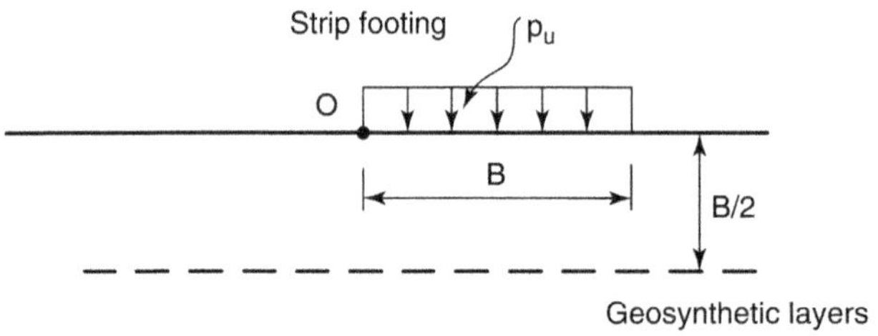

Figure 25.7s Illustration of foundation failure with geogrid.

In this problem, $p_L = 500$ kPa, B = 2 m, and T = 100 kN/m; hence:

$$p_{u1} = P_L = 500 \text{ kPa}$$

$$p_{u2} = P_L + \frac{2T}{B} = 500 + \frac{2 \times 100}{2} = 600 \text{ kPa}$$

Using the geogrid improves the ultimate bearing capacity by 20%.
The required length of geogrid can be calculated:

$$F_{us} = 2L_e K \sigma'_v \tan\phi'$$

Assume that the soil unit weight is 20 kN/m^3. The geogrid is buried 1 m beneath the foundation; therefore:

$$\sigma'_v = \gamma h = 20 \times 1 = 20 \text{ kPa}$$

$$L_e = \frac{F_{us}}{2K\sigma'_v \tan\phi'} = \frac{100}{2 \times 0.3 \times 20 \times \tan 30^\circ} = 14.4 \text{ m}$$

So, the length of geogrid required to safely anchor the geogrid on each side of the footing is 14.4 m.

Problem 25.8

A 30 m wide, 7 m high embankment is placed on soft clay with a geotextile between the surface of the soft clay and the embankment fill. The purpose of the geotextile is to increase the bearing capacity and reduce the settlement reduction, but it is also used for separation, drainage, and filtering. The bottom of the embankment settles along an arc of circle with 1 m of settlement at the center and a negligible amount at the edges. What will be the tension load in the geotextile if its modulus is 500 kN/m?

Solution 25.8

$$\text{Modulus, } E = 500 \text{ kN/m}$$

$$\text{Settlement, } s = 1 \text{ m}$$

$$\text{Width of embankment, } L = 30 \text{ m}$$

Radius R of the circle is:

$$R = \frac{L^2}{8s} = \frac{30^2}{8 \times 1} = 112.5 \text{ m}$$

Deformed length L′:

$$L' = 2RArc\sin\frac{L}{2R} = 30.09 \text{ m}$$

Strain ε in the geosynthetic:

$$\varepsilon = \frac{Arc\sin\left(\frac{4s}{L}\right)}{\frac{4s}{L}} - 1 = \frac{L' - L}{L} = \frac{30.09 - 30}{30} = 0.003$$

Tension T equal to:

$$T = E\varepsilon = 500 \times 0.003 = 1.5 \text{ kN/m}$$

Problem 25.9

A geotextile has an ultimate tensile strength of 100 kN/m and a maximum flow rate capacity of 8×10^{-7} m^3/s per meter of geotextile. What are reasonable values of the allowable tensile strength and allowable flow rate for this geotextile?

Solution 25.9

$$\text{Ultimate tensile strength, } T_{ult} = 100 \text{ kN/m}$$

$$\text{Maximum flow rate, } q_{ult} = 8 \times 10^{-7}\text{m}^3/\text{s per meter of geotextile.}$$

The strength reduction factors take into account installation damage ID, creep CR, and chemical and biological degradation CBD. They are RFID, RFCR, and RFCBD. They average respectively 1.55, 2.15, and 1.32:

$$T_{allow} = \frac{T_{ult}}{RF_{ID} \times RF_{CR} \times RF_{CBD}} = \frac{100}{1.55 \times 2.15 \times 1.32} = 23.05 \text{ kN/m}$$

The flow reduction factors take into account soil clogging and blinding, creep reduction of void space, adjacent materials intruding into the geotextile void space, chemical clogging, and biological clogging. They are RF_{SCB}, RF_{CR}, RF_{IN}, RF_{CC}, and RF_{BC}. Their respective average values are: 4.41 (RF_{SCB}), 1.83 (RF_{CR}), 1.1 (RF_{IN}), 1.25 (RF_{CC}), and 2.2 (RF_{BC}).

$$q_{allow} = \frac{q_{ult}}{(RF_{SCB} \times RF_{CR} \times RF_{IN} \times RF_{CC} \times RF_{BC})} = \frac{8 \times 10^{-7}}{4.41 \times 1.83 \times 1.1 \times 1.25 \times 2.2} = \frac{8 \times 10^{-7}}{24.41}$$
$$= 0.33 \times 10^{-7}\text{m}^3/\text{s per meter of geotextile}$$

Problem 25.10

A construction site has a 30 m long erodible slope with an angle of 6°. Silt fences are required.

a. How many silt fences are needed?
b. If the 10-year rainstorm generates 100 mm/hr, what is the flow rate to be handled per meter of width of the fence?
c. Calculate the height of the fence so that it can safely handle two 10-year storms each lasting 3 hours.
d. Posts are placed every 3 m and the fence is allowed to deflect 0.2 m at its center. Estimate the tension in the fence fabric.

Solution 25.10

a. Number of silt fences:

$$L_{\max}\,(\text{m}) = 36.2e^{-11.1\tan\alpha}$$

$$\alpha = 6^\circ \Rightarrow L_{\max} = 11.3\text{ m} \Rightarrow \text{For 30 m long slope, 3 silt fences are needed}$$

b. Flow rate per meter of fence:

$$Q(\text{m}^3/\text{hr}) = C \times I\,(\text{m/hr}) \times A(\text{m}^2)$$

$$\left.\begin{aligned} C &= 0.5 \\ I &= 0.1 \\ A &= 1 \times 11.3 \end{aligned}\right\} \Rightarrow Q = 0.565\,(\text{m}^3/\text{hr/m of fence})$$

c. Height of fence:

$$V(\text{m}^3) = Qt = H\left(\frac{H}{tan\alpha}\right) \times 1(\text{m})$$

The time t (duration of 10 yr rain storm) is 3 hours

$$0.565 \times 3 \times 2 = H\left(\frac{H}{\tan 6^\circ}\right) \Rightarrow H = 0.6\text{ m}$$

d. Tension in the fence geosynthetic fabric:

$$\frac{0.2}{3} = 0.067, \text{ small enough to use the simplified equation :}$$

$$T = \frac{\gamma_w H L^2}{16s} \Rightarrow T = \frac{9.81 \times 0.6 \times 3^2}{16 \times 0.2} = 16.55\text{ kN/m}$$

Problem 25.11

Derive Eq. 25.56 for silt fences.

Solution 25.11

The average pressure on the fence is:

$$p = \frac{1}{2}\gamma_w H^2$$

The corresponding load on the fence is pL, where L is the length between posts. The resistance comes from the tension T in the fence geosynthetic. The component of T in the direction of the load is $T\sin\alpha$ (Figure 25.8s). For equilibrium:

$$pL = 2T\sin\alpha$$

In triangle OAC (Figure 25.8s), $\sin\alpha$ is given by:

$$\sin\alpha = \frac{L/2}{R}$$

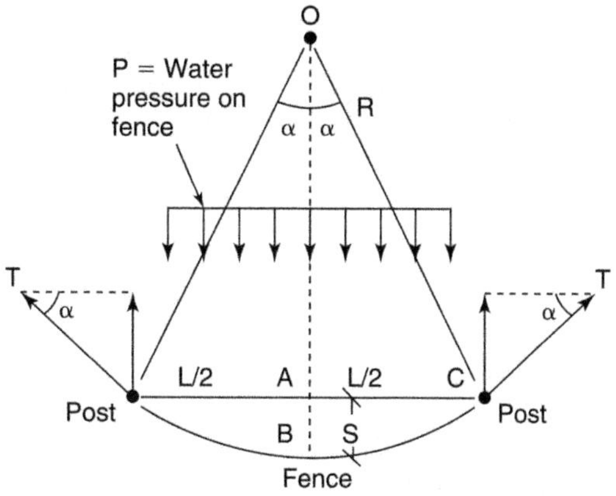

Figure 25.8s Plan view of deformed silt fence.

So,

$$T = pR$$

The radius R is given by using triangle OAC (Figure 25.8s):

$$R^2 = (R - s)^2 + \left(\frac{L}{2}\right)^2$$

Because s is small compared to R and L, this gives:

$$R = \frac{L^2}{8s}$$

Then:

$$T = p\frac{L^2}{8s} = \frac{\gamma_w H L^2}{16s}$$

Problem 25.12

A 7 m high, 60 m wide embankment is to be built on a layer of soft clay with a water table at the ground surface. The soft clay is 5 m thick and the increase in stress in the clay layer can be taken as the pressure under the embankment because the clay layer is thin compared to the width of the embankment. The clay layer has the following consolidation characteristics: $e_o = 1.1$, $\gamma = 19$ kN/m^3, $C_c = 0.5$. Two options are considered for the embankment fill: soil fill and geofoam fill. The soil fill has a unit weight of 20 kN/m^3 and the geofoam fill 2 kN/m^3. What will be the settlement of the embankment in each case? Which fill type will have the shortest time to reach 90% consolidation?

Solution 25.12

Using the consolidation theory:

$$s = H\frac{C_c}{1 + e_0}\log\frac{\sigma'_{ov} + \Delta\sigma_v}{\sigma'_{ov}}$$

a. Option 1: Soil fill.
 The increase in stress in the clay layer is:

$$\Delta\sigma_v = 20 \times 7 = 140\,\text{kPa}$$

The initial effective stress in the middle of the clay layer is:

$$\sigma'_{ov} = 2.5 \times 19 - 2.5 \times 9.81 = 23\,\text{kPa}$$

Therefore:

$$s = 5\frac{0.5}{1+1.1}\log\frac{23+140}{23} = 1.01\text{ m}$$

b. Option 2: Geofoam.
   The increase in stress in the clay layer is:

$$\Delta\sigma_v = 2 \times 7 = 14\text{ kPa}$$

The initial effective stress in the middle of the clay layer is still:

$$\sigma'_{ov} = 2.5 \times 19 - 2.5 \times 9.81 = 23\text{ kPa}$$

Therefore:

$$s = 5\frac{0.5}{1+1.1}\log\frac{23+14}{23} = 0.25\text{ m}$$

So, the settlement is reduced by a factor of 4 but the time to reach 90% consolidation is unchanged; the time required for the settlement to take place does not depend on the stress level, but rather on the drainage length and the properties of the compressing layer:

$$t = T_v\frac{H_{dr}^2}{c_v}$$

Problem 25.13

A building refrigerated at −5°C is being designed on a soil with a high water table. The concern is the cost of the power (watts) to maintain the difference in temperature across the foundation. Two alternatives are considered. The first consists of a relatively inexpensive 100 mm thick concrete slab on grade on top of the soil. The second one consists of the same slab on grade on top of a 150 mm thick geofoam. The concrete has a thermal resistance R value equal to 0.9 m^2.C/W per meter of thickness and the geofoam 35 m^2.C/W per meter of thickness. Calculate the amount of power required in each case to maintain the difference in temperature at −5°C above the slab and 0°C on top of the soil.

Solution 25.13

The heat flow is defined as:

$$\frac{\Delta Q}{\Delta t}(W) = kA\frac{\Delta T}{\Delta x}$$

Where:

k = Material thermal conductivity (W/m.°C)

A = Cross-sectional area

$\Delta T/\Delta x$ = Temperature gradient

In terms of thermal resistance, the preceding equation can be written as:

$$\frac{\Delta Q}{\Delta t}(\text{W}) = \frac{A\Delta T}{R}$$

where R (m^2.°C/W) is the thermal resistance per meter thickness of the material. Note that the thermal resistance of a layered system is equal to the sum of each layer's thermal resistance. Assuming a unit area of the slab ($A = 1\text{ m}^2$), for the case of the concrete slab only, the heat flow is:

$$\frac{\Delta Q}{\Delta t}(\text{W}) = \frac{1 \times 5}{0.9 \times 0.1} = 55.5\text{ Watt}$$

For the case of the concrete slab + 150 mm of geofoam:

$$\frac{\Delta Q}{\Delta t}(\text{W}) = \frac{1 \times 5}{(0.9 \times 0.1 + 35 \times 0.15)} = 0.93\text{ Watt}$$

The use of 150 mm of geofoam can reduce the power usage by 98.3%.

Problem 25.14

A Styrofoam coffee cup holds coffee at 80°C. Your hand holding the coffee cup is at 30°C. Assuming a steady-state heat transfer in the cup wall, what is the R rating per meter of the Styrofoam if the amount of heat released from the coffee cup through the wall of the cup is 45 W? What is the thermal conductivity of the Styrofoam if the cup wall is 1.5 mm thick?

Solution 25.14

Assume a coffee cup with an internal radius $r_1 = 40$ mm, an external radius $r_2 = 41.5$ mm and a height of 160 mm.

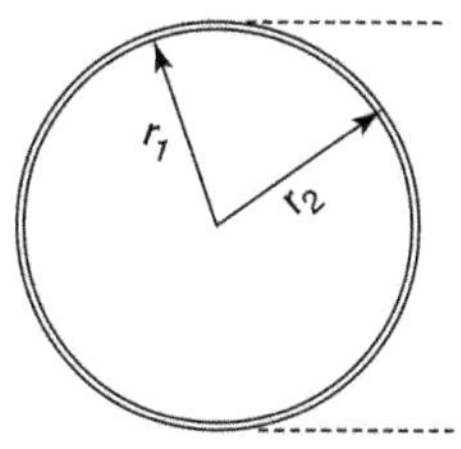

Figure 25.9s Coffee cup dimensions.

At steady state, the amount of heat Q (W) released from the coffee cup, assuming that it is a long hollow cylinder, can be calculated as follows:

$$Q(\mathrm{W}) = -kA\frac{\Delta T}{r_1 \times \ln\left(\frac{r_2}{r_1}\right)} = k2\pi r_1 L\frac{\Delta T}{r_1 \times \ln\left(\frac{r_2}{r_1}\right)}$$

with A (m^2) $= 2\pi r_1 L$ as the internal surface area of the cup.

k (W/m°C) is the thermal conductivity of the cup wall.

ΔT is the temperature difference between the inside and outside faces of the cup wall.

The preceding equation can be rewritten in terms of thermal resistance:

$$Q(\mathrm{W}) = -2\pi L\frac{\Delta T}{R}$$

where R (m^{2}°C/W/m) is the thermal resistance per meter of the cup wall and is equal to:

$$R = \frac{\ln\left(\frac{r_2}{r_1}\right)}{k}$$

Based on the data given in the problem statement, the thermal resistance of the Styrofoam is:

$$R = -2\pi L\frac{\Delta T}{Q} = -2\pi \times 0.16 \times \frac{(30-80)}{45} = 1.117\ \mathrm{m^2.^\circ C/W/m}$$

The thermal conductivity of the Styrofoam is then:

$$k(\mathrm{W/m.^\circ C}) = \frac{\ln\left(\frac{r_2}{r_1}\right)}{R} = \frac{\ln\left(\frac{0.0415}{0.04}\right)}{1.117} = 0.033$$

CHAPTER 26

Soil Improvement

Soil improvement is an alternative considered when the natural soil does not meet the engineering requirements for a project. As an example, if the soil is too weak to carry the structure on a shallow foundation, two alternatives exist: deep foundations or soil improvement plus a shallow foundation. A soil improvement technique is sought that would make a shallow foundation feasible. If the deep foundation will cost \$1,000,000, while the soil improvement will cost \$250,000 and the shallow foundation \$500,000, then the soil improvement alternative becomes attractive. Typically in this case the soil improvement technique is verified by in situ testing to demonstrate that a sufficiently improved soil strength and soil modulus can be reached so that a shallow foundation is viable.

A very large number of methods are aimed at soil improvement; this chapter summarizes the main methods. For additional information, the following excellent references can be consulted: the state-of-the-art report, published by the ISSMGE Technical Committee on ground improvement and presented at the 2009 International Conference on Soil Mechanics and Geotechnical Engineering (Chu et al. 2009); the book by Moseley and Kirsch (2004); the NHI manual (Elias et al. 2006); and the web site www.geotechtools.org (Schaefer 2013).

26.1 OVERVIEW

Over the past 50 years, many different soil improvement techniques have been developed, and they continue to be developed and revised as the space available for human activities decreases. These methods have been classified by the ISSMGE Technical Committee on Ground Improvement as shown in Table 26.1. The word *ground* is used in that classification because it can incorporate rock, but because this book is limited to soil, the term *soil improvement* is used here. There are five major categories of soil improvement methods.

1. Soil improvement without admixture in coarse-grained soils
2. Soil improvement without admixture in fine-grained soils
3. Soil improvement with replacement
4. Soil improvement with grouting and admixtures
5. Soil improvement with inclusions

26.2 SOIL IMPROVEMENT WITHOUT ADMIXTURE IN COARSE-GRAINED SOILS

26.2.1 Compaction

Compaction in this instance refers to roller compaction for shallow densification of soil deposits. The rollers used are static rollers, such as sheep-foot rollers for fine-grained soils or vibratory rollers for coarse-grained soils. Most rollers are cylindrical, but some are uneven rollers. The depth of compaction is at most 1 m and is highest near the surface. Compaction is used to prepare pavement layers, retaining wall backfills, and embankment fills. This topic is covered in Chapter 20.

26.2.2 Dynamic Compaction

Because of the limited depth of conventional compaction techniques and the need to compact natural soils at larger depths, the idea of dropping a heavy weight from a height onto the soil surface was pioneered by Louis Menard (Menard and Broise 1975). A typical combination would be a 20-ton weight dropping from a height of 20 m. This technique is best suited to compaction of coarse-grained soils. This topic, including the depth that can be reached and the improvement ratio versus depth, is covered in Chapter 20.

26.2.3 Vibrocompaction

The vibrocompaction method consists of lowering a cylindrical vibrator from a crane into the soil to densify the soil (Figure 26.1). A grid of 3 to 4 meters center to center is common. The vibrator is 2 to 5 m long and 0.3 to 0.5 m in diameter, and weighs 15 to 40 kN. The vibrations are generated in the horizontal direction by rotating eccentric masses. The frequency of vibration is in the range of 25 to 35 Hz with amplitudes between 10 to 30 mm. The vibrator typically

Table 26.1 Classification of Soil Improvement Methods (Chu et al. 2009)

Category	Method	Principle
A. Ground improvement without admixtures in noncohesive soils or fill materials	A1. Dynamic compaction	Densification of granular soil by dropping a heavy weight from air onto ground.
	A2. Vibrocompaction	Densification of granular soil using a vibratory probe inserted into ground.
	A3. Explosive compaction	Shock waves and vibrations generated by blasting cause granular soil ground to settle through liquefaction or compaction.
	A4. Electric pulse compaction	Densification of granular soil using the shock waves and energy generated by electric pulse under ultra-high voltage.
	A5. Surface compaction (including rapid impact compaction)	Compaction of fill or ground at the surface or shallow depth using a variety of compaction machines.
B. Ground improvement without admixtures in cohesive soils	B1. Replacement, displacement (including load reduction using lightweight materials)	Remove bad soil by excavation or displacement and replace it by good soil or rocks. Some lightweight materials may be used as backfill to reduce the load or earth pressure.
	B2. Preloading using fill (including the use of vertical drains)	Fill is applied and removed to preconsolidate compressible soil so that its compressibility will be much reduced when future loads are applied.
	B3. Preloading using vacuum (including combined fill and vacuum)	Vacuum pressure of up to 90 kPa is used to preconsolidate compressible soil so that its compressibility will be much reduced when future loads are applied.
	B4. Dynamic consolidation with drainage (including the use of vacuum)	Similar to dynamic compaction except that vertical or horizontal drains (or together with vacuum) are used to dissipate pore pressures generated in soil during compaction.
	B5. Electro-osmosis or electrokinetic consolidation	DC current causes water in soil or solutions to flow from anodes to cathodes installed in soil.
	B6. Thermal stabilization using heating or freezing	Change the physical or mechanical properties of soil permanently or temporarily by heating or freezing the soil.
	B7. Hydro-blasting compaction	Collapsible soil (loess) is compacted by a combined wetting and deep explosion action along a borehole.
C. Ground improvement with admixtures or inclusions	C1. Vibro replacement or stone columns	Hole jetted into soft, fine-grained soil and backfilled with densely compacted gravel or sand to form columns.
	C2. Dynamic replacement	Aggregates are driven into soil by high-energy dynamic impact to form columns. The backfill can be either sand, gravel, stones, or demolition debris.
	C3. Sand compaction piles	Sand is fed into ground through a casing pipe and compacted by vibration, dynamic impact, or static excitation to form columns.
	C4. Geotextile confined columns	Sand is fed into a closed-bottom, geotextile-lined cylindrical hole to form a column.
	C5. Rigid inclusions	Use of piles, rigid or semirigid bodies, or columns that are either premade or formed in situ to strengthen soft ground.
	C6. Geosynthetic-reinforced column or pile-supported embankment	Use of piles, rigid or semirigid columns/inclusions, and geosynthetic girds to enhance the stability and reduce the settlement of embankments.
	C7. Microbial methods	Use of microbial materials to modify soil to increase its strength or reduce its permeability.
	C8. Other methods	Unconventional methods, such as formation of sand piles using blasting, and the use of bamboo, timber, and other natural products.

(*Continued*)

Table 26.1 *(Continued)*

Category	Method	Principle
D. Ground improvement with grouting-type admixtures	D1. Particulate grouting	Grout granular soil or cavities or fissures in soil or rock by injecting cement or other particulate grouts to either increase the strength or reduce the permeability of soil or ground.
	D2. Chemical grouting	Solutions of two or more chemicals react in soil pores to form a gel or a solid precipitate to either increase the strength or reduce the permeability of soil or ground.
	D3. Mixing methods (including premixing or deep mixing)	Treat the weak soil by mixing it with cement, lime, or other binders in situ using a mixing machine or before placement.
	D4. Jet grouting	High-speed jets at depth erode the soil and inject grout to form columns or panels.
	D5. Compaction grouting	Very stiff, mortar-like grout is injected into discrete soil zones and remains in a homogenous mass to densify loose soil or lift settled ground.
	D6. Compensation grouting	Medium- to high-viscosity particulate suspensions are injected into the ground between a subsurface excavation and a structure to negate or reduce settlement of the structure due to ongoing excavation.
E. Earth reinforcement	E1. Geosynthetics or mechanically stabilized earth (MSE)	Use of the tensile strength of various steel or geosynthetic materials to enhance the shear strength of soil and stability of roads, foundations, embankments, slopes, or retaining walls.
	E2. Ground anchors or soil nails	Use of the tensile strength of embedded nails or anchors to enhance the stability of slopes or retaining walls.
	E3. Biological methods using vegetation	Use of the roots of vegetation to create and improve stability of slopes.

Figure 26.1 Example of vibrocompactor. (Courtesy of Earth Tech, LLC.)

reaches depths of 20 to 30 m, with 60 m being rare. Pipes go through the body of the vibrator and can supply water or air to the bottom of the vibrator to help with penetration if necessary.

The soils best suited to use of this technique are clean sands. If the fine content becomes higher than 10 to 15%, the vibrocompaction process becomes much less efficient (Mitchell and Jardine 2002). Massarsch (1991) proposed a CPT-based chart indicating which soils are most applicable to vibrocompaction (Figure 26.2).

26.2.4 Other Methods

Other compaction methods include rapid impact compaction (Figure 26.3), explosive compaction, and electric pulse compaction. In rapid impact compaction (Watts and Charles 1993), a tamper is pounded repeatedly on the ground surface. The weight is lifted about 1 m up in the air and dropped at a rate of around 40 drops per minute. The hammer weighs about 100 kN and has a diameter between 1.5 and 1.8 m. This technique is best for sands and gravels and is not suited for saturated silts and clays.

Explosive compaction consists of setting a series of detonation charges in the deposit. These detonations create waves that propagate in the soil and compact it. This technique is not commonly used, but has the advantage of being relatively inexpensive. Electric pulse compaction consists of lowering a probe into the soil and discharging high voltage sparks at a rate of about 10 per minute. This recent method is as yet unproven.

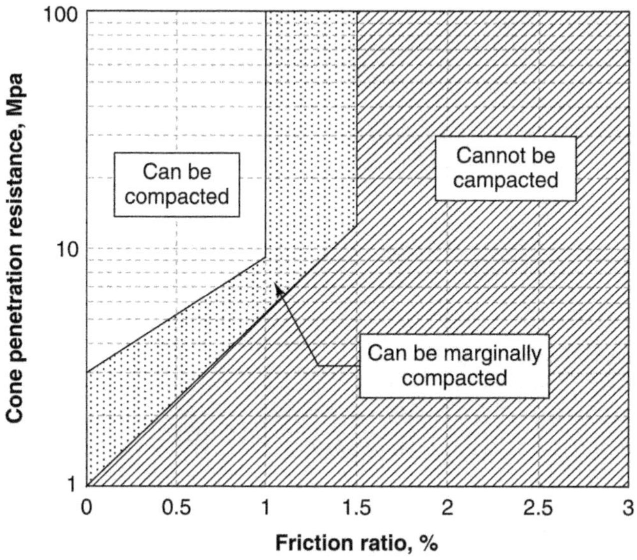

Figure 26.2 Soil suitability for vibrocompaction based on CPT. (Courtesy of Dr. Rainer Massarsch)

Figure 26.3 Example of rapid impact compactor. (Courtesy of Menard Bachy, Inc.)

26.3 SOIL IMPROVEMENT WITHOUT ADMIXTURE IN FINE-GRAINED SOILS

26.3.1 Displacement–Replacement

The displacement-replacement technique consists of simply excavating the weak soil (say, $s_u < 20\,\text{kPa}$) and replacing it with stronger soil. Excavation depths beyond 8 m are uncommon; the method can be costly and environmentally unfriendly because of the amount of spoil to be disposed of. In the case of peat bogs, the backfill, which may be twice as heavy as the natural soil, can create very large settlements. Sometimes the backfill is made of lightweight material such as geofoam blocks (see section 25.3.5) to avoid excessive settlement and bearing capacity issues.

26.3.2 Preloading Using Fill

The technique of preloading using fill consists of loading the soil surface with a fill, as in the case of an embankment and a surcharge fill. It has been used for many years to shorten the time required to reach a certain settlement under the design load. Once the settlement is reached, the surcharge is withdrawn and the road can be paved, for example. It is important to note, in this respect, that the time t_U to reach U percent of consolidation depends not on the height of the fill but on the drainage length H and the soil coefficient of consolidation c_v. In other words, if it takes 5 years to reach 90% of the final settlement under a 5 m high embankment, it will also take 5 years to reach 90% of the final settlement under a 10 m embankment. However, if it takes 5 years to reach 90% of the final settlement under a 5 m high embankment, it will take a lot less time to reach that same settlement under a 10 m high embankment. To find out what height h_s must be added as a surcharge on top of an h_e high embankment to reach, say, 90% of the settlement of the embankment within a target time t_t, use the following steps (Figure 26.4):

1. Calculate the maximum settlement of the embankment $s_{\max(emb)}$. For a normally consolidated clay, the following equation can be used (see section 17.8.9):

$$s_{\max(emb)} = h_o \frac{C_c}{1+e_o} \log\left(\frac{\sigma'_{ov} + \Delta\sigma'}{\sigma'_{ov}}\right) \tag{26.1}$$

   where h_o is the height of the soft clay layer, C_c is the compression index from consolidation tests, e_o is the initial void ratio of the soft clay layer, σ'_{ov} is the initial effective stress in the middle of the soft clay layer, and $\Delta\sigma'$ is the increase in stress in the middle of the soft clay layer.
2. Choose the target time t_t to reach $s_{\max(emb)}$.
3. Knowing the target time t_t and the coefficient of consolidation c_v of the soft clay layer, calculate the time factor T_U corresponding to t_t using the equation (see section 17.8.10):

$$T_U = \frac{t_t c_v}{h_d^2} \tag{26.2}$$

   where h_d is the drainage length.

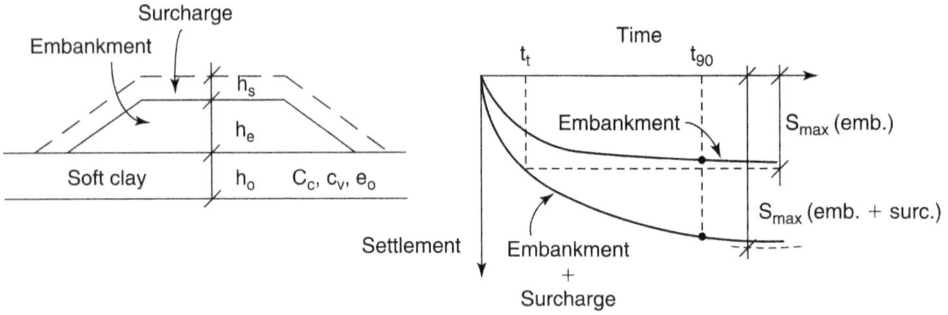

Figure 26.4 Surcharge to accelerate embankment settlement.

This drainage length is equal to the soft clay layer thickness if the water can only drain through the top or the bottom of the layer; equal to one-half of the layer thickness if the water can drain through the top and the bottom of the layer; and equal to the horizontal distance between vertical drains if such drains are installed.

4. Then find the average percent consolidation U corresponding to the time factor T_U by using the curve that links both parameters (Figure 26.5). Note that U is equal to:

$$U = \frac{s(t)}{s_{max}} \quad (26.3)$$

where $s(t)$ is the settlement after a time t and s_{max} is the final settlement.

5. Knowing U, use Eq. 26.3 to calculate the maximum settlement $s_{max(emb+surch)}$ under the embankment plus the surcharge. In Eq. 26.3, U is known and $s(t)$ is equal to the settlement under the embankment plus the surcharge after a time equal to the target time $s(t_t)$. By design, this settlement is equal to the maximum settlement under the embankment only, $s_{max(emb)}$:

$$s(t_t) = s_{max(emb)} \quad (26.4)$$

$$s_{max(emb+surch)} = \frac{s_{max(emb)}}{U} \quad (26.5)$$

6. Once the maximum settlement under the embankment and the surcharge $s_{max(emb+surch)}$ is known, Eq. 26.1 can be used to back-calculate the value of $\Delta\sigma'$ induced by the surcharge:

$$\Delta\sigma' = \sigma'_{ov}\left(10^{\left(\frac{(1+e_o)s_{max(emb+surch)}}{h_o C_c}\right)} - 1\right) \quad (26.6)$$

7. Finally, the height of the surcharge h_s is the height that generates an increase in effective stress in the soft clay layer equal to $\Delta\sigma'$. Often, if the soft clay layer is not very thick compared to the width of the embankment, the increase in stress is equal to the pressure generated by the surcharge at the ground surface and the height of the surcharge is:

$$h_s = \frac{\Delta\sigma'}{\gamma_s} \quad (26.7)$$

where γ_s is the unit weight of the surcharge soil.

8. Note that if the surcharge is too high, a slope stability or bearing capacity problem arises for the side of the

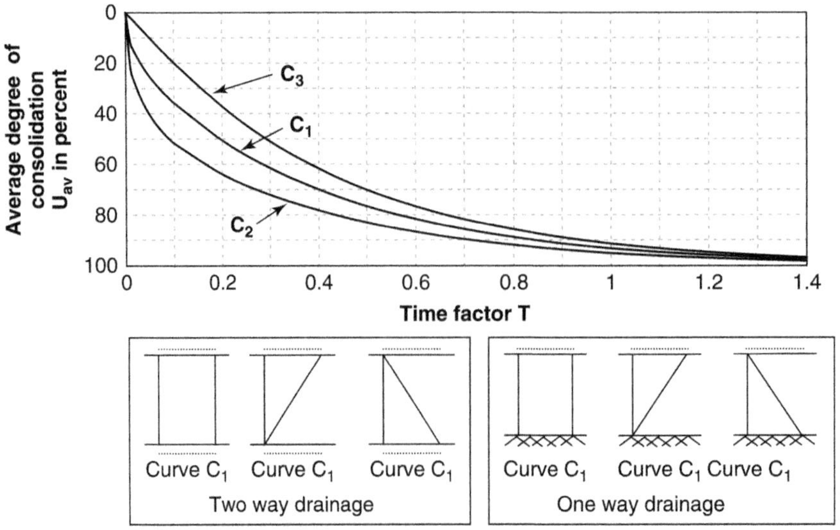

Figure 26.5 Average percent consolidation U versus time factor T_U.

embankment. In that regard, the height of the surcharge $h_{s\,max}$ that would generate a bearing capacity failure in a clay of undrained shear strength s_u can be estimated by:

$$h_{s\,max} = \frac{5.14 s_u}{\gamma_s} \tag{26.8}$$

26.3.3 Prefabricated Vertical Drains and Preloading Using Fill

The technique of using vertical drains and preloading consists of loading the soil surface with a fill while accelerating the consolidation process by installing prefabricated vertical drains (PVDs) or sand drains. Prefabricated vertical drains are also called *wick drains* or *band drains*. They are installed to decrease the drainage length h_d, thereby reducing the time necessary for the consolidation settlement to take place. The drainage is then shifted from a vertical drainage problem involving the vertical hydraulic conductivity k_v to a horizontal drainage problem between PVDs involving the horizontal hydraulic conductivity k_h. For example, if a soft clay layer is 10 m thick and has one-way drainage, the drainage length will be 10 m and the time required for 90% of the final settlement will be t_1. If PVDs are installed on a grid with a center-to-center spacing equal to 2 m, the drainage length is controlled by the horizontal spacing and becomes much shorter, so the time t_2 for 90% of the final settlement is dramatically reduced compared to t_1. Note that the ratio of the two times requires comparing the solution of the one-dimensional consolidation problem for the embankment on top of the layer without PVDs (see section 11.4.6) to the solution for the drainage around a grid of drains; this is often approximated by the radial consolidation problem (Moseley and Kirsch 2004).

PVDs are typically made of a filter material covering both sides of a corrugated plastic shell (Figure 26.6). The width may be 100 mm, the thickness 3 to 4 mm, and the installed length can be 30 m. The flow rate out of such drains is in the range of 2 to 8 liters per minute, but may decrease with time because of siltation, for example. At the same time, the actual flow through the PVD decreases with time as consolidation takes place. Installation of PVDs is done by tying one end of the PVD with an anchor inside a small-diameter pipe called a *mandrel* and pushing the mandrel vertically into the soil and dragging the PVD with it (Figure 26.7). Once at the required depth, the mandrel is withdrawn and the PVD is left in place. PVDs can be placed to depths of several tens of meters on a grid with spacing in the range of 1 to 2.5 m. The tops of the drains are bound by a drainage layer or drainage blanket (0.5 to 1 m thick) made of clean sand, and the water is pumped away from the site. The drainage blanket is often placed before the PVDs are installed and serves as a work platform for the equipment. One issue associated with the placement of PVDs is the development of a "smear zone" in soft clays at the boundary between the soil and the PVD. This smear zone

Figure 26.6 Prefabricated vertical drain. (Courtesy of Layfield Environmental Systems, Layfield Group Limited, 11120 Silversmith Place, Richmond, British Columbia, Canada V7A 5E4.)

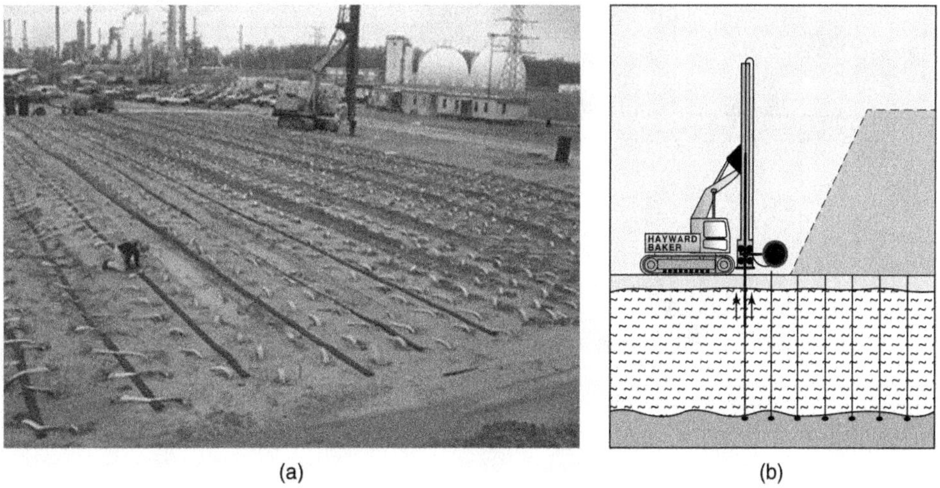

(a) (b)

Figure 26.7 Installation of prefabricated vertical drains. (Courtesy of Hayward Baker Geotechnical Construction.)

is a few PVD diameters thick and can reduce the permeability of the interface.

The purpose of PVDs is to minimize the consolidation time t needed to reach a given percent consolidation U_h taken as a ratio, not a percent, in Eq. 26.9. This time t can be calculated by using the Barron-Hansbo formula (Barron 1948; Hansbo 1981):

$$t = \frac{d_w^2}{8c_h}\left(Ln\left(\frac{d_w}{d_e}\right) - 0.75 + F_s\right)Ln\left(\frac{1}{1-U_h}\right) \quad (26.9)$$

where d_e is the equivalent diameter of the PVD defined in Eq. 26.10, c_h is the horizontal coefficient of consolidation, d_w is the well influence diameter (taken as 1.05 s for an equilateral triangle spacing pattern and 1.13 s for a square spacing pattern) where s is the spacing between PVDs, and F_s is a soil disturbance factor (taken as 2 for highly plastic sensitive soils but taken as zero if c_h has been conservatively estimated or accurately measured):

$$d_e = \frac{2(a+b)}{\pi} \quad (26.10)$$

where a is the PVD thickness and b is the PVD width.

26.3.4 Preloading Using Vacuum

Sometimes the soil is so soft that a surcharge fill cannot be placed to a sufficient height to be useful. In this case, preloading by vacuum is an alternative. The method consists of applying a vacuum, thereby decreasing the water stress, increasing the effective stress, and compressing the soil. A vacuum of 0.8 atmosphere is commonly applied and is equal to about 4 m of soil surcharge. One difference between this method and a fill method is that for vacuum preloading, the increase in effective stress is applied isotropically, as opposed to anisotropically for the fill. Figure 26.8 shows the difference in effective stress path between a surcharge fill and vacuum preloading.

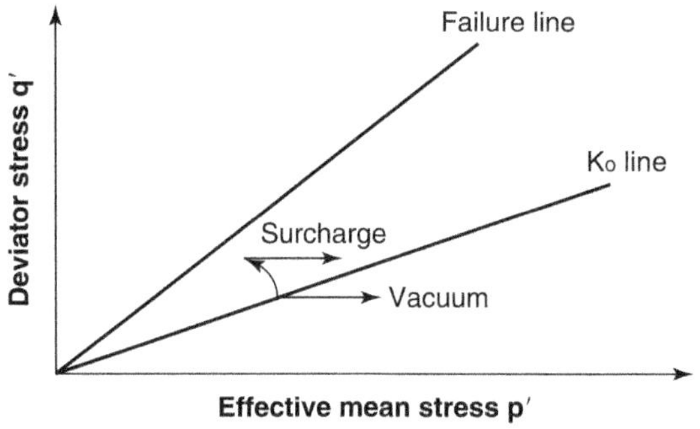

Figure 26.8 Stress path comparison between surcharge preloading and vacuum preloading. (After Chu et al. 2009.)

The construction sequence consists of constructing a 0.3 m thick sand blanket on the site, installing prefabricated vertical drains on a square grid (say, 1 m center to center), laying down a grid of geotextile-covered perforated pipes in the sand blanket to connect the PVDs to the vacuum pump, and covering the ground surface with a geomembrane to seal the soil volume. The vacuum pump is turned on and vacuum consolidation takes place. A variant of this process is shown in Figure 26.9.

The vacuum preloading method works well when the soil is soft, low permeability, and relatively homogeneous. If clean

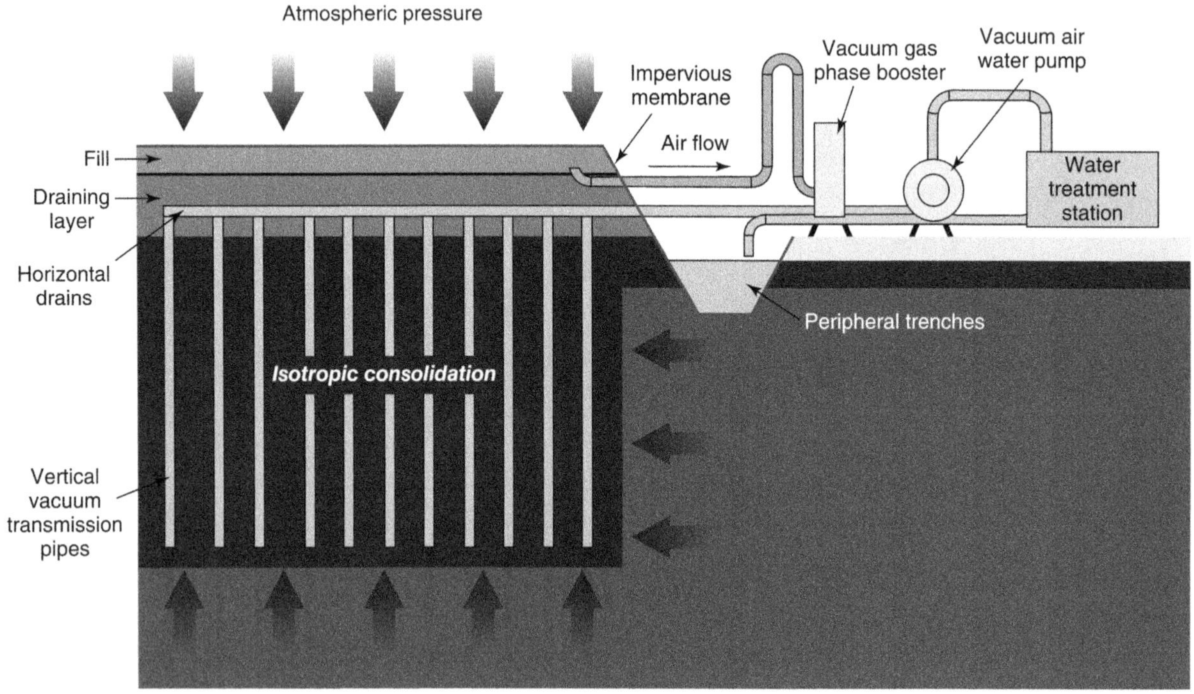

Figure 26.9 Menard vacuum consolidation. (Courtesy of Menard, Bridgeville, PA; www.menard usa.com)

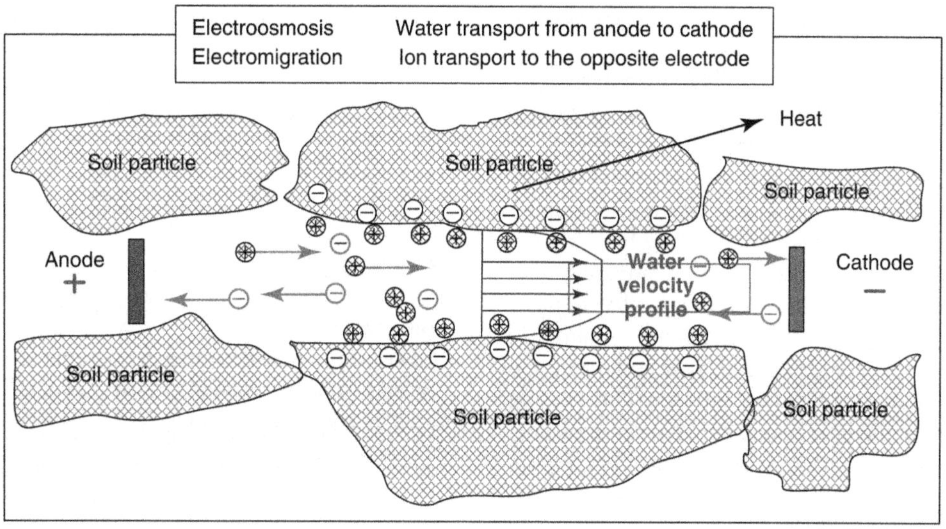

Figure 26.10 Electro-osmosis in clays. (Courtesy of C.J. Athmer—Terran Corporation.)

sand layers are interbedded in a deposit of soft clay, the efficiency of the process decreases unless cutoff walls can be installed first. Also, because vacuum preloading is isotropic, compression occurs in all direction equally and horizontal shortening takes place. This leads to vertical cracks in the soil mass.

26.3.5 Electro-osmosis

The *electro-osmosis* process was discovered in the early 1800s and applied to soils by Leo Cassagrande in the early 1940s. It is based on the fact that when a DC electrical current is established between two electrodes (e.g., steel bars) driven into fine-grained soil, the water flows from the anode (positive charge) to the cathode (negative charge) (Figure 26.10). The reason for this water movement is as follows. Clay particles are negatively charged and as such attract cations (positively charged) such as sodium, calcium and magnesium to their surfaces. When a DC current is established between two metal rods, the cations that line the surface of the clay particles start sliding toward the cathode by electrical attraction. The movement of this boundary layer of cations drags the bulk soil water with it. The water that accumulates at the cathode is drained away and the water content of the clay decreases, with an associated increase in strength and stiffness.

26.3.6 Ground Freezing

The technique of ground freezing (Figure 26.11) consists of freezing the soil by installing a network of steel pipes and circulating either brine water or liquid nitrogen. The temperature of circulating brine water is typically −20C; liquid nitrogen is much colder, at around −200C. Brine is much less expensive, but nitrogen takes a lot less time to freeze the ground. As a result, brine is used for large projects, whereas nitrogen may be economical when time is more important than cost savings. The advantage of ground freezing is that it is applicable to almost all soil conditions as long as the soil is saturated. Recall, however, that when water turns to ice, it expands by 10%. Applications include tunneling, retaining walls, cutoff walls, and contamination remediation.

Figure 26.11 Ground freezing. (Courtesy of British Drilling and Freezing Co. Ltd.)

26.3.7 Hydro-Blasting Compaction

The hydro-blasting compaction technique is particularly well suited to the treatment of collapsible soils. It consists of wetting the soil to induce collapse and then detonating explosives in sequence to shake the soil into a more compact arrangement.

26.4 SOIL IMPROVEMENT WITH REPLACEMENT

26.4.1 Stone Columns without Geosynthetic Sock

Stone columns, also called *aggregate columns* (Figure 26.12), are constructed by opening holes in the soil to be improved (say, 1 m diameter) down to a chosen depth (say, 10 m) and backfilling them with aggregates or crushed stones. Opening of the hole in which to place the stones is done by vibration or by jetting. In the vibration technique, a vibrating cylinder is used (section 26.2.3) and the stones are placed upon withdrawal and are compacted using the same vibrator. In the jetting technique, the hole is created by a probe inserted to the chosen depth and rotated out of the hole while jetting horizontally to enlarge the hole before the stones are placed. A third technique, called the *rammed aggregate pier method*, consists of opening a hole with an auger and compacting the stones in the open hole in 0.3 m thick lifts.

In all cases a stone column is placed in the soil to reinforce it vertically. This column can carry vertical compression load, but very little uplift load and horizontal load. It can also carry shear load, as required for the stabilization of unstable slopes. This latter case is handled as a slope stability problem. The rest of this section deals with the vertical compression capacity and settlement of stone columns.

The column can be considered as a large sample of gravel loaded in a manner similar to a triaxial test. Therefore, at failure of the column, the ratio between the vertical effective stress σ'_1 and the horizontal effective stress σ'_3 is given by:

$$\sigma'_1 = K_p \sigma'_3 \tag{26.11}$$

where K_p is the coefficient of passive earth pressure.

In this large-scale triaxial test, σ'_3 is limited by the maximum horizontal pressure that the soil can resist. This is given by the effective stress limit pressure p'_L of the pressuremeter test. The value of p'_L can be obtained by performing a drained pressuremeter test (pressure steps lasting until the probe volume stabilizes) and assuming that the water stress u_w is equal to the hydrostatic pressure:

$$p'_L = p_L - u_w \tag{26.12}$$

Figure 26.12 Stone column construction. (Courtesy of Menard Bachy, Inc.)

Therefore, the drained ultimate load on the stone column is:

$$Q_u = K_p (p_L - u_w) A \tag{26.13}$$

where Q_u is the ultimate load on the stone column, p_L is the limit pressure from a drained pressuremeter test, u_w is the hydrostatic pressure at the PMT testing depth, and A is the cross-sectional area of the stone column. Of course, there is a beneficial effect that increases when the spacing between stone columns decreases; this observation makes Eq. 26.13 conservative.

The settlement can also be estimated using pressuremeter data. The horizontal relative expansion of the column is considered to be equal to the relative expansion of the pressuremeter for the same horizontal pressure:

$$\frac{\Delta B}{B} = \frac{\Delta R}{R} \tag{26.14}$$

where B and ΔB are the initial diameter and increase in diameter of the stone column respectively, and R and ΔR are the radius and increase in radius of the pressuremeter probe at a pressure corresponding to p_L divided by a chosen factor of safety against horizontal expansion failure. Therefore, ΔB can be obtained from Eq. 26.14. The volume involved in the barrel-like deformation shown in Figure 26.13 extends to a depth equal to about 2 times the diameter of the stone column (Hughes and Withers 1974). Thus, the initial volume involved in the deformation is:

$$V_o = 2B \frac{\pi B^2}{4} \tag{26.15}$$

If, during the deformation of the column under load, the volume of stone experiences a volume change ΔV, then the volume V of the deformed column under load will be:

$$V = V_o + \Delta V \tag{26.16}$$

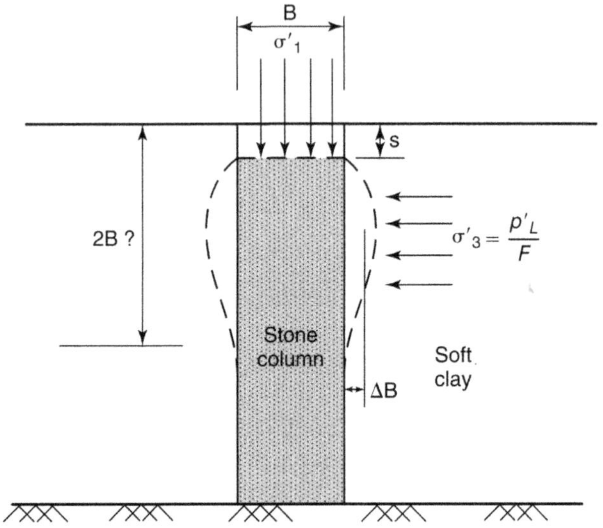

Figure 26.13 Expansion of a stone column under load.

The deformed volume V is also equal to:

$$V = (2B - s)\frac{\pi}{4}(B + \Delta B)^2 \quad (26.17)$$

where s is the settlement of the stone column. This settlement s is then given by:

$$s = 2B\left(1 - \frac{\left(1 + \frac{\Delta V}{V_o}\right)}{\left(1 + \frac{\Delta B}{B}\right)^2}\right) \quad (26.18)$$

The relative increase in stone column diameter $\Delta B/B$ is obtained from Eq. 26.14 using a ratio $\Delta R/R$ from a pressuremeter test at a pressure corresponding to p_L divided by a chosen factor of safety against horizontal expansion failure. The relative change in volume $\Delta V/V$ can be obtained from a triaxial test on the stone column material. The value of $\Delta V/V$ is the one that corresponds to a vertical stress σ'_1 applied at the top of the stone column. Therefore, the settlement s corresponds to a top load Q equal to:

$$Q = \sigma'_1 \frac{\pi B^2}{4} \quad (26.19)$$

If ΔV is 0 and if $\Delta R/R$ is small, then Eq. 26.18 reduces to:

$$s = 4B\frac{\Delta R}{R} \quad (26.20)$$

Another possible mode of failure is sliding along the sides of the column as a pile. The rules of design for piles can be used in this case, assuming that the failure will take place in the soft clay rather than the stone column at the vertical friction interface.

Besides strengthening the soft soil, stone columns act as large drains. When the surface is loaded, the water squeezes out of the soil horizontally (because the drainage length is shorter in that direction), drains into the stone column, and is collected at the surface. The design of stone columns as drains follows the same process as for prefabricated vertical drains (section 26.3.3).

26.4.2 Stone Columns with Geosynthetic Encasement

More recently, geosynthetic encasement, in the form of a large sock (Figure 26.14), has been used to increase the horizontal resistance and therefore vertical capacity of the stone column. Because improved horizontal drainage is also an attribute of stone columns, the geosynthetic used is a geotextile that provides a filter between the native soil (often soft clay) and the stone column material.

The critical factors for the encasement are the tensile capacity of the geotextile T_u (kN/m) and its modulus E (kN/m). The value of T_u ranges from 25 to 60 kN/m and that of E from 30 to 150 kN/m. The modulus E is defined as:

$$E = \frac{T}{\varepsilon} \quad (26.21)$$

Figure 26.14 Stone column with geotextile encasement. (Courtesy of HUESKER Inc.)

where T is the force applied per meter of fabric and ε is the corresponding tensile strain.

Note that for geotextiles, the tensile strain at failure ε_f is very large, in the range of 25 to 70%. Therefore, it is likely that the soil would fail before the geotextile sock did. The failure mechanism may involve failure of the column aggregate, failure of the geotextile encasement, or failure of the soil laterally. Because of the large strains required for the geotextile to fail, this failure mechanism is not likely. The ultimate pressure that can be placed at the top of the encased stone column is given by:

Soil fails laterally $\quad p_{u1} = k_p\sigma'_3 = k_p(p'_L + p_{geo}) \quad (26.22)$

Geotextile fails in hoop tension

$$p_{u2} = k_p\sigma'_3 = k_p\left(2G\frac{\Delta r}{r_o} + p_{geo\,f}\right) \quad (26.23)$$

where p_{u1} is the ultimate pressure that can be placed at the top of the stone column if the soil fails first by reaching the soil effective stress horizontal limit pressure p'_L, k_p is the coefficient of passive earth pressure of the soil, σ'_3 is the horizontal stress generated by the combination of geotextile and soil, p_{geo} is the pressure contributed by the geotextile when stretched at $\Delta r/r_o$, $p_{geo\,f}$ is the pressure contributed by the geotextile at failure of the geotextile, G is the shear modulus of the soil outside the geotextile (soil being improved),

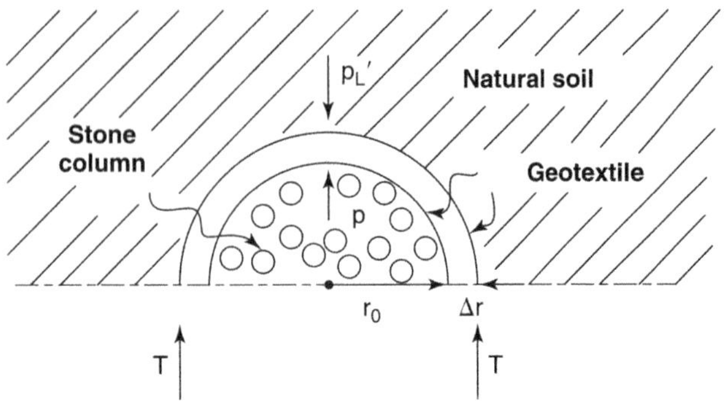

Figure 26.15 Pressure and tension in the geotextile encasement.

and $\Delta r/r_o$ is the relative increase in radius of the stone column. The expression $2G\Delta r/r_o$ is the pressure contributed by the soil outside the geotextile for a radial strain $\Delta r/r_o$.

Because the geotextile strain at failure is often very large, p_{u1} is likely to control. The pressure p_{geo} contributed by the geotextile at a relative increase in radius of the stone column equal to $\Delta r/r_o$ is given by (Figure 26.15):

$$p_{geo}2r = 2T = 2E\varepsilon = 2E\frac{\Delta r}{r_o} \quad \text{or} \quad p_{geo} = E\frac{\Delta r}{r_o^2} \quad (26.24)$$

Then the ultimate load on the encased stone column corresponding to this failure mechanism is:

$$Q_{u1} = k_p\left(p_L' + E\frac{\Delta r}{r_o^2}\right)\pi r_o^2 \quad (26.25)$$

The value of k_p is obtained from the friction angle of the stone column material, p_L' from a pressuremeter test in the natural soil, E from the geotextile material, r_o from the size of the stone column, and $\Delta r/r_o$ as 0.41 to correspond with the strain at failure for the limit pressure. Note that the product E $\Delta r/r_o$ cannot be larger than the tensile capacity T_u of the geotextile.

The settlement calculations become rather cumbersome in close form and are best handled by numerical simulations starting with elasticity. Alexiew et al. (2003) and Raithel et al. (2005) proposed simplified method for hand calculations.

26.4.3 Dynamic Replacement

Dynamic replacement (DR) starts by placing a blanket of aggregates on top of the soil to be improved. Then a dynamic compaction (DC) operation is performed, creating craters that are filled with aggregates to form a plug. More pounding takes place on top of the plug at the same locations; the craters deepen and more aggregates are placed in the open hole. The process is repeated until the crater decreases in depth. In this fashion a column of compacted aggregates is formed in place (Figure 26.16). The same range of weight, drop height, and pounder diameter are used for both DC and DR. If the energy used is high (200 to 400 kJ/m^3) and the soil is softer (PMT limit pressure 100 to 400 kPa), then the craters are deep, DR takes place, and the degree of improvement is high. In contrast, if the energy is lower (50 to 250 kJ/m3) and the soil is stronger (PMT limit pressure 250 to 700 kPa), then the craters are limited in depth, DC take place, and the degree of improvement is lower.

26.5 SOIL IMPROVEMENT WITH GROUTING AND ADMIXTURES

You might have heard the words *grout, concrete, cement*, and *mortar*: what are they, and what is the difference?

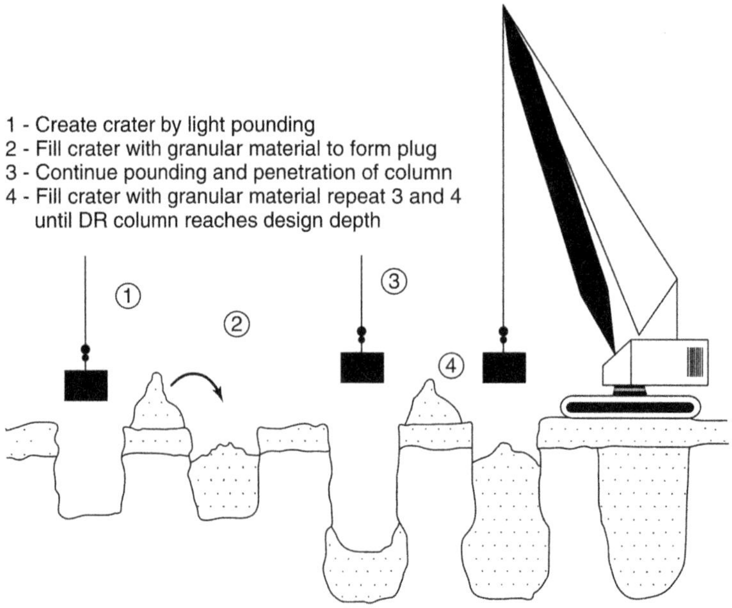

Figure 26.16 Dynamic replacement method (After Chu et al. 2009).

Cement is made of calcium and silicon. If you want to make cement in your kitchen, you mix powdered limestone (calcium carbonate, $CaCO_3$) and powdered clay (mostly silica, SiO_2) and heat it to 1450°C,; you will get a hard piece of rock out of the oven. (Note that the oven in your kitchen is very unlikely to be able to reach such high temperatures.) If you then grind that piece of rock into a very fine powder, you will have a crude cement. When you add water to that very dry cement powder, an exothermic reaction (generates heat) called *hydration* takes place and produces calcium silicate hydrate, which is the main source of cement strength. Cement is the binder in concrete, mortar, and grout. *Concrete* is the combination of cement, water, sand, gravel, and even larger aggregates. *Mortar* and *grout* are combinations of cement, water, and fine sand. The difference between mortar and grout is that typically grout will be more fluid than mortar. Sometimes grout is simply cement and water.

Different grouting techniques are used depending on the type of soil to be improved (Warner 2004). For gravels and coarse sands, the grout is injected by gravity or under pressure and fills the soil voids; the smaller the D_{50} of the soil, the finer, the more fluid, and the less viscous the grout has to be. These techniques include particulate grouting and chemical grouting. For fine sands and fine-grained soils, the grout is placed in a hole made in the soil to be improved. These techniques include jet grouting, compaction grouting, and compensation grouting. Also, for fine-grained soils, the soil can be mixed with grout that acts as a drilling fluid; this is *soil mixing*. Figure 26.17 shows the range of applicability of various grouting techniques.

26.5.1 Particulate Grouting

Particulate grouting refers to grouting coarse-grained soils by injecting the grout under gravity or under pressure into the soil voids. It also refers to grouting fissures in rocks and cavities such as sinkholes. Particulate grouting consists of opening a borehole down to the desired depth, sealing it, and then injecting the grout. The spacing between boreholes is in the range of 1 to 2 m and the hydraulic conductivity of the soil for which this technique is applicable is 10^{-2} to 10^{-5} m/s. The tube a manchettes (TAM) technique can be used to inject the grout into the soil under pressure. The TAM consists of a casing with holes at regular intervals (say, 0.5 m) covered by rubber sleeves. Two packers inside the TAM casing are inflated, one above the holes and one below; then the grout can be injected through that hole to force the rubber sleeve to lift off and allow the grout to flow into the adjacent soil under pressure. The pumping rate for particulate grouting can vary from 0.1 to 25 m^3/hr under a pressure of 0.5 to 10 MPa.

The groutability of soils is often evaluated through a ratio *N* of the soil grain size to the grout grain size. For example:

$$N_1 = \frac{D_{10(soil)}}{D_{65(grout)}} \quad \text{or} \quad N_2 = \frac{D_{10(soil)}}{D_{95(grout)}} \tag{26.26}$$

where $D_{10(soil)}$ is the grain size of the soil corresponding to 10% fines, and $D_{65(grout)}$ and $D_{90(grout)}$ are the grain size of the grout corresponding to 65 and 90% fines respectively.

Mitchell and Katti (1981) state that grouting is feasible if $N_1 > 24$ and not feasible if $N_1 < 11$. Karol (2003) states that grouting is feasible if $N_2 > 11$ and not feasible if $N_2 < 6$. Groutability also depends on how fluid the grout is and what injection pressure is applied. Akbulut and Saglamer (2002) proposed a more complete expression that reflects the influence of these parameters:

$$N_3 = \frac{D_{10(soil)}}{D_{90(grout)}} + k_1 \frac{w/c}{FC} + k_2 \frac{P}{D_r} \tag{26.27}$$

where k_1 and k_2 are soil-specific factors (0.5 and 0.01 for the soil tested by Akbulut and Saglamer), w/c is the water-to-cement ratio of the grout, *FC* is the fine content, *P* is the grout pressure in kPa, and D_r is the soil relative density. The soil is considered groutable if N_3 is larger than 26.

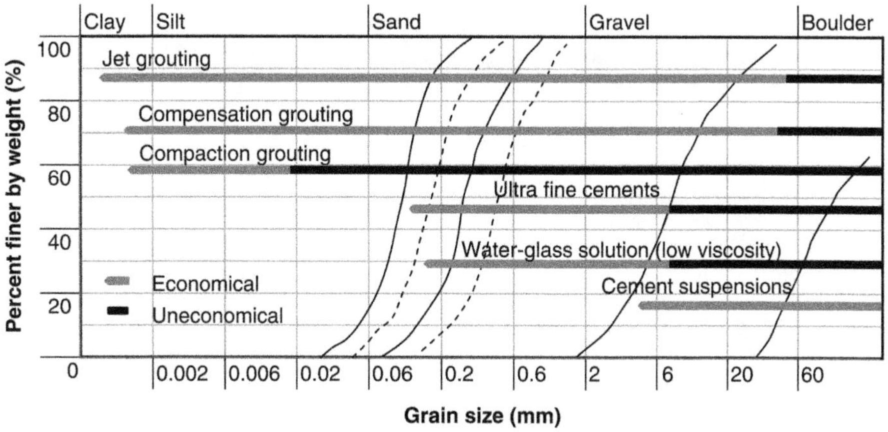

Figure 26.17 Range of application of grouting techniques (After Keller 2012; www.kellergrundbau.com/download/pdf/en/Keller_66–01E.pdf).

26.5.2 Chemical Grouting

Chemical grouting makes use of any grout that is a pure solution with no particles in suspension. Because it does not have any solids in suspension, it can penetrate finer soils. Whereas the groutability of soils by particulate grouts depends on the grain size of the solids in the grout, the groutability of chemical grouts depends on their viscosity. Chemical grout can be used in soils as fine as coarse silt.

26.5.3 Jet Grouting

Particulate and chemical grouts permeate the soil and fill the voids with grout. These techniques apply mostly to coarse-grained soils. For fine-grained soils, it is not possible for the grout to penetrate the voids, because they are too small. Instead, the approach consists of creating columns of grout in place. This is done by jet grouting, or compaction grouting, or compensation grouting, or soil mixing. Note that these techniques are also applicable to coarse-grained soils (Figure 26.17).

Jet grouting consists of drilling a borehole down to the desired depth. The drill bit has a diameter in the range of 100 to 150 mm. Once the required depth is reached, a horizontal high-pressure jet ($\sim$20 MPa) is generated to erode the soil laterally. The rod is withdrawn while rotating (Figure 26.18). This erosion process generates a larger-diameter hole (1 to 1.5 m) that is then filled with grout.

26.5.4 Compaction Grouting

Compaction grouting (Figure 26.19; Al-Alusi 1997) consists of drilling a hole with a small casing to the depth where grouting is to start, and then injecting very stiff grout with 25 mm slump or less (decrease in height in a standard cone test) under 3 to 7 MPa pressure while withdrawing the grout casing. The grout injection is performed at discrete locations and forms bulbs of grout that are 0.3 to 0.6 m thick. The grout does not penetrate the soil voids, but instead displaces and densifies the soil around the bulb. A sudden drop in pressure often indicates soil fracture. The spacing between grouting holes is in the range of 2.5 to 3.5 m center to center.

26.5.5 Compensation Grouting

Compensation grouting is used to minimize the amount of soil deformation potentially created by excavation and tunneling. It consists of injecting a volume of grout that compensates for the volume of soil displaced so that the adjacent ground surface or buildings do not deflect excessively. The grout can be injected by intrusion grouting, fracture grouting, or compaction grouting. The method is used in many different types

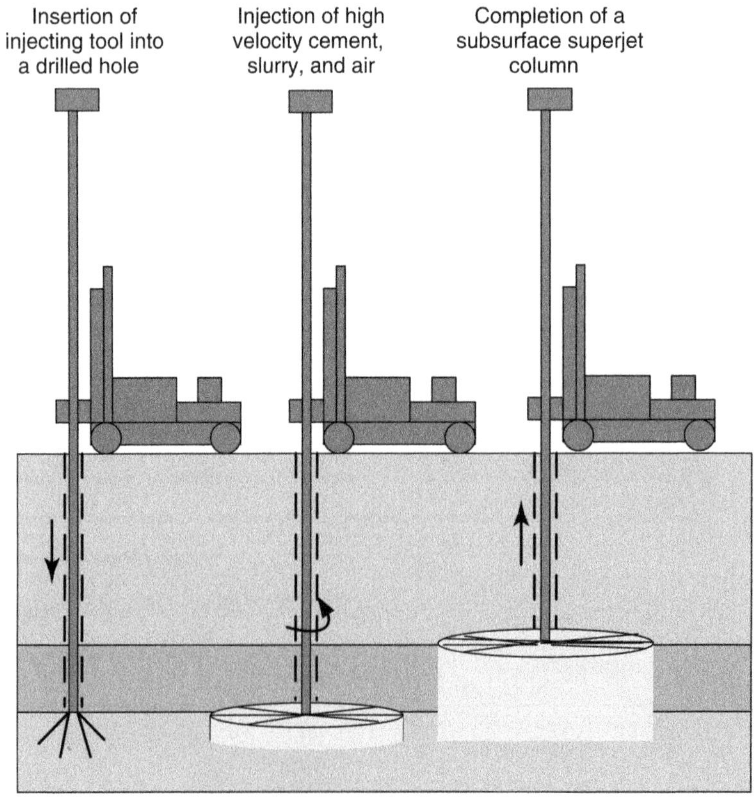

Figure 26.18 Jet grouting (After Hayward Baker, Inc.).

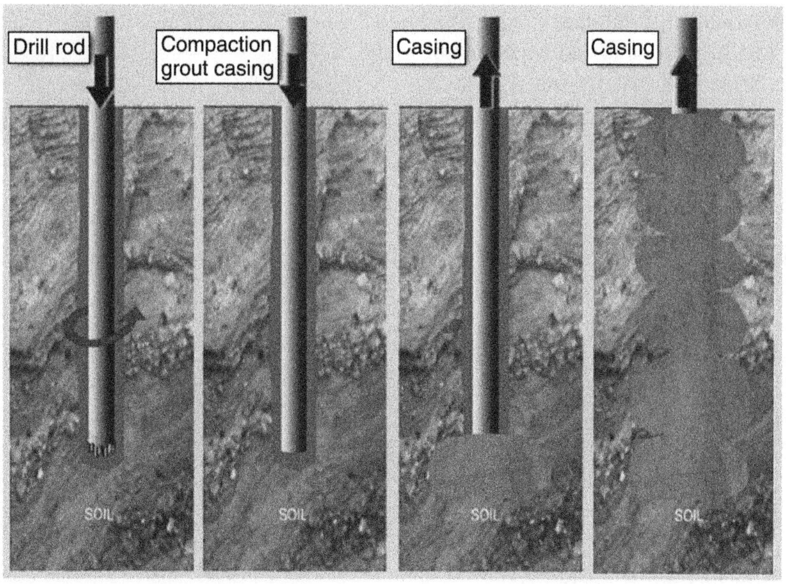

Figure 26.19 Compaction grouting. (Courtesy of Arizona Repair Masons Inc.)

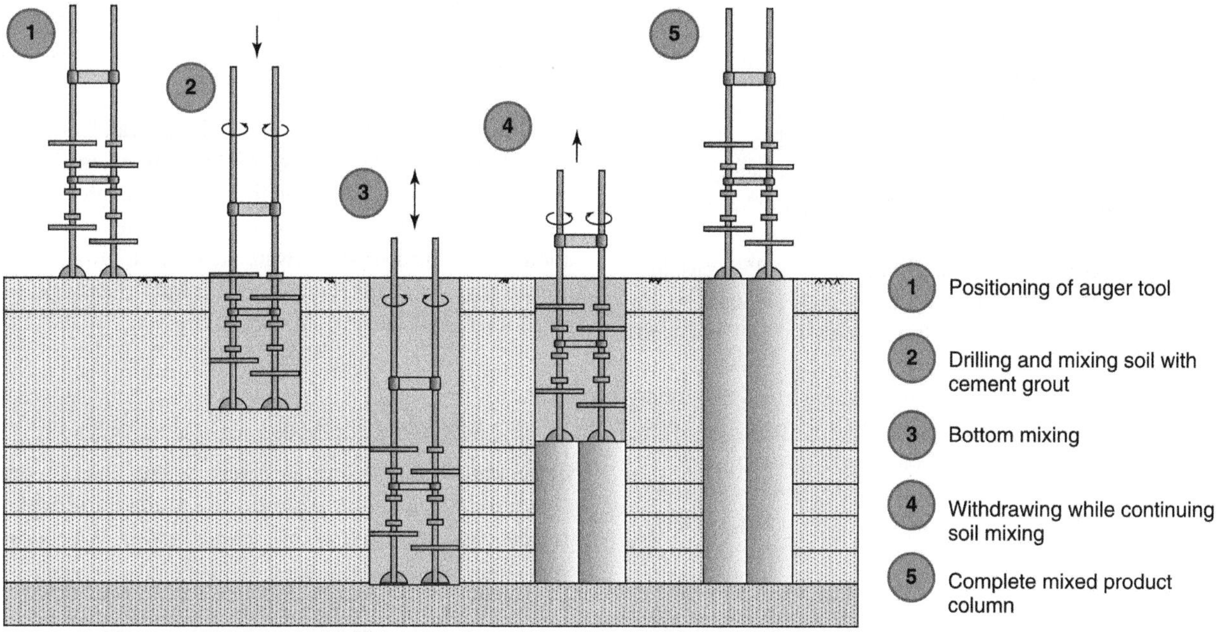

Figure 26.20 Example of SCM excavation support construction sequence. (Courtesy of JAFEC USA, Inc. Geotechnical Constructors)

of soils, but mostly in fine-grained soils, although difficulties have been encountered in soft clays (Chu et al. 2009).

26.5.6 Mixing Method

The mixing method consists of mixing soil with grout in place. The grout serves as the slurry for the drilling process and the soil-grout mixture creates a strengthened column in situ. The technique is called *deep soil mixing* (DSM) or *deep cement mixing* (DCM) or *soil cement mixing* (SCM). The drilling tool is usually a paddle auger (Figure 26.20) about 1 m in diameter; several side-by-side augers can be used at one time. Examples of construction with SCM include walls for deep excavation in soft clays, flow barriers, or simply forming a block of strengthened soil mass. The ratio of grout to soil varies from 0.15 to 0.4. Soils usually have compressive strengths less than 200 kPa and concrete more than 20,000 kPa; in SCM the soil-cement mixtures have unconfined compressive strength in the 2000 kPa range. The modulus of deformation of SCM varies in the range of 100 to 1000 MPa and can be estimated by the following equation (Briaud and Rutherford 2010):

$$E_{Soil\ Cement}\ (\text{kPa}) = 12{,}900(f_c'\ (\text{kPa}))^{0.41} \qquad (26.28)$$

26.5.7 Lime Treatment

If you want to make lime, you take a piece of natural limestone rock ($CaCO_3$); heat it to about 1000°C, which drives the carbon (CO_2) out of the limestone, and then grind the leftover piece of rock. You will have a white powder called *lime* or *calcium oxide* (CaO). If you then mix this white powder with a wet clay, it will hydrate, reabsorb carbon dioxide, and turn back into limestone. The difference between cement and lime is that lime does not strengthen as rapidly as cement; also, it is not as strong and more brittle than cement. The strengthening of the lime-soil mixture is accompanied by a decrease in water content of the clay, an increase in pH (more alkaline), an increase in plastic limit, a decrease in plasticity index, and a decrease in shrink-swell potential. The lime affects the electrostatic field around the clay particles, which tend to flocculate and assume a more granular structure. The typical amount of lime added to a clay ranges between 2 and 8%. The design of the mix and the impact on the soil properties are given in Little (1999). The clay-lime mixture has unconfined compression strengths between 700 and 1400 kPa and moduli between 200 and 3000 MPa. Lime treatment is often used to stabilize pavement foundation layers, and works best when the soil has at least 25% passing the 75 micron sieve and a plasticity index (PI) of at least 10. In the field, the lime is mixed with the surface soil and hydrated (Figure 26.21).

There is one case in which using lime can be very counterproductive: This is the case where the soil to be stabilized contains a certain amount of sulfate in the form of gypsum ($CaSO_4\ 2H_2O$). The addition of lime (CaO) and water (H_2O) to this type of soil will form ettringite ($Ca_6Al_2(SO_4)_3(OH)_{12}$

After applying lime slurry to prepared soil, the machine is run in reverse to ensure thorough mixing to the specified depth.

Figure 26.21 Lime stabilization of pavement layers. (Photo by James Cowlin/Asphalt Busters, Phoenix, AZ.)

Figure 26.22 Ettringite crystals. (Courtesy of www.sciencedirect.com/science/article/pii/S0008884698001379)

$26H_2O$), which is a highly expansive mineral. Ettringite crystals are needle like (Figure 26.22) and when mixed with water can swell to 250% of their initial height and destroy pavements. If the total soluble sulfate level is greater than about 0.3% in a 10-to-1 water-to-soil solution, additional precautions to guard against this sulfate reactions, such as swell tests, may be warranted (Little 1999).

26.5.8 Microbial Methods

Certain naturally occurring bacteria are able to generate material that can either plug the soil voids (bio-plugging) or cement particles together (bio-cementation). Water-insoluble microbial slime is produced by facultative anaerobic and microaerophilic bacteria to plug the soil voids. Bio-plugging can decrease the hydraulic conductivity of the soil by a factor of 2 (Ng et al. 2012). Calcite is produced by ureolytic bacteria that precipitate calcium carbonate.

Bio-cementing increases the shear strength of the soil by cementing the particles together. This process is called microbial-induced calcite precipitation (MICP) and works best with sand particles. It takes place when the urease enzyme produced by bacteria such as *Bacillus megaterium* decomposes urea by hydrolysis and produces ammonium. In turn, ammonium increases the pH and starts the precipitation of calcium carbonate. Calcium carbonate is the glue that cements the soil grains together (Figures 26.23 and 26.24) and can increase the shear strength of the sand by a factor of 2 (Ng et al. 2012).

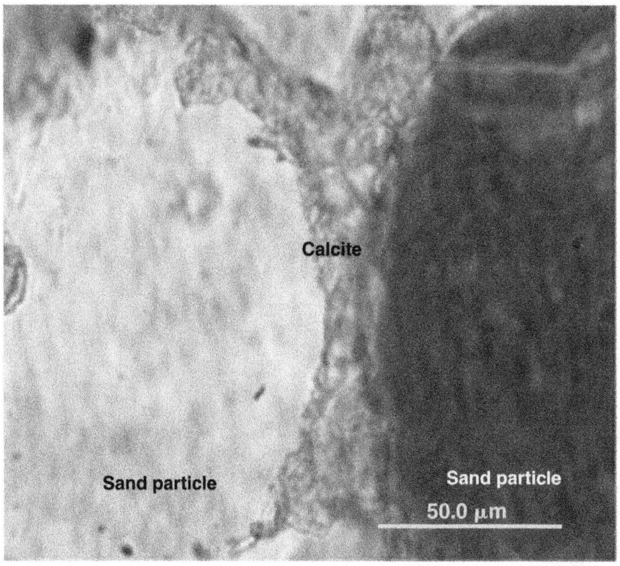

Figure 26.23 Light microscopic image of calcite crystals, produced by ureolytic bacteria, cementing two sand particles. (Courtesy of Salwa Al-Thawadi.)

26.6 SOIL IMPROVEMENT WITH INCLUSIONS

26.6.1 Mechanically or Geosynthetically Stabilized Earth

Mechanically stabilized earth (MSE) walls are covered in section 21.10. Geosynthetically stabilized earth (GSE) walls are covered in section 25.6.2.

26.6.2 Ground Anchors and Soil Nails

Ground anchor walls are covered in section 21.12 and soil nail walls are covered in section 21.13.

26.6.3 Geosynthetic Mat and Column-Supported Embankment

Geosynthetic mat and column-supported embankments (GM-CSs) (Figure 26.25) are increasingly being used as a way to rapidly construct or widen embankments on soft soils. The construction proceeds by first constructing the columns to the required depth, and preferably to a strong layer; then covering them with a bridging layer made of interbedded select fill and geosynthetic layers (say, 1 m thick); and then completing the embankment to the design height. The design process proceeds as follows (Smith 2005; Schaefer 2013):

1. Investigate the site to collect the properties of the natural soil.
2. Choose the depth and spacing of the columns. Identify the repeatable shape in plan view of the group of columns called the *unit cell*. The depth should be chosen such that the columns reach a strong layer. The spacing s should be smaller than the following values:

$$s \leq 0.67H + a$$
$$s \leq 1.23H - 1.2a$$
$$s \leq a + 3 \text{ meters} \qquad (26.29)$$

where H is he height of the embankment, a is the side of the individual square cap on top of the column. If there is no cap, a is taken as 0.89 times the diameter of the column (equivalent areas).

Center-to-center spacings of between 2 and 5 column diameters are common. The conditions placed on the spacing (Eq. 26.29) are set to ensure that proper arching will develop in the embankment through the bridging

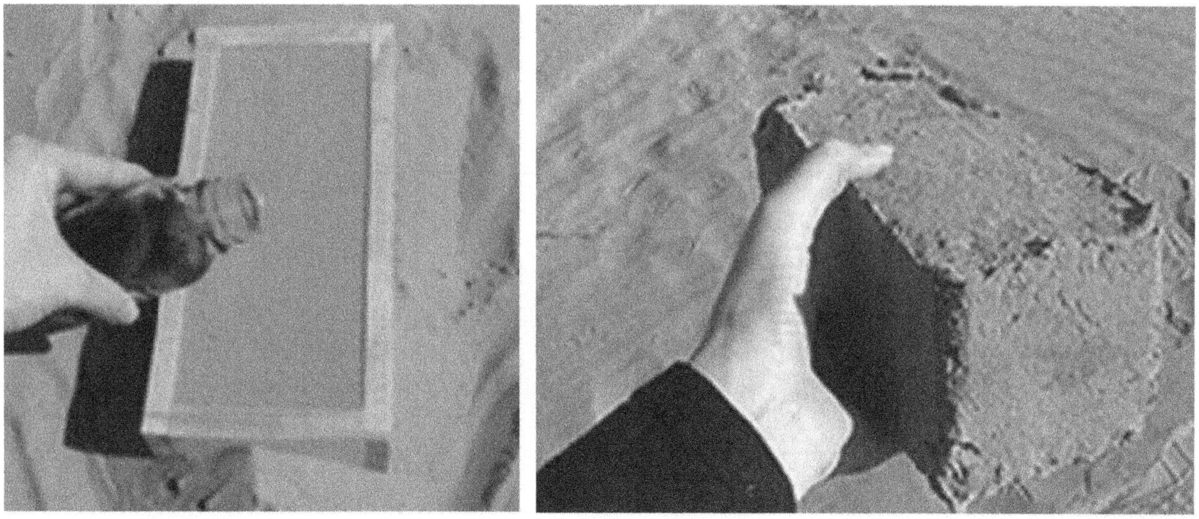

(a) (b)

Figure 26.24 Microbial-induced calcite precipitation: (a) Bacteria and calcium chloride. (b) Brick-like product. (Courtesy of Ginger Krieg Dosier, bioMASON Inc.)

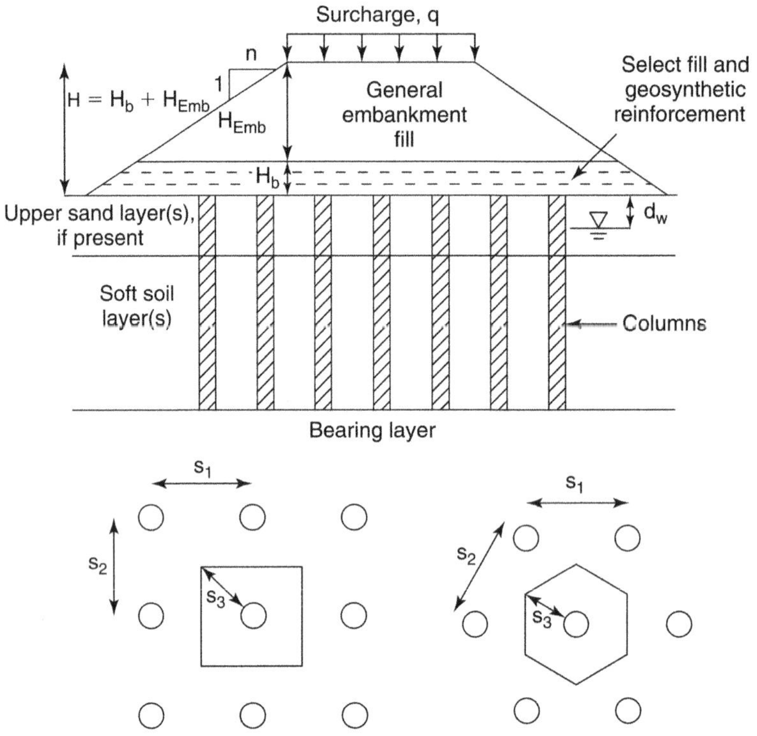

Figure 26.25 Geosynthetic reinforced column supported embankment. (Courtesy of Professor Vernon Schaefer, vern@iastate.edu)

layer to transfer all the embankment weight to the columns. Arching is also ensured by selecting a material for the bridging layer that satisfies a number of criteria (Schaefer 2013).

3. Determine the column load Q_{col} knowing the height of the embankment:

$$Q_{col} = (\gamma H + q)A \qquad (26.30)$$

where γ is the unit weight of the embankment fill, H is the height of the embankment, q is the traffic surcharge, and A is the tributary area of the column or unit cell (Figure 26.25).

4. Design the piles to safely carry Q_{col}. See sections 18.4 and 18.5.
5. Calculate the tension load in the geosynthetic layer. This tension load has two components: the tension load T_1 due to the vertical load transferred from the embankment to the columns through the bridging layer, and the tension load T_2 due to the tendency of the embankment to spread laterally. Filz et al. (2012) recommend the following expression for calculating the value of T_1:

$$6T_1^{\,3} - (6T_1 - E_{GS})\left(\frac{\sigma_{net}A_{soil}}{p}\right) = 0 \qquad (26.31)$$

where T_1 is the tension load per unit length of embankment in the geosynthetic due to the embankment vertical load, E_{GS} is the modulus of the geosynthetic layers (kN/m), σ_{net} is the difference between the soil pressure above the geosynthetic and below the geosynthetic, A_{soil} is the area within the unit cell underlain by soil, and p is the column or pile cap perimeter.

The tension load T_2 is obtained from:

$$T_2 = \frac{1}{2}K_a\gamma H^2 + qK_aH \qquad (26.32)$$

where T_2 is the tension load per unit length of embankment in the geosynthetic due to the embankment tendency to spread laterally, K_a is the coefficient of active earth pressure of the embankment soil, H is the embankment height, and q is the traffic surcharge.

6. Select a suitable geosynthetic. Geogrids are most commonly used for this application. The geosynthetic has two strengths that must be checked: the creep-limited strength at 5% strain and the allowable tensile strength (see section 25.3.3).
7. Calculate the embedment length L_e of the geosynthetic layer:

$$L_e \geq \frac{(T_1 + T_2)F}{\gamma H(\tan\delta_1 + \tan\delta_2)} \qquad (26.33)$$

where F is the factor of safety, γ is the unit weight of the embankment soil, H is the height of the embankment, and $\tan\delta_1$ and $\tan\delta_2$ are the coefficients of friction between the geosynthetic and the soil above and below in the bridging layer.

8. Calculate the total settlement s of the embankment, which includes the compression of the embankment soil under its own weight, the compression of the columns under load, and the settlement of the group of columns (see section 18.5). If the settlement is excessive, the spacing between columns can be reduced and the columns can be lengthened.
9. The lateral extent of the group of columns should be decided by stability analysis of the embankment slope reinforced by the columns (see section 19.14). A factor of safety of 1.3 to 1.5 is common.

26.7 SELECTION OF SOIL IMPROVEMENT METHOD

Considering how many different methods exist for soil improvement, it is important to have a tool that can optimize the choice of method for the given situation. The factors to be considered include the soil type; the fine content and size; the soil strength and compressibility; the area and depth of treatment; the proposed structure; the settlement criteria; the availability of skills, equipment, and materials; and the cost of the possible techniques. Sadek and Khouri (2000) proposed a software product called Soil and Site Improvement Guide to optimize the choice. More recently, Schaefer (2013) optimized the decision process through freeware available at www.geotechtools.org.

PROBLEMS

26.1 Three soils have the following CPT characteristics. Can they be vibrocompacted?

Soil	Point Resistance	Friction Ratio
Soil 1	10 MPa	1.2%
Soil 2	8 MPa	0.5%
Soil 3	15 MPa	2%

26.2 An embankment was built as shown in Figure 26.1s.

a. What is the maximum settlement of the embankment?
b. How much time is required for 90% of that settlement to occur?
c. How much surcharge is required to get the maximum settlement in 6 months? (Assume that the stress increase in the clay layer is equal to the stress at the bottom of the embankment.)

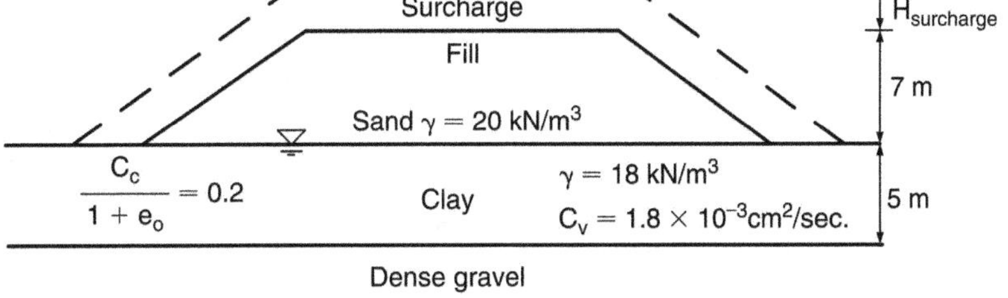

Figure 26.1s Highway embankment for preloading problem.

26.3 A highway embankment is to be built on a soft clay layer. What is the spacing of prefabricated vertical drains necessary to obtain 90% consolidation in 12 months? The PVDs are 100 mm wide and 4 mm thick, and are constructed on a square grid. The soil data are shown in Figure 26.2s.

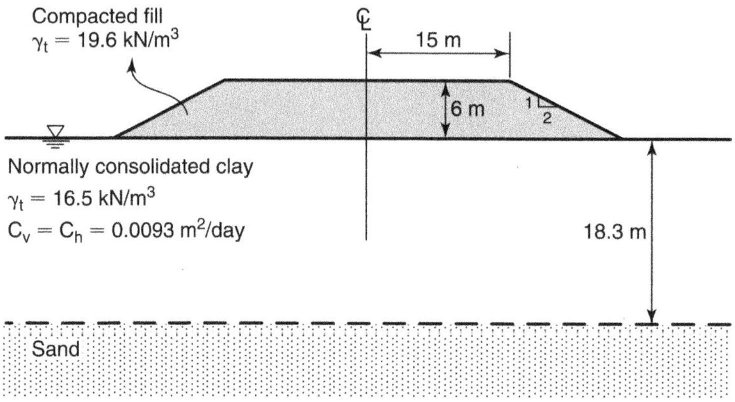

Figure 26.2s Highway embankment for prefabricated vertical drain problem.

26.4 A docking facility needs to have a 5 m high fill built on top of a 10 m thick soft silt layer underlain by dense sand. The groundwater level is at the ground surface. Stone columns 1 m in diameter and 10 m long are built. Long-term drained pressuremeter tests are performed and a conservative value of the limit pressure and modulus are 150 kPa and 1700 kPa respectively. The friction angle of the gravel used for the columns is 38°. Calculate the load that can be safely carried by one column and the settlement of the top of the column under that load. Assume that no volume change takes place in the column.

26.5 Repeat problem 26.4 but this time the stone columns are encased in a geotextile with a stiffness E equal to 150 kN/m and a tensile strength of 60 kN/m.

26.6 How would you make cement? How do you make lime? What is the difference between cement and lime?

26.7 A soil has a D_{10} equal to 2 mm. The grout used to strengthen it has a D_{65} of 60 μm and a D_{95} of 130 μm. Can particulate grouting be successful?

26.8 A geosynthetic mat and column-supported embankment (GMCS) is used to build an embankment on soft clay. The embankment is 7 m high, built with a fill with a compacted unit weight of 20 kN/m^3 and a friction angle of 32°. The columns are 1 m in diameter with no pile cap and are placed on a square 2 m center-to-center grid. The bridging layer is 1 m thick with two layers of geosynthetic. Calculate the load per column, the tension in the geosynthetic layers due to spanning across the columns (T_1), and the tension due to lateral spreading (T_2). Assume that the net difference in stress on either side of the geosynthetic layer is 80% of the pressure under the embankment and that the geosynthetic has a modulus equal to 60 kN/m.

Problems and Solutions

Problem 26.1

Three soils have the following CPT characteristics. Can they be vibrocompacted?

Soil	Point Resistance	Friction Ratio
Soil 1	10 MPa	1.2%
Soil 2	8 MPa	0.5%
Soil 3	15 MPa	2%

Solution 26.1

According to Massarsch guidelines, and Figure 26.2:

- Soil 1—Marginal
- Soil 2—Yes
- Soil 3—No

Problem 26.2

An embankment was built as shown in Figure 26.1s.

a. What is the maximum settlement of the embankment?
b. How much time is required for 90% of that settlement to occur?
c. How much surcharge is required to get the maximum settlement in 6 months? (Assume that the stress increase in the clay layer is equal to the stress at the bottom of the embankment.)

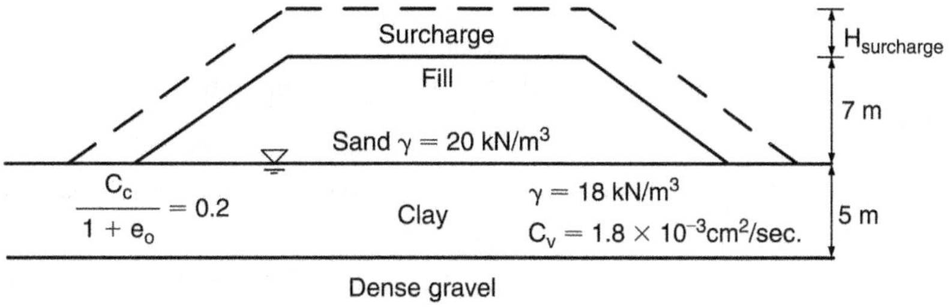

Figure 26.1s Highway embankment for preloading problem.

Solution 26.2

a. Maximum Settlement of the Embankment

$$S_{max} = H\frac{C_c}{1+e_o}\log\frac{\sigma'_{ov}+\Delta\sigma'}{\sigma'_{ov}}$$

$$\sigma'_{ov} = 2.5 \times 18 - 2.5 \times 9.81 = 20.48 \text{ kPa}$$

$$\Delta\sigma' = 20 \times 7 = 140 \text{ kPa}$$

$$\therefore S_{max} = 5 \times 0.2 \times \log\frac{20.48+140}{20.48} = 0.89 \text{ m}$$

b. Time Required for 90% Settlement

$$T_v = C_v\frac{t}{H_{dr}^2} \qquad \therefore t = T_v\frac{H_{dr}^2}{C_v}$$

when $U = 90\%$, $T_v = 0.848$

$$\therefore t = 0.848\frac{2.5^2}{1.8 \times 10^{-3} \times 10^{-4}} = 2.94 \times 10^7 \text{ sec.} = 341 \text{ days}$$

c. Surcharge Needed to Reach Maximum Settlement in 6 Months

The surcharge needed to reach the maximum settlement in 6 months is calculated as follows:

$$T_v = 1.8 \times 10^{-7} \times \frac{6 \times 30 \times 24 \times 3600}{2.5^2} = 0.448$$

When $T_v = 0.448$, $U = 73.2\%$

$$U = \frac{S_{6\ months\ \text{(fill+surcharge)}}}{S_{max(fill+surcharge)}}$$

Because we want $S_{6\text{ months}}$ (fill + surcharge) = S_{max}(fill), then:

$$U = \frac{S_{max\ (fill)}}{S_{max(fill+surcharge)}}$$

$$S_{max}(F+S) = \frac{0.89}{0.732} = H_o\frac{C_c}{1+e_o}\log\frac{\sigma'_{ov}+\Delta\sigma'}{\sigma'_{ov}} = 5 \times 0.2 \times \log\left(\frac{20.48+\Delta\sigma}{20.48}\right)$$

$$1.216 = \log\left(\frac{20.48 + \Delta\sigma}{20.48}\right)$$

$$\Delta\sigma = 316.3\text{ kPa} = \Delta\sigma(F) + \Delta\sigma(S)$$

$$316.26\text{ kPa} = 20 \times 7 + 20 \times h$$

$$h = 8.81\text{ m}$$

Problem 26.3

A highway embankment is to be built on a soft clay layer. What is the spacing of prefabricated vertical drains necessary to obtain 90% consolidation in 12 months? The PVDs are 100 mm wide and 4 mm thick, and are constructed on a square grid. The soil data are shown in Figure 26.2s.

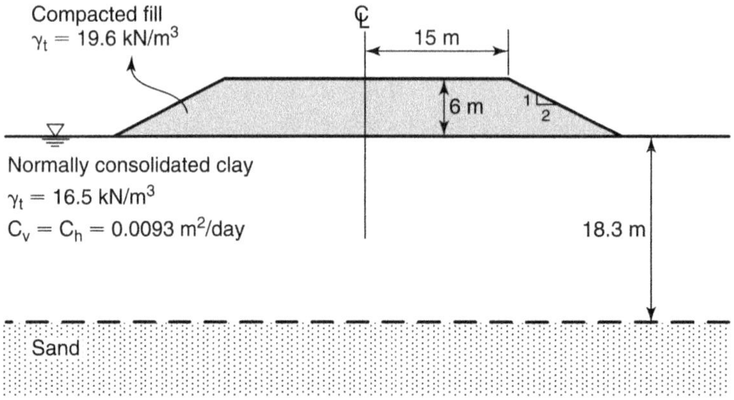

Figure 26.2s Highway embankment for prefabricated vertical drain problem.

Solution 26.3

The equation that gives the spacing is:

$$t = \frac{d_w^2}{8c_h}\left(Ln\left(\frac{d_w}{d_e}\right) - 0.75 + F_s\right) Ln\left(\frac{1}{1 - U_h}\right)$$

where d_e is the equivalent diameter of the PVD defined as $2(a + b)/\pi$, a and b are the width and thickness of the PVD, c_h is the horizontal coefficient of consolidation, d_w is the well influence diameter (taken as 1.05 s for an equilateral triangle spacing pattern and 1.13 s for a square spacing pattern) where s is the spacing between PVDs, and F_s is a soil disturbance factor (taken as 2 for highly plastic sensitive soils but taken as zero if c_h has been conservatively estimated or accurately measured). Using the parameters given in the problem, we get:

$$12 \times 30 = \frac{(1.13s)^2}{8 \times 0.0093}\left(Ln\left(\frac{1.13s}{2\,(0.1 + 0.004)\,/\pi}\right) - 0.75 + 1\right) Ln\left(\frac{1}{1 - 0.9}\right)$$

Note that an average value of $F_s = 1$ is used in this case:

$$360 = \frac{1.277s^2}{0.0744}(Ln(17.07s) - 1.75) \times 2.303$$

$$9.108 = s^2(Ln(17.07s) - 1.75)$$

This equation is solved by trial and error and gives a center-to-center spacing of $s = 2.2$ m.

Problem 26.4

A docking facility needs to have a 5 m high fill built on top of a 10 m thick soft silt layer underlain by dense sand. The groundwater level is at the ground surface. Stone columns 1 m in diameter and 10 m long are built. Long-term drained pressuremeter tests are performed and a conservative value of the limit pressure and modulus are 150 kPa and 1700 kPa respectively. The friction angle of the gravel used for the columns is 38°. Calculate the load that can be safely carried by one column and the settlement of the top of the column under that load. Assume that no volume change takes place in the column.

Solution 26.4

$$Q_u = K_p(p_L - u_w)A$$

$$K_p = \frac{1+\sin 38}{1-\sin 38} = 4.2$$

$$Q_u = 4.2 \times (150 - 5 \times 9.8) \times 0.5^2\pi = 333 \text{ kN}$$

$$Q_{allowable} = \frac{Q_u}{F.S.} = \frac{333}{2} = 166.6 \text{ kN}$$

$$s = 4B\frac{\Delta R}{R}$$

$$\text{E} = 1700 \text{ kPa and } v = 0.35$$

$$\frac{\Delta R}{R} = \frac{P}{2G} = \frac{1+v}{E}\frac{Pl}{F.S.}$$

$$\frac{\Delta R}{R} = \frac{1.35}{1700}\frac{150}{2} = 0.06$$

$$s = 4 \times 0.06 = 0.24 \text{ m}$$

Problem 26.5

Repeat problem 26.4 but this time the stone columns are encased in a geotextile with a stiffness E equal to 150 kN/m and a tensile strength of 60 kN/m.

Solution 26.5

Figure 26.3s illustrates this problem.

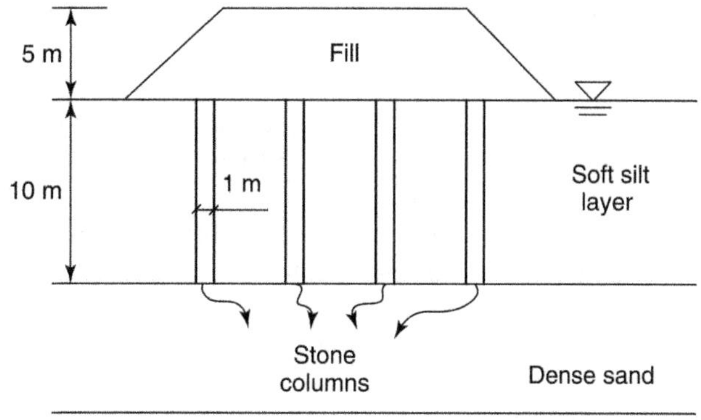

Figure 26.3s Illustration of the soil profile.

Failure mechanism 1: Soil fails laterally

The ultimate pressure the stone column can carry is calculated as:

$$p_{u1} = k_p(p'_L + p_{geo})$$

where k_p is the passive earth pressure coefficient of the soil, p'_L is the drained pressuremeter limit pressure, and p_{geo} is the lateral confinement pressure generated by the geotextile at the strain corresponding to failure of the soil.

$$k_p = \frac{1+\sin\varphi'}{1-\sin\varphi'} = \frac{1+\sin 38}{1-\sin 38} = 4.2$$

$$p'_L = p_L - u_w = 150 - 5 \times 9.81 = 101 \text{ kPa}$$

The limit pressure is associated with a radial strain or hoop strain equal to 41%. We use this strain to calculate p_{geo}. Therefore, the deformation of the geotextile at failure is:

$$\Delta r = r_o \times \varepsilon = 0.5 \times 0.41 = 0.205 \text{ m}$$

$$p_{geo} = E\frac{\Delta r}{r_o^2} = 150 \times \frac{0.205}{0.5^2} = 123 \text{ kPa}$$

Hence, the ultimate load the stone column can carry is:

$$Q_{u1} = k_p(p'_L + p_{geo})\pi r_o^2 = 4.2 \times (150 - 5 \times 9.81 + 123) \times \pi \times 0.5^2 = 739 \text{ kN}$$

As can be seen, the geotextile encasement more than doubles the ultimate load the stone column can carry.

Failure mechanism 2: Geotextile fails in hoop tension

The ultimate pressure the stone column can carry is calculated as:

$$p_{u2} = k_p\left(2G\frac{\Delta r}{r_o} + p_{geo\ f}\right)$$

where k_p is the passive earth pressure coefficient of the soil, G is the shear modulus of the soil outside the geotextile, $\Delta r/r_o$ is the relative increase in radius of the stone column, and $p_{geo\ f}$ is the confining pressure generated by the geotextile at failure. In this failure mechanism, the tensile strain of the geotextile at failure is calculated as:

$$\varepsilon = \frac{T}{E} = \frac{60 \text{ kN/m}}{150 \text{ kN/m}} = 0.4$$

Note that when the hoop strain of the geotextile is 0.4, the soil is approximately at the limit pressure (radial or hoop strain of 0.41), so in this fortuitous case, failure mechanisms 1 and 2 are the same.

Therefore, the ultimate load per stone column is 739 kN.

Problem 26.6

How would you make cement? How do you make lime? What is the difference between cement and lime?

Solution 26.6

Cement is made of calcium and silicon. To make cement in your kitchen, you mix powdered limestone (calcium carbonate, $CaCO_3$) and powdered clay (mostly silica SiO_2) and heat it to 1450°C; you will get a hard piece of rock. (Note that the oven in your kitchen is very unlikely to be able to generate this high a temperature.) If you then grind that piece of rock into a very fine powder, you will have a crude cement. When you add water to that very dry cement powder, an exothermic reaction (generates heat) called hydration takes places and produces calcium silicate hydrate, which is the main source of cement strength. Cement is the binder in concrete, mortar, and grout.

To make lime, take a piece of natural limestone rock ($CaCO_3$), heat it to about 1000°C to drive the carbon (CO_2) out of the limestone, and then grind the leftover piece of rock; you will have a white powder called lime or calcium oxide (CaO). If you then mix this white powder with a wet clay, it will hydrate, reabsorb carbon dioxide, and turn back into limestone.

The difference between cement and lime is that lime does not strengthen as rapidly as cement. It is weaker and more brittle than cement.

Problem 26.7

A soil has a D_{10} equal to 2 mm. The grout used to strengthen it has a D_{65} of 60 μm and a D_{95} of 130 μm. Can particulate grouting be successful?

Solution 26.7

$$N_1 = \frac{D_{10(soil)}}{D_{65(grout)}} \quad or \quad N_2 = \frac{D_{10(soil)}}{D_{95(grout)}}$$

According to one theory, grouting is feasible if $N_1 > 24$ and not feasible if $N_1 < 11$. According to another theory, grouting is feasible if $N_2 > 11$ and not feasible if $N_2 < 6$.

In this problem, grouting is feasible because:

$$N_1 = \frac{D_{10(soil)}}{D_{65(grout)}} = \frac{2 \text{ mm}}{60 \ \mu\text{m}} = 33.3 > 24$$

$$N_2 = \frac{D_{10(soil)}}{D_{95(grout)}} = \frac{2 \text{ mm}}{130 \ \mu\text{m}} = 15.4 > 11$$

Problem 26.8

A geosynthetic mat and column-supported embankment (GMCS) is used to build an embankment on soft clay. The embankment is 7 m high, built with a fill with a compacted unit weight of 20 kN/m^3 and a friction angle of 32°. The columns are 1 m in diameter with no pile cap and are placed on a square 2 m center-to-center grid. The bridging layer is 1 m thick with two layers of geosynthetic. Calculate the load per column, the tension in the geosynthetic layers due to spanning across the columns (T_1), and the tension due to lateral spreading (T_2). Assume that the net difference in stress on either side of the geosynthetic layer is 80% of the pressure under the embankment and that the geosynthetic has a modulus equal to 60 kN/m.

Solution 26.8

The load per column is:

$$Q_{col} = (\gamma H + q)A$$

$$Q_{col} = (20 \times 7 + 0) \times 2 \times 2 = 560 \text{ kN}$$

The tension T_1 due to the bridging effect between columns is given by:

$$6{T_1}^3 - (6T_1 - E_{GS})\left(\frac{\sigma_{net} \times A_{soil}}{p}\right) = 0$$

$$6{T_1}^3 - (6T_1 - 60)\left(\frac{0.8 \times 20 \times 7 \times (4 - \pi \times 0.5^2)}{\pi \times 1}\right) = 0$$

$${T_1}^3 - 114.6(T_1 - 10) = 0$$

which gives a tension T_1 equal to:

$$T_1 = -14 \text{ kN/m}$$

The tension T_2 due to the lateral spread of the embankment is given by:

$$T_2 = \frac{1}{2}K_a \gamma H^2 + qK_a H$$

$$K_a = \frac{1 - \sin\varphi}{1 + \sin\varphi}$$

$$K_a = \frac{1 - \sin 30}{1 + \sin 30} = \frac{0.5}{1.5} = 0.333$$

$$T_2 = \frac{1}{2} \times 0.333 \times 20 \times 7^2 + 0 = 163 \text{ kN/m}$$

So, the lateral spreading effect is much more severe than the bridging effect in this case.

CHAPTER 27

Technical Communications

27.1 GENERAL

The most important concepts in technical communications are:

1. Be brief
2. Be clear
3. Be right technically
4. Be correct from the communication point of view

"Be brief" is essential, as you may lose your reader or listener if your statement drags on. It is often better to be too short and entice your audience to come back to you for more than to be too long and boring. Of course, in the end you wish to hit exactly the right length. In technical writing, 10-word sentences are about the best length. In oral communications, you have the choice between the 15-second sound bite, the 2-minute exposé, the 10-minute discourse, and the 1-hour lecture. Think about which one is right for the situation.

"Be clear" requires that you put yourself in the shoes of your readers or listeners and aim at the sophistication level that most closely corresponds to their background. If you are unsure, assume a lower level and gradually increase the sophistication of the message. This sophistication level refers to the sophistication of for the vocabulary as well as the sophistication of the technical content and thought process.

"Be right technically" is crucial in our field; it requires that any statement made be based on prior work by others or your own work. If not, it is necessary to acknowledge that your statement is based on your intuition or experience. If you use prior work in your statement, you must quote the source and respect intellectual property.

"Be correct from the communication point of view" requires proper vocabulary, grammar, and diction, including being politically correct. Make sure to proofread your written work. Don't forget that your listener may not speak your language, so be prepared to speak slowly and exercise patience when you get indications that you have lost contact.

27.2 E-MAILS

E-mail has become a huge part of daily communications, because these messages are very convenient and time efficient. They include the distribution lines (To, Cc, Bcc), the title, and the body of the message. In the distribution line, make sure that you copy those who truly need to see your message—and no more. The Bcc can be dangerous, as you are obviously hiding something from someone. Remember this golden rule of communication: It is always best to communicate in such a way that if your message were published on the front page of a major newspaper, you would not be embarrassed.

Sometimes you will receive an unpleasant message. When you do, please follow this other golden rule: It is best not to answer unpleasant messages right away. In fact, it is often best not to answer at all. Answering right away with another unpleasant message may give you a few seconds of pleasure, but days of agony later on. Unpleasant messages are best left to simmer for a few days (and it is often disturbing to the sender when such messages remain unanswered).

An email signature with your complete title and contact information is important and convenient. It allows your reader to know who you are and gives your contact information in case a phone call is more appropriate as a response. However, if you do not wish to be contacted, or if your reader knows you well, these items are not useful and may convey a message of misplaced egocentric pride. When you write your name in your signature, write your first name in lower case and your family name in capital letters; you may be sending an email to someone in a country where it is not obvious which is your first name and which is your last name. Another problem may be that, in that other country, names are so different that your gender is not obvious. One trick is to answer by saying "Dear Dr. Something": that way you do not have to decide. By the way, make sure that you include your country as part of your signature contact information.

27.3 LETTERS

Letters are no longer very common. They are used for extraordinary and more formal occasions. Letter formats vary, but, generally speaking, you will have the letterhead at the top or bottom of the page (or both) with the text of your message in between. The letterhead has the name of your organization and general contact information. Start by indicating the date of the letter, then follow with the name, title, affiliation, and address of the person you are writing to. The greeting line should be Dear Mr. X, Dear Mrs. X, Dear Ms. X (for women not married or if you are not sure of their marital status), Dear Dr. X, Dear Professor X, and so on. If you have a choice of two titles, it is always best to choose the title indicating higher rank. Note that one uses Mr. President but Madame President, not Mrs. President. The text of your letter follows. This text should be two pages or less; letters that must be longer probably should be reports. Letters rarely have attachments unless they are cover letters.

27.4 GEOTECHNICAL REPORTS

A geotechnical design report communicates the site conditions, design, and construction information to the owner of the project, or to the owner's representative. It is an essential part of the construction process and is used at the design stage, the construction stage, and after construction if there are claims. In any case, it should be clear, concise, and accurate. Although the content of a geotechnical report will vary depending on the type and complexity of the project, a typical report will contain at least the following information:

1. Summary of all site investigation data, including layering, groundwater conditions, and variability based on borings
2. Geologic interpretation
3. In situ test results
4. Laboratory test results
5. Interpretation and analysis of the subsurface data
6. Predictive analyses, design
7. Engineering judgment and recommendations, including solutions for possible problems
8. Recommended special provisions and limiting conditions

The detailed data usually appears in appendices with all figures, including boring logs, soil profiles, and test results. Remember that when you have only a few borings at some distance from each other, it is unwise to infer the layering in between the borings unless confirmed through the geology of the site, geophysical methods, or other evidence. Though it is tempting to draw a continuous layering graph, it is best to use question marks between borings when suggesting a layering profile.

The report is written to help in the design of the project and so must be helpful to the person who will use it. At the same time, you have to be careful not to make statements that may hurt you or your company in the future. Although detailed calculations are not included in a geotechnical report, it is important to keep all your calculations, as you may have to go back to them later on. This is why it is important, when you make calculations, to clearly document the steps you took, why you assumed some values, how you came to a conclusion, and what published references you used. Once the geotechnical report is ready, have it proofread by a senior and experienced engineer.

27.5 THESES AND DISSERTATIONS

A *thesis* is usually required for a master's degree, whereas a *dissertation* is produced by a candidate for a PhD degree. Both have the same typical organization.

1. *Title.* The title should reflect as precisely as possible the content of the work—no more, no less. It is better to have a longer, more descriptive title than a short and misleading or vague title. Overall, titles of about 50 to 75 characters (5 to 10 words) are best.

2. *Cover page.* The cover page should include the title, the author' (or authors') name(s), and the name of the institution. The date should also appear on the cover page.

3. *Dedication.* You may wish to dedicate your work to someone who is important to you.

4. *Acknowledgments.* This is where you thank those who have contributed to the work but are not the author(s). Don't forget the name of the sponsoring organization and anyone in that organization who helped you in some fashion.

5. *Table of contents (TOC).* This is the first major step in writing a thesis or dissertation. The more time you spend on the table of contents, the less time you will have to spend writing and iterating. Start with major section and subsection titles. As you do so, think about the natural flow of the work. Then, for each subsection, write notes to yourself in bullet form identifying what you will talk about in each paragraph. The more detailed the TOC, the easier the body of the paper will be to write. A skimpy or poorly organized table of contents leads to many rewrites, frustration, and a feeling of not making progress.

6. *Executive summary, abstract, or summary.* Summaries or abstracts are a very important part of a thesis or dissertation, as people often do not take the time to read the details of your work. Describe the problem, summarize the important findings of each section (in order), and briefly state the most important conclusions. Usually there are no figures, tables, or photos in this part.

7. *Introduction.* The purpose of the introduction is to answer the following questions: what, why, how, where, by whom, and for whom. Once these questions are answered, you can present a narrative outline of the thesis or dissertation.

8. *Review of existing knowledge (literature review).* It is important to collect and study existing information so you

do not repeat work that has already been well established. It is sometimes good to duplicate important experiments done by others, especially if there is some level of controversy regarding the techniques or findings, but overall this is not a great way to progress. Once you have summarized the existing knowledge, take the time to synthesize that knowledge, give your opinion and point out why your work was necessary or how it built on or extended previous work.

9. *Experiments*. A dimensional analysis is always a helpful initial step. If the experiment is a small-scale version of the full-scale prototype, scaling laws must be addressed and extrapolation of the results to full scale explained. Experiments should be reported by first explaining what the purpose of the experiment was, then the design of the experiment, the description of the mechanical and electronic parts, the test procedure, the data acquisition, and the results. If the project included a large number of experiments, a table listing all the experiments should be presented. If there are too many parameters to report for each experiment, a number designation (e.g., T46) should be attributed to each one and the table should give all parameters. If there are too many results or figures to present in the main text, present a few strong examples in the main text and put all the results in an appendix. A summary table should be the first page in the appendix. The analysis of the results can appear here or in a separate section.

10. *Numerical simulations.* The motivation behind running numerical simulations should be outlined. The mesh size should be discussed first, demonstrating the reason for choosing the distance to the boundaries. The boundary conditions should be explained. The selection of the soil model and of the input parameters should be discussed next. A table summarizing the number of simulation cases helps readers understand the extent of the work and identify which parameters were varied. If the number of simulations is not too large, the results can be presented in the main text. If not, put the results in the appendix that starts with a summary table. The analysis of the results can be done here or in a separate section.

11. *Analysis of data.* This section makes use of all data accumulated to formulate a solution to the problem posed. Theory, measurements, engineering judgment, logic, and common sense all contribute to making the outcome and results as simple, sound, and useful as possible. It often takes a lot of effort to reach the optimum threshold of simplicity.

12. *Conclusions.* This is where you demonstrate your contribution to new knowledge in a succinct way. It is often convenient to go chapter by chapter and collect your conclusions from each part, arranging them in a consistent framework that shows progress in geotechnical engineering. I am reminded of two comments I received on my research in the early 1980s, one from my father and one from Geoff Meyerhof. My father, after patiently listening to my research work, looked straight at me and said: "So what?!" Meyerhof, after reading my early work on laterally loaded piles, said: "Too complicated!" So, while you have to deal with great complexity to solve the problem, in the end your goal should be to develop something "useful and simple."

13. *References.* The purpose of a reference is to acknowledge the work of others and support your statement. Remember that in engineering, when you make a statement, you must have proof (experiment, theory, simulation, reference) or at least a factual basis for that statement; you may need to say that your statement is based on your experience or common sense or engineering judgment. The best way to quote a reference in the text is according to your institution's mandated or preferred system; most use the author-date system, In this system, you use the name of the first author, followed by the name of the second author if there are only two authors, or by "et al." if there are more than two authors, and followed by the year of publication. In the reference list, the full citation information for each source is given, organized in such a way that readers can easily track down and obtain the referenced publication. A typical presentation is

- Last name and initials of all authors, year of publication (in parentheses); title of paper, report, or book chapter; title of periodical, proceedings, or book (usually in italics); volume number and issue number; name of publisher and the publisher's location; inclusive page numbers.

If the reference is a web site address (URL), the reference is organized as follows:

- Author if any is credited, copyright or posting date, title, the address/URL of the web site from which the piece was retrieved, and the date the material was accessed or downloaded.

If the reference is a CD, the reference citation is organized as follows:

- Author(s), copyright date, title, medium, and producer/publisher and publisher's location.

14. *Appendices.* The bulk of your data should appear in an appendix; you may need to use more than one. The front page of each appendix should explain what is in that appendix. This is where a summary table of tests or simulations becomes most useful.

27.6 VISUAL AIDS FOR REPORTS

Visual aids for reports may include figures, tables, and photographs. Figures showing graphs of data should have the two axes labeled with the spelled-out name of the variable, the letter symbol, and the unit in parentheses. The axes should have scales with about 20 tick marks and 4 or 5 numbers on the scale. The data points should be clearly identified. If you have more than one set of data points, use different symbols. Although color graphs are more appealing and easier to read, remember that a color graph created electronically may end up printed in black and white. In this case, if you have used

the same symbols but different colors, the data points will be undistinguishable between sets. If a set of data points leads to a recommended design line, leave the data points with the recommended line; this will help the user gauge the extent of the scatter and select a different value than your recommendation if necessary. If a regression line is drawn, indicate the equation of the line and the value of the coefficient of regression R^2. The size of the letters or numbers in a graph should be such that the graph can be easily read; the minimum size to ensure easy readability is about 1/20 of the size of the graph.

Tables should have the name, symbol, and unit of the parameters at the top of each column or the beginning of each row. The caption of a table precedes the body of the table, whereas the caption of a figure or photo goes below the artwork (don't ask me why!). For best results, photos must be sharp and high resolution. The rules about using visual aids that are not yours, or that are yours but that you have signed over to a publisher by signing a copyright agreement, vary from one source to another. For noncommercial purposes, the general rule is that the source of each visual aid must be acknowledged unless it is your own original work. The acknowledgment may be made simply by placing in the caption the name of the author and date of the publication where the visual aid was found (essentially, giving the reference citation. For commercial purposes, written permission must be obtained from the publisher of the visual aid, and that permission or credit line must be mentioned along with the acknowledgment of the source. Student work is noncommercial, but it is essential to get into the good habit of acknowledging any intellectual property you use that is not your own. People always appreciate when they are recognized and get upset when they are not.

27.7 PHONE CALLS

In all cases, one should prepare for a phone call—if nothing else, to anticipate questions and minimize cost. Know what you wish to achieve and have a plan on how to maneuver if the conversation goes in a different direction than you anticipated.

E-mail or phone call: that is the question! Most of the time, e-mails are very efficient, but there are some situations in which they are very dangerous, misleading, and inefficient. It is amazing to see how many different ways a given e-mail may be interpreted by different people. There is a big difference between the written word and the spoken word. For example, if an interaction might be contentious, it is best to pick up the phone. People tend to understand much better when spoken to than when written to. Reading an e-mail can lead to a serious misunderstanding and an escalating response; it is often much easier to diffuse a misunderstanding on the telephone.

There are also times when you simply have to be courageous enough to call the person rather than hiding behind an impersonal and cold e-mail. Most people appreciate being told unpleasant truths "in person" by a telephone call rather than reading them in e-mail. One might argue that some things must be in writing, and that is true. However, the best approach in those cases is to talk on the telephone and explain that the conversation will be followed by a follow-up e-mail to restate and formally memorialize the points covered in the conversation.

27.8 MEETINGS

Three of the most important rules for efficient meetings are:

1. Do not interrupt anyone
2. Be brief
3. Be professional in your attitude toward your colleagues

Interrupting people when they speak is rude, but they have to respect your right to contribute as well by being brief. From time to time, someone may get under your skin, but it is important to remain calm under fire and concentrate on facts, data, logic, analysis, and reasoning to win your arguments rather than shouting or attacking someone personally. Accept that sometimes your point of view is not the view of the majority and that you are only a member of the team. In many situations, it is important to have the courage to change the things you can change, accept those that you cannot change, and have the wisdom to know the difference.

If you are a participant in a meeting, speak up only when you really have something important to say—something that will advance the process. If you are presiding at the meeting, keep in mind the time allotted for each item on the agenda, have a plan if a discussion drags on for good reasons, cut off any unnecessary chat, and help the group stay focused on the topic by repeating during the discussion the original problem to be solved or question to be answered. Also, as the leader of the meeting, start by establishing some initial rules about cell phone use, side chats, and texting or answering e-mails during the meeting. All are distracting (and discourteous) and should not be allowed during meeting.

Motions and votes are very valuable because the decision becomes extremely clear. It takes place by

1. A motion is proposed by someone.
2. The motion is seconded by a second person. If not seconded, the motion dies.
3. Once the motion is seconded, a discussion period takes place during which you try to convince your colleagues that they should vote in a particular way.
4. The person presiding calls for the vote when the discussion is over and then the votes are recorded. This can be done by show of hands, voice call, or written ballots. The choices are yes, no, or abstain. For delicate matters, the vote may be secret, depending on the rules of the organization.

Although motions may seem cumbersome at times, they are very useful in case of arguments after the fact. Remember

that after any motion passes, there is usually a need for an action item: who will do what to implement the decision. Any action items or assignments should be included in the recorded minutes of the meeting. For further help in running meetings efficiently, consult *Roberts' Rules of Order*.

27.9 PRESENTATIONS AND POWERPOINT SLIDES

If you are going to give an important lecture in front of many people, make sure that nothing is left to chance. The best presentations can be ruined if something fails to work. Here is a helpful checklist for presentation success:

1. Hook up your laptop to the local projector and check that it works properly. If you have to use someone else's laptop, make sure you know how to use the basic functions on that laptop.
2. Check all your slides to make sure that they are exactly what you expect (equations are not changed, movies are working, and so on).
3. Bring a pointer or find out if one is available to borrow.
4. Know who will advance the slides. If you are not doing this yourself, what will be the signal to advance the slides? Constant use of a "next slide" request is not elegant; a sign of some sort between you and the projectionist is best, including when to start any movies or animations.
5. Keep an eye on time and pace yourself. It is best to practice the full presentation ahead of time and under "field" conditions to measure the time required.
6. The average time per slide is one minute; however, slides with only photos will go faster and slides with sample calculations will go slower.
7. Have a back-up plan if something fails to work. Can you project your voice without the microphone? Can you complete the presentation without slides, for example? Develop the talent of not requiring slides to guide your thoughts.
8. Have a special title slide and final slide that set your desired tone and reflect your personality.
9. Keep an eye on your audience to see if you are getting blank stares or interested looks. Adjust accordingly.
10. If there is no podium to lean on, you may find yourself on an open stage not knowing what to do with your hands and being self-conscious. A good trick in this situation is to grab a pen or a pointer. Both hands will naturally join to hold it, and you will not think about that any more.

PowerPoint presentations are subject to a fair amount of personal taste with regard to color, background, animations, and so on. However, there are some fundamental rules:

1. Do not put too much information on the slide. Four bullets, or one graph (possibly two graphs) with explanation, or a couple of photos is a maximum.
2. Graphs speak well to an engineering audience. These graphs follow the same rules as the figures of a report, thesis, or dissertation (see section 27.6). With PowerPoint, the lettering should be even larger than for figures in a printed report.
3. An audience cannot absorb tables with more than 10 numbers in them at the normal rate of presentation. Generally speaking, tables are not a good way to convey an idea or a result in PowerPoint.
4. Equations may be necessary, but should be limited in length and complexity unless the audience is well versed in that aspect of the work.
5. The use of movies is entertaining and holds the audience's attention. If you intend to use movies, make sure that they work and double-check them right before your presentation.

27.10 MEDIA INTERACTION

The media has essentially three forms: the written press, the audio press, and the video press. In all cases, the most likely interaction will be an interview, although a written communication may be involved as well. This written part may be a press release or a letter to the editor sent to newspapers. In this case, you will have time to prepare and proofread your statements. Interviews for the written press are conducted in an informal setting, often by telephone, and are less stressful than audio and video interviews. For the written press, note that saying "off the record" is best avoided, as you have no insurance that your request will be honored. Always only say what you do not mind seeing printed.

For the video press, you can have either a taped interview (that may be edited) or a live interview. While the possibility of having your statements edited may give you some level of confidence against mistakes, you should not behave differently. Remember also that you most likely will not have an opportunity to edit your statements. Video editing is very time consuming and not as easy as text editing.

In preparing for the interview, take the time to review your notes and check your appearance in a mirror. Before the interview begins, ask the reporter about the line of questioning, including typical questions, and make sure the reporter knows how to pronounce your name and affiliation correctly. Find a way to be comfortable in front of the camera; the best way to do that is to ignore it completely. Just talk to the reporter as if you were chatting with him or her at the kitchen table. Overall, speak your mind, but do not say anything that you are not very sure of. Live TV interviews are an exercise in fast thinking and right thinking. Remember that the 15-second sound bite dominates the TV market. If you are uncomfortable with a question, find a way to answer by talking about what you really wish to talk about. To minimize errors in your answer and give yourself time to think, take a second before answering to look in the distance or at the

ceiling; then start with the obvious while formulating the rest of your answer in your mind.

27.11 ETHICAL BEHAVIOR

In the end, you have to answer to yourself and the dictates of your conscience. In any decision process, you are always free to choose what is right for you. Regardless of your decision, you will also have to face the consequences of that decision. You may get by with a few lucky ones, but you may also find yourself implicated in undeserved conflicts. There are close to 9 billion people on our planet and each one thinks differently—yet everyone thinks that they are right. It often makes it smooth interaction very difficult. Nevertheless, there are reasonable guidelines governing ethical behavior.

As an engineer making a decision, remember the following:

- As engineers, we must uphold, as the highest priority, the safety of the general public within reasonable economic constraints.
- If you are unsure about something, get advice from people whom you respect and who have a proven track record.
- If at all possible, do not rush the decision.
- In the process of deciding, reverse the roles; put yourself in the other person's shoes and treat people the way you would like to be treated.

Whatever you decide after reasonable thought, remember that you have done your best and you should not feel badly about it. If the outcome is unpleasant, do not quit: Keep fighting for what you think is right until it becomes right or you run out of energy. In any case, worrying and stressing are useless (and actually harmful to your health)—but that does not mean that you should take everything lightly. So, don't stress and don't worry; just prepare, plan, and concentrate. Easy to say but hard to do!

27.12 PROFESSIONAL SOCIETIES

In your life, you have two families: your blood family and your professional family. It is important to support your professional family by belonging to your professional organization. In the United States, it is the Geo-Institute (http://www.asce.org/geo/). For the world scene it is the International Society for Soil Mechanics and Geotechnical Engineering (http://www.issmge.org/). By belonging to and being active in your professional society, you will participate in the work of technical committees, contribute to national decisions, and more generally strengthen and advance the practice of geotechnical engineering. In your work as a volunteer, you will be interacting and socializing with your peers; you will learn from them and you will teach them. You will also improve your technical communication skills, as you will naturally find yourself engaging in various types of communication. Being a member of your professional society ranks at the level of a family obligation; you should ask how you can help your professional society rather than ask what it can do for you.

27.13 RULES FOR A SUCCESSFUL CAREER

A successful career is built on a series of demonstrated successes by an individual, either alone or as part of a team. In the performance of your job, remember when you make a decision of any sort that it will take ten successes to erase one mistake from the minds of your peers. This is why it is always important to concentrate and plan. Also, before a challenging moment, remember that you may have been through similar tough moments before and done well; this recollection will give you added confidence and lower the stress.

The following "Top Ten" are some thoughts on what is important in a career. They have been inspired by discussion with many engineers over time, including Clyde Baker, and personal experiences as well:

10. Choose the relentless pursuit of excellence as a way of life.
9. Be curious. The discovery process is a fountain of youth.
8. Work hard but balance your interests (fun, family, sport, art, world news).
7. Make lots of friends. Nurture your public relations.
6. Look for solutions and not who is to blame. Leave that to the judge.
5. Be firm in your decisions, but always be fair and polite.
4. Treat others as you wish to be treated, and you will lead by example.
3. Communication is the best way to solve problems. Convince through logic and data.
2. Surround yourself with smart people and positive role models.
1. Pursue your dreams with vision and perseverance.

REFERENCES

Aas, G., S. Lacasse, T. Lunne, and K. Hoeg. 1986. "Use of In Situ Tests for Foundation Design on Clay." In *Proceedings of In Situ '86, Use of In Situ Tests in Geotechnical Engineering*, 1–30. Blacksburg, VA: ASCE GSP 6.

AASHTO (American Association of State Highway and Transportation Officials). 2007. "SI Units." In *LRFD Bridge Design Specifications*, 5th ed. Washington, DC: Transportation Research Board, Academy of Sciences.

Abdelmalak, R. 2007. "Soil Structure Interaction for Shrink-Swell Soils: A New Design Procedure for Foundation Slabs on Shrink-Swell Soils." PhD dissertation, Department of Civil Engineering, Texas A&M University.

Abramson, L. W., T. S. Lee, S. Sharma, G. M. Boyce. 1996. *Slope Stability and Stabilization Methods* (1st ed.). New York: John Wiley & Sons.

Abramson, L. W., T. S. Lee, S. Sharma, and G. M. Boyce. 2002. *Slope Stability and Stabilization Methods*. New York: John Wiley & Sons.

ADSC-DFI. 2004. *Drilled Shaft Inspector's Manual*, 2nd ed. Dallas, TX: International Association of Foundation Drilling.

Akbulut, S., and A. Saglamer. 2002. "Estimating the Groutability of Granular Soils: A New Approach." *Tunneling and Underground Space Technology* 7(4): 371–80.

Al-Alusi, H. R. 1997. "Compaction Grouting: From Practice to Theory." *Geo-Logan 1997, GSP-66: Grouting: Compaction, Remediation, and Testing*. (July): 43–54.

Alexiew, D., G. J. Horgan, and D. Brokemper. 2003. "Geotextile Encased Columns (GEC): Load Capacity and Geotextile Selection." In *BGA International Conference on Foundations: Innovations, Observations, Design and Practice* 81–90.

Alonso, E. E., A. Gens, and A. Josa. 1990. "A Constitutive Model for Partially Saturated Soils." *Géotechnique* 40(3): 405–30.

Al-Thawadi, S. M. 2008. "High Strength In-Situ Biocementation of Soil by Calcite Precipitating Locally Isolated Ureolytic Bacteria." PhD diss., Murdoch University, Western Australia.

Anderegg, R. 1997. "Nichtlineare schwingungen bei dynamischen bodenverdichtern." Diss. 12419, Eidgenössische Technische Hochschule Zürich, Schweiz, Switzerland.

Andersen, K. H., H. P. Jostad, and R. Dyvik. 2008. "Penetration Resistance of Offshore Skirted Foundations and Anchors in Dense Sand." *Journal of Geotechnical & Geoenvironmental Engineering* 134(1): 106–16.

Andrus, R. D., and K. H. Ii Stokoe. 2000. "Liquefaction Resistance of Soils from Shear-Wave Velocity." *ASCE Journal of Geotechnical & Environmental Engineering* 126(11): 1015–22.

Annandale, G. W. 1995. "Erodibility." *Journal of Hydraulic Research* 33(4): 471–94.

API (American Petroleum Institute). 2000. *Recommended Practice for Planning, Designing and Constructing Fixed Offshore Platforms: Working Stresses Design* (Recommended Practice 2A). Washington, DC: Author.

_____. 2012. *Design and Analysis of Station Keeping Systems for Floating Structures* (Recommended Practice 2SK). Washington, DC: Author.

ARA-ERES. 2000. "Appendix DD-1: Resilient Modulus as Function of Soil Moisture—Summary of Predictive Models." In *Guide for Mechanistic-Empirical Design of New and Rehabilitated Pavement Structures* (NCHRP Report on Project 1-37a). Washington, DCc: Transportation Research Board.

ASCE (American Society of Civil Engineers) 7. 2011. *Minimum Design Loads for Buildings and Other Structures*. Washington, DC: Author.

ASTM (American Society for Testing and Materials). 1995. *Standard Guide for Risk-Based Corrective Action Applied at Petroleum Release Sites* (E1739). West Conshohoken, PA: Author.

ASTM (American Society for Testing and Materials) International. 2013. "Section 04 Construction." In *Annual Book of ASTM Standards*. Philadelphia, PA: ASTM.

Athanasopoulos, C. A. 2007. "Reducing the Seismic Earth Pressures on Retaining Walls by EPS Geofoam Buffers: Numerical Parametric Analyses." In *Proceedings of Geosynthetics 2007*. Washington, DC: IFAI Publication.

Aubeny, C. P., and R. L. Lytton. 2004. "Shallow Slides in High Plasticity Clay Slopes." *ASCE Journal of Geotechnical & Environmental Engineering*130(7): 717–27.

Austin, D. N., K. J. Wu, and D. F. White. 1993. "The Influence of Test Parameters and Procedures on the Tensile Modulus of Stiff Geogrids." In *Geosynthetic Soil Reinforcement Testing Procedures* (ASTM STP 1190). Philadelphia, PA: ASTM.

Australian Standard. 1996. *Residential Slabs and Footings* (AS 2870). Sydney, Australia: Standard House.

Averjanov, S. F. 1950. "About Permeability of Subsurface Soils in Case of Incomplete Saturation." *Engineering College* vol. 7.

Baguelin, F., R. Frank, and J.-F. Jezequel. 1982. "Parameters for Friction Piles in Marine Soils." Second International Conference in Numerical Methods in Offshore Piling, Department of Civil Engineering, University of Texas, Austin, TX.

Baguelin, F., J.-F. Jezequel, and D. H. Shields. 1978. *The Pressuremeter and Foundation Engineering*. Clausthal-Zellerfeld, West Germany: Trans Tech Publications.

Baker, C. N., E. E. Drumright, J.-L. Briaud, F. D. Mensah, and G. Parikh. 1993. *Drilled Shafts for Bridge Foundations* (Publication No. FHWA-RD-92-004). Washington, DC: Federal Highway Administration.

Baligh, M. M., V. Vivatrat, and C. C. Ladd. 1980. "Cone Penetration in Soil Profiling." *Journal of the Geotechnical Engineering Division* 106(4): 447–61.

Barker, R. M., J. M. Duncan, K. B. Rojiani, P. S. K. Ooi, C. K. Tan, and S. G. Kim. 1991. *Manuals for the Design of Bridge Foundations* (NCHRP Report 343). Washington, DC: Transportation Research Board, National Research Council.

Barron, R. A. 1948. "Consolidation of Fine-Grained Soils by Drain Wells." *Transactions ASCE* 113 (Paper 2346): 718–24.

Bath, M. 1966. "Earthquake Energy and Magnitude." In *Physics and Chemistry of the Earth*, vol. 7, edited by L. H. Ahrens, F. Press, S. K. Runcorn, and H. C. Urey, 117–65. New York: Pergamon Press.

Bell, F. G. 2000. *Engineering Properties of Soils and Rocks* (4th ed.).,Malden, MA: Blackwell Science Inc.

_____. 2007. *Engineering Geology*, 2nd ed. Burlington, UK: Elsevier.

Bermingham, P., and M. Janes. 1989. "An Innovative Approach to Load Testing of High Capacity Piles." In *Proceedings of the International Conference on Piling and Deep Foundations*, vol. 1, 409–13. Rotterdam: A.A. Balkema Publishers.

Bieniawski, Z. T. 1989. *Engineering Rock Mass Classifications*. New York: John Wiley & Sons.

Bishop, A. W. 1955. "The Use of the Slip Circle in the Stability Analysis of Slopes." *Géotechnique* 5(1): 7–17.

Bishop, A. W., and D. J. Henkel. 1962. *The Measurement of Soil Properties in the Triaxial Test*, 2nd ed. London: Edward Arnold.

Bjerrum, L. 1972. "Embankments on Soft Ground." In *Proceedings of the ASCE Conference on Performance of Earth-Supported Structures*, vol. 2, 1–54. Purdue University.

Bjerrum, L., and A. Landva. 1966. "Direct Simple-Shear Test on a Norwegian Quick Clay." *Géotechnique* 16(1): 1–20.

Blight, G. E. 1971. "Flow of Air through Soils." *ASCE Journal of the Soil Mechanics and Foundations Division* 97 (Sm4): 607–24.

Bobet, A. 2010. "Numerical Methods in Geomechanics." *Arabian Journal for Science & Engineering* 35 (1B 2).

Bolduc, L. C., P. Gardoni, and J.-L. Briaud. 2008, February. "Probability of Exceedance Estimates for Scour Depth around Bridge Piers." *Journal of Geotechnical & Geoenvironmental Engineering* 134(2):.

Bollaert, E. 2002. *Transient Water Pressures in Joints and Formation of Rock Scour Due to High-Velocity Jet Impact* (Communication No. 13). Lausanne, Switzerland: Laboratory of Hydraulic Constructions, École Polytechnique Federale de Lausanne.

Bonaparte, R., D. E. Daniel, and R. M. Koerner. 2002. Assessment and Recommendations for Improving the Performance of Waste Containment Systems (EPA/600/R-02/099). Washington, DC: U.S. Environmental Protection Agency.

Boore, D. M., W. B. Joyner, and T. E. Fumal. 1997. "Equations for Estimating Horizontal Response Spectra and Peak Acceleration from Western North American Earthquakes: A Summary of Recent Work." *Seismic Research Letters* 68: 128–53.

Boutwell, G. P., and R. K. Derick. 1986. *Groundwater Protection for Sanitary Landfills in the Saturated Zone*. Paper presented at Waste Tech '86, National Solid Waste Management Association, Chicago, IL.

Bowders, J. J., and D. E. Daniel. 1987. "Hydraulic Conductivity of Compacted Clay to Dilute Organic Chemicals." *Journal of Geotechnical Engineering* 113(12): 1432–48.

Bowles, J. E. 1996. *Foundation Analysis and Design*. New York: McGraw-Hill.

Brandimarte, L., A. Montanari, J.-L. Briaud, and P. D'Odorico. 2006. "Stochastic Flow Analysis for Predicting Scour of Cohesive Soils." *Journal of Hydraulic Engineering* 132(5):

Bray, J. D., R. B. Sancio, L. F. Youd, C. Christensen, O. Cetin, A. Onalp, T. Durgunoglu, J. P. C. Stewart, R. B. Seed, M. B. Baturay, T. Karadayilar, and C. Oge. 2001, February. "Documenting Incidents of Ground Failure Resulting from the August 17, 1999 Kocaeli, Turkey Earthquake." Available at Pacific Earthquake Engineering Research Center, http://peer.berkeley.edu/turkey/adapazari/index.html.

Brebbia, C. A., J. C. F. Telles, and L. C. Wrobel. 1984. *Boundary Element Techniques*. Berlin, Germany: Springer-Verlag.

Briaud, J.-L. 1992. *The Pressuremeter*. Oxford: Taylor and Francis.

_____. 1997. "SALLOP: Simple Approach for Lateral Loads on Piles." *Journal of Geotechnical & Geoenvironmental Engineering* 123(10): 958–64.

_____. 2006a. "Bridge Scour." *Geotechnical News* 24 (3, September):

_____. 2006b. *Erosion Tests on New Orleans Levee Samples* (Internal Report). College Station, TX: Zachry Department of Civil Engineering, Texas A&M University.

_____. 2007. "Spread Footings in Sand: Load Settlement Curve Approach." *Journal of Geotechnical & Geoenvironmental Engineering* 133 (8, August).

_____. 2008. "Case Histories in Soil and Rock Erosion: Woodrow Wilson Bridge, Brazos River Meander, Normandy Cliffs, and New Orleans Levees" (The 9th Ralph B. Peck Lecture). *Journal of Geotechnical & Geoenvironmental Engineering* 134(10).

_____. 2012. *Summary Report*. Available at http://ceprofs.tamu.edu/briaud/, under "research," then under "scour."

Briaud, J.-L., R. Abdelmalak, and X. Zhang. 2010. "Design of Stiffened Slabs on Grade on Shrink Swell Soils" (Keynote Lecture). In *Proceedings of the International Conference on Unsaturated Soils [UNSAT 2010]*, Barcelona, Spain. : CRC Press-Balkema-Taylor and Francis Group.

Briaud, J.-L., M. Ballouz, and G. Nasr. 2000. "Static Capacity Prediction by Dynamic Methods for Three Bored Piles." *Journal of Geotechnical & Geoenvironmental Engineering* 126(7): 640–49.

_____. 2002. "Defect and Length Prediction by NDT Methods for Nine Bored Piles." In *Proceedings of the International Deep Foundation Congress* (GSP 116), Orlando, Florida, February 14-16, 2002.

Briaud, J.-L., M. Bernhardt, and M. Leclair. 2011. "The Pocket Erodometer: A Simple Device to Estimate Soil Erodibility." *ASTM Geotechnical Testing Journal*.

_____. 2012. "The Pocket Erodometer Test: Development and Preliminary Results." *ASTM Geotechnical Testing Journal*, March 2012, ASTM

Briaud, J.-L., L. Brandimarte, J. Wang, and P. D'Odorico. 2007. "Probability of Scour Depth Exceedance due to Hydrologic Uncertainty." *Georisk Journal for Assessment and Management of Risk for Engineered Systems and Geohazards* 1 (2, March): 77–88.

Briaud, J.-L., and A. Chaouch. 1997. "Hydrate Melting around Hot Conductor." *Journal of Geotechnical & Geoenvironmental Engineering* 123(7): 645–53.

Briaud, J.-L., H.-C. Chen, K.-A. Chang, Y.-A. Chung, N. Park, W. Wang, and P.-H. Yeh. 2007. *Establish Guidance for Soil Properties-Based Prediction of Meander Migration Rate* (Report FHWA/TX-07/0-4378-1). College Station, TX: Zachry Department of Civil Engineering, Texas A&M University.

Briaud, J.-L., H.-C. Chen, P. Nurtjahyo, and J. Wang. 2004. *Pier and Contraction Scour in Cohesive Soils* (NCHRP Report 516). Washington, DC: Transportation Research Board.

Briaud, J.-L., H. C. Chen, K. Kwak, S.-W. Han, and F. Ting, 2001. "Multiflood and Multilayer Method for Scour Rate Prediction at Bridge Piers." *Journal of Geotechnical & Geoenvironmental Engineering* 127(2): 114–23.

Briaud, J.-L., H. M. Coyle, and H. M. Tucker. 1990. "Axial Response of 3 Vibratory and 3 Impact Driven H Piles in Sand." *Transportation Research Record* (January).

Briaud, J.-L., and G. Y. Felio. 1986. "Cyclic Axial Loads on Piles: Analysis of Existing Data." *Canadian Geotechnical Journal* 23(3): 362–71.

Briaud, J.-L., and E. E. Garland. 1985. "Loading Rate Method for Pile Response in Clay." *Geotechnical Engineering Journal* 111(3): 319–35.

Briaud, J.-L., and R. M. Gibbens. (1994). "Test and Prediction Results for Five Large Spread Footings on Sand." In *FHWA Prediction Symposium* (ASCE Geotechnology Special Publication No. 41), 92–128. New York: ASCE.

_____. 1999. "Behavior of Five Large Spread Footings in Sand." *Journal of Geotechnical & Geoenvironmental Engineering* 125(9): 787–97.

Briaud, J.-L., L. G. Huff, L. M. Tucker, and H. M. Coyle. 1984. *Evaluation of In Situ Test Design Methods for Vertically Loaded H Piles at Lock & Dam No. 26 Replacement Site* (Research Report 4690, USAE, Waterways Experiment Station).

Briaud, J.-L., and N. K. Kim. 1998. "Beam Column Method for Tieback Walls." *Journal of Geotechnical & Geoenvironmental Engineering* 124(1): 67–69

Briaud, J.-L., Y. Li, and K. Rhee. 2006. "BCD: A Soil Modulus Device for Compaction Control." *Journal of Geotechnical & Geoenvironmental Engineering* 132(1): 108–115.

Briaud, J.-L., and Y. Lim. 1997. "Soil-Nailed Wall under Piled Bridge Abutment: Simulation and Guidelines." *Journal of Geotechnical & Geoenvironmental Engineering* 123(11): 1043–50.

_____. 1999. "Tieback Walls in Sand: Numerical Simulation and Design Implications." *Journal of Geotechnical & Geoenvironmental Engineering* 125(2): 101–10.

Briaud, J.-L., M. Meriwether, and H. Porwol. 1983. *Pressuremeter Design of Retaining Walls* (Texas Department of Transportation Report No. FHWA/TX-84/54+340-4f).

Briaud, J.-L., and J. Miran. 1992a. *The Cone Penetrometer Test* (Publication No. FHWA-SA-91-043). Washington, DC: U.S. Department of Transportation.

_____. 1992b. *The Flat Dilatometer Test* (Publication No. FHWA-SA-91-044). Washington, DC: Federal Highway Administration, Office of Technology Applications.

Briaud, J.-L., J. Nicks, K. Rhee, and G. Stieben. 2007. "The San Jacinto Monument Case History." *Journal of Geotechnical & Geoenvironmental Engineering* 133(11): 1337–51.

Briaud, J.-L., and C. J. Rutherford. 2010. "Excavation Support Using Deep Mixing Technology" (Keynote Lecture). In *Proceedings of the International Geotechnical Conference on Geotechnical Challenges in Megacities, Moscow, Russia, 7–10 June 2010*. St Petersburg, Russia: GRF.

Briaud, J.-L., B. Smith, K.-Y. Rhee, H. Lacy, and J. Nicks. 2009. "The Washington Monument Case History." *International Journal of Geoengineering Case Histories* 1(3): 170–88. Available at http://casehistories.geoengineer.org

Briaud, J.-L., F. Ting, H. C. Chen, Y. Cao, S.-W. Han, and K. Kwak. 2001. "Erosion Function Apparatus for Scour Rate Predictions." *Journal of Geotechnical & Geoenvironmental Engineering* 127(2): 105–13.

Briaud, J.-L., F. C. K. Ting, H. C. Chen, R. Gudavalli, S. Perugu, and G. Wei. 1999. "SRICOS: Prediction of Scour Rate in Cohesive Soils at Bridge Piers." *Journal of Geotechnical & Geoenvironmental Engineering* 123(4): 237–46.

Briaud J.-L., and L. M. Tucker. 1984a. "Piles in Sand: A Method Including Residual Stresses." *Journal of Geotechnical Engineering* 110(11): 1666–1680.

_____. 1984b. "Residual Stresses in Piles and the Wave Equation." Presented at ASCE Symposium on Deep Foundations, San Francisco, October.

_____. 1987. "Horizontally Loaded Piles Next to a Trench." In *Proceedings of an ASCE Session on Foundations for Transmission Towers*, Atlantic City, NJ, April 1987.

_____. 1988. "Measured and Predicted Axial Response of 98 Piles." *Journal of Geotechnical Engineering* 114(9): 984–1001.

_____. 1997. *Design and Construction Guidelines for Downdrag on Uncoated and Bitumen-Coated Piles* (NCHRP Report 393). Washington, DC: Transportation Research Board, National Academy Press.

Briaud, J.-L., L. M. Tucker, R. L. Lytton, and H. M. Coyle. (1985). *Behavior of Piles and Pile Groups in Cohesionless Soils* (FHWA-IRD-831038). Washington, DC: Federal Highway Administration.

Briaud, J.-L., L. M. Tucker, and E. Ng. 1989. "Axially Loaded 5 Pile Group and Single Pile in Sand." In *Proceedings of the 12th International Conference on Soil Mechanics and Foundation Engineering* (Rio de Janeiro), vol. 2, 1121–24. Rotterdam: A. A. Balkema.

Briaud, J.-L., B. Smith, K.-Y. Rhee, H. Lacy, and J. Nicks. 2009. "The Washington Monument Case History." *International Journal of Geoengineering Case Histories* 1(3): 170–188. Available at http://casehistories.geoengineer.org.

Briaud J.-L., X. Zhang, and S. Moon. 2003. "Shrink Test: Water Content Method for Shrink and Swell Predictions." *Journal of Geotechnical & Geoenvironmental Engineering* 129(7): 590–600.

Brice, J. C. 1974. "Evolution of Meander Loops." *Geological Society of America Bulletin* 85: 581–86.

Brooks, R. H., and A. T. Corey. 1964. *Hydraulic Properties of Porous Media* (Hydrology Paper 3). Fort Collins, CO: Colorado State University.

Brown, D. A., J. P. Turner, and R. J. Castelli. 2010. *Drilled Shafts: Construction Procedures and LRFD Design Methods* (NHI Course 132014, Geotechnical Engineering Circular no. 10, Federal Highway Administration Report no. FHWA NHI-10-016). Washington, DC:

Brunner, G. W. 2002. *HEC-RAS River Analysis System Hydraulic Reference Manual* (Version 3.1) (Report No. CPD-69). Davis, CA: U.S. Army Corps of Engineers, Institute for Water Resources, Hydrologic Engineering Research Center.

Building Research Advisory Board (BRAB). 1968. *National Research Council Criteria for Selection and Design of Residential Slabs-on-Ground* (U.S. National Academy of Sciences Publication 1571). Washington, DC: National Academy Press.

Bulut, R. 2003. "Measurement of Soil Suction." Presentation at the Foundation Performance Association, Houston, TX, August 20.

Bulut, R., S. M. Hineidi, and B. Bailey. 2002. "Suction Measurements: Filter Paper and Chilled Mirror Psychrometer." In *Proceedings of the Texas ASCE 2002 Fall Meeting*, Waco, TX.

Bulut, R., R. L. Lytton, and W. K. Wray. 2001. "Soil Suction Measurements by Filter Paper." In *Expansive Clay Soils and Vegetative Influence on Shallow Foundations* (ASCE Geotechnical Special Publication no. 115), 243–261.

Bulut, R., and W. K. Wray. (2005). "Free Energy of Water—Suction—in Filter Papers." *ASTM Geotechnical Testing Journal* 28(4): 355–64.

Burenkova, V. V. 1993. "Assessment of Suffusion in Noncohesive and Graded Soils." In *Proceedings of the First International Conference on Geo-Filters, Karlsruhe, Germany*, 357–60. Rotterdam: Balkema.

Burns, S. E., and P. W. Mayne. 1998. "Monotonic and Dilatory Pore Pressure Decay during Piezocone Tests." *Canadian Geotechnical Journal* 35(6): 1063–73.

Byrne, R. J., D. Cotton, J. Porterfield, C. Wolschlag, and G. Ueblacker. 1998. *Manual for Design and Construction Monitoring of Soil Nail Walls* (Report FHWA-SA-96-69R). Washington, DC: Federal Highway Administration.

CALTRANS. 1991. *A User's Manual for the SNAIL Program, Version 2.02—Updated PC Version*. Sacramento, CA: California Department of Transportation, Division of New Technology, Material and Research, Office of Geotechnical Engineering.

Canadian Foundation Engineering Manual (4th ed.). 2007. Richmond, BC: Canadian Geotechnical Society/BiTech.

Cao, Y., J. Wang, J.-L. Briaud, H. C. Chen, Y. Li, and P. Nurtjahyo. 2002. "EFA Tests and the Influence of Various Factors on the Erodibility of Cohesive Soils." In *Proceedings of the First International Conference on Scour of Foundations*. College Station, TX: Texas A&M University, Department of Civil Engineering.

Cardimona, S. 1993. *Electrical Resistivity Techniques for Subsurface Investigation* (Internal Document). Rolla, MO: Department of Geology and Geophysics, University of Missouri-Rolla.

Carman, P. C. 1938. "The Determination of the Specific Surface of Powders." *Journal of the Society for Chemical Industrial Transport* 57: 225.

Carrier, D. W., III. 2003. "Goodbye, Hazen; Hello, Kozeny-Carman." *Journal of Geotechnical & Geoenvironmental Engineering* 129(11): 1054–1056.

Carrier, W. D., and J. F. Beckman. 1984. "Correlations between Index Tests and the Properties of Remolded Clays." *Géotechnique* 34(2): 211–28.

Carroll, R. G., Jr. 1983. "Geotextile Filter Criteria." In TRR 916, *Engineering Fabrics in Transportation Construction*, 46–63. Washington, DC.

Carslaw, H. S., and J. C. Jaeger. 1947. *Conduction of Heat in Solids*, 2nd ed. Oxford, UK: Oxford University Press.

Cassagrande, A. 1938. "Notes on Soil Mechanics—First Semester." Cambridge, MA: Harvard University (unpublished).

Cazzuffi, D., and S. Venesia. 1986. "The Mechanical Properties of Geotextiles: Italian Standard and Interlaboratory Test Comparison." In *Proceeding of the Third Conference on Geotextiles*, IFAI.

Cedergren, H. 1967. *Seepage, Drainage, and Flownets*. New York: McGraw-Hill.

Chapuis, R. P., and T. Gatien. 1986. "An Improved Rotating Cylinder Technique for Quantitative Measurements of the Scour Resistance of Clays." *Canadian Geotechnology Journal* 23: 83–87.

Chen, H.-C. 2002. "Numerical Simulation of Scour around Complex Piers in Cohesive Soil." In *Proceedings of First International Conference on Scour of Foundations* (November 17-20, 2002, College Station, TX), 14–33.

Chen H.-C., V. C. Patel, and S. Ju. 1990. "Solution of Reynolds-Averaged Navier Stokes Equations for Three-Dimensional Incompressible Flows." *Journal of Computational Physics* 88(2): 305–36.

Chen, T., and Y. Fang. 2008. "Earth Pressure due to Vibratory Compaction." *Journal of Geotechnical & Geoenvironmental Engineering* 134(4): 437–44.

Chow, V. T., D. R. Maidment, and L. W. Mays. 1988. *Applied Hydrology*. New York: McGraw-Hill.

Christian, J. T., and W. D. Carrier III. 1978. "Janbu, Bjerrum, and Kjaernsli's Chart Reinterpreted." *Canadian Geotechnical Journal* 15: 127.

Chu, J., S. Varaksin, U. Klotz, and P. Menge. 2009. "Construction Processes." In *Proceedings of the International Conference on Soil Mechanics and Geotechnical Engineering*, Alexandria, Egypt: IOS Press.

Chua, K. M., S. Gardner, and L. L. Lowery. 1987. "Wave Equation Analysis of a Vibratory Hammer-Driven Pile." Presented at Offshore Technology Conference, 27-30 April 1987, Houston, TX.

Clayton, C. R. I., N. E. Simons, and M. C. Matthews. 1982. *Site Investigation: A Handbook for Engineers*. Halsted Press.—a Division of John Wiley & Sons Inc.

Clough, G. W., and T. D. O'Rourke. 1990. *Construction Induced Movements of In Situ Walls: Design and Performance of Retaining Structures* (Geotechnical Special Publications No. GSP 25), 439–70. : ASCE.

Clough, R. W. 1960. "The Finite Element in Plane Stress Analysis." In *Proceedings of the Second ASCE Conference on Electronic Computation*, Pittsburgh, PA, 345–78.

Cornell, C. A., H. Banon, and A. F. Shakal. 1979. "Seismic Motion and Response Prediction Alternatives." *Earthquake Engineering & Structural Dynamics* 7(4): 295–315.

Couroyer, C. 2000. *Attrition of Alumina Catalyst Carrier Beads*. PhD thesis, University of Surrey, Guildford, Surrey, UK. Available at www.engineering.leeds.ac.uk/ipse.old/research/ghadiri/ghadiri009.shtml

Cox, W. R., D. A. Dixon, and B. S. Murphy. 1983. *Lateral Load Tests on 25.4 mm Diameter Piles in Very Soft Clay in Side by Side and In Line Groups* (ASTM Special Technical Publication no. STP 835), 122–40.

Coyle, H. M., R. E. Bartoskewitz, and W. J. Berger. 1973. *Bearing Capacity by Wave Equation Analysis—State of the Art*. College Stations, TX: Texas Transportation Institute, Texas A&M University.

Coyle, H. M., and G. C. Gibbson. 1970. "Empirical Damping Constants for Sand and Clay." *Journal of the Soil Mechanics & Foundation Division, ASCE . .*

Crank, J. 1956. *The Mathematics of Diffusion*. New York: Oxford University Press.

Croad, R. N. 1981. *Physics of Erosion of Cohesive Soils*. PhD thesis, Department of Civil Engineering, University of Auckland, Auckland, New Zealand.

Crouch, S. L., and A. M. Starfield, 1983. *Boundary Element Methods in Solid Mechanics*. London: Allen and Unwin.

Cundall, P. A., and O. D. L. Strack. 1979. "A Discrete Numerical Model for Granular Assemblies." *Géotechnique* 29(1): 47–65.

Dalrymple, G. B. 1994. *The Age of the Earth*. Palo Alto, CA: Stanford University Press.

Daniel, D. E. 1989. "In Situ Hydraulic Conductivity Tests for Compacted Clay." *Journal of Geotechnical Engineering* 115(9): 1205–1226.

Daniel, D. E., J. J. Bowders, and R. B. Gilbert. 1997. "Laboratory Hydraulic Conductivity Testing of GCLs in Flexible Wall Permeameters." In *Testing and Acceptance Criteria for Geosynthetic Clay Liners* (ASTM STP 1308), edited by Larry W. Well, 208–28. ASTM.

Davis, R. O., and A. P. S. Selvadurai. 2002. *Plasticity in Geomechanics*. Cambridge, UK: Cambridge University Press.

De Alba, P., H. B. Seed, and C. K. Chan. 1976. "Sand Liquefaction in Large Scale Simple Shear Tests." *Journal of the Geotechnical Engineering Division, ASCE* 102 (GT9): 909–27.

Dejong, J., and R. W. Boulanger. 2000. *An Introduction to Drilling and Sampling in Geotechnical Practice*—2nd ed. [instructional video]. Department of Civil and Environmental Engineering, University of California-Davis.

De Moor, J. J., R. T. Van Balen, and C. Kasse. 2007. "Simulating Meander Evolution of the Geul River (the Netherlands) Using a Topographic Steering Model." *Earth Surface Processes and Landforms* 32(7): 1077–1093.

Den Hoedt, G. 1986. "Creep and Relaxation of Geotextile Fabrics." *Journal of Geotextiles & Geomembranes* 4(2): 83–92.

Denk, E. W., W. A. Dunlap, W. R. Bryant, L. J. Milberger, and T. J. Whelan. 1981. "A Pressurized Core Barrel for Sampling Gas Charged Sediments." In *Proceedings of the 13th Offshore Technology Conference*, 43–52, Richardson, TX

Desai, C. S., and J. F. Abel. 1972. *Introduction to the Finite Element Method: A Numerical Method for Engineering Analysis* (10th reprint). New York: Van Nostrand Reinhold.

De Wilde, P. 1996. *Neural Network Models: An Analysis*. London, UK: Springer-Verlag.

Donald, I. B. 1961. *The Mechanical Properties of Saturated and Partly Saturated Soils with Special Reference to Negative Pore Water Pressure*. PhD diss., University of London.

Drucker, D. C., and W. Prager. 1952. "Soil Mechanics and Plastic Analysis or Limit Design." Quarterly Journal of Applied Mathematics 10(2): 157–65.

Duncan, J. M., and C. Y. Chang. 1970. "Non-Linear Analysis of Stress and Strain in Soils." *Journal of Soil Mechanics* 96 (Sm5): 1629–53.

Duncan, J. M., and R. B. Seed. 1986. "Compaction-Induced Earth Pressures under Ko-Conditions." *Journal of Geotechnical Engineering* 112(1): 1–21.

Duncan, J. M., and S. P. Wright. 2005. *Soil Strength and Slope Stability*. Hoboken, NJ: John Wiley & Sons.

Elias, V., and I. Juran. 1991. *Soil Nailing for Stabilization of Highway Slopes and Excavations* (Publication FHWA-RD-89-198). Washington, DC: Federal Highway Administration.

Elias, V., J. Welsh, J. Warren, R. Lukas, J. G. Collin, and R. R. Berg. 2006. *Ground Improvement Methods* (NHI-06-020). Washington, DC: Federal Highway Administration.

Elton, D. J. 1999. *Soils Magic* (Geotechnical Special Publication no. 114). Reston, VA: American Society of Civil Engineers.

Escario, V., J. Juca, and M. S. Coppe. 1989. "Strength and Deformation of Partly Saturated Soils." In *Proceedings of the 12th International Conference on Soil Mechanics and Foundation Engineering, Rio de Janeiro*, vol. 3, 43–46.

Eurocode 7. 2011. *Geotechnical Design. Part 1: General Rules, Part 2: Ground Investigation and Testing* (European Standard EN-1997). Brussels, Belgium: European Committee for Standardization.

European Committee for Standardization. 2004. *Geotechnical Investigation and Testing—Laboratory Testing of Soil Part 6: Fall Cone Test (ISO/TS 17892-6:2004)*. Paris, France: Author.

Fang, H. Y., ed. 1991. *Foundation Engineering Handbook*. New York: Van Nostrand Reinhold.

Fascicule 62. 1993. *Regles techniques de conception et de calcul des foundations des ouvrages de genie civil* (Ministere de l'equipement, du logement, et des transports). Paris, France: Publications Eyrolles.

Feda, J. 1966. "Structural Stability of Subsident Loess Soil from Prahadejuice." *Engineering Geology* 1: 201–19. [Available from Elsevier Science Publishers B. V., Box 211, 1000 AE, Amsterdam, The Netherlands.]

Federal Highway Administration. XXXX *See* FHWA.

Fell, R., and J.-J. Fry. 2005. "The State of the Art of Assessing the Likelihood of Internal Erosion of Embankment Dams, Water Retaining Structures and Their Foundations." In *Internal Erosion of Dams and their Foundations*, edited by R. Fell and J.-J. Fry,. London, UK: Taylor and Francis Group.

Fell, R., P. MacGregor, D. Stapledon, and G. Bell. 2005. *Geotechnical Engineering of Dams*. Leiden, Germany: Balkema.

Fell, R., and C. F. Wan. 2005. "Investigation of Internal Erosion by the Process of Suffusion in Embankment Dams and Their Foundations." In *Internal Erosion of Dams and their Foundations*, edited by R. Fell and J.-J. Fry, 219–34. London, UK: Taylor and Francis Group.

Fellenius, W. 1927. *Erdstatische berechnungen mit reibung and kohaesion*. Berlin: Ernst.

FERUM. 2001. "Finite Element Reliability Using Matlab." Available at http://www.ce.berkeley.edu/projects/ferum/index.html.

Fetter, C. W. 1992. *Contaminant Hydrogeology*. New York: Macmillan.

FHWA (Federal Highway Administration). 1984. *Permanent Ground Anchors* (Report FHWA-DP-68-1R). Washington, DC: Author.

_____. 1998. *Geotechnical Earthquake Engineering* (FHWA HI-99-012). Washington, DC: Author.

_____. 2006. "Geotechnical Inputs for Pavement Design." In *Geotechnical Aspects of Pavements Reference Manual* (U.S. Department of Transportation Publication No. FHWA NHI-05-037, NHI Course No. 132040), ch. 5. Washington, DC: Author.

_____. 2010. *Drilled Shaft Manual* (Publication No. FHWA-NHI-10-016). Washington, DC: Author.

Filz, G., J. Sloan, M. P. McGuire, J. Collin, and M. Smith. 2012. "Column-Supported Embankments: Settlement and Load Transfer." In *Geotechnical Engineering State of the Art and Practice: Proceedings of the Geo-Institute Congress*, Oakland, CA.

Finno, R. J., and S. L. Gassman. 1998. "Impulse Response Evaluation of Drilled Shafts." *Journal of Geotechnical & Geoenvironmental Engineering* 124(10): 965–75.

Floss, R., and H.-J. Kloubert. 2000. "Newest Innovations in Soil and Asphalt Compaction Technology." Presented at International Workshop on Compaction of Soil, Granulates and Powders, Innsbruck, February 2000. Rotterdam: A.A. Balkema-Verlag.

Focht, J. A., Jr., F. R. Khan, and J. P. Gemeinhardt. 1978. "Performance of One Shell Plaza Deep Mat Foundation." *Journal of the Geotechnical Engineering Division* 104 (GT5): 593–608.

Foster, M., R. Fell, and M. Spannagle. 2000a. "The Statistics of Embankment Dam Failures and Accidents." *Canadian Geotechnical Journal* 37(5): 1000–24.

_____. 2000b. "A Method for Assessing the Relative Likelihood of Failure of Embankment Dams by Piping." *Canadian Geotechnical Journal* 37(5), 1025–61.

Foster, R. J. 1988. *Physical Geology*, 5th ed. New York: Prentice Hall.

Fourier, J. B. 1822. *Théorie Analytique de la Chaleur*. Translated by A. Freeman. New York: Dover.

Fredlund, D. G., and H. Rahardjo. 1993. *Soil Mechanics for Unsaturated Soils*. New York: John Wiley & Sons.

Fredlund, D. G., and Xing Anqing. 1994. "Equations for the Soil-Water Characteristic Curve." *Canadian Geotechnical Journal* 31: 521–32.

Fredlund, D. G., A. Xing, and S. Huang. 1994. "Predicting the Permeability Function for Unsaturated Soils Using the Soil-Water Characteristic Curve." *Canadian Geotechnical Journal* 31: 533–46.

Gardner, W. R. 1958. "Some Steady State Solutions of the Unsaturated Moisture Flow Equation with Application to Evaporation from a Water Table." *Soil Science* 85(4): 228–232..

Garner, D. 2002. "*Comparison of Suction and Water Content Determination.*" Personal communication.

Ghaboussi, J., and R. Barbosa. 1990. "Three-Dimensional Discrete Element Method for Granular Materials." *International Journal of Numerical & Analytical Methods in Geomechanics* 14: 451–472.

Gibbs, H. J., and Bara, J. P. 1962. *Predicting Surface Subsidence from Basic Soil Tests* (Field Testing of Soils, Special Technical Publication No. 322), 231–247. Available from American Society for Testing and Materials, 1916 Race Street, Philadelphia, PA 19103.

Giroud, J.-P. 1981. "Designing with Geotextiles." *Materiaux et Construction (Paris)* 14(82): 257–252.

_____. 2010. "Development of Criteria for Geotextiles and Granular Filters" (Prestigious Lecture). In *Proceedings of the 9th International Conference on Geosynthetics, Guarujá, Brazil* 1, 45–64.

Giroud, J.-P., and R. Bonaparte. 1989. "Leakage through Liners Constructed with Geomembranes: Geomembrane Liners." *Geotextiles & Geomembranes* 8(1):.

Goble, G. G., F. Rausche, and G. E. Likins. 1993. *Case Pile Wave Analysis Program—CAPWAP Manual*. Cleveland, OH: GRL.

Golder Associates. 1993. *GOLDNAIL Soil Nailing Design Program*. Seattle, WA: Author.

Goodman, R. E. 1989. *Rock Mechanics*, 2nd ed. New York: John Wiley & Sons.

Griffiths, D. V., and P. A. Lane. 1999. "Slope Stability Analysis by Finite Elements." *Géotechnique* 49(3): 387–403.

GRLWEAP. 2012. *Goble Rausche Likins Wave Equation Analysis Program—User's Manual*. Cleveland, OH:Pile Dynamics Inc.

Gutenberg, B., and C. F. Richter. 1944. "Frequency of Earthquakes in California." *Bulletin of the Seismological Society of America* 34(4): 1985–1988.

Haberfield, C. 2010. "Engineering the Foundations for the Nakheel Tower." In *Piling and Deep Foundations Middle East 2010*. Dubai, UAE: Deep Foundation Institute.

Hall, W. J., and N. M. Newmark. 1977. "Seismic Design Criteria for Pipelines and Facilities." In *The Current State of Knowledge of Lifeline Earthquake Engineering, Proceedings of the ASCE*, 18–34. New York: ASCE.

Handy, R. L. 1975. "Measurement of In-Situ Shear Strength." In *Proceedings of the Conference on In Situ Measurement of Soil Properties* 2, 143–149. New York: ASCE.

_____. 1986. "Borehole Shear Test and Slope Stability." In *Use of In Situ Tests in Geotechnical Engineering*, edited by Samuel P. Clemence, 161–175. New York: ASCE.

Hannigan, P. J. 1990. "Dynamic Monitoring and Analysis of Pile Foundation Installations." In Deep Foundations Institute short-course text, 1st ed., 69.

Hannigan, P. J., G. G. Goble, G. Thendean, G. E. Likins, and F. Rausche. 1998. *Design and Construction of Driven Pile Foundations*, Vols. 1 and 2 (FHWA Publication HI 97-014, NHI course nos 13221 and 13222). Washington DC: FHWA.

Hansbo, S. 1981. "Consolidation of Fine-Grained Soils by Prefabricated Drains." In *Proceedings of the 10th International Conference on Soil Mechanics and Foundation Engineering*, Vol. 3, Stockholm:

Hanson, G. J. (1991). "Development of a Jet Index to Characterize Erosion Resistance of Soils in Earthen Spillways." *Transactions of ASAE* 34(5): 2015–2020.

Hardin, B. O. (1978) "The Nature of Stress-Strain Behavior for Soils." *In Proceedings of the ASCE Geotechnical Engineering Division Specialty Conference on Earthquake Engineering and Soil Dynamics*, Vol. 1, 3–89.

Hardin, B. O., and V. P. Drnevich. 1972. "Shear Modulus and Damping in Soils: Design Equations and Curves." *Journal of Soil Mechanics, Foundations Division* 98 (SM7): 667–692.

Harland, W. B., A. V. Cox, P. G. Llewellyn, C. A. Pickton, A. G. Smith, and R. Walters. 1990. *A Geologic Time Scale*. New York: Cambridge University Press.

Hart, R., P. A. Cundall, and J. Lemos. 1998. "Formulation of a Three-Dimensional Distinct Element Model—Part II. Mechanical Calculations for Motion and Interaction of a System Composed of Many Polyhedral Blocks." *International Journal of Rock Mechanics & Mining Sciences* 25(3): 117–125.

Hazen, A., 1892. *Some Physical Properties of Sands and Gravels, with Special Reference to Their Use in Fltration* (24th Annual Report, Massachusetts State Board of Health, Pub. Doc. No. 34), 539–556.

Heibaum, M. W. 2004. "Geotechnical Filters: The Important Link in Scour Protection." In *Proceedings of the 2nd International Conference on Scour and Erosion*, edited by Y. M. Chiew et al., 13–28. Singapore: Nanyang Technology University, Maritime Research Center.

Hemond, H. F., and E. J. Fechner-Levy. 2000. *Chemical Fate and Transport in the Environment* (2nd ed.). San Diego, CA: Academic Press.

Hermana, R. 2001. "An Introduction to Electrical Resistivity in Geophysics." *American Journal of Physics* 69(9).

Hertlein, B. H. 2009. "Analysis Collateral: Information Needed for Foundation Integrity Test Data Analysis." In *2009 International Foundation Congress and Equipment Expo*, Orlando, Florida, 528–535. : ASCE.

Hickin, E. J., and G. C. Nanson. 1984. "Lateral Migration Rates of River Bends." *Journal of Hydraulic Engineering* (ASCE) 110: 1557–1567.

Hjulström, F. 1935. "The Morphological Activity of Rivers as Illustrated by River Fyris." *Bulletin of the Geological Institute Uppsala ch.* 3(23): 221.

Hoffmans, G. J. C. M., and H. J. Verheij. 1997. *Scour Manual*. Rotterdam: Balkema.

Hofland, B., J. A. Battjes, and R. Booij. 2005. "Measurement of Fluctuating Pressures on Coarse Bed Material." *Journal of Hydraulic Engineering* 131(9): 770–781.

Hoek, E. 1994. "Strength of Rock and Rock Masses." *ISRM News Journal* 2: 4–16.

Hoit, M., C. Hays, and M. Mcvay. 1997. "The Florida Pier Analysis Program: Methods and Models for Pier Ananlysis and Design." *Journal of the Transportation Research Board* 1569: 1–7.

Hollema, D. A., and L. D. Olson. 2002. "Crosshole Sonic Logging ND Tomography Velocity Imaging of a New Bridge Foundation." Paper presented at Structural Materials Technology Conference, American Society for Non-Destructive Testing, Cincinnati, OH, 10-13 September 2012.

Holtz, R. D., W. D. Kovacs, and T. C. Sheahan. 2011. *An Introduction to Geotechnical Engineering*, 2nd ed. Upper Saddle River, NJ: Pearson-Prentice Hall.

Hong, G. T., C. P. Aubeny, and R. L. Lytton. 2010. "Lateral Earth Pressure on a Wall in Expansive Soils." In *Proceedings of the International Conference on Unsaturated Soils* (UNSAT 2010), September 2010, Spain.: CRC Press-Balkema-Taylorand Francis Group.

Hooke, J. M. 1984. "Changes in River Meanders: A Review of Techniques and Results of Analysis." *Progress in Physical Geography* 8: 473–508.

Horvath, J. S. 1991. "Using Geosynthetics to Reduce Earth Loads on Rigid Retaining Structures." In *Proceedings Geosynthetics '91*, 409–423. Washington, DC: IFAI Publications.

Horvath, J. S., ed. 1994. "Polystyrene Foam in Below Grade Applications." In *Proceedings of International Symposium*, New York: Manhattan College.

Hossain, K. M. 1996. *Load Settlement Curve Method for Footings in Sand at Various Depths, under Eccentric or Inclined Loads, and Near Slopes*. PhD diss., Texas A&M University, Department of Civil Engineering, College Station, TX.

Houlsby, G., 1991. "How the Dilatancy of Soils Affects Their Behavior." Invited Lecture, 10th European Conference on Soil Mechanics and Foundation Engineering, Florence, Italy.

Hughes, J. M. O., and N. J. Withers. 1974. "Reinforcing of Soft Cohesive Soils with Stone Columns." *Ground Engineering* (May): 42–49.

Hunt, B. 2001. "Measured Scour Depths and Velocities at the Old Woodrow Wilson Bridge in Washington DC." Personal communication.

Hunt, R. 1984. *Geotechnical Engineering Investigation Manual*. New York: McGraw Hill.

_____. 1986. *Geotechnical Engineering Techniques and Practices*. New York: McGraw Hill.

_____. 2005. *Geotechnical Investigations Methods: A Field Guide for Geotechnical Engineers*. Boca Raton, FL: CRC Press/Taylor and Francis.

Hvorslev, M. J. 1949. Subsurface Exploration and Sampling of Soils for Civil Engineering Purposes (Waterways Experimental Station, Vicksburg, MS).

_____. 1951. "Time Lag in the Observation of Ground-Water Levels and Pressures (U.S. Army Engineers Waterways Experiment Station, Vicksburg, Miss.)." *In International Congress of Soil Science* 2: 37–48.

Idriss, I. M. 1990. "Response of Soft Soil Sites During Earthquakes." In *Proceedings of the H. Bolton Seed Memorial Symposium*, J. M. Duncan (ed.), *BiTech Publ.*, vol. 2, 273–290.

Idriss, I. M., and R. W. Boulanger. 2008. *Soil Liquefaction during Earthquakes* (Monograph MNO-12). Oakland, CA: Earthquake Engineering Research Institute.

Idriss, I. M., and Sun, J. I. 1992. *SHAKE91: A Computer Program for Conducting Equivalent Linear Seismic Response Analyses of Horizontally Layered Soil Deposits*. Davis, CA: Center for Geotechnical Modeling, Department of Civil and Environmental Engineering, University of California at Davis.

Iskander, M., S. Kelley, C. Ealy, and D. Roy. 2001. *Class-A Prediction of Construction Defects in Drilled Shafts* (TRB ID No. 01-0308). Washington, DC: Transportation Research Board.

ISSMGE Technical Committee on Offshore Geotechnics. 2005. "Geotechnical and Geophysical Investigations for Offshore and Nearshore Developments." Retrieved from http://www.ISSMGE.rg/

Jaky, J. 1944. "A nyugalmi nyomas tenyezoje (The coefficient of earth pressure at rest)." Magyar Mernok es Epitesz-Egylet Kozlonye, Hungary: 355–358 [in Hungarian].

Jamiolkowski, M. 2001. "The Leaning Tower of Pisa, End of an Odyssey". In *Proceedings of the 15th International Conference on Soil Mechanics and Geotechnical Engineering*, Vol. 4, 2979–2996. Rotterdam: Balkema.

Jamiolkowski, M., S. Leroueil, and D. C. F. Lo Presti. (1991). "Design Parameters from Theory to Practice" (Theme Lecture). In *Proceedings of Geo-Coast* '91, 1–41.

Janbu, N. 1954. *Stability Analysis for Slopes with Dimensionless Parameters* (Harvard University Soil Mechanics Series No. 46). Doctor of Sciences thesis in the Field of Civil Engineering.

_____. 1968. *Slope Stability Computations* (Soil Mechanics and Foundation Engineering Report). Trondheim, Norway: Technical University of Norway.

_____. 1973. "Slope Stability Computations in Embankment-Dam Engineering." In , edited by R. C. Hirshfeld and R. J. Poulos, 47–86. New York: John Wiley & Sons.

Jeanjean, P. 1995. Load Settlement Curve Method for Spread Footings on Sand from the Pressuremeter Test. PhD diss., Texas A&M University, Department of Civil Engineering, College Station, TX.

Jeanjean, P. 2012. "State of Practice: Offshore Geotechnics Throughout the Life of an Oil and Gas Field." In *Proceedings of the Geo-Institute Congress* (Oakland, March 2012), . Reston, VA: ASCE.

Jefferis, D. 2008. *Exploring Our Solar System: The Earth, Our Home Planet*. New York: Crabtree Publishing.

Jennings, J. E., and K. Knight. 1975. "A Guide to Construction on or with Materials Exhibiting Additional Settlement Due to Collapse of Grain Structure." *In Sixth Regional Conference for Africa and Soil Mechanics and Foundation Engineering*, 99–105.

Jeong, S., and Briaud J.-L. 1994. "Nonlinear Three-Dimensional Analysis of Downdrag on Pile Groups." Paper presented at ASCE Specialty Conference, "Settlement 94," Texas A&M University, June 1994.

Jing, L., and J. A. Hudson. 2002. "Numerical Methods in Rock Mechanics." *International Journal of Rock Mechanics & Mining Sciences* 39: 409–427.

Jumikis, A. R. 1977. *Thermal Geotechnics*. New Brunswick, NJ: Rutgers University Press.

Kanamori. H. 1977. "The Energy Release in Great Earthquakes." *Journal of Geophysical Research* 82: 2981–2987.

Karlsrud, K., R. Nadim, and T. Haugen. 1986. "Piles in Clay under Cyclic Axial Load: Field Tests and Computational Modeling." In *Proceedings of the Third International Conference on Numerical Methods in Offshore Piling, Nantes, France*, 165–190.

Karol, R. H. 2003. *Chemical Grouting and Soil Stabilization* (3rd ed.). Boca Raton, FL: CRC Press.

Karzenbach, R. 2012. *International CPRF Draft Guidelines* (ISSMGE Technical Committee on Deep Foundations). London: ISSMGE.

Kavazanjian, E. 1999. "Seismic Design of Solid Waste Containment Facilities." In *Proceedings of the 8th Canadian Conference on Earthquake Engineering*, Vancouver, B.C., Canada, (June): 51–89.

Kavazanjian, E., N. Matasovic, T. Hadj-Hamou, and P. J. Sabatini. 1997. *Geotechnical Earthquake Engineering for Highways: Vol 1—Design Principles and Vol 2—Design Example* (Geotechnical Engineering Circular No. 3, FHWA SA-97-076). Washington, DC: FHWA.

Kavazanjian, E., J.-N. J. Wang, G. R. Martin., A. Shamsabadi, I. P. Lam, S. E. Dickenson, and C. J. Hung. 2011. LRFD *Seismic Analysis and Design of Transportation Geotechnical Features and Structural Foundations* (NHI Course no. 130094 reference manual, Geotechnical Engineering Circular no. 3, FHWA-NHI-11-032). Washington, DC: FHWA.

Keller . 2012. "Range of Application of Grouting Techniques." Retrieved from http://www.kellergrundbau.com/download/pdf/en/Keller_66-01E.pdf

Kenney, T. C., and D. Lau. 1986. "Internal Stability of Granular Filters." *Canadian Geotechnical Journal* 22(2): 215–223.

Kerisel, J. (1985). "The History of Geotechnical Engineering up until 1700." In *Proceedings of the Eleventh International Conference on Soil Mechanics and Foundation Engineering* (San Francisco, Golden Jubilee Volume), 3–93. Rotterdam: A. A. Balkema.

Khalili, N., and M. H. Khabbaz. 1998. "Unique Relationship for the Determination of the Shear Strength of Unsaturated Soils." *Géotechnique* 48(5): 681–687.

Kim, D.-S., W.-S. Seo, and S.-H. Lee. 2006. "Development of Modulus-Soil Moisture Model for Subgrade Soils Using Soil Suction Control Testing System." In *Pavement Mechanics and Performance* (GSP 154), 256–263. Washington, DC: ASCE.

Kim, K. 2010. *Numerical Simulation of Impact Roller for Estimating the Influence Depth on Soil Compaction*. Master's thesis, Department of Civil Engineering, Texas A&M University, College Station, TX.

Kirsten, H. A. D., J. S. Moore, L. H. Kirsten, and D. M. Temple. 1996. "Erodibility Criterion for Auxiliary Spillways of Dams." Paper presented at ASAE International Meeting, Phoenix, AZ (Paper No. 962099).

Koerner, R. M. 2012. *Designing with Geosynthetics* (6th ed.), vols. 1 and 2. Bloomington, IN: Xlibris Corp.

Koerner, R. M., and T.-Y. Soong. 2001. "Geosynthetic Reinforced Segmental Retaining Walls." *Geotextiles & Geomembranes* 19(6): 359–386.

Komine, H., and N. Ogata. 2004. "Predicting Swelling Characteristics of Bentonites." *Journal of Geotechnical & Geoenvironmental Engineering* 130(8): 818–829.

Kozeny, J. 1927, "Ueber kapillare Leitung des Wassers im Boden." *Wien Akademie Wiss.* 136(2a): 271.

Kramer, S. L. 1996. *Geotechnical Earthquake Engineering*. New York: Prentice Hall.

Kulhawy, F. H., and J.-R. Chen. 2007. "Discussion of 'Drilled Shaft Side Resistance in Gravelly Soils' by Rollins K. M., Clayton R. J., Mikesell R. C., Blaise B. C." *Journal of Geotechnical & Geoenvironmental Engineering* 133(10): 1325–1328.

Kulhawy, F. H., and C. S. Jackson. 1989. "Some Observations on Undrained Side Resistance of CIDH Piles." In *Proceedings, Foundation Engineering: Current Principles and Practices*, Vol. 2, 1011–1025. Washington, DC: ASCE.

Kulhawy, F. H., and P. W. Mayne. 1990. *Manual on Estimating Soil Properties for Foundation Design* (Electric Power Research Institute Report EL-6800). Palo Alto, CA: Electric Power Research Institute.

Kwak, K., J.-L. Briaud, Y. Cao, M.-K. Chung, B. Hunt, and S. Davis. 2002. "Pier Scour at Woodrow Wilson Bridge and SRICOS Method." In *Proceedings of the First International Conference on Scour of Foundations*, Department of Civil Engineering, Texas A&M University, College Station, TX.

Kwak, K., J.-L. Briaud, and H.-C. Chen. 2001. "SRICOS: Computer Program for Bridge Pier Scour." In *Proceedings of the 15th*

International Conference on Soil Mechanics and Geotechnical Engineering, Vol. 3, 2235–2238 Rotterdam: A.A. Balkema.

Ladd, C. C. 1991. "Stability Evaluation during Staged Construction: 22nd Terzaghi Lecture." *Journal of Geotechnical Engineering* 117(4): 537–615.

Ladd, C. C., and R. Foott. 1974. "New Design Procedure for Stability of Soft Clays." *Journal of Geotechnical Engineering* 100 (GT7): 763–786.

Ladd C. C., R. Foott, K. Ishihara, F. Schlosser, and H. Poulos. 1977. "Stress-Deformation and Strength Characteristics"(State of the Art Report). In *Proceedings of the 9th International Conference on Soil Mechanics and Foundation Engineering, Tokyo*, vol. 2, 421–494.

Lagasse, P. F., P. E. Clopper, J. E. Pagán-Ortiz, L. W. Zevenbergen, L. A. Arneson, J. D. Schall, and L. G. Girard. 2009. *Bridge Scour and Stream Instability Countermeasures: Experience, Selection and Design Guidance* (Hydraulic Engineering Circular no. 23, 3rd ed., vols. 1 and 2). Washington, DC: Federal Highway Administration.

Lagasse, P. F., J. D. Schall, and E. V. Richardson. 2001. *Stream Stability at Highway Bridges* (Hydraulic Engineering Circular no. 20, 3rd ed.). Washington, DC: Federal Highway Administration.

Laliberte, G. E., and A. T. Corey. 1966. *Hydraulic Properties of Disturbed and Undisturbed Clays* (ASTM STP No. 417). West Conshohocken, PA: ASTM.

Lambe, T. W., and R. V. Whitman. 1979. *Soil Mechanics—SI Version*. New York: John Wiley & Sons.

Landva, A. O., and J. L. Clark. 1990. *Geotechnics of Wastefills: Theory and Practice* (ASTM STP 1070), 86–103. West Conshohocken, PA: ASTM.

Lawson, W. D., P. W. Jayawickrama, T. A. Wood, and J. G. Surles. 2011. *MSE Pullout Testing for RECO HA Ladder and HA Strip Reinforcements* (Research Report for The Reinforced Earth Company, Vienna, VA). Texas Tech University, Department of Civil Engineering.

Lay, T., H. Kanamori, C. J. Ammon, M. Nettles, S. N. Ward, R. C. Aster, . . . and S. Sipkin. 2005. "The Great Sumatra-Andaman Earthquake of 26 December 2004." *Science* 308: 1127–1133.

Lazarte, C. A., V. Elias, R. D. Espinoza, and P. J. Sabatini. 2003. *Geotechnical Engineering Circular No. 7: Soil Nail Walls* (Report No. FHWA0-IF-03-017). Washington, DC: Federal Highway Administration.

Lemoine, B. 2006. *La Tour de 300 Metres*. Paris: Societe des Imprimeurs Lemercier.

Li, X., L. M. Zhang, and D. G. Fredlund. 2009. "Wetting Front Advancing Column Test for Measuring Unsaturated Hydraulic Conductivity." *Canadian Geotechnical Journal* 46(December 2009): 1431–1445.

Likins, G. E., and M. Hussein. 1988. "A Summary of the Pile Driving Analyzer Capacity Methods: Past and Present." Paper presented at the 11th Pile Driving Analyzer's Seminar, Cleveland, OH.

Lisyuk, M., and V. Ulitsky. 2012. "Soil Structure Interaction." In *Proceedings of the Russia-USA Geotechnical Summit* (March) Oakland, CA.

Little, D. N. 1999. "Evaluation of Structural Properties of Lime Stabilized Soils and Aggregates." Retrieved from www.lime .org/documents/publications/free_downloads/soils-aggregates-vol1.pdf.

Lowe, J., and L. Karafiath. 1960. "Stability of Earth Dams upon Drawdown." In *Proceedings of the First Pan American Conference on Soil Mechanics and Foundation Engineering*, Mexico City, 537–552.

Lowery, L. L. 1993. *Pile Driving Analysis by the Wave Equation—Microwave Manual*. Bryan, TX: Wild West Software.

Lowery, L. L., T. J. Hirsch, and C. H. Samson, Jr. 1967. *Pile Driving Analysis: Simulation of Hammers, Pile and Soils* (Research Report 33-9, Project 2-5-62-33, Piling Behavior). College Station, TX: Texas Transportation Institute, Texas A&M University.

Lu, N., and W. J. Likos. 2004. *Unsaturated Soil Mechanics*. Hoboken, NJ: John Wiley & Sons.

Luettich, S. M., J. P. Giroud, and R. C. Bachus. 1992. "Geotextile Filter Design Guide." *Journal of Geotextiles & Geomembranes* 11(4-6): 19–34.

Lukas, R. G. 1995. *Dynamic Compaction* (Geotechnical Engineering Circular No.1, FHWA-SA-95-037). Washington, DC: Federal Highway Administration.

Lundberg, G. 1939. "Elastische beruehrung zweier halbraeume." *Forsch. Geb. Ingenieurwes.*, 10–5:. 201–211.

Lunne, T., and A. Kleven. 1981. "Role of CPT in North Sea Foundation Engineering." In *Symposium on Cone Penetration Engineering Division*, 49–75. ASCE.

Lupini, J. F., A. E. Skinner. and P. R. Vaughan. 1981. "The Drained Residual Strength of Cohesive Soils." *Géotechnique* 31(2): 181–213.

Lytton, R. L. 1994. "Prediction of Movement in Expansive Clays." In *Proceedings of the Settlement Conference at Texas A&M University* (ASCE Geotechnical Special Publication no. 40), vol. 2, 1827–1845.

Machan, G., and V. Bennett. 2008. *Use of Inclinometers for Geotechnical Instrumentation on Transportation Projects: State of the Practice* (Transportation Research Circular No. E-C129, October 2008).

Makdisi, F. I., and H. B. Seed. 1978. "Simplified Procedure for Evaluating Dam and Embankment Earthquake Induced Deformation." *Journal of the Geotechnical Engineering Division* 104 (GT7): 849–867.

Manso, P. F. 2006. *The Influence of Pool Geometry and Induced Flow Patterns in Rock Scour by High-Velocity Plunging Jets*. PhD thesis, Laboratory of Hydraulic Constructions, École Polytechnique Federale de Lausanne, Switzerland.

Mansur, C. I., and R. I. Kaufman. 1962. "Dewatering." In *Foundation Engineering*, edited by G. A. Leonards, ch. 3. New York: McGraw-Hill.

Marchetti, S. 1975. "A New In Situ Test for the Measurement of Horizontal Soil Deformability." In *Proceedings of the Conference on In Situ Measurement of Soil Properties* (ASCE Special Conference, Raleigh, NC), vol. 2, 255–259.

Marcuson, W. F., III. 2001. "Construction of the Panama Canal." In *Proceedings of the 15th International Conference on Soil Mechanics and Geotechnical Engineering*, vol. 3, 2375–2390. Rotterdam: Balkema.

Marcuson, W. F., and A. G. Franklin. 1983. "Seismic Design, Analysis, and Remedial Measures to Improve the Stability of Existing Earth Dams—Corps of Engineers Approach." In *Seismic Design of Embankments and Caverns*, T. R. Howard (ed.). . New York: ASCE.

Marinos, P., and E. Hoek. 2000. "GSI—A Geologically Friendly Tool for Rock Mass Strength Estimation." In *Proceedings of the GeoEng2000 Conference*, Melbourne, Australia, 1422–1442.

Massarsch, K. R. 1991. "Deep Soil Compaction Using Vibratory Probes." In *Proceedings of Symposium on Design, Construction, and Testing of Deep Foundation Improvement: Stone Columns and Related Techniques* (ASTM Special Technical Publication, STP 1089), edited by R. C. Bachus, 297–319. Philadelphia: ASTM.

Matasovic, N. 1993. *Seismic Response of Composite Horizontally-Layered Soil Deposits*. PhD diss., Civil Engineering Department, University of California-Los Angeles.

Mayne, P. W. 1985. "Ground Vibrations During Dynamic Compaction." In *Vibration Problems in Geotechnical Engineering*, 247–265. New York: ASCE.

_____. 2007a. *Cone Penetration Testing State-of-Practice* (NCHRP Project 20-05; Task 37-14: Synthesis on Cone Penetration Test). Washington, DC: National Cooperative Highway Research Program, National Academy of Sciences.

_____. 2007b. *Cone Penetration Testing* (NCHRP Synthesis 368). Washington, DC: Transportation Research Board, National Research Council.

Mayne, P. W., B. Christopher, R. Berg, and J. DeJong. 2002. *Subsurface Investigations—Geotechnical Site Characterization* (Publication No. FHWA-NH-01-031). Washington, DC: National Highway Institute, Federal Highway Administration.

Mayne, P. W., M. R. Coop, S. M. Springman, A. Huang, and J. G. Zornberg. 2009. "Geomaterial Behavior and Testing." In *Proceedings of the 17th International Conference on Soil Mechanics and Geotechnical Engineering*, 2777–2872. London: ISSMGE.

Mayne, P. W., and G. J. Rix. 1993. "Gmax—qc Relationships for Clays." *Geotechnical Testing Journal* 16(1): 54–60.

McDowell, C. 1956. "Interrelationships of Loads, Volume Change, and Layer Thickness of Soils to the Behavior of Engineering Structures." In *Highway Research Board, Proceedings of the Thirty Fifth Annual Meetings* (Publication No. 426), 754–772. Washington, DC: Transportation Research Board.

Menard, L. 1963a. "Calcul de la force portante des fondations sur la base des resultats des essais pressiometriques." *Sols-Soils* 2(5 & 6), Techniques Louis Menard.

_____. 1963b. "Calcul de la force portante des fondations sur la base des resultats des essais pressiometriques: seconde partie, resultats experimentaux et conclusions." *Sols-Soils* 2(6), Techniques Louis Menard.

Menard, L., and Y. Broise. 1975. "Theoretical and Practical Aspects of Dynamic Consolidation." *Géotechnique* 25(1): 3–18.

Mercalli, G. 1883. "Vulcani e fenomeni vulcanici in Italia" In *Geologia d'Italia*, edited by G. Negri, A. Stoppani, and G. Mercalli, 217–218. : Vallardi.

Mesri, G., and M. Abdelghafar. 1993. "Cohesion Intercept in Effective Stress Stability Analysis." *Journal of Geotechnical Engineering* 99(119): 1229–1249.

Meyerhof, G. G. 1951. "The Ultimate Bearing Capacity of Foundations." *Géotechnique* 2(4): 301–332.

_____. 1953. "The Bearing Capacity of Foundations under Eccentric and Inclined Loads." In *Proceedings of the 3rd International Conference on Soil Mechanics and Foundation Engineering, Zurich*, vol. 1, 440–445.

_____. 1955. "Influence of Roughness of Base and Ground Water Conditions on the Ultimate Bearing Capacity of Foundations." *Géotechnique* 5: 227–242.

Mindlin, R. D. 1936. "Force at a Point in the Interior of a Semi-Infinite Solid." *Physics* 7: 195.

Mitchell, J. K. 1976. *Fundamentals of Soil Behavior*. New York: John Wiley & Sons.

Mitchell, J. K., and R. K. Katti. 1981. "Soil Improvement—State of the Art Report." 10th ICSMFE, Stockholm, 4: 509–565.

Mitchell, J. K., and F. T. Madsen. 1987. "Chemical Effects on Clay Hydraulic Conductivity." In *Geotechnical Practice for Waste Disposal '87* (Geotechnical Special Publication No. 13), 87–116. Reston, VA: ASCE.

Mitchell, J. K., and K. Soga. 2005. *Fundamentals of Soil Behavior*, 3rd ed. Hoboken, NJ: John Wiley & Sons.

Mitchell, J. M., and F. M. Jardine. 2002. *A Guide to Ground Treatment*. London: CIRIA.

Monaco, P., S. Marchetti, G. Totani, and M. Calabrese. 2005. "Sand Liquefiability Assessment by Flat Dilatometer Test (DMT)." In *Proceedings of XVI ICSMGE, Osaka*, 4, 2693–2697.

Mononobe, N., and H. Matsuo. 1929. "On the Deformation of Earth Pressure During Earthquakes." In *Proceedings of the World Engineering Conference*, vol. 9, 177.

Moody, L. F. 1944. "Friction Factors for Pipe Flow." *Transaction of the American Society of Civil Engineers* 66: 671–684.

Mooney, M. A., and P. K. Miller. 2009. "Analysis of Lightweight Deflectometer Test Based on In Situ Stress and Strain Response." *Journal of Geotechnical & Geoenvironmental Engineering* 135(2): 199–208.

Morgenstern, N. R. 1963. "Stability Charts for Earth Slopes During Rapid Drawdown." *Géotechnique* 13(1): 121–131.

Morgenstern, N. R., and V. E. Price. 1965. "The Analysis of the Stability of General Slip Surfaces." *Géotechnique* 15(1): 77–93.

Moseley, M. P., and K. Kirsch. 2004. *Ground Improvement*. Oxford: Spon Press.

Moulton L. K., V. S. G. Hota, and G. T. Halvorsen. 1985. *Tolerable Movement Criteria for Highway Bridges* (FHWA/RD-85/107). Washington, DC: Federal Highway Administration.

National Cooperative Highway Research Program (NCHRP). 2008. *NCHRP Project 12-70: Seismic Analysis and Design of Retaining Walls, Buried Structures, Slopes and Embankments, Recommended Specifications, Commentaries and Example Problems* (NCHRP Report 611). Washington, DC: Transportation Research Board.

Negussey, D. 1997. "Properties and Applications of Geofoam." Washington, DC: Society of the Plastics Industry.

Nenad Bicanic. 2007. "Discrete Element Methods." *Encyclopedia of Computational Mechanics*, Vol. 1: Fundamentals, edited by Erwin Stein, Rene de Borst, and Thomas J. R. Hughes, ch. 11. London: John Wiley & Sons, Ltd.

Newmark, N. M. 1965. "Effects of Earthquakes on Dams and Embankments." *Géotechnique* 15(2): 129–160.

Ng, E., J.-L. Briaud, and L. M. Tucker. 1988. *Field Study of Pile Group Action in Sand* (Research Report FHWA-RD-88-QB2). Washington, DC: FHWA.

Ng, W.-S., M.-L. Lee, and S.-L. Hii. 2012. "An Overview of the Factors Affecting Microbial-Induced Calcite Precipitation and Its Potential Application in Soil Improvement." *World Academy of Science, Engineering and Technology* 62: 723–729.

Nurtjahyo, P. Y. 2003. *Chimera RANS Simulations of Pier Scour and Contraction Scour in Cohesive Soils*. PhD diss., Zachry Department of Civil Engineering, Texas A&M University, College Station, TX.

Ohta, Y., and N. Goto. 1976. "Estimation of S-Wave Velocity in Terms of Characteristic Indices of Soil." *Butsuri-Tanko* 29(4): 34–41.

Okabe, S. 1926. "General Theory of Earth Pressure." *Journal of the Japanese Society of Civil Engineers* 12(1): 123–134.

Okabe, T. 1977. "Large Negative Friction and Friction-Free Pile Methods." In *Proceedings of the 9th International Conference on Soil Mechanics and Foundation Engineering*, 679–682. Tokyo:

Olson, S. M., and T. D. Stark. 2002. "Liquefied Strength Ratio from Liquefaction Flow Failure Case Histories." *Canadian Geotechnical Journal* 39: 629–647.

O'Neill, M. W. 1983. "Group Action in Offshore Piles." Presented at ASCE Specialty Conference on Geotechnical Engineering in Offshore Engineering, Austin, TX.

O'Neill, M. W., and L. C. Reese. 1999. *Drilled Shafts: Construction Procedures and Design Methods* (Publication FHWA-If-99-025). Washington, DC: Federal Highway Administration.

Osterberg, J. O. 1984. "A New Simplified Method for Load Testing of Drilled Shafts." *Foundation Drilling Magazine* 23(6): 9–11.

Othman, M. A., R. Bonaparte, and B. A. Gross. 1997. "Preliminary Results of Study of Composite Liner Field Performance." In *Proceedings of the Gri-10 Conference*, 115–142. Folsom, PA: GII Publishing.

Paikowsky, S. G. 2004. *Load and Resistance Factor Design (LRFD) for Deep Foundations* (NCHRP Report 507). Washington, DC: Transportation Research Board, National Academy of Sciences.

Paikowsky, S. G., M. C. Canniff, N. Chelmsford, K. Lesny, A. Kisse, S. Amatya, and R. Muganga. 2010. *LRFD Design and Construction of Shallow Foundations for Highway Bridge Structures* (NCHRP Report 651). Washington, DC: Transportation Research Board, National Academy of Sciences.

Parez, L., and R. Fauriel. 1988. "Le piezocone: Ameliorations apportees a la reconnaissance de sols." *Révue Française de Géotechnique* 44: 13–27.

Park, N. 2007. *A Prediction of Meander Migration Based on Large-Scale Flume Tests in Clay*. PhD diss., Zachry Department of Civil Engineering, Texas A&M University, College Station, TX.

PDI. 2012. *CAPWAP (Case Pile Wave Analysis Program)*. Cleveland, OH: Pile Dynamics Inc.

Peck, R. B. 1969. "Deep Excavations and Tunneling in Soft Ground" (State of the Art Report). In *Proceedings of the 7th International Conference on Soil Mechanics and Foundation Engineering*, Mexico City, 215–290.

_____. 1985. "The Last Sixty Years." In *Proceedings of the XI International Conference on Soil Mechanics and Foundation Engineering, San Francisco* (Golden Jubilee Volume), 123–133. Rotterdam: A. A. Balkema.

Pereira, J. H. F., and D. G. Fredlund. 2000. "Volume Change Behavior of Collapsible Compacted Gneiss Soil." *Journal of Geotechnical & Geoenvironmental Engineering*, 126(10): 907–916.

Perzlmaier, S. 2005. "Hydraulic Criteria for Internal Erosion in Cohesionless Soils." In *Internal Erosion of Dams and Their Foundations*, edited by R. Fell and J.-J. Fry, 179–190. London: Taylor and Francis Group.

Philiponnat, G. 1986. "Le Phicomètre: Essai de cisaillement direct in situ." *Révue Française de Géotechnique* 35: 43–60.

Philipponnat, G., and M. Zerhouni. 1993. "Interprétation de l'essai au Phicomètre." *Révue Française de Géotechnique* 65: 3–28.

Polshin, D. E., and R. A. Tokar. 1957. "Maximum Allowable Nonuniform Settlement of Structures." In *Proceedings of the 4th International Conference, Soil Mechanics and Foundation Engineering*, 402–406. London: Butterworth Heinemann.

Post-Tensioning Institute (PTI). 2004. *Design and Construction of Post-Tensioned Slabs-on-Ground* (3rd ed.). Phoenix, AZ: Post-Tensioning Institute.

Potts, D. M., and L. Zdravkovic. 1999. *Finite Element Analysis in Geotechnical Engineering*. London: Thomas Telford Ltd.

Poulos, H. G. 1988. *Marine Geotechnics*. London: Unwin Hyman Ltd.

_____. 2012. *User's Guide to DEFPIG: Deformation Analysis of Pile Groups*. Sydney, Australia: Civil Engineering Department, University of Sydney.

Poulos, H. G., and E. H. Davis. 1980. *Pile Foundation Analysis and Design*. New York: John Wiley & Sons.

Priest, S. D. (1993). *Discontinuity Analysis for Rock Engineering*. New York: Chapman & Hall.

Raithel, M., A. Kirchner, C. Schade, and E. Leusink. 2005. "Foundation of Constructions on Very Soft Soils with Geotextile Encased Columns—State of the art." *Geotechnical Special Publication: Innovations in Grouting and Soil Improvement* 136: 1867–1877.

Randolph, M. F. 1980. *PIGLET: A Computer Program for the Analysis and Design of Pile Groups under General Loading Conditions* (Soil Report TR91, CUED/D). Cambridge, UK: University of Cambridge.

Randolph, M. F., and B. S. Murphy. 1985. "Shaft Capacity of Driven Piles in Clay." In *Proceedings of the Offshore Technology Conference*, Houston, TX. Paper 4883, vol. 1, pp. 371–378.

Randolph, M. F., and C. P. Wroth. 1978. "Analysis of Deformation on Vertically Loaded Piles." *Journal of Soil Mechanics & Foundation Engineering* 2 (GT12): 1465–1488.

Raudkivi, A. J. 1998. *Loose Boundary Hydraulics*. Rotterdam: Balkema.

Rausche, F. 2002. "Modeling of Vibratory Pile Driving." *In Proceedings of the International Conference on Vibratory Pile Driving and Deep Soil Compaction, Transvib 2002*, 21–32.

Reese, L. C., and M. W. O'Neill. 1988. *Drilled Shafts: Construction and Design* (FHWA Publication No. HI-88-042). Washington, DC: Federal Highway Administration.

Reese, L. C., F. Touma, and M. W. O'Neill. 1976. "Behavior of Drilled Piers under Axial Loading." *Journal of the Geotechnical Engineering Division, ASCE* 102 (GT5): 493–510.

Richards, A. 1988. *Vane Shear Strength Testing of Soils: Field and Laboratory Studies* (ASTM STP 1014). Philadelphia, PA: American Society for Testing and Materials.

Richards, A. F., and H. M. Zuidberg. 1985. "In Situ Determination of the Strength of Marine Soils." In *Strength Testing of Marine Sediments: Laboratory and In-Situ Measurements* (ASTM STP 883), 11–40. West Conshohocken, PA: American Society for Testing & Materials.

Richardson, E. V., and S. M. Davis. 2001. *Evaluating Scour at Bridges* (Publication No. FWHA-IP-90-017, Hydraulic

Engineering Circular No. 18). Washington, DC: Federal Highway Administration.

Richter, C. F. 1935. "An Instrumental Earthquake Magnitude Scale." *Bulletin of the Seismological Society of America* 25(1): 1–32.

Riessner, H. 1924. "Zum Erdruck Problem" (Concerning the earth pressure problem). In *Proceedings of the First International Congress of Applied Mechanics*, 295–311. Delft, The Netherlands:

Rix, G. J., and K. H. Stokoe. 1991. "Correlation of Initial Tangent Modulus and Cone Resistance." In *International Symposium on Calibration Chamber Testing*, 351–362. New York: Elsevier.

Robertson, P. K., and R. G. Campanella. 1983. "Interpretation of Cone Penetration Tests: Sands." *Canadian Geotechnical Journal* 20(4): 719–733.

Robertson, P. K., R. G. Campanella, D. Gillespie, and J. Greig. 1986. "Use of Piezometer Cone Data." In *Use of In-Situ Tests in Geotechnical Engineering* (GSP 6), 1263–1280. Reston, VA: American Society of Civil Engineers.

Robertson, P. K., and C. E. Wride. 1998. "Evaluating Cyclic Liquefaction Potential Using the Cone Penetration Test." *Canadian Geotechnical Journal* 35: 442–459.

Roscoe, K. H., and J. B. Burland. 1968. "On the Generalized Behaviour of 'Wet' Clay." *Engineering Plasticity* 48: 535–609.

Roscoe, K. H., A. N. Schofield, and C. P. Wroth. 1958. "On the Yielding of Soils." *Géotechnique* 8(1): 22–52.

Rutherford, C., G. Biscontin, and J.-L. Briaud. 2005. *Design Manual for Excavation Support Using Deep Mixing Technology* (Internal Report). Department of Civil Engineering, Texas A&M University, College Station, TX.

Sadek, S., and G. Khoury. 2000. "Soil and Site Improvement Guide: An Educational Tool for Engineered Ground Modification." *International Journal of Engineering Education* 16(6): 499–508.

Safir, L., and W. Safire. 1982. *Good Advice*. New York: Times Books.

Samson, C. H., T. J. Hirsch, and L. L. Lowery. 1963. "Computer Study of Dynamic Behavior of Pilings." *Journal of the Structural Division, ASCE* 89 (ST4): Proceedings Paper 3608.

Sandström, Å. (1994). *Numerical Simulation of a Vibratory Roller on Cohesionless Soil* (Internal Report). Stockholm, Sweden: Geodynamik.

Sarma, S. K. 1973. "Stability Analysis of Embankments and Slopes." *Géotechnique* 23(3): 423–433.

Sarsby, R. W. 1985. "The Influence of Aperture Size and Particle Size on the Efficiency of Grid Reinforcement." In *Proceedings of the 2nd Canadian Symposium on Geotextiles and Geomembranes*, 7–12. Edmonton, Canada: The Geotechnical Society of Edmonton.

Saye, S. R., J. C. Volk, and P. C. Gerhart. 2000. "Design-Built I-95 Highway Reconstruction." : GeoStrata, Geo-Institute, ASCE.

Schaefer, V. R., and R. R. Berg. 2013. "*SHRP 2 R02: Geotechnical Solutions for Soil Improvement, Rapid Embankment Construction, and Stabilization of the Pavement Working Platform—Final Phase 2 Summary Report*" (The Strategic Highway Research Program 2). Transportation Research Board of The National Academies. Web-based system available at www.GeoTechTools.org

Schalkoff, R. J. 1997. *Artificial Neural Networks*. New York: McGraw-Hill.

Schmertmann, J. H. 1970. "Static Cone to Compute Static Settlement over Sand." *Journal of Soil Mechanics & Foundations Division* 96(3): 1011–1043.

_____. 1975. "Measurements of In Situ Strength." In *Proceedings of the ASCE Specialty Conference on In Situ Measurements of Soil Properties*, vol. 2, 57–138.

Schmertmann, J. M., J. P. Hartman, and P. R. Brown. 1978. "Improved Strain Influence Factor Diagram." *Journal of the Geotechnical Engineering Division* 104: 1134.

Schnabel, P. B., J. Lysmer, and H. B. Seed. 1972. *SHAKE: A Computer Program for Earthquake Response Analysis of Horizontally Layered Sites* (Report No. EERC 72-12). Berkeley: Earthquake Engineering Research Center, University of California.

Schofield, A. N., and C. P. Wroth. 1968. *Critical State Soil Mechanics*. New York: McGraw-Hill.

Schofield, R. K. 1935. "The pF of the Water in Soil." In *Transactions, 3rd International Congress of Soil Science*, vol. 2, pp. 37–48.

Schwartz, D. P., and K. J. Coppersmith. 1984. "Fault Behavior and Characteristic Earthquakes: Examples from the Wasatch and San Andreas Fault Zones." *Journal of Geophysical Research* 89(B7): 5681–5698.

Seed, B., R. T. Wong, I. M. Idriss, and K. Tokimatsu. 1986. "Moduli and Damping Factors for Dynamic Analyses of Cohesionless Soils." *Journal of Geotechnical Engineering* 112 (GT11): 1016–1032.

Seed, H. B., and I. M. Idriss. 1970. *Soil Moduli and Damping Factors for Dynamic Response Analyses* (Earthquake Engineering Research Center, Report No. EERC 70-10). Berkeley: University of California.

_____. 1971. "Simplified Procedure for Evaluating Soil Liquefaction Potential." *Journal of the Soil Mechanics & Foundation Division, ASCE* 107 (SM9): 1229–1274.

Seed, R. B., and L. F. Harder. 1990. "SPT-Based Analysis of Cyclic Pore Pressure Generation and Undrained Residual Strength." In *H. B. Seed Memorial Symposium*, vol. 2, 351–376. Berkeley, CA: BiTech Publishing, Ltd.

Shackelford, C. D. 1994. "Waste-Soil Interactions That Alter Hydraulic Conductivity." In Hydraulic Conductivity and Waste Contaminant Transport in Soil (ASTM STP 1142), edited by S. J. Trautwein and E. D. Daniel, . Philaelphia, PA: ASTM.

Shackelford, C. D., C. H. Benson, T. Katsumi, T. B. Edil, and L. Lin. 2000. "Evaluating the Hydraulic Conductivity of Gcls Permeated with Non-Standard Liquids." *Geotextiles & Geomembranes* 18: 133–161.

Sharma, H. D. 2000. "Solid Waste Landfills: Settlements and Post-Closure Perspectives." In *Proceedings of the ASCE National Conference on Environmental and Pipeline Engineering*, edited by R. Y. Surampali, 447–455.

Sharma, H. D., D. E. Hullings, and F. R. Greguras. 1997. "Interface Strength Tests and Application to Landfill Design." In *Proceedings of the Conference on Geosynthetics '97*, 913–924. Longbeach, CA: IFAI.

Sharma, H. D., and K. R. Reddy. 2004. *Geoenvironmental Engineering*. Hoboken, NJ: John Wiley & Sons.

Sherard, J. L. 1979. "Sinkholes in Dams of Coarse, Broadly Graded Soils." In *Proceedings of the 13th International Congress on Large Dams, New Delhi*, vol. 2, 23–35. Paris: International Commission on Large Dams.

Shields, A. 1936. "Anwendung der Aehnlichkeitsmechanik und der Turbulenzforschung auf die Geschiebebewegung." Doktor-Ingenieurs diss., Technischen Hochschule, Berlin [in German].

Sjoberg, J. 1997. *Estimating Rock Mass Strength Using the Hoek-Brown Failure Criterion and Rock Mass Classification: A Review and Application to the Aznalcollar Open Pit* (Internal Report BM 1997-02). Lulea, Sweden: Lulea University of Technology.

Skempton, A. W. 1951. "The Bearing Capacity of Clays." In *Proceedings of the Building Research Congress*, vol. 1, 180–189.

_____. 1954. "The Pore Pressure Coefficients A and B." *Géotechnique* 4(4): 143–147.

_____. 1985. "A History of Soil Properties, 1717–1927." In Proceedings, XI International Conference on Soil Mechanics and Foundation Engineering, San Francisco (Golden Jubilee Volume), 95–121. Rotterdam: A. A. Balkema.

Skempton, A. W., and D. H. MacDonald. 1956. "The Allowable Settlement of Buildings." *Proceeding of the Institution of Civil Engineers* 3(5): 727–768.

Smith, E. A. L. 1960. "Pile Driving Analysis by the Wave Equation." *Journal of Soil Mechanics & Foundation Engineering, ASCE*, 86 (SM4): 35–61. (Discussions in 87 (SM1): 63–75.)

Smith, M. E. 2005. Design of Bridging Layers in Geosynthetic Reinforced Column Supported Embankments. PhD diss., Virginia Tech.

Sorrell, C. A., and G. F. Sandström. 2001. *Rocks and Minerals: A Guide to Field Identification*. New York: Macmillan.

Sowers, G. F. 1979. *Introductory Soil Mechanics and Foundations: Geotechnical Engineering*, 4th ed. New York: Macmillan.

Spencer, E. 1967. "A Method of Analysis of the Stability of Embankments Assuming Parallel Inter-Slice Forces." *Géotechnique* 17: 11–26.

Stark, T. D. 2003. "Three-Dimensional Slope Stability Methods in Geotechnical Practice." In *Proceedings of the 51st Annual Geotechnical Engineering Conference*, University of Minnesota, St. Paul.

Stark, T. D., and H. T. Eid. 1994. "Drained Residual Strength of Cohesive Soils." *Journal of Geotechnical Engineering* 120(5): 856–871.

Stokoe, K. H., S. J. Joh, and R. D. Woods. 2004. "Some Contributions of In Situ Geophysical Measurements to Solving Geotechnical Engineering Problems." Presented at International Conference on Site Characterization (ISC-2), Porto, Portugal, September 19–22.

Styropor. 1993. "Construction, Highway Construction, and Ground Insulation" (Technical Information Bulletin No. 1-800e). Ludwigshafe, Germany: BASF AG.

Tand, K. E., E. G. Funegard, and J.-L. Briaud. 1986. "Bearing Capacity of Footings on Clay: CPT Method." Presented at ASCE Specialty Conference, Use of In Situ Tests in Geotechnical Engineering, Vicksburg, MS.

Taylor, D. W. 1948. *Fundamentals of Soil Mechanics*. New York: John Wiley & Sons.

Temple, D. M., and J. S. Moore. 1994. "Headcut Advance Prediction for Earth Spillways." In *Proceedings of the ASAE International Winter Meeting*, Atlanta, Georgia, Paper No. 942340.

Terzaghi, K. 1943. *Theoretical Soil Mechanics*. New York: John Wiley & Sons.

Terzaghi, K., and R. B. Peck. 1963. *Soil Mechanics in Engineering Practice*. New York: John Wiley & Sons.

_____. 1967. *Soil Mechanics in Engineering Practice* (2nd ed.). New York: John Wiley & Sons.

Terzaghi, K., R. B. Peck, and G. Mesri. 1996. *Soil Mechanics in Engineering Practice* (3rd ed.). . New York: John Wiley & Sons.

Tetens, O. 1930. "Uber einige meteorologische Begriffe." *Zeitschrift Geophysic* 6: 297–309.

Theisen, M. S. 1992. "The Role of Geosynthetics in Erosion and Sediment Control: An Overview." *Journal of Geotextiles & Geomembranes* 11(4-6): 199–214.

Thompson, M. J., and D. J. White. 2008. "Estimating Compaction of Cohesive Soils from Machine Drive Power." *Journal of Geotechnical & Geoenvironmental Engineering* 134(12): 1771–1777.

Thurner, H., and A. Sandstrom. 1980. "Continuous Compaction Control, CCC." In *Proceedings of the International Conference on Compaction*, 237–245. Paris:

TNO. 2012. *DLTWAVE—Dynamic Load Testing WAVE Program*. The Hague: Toegepast Natuurwetenschappelijk Onderzoek.

TNOWAVE. 2012. *Toegepast Natuurwetenschappelijk Onderzoek Wave Equation Analysis Program*. The Hague: TNO.

Towhata, I. 2008. *Geotechnical Earthquake Engineering*. Berlin/Heidelberg: Springer.

Trautwein, S. J., and E. D. Daniel. 1994. *Hydraulic Conductivity and Waste Contaminant Transport in Soil* (ASTM STP 1142). Philadelphia, PA: American Society for Testing and Materials.

Tucker, L. M., and J.-L. Briaud. 1988. "Analysis of the Behavior of a 5 Pile Group and a Single Pile in Sand at Hunter's Point" (Research Report 7065 2 to GeoResource Consultants and FHWA, Civil Engineering). College Station: Texas A&M University.

U.S. Army Corps of Engineers (USACE). 1970, 1982. *Slope Stability Manual* (EM-1110-2-1902). Washington, DC: Department of the Army, Office of the Chief Engineer.

_____. 1990. *Engineering and Design: Settlement Analysis* (EM 1110-1-1904). Washington, DC: Department of the Army, Office of the Chief Engineer.

_____. 1991. *Hydraulic Design of Flood Control Channels* (EM 1110-2-1601). Washington, DC: Department of the Army, Office of the Chief Engineer.

_____. 2003. "Recommendations for Seepage Design Criteria, Evaluation, and Design Practice" (Report prepared by the Corps of Engineers, Sacramento District). Washington, DC: Department of the Army.

U.S. Environmental Protection Agency (USEPA). 1991. *Subsurface Contamination Reference Guide* (Publication EPA/540/2-90/011). Washington, DC: US Environmental Protection Agency.

_____. 2001. *Risk Assessment Guidance for Superfund*, Vol. II, Part A: "Process for Conducting Probabilistic Risk Assessment" (Publication EPA/540/R-02/002). Washington, DC: Office of Emergency and Remedial Response.

U.S. Navy. 1982. *Foundations and Earth Structures* (NAVFAC Design Manual DM-7.2). Washington, DC: US Government Printing Office.

Vanapalli, S. K., W. S. Sillers, and M. D. Fredlund. 1998. "The Meaning and Relevance of Residual State to Unsaturated Soils." In *Proceedings of the 51st Canadian Geotechnical Conference*, Edmonton, Alberta: Canadian Geotechnical Society.

Van Genuchten, M. (1980). "A Closed-Form Equation for Predicting the Hydraulic Conductivity of Unsaturated Soils." *Soil Science Society of America Journal* 44: 892–898.

Van Schalkwyk, A., J. M. Jordaan, and N. Dooge. 1995. *The Erodibility of Different Rock Formations* (Water Research Commission Report No. 302/1/95). Pretoria, South Africa:

Van Susante, P. J., and M. A. Mooney. 2008. "Capturing Nonlinear Vibratory Roller Compactor Behavior through Lumped Parameter Modeling." *Journal of Engineering Mechanics* 134(8): 684–693.

Vucetic, M., and R. Dobry. 1991. "Effect of Soils Plasticity on Cyclic Response." *Journal of Goetechnical Engineering* 117(1). 898–907.

Wahls, H. E. 1994. "Tolerable Deformations." In *Geotechnical Special Publication No. 40*, edited by A. T. Yeung and G. Y. Félio, 1611–1628. New York: ASCE.

Waltham, A. C. 1994. *Foundations of Engineering Geology*. New York: E&FN SPON.

Wan, C. F., and R. Fell. 2004. "Investigation of Rate of Erosion of Soils in Embankment Dams." *Journal of Geotechnical & Geoenvironmental Engineering* 130(4): 373–380.

Wang, W. 2006. *A Hydrograph-Based Prediction of Meander Migration*. PhD diss., Zachry Department of Civil Engineering, Texas A&M University, College Station, TX.

Warner, J. 2004. *Practical Handbook of Grouting: Soil, Rock and Structures*. New York: John Wiley & Sons.

Warrington, D. C. 1992. "Vibratory and Impact-Vibration Pile Driving Equipment." (October) *Pile Buck* magazine, 2A-28A.

Watts, K. S., and J. A. Charles. 1993. "Initial Assessment of a New Rapid Impact Ground Compactor." In *Proceedings of the Conference on Engineered Fills '93*, 399–412.

Welsh, J. P., and G. K. Burke. 2000. "Advances in Grouting Technology." In *Proceedings of Geoengineering 2000*. Melbourne, Australia:

Westergaard, H. 1931. "Water Pressure on Dams During Earthquakes." *Transactions of ASE*, (Paper No. 1835): 418–433.

White, D., and M. Thompson. 2008. "Relationships between In Situ and Roller-Integrated Compaction Measurements for Granular Soils." *Journal of Geotechnical & Geoenvironmental Engineering* 134(12): 1763–1770.

White, D. J., E. J. Jaselskis, V. R. Schaefer, and E. T. Cackler. 2005. "Real-Time Compaction Monitoring in Cohesive Soils from Machine Response." *Transportation Resources Record* 1936: 171–180.

Whittle, A. J. 1993. "Evaluation of a Constitutive Model for Overconsolidated Clays." *Géotechnique* 43(2): 289–313.

Williamson, D. A. 1984. "Unified Rock Classification System." Bulletin of the Association for Engineering Geology 21: 345–354.

Winter, T. C., J. W. Harvey, O. L. Franke, and W. M. Alley. 1999. *Ground Water and Surface Water: A Single Resource* (U.S. Geological Survey Circular 1139). Reston, VA: US Geological Survey (available at http://pubs.usgs.gov/circ/circ1139/#pdf).

Wire Reinforcement Institute (WRI). 1981. *Design of Slab-on-Ground Foundations*. Hartford, CT: Wire Reinforcement Institute.

Wischmeier, W. H., and D. D. Smith. 1960. "A Universal Soil-Loss Equation to Guide Conservation Farm Planning." In *Proceedings of the 7th International Conference on Soil Science*. Soil Science Society of America.

Wood, D. M. 1990. *Soil Behaviour and Critical State Soil Mechanics*. Cambridge: Cambridge University Press.

Yao, C. 2013. *LRFD Calibration and Risk Analysis for Bridge Foundations Subjected to Scour*. PhD diss., Zachry Department of Civil Engineering, Texas A&M University, College Station, TX.

Yeh, P.-H. 2008. *Physical Models of Meander Channel Migration*. PhD diss., Zachry Department of Civil Engineering, Texas A&M University, College Station, TX.

Youd, T. L., and I. M. Idriss. 1997. *Proceedings of the NCEER Workshop on Evaluation of Liquefaction Resistance of Soils, Salt Lake City, UT, January 5-6, 1996* (Technical Report NCEER-97-0022). Buffalo, NY: National Center for Earthquake Engineering Research, University at Buffalo.

Zhang, L., and A. M. Y. Ng. 2007. *Limiting Tolerable Settlement and Angular Distortion for Building Foundations* (GSP 170, Probabilistic Applications in Geotechnical Engineering). ASCE.

Zienkiewicz, O. C., R. L. Taylor, and J. Z. Zhu. 2005. *The Finite Element Method* (6th ed.). Oxford: Butterworth-Heinemann.

INDEX

D

M

Q

R

S

T

Y

Z